Sports Palace Sant Jordi, Barcelona, Spain

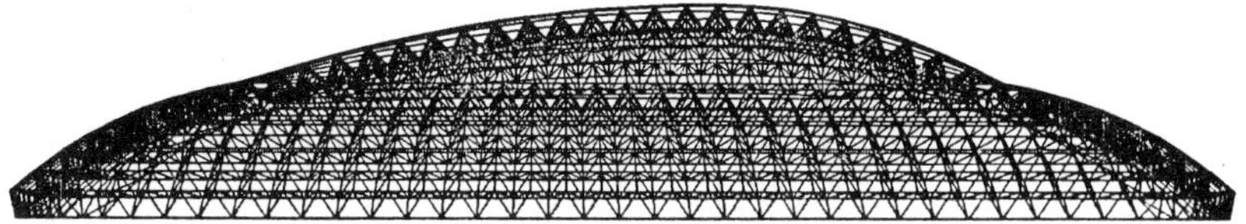

Architect: Arata Isozaki

This double-layer dome structure has a length of 127.87 m and a width of 105.60 m, covering an area of 13 462 m^2.

The Sports Palace was assembled from a total of 9070 bars and 2343 nodes.

This construction was the main arena for the indoor sports at the 1992 Olympic Games in Barcelona.

Space structures 4

Edited by G. A. R. Parke
and C. M. Howard

Volume 1

This volume contains the papers presented at the Fourth International Conference on Space Structures, held at the University of Surrey, Guildford, UK, on 5 - 10 September 1993

Organising committee: Dr G. A. R. Parke, Mr C. M. Howard, Dr P. Disney, Dr H. Nooshin, Professor J. E. Harding and Mr H. Wickens

Sponsors: Institution of Civil Engineers, UK; Institution of Structural Engineers, UK; Royal Institute of British Architects; IASS; Orona Space Structures, Spain; Tomoe Corporation, Japan; Kawatetsu Steel Products, Japan; Taiyo Kogyo Corporation, Japan; Mero, Germany

Published by Thomas Telford Services Ltd, 1 Heron Quay, London E14 4JD

First published 1993

Distributors for Thomas Telford books are
USA: American Society of Civil Engineers, Publications Sales Department, 345 East 47th Street, New York, NY 10017-2398
Japan: Maruzen Co Ltd, Book Department, 3-10 Nihonbashi 2-chome, Chuo-ku, Tokyo 103
Australia: DA Books and Journals, 11 Station Street, Mitcham 2131, Victoria

A catalogue record for this book is available from the British Library

Classification
Availability: Unrestricted
Content: Collected papers
Status: Invited authors' opinion
User: Structural engineers

ISBN: 0 7277 1968 8

Printed and bound in Great Britain

Preface

The diversity of constructions included in this compilation on space structures ranges from antenna reflectors and masts positioned in space, to equally exciting terrestrial structures, notably large-span domes, barrel vaults, multi-layered grids, cable and membrane systems, and pneumatic structures. This collection of more than two hundred and twenty papers, presented in two volumes, is the work of leading international experts for presentation at the Fourth International Conference on Space Structures held at the University of Surrey, UK, on 5-10th September 1993.

These two volumes contain a prodigious amount of original and innovative information on space structures that will be of especial interest to engineers, architects and other professionals engaged in the planning, design, fabrication and erection of novel constructions.

The editors are indebted to the many engineers and architects who have presented their work for this publication, and also to the members of the Technical Committee who gave their time and valuable advices on the selection and editing of the published material. Thanks are also due to the conference sponsors, whose support and encouragement was essential in facilitating the interchange of ideas and innovative concepts that was a prominent feature of the conference.

The editors and conference organisers are staff of the Space Structures Research Centre at the University of Surrey, Guildford. This centre was founded in 1964 by Professor Z. S. Makowski, an eminent leader in this field, who pioneered many programmes of teaching and research into space structures behaviour.

G. A. R. Parke
C. M. Howard

Contents

1. Space structures - a review of the developments within the last decade

Z. S. MAKOWSKI, University of Surrey, UK

SYNOPSIS. Once regarded as exotic and unconventional, space structures are now universally accepted as economical and aesthetically pleasing. During the past decade advances in computerisation have resulted in major changes in space structure design, and in the analysis of their stability and plastic behaviour. Membrane structures, many of polymer-covered glass fibre have seen remarkable development, and offer architects immense potential in novel air-supported and air-inflated structures. Among space structures recently erected are the Toronto Sky Dome with a retractable roof, the Fukoka Dome (1993) which will be the world's largest retractable steel structure, and the Olympic Sports Hall, Barcelona (1992). Although these structures are steel, wood also has been widely used, e.g. the wood lattice dome for Expo'90 Osaka. All evidence shows that development in space structure design, especially light-weight construction, is accelerating rapidly.

INTRODUCTION.

1. During the last decade the world witnessed fundamental changes affecting all aspects of our life. Great political changes, which could not be foreseen a few years ago, have taken place. The scientific revolution brought by the phenomenal developments in electronic computation greatly influenced all branches of engineering architecture. At the same time the economic depression all over the world is affecting all countries.

2. These changes are also visible in the field of three-dimensional structures, the subject which will be discussed during the Fourth International Conference on Space Structures. In fact the methods of analysis and design of these structures have now changed beyond recognition.

3. The 1st Space Structures Conference arranged by the Space Structures Research Centre in London at the Battersea College of Advanced Technology in 1966, was followed at nine years intervals by the 2nd and 3rd International Space Structures Conferences in Guildford in 1975 and 1984, and now in 1993 today we open the 4th International Conference, again at Guildford at the University of

Surrey. This Conference gives us the welcome opportunity to survey the important changes which occurred during the last few years.

4. The visible progress in the development of space structures is extremely significant emphasising continuous achievement of creative designers, engineers and architects alike.

5. Whereas in 1966 space structures were often referred to as "interesting but exotic, non-conventional structures", in 1993 these structures are now accepted all over the world.

6. An increasing number of universities, schools of engineering and architecture, include now the study of space structures in their curricula. Space structures have became an established discipline.

7. Over the last century engineers have been called upon to build larger and larger structures, all of them with less material and less cost. The best engineers matured under the discipline of extreme economy. Their ideas and their styles developed under competitive cost control. The last decade brought tremendous changes through greatly improved theoretical advances in the knowledge of elastic and non-elastic behaviour under applied loading. New theoretical methods and introduction of improved materials has had far-reaching influence on structural steel design. Architectural concepts of aesthetics are changing all the time, leading to the introduction of new structural systems.

8. Designers agree that for large spans conventional beam and truss structures prove to be uneconomical. The advantages of three-dimensional structures have been known for many years. Engineers always appreciated the inherent rigidity of three-dimensional structures and their ability to cover large spans with minimum weight. However the difficulty of the complicated stress analysis of such systems originally contributed to their limited use. The introduction of electronic computers and their present wide-spread use changed this picture. Architects started to experiment with new shapes and some fifty years ago reinforced concrete shells entered the market. The visual beauty of these shells appealed to architects.

9. However, there is no ideal structural material or form, all of them have their advantages and disadvantages. Soon progressive designers realised the limitation of reinforced concrete shells and turned their attention to three-dimensional skeleton frameworks. The activities of the great American designer - Richard Buckminster Fuller (1895-1981), and the original French mathematician and engineer, Robert Le Ricolais (1894-1977), each in a different way contributed greatly to the present acceptance of space structures. Le Ricolais will be remembered as one of the first few, who drew our attention to the "organic architecture" and to the exceptional rigidity and lightness of three-dimensional structures created by Nature. It was Le Ricolais who influenced designers to look closely at the double-layer grid configuration - an area of extreme importance in the design of large span space frames. These structures are now universally used for public assembly buildings, huge sports arenas, exhibition halls, industrial factories etc; - they have a biological equivalent in the skeletons of radiolaria. It is interesting to note that the ideas of this great French designer were originally

completely ignored in his own country, France, and only in the USA did he find the opportunity to continue his research.

10. The proceedings of the recent conferences organised by the International Association for Shell and Spatial Structures (I.A.S.S.) contain a vast number of valuable scientific publications contributing to further developments of three-dimensional systems.

11. The following areas should receive our particular attention:

(a) the most important impact upon the development of space structures is provided by electronic computation. In this area the progress is almost unbelievable. The optimisation studies carried out using the computer, investigating the least weight and minimum cost of space structures continue. One must remember that minimum economy does not necessarily mean minimum material cost. The introduction of computer aided design (CAD), has greatly affected the design of space structures within the last decade. The original work of Dr H Nooshin at the University of Surrey on the Formex configuration processing, provided a truly invaluable basis for solution of problems of data preparation and graphics in the computer-aided design processes. Formex Algebra proved to be an extremely efficient tool dealing with complex configurations of three-dimensional structures. Dr Nooshin is also responsible for the development of a programming language called Formian, for a comprehensive coverage of his Formex complex. Using Formian the information generated about any space structure may be rapidly deployed for graphic visualisation or may be submitted as input data to analysis packages. It should be noted that the introduction of computer-aided design brought a major change for prefabricated space structures. Whereas in the early periods of development the main emphasis was put towards standardisation and use of identical modular units, now with the vastly improved computer techniques, designers are realising the advantages of one-of-a-kind configurations.

(b) the development of light-weight structures greatly changed the engineering approach to design. In the past, as a rule, only the heavy gravity load had to be considered during the analysis - the wind loading rarely had any significant influence. In light-weight structures the uplift due to wind started to change the stresses as the live loading acting upon the structure often exceeds the gravity load. The lightness of space structures created a problem of instability. Within the last decade we have seen increased research into the plastic behaviour of space structures - the number of research investigations into the instability of space structures and characteristics of failure under excessive load, have increased at universities at a phenomenal rate. The collapse analysis of various types of space structures has been carried out by numerous investigators. The double-layer space frames have received within the last decade special attention (Schmidt, 1979, Collins, 1981 etc.) with reference to the brittle load-displacement response of compression members in such systems. It is of interest to note the development of a novel technique of including ductility into the load-displacement response of double-layer grids through the incorporation of soft members into the top chord compression zones (G A Parke, 1988). Whereas previous researchers tried to obtain a ductile load-displacement in space frames through construction of weak members in the tensile chords, Dr Parke proved that one can obtain a significant improvement of the load-carrying capacity of space structures by introduction of specially constructed soft compression members designed to act both as a soft member with reduced axial stiffness and as a load-limiting device. The specially valuable publication of the Analysis, Design and Realisation of Space Frames, issued in 1984, by the IASS dealt with the buckling behaviour of braced domes - it contains an extremely useful discussion of investigations carried out in comparative studies of the accuracy of several approximate approaches as well as the perturbation method, the Newton-Raphson and non-linear programming techniques. The excellent review of buckling of space frames by Kashani & Croll (1987) could be of real interest to practising designers. One should also note the proceedings of an IASS conference on domes, held in Istanbul 1988 which contains many valuable papers on the same topic presenting the very recent research investigations carried out at Japanese universities.

(c) Rapid improvement in structural textiles has lead within the least decade to the dramatic development of novel structural forms; tension cable structures, membranal and pneumatic systems, rarely attempted three decades ago, are now accepted forms of space structures used frequently by progressive designers. Membranal systems are kinematically indeterminate and they only acquire their rigidity by the use of prestress. Such systems can be generated on computer, using linear, triangular and higher order elements for surface discretisation. During the last decade several new techniques have been developed of dynamic relaxation with kinetic damping. The research publications of M R Barnes (1982), W Y Lewis (1989), P D Gesling (90-92) must be mentioned as well as very recent original studies at University of Montpellier from R Motro's research school. New techniques of membrane form finding have been developed within the last few years. It is obvious that membranal structures constitute a field of remarkable development. The original work of Frei Otto and René Sarger has been considerably extended by numerous researchers. The activities of Mamoru Kawaguchi and his Japanese colleagues are receiving now universal acclamation and are bringing to the attention of architects the immense potential of this form of construction which dramatically reversed the previous popularity of shell structures. Tensile stresses are produced in the membranes either by pulling the membrane at its ends (so called "pulled membranes") or by pressurising the membranes by air (so called "pneumatic membranes"). There is no doubt that the recent rapid acceptance all over the world of air-supported or air-inflated structures is mainly due to the work of Japanese and American engineers. The large space structures constructed by David Geiger or Ryotaro Nohmura, of the Taiyo Kogyo Co., demonstrate in a very convincing way the great potential of permanent, non-combustible, low-profile domical structures using greatly improved skin material - usually fibre-glass fabric coated with Teflon resin. It should be noted that in the West only a very few metal membranal structures have been built. Those built in the U.S.A. or Canada do not seem to be very economical. Also, the technical difficulties at the connection of the constituent strips are considerable. It is most interesting to note that the Russian designers seem to be more adventurous in their designs of thin-sheet metal membranal shells, especially V. Trofimov and P. E. Eremejev. They have been responsible for the wide use of metal suspended membrane roofs in the former U.S.S.R. Among the constructed buildings one must mention the world largest metal suspended roof stadium, erected in 1980 for the summer Olympic Games in Moscow, the "Olympijskij" arena of 224m by 183m, the cycle track "Krylatskoje" spanning over an area of 168m by 138m, or the sports palace "Izmailovo", with a diameter of 160m. According to Russian designers the use of membrane roofs leads to lower cost. Such structures have been used for large markets, exhibition halls and sports buildings. As a rule, 12m wide steel sheets are manufactured at the factory and delivered to the site in coils. The membrane's sheets, 5mm thick, are lap welded to each other using a continuous fillet weld. During erection penetration spot welds are used, or high strength bolts to connect the sheets. Practising designers will find the proceedings of a conference on shells, membranes and space frames, held in 1986 in Osaka edited by Koichiro Heki, a real mine of extremely useful information on analysis and design of membranal and cable structures.

(d) Topological investigations in structural configuration of space structures received a great deal of attention within the last decade. Special attention must be referred to the proceedings of an international seminar on structural morphology held in 1992 at the University of Montpellier, edited by René Motro and Ture Wester. Morphology is a study of fundamental properties of configuration, such as combination, connectivity of geometrical three-dimensional forms and generation of complex shapes. The proceedings of this seminar will be of particular value to architects. They contain the results of investigations of D G Emmerich, T Tarnai, H Lalvani, J F Gabriel and R Motro, just to mention the main contributors. It is obvious that these researchers are not advocating a capricious use of arbitrary though aesthetically very pleasing forms merely for the sake of novelty, but they show that three-dimensional structures produce a real wealth of possibility to the architect and it is up to him to take up the challenge.

(e) Some forty years ago, R Buckminster Fuller drew attention to the novel concept of tensegrity structures. The initial concept has been extended slowly over the years by numerous American, Japanese, Spanish and now by French engineers. The last decade, provided very significant advances and extension of the tensegrity concept for various forms of double-layer grids and braced domes, mainly through the work of Geiger, Emmerich, Vilnay, Hanaor and now by Motro. As is well known, a tensegrity structure is a free-standing prestressed cable network in which the tendons (under the action of tensile forces) are prestressed against a discontinuous system of compressive bars. Originally this concept was used for free-standing masts, but recent work shows the exceptional potential of tensegrity systems in light-weight domes. The huge tensegrity domes built by Geiger in Korea for the recent Korean Olympics in Seoul, 1986, diameter 120m, and in the United States, in Florida, diameter 210m, in 1989, highlight the really remarkable economy of this novel form of construction.

Another area of rapid recent development is in the deployable-collapsible articulated grid structures. The original expandable two way double-layer grid structures have been proposed and many constructed by the Spanish architect Emilio Perez Piñero for travelling theatres and public exhibition buildings in Madrid in 1961. Since that time several designers improved gradually the original concept (T R Zeigler, 1974, D L Derus, 1985, R C Clarke, 1984, F Escrig, 1985, S T Krishnapillai, 1986 and Y Rosenfeld and D Logcher, 1987). The publication of these researchers has attracted recently a great deal of attention and has stimulated commercial development of this promising field of deployable-collapsible systems. There are already several commercial firms manufacturing these structures. It is hoped that such systems will become popular as large-size tents, medical shelters and maintenance sheds. There is also a growing interest in retractable roof structures suitable for large stadia, swimming pools, etc. These structures may be designed to cover very large areas. A typical example is the very recently completed Toronto SkyDome though this illustrates rather a heavy structure. Japanese firms concentrate their interest on research into lightweight retractable structures consisting of three-dimensional movable frameworks and covering their structures with membranal fabric material which is light and in most cases translucent. The Ariake Colosseum (built in 1991) is a typical example of a sliding retractable roof (consisting of double-layer braced barrel vaults and spanning 136m.). The Fukoka Dome (completed in 1993) covers a stadium having a clear span of 222m with a retractable roof. It provides seating capacity for 40,000 people. It is interesting to note that despite the enthusiastic predictions in 1970s and 1980s about the future of structural applications of plastics, the promising developments of space structures in glass fibre reinforced polyester stopped after the oil crisis in 1973 due to financial reasons. The sharp increase in the price of polyester (a by-product of oil) prevented the use of this material in three-dimensional structures and present research in this area is unfortunately minimal. Also the futuristic predictions about the structures in outer space have not been realised at present. We still have to wait to see the practical applications of space structures for manned space stations. Though there is enough theoretical evidence proving that the construction of large space geostationary communication platforms, or three-dimensional mobile satellite antennas, is entirely feasible, the cost at this stage is still prohibitive, even for world powers such as the USA.

(f) The best proof of general acceptance of space structures is the ever increasing number of applications of such structures in practice. Many thousands of such structures have been designed and built during the last decade in many countries. One will be very selective if only a few of outstanding recent examples are mentioned. The Toronto SkyDome opened in 1989 in Canada is at present the world's largest retractable entertainment/exhibition centre. The seating capacity of this huge structure is up to 68,000 people. The roof, when closed, creates an enclosed fully air-conditioned stadium. It consists of four separate steel parts, three of which move. The roof, when retracted, exposes 91% of the seats to the open sky. The middle panels are in the form of barrel vaults, and the two end panels are in the form of quarter domes. The covered circular area has a diameter of 203m. The structure used approximately 10,000 tons of structural steel. Several other, equally outstanding, examples of retractable roofs over stadia have been provided by Japanese designers. The already mentioned Fukoka Dome, when completed in 1993 will become the largest retractable steel structure, diameter 222m. It has been designed by Takenaka and Maeda Corporations. This most ingenious design consists of three panels making a fan shape of 125 degrees at the open position. The lower panel over the inner field is fixed and the middle panel and the upper panel are movable on circular guide tracks. Each panel is self-supporting and has a configuration of a triangulated three-way grid of lamella trusses. American designers can be justly proud of the huge space structures over the Anaheim Arena in California (just finished in 1993) covering an area of 408 feet, by 536 feet as well as the giant new Chicago Stadium (now under construction). The most interesting and highly economical examples of Geiger tensegrity domes are provided by the Florida Suncoast Dome (built in 1988), span 210m, as well as the Heerenveen Skating Rink (built in 1986) over-all area of 90m x 200m, or the Köln Arena ellipsoidal in shape, having a length of 160m and a width of 140m. Although steel is the predominant structural material for such applications, one has to remember the famous laminated timber dome of the parallel lamella type, so far the biggest timber dome in the USA, completed in 1983 in Tacoma, Washington, having a clear span of 162m and a rise of 48m. American designers stress that wood is a natural insulator and the timber roof deck fixed to a glue-laminated timber structure provides natural insulation and complete environmental control. Another equally convincing example of a laminated timber dome is the Round Valley Ensphere, completed in 1991, of clear span of 132m. According to the designer the cost of this structure proved to be almost 50% cheaper than in steel. One should also mention the beautiful wood lattice dome constructed for Expo'90 in Osaka. Another Japanese design the IZUMO timber dome has a clear span of 140m and a height of 49m. One of the most elegant space frames is, no doubt, the huge steel tubular structure erected for the Olympic Games 1992 in Barcelona over the Olympic Sports Hall, "Palau d'esports Sant Jordi"; it covers an area of 106m x 128m. It has been designed by Arata Isozaki with Professor Mamoru Kawaguchi, the inventor of the ingenious Pantadome erection technique for space frames, as technical advisor. One of the important features of the Pantadome system is that the central dome during erection has a freedom of movement only in the vertical direction, at the same time being stable in lateral directions. The first application of this technique by Kawaguchi was during the construction of the World Memorial Hall in Kobe. The third recent application of this technique is the recently constructed Singapore National Indoor Stadium covering a rhombic plan of 200m by 120m. There are numerous recent examples of large span double-layer grids, though the most remarkable one is the Javits Centre in New York, covered by a huge steel space frame with a plan area of over 53,000m^2, formally completed in 1986. One must mention also the huge hyperbolic paraboloid tubular steel structure erected in 1988 by MERO for the Islamic University in Riyadh in Saudi Arabia. The clear span of this remarkable structure is 150m with an overhang of 55m. Though the dimensions are smaller, one must refer also to the two fan-shaped hyperbolic paraboloids, each supported by 3 columns and having a three-way double-layer grid configuration, for the International Exposition in Tsukuba, Japan, erected in 1985. The design is by Dr Fujio Matsushita of the Tomoegumi Iron Works Ltd of Tokyo. The distance between the columns is 54m and the total length of each fan is 80m.

CONCLUSIONS

12. The pace of development in the field of three-dimensional structures is clearly and rapidly accelerating. The emphasis of designers has been to reduce the weight of their structures hence the great interest in the light-weight construction.

13. Within the last decade numerous designers have contributed to this stage of development though the main contributions come from a rather limited number of distinguished engineers and architects belonging to the younger generation. One must mention especially the names of: M. Kawaguchi, D. Richter, M. Eekhout, F. Gabriel, H. Lalvani, R. Motro, M. Majowiecki, P. Huybers, T. Tarnai, A.A. Noble, F. Rebielak, G. N. Pavlov, just to mention some who contributed most.

14. At the same time it is sad to note that the space structures fraternity has suffered some terrible losses. Four distinguished men of outstanding abilities and remarkable imagination have left us forever.

(a) Dr Max Mengeringhausen (born in 1903), the original developer of the MERO system, died on 13 April 1988.

(b) Professor Yoshikatsu Tsuboi (born in 1907), the past president of IASS, widely recognised as the outstanding authority and designer of several large span structures, died on 6 December 1990.

(c) Mr Harold Gordon Fentiman (born in 1921) the originator of the TRIODETIC system, died on 14 May 1986.

(d) Professor Jose Mirafuentes Galvan (born in 1934) died on 28 May 1991.

15. In recognition of the contributions to the development of space structures the first three colleagues received the special PIONEER Award during the 3rd International Conference on Space Structures in Guildford at the University of Surrey in September 1984.

16. The contributions to the development of space structures of these four distinguished designers will be remembered for many years to come.

2. Post-buckling behaviour of perfect and randomly imperfect grid shell structures

O. BACCO and C. BORRI, Universita Degli Studi di Firenze, Italy

SYNOPSIS. A new approach to the study of the effects of randomly distributed imperfections on grid-shell structures is presented. The simulation of randomly distributed out-of-scheme imperfections has been performed, introducing a kind of spatial correlation in order to avoid events that even if statistically possible, has to be considered not achievable from a pragmatic point of view. The fields of imperfections take into account the initial stress state of the imperfect structure that may appear as a consequence of the geometric perturbation. Some computational tests have been finally carried out on a sample structure.

INTRODUCTION

1. The lightness of grid-shell structures (Otto et al. ref. 1) and the prevailing compression stress due to peculiar load conditions, in the elements of these structures, requires that an investigation into the *"sensitivity"* to structural imperfections be undertaken, devoting special attention to geometric imperfections, i.e. deviations of the actual geometry of a structure from the theoretical scheme (Bacco, ref. 2).

2. Investigations of the effects of geometrical imperfections on reticulated shell structures have been made in the recent past: they revealed a high sensitivity of this structural typology to the impressed distortions (Borri, ref. 3). The aim of this paper is to analyse the case of grid-shells, introducing some innovations on the way to *"manage"* the imperfections and introducing a given space correlation among the imperfections themselves.

STOCHASTIC IMPERFECTIONS

3. Imperfections may influence the whole static behaviour of imperfect structures, as they may alter them, not only from a geometric point of view, but also from a tensional one, with the appearance of an initial state of internal constraints. So in order to conduct the analyses of sensitivity of grid-shells to imperfections in a correct way, a first basic aspect has to be considered, concerning the way to "manage" the geometrical imperfections.

4. Besides, the random character of the distribution of imperfections on a structure makes a stochastic approach necessary for the analysis of imperfect structures. This casual character, may be simulated modelling the output of a random generator, guided (by gaussian bimodal frequencies histograms) on real values included between the maximum admissible values (ρ and $-\rho$, where ρ is a characteristic parameter of the section of the elements) that can be attained from the imperfections.

5. The gaussian bimodal distribution is to be considered very pessimistic if compared to other distributions such as the constant or the simple gaussian ones. So the analysis based on the hypothesis of a bimodal distribution of the imperfections will lead to very heavy conditions for a structure, which has to be considered as the extreme situation (Borri and Spinelli, ref. 4).

6. The stochastic generation of the perturbations impressed to a perfect structure doesn't avoid those situations, which even if possible from a probabilistic point of view, are not really admissible, as for example, all those cases in which the pattern of imperfections includes two opposite maximum values in two neighbouring nodes. It is difficult to envisage that such a condition can be realised in the constructive reality of the structure. Then, a second relevant aspect to be considered is the introduction of a space correlation between imperfections; correlation that may strongly influence the reliability of results.

7. The field of stochastic, spatially correlated, imperfections can be generated starting from a system of primary perturbations assigned in a restricted area of the structure, chosen in some characteristic zones in which one can imagine subdivision of the perfect structure itself. Under the action of the impressed imperfections, the structure finds a new equilibrated configuration characterised by a geometry altered by zones and by an initial internal stress state.

NON-LINEAR STABILITY ANALYSES AND TRACING OF SECONDARY EQUILIBRIUM PATHS.

8. The linear stability analyses give, as is well known, only qualitative information over the presence and the nature of critical states of equilibrium. In order to establish the *"exact"* value of the critical threshold, it is necessary to remove the hypothesis of negligible initial displacements, on which the linear theory is based, and to carry out a non-linear stability analysis. The condition of indifferent equilibrium connected to a critical state is expressed by the eigenvalue problem:

$$\det\left([K_E] + \lambda_{cr}[K_{Go}]\right) = 0 \qquad (1)$$

where K_E and K_{Go} represent respectively the linear elastic stiffness matrix and the geometrical one. Removing the hypothesis of negligible initial displacements and taking into account the non-linear terms of stiffness, the eigenvalue problem becomes:

$$\det([K_E] + \bar{\lambda}_{cr}[K_{GLo} + K_{UL}] + \bar{\lambda}_{cr}^2 [K_{GNLo} + K_{UNL}]) = 0 \tag{2}$$

where K_{UL} and K_{UNL} are respectively the linear and non-linear terms of the stiffness due to initial displacements; the quadratic problem represented by Eq. 2, may be solved with a step-by-step linearisation. This process will end when the value $\bar{\lambda}_{cr} = 1$ is achieved, as for that value the solution would coincide with the exact one (Eckstein, ref. 5; Borri and Spinelli, ref.4; Borri and Chiostrini, ref. 6).

9. In the presence of bifurcation points it may be suitable to find eventual secondary decreasing equilibrium paths. One of the commonly used methods of bifurcation post-buckling behaviours, is the perturbation method.

10. The perturbation, in terms of displacements, may be expressed as suggested by the classical post-buckling theory:

$$U_s = U_c + \varepsilon U_a \tag{3}$$

where U_s is the displacement vector in the new state, U_c is the vector of nodal displacements in the critical state, U_a is the normalised eigenvector that is associated to the critical eigenvalue and finally ε is a factor chosen in order that the product εU_a is little if compared with the vector of the effective displacements of the bifurcation critical state.

AN APPLICATION

11. During this research work a grid-shell test structure, with an octagonal base of 60m span and 16m height (Fig. 1), has been investigated. In order to determine the expected value of the variation of the ultimate bearing load factor (U.B.L.F.) of the test structure due to the presence of imperfections, analysis on the perfect structure and on that affected by stochastic geometrical imperfections has been carried out.

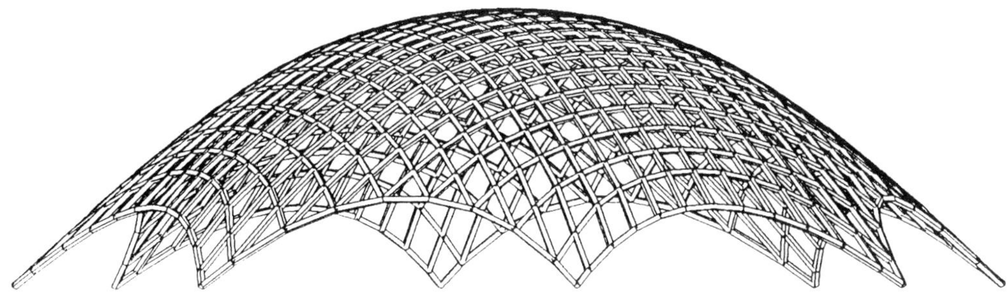

Fig. 1. Grid-shell structure with octagonal base.

12. For the execution of the analyses the iterative methods of Newton-Raphson and Riks-Wempner-Wessels, supported by incremental procedures, have been used. In order to study the sensitivity to geometrical imperfections of the test grid-shell structure, this has been subjected to different fields of perturbations and the eventual variation of the "critical bearing load" factor of the structure has been estimated, i.e. the ultimate load factor for which the displacements corresponding to small load increments, are not admissible any more.

13. The structure has been affected by two different kinds of imperfections, so that it has been possible to verify the influence of the initial stress state induced by geometrical imperfections:

- -type 0: perturbation of the geometry only, by means of perturbating the three co-ordinates of the nodes of the structure;
- -type 1: geometrical perturbations of the three local co-ordinates of the nodes, considered as actual impressed displacements, which produce a variation of the geometry and an internal stress state in the structure.

14. Eight tests have been carried out for each kind of imperfection, i.e. eight fields of perturbations of the co-ordinates have been generated and applied to the perfect structure. A static non-linear analysis of the imperfect structure has been considered sufficient to trace the primary equilibrium path and to point out the eventual variation of the bearing load factor, as for this specific case the variation of the values of the critical loads are not significant.

15. Imperfections type 1 have been spatially correlated and applied in a restricted area, i.e. 32 nodes chosen in some zones delimited by the symmetry axes of the structure. From the non-linear static analyses of the structures affected by the two systems of imperfections, the load displacement curves for some degrees of freedom have been computed, for example the vertical displacement of a node belonging to the border (node 692) and of an internal node of the shell (node 306) (Figs 2 and 3).

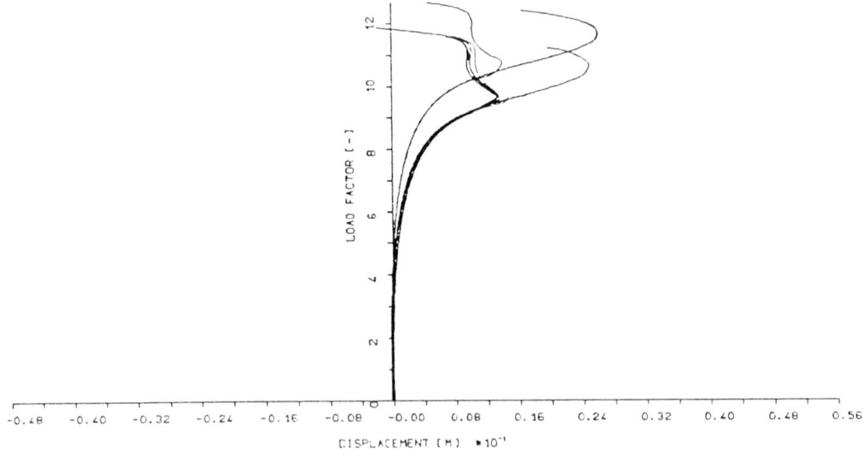

Fig. 2. Load-displacement curve of the vertical d.o.f. of node 692.

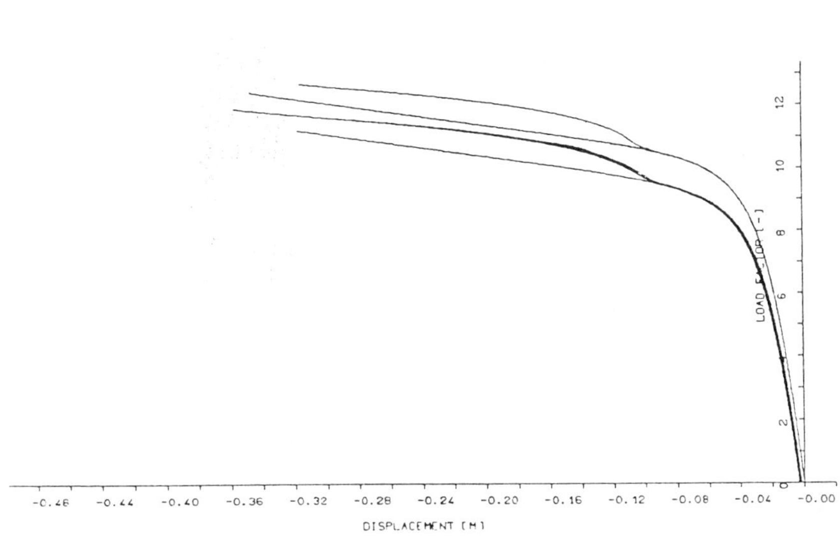

Fig. 3. Load-displacement curve of vertical d.o.f. of node 306.

16. The structures affected by the two kinds of imperfections behave in a really different way, though the value of the maximum imperfection has been maintained as constant for both typologies; the equilibrium paths of the eight structures affected by the fields of type 0, are very close to the principal one and to the secondary branch of the "perfect" structure, as shown in Table 1 in which the values of the percentage variation of the bearing load factor corresponding to three of the analysed node, are reported.

Table 1. The percentage variation of the bearing load factor - Group I.

Group I (Type of imperfections 0)								
Test ***Node***	1	2	3	4	5	6	7	8
706	+1.0	+1.9	+0.9	-0.2	-1.7	+1.7	-2.2	0.0
306	-1.2	+0.8	+2.2	-2.8	+0.8	+1.0	-4.3	0.0
692	+3.4	0.0	+3.2	+5.4	-3.5	-5.9	+4.4	+3.9

17. With reference to the results concerning the structures affected by imperfections of type 1, a reduction of the load factor has been recorded and it attains, in some cases, a value of 12 % (Table 2).

18. The frequency histograms of the percentage variation of the U.B.L.F., referred to a field of random imperfections of types 1 and 0 on the octagonal shell, are shown in Fig. 4.

Table 2. Percentage variation of the bearing load factor - Group II.

Group II (Type of imperfections 1)								
Test ***Node***	1	2	3	4	5	6	7	8
706	-9.8	-6.0	+1.4	-9.45	-10.2	-10.5	-12.9	-10.9
306	-12.1	-12.4	-6.0	-12.5	-12.	-11.8	-12.13	-12.8
692	-10.3	-11.1	-6.0	-12.7	-12.2	-11.2	-12.09	-12.0

CONCLUDING REMARKS

19. The results of the repeated tests show that the effects of the impressed imperfections on the considered grid-shell structure are not very relevant when the imperfections alter only the geometry of the structure, at least until the maximum value of imperfections does not exceed the values chosen for the tests, which are acceptable from a technological point of view.

20. The factor which influences the variation of the U.B.L.F., of the structure, is the initial stress state induced by the imperfections themselves.

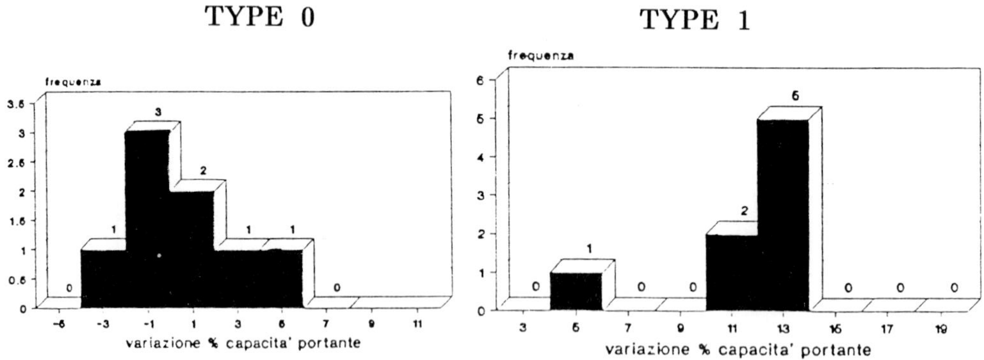

Fig. 4. Frequency histograms of the U.B.L.F. percentage variation referred to node 306.

21. Useful and relevant information can also be obtained, as a concluding aspect, from the comparison with the results obtained by previous investigations on reticulated shells, i.e. another kind of single-layer light-weight structures.

22. Considering the analyses carried out by Borri (ref. 3), and Borri, Spinelli (ref. 4) on the sensibility of such shells to stochastic out-of-scheme imperfections, the critical load reduction due to stochastic imperfections with bimodal gaussian distribution achieves an expected value of 615. It therefor becomes evident that reticulated shells show a higher degree of sensitivity to geometric imperfections if compared with grid-shells, given that the maximum value of the imperfection parameter has been assumed with the same criterion.

REFERENCES

1. OTTO F. Ed., IL-10, Gitterschalen-Gridshells, Institut für leichte Flächentragwerke Univ. Stuttgart, 1974.
2. BACCO O., Analisi non lineare di strutture grigliate a guscio, Tesi di Laurea, Università di Firenze, Facoltà di Ingegneria, 1992.
3. BORRI C., Sul comportamento post-critico di gusci reticolari monostrato in presenza di imperfezioni, Meccanica Teorica ed Applicata, Atti dell' VIII Congresso A.I.ME.T.A., Torino, Italia 29/9-3/10, 1986.
4. BORRI C., SPINELLI P., Buckling and post-buckling behaviour of single layer reticulated shells affected by random imperfections, Computers & Structures, 30, 4, 937-943, 1988.
5. ECKSTEIN U., Nichtlineare Stäbiltatsberechnung elastischer Schalentragwerke, Techn. Rept. 83-3, Inst. für konstr. Ingenieurbau, Ruhr-Univ., Bochum, 1983
6. BORRI C., CHIOSTRINI S., A non-linear approach to the stability analysis of space beam structures, Int. J. of Space Structures, 4, 193-217, 1989.

ACKNOWLEDGEMENTS

The Authors thank the financial contribution of C.N.R. (Progetto di ricerca internazionale bilaterale, Contr. n. 91.02762.CT07); moreover thanks are due to Dott. Ing. W. Zahlten and Dott. Ing. F. Crocchini for the precious help given in the computational phase of the research.

3. Collapse studies of the effects of eccentricity in compression Tee-section chords in double-layer space trusses

J. A. MWAKALI, Makerere University, Uganda

SYNOPSIS. Results of a recent investigation (ref. 1) on eccentricity effects on the behaviour of mild steel tee-section compression members loaded into the plastic region are presented. Tee-sections were also used to study the collapse behaviour of square-on-square double-layer space trusses with the aim of establishing the amount of ductility that results from using eccentric, in lieu of concentric, top-layer chord members. The study covered both theoretical and experimental work.

INTRODUCTION

1. The design of space trusses is still limited to the linear elastic range of behaviour using the "allowable stress" method. In contrast "limit state" design has virtually superseded the allowable stress design method in more conventional structural forms such as steel and reinforced concrete frames. The main reason for this transition has been the desire to adopt a more rational explanation of the behaviour and a more efficient use, of structural steel. There is growing need to develop comparable methods for the design of space trusses. "Collapse" methods have been suggested in which the structure is loaded beyond the elastic capacity of the members. It is assumed that the failure of one or more members may still leave the structure able to perform its intended use.

2. The main drawback to designs based on collapse analysis lies in the fact that the constituent members of space trusses are ordinarily centrally loaded and their failure in compression is sudden, brittle and dynamic when they fall in the economic range of slenderness ratio. The failure of even one such member could precipitate the collapse of the entire structure. Hence, the benefits that may otherwise be derived from a collapse design are not realised. To overcome this problem most investigators suggest special design details to reduce the brittleness of the critical members of the structure. Among these designs is the introduction of eccentricity in the members. The use of eccentricity for brittleness reduction is based on the fact that although the ultimate strength of a member under axial compression is significantly reduced if the member is at the same time subjected to bending moments, a much smoother characteristic is displayed at the maximum

load and the post buckling drop-off of load capacity is less severe than that exhibited by a member subjected solely to concentric loading.

3. Most collapse methods assume that the relationship prior to the attainment of maximum load is linear, and that the post-buckling behaviour controls the analysis. The first step in the analysis requires determining a mathematical formulation of the full loading behaviour of the members.

BEHAVIOUR OF COMPRESSION MEMBERS

4. A space truss is considered as a collection of finite two-noded straight members joined together at their ends with "pins". A pin-jointed member can sustain a single type of loading only, namely, a uniform axial force which may be tensile or compressive. The behaviour of steel members in tension is a property of the material and is well understood and fairly predictable. Compression element behaviour, on the other hand, not only is a property of the material and geometry of the member but is also greatly influenced by "imperfections" that may be present in the member such as residual stresses, crookedness, eccentricity and end-restraints. Many formulae have been devised over the years to incorporate these phenomena but most compression member designs are based on the Perry Robertson formula.

5. Although eccentricity in compression members is usually small, accidental and undesirable, there are cases when the eccentricity may be introduced intentionally. In this paper one beneficial application of eccentric compression members in space trusses is discussed. The load deformation relationships for an eccentrically loaded member can be derived using a procedure suggested by Jezek (ref. 2). The material is assumed to be ideal elastic-plastic. Consider a pin-ended straight column, length l, of rectangular cross-section which is axially loaded by the force P at an eccentricity of e (Fig. 1).

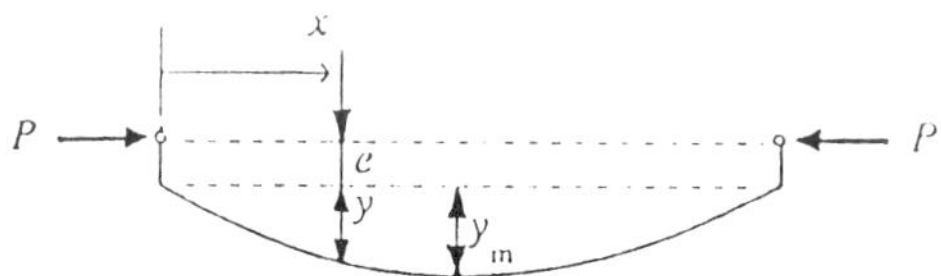

Fig. 1. Eccentrically loaded compression member.

At a cross-section x the column is subjected to a lateral deflection y and total bending moment M given by:

$$M = P(y+e) = Py+Pe \qquad (1)$$

Where Py is the moment due to flexure of the column and Pe is the moment due to the applied eccentricity. If it is assumed that the axis of the deflected column is a half-wave of a sine curve, then:

$$y = y_m \sin(\pi x/l) \tag{2}$$

where y_m is the maximum deflection at the centre of the column. Curvature is defined as the reciprocal of the radius of curvature f that is:

$$\frac{1}{f} = \frac{-d^2y/dx^2}{(1+(dy/dx)^2)^{3/2}} \tag{3}$$

For small deflections $(dy/dx)^2$ is very small compared with unity. Hence, the curvature l/f_m at the column mid-height is given approximately, by:

$$\frac{1}{f_m} = -d^2y/dx^2\Big|_{x=\ell/2} = (\pi^2/\ell^2)y_m \tag{4}$$

The expression for curvature can also be derived from the distribution of stress at the column mid-height, and by equating the expression, so obtained with Eq. 4, a relation is obtained between y_m and σ_0, the average axial stress. The distribution of stress depends on the degree of deformation of the column. Fig. 2 shows stages through which the column passes as it is bent from the straight position in (a) until the cross-section at the column mid-height is fully plastic in (e). In general, therefore, at mid-section two different distributions of stress are possible, as illustrated in Fig. 3.

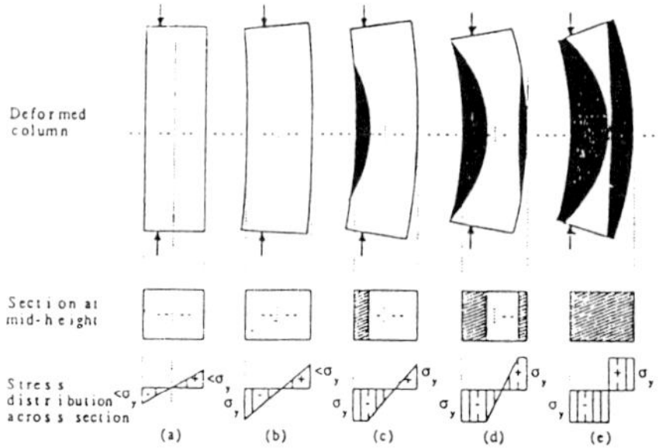

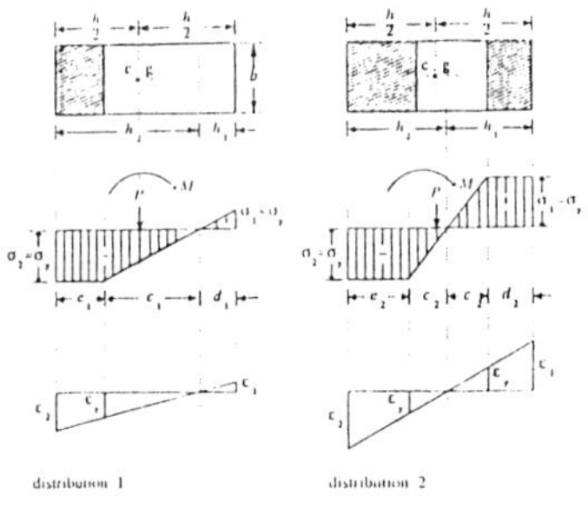

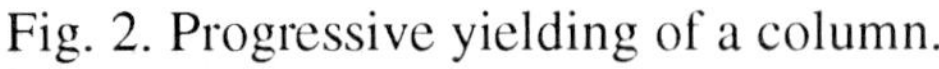

Fig. 2. Progressive yielding of a column.

Fig. 3. Stress and strain in a section.

6. Stress distribution 1. In this case the yield point is reached on the concave side of the bent column, while the tensile stresses on the convex side still remain

in the elastic range. The conditions of equilibrium between the internal and external forces at the cross-section at mid-height of the column are:

$$P = \int_A \sigma dA = b\int_{h_1}^{h_2} \sigma dz \tag{5}$$

and

$$M = M_m = \int_A \sigma z dA = b\int_{h_1}^{h_2} \sigma z dz \tag{6}$$

Where z is the fibre distance from the σ_o-axis, σ is the variable stress across the section and M_m is the total bending moment at mid-height. Eqs 5 and 6 are expanded to give, respectively:

$$P = b_1 e_1 \sigma_y + \tfrac{1}{2} b c_1 \sigma_y - \tfrac{1}{2} b d_1 \sigma_1 \tag{7}$$

$$M_m = b\sigma_y\left[e_1(h/2 - e_1/2) + \frac{(h/2 - e_1)^2 (h/2 - d_1)}{2c_1} + \frac{(h/2 - e_1)}{3c_1} + \frac{(h/2 - d_1)}{6c_1}\right]$$

$$+\, b d_1 (h/2 - d_1/3)\sigma_1 \tag{8}$$

From Fig. 3 linear proportioning gives $\sigma_1 = (d_1/c_1)\sigma_y$. Hence Eqs 7 and 8 become:

$$P = b\sigma_y(e_1 + c_1/2 - d_1^2/2c_1) \tag{9}$$

$$M_m = b\sigma_y/2\Big[e_1(h - e_1) + (h/2 - e_1)^2(5h/6 - d_1 - 2e_1/3) - \frac{(h/2 - d_1)^3}{3c_1} + \frac{d_1^2(h/2 - d_1/3)}{c_1}\Big] \tag{10}$$

Noting that $e_1+c_1+d_1=h$, d_1 may be eliminated from Eqs 9 and 10 to give:

$$P = b\sigma_y\left[e_1 + c_1/2 - \frac{(h - e_1 - c_1)^2}{2c_1}\right] \tag{11}$$

$$M_m = b\sigma_y[e_1(h-e_1)+(\frac{h}{2}-e_1)^2(\frac{-h}{6}+\frac{e_1}{3}+c_1)-\frac{(-h/2+e_1-c_1)^3}{3c_1} + \frac{(h-e_1-c_1)^2(h/2-e_1+c_1)}{3c_1}] \quad (12)$$

Introducing $\sigma_0 = P/A = P/bh$ in Eq. 11 gives:

$$\frac{\sigma_0 h}{\sigma_y} = \frac{2e_1c_1 + c_1^2 - (h-e_1-c_1)^2}{2c_1} \quad (13)$$

Solving for e_1 and ignoring the positive square root (since e_1 cannot be greater than h) yields:

$$e_1 = h - \sqrt{2hc_1(1-\sigma_0/\sigma_y)} \quad (14)$$

Substituting back into Eq. 12 and rearranging gives:

$$c_1/h = \frac{9(\sigma_y - \sigma_0 - 2M_m/bh^2)^2\sigma_y}{8(\sigma_y - \sigma_0)^3} \quad (15)$$

From Fig. 3, $(\varepsilon_2 - \varepsilon_1)/h = 1/f_m$ Since $\varepsilon_y = \sigma_y/E$, it follows that:

$$1/f_m = \sigma_y/Ec_1 \quad (16)$$

Eliminating c_1 between Eqs 15 and 16 gives the expression for the curvature at the column mid-height as:

$$\frac{1}{f_m} = \frac{2\sigma_o(\sigma_y/\sigma_o - 1)^3 h}{9E[h/2(\sigma_y/\sigma_o - 1) - M_m/P]^2} \quad (17)$$

If M_m is replaced by (Pe+Py) in the above expression and the resulting expression equated with Eq. 4, the following relationship between mid-column lateral deflection y_m σ_o and e is generated:

$$9Ey_m[h/2(\sigma_y/\sigma_0 - 1) - y_m - e]^2 - 2\sigma_0 h(\sigma_y/\sigma_0 - 1)^3 l^2/\pi^2 = 0 \quad (18)$$

The critical buckling stress σ_{cr}, is the maximum value of σ_0 defined by $d\sigma_0/dy_m=0$. Differentiating and solving for the critical values of y_m and σ_0 and substituting $r^2 = I/A = h^2/12$ gives:

$y_{cr} = 1/3(h/2(\sigma_y/\sigma_0 - 1) - e)$

$$\sigma_{cr} = \frac{\pi^2 E}{(1/r)^2}\left[\frac{\sigma_y/\sigma_{cr} - 1 - 2e/h}{\sigma_y/\sigma_{cr} - 1}\right]^3 \tag{19}$$

If h is replaced by the core radius $e_r = h/6$ for a rectangle, Eq. 19 may be applied to other sections by being rewritten as:

$$\sigma_{cr} = \frac{\pi^2 E}{(1/r)^2}\left[\frac{3(\sigma_y/\sigma_{cr} - 1 - k}{3(\sigma_y/\sigma_{cr} - 1}\right]^3 \quad \text{for} \quad (1/r)^2 - \frac{\pi^2 E k^3}{9\sigma_y(3-k)} > 0 \tag{20}$$

where the eccentricity ratio $k = e/e_r$

7. Stress Distribution 2. In this case the yield stress is reached at both sides of the cross-section (Fig. 3). Using the same procedure as for stress distribution 1, the following y_m-σ_0 relationship is derived:

$$y_m\sqrt{\frac{h\sigma_y}{4\sigma_0}(1-\sigma_0^2/\sigma_y^2) - y_m - e} - \frac{1^2}{\pi}\sqrt{\frac{\sigma_y^3}{3hE^2\sigma_o}} = 0 \tag{21}$$

The critical values of y_m and σ_0 are given by:

$$y_{cr} = \frac{h\sigma_y}{6\sigma_0}(1-\sigma_0^2/\sigma_y^2) - \frac{2e}{3} \tag{22}$$

$$\sigma_y = \frac{\pi^2 E}{(1/r)^2}\sqrt{\frac{\sigma_{cr}}{\sigma_y}(\frac{\sigma_y}{\sigma_{cr}} - \frac{\sigma_{cr}}{\sigma_y} - \frac{4e}{h})^3} \tag{23}$$

or, in general,

$$\sigma_y = \pi^2 E/(1/r)^2\sqrt{\sigma_{cr}/\sigma_y(\sigma_y/\sigma_{cr} - \sigma_{cr}/\sigma_y - \frac{2}{3}k)^3}$$
$$\text{for} \quad (1/r)^2 - \pi^2 E k^3/\left[9\sigma_y(3-k)\right] < 0 \tag{24}$$

When the eccentricities are large $k \geq 3$, stress distribution 2 and Eqs 21 to 24 apply. Eqs 18 and 21 may also be written in terms of the axial deformation δ instead of the central lateral deformation y_m by using the relationship:

$$\delta = Pl / AE + \pi^2 y_m^2 / 4l \tag{25}$$

Fig. 4 shows the P-δ curves for various eccentricities of the tee-section whose geometrical and mechanical properties are as indicated. The curves show that:

(i) the maximum load supported by the column decreases with increasing eccentricity;

(ii) the non-linearity of the P-δ curve near the maximum load increases with increasing eccentricity;

(iii) the post-critical portion of the P-δ curves for the centrally loaded ($e = 0$) member lies below those for eccentricities of 5, 10, 15 and 20mm.

This is because the centrally loaded column is presumed to fail by buckling about the weak (y-y) axis whereas the eccentrically loaded columns fail about the strong (x-x) axis.

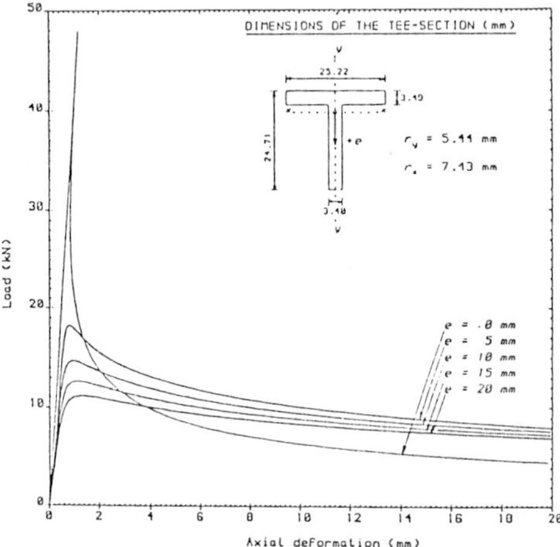

Fig. 4. Load-axial deformation curves for an eccentric column 800mm long.

Experimental work

8. Tests were undertaken on pin-ended compression tee-sections loaded at different eccentricities. The eccentricity was measured in the plane of the weld, that is the symmetry axis. Fig. 5 shows results of tests for two different compression member lengths. A comparison of the experimental and theoretical curves reveals reasonable agreement in their shapes. Experiments showed that the buckling failure of the centrally loaded columns about the minor y-y axis occurred suddenly and dynamically, with a loud "bang", with a big fall in the load. The

eccentrically loaded columns in contrast failed more gently about the strong x-x axis. The brittle failure of concentrically loaded columns at slenderness ratio greater than the transition slenderness ratio is a common phenomenon.

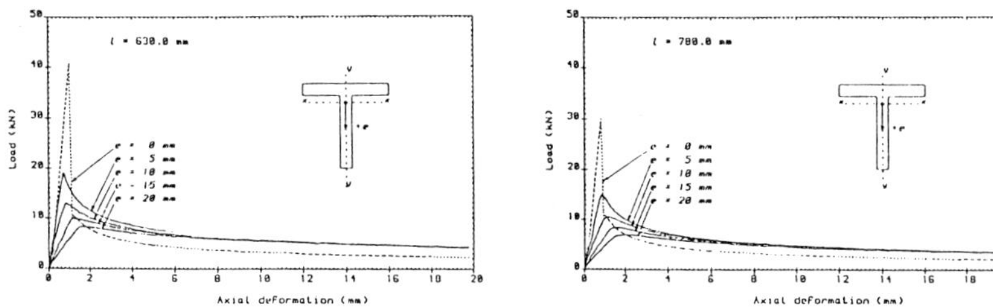

Fig. 5. Results of compression tests on eccentrically loaded tee-sections.

COLLAPSE BEHAVIOUR OF DOUBLE LAYER SPACE TRUSSES

9. Having established that eccentrically loaded column members failed less brittley than concentrically loaded ones, their effects on the overall performance of a space truss were investigated. The theoretical investigations were carried out using a FORTRAN 77 computer program based on the stiffness method of structural analysis (ref. 3). Since large deformations were involved and the loading was taken beyond the linear elastic region, the behaviour of the structure was non-linear. The solution of the problem was simplified by approximating the load-displacement curves of the members by a series of straight lines. The solution involved developing incremental equations of static compatibility in discrete form, whereby the reference configuration was continuously updated. The solution was obtained by considering successive collapse cycles, each cycle being treated as a small-displacement linear problem. The element force resultants were computed on the basis of the geometry at the beginning of the cycle. The geometry was then adjusted at the end of the cycle. Fig. 6 shows the shape of the idealised load-deformation relationship used in the computer analysis and how the various line segments or "stiffness states" are inter-related. The derivation of the lines is described in (ref. 1).

10. Fig. 7 shows the two configurations of the double-layer space trusses analysed for eccentricities of 0, 5, 10, 15 and 20 mm in the top-layer chords. Fig.8 shows the load-central node deflection relationships obtained in the analysis. These relationships show that the eccentricity has a profound effect not only on the ultimate load but also on the collapse pattern of the structure. The point of initial collapse drops drastically when an eccentricity is introduced in the top-layer members, but for equal increments of eccentricity the drop becomes less pronounced. The collapse patterns for the type 1 trusses also reveal that: (i) the failure of some of the members does not necessarily lead to the complete collapse of the structure; the structure may carry more load above that which caused first collapse; (ii) the introduction of an eccentricity results in a more gradual attainment of the ultimate load of the structure, and a more gradual falling off of

load in the post-ultimate region. Because the type 2 truss is not very redundant, the failure of the top-layer chords leaves the structure with very little stiffness, resulting in very large deformations. Hence the theoretical post-ultimate characteristics do not offer much useful information.

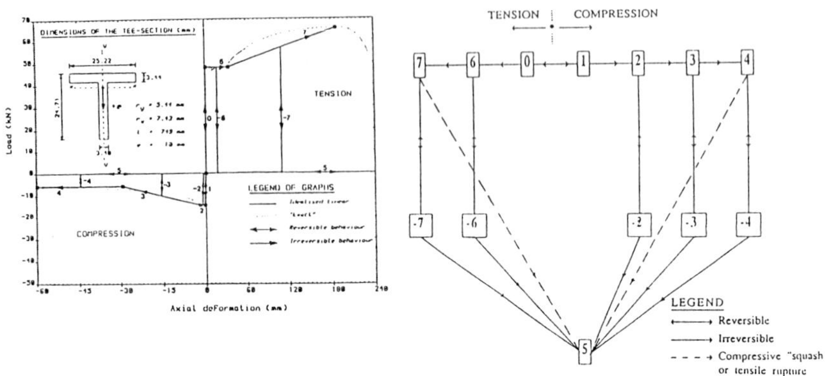

Fig.6. Idealised load-deformation relationship and its stiffness states.

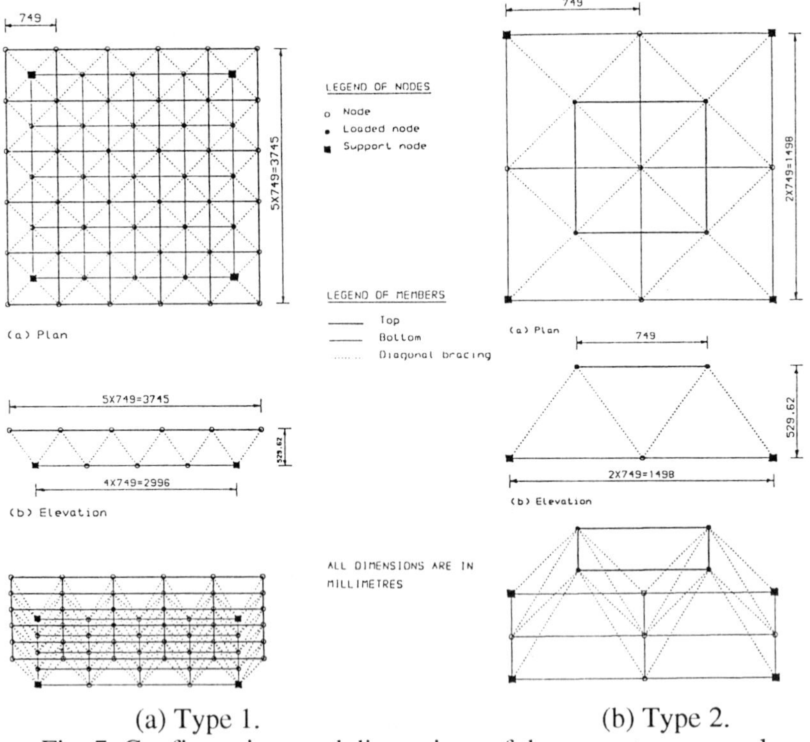

(a) Type 1. (b) Type 2.

Fig. 7. Configurations and dimensions of the space trusses analysed.

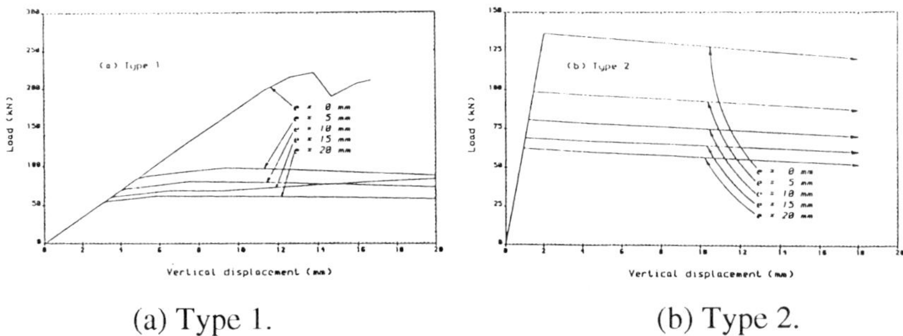

(a) Type 1. (b) Type 2.

Fig. 8. Load deflection relationships for the trusses analysed.

11. In the experiments on space truss models, two types of structural configuration were used: Type 1 and Type 2 with dimensions, loading and boundary conditions as shown in Fig. 7. One Type 1 and six Type 2 models were tested under displacement-controlled loading. The trusses were constructed from tee-sections of the batch used in compression tests except that to preclude premature failure of the four highly stressed corner diagonal members mild steel one-inch-diameter round bars were used in these positions. All joints in the Type 1 model were formed by welding the members on specially designed Triodetic-type nodes, with all the top-layer members welded at an eccentricity of 5mm and the rest concentric with respect to the nodal points. The top-layer members, however, performed as would be expected for concentrically rather than eccentrically loaded members. Two theories were suggested to explain this discrepancy: (i) The top-layer members were probably not subjected to an eccentricity because of possible undercutting and weld defects, causing distortion in the alignment of the lines of action of the member forces; (ii) While the theory of eccentric buckling discussed in paragraphs 5-8 dealt with "pinned" columns, in actual structures the compression members usually are framed into the adjacent members of the structure and may be regarded as columns having restrained ends. Some studies have shown that continuity, joint detail and adjoining members reduce the effect of the movement of eccentricity upon the buckling load of the eccentrically loaded members (refs. 4).

12. In view of the unsatisfactory performance of welded eccentrically loaded members in the Type 1 truss model, in the subsequent Type 2 models eccentricity in the top-layer members was achieved using bolted joints on a specially designed node. Fig. 9 shows some of the results obtained from the tests. The results show that increasing the eccentricity in the top-chord members resulted in a fall in the maximum load supported by the truss but the collapse failure of structure became less sudden and less steep. The graphs also show that on unloading the structures, the recovery lines were practically parallel to the pre-ultimate loading lines, hence confirming the assumptions made in the computer analysis. The residual unrecoverable plastic deformation decreased with increasing eccentricity showing that the shock in the structure at initial collapse reduced when eccentricity was introduced in the top-chord members.

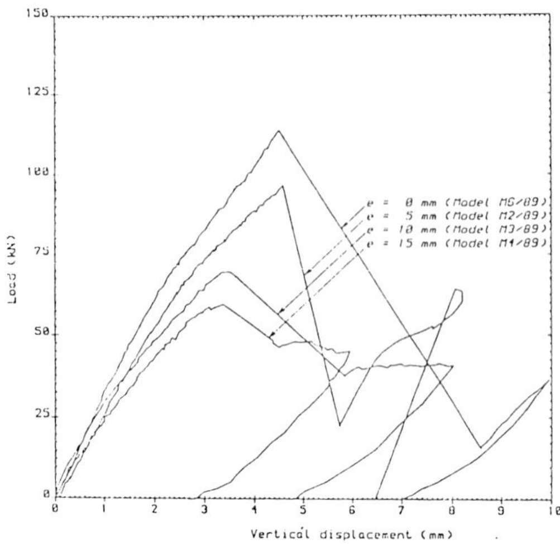

Fig. 9. Load-displacement relationships for four of the Type 2 models.

CONCLUSION

13. It has been demonstrated in the studies described in this paper that eccentricity of compression chords in a space truss can be used to regulate its failure characteristics. Eccentricity in a compression member affects its behaviour by introducing non-linearity and a gentler collapse mode, albeit at the expense of the maximum load supported by the member. The load capacity of many space trusses is usually controlled by compression members. The failure of such compression members, when concentrically loaded, is usually brittle and the failure of only a few members can precipitate a chain reaction in the entire structure. Introduction of eccentricity in some carefully selected compression members reduces brittleness in the structure. The increased ductility of the structure is a valuable property when, for instance, the safety of the occupants of the building subject to a sudden shock such as an earthquake is considered. It is possible to determine, by minimisation-maximisation linear programming techniques, optimum values of eccentricity and the best compression members in which to apply the eccentrities for any given configuration of space truss. This would ensure reduction of brittleness is achieved without much loss of maximum load. This paper has not covered this area. The studies showed that the benefits of eccentricity in compression members are best realised in highly redundant structures where the failure of some members will generally not lead to complete loss of the integrity of the structure.

REFERENCES

1. MWAKALI J. A. The collapse behaviour of double-layer space trusses incorporating eccentrically loaded tee-section members. PhD Thesis, University of Surrey, November, 1990.
2. JEZEK K. Die Festigkeit von Druckstaben aus Stahl. Julius Springer, Vienna, 1937.
3. LIVESLEY R. K. Matrix methods of structural analysis. Second edition, Pergamon Press Ltd, Oxford, 1975.

4. SCHMIDT L. C., MORGAN P. R. and COULTHARD B. R. The influence of eccentricity and continuity on the inelastic behaviour of a space truss. Proceedings of the Sixth Australasian Conference on the mechanics of structures and materials. University of Canterbury, Christ Church, New Zealand, August 22-24, 1977, 274-281.

4. Path-switching in bifurcation analysis by using line search and generalized inverse

K. K. CHOONG and Y. HANGAI, University of Tokyo, Japan

SYNOPSIS. A path-switching strategy obtained by combining the use of generalised inverse and line search is proposed. In this strategy, a reliable predictor for the bifurcation mode at a bifurcation point, is first computed by using the generalised inverse approach. A search in the direction of maximum potential gradient at the point of intersection between the above predictor for the switching direction and a constant loading plane located in the vicinity of the bifurcation point is initiated in order to obtain an improved approximation for a point on the bifurcation path.

INTRODUCTION

1. Path-switching is used to refer to the process of bringing a point lying initially on a primary path of an engineering structure on to the bifurcation path. It is necessary when one wishes to investigate the structural behaviour after bifurcation buckling. Assuming that the position of the bifurcation point lying on the equilibrium path has been determined, then the associated bifurcation path could be traced as long as the direction where the path is extending is known. The evaluation of this so-called branching-direction is one of the core issues in any attempt for path switching. Another major difficulty encountered is the task of deciding how far away from the bifurcation point should one proceed along the branching-direction in order to obtain a reasonably good approximation for an actual point lying on the bifurcation path (Fig. 1).

2. Fujii and Choong (refs 1,2) have proposed a line search process for locating an initial approximation for a point on the bifurcation path. Provided that the search direction is appropriately defined, this line search approach will result in a point lying in the vicinity of the bifurcation path being detected. On the other hand, by using a generalised inverse approach and the existence condition of a solution at the bifurcation point, Hangai et. al. (refs 3-5) have shown that a reliable predictor for path switching direction could be obtained in a computationally neat way.

3. In this paper, the possibility of combining both the generalised inverse and line search approach to form an alternative path-switching strategy has been studied. In this strategy, a reliable path-switching direction will first be computed

by using the generalised inverse approach. A search in the direction of the maximum potential gradient at the intersection point of the above path-switching direction with a constant loading plane, introduced in the vicinity of the bifurcation point is then carried out. By doing this, an improved approximation for a point on the bifurcation path could be obtained. In the following sections, after a brief review of the available path-switching methods, the proposed strategy will be described. This will then be followed by the analysis results obtained for two space truss models and a conclusion. In this study, only the bifurcation problem of geometrically non-linear elastic structures subjected to conservative proportional loading will be treated.

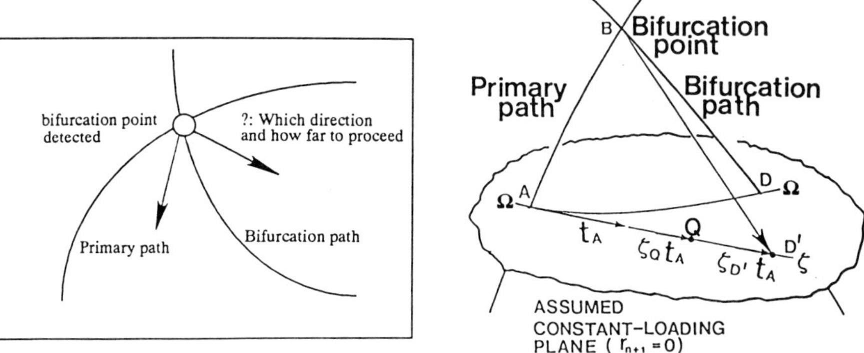

Fig. 1 Switching from primary to bifurcation path. Fig. 2 Concept of line search.

PATH-SWITCHING IN BIFURCATION ANALYSIS

General review

4. An initial approximation for a point lying on the bifurcation path could be obtained by adding a suitably chosen approximate bifurcation mode with appropriate magnitude to the detected bifurcation point as follows,

$$\dot{\mathbf{x}}_{II}=(\alpha_1\dot{\mathbf{x}}_I+\alpha_2\mathbf{b})/\|\alpha_1\dot{\mathbf{x}}_I+\alpha_2\mathbf{b}\|,\quad \dot{\mathbf{x}}_{II}=\{\dot{\mathbf{v}}_{II},\dot{\lambda}_{II}\}^T \tag{1}$$

$$\mathbf{X}_{II}=\mathbf{X}_B+\Delta\eta\dot{\mathbf{x}}_{II} \tag{2}$$

where $\dot{X}_{II}$ = linear approximation for branching direction, $\dot{X}_I$ = tangent vector to the primary path, b = approximate bifurcation mode, α_1, α_2 = magnitude of $\dot{X}_I$ and b respectively, X_{II} = initial approximation for a point on the bifurcation path, $\Delta\eta$ = magnitude of $\dot{X}_{II}$ and superscript T = transposition. The eigenvector corresponding to the smallest eigenvalue is used in most of the methods proposed in order to construct the approximate branching direction $\dot{X}_{II}$ (refs 6-9). In refs 1 & 2, alternative strategies whereby no eigenproblem is involved have been proposed. Methods where second order derivatives of equilibrium equations are used in the determination of $\dot{X}_{II}$ (refs 3-5, 12, 13) are few compared with those that make use of only linear incremental stiffness equations together with additional constraints (refs 1, 2, 10, 14). As for the determination of $\Delta\eta$, a trial-and-error method is the most commonly used method. Methods whereby the magnitude of the branching direction could be computed directly have been proposed in (refs 1, 2, 10). For attaining a point on the bifurcation path from the initial approximation, an iterative process with various special constraints introduced for the purpose of ensuring convergence onto a path other than the already known primary path, is often used (refs 6, 8, 10, 14, 17).

Path-switching strategy using generalised inverse (refs 3-5)

5. The set of equilibrium equations for a discretised structural system with N degrees of freedom could be generally written as:

$$\mathbf{r}(\mathbf{v},\lambda)=\mathbf{f}(\mathbf{v})-\lambda\mathbf{p}_o=\mathbf{0} \tag{3}$$

where f = internal resisting force, v = displacement vector, λ = loading parameter and $\mathbf{p}_o$ generalised loading pattern. Eq. 3 could be repeatedly differentiated with respect to a path parameter yielding the following sets of equations:

$$r_{i,j}\dot{v}_j+r_{i,\lambda}\dot{\lambda}=0 \tag{4}$$

$$r_{i,j}\ddot{v}_j+r_{i,\lambda}\ddot{\lambda}+r_{i,jk}\dot{v}_j\dot{v}_k+2r_{i,j\lambda}\dot{v}_j\dot{\lambda}+r_{i,\lambda\lambda}\dot{\lambda}^2=0 \tag{5}$$

where ( ), j, ( ), λ = partial differentiation with respect to v_j and λ respectively, and $(\dot{\ })$ = d( )/dη. By using matrix notation, Eq. 4 can be rewritten as:

$$\mathbf{K}\dot{\mathbf{v}}-\dot{\lambda}\mathbf{p}_o=\mathbf{0} \tag{6}$$

where $\mathbf{K}$ = NxN tangent stiffness matrix($K_{ij} = r_{ij}$) and $\mathbf{p}_o$ = -$\mathbf{r}$, λ. Hereafter, the summation convention will be assumed for repeated indices and they will vary from 1 to N. Making use of the generalised inverse and existence condition for the solution of Eq. 6;

$$[\mathbf{I}-\mathbf{K}\,\mathbf{K}^-]\dot{\lambda}\mathbf{p}_o=\mathbf{0} \tag{7}$$

incremental solution at a bifurcation point could be represented in general form as:

$$\dot{\mathbf{v}}=\dot{\lambda}\mathbf{K}^-\mathbf{p}_o+[\mathbf{I}-\mathbf{K}^-\mathbf{K}]\dot{\boldsymbol{\alpha}} \tag{8}$$

where $\mathbf{I}$ = identity matrix, K^- = Moore-Penrose generalised inverse matrix of $\mathbf{K}$ and $\dot{\alpha}$ = arbitrary vector composed of the scaling factors for the linearly independent column vectors of matrix [$\mathbf{I}$- K^- $\mathbf{K}$]. At a distinct bifurcation point, since the rank of [$\mathbf{I}$- K^- $\mathbf{K}$] = 1, Eq. 8 will become:

$$\dot{v}=\dot{\lambda}K^{-}p_o+\dot{\alpha}a \tag{9}$$

where a = linearly independent vector of [**I**- K^{-} **K**]. Making use of the existence condition for solution of Eq. 5:

$$[\mathbf{I}\text{-}\mathbf{K}\,\mathbf{K}^{-}]\mathbf{h}=0 \tag{10}$$

where $h_i = r_{ijk}\,\dot{v}_j\,\dot{v}_k$, a quadratic equations

$$A\kappa^2+2B\kappa+C=0 \tag{11}$$

could be obtained where $\kappa = \dot{\lambda}/\dot{\alpha}$ and A, B, C = computed constants. With values of κ determined from Eq. 11, the initial predictor for the branching direction will take the following form:

$$\{\dot{v},\dot{\lambda}\}^T=(\kappa\{\mathbf{K}^{-}\mathbf{p}_o\,,1\}^T+\{\mathbf{a},0\}^T)\dot{\alpha} \tag{12}$$

Line search

6. Fujii and Choong (refs 1, 2) have proposed a process called line search to be carried out in the vicinity of the bifurcation point on a constant loading plane along the direction of a tangent vector $\mathbf{t}_A$ of curve Ω. This curve Ω is obtained by replacing equation $r_i = 0$ (i=1 to N) with r_{N+1} ($= \lambda - \lambda_A$) = 0. The purpose of this search is to locate a point D' which lies sufficiently close to point D as shown in Fig. 2. The position of point D' corresponds to the stationary point of potential Π along search direction $\mathbf{t}_A$. With point D' determined, iteration carried out on the constant loading plane to obtain point D could then be started by using;

$$\mathbf{X}_{D'}=\mathbf{X}_A+\zeta_{D'}\mathbf{t}_A \tag{13}$$

as the first approximation, where $\mathbf{X}_{D'}$, $\mathbf{X}_A$ = position vector of point D' and A respectively.

PATH-SWITCHING STRATEGY BY USING GENERALISED INVERSE AND LINE SEARCH

7. By using a generalised inverse approach, one obvious advantage is that numerical trouble associated with the near-critical condition of the tangent stiffness matrix when operating near the bifurcation point could be avoided. Consequently a very reliable linear approximation for path-switching direction

established in the very vicinity of the bifurcation point could be obtained. By assigning an appropriate magnitude to this switching direction predictor, a point on the bifurcation path could then be determined. Here lies a possible source of numerical trouble as there exists no criteria with which the selection of $\dot{\alpha}$ in Eq. 12 could be made. The choice of $\dot{\alpha}$ has to be made on a trial-and-error basis. On the other hand although the magnitude of the switching direction vector could be obtained through computation when a line search is used, there exists an uncertainty in the selection of the search direction vector $\mathbf{t}_A$. In this paper, the simplicity of the former approach in evaluating a reliable switching direction and the characteristics of the latter approach in searching for a stationary point of potential is combined to form an alternative path-switching strategy as described in the following section.

8. The path-switching direction s extending from bifurcation point B is first obtained through the use of the generalised inverse approach (Fig. 3). The still undetermined magnitude $\dot{\alpha}$ is then computed by first extending the vector **s** until its intersection point D' with a constant loading plane introduced in the vicinity of the bifurcation point. Since $\kappa = \dot{\lambda}/\dot{\alpha}$ and by approximating $\dot{\lambda}$ as λ_A-λ_B, $\dot{\alpha}$ could be computed as:

$$\dot{\alpha}=\Delta\lambda_{BA}/\kappa, \quad \kappa\neq 0 \tag{14}$$

where $\Delta\lambda_{BA} = \lambda_A - \lambda_B$. In cases where $\kappa = 0$ which corresponds to a symmetric bifurcation point, $\dot{\alpha}$ could be specified to be the same as the size of the path parameter used just prior to reaching the bifurcation point during the path tracing process. Using point D' as the first approximation for a point lying on the bifurcation path, a search for a better approximation D" is then initiated by carrying out a line search in the direction of $\mathbf{t}_{D'}$. The proper choice of search direction is important in ensuring that point D could be approximated as closely as possible. In this study, a search direction corresponding to the maximum gradient of potential at point D' is used. This maximum gradient of potential $\mathbf{r}_{D'}$ will be given by the residual vector of Eq. 3. A displacement perturbation pattern $e_{D'}$ will be first computed and normalised to give the search direction $\mathbf{t}_{D'}$ as follows:

$$\mathbf{e}_{D'}=-\mathbf{K}_{D'}^{-1}\mathbf{r}_{D'}, \quad \mathbf{t}_{D'}=\mathbf{e}_{D'}/\|\mathbf{e}_{D'}\| \tag{15}$$

A series of points Q lying along the direction of $t_{D'}$ computed as follow:

$$\mathbf{X}_Q=\mathbf{X}_{D'}+\zeta_Q\mathbf{t}_{D'} \tag{16}$$

will then be examined in turn in order to obtain a better approximation for point D. As a criteria for judging the encounter of such a better approximation (point D" in Fig. 3) the stationary condition of potential Π along search direction $t_{D'}$ is used.

The derivative of potential Π along search direction $t_{D'}$ (ref. 1) could be expressed as:

$$d\Pi/d\zeta = \mathbf{t}_{D'}^{T}\mathbf{r}_{Q} \tag{17}$$

Therefore during the line search the vanishing of $\mathbf{t}_{D'}^{T}\mathbf{r}Q$ is used as the judging criteria for determining the position of point D". The detected point D" could then be used as the first approximation necessary for the subsequent pursuing of bifurcation path.

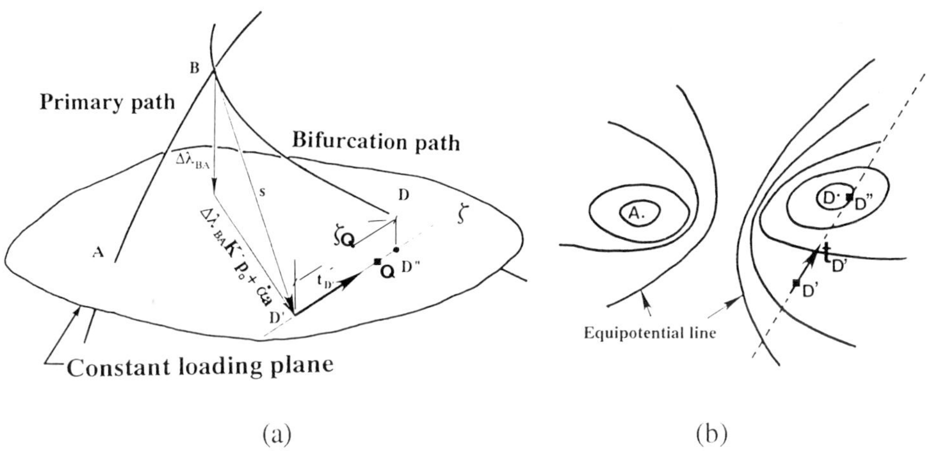

Fig. 3a. The proposed path-switching strategy. Fig. 3b. The search for a stationary point of potential D" in the vicinity of bifurcation point B.

NUMERICAL EXAMPLES

9. A three-bar space truss and a reticulated truss dome (Fig. 4) have been analysed in order to test the validity of the above proposed path-switching strategy. The occurrence of a bifurcation point are judged by monitoring the change in the number of negative diagonal elements of **D** matrix during the decomposition of **K**, where $\mathbf{K}=\mathbf{LDL}^{T}$ and **L** is a lower triangular matrix. Fig. 5 shows the computed primary and bifurcation path for the two models.

Three-bar space truss

10. This example has been analysed by Nishino (ref. 12). Since two negatives diagonal elements are detected after traversing the first bifurcation point, B1 is a compound bifurcation point with the first and second eigenmodes corresponding to a symmetric and asymmetric bifurcation modes respectively. In this study, only the switching to the more general asymmetric bifurcation path is attempted. The computed loading parameter λ_B and λ_A are 2.282 and 2.096 respectively. From Eq. 11 two values of κ are found to be κ_1 = infinity and $\kappa_2 = 0.331$ respectively.

The case of κ_1 = infinity indicates that $\dot{\alpha}_1 = 0$. This represents the tangent vector to the primary path. Using κ_2, and $\Delta\lambda = -0.062$ (= λ_A-λ_B), $\dot{\alpha}_2$ is computed to be -0.188. Fig. 6 shows the search direction mode used. It can be seen that it resembles closely the asymmetric bifurcation mode. A stationary point is detected at ζ, = -0.103 along this search direction. As a way of checking the accuracy of this detected point D", the maximum absolute norm of residual vector $\mathbf{r}_{D''}$ has been computed and its value is 3×10^{-4}. From this, it can be concluded that point D" approximates the actual point on the asymmetric bifurcation path with good accuracy. The variation of potential gradient along the search direction is illustrated in Fig. 7.

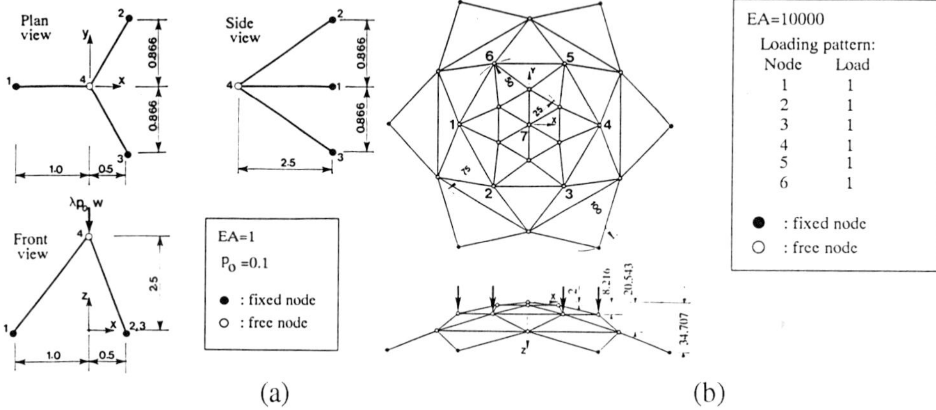

(a) (b)

Fig. 4. Numerical models: a) Three bar space truss; b) Reticulated truss dome.

Reticulated truss dome

11. Three bifurcation points B1, B2 and B3 are detected before the occurrence of the first limit point L1. Since the number of negative diagonal elements changes in the sequence of 0-1-2-3 when traversing the three bifurcation points, all three of them are distinct bifurcation points. By checking the computed value of κ_1 and κ_2 (Table 1), it can be concluded that both point B1 and B3 are symmetric bifurcation points whilst point B2 is an asymmetric one. Table 1 lists the values of ζ, where stationary points are encountered in all three attempts of line search. Using the detected point D" as a first approximation, the bifurcation path extending from B1 and B2 could be followed successfully. Tracing by using the point D" detected for the case of B3 is however not successful and the path traced is the same as the bifurcation path extending from B2. This failure could be possibly due to both B2 and B3 being very closely spaced, as can be judged from their respective loading parameter (Table 1). Path-switching in the case of such a type of bifurcation points requires further research. Fig. 8 shows an example of the variation of potential gradient along the search direction for the case of a line search in the vicinity of B1.

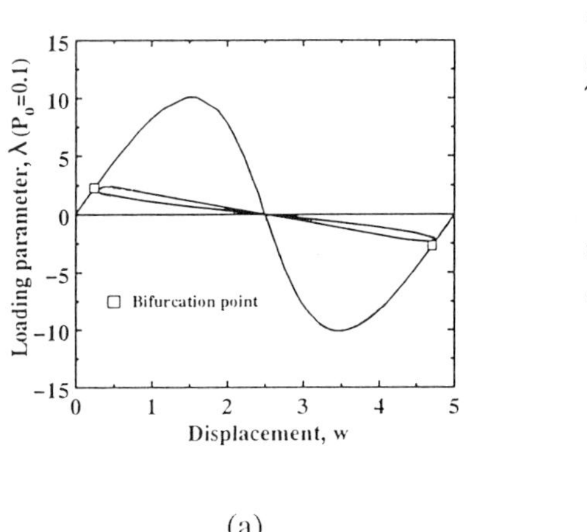

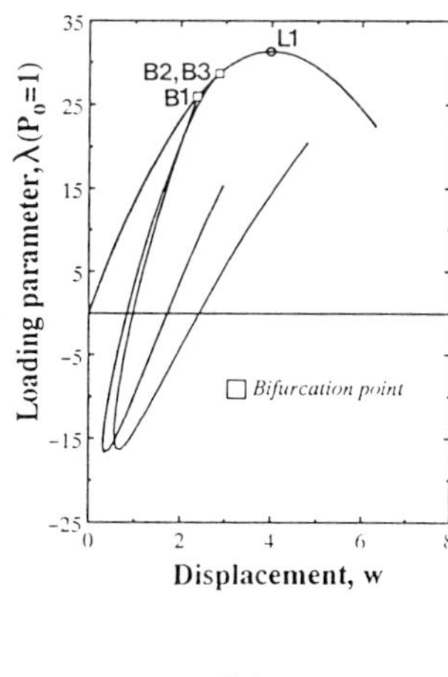

(a) (b)

Fig. 5. Load-deflection curves for:
a) Three bar space truss (w = vertical displacement of node 4);
b) Reticulated truss dome (w = vertical displacement of node 7).

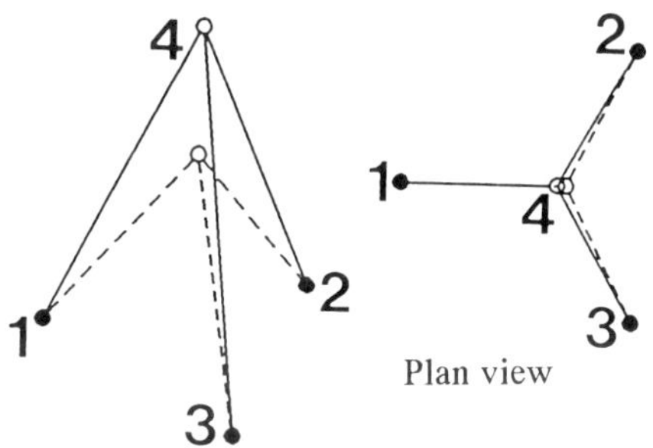

Fig. 6. Perturbation mode **t**D' used in line search for three-bar space truss.

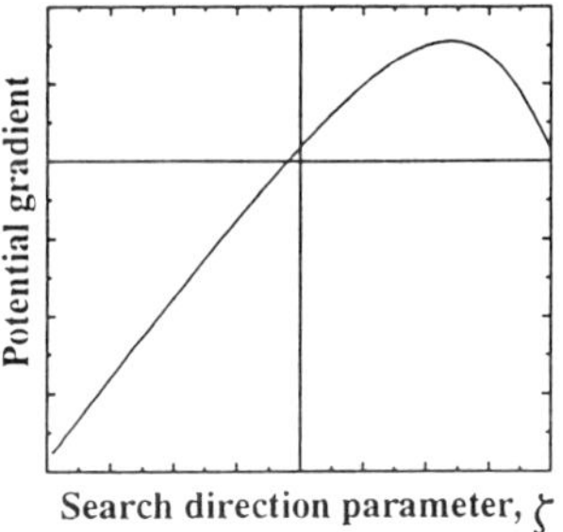

Fig. 7. Variation of $d\Pi/d\zeta$, along **t**D'

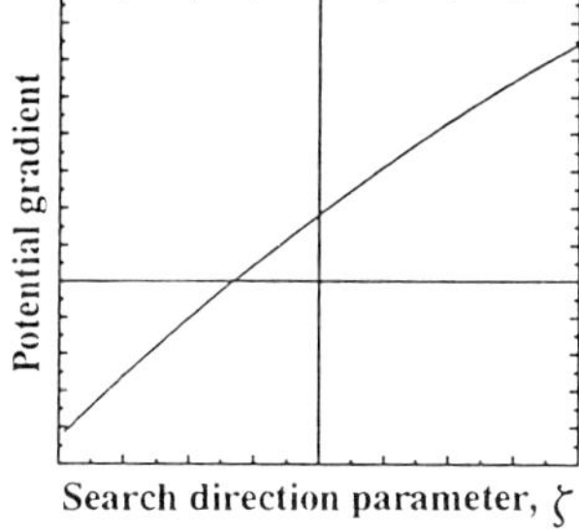

Fig. 8. Variation of $d\Pi/d\zeta$, along $\mathbf{t}_{D'}$ during line search in the vicinity of B1.

Table 1. Results of line search.

B.P.	κ_1	κ_2	λ_A	λ_B	ζ	$\dot{\alpha}$
B1	402519	0.000	25.937	25.957	-0.658	0.050
B2	-82.354	0.928	28.695	28.718	-0.908	-0.025
B3	∞	0.000	28.745	28.770	-0.129	0.050

Note: 1.B.P.=Bifurcation point
2.∞=very big value

CONCLUSION

12. A path-switching strategy by combining the approach of generalised inverse and line search has been proposed and tested on two simple truss structures. From the results obtained, it can be said that the strategy works quite satisfactorily. Cases where bifurcation points are closely spaced together or coincident will need further research. Although the first derivatives of the tangent stiffness matrix are required when computing the approximate bifurcation mode, a reliable predictor for path-switching direction extending from a bifurcation point could be obtained. Furthermore by examining the computed constant such as C in Eq. 11, the type of bifurcation point could be determined. This will provide very useful information for planning the path-switching analysis. By searching in the direction of maximum gradient at the first approximate location of a bifurcation point, a stationary point of potential, lying in the vicinity of the bifurcation path could be obtained. Using this improved location as an initial approximation, the bifurcation path could be pursued with a higher rate of success.

REFERENCES

1. FUJII F. and CHOONG K.K. Branch-switching in Bifurcation Problem of Structures. J. of Engineering Mechanics, ASCE, 1992, Vol.118, No.8, Aug., 1578-1596.
2. FUJII F. and CHOONG K.K. Branch-switching in Symmetric / Asymmetric Bifurcation Problems of Elastic Structures. Journal of Structural Engineering, Japanese Association of Civil Engineering, 1991, Vol.37A, March, 343-352.
3. HANGAI Y. Application of the Generalised Inverse to the Geometrically Non-linear Problem. SM Archives, 1981, Vol.6, Issue 1, Jan., 129-165.
4. HANGAI Y and LIN X.G. Geometrically Non-linear Analysis in the Vicinity of Critical Points by the Generalised Inverse. Int.J. of Space Structures, 1989, Vol.4, No.4, 181-191.
5. HANGAI Y., "Numerical Analysis in the Vicinity of Critical Points by the Generalised Inverse", Bulletin of the International Association for Shell Structures, 1987, Vol.XXVIII-3, No.95, 23-26.
6. SKEIE G. and FELIPPA C.A. Detecting and Traversing Bifurcation Points in Non-linear Structural Analysis. Int. J. of Space Structures, 1991,Vol.6, No.2, 77-98.
7. WAGNER W. and WRIGGERS P. A Simple Method for the Calculation of Post critical Branches. Eng. Comput., 1988, Vol.5, June,103-109.
8. KOUHIA R. and MIKKOLA M. Tracing the Equilibrium Path Beyond Simple Critical Points. Int. J. for Numer. Methods in Engrg., 1989, Vol.28, 2923-2941.
9. STEIN E., WAGNER W. and WRIGGERS P. Non-linear Stability-Analysis of Shell and Contact-Problems including Branch-Switching. Computational Mechanics, 1990, 5, 428-446.
10. KROPLIN B., DINKLER D. and HILLMAN J. An Energy Perturbation Applied to Non-linear Structural Analysis. Computer Methods in Applied Mechanics and Engineering, 1985, 52, 885-897.

11. NISHINO F., IKEDA K., SAKURAI T. and HASEGAWA A. A Total Lagrangian Non-linear Analysis of Elastic Trusses. Proc. of JSCE, No.344/I-1 (Structural Eng./Earthquake Eng.), 1984, April, 39-53.
12. NISHINO F., HARTONO W., FUJIWARA O. and KARASUDHI P. A Study of the Stability and Behaviour at the Critical Point by the Taylor Expansion. Structural Engrg./Earthquake Engrg., Proc. of JSCE, 1987, Vol.4, No.1, April, ls-9s.
13. HARTONO W., NISHINO F., FUJIWARA O. and KARASUDHI P. On Tracing Bifurcation Equilibrium Paths of Geometrically Non-linear Structures. Structural Engrg./Earthquake Engrg., Proc. of JSCE, Vol.4, No.1, No.380/I-7, 11s-17s.
14. RIKS E. An Incremental Approach to the Solution of Snapping and Buckling Problems. Int. J. Solids Structures, 1979, Vol.15, 529-551.
15. RIKS E. Bifurcation and Stability, a Numerical Approach. Proceedings of WAM Symposium on Innovative Methods for Non-linear Problems, New Orleans, Nov., 1984.
16. RIKS E. "Some Computational Aspects of the Stability Analysis of Non-linear Structures", Computer Methods in Applied Mechanics and Engineering, 1984, 47, 219-259.
17.RHEINBOLDT W.C. Numerical Methods for a Class of Finite Dimensional Bifurcation Problems. SIAM J.Numer. Anal., 1978, 15, 1-11.
18. CHOONG K.K., FUJII F. and KITAGAWA T. Access to Bifurcated and Isolated Equilibrium Branches of Structures. Proc. of Pacific Conference on Computational Mechanics. 1991, Vol.1, 463-468.

5. Multidisciplinary modelling and analysis of aerospace structures

K. K. GUPTA, NASA Ames Research Center, USA

SYNOPSIS. This article is concerned with the modelling and simulation of modern complex aerospace vehicles that involve interaction of a number of major technical disciplines such as structures, aerodynamics, propulsion, heat transfer and control engineering, among others. In this connection the pertinent details of the development and application of an integrated multidisciplinary finite element analysis tool that include relevant numerical formulations are presented. Some numerical examples pertaining to a number of practical problems are also presented, in some detail, that testifies to the efficacy of the currently evolved solution techniques.

INTRODUCTION

1. Modem high-performance aerospace craft are characterised by unprecedented levels of multidisciplinary aero-structural-controls-propulsion interactions that tend to impose considerable constraint on dynamic stability and controls performance margins required for flight safety. An accurate prediction of such flight characteristics, prior to flight envelope expansion, is of vital interest to ensure safety of the vehicle. NASA Dryden's multidisciplinary research is focused on the continued development of an advanced, effective prediction analysis capability, and also simultaneous application of the related tool for the solution of the ongoing NASA projects related problems.

2. Associated analysis methodology is essentially based on the unifying finite element technology approaches within each critical discipline area, enabling effective integration in a natural and consistent fashion and yielding a compact and coherent analysis framework. In the area of structural analysis the relevant general purpose computer code is entirely based on standard finite element (FE) procedures and is capable of modelling a wide variety of structural materials. Likewise, the heat transfer analysis module is also based on the FE procedure and may be conveniently utilised or be the solution of a wide variety of problems. In the area of computational fluid dynamics (CFD) analysis, the relevant matrix equations derived from finite element discretisation of Euler or Navier-Stokes formulations are solved by either an explicit or an alternative quasi-implicit

solution technique. Significant acceleration in solution convergence, in this connection, has been achieved by developing an Aitken solution technique as well as by implementing an adaptive mesh generation capability. Both 3-D external aerodynamic as well as internal propulsion flow may be simulated by the CFD code. An associated aeroservoelastic analysis procedure integrates each of the critical discipline areas in an appropriate manner (Fig. 1) to yield the desired static and dynamic stability derivatives that are critical for effective hypersonic as well as other relevant simulation analyses. The associated multidisciplinary finite element computer program STARS (ref. 1) (STructures, Aerodynamics and Related Systems analysis program), developed in this connection, integrates the various individual disciplinary codes in an efficient manner and is capable of effective modelling and simulation of advanced aerospace vehicles (refs 2 and 3).

3. This paper provides some pertinent details of the numerical formulations adapted for the current analysis effort. A number of example problems are also presented that demonstrate the efficacy of the presently developed numerical algorithms and solution schemes.

DEVELOPMENT OF NUMERICAL FORMULATIONS

4. To ensure accurate interaction of various disciplines for multidisciplinary modelling and simulation of advanced aerospace vehicles, the finite element method has been commonly used for discretisation of both solids and fluids continua. Thus, general purpose finite element codes have been developed in each critical discipline area which were then appropriately combined to yield an integrated multidisciplinary analysis tool. The various modules of the associated STARS computer program are depicted in Fig. 2, and they are capable of modelling complex aerospace vehicles. Summaries of relevant numerical finite element formulations pertaining to each major discipline area as well as the integrated systems simulation approach are presented next, the current effort being an extension of work performed earlier (refs 2 and 3).

Finite Element Structural Analysis

5. This module of the STARS program is capable of modelling and simulation of any complicated 3-D structural system with arbitrary loading and boundary conditions. The program caters for general anisotropic material properties and the extensive element library includes improved composite elements (ref. 4). A typical matrix numerical formulation encompassing a broad class of structural problems may be written as below:

$$[\mathbf{K}(1 + ig) + \mathbf{K}_G + \mathbf{K}']\mathbf{U} + [\mathbf{C}_c + \mathbf{C}_D]\,\dot{\mathbf{U}} + \mathbf{M}\ddot{\mathbf{U}} = \mathbf{F}(t) \quad (1)$$

that pertains to the dynamic response analysis of structures including spinning ones; relevant definitions of associated matrices and parameters are as follows:

$\mathbf{K}, \mathbf{K}_G, \mathbf{K}'$ = Elastic, geometric and centrifugal stiffness matrices, respectively

$\mathbf{C}_c$, $\mathbf{C}_D$,	=	Coriolis and viscous damping matrices
$\mathbf{M}$	=	Inertia matrix
$\mathbf{U}$	=	Unknown deformation vector
$\mathbf{F}(t)$	=	Dynamic forcing function
g	=	Structural damping

Such matrix equations pertain to static, buckling, vibrational and dynamic response problems of usual and spinning structural systems, that tend to be sparse in nature. The numerical analysis sub-module of the program exploits fully such matrix sparsity and is capable of implementing efficient solution of the associated large order simultaneous equations as well as eigenvalue problems.

Finite Element Heat Transfer Analysis

6. This solution module is capable of performing both steady state and transient analyses pertaining to heat conduction problems. The element library includes 1-, 2- and 3-D finite elements including composites. Nonlinear effects such as radiation boundary conditions as well as temperature dependent material properties are currently being implemented in the STARS program. Associated numerical formulation for the general nonlinear transient analysis may be written as below:

$$[\mathbf{K}_c(T) + \mathbf{K}_r(T) + \mathbf{K}_h(T)]\,\mathbf{T}(t) + \mathbf{C}(T)\,\dot{\mathbf{T}}(t) = \mathbf{R}_Q(T,t) + \mathbf{R}_q(T,t) + \mathbf{R}_h(T,t) + \mathbf{R}_r(T,t) \qquad (2)$$

in which,

$\mathbf{K}_c$, $\mathbf{K}_r$, $\mathbf{K}_h$	=	conductance matrices relating to conduction, radiation and convection, respectively.
$\mathbf{C}$	=	capacitance matrix.
$\mathbf{R}_Q$, $\mathbf{R}_q$, $\mathbf{R}_h$, $\mathbf{R}_r$	=	heat load vectors pertaining to internal heat generation specified surface heating, surface convection, and incident surface radiant heating, respectively.
$\mathbf{T}$, T, t	=	unknown nodal temperature vector, nodal temperature and time.

The modal method is employed for the linear transient analysis whereas for the nonlinear case a time integration scheme in conjunction with a Newton Raphson iterative solution procedure is adopted for the relevant analysis. Nonlinearities caused by temperature-dependent properties and heating as well as radiation heat transfer are currently being implemented in the STARS program.

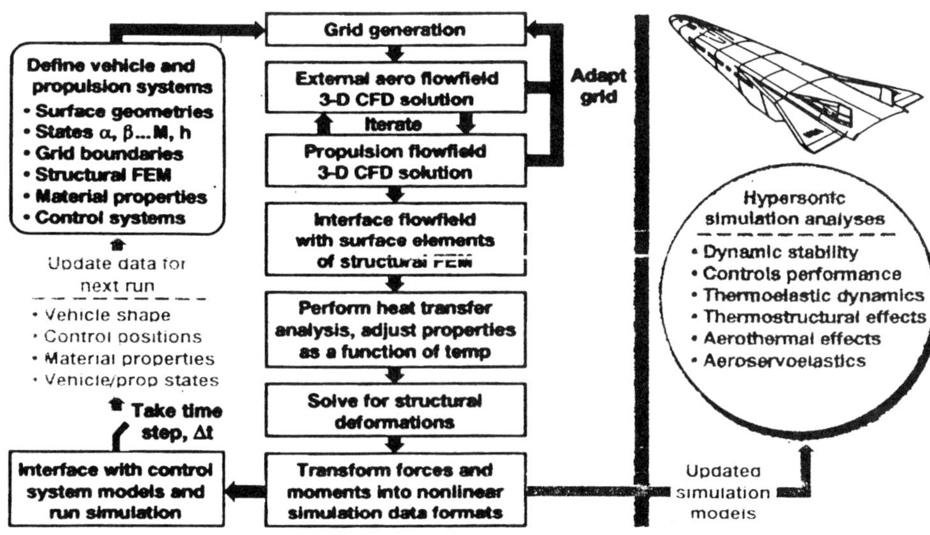

Fig. 1. Nonlinear multidisciplinary simulation approach.

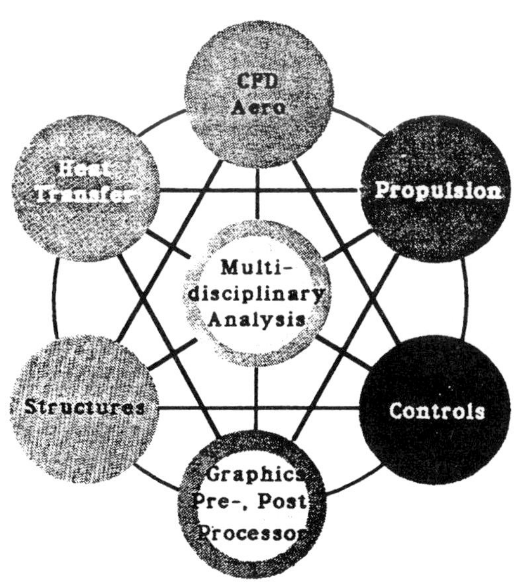

Fig. 2. STARS multidisciplinary finite element analysis modules.

Finite Element CFD Analysis

7. The dynamic equation for a viscous, heat-conducting compressible fluid obeying conservation of mass, momentum, and energy can be expressed by a set of partial differential equations:

$$\frac{\partial \mathbf{V}}{\partial t}+\frac{\partial \mathbf{F}_i}{\partial x_i} = \mathbf{f}_b, \qquad i=1,2,3 \tag{3}$$

where the solution, flux and body forces column vectors, as well as the viscous stress tensor, are defined as below:

$$\mathbf{V} = \{\rho \quad \rho u_j \quad \rho E\} \tag{4}$$

$$\mathbf{F}_i = \left\{\rho u_i \quad \rho u_i u_j + p\delta_{ij} + \sigma_{ij} \quad u_i(\rho E + p) + u_l \sigma_{li} + k\frac{\partial T}{\partial x_i}\right\} \tag{5}$$

$$\mathbf{f}_b = \{0 \quad \mathbf{f}_{b_j} \quad u_l \mathbf{f}_{b_l}\} \tag{6}$$

$$\sigma_{ij} = -\frac{2}{3}\mu\frac{\partial u_k}{\partial x_k}\delta_{ij} + \mu\left(\frac{\partial u_i}{\partial x_j}+\frac{\partial u_j}{\partial x_i}\right) \tag{7}$$

in which ρ, p, E are the density, average pressure intensity and total energy, respectively, δ_{ij} being the Knonecker delta; u_j is the velocity component in the direction x_j of a cartesian coordinate system, μ the viscosity and k is the thermal conductivity, $\mathbf{f}_b$ being the body forces. The above equations are supplemented with state equations:

$$p = (\gamma-1)\rho\left[E-\frac{1}{2}u_i u_i\right] \tag{8}$$

$$T = \left[E-\frac{1}{2}u_i u_i\right]c_v \tag{9}$$

for a complete solution, in which γ is the ratio of specific heats and c_v is the specific heat at constant volume, such a formulation being valid for a perfect gas.

8. Solution of the non-viscous form of Eq. 1 may be achieved by first obtaining a Taylor series expansion of **V** in time domain. The spatial domain Ω is next discretized by unstructured meshes consisting of 3-D tetrahedron elements. Using linear finite element approximations $\mathbf{V} = a\hat{\mathbf{V}}, \hat{\mathbf{V}}$ being nodal variable values, and employing Galerkin weighted residual procedure, a time-dependent form of the governing equations may be obtained as below:

$$\mathbf{M}\delta\hat{\mathbf{V}} = -\Delta t\left[\mathbf{C}\hat{\mathbf{V}}\right] + \mathbf{R} \tag{10}$$

in which **R** includes artificial viscosity effects essential for capturing shocks. Solution of Eq. 10 is effected by advancing this time-dependent form until steady conditions are obtained; an explicit time-stepping iterative scheme (ref. 5) as well as an alternative quasi-implicit solution scheme has been implemented in the STARS program to that effect.

9. An accelerated Euler solution procedure based on the Aitken acceleration technique has recently been implemented (ref. 3) that effects considerable improvement in solution convergence rate. The associated 3-D automated unstructured mesh generation code is based on an advancing front technique (ref. 5) and has proved to be rather versatile for modelling of any practical CFD solution domain around a complex structural form such as an aircraft. However, this procedure involves a rather extensive search for nodes and faces on the front and as such the grid generation time tends to be rather large for complex configurations. Significant savings in solution time has been achieved by first generating a grid whose cells have linear dimension about twice the required size, and then reducing each cell locally to reach desired cell sizes.

Controls Engineering

10. A control input is treated as finite displacement rate, which is essentially equivalent to the loading term on the right hand side in Eq. 1. A step-by-step small displacement input is utilised for the overall solution strategy.

Materials

11. Layered, anisotropic, advanced composite finite elements (ref. 1) are utilised for the design of modern aerospacecraft. Since the temperature-dependent anisotropic material properties of such an element augments the stiffness matrix of Eq. 1, a step-by-step solution approach is considered necessary.

Propulsion Technology

12. The finite element CFD program, described earlier, is being used to solve the propulsion related problems. Additional features, as below, are currently being augmented to the FEM CFD code:

(a) Turbulence - a Baldwin-Lomax model to be used for near-wall turbulence.

(b) Combustion - a finite rate combustion model with two-step kinetics to be used for the solution; the CFD code will utilise a MacCormick algorithm with point implicit treatment of the chemistry source terms.

Flutter and Aeroservoelastic Analysis (ASE)

13. A flutter analysis procedure starts with the finite element structural model and subsequently computes the natural frequencies (ω) and modes (ϕ) that consist of rigid body, elastic, and control surface motions, by solving:

$$\mathbf{M}\ddot{\mathbf{u}} + \mathbf{K}\mathbf{u} = \mathbf{0} \tag{11}$$

in which **u** is the displacement vector. An efficient block Lanczos procedure, that fully exploits matrix sparsity (refs 6 and 7), is utilised for such a solution. Next, a steady-state Euler solution is effected through an explicit or an alternative quasi implicit local time-stepping solution procedure that employs a residual smoothing strategy. The resulting vehicle equation of motion may be cast into the frequency domain as follows:

$$\hat{\mathbf{M}}\ddot{\mathbf{q}} + \hat{\mathbf{C}}\dot{\mathbf{q}} + \hat{\mathbf{K}}\mathbf{q} + \hat{\mathbf{f}}_a(t) + \hat{\mathbf{f}}_I(t) = \mathbf{0} \tag{12}$$

in which the generalised matrices and vectors are as follows:

$\hat{\mathbf{M}}$ = inertia matrix ($= \Phi^T \mathbf{M} \Phi$), and, similarly,

$\hat{\mathbf{K}}, \hat{\mathbf{C}}$ = stiffness and damping matrices

q = displacement vector ($= \Phi^T \mathbf{u}$)

$\hat{\mathbf{f}}_a(t)$ = aerodynamic (CFD) load vector ($= \Phi^T pA$), p is the Euler pressure A is the appropriate surface area

$\hat{\mathbf{f}}_I(t)$ = impulse force vector ($= \Phi^T \mathbf{f}_I$)

where $\mathbf{f}_I$ is the user input hat contains a number of modes of interest. A state-space matrix equation form (ref. 2) of Eq. 12 may be formulated as:

$$\dot{\mathbf{X}} = \mathbf{A}\mathbf{X} + \mathbf{b}_a(t) + \mathbf{b}_I(t) \tag{13}$$

where:

$$\mathbf{X} = \begin{bmatrix} \mathbf{q} \\ \dot{\mathbf{q}} \end{bmatrix}$$

$$\mathbf{A} = \begin{bmatrix} \mathbf{0} & \mathbf{I} \\ -\hat{\mathbf{M}}^{-1}\hat{\mathbf{K}} & -\hat{\mathbf{M}}^{-1}\hat{\mathbf{C}} \end{bmatrix}$$

$$\mathbf{b}_a(t) = \begin{bmatrix} \mathbf{0} \\ -\hat{\mathbf{M}}^{-1}\hat{\mathbf{f}}_a(t) \end{bmatrix}$$

$$\mathbf{b}_I(t) = \begin{bmatrix} \mathbf{0} \\ -\hat{\mathbf{M}}^{-1}\hat{\mathbf{f}}_I(t) \end{bmatrix}$$

and a time response solution of Eq. 13 in an interval Δt (= t_{n+l}-t_n) is obtained as:

$$\mathbf{X}_{n+1} = e^{\mathbf{A}\Delta t}\mathbf{X}_n + \mathbf{A}^{-1}\left[e^{\mathbf{A}\Delta t} - \mathbf{I}\right]\left[\mathbf{b}_a(t_n) + \mathbf{b}_I(t_n)\right] \tag{14}$$

which are stored for later processing.

14. The structural deformations $\mathbf{u}$ and velocities $\dot{\mathbf{u}}$ are then computed from $\mathbf{q}$ and $\dot{\mathbf{q}}$, respectively, and the aerodynamic mesh is updated only if large motions are encountered. Such values are introduced into the CFD code to change velocity boundary conditions at the solid boundary. This is followed by a one-step Euler solution using a global time-stepping scheme, and the entire solution process is repeated for the specified number of time steps.

15. The response data, as above, may be resolved into modal components utilising an FFF, as below:

$$\mathbf{q} = \sum_{m=1}^{p} e^{\zeta_m t}(a_m \cos\omega_m t + b_m \sin\omega_m t) \tag{15}$$

to yield the damping (ζ) and frequency (ω) values. This procedure is repeated for a number of dynamic pressure values, $\bar{q} = \frac{1}{2}\rho V^2$ and the (ζ) and (ω) values plotted against $\bar{\mathbf{q}}$ or Mach number. Such a plot depicting stability characteristics of the vehicle enables prediction of onset of flutter or divergence occurring within the entire flight regime.

16. For aeroservoelastic analysis, when the control derivatives are not known *a priori*, since the non-linear CFD analysis has been used, an autoregression procedure may be used to reconstruct a model based on the past history of aircraft input and output information. Typically, for a small incremental motion of a control surface, the vehicle body forces and moments are first calculated from the computed surface pressure distribution. This is followed by an estimation of such parameters as α, side slip (β), control surface deflections and hinge moments, as well as roll, pitch, and yaw rates employing $\mathbf{q}$ and $\dot{\mathbf{q}}$ values computed from the

appropriate equations of motion. These calculations are to be performed for a large number of time steps that represent the entire range of control surface motion, and the resulting data are then employed to obtain static and dynamic stability derivatives for the hypersonic simulation analysis.

NUMERICAL EXAMPLES

17. The STARS program is routinely used to solve on-going NASA project related problems. A number of such representative examples are presented herein.

Generic Hypersonic Vehicle - vibration analysis

18. A complete generic hypersonic vehicle was chosen to perform some representative free vibration and subsequent aeroservoelastic analyses. The finite element model developed in this connection has the following relevant details:

No. of elements	=	4,860
No. of nodes	=	2,800
No. of degrees of freedom	=	16,800

A free vibration analysis of the problem was performed using the Lanczos solution option and the desired first twenty five eigenvalues and vectors, including six rigid body modes, were computed by the program. The total CPU time for the analysis, employing an IBM RISC 6000/580 computer, amounted to 2 hr 30 min. Table 1 presents a summary of such results, whereas Fig. 3 depicts a few mode shapes.

Table 1. Generic hypersonic vehicle - STARS free vibration analysis.

Mod No.	Frequency (Hz)	Description
1-6	0.0	rigid body modes
7	3.05	fuselage 1st bending (S)
8	4.01	wing 1st bending (S)
9	4.03	wing 1st bending (A/S)
10	5.21	fuselage 1st bending (A/S)
11	6.92	wing 1st torsion
12	7.16	fuselage 2nd bending (S)
13	9.40	fin 1st bending (S)

X-29A aircraft - linear aeroservoelastic(ASE) analysis

19. The X-29A is a canard-configured research aircraft that possesses such unique features as a very thin and flexible forward swept-wing, aeroelastically tailored with advanced composite wing covers designed to elastically eliminate structural divergence without a weight penalty. Other critical features include supercritical airfoil, variable-incidence canards operating along with full-span variable camber flaperons and strake flaps. The vehicle is up to 35% statically

unstable requiring appropriate feedback flight controls for augmented static stability. Therefore, an appropriate ASE analysis was of vital importance to ensure flight safety; the STARS program was effectively utilised to that effect. Fig. 4 depicts the key aspects of this multidisciplinary analysis that involves structural, aerodynamic and controls modelling and Fig. 4(e), in particular, provides a comparison of analysis and flight test results pertaining to damping characteristics.

Oscillating double wedge airfoil - non-linear ASE analysis

20. The STARS ASE module was utilised to generate unsteady aerodynamic forces in the airfoil (ref. 3), undergoing pitching motion along the trailing edge and oscillating at a frequency of 670 rad/sec. Such forces were computed for Mach 3.0, the maximum pitching angle being 0.1 rad., and correlated with those computed by simple piston theory; Fig. 5 shows such comparison.

CONCLUDING REMARKS

21. An integrated, finite element based numerical methodology and associated computer code developed for the modelling and simulation of advanced aerospace vehicles has been presented in the paper. Numerical results, pertaining to a number of example problems have also been presented that testify to the efficiency of the presently developed solution algorithms and the associated computer program.

REFERENCES

1. GUPTA K. K. STARS - An integrated general-purpose finite element structural, aeroelastic, and aeroservoelastic analysis computer program. NASA TM-101709, June 1990.
2. GUPTA K. K., BRERMER M. J. and VOELKER L. S. Integrated aeroservoelastic analysis capability with X-29A comparisons. J. Aircraft, 26, 1989, pp. 84-90.
3. GUPTA K. K., PETERSEN K. L. and LAWSON C. L. On some recent advances in multidisciplinary analysis of hypersonic vehicles. Proceeding of the 4th AIAA International Aerospace Planes Conference, December 24, 1992, Orlando, FL, Paper No. AIAA -92-.
4. LUNG S. F., MARTIN C. W. and GUPTA K. K. Three-node C° element for analysis of laminated composite sandwich shells. NASA TM 4125, 1989.
5. MORGAN K., PERAIRE J., PEIRO J. and HASSAN O. The computation of three-dimensional flow using unstructured grids. Comput. Methods Appl. Mech. Engrg., 87, 1991, pp. 335-352.

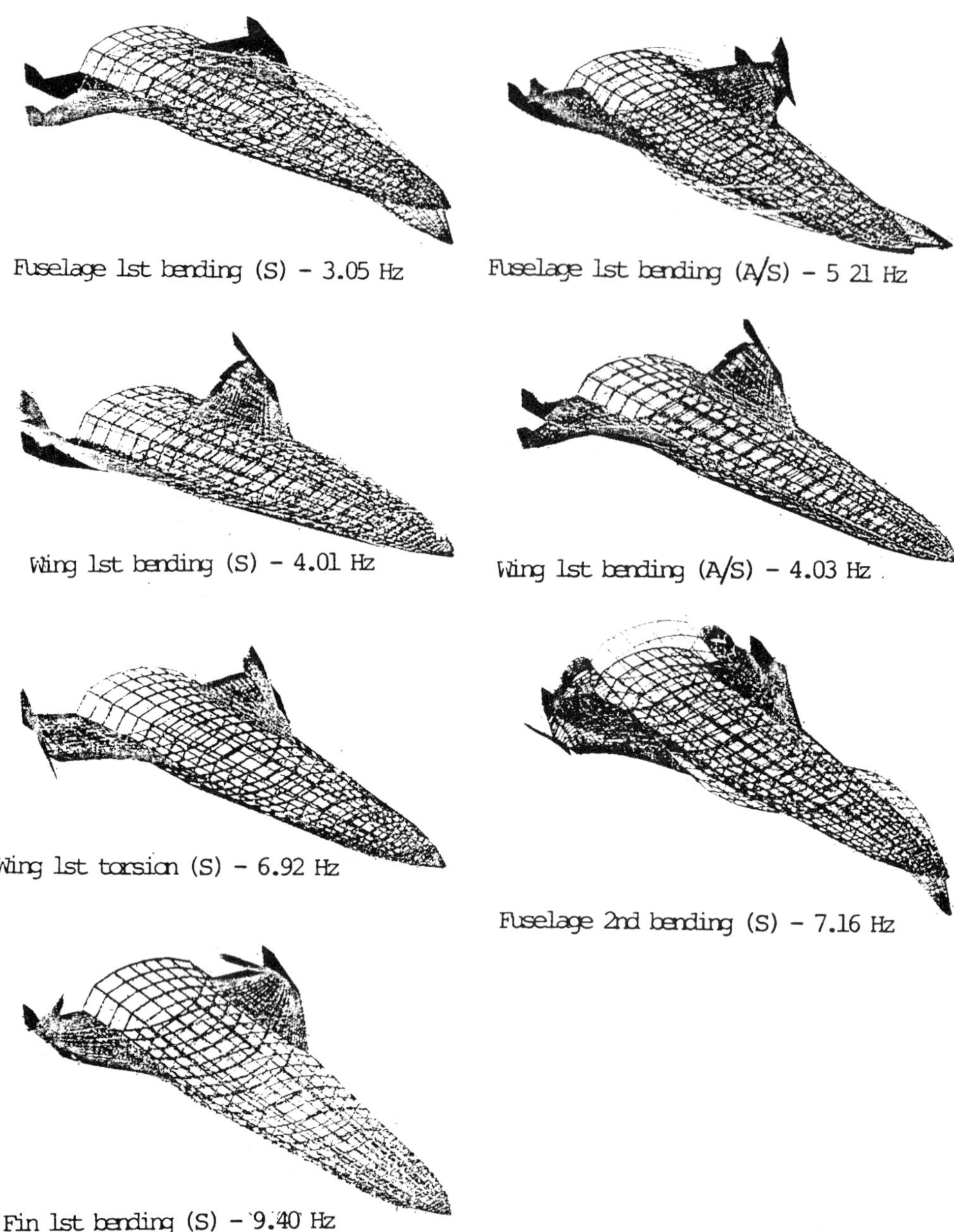

Fig. 3. Generic hypersonic vehicle - STARS FEM elastic mode shapes. (Symmetric (S) and Anti-symmetric (AS)).

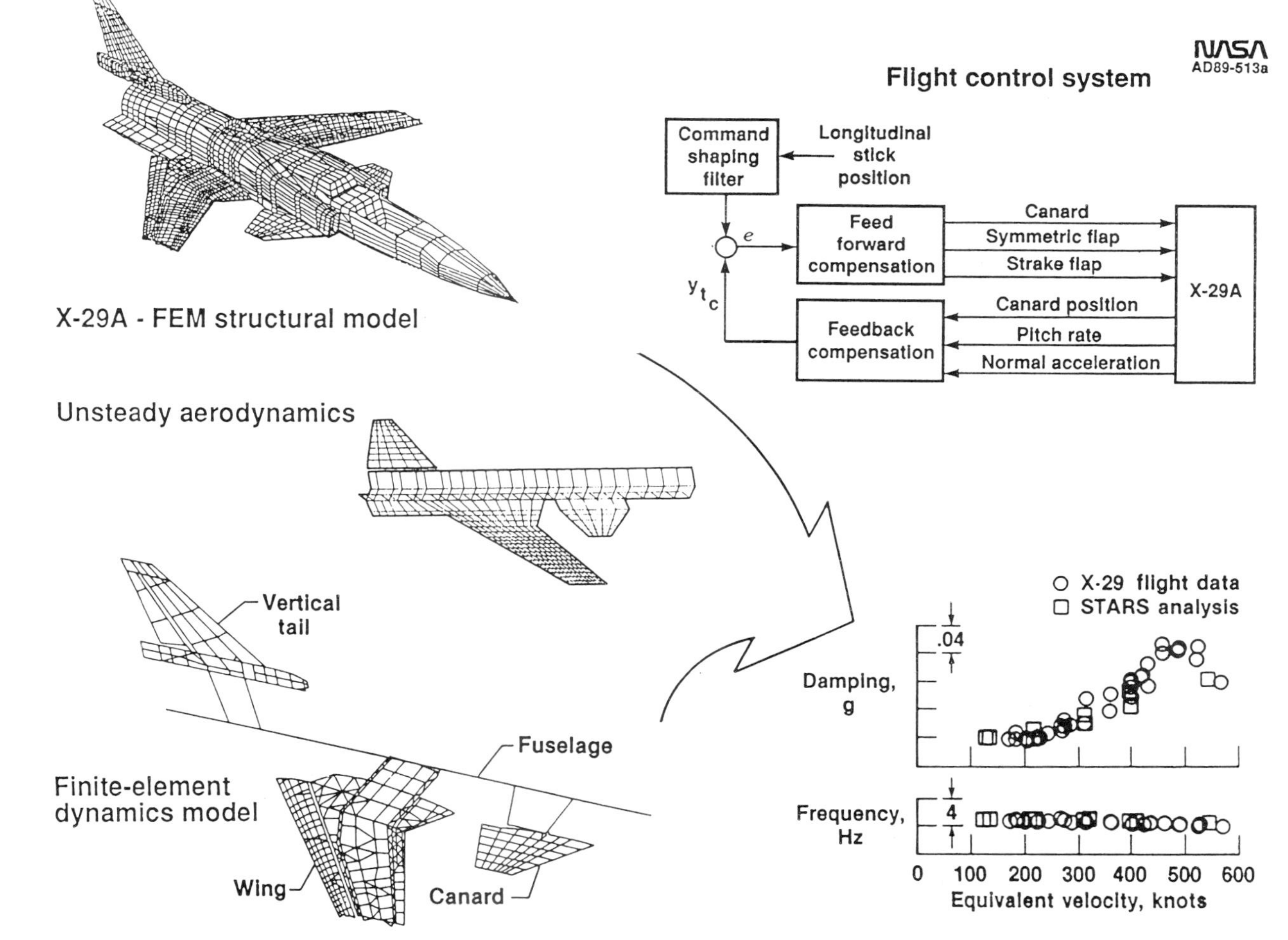

Fig. 4. X-29A aeroservoelastic (ASE) analysis.

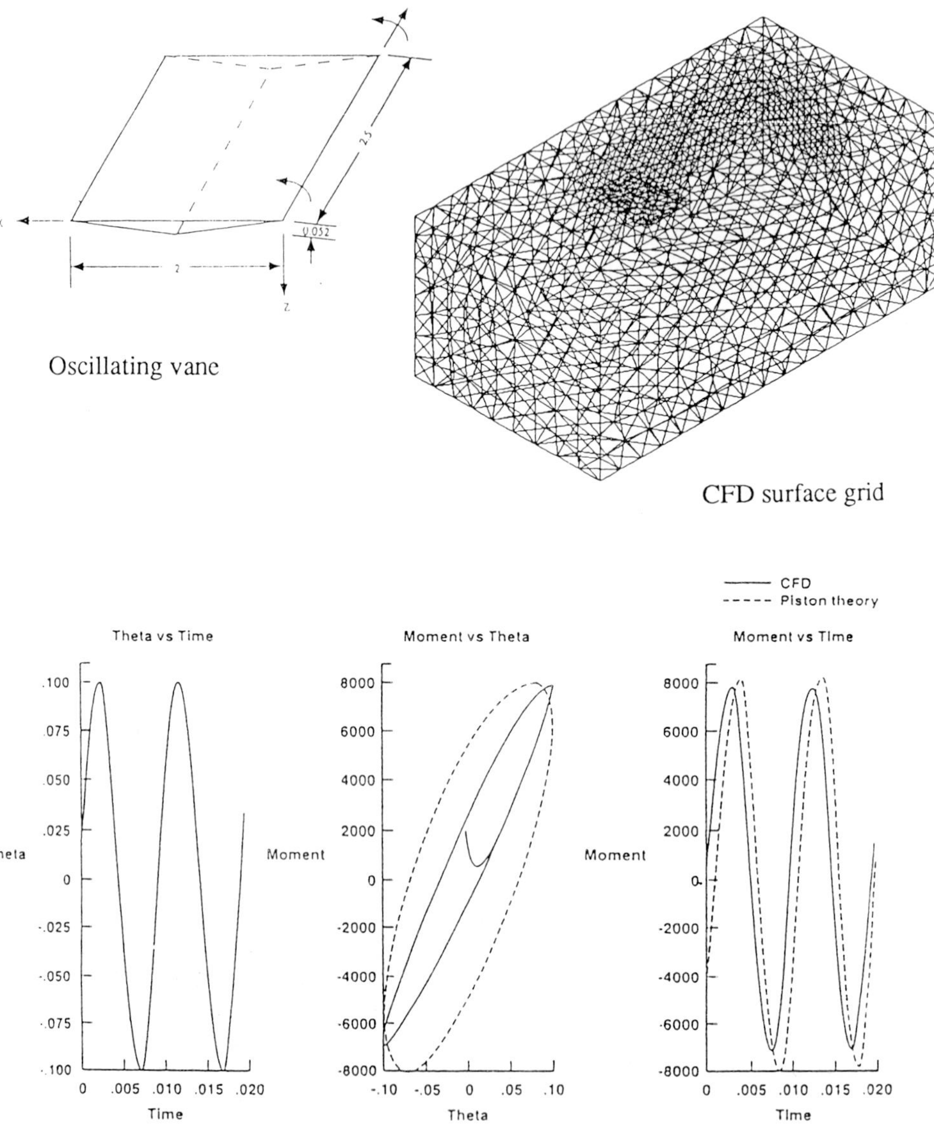

Fig. 5. Double wedge airfoil - comparision of unsteady aeodynamic forces obtained from CFD and Piston theory solution.

6. Inelastic buckling in three-dimensional steel structures

J. W. BUTTERWORTH and M. J. BEAMISH, University of Auckland, New Zealand

SYNOPSIS. This paper describes a technique for modelling the complex behaviour associated with the large, cyclic, inelastic deformations that steel framed structures may undergo during severe earthquakes. The finite element method was used with a hybrid thin-walled beam element in which rectangular shell elements were linked to the beam axis, thereby preserving the advantage of a one dimensional element. Large displacements were incorporated within a total Lagrangian formulation with axis updating and second order displacement equations. The problem of rotational degrees of freedom associated with large deformations was overcome by the use of the semi-tangential definition of moments and finite rotations. A comparison with the results of a physical test involving cyclic loading, local and lateral-torsional buckling of part of a steel frame showed good prediction of the general behaviour.

INTRODUCTION

1. Efficient design of steel framed structures generally leads to the use of slender, thin-walled members able to exploit the high strength and stiffness of the material. As a result, stability considerations, such as local, Euler and lateral-torsional buckling will frequently curtail the capacity of a member or assemblage despite stress levels which are well within the elastic range. Elastic design procedures are therefore commonly used. However, in seismically active regions of the world it is uneconomic and impractical to design structures to respond elastically to earthquake forces and it is the practice in New Zealand and a number of other countries to design structures to withstand moderate earthquakes with only minor damage and to avoid collapse in major earthquakes by incorporating the ability to dissipate energy by cyclic, inelastic deformations in selected components without significant loss of strength. The energy dissipating mechanism (comparable to the plastic collapse mechanism of plastic design) limits the maximum forces in other parts of the structure (typically the major gravity load carrying elements such as the columns), lengthens the natural periods and increases the level of damping, resulting in an attenuation of the response.

2. When applying this design procedure it is obviously important that the ductile components have predictable post-elastic response, with sufficient strength retention to allow them to safely undergo repeated cyclic excursions into the inelastic regime. Premature loss of strength due to local or lateral torsional buckling, both of which are exacerbated by cyclic loading, must be avoided. It is equally important to ensure that the intended ductile components are not over strong, forcing unintentional yielding into the non-ductile members. For example, in multi storey moment resisting frames, NZS 4203 (ref. 1) requires a mechanism where plastic hinges form in the beams rather than the columns to avoid the possibility of a soft storey collapse. The considerable flange and web distortions that can occur at plastic hinge sites during seismically induced cyclic deformations are shown in Fig. 1. The figure is taken from ref. 2 which describes simulated seismic loading of portal frame knee joints.

3. Assessing the performance of members and joints in the inelastic, post-buckled regime has most often been achieved by physical testing. Ref. 3 describes an attempt at the theoretical modelling of this difficult problem. The approach taken was to develop a special purpose finite element incorporating local and lateral buckling deformations, large displacements, finite rotations, residual stresses, eccentrically applied loads and discrete and continuous elastic restraints. The element was then embedded in an existing non-linear finite element analysis program, ANSR III (ref. 4), which provided a material model for the cyclic stress strain behaviour of steel as well as appropriate solution strategies for controlling the overall analysis. Due to space limitations the present paper can provide only a summary view of some aspects of the development and application of the finite element. Full details may be found in ref. 3.

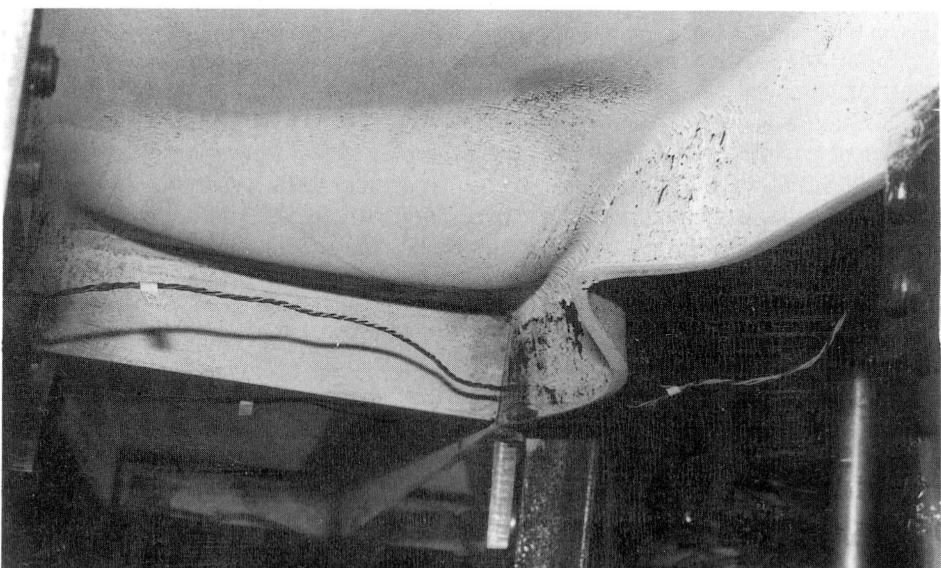

Fig. 1. Section distortion after simulated seismic loading.

THIN WALLED BEAM ELEMENT FORMULATION

Modelling approach

4. Accurate analysis of thin walled beams including local deformations of the cross section as well as lateral-torsional buckling deformations requires the use of three dimensional shell theory. A direct approach to the problem using 3-D quadrilateral shell elements was rejected as cumbersome, especially in the areas of data preparation and the interpretation of results in terms of familiar stress resultant quantities such as shear forces and bending moments. Demands on computational power were also thought likely to be excessive.

5. The approach adopted used Rajasekeran and Murray's (ref. 8) description of the displacement field as a set of local buckling deformations superimposed on the standard thin-walled beam displacements, to develop a general purpose thin-walled beam finite element. A second element that did not allow local deformation was also developed.

Objectives

6. The primary objectives for the new element were as follows:
(a) A true large displacement formulation, including large rotations, based on consistent continuum mechanics principles.
(b) Both lateral-torsional and local buckling displacement modes to be included.
(c) Shear deformations to be included.
(d) An appropriate constitutive model for cyclic deformations in steel.
(e) Inclusion of residual stresses and initial deformations.
(f) Provision for eccentrically applied loads.
(g) Discrete and continuous elastic lateral restraints.
(h) Results in terms of member axis displacements and stress resultants.
(i) Compatibility with an existing non-proprietary finite element package.
(j) Ease of use and flexibility.

Idealisation

7. The member was idealised as an assemblage of flat rectangular strips rigidly connected along the longitudinal edges (Fig. 2), with the sub-elements treated as physically separate shell finite elements connected by rigid links to the member axis. A single longitudinal reference axis replaced the shear centre axis plus centroidal axis used in other formulations. Displacements and rotations of the member axis cause mainly membrane type displacements in the sub-elements due to the generally small transverse curvatures. Practical plate slenderness ratios are such that thin plate theory gives acceptable results, so the plate displacement field in the normal direction could be based on a thin plate finite element, but the in-plane displacement field had to include shear deformations. The displacements of any point in the member were considered to be the sum of the displacements assuming no distortion of the cross section and the displacements due to flexure of the component plates of the section. This allowed the use of the standard 14 nodal degrees of freedom for a thin-walled beam element, and the corresponding forces and moments at the ends.

8. Vlasov's assumptions (ref. 3) were adopted as the basis for the displacement field but were modified to include shear deformations by assuming that shear strains in the element mid-surface were constant over the element length. Cubic interpolation functions were used for the transverse displacements and axial twist so that the required nodal generalised displacements included the transverse displacement (or twist) and the axis slope (or rate of twist). If warping and local deformations were excluded the resulting displacement field would be the same as that of a Timoshenko beam. To consider the large transverse curvatures arising from local buckling, a bi-cubic field of displacements relative to the displaced position of the member cross section was assumed. This required additional degrees of freedom at the member ends. These latter displacements were assumed to be fully continuous along the longitudinal junctions of the sub-elements. A total Lagrangian formulation was used with optional axis updating as noted in §11.

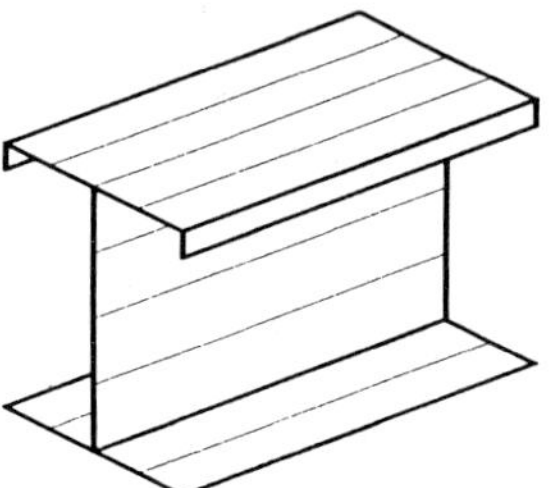

Fig. 2.Thin walled beam discretisation.

Large rotations

9. Large displacement analysis of three dimensional structures with rotational degrees of freedom requires a suitable description of finite rotations at element node points to uniquely define the motion of the body. It is well known that a series of finite rotations defined with respect to a set of fixed axes are not commutative and that this leads to non-unique strain energy functionals and non-symmetric tangent stiffness matrices (ref. 5). Argyris (ref. 6) has presented a comprehensive treatment of the matrix formulation of large rotations and the present treatment stems from that work.

10. Quasi and semi-tangential moments. Moments that are applied about a fixed axis or about a follower axis that rotates with the body can be shown to be non-conservative as their work is path dependent. Quasi-tangential moments (ref. 7) are conservative moments formed by a single pair of opposite forces with fixed direction acting on a rigid lever, such as bending moments in a beam. Semi-tangential moments are formed by two or more equal quasi-tangential moments about the same axis applied via levers spaced apart by equal angles. The St. Venant torque in a beam is an example of a semi-tangential moment, while the warping torque can be quasi-tangential, semi-tangential or a combination. Argyris (ref. 5) showed that the use of either semi-tangential or quasi-tangential moments as generalised nodal forces ensures symmetry of the element geometric stiffness matrices. However, quasi-tangential moments do not follow standard co-ordinate

transformation rules causing problems when two members connect at an angle. The consequences include the calculation of incorrect critical loads and convergence difficulties in incremental large displacement analysis. The semi-tangential moment definition on the other hand, obeys the standard co-ordinate transformation rules and does not require knowledge of the type of moment at the member ends to establish joint equilibrium equations. The semi-tangential moment definition was adopted and used exclusively for element end moments and applied loads.

11. Semi-tangential rotations. The use of semi-tangential moments as generalised forces in an energy based finite element formulation requires the use of corresponding generalised displacements, known as semi-tangential rotations, that are conjugate in the energy sense. Argyris (ref. 5) provides the necessary definitions and these were adopted along with the rules of transformation necessary to deal with changes of reference axis system (such as from local element co-ordinates to the global axes), sequences of semi-tangential rotation increments and rotations of the structure. An option to update element reference axes was included to allow for truncation of higher order terms in the rotation matrices and the neglecting of some of the less significant strain energy terms.

Element excluding local deformations

12. This element provided an efficient model for cases where lateral buckling was the dominant effect. Seven degrees of freedom (three translations, three semi-tangential rotations and the warping function) were used at the end nodes with cubic interpolations for transverse displacement and axial twist. Shear deformations were provided for by keeping rotation variables independent of the transverse displacement variables. The chosen interpolation functions required constraints using three shear strain parameters to relate the cross section rotations and the warping parameter to the axis slopes and rate of twist. Four additional internal degrees of freedom were required for the shear strain parameters plus an additional axial incompatible displacement. The total of three axial displacement degrees of freedom allowed quadratic axial displacement interpolation, avoiding coupling with flexural displacements when the member reference axis was not located at the centroid.

Element including local deformations

13. This element superimposed local deformations of the sub-elements on the deformations described in the preceding section. The local displacement field used conforming bicubic functions for the transverse displacement, while the membrane displacement interpolations were cubic in the longitudinal direction but only linear in the local orthogonal in-plane direction. Additional degrees of freedom were needed in order to retain the local deformation relative displacements at the member ends. The number of these depended on the number of local nodes used in the cross section model.

Material model

14. For elastic material response a plane stress state was assumed with the standard constitutive matrix relating 2nd Piola-Kirchhoff stresses and Green-Lagrange strains. Modelling of the behaviour of steel subject to cyclic inelastic deformation requires a constitutive model that will account for both strain hardening effects and the Bauschinger effect. Mroz (ref. 9) used several congruent yield surfaces of the Von Mises type, each of which may translate and/or change size according to a mixed isotropic/kinematics hardening rule. Each yield surface has a different tangent plastic modulus associated with its motion in stress space, derived from a multilinear representation of a monotonic curve. The inner yield surface which contains the current stress point translates until the next yield point is engaged with a corresponding change in plastic modulus and so on. The material model used in the ANSR-III finite element package in which the new element was implemented was an extension the basic model just described and included cyclic plasticity and strain rate dependency. The virgin and fully cycled (or saturated) stress - plastic strain relations were represented by multilinear curves, where each linear segment was associated with a yield surface as for the standard Mroz model.

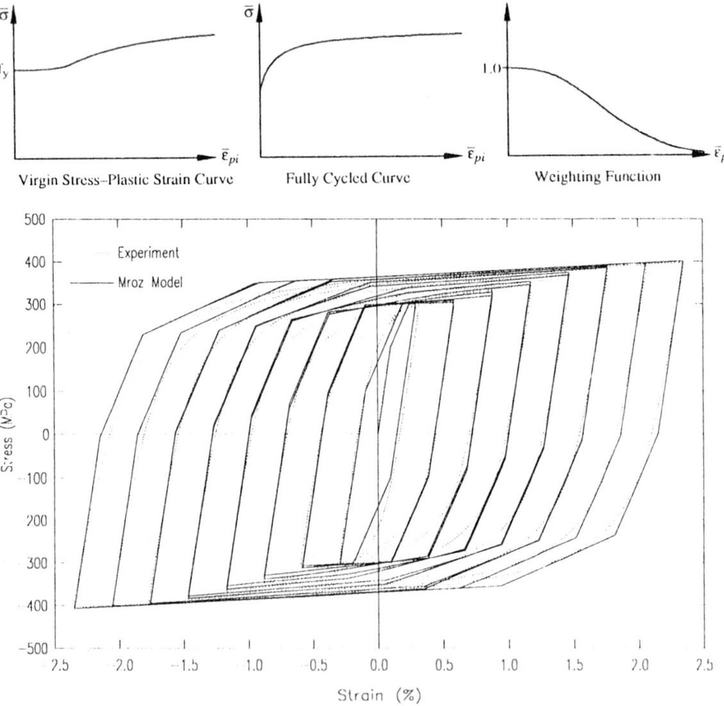

Fig. 3. Comparison of experimental cyclic stress-strain curve with extended Mroz model.

The saturated curve could be obtained from cyclic stress-strain data by taking one leg of a stabilised loop and halving the values on both axes. The tangent plastic modulus and yield stress for each yield surface were determined as a weighted average of the virgin and saturated values. The weighting function depended only on the accumulated effective plastic strain. Fig. 3, shows sample experimental data curves together with the model results.

Residual stresses

15. Residual stresses were provided for by specifying the residual stress at evenly spaced discrete points over the middle surface of each sub-element. Interpolation was by standard Lagrangian interpolation functions up to fourth order. For example, Fig. 4 shows a residual stress pattern with parabolic flange and quartic web distributions specified by three residual stresses in each flange outstand and five in each half of the web. The residual stress variation along the length of the element was linear between the specified end values. In the actual formulation, no change to the model was required as the residual stresses were incorporated by simply updating the initial material stress state and setting the corresponding strains to zero.

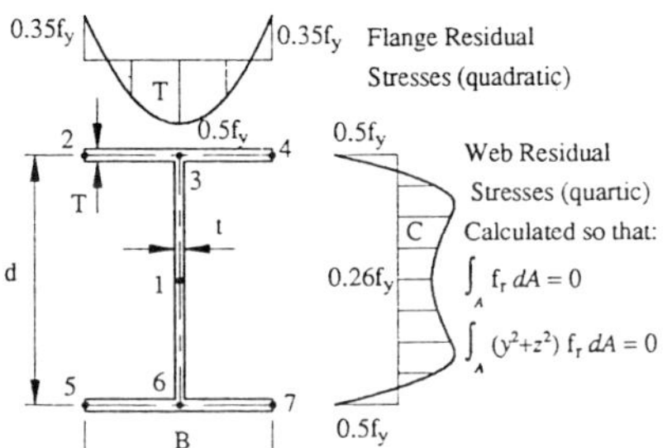

Sample Residual Stress Pattern Specification							
Dimensions: d=400, B=160, T=12, t=8							
Sub-Elem	End Nodes		Residual Stresses (f_r/f_y)				
			1	2	3	4	5
1	2	3	-0.35	0.29	0.50		
2	3	4	0.50	0.29	-0.35		
3	5	6	-0.35	0.29	0.50		
4	6	7	0.50	0.29	-0.35		
5	3	1	0.50	-0.30	-0.43	-0.32	-0.26
6	1	6	-0.26	-0.32	-0.43	-0.30	0.50

Fig. 4. Residual stress pattern example.

Initial deformations

16. Initial deformations were specified by assigning initial values to the nodal degrees of freedom and initialising the total displacement vector accordingly. To permit different initial local buckling deformations in adjacent sub-elements, the initial nodal local deformations for each sub-element were specified separately. The strain distribution was calculated at the beginning of an analysis and the material state at all points updated accordingly with the corresponding stress increment set to zero. As with the residual stresses, no changes to the basic formulation were required in an incremental analysis.

Eccentric loading

17. Provision for the application of point or distributed loads at points other than the element axis line was achieved by including the virtual work associated with the displacement of the load points in the element work expression enabling the calculation of the additional load vector (forces and semi-tangential moments) to

be applied at the element nodes. There was also a contribution to the geometric stiffness matrix which accounted for the de-stabilising effect of loads applied above the element axis.

Elastic lateral restraints

18. Discrete restraints. Elastic (or inelastic) restraints at discrete points were modelled by three linear (or non-linear) translational and three rotational springs at the specified point in the plane of the member's end. Moments in rotational springs eccentric to the member axis also have a bimoment component. For example, if one flange of an I-beam is restrained (as in a connection with a top flange plate) then warping displacements as well as the minor axis rotation are partially restrained. A warping restraint spring was also provided, allowing a wide range of restraint conditions to be modelled. Inclusion of appropriate virtual work terms permitted the calculation of the contribution to the element stiffness matrix. The geometric stiffness contribution arising from the eccentrically applied translational spring forces was identical to that of the eccentric point loads referred to in §17. Non-linear springs were accommodated in the incremental formulation by substituting the tangent spring stiffnesses.

19. Discrete restraints applied at element node points. If the elastic restraint springs act eccentrically at a sub-element node point, the restraint forces in the real situation may cause local distortion of the element cross section if no stiffening is provided. To account for this, the restraints were applied at the sub-element node point directly and the contribution to the element stiffness calculated..

20. Continuous elastic restraints. Elastic lateral restraints acting continuously over the full length of the element can be used to simulate cladding, floor slabs or continuous warping stiffeners. Any continuous translational springs fixed to the element will rotate with the element about the local transverse reference axes but should not follow the element twist. Hence the usual second order displacement expression is simplified, eliminating all second order terms not associated with the axial twist. Continuous restraints can also be applied at a sub-element edge to model situations such as a slab on the top flange of a beam where cross section distortions are allowed. The displacement equation modifications are the same as for the discrete restraint case except for the axial strain.

Local buckling restraints

21. Local buckling in a steel beam can be restrained by stiffener plates in a variety of configurations or by the presence of floor slabs, walls, concrete encasement, etc. Continuous constraints on local buckling deformation can be achieved by deleting or reassigning the applicable local relative displacement degrees of freedom at the element level, while point constraints on relative displacements at the element ends could be applied to the global displacement vector. Flexible local deformation restraints or more complex internal constraints can be applied by modelling them as simple elastic springs connecting the restrained relative degrees of freedom to a fixed point or to another local degree of freedom.

IMPLEMENTATION

22. Theoretical formulations for the elements described in the preceding sections were coded in FORTRAN V and added to the element library of the finite element package ANSR-III (ref. 4). Some further modifications were made to ANSR-III, for example, to accommodate the larger number of nodal degrees of freedom required by the thin-walled beam elements and to provide some post-processing capabilities. ANSR-III provided the necessary infrastructure, including non-linear solution strategies for dealing with bifurcation, snap buckling, etc., and the extended Mroz material model for cyclic deformation of steel members. A number of versions of the program were produced, tailored to run on machines such as IBM3081, Silicon Graphics 4D/240S and IBM PC.

EXAMPLE

23. The performance of the element was verified over a range of problems which had known theoretical solutions or previously published analysis results (ref. 3). The example given here is an analysis of a steel cantilever I-beam with geometry and material properties based on experimental measurements taken from laboratory tests on a portal frame knee joint sub-assemblage under cyclic loading (ref. 2). The section of the portal frame modelled extended from the knee joint, through the plastic hinge zone and further along the rafter. This was modelled as a cantilever with the end load inclined to provide similar loading conditions to the portal frame. Fig. 5, shows the discretisation used, the model dimensions and the assumed residual stress distribution (fig. 3, shows the cyclic stress-strain curve for the material).

24. Fig. 6 shows the load-deflection plot for the cantilever. When compared with the measured experimental values (ref. 3) the plot showed many similar characteristics but with some over-estimation of the loads. Local buckling occurred at a similar displacement value but in the second cycle rather than the first and strength degradation was not as severe as in the test. The buckled shape, shown in Fig. 7, was similar to the observed form but with lower amplitude as would be expected given the crudeness of the model and the low order membrane displacement interpolation.

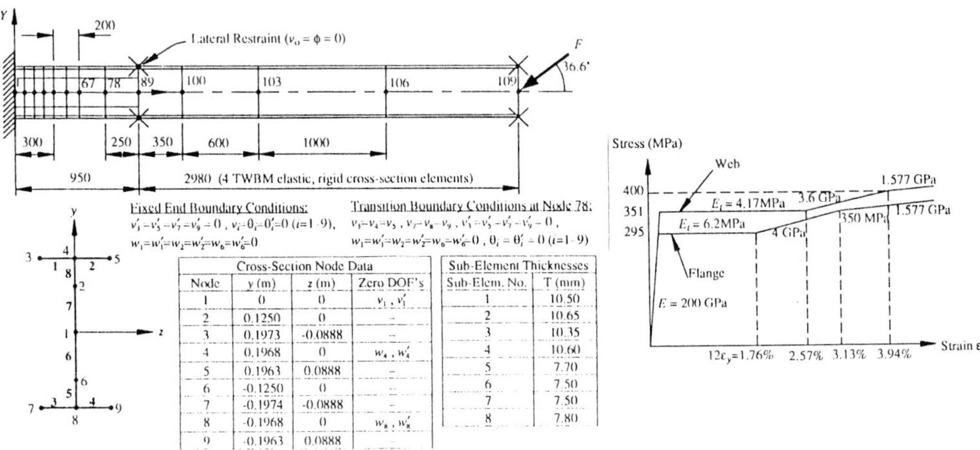

Cross-Section Node Data			
Node	y (m)	z (m)	Zero DOF's
1	0	0	v_1, v'_1
2	0.1250	0	-
3	0.1973	-0.0888	-
4	0.1968	0	w_4, w'_4
5	0.1963	0.0888	-
6	-0.1250	0	-
7	-0.1974	-0.0888	-
8	-0.1968	0	w_8, w'_8
9	-0.1963	0.0888	-

Sub-Element Thicknesses	
Sub-Elem. No.	T (mm)
1	10.50
2	10.65
3	10.35
4	10.60
5	7.70
6	7.50
7	7.50
8	7.80

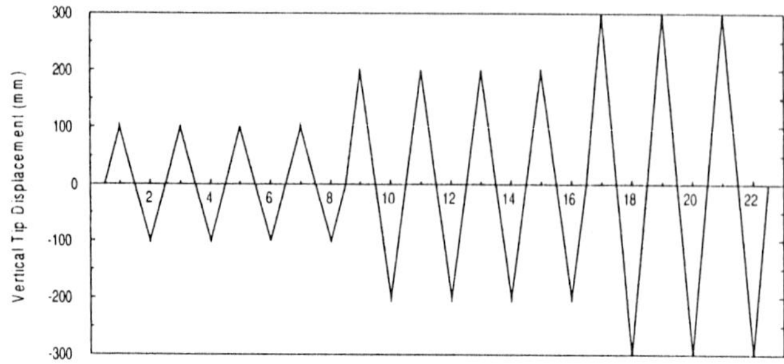

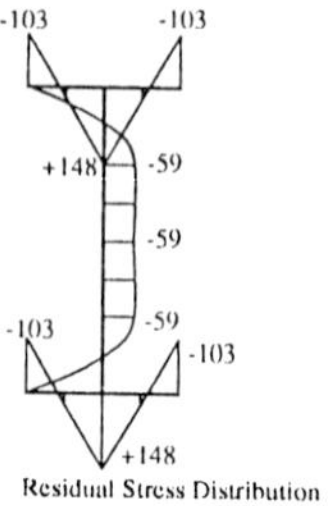

Fig. 5. Details of cyclically loaded I-beam model.

CONCLUSIONS

25. An inelastic, non-linear, thin-walled beam finite element has been developed and implemented in a general purpose finite element package. Analyses of the cyclic loading of a sub-assemblage in the plastic hinge zone adjacent to a portal frame knee joint showed good general agreement with experimentally observed behaviour.

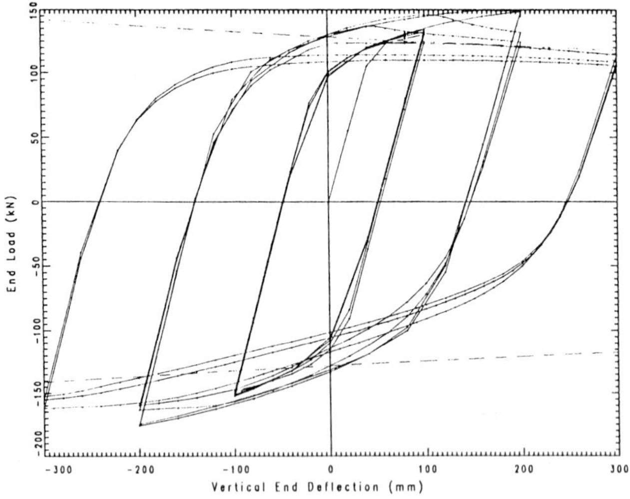

Fig. 6. Load-deflection results for cyclic analysis.

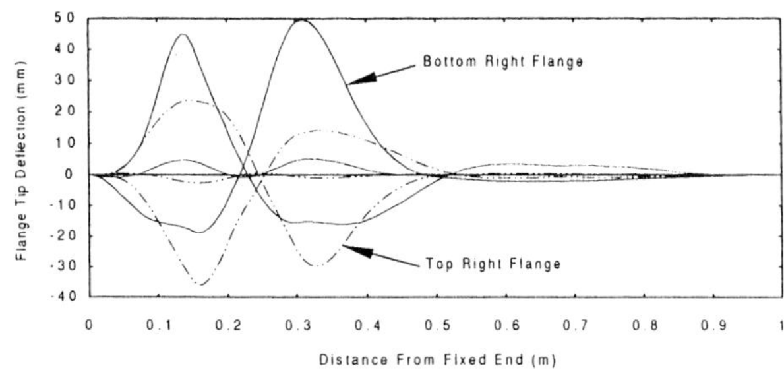

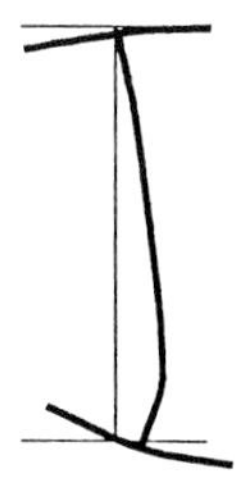

Fig. 7. Local deformation in flanges and section distortion.

REFERENCES

1. STANDARDS ASSOCIATION OF NEW ZEALAND, Code of practice for general structural design and design loadings for buildings, NZS 4203: 1976.

2. BEAMISH, M.J., Seismic behaviour of steel portal frame knee joints, University of Auckland School of Engineering Report No. 348, 1984.

3. BEAMISH, M.J., and BUTTERWORTH, J.W., Inelastic local and lateral buckling of thin-walled steel members, University of Auckland School of Engineering Report No. 494, 1991.

4. OUGHOURLIAN, C.V., and POWELL, G.H, ANSR-III general purpose computer program for non-linear structural analysis, Report No. UCB/EERC-82/21, Earthquake Engineering Research Centre, University of California, Berkeley, 1982.

5. ARGYRIS, J.H., DUNNE, P.C. and SCHARPF, D.W.,On large displacement-small strain analysis of structures with rotational degrees of freedom, Computer Methods in Applied Mechanics and Engineering, Vol. 14, 1978, pp 401-451.

6. ARGYRIS, J., An excursion into large rotations, Computer Methods in Applied Mechanics and Engineering, Vol. 32, 1982, pp 85-155.

7. ZIEGLER, H., Principles of structural stability, Blaisdell, Waltham, Massachusetts, 1968.

8. RAJASEKERAN, S., and MURRAY, N., Coupled local buckling in wide-flange beam-columns, Journal of the Structural Division, ASCE, Vol. 99, No. ST6, June 1973 pp 1003-1023.

9. MROZ, Z., An attempt to describe the behaviour of metals under cyclic ioads using a more general work hardening model, Acta Mechanica, Vol. 7, 1969.

7. Stability behaviour investigation of braced barrel vaults

DU SHOUGUN, SUN JIANHENG and XIA HENGXI, Agricultural University of Hebai, P. R. China

SYNOPSIS. In this paper, based on the geometrically non-linear finite element method and the post-buckling analysis technique, a large amount of numerical work has been done to investigate the stability behaviour of barrel vaults. The influence factors investigated include: (a) four configurations, (b) ratio of longitudinal span to transverse span, and (c) height to transverse-span ratio. By comparing the equilibrium paths and buckling load, some suggestions are presented to aid structural design.

INTRODUCTION

1. Braced barrel vaults, categorised as large span space structures, have been widely used in the past decades. In comparison with double layer barrel vaults, much attention must be paid to the stability behaviour of single layer barrel vaults. In recent years, a lot of numerical work has been undertaken on the non-linear stability of braced domes (ref. 1-4), but the stability of single layer braced barrel vaults has received only limited attention (refs 5,6). Also, most of the previous research focused mainly on the non-linear finite element method and the post-buckling analysis technique.

2. Based on the theory presented in ref. 4, a large amount of numerical work has been undertaken for this paper. All the braced barrel vaults investigated are transversely spanning, and are supported along only the horizontal longitudinal edges with all the translational movements of the boundary joints restrained. All members of the vaults are circular hollow section and have the same outer diameter of 140mm and wall thickness of 3mm. The investigated factors which affect the stability behaviour of the vaults include: (a) four configurations of vault, (b) ratio of longitudinal-span to transverse-span, and (c) ratio of height to transverse-span.

CONFIGURATION

3. To enable a comparison to be made between the stability behaviour of various configurations, four single layer barrel vaults shown in Fig. 1 are considered. The vaults have the same transverse-span of 18m, longitudinal-span

of 30m and ratio of height to transverse-span of 0.22. Vault A is a rectangular configuration without diagonal bracing, while B and C are with diagonal bracing. Vault D is a isosceles triangle configuration. The load-deflection paths of the four vault types are shown in Fig. 2.

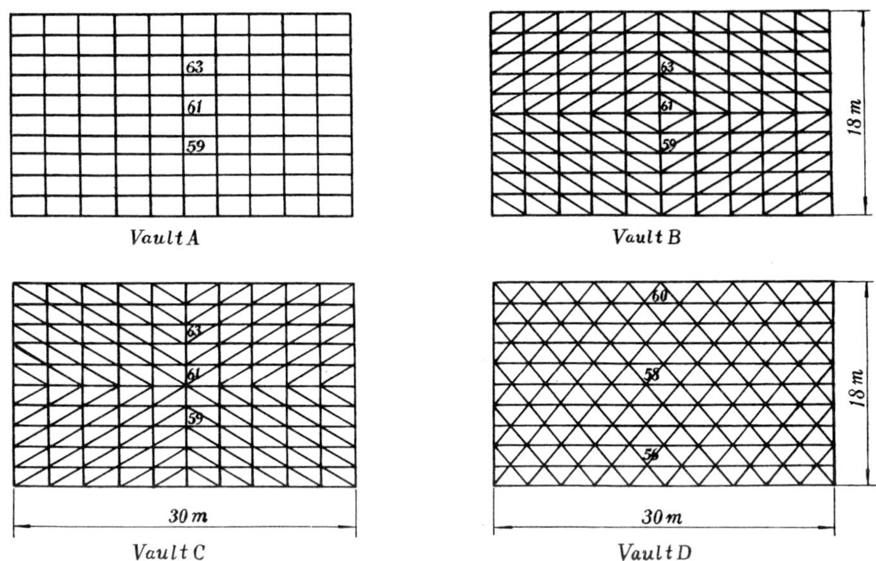

Fig. 1. Barrel vaults of four different configurations.

4. It is seen from Fig. 2 that for vault A, the critical load and the maximum deflection (joint 61) at the buckling load is very small. For vaults B and C, although the different arrangements of diagonal bracing cause the diagonal members of vault B to be in tension while the diagonal members of vault C are in compression, the load-deflection paths of the two vaults are very similar. In comparison with vaults A, B and C; vault D, the isosceles triangular configuration, has a much higher critical load, and its deflection at the buckling load has greatly exceeded the normal engineering tolerances.

5. What is interesting is that the steel requirements of vault B and C is about 1.7 times of that of vault A, but their critical loads are only a little higher than vault A. Also the steel requirements of vault D is only 1.6 times higher than that of vault A, and its critical load is 4.9 times greater than the critical load of vault A, and 3.7 times that of vault B and C.

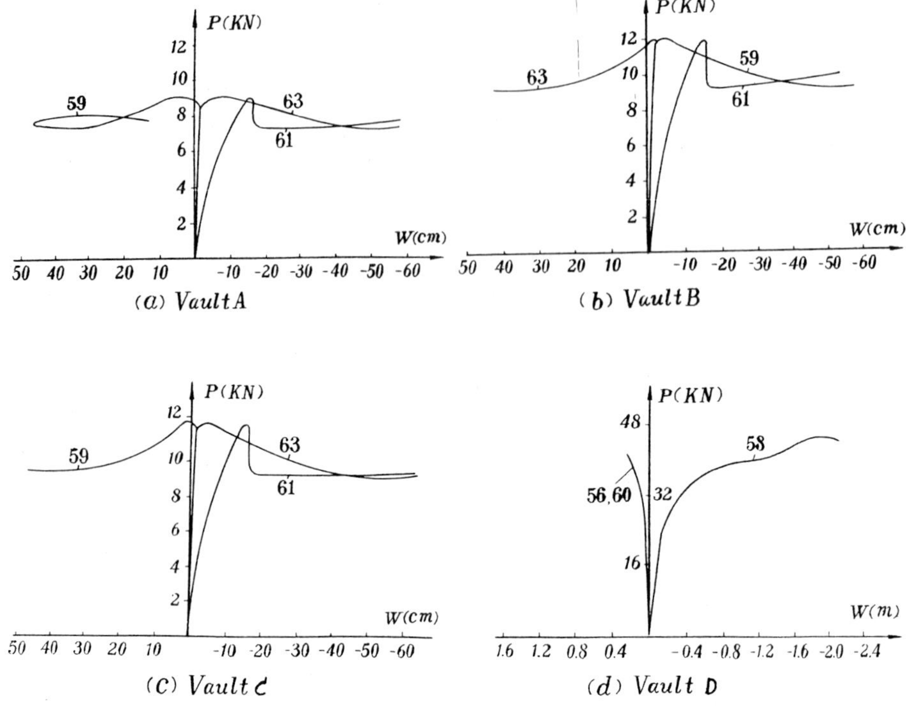

Fig. 2. Load-deflection curve for vaults of different configurations.

Table 1. Critical loads and steel requirements for different configurations.

Barrel vault	A	B	C	D
Critical load (KN / m^2)	1.72	2.26	2.28	8.4
Steel consumption (Kg / m^2)	10.36	17.15	17.15	16.39

6. As for the buckling mode, vaults A, B and C are unsymmetrical buckling (Fig. 3a). At the bifurcation point, the deformation of the structures become unsymmetrical, and the loads decrease abruptly, then the deformation increases rapidly when the loads remain almost unchanged. Vault D exhibits symmetrical buckling (Fig. 3b) and the buckling point (limit point) occurs at a much higher load level.

Fig. 3. Unsymmetrical and symmetrical buckling modes.

RATIO OF LONGITUDINAL SPAN TO TRANSVERSE SPAN

7. Let L represent the longitudinal span and S the transverse span of the barrel vault. In the analysis of the effect of ratio L/S to the stability behaviour, the vault formed from isosceles triangles, shown in Fig. 4, is considered. The ratio L/S varies with the increase in the longitudinal span while the transverse span and the ratio of height to transverse span remain constant at 18m and 0.22 respectively. The pre-and post-buckling load-deflection curves of vaults of various ratios of L/S are shown in Fig. 5a. The relationship between the upper buckling loads and the ratio L/S is shown in Fig. 5b.

8. Figs 5a and 5b show that when $L/S < 1.0$, the critical loads increase rapidly with the decrease in the ratio of longitudinal span to transverse span, and when $1.0 < L/S \leq 1.67$ the change of critical loads is not obvious. This region, $1.0 < L / S < 1.67$ is termed the insensitive region of the critical load. In barrel vault structures stiffening ribs spanning transversely are often needed to improve the stability behaviour of single layer structures. Consequently, this paper suggest that the distance between stiffening ribs or the distance between the stiffening rib and the end bracing frame should be smaller than the transverse span of the vault, otherwise the stiffening ribs can not play their full role.

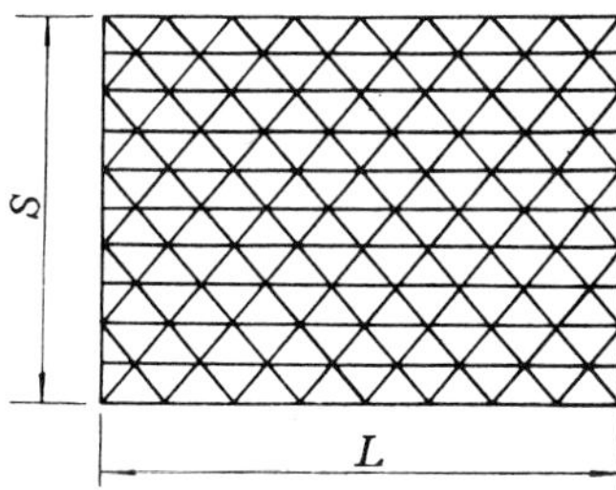

Fig. 4. Vault of isosceles triangle configuration.

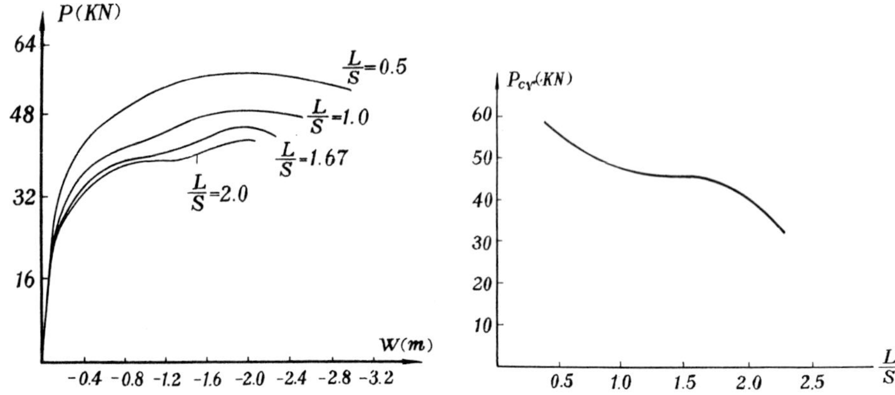

Fig. 5a. Load-deflection curve for various ratios L/S.

Fig. 5b. Pcr vs L/S for vaults of isosceles configuration.

RATIO OF HEIGHT TO TRANSVERSE SPAN

9. In this section, the vault formed from isosceles triangles shown in Fig. 4 is considered again. In order to have a clear understanding of how the ratio of height to transverse span affects the stability behaviour of barrel vaults, the span in both directions are 18m and remain unchanged while the height gradually increases. Six vaults of different height are analysed and their load-deflection curves are shown in Fig. 6.

10. By comparison of the load-deflection curves shown in Fig. 6, it is evident that when the ratio H/S is relatively small, the critical load will increase rapidly with increasing height of the vault, and for normal loading condition, the vault has enough stiffness. Especially when H/S = 0.17, the vault has the greatest structural stiffness. This is clearly seen in Fig. 7 which represents the maximum deflection of different H/S under the load of 2.8kN/m^2. This conclusion is totally coincident with the point of view presented in ref. 6 that the height-to-span ratio of 0.17 gives the minimum structural weight of vaults. To compare with other different height-to-span ratios, at a ratio of 0.17, the members of vault exhibit the maximum axial forces and minimum bending moments, which result in the greatest stiffness and minimum structural weight.

11. When the height-to-span ratio is relatively large, such as H/S = 0.285, H/S = 0.4 and H/S = 0.5 as shown in Fig. 6, the instability doesn't occur throughout the load-deflection curve, but the structure exhibits greater flexibility. This behaviour becomes more and more obvious with the increase in height of the vaults. Generally the design load for barrel vaults are much smaller than their critical load. Too large height-to-span ratio may cause structural deflection that

can not satisfy design requirements, although the stability behaviour has improved. Taking this behaviour into consideration it is suggested that the height-to-span ratio of single layer barrel vaults should not be greater than 0.3, if there are no special architectural requirements.

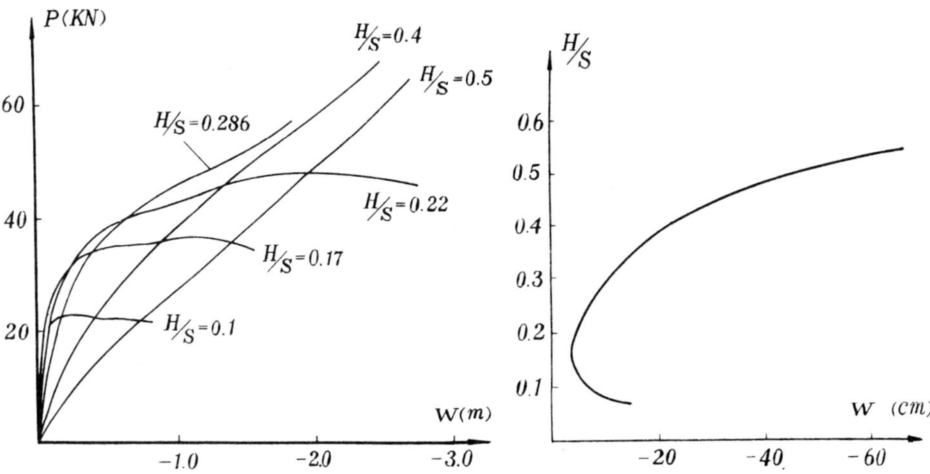

Fig. 6. Load-deflection curves of vault with different height-to-span ratio H/S.

Fig. 7. Maximum vertical deflection versus height-to-span ratio H/S.

SUMMARY AND CONCLUSION

12. The stability capacity of barrel vaults of rectangular configuration is very low. For the vault of rectangular configuration with diagonal bracing, although the steel requirement increases, the stability behaviour does not improve significantly whether the diagonal bracings are in tension or compression. The vaults formed from isosceles triangles exhibit superior stability behaviour compared with the other three types of configurations of vaults and its structural weight also is the lowest (ref. 6). Another advantage of this type of configuration is that it is possible to construct the vault from equal length members which eases construction of the structure.

13. Reducing the ratio of longitudinal span to transverse span and increasing the height-to-span ratio are effective ways to improve the stability of single layer barrel vaults. When the longitudinal span is much larger than the transverse span, the distance between the stiffening ribs, if they are used, should be less than the transverse span. A favourable height-to-span ratio, which will allow the barrel vault to exhibit superior stability behaviour and at the same time having enough stiffness to meet the limitation of structural deflection, should be between 0.17 and 0.3.

REFERENCES

1. J.L.MEEK and H.S.TAN, Geometrically non-linear analysis of space frames by an incremental iterative technique, Computer method in applied mechanics and engineering, VOL.47, 1984.

2. J.L.MEEK and S.LOGANATHAN, Large displacement analysis of space frame structures, Computer method in applied mechanics and engineering, VOL.72, 1989.

3. SUN JIANHENG and ZHAO HUILIN, Snap-through instability analysis and experimental Studies of single layer braced dome, Proceedings of fourth international colloquium on structural stability, International academic publisher, 1989.

4. SUN JIANHENG, Modified tangent stiffness matrix for non-linear stability analysis of single layer braced domes, Journal of structural engineering VOL.3,4, 1990 (in Chinese).

5. SHU GANGPING and WU LAN, A theoretical analysis and experimental study on the non-linear stability of a single layer reticulated barrel vault, Symposiums on space gird structures and reticulated shells, Tongji University Publisher, 1991, (in Chinese).

6. Z.S.MAKOWSKI, Analysis, design and construction of braced barrel vaults, Elsevier Applied Science Publishers, 1985.

8. Shape creation and erection of metal space structures by means of post-tensioning

L. C. SCHMIDT and G. DEHDASHTI, University of Wollongong, Australia

SYNOPSIS. The paper presents a brief review of the concept of the use of post tensioning as a means of creating the final shape of a curved space truss. The concept is basically one where the whole structure is assembled initially at ground level in an essentially flat or planar condition. By the use of post-tensioning, in an appropriate way, the space structure can be erected by deflection into its final shape and into position. Potentially, the procedure can offer economies over traditional methods of construction of curved space trusses. Two series of barrel vaults are described for their shape formation and subsequent testing to destruction. Linear elastic analysis is seen to furnish conservative results for the ultimate load-carrying capacity.

INTRODUCTION

1. Large clear spans are required for a variety of uses, such as sports stadia, industrial buildings, warehouses, aircraft hangars, supermarkets, leisure centres and swimming pools. The provision of these structures is a challenge for engineers and architects to achieve economical and aesthetically pleasing forms.

2. Eberlein (ref. 1) has referred to the erection of space structures used for such purposes. He considered three methods of erection: the cantilever method, the lift span method, and a combination of these two. He indicated that the third method is the most common, where units of a structure are assembled at ground level, are lifted into position, and are then connected to other units in the air by individual members placed separately. Usually, flat-decked space structures are erected in this way. Depending on the capacity of the lifting plant available, so the size of the structural units assembled at ground level will vary. With recent developments in mobile cranes, increased sizes of units are being handled. However, construction practices are always under review.

3. A purpose of the present work is to investigate the possibility of generalising the concepts introduced by Strarch (ref. 2) and Ellen (ref. 3), and to see whether planar space trusses can be formed or folded into shapes that are of architectural interest, such as barrel vaults. The folding or shaping is integral with the erection process.

4. Barrel vaults are usually constructed in their final shape *in situ*, in either single-or double-layer forms. While a single-layer chord system, even with in-plane shear stiffness, can be readily folded or shaped in single curvature, instability under load in its final form can quickly become a problem owing to the low out of-plane flexural stiffness. Double-layer truss systems, on the other hand, rarely suffer this problem, but they are unable to be folded or shaped from a flat position. In this paper, however, attention is focused on a space truss consisting of a single layer of chords together with a system of out-of-plane web members. Such a system (referred to as a single-chorded space truss, SCST, Fig. 1), even when pin-connected, is intermediate in out-of-plane flexural stiffness between the single-and double-layer planar space-truss systems (ref. 4). This system has the advantage of the single-layer system in that it can be folded or shaped in various ways. After erection, however, the system depends for its structural behaviour on the axial stiffness of the members and not on their flexural stiffness.

5. The out-of-plane stiffness of the barrel vault SCST will increase with the height of the pyramidal web system apex from the plane of the chords (ref. 5). For the cases investigated herein, the ratio of the distance between parallel chords, or pyramid base width, and the height of the pyramids was chosen to be 1.42.

6. The work described herein refers to two sets of articulated barrel vaults formed from the planar state, and to the results of their subsequent tests to destruction.

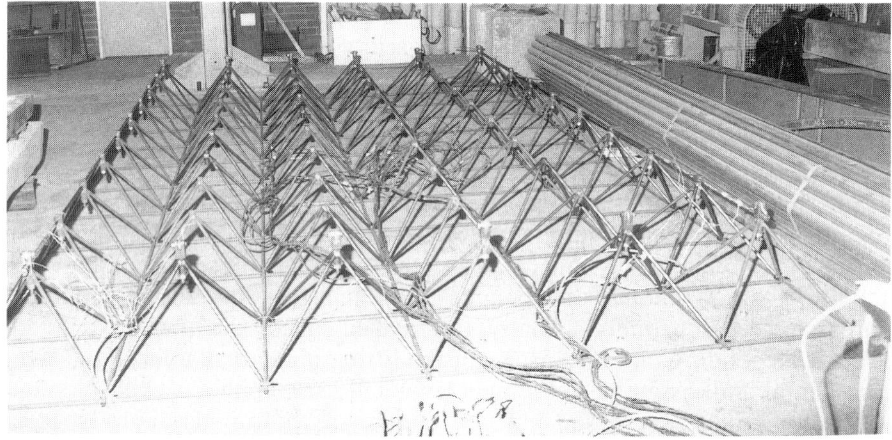

Fig. 1. Single-chorded space truss in initial form.

TEST TRUSSES

7. The dimensions of the two model types are shown in Figs 2a and 2b, in developed or planar form. The 6 x 6 mesh panel truss had a 5.2m radius of curvature, while the 10 x 6 mesh panel truss had a smaller radius at 1.18m. Figs

3 and 4 show the curved 6 x 6 mesh truss and the more sharply curved 10 x 6 mesh truss, respectively, after post-tensioning.

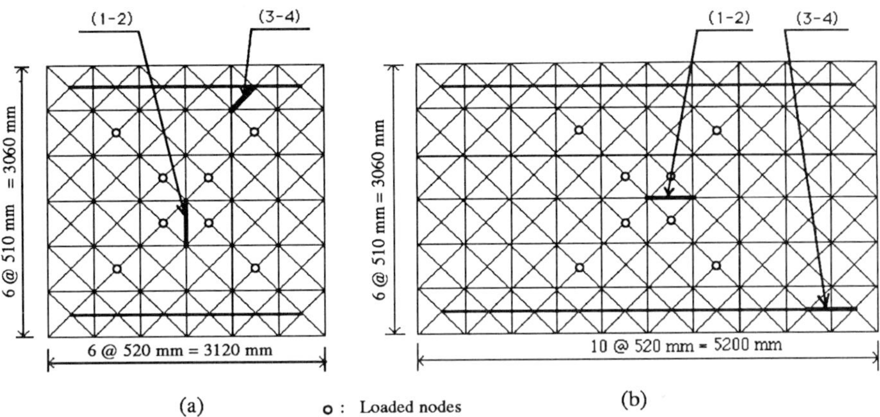

Fig. 2. Dimensions of two space truss types, and positions of loads and failed members.

8. The basic fabrication consisted of 13 x 13 x 1.8mm SHS steel chords, with steel web members of 13.5 x 2.3mm CHS. These sizes were kept constant for all six trusses tested. The depth of the apices of the pyramids formed by the web members were 365mm below the plane of contact of the orthogonal chord members. The nominal slenderness ratio of the chords, between panel points, in the flat position, was 121.

Fig. 3 Shape of vaults with 5.2m radius of curvature.

9. One layer of chords was placed at right angles to the other layer and bolted together (6mm HT cap screws), in order to form a rectangular mesh. The chords were continuous, therefore, through the joints in the direction of each set of chords. Eccentricities existed at the joints, between the centrelines of the orthogonal SHS chords, of value equal to the depth of the square section. Previous work (refs 6-9) has shown that such eccentricities are not as detrimental to the load-carrying capacity as might be thought at first. In regions of the truss system where the chord forces are high, the shear contribution to the axial chord forces from the web members is small; therefore moments caused by the eccentric shear forces are small (ref. 6). In regions of the truss system near the points of support, where the chord forces are small, the shear contributions from the web members are high.

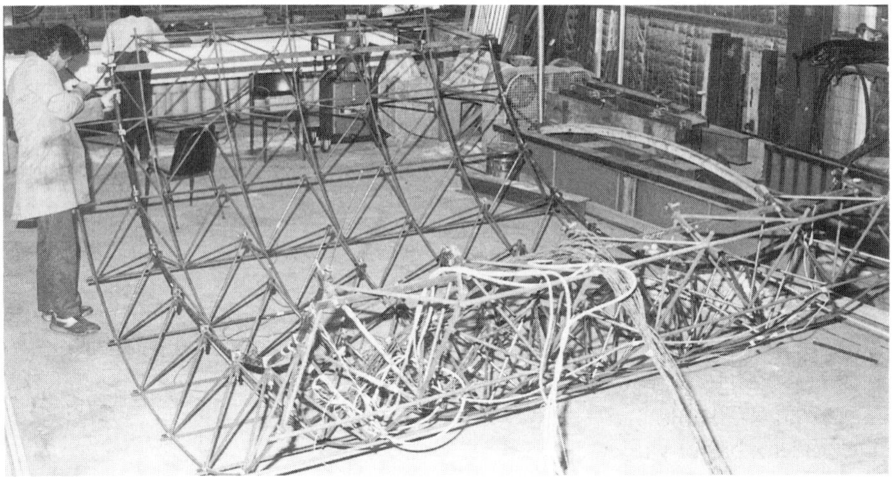

Fig. 4. Shape of vaults with 1.18m radius of curvature.

Because of the small axial forces in the chords in these locations, there is a margin of strength available to withstand the bending moments caused by the eccentric shear forces. This argument assumes that the chord members are essentially of the same size throughout the truss system.

10. The simplicity of fabrication (and therefore lower cost) using continuous chords, which furnishes member continuity through the joint, helps to offset any small loss of strength (and therefore increased cost) due to the eccentricity. Additionally, any potential loss of strength caused by a joint instability effect (ref. 7), due to the use of discrete members and nodes (hubs or connectors), is avoided.

11. Although dealing with laboratory scale barrel vaults, the authors were able to compare, in round terms, the time of fabrication of a discrete-member welded vault (ref. 5) with the similarly-sized continuous-chorded trusses of the investigation herein. The ratio is of the order of 4 to 1 in favour of the currently proposed procedure for fabrication. The fabrication tolerances were also able to be compared, and, again, were in favour of the continuous-chorded trusses.

SHAPES FORMED

12. Six model trusses were fabricated and shaped before testing. All models were constructed in the flat single-layer condition, as shown in Fig. 1. A second layer of chords, at the depth of the web apices, was not used, as indicated earlier, so as to allow each structure to be readily formed into a barrel vault with an appropriate radius of curvature. Two values were chosen, as indicated earlier. The first set of three identical trusses was formed from a flat 6 x 6 mesh (Fig. 3). The second set of almost identical vaults was formed from a flat 10 x 6 mesh (Fig. 4).

13. It is worthy to note that for a space truss to develop plate or shell-like action (ie. to develop two-way action in order to transmit forces and moments in two directions to supports), the edges of the truss need to have significant stiffness compared with that at the interior regions of the space truss. Consequently, as each of the barrel vaults was formed from a single-chorded planar space truss, all four edges had to be stiffened or supported in some way.

14. Two opposite sides were supported on pins at each lower web node. The remaining opposite two edges had tubular members located at lower chord level with a predetermined lack of fit. This lack of fit was calculated so that, after stressing of a 5.05mm diameter high tensile steel strand passing through the lower chord tubes, a curved vault of the desired radius of curvature was formed (Figs. 3 and 4). Table 1 lists the radius of curvature and the rise/span ratio for each of the six barrel vaults tested.

Table 1. Test program.

No	No of Panels	Radius of Curvature (m)	Post-tensioning force (kN)	Rise/Span Ratio
1	6 x 6	5.2	9.5	0.07
2	6 x 6	5.2	9.5	0.07
3	6 x 6	5.2	13.5	0.07
4	10 x 6	1.18	11.3	0.44
5	10 x 6	1.18	18.2	0.44
6	10 x 6	1.18	18.2	0.44

15. The values of the post-tensioning forces used in the prestressing strands are also given in Table 1. Different post-tensioning forces were used for the various cases so as to alter the internal load at which separation would occur in the edge lower chord members, due to the additional tensile force equalling and overcoming the initial compressive force in the member.

16. For trusses 1 - 4, the lower edge chords were of size 13.5 x 2.3mm CHS tubes. With increasing post-tensioning forces the size of the CHS used for truss 5 was 17.1 x 3.2mm, and for truss 6, 26.7 x 5.6mm.

17. The four edges of the single-chorded space truss were, therefore, stiffened. The structural action expected was that vertical internal distributed load would be transferred by shell-like action directly to the side supports, or transversely to the edge trusses and then to the foundation.

TEST RESULTS

Member Tests

18. The material yield stress was taken from three tests on the SHS tubes used for the chords, and gave a value of 400MPa. Likewise, three tests on the CHS tubes used for the web members gave a yield stress of 445MPa

19. Pin-jointed strut tests (loaded through bolts) were carried out on straight sections of the top chord tube with slenderness ratios equal to those used in the model vaults. The average of five tests furnished a value of 10.8kN, which was used subsequently in the simplified theoretical predictions of the ultimate load capacities of the various vault tests.

Ultimate Load Tests

20. Table 2 lists significant test results. The trusses were loaded at the bases of the pyramidal web system in patterns as indicated in Figs 2a and 2b. A whiffletree arrangement ensured equal load values at all loaded points.

Table 2. Test and computer results.

No	Load at separation (kN)	Initial compressive chord force (kN)	Ultimate load capacity (kN)		Test failure position
			Test	Computer	
1	18	---	52.2	36.0	Top chord
2	10	---	45.7	36.0	Top chord
3	20	0.2	51.7	---	Web
4	10	---	36.7	---	Bottom chord
5	30	---	36.7	---	Bottom chord
6	35	0.0	49.0	38.0	Top chord

21. The external load that caused the compressive force to be reduced to zero in the lower chord member on the transverse centreline, in each truss, is given in the Table. At this stage the tensile force in the member is carried by the high tensile

strand on the tube centreline. The loads to cause separation are less than 50% of the ultimate loads for tests 1 - 4. The higher post-tensioning force in tests 5 and 6 caused separation to occur later in the loading history. A loss of vertical stiffness is expected after separation, but this is not apparent in the load versus deflection results shown in Fig. 5 and 6.

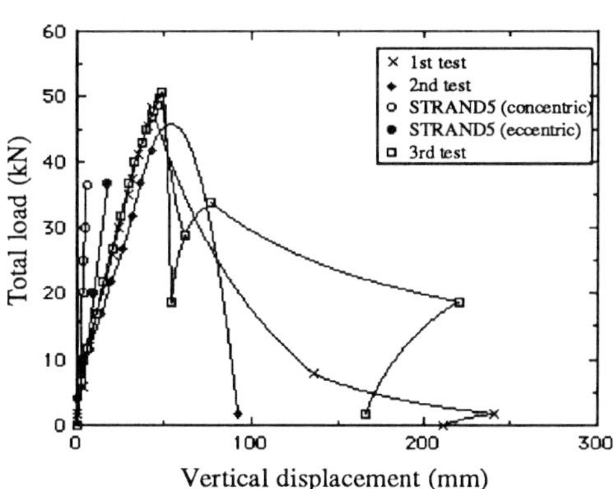

Fig. 5. Experimental and theoretical results for tests 1, 2 and 3.

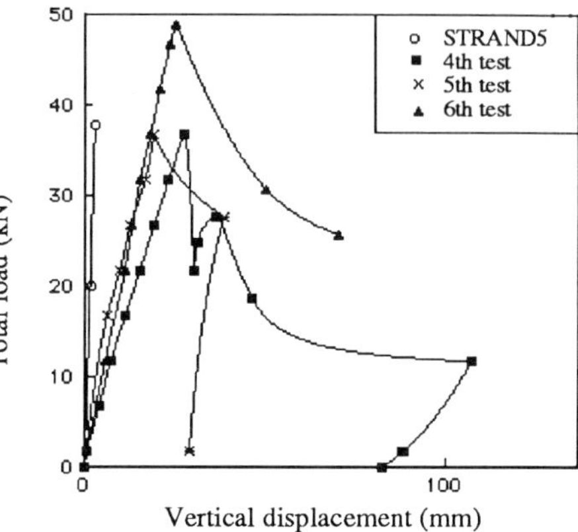

Fig. 6. Experimental and theoretical results for tests 4, 5 and 6.

22. The member failures varied between the trusses according to the level of the post-tensioning force. For example, test 3 failed initially by web failure (member 3 - 4, Fig. 2a), although the external load reached was almost as great as that in test 1, when the top chord member 1 - 2 (Fig. 2a) failed. With the greater curvature tests (tests 4 - 6), it is noted that a bottom chord member (member 3 - 1, Fig. 2b), adjacent to the support, failed in compression. The lower chord member sizes were increased to CHS 19.1 x 3.25mm, as was the post-tensioning force. Again the bottom chord member in the same position failed. Increasing the member sizes again (to CHS 26.7 x 5.6mm) caused top chord compressive failure at a load of 49.0kN (member 1 - 2, Fig. 2b).

23. In the cases where top chord member failure occurred (tests 1, 2 and 6), the failure was dramatic, with subsequent sway deflections of the adjoining chord members. As the load versus deflection curves show (Figs. 5 and 6), the ultimate load was reached when the first member collapsed; no higher peak load was reached.

24. The initial top chord axial forces in the members expected to fail were small (Table 2), and played no significant role in the failure pattern. The maximum top chord member force for truss 6 (Table 2) was recorded by strain gauges as having a level greater than the pin-ended value (10.8kN), indicating that the elastic restraint provided by the adjacent spans increased the load capacity.

COMPUTER ANALYSIS

25. Linear-elastic computer analyses (ref. 10) of a pin-jointed theoretical model were carried out for the two sets of trusses, assuming that the maximum compressive chord force that could be achieved was 10.8kN. The theoretical maximum load-carrying capacities for trusses 1, 2 and 6, where compressive chord failure occurred were 69, 79 and 78 percent, respectively, of the experimental results (Table 2). The results are plotted in Figs 5 and 6.

26. This comparison reflects the conservative nature of the assumption of pin ended strut capacity as the governing criterion for compressive chord failure. However, this assumption is a common one adopted in many design codes throughout the world.

CONCLUSIONS

On the basis of the model studies carried out herein, the following conclusions can be drawn.

27. It appears that curving or forming a barrel vault by the post-tensioning procedure furnishes a far more economical structure than by constructing the same vault, composed of discrete members, in its final position.

28. The tolerances achieved in the formation of the barrel vaults were equal to or better than, those obtained by conventional methods, such as welding.

29. After web-or lower chord-failure was prevented, the ultimate load capacity of each vault tested was determined by the axial load capacity of the top chord critical compression member between panel points. The out-of-plane web members prevented an initial overall instability of the chord members.

30. A local zone of collapsed members formed in each case after the first compression element buckled.

31. The pin-ended member load capacities, together with linear analyses, enabled conservative estimates to be made of the ultimate load capacities of the vaults in each case where failure occurred in the top chord compression members. The theoretical estimates of the ultimate loads were approximately 30 percent below the experimental values.

32. The observed vertical central stiffnesses of the eccentrically connected vaults were about seven times less than those derived from linear analyses of the simple pin-connected theoretical models adopted.

REFERENCES

1. EBERLEIN H. Latticed Space Structures - a Comprehensive Survey, Acier Stahl - Steel, 2 1975, 50 - 66.
2. STRARCH INTERNATIONAL (Singapore) Pty Ltd., 1990.
3. ELLEN P.E. The development of membrane structures 25 year progress report. IASS Congress, Madrid, Vol. 3, 11 - 15 Sept, 1989.
4. SCHMIDT L.C. Erection of space trusses of different forms by tensioning of near mechanisms. IASS Congress, Vol. 4, Madrid, Spain, 11 - 15 Sept, 1989.
5. TAN T.H. and SCHMIDT L.C. Ultimate load behaviour of a barrel vault space truss. Intl. J. Space Structures, 2, 1986/87, 1 - 10.

6. SCHMIDT L.C., MORGAN P.R. and COULTHARD B. The Influence of eccentricity and continuity on the inelastic behaviour of a space truss. Sixth Austin. Conf. Mech. Str. and Matls., Univ of Canterbury, Christchurch, NZ, Aug, 1977.

7. SCHMIDT L.C. and MORGAN P.R. Behaviour of some joint designs in space trusses. Civ. Eng. Trans., I.E. Aust., Vol. CE23, No. 3, Aug 1981, 172-177.

8. SCHMIDT L.C., MORGAN P.R. and PHANG P.W. Influence of joint eccentricity and rigidity on the load capacity of a space truss sub-assemblage. J. Cons. Steel Res., Sept, 1981, 16 - 22.

9. SCHMIDT, L.C. The use of continuous chords in plate-like space trusses. IASS Congress, Moscow, vol. IV - B, 1985, 272 - 285.

10. STRAND 5. User Manual. G + D Computing, Sydney, 1989.

9. Stability of a space truss composed of square pyramid units

T. SUZUKI and T. OGAWA, Tokyo Institute of Technology, Japan, and Y. TSUSHIMA, M. KIMURA and H. KANEKO, Takenaka Technical Research Laboratory, Japan

SYNOPSIS. In this paper, the square pyramid space truss, one of the most efficient truss construction forms, is reviewed, and an analytical investigation has been made on the stability of a structure to which a vertical load and a horizontal load are simultaneously applied. Because the pin joint is unstable, various methods have been used to make this space truss a stable structure. From the maximum strength onwards, deformation capacity becomes important as a latent safety factor. A method has been proposed, which is suitable for securing deformation capacity, by taking into consideration individual member buckling.

INTRODUCTION

1. The space truss, as a large-span roof structure, has been designed by Allowable Stress Design mainly under vertical loading. The joints of the space truss are frequently made into a ball joint of system trusses, taking into consideration its workability, which is dynamically evaluated as a pin joint.

2. The square pyramid space truss becomes an unstable structure in the case of the pin-joint due to the relative rotation of the square pyramid as shown in Fig. 1 (ref. 1). To make the square pyramid space truss into a stable structure, the following conditions should be satisfied:

(1). To make the periphery a simple support.

(2). To stop the entire in-plane motion parallel to the horizontal plane.

(3). To stop the relative rotation of the square pyramid.

3. This paper is concerned with the investigation of the square pyramid space truss. Various methods have been undertaken to render the space truss stable, together with the assessment of its structural stability in the plastic region, which is accompanied with individual member buckling. Both horizontal and vertical loading is used.

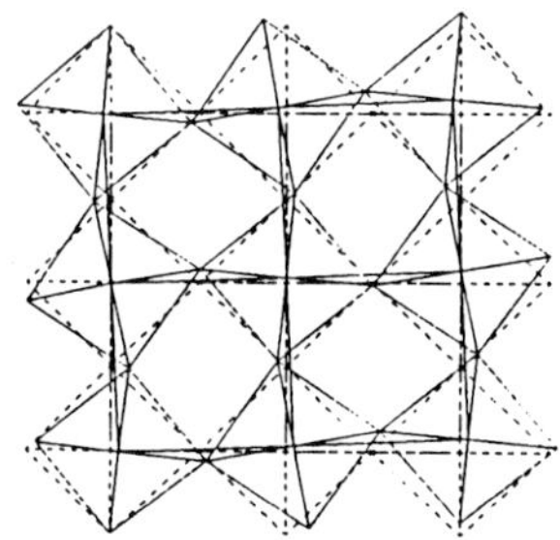

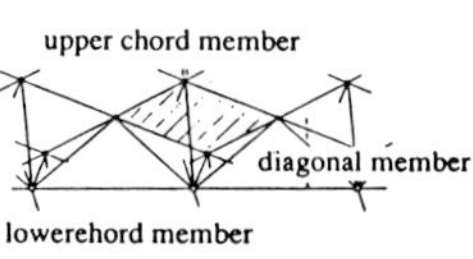

(a) Relative rotation. (b) Unit of square pyramid.

Fig. 1. Unstable behaviour of the square pyramid space truss with pin-joint.

MODELLING AND OUTLINE OF ANALYSIS

Analysis Model

4. Fig. 2 shows the shape and dimensions of the analysis model. The grid of 2 x 2 bottom chord members is made into the composition unit of the space truss, and the sectional shape of the truss member is made constant. Characteristics of the truss members are shown in Table 1. The nominal yield stress for the steel members is 235 MPa. The buckling stress in the table is the elasto-plastic flexural buckling stress of truss members, both ends of which are simply supported, having the nodal point distance as the buckling length.

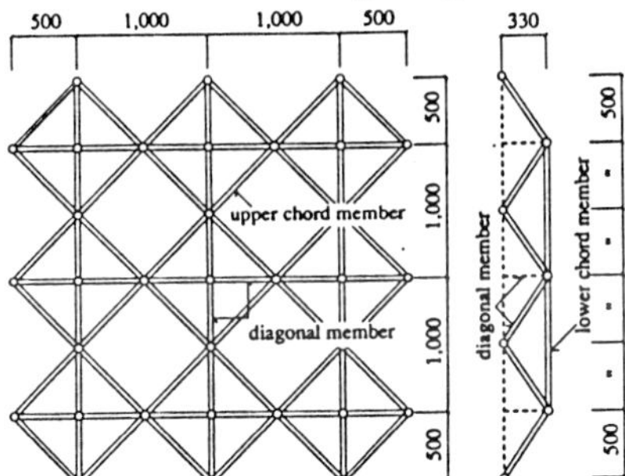

Fig. 2. Shape and dimensions of analysis model.

Table 1. Characteristics of truss members.

member	cross section	nodal point distance L (mm)	slenderness ratio λ	yield stress F (MPa)	buckling stress σcr (MPa)
upper chord	O-27.2×2.6	599.1	68.5	235	178
lower chord	〃	1000.0	114.3	〃	107
diagonal	〃	707.1	80.8	〃	161

$$\sigma cr = \frac{\left\{1-0.4\left(\frac{\lambda}{\Lambda}\right)^2\right\}F}{1+\frac{4}{9}\left(\frac{\lambda}{\Lambda}\right)} \qquad \Lambda = \sqrt{\frac{\pi^2 E}{0.6\,F}}$$

5. Type 1 achieves structural stability by constraining relative rotation of the pin jointed square pyramid by installing a peripheral beam at the boundary of the upper chord members. For the peripheral beam, two kinds of members are selected, i. e. (1) B-1: the same member as that of the truss member and (2) B-2: a member with greater rigidity. Fig. 3 shows the specimen model, and Table 2 shows the sectional characteristics of the peripheral beams.

Table 2. Sectional characteristics of the peripheral beams.

beam	cross section	sectional area (cm^2)	geometrical moment of inertia Iy (cm^4)	radius of gyration iy (cm^4)
B − 1	ϕ-27.2×2.6	2.009	1.537	0.875
B − 2	H-100×100×6×8	21.90	134	2.47

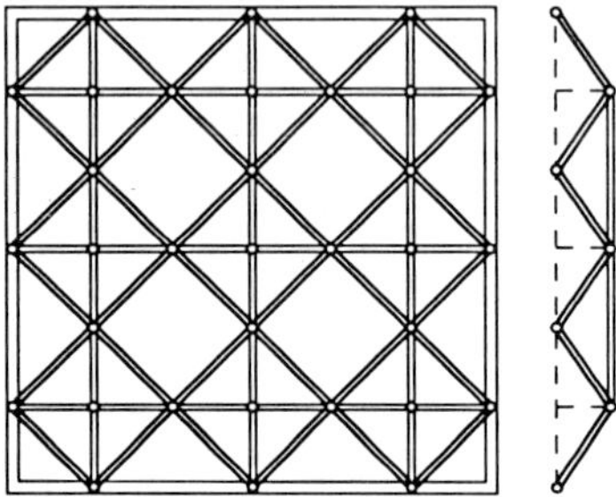

Fig. 3. Model with peripheral beams.

6. Type 2 is a model which achieves structural stability by constraining the relative rotation of the square pyramid through the addition of a member thus forming a triangle . For the analysis the number of inserted members is adopted as a parameter. The inserted members are the same as the truss members. Fig. 4 shows the specimen.

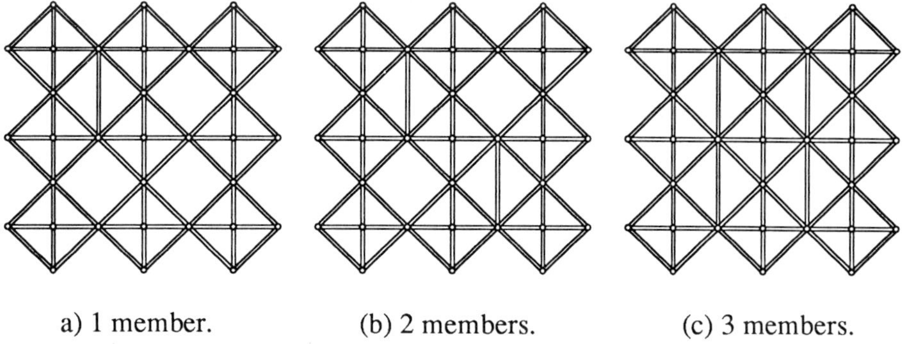

a) 1 member. (b) 2 members. (c) 3 members.

Fig. 4. Model with inserted members in the square of the upper chord member.

7. Type 3 achieves the structural stability by making the relative rotation of the square pyramids difficult through additional rotational rigidity of the joints. As shown in Fig. 5, the nodal point of the truss and the member are connected by a spring element. The geometrical moment of inertia of the member is denoting by I_o and that of the spring element is changed to I_o, $I_o/2$, $I_o/4$ and $I_o/16$. In this case, the spring element is assumed to transmit the axial force of the truss member, and the yield stress is assumed to be $\sigma_y = 819$ MPa.

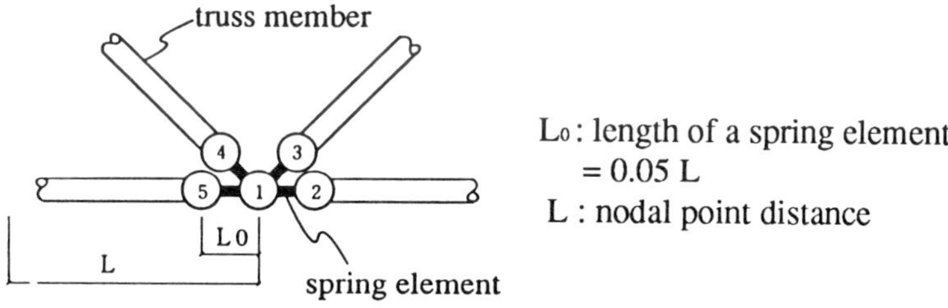

Fig. 5. Modelling of the joints.

Boundary Conditions and Loading Conditions

8. The boundary conditions which do not constrain the relative rotations, as shown in Fig. 6, are used. u, v, and w as well as θ_x, θ_y, and θ_z are translational displacements and rotational displacements in the X-, Y-, and Z-directions, respectively. Numerals I, II, and IV are kinds of boundary condition and are given at all nodal point positions of the periphery of the upper chord members.

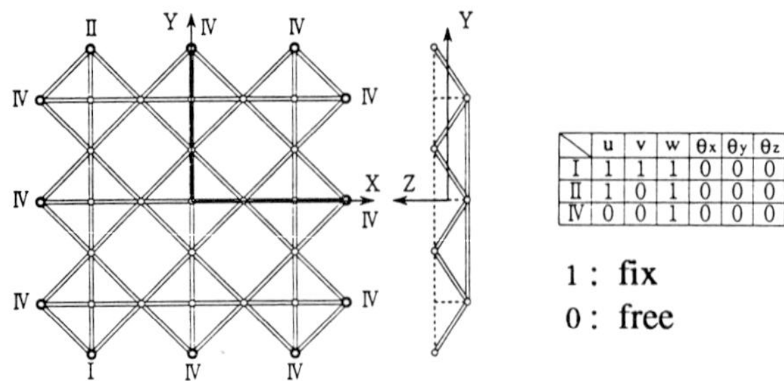

	u	v	w	θ_x	θ_y	θ_z
I	1	1	1	0	0	0
II	1	0	1	0	0	0
IV	0	0	1	0	0	0

Fig. 6. Boundary conditions.

9. To apply vertical and horizontal loads simultaneously, the vector of the load is set up (Fig. 7). Concentrated loads are applied at all nodal points, except the periphery of the upper chord members. Components of the load vector are set to a ratio of 1: 2: -3 in the horizontal X- and Y-directions and in the vertical Z-direction, and loads are applied under loading control.

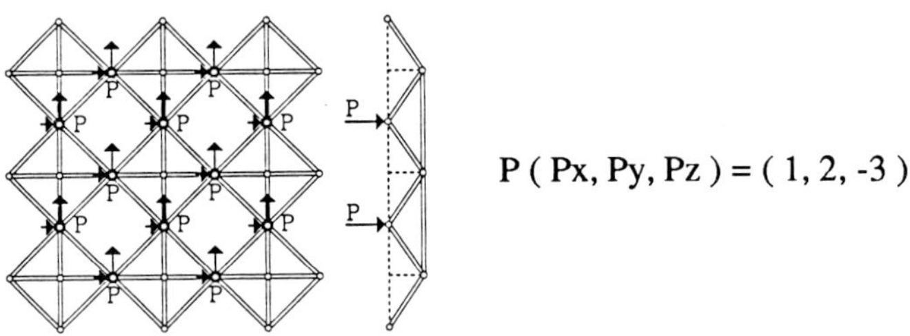

Fig. 7. Loading points and components of the load vector.

Analysis Sequence Taking into Consideration Individual Member Buckling

10. The Analysis was performed using the universal structural analysis program "NASTRAN". An individual member between nodal points of the space truss corresponds to a single finite element. At the instant when the member has reached the compressive stress corresponding to elasto-plastic flexural buckling strength, the member is judged to have buckled. The member which has buckled is removed, and in the new space truss composition, an analysis is again carried out from the starting point. The fact that the buckled member is removed is considered due to sudden development of a strength drop by buckling and that it is unable to transmit further compressive force. On the other hand, tension is

considered to be perfect elasto-plastic. Fig. 8 shows skeleton curves of the tensile and compressive members.

11. The post-elasto-plastic buckling behaviour is expressed by enveloping the point of the load and the displacement at the individual member buckling generating point in the new space truss composition (Fig. 9). Provided, however, that at the initial period, the load-displacement relationship includes the point which has caused tensile yielding.

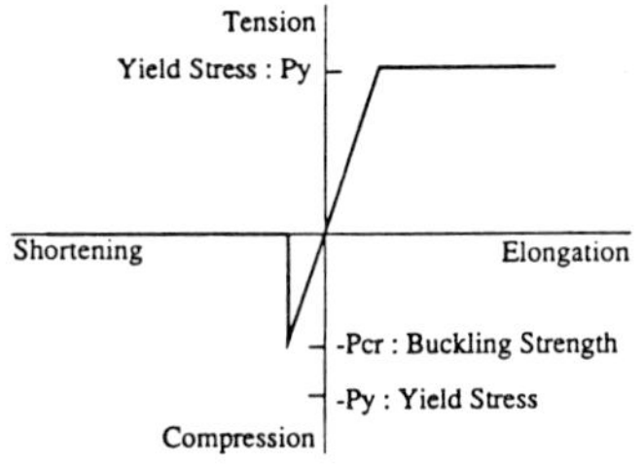

Fig. 8. Skeleton curves of the tensile and compressive members.

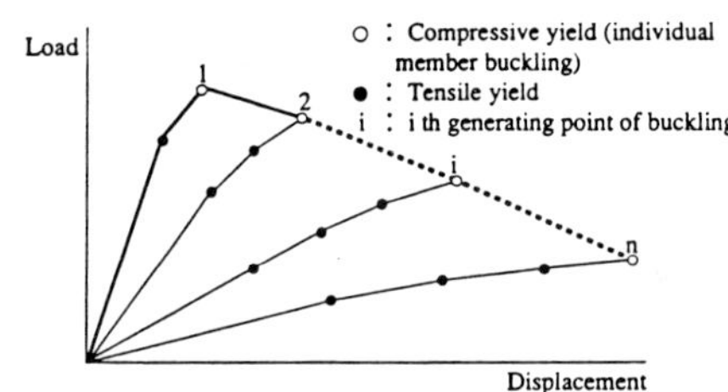

Fig. 9. Load-displacement curves of space truss accompanied with flexural buckling.

STABILITY OF SPACE TRUSS

12. First, the structural stability and instability is classified by the assessment of statically-determinacy or indeterminacy. In general, the order of statically-indeterminacy can be obtained from the number of nodal points, joining conditions, number of members, number of reactions and supporting conditions. From the viewpoint of the sufficient conditions, the value of the determinant of the stiffness matrix is obtained the numerical analysis, and when this value is evaluated as zero, the structure can be considered as being instable.

13. Fig. 10 shows the relationship between the number of inserted members of Type 2 and value of the determinant $\overline{D}$ of the stiffness matrix. The value of $\overline{D}$ of the axis of ordinates has been made dimensionless using the value of $\overline{D}$, when the number of inserted members is 2. When a single member is inserted, the value of $\overline{D}$ is increased by a 2-digit exponent with respect to the zero number of the inserted members. Further, as the number of inserted members is increased, the value of $\overline{D}$ increases by a multiple of nearly one digit, and gradually tends to converge. The method of inserting a truss member diagonally on the square is effective in stabilising the pin-connected square pyramid space truss, thereby indicating effectiveness which is obtained simply by inserting a single member.

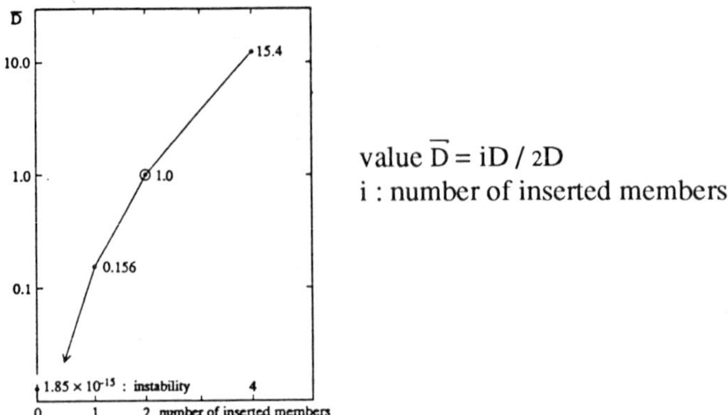

Fig. 10. Relationship between the number of inserted members and value of D̄.

STABILITY IN ELASTO-PLASTIC REGION

14. Evaluation of the deformation capacity of the space truss is important in applying the space truss to a structure that can resist seismic force. Regarding three cases for stabilising the pin-jointed square pyramid space truss, which is an unstable structure, the examination results of elasto-plastic behaviours are shown below.

In the case of Type 1

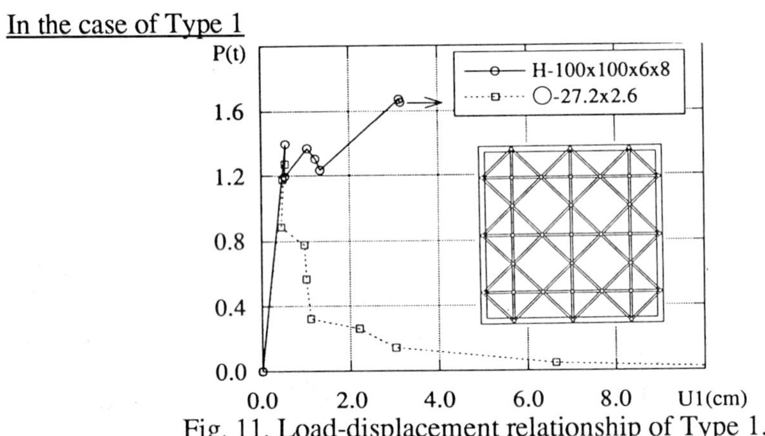

Fig. 11. Load-displacement relationship of Type 1.

15. Fig. 11 shows the load-displacement relationship, when a periphery beam is provided at the boundary. The load at the axis of ordinates is a value of concentrated force on one of the nodal points, and expressed in the magnitude of vector (1:2:-3). Displacement Ul of the axis of abscissas is the nodal-point displacement at the centre of the bottom chord, and is expressed by the magnitude of the vector from the original co-ordinate point. When the peripheral beam (□ : B - 1) is made the same size as the truss members, the structure is

stabilized, but as displacement progresses after the maximum strength, a sudden drop in strength is caused when the structure becomes unstable. When the periphery beam (o: B - 2) is made to have a cross section about 10 times that of the truss member, the effect of the periphery beam becomes large due to the small number of grids, thereby stabilising the structure even after maximum strength.

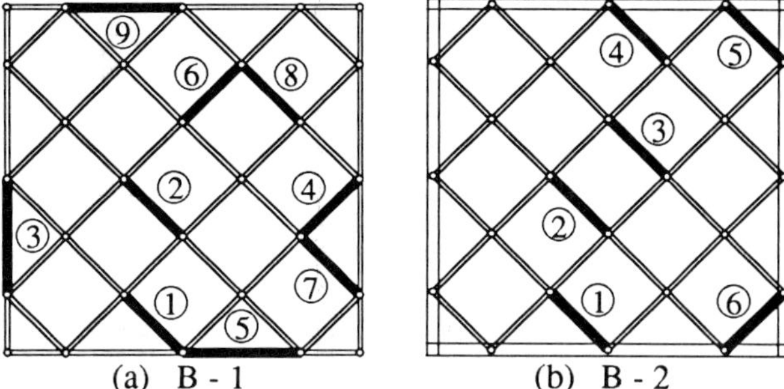

Fig.12. Generating position and sequence of individual member buckling.

16 . Fig. 12 shows the position and sequence of individual member buckling of the upper chord members. The object member is expressed by a black colour. Numerals show the sequence of buckling generation. No individual member buckling occurred on the bottom chord member and diagonal member. The peripheral beam B-1 has individual member buckling from the early stage, but the peripheral beam B-2 has no individual member buckling.

17. Consequently, when the peripheral beam has a proportion of developing individual member buckling, the peripheral beam becomes unable to constrain the relative rotation of the square pyramid, and the square pyramid space truss becomes an unstable structure. However, in the case of the proportion in which the periphery beam will not develop individual member buckling, the relative rotation of the square pyramid is constrained and thus the square-pyramid space truss has deformation capacity.

<u>In the case of Type 2</u>

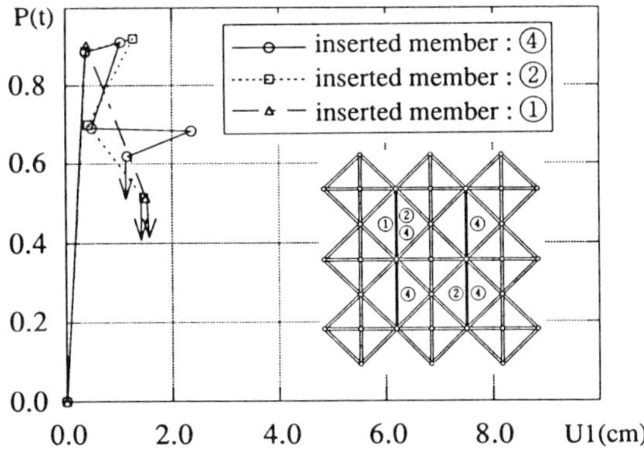

Fig. 13. Load - displacement relationship of Type 2.

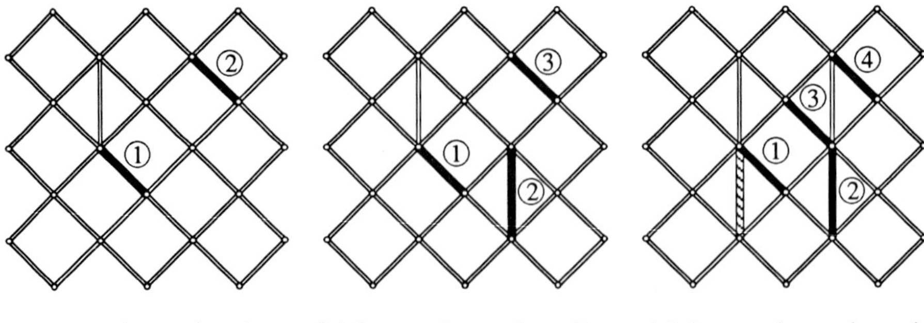

(a) Inserted member 1 (b) Inserted members 2 (c) Inserted members 4

Fig. 14. Position and sequence of individual member buckling.

18. Fig. 13 shows the load-displacement relationship when a member having the same dimension as the truss member is inserted into the square of the upper chord surrounded by the square pyramid. Numbers of inserted members 1, 2, and 4 are indicated by Δ, □, and o respectively. Figs 14 (a)~(c) show the positions and sequences of individual member buckling, when the number of inserted members is 1, 2 and 4. In this case also, no individual member buckling occurs at the bottom chord member and diagonal member.

19. When a single member is inserted, the pin-jointed square pyramid space truss is a stable structure and has load carrying capacity. When the first member of the upper chord shown in Fig. 14(a) buckles, the strength suddenly drops. Further, when the second member buckles, the load carrying capacity is lost and displacement becomes infinite, thereby making the square pyramid space truss an unstable structure. Even if the number of inserted members are increased to 2 and 4, the inserted members buckle, and the pyramid space truss becomes an unstable structure. In the truss structure shown here, even if only a single member is inserted, it will contribute to the stability of the structure, but regardless of the number of inserted members, no deformation capacity can be expected.

In the case of Type 3.

20. Fig. 15 shows the load-displacement relationship. Plots Δ, □,◇ and o in the Figure indicate the cases where the geometrical moments of inertia of spring elements are $I_o/16$, $I_o/4$, $I_o/2$, and I_o, respectively. When the rigidity of the spring element is increased, the Rahmen effect can be expected with respect to the deformation of the rhomboid of the square, and the space truss action is stabilised. However, until the geometrical moment of inertia of the spring element becomes similar to that of the truss member, plasticization of the spring element progresses due to relative rotation of the square pyramid; hence axial force acting on the truss member does not reach buckling strength. In such a case, for securing both strength and deformation capacity, the use of a rigid joint can be

said to be most suitable. Provided, however, that the spring element is to form a plastic hinge.

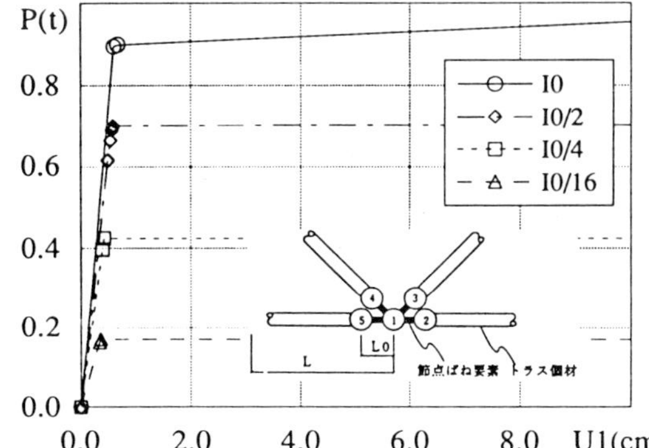

Fig. 14. Load - displacement relationship of Type 3.

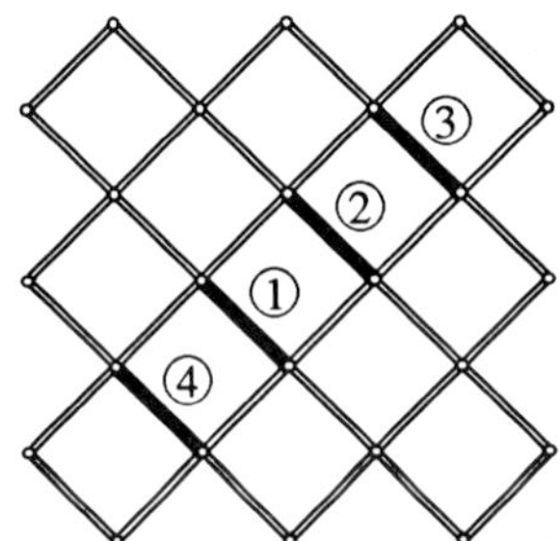

Fig. 16. Position and sequence of individual member buckling (rigid joint).

21. Fig. 16 shows the position and sequence of individual member buckling of the upper chord members in the case of the rigid-joint. Four members buckle, but even at this stage, no drop in strength has occurred.

CONCLUSION

22. Various methods of stabilising a pin-jointed square pyramid space truss have been investigated, and the effect of the respective methods on the rigidity, strength, and displacement behaviour after individual member buckling have been studied.

23. The method (Type 1) of installing the peripheral beam and the method (Type 2) of inserting a member give nearly the same elastic rigidity, because the relative rotation of the square pyramid is constrained. However, Type 1, which uses more members, has higher strength than that of Type 2.

24. The method (Type 3) of employing a rigid-joint cannot completely constrain the relative rotation of the square pyramid; hence, the elastic rigidity is less than

that of the other two types. However, the strength of Type 3 is similar to that of Type 2, since strength is determined by individual member buckling.

25. To secure the deformation capacity of the square pyramid space truss, the method of employing a rigid-joint is effective.

REFERENCES

1. KATO B ., TAKANASHI Y., TSUSHIMA Y., and HIRATA Y. The analysis of a space truss composed of square pyramid units. The international Conference on Space Structures, University of Surrey, 1966, September, 201-212.

2. ONO T., KIMURA., KANEKO H., IWATA M., MUTO Y., OKADA A., and ISHIDA K. On the behaviour of rigid joint space truss under vertical and lateral loading. International Association for Shell and Spatial Structures, Symposium Spatial Structures, Dresden and Cottbus, 1990, September, 223-232.

3. KONISHI H., SAKA T., and TANIGUCHI Y. The effect of boundary members on the elastic buckling strength of "square-and-diagonal" double-layer grid. Architectural Institute of Japan, Summaries of Technical Papers of Annual Meeting, 1991, September, 1259-1260.

10. Buckling behaviour of space trusses constructed by a bolted jointing system

Y. TANIGUCHI and T. SAKA, Osaka City University, Japan, and
Y. SHUKU, Taiyo Kogyo Corporation, Japan

SYNOPSIS. The bending moment-rotation relationship and axial rigidities, determined by loading tests on members, including the joints, are given for two typical bolted jointing systems . The member model, which consists of a uniform member, rigid end parts and rotational springs, is applied to the practical bolted jointing systems. The buckling loads and post-buckling behaviour of simple space trusses are predicted by theoretical analyses, and compared with the corresponding experimental results, to study the validity of the member model.

INTRODUCTION

1. Space trusses which are composed of similar structural units are often used as space structures of varied configurations, such as grids, spherical domes and cylindrical shells. In such a case, it may be desirable that constituent members can be simply connected to a node in any direction and that the jointing system can be industrialised. In order to satisfy these requirements, the bolted jointing systems, such as TM truss, are widely used.

2. In analysing the bolted jointing systems, the second author presented a member model which consisted of a uniform member, rigid end parts and rotational springs at both ends. In considering the rigidity of the joints, the elastic buckling load of a space truss usually becomes greater than that of the corresponding pin-jointed model. However, if the rotational spring rigidities of joints with finite sizes are too small, the elastic buckling load becomes smaller than that of the pin-jointed case (ref. 1).

3. For the post-buckling behaviour of space trusses with a great number of members, some simplified non-linear analyses are used (refs 2, 3 and 4). They are full stage-by-stage analyses, where the characteristics of compressive members are idealised by their axial force-axial deformation relationships. Other methods, allow for geometrically and materially non-linear analyses using plastic hinge theory or the finite element method (refs 5, 6).

4. The purpose of this paper is to study the application of the member model to practical jointing systems and to present a numerical method to model the post-buckling behaviour of space trusses. Firstly, for two typical bolted joint systems,

bending tests on members including the joints were carried out in order to determine the moment - rotation relationships of the connections. The axial rigidities of the connections were also obtained by compressive and tensile loading tests. Most bending tests on the connections have been carried out by treating the test specimen as a simple beam where the centre node was subjected to a concentrated load. In such a case, there are shear forces as well as bending moments acting at the connection. The estimation of bending moments at the connection, therefore, must be carefully performed. Secondly, the lengths of the rigid end parts, the rotational spring rigidities and the effective axial rigidity of the member models are estimated from the experimental results. Moreover, for some simple space system trusses the buckling loads and post-buckling behaviour are estimated by a non-linear analysis using the member model. These results are compared with the experimental results in order to confirm the justification of the member model. It is noted that the rotational spring characteristics of the member model are modelled by using the moment rotation relationships obtained by the bending tests and that the axial rigidity of the connection is taken into account as an effective axial rigidity of the member model. Finally as a numerical example, the buckling load and post-buckling behaviour of a double-layer truss are investigated using non-linear analysis, together with the test results of the connections and members.

LOADING TESTS ON SPACE TRUSS CONNECTIONS

Specimens

5. The two typical bolted jointing systems, shown in Fig. 1, were tested. The sizes of the parts of the jointing systems are shown in Tables 1 and 2.

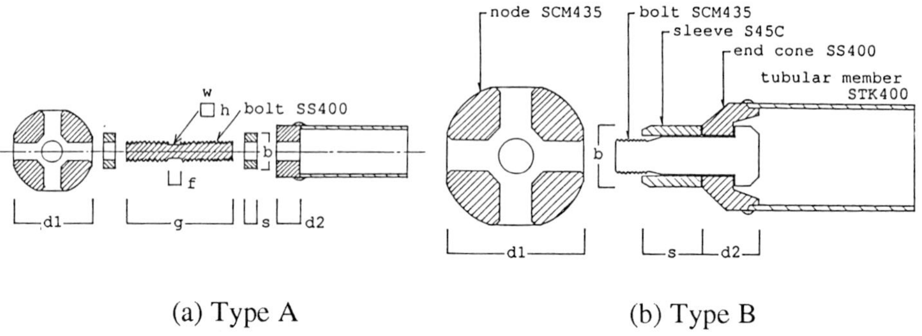

Fig. 1. Details of the connections.

6. Type A consists of a node, a screw bolt and a tubular member. It is the system which resists both compressive and tensile loads, in the same mechanism. Type B consists of a node, a high tensile bolt and a sleeve covering the bolt. The system resists both compressive and tensile loads using different mechanisms. Two kinds of type A and three kinds of type B were tested.

Table 1. Sizes of type A.

(mm)

	d1	d2	s	b	w x h Bending	w x h Compression	w x h Tension	f	g	Bolt	Member φ x t
A22a	46	14.0	6.7	18.7	8.8 x 8.8(L) 8.8 x 8.8(R)	9.2 x 8.9	8.6 x 8.6	6.1	66.2	M12	21.78 x 1.70
A22b	46	14.0	6.7	18.7	8.6 x 8.6(L) 8.6 x 8.6(R)	8.8 x 8.8	8.7 x 8.7	6.1	66.2	M12	21.78 x 1.70
A22c	46	14.0	6.7	18.7	8.6 x 8.6(L) 8.6 x 8.6(R)	8.8 x 8.8	8.7 x 8.7	6.1	66.2	M12	21.78 x 1.70
A60a	103	23.8	13.7	35.6	19.9 x 19.9(L) 20.0 x 19.9(R)	19.9 x 19.9	19.3 x 19.3	12.1	132	M24	60.63 x 3.09
A60b	103	23.8	13.7	35.6	19.6 x 19.9(L) 20.0 x 20.0(R)	19.8 x 19.8	19.7 x 19.7	12.1	132	M24	60.63 x 3.09
A60c	103	23.8	13.7	35.6	19.2 x 19.4(L) 19.8 x 19.9(R)	19.8 x 19.8	19.9 x 19.7	12.1	132	M24	60.63 x 3.09

Table 2. Sizes of type B. (mm)

	d1	d2	s	b	Bolt	Member φ x t
B34	78	18	30	24	M16	34.03x2.30
B42	78	18	30	24	M16	42.66x2.05
B60	78	33	30	24	M16	60.51x2.98

Test methods

7. Bending tests. The loading system shown in Fig. 2 was chosen in order to apply only bending moments onto the connections. The specimens are simply supported by thin plates and loaded by a tournament system. The relative deflections between the centre of the node and the two points on the tubular members are measured with a strain gauge type transducer. The strain gauges are attached on the tubular members in order to confirm the uniformity of bending moments. The load is measured by a load cell.

8. Compressive and tensile loading tests. As shown in Fig. 3, the relative deformations between the centre of a node and a point on the tubular member are measured with three strain gauge type transducers. Four strain gauges are attached on a tubular member in the same level. This test method can measure the axial rigidities of the connection and the tubular member at the same time.

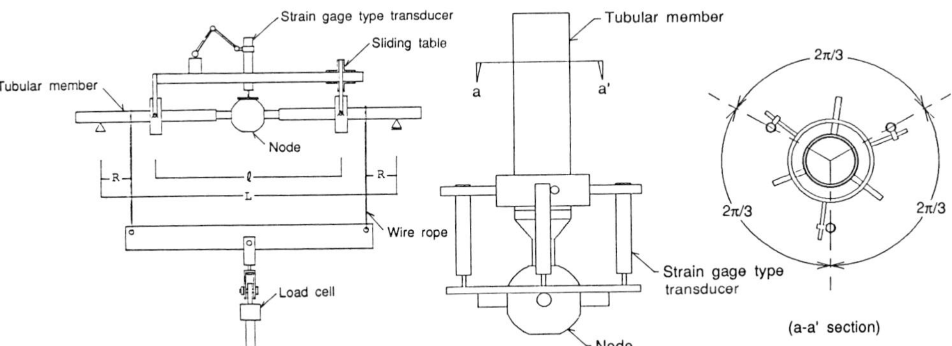

Fig. 2. Loading system for bending tests.

Fig. 3. Loading system for compressive and tensile tests.

<u>Bending test results</u>

9. The bending moments M - rotation θ relationships obtained from testing the connections are estimated from the experimental results. The survey of the residual deformations of the connections gives the centre of rotation as shown in Fig. 4. The rotational spring of the model is, therefore, assumed to be located in this position. The bending moment M - rotation θ relationships of the connections are estimated by using the member model (as shown in Fig. 5). The results obtained for type A and type B (ϕ60) are shown in Figs 6 and 7.

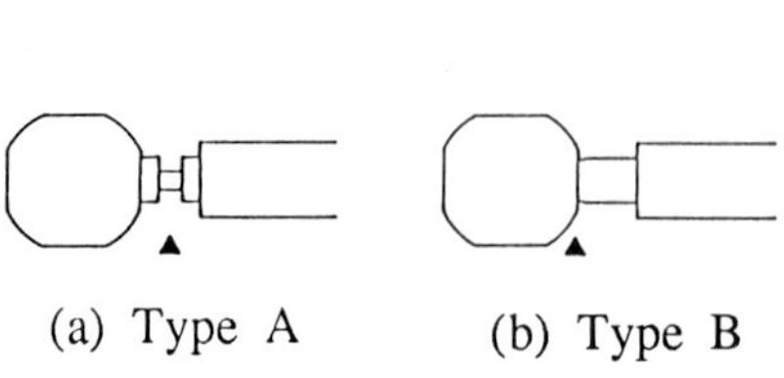

Fig. 4. Rotational centre estimated from the residual deformation.

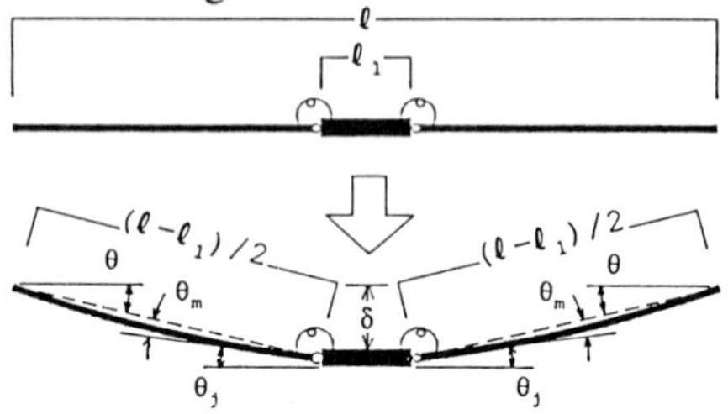

Fig. 5. Member model for the estimation of bending moment M-rotation θ.

10. The moment - rotation relationships of type A are close to a bi-linear behaviour. However type B shows remarkable non-linearity in the moment - rotation relationship. The rigidities of the unloading range are almost constant for each case. As for type B, the rigidities of the unloading range become smaller in the range of small bending moments. This may be due to the fulcrum point of the sleeve moving during the unloading range.

11. <u>Regression curve of bending moment-rotation relationships.</u> The regression curve of bending moment-rotation relationships is estimated from the experimental results. In this paper, the regression curve is assumed to be an exponential function given by Eq. 1 (ref. 7). The rigidity of the springs is given by differentiating Eq. 1.

$$M = \sum_{i=1}^{m} A_i \left(1 - \exp\left(-\frac{|\theta|}{2i\alpha}\right)\right) + \beta\,|\theta| \qquad (1)$$

12. The coefficients of Eq. 1 are determined as follows:

i) $m=3$, $\beta \geq 0$.

ii)The mean value of unloading rigidities is taken as an initial rotational rigidity (shown in Table 3).

iii)The two points in the moment - rotation relationships are taken from the experimental results.

iv)The coefficients are determined from the experimental results by the method of least squares. The results obtained are shown in Table 4. The regression curves obtained for type A and type B (φ60) are shown in Figs 6 and 7 with the experimental results. The dashed and dotted line in Fig. 6 is the regression curve for the case of $\beta = 0$ for type A.

Table 3. Elastic rotational spring rigidity.

Elastic rotational spring rigidity	A22	A60	B34, 42	B60
C_{ji} (kNcm/rad)	478.8	3907.5	3018.8	2838.6

Table 4. Coefficient values of Eq. (1).

	A22	A60	B34, 42	B60
A_1	-0.6922	-25.59	-8.946	-10.59
A_2	3.405	102.7	53.04	54.12
A_3	2.359	-6.895	39.90	42.76
α	0.00188	0.00305	0.00514	0.00546
β	0.775	65.7	15.40	24.9

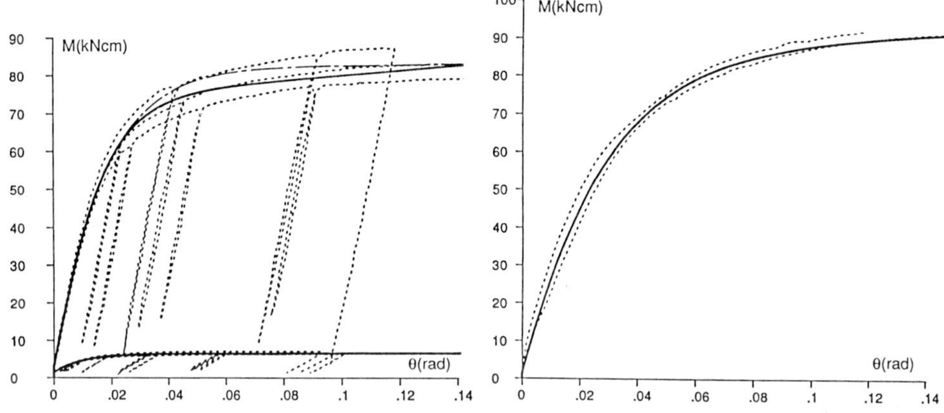

Fig. 6. Bending moment (M)-rotation (θ) relationship for type A.

Fig. 7. Bending moment (M)-rotation (θ) relationship for type B (φ60).

Compressive and tensile loading test results

13. The axial rigidities of the connections are estimated from the load-deformation relationships which are given by taking away the axial deformations of the tubular member from the total deformations. The results obtained are shown in Tables 5 and 6. It is noted that the connection includes a sleeve and the end cone of a tubular member, and then the length of the connection is (s + d2), as shown in Fig. 1.

Table 5. Elastic axial rigidity (type A).

Elastic axial rigidity	Compression (kN)	Tension (kN)
A22	9.320×10^3	8.519×10^3
A60	4.025×10^4	4.881×10^4

Table 6. Elastic axial rigidity (type B).

Elastic axial rigidity	Compression (kN)	Tension (kN)
B34	4.164×10^4	1.962×10^4
B42	3.660×10^4	2.338×10^4
B60	4.442×10^4	2.206×10^4

14. For type A connection, the compressive axial rigidity and tensile rigidity are close to each other. This is presented in ref. 8 where a similar connection was investigated. As for type B, the axial tensile rigidity is about 60% of the compressive rigidity. It can be concluded therefore there are large differences between the compressive and tensile axial rigidities for the connections type B, which resists axial loads by different mechanisms.

BUCKLING BEHAVIOUR OF SIMPLE SPACE TRUSSES CONSTRUCTED BY A BOLTED JOINTING SYSTEM

Forms of Octahedral Space Trusses

15. Octahedral space trusses constructed with type B connections, as shown in Fig. 8, are treated as simple space trusses. The numerical analyses and experimental tests are performed in tension with both joints in the diagonal line. The initial deflections for compression member specimens are shown in Table 7.

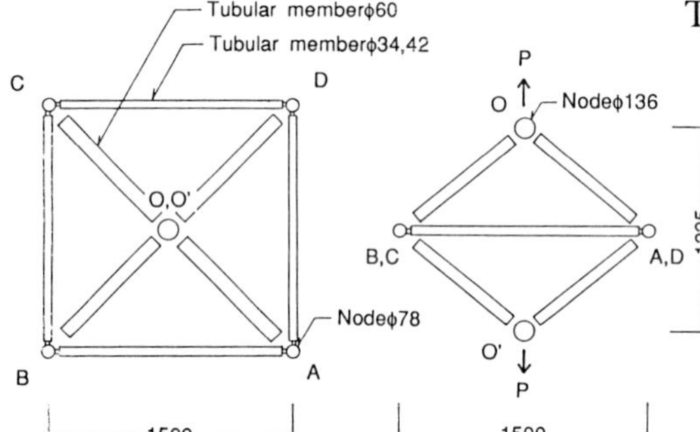

Fig. 8. Octahedral space truss.

Table 7. Initial lateral deflection of compression members.

	Octa34		Octa42	
	θ (deg.)	δ (mm)	θ (deg.)	δ (mm)
A - B	180	1.0	180	1.5
B - C	90	1.0	180	1.5
C - D	0	0	70	1.6
D - A	45	0.7	135	0.7

Theoretical buckling load and post-buckling behaviour

16. Elastic buckling loads are estimated by the elastic buckling analysis using the slope-deformation equations modified for an axial force in the member model (ref. 1).

17. The post-buckling behaviour is estimated by a non-linear analysis which considers geometrical and material non-linearity. The member model of the non-linear analysis is shown in Fig. 9. In the non-linear analysis, the elastic stiffness matrix of a uniform member (a-m-b) is the tangent stiffness matrix presented in ref. 9. The material non-linearity are taken into account by the method of generalised plastic hinge theory (ref. 10). In this paper, plastic hinges are considered at both the ends (a, b) and in the middle (m) of the uniform member. The yield function of a uniform member is represented in terms of the axial force N and the bending moment M as follows:

$$F=\sqrt{\left(\frac{M}{M_Y}\right)^2}-\cos\left(\frac{\pi}{2}\frac{N}{N_Y}\right)=0, \qquad M=\sqrt{M_y^2+M_Z^2} \tag{2}$$

18. The moment - rotation relationship for the rotational springs follows the preceding regression curve obtained from the bending tests. The yield axial forces N_y and yield bending moments M_Y obtained from the loading tests are shown in Table 8.

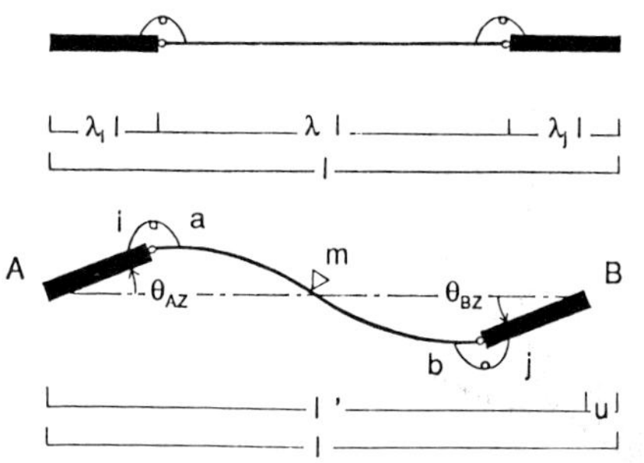

Fig. 9. Member model.

Table 8. Yield axial forces and yield bending moments.

	N_Y(kN)	M_Y(kNcm)	Strain hardening ratio ν
ϕ34	95.71	96.79	0.0380
ϕ42	83.65	108.27	0.0333

19. The non-linear analysis takes into account the initial lateral deflection of members. Since tensile axial forces may reduce the rotational spring rigidity, the numerical analysis is also carried out for the case that the rotational spring rigidities of web members are reduced by 20%.

20. The elastic buckling loads and load-carrying capacities obtained by the analyses are shown in Table 9 together with the experimental results. The experimental load (P)-deformation (δ) relationships for the octahedral space trusses are shown in Figs 10 and 11.

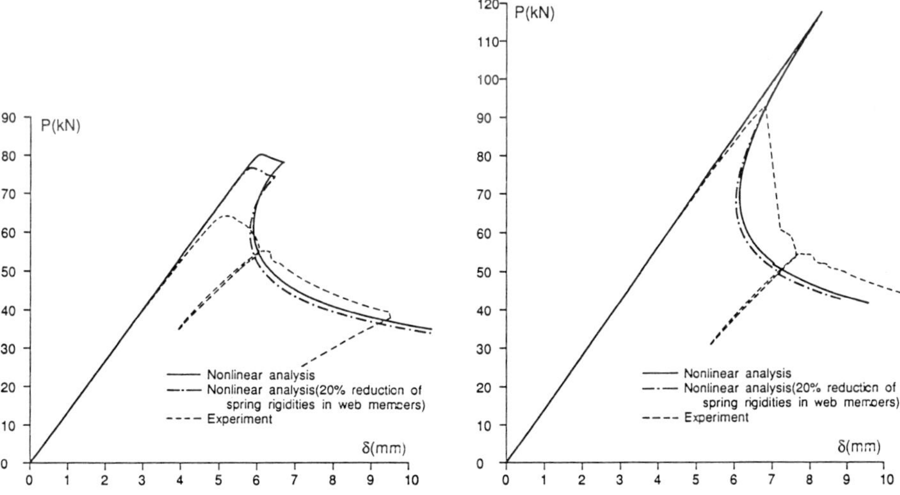

Fig. 10. Load (P) - deformation (δ) of octa34.

Fig. 11. Load (P) - deformation (δ) of octa42.

Table 9. Elastic buckling loads and load-carrying capacities.

	Octa34		Octa42	
Theoretical elastic buckling load Pe (kN)	82.96 (128%)	*79.14 (123%)	128.0 (139%)	*123.4 (134%)
Theoretical load carrying capacity Pmax (kN)	79.92 (124%)	*76.69 (119%)	116.7 (127%)	*115.7 (125%)

* represents the 20% reduction of spring rigidities in web members.
( ) represents the ratio of the value to the experimental result.

Comparison between theoretical and experimental results

21. The load-carrying capacities obtained from the non-linear analysis are 25% higher than the experimental results. With the 20% reduction in the rotational spring rigidities of the web members, the load-carrying capacities decrease 5% for octa34 and 3% for octa42. However the post-buckling behaviour of each case is in agreement with the experimental results. It can be seen that when the type B connections are subjected to tensile forces, the rotational rigidities decrease. From the theoretical results, the sensitivity of load-carrying capacities to initial imperfections is large.

22. Since the non-linear analysis can predict the post-buckling behaviour of space trusses, the application of the member model to a practical bolted jointing system is suitable.

BUCKLING BEHAVIOUR OF A DOUBLE-LAYER TRUSS CONSTRUCTED USING A PRACTICAL BOLTED JOINTING SYSTEM

23. A square-on-square double-layer grid constructed with type B (ϕ42) connection is considered. The grid consists of 6 x 6 structural units (12 x 12m). The geometrical and mechanical data of the grid are shown in Table 10. The grid is simply supported at all the perimeter nodes and subjected to a uniform vertical load on the upper nodes. In the non-linear analysis, the initial lateral deflections of the centre compressive members are 2.5mm in the middle of the members. It is 1/800 of the length between the two nodes.

Table 10. Geometrical and mechanical data of the grid.

Length of chord members ℓ_s;	200.0 (cm)	Diameter of ball joints;	7.8 (cm)
Rotational spring rigidity C_j;	3.02 x 10^3 (kNcm/rad)	Tubular member	ϕ 42.7 x t 2.1 (mm)
		Section area A;	2.62 (cm^2)
Modulus of elasticity E;	2.145 x 10^4 (kN/cm^2)	Moment inertia I;	5.41 (cm^4)

24. The total load (P) - deflection(δ) relationship of the centre node is shown in Fig. 12. The elastic buckling load is 445kN. The first peak load of Fig. 12 is 445kN, which is the load-carrying capacity of the grid. In the case of the slenderness ratio of the members of the grid, the elastic buckling load is equal to the first peak load which is determined by the yield of the centre compression members. After the initial yield of the centre compression members, the load-bearing capacity decreases. However, the load increases again until the next upper

members yield. The present grid has, therefore, the deformation capacity in the post initial yield state. This characteristic of the grid corresponds to that of the model tests (ref. 1)

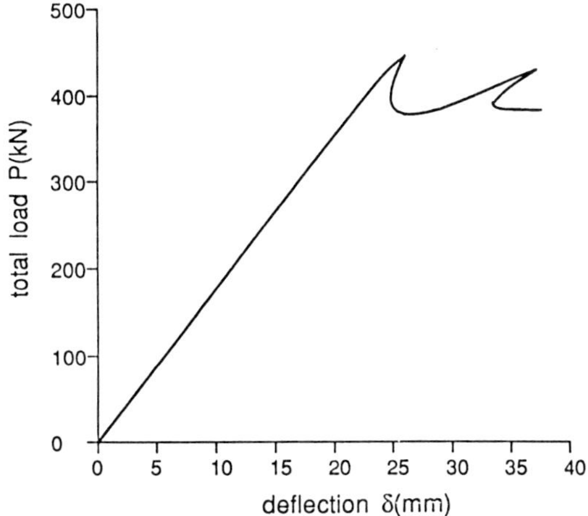

Fig. 12. Total load (P) - deflection (ϕ) relationship of the grid.

CONCLUSIONS

25. Loading tests were undertaken on two typical bolted jointing systems. In conclusion, for the bolted jointing system (type A) which resists tensile axial and compressive forces by the same mechanism, the bending moment-rotation relationship is close to a bi-linear behaviour. The compressive and tensile axial rigidities are almost equal to each other. On the other hand, for the bolted jointing system (type B) which resists the two axial forces differently, the bending moment-rotation relationship is non-linear. The compressive and tensile axial rigidities are different from each other. From the experiments on some simple space trusses, the application of the member model, consisting of rigid end parts, rotational springs and a uniform member, to a practical bolted jointing system is suitable. For a square-on-square double-layer grid constructed with a practical bolted jointing system, the elastic buckling load and post-buckling behaviour are predicted. The deformation capacity of the grid is followed into the non-linear region.

ACKNOWLEDGEMENTS

The authors wish to thank Mr. K. Yamaguchi for his valuable work during the loading tests.

REFERENCES

1. SAKA T. and HEKI K. The effect of joints on the strength of space trusses. Proceedings of Third International Conference on Space Structures, Edited by H. Nooshin, Elsevier Applied Science Publishers, London, 1984, 417-422.

2. SUPPLE. and COLLINS I. Limit state analysis of double-layer grids. Analysis, Design and Construction of Double-Layer Grids, Applied Science Publishers, London, 1981, 93-117.

3. SMITH E.A. Space truss non-linear analysis. Journal of Structural Engineering, ASCE, 1984, Vol. 110, 688-705.

4. SAKA T. and TANIGUCHI Y. Post-buckling behaviour of square-on-diagonal double-layer grids constructed by a bolted jointing system. Proceedings of the International IASS Symposium, Spatial Structures at the Turn of the Millennium, 1991, Vol. 3, September, 199-206.

5. INOUE K., YAMAMOTO K., MATSUMOTO T. and WAKIYAMA K. Non-linear analysis of tubular truss towers subjected to cyclic horizontal force. Proceedings of Safety Criteria in Design of Tubular Structures, Edited by Y. Kurobane and Y. Makino, 1987, 47-59.

6. FUJIMOTO M., WADA A., IWATA M. and NAKATANI F. Non-linear three dimensional analysis of steel frame structure. Transactions of the Architectural Institute of Japan, 1975, No. 227, January, 75-90.

7. LUI E.M. and CHEN W.F. Behaviour of braced and unbraced semi-rigid frames. International Journal of Solids and Structures, 1988, Vol. 24, No. 9, 893-913.

8. KUBODERA I., KATO H. and ISHIZU H. Experimental study on strength and stiffness of nodes in new space frames. Proceedings of Safety Criteria in Design of Tubular Structures, Edited by Y. Kurobane and Y. Makino, 1987, 373-380.

9. ORAN C. Tangent stiffness in space frames. Journal of the Structural Division, ASCE, 1973, Vol. 99, No. ST6, 987-1001.

10. MORRIS G.A. and FENVES S.J. Elastic-plastic analysis of frameworks. Journal of the Structural Division, ASCE, 1970, Vol. 96, No. ST5, May, 931-946.

11. Effect of initial imperfections on the buckling of space truss structures

F. TRENTADUE, Politecnico di Bari, Italy

SYNOPSIS. In this work the effect of arbitrary imperfections on the bearing capacity of space truss structures subject to elastic buckling is examined. A perturbation technique for studying imperfection sensitivity of general elastic structures is presented. This technique is applied here to space truss structures considering various kinds of imperfection variables, including load vectors, nodal initial position vectors, initial lengths of the members, and the product of Young's modulus and the cross-sectional area. Previous work related to the same subject, treat only structural "small" imperfections. Here moderately large imperfections are considered and the imperfections of the loads are included in this analysis.

GENERAL THEORY

1. Consider a family of elastic structures under proportional loading, and writing the geometrically non-linear matrix equilibrium equation:

$$\mathbf{F}(\mathbf{U}, \Lambda\mathbf{p}, \varepsilon\mathbf{b}) = 0 \qquad (1)$$

where $F: R^n x R^n x R^p \Rightarrow R^n$ is a sufficiently smooth (analytical) function. The term $\varepsilon b \in R^p (\varepsilon \in R)$ is the imperfection vector; $\Lambda p \in R^n$ is the load vector; $\Lambda \in R$ is the load factor. $U \in R^n$ is the position vector.

2. Defining the *perfect* structure as giving $\varepsilon b = 0$. The solution of Eq. 1, which consist of $(U, \Lambda) \in R^n x R$ satisfying the equation, allows the formation of equilibrium paths. The load bearing capacity of the perfect system is supposed to be governed by a simple critical point (U_{cr}, Λ_{cr}). At this point the (symmetric) tangent stiffness matrix $K = F_{100}(U_{cr}, \Lambda_{cr} p, 0)$ is singular, giving:

$$\mathbf{F}^c = \mathbf{F}(\mathbf{U}_{cr}, \Lambda_{cr}\mathbf{p}, 0) = 0 \qquad (2.1)$$

$$\mathbf{F}^c_{100}\,\mathbf{e} = \mathbf{F}_{100}(\mathbf{U}_{cr}, \Lambda_{cr}\mathbf{p}, 0)\mathbf{e} = 0 \qquad (2.2)$$

where $e \in R^n$ is the critical eigenvector of the perfect structure and the superscript c denotes that the corresponding operator[1] is evaluated at the critical point of the perfect system.

3. The critical state of an arbitrary imperfect system is now considered. The following equations hold:

$$F(U_{cr} + u,\ (\Lambda_{cr} + \lambda)p\ ,\ \varepsilon b) = 0 \tag{3.1}$$

$$F_{100}(U_{cr} + u,\ (\Lambda_{cr} + \lambda)p,\ \varepsilon b)e^{(imp)} = 0 \tag{3.2}$$

where $e^{(imp)} \in R^n$ is the critical eigenvector of the imperfect structure. It can be shown by means of the Implicit Function Theorem that the solution of the system of Eq. 3 can be expanded in a power series of the parameter s:

$$u(\varepsilon b) = \sum_{i=1}^{\infty} \frac{1}{i!} s^i\, v_i(b)\, e + \sum_{i=1}^{\infty} \frac{1}{i!} s^i\, w_i(b) \tag{4.1}$$

$$\lambda(\varepsilon b) = \sum_{i=1}^{\infty} \frac{1}{i!} s^i\, \lambda_i(b) \tag{4.2}$$

where $s = \varepsilon$ for the limit point, $s = |\varepsilon|^{1/2}$ for the asymmetric bifurcation point, $s = \varepsilon^{1/3}$ for the symmetric bifurcation point.

4. Adopting the Lyapunov-Schmidt decomposition, it is required that the functions w_i (b) satisfy the conditions:

$$e^T w_i(b) = 0 \tag{5}$$

5. The following perturbation technique can be used for finding the functions vi(b), wi(b) and λi(b). Differentiating first Eq. 3.1 with respect to s gives:

$$F^c_{100}\left(v_1 e + w_1 \right) + F^c_{010}\, \lambda_1 + \varepsilon_1\, F^c_{001} b = 0 \tag{6}$$

1--------------------------------

The present notation is adopted:

$$F_{rst}(U,\ \Lambda p,\ b)\ u^{(1)}u^{(2)}...\ u^{(r)}\ \lambda^{(1)}\lambda^{(2)}...\ \lambda^{(s)}\ b^{(1)}b^{(2)}...\ b^{(t)} =$$

$$= \sum_{\substack{i_1, i_2, .., i_r = 1,2,..,n \\ j_1, j_2, .., j_t = 1,2,..,p}} \frac{\partial^{(r+s+t)} F(U,\ \Lambda p,\ b)}{\partial U_{i_1} \partial U_{i_2} .. \partial U_{i_r}\ \partial \Lambda^s\ \partial b_{j_1} \partial b_{j_2} .. \partial b_{j_t}} u^{(1)}_{i_1} u^{(2)}_{i_2} ...\ u^{(r)}_{i_r}\ \lambda^{(1)}\lambda^{(2)}...\ \lambda^{(s)}\ b^{(1)}_{j_1} b^{(2)}_{j_2} ...\ b^{(t)}_{j_t}$$

where vi(b), wi(b) and λi (b) are denoted by vi, wi and λi. Pre-multiplying Eq. 6 by the projection matrix ee^T gives:

$$\mathbf{e}\left(\mathbf{e}^T\mathbf{F}^c_{100}\left(v_1\mathbf{e} + \mathbf{w}_1\right) + \lambda_1\mathbf{e}^T\mathbf{F}^c_{010} + \varepsilon_1\mathbf{e}^T\mathbf{F}^c_{001}\mathbf{b} \right) = 0 \qquad (7)$$

from which the following scalar equation is obtained:

$$\left(L_{100}v_1 + \lambda_1 L_{010} + \varepsilon_1 L_{001}(\mathbf{b}) \right) = 0 \qquad (8.1)$$

6. First a system of two perturbation equations is formed, by associating to this equilibrium equation a critical state equation. It can be shown that the critical state equations can be obtained simply by partially differentiating with respect to vi the corresponding equilibrium equations. Therefore the first critical state equation (that is identically satisfied) is:

$$\mathbf{e}^T\mathbf{F}^c_{100}\mathbf{e} = L_{100} = 0 \qquad (8.2)$$

7. Pre-multiplying Eq. 6 by the projection matrix $(I - ee^T)$ gives:

$$(\mathbf{I} - \mathbf{e}\mathbf{e}^T)\left(\mathbf{F}^c_{100}(v_1\mathbf{e} + \mathbf{w}_1) + \mathbf{F}^c_{010}\lambda_1 + \varepsilon_1\mathbf{F}^c_{001}\mathbf{b} \right) = 0 \qquad (9)$$

The system of n scalar equations represented by Eq. 9 can be solved using the first of the conditions Eq. 5, (where i = 1) to obtain:

$$\mathbf{w}_1 = \mathbf{w}_{010}\lambda_1 + \varepsilon_1\mathbf{w}_{001}(\mathbf{b}) \qquad (10)$$

8. Differentiating Eq. 7 with respect to s and using Eq. 10, gives the second order equilibrium equation:

$$\begin{aligned}
\mathbf{e}\Bigg(& \left(\mathbf{e}^T\mathbf{F}^c_{200}\mathbf{e}\,\mathbf{e}\right)v_1^2 + 2\left(\mathbf{e}^T\mathbf{F}^c_{200}\mathbf{e}\,\mathbf{w}_{010} + \mathbf{e}^T\mathbf{F}^c_{110}\mathbf{e}\right)v_1\lambda_1 + \\
&+ 2\left(\mathbf{e}^T\mathbf{F}^c_{200}\mathbf{e}\,\mathbf{w}_{001}(\mathbf{b}) + \mathbf{e}^T\mathbf{F}^c_{101}\mathbf{e}\,\mathbf{b}\right)v_1\varepsilon_1 + \\
&+ \left(\mathbf{e}^T\mathbf{F}^c_{200}\mathbf{w}_{010}\mathbf{w}_{010} + 2\,\mathbf{e}^T\mathbf{F}^c_{110}\mathbf{w}_{010} + \mathbf{e}^T\mathbf{F}^c_{020}\right)\lambda_1^2 + \\
&+ 2\left(\mathbf{e}^T\mathbf{F}^c_{200}\mathbf{w}_{001}(\mathbf{b})\mathbf{w}_{010} + \mathbf{e}^T\mathbf{F}^c_{110}\mathbf{w}_{001}(\mathbf{b}) + \mathbf{e}^T\mathbf{F}^c_{011}\mathbf{b} + \mathbf{e}^T\mathbf{F}^c_{101}\mathbf{w}_{010}\mathbf{b}\right)\lambda_1\varepsilon_1 \\
&+ \left(\mathbf{e}^T\mathbf{F}^c_{200}\mathbf{w}_{001}(\mathbf{b})\mathbf{w}_{001}(\mathbf{b}) + 2\,\mathbf{e}^T\mathbf{F}^c_{101}\mathbf{w}_{001}(\mathbf{b}) + \mathbf{e}^T\mathbf{F}^c_{002}\mathbf{b}\,\mathbf{b}\right)\varepsilon_1^2 + \\
&+ \mathbf{e}^T\mathbf{F}^c_{100}(v_2\mathbf{e} + \mathbf{w}_2) + \lambda_2\mathbf{e}^T\mathbf{F}^c_{010} + \varepsilon_2\mathbf{e}^T\mathbf{F}^c_{001}\mathbf{b} \Bigg) = 0
\end{aligned} \qquad (11)$$

rearranging gives:

$$L_{200}v_1^2 + 2\,L_{110}v_1\lambda_1 + 2\,L_{101}(\,b\,)v_1\varepsilon_1 + L_{020}\lambda_1^2 + 2\,L_{011}(\,b\,)v_1\lambda_1 +$$

$$+ L_{002}(\,b\,)\varepsilon_1^2 + L_{010}\,\lambda_2 + L_{001}(\,b\,)\varepsilon_2 \quad = \;0 \qquad (\;12.1\;)$$

9. The second critical state equation is obtained by differentiating Eq. 12.1 with respect to v_1:

$$L_{200}v_1 + L_{110}\lambda_1 + \varepsilon_1 L_{101}(\,b\,) = 0 \qquad (\;12.2\;)$$

10. The Eqs 12.1 and 12.2 constitute the second system of perturbation equations. Following the procedure above, gives a series of two scalar equations in the unknowns vi(b) and λi(b).

11. Here the expressions of λ1(b) and λ2(b) are presented in the following cases:

LIMIT POINT

In this case, the following relationships result: $\varepsilon_1 = 1$; $\varepsilon_i = 0$ if $i > 1$.
From the system of equations represented by Eq. 7 the following is obtained:

$$\lambda_1(\,b\,) = -\,L_{001}(\,b\,)/\,L_{010} \qquad (\;13\;)$$

and from Eqs 12.1 and 12.2:

$$\lambda_2(\,b\,) = \left[\left(\lambda_1(\,b\,)\right)^2\left(L_{110}^2 - L_{200}L_{002}(\,b\,)\right) + 2\lambda_1(\,b\,)\left(L_{110}L_{101}(\,b\,) -\right.\right.$$

$$\left.\left. - L_{200}L_{011}(\,b\,)\right) + \left(L_{101}^2(\,b\,) - L_{200}L_{002}(\,b\,)\right)\right]/\left(L_{010}L_{200}\right) \qquad (\;14\;)$$

ASYMMETRIC BIFURCATION POINT

In this case, the following relationships result: $L_{010} = 0$; $\varepsilon_1 = 0$; $\varepsilon_2 = 2\mathrm{sgn}(\varepsilon)$.
The system represented by Eq. 7 is identically satisfied. From Eq. 12:

$$\lambda_1(\,b\,) = \pm\left(2\mathrm{sgn}(\,\varepsilon\,)\;L_{001}(\,b\,)L_{200}\;/(L_{110}^2 - L_{020}L_{200})\right)^{1/2} \qquad (\;15\;)$$

and for the third system of perturbation equations (not reported here):

$$\lambda_2(\,b\,) = \left[\left(\lambda_1(\,b\,)\right)^2\left(-L_{300}L_{110}^3/(3L_{200}^2) + L_{210}L_{110}^2/L_{200} - L_{120}L_{110} +\right.\right.$$

$$\left.\left. + L_{030}/(3L_{200})\right) + 2\mathrm{sgn}(\,\varepsilon\,)\left(L_{011}(\,b\,)L_{200} - L_{110}L_{101}(\,b\,)\right)\right]/$$

$$/(L_{110}^2 - L_{020}L_{200}) \qquad (\;16\;)$$

APPLICATION TO SPACE TRUSS STRUCTURES.

12. Consider a space truss structure with M members, Nj1 joints free and Nj2 joints fixed. N = 3Nj1 is the number of degrees of freedom of the structure. Let Nj = Nj1 + Nj2 be the total number of joints. For sake of simplicity suppose that the indices a of the joints satisfy the following conditions:

$1 \le a \le Nj1$ if the joint is free.

$Nj1 < a \le Nj$ if the joint is fixed.

Denoting by a_i the index of the i-th (i =1,2,3) co-ordinate x_{ai} of the joint of index a and hence:

$$a_i = 3(a - 1) + i$$

13. Let Lh be the deformed length of the member of index h connecting the nodes of index h_1 and h_2 and let Loh be the initial length of the same member:

$$L_h = \left(\left(\sum_{i=1}^{3}(x_{h1_i} - x_{h2_i})^2\right)\right)^{1/2} ; \qquad L_{0h} = \left(\left(\sum_{i=1}^{3}(x_{0h1_i} - x_{0h2_i})^2\right)\right)^{1/2} \qquad (17)$$

(the subscript o indicates the initial value of the corresponding variable). Writing the equilibrium equations of the perfect structure in the following form:

$$\mathbf{F}_{a_i} = \sum_{h=1}^{M} \left(\delta_{a,h1} - \delta_{a,h2}\right) \frac{(EA)_h}{L_{0h}} \left(L_h - L_{0h}\right) \frac{\left(x_{h1_i} - x_{h2_i}\right)}{L_h} - \Lambda \mathbf{p}_{a_i} =$$

$$= \sum_{h=1}^{M} \left(\delta_{a,h1} - \delta_{a,h2}\right)(EA)_h \left(1/L_{0h} - 1/L_h\right)\left(x_{h1_i} - x_{h2_i}\right) - \Lambda \mathbf{p}_{a_i} = 0$$

$$(1 \le a_i \le N) \qquad (18)$$

where δ is the Kroenecker symbol. Now the expressions of the operators F100, F010, F001 can be written. The tangent stiffness matrix F100 is obtained by differentiating Eq. 18 with respect to the position vector's components x_{b_j}

$$\left(\mathbf{F}_{100}\right)_{a_i b_j} = \sum_{h=1}^{M} \left(\delta_{a,h1} - \delta_{a,h2}\right)\left(\delta_{b,h1} - \delta_{b,h2}\right)(EA)_h \left[\left(1/L_{0h} - \right.\right.$$

$$\left.\left. - 1/L_h\right)\delta_{i,j} + \left(1/L_h^3\right)\left(x_{h1_i} - x_{h2_i}\right)\left(x_{h1_j} - x_{h2_j}\right)\right]$$

$$(1 \le a_i \le N) ; \quad (1 \le b_j \le N) \qquad (19)$$

F_{010} is obtained by differentiating Eq. 18 with respect to Λ:

$$\left(\mathbf{F}_{010}\right)_{a_i} = -\mathbf{p}_{a_i} \qquad (1 \le a_i \le N) \qquad (20)$$

The following expression of the imperfection vector is assumed:

$$\mathbf{b}^T = \left\{\Delta x_{01}, .., \Delta x_{0(3xNj)}, \ \Delta(EA)_1, .., \Delta(EA)_M, \ \Delta L_{01}, .., \Delta L_{0M}, \ \Delta p_1, .., \Delta p_N\right\} =$$

$$= \left\{ \underset{\substack{3Nj \\ \text{columns}}}{\mathbf{b}_X^T}, \ \underset{\substack{M \\ \text{column}}}{\mathbf{b}_{EA}^T}, \ \underset{\substack{M \\ \text{columns}}}{\mathbf{b}_L^T}, \ \underset{\substack{N \\ \text{column}}}{\mathbf{b}_f^T} \right\}$$

where bx is the vector of the variations of the initial positions of the joints. The value bEA is the vector of the variations of the section properties. The expression bL is the vector of the variations of the initial lengths of the members. (This kind of imperfection produces an initial state of stress).The value bf is the vector of the variation of the forces.

14. The first 3Nj columns of the sensitivity matrix F001 are obtained by partially differentiating with respect to the initial position vectors components $x_{o\,\ell_j}$ of Eq. 18 :

$$\left(\mathbf{F}_{001}\right)_{a_i \ell_j} = \sum_{h=1}^{M} \left(\delta_{a,h1} - \delta_{a,h2}\right)\left(\delta_{\ell,h1} - \delta_{\ell,h2}\right)(EA)_h \Big[-$$

$$- \left(1/L_{0h}^3\right)\left(x_{h1_i} - x_{h2_i}\right)\left(x_{oh1_j} - x_{oh2_j}\right) + \chi_\ell\left(1/L_{0h} - 1/L_h\right)\delta_{i,j} +$$

$$+ \chi_\ell\left(1/L_h^3\right)\left(x_{h1_i} - x_{h2_i}\right)\left(x_{h1_j} - x_{h2_j}\right)\Big] \qquad \begin{matrix}(1 \le a_i \le N) \\ (1 \le \ell_j \le 3Nj)\end{matrix}$$

$$\text{where} \quad \chi_\ell = \begin{cases} 1 & \text{if} \quad Nj1 < \ell \le Nj \\ 0 & \text{if} \quad 1 \le \ell \le Nj1 \end{cases} \qquad (22)$$

15. The columns of the sensitivity matrix of indices k = 3Nj + h (with 1 ≤ h ≤ M) are obtained by partially differentiating Eq. 18 with respect to EA_h:

$$\left(\mathbf{F}_{001}\right)_{a_i,3Nj+h} = \left(\delta_{a,h1} - \delta_{a,h2}\right)\left(1/L_{0h} - 1/L_h\right)\left(x_{h1} - x_{h2}\right) \qquad (23)$$

$$(1 \le a_i \le N); \qquad (3Nj < 3Nj+h \le 3Nj+M)$$

16. The columns of indices k = 3Nj+h+M (with 1 ≤ h ≤ M) are obtained by partially differentiating Eq. 18 with respect to L_{0h}:

$$\left(F_{001}\right)_{a_i,3Nj+M+h} = -\left(\delta_{a,h1} - \delta_{a,h2}\right)\left(1/L^2_{0h}\right)\left(x_{h1_i} - x_{h2_i}\right) \qquad (24)$$

$(1 \le a_i \le N)$; $(3Nj+M < 3Nj+h+M \le 3Nj+2M)$

17. The last columns of indices k = 3Nj+ 2M+ b_j (with $1 \le b_j \le N$) have the following expression:

$$\left(F_{001}\right)_{a_i,3Nj+2M+b_j} = -\delta_{a_i,b_j} \qquad (25)$$

$(1 \le a_i \le N)$; $(3Nj+2M < 3Nj+2M+b_j \le 3Nj+2M+N)$

18. The operators Frst of higher order (for brevity, not reported here) can be obtained using the same technique outlined previously.

NUMERICAL EXAMPLE.

The effect of various kind of imperfections on the reticulated dome shown in Fig. 1 is examined.

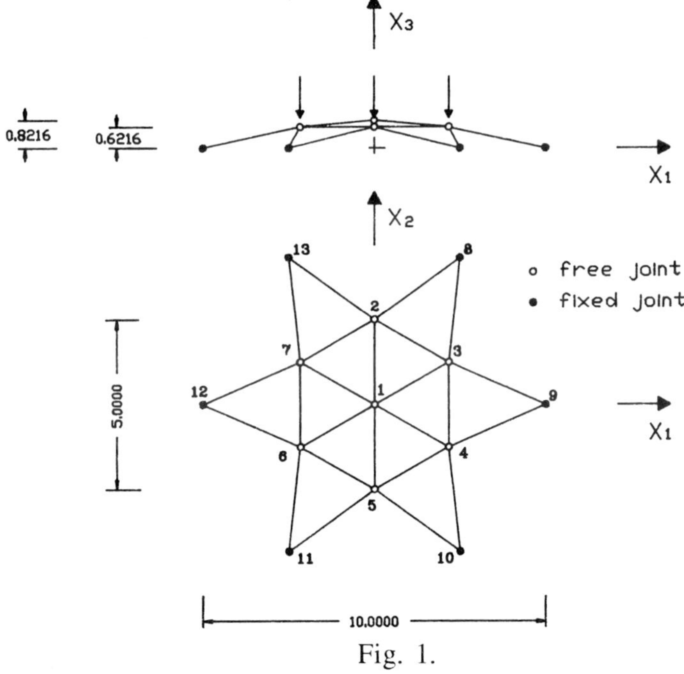

Fig. 1.

19. The *perfect* loads acting on this structure consist of the system of vertical loads Λp_{a_3}:

$$P_{a_3} = -1(a = 1,2,....,7) \qquad (26)$$

All the members have the same sections and results: $EA_h = 1000$, $(h = 1,....,24)$. The load bearing is governed by a limit point with $\Lambda_{cr} = 0.7658$. Consider first the effect of two "imperfections" of a system of horizontal loads $\varepsilon \Delta p_{a_1}^{(1)}$:

$$\Delta p_{a_1}^{(1)} = 1 \quad (a = 1,2,...,7) \tag{27}$$

The second imperfection consists of a system of loads $\varepsilon \Delta p_{a_1}^{(2)}$:

$$\Delta p_{a_1}^{(2)} = 1 \quad (a = 1,2,3,4) \tag{28}$$

20. In the table below, for various values of imperfection magnitude ε, a comparison is found between the critical load factor estimated using the first two terms of the series represented by Eq. 4.2:

$$\Lambda_{cr}(\varepsilon \mathbf{b}) = \Lambda_{cr} + \varepsilon \lambda_1(\mathbf{b}) + \varepsilon^2 \lambda_2(\mathbf{b})/2 \tag{29}$$

and the exact value $\Lambda_{cr}^{(imp)}$ obtained following the equilibrium paths of the imperfect structures:

ε	Imperfection 1		Imperfection 2	
	$\Lambda_{cr}(\varepsilon \mathbf{b})$	$\Lambda_{cr}^{(imp)}$	$\Lambda_{cr}(\varepsilon \mathbf{b})$	$\Lambda_{cr}^{(imp)}$
0		0.7685		0.7685
0.2	0.7660	0.7660	0.7118	0.7106
0.4	0.7584	0.7590	0.6675	0.6597
0.6	0.7485	0.7502	0.6355	0.6134
0.8	0.7282	0.7819	0.6159	0.5702
1.0	0.7055	0.7025	0.6087	0.5295

21. Finally, choosing a third type of imperfection consisting of a reduction $\varepsilon \Delta(EA)_1$ of the product EA for the member of index 1 connecting the nodes of indices 1 and 7, results in:

$$\Delta(EA)_1 = -(EA)_1 \tag{30}$$

and fourth type of imperfection consisting of variations in the initial position of the nodes of indices 3, 4:

$$\varepsilon \Delta x_{0a_1} = 0.5\varepsilon \; (a = 3,4) \tag{31}$$

22. In the table below the last two types of imperfection $\Lambda_{cr}(\varepsilon b)$ and $\Lambda_{cr}^{(imp)}$ are compared:

ε	Imperfection 3		Imperfection 4	
	$\Lambda_{cr}(\varepsilon \mathbf{b})$	$\Lambda_{cr}^{(imp)}$	$\Lambda_{cr}(\varepsilon \mathbf{b})$	$\Lambda_{cr}^{(imp)}$
0		0.7685		0.7685
0.2	0.7060	0.7049	0.7577	0.7577
0.4	0.6385	0.6329	0.7477	0.7475
0.6	0.5660	0.5320	0.7384	0.7379
0.8	0.4886	0.3802	0.7300	0.7288
1.0	0.4062	0.0691	0.7224	0.7202

23. These tables confirm the effectiveness of the use of Eq. 29 in estimating the effect of the imperfections considered in this example on the critical load factor.

CONCLUSIONS.

24. The analysis presented in this work permits the examination of the effect of arbitrary imperfections on the critical load factor of space truss structures. The expressions of the second order terms $\lambda_2(b)$ of the series given by Eq. 4.2, obtained using a new perturbation technique, are presented. The numerical results show the effectiveness of the proposed technique for "moderately large" values of the imperfections.

ACKNOWLEDGEMENT

This research has been executed by the contribution of the Ministero dell'Universtita e della Ricerca Scientifica.

REFERENCES

1. KOITER W. T. On the stability of elastic equilibrium. English translation: NASA Tech. Trans., F10, 1-833 1967.
2. THOMPSON J. M. T. and HUNT G. W. A general theory of elastic stability, John Wiley & Sons, 1973.
3. IKEDA K. and MUROTA K. Computation of critical initial imperfection of truss structures, ASCE J. of Engineering Mech., October 1990, Vol. 116, No 10.
4. ARBOCZ J., POTIER-FERRY M., SINGER J. and TVERGAARD V. Buckling and post-buckling. Lecture notes in physic, Vol. 288, Springer Verlag, Berlin, 1985.
5. BEN-HAIM Y. and ELISHAKOFF I., Convex model of uncertainty in applied mechanics, Elsevier, New York 1990
6. VAINBERG M. M. and TRENOGIN V. A., Theory of branching of solutions of non-linear equations, Noordhoff International Publishing, 1974.

12. Stability analysis of space truss systems under random element imperfections

O. M. SADIQ, Ahmadu Bello University, Nigeria, and A. O. ABATAN, Virginia Polytechnic and State University, USA

SYNOPSIS. An investigation into the correspondence between imperfection statistics and statistical properties of the critical load of space truss systems is presented. Our analysis shows that the normal distribution is a good enough model for the critical load; the log-normal distribution however appears to be better when the random variable is associated with a relatively high value of standard deviation. The mean critical load was found not to be significantly affected by a variation in imperfection standard deviations. A rising trend was found in the case of the '95% confidence load' for decreasing values of both mean and standard deviations.

INTRODUCTION

1. The stability problem of reticulated or lattice space structures in general has received considerable attention in the past few years. The snap-through phenomenon in particular has been widely researched. Coan and Plaut (ref. 2) investigated the dynamic response of a lattice dome modelled as a 24-bar space truss and subjected to three independent sets of impulse loads. Holzer and Plaut (ref. 4) also considered the stability problem in terms of multiple independent loading parameters as opposed to a single parameter-based analysis common among analysts.

2. The effects of the presence of imperfections on the stability behaviour of imperfection-sensitive structures has been investigated from a statistical view point by Rhoda (ref. 7). He presented a mathematical model for estimating the statistics of the critical load in general. Rhoda's work is however based on a pre-determined deterministic relation between imperfection and critical load which limits its application to systems where this is known.

3. Statistical investigations have been conducted in the literature by methods other than purely analytical. In particular, Sadeeq (ref. 10) and Warmer (ref. 13) employed the Monte-Carlo technique. Warmer investigated the variability in the ultimate strength of a short reinforced concrete column under axial load, while Sadeeq's work is an M.Sc. on the variability of reinforced concrete beams and columns.

4. Terry (ref. 11) undertook the study of the stability of space trusses with initial element imperfections. His work involved the tracing of the equilibrium paths for geometrically non-linear systems with local imperfections in the form of member curvatures. One or two members were selectively given a single half sine wave and the structure analysed for various load cases. He employed the formulations of Rosen and Schmidt (ref. 8) and his results indicated a general reduction in the critical load with increasing magnitude of imperfection. His work also revealed that deposition of the imperfect elements, with respect to the perfect ones, has a bearing on the value of the critical load. However, the snag with Terry's work is that imperfection was modelled wholly in a deterministic manner. Element imperfections are the result of manufacturing, testing and handling processes which by their very nature are subject to a large degree of randomness. Properly therefore, imperfection analysis belongs to the realms of probability and statistics.

SOLUTION TECHNIQUE

Monte-Carlo Simulation

5. In principle, the Monte-Carlo simulation is a procedure for solving a deterministic analytical problem, by converting it into a probabilistic analogue having the same mathematical formulation. Random sampling techniques are then employed in finding a solution to the problem.

6. The fundamental step in a Monte-Carlo analysis is the generation of random members. Most of todays MICROS and MACROS have in-built library routines for this purpose. In this paper we found it necessary to adapt a standard routine (ref. 6). With this adaptation, it became possible and convenient to vary the initial seeding as necessary to achieve the purpose of this work.

Random Numbers

7. The use of an algorithm to generate random numbers seems to violate the basic principles of randomness as the process involved is deterministic. Sequences of software-generated random numbers are therefore referred to as synthetic quasi-random or pseudo-random numbers. The generated sequences are found however to meet most criteria for randomness, always starting with an initial value called 'seed' and proceeding in a completely deterministic, repeatable fashion. Basic requirements for pseudo-random sequences are that the uniform distribution is followed and that the sequence length or repeatability cycle is long enough.

8. In this paper, imperfection is assumed to be normally distributed. This may be justified by considering that the total imperfection in a specimen is a consequence of a large number of independent additive factors in the manufacturing process. By the central limit theorem, therefore, a summation of a large number of independent random variables will be approximately normally distributed almost regardless of their individual distributions.

9. A number of algorithms are available for generating random samples following the standardised normal distribution from pseudo-random (uniform)

sequences. The Box-Mueller transformation (ref. 8), is a common one and is adopted here. The transformation uses the relation

$$Z = \left(-21\eta R_1\right)^{1/2} \cos\left(2\pi R_2\right) \qquad (1)$$

where R_1, R_2 are two uniform random numbers, and Z is standard normal variate. For a specific value, normal variates are obtained from a further transformation:

$$X = 26 + \mu \qquad (2)$$

10. A variate of the Box-Muller transformation, the so-called Polar method (ref. 8) exists. The method generates two standard normal random numbers from two independent pseudo-random numbers.

11. There is also the generalised Lambda transformation (ref. 12). For N, a uniform random number, y, a number from an arbitrary (generalised Lambda) distribution is such that:

$$y = \frac{A + \left[N^C - (1-N)^D\right]}{B} \qquad (3)$$

12. By suitably choosing values for A, B, C and D various distributions can be simulated. In particular for the standard normal case, A = 0, B = 0.1975, C = D = 0.135.

EXAMPLES, ANALYSIS AND DISCUSSIONS

13. A computer model was developed for the stability analysis of any pin-connected space framework within its capacity. In this paper we present the results of a 24-bar truss system with 21 degrees of freedom shown in Fig. 1. The truss spans 86.6cm with a rise of 8.161cm giving a rise-to-span ratio of 0.0942.

14. The material of the structure is considered to be steel with modulus of elasticity, E = 21 x 10^5 kN/ cm^2. The truss has a bar diameter of 1.382cm with a cross-sectional area of 1.50cm^2 . It is loaded in the positive y - direction at points 3, 4, 6, 7, 8, 10 and 11. The crown and ring loads are applied simultaneously.

15. A total of 1200 imperfect truss systems were analysed, and twelve different imperfection statistics were considered, for which 100 systems were analysed each with a different sets of element imperfections, the various sets all conforming with the same probability distribution. Estimates were also made of a '95% confidence load' as well as a factor of safety.

16. Two charts, Figs 2 and 3 have been produced from the information obtained from our analysis. Fig. 2. is a chart of the normalised mean critical load versus the imperfection statistics, while Fig. 3. is a corresponding chart for the 95% confidence load (normalised).

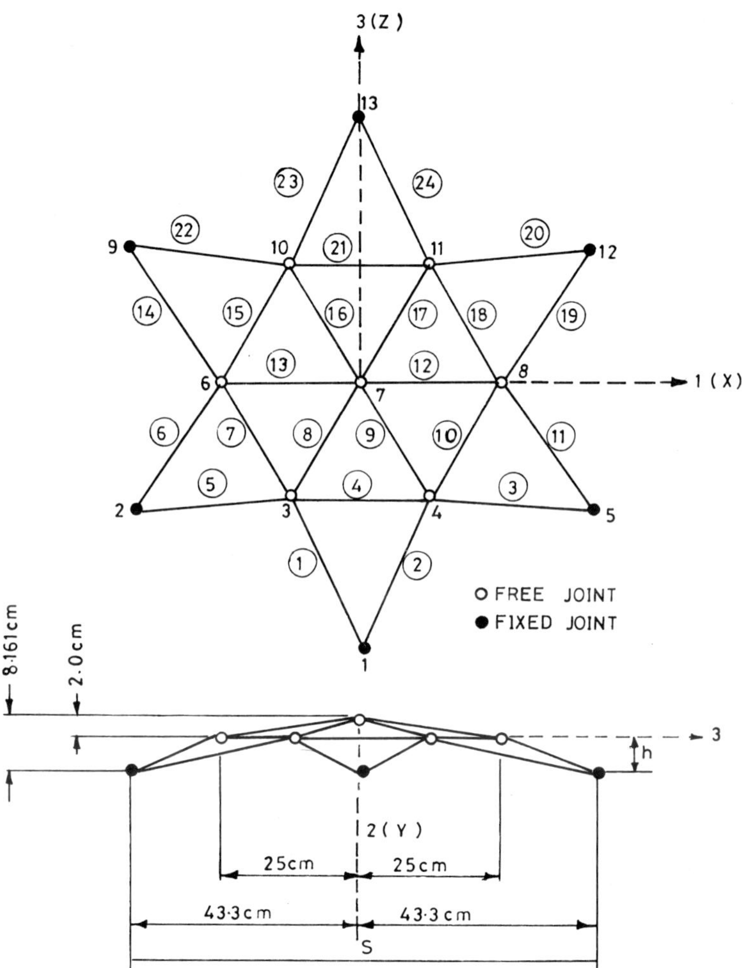

Fig. 1. 24-bar space truss (21 degrees of freedom).

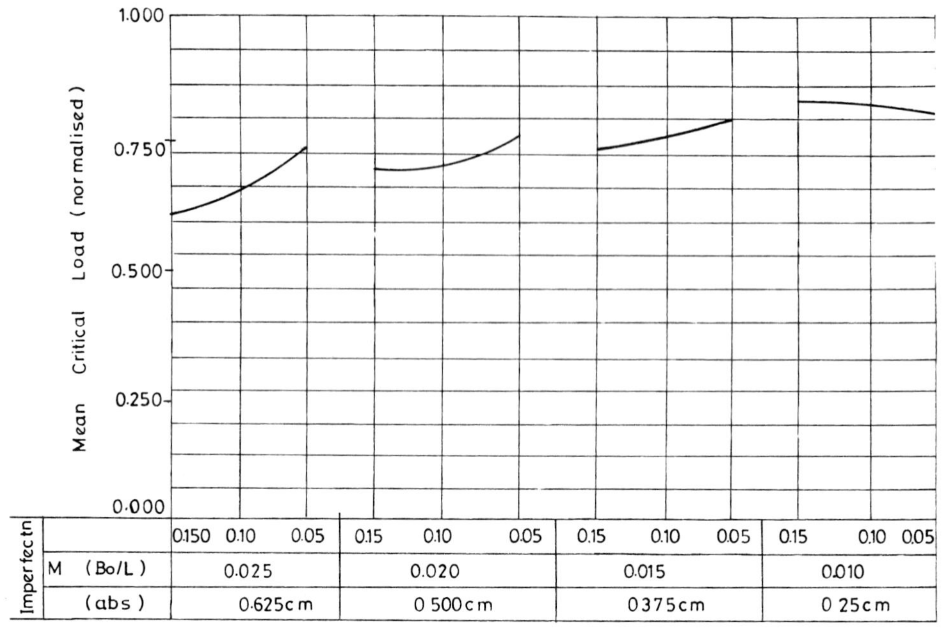

Fig. 2. Mean critical load vs imperfection statistics.

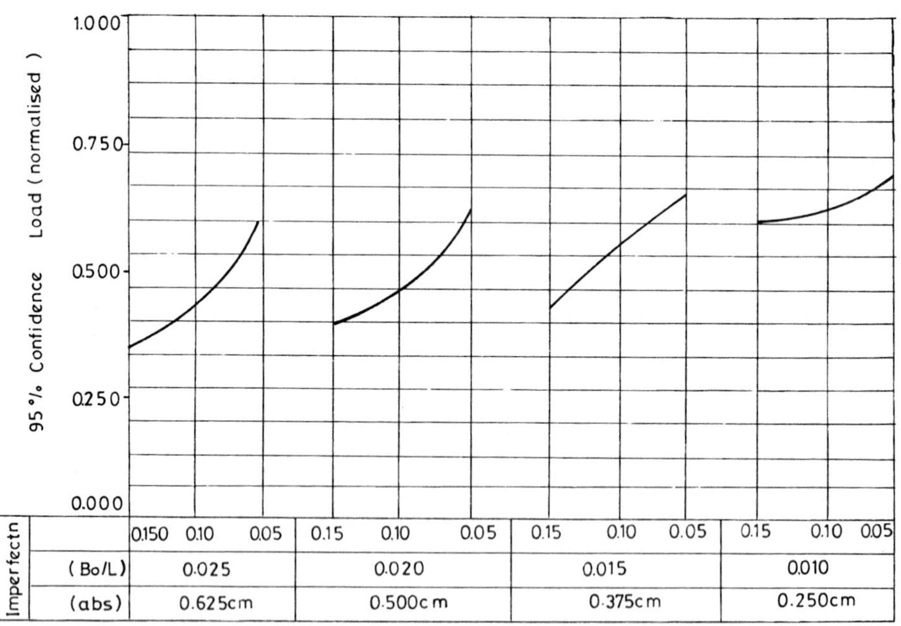

Fig. 3. 95% confidence load vs imperfection statistics.

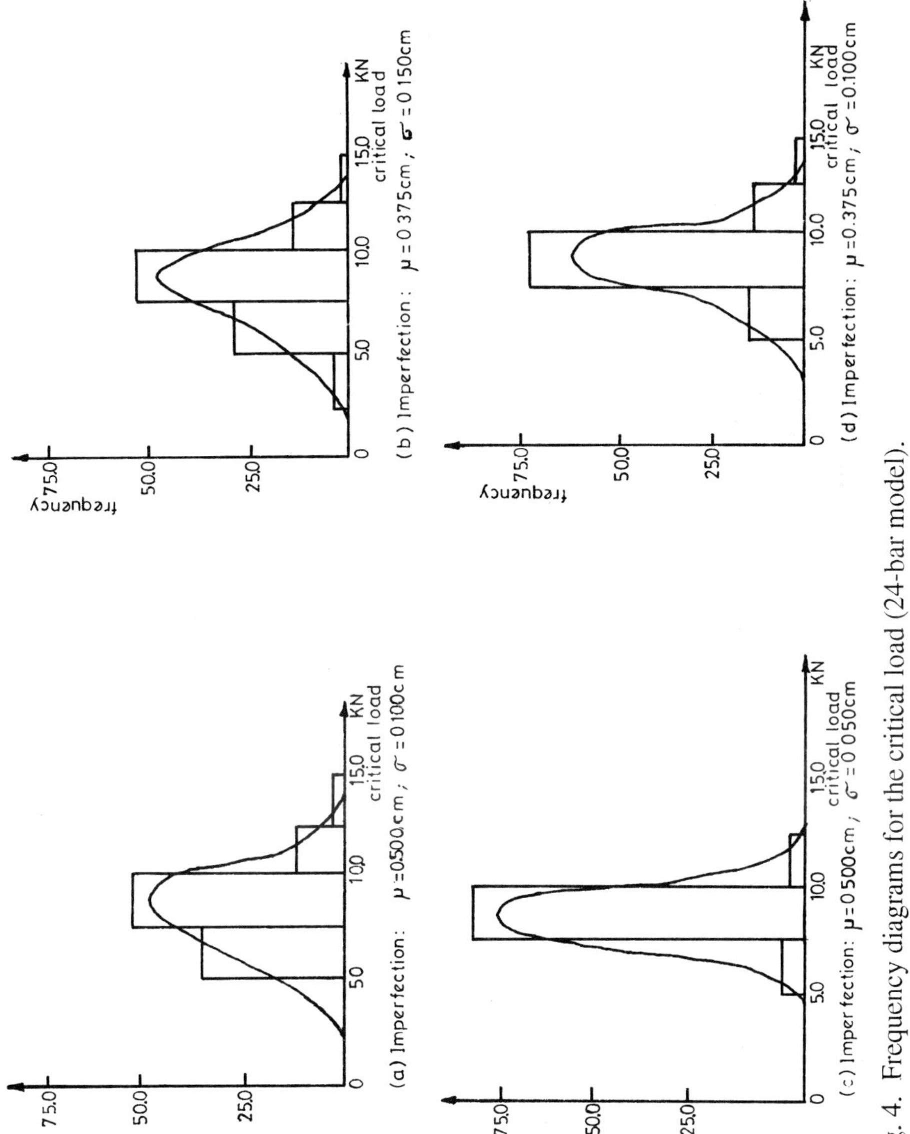

Fig. 4. Frequency diagrams for the critical load (24-bar model).

17. It is observed from Fig. 2. that for a given imperfection mean, the variation in the standard deviation has very little effect on the mean critical load. Fig. 2. also indicates a general rise in the mean critical load values with a fall in the mean value of the imperfection. The implication of this is that for a given mean value of imperfection, the mean value of the critical load would be expected to remain relatively the same over a range of imperfection standard deviations. However, a progressive variation of the imperfection mean would cause a corresponding variation in the value of the mean critical load, but in the opposite direction.

18. On the other hand, Fig. 3. indicates a definite rising tendency of the 95% confidence load both with decreasing mean values of the imperfection and with a lowering of the standard deviation for the same mean. A low value of standard deviation implies a limited amount of uncertainty in the mean value assumed or prescribed. It then follows that at a given load level, the space truss would be expected to be more stable the lower the value of standard deviation associated with the imperfection parameter.

19. Fig. 4. shows some of the frequency diagrams for the space truss.

REFERENCES

1. ABATAN A.O. and HOLZER S.M. Degree of stability of geodesic domes with independent loading parameters. Computer and structures, 1978, Vol. 9, p. 43 - 51.

2. COAN C.N. and PLAUT R.H. Dynamic stability of a lattice dome. Earthquake Engineering and structural Dynamics 1983, Vol. 11, p. 269 - 274.

3. GUTTMAN A.J. Programming and Algorithms: An introduction. Heinemann Education. Books, London, 1977.

4. HOLZER S.M., PLAUT R.H. and SOMERS A.E. Jr. Stability of lattice structures under combined loads. Journal of Engineering Mech. Div. A.S.C.E., No. 106(2), p. 289 - 305, 1980.

5. MAKOWSKI Z.S. A world wide review of space structures in sports buildings. Bulletin IASS Vol. XXIX - 2(97), p. 11 - 40, 1988.

6. MERCHANT M.T. The ABC of FORTRAN Programming. Wadsworth Publishing Company, California, U.S.A., 1979.

7. RHODA J. Some statistical aspects of the buckling of imperfection sensitive structures. Journal Mechanics and Physics of Solids, vol. 17, p. 111 - 123, 1969.

8. ROSEN J.A. and SCHMIDT L.A. Jr. Design oriented analysis of imperfect truss structure - part 1 - Accurate analysis. International Journal of Numerical Methods in Engineering. Vol. 14, p. 1309 - 1321, 1979.

9. ROTHERT H. Snap-through buckling of reticulated space trusses. Journal of Structural Div. ASCE, Vol. 107, p. 129 - 143, 1981.

10. SADEEQ J.A. Monte-Carlo Study of variability in strength of reinforced concrete beams and columns. M.Sc. Thesis Ahmadu Bello University, Zaria, Nigeria 1988.

11. TERRY W.C. Stability investigations of space trusses with element imperfection. M.Sc. Thesis, Virginia Polytechnic Institute and State University Blacksburg Virginia, U.S.A., 1982.
12. WALKER H.M. Problems for computer solutions using FORTRAN. Winthop Computer System Series, Cambridge, Massachusetts, 1980.
13. WARMER R.F. and KABAILA A.P. Monte-Carlo study of structural safety. Journal of Structural Div. ASCE, Vol. 94, p. 2847 - 2859, 1968.

13. A novel soft member for space trusses

G. A. R. PARKE, University of Surrey, UK

SYNOPSIS. A novel 'soft' compression member has been designed and tested which exhibits an initial reduced axial stiffness, followed by a steady load plateau and subsequent enhanced stiffness and strength, before final failure. The member has been designed with the aim of introducing force redistribution and post-buckling ductility into selected double-layer space trusses.

INTRODUCTION

1. Double-layer space trusses are one of the most frequently used forms of structure capable of providing large column free cover, required for a diverse range of social and commercial activities. The strength, or load-carrying capacity of a double-layer space truss is normally assessed by considering the elastic load-displacement response of the structure up to the point where the first compression member fails. Normally, no account is taken of any post-buckling reserves of strength which the double-layer space truss may possess. The inelastic behaviour of the structure is usually ignored in the design process because it is considered that, due to their high degree of statical indeterminacy, double-layer space trusses, fabricated from ductile materials, will possess reserves of strength in excess of their elastic capacity. However, this is not necessarily true.

2. The complete load-displacement behaviour of a space truss depends on the topology of the structure, the support conditions and the load-displacement response of the individual members forming the structure. Post-elastic reserves of strength will only exist in a double-layer space truss if after the failure of the first compression member the adjacent members have sufficient strength to carry the additional load transferred into the structure from the buckled member.

3. Practical considerations usually dictate that the compression members used in a double-layer space truss have intermediate slenderness and consequently buckle inelastically. The inelastic or plastic buckling of a compression member is a dynamic process resulting in a sudden loss of both member stability and load-carrying capacity. If the double-layer space truss encompassing the failed member can absorb this release of energy then the structure may also be able to support an increase in the imposed loading. However, if the force redistribution resulting

from the failed member causes further members to fail, then it is likely that progressive inelastic buckling of the compression members will occur, resulting in a complete collapse of the structure.

4. The major objective of the present study was to investigate means of inducing ductility into the load-displacement response of double-layer space trusses. The method considered here, involved the development of a novel 'soft' member which, when loaded in compression, exhibits an elastic-plastic load-displacement response. It is envisaged that when the most heavily stressed compression members in a double-layer space truss are replaced by the novel 'soft' members, a redistribution of forces will occur within the structure, leading to an enhanced load carrying capacity and improved post-elastic ductility.

FORCE-LIMITING DEVICES

5. In order to improve the load carrying capacity and post-buckling behaviour of compression-chord-critical double-layer grids, several engineers have investigated the possibility of introducing artificial ductility into compression member behaviour (refs 1 and 2). This ductility will, in turn, introduce ductility into the structure and decrease the possibility of progressive collapse occurring after failure of the most heavily stressed compression member. Artificial ductility can be created in compression member behaviour, by incorporating into the member a force-limiting device. This device would, ideally, limit the compression force in the member to a pre-determined level which would remain constant under increasing deflection. Consequently a compression member, protected by a force-limiting device, would exhibit the elastic-plastic load deflection characteristics, shown in Fig. 1, instead of the highly unstable, brittle, post-buckling characteristics typical of transition length compression members. The value of the load plateau set by the force-limiting-device must be lower than the average compression member buckling load to ensure that the device becomes operative before the member buckles. However, it should be appreciated that the introduction of the force-limiting-device itself, may alter the buckling load of the protected member.

6. The characteristics required from the force-limiting devices in addition to the ideal elastic-plastic behaviour are reliability and repeatability. The device must be capable of providing a constant limit force with a load plateau of sufficient length to allow redistribution of member forces to occur within the encompassing space structure. In addition, the device should function with the minimum of maintenance and the behaviour characteristics should be independent of loading sequence and time. To a large extent, these operating requirements are shared by energy absorbing devices, and the design concepts used in energy absorbers can be modified where necessary for use in force-limiting devices. Energy absorbers are rated on both their energy absorptions per unit weight, and their ratio of stroke length to device length. These parameters are generally not critical in assessing force-limiting devices in which acceptable load-displacement requirements predominate. Several different types of energy absorbers are available, all of which may be considered for use as force-limiting-devices. To assist in their

comparison, they have been classified according to their mode of operation which may involve material deformation, extrusion or friction (ref. 3).

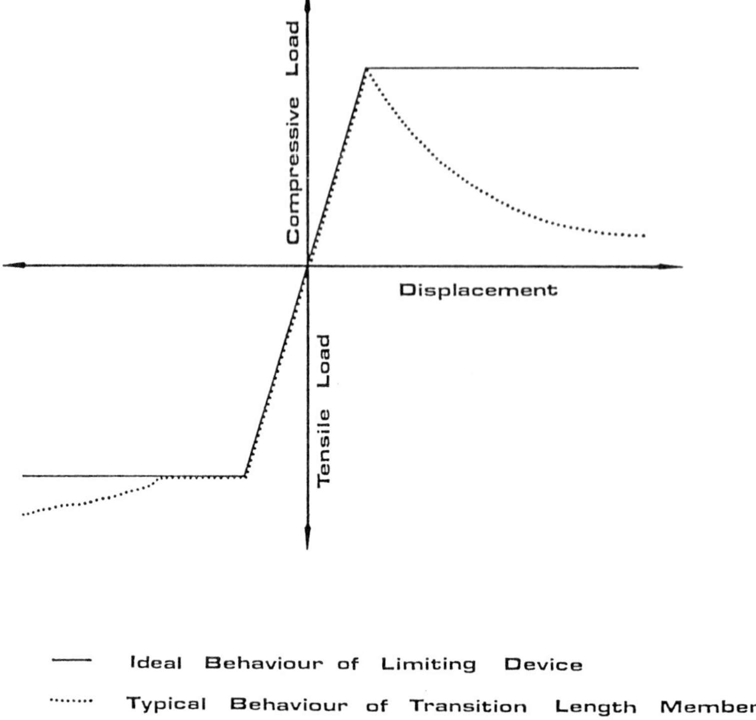

Fig. 1. Ideal behaviour of a force limiting device.

NOVEL FORCE-LIMITING DEVICE

7. One of the recurring problems arising from the use of force-limiting devices, is the general unsuitability and lack of control of their load-displacement characteristics. Several of the devices considered showed either large fluctuations in their working limit loads, or a high initial load occurring before the load decreased and stabilised at the working limit load (refs 4 to 7). In an attempt to overcome these problems, a novel compression member has been designed to act both as a 'soft' member, with reduced axial stiffness, and as a load-limiting device (ref. 8).

Mode of operation of soft members

8. Details of a full size soft member are shown in Fig. 2. The compression member consists of two square hollow section tubes and four rectangular strips. The two hollow section tubes are proportioned so that the smaller tube plus strips just fit inside the large tube. Fig. 3 shows the load-strain relationship obtained for the soft member loaded in compression. Under a compression load the soft member initially behaves elastically, but with a reduced elastic stiffness (A to B,

Fig. 3). At a pre-determined compression load acting on the soft member, the four middle strips yield in tension and provide a load plateau for the complete soft member (B to C, Fig. 3). At this stage, the member is only capable of supporting a constant compression force. Under this steady compression, yielding of the middle strip will continue until the outer and inner tubes mate simultaneously at both ends of the soft member, represented by point C in Fig. 3. When this has occurred, the stiffness of the soft member increases and the compression member acts as a pre-stressed column which is capable of supporting a further increase in compression loading (C to D, Fig. 3). At point D, the column buckles exhibiting a 'brittle type' strut buckling path.

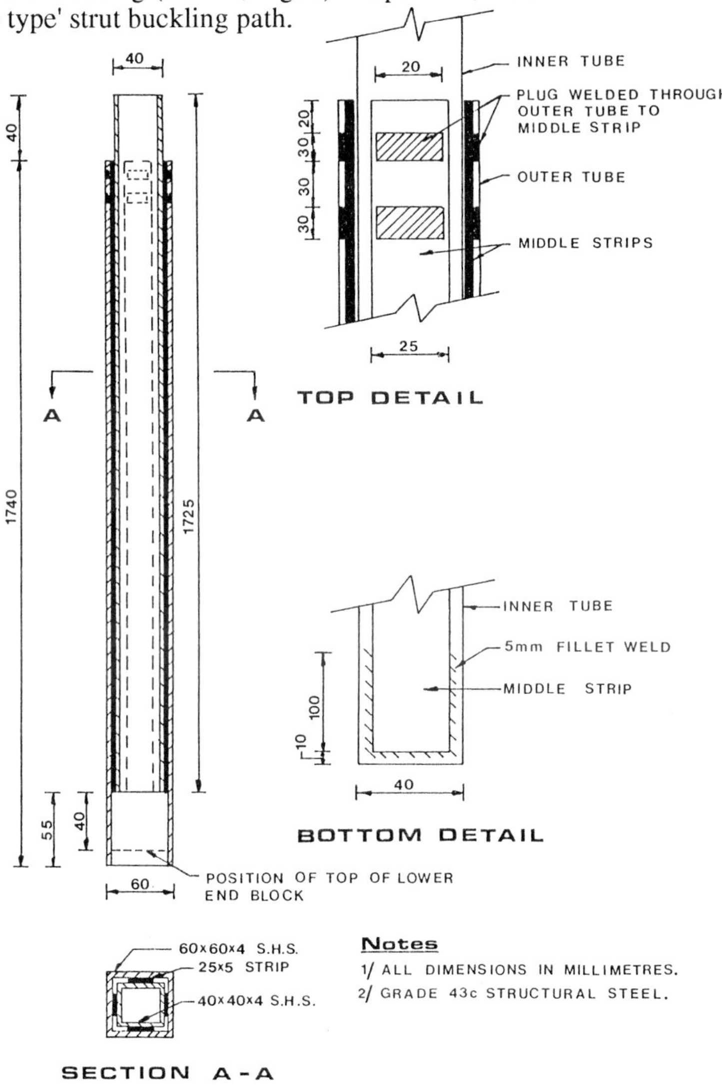

Fig. 2. Full size soft member.

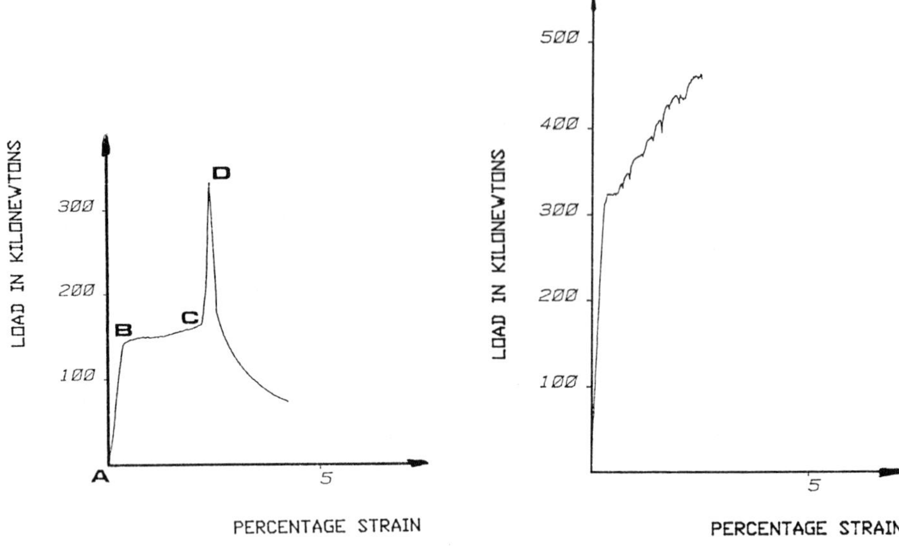

Fig. 3. Compressive load-strain relationship for full size soft member.

Fig. 4. Tensile load-strain relationship for full size soft member.

9. The characteristics of the soft member shown in Fig. 3, can be significantly altered by varying both the length and cross-sectional area of the four steel strips. The initial stiffness of the system can be increased by shortening the length of the four strips, and the magnitude of the load plateau can be enhanced by increasing the cross-sectional area of the strips. In addition, the length of the load plateau can be controlled by varying the end distance occurring between the inner and outer square tubes. Fig. 4 shows the tensile load-strain behaviour obtained from the soft member. If the soft member is loaded in tension, then the four steel strips are subjected to compression forces, and both the inner and outer square tubes are stressed in tension. If the gap between the strips and tubes is very small, then the steel strips will yield and squash in compression. If this is not the case, the strips will buckle and deflect until they are restrained by the inner and outer tubes, whereupon the member will carry additional load, until the strips yield in compression and deform plastically. When the strips have yielded, the soft member will carry an almost constant tensile load until the strip material strain hardens, and the member supports an additional tensile load. The tensile load carried by the member will increase, until either the end welds fail, or the inner or outer square tubes yield, and finally rupture in tension.

10. Several improvements to the type of soft member shown in Fig. 2 have been made. The relative maximum load capacity of the device was enhanced, by changing the two square tubes to round tubes, and more significantly, by replacing the four separate steel strips also by a round steel tube. The three tubes were fitted closely one inside the other, and the assembly was fabricated using a similar procedure to that adopted for the square soft members. These triple-tube

soft members have a simple mode of operation, but the ultimate capacity of the novel triple-tube compression member is difficult to assess, due to the interdependence of the three tubes .

Theoretical behaviour and ultimate capacity of triple-tube soft members

11. The theoretical response of a perfect soft member supporting a compressive load, can be investigated by considering the stiffness of the individual tubes and their joint interaction. Fig. 5 shows the changes in stress occurring within the three tubes of the soft member as the external compressive load is increased. Before the soft member has closed, each of the three tubes will carry a force equal in magnitude to the external load. This will result in a compressive stress in the inner and outer tubes, and a tensile stress in the middle tube (Fig. 5A). As the external load is increased, the force carried by each of the three tubes will increase by the same amount. Provided that the compression force in the inner and outer tubes is always insufficient to cause premature elastic buckling of both of these members, the tensile force in the middle tube will continue to increase linearly. When the tensile force acting on the middle tube is sufficient to cause this tube to yield, plastic flow occurs within this member and the soft member will not support any increase in external load (Fig. 5B). The middle tube continues to yield until the three tubes close up and form a pre-stressed triple tube compression member.

12. As the external load is increased, the force within each of the three tubes now alters. The middle tube unloads elastically, while the compressive stress in the inner and outer tubes increases (Fig. 5C). A further increase in the external loading will cause a corresponding increase in the compressive stress in the inner and outer tubes, and consequently increase their tendency to buckle. However, because the three tubes are in very close proximity to each other, premature elastic buckling of either the inner or outer tubes will not occur due to the restraint provided by the middle tube still in tension. As the load is further increased, the compressive stress in the inner and outer tubes will approach yield and the middle tube will continue to unload elastically.

13. If perfect elastic-plastic material behaviour is assumed, then when the inner and outer tubes yield in compression they are unable to support any further increase in external load (Fig. 5D). In addition, when the inner and outer tubes have yielded, the bending stiffness of the triple column will have decreased to a value equal to the bending stiffness of the middle tube only. Any further increase in the external load must now be supported only by the middle tube. Consequently, the stress in the middle tube will change from tensile to compressive, and will increase until the tube becomes unstable, and the middle portion of the tube begins to displace horizontally. This horizontal displacement will occur with buckling at the Euler buckling load, if the middle tube fails by elastic buckling, or will occur at the tangent-modulus load if the tube is prone to failure by plastic buckling.

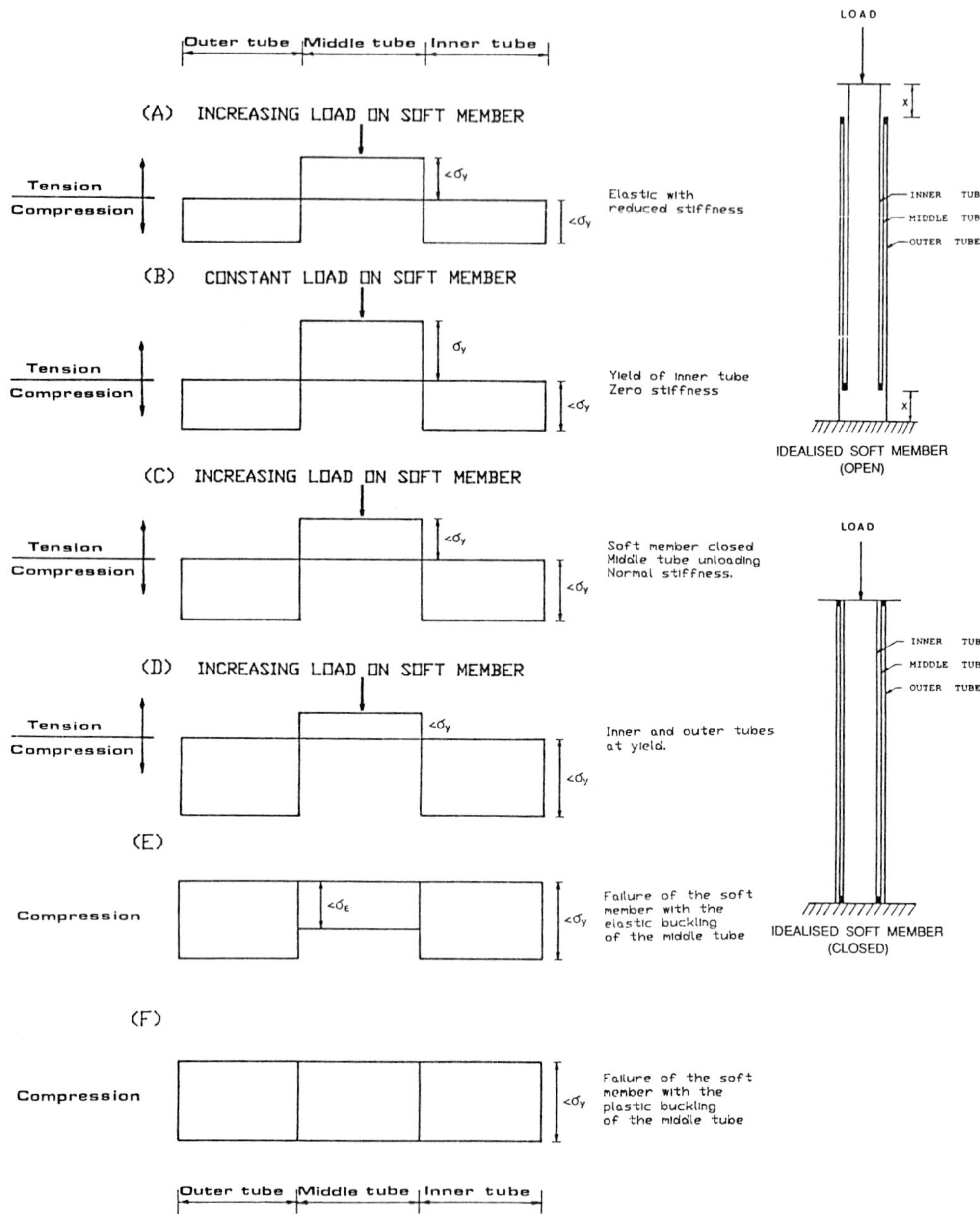

Fig. 5. Stress diagrams showing the theoretical response of a perfect soft member supporting an axial compression load.

14. If elastic buckling of the middle tube occurs, the bending stiffness of the member has been decreased to zero, and all lateral restraint provided to both the inner and outer tubes by the middle tube vanishes, causing the complete soft member to buckle (Fig. 5E). If the middle member fails by plastic buckling it will support a load greater than the tangent modulus load but less than the double-modulus load. However, buckling of the middle tube will still cause buckling of the complete soft member (Fig. 5F). Consequently, the buckling load of the perfect soft member will be controlled by the buckling load of the middle tube. If the middle tube fails by elastic buckling, a perfect soft member will fail at a theoretical load, equal to the sum of the squash loads of the inner and outer tubes, plus the Euler critical buckling load of the middle member. However, if the middle tube fails by plastic buckling, a perfect soft member will fail at a theoretical load equal to the sum of the plastic buckling load of all three tubes.

15. If perfect elastic-plastic material behaviour is not assumed, then material strain hardening will allow the inner and outer tubes to sustain a compressive stress greater than the yield stress, provided the stress in the middle tube remains below its critical buckling value. In addition, because the middle tube has been extensively strained in tension before being stressed in compression, both the compressive yield stress and the member stiffness in compression will be decreased as a direct result of the Bauschinger effect. These small decreases in compression stiffness and yield stress, will decrease both the elastic and plastic buckling loads of the middle tube, which will in turn reduce the collapse load of the complete soft member.

16. The presence of imperfections will also decrease the collapse load of the soft member. If both the inner and outer tubes yield in compression before the middle member fails, it will be imperfections occurring within the middle member which will have the dominating influence on the collapse load of the entire soft member.

EXPERIMENTAL

17. An experimental investigation has been undertaken to assess the behaviour of soft members loaded in both tension and compression. Accurate and reliable values are required for the initial stiffness, yield load, collapse load and the post-buckled response of the soft member. This data was used to compare the theoretical and experimental behaviour of the soft member in addition to obtaining the theoretical response of space trusses incorporating soft members. Several groups of soft members have been tested. The load-displacement behaviour has been obtained for both model soft members and a limited number of full size soft members.

Full size soft members

18. Six full size soft members have been fabricated and tested. Five have been loaded in compression and only one member tested in tension. During this investigation modifications were made to improve the behaviour of the soft

member. Changes were made to both the connection details and also to the lengths of the individual components.

19. Fabrication of members. All six of the soft members were fabricated from grade 43C structural steel (ref. 9). This grade of steel was chosen because of its high ductility and good weldability. The first three members SM1, SM2 and SM3 were fabricated from 100 x 100 x 4mm and 80 x 80 x 5mm thick, square hollow section and four, 50 x 5mm thick flat steel strips. The second group of soft members SM4, SM5, and SM6 were fabricated from 60 x 60 x 4mm and 40 x 40 x 4mm thick, square hollow section and four, 25 x 5mm thick flat steel strips. All of the square hollow sections were hot finished seamed tubes.

Model soft member tests

20. Fig. 6 shows typical details of the model triple-tube soft members which were fabricated to be incorporated into steel model double-layer space trusses. In order to assess their typical load-displacement responses, twenty eight, nominally identical, soft members were tested in tension and sixteen tested in compression.

21. Fabrication of members. All forty four soft members were fabricated from annealed cold-drawn seamless steel tubes with the diameter and wall thickness of the outside, middle and inner tubes being, 9.52 x 0.82; 7.94 x 0.5 and 6.35 x 1.2mm respectively. The three tubes were carefully proportioned so that they would just fit inside each other as shown in Fig. 6.

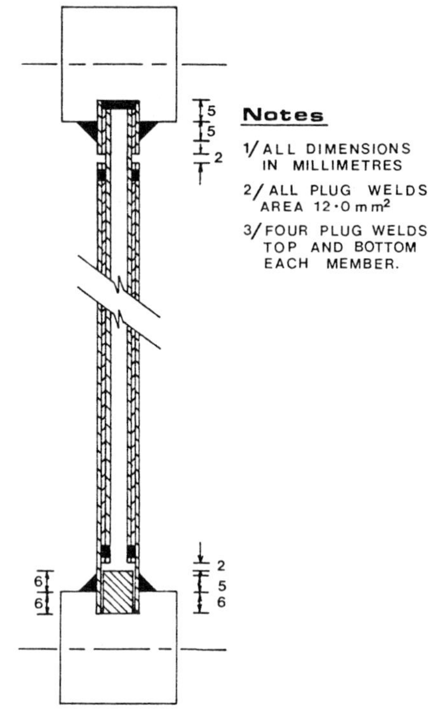

Fig. 6. Model soft member.

RESULTS

Full Size Soft Members

22. Fig. 3 shows the typical compressive load-displacement response for the full size soft members. The members exhibited an initial stiffness of 27.4kN/mm followed by an average yield plateau load of 150kN. Table 1 gives both the actual and theoretical ultimate buckling loads for the soft members SM5 and SM6.

Table 1. Comparison of experimental and theoretical flexural buckling loads for full size soft members.

Theoretical flexural buckling load allowing for the measured bow.	Theoretical flexural buckling load calculated using Table 27a BS5950.	Experimental flexural buckling load.
351.9kN	310.23kN	331.58kN

Model soft members

23. Figs 7 and 8 show the tensile and compressive load-displacement response respectively for a typical model soft member. In compression the member exhibits an initial elastic stiffness of 3259N/mm. The member yields under an average force of 3916N and buckles under a force of 7739N.

24. Table 2 gives the theoretical and measured buckling loads for three sets of the model members, each set being fabricated from a particular batch of material. It can be seen from the experimental results obtained, that the behaviour of the triple-tubular soft members is governed by the buckling capacity of only the outside tubes.

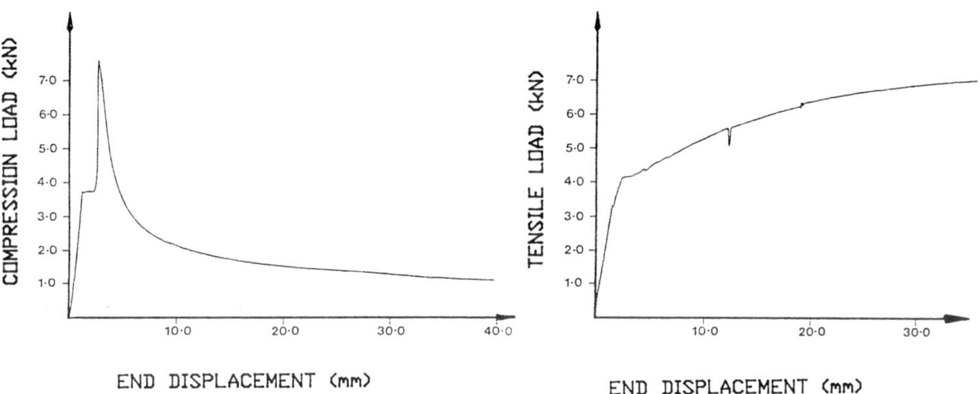

Fig. 7. Load-displacement relationship for a compression model soft member.

Fig. 8. Load-displacement relationship for a tensile model soft member.

Model soft member classification.	Average experimental buckling load kN.	Theoretical buckling load kN (a)	Theoretical buckling load kN (b)	Theoretical buckling load kN (c)	Theoretical buckling load kN (d)
White	7.556	7.232	6.281	13.161	13.049
Yellow	8.109	7.289	6.376	13.731	13.619
Red	7.553	7.222	6.217	12.763	12.667

(a) Assuming soft member fails when the outside tube buckles based on average measured bow of 1/2270.

(b) Assuming soft member fails when the outside tube buckles based on Table 27a BS5950 with a bow of 1/1000.
(c) Assuming soft member fails when the middle tube buckles based on average measured bow of 1/2270.
(d) Assuming soft member fails when the middle tube buckles based on Table 27a BS5950 with a bow of 1/1000.

CONCLUSIONS

25. The experimental investigations described show that a compression member has been constructed capable of exhibiting a steady load plateau followed by a reserve of strength before failure.

26. Initial theoretical investigations into the stability of the triple-tube members indicated that their ultimate capacity would be governed by the capacity of the middle tube, which first yielded in tension and then subsequently loaded in compression. However, the limited experimental investigation has shown the ultimate capacity of the triple-tube soft member to be dependent on the stability of the outside tube, which may obtain restraint from the middle tube, depending on the tolerance in fit occurring between the three individual component tubes of the member.

27. Additional studies are now required to investigate the response of double-layer space trusses incorporating the new soft members. These investigations should consider optimising the ratio of yield load to member buckling for the soft members and where best to incorporate them into a structure to achieve maximum benefit in both load carrying capacity and post buckling ductility.

REFERENCES

1. SCHMIDT L. C. and HANAOR A., Force limiting devices in space trusses. ASCE J. Struct. Div., 105 (ST5), 939-951, (1979).
2. HANAOR A. and SCHMIDT C., Space truss studies with force limiting devices ASCE J. Struct. Div., 106 (ST11), 2313-2329, (1980).
3. EZRA A. A. and FAY, R. J., An assessment of energy absorbing devices for prospective use in aircraft impact situations. Dynamic Response of Structures, Eds. Herrmann, G., Perrone, N., Pergamon Press, 225-246, (1972).
4. JOHNSON W., Impact Strength Of Materials, London Arnold, (1972).
5. JOHNSON W., REID S.R. and SINGH L. B. Experimental study of the rolling torous load limiter, Int. J. Mech. Sci., 19, 595-602, (1975).
6. HANAOR A., Force limiting devices in space trusses. PhD Thesis, University of Melbourne, (1979).
7. ROBINSON W. H. Properties of an extrusion damper, Sixth Australasian Conference on the Mechanics of Structures and Materials. University of Canterbury, Christchurch, New Zealand, 372-375, (1977).
8. PARKE G A. R., The behaviour of space trusses incorporating novel compression members. PhD Thesis, University of Surrey, (1988).
9. BRITISH STANDARDS INSTITUTION Specification of weldable structural steels, BS4360, London, BSI, (1986).

14. A method to calculate the critical loads of single layer shallow lattice domes with initial imperfections

H. L. ZHAO, W. M. HUANG and C. Q. ZHAO, Southeast University, P. R. China

SYNOPSIS. An iterative approach is suggested to calculate the critical loads of single-layer shallow domes with initial stochastic geometrical imperfections. The tangent stiffness matrix of an element with a small initial curvature is also presented in this paper.

INTRODUCTION

1. Single-layer lattice domes are very sensitive to initial imperfections. The critical load calculated from an imaginary perfect model is always much greater than that of the real structure. To estimate accurately the critical load of such a dome it is essential to take initial imperfections into account.

2. From previous works, such as (ref. 1), it is evident that geometric imperfections are the most important factor in all kinds of imperfections which affect the critical loads of single-layer shallow lattice domes (SLSLD). Here two major types of geometric imperfections are considered namely the initial deviation of the nodes from an imaginary perfect position and initial member curvature.

3. Most of the previous work on this topic only deals with the effects of some assumed modes of imperfection, such as (ref. 2), but in real structures, imperfections are distributed stochastically. What is evident from statistics and experience is the extent of such imperfections.

4. In this paper, an iterative method is presented to calculate the critical load of SLSLD with stochastical geometrical imperfections. This is a different approach to most previous work which only take initial node deviation into account. Initial member curvature is also included here in the imperfection mode.

SOLUTION APPROACH

5. It is known that for a given range of imperfections, there is a critical initial imperfection mode which corresponds to the minimum critical load. The work of D. Ho (ref. 3) has shown that if the post-buckling path is steep, reduction in buckling load will be large. If the imperfection mode is parallel to the steepest post-buckling path, the minimum critical load is obtained. Generally, it is very hard to obtain the actual minimum critical load, but it is possible to use an iterative

approach to gain the approximate least critical load, i.e. nominal least critical load of SLSLD with stochastic initial imperfections. The steps of such an iterative approach are:

(a) calculating the buckling mode of the perfect structure;
(b) adding the buckling mode to the perfect structure;
(c) calculating the buckling mode and buckling load of the modified structure;
(d) returning to (b) until the difference in buckling loads between two adjacent iterative times is sufficiently small . The buckling load at that time is called the nominal least critical load.

6. However, it is very conservative if structures are designed according to this value. The most important value for an engineering designer is the critical loads of the structure which meets given reliability demands. The reliability of the critical loads obtained by such an iterative approach depends on the number of iterations. If only one iteration is undertaken, it will be a sort of assumed mode method which takes the critical mode corresponding to the lowest critical eigen-value of the perfect structure as the imperfect mode.

TANGENT STIFFNESS MATRIX FOR AN ELEMENT WITH A SMALL INITIAL CURVATURE

7. In real SLSLD, the initial curvature of the member is very small. Here, it is assumed that the initial curvature is cubic in each of the two principal directions. So the initial curvature could be depicted by initial nodal angles. By applying the principle of minimum potential energy, the element tangent stiffness matrix can be obtained in the local co-ordinate system as shown in Figs 1(a) and 1(b).

$$[\overline{K}] = [A][K]^{o}[A]^{T} + [D] \tag{1}$$

Here, $[K]^{\circ} = [K] + [K]'$, [A], [D], and [K] are the same as in ref. 4. (M_{13}, M_{23}, M_{12}, M_{22}, M_t and P are as shown below). The non-zero elements in [K]' are:

$K_{11} = -4C$, $K_{12} = K_{21} = C$, $K_{22} = -4C$, $K_{33} = -4C$, $K_{34} = K_{43} = C$,

$K_{44} = -4C$, $C = EA \times l \times T / 900$

$\theta_{ij} = \overline{\theta}_{ij} + \tilde{\theta}_{ij}$ (i = 1,2; j = 2,3)

θ_{ij} is the total rotation of node;

$\overline{\theta}_{ij}$ is the initial angle of node;

$\tilde{\theta}_{ij}$ is the actual rotation of node.

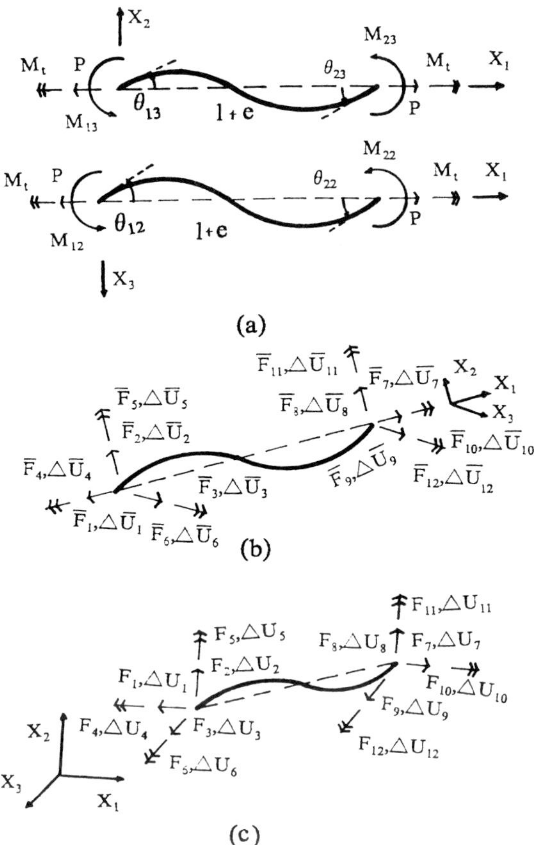

Fig. 1. Member deformations and associated forces.

$$M_{13} = \frac{4EI_3}{\ell}\tilde{\theta}_{13} + \frac{2EI_3}{\ell}\tilde{\theta}_{23} + \frac{P\ell}{30}(4\theta_{13} - \theta_{23})$$

$$M_{23} = \frac{4EI_3}{\ell}\tilde{\theta}_{23} + \frac{2EI_3}{\ell}\tilde{\theta}_{13} + \frac{P\ell}{30}(4\theta_{23} - \theta_{13})$$

$$M_{12} = \frac{4EI_2}{\ell}\tilde{\theta}_{12} + \frac{2EI_2}{\ell}\tilde{\theta}_{22} + \frac{P\ell}{30}(4\theta_{12} - \theta_{22})$$

$$M_{22} = \frac{4EI_2}{\ell}\tilde{\theta}_{22} + \frac{2EI_2}{\ell}\tilde{\theta}_{12} + \frac{P\ell}{30}(4\theta_{22} - \theta_{12})$$

$$M_t = \frac{GJ}{\ell}\theta_t$$

$$P = EA\left[\frac{e}{\ell} + \frac{1}{30}\left(2\theta_{12}^2 + 2\theta_{22}^2 - \theta_{12}\theta_{22} + 2\theta_{13}^2 + 2\theta_{23}^2 - \theta_{13}\theta_{23}\right) - \frac{T}{30}\right]$$

$$T = 2\overline{\theta}_{12}^2 + 2\overline{\theta}_{22}^2 + 2\overline{\theta}_{13}^2 + 2\overline{\theta}_{23}^2 - \overline{\theta}_{12}\overline{\theta}_{22} - \overline{\theta}_{13}\overline{\theta}_{23}$$

8. EI_2 and EI_3 are the flexural rigidities about the X_2 axis and X_3 axis respectively. The value EA is axial rigidity and GJ is the torsional rigidity.

9. To transform Eq. 1 into global co-ordinates, and using the matrix [R] as:

$$[R] = [[r],[r],[r],[r]]$$

Where [r] is matrix of direction cosines of the local member axes with the global axes (Fig. lc), then:

$$[K_G] = [R]\left[\overline{K}\right][R]^{\mathrm{T}} \tag{2}$$

and the non-linear equilibrium equation is:

$$\{\Delta F\} = [K_G]\{\Delta U\} \tag{3}$$

Where:

$\{\Delta F\}^{\mathrm{T}} = \{\Delta F_1\ \Delta F_2\ \Delta F_3\ \Delta F_4\ \Delta F_5\ \Delta F_6\ \Delta F_7\ \Delta F_8\ \Delta F_9\ \Delta F_{10}\ \Delta F_{11}\ \Delta F_{12}\}$ are increments of load and

$\{\Delta U\}^{\mathrm{T}} = \{\Delta U_1\ \Delta U_2\ \Delta U_3\ \Delta U_4\ \Delta U_5\ \Delta U_6\ \Delta U_7\ \Delta U_8\ \Delta U_9\ \Delta U_{10}\ \Delta U_{11}\ \Delta U_{12}\}$ are increments of displacement.

10. If there is no initial curvature $\left(\overline{\theta}_{ij} = 0, i = 1,2; j = 2,3\right)$, the tangent stiffness matrix above is the same as the tangent stiffness matrix in (ref. 4).

NUMERICAL EXAMPLES

Example l

11. Fig. 2 shows a 42-member, 19 node shallow dome (ref. 5). The nodal co-ordinates are given in Table 1. At the six support nodes (1, 3, 8, 12, 17 and 19) all six degrees of freedom are restrained. The load case is shown in Fig. 3. The member cross-section is a 4 x 4mm square mild steel bar which has an E = 211.7kN/mm^2, and G = 80kN/mm^2. The load displacement relationship for nodes 10 and 2 in the vertical direction are shown in Fig. 4.

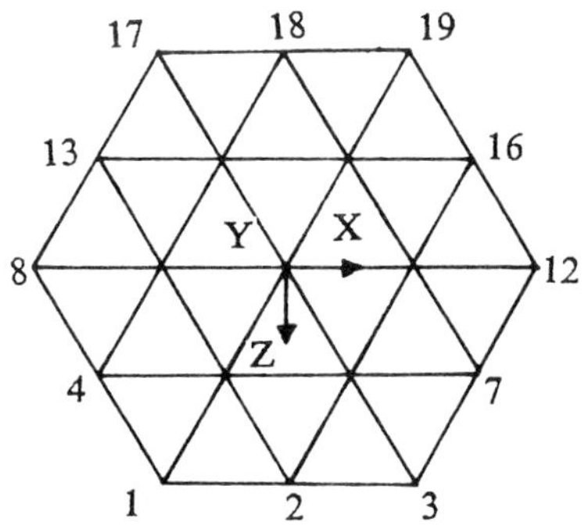

Fig. 2. Model for Example l.

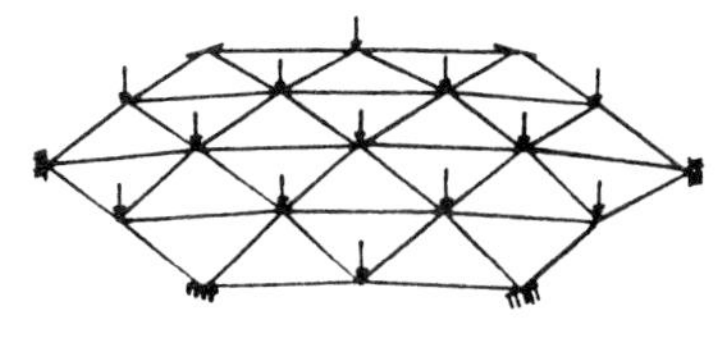

Fig. 3. Loading case.

Table 1. Nodal co-ordinates of the 42-member dome.

Node	X(mm)	Y(mm)	Z(mm)
1	−250.0	0.00	433.01
2	0.0	10.05	433.01
3	250.0	0.00	433.01
4	−375.0	10.05	216.51
5	−125.0	30.05	216.51
6	125.0	30.05	216.51
7	375.0	10.05	216.51
8	−500.0	0.00	0.00
9	−250.0	30.05	0.00
10	0.0	40.00	0.00
11	250.0	30.05	0.00
12	500.0	0.00	0.00
13	−375.0	10.05	−216.51
14	−125.0	30.05	−216.51
15	125.0	30.05	−216.51
16	375.0	10.05	−216.51
17	−250.0	0.00	−433.01
18	0.0	10.05	−433.01
19	250.0	0.00	−433.01

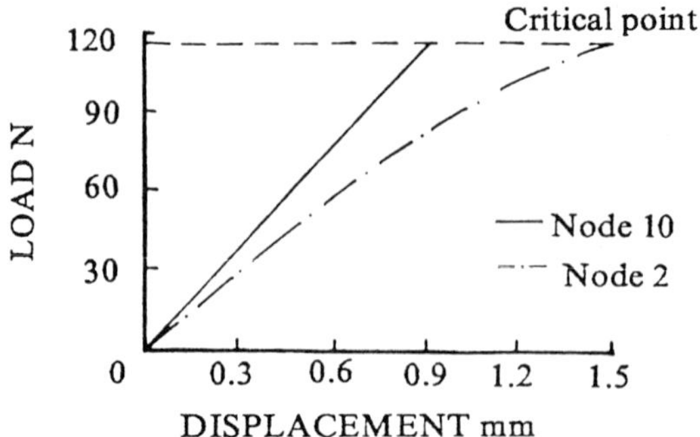

Fig. 4. Load-displacement results.

Table 2. Critical loads and statistical results (Unit: N).

$\alpha=\pi/1000$	$\triangle=0.15$mm	$\triangle=0.30$mm	$\triangle=0.45$mm
1	103.7	92.6	95.4
2	103.5	91.9	80.2
3	96.2	92.1	80.7
4	91.2	87.5	72.7
5	103.2	90.9	78.8
6	103.6	100.5	94.5
7	102.3	100.8	80.6
8	95.7	91.6	79.9
9	103.6	92.0	80.6
10	102.9	89.7	76.8
11	101.9	86.1	70.6
12	103.2	90.7	78.5
13	102.6	88.7	87.9
14	100.1	94.3	83.7
15	103.4	91.4	79.6
16	103.2	90.8	92.0
17	97.4	94.3	98.7
18	98.7	101.5	93.7
19	103.5	80.2	69.5
20	95.0	92.0	80.7
21	99.2	91.8	94.2
22	102.3	90.4	78.6
23	102.2	100.7	98.6
24	101.4	87.3	72.8
25	103.0	84.6	79.6
P^{cri}_{min}	91.2	80.2	69.5
P^{cri}_{ave}	100.9	91.8	83.2
σ	3.3	5.0	8.6

12. It is assumed that the initial deviation of the nodes and the initial angles of the elements in a real structure can be modelled by a normal probability density function within two standard deviations of the mean. The terms Δ and α represent the maxima of the initial deviation of the nodes in three dimensions and the initial angles of the elements respectively. When $\alpha = \pi/1000$, Δ equals 0.15mm, 0.3mm and 0.45mm respectively. A total of twenty five calculations were undertaken. The critical loads $\left(P^{cri}\right)$, minimum $\left(P^{cri}_{min}\right)$, average $\left(P^{cri}_{ave}\right)$ and the mean square deviation (σ) of each group are shown in Table 2.

13. As the imperfection is distributed normally, the critical loads can be assumed to be distributed normally. If the distribution is normal, the probability within one, two and three times of σ at P^{cri}_{ave} are 0.6827, 0.9543 and 0.9973 respectively.

14. In previous work, the imperfection mode is assumed to be the buckling mode of the perfect structure. In Table 3, it is clear that this assumption could give an acceptable result only when the imperfection is small ($\Delta = 0.15$mm). If the perfection increases ($\Delta = 0.3$mm and 0.45mm), the present method will give a reliable result.

Table 3. Iterative results (Unit: N).

$\alpha = \pi / 1000$		$\Delta = 0.15$mm	$\Delta = 0.3$mm	$\Delta = 0.45$mm
P^{cri}_{ave}		100.9	91.8	83.2
$P^{cri}_{ave} - \sigma$		97.6	86.8	74.6
$P^{cri}_{ave} - 2\sigma$		94.3	81.8	66
$P^{cri}_{ave} - 3\sigma$		91	76.8	57.4
Itera-tive results	First	93.1	82.2	76.5
	Second	92.5	83.1	70.7
	Third	89.8	77.8	63.8

Example 2

15. Two test models have the same dimensions as shown in Fig. 5. All nodes are designed to be at the surface of a sphere with a radius 1500mm. The model is 200mm high with a vertical load at the central node (node 19). E = 2.06 x 10^5Mpa, and G = 8.05 x 10^4Mpa. By measurement of the real models, it is found that the nodal deviation is distributed normally and the maximum deviation is 5.6mm. Both models failed by local buckling in the central area as shown in Fig. 6. Test results and theoretical results are presented in Fig. 7.

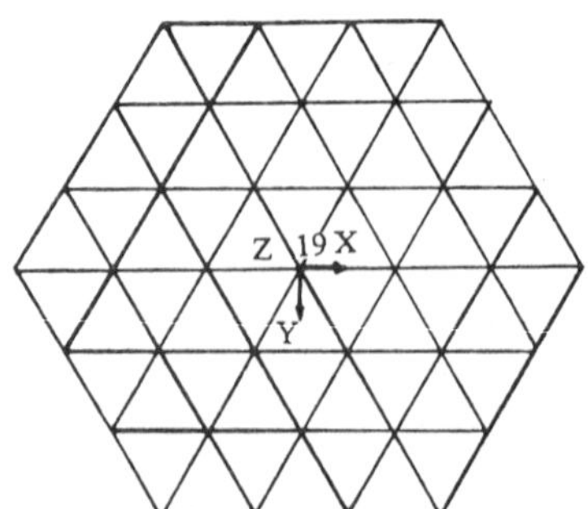

Fig. 5. Model of test.

(a) Buckling shape.

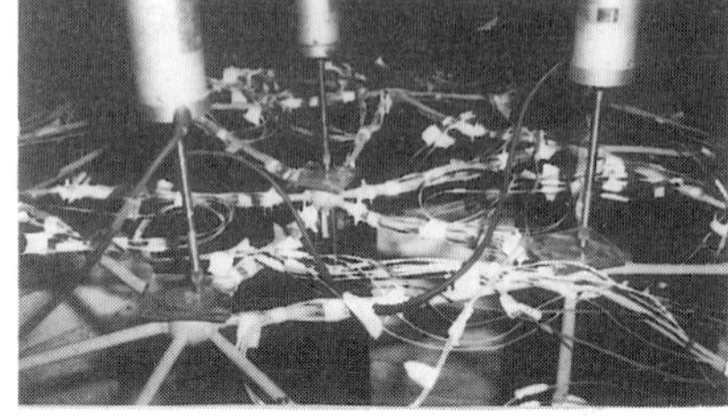

(b) Local buckling area.

Fig. 6. Buckling shape of model.

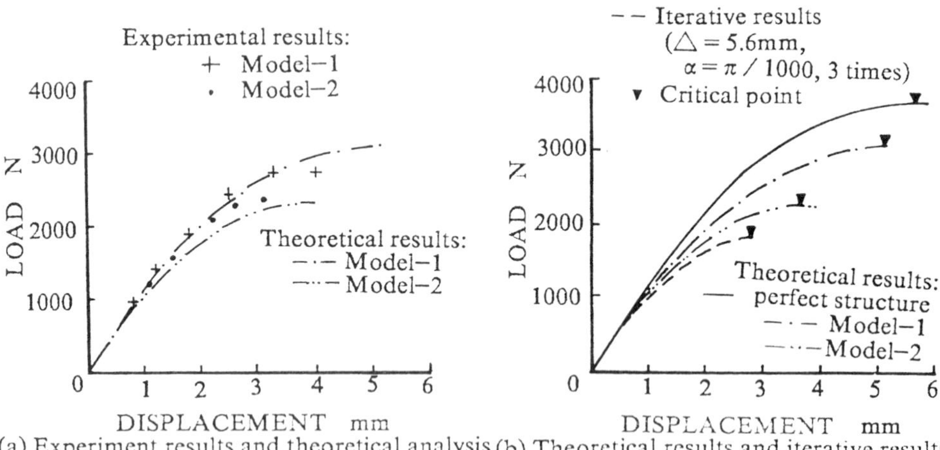

(a) Experiment results and theoretical analysis (b) Theoretical results and iterative results

Fig. 7. Test results and theoretical results (node 19).

16. It is very clear that to accurately estimate the critical loads and provide reliable information about the load carrying capacity of SLSLD, it is essential to take initial geometric imperfections into consideration. The result obtained by the iterative method compare favourably with the test result.

CONCLUSION

17. Initial node deviation and initial member curvature are important factors which greatly decrease the critical loads of SLSLD.

18. The method which takes imperfection into consideration by assuming that an imperfect mode is a buckling mode of the perfect structure is not suitable for SLSLD with relatively large imperfections. The present method is reliable for this case.

19. The tangent stiffness matrix of an element with initial curvature presented here is suitable for structures with bars with small initial curvature. Sub-division is avoided and expenditure on calculations are reduced.

20. The deduction of the SLSLD's critical load is dependent on the extent of the imperfection. Just dividing the critical load of a perfect structure by an invariant factor is not appropriate.

REFERENCES

1. BABCOCK C. D. Shell stability, Journal of Applied Mechanics, 1983,50(46), 935-940.

2. SEE T. and McCONNEL R. E. Large displacement elastic buckling of space structure, J. of Structural Engineering, 1986, 112(5), 1052-1069.

3. HO D. Buckling load of non-linear systems with multiple eigen values, International Journal of Solids Structures, 1974, 10, 1315-1330.

4. MEEK J. L. and TAN H. S. Geometrical non-linear analysis of space frames by an incremental iterative technique, Computer Methods in Applied Mechanics and Engineering, 1984, 47, 261-282.

5. KANI J. M. and McCONNEL R. E. Single layer shallow lattice domes: analysis, general behaviour and collapse, Int. Journal of Space Structure, 1988, 3(2),64-73.

15. Combined non-linear analysis of shallow lattice domes using a beam super element

I. M. KANI, Tehran University, Iran

SYNOPSIS. Based on a fast and efficient non-linear two-noded isoparametric beam element developed by the author, a super element is devised consisting of four to eight of these two-noded sub-elements. The internal nodes are condensed out so that only two nodes suffice to describe the element. Compared with the rigorous three-noded isoparametric beam element used by the author, this super element provides striking reductions in CPU time , and maintains comparable accuracy. Two shallow lattice domes are analysed using the super element and the results are compared with previous observations.

INTRODUCTION

1. Based on Timoshenko beam theory a simple and efficient two-noded beam element was presented in ref. 1 for the combined non-linear analysis of frameworks. Due to the simplifying assumptions involved in the formulation of this element it was necessary to use four to eight elements to model each member of the framework for accurate results in the presence of severe non-linearities.

2. To get round this problem, which makes data file preparation a very time consuming task, a super element consisting of a variable number of the above mentioned two-noded beam elements is developed in the following sections. The static condensation technique is used to condense out the internal degrees of freedom taking into account the non-linear nature of the problem. Therefore, only the end nodes are required to describe the element in space.

THE TWO-NODED BEAM SUB ELEMENT WHICH BUILDS UP THE SUPER ELEMENT

3. This is an isoparametric element with six degrees of freedom per node as shown in Fig. 1.

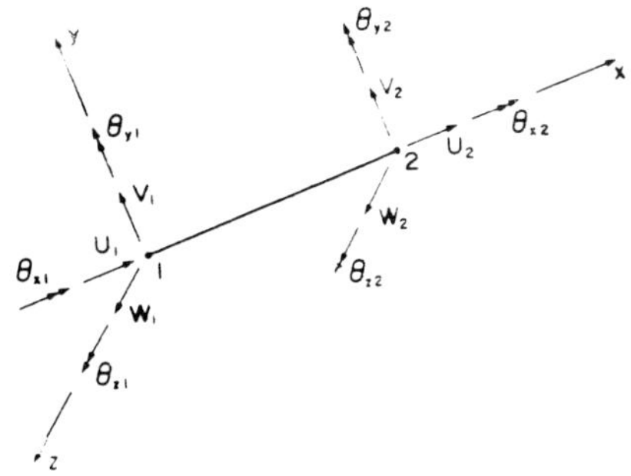

Fig. 1. Element degrees of freedom.

The element displacements and rotations are interpolated using the corresponding nodal values shown in Fig. 1 and shape functions h_1 and h_2:

$$
\begin{aligned}
u &= h_1 \cdot U_1 + h_2 \cdot U_2 \\
v &= h_1 \cdot V_1 + h_2 \cdot V_2 \\
&\vdots \\
\theta_z &= h_1 \cdot \theta_{z1} + h_2 \cdot \theta_{z2},
\end{aligned} \tag{1}
$$

where $$h_1 = 1 - \beta \tag{2}$$

$$h_2 = \beta$$

and $$\beta = x/l. \tag{3}$$

where l is the length of the beam element and x is measured along the element from node 1 to node 2 as shown in Fig. 2 where y and z are the principal axes of the section.

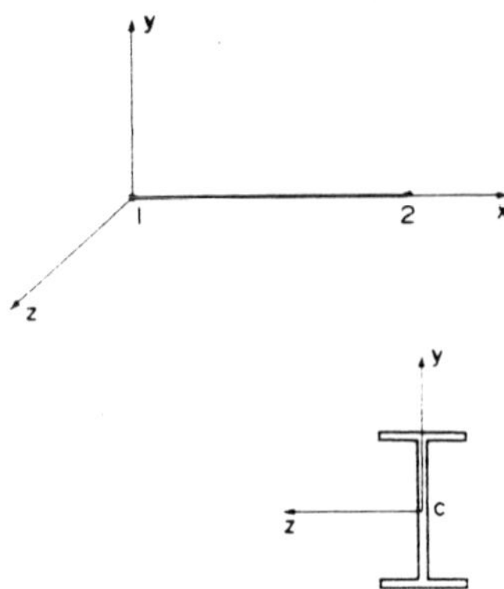

Fig. 2. Element local co-ordinate system.

4. The element stiffness matrix in the small deflection elasto-plastic range. As described in detail in ref. 2, the principle of virtual work can be used to derive the element tangent stiffness matrix k_m which includes material non-linearity:

$$\mathbf{k}_m = \int_V (\mathbf{B}_1 \cdot E_T \cdot \mathbf{B}_1^T + \mathbf{B}_2 \cdot G \cdot \mathbf{B}_2^T + \mathbf{B}_3 \cdot G \cdot \mathbf{B}_3^T)\, dV, \tag{4}$$

where the vectors B_1,B_2 and B_3, satisfy the following incremental strain displacement relations:

$$\dot{\varepsilon}_x = \mathbf{B}_1^T \cdot \dot{\mathbf{U}}_n$$

$$\dot{\gamma}_{xy} = \mathbf{B}_2^T \cdot \dot{\mathbf{U}}_n$$

$$\dot{\gamma}_{xz} = \mathbf{B}_3^T \cdot \dot{\mathbf{U}}_n. \tag{5}$$

The quantities $\dot{\varepsilon}_x, \dot{\gamma}_{xy}$ and $\dot{\gamma}_{xz}$ are the increments in the axial and shear strains at any point of the beam element as a result of the increment $\dot{U}_n$ in the element nodal degrees of freedom. E_T and G are the tangent and shear moduli in the stress-strain relations, respectively. The matrix k_m in Eq. 4 relates the increment in nodal forces $\dot{f}$ to the corresponding increment in nodal displacements $\dot{U}_n$, in the element local directions:

$$k_m . \dot{U}_n = \dot{f} \tag{6}$$

5. The integral in Eq. 4 is evaluated over the volume of the element. In practice, the integral along the length is calculated by Gaussian quadrature using one Gauss point (at $\beta = 1/2$) to avoid shear locking (ref. 2). The only non-linearity is due to plasticity (effects of shear stresses on plasticity are ignored). In order to detect the growth of plasticity over the cross-section of the element the cross-section is divided into a number of sampling points (Fig. 3.), and the integrand in Eq. 4 is sampled at all these points at the integration station. When the element is elastic, k_m can be derived explicitly as presented in refs 1, 2.

INCLUSION OF GEOMETRICALLY NONLINEAR EFFECTS IN THE SUB ELEMENT STIFFNESS MATRIX

6. Eq. 4 yields the sub-element stiffness matrix in its local directions which are assumed to remain unchanged in any small deflection analysis. For large deflection analysis, the equilibrium equations of the element should be written in its deformed configuration. In order to tackle instability problems, the effect of element stresses on its stiffness should also be considered.

7. To incorporate the large deflection effects it is necessary to update the element geometry. Instability effects can be incorporated by augmenting the usual element stiffness matrix with a geometric stiffness matrix.

8. <u>A simplified geometric stiffness matrix for the beam sub-element</u>
The only instability effect that is considered in this study is that due to axial force. The form of the corresponding geometric stiffness matrix can be generalised to three dimensions from that of a two-dimensional truss element (ref. 3) as follows:

$$\mathbf{k}_g = (p/l) \begin{bmatrix} 1 & 0 & 0 & 0 & 0 & 0 & -1 & 0 & 0 & 0 & 0 & 0 \\ & 1 & 0 & 0 & 0 & 0 & 0 & -1 & 0 & 0 & 0 & 0 \\ & & 1 & 0 & 0 & 0 & 0 & 0 & -1 & 0 & 0 & 0 \\ & & & 0 & 0 & 0 & 0 & 0 & 0 & 0 & 0 & 0 \\ & & & & 0 & 0 & 0 & 0 & 0 & 0 & 0 & 0 \\ & & & & & 0 & 0 & 0 & 0 & 0 & 0 & 0 \\ & & & & & & 1 & 0 & 0 & 0 & 0 & 0 \\ & \text{Symm.} & & & & & & 1 & 0 & 0 & 0 & 0 \\ & & & & & & & & 1 & 0 & 0 & 0 \\ & & & & & & & & & 0 & 0 & 0 \\ & & & & & & & & & & 0 & 0 \\ & & & & & & & & & & & 0 \end{bmatrix} \tag{7}$$

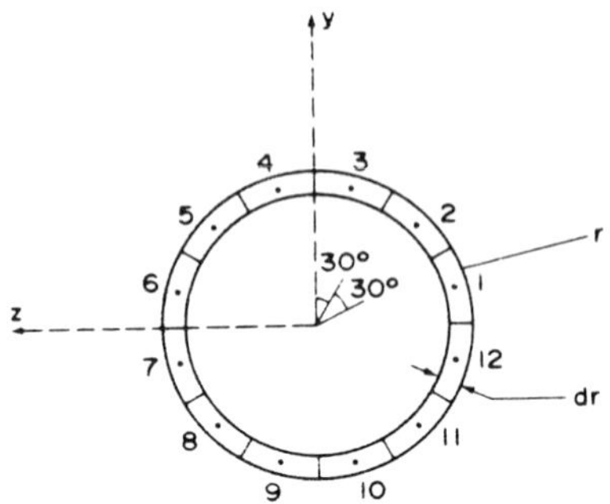

Fig. 3. A reasonable distribution of sampling points.

where l is the element length and p is the element axial force which is positive when tensile. The matrix k_m of Eq. 4 and k_g of Eq. 7 are calculated in element local system and added together (ref. 3):

$$k_l = k_m + k_g \tag{8}$$

The resulting element total stiffness matrix k_l is then transformed to the global system of axes, using the element transformation matrix, which is a function of the element displaced position in the large deflection regime.

9. Updating the sub-element transformation matrix for large deflections. Fig. 4 shows the beam sub-element in two configurations at time 0 and at time t, where time here represents a stage in the loading process. The system r,s,t in this figure is a non-dimensional moving co-ordinate system. The non-dimensional co-ordinates r, s and t vary between -1 and 1 in the element. The unit vectors V_s and V_t, are normal to each other and fixed to the element nodes at the centroids. These vectors are in the s and t directions respectively. Assuming a_k and b_k are representative cross-sectional dimensions in the direction of vectors V_s, and V_t, respectively, the Cartesian co-ordinates of a point in the element at time t is given in Eq. 9, (ref. 4).

$$ {}^t x_i = \sum_{k=1}^{2} h_k \cdot {}^t x_i^k + s \sum_{k=1}^{2} h_k \cdot \frac{a_k}{2} \cdot {}^t V_{si}^k + t \sum_{k=1}^{2} h_k \cdot \frac{b_k}{2} \cdot {}^t V_{ti}^k \quad (i = 1, 2, 3) \tag{9} $$

where:

k: node number

${}^t x_i^k$: Cartesian coordinates of node k at time t

h_k: shape function corresponding to node k, which is a function of r

${}^t V_{si}^k$: components of vector $\mathbf{V}_s$ at node k and at time t

${}^t V_{ti}^k$: components of vector $\mathbf{V}_t$ at node k and at time t.

10. The natural co-ordinate r is -1 at node 1, and +1 at node 2. The other natural co-ordinates s and t are zero at the centroid of any cross-section of the beam and vary between -1 and 1 over the cross-section. In terms of r the relations Eq. 2 can be written as:

$$ h_1 = (1 - r)/2; \quad h_2 = (1 + r)/2 \tag{10} $$

A Jacobian matrix can be defined at any material point of the element as follows:

$$ \mathbf{J} = \begin{bmatrix} \dfrac{\partial x_1}{\partial r} & \dfrac{\partial x_2}{\partial r} & \dfrac{\partial x_3}{\partial r} \\ \dfrac{\partial x_1}{\partial s} & \dfrac{\partial x_2}{\partial s} & \dfrac{\partial x_3}{\partial s} \\ \dfrac{\partial x_1}{\partial t} & \dfrac{\partial x_2}{\partial t} & \dfrac{\partial x_3}{\partial t} \end{bmatrix} \tag{11} $$

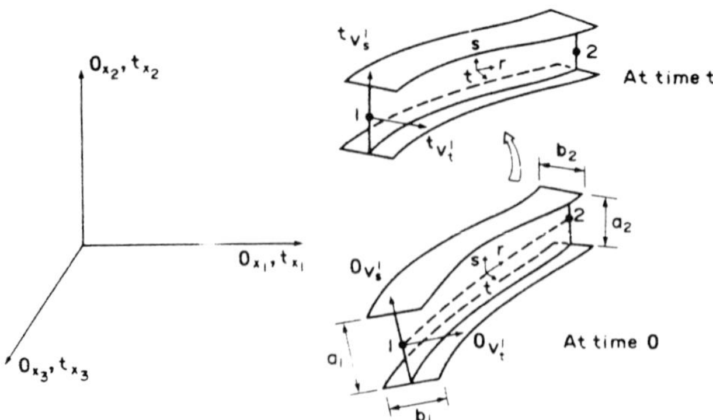

Fig. 4. Different configurations of the beam element.

11. From the first principles of space curve geometry it can be shown that the rows of J in Eq. 11 represent vectors parallel to the r, s and t axes respectively. Assuming that cross-sectional dimensions do not change along the element we can write:

$$
\begin{aligned}
{}^tx_{i,r} &= \sum_{k=1}^{2} h_{k,r} \cdot {}^tx_i^k + \frac{sa}{2} \sum_{k=1}^{2} h_{k,r} \cdot {}^tV_{si}^k + \frac{tb}{2} \sum_{k=1}^{2} h_{k,r} \cdot {}^tV_{ti}^k \\
{}^tx_{i,s} &= \frac{a}{2} \sum_{k=1}^{2} h_k \cdot {}^tV_{si}^k \quad (i = 1, 2, 3) \\
{}^tx_{i,t} &= \frac{b}{2} \sum_{k=1}^{2} h_k \cdot {}^tV_{ti}^k .
\end{aligned}
\tag{12}
$$

12. In the process of deformation it is assumed that the cross-sections are not distorted, so the vectors, ${}^tX_{i,s}$ and ${}^tX_{i,t}$, which are parallel to the s and t axes respectively remain normal to each other. These vectors are calculated at the Gauss integration point (where r=s=t=0), and normalised to unit length. The vector ${}^tX_{i,r}$, may not be normal to the above normalised vectors due to shear deformations. So, a third vector is calculated as the cross product of the normalised vectors ${}^tX_{i,s}$ and ${}^tX_{i,t}$. The resulting system of orthonormal vectors constitute a local co-ordinate system with respect to which the sub-element stiffness matrix k_t of Eq. 8 is calculated before it is transformed to the global system.

CONSTRUCTION OF A BEAM SUPER ELEMENT

13. The two-noded beam sub-element described above can be used to solve highly non-linear problems as demonstrated by two cases in ref. 1. However, due to the simplifying assumptions involved in its formulation, it is necessary to use four to eight elements to model each member of the framework for accurate results in the presence of severe non-linearities.

14. This type of modelling clearly becomes impractical for the non-linear analysis of lattice domes where a lot of members are involved. To get round this problem, a super element consisting of a variable number of the two-noded sub-elements was developed. For those members where the axial load will become close to the Euler buckling load the use of eight sub-elements is recommended and for the rest the use of four sub-elements is quite sufficient. The subdivision of the member into sub-elements is carried out automatically by the program and the internal nodes are eliminated by static condensation. In the presence of non-linearities a special technique should be used for static condensation which is fully described in refs 2,5, together with the solution of incremental/iterative equations.

SOME EXAMPLES

15. The above non-linear two-noded beam super element was incorporated in an incremental/iterative stiffness based computer program. Many numerical tests were carried out which demonstrated that the super element is extremely fast and accurate. The following two tests are reported here.

16. Nonlinear elastic analysis of a 12 member model dome. This space frame is illustrated in Fig. 5. The Young's modulus, shear modulus and cross-sectional area of each member are E=439.8 ksi, G=159.0 ksi and A=0.494 in^2 respectively. Each member is modelled by a super element (consisting of 8 sub-elements).

17. The load-deflection curve of this dome is depicted in Fig. 5. The results obtained are compared with those of Papadrakakis (ref. 6). The first limit point load obtained by the current program is 56.2lb compared with the experimental value of 56.5lb.

18. In Fig. 6 the results of the current program are compared with experimental results and those in refs 7-9.

19. Nonlinear analysis of a 42 member dome. The geometry and loading of this dome are illustrated in Figs 7, 8 and in Table 1. All the members are of $4x4mm^2$ cross-section made from mild steel bar with mechanical properties of E=210 kN/mm^2, σ_y = $0.25kN/mm^2$ and $E_T = 0.00$ where E, σ_y and E_T are the elastic modulus, yield stress and strain hardening tangent modulus respectively; each member is modelled by a super element (consisting of 8 sub-elements and 9 sampling points at each Gauss station).

20. The resulting load-deflection curve of the dome is illustrated in Fig. 9. This curve is compared with the corresponding result obtained using a rigorous non-linear three-noded beam element (in which the middle node is condensed), coded in (ref. 2) The central 6 members were modelled using two rigorous elements and the rest using one element each. The two sets of results are very close to each other in most parts.

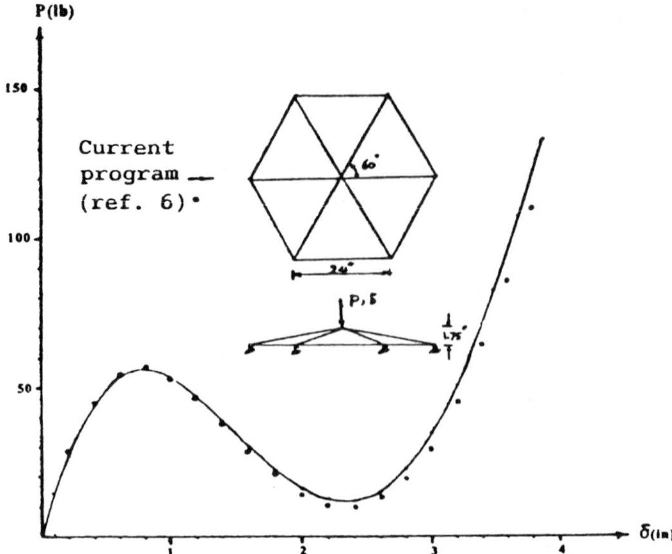

Fig. 5. Load-deflection curve of the 12 member model dome.

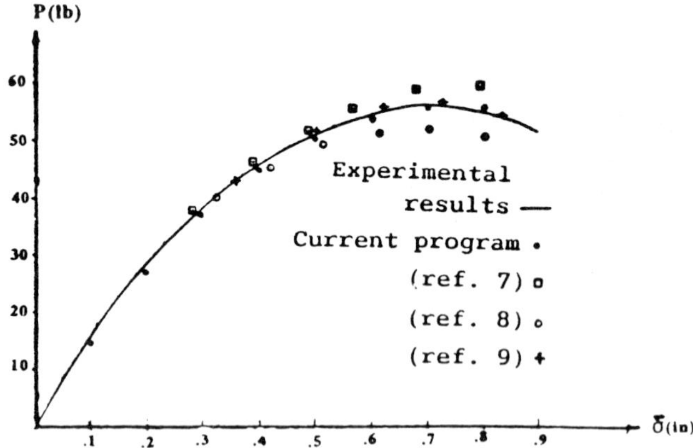

Fig. 6. Comparison of various results for the 12 member model dome.

CONCLUSION

21. The two-noded three-dimensional non-linear beam super element presented in this paper provides a fast and accurate numerical model for collapse and post-collapse analysis of shallow lattice domes. Only the two end nodes are required to specify the super element. The number of sub-elements included in the super element is chosen by the user. Four to eight sub-elements are sufficient for most applications. In the bench mark tests, carried out using the super element, up to 85% reduction in CPU time has been observed compared with the rigorous non-linear beam element coded in (ref. 2). The accuracy observed was the same if not better.

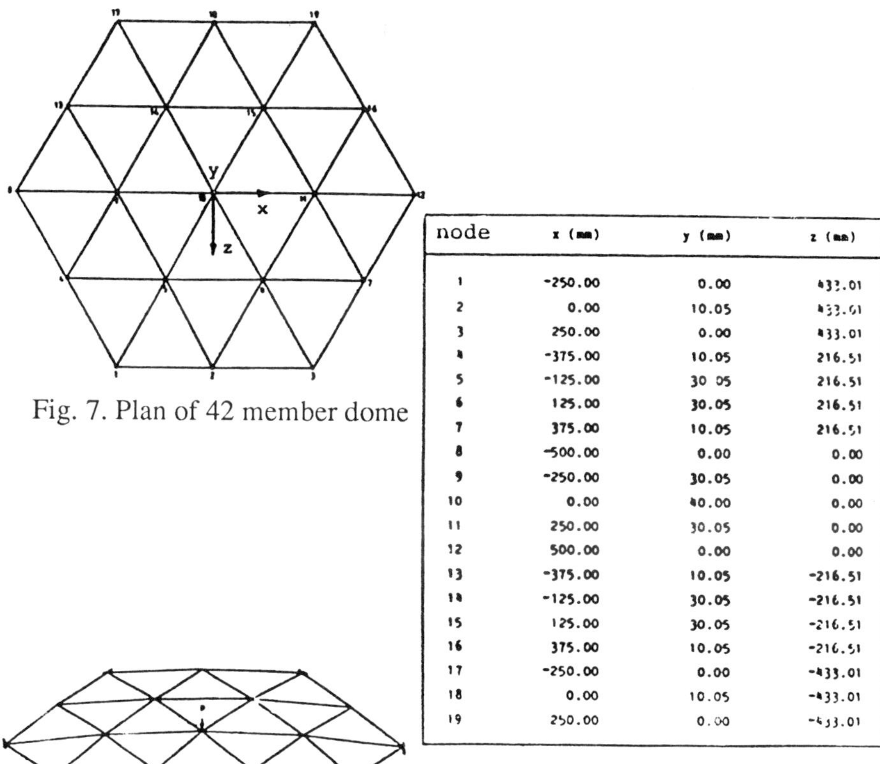

Fig. 7. Plan of 42 member dome

node	x (mm)	y (mm)	z (mm)
1	-250.00	0.00	433.01
2	0.00	10.05	433.01
3	250.00	0.00	433.01
4	-375.00	10.05	216.51
5	-125.00	30.05	216.51
6	125.00	30.05	216.51
7	375.00	10.05	216.51
8	-500.00	0.00	0.00
9	-250.00	30.05	0.00
10	0.00	40.00	0.00
11	250.00	30.05	0.00
12	500.00	0.00	0.00
13	-375.00	10.05	-216.51
14	-125.00	30.05	-216.51
15	125.00	30.05	-216.51
16	375.00	10.05	-216.51
17	-250.00	0.00	-433.01
18	0.00	10.05	-433.01
19	250.00	0.00	-433.01

Table 1. Joint coordinates of 42 member dome.

Fig. 8. Loading and fixed supports of 42 member dome.

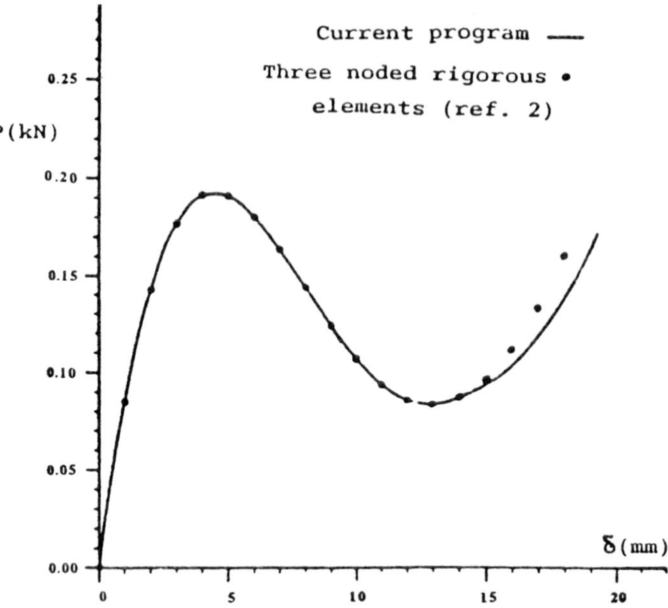

Fig. 9. Load-deflection curve of the 42 member dome.

ACKNOWLEDGEMENT

22. I would like to thank my research student Mr M.R. Moarefzadeh who helped in coding the super element and running the examples on a PC .

REFERENCES

1. KANI I.M. and McCONNEL R.E. A simple and efficient beam element for the combined non-linear analysis of frameworks. Computers and structures, 1987, Vol. 25, No. 3, 457-462.

2 . KANI I . M . A theoretical and experimental investigation of the collapse of shallow reticulated domes. Ph.D. Thesis, University of Cambridge, U.K., 1986.

3 . MARTIN H.C. and CAREY G.F. Introduction to finite element analysis, McGraw-Hill, New York, 1979.

4. BATHE K.J. Finite element procedures in engineering analysis. Prentice-Hall, Englewood cliffs, NJ, 1982.

5. KANI I.M. and McCONNEL R.E. Single layer shallow lattices domes: analysis, general behaviour and collapse. International Journal of Space Structures, 1988, Vol. 3, No. 2, 64-73.

6. PAPADRAKAKIS M. Post buckling analysis of spatial structures by vector iteration methods. Computers and structures, 1981, Vol. 14, No. 5-6, 393-402.

7. CHU K.H. and RAMPETSREITER R.H. Large deflection buckling of space frames. ASCE, Structural Division, 1972, Vol. 98, No. ST12, 2701- 2722 .

8. CONNOR J.J., LOGCHER D. and CHAN S.C. Nonlinear analysis of elastic framed structures. ASCE, Structural Division, 1968, Vol. 94, No. ST6, 1525-1548.

9. BRENDEL B. and RAMM E. Stabilitatsunter suchungen weitgespannter tragwerke mit der methode der finiten element. Proc. of Int. Symp. on Wide Span Surface Structures, Stuttgart, 1976.

16. Propagation of local buckling in reticulated shells

V. GIONCU, Building Research Institute, Romania, and P. LENZA, University of Naples, Italy

SYNOPSIS. The local buckling of a node is one of the most dangerous forms of instability in reticulated shells. This phenomenon can be initiated by some deviation in the geometry of the ideal shell or by the Eulerian buckling of a single member. It is very difficult for the designer to predict the consequence of this phenomenon because the local loss of stability has a dynamic effect, produced by a snap-through which can cause the collapse of a large zone of the reticulated lattice structure. The effect of this dangerous event should be carefully analysed. Using complex and time-expensive numerical analyses the authors think that it is possible to give designers some useful guidelines, derived from general theoretical considerations and from the results of selected numerical analyses and experimental tests on models.

DYNAMIC EFFECTS PRODUCED BY NODAL SNAP-THROUGH

1. Single layer lattices can be classified according to their Gaussian curvature. The topology of positive Gaussian curvature (elliptic paraboloid) is characterised by a very rigid shape (ref. 1). It is impossible to develop this surface onto a plane without distorting it. A nodal snap-through requires an inversion in local curvature and so a violent dynamic effect has to be considered. Usually the members of lattice structures are arranged in a single (or double) diagonal-square pattern, where both longitudinal and transverse bars may be in compression. The load capacity of the nodes decrease following a softening behaviour culminating in a limit point.

2. Barrel vaults (zero Gaussian curvature) are formed from transverse arches usually with bending rigidity and longitudinal bars spanning between the arches (ref. 2). This surface can be developed onto a plane, therefore its shape is not as rigid as in the previous cases, so a nodal collapse should produce less violent effects.

3. Hypar shells (negative Gaussian curvature) are characterised by lines with curvatures of different sign. This causes tension or compression in the members situated along these lines resulting in their corresponding hardening and softening behaviour. The surface shape is less rigid because a large radius of curvature is

usually adopted. In this case the hardening behaviour of the tension members could avoid any dynamic effect caused by a nodal failure.

AN ELEMENTARY TRUSS

4. The previous qualitative predictions can be confirmed by the analysis of the simple truss shown in Fig. 1. The non-linear behaviour of this single degree of freedom scheme is well known. When the external load P reaches the limit value the geometry suddenly changes with dynamic effects. The segment "b" represents the gap between the applied load P and the possible elastic reaction R during the collapse. To comply with D'Alembert's principle "b" has to be equal to the inertial action (neglecting damping). The area "A" represents the kinetic energy gained by the system. This energy transforms itself in an impulsive action applied to nearby nodes when the snap-through has occurred.

5. Consider now, the same elementary truss with a linear spring of stiffness K at the node. Fig. 2 shows how the behaviour of the truss varies when the rigidity of K increases. Several relationships can be compared by considering the dimensionless parameter α that expresses the spring to member rigidity ratio. The figure shows that by increasing α the limit load increases, while the kinetic energy associated with the snap-through and measured by the area "A" greatly decreases. Increasing still further the value of α and the limit point degenerates into a point of contra-flexure and no kinetic energy arises if the load is slowly increased.

6. Analysing the dynamic effects of the nodal collapse, consider the system to be in equilibrium in the deformed situation, when P is equal to P_{lim}. By increasing P by a small quantity it is possible to analyse step by step, the development of the dynamic snap-through; determining at each instant t the displacement, velocity and acceleration values. Table 1 gives the maximum acceleration values obtained as a function of α.

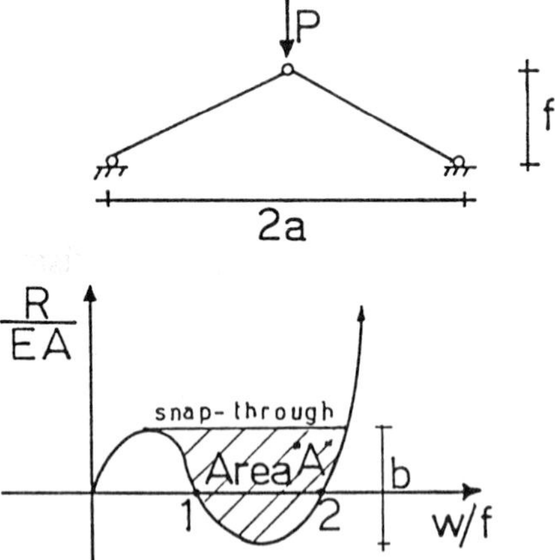

Fig. 1. The classic snap-through of an elementary truss.

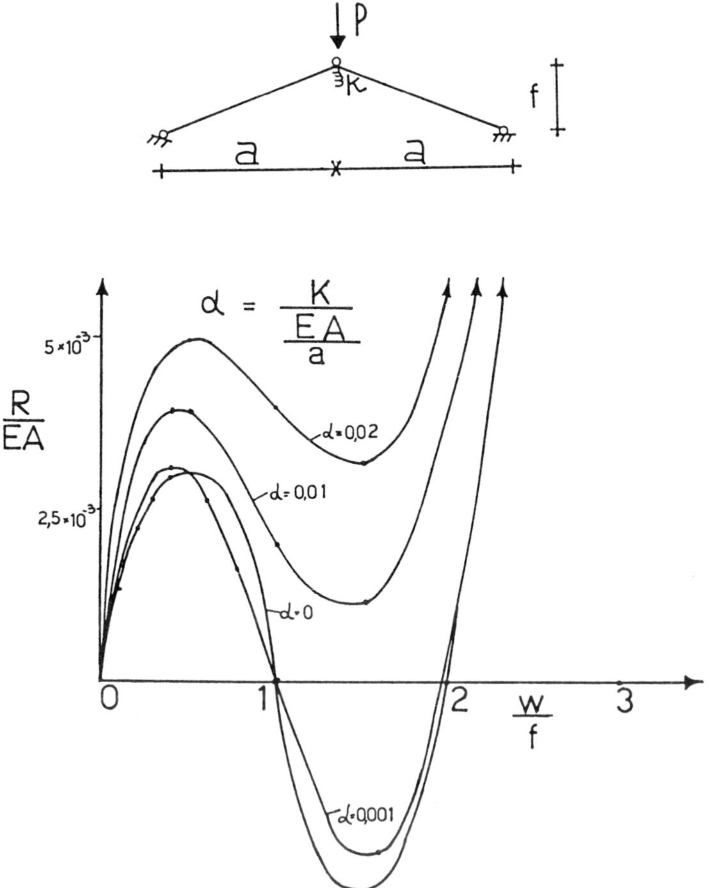

Fig. 2. Different behaviour determined by a linear spring whose rigidity is K. (a = 50cm; EA = 4.1 x 10^7kg).

Table 1.

α	**amax/g**
0	7.62
1×10^{-3}	7.13
1×10^{-2}	3.77
2×10^{-2}	1.58

7. The values given in Table 1 confirm the relationships between the kinetic energy gained in the nodal collapse and the maximum inertia action produced by the snap-through.

8. A technical valuation of the inertia action can also be reached by considering the principle momentum, hence:

$$M.Dv = P.Dt \qquad (1)$$

9. Where M is the mass of the system; Dv is the increment of velocity (from zero in this case); P is the impulsive force (to be predicted) and Dt the duration over which the force acts. Dv can be easily derived from the value of kinetic energy (area A obtained from the behaviour curve) while only an approximate value of Dt can be proposed, related to the natural period of the system in the strained configuration.

10. The authors consider that the case with $\alpha = 0$ could simulate the behaviour of a negative Gaussian lattice structure characterised by a violent dynamic snap-through effect. The spring with its linear rigidity K could represent the contribution of longitudinal bars in the barrel vaults, which exhibit a less violent snap-through. Increasing the value of α could simulate the behaviour of the hypar shells where the contribution of the tension members strongly reduce (or negate) the kinetic energy of the system during nodal collapse.

RETICULATED SHELLS

11. A reticulated shell can be schematised as a system composed of many elementary trusses similar to the structures analysed previously. A nodal scheme is an asymmetrical spatial structure with several degrees of freedom. The incremental analysis of the dynamic collapse of three elementary cells, where the bordering adjacent nodes are restrained only in the vertical direction, confirms the behaviour predicted according to the Gaussian curvature (Fig. 3). Unfortunately, the impulsive action indicated by this simplified local analysis has to be considered to determine the effective local limit load. The analysis of the isolated cell does not consider the elastic reactions of the bordering surface and also neglects the stresses produced in the members by loads acting on the other nodes.

12. Only the analysis of the whole reticulated surface makes it possible to predict rigorously the dynamic effects of a nodal collapse. With reference to the reticular systems having many degrees of freedom, the unknowns can be taken as the node co-ordinates in the strained configuration as a function of which it is possible to express both the beam stresses and the changed direction cosines.

13. The dynamic equilibrium equations relative to each degree of freedom, can be solved by a step by step iterative incremental analysis using Newmark's procedure. The first results of such an analysis were presented in ref. 3. They refer to a nodal collapse obtained on a doubly curved model with positive Gaussian curvature. This analysis emphasised the large inertia actions which rapidly propagates the collapse. The maximum vertical acceleration value was in that case, about 16g, neglecting any damping effect (Fig. 4).

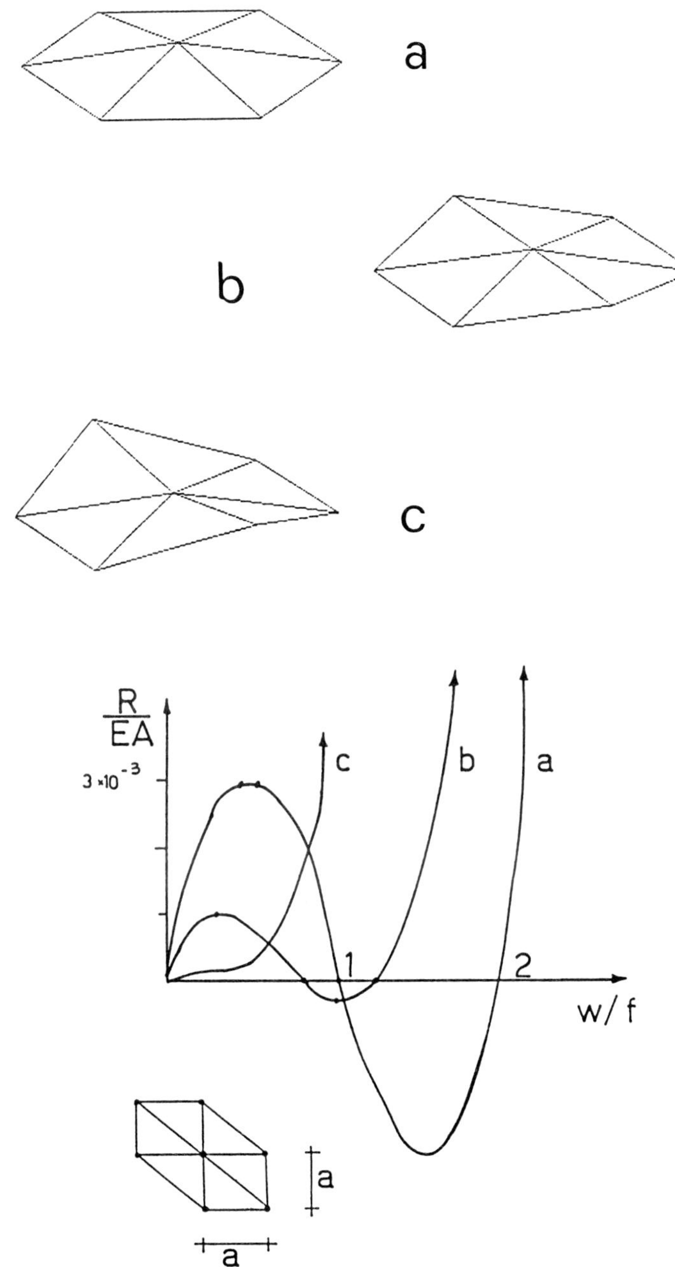

Fig. 3. An elementary cell with negative, zero and positive Gaussian curvature which determines different collapse behaviours.
(a = 50cm; EA = 4.1 x 10^7 kg; f = 10cm)

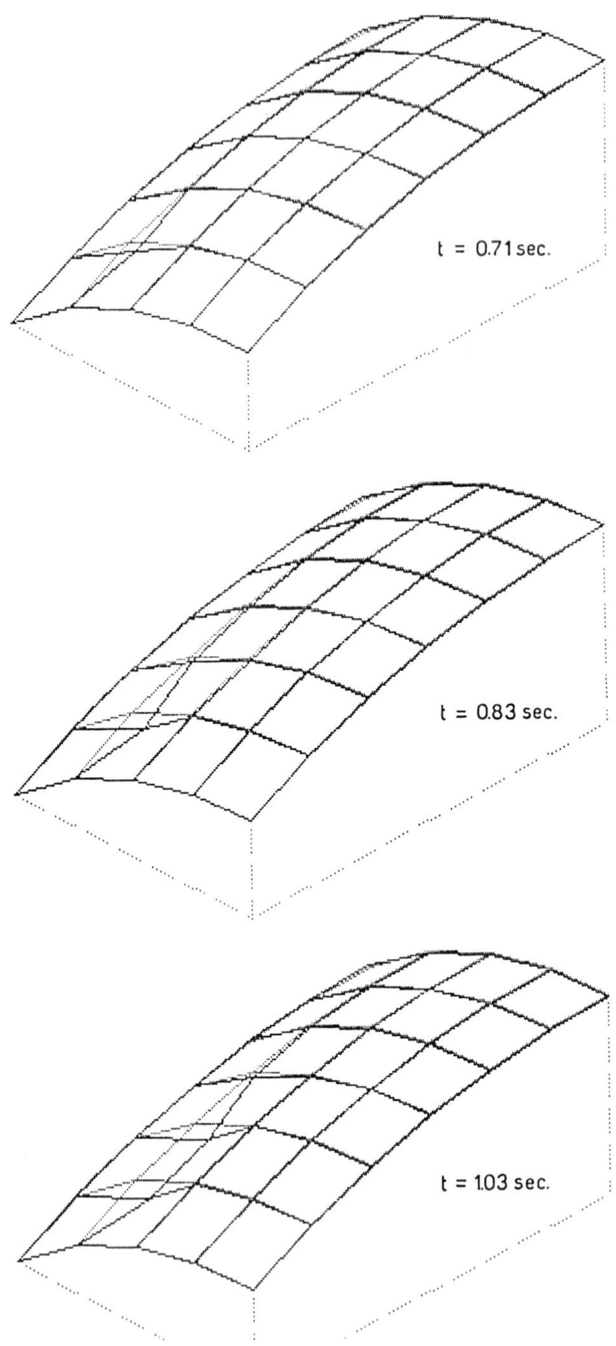

Fig. 4. The dynamic propagation of local collapse predicted in a doubly curved model (ref. 3). Displacements are drawn without any amplification scale.

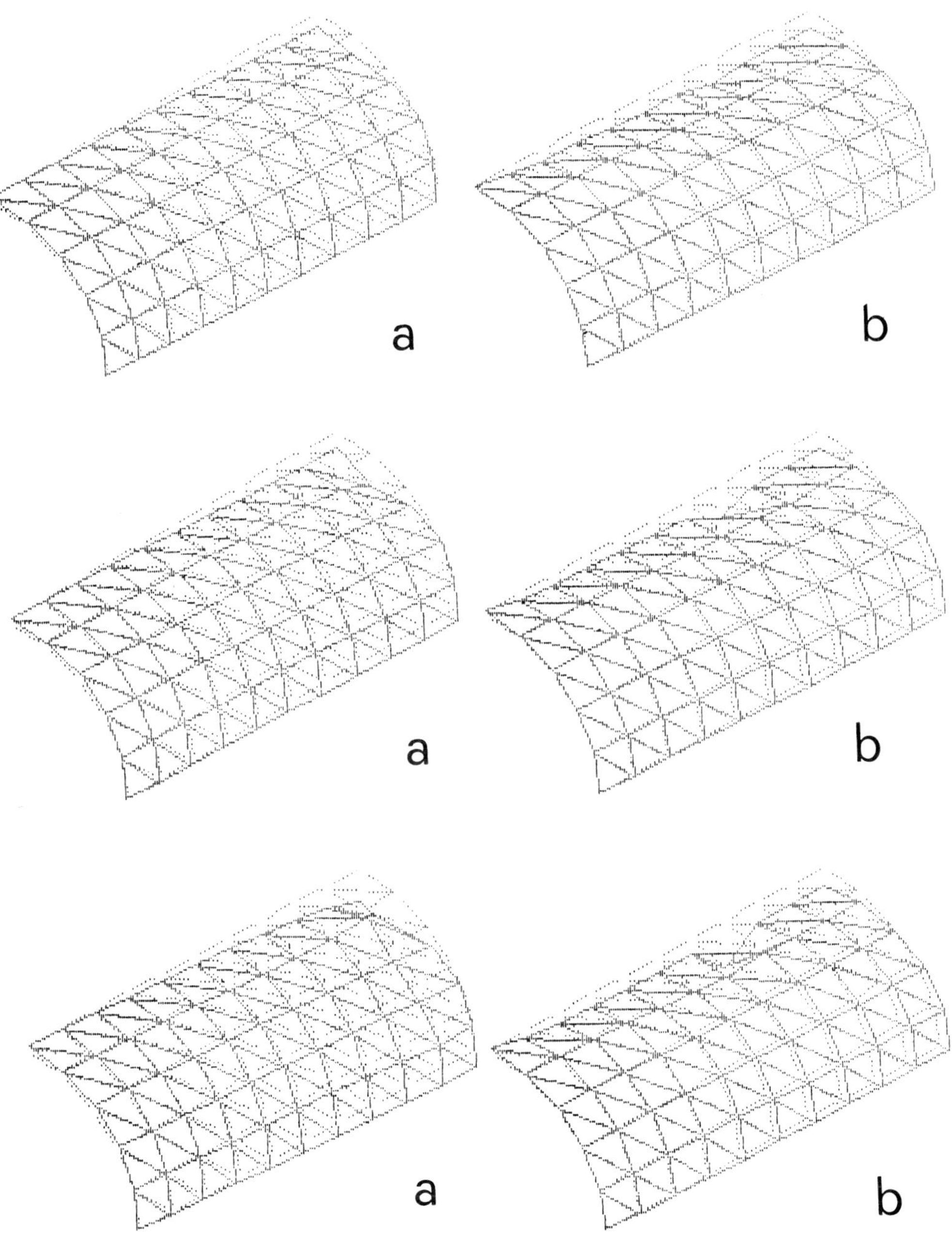

Fig. 5. Local collapse and general collapse obtained in a barrel model for two different load cases. The adopted scale amplifies only twice the node displacements.

14. In this paper the authors compare those results (ref. 3) with the dynamic analysis of a barrel vault subjected to experimental tests (ref. 4). The reticulated model has a semi-circular directrix (R = 13.5cm), is free on the lateral edges and is supported by two end frames which are stiff in their plane. The vault consists of a lattice of circular section steel beams (D = 2mm) forming a square grid mesh (4 x 4cm), each cell stiffened by one diagonal.

15. The transverse arches were considered as beam elements (with bending rigidity). The longitudinal and diagonal members were considered to have only axial stiffness. Two load conditions have been considered:

(a) vertical loads at all the nodes (up to 1.7kg);

(b) vertical loads only at the nodes of the crown generatrix (up to 9.7kg).

16. In both cases the loads have been increased gradually in order to start the nodal collapse without the dynamic effect. When the nodal collapse starts the loads are kept constant and the phenomenon analysed. As in the analysis of doubly curved vaults, the nodal failure is initiated by member buckling, giving a classic case of coupled instability with consequent dynamic effects.

17. In comparison with the phenomenon examined in the doubly curved model, the dynamic analysis of the barrel vault model gave acceleration values of about 6g in case (a) and 1.8g in case (b) (Fig. 5). for the latter case as the failure was less violent but also failure of the entire structure followed nodal collapse (Fig. 5). Other differences, more important than the inertial action, have to be emphasised. In fact for the doubly curved model the snap-through occurred when only one bar was in its post critical stage. For the case of the barrel vault model many longitudinal members had already buckled before the node collapse occurred so that general instability, or more properly, a line instability occurred. In fact the values of the vertical acceleration of several nodes were consistent and a large portion of the lattice was already in motion when the node collapsed. Consequently the inertial actions produced by the nodal snap-through gave to the structure only a "coup de grâce".

CONCLUDING REMARKS

18. The dynamic propagation of local collapse is certainly an insidious phenomenon in the field of single-layer reticulated vaults for both positive and zero Gaussian curvature. If a local phenomenon such as nodal snap-through can cause the general collapse of a structure, the influence of local geometrical imperfections become very important because these can lower the local limit load and so also initiate the collapse propagation.

19. Obviously, the danger is more significant in the case of mass action (snow loads) than in the case of surface actions (wind pressure) because the entity of the mass has a great influence on the phenomenon.

20. The first results of this study show how this event can assume different aspects according to the Gaussian curvature, but also how it is difficult in the present stage, to define a general and simple model, in order to predict the entity of inertial actions. It is hoped that the rapid improvement in computer

performance can facilitate the non-linear dynamic analysis of the whole reticular surface.

REFERENCES

1. PAGANO M. and LENZA P. Theoretical Studies and Experimental Research on Structural Behaviour of Reticulated Vaults. Analysis Design and Construction of Braced Barrel Vaults. Edited by Makowski Z.S. chapt. 19, Elsevier Science Publisher 1985.
2. GIONCU V., et al. Instability Behaviour of Triangulated Barrel Vaults. Analysis Design and Construction of Braced Barrel Vaults. Edited by Makowski Z.S. chapt. 9, Elsevier Science Publisher, 1985.
3. LENZA P. Some Development in Shell Structural Analysis. Proc. International IASS Congress - Madrid, 1989, Vol 1.
4. LENZA P. Non Linear Behaviour of Reticulated Cylindrical Vaults. Proc. International IASS Symposium, Vol. III, Copenhagen, 1991.

17. Geometrically non-linear and experimental studies of single-layer aluminium braced domes

XI-LIANG LIU, Tianjin University, P. R. China, and YAN ZHENG, Ministry of Machinery and Electronics, P. R. China

SYNOPSIS. This paper discusses the feasibility of using aluminium in single layer reticulated domes. Based on the beam-column theory, the accurate tangent stiffness matrix of a beam member in space is deduced. A finite element method computer program for the stability analysis of single layer domes, with consideration of large displacements, has been completed. In this program, the method of automatic solution is used to trace the equilibrium configuration of the structures. To verify the reliability of the analytical results, an experiment on the stability of an aluminium dome has been carried out. The results of the experiment are basically consistent with that predicted by the theory.

ALUMINIUM-ALLOY

Properties of aluminium-alloy

1. Aluminium is the most abundant metallic element, estimated to form eight percent of the solid portion of the earth's crust.

2. Aluminium alloy is composed of pure aluminium, copper, magnesium, manganese, silicon and other metals. It has properties of low density (about 1/3rd of that of steel), good corrosion resistance and good machinability. Its strength is equal to, or may exceed, that of carbon steel (about 700-800 MPa). Moreover, as temperature decreases its strength increases. This is why aluminium alloys have been widely used in structural engineering.

Problems of aluminium-alloy used in space structures

3. When aluminium alloy is compared with steel, which plays a leading role in space structures, the following characteristics may be found:

(a). Since its density is about 1/3rd of that of steel, the self weight of an aluminium-alloy structure can be a great deal less than that of an equivalent steel structure.

(b). Its modulus of elasticity is relatively low (about 1/3rd of that of steel) and therefore the elastic flexibility of the structure becomes a problem in design.

Space structures 4. Thomas Telford, London, 1993

(c). Although its thermal coefficient of dilatation is twice as much as that of steel, this is compensated by the low modulus of elasticity, giving thermal stresses of only 2/3rd of that of steel.

In space structures, problems of elastic stability must be taken into account, especially in the design of single layer domes. As we know, the elastic stability of a structure relates to the modulus of elasticity and the torsional elasticity. The ultimate load bearing capacity will decrease when aluminium-alloy, which has a relatively low modulus of elasticity, is used. So it is necessary to carry out stability calculations. The physical properties of aluminium and steel are shown in Table 1.

Table 1. The physical properties of aluminium and steel.

Property	Density (t/mm)	Modulus of Elasticity (N/mm)	Thermal coefficient of dilatation (1/ C)	Thermal coefficient of conduction
Aluminium-alloy	2.7	70,000	23.5*10	0.52
Steel	7.87	210,000	12.1*10	0.082

4. There are also large differences between the mechanical properties of aluminium-alloy and steel, which are described by the shape of the stress strain diagram (also called σ and ε curve, see Fig. 1). Fig. 1 shows that when steel is tested there is an evident transition-stage between elastic and plastic behaviour, and that there are high and low yield points over the first yielding point. This phenomenon is rarely seen for aluminium-alloy. Aluminium-alloy exhibits a more gradual transition from the linear to the nonlinear region. Therefore, the yield strength is defined as the stress that will induce a specified permanent set, usually 0.2%, which is equivalent to a plastic strain of 0.002. This yield is also called condition yield strength denoted by 0.2.

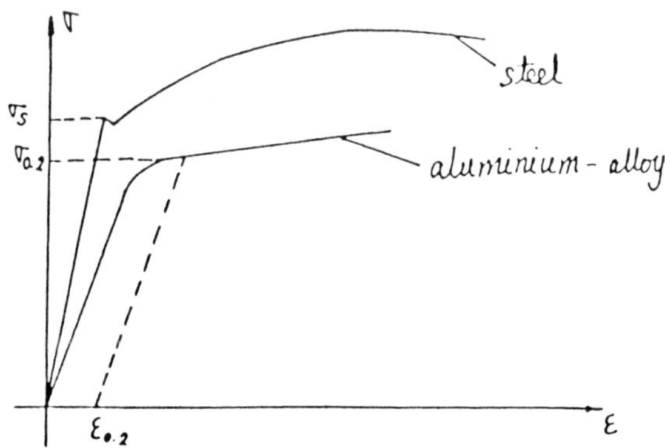

Fig. 1. Stress-strain relationship diagram.

5. The properties of aluminium-alloy define the range of its usage. It is uneconomical to use aluminium-alloy for structures which which are subjected to high loads. However, there are many advantages when it is used in large span structures which are relatively lightly loaded.

6. In the design of structures, the properties of aluminium-alloy must be considered comprehensively in order to find the suitable kind of alloy. Aluminium-alloys 6061 and 6063 are often used as structural materials.

THEORETICAL ANALYSIS ON THE NONLINEAR STABILITY OF SINGLE-LAYER BRACED DOMES

Deducing the accurate tangent stiffness matrix

7. On the basis of the beam-column theory, a theoretical analysis is derived from the nonlinear stability of single-layer braced domes. By adopting the achievements in the scientific research of C.Oran and considering the coupling effect of bending in the two principal directions, the accurate tangent stiffness matrix is obtained.

8. The basic assumptions in this paper are listed as follows:

(a). Initially, the structural members are straight. The cross-sections are doubly symmetric and constant throughout the their length. At the joints,there should be rigid coupling and no eccentricity.

(b). The external loads are conservative and applied at the joints only.

(c). Transverse shear deformations and warping effects are neglected.

(d). The material is linearly elastic.

(e). The displacements and rotations of the structural joints can be quite large, but the deformations of structural members are very small.

9. For convenience, two types of coordinate systems are employed. One is a fixed global set of coordinates (X,Y,Z),and there is a local convective system (x,y,z) which rotates and translates with the structural member. The two systems are both right-handed coordinate systems. The member coordinate system is defined as follows:The X-axis passes through the centroids of the member's end sections, the Y- and Z-axes are parallel to the principal directions of the cross section. When the relative deformations of the members take place, the two end sections are generally not parallel to each other. There the y- and z-axes are defined as the average of the principal directions of the two end sections.

10. According to ref. l, the force-deformation relations have been set up:

$$M_{in} = \frac{EI_n}{L}\left(C_{1n}\theta_{in} + C_{2n}\theta_{jn}\right) \qquad (n = 2,\ 3) \tag{1}$$

$$M_{jn} = \frac{EI_n}{L}\left(C_{2n}\theta_{in} + C_{1n}\theta_{jn}\right) \tag{2}$$

$$M_t = \frac{GJ}{L}\varphi_t \tag{3}$$

$$P = EA\left(\frac{\mu}{L} - C_{b3} - C_{b2}\right) \tag{4}$$

where, θ_{mn} are end angles; u is the relative axial displacement; C_{1n} and C_{2n} are the stability functions associated with bending moments about the y- and z-axes (ref. 2); C_{bn} is the length correction factor resulting from flexural deformations about the y- and z-axes. As a rule, tension is defined as positive, and compression as negative.

11 . C_{1n}, C_{2n} can be obtained from Eq. 5:

$$q_n = \frac{PL^2}{\pi^2 EI_n} \tag{5}$$

For qn<0 (compression):

$$C_{1n} = \frac{\varphi(\sin\varphi - \varphi\cos\varphi)}{2(1-\cos\varphi) - \varphi\sin\varphi} \tag{6}$$

$$C_{2n} = \frac{\varphi(\varphi - \sin\varphi)}{2(1-\cos\varphi) - \varphi\sin\varphi} \tag{7}$$

$$\varphi^2 = -\frac{PL^2}{EI_n} = -\pi^2 q_n \tag{8}$$

For qn<0 (no axial force):

$$C_{1n} = 4, \quad C_{2n} = 2. \tag{9}$$

For qn>0 (tension):

$$C_{1n} = \frac{\psi(\psi\cosh\psi - \sinh\psi)}{2(1-\cosh\psi) + \psi\sinh\psi} \tag{10}$$

$$C_{2n} = \frac{\psi(\sinh\psi - \psi)}{2(1-\cosh\psi) + \psi\sinh\psi} \tag{11}$$

$$\psi^2 = \frac{PL^2}{EI_n} = \pi^2 q_n \tag{12}$$

C_{bn} is given by:

$$C_{bn} = b_{1n}(\theta_{in} + \theta_{jn})^2 + b_{2n}(\theta_{in} - \theta_{jn})^2 \tag{13}$$

$$b_{1n} = \frac{(C_{1n} + C_{2n})}{8\pi^2 q_n}(C_{2n} - 2) \tag{14}$$

$$b_{2n} = \frac{C_{2n}}{8(C_{1n} + C_{2n})} \tag{15}$$

Let: $\xi_n = \frac{I_n}{I} \quad \eta = \frac{GJ}{EI} \quad \lambda = \frac{L}{\sqrt{I/A}} \quad q = \frac{PL^2}{\pi^2 EI} = \xi_n q_n$ (16)

12. Let {S} be the relative force vector of member ends. $\{S\}^T = \{M_{13}, M_{23}, M_{12}, M_{22}, Mt, P\}$; {V} is the relative displacement vector of members. $\{V\}^T = \{\theta_{13}, \theta_{23}, \theta_{12}, \theta_{22}, \phi_t, u\}$.

The relation between incremental vectors $\{\Delta S\}$ and $\{\Delta V\}$ can be expressed as:

$$\{\Delta S\} = [t]\{\Delta V\} \tag{17}$$

where:

$$t_{ij} = \frac{\partial S_i}{\partial V_j} + \frac{\partial S_i}{\partial q}\frac{\partial q}{\partial V_j} \quad (i,\ j = 1,\ 2 \cdots 6) \tag{18}$$

[t] tangent stiffness matrix for relative deformations can be written:

$$[t] = \frac{EI}{L}\begin{bmatrix} \xi_3 C_{13} + \frac{G_{13}^2}{\pi^2 H} & \xi_3 C_{23} + \frac{G_{13}G_{23}}{\pi^2 H} & \frac{G_{13}G_{12}}{\pi^2 H} & \frac{G_{13}G_{22}}{\pi^2 H} & 0 & \frac{G_{13}}{HL} \\ \xi_3 C_{23} + \frac{G_{23}G_{13}}{\pi^2 H} & \xi_3 C_{13} + \frac{G_{23}^2}{\pi^2 H} & \frac{G_{23}G_{12}}{\pi^2 H} & \frac{G_{23}G_{22}}{\pi^2 H} & 0 & \frac{G_{23}}{HL} \\ \frac{G_{12}G_{13}}{\pi^2 H} & \frac{G_{12}G_{23}}{\pi^2 H} & \xi_2 C_{12} + \frac{G_{12}^2}{\pi^2 H} & \xi_2 C_{22} + \frac{G_{12}G_{22}}{\pi^2 H} & 0 & \frac{G_{12}}{HL} \\ \frac{G_{22}G_{13}}{\pi^2 H} & \frac{G_{22}G_{23}}{\pi^2 H} & \xi_2 C_{22} + \frac{G_{22}G_{12}}{\pi^2 H} & \xi_2 C_{12} + \frac{G_{22}^2}{\pi^2 H} & 0 & \frac{G_{22}}{HL} \\ 0 & 0 & 0 & 0 & \eta & 0 \\ \frac{G_{13}}{HL} & \frac{G_{23}}{HL} & \frac{G_{12}}{HL} & \frac{G_{22}}{HL} & 0 & \frac{\pi^2}{HL^2} \end{bmatrix} \tag{19}$$

in which:

$$G_{1n} = C'_{1n}\theta_{in} + C'_{2n}\theta_{jn} \tag{20}$$

$$G_{2n} = C'_{2n}\theta_{in} + C'_{1n}\theta_{jn} \tag{21}$$

$$H = \frac{\pi^2}{\lambda^2} + \sum_{n=3,2} \frac{1}{\xi_n}\left[b'_{1n}(\theta_{in}+\theta_{jn})^2 + b'_{2n}(\theta_{in}-\theta_{jn})^2\right] \tag{22}$$

In each equation above C'_{1n}, C'_{2n}, b'_{1n}', b'_{2n} indicate differentiation with respect to q_n.

13. The relative member end forces and displacements are transformed to member end forces and displacement in member coordinates by using [B], the instantaneous local static matrix:

$$\{\bar{F}\} = [B]\{S\} \tag{23}$$

$$\{\Delta V\} = [B]\{\Delta \bar{u}\} \tag{24}$$

in which, {F} represents the 12 member end forces with respect to member coordinates; $\{\Delta\bar{u}\}$ represents the 12 member end displacements with respect to member coordinates.

14. The incremental equation form of Eq. 23 is:

$$\{\Delta\bar{F}\} = [B]\{\Delta S\} + [\Delta B]\{S\} \tag{25}$$

In another form, Eq. 25 can be written as:

$$\{\Delta\bar{F}\} = ([B][K][B]^T + [G])\{\Delta\bar{u}\} \tag{26}$$

in which:

$$[G] = \begin{bmatrix} [g] & [0] & -[g] & [0] \\ [0] & [0] & [0] & [0] \\ -[g] & [0] & [g] & [0] \\ [0] & [0] & [0] & [g] \end{bmatrix} \tag{27}$$

$$[g] = \begin{bmatrix} 0 & (M_{13}+M_{23})/L^2 & -(M_{12}+M_{22})/L^2 \\ (M_{13}+M_{23})/L^2 & P/L & 0 \\ -(M_{12}+M_{22})/L^2 & 0 & P/L \end{bmatrix} \tag{28}$$

Let {ΔF} be the member end force vector with respect to global coordinates:

$$\{\Delta F\}=[R]([B][K][B]^T+[G])[R]^T\{\Delta u\}=[K]\{\Delta u\} \tag{29}$$

in which, [K] is tangent stiffness matrix with respect to global coordinate for the space beam member, and [R] is transformation matrix between local and global coordinates.

The amendment of the displacement and transformation matrix in the case of large rotation

15. Since the joint incremental rotations are far from small, the theorem of superposition of angular rotations is no longer applicable. Hence, the angular rotation should be amended.

16. This paper introduces the joint orientation matrix [α], which is an orthogonal matrix arranged as the columns by the direction cosines of three mutually perpendicular lines rigidly connected to the joint in the whole coordinate axes.

17. Let θ_x^i, θ_y^i, θ_z^i be rotation angles about three coordinate axes for one joint. According to the rotation theory, we know:

$$[\alpha_i] = \begin{bmatrix} 1 & 0 & 0 \\ 0 & 1 & 0 \\ 0 & 0 & 1 \end{bmatrix} \cos\delta^i + \begin{bmatrix} C_1^2 & C_1C_2 & C_1C_3 \\ C_2C_1 & C_2^2 & C_2C_3 \\ C_3C_1 & C_3C_2 & C_3^2 \end{bmatrix} (1-\cos\delta^i) + \begin{bmatrix} 0 & -C_3 & -C_2 \\ C_3 & 0 & -C_1 \\ -C_2 & C_1 & 0 \end{bmatrix} \sin\delta^i \quad (30)$$

where,

$$\delta^i = \sqrt{\theta_x^{i2} + \theta_y^{i2} + \theta_z^{i2}} \quad C_1 = \frac{\theta_x^i}{\delta^i} \quad C_2 = \frac{\theta_y^i}{\delta^i} \quad C_3 = \frac{\theta_z^i}{\delta^i} \quad (31)$$

18. After member deformation, the incremental rotations $\Delta\theta_x^{i+1}$, $\Delta\theta_y^{i+1}$, $\Delta\theta_z^{i+1}$ are created. Therfore, $[\Delta\alpha_{i+1}]$is given by:

$$[\alpha_{i+1}] = [\Delta\alpha_{i+1}][\alpha_i] \quad (32)$$

θ_x^{i+1}, θ_y^{i+1}, θ_z^{i+1} can also be deduced from Eqs 30 and 31.

19. The transformation matrix at the initial stage is:

$$[R_O]=\text{diag}\{[ro], [ro], [ro], [ro]\} \quad (33)$$

On account of the relative deformations of members, the principal directions will vary slightly from one cross-section to another. The new principal directions are defined for the average cross-section. Let $[p^{(1)}]$, $[p^{(2)}]$ be the end section orientation matrix:

$$[P^{(1)}]=[\alpha^{(1)}][r_O] \quad [P^{(2)}]=[\alpha^{(2)}][r_O] \quad (34)$$

Since relative angular rotations are very small:

$$\tan\theta_{mn} \cong \sin\theta_{mn} \cong \theta_{mn} \quad (35)$$

Hence, θ_{mn} can be calculated from the following equation:

$$\begin{Bmatrix} 1 \\ -\theta_{13} \\ \theta_{12} \end{Bmatrix} = [P^{(1)}]^T \{r_1\}, \quad \begin{Bmatrix} 1 \\ -\theta_{23} \\ \theta_{22} \end{Bmatrix} = [P^{(2)}]^T \{r_1\} \quad (36)$$

$$\varphi_1 \doteq - \{P_2^{(1)}\}^T \{P_3^{(2)}\} \doteq \{P_3^{(1)}\}^T \{P_2^{(2)}\}$$

The second and third columns of [r] can be determined by:

$$[e^{(1)}] = \begin{bmatrix} 1 & \theta_{13} & -\theta_{12} \\ -\theta_{13} & 1 & 0 \\ \theta_{12} & 0 & 1 \end{bmatrix}, \quad [e^{(2)}] = \begin{bmatrix} 1 & \theta_{23} & -\theta_{22} \\ -\theta_{23} & 1 & 0 \\ \theta_{22} & 0 & 1 \end{bmatrix} \tag{37}$$

The new member orientation matrix is:

$$[r] = \frac{1}{2}[P^{(1)}][e^{(1)}] + \frac{1}{2}[P^{(2)}][e^{(2)}] \tag{38}$$

Equilibrium configuration tracing

20. Among the methods of equilibrium configuration tracing, the most effective ones are arc-length methods. However, in this kind of method each control parameter should be selected by experience and be fixed gradually by calculation.

21. When the structure is loaded, the equilibrium equations can be written as:

$$^{t+\Delta t}R - {}^{t+\Delta t}F = 0 \tag{39}$$

where, $^{t+\Delta t}R$ is a vector of externally applied nodal point loads, and $^{t+\Delta t}F$ is a vector of nodal point forces equivalent (in the virtual work sense) to the internal element stresses, both being evaluated at time t+Δt. Let λ be load factor and R be load vector, then the incremental form of Eq. 39 is as follows:

$$^{t}K\Delta U^{(1)} = {}^{t+\Delta t}\lambda^{(1)}R - {}^{t+\Delta t}F^{(i-1)} \tag{40}$$

In which, K is tangent stiffness matrix.

The iterative displacement $\Delta U^{(i)}$ is:

$$\Delta U^{(i)} = \Delta U_r^{(i)} + \Delta\lambda^{(i)}\Delta U_e^{(i)} \tag{41}$$

where:

$$^{t}K\Delta U_r^{(i)} = {}^{t+\Delta t}\lambda^{(i-1)}R - {}^{t+\Delta t}F^{(i)} \qquad {}^{t}K\Delta U_e^{(i)} = R \tag{42}$$

The incremental displacement up to the ith iteration $U^{(i)}$ is given by:

$$U^{(i)} = U^{(i-1)} + \Delta U^{(i)} \tag{43}$$

The displacement up to the ith iteration $^{t+\Delta t}U^{(i)}$ in an incremental process is:

$$^{t+\Delta t}U^{(i)} = {}^{t+\Delta t}U^{(i-1)} + \Delta U_r^{(i)} + \Delta\lambda^{(i)}\Delta U_e^{(i)} \tag{44}$$

In this case, the arc-length equation is:

$$U^{(i)T}U^{(i)}+\eta\ (\lambda^{(i)})^2=\Delta l_i^2 \tag{45}$$

then:

$$a\Delta\lambda^{(i)^2}+b\Delta\lambda^{(i)}+c=0 \tag{46}$$

in which:

$$a=\eta+\Delta U_c^{(i)T}\Delta U_c^{(i)};$$

$$b=2\ [\ (\Delta U_r^{(i)}+U^{(i-1)}]^T\Delta U_c^{(i)}+\eta\lambda^{(i-1)};$$

$$c=\ (U^{(i-1)}+\Delta U_r^{(i)})^T\ (\Delta U_r^{(i)}+U^{(i-1)})\ +\eta\lambda^{(i-1)^2}-\Delta l_i^2.$$

$\Delta\lambda^{(i)}$ is the real root of Eq. 46. The selection of an appropriate root was suggested by Crisfield [ref. 3] to avoid doubling back on the solution path.

22. The convergence in the iteration is measured as in Eq. 47:

$$\frac{(\Delta U^{(i)})^T\ ({}^{t+\Delta t}\lambda^{(i-1)}R-{}^{t+\Delta t}F^{(i-1)})}{\Delta U^{(1)}\ (\Delta\lambda^{(1)}R)}\leqslant\varepsilon \tag{47}$$

where ε is an energy convergence tolerance.

THE STABILITY MODEL EXPERIMENT OF AN ALUMINIUM DOME

23. In order to verify the reliability of the results of the theoretical analysis and the computer program, a stability model experiment on an aluminium-dome has been undertaken. This model, a K6 single layer aluminium dome, is 1500mm in span, and 100mm in height (Fig. 3). It consists of 37 joints and 70 members which are made of aluminium tubes $\phi 10*1$. The boundary nodes are restrained against transitional movement in all directions. The model is welded on a steel plate which has high rigidity.

24. As the experiment was carried out under a restricted condition, a method of hanging weights and simultaneous loading was adopted. As a result, the load-deflection curve of pre-buckling and the ultimate load for brace-dome were obtained.

25. Compared with the ultimate load of 15.2kg/joint found from the theoretical calculation, the ultimate load obtained from the experiment was 12.4kg/joint, that is, 81.6% of theoretical result. This difference in the results was attributed to the errors of measurement and the initial imperfection.

CONCLUSION

26. Although aluminium-alloy can be used in space structures, the determination of proper patterns of the structure and construction procedures are very important. In the structural design of a single layer aluminium dome, non-linear stability analysis must be undertaken. The results of theoretical analysis, calculation, and experiment illustrate that the accurate tangent stiffness matrix is essential in the non-linear stability analysis of single layer domes.

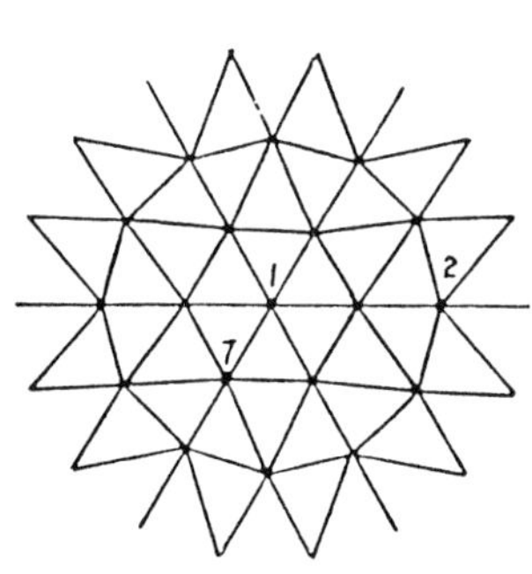

Fig. 2. The model of aluminium-alloy dome.

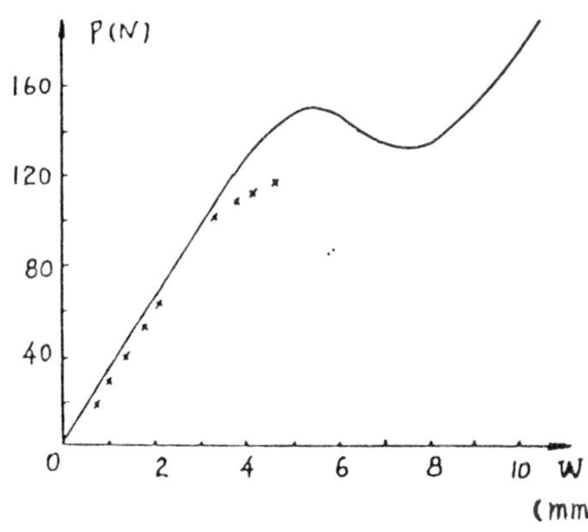

Fig. 3. Load-deflection curve (joint 1).

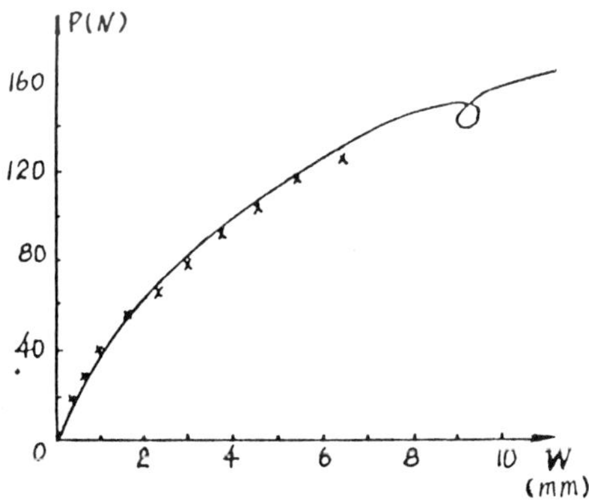

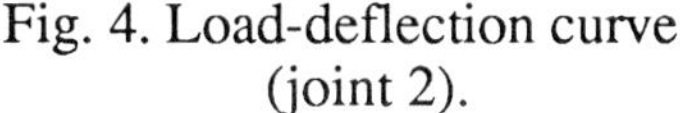

Fig. 4. Load-deflection curve (joint 2).

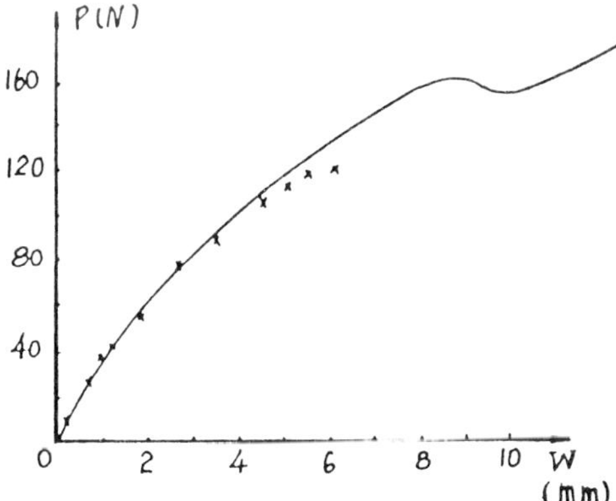

Fig. 5. Load-deflection curve (joint 7).

REFERENCES

1. ORAN, C. "Tangent Stiffness in Space Frames", Journal of the Structural Division ASCE, vol.99, No.ST6, pp 987-1001, 1973.
2. ORAN, C. "Tangent Stiffness in Space Frames", Journal of the Structural Division ASCE, vol.99, No. ST6, pp 973-985, 1973.
3. CRISFIELD, M.A."A Fast Incremental/Iterative Solution Procedure that Handles 'Snap-Through'", Computers and Structures, vol. 13 pp 55-62, 1981.
4. BATHE,Klaus-Jürgen and DVORKIN, Eduardo N. "On the Automatic Solution of Non-linear Finite Element Equations", Computers and Structures, vol.17, No.5-6, pp.871-879, 1983.
5. MEEK,J.L. and LOGANATHAN, S. "Large Displacement Analysis of Space-Frame Structures", Comp. Meths. Appl. Mech. Eng., vol.72, pp.57-75, 1989.
6. MEEK, J.L and TAN, Hoon Swee "Geometrically Non-linear Analysis of Space Frames by and Incremental Iterative Technique", Comp. Meth. Appl. Mech. Eng., vol.47, pp.216-282, 1984.
7. BATOZ, Jean-louis and DHATT, Gouri "Incremental Displacement Algorithms for Non-linear Problems", Int. J. Num. Meth. Eng., vol.14, pp.1262-1267, 1979.
8. 陈昕，"空间网格结构全过程分析及单层鞍形网壳的稳定性"，博士学位论文，哈尔滨建筑工程学院，1990
9. 钱若军，"弹性结构非线性稳定全过程分析述评"，《空间网格结构论文集》，pp.1-18，同济大学出版社，1991.4

18. Study of the elasto-plastic buckling of single-layer domes

F. MATSUSHITA, T. OHYA, Y. MUKAIYANA and Y. MATSUE,
Tomoe Corporation, Japan

SYNOPSIS. The present paper investigates the elasto-plastic buckling loads of single layer domes on a circular plane. The main analytical parameter is the type of mesh. Also, consideration is given to the mesh size, the rise of dome, boundary conditions, and the nodal joint system. The mesh types compared are the triangular and quadrilateral configurations. The buckling analyses performed are linear buckling and elasto-plastic buckling, taking into account geometric and material non-linear characteristics. Based on the comparison and summary of the results, this paper presents an effective method to examine the ultimate strength of the domes of triangular and quadrilateral mesh forms.

INTRODUCTION

1. In most cases, a triangular unit is used as the basic unit of the mesh for single layer domes. A great deal of research has been performed on single layer domes with a triangular mesh, and several effective estimation methods have been proposed for the ultimate strength of these domes (refs 1,2,3).

2. In this study, the quadrilateral mesh is chosen for comparison with a triangular mesh, because the former mesh is often adopted in spite of its small in-plane shear stiffness. The strengths are calculated for both meshes based on linear and elastoplastic analysis. By comparing the buckling behaviours, the characteristics of the quadrilateral mesh are determined and an effective method for estimating buckling is presented, in terms of column strength curves, similar to the domes formed from a triangular mesh (ref. 3).

MODELS AND METHOD OF ANALYSIS

Configuration of the domes

3. The domes are shown in Fig. 1, and each is a shaped spherical surface on a circular plane. In order to compare the triangular mesh with the quadrilateral mesh, the division number n is assumed to be 6,7,9 or 10 for the quadrilateral mesh and 3,4,5 or 6 for the triangular mesh. The total length (weight) of the members in each comparable pair, for example 10 for the quadrilateral mesh and 6 for the triangular one, is almost identical. Table 1 shows the division number n,

average member length ℓ_0 and the total length ℓ_t, of the members for each model, where $n \cdot \ell_0$ is set equal to $R \cdot \theta$.

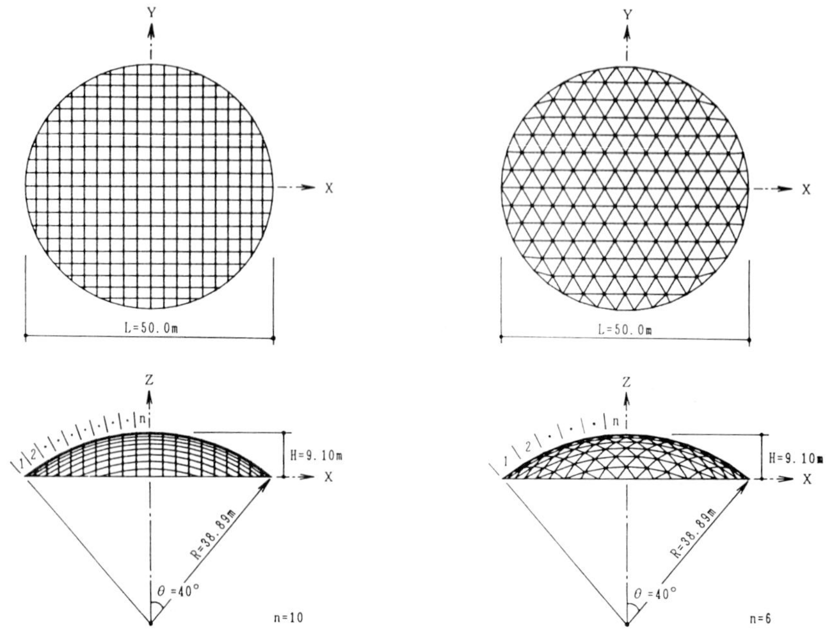

(a) Quadrilateral mesh.

(b) Triangular mesh.

Fig. 1. Configuration of domes.

Table 1. Properties of models.
(θ = 40 deg., H/L = 0.182)

mesh type	n	ℓ_0 (m)	ℓt (m)
Quadrilateral (Q-mesh)	6	4.525	960.2
	7	3.879	1125.7
	9	3.017	1455.8
	10	2.715	1620.5
Triangular (T-mesh)	3	9.051	905.8
	4	6.788	1183.9
	5	5.430	1458.5
	6	4.525	1713.3

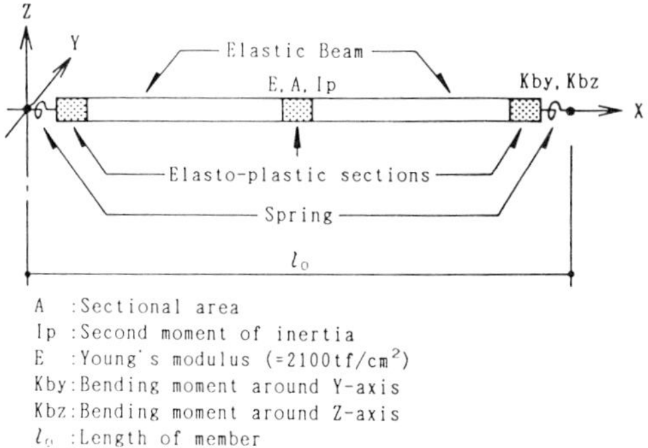

Fig. 2. Geometry of a constitutive line element.

Constitutive members

4. Each component, considered to be one member in the single layer dome, consists of two rotational springs and a general beam element as shown in Fig. 2. Steel tubes of rectangular section are assumed for members as shown in Table 2. According to finite deflection theory, the present analysis deals with geometrical non-linearity, composed partly of member buckling and partly of nodal displacements. The stiffness matrix for elastic members is formulated by the slope deflection method and takes into account member buckling. To take into consideration material non-linearity, elasto-plastic sections are placed at both ends and at the centre of the general beam elements. The two beam sections in one general element, are treated as elastic. The mechanical behaviour of the elasto-plastic section is assumed elastic before yielding, however, after yielding the axial force and the bending moments move on the yield surface defined by Eq. 1. The yield stress of steel tubes σy is assumed to be $2.4 tf/cm^2$. Table 2 shows the properties of the members.

$$f = \left(\frac{N}{Np}\right)^2 + \sqrt{\left(\frac{My}{Mp}\right)^2 + \left(\frac{Mz}{Mp}\right)^2} - 1 = 0 \qquad (1)$$

$Np = \sigma y \cdot A, \quad Mp = \sigma y \cdot Zp$

where N : axial force

My : bending moment around Y-axis

Mz : bending moment around Z-axis

A : sectional area

Zp : plastic modulus of section

Table 2. Member properties.

section	$E(tf/cm^2)$	$A(cm^2)$	$Iy(cm^4)$	$Iz(cm^4)$	Np(tf)	Mp(tf·cm)
2x□-175x175x6.0	2100.0	79.26	3720.0	3720.0	190.22	12334.35
H-800x300x14x26	2100.0	267.40	$2.92x10^5$	$1.17x10^4$	·······	·······

Boundary conditions

5. Two types of boundary conditions are adopted; one is a pin-support and the other is a roller-support. For the pin-support condition, the nodal displacements on the boundary are completely restrained against translation without any restrictions on rotation. For the roller-support condition, the vertical displacements at the nodal points on the boundary are fixed, and the rest are free on the X-Y plane. An additional assumption is made that the peripheral ring does not yield.

Joint system for members

6. This analysis considers two conditions for the joint rigidity at nodes; pin-jointed and rigid-jointed. In the case of the quadrilateral mesh, the pin-joint means that the two intersecting members are connected by a pin without any loss of bending rigidity in the members. However, in the case of the triangular mesh, the pin-joint means that both ends of members are considered to be free from bending moments.

Load distribution

7. The load is assumed to be vertical and uniform over the dome surface. The force at each node is calculated to be in proportion to the area occupied by each node.

Method of analysis

8. The static buckling loads of the domes under vertical loading are examined in two separate ways using a linear buckling analysis and an elasto-plastic buckling analysis. For linear buckling, the axial forces are assumed to be proportional to those obtained from a linear-elastic analysis. For the elasto-plastic buckling analysis, the Newton Raphson method is used, along with the load incremental method or the displacement incremental method. In this paper, the elasto-plastic buckling load is defined as the maximum load obtained by the elasto-plastic buckling analysis.

RESULTS OF THE ANALYSIS

Effect of the division number n

9. Four models for each mesh type were used to determine the effect of the division number. The other parameters are as follows: 40 degrees for the half open angle θ; pin-supports for the boundary condition; and rigid joints at the nodes. Fig. 3 shows the relationship between the elasto-plastic buckling loads Pcr and the division number n. In this figure, the axis of abscissa is weight of steel per unit surface, and the axis of the ordinate is the buckling load per unit surface. The buckling load varies almost linearly with the weight of steel not depending on the mesh form. In addition, the buckling strength of the triangular mesh is twice that of the quadrilateral mesh for the same weight of steel.

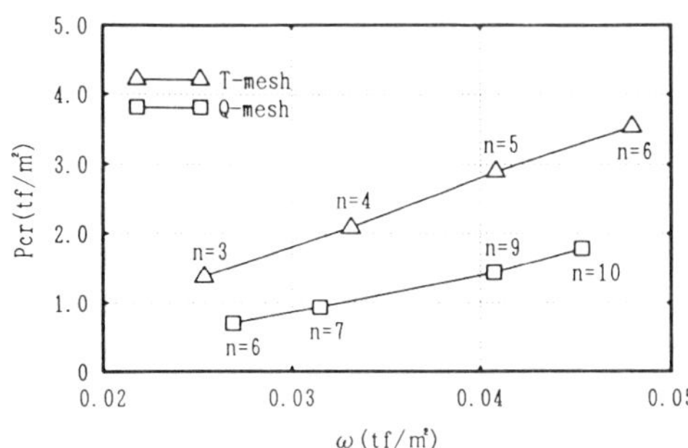

Fig. 3. Relationship between Pcr and n.

Effect of the rise of the domes

10. Four models for each mesh type were used to determine the influence on the elastic-plastic buckling load of the rise of the domes. The half open angle corresponding to the dome rise is set to 20,40,60 and 80 degrees. The other parameters are as follows: 10 and 6 as the division number for the quadrilateral and the triangular mesh, respectively; pin-supports for the boundary condition; rigid-joints at the nodes. The relationship between the half open angle and the elasto-plastic buckling load is shown in Fig. 4. This figure shows that for both the quadrilateral and triangular mesh models, the buckling strength increases as the inclination angle increases up to 60 degrees. This tendency may be attributed to the fact that member stresses decrease as θ becomes higher up to an inclination of 60 degrees. With a half open angle of 80 degrees, however, the dome strength is lower than that obtained for 60 degrees, as the inclination angle exceeds 60 degrees the surface area rapidly increases and the members become elongated.

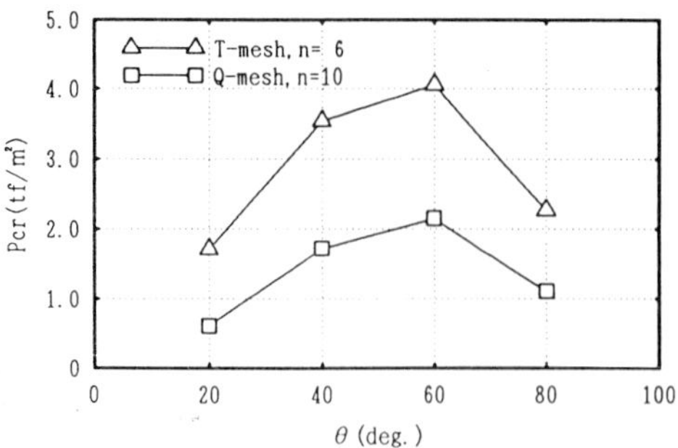

Fig. 4. Relationship between Pcr and θ.

Non-dimensional buckling strength η and normalised slenderness ratio Λ

11. To compare the linear buckling loads with the elasto-plastic buckling loads, the non-dimensional buckling strength η and the normalised slenderness ratio Λ are defined as follows.

a) For a particular member, which has the largest elastic axial force for a design load P_d ; the maximum force is assigned the value N_d ;

b) The normalised slenderness ratio Λ is calculated using Eq. 3 where Ncr^{lin} is the linear buckling axial force.

c) The maximum axial force Ncr is estimated using Eq. 4, where it is assumed that Ncr is proportional to the elasto-plastic buckling load Pcr.

$$Ncr^{lin} = N_d \cdot \frac{Pcr^{lin}}{P_d} \tag{2}$$

$$\Lambda = \sqrt{\frac{Np}{Ncr^{lin}}} \tag{3}$$

$$Ncr = N_d \cdot \frac{Pcr}{P_d} \tag{4}$$

$$\eta = \frac{Ncr}{Np} \tag{5}$$

where P_d : design load (100kgf/m^2)
Pcr^{lin}: linear buckling load
Pcr : elasto-plastic buckling load
N_d : maximum axial force under P_d by linear analysis

Effect of the boundary condition

12. The results of the analyses depending on the different of boundary conditions are explained in terms of η and Λ, which are shown in Fig. 5. A solid line represents linear buckling (LB) for Λ greater than 1.0, and also represents the yield strength of material Ncr/Np=1.0 for values of Λ less than 1.0. The alternate long and short dashed line represents the modified Dunkerley formulation (MDF) as shown by Eq. 6, (ref. 5).

$$\left|\frac{Ncr}{Ncr^{lin}}\right| + \left(\frac{Ncr}{Np}\right)^2 = 1 \tag{6}$$

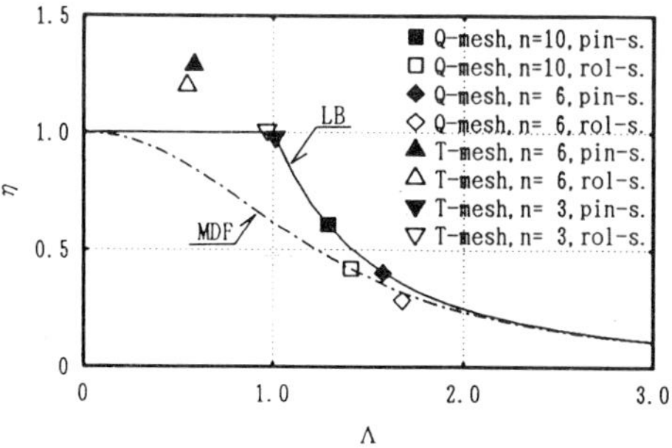

Fig. 5. Effects of boundary conditions.
(θ = 40 deg., rigid-joint)

13. From this figure, the strength of the roller-supported domes are almost equal to that of pin-supported domes. With the case of the quadrilateral mesh structures, the plots of the ultimate strength fall almost on the linear buckling line, $\eta=1/\Lambda^2$, irrespective of the boundary conditions; however, a small decrease in buckling loads ocurs in the roller support condition compared with the pin-support condition. On the contrary, with the case of the triangular mesh, the plots fall near the line of $\eta=1$. The difference due to the boundary conditions seems very small in the triangular mesh structures possibly because the half open angle θ is sufficiently large so that the members reach their elastic or plastic buckling capacity without being subjected to bending moments.

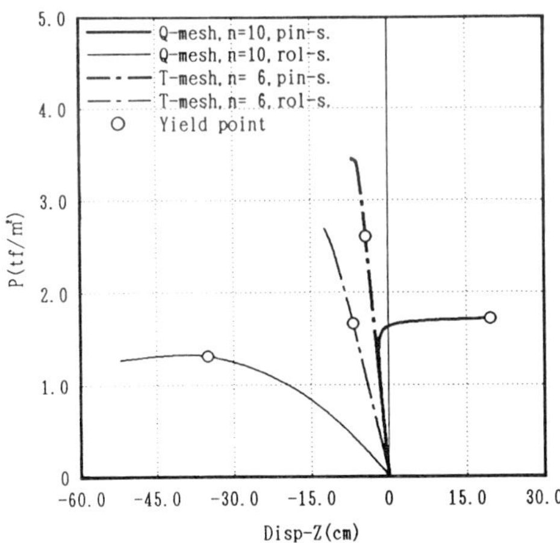

Fig. 6. Load-displacement curve at the apex.

Effect of joint rigidity

14. The results related to the difference in joint rigidity are expressed in terms of η and Λ as shown in Fig. 7. In the case of the quadrilateral mesh, the ultimate

strength of the domes are almost similar to that obtained for linear buckling, because the plots fall on the curve of $1/\Lambda^2$. The effect of joint rigidity can be clearly seen for each number of mesh division. On the other hand, for the triangular mesh structures, the ultimate strength is almost determined by the plastic buckling capacity of the members except for the case of n=3. If an actual mesh division is considered, the case of n=3 would be rare. Thus, when the half open angle is large, the ultimate strength of the domes may be almost the same irrespective of the joint rigidity. The reason may be that the members reach their plastic buckling capacity as one member without nodal displacements or bending moments at either or both ends.

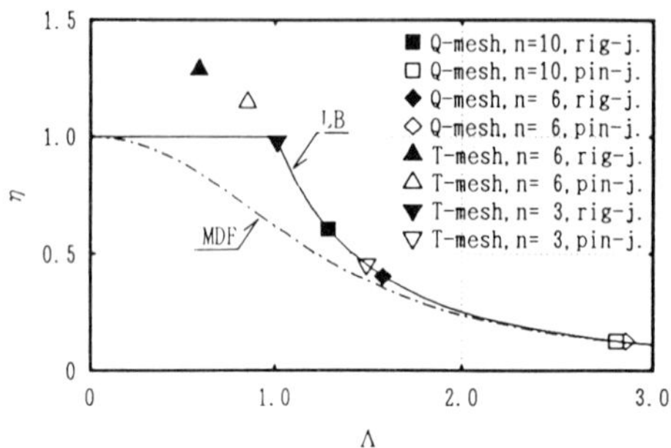

Fig. 7. Effects of joint rigidity.
(θ = 40 deg., pin-support)

15. The buckling modes are shown in Fig. 8. From this figure, it can be seen that the domes formed from a quadrilateral mesh have a large elastic buckling wave length, while the domes formed from a triangular mesh are not subjected to nodal displacements resulting in a short wave length indicating that the domes are subjected to plastic axial yield of members.

Estimation of the ultimate strength of the domes

16. The results of the investigation lie almost along the solid line shown in Figs 5 and 7, which are above the modified Dunkerley formulation. Accordingly, the ultimate strength of the domes may be estimated using the following method which is independent of mesh form, boundary conditions, and joint rigidity.

17. First, using linear buckling analysis, the normalised slenderness ratio Λ is calculated by using Eqs 2 and 3. Subsequently, the non-dimensional buckling strength η is obtained by using the modified Dunkerley formulation. Then the elasto-plastic buckling load is estimated using Eqs 4 and 5.

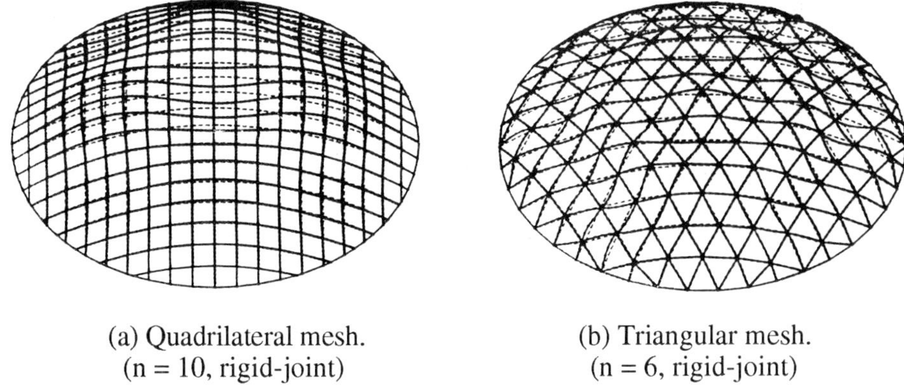

(a) Quadrilateral mesh.
(n = 10, rigid-joint)

(b) Triangular mesh.
(n = 6, rigid-joint)

Fig. 8. Buckling modes by linear buckling analysis.

CONCLUSIONS

18. It has been shown by the authors that the buckling loads of single layer domes under uniform loading can be effectively estimated for any mesh type by using the modified Dunkerley formulation. Future work, will investigate the effects of load distribution, geometric imperfections and other parameters. For quadrilateral meshed domes, the effect of reinforcement using wires or slender rods in the structure to enhance in-plane shear rigidity will also be considered.

REFERENCES

1. KATO S., SHIBATA R., TAKASHIMA H. and UEKI T. A computational procedure to evaluate buckling strength of reticular domes based on a concept of column buckling, Proceedings of IASS Congress, Toronto, Vol. 2, 1992, pp.552-563.

2. MUTOH I., UEKI T. and KATO S. On estimation of the buckling load of lattice domes by generalised slenderness, Proceedings of IASS Congress, Toronto, Vol. 2, 1992, pp.564-575.

3. KATO S., SHIBATA R. and UEKI T. An estimation procedure of the buckling load of a rigidly jointed single layer reticular dome, Journal of structural and construction engineering, Transactions of AIJ, No. 436, 1992, pp.91-103.

4. UEKI T., MUKAIYAMA Y., SHOMURA M. and KATO S. Loading test and elasto-plastic buckling analysis of a single layer latticed dome, Journal of structural and construction engineering, Transactions of AIJ, No. 421, 1991, pp.117-128.

5. KOLLAR L. and DULACSKA E. Buckling of shells for engineers, John Wiley & Sons, Chichester, 1984, pp.260-265.

19. Comparison of buckling loads between single-layer lattice domes and spherical shells

I. MUTOH and S. KATO, Gifu National College of Technology, Japan

SYNOPSIS. According to the equivalent shell analogy approach the buckling strength of equilateral triangulated lattice domes are made in relation to the spherical shell buckling by using the same abscissa of the shell slenderness parameter. Since the generalised slenderness that is determined by the eigen-value analysis, for buckling of lattice domes is used as the abscissa in an earlier study, conversions of this into shell-like slenderness are carried out by means of evaluating the numerical results of lattice domes by buckling criteria and by comparing the buckling wave-length ratio, derived from classical shell buckling theory, with the generalised slenderness through characteristic shape parameters for lattice domes.

INTRODUCTION

1. In this paper, a direct comparison between the buckling strength of a single-layer lattice dome and the equivalent spherical shell is made. A shell analogy is used with the aid of slenderness parameters, on the buckling curves.

2. Design curves for spherical shells consist of the strength measure and shell slenderness parameters $\bar{\lambda}_s$ in view of reduction factors which are the knockdown factor, ρ and the plasticity effect, γ (e.g. refs 1-3).

3. Buckling strengths of single-layer lattice domes have been discussed from the viewpoint of the equivalent shell analogy (refs 4-5) in order to relate these buckling characteristics to that of spherical shells considering specific shape parameters related to the geometrical parameters of the shells.

4. A procedure (ref. 6) to estimate the buckling strength of lattice domes was presented considering linear buckling for constituent members. The generalised slenderness parameter $\Lambda = \sqrt{N_y / N_{cr}^{lin}}$ is calculated from the member critical axial force N_{cr}^{lin} using the linear eigen-value analysis and the yield axial force N_y. Using semi-empirical buckling curves, such as column strength curves and Dunkerley-like strength curves, the normalised buckling strength may be determined using the generalised slenderness parameter Λ as abscissa.

5. Discussion of the estimation for spherical shell buckling and lattice dome buckling was made by using the same abscissa as the slenderness parameters, Λ

Space structures 4. Thomas Telford, London, 1993

and $\overline{\lambda}_s$, including the effects of geometrical imperfections (ref. 7). Fig. 1 shows, for example, the results of comparisons between shell buckling curves, denoted by (1), (2) and (3), and pin-supported lattice domes, denoted by empty circles and solid hexagons.

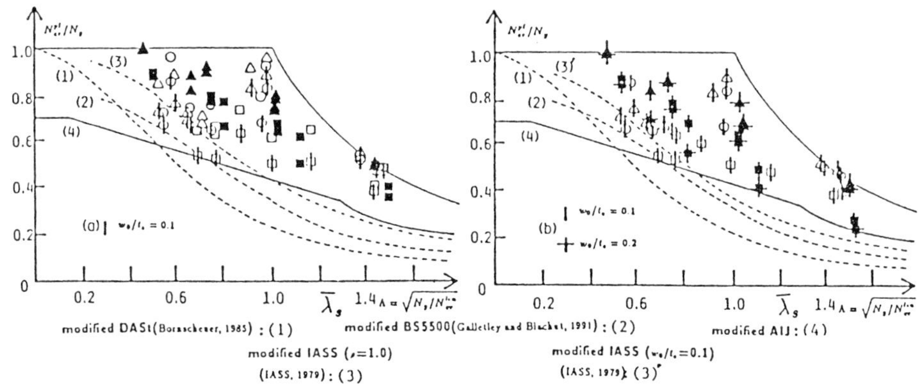

Fig. 1. Comparison of numerical results for lattice domes and design curves of spherical shells (ref. 7).

6. Unfortunately, since the notation for abscissa are different, it is difficult to compare directly the characteristic dependence of the strength, on the slenderness parameters. Hence we infer the buckling strength of lattice domes by applying a shell analogy approach to convert Λ to $\overline{\lambda}_s$, considering the influence of characteristic shape parameters, member slenderness λ_0, the subtended half-open angle of the member at apex θ_0 and plan forms, on the buckling loads.

EVALUATION OF BUCKLING LOADS BY MEANS OF SHELL ANALOGY

7. Representative shape parameters for relatively shallow lattice shells are the member slenderness $\lambda_0 = l_0 / i_0$ where l_0 and i_0 are the member length and the radius of gyration, and the half-open angle of subtended members at the apex of the dome $\theta_0 = l_0 / 2R$. Both λ_0 and θ_0 relate to the characteristic slenderness, which corresponds to the reduced slenderness ratio, $\overline{\lambda} = \lambda_0 / \pi\sqrt{\sigma_y / E}$, for columns, defined by $\alpha = l_0^2 / Ri_0$ (where R is the radius of curvature) after Lind (ref. 5). Also the differences in plan forms (circular, hexagonal, rectangular), supporting conditions (pin-support, roller-support), and number of subdivisions (n) may result in significant buckling behaviour of lattice shells.

Equivalent buckling loads and characteristic shape parameters

8. In view of Wright's buckling criteria (ref. 4) buckling loads and characteristic shape parameters are derived using the notation, η versus ξ by one of the authors.

$$\eta = \frac{P_{cr}}{EA\theta_0};\quad \xi = \frac{12\sqrt{2}}{\lambda_0\theta_0\sqrt{1+2/\kappa}} \tag{1}$$

Focusing on the buckling loads of the rigid-jointed lattice domes, the characteristic parameter ξ is given as $\xi_0 = 12\sqrt{2}/(\lambda_0\theta_0)$ setting the term $\sqrt{1+2/\kappa}$ equal to one. Also, Heki gives the non-dimensional buckling load based on the effective strength as $\varsigma = P_{cr}/71EA\theta_0/\lambda_0^2$ (ref. 8).

Table 1. Buckling loads and characteristic parameters for equilateral triangulated lattice domes.

Shape Parameter (1)	General Buckling (2)	Member Buckling (3)	Snap-Buckling (4)	Ref. (5)
$\alpha = \frac{\ell_0^2}{i_0 R}$	$C_G = \frac{8\pi^2}{8\pi^2+\alpha^2}(\frac{0.865B}{\alpha} + 0.136\alpha)$	$C^{[a]}_{Euler} = \frac{8.55}{\alpha}$	$C^{[b]}_{snap} = \frac{0.136\alpha}{1+\alpha^2/8\pi^2}$	*Lind*
$S = \frac{\ell_0}{\sqrt{R}}\left(\frac{D}{K}\right)^{1/4}$	$\alpha_G = 1.0$	$\alpha_m = \frac{7.33}{S^2}$	$\alpha_1 = 0.045S^2$	*Yamada*
$\xi = \frac{12\sqrt{2}}{\lambda_0\theta_0\sqrt{1+2/\kappa}}$	$\eta_G = \xi$	$\eta^{[c]}_{Euler} = 0.41\xi^2$	$\eta^{[d]}_{snap} = \frac{2}{\sqrt{3}}$	*Kato*

Note: [a],[c]:eq.(25) in *Wright,1965*, [b]:pin-jointed dome by setting $B=0$ from C_G, [d]:eq.(29) in *Wright,1965*

ℓ_0:member length, i_0:radius of gyration, R:radius of curvature, B:nondimensional bending stiffness of member end, K:equivalent extensional stiffness, D:equivalent bending stiffness, λ_0:member slenderness ratio, θ_0: subtended half-open angle of apex member, κ:nondimensional rotational rigidity of joint

9. Evaluation of buckling loads, with respect to the key-parameter of $\eta - \xi_0$ notation given by Eqs 2 and 3 adopting results of the numerical analyses (refs 6, 8-10), are made based on the values given in Table 1 and are shown as in Fig. 2. These results were obtained from numerical analyses for lattice domes with the pin-support boundary under the following conditions:

a) hexagonal plan form (ref. 6) with λ_0 =60, 100 and 140, θ_0 = 2 and 4° and n = 4;

b) hexagonal plan form (ref. 8) with λ_0 = 100, 150 and 200, θ_0 = 2, 3, 4 and 5° and n = 5;

c) circular plan form (ref. 9) with λ_0 = 52.7, 62.5, 98.4 and 145.6, θ_0 = 2, 3 and 4°, and number of sub-divisions n = 6;

d) rectangular plan form (ref. 10) with λ_0 =54, 64, 78, 100, 148 and 186, θ_0 = 2 and 4° and n = 4.

The member length l_0 is 3m for cases a),b),d), and for case c) l_0 = 3.1m.

$$C = \eta\frac{1.22}{\xi_0};\quad \alpha(\text{Yamada}) = \eta\frac{1.175}{\xi_0};\quad \varsigma = \eta\frac{4.06}{\xi_0^2} \tag{2}$$

$$\xi_0 = \frac{12\sqrt{2}}{\lambda_0\theta_0}; \quad \alpha(\text{Lind}) = \frac{24\sqrt{2}}{\xi_0}; \quad S = \frac{5.51}{\sqrt{\xi_0}} \tag{3}$$

10. In the case of linear buckling (denoted by empty marks), an almost linear relationship given by $\eta_{cl} = \xi_0$, appears in Fig. 2(a) above the value of $\xi_0 \approx 4$. The coupling behaviour of general buckling and member buckling may be observed over the range of ξ_0, 3 to 4 in Fig. 2(a).

11. The results for elasto-plastic buckling are shown in Fig. 2. The differing values of θ_0 and λ_0 for the circular plan form are denoted by numerals 1, 2 and 3 and a, b and c in the figure. The effect of shape perameters on the buckling loads in elasto-plastic buckling is weak for values of λ_0 greater than 100. However for lower values of λ_0 the member buckling criterion might become dominant.

Relationships between shape parameters and slenderness parameters

12. In order to investigate the buckling criteria in relation to the spherical shell buckling, discussions of buckling wavelength are available in view of the buckling length ratio $\bar{l}_{cr}$, namely $\bar{l}_{cr} = 2l_0 / l_{cr0}$. Here l_{cr0} for equivalent spherical caps is assumed as $3.5\sqrt{Rt}$ from the theoretical wavelength of a complete sphere. In Fig. 3(a) the buckling length ratios, $\bar{l}_{cr}$ are plotted against the shape parameters given by $\sqrt{\xi_0}$ including a dotted curve obtained from the shell analogy with a value of $\bar{l}_{cr} = 1.36/\sqrt{\xi_0}$

13. Equivalent shell slenderness parameter $\bar{\lambda}_s$, which is defined by Eq. 4-1 below, and aforementioned generalised slenderness Λ from numerical analyses, are compared as shown in Figs 3(b) and 3(c). Lines with numbers (2),(3),(4) and (5), drawn for the shell analogy, are the cases where θ_0 = 2, 3, 4 and 5° respectively.

$$\bar{\lambda}_s \approx \frac{0.083}{\theta_0\sqrt{\xi_0}}\left(\approx \frac{0.014\alpha}{\theta_0} \approx \frac{0.0151S}{\theta_0}\right) \tag{4-1}$$

14. If the buckling criteria for lattice domes is that of a member buckling, there might be linear dependence of $\bar{\lambda}_s$ to $\bar{\lambda}$, as shown in Fig. 4. Fig. 4.(b) shows the relationship of Λ to $\bar{\lambda}$. Fig. 4(a) shows the equivalent shell slenderness $\bar{\lambda}_s$ plotted against $\bar{\lambda}$.

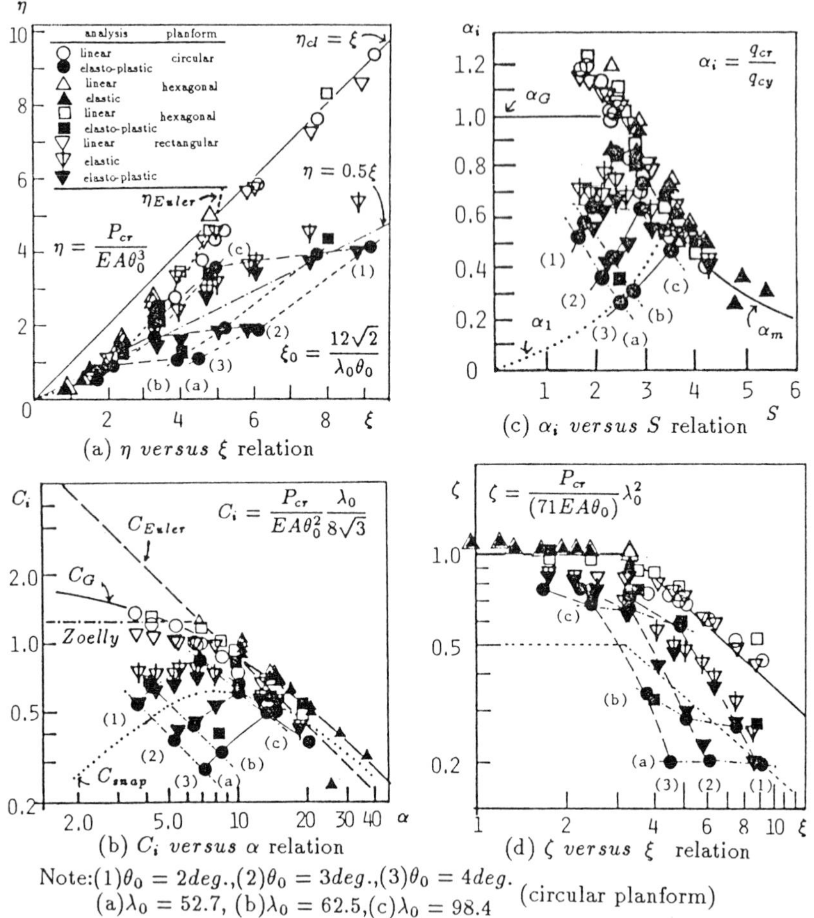

Fig. 2. Evaluation of buckling loads with characteristic shape parameters.

Here Λ is redefined by the reduced slenderness ratio $\bar{\lambda}$ and given by:

$$\Lambda = \sqrt{\frac{N_y}{N_{cr}^{lin}}} = \bar{\lambda}\sqrt{\frac{N_{Euler}}{N_{cr}^{lin}}} \tag{4-2}$$

where N_{cr}^{lin} is the critical axial force for the member determined by the linear buckling load through an eigen-value analysis. Also Λ is related to $\bar{\lambda}_s$, due to $\bar{\lambda} \propto 1/(\theta_0 \xi_0)$ using Eqs 4-1 and 4-2. Finally, Λ can be described in relation to $\bar{\lambda}_s$ as follow:

$$\Lambda \propto \frac{\bar{\lambda}_s}{\sqrt{\xi_0}}\sqrt{\frac{N_{Euler}}{N_{cr}^{lin}}} \tag{4-3}$$

In Fig. 3(c) solid curves approximately correspond to the results calculated by Eq. 4-3 with a specific value of $\varsigma = \sqrt{N_{Euler} / N_{cr}^{lin}}$ dependent on ξ_0. And from Fig. 4(a) when $\bar{\lambda} = 1$, Λ coincides with ς according to Eq. 4-2. From Fig. 4(b) non-linear dependence of $\bar{\lambda}_s$ against $\bar{\lambda}$ appears, since $\bar{\lambda}_s \propto \sqrt{\xi_0}\,\bar{\lambda}$ is obtained from a shell analogy.

BUCKLING STRENGTH CURVES FOR LATTICE DOMES

15. Here comparisons of the buckling strength of spherical shells and lattice domes are investigated by means of normalised strengths P_{cr} / P_y, plotted against equivalent shell slenderness parameters $\bar{\lambda}_s$. In so far as the linear relation between loads and axial forces can be assumed, these plots might be implicated with the estimation of buckling stresses available on the same diagram in a similar manner to design codes for shells (ref. 11).

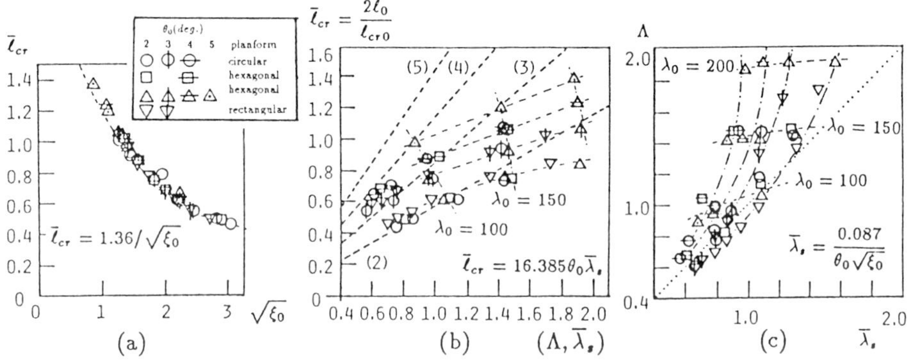

Fig. 3. Relationships between buckling length ratio and shape parameters.

Buckling loads of equivalent spherical shells and geometrical parameters

16. Numerical results adopted in this study are normalised to the classical buckling loads P_{cl} for equivalent spherical shells depicted as shown in Fig. 5. The geometrical parameter is approximately given by a shell analogy as:

$$\lambda\left(= \left[12\left(1 - \nu^2\right)\right]^{1/4} \beta\sqrt{\frac{R}{t}}\right) \approx 1.37n\sqrt{\lambda_0\theta_0} \tag{5}$$

where n is the number of subdivision of the lattice dome and is related to a half open angle of the equivalent spherical shell, β or $2n\theta_0$. As may be seen in Fig. 5, the range of λ for numerical results is from 8 to 30 approximately. Also both P_{cr}^{lin}/P_{cl} and P_{cr}^{pl} / P_{cl} are very dependent to θ_0 and λ_0, namely, the shape parameter, ξ_0. In the case of a constant member slenderness, $\bar{\lambda} \approx 1.0$, strength reductions can be seen from the results of elasto-plastic buckling analyses regardless of the plan forms.

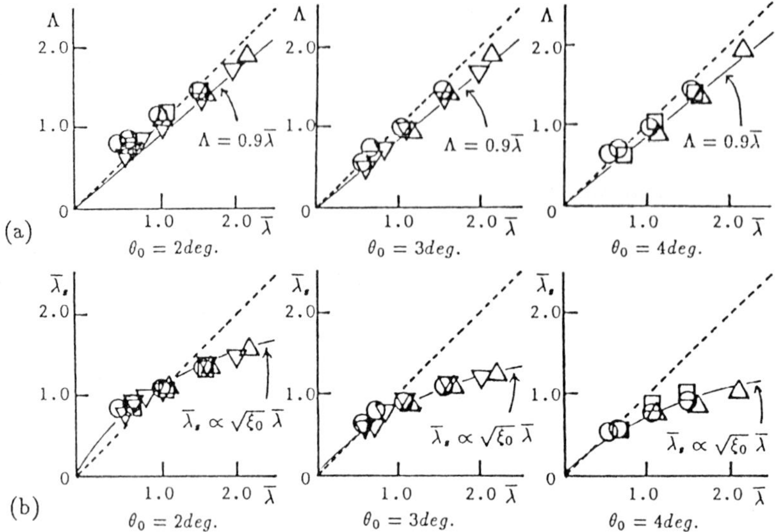

Fig. 4. Relationships in-between slenderness parameters ($\bar{\lambda}, \Lambda, \bar{\lambda}_s$).

Proposal of buckling curves for lattice domes

17. Buckling strength curves for continuum shells may be described by $\bar{P}_u(= P_{cr} / P_y) = \phi\left(\sqrt{P_y / P^{el}_{cr}}\right)$ and $\gamma = \Delta\left(P_y / \rho P_{cl}\right)$. We use the buckling strength curves using the notation based on the yield strength in contrast with the elastic buckling P^{el}_{cr}, namely the relationship between P_{cr}/P_y and $\bar{\lambda}_s$, given by $\phi(\bar{\lambda}_s)$. Such a concept for evaluation of the ultimate strength may be found elsewhere, for instance, the buckling curves for columns, plates and cylindrical shells under axial loads (ref. 12). In this study, the Merchant-Schwartz-Rankine formula, given by Eq. 6, for the buckling strength curve is proposed in order to discuss the evaluation of buckling loads for lattice domes by a shell analogy;

$$\frac{P_{cr}}{P_y} = k_1 \frac{1}{\left[k_2\bar{\lambda}_s^{\,4} + k_3\bar{\lambda}_s^{\,2} + 1.0\right]^{1/2}} \qquad (6)$$

where P_y is the yield strength, calculated by $P_y = 6\theta_0 N_y$, and coefficients $k_1 = 1/1.25$, $k_2 = 0.75$ and $k_3 = 1.0$.

18. In the case of lattice dome buckling, elastic buckling of members may be dominant due to the combination of θ_0 and λ_0 aforementioned in Fig. 4. Hence the buckling strength is discussed by comparison of the linear buckling between the numerical results and the $\eta_{cl} = \xi_0$ curve for equivalent shells as shown in Fig.

6(a). Bold curves with notations (1)(2)(3) and (4) inserted in this figure indicate the case of member buckling criteria as illustrated in Fig. 2. Curves are obtained from the relation as shown in Table 1, $\eta_{Euler} \propto \xi_0^{\ 2}$ and Eq. 4, $1/\overline{\lambda}_s^{\ 4} \propto \theta_0^{\ 4}\xi_0^{\ 2}$ as in $P_{Euler} / P_y \propto 1/\left(\theta_0^{\ 2}\overline{\lambda}_s^{\ 4}\right)$. Here notations (1), (2), (3) and (4) represent $\theta_0 = 2$, 3, 4 and 5°, respectively.

19. It is noted that in the case of $\theta_0 = 3, 4, 5°$ with $\lambda_0 > 100$ plots appear along the curves, (2)(3) and (4), below $P_{cr}/P_y = 0.5$. The dot-dashed line inserted in this figure indicates the bounds of the buckling criteria which are estimated by the relationship of ratios of apparent effective length to member length against ξ_0 for numerical results for hexagonal lattice domes where the limiting value of ξ_0 is approximately 2.36. As can be seen in Fig. 2(a), this corresponds to the coincidence of curves, η_{Euler} and $\eta = 0.5\xi_0$, around $\xi_0 = 2.5$. Although an apparent effective slenderness is proportional to $\sqrt{\xi_0}$, departure from such relation may be observed around the range of ξ_0 from 4.0 to 2.36.

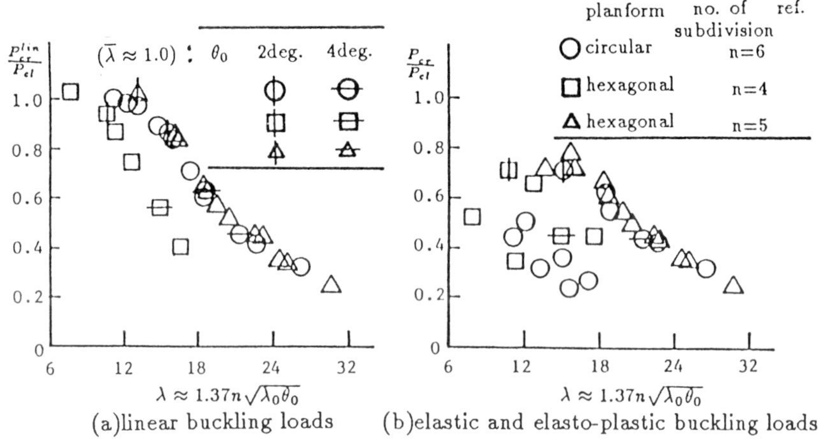

Fig. 5. Normalised buckling loads versus geometric parameter of equivalent shell.

20. In Fig. 6(b) buckling strength curves denoted by the bold curve (a) and dot-dashed curve (b) are compared with numerical results obtained from the elastic buckling analyses and elasto-plastic buckling analyses. Curve (b) means the ultimate strength estimation given by Eq. 6. However, curve (b), given by $0.5/\overline{\lambda}_s$, equivalent to $\eta = 0.5\xi_0$, could correspond to the dot-dashed line in Fig. 6(a) as mentioned earlier. It might be postulated that curve (a) shows the lower bound of imperfection sensitivity in elastic buckling of lattice domes from the viewpoint of equivalent spherical shells (ref. 7). When the focus of elasto-plastic buckling is attained in this figure, the proposed buckling curve, below the range of $\overline{\lambda}_s \approx 1.0$, may become an appropriate candidate for estimation of the buckling strength of lattice domes adopted in this study.

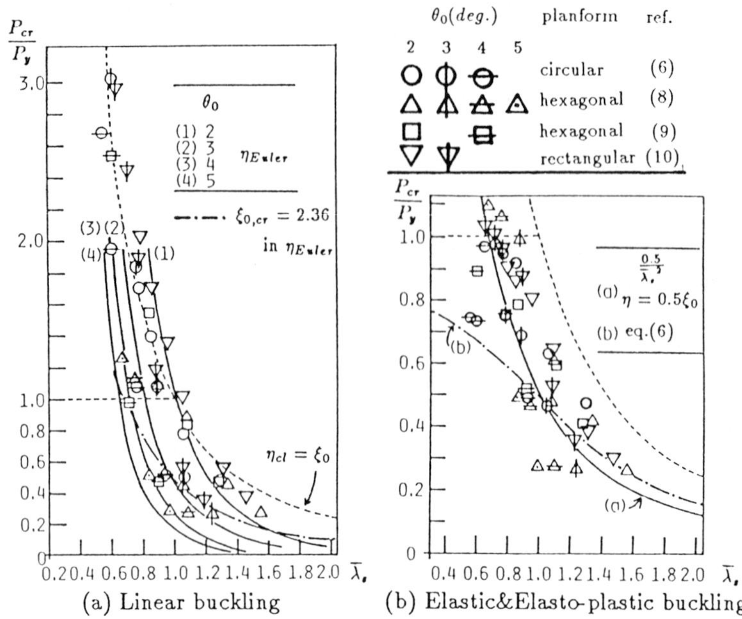

Fig. 6. Evaluation of buckling strength with equivalent shell slenderness, $\overline{\lambda}_s$.

CONCLUSIONS

21. It is concluded that the buckling strengths may be estimated both in the elastic and elasto-plastic range of $\overline{\lambda}_s$ in relation to the design curves for spherical shells as follow:

i) In the elastic range, the buckling strengths are given by $0.5/\overline{\lambda}_s$ and this curve corresponds to the bounds of the buckling criteria, general buckling and member buckling, which are illustrated by the relationships between the ratios of the Euler buckling loads to the linear buckling loads and the shape parameters, ξ_0.

ii) In the elasto-plastic range, Eq. 6 given in the manner of the Merchant-Schwartz-Rankine formula represents a lower bound to the numerical results. This curve is also comparable to some design curves for spherical shells. A further insight into the buckling strength estimation for the overall range of slenderness parameters might recall by means of the unification of curve (a) and curve (b) in a practical way considering the effects of initial imperfections on the elasto-plastic buckling. Moreover, the effects of mesh patterns and roller-supported condition on the buckling strengths of lattice domes are challenging subjects for the future. Also detailed research will be necessary for discussion of the buckling criteria, in particular, the boundary of member buckling and general buckling.

REFERENCES

1. ECCS The buckling of shells, European Recommendations for Steel Construction, Brussels, Publication 29,1983.

2. DASt Beulsicherheitsnachweise fur Schalen, Richt-linieO13, Deutscher Ausschuss fur Stahlbau, Cologne, Germany, 1980.
3. BS5500 Specification for Unfired Fusion Welded Pressure Vessels, British Standards Institution, London, 1984.
4. WRIGHT D.T. Membrane forces and buckling in reticulated shells, Journal of the Structural Division, ASCE, 1965, vol.91, No.ST1, 173-201.
5. LIND N.C. Local instability analysis of triangulated dome framework, The Structural Engineering, 1969, vol.47, No.8, 317-324.
6. KATO S. et al. Study on the buckling stress of a rigidly jointed single-layer reticular dome, Journal of Struc. Constr. Engng, AIJ, 1991, No.428, 97-105.(in Japanese).
7. MUTOH I. et al. On estimation of the buckling load of lattice domes by generalised slenderness, Proceedings of the IASS-CSCE Int. Congress (Toronto Canada), 1992, vol.2, 564-575.
8. MURAKAMI M. and HEKI K. On the analysis of elastic buckling of single-layer latticed domes with regular hexagonal plan under gravity load, Proceedings of Int. IASS Symp. (Copenhagen Denmark), 1991, vol.III, 101-108.
9. KATO S. et al. Estimation of elasto-plastic buckling loads for reticular domes on a circular plan, Journal of Struct. Constr. Engng, AIJ, 1992, No.439, 111-119.(in Japanese).
10. UEKI T. et al. Study on the elastic and elasto-plastic buckling behaviour of single layered domes composed of members having axial and bending springs at both ends, Proc of Int. IASS Symp. (Copenhagen), 1991, vol.III, 93-100.
11. GALLETLY G.D. and BLACHUT J. Buckling design of imperfect welded hemispherical shells subjected to external pressure, Proceedings of Instn Mech. Engrs. Part C, 1991, vol.205, 175-188.
12. BORNSCHEUER B.-F. Stabilitatsnachweise fur Platten und Shcalen in gleicher Darstellung wie fur Stabe, Der Stahlbau, 1985, 54, H.12, 364-368.

20. Space structures and their topographical invariants

A. KAVEH, University of Science and Technology, Iran

SYNOPSIS. In this paper topological properties of the graph models of space structures are studied. Four topological invariants of the graphs, namely the first Betti number, crossing number, genus and thickness of graphs are considered and their applications to the force method of structural analysis are described. It is shown that such study simplifies many problems involved in the analysis of space structures.

INTRODUCTION

1. Analysis of skeletal structures involves the study of three different properties; topological, geometrical and material. For an optimal analysis and design, topology plays an important role, because once the topology of a structure is established, the necessary geometry can be assigned and appropriate material can be decided upon in the course of analysis and design .

2. The simplicity of some topological concepts have perhaps been responsible for the lack of attention to such an important property of the structures. As an example, the natural existence of a co-cycle basis of the graph model of a structure, corresponding to a kinematical basis leading to a rather sparse stiffness matrix, allows the analyst to use the displacement method without recourse to the topology of a structure in an explicit form. Such a property can not be ignored when force method has been employed as will become obvious from the content of this paper. Even in the displacement analysis of a large structure, topology is involved in other forms such as configuration processing, sub-structuring, ordering for efficient solution of equations and dynamics. For an optimal solution, where the formation of the most sparse stiffness matrix of a structure is required, considerable information from topology will be necessary to generate an optimal cut-set basis leading to a matrix of this quality.

3. In this paper only those topological properties of structures are studied, which are relevant to the force method of their analysis. The mathematical terminology and definitions used here may be found in ref. 1.

Space structures 4. Thomas Telford, London, 1993

FORCE METHOD OF STRUCTURAL ANALYSIS

4. The stress resultant distributions due to a given loading on a skeletal structure, S, obtained by the force method is given by:

$$r = B_0 p + B_1 q \tag{1}$$

where B_0p is a particular solution corresponding to applied load p, and B_1q is a complementary solution related to redundants q.

5. For rigid-jointed skeletal structures , B_0 can easily be generated on a shortest route tree (forest), (ref. 1). Matrix B_1 has its columns as a maximal set of independent self-equilibrating stress systems (SESs), known as a *statical basis*. The cardinality of such a basis is the same as the degree of statical indeterminacy (DSI) of the structure, denoted by $\gamma(S)$. Considering the compatibility conditions, and employing Eq.(1) leads to:

$$r = \left[B_0 - B_1 \left(B_1^t F_m B_1 \right)^{-1} B_1^t F_m B_0 \right] p \tag{2}$$

where F_m contains the flexibilities of the individual members of the structure.

6. For an efficient force method, the flexibility matrix $G = B_1^t F_m B_1$ should be sparse, properly structured (e.g. banded) and well conditioned. Analysis employing matrices with these properties is called an optimal analysis. The properties of G depend on those of B_1, since F_m have a definite pattern for the members of a structure. A statical basis corresponding to the sparsest B_1 is called a minimal statical (null) basis, and the one leading to the sparsest matrix G, is defined as an optimal statical basis.

7. The mathematical model of a frame can be considered as a simple graph. A graph is defined as a set of vertices and a set of edges together with a relation of incidence which associates two distinct vertices with each edge. There is a 1-1 correspondence between members (nodes) of a structure and edges (vertices) of its graph model, except support nodes which are taken support (ground) vertex, (Fig. 1).

8. A vector space can be associated with a graph, known as a cycle space. A basis of this space consisting $b_1(S)$ mod 2 independent cycles is known as a cycle basis. The cardinality of this basis is given by:

$$b_1(S) = M(S) - N(S) + b_0(S) \tag{3}$$

where $b_1(S)$ and $b_0(S)$ are the first and zero Betti numbers of S, and M(S) and N(S) are the numbers of members and nodes of S, respectively.

9. A cycle basis corresponding to maximal sparsity of cycle - edge incidence matrix C is called minimal and if corresponds to maximal sparsity of cycle adjacency matrix $D = CC^t$, it is known as optimal cycle basis. The sparsity of a matrix is taken as the number of its non - zero entries. If the sparsities X(C) and

X(D) are near minimum, then such bases are called sub-minimal and sub-optimal, respectively.

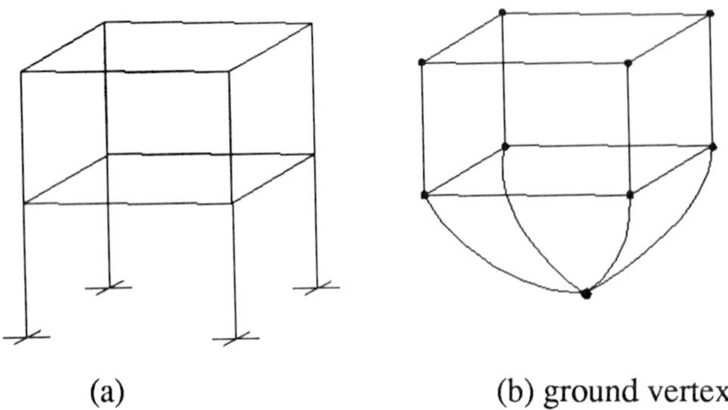

(a) (b) ground vertex

Fig. 1. A space structure and its graph model S.

10. The formation of a statical basis can be decomposed into two steps. In the first step a cycle basis of S is formed, and in the second step, 3 or 6 SESs are constructed on each selected cycle, depending on the corresponding structure being a planar or space frame. The latter step is a straight forward process and any improvement on statical basis will be the consequence of selecting a better cycle basis. Therefore the problem of forming a minimal or and optimal statical basis is transferred to the construction of a minimal or and optimal cycle basis of the structural model, respectively.

DSI OF SPACE STRUCTURES - CROSSING NUMBER OF S

11. The first step of analysis by the force method consists of determining the DSI of a structure. In the following, two theorems are presented for calculating the DSI of space structures, using the crossing numbers of their planar drawings.

Definitions

12. A drawing S^P of S is a mapping of S into a surface. The nodes of S go into distinct nodes of S^P. A member and incident nodes map into a homeomorphic image of the closed interval [0,1] with the relevant nodes as end points and the interior as a member, containing no nodes. An admissible drawing is one in which no two members incident with a common node have a common point, and

no two members have more than one point in common. A common point of two members is a crossing. An optimal drawing in a given surface is one which exhibits the least possible crossings. The number of crossing points of S after drawing on a plane or sphere S^P , is denoted by $\nu(S^P)$. For cases when the drawing is optimal, $\nu(S^P)$ becomes the crossing number of S.

THEOREM A - Let S be a space rigid-jointed frame. Then:

$$\gamma(S) = 6\, b_1(S) = 6\, [R_i(S^P) - \nu(S^P)] \tag{4}$$

where $R_i(S^P)$ is the number of internal regions of S^P, i.e. $R_i(S^P) = R(S^P) - 1$.

THEOREM B - Let S be a ball-jointed space truss. Then:

$$\gamma(S) = \nu(S^P) - M_C(S^P) \tag{5}$$

where $M_C(S^P)$ is the number of members required for full triangulation of S^P. Simple proofs of the above theorems, using Euler's polyhedron formula, are given in ref. 1.

EXAMPLE 1 : Let S be the graph model of a space frame as shown in Fig. 2a. A drawing S^P of S is shown in Fig. 2b for which $\nu(S^P) = 8$. Using Eq.(4) leads to:

$$\gamma(S) = 6[41 - 8] = 198. \tag{6}$$

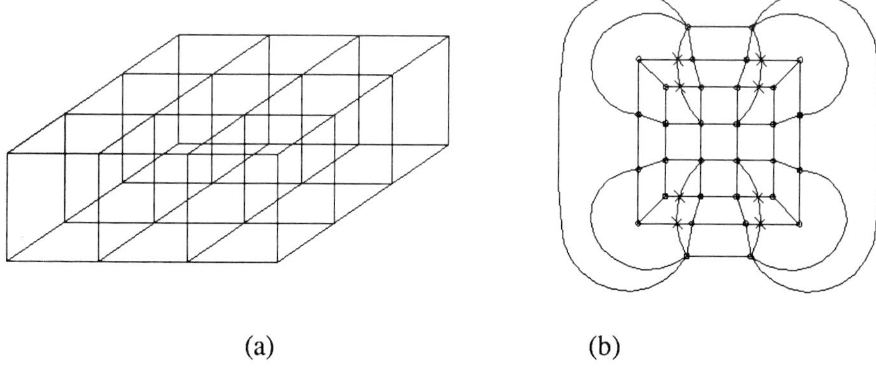

(a) (b)

Fig. 2. A space frame and its drawing.

EXAMPLE 2 : S is the graph model of a double layer grid a drawing of which is shown in Fig.3. For this drawing $\nu(S^P) = 60$, and considering 21 members for triangulation of the outer polygon, Eq.(5) leads to:

$$\gamma(S) = 60 - 21 = 39 \qquad (7)$$

13. Obviously it is advantageous to have an optimal drawing of S, in order to reduce the number of countings for determining $\nu(S)$. However, in general, little is known about the crossing number of S, (ref.2). Further research is required to develop a systematic approach for constructing an optimal or at least a sub-optimal drawing of S. Notice that for the case $\nu(S^P) = 0$, the corresponding mesh basis of S^P forms a suitable cycle basis of S.

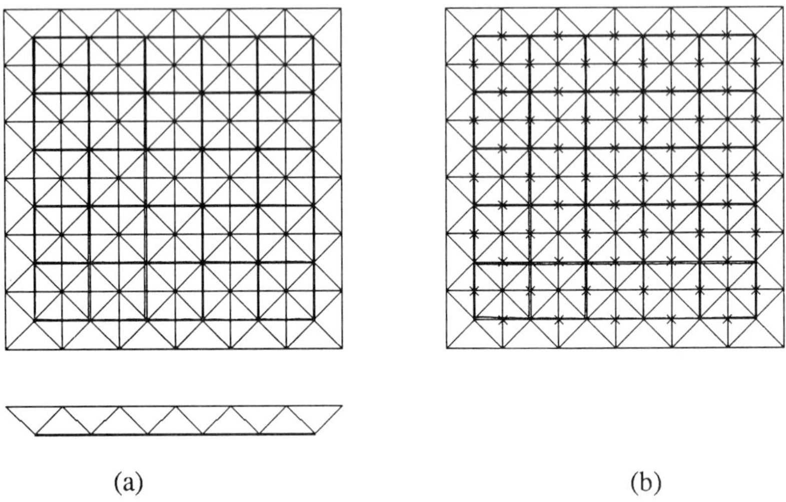

Fig. 3. A double layer grid with an arbitrary drawing.

EMBEDDING FOR CYCLE BASES SELECTION : GENUS OF S

14. The quality of a mesh basis consisting of the cycles bounding the finite regions of a planar graph embedded in $\mathbb{R}^2$, suggests the extension of this approach to general non-planar graph. Every graph can be embedded on an orientable surface. Such a surface may be viewed as a sphere with "h" handles, h being called the genus of the surface.

15. The genus g(S) of S is the minimum genus of the surface on which it can be embedded. An embedding is minimal if the genus of surface is the same as g(S). Although Edmond's permutation technique provides a method for embedding S in an orientable manifold, it is lengthy and non-practical. Therefore only intuitive

embeddings have to be employed which limits its application to simple structural models.

16. The regional cycles obtained by embedding S on a manifold M, are not independent and embedding should be modified to provide the independence of the selected cycles. For each handle of M two fillings should be added, and one perforation should be made on the sphere.

EXAMPLE: Let S be a 3-storey frame as shown in Fig. 4a. The first storey can be embedded on a torus and the third storey on a second torus. Four tubes can be added to embed the columns of the second storey. Identifying these toruses through one of the tubes, results in a sphere with five handles as depicted in Fig. 4b. Once the embedding is obtained, modifications can easily be made. Fillings are shaded and the perforation is shown by P, Fig. 4b.

17. It is ideal to embedded S on a manifold with minimum number of handles, since the fillings, in general, increase the overlaps the cycles. However, a much more efficient method is required for embedding before such an approach can be used in practice.

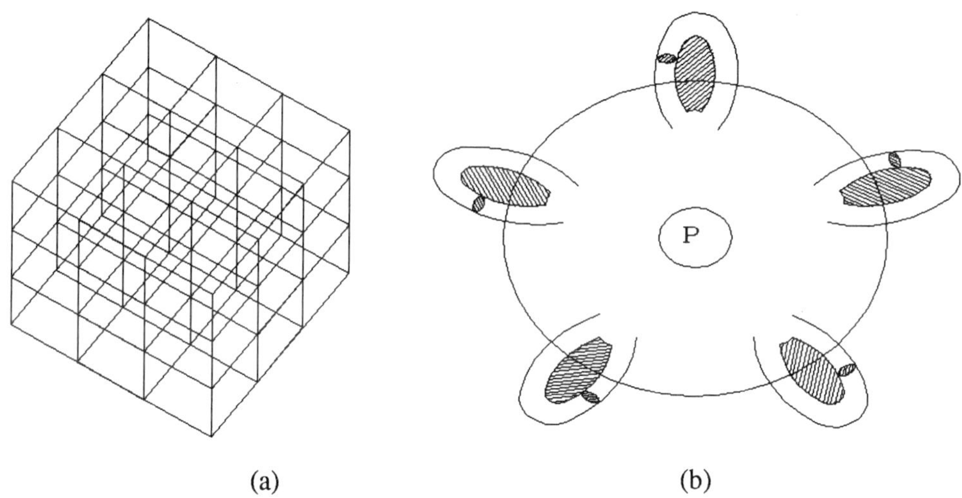

Fig. 4. Manifold on which S is embedded, and its modifications.

EMBEDDING ON A UNION OF DISKS : THICKNESS OF S

18. A graph S can be considered as the union of some planar sub-graph of S. The regioned cycles can then be selected as a cycle basis, provided some considerations are taken into account, as describe below :

19. Identify a planar sub-graph S, and embed it on a disk d_1 whose dissection K_1 is isomorphic to S_1. The second sub-graph S_2 is identified such that the corresponding K_2 has a 2 - cell with a free 1-face and $K_1 \cap K_2$ is a connected

subspace of the frontier of d_l . This process is continued and at its step K_i is joined to $K^{i-1} = \cup_{j=1}^{i-1} K_j$ with K_i having a free 1-face and $K^{i-1} \cap K_i$ is a connected subspace of the frontier of d_i. This process terminates when all members of S are embedded in $K = \cup_{j=1}^{q} K_j$. The regional cycles of K are independent and form a cycle basis of S, since K is collapsible into a point, (ref. 3).

EXAMPLE: The space frame of Fig. 2a is embedded on three disks as shown in Fig. 5.

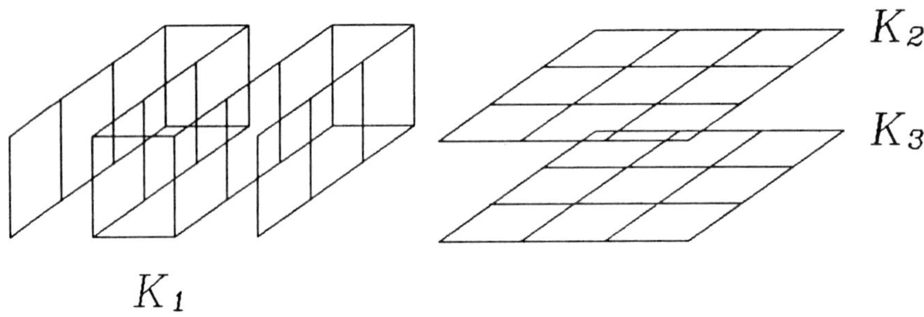

Fig. 5. S embedded on the union of three disks.

20. It is ideal to embed S in a K with minimum number of disks, in order to have smaller overlaps for the cycles. This number is known as the thickness of S, denoted by t(S). Again only bounds are known for thickness of special graphs.

AN EXPANSION PROCESS FOR CYCLE BASES SELECTION

21. Embedding graphs on an orientable surface and union of disks are theoretically attractive, however, in practice the computer automation of these approaches is not simple, if not impossible. In order to overcome the difficulties involved, an expansion process has been adopted (ref. 4), which starts from a specified node and generates connected cycles of the least length on unused members. Admissibility of a typical cycle C_{k+1} is provided by imposing restrictions on $C^k \cap C_{k+1}$ being a path, here $C^k = \cup_{i=1}^{k} C_i$. This algorithm is order dependent like many other combinatorial algorithms. Further modifications have been made by allowing the formation of disjoint cycles, (ref.5). In this method the following expansion process is employed:

$$C_1 = C^1 \rightarrow C^2 \rightarrow C^3 \rightarrow \ldots \rightarrow C^{b_1(S)} = S \tag{8}$$

A cycle C_{k+1} is called an admissible cycle if for $C^k \rightarrow C^k \cup C_{k+1}$,

$$b_1(C^{k+1}) = b_1(C^k) + 1 \tag{9}$$

The process terminates when $b_1(S)$ independent cycles are formed. This method has been implemented and efficient generation of minimal cycle bases has become feasible, (ref. 6).

EXAMPLE: Consider S as shown in Fig. 6, for which $b_1(S) = 8$. The following cycles are selected as a basis:
C_1 = (1,2,3), C_2= (1,8,9), C_3= (2,6,3), C_4= (2,5,6), C_5=(1,4,5,2) C_6= (1,7,5,2), C_7 (8,6,2,1), and C_8= (10,8,6,3,11). No embedding could be considered to lead to such a cycle basis with length equal to 29. Although a polynomial time algorithm has recently been developed for the formation of minimal cycle bases (ref. 7). However, the complexity of this method is much higher than that of ref. 6.

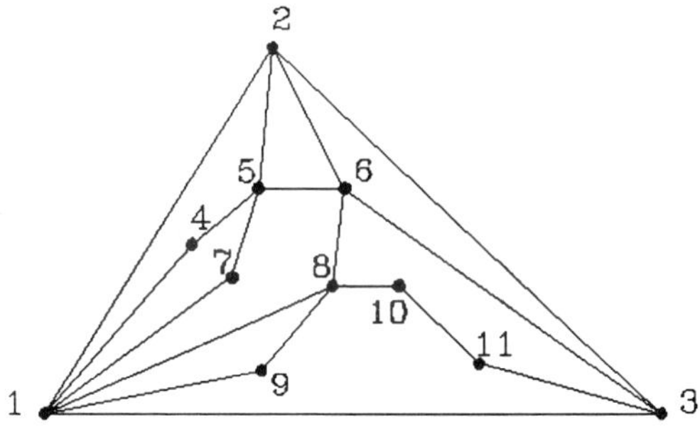

Fig. 6. A simple graph S.

22. In the above methods, the total length of the cycles of a basis is reduced in order to decrease the overlaps of the cycles. However, such a basis, in general, does not lead to a cycle adjacency matrix D of maximal sparsity. A direct method for maximising the sparsity of D has been developed in refs 8-9. For this purpose two weights are assigned to the members of S corresponding to the length and the number of minimal cycles which can be generated on each member of S. In the course of selecting cycles of minimal length, additional restrictions are imposed in order to generate cycles of smallest overlaps. In this method

additional measures can be assigned to the members of the structure, representing their flexibility or stiffness, and minimal measure cycles can then be selected to improve the conditioning of the flexibility, (ref. 10).

23. Once an optimal or sub-optimal cycle basis of S is generated, matrix B_1 corresponding to highly sparse flexibility matrix G can be formed. The bandwidth of G can also be reduced by ordering the selected cycles using the algorithms of ref. 1.

CONCLUDING REMARKS

24. It is shown how four topological invariants, namely the first Betti number, crossing number, genus and thickness of a graph are involved in the force method of structural analysis. Any advances in the direction of systematic computation of these invariants in the field of graph theory will naturally influence the efficiency of the force method. The advantage of the author's expansion process lies in its simplicity for computer programming and capability of including material and metrical properties, for improving the conditioning as well as the sparsity of flexibility matrices.

REFERENCES

1. KAVEH A. Structural mechanics : graph and matrix methods. Research Studies Press, Exeter, 1992.

2. WHITE A.T. and BEINEKE L.W. Topological graph theory. In selected topics in graph theory (eds Beineke and Wilson), 1978.

3. HENDERSON J. C. de C. and MAUNDER E.A.W. A problem in applied topology. J. Inst. Math. Applics, 1969, vol.5, 254-269.

4. KAVEH A. Application of topology and matroid theory to the analysis of structures. Ph.D. thesis, IC, Lond. Univer. 1974.

5. KAVEH A. Improved cycle bases for the flexibility analysis of structures. Computer Meth Appl. Mech. Engng. 1976,vol.9,267-272.

6. KAVEH A. An efficient program for generating sub-minimal cycle bases for the flexibility analysis of structures. Commun. Appl. Numer. Methods. 1986, vol.2, 339-344.

7. HORTON J.D. A polynomial time algorithm to find the shortest cycle basis of a graph, SIAM J. Comput., 1987, vol.16, 358-366.

8. KAVEH A. Sub-optimal cycle bases of graphs for the flexibility analysis of skeletal structures. Comput.. Meths Appl. Mech. Engng. 1988, vol.71, 259-271.

9. KAVEH A. Recent developments in the force method of structural analysis, Appl. Mech. Review, 1992, vol. 45, 401-418.

10. KAVEH A. Optimising the conditioning of structural flexibility matrices, Comput. Struct., 1991, vol.41, 1217-1223.

21. On parameter investigations of instabilities in reticulated space frames

A. ERIKSSON and C. PACOSTE, Royal Institute of Technology, Sweden

SYNOPSIS. Reticulated space frames are typical examples of optimally designed structures with respect to load carrying capacity. In connection with such systems the main problems seem to be related to the instability phenomena, which may occur due to their extreme slenderness and are directly associated with structural failure. Traditional analyses of these structures are based on continuation methods with respect to the main load parameter. The obtained equilibrium path corresponds to a certain section of a much more complex equilibrium manifold in a multi-parametric space and hence, is of limited relevance for an accurate prediction of the safety margin against an instability failure. The proper analysis procedure must include the parameter sensitivity aspects, preferably in an automatic manner. A numerical technique for this purpose can be based on systematic cutset evaluations, with particular emphasis on finding fold lines for the given problem. As a result, a good image of the dependence of critical behaviour on a chosen set of geometrical, mechanical and loading parameters will be obtained. Compared to traditional methods, this will improve the understanding of structural behaviour in both qualitative and quantitative terms. In the course of this paper, the usefulness of the proposed concept and method will be illustrated for the case of node instability in triangulated barrel vaults.

NODE INSTABILITY IN TRIANGULATED BARREL VAULTS

1. The simplified model used in order to investigate the node instability in triangulated barrel vaults is obtained by isolating a hexagonal cell from the real structure (refs 1 and 2), as shown in Fig. 1a. The connections of the bars in the nodes of the cell are assumed to be perfect hinges and the angle α is assumed to be small enough, such that: $\sin \alpha = \tan \alpha = \alpha$ and $\cos \alpha = 1$. In order to determine the equation which describes the behaviour of the model, the loads are assumed to be applied in two steps. The first loading step (a) is defined in Fig. 1c. The load P_r, which acts in the central node, is determined in such a way that the model undergoes radial displacements only. The equation connecting P and P_r may be obtained by expressing the equilibrium condition at node (1) in the radial direction, which gives:

$$P_r = P\frac{f - w_i}{f + w_i} \tag{1}$$

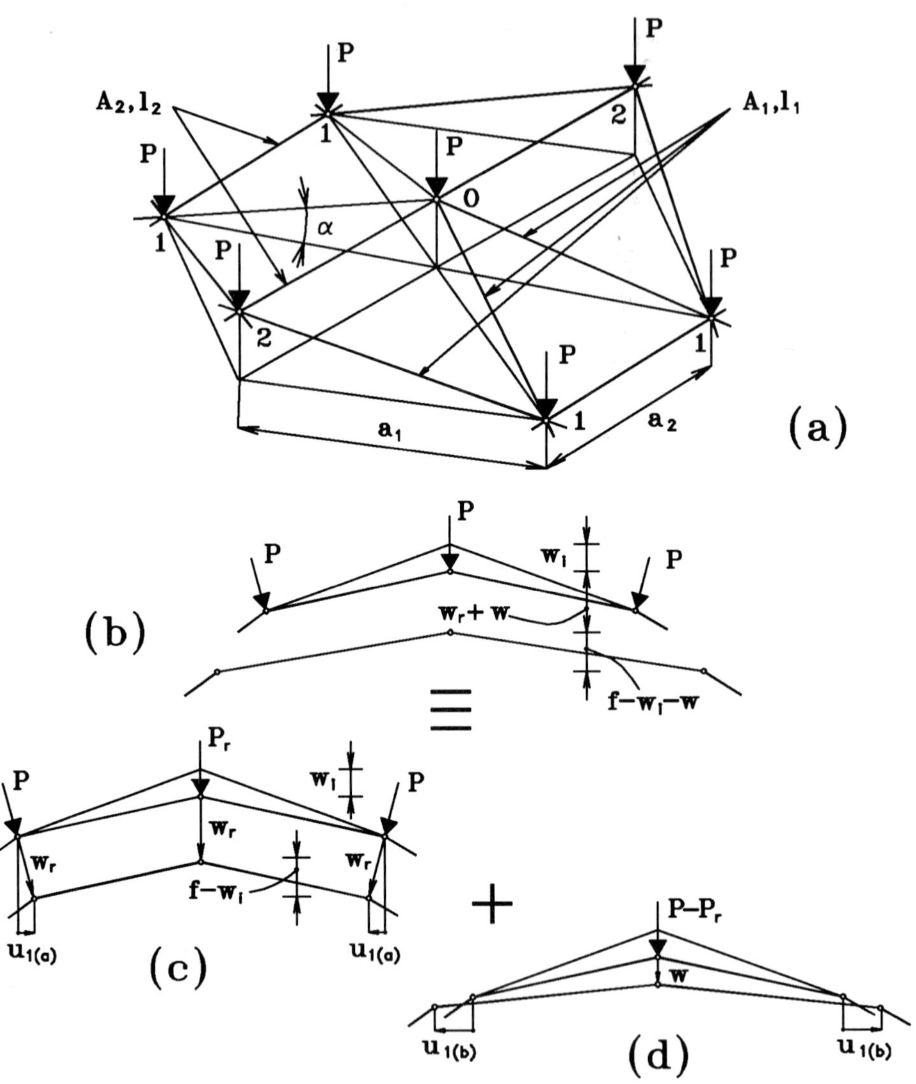

Fig. 1. Geometrical model for analytical treatment of node instabilities.

Further, the equilibrium equation at node (O) (radial direction) yields the following result:

$$\Delta l_{1(a)} = u_{1(a)} = -\frac{Pl_1^2}{4EA_1(f+w_i)} \tag{2}$$

where $l_{1(a)}$ denotes the axial shortening of the bars 0-1 and $u_{1(a)}$ denotes the horizontal displacements of nodes (1), both quantities corresponding to the loading step (a).

2. The second loading step (b) is defined in Fig. 1d. The load P - P_r, applied in the central node will generate the horizontal displacements $u_{1(b)}$ and $u_{2(b)}$ at nodes (1) and (2) and the radial displacement w at node (O). The horizontal displacements $u_{1(b)}$ and $u_{2(b)}$ depend on the relative stiffness of the hexagonal cell with respect to the surrounding structure. This dependence may be expressed by means of the coefficients:

$$k_1 = -\frac{u_{1m}}{u_{1s}} \quad \text{and} \quad k_2 = -\frac{u_{2m}}{u_{2s}} \tag{3}$$

where the meaning of the quantities u_{1m}, u_{1s}, u_{2m} and u_{2s} is shown in Fig. 2.

3. The axial shortening of the bars 0-1 in the loading step (b) is given by:

$$\begin{aligned} \Delta l_{1(b)} &= \sqrt{a_1^2 + (f - w_i - w)^2} - l_1 + u_{1(b)} \simeq \\ &\simeq -\frac{l_1}{2}\frac{w(2f - 2w_i - w)}{a_1^2} - \frac{\Delta l_{1(b)}}{k_1} \end{aligned} \tag{4}$$

and further:

$$\Delta l_{1(b)} = -\frac{1}{1+\frac{1}{k_1}}\frac{l_1}{2}\frac{w(2f - 2w_i - w)}{a_1^2} \tag{5}$$

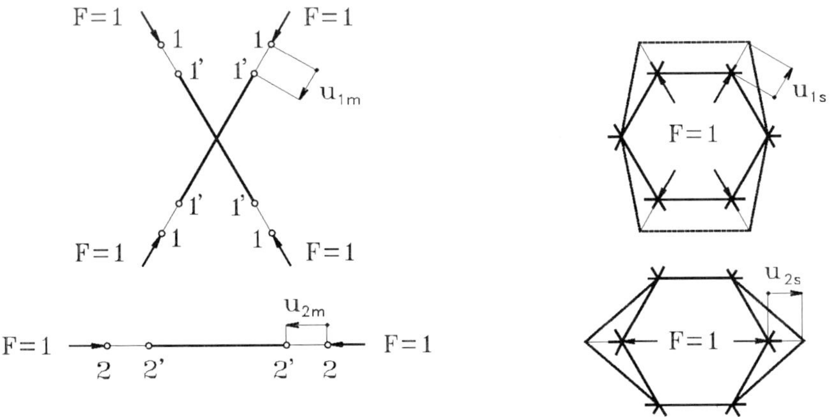

Fig. 2. Displacement quantities in derivation of hexagonal cell behaviour.

4. In a similar manner, the axial shortening for the bars 0-2 is obtained in the form:

$$\Delta l_{2(b)} = \frac{1}{1+\frac{1}{k_2}} \frac{l_2}{2} \frac{w^2}{a_2^2} \tag{6}$$

The total axial forces in the bars 0-1 are given by:

$$N_{01} = N_{01(a)} + N_{01(b)} = \frac{EA_1}{1+\frac{1}{k_1}} \frac{w(2f-2w_i-w)}{2a_1^2} - \frac{Pl_1}{4(f+w_i)} \tag{7}$$

and for the bars 0-2 by:

$$N_{02} = N_{02(a)} + N_{02(b)} = \frac{EA_2}{1+\frac{1}{k_2}} \frac{w^2}{2a_2^2} \tag{8}$$

The equilibrium condition at node (0) expressed in the radial direction gives:

$$P = -4N_{01}\frac{f-w_i-w}{a_1} + 2N_{02}\frac{w}{a_2} \tag{9}$$

which, by making use of Eqs 7 and 8, becomes:

$$2\frac{EA_{1r}}{a_1^3} f^3 \bar{w}(1-\bar{w}_i-\bar{w})(2-2\bar{w}_i-\bar{w}) + \frac{EA_{2r}}{a_2^3} f^3(\bar{w})^3 - P(2\bar{w}_i+\bar{w}) = 0 \tag{10}$$

where:

$$EA_{1r} = \frac{EA_1}{1+\frac{1}{k_1}} \; ; \; EA_{2r} = \frac{EA_2}{1+\frac{1}{k_2}} \; ; \; \bar{w} = \frac{w}{f}$$

For the case of a perfect model ($w_i = 0$), the equilibrium path exhibits an asymmetric bifurcation point while the imperfect systems ($w_i \neq 0$) lose their stability at limit points, as illustrated in Fig. 3a ($\bar{P} = P / \left(\left(EA_{1r}/a_1^3\right)f^3\right)$).

5. If the loads at nodes (1) and (2) are maintained at a constant level $P_0 < P_{cr}$ and the load in the central node (O) is assumed as a variable (ref. 3); Eq. 10 becomes:

$$2\frac{EA_{1r}}{a_1^3} f^3 \bar{w}(1-\bar{w}_i-\bar{w})(2-2\bar{w}_i-\bar{w}) + \frac{EA_{2r}}{a_2^3} f^3(\bar{w})^3 + P_0(1-2\bar{w}_i-\bar{w}) - P = 0 \tag{11}$$

For various values of the parameter P_0, Eq. 11 yields the equilibrium curves in Fig. 3b, while the equilibrium curves in Fig. 3c are obtained for a fixed P_0 on varying the values of the ratio:

$$\frac{EA_{1r}}{\;_{2r}}\frac{a_1^3}{a_2^3}$$

Finally, an interact... ...nd P is represented in Fig. 3d.

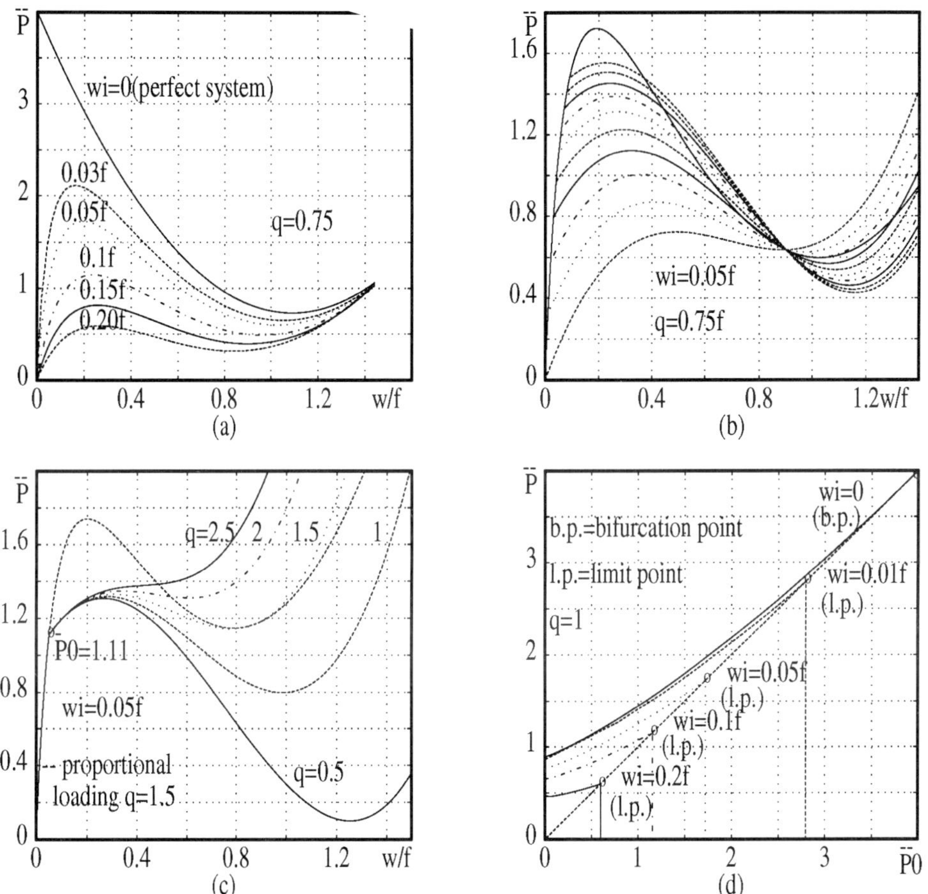

Fig. 3. Equilibrium paths for the geometrical model.
a) Proportional loading. b) Independent load variation. c) Influence of the longitudinal bars. d) P-P_0 interaction diagram.

FOLD LINE CONCEPT

6. Although the analysis of node instability, put forth in the previous section, offers a good qualitative description of the phenomenon, its relevance for full

scale structures is certainly reduced from a quantitative point of view. In order to overcome this inconvenience, the adoption of a numerical approach is necessary. Since such an approach must include the parameter sensitivity aspects outlined in Fig. 3, it follows that a traditional continuation method is not suitable. Instead, an approach based on the fold line concept will be introduced in this section.

7. Fold line evaluation is a method to quantitatively describe the dependence of critical equilibrium states on the variations of the parameters defining the analysis model. In order to obtain this more complex model description, a formulation is used according to

$$\underline{F}(\underline{d}, \underline{\Lambda}) = \underline{o} \tag{12}$$

introducing p parameters $\underline{\Lambda} = (\lambda_1,...\lambda_p)^T$, rather than just one load factor. The added parameters can be load factors or geometrical parameters. This means that the equilibrium equations can be described in either of the forms;

$$\underline{f}(\underline{d}) - \underline{p}(\underline{\Lambda}) = \underline{o} \tag{13a}$$

$$\underline{f}(\underline{d}, \underline{\Lambda}^1) - \underline{p}(\underline{\Lambda}^2) = \underline{o} \tag{13b}$$

relating internal and external forces. In the latter case, the parameters $\underline{\Lambda}$ are split into two sets. The description in Eq. 12 will lead to higher order solution spaces, for instance, with two control parameters, to continuous surfaces in an $(n + 2)$ dimensional space. With the terminology from Sewell (ref. 6), the total solution space will be a p-dimensional 'equilibrium surface'.

8. In the obtained equilibrium surface, certain sections are interesting. A common continuation algorithm will, for instance, find the set of states on the equilibrium surface, that have certain values for the added parameters. Also other cutsets of the total equilibrium space are, however, interesting. The formulation for the sought cutset can be written, (ref. 5), as:

$$\underline{G}(\underline{d}, \underline{\Lambda}) = \underline{o} \tag{14a}$$

where the set

$$\underline{G}(\underline{d}, \underline{\Lambda}) \equiv \begin{pmatrix} \underline{F}(\underline{d}, \underline{\Lambda}) \\ \underline{g}(\underline{d}, \underline{\Lambda}) \end{pmatrix} \tag{14b}$$

demands fulfilment of both equilibrium and some added relations. If p - 1 such relations are added, the solution to Eq. 14 will be a set of curve segments in the (n + p), dimensional space.

9. Fold lines correspond to such curve segments, connecting all critical states in the structural behaviour, for varying values of the parameters. The fold lines are the intersections of the equilibrium subspace, and the subspace of states where the

tangential stiffness matrix is singular. Introducing the augmented problem into a continuation algorithm, will give a method that follows the critical boundaries of the problem, in the higher dimensional space. As these boundaries will include the stability boundaries (ref. 6), the additional information concerning the studied model, is valuable for the analysis of imperfection sensitivity. By evaluation of the fold lines, the critical combinations of parameters will thus be obtained. In the present context, the most interesting result is a description $\lambda_1^c = \lambda_c^1(\lambda_2, \ldots \lambda_p)$ giving the critical load level in the main load case as a function of the added parameters. In particular, turning points of the obtained fold lines, will show values for the added parameters where certain critical states disappear, or where several criticalities interact.

10. A fold line can be defined in several ways, using the equations $\underline{\mathbf{g}}$ to define criticality. Many such definitions have been proposed for the general case. The methods differ in the used characterisation of the criticality, and also in the dimension of the extended systems. A few of these, relevant for the studied class of problems, are discussed and evaluated in ref. 7; a method by Dai (ref. 8), is seen to be a good formulation.

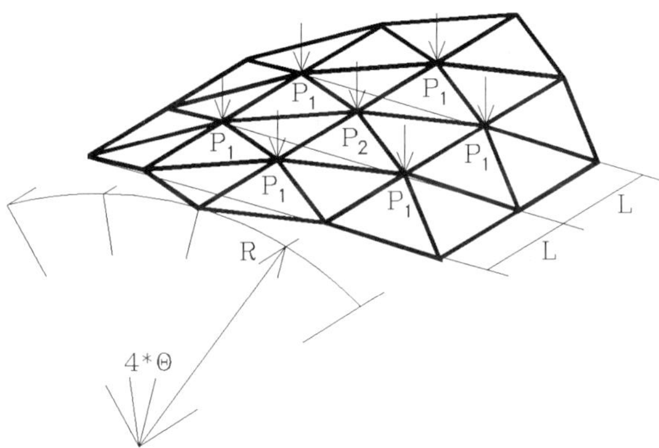

Fig. 4. Simplified model for analysis of node instability in a reticulated vault frame. The figure corresponds to Model 2; Model 1 contains the middle seven nodes.

ILLUSTRATIVE EXAMPLES

11. The case in Fig. 2 has been studied with a fold line formulation. A number of calculations have been performed; in this paper only two illustrative examples are included. The calculations have been based on models according to Fig. 4. Model 2 is the one shown in the figure, with fixed supports at the 12 outer nodes. Model 1 is the inner part, supported at the six nodes, marked P1 in the figure.

12. The main load case is chosen as a case of equal vertical loads in all free nodes. The basic model was described by the data: L = 3m, R = 11.417m, θ =

13.16°, $E = 2.06x10^{11}N/m^2$, $A_l = 12.3x10^{-4}m^2$. In the demonstrated case, only the area A_2 of the longitudinal bars was varied. The results are shown in Fig. 5, model 1, and Fig. 6, model 2.

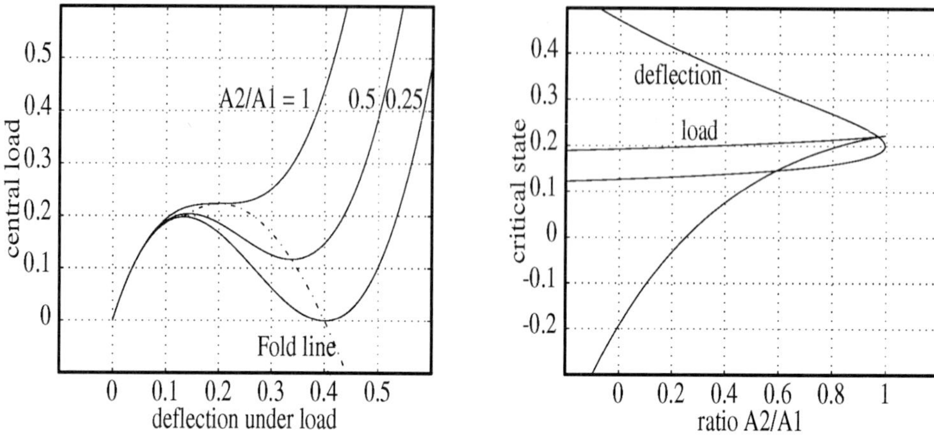

Fig. 5. Results from the case in Fig. 4; Model 1. Results for central vertical load, with various ratios between longitudinal and transversal member areas. a) Vertical load vs deflection, b) critical states vs area ratio.

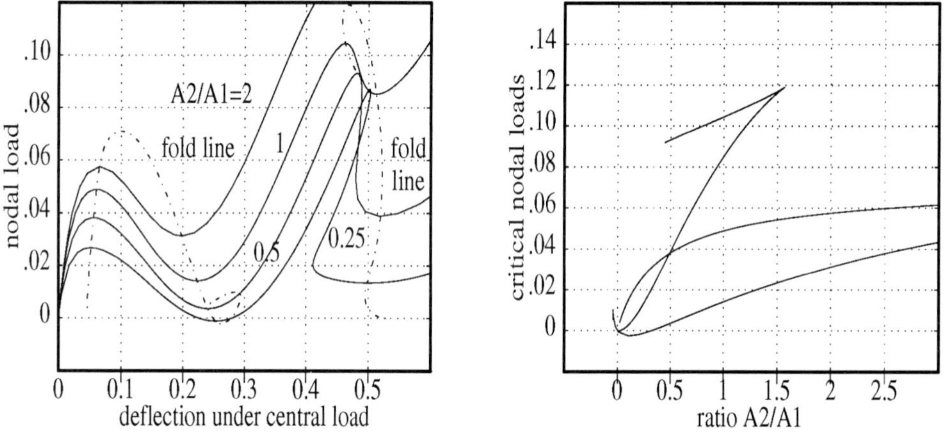

Fig. 6. Results from the case in Fig. 4; Model 2. Results for 7 equal vertical loads, with various ratios between longitudinal and transversal member areas. a) Vertical load vs deflection, b) critical loads vs area ratio.

13. Fig. 5 shows that for small values of A_2 the model will show a snapping behaviour, with one maximum and one minimum load. For a larger value, this snapping behaviour will disappear. The positions of the critical states are shown

in the left sub-figure. The loads and deflections of the critical states are shown as functions of the ratio A_2/A_1 in the right sub-figure. By the fold line evaluation, the snapping behaviour can be shown to disappear at an area ratio of 0.9941.

14. For model 2, similar results are shown in Fig. 6. With seven equal vertical loads, two snapping states are shown to exist for small area ratios. The fold line evaluations show that the second snapping state disappears for a ratio of 1.566. As seen from the left sub-figure, also the first snapping state will disappear for a certain area ratio; this ratio is found to be 7.912. For extreme values of the ratios, the fold lines show very complex shapes; this corresponds to complex instability behaviour, without practical relevance.

CONCLUSIONS

15. The diagrams plotted in Figs 3, 5 and 6 outline the following conclusions. The effect of geometrical imperfections on the values of the critical load is significant (Fig. 3a). Moreover, the effect of geometrical imperfections must not be accounted for by proportionally increasing the loads and performing a stability test for the perfect structure at the new loading level. As shown in Figs 3b and d such a procedure may lead to incorrect conclusions.

16. The bars in the longitudinal direction have a significant influence on the stability of the node, as illustrated in Figs 3c and 6. The simplified analysis suggests that, for a value of the ratio $A_2/A_1 > 2.5$, an unstable behaviour can be avoided. However, the numerical analysis shows that this prediction is too optimistic and just for much higher ratios A_2/A_1 the snapping state will disappear. This discrepancy emphasises the necessity of a numerical approach whenever reliable numerical conclusions are required.

17. This type of numerical calculations can be based on a fold line evaluation method. This demands a numerical formulation based on a multi-parametric structural model. The fold lines can be followed by use of common continuation algorithms; more efficient algorithms can, however, be developed by utilising the special properties of the equations.

18. In addition to giving imperfection sensitivity measures, the fold lines can show parameter ranges where certain instabilities exist, and parameter combinations where instabilities interact. However, this does not reduce the necessity for a theoretical analysis in order to decide on the relevant parameters. Research is in progress concerning methods to find these parameters from, for instance, catastrophe theoretical ideas.

REFERENCES

1. LIND, N. C. Stability criteria for reticulated shells, Proc. IASS Congress, Leningrad, 1, 444-460 (1966).

2. JURAVLEV, A. A. O mestnoi ustoicivosti setciatih kupolov s trengolnii resetkoi, IVUZ, Stroit, i arh, 5, 77-100 (1971).

3. PACOSTE, C. C. Noi tipuri de solutii constructive pentru structuri reticulate planare si curbe, Dept. Theoret. and App. Mech., Civil Eng. Inst., Bucharest, 1989.

4. PACOSTE, C. C. On the Application of Catastrophe Theory to Instability Analyses of Elastic Structures, Dr. thesis, Dept. Struct. Engng, Royal Inst. Techn., Stockholm, 1993.
5. ERIKSSON, A. Cutsets and augmented equilibrium equations for multi-parametric structural analysis models, Dept. Struct. Mech., Royal Inst. Techn., Stockholm, 1993.
6. SEWELL, M. J. On the connection between stability and the shape of the equilibrium surface, J. Mech. Phys. Solids, 14, 203-230 (1966).
7. ERIKSSON, A. Fold lines for sensitivity analyses in structural instability, Dept. Struct. Engng, Royal Inst. Techn., Stockholm, 1992.
8. DAI, R.-X. Characterisation and Computation of Fold sets for Parameter dependent Equations, Ph.D. thesis, University of Pittsburgh, 1988.

22. Load-carrying capacity of space trusses under the influence of imperfections

M. TADA and K. WAKIYAMA, Osaka University, Japan

SYNOPSIS. Some imperfections such as errors in setting levels of the supports may cause a reduction in the load-carrying capacity of ordinary double-layer space trusses. Schmidt *et al* have already proposed the idea that the constant load-carrying capacity can be obtained by incorporating force limiting devices (FLDs) into trusses (ref. 1). This paper deals with the combined non-linear analysis to evaluate the reduction of the load-carrying capacity for ordinary trusses. Moreover, it is confirmed that the constant load-carrying capacity can be obtained for FLD trusses regardless of the influence of imperfections.

INTRODUCTION

1. The ultimate load-carrying capacity of double-layer space trusses is generally determined by the buckling of a compressive member. Therefore, the imperfections during the construction, such as errors in the member-length or in setting levels of supports, may cause a reduction in the ultimate load-carrying capacity of the truss (ref. 2). In this respect, Schmidt *et al* have already proposed the idea that a constant load-carrying capacity can be obtained by incorporating force limiting devices (FLDs), which would yield under a predetermined axial member-force (Fig. 1), into the compressive members (ref. 1). A study on force limiting devices has been developed by Parke (ref. 3).

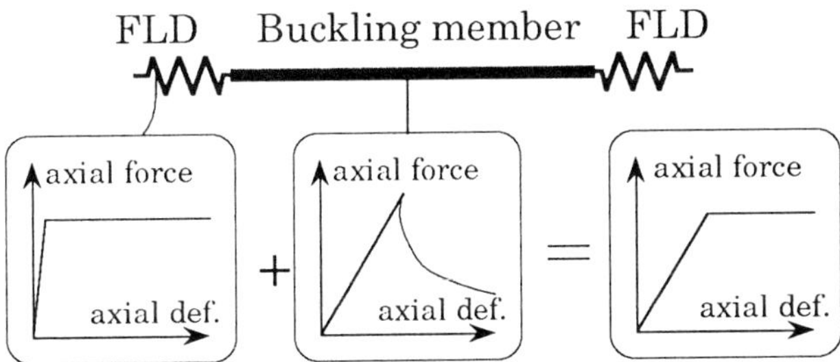

Fig. 1. Explanation of FLD.

However, a quantitative consideration related to the amount of the reduction has not been extensively discussed.

2. This paper deals with the combined non-linear analysis of a double-layer space truss undertaken to evaluate the reduction of the load-carrying capacity caused by the influence of imperfections. The imperfections applied are the length of members, the setting level of supports and the combination of such imperfections. The trusses incorporating force limiting devices are also analysed to confirm that a constant load carrying capacity can be obtained regardless of the influence of imperfections.

DOUBLE-LAYER SPACE TRUSS TO BE CONSIDERED

3. The double-layer space truss to be considered is simply supported at all nodes along the perimeter (Fig. 2). The number of bays is seven by seven. The members are pin-connected circular tubes with a uniform length. The members are so designed that the normal stress may not exceed the temporary allowable stress (235MPa) under the vertical load of 4.80kN/m^2 (permanent load: 0.39kN/m^2, snow load: 4.41kN/m^2).

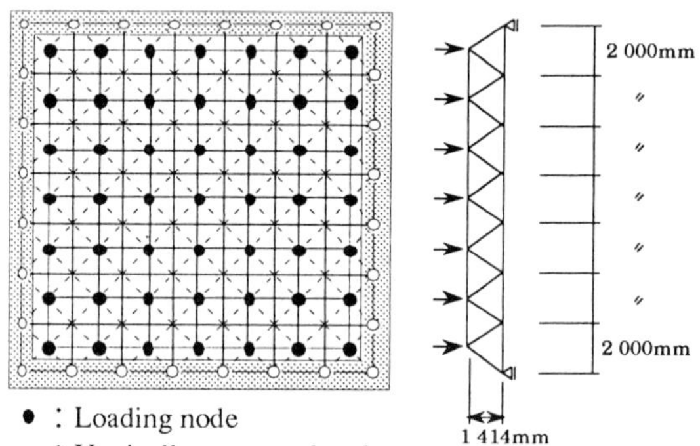

Fig. 2. Space truss to be considered.

Table 1. Schedule of members.

Sign	$D \times t$ (mm)	A (cm^2)	I (cm^4)	λ	$_cN_A'$ (ton)	N_k^J (ton)	N_k (ton)
– – –	56.8ϕ x 1.71	2.956	11.21	103	3.78	4.93	4.99
——	66.6ϕ x 2.01	4.07	21.26	88	6.22	7.60	8.14
▬ ▬ ▬	74.9ϕ x 2.25	5.146	33.99	78	8.65	10.2	11.0
▬▬▬	82.3ϕ x 2.48	6.205	49.41	71	11.1	12.7	13.6

D : Diameter, t : Thickness, A : Sectional area, I : Geometrical moment of inertia,
λ : Slenderness ratio, $_cN_A'$: Temporary allowable compressive force,
N_k^J : Johnson's buckling force, N_k: Buckling force under the initial deflection of l / 1000

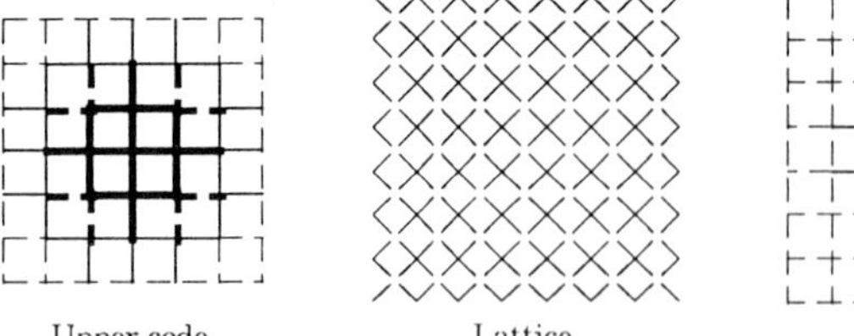

Fig. 3. Layout of members.

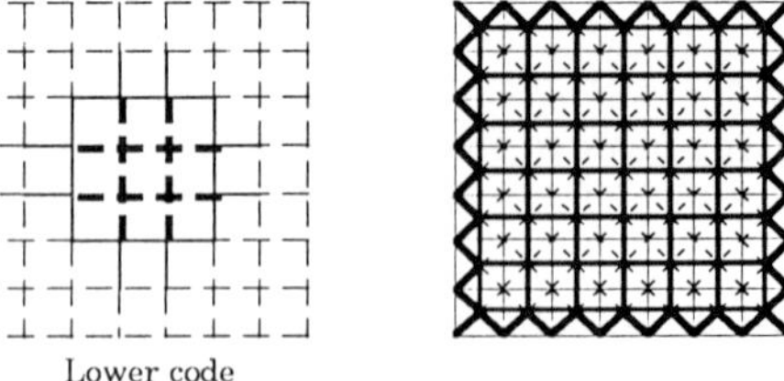

Fig. 4. Layout of FLDs.

The calculation of the allowable stress conforms to the standard of the Architectural Institute of Japan (ref. 4). The members are divided into four groups depending on the level of the stress. The layout of the members is shown in Fig. 3. The schedule of the members is shown in Table 1.

4. All members of the FLD truss are identical with those of the ordinary truss. The FLDs are incorporated into the members whose compressive force is dominant; these are all upper code-members and the lattice-members at the perimeter zone (the members shown by bold lines in Fig. 4). As the FLD should exhibit a stable yield to prevent the member from buckling, the limiting force of the FLD should be sufficiently small in view of the influence of the initial imperfections of the member and in view of the fluctuation of the limiting force of the FLD. But too conservative a setting of the limiting force is clearly uneconomical. In this paper, it is postulated that the members are considered not to buckle up to the temporary allowable compressive force and that the fluctuation of the limiting force of the FLD is negligible. From the above postulation, the limiting force of the FLD is set at the level of the temporary allowable compressive force.

5. The combined non-linear analysis using the strain hardening general yield hinge model and using the incremental perturbation technique (ref. 5) is adopted. Vertical loads of the same magnitude are proportionally applied to all the upper nodes. The plasticity of the section is considered to occur only at the mid point of each member. The post-buckling behaviour can be analysed through the accuracy of the geometrical non-linearity. The initial transverse deformation of the member is given as an amount equivalent to 1/1000 of the member-length (ℓ), because the buckling force of the member calculated under the initial deformation of $\ell/1000$ corresponds well with that given by Johnson's estimation (compare N_k^J and N_k in Table 1). The bi-linear stress strain relation is postulated in the analysis. The yield stress is 235Mpa and the strain-hardening coefficient is 1%. The perfectly rigid-plastic relation is postulated for the FLDs.

IMPERFECTIONS TO BE CONSIDERED

6. Errors in the setting level of supports and errors in the member-length are introduced as the imperfections during the construction. A level error in the support is introduced at support A in Fig. 5 to which the member having the biggest compressive force of all lattice-members is connected. Errors of the member-length are introduced to the upper chord member B and B', which have the biggest compressive forces of all the members, as well as into member C which is located next to the member B.

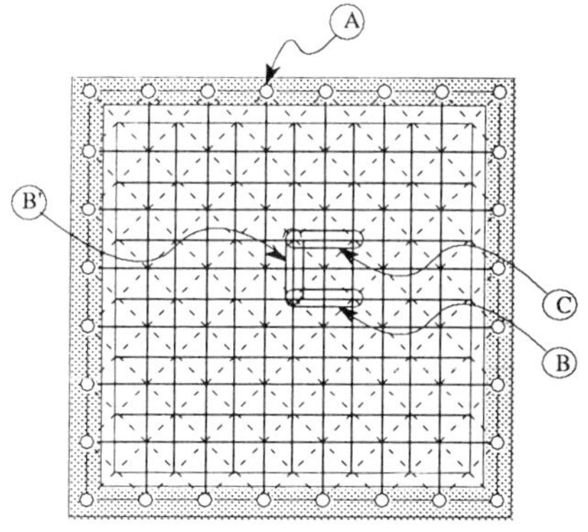

Fig. 5. Explanation of imperfections.

The imperfections are:

A :the case where the support A is set at a level 3mm higher than the level of the other supports;

B :the case where the member B is 3mm longer than the standard length;

B' :the case where the member B' is 3mm longer than the standard length;

C :the case where the member C is 3mm longer than the standard length;

The maximum allowable error in the formation level of a column base plate and the maximum allowable error in the length of the beam and column are specified as 3mm, according to Japanese Architectural Standard Specification for Steelwork (ref. 6). The amount of 3mm is determined by these specifications.

7. In the case where no imperfection exists, the axial forces in the members, which are located at the symmetrical positions, simultaneously reach their buckling forces because symmetrical loads are applied to the symmetrical truss. In practice, however, a perfectly symmetrical state does not exist and a member whose stress is more or less bigger than another initiates the buckling. In the present analysis, when the stresses of some of the members simultaneously reach the yield stress, the equilibrium path is calculated by introducing plastic hinges, one by one into these members. It is therefore to be noted that "the perfect truss" in this paper does not strictly mean "perfect".

RESULT OF ANALYSIS AND DISCUSSION

8. The analytical results of the load-deflection relations are shown in Figs 6a to 6g. The figures show the cases where no imperfection exists, where the single imperfection A, B or C exists and where the combined imperfection A+B, A+B' or B+C exists. The solid lines show the results for the ordinary truss and the dotted lines show the results for the FLD truss. The ordinates (V_n) show the vertical load divided by the design-load. The abscissas (D_n) show the average vertical deflection of all upper nodes divided by the deflection when the design-load is applied.

9. The load-deflection relations for the ordinary truss show zigzag lines. Each peak corresponds to a time when a member buckles. Both the load and the deflection decrease after the buckling. The reason is that the analysis here deals with the static equilibrium. In practice, if more load is applied or if the deflection is continuously increased, the truss will be dynamically balanced by the introduction of the inertia force.

10. The values denoted by ∇ in Fig. 6 show the maximum loads for the ordinary trusses and the loads at the end of the analysis ($D_n = 2.0$) for the FLD trusses. The load-carrying capacities of the perfect ordinary truss and the perfect FLD truss are almost the same as each other (Fig. 6a). Although Schmidt *et al* mentioned that a considerable increase in ultimate load-carrying capacity is expected using FLD (ref. 1), only a little increase is obtained in this example. In fact, the amount of increase depends on the limiting forces of the FLDs and the amount of allowance of the ultimate axial force of each member.

11. In the cases where the imperfections exist, the load-carrying capacities of the ordinary trusses are reduced by 10 to 15% from that of the perfect truss.

Although the load-levels, when a member initially buckles, are different from each other among the cases, the maximum load-levels are almost the same. Moreover, the differences of the maximum load-levels are not recognised between the cases of the single imperfection and those of the combined imperfections. The constant load carrying capacity is obtained for the FLD trusses regardless of the influence of imperfections. The efficiency of the FLD truss that Schmidt et al. proposed is evaluated quantitatively by these results.

12. Fig. 7 shows the maximum deformation ($_mu^P$) of the FLD at the end of the analysis. The required deformation of the FLD cannot be generalised from these results. As far as these examples are concerned, it can be said that about 10mm deformation is sufficient as a required deformation for the FLD.

13. Figs 8a to 8g show the plastic members at the maximum load and at the end of the analysis. The bold solid lines show the buckled members and the bold dotted lines show the tensile yielding members.

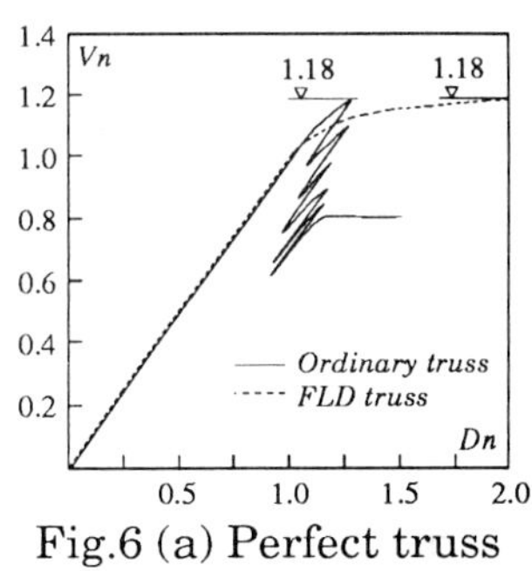

Fig.6 (a) Perfect truss

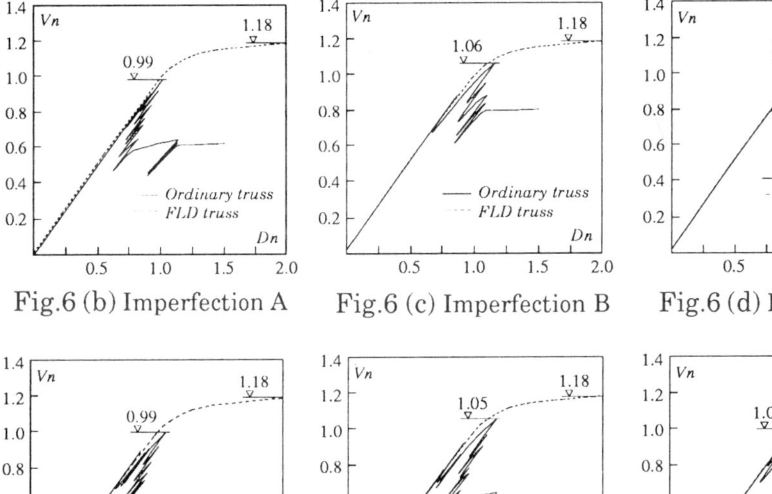

Fig.6 (b) Imperfection A

Fig.6 (c) Imperfection B

Fig.6 (d) Imperfection C

Fig.6 (e) Imperfection A+B

Fig.6 (f) Imperfection A+B'

Fig.6 (g) Imperfection B+C

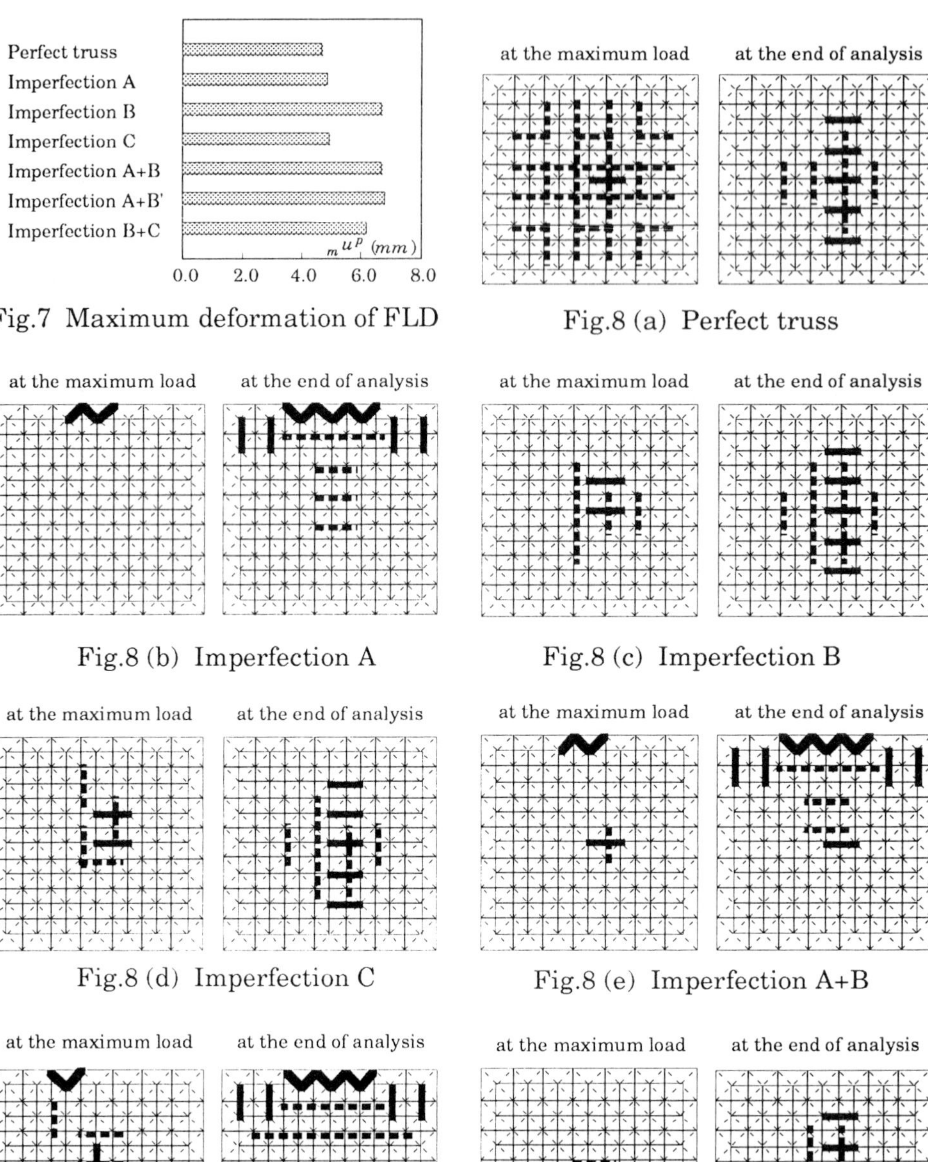

Fig.7 Maximum deformation of FLD

Fig.8 (a) Perfect truss

Fig.8 (b) Imperfection A

Fig.8 (c) Imperfection B

Fig.8 (d) Imperfection C

Fig.8 (e) Imperfection A+B

Fig.8 (f) Imperfection A+B'

Fig.8 (g) Imperfection B+C

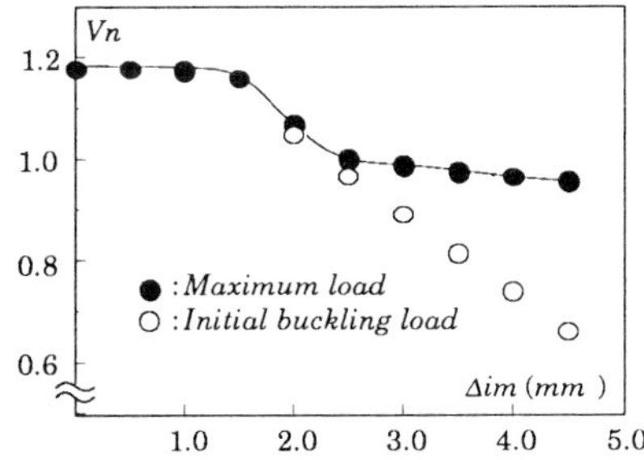

Fig. 9a. Imperfection A.

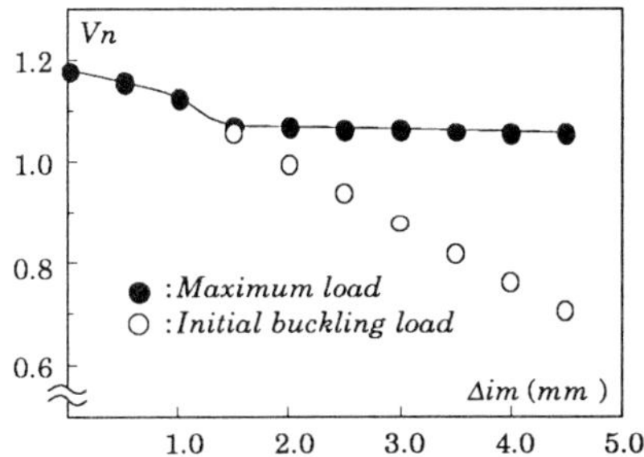

Fig. 9b. Imperfection B.

In the case of the perfect truss, the collapse mechanism is characterised by the formation of the yield line in the upper chord members across the truss (Fig. 8a). While the collapse mechanisms of the trusses including the imperfection B or C are the same with that of the perfect truss (Figs 8c, 8d and 8g), the imperfection A leads to a change in the mechanism (Figs 8b, 8e and 8f).

14. Figs 9a and 9b show the relation of the maximum load and initial buckling load versus the amount of the imperfection (Δ_{im}) for the ordinary trusses including the imperfection A or B. The sign ● denotes the maximum load and the sign O denotes the initial buckling load. The ordinate represents the non-dimensional load V_n. In the case of the imperfection B, even a small imperfection reduces the maximum load. But in the case of the imperfection A, the maximum load remains unaffected by the small imperfection (Δ_{im} = 0 - 1.5mm). This is because the member B is critical under the design load and the members connecting to support A have some amount of allowance. While the initial buckling load is proportionally reduced as the amount of the imperfection increases, the maximum load ceases to decrease at a certain level. It is because the axial force of the member suddenly weakens after buckling and it gradually approaches the constant post-buckling strength.

CONCLUSIONS

15. This paper deals with the combined non-linear analysis of double-layer space truss in order to evaluate the load-carrying capacity of the ordinary or FLD space truss under the influence of the imperfections. The primary conclusions obtained are as follows.

- The load-carrying capacity of the ordinary double-layer space truss is reduced by 10 to 15% under the influence of imperfections which are generally accepted in JASS6 (ref. 6).
- The idea of Schmidt *et al* that a constant load-carrying capacity can be obtained regardless of the influence of imperfections if FLDs are incorporated into ordinary trusses, is confirmed by the numerical analysis.
- The difference in the load-carrying capacity of the ordinary space truss is not dependent on a particular single or combined case of imperfection.
- The reduction in the load-carrying capacity of the ordinary space truss is not linearly dependent on the size of the imperfection.

ACKNOWLEDGEMENTS
The authors are grateful to Dr. Katsuhiko Imai of Kawatetsu Steel Products corporation for his discussion and to Graduate Student of Osaka University, Miss Mariko Hayashi for her help in the numerical analysis.

REFERENCES

1. SCHMIDT L. C. and HANAOR A. Force limiting devices in space trusses. Journal of the Structural Division ASCE, 1979, vol. 105, No. ST5, May, 939-951.
2. SCHMIDT L. C., MORGAN P. R. and CLARKSON J. A. Space trusses with brittle type strut buckling. Journal of the Structural Division ASCE, 1976, vol. 102, No. ST7, July, 1479-1492.
3. PARKE G. A. R. The behaviour of space trusses incorporating novel compression members. Thesis for the degree of Doctor of philosophy, University of Surrey, May 1988.
4. Design standard for steel structures, Architectural Institute of Japan, 1970.
5. TADA M., INOUE K. and KUWAHARA S. Application of perturbation techniques in the analysis of braced frames, sixth Canadian Conference on Earthquake Engineering, June 1991, 77-84.
6. Japanese architectural standard specification JASS6 Steel work, Architectural Institute of Japan, 1988.

23. Formulae for grids with warping torsion

D. L. DEAN, Dean & Associates, USA

SYNOPSIS. The rectangular grid is the simplest and most frequently used space structure both alone and as a component of a more complex structure, such as a waffle plate. Closed formulae have been derived for torsionless grids and for grids with simple St. Venant torsion by the micro or difference equation approach. Such an approach could also be used for the grid with warping torsion also, but is not available at this time. Here the macro or summation equation approach is used to write less complex and more useful formulae for the subject structure with both simple and flexible supports.

INTRODUCTION

1. The plane flexural grid with out-of-plane loading is the simplest structure within the space structures category and may be considered geometrically dull. However, it is frequently used for a wide variety of purposes including function as a component of a more complicated unit such as a waffle plate. It is usually regular and thus has attracted the attention of analysts deriving closed formulae for an arbitrary number of nodes in both directions. To date this treatment has been limited to torsionless grids and grids with simple St. Venant torsion, which omits the major torsional effects for many members. Existing grid formulae were derived by the micro discrete field approach. That is, general boundary solutions for the grid elements were used to write a difference equation model which, in turn was solved in closed form. Such an approach can also be used for the grid with warping torsion, but it is much more complicated and awkward to use than the alternate macro discrete field approach. That is, general load solutions for the grid beams (full length not just element between nodes) are used to write a summation equation model which is solved in a unified form with both continuous and discrete independent variables. The macro field method solves many problems intractable by the micro field method, e.g. those structures with elements of more than one dimension. A secondary goal here is to interest other analysts in applying this powerful, but rarely used method for structural analysis.

2. Because the macro field approach is less well known than alternate methods, the basic formulae are presented in the appendix. A simple extension of these

basic one-dimensional formulae and macro field concepts permits the rational model for the subject structure to be written almost by inspection, as will be demonstrated below.

DERIVATION OF GRID MODEL

3. A flexibility approach will be used. Three sets of interactive forces between the two sets of beams will be found. They are: 1) the out-of-plane interactive node forces, R(r,s), (acting up on beams in x and r direction and down on the beams in y and s direction); 2) the node couples, T(r,s), acting in the negative direction on the beams in x and r direction (and negative torques on beams in the y and s direction); and 3) the node couples, $\underline{T}$(r,s), acting in the negative direction on beams in the y and s directions (and positive torques on beams in x and r direction). The series for these forces are:

$$R(r,s) = \sum_{k=1}^{m-1} \sum_{l=1}^{n-1} R_{kl} \sin \lambda_k r \sin \lambda'_l s \qquad \lambda'_l = l\pi/n \tag{1a,b}$$

$$T(r,s) = \sum_{k=0}^{m} \sum_{l=1}^{n-1} T_{kl} \cos \lambda_k r \sin \lambda'_l s \tag{2}$$

$$\underline{T}(r,s) = \sum_{k=1}^{m-1} \sum_{l=0}^{n} \underline{T}_{kl} \sin \lambda_k r \cos \lambda'_l s \tag{3}$$

4. Thus, referring to Eqs A.4, A.6a and A.16a, one can easily write the continuous deflections for the beams in the x direction as:

$$w(x,s) = \sum_{i=1}^{\infty} \sum_{l=1}^{n-1} \left[(q_{il} - \frac{m}{a} R_{il}) \overset{*}{C}{}^4_i - \frac{m}{a} T_{il} \overset{*}{C}{}^3_i \right] \sin \alpha_i x \sin \lambda'_l s \tag{4}$$

in which R_{il} and T_{il} are sine-wise and cosine-wise cyclic with respect to the first index respectively. Then, referring to Eqs. A.8 and A.16, one can similarly write the node deflection for the beams in the r direction as:

$$w(\frac{a}{m} r,s) = \sum_{k=1}^{m-1} \sum_{l=1}^{n-1} \left[W^0_{kl} - R_{kl} C^4_k - T_{kl} C^3_k \right] \sin \lambda_k r \sin \lambda'_l s \tag{5}$$

in which the coefficients, W^0_{k1}, of the node deflections for the unrestrained beams are given by:

$$W^0_{kl} = \sum_{I=-\infty}^{+\infty} q_{2Im+k,l} \overset{*}{C}{}^4_{2Im+k} \tag{6}$$

Similar expressions for the continuous and discrete deflections, slopes, and twists of the beams in the y and s direction can be written by induction.

5. Simple algebraic equations for the unknown force coefficients, R_{kl}, T_{kl} and $\underline{T}_{kl}$ result from substitution of the deflection expressions into the following compatibility relations:

$$w(\tfrac{a}{m}\,r,s) = \underline{w}(r,\tfrac{b}{n}\,s),\quad \theta(\tfrac{a}{m}\,r,s) = -\underline{\varphi}(r,\tfrac{b}{n}\,s),\quad \underline{\theta}(r,\tfrac{b}{n}\,s) = \varphi(\tfrac{a}{m}\,r,s) \qquad (7a,b,c)$$

The resulting equations are:

$$\begin{bmatrix} A_{ij} \end{bmatrix} \begin{bmatrix} R_{kl} \\ T_{kl} \\ \underline{T}_{kl} \end{bmatrix} = \begin{bmatrix} W^{0}_{kl} - \underline{W}^{0}_{kl} \\ \theta^{0}_{kl} \\ \underline{\theta}^{0}_{kl} \end{bmatrix} \qquad (8a,b,c)$$

$$\begin{bmatrix} A_{ij} \end{bmatrix} = \begin{bmatrix} C^{4}_{k} + \underline{C}^{4}_{l} & C^{3}_{k} & -\underline{C}^{3}_{l} \\ C^{3}_{k} & C^{2}_{k} + \underline{C}^{t}_{l} & 0 \\ -\underline{C}^{3}_{l} & 0 & \underline{C}^{2}_{l} + C^{t}_{k} \end{bmatrix} \qquad (8d)$$

Eqs 8 complete the closed analysis of a simply-supported grid for both continuous and discrete independent variables using elementary beam bending and restrained warping torsion theories.

EXTENSION TO THE CASE WITH BOUNDARY DISPLACEMENTS

6. For an analysis to be non-trivial, it should be feasible to extend coverage to different boundary conditions. With a Fourier series, it is desirable to have techniques available to extend the solution to in homogeneous boundary conditions or unnatural homogeneous boundary conditions. This is usually possible with the macro field analysis. Space limitations limit coverage to a given arbitrary boundary displacement. Extension to an unnatural homogeneous boundary condition, say spring supports, is routine. Forces on the boundary beam due to the particular and homogeneous solutions are reconciled with an analysis of the boundary beam to find the arbitrarily imposed boundary displacements assumed in the homogeneous solution.

Homogeneous solution

7. The objective here is analysis of the subject grid with no interior loads and an arbitrary set of boundary node displacements along r = m. The displacements will be given as coefficients of a finite series, i.e.,

$$w^{h}(a,s) = \sum_{l=1}^{n-1} \underline{W}^{b}_{l} \sin \lambda'_{l} s \qquad (9)$$

The solution method is to write the homogeneous solution as the same series used for the particular solution above plus a simple function which satisfies the boundary condition. This modified form is then substituted into the original compatibility equations with the boundary function expanded as a series similar to the loads in the particular solution. Matching coefficients as before gives the coefficients of the homogeneous series in terms of $\underline{W}_l^b$ as if it were a loading term.

8. The boundary function used here is a simple finite conoid with zero deflection along s = 0, s = n, and r = 0 and $w^b(s)$ along r = m. In this case, the assumed boundary function is the deflection of the released grid, i.e., the interactive forces equal zero, but this need not be the case for a suitable boundary function. Thus, the homogeneous analogues to Eqs 4 and 5 are:

$$w^h(x,s) = \sum_{l=1}^{n-1}\left[\frac{x}{a}\underline{W}_l^b - \frac{m}{a}\sum_{i=1}^{\infty}(R_{il}^h\overset{*}{C}{}_i^4 + T_{il}^h\overset{*}{C}{}_i^3)\sin\alpha_i x\right]\sin\lambda_l' s \tag{10}$$

$$w^h\left(\frac{a}{m}r,s\right) = \sum_{l=1}^{n-1}\left[\frac{r}{m}\underline{W}_l^b - \sum_{k=1}^{m-1}(R_{kl}^h C_k^4 + T_{kl}^h C_k^3)\sin\lambda_k r\right]\sin\lambda_l' s \tag{11}$$

In order to match coefficients of the double finite series in the compatibility equations, the linear term, r/m, in Eq. 11 must be expressed as a finite sine series valid at the interior nodes.

$$\frac{r}{m} = \sum_{k=1}^{m-1} C_k^f \sin\lambda_k r \qquad C_k^f = -\frac{\cos k\pi}{m}\,\frac{\sin\lambda_k}{\gamma_k} \tag{12a, b}$$

Thus the equations for the coefficients of the homogeneous solution are:

$$\left[A_{ij}\right]\begin{bmatrix} R_{kl}^h \\ T_{kl}^h \\ \underline{T}_{kl}^h \end{bmatrix} = \begin{bmatrix} -C_k^f W_l^b \\ 0 \\ 0 \end{bmatrix} \tag{13}$$

9. Eq. 13 completes the objective homogeneous solution. Other or multiple homogeneous solutions, which are either end goals or intermediate steps toward other unnatural homogeneous solutions, can be routinely written in a similar fashion.

CONCLUSIONS

10. Compact formulae were given for the bending and warping torsion analysis of regular grids simply-supported along a rectangular boundary. As a heuristic example toward extending the analysis to arbitrary boundary conditions, formulae were also given for an arbitrary set of imposed out-of-plane displacements along one side of the boundary. The macro field approach is usually applied to more

complex structures. A secondary goal of its use, for a system that could be solved by other field or open form methods, is to introduce this powerful method to a wider audience of structural analysts than has been the case to date.

APPENDIX: BEAM FORMULAE

A1. Classical beam theory is well known and not normally reviewed in an international journal paper. However, application to macro discrete field analysis is so rare that a review is needed to follow the condensed presentation in the paper. For the better developed micro discrete techniques, general boundary solutions for system elements are used to write the mathematical models as difference equations. For macro discrete field techniques, general load solutions for the system components (with dimensions equal those of the system) are used to write the mathematical models as summation equations. Specifically, component solutions for movable unit impulse loads (kernel functions) which satisfy the given system boundary conditions yield unified solutions valid for both continuous and discrete system co-ordinates.

Impulse load

A2. As a first example, consider the simple bending analysis of a grid beam parallel to the x axis. The kernel function model for deflection due to a unit impulse load is:

$$B\, D_x^4\, K^{wq}(x,\xi) = \delta(x-\xi)\ , \qquad K^{wq}\left({}^{0}_{a},\xi\right) = D_x^2\, K^{wq}\left({}^{0}_{a},\xi\right) = 0 \qquad \text{(A.1a,b,c)}$$

The classical half range sine load and solution series are:

$$\begin{bmatrix} \delta(x-\xi) \\ K^{wq}(x,\xi) \end{bmatrix} = \frac{2}{a}\sum_{i=1}^{\infty}\begin{bmatrix} 1 \\ \overset{\star}{C}{}_i^4 \end{bmatrix} \sin\alpha_i\xi\, \sin\alpha_i x \qquad \text{(A. 2a b)}$$

The superscript, ^{4}, in the kernel function coefficient denotes the exponent of the series index, i.e.,

$$\alpha_i = i\pi/a\ , \qquad \overset{\star}{C}{}_i^4 = \frac{1}{B\alpha_i^4} \qquad \text{(A.3a,b)}$$

Distributed load

A3. For load, q(x), with infinite series coefficients q_i, the continuous deflection field is:

$$w(x) = \sum_{i=1}^{\infty} q_i\, \overset{\star}{C}{}_i^4 \sin\alpha_i x \qquad \text{(A.4)}$$

Discrete loading

A4. An important concept in macro field analysis is the expression of beam quantities, continuous and discrete, due to a discrete field of impulse loads.

Consider the following set of concentrated loads, spaced at a/m, which are expressed in terms of a finite sine series:

$$P(r) = \sum_{k=1}^{m-1} P_k \sin \lambda_k r \ , \quad P_k = \frac{2}{m}\sum_{k=1}^{m-1} P(r) \sin \lambda_k r \ , \quad \lambda_k = k\pi/m \qquad \text{(A.5a,b,c)}$$

Summing the kernel function for $\xi = ar/m$, results in the following expression for the continuous deflection field in terms of the discrete loading coefficients:

$$w(x) = \frac{m}{a}\sum_{i=1}^{\infty} P_i \overset{*}{C}{}_i^4 \sin \alpha_i x \quad \text{in which } P_k = \pm P_{2Im\pm k} \qquad \text{(A.6a,b)}$$

Use of the sine-wise cyclic nature of the m-1 distinct values of P_i in Eq. A.6a gives exact results and the form is convenient for derivations. For programming, however, it is more convenient to use:

$$w(x) = \frac{m}{a}\sum_{k=1}^{m-1} P_k \sum_{I=-H}^{+H} \overset{*}{C}{}_{2Im+k}^4 \sin(\pi\frac{x}{a}(2Im+k)) \qquad \text{(A.7)}$$

in which H is selected to provide sufficient accuracy for this rapidly converging series, e.g. H = 10.

<u>Node quantities</u>

A5. Eq. A.7 can be used to determine node quantities such as $W(a/m)r$, but to write closed form solutions one requires finite series expressions for such discrete fields, i.e.,

$$w(\frac{a}{m} r) = \sum_{k=1}^{m-1} P_k C_k^4 \sin \lambda_k r \ , \qquad C_k^4 = \frac{m}{a}\sum_{I=-\infty}^{\infty} \overset{*}{C}{}_{2Im+k}^4 \qquad \text{(A.8a,b)}$$

The infinite sum in Eq. A.8b, as with most sums required for structural systems, is available in closed form as:

$$C_k^4 = \frac{1}{12B}\left(\frac{a}{m}\right)^3 \frac{3-\gamma_k}{\gamma_k^2} \ , \qquad \gamma_k = 1 - \cos \lambda_k \qquad \text{(A.9a,b)}$$

<u>Other sums</u>

A6. Similar formal sums are required to get finite series expressions for node quantities resulting from taking the first, second and third derivatives of formula A.6, i.e., slope, moment and shear. One should recall that infinite Fourier series converge to the mean of a function with a "jump" discontinuity at a node.

$$C_k^3 = \frac{m}{a}\sum_{I=-\infty}^{\infty} \overset{*}{C}{}_{2Im+k}^3 = \frac{1}{4\ B}\left(\frac{a}{m}\right)^2 \frac{\sin \lambda_k}{\gamma_k^2} \qquad \text{(A.10a,b)}$$

$$C_k^2 = \frac{m}{a}\sum_{I=-\infty}^{\infty} \overset{*}{C}{}_{2Im+k}^2 = \frac{1}{2\ B}\left(\frac{a}{m}\right) \frac{1}{\gamma_k} \qquad \text{(A.11a,b)}$$

$$C^1_k = \frac{m}{a}\sum_{I=-\infty}^{\infty} \overset{*}{C}{}^1_{2Im+k} = \frac{1}{2\,B}\;\frac{\sin\lambda_k}{\gamma_k} \tag{A.12a,b}$$

Impulse shear

A7. The second required bending analysis is for a beam loaded with a movable impulse shear in contrast to the impulse load above. This loading equals the derivative of δ(x-ξ) with respect to x or minus the derivative with respect to ξ-- herein denoted by δ'(x-ξ) -- and is physically interpreted as a negative unit couple (as imparted by the twisting of an intersecting beam).

$$B\,D^4_x\,K^{wv}(x,\xi) = \delta'(x-\xi) = \frac{-2}{a}\sum_{i=1}^{\infty}\alpha_i\,\cos\alpha_i\xi\,\sin\alpha_i x \tag{A.13a,b}$$

$$K^{wv}(x,\xi) = \frac{-2}{a}\sum_{i=1}^{\infty}\overset{*}{C}{}^3_i\,\cos\alpha_i\xi\,\sin\alpha_i x \tag{A.14}$$

Discrete field of couples

A8. Use of Eq. A.14 to find the continuous response due to a discrete field of impulse shears (negative couples) requires that the couples be expressed as a finite cosine series, whose orthogonality range includes the boundary nodes. Symmetry requires the introduction of a weighting equal to one, at interior, and one-half, at boundary nodes. i.e.,

$$\frac{1}{\omega_r}\,T(r) = \sum_{k=0}^{m} T_k\,\cos\lambda_k r \qquad T_k = \frac{2\omega_k}{m}\sum_{r=0}^{m} T(r)\,\cos\lambda_k r \tag{A.15a,b}$$

in which

$$\omega_k = 1 - \frac{1}{2}(\delta^0_k + \delta^m_k) \tag{A.15c}$$

Summing Eq. A.14 for ξ = (a/m)r gives the continuous (and differentiable) deflection field for a discrete field of couples. Then, as above, summing that expression gives the node deflections, i.e.,

$$w(x) = \frac{-m}{a}\sum_{i=1}^{\infty} T_i\,\overset{*}{C}{}^3_i\,\sin\alpha_i = \frac{-m}{a}\sum_{k=0}^{m}\omega_k T_k\sum_{I=-H}^{+H}\overset{*}{C}{}^3_{2Im+k}\,\sin(\pi\tfrac{x}{a}(2Im+k)) \tag{A.16a,b}$$

$$w(\tfrac{a}{m}\,r) = -\sum_{k=1}^{m-1} T_k\,C^3_k\,\sin\lambda_k r \tag{A.16c}$$

The indeterminate term k = I = 0 in Eq. A.16b, if it occurs, is set to zero. Eqs 15 and 16 provide for an arbitrary field of forces, T(r), with m+l degrees of freedom. For applications here, the couples are moment spring reactions to a sine series, loading with m-l degrees of freedom. While the field extends to the boundary nodes, there are only m-l degrees of freedom and the sums are from k =

1 to m-l. The moment spring constant at the boundaries is one-half that at the interior nodes.

Beam torsion with warping restraint

A9. As with the bending analysis, one requires the solution for the angle of twist, φ, due to a movable impulse torque. The model for this kernel function, with simple end restraints, is (ref. 2):

$$(EI_w D_x^4 - GK_t D_x^2)K^{\varphi t}(x,\xi) = \delta(x-\xi), \quad K^{\varphi t}({}^0_a,\xi) = D_x^2 K^{\varphi t}({}^0_a,\xi) = 0 \qquad \text{(A.17a,b)}$$

In which K_t is the standard St. Venant torsional constant (equal to the sum of $bt^3/3$ for plates in a thin open section) and I_w for a WF section is the stiffness against twist offered by in-plane shear in the flanges with a couple with a distance equal to the beam depth, d. That is, $I_w = b^3td^2/24$. As above, the kernel function can be written as:

$$K^{\varphi t}(x,\xi) = \frac{2}{a}\sum_{i=1}^{\infty} \overset{*}{C}{}_i^t \sin\alpha_i\xi \sin\alpha_i x, \quad \overset{*}{C}{}_i^t = \frac{1}{\alpha_i^2(B_w\alpha_i^2 + B_t)} \qquad \text{(A.18a,b)}$$

Discrete field of torques

A10. As in the bending cases, one needs to express the twist angles, φ, in terms of a discrete field of torques, T(r), expressed as a finite series as:

$$\underline{T}(r) = \sum_{k=1}^{m-1} \underline{T}_k \sin\lambda_k r, \quad \varphi(\tfrac{a}{m}r) = \sum_{k=1}^{m-1} \underline{T}_k C_k^t \sin\lambda_k r \qquad \text{(A.19a,b)}$$

$$C_k^t = \frac{m}{a}\sum_{I=-\infty}^{+\infty} \overset{*}{C}{}^t_{2Im+k} = \frac{1}{2B_w}\left(\frac{a}{\pi}\right)^3 \frac{1}{m\rho^2}\left[\frac{\pi}{\gamma_k} - \frac{m}{\rho}\,\frac{\sinh(\rho\pi/m)}{(\cosh(\rho\pi/m) - \cos\lambda_k)}\right] \qquad \text{(A.19c,d)}$$

in which

$$\rho = \frac{a}{\pi}\sqrt{B_t/B_w} \qquad \text{(A.19e)}$$

REFERENCES

1. DEAN D. L. Discrete Field Analysis of Structural Systems, Springer Verlag, Wien, 1976.

2. GALAMBOS T. V. Structural Members and Frames, Prentice Hall, Englewood Cliffs, N.J., USA, 1968.

24. Interpolation of thermal gradients based on the prescribed average temperature method

S. APPEL, Fokker Space and Systems B.V., The Netherlands

SYNOPSIS. For the purpose of thermo-elastic deformation and stress analysis of a mechanical structure, the temperatures on the nodes of the finite element model are required. Thermal states are determined by thermal analyses in which the mesh used is often much coarser than that of the structural model. The Prescribed Average Temperature (PAT) method has been developed to interpolate the temperatures represented at the thermal nodes to those at the nodes of the structural finite element model. While the basic implementation of the PAT-method results only in structural node temperatures, it is unable to determine thermal gradients across sections of plates and beams. A method, which is based on the PAT-method, to interpolate these cross-sectional thermal gradients, is described.

PROBLEM DEFINITION

1. Thermal loads, which represent one of the important loading conditions for a spacecraft, need to be applied to a structural finite element model. The resulting thermally induced deformation and stresses can then be determined.

2. The thermal states are calculated by thermal analyses, which result in temperatures at thermal nodes. While the nodes of the finite element model, the "structural" nodes, describe the condition at a point on the spacecraft, a thermal node does this for a surface or a volume of the same structure (ref. 1). For this reason, the finite element mesh is in general much finer than that of the corresponding thermal model. One thermal node may span, for instance, 10 or more finite elements. To apply the temperatures determined by the thermal analyses to the structural nodes of the finite element model, an interpolation method must be used.

3. At Fokker Space & Systems the Prescribed Average Temperature method (PAT-method) has been developed to interpolate the thermal node temperatures to those at the structural nodes. This method has been implemented in the computer program SINAS II (ref. 2) and has been applied successfully to several projects: ERS, ARIANE 5 engine frame, Aristoteles and Cluster, to name but a few. The PAT-method is briefly described further in this paper.

4. As described later, the basic implementation of the PAT-method results in structural node temperatures. These temperatures are assumed to be constant over the cross-section of a plate or a beam. In general this information is accurate enough to compute the distortion of a structure due to temperature loads. However, in the case of accurate pointing devices, such as antennae, knowledge of the temperature distribution across the cross-section of plates and beams then becomes essential for the prediction of distortions due to temperature fields. To satisfy this need, a method for the interpolation of these cross-sectional thermal gradients has recently been developed. This method, which is based on the PAT-method, is explained in this paper.

THE PRESCRIBED AVERAGE TEMPERATURE METHOD

5. A thermal node represents the thermal behaviour of a region of a structure. The PAT-method uses the same assumption, as made in the derivation of the thermal model, namely that the temperature is constant over the part of the structure covered by the thermal node. This assumption is translated to the following requirement, with regard to the structural nodes: the average temperature of the structural nodes, spanned by a thermal node, must be equal to the temperature of the thermal node itself. This requirement can be written as:

$$\sum_{i=1}^{n} a_{ij} T_i^s = T_j^t \qquad (1)$$

and

$$\sum_{i=1}^{n} a_{ij} = 1 \qquad (2)$$

in which T_i^s are the temperatures of the structural nodes spanned by thermal node j, which has a temperature T_j^t. The coefficients a_{ij} are weighting factors.

6. Physically, the coefficients a_{ij} represent the relative amount of material related to structural node i. These values can be determined using the shape functions of the finite elements spanned by the thermal node. This can be done through the application of a constant heat flux per unit volume to each such finite element. The finite element package distributes these heat fluxes to the structural nodes. To satisfy the requirement of Eq. 2, the vector with nodal heat fluxes must be normalised to unit length.

7. For a finite element model spanned by several thermal nodes, Eq. 1 can be written in matrix form as:

$$AT^s = T^t \qquad (3)$$

Where the matrix A contains the coefficients a_{ij}, and the vectors T^s and T^t contain the structural node and thermal node temperatures respectively.

8. If it is assumed that the structure is in a stationary state and, that the thermal node defines a side constraint, as described by the PAT-relation in Eq. 3, the following energy functional can be written for the finite element model.

$$U = \frac{1}{2}T^{s^T}CT^s + q^T(AT^s - T^t) \tag{4}$$

9. In this functional U, C is the conduction matrix of the finite element model and q is a vector containing the Lagrange multipliers for the constraint relation as defined by Eq. 3. These Lagrange multipliers can be interpreted as the heat flow into each thermal node.

10. Applying the principle of minimum energy, the following relation is obtained:

$$\begin{bmatrix} C & A^T \\ A & 0 \end{bmatrix} \begin{bmatrix} T^s \\ q \end{bmatrix} = \begin{bmatrix} 0 \\ T^t \end{bmatrix} \tag{5}$$

11. This system can efficiently be solved by the program SINAS II (ref. 2). The resulting vector T^s can be used for the distortion analysis of the structure, using the finite element method.

12. A more detailed description of the PAT-method can be found in ref. 3.

THERMAL GRADIENTS

13. If in the thermal model, thermal nodes are defined at both sides of a plate, then different temperatures can be determined at the two sides. This leads to thermal gradients across the thickness, which results in a curvature of the plate concerned. Similarly, four different temperatures at the sides of a beam specify thermal gradients across the beam section.

14. To prevent confusion about the term "thermal gradients" it is here explicitly defined as being the temperature gradients across the thickness of a plate or across the cross-section of a beam.

15. A region of a structure subjected to thermal gradients is referred to as "gradient area". Two types of gradient area can be distinguished: the plate type gradient area and the beam type. The region of a structure not subjected to thermal gradients is termed the "no-gradient area". In Fig. 1, the two types of gradient areas are illustrated.

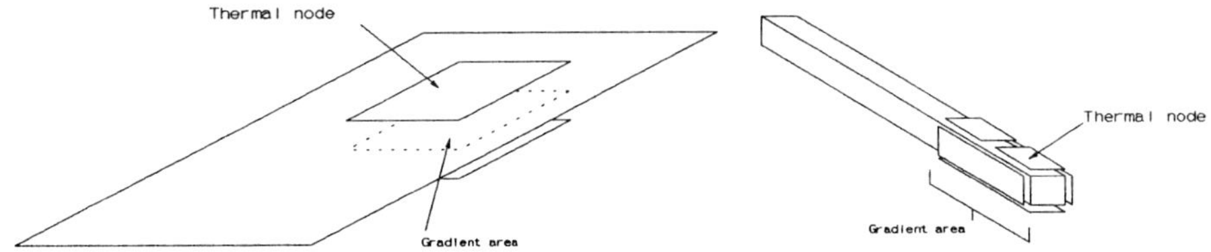

Fig. 1. The two types of gradient areas, with thermal nodes at different sides of a plate and a beam.

STRUCTURAL NODE LAYERS

16. To determine the curvature of plates and beams due to thermal gradients, the temperatures at both sides of all plates and at all cross sectional edges of all beams, must first be known. This implies that the thermal node temperatures must be interpolated to the two sides of a plate or the four sides of a beam cross-section.

17. If we consider a plate subjected to thermal gradients, the temperatures of a single side can be interpolated using Eq. 5. The same equation can then be used for the temperature interpolation of the opposite side. Such an approach can be considered to be the interpolation of two separate plates, for each of which Eq. 5 can be used. To ensure that the total conduction of these two plates is equal to that of the original plate, the sum of the thicknesses of the two must be equal to the thickness of the original.

18. If this is related to the finite element model of the same plate, then the original finite elements and structural nodes are replaced by two layers of new elements and corresponding new structural nodes. The conduction matrices of these layers can then be derived from the conduction matrix of the original finite element model of the plate, using:

$$\begin{aligned} C^{+N} &= f^{+N} C \\ C^{-N} &= f^{-N} C \end{aligned} \qquad (6)$$

where C is the conduction matrix of the part of the plate subjected to thermal gradients. The matrices C^{+N} and C^{-N} represent the conduction of the top and bottom layers respectively. To ensure that the total conduction of these layers is equal to that of the original model, the proportions of the total conduction provided by the top and bottom layers, f^{+N} and f^{-N}, must fulfil the following requirement:

$$f^{+N} + f^{-N} = 1 \qquad (7)$$

19. Since the conductive information of the elements is contained in the conduction matrix, the layers can then be represented by a combination of the new structural nodes and the conduction matrix describing the conduction relation between these new structural nodes. For this reason the name "structural node layer" is adopted in relation to such a layer.

20. In Fig. 2 a part of a plate is subjected to thermal gradients. This part is split into two structural node layers, referred to as "new" structural nodes, because they replace, for the purpose of interpolation, the original nodes. For convenience of implementation, the structural nodes of the no-gradient area are also replaced by new structural nodes.

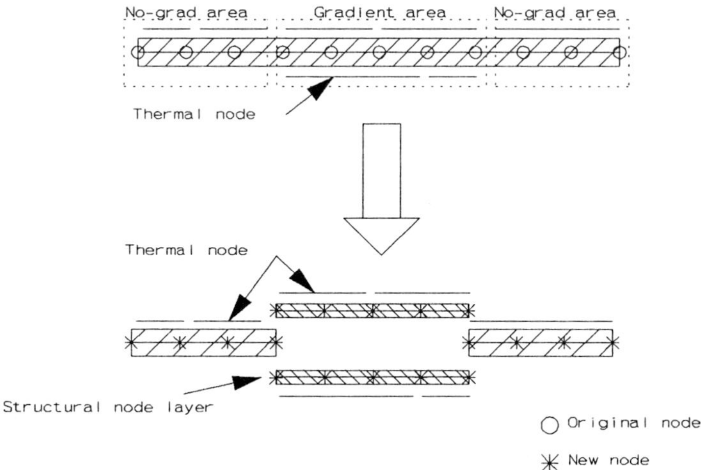

Fig. 2. Structural node layers for a plate type gradient area.

21. In most finite element packages, the temperature variation over the cross-section of a beam can be defined by the temperatures at four distinct positions within the cross-section. These positions are referred to as "temperature points". With these four temperatures, a bilinear distribution can then be defined. In a manner similar to that of the plate type gradient area in Fig. 2, a beam type gradient area can be split into four structural node layers. The new structural nodes of the structural node layers are then located at the temperature points. Each temperature point within a section is connected to a different structural node layer. When the temperatures at the thermal nodes are interpolated to those at the new structural nodes, the temperature variation across each beam section is then known.

COMPATIBILITY RELATIONS

22. As is shown in Fig. 2, the structural node layers are not connected to each other. If Eq. 5 is solved independently for each structural node layer, the influence of neighbouring gradient areas and the no-gradient area is neglected. For this reason, the structural node layers of the different gradient areas should be connected to each other. This connection is achieved mathematically through the satisfaction of certain compatibility relations.

23. The basis for these compatibility relations is the requirement that only the distribution of the average temperatures within a gradient area may have influence on the temperature distribution outside the gradient area. This implies that the average temperatures at both sides of an interface between two gradient areas or between a gradient area and the no-gradient area must be the same.

24. If we consider the interface between a plate type gradient area and the no-gradient area, the following relation can be written:

$$T^s - \frac{1}{2}(T^{+N} + T^{-N}) = 0 \qquad (8)$$

in which T^s is the temperature of the new structural node at the no-gradient area side of the interface, T^{+N} and T^{-N} are the temperatures of the new structural nodes at the edges of the structural node layers at the top and bottom faces of the plate. These three new structural nodes all derive from the same original structural node at the interface between the no-gradient area and a plate type gradient area.

25. A similar relation can also be derived for a beam type gradient area and the no-gradient area, or if two gradient areas interface with each other.

26. All these compatibility relations can be written in matrix form as:

$$M^* T^{s*} = 0 \qquad (9)$$

in which T^{s*} is the temperature vector corresponding to the new structural nodes and M^* is the matrix containing the coefficients of the compatibility relations. A characteristic of this matrix is that the sum of each row is equal to zero.

THE FINAL SYSTEM TO BE SOLVED

27. In Eq. 9 the vector T^{s*} is introduced. This vector contains the temperatures of the new structural nodes. According to the sequence of these nodes a conduction matrix of the total system, C^*, can be built up from the conduction matrices of all the structural node layers and the no-gradient area. In addition, based on the order of the temperatures of the new structural nodes in the vector T^{s*}, the PAT-relation, Eq. 3, can be rewritten as:

$$A^* T^{s*} = T^{t*} \qquad (10)$$

where matrix A^* contains the coefficients of the PAT-relation corresponding to the order of the new structural node temperatures in the vector T^{s*}. The vector T^{t*} is the vector of thermal node temperatures, which corresponds to the order of the rows within matrix A^*.

28. If the compatibility relations are considered as an additional constraint, the energy functional for the system with respect to the new structural nodes can be written as:

$$U = \frac{1}{2} T^{s*T} C^* T^{s*} + q_A^{*T} \left(A^* T^{s*} - T^{t*} \right) + q_M^{*T} M^* T^{s*} \qquad (11)$$

In this relation, two vectors with Lagrange multipliers are used. The Lagrange multipliers for the PAT-relations are represented by the vector $\overset{*}{q}_A$. Those for the compatibility relations are represented by the vector $\overset{*}{q}_M$.

29. Using the principle of minimum energy the following relation is obtained:

$$\begin{bmatrix} C^* & A^{*T} & M^{*T} \\ A^* & 0 & 0 \\ M^* & 0 & 0 \end{bmatrix} \begin{bmatrix} T^{s*} \\ q_A^* \\ q_M^* \end{bmatrix} = \begin{bmatrix} 0 \\ T^{t*} \\ 0 \end{bmatrix} \tag{12}$$

This system has the same form as that of Eq. 5. For this reason, Eq. 12 can be solved by the same algorithm as used for Eq. 5.

30. With the solution of the system as represented by Eq. 12, the thermal node temperatures are then interpolated to the new structural nodes. In the next section, it is explained how these temperatures can be used to apply the thermal gradients to the structural finite element model.

APPLICATION OF THERMAL GRADIENTS TO THE FINITE ELEMENT MODEL

31. In most finite element packages, thermal gradients can be applied to plate elements through the specification of the temperatures at the two sides of the element. To apply the thermal gradients to beam elements, the temperatures at the temperature points, as described above, must be specified.

32. If we consider a plate, structural node layers are defined at the top and bottom sides. However, a decision has to be made, as to which side of the plate is to be considered as the "top" side and which the "bottom". As a result, the temperatures at the new structural nodes are temperatures at the geometrical top and bottom sides of the plate. If these temperatures are to be applied to the finite elements, then we must know which side of the element corresponds to which side of the plate.

33. The orientation of each of the elements can be derived from the input data describing the finite element model. In addition to the orientation of the elements themselves, the relative orientation with respect to the geometry of the plate can also be deduced. Thus, all the necessary information is available for the assignment of the temperatures of the new structural nodes to the plate elements.

34. In the case of beam elements, the situation is more complicated. The structural node layers connect each temperature point between beam elements, as is illustrated in Fig. 3.

35. For the case of a certain element in a beam type gradient area, structural node layer 1 can be related to temperature point 1 in the cross-section of both ends of the beam element. Similarly, structural node layer 2, 3 and 4 can be related to temperature points 2, 3 and 4 respectively. This element is termed the "reference" element.

36. As the structural node layers now link with the element connected to this reference element, the problem now arises whereby the position of the temperature points can be specified by the user for each beam element. This implies that both the order in which the temperature points are defined and the relative position of the temperature points may differ from element to element.

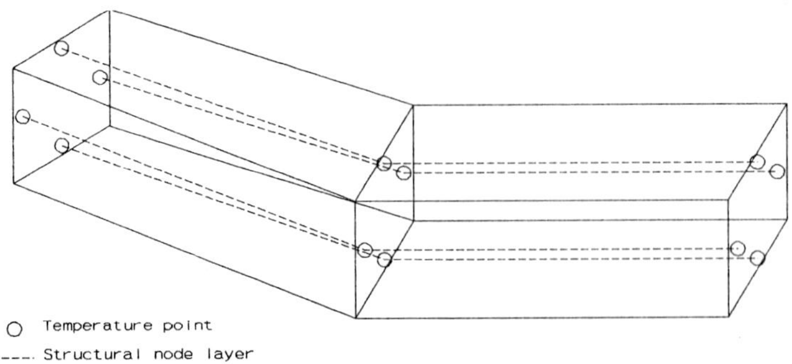

Fig. 3. Structural node layers connecting temperature points.

37. If we consider an element connected to the reference element, we need to determine which temperature point of the reference element corresponds to which temperature point of the element considered. In ref. 4, a means is explained in which this can be achieved using the information contained within the input data, describing the finite element model.

38. The temperatures of the thermal nodes can now be interpolated to those at the new structural nodes, and as the relation between the position of the structural node layers and the finite elements is known, the temperature load data can be generated for the finite element analysis. The stresses and deformations caused by temperature fields considered can now be determined.

CONCLUSIONS

39. The method for the interpolation of thermal gradients has recently been implemented in a computer program. It has been tested using different finite element models and the results obtained have been found to be good.

ACKNOWLEDGEMENTS

40. The method as described above was developed under a contract from the European Space Agency (ESA). The author wants to thank Mr. M. Klein, for his guidance and helpful comments.

REFERENCES

1. ESA/ESTEC and Engineering Research Centre of GEC Alsthom, ESATAN Engineering Manual, EM-ESATAN-056
2. Fokker Space & Systems, SINAS II documentation
3. M.H.M. Ellenbroek, Temperature Interpolation, SINDA-MSC/NASTRAN interface, Proceedings of the MSC/NASTRAN European User's Conference, 1987.
4. APPEL, S. Theory Applied for the interpolation of thermal gradients. Report FSS-R-91-0336.

25. Formfinding, analysis and construction of a synclastic curved glass dome

M. MOLLAERT and E. MAS, Portal Europe, Belgium

SYNOPSIS. The basic idea is to build a synclastic glass roof as transparent and light as possible. The dome is realised with a bi-directional network of bars. The network is a discretisation (i.e. discrete elements) of a shell in pure compression. The theoretical co-ordinates are calculated according to this principle. To assure stability under an asymmetric snow load or wind pressure, cables are added along the diagonals of the quadrilateral meshes. The structural analysis was performed with the program ROBOT. A prototype was constructed and tested.

THE DEFINITION OF THE FORMS

1. To develop the structural system, a prototype with a regular shape was manufactured. Its base is a circle of 10m diameter, its height in the middle is 1m. The height to span ratio is thus only 0.1.

2. To ensure that compression is the principal action in the structural bars, the form is built on the mathematical principle of two reversed chain curves. The chain curves are approximated by rectilinear bars. From the main axes, flat glazing panels with equal length sides, are calculated with the main grid members.

Bar

3. Each bar has two connection pieces at its ends. Four connection pieces are assembled in one node (Fig. 1).

a = bar length; b = length of connection piece; α = angle of connection piece.

Fig. 1. Typical bar element.

Each connection piece can initially freely rotate (α) around a horizontal axis in the bar, and is fixed afterwards by two bolts.

4. For the prototype there are four types of bars (Fig. 2).

Position of bar	α (degrees)
1	2.800
2	2.773
3	2.697
4	2.579

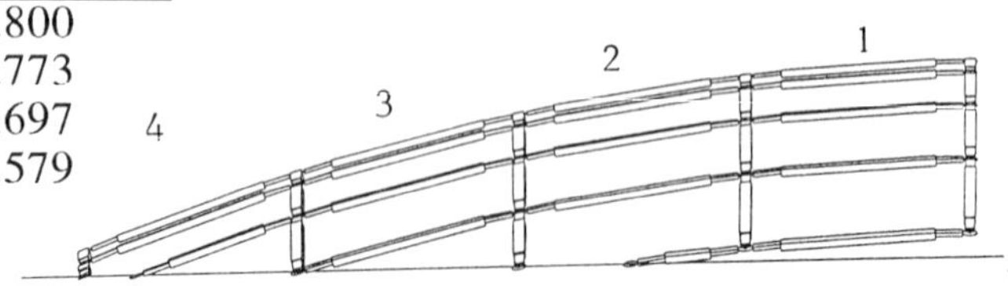

Fig. 2. The corresponding α value for each bar position.

Node

5. There are two handed types of connection pieces. Each of them is positioned one time upright, and one time inverted into each node (Fig. 3).

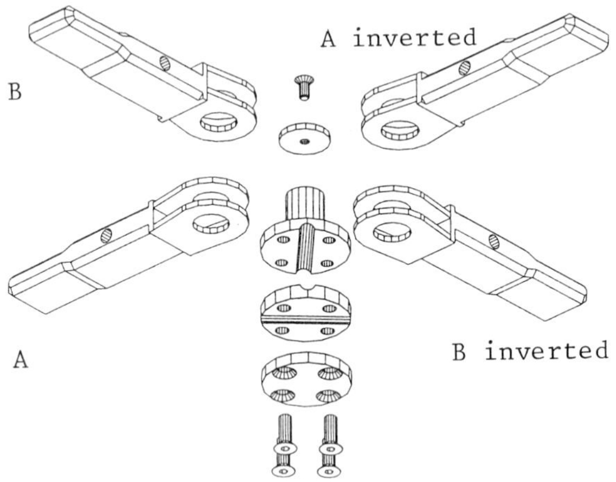

Fig. 3 . Exploded view of the node.

For each node in the dome there are seven variable values:

- $\alpha 1$, $\alpha 2$, $\alpha 3$ and $\alpha 4$: the slope of the four connection pieces,
- β, γ, δ: the angles between the connection pieces in the tangential plane (Figs 4 and 5).

6. In the central element of the node, which connects the four bars, provision is also made to retain two cables at two drilled holes.

7. All pieces of the node are assembled by countersunk headed bolts (Fig. 3). If the length of the bars (a), the length of the connection pieces (b), the angle at the top ($\alpha 0$), and the number of required bars in the x- and the y-directions are specified, then the form of the chain curves can be calculated and the necessary

theoretical co-ordinates of the dome can be obtained. The algorithm has been implemented in a LISP-program in AUTOCAD.

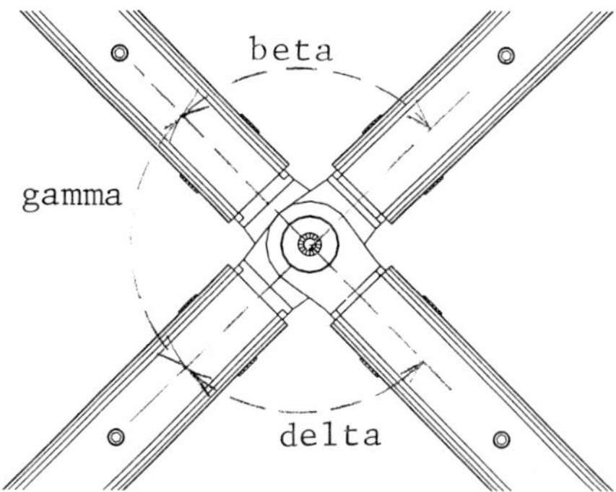

Fig 4. Plan view of the node.

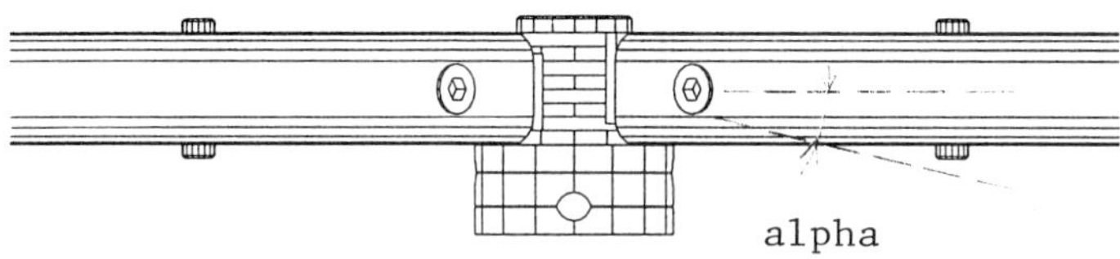

Fig. 5. Front view of the node.

Meshes

8. The glazing sheets are flat, but their rhomboidal form varies for different positions in the dome. In the middle the angles of the rhombus are 89.86 and 90.14 degrees; at the border 86.77 and 93.23 degrees (Fig 6).

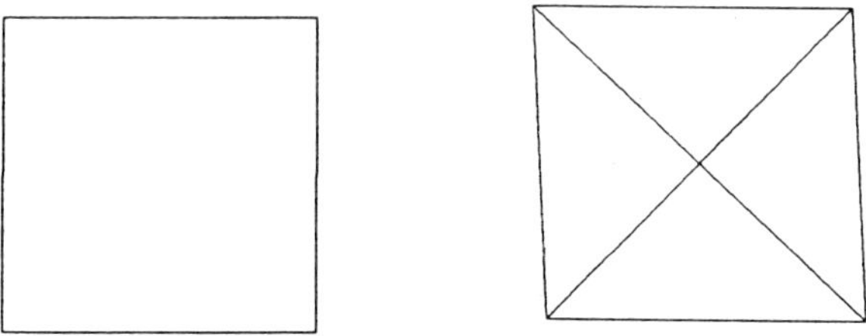

Fig. 6. The 'almost' square and 'most' rhombus.

SPECIFICATION OF THE BOUNDARY AND THE SUPPORTS

9. In a preliminary test it became clear that the precise position of the supports is a prerequisite to build the designed form. For this reason special attention was drawn to these supports, which were designed to be adjustable:

- in their x, y and z co-ordinates (position of the base);
- and in their α ,β, γ angles and L length (position of the endpoint), (Fig. 7).

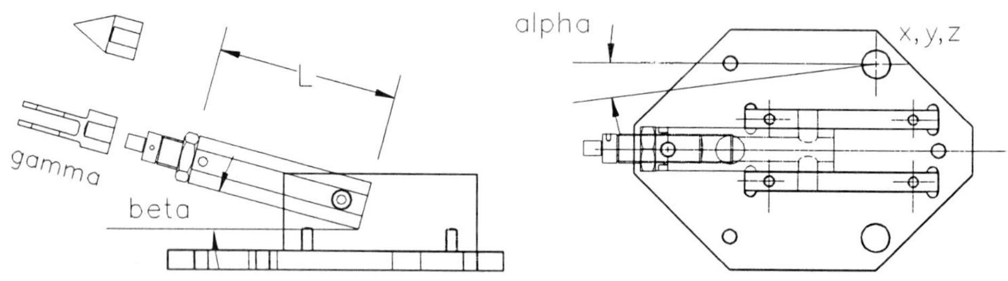

Fig. 7. Parameters to be adjusted at each support.

10. All the supporting elements have to be measured in the theoretical model, and they have to be manufactured and positioned with high precision.

STRUCTURAL CALCULATIONS

11. With the geometry defined in the previous steps, a calculation model was prepared. The structural analysis was performed with a computer program ROBOT.

<u>Sections (mm)</u>

- Principal bars: 60 by 40 by 5mm, for each bar an individual rotation angle had to be specified (Fig. 8).
- Connection pieces: two flat steel plates 50 by 5mm, at a distance of 15mm.
- Cables: diameter 12mm, cables only act in tension.
- Cylinder in between the connection pieces and the cables: diameter 30mm.
- The whole prototype was constructed on a ringbeam 150 by 150 by 8mm.

<u>Degrees of freedom</u>

12. The bars are free to rotate around Z axis (perpendicular to the surface) at their extremities, cables only transmit axial forces.

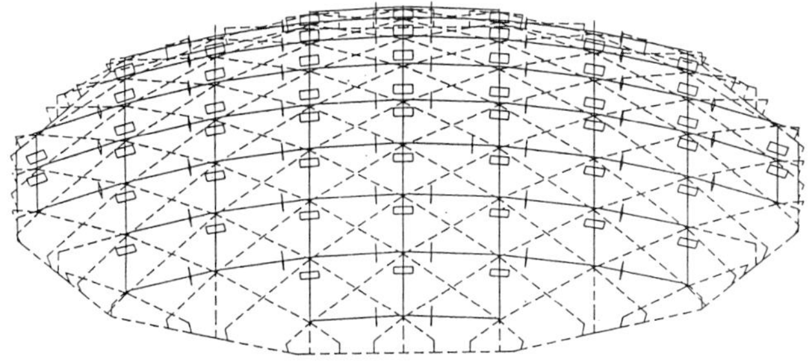

Fig. 8. Visualisation of the orientation of the bars.

Linear calculation

13. To check the geometry of the construction before loading, the forces in the structure were calculated under its self weight loading. Only the bars were considered. The maximum deflection was 1.63mm, and the maximum compression in the bars is 2320N. The difference between the theoretical position of the nodes and the measured values was only 8.0mm.

14. The proposed structural system is very flexible, and non-linear calculations have to be considered.

Non-linear analysis for dead load + symmetric snow load

15 Cables are taken into account if they are in tension. The glass (10/12/6.6.2) load is 55ON/m^2, the snow load 5OON/m^2. For a mesh of 123cm by 123cm, the load per node is 1600N. The structure has been calculated with a load of 1800N per node.

16. For the principal bars and connection pieces the maximum compression is 16835N, the maximum bending moment 140000Nmm. The maximum cable tension is 7970N. For the cylinder in between the connection pieces and the cables, the maximum tension is 2270N, the maximum shear 7439N and maximum bending moment 261225Nmm. The maximum deflection is 6.95mm.

THE BUCKLING PROBLEM

17. For a part of the structure shown in Fig. 9, consisting of four bars and the surrounding cables, placed isostatically on external supports, the critical load at the top can be calculated according to the following formula:

$$F = \frac{4(2)^{0.5}\,Eb\;Ab\;Ec\;Ac\;tg\beta\left(\cos\beta - \cos(\beta 0)\right)}{Eb\;Ab\;\cos\beta + Ec\;Ac(2)^{0.5}} \qquad (1)$$

with $Eb = Ec = 2.1 \times 10^5 N/mm^2$
$\beta 0 = 2.579$ degrees
$Ab = 900mm^2$, $Ac = 113mm^2$

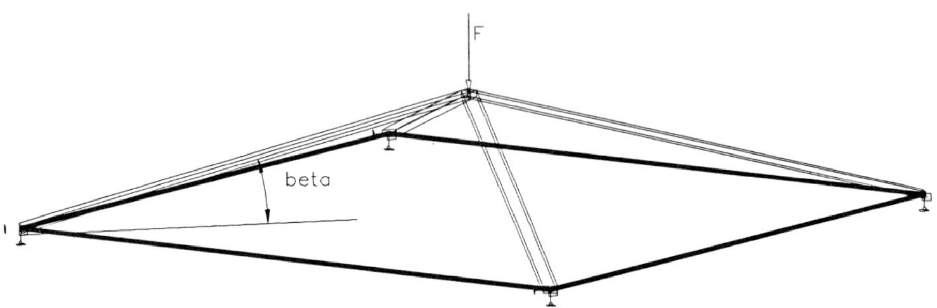

Fig. 9. A typical pyramid in the dome.

18. One must also take into account the fact that the critical load is 2000N, and the corresponding value of beta is 1.5 degrees. The extreme value for the axial force is 19100N, while the maximum value under dead load + snow was 16835N. If Ec is only $1.47 \times 10^5 N/mm^2$ (70 %), the critical load is reduced to 1467N.

19. The calculated critical load is a worst value, since the four boundary nodes are partially held by the surrounding structure. The top node in the calculation model is a spherical pin, while in the prototype only the rotation around the Z-axis is possible.

CONSTRUCTION

20. Each bar is connected to two connection pieces, and the angles at the extremities are adjusted and definitively fixed. The supporting elements are positioned on the ring beam and adjusted in their x, y and z co-ordinates at their base and at their endpoint. From the boundary, further grid meshes are formed by adding the bars two by two. At this assembly stage the structure is very flexible. It is only after the cables are attached across from a boundary node, through the internal nodes, and across to their opposite boundary node, that the structure achieves some rigidity.

21. By the subsequent introduction of an initial uniform pretension, it becomes possible to withstand both compression and tension in the cables (Fig. 10).

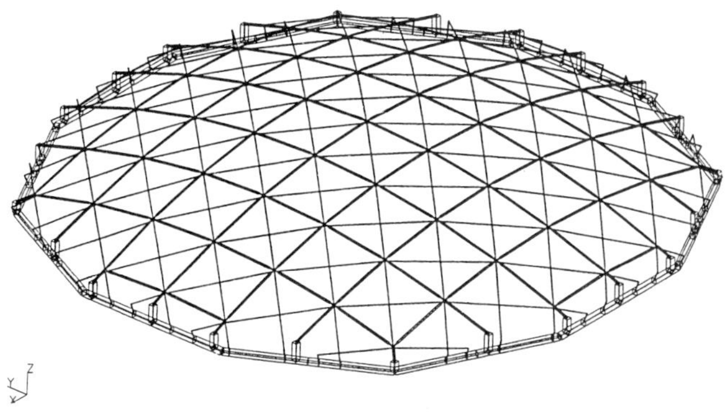

Fig. 10. General view of the structure.

EXPERIMENTAL TESTING

22. The structure with cables was loaded at the principal nodes. Loading was increased incrementally by adding 500N per node at each step. Deformations and constraints were measured by strain gauges.

23. For a loading of 1000N/node, a maximum deformation of 0.754mm was obtained (the calculated value was 0.835mm), and the maximum compression in the bars was 8380N (calculated value 5670N).

24. The structure collapsed under a load of 3250N/node, which is more than twice the maximum estimated symmetric loading.

25. Although the calculation program ROBOT was not able to cope with large deformations, the calculated shape at the first buckling mode did correspond with the form obtained at failure (Fig. 11).

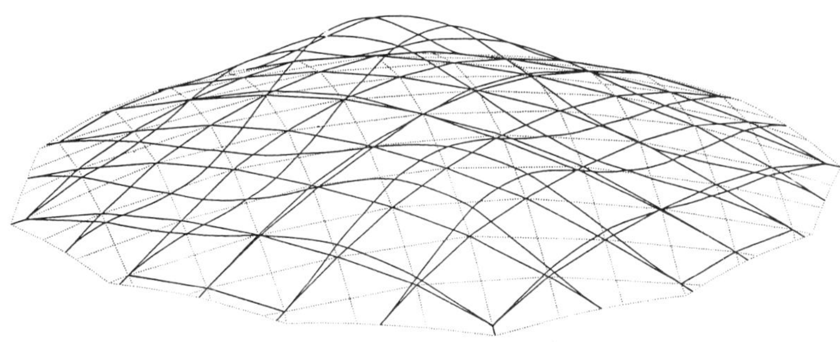

Fig. 11. First buckling mode.

CONCLUSION

26. The construction and testing of the prototype did highlight the importance of the installation precision necessary at the supports of the structure, and clearly demonstrated the buckling behaviour of this single layer dome.

27. Further non-linear calculations, including the proper modelling of large deformations are to be performed. Correspondance with the experimental results, both for symmetric and asymmetric loading should be obtained, to allow the further use of the software, to dimension and analyse, this type of transparent and light weight structure.

REFERENCES

1. MAKOWSKI Z.S. Steel braced barrel vaults. Space Structures Research Centre, University of Surrey, Building specification, October 1977.
2. RAMMERSTORFER F.G. and FISCHER D.F. Non-linear elastic-plastic stability and contact problems concerning the straightening of a wave-like deformed strip plate. Numerical methods for non-linear problems, vol. 1, p.307-315.
3. McCONNEL R.E. and KLIMKE H. Geometrically non-linear pinjointed space frames. Numerical methods for non-linear problems, vol. 1, p.333-342.
4. VAN DE PITTE D. Berekening van constructies bouwkunde en civiele techniek, Deel 2, 1981, Gent.
5. SAITOH M. Buckling loads of reticulated single-layer domes. Proceedings of the IASS symposium on membrane structures and space frames, vol.3, p. 121 - 128, September, 1986.
6. LENZA P. The collapse of doubly curved vaults. (ed.), p.169-176.
7. KOLOSOWSKI J. A method of the subdivision of braced spherical domes with the maximum number of members of equal length. (ed.), p.253-260.
8. KITAMURA S. The practical design of the single layered steel space frame roof of a gymnasium. (ed.), p.381-386.

26. Design procedure for large single-layer latticed domes

T. UEKI and F. MATSUSHITA, Tomoe Corporation, Japan, and R. SHIBATA and S. KATO, Toyohashi University of Technology, Japan

SYNOPSIS. The present paper discusses the applicability and effectiveness of the structural design procedure which has been developed from a project where the authors considered a particular single-layer latticed dome in Japan, by evaluating the ultimate strength of the structure. The steps leading from the experiments to design practice, are introduced by focusing on the loading tests for strength and rigidity at nodes and by comparing the experiments on domes with the elasto-plastic buckling analyses, and further, by extending the results to a design formula based on the concept of column strength curves.

INTRODUCTION

1. The bending rigidity and strength at the joints has a large effect on the buckling of single-layer latticed domes especially in the case of ball joints. The present screw-joint system, as shown in Fig. 1 and Photo. 1, which has been developed as a semi-rigid connection for tubular members, was investigated in a series of loading tests to find the bending characteristics; the bending rigidity and strength for the springs at both ends of a member. In the analysis, the members are modelled as illustrated in Fig. 7. The data from experiments are utilised in the numerical modelling of the members with springs, therefore, the accuracy in modelling corresponds to that in the numerical estimation of the buckling strength of reticular domes.

2. The initial part of this paper puts into focus the experiments, undertaken for the above project, using the present screw joints in a four bay dome, as shown in Photo. 2. The experimental investigation was undertaken to investigate the validity of the elasto-plastic geometrically non-linear analysis procedure. The middle part of the paper describes the structure, and the latter part of the work covers the analysis and design using the concept of column strength curves.

TESTING OF JOINTS

3. Fig. 1 and Photo. 1 shows the ball joint of the truss system named, T.U.T.-S, which was developed to form a new space frame system for single-layer domes with sufficient rigidity and strength against buckling. The joint has semi-

rigid bending stiffness. A schematic test was conducted on the joints to prove that the joints have sufficient rigidity and strength for the domes to function as required in the design (ref. 1).

Initial rigidity testing of joints

4. Fig. 2 illustrates the loading test of a joint. The specimen consisted of two members with one ball between them and two hydraulic jacks positioned to apply an axial force and bending moments to the joint simultaneously or separately depending on the requirements. Four types of test were conducted to find the initial rigidity under moderate loading: (a) cyclic axial loading [+N] for axial rigidity; (b) cyclic bending [+M], (c) cyclic bending [+M] under constant tension [+N] and; (d) cyclic bending [+M] under constant compression [-N] for bending rigidity.

Photo. 1. Ball-joint of T.U.U.-S system.

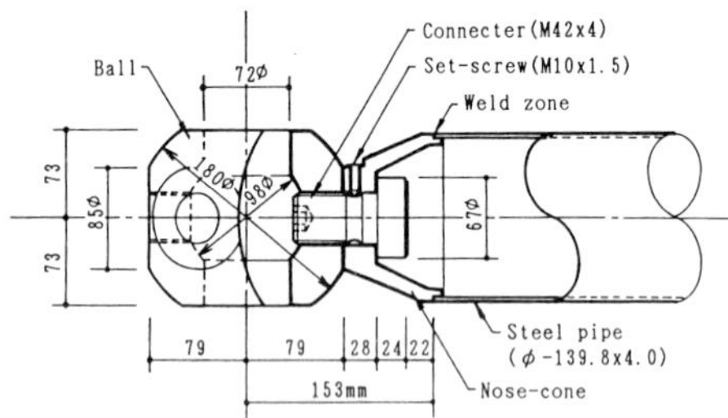

Fig. 1. Detail of joint.

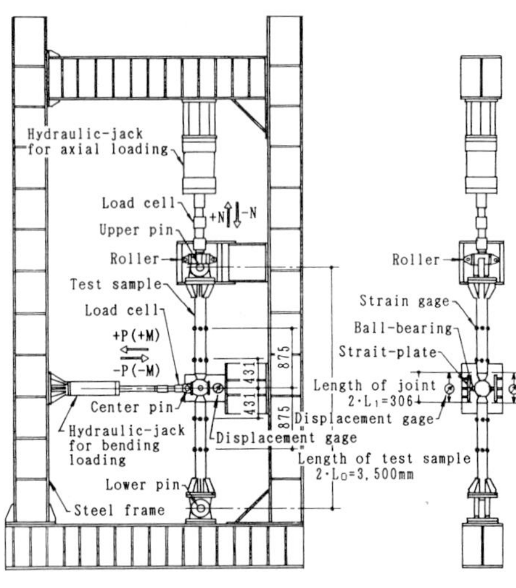

Fig. 2. Schematic joint test.

5. Axial rigidity test (a) Fig. 3 gives information on the axial rigidity of the joints under cyclic axial load. Although the rigidity is higher in compression and smaller in tension than that of bare steel tubes, the difference has little effect on the global equivalent axial rigidity, because the length of joints is much smaller than the length of tubular members. Thus, we may approximately adopt the net rigidity of tubular members for design.

6. Bending rigidity tests (b, c and d). Fig. 5 shows the difference in bending rigidity depending on constant tension or compression. The rigidity increases under compression and the joint behaves in a similar manner to a perfectly rigid connection in compression. On the other hand, the rigidity reduces under tension. However, the effects of low rigidity under tension may be insignificant as no

dome buckles in tension (ref. 2). From a series of these experiments, the authors conclude that the bending rigidity of the present system is sufficient for single-layer domes.

Strength testing of joints

7. Several kinds of strength tests were conducted by axial loading to each part of the joint, such as connectors, tubes, balls and nose-cones. The results have revealed that the global strength of joints is higher than that of tubular members. Several more loading tests were also performed to investigate the bending strength. Some of those results are shown in Fig. 4. With the bending moment increasing at the joint, the end part of the tubular members was found to be yielding locally near the nose-cone and every result has proved that the bending strength is higher than that of the tubular members. From both the axial and bending tests undertaken, the authors may conclude that the joints have sufficient ductility under cyclic loading and that the tubular members yield or buckle before the axial force and bending moment acting on the joints reach their ultimate strength.

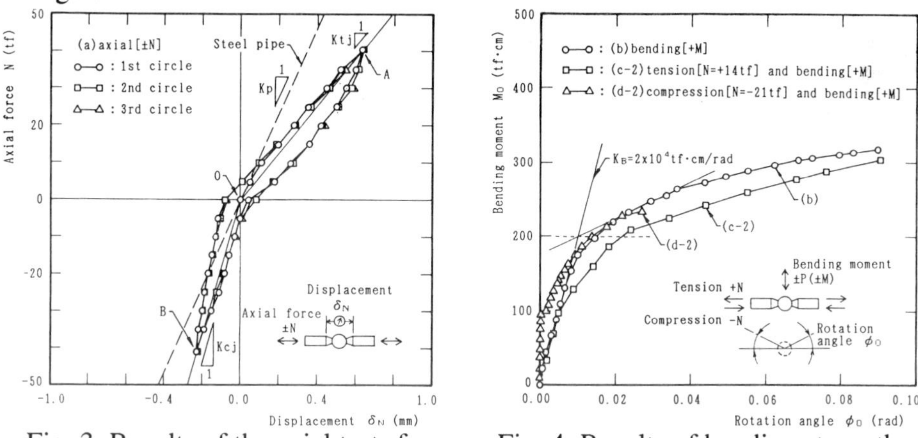

Fig. 3. Results of the axial tests for the initial rigidity of joints.

Fig. 4. Results of bending strength tests on joints.

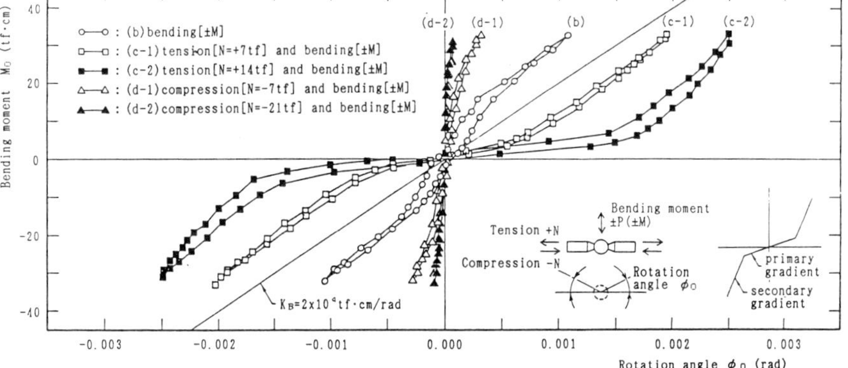

Fig. 5. Results of bending tests for the initial rigidity of joints.

LABORATORY EXPERIMENT ON A FOUR SPANNED DOME

8. Several static loading tests were performed on the domes to prove their high rigidity and strength against buckling as well as to ensure the accuracy of the material and geometric non-linear analysis developed for the present project (ref. 1).

Method of loading tests

9. Fig. 6 and Photo. 2 show a four bay dome with a diagonal span of 12m and a rise of 0.63m. The half open angle θ_0 is 3 degrees. It was part of a dome constructed using tubular members of diameter 139.8mm and wall thickness 4.0mm and the same size of balls joints as used in the previous tests. The inner seven nodes were unsupported with the outer twelve nodes vertically supported by horizontal rollers. The load was applied to node 15.

Method of analysis

10. Each member is assumed to be as shown in Fig. 7. It is composed of two rigid balls, bending springs at both ends and a beam element. However, the torsional rigidity is assumed to be negligible. Plasticity is assumed at both the centre and the ends of each member, to include the material non-linearity and the other sections are treated as materially linear. The plasticity is treated by the following strength interaction.

$$f = \left(\frac{N}{Ny}\right)^2 + \sqrt{\left(\frac{M_Y}{My}\right)^2 + \left(\frac{M_Z}{My}\right)^2} - 1 = 0 \qquad (1)$$

where N is the axial force; Ny the yield axial force; My the yield bending moment M_Y the bending moment around the Y-axis and M_Z the bending moment around the Z-axis.

In this analysis, the yield strength for axial force Ny, and yield strength for bending moment My, were values obtained by the strength tests on the joints. The geometrical non-linearity is included by using the slope deflection method for general member buckling for the beams, using a total Lagrangian approach.

Comparison of the experimental and analytical results

11. The experimental results are compared with the analytical results given in Figs. 8 and 9. The former figure is for the relationship between the applied load P and the downward deflection δ_Z at node 15; the latter is the interaction diagram between the axial force and the bending moments in member 'A', beside node 15. Notice that the bending rigidity of $K_B = 2 \times 10^4$tf cm/rad corresponds to the measured value obtained by testing the joints, and the strength for both the joints and members are assumed according to the values obtained from the same experimental tests. Judging from the agreement of the analytical paths with those from the experimental tests, the procedure given by the present non-linear

analysis is sufficiently reasonable to be used as a design tool for ultimate strength calculations.

Photo. 2. Dome under loading.

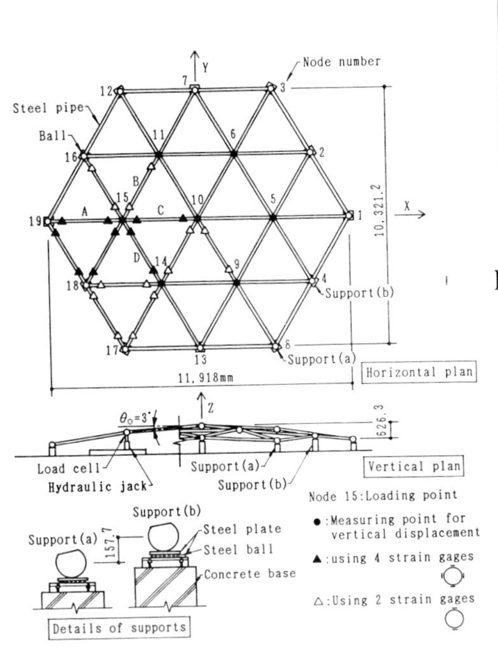

Fig. 6. Schematic dome test.

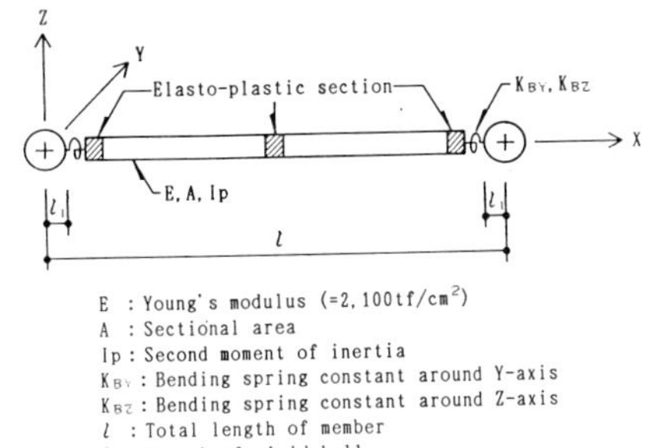

Fig. 7. Analytical model of member.

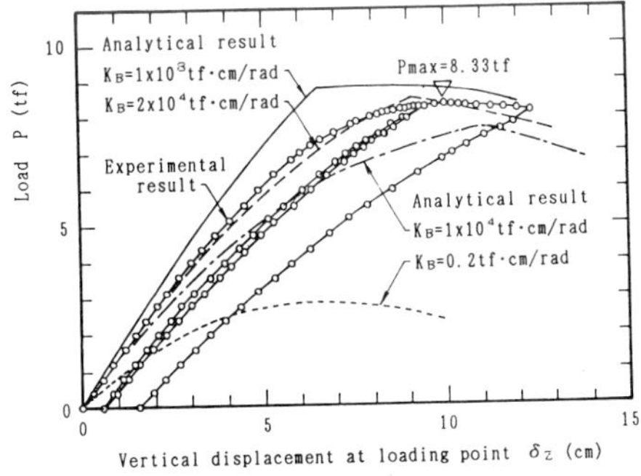

Fig. 8. Results of the dome test. Relation between P and δ_Z.

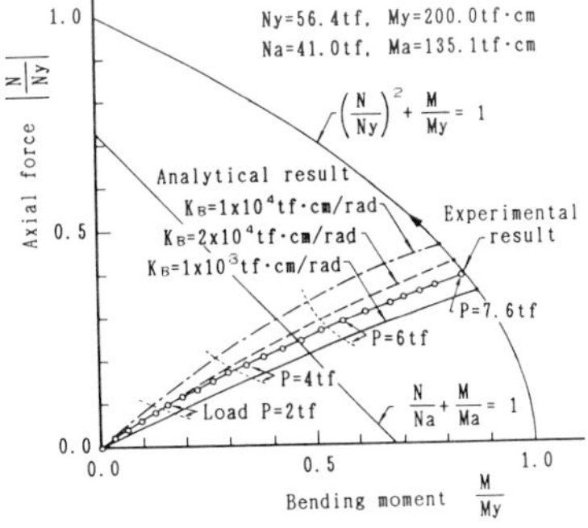

Fig. 9. Results of dome test. Relation between N and M (Me.A-No.15).

12. In a typical dome, there is some difference between the theoretical behaviour given by the present analysis and that of the actual dome, because the bending spring constant of the joint will be altered by a change of axial force caused by an increase in load. However, the behaviour of an actual dome can be estimated almost exact, by the present method of analysis.

BUCKLING BEHAVIOUR OF SINGLE-LAYER LATTICED DOMES

13. The present and next sections introduce the single-layer latticed dome and arch constructed using the T.U.T.-S system and explain several results on the non-linear analysis. Fig. 9 illustrates the configuration of a roof in the T.U.T-S system, approved by Japan Architecture Centre, as the first structure constructed adopting this system. The global roof is composed of two roofs of the same structural shape. Each one has a rectangular plan of 18.9m x 19.5m with a radius of 24.0m as a spherical shell. The design was based on the aforementioned experiments and the method of non-linear analysis. The analyses performed for the roof are summarised in the following paragraphs.

Model for the analysis

14. The configuration of the model for the analytical studies is shown in Fig. 11 (ref. 3). The shape is of a rectangular spherical segment with the half open angle θ_o varying from 1, 2 and 3 degrees, because this angle is one of the dominant structural factors. The models numbered as Model 1, 2 and 3 depend on the value of the angle. The same size of steel tube is assumed for every member. The bending spring constant K_B of the member ends, described with a dimensionless bending constant κ in Eq. 2, is set to the value $\kappa = 1.0$ and $\kappa = 1000$. The value of κ=1.0 is assumed for the semi-rigid joint connection and $\kappa = 1000$ is assumed for a fully rigid connection.

$$\kappa = K_{BY} \cdot l_o/(E \cdot I_P) = K_{BZ} \cdot l_o/(E \cdot I_p)$$

15. Two types of boundary conditions, B1 and B2, were used. The former is a pin-support on the entire periphery, and the latter is a combination of pin-supports at four corners and roller-supports along the four sides.

16. With these boundary condition, two kinds of analysis were performed; linear eigen-value analysis, and material and geometric non-linear analysis. Plasticity is assumed only for tubular members, and Ny and My are calculated by Eq. 3, where the yield strength σy is assumed to be 2.4tf/cm^2.

$$Ny = sy \cdot A \quad My = sy \cdot Z_B \tag{3}$$

where A is the cross-sectional area for the steel tube.

Z_B is the plastic modulus for the steel tube.

Expression in terms of column strength curve

17. Fig. 12 expresses the ultimate strength of domes in terms of the dimensionless ultimate strength n_{cr} and the normalised slenderness ratio L,

similar to straight columns. The values of n_{cr} and Λ are defined for the particular member which is the most relevant to the ultimate behaviour of domes. They are calculated as described by Eq. 4. P_{cr}^{lin} is the linear buckling load determined by a linear eigen-value analysis and P_{cr}^{pl} is the elasto-plastic buckling load determined by material and geometric non-linear analysis.

$$\begin{aligned} &Ny = \sigma y \cdot A_{(m)}, \quad N_{cr} = N \cdot P_{cr}^{pl} / Pd, \quad N_{cr\,(m)}^{lin} = N\, P_{cr}^{lin} / Pd \\ &n_{cr} = N_{cr} / Ny, \quad \Lambda = \sqrt{Ny / N_{cr(m)}^{lin}} \end{aligned} \tag{4}$$

where N is the axial force for the particular member under design load and Pd is the design load per node.

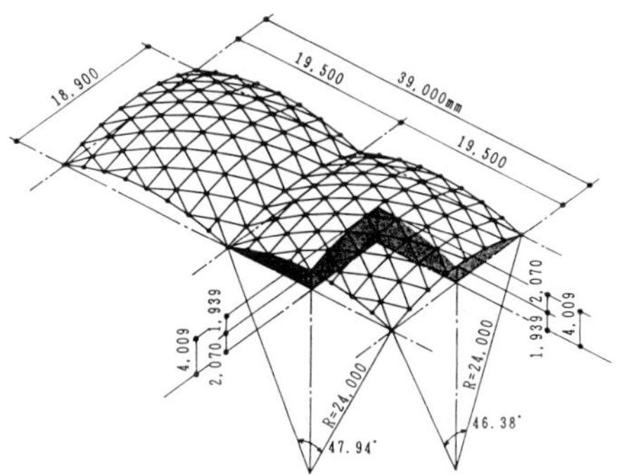

Fig. 10. Configuration of roof for T.U.T. works.

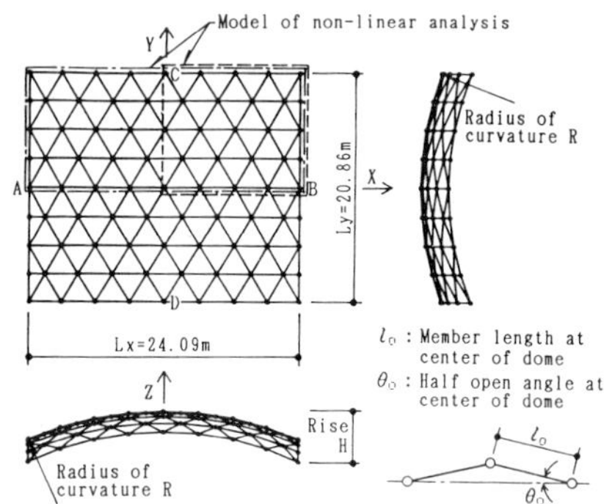

Fig. 11. Analytical model of single-layer latticed dome.

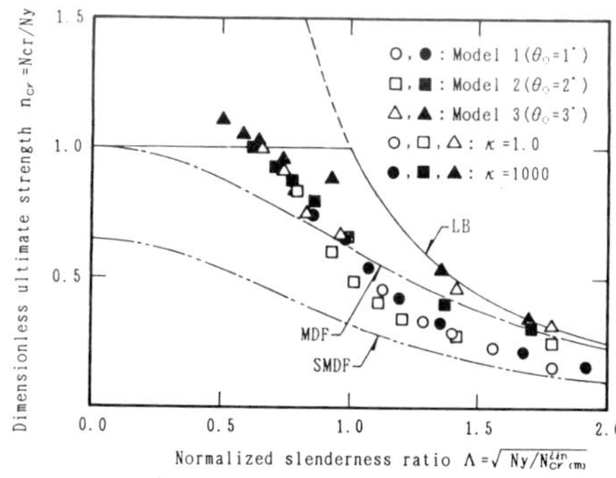

(a) B1 boundary condition.

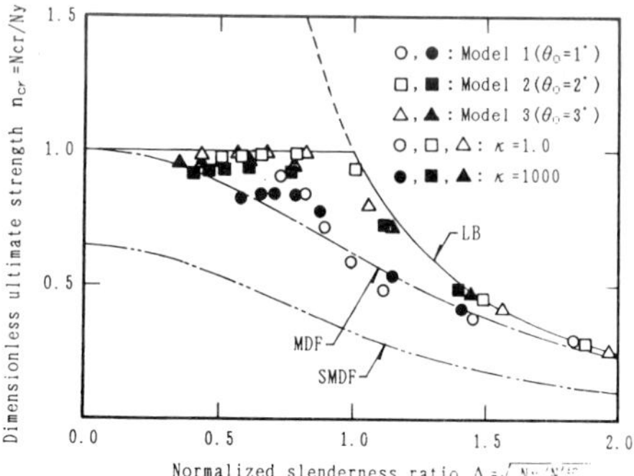

(b) B2 boundary condition.

Fig. 12. Results of analysis. Relationship between n_{cr} and Λ for the single-layer latticed dome under uniform loading.

Results of the analysis

18. In Fig. 12, the solid line is composed of two lines, based on the linear buckling (LB) for the parameter Λ being greater than 1 and full axial strength for L being less than 1. The alternate long and short dash line represents a curve of the modified Dunkerley formulation (MDF) as shown in Eq. 5, and the alternate long and two short dashes line SMDF, Eq. 6, is the curve using safety factors K_{el} and K_{pl} respectively for elastic buckling and plastic failure. In the present case, 2.30 and 1.55 are adopted for K_{el} and K_{pl} (ref. 4).

$$\left|N_{cr}/N_{cr(m)}^{lin}\right|+\left(N_{cr}/N_y\right)^2=\Lambda^2\left|N_{cr}/N_y\right|+\left(N_{cr}/N_y\right)^2=1 \qquad (5)$$

$$\Lambda^2\left|K_{el}\bullet N_{cr}/N_y\right|+\left(K_{pl}\bullet N_{cr}/N_y\right)^2=1 \qquad (6)$$

19. In the case of the B1 boundary condition, the numerical results distribute, irrespective of the models, between MDF and SMDF for Λ being greater than 1 and between MDF and $n_{cr}=1$ for Λ being less than 1. On the other hand, in the case of B2 boundary conditions, the results of Model 1 are almost similar to the curve of MDF. Also the results of the Models 2 and 3 almost correspond to LB for Λ being greater than 1, and between MDF and $n_{cr}=1$ for Λ being less than 1. The effect of member crookedness, initial nodal imperfections and load distribution were examined, however, those results are not presented in the present paper.

BUCKLING BEHAVIOUR OF FOLDED PLATE-LIKE RETICULAR ARCHES

20. Fig. 13 represents the shape of a roof constructed for a gymnasium in the T.U.T.-S system. The roof has a rectangular plan of 35.4m x 43.04m like a series of folded plate-like arches, with repetition at an interval of 3.91m.

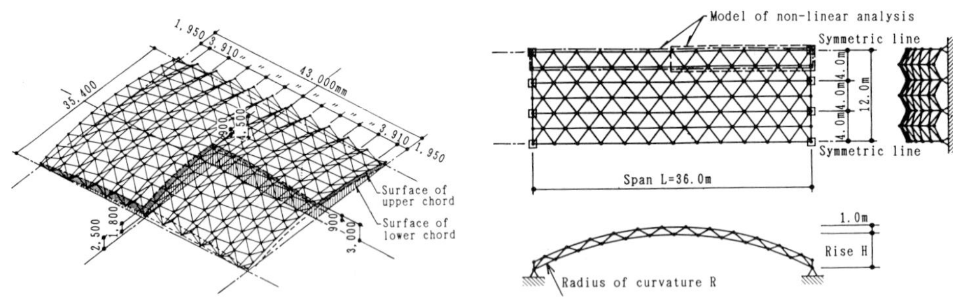

Fig. 13. Configuration of the roof for the Tachikawa gymnasium.

Fig. 14. Analytical model of folded plate-like reticular arch.

Analysis model

21. The configuration of the analytical model is shown in Fig. 14 (ref. 5). The rise H of arch is assumed to be 4.5, 3.0 and 2.0m respectively for arches 1, 2 and 3, because of the need to investigated the effect of the rise. Pin-supports are adopted at the ends of the lower chords, marked as squares. The value of the bending rigidity for joints is set to $\kappa = 0.01$ for pin-joints and $\kappa = 1000$ for rigid-joints.

Results of analysis

22. The results for uniform loading are shown in Fig. 15 in terms of n_{cr} and Λ. The results resemble each other irrespective of rise H. Moreover, n_{cr} corresponds almost to the linear buckling LB for Λ being greater than 1 and is nearly in agreement with $n_{cr} = 1$ for Λ being less than 1. The effects of member crookedness, initial nodal deflections, load distribution and boundary conditions were examined.

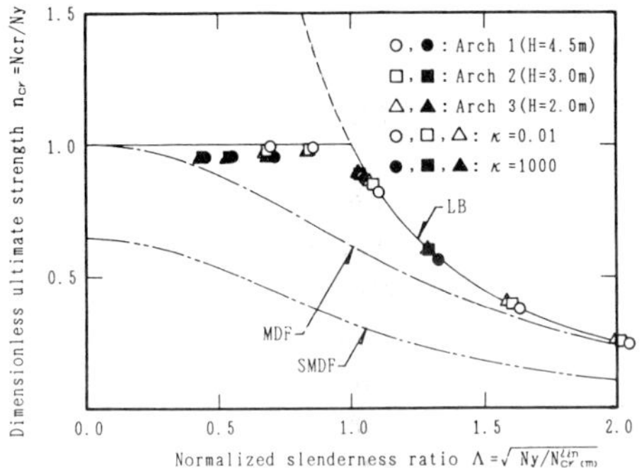

Fig. 15. Results of the analysis. Relationship between n_{cr} and Λ for folded plate-like arch.

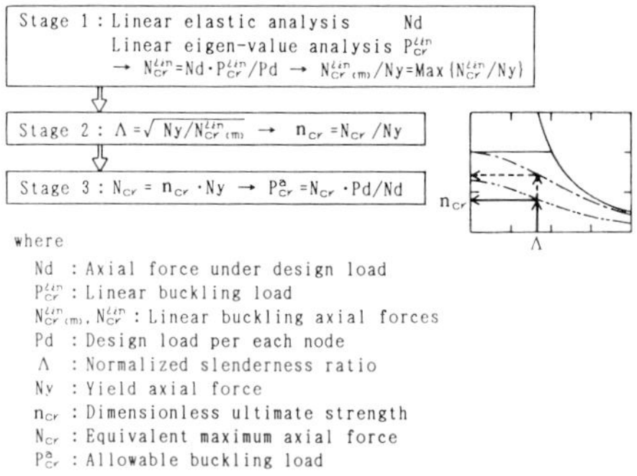

Stage 1 : Linear elastic analysis Nd
Linear eigen-value analysis P^{lin}_{cr}
$\rightarrow N^{lin}_{cr} = Nd \cdot P^{lin}_{cr}/Pd \rightarrow N^{lin}_{cr(m)}/Ny = Max\{N^{lin}_{cr}/Ny\}$

Stage 2 : $\Lambda = \sqrt{Ny/N^{lin}_{cr(m)}} \rightarrow n_{cr} = N_{cr}/Ny$

Stage 3 : $N_{cr} = n_{cr} \cdot Ny \rightarrow P^{a}_{cr} = N_{cr} \cdot Pd/Nd$

n_{cr} Λ

where

Nd : Axial force under design load
P^{lin}_{cr} : Linear buckling load
$N^{lin}_{cr(m)}$, N^{lin}_{cr} : Linear buckling axial forces
Pd : Design load per each node
Λ : Normalized slenderness ratio
Ny : Yield axial force
n_{cr} : Dimensionless ultimate strength
N_{cr} : Equivalent maximum axial force
P^{a}_{cr} : Allowable buckling load

Fig. 16. Estimation method for the allowable buckling load.

ESTIMATION METHOD FOR THE ULTIMATE STRENGTH OF LATTICED DOMES AND ARCHES

23. As a consequence of comparing results from the non-linear analysis with the experimental results, the proposed non-linear analysis and also the expression for the ultimate buckling loads in terms of column strength curves have proved to be applicable and efficient in design practice. Thus, the buckling of single-layer latticed structures adopted in this paper may be estimated based merely on the relationship between n_{cr} and Λ for the particular member; by taking into consideration the imperfection sensitivity as recommended in many shell codes. The steps for estimation is shown in Fig. 16. When more precise attention is required to imperfections, the foregoing non-linear analysis program can be efficient.

Photo. 3. Interior of single-layer latticed dome for T.U.T. works.

Photo. 4. Shape of folded plate-like reticular arch for the Tachikawa gymnasium.

CONCLUSION

24. The method of analysis using the proposed element, is applicable for estimating the buckling behaviour of single-layer domes composed of semi-rigid joints, because the experimental results for loading test on domes are in agreement with the analytical results. The ultimate strength results for single-layer domes and folded plate-like reticular arches could be estimated effectively by using the concept of column strength curves considering appropriate safety factors.

REFERENCES

1. UEKI T., KUBODERA I., MATSUSHITA F. and KATO S. A new approach to the design of a single layer reticulated dome. Testing of ball-type rigid joints, and construction of a dome, Proceedings of IASS congress, Madrid, Vol.4, 1989.
2. UEKI T., MUKAIYAMA Y., KUBODERA I. and KATO S. Buckling behaviour of single layered domes composed of members with axial and bending springs at both ends, Proceedings of IASS congress, Madrid, Vol. 4, 1989.
3. UEKI T., KUBODERA I., MATSUSHITA F. and KATO S. Study on the elastic and elasto-plastic buckling behaviour of single layered domes composed of members having axial and bending springs at both ends, Proceedings of IASS symposium, Copenhagen, Denmark, Vol. 3, 1991, pp. 93-100.
4. KOLLAR L. and DULACSKA E., Buckling of shells for engineers, John Wiley & Sons, Chichester, 1984, pp. 260-265.
5. UEKI T., MUKAIYAMA Y., KUBODERA I. and KATO S. Estimation of ultimate strength of folded plate-like reticular arch based on linear eigen value analysis, Proceedings of the third of the summer colloquium on shell and spatial structures, Taegu, Korea, 1990, pp. 493-500.

27. The design, analysis and fabrication of a glass-reinforced domed structure for use as a flight simulator

J. D. CLINTON and M. PAGNOTTA, 3D Structures, USA

SYNOPSIS. The commercial world is often a harsh environment when compared to the scientific or theoretical world. The beauty of geometry, architecture, and engineering theory give way to constraints and requirements, most notably cost and time. The structure examined in this paper, a UH-lN helicopter flight simulator, is a blend of theory and practicality, using a variety of materials, geometric shapes, and manufacturing methods.

OVERVIEW

1. The design of a flight simulator is dependent upon the design of the aircraft and the desired training exercises for that aircraft. The design of the UH-lN helicopter flight simulator is no exception. The UH-lN helicopter is a wide body aircraft with the pilot and co-pilot seated side by side and having a field of view of approximately 110 degrees in azimuth and +20 to -50 degrees in elevation. The trainer consists of a 7.3 meter diameter spherical projection screen, 5 computer controlled video projectors for real time combat simulation, a display support structure (DSS), and an instructor station, all on a motion base. The choice and placement of projectors, the type of projection surface, the location of the pilot's and co-pilot's eye points relative to the spherical screen centre, and the centre of gravity and weight of the system all are critical factors to be considered in the design of the trainer. The image brightness, sharpness, contrast, and manoeuvrability are influenced by the choices made during the design process. The main objective of the system is to provide as realistic a training environment as possible within a reasonable cost. This paper will concentrate on the engineering and structural design of the UH-lN simulator. The main structural elements of the UH-lN simulator are the dome, the DSS, and the projector truss.

UNIQUE DESIGN FEATURES

2. There is nothing unique about the spherical dome used as the viewing area of the simulator; most flight simulators use either full or partial spheres based on the amount of viewing area required. What is unique about the UH-lN simulator is that the spherical shape was kept as much as possible in order to take advantage of the inherent stiffness of a sphere as compared to a flat surface. Only a technical requirement to have an opening large enough to install the helicopter cockpit

prevented the design from being a full sphere cut off at 50 degrees below the equator.

3. Another feature of this simulator that has not been used in flight simulators built to date is that the dome was used as structural support for the heaviest items in the system, the projectors and related electronics. Typically, the domes currently used on motion based flight simulators react only their own weight. The UH-1N simulator sphere is constructed with lightweight composite material which can support the projector system in an environment which imparts 8 times the force of gravity in any direction.

4. The projector support frame was designed with more conventional materials. Aluminium tubing was used to form a fully triangulated three dimensional space truss. While the space truss could not be designed as a pure octahedral or octahedral/tetrahedral truss because of the fixed projector locations, the resulting truss is sufficiently stiff and strong so the dome, projector, and display support structure behave as a single unit. That is, there is very little relative deformation between the three main structural components while in motion or at the primary resonance mode shapes. This feature is important in maintaining image sharpness during training exercises. Fig. 1 shows an isometric view of the exterior of the structure and Fig. 2 shows the same view with the upper dome panels removed, revealing the projector truss.

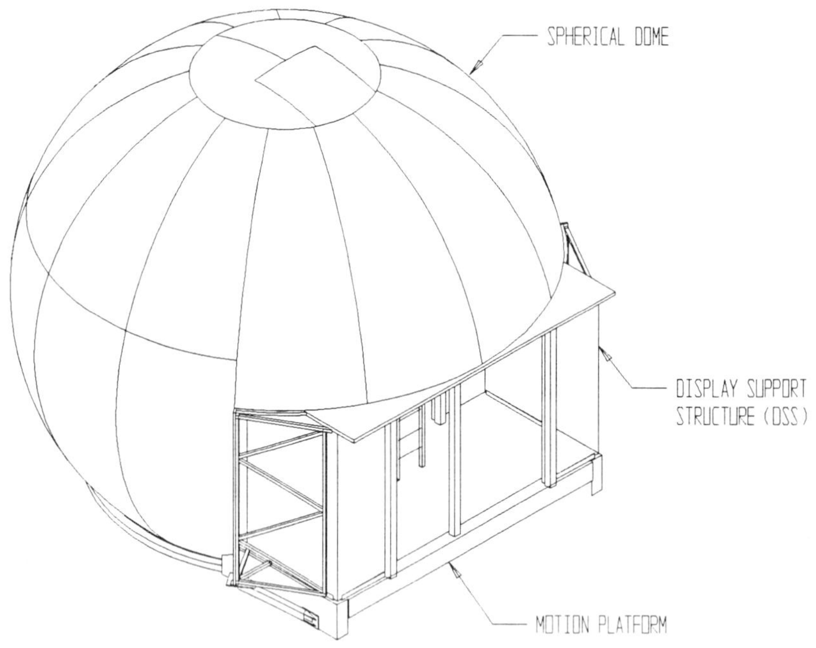

Fig. 1. Isometric view of UH-1N simulator exterior.

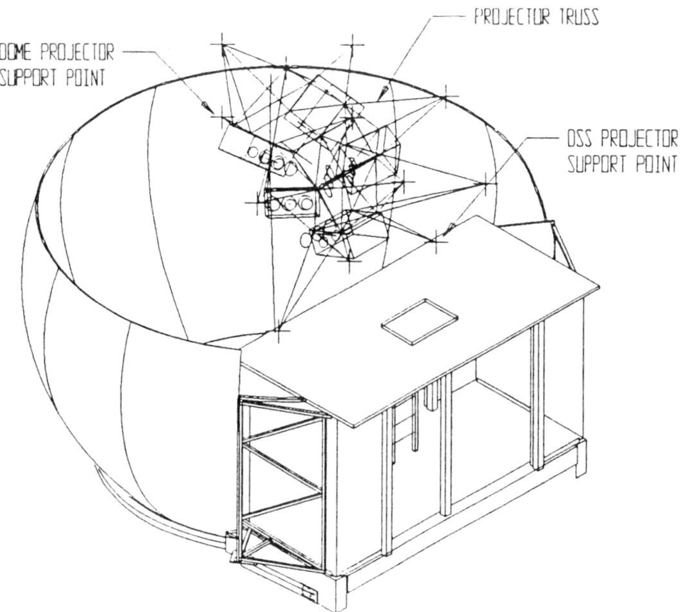

Fig. 2. Isometric view of UH-1N simulator with upper dome panels removed.

DESIGN DECISIONS

5. The key technical drivers in the design of the UH-lN simulator were:
   a) Minimum fundamental frequency of 15 Hz for the structure.
   b) The ability to withstand 8 G's without permanent deformation.
   c) Mass moment of inertia limits to ensure performance of the motion system.
   d) Surface quality of the viewing surface.

It was decided to design the structure to meet the frequency requirement first. Once this was accomplished, weight and mass moment of inertia estimates could be made and the detailed structural analysis could be performed with the knowledge of what, if any, additional weight could be added to meet strength criteria and yet not compromise the other requirements. Since the performance of the motion system will decline with excess weight, a graphite composite structure is the best choice for stiffness vs. weight.

6. For economic reasons, the dome was constructed using a glass weave with vinyl ester resin as the skins on a 2.5 cm foam core. The foam core, thicker than what is typical for similar domes, was chosen with the knowledge that there would be point loading on the spherical panels from the projector truss attachments. The thicker foam provides superior bending strength and rigidity in point loading applications. The use of the thicker foam throughout the sphere, even in lightly loaded areas, was an acceptable weight penalty since the foam is a small percentage of the total panel area density. Since the viewing surface quality is so important to the overall quality of the simulator, the dimensional tolerances on the viewing area are very strict. In addition, the viewing surface must be very smooth and uniform in colour to allow an optical finish of either a lenticular material or a paint.

7. When using composites, it is usually desirable to form the skins in such a way that the mechanical properties are uniform (isotropic) regardless of the direction of the loading, as is the case with aluminium and other metals.

Manufacturing limitations dictated that this could not be the case with the UH-lN simulator sphere. Therefore, the anisotropic behaviour of the sphere had to be considered in any finite element modelling and analysis performed on the sphere. While this increased the complexity of the analytical tasks, it was not a controlling factor in the finished product.

8. The main driver in meeting the frequency requirement was reinforcing the opening in the sphere. The large opening considerably softened and weakened the total structure, so much so that the housing which replaced the opening, the display support structure, or DSS, had to be constructed with graphite reinforced vinyl ester skins on 5.0 cm foam core. Depending on the specific panel, either two layers of graphite (1.12 mm) or three layers (1.68 mm) were used for the skins. This change, along with additional reinforcing at the openings of the DSS, resulted in a structure which was able to meet the frequency requirement. In addition, the interface connections of both the dome and DSS to the motion base had to maintain the structural stiffness and transfer the imposed loads to the motion platform. In the case of the dome connection, the structure could not exist on the dome interior as the viewing area of the dome extended to the bottom.

9. The aluminium tube space truss was chosen as the projector support for a variety of reasons. First, four specific attachment locations had to be provided for the five projector assemblies for a total of twenty pickup points. Second, the projector area had to be kept as open as possible to allow maintenance to be performed on the projectors without removal of structure. Third, since the dome was being used to support the projector truss, it was desirable to spread the load to a number of points which cover a large area of the dome non-viewing surface. In this way, the point loads could be kept to a minimum and the dome geometry could be used to its fullest advantage structurally. Finally, the tube design allowed for considerable tolerance adjustment. Dimensional variations in both the dome and DSS can be taken by adjusting the tube lengths without affecting the fixed projector pickup points.

ANALYTICAL ASPECTS

10. The first task in performing the structural analysis of the UH-1N simulator was to establish a material data base for both the glass/vinyl ester and the graphite/vinyl ester. Sample panels of both materials were fabricated and tested for both strength and modulus of elasticity. With this baseline data, the panel characteristics could be determined for incorporation into a finite element model using classical lamination theory. Therefore, in the case of the dome panels, the anisotropic nature of the skins was represented in the model. With this data, the basic frequency, deflection, and overall strength analysis could be performed directly with the finite element model. It should be noted that much of the design work and all of the finite element modelling was performed on a personal computer. No mainframes or workstations were required. However, the detailed connections were analysed by hand, as were buckling and local instability checks. As is often the case with structures, it was the detailed connections that required the most design and analytical effort.

Panel to panel joint.

11. The sphere was constructed in two main sections, the viewing area and the projector support area. The viewing area consists of nine panels (thirteen panels

close the section) from the bottom to 20 degrees above the equator. The projector support consists of thirteen panels from 20 degrees above the equator to 70 degrees above the equator, with a two piece cap forming the top. A key concern was the panel to panel joint, which had to be as strong as the basic dome panel and also had to withstand local bending and membrane forces from the projector truss attachment points. The final design, shown in Fig. 3, is a dense build-up of glass at the panel edges. The mating surface is the panel mid plane. Fasteners are used at 10.0 cm intervals into inserts in one of the mating parts. This joint is at all panel to panel vertical seams and the horizontal seam at 20 degrees above the equator. While this joint adds considerable weight to the structure, test results verified that the joint is at least as strong as the basic dome panel, which allows it to act as a natural projector truss pickup location.

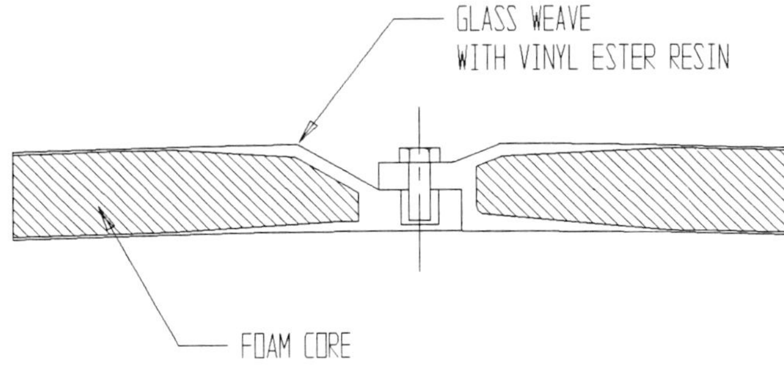

Fig. 3. Cross-section of dome panel to panel joint.

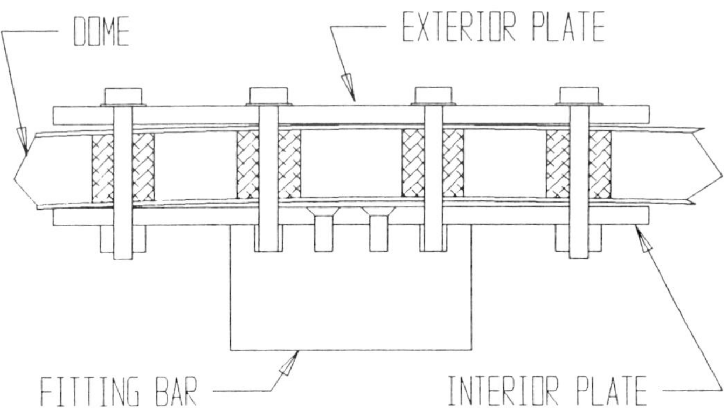

Fig. 4. Cross-section of projector truss to dome joint.

<u>Projector truss to dome joint.</u>

12. The projector truss tubes were designed to have their lines of action coincide at the panel midpoint. This would naturally result in very high local stresses, so a connection has to be designed which would locally distribute the load from the tubes. The design utilised, shown in Fig. 4, consists of a pair of aluminium

plates, one on the dome interior and one on the dome exterior. The plates are attached to each other by a combination of through bolts which pass through the core, and bolts through the panel to panel joint. The tube end fittings are attached to a fitting bar which is in turn attached to the interior plates.

DSS to motion platform joint.

13. The connection at the base of the DSS panels had a space restriction to avoid interference with the systems electronics. It was desired to transfer loads from both panel skins directly to the motion platform. Fig. 5, shows the design, which is a custom extruded aluminium channel which captures the lower portions of the DSS walls with through bolts. The dimensional tolerance of both the DSS walls and the channel interior were controlled in order to minimise the stresses imposed on the channel due to assembly bolt torque.

Dome to motion platform joint.

14. The dome to motion platform connection had to be made on the exterior surface only due to the viewing area requirements. The connection used is a glass/vinyl ester angle build-up which conforms to the dome surface. This double curvature gives the angle added strength and stiffness as compared with segmented flat angles. To connect to the dome, inserts were co-cured into the dome core in resin plugs. The angle is then bolted to the dome exterior. In addition, a field bond is used for added strength. The leg of the angle which attaches to the motion platform is bolted using threaded rods and threaded shoulder bushings to level the dome.

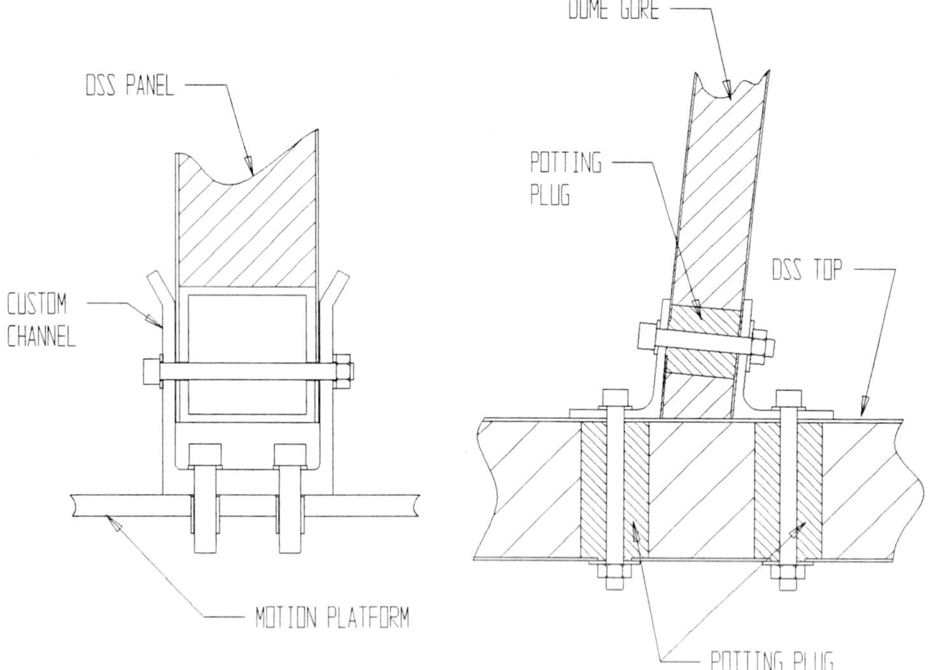

Fig. 5. DSS to motion platform joint.

Fig. 6. Typical connection joint.

Other panel to panel connections.

15. With the vast number of connections on this structure, custom connections such as those described in paragraphs 11, 13, and 14 had to be kept to a minimum for economic reasons. A typical joint which used common sections is shown in Fig. 6. This cross section is the region where the dome meets the DSS horizontal top surface. Aluminium angles are used to provide a two sided connection from the dome panel to the DSS top. In order to avoid local core crushing and to provide in plane load transfer capability, plugs of resin are inserted into both the dome and DSS panels at the bolt locations. Considerable effort was spent in determining the proper combination of material and process which gave high reliability and high strength.

DESIGN AND ANALYSIS SUMMARY

16. Each of the main components of the UH-1N flight simulator is unique in both geometric shape and materials used. Table 1 summarises the primary features of these components.

Table 1

Component	Geometric features	Materials utilised
DOME	Spherical	Glass/VE, foam core
DSS	Rectangular Prism	Graphite/VE, foam core
PROJECTOR SUPPORT	Space Truss	Aluminium tubing

The entire project is just a small indication of the vast choices that exist for the modern Design and Structural Engineer. One can take the path of simplicity, cost effectiveness, pure theory, or state of the art technology. Or as is the case with the UH-lN flight simulator, a portion of each was chosen and blended to produce a structure which has both its common and unique aspects, and most importantly meets the technical requirements.

28. A new method for the subdivision of braced domes

J. KOLOSOWSKI, Calgary, Canada

SYNOPSIS. The proposed grid configuration for braced domes combines some advantages of the following ideas and practical aspects: easy calculation of nodal co-ordinates; has small number of different members, joints and roof covering panels as possible; members interconnected at the same angles; easy fabrication and erection with the possibility of eliminating inaccuracies of fabrication; and finally, the possibility of pre-stressing.

INTRODUCTION

1. The optimisation of braced domes is a very complex function of many parameters such as span, support condition, cross-sectional properties of members with their mutual ratios, type of joint, type of roof covering, fabrication, erection techniques, mass production and many others. It is very difficult, if at all possible, to include all of them into any theoretical analysis, which should always be preceded by the selection of the type of bracing, the latter has a fundamental influence upon the load-carrying capacity and its deflection characteristics.

2. The objective of this paper is to present a new method for the subdivision of braced domes - taking into account as many parameters as possible. Further improvements of the proposed method are under way.

PROCEDURE

3. It is proposed to introduce a surface obtained by rotating a small circle 'A' along two much larger quarter circles with a horizontal part closing the gap between the quarter circles. If two cuts along the lines B-B and C-C with two other cuts $0'_1$-D are applied, then the remaining surface represents two opposite segments of the proposed dome as shown in Figs 1,2.

4. The convenient properties of the proposed shape of domes are:

a) any section through $0'_1$ perpendicular to the plane XZ_1 leaves a circular trace on its surface of a constant radius r, and

b) any section parallel to the XZ_1 plane is represented by a circular arc with the centre on the Y_1 axis.

Space structures 4. Thomas Telford, London, 1993

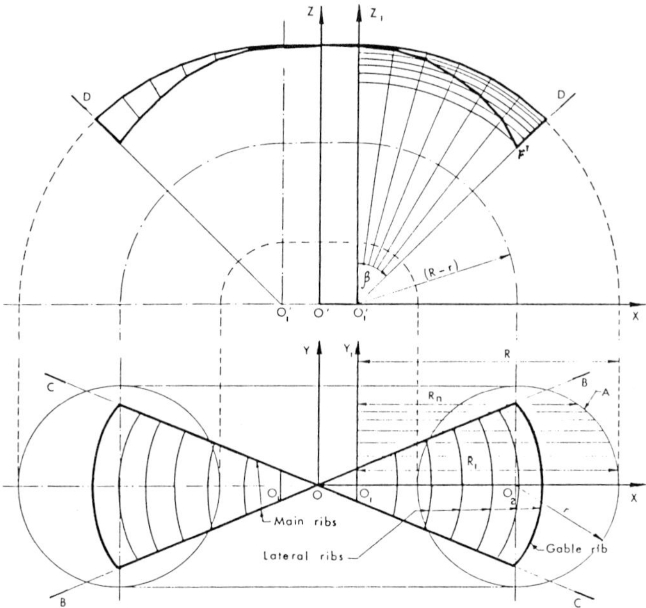

Fig. 1. Surface of segments.

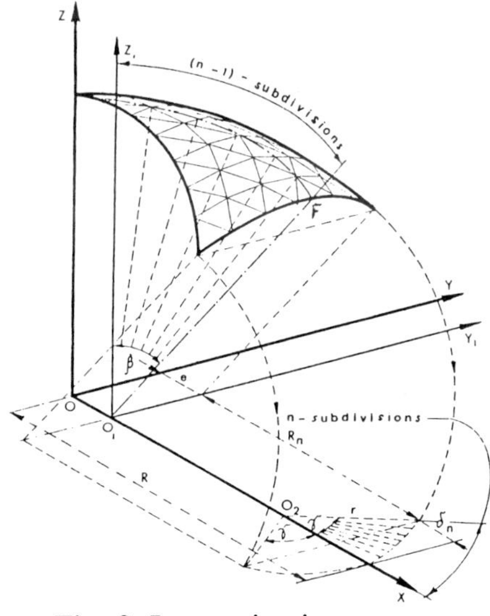

Fig. 2. Isometric view.

5. Finally, the surface of the proposed dome is composed of 'm' identical segments meeting along the valleys between them. Thus, the lateral ribs between the main ribs represent closed rings composed of circular arcs of the constant

curvature l/r and variable span, as it can be seen in Figs 5,6. All the joints of each segment are positioned on these arcs.

6. There are two important advantages of this arrangement of arches:
a) the horizontal perpendicular to the main ribs' components of the reactions of these arches meet on the main ribs and cancel each other for any symmetrical load, and b) any point or irregular load on a part of the dome is transferred to neighbouring parts of the dome and thus is distributed over a larger portion of the dome.

7. The angle 2γ subtended by the gable rib about the centre 0_2 must be subdivided into 2n parts. The planes parallel to the XZ_1 plane, through the points of subdivision on the gable rib, intersect the considered surface along the circular arcs with their centres on the Y_1 axis. The radii of these arcs are gradually reduced by δ_n from the value of R at the centre line of the segment to R_n at the support F, as shown in Figs 1,2. The same subdivision of the angle γ refers to all the lateral ribs. Therefore, the y co-ordinates of the joints on the lines parallel to the X axis are the same.

8. The angle β subtended between the YZ plane and the plane of the gable rib, i.e. $\measuredangle Z_1 0'_1 F'$ must be subdivided into (n-l) equal parts as shown in Fig. 2. The planes of these subdivisions represent the planes of the lateral ribs between the main ribs. If these planes are rotated about the axis Y_1 by the angles $(90^o-\beta_n)$ until they reach the XY plane, then all of them will coincide with the circle of the radius r in the plan. They represent the true shape of the lateral ribs.

PREVIOUS THEORETICAL AND EXPERIMENTAL INVESTIGATIONS ON DOMES

9. The most important reference is made to the results of extensive feasibility studies of spherical domes and domes composed of barrel-vaults carried out in the Department of Civil Engineering of the University of Surrey, Guildford, UK. Analytical studies for the linear analysis of stiff-jointed skeletal structures based on stiffness method have been augmented by experiments on small-scale models of domes (refs 1-4). The conclusions from these studies support the advantages of the suggested subdivision based originally on just intuition.

10. The next reference is made to Timoshenko and Gere (ref. 5) who dealt with the problem of buckling of a circular arch under a uniform pressure and found that for practical use, it is convenient to represent the critical pressure as a function of the span ℓ and the rise h of the arch. The formula for the critical pressure takes the form shown in Fig. 3. The numerical values of the coefficient γ_2 given by Timoshenko are represented on the graph for the following cases: no hinges; one hinge; two hinges; and three hinges.

11. The following conclusions can be drawn from this graph:
a) the maximum value of γ_2 for all the cases occurs when:

$$h : \ell = 0\ 25 \tag{1}$$

b) the smaller the number of hinges the greater the critical pressure.

12. Eq. 1 applies to circular arches under a uniform pressure acting radially with no translations at the ends; while in the proposed system of domes loads are acting mainly vertically, the ends are partially restrained but not rigidly and the arches are inclined by some angles to the vertical plane. Nevertheless, it is felt that the behaviour of these arches should be similar to the behaviour of the arches considered by Timoshenko. Therefore, for a preliminary design of domes it should be reasonable to assume that the ratio of rise to span, $h:\ell = 0.25$ and that the joints are rigid.

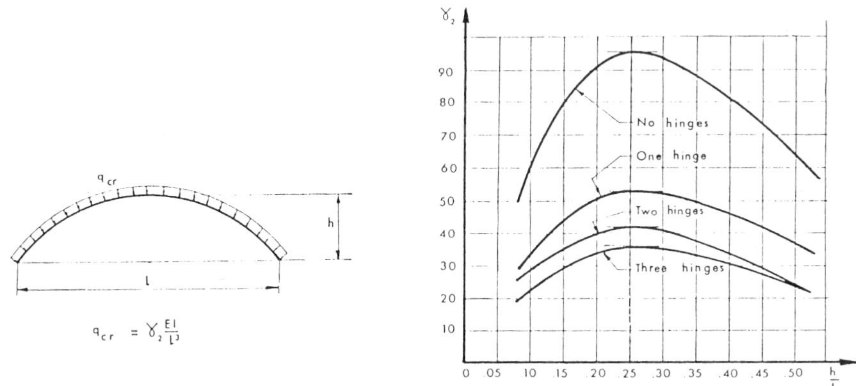

Fig. 3. The graphical representation of the critical pressure of circular arches, (ref. 5).

DERIVATION OF THE FORMULAS FOR CALCULATION OF NODAL CO-ORDINATES

13. The span of the dome 'ℓ' is usually given. The rise of the dome 'h' can be determined from Eq. 1. The relation between the main radius of dome R, the rise and the span of dome can be determined from the property of the circle:

$$(\ell /2)^2 = h(2R-h)$$

therefore: $R = \ell^2/8h + 1/2h$ (2)

14. Referring to Fig. 4, the angle α subtended by the main rib can be found from the triangle E"F"O" as:

$\sin\alpha = E''F'':F''O'' = \ell:2R$ (3)

or from the triangle E"F"G"

$\tan 1/2\alpha = E''G'':E''F'' = 2h:\ell$ (4)

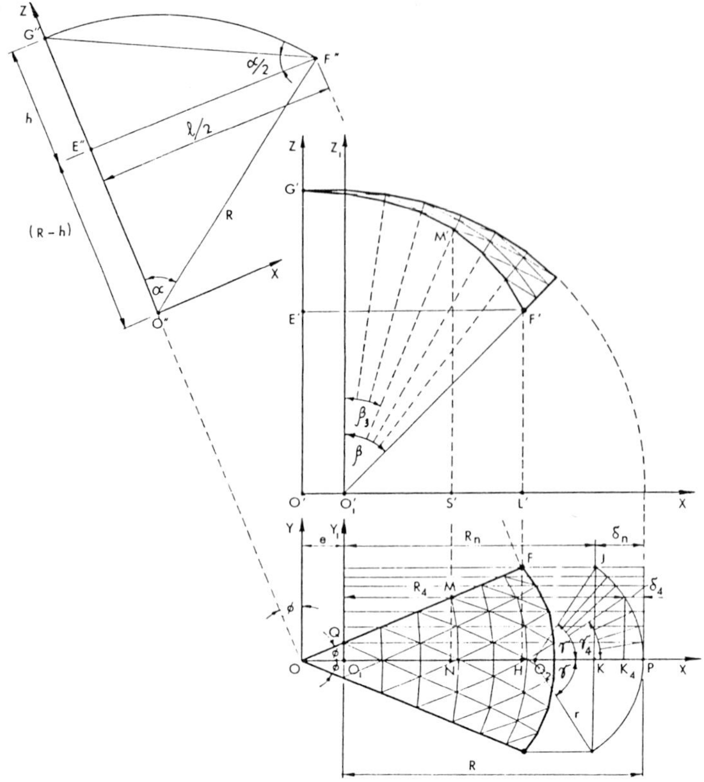

Fig. 4. Geometry of segments.

15. The number of segments 'm' and the number of subdivisions in the gable ribs 'n' can be selected to suit best the size of dome and architectural requirements. Then the angle subtended by one segment in plan 2ϕ can be determined from:

$$2\phi = 360^o : \text{(number of supports)} \tag{5}$$

16. The co-ordinates of the support F should be selected so that the support F lies on the considered surface. The y co-ordinate of the support F can be found from the triangle JKO_2 and from the triangle FHO:

$$r \sin\gamma = R \sin\alpha \ \sin\phi \tag{6}$$

17. If we assume the rise-to-span ratio is the same for both the main and lateral ribs, then $\alpha = y$ and Eq. 6 can be simplified to:

$$r = R \sin\phi \tag{7}$$

18. The z co-ordinate of the support F can be determined from the triangle E"F"O" and the triangle F'L'O'$_1$; $R\cos\alpha = R_n\cos\beta$

$$\cos\beta = R\cos\alpha : R_n \tag{8}$$

19. The x co-ordinate of the support F with respect to the origin O can be found from the triangle FOH:

$$x_F = FH : \tan\phi = r \sin\alpha : \tan\phi \tag{9}$$

20. If the lateral ribs are rotated about the Y_1 axis by the angles (90° - β_n), as explained in §7, the radii described by the joints can be evaluated from:

$$R_n = R - \delta_n \tag{10}$$

where, for example, for the joint M, $R_4 = (R-\delta_4)$ and:

$$\delta_4 = O_2P - O_2K_4 = r - r\cos\gamma_4 = r(1-\cos\gamma_4) \tag{11}$$

21. Now the eccentricity 'e', or the shift of the origin O_1 from the centre of the dome, $e = OO_1, = HO - HO_1$, where HO and HO_1 can be determined from the triangles FHO and F'L'O'$_1$ respectively

$$e = R.\sin\alpha \cos\gamma - R_n.\sin\beta \tag{12}$$

22. There are two alternative ways for the subdivision of the gable rib as shown on the numerical examples in Figs 5,6.
a) The alternative subdivision A when the angle γ subtended by the arc JP is divided into 'n' equal parts, and then certain gaps between the end joints of the lateral ribs remain.
b) The alternative subdivision B when the angles γ_n are adjusted so that the end joints of the lateral ribs meet each other along the straight line between two neighbouring segments in plan.

23. For the alternative subdivision B the y co-ordinate of the first joint from the apex of the dome can be found from the triangle OO_1Q; $y_1 = O_1Q = e \tan\phi$ but $O_1Q = r\sin\gamma_1$,

$$\sin\gamma_1 = e.\tan\phi : r \tag{13}$$

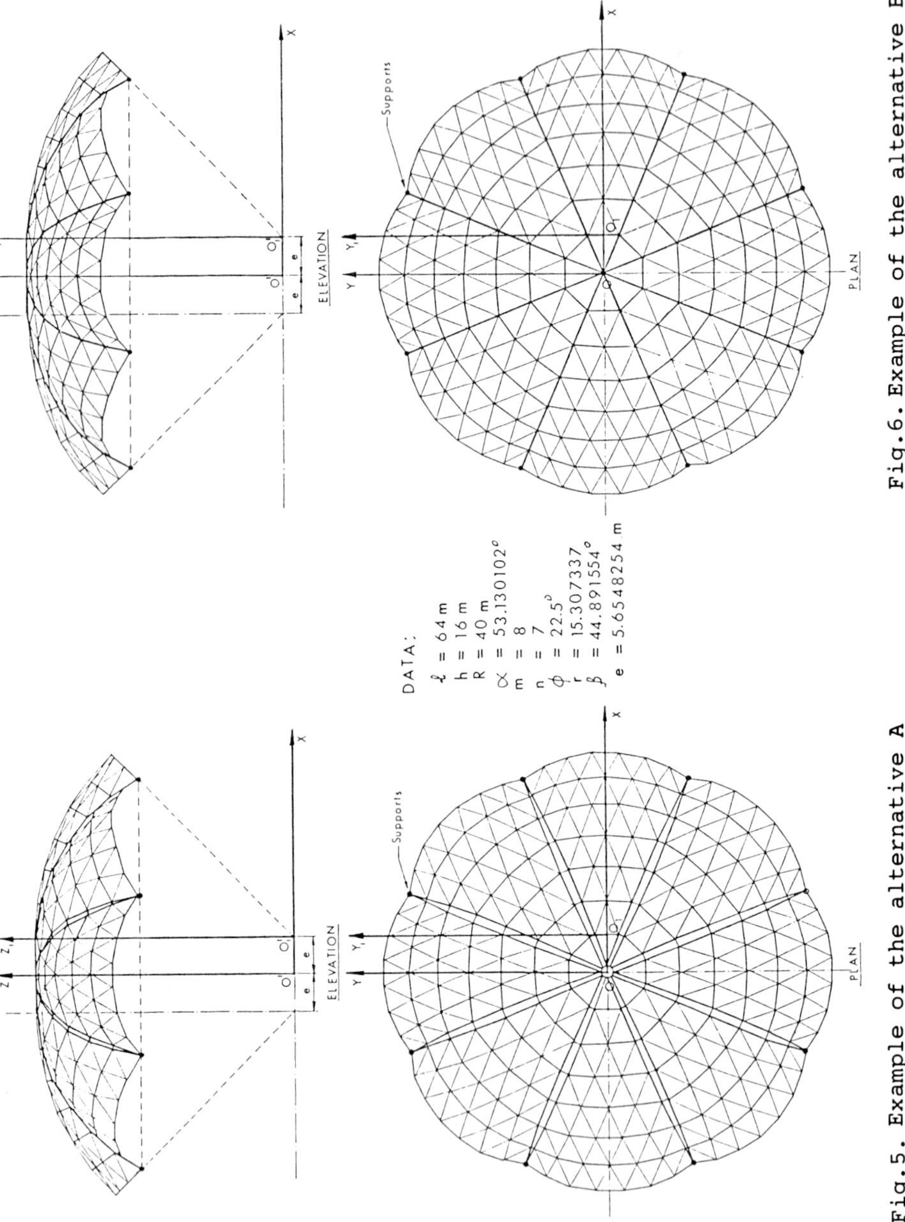

Fig.5. Example of the alternative A

Fig.6. Example of the alternative B

24. The x co-ordinates of other joints, for example of the joint M, can be determined from the triangle OMN:

$$x_4 = y_4 : \tan\phi = r.\sin\gamma_4 : \tan\phi \qquad (14)$$

Also, $x_4 = OO_1, + O_1N_1$ where O_1N can be found from the triangle $O'_1,M'S'$:

$$x_4 = e + R_4.\sin\beta_3 \qquad (15)$$

By equating Eqs 14,15 and substituting the values of R and δ_4 from Eqs 10,11 respectively, the following formula is derived:

$$r.\sin\gamma_4 : \tan\phi = e + [R - r(1-\cos\gamma_4)]\ \sin\beta_3 \qquad (16)$$

25. From this relation the angle γ_4 can be evaluated. Similarly, the angles γ_n for other joints can be determined and the co-ordinates of joints for both alternative subdivisions A and B with respect to the origin O_1 can be expressed, as shown in Fig. 7.

$$\begin{aligned} x_n &= R_n.\sin\beta_{n-1} \\ y_n &= r.\sin\gamma_n \\ z_n &= R_n.\cos\beta_{n-1} \end{aligned} \qquad (17)$$

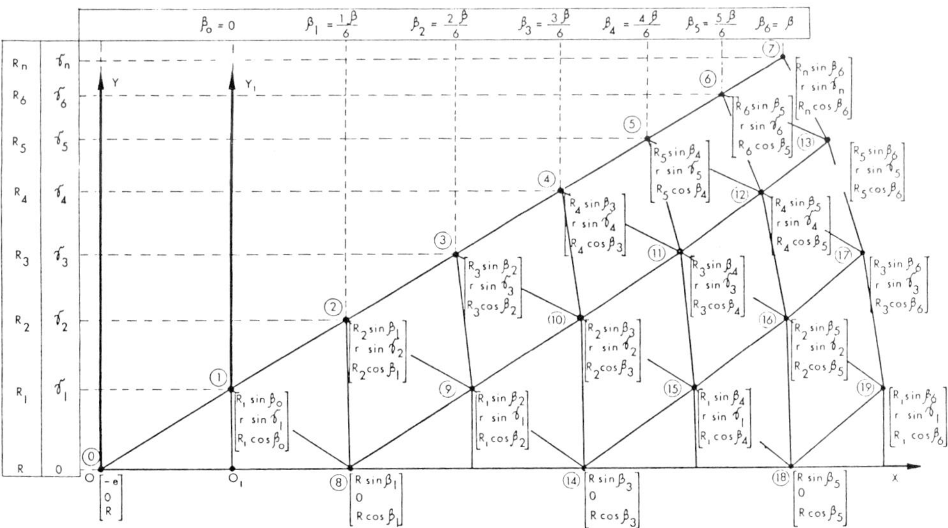

Fig. 7. Co-ordinates of joints.

26. The main ribs should be designed as tapering trusses, as shown in Fig. 8. The joints of the bottom chords of these trusses for the alternative B coincide with the end joints of the lateral ribs. The slopes of columns of these trusses to the XY plane are determined as shown in Fig. 8.

$$\tan\psi_n = x_n : (z_n \cos\phi) = \tan\beta_n : \cos\phi \qquad (18)$$

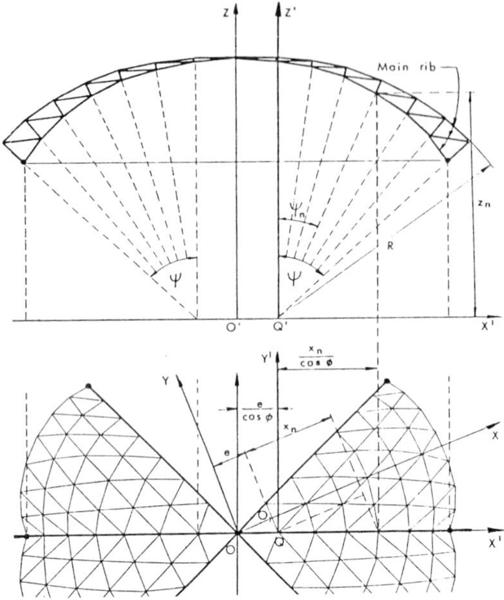

Fig. 8. Main rib.

DISCUSSION

27. From the previous theoretical and experimental investigations it may be concluded that a double curvature is preferred to a single one and, similarly, a large curvature to a small one. The proposed grid configuration is a practical example of braced domes following these guides and is adapted to satisfy the following points as far as possible:

a) to facilitate the calculation of co-ordinates of nodal joints,
b) to extend the mass production,
c) to simplify the method of erection and eliminate inaccuracies of fabrication and initial lack of fit,
d) to simplify the method of welding joints.

28. One of the most important problems facing a designer is to find the grid configuration with the smallest number of different members, joints and panels for the roof covering.

29. Fig. 9 shows that in lines parallel to the X axis:

a) all the panels for the roof covering are identical but left and right;
b) the lengths of members are the same;

c) the angles between any two corresponding members are also the same.
Owing to the above characteristics of the suggested method for the subdivision, the method of fabrication, erection and welding of joints can be simplified substantially. The elements between the main ribs can be fabricated on one construction jig.

30. There is a possibility of eliminating inaccuracies of fabrication and initial lack of fit by jacking up and down any particular joint.

31. For medium and large domes the main ribs constructed as tapering trusses can be, if necessary, pre-stressed by tension cables hidden inside the tubes of the top chords. In this way the stiffness of domes would be increased substantially.

32. The suggested method for the subdivision of braced domes can be applied to a large range of sizes of domes because the number of supports and subdivisions can be selected independently.

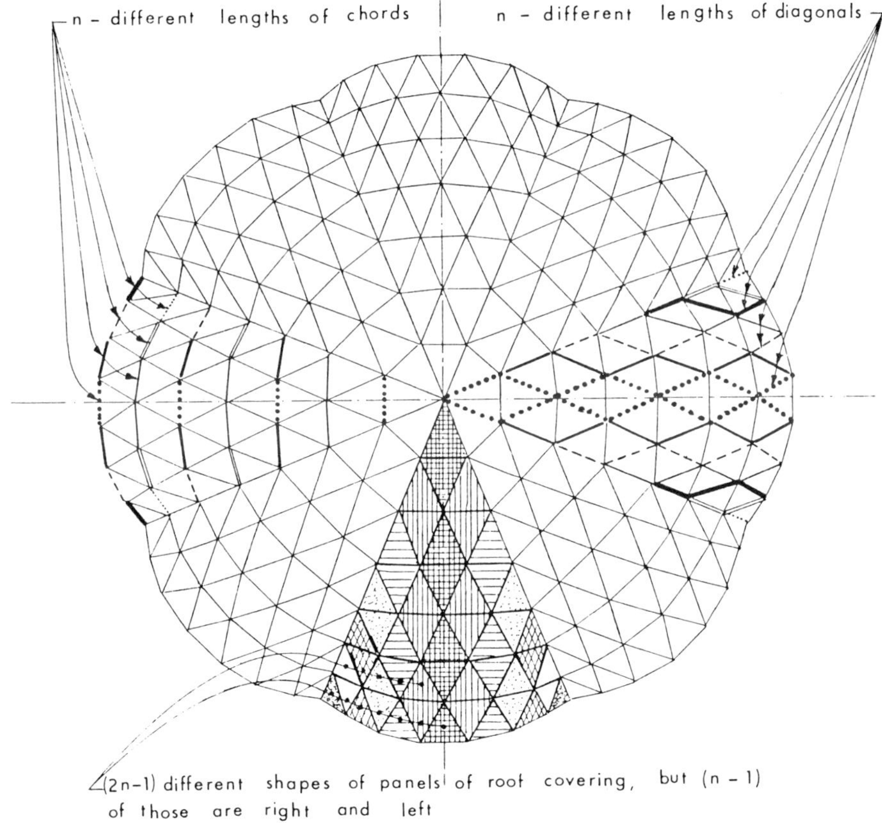

Fig. 9. Properties of the proposed subdivision.

REFERENCES

1. KOLOSOWSKI J. An economic method for the subdivision of braced domes with some recommendations for their design. In: Analysis design and construction of braced domes, by Z.S. Makowski, Granada, London, Toronto, Sydney and New York, Chapter 12, pp. 337-383, 1984.
2. KOLOSOWSKI J. A method of the subdivision of braced domes with the maximum number of members of equal length. International Symposium on Membrane Structures and Space Frames, Osaka, Japan, September 15-19, 1986.
3. MAKOWSKI Z.S. A history of the development of domes and a review of recent achievements world-wide. In: Analysis design and construction of braced domes, by Z.S. Makowski, Granada, London, Toronto, Sydney and New York, Chapter 1, pp. 1-85, 1984.
4. PANICKER K.S. An attempt at economical design of a dome, MSc Thesis, University of Surrey, Guildford, Surrey GU2 5XH, UK., 1977.
5. TIMOSHENKO P.S. and GERE J.M. Theory of elastic stability, McGraw-Hill Book Company, Chapter 7, pp. 299-305, 1961.

29. An analysis of braced domes

V. QUINTAS and J. M. AVILA, Universidad Politecnica de Madrid, Spain

SYNOPSIS. An approximate method of analysis, that makes it possible to obtain directly member forces in braced domes, is presented in this paper. It is based on obtaining equilibrium between global forces acting on the dome, and forces in the members. For non-symmetric forces the structure is not statically determinate and Bernouilli's hypothesis is used to take this into account. A comparison between the results obtained by this method, and those given by means of a matrix method of analysis, completes this work.

INTRODUCTION

1. A braced dome may be defined as a structure generated by a family of polygonal arches crossing at the crown of the dome, joined by horizontal rings and triangulated by diagonals elements. Any other type of braced dome can be analysed, by considering the behaviour of this basic type.

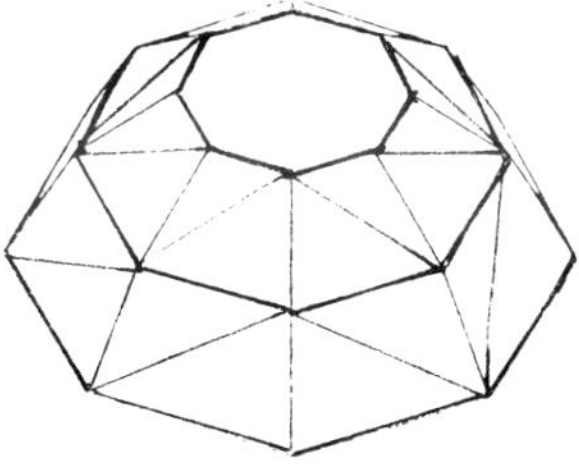

Fig. 1.

2. If the global forces acting on the dome are known, it is possible to deduce forces in the bars, following the analysis developed by F. Dischinger (ref. 1) for continuous domes.

GLOBAL FORCES

3. Suppose a braced dome is supported by a ring and submitted to any kind of vertical and horizontal loads, P_v and P_h. At a horizontal cross-section, struts of the dome will be subjected to three global forces; a vertical resultant, $Q_m = \Sigma P_{vm}$,

a horizontal resultant, $T_m = \Sigma P_{hm}$, and a global bending moment, $M_m = \Sigma P_{vm} x_m + \Sigma P_{hm} y_m = M_{vm} + M_{hm}$. The member forces may be obtained only as functions of Q_m and M_m, as follows:

$$T_m = \frac{\Delta M_{hm}}{\Delta y_m} \quad (1)$$

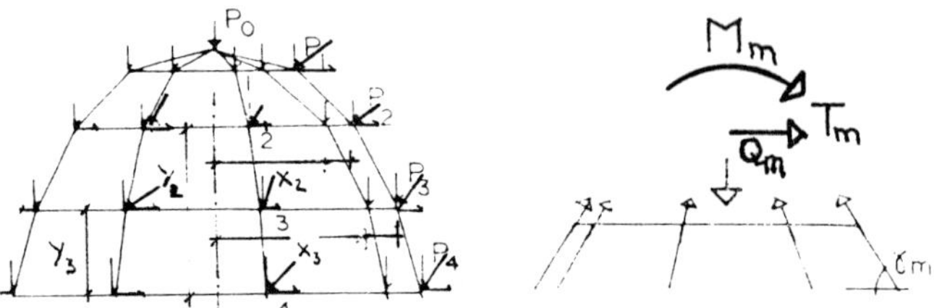

Fig. 2.

FORCES DUE TO VERTICAL RESULTANT Q_m

4. As the load and structure have axial symmetry, all member forces will also be symmetrical. Consequently, it can then be deduced that the diagonals members are redundant.

Forces in arches

5. The vertical component of forces in an arch at a cross section must equal the vertical resultant. So: $N\gamma_m \, n_a \sin \gamma_m = Q_m$ and:

$$N\gamma_m = -\frac{Q_m}{n_a \sin \gamma_m} \quad (2)$$

if the loads act downwards, $N\gamma_m$ are the forces in the arches, n_a is their number and γ_m their slope.

Forces in rings

6. Horizontal components of $N\gamma_m$, $H\gamma_m = N\gamma_m.\cos \gamma_m$, are not in equilibrium at any node. Then, acting at the rings the non-equilibrated forces are:

$$\Delta H\gamma_m = H\gamma_{m-1} - H\gamma_m \quad (3)$$

7. The ring must counteract these forces by acting in tension or compression giving $N\theta_m$ forming a force polygon similar to the triangle that forms with the length of a bar of the ring Δd_m and its radius r_m. So, it can be stated that:

$$\frac{\Delta H\gamma_m}{\Delta d_m} = \frac{N\theta_m}{r_m} \tag{4}$$

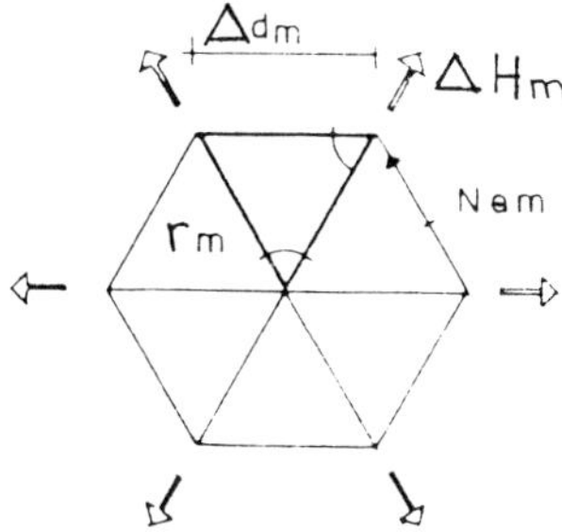

Fig. 3.

$$\text{and;}\quad N\theta_m = \frac{\Delta H\gamma_m}{\Delta d_m} r_m \tag{5}$$

$$\text{As;}\quad \sin\frac{\pi}{n} = \frac{\Delta d_m}{2r_m} \tag{6}$$

the following can be obtained:

$$N\theta_m = \frac{\Delta H\gamma_m}{2\sin\dfrac{\pi}{n}} \tag{7}$$

8. These forces can be obtained by means of Dischinger's diagram (ref. 2) which for this particular case is exact.

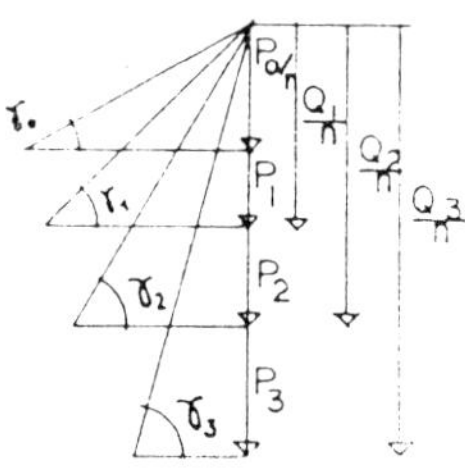

Fig. 4.

FORCES DUE TO GLOBAL BENDING MOMENTS M_m

9. If a horizontal cross-section is subjected to a bending moment, it will be counteracted by forces in the arches and in the diagonals. Neglecting the contribution of forces in the diagonals, the equilibrium of forces gives:

$$\sum_{0}^{n_a} N\gamma_{m,n} \sin\gamma_m x_{m,n} = M_m \tag{8}$$

10. To obtain $N\gamma_{m,n}$, an additional hypothesis has to be assumed. It has been outlined in previous work (ref. 3), Bernouilli 's hypothesis, that plane sections remain plane, is taken as valid. Then, the elongation of a bar of an arch U_{mn} follows:

$$U_{m,n} = K_e x_{m,n} \tag{9}$$

11. If as usual, all the areas of the bars A in a cross-section are equal, and the elements are made from the same material:

$$U_{m,n} = e_{m,n}\, \Delta S_m = \frac{N\gamma_{m,n}}{AE} \Delta S_m \tag{10}$$

if ΔS_m is the length of the bars in the arches. Then:

$$N\gamma_{m,n} = \frac{AE}{\Delta S_m} K_e x_{m,n} \tag{11}$$

$$K = \frac{AE}{\Delta S_m} K_e \tag{12}$$

K is also a constant, so $N\gamma_m$ obeys the simple law: $N\gamma_m = K.x_{mn}$ Eq. 13 that substituted into Eq. 8, allows the value of K to be determined as:

$$K = \frac{M_m}{\sin\gamma_m \sum_{0}^{n_a} x^2_{m,n}} \tag{14}$$

$$\sum_{0}^{n_a} x^2_{m,n} = I_{0m} \tag{15}$$

I_{om} is the moment of inertia of the bars of the arch at a cross-section, if all the areas are equal to unity. So Eq. 13 is transformed into:

$$N\gamma_{m,n} = \frac{M_m}{\sin\gamma_m I_{0m}} \quad x_{m,n} \tag{16}$$

Forces in diagonals

12. Equilibrium of an isolated portion of the structure between two arches and two rings, allows the determination of the two forces $Nx\gamma_{m,n}$ that will be supported by the diagonals members:

$$2Nx\gamma_{m,n} = \sum_{0}^{n} \Delta N\gamma_{m,n} \sin\gamma_m \tag{17}$$

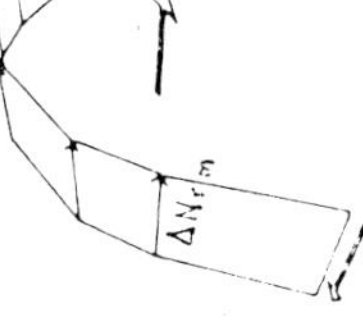

Fig. 5.

As $\Delta N\gamma_{m,n} = N\gamma_{m,n} - N\gamma_{m,n-1}$, using Eq. 16 it is possible to obtain:

$$Nx\gamma_{m,n} = \frac{1}{2}\left[\frac{M_m}{I_{0m}}\sum_{0}^{n} x_{m,n} - \frac{M_{m-1}}{I_{0m-1}}\sum_{0}^{n} x_{m-1,n}\right] \tag{18}$$

31. The expression given below;

$$\sum_{0}^{n} x_{m,n} = S_{0xm,n} \tag{19}$$

is the static moment of the arches areas of the isolated portion, if they are unity. A relation can be obtained between $S_{0xm,n}$ and $S_{0xm-1,n}$. As: $x_{n,m} = \Delta Y_m.\tan\beta_n + x_{n,m-1}$, and:

$$\tan\beta_n = \frac{x_{m,n}}{Z_m} \tag{20}$$

$$x_{n,m-1} = \left[1 - \frac{\Delta Y_m}{Z_m}\right] x_{n,m} \tag{21}$$

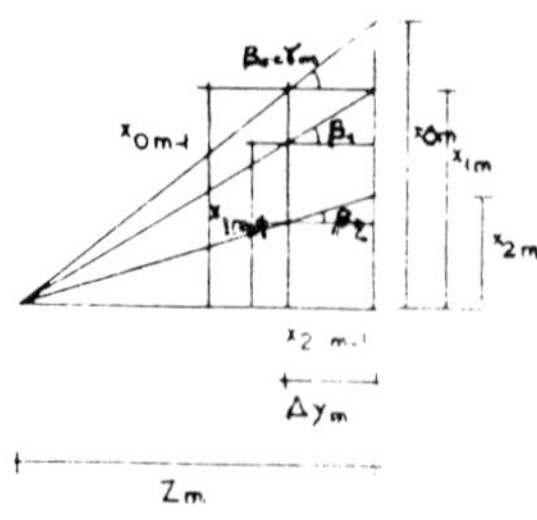

Fig. 6.

and the relations:

$$S_{0xn,m-1} = \left(1 - \frac{\Delta Y_m}{Z_m}\right) S_0 x_{n,m} \qquad (22)$$

$$I_{0m-1} = \left(1 - \frac{\Delta Y_m}{Z_m}\right)^2 I_{0m} \qquad (23)$$

14. Eq. 21 is then transformed into:

$$N x\gamma_m = \frac{\left(\frac{\Delta M_m}{\Delta Y_m} - \frac{M_m}{Z_m}\right)\Delta Y_m}{2 I_{0m-1}} S_{0xn,m-1} \qquad (24)$$

and: $Z_m = r_m \tan \gamma_m$

15. The first term is the global shear force T_m, that was obtained from Eq. 1 if $M_{vm} = 0$, that is, there are no vertical loads or these loads have axial symmetry. The second term is the contribution of forces $N\gamma_m$ to balance T_m, as found by Dischinger (ref. 1). This will be negative if the cross-section of the dome increases as the bending moment grows and positive if the contrary is true.

16. The existence of vertical shear forces $Nx\gamma_{m,n}$ compels that of horizontal shear forces $N\alpha\beta_{m,n}$. Equilibrium at a portion of the structure between two consecutive rings and arches gives:
$N\alpha\beta_{m,n}\, \Delta Y_m = Nx\gamma_{m,n}\, \Delta d_{m-1}$, and the expression for $N\alpha\beta_{m,n}$ is obtained which is more useful than $Nx\gamma_{m,n}$:

$$N\alpha\beta_{m,n} = \frac{\dfrac{\Delta M_m}{\Delta Y_m} - \dfrac{M_m}{Z_m}}{2I_m - 1} \Delta d_{m-1}\, Sx_{n,m-1} \qquad (25)$$

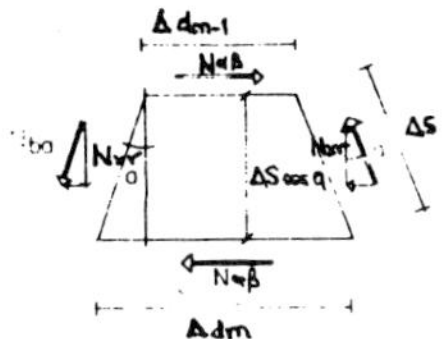

Fig. 7.

17. The forces in the diagonals $Nd_{m,n}$ can then be obtained readily from:

$$Nd_{m,n} = \frac{N\alpha\beta_{m,n}}{\cos\alpha_m} \qquad (26)$$

if there is only one diagonal:

$$Nd_{m,n} = \frac{N\alpha\beta_{m,n}}{2\cos\alpha_m} \qquad (27)$$

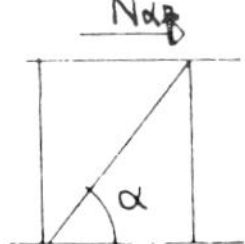

Fig. 8.

Forces in rings

18. For the vertical resultant, the Q_m forces, can be produced by the horizontal components of $N\gamma_m$ with the only difference that these forces are not equal at every bar of the ring. This non-equilibrium is redressed by the force components in the rings, from the diagonals, that are precisely $N\alpha\beta$. Calling $N\alpha\beta'$, the least value of the two forces $N\alpha\beta$ that act at a bar in a ring, the force $N\theta_{n,m}$ at a ring is:

$$N\theta_{m,n} = \frac{\Delta H_{m,n}}{2\sin\frac{\pi}{n}} + N\alpha\beta^{\ell} \tag{28}$$

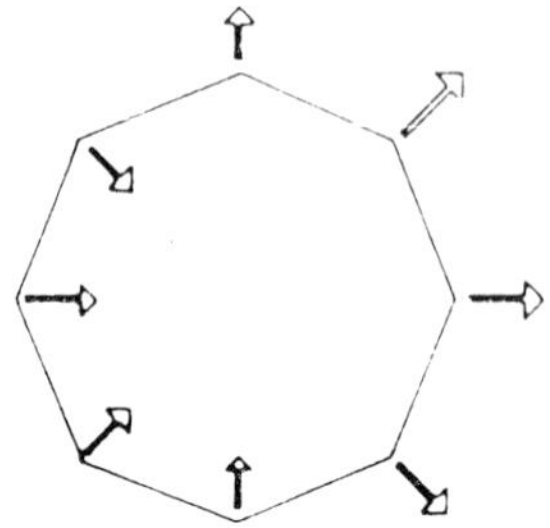

Fig. 9.

FORCES DUE TO LOADS

19. A correction shall be made to the forces in the arches for vertical loads, and forces in the rings for horizontal loads, in order to equilibrate the actual loads that act at the nodes of the dome. That modification does not alter significantly the forces in the bars.

STRESSES IN BARS

20. The final forces will be the sum of those due to Q_m and that due to M_m:

$$N\gamma_{m,n} = -\frac{1}{\sin\gamma_m}\left[\frac{Q\gamma_m}{n_a} - \frac{M_m}{I_{om}}x_{m,n}\right] \tag{29}$$

$$N\alpha\beta_{m,n} = \frac{1}{2I_{0m}}\left[\frac{\Delta M_m}{\Delta Y_m} - \frac{M_m}{Z_m}\right]\Delta d_{m-1}\ Sx_{n,m-1} \tag{30}$$

$$N\theta_{m,n} = \frac{\Delta H_{m,n}}{2\sin\frac{\pi}{n}} + N\alpha\beta^{\ell}_{m,n} \tag{31}$$

21. The forces in bars can then be obtained as:

$$N_a = N\gamma_{m,n} \tag{32}$$

$$N_d = \frac{N\alpha\beta_{m,n}}{2\cos\alpha_m} \tag{33}$$

$$N_r = N\theta_{m,n} \tag{34}$$

Where N_a are the forces in the arches, N_d are the forces in the diagonals and N_r are the forces in the rings, respectively.

22. If the arches of the dome are replaced by two diagonals, the well known transformation can be used:

$$Nd_{m,n} = \frac{N\gamma_{m,n}}{2\sin\alpha_m} \pm \frac{N\alpha\beta_{m,n}}{2\cos\alpha_m} \tag{35}$$

$$Nr_{m,n} = N\theta_{m,n} \tag{36}$$

23. In domes in which the triangles form a variable angle with the arches such as in geodesic domes, equilibrium of an isolated triangle should be considered with forces $N\gamma$ and $N\theta$ and $N\alpha\beta$, in order to obtain the forces in bars.

COMPARISON OF RESULTS.

24. A comparison has been undertaken between the forces obtained from the approximate method and those obtained by the standard matrix method, using a computer program prepared for this work by the second author.

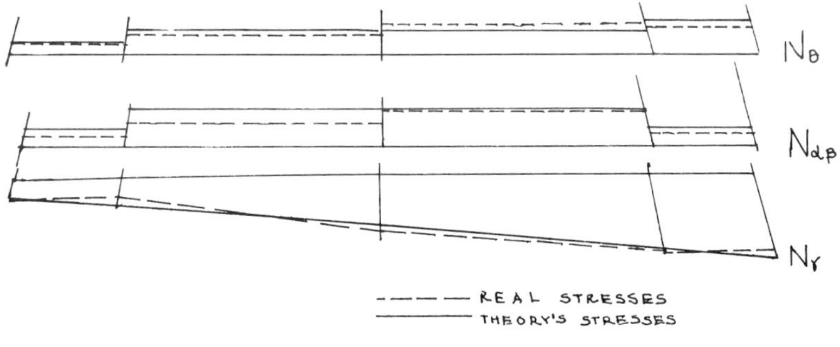

Fig. 10.

25. The results are very similar to those obtained for braced barrel vaults (ref. 3). Using the approximate method of analysis the forces $N\gamma$ are slightly greater than the values obtained from the matrix approach. The forces $N\alpha\beta$ and $N\theta$ obtained from the approximate method are slightly smaller in some cases than their corresponding values given by the matrix method of analysis.

CONCLUSIONS

26. As a general conclusion the approximate method outlined in this paper gives a simple means of obtaining member forces that can be used at the design stage for projects using braced domes.

27. It also gives a simple insight into how the dome works, which is very useful for the general design of braced domes.

REFERENCES

1. DISCHINGER F. Die Rotationsschalen mit unsymmetrischer Form und Belastung. Der Bauingenieur. Vol. 16. 1935. pp. 374-381, 393-398.

2. DISCHINGER F. Handbuch fiir Eisenbetonbau. Vol. XII. Chapter II. Berlin. Ernst. 1928.

3. QUINTAS V., AVILA J.M., and ANTUNA J. An Analysis of braced barrel vaults. Bulletin of the International Association for Shell and Spatial Structures. 1990. Vol.31. April-Aug. pp. 102-103.

30. Experimental study of a reticular dome constructed from aluminium alloy sections

R. I. MOLEVA, Nizhny Novgorod, Russia

SYNOPSIS. This paper analyses the behaviour under static load of members and joints of a reticular dome construction of aluminium alloy. The design of test specimens permit checking of the behaviour of supporting and common joints, as well as bearing bars of the frame, developing both tensile and compressive stresses. In the course of the experiment, breaking load values were recorded and stress-strain conditions in the bars and joints were analysed. The experimental data obtained allow for further improvement in structural design of reticular domes.

OBJECTIVES AND TASKS OF THE STUDY

1. Much experience has been accumulated in designing the dome type roofs at present. To substantiate theoretical assumptions, and study further the behaviour of their elements and units under various conditions of loading, it is necessary to have adequate experimental data.

2. The issues considered in the course of this experimental study included:

(a) the actual behaviour of main bearing members of the dome structure under static loading with maximum allowance made for structural features;

(b) the assessment of the stress condition in the bearing bars and connecting joints by measured strain values;

(c) the investigation of specific features in the behaviour of the units in an elastic stage, spread of plastic deformations in the unit and its transition to the limiting state condition;

(d) the analysis of behaviour under load of the joints of various design for the purpose of selecting the most appropriate one.

TEST SPECIMENS

3. The experiments were carried out on a reticular dome roof (Fig. 1). The skeleton of the dome designed to cover an area of 24-60m in diameter consisted of closed-box section bars ($A = 19.8\ cm^2$, $r_x = 5.19$ cm, $r_y = 1.87$ cm, $I_x = 535\ cm^4$, $I_y = 61.1\ cm^4$) connected in the joints by means of star-shaped elements with six points.

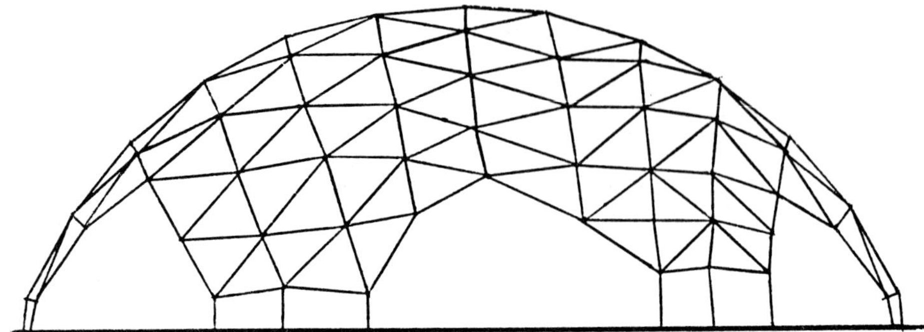

Fig. 1. Reticular dome structure from aluminium alloy; d = 24m.

4. Several types of specimens were tested:
(a) A-series: fragments of the dome roof made to full scale from 5 bearing bars connected by joint units to form a rhombus (Fig. 2). The triangular roof panels from sheet aluminium 0.8mm thick were placed onto the skeleton bars and attached to them by joint plates and self-tapping screws. The joints in this series were of two types, viz. joints with deformed bar ends to match the overall dimensions of the joint (A-series) and those with additional inserts in the connection zone intended to strengthen the section without developing a complicated stress condition in the web during assembling (AU-series),(Fig. 3);
(b) B-series: test specimens of this series consisted of a skeleton bar and two joint elements tested for central compression;
(c) C-series: specimens intended for the test in compression of two supporting units connected in pair in a "triangle". The length of the members was selected so as to provide for their stability in the whole range of loading;
(d) D-series: specimens intended for the test in tension and made of a joint unit and a length of the main supporting bar connected by high strength bolts (Fig. 4). Special type washers (cone-type, shaped, fitted with additional intermediate elements made from high strength wire in the form of a star and put between the contact surfaces in the assembly) were used in addition to common ones to enhance the bearing capacity of the joint;
(e) E-series: specimens of the D-series type assembled with steel bolts made from steel of the S-235 grade.

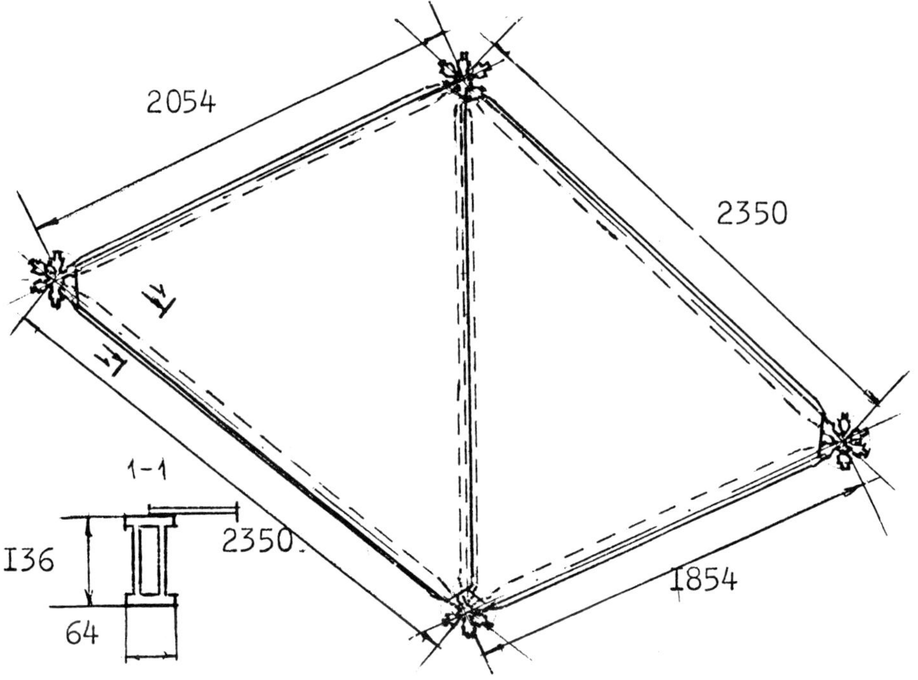

Fig. 2. Specimen of the A-series.

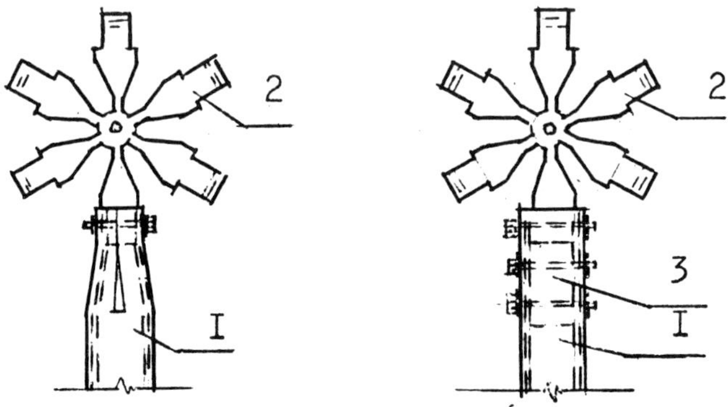

Fig. 3. Design variants for a joint of the reticular dome roof: a) A-series; b) AU-series 1-bar, 2-joint element, 3-strengthening element.

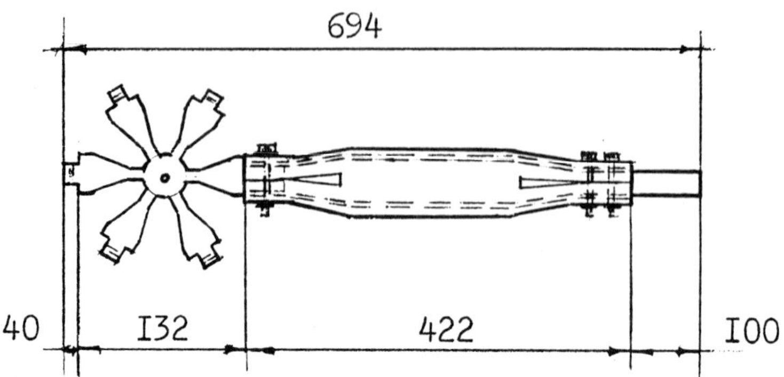

Fig. 4. Specimen of the D-series.

5. All the bars intended for testing in compression were to suit special requirements concerning their straightness. They were checked for initial distortion which did not exceed L/750 in all the bars.

6. The material used in the specimens was aluminium alloy AD31Tl. The mechanical characteristics were determined for each type of the material used in the experiment. The techniques applied for evaluation of the tensile properties of the aluminium alloy were in compliance with state standard GOST 1497-73, the compressive strength being assessed according to recommendations in ref. 1. The statistical processing of the data gave the values for proportionality limits $\sigma_{0.05}$, yield limit $\sigma_{0.2}$ and ultimate strength σ_5. For the material of the bearing bars $\sigma_{0.2}$ = 120 - 135 MPa, and for joint elements $\sigma_{0.2}$ = 122 - 200 MPa.

METHODS OF EXPERIMENTAL STUDY

7. The test specimens of the A- and B- series were tested in compression on a special installation. The load was produced by a 100 ton hydraulic jack. The pressure from the jack was transmitted directly to the supporting members fitted with high strength pins (σ_b = 1,100 MPa) and passed through the central holes in the joints. This mode of transmitting the load most closely corresponded to the theoretical premises (bar skeleton of the dome with hinged joints).

8. The supporting units of the C-series were tested in compression by means of UIM-500 hydraulic press. The loads were transmitted through the centre of the supporting unit shoe by special hinged plates.

9. The loading in all the tests was a step-by-step procedure, each step amounting to 15% of the design load . As the critical value was approached the magnitude of the step load was reduced. The deformations in the bars and joints

were recorded from resistive-strain sensors with a base of 10mm and 5mm and a digital CTM-3 strain-gauge bridge. The sensors were located in the zones of most intensive stress of the specimens. The overall number of sensors was 3,000. To measure lateral deflection in the bar web, deformation in the supports and vertical movements of bars in compression clock-type indicators were used with a scale division of 0.01mm. Indicators of the same type fitted with a special device were used to record reciprocal displacement of elements in the bolted joints in tension.

ANALYSIS OF RESULTS

10. The bearing capacity of the members in compression was limited either by the bearing capacity of the joint element neck or the bar capacity for stability, the adopted design scheme of the unit being composed of a bar with hinge-supported ends. The theoretical bearing capacity value for the specimens calculated with allowance for the forecast characteristics based on the actual data are given in Table 1. The values of experimental breaking loads applied to the test specimens are shown in Table 1 as well. These data clearly indicate an appreciable increase in the bearing capacity of the specimens fitted with additional strengthening element (AU-series). The analysis of stress-strain condition of elements in compression showed that both in the joint section and in the bar length the material behaved elastically in the whole range of the loads applied. The distribution of normal longitudinal stresses in the bar is approximately uniform (Fig. 5). Bending stress was a small portion of the overall stress, not more than 16% in the A-type bars, and 10% in the AU-type ones. The numerical values of axial stresses in the AU-type bars were considerably lower than in the bars of the A-series. The most difficult condition was created for the behaviour of the joint element neck and the main bearing member web length between the bolts in the support zone. These areas were studied with special care while analysing the stress-strain condition in the specimens. The axial stress values in this area exceeded the yield limit of the material. In the joint element sections the proportion of bending stresses is appreciably higher than the measured values (between 25% and 84% in the A-type joints and 9% - 31% in the AU-type joints). In specimens with joints of the AU-type there was a considerable reduction in concentration of tangential and normal stresses in the transverse direction (3 - 4 times less than in the joints of the A-series.) The eccentricity e_{oy}, e_{ox} and relative eccentricity values m_{oy} and m_{ox} in the specimens of the AU-series (Table 2), were also considerably lower. The destruction of specimens of both the A-series and AU-series occurred due to the general buckling of the bar accompanied by local bar web buckling in the end zone at breaking load values given in Table 1.

11. The bars of the B-series behaved like those of the A-type.

12. During the test of supporting units of the C-series the behaviour of supporting joint elements and the shoes was analysed in compression. Processing the strain-gauge data permitted identification of the zones of the most intensive stress and gave the numerical values for the normal and tangential stresses in these areas.

Table 1. Bearing capacity of membrane.

Specimen type	Theoretical bearing power, kN		Experimental value of breaking load kN
	bar stability	joint strength	
A - 1	199.4	285.0	183.0
A - 3	200.1	172.9	150.0
A - 4	200.1	172.9	140.0
AU- 1	199.4	285.0	280.0
AU- 2	199.4	285.0	285.0
AU- 3	199.4	285.0	255.0

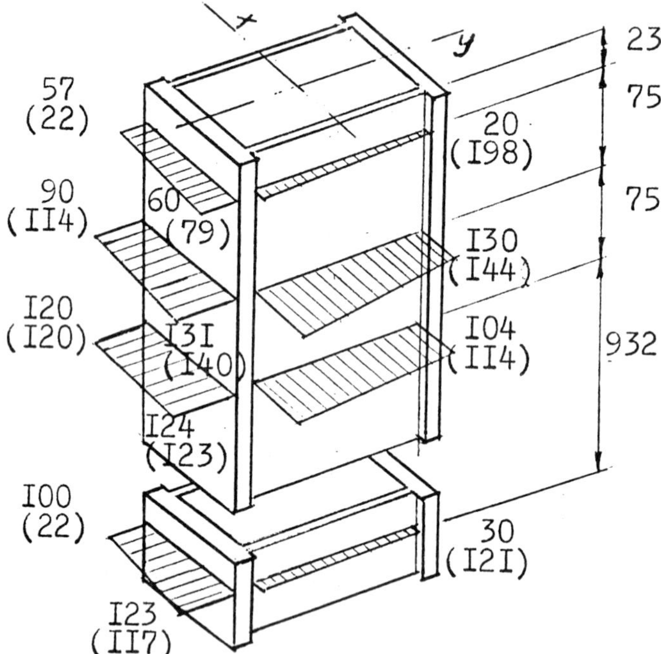

Fig. 5. Distribution of direct longitudinal stresses in the end zone of the AU-series bar at a load of 180 kN (in brackets: stress values in the A-type bar at the same load).

Table 2. Eccentricity values e_{oy}, e_{ox}, m_{oy}, and m_{ox}.

Speci-men type	e_{oy}		$m_{oy}=e_{oy}/\beta_y$		e_{ox}		$m_{ox}=e_{ox}/\beta_x$	
	bar	joint	bar	joint	bar	joint	bar	joint
A - 1	0.824	1.603	0.065	0.802	1.407	7.773	0.035	0.395
A - 3	o.500	0.180	0.040	0.090	4.800	5.950	0.120	0.302
A - 4	1.000	0.270	0.080	0.135	6.370	14.20	0.160	0.721
AU- 1	0.096	0.313	0.008	0.157	2.067	2.520	0.052	0.128
AU- 2	0.100	0.200	0.008	0.100	1.800	1.580	0.045	0.080
AU- 3	0.422	0.034	0.087	0.044	2.020	0.236	0.051	0.012

13. The behaviour of the specimens in tension corresponded to the eccentric tension pattern. The additional bend affected the form of the distribution diagram of the normal longitudinal stresses. The most non-uniformly distributed stresses were located in the section weakened by the bolt holes. The maximum values of normal and tangential stresses were also registered in the same section of the bar web. The destruction of all specimens of the D- and E-series was caused by the bar web tear-out between the bolt hole and the end of the bar. The shearing moment was recorded in all specimens at loads equal to 40-50% of the design load value. Setting the bolts close to the edge of the bar, deforming the bar web in assembling, causes failure of the elements at loads barely reaching 0.7-0.75 of the design load. Attempt was made to improve the abutting joint design by application of special shape washers. However, the bearing capacity increase obtained was hardly adequate to take the maximum tensile stress. The best results can be achieved in the design with strengthening elements.

CONCLUSIONS

14. The final results of testing the specimens to the point of their failure and the respective stress distribution diagrams were taken as a basis for analysing the stress-strain condition in the reticular dome members.

15. The bearing capacity of the joint is a factor which determines the bearing capacity of the dome roof as a whole.

16. The test permitted identification of the dangerous stress condition locations and prompted introduction of strengthening elements in the dome roof design.

17. The use of the strengthening elements allows for a marked increase in the bearing capacity of members and joints of the reticular aluminium dome structure

by reducing the eccentricity which occurs in transmitting the load from the joint to the bar member and reducing the concentration of normal and tangential stresses in the bar web and end zone as in well as components of the joint.

18. The analyses of the distribution of stresses show the relation between the stress values in the web and those in the flanges. The highest values of σ and τ are registered in the web, the flanges being under-loaded.

REFERENCES

1. SHELESTENKO L.R., NACHEVICH Yu.M. Study of mechanical and physical properties of aluminium alloys of the D l-T, D 16-T, AMr 61, D 16-A (l/k) type, Aluminium alloy structures, pp.57-77., Stroyizdat, Moscow, 1962.

31. Elastic stability of shallow arches

H. MATSUNAGA, Setsunan University, Japan

SYNOPSIS. This paper presents a numerical method used to obtain the values of limit points of load and bifurcation by pursuing the non-linear equilibrium path of shallow arches. A set of non-linear equilibrium equations which are transformed into the set of algebraic equations by the Galerkin method. An iteration method based on the secant stiffness relations of total equilibrium is used to obtain the non-linear solutions with controlling loads or displacements at a specified point on the equilibrium path. Fundamental and branching paths of simply supported shallow arches subjected to uniformly distributed loads are pursued numerically.

INTRODUCTION

1. The shallow arch is one of the simplest structures which show instability phenomena such as a symmetric buckling at the highest point in the equilibrium path and a bifurcation into an asymmetric buckling mode. The stability of shallow arches subjected to symmetrical loading has been the object of a number of investigations (refs 1 - 4). One of the simplest methods of estimating the buckling loads of shallow arches is to compute their Euler loads from an assumption of linear pre-buckling behaviour. In general, however, the Euler loads do not always agree with the experimental buckling loads. It is necessary to obtain complete solutions by analysing a set of governing non-linear equations for shallow arches.

2. Incremental formulations are widely used in numerical techniques for the solution of non-linear structural problems. The major difficulty associated with the incremental solution scheme is the accuracy of solutions due to numerical convergence (ref. 5). In order to ensure the accuracy of solutions, it is necessary to check the total equilibrium conditions at each step of the analyses.

3. Along the load-deflection curves, limit points of loads and bifurcation points may appear and instability characteristics will occur. Usual methods of numerical computations by controlling only the load components cannot pursue the load deflection curves beyond such points of instability.

4. This paper presents a numerical method to obtain accurately the non-linear load-deflection curves of shallow arches. Based on the shallow arch theory, the governing non-linear equations of total equilibrium are expressed in terms of the

two displacement components. The displacement components may be expanded in trigonometric series satisfying the prescribed boundary conditions. A set of the governing differential equations can be transformed into a set of algebraic equations by the Galerkin method. For the set of algebraic equations, an iterative procedure is applied, for each increment of loads or displacements at a specified point. An iteration method based on the secant stiffness relations by controlling loads or displacement components in accordance with the load-deflection response is applied to solve the set of non-linear equations. (see, refs 6 - 8).

5. The non-linear load-deflection curves for simply supported shallow arches are clarified in considerable detail. Numerical examples of shallow arches exhibiting load limit point and bifurcation point behaviour are solved and load-deflection curves of several types are obtained. The convergence studies of present numerical results, are carefully examined. The determinant of the secant stiffness matrix is computed along the load-deflection curves. It is noticed that the sign of the determinant changes at the unstable points on the load-deflection curves. Through the numerical solution of non-linear load-deflection curves for shallow arches, limit loads and bifurcation loads are accurately determined.

FUNDAMENTAL EQUATIONS OF SHALLOW ARCHES

6. The geometrical configuration of a shallow arch.

$$z = \frac{1}{2A}(x-a)^2, \qquad z_{,xx} = \frac{1}{A} \tag{1}$$

where a comma denotes differentiation with respect to the co-ordinate subscripts that follow.

7. Strain-displacement relations.

$$\gamma_x = u_{,x} - z_{,xx}w + w_{0,x}w_{,x} + \frac{1}{2}w_{,x}^2, \qquad \kappa_x = -w_{,xx} \tag{2}$$

where w_0 is initial deflections of the arch.

8. Equilibrium equations. When tangential loads are zero, the equilibrium equations of shallow arches are expressed as:

$$N_{x,x} = 0, \qquad M_{x,xx} + (z + w + w_0)_{,xx}N_x + p_z = 0 \tag{3}$$

9. Stress resultants.

$$N_x = Et\gamma_x = Et(u_{,x} - z_{,xx}w + w_{0,x}w_{,x} + \frac{1}{2}w_{,x}^2), \qquad M_x = EI\kappa_x = -EIw_{,xx} \tag{4}$$

where E is Young's modulus and t is the thickness of arch, and $I = t^3/12$.

10. Dimensionless quantities.

$$x = a\xi; \qquad u = tU, \quad w = tW, \quad w_0 = tW_0; \qquad P_z = \frac{p_z a^4}{EIt} \tag{5}$$

where a is half span length of arch.

11. Equilibrium equations in terms of dimensionless displacements. The equilibrium equations in terms of the dimensionless displacement components can be written as:

$$f_x(U,W) = U_{,\xi\xi} - \left(\frac{a}{A}\right)W_{,\xi} + \left(\frac{t}{a}\right)\left\{(W+W_0)_{,\xi}(W+W_0)_{,\xi\xi} - W_{0,\xi}W_{0,\xi\xi}\right\} = 0 \quad (6)$$

$$f_z(U,W) = W_{,\xi\xi\xi\xi} - 12\left(\frac{a}{t}\right)\left\{\left(\frac{a}{t}\right)\left(\frac{a}{A}\right) + (W+W_0)_{,\xi\xi}\right\}$$
$$\times\left[U_{,\xi} - \left(\frac{a}{A}\right)W + \frac{1}{2}\left(\frac{t}{a}\right)\left\{(W+W_0)^2_{,\xi} - W^2_{0,\xi}\right\}\right] = P_z \quad (7)$$

DISCRETIZATION OF FUNDAMENTAL EQUATIONS

12. The equilibrium equations are non-linear differential equations for which it is difficult to obtain the analytical solutions. It is possible to transform these equations into algebraic equations which may be solved using numerical procedures. For this purpose, the Galerkin method is applied here. If the boundary conditions are assumed to be simply-supported and then expressed as:

$$U = W = W_{,\xi\xi} = 0, \qquad at \ \ \xi = 0,\ 2 \quad (8)$$

13. Series expansion of load and displacement components. The load and displacement components are expanded into the trigonometric series satisfying the prescribed boundary conditions as follows:

$$P_z = \sum_r P_{zr}\sin\lambda_r\xi, \qquad W_0 = \sum_r W_{0,r}\sin\lambda_r\xi \quad (9)$$

$$W = \sum_r W_r\sin\lambda_r\xi, \qquad U = \sum_r U_r\sin\lambda_r\xi \quad (10)$$

where $\lambda_r = r\pi/2$; $r = 1, 2, 3, \ldots, M$. M is the termination number of the truncated trigonometric series. U_r and W_r are the unknown displacement coefficients to be determined. $W_{0,r}$ is given by the distribution of initial deflections. The undetermined coefficients are defined as follows:

$$U_r = W_r = 0, \qquad if \ \ r > M \ \ or \ \ r \leq 0 \quad (11)$$

14. Equilibrium equations in terms of generalised displacements. By applying the Galerkin method to Eqs 6 and 7, the following equations result:

$$\int_0^2\int_0^2 f_x(U,W)\sin\lambda_r\xi d\xi = 0, \qquad \int_0^2\int_0^2 f_z(U,W)\sin\lambda_r\xi d\xi = 0. \quad (12)$$

With the use of the addition theorem and the orthogonality property of trigonometric functions, the equilibrium equations may be represented as follows:

$$\lambda_r^2 U_r + \left(\frac{2}{\pi}\right)\left(\frac{a}{A}\right)\sum_m W_m \lambda_m g(r,m) + \frac{1}{2}\left(\frac{t}{a}\right)\sum_m W_m \lambda_m \Big[\big\{(W_{r-m} + W_{0,r-m})\lambda_{r-m}^2$$

$$-(W_{m-r} + W_{0,m-r})\lambda_{m-r}^2 + (W_{r+m} + W_{0,r+m})\lambda_{r+m}^2\big\}$$

$$+\lambda_m(W_{0,r-m}\lambda_{r-m} + W_{0,m-r}\lambda_{m-r} - W_{0,r+m}\lambda_{r+m})\Big] = 0 \qquad (13)$$

$$\lambda_r^4 W_r - 12\left(\frac{a}{t}\right)^2\left(\frac{a}{A}\right)\Big\{\left(\frac{2}{\pi}\right)\sum_m U_m \lambda_m g(r,m) - \left(\frac{a}{A}\right)W_r\Big\}$$

$$+6\left(\frac{a}{t}\right)\Big[\sum_m U_m \lambda_m \big\{(W_{r-m} + W_{0,r-m})\lambda_{r-m}^2$$

$$-(W_{m-r} + W_{0,m-r})\lambda_{m-r}^2 + (W_{r+m} + W_{0,r+m})\lambda_{r+m}^2\big\}$$

$$+\left(\frac{2}{\pi}\right)\left(\frac{a}{A}\right)\sum_m\sum_i W_m(W_i + W_{0,i})\lambda_i^2\{g(r,m+i) - g(r,m-i)\}$$

$$-\left(\frac{1}{\pi}\right)\left(\frac{a}{A}\right)\sum_m\sum_i W_m(W_i + 2W_{0,i})\lambda_m\lambda_i\{g(r,m+i) + g(r,m-i)\}$$

$$+\frac{1}{4}\left(\frac{t}{a}\right)\sum_m\sum_i W_m(W_i + 2W_{0,i})\lambda_m\lambda_i \sum_{\sigma=1}^{8}(-1)^\sigma(W_{p(\sigma)} + W_{0,p(\sigma)})\lambda_{p(\sigma)}^2\Big] = P_r \qquad (14)$$

where the function g is defined as:

$$g(a,b) \equiv \begin{cases} [1-(-1)^{a+b}]\frac{a}{a^2-b^2}, & (|\,a\,| \neq |\,b\,|) \\ 0, & (|\,a\,| = |\,b\,|) \end{cases} \qquad (15)$$

and the lower subscript $p(\sigma)$takes the value of the following eight combinations of plus and minus signs.

$$p(\sigma) = \begin{cases} +(r \pm m \pm i), & \sigma = 2,4,6,8 \\ -(r \pm m \pm i), & \sigma = 1,3,5,7 \end{cases} \qquad (16)$$

Eqs 13 and 14 are the non-linear equations with respect to the undetermined coefficients U_r and W_r, and may be solved numerically by iterative procedures.

METHOD OF NUMERICAL SOLUTIONS

15. In order to ensure the accuracy and reliability of the solution, the total equilibrium equations should be satisfied to within a specified tolerance at each point along the equilibrium path. For the set of algebraic equations, an iterative method based on the secant stiffness relations is applied for each step of total equilibrium by controlling the loads or displacements in accordance with the load-deflection response. Eqs 13 and 14 may be represented in the form:

$$[K_{stiff}] \cdot \{D\} = \{P\} \tag{17}$$

where $[K_{stiff}]$ is the secant stiffness matrix composed of the coefficients in Eqs 13 and 14 and involves partially the undetermined total displacement coefficients U_m and W_m. $\{D\}$ is denoted as a column matrix:

$$\{D\}^T = \{U_m;\ W_m\} \qquad (m = 1, 2, 3, \ldots, M) \tag{18}$$

and $\{P\}$ is denoted as a column matrix composed of the known constants P_m of the total load terms.

16. In the case of the load-controlling procedure, the above Eq. 17 remains to be solved. However, in the case of displacement-controlling procedure, Eq. 17 may be rewritten by interchanging a column matrix $\{P\}$ with a column of the stiffness matrix $[K_{stiff}]$ which corresponds to the specified displacement coefficient to be controlled. For these two cases of controlling procedures, Eq. 17 can be expressed as a generalised form:

$$[K] \cdot \{X\} = \{Y\} \tag{19}$$

In order to find the numerical solutions of the non-linear Eqs 19, an iterative procedure is applied for each incrementally increased step of loading or displacement component $\{Y\}$. The iteration procedure for this problem will be stated in the following.

17. Consider a basic state at which Eq. 19 is satisfied for a certain load or displacement intensity Y and an adjacent state at which an incremental load or displacement ΔY is applied to the basic state. Then, in the adjacent state, Eq. 19 is solved to obtain the first-step solution$\{X^{(1)}\}$ with using the matrix $[K^{(0)}]$ at the basic state and the matrix $[K]$ is evaluated as $[K^{(1)}]$ to obtain the second-step solution $\{X^{(2)}\}$ from the following equation:

$$[K^{(1)}] \cdot \{X^{(2)}\} = \{Y + \Delta Y\} \tag{20}$$

From this result, the matrix $[K]$ is evaluated at $\{X^{(2)}\}$ as $[K^{(2)}]$. Then, the iterative procedure may be generalised for the ith step to obtain the solution $\{X^{(i)}\}$ and to evaluate the matrix $[K^{(i)}]$ for $\{Y+\Delta Y\}$ as follows:

$$[K^{(i-1)}]\cdot\{X^{(i)}\} = \{Y + \Delta Y\} \tag{21}$$

and the $(i + 1)$th step solution $\{X^{(i+1)}\}$ can be obtained from the equation:

$$[K^{(i)}]\cdot\{X^{(i+1)}\} = \{Y + \Delta Y\} \tag{22}$$

18. This iteration procedure will be repeated to the negligible order of magnitude of the load or displacement coefficients. The convergence criterion may be stated as follows:

$$\left|\frac{X^{(i+1)} - X^{(i)}}{X^{(i+1)}}\right| \leq \delta \tag{23}$$

where δ is the desired accuracy. If the desired accuracy is once attained for the current level of load or displacement step $\{Y + \Delta Y\}$, the calculated state $\{X^{(i+1)})\}$ is then regarded as a basic state for the next step and the iteration procedure stated above will be repeated. It is very important to accumulate the perfectly accurate solutions at each step of the equilibrium path of non-linear problems. By introducing small values of asymmetric displacement components on the symmetric equilibrium path, the bifurcation point can be obtained.

NUMERICAL EXAMPLES AND RESULTS

19. A simply-supported shallow arch subjected to uniformly distributed loads is analysed. The slenderness parameter is fixed to be $a/t = 100$ and the curvature parameter a/A is varied as follows:

$$a/A = 0.010,\quad 0.020,\quad 0.025,\quad 0.030,\quad 0.040,\ldots,0.100 \tag{24}$$

In order to plot the equilibrium path of the, multi-degree of freedom system, a mean deflection defined in the following equation is introduced.

$$W_{av} = \frac{1}{2}\int_{-1}^{+1} W d\xi \tag{25}$$

20. The convergence properties of the equilibrium paths according to the terminal number M are examined to verify the accuracy of the numerical

solutions. The numerical results for M = 11 which are converged accurately enough along the equilibrium paths are shown in the following. The desired accuracy of the solution at each step of the equilibrium path is 10^{-5}.

21. Fig. 1 shows the equilibrium paths for shallow arches with the curvature parameter (i) a/A = 0.025 and (ii) a/A = 0.030. In the figures, "A11025", for instance, denotes the case of M = 11 and a/A = 0.025. The relations between the load and the mean deflection are plotted for symmetric and asymmetric equilibrium paths. Instability phenomena of shallow arches are noticed at the critical buckling and bifurcation points on the equilibrium paths.

22. By controlling an appropriate displacement coefficient, the present method can pursue the equilibrium path beyond the unstable points such as critical points and bifurcation points. The critical buckling and bifurcation point loads are also obtained accurately.

23. In order to estimate the stability phenomenon of the equilibrium path the ratio of the determinant of secant stiffness matrix to that of the initial state is computed at each point on the equilibrium path and has plotted in Fig. 2. It is noticed that the sign of the determinant is changed on passing through the unstable points. Figs 3 and 4 shows the deflection curves for several load levels.

DISCUSSIONS AND CONCLUSIONS

24. Based on the shallow arch theory, the governing non-linear equations of total equilibrium are transformed into algebraic equations by the Galerkin method. For the set of algebraic equations, an iteration method based on the secant stiffness relations is applied for each increment of controlled displacements at a specified point.

25. Although the incremental formulation is widely used for the geometrically non-linear analyses of structural problems, this procedure has a serious disadvantage in that no estimate of the accuracy of the solution is generally available. Incremental methods require that the controlling load or displacement increments must be kept small so that the solution does not deviate too far from the satisfaction of equilibrium. The purely incremental method is characterised by the continuous accumulation of truncation errors. In order to ensure the accuracy and reliability of the solution, the total equilibrium equations should be satisfied to within a specified tolerance along the equilibrium paths. It is very important to accumulate the perfectly accurate solutions at each step of the equilibrium path of non-linear problems. Through the equilibrium iterations for the governing equations of total equilibrium, any desired degree of the solution can be obtained.

26. For shallow arches, limit points of loads may appear on the load-deflection curves and instability characteristics will occur. Therefore usual methods of numerical computations by controlling only the load components cannot pursue the load-deflection curves beyond a load limit point. By selecting a monotonously increasing displacement component as the controlling parameter, limit points of loads on the load-deflection curves can be passed without difficulty.

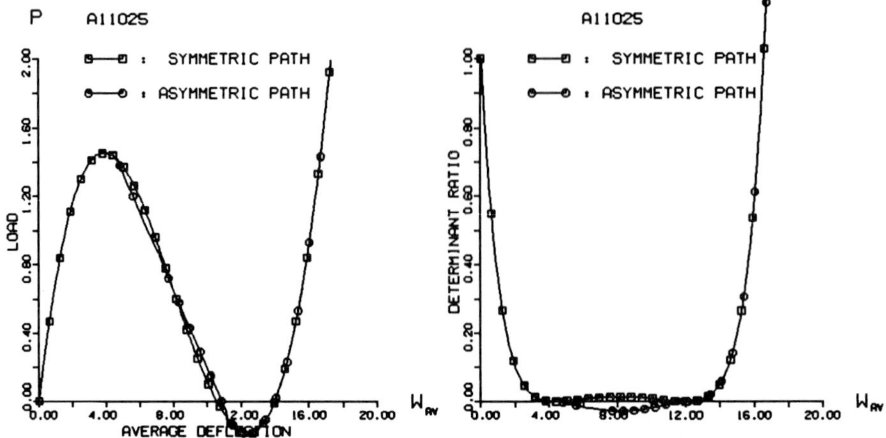

Fig. 1-i. Equilibrium path.

Fig. 2-i. Determinant ratio of stiffness matrix.

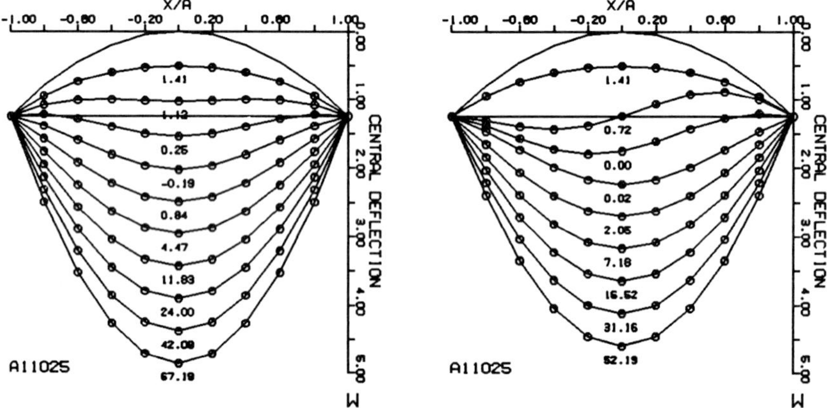

Fig. 3-i. Deflection curves (symmetric).

Fig. 4-i. Deflection curves (asymmetric).

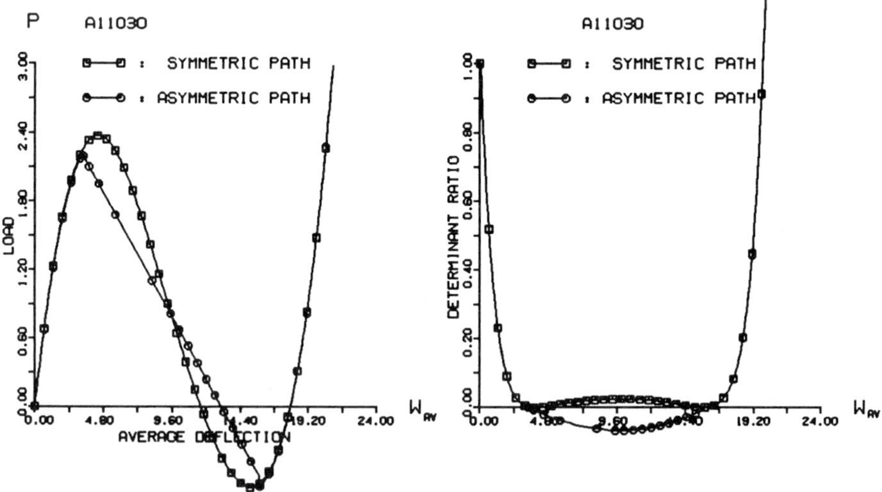

Fig. 1-ii. Equilibrium path.

Fig. 2-ii. Determinant ratio of stiffness matrix.

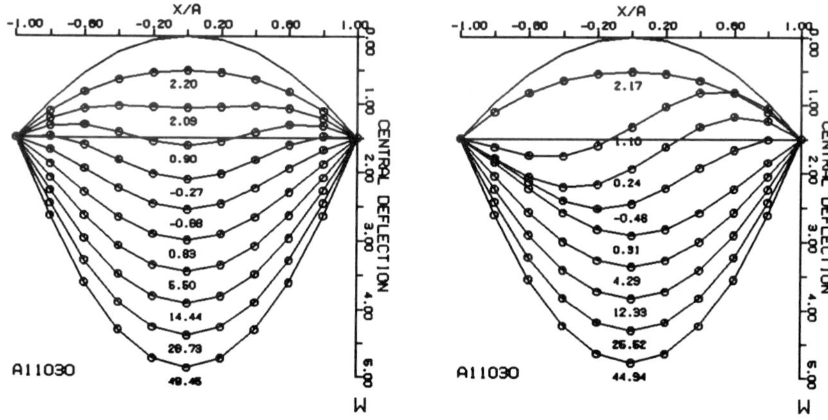

Fig. 3-ii. Deflection curves (symmetric).

Fig. 4-ii. Deflection curves (asymmetric)

27. Since it is most important and essential to prove the accuracy of the non-linear solutions, convergence studies on the numerical results have been carefully examined. The benchmark load-deflection curves for shallow arches have been followed for considerably large deflections. The buckling loads at critical points on the load-deflection curves have been obtained easily and accurately. Through the numerical solutions of non-linear load-deflection curves of shallow arches, limit loads and bifurcation loads have been accurately determined.

REFERENCES

1. SCHREYER H. L. and MASUR E. F., Buckling of shallow arches, Journal of the Engineering Mechanics Division, ASCE, 1966, Vol. 92, No. EM4, August, 1-19.

2. DICKIE, J. F. and BROUGHTON, P., Stability criteria for shallow arches, Journal of the Engineering Mechanics Division, ASCE, 1971, Vol. 97, No. EM3, June, 951-965.

3. DIM, C. L., Bifurcation analyses for shallow arches, Journal of the Engineering Mechanics Division, ASCE, 1973, Vol. 99, No. EM2, April, 287-301.

4. DYM C. L., Stability theory and its applications to structural mechanics, Mechanics of Elastic Stability, Noordhoff International Publishing, 1974.

5. BERGAN P. G., HORRIGMOE G., KRAKELAND B. and S0REIDE T. H., Solution techniques for non-linear finite element problems, Int. J. for Num. Meth. in Engng, 1978, Vol. 12, 1677-1696.

6. MATSUNAGA, H., Non-linear analysis of equilibrium path of shallow arches, Proc. Annual Meeting A.I.J., 1991, 1173-1174 (in Japanese).

7. MATSUNAGA H., Non-linear analysis of shallow hyperbolic paraboloidal shells by the Galerkin method, Proc. 10 Years of Progress in Shell and Spatial Structures, 30 Anniversary of IASS, Madrid 1989, Vol. 1, Shells (1).

8. MATSUNAGA H., Non-linear load-deflection curves of shallow shells with non-negative Gaussian curvatures, Proc. IASS/CSCE International Congress, Innovative Large Span Structures, Toronto 1992, Vol. 2, 758-769.

32. A study of instability mechanisms of reticulated shells and analytical model

R. J. QIAN and Z. Y. SHEN, Tongji University, P. R. China, and
S. H. XIA, Hebai University, P. R. China

SYNOPSIS. The Auto Increment procedure has been widely adopted for the nonlinear stability analysis of reticulated shells and has been successfully used for tracing the pre-buckling path and determining critical loads. However, in the process of forming a secondary stable equilibrium state from the buckled system in the region of the critical point, not all these techniques are effective in actual implementation. It is therefore necessary to carry out an investigation into the unstable mechanisms of reticulated shells and their analytical models. In this paper, the authors present a definition of the instability of a structure and the secondary state, discuss the factors which could make a notable difference between theoretical and experimental critical loads and provide a nonlinear analytical model. The authors present the assumptions regarding to the unstable mechanisms of reticulated shells, and the forming of the secondary state. Some numerical examples are provided for verification.

INTRODUCTION

1. Since the 1960's, nonlinear theory has been widely used in structural stability analysis. The increment iteration procedure, which is the basis of the auto-analysis technique is accepted as the main method for tracing the structural equilibrium configuration. The size of the loading increment is effectively controlled, the pre-buckling path is traced, the behaviour of a structure in the vicinity of a critical point on equilibrium path is detected, the critical load is determined and the post-buckling path is further traced in the duration of auto increment iteration process. However, there are many problems unsolved especially in a system with multi-degree of freedom like a reticulated shells. It is considered if it is possible to satisfy these requirements with classical stability theory and traditional methods. There is no doubt that the methods mentioned in ref. 1 can be employed in pre-buckling analysis of reticulated shells, but in further analysis. Therefore, an investigation on the definition of structural stability and the unstable mechanisms of reticulated shells is required. In this paper, the authors firstly present the definition of structural instability and the secondary state, then, discuss and explain the differences between the theoretical analysis

and test results in the study on structural stability. Having compared two analytical models, the authors provide a nonlinear analytical model and numerical examples. Finally, the authors present the unstable mechanisms of reticulated structures.

DEFINITION OF INSTABILITY

2. Because of the lack of attention paid to the familiar sight of stable phenomenon, instability is used as a measure for analysing structural stability. There has been no unique definition of structural stability since the beginning of research on the structural unstable phenomenon. It depends on the theory that is adopted. At present, both mechanics and energy principles are used in describing and defining the unstable phenomenon, The critical state is mentioned as a transforming state. The transformation of the equilibrium state of a structure and the form of energy in a structure occurs as soon as the structure buckles. It can also be said that unstable phenomenon is a sort of transformation of deformation. The elastic or plastic deformation occurs before structure buckling, however, some geometric deformation of a structure would occur after it goes over the critical point. The elastic or plastic deformation may be determined. There is much in homologue between the deformation and load, also, between the deformation and stress in the members. On the contrary, there is no homologous relationship between geometric deformation and load. The geometric deformation is in fact a rigid-body motion. The structure deformed in its geometry is impossible to return to the original shape even if the load is moved, consequently, there is no relationship between geometric deformation and stress in a structure. There is a transformation from elastic or plastic deformation to geometric deformation when the structure buckles. Before buckling a structure is in the primary state, and enters a secondary state after buckling. The primary state is also called the state of fundamental equilibrium or pre-buckling state. The stress may be found from the strain in a structure both in the primary and secondary states only if the equilibrium is stable. During the transformation of the state the structure exhibits a unstable response until it arrives at a stable secondary state. The primary state corresponds to the original geometry, physical and loading conditions, while the initial condition of the secondary state corresponds to the critical state. The geometry of the structure has been changed, the stress and loading corresponding to the critical point are distributed in and applied to the structure at the beginning of the secondary state. Therefore, a full trace of the equilibrium configuration of the structure has to consist of the trace of the primary equilibrium state, determination of the secondary state, and trace of the stable secondary path. Some techniques and processes have been provided for tracing the equilibrium configuration (refs 1,2). All these methods may be used in effectively analysing the stable path. A technique for shape finding may be introduced in determining the initial geometry of the secondary state. Therefore, it is very important to predict the stability of the secondary path by means of estimating the buckling mode of a structure, that is bifurcation or limit point buckling (ref. 8).

THE INVESTIGATIONS ON STABILITY THEORY

3. It may be observed from the test results on the critical load shells and reticulated shells, that an obvious difference exists between theoretical and experimentally measured critical loads (ref. 3). However, an exception to this is reported by McConnel (ref. 4). He undertook an experiment with two spherical domes which were meticulously made. The models had 19 nodes and 42 members. The comparison between theoretical analysis and experimental results showed that if only one concentrated load was applied to the top of the dome, the reduction of actual buckling load is about 4% of theoretical one, because of geometric errors in the model. However, a quite different result was derived when several loads were simultaneously applied to the central nodes, the actual critical load rapidly reduced by 40% due to the influence of imperfections in the dome. Except for the geometric error none of other imperfections was taken into account in his research. It can be seen that there are many reasons for the difference between theoretical and experimental critical loads, when all the theoretical and experimental results, provided from different countries during the last several decades, are taken together and studied. The main factors influencing the reduction in critical load include:

1) the theoretical analysis model;
2) the algorithm for evaluation of nonlinear equations;
3) the initial shape of the structure;
4) the type of load applied to the structure;
5) the boundary condition and actual restriction of the structure;
6) imperfection.

As mentioned above, further studies on the mathematics-mechanics model and unstable mechanism of reticulated shells are obviously needed.

MATHEMATICS-MECHANICS MODEL

4. So far, the widely used analysis models have been the one presented by Oran (ref. 5) and the other based on the nonlinear finite element method (FEM). The influence of axial deformation and slenderness change in length of the members are taken into account in the Oran and Safan model, while, only a contribution of axial force is in consideration and the rotations in two planes are incompatible. However, more than that mentioned above is taken into account in FEM. However, the static displacement model is adopted and the change in length of the members is ignored.

5. The expression of strain in a beam element is given by the Green strain tensor. The strain vector consists of linear and nonlinear terms, i. e:

$$\Delta\varepsilon = \Delta\varepsilon_L + \Delta\varepsilon_{NL} \tag{1}$$

The strain vector can also be expressed in terms of a nodal displacement vector, as:

$$\Delta\varepsilon = (B_L + \widetilde{B_{NL}})\Delta u_e \tag{2}$$

The virtual strain, namely, the variation of the strain is:

$$\delta\Delta_e = (B_L + B_{NL})\Delta u_e \tag{3}$$

The stress in a beam element is:

$$\sigma = \sigma^0 + \Delta\sigma \tag{4}$$

in which, σ^0 and $\Delta\sigma$ are the initial stress and increment of stress in the elements, respectively. From the virtual work principle:

$$\int_V \delta\Delta\varepsilon^T \sigma dv = \int_V \delta\Delta u^T x dv + \int_s \delta\Delta u^T \bar{x} ds$$

the incremental finite element fundamental equations may be derived as below:

$$k_e \Delta u_e = p_e - f_e \tag{5}$$

in which k_e is the element stiffness matrix defined in local the co-ordinates system. It consists of the elastic stiffness matrix k_E, geometry stiffness matrix k_G and initial strain matrix k_e, i. e.

$$k_e = k_E + k_G + k_\varepsilon \tag{6}$$

here

$$k_G = k_{GN} + k_{GM} + k_{G\Delta} \tag{7}$$

and

$$k_\varepsilon = \frac{1}{2}\int_V B_L^T D B_{NL} dV + \int_V B_{NL}^T D B_L dV + \frac{1}{2}\int_V B_{NL}^T D B_{NL} dV \tag{8}$$

fe is the load vector equivalent to the forces in members, given by:

$$f_e = \int_V B_L^T \sigma^0 dV \tag{9}$$

6. The terms with high order in the expression of strain have to be taken into account in deriving the formulas. From the definition of the Green strain tensor, the axial strain may be written as:

$$\varepsilon_{xx} = \frac{\partial u}{\partial x} + \frac{1}{2}\left[\left(\frac{\partial u}{\partial x}\right)^2 + \left(\frac{\partial v}{\partial x}\right)^2 + \left(\frac{\partial w}{\partial x}\right)^2\right]$$

7. Large errors may result if the second order terms are ignored. Also, the terms with high order should be considered in the flexural strain. Although, the influence of the high order terms reduce rapidly.

8. Besides the contribution of axial force in a member, both flexural stress and p-Δ effects are taken into account in expressing the stress. The p-Δ effect has little influence if no transverse load is applied to the members of a reticulated shell.

9. The terms k_{GN}, k_{GM} and $k_{G\Delta}$ in Eq. 7 are the geometric stiffness matrices, corresponding to the contribution of axial force, flexural stress and the p-Δ effect, respectively.

10. It should be noted that the static deformation model is employed. This model is available for analysing the strength and also tracing the stable equilibrium path in the load-displacement space.

INVESTIGATION ON UNSTABLE MECHANISM

11. It is important to find unstable mechanism in order to determine the buckling load, bearing capacity of a structure and secondary state. There are some factors inherent in the unstable mechanism of reticulated shells. It may be seen from results derived by tracing analyses that the stability of a reticulated shell is closely concerned with the distribution and value of initial stress which exists in the members. In general, if there is a set of unbalanced initial stress in the flexible structure, it may maintain equilibrium by means of adjusting the geometry configuration and changing the distribution of initial stress. The shape of a flexible structure depends on the distribution of initial stress. However, the rigid structure lacks this self-adjusting capacity for distributing the unbalanced forces. A potential energy gathered by the unbalanced force makes reticulated shells unstable. The structure forms a new shape after buckling, so that the forces are redistributed. Numerical examples show that the stresses in members of a structure are high at the vicinity of limit point in the primary state. Unfortunately, the solution is pseudo and the values of the initial stresses in members are inaccurate.

12. This also depends on the transformation of the type of strain energy and force elements for the stability of the reticulated shells. The transformation would be performed from the state mainly in membrane stress to that of flexural stress or the state that both membrane and flexural stress exist, because of the variation in structural stiffness. Therefore, flexural stresses in the reticulated shells would reduce the bearing capacity of the structure, and the imperfection, both in the initial geometry and stress or flexure, would result in flexural stress.

NUMERICAL EXAMPLES

13. Fig. 1 shows a parallel Lamella dome with a span of 24.0m and a rise of 2.4m and 4.0m, respectively. The ratio of rise and span is 1/10 and 1/6 respectively. A dead load of $1kN/m^2$ and a live load of $0.5kN/m^2$ are considered as reference loads. The dome is elastically restricted normal to the boundary, the spring constant of a elastic support is 30kN/cm. The following five stiffness

matrices were taken into account for comparing the analytical model, i. e: mk= 1, implies that only elastic stiffness matrix k_E is taken into account; mk= 2, implies both k_E and k_{GN} were taken into consideration; mk= 3, implies that k_E, k_{GN} and k_{GA} were taken into consideration; mk= 4, implies that k_E, k_{GN} and k_{GM} were taken into consideration; and mk = 5, implies that k_E, k_{GN}, k_{GM} and $k_{G\Delta}$ were taken into consideration. The result show in Figs 2 and 3 and Table 1.

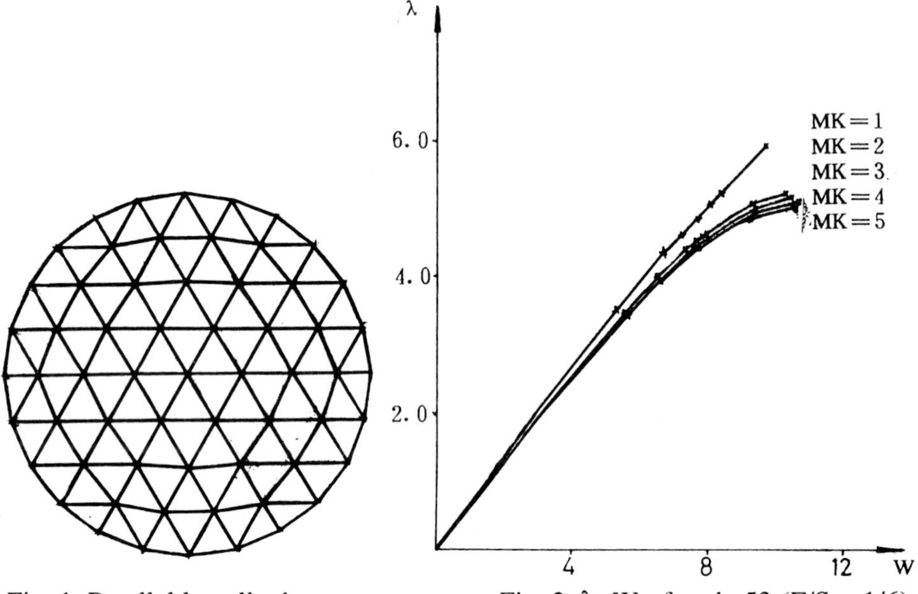

Fig. 1. Parallel lamella dome.

Fig. 2. λ–W of node 53 (F/S = 1/6).

14. In order to establish the influence of imperfection on the critical load, a pre-buckling analysis was carried out, in which, a deviation of the central node coordinates of 1.0cm, 1.0cm and 3.0cm along X ,Y and Z directions respectively, and the initial axial force N^0 and moment M^0 equivalent to initial strain of 0. 001 are taken into account. From the numerical examples, it can be seen that in a full-tracing analysis, the nonlinear stiffness matrix has to be taken into account (Fig. 4) otherwise, the critical load is exaggerated unrealistically. The bifurcation load is equal to 5.2, and the limit load equals 1.5, as the ratio of rise to span are 1/6 and 1/10, respectively. The results show that if only the vertical concentrated loads are apllied to the nodes of the dome, acceptable errors in pre-buckling analysis and the nonlinear geometry stiffness matrices k_{GM} and $k_{G\Delta}$, are ignored. In contrast, the contribution of axial force to the stiffness is the most important.

15. The imperfection in a local region in a reticulated shell does not seem to have much effect on the buckling load. The imperfection would not change the buckling load of a structure, which exhibits the limit point buckling, by very much . A further study needs to be performed to establish if the type of buckling would be changed if the limit point buckling occurred ahead of the bifurcation

point buckling for a structure that exhibits bifurcation buckling. The nodes located in the buckled region of a structure would move in rigid body deformation as soon as the structure buckled. A large out of balanced force exist at the buckled nodes, meanwhile, the force in the members connected to the buckled node are unchanged. For example, as the dome buckled, the unbalanced forces at the node 56 are as below:

$$F_{x56} = -607.723^{KN},\ F_{y56} = 105.417^{KN},\ F_{z56} = 196.628^{KN},$$
$$M_{x56} = 196.628^{KN-cm},\ M_{y56} = 106.395^{KN-cm},\ M_{z56} = -1.224^{KN-cm}$$

which the deformation of the node are:

$$U_{56} = -0.0816^{cm},\ V_{56} = 0.1307^{cm},\ W_{56} = 97.0486^{cm},$$
$$\theta_{x56} = 0.011 \quad \theta_{y56} = 0.0053 \quad \theta_{z56} = 0.0002$$

However, for instance, the forces in the member which is between nodes 56 and 60 are:

$$N = 105.7^{KN},\ Q_y = 0.0^{KN},\ Q_z = -3.1^{KN},$$
$$M_x = 4.5^{KN-cm},\ M_{yi} = 571.2^{KN-cm}$$
$$M_{yj} = 520.7^{KN-cm},\ M_{zi} = 0.5^{KN-cm},\ M_{zj} = -1.5^{KN-cm}$$

16. It obviously does not make any numerical sense to evaluate the nodal deformations at a critical point using the methods provided so far.

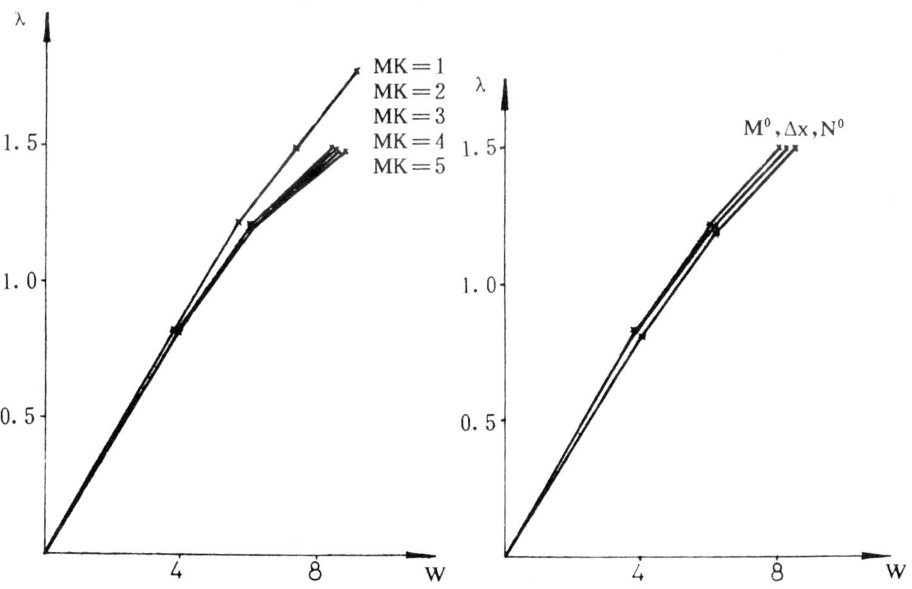

Fig. 3. λ–W of node 53 (F/S = 1/10).

Fig. 4. λ–W of node 53 (F/S = 1/10).

Table 1. η_c and η_t refer to ref. 8

	F/S=1/6					F/S=1/10				
	1	2	3	4	5	1	2	3	4	5
W (cm)	8.41	10.15	10.16	9.81	10.55	9.14	8.40	8.40	8.40	8.40
k_s (10^{-3}	0.50	−0.22	−0.37	−0.38	−0.37	0.18	0.13	0.12	0.13	0.13
η_c	−1.16	−1.13	−1.13	−1.13	−1.13	−1.35	−1.34	−1.34	−1.34	−1.34
η_t	1.05	1.12	1.12	1.12	1.12	1.08	1.12	1.12	1.12	1.12

CONCLUSION

17. The buckling of a structure is in reality a sort of transformation of equilibrium state, or the form of strain energy, or the force elements and the transformation of deformation.

18. The structure has an inherent buckling behaviour, which is characterised by no change in stress during infinite deformation in the geometry softening region.

19. The primary state is based on initial geometry, load and stresses, and the secondary state depends on the geometry, load and stresses corresponding to critical point. The secondary state may be stable or unstable.

20. The analytical models and algorithms provided so far give a stable equilibrium path but not the process of buckling.

21. In this paper the analytical model presented by the authors is a relatively complete model.

22. The technology of Form Finding may to be employed for determining the secondary state.

REFERENCES

1. RUO JUN QIAN, A comment on nonlinear stability analysis of elastic structure, The proceedings on space grid structures and reticulated shells, Tongji University Press, 1991, 4.

2. RUO JUN QIAN and YA LING LI, A study on nonlinear stability analysis of reticulated shell structure, The proceedings on space grid structures reticulated shells, Tongji University Press, 1991, 4.

3. L. KOLLAR and DULACSKA, Buckling of shells for engineers, a kademinal Kiado, Budapest, 1984

4. IRADJ M KANI and RICHARD E. MCCONNEL, Single layer shallow lattice dome analysis, General behaviour and collapse. Int.. J. of Space Structures, vol. 3, No. 2, 1988.

5. CENAP QRAN, Tangent stiffness in space frames. Journal of the Structural Division ASCE,1973

6. Z. S. MAKOWSKI, Analysis, design and construction of braced domes

7. C. BORRI and H. HUFENDIEK, Geometrically-nonlinear behaviour of grid shells. The Proceedings of the Third International Conference on Space Structures, Surrey, 1984.
8. SHAO HUA XIA and RUO JUN QIAN, Judging the buckling type of structure. The Proceedings of the Sixth National Conference on Space Structure, Earthquake Engineering Press, Beijing, 1992.

ACKNOWLEDGEMENT
This paper is granted by the China Natural Science Fund.

33. Non-linear buckling analysis of double-layer latticed domes and application of modal analysis

S. E. HAN and H. KUNIEDA, Disaster Prevention Research Institute, Japan

SYNOPSIS. The primary objective of this paper is to research the characteristics of geometrically non-linear buckling behaviour, and to discuss the application of modal analysis, of double-layer latticed spherical domes. Modal analysis is applied to the derivation of the system matrices transforming variables into generalised co-ordinates which are coefficients multiplied by independent and orthogonal eigen-modes. By using the modal analysis method, it is expected to reduce the size of the stiffness matrices to that of the generalised co-ordinates. It can be shown that a decrease in the possibility of divergence in the calculation of post-buckling path can be achieved. The Arc-Length method and Newton-Raphson iteration method are used to obtain the geometrically non-linear equilibrium path.

INTRODUCTION

1. Because it is difficult to use models for complex structural systems encountered in physical situations, mathematical models with a large number of degrees of freedom are usually used. Very large numbers of degrees of freedom are necessary to study the mechanical behaviour of structures such as double-layer latticed domes. However, to solve the large scale stiffness matrices in the normal way, is not efficient or economical. This problem is particularly relevant in non-linear analysis. Therefore, there is a necessity for the development of new analytical schemes for the analysis of double-layer latticed domes that have a high number of degrees of freedom. This paper reports a study of the characteristics of geometrically non-linear buckling behaviour of double-layer latticed spherical domes using the arc-length method and the Newton-Raphson iteration method. Comparison and discussion on the theoretical process for the application of a new approach using modal analysis is given. The analysis is shown to be made easier by the reduction in the number of variables to examine the structural characteristics.

2. Modal analysis derives the small system matrices of simultaneous differential equations, representing the displacements as linear combinations of several Eigen-modes which are independent and orthogonal to each other and regarding the

generalised co-ordinates, which are coefficients multiplied by Eigen-modes (about twenty eigen-modes are used in this paper), as independent variables. These Eigen-modes are calculated from the global system matrices. Alternatively, the Eigen-modes of free vibration of the same type for continuous shells may be obtained. Although it is difficult to select the necessary eigen-modes from the global system matrices, twenty natural modes related to the 1st to the 20th eigen-values are used in this paper. By using the modal analysis method in the buckling analysis of double-layer latticed domes, a reduction is expected in the large stiffness matrices of spatial displacement variables of the global system.

3. The merit of use of modal analysis is to avoid the degeneration of system matrices in multi-degrees on bifurcation points. We can decrease the possibility of divergence in the procedure of numerical calculation for the direction of the post-buckling path. Usefulness of the application of modal analysis has already become clear in the buckling analysis of continuous domes using several modes represented by Legendre polynomials (ref. 1).

FORMULATION OF MEMBER STIFFNESS MATRIX

Pin nodal model

4. The member stiffness matrix related to the deformed state is derived from the co-ordinates of the undeformed state. The member stiffness matrix is as follows:

$$[f_{i,j}] = \left[\frac{\partial f_i}{\partial u_j}\right] = \begin{bmatrix} [a] & [a] \\ [a] & [a] \end{bmatrix} \tag{1}$$

$$[a] = EA\begin{bmatrix} \frac{1}{l_0} - \frac{1}{l^3}\left(3u^3 - \frac{v^2+w^2}{2}\right) & \frac{v}{l^2}\left(1+\frac{u}{l}\right) & \frac{w}{l^2}\left(1+\frac{u}{l^2}\right) \\ & \frac{1}{l_0} - \frac{1}{l} + \frac{u}{l^2} - \frac{u^2}{l^3} + \frac{3v^2+w^2}{2l^3} & \frac{vw}{l^3} \\ Sym & & \frac{1}{l_0} - \frac{1}{l} + \frac{u}{l^2} - \frac{u^2}{l^3} + \frac{v^2+3w^2}{2l^3} \end{bmatrix} \tag{2}$$

where $l_0$is the member length in the undeformed state; l is the member length in the deformed state; and u, v, w are displacement components in the x, y, z directions respectively.

Rigid nodal model

5. It is necessary to derive the member stiffness matrix with consideration of geometrical non-linearity of rigid nodal latticed domes. Joint nodes consist of an undeformable rigid body. In this study, member elastic properties are assumed to be concentrated along their axes. External loads are applied only at the joints. When a member deformation is very small and the nodal displacement is very large the total potential energy can be calculated as:

$$\Pi = \int_0^{\ell} \left\{\frac{EA\,\varepsilon_N^2}{2} + \frac{EI\,\kappa_y^2}{2} + \frac{EI\,\kappa_z^2}{2}\right\} dx - \{\tilde{P}\}^T\{\tilde{x}\} \tag{3}$$

where

$$\varepsilon_N = \frac{du}{dx} + \frac{1}{2}\{(\frac{dv}{dx})^2 + (\frac{dw}{dx})^2\},\ \kappa_y = -\frac{d^2w}{dx^2},\ \kappa_z = -\frac{d^2v}{dx^2}$$

6. In order to obtain the member stiffness matrix, we must solve the variational problem regarding Π as a functional. The equilibrium equation of internal and external forces of the member can be obtained as the simultaneous non-linear differential equation of u, v, w from the stationary conditions of Π. Transverse displacements v and w are assumed to be constrained to vary cubically with the distance along the member chord, and the longitudinal displacement, u, to vary linearly as shown in Fig. 1.

Fig. 1. Deformation of unit member model.

In view of the above, we can derive the < Stiffness Matrix 1 > (ref. 2). Differentiating Eq. 4 for each displacement, the tangential stiffness matrix can be obtained.

$$\begin{Bmatrix} P \\ M_a \\ M_b \end{Bmatrix} = \begin{bmatrix} \dfrac{EA}{l} & \dfrac{EA(4\theta_a-\theta_b)}{60} & \dfrac{EA(4\theta_b-\theta_a)}{60} \\ \dfrac{EA(4\theta_a-\theta_b)}{30} & \dfrac{4EI}{l}+\dfrac{EAl}{140}(4\theta_a^2-\theta_a\theta_b+\theta_a^2) & \dfrac{2EI}{l}-\dfrac{EAl}{280}(\theta_a^2+\theta_b^2) \\ \dfrac{EA(4\theta_b-\theta_a)}{30} & \dfrac{2EI}{l}-\dfrac{EAl}{280}(\theta_a^2+\theta_b^2) & \dfrac{4EI}{l}+\dfrac{EAl}{140}(4\theta_b^2-\theta_a\theta_b+\theta_a^2) \end{bmatrix} \begin{Bmatrix} \bar{u} \\ \theta_a \\ \theta_b \end{Bmatrix} \tag{4}$$

When the rotation angles are sufficiently small, higher order terms can be neglected. Hence, Eq. 4 reduces to Eq. 5 as < Stiffness Matrix 2 >.

$$\begin{Bmatrix} P \\ M_a \\ M_b \end{Bmatrix} = \begin{bmatrix} \dfrac{EA}{l} & 0 & 0 \\ 0 & \dfrac{4EI}{l}+\dfrac{2EA\bar{u}}{15} & \dfrac{2EI}{l}-\dfrac{EA\bar{u}}{30} \\ 0 & \dfrac{2EI}{l}-\dfrac{EA\bar{u}}{30} & \dfrac{4EI}{l}+\dfrac{2EA\bar{u}}{15} \end{bmatrix} \begin{Bmatrix} \bar{u} \\ \theta_a \\ \theta_b \end{Bmatrix} \tag{5}$$

Generally, < Stiffness Matrix 3 > is used for the slope-deflection method in the case of finite deformation. The tangential stiffness matrix is given by Eq. 6.

$$\begin{Bmatrix} P \\ M_a \\ M_b \end{Bmatrix} = \begin{bmatrix} \dfrac{EA}{l} & 0 & 0 \\ 0 & \dfrac{EI\alpha(\sinh\alpha l - \alpha l\cosh\alpha l)}{2(\cosh\alpha l - 1) - \alpha l\sinh\alpha l} & \dfrac{EI\alpha(-\sinh\alpha l + \alpha l)}{2(1 - \cosh\alpha l) + \alpha l\sinh\alpha l} \\ 0 & \dfrac{EI\alpha(-\sinh\alpha l + \alpha l)}{2(1 - \cosh\alpha l) - \alpha l\sinh\alpha l} & \dfrac{EI\alpha(\sinh\alpha l - \alpha l\cosh\alpha l)}{2(\cosh\alpha l - 1) - \alpha l\sinh\alpha l} \end{bmatrix} \begin{Bmatrix} \overline{u} \\ \theta_a \\ \theta_b \end{Bmatrix} \tag{6}$$

ARC-LENGTH METHOD AND DETERMINATION OF CRITICAL POINT

7. The Arc-Length method, proposed by E. Riks (refs 3, 4), is used to obtain the solution of the non-linear global stiffness matrices. Fig. 2 shows the process of the Newton-Raphson iteration and the determination of the critical point.

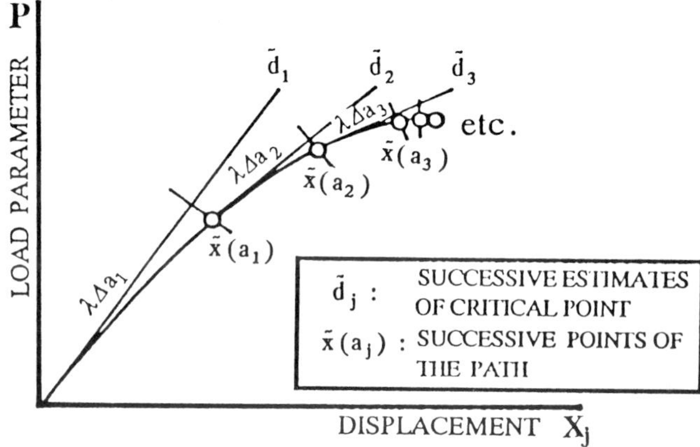

Fig. 2. Process for the decision of a critical point in the load-displacement space.

8. Firstly, suppose that the exact solution vector $\tilde{x}(a_j)$ is within a certain neighbourhood of the critical point. Approximate solution $\tilde{d}^{(j)}$ can conveniently be calculated with the basic procedure of the Newton-Raphson method.

$$\tilde{d}^{(j)} = \tilde{x}(a_j) + \lambda\, \Delta a^{*}\, \dot{\tilde{x}}(a_j) \tag{7}$$

In this expression λ (= 0.3 in this paper) is the control parameter to serve as a means of keeping the stepsize $\Delta a = \lambda \Delta a^*$ within acceptable bounds. This approximate solution $\tilde{d}^{(j)}$ can be improved by the successive solution of:

$$f_{\alpha,\beta}(\tilde{d}^{(J)})\Delta d_{\beta}^{(J)} + f_{\alpha}(\tilde{d}^{(J)}; a_J + \Delta a) = 0 \tag{8}$$

$$\tilde{d}^{(J+1)} = \tilde{d}^{(J)} + \Delta\tilde{d}^{(J)}$$

In this process, when the minimum eigen-value of the tangential stiffness matrix is equal to zero its point a^* is judged to be the critical point. By Taylor expansion around the critical point a^*:

$$\dot{\omega}(a^*) = \omega(a_k) + \dot{\omega}(a_k) \cdot da_k = 0 \tag{9}$$

Eq. 9 can be rewritten as:

$$\begin{aligned} da_k &= -\omega(a_k) / \dot{\omega}(a_k) \\ &\approx -\omega(a_k) \cdot da_{k-1} / \{\omega(a_k) - \omega(a_{k-1})\} \end{aligned} \tag{10}$$

If a_k approaches a^* we can decide such a point (= a_k) as the critical point (= a^*) because the arc-length da_k becomes zero at that point.

$$da_k \leq \varepsilon \cdot da_1 \tag{11}$$

where ε is a constant decided according to the model (ε = 0.05 in this paper).

ANALYTICAL MODEL

9. A pin-jointed spherical dome with 8 segments in the meridianal direction, radius R=28m, and span L=27.5m (analytical model 1 (T)) and a rigid-jointed model with 6 segments, radius R=30m and span L=25.4m (analytical model 2 (F)) are used as the analytical models of double-layer latticed domes. Boundary conditions were assumed to be both pinned and fixed supports. The classification of analytical models is as follows:

10. For the consideration of the feasibility of the numerical calculation, physical quantities are given as dimensions of structural members. Unit members composing the double-layer latticed dome are elastic straight pipes and the ratio of cross-sectional area to moment of inertia is set as 17 (referred to Saka's data). In the case of the rigid-jointed model, the cross-sectional area of the upper chord members is 12.18cm^2, buckling length is 387cm, and slenderness ratio is 100. In the case of the pin-jointed model, the cross-sectional area is 12.18cm^2, member length is 390cm, and slenderness ratio is 100.

Table 1. Classification of analytical model 1.

CONTENTS OF CLASSIFICATION			NO.
Half Open Angle (A)		32°	1
		28°	2
		24°	3
Load (P)	Model 1	All , 2/4 , 3/4	1, 2, 3
	Model 2	Half, 6 Division	4, 5
Slenderness Ratio (S)		100	1
		75	2
		130	3

Table 2. Classification of analytical model 2.

CONTENTS OF CLASSIFICATION		NO.
Stiffness Matrix (K)	Stiffness Matrix 1	1
	Stiffness Matrix 2	2
	Stiffness Matrix 3	3
Node (N)	Pin Node	2
	Rigid Node	1
Support (B)	Pin Support	2
	Fixed Support	1

APPLICATION OF THE MODAL ANALYSIS

11. Let us consider the structural systems consisting of n-degrees of freedom represented by global co-ordinates. We can dispense with the vector notation, and denote the spatial displacements by {D}

$$^{n}\{\overset{1}{dF}\} = {}^{n}[\overset{n}{K(N)}]\ {}^{n}\{\overset{1}{dD}\} \qquad (12)$$

where {dF}, [K(N)], {dD} are incremental loading, global tangential stiffness matrix, and spatial incremental displacement respectively. Eq. 12 represents a set of n-simultaneous partial differential equations of the incremental system in the non-linear buckling problem. In general, to obtain the solution to Eq. 12 is not straight forward, since n is usually very large. Therefore, there is a need for a new analytical method to facilitate this. In this paper, we develop the modal analysis as a new solution. In view of the above, we derive the new stiffness matrices with respect to a different set of generalised co-ordinates Λ_J (j = 1, 2,

..., m) such that spatial displacement vector {D} can be represented by a linear combination of the Eigen-modes. Let us consider the linear transformation:

$$ {}^{n}\{\overset{1}{D}\} = \Lambda_1 {}^{n}\{\overset{1}{D_1}\} + \Lambda_2 {}^{n}\{\overset{1}{D_2}\} + \cdots + \Lambda_m {}^{n}\{\overset{1}{D_m}\} \tag{13} $$

$$ {}^{n}\{\overset{1}{dD}\} = d\Lambda_1 {}^{n}\{\overset{1}{D_1}\} + d\Lambda_2 {}^{n}\{\overset{1}{D_2}\} + \cdots + d\Lambda_m {}^{n}\{\overset{1}{D_m}\} \tag{14} $$

in which Λ_J is the generalised co-ordinate associated with the basis Eigen-modes $\{D_1\}, \{D_2\}, ..., \{D_m\}$ where $\{D_j\}$ takes a constant value. Member forces can be calculated from the physical displacements:

$$ {}^{k}\{\overset{1}{N}\} = {}^{k}[\overset{n}{C}]\, {}^{n}\{\overset{1}{D}\} \tag{15} $$

where {N} is a internal force vector and [C] is a stiffness matrix. Substituting Eq. 13 into Eq. 15 gives:

$$ \{N\} = \Lambda_1\{H_1\} + \Lambda_2\{H_2\} + \cdots + \Lambda_m\{H_m\} \tag{16} $$

where

$$ {}^{k}\{\overset{1}{H_J}\} = {}^{k}[\overset{n}{C}]\, {}^{n}\{\overset{1}{D_J}\} = \text{const.} \tag{17} $$

Therefore, elements of the internal forces can be rewritten as

$$ N_{\alpha\beta} = \Lambda_1 h_{1\alpha\beta} + \Lambda_2 h_{2\alpha\beta} + \cdots + \Lambda_m h_{m\alpha\beta} \tag{18} $$

in which $h_{j\alpha\beta}$ is obtained from Eq. 17 .
Associated with elements k_{ij} of [K(N)] of Eq. 12

$$ \begin{aligned} k_{IJ}(N_{\alpha\beta}) &= k_{IJ}(\Lambda_1 h_{1\alpha\beta} + \Lambda_2 h_{2\alpha\beta} + \cdots + \Lambda_m h_{m\alpha\beta}) \\ &= \tilde{k}_{IJ}(\Lambda_1, \Lambda_2, \cdots, \Lambda_m) \end{aligned} \tag{19} $$

If $\Lambda_1 \sim \Lambda_m$ are given we can calculate the elements $\tilde{k}_{ij}$ of Eq. 19. That is to say, $\tilde{k}_{ij}$ is the function of generalised co-ordinates. Eq. 12 can be re-written in the form:

$$
{}^{n}\overset{1}{\{dF\}} = {}^{n}\overset{n}{[K(\Lambda_1,\Lambda_2,\cdots,\Lambda_m)]}\;{}^{n}\overset{1}{\{dD\}}
$$

$$
= [\{B_1\},\{B_2\},\cdots,\{B_m\}]\begin{Bmatrix} d\Lambda_1 \\ \cdot \\ d\Lambda_m \end{Bmatrix} \tag{20}
$$

$$
= {}^{n}\overset{m}{[B]}\{d\Lambda\} \tag{21}
$$

where

$$
{}^{n}\overset{1}{\{B_j\}} = [K(\Lambda_1,\Lambda_2,\cdots,\Lambda_m)]\{D_j\} \tag{22}
$$

Applying Galerkin's method to Eq. 21, we can obtain the final incremental global stiffness matrix with respect to generalised co-ordinates.

$$
\{D_j\}^T\{dF\} = \{D_j\}^T[B]\{d\Lambda\} \qquad (j=1,2,\cdots,m) \tag{23}
$$

or re-writing as:

$$
{}^{m}\overset{1}{\{dP\}} = {}^{m}\overset{m}{[K]}\;{}^{m}\overset{1}{\{d\Lambda\}} \tag{24}
$$

in which:

$$
\{dP\} \equiv \begin{Bmatrix} \{D_1\}^T\{dF\} \\ \bullet \\ \{D_m\}^T\{dF\} \end{Bmatrix} \tag{25}
$$

$$
{}^{m}\overset{m}{[K]} \equiv \begin{bmatrix} \{D_1\}^T[K]\{D_1\} & \cdots & \{D_1\}^T[K]\{D_m\} \\ \bullet & \bullet & \bullet \\ \{D_m\}^T[K]\{D_1\} & \cdots & \{D_m\}^T[K]\{D_m\} \end{bmatrix} \tag{26}
$$

The geometrically non-linear buckling problems can be analysed by calculating Eq. 26 by means of the Arc-Length method.

EXAMPLES OF EIGEN-MODES AND NUMERICAL RESULTS

12. Fig. 3 shows 3 eigen-modes (1st, 10th and 20th) of the 20 eigen-modes of analytical model 2 used in the analysis. Although it is difficult to select the necessary eigen-modes from the global system matrices, twenty natural modes related to the 1st to 20th eigen-values are used in this paper.

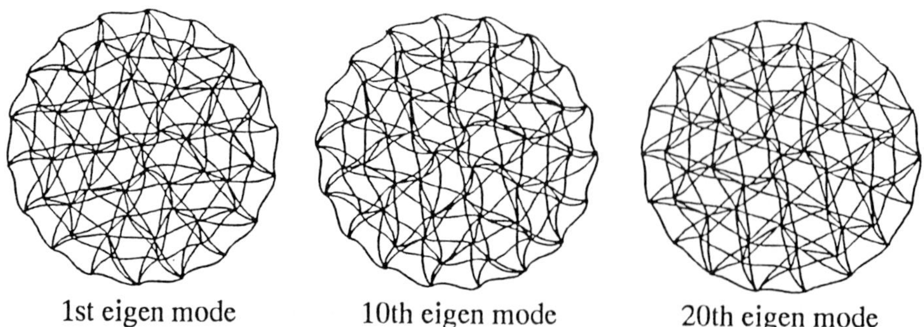

1st eigen mode 10th eigen mode 20th eigen mode

Fig. 3. Examples of eigen-modes for analytical model 2.

13. We introduce T-NBKH and F-APS as symbols to represent the analytical models 1 and 2 according to Table 1 and 2 respectively. Fig. 4 shows the distribution of member strain energy for model T-121, and this becomes one index to judge the member that is most likely to buckle.

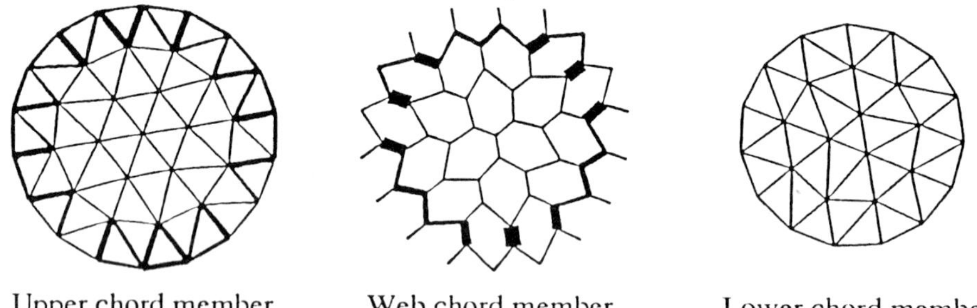

Upper chord member Web chord member Lower chord member

Fig. 4. Distribution of member strain energy of analytical model T 121.

14. We know that the buckling behaviour is affected more by the half-open angle than other factors and bifurcation buckling rather than snap-through occurs in the large portion of models. Fig. 6 shows the buckling paths in the three-dimensional space in case of model 1 according to the variation of boundary conditions.

15. It is impossible to obtain the bifurcation paths after the critical point in the region of degeneration in multi-degrees and linear behaviour, as shown in Fig. 7. This is because there were some problems in the process of our numerical calculation. Numerical results of modal analysis could not be included in the paper. Therefore, the numerical results will be presented at the conference.

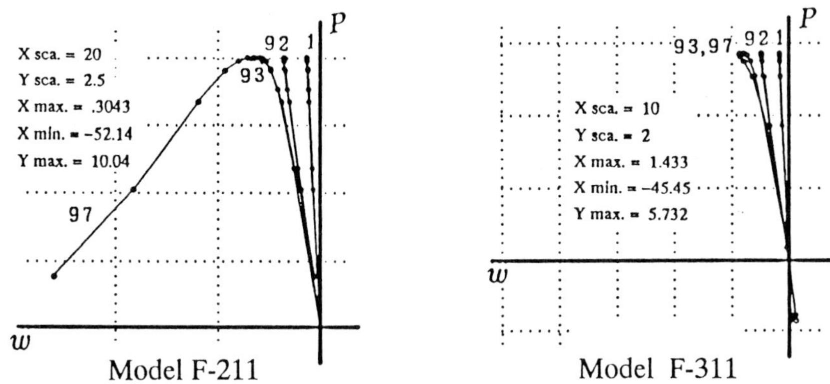

Fig. 5 Influence of half-open angle on buckling behaviour.

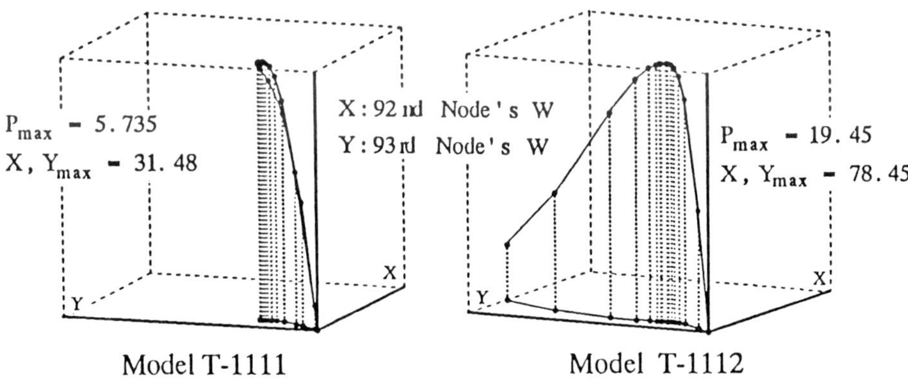

Fig. 6. Three-dimensional presentation of buckling paths by boundary conditions.

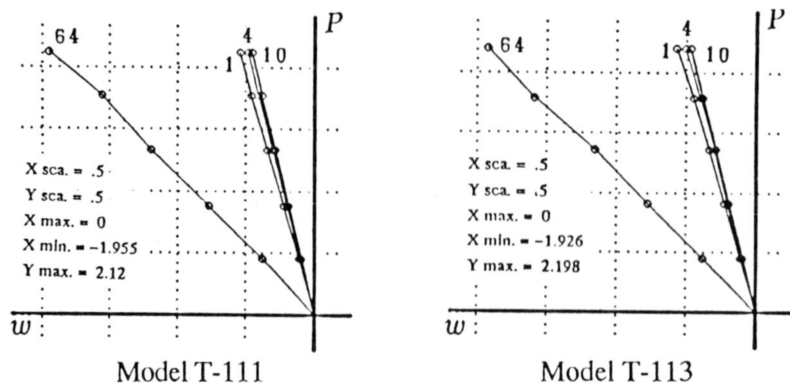

Fig. 7. Influence of Stiffness matrix 1 an 3 in the analytical model 2.

REFERENCES

1. KUNIEDA H. Classical buckling load of spherical domes under uniform pressures. Proc. ASCE. Vol.118, No.EM8, August, 1992, pp.1513 - 1525.
2. POWELL G.H.. Theory of non-linear elastic structures. Proc. ASCE, Vol.95, No.ST12, December, 1969, pp. 2687 - 2701.
3. RIKS E. An incremental approach to the solution of snapping and buckling problems. Int. J. of Solids and Structures, Vol.15, 1979, pp. 529 - 551.
4. RIKS E. Some computational aspects of the stability analysis of non-linear structures. Comput. Meths. Appl. Mech. Engrg., Vol.47, 1984, pp. 219 - 259.

34. Shape generation and stability analysis of funicular domes

S. RAJASEKARAN and K. MUDUMBI, P. S. G. College of Technology, India

SYNOPSIS. Contrary to the usual procedure, a desired state of stress is assumed and the corresponding funicular shell shape is obtained. The domes thus developed from these equivalent shells are called Funicular domes. For shallow shells, the governing equation is a Poisson's equation and for deep shells the governing equation is a non-linear partial differential equation. The boundary integral element method (ref. 1) is applied to obtain shapes of the funicular domes. Static and non-linear stability analysis is carried out to investigate the snap through buckling and load deformation response.

INTRODUCTION

1. Domes are of special interest to engineers and architects as they enclose a maximum amount of space with a minimum surface area and hence have proved to be economical in the use of construction materials. Recently, skeletal structures such as Geodesic domes and Schwedler domes have been developed.

FUNICULAR CONCEPT

2. This concept was first introduced by Ramaswamy (ref. 5) in the development of funicular shells. A cable carrying a uniformly distributed load/unit length of horizontal projection will develop tension throughout and take the form of a parabola which is the funicular polygon for the applied loads. If the cable is frozen in this position by some means and inverted as shown in Fig. 1 it will continue to be the funicular polygon but the loads will now be carried by pure compression instead of tension. There will not be any bending moment in the arch. The same concept is applied to funicular shells.

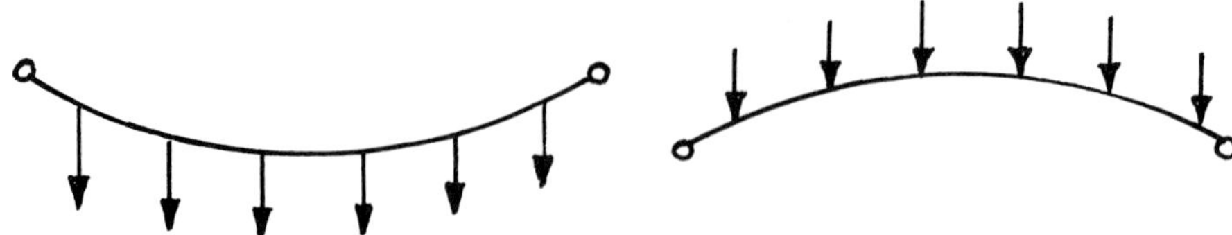

a) Funicular polygon for udl.

b) Funicular arch for udl.

Fig. 1.

EQUIVALENT SHELL THICKNESS

3. In order to generate the form of the funicular dome, first the skeletal structure (funicular dome) must be converted to an equivalent shell and the thickness of the equivalent shell is given by Wright (ref. 6) as:

$$t = 2\sqrt{3}\sqrt{(I / A)} = 2\sqrt{3}.r_g \qquad (1)$$

4. For hollow circular members $r_g \geq d / 3$ in which d is the outer diameter and the effective thickness of the shell is given by:

$$t \geq 2d / \sqrt{3} \qquad (2)$$

The force per unit length of the equivalent shell is obtained by multiplying the allowable stress in the membrane of the dome with the equivalent shell thickness.

GOVERNING EQUATION

5. If it is assumed that the desirable optimum stress distribution is one of uniform compression ($N\theta = N\phi = -S = \sigma$ t and $N\theta\phi = 0$) then the equilibrium equation for the shell is given by Harrenstein (ref. 2) as:

$$1/R_1 + 1/R_2 = P_3 / S \qquad (3)$$

where P_3 is the normal pressure on the dome.

6. For shallow shells, Eq. 3 reduces to :

$$\partial^2 Z/\partial X^2 + \partial^2 Z/\partial Y^2 = P_3 / S \qquad (4a)$$

and for deep shells:

$$K_X \partial^2 Z/\partial X^2 + K_Y \partial^2 Z/\partial Y^2 = P_3 / S \qquad (4b)$$

where,

$$K_X = 1/\left(1+(\partial Z/\partial X)^2\right)\sqrt{\left(1+(\partial Z/\partial X)^2+(\partial Z/\partial Y)^2\right)} \quad (4c)$$

$$K_Y = 1/\left(1+(\partial Z/\partial Y)^2\right)\sqrt{\left(1+(\partial Z/\partial X)^2+(\partial Z/\partial Y)^2\right)} \quad (4d)$$

BOUNDARY INTEGRAL ELEMENT METHOD FOR SHALLOW FUNICULAR SHELLS (see ref. 1,)

7. The governing equation is Poisson's equation:

$$v^2 Z = P_3 / S = C \quad (5)$$

which is used in the domain Ω with boundary conditions (see Fig. 2)

$$Z = Z \text{ on } \Gamma_1 \quad (6a)$$

$$\delta z/\delta n = V \text{ on } \Gamma 2 \quad (6b)$$

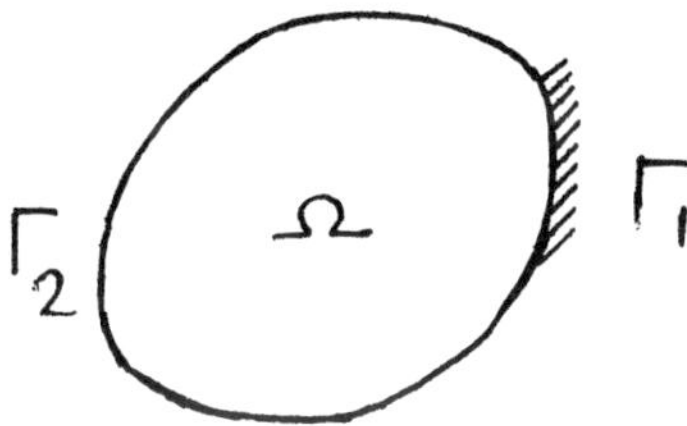

Fig. 2. Region and Boundary.

where C is a known function and Z is the vertical ordinate and $\delta z/\delta n$ is the slope. Applying the method of weighted residual to Eq. 5 and integrating by parts twice, gives on the boundary:

$$Z_i / 2 + \sum_{j=1}^{n} \int_{\Gamma_j} Z_j \partial W / \partial n . d\Gamma + \int_A CW.dX.dY = \sum_{j=1}^{n} \int_{\Gamma_j} V_j W.d\Gamma \quad (7)$$

The boundary can be idealised into either constant, linear or quadratic elements as in the finite element method. Assuming constant elements Eq. 7 reduces to:

$$Z_i/2+\sum_{j=1}^{n}\hat{H}_{ij}Z_j=\sum_{j=1}^{n}G_{ij}V_j-\int C\,W.dX.dY \qquad (8)$$

where

$$G_{ij}=\int_{\Gamma j}W.d\Gamma_j \qquad (9a)$$

$$H_{ij}=\int_{\Gamma j}\partial W/\partial n.d\Gamma_j \qquad (9b)$$

The fundamental solution for the problem is given by:

$$W=LOG\,(1/r)/(2\pi) \qquad (10)$$

The second term of the right hand side of Eq. 8 is a region integral which can be evaluated as:

$$Q_i+\int C\,W.dX.dY=\sum_{e=1}^{M}\sum_{K=1}^{e}P_K(CW)_K\Omega \qquad (11)$$

where e denotes the cell varying from 1 to M and M being the total number of elements in the region and P_K the integration weights. Eq.8 on the boundary reduces to:

$$\begin{vmatrix}H_{11} & H_{12}\\ H_{21} & H_{22}\end{vmatrix}\begin{Bmatrix}\overline{Z}\\ Z\end{Bmatrix}=\begin{vmatrix}G_{11} & G_{12}\\ G_{21} & G_{22}\end{vmatrix}\begin{Bmatrix}V\\ \overline{V}\end{Bmatrix}-\{Q\} \qquad (12)$$

where Z is known on the boundary Γ consisting of n_1 elements and V is known on the boundary Γ consisting of n_Z elements. Rearranging the equations, Z and V on the boundary may be evaluated. Once Z and V on the boundary are known, Z at the interior points may be evaluated as:

$$Z_i=\sum_{j-1}^{n}G_{ij}V_j-\sum_{j=1}^{n}\hat{H}_{ij}Z_j-Q_i \qquad (13)$$

and hence the shape of the shallow funicular dome.

BIE METHOD FOR A DEEP FUNICULAR DOME

8. Eq. 4b, is the governing equation for the shape generation of a deep funicular dome. By using the transformations:

$$X_n=X/\sqrt{K_X} \qquad\qquad Y_n=Y/\sqrt{K_Y} \qquad (14)$$

Eq. 4b is now transformed into:

$$\partial^2 Z/\partial X_n^{\ 2} + \partial^2 Z/\partial Y_n^{\ 2} = p_3 / S = C \tag{15}$$

and the fundamental solution is:

$$W = LOG\,(1/r) / \left(2\Pi\sqrt{K_X}\sqrt{K_Y}\right) \tag{16}$$

Eq. 4b has to be solved iteratively by using:

$$\partial Z/\partial X = 0 \qquad \partial Z/\partial Y = 0 \qquad \text{i.e.: } K_X = K_Y = 1$$

solve as discussed, for shallow shells as:

$$[G]_1\{V\}_0 - [H]_1\{Z\}_0 = \{Q\}_0 \tag{17}$$

On the boundary $\{Z\}_0 = \{0\}$ $\{V\}_0$ can be computed as:

$$\{V\}_0 = [G]_1^{-1}\{Q\}_0 \tag{18a}$$

or in general

$$\{V\}_{K+1} = [G]_K^{-1}\{Q\}_K \tag{18b}$$

At any step knowing the values of Z, $\partial Z/\partial X$, $\partial Z/\partial Y$; K_X and K_Y are evaluated and hence W. Knowing $[G]_K$, the value of $\{V\}_{K+1}$ on the boundary can be evaluated. Iterations are repeated until the desired convergence is reached as:

$$\{\Delta Z\}^T\{\Delta Z\} / \{Z\}^T\{Z\} \le 0.0001 \tag{19}$$

STATIC AND STABILITY ANALYSIS

9. All ductile structures respond in a non-linear manner immediately prior to and during collapse. In order to predict the maximum load carrying capacity it is therefore necessary to be able to assess the effect of the non-linear response. Using the approach of Rajasekaran and Murray (ref. 4) the total and incremental equilibrium equations are written as:

$$([K] + 1/2\ [N_1] + 1/3\ [N_2])\{u\} = \{F\} \tag{20.a}$$

$$([K] + [N_1] + [N_2])\{u\} = \{\Delta u\} = \{\Delta F\} \tag{20.b}$$

The non-linear load deformation response is traced by incrementing the displacement at the centre node and using the Newton Raphson or Modified Newton Raphson method. In addition to the above, the bifurcation load is determined by carrying out the classical eigen value analysis as:

$$[K]\{\Delta u\} = \mu([N_1]+[N_2])\{\Delta u\} \tag{21}$$

The classical eigen value analysis ignores all the pre-buckling deformations prior to the attainment of the buckling load. In case of structures such as domes, which exhibit snap through phenomenon, the eigen value analysis is not realistic and a complete non-linear load deformation response is needed to find the stability load.

NUMERICAL EXAMPLES

Example 1

10. A circular dome of 30m diameter is subjected to a uniformly distributed load of 2kN/sq.m. Assuming σ_{all} (comp)=150MPa the shape of the funicular dome is obtained as shown in Fig. 3. The stress analysis is carried out and it is found that the stresses in the meridional and circumferential members vary from 80 to 140 MPa and the stresses in the diagonal members are very small. Buckling analysis of this dome is carried out by incrementing the displacement of the central node in the Z direction and the corresponding load is determined. The load deformation response is shown in Fig. 4 and the buckled shape is given in Fig. 5. The buckling load obtained from the load deformation response is 19.063kN as compared to classical eigen value analysis of 31.47kN.

Example 2

11. A square based dome of 20m sides is subjected to a uniform pressure load of 2kN/sq.m. The funicular shell obtained on this plan area for the given load is shown in Fig. 6. and it is found that the compressive stress in the members are of the order of 102 to 138 MPa. The buckled shape is also shown in Fig. 7.

Example 3

12. A circular dome of 20m diameter is subjected to a concentrated load of 100kN. The funicular dome obtained is shown in Fig. 8. The stresses in the compression members are of the order of 90 to 130 MPa. The results obtained by the BIE solution agree with the closed form solution. More examples are found in (ref. 3).

CONCLUSION

13. Optimal shapes of funicular domes on different ground plans subjected to different loading conditions are obtained using BIEM. Non-linear stability analysis for these domes is carried out to investigate snap through buckling and load deformation response. Static analysis for these funicular domes are carried out to check whether the stresses are within the allowable limit.

ACKNOWLEDGEMENTS

14. The authors thank the management and Prof. S. Subramanyan, Principal, P.S.G.College of Technology for giving the necessary facilities to carry out the research work reported in this paper.

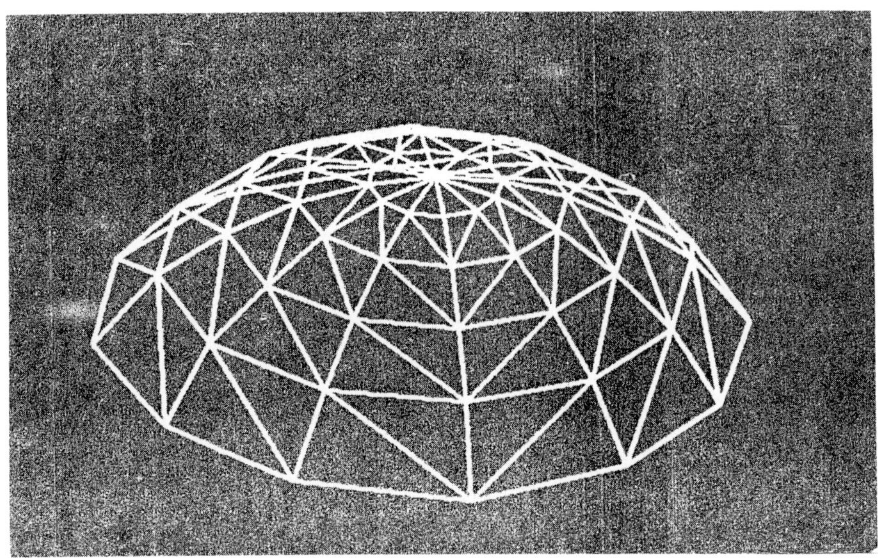

Fig. 3. Funicular dome on a circular plan.
(d = 30m; udl = 2kN/m^2)

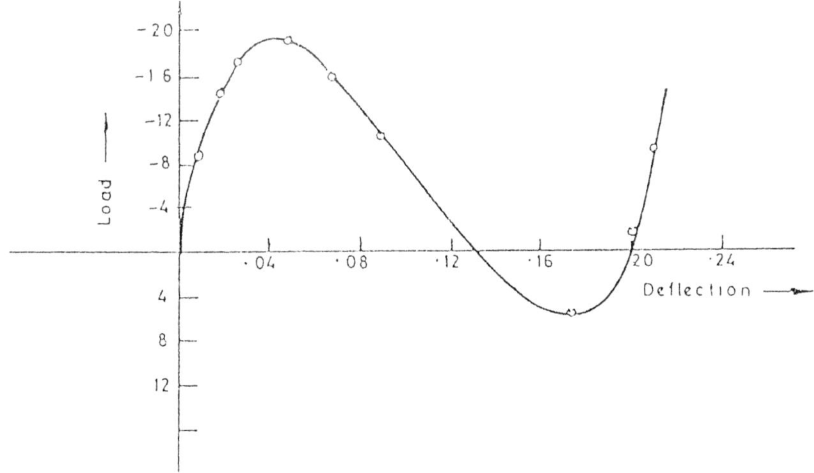

Fig. 4. Load deformation response of a circular dome.

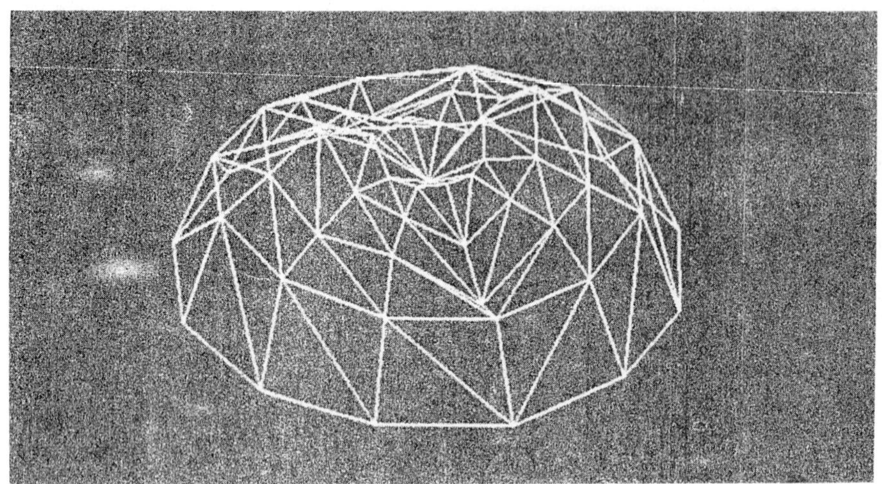

Fig. 5. Buckled shape of the dome on a circular plan.

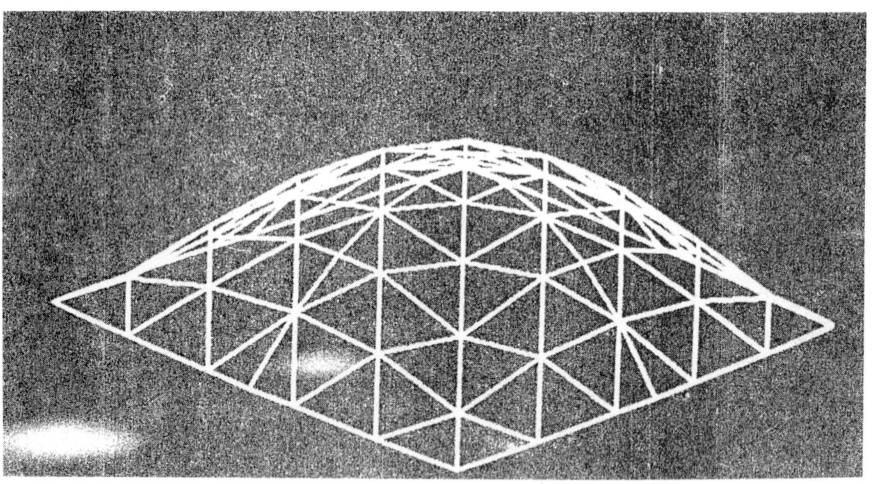

Fig. 6. Funicular dome on a square plan.
(a = 20m; udl = 2kN/m^2)

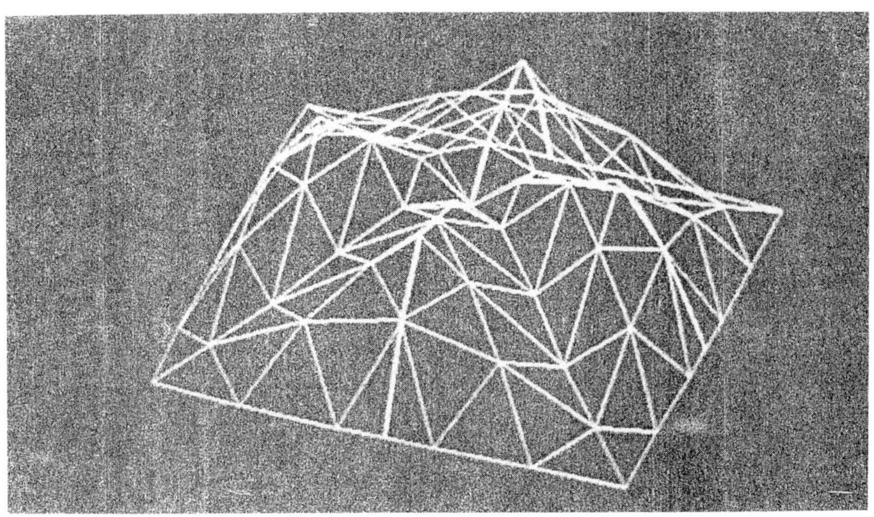

Fig. 7. Buckled shape of the dome on a square plan.

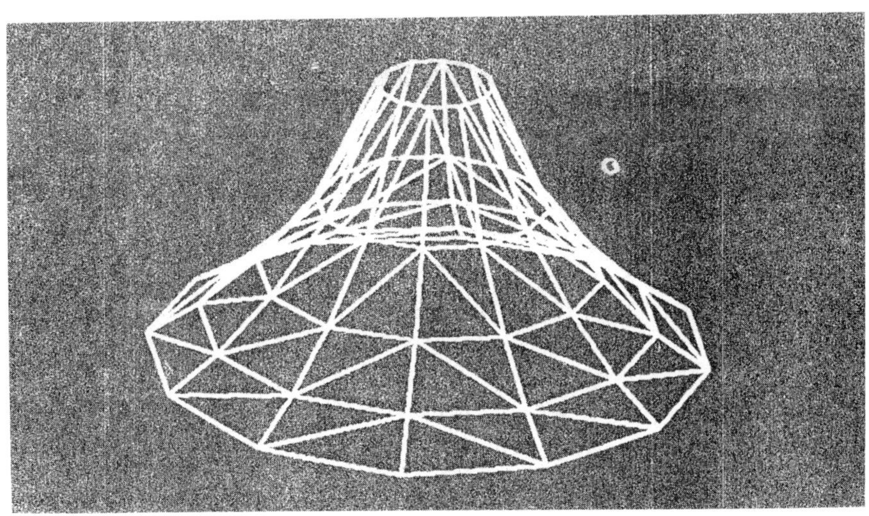

Fig. 8. Funicular dome on a circular plan.
(d = 20m; P = 100kN

REFERENCES

1. BREBBIA, C.A. The Boundary Element Method for Engineers, Pentech Press, London, 1978.
2. HARRENSTIEN, H.P. Configuration of Shell structures for optimal stress. Proceedings of the Symposium on shell research, Aug.30-Sept2 1961.
3. KARUNA, M. Shape Generation, Static and Instability Analysis of Funicular Domes. Thesis submitted to the Bharathlar University as a partial fulfilment of M.E. degree, 1992.
4. RAJASEKARAN, S., and MURRAY, D.W. Incremental Finite Element Matrices. ASCE. Journal of Structural Division, Vol.99, Dec.1973.
5. RAMASUAMY, G.S. The theory of a New Funicular Shell of Double Curvature, Indian Concrete Journal, Sept.1960.
6. WRIGHT, T. Membrane forces and Bucking in Reticulated Shells, ASCE. Journal of Structural Division, Vol.91, No.ST.1 Feb. 1965.

35. Stability of braced domes under dynamic loads

S. JIANHENG, Hebei Agricultural University, P. R. China

SYNOPSIS. In this paper, the large deflection stability behaviour of two well known domes subjected to dynamic loads are examined by the geometrically non-linear finite element method based on the Updated Lagrangian co-ordinate. Three types of time-dependent loading cases are considered, namely, step load and sinusoidal load of infinite duration and triangular load of finite duration. In the numerical solution, the Newmark method is used to satisfy the equations of motion.

INTRODUCTION

1. The stability of braced domes under static loads has been subject of considerable research in the past decades, and a lot of useful results have been obtained (refs 1-4). However, little attention has been paid to the non-linear stability of braced domes under dynamic loads, although the non-linear dynamic stability of pin-jointed truss structures has been studied (ref. 5). For clear understanding of the stability behaviour of braced domes under various loading conditions, the dynamic stability analysis of these structures is very important.

2. In this paper, based on the Updated Lagrangian co-ordinate, a geometrically non-linear tangent matrix is introduced, and two well-known braced domes are examined. In the numerical analysis, three types of dynamic loads are considered, namely, step load and sinusoidal load of infinite duration, and triangular load of finite duration. The Newmark method is used in the step-by-step integration procedure to satisfy the equation of motion. The load-deflection curves for the two braced domes subjected to three different dynamic loads are presented. The effects of the duration of the triangular load and the cycle of sinusoidal load are also presented for a 12-member braced dome.

NON-LINEAR EQUATION AND ANALYSIS PROCEDURE

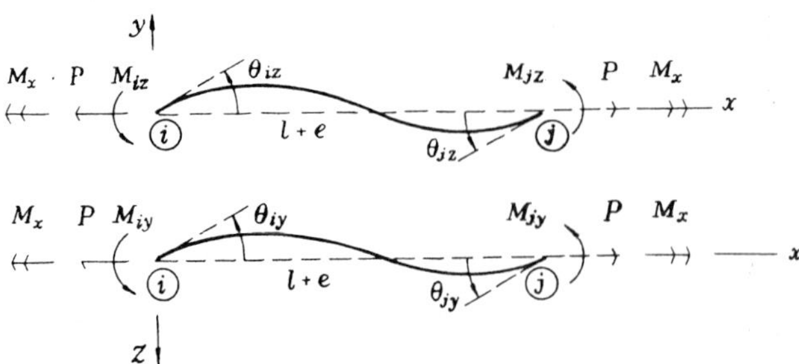

Fig. 1. Member basic displacements and forces.

3. As shown in Fig. 1, in the derivation of the geometrically non-linear tangent stiffness matrix for a three dimensional frame element, it is assumed that the deflection curves are cubic in two principal directions of the cross section of the member, and that the axial displacements vary linearly along the member. By the principle of minimum potential energy the basic member force-deformation relation can be expressed as follows:

$$M_{iz} = EA\{ [\frac{4e}{30} + \frac{1}{2}(\frac{2l}{35}C_1 + \frac{l}{210}C_2 - \frac{2l}{140}C_3)]\ \theta_{iz} - [\frac{e}{30} + \frac{1}{2}(\frac{l}{140}C_1 + \frac{l}{140}C_2 - \frac{2l}{210}C_3)\theta_{jz}\} + \frac{4EI_z}{l}\theta_{iz} + \frac{2EI_z}{l}\theta_{jz}$$

$$M_{jz} = EA\{ [\frac{4e}{30} + \frac{1}{2}(\frac{l}{210}C_1 + \frac{2l}{35}C_2 - \frac{2l}{140}C_3)]\ \theta_{jz} - [\frac{e}{30} + \frac{1}{2}(\frac{l}{140}C_1 + \frac{l}{140}C_2 - \frac{2l}{210}C_3)\theta_{iz}\} + \frac{4EI_z}{l}\theta_{jz} + \frac{2EI_z}{l}\theta_{iz}$$

$$M_{iy} = EA\{ [\frac{4e}{30} + \frac{1}{2}(\frac{2l}{35}C_1 + \frac{l}{210}C_2 - \frac{2l}{140}C_3)]\ \theta_{iy} - [\frac{e}{30} + \frac{1}{2}(\frac{l}{140}C_1 + \frac{l}{140}C_2 - \frac{2l}{210}C_3)\theta_{jy}\} + \frac{4EI_y}{l}\theta_{iy} + \frac{2EI_y}{l}\theta_{jy}$$

$$M_{jy} = EA\{ [\frac{4e}{30} + \frac{1}{2}(\frac{l}{210}C_1 + \frac{2l}{35}C_2 - \frac{2l}{140}C_3)]\ \theta_{jy} - [\frac{e}{30} + \frac{1}{2}(\frac{l}{140}C_1 + \frac{l}{140}C_2 - \frac{2l}{210}C_3)]\ \theta_{iy}\} + \frac{4EI_y}{l}\theta_{iy} + \frac{2EI_y}{l}\theta_{iy}$$

$$P = EA\{ [\frac{e}{l} + \frac{1}{30} (2\theta_{iz}^2 - \theta_{iz}\theta_{jz} + 2\theta_{jz}^2) + \frac{1}{30} (2\theta_{iy}^2 - \theta_{iy}\theta_{jy} + 2\theta_{jy}^2)]$$

$$M_x = \frac{GJ}{l}\theta_x \tag{1}$$

In which $C_1 = \theta_{iy}^2 + \theta_{iz}^2, = \theta_{jy}^2\theta_{jy}^2 + \theta_{jz}^2, C^3 = \theta_{iy}\theta_{jy} + \theta_{iz}\theta_{jz}$, EA = axial rigidity, EI_y and EI_z are the flexural rigidities about the y-axis and the z-axis, GJ = torsional rigidity, and l is the initial length of the undeformed members. By partial differentiation of Eq.1 with respect to each of the basic displacements, the increments of basic member forces can be obtained. Relating these with the increments of nodal forces in local co-ordinates will yield the geometrically non-linear tangent stiffness matrix. The detailed derivation and results can be found in refs 1-4.

4. In the present paper, the mass of each structural member is lumped at the joints at the member ends and the effect of damping is neglected. The equation of motion of the structure as a whole can be written as follows:

$$[M]\{\ddot{V}\} + \{F\} = \{P\} \tag{2}$$

in which {P} = dynamic load acting at the joints of structure, {F} = resultant internal forces, $[M]$ = diagonal mass matrix, and $\{\ddot{V}\}$ = acceleration of joints in globe co-ordinate. A dot on the displacement V represents a differentiation with respect to time.

5. For the non-linear numerical analysis, Eq. 2 should be rewritten into the incremental form as follows:

$$[M]\{\Delta\ddot{V}\} + [K]\{\Delta V\} = \{\Delta P\} \tag{3}$$

in which $\{\Delta V\}$ = increments of displacements in global co-ordinates, and [K] = geometrically non-linear tangent stiffness matrix for the whole structure, and $\{\Delta\ddot{V}\}$ = increments of acceleration.

6. In this study, a step-by-step integration technique utilising the Newton-Raphson iteration procedure is used to satisfy the equations of motion (Eq. 2 and Eq. 3) to obtain the dynamic response of the braced domes, The Newmark method (ref. 6) is employed in each of the iteration cycles.

7. Assuming that the configuration of the structure has been known at time t_k, when $\gamma = 1/2$ and $\beta = 1/4$, the Newmark method is unconditionally stable, and the increments of displacements and velocities between t_k and t_{k+1} can be written as follows:

$$\{\triangle V\} = (\triangle t)\ \{\dot{V}^{(k)}\} + \frac{1}{4}(\triangle t)^2\{\ddot{V}^{(k)}\} + \frac{1}{4}(\triangle t)^2\{\ddot{V}^{(k+1)}\} \tag{4}$$

$$\{\triangle \dot{V}\} = \frac{1}{2}\ (\triangle t)\ \{\ddot{V}^{(k)}\} + \frac{1}{2}(\triangle t)\{\ddot{V}^{(k+1)}\} \tag{5}$$

in which $\{\dot{V}\}$ = velocity vector, $\Delta t = t_{k+1} - t_k$ is the time increment. Substituting $\{\ddot{V}^{(k+1)}\} = \{\ddot{V}^{(k)}\} + \{\Delta\ddot{V}\}$ into Eq. 4, gives the increment of acceleration Eq. 6.

$$\{\triangle \ddot{V}\} = \frac{4}{(\triangle t)^2}\{\triangle V\} - \frac{4}{(\triangle t)}\{\dot{V}^{(k)}\} - 2\{\ddot{V}^{(k)}\} \tag{6}$$

Substituting Eq. 6 into the incremental equation of motion, Eq. 3, and re-arranging, gives:

$$[K]\ \{\triangle V\} + \frac{4}{(\triangle t)^2}\ [M]\ \{\triangle V\} = \{\triangle P\}$$
$$+ \frac{4}{\triangle t}\ [M]\ \{\ \dot{V}^{(K)}\} + 2\ [M]\ \{\ddot{V}^{(K)}\} \tag{7}$$

Solving Eq. 7 can obtain the approximate configuration of structure at time t_{k+1}, and then the Newton-Raphson iteration procedure is used to correct this approximate configuration until the equation of motion (Eq. 2) is satisfied. Details of the Newton—Raphson iteration procedure can be found in ref. 5.

NUMERICAL ANALYSIS OF DYNAMIC STABILITY

8. In the stability analysis of structures it is very important to choose the criterion by which the critical load is determined. In the present study, the Budinsky-Roth criterion is used to determined the dynamic critical load of structures. By this criterion, the dynamic response of structures is monitored under increasing load level, and a sudden increase in response due to a very small increase in the magnitude of load is considered as an indication of instability (ref. 5).

9. In order to have a clear understanding of the stability behaviour of braced domes under various dynamic loading, three types of dynamic loads are considered. They are the step load and the sinusoidal load of infinite duration and the triangular load of finite duration. These loading functions are shown in Fig. 2.

12-Member hexagonal dome

10. The stability of this dome under static loading (Fig. 3) has been studied by various authors (ref. 4). In the present analysis, the six boundary nodes are assumed to be restrained against translational movement in all directions, and a concentrated vertical load acts at the central node. The load-deflection curves for

pre-buckling (ref. 4) are shown in Figs 4-6. This dome has a buckling point of typical limit type, and its critical load is 97.8 lb (435.2N), (ref. 4).

11. The results of the numerical analysis for the 12-member hexagonal dome under dynamic load are summarised in Figs 4-7. The load versus maximum deflection of joint 1 due to step loading is shown in Fig. 4. The critical load under the step loading is 53.81b (239.4N), a reduction of 45% occurring at the critical load point.

12. Figs 5-6 show some typical curves of load against maximum deflection for joint 1 due to the triangular load and the sinusoidal load respectively, and Fig. 7 shows how the critical loads vary with the impulse duration of the triangular load and the cycle of sinusoidal load. For the triangular load, the variation of critical load with the impulse duration is very similar to the results obtained by Kassimali (ref. 5) for a two member truss and the pin-jointed geodesic dome. For short impulse duration, the critical load is higher than that for a static load. The critical loads decrease as the impulse duration increases, and for long impulse duration it approaches the critical value for step loading.

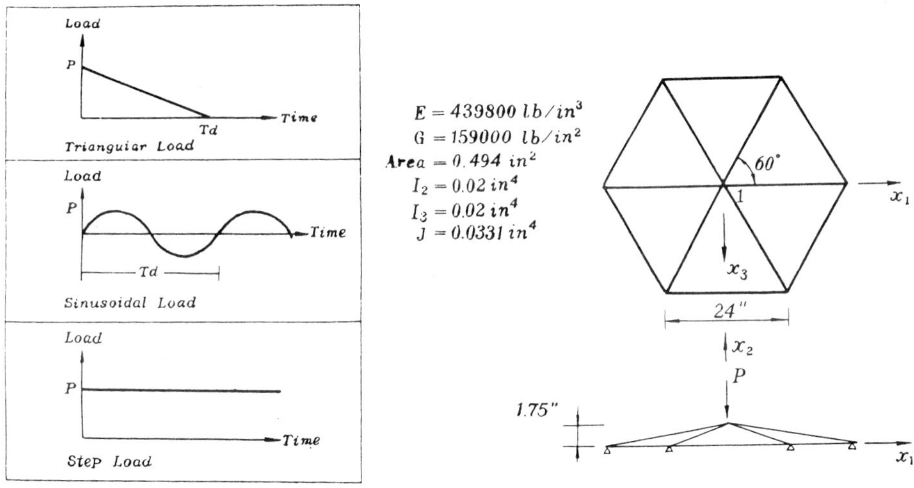

Fig. 2. Dynamic loading functions.

Fig. 3. 12 - member hexagonal dome, (1inch=25.4m, 1lb=4.45N).

13. For the sinusoidal loading case, the variation of critical load with the cycle of sinusoidal load is very different form the results obtained by Kassimali (ref. 5) for a two member truss and a pin-jointed geodesic dome, except for the resonance effect. The critical load is always higher than that for a static case, except at the resonance point. As the cycle increases, it approaches the static critical value.

Geodesic dome

14. The static stability behaviour of this dome (Fig. 8) has been a subject of much research (refs 1,4). Kassimali (ref. 5) conducted a detailed investigation into the dynamic stability behaviour of a space truss. In the present paper, it is re-analysed as a space frame with a vertical concentrated load at the central node. Its static critical load is 571,7N (ref. 4). The results for the three different loadings are shown in Figs 8-11.

15. Fig. 8 shows that for step loading, the critical load is 305N, indicating a reduction of 41.1%. For the triangular loading case, Fig. 10 shows that with increasing impulse duration the critical load decreases. In Fig. 11, when the cycle of sinusoidal loading Td = 0.03 sec, the critical load is very low (25.5N), only 4.5% of that for the static load, when Td = 0.05 sec, the critical load is 35.7% higher than that for the static load.

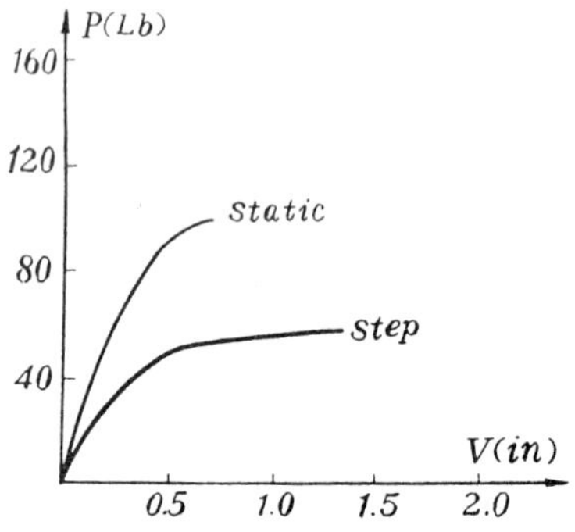

Fig. 4. Load-deflection curve for step loading (joint 1), (1lb=4.45N, 1inch=25.4mm).

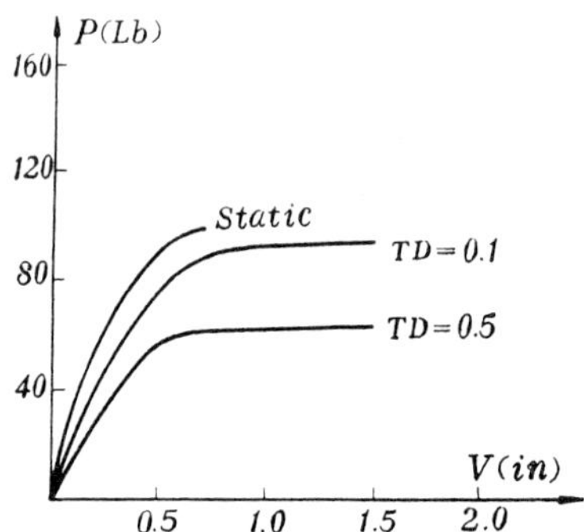

Fig. 5. Load-deflection curve for triangular loading (joint 1), (1lb=4.45N, 1inch=25.4mm).

CONCLUSION.

16. Based on the geometrically non-linear analysis method accounting for large rotation and displacements of joints in Updated Lagrangian co-ordinates, the large deformation response and stability behaviour of two well known frames, a 12-member hexagonal dome and a 24-member geodesic dome, under dynamic load are examined. Three loading cases, namely, step loads and sinusoidal loads of infinite duration, and triangular loads of finite duration are considered. The load-deflection curves are presented. For the critical loads of the 12-member hexagonal dome, the effects of the duration of triangular load and the cycle of sinusoidal load are investigated. The presented results will be helpful for a clear understanding of the dynamic stability behaviour of braced domes.

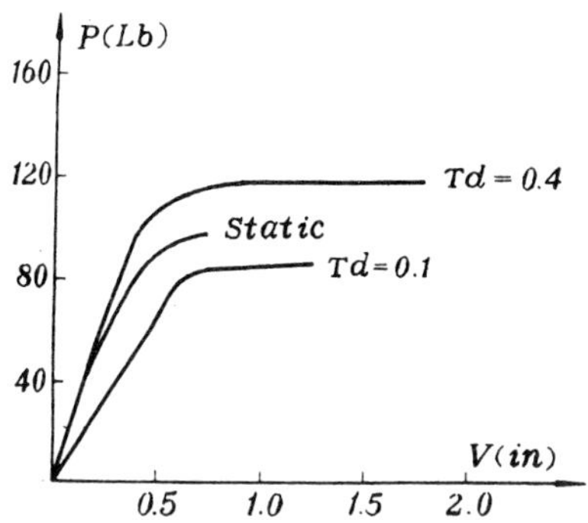

Fig. 6. Load-deflection curve for the sinusoidal loading, (1lb=4.45N, 1inch=25.4mm)

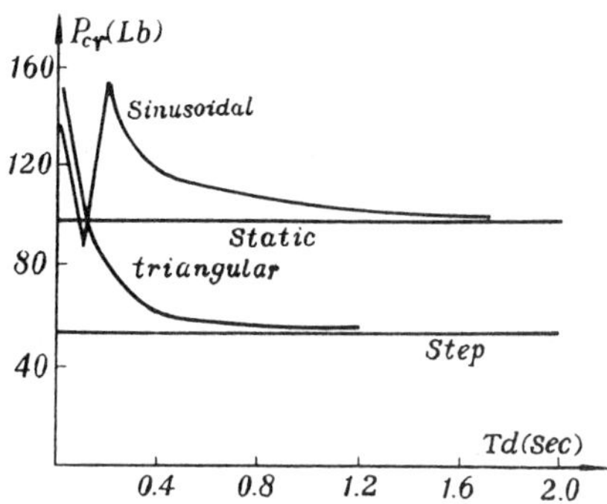

Fig. 7. P_{cr} vs Td for the 12-member hexagonal dome, (1lb=4.45N, 1inch=2.45mm)

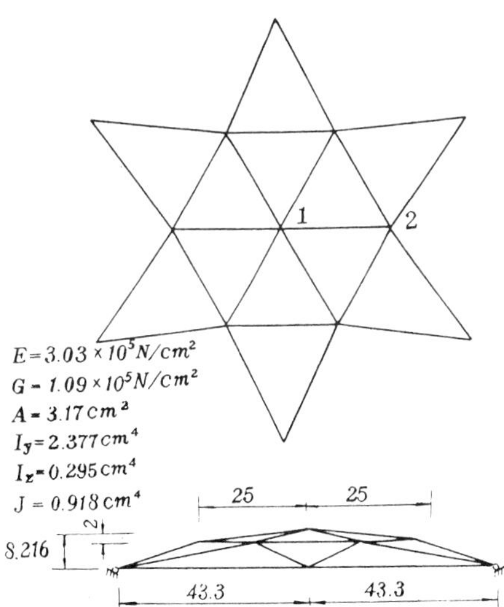

Fig. 8. Geodesic dome: geometry and properties.

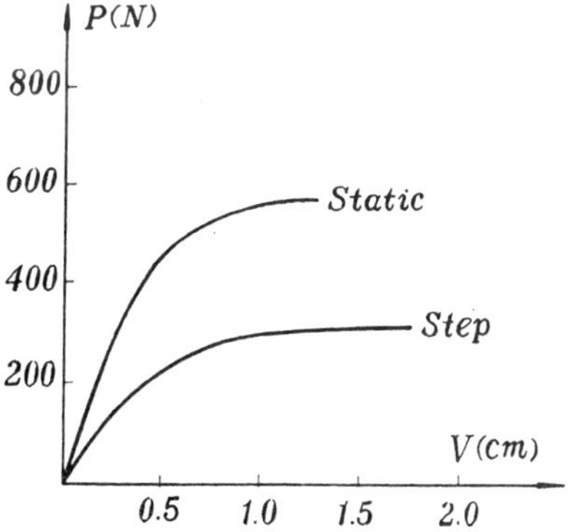

Fig. 9. Load-deflection curve for the step loading (joint 1).

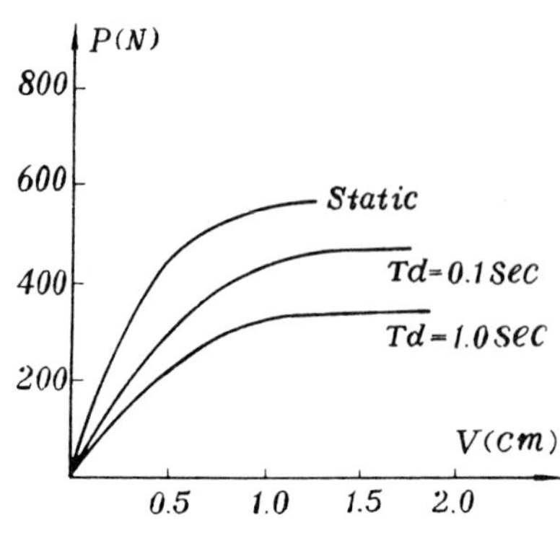

Fig. 10. Load-deflection for the triangular loading (joint 1).

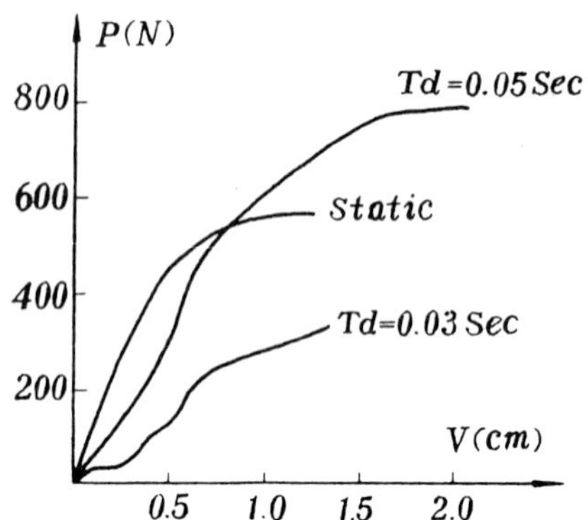

Fig. 11. Load-deflection curve for the sinusoidal loading (joint 1).

REFERENCES

1. MEEK, J.L. and TAN, H.S. Geometrically non-linear analysis of space frames by an incremental iterative technique, Computer methods in applied mechanics and engineering, Vol.47, 1984.
2. MEEK, J.L. and LOGANATHAN, S. Large displacement analysis of space frame structures, Computer methods in applied mechanics and engineering, Vol.72, 1989.
3. JIANHENG, S. and HUILIN, Z. Snap-through instability analysis and experimental studies of single layer braced domes, Proceedings of fourth international colloquium on structural stability, International academic publisher, 1989.
4. JIANHENG, S., Modified tangent stiffness matrix for non-linear stability analysis of single layer braced domes, Journal of structural engineering Vol.3, 4, 1990, (in Chinese).
5. KASSIMALI, A. and BIDHENDI, E. Stability of trusses under dynamic loads, Computers & structures, Vol.29, 1988.
6. NEWMARK, N.M. A method of computation for structural dynamic, Journal of engineering division, ASCE85, 1959.

36. A buckling analysis for 3D beams

K. M. HSIAO and R. T. YANG, National Chiao Tung University, Taiwan

SYNOPSIS. A co-rotational total Lagrangian finite element formulation for the buckling analysis of 3-D elastic Euler beams is presented. It is assumed that the pre-buckling displacement and rotations are small but not negligible, and are linearly proportional to the external loading. The zero value of the tangent stiffness matrix determinant is used as the criterion of the buckling state. A bisection method is used to find the buckling load. Numerical examples are presented to investigate the effect of the pre-buckling displacements on the buckling load.

INTRODUCTION

1. In the usual treatment (ref. 1) of the lateral buckling problem, the pre-buckling deflections of the beam are assumed small enough to be neglected. This assumption may be sufficiently accurate when the ratio of the minor axis flexural rigidity to the major axis flexural rigidity is much smaller than unity. However, the pre-buckling deflection may become quite important when this ratio is not small. In the extreme case when this ratio is equal to unity, the beam does not buckle (ref. 2).

2. The object of this paper is to investigate the effect of the pre-buckling displacements on the buckling load. In ref. 3, Hsiao presented a co-rotational total Lagrangian formulation of a beam element for the non-linear analysis of 3-D beam structures with large rotations but small strains. The non-linear coupling between the bending, twisting, and stretching deformations is considered. It seems that this element can be applied for the buckling analysis of beam structures with consideration of the pre-buckling displacements. Thus, the beam element presented in ref. 3 is employed here. The pre-buckling displacements and rotations are assumed to be linearly proportional to the external loading in this study.

FINITE ELEMENT FORMULATION

3. In the following only a brief description of the beam element is given. The more detailed description may be obtained from ref. 3.

Basic assumptions

4. The following assumptions are made in derivation of the beam element behaviour.
(i) The beam is prismatic and slender, and the Euler-Bernoulli hypothesis is valid.
(ii) The centroid and the shear centre of the cross-section are coincident.
(iii) The unit extension and the twist rate of the centroid axis of the beam element are uniform.
(iv) The cross-section of the beam element does not deform in its own plane and strains within this cross-section can be neglected.
(v) The out-of-plane warping of the cross-section is the product of the twist rate of the beam element and the Saint Venant warping function for a prismatic beam of the same cross-section.
(vi) The deformations of the beam element are small.
(vii) The pre-buckling displacements and rotations are linearly proportional to the external loading.

Co-ordinate systems

5. In this paper, a co-rotational total Lagrangian formulation is adopted. In order to describe the system, following ref. 3, we define three sets of co-ordinate systems (see Fig. 1):
(i) A fixed global set of co-ordinates, $X_i, i = 1,2,3$. The nodal co-ordinates, displacements, and rotations, and the stiffness matrix of the system are defined with respect to these co-ordinates.
(ii) Element co-ordinates, $x_i, i = 1,2,3$; a set of element co-ordinates associated with each element. The origin of this co-ordinate system is located at node 1, and the x_1 axis is chosen to pass through two end nodes of the element. The deformations and stiffness matrices of the elements are defined in terms of these co-ordinates.
(iii) Element cross-section co-ordinates, $x_i^s, i = 1,2,3$; a set of element cross-section co-ordinates is associated with each cross-section of the beam element. The origin of this co-ordinate system is rigidly tied to the shear centre of the unwarped cross-section. The x_1^s axes are chosen to coincide with the normal of the corresponding unwarped cross-section and the x_2^s and x_3^s axes are chosen to be the principal directions of the unwarped cross-section. In this paper the element deformations are determined by the rotation of these co-ordinate systems relative to the element co-ordinate system.

Rotation vector and rotation parameters

6. For convenience of the later discussion, the term 'rotation vector' is used to represent a finite rotation. Fig. 2 shows that a vector **b** which as a result of the application of a rotation vector $\phi\mathbf{a}$ is transported to the new position **b'**. The relation between **b** and **b'** may be expressed as (ref. 4)

$$\mathbf{b}' = \cos\phi\mathbf{b} + (1 - \cos\phi)(\mathbf{a}\cdot\mathbf{b}) + \sin\phi(\mathbf{a}\times\mathbf{b}) \qquad (1)$$

where ϕ is the angle of counter clockwise rotation, and a is the unit vector along the axis of rotation.

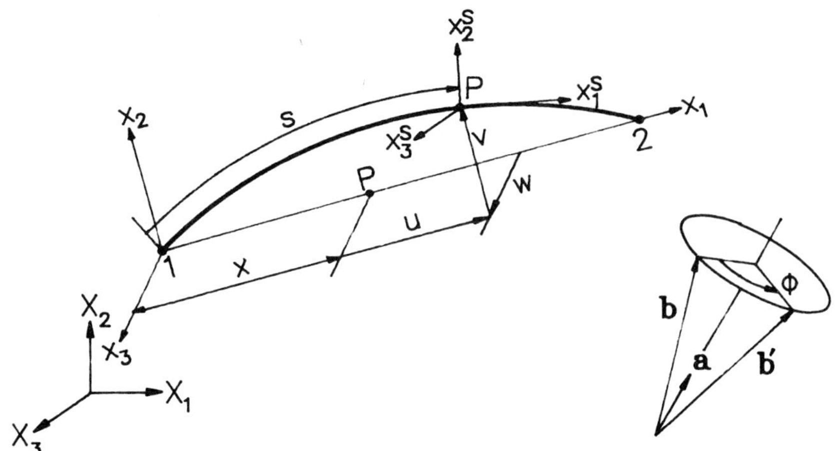

Fig. 1. Co-ordinate systems. Fig. 2. Rotation vector.

7. Let $\mathbf{e}_i$ and e_i^s (i = 1,2,3) denote the unit vectors associated with the x_i and x_i^s axes, respectively. Here, the traid $\mathbf{e}_i^s$ in the deformed state is assumed to be achieved by the application of the following two rotation vectors to the traid $\mathbf{e}_i$:

$$\theta_n = \theta_n \mathbf{n} \tag{2}$$

and

$$\theta_t = \theta_t \mathbf{e}_1^s \tag{3}$$

where $\mathbf{n}$ is the unit vector perpendicular to the vectors $\mathbf{e}_1$ and $\mathbf{e}_1^s$, θ_n is defined by

$$\cos\theta_n = (1 - \theta_2^2 - \theta_3^2)^{1/2}$$
$$\theta_2 = -\frac{dw(s)}{ds} \tag{4}$$
$$\theta_3 = \frac{dv(s)}{ds}$$

in which $v(s)$ and $w(s)$ are the lateral deflections of the centroid axis of the beam element in the x_2 and x_3 directions, respectively, and s is the arc length of the centroid axis.

8. The rotation vectors θ_n and θ_t are determined by θ_i (i = 1,2,3). Thus, θ_i are called rotation parameters in this study.

Nodal parameters

9. The global nodal parameters for the system of equations corresponding to the element local nodes j(j= 1,2) are ΔU_{ij}, the X_i (i = 1,2,3) components of the incremental translation vectors ΔU_j at node j and $\Delta\Phi_{ij}$, the X_i (i = 1,2,3) components of the incremental rotation vectors $\Delta\Phi_j$ at node j.

10. The element employed here has six degrees of freedom per node. Two sets of element nodal parameters termed 'explicit nodal parameters' and 'implicit nodal parameters' are employed. The explicit nodal parameters of the element are used for the assembly of the system equations from the element equations. Thus, they should be consistent with the global nodal parameters, and are chosen to be Δu_{ij}, the x_i (i = 1,2,3) components of the incremental translation vectors $\Delta\mathbf{u}_j$ at node j (j= 1, 2) and $\Delta\phi_{ij}$, the x_i (i = 1,2,3) components of the incremental rotation vectors $\Delta\phi_j$ at node j. The implicit nodal parameters of the element are used to determine the deformation of the beam element. They are chosen to be u_{ij}, the x_i (i = 1, 2,3) components of the total displacement vectors $\mathbf{u}_j$ at node j and θ_{ij}, the nodal values of the rotation parameters θi (i = 1, 2, 3) at node j.

11. The relations between the variation of the implicit and explicit nodal parameters may be expressed as, (ref. 3):

$$\delta\mathbf{q}_\theta = \begin{Bmatrix} \delta\mathbf{u}_1 \\ \delta\theta_1 \\ \delta\mathbf{u}_2 \\ \delta\theta_2 \end{Bmatrix} = \begin{bmatrix} \mathbf{I} & \mathbf{0} & \mathbf{0} & \mathbf{0} \\ \mathbf{0} & \mathbf{T}_1^{-1} & \mathbf{0} & \mathbf{0} \\ \mathbf{0} & \mathbf{0} & \mathbf{I} & \mathbf{0} \\ \mathbf{0} & \mathbf{0} & \mathbf{0} & \mathbf{T}_2^{-1} \end{bmatrix} \begin{Bmatrix} \delta\mathbf{u}_1 \\ \delta\phi_1 \\ \delta\mathbf{u}_2 \\ \delta\phi_2 \end{Bmatrix} = \mathbf{T}_{\theta q}\delta\mathbf{q}, \tag{5}$$

where:

$$\mathbf{T}_j^{-1} = \begin{bmatrix} 1 & \frac{\theta_{3j}}{2} & \frac{-\theta_{2j}}{2} \\ -\theta_{3j} & 1 & 0 \\ \theta_{2j} & 0 & 1 \end{bmatrix}, \tag{6}$$

I and **0** are the identity and zero matrices of order 3 x 3, respectively.

Kinematics of beam element

12. The deformations of the beam element are described in the current element co-ordinate system. From the kinematic assumptions made in this paper, the deformations of the beam element may be determined by the displacements of the centroid axis of the beam element, orientation of the cross-section (element cross-section co-ordinates), and the out-of-plane warping of the cross-section. Here, the lateral deflections $v(s)$ and $w(s)$ are assumed to be the Hermitian polynomials

of s, the arc length of the centroid axis, and $\theta_1(s)$ (Eq. 3) is assumed to be linear polynomials of s. The axial displacements of the centroid axis may be determined from the lateral deflections and the unit extension of the centroid axis.

Determination of element co-ordinates and element rotation parameters

13. Assume that the external loading is proportional to a loading parameter λ. Thus the linear solution is proportional to λ. In this study the linear solution is assumed to be the pre-buckling displacements.

14. Let $\mathbf{u}_j^L$ and ϕ_j^L ($j = 1,2$) (referred to the initial element co-ordinates) denote displacement and rotation vectors of an element at node j, which are extracted from the linear solution for loading parameter λ and then transformed from the global co-ordinates to the initial element co-ordinates. The current element co-ordinates corresponding to $\mathbf{u}_j^L$ and ϕ_j^L may be obtained by the application of a rotation vector ϕ^R to the initial element co-ordinates. Here ϕ^R is assumed to be proportional to the linear solution, and the components of ϕ^R referred to the initial element co-ordinates may be defined by:

$$\begin{aligned} \phi_1^R &= \frac{1}{2}(\phi_{11}^L + \phi_{12}^L) \\ \phi_2^R &= \frac{1}{L}(u_{32}^L - u_{31}^L) \\ \phi_3^R &= \frac{1}{L}(u_{22}^L - u_{21}^L) \end{aligned} \qquad (7)$$

where L is the initial length of the beam element.

15. Here the element rotation parameters are also assumed to be proportional to the linear solution. Thus the vector of the rotation parameters at node j ($j = 1, 2$) may be determined by:

$$\theta_j = \phi_j^L - \phi^R \qquad (8)$$

Element stiffness matrix

16. The element tangent stiffness matrix referred to the global co-ordinate system is first calculated in the current element co-ordinate system and then transformed to the global co-ordinate system using the standard co-ordinate transformation. The derivation of the element stiffness matrix corresponding to the explicit nodal parameters was given in ref. 3, and is not repeated here. If the element sizes are chosen to be small enough, the assumption of small deformation of the beam element in the element co-ordinates can always be valid. Thus, only the terms up to the first order of nodal parameters are retained in the element stiffness matrix (ref. 3). The tangent stiffness matrix (referred to the current element co-ordinates) may be written as:

$$\mathbf{k}_t = \mathbf{k}_0 + \mathbf{k}_1 \qquad (9)$$

where $\mathbf{k}_0$ is the linear stiffness matrix, and $\mathbf{k}_1$ is the geometric stiffness matrix which is linearly proportional to the loading parameter. Note that even the element tangent stiffness matrix referred to the current element co-ordinates can be approximated by a linear function of the loading parameter, it is a non-linear function of loading parameter when transformed to the global co-ordinates.

Calculation of the buckling load

17. Here, the zero value of the tangent stiffness matrix determinant is used as the criterion of the buckling state. Let $\mathbf{K}_T(\lambda)$ denote the tangent stiffness matrix of the structure corresponding to the loading parameter. The criterion of the buckling state may be expressed as:

$$D(\lambda) = \det\left|\mathbf{K}_T(\lambda)\right| = 0 \tag{10}$$

18. The minimum root of Eq. 10 is the buckling load. Here, a bisection method (ref. 5) is used to find the minimum root of Eq. 10. The convergence criterion employed for the bisection iterations is given by:

$$e = \frac{|D(\lambda)|}{|D(0)|} \le e_{tol}, \tag{11}$$

where e_{tol} is a prescribed value of error tolerance.

NUMERICAL STUDIES

19. The example considered here is a simply-supported beam under uniform moment as shown in Fig. 3. The ends of the beam are free to rotate with respect to the X_2 and X_3 axes, but are rigidly held against rotation about the X_1 axis. Twenty elements are used for discretization. The error tolerance is set to 10^{-8} for bisection iterations.

20. The buckling moments for different $\alpha = I_2/I_3$ and $\beta = GJ/EI_3$ are shown in Fig. 4. It is seen that when α and β are much smaller than unity, the buckling moments converge to the classical buckling moment as expected; when α and/or β is not small, there are discrepancies between the present results and the classical solutions.

CONCLUSIONS

21. This paper has described a co-rotational finite element formulation for the buckling analysis of 3-D beam in which the effect of the pre-buckling displacement is considered. The pre-buckling displacements are assumed to be linearly proportional to the external loading. A practical procedure is proposed to determine the element co-ordinates and nodal parameters of the beam element. In conjunction with the co-rotational formulation, the higher order terms of nodal parameters in element stiffness matrix are consistently dropped.

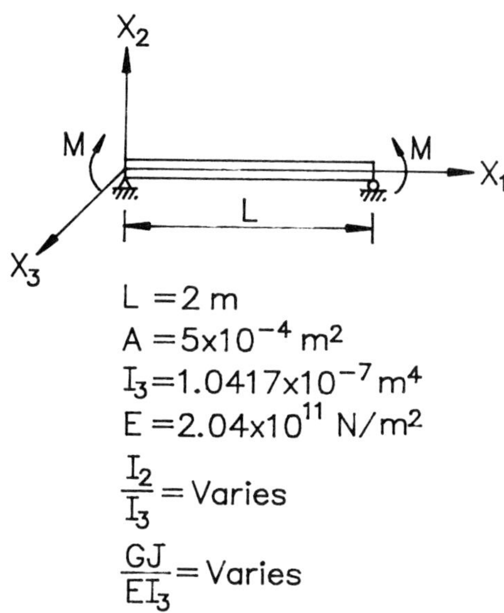

Fig. 3. Simply supported beam.

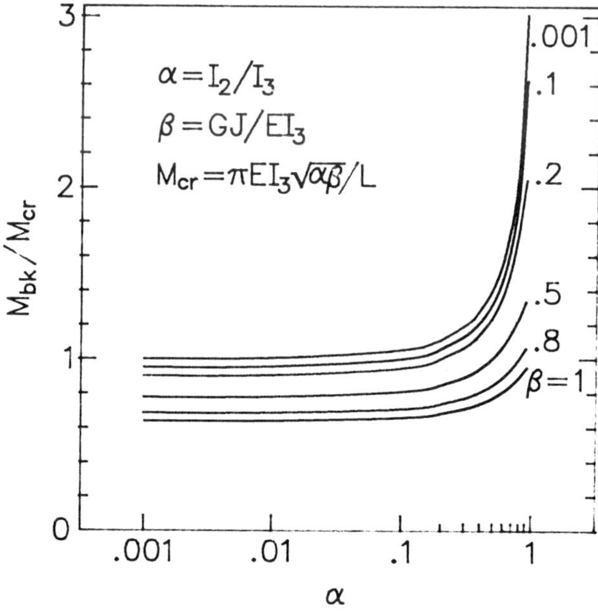

Fig. 4. Buckling moment of simply-supported beam.

22. Results obtained for simply-supported beam subjected to uniform moment demonstrated clearly the influence of the pre-buckling displacements on the buckling load.

REFERENCES

1. TIMOSHENKO, S.P. and GERE, J.M. Theory of Elastic stability, 2nd ed., McGraw-Hill, New York, 1 970.
2. TRAHAIR, N.S. and WOOLCOCK, S.T. Effect of major axis curvature on I-beam stability, Journal of the engineering mechanics division, ASCE, 1973, vol. 99, February, 85-98.
3. HSIAO, K.M. Co-rotational total Lagrangian formulation for three-dimensional beam element, AIAA Journal, 1992, vol. 30, March, 797-804.
4. GOLDSTEIN, H. Classical Mechanics, Addision-Wesley, Reading, MA, 1980.
5. HILDEBRAND, F.B. Introduction to numerical analysis, 2nd ed., McGraw-Hill, New York, 1974.

37. Some stability aspects of reticulated shells

J. SUMEC, Slovak Technical University, Slovakia

SYNOPSIS. In this paper some stability problems of reticulated shells are studied. Adequate mathematical models for the analysis of shell instability, using a direct method and a shell analogy method, are presented. It is assumed that the members of the analysed shells consist of a homogeneous, isotropic elastic material. The theoretical results, obtained from some numerical examples, are also presented .

INTRODUCTION

1. Framed space structures, which also include regular lattice plates and shells, constitute an additional developmental stage of classic trusses.

2. In addition to the static analysis of the bar-systems, their stability analysis require, the following problems to be solved: a) bar stability (refs 2 and 3); b) local stability (refs 1, 3-5) and; c) overall stability (refs 1-3, 6-9, 11-16).

3. The stability of the reticulated shell is also conditioned by the topology and geometrical shape, by material properties and the supporting mode, by the degree of static indeterminacy and the stiffness of joints. Already for the single-layer reticulated shells mentioned above, the dominant factor of maintaining the stability of shape is its curvature. Finally, experimental evidence (refs 2, 10, and 14) suggests that buckling is initiated by local dimpling.

BAR STABILITY

4. The critical load F_{cr} of a straight prismatic bar element of a reticulated shell is given by the well-known relation:

$$F_{cr} = k\pi^2 E_{ef} J_{min} / \left(\ell^2\right) \tag{1}$$

where $k = k(a_i, a_j, w_0, \xi_0, s)$ is a parameter depending on the rotational stiffnesses (a_i, a_j), of the terminal joints (i), (j), the initial geometrical imperfection w_0, the eccentricity ξ_0, the compression force and the magnitude of shear forces and bending moments acting on the end sections of the bar s, the minimal moment of inertia of the bar section, J_{min} and the effective modulus of elasticity E_{ef} . A schematic illustration of the bar and the meaning of the individual parameters is apparent from Fig. 1.

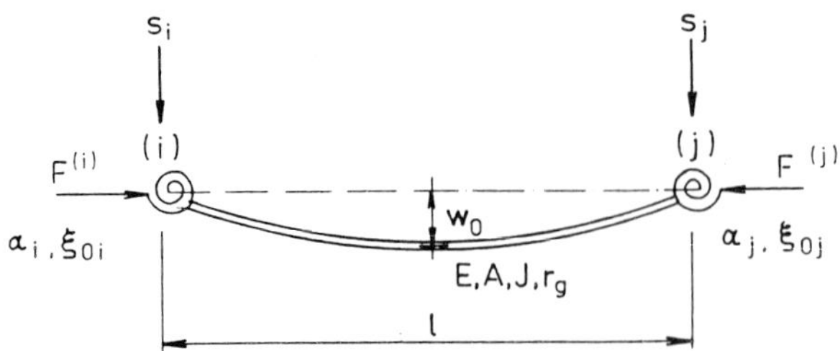

Fig. 1. Calculating critical bar loading.

5. If $k = k(a_i, a_j)$, then the parameter k has already been studied, for example by Sandorf (ref. 3). The behaviour of the parameter $k = k(a_i, a_j)$ is given in Fig. 2.

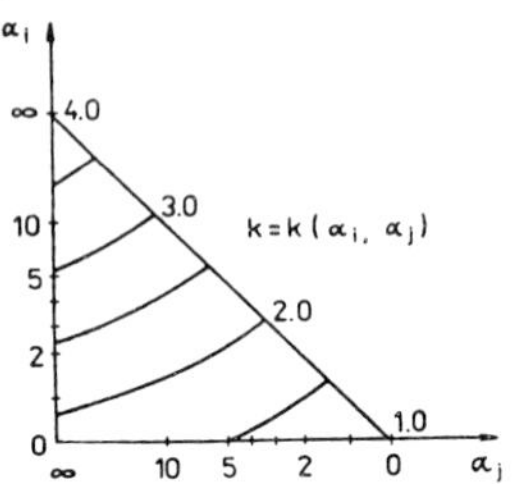

Fig. 2. Dependence of $k = k(a_i, a_j)$ on the parameters a_i, a_j .

6. For the sake of security, a value of the parameter k is sometimes chosen that corresponds to pinned joints (i.e. the Euler case). Then the corresponding stress σ_{cr}, is determined from the relationship:

$$\sigma_{cr} = k\pi^2 E / (\Psi^2) \qquad (2)$$

where $\Psi = \ell r_g^{-1}$ is the slenderness of the bar, and r_g its radius of gyration. What corresponds to the critical proportional limit stress is the slenderness ratio:

$$\Psi_p = \sqrt{k\pi^2 E / \sigma_p} \qquad (3)$$

<u>Character of buckling</u>

7. With respect to the parameter Ψ, three cases are distinguished, in general: elastic buckling ($\Psi >> \Psi_p$); elastic-plastic buckling ($\Psi = \Psi_p$); and failure at the yield point ($\Psi << \Psi_p$). Lattice shells are usually characterised by large numbers of framed elements; hence it is rather complicated to estimate the parameter k because of being heavily influenced by the ambient environment. In the treatise

given in ref. 3, it is recommended that the value $k \in$ <1.5; 2.5> is chosen, whence an estimation for the bar slenderness is obtained at the limits:

$$\sqrt{\pi^2 E / \left(0.67\sigma_p\right)} \le \Psi_p \le \sqrt{\pi^2 E / \left(0.4\sigma_p\right)} \tag{4}$$

and for the limiting case:

$$\sqrt{\pi^2 E / \sigma_p} \le \Psi_p \le \sqrt{\pi^2 E / \left(0.25\sigma_p\right)} \tag{5}$$

For a bar of rectangular cross-section, the relationship $\sigma_{cr} \sim \Psi$ is given in Fig. 3.

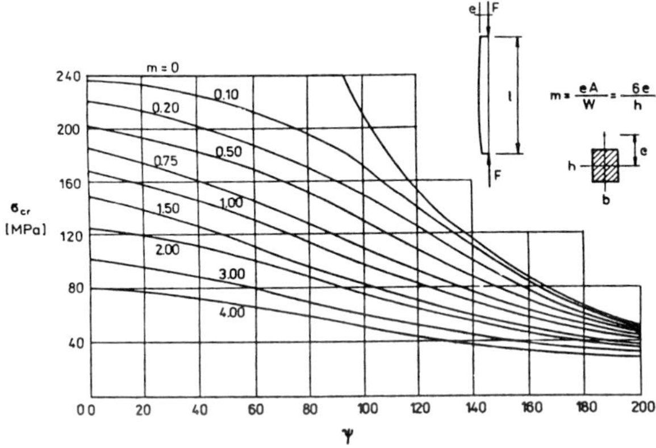

Fig. 3. Diagram of σ_{cr} (Ψ) for the eccentrically compressed steel bar.

LOCAL STABILITY

8. Examination of the loss of local stability in reticulated shells as space structures appears to be a rather sophisticated problem because it requires keeping track of the response of the entire structure. This phenomenon focused upon the node area with the neighbouring bars shown in Fig. 4, and can be analysed in a simplified way, as pointed out in (refs 1, 4 and 5) for a shell with an "isotropic" network.

9. Consider a sub-domain of the reticulated shell whose bars form a grid of equilateral triangles with constant parameters E, A, ℓ, R, where A is the cross-sectional area of the bar and R is the radius of curvature of the analysed shell. From the condition of the equilibrium of forces in node <2> (Fig. 5) and maximising F with respect to f, gives the upper constraint of critical load:

$$^{u}F_{cr} = 2EAf^3 / \left(\sqrt{3}l^3\right) \tag{6}$$

and in a similar way for the lower constraint:

$$ {}^{\ell}F_{cr} = EAf^3 / \left(\sqrt{3}l^3\right) \tag{7} $$

Then the actual uniform critical load P_{cr} subsequently lies within the bounds:

$$ EA\ell / \left(12R^3\right) \leq P_{cr} \leq EA\ell / \left(6R^3\right) \tag{8} $$

<u>Including the bending strength</u>

10. If, in addition, the bending stiffness of nodes are taken into consideration, then for the upper estimation:

$$ {}^{u}F_{cr} = 3EAf'\left(f^2 - f'^2\right) / \ell^3 + 44EJ(f - f') / \ell^3 \tag{9} $$

and for the lower estimate:

$$ {}^{l}F_{cr} = 3EAf'\left(f^2 - f'^2\right) / \left(2\ell^3\right) + 44EJ(f - f') / \ell^3 \tag{10} $$

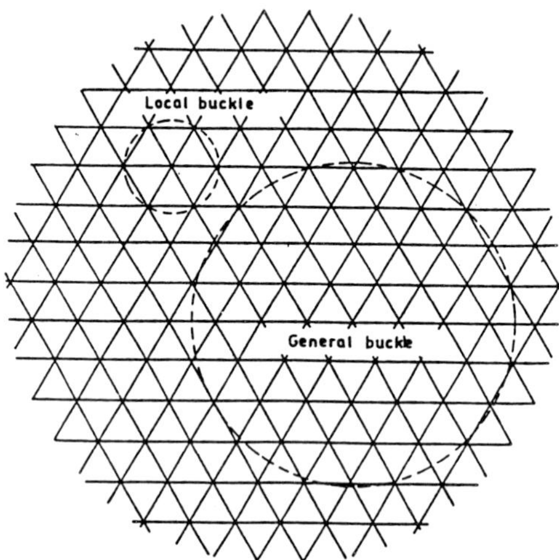

Fig. 4. Domains of local and global loss of stability.

11. Lind (ref. 4) introduces an approximate value for the buckling nodal load of the spherical lattice shell with an "isotropic" grid in the form:

$$ F_{cr} = \frac{E_T}{1 + \tilde{\alpha}^2 / \left(8\pi^2\right)} \left(0.47A\ell^3 / R^3 + 3BJ / (R\ell)\right) \tag{11} $$

where E_T is the tangent modulus, $\tilde{\alpha} = \ell^2 / (r_g R)$ and B is the dimensionless bending stiffness quantity of the ambient structure, depending on the parameter α. For the snap through of a joint in a regular dome supporting a concentrated load:

$$F_{cr} = \frac{3EAf^3}{\ell^3}\left[\frac{8B}{\tilde{\alpha}^2} + 0.241\left[1 - 1.595\frac{8B}{\tilde{\alpha}^2}\right]\right] \tag{12}$$

or for reticulated shells with pinned nodes:

$$F_{cr} = 0.0905\ EA\ell^3 / R^3 \tag{13}$$

If the slenderness ratio of bars is low, it is recommended to analyse the loss of local stability, not in the elastic domain but in the elastic-plastic area.

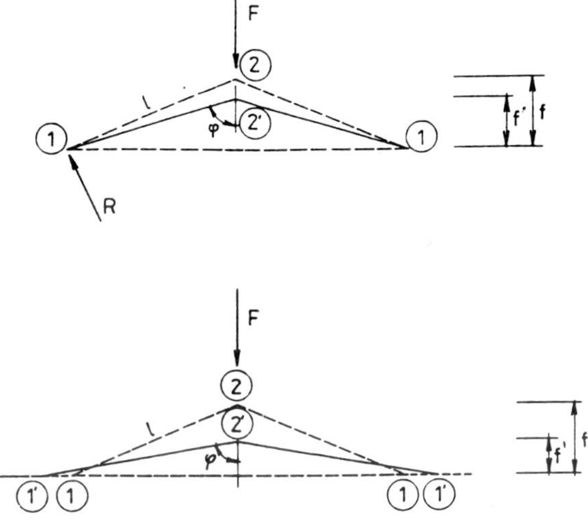

Fig. 5. Segment diagram for calculating the upper (lower) estimate of the critical load.

OVERALL STABILITY

12. The domain of overall stability loss of the reticulated shell structure is indicated in Fig. 4. In this part, the loss of overall stability is analysed by the continuum analogy method. The direct method of analysis of this problem is demonstrated in detailed in refs 2 and 7. The measure of the approximation accuracy of the solution is conditioned by the "density" of the lattice mesh, and by sufficient smoothness of the "equivalent" central surface, as well as by the type of structure support. As experimental measurements show (refs 2, 7, 10, and 14),

the overall loss of stability in these structures is due to the formation of a local deformation whose dimensions are small in comparison with the radius of curvature of the shell.

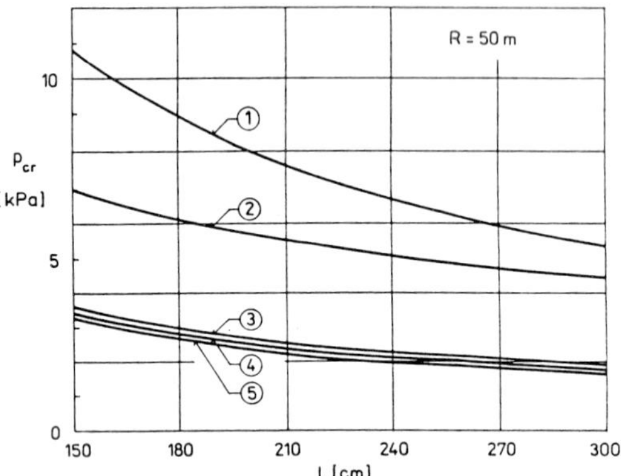

Fig. 6. Dependence of the values p_{cr} on the radius of curvature R and the bar lengths ℓ.
(1 Klöppel, Ross; 2 Beles, Soare; 3 Wright; 4 Buchert; 5 Author.)

13. Starting from strain energy and the potential energy of the external load, and using the Ritz variational method to determine the uniformly distributed critical load of the shell, gives:

$$p_{cr} = \tilde{\beta}^* E(A/(Ra))^2 = \lambda_{cr} p \quad (14)$$

where β^* is the dimensionless parameter (ref. 7); a is the distance between bundles of bars; λ_{cr} is the critical load parameter and p is the external load on the shell.

Solution by other authors

14. Buchert (ref. 6), Crawford and Schwartz (ref. 11), Klöppel and Ross (ref. 12) suggested for p_{cr} the relationship:

$$p_{cr} = CE(t_m / R)^2 (t_b / t_m)^{3/2} \quad (15)$$

where C is the constant, t_m is the effective "membrane" thickness of the shell, and t_b is the effective "bending" thickness, such that:

$$t_m = 2A/(\sqrt{3}\ell), \quad t_b = (9\sqrt{3}J/\ell)^{1/3}$$

15. The value of the constant C follows from the method of derivation applied by the individual authors. Buchert proposed the value C = 0.36. Klöppel and Ross suggested the value C = 1.16, just as Crawford and Schwartz, Gerard and Lakshmikantham (ref. 13). Model measurements on the given type of shells were made by Matsushita *et al.* (ref. 10), and Okumura *et al.* (ref. 14), who recommended the value C = 0.6-0.92 with respect to the initial imperfection and plasticity effect. Beles and Soare (ref. 2) proposed the expression:

$$p_{cr} = 0.125E\,(t/R)^2 \tag{16}$$

where $t = (24J/\ell)^{1/3}$. By the underlying methods of continuum modelling Wright (ref. 1) derived the relation:

$$p_{cr} = 1.60E\sqrt{AJ}/(R^2\ell) \tag{17}$$

From the numerical solution of the problem the author proposes the value C = 0.3506 in Eq. 15.

16. As a comparative example, choosing the calculation of the critical load of reticulated shells with an "isotropic" trelliswork whose material and geometric parameters are as follows: E = 210 GPa, $\overline{\mu} = 0.77$, $r_1 = 0.0415$m, $r_2 = 0.0355$m (external and internal radius of tube cross-section), R = 50m, A = 0.0014914m^2, a =0.86602ℓ, J = 0.10822x10^{-5}m^4, ℓ = 1.5, 1.8, 2.1, 2.4, 2.7, 3.0m. The behaviour of p_{cr}, is given in Fig. 6.

COMPARISON OF TWO STABILITY ANALYSIS METHODS

17. In this part the variation of the critical load parameters λ_{cr}, which depends on the initial changes in shape geometry, is investigated. The treatise (ref. 16) presents the derivation of the formula which allows consideration of the approximate effect which the variation of the initial shell shape geometry has on its critical load, provided that the critical load for the shell with an ideal geometry is known in advance. The reduced parameter of the critical load $\lambda_{cr,red}$ is determined from the relationships:

$$\lambda_{cr} = K_{red}\lambda_{cr} \tag{18}$$

where;

$$k_{red} = \overline{k}_1^2 \Big/ \left\{1 - \frac{1-\sqrt{\overline{k}_1}}{1-\overline{k}_1}\left[(1-\overline{k}_1)\left(1+0.261\sqrt{\overline{t}}/\overline{c}+\overline{k}_1\overline{f}\right)\right]\sqrt{1-\overline{c}^2\overline{k}_1^2}\right\} \tag{19}$$

$$\overline{k}_1 = 2\left(1-\overline{f}-\sqrt{1-\overline{c}^2}\right)/\left[2\left(1-\overline{f}\right)\left(1-\sqrt{1-\overline{c}^2}\right)+\overline{f}^2\right] \tag{20}$$

t = t/R; c = c/R; f = f/R; f in this case signifies the settlements of the shell vertex, c is the radius of the domain with initial shape imperfection and t is the effective "thickness" of the continuum model.

Example

18. The shell Fig. 7 has a span of 50m and a rise of 6.7m. Additional material and sectional characteristics are as follows: E = 210 GPa, A = $0.0001131m^2$, R = 50m, r_g = 0.0439m, J = $0.218x10^{-5}m^4$, ℓ = 2.53-2.70m, t = 0.152m.

19. The problem is solved for the following values of parameters f and c: f = 0, 0.005, 0.10, 0.20, 0.25, 0.30m; c = 5, 10, 15m;

20. The reference value of the external uniform load is p= $900N/m^2$. The value of the corresponding critical load parameter is λ_{cr}, = 6.35, which has been calculated by the direct method (FEM) and applies to the ideal (designed) shell shape. The reduced parameter values of the critical load $\lambda_{cr,red}$ are then calculated in the terms of Eq. 18. In Fig. 8 the values $\lambda_{cr,red}$ are plotted, having been calculated according to both methods, depending on the dimensionless parameters $\left(\overline{R}-R\right)/R$, where $\overline{R}$ is the radius of curvature of the shell section with the changed geometry. This radius is calculated from the relationship:

$$\overline{R} = \frac{2(f-R)\left(\sqrt{R^2-c^2}-R\right)+f^2}{2\left(R-f-\sqrt{R^2-c^2}\right)} \tag{21}$$

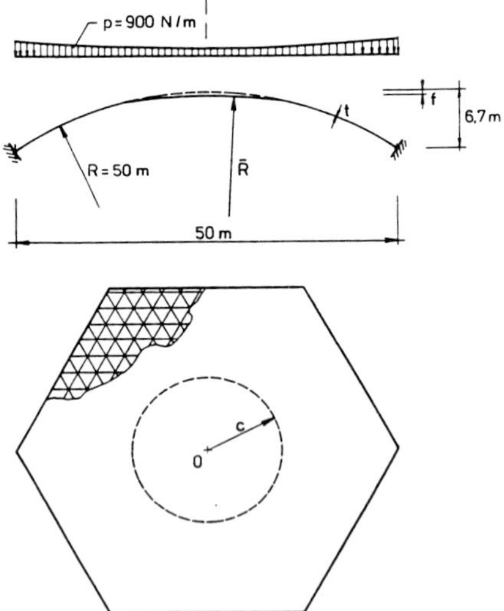

Fig. 7. Diagram of the shell analysed.

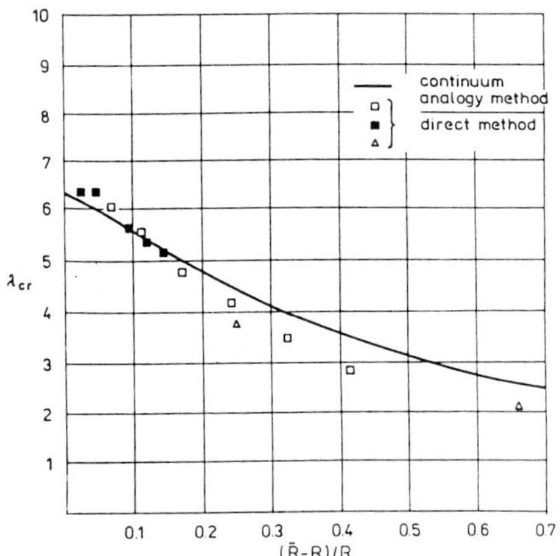

Fig. 8. Critical load factor versus the parameter $\left(\overline{R}-R\right)/R$.
Δ – c = 5m,; ❑ – c= 10m; ■ – c= 15m.

REFERENCES

1. WRIGHT D.T. Membrane forces and buckling in reticulated shells. J. of Struct. Div., Proc. of ASCE, 1965, vol.91, STl, February, 177-201.
2. DAVIES R.M. (ed.) Space structures, A study of methods and developments in three dimensional construction resulting from the International Conference on Space Structures, p.1233. Wiley, New York, 1966.
3. TSUBOI, Y.(Chairman) Analysis, design and realisation of space frames. Bull. IASS, 1984, no.84/85, p.106.
4. LIND N.C. Local instability analysis of triangular dome frameworks. The Structural Engineer, 1969, vol.47, no.8, August, 317-324.
5. JAGANATHAN S., EPSTEIN H. I. and CHRISTIANO P. Non-linear analysis of reticulated space trusses. J. Struct. Div., ASCE, 1975, vol.101, ST12, December, 2641-2673.
6. BUCHERT K. P. Buckling of shell and shell-like structures.(Internal Report). Columbia, University of Missouri, 1973.
7. SUMEC J. Regular lattice plates and shells, p.554. Elsevier, Amsterdam, 1990.
8. NAYFEH A. H. Continuum modelling of the mechanical and thermal behaviour of discrete large structures. AIAA Journal, 1981, vol.19, June, 766-773.
9. IKEDA K. and TORII K. Bifurcation behaviour of an octagonal truss dome with imperfections. Proc. of JSCE, 1987, vol.4, April, 225-228.
10. MATSUSHITA F., SATO M. and HAYASHI T. An experimental study of non-uniformly trussed steel shells consisting of triangular elements of varying

dimensions. In: Symposium on Steel Structures, Japan Society for the Promotion of Science, Tokyo, 1961.
11. CRAWFORD R.F. and SCHWARTZ D.B. General instability and optimum design of grid-stiffened spherical dome. ALAA Jounal, 1965, vol.3, no.3, March, 511-543.
12. KLÖPPEL K. and ROSS E. Beitrag zum Durschlagsproblem dünnwandiger versteifter und unversteifter Kugelschalen für voll-und halbseitige Belastung. Der Stahlbau, 1956, vol.25, no.3, Martz, 1-25.
13. GERARD G. and LAKSHMIKANTHAM C. Elastic and plastic stability of geometrically orthotropic spherical shells.(Intemal Report). Montreal, Concordia University, 1964.
14. OKUMURA T. *et al.* Experimental research on load distribution of grid structure whose main girders have each different rigidity. In: Symposium on the New Ideas in Structural Design. Japan Society for the Promotion of Science, Tokyo, 1963, 146-171.
15. NOOR A. K. Continuum modelling for repetitive lattice structures. Appl. Mech. Rev., 1988, vol.41, no.7, July, 285-296.
16. SUMEC J. and ZINGALI A.E. A study of the influence of initial shape imperfections on the stability of lattice shells by direct and shell analogy methods. Int. J. of Space Struct., 1986/87, vol.2, no.4, 223-230.

38. Load-bearing capacity of square-on-square double-layer braced domes

T. SAKA, Y. TANIGUCHI and H. KONISHI, Osaka City University, Japan

SYNOPSIS. The effective strength of square-on-square double-layer grids under equivalent stress resultants and stress couples, are derived. For square-on-square double-layer braced domes with a square plan, the approximate elastic buckling loads are estimated from the effective strength and the stress analysis of space trusses. The elastic buckling loads are also calculated by a discrete element method. The post-buckling behaviour is predicted by a limit state analysis, which is a full stage-by-stage analysis. The member model used in the analyses consists of rigid end parts, a uniform member, and rotational springs.

INTRODUCTION

1. Square-on-square double-layer grids are widely used as plates, cylinders, domes, etc. There are several theoretical methods which estimate the load-bearing capacities of space grids. To study the load-bearing capacities of space grids, elastic buckling loads and post-buckling behaviour of grids must be made clear. Among these methods, there are the approximate method which use the effective strength expressed in terms of stress resultants and stress couples of continua (refs 1 and 2). The effective strength of the grids under biaxial equivalent bending moments, M^{11}, M^{22}, without equivalent twisting moments, M^{12}, M^{21}, were already derived. When braced domes are treated, the effective strengths under not only biaxial bending moments but also normal, shearing forces, N^{11}, N^{22}, $N^{12}(=N^{21})$, and transverse shear forces, Q^1, Q^2, are necessary.

2. In this paper, the effective strengths of the grids subjected to equivalent stress resultants and stress couples, i.e. normal, shearing, transverse shear forces and the bending moments, are derived. For square-on-square double-layer braced domes with a square plan, the approximate elastic buckling loads are estimated from the effective strength and the results of the elastic stress analysis, by a discrete element method. The exact elastic buckling loads of the above braced domes are also calculated by a discrete element method. Further, the post-buckling behaviours of some braced domes are estimated by using a simplified non-linear analysis (refs 3-5), where the member non-linear characteristics are idealised as the axial force-axial deformation relationships. The buckling strengths of compression members are estimated from the effective strength, and the

member characteristics after elastic buckling are estimated by assuming the mode of the buckled member (ref. 6). In these theoretical analyses, the member model, which was presented by the author and consists of a uniform member, rigid end parts, and rotational springs connecting them, are utilised in order to consider both the effects of sizes and connecting rigidities of joints. As parameters, the rigidity of rotational springs, the ratio of the flexural rigidity of chord members to web members are varied. The curvature and boundary conditions of domes are also varied in the analyses of the domes.

EFFECTIVE STRENGTH OF SQUARE-ON-SQUARE DOUBLE-LAYER GRIDS

Form of a structural unit and a constituent member model

3. The square-on-square double-layer grids and their equivalent stress resultants and stress couples are shown in Fig. 1. The member consists of a uniform member, rigid end parts and rotational springs as shown in Fig. 2.

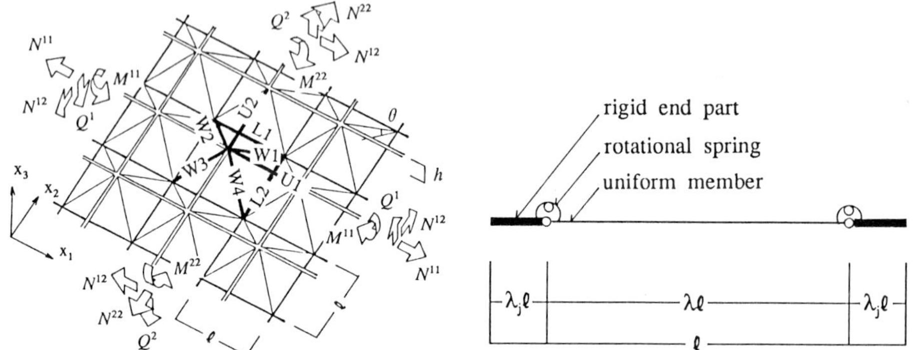

Fig. 1. Form of a unit.

Fig. 2. Model of constituent member.

Relation among the equivalent stress resultants and stress couples and axial forces of the member

4. The axial forces of a structural unit are derived from the equivalent stress resultants and stress couples as follows:

$$\begin{Bmatrix} N^{11} \\ N^{22} \\ M^{11} \\ M^{22} \\ N^{12} \\ Q^1 \\ Q^2 \end{Bmatrix} = \frac{1}{\ell} \begin{bmatrix} 1 & 0 & 1 & 0 & & & \\ 0 & 1 & 0 & 1 & & [0] & \\ \frac{h}{2} & 0 & -\frac{h}{2} & 0 & & & \\ 0 & \frac{h}{2} & 0 & -\frac{h}{2} & & & \\ & & & & \frac{\sqrt{2}\cos\theta}{2} & 0 & \frac{\sqrt{2}\cos\theta}{2} \\ & [0] & & & 0 & \sin\theta & \sin\theta \\ & & & & -\sin\theta & -\sin\theta & 0 \end{bmatrix} \begin{Bmatrix} N_{U1} \\ N_{U2} \\ N_{L1} \\ N_{L2} \\ N_{W1} \\ N_{W2} \\ N_{W3} \end{Bmatrix} \quad (1)$$

5. The subscripts of Eq. 1 represent the location of members as shown in Fig. 1 The chord members resist equivalent normal forces(N^{ii}) and equivalent bending moments(M^{ii}). The web members resist equivalent shearing forces(N^{12}) and equivalent transverse shear forces(Q^{i}). For the grids, equivalent twisting moments(M^{12}) are neglected because the twisting rigidity is very small.

Effective strengths of (N^{ii}, M^{ii}) stress states

6. In this paper, the ratio of rigid end parts to the length of the members is 0.04. The torsional rigidities of members are neglected. The angle θ shown in Fig. 1. is equal to π/4. The effective strengths of M^{ii}, N^{ii} are shown in Fig. 3.

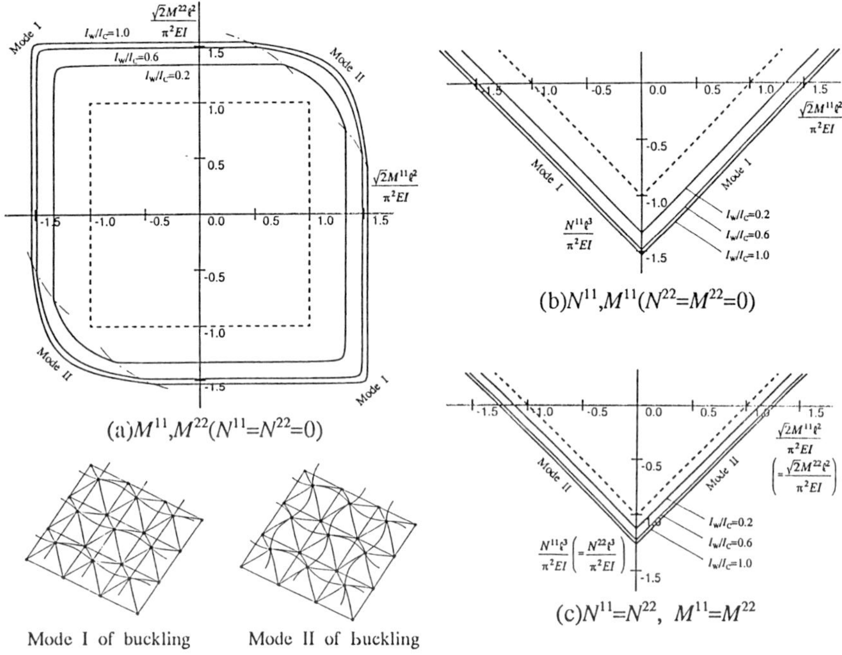

Fig. 3. The effective strength of (N^{ii}, M^{ii}) stress states (c_{jc} = 2.0).

Effective Strength of (N^{12}, Q^{i}) Stress States

7. The effective strengths of (N^{12}, Q^{i}) stress states on the pin-jointed case are shown in Fig. 4. In the figure, N_e is the Euler's buckling load and N_y is a yield tensile axial force. In the case of $|N_y|=|N_e|$ the form of the effective strength is a octahedral surface. In the case of $|N_y|>3|N_e|$ the form of the effective strength is a tetrahedral surface.

8. The effective strengths of (N^{12}, Q^{i}) stress states of the grids are shown in Figs 5 (a to e). The buckling modes of those effective strengths are shown in Figs 5.(f to g). When $|N_y|$ is small, the surface determined by the yield of tensile members appear at the vertex of the tetrahedral surface, such as the pin-jointed case.

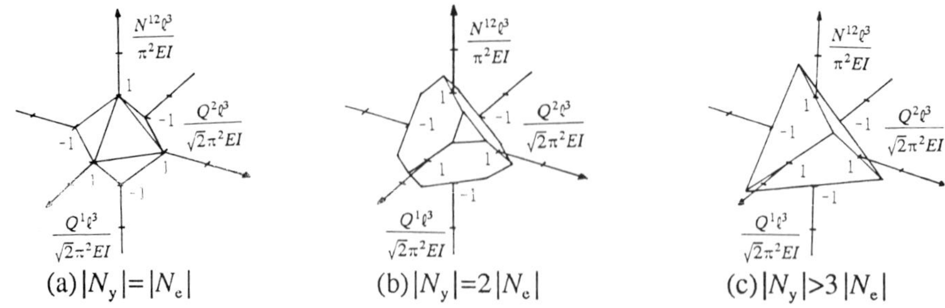

(a) $|N_y|=|N_e|$ (b) $|N_y|=2|N_e|$ (c) $|N_y|>3|N_e|$

Fig. 4. The effective strength of (N^{12}, Q^i) stress states on the pin-jointed case.

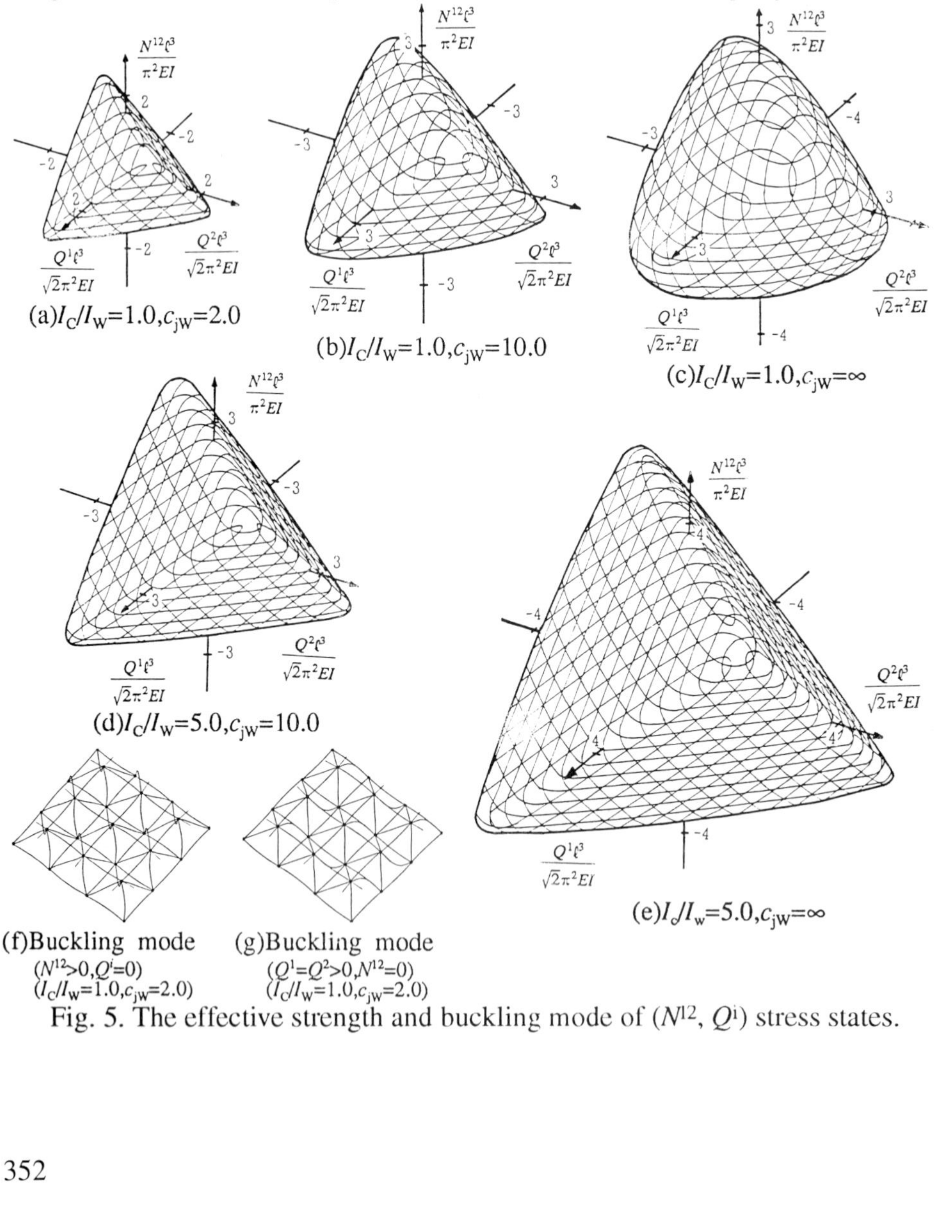

(a) $I_C/I_W=1.0, c_{jW}=2.0$

(b) $I_C/I_W=1.0, c_{jW}=10.0$

(c) $I_C/I_W=1.0, c_{jW}=\infty$

(d) $I_C/I_W=5.0, c_{jW}=10.0$

(e) $I_C/I_W=5.0, c_{jW}=\infty$

(f) Buckling mode ($N^{12}>0, Q^i=0$) ($I_C/I_W=1.0, c_{jW}=2.0$)

(g) Buckling mode ($Q^1=Q^2>0, N^{12}=0$) ($I_C/I_W=1.0, c_{jW}=2.0$)

Fig. 5. The effective strength and buckling mode of (N^{12}, Q^i) stress states.

ELASTIC BUCKLING LOADS OF SQUARE-ON-SQUARE DOUBLE-LAYER BRACED DOMES

Configuration, Loading and Boundary Conditions

9. The square-on-square double-layer braced dome with a square plan (30m x 30m) is shown in Fig. 6, where θ is the angle subtended by one half of the span, R_U is the radius of curvature of the upper chord spherical surface and R_L is that of the lower chord spherical surface. The double-layer braced domes are made up from 10 units x 10 units. One kind of grid (θ = 0°) and two kinds of domes (20°, 40°) are treated and the length of the upper chord member is equal to the length of the web member at the centre of the dome. The forms of a grid and domes are shown in Table 1. The grid and domes are subjected to a uniform vertical load at the upper nodes.

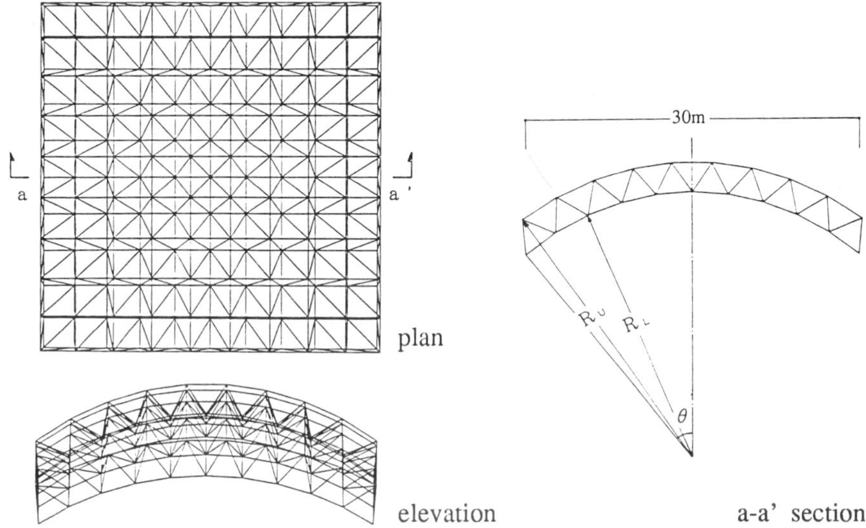

Fig. 6. Square-on-square double-layer braced dome.

Table 1. Form of Domes.

Analytical model	θ(°)	Rise-to-span ratio	Radius of curvature			Member's length (m)		
			R_L(m)	R_U(m)		Lower chord member	Upper chord member	Web member
DM00	0.00	0.0	(∞)			3.00	3.00	3.00
DM20	20.0	0.088	43.85	46.19	standard minimum maximum	3.06 3.05 3.06	3.22 3.21 3.22	3.22 3.16 3.34
DM40	40.0	0.182	23.34	26.02	standard minimum maximum	3.26 3.01 3.26	3.63 3.45 3.63	3.63 3.37 4.16

10. The structures are supported by three support conditions. Type (α) denotes the condition that the grid and domes are simply supported at the perimeter nodes.

Type (β) denotes the condition that the grid and domes are pin-supported at the all perimeter nodes. Type (γ) denotes the condition of the four corner being pin-supported.

Mechanical properties of a member and a joint

11. The grid and domes consist of the same member model. Mechanical and section properties of the constituent member and joints are given as shown in Table 2.

Table 2. Mechanical and section properties of the constituent member.

Member		Rigid part and rotational spring	
Section area A	5.76cm^2	Dimension of rigid part $\lambda_j \ell$	12.00cm
(Web member in model -W2)	3.05cm^4	(Rate of dimension of rigid part to	
Moment Inertia I	23.7cm^4	Standard member's length λ	0.04)
(Web member in model -W2)	4.74cm^4	Rotational spring rigidity C_j	3.25×10^{3}kN·cm
Modules of elasticity E	2.05910^{4}kN	(Nondimensional rotational spring	
Yield bending moment M_p	247.1kN·cm	rigidity c_j of Standard member	2.0)
Yield tensile strength N_p	178.6kN	Yield moment of rotational spring M_{jp}	123.6kN·cm

Approximate elastic buckling loads and exact elastic buckling loads

12. The approximate elastic buckling loads are estimated from the result of elastic stress analysis and the effective strength. The exact elastic buckling loads are estimated by a discrete element method. The results obtained are shown in Table 3. In the table, A, B-G denote the locations of buckling members as shown in Fig. 7.

Table 3. Total buckling loads and buckling members.

Analytical model	Approximate elastic buckling analysis (P_c') (kN)		Elastic buckling analysis (P_c) (kN)		P_c'/Pe(%)
DM00-α -W1	636.1	A	677.7	A	95.9
DM00-α -W2	597.2	A	620.4	A	96.2
DM00-β -W1	705.7	A	735.7	A	95.9
DM00-β -W2	602.6	A	647.7	A	93.1
DM00-γ -W1	264.8	B	212.3	B	124.7
DM00-γ1-W1	245.7	C	238.9	C	102.8
DM20-α -W1	696.8	A	724.8	A	96.1
DM20-α -W2	608.8	A	656.6	A	92.7
DM20-β -W1	1558.1	A	1612.1	A	96.5
DM20-β -W2	1635.3	E	1358.9	D	120.3
DM20-β1-W2	1409.4	A	1501.5	A	93.9
DM20-γ -W1	314.0	F	302.3	F	103.9
DM40-α -W1	928.9	A	965.7	A	96.2
DM40-α -W2	792.4	G	797.8	G	99.3
DM40-β -W1	2246.1	E	2198.3	E	102.2
DM40-β -W2	1902.8	E	1561.6	D	121.9
DM40-β1-W2	1948.5	E	2004.5	E	95.3
DM40-γ -W1	396.5	F	382.2	F	103.9

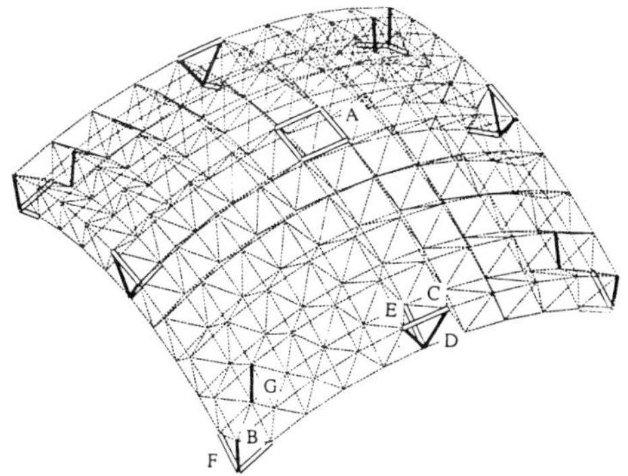

Fig. 7. Locations of buckled members.

Table 4. Section properties and locations of replaced web members.

Analytical model	Replaced members	Section Properties
DM00-γ1-W1	web members in four corner	A=11.8(cm^2) I=118.5(cm^4)
DM20-β1-W2 DM40-β1-W2	all boundary web members	A=5.76(cm^2) I=23.7(cm^4)

13. In Table 3, the approximate elastic buckling loads are in good agreement with the exact buckling loads except for the case where boundary web members buckle. For the three cases of boundary web members buckling, the approximate elastic buckling loads are about 22% higher than the exact buckling loads. However when the rigidities of the boundary web members are larger (see Table 4), the web members do not buckle and the approximate elastic buckling loads are closer to the exact buckling loads. When the boundary web members have buckled, another effective strength, taking into account the boundary conditions, must be derived. In practical design, it is desirable to avoid the buckling of boundary web members.

POST-BUCKLING BEHAVIOUR OF SQUARE-ON-SQUARE DOUBLE-LAYER BRACED DOMES

Analytical method

14. The post-buckling behaviours of the grid and domes are predicted by a simplified non-linear analysis with the effective strength (ref. 6). The procedure is briefly described as follows:

(a) At each step, the equivalent sectional forces corresponding to plate-like continuum at each constituent member are estimated by the axial forces in the relevant member and in adjoining members.

(b) When the equivalent sectional forces initially reach the corresponding effective strength, the axial force in the member is considered to be the buckling strength in the member.

(c) The non-linear characteristic of the member after elastic buckling, is derived by assuming the mode of the deformed member. In the paper, the buckling strength of the member is reserved until full yield occurs in the middle of the member and at the rotational springs. Plastic hinges are formed at these points.

(d) A tensile member is an elastic perfectly plastic body.

15. The present non-linear analysis can take into account both the rigidities and the stress states of adjoining members.

Analytical results

16. The post-buckling behaviours of the grid (DM00-A-W1) and two braced domes (DM20-A-W1, DM40-A-W1) are predicted. The total load-deflection relationships at the centre node are shown in Fig. 8. The solid line represents the result of the dome (DM40), the dashed-and-dotted line represents that of the dome (DM20). The broken line is the case of the grid (DM00).

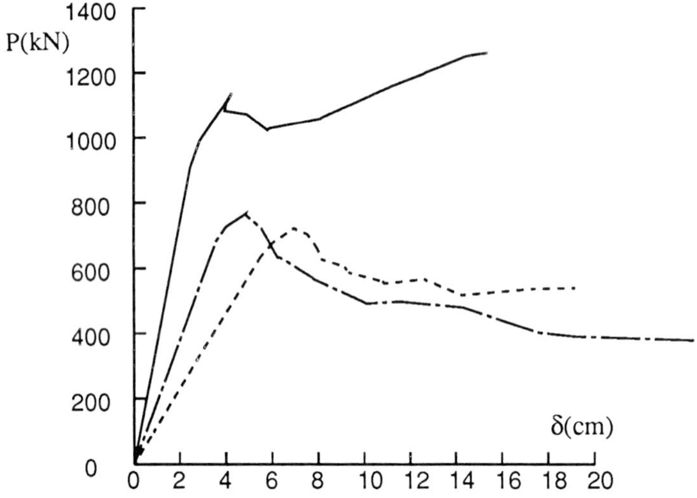

Fig. 8. Total load-deflection relationships of the double-layer braced domes.

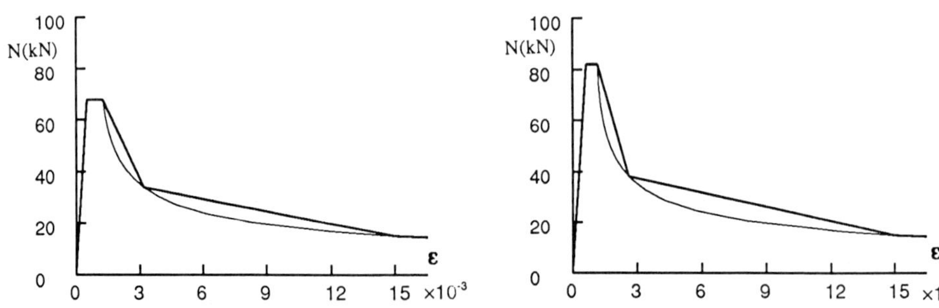

Fig. 9. Member characteristic at the lowest buckling strength.

Fig. 10. Member characteristic at the highest buckling strength.

17. In the analysis, the non-linear characteristic of each member undergoes a change by the stress state of the grid. The non-linear characteristics of the

members which buckle at the highest and lowest axial forces are shown in Fig. 9 and Fig. 10 respectively. The non-linear characteristics of the other buckled members exist between both the characteristics.

18. In Fig. 8 the stiffness of the domes are different from each other because of the curvature. The load-carrying capacity increases, with increasing stiffness of the domes. The load-carrying capacity of the dome (DM20) is larger than that of the grid (DM00). However the load-bearing capacity of the dome (DM20) after the peak load, is smaller than that of the grid (DM00). For the dome (DM40), the remarkable decrease in the load-bearing capacity has not appeared. Deformation capacity in the ultimate state is expected. The collapse modes of the grid (DM00) and the dome (DM20) are cross-type collapse patterns. For the dome (DM40), a cross-type collapse pattern appeared in the centre part of the grid. After the collapse pattern appeared, the web members near the four corners begin to buckle, and the load-bearing capacity increase gradually. However, if the axial forces of the buckled web members begin to decrease, the load-bearing capacity may decrease remarkably.

CONCLUSIONS

19. The effective strengths' in terms of equivalent stress resultants and stress couples, are derived for square-on-square double-layer grids constructed by a bolted jointing system. The practical applications of the effective strength are presented.

20. For double-layer braced domes, the approximate elastic buckling loads, which are estimated from the effective strength and the results of the elastic stress analysis of space trusses, are estimated, and the results are in good agreement with the accurate buckling loads estimated by a discrete element method, except for the case where boundary web members buckle.

21. The post-buckling behaviour of braced domes is predicted by a simplified non-linear analysis which can take into account the rigidity and stress state of constituent members by using the effective strength.

ACKNOWLEDGEMENT The authors wish to thank Prof. K. Heki for his valuable advice.

REFERENCES

1. SAKA T. and HEKI K. Limit analysis of lattice plates. Proc. 1971, IASS Pacific Symposium Part II on Tension Structures and Space Frames, Architectural Institute of Japan, 1972, 675-690.

2. SAKA T. and HEKI K. The effect of joints on the strength of space trusses. Third International Conference on Space Structures, Edited by H. Nooshin, Elsevier Applied Science Publishers, London, 1984, 417-422.

3. SCHMIDT, L. C., MORGAN, P.R. and CLARKSON, J. A. Space trusses with brittle-type strut buckling. Journal of the Structural Division, ASCE, Vol.102, No.ST7, 1976, 1479-1492.

4. SUPPLE L. C. and COLLINS, I. Limit state analysis of double layer grids. Analysis, Design and Construction of Double-Layer Grids, Applied Science Publishers, London, 1981, 93-117.
5. SMITH, E. A. Space truss non-linear analysis. Journal of Structural Engineering, ASCE, 110, 1984, 688-705.
6. SAKA T. Approximate analysis method for post-buckling of double-layer Ssace grids constructed by a bolted jointing system. Proc of 10 Years of Progress in Shell and Spatial Structures, IASS, Madrid, Vol.4, 1989.9, 387-402.

39. Comparison of the behaviour of three types of braced domes

P. D. PAKANDAM and B. SARSHAR, Ontario, Canada

SYNOPSIS. Three types of braced domes, i.e. geodesic, Schwedler and three-way grid domes, are investigated with a view to understand their behaviour. The behaviour of these domes is studied under four load cases with two boundary conditions and two height to span ratios. Furthermore, a comparison is made of the behaviour of these domes. For this purpose, a computer package, LUSAS (based on finite element method of analysis) is used assuming pin-connected conditions. The data for the analysis are prepared with the help of an automatic data generation system using formex algebra. Computer programs using FORTRAN77 were developed to present the results graphically.

INTRODUCTION

1. Various type of bracing system are used for the construction of skeletal domes. Professor Makowski in his book (ref. 1) discusses the development of the different types of bracing most commonly used in domes.

2. In our proposed paper the structural behaviour of three types of braced domes are studied under four load cases with two height to span ratios. The search has been carried out for a single-layer 50 meter diameter braced dome of minimum weight.

OUT LINE OF THE RESEARCH STUDIES

3. The investigation carried out to determine a single-layer 50m diameter braced dome with minimum weight, involved the following structures:

(a) three types of braced domes, i.e. geodesic, Schwedler and three-way grid domes Figs 1, 2 and 3;

(b) two height to diameter ratios (1/3 and 1/5) which are referred to as configuration 1 and 2 respectively;

(c) two types of boundary conditions are considered.

Support Type Sl - All the boundary nodes of the dome are supported, five or six of them are restrained against complete translation movements and are vertically constrained.

Support Type S2 - All the boundary nodes of the dome are restrained against translation movements, both in the horizontal and vertical planes of the structure. All the support points are considered free to rotate in any direction.

(d) Four load cases:

i	- Dead load and imposed load	LA
ii	- Dead load and unsymmetric snow load	LB
iii	- Dead load and wind load	LC
iv	- Dead load, snow load and wind load	LD

4. A total of 48 analyses were carried out.

5. Space limitation allows only some of the graphical results of configuration 1 to be presented. However, the maximum values of the axial forces, deflections and support reactions are tabulated for both of the configurations, two types of supports and all the load cases.

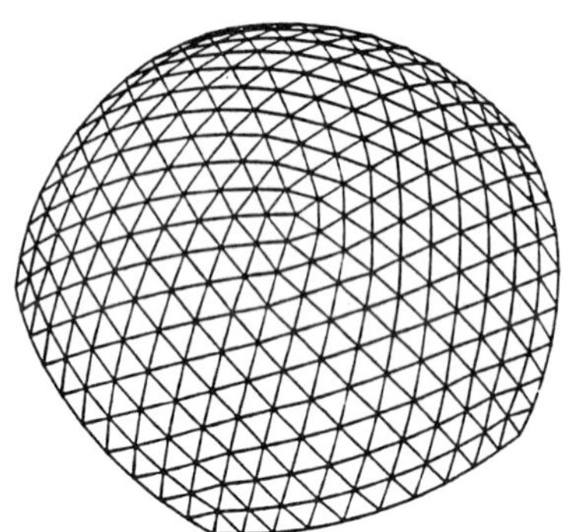

Geodesic Dome.

Fig. 1.

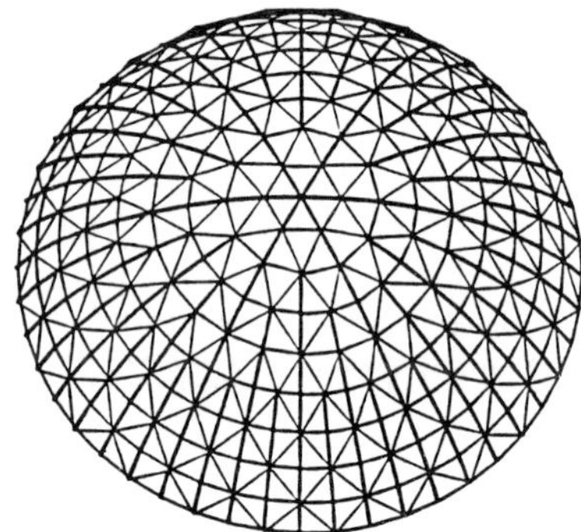

Schwedler Dome.

Fig. 2.

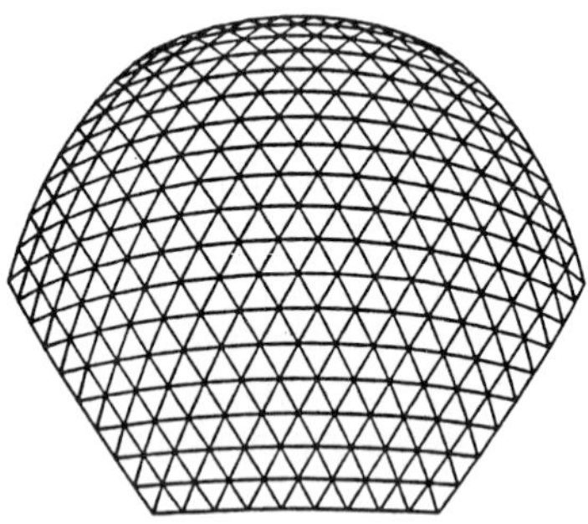

Three - way grid dome.

Fig. 3.

DATA GENERATION

6. Formex algebra and its interactive programming language, Formian, have been used to generate topology and nodal co-ordinates required for the computer analysis.

DETERMINATION OF LOAD CASES

7. The prime loading is obtained from BS6399, 1984, which gives the value for dead and superimposed loads, and BSI CP3, 1972, which provides information on the determination of the design wind velocities in relation to the surrounding terrain and the height of the structure.

8. The unsymmetric snow load is obtained from the proposal given in ISO/DIS/4355. The coefficients of the wind loads are determined from ECCS, 1980. The description of load case LD is illustrated in Fig. 4.

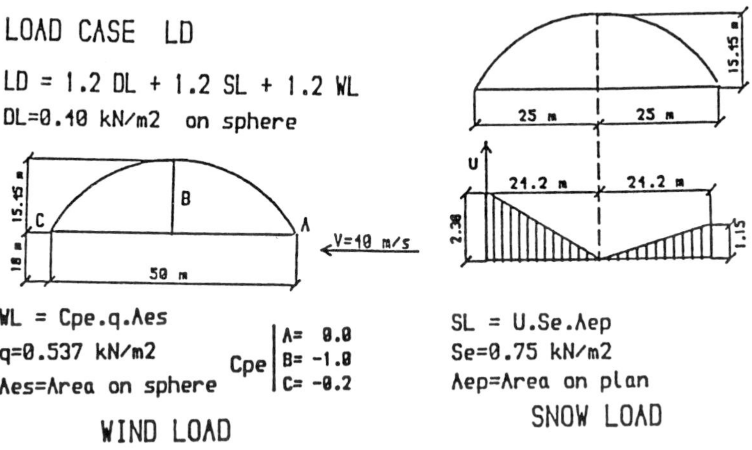

Fig. 4.

9. The loading on the dome is transferred to the nodes of the structure for the analysis.

10. A computer based technique has been devised to find the accurate projected area for each node, and the value of dead, imposed, snow and wind loads at each node is determined by using the corresponding surface area.

STRUCTURAL ANALYSIS AND DESIGN

11. The preliminary analysis was carried out assuming identical cross-sectional area for members and pin-connected end conditions, under the influence of symmetrical uniformly distributed load LA and also unsymmetric snow load LB. Having obtained the member forces, design was done by considering the minimum number of various cross-sectional areas of the members.

12. With the approximate sizes of the members, an exact linear elastic analysis of the skeletal structure was carried out under the four load cases. The result of each analysis were checked for the tension capacity and compression resistance.

COMPARISON OF RESULTS

13. To simplify the comparison of member forces, nodal displacements and support reactions obtained from each analysis, the results are represented in graphical form. The discussion of axial forces, deflections and support reactions is made below.

A) Axial forces

14. The distribution of the force is shown in the form of coefficients recorded along the lengths of the corresponding members. The value of a member force can be obtained by multiplying the coefficient of a member with the given factor on each of the figures. Also, the thickness representing a member is proportional to the magnitude of the force in it.

15. The compressive axial force in a member can be identified by a negative sign proceeding its coefficient. In addition, the members with compressive force are represented by firm lines as against dotted for the representation of a member with tensile axial forces. The comparison of forces in the geodesic dome for load case LB with two type of supports is made in Fig. 5.

B) Deflections

16. The nodal deflection in configuration GD1 for load case LB and both support conditions are shown in Fig. 6.

17. The deflections are shown in the form of dots. The size of dot is related to the magnitude of the corresponding nodal deflection in the vertical direction. A value is also recorded adjacent to each dot, which can be used to determine the magnitude of the deflection by multiplying it with the given factor. A negative value indicates a downward deflection.

C) Support reactions

18. A vertical reaction is shown by a line directed from the support point to the centre of the plan view of the structure. The length of such line is proportional to the magnitude of reaction. The greater the magnitude of the reaction, the longer line will be. In addition, the value of each reaction is also given in these figures. The distribution of vertical reactions is illustrated by means of dotted lines.

19. The resultant of horizontal reaction, if any, for the support point is given by a dashed line whose length is again proportional to its magnitude. The distribution of support reactions in configuration GD1 for load case LB and two support conditions are displayed in Figs 7 and 8.

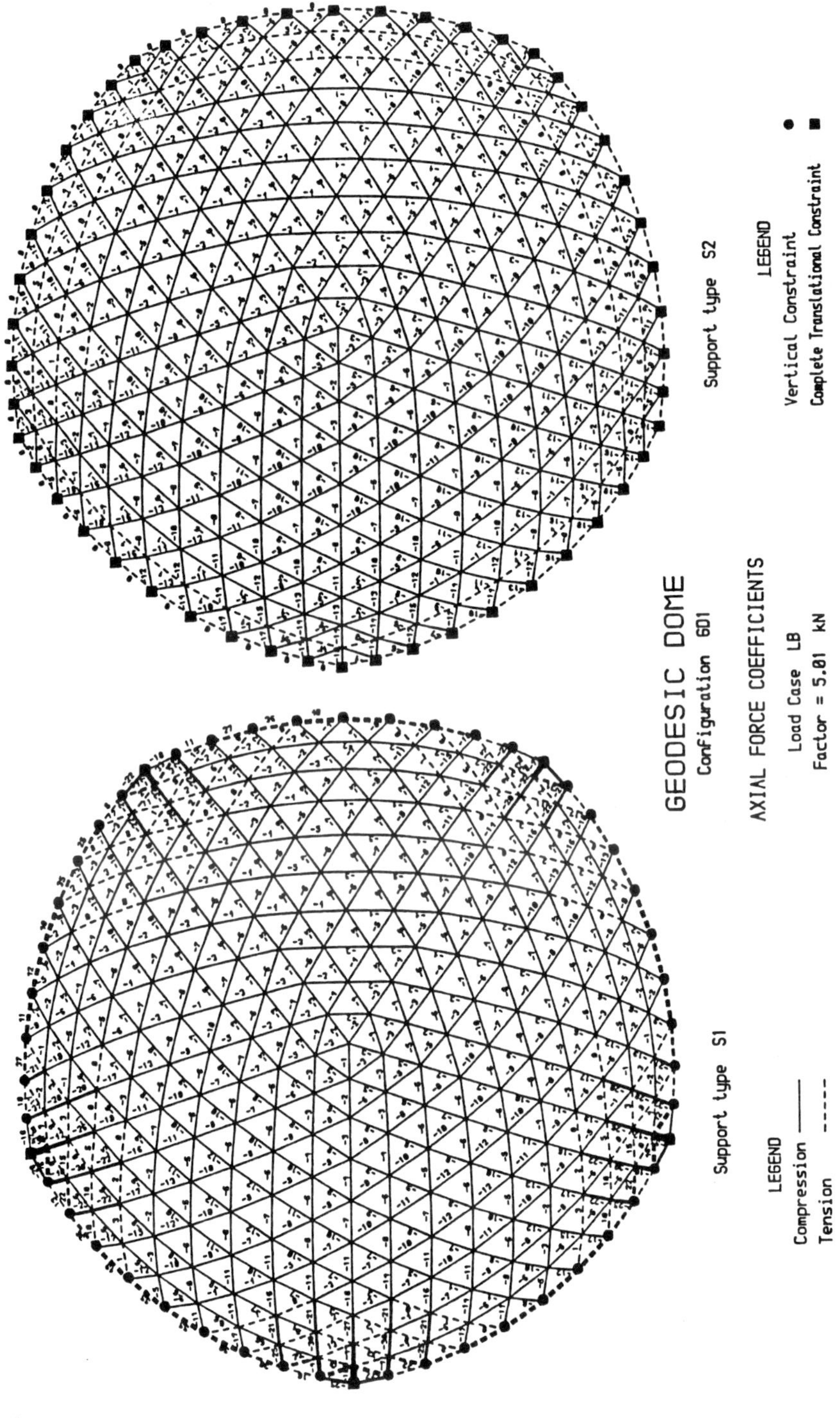
GEODESIC DOME
Configuration GD1
AXIAL FORCE COEFFICIENTS
Load Case LB
Factor = 5.01 kN
Support type S1
LEGEND
Compression
Tension
Support type S2
LEGEND
Vertical Constraint
Complete Translational Constraint

Fig. 5.

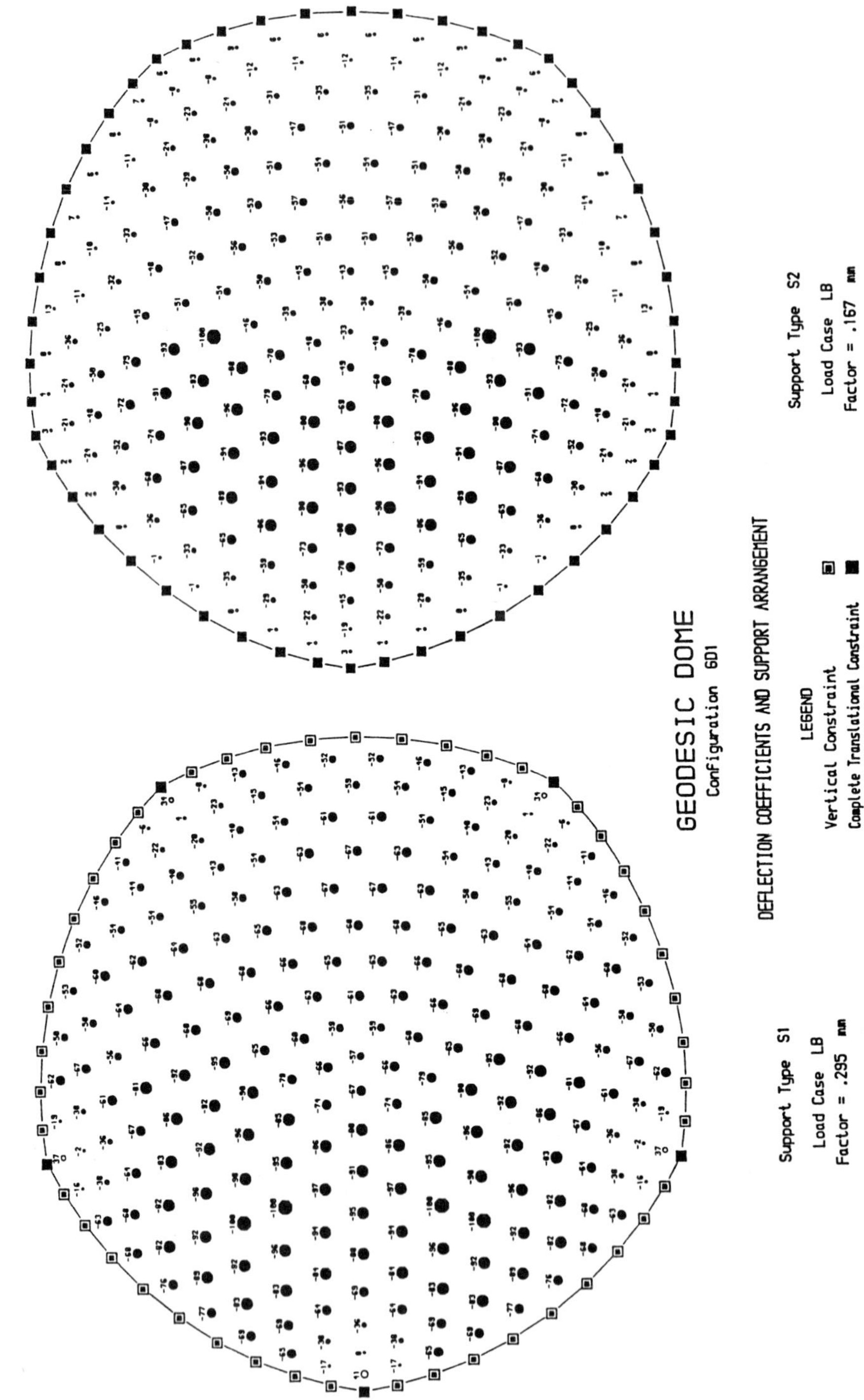

Fig. 6.

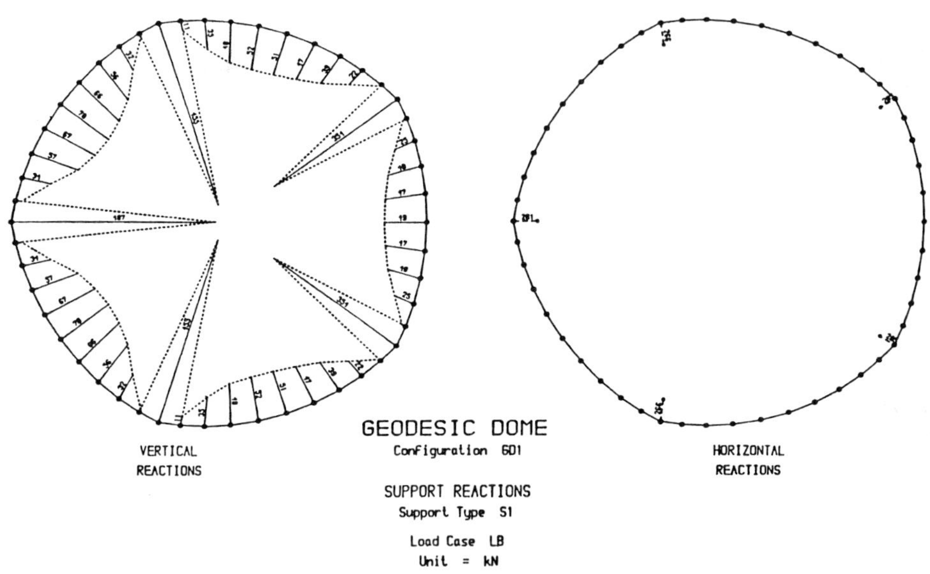
VERTICAL
REACTIONS
GEODESIC DOME
Configuration GD1
SUPPORT REACTIONS
Support Type S1
Load Case LB
Unit = kN
HORIZONTAL
REACTIONS

Fig. 7.

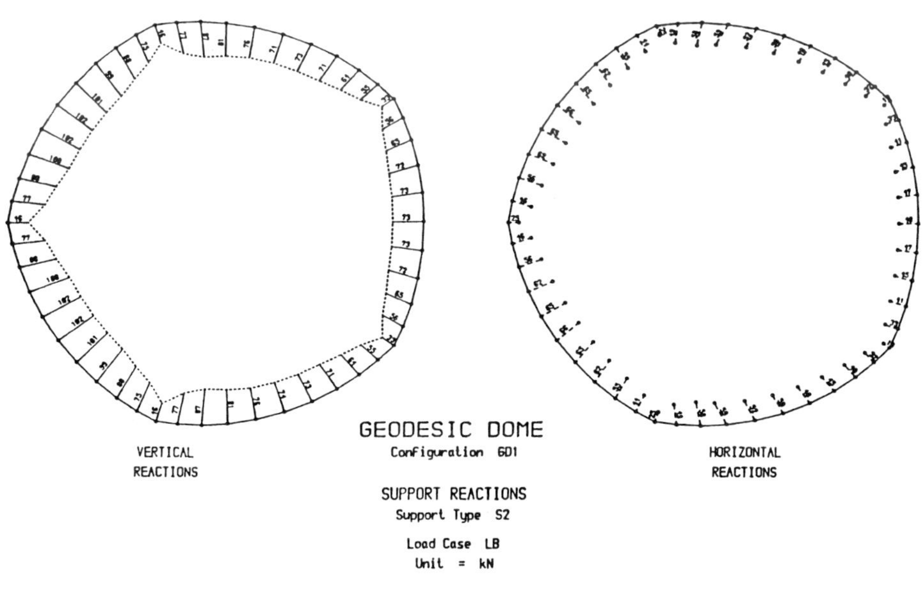
VERTICAL
REACTIONS
GEODESIC DOME
Configuration GD1
SUPPORT REACTIONS
Support Type S2
Load Case LB
Unit = kN
HORIZONTAL
REACTIONS

Fig. 8.

Table 1.

COMPARISON OF AXIAL FORCES

Support Type		Load Cases	GEODESIC DOMES GD1	GD2	SCHWEDLER DOMES SD1	SD2	THREE-WAY GRID DOMES TW1	TW2
S1	maximum Compressive AXIAL FORCE kN	LA	397	494	370	603	310	423
		LB	501	598	481	746	397	520
		LC	267	300	228	355	206	259
		LD	473	568	441	691	381	494
	maximum Tensile AXIAL FORCE kN	LA	226	242	178	241	167	207
		LB	276	275	212	291	206	239
		LC	150	144	123	146	138	122
		LD	251	261	212	283	190	222
S2	maximum Compressive AXIAL FORCE kN	LA	75	76	75	83	59	62
		LB	95	100	95	114	79	102
		LC	54	47	57	53	44	41
		LD	85	86	94	100	80	92
	maximum Tensile AXIAL FORCE kN	LA	42	0	47	0	44	0
		LB	35	0	42	0	26	0
		LC	25	0	11	0	16	0
		LD	32	0	37	0	25	0

Table 2.

COMPARISON OF NODAL DEFLECTIONS

Support Type		Load Cases	GEODESIC DOMES GD1	GD2	SCHWEDLER DOMES SD1	SD2	THREE-WAY GRID DOMES TW1	TW2
S1	maximum Downward DEFLECTION mm	LA	37.7	47.6	23.9	34.8	29.5	35.5
		LB	29.5	42.5	25.3	44.3	25.6	41.5
		LC	22.8	27.6	13.8	21.6	16.9	22.0
		LD	29.5	39.2	24.2	42.3	25.5	41.5
	maximum Upward DEFLECTION mm	LA	11.3	18.6	6.7	13.6	10.6	16.3
		LB	12.1	14.9	7.1	12.0	11.5	12.9
		LC	6.8	10.8	3.2	8.0	5.2	9.5
		LD	11.2	14.1	6.3	10.6	10.5	12.5
S2	maximum Downward DEFLECTION mm	LA	26.8	29.5	17.1	24.0	18.3	23.1
		LB	16.7	19.3	17.3	25.0	12.9	20.6
		LC	16.3	17.1	9.5	13.7	10.3	13.1
		LD	18.3	17.3	16.7	24.6	13.6	20.3
	maximum Upward DEFLECTION mm	LA	2.3	0	2.7	1.1	1.7	0
		LB	2.2	1.0	2.3	11.3	1.7	6.0
		LC	1.6	0	1.6	0.8	1.1	0
		LD	2.2	0	2.5	5.2	1.5	0

Table 3.

COMPARISON OF SUPPORT REACTIONS

Support Type	Reaction (kN)	Load Cases	GEODESIC DOMES		SCHWEDLER DOMES		THREE-WAY GRID DOMES	
			GD1	GD2	SD1	SD2	TW1	TW2
S1	maximum Vertical SUPPORT REACTION kN	LA	390	466	303	385	405	466
		LB	487	542	395	480	505	535
		LC	248	275	185	227	251	275
		LD	444	504	360	444	466	513
	maximum Horizontal SUPPORT REACTION kN	LA	224	546	148	436	205	545
		LB	281	630	184	541	257	626
		LC	176	315	77	256	130	321
		LD	255	587	166	500	236	597
S2	maximum Vertical SUPPORT REACTION kN	LA	80	66	104	95	78	67
		LB	102	80	130	128	95	75
		LC	58	41	80	62	54	41
		LD	95	78	134	124	95	74
	maximum Horizontal SUPPORT REACTION kN	LA	53	87	49	101	53	88
		LB	66	105	67	138	63	100
		LC	36	51	36	64	33	53
		LD	61	98	64	129	61	96

CONCLUDING REMARKS

20. The maximum axial forces, maximum deflections and maximum vertical and horizontal support reactions in both the configurations for the four load cases and two boundary conditions are recorded in Tables 1, 2 and 3 respectively.

21. The configurations type 1, which have a relatively greater rise, deflect less than the configuration type 2. Similarly, the forces in configurations 1 are considerably smaller than the other configuration. Furthermore, the horizontal reactions in configuration 1 are significantly smaller than those in configuration 2.

22. The stress distributions in all the domes show that most of the members are carrying compressive axial forces, whilst the extent of tensile forces is limited to the areas near the supports. The stress distribution in the domes for all load cases, with both the support types are almost uniform. The tensile forces are generally in the members in the two or three outer rings near the vertically constrained nodes.

23. With support type S1 for the domes, the nodes next to fully constrained points lift vertically, under the four load cases. However, the upward deflections are small and are not significant. In configuration 1 of the domes, with support type S2 for all the load cases, the nodes next to the support nodes have a slight upward deflections.

24. The vertical reactions for all the domes are directed upward, with the exception of the three-way grid dome with support type S1. In the three-way grid and the configuration GD2 of geodesic domes, the nodes next to the fully constrained points have small negative reactions (downward) for all the load cases. The horizontal reactions on all the support points are directed inward.

25. In both the configurations the application of load case LA, which is the combination of dead and imposed loads, produces symmetrical stress distributions, nodal deflections and support reactions.

26. In load cases LB and LD because of the high intensity of snow loading on one side of the structure, the forces in the members, deflections and support reactions are larger than those of the other side.

27. The member forces, nodal deflections and reactions for load case LC are significantly smaller than those for the other load cases.

REFERENCE

1. MAKOWSKI Z.S., Analysis, Design and Construction of Braced Domes, Granada Publishing, London, 1984.

2. NOOSHIN H., Formex Configuration Processing in Structural Engineering, Elsevier Applied Science Publishers, London, 1984.

40. Large space frames recently constructed in diamond truss and Tomoe Unitruss systems

H. KATO, T. KAWAZOE, I. KUBODERA and F. MATSUSHITA,
Tomoe Corporation, Japan

SYNOPSIS. This paper introduces the comparatively large and unique space structures manufactured and erected recently by the TOMOE CORPORATION. The space frame systems developed by the TOMOE CORPORATION were adopted for all eight buildings reported in this paper; the Diamond Truss system for three buildings and the Tomoe Unitruss system for the remaining five buildings. The paper describes their outlines, as well as the features of the space structures, including their structural design, and erection methods.

INTRODUCTION

1. The Diamond Truss system is a space structure which was developed by TOMOE CORPORATION about 60 years ago. This structure is composed of prefabricated flat trusses, which are further three-dimensionally fabricated into diamond or triangular shapes. The unit truss is composed, in principle, of top and bottom chord members arranged in parallel with each other and is coupled by vertical or diagonal members. Therefore, the system exhibits high rigidity as a double-layer three-dimensional grid, and its structure can be analysed as a single-layer grid.

2. On the other hand, the Tomoe Unitruss system is a ball joint type space frame system made of steel tubes, and about 10 years have passed since its commercialisation. The system is featured by the continuous utilisation of computers used for design through to manufacture. The standard component parts of the system consist of 10 kinds of balls and 12 kinds of tubular members. Space frames that consist of basic units, namely, trigonal and quadrangular pyramids, are composed of these parts. In addition to the pin-jointed type that has been used for double-layer three-dimensional structures, the TOMOE CORPORATION has developed a rigid-joint type for single-layer truss shells.

3. Introduced on the following pages are the large buildings in which the features of the Diamond Truss or Tomoe Unitruss system has been exemplified.

DIAMOND TRUSS SYSTEM

Fantasy Dome

Architectural firm : Ishimoto Architectural & Engineering Firm, Inc.
Contractor : Shimizu-Iwakura Joint Venture
Completion : 1990

4. This building planned as a leisure facility with roller coasters, merry-go-rounds, and many other amusement facilities, was constructed in Hokkaido, a northern island of Japan. The roof is almost flat, measuring 76.05m by 92.95m. and is about 43m above the ground.

5. The Diamond Truss system, with diagonally crossing beams, was adopted for the roof structure. With the truss depth being 4.5m, this building is one of the largest flat Diamond Truss structures. In the centre of the roof the main trusses were replaced by the Tomoe Unitruss system over an area of about 928m^2 to allow natural lighting. The construction site is in a snow region, which requires a design snow loading of 0.98kN/m^2. The stainless steel sheet roof, with fireproof cemented excelsior board backing, is lightweight at 0.29kN/m^2. Because of this construction, no special steel members were used as truss members and the details of joints are those of an ordinary Diamond Truss system, although the roof structure is very large.

6. Prior to the erection of the roof structure, the erection method was examined in terms of the construction period, safety, costs, and many other factors. As a results the lift-up method was adopted because the roof structure had a large area and was located at a high elevation. To raise the roof, six temporary supports were erected inside the roof, and two centre-haul jacks were installed for each support, to hoist the roof. To ease the connection of the roof to the side columns after the structure was raised, the temporary lifting supports were positioned at locations in the roof having almost equal deflections.

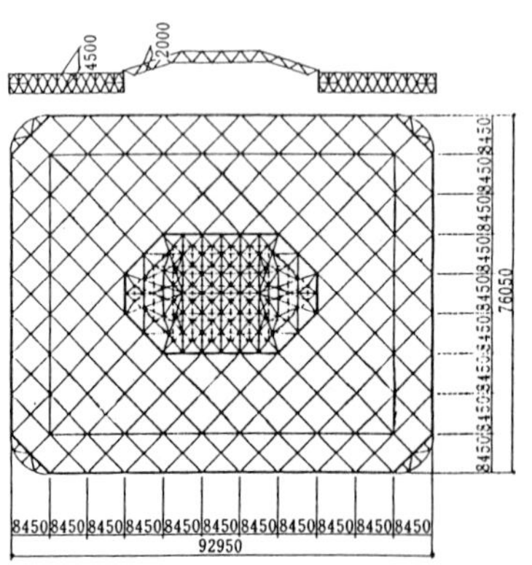

Fig. 1. Plan and Section.

Fig. 2. Erection.

Yamagata Pref. General Gymnasium

Architectural firm : Yamashita Sekkei Inc.
Contractor : Kajima-Tonan Joint Venture
Completion : 1990

7. This building provides multi-purpose gymnastic facilities and was established for the National Athletic Meeting held in Yamagata Prefecture in 1992. Yamagata Prefecture is located in the northern part of Japan and has a heavy snowfall exceeding 100cm. The roof of the large arena is cylindrical in the centre and flat in its periphery. The entire facilities measure 72.0m in the span direction and 75.0m in the cross-beam direction.

8. The Diamond Truss system was adopted for this large roof structure. The roof structure is composed of three structural elements. One is the cylindrical shell in the centre, measuring 48.0m in the span direction, 56.0m in the cross-beam direction, 9.5m in arch rise, and 2.2m in truss depth. The other two elements are the diagonally crossing truss on the base side of the cylindrical shell and the parallel truss on the gable side arranged in the flat plate portion at the periphery. The depth of both trusses is 3.0m.

9. The supporting points of the cylindrical shell at each of the four corners divide the force transmitted by the entire roof structure into two. In the cross-beam direction, force transmission is achieved by the structure behaving like a simple beam when the structure is viewed macroscopically, and in the span direction, force transmission is achieved by bending of the truss beams. To make the best use of this behaviour, the rib arch of the cylindrical shell, including tie beams, were made very rigid. As a result, the four corner supporting points of the cylindrical shell carry 60% of the roof load.

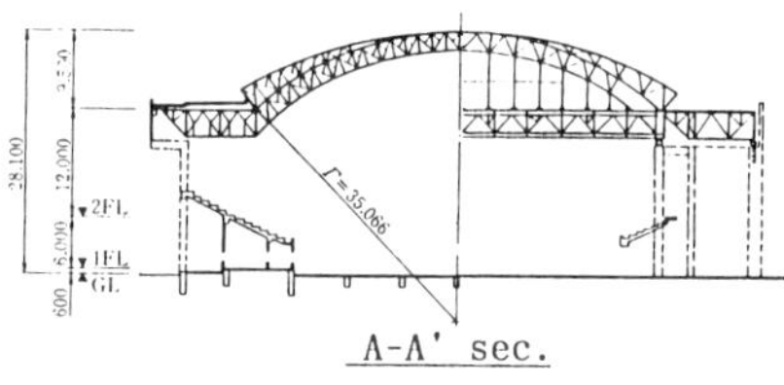

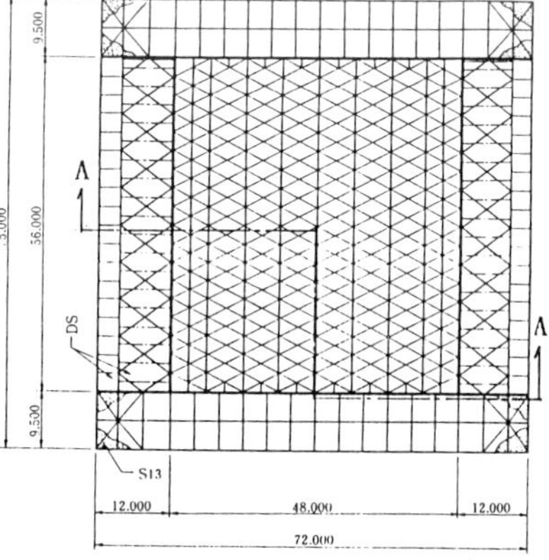

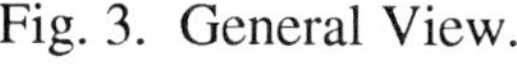

Fig. 3. General View.

Fig. 4. Plan and Section.

Shiga Pref. Nagahama Dome

Architectural firm : Kume Architects-Engineers
Contractor : Kumagai-Zaiko Joint Venture
Completion : 1992

10. This huge dome constructed near Lake Biwa, the largest lake in Japan, is one of the prominent multi-purpose gymnastic facilities which are available throughout the year irrespective of the weather conditions. The pseudo elliptical plane is composed of arcs, measuring 107.5m in the short span direction, 162.5m in the long span direction, and 32.2m in height.

11. The Diamond Truss system was adopted for building the structure. Fabricated in triangular grids, the space frame has an extremely high in-plane rigidity. In terms of the roof shape, the structure can be called a shell structure. A flat portion in the centre of the roof causes local bending moments, and especially when the load is asymmetric, larger bending moments are generated. Therefore the truss is 2.4m deep, which is about 1/25 of the radius of the roof curvature, a very large value in comparison with other shell structures.

12. The roof structure is supported on its periphery by 60 reinforced concrete columns. To cope with the horizontal reaction in the superstructure, prestressed concrete ring beams are formed at the top and bottom of these columns. Furthermore, opposing columns are linked by underground tie beams.

13. The roof structure was erected by the block method. The structure is composed of about 36,000 pieces, including 3,000 main trusses. It was judged, therefore, that the block method, which allowed the greater part of member erection work to be done on the ground, was the optimum method in view of the required accuracy, work efficiency, and safety. Truss blocks fabricated on the ground were raised by a crane from the periphery of the structure to the inside and were supported at critical locations by temporary supports.

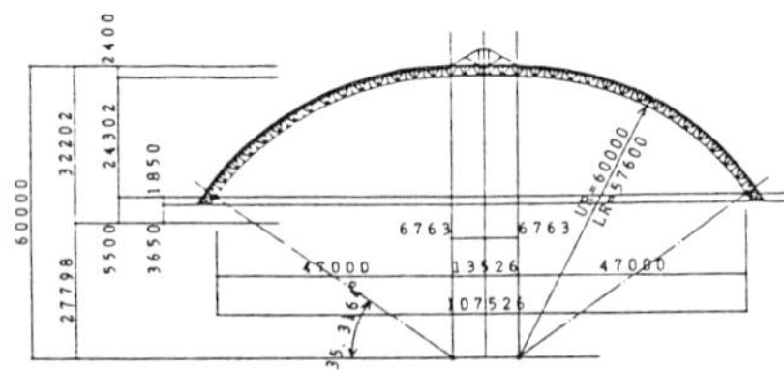

Fig. 5. Section through the Nagahama Dome.

Fig. 6. General and interior views.

TOMOE UNITRUSS SYSTEM

Tokyo Metropolitan Art Space

Architectural firm : Ashihara. Architect & Associates
Takumi Orimoto-Engineer (Structural Design)

Contractor : Taisei et al. Joint Venture

Completion : 1990

14. Tokyo Metropolitan Art Space provides art and cultural facilities and was constructed by the Tokyo Metropolitan Government at the front of Ikebukuro Station West Gate. It has an atrium, which is situated on the front of the building. The atrium is 28m high and covers an area of 1,802m^2. It is constructed with external walls and a roof made of trusses most of which are covered with glass.

15. The Tomoe Unitruss system was adopted for the roof and wall structures of the atrium. The basic grid of the truss is a quadrangular pyramid for the roof and a trigonal pyramid for the wall. The grid measures 3.0m by 3.0m in the flat portion of the roof and 2.2m in height (truss depth). The inclined surface measures about 3.56m on one side of a rhombus and is about 2.15m high. The wall is shaped into an equilateral triangle with one side measuring about 3.56m and is 2.0m high.

16. In Japan where earthquakes abound, the condition for joining the atrium frame to the main structure is one of the most important points in the structural design. In this building, the roof portion of the space frame atrium is pin-supported on two sides by brackets extending from the upper portion of the main building. The atrium wall is one-way or two-way roller-supported by a large reinforced concrete beam structure which surrounds the periphery of the building at the second floor level. Three supporting points are provided on the large beam with the taller walls supported on one-way roller-supported, while the supporting points for the lower walls provide two-way roller-supports. Thus, the horizontal constraint between the atrium and the frame of the main structure is weakened so that the atrium frame will not become an aseismic element of the entire building.

Fig. 7. General view.

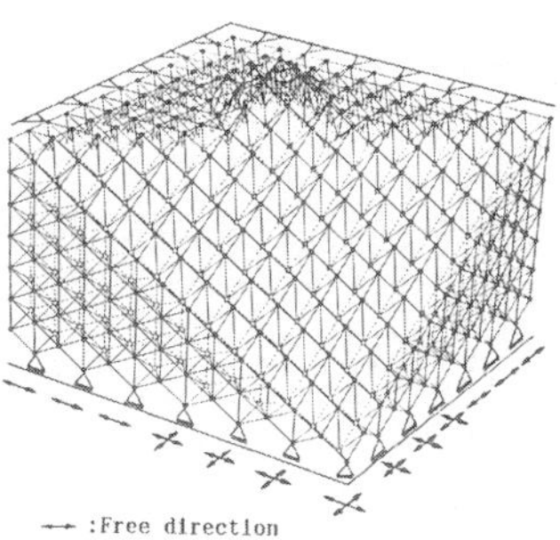

Fig. 8. Analysis model.

Asakita-ward Sports Centre of Hiroshima City

Architectural firm : Kume Architects-Engineers
Contractor : Sumitomo et al. Joint Venture
Completion : 1991

17. Hiroshima City is promoting the construction of a sports centre in every administrative district, and the present structure is one of them. This sports centre is planned to be used for the Asian Games to be held in 1994 and many other international sports meeting. The roof of the large gymnasium is shaped into a characteristic dome.

18. The Tomoe Unitruss system is used for the roof structure. This space frame features load transmission through almost four keels alone to the lower main structure. The keel span is 63.6m in width and 66.8m in length, while the radius of curvature is about 60m on the top chord surface in both directions. The eight keel supporting points are in turn supported by large cantilever columns made of prestressed concrete. The deformation and stress in the space truss changes greatly according to the degree of rigidity. Accordingly, the space truss was designed by taking these important effects into consideration.

19. The stress of the space truss was analysed by dividing the bearing conditions of the keel points into two cases. These are when the supporting points will not move (Case 1) and where the supporting columns for the keels move due to the reaction forces, (Case 2). In Case 1, the rigidity of the lower structure was assumed to be infinite, and displacements other than rotation were constrained at all keel supporting points.
In Case 2, the horizontal rigidities of the columns were replaced by spring members for the stress analysis. The cross sections of the truss members were determined according to the maximum stress obtained from the above two cases.

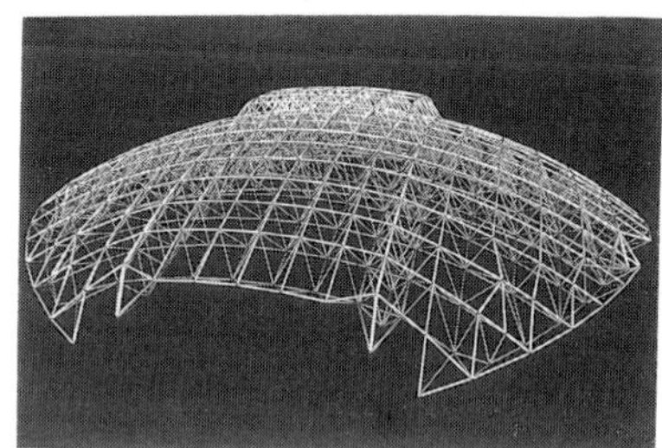

Fig. 9. Skeleton.

Fig. 10. General and interior views.

Indoor Swimming Pool of Nagoya City General Gymnasium

Architectural firm : Azusa Sekkei Co. Ltd. and Kume Architects-Engineers
Contractor : Ohbayashi et al. Joint Venture
Completion : 1991

20. This building with a main swimming pool suitable for international swimming competitions was built in Nagoya City. The building measures, 78.0m in the span direction and 111.648m in the ridge direction. The roof is formed by combining three different radiused spheres (R = 481.254m, 443.717m, and 459.547m) making the top view of the building almost elliptical.

21. A composite structure was constructed, in which the Tomoe Unitruss system and a tension cable structure were combined to form the roof structure. The structure was designed as a space frame without supporting points on the gable side. Seventeen cables were used across the width of the structure and fifteen cables were used for the middle section. The cables were positioned under the space frame and attached to struts hung from the bottom chord members. A tensile force introduced into the cable generates a hogging bending moment in the structure opposite to the sagging bending moment generated in the beam elements due to live load and self-weight, thus reducing the final sagging bending moment in the beam elements.

22. The major point in erecting the roof structure was determining the value of the tensile force to be introduced into the cables and also the optimum time for tensioning. The tensile force introduced was determined in such a manner so that the bending moment generated in the frame would approach zero and a uniform force in the axial direction would be generated in all members. The tensioning was undertaken at the earliest possible time in the construction period, when the fluctuations in the roof load were small, allowing work on the ceiling and roof cladding to proceed.

Fig. 11. General and interior views

Fujikura Futtsu Factory
Architectural firm : Nissoken Architects-Engineers
Contractor : Ando Construction Co., Ltd.
Completion : 1991

23. This building was constructed in Futtsu City, Chiba Prefecture, which is located to the Southeast of the metropolitan area and is one of the most advanced power cable factories in the world, furnished with two 50-ton cranes. The roof is an almost flat cylinder with a radius of curvature of about 671m, measuring 360m in the ridge direction, 80m (40m by 2) in span, covering a total floor area 24,000m^2.

24. The Tomoe Unitruss system was adopted for the roof structure. The standard grid of the space frame is 5.0m by 5.0m, the truss depth is 3.5m, and the eaves height is 21.23m. The structure is composed of 7,936 steel tubes and 2,120 joint balls. Expansion joints provided in three locations in the middle part divide the roof into four blocks, to alleviate the stress generated in the members as the temperature changes.

25. The sliding method was adopted for erection of the roof structure, because a long and narrow plane extended in the ridge direction, the simple shape was almost like a plate. Rationalisation of the erection technique minimised construction time. The truss was fabricated on staging installed throughout the section between line 0 to line 4 (40m) in order to ensure safe working. First, a truss for two grids (l0m) and secondary members were fabricated. Then they were pulled l0m by wires and a winch from line 0 to line 36 in the ridge direction, then the next l0m truss was fabricated. During that time, the painter's work, roof finishing, and electrical equipment work were completed. This operation was repeated 21 times, and a 225m truss (weighing about 990 tons) was slid.

Fig. 12. Erection.

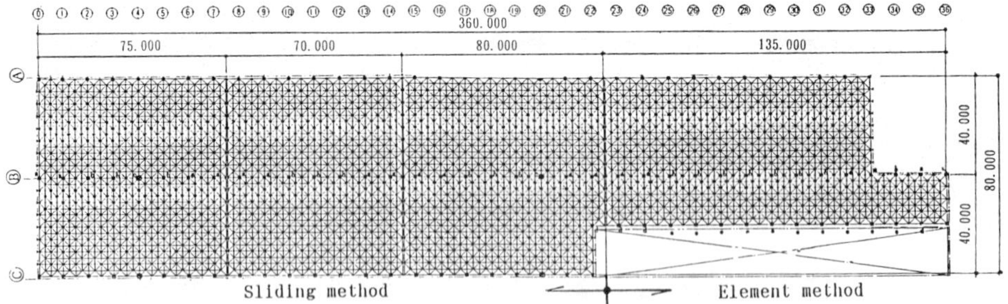

Fig. 13. Plan.

Tatsumi Indoor Swimming Pool

Architectural firm	: Senda & Environment Design Institute Kozo Keikaku Engineering Inc.
Contractor	: Shimizu et al. Joint Venture
Completion	: 1993 (scheduled)

26. This building constructed in Koto-ku, Tokyo is an indoor swimming hall equipped with a main swimming pool, diving event pool, an auxiliary swimming pool, seats for spectators, and an arena. This large building measures about 100m in the maximum span direction, has a maximum height of 36.5m and provides cover to about 12,400m^2.

27. The Tomoe Unitruss system was adopted for the large span roof structure. The basic unit of the truss is a quadrangular pyramid. The sub-structure is made of reinforced concrete. As clearly shown in the shape, the bases of columns supporting the space frame are located in limited positions in the centre and edge portions where the separate roofs meet. Since the structure had few supporting points within a large space and the self-weight was comparatively small, deformation due to wind-load became a grave problem in designing the space frame. To solve the problem, the rigidity of the entire roof was increased, while the following points were taken into consideration.

a) The portion where roofs overlapped were linked with members to form an integral frame, called a rib arch.

b) The rising portion of the rib arch was made into a solid web, fabricated from plates.

c) The centre of the roof structure was composed of a system truss with bracing members inside.

d) The standard unit of the column base was changed from a quadrangle pyramid to a trigonal pyramid.

28. The element method was adopted for erection, because the shape of the roof was complicated. Although the frame had large spans and was complicated in shape, the desired accuracy was obtained both during the manufacture of the truss and also during erection on site.

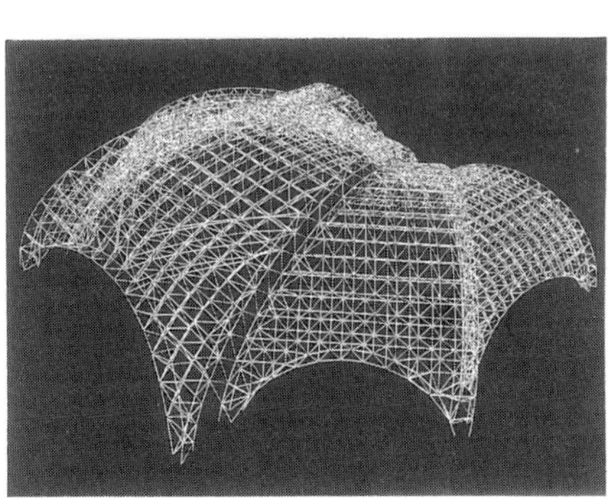

Fig. 14. Skeleton.

Fig. 15. Erection.

41. The design and construction of the domes for the Cairo stadium

S. SRINIVASAN, Dar Al-Handasah, UK, and A. AYOUB and A. ELSHERIF, The Arab Contractors, Egypt

SYNOPSIS. This paper describes the design and construction of tensioned cable dome structures of varying diameters from 40m to 120m as part of a fast track design and build project with Arab Contractors in Cairo. The domes form the roofs to the covered stadium for the African Games in Cairo, Egypt. The feature of the design has been to provide a structure that can be built safely and quickly without interruption to other construction activities below in the stadium. The whole construction was carried out in record time. The resulting structure is an example of a state of the art structure built simply with little sophistication.

INTRODUCTION

1. Following the selection of Cairo as the venue for the 1991 African games, Dar Al Handasah Consultants adopted a design and construct strategy with the full co-operation of the Arab Contractors of Egypt who were awarded the task of completing the project in 18 months.

2. The stadium complex consists of four sports halls. The main arena, with a 55m diameter playing field and a seating capacity of 20,000 spectators, is 120m in diameter with a maximum internal height of 29.8m. A secondary hall seating approximately 1500 spectators, has an internal diameter of 60m and an internal height of 24m. Two smaller halls seating 750 spectators each are 50m in diameter and a maximum internal height of 19m.

DESIGN REQUIREMENTS

3. Two overriding constraints guided the design of the roof structures for the Cairo Indoor Stadium. First, a concept had to be developed that would enable the structure to be built in a short period of time and since the overall time for the construction of the entire complex was also limited, the construction of the roof had to proceed with out hindrance to the construction activities below the roof. Finishing and furnishing of the interior had to proceed, while the roof was being constructed. Safety of the workers below the roof was absolutely necessary and this was further accentuated by the fast pace at which construction was proceeding in all these areas. Simplicity both in concept and in construction

Space structures 4. Thomas Telford, London, 1993

procedures was essential to minimise risks to all involved in the construction of the structures.

4. Long span roof structures designed and built in recent times have been cable supported type of structures because of the freedom of support it offers and the lightness of the structures. They have however, tended to be rather involved in construction. The challenge therefore was to conceive a structure that could be quickly and easily built, without interference to the work being carried below.

5. The choice was made against conventional concrete or steel space structure for the reason already stated. Aesthetics was an important requirement in the criteria.

THE CONCEPT

6. Several concepts were examined for a roof that is circular in plan and is dome shaped. Cable supported roofs have been attempted before but there have always been problems in building them. The construction was complicated and involved falsework to build the structure. The sophistication in form and function had a penalty in construction cost.

7. The concept of the roof for the stadium is very simple and avoids all problems related to construction. This has considerably simplified the whole process thus effecting considerable saving in time and cost. There are two variations on the theme.

The Double Layered Cable system

8. This scheme consists of a ring beam supported on bearings which in turn is supported on discrete columns. The lower layer of cables is draped across the ring beam at a predetermined offset to the diameter to form a symmetrical layout. Steel trusses of specific height and diameter are placed on this layer to obtain the desired curvature of the dome. The second layer of cables is similarly offset on the diameter passed over the tops of the trusses and stressed to the required force to put the structure in tension for any system of loading. The top layer of cables are covered with a membrane or any desired cladding.

9. This system is suitable only for small diameter roofs as with increase in the span the cable sag is visually intrusive.

The Multi- Layered Cable system

10. This concept is used for large diameter roofs where the sag in the cable becomes significant. Reducing the plan area of the cables also reduces the span lengths of the cables and thus their sag. This scheme therefore consists of the lowest layer of cables anchored to the ring beam and supporting the first truss. The second layer of cables is anchored to the ring beam and passed over the first truss. This layer supports the second truss and so on. The top-most layer passes over the top flanges of all the trusses and is stressed down to the ring beam to keep the whole system in tension. The membrane or cladding is fixed on to the trusses.

11. The most important feature of the concept is that it has a very large safety factor which is intrinsic to this design and is thus free from progressive failure. The tangential layout of the cables is a stable system in itself which can accept vertical loads from the truss. Likewise the trusses are in self equilibrium and are supported on the cables. This makes it very simple to construct without the necessity for staging or falsework.

DESCRIPTION OF THE STRUCTURE

12. There are four domes in all, of which two are 40m in diameter, double-layer configuration, and the 60m and 120m diameter domes are of the multi-layer type.

The Small Dome

13. The roof to this hall consists of a concrete ring beam of 41.6m diameter measured on the centre line of the ring beam. This beam is supported on twin 25 tonne capacity, laminated bearings, on each of the 18 columns equally spaced along the circumference of the hall. Eighteen cables made up of six, 15.7mm strands each, is anchored with a predetermined sag and offset from the diameter by connecting every 8th point on the circumference to give a symmetrical layout of cables. The diameter and the heights of the trusses were determined from the profile of the dome surface. The top layer of the cables, made up of two, 15.7mm diameter strands, each pass over the top flanges of the two trusses and follow the configuration of the bottom layer of cables. These are stressed to a designed force to keep the system in tension for all load conditions.

14. Specially designed saddles are provided on the flanges of the trusses to seat the cables. These saddles are welded to the flanges of the truss and the cables are clamped to the saddles so that they are prevented from displacement.

15. Rafters and purlins are placed on the trusses to obtain the curvature of the dome. This is then covered with aluminium cladding. This alternative was preferred to a membrane because of the time constraint and also because it appears visually more permanent.

The Large Dome

16. The ring beam for the large hall is 120m in diameter on the centre line axis of the beam. The ring beam is supported on 36 equally spaced points on the circumference. There are four layers of cables and three trusses of 9.5m height and 85.9m diameter, 6.0m height and 58.2m diameter, and 4.0m height and 26.85m diameter. The bottom layer consists of 36 cables of twenty six, 15.7mm diameter strands each. These are anchored to the ring beam connecting every tenth point on the circumference as shown in Fig. 1. The truss T1 with the largest diameter is placed on this layer of cables. The second layer of 36 cables made up of nine 15.7mm diameter strands each is anchored to the ring beam and passed over the first truss at an offset to the diameter and connected at every 13th point on the circumference as shown in Fig. 2. The truss T2 is assembled on this layer. The third layer of 18 cables consisting of seven 15.7mm diameter strands are passed over the top flanges of the trusses and tensioned to the designed force.

Truss T3 is erected on this layer as shown in Fig. 3. The top-most layer of cables consisting of nine 15.7mm diameter strands passes over all the trusses and is tensioned to put the whole system in tension as in Fig. 4. The cables in every layer are stressed to a designed value of force and sag. The offsets of the cable layers are dependent on the diameter of the trusses used. The heights of the trusses are designed to give the desired profile of the dome. The positioning of the truss on the cable layer is always at a distance from the inner edge of the cable layer so that the truss bears on no more than two thicknesses of cables. This gives the difference in heights of saddles at an intersection of not more than one cable diameter.

17. The prestressing strands are polyethylene coated and housed in H.D.P.E. tubes. The cables are housed in oversize ducts in concrete to allow for variations in the cable deflections of the structure in service and also to prevent crimping of the cable against the concrete at the beam edge.

ANALYSIS

18. The following assumptions were used in simplifying the analysis:
The cable system and the ring beam are uncoupled. The ring trusses have negligible bending and shear stiffness. The loads from the upper level cables are transferred to the lower level by struts separating the two layers.

Parameters used in design

19. Loading;

Imposed loading 0.6 kN/sq.m.
Local patch load 1.0 kN/sq.m.
Temperature range ± 25°C

Load Combinations

20. Main combination

$$\left.\begin{array}{l}\text{PAL } t = 0 \text{ or} \\ \text{PAL } t = \infty\end{array}\right\} + \text{ Imposed load or Wind load or Temporary load.}$$

Secondary combination

$$\left.\begin{array}{l}\text{PAL } t = 0 \\ \text{PAL } t = \infty\end{array}\right\} + \text{ Imposed load or Wind } + \text{ 0.5 Temporary load.}$$

Cable Stress limits

21. Combinations Main combination. 0.1 fpu < fp < 0.45 fpu.
Secondary combination. 0.1 fpu < fp < 0.56 fpu.

22. The structures being axi-symmetric the analysis was considerably simplified by being able to simulate the structure as a two dimensional model. Non-linear analysis of the structure to simulate the stages of construction were carried out. From the analysis the final force in cables in various layers and their corresponding deflections were obtained.

23. The determination of the stressing load for the cables and their sequence for each layer was much more elaborate and entailed iterative procedures to arrive at the forces.

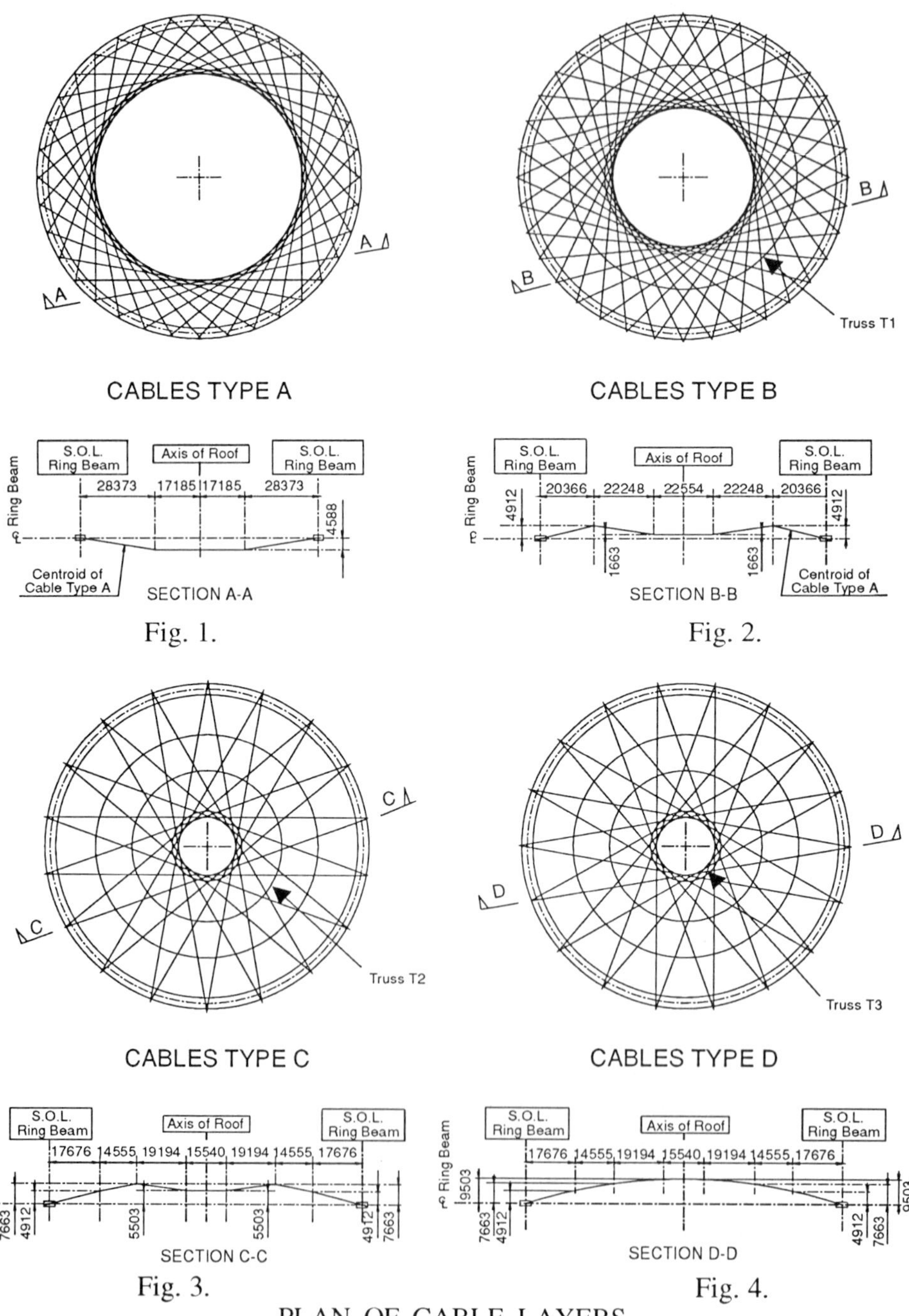

Fig. 1.

Fig. 2.

Fig. 3.

Fig. 4.

PLAN OF CABLE LAYERS

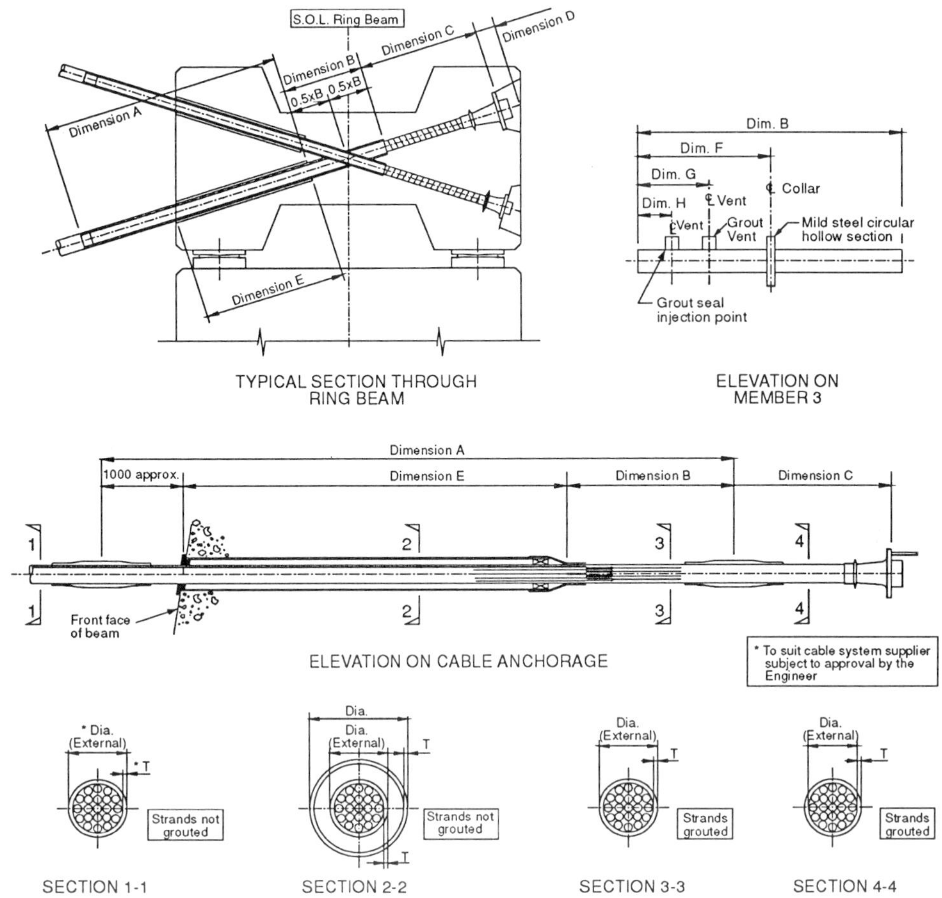

S.O.L. Ring Beam
Dimension A
Dimension B
0.5xB 0.5xB
Dimension C
Dimension D
Dimension E

TYPICAL SECTION THROUGH RING BEAM

Dim. B
Dim. F
Dim. G
Dim. H
℄ Vent
℄ Vent
Grout Vent
℄ Collar
Mild steel circular hollow section
Grout seal injection point

ELEVATION ON MEMBER 3

Dimension A
1000 approx.
Dimension E
Dimension B
Dimension C
Front face of beam
* To suit cable system supplier subject to approval by the Engineer

ELEVATION ON CABLE ANCHORAGE

* Dia. (External)
* T
Strands not grouted

SECTION 1-1

Dia.
Dia. (External)
T
Strands not grouted

SECTION 2-2

Dia. (External)
T
Strands grouted

SECTION 3-3

Dia. (External)
T
Strands grouted

SECTION 4-4

ANCHORAGE DETAILS

Construction phase of the small roof showing weights hung to stress top cables.

View of completed dome.

View of the large dome during construction before the installation of cladding.

CONSTRUCTION

24. The construction of the dome starts with the concrete ring beam which in all the roofs is supported on bearings on columns. The anchorages are cast in the beam on the cable alignments. The ring beam being always in compression is designed as a reinforced concrete member.

25. The first layer of cables is installed, one by one, to the required dip and the calculated force in the tendon for the condition of no load on the cables. The first truss is erected on this layer of cables and the tension in the cables checked to be the same value in all of them.

26. The second layer of cable is draped over the truss and anchored to the ring beam to give the correct dip consistent with the force in the cable. This is achieved through iterative stressing of the cables to obtain the same force in all them in the second layer with the same dip.

27. The second truss is erected on this layer and the cable force in the first and second layer checked to give equal forces in the same layers. This operation is repeated with the third layer and the third truss erected thereon.

28. The fourth layer of cables is passed over all the trusses and tensioned across the ring beam to the required design force. This is again subject to the iterative procedure to equalise the forces in all cables in any one layer.

29. The rafters and purlins are added and the aluminium cladding installed.

Construction of the Small Dome

30. The construction of this dome was considered as a prototype for the execution of the other more involved types to follow. The sequence of construction was as follows.

31. The concrete ring beam was cast with all the prestressing anchorage accurately assembled. The lower layer consisted of six 15.7mm diameter strands and the upper layer consisted of two 15.7mm diameter strands

32. After the concrete in the beam had gained strength, H.D.P.E pipes cut to the correct lengths were hung in place with the aid of a tower crane. The strands were pulled through the pipes. After accurate positioning of each cable in the pipes they were fixed by the wedges in the prestressing anchorages and tensioned by jacks to give the designed sag at the middle of each cable.

33. The truss T1 was erected in 9 segments and the prestressing force in the cables adjusted to give the calculated deflection with the conforming force. The truss T2 was assembled next and the forces and sag adjusted to conform to the design values. The readings at this stage agreed with the calculated values.

34. The top layer of cables was assembled as before and located at their anchorages. Instead of applying sequential stressing, calculated weights were hung from the trusses T1 and T2 to give the trusses the required downward displacement as when fully stressed by the top cables. This uniform load was suspended from 36 points around the circumference of the trusses. The cables in layer B were then stressed to a nominal load to take out the slack.

35. The suspended loads were then progressively released and tension in the top cables checked to conform to the design values.

36. The final checking of the loads in the cables confirmed good agreement with the design values.

37. The steel purlins were erected on the trusses and the aluminium cladding with the glass wool insulation were installed.

38. The epoxy resin plugs were first formed in the pipes cast in the ring beam, to prevent the grout flowing into the cables. The anchorages were grouted with anti-shrink cement grout and the construction completed.

Construction of the Large Dome

39. The construction of this dome could not follow exactly the same procedure as the smaller one because the suspended load weights were too large to adopt. Instead the cables had to be stressed sequentially, and to progressively decreasing load, so that at the end of the operation all the cables would have the same value of force in any one layer.

40. Steel shuttering was erected for the ring beam to help in placing the ends of all the cables accurately through the ring beam and the ring beam was concreted, supported on 36 bearing points.

41. A special truss was used to erect the layer A cables. After all the cables were in position they were checked for sag and tension. A check was carried out using multi-strand jacks to make sure that all the strands were stressed to the same value. A final check using a multi-strand jack to obtain the exact force coincident with the sag, was carried out to a tolerance of ± 3%. The first truss was assembled on this layer in six days.

42. The force in the cables at this stage had to be adjusted to bring it down to the design value. Deflection measurements were taken on the bottom of the truss at control positions to make sure that the sag was uniform The second layer of cables were installed and the cables stressed in sequence, to obtain the resultant force in all cables. The truss deflections were checked and random checks carried out on layer A cables. The process was repeated with trusses T2, T3 and layer C and D cables. After each operation, random checks were made on all cables installed, to make sure that they had approximately the design forces in them.

43. The steel purlins, cladding installation and grouting of the anchorages followed as per the smaller dome construction.

44. The entire construction of the main hall, from commencement of laying cable layer A to completion of the cladding was accomplished in 43 days.

LARGE DOME - VALUES OF FORCE-DEFLECTION FOR CABLES.

PHASE	LEVEL MEASURED FROM ℄ OF BEAM						CABLE FORCES (KA)			
	BOTTOM OF T1	BOTTOM OF T2	BOTTOM OF T3	MID. POINT OF CABLE A	MID. POINT OF CABLE B	MID. POINT OF CABLE C	CABLE A	CABLE B	CABLE C	CABLE D
ERECT / STRESS CABLES A	-	-	-	-3413	-	-	94	-	-	-
ERECT TRUSS T1	-3178	-	-	-3378	-	-	233	-	-	-
ERECT / STREE CABLES B	-3958	-	-	-3978	+5473	-	1792	818	-	-
ERECT TRUSS T2	-3978	+5143	-	-3998	+5133	-	1883	807	-	-
ERECT STRESS CABLES C	-4378	+2443	-	-4398	+2433	+8413	2828	1040	960	-
ERECT TRUSS T3	-4388	+2423	+8283	-4398	+2423	+8283	2836	1043	949	-
STRESS CABLES D	-4563	+1745	+5693	-4576	+1739	+5683	3300	1150	782	737
ERECT RAFTER + CLADDING	-4588	+1663	+5503	-4606	+1653	+5503	3378	1154	754	610

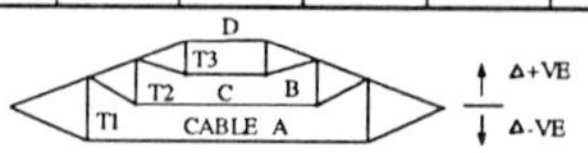

D
T3
T2 C B
T1 CABLE A
Δ+VE
Δ-VE

42. Experimental study on the ultimate strength of single-layer reticular domes

R. SHIBATA, S. KATO and S. YAMADA, Toyohashi University of Technology, Japan, and T. UEKI, Tomoe Corporation, Japan

SYNOPSIS. The present paper describes the experimental investigation into the effects of the joint properties on the ultimate strength of single layer reticular domes. In the present tests, the domes are loaded into the post buckling region by controlling displacement at the central node. The experimental results are compared with analytical ones obtained from the finite element method, considering the semi-rigidity of the joint system as an elastic plastic spring. According to the results obtained experimentally and analytically, the present study has confirmed that the semi-rigidity of the joint system remarkably affects the ultimate strength.

INTRODUCTION

1. Single layer reticular domes generally consist of straight members jointed with screw connectors to a nodal rigid ball. The characteristics of these joint systems have remarkable effects on both the load carrying capacity and the buckling behaviour. If the joint system cannot provide sufficient bending rigidity, the dome has a great possibility of local buckling at a lower critical load as compared with the rigidly jointed reticular domes. Several experimental and analytical works have already been reported by researchers (refs 1-3) on the properties of the joint system. However, it is not enough to understand about the ultimate strength and the post buckling behaviour of single layer reticular domes through experiments using actual construction members.

2. In the present experiments, a semi-rigid joint system is considered. First, bending tests of the joint between members are performed to confirm the bending properties and bending capacity. Secondly, loading tests of the unit reticular dome as a part of a large dome, are carried out to obtain the load carrying capacity. Three types of joint system are used in order to compare their influence on the responses. Eight domes are tested, considering four parameters as follows: (a) length of member ℓ_0, (b) subtended half angle θ_0, (c) diameter of member d_0 and (d) joint type.

3. The load and displacements at the crown of the dome are measured. The ultimate strength and the hysteresis characteristics of the specimens are discussed

by comparing their responses. The stiffness of the elastic plastic springs connected to both ends of the member is investigated using the results of experimental tests.

THE PROPERTIES OF THE THREE TYPES OF JOINT SYSTEM

4. In order to establish the bending rigidity for an estimation of the ultimate strength of the test domes, bending tests are carried out with jointed, simply supported beams.

Joint type

5. Three types of the joint system used in the test domes are shown in Fig. 1. The configurations of these joint systems are described as follows.

(a) S14 type:- The nose cone is in contact with the ball surface occupying a relatively large metal contact area. Thus, this joint system is assumed to have the largest bending rigidity of the three types.

(b) D14 type:- In this joint system, the nose cone is connected to the ball with the connector, being not in contact between the nose cone and the ball.

(c) D06 type:- This joint system is similar to the D14 type except that a smaller connector and a slender steel pipe are used.

6. The element size and material properties for the three types of the joint system, are shown in Table 1.

Bending test of the joint system

7. Fig. 2 shows the experimental set-up and the loading device. Each tested piece is a simple beam consisting of two steel pipes connected to a rigid ball with either of the three types. Two weights are hung near both ends of the beam to cancel the initial bending moment, caused by the dead load, at the central joint. The load is applied on the top of the ball using a hydraulic jack. The vertical displacement of the ball is measured by two displacement gauges. From measurement, the relations between the bending moment and the rotation angle are obtained and plotted in Fig. 3. The bending moment is calculated at the centre of the simple beam subjected to a concentrated centric load. The rotation angle is calculated by subtracting the elastic deformation of the steel pipe from the measured deformation.

8. In the case of the S14 type, the bending moment is linearly proportional to the rotation angle at the initial loading stage because the semi-rigid joint and the bending stress, 4.3*tf* in the connector, occurring between the ball and the nose cone. In contrast, in the D14 and D06 types the rotation angle grows without any bending resistance moment due to the looseness of the screw. The bending rigidity K_B and yield bending moment M_y of the three joint svstems are summarized in Table 2.

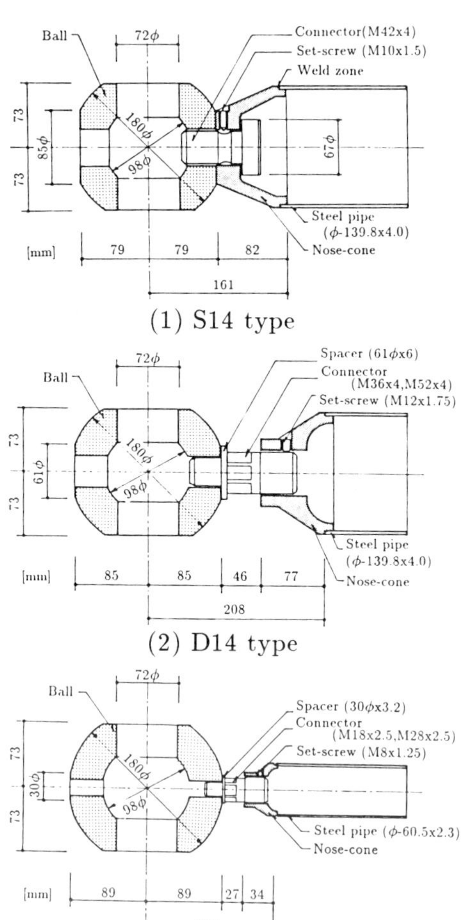

Fig. 1. Details of three joint systems

Table 1. Mechanical properties of elements in the joint system
Y.S. : yield stress
T.S. : tensile strength
El. : elongation

element	size	Y.S. $\left(\frac{tf}{cm^2}\right)$	T.S. $\left(\frac{tf}{cm^2}\right)$	El. (%)
ball	180φ	5.20	7.00	25
nose cone	(S14)	3.17	4.96	38
	(D14)	3.27	5.00	34
	(D06)	3.27	4.80	36
connector	M42x4	9.96	10.95	18
	M36x4	9.41	10.23	21
	M18x2.5	9.93	10.63	21
steel pipe	φ-139.8x4.0	3.31	4.10	36
	φ-60.5x2.3	4.39	5.14	51

Table 2. Results of bending test
K_B : bending stiffness
M_y : yield bending moment
ϕ_s : looseness of joint
ϕ_y : rotation angle at yielding

joint system	K_B $(tf{\cdot}cm/rad)$	M_y $(tf{\cdot}cm)$	ϕ_s (rad)	ϕ_y (rad)
S14	20000	200	0.0000	0.0100
D14	3800	100	0.0025	0.0288
D06	580	13	0.0108	0.0322

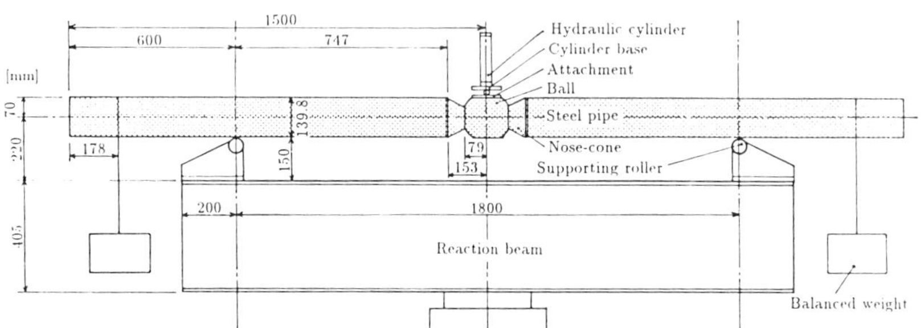

Fig. 2. Schematic bending test

LOADING TEST OF UNIT RETICULAR DOMES

Tested specimens and loading device

9. Fig. 4 shows the experimental set-up and the loading device. Each specimen is a single layer unit dome on a hexagonal plan. The parameters considered are the length of the member ℓ_0, subtended half angle θ_0 and slenderness ratio of the member λ_0, summarised in Table 3. The number of each specimen indicates, in the order of letters, the joint system, length of the member and subtended half angle, respectively. In each dome, the same size of member and the same joint system are used, which are arranged as illustrated in Fig. 4. The dome has seven joints and twelve members; the node number 4, at the crown moves freely, the nodes numbers 3, 5 are roller supported in the X direction, the nodes numbers 1, 2, 6, 7 are roller supported in both the X and Y directions. The detail of the support is illustrated in Fig. 5.

10. The vertical load is applied to the ball with node number 4 (crown) by using a hydraulic jack and spherical support. The load level is measured by a load cell and implemented by controlling the displacement at the crown.

Testing procedure and Measuring methods

11. The loading is programmed as follows:

(a) Loads are cyclically applied within the elastic range.

(b) After cyclic loading, loading is continued until the displacement at crown passes over the rise of the dome.

12. The strains of the steel pipes are measured by a pair of gauges or a set of four gauges in the specified section of steel pipes, and their positions are indicated in Fig. 4. Axial force and bending moment are calculated using the values obtained from the gauges.

RESULTS AND COMMENTS ABOUT LOADING TESTS ON DOMES

13. From the present experiments, summaries are made on the following characteristics:

(a) ultimate strength of the dome;

(b) load-displacement relationship;

(c) axial force and bending moment of members.

Numerical analyses are also performed to be compared with the experimental results. The analysis procedure is based on the finite element method with the Newton-Raphson procedure considering geometrical and material non-linearities. Since the bending rigidity and capacity of the springs at the ends of members give a great deal of effects to responses, the values K_B and M_y for the bending springs are assumed based on the above mentioned bending tests. The procedure to trace the equilibrium path of the post buckling region is abbreviated in the present paper owing to the fact that the procedure is well known and reported in ref. 4.

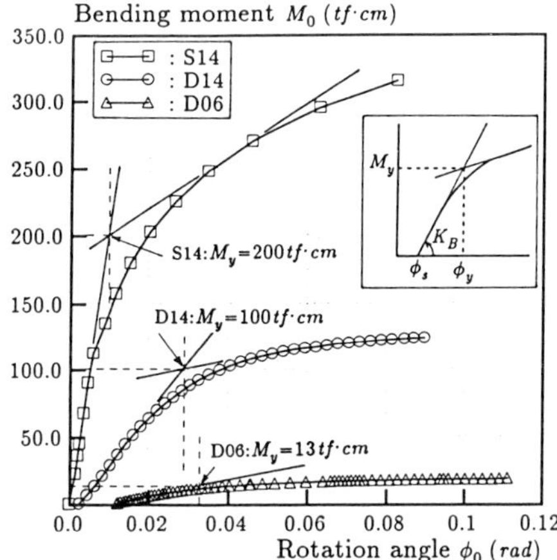

Fig. 3. Initial rigidity of three joint systems

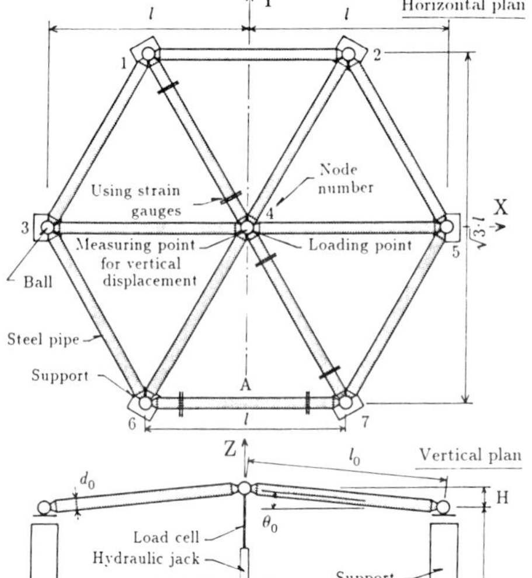

Fig. 4. Schematic loading test

Table 3. Configuration of eight tested domes

l : length of member A
θ_0 : subtended half angle
H : rise of dome
λ_0 : slenderness ratio of member A

specimen's name	l (cm)	θ_0 $(degree)$	H (cm)	λ_0 (cm)
S14-30-1	300.0	0.80	4.2	62
S14-30-3	300.0	2.88	15.1	62
D14-30-1	300.0	0.86	4.5	62
D14-30-3	300.0	2.88	15.1	62
D06-30-1	300.0	0.84	4.4	146
D06-30-3	300.0	2.88	15.1	146
D06-15-1	150.0	0.84	2.2	73
D06-15-3	150.0	2.82	7.4	73

Table 4. Results of loading test

P_{max} : ultimate strength
δ_{zp} : vertical dispacement at ultimate strength
H : rise of dome

specimen's name	P_{max} (tf)	δ_{zp} (cm)	H (cm)	δ_{zp}/H
S14-30-1	-	-	4.2	-
S14-30-3	5.91	9.32	15.1	0.62
D14-30-1	-	-	4.5	-
D14-30-3	3.10	9.01	15.1	0.60
D06-30-1	0.03	2.00	4.4	0.45
D06-30-3	0.68	6.25	15.1	0.41
D06-15-1	-	-	2.2	-
D06-15-3	0.89	4.20	7.4	0.56

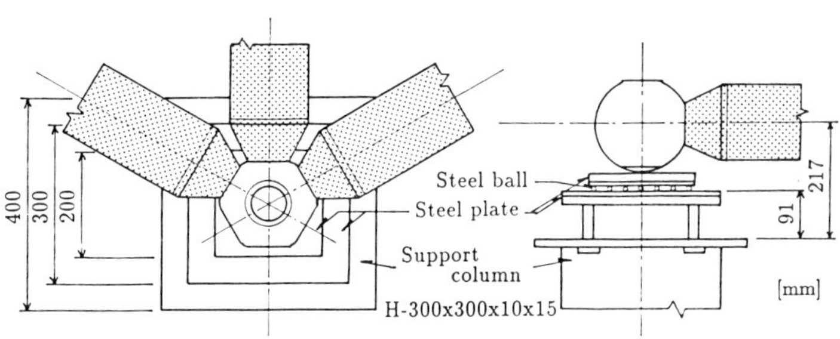

Fig. 5. Details of support

Ultimate strength and displacement

14. The ultimate strength and displacements of the test domes are summarised in Table 4. For the shallow domes (S14-30-1, D14-30-1, D06-15-1), the ultimate strengths are not obtained from the tests, because these domes could sustain loads after inverting.

15. The influence of the joint system on the ultimate strength. As compared between S14-30-3 and D14-30-3, the S14 type has an ultimate strength twice that of the D14 type. Thus, it is confirmed that the properties of the joint system severely affects the ultimate strength of the dome.

16. The influence of the subtended half angle (rise) on the ultimate strength. As for D06-30-3 compared with D06-30-1, the rise provides a remarkable improvement in the load carrying capacity.

Load displacement relation

17. The measured load-displacement relations are plotted using squared dots in Fig. 6. In this figure, three solid lines denote the results obtained by the numerical analysis, and the vertical broken line indicates the height of each dome.

18. In consequence of adopting the bending rigidity based on the experiments, most of the numerical results are in agreement with the experimental results.

19. Influence of the joint system on the equilibrium path. As compared between S14-30-3 and D14-30-3, the initial rigidity of the S14 dome types is larger than that of the D14 type. The bending rigidity of the joint system considerably effects the load carrying capacity of the dome because the ultimate load of S14-30-3 is almost twice that of D14-30-3. In the hysteresis loop of dome S14-30-3, residual displacements are observed under loading exceeding a limit point.

20. Influence of the subtended half angle (rise) on the equilibrium path. In the equilibrium path of dome D06-30-1, snap through is observed under a small loading level. So, it is confirmed that the ultimate strength of the dome is severely influenced by both the rise (subtended half angle θ_0) and the bending rigidity. In the equilibrium path of D06-30-3, elastic buckling of a member is observed after a limit point, as shown in Fig. 6 (D06-30-3). Similar behaviour to the member buckling is also found in the numerical results. In the case of D06-30-3, the dome markedly loses stiffness in its post-buckled state; consequently the load carrying capacity decreases more suddenly compared with the other domes.

Distribution of stresses of members

21. Under half the maximum load, the axial forces and bending moments, obtained experimentally and analytically, are illustrated in Fig. 7. In all cases, both the measured axial forces and bending moments are judged to be similar to the analysed results, but with some errors. The numerical procedure may simulate the stress distributions.

22. Influence of the joint system on the stresses of members. Examining the ratio of the bending moment to the axial force, S14-30-3 tends to be subjected to larger bending moments than D14-30-3. The reason is that the rigidity of the bending springs contribute to the load carrying mechanism in reticular domes.

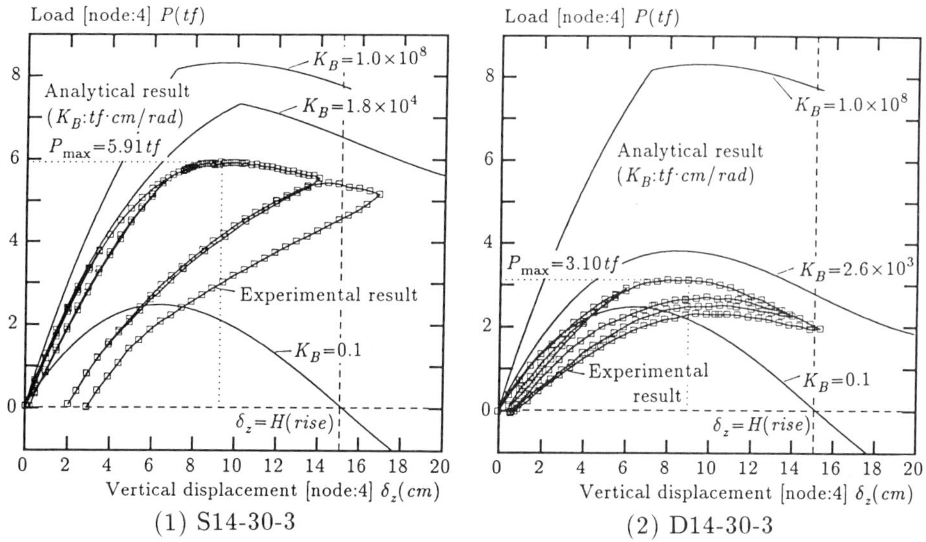

Influence of the joint system

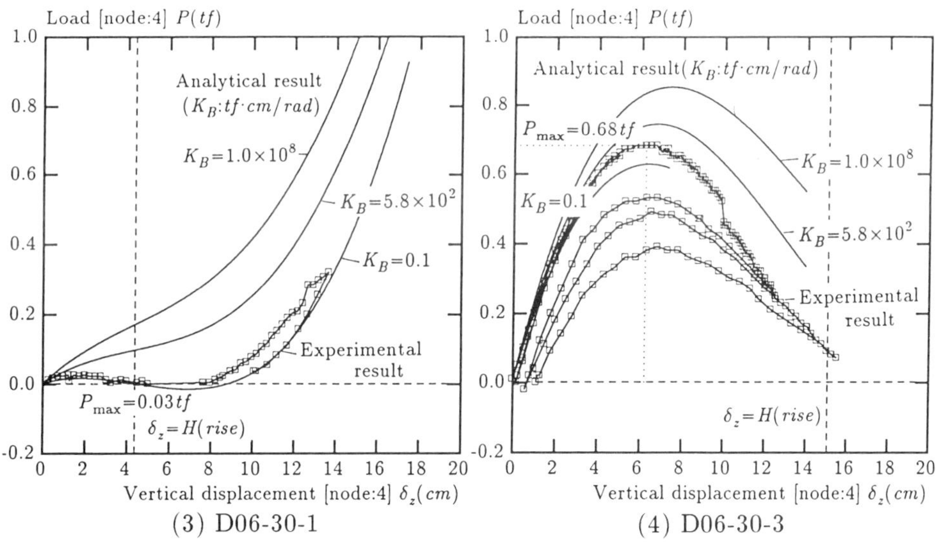

Influence of the subtended half angle θ_0

Fig. 6. Relations between load P and Displacement δ_z at crown

CONCLUSIONS

23. According to the results obtained, both from experiments and analyses, the conclusions drawn are as follows:

(a) It is confirmed that the joint system plays an important role on the ultimate strength of the domes. With increment in the bending rigidity, the domes obtain a large ultimate strength. In the cases of the D14 and D06 types, the looseness of connectors causes a considerable decrease in the ultimate strength of shallow domes ($\theta_0 = 1°$).

(b) For domes consisting of slender members such as type of D0630-3 with a relatively high rise of $\theta_0 = 3°$, a large decrement in the load carrying capacity is observed in the post buckling equilibrium path.

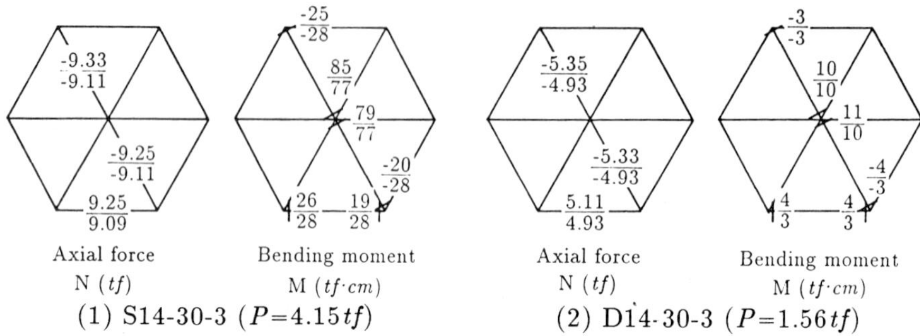

(1) S14-30-3 (P=4.15tf) (2) D14-30-3 (P=1.56tf)

Influence of the joint system

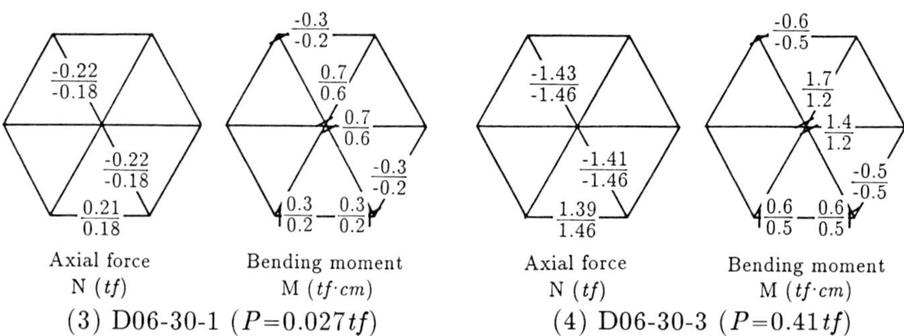

(3) D06-30-1 (P=0.027tf) (4) D06-30-3 (P=0.41tf)

Influence of the subtended half angle θ_0

Fig. 7. Distribution of axial force N and bending moment M: $\frac{Experimental-result}{Analytical-result}$

24. Through the present loading tests, it is clear that the influence of the bending rigidity on the buckling behaviour is noticeable. However, studies are still needed on both the buckling analysis and loading test of actual scale specimens, to establishing an estimation procedure (ref. 5) for the ultimate strength of single layer reticular domes.

REFERENCES

1. UEKI T., MATUMOTO K., KUBODERA I., MATSUSHITA F. and KATO S. Experiment and realisation of single layer space structure. Proceeding of 4th International Conference on STEEL STRUCTURES & SPACE FRAMES, 1990, February, 223-230.

2. SUZUKI T., OGAWA T., KUBODERA I. and IKARASHI K. Experimental and theoretical study of a single layer reticulated dome. Proceedings of the IASS symposium, Volume III, 1991, September, 85-92.

3. MEEK J.L. and LOGANATHAN S. Geometric and material non-linear behaviour of structures comprising circular and rectangular hollow sections. Proceedings of the IASS symposium, Volume 4, 1989, September, 223-244.

4. UEKI T., KATO S., KUBODERA I. and MUKAIYAMA Y. Study on the elastic and elasto-plastic buckling behaviour of single layered domes composed of members having axial and bending springs at both ends. Proceedings of the IASS symposium, Volume III, 1991, September, 93-100.

5. KATO S., SHIBATA R., TAKASHIMA H. and UEKI T. A computational procedure to evaluate buckling strength of reticular domes based on a concept of column buckling. Proceedings of the IASS symposium, Volume 2, 1992, July, 552-563

43. Experiment and numerical analysis of the buckling of a single-layer latticed dome

M. FUJIMOTO and T. SAKA, Osaka City University, Japan, and
K. IMAI and T. MORITA, Kawatetsu Steel Products Corporation, Japan

SYNOPSIS. This paper reports an experiment on the buckling of a shallow single-layer latticed dome under nodal gravity loading at the crown. The effect of a small increase of rise of the latticed dome on the buckling behaviour is examined. The discrete treatment method presented in this paper is used to calculate the load-deflection relationship of the dome. It is confirmed that the small increase of rise increases the buckling load. The validity of the method is shown by the comparison of numerical and experimental results for the latticed dome undergoing snap-through buckling.

INTRODUCTION

1. Several experimental and numerical studies report the buckling behaviour and design of single-layer latticed domes (refs 1-5). These studies focus on the mesh type, the rise-to-span ratio, the division number, the member slenderness ratio, the half open angle subtended by member, the initial imperfections, and the load distribution pattern. However, little work is reported on single-layer latticed domes composed of a standardised truss system.

2. In this paper, single-layer latticed domes composed of a standardised truss system are investigated. Experiments on the buckling of a single-layer latticed dome under nodal gravity loading at the crown are reported. Two latticed domes are adopted: Dome-I - all nodes are expected to be situated on a spherical surface; and Dome-II - the length of members connected with the crown of Dome-I are slightly increased and the rise of the dome is slightly greater than that for Dome-I. The effect of small increase in the rise of the dome on the buckling behaviour is examined (refs 6,7).

3. The discrete treatment method, based on the geometrical and material non-linearity, is presented. The characteristic of this method is that joints of the dome are modelled as rigid zones with rotational springs. The arc length method, a numerical technique, is used to trace the equilibrium path, and the skyline method is applied to the solution of the linear simultaneous equations.

4. For Dome-I and Dome-II, the relation of load to nodal gravity displacement, the relation of load to member axial strain, the elasto-plastic buckling load, and

the buckling mode are calculated by the present method. Results of numerical analysis are compared with those of experiment. The validity of the discrete treatment method, presented in this paper, is discussed in the calculation of elasto-plastic behaviour of a shallow single-layer latticed dome subjected to snap through buckling.

EXPERIMENT ON THE LATTICED DOME

Test Dome

5. The elevation of the dome and the overall configuration of the dome are shown in Figs 1 and 2, respectively. The parallel lamella dome is used as the mesh type. For Dome-I all nodes are expected to be on a spherical surface, and for Dome-II the rise of the dome is about 12 mm greater than that of Dome-I. The member slenderness ratio in the meridian direction from the crown of dome, λ, the half open angle subtended by that member, ϕ, and the division number of dome, n, are used to determine the dimension of the dome. In this study, $\lambda=100$, $\phi=3^0$, and n=3 are adopted.

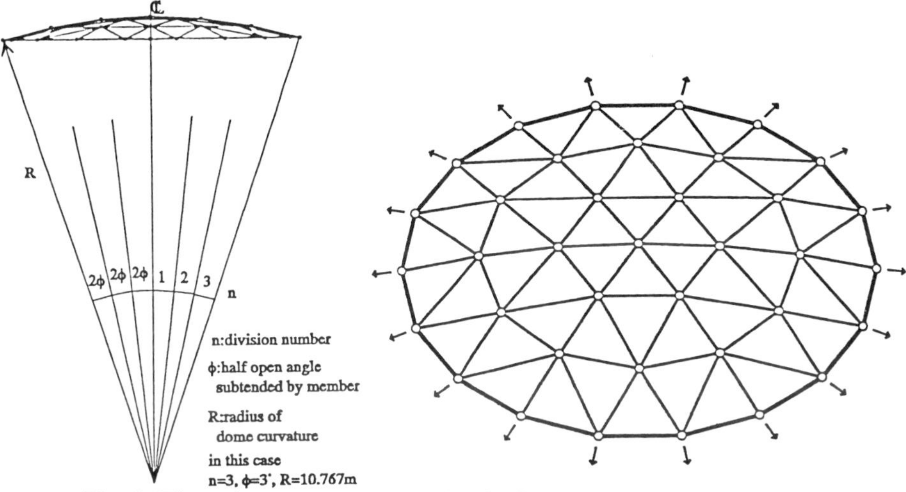

Fig. 1. Elevation.

Fig. 2. Overall configuration.

6. Circular hollow tubes are used as constituent members of the dome. The diameter of the members in the peripheral ring is 60.5 mm, and that of all other members is 34 mm. The sectional properties of the members are shown in Table 1. The slenderness ratio for members connected with the crown of the dome is about 100, the length of other members is determined. The material properties of members are shown in Table 2. The joint used, the KT truss system, is composed of a solid spherical node and joint assembly (ref. 8). The material properties of the joint are shown in Table 3.

7. The support condition of the dome is a one way roller in the meridian direction. The meridian direction is shown in Fig. 2. The dome is placed on the supporting unit via the one way roller set-up.

Method of loading

8. The control parameter of the experiment is the gravity displacement of the crown of the dome. A static gravity load is applied to the crown of the dome. The loading arrangement consists of a hydraulic jack, load cell, loading set-up, and PC bar. The loading arrangement is shown in Fig. 3.

Table 1. Sectional properties of member.

ϕ (mm)	t(mm)	A(cm^2)	Z_P(cm^3)	I(cm^4)
34	2.0	2.01	2.05	2.58
60.5	3.2	5.76	10.5	23.7

Table 2. Material properties of member.

	(KN/cm^2)
Elastic modulus	2.07x10^4
Yield stress	39.0
Tensile strength	42.6

Table 3. Material properties of joint.

	(KN·cm)
Coef. rotational spring	9.34x10^3
Maximum bending moment	71.1

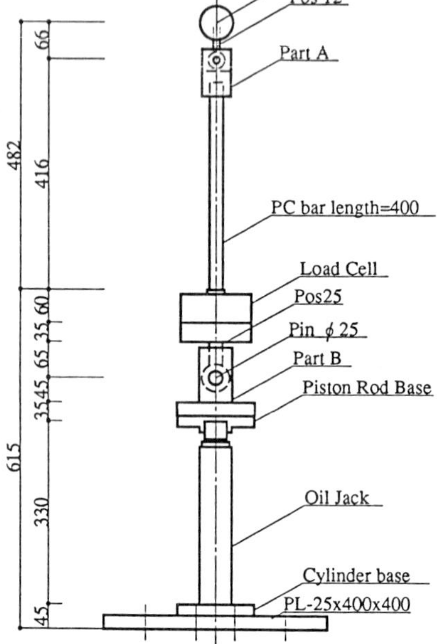

Fig. 3. Loading arrangement.

Method of measurement

9. The gravity load at the crown of dome is measured by the load cell attached to the head of oil jack. The gravity displacement of the crown of the dome and that of the node in first ring and second ring are measured by the linear transducer (stroke 100mm) and by the dial gauge type linear transducer (stroke 50mm), respectively. The horizontal displacement of the node in the peripheral ring is measured by the linear transducer (stroke 2mm). Strain gauges are used to measure the axial strains and the bending strains. The bending strain of the member connected with the crown node of the dome is measured at the inner end of the member by strain gauges attached to the top and bottom. The axial strain of the member is measured at middle point of the member, gauges are attached to the both sides of the member. Items and nodes of measurement are shown in Fig. 4.

10. The gravity load and the gravity displacement of the crown of the dome are measured by a dynamic strain meter. Other nodal displacements and the strain of members are measured using a static strain meter.

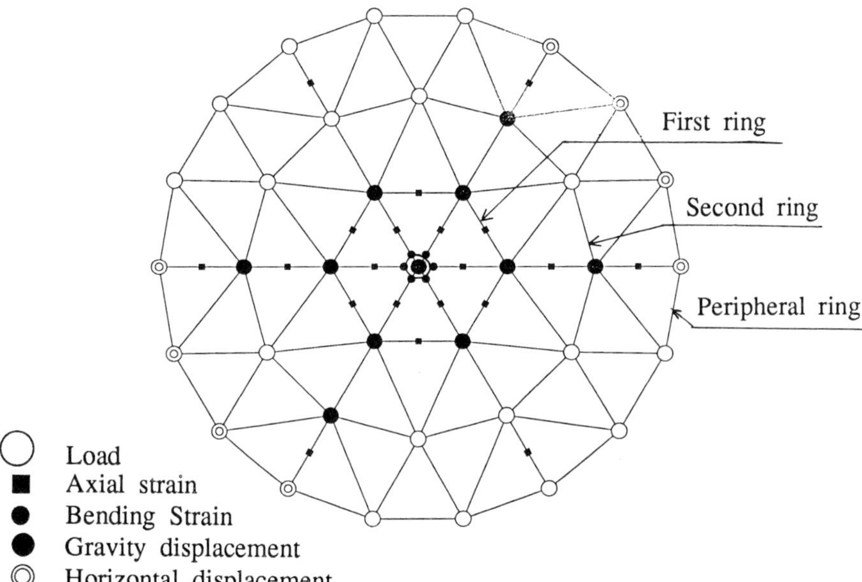

Fig. 4. Items and nodes of measurement.

NUMERICAL ANALYSIS METHOD BASED ON DISCRETE TREATMENT

Method of analysis

11. The discrete treatment based on the geometrical and material non-linearity is used to trace the buckling behaviour of a single-layer latticed dome. To consider both the size and the rigidity in the joint, the ball joint and the junction between the tubular member and the ball joint are modelled as a rigid zone and a rotational spring, respectively (ref. 9). The member model is shown in Fig. 5.

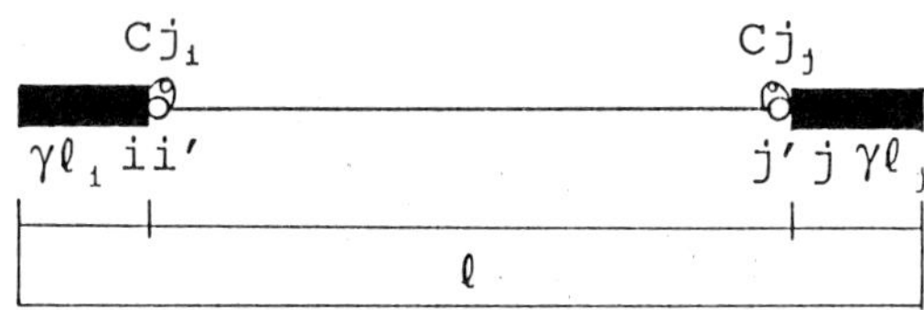

Fig. 5. Member model.

12. The finite displacement of the node, the change in flexural rigidity with axial force, and the change in member end distance with bending deformation of member, the so-called bowing effect, are treated as the geometrical non-linearity. The tangent stiffness matrix (refs 10,11), based on the slope-deflection method using stability functions, is used as the incremental elastic stiffness matrix for the

tubular members. The incremental elasto-plastic stiffness matrix for the tubular members is determined by the generalised hardening hinge method (ref. 12).

13. To consider both the compound non-linearity and the effect of the joint simultaneously, the member model (Fig. 5) and the incremental elasto-plastic stiffness matrix (ref. 13) are adopted in this study.

14. To make the numerical calculation easy, Eq. 1 is used as the yield function of the tubular members.

$$F = \left(\frac{N}{N_p}\right)^2 + \left(\frac{M_y}{M_{yp}}\right)^2 + \left(\frac{M_z}{M_{zp}}\right)^2 - 1, \quad F = 0 \tag{1}$$

in which, N is the axial force, M_y and M_z are the bending moments in the tubular member about the y-axis and z-axis, respectively and N_p is the yield axial force, M_{yp} and M_{zp} are the standard bending moments to judge the plastification of the tubular member.

15. For rotational spring, the relation of moment to the rotational angle is assumed to be bilinear. The plastification of the rotational spring is judged from the interaction between the bending moment about the y-axis or z-axis and the axial force in rotational spring. The yield function of the rotational spring is shown in Eq. 2.

$$F = \left(\frac{N}{N_{ps}}\right)^2 + \left(\frac{M}{M_{ps}}\right) - 1, \quad F = 0 \tag{2}$$

where, N_{ps} is the yield axial force of the rotational spring and M_{ps} is the maximum bending moment about the y-axis or z-axis.

16. The arc length method is used to trace the equilibrium path (ref. 14). The skyline method is used for the numerical solution of the simultaneous equations (ref. 15).

Analytical modelling of the dome

17. For the nodal co-ordinates of the analysed latticed dome, the design value is used as the horizontal component, and the measured value, obtained from the test dome was used as the vertical component.

18. For the material and geometric properties of tubular members and joints, the values obtained from the experiment are adopted. Where, the fully plastic moment or the yield moment is used as the standard moment, and the ratio of the coefficient of second stiffness of rotational spring to that of first stiffness is 0.01 or 0.05. The yield axial force of the rotational spring is 49 kN.

19. For the latticed dome analysed, the applied load is assumed to be the nodal gravity load acting on the crown of the dome, and the support condition of nodes on the peripheral ring is assumed to be a one way roller in meridian direction.

RESULTS

20. The buckling loads obtained from the experiments and the numerical analyses are shown in Table 4. In this table and following figures, MP and MY mean full plastic moment and yield moment, used as the standard bending moment to judge the plastification of tubular members. The numbers 1 and 5 mean the percentage of second stiffness coefficient of rotational spring to the first stiffness coefficient.

Table 4. Buckling loads.

Dome	Experiment(KN)	Numerical analysis(KN)			
		MY1	MY5	MP1	MP5
Dome-I	8.27	9.18	9.18	9.18	9.18
Dome-II	9.80	10.19	10.40	10.19	10.40

21. To exhibit the effect of small growth of rise of single-layer latticed dome on the buckling behaviour, the relation of load to gravity displacement of the crown of the dome and that of the average gravity displacement of the node on the first ring, obtained by experiments are shown in Fig. 6(a) and Fig. 6(b), respectively. In these figures, the broken lines and the dashed-and-solid lines mean the experimental results of Dome-I, and those of Dome-II, respectively. The open squares and the closed squares denote the experimental results of Dome-I and those of Dome-II, respectively.

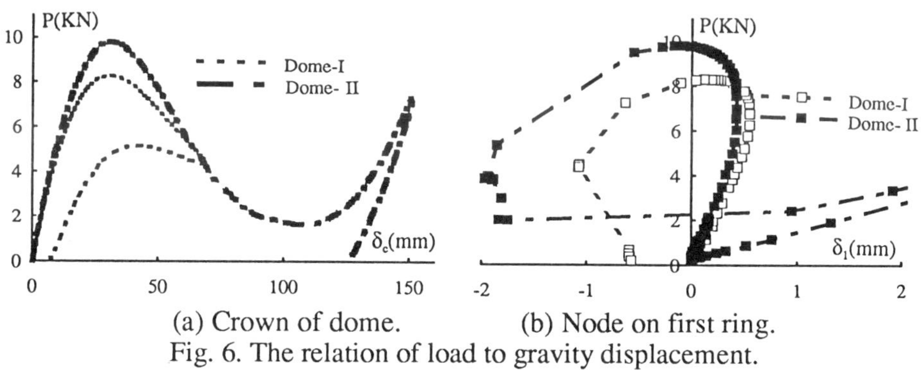

(a) Crown of dome. (b) Node on first ring.

Fig. 6. The relation of load to gravity displacement.

22. To compare the results of experiments with those of numerical analysis, the relation of load to gravity displacement of the crown of the dome is shown in Fig. 7. The relation of load to average gravity displacement of node in the first ring is shown in Fig. 8. The relation of load to average axial strain of the member connected with the crown of the dome is shown in Fig. 9. In these figures, the solid lines without marks indicate the results of the numerical analyses.

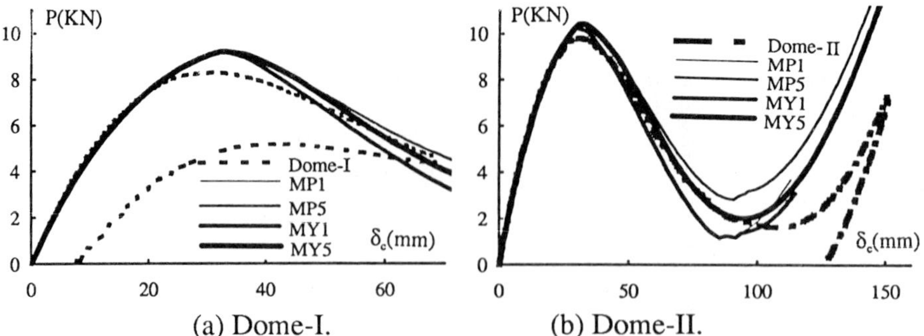

(a) Dome-I. (b) Dome-II.

Fig. 7. The relation of load to gravity displacement of crown of dome.

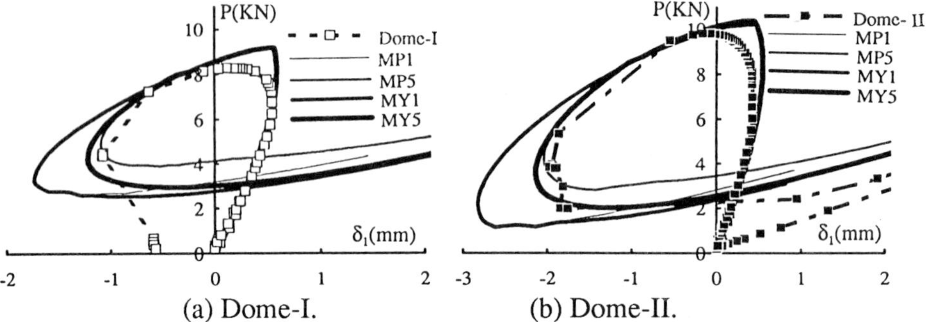

(a) Dome-I. (b) Dome-II.

Fig. 8. The relation of load to gravity displacement of node in first ring.

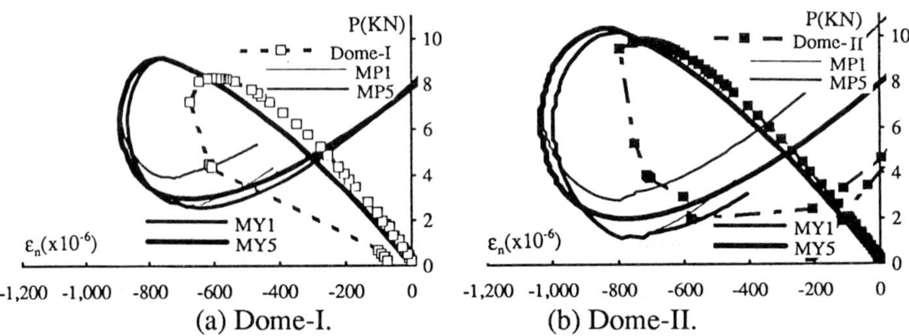

(a) Dome-I. (b) Dome-II.

Fig. 9. The relation of load to axial strain of member.

23. The final deformed configuration of the dome and the buckling mode obtained by the numerical analysis, presented by this paper, are shown in Photo. 1, and, Fig. 10, respectively.

Photo l. Final deformation of dome.

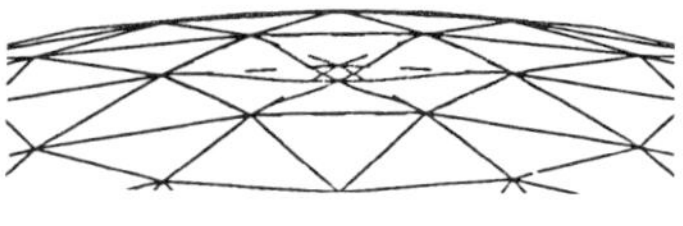

Fig. 10. Buckling mode of dome.

CONSIDERATION

Buckling load

24. For the buckling load obtained by experiment, the value of Dome-II is about 15% larger than that of Dome-I. Similarly, for the buckling load obtained by numerical analysis, the former is 10% to 13% larger than the latter. This shows that the small growth of rise of latticed dome increases the buckling load in this case. The effect of small growth of rise on the buckling load (refs 6,7) is confirmed experimentally and numerically.

25. The ratio of the buckling load of numerical analysis to that of experiment is about 1.1 for Dome-I and about 1.05 for Dome-II. This shows that the discrete method in this paper is applicable to the calculation of buckling load determined by snap-through of shallow latticed domes. These slight differences in buckling load between the experiments and numerical analyses result from the assumption that the relation of moment to rotational angle for the rotational spring is bilinear.

The relation of load to gravity displacement

26. The gravity displacement of the crown of the dome and the equilibrium paths of both domes show that the rigidity of the latticed dome decreases gradually, by geometrical non-linearity, as the applied load increases and the buckling point appears as the limit point of load. The equilibrium path denotes the snap-through phenomenon. For the equilibrium path in both the range of loading and that of post-buckling, the slope of Dome-II is larger than that of Dome-I. For the gravity displacement of the node on the first ring, the equilibrium paths of both domes show that the increment of the displacement decreases gradually as the applied load increases, the inflection point of the displacement appears; then, the limit point of load appears.

27. For the equilibrium path from the initial state to the range of snap-through, the results of the numerical analysis almost agree with those of the experiment. The buckling point of the latticed dome by the numerical analysis as well as that by experiment is the limit point of load. These show that the discrete treatment method is applicable to calculate the equilibrium path involving snap-through for the shallow single-layer latticed domes.

Axial strain

28. For the relation of load to axial strain obtained by the experiment, the result of Dome-I is in good agreement with that of Dome-II.

29. The axial strain before the buckling point obtained by the numerical analysis is in agreement with that by experiment. But, the axial strain after the buckling point by the numerical analysis is larger than that of the experiment.

The final deformation and the buckling load

30. The final deformation of each dome is the snap-through mode of the crown of the dome after the limit point appears. The buckling mode by the numerical analysis is in good agreement with the final displacement from the experiment.

CONCLUSIONS

31. In this paper, the shallow single-layer latticed dome composed of a standardised truss system was investigated. Experiments of the buckling of the dome under nodal gravity load at the crown were reported. The effect of small increase in the rise of dome on the buckling behaviour and the validity of the discrete treatment, presented in this paper, were examined. The following conclusions were drawn:

(a). As the result of the comparison of Dome-I and Dome-II for the buckling load, it was confirmed that the buckling load increased by the effect of small increase in the rise of dome on the buckling behaviour, as is shown in the numerical studies based on the geometrical non-linearity.

(b). Results of the numerical analyses obtained by the discrete treatment method were compared with those of experiments for the buckling behaviour of the dome. The validity of the discrete treatment method was shown by the comparison of the numerical results with the experimental results of shallow single-layer latticed dome subject to snap-through buckling.

ACKNOWLEDGEMENT

The authors wish to thank Mr. Honbo and Mr. Matsuura for their valuable assistance during the experimental work.

REFERENCES

1. ARCHITECTURAL INSTITUTE OF JAPAN. Stability of single-layer latticed dome (in Japanese), Edited by HEKI K., 1989.

2. MEEK J. L. and LOGANATHAN S. Theoretical and experimental investigation of a shallow geodesic dome, Int. J. Space Structures, 1989, Vol.4, No.2, 89-105.

3. MCCONNEL R. E., FATHELBAB F. A., and HATZIS D. The buckling behaviour of some single-layer, shallow lattice domes, Shells, Membranes and Space Frames, Proc. IASS Symp., Osaka, 1986, Vol.3, 97-104.

4. SUZUKI T., OGAWA T., KUBODERA I., and IKARASHI K. Experimental and theoretical study of a single-layer reticulated dome, Proc. Int. IASS Symp., Copenhagen, 1991, Vol.3, 85-92.

5. UEKI T., MUKAIYAMA Y., SHOMURA M., and KATO S. Loading test and elasto-plastic buckling analysis of a single-layer latticed dome (in Japanese), Journal of Struct. Constr. Engng. AIJ, 1991, No.421, Mar., 117-128.
6. MORITA T., IMAI K., YAMAOKA Y., and TAGAMI K. Advanced formian method considering the reasonability of structural mechanics (in Japanese), Summaries of Technical Papers of Annual Meeting of AIJ, 1991, 1271-1272.
7. KASHANI M. and CROLL J. G. A. Non-linear buckling response of spherical space domes, Int. Conf. IASS, Madrid Spain, 1989, Vol.5.
8. IMAI K., WAKIYAMA K., TSUJIOKA S., and YAMADA Y. Proposing a new joint system (KT-system) of space frame with threaded spherical nodes and its fatigue characteristics, Int. Conf. IASS, Madrid Spain, 1989, Vol.4.
9. SAKA T. and HEKI K. The effect of joints on the strength of space trusses, Proc. 3rd International Conference on Space Structures, Guildford, 1984, 417-422.
10. ORAN C. Tangent stiffness in space frame, Proc. of ASCE, 1973, Vol.99, No. ST6, 987-1001.
11. KASSIMALI A. Large deformation analysis of elastic plastic frames, Proc. of ASCE, 1983, Vol.109, No.8, 1869-1886.
12. INOUE K. and OGAWA K. A study on the plastic design of braced multi-storey steel frames Part 2 On the overall static and dynamic behaviours of plastically designed multi-storey braced frames (in Japanese), Trans. of AIJ, 1978, No.268, 87-98.
13. MURAKAMI M. Numerical analysis of elastic buckling of single-layer latticed domes under gravity load, IASS-CSCE International Congress 1992, Vol.2, 576-586.
14. RIKS E. An incremental approach to the solution of snapping and buckling problem, Int. J. Solids and Structures, 1978, Vol.15, 529-551.
15. BATHE K. and WILSON E. L. Eigen-solution of large structural systems with small bandwidth, Proc. of ASCE, 1973, Vol.99, No.EM3, 467-479.

44. Non-linear dynamic analysis of roof trusses connected by rigid, pinned and semi-rigid joints

S. L. CHAN and P. P. T. CHUI, Hong Kong Polytechnic, Hong Kong

SYNOPSIS. The dynamic non-linear behaviour of a 68.58m span roof truss composed of various types of joint details is studied using an incremental iterative numerical procedure embedded in the program, NAF-SR-D (Non-linear Analysis of Frames composed of Semi-rigid connections including the Dynamic effects). The importance of including the flexibility of connections in studying the response of the truss subjected to a sinusoidal applied load is investigated via various assumed joint stiffnesses.

INTRODUCTION

1. In many national design codes for space frames and steel trusses such as the Chinese national design code for space frames (ref. 1) and the British Standard 5950 (ref. 2), the effective length of a structural member in a truss is not necessarily equal to unity, implying that a certain rotational restraint exists at the connections. This leads to the inconsistency between the design and the analysis if the designer is not using a computer program allowing for the effect of semi-rigid connections.

2. In a previous paper by Chan and Yau (ref. 3), the static non-linear behaviour of trusses composed of various types of connections was studied, the pre- and post-buckling paths were plotted and the maximum loads were also located. It was concluded that the static capacity of a truss will be underestimated by some 18% if all joints are assumed pinned. However, if the top and bottom chords are assumed to be continuous and the web members are assumed to be pinned, the behaviour of the truss is very similar to the analysis based on the rigid joint assumption. The assumption for the connection stiffness of the joints between the web to chord members was noted to be unimportant in affecting the overall behaviour of the structure.

3. In this paper, the work is extended to the dynamic analysis of arch trusses subjected to a harmonic point load at the top. Three connection idealisations are considered, which are the rigid, pinned and semi-rigid connections. The transient response of the truss under these assumptions are plotted and compared.

4. Previous work on the non-linear transient response of space frames and trusses allowing for semi-rigid connection behaviour are limited. Most of the

previous studies on space frames are focused on the analysis of frames with idealised joint details such as the space truss by Peter and Noor (ref. 4), space frames by Remseth (ref. 5) and many others. Studies on space frames allowing for semi-rigid connection were conducted by Shi and Atluri (ref. 6) using the complementary energy principle. However, the work approach may prove difficult to incorporate into the commonly used displacement based analysis program. Further, the formulation of a consistent mass matrix has not been detailed.

5. In this paper, the simple but efficient displacement based computer method for non-linear analysis of space frames is applied to study the large deflection and dynamic response of a arch truss composed of various types of connections. Using the concept proposed in this paper and the one previously presented by Chan and Yau (ref. 3), a typical computer program for the analysis of space frames can be easily modified to account for non-linear effects and to study its dynamic response using a micro-computer.

FORMULATION

6. Various methods are available for the formulation of linear stiffness, geometric stiffness and mass matrices. However, an approach of adopting a displacement function allowing for springs at both ends is proposed in this paper for consistency and clarity. Once the displacement function is verified against its capacity of modelling the connection stiffness, the following analysis procedure becomes standard.

7. The transverse displacement along the axis passing through the two ended nodes of an element, v, can be expressed in terms of the two ended rotations, θ_1 and θ_2, by imposing the condition that the displacement at the two nodes are equal to zero and its derivative equal to the nodal rotations as,

$$v = \left(\rho_1^2 \rho_2 L \bar{\theta}_1 - \rho_1 \rho_2^2 L \bar{\theta}_2 + \rho_1 v_1 + \rho_2 v_2 \right) \quad (1)$$

8. Expressing the displacement in terms of the rotations allowing for the flexibility of joints or of the connection spring element (Fig. 1), gives:

$$v = \frac{[\rho_1^2 \rho_2 L \quad -\rho_1 \rho_2^1 L] \begin{bmatrix} r_2 + \frac{4EI}{L} & -\frac{2EI}{L} \\ -\frac{2EI}{L} & r_1 + \frac{4EI}{L} \end{bmatrix} \begin{bmatrix} r_1 & 0 \\ 0 & r_2 \end{bmatrix} \begin{bmatrix} 1 & \frac{1}{L} & 0 & \frac{1}{L} \\ 0 & -\frac{1}{L} & 1 & -\frac{1}{L} \end{bmatrix} \begin{bmatrix} \theta_1 \\ v_1 \\ \theta_2 \\ v_2 \end{bmatrix}}{(r_1 + \frac{4EI}{L})(r_2 + \frac{4EI}{L}) - 4\left(\frac{EI}{L}\right)^2}$$

$$+ \rho_1 v_1 + \rho_2 v_2 \quad (2)$$

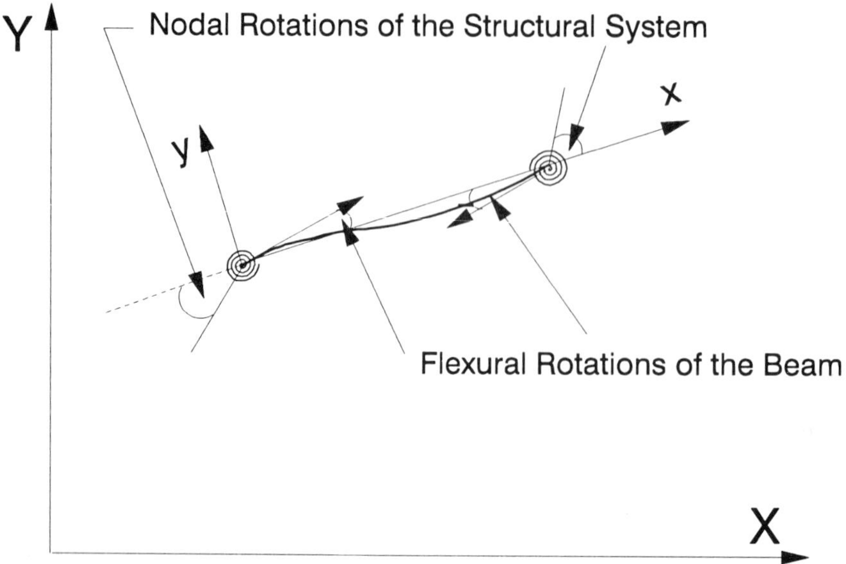

Fig. 1. Nodal rotations of a Hybrid Element.

9. The expression in Eq. 2 will convert to a linear function when the connection stiffness is equal to zero, which is the case for a truss element. It becomes a conventional beam element when the connection stiffness is indefinite. Note that the formulation is not limited to cubic Hermite function and a higher order element can be modified in the same procedure. Also, in the case for a truss element, the proposed approach can be used to consider the buckling of an individual element which is not possible for a conventional truss element having only axial stiffness. In actual computer coding, the spring stiffness is set to 1×10^{-10} of the flexural stiffness (EI/L) for the pin-jointed case and to $1 \times 10^{+10}$ of the flexural stiffness (EI/L) for the rigid jointed case.

FORMULATION OF THE LINEAR, MASS AND GEOMETRIC STIFFNESS MATRICES

10. Once the shape function is obtained, the flexural or bending linear, the mass and the geometric stiffness matrices can be obtained directly as,

$$[\,K_L\,] = EI \int_0^L \left(\frac{\partial^2 v}{\partial x^2}\right)\left(\frac{\partial^2 v}{\partial x^2}\right)^T dx \qquad (3)$$

$$[\,M\,] = \int_0^L \bar{m}\; v\; v^T dx \tag{4}$$

$$[\,K_G\,] = \frac{P}{2}\int_0^L \left(\frac{\partial v}{\partial x}\right)\left(\frac{\partial v}{\partial x}\right)^T dx \tag{5}$$

in which $[K_L]$ $[M]$ and $[K_G]$ are the linear stiffness, the consistent mass and the geometric stiffness matrices for an element with end springs. The term $\overline{m}$ is the mass per unit length and x is the distance along the element axis.

11. Alternative simpler formulations are available for the formation of the stiffness matrices. This includes the use of the lump mass and the addition of a matrix accounting for the work done due to an initial load and nodal displacements (ref. 7) as the geometric stiffness matrix.

NUMERICAL PROCEDURE FOR LARGE DEFLECTION TRANSIENT ANALYSIS

12. The present paper uses the Newmark numerical integration scheme to relate acceleration and velocity to displacements at time t and $t + \Delta t$. The constant-average-acceleration assumption is made to obtain the values of α and β, as 0.5 and 0.25 respectively. Thus, the incremental equilibrium equation can be written as:

$$\left[[K]_L + [K]_G + \frac{[M]}{\alpha(\Delta t)^2} + \frac{\beta[C]}{\alpha\Delta t}\right][\Delta u] = [\Delta F] + [M]\left[\frac{{}^t\dot{u}}{\alpha\Delta t} + \frac{{}^t\ddot{u}}{2\alpha}\right] +$$

$$[C]\left[\frac{\beta\;{}^t\dot{u}}{\alpha} + \frac{\Delta t}{2}\left(\frac{\beta}{\alpha}-2\right)\;{}^t\ddot{u}\right] \tag{6}$$

in which Δt is the time increment, Δu is the displacement increment and ΔF is the incremental force. Each super dot represents a differentiation with respect to the time.

13. Once the displacement increment in Eq. 6 is solved, it will be used to update the geometry of the structure by adding the displacement to the nodal co-ordinates, to update the acceleration and velocity as:

$$ {}^{t+\Delta t}[x] = {}^{t}[x] + [\Delta u] \qquad (8) $$

$$ {}^{t+\Delta t}[u] = {}^{t}[u] + [\Delta u] \qquad (10) $$

$$ {}^{t+\Delta t}[\ddot{u}] = \frac{1}{\alpha\ (\Delta t)^2}[\Delta u] - \frac{1}{\alpha\ \Delta t}\ {}^{t}[\dot{u}] - \left(\frac{1}{2\alpha} - 1\right)\ {}^{t}[\ddot{u}] \qquad (11) $$

$$ {}^{t+\Delta t}[\dot{u}] = {}^{t}[u] + \Delta t(1-\beta)\ {}^{t}[\ddot{u}] + \beta\,\Delta t\ {}^{t+\Delta t}[\ddot{u}] \qquad (12) $$

For the checking of equilibrium, the following equations can be adopted:

$$ [\Delta F] = [F_a]_{i+1} - [R]_i - [M][\ddot{u}]_i - [C][\dot{u}]_i \qquad (13) $$

$$ [\bar{K}][\Delta u] = [\Delta F] \qquad (14) $$

in which $[F_a]$ is the applied nodal force and [R] is the resistant force vector of the structure and can be computed in the procedure as in the static case (ref. 8). The symbol i represents the number of iterations in a particular time increment.

14. The residual displacements obtained from Eq. 14 are then used to update the geometry, displacement, velocity and acceleration as:

$$ {}^{t}[x]_{i+1} = {}^{t}[x]_i + [\Delta u]_i \qquad (15) $$

$$ {}^{t}[u]_{i+1} = {}^{t}[u]_i + [\Delta u] \qquad (16) $$

$$ {}^{t}[\dot{u}]_{i+1} = {}^{t}[\dot{u}]_i + \frac{\beta}{\alpha\,\Delta t}[\Delta u]_i \qquad (17) $$

$$ {}^{t}[\ddot{u}]_{i+1} = {}^{t}[\ddot{u}]_i + \frac{1}{\alpha(\Delta t)^2}[\Delta u]_i \qquad (18) $$

NUMERICAL EXAMPLES

15. The 68.58m span arch truss shown in Fig. 2 is analysed dynamically using the proposed procedure. Three cases are considered in total in order to investigate the importance of joint stiffness on the behaviour of the truss. The geometrical and material properties of the structure are also shown in Fig. 2. The material density is assumed to be 2700kg/m^3. The applied load is sinusoidal with amplitude of 20kN and angular velocity equal to the natural frequency of the truss using the linear assumption, which are equal to 5.124, 5.462 and 5.402 radian per second respectively for pinned, rigid and semi-rigid cases.

16. It can be seen in Fig. 3. that the response of the three types of trusses are smaller than the static case by Chan and Yau (ref. 3). It is expected since the mass of all three trusses are the same whilst the resistance against applied load is composed of its stiffness as well as the inertia force. The first maximum displacements are smaller when the truss stiffness is larger.

17. From the example, it can be seen that the effects of semi-rigid connections is mainly related to the stiffness of the structure as a whole. In the present example, the stiffness is dominant by the axial stiffness and the arch effect. Therefore, the effects of semi-rigid connections are minimal. It is expected that the argument may not be applicable to beam-column steel frames of which the investigation is being undertaken by the authors.

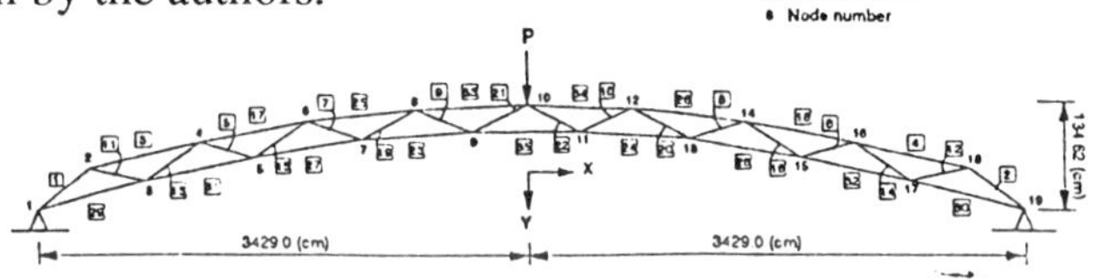

Node Number	y coordinate	x coordinate
1,19	± 3429	0.00
2,18	± 3048	50.65
3,17	± 2667	34.75
4,16	± 2286	83.82
5,15	± 1905	65.30
6,14	± 1524	110.85
7,13	± 1143	87.99
8,12	± 762	128.50
9,11	± 381	100.05
10	0	134.60

Coordinates of the Arch Truss

Member Number (see Figure 2)	Cross Sectional Area (cm^2)
1 to 10 and 35	51.61
11 , 12	64.52
13 to 16	83.87
17 , 18	96.77
19 to 22	103.23
23,24	161.29
25 , 26	193.55
27 , 28	258.06
29 to 32	290.32
33 , 34	309.68

Cross Sectional Area of the Members

Fig. 2. Properties of the arch truss.

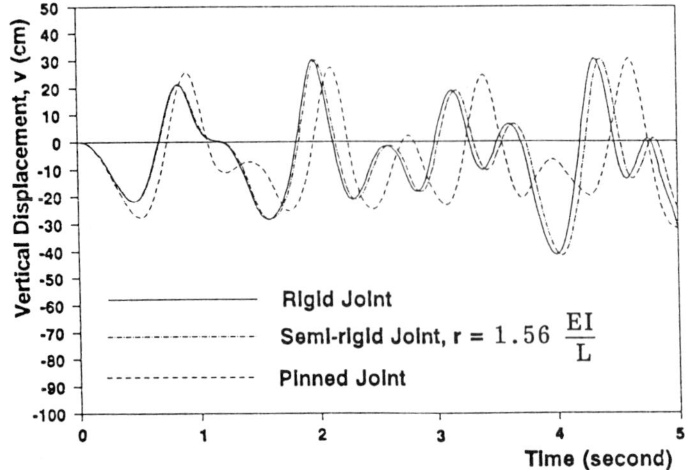

Fig. 3. Transient response of the arch truss.

ACKNOWLEDGEMENT

The authors thank Mr. Chun Yee YAU for the valuable help given in preparing the manuscript of this paper and also his comments on the extension of his work in static problems to dynamic cases.

REFERENCES

1. Code of Practice for Design and Construction of Space Structures (JGJ 780), Beijing, China,1981 (Chinese).

2. BS5950, Structural use steelwork in building, British Standards Institution, London, 1985.

3. CHAN S.L. and YAU C.Y., Non-linear analysis of roof truss composed of pinned, rigid and semi-rigid jointed members, Proceeding of International Congress on Innovative Long Span Structures, Toronto, Canada, 13-17 July, 1992, pp. 175-185.

4. NOOR A.K. and PETERS J.M., Non-linear dynamic analysis of space trusses, Computer Methods in Applied Mechanics and Engineering, 21, 1980, pp.131-151.

5. REMSETH S.N., Non-linear static and dynamic analysis of framed structures, Computers and Structures, 10, 1979, pp. 879-897.

6. SHI G. and ATLURI S.N., Static and dynamic analysis of space frames with non-linear flexible connections, International Journal for Numerical Method in Engineering, 28, 1989, pp.2635-2650.

7. HO W.M.G. and CHAN S.L., Semi-bifurcation and bifiurcation analysis of flexibly connected steel frames, Journal of Structural Engineering, ASCE, 117(8), 1991,pp.2299-2319.

8. CHAN S.L., Inelastic post-buckling analysis of tubular beam-columns and frames, Journal of Engineering Structures, 11,1989, pp.23-30.

45. A method to determine dynamic response of truss structures during sudden consecutive member failure

R. MALLA and B. WANG, University of Connecticut, USA

SYNOPSIS. Damage to one individual or a group of members can initiate a chain reaction of member failure in truss-type structures, which may, in turn, lead to total collapse of the structures. This paper presents a systematic methodology to determine the dynamic response of a truss structure during the event of the progressive member failure. Members are considered to lose their load carrying capacities in a sudden fashion. Emphasis has been given to simulate the dynamic nature of the event. The central point in the approach developed is that a sudden member failure is represented by providing appropriate external forcing functions at the joints where the member is connected to the rest of the structure. The method has been applied to determine the response of a redundant planar cantilever truss. Results presented include natural frequencies, member stresses and joint displacements.

INTRODUCTION

1. Sudden damage to an individual member or a portion of a structure in service environments can come from a variety of causes, such as, material defect, fabrication errors, construction flaws, impact, extreme temperature, and abnormal loads. Moreover, member behaviour such as snap-through buckling, dynamic jump, and brittleness can also give rise to such failure. Member failure affects a small portion of the structure initially, but has potential for propagating to other parts of the structure and may ultimately cause total collapse of the structure ("progressive collapse phenomenon,") (ref. 1). This mechanism of failure is widely observed in a truss-type structure.

2. Truss-type structures, because of their lightweight and high stiffness are the preferred candidates for various large scale construction projects on earth as well as in space. In practice, truss-type structures are designed to have a high number of redundant members. However, collapse of such structures is not uncommon and can have catastrophic consequences as exemplified by numerous lattice roof collapses on earth (refs 2 and 3).

3. There are techniques developed to evaluate the structural integrity and detect damage of structures using vibration measurements. However, it may not be possible to detect and then send remedial measures to guard against progressive-

type collapse in a lattice structure, because the collapse usually lasts for a short duration. This increases the importance of understanding thoroughly, in advance, the characteristics and response of the structure during member failure, for safe design.

4. The loss or failure of a member of a truss-like structure causes force redistribution among the remaining members. The force redistribution may cause these remaining members to carry greater loads forcing them to fail. Static analyses of terrestrial double-layer grids have shown that the loss of only one critical member can lead to total collapse at full service loadings (ref. 4). The failure of members in a lattice truss, however, has been found to be dynamic in nature (refs 4-6). It is, therefore, important to include such dynamic effects in the evaluation of response of these structures. An attempt has been made recently to represent a sudden member failure by using energy principles (ref. 7).

5. The objective of this study is to develop a method to investigate the dynamic response of a truss structure during sudden progressive failure of members. Emphasis has been given to account for the dynamic nature of the member failure. Once damaged the member is assumed to have no load carrying capacity. This paper also presents results obtained from applying the methodology to a two-dimensional cantilever truss. The results presented include static displacements and stresses, natural frequencies, time variations of member stresses and nodal displacements. The structure is assumed to have material linearity. Damping is ignored. Member buckling, dynamic jump or snap-through phenomenon were not considered.

ANALYSIS TECHNIQUE

6. The dynamic (sudden) nature of member failure is represented by first replacing the member by its equivalent force applied at its end joints and then abruptly dropping the force to zero or reduced value. Thus, the dynamic response of the truss structure is that due to this suddenly removed force.

7. The first step in the approach, reported in this paper to study the dynamic response, is to determine accurately the forces to be applied at end joints of the members which are to fail. These forcing functions must be such that when they are applied at the appropriate joints replacing the members, the response of the structure determined at a time before the failure of the members should remain unchanged from that when the members were physically present. The response of the structure can then be obtained by using well established regular dynamic analysis techniques. The method used for the determination of the forcing functions is outlined below.

8. Fig. 1(a). shows a pin-pin supported truss under a uniformly distributed static load of intensity p. The structure deflects to position 1 Fig. 1(b) under the static load. At this position, members of the truss will acquire certain axial internal forces. The first member damage occur at this position initiating the force redistribution and hence, the possible chain reaction of member failure. Curves 2 and 3 in Fig. 1(b) represent respectively the deflected positions of the structure

after the first member i and then the second member j fail. The goal here is to simulate this physical scenario of progressive member failure analytically.

First member failure

9. The deflected position 1 Fig. 1(b) of the intact structure under the applied static load can be obtained analytically by applying the load p quasi-statically as shown in Fig. 2(a). This position at which the first member i is to damage can also be obtained analytically replacing the member by its member force f_i applied at the two end nodes of the member in the original (undeformed) position of the structure Fig. 2(b). In order to facilitate the subsequent dynamic analysis, the static load p and the member force f_i are to be applied in a quasi-static fashion. To obtain the adequate static effect of the loads thus applied, the rise time t_r for the loading versus time curves must be greater than three times the natural period of the structure (refs 8 and 9).

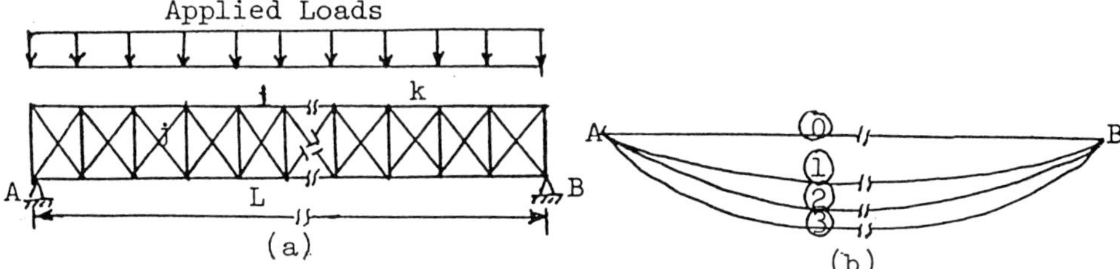

L: length of the strucutre
⓪ :intact undeformed position
① : intact deformed position under the applied loads but before member i breaks at time t_1
② : deformed position after the first member i fails but before the second member j fails at time t_2
③ : deformed position after members i and j fail but before the third member k fails at time t_3

Fig. 1. Physical member failure sequence: (a) a truss structure under loading; (b) deformed positions of the structure.

10. (a) For the structure with all the members intact, the equation of motion for undamped vibration can be written in matrix form as:

$$[M]\{\ddot{y}\}+[K]\{y\}=\{p\} \qquad (1)$$

where, $\{y\}$ and $\{\ddot{y}\}$ are displacement and acceleration vectors, respectively. t indicates time. [M] and [K] represent mass and stiffness matrices of the structure. The static nodal load matrix {p} applied quasi-statically at each joint on the structure is given by:

$$\{p\}=\{P_0\}t \text{ for } 0\le t\le t_r \text{ and } \{P_0\}\, for\ t\ge t_r \qquad (2)$$

where $\{P_o\}$ is the magnitude of the static load applied at each node, and t_r the rise time for the loads.

11. (b) For the structure with member i replaced by its internal force, the equation of motion becomes:

$$[M]\{\ddot{y}\}+[K_I]\{y\}=\{p\}+\{f_i\} \tag{3}$$

Here, $[K_I]$ represents stiffness matrix of the structure without the contribution from member i. Note that the mass matrix [M] is left unaltered, which indicates that the damaged member does not have load carrying capacity, but is still a part of the structure. The member force $\{f_i\}$ Fig. 2(b) is given by:

$$\{f_i\}=\{F_{0i}\}t \text{ for } 0\leq t\leq t_r \text{ and } \{F_{0i}\} \text{ for } t\geq t_r \tag{4}$$

$\{F_{oi}\}$ is the magnitude of the force in member i due to the static load $\{p\}$ before it fails. It is applied quasi-statically at the end nodes of member i in the original (undeformed) position of the structure Fig. 2(b).

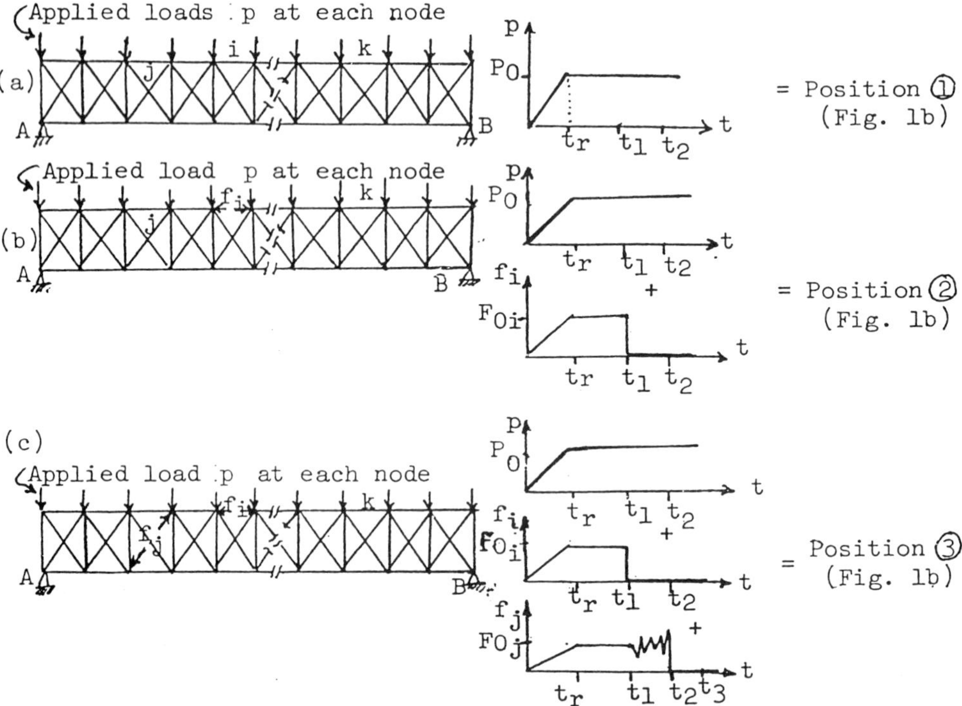

Fig. 2. Analytical representation of member failure.

12. Both of the above formulations, (a) and (b), represent the intact structural response. The sudden damage of member i at time t_l ($t_l > t_r$) now can be represented by dropping the member force $\{f_i\}$ to zero at this time (Fig. 2(b),

portion for $t > t_1$). This means that the complete distribution of force $\{f_i\}$ with time is given by:

$$\{f_i\} = \{F_{0i}\}t \text{ for } 0 \le t \le t_r; \{F_{0i}\} \text{ for } t_r \le t \le t_1; \text{ and } \{0\} \text{ for } t > t_1 \qquad (5)$$

Second member failure

13. The determination of the forcing function to represent the failure of the second member is more involved. This is mainly because the force in the second member after the first one is damaged, is a dynamic force and this aspect should be taken into consideration. The equation of motion for this case can be written as:

$$[M]\{\ddot{y}\} + [K_{II}]\{y\} = \{p\} + \{f_i\} + \{f_j\} \qquad (6)$$

where, $[K_{II}]$ represents the stiffness matrix of the structure without the contribution from the first and the second members i and j. Forces $\{p\}$ and $\{f_i\}$ are given by Eqs 2 and 5, respectively. The force $\{f_j\}$ in the second member j (Fig. 2c) is given by:

$$\{f_j\} = \{F_{0j}\}t \text{ for } 0 \le t \le t_r; \{F_{0j}\} \text{ for } t_r \le t \le t_1; \{f_{dj}\} \text{ for } t_1 \le t \le t_2; \text{ and } \{0\} \text{ for } t > t_2 \qquad (7)$$

where $\{F_{oj}\}$ is the magnitude of the internal force in member j due to the applied static load before the first member fails at time t_1. $\{f_{dj}\}$ is the time varying force in the second member between time t_1 and t_2, where t_2 is the time at which the second member (j) fails. This portion of force $\{f_j\}$ is obtained from the dynamic analysis of the structure with the first member damaged according to the steps outlined above. The sudden failure of the second member is represented by dropping the force to zero at time t_2 (Fig. 2c, portion $t>t_2$). Again, the force $\{f_j\}$ is applied only at the end nodes of member j in the direction of the member in the original (undeformed) position of the structure.

Third and subsequent member failure

14. A procedure similar to that used for the second member failure can be used to determine the forcing function to represent the failure of the third and subsequent members. Force terms corresponding to each additional damaged members are to be added on the right side of the corresponding equation of motion.

APPLICATION

15. The approach outlined in the previous section for determining the dynamic response of a truss structure during sudden progressive member damage had been applied to a 2-dimensional 3-bay, 9 m-long aluminium cantilever truss. Fig. 3. shows the planar cantilever truss and its dimensions and material properties. All members of the structure are considered to have a circular tubular cross-section. The structure is under a total static load of 120N amplitude applied equally at the 8 nodes in the vertical direction. The structure is considered initially designed safe

against static progressive member damage both in tension and compression. The structure is such that all of its members are safe against yielding throughout the total period of the progressive failure. Therefore, the progressive damage in the structure occurs by failure of compressive members. Since the objective of the study is to present the methodology to study the progressive failure of members of the structure, the case of only two consecutive member failures is investigated.

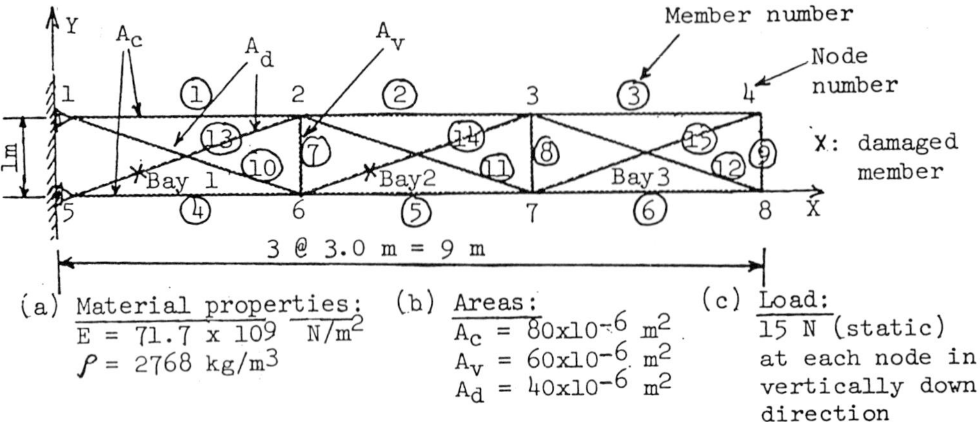

Fig. 3. A planar cantilever truss.

16. To initiate the progressive member failure, compressive diagonal member 13 is damaged first at a stress level somewhat below its critical load. In doing so, it is assumed that there exists certain unusual circumstances such as fabrication errors and material defects in the member warranting the failure. Next, compressive diagonal member 14 is assumed to fail by exceeding its Euler critical load. Each damaged diagonal is considered to fail in a sudden fashion and not have any load carrying capacity after it fails, but remains physically attached to the structure.

17. The analysis to determine the static and dynamic responses of the structure were performed with the help of COSMOS finite element software package for a Sun computer system. The dynamic analysis was performed using the non-linear module available in the software package. The analysis is based on a step-by-step direct time integration using the Newmark-Beta method. The computation algorithm utilises the modified Newton-Raphson iteration procedure. The present study aims at determining the maximum response and therefore, deals with the undamped vibration situation. Material behaviour is linear elastic.

RESULTS AND DISCUSSION

Static analysis

18. Static analysis of the structure in intact and damaged configurations were performed under the total distributed load of magnitude 120N. Nodal displacements and member stresses were obtained for the intact structure as well as the damaged ones as obtained by failing the two compressive diagonals (13

and 14) one after another. The largest vertical displacements at the free end are -0.0051m for the intact structure, -0.0058m when member 13 is damaged, and -0.0063m when members 13 and 14 are damaged. It is observed that the largest horizontal displacement of each node is less than one-tenth of the corresponding vertical displacements of the joint.

19. The static analysis results show that damage of a diagonal member significantly affects stresses only in members in the corresponding bay containing the damaged member. The results also show that when a compressive diagonal member fails the magnitude of the stress in the top chord member of the corresponding bay is reduced, where as they are increased in the bottom, other diagonal, and vertical chord members. For example, when member 13 in bay 1 was damaged, the stress magnitude in top chord member 1 decreased from 5.0625MPa (tension) to 3.375MPa (tension), in bottom chord member 4 increased from 5.0625MPa (compression) to 6.750MPa (compression), in diagonal member 10 from 3.558MPa (tension) to 7.115MPa (tension), and in vertical member 7 from 0 MPa to 0.745MPa (compression). Similar trends can be observed when the second diagonal (member 14 in bay 2) fails following the failure of member 13.

Dynamic analysis

20. The analysis was performed considering the structure to be a lumped mass model. Twelve mode shapes and frequencies were obtained for the intact and damaged structures. The frequencies vary from 11.40Hz (period = 0.877s) to 611.94Hz (period = 0.0016s) for the intact structure, from 10.45Hz to 611.94Hz when member 13 is damaged, and from 9.99Hz to 611.91Hz when members 13 and 14 are damaged. The 1st, 2nd, and 3rd frequencies for these three cases respectively are 11.40Hz, 44.22Hz, and 79.37Hz for the intact structure; 10.50Hz, 33.99Hz, and 74.70Hz for member 13 damaged; and 9.99Hz, 33.80Hz, and 58.05Hz for members 13 and 14 damaged. Based on the period of the structure obtained above, the time step size for the dynamic analysis is chosen to be equal to 0.001s. The rise time t_r for the loads/forces applied in order to obtain quasi-static response is taken equal to 1.5s.

21. From the static analysis of the intact structure the forces in members 13 and 14 are determined to be -142.30N (F_{oi}) and -94.868N (F_{oj}), respectively. The forcing functions to facilitate the dynamic analysis due to the damage of members were determined using the procedure outlined under the section entitled "Analysis Technique." These forcing functions are shown in Fig. 4. The responses during the time period t_r (1 .5s) $\leq t \leq t_1$ (2.0s) compare closely with the corresponding static values as expected. The dynamic analysis results with member 13 damaged (sudden drop of the force magnitude to zero) show that the stress in the second diagonal (No. 14) exceeds its critical value at time 2.055s (t_2). Therefore, member 14 is damaged at 2.055s. The time varying stress in member 14 during the period from 2.0s (time at which member 13 fails) to 2.055s (time at which member 14 fails) is also shown in Fig. 4. A comparison between the stresses and displacement at time 2.055s obtained from the two situations, one, member 13

damaged with member 14 left intact and the other, member 13 damaged with member 14 replaced by the forcing function, show a close match, indicating that the forcing functions used to replace the damaged members are sufficiently accurate. The dynamic response of the entire structure during the progressive failure of member 13 followed by member 14 is obtained by performing the dynamic analysis of the structure with these forcing functions (Fig. 4) applied at appropriate joints in place of the members.

22. Fig. 5 shows time variations of vertical displacements of joint 4. Fig. 6 shows time variations of stresses in some selected members. Also shown on these plots are corresponding member stresses from the static analysis. The results show that the dynamic nature of member failure affects the behaviour of members also in bays other than that of the damaged member.

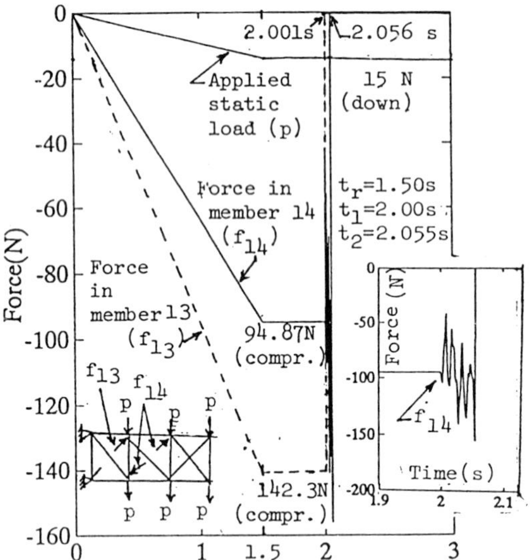

Fig. 4. Forcing functions for dynamic analysis.

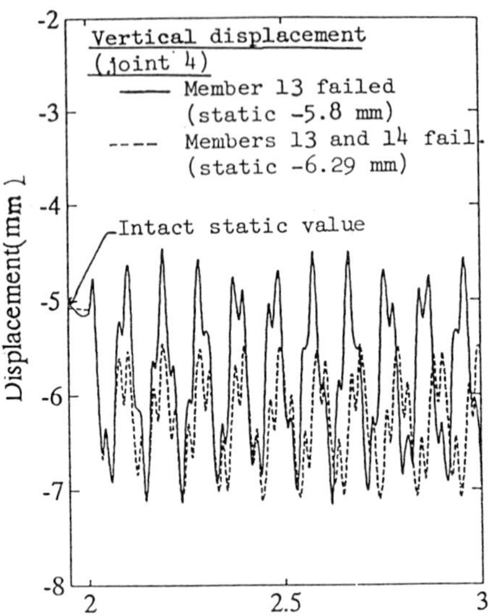

Fig. 5. Time variation of joint displacement (joint no. 4)

CONCLUSIONS

23. A systematic analysis technique has been presented to evaluate the dynamic response of truss structures during sudden progressive member failure. The members are assumed to lose all load carrying capacity once they are damaged. However, they remain physically attached to the structure. The dynamic nature of member failure is accounted for. The method was employed for a planar cantilever truss subjected to sudden damage of diagonal compressive members. The results presented include static displacements and stresses, frequencies, and dynamic displacements and member stresses of the truss. The results show that static progressive member failure has localised effects, whereas the dynamic

consideration of progressive member failure has wide spread effects in the structure. It is expected that this technique will be suitable to represent the snap-through type of member failure in a truss-type structure. In depth studies accounting for buckling, dynamic jump or snap-through phenomenon, material and deformation, non-linearity effects are in progress.

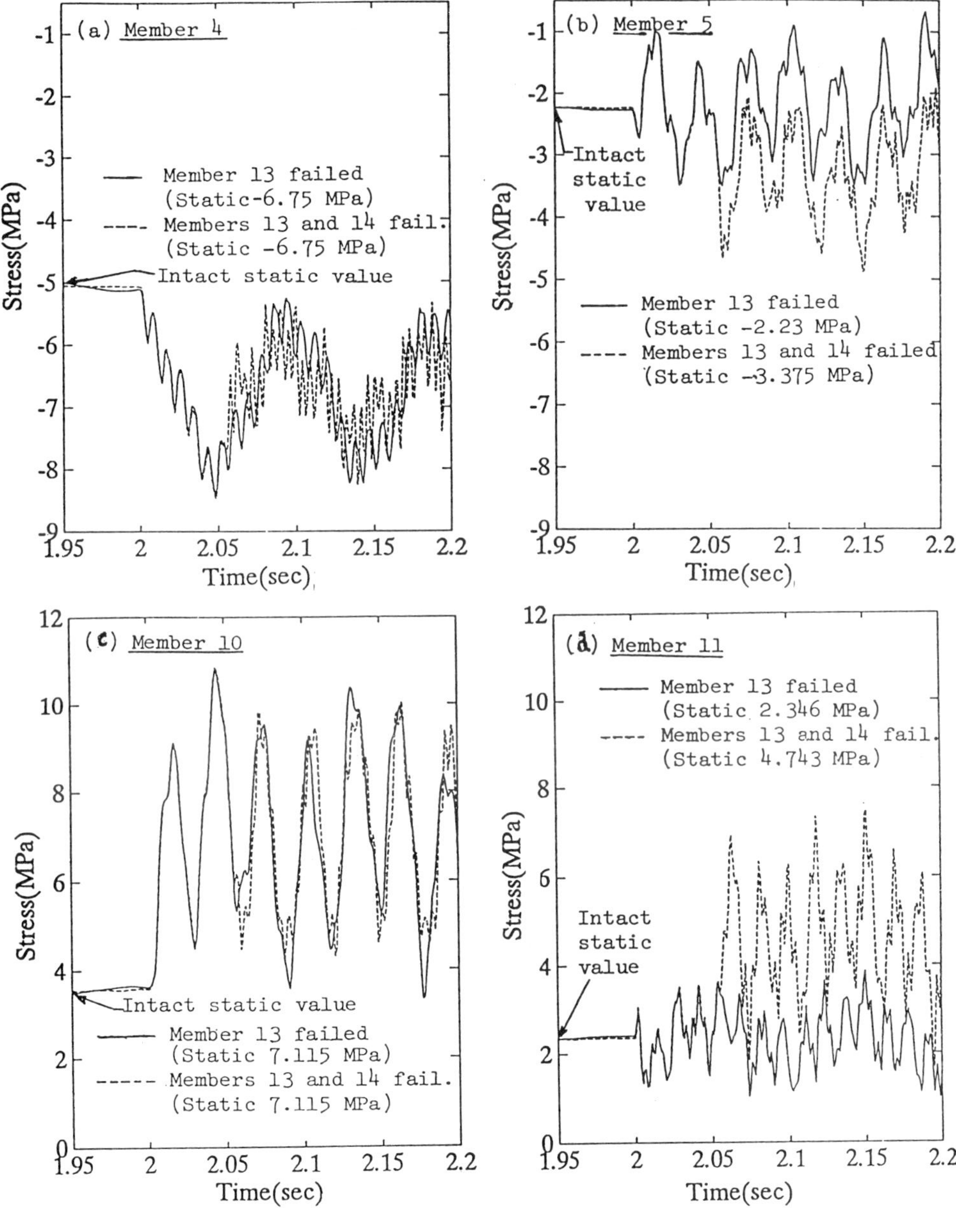

Fig. 6. Variations of member stresses with time: (a) member no. 4, (b) member no. 5, (c) member no. 10, and (c) member no. 11.

ACKNOWLEDGEMENT

24. The authors acknowledge the financial support obtained from the National Science Foundation, Washington, D.C. under Research Initiation Award (Grant No. MSS-9110900) and The University of Connecticut, Storrs, CT during this work.

REFERENCES

1. BREEN J. Summary report of the research workshop on progressive collapse of building structures. Univ. of Texas at Austin, Tx, 1975.
2. TASK COMMITTEE on Latticed Structures under Extreme Loads of the Committee on Special Structures of the ASCE-Structural Division (Chair, S. Holzer.) Dynamic consideration in latticed structures. J. of Structural Engineering, Vol. 110, no 10, Oct. 1984, 2547-2551.
3. SMITH E. and EPSTEIN, H. Hartford coliseum roof collapse: structural sequence and lesson learned. Civil Engineering, ASCE, Vol. 52, No. 4, April, 1980, 59-62.
4. MURTHA-SMITH E. Alternate path analysis of space trusses for progressive collapse. J. of Structural Engineering, Vol. 114, no 9, Sept., 1988, 1978- 1999.
5. DAVIES G. and NEAL B. The dynamical behaviour of a strut in a truss framework. Proc. Royal Soc., A253, 1959, 542-562.
6. SCHMIDT L., MORGAN P. and CLARKSON J. Space trusses with brittle type strut bucking. J. of the Structural Division, ASCE, Vol. 102, No. ST7, July 1976, 1479-1492.
7. MALLA R. and WANG B. Response of space structures under sudden local damage. Engineering, Construction and Operations in Space III (Vol. I.) Proceedings of the SPACE 92 conference, ASCE, New York, May, 1992, 909-920.
8. WEAVER W., TIMOSHENKO S. and YOUNG D. Vibration problems in engineering. 5th. Ed., John Wiley & Sons, New York, 1990, 103-105.
9. CRAIG R. Structural dynamics - an introduction to computer methods. John Wiley & Sons, New York, 1981, 115-117.

46. Natural frequencies of rotationally or linearly repetitive space frames

F. W. WILLIAMS, D. KENNEDY, L. M. SONG and A. K. JEMAH,
University of Wales College of Cardiff, UK

SYNOPSIS. Existing theory enables natural frequencies of space frames which are repetitive in up to three Cartesian directions, or which are rotationally periodic and optionally also repetitive along the axis of periodicity, to be calculated by analysing only one repeating portion and with the distributed mass of members treated exactly. New theory presented uses Lagrangian multipliers to couple different response wavelengths, allowing supports to be present in only some of the repeating portions, while still only analysing one of them. Illustrative examples include a double-layer roof grid and a dome with 36 identical portions and four, nine or twelve supports.

DEFINITION OF THE PROBLEM AND OF THE EXAMPLES USED

1. The large space frame computer program BUNVIS-RG (ref. 1) enables natural frequencies of undamped vibration to be found, while allowing exactly for the distributed mass of members so that it contrasts with the approximate lumped mass or consistent mass matrix techniques of traditional finite element analyses. The program permits the exact analysis of structures which are rotationally periodic, or repetitive in one or more Cartesian directions, by modelling only a single repeating portion. The present paper gives the key features of theory currently under development, which uses Lagrangian multipliers to extend BUNVIS-RG to include point supports which are repetitive, but at less frequent intervals than the single repeating portion of the structure that is modelled. Thus if b and ℓ (= kb, where k is an integer) represent the bay lengths over which, respectively, the structure and its support system are repetitive in any direction.

2. The dome of Fig. 2(a) consists of 36 identical portions of the type shown in Fig. 1(a). For such rotationally periodic structures it is convenient to express the structure bay length b as the subtended angle, i.e. 10°, so that for the supports shown on Fig. 2(c) the support bay length becomes 30° and k = 3. The beam of Fig. 3(a), which repeats in one Cartesian direction, involves only analysis of the repeating portion shown in Fig. 1(b), so that b = 2m. The support pattern repeats over a length ℓ = 24b = 48m. Finally, the double-layer grid of Fig. 4 involves repetition in the two Cartesian directions x and y, with $b_x = b_y$ = 2m. Note that,

as in Fig. 3, the point supports are not uniformly spaced so that ℓ_x and ℓ_y are the lengths over which the support pattern repeats.

3. For the beam and grid portions of Fig. 1, the nodes denoted by circles and squares lie respectively in upper and lower horizontal planes which are 2m apart. Only the numbered nodes lie in the repeating portion, the other nodes belonging to adjacent portions. Therefore, since there are six degrees of freedom at each node, the analysis uses a stiffness matrix of order only 18 for the beam and 12 for the grid. Similarly, only 29 nodes belong to the repeating portion of the dome shown in Fig. 1(a) and so its stiffness matrix has order 174. The dome is of parabolic profile and constructed of straight members, with the horizontal rings of members being regular polygons of radius 5m, 10m, 15m, 20m, 25m, 30m and 35m.

4. The results given later are all for the dome, beam and grid of Figs 2-4, with rigid joints assumed and with all of the members being thin circular steel tubes having E = 210GN/m^2, density = 7,000kg/m^3, Poisson's ratio = 0.3, EA = 322.6MN and EI = 496.3kNm^2, so that their radius of gyration is 39.2mm and their outer diameter and thickness are approximately 115mm and 4.4mm.

THEORY

5. There is only space here to give a brief account of the theory, which is currently being further developed. The key points which follow are restricted to repetition in up to three Cartesian directions, but the full theory also handles rotational periodicity with the option of linear repetition parallel to the axis of periodicity.

6. The frame is of infinite size and so its stiffness matrix [K] would be of infinite order, with the natural frequencies and modes given by:

$$[K]\,[D] = [0] \tag{1}$$

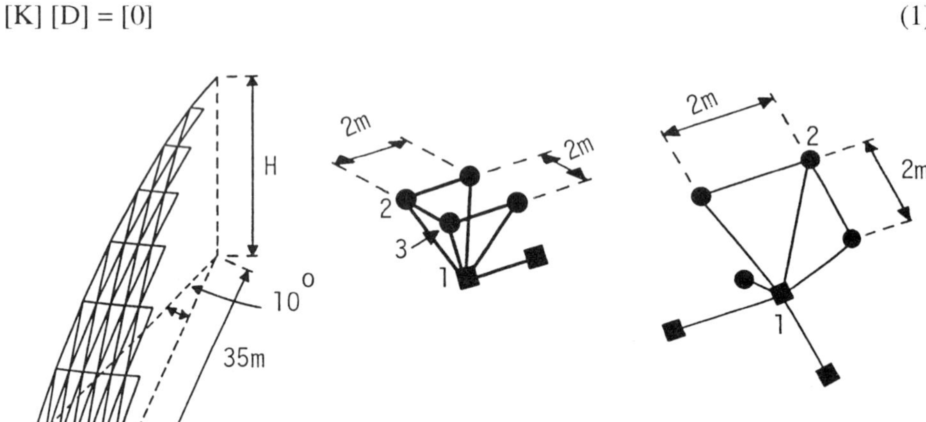

Fig. 1. Repeating portions of structure for: (a) the rotationally periodic dome of Fig. 2; (b) the truss-type beam of Fig. 3 and; (c) the double-layer grid of Fig. 4.

noting that [K] is a dynamic stiffness matrix, i.e. its elements are transcendental functions of frequency, because distributed member mass is treated exactly.

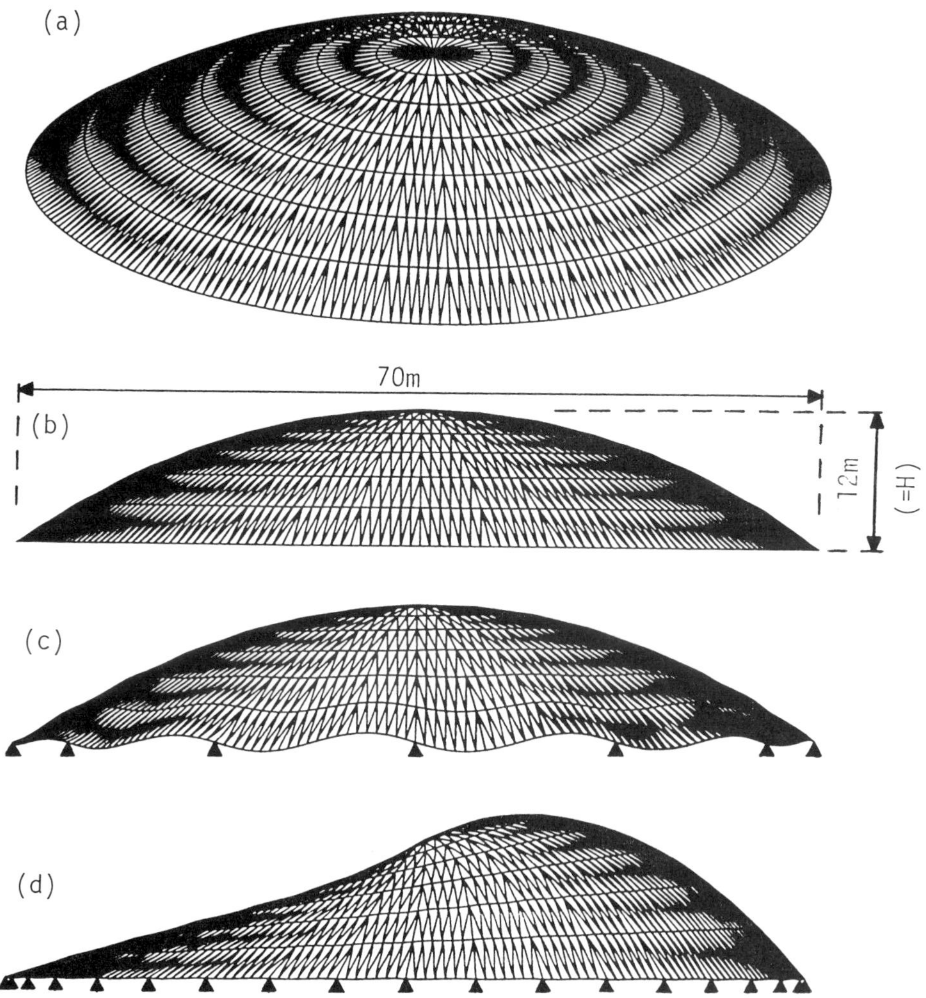

Fig. 2. (a) Isometric view of dome; (b) side view of dome; (c) fundamental mode for 12 supports case and; (d) fundamental mode for 36 supports case.

When k=1, it has been shown (ref. 1) that deflections vary sinusoidally such that if $[D_o]$ are those components of [D] which correspond to a datum repeating portion and$[D_j]$ are the corresponding components for a repeating portion which is located j_x, j_y and j_z bays from it in the x, y and z directions, then:

$$[D_j] = [D_0]\exp\{2i\pi(n_x j_x / N_x + n_y j_y / N_y + n_z j_z / N_z)\} \tag{2}$$

where $i = \sqrt{-1}$ and the N_α (α = x,y or z) are the number of structure bay widths over which the mode is assumed to repeat. The N_α can be chosen freely and computations can be repeated for several combinations. In this extension of the theory, N_α will usually be a multiple M_α of the support system bay length, so that $N_\alpha = M_\alpha k_\alpha$. Good results are often obtained with $M_\alpha = 2$, which is the value used for the beam and grid results given in Fig. 3 and Table 2. The values of n_x, n_y and n_z define the wavelengths of sinusoidal response of the mode in the x, y and z directions, being the number of complete waves in the lengths N_x,b_x, $N_y b_y$ and $N_z b_z$ of the frame. They must cover the range:

$$n_\alpha = 0, \pm 1, \pm 2, \ldots, \pm \text{integer}(N_\alpha/2) \qquad (\alpha = x, y \text{ or } z) \tag{3}$$

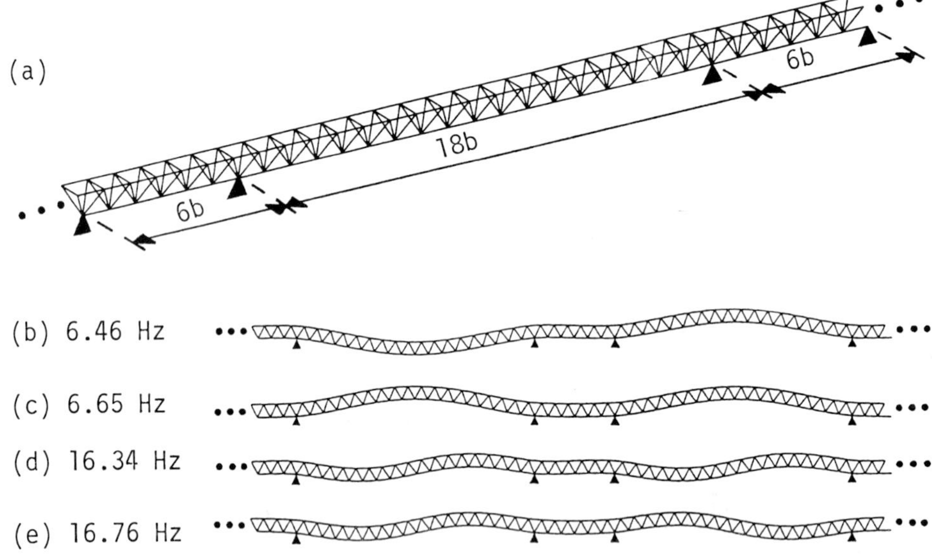

Fig. 3. Truss-type beam, showing as (a)-(e) its undeflected shape and its first four modes (with M=2) which are symmetric with respect to its vertical longitudinal plane of symmetry. The supports shown prevent all translational freedoms.

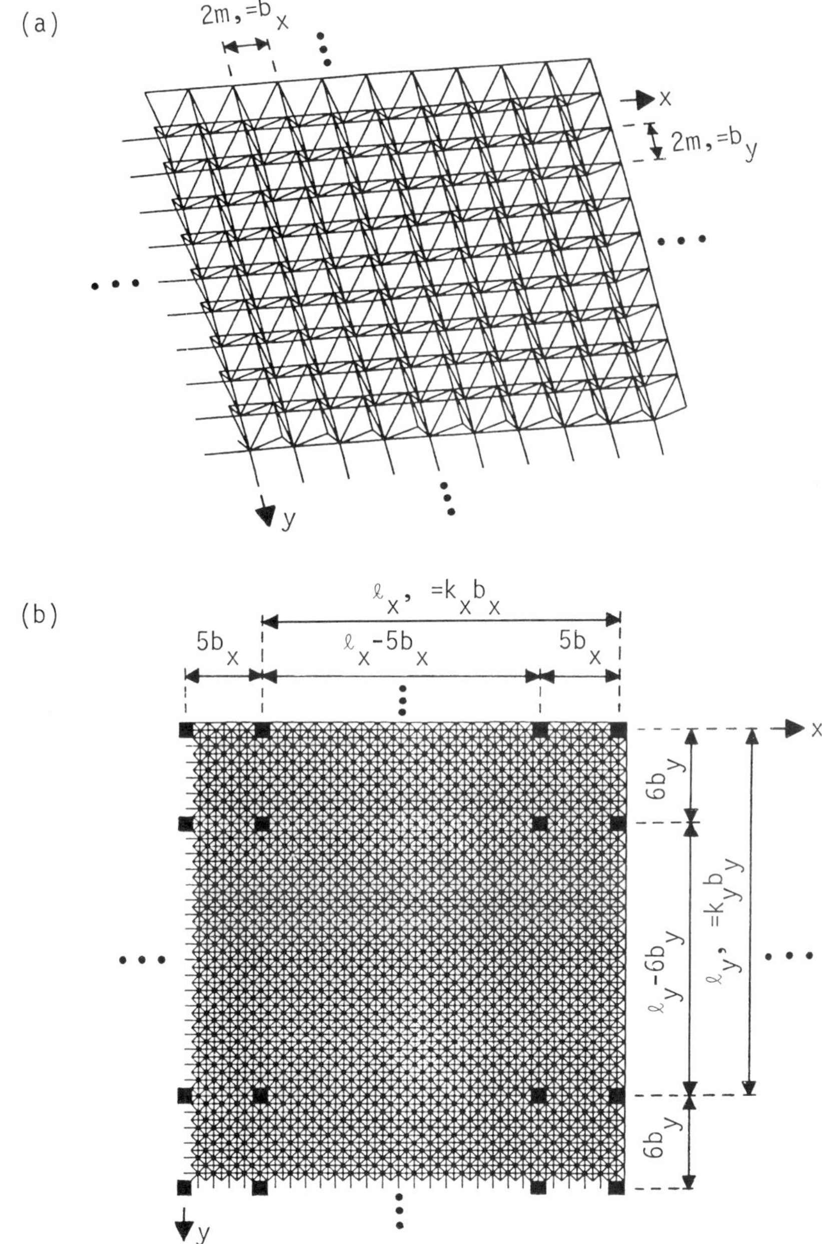

Fig. 4. (a) Isometric view of double-layer grid. (b) Plan view, showing positions of supports which prevent all three translational freedoms of bottom layer nodes.

In this series - integer ($N\alpha/2$) can be omitted when $N\alpha$ is even, and all the negative signs can be omitted for any one of x, y or z without affecting the answers obtained.

7. Substituting from Eq. 2 reduces Eq. 1 to:

$$[K_{0m}][D_{0m}] = [0] \qquad (m = 1,2,....) \qquad (4)$$

where $[K_{0m}]$ is a complex dynamic stiffness matrix of order corresponding to the size of only a single repeating portion of the frame and m denotes a particular combination of n_x, n_y and n_z. $[K_{0m}]$ can be assembled directly from the repeating portion of the frame, e.g. from the portions of Fig. 1.

8. For a given choice of the N_α, Lagrangian multipliers can be used to couple the stiffness matrices $[K_{0m}]$ for m = 1,2,.... so as to satisfy a support pattern repeating at intervals $\ell_\alpha = k_\alpha b_\alpha$, i.e. less frequently than the repetitions of the structure. In effect, this forms a Fourier series analysis which couples the responses with an appropriate set of wavelengths in each of the directions x, y and z, such that the resulting displacement is zero at each of the supports. Such theory has been given in a refined form elsewhere (ref. 2) for prismatic stiffened plate structures with an infinite number of bays in the direction parallel to the stiffeners. The principal refinements are equivalent to showing that only subsets of the n_α given by Eq. 3 can couple with each other. As a consequence, only a single repetition of the support pattern need be included in the analysis instead of all the supports in an interval over which the mode repeats. In the present context, this means that only the supports within the volume $(k_x b_x)(k_y b_y)(k_z b_z)$ need be included rather than the larger number in the volume $(N_x b_x)(N_y b_y)(N_z b_y)$. For example in Fig. 4(b) only four of the supports shown need be modelled, e.g. those in the top left-hand corner.

9. The analogy with stiffened plate structures is incomplete, both because plate structures are continuous rather than discrete in the x direction and also because for them Lagrangian multipliers model supports repeating in only the x direction, rather than in all three Cartesian directions. A formal derivation of the equivalent theory for the frames of the present paper would be far too long to present here, but the key ideas are summarised below and Tables 1 and 2 show which wavelengths of response were coupled for each of the natural frequencies presented.

10. The supports are rigid and can be represented by the constraint equations

$$\sum_m [E_m][D_{0m}] = [0] \qquad (5)$$

where the number of matrix products summed is equal to the number of possible combinations of the values of nx, ny and nz given by Eq. 3 (e.g. 144 when $N_x = N_y = N_z = 6$) in the absence of the refinements referred to above, but is much smaller when the refinements are included. Here each scalar row of Eq. 5 has the form:

(6)

where d_{0m} is the amplitude of one of the sinusoidal components of the response of the constrained freedom of $[D_{0m}]$ and j_x^*, j_y^* and j_z^* locate the repeating portion in which the support lies. Thus for the origin of co-ordinates shown in Fig. 4(b) the four supports to be included are those for which (j_x^*, j_y^*) take the values (0,0), (0,6),(5,0) and (5,6).

11. Eqs 4 and 5 combine to give the following equation, from which all the eigenvalues can be extracted.

$$\left[\begin{array}{cccccc|c} [K_{01}] & [0] & \cdots & [0] & \cdots & & [E_1]^H \\ [0] & [K_{02}] & \cdots & [0] & \cdots & & [E_2]^H \\ \vdots & \vdots & \ddots & \vdots & & & \vdots \\ \vdots & \vdots & & [K_{0m}] & \cdots & & [E_m]^H \\ \vdots & \vdots & & \vdots & \ddots & & \vdots \\ \hline [E_1] & [E_2] & \cdots & [E_m] & \cdots & & [0] \end{array}\right] \left[\begin{array}{c} [D_{01}] \\ [D_{02}] \\ \vdots \\ [D_{0m}] \\ \vdots \\ \hline [\gamma] \end{array}\right] = \left[\begin{array}{c} [0] \\ [0] \\ \vdots \\ [0] \\ \vdots \\ \hline [0] \end{array}\right] \qquad (7)$$

where $[\gamma]$ are the Lagrangian multipliers, H denotes Hermitian transpose (i.e. the complex conjugate of the transpose) and the entire left-hand side matrix is Hermitian.

12. The extraction of the natural frequencies from Eq. 7 without the chance of any being missed can be accomplished by using an existing algorithm (ref. 3), in an extended form (ref. 4) which includes Lagrangian multipliers. Again, the detailed application of this algorithm involves theory which is too detailed to present here, but which is analogous to that previously used for stiffened plate structures (ref. 2).

RESULTS

13. Several natural frequencies and the associated mode shapes of the continuous beam are shown in Fig. 3. The fundamental natural frequencies of the dome and of the double-layer grid are given in Tables 1 and 2, with two fundamental modes of the dome illustrated in Fig. 2.

Table 1. Fundamental natural frequency for variants of the dome of Fig. 2. Each variant has 36 repeating portions of the type shown in Fig. 1(a) and so the column headed 36/k gives the number of point supports.

H (m)	36/k	Nat. freq. (Hz)	Dominant half-wavelengths, expressed as number of complete waves round circumference (with % they contribute to the fundamental mode in brackets)
9	12	2.050	1 (23%), 11 (29%), 13 (48%)
12	12*	2.131	1 (9%), 11 (42%), 13 (49%)
15	12	2.068	1 (6%), 11 (44%), 13 (50%)
12	36	2.517	1 (100%)
12	12*	2.131	1 (9%), 11 (42%), 13 (49%)
12	9	1.409	1 (5%), 8 (42%), 10 (49%), 17 (4%)
12	4	0.387	0^+ (5%), 4 (94%), 8 (1%), ...

Table 2. Fundamental natural frequency for the double-layer grid of Fig. 4 for various combinations of ℓ_x and ℓ_y, for M_x=M_y=2.

k_x $=\ell_x/b_x$	k_y $=\ell_y/b_y$	Nat. freq. (Hz)	Dominant half-wavelength combinations expressed as number of complete waves in length $2\ell_x/2\ell_y$ (with % they contribute to the fundamental mode in brackets)
19	24	4.068	0/1(26%), 2/3(15%), 2/1(14%), ...
23	24	3.243	1/1(59%), 3/3(7%), 1/3(5%), ...
27	24	2.700	1/1(61%), 3/3(6%), 3/1(5%), ...
23	28	2.702	0/1(27%), 2/3(16%), 2/1(14%), ...
27	28	2.241	1/1(63%), 3/3(6%), 1/3(4%), ...
31	28	1.927	1/1(65%), 3/3(5%), 3/1(4%), ...

ACKNOWLEDGEMENTS

14. The authors are grateful for the support of both the Science and Engineering Research Council and the British Council.

REFERENCES

1. ANDERSON, M.S. and WILLIAMS, F.W. BUNVIS-RG: Exact frame buckling and vibration program, with repetitive geometry and substructuring. Journal of Spacecraft and Rockets, 1987, vol. 24, no. 4, 353-361.

2. ANDERSON, M. S., WILLIAMS, F.W. and WRIGHT, C.J. Buckling and vibration of any prismatic assembly of shear and compression loaded anisotropic plates with an arbitrary supporting structure. International Journal of Mechanical Sciences, 1983, vol. 25, no. 8, 585-596.

3. WITTRICK, W.H. and WILLIAMS, F.W. A general algorithm for computing natural frequencies of elastic structures. Quarterly Journal of Mechanics and Applied Mathematics, 1971, vol. 24, part 3, 263-284.

4. WILLIAMS, F.W. and ANDERSON, M. S. Incorporation of Lagrangian multipliers into an algorithm for finding exact natural frequencies or critical buckling loads. International Journal of Mechanical Sciences, 1983, vol. 25, no. 8 579-584.

47. The modal analysis of a unit building block for a 5.0 m deployable antenna for space applications

P. FANNING and L. H. HOLLAWAY, University of Surrey, UK

SYNOPSIS. The volume constraints imposed by current launch modules and the desirability of large reflectors and communication booms for space applications has given rise to active research in the field of deployable systems. These systems are stowed in a compact package for launch before being deployed into their operating configurations in orbit. The main design drivers for deployable systems are that they deploy reliably, have a minimised mass and part count and that they are dynamically isolated from their supporting spacecraft. A new concept for a deployable antenna is presented. Modal analyses of a unit building block of the complete structure were undertaken to validate the modelling techniques to be used for the complete structure. The numerical results for the unit building block are compared to those obtained experimentally.

DEPLOYABLE SYSTEMS

1. The *Composite Structures Research Unit (C.S.R.U.)* in the Department of Civil Engineering, has, over a period of years, been examining the potential of large skeletal systems for use in space. The inherent stiffness of skeletal systems allied with the high specific stiffness of composite materials produces a structure of high stiffness and low mass. It is envisaged that composite skeletal systems will be used for communication booms, and integral parts of future space stations. In the near term their most important application is likely to be as backing structures for large reflecting antennas.

2. Large antennas concentrate the reflected beam over a smaller area allowing the use of smaller, less expensive and lower-powered ground and space based receivers. These types of receivers are particularly attractive to mobile ground users. The continuing trend to increase the size of reflector antennas is demonstrated by considering that the Orbital Test Satellite (OTS), launched in 1979, for the European telecommunications antenna was 0.6m in diameter. The Maritime Communications Satellite (MARECS), launched in 1983, for the European Space Agency (ESA) had a single dish of 2.0m; most of the current Eurostar systems use 2.5m antennas. The antennas used for Intelsat Vl, launched in 1987, are 3.6m in diameter and are likely to be the largest solid reflectors

capable of being launched because of volume constraints imposed by current launch vehicles, (ref. 1).

3. As the size of a reflector increases a reduction in the natural frequency, due to the greater flexibility of the configuration, results. To ensure that the antenna is not excited by one of the control frequencies of its supporting spacecraft its first natural frequency should be above a certain minimum threshold level, specified as 1.5Hz in (ref. 2). Freeland, (ref. 3), Garrett and Ferebec, (ref. 4) and Anderson, Garrett and Calleston, (ref. 5) have carried out surveys on different deployable antenna concepts. Deployable skeletal systems have a number of advantages, namely, high structural strength and stiffness due to the truss depth, nodal joints which provide locations for mounting subsystems, easy division of the truss into modules, high degree of redundancy which minimises the effect of single member failure, and a number of hard points for locating a suitable reflecting surface. These advantages make the use of deployable skeletal systems more attractive than other deployable concepts such as, inflatable antennas, hoop column antennas, the wrap rib antennas or the spin stiffened membrane antennas, (refs 3-5).

THE DEPLOYABLE CONCEPT

4. The deployable antenna is a skeletal system comprising carbon fibre epoxy resin structural members joined together at aluminium node assemblies and deployed by a series of energy loaded joints located at the ends of each of the structural members, at the centres of the inner ring members and long radial members. The deployed antenna and joint locations are shown in Fig. 1.

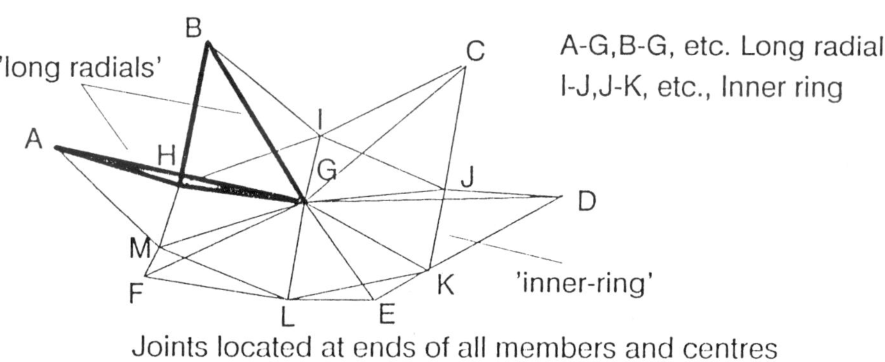

Fig. 1. The deployed antenna illustrating the unit building block (UBB).

5. The deployment sequence, driven by the energy loaded joints, has been obtained by numerical analysis, (ref. 6), and is illustrated in Fig. 2. The potential energy stored in the joints is transferred to the structure as kinetic energy thus

causing it to deploy. The structure is a one degree of freedom system and once the antenna has reached its deployed configuration the joints are rigidly locked.

6. The proposed concept is shown in ref. 6 to offer several important advantages over similar existing deployable antennas. It possesses fewer structural members and mechanical joints, thereby reducing the part count and manufacturing costs. The relatively small number of mechanical joints, with respect to similar skeletal systems is particularly attractive to the aerospace industry which favours simpler and therefore more reliable deployable mechanisms.

Structural Members

7. The tube members which were used for the development model were manufactured from carbon fibre/epoxy resin by the 'pullwinding' technique, (ref. 7); this is a fabrication process developed by Pultrex Ltd., Clacton-on-Sea. The tube members have a nominally 25mm internal diameter with 1.2mm wall thickness. In addition to the current work these tubes are being evaluated for use in space, by British Aerospace, as part of the model development. Three filament wound tubes with 50mm internal diameter and 4.0mm wall thickness were also manufactured; these members were located at the point of connection to the spacecraft.

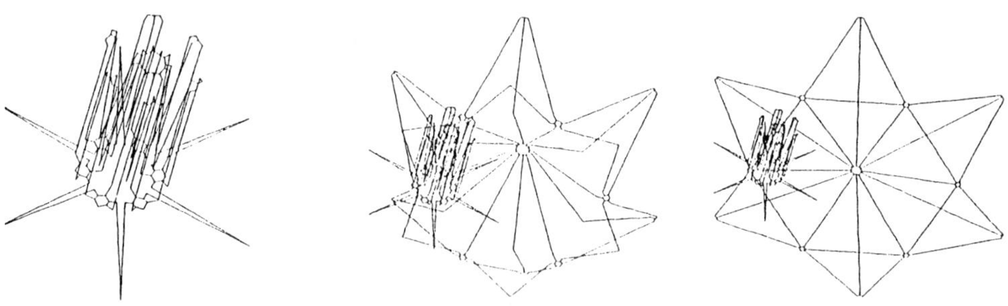

Fig. 2. The deployment sequence.

Energy loaded joints

8. A typical energy loaded joint is shown in its stowed, partially deployed and deployed configurations in Fig. 3. The joint constituents are, two component inserts bonded into the carbon fibre struts which are to be joined, a link plate connecting these two components, a sliding sleeve and a spring. In its stowed configuration the compressed spring exerts a force on the sliding collar which in turn generates a moment about the link pin causing the joint to deploy.

9. The joint inserts for the development model were manufactured in titanium. Generally aluminium is only used for non-moving components. It is well known that two aluminium surfaces in contact with, or sliding over, each other can become cold welded together. This would render a mechanical joint inoperative

and hence titanium, although heavier, is preferred. The joint is lubricated using powder graphite to minimise the frictional forces as the joint deploys.

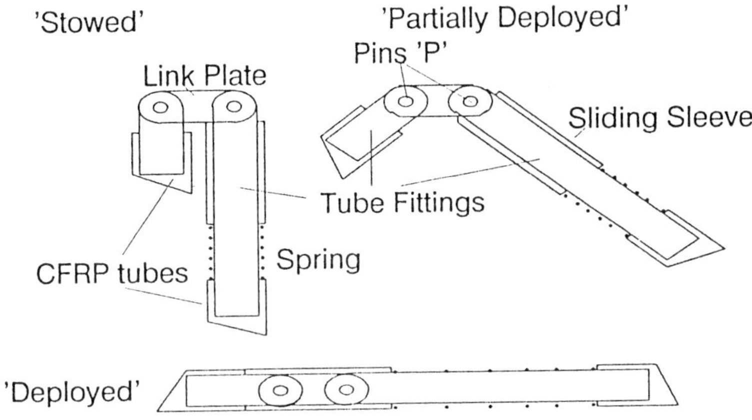

Fig. 3. The energy loaded joint.

Node assemblies and member end connectors

10. Typical node assemblies with member end connectors, manufactured from aluminium and titanium respectively, for the full scale model are shown in Fig. 4. The member end connectors were fixed to the nodes using a single bolt and a rivet to prevent rotation, Fig. 5. (For a flight model it is envisage that they would be bonded into position and secured with two bolts.)

THE UNIT BUILDING BLOCK, (UBB)

11. The constraint on minimum frequency of the deployable system means that the numerical modelling techniques for the complete structure must be carefully validated prior to its manufacture to ensure this constraint is satisfied. A unit of the complete structure was therefore manufactured and tested to validate the numerical modelling techniques which can then be used to qualify the dimensions of the complete structure. The unit building block comprises one partial tetrahedron extracted from the full scale composite antenna as shown in Fig. 1. The node assemblies used in its manufacture are similar to those used in the full scale structure.

Experimental modal analysis of the UBB

12. Modal analyses, (ref. 8), were undertaken for this structure to determine its natural frequencies and modes of vibration and thereby validate the numerical modelling techniques that were employed for the complete antenna.

13. The unit was considered initially with its full complement of energy loaded joints and its response was seen to be non-linear due to the small clearance between the joint collars and fittings, (ref. 6). The structure was linearised by replacing the articulated members with continuous carbon fibre tubes.

14. The structure was suspended from a rigid support using 4.0m lengths of nylon string in series with soft springs at each of its node points. The unit was excited using a random noise energy source via a shaker and with an impact hammer. Measurements were taken at the ends of each member, at the nodes, and additionally at the third points in three orthogonal directions to determine the mode shapes.

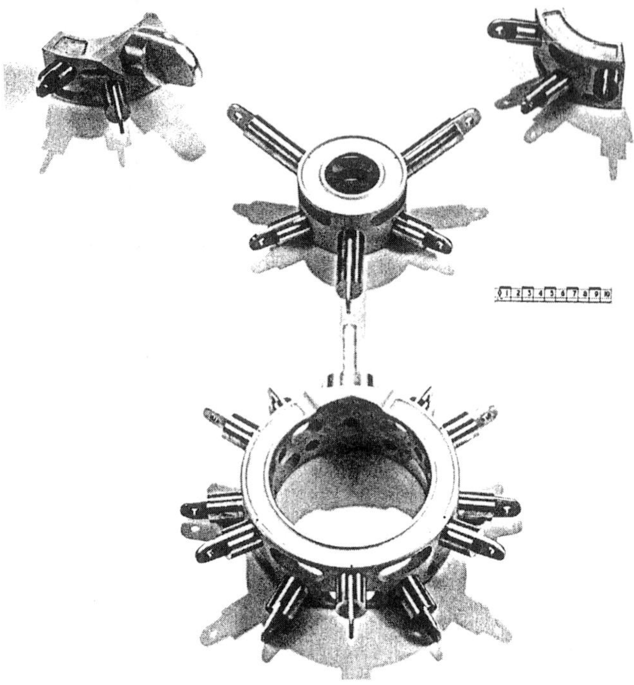

Fig. 4. Node assembly.

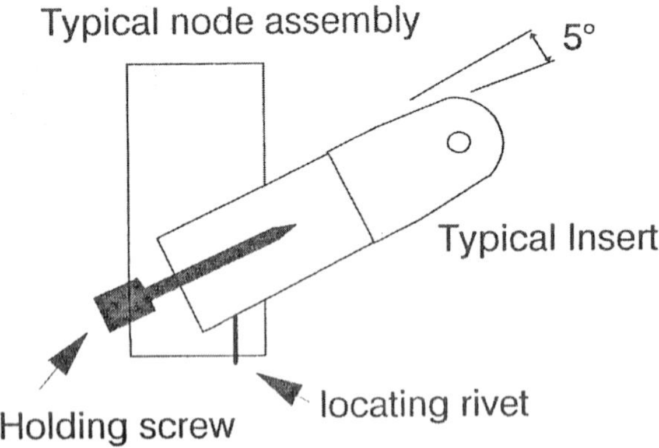

Fig. 5. Section through node assembly showing fixing of member end connectors.

15. The first five modes of the structure were determined experimentally. The natural frequencies for the two experimental analyses are given in Table 1, columns one and two. The difference between these two values is observed to be less than 2.5% for each of the measured modes using the two different excitation techniques.

Table 1. Natural frequencies for the UBB.

Mode	Frequency (Hz)		
	Impact Hammer	Vibrator	Numerical
1	7.42	7.58 (+2)	8.22 (+10)
2	18.26	18.62 (+2)	20.03 (+9)
3	24.02	23.59 (-2)	24.9 (+11)
4	24.61	24.9 (+1)	27.1 (+10)
5	28.61	28.52 (-0)	32.2 (+9)

16. The mode shapes obtained using the mechanical vibrator and the impact hammer were normalised, with respect to the mass matrix of the structure, to examine their respective orthogonality properties, (ref. 9). Targoff suggests that off diagonal terms for normalised mode shapes of 0.1 or less indicate a good set of measured modes. The orthogonality matrices for the two analyses are given in Tables 2 and 3. The off diagonal terms in the case of external excitation using the shaker are significantly higher than those for the analysis undertaken with the impact hammer. This suggests that the response of the structure was affected by the attachment of the shaker. The off diagonal terms for each of the modes examined with the impact hammer are small except for mode number four where there is coupling with mode two and to a lesser degree mode number three and it is concluded that experimentally measured mode number four may not have been an accurate representation of that mode. Plan and three dimensional views of the first five modes measured using the impact hammer are shown in Fig. 6.

Table 2. Orthogonality of modes measured using the impact hammer.

Mode	1	2	3	4	5
1	1.0	0.05	0.14	-0.04	-0.04
2		1.0	-0.04	0.49	-0.02
3			1.0	0.19	0.03
4	symmetric			1.0	0.0
5					1.0

Table 3. Orthogonality of modes measured using the mechanical vibrator.

Mode	1	2	3	4	5
1	1.0	-0.13	0.08	-0.02	0.04
2		1.0	-0.23	-0.25	-0.14
3			1.0	0.49	-0.05
4	symmetric			1.0	-0.43
5					1.0

Numerical Analysis of the UBB

17. Finite element eigenvalue extraction analyses were undertaken using the ABAQUS finite element software package to predict the natural frequencies and modes of vibration of the UBB in a free-free condition. The structure was modelled using the lower order continuous beam element implemented by ABAQUS. This element includes rotary inertia terms in its mass matrix and is capable of transverse shear deflections. The lower order elements are shown to give better eigenvalue for a given mesh density, (ref. 10).

18. The node assemblies, member end connectors and composite members were modelled using beam elements with isotropic material definitions. The material properties of the carbon fibre tubes were orthotropic, (ref. 11) but the dynamic response of the lower modes of these tubes can be shown to be accurately represented by a simplified isotropic material model, (ref. 6). The carbon fibre tubes for the UBB are 25.0mm internal diameter with 1.2mm wall thicknesses. Their respective material properties, used for the isotropic definition, calculated from static stiffness tests, (ref. 6), are E= 118.6GN/m^2 and G= 15GN/m^2. The mechanical properties used for aluminium and titanium are E=68.0GN/m^2, G=26.0GN/m^2 and E=110.0GN/m^2, G=42.0GN/m^2 respectively. The composite tubes are bonded to the member end connectors using an araldite adhesive and complete transfer of moments and forces across these bonds were assumed for the numerical model. The first five natural frequencies predicted by the numerical model are given in the third column of Table 1.

Comparison of experimental and numerical results

19. The frequencies of the unit predicted by numerical analyses were observed to be consistently about 10% greater than those measured experimentally. The analysis of single carbon fibre tubes experimentally and modelled numerically using an isotropic material definition, in (ref. 6), were seen to vary by 3% only. The difference in results was attributed to the difficulty in accurately modelling the semi-rigid fixing of the two. The difference between these two values is observed to be less than 2.5% for each of the measured modes using the two different excitation techniques.

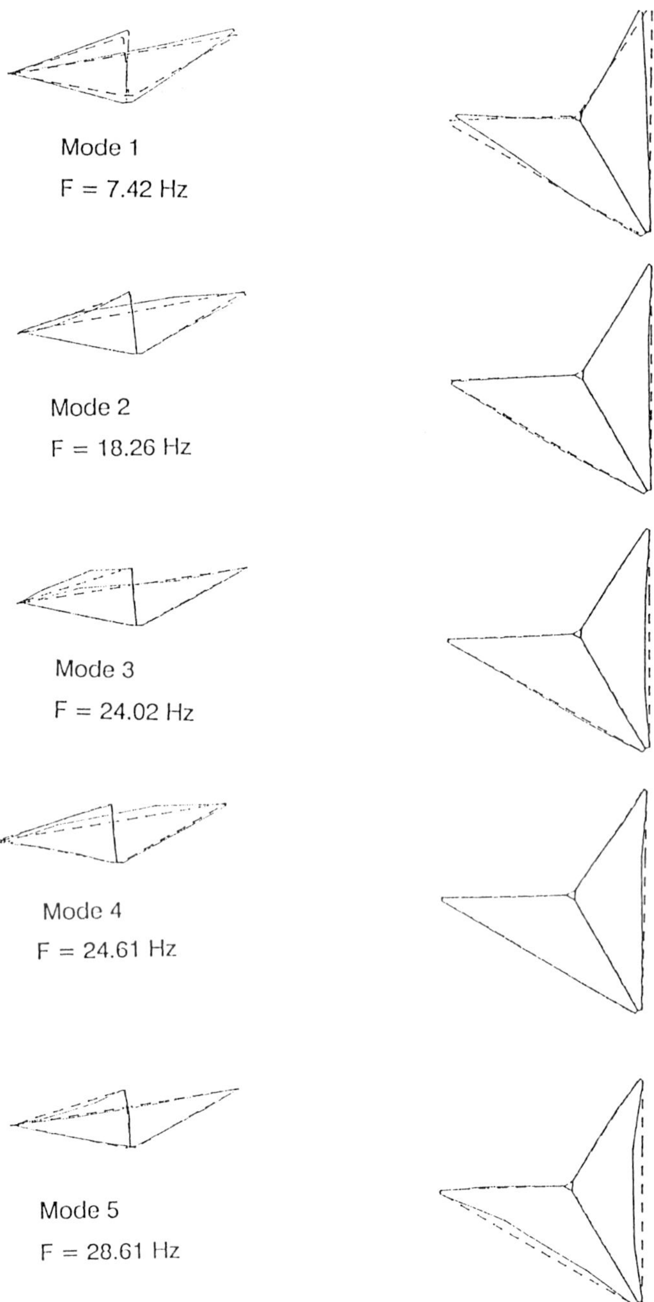

Fig. 6. Three dimensional and plan views of the first five mode shapes of the UBB obtained experimentally.

20. Table 4 shows the orthogonality checks for the numerical modes compared to those obtained using the impact hammer. The off axis terms for mode numbers one, two and five are low, <0.1, and the main diagonal terms approach unity indicating a good numerical representation of these modes. The close proximity of mode numbers three and four and the similarity of their mode shapes made it difficult to identify each mode accurately using single point excitation.

Table 4. Impact hammer modes compared with numerical modes.

	Mode	1	2	3	4	5
	1	0.96	0.01	0.04	0.03	0.06
E	2	0.04	0.93	-0.13	-0.19	-0.0
X	3	0.12	-0.08	0.52	0.58	0.03
P	4	-0.08	0.4	-0.75	0.07	-0.08
	5	0.02	-0.01	0.04	-0.0	0.97

21. Table 5 shows the comparison between the experimental modes measured using the mechanical shaker and the numerical analyses. The coefficients on the main diagonal, for mode numbers one, two and five, are lower than those achieved for comparison with the impact hammer results.

Table 5. Mechanical vibrator modes compared with numerical modes.

	Mode	1	2	3	4	5
	1	0.75	0.05	-0.09	0.01	0.05
E	2	0.08	0.86	-0.12	0.19	-0.15
X	3	0.08	0.06	0.17	0.89	0.22
P	4	-0.01	0.13	0.46	0.7	0.18
	5	-0.02	0.13	-0.07	0.32	0.82

CONCLUSIONS

22. The choice of external excitation was seen to influence the quality of the experimental results and subsequent comparison with numerical predictions. The virtually non-intrusive nature of the impact testing techniques was shown to produce a better set of experimental modes than the shaker test for this rather flexible structure.

23. The comparison of the experimental and numerical results of the UBB, manufactured from aluminium, titanium and composite elements, using a number of differing techniques to join these materials together, showed that the natural frequencies could be reliably predicted with good mode shape representation.

REFERENCES:

1. HOLLAWAY, L., THORNE A., SPARRY D., O'NEILL M., and YORK D. "Large space structures - their implications and requirements", International Journal of Space Structures, Vol. 6, No. 1, 1991, pp 1-10.

2. RUSSELL A., "Development of a large deployable space structure", Space Structures, Design, Deployment and Stability Conference, 6th November, 1991, Institution of Mechanical Engineers, 1 Bird Cage Walk, Westminster, London SW1 H9JJ.

3. FREELAND R., "Survey of deployable antenna concepts", in: Satellite Communication Antenna Technology, Elsevier Science Publishers B.V. (North Holland), 1983, pp 613-652.

4. GARRET L.B. and FEREBEC M.J., "Systems Design and Comparative Analysis of Large Antenna Concepts", Large Space Antenna Systems Technology, NASA CP2269, Part 1, 1982.

5. ANDERSON G.C., GARRET L.B. and CALLESTON R.E. "Comparative analysis of on-orbit dynamic performance of several large antenna concepts", Proceedings of the AIAA/ASME/ASCE/AHS 26th Structures, Structural Dynamics and Materials Conference, 1985, pp 707-721.

6. FANNING P., "Development and analyses of a deployable skeletal reflector for spacecraft antennas", PhD Thesis to be presented in May, 1993.

7. HOLLAWAY L., "Pultrusions", Chapter 1, 'Developments in Plastics Technology III', ed. A. Whelan & J.L. Craft, Pub., Elsevier Science Publishers, 1986.

8. EWINS D.J., "Whys and wherefores of modal testing", Jnl. Soc. Env. Engrs., 1979, V18, pt.3, pp 3-15.

9. TARGOFF W. P., "Orthogonality check and correction of measured modes", AIAA Journal, Vol.14, No.2, 1976, pp 164-167

10. HIBBIT, KARLSSON and SORENSON Inc., "ABAQUS", Version 4.5, User's Manual, 1985.

11. TSAI S.W., "Introduction to Composite Materials", 1980, Technomic Publ. Co., Inc.

48. The reliability and sensitivity of space truss structures under time-variable loads

P. SNIADY and S. ZUKOWSKI, Technical University of Wroclaw, Poland

SYNOPSIS. The problem of reliability and sensitivity of space truss structures, subjected to time-varying loads, is considered. Taking into account the possible operation of the structure in the range of elastic-plastic strains it is proposed to evaluate its reliability and sensitivity in terms of the shakedown theory for different stochastic models of loading.

INTRODUCTION

1. A considerable number of papers have been published in recent years which are devoted to the evaluation of structure reliability taking account of plastic strains (refs 5-7). In most cases, the problem is considered in respect to the limiting of load capacity. In the case of time variable loads these analyses are inadequate or unsatisfactory. When dealing with a load which is changing in time it is necessary to consider the reliability problem in the elastic range of materials or in terms of the shakedown theory (refs 2,3).

2. In this paper, on the basis of the Melan's shakedown theorem, the reliability and sensitivity problem is solved. The presented relations allow, through the development of computer software, for the successful computation of reliability and sensitivity, using procedures or programs for the deterministic analysis of trusses.

THE PROBABILISTIC CHARACTERISTICS OF EXTREME AXIAL FORCES IN BARS

3. The maximum and minimum axial forces in bars of the elastic system may be determined by the relationships

$$\begin{aligned} \max \mathbf{N} &= \mathbb{N}^T \cdot (\Phi \cdot \mathbf{P}^{\wedge} + \Psi \cdot \mathbf{P}^{\vee}) \\ \min \mathbf{N} &= \mathbb{N}^T \cdot (\Psi \cdot \mathbf{P}^{\wedge} + \Phi \cdot \mathbf{P}^{\vee}) \end{aligned} \qquad (1)$$

where: ^{T} denotes the transposition of a matrix,
max $\mathbf{N} = [\max N_1,\dots, \max N_n]^T$ is a vector of components being the maximum axial forces in bars,
min $\mathbf{N} = [\min N_1,\dots, \min N_n]$ is a vector of components being the minimum axial forces in bars,

$$\mathbb{N} = \begin{bmatrix} N_1^1, \dots, N_1^m \\ \vdots \quad\quad \vdots \\ N_n^1, \dots, N_n^m \end{bmatrix}$$

is a matrix, elements of which are bar axial forces caused by unity forces acting on subsequent nodes,

$$\Phi = \begin{bmatrix} \varphi_1^1, \dots, \varphi_1^m \\ \vdots \quad\quad \vdots \\ \varphi_n^1, \dots, \varphi_n^m \end{bmatrix}, \quad \Psi = \begin{bmatrix} 1-\varphi_1^1, \dots, 1-\varphi_1^m \\ \vdots \quad\quad \vdots \\ 1-\varphi_n^1, \dots, 1-\varphi_n^m \end{bmatrix}$$

are deterministic matrices, elements of which are determined by the parameters

$$\varphi_i^j = \begin{cases} 0 \text{ for } N_i < 0 \\ 1 \text{ for } N_i^j > 0 \end{cases},$$

$$\mathbf{P}^\wedge = [P_1^\wedge, \dots, P_m^\wedge]^T, \quad \mathbf{P}^\vee = [P_1^\vee, \dots, P_m^\vee]^T$$

are vectors of components being the upper and lower boundaries of load parameters, respectively, n is the number of bars, m is the number of parameters determining the load.

The expressions for random fluctuations of the extreme axial forces are:

$$\begin{aligned} \max{}^\sim\mathbf{N} &= \mathring{\mathbb{N}}^T \cdot (\Phi \cdot \tilde{\mathbf{P}}^\wedge + \Psi \cdot \tilde{\mathbf{P}}^\vee) + \tilde{\mathbb{N}}^T \cdot (\Phi \cdot \mathring{\mathbf{P}}^\wedge + \Psi \cdot \mathring{\mathbf{P}}^\vee) \\ \min{}^\sim\mathbf{N} &= \mathring{\mathbb{N}}^T \cdot (\Psi \cdot \tilde{\mathbf{P}}^\wedge + \Phi \cdot \tilde{\mathbf{P}}^\vee) + \tilde{\mathbb{N}}^T \cdot (\Psi \cdot \mathring{\mathbf{P}}^\wedge + \Phi \cdot \mathring{\mathbf{P}}^\vee) \end{aligned} \tag{2}$$

where the marks o and ~ over symbols denote the mean value and the random fluctuation of the given item, respectively. The axial forces in the elastic truss due to a unit load $P_j = 1$ may be obtained by the stiffness method or by the flexibility method for the case of the statically indeterminate system.

4. By the linearization method the equation of the stiffness method becomes (ref. 1):

$$\mathring{\mathbb{K}} \cdot \mathring{\mathbb{Q}} + \mathring{\mathbb{K}} \cdot \tilde{\mathbb{Q}} + \tilde{\mathbb{K}} \cdot \mathring{\mathbb{Q}} = \mathring{\mathbb{F}} \tag{3}$$

where:

is a stiffness matrix,
is a matrix, elements of which are vectors of displacements caused by unit loads $P_j = 1$,
is a deterministic matrix, elements of which are load vectors in the global base, specified by unit value parameters Pj.

From the relationship given in Eq. 3 one may derive the matrix of mean displacements:

$$\mathring{\mathbb{Q}} = \mathring{\mathbb{D}} \cdot \mathring{\mathbb{F}} \tag{4}$$

and the matrix of the random displacement fluctuations:

$$\tilde{\mathbb{Q}} = - \mathring{\mathbb{D}} \cdot \tilde{\mathbb{K}} \cdot \mathring{\mathbb{Q}} \tag{5}$$

where:

$$\mathring{\mathbb{D}} = \mathring{\mathbb{K}}^{-1} .$$

The matrices of mean axial forces and their random fluctuations may be rewritten as:

$$\mathring{\mathbb{N}} = \mathring{\mathbb{H}} \cdot \mathring{\mathbb{Q}}$$

$$\tilde{\mathbb{N}} = \mathring{\mathbb{H}} \cdot \tilde{\mathbb{Q}} + \tilde{\mathbb{H}} \cdot \mathring{\mathbb{Q}} \tag{6}$$

where $\mathbb{H}$ is a transformation matrix of node displacements onto bar axial forces.

5. The alterations of the direction cosines due to small random fluctuations of structural dimensions are small with respect to this fluctuation and may be disregarded. The stiffness matrix and the node displacement to axial forces transformation matrix are functions of the Young's moduli E_i, bar cross-sections A_i and bar lengths L_i. The random fluctuations of the matrices may be represented as:

$$\tilde{\mathbb{K}} = \sum_{i=1}^{n} \mathring{\mathbb{K}}^{i} \cdot (\tilde{e}_i + \tilde{a}_i - \tilde{l}_i)$$

$$\tilde{\mathbb{H}} = \sum_{i=1}^{n} \mathring{\mathbb{H}}^{i} \cdot (\tilde{e}_i + \tilde{a}_i - \tilde{l}_i) \tag{7}$$

where:

$$\mathring{\mathbb{K}}^i = \frac{\partial \mathbb{K}}{\partial E_i}\Bigg|_{A_i=\mathring{A}_i,\, L_i=\mathring{L}_i} \cdot \mathring{E}_i = \frac{\partial \mathbb{K}}{\partial A_i}\Bigg|_{E_i=\mathring{E}_i,\, L_i=\mathring{L}_i} \cdot \mathring{A}_i = -\frac{\partial \mathbb{K}}{\partial L_i}\Bigg|_{E_i=\mathring{E}_i,\, A_i=\mathring{A}_i} \cdot \mathring{L}_i ,$$

$$\mathring{\mathbb{H}}^i = \frac{\partial \mathbb{H}}{\partial E_i}\Bigg|_{A_i=\mathring{A}_i,\, L_i=\mathring{L}_i} \cdot \mathring{E}_i = \frac{\partial \mathbb{H}}{\partial A_i}\Bigg|_{E_i=\mathring{E}_i,\, L_i=\mathring{L}_i} \cdot \mathring{A}_i = -\frac{\partial \mathbb{H}}{\partial L_i}\Bigg|_{E_i=\mathring{E}_i,\, A_i=\mathring{A}_i} \cdot \mathring{L}_i ,$$

$$\tilde{e}_i = \tilde{E}_i / \mathring{E}_i , \quad \tilde{a}_i = \tilde{A}_i / \mathring{A}_i , \quad \tilde{l}_i = \tilde{L}_i / \mathring{L}_i$$

are normalised random fluctuations of parameters E_j, A_i, L_i, respectively. These parameters are mutually independent variables. For different bars i and j, the variables may be inter-correlated in groups, i.e. E_i and E_j, A_i and A_j, or L_i and L_j.

After substituting Eq. 7 into Eqs 5 and 6 the fluctuation matrices of random displacements and axial forces caused by unit loads $P_j = 1$ become:

$$\begin{aligned} \tilde{\mathbb{Q}} &= -\sum_{i=1}^{n} \mathring{\mathbb{D}} \cdot \mathring{\mathbb{K}}^i \cdot \mathring{\mathbb{Q}} \cdot (\tilde{e}_i + \tilde{a}_i - \tilde{l}_i) \\ \tilde{\mathbb{N}} &= \sum_{i=1}^{n} (\mathring{\mathbb{H}}^i - \mathring{\mathbb{H}} \cdot \mathring{\mathbb{D}} \cdot \mathring{\mathbb{K}}^i) \cdot \mathring{\mathbb{Q}} \cdot (\tilde{e}_i + \tilde{a}_i - \tilde{l}_i) \end{aligned} \tag{8}$$

After inserting Eq. 8 into Eq. 2 the expressions determining the random fluctuations of the extreme axial forces are as follows:

$$\begin{aligned} \max \tilde{\mathbf{N}} &= \sum_{i=1}^{n} \max \mathring{\mathbf{N}}^i \cdot (\tilde{e}_i + \tilde{a}_i - \tilde{l}_i) + \\ &\quad + \sum_{j=1}^{m} \max \mathring{\mathbf{N}}_j^{\wedge} \cdot \tilde{p}_j^{\wedge} + \sum_{j=1}^{m} \max \mathring{\mathbf{N}}_j^{\vee} \cdot \tilde{p}_j^{\vee} , \\ \min \tilde{\mathbf{N}} &= \sum_{i=1}^{n} \min \mathring{\mathbf{N}}^i \cdot (\tilde{e}_i + \tilde{a}_i - \tilde{l}_i) + \\ &\quad + \sum_{j=1}^{m} \min \mathring{\mathbf{N}}_j^{\wedge} \cdot \tilde{p}_j^{\wedge} + \sum_{j=1}^{m} \min \mathring{\mathbf{N}}_j^{\vee} \cdot \tilde{p}_j^{\vee} , \end{aligned} \tag{9}$$

where
$$\max\mathring{\mathbf{N}}^{i} = \mathring{\mathbb{Q}}^{T}\cdot(\mathring{\mathbb{H}}^{i} - \mathring{\mathbb{H}}\cdot\mathring{\mathbb{D}}\cdot\mathring{\mathbb{K}}^{i})^{T}\cdot(\Phi\cdot\mathring{\mathbf{P}}^{\wedge} + \Psi\cdot\mathring{\mathbf{P}}^{\vee}),$$
$$\min\mathring{\mathbf{N}}^{i} = \mathring{\mathbb{Q}}^{T}\cdot(\mathring{\mathbb{H}}^{i} - \mathring{\mathbb{H}}\cdot\mathring{\mathbb{D}}\cdot\mathring{\mathbb{K}}^{i})^{T}\cdot(\Psi\cdot\mathring{\mathbf{P}}^{\wedge} + \Phi\cdot\mathring{\mathbf{P}}^{\vee}),$$
$$\max\mathring{\mathbf{N}}_{j}^{\wedge} = \mathring{\mathbb{N}}^{T}\cdot\Phi\cdot\mathring{\mathbf{P}}_{j}^{\wedge}, \qquad \min\mathring{\mathbf{N}}_{j}^{\wedge} = \mathring{\mathbb{N}}^{T}\cdot\Psi\cdot\mathring{\mathbf{P}}_{j}^{\wedge},$$
$$\max\mathring{\mathbf{N}}_{j}^{\vee} = \mathring{\mathbb{N}}^{T}\cdot\Psi\cdot\mathring{\mathbf{P}}_{j}^{\vee}, \qquad \min\mathring{\mathbf{N}}_{j}^{\vee} = \mathring{\mathbb{N}}^{T}\cdot\Phi\cdot\mathring{\mathbf{P}}_{j}^{\vee},$$
$$\tilde{p}_{j}^{\wedge} = \tilde{P}_{j}^{\wedge}/\mathring{P}_{j}^{\wedge}, \qquad \tilde{p}_{j}^{\vee} = \tilde{P}_{j}^{\vee}/\mathring{P}_{i}^{\vee}$$

are normalised random fluctuations of lower and upper bounds of the load parameters, respectively.

SHAKEDOWN

6. In the case of a structure fabricated from an elastic-plastic material and loaded with alternating load, the safe usage of the plastic reserve is allowed by the shakedown theory.

7. Following the Melan's shakedown theorem the truss shakedown problem may be described as a linear programming task:

find max μ

while:
$$\begin{aligned} \mathbf{n}.\mathbf{X}^{R} + \max\mathbf{N}.\mu &\le \mathbf{N}_{0} \\ -\mathbf{n}.\mathbf{X}^{R} - \max\mathbf{N}.\mu &\le \mathbf{N}_{0} \\ \mu &\ge 0 \end{aligned} \tag{10}$$

where: $\mathbf{N}_{0} = \mathbf{N}_{e}$ is a vector of the plastic and elastic load capacity of bars under tension,
n is a deterministic influence matrix of the bar axial forces for the unit hyperstatic loads,
$\mathbf{X}^{R}$ is a vector of the residual hyperstatic forces,
μ is the load bounds multiplier.

The solutions of the problem Eq. 10 are: the load bounds multiplier μ, the residual hyperstatic forces $\mathbf{X}^{R}$ and the dual variables q^{+}, q^{-}. The latter are the velocities of plastic longitudinal strains of bars which specify the mechanism of plastic flow in the unshakedown state corresponding to the optimum solution. For the optimum solution, some of the constraints are satisfied as equalities, and the rest of them as inequalities. The equality-satisfied constraints (the number of which is equal to the degree of the statical indeterminacy of the structure plus one) lead to the linear equation system:

$$\mathbb{A}\cdot\mathbf{Y} = \mathbf{B} \tag{11}$$

where:

$$\mathbb{A} = \begin{bmatrix} \mathbf{n}^*, & \max \mathbf{N}^* \\ -\mathbf{n}^*, & -\min \mathbf{N}^* \end{bmatrix} = [\mathbf{u}, \ \mathbf{U}], \quad \mathbf{Y} = \begin{bmatrix} \mathbf{X}^R \\ \mu \end{bmatrix}, \quad \mathbf{B} = \begin{bmatrix} \mathbf{N_0}^* \\ \mathbf{N_0}^* \end{bmatrix},$$

$\mathbf{n}^*$, max $\mathbf{N}^*$, min $\mathbf{N}^*$, $\mathbf{N_0}^*$ are the matrices $\mathbf{n}$, max $\mathbf{N}$, min $\mathbf{N}$, $\mathbf{N_0}$, respectively, in which appropriate rows were erased .

8. The random fluctuations of the matrices $\mathbb{A}$ and $\mathbf{B}$ are of the form:

$$\tilde{\mathbb{A}} = [0, \ \tilde{\mathbf{U}}] = \sum_{i=1}^{n} [0, \ \mathring{\mathbf{U}}^i]\cdot(\tilde{e}_i + \tilde{a}_i - \tilde{I}_i) + \sum_{j=1}^{m} [0, \ \mathring{\mathbf{U}}_j^\wedge]\cdot\tilde{p}_j^\wedge + \sum_{j=1}^{m} [0, \ \mathring{\mathbf{U}}_j^\vee]\cdot\tilde{p}_j^\vee, \tag{12}$$

$$\tilde{\mathbf{B}} = \sum_{i=1}^{n} \mathring{\mathbf{B}}^i\cdot\tilde{n}_i \tag{13}$$

where the matrices:

$$\mathring{\mathbf{U}}^i, \quad \mathring{\mathbf{U}}_j^\wedge, \quad \mathring{\mathbf{U}}_j^\vee$$

are defined by Eq. 9.

$$\mathring{\mathbf{B}}^i = \frac{\partial \mathbf{B}}{\partial N_{0\,i}}\cdot \mathring{N}_{0\,i}, \qquad \tilde{n}_i = \tilde{N}_{0\,i} / \mathring{N}_{0\,i}.$$

9. The random fluctuation of the vector of unknown values $\mathbf{Y}$ is determined as

$$\mathring{\mathbb{A}}\cdot\tilde{\mathbf{Y}} = \tilde{\mathbf{B}} - \tilde{\mathbb{A}}\cdot\mathring{\mathbf{Y}} \tag{14}$$

and from the above relationship the random fluctuation of the multiplier μ, which specifies the load level corresponding to the load capacity, may be represented as follows:

$$\tilde{\mu} = \mathring{\mathbf{G}}\cdot(\tilde{\mathbf{B}} - \tilde{\mathbf{U}}\cdot\mathring{\mu}) = \sum_{i=1}^{n} \mathring{\mathbf{G}}\cdot\mathring{\mathbf{B}}^i\cdot\tilde{n}_i + \sum_{i=1}^{n} \mathring{\mathbf{G}}\cdot\mathring{\mathbf{U}}^i\cdot(\tilde{I}_i - \tilde{e}_i - \tilde{a}_i) - \sum_{j=1}^{m} \mathring{\mathbf{G}}\cdot\mathring{\mathbf{U}}_j^\wedge\cdot\tilde{p}_j^\wedge - \sum_{j=1}^{m} \mathring{\mathbf{G}}\cdot\mathring{\mathbf{U}}_j^\vee\cdot\tilde{p}_j^\vee, \tag{15}$$

where:

$$\mathring{G} = [\mathring{G}_{k1}, \ldots, \mathring{G}_{kk}]$$

is the last row of the matrix:

$$\mathring{G} = \mathring{A}^{-1}.$$

The variance of parameter μ is given as:

$$\begin{aligned} Var(\mu) = E[\tilde{\mu}\cdot\tilde{\mu}^T] = \sum_{i=1}^{n}\sum_{j=1}^{n} \Big(&\mathring{G}\cdot\mathring{B}^i\cdot(\mathring{G}\cdot\mathring{B}^j)^T\cdot n_{ij} + \\ &+ \mathring{G}\cdot\mathring{U}^i\cdot(\mathring{G}\cdot\mathring{U}^j)^T\cdot(e_{ij} + a_{ij} + 1_{ij})\Big) + \\ + \sum_{i=1}^{m}\sum_{j=1}^{m}\Big(&\mathring{G}\cdot\mathring{U}_i^{\wedge}\cdot(\mathring{G}\cdot\mathring{U}_j^{\wedge})^T\cdot p_{ij}^{\wedge\wedge} + \mathring{G}\cdot\mathring{U}_i^{\vee}\cdot(\mathring{G}\cdot\mathring{U}_j^{\vee})^T\cdot p_{ij}^{\vee\vee} + \\ &+ 2\cdot\mathring{G}\cdot\mathring{U}_i^{\wedge}\cdot(\mathring{G}\cdot\mathring{U}_j^{\vee})^T\cdot p_{ij}^{\wedge\vee}\Big) \end{aligned} \tag{16}$$

where E[·] denotes the expected value operator,

$$n_{ij} = E[\tilde{n}_i\cdot\tilde{n}_j] = r_{NN}^{ij}\cdot V_N^i\cdot V_N^j, \quad e_{ij} = E[\tilde{e}_i\cdot\tilde{e}_j] = r_{EE}^{ij}\cdot V_E^i\cdot V_E^j,$$

$$a_{ij} = E[\tilde{a}_i\cdot\tilde{a}_j] = r_{AA}^{ij}\cdot V_A^i\cdot V_A^j, \quad p_{ij}^{\wedge\wedge} = E[\tilde{p}_i^{\wedge}\cdot\tilde{p}_j^{\wedge}] = r_{\hat{P}\hat{P}}^{ij}\cdot V_{\hat{P}}^i\cdot V_{\hat{P}}^j,$$

$$p_{ij}^{\wedge\vee} = E[\tilde{p}_i^{\wedge}\cdot\tilde{p}_j^{\vee}] = r_{\hat{P}\check{P}}^{ij}\cdot V_{\hat{P}}^i\cdot V_{\check{P}}^j, \quad p_{ij}^{\vee\vee} = E[\tilde{p}_i^{\vee}\cdot\tilde{p}_j^{\vee}] = r_{\check{P}\check{P}}^{ij}\cdot V_{\check{P}}^i\cdot V_{\check{P}}^j,$$

$$r_{NN}^{ij},\ r_{EE}^{ij},\ r_{AA}^{ij},\ r_{\hat{P}\hat{P}}^{ij},\ r_{\hat{P}\check{P}}^{ij},\ r_{\check{P}\check{P}}^{ij}$$

are the correlation coefficients of the parameters No_i and No_j, E_i and E_j, A_i and A_j, $P_i^{\wedge}$ and $P_j^{\wedge}$, $P_i^{\wedge}$ and $P_j^{\vee}$, $P_i^{\vee}$ and $P_j^{\vee}$, respectively and V_N^i, V_N^j, V_E^i, V_E^j, V_A^i, V_A^j, $V_{\hat{P}}^i.V_{\hat{P}}^j$, $V_{\hat{P}}^i.V_{\check{P}}^j$, $V_{\check{P}}^i.V_{\check{P}}^j$, are the variation coefficients for the parameters mentioned.

RELIABILITY INDEX

10. Denoting by S the worst-case load of the structure corresponding to the assumed load variation program, the load capacity (R) of the structure may be presented as:

$$R = S.\mu \tag{17}$$

and the normalised load capacity reserve as:

$$z = (R - S)/S = \mu - 1. \tag{18}$$

The reliability index is then:

$$\beta = \frac{\mathring{\mu} - 1}{\sqrt{\mathrm{Var}(\mu)}} . \tag{19}$$

SENSITIVITY

11. The sensitivity measure for a given quantity due to alteration of any deterministic or probabilistic parameter was assumed (ref. 4).

12. By virtue of Eq. 11 and taking into account the character of the matrix **s** the sensitivity of the mean value of the load capacity parameter μ due to variations of any parameters α may be shown as:

$$\mathring{\mu},_{\alpha} = \mathring{\mathbf{G}}\cdot(\mathring{\mathbf{B}},_{\alpha} - \mathring{\mathbf{U}},_{\alpha}\cdot\mathring{\mu}), \tag{20}$$

The sensitivity of the reliability index β is:

$$\beta,_{\alpha} = \mathring{\mu},_{\alpha}/\sqrt{\mathrm{Var}(\mu)} - \beta\cdot\mathrm{Var},_{\alpha}(\mu)/\mathrm{Var}(\mu) \tag{21}$$

The sensitivity of the variation of the multiplier μ, $(\mathrm{Var},_{\alpha}(\mu))$ is determined on the basis of Eq. 14. Right-side multiplication of the left and right side of Eq. 14 by:

$$(\mathring{\mathbf{A}}\cdot\tilde{\mathbf{Y}})^{T} \quad \text{and} \quad (\tilde{\mathbf{B}} - \tilde{\mathbf{A}}\cdot\mathring{\mathbf{Y}})^{T},$$

respectively, and expected value operation leads to:

$$\mathring{\mathbf{A}}\cdot C_{YY}\cdot\mathring{\mathbf{A}}^{T} = C_{BB} + \mathring{\mu}^{2}\cdot C_{UU} \tag{22}$$

where: $C_{YY} = E[\tilde{\mathbf{Y}} \cdot \tilde{\mathbf{Y}}^T],$

$$C_{BB} = E[\tilde{\mathbf{B}} \cdot \tilde{\mathbf{B}}^T] = \sum_{i=1}^{n} \sum_{j=1}^{n} \mathring{\mathbf{B}}^i \cdot (\mathring{\mathbf{B}}^j)^T \cdot n_{ij},$$

$$C_{UU} = E[\tilde{\mathbf{U}} \cdot \tilde{\mathbf{U}}^T] = \sum_{i=1}^{n} \sum_{j=1}^{n} \mathring{\mathbf{U}}^i \cdot (\mathring{\mathbf{U}}^j)^T \cdot (e_{ij} + a_{ij} + l_{ij}) +$$

$$+ \sum_{i=1}^{m} \sum_{j=1}^{m} [\mathring{\mathbf{U}}_i^{\wedge} \cdot (\mathring{\mathbf{U}}_j^{\wedge})^T \cdot p_{ij}^{\wedge\wedge} + \mathring{\mathbf{U}}_i^{\wedge} \cdot (\mathring{\mathbf{U}}_j^{\vee})^T \cdot p_{ij}^{\wedge\vee} + \mathring{\mathbf{U}}_i^{\vee} \cdot (\mathring{\mathbf{U}}_j^{\vee})^T \cdot p_{ij}^{\vee\vee}].$$

On the basis of Eq. 22 the formulae to calculate the sensitivity of variation of the multiplier μ, which is the last element of the vector $\mathbf{Y}$, has the form:

$$\mathrm{Var},_{\alpha}(\mu) = \mathring{\mathbf{G}} \cdot [C_{YY},_{\alpha} = \mathring{\mathbf{A}}^{-1} \cdot (C_{BB},_{\alpha} + 2 \cdot \mathring{\mu} \cdot \mathring{\mu},_{\alpha} \cdot C_{UU} +$$

$$+ \mathring{\mu}^2 \cdot C_{UU},_{\alpha} - \mathring{\mathbf{A}},_{\alpha} \cdot C_{UU} \cdot \mathring{\mathbf{A}}^T + \mathring{\mathbf{A}} \cdot C_{UU} \cdot \mathring{\mathbf{A}}^T,_{\alpha}] \cdot \mathbf{G}^T. \tag{23}$$

FINAL REMARKS

13. The relations derived allow the determination of the first and second order probabilistic characteristics of the load capacity of the truss with respect to the shakedown, the structure reliability index, and the sensitivities of the quantities due to variations of any deterministic and probabilistic parameters.

14. The probabilistic characteristics of the investigated quantities and their sensitivities are determined by solving the deterministic problem and by the probabilistic characteristics of the basic physical and geometrical parameters of the structure and the bounds of load variation. These relations allow the creation of computer software for the successful computation of reliability and sensitivity, using procedures or programs for the deterministic analysis of trusses.

15. The numerical test calculations show that the algorithm introduced here allows the effective reliability and sensitivity analysis of plane and space trusses in the scope given.

ACKNOWLEDGEMENT

This research was supported by the Scientific Research Committee in Varsov under grant number 772309203.

REFERENCES

1. SNIADY P.and ZUKOWSKI S. Probabilistic analysis of trusses, Archiwum Inzynierii Ladowej, Part 2, 1988, 175-189 (in Polish).

2. KONIG J.A. Shakedown of elastic-plastic structures, PWN, Warszawa, 1987.

3. SNIADY P.and ZUKOWSKI S. The reliability of bar structures in the light of the shakedown theory, Proc.of the 4th IFIP WG 7.5 Conf. on Reliability and optimization of structural systems, Munich, September 11-13, 1991, Springer-Verlag, IFIP 76, 351-362.
4. HAUNG E.J., CHOI K.K.and KOMKOW V. Design sensitivity analysis of Structural Systems, Acad.Press, 1985.
5. MUROTSU Y. Development in structural systems reliability theory. Transactions 8th Inter. Conf. on Struct. Mech. in Reactor Technology 1985, M1, 7-14.
6. MELCHERS R. E. and TANG L. K. Dominant failure modes in stochastic structural systems, Structural Safety, 1984, 2, 127-143.
7. RANGANATHAN R.and DESHANDE A.G. Generation of dominant modes and reliability analysis of frames, Structural Safety, 1987, 4, 217-288.

49. New knowledge of the fatigue behaviour of steel structures and especially of hollow section structures under different load spectra

O. BUCAK and F. MANG, Universitat Karlsruhe, Germany

SYNOPSIS. Many constant amplitude (CA) tests with different notch cases and types of joints (T-, X-, N-, K,...) made from hollow sections, have been performed and the results were made available to the user as S-N-line responses (ref. 2) some years ago. Based on these investigations, design rules for the standard specification Eurocode 3 (ref. 1) have been written. It is possible to design a structure under fatigue loading economically, if a realistic load spectra and cumulative damage assessment is used. In the following, the results of the fatigue investigations under variable amplitude on hollow section joints are given, and the application of the Miner-rule for structures under variable loads is shown.

SYMBOLS AND DEFINITIONS

CA	Constant Amplitude	d	Diameter of the CHS
VA	Variable Amplitude	d_0	Diameter of the chord member CHS
CHS	Circular Hollow Sections	b	Width of the RHS
RHS	Rectangular Hollow Sections	b_0	Width of the chord member RHS
R	Stress ratio R = max σ/min σ	t	Wall thickness
C	Constants	t_0	Wall thickness of the chord member
S	Miner-sum	m	Slope of the S-N-line
S_R	Stress range [N/mm^2]; $S_R = (\sigma_{max} - \sigma_{min})$	σ	Stress [N/mm^2]
		σ_m	Maximum stress of a load sequence [N/mm^2]
Pü	Probability of survival	c	Safety factor
GL	Germanisch Lloyd	WG3	ECCS-Working Group 3
IABG	Industrieanlagen-Betriebs-gesellschaft, Ottobrunn	LBF	Laboratorium für Betriebs-festigkeit Darmstadt

f_E Increase factor for the stress range $f_E = \dfrac{\Delta\sigma_{var.ampl.}}{\Delta\sigma_{const.ampl.}}$

f_A Load increment coefficient for the collective A (similar to f_E).

f_B load increment coefficient for the collective B (similar to f_E).

V_A Fullness of the load sequence A. V_B Fullness of the load sequence B.

Space structures 4. Thomas Telford, London, 1993

TESTING PROGRAMME

1. Systematic tests on hollow section joints subjected to spectrum loading have been carried out at first with X- and K-type test specimens (Fig. 1) under axial load, applying load spectra known from literature (refs 4, 7, 8, 11 etc.).

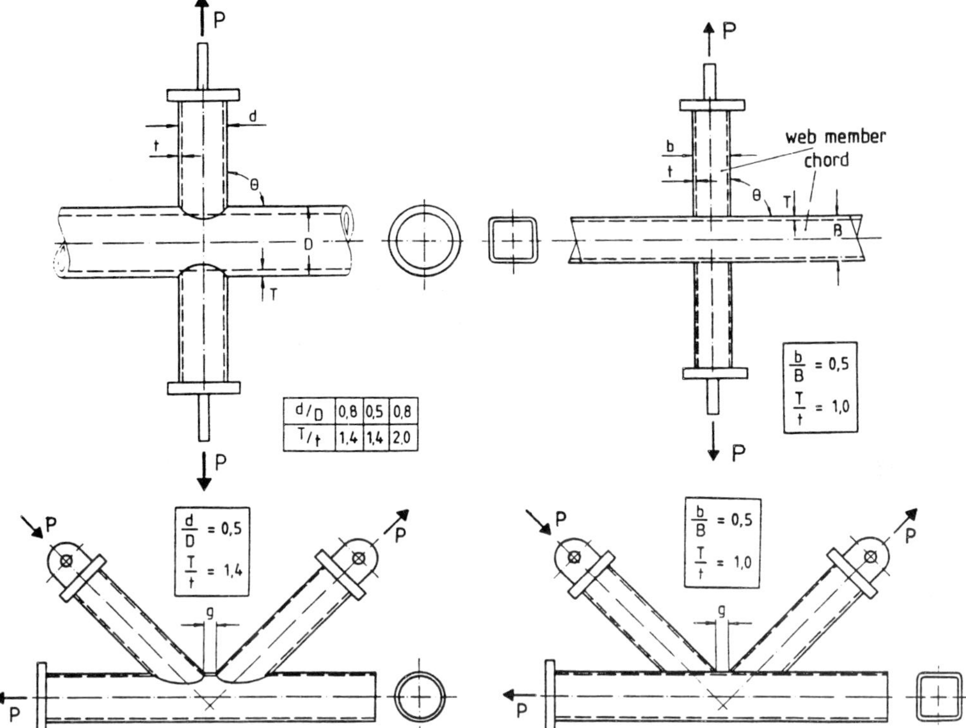

Fig. 1. X- and K-type joints made from hollow sections (circular and rectangular) for the tests carried out in Karlsruhe.

2. The basis for this work are fatigue tests carried out under constant amplitude (S-N lines). Based on the CA-tests, block tests were performed under the load sequence P = 1/3 and P = 2/3 (see Fig. 2), which are known to be necessary for cranes and crane railways (refs 7, 8, 11).

3. By means of the load spectra 'S-III Laplace' (ref. 7), 'North Sea' (ref. 7) and 'North Sea Sequence' (ref. 12), a connection to the European Offshore Research was established. Two further load spectra, one structure subjected to wind (mast, 200m high) the other one to the wave loads of the Gulf of Mexico completed the testing programme.

4. Fig. 2 shows a compilation of the applied load spectra. The test values under load spectra have been compared with those of the Wöhler curves (S-N-lines) and the applicability of the Miner-rule has been checked.

5. Endurance tests will be evaluated by means of the Miner-rule whose accuracy and applicability will be examined in the course of the present investigations with regard to the treated specimen form and the load spectra applied.

Table 1. Data of the test specimens.

Type of joints		chord member	bracing	material	R
CHS	X	Ø101.6 x 5.6 Ø101.6 x 5.6 Ø101.6 x 8.0	Ø51 x 4.0 Ø82 x 4.0 Ø82 x 4.0	Fe 360	+0.1 +0.1 -1.0
	K	Ø101.6 x 5.6	Ø51 x 4.0		+0.1
RHS	X	100x100x4.0 100x100x6.0 100x100x4.0 100x100x6.3	50x50x4.0 90x90x3.0 40x40x4.0 100x100x5.0	Fe 360 FeE 460 FeE 460	+0.1 +0.1 +0.1 +0.1
	K	100x100x4.0	50x50x4.0	Fe 360	+0.1
K-joints special		100x150x4.0	Ø57x4.0 with flattened ends	Fe 360	+0.1

6. The investigations of CHS X-Joints were carried out with test specimens of three different dimensions (2 diameter ratios - d/d_0 = 0.5 and 0.8; and two wall thickness ratios - t_0/t = 1.4 and 2.0). Using these specimens, the dependence of such hollow section joints on the existing geometric parameter ratios could be considered. The given parameter ratios leads to three different types of failure:

a) Joint failure in chord shear (see Fig. 3).

b) Crack starting from the weld toe of the web member (see Fig. 4).

c) Crack starting from the inside surface of the tube in the non-welded area of the chord (see Fig. 5)

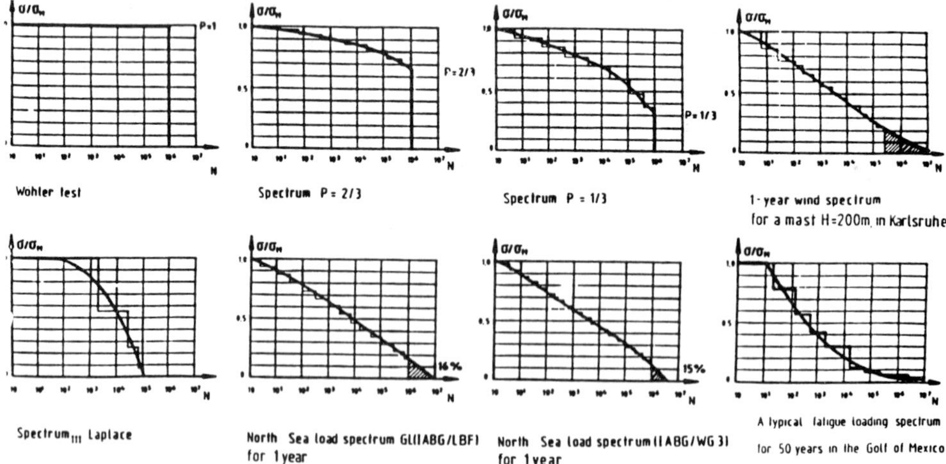

Fig. 2. Compilation of spectra applied.

7. A detailed report has been given in (refs 4, 5 and 6). In the following, further investigations will be presented. The failure mode for the rectangular hollow section (RHS) joints is identical to that of the circular hollow section (CHS) joints. Fig. 6 shows the test specimen with the geometric parameter $b/b_0 = 0.5$ and $t/t_0 = 1.0$ after the failure.

8. Fig. 7 shows the result of the tests on the X-type CHS-joints, and from Figs 8 to 10, the results of X-type RHS joints can be seen. The stress range ($S_R = \Delta\sigma$) is plotted in the vertical axis of these diagrams. The test results underload sequence have been plotted on the level of the maximum stress range in a group. The tests were performed with the same testing machines, under the same test conditions without corrosion impact.

Fig. 3. Joint failure in chord shear (CHS; $d/d_0 = 0.5$ and $t_0/t = 1.4$).

Fig. 4. Joint failure from the weld toe of the web member (CHS; $d/d_0 = 0.8$ and $t_0/t= 1.4$).

Fig. 5. Joint failure from the inside surface of the tube in the non-welded area of the chord (CHS; $d/d_0 = 0.8$ and $t_0/t = 2.0$).

Fig. 6. Joint failure in the chord shear.
(RHS; $b/b_0 = 0.5$ and $t_0/t = 1.58$, Fe E 460).

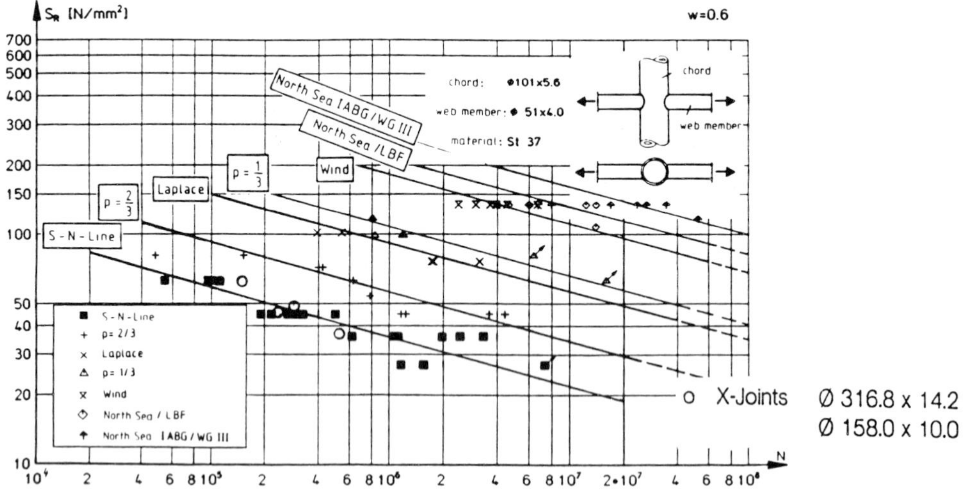

Fig. 7. Results of the fatigue investigations on CHS X-type joints under different load spectrum (CHS; $d/d_0 = 0.5$ and $t_0/t = 1.4$).

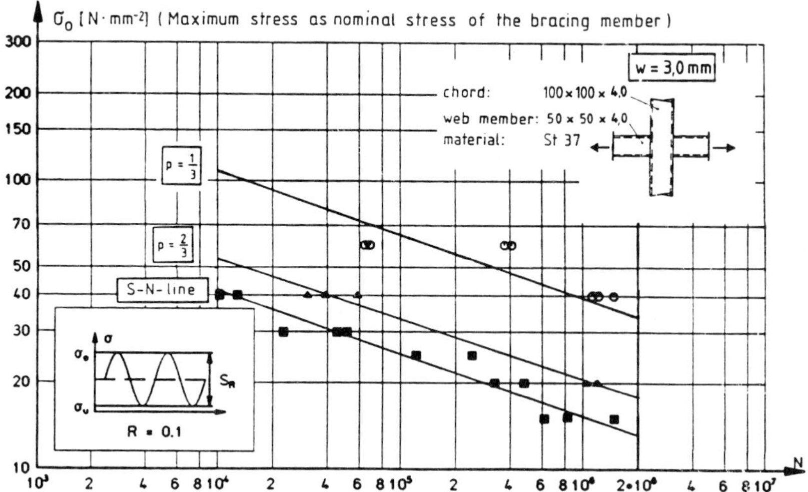

Fig. 8. Results of the fatigue investigations on RHS X-type joints

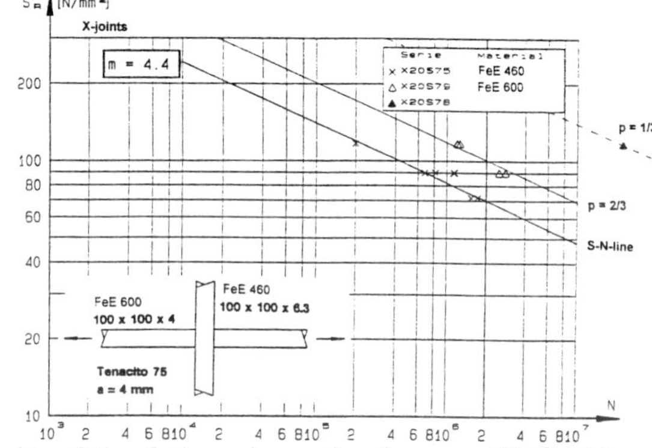

Fig. 9. Results of the fatigue investigations on RHS X-type joints under a different load spectrum ($b/b_0 = 1.0$ and $t_0/t = 1.58$; material Fe E460).

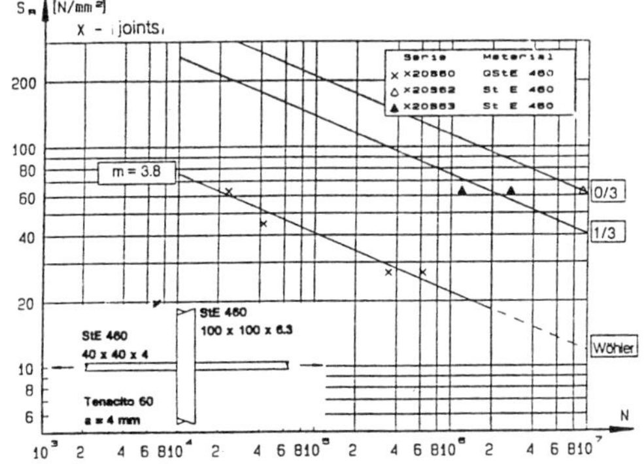

Fig. 10. Results of the fatigue investigations on RHS X-type joints loaded under a different load spectrum; material Fe E 460.

9. The determination of the S-N lines for different load sequences has the same slope as the S-N-lines for constant amplitude tests. After completion of the first tests under the collectives P = 2/3, Laplace S_{III} and P = 1/3, the authors found that the slope of each S-N-line is the same. Therefore, it was decided to perform the tests under load collectives on one stress range level, and to draw a fatigue life line with the same slope as found under constant amplitude, by means of the mean value of the endurance of the test. From tests on small specimens and L-type specimen tests, it is known that the slope can become smaller, e.g. bigger values for m. In favour of considering the scatters, this positive phenomenon has been ignored.

10. An important question to be answered concerns the size of the crack, when a joint type loses its serviceability. Since the definition of the bearing capacity of the applied testing load proved to be only a bulk value, measurements of the crack growth were conducted. As a criterion for disconnecting the testing machine, an elongation of the test pieces from 0.1 up to 3mm, divided into various stages, was fixed. The limitation of the maximum elongation of the test pieces to 3mm proved to be sufficient, since the crack reached over half of the width of the web chord when using this value. Nevertheless, even in this failure condition, the test piece could still bear the maximum testing force. The EGKS-standards (ref. 7) indicate four different load cycles or failure criteria for such investigations.

11. In the figure, each stage corresponds to an elongation of the test specimen by 0.1mm. In order to clarify the effect of the shape influence on the service fatigue strengths, besides X-type joints, K-type joints made of circular and rectangular hollow sections were also tested under identical loading collectives up to now, according to Fig. 2. A test rig, which has been used for previous investigations of K-Joints, was applied in order to achieve truss forces as they occur for lattice structures.

12. The results of the investigations on the fatigue strength of K-type truss joints are presented in Fig. 12. Fig. 13 shows one of the specimens after failure. The type of failures is identical to that of X-type joints according to Figs 3 and 6.

N1 Load cycle by a strain reduction of 15% (ε-15% = 0.85ε) close to the first crack.

N2 Load cycle when the first discernible crack occurs.

N3 Load cycle through wall cracking (crack goes through the whole wall thickness).

N4 Load cycle at the end of the test.

Where the load cycles N1 and N2 only differ slightly from each other.

13. Before defining the cut-off criteria, the relation between the load cycle numbers has been determined by means of single tests. One of the test results is given in Fig 11.

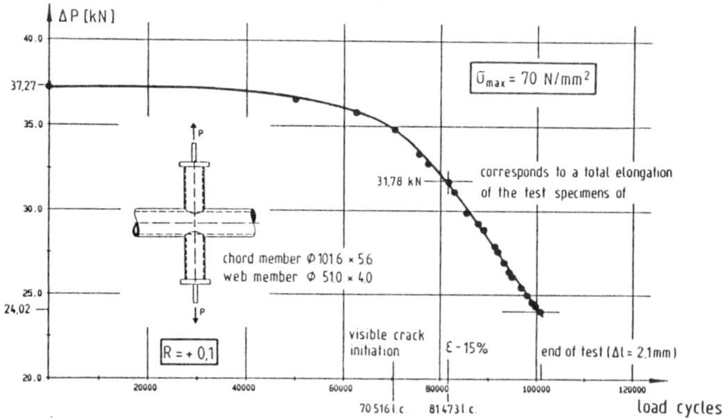

Fig. 11. Reduction of the upper loads (simultaneous increase of the lower load) when testing a hollow section joint.

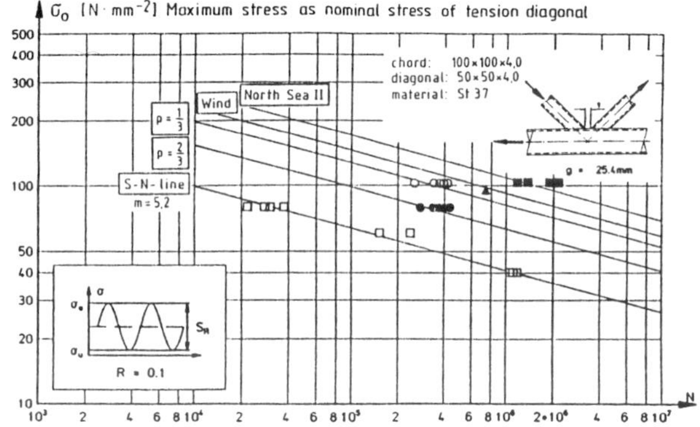

Fig. 12. Results of the fatigue investigations (RHS K-type joints) under different load spectra ($b/b_0 = 0.5$; $T/t = 1.0$).

Fig. 13. Test specimen after failure; K-type joint (b/bo = 0 5).

14. Fig. 14 shows the results of the special K-Joint for a crane structure and Fig. 15 the form and failure mode of the specimens.

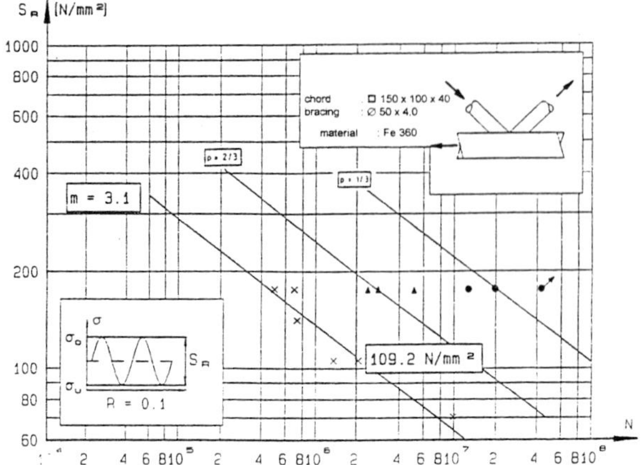

Fig. 14. Results of the fatigue investigations (special K-type joints under different load spectra).

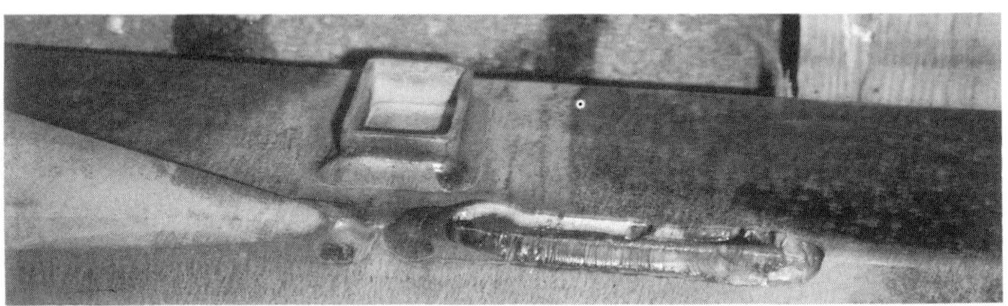

Fig. 15. Test specimen dimensions according to Fig. 4 and test specimens after failure.

PROPOSAL FOR THE UTILISATION OF TEST RESULTS

15. Through the comparison of stress ranges for a given load cycle, increase factors are obtained from the test values for the stress range dependent on the load sequence.

$$f_E = \frac{\Delta_{\sigma \text{Collective}}}{\Delta_{\sigma \text{Wöhler}}}$$

For sequence types:

2/3	=>	f_E=1.3
1/3	=>	f_E=1.8
North Sea LBF	=>	f_E=3.0
North Sea WG3	=>	f_E=3.4
Wind	=>	f_E=3.0

APPLICATION OF THE PALMGREN-MINER-RULE

16. Most of the new design rules recommend the application of the Miner-rule for the consideration of varying load (load sequence), e.g. for the proof of fatigue strength. For this reason, the meaningfulness of the Miner-rule has been checked in the scope of this work. The results of the Miner evaluation are presented graphically by means of Fig. 16.

17. The application of the Palmgren-Miner-Rule did not result in a good accordance in the field investigated (Fig. 16).

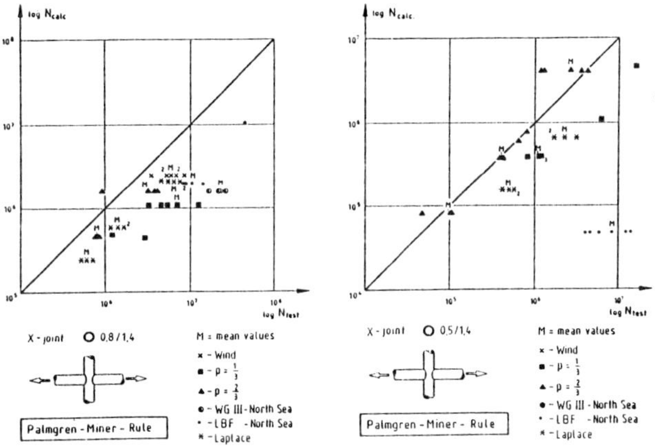

Fig. 16. Graphical illustration of the Miner results.

18. This statement is applicable to all possibilities of modified S-N-lines (different cut-off limits; different slope etc.) (ref. 5).

19. From Fig. 17, the frequency of the Miner sum, determined from the tests can be seen. The results of the investigations made in Karlsruhe are recorded comparatively into the diagram according to Schütz/Zenner (ref. 3).

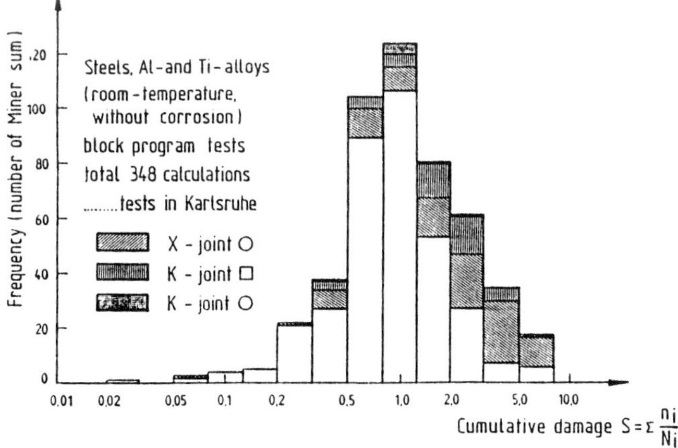

Fig. 17. Failure frequency; presentation according to (ref. 3) supplemented by tests carried out in Karlsruhe.

20. It can be concluded from available tests that the cumulative damage results in values below 1.0 for load collectives with a higher fullness, as for example 2/3, and in values between 3.0 and 4.0 for a smaller fullness.

21. For the practical application of the Miner-rule, the cumulative damage indicated in Table 2 is recommended for different types of collectives.

Table 2. Cumulative damages in dependence on the type of collective; S_{min} is the lowest value determined during the tests in Karlsruhe; $S_{proposed}$ is the sum of damages to be used for the calculation. The poor single results due to weld defects and roughness of the specimens were not considered.

Form of load spectrum Collective	Damage accumulation All test results (All types of failure) for all R-values		Damage accumulation non considering the series with the crack on the inside of the tube and $R \geq 0.1$	
	S_{min}	$S_{proposed}$	S_{min}	$S_{proposed}$
2/3	0.27	0.3	0.27	0.5
1/3	0.34	0.5	2.11	3.0
Laplace	1.87	----	1.87	2.5
North Sea (LBF-IABG)	0.48	1.0	3.18	4.0
North Sea (LBFWG III)	0.56	1.0	1.59	4.0
Wind	0.33	1.0	1.47	2.8

22. These indications are valid for structures with small tube dimensions as they occur for crane systems, masts etc. Presently, experimental investigations are being carried out with hollow sections of a bigger dimension and bigger wall thickness. The results of the big-size joint tests (X-joints) with the dimensions of the chord member ϕ 316.8 x 14.2 and web member ϕ 158 x 10.0 are plotted in the diagram shown in Fig. 7.

PROPOSAL FOR THE DESIGN OF HOLLOW SECTION CONSTRUCTIONS UNDER LOADING SPECTRUM

23. As the design calculation using the Miner-rule or the modified Miner-rule did not give good results in the past, new design methods were sought for practical applications.

24. While considering the load spectrum with the area under the corresponding staircase curves (blocks), and the fatigue life lines for these collectives determined by tests, according to Figs 7, 10 and 12 as well as the test result given in (ref. 5), it occurs to one that the maximum stresses or alternatively the endurance (load cycles) for certain stress levels, increase with the decreasing fullness coefficient.

The fullness of a load spectrum is an integration of each level over the number of cycles (N).
(The following hypothesis is formulated using this reverse effect:
"The ratio of the fullness (area) of two load spectra behaves inversely proportional to the load increment coefficients of these load spectra".
According to the hypothesis, the following is valid for the load spectra A and B:

$$\frac{V_A}{V_B} = \frac{f_B}{f_A}$$

V_A	fullness of the load spectrum (collective) A
V_B	fullness of the load spectrum (collective) B
f_A	load increment coefficient for the load spectrum (collective) A
f_B	load increment coefficient for the load spectrum (collective) B

The load increment coefficient is defined as follows:

$$f_i = \frac{\text{Max. stress for a joint under load spectrum i (VA)}}{\text{Max. stress for the same joint according to the Wöhler tests (CA)}}$$

Fig. 18 shows the determination of f_i-values (schematically). In general, the above mentioned hypothesis can be expressed for any load spectrum i as follows:

$V_i.f_i = C$ where C is a constant, which has to be determined empirically by tests.

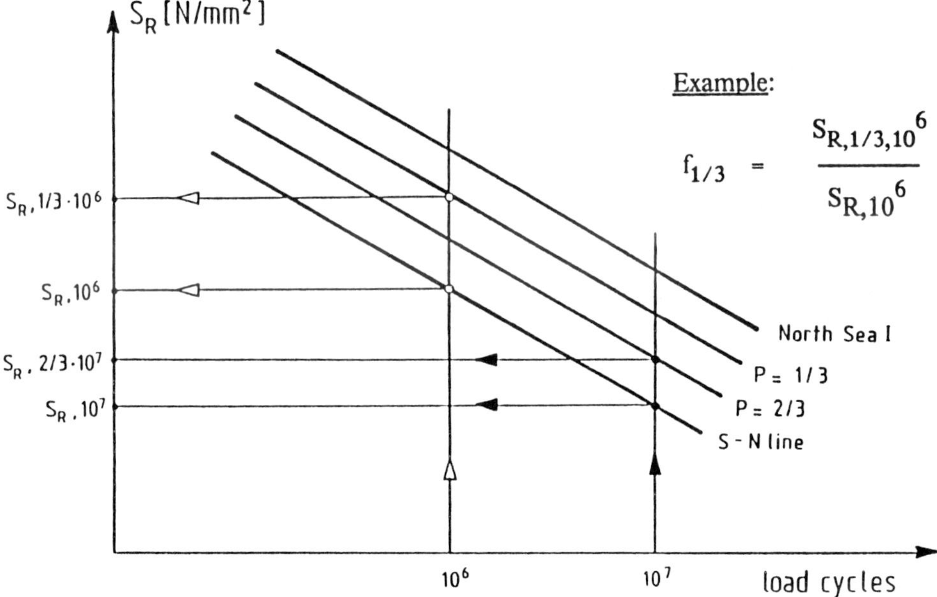

Fig. 18. Determination of the load factors (schematically).

25. The argument for the above hypothesis and the advantages and disadvantages of using this method are given in (ref. 5). This hypothesis was checked for various types of load spectra (collectives) by means of extensive investigations.

26. Table 3 contains the maximum and minimum values of the load increment coefficient calculated using the test data and the measuring number for the design according to the Figs 5, 6, 7, and 9 and other test values according to (ref. 5).

27. A constant close to $7.5x10^7$ for a collective range of 10^6 loading cycles is obtained by multiplying the corresponding fullness coefficients with the measuring values on the right hand column of the table. This value is, however, only an approximate value, that represents an acceptable calculation value for all described load collectives.

28. If the fullness coefficient for a new collective is obtained by integrating the staircase curve, the load increment coefficient can be calculated by determining the ratio of the limiting value $7.5x10^7$ to the given fullness coefficient.

Table 3. $V_{colletive}$, $f_{collective}$ -design for different load spectra.

Definition of the load spectrum (Block collective)	Fullness[1)] $V_{collective}$	Load increment coefficient Min. $f_{coll.min}$	Max. $f_{coll.max}$	Proposed ratio value[2)] for the design $f_{coll.design}$
2/3	$6.586 \cdot 10^7$	1.31	1.68	1.3
1/3	$4.395 \cdot 10^7$	1.82	2.98	1.8
North Sea LBF	$2.330 \cdot 10^7$	3.15	5.60	3.0
North Sea WGIII	$2.164 \cdot 10^7$	3.40	6.80	3.4
Wind	$2.201 \cdot 10^7$	2.87	4.80	2.8

1) Determined for a maximum stress range of $S_r = 100\ N/mm^2$ and a collective range of 10^6 cycles.

2) Due to the small number of test data, one starts from the lowest value, while the minimum value is rounded off to the lower side (presently, $f_{coll.design}$, is similar to the increase factor f_E).

REFERENCES

1. N.N.: Eurocode 3. Design of Steel Structures, August 88.

2. MANG, F., BUCAK, Ö. I. Wöhlerlinien-Katalog für Hohlprofilverbindungen (S-N-Line Catalogue for Hollow Section Joints). Studiengesellschaft für Anwendungstechnik von Eisen und Stahl e.V., Düsseldorf, Juni 1987, Januar 1988.

3. SCHÜTZ, ZENNER. Schadensakkumulationshypothesen zur Lebensdauervorhersage bei schwingender Beanspruchung (Damage Accumulation Hypotheses for the Fatigue Life at Variable Load). Zeitschrift für Werkstofftechnik 1973 Part 1, No.l, p.25-33, Part 2, No.2, p.97-102.

4. MANG, F., BUCAK, Ö. Investigations into the Fatigue Behaviour of Hollow Section Joints Subjected to Spectrum Loading, ISOPE '91, Edinburgh 11.-15.08., 1991

5. BUCAK, Ö. Bewertung des Ermudungsverhaltens von Hohlprofilkonstruktionen (Evaluation of the Fatigue Behaviour of Hollow Section Joints), PhD Thesis, University of Karlsruhe, May, 1990.

6. BUCAK, Ö., MANG, F. Investigations into the Fatigue Behaviour of CHS-Joints Subjected to Spectrum Loading. Third International Symposium on Tubular Structures, Lappeenranta, September,1989.

7. N.N.: Stahl in Meeresbauwerken (Steel in Marine Building Structures), Proceedings of the International Conference in Paris, Oct. 15-18, 1981, EURBericht Nr. 7347 (1981), p.439-483 and Plenary Session No. 5.

8. BIERETT, G. Einige wichtige Gesetze der Betriebsfestigkeit geschweißter Bauteile aus Stahl (Some Important Rule of the Fatigue Strength of Welded Structural Members Made of Steel), Schweißen und Schneiden, Heft 11,1072, p.429-434.

9. HAIBACH, E. Modifizierte lineare Schadensakkumulationshypothese zur Berücksichtigung des Dauerfestigkeitsabfalls mit Fortschreiten der Schädigung (Modified Linear Damage Accumulation Hypothesis for the Consideration of the Fatigue Strength Reduction with an Increase of the Damage), LBF Darmstadt, TM 59/70.

10. LIPP, W. Zur Lebensdauerabschätzung mit dem Blockprogramm und Zufallslastenversuch (On the Estimation of the Service Life with the Block Program and Load Test at Random) LBF Darmstadt, TM 76/80.

11. DIN 15018. Krane, Grundsätze für Stahltragwerke; Berechnungen (Cranes, Fundamentals for Steel Supporting Frameworks; Calculations), Ausgabe, April, 1974.

12. SONSINO, C.M., KLÄTSCHKE, H., SCHÜTZ, W., HÜCK, M.: Standardised Load Sequence for Offshore Structures. WASH-1 LBF Report No. FB 181 (1988) IABG-Report No. TF 2347 (1988).

13. SONSINO, C.M., LIPP, K: Übertragbarkeit des an Winkelproben ermittelten Betriebsfestigkeitsverhaltens auf große Rohrknoten für die Offshore Technik (Transferability of the Fatigue Strength Behaviour Determined on Wedge Tests to Big Tube Joints for Offshore Technology), 5. LBF Kolloquium in Darmstadt, March 8-9,1988. Report No. TB-180 (1988)

14. BUCAK, Ö., MANG, F. New Knowledge on the Fatigue Behaviour of Steel Structures and Especially on Hollow Section Structures under Different Load Spectra. 1st World Conf. "Constructional Steel Design", Acapulco Nov.-Dec. 92.

50. Dynamic buckling behaviour of single and double layer latticed domes due to vertical earthquake motions

K. ISHIKAWA, Ishikawa College of Technology, Japan, and S. KATO, Toyohashi University of Technology, Japan

SYNOPSIS. The present paper deals with the dynamic buckling behaviour and the earthquake resistance of single and double-layer pin-jointed latticed domes under vertical seismic motions. Elastic-plastic, static and dynamic analysis, allowing for both geometric and material non-linearity has been considered. Based on the results of the above analyses the earthquake resistant capacity is evaluated by using the maximum amplitude of the acceleration of the input earthquake wave.

INTRODUCTION

1. Structural and vibrational characteristics of a shell structure vary according to the configuration of the structure. As a result, the vibrational analysis of each case may be needed in order to estimate the response of the structure subject to earthquake motions.

2. The present work described in this paper is as follows:

(a) Investigation of the difference between the buckling collapse behaviour both of double and single-layer latticed domes under vertical earthquake motions investigated by an elastic-plastic analysis considering geometrical and material non-linearities.

(b) Mechanisms related to the dynamic collapse of both domes are considered using the results of the analysis.

(c) The vertical displacement at the central point of the dome can be used as a guide to determine the magnitude of the dynamic collapse for double-layer latticed domes.

(d) For a single-layer latticed dome the maximum acceleration of an input wave which induces the dome to collapse dynamically can be estimated by using linear dynamic analysis.

LATTICED DOMES FOR ANALYSIS

Single-layer latticed dome

3. Global configuration. The single layer latticed dome analysed had the following geometry: span L = 3287cm; the radius R = 2456cm; the rise H = 631cm and the half open angle $\Phi_0 = 42^0$. The structure is shown in Fig. la. The

angle ϕ_1 may be called the half open angle of the members as shown in Fig. lb. For the present investigation ϕ_1 = 3.5^0. The member length, denoted by an arrow OI in Fig. la, is 300cm and is almost identical for every member in the structure.

4. As for the boundary condition, the mark O in Fig. la indicates roller supports. On the rollers the nodes can move horizontally in every direction in the horizontal plane. At the six points, A, B, C, D, E and F each support can move only in the direction given by a line connecting the centre point C with the node in the horizontal plane.

Structural characteristic of members

5. Under the present assumption a member can buckle due to a compression and can yield in tension. Each member is pin connected to others and the elastic-plastic relation between the average stress and strain is based on a result obtained by an elastic-plastic buckling analysis of one member. The curves in Fig. 2 are the hysteretic curves used in the present analysis. The curves adopted correspond to members with slenderness ratios of λ = 60, 80, and 100 respectively. The yield stress σ_y and Young's modulus E are set at 2400 kgf/cm^2 and 2.1x10^6 kgf/cm^2 respectively. The maximum compression stress σ_{cr} is determined using the following equations considering initial geometric imperfections and residual stress present in a member.

$$\frac{\sigma_{cr}}{\sigma_y} = \frac{1-0.4(\lambda/\Lambda)^2}{1+(2\lambda/3\Lambda)^2} \quad \text{: in the case of } \lambda \leqq \Lambda$$

$$\frac{\sigma_{cr}}{\sigma_y} = \frac{1}{1.44}\frac{\pi^2 E}{\lambda^2} \quad \text{: in the case of } \lambda > \Lambda \qquad (1)$$

Where Λ is the critical slenderness ratio.

Geometric imperfections

6. The load carrying capacity of pin-jointed single-layer latticed domes is largely affected by the level of geometric imperfection (ref. 3).

7. In the present study a very simple and intrinsic imperfection is applied. The nodal imperfection is defined as a deviation g of a nodal co-ordinate at a single node as shown in Fig. 3.

$$_g\varepsilon = g/h \qquad (2)$$

Where h is the height of the unit hexagonal dome (Fig. lb) and g is the reduction of the height toward the centre of the sphere. Accordingly the value (R-g) is the reduced radius.

Analysed model (Dome S)

8. The analysed model is shown in Fig. 4. The member slope ϕ_l of the dome is 3.5^0 for Dome S. For determining member sizes, a uniformly distributed design load of 120kgf/cm^2 was used.

9. In Dome S (Fig. 4), several different members were used to investigate the member buckling by using the method of estimating the elastic-plastic buckling load (ref. 3). In the numerical computation, the model is analysed assuming symmetry about the X-X axis (Fig. 1a).

Double-layer dome

10. Global configuration. The double-layer latticed dome analysed had a span L = 4800 cm, a rise H = 1410cm, a depth h = 120cm, the radius R1 = 4800 cm, R2 = 4680 cm for the upper and lower surface respectively, the half open angle ϕ_1 = 30^0. The dome is shown in Fig. 5.

11. The member length of the upper and lower chords is almost 400cm. The length of the web members is 150cm.

12. As for the boundary condition, the mark O in Fig. 5 shows the roller supports in the same position as the roller supports of Dome S. The mark ● shows the pin supports for which no translation is permitted.

Structural characteristic of members and analysed model (Dome D)

13. The analysed model is shown in Fig. 6. The design load used to determine the member characteristics is 60kgf/cm^2 treated as an uniformly distributed load. The cross-sectional area of the members is shown in Table 1. The cross-sectional areas and the slenderness ratios of the members in Dome D are determined using the following procedure:

(a) The slenderness ratio λ for all the upper and the lower chords was equal to 80. The value of λ for all the web members was 100.

(b) The material yield stress σ_y was 2400kgf/cm^2. The plan of the dome was divided into three parts as shown in Fig. 6. The same mechanical properties for the member were assumed for use in each part.

(c) The cross-sectional area of the member in each part was obtained by considering the maximum axial force in one member, in each of the three segments and dividing by the allowable unit stress f_c (= σ_{cr} /1 5) where σ_{cr} is the maximum compressive stress (Eq. 1).

STATIC ELASTIC-PLASTIC ANALYSIS

14. Before undertaking the dynamic analysis the static elastic-plastic buckling analysis, allowing for both geometric and member buckling non-linearities, was undertaken for both Dome S and D to investigate their static vertical load carrying capacity.

Single-layer dome (Dome S)

15. The results of the static analysis on the single-layer dome gave the ratio of $P_{cr}^A/P_d = 3.81$, where P_{cr}^A is the buckling load applied to a nodal point and P_d is the design load for a nodal point.

Double-layer dome (Dome D)

16. The static analysis of Dome D, gave the relationship between the nodal load P and the vertical displacement of the central point on the plan W (node D) as shown in Fig. 7, where P_s is the design load, P_p is the load at which member buckling first occurred, W_p^s is the vertical displacement under the load P_p and W_f^s is the vertical displacement at which the load carrying capacity of the dome is just P_s after decreasing from the maximum load P_{max}.

17. From these results it can be seen that the increase in the load carrying capacity from P_p to the maximum load P_{max}, is only an increase of 10% of P_p. Also after decreasing from the maximum load P_{max}, the equilibrium path shows load-shedding behaviour.

DYNAMIC ELASTIC-PLASTIC ANALYSIS

18. In this method both geometric and material non-linearities are considered. This means a large displacement analysis considering member buckling.

19. The generalised Rayleigh damping matrix is used in this analysis. The damping factor h = 0.02 is used for all the vibration modes in the damping matrix.

20. In the present study the earthquake resistant capacity of the domes is evaluated by investigating the maximum response of displacement W_{max} at the central node of the dome corresponding to the peak acceleration A_{max} of the input wave. From Fig. 9, which shows the relationship between W_{max} and A_{max} the dynamic buckling collapse can be defined when the dome vibration behaviour becomes divergent at certain value of A_{max}, called the collapse acceleration A_f.

Single-layer dome (Dome S)

21. From the results of Dome S, without the imperfection, the following results are summarised. The total number of natural periods of Dome S is 176.

(a) The natural periods of the l'st, 2'nd, 3'rd and 176'th modes are 0.22974 sec, 0.22949 sec, 0.22650 sec and 0.0030675 sec respectively.

(b) The the l'st, 2'nd and 3'rd mode shapes could be recognised due to the vertical movement of the several node points.

22. The results of Dome S with an imperfection are summarised as follows:

(a) As for the first mode shape only the node including the imperfection moves extensively downward. The 2'nd and 3'rd mode shapes are similar to the l'st and 2'nd mode of Dome S without the imperfection.

(b) The natural periods change according to the amount $_g\varepsilon$ of the imperfection. In the case of $_g\varepsilon = 0.1$, the natural periods of 1'st, 2'nd, 3'rd and 176'th modes are 0.24798 sec, 0.22939 sec, 0.22915 sec and 0.0030675 sec respectively. In the case of $_g\varepsilon$ = 0.2, the natural periods of l'st, 2'nd, 3'rd and 176'th modes are 0.30260 sec, 0.22933 sec, 0.22923 sec and 0.0030675 sec respectively.

23. Several dynamic analyses for the domes were performed, both with and without the imperfection at the node point 27. The domes were analysed subjected to the vertical earthquake motions together with the uniformly distributed dead load allowing the determination of A_f and the buckled members. Node point 27 was chosen for the imperfection because the static buckling load is most sensitive to an imperfection at this node, determined from the result of the sensitive analysis. Taft UD, El-Centro UD and Hachinohe UD are used as the input waves which have a duration of 5 seconds. The acceleration linear response for each input wave are shown in Fig. 8.

24. The relationship between W_{max} at the specified nodes and A_{max} in the case of the dome without the imperfection under Taft UD is shown in Fig. 9. In the case of the other waves and the dome with the imperfection, the results of A_f is shown in Table 2.

25. From the results the following points are noted:

a). The value A_f of Dome S without the imperfection under Taft UD is approximately determined as 800cm/sec^2 from Fig. 9. The dynamic buckling behaviour under Taft UD with A_f = 800cm/sec^2 can be observed from Fig. 10.

b) The A_f value from Dome S with $_g\varepsilon$ = 0.1 is derived as 780cm/sec^2 and the A_f value of Dome S with $_g\varepsilon$ = 0.2 is derived as 740 cm/sec^2. Consequently increasing the amount of $_g\varepsilon$, A_f decreases but the difference between A_f corresponding to $_g\varepsilon$ =0, 0.1 and 0.2 is small. From this result it can be seen that the value of imperfection $_g\varepsilon$ has little effect on the dynamic buckling collapse load. The dynamic buckling configurations of the dome with $_g\varepsilon$ = 0.2 are shown in Fig. 11.

Double-layer dome (Dome D)

26. The model for the eigenvalue and dynamic analysis of Dome D was a quarter of the dome assuming symmetrical conditions. Consequently, the total degrees of freedom of the model is 218.

(a) As for the l'st and 2'nd mode shape the nodes of the central part of the dome move downward.

(b) The natural periods of l'st, 2'nd, 3'rd and 218'th modes are 0.419 sec, 0.321sec, 0.317sec and 0.0079sec respectively.

27. The relationship between W_{max} at the central point node and A_{max}, is shown in Fig. 12. From the result the followings have been observed:

(a) The value of A_f of Dome D was determined as 120cm/sec^2 . The relationship from A_{max} = 0 to 115cm/sec^2 is linear. The value of W_{max} is taken to be 6.9cm at the turning point when A_{max} = 115cm/sec^2 . The present study focuses on this turning point in order to clarify the frequency of occurrence of the dynamic buckling collapse of Dome D. From Fig. 7, which shows the results of the static analysis, the vertical displacement $W_f{}^s$ at the central node, corresponding to the design load P_s after exceeding the maximum load P_{max}, is observed to be 6.8cm. Accordingly it may be considered that the dynamic buckling collapse occurs when the response displacement at the central node exceeds W_f^s. This means that the collapse occurs before the load carrying capacity reaches the dead load.

(b) The dynamic buckling behaviour and the deformed configurations of collapse of the dome due to Taft UD with A_{max} = 120cm/sec^2 are show in Fig. 11 at the time t = 1.0, 4.75 and 5.0 sec. From Fig. 13 it could be observed that the large vertical displacement around the central part is induced by the sequence in which members buckled. The time history diagrams of the acceleration, the velocity and the displacement at the central node point of the dome due to Taft UD with A_f^s = 120cm/sec^2 are shown in Fig. 14. The response of the acceleration shows that the vibration period after 4 seconds is shorter than that before. The response of the displacement shows that the response curve obtained vibrate after the displacement exceeds W_f^s at a period of around 4 seconds.

DISCUSSIONS OF EARTHQUAKE RESISTANT CAPACITY

28. Based on the results of this investigation of the static and the dynamic buckling analysis of domes, the earthquake resistant capacity of Domes S and D evaluated by the proposed method.

Single-layer dome (Dome S)

29. The dynamic collapse of Dome S is induced by the large vertical displacement occurring at several specified nodes in connection with the buckled members. Accordingly it can be considered that A_f of Dome S is predicted by using linear dynamic analysis. This means that $A_{max}{}^L$ given in Fig. 9, may be determined by a linear dynamic analysis and a simple calculation using the linear relationship between member stress and A_{max} . The validity of this method is shown by considering the small difference between A_f and $A_{max}{}^L$ as shown in Fig. 9.

Double-layer dome (Dome D)

30. Even if the load carrying capacity of Dome D under the dead load P_s is lost due to several members buckling, the dome can endure both P_s and the vertical earthquake motions due to the redundancy of the other members.

31. Now A_p defines the maximum acceleration of an input wave due to which the maximum vertical displacement response at the central node point W_{max} is W_p^s corresponding to the static load P_p . If it can be assumed that the dome has an elastic response behaviour due to the input wave with a value greater than A_p, the maximum elastic vertical displacement response W_{ef} corresponding to A_f can be estimated using the following equation.

$$W_{ef} = W_S + \frac{A_f}{A_p}\left(W_P^S - W_S\right) \quad (3)$$

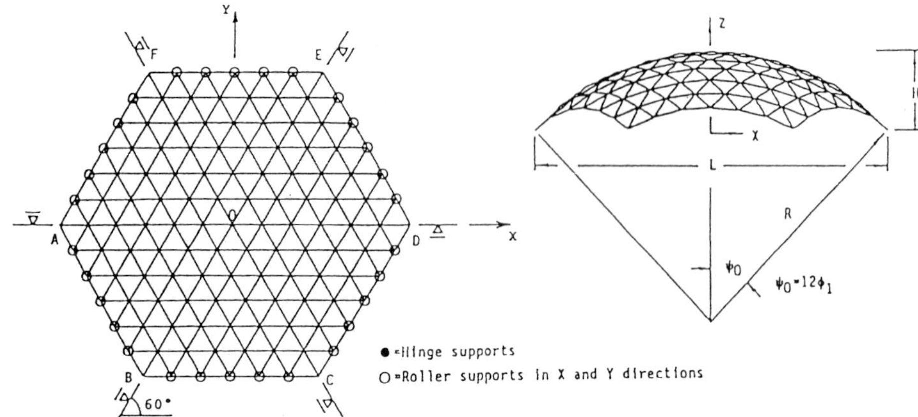

Fig. 1-(a). Dome geometry and boundary condition.

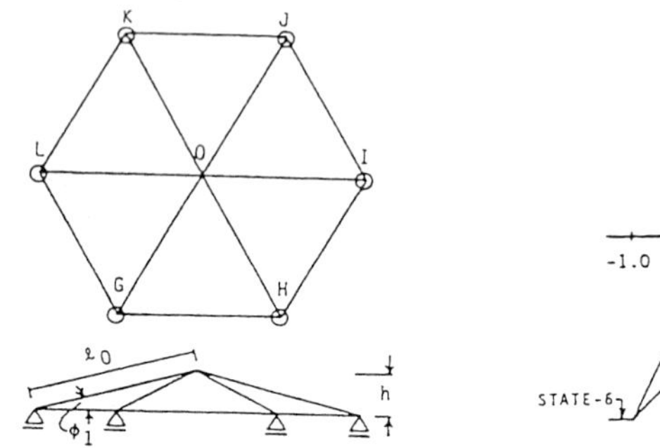

Fig. 1-(b). Unit dome and half open angle.

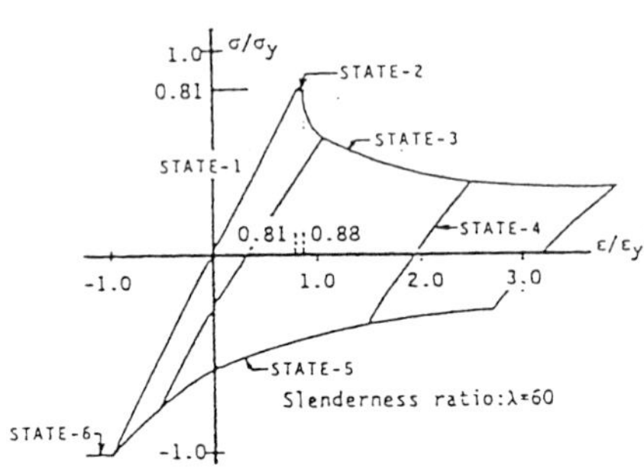

Fig. 2. Hysteresis curve of a pin-end member.

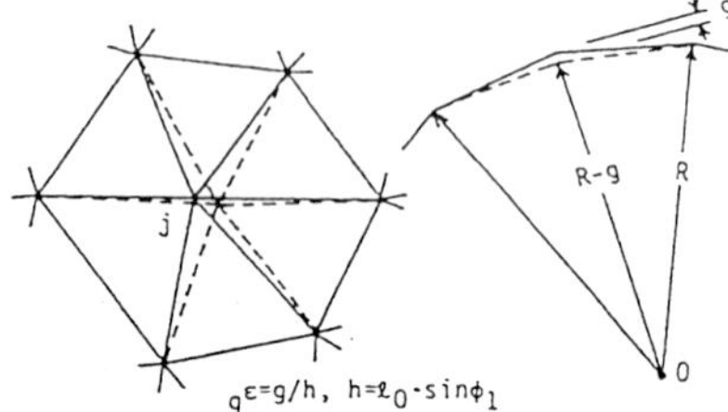

Fig. 3. A geometric imperfection.

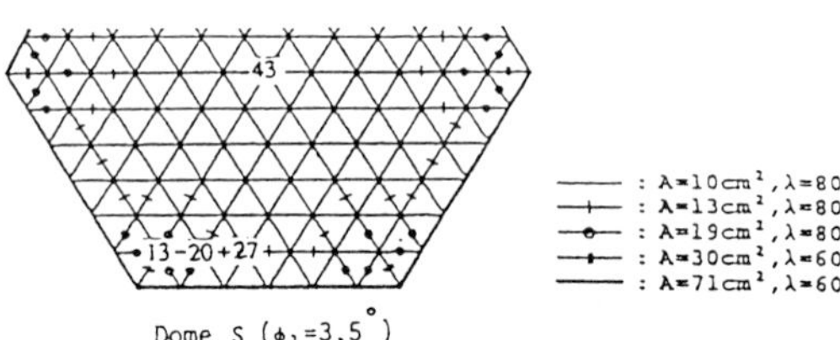

Fig. 4. Analyzed Dome S and arragement of members.

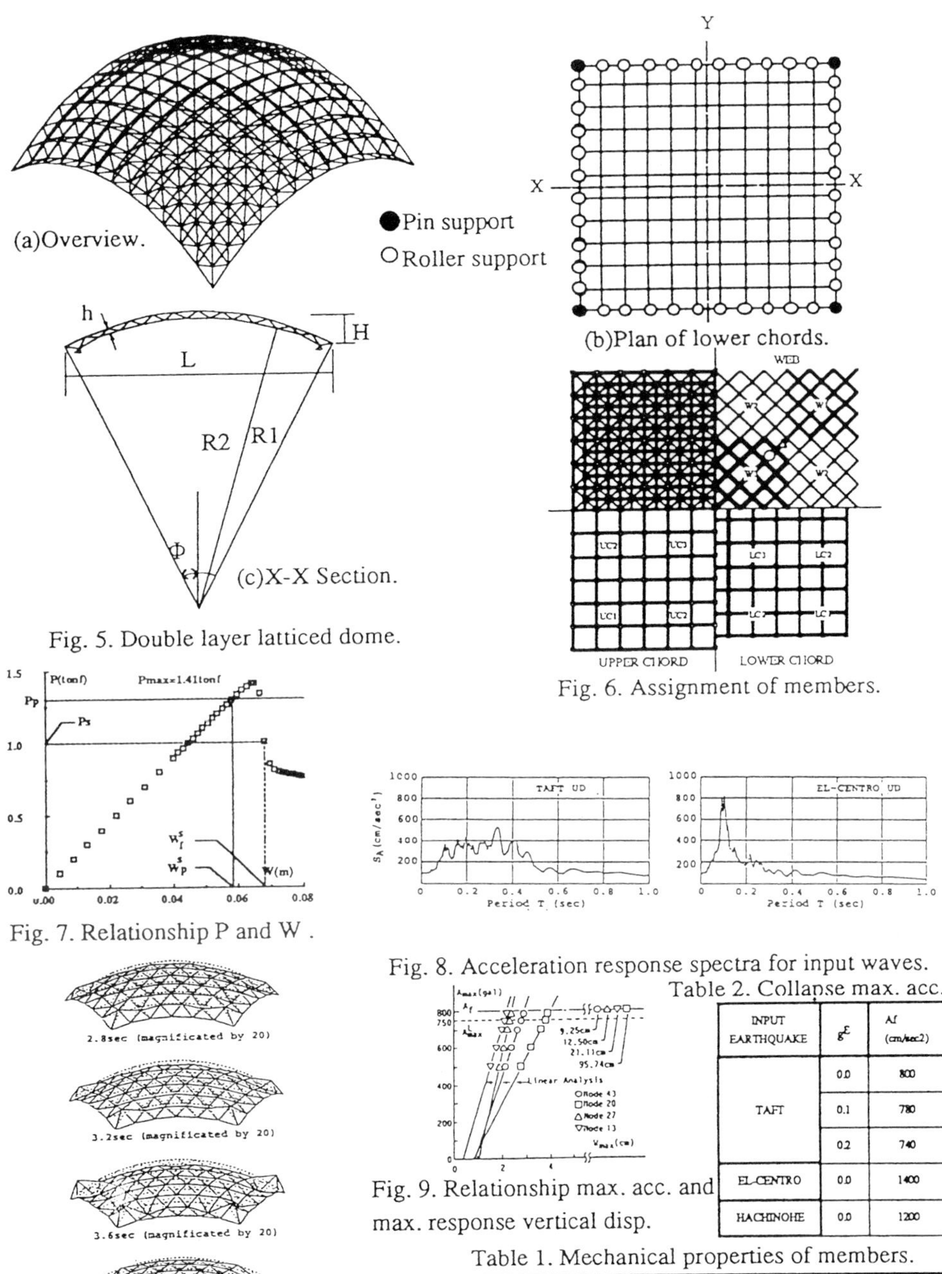

Fig. 5. Double layer latticed dome.

Fig. 6. Assignment of members.

Fig. 7. Relationship P and W .

Fig. 8. Acceleration response spectra for input waves.

Fig. 9. Relationship max. acc. and max. response vertical disp.

Fig. 10. Dynamic buckling configuration in case of Taft UD(A_{max}=800gal).

Table 2. Collapse max. acc.

INPUT EARTHQUAKE	g^{ε}	Af (cm/sec2)
TAFT	0.0	800
	0.1	780
	0.2	740
EL-CENTRO	0.0	1400
HACHINOHE	0.0	1200

Table 1. Mechanical properties of members.

MEMBER	UPPER CHORDS			LOWER CHORDS			WEBS		
SYMBOL	UC1	UC2	UC3	LC1	LC2	LC3	W1	W2	W3
λ	80	80	80	80	80	80	100	100	100
AREA (cm2)	0.813	2.387	2.867	2.699	4.179	2.208	2.406	2.361	0.866

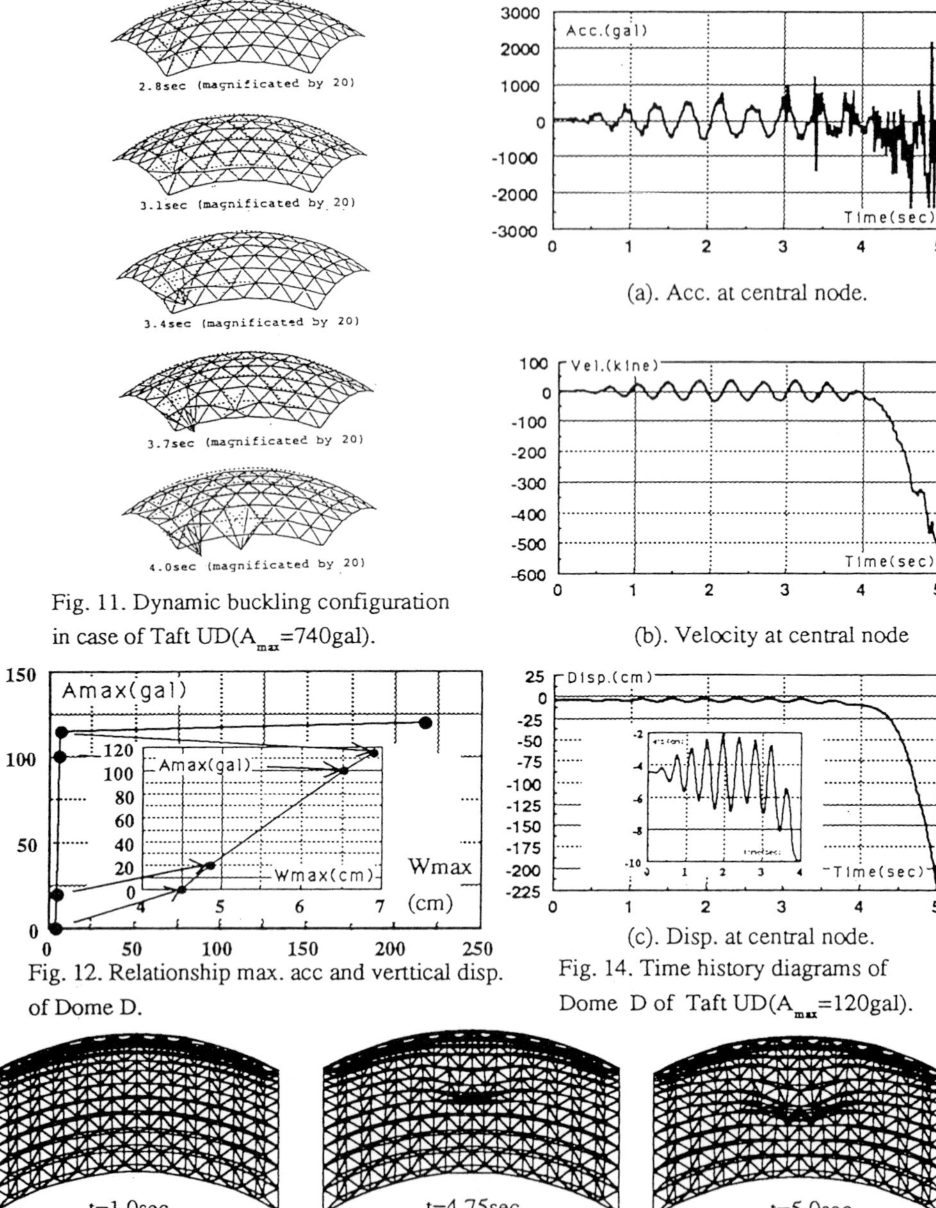

Fig. 11. Dynamic buckling configuration in case of Taft UD(A_{max}=740gal).

Fig. 12. Relationship max. acc and verttical disp. of Dome D.

Fig. 14. Time history diagrams of Dome D of Taft UD(A_{max}=120gal).

Fig. 13. Dynamic buckling configuration of Dome D in case of Taft UD(A_{max}=120gal).

Table 3. Numerical values of results of analysis of Dome D.

P_s (tonf)	P_p (tonf)	W_s (cm)	S_D (cm)	A_p (cm/sec^2)	A_f (cm/sec^2)	A_f^D (cm/sec^2)	A_f/A^p	ν_D
1.00	1.298	4.45	2.05	65	120	119	1.85	0.55

Where W_s is the vertical displacement due to the dead load P_s.

32. If it can be assumed that the dome collapse occurs when the maximum elastic response displacement due to A_f^D reaches W_{ef}, the following relationship can be stated.

$$\frac{A_f^D}{A_k} \cdot S_D(A_k) + W_s = W_{ef} \tag{4}$$

Where A_f^D is the estimated value of the collapse maximum acceleration, $S_D(A_k)$ is a displacement response spectrum for an earthquake wave with a peak acceleration A_k (cm/sec^2).

33. Substituting Eq. 4 into Eq. 3 and considering $W_p^s = (P_p/P_s) \times W_s$, the following equation can be given.

$$A_f^D = \nu_D \times \frac{A_k}{S_D(A_k)} \times W_s \tag{5}$$

where $\nu_D = \tilde{\nu}_D \left(\frac{P_p}{P_s} - 1 \right), \ \tilde{\nu}_D = \frac{A_f}{A_p}.$

34. The term υ_D may be called the static collapse seismic coefficient because of the following relationship that $\upsilon_D \times W_s + W_s = W_{ef}$.

35. The values of A_f^D and υ_D are shown in Table 3. From this result good agreement is obtained A_f and A_f^D. The value of $\upsilon_D = 0.55$ is derived from this method. This means that the earthquake resistant capacity, corresponding to the dynamic collapse of the dome due to Taft UD with A_f = 120cm/sec^2 is equivalent to both the dead load P_s and the static seismic load 0.55 x P_s.

REFERENCES

1. KUNIEDA H. Solutions of free vibrations of spherical shells, AIJ, No.325, 1983, 57-66.

2. KATO S., ISHIKAWA K. Materially non-linear stability of a double-layer grid due to vertical earthquake motions, Proceedings IASS Symposium, Osaka, 1986, Vol. 3, 9-16.

3. KATO S., ISHIKAWA K. On elastic-plastic buckling of a spherical reticular dome of single layer on a hexagonal plan, Proceedings IASS-MSU Symposium, Istanbul, 1988, 515-526.

51. Seismic design of double-layer space grids and their supports

A. S. KARAMANOS, EPPO, Greece, and S. A. KARAMANOS,
University of Texas, USA

SYNOPSIS. Strong earthquakes have a major effect, referred to as seismic actions on structures. This paper investigates the effect of horizontal seismic actions on the most used type of space frame, the double-layer space grid, as well as providing an outline of the main features of earthquake resistant design for the grid structure and its supports, in accordance with modern seismic design codes, the available literature and the authors design experience.

INTRODUCTION

1. Space structures erected in seismic areas of earthquake prone countries must be designed for the so called seismic actions, i.e. for the effects imposed on them by strong earthquakes, in addition to their design for usual loads. Seismic action is a special type of loading characterised by low probability of occurrence, high intensity and a number of independent parameters such as maximum ground acceleration, frequency content, duration, etc, which vary significantly among different earthquakes and have a direct impact on structural design.

2. Seismic action is due to strong ground motion during an earthquake, and essentially consists of forced motion of the base of the structure, which, due to the inertia of the various masses, results in a forced deformation of the structure and its members. Therefore, the system of seismic forces (or loads) acting on a structure is a consequence of the base excitation (inertia forces). For practical reasons this system is usually considered as an equivalent system of external loads representing seismic action.

3. Although extensive theoretical and experimental research on the behaviour of space frames has been reported, there is very limited information available concerning their seismic response and design. More specifically, effort has been devoted on the analytical investigation of space frame stability under static vertical load, using computational techniques for tracing the post-buckling equilibrium paths (refs 2,8 and 10) or determining the ultimate load (ref. 3). In addition, for the same type of loading, several experimental and analytical works concerning the ductility of such structures have been published (refs 4, 5 and 13).

4. The influence of the member connection details on the frame ductility has also been examined in the experimental work of Schmidt and Morgan (ref. 9) by single cycle static loading of specimens. On the other hand, the cyclic behaviour of tubular members (widely used in space frames), has been experimentally investigated by Zayas et al. (ref. 6) and Black et al. (ref. 7), shows the superiority of such sections over other structural sections.

5. The effects of an earthquake excitation on a double grid has been investigated by Kato et al. (refs 14 and 15), including geometric and material nonlinearities, whereas a simplified procedure has been proposed by Robertson (ref. 12). However, there have been no reports of any specific guidelines for the earthquake design of such structures. Of course, the development of such design aids requires extensive experimental and analytical research, and is beyond the scope of this paper. Nevertheless, based on the available literature, as well as on the design experience of the authors, the present work aims at presenting some basic features concerning the behaviour, and safe design, of double-layer space grids under earthquake loading. Finally, particular emphasis is given to the important issue of frame support design in order to optimise the earthquake frame response.

CONSIDERATIONS ON SPACE GRID SEISMIC BEHAVIOUR

6. Conceptually, earthquake resistant steel structures may be divided into two categories, non-dissipative and dissipative (ref. 16). For the first category, the seismic action effects (axial forces, moments and shears) are calculated on the basis of an elastic analysis without considering material non-linear behaviour. The strength of members and connections in resisting the design seismic loading is evaluated without any ductility requirements, i.e. behaviour factor $q= 1$. On the contrary, dissipative structures are designed so that during the earthquake some of their parts, referred to as dissipative zones, enter the inelastic region and undergo large inelastic deformations, dissipating seismic energy by means of ductile hysteretic behaviour. This ductile capacity of the structure, allows the reduction of design seismic loading by applying a reduction factor of $1/q$ of that for an elastic response, with obvious cost benefits. To provide adequate ductility for the structure, a reliable plastic mechanism without the danger of a brittle and/or progressive collapse must be ensured. This is achieved by means of capacity design and avoidance of all forms (overall and local) of early buckling phenomena (refs 16-18).

7. Space grids are structures consisting of axially loaded members only. Therefore, ductility requirements must take account of the axial hysteretic behaviour of members, as well as the increased danger of buckling of the numerous members in compression. In the presence of significant vertical load, additional members loading, due to an earthquake, may cause buckling of a particular member with possible catastrophic consequences for the structure. It has been shown analytically (refs 4 and 11) and experimentally (refs 4 and 5), that double-layer grid frameworks exhibit a brittle mode of progressive collapse when individual members buckle, due to extensive compressive force. When a compression member fails its contribution to the frame stiffness is negligible. In

this case, a collapse mechanism is developed, internal forces cannot be redistributed and the truss collapses. A detailed investigation of possible collapse mechanisms is provided by Affan and Calladine (ref. 1). A certain ductility of these structures has been observed when extensive tensile chord yield is involved prior to compression bar buckling. However, it has been established experimentally that, even for this case, internal force redistribution causes individual compression member buckling and leads again to brittle-like collapse (ref. 5). It is obvious that this phenomenon is more likely to occur when the bars are subjected to large axial loads, i.e. when the seismic energy is dissipated mainly by the space grid itself, and therefore, seismic response depends on the nature of the grid supporting substructure (flexible or rigid).

8. It should be noted that the aforementioned ductility considerations have been based on the response of grid frameworks to vertical and not seismic loads. Of course, during an earthquake, vertical loads are always present and, upon buckling of a compression bar, the collapse mechanism may be activated. However, these ductility considerations can only qualitatively be extended to the seismic loading case. The nature of earthquake actions requires a profound investigation of the phenomenon; this is, so far, not available in the literature. As mentioned in the introduction, most of the research on these structures has been conducted for the vertical static load in order to determine the equilibrium path (refs 2, 8 and 10).

9. The cyclic-load response of tubular members which are almost exclusively used in space frames, provides a significant advantage of these structures over ordinary structures composed of W-shaped, tee or double-angle members. More specifically, it has been shown experimentally that tubular members are superior to the above sections, in terms of their hysteretic envelops, when subjected to cycling loading (refs 6 and 7). This is explained by the double symmetry of the tube which eliminates the lateral-torsional buckling and results in member-buckling in a purely flexural mode. Moreover, it has been observed that double angles, as well as tees, exhibit a more pronounced local buckling behaviour which further reduces the cyclic-load capacity.

OUTLINE OF SPACE GRID SEISMIC DESIGN

10. The standard design procedure allowed by modern seismic codes (refs 16-18), is the Response Spectrum Method of Analysis. According to this method, the seismic design loading is determined by using a Design Response Spectrum in conjunction with a number of parameters concerning the seismicity of the area (zone), the importance of the structure, its dynamic characteristics (vibration period T), and its ductility (behaviour factor q). Under certain favourable conditions (regularity etc), the above dynamic method of analysis can be reduced to a simplified procedure called Equivalent Static Force Analysis, which is an approximate, yet adequate, method for most practical cases, including the seismic design of double-layer grid frameworks of usual dimensions and regular shape and geometry.

11. With regard to such a type of structure, two different cases must be examined, depending on whether the grid is supported on relatively flexible supports (e.g. slender columns), or on a stiff substructure (e.g. massive reinforced concrete).

Space grid on flexible supports

12. Common layouts of this case are shown in Figs la and lb. Vertical bracing may be needed in order to reduce the horizontal drift. It is recommended that the bracing be placed at or near to the centre of the structure to reduce the self-strained forces resulting from temperature changes. It should be noted that in this type of structure, the connections between space grid and columns are generally laterally fixed and rotationally free (Fig. 2). Therefore, the columns provide an elastic lateral restraint to the grid, resulting in the development of self-strained forces in the grid members. These forces, in the case of flexible columns, are of small magnitude and consequently do not have a serious impact on the dimensioning of the space grid members.

13. In the case shown in Fig. 1a, resistance to horizontal seismic loads is provided by the columns acting as cantilevers, transmitting forces between them through the space grid, while in the case of Fig. 1b, resistance is provided by the vertical bracing. It should be noted, that in this type of structure the total mass is almost that of the space frame, the mass of columns being very small. Secondly, the space grid has enough in-plane stiffness, that it can be considered as a diaphragm, distributing the total seismic shear to the columns in accordance with their flexural stiffness.

14. The total seismic shear load (base shear) V is generally given by codes (refs 16 and 17), in the following form:

$$V = \frac{Z\,I\,C}{q}.W \qquad (1)$$

where:

Z = Seismic zone factor given by the codes.
I = Importance factor also given by the codes.
C = Numerical coefficient depending on the soil characteristics and the fundamental period of vibration of the structure, given by the codes (design response spectrum or formula).
W = The total seismic dead load.
q = The behaviour factor, given by the codes for many types of dissipative structures, but not clearly for the type of structure under examination.

Eurocode 8 (ref. 16), for steel structures, consisting of a rigid roof over flexible columns, where the dissipative zones are concentrated at the columns, proposes a value of q between 1.5 and 2.0 (depending on the degree of regularity), which in the authors opinion, is rather conservative. The reason is that this type of structure is classified by the Eurocode in the same category as that of a simple cantilever.

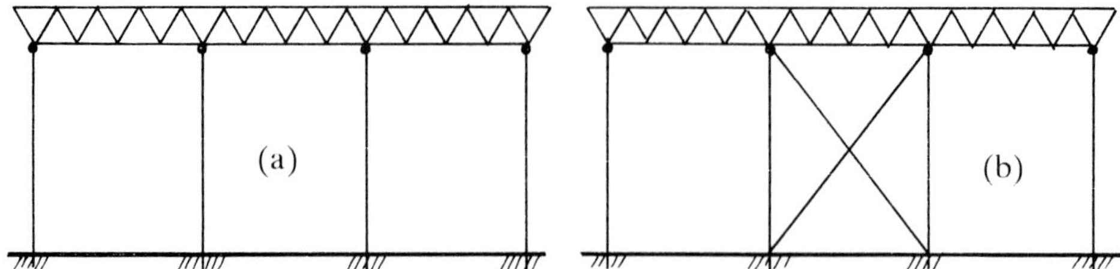

Fig. 1. Layout examples of space grids on flexible supports.

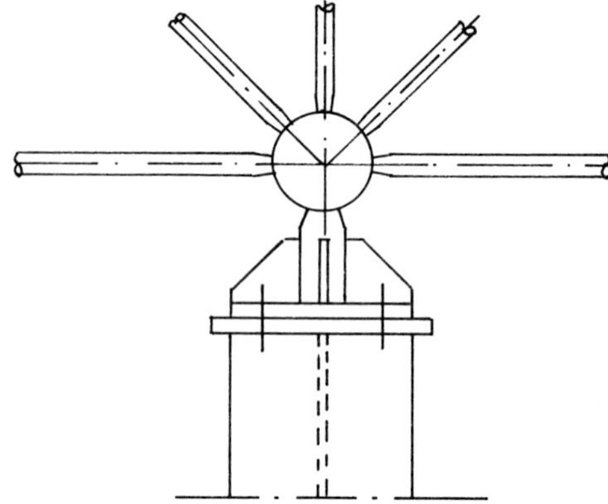

Fig. 2. Connection example between space grid and flexible column.

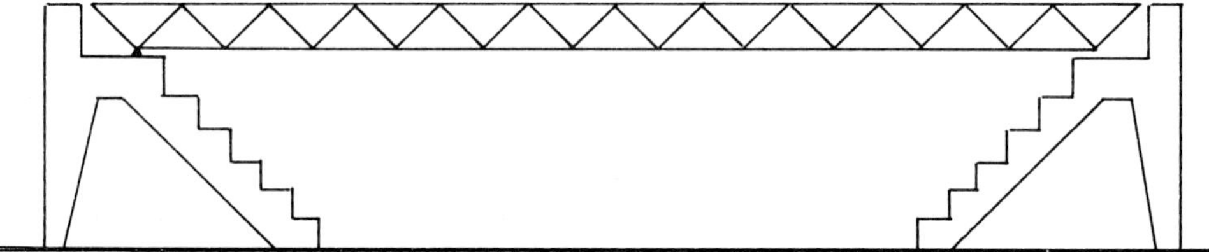

Fig. 3. Schematic view of a space grid covering a stadium or arena.

This, of course, is a conservative approximation, because the system of space grid and columns behave with frame action, which may be quite important when the structure enters the inelastic range. It should be noted that the codes give q values for ordinary frame behaviour i.e. without special ductility provisions, ranging between 4.0 and 6.0 (refs 16 and 17). Therefore it is necessary to conduct analytical and experimental research in order to establish more realistic "q" values for the type of structures under consideration. As for the determination of period T, any accurate or approximate method can be used, e.g. the formula:

$$T = 2\pi\sqrt{\frac{M}{K}} \tag{2}$$

where M is the total vibrating mass (space frame and half height of columns), and K is the total stiffness of resisting columns in the direction under examination.

15. The axial forces of grid members resulting from the analysis under the design seismic force, are combined with those of other loadings in order to provide the most unfavourable values for dimensioning. This combination is referred to as the *seismic design combination*. The other loadings are the permanent loads and a percentage of live loads specified by the code. Effects due to temperature changes can be neglected in the above combination.

16. The verification of all critical members of the space grid must satisfy the basic relation of structural safety $S_d =< R_d$, where:

S_d = member axial forces due to the seismic design combination.

R_d = member design resistances (calculated according to relevant steel design codes with values of partial safety factors γ_m valid for the basic usual loading combinations (ref. 18).

There is need for special ductility requirements at the columns, where almost the total seismic energy is dissipated by the formation of plastic hinges. In order that this mechanism be reliable, adequate redundant strength must be ensured, so that the formation of plastic hinges occur at the expected places (columns), and to enable the redistribution of forces. The latter is dictated from the fact that all columns do not enter the inelastic range simultaneously. This requirement can be fulfilled provided that the space grid members and the connections to the columns are designed with the member axial forces, S_d, multiplied by a magnification factor of the order of 1.20 to 1.50.

17. As mentioned above, the columns are the principal elements participating in the inelastic response of the structure. They must be designed using moment Ms and axial force Ns derived from the seismic design combination. The steel section must satisfy the limiting values of width-to-thickness ratio, b/t, for their compression parts, in order to avoid early local buckling and to allow for inelastic deformation assumed through the q factor in the design. Limiting values b/t can be found in the relevant seismic design codes. In the case of layout shown in Fig. 1b, where a vertical bracing is the resisting system, the inelastic response can be limited to the tension diagonals. Their design, as well as that of their end connections, shall follow the specific application rules provided by the relevant codes (refs 16 and 18). For example, limits are specified for the slenderness of the diagonal, and the redundant strength factor must be considered for the resistance of end connections over the plastic resistance of the member.

18. This is the case where a space grid is erected to cover stadiums and arenas (Fig. 3), atriums in commercial centres (Fig. 4) and in general areas where the

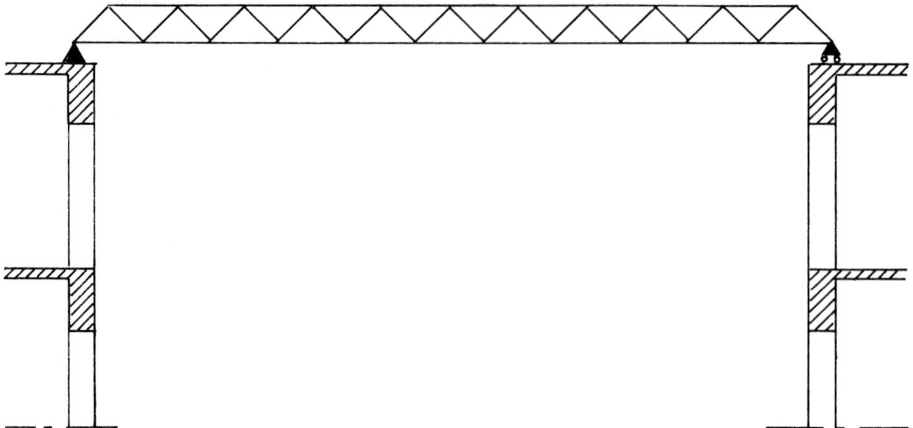

Fig. 4. Usual layout of a space grid covering an atrium.

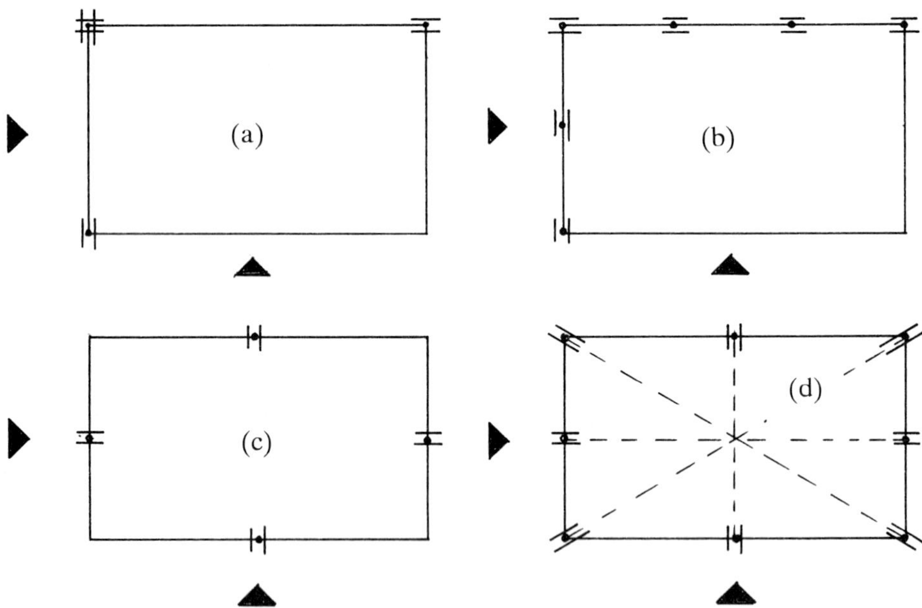

Fig. 5. Common arrangements of grid connections with substructure to resist horizontal loads.

supporting substructure is very stiff and massive compared to the space grid. In these cases, the space grid cannot be considered as a stiff diaphragm acting between the stiff vertical supports.

19. The design of connections between the space grid and the substructure becomes significant. Due to the stiffness of the substructure, the connections cannot be made laterally fixed, because high self-strained forces are developed in the grid. As a consequence, their selection (layout and type) must satisfy two opposing requirements, i.e. the need to avoid connection restraints which induce self-strained forces in the grid due to temperature changes, and on the other hand, the need to have a number of fixed connections to resist the horizontal seismic forces in the two main directions. Some arrangements which satisfy the above two requirements are shown in Fig. 5. They all permit the free movement of the space grid under temperature changes and, at the same time, provide the required resistance to seismic forces. It is evident that in these cases the space grid does not contribute as a diaphragm to the resisting system, as far as the seismic loads are concerned. These loads are finally resisted by the substructure alone, and it is there that ductility requirements should be directed. The space grid simply conveys its reactions to the substructure, where the dissipative plastic mechanism must be developed. For the design of connections between grid and substructure it is recommended that the redundant strength factor between 1.20 and 1.50 be taken, to ensure that these will not fail before the activation of the plastic mechanism of the substructure.

CONCLUSIONS

20. In this paper, some basic aspects concerning the behaviour of double-layer space grids under earthquake loading are presented. It is important to note that grid frameworks are parts of larger structural systems and, therefore, their seismic response depends on their interaction with the substructure.

21. Their design must primarily ensure the seismic energy dissipation through the formation of dissipative zones. A crucial point is the design of connections between the grid and the substructure. More specifically, the connections must satisfy the over strength requirement in order to allow activation of the dissipative plastic mechanism. On the other hand, they should not induce any significant self-strained forces due to temperature changes. Some usual support arrangements are indicated for the case of a rigid substructure, where the problem is more pronounced.

22. Furthermore, in order to improve the hysteretic response of grid members the use of tubular members is recommended. Nevertheless, it should be emphasised that research concerning the seismic design of structural systems, which include grid frameworks, is very limited. Therefore, extensive experimental and analytical research is necessary to investigate the ductility of these structures, as well as several important parameters, such as the effects of substructure rigidity, support connection adequacy, or ductility of the space grid itself.

REFERENCES

1. AFFAN A. and CALLADINE C.R., "Structural Mechanics of Double-Layer Space Grids", Shells, Membranes and Space Frames, Proceedings IASS Symposium, Osaka, Japan.
2. PAPADRAKAKIS M., "Analysis Methods for Spatial Structures", Shells and Spatial Structures; Computational Aspects", Proc. Int. Symp., Leuven, 1986.
3. SUPPLE W.J. and COLLINS I., "Limit State Analysis of Double-Layer Grids", in Analysis, Design and Construction of Double-Layer Grids by Z.S.Makowski, John Wiley, London 1981.
4. SCHMIDT L.C., MORGAN P.R. and CLARKSON J.A., "Space Trusses with Brittle-Type Strut Buckling", J.Str.Div., ASCE, Vol. 102, pp. 1479-1492, July 1976.
5. SCHMIDT L.C., MORGAN P.R. and HANAOR A., "Ultimate Load Testing of Space Trusses", J. Str. Div., ASCE, Vol. 108, pp. 1324-1335, June 1982.
6. ZAYAS V.A., POPOV E.P. and MAHIN S.A., "Cyclic inelastic buckling of Tubular Steel Braces", UCB/EERC-80/16, Berkeley, CA, June 1980.
7. BLACK R.G., WENGER W.A.B. and POPOV E.P., "Inelastic buckling of Steel Struts under Cyclic Load Reversals", UCB/EERC-80/40, Berkeley, CA, October 1980.
8. HANAOR A. "Analysis of Double-Layer Grids with Material Non-linearities-a practical approach", Space Structures, Vol. 1, pp.33-40, 1985.
9. SCHMIDT L.C. and MORGAN P.R., "Member Ductility and Design Detail of Some Welded Joints", Space Structures, Vol.2, pp. 139-145, 1986/87.
10. KANI I.M. and McCONNEL R.E., "Single Layer Shallow Lattice Domes: Analysis, General Behaviour and Collapse", Space Structures, Vol.3, pp. 64-73, 1988.
11. HANAOR A. and ONG A.F., "On Structural Redundancy in Space Trusses", Space Structures, Vol.3, pp. 237-241, 1988.
12. ROBERTSON S.R., "Seismic Analysis of a Space Frame", 3rd Int.Conf.Space Structures, Guildford, UK, 1984.
13. SMITH E.A., "Ductility in Double-Layer Grid Space Trusses", 3rd Int. Conf. Space Structures, Guildford, UK, 1984.
14. KATO S. and ISHIKAWA K., "Materially Non-Linear Stability of a Double-Layer Grid due to Vertical Earthquake Motions", Shells, Membranes and Space Frames, Proceedings IASS Symposium, Osaka, Japan, 1986.
15. KATO S., ISHIKAWA K. and YOKOO Y., "Earthquake Resistant Capacity of Long Span Trussed Structures", Trans. AIJ, No. 360, 1986.
16. Eurocode No 8, "Design of Structures in Seismic Regions", May 1988.
17. SEAOC, "Recommended Lateral Force Requirements and Commentary", 1990.
18. New Seismic Design Code of Greece, Athens, August 1992.
19. ANASTASIADIS K., "Seismic Resistant Structures", Vol.1, Thessaloniki 1991.
20. ROSENBLUETH E., "Design of Earthquake Resistant Structures", Pentech. Press, 1980.

52. Comparative investigation of the fatigue behaviour of uniplanar and multiplanar K-joints with gap

O. BUCAK, F. MANG and S. HERION, Universitat Karlsruhe, Germany

SYNOPSIS. In the past years, extensive evaluation and practical tests were performed to investigate the fatigue behaviour of uni-planar K-type square hollow section joints leading to the design formulae for this type of joint at the University of Karlsruhe. A major part of these research results has been incorporated into Eurocode 3 "Design of Steel Structures" initiated by the European Community. Regarding multi-planar K-joints, to the authors' knowledge no experimental work has been reported in literature. In almost all the cases, the tests were carried out under static load. The first fundamental investigations on multi-planar joints made of circular and square hollow sections under fatigue load are being carried out in the Netherlands (Delft University of Technology and TNO Building and Construction Research, Rijswijk) and in Germany (Karlsruhe University) sponsored by ECSC (refs 1, 2 and 3). This paper describes the developments achieved in the field of uni-planar K-type joints, as well as the test series initiated at the University of Karlsruhe, to investigate the fatigue behaviour of isolated multi-planar KK-joints made of square hollow sections. In order to determine the influence of combined axial force and in-plane and/or out-of-plane bending moment on the stress concentrations in multi-planar KK-joints, static tests with strain measurements using strain gauges and chains were performed. Following the static procedure, fatigue tests on joints with various geometrical parameters were made plotting S_r (stress range) verses N_f (loading cycles to failure) diagrams. Furthermore, fatigue tests were done on the cracked joints after repair to find out how they can be repaired efficiently.

INTRODUCTION

1. It is well-known that stress or strain concentrations occur in a loaded construction at the places of discontinuity of form and cross-section causing localised peak stresses, which can be many times higher than the nominal stresses calculated according to simple elastic theory. This phenomenon is also observed at the welds, where the stress or strain concentration takes place due to the change over from basic to weld material at the weld toe depending on the shape of the weld toe, the weld angle and the weld defects (undercut, excess weld metal,

overlap etc.). In short, the localised peak stresses appear in welded structures caused by: a) geometrical stress concentration, and b) stress concentration due to weld.

2. While the role played by the local stress or strain concentration is of minor importance for statically loaded joints due to redistribution of stresses by local yielding of the material, it is vital for joints under fatigue loading. The initiation of fatigue cracks occurs at the location of largest, so called "hot spot" stresses, which, therefore, must be taken into account for fatigue design.

3. The phenomenon of stress concentration is especially significant for welded hollow section joints in lattice structures, where bracings and chords of different stiffnesses intersect and the loading direction changes depending on the joint configuration. In order to incorporate stress concentration into the hollow section fatigue design, it was necessary to carry out fatigue tests and plot curves with the measured hot spot stress ($S_{r,h.s}$) or strain ($\varepsilon_{r,h.s.}$) ranges against the loading cycles of failure (N_f). For practical design of hollow section joints, only the effect of the geometrical stress concentration, which depends on the various geometrical parameters, such as α, β, γ, τ, ξ, and θ, has to be known, as the basic $S_{r,h.s}$-N_f curves are determined for specific details. The measurement of hot spot strains are made in a manner that the local weld effects are not included.

4. In order to simply calculate the hot spot stresses of lattice girder joints in hollow sections, it is useful to relate the hot spot stress in the joint at the weld toe to the nominal stress in bracing. The factor of relationship, known as the strain concentration factor SNCF, is defined as:

$$SNCF = \frac{\varepsilon_{max,\ weld\ toe}}{\left(\varepsilon_{ax}+\varepsilon_{ipb}+\varepsilon_{opb}\right)\ \text{brace in tension}} \quad (1)$$

Further,

$$SCF = 1.1 \cdot SNCF$$

5. The calculated SNCF values depend on the type of joints (such as T, X, K), the type of loading (axial force, in-plane and out-of-plane bending moment) and geometrical parameters (α, β, γ, τ, ξ, and θ) as shown in Fig. 1.

6. The hot spot stress range at a particular location is calculated according to the following equation:

$$S_{r,h.s} = SCF_{ax.} \cdot S_{r,ax,nom.} + SCF_{ipb} \cdot S_{r,ipb,nom} + SCF_{opb} \cdot S_{r,opb,nom} \quad (2)$$

7. The determination of hot spot stresses at the weld toe of the intersection of bracing and chord is made either experimentally by means of strain gauge measurements or theoretically using the Finite Element Method.

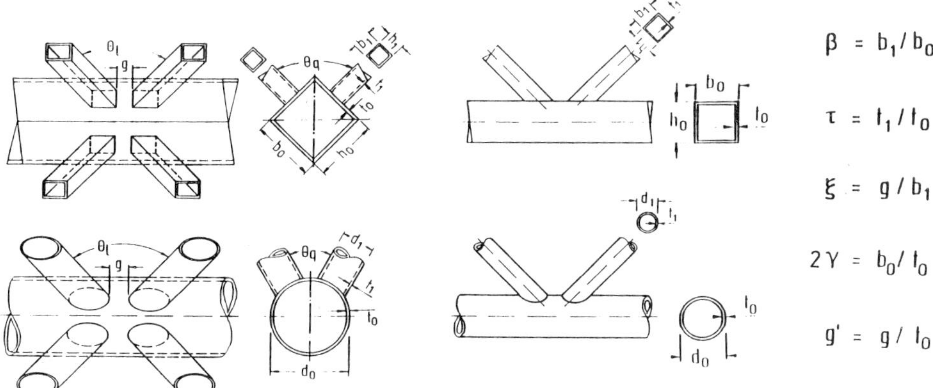

Fig. 1. Joint types and geometrical parameters.

8. The measurement of the geometrical strains at the weld toe with strain gauges were made using the method of extrapolation as shown in Fig. 2. Between the linear and quadratic extrapolation the latter was chosen for rectangular hollow section joints because of their highly non-linear stress gradients.

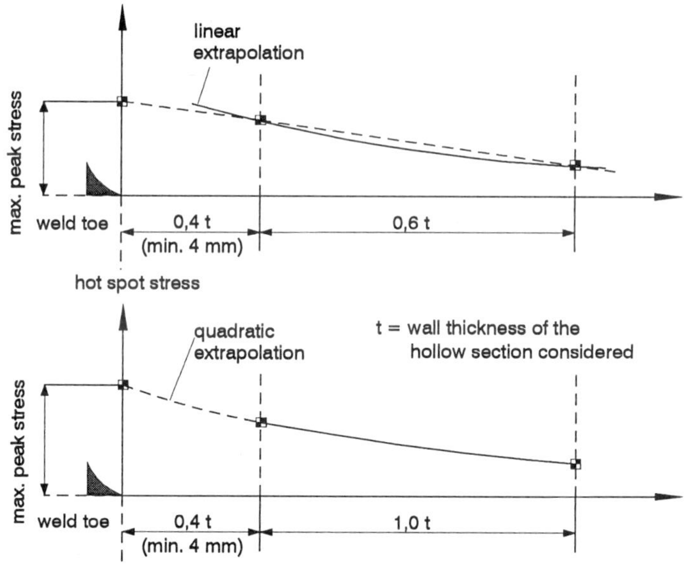

Fig. 2 Methods of extrapolation.

9. For modelling the joint using Finite Elements, shell elements and solid elements are most suitable. By means of these elements, two models should be used:

- model I using solid elements only;
- model II with shell elements and solids elements in the area of the welds;

(the connection of both element types was done by constraint equations).

10. In both cases, special work is caused by the simulation of the welds or the connection of the shells to the solids. Figs 3a and 3b show the problem in the area of the welds.

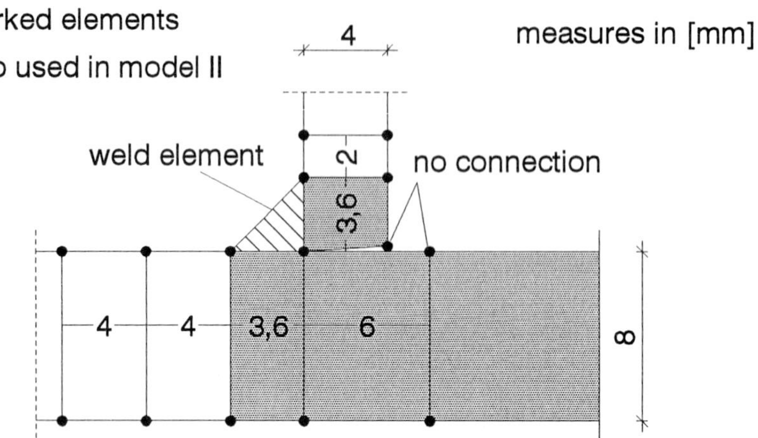

Fig. 3a. Weld area of model I.

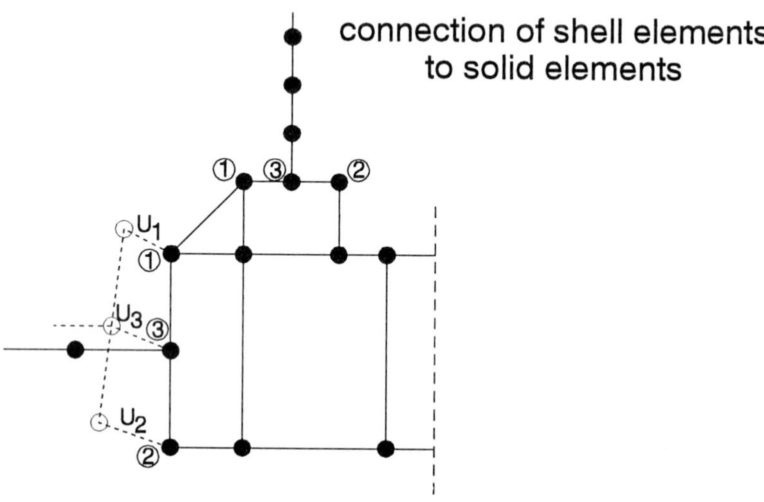

Fig. 3b. Weld area of model II.

11. The corner fillets of the hollow sections can be approximated with two or three elements respectively in dependence on the wall thickness as is shown schematically in Fig. 4 .

12. The sizes of the elements result from the geometry such as corner fillets, width ratio $\beta = b_1/b_0$ and the size of the gap g. In addition, the element sizes are

to be selected in a way that sufficient information is available for carrying out quadratic extrapolations.

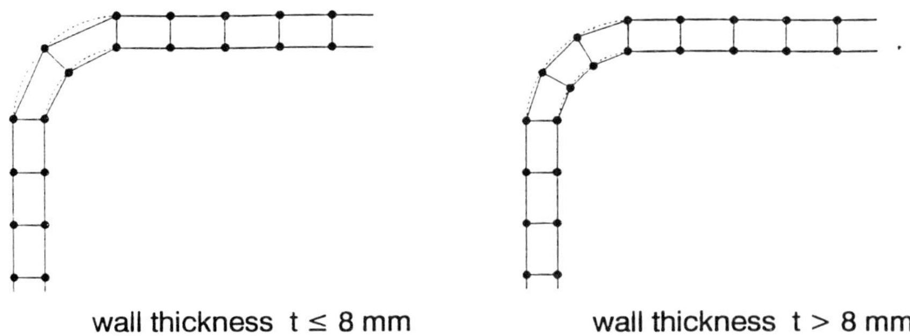

Fig. 4. Finite Element model in the corner area.

13. In order to obtain a simple means to calculate the hot spot stresses, parametric formulae respective design curves for SCF were developed using regression analyses based on an extensive parameter study with a computer program.

14. The ultimate aim is to give to the designers in practice a calculation method for RHS joints under fatigue load, as described in Fig. 5.

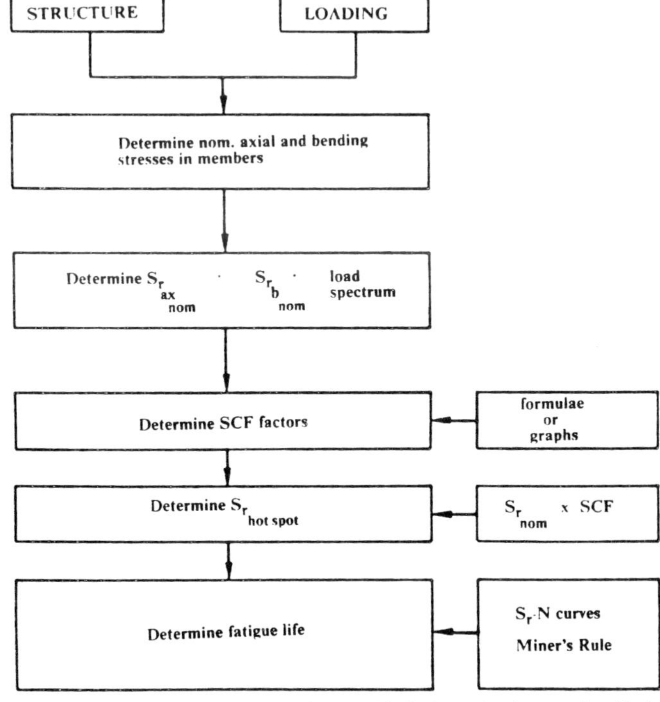

Fig. 5. Design procedure for RHS-joints (schematically).

UNI-PLANAR K-TYPE JOINTS OF SQUARE HOLLOW SECTIONS

15. Sponsored by the European Community, Brussels and CIDECT (Comité International pour le Développement et l'Etude de la Construction Tubulaire), Düsseldorf, an extensive research programme was conducted to investigate the fatigue behaviour of uni-planar square hollow section joints of T-, X- and K-types. This work was performed jointly by the Delft University of Technology and TNO-IBBC, Delft (T- and X-type joints) and the University of Karlsruhe (K-type joints with gap and overlap).

16. The experimental part of the investigations on K-joints with overlap consisted of four series with a total of 12 joints, and of K-joints with a gap, five series with a total of 24 joints, whereby the joints with a stress ratio of R = +0.1 have been checked at various stress levels. The test rig and manufacturing procedure are the same as for the previous investigations of K- and N-joints also sponsored by ECSC (ref. 4).

17. The objective of the experimental investigations was to measure the strain concentration in the critical areas of the joint to make comparisons with the results of the numerical investigations, using the Finite Element Method. In addition, fatigue tests were necessary for the determination of S-N-curves, in order to be able to check the design recommendations given by Eurocode 3 and others.

18. The maximum strains were determined in six and fourteen extrapolation lines respectively, placed near the corners of a joint (see Fig. 6). In addition, one specimen from each series was equipped with a strain measuring chain. Since the influence of the strains directly at the toe of the welds could not be determined due to local weld notch, extrapolation methods had to be carried out, following the agreement made by the experts involved in the European Offshore Programme.

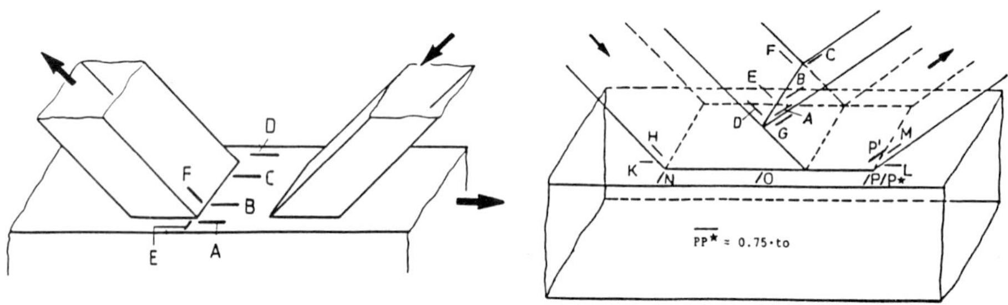

Fig. 6. Positions of the strain gauges.

19. As the axial force and in-plane bending components were not determined separately, the bending components were indirectly included, which means that the SCFs were calculated as follows:

$$SCF = \frac{S_{r,h.s.}}{S_{r,ax,nom}}$$ and not as mentioned in Eq. 1.

20. In order to take account of the secondary bending moments in the joints, caused by the actual bending stiffness of the joint and joint eccentricities, the recommendation of the fatigue rules of the IIW-Subcommission XV E (ref. 5) has to be applied. This recommendation is to multiply the stress range due to axial loading by the factors in Table 1, when in-plane bending stress is not known, implying that the nominal bending stress due to secondary moments is generally about 50% of the nominal axial stress (or 33% of the nominal total stress).

21. Basing on a statistical analysis of all relevant experimental data (T-, X-, K-Joints) design curves $S_{r,h.s}$ -N_f for square hollow section joints (also applicable to rectangular hollow section joints) were derived as shown in Fig. 7.

Table 1. Factors to be applied to K-joints in square hollow sections, which take secondary bending moments into account.

Type of joints	chords	braces
gap joints	1.5	1.5
overlap joints	1.5	1.3

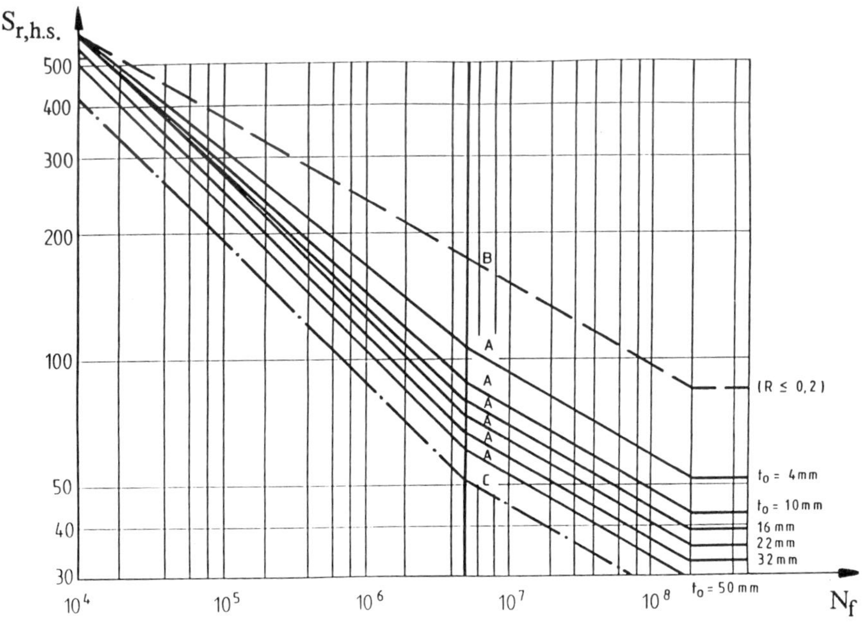

Fig. 7. $S_{r,h.s}$ - N_f design curves.

22. Parallel to the tests, parametric studies were carried out by using the Finite Element method. After it was guaranteed that the Finite Element model, which has been developed, provided reliable results, the parametric studies could be started.

23. In total, 103 analyses have been carried out using the Finite Element Method, 48 for K-joints with overlap and 55 for K-joints with gap. With this, the governing parameters have been varied over wide ranges. The angle u between the brace and the chord axis has been kept constant at 45°; but the range of validity can be increased from 35° to 60° according to indications in literature (ref. 6).

24. These studies resulted in formulae as given in Tables 2 and 3.

25. The design calculation of uni-planar RHS joints according to Fig. 5 can now be carried out, where required data can be obtained as follows:

SCF from Tables 2 and 3
$S_{r,h.s}$ - N_f design curves from Fig. 7

Table 2. SCF formulae for K-joints with a gap in square hollow sections subjected to balanced axial forces and secondary moments.

Member	Formulae for SCFs at weld toe in brace and chord Parametric range: $0.35 \le \beta \le 1.0$, $12.5 \le 2\gamma \le 25.0$, $0.25 \le \tau \le 1.0$, $0.25 \le \xi \le 0.75$, $1.5 \le g' \le 7.0$, $35° \le \theta \le 55°$
Brace (maximum value)	$SCF = 3.63\,\tau\,(2-\tau) + 0.336\,\xi\,\gamma^2\,(0.3 - 0.01\,\xi\,\gamma) + 0.044\,\gamma\,\beta\,(6.38 - \gamma\,\beta^2) - 4.81\,(\frac{\gamma g'}{100})^2 - 2.2$ where: $\xi = g' / (2\,\gamma\,\beta)$ $(= g / b_1)$
chord (maximum value)	$SCF = 1.1\,\tau\,(0.00288\,\gamma^3 + g') + 5.73\,\xi\,(1 - 0.178\,\xi^2\,g') - 0.166\,\xi^3\,g'^2 - 1.73$ where: $\xi = g' / (2\,\gamma\,\beta)$ $(= g / b_1)$

MULTI-PLANAR DOUBLE K-JOINTS OF SQUARE HOLLOW SECTIONS

26. An additional research programme dealing with the fatigue behaviour of multi-planar double K-Joints (Fig. 1) as are found in triangular girders, has just been finished (ref. 2). Recent strain measurement results under static loading as well as fatigue tests are presented in the following. The aim of the investigation was to determine a design method for multi-planar double K-joints following a similar procedure to that for uni-planar joints described in the first part of this paper.

27. Tables 4 and 5 show results of the fatigue tests together with the determined SCF factors obtained through the extrapolated hot spot strains ε_{max} at the weld

toe of the bracings. Included are some test results of uni-planar joints with the same dimensions. The layout of the strain strips can be seen in Fig. 8.

Table 3. SCF formulae for K-joints with overlap square hollow sections, subjected to balanced axial forces and secondary moments.

Member	Formulae for SCFs at weld toe in brace and chord
	Parametric range: $0.35 \le \beta \le 0.7$, $12.5 \le 2\gamma \le 25.0$, $0.4 \le \tau \le 1.0$, $-0.40 \le \xi \le -1.0$, $-17.0 \le g' \le -2.5$, $35° \le \theta \le 55°$
Brace (maximum value)	$SCF = 0.144\,\beta\,\gamma^2\,(1 - 0\,813\,\beta^2) + 1.84\,\beta\,\tau + 3.23\,\xi^2\,(1.94\,\tau^2 - 1.9\,\tau^3 - (\frac{\gamma}{10})^3) - 0.26$ where: $\xi = g' / (2\,\gamma\,\beta)$ $(= g / b_1)$
chord (maximum value)	$SCF = -40.22\beta\,\gamma^2\,(1 - 0.59\,\xi^2) + 0.028\,\gamma^2\,(8.9\,\beta + \tau) - 5.41\,\gamma\,\beta^3 - 0.008\,\xi^2\,\gamma^3 + 2.109\,\xi^6 - 4.24$ where: $\xi = g' / (2\,\gamma\,\beta)$ $(= g / b_1)$

Table 4. Test results for $\beta = 0.4$.

test specimen *)	dimensions [mm]	type of failure	S_R [N/mm^2]	SCF	$S_{R,h.s.}$ [N/mm^2]	$S_{R,h.s.}$ (t = 16 mm) [N/mm^2]	$N_{failure}$
R10S01-1	200x200x8.0	brace	90.0	3.72 1)	334.8	144.6	320000
	80x 80x4.0						
R10S01-2	g = 25.4	weld	63.0	3.72 1)	234.4	156.5	197000
R10S01-3		chord	63.0	3.91	246.3	163.6	231700
R10S01-4		chord	45.0	3.52	158.4	101.5	684900
R10S04-1	200x200x8.0	chord	36.0	4.26	153.4	93.8	2854000
	80x 80x4.0						
R10S04-2	g = 87.0	chord	67.0	4.26 1)	285.4	193.8	119500
K20S51-1	200x200x8.0	chord	54.0	4.22	227.9	146.9	572500
	80x 80x4.0						
K20S51-2	g = 25.4	chord	76.5	4.22	322.8	223.0	177840
K20S51-3		-	45.0	4.22	189.9	-	6365300
K20S51-4		brace	90.0	4.22	379.8	160.6	442000

1) mean taken from the "measured" SNCF-values of each series

*) R: multiplanar, K: uniplanar

Table 5. Test results for $\beta = 0.6$.

test specimen *)	dimensions [mm]	type of failure	S_R [N/mm²]	SCF	$S_{R,h.s.}$ [N/mm²]	$S_{R,h.s.}$ (t = 16 mm) [N/mm²]	$N_{failure}$
R10S02-1	200x200x8.0	chord	90.0	3.67 1)	330.3	229.3	61500
	120x120x4.0						
R10S02-2	g = 25.4	chord	63.0	3.89	245.1	161.2	315950
R10S02-3		chord	45.0	3.45	155.3	95.9	2109800
R10S02-4		chord	54.0	3.67 1)	198.2	124.4	1280600
K20S52-1	200x200x8.0	chord	60.0	3.93 2)	235.6	151.9	571068
	120x120x4.0						
K20S52-2	g = 25.4	chord	67.5	3.93 2)	241.0	160.0	235440

1) mean taken from the "measured" SNCF-values of each series
2) SCF from formula

*) R: multiplanar, K: uniplanar

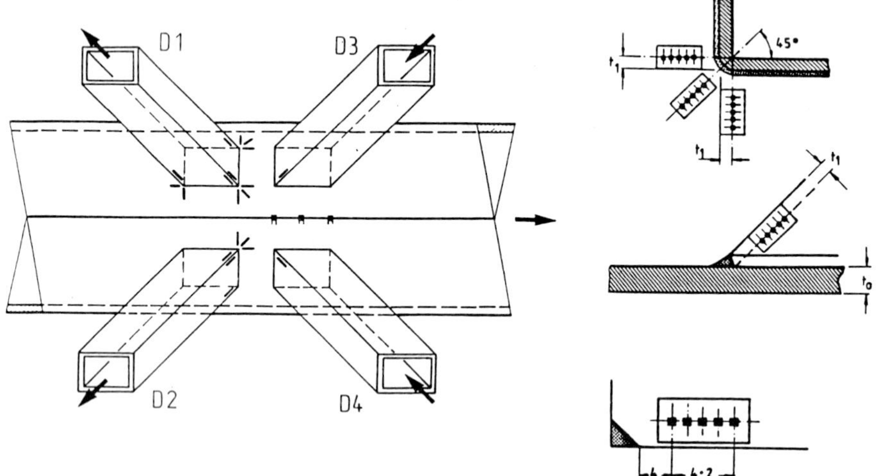

Fig. 8. Position of the strain gauges.

28. Finally, all the test results in the fatigue tests on multi-planar K-joints have been plotted in one diagram, without considering any wall thickness effect (Fig. 9), together with some data for uni-planar joints using the SCF formula as mentioned above. In addition to this, the scatter band for uni-planar K-joints has also been plotted considering the wall thickness effect.

29. It can be seen that the uni-planar K-joints K20S51 and K20S52 are well within the scatter band, if the wall thickness effect has been taken into account. It becomes also evident that the test results for the multi-planar joints, without considering the wall thickness effect, are well within the scatter band of the uni-planar joints.

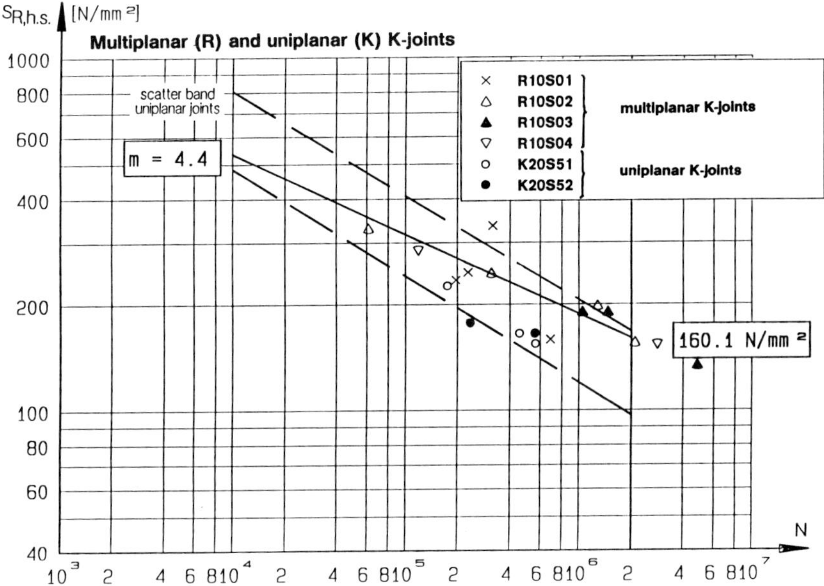

Fig. 9. $S_{r,h.s}$ - N_f lines without thickness correction.

30. Parallel to the procedure for uni-planar joints, parametric studies for the determination of SNCF for multi-planar joints are being performed at present, (refs 7, 8 and 9). Fig. 10 shows the first results of these studies. The publication of these investigations, considering the influences of pure axial load as well as moments, from in-plane bending and out-of-plane bending (IPB and OPB) separately, is foreseen at the beginning of 1994.

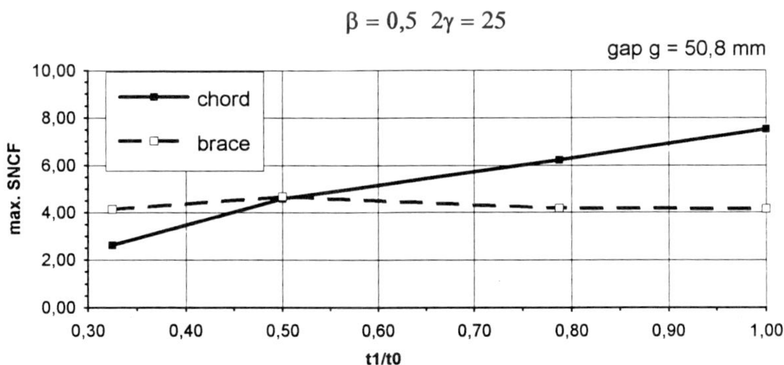

Fig. 10. SNCF versus τ for $\beta = 0.5$ and $g = 50.8$ mm.

REFERENCES

1. PUTHLI, R., WARDENIER, VAN WINGERDE, and VERHEUL, C.H.M. ECSC research programme 7210/-SA 114. Fatigue Behaviour of Multiplanar Welded Hollow Section Joints and Reinforcement Measures for Repair. Interim Reports, TNO Delft, (1988 1991).
2. MANG, F., HERION, S., and BUCAK, Ö. ECS research programme 7210-SA/114. Zeit- und Dauerfestigkeit von räumlichen und geschweiβten, unversteiften Rund- und Rechteckhohlprofilverbindungen sowie Verstärkungsmaβnahmen (Fatigue Strength of Multiplanar Welded, Unstiffened Circular and Rectangular Hollow Sections as well as Reinforcement Measures=. Zwischenberichte, TU Karlsruhe, Versuchsanstalt fiir Stahl, Holz und Steine, (1991-1).
3. MANG, F:, HERION, S.; and BUCAK, Ö. Fatigue Behaviour of Multiplanar K-Joints made of Hollow Sections, First International Offshore and Polar Engineering Conference (ISOPE 91), (1991-1).
4. MANG, F. Zeit- Und Dauerfestigkeit von einfachen und geschweiβten Fachwerkknoten aus Rundhohlprofilen. ECSC Research 6210-KD/1/103, April 1980.
5. DIJKSTRA, O.D., and DE BACK, J. Fatigue Strength of Tubular X- and T-Joints, Offshore Technology Conference , paper no. 3696, Houston 6. DUTTA, D., MANG, F., WARDENIER, J. Fatigue Behaviour of Welded Hollow Section Joints. CIDECT MONOGRAPH No. 7, (1980).
7. MANG, F., and HERION, S. Deutsche Forschungsgemeinschaft DFG Research Programme MA 761-6-1: Räumliche Hohlprofilverbindungen (Multiplanar Hollow Section Joints), TU, Karlsruhe, Versuchsanstalt für Stahl, Holz und Steine, (1991-2).
8. MANG, F., HERION, S. Comparative Investigation on the Fatigue Behaviour of Uniplanar and Multiplanar K-Joints with Gap. Second International Offshore and Polar Engineering Conference (ISOPE 92), (1992).
9. HERION, S. Untersuchungen über SNCF Faktoren an Konstruktionen aus Hohlprofilen (Investigations on SNCF factors on Constructions made of Hollow Sections), PhD thesis, University of Karlsruhe, in preparation.

53. Earthquake-strain constrained design for large space frames

TSUNEYOSHI NAKAMURA and MAKOTO OHSAKI, Kyoto University, Japan

SYNOPSIS. A practically efficient method of computing a sequence of seismic strain-constrained designs of a large space frame for a design strain range is presented. The stiffnesses of all the members of the frame are adjusted or controlled in such a way that the mean maximum response strain in every member, due to a set of spectrum-compatible earthquake motions, would not exceed the specified design upper bound. The effect of rigid connections on the designs is discussed and compared with the designs for corresponding idealised trusses.

INTRODUCTION

1. The problem considered in this paper is to find a sequence of seismic-strain controlled designs of a large space frame for a design strain range. The mean maximum response strains of all the members to a set of design earthquake motions compatible with a prescribed design response spectrum are adjusted or controlled in terms of the member stiffnesses. Any previous computational method (refs 1-8) is based on a mathematical programming method together with large-sized sensitivity matrices. Recently, an efficient method, which does not resort to the latter, has been proposed by the present authors for the problem of designing idealised space trusses (refs 9-11).

2. The purpose of this paper is to extend the authors' sequential generator technique to rigidly-jointed space frames. Here the location of the member-end fibre in which the maximum extensional strain occurs is also a function of the set of member stiffnesses. A sequence of seismic strain-constrained designs of a rigidly-jointed space frame is compared with that for an idealised space truss (ref. 11) with the same centreline dimensions and support conditions. The effect of the different jointing on the designs is evaluated.

PROBLEM OF SEISMIC STRAIN-CONSTRAINED DESIGNS

3. Consider a space frame as shown in Fig. 1. The centreline dimensions of the members and the support conditions are to be prescribed. All the members of the frame are made of a linear elastic material and are rigidly-jointed. For simplicity, it

is assumed that each member is a pipe whose middle surface is a circular cylinder with a radius R_i ,as shown in Fig. 2, and that the thickness t_i is proportional to the radius, i.e. $t_i = \tau R_i$, where τ denotes a common scale factor.

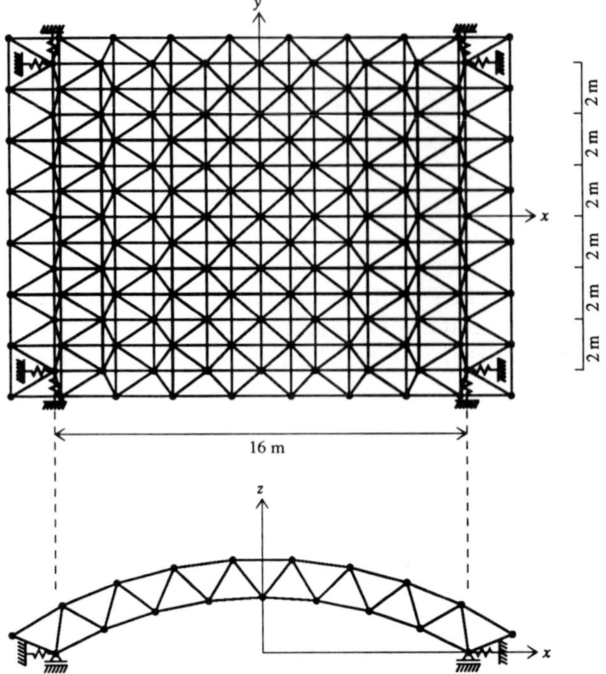

Fig. 1. A 504-bar circular cylindrical space frame with rigid joints.

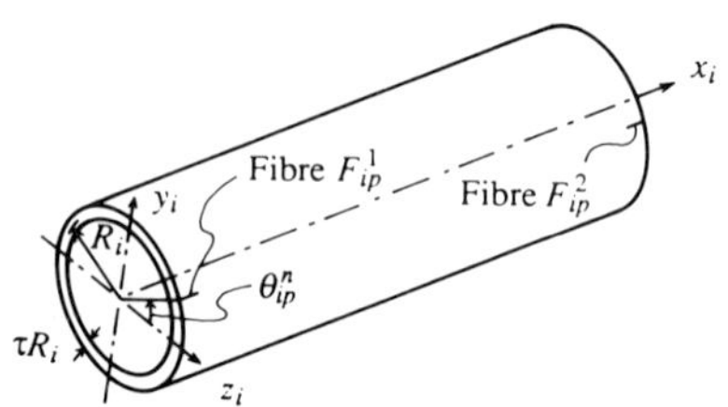

Fig. 2. A tubular member and the member co-ordinates.

The cross-sectional area A_i , moment of inertia I_i^B and polar moment of inertia I_i^P may be written as follows:

$$A_i = 2\pi\tau R_i^2, \qquad I_i^B = \frac{\pi}{4}\tau(4+\tau^2)R_i^4, \qquad I_i^P = \frac{1}{2}I_i^B \tag{1}$$

The frame is defined completely with a set of design variables $\{R_i\}$ which is denoted by a vector $\boldsymbol{R}$. A set of lower bounds of the design variables denoted by $\overline{\boldsymbol{R}} = \{\overline{R}_i\}$ is also prescribed.

4. The frame is to be designed so that the mean value of the peaks of representative seismic strain in each member, due to a set of spectrum compatible design earthquakes, is within a prescribed limit. Let $S_D(\Omega)$ denote a prescribed design displacement response spectrum (e.g. ref. 12) with respect to the eigen-value of the frame. For the sake of brevity, the frame is to be subjected only to a set of vertical ground motions.

5. For each member, a cartesian co-ordinate system (x_i, y_i, z_i) is defined as shown in Fig. 2. The co-ordinate i is taken along the axis of the member i. The location of member-end fibre F_{ik}^n is defined with respect to the angular co-ordinate θ_{ik}^n, where the subscript k denotes the number of equally spaced points on an end circle n (= 1, 2) on the exterior circular cylindrical surface of member i. The representative strain of member i is the maximum among the mean seismic extensional strains $\{\varepsilon_{ik}^n\}$ in fibres $\{F_{ik}^n | \ n = 1,2,; \ k = 1,2,...\}$. Let ε_i^x denote the extensional strain of the axis of member i due to the axial force. Let k_i^{yn} and k_i^{zn} denote the components of the curvature of the deformed axis related respectively with a negative sign to the second derivatives of the deflection functions $v_i(x_i)$ and $w_i(x_i)$ in the (x_i, y_i)—plane and in the (x_i, z_i)—plane. Both $v_i(x_i)$ and $w_i(x_i)$ are expressed in terms of cubic functions. The total extensional strain in fibre F_{ik}^n may be written as

$$\varepsilon_{ip}^n = \varepsilon_i^x + (1 + 0.5\tau)R_i\{\kappa_i^{zn}\cos\theta_{ik}^n + \kappa_i^{yn}\sin\theta_{ik}^n\} \tag{2}$$

The longitudinal displacement and the angle of twist are linear functions of x_i

6. Let $\boldsymbol{K}(\boldsymbol{R})$ $\boldsymbol{M}_D(\boldsymbol{R})$ and $\boldsymbol{M}_J$ denote the elastic stiffness matrix, the consistent mass matrix due to member masses, and the mass matrix due to lumped masses at joints, respectively, of the frame of design $\boldsymbol{R}$ with respect to a structure co-ordinate system. The rth eigen-value and the corresponding eigen-vector are denoted by $\Omega_r(\boldsymbol{R})$ and $\Phi_r(\boldsymbol{R})$ respectively. The equation of eigen-vibration in the rth eigen-mode may be written as

$$\boldsymbol{K}(\boldsymbol{R})\boldsymbol{\Phi}_r(\boldsymbol{R}) = \Omega_r(\boldsymbol{R})\{\boldsymbol{M}_D(\boldsymbol{R}) + \boldsymbol{M}_J\}\boldsymbol{\Phi}_r(\boldsymbol{R}) \qquad (r = 1, 2, \ldots, s) \tag{3}$$

The lowest s modes are taken into consideration in evaluating the seismic strain. The rth mode strain in fibre F_{ik}^n corresponding to Φ_r is denoted by $e_{ik}^{n(r)}$.

7. Let $\boldsymbol{I}$ denote a vector in which elements corresponding to vertical displacements are equal to unity and the remainders are null. The participation factor β_r of the rth mode may be written, in terms of the normalised eigen-vectors $\{\Phi_r\}$ as:

$$\beta_r = \boldsymbol{\Phi}_r^T[\boldsymbol{M}_D + \boldsymbol{M}_J]\boldsymbol{I} \qquad (r = 1, 2, \ldots, s) \tag{4}$$

8. The mean maximum responses are evaluated by means of the complete quadratic combination (CQC) method (refs 13,14). The mean maximum seismic extensional strain ε_{ik}^{vn} may be evaluated by:

$$\varepsilon_{ik}^{vn} = \left[\sum_{p=1}^{s}\sum_{q=1}^{s}\left\{(e_{ik}^{n(p)}\beta_p S_D(\Omega_p))\rho_{pq}(e_{ik}^{n(q)}\beta_q S_D(\Omega_q))\right\}\right]^{\frac{1}{2}} \tag{5}$$

where ρ_{pq} denotes the cross-modal correlation coefficient between the pth and qth modes (refs 10, 11).

9. Let ε_{ik}^{sn} denote the static extensional strain in fibre F_{ik}^{n} due to dead loads. The total mean maximum seismic extensional strain ε_i in member i is defined by:

$$\varepsilon_i = \max_{k,n}\{|\varepsilon_{ik}^{sn}| + \varepsilon_{ik}^{vn}\} \qquad (i = 1, 2, \ldots, m) \tag{6}$$

SEQUENCE OF SEISMIC STRAIN CONSTRAINED FRAMES

10. Let Λ_i denote the design strain for member i. The problem RSCD of finding a seismic response strain-constrained design may be stated as follows: find the design $\tilde{\boldsymbol{R}}$ for a specified set of design strains $\{\Lambda_i\}$ such that the constraints, $\varepsilon_i(\tilde{\boldsymbol{R}}) = \Lambda_i$ for $\tilde{R}_i \geq \bar{R}_i$ and $\varepsilon_i(\tilde{\boldsymbol{R}} \leq \Lambda_i) < \Lambda_i$ for $\tilde{R}_i = \bar{R}_i$ $(i = 1, 2, \ldots, m)$ are all satisfied.

11. Consider next a sequence of design strain sets defined by $\{\Lambda_i\} = \xi\{\Lambda_i^b\}$ (Λ_i^b prescribed) for a range $\xi_L \leq \xi \leq \xi_U$. Then the corresponding sequence of the solutions $\tilde{\boldsymbol{R}}$ to the sequence of problems RSCD may be conceived as a function of ξ and is denoted by $\hat{\boldsymbol{R}}(\xi)$. All the corresponding response quantities of the frames of design $\hat{\boldsymbol{R}}(\xi)$ may then be regarded as functions of ξ. Each response quantity is denoted by the same symbol but is distinguished by a circumflex, for instance, $\varepsilon_i(\tilde{\boldsymbol{R}}(\xi)) \equiv \hat{\varepsilon}_i(\xi)$ It is assumed here that $\hat{\varepsilon}_i(\xi), \hat{\boldsymbol{R}}(\xi)$ and other response quantities as functions of ξ are n times continuously differentiable.

12. The problem of finding the sequence of solutions $\hat{\boldsymbol{R}}(\xi)$ to RSCD problems may be stated as follows:

PROBLEM SRSCD: For $\xi_L\{\Lambda_i^b\} \leq \xi\{\Lambda_i^b\} \leq \xi_U\{\Lambda_i^b\}$, find $\hat{\boldsymbol{R}}(\xi)$ such that:

$$\hat{\varepsilon}_i(\xi) = \xi\Lambda_i^b \quad \text{for} \quad \hat{R}_i(\xi) > \bar{R}_i \tag{7a}$$

$$\hat{\varepsilon}_i(\xi) \leq \xi\Lambda_i^b \quad \text{for} \quad \hat{R}_i(\xi) = \bar{R}_i \tag{7b}$$

where $\{\hat{\varepsilon}_i(\xi)\}$ are governed by Eqs 1-6, equations of design response spectrum and other pertinent equations (refs 10, 11).

13. If the solution to an RSCD problem is found, then the solution to PROBLEM SRSCD may be found by means of the sequential generator technique (refs 9-11). Consider therefore the design $\overline{\boldsymbol{R}}$. Let $\bar{\xi}$ denote $\max\{\varepsilon_i(\overline{\boldsymbol{R}})/\Lambda_i^b\}$ (i = 1,2,...,m). Obviously $\overline{\boldsymbol{R}}$ is the solution to PROBLEM RSCD for $\bar{\xi}\{\Lambda_i^b\}$. $\overline{\boldsymbol{R}}$ may always be scaled down so as to satisfy $\xi_U \leq \bar{\xi}$. The sequential generator may be formed with $\overline{\boldsymbol{R}}$ as the initial solution. Rename the member number in such a way that $\varepsilon_1(\overline{\boldsymbol{R}})/\Lambda_1^b = \bar{\xi}$. As ξ is decreased from $\bar{\xi}$, R_1 must increase while every other R_1 remains at $\overline{\boldsymbol{R}}_i$. All the governing equations are differentiated with respect to ξ to the desired order. Note that $\hat{R}_i' = 0$ for $i \geq 2$. Each set of differentiated equations are evaluated at $\bar{\xi}$. Then a set of $(2s^2 + f + fs + 3s + 4)$ simultaneous linear equations of the form $\boldsymbol{Bx}' = \boldsymbol{b}_1$ is derived and solved for the vector $\boldsymbol{x}'$ of unknown first differential coefficients. The set of equations for higher differential coefficients is reduced to the form $\boldsymbol{Bx}^{(j)} = \boldsymbol{b}_j$ after all the known lower differential coefficients have been incorporated.

14. The resulting Taylor series representation is applicable and ξ may be decreased until one of the following terminal point conditions is satisfied: (T1) - another member strain reached $\xi\Lambda_i^b$; (T2) - discontinuous change of $\boldsymbol{S}'_D(\Omega_p(\xi))$; (T3) - the sign of some $\hat{\varepsilon}_i^s$ changes; (T4) - truncation error limit, and; (T5) - some $\hat{R}_i$ (> $\overline{\boldsymbol{R}}_i$) returns to $\overline{\boldsymbol{R}}_i$. The next step, similar to the first, may then be initiated at the previous terminal point state and so on.

EXAMPLE

15. Consider a 504-bar rigidly-jointed double-layer space frame as shown in Fig. 1. The upper and lower joints are on circular cylindrical surfaces whose radii are 16 m and 17.5 m, respectively. The frame is roller-supported with horizontal springs whose extensional stiffness is 1.029 x 10^5 N/m. Each upper joint is to carry a non-structural mass of 800 kg. Each member is a pipe made of steel with mass density 7.86 x 10^3 kg/cm and with Young's modulus of 2.058 x 10^{11} Pa. The ratio $\tau\ (= t_i / R_i)$ is 0.07. The minimum radius $\overline{\boldsymbol{R}}_i$ is 3.37168 cm for all the members and corresponds to the cross-sectional area of 5.0 cm^2. Let C_A C_V and C_D denote the maximum acceleration, the maximum velocity and the maximum displacement, respectively, of the vertical ground motions. The following design response spectra proposed by Newmark and Hall (ref. 12) is adopted:

$$
\begin{aligned}
S_D(\Omega) &= S_D^{(1)}(\Omega) = C_A/\Omega \quad \text{for} \quad \Omega^{(1)} \leq \Omega \\
S_D(\Omega) &= S_D^{(2)}(\Omega) = 16.2C_A\{3.21 - 0.68\ln(100h)\}\Omega^{-1.36} \quad \text{for} \quad \Omega^{(2)} \leq \Omega < \Omega^{(1)} \\
S_D(\Omega) &= S_D^{(3)}(\Omega) = C_A\{3.21 - 0.68\ln(100h)\}/\Omega \quad \text{for} \quad \Omega^{(3)} \leq \Omega < \Omega^{(2)} \\
S_D(\Omega) &= S_D^{(4)}(\Omega) = C_V\{2.31 - 0.41\ln(100h)\}/\sqrt{\Omega} \quad \text{for} \quad \Omega^{(4)} \leq \Omega < \Omega^{(3)} \\
S_D(\Omega) &= S_D^{(5)}(\Omega) = C_D\{1.82 - 0.27\ln(100h)\} \quad \text{for} \quad \Omega < \Omega^{(4)}
\end{aligned}
\tag{8a-d}
$$

where $\Omega^{(1)}$, $\Omega^{(2)}$, $\Omega^{(3)}$ and $\Omega^{(4)}$ denote the transition eigen-values. The modal damping ratio h = 0.02 has been adopted for all the s modes. The following values are adopted for the vertical ground motions: C_A = 134.0 cm/sec^2, C_V = 16.7 cm/sec, C_D = 12.5 cm. These data except $\overline{\boldsymbol{R}}_i$ are exactly the same as for the space truss in ref. 11 so that the effect of rigid joints can be directly observed.

16. The lowest four symmetric modes are taken into consideration. The lowest eigen-value of the frame of $\overline{\boldsymbol{R}}$ is 306.688 rad^2/sec^2 which is only slightly greater than 299.556 rad^2/sec^2 for the corresponding truss. The effect of rigid joints and of strain due to bending manifests itself in the maximum total strain of 1.60276 x 10^{-3} as compared to 1.16480 x 10^3 for the truss, which are values for the two initial designs.

17. Fig. 3a-d shows the radii and cross-sectional areas of member pipes at four values of: (a) 6.54409 x 10^{-4}, (b) 4.41556 x 10^{-4}, (C) 3.41665 x 10^{-4} and (d) 2.73170 x 10^{-4} for one of the four symmetric parts. It can be observed from these figures that the proposed sequential generator algorithm works successfully so that primary inertia carrying members are identified and generated with necessary cross-sectional areas as ξ is decreased. In particular an edge beam consisting of members on a pair of lower and upper chord lines or an edge arch consisting of members on a pair of lower and upper chord arch lines is generated naturally between a pair of adjacent supports. Variation of the response strains in, and the cross-sectional areas of the members designated by the numbers shown in Fig. 1, are plotted with respect to ξ in Fig. 4 and Fig. 5, respectively. Fig. 4 indicates that the maximum strain in any of these members is decreased as soon as it coincided with the current ξ value. At that same value of ξ,, the cross-sectional area of that member is increased as shown in Fig. 5. It is noted that the member numbers in Figs 4, 5 and 2 are different from the numbers of those members whose cross-sectional areas have been consecutively increased from their uniform lower bound.

18. While direct verification of the present numerical results needs to be made via time history analysis, it is noted that very good agreement has been demonstrated between the theoretical results on two double-layer space trusses and the results of the corresponding time history analysis on those trusses subjected to 20 spectrum compatible artificial earthquakes.

CONCLUSIONS

19. The following conclusions can be drawn from the results:

(i) The concept of "an ordered set of seismic response constrained designs with respect to a design strain parameter" is useful and an efficient algorithm for generating a sequence of such designs and can be formed for space frames. The sequential generator is efficient for large space frames.

(ii) The process in which a pair of edge beams and a pair of arches, between adjacent supports, are formed in a space frame as the design strain level is decreased, is quite similar to that for the corresponding space truss. The effect of rigid-joints on the nature of the sequence of seismic response-constrained designs is only slight. The effect of strain due to bending results in a frame of slightly larger cross-sectional areas as compared with those of the corresponding truss for the same design strain level.

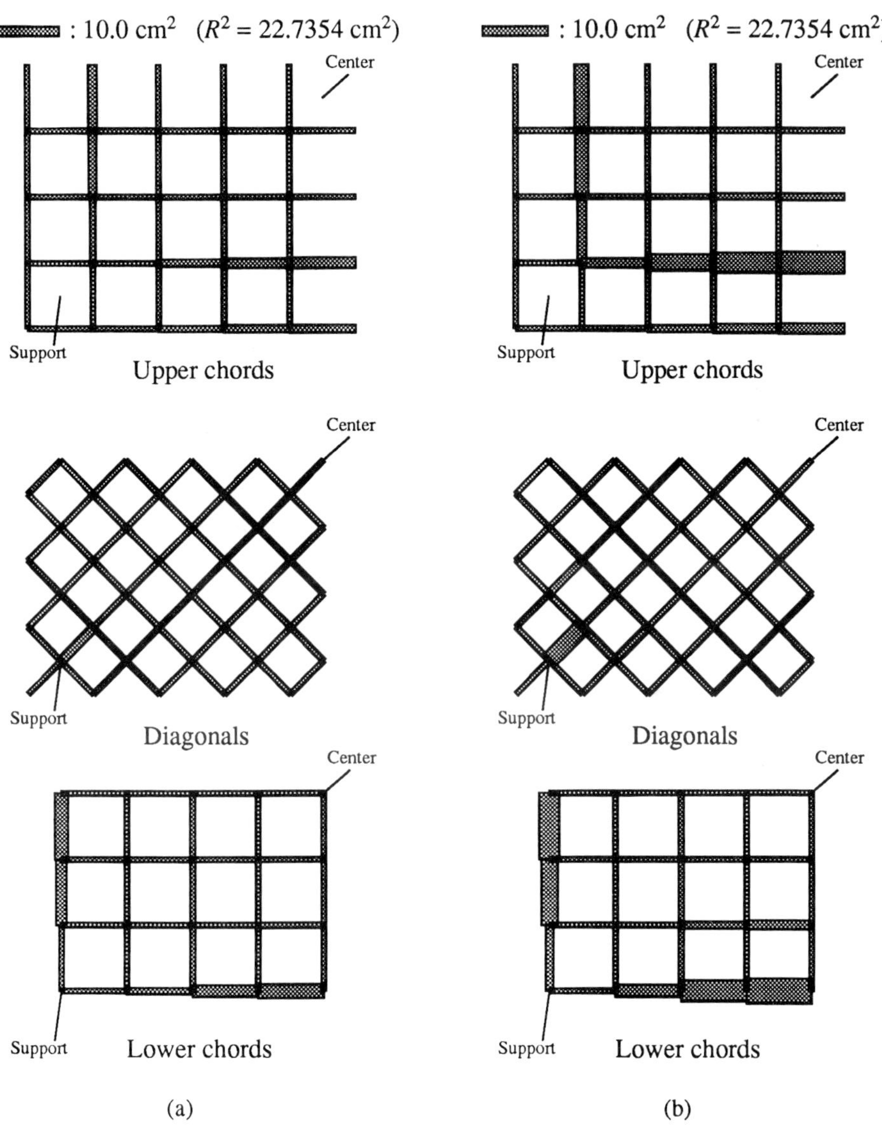

Fig. 3(a), (b). Cross-sectional area of the seismic strain-constrained frames (a) $\xi = 6.54409 \times 10^{-4}$ and (b) $\xi = 4.41556 \times 10^{-4}$.

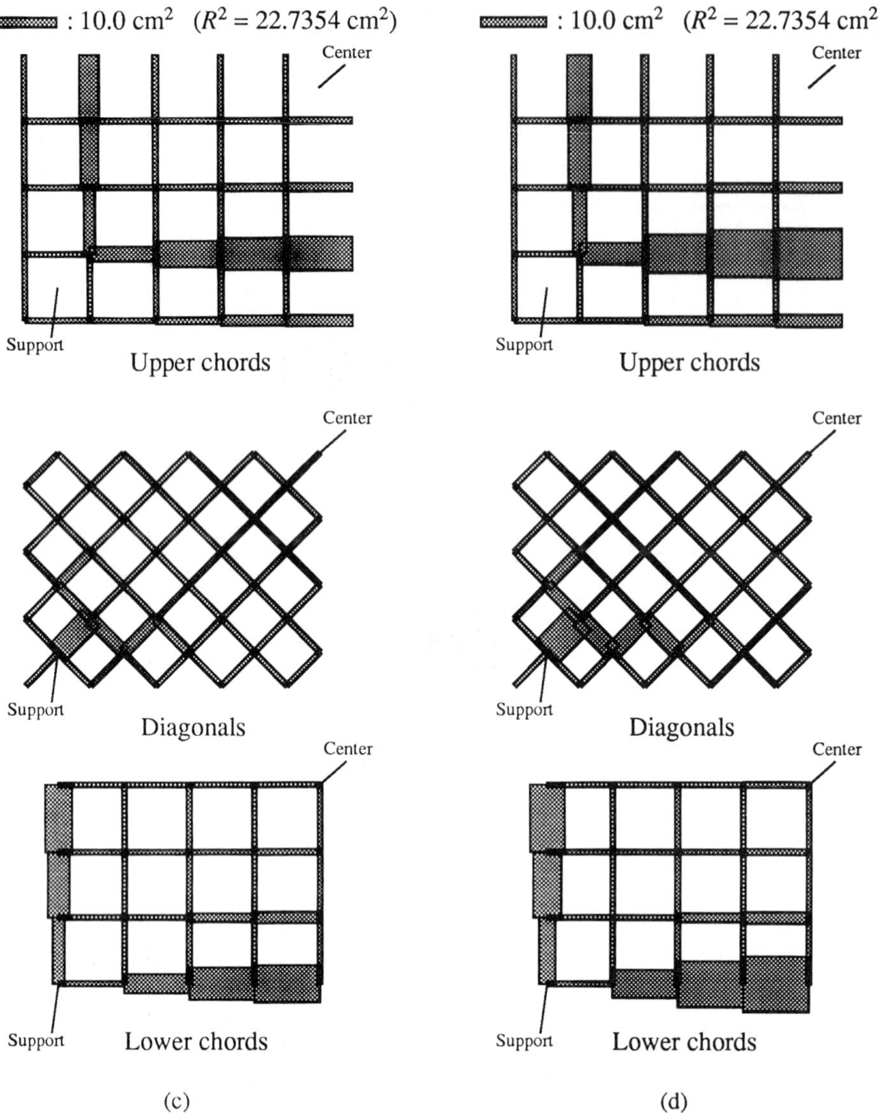

Fig. 3(c), (d). Cross-sectional area of the seismic strain-constrained frames (a) $\xi = 3.41665 \times 10^{-4}$ and (b) $\xi = 2.73170 \times 10^{-4}$.

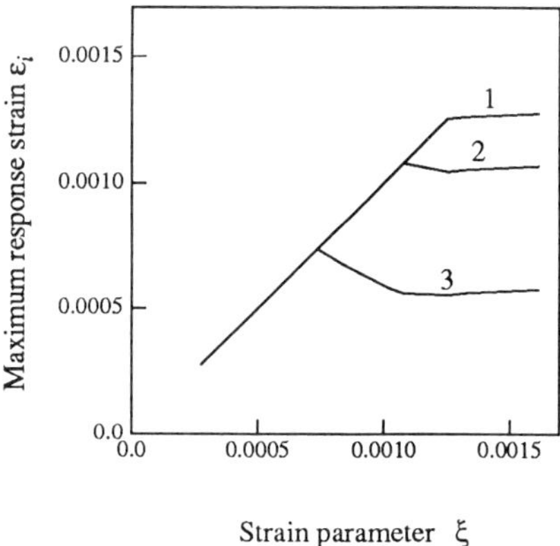

Fig. 4. Variation of the response strains in the members designated in Fig. 1.

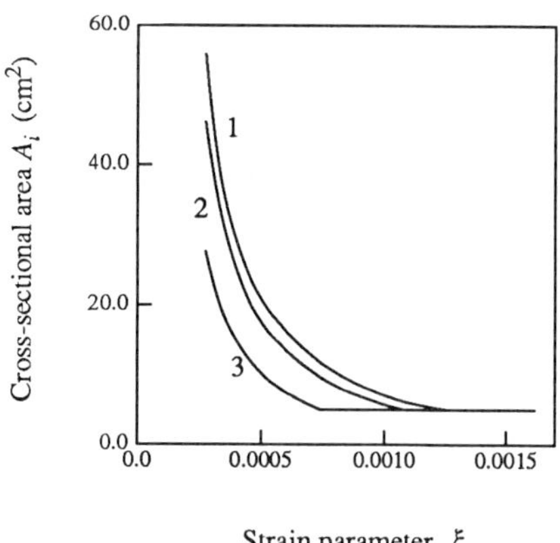

Fig. 5. Variation of the cross-sectional areas of the members designated in Fig. 1.

REFERENCES

1. FOX, R.L. and KAPOOR, M.P. Structural optimisation in the dynamic response regime; A computational approach, AIAA J., Vol.8, pp.l798-1804, 1970.

2. FENG, T., ARORA, J.S. and HAUG, E.J. Optimal structural design under dynamic loads, Int. J. Numer, Mech. Engng, Vol.ll, pp.39-52, 1977.

3. ARORA, J.S. and Haug, E.J. Optimum structural design with dynamic constraints, J. Struct. Div., ASCE, Vol.103, pp.2071-2074, 1979.

4. HAUG, E.J. and ARORA, J.S. Applied optimal design, John Wiley, New York, 1979.

5. BHATTI, M.A. and PISTER, K.S. A dual criteria approach for optimal design of earthquake-resistant structural systems, Earthquake Engng Struct. Dyn., Vol.9,pp.557-572,1981.

6. BALLING, R.J. PISTER, K.S. and CIAMPI, V. Optimal seismic-resistant design of a planner steel frame, Earthquake Engng Struct. Dyn., Vol.ll, pp.541-556, 1983.

7. YAMAKAWA, H. Optimum structural designs for dynamic response, New Directions in Optimum Structural Design, (Eds. Etres, E., Gallagher, R.H., Ragsdell, E.M. and Zienkiewicz, O.C.), pp.249-266, John Wiley, New York, 1984.

8. HSIEH, C.C. and ARORA, J.S. Design sensitivity analysis and optimisation of dynamic response, Comput. Mech. Appl. Mech. Engng, Vol.43, pp.195-219, 1984.

9. NAKAMURA TSUNEYOSHI and OHSAKI, M. Sequential optimal truss generator for frequency ranges, Comput. Mech. Appl. Mech. Engng, Vol.67, pp.189-209, 1988.

10. NAKAMURA TSUNEYOSHI and OHSAKI, M. Sequential generator of earthquake response constrained trusses for design strain ranges, Int. j. Comput. & Struct., Vol.33(6), pp.1403-1416, 1989.
11. NAKAMURA TSUNEYOSHI and OHSAKI, M. Sequential generator of earthquake strain constrained curved space trusses for design strain ranges, In: Optimisation of Structural Systems and Industrial Applications, Proc. OPTI91, Cambridge, MA, pp233-244, 1991.
12. NEWMARK, N.M. and HALL, W.J. Earthquake spectra and design, Earthquake Engineering Research Institute, Berkeley, CA, 1982.
13. KIUREGHIAN, A.D. On response of structures to stationary excitation, Report No. EERC 79-32, Earthquake Engineering Research Centre, University of California, Berkeley, CA, 1979.
14. WILSON, E.L., KIUREGHIAN, A.D. and BAYO, E.P. A replacement for the SRSS method in seismic analysis, Earthquake Engng Struct. Dyn., Vol.9, pp.187-194,1981.

54. Dynamic stability of spherical domes subject to earthquakes

H. KUNIEDA, Kyoto University, Japan

SYNOPSIS. The purpose of this paper is to provide the final non-linear governing equations of spherical domes subjected to vertical earthquake loading in an analytical form. The existence of dynamically unstable responses for a real earthquake data, recorded through numerical calculation of these equations in single and multi-modal states, is also discussed.

INTRODUCTION

1. It has been shown by the author that extremely high level response acceleration is induced in spherical domes when subjected to a real vertical earthquake loading and when the geometrical and material parameter A (= ma^2/Egh) takes the value in a specific range (ref. 1). This result implicates the response behaviour of domes to be examined as a dynamic instability problem. It is therefore an urgent issue to confirm the occurrence of the large acceleration response for shells under finite deformation treatment and examine the possibility of the occurrence of dynamic instability behaviour under real earthquake loading.

2. The purpose of this paper is to provide a useful expression, for the numerical calculation, of the governing equation for the non-linear response of spherical domes in the axi-symmetric state and elucidate the tendency of the dynamic stability limit intensity of vertical earthquakes. The modal analysis method is introduced to deduce the finally governing non-linear equations for the analysis of responses, using the approximate eigen-modes (ref. 2) of the axi-symmetric flexural vibration of domes. Since each eigen-mode used in this paper is represented as a finite series of the Legendre polynomials, all integrals with respect to a spatial variable, involved in the procedure of the modal analysis, can be reduced to analytical functions of closed form, even in non-linear terms. Numerical calculation, therefore, of these functions can be executed very easily. This is one of the most important merits of this paper.

3. Many references are available for the criteria of dynamic instability and many papers have been published on the analysis of dynamic instability of spherical domes. But, they are limited in the case of conservative loading or statically non-conservative loading as reviewed in refs 3 and 4. For non-conservative dynamic

excitations, mathematical prediction of the stability critical intensity of excitations has not yet been put into practical use, even for a trigonometric excitation and a single-degree-of-freedom system (refs 5-6). Since a spherical dome is a continuum and has an infinite number of eigen-modes, the effect of modal coupling must be taken into account. That is, multi-modal states must be postulated for the examination of the dynamic instability in responses. In addition, excitations considered here are due to earthquakes. There is no suitable mathematically analytical method of determination of stability limit available for multi-degrees-of-freedom systems and for earthquake excitations, except direct numerical calculations of responses and visual judgement.

4. Finally, derived non-linear simultaneous differential equations with all coefficients, represented analytical functions of closed form, are solved numerically for an earthquake excitation and the stability critical intensity of the acceleration of the referred earthquake is elucidated and examined in coupled modal, as well as, single modal.

REDUCTION OF FINAL GOVERNING EQUATIONS

5. Applying the D'Alembert principle to the Sanders' equations (ref. 7), the kinematic non-linear equations of domes in the axi-symmetric state can be derived as:

$$(sin\phi N_\phi)^o - sin\phi Q_\phi - cos\phi N_\theta - (ma/g)\ sin\phi\ (\ddot{u} - \ddot{U} sin\phi) = 0 \tag{1}$$

$$(sin\phi Q_\phi)^o - sin\phi(N_\phi + N_\theta) + (sin\phi N_\phi w^o)^o\ /\ a - (ma/g) sin\phi(\ddot{w} - \ddot{U}\ cos\phi) = 0 \tag{2}$$

where U is the vertical earthquake motion; m is the dome weight per unit area; a is the dome radius and h is the dome thickness. The strain-displacement relations used in this paper are:

$$ae_\phi = u^o - w + (w^o)^2/2a\ , \quad ae_\theta = ucot\phi - w$$
$$a^2\kappa_\phi = w^{oo} + u^o, \quad a^2\kappa_\theta = (w^o + u)\ cot\phi\ , \quad B^o \equiv dB/d\phi \tag{3}$$

in which the non-linear term, with respect to ω only, is taken into consideration. Dividing strains into two parts, linear and non-linear parts, and introducing stress function ψ, which is relevant to the linear terms, so as to satisfy Eq. 1 as:

$${}_LN_\phi = [\psi^o\ cot\phi + \psi - (D/a)\ H_2(w)]/a^2 + (ma/g)\ \ddot{U}cos\phi \tag{4}$$

$${}_LN_\theta = [\psi^{oo} + \psi - (D/a)\ H_2(w)]/a^2 + (ma/g)\ \ddot{U}cos\phi \tag{5}$$

$$H_2(B) \equiv B^{oo} + B^o cot\phi + 2B$$

and substituting these into Eq. 2, we can rewrite Eq. 2 in the form of the equations shown below:

$$H_2\,H_2(w) - \frac{a}{D}\,H_2(\psi) + \frac{ma^4}{gD}\ddot{w} - \frac{a^2}{D sin\phi}\,(sin\phi\,{}_L\overline{N}_\phi w^o)^o$$
$$- \frac{a^3}{D sin\phi}[\,sin\phi({}_NN_\phi + A_N N_\theta) + \frac{1}{a}(sin\phi\,{}_NN_\phi\,w^o)^o\,]$$
$$= 3\,\frac{ma^4}{gD}\,\ddot{U}cos\phi + \frac{ma^3}{gD}\,\ddot{U}\,[\,cos\phi\,H_0(w) - w^o\,sin\phi\,] \qquad (6)$$

in which $I_L\overline{N}_\phi$ denotes the rest part of Eq. 4 without the term containing U. Non-linear terms of stress resultants, $I_N N_\phi$ and $I_N N_\theta$, are calculated directly from the relevant non-linear part of strains, which can be determined if ω is given.

6. The following approximate representation of eigen-modes of axi-symmetric state (ref. 2) is used.

$$w = \sum_{s=1}^{M} W_s(\phi)\;T_s(t)\;,\quad W_s(\phi) = \sum_{j=1}^{N} G_{sj}\;P_j(cos\phi) \qquad (7\text{-a})$$

$$\psi = \frac{D}{a}\sum_{s=1}^{M}\sum_{j=1}^{N} G_{sj}\,\lambda_j\,P_j\,(cos\phi)\;T_s\,(t)\;, \qquad (7\text{-b})$$

$$\lambda_j = \frac{(1-\nu)[2-j(j+1)]-\kappa}{1-\nu-j(j+1)}\;,\quad \kappa = \frac{12(1-\nu^2)\;a^2}{h^2}$$

in which

M_{total} : *total number of modes at making eqs(7)*

M : *number of modes used here* $(M \le M_{total})$

$N \equiv K_{M_{total}}$, $K_{M_{total}}$: *see ref.2*

G_{sj} : *replaced* G_{ij} *of* M_{total} *according to* M

7. Applying the Galerkin's procedure with Eqs 7, simultaneous non-linear differential equations with respect to a time variable can be obtained as the final governing equations of kinematics of a dome.

$$\ddot{T}_s + \Theta_{1s}T_s + \sum_{p=1}^{M}\sum_{q=1}^{M}\Theta_{2s}(p\;,\;q)\;T_p\;T_q$$
$$+ \sum_{p=1}^{M}\sum_{q=1}^{M}\sum_{r=1}^{M}\Theta_{3s}(p,q,r)\;T_p\;T_q\;T_r = \ddot{U}(\Theta_{4s} + \sum_{p=1}^{M}\Theta_{5s}(p)\;T_p) \qquad (8)$$

8. Compatibility relations are all satisfied by use of Eqs 7a and 7b, since φ in Eq. 7b has been determined to satisfy the linear compatibility relations and non-linear part of strains is determined directly from ω. Since eigen-modes Eqs 7a and 7b are used, all coefficients $\Theta_{1s} - \Theta_{5s}$ in these equations are represented as analytical functions of closed form as below:

$$\Theta_{1s} = \overline{\Theta}_{1s}/(A\Theta_{0s}) , \quad \Theta_{2s} = \left[\frac{h}{a}\right] \overline{\Theta}_{2s}/(A\Theta_{0s})$$

$$\Theta_{3s} = \left[\frac{h}{a}\right]^2 \overline{\Theta}_{3s}/(A\Theta_{0s}) , \quad \Theta_{4s} = \frac{1}{h} \overline{\Theta}_{4s}/\Theta_{0s} ,$$

$$\Theta_{5s} = \frac{1}{a} \overline{\Theta}_{5s}/\Theta_{0s} , \quad \Theta_{0s} = \sum_{j=1}^{N} \sum_{i=1}^{N} G_{sj} \, G_{si} \, S(j \, , i), \quad A = \frac{ma^2}{Ehg}$$

$$\overline{\Theta}_{1s} = \frac{1}{\kappa} \sum_{j=1}^{N} \sum_{i=1}^{N} G_{sj} \, G_{si} \, [2 - j(j+1)][2 - j(j+1) - \lambda_j] \, S(j \, , i)$$

$$\overline{\Theta}_{2s}(p,q) = - \frac{a^3}{\kappa D} \left(sin\phi \; W_p{}^o \; {}_L\overline{N}_{\phi q} \; W_s\right)|_{\phi_0}$$

$$+ \frac{1}{\kappa} \sum_{j=1}^{N} \sum_{k=0}^{\overline{N}} G_{qj} C_{kps} \left[\lambda_j \, S_2(k,j) + (\lambda_j - 2 + j\langle j+1 \rangle) \, S(k,j) \right]$$

$$- \frac{1}{2(1-\nu)} \sum_{j=1}^{N} \sum_{k=0}^{\overline{N}} G_{sj} \, C_{kpq} \, S(k \, , j)$$

$$\overline{\Theta}_{3s}(p,q,r) = - \frac{1}{2(1-\nu^2)} \left(sin\phi \, W_p{}^o \, W_q{}^o \, W_r{}^o \, W_s \right)|_{\phi_0}$$

$$+ \frac{1}{2(1-\nu^2)} \sum_{k=0}^{\overline{N}} \sum_{l=0}^{\overline{N}} C_{kpq} C_{lrs} \, S(k \, , l)$$

$$\overline{\Theta}_{4s} = 3 \sum_{j=1}^{N} G_{sj} \, S\,(j \, , 1)$$

$$\overline{\Theta}_{5s}(p) = \sum_{j=1}^{N} \sum_{i=1}^{N} G_{pj} G_{si} \left[(j+2)S(j+1,i) + (j-1)S(j-1,i)\right] \left\{- \frac{j(j+1)}{2j+1}\right\}$$

$\overline{N} = 2N \; (\equiv 2K_{M_{total}})$ $\quad {}_L\overline{N}_{\phi p}$: *the part of* ${}_LN_\phi$ *corresponding to* W_p

$$S(r,s) = \int_0^{\phi_0} P_r \, (cos\phi) \; P_s \, (cos\phi) \; sin\phi \; d\phi$$

$$= \frac{1}{2r+1} \left\{ 1 - (P_r P_r cos\phi)|_{\phi=\phi_0} - 2 \sum_{p=1}^{[r/2]} (2r+1-4p) S(r,r-2p) \right\} , \quad r = s$$

$$= \frac{1}{2(s+1) - r(r+1)} \left\{ (r-s) cos\phi P_r P_s + s \; P_r P_{s-1} - r \; P_{r-1} P_s \right\}|_{\phi=\phi_0} , \quad r \neq s$$

$$S_2(r\,,\,s) = -\int_0^{\phi_0} P_r\,(cos\phi)P^1{}_s(cos\phi)\;cos\phi\;d\phi$$

$$= -\,s\;S(r,s) - \sum_{p=1}^{[s/2]} (2s+1-4p)\;S(r,\,s-2p)$$

$[\,n\,]$: *Gaussian symbol*

$$(w^o)^2 = \sum_{s=1}^{M}\sum_{t=1}^{M}\sum_{j=1}^{N}\sum_{i=1}^{N} G_{sj}\;G_{ti}\;P^1{}_j(cos\phi)\;P^1{}_i(cos\phi)\;T_s\;T_t$$

$$= \sum_{s=1}^{M}\sum_{t=1}^{M}\sum_{k=0}^{\overline{N}} C_{kst}P_k(cos\phi)\;T_s\;T_t$$

$$C_{kst} = \sum_{p=1}^{N}\sum_{q=1}^{N} G_{sp}\;G_{tq\;\;p,q}D_k$$

$${}_{p,q}D_k = \frac{1}{2}\left\{p\;(p+1) + q\;(q+1) - k\;(k+1)\right\}\;{}_{p,q}\overline{C}_k$$

$${}_{p,q}\overline{C}_k = \frac{2k+1}{p+q+k+1}\;\frac{R\left(\dfrac{p-q+k}{2}\right)R\left(\dfrac{q-p+k}{2}\right)R\left(\dfrac{p+q-k}{2}\right)}{R\left(\dfrac{p+q+k}{2}\right)}$$

$$R(n) = \frac{(2n-1)!!}{n!}\;,\quad (2n-1)!! = 1\cdot3\cdot5\cdots\cdot(2n-3)(2n-1)$$

both $|p-q|$, k : *even* $\quad {}_{p,q}D_k = 0$
both $|p-q|$, k : *odd* $\quad {}_{p,q}D_k = 0$

9. If we use exact eigen-modes, which are represented by Legendre functions with fractional and complex orders, all integrals involved in the modal analysis can not be represented in analytical functions of closed form.

10. The non-linear response behaviours of domes can be determined by solving Eq. 8 under given excitations. If excitations are given in the form of numerical data, it is not possible to obtain an analytical solution to Eq. 8 and a numerically solving method must be applied.

NUMERICAL RESULTS AND CONCLUSIONS

11. As an example, the El'Centro '40 vertical earthquake, of which the absolute maximum acceleration is 206.3 gals, will be used as the real earthquake in this paper. The digitised record of the accelerations of this earthquake is given in ref. 8.

Table 1. Λ_0 in single modal state, *a/h = 500, a = 40m, zeta = 0.01, fixed end.*

(*a*) *thecase*: ϕ_0 = 30 *degrees*

mode	A = 2.0E-4			A = 6.5E-4		
	a/h=300	a/h=500	a/h=1000	a/h=300	a/h=500	a/h=1000
1	—	—	—	—	—	6.8437
2	—	—	—	—	—	—
3	—	—	—	—	—	5.1796
4	—	5.0078	1.7109	3.1797	2.3750	—
5	6.0156	0.7949	1.0937	3.1797	1.5937	1.3750
6	—	1.3047	1.8437	3.0390	2.3750	2.8750
7	7.5312	1.8750	1.0937	3.3359	0.5937	1.3750

(*b*) *thecase*: ϕ_0 = 45 *degrees*

mode	A = 2.0E-4			A = 6.5E-4		
	a/h=300	a/h=500	a/h=1000	a/h=300	a/h=500	a/h=1000
1	—	—	1.5178	—	3.5937	1.2734
2	—	—	—	—	—	6.5625
3	—	—	—	—	—	—
4	—	—	—	4.2109	7.3750	6.6562
5	3.1953	4.7031	3.8750	1.70313	2.7187	2.3750
6	1.3750	2.8750	2.0078	1.0234	1.0937	0.8750
7	1.2109	2.7187	0.8750	2.3750	0.7187	2.3750
8	1.3750	1.0859	1.3437	—	0.5234	3.3750
9	0.5937	1.5937	2.7187	—	1.3750	1.0937
10	—	1.3750	1.0937	1.8750	—	1.3750

(*c*) *thecase*: ϕ_0 = 60 *degrees*

mode	A = 2.0E-4			A = 6.5E-4		
	a/h=300	a/h=500	a/h=1000	a/h=300	a/h=500	a/h=1000
1	—	—	—	—	—	0.7089
2	—	—	—	—	—	—
3	—	—	—	—	—	—
4	—	—	—	—	—	—
5	—	—	—	—	—	—
6	7.8437	—	—	4.3437	—	—
7	1.0234	—	—	0.8750	—	—
8	4.0937	—	3.6953	0.5234	—	1.7656
9	—	—	6.3437	6.5625	6.0547	2.2734
10	1.8750	4.8750	1.5176	1.0937	1.7734	0.8437

12. The dynamic stability critical intensity, if it exists, of this factor is calculated by multiplying the attenuation or amplification factor to this record and substituting this into $U(t)$. This factor is denoted as Λ and called the load factor in the following discussion. The case $\Lambda = 1$ implies the original earthquake state. For the amplified or attenuated El'Centro '40 earthquake in accordance with the stability critical intensity of the load factor, response of domes show unstable behaviour.

13. Since a real earthquake record is applied, to Eq. 8, to obtain mathematical closed form solutions is difficult; the Runge- Kutta-Gill method was therefore used. The numerical calculation was executed to solve Eq. 8. Because the earthquake record is given in real physical quantities, terms in Eq. 8 should also be calculated using physical quantities. Therefore, the radius is set as $a = 40$ meters for an example in this paper. A velocity proportional damping term is added to Eq. 8, taking the damping ratio of *0.01*.

Table 2. Λ_0 in multi-modal state, $\phi_0 = 45$ *degrees*, $a/h = 500$, $a = 40m$, $A \equiv ma^2/(Ehg) = 6.5E\text{-}4$, $\varsigma = 0.01$.

coexi.modes	Λ_0	coexi.modes	Λ_0
1, 2	2.67969	1, 2, 6	2.37500
1, 3	2.27344	1, 3, 5	0.78711
1, 5	0.87500	1, 3, 6	1.14844
2, 3	—	2, 3, 5	1.37500
1, 2, 3	1.82813	2, 3, 6	2.87500
1, 2, 5	0.83008	1, 2, 5, 6	1.87500
1,2,3,4,5	1.08590	1,2,3,4,5,6	1.37500

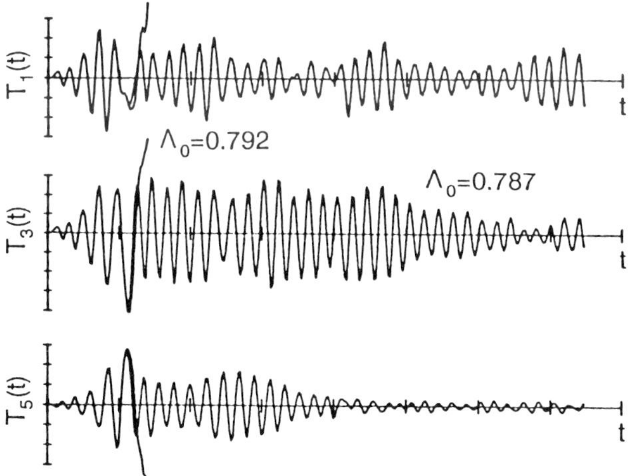

Fig. 1. Time history of response $T_i(t)$, $\phi_0 = 45$ *degrees*, $a/h = 500$, $a = 40m$, $A \equiv ma^2/(Ehg) = 6.5E\text{-}4$, $\varsigma = 0.01$.

14. Eq. 8 contains the non-linearity of both quadratic and cubic terms, and the coefficient of the cubic term in the single modal state usually takes a negative

value as shown in Table 4. It is, therefore, pertinent to presume the appearance of a dynamic instability in response in the single modal state. It is apparent that dynamic instability may appear in response in the multi-modal state, if dynamic instability exists in the single modal state.

15. In order to determine a stability limit, the instability phenomenon was found by a sudden increase of the absolutely maximum deflection with a gradually increasing load factor. The critical load factor Λ_0 was determined here in the single modal and multi-modal states as given in Tables 1 and 2, respectively. In this paper, the upper limit of Λ is set as $\Lambda = 8.5$ for the calculation of response, avoiding impractical examination. The case of $\Lambda > 8.5$ has not been examined in this paper. The effect of damping on Λ_0 will also be discussed later.

Table 3. Influence of the damping ratio and *a/h* on Λ_0
$\phi_0 = 45$ *degrees*, $a = 40m$, $A \equiv ma^2 / (Ehg) = 6.5E\text{-}4$.

(*a*) *the case*: $\zeta = 0.01$

coexi.modes	a/h	Λ_0	a/h	Λ_0
1,2,3,4,5,6	300	2.37500	700	1.37500
	400	1.09375	800	0.52344
	500	1.37500	900	0.71875
	600	3.37500	1000	1.87500

(*b*) *the case*: $\zeta = 0.05$

coexi.modes	a/h	Λ_0	a/h	Λ_0
1,2,3,4,5,6	300	1.84375	700	1.37500
	400	0.71094	800	2.87504
	500	1.21875	900	2.37505
	600	3.37500	1000	1.87500

16. In the tables given in this paper, the symbol "-" implies the case that Λ_0 can not be found within the range $\Lambda = 0$ to 8.5. Λ_0 for domes of $\phi_0 = 30$, 45 and 60 degrees (ϕ_0 is the half open angle; $\phi_0 = 90$ degrees in a hemi-spherical dome) in the single mode is given in Table 1. When several modes co-exist, the phenomenon of instability response emerging in one mode, and not in another mode was not found. Even if the instability response of one mode triggers instability, responses of all modes show instability almost simultaneously as depicted in Fig. 1. Λ_0 of multi-modal states is also obtained in Table 2, showing that Λ_0 varies considerably according to the selected modes. This table also shows that the fact that lower Λ_0 is not necessarily obtained from a number of modes co-existing. Since an infinite number of modes co-exist in a real dome, it is not yet apparent how to estimate the effect of co-existence of modes on Λ_0 and how to select the necessary modes. The relation of Λ_0 of the single mode with that of multi-modes has not manifested itself.

17. Values of load factor Λ_0 were determined as that which caused a sudden five fold increase in the deflections. However, for a two fold increase in deflections, the value of Λ_0 varies slightly, however, there is no serious

difference in most Λ_0 values. Λ_0 values vary a little, depending on the cut off value of computation, to avoid overflow calculation in the vicinity of stability limit. This implies that values given in Tables 1 and 2 approximate, since direct numerical calculation and visual judgement are employed. The precise value of Λ_0 should be determined by mathematical treatment, if possible.

Table 4. Examples of coefficients of Eq. 8 in the single mode
ϕ_0 = 45 degrees, a/h = 500, A = 6.5E-4.

MODE	Θ_1	Θ_2	Θ_3	Θ_4	Θ_5
1	1.469+3	2.056+0	-2.754+2	-8.968+0	-8.180-2
2	1.527+3	-1.029+1	-8.503+1	-1.634+0	-2.146-1
3	1.561+3	1.216+1	-1.812+2	-2.125+0	-4.198-1
4	1.617+3	-6.783+0	-5.047+2	-3.656+0	-6.748-1
5	1.714+3	-8.826+0	-1.016+3	7.596+0	-9.833-1
6	1.872+3	2.830+0	-2.050+3	1.533+1	-1.347+0
7	2.115+3	-6.264+0	-1.963+4	2.888+1	-1.767+0
8	2.467+3	7.744-1	-1.274+5	6.931+1	-2.236+0
9	2.952+3	5.583+0	-6.441+4	-1.389+2	-2.736+0
10	3.589+3	2.754+0	-7.267+4	-4.188+2	-3.104+0

18. Influence of the ratio of radius to thickness and damping ratio on Λ_0 is shown in Table 3. It is known from this table that the influence of these ratios on Λ_0 is not simple, due to the parameter A taking on a constant value and, therefore, mutual relation between frequency characteristics of the earthquake and natural frequencies of domes is involved. This points to the important conclusion greater values of Λ_0 are not necessarily achieved by an increase in the damping ratio.

19. Finally, values of some coefficients of Eq. 8 in the single mode are tabulated in Table 4. All coefficients of a cubic term exemplified in this table are negative. This fact is very important for the calculation of the dynamic response of domes.

REFERENCES

1. KUNIEDA, H., Responses of spherical domes subjected to vertical earthquake, Proc.IASS Osaka JAPAN, 1986, pp.33-40.
2. KUNIEDA, H., Classical buckling loads of spherical domes under uniform pressure, Jour. EM, ASCE, Vol.118, No.8, 1992, pp.1513-1525.
3. Recent advancement of structural mechanics in building engineering (in Japanese. Partly written by the author), AIJ, 1987.
4. Vibrations of shells and latticed structures (in Japanese. Edited and partly written by the author), AIJ, 1993.
5. KUNIEDA, H., On dynamic instability of structures subject to dynamic excitation: a trial regarding prediction on incipient instability, Proc. 9WCEE Tokyo JAPAN, 1989, Vol.5, pp.1-6.
6. KUNIEDA, H., Prediction of critical force utilising the solution of autonomous system, Proc. Annual Meeting of AIJ, 1991, pp.1179-1180.

7. SANDERS, J. L. Jr., Non-linear theories for thin shells, Quart.Appl.Mech., Vol.XXI, No.1, 1963, pp.21-36
8. Strong motion earthquake accelerograms, CALTEC EERL 71-21, 1971.

55. Mine shaft tower structure to accommodate rock movement

G. J. KRIGE, University of the Witwatersrand, South Africa, and
S. D. DEVY, Samar Consulting Engineers, South Africa

SYNOPSIS. For various reasons relating to geological and mining factors, deep mine shafts are commonly subjected to high strains of the surrounding rock at certain places. Guide steelwork is usually fixed to the sides of the shaft at regular intervals, but the high strains are likely to preclude this. The solution which has been adopted is the provision of a "guide tower" in the shaft. These guide towers are designed as space frames, consisting of the guide rails and such other steelwork as may be required. The towers are required to accommodate substantial vertical movement by means of expansion joints, and lateral movement by means of deformation of the tower into a smooth curve, which allows smooth and high speed hoisting. Their dynamic behaviour is modelled by a simplified computer model which predicts the magnitude of displacements and forces developed. Modular construction of the towers, to allow for ease of installation, without disrupting shaft production, must also be achieved.

INTRODUCTION

1. South Africa has large known gold reserves at great depths, in excess of 4000m below surface. There are many additional engineering problems associated with mining at such depths, which must be solved if the gold is to be profitably exploited. Several of these additional problems involve the mine shaft, through which all access to the mine for personnel and equipment as well as removal of ore, must pass. The present paper explores an innovative solution to one of these problems, high rock strains in the vicinity of the reef being mined, which would otherwise cause unacceptable distortion to the equipment in the shaft.

ROCK STRAINS

2. The virgin stress in the rock at approximately 3000m below the surface in the deep gold mines in South Africa, is of the order of 80MPa. In the virgin condition, the rock is triaxially contained. Any mining activity disturbs the triaxial containment, and leads to straining of the rock. In particular, the extraction of the reef leaves a plane which must have zero stress. The change from a high virgin stress to a zero stress leads to large strains in a direction perpendicular to the reef, as shown in Fig. 1. Valente (refs 1 and 2) assuming that the rock may typically

have a low elastic modulus of the order 17×10^3MPa, shows that this strain will be a tensile strain which may be in the region of 1.5×10^{-3}. The closure across the reef depends on the extent of removal of reef prior to any backfilling, but may typically be up to 700mm. Once backfilling has been completed, and mining of the reef continues further away from the shaft, the compressive stress is again introduced into the rock immediately above and below the reef as the rock mass above the reef moves downwards. The downwards movement of the overlying rock mass leads to an upward movement of the tensile strain towards the surface. This strain of the rock mass leads to vertical strain of the shaft, and where the reef is inclined, a concurrent shear strain. These can be seen in Fig. 1. Typical strain values according to Valente (ref. 2) are given in Table 1.

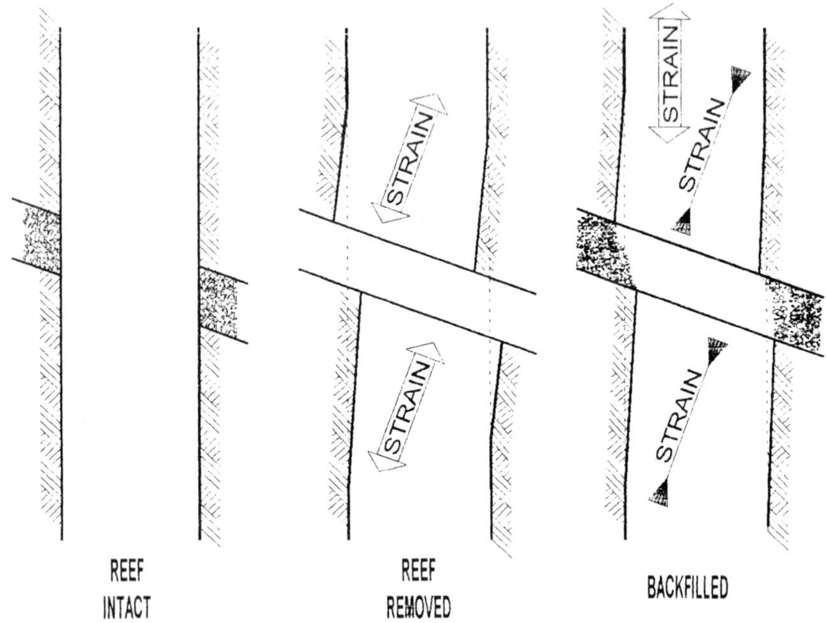

Fig. 1. Typical rock strains in the vicinity of the reef.

3. Traditional mining methods used a "shaft pillar" to support the shaft, and reduce these strains. This is a portion of rock immediately surrounding the shaft which is left intact until the end of the operational life of the shaft. It typically has a radius which is about 100m. When this is finally removed, the shaft and all equipment initially experience large tensile and shear strains, followed later by compressive strains if mining continues. More recently, mining practice has been to mine the reef in the vicinity of the shaft before the installation of any permanent equipment in the shaft. The initial tensile strain thus has no effect on any shaft equipment, but the following compressive strain must be accommodated by the shaft equipment.

Table 1. Typical shaft strain and deformation values.

Position	Type of deformation		Strain or deformation	
			Before backfill	After backfill
Near reef	Closure across reef		700 mm	800mm
	Shear movement across reef		80mm	100mm
	Vertical strain	($x10^{-3}$)	+1.50	+0.40
	Shear strain	($x10^{-3}$)	-0.50	-0.25
100m above reef	Vertical strain	($x10^{-3}$)	+1.20	+0.60
	Shear strain	($x10^{-3}$)	+0.20	-0.10
200m above reef	Vertical strain	($x10^{-3}$)	+0.60	+0.60
	Shear strain	($x10^{-3}$)	+0.25	-0.00
300m above reef	Vertical strain	($x10^{-3}$)	+0.40	+0.60
	Shear strain	($x10^{-3}$)	+0.20	+0.10

4. The shaft equipment, which is most severely influenced by straining of the shaft, is the guide steelwork. The conveyances, both cages and ore skips, are typically guided in the shaft by rails which run the full length of the shaft, and are attached at regular intervals, of 4m to 5m, to sides of the shaft by horizontal beams, called "buntons", as shown in Fig. 2. These rails and buntons make up the guide steelwork. The tensile strains which may be experienced can lead to stresses in the rails which approach the yield stress, and resultant high bending moments in the buntons. The compressive strains may lead to buckling of the rails which deflects them and reduces their stiffness, both of which have serious implications for the operation of the conveyances.

GUIDE TOWER CONCEPT

5. The high strains which occur in the vicinity of the reef thus preclude the use of the traditional buntons, attached at regular intervals to the sides of the shaft. The problem faced by shaft engineers is to find some manner in which the guide steelwork can reliably guide conveyances through any portion of the shaft where large horizontal and vertical strains are experienced. Early solutions to this problem concentrated on adjustable connections between the buntons and the sides of the shaft. A sophisticated arrangement of this type is to support the buntons vertically on springs, and horizontally by a screw-adjustable arrangement which can be adjusted to accommodate any shaft strains which may take place.

This is shown in Fig. 2. This arrangement is known to work effectively, but leads to rather high maintenance demands, as adjustment of the buntons must be done at approximately 6 monthly intervals and is rather time consuming.

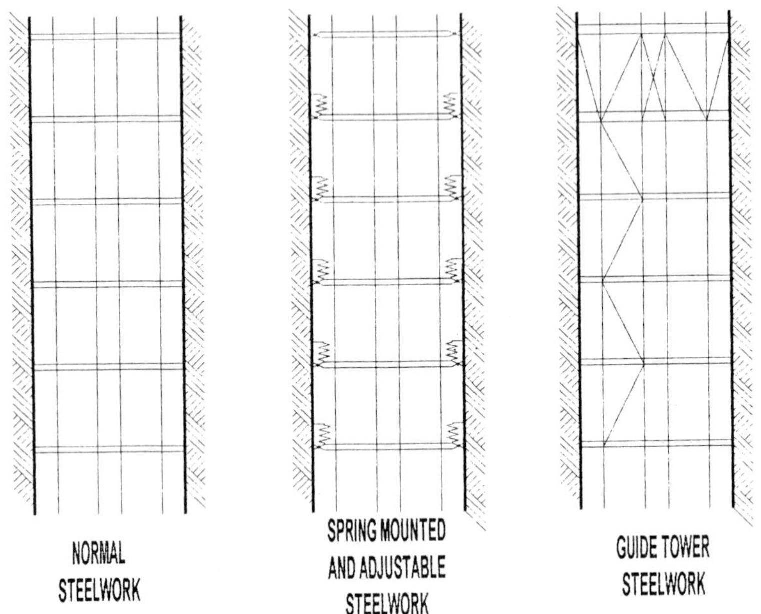

Fig. 2. Typical guide steelwork layouts.

6. The solution now being adopted is that of using a guide tower, which is a space structure and is completely isolated from the sides of the shaft in the vicinity of the reef. Typical guide towers may be up to 80m, or even 100m, long. The highest strains are thus avoided, but a certain amount of strain must nevertheless be accommodated by the tower. The length of the tower however, enables it to accommodate quite high levels of strain without high local deformations which most adversely influence the operation of conveyances.

REQUIREMENTS OF THE GUIDE TOWER

7. The chief function of the guide tower is to locate the guide rails adequately, in order to allow safe, full speed hoisting of conveyances in the shaft. Krige and Devy (ref. 4) have defined the most important requirements of a guide tower to enable it to perform its function.

The tower must accommodate vertical strains

8. The shaft strain leads to relative vertical movements between the ends of the tower. These may be either shortening or lengthening, depending on the mining

procedures used. The simplest manner in which these strains may be accommodated is the provision of a vertical expansion at the lower end of the tower. This joint should typically allow for approximately 1m of closure, or 0.2m opening.

<u>The tower must accommodate horizontal strains</u>

9. Shear strain of the shaft must be taken up by a smooth curve along the length of the tower. Full speed hoisting usually implies about 15m/s, so that abrupt changes in direction or local deformations cannot be allowed. This requires that the tower must be fixed against rotation at both ends, to prevent any abrupt changes of direction at these points. This is quite easily achieved at the top, by supporting the tower from at least two levels of heavy buntons, but it requires careful detailing at the expansion joint at the bottom.

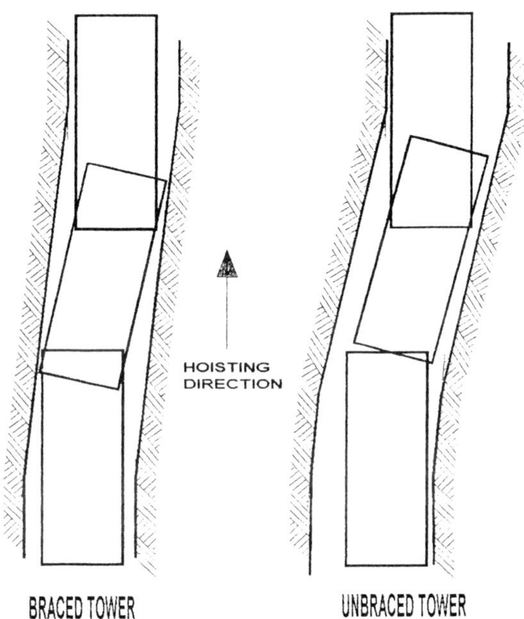

Fig. 3. Conveyance negotiating guide tower.

10. Guide towers may be either braced or unbraced. Fig. 3 shows a cage or skip negotiating the tower. The deflection of the cage or skip in a braced tower is less than that in an unbraced tower, but the forces generated by shear strains are higher. An unbraced tower has quite small forces induced by shear strain of the shaft.

11. The forces generated between the conveyance and the guide tower, in guiding the conveyance past the shear displacement at the reef, may also lead to

torsion of the tower if the conveyance is located at the side of the shaft, as shown in Fig. 4. This must be catered for in the tower design.

The tower must resist certain vibrations

12. The dynamic behaviour of the guide tower must be defined for two reasons. The first is that large conveyances which travel at high speeds must be adequately guided. The buntons, or horizontal beams in the towers are spaced at regular intervals, leading to a regular variation of stiffness. This in turn has the potential to develop regular lateral excitation of the conveyances, leading to possible resonance difficulties. The second reason is that mining induced earth tremors are quite frequent, and cannot be allowed to lead to accidents in the shaft. The important frequencies are the frequency of passing buntons and half this frequency, as well as earth tremor frequencies. Typical values of these frequencies are given in Table 2.

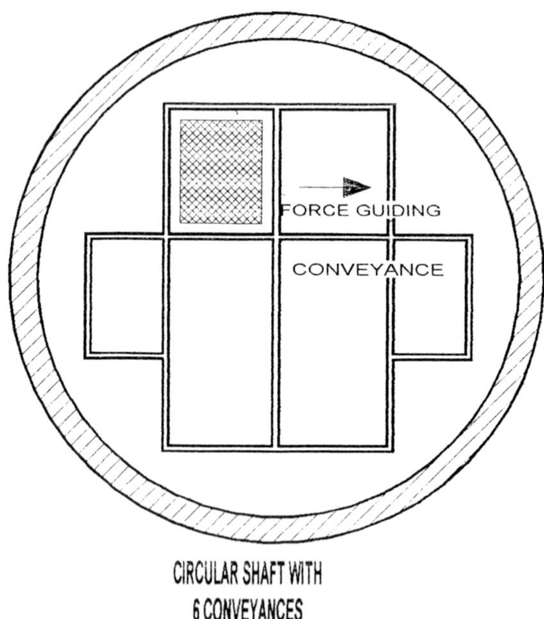

Fig. 4. Typical shaft layout indicating torsion on guide tower.

13. Braced towers are stiffer, and thus have higher natural frequencies as is indicated in Table 2. Unbraced towers rely for their dynamic performance on the stiffness of the guides and the mass of the buntons. They are more flexible, so have lower natural frequencies which make them more susceptible to dynamic magnification at resonance. Greater deflections also result from any lateral motion of the conveyances. Unbraced towers must thus always be carefully analysed to

confirm adequate dynamic behaviour. Krige *et al* (ref. 6) describe an analysis of the dynamic behaviour of an unbraced guide tower as conveyances traverse it. This is based on a simplified mass and stiffness model of the tower, as shown schematically in Fig. 5.

Table 2. Typical natural and excitation frequencies for guide towers.

Structure	Cause of excitation	Frequency (Hz)
80m braced tower		5.0 to 7.0
80m unbraced tower		0.2 to 0.5
	Passing buntons	2.5 to 4.5
	Earth tremor	10 to 15

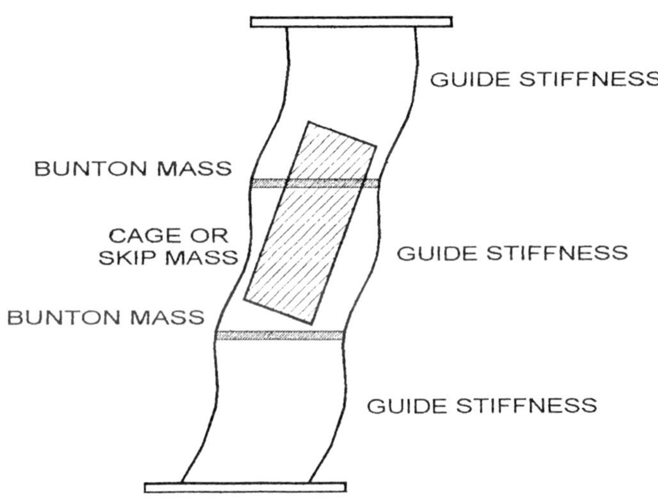

Fig. 5. Schematic layout and analytical model of guide tower.

14. The geometric properties and mass of the tower, as well as the mass and hoisting velocity of the conveyance in the shaft, are all included in this model, which predicts the forces and displacements as the cage or skip traverses the tower. Specific maximum values of these forces and displacements are given in Table 3 for specific parameters of a 40m long tower.

Table 3. Predicted displacements and forces in an unbraced guide tower.

Conveyance mass (kg)	Shear displ. of tower (m)	Max. displ. (m)	Max. force (kN)	Comments
10000	0.10	0.052	17.7	OK
10000	0.05	0.011	4.5	OK
30000	0.10	0.226	121.7	Unacceptable
30000	0.05	0.081	28.1	OK

15. Where the behaviour of an unbraced tower is unacceptable, a braced tower must be used. This is typically required for long towers, high values of tower shear, and large conveyances.

Modular construction

16. If the guide tower is installed at an early stage in the life of the mine, there are not likely to be any extra difficulties involved with the construction of the tower. However, when the guide tower is installed later on in the life of the mine, which is commonly the case, it must not adversely influence the production of the mine. This commonly introduces a requirement that construction be carried out during periods when the shaft is not normally used. Typical production schedules only allow engineering work in the shaft to be carried out over one and a half days each weekend. Krige *et al* (ref. 6) describe one such exercise, where a modular solution had to be adopted. This consisted of installing a working platform, removing a section of existing guide steelwork, fitting the equivalent section of the new guide tower, aligning the guides, and removing the working platform. All of this had to be achieved for each module within a forty hour period.

CONCLUSION

17. The use of a space tower in a mine shaft is shown to solve the problem of guiding conveyances through portions of the shaft which experience large rock strains due to mining activity. The major requirements of the tower have been defined, and an analytical investigation of the performance of an unbraced tower has shown that it can perform adequately under certain circumstances, but that a braced tower must be used in other cases.

REFERENCES

1. VALENTE A. Rock mechanics design of a deep level, mechanised gold mine with high stopping widths. International deep mining conference. South African Institute of Mining and Metallurgy, Johannesburg, 1990, 1093-1108.

2. VALENTE A. The South Deep project. Elastic analysis of induced deformation and stress for shaft steelwork design. Internal report, Johannesburg Consolidated Investments, 1988.
3. THOMAS G. Design guidelines for the dynamic performance of shaft steelwork and conveyances. Prepared for COMRO, Johannesburg, 1990.
4. KRIGE G.J. and DEVY S.D. Specifying shaft steelwork at reef intersection in deep mines. International deep mining conference. South African Institute of Mining and Metallurgy, Johannesburg, 1990, 903-915.
5. SPOTTISWOODE S.M. Source mechanism studies on Witwatersrand seismic events. Ph.D thesis, University of the Witwatersrand, Johannesburg, 1980.
6. KRIGE G.J., DEVY S.D., NAPIER L.G.D. and READ H.J. Western Areas Gold Mine north shaft - Shaft refurbishment through weathered and fragmented lava stratum. Colloquium on shaft sinking, South African Institute of Mining and Metallurgy, Johannesburg, 1989.

56. A simple estimation of free vibration behaviour of latticed cylindrical panel structures

S. YAMADA and A. TAKEUCHI, Toyohashi University of Technology, Japan

SYNOPSIS. A simple analytical procedure is presented for the estimation of the free vibration frequency of shells or shell-like structures in consideration of the effects of initial stresses. Then it is applied to latticed cylindrical panel structures through orthotropic shell analogy, and the effects of geometric, stiffness and loading parameters on the dynamic behaviour are presented.

INTRODUCTION

1. Latticed cylindrical space frames have very simple member arrangements over that of latticed spherical dome structures. However it has not been enough to study the mechanical characteristics. The first author and his co-worker (ref. 1) have carried out the static buckling analyses of monocoque cylindrical panels under external pressure, and demonstrated that an idealised classical buckling pressure is shown to provide unreliable upper bounds to the buckling of imperfect cylindrical panels. However, a reduced stiffness critical pressure is shown to provide convincing lower bounds to the imperfection sensitivity. The first author (ref. 2) has confirmed this for latticed space frames, and has also shown the good agreement between a finite element model and an orthotropic continuum shell model.

2. For the design of shell-like space frames, it is of primary importance to ensure adequate resistance to buckling. In addition however, it is necessary to give considerable attention to the dynamic behaviour. This interest has been particularly stimulated by the many constructions of large-spanned space frames in the deep snow region in Japan where a seismically-induced response cannot be ignored in design. A simple, but effective, estimation method for dynamic characteristics related to the static buckling load, would be very convenient for a designer who wants to take advantage of the possibilities offered by optimisation.

3. The present study proposes a simple estimation method for the free vibration frequency based upon the mode similarity to the static buckling mode. The method is then applied to the orthotropic continuum cylindrical panels modelled from latticed space frames, and the effects of geometric parameters, initial

pressure loading factor, and effective stiffness ratio to the dynamic behaviour, qualitatively and quantitatively clarified.

METHOD

4. As the energy criterion for static stability is based on potential energy consideration (refs 3 and 4), the classical bifurcation buckling condition of a shell-like elastic structure may be written as:

$$U_{2b} + U_{2m} + V_{2m} = 0 \tag{1}$$

for each critical load q_{ck} and the associated critical mode $\{u\}_k$. In this expression, U_{2b} and U_{2m} are the component parts of the quadratic form for incremental energy related to the usual linear bending and membrane strain energy terms, respectively. The last term V_{2m}, which is called the "linearised" membrane energy, represents the interaction between the fundamental equilibrium state (affected by so-called "initial stresses") and the non-linear incremental membrane action. Since the fundamental initial compressive stress state is assumed to be a linear function of the load parameter Q, the term V_{2m} can be rewritten as:

$$V_{2m} = -Q_{ck}\alpha_m \tag{2}$$

where $Q_{ck} = q_{ck} / q_d$ and α_m = known constants. This load normaliser q_d can be alternatively defined as, for instance, a design load or the minimum critical buckling load.

5. We are concerned with small-amplitude linear undamped vibration about an unbuckled equilibrium state applied to the static load, $q = Qq_d$. When an incremental vibration mode is approximately equal to the one of the static buckling critical modes, from the well-known Lagrange equations we obtain the following decoupled equations:

$$T_2 - (U_{2b} + U_{2m} - Q\alpha_m) = 0 \tag{3}$$

where the first term T_2 is the quadratic form of kinetic energy in terms of the incremental velocity, $\{\dot{u}\}_k\{\partial u / \partial \tau\}_k$, and τ represents the time. For each mode of $\{u\}_k$ and the associated circular frequency of vibration ω_k, the term T_2 is written as:

$$T_2 = \omega_k^2 \alpha_t \tag{4}$$

where α_t = known constants

6. In general, a free vibration mode is not necessarily equal to a static critical mode. But as the static load Q, related to the initial stresses, increases and approaches Q_{ck} both modes become consistent. In addition the vibration mode would approximately be the sum of the static critical modes, and the dominant

wavelength of the vibration mode would be very close to the length specific. From Eqs 1-4, we derive:

$$\omega_k^2 = (Q_{ck} - Q)\alpha_m/\alpha_t \tag{5}$$

7. In shell structure design it is most important to ensure adequate resistance to buckling, and from this view point it is very useful to analyse the energy of the resistance to deformations associated with each of the various critical modes, this is, to obtain the critical load spectrum (ref. 1). In the successive design procedure, it is very easy to derive the frequency spectrum of free vibration for the same modes from Eq. 5. Consequently, it would be possible to understand the interaction between the static buckling behaviour and the dynamic fundamental response. In the next section we will show the result of the application of Eq. 5 to latticed cylindrical panel structures.

APPLICATION OF ORTHOTROPIC SHELL ANALOGY

8. Let us consider a latticed cylindrical panel of length L, radius R and central angle ϕ as shown in Fig. 1. The latticed structure can be approximately replaced by an orthotropic panel through the application of the well-known continuum analogy. The initial stresses as membrane fundamental solution under pressure loading q are approximately given as (N_x^F, N_y^F, N_{xy}^F) = (0, -qR, 0) except for nearly-flat and/or thick panels (ref. 1). The constitutive relations are:

$$M_x = D_x\rho_x + D_p\rho_y\ , \quad M_y = D_p\rho_x + D_y\rho_y\ , \quad M_{xy} = D_{xy}\kappa_{xy} \tag{6a}$$

$$N_x = C_x\varepsilon_x + C_p\varepsilon_y\ , \quad N_y = C_p\varepsilon_x + C_y\varepsilon_y\ , \quad N_{xy} = C_{xy}\gamma_{xy} \tag{6b}$$

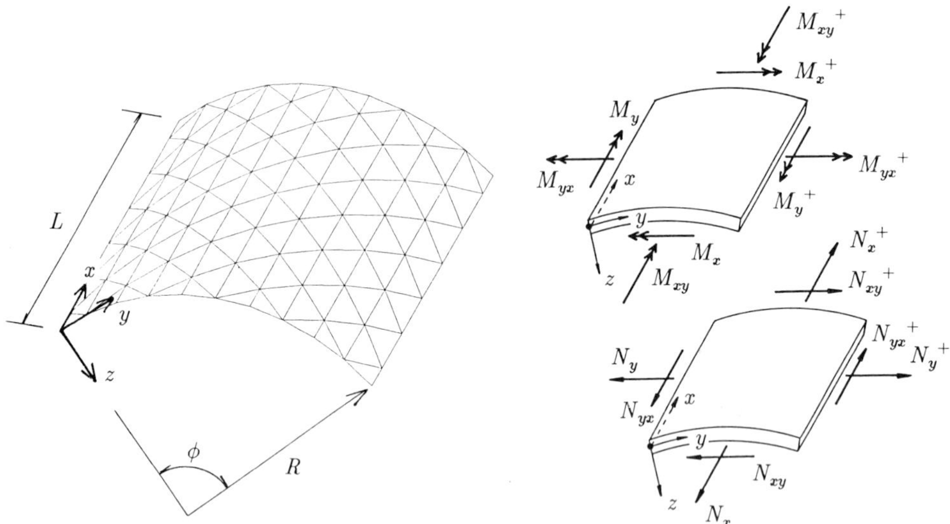

Fig. 1. Geometry of a latticed cylindrical panel and the equivalent stress and moment resultants of its continuum analogy.

where (M_x, M_y, M_{xy}) and (N_x, N_y, N_{xy}) are the bending and membrane stress resultants; (ρ_x, ρ_y, κ_{xy}) and (ε_x, ε_y, γ_{xy}) are the bending and membrane strain components; (D_x, D_p, D_y, D_{xy}) and (C_x, C_p, C_y, C_{xy}) are the bending and membrane stiffness components computed by Heki's method (ref. 5).

9. The static buckling problem for the present structure assuming Donnell-type formulation and classical simple supported boundary condition (so-called SS3) has been solved by Yamada (ref. 2); the critical loads q_c are:

$$Q_c \equiv q_c/q_d = Q_b + Q_m \tag{7}$$

For this result as well as all subsequent results, the normaliser q_d was chosen as:

$$q_d \equiv \pi^2 D_y/(RL^2) \tag{8}$$

The associated eigen-modes, having longitudinal half-wave number m and circumferential half-wave number n, are:

$$u = \overline{u}_{mn}\,(Lt/R)\cos(m\pi x/L)\sin(n\pi y/R\phi) \tag{9a}$$

$$v = \overline{v}_{mn}\,(Lt/R)\sin(m\pi x/L)\cos(n\pi y/R\phi) \tag{9b}$$

$$w = \overline{w}_{mn}\,t\sin(m\pi x/L)\sin(n\pi y/R\phi) \tag{9c}$$

In the expression of Eq. 7, we use the following forms;

$$Q_b = B^2+2(D_2+2D_3)m^2+D_1m^4B^{-2} \tag{10}$$

$$Q_m = 12(C_1-C_2^{\,2})C_3\beta^2m^4/(\pi^4B^2S) \tag{11}$$

$$B = nL/(R\phi) \tag{12}$$

$$\beta = L^2/(Rt) \tag{13}$$

$$S = C_3B^4+(C_1-C_2^{\,2}-2C_2C_3)B^2m^2+C_1C_3m^4 \tag{14}$$

$$t = (12D_y/C_y)^{1/2} \tag{15}$$

$$D_1 = D_x/D_y\,,\quad D_2 = D_p/D_y\,,\quad D_3 = D_{xy}/D_y \tag{16}$$

$$C_1 = C_x/C_y\,,\quad C_2 = C_p/C_y\,,\quad C_3 = C_{xy}/C_y \tag{17}$$

$$\overline{u}_{mn} = \overline{w}_{mn}(A_5A_2-A_3A_4)/(A_1A_4-A_2^{\,2}) \tag{18a}$$

$$\overline{v}_{mn} = \overline{w}_{mn}(A_2A_3-A_1A_5)/(A_1A_4-A_2^{\,2}) \tag{18b}$$

where,
$A_1 = C_1m^2+C_3B^2$, $A_2 = (C_2+C_3)mB$, $A_3 = C_2m/\pi$, $A_4 = C_3m^2+B^2$, and $A_5 = B/\pi$.

10. We assume that the undamped free vibration in the following mode having m axial halfwaves and n circumferential halfwaves (refer to Eq. 9) becomes:

$$(u_d\,,\,v_d\,,\,w_d) = (u\,,\,v\,,\,w)\,\sin(\omega\tau) \tag{19}$$

In the present problem, it is easy to recognise Eq. 19 to be exact solution. On account of the fit with the reference level of the static strain energy components, the quadratic form of the kinematic energy is defined as:

$$T_2 = \frac{2}{s}\int_0^s \frac{q_w}{g}\int_0^{R\phi}\int_0^L \frac{1}{2}\left\{\left(\frac{\partial u_d}{\partial \tau}\right)^2 + \left(\frac{\partial v_d}{\partial \tau}\right)^2 + \left(\frac{\partial w_d}{\partial \tau}\right)^2\right\} dx\,dy\,d\tau \tag{20}$$

in this expression, $s \equiv 2\pi/\omega$ = the natural periods, g = the acceleration due to gravity, and q_w = weight per square arising from self-weight and a dead load. Through defining the energy normaliser, $E_o \equiv 12 D_y L\phi(\overline{\omega}_{mn}{}^2)/R$, and the load parameter, $Q_w \equiv q_w / q_d$ we obtain α_t in Eq. 5 from Eqs 18 and 20, as:

$$\alpha_t = Q_w E_0(\alpha_{tu} + \alpha_{tv} + \alpha_{tw}) \tag{21}$$

where α_{tu}, α_{tv} and α_{tw} are the integral products of Eq. 20 associated with the inertia force in the x, y and z directions, respectively. According to our preliminary study, α_{tu} and α_{tv} are so small that they may be neglected in this problem, and then we derive:

$$\alpha_t \approx E_0\alpha_{tw}Q_w = E_0\pi^2 t Q_w/(96 g\beta) \tag{22}$$

11. The quadratic form of the linearised membrane strain energy is given as:

$$V_{2m} = \frac{2}{s}\int_0^s (-qR)\int_0^{R\phi}\int_0^L \frac{1}{2}\left(\frac{\partial w_d}{\partial y}\right)^2 dx\,dy\,d\tau \tag{23}$$

in this expression the external pressure q would be assumed to be the sum of the selfweight load q_w and a hydrostatic load arising from wind induced pressure q_h,

$$q = q_w + q_h = Q q_d \tag{24}$$

Consequently we obtain α_m in Eq. 5, as:

$$\alpha_m = E_0\pi^4 B^2/(96\beta^2) \tag{25}$$

12. By substituting Eqs 22 and 25 into Eq. 5, the circular frequencies of vibration ω associated with the mode (m, n) are obtained as:

$$\omega^2 = g\pi^2 B^2(Q_c - Q)/(\beta t Q_w) \tag{26}$$

When we define the following non-dimensional forms;

$$F \equiv (\omega/2\pi)(t/g)^{1/2} \quad \text{(non-dimensional frequency)}$$
$$h \equiv q_h/q_w = Q/Q_w - 1 \quad \text{(hydrostatic pressure parameter)}$$
$$\xi \equiv Q/Q_c \quad \text{(initial stress factor)}$$

the frequency spectrum can be obtained from the simple expression:

$$F^2 = (\xi^{-1} - 1)(1+h)B^2/(4\beta) \tag{27}$$

RESULTS

13. In Fig. 2 the solid lines represent the present estimates by using Eq. 27 for Yamada's space frame (ref. 2) having the geometric parameter $\beta = 96.6$, the minimum buckling load parameter $Q_{c0} = 11.42$, the associated mode $(m, B_{c0}) = (1, 2.62)$, the stiffness $C_1 = 1$, $C_2 = C_3 = 1/3$, $D_1 = 1$, $D_2 = 0.061$, $D_3 = 0.469$, and the loading condition $h = 0$. Care must be taken that a loading ratio λ is defined as $\lambda \equiv Q_w / Q_{c0}$. The closed dots represent the minimum value of the non-dimensional frequencies, F_0. The open dots are the eigenvalue analytical predictions for the finite element modelling having the aspect ratio $a = L/R\phi = 0.693$. Good agreement is obtained even though these two analytical procedures are quite different. From this figure it is seen that, as λ increases, F_0 decreases severely, however, the associated circumferential wave number parameter B_0 approaches to the static buckling mode B_{c0} through values smaller than B_{c0}.

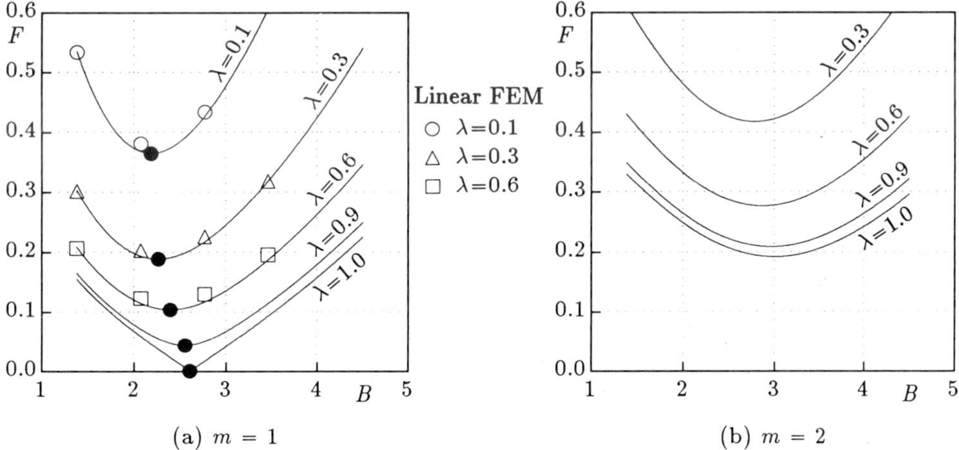

Fig. 2. Non-dimensional frequency F versus circumferential wave parameter B curves for Yamada's space frame (ref. 2) in the case of h=0.

14. Fig. 3 indicates the effects of the geometric parameter , the stiffness ratio C_1 and the hydrostatic pressure parameter h on the minimum frequency F_0. From

Eqs 7 and 11 we find that the minimum buckling load Q_{c0} decreases as C_1 decreases. However, by considering an advantage in design, the normaliser of , Q_{c0}, was taken to be that for the isotropic stiffness ($C_1 = 1$) in the present paper. In our analytical conditions, F_0 was found to be highly sensitive to the parameters C_1 and λ. However, the frequency F_0 varies from 4.0% to 31.8% as $h = 0 \rightarrow 1$.

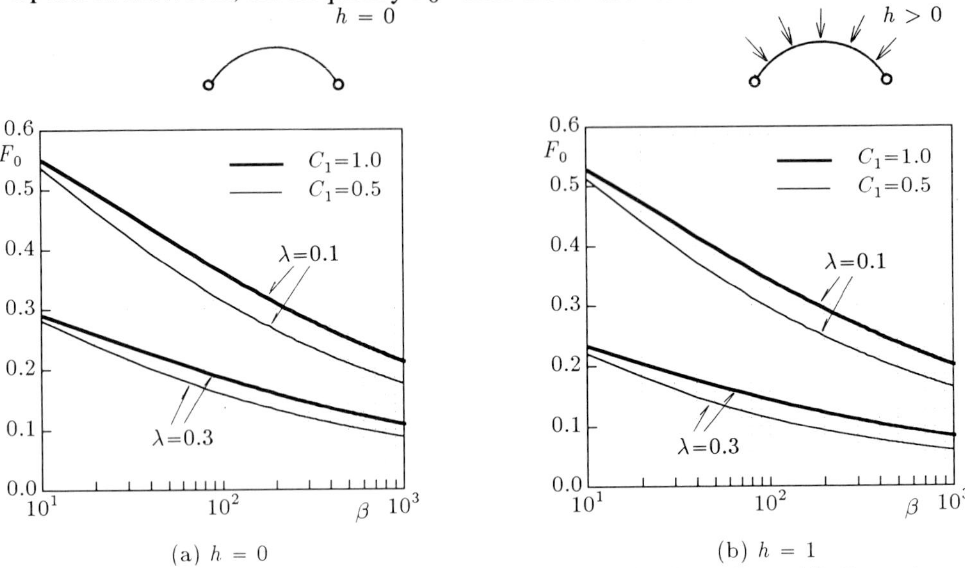

Fig. 3. Parametric plots of minimum frequency F_0 for $C_2 = C_3 = 1/3$, $D_1 = 1$, $D_2 = 0.061$ and $D_3 = 0.469$.

EFFECTS OF THE NON-LINEARITY OF INITIAL STRESSES

15. In the buckling problem of cylindrical barrel vault roof structures, the prebuckling non-linearity by geometric and/or loading imperfections cannot be neglected (ref. 1). This is typically rehearsed in Fig. 4(a) which shows the load versus maximum deflection curve obtained by non-linear finite element analysis for a geometrically perfect space frame with roller supports (ref. 2). Generally the effect of the above imperfections will reveal the non-linearity of initial stresses. Now consider the imperfection amplitude of mode(m, n) involved between geometric and loading, w^0_{mn}. The associated displacement amplitude w^r_{mn} would be:

$$w^r{}_{mn} = w^0{}_{mn} Q/(Q_r - Q) \tag{28}$$

where Q_r is the associated buckling load of the imperfect model. This displacement modelling is a quasi-inextentional deflection one treated by Croll (ref. 6). Its stiffness at a given load and a given imperfection form and amplitude is always lower than that for the exact non-linear solution, therefore at a given load level it will be an upper bound to the corresponding exact deformation.

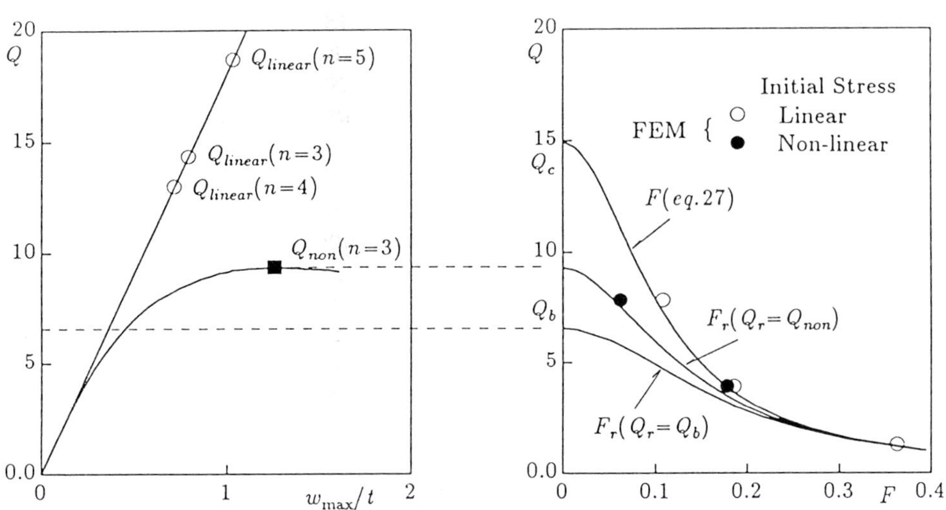

(a) Static buckling analysis by finite element method.

(b) Load versus free vibration frequency for (m, n) = (1, 3).

Fig. 4. Static and dynamic behaviour of the latticed cylindrical panel having $\beta = 96.6$, $a = 0.693$, $C_1 = 1$, $C_2 = C_3 = 1/3$, $D_1 = 1$, $D_2 = 0.061$ and $D_3 = 0.469$.

16. A recent study performed by Virgin (ref. 7) has shown that in the non-linear bending equilibrium state of bars, the frequency squared is in proportion to the deflection amplitude squared (see Eq. 41 in ref. 7). In the present problem, it would be possible to assume that the free vibration frequency is affected by the non-linearity of initial stresses, F_r, as follows:

$$F_r^2 = a_0 + a_1(w^r{}_{mn})^2 \tag{29}$$

where a_0 and a_1 are unknown coefficients. By substituting Eq. 28 into Eq. 29 and by taking the following conditions:

$$F_r = F \quad \text{for } Q = 0\,, \qquad F_r = 0 \quad \text{for } Q = Q_r \tag{30}$$

we derive

$$F_r^2 = F^2\{1 - (Q/Q_r)^2\} \tag{31}$$

17. Fig. 4(b) shows the variation of the free vibration frequency associated (m, n) = (1, 3) mode in the case of Yamada's space frame (ref. 2) and h=0. In this figure, the open and closed dots indicate the results obtained by the finite element analysis; the initial stress state of the open dots is linear, and that of the closed dots non-linear. It is shown that the present simple result adopting the buckling load as shown in Fig. 4(a), Q_{non} as Q_r in Eq. 31, is good in agreement with the associated, fully analytical results of the closed dots. With both the shapes and magnitudes of geometric imperfections unknown at the stage of design, the static critical load Q_r cannot be realistically estimated. But ref. 1 has shown that its lower bound can be provided by the so-called reduced stiffness method. By adopting the reduced stiffness load Q_b obtained in Eq. 10 as Q_r in Eq. 31, we would predict an alternative lower bound for the free vibration frequency of latticed cylindrical panel structures.

CONCLUSIONS

18. An alternative method on the free vibration estimations of shell-like structures having initial stresses has been developed. It has been applied to latticed cylindrical panel structures. The present frequency estimate, Eq. 27, is shown to provide an upper bound to the free vibration frequency. However, on account of the high levels of geometric and intrinsic loading imperfections, the present lower bound estimate using $Q_r = Q_b$ in Eq. 31 will be a realistic representation in design.

REFERENCES

1. YAMADA S. and CROLL J.G.A. Buckling behaviour of pressure loaded cylindrical panels. J. Engrg. Mech., ASCE, 1989, vol.115, 327-344.
2. YAMADA S. Relationship between non-linear numerical experiments and a linear lower bound analysis using finite element method on the overall buckling of reticular partial cylindrical space frames. Computer Applications in Civil and Building Engineering, Kozo System Inc., Tokyo, 1991, 259-266.
3. CROLL J.G.A and ELLINAS C.P. Reduced stiffness axial load buckling of cylinders. Int. J. of Solids and Structures, 1983, vol.19, 461-477.
4. YAMADA S. and CROLL J.G.A. Non-linear buckling response of pressure loaded cylindrical panels and its interpretation for design. The Technology Reports of the Tohoku University, 1988, vol.52, 71-95.
5. HEKI K. On the effective rigidities of lattice plates. Recent Researches of Structural Mechanics, Uno Shoten, Tokyo, 1968, 31-46.
6. CROLL J.G.A. Elasto-plastic buckling of pressure and axial loaded cylinders. Proc. Instn Civil Engrs, 1982, vol.73, 633-652.
7. VIRGIN L.N. Postbuckling dynamics of struts as related to their loading devices. Engrg Struct., 1986, vol.8, 127-133.

57. Vibration analysis of cylindrical curved panels using Kantorovich integral method

H. A. HADID and D. A. HASSON, University of Mosul, Iraq

SYNOPSIS. Free vibration characteristics of cylindrical curved panels are obtained by solving the partial differential equations of motion. The method of Kantorovich is used to reduce these equations to a set of ordinary differential equations. These equations are then solved using an integral equation technique with a cubic spline function. The results obtained were compared with other theoretical solutions and experimental results.

INTRODUCTION

1. The accurate prediction of the natural frequencies of the transverse vibrations of cylindrically curved panels are important for designing panels of aircraft and spacecraft against acoustic fatigue .

2. Sewall (ref. 9), used the energy method employing the Rayleigh-Ritz procedure to solve the free vibration of a cylindrically curved panel with simply supported and clamped edges . The same approach was used for the free vibration of cylindrical shell panels by Blevins (ref. 2). Webster (ref. 11) applied a double-power series method to solve the vibration of shallow clamped shell. Petyt and Deb Nath (ref. 8) used the Kantorovich method to obtained the free vibration characteristics of singly-curved rectangular plates. Mizusawa (ref. 6) used the spline strip method to solve the free vibration of open cylindrical shells. The finite element method was used for dynamic analysis of a shallow shell by Olson and Lindberg (ref. 7).

3. In this paper the free vibration characteristics of clamped, free, and simply-supported boundary conditions with different values of curvature are studied.

DETERMINATION OF THE NATURAL FREQUENCY

4. The non-dimensional, governing, partial differential equations for a circular cylindrical thin shell panel, as derived by Ugral (ref. 10), are:

$$\frac{\partial^2 u}{\partial x^2} + \left(\frac{1-\upsilon}{2}\right) r^2 \frac{\partial^2 u}{\partial y^2} + \left(\frac{1+\upsilon}{2}\right) r \frac{\partial v^2}{\partial x \partial y} - \upsilon \frac{\ell}{a} \frac{\partial w}{\partial x} = \lambda^\circ u$$

$$\left(\frac{1+\upsilon}{2}\right)r\frac{\partial^2 u}{\partial x\partial y}+\left(\frac{1-\upsilon}{2}\right)\frac{\partial^2 v}{\partial x^2}+r^2\frac{\partial^2 v}{\partial y^2}-\frac{r\ell}{a}\frac{\partial w}{\partial y}=\lambda^\circ v$$

$$\frac{\upsilon\ell}{a}\frac{\partial u}{\partial x}+\frac{r\ell}{a}\frac{\partial v}{\partial y}-\frac{\ell^2}{a^2}w-\frac{h^2}{12\ell^2}\left(\frac{\partial^4 w}{\partial x^4}+2r^2\frac{\partial^4 w}{\partial x^2\partial y^2}+r^4\frac{\partial^4 w}{\partial y^4}\right)=\lambda^\circ w \qquad (1)$$

where $\lambda^\circ = -m\omega^2\ell^2(1-\upsilon)/Eh$, $r = \ell/b$, m = mass density per unit area, ν = Poisson's ratio, E = modulus of elasticity, ω is the circular frequency of vibration. ℓ,b,a and h are length, width, radius and thickness of the shell respectively. u, v and w are displacements in the x, y and z directions respectively. For a clamped cylindrical shell panel in the x-direction the partial differential Eq. 1 may be written in the following form:

$$\begin{aligned} L_1(u,v,w) - \lambda^\circ Iu &= 0 \\ L_2(u,v,w) - \lambda^\circ Iv &= 0 \\ L_3(u,v,w) - \lambda^\circ Iw &= 0 \end{aligned} \qquad (2)$$

5. The modes of vibration along the straight line generators of a cylindrical shell panel can be closely approximated by a straight beam eigen function. So that, the solution of Eq. 2 takes the following form:

$$\begin{aligned} u &= \frac{1}{\lambda_m}F'_m(x)\cdot g_1(y) \\ v &= F_m(x)\cdot g_2(y) \\ w &= F_m(x)\cdot g_3(y) \end{aligned} \qquad (3)$$

λ_m and $F_m(x)$ are mth eigenvalues and eigen functions of a straight beam, $g_1(y)$, $g_2(y)$ and $g_3(y)$ are functions to be determined.

where:

$$F_m(x) = \cosh(\lambda_m x)-\cos(\lambda_m x)-\sigma_m(\sinh(\lambda_m x)-\sin(\lambda_m x)), 0\le x\le 1$$

$$\lambda_m = (4.73, 7.8532, 10.996, \ldots, (2_m+1)/\pi, m=1,2,\ldots,\infty)$$

$$\sigma_m = (\cosh\lambda_m-\cos\lambda_m)/(\sinh\lambda_m-\sin\lambda_m)$$

6. Substituting Eq. 3 into Eq. 2 and multiplying the equation by $1/\lambda_m F_m'(x)$, $F_m(x)$ and $F_m(x)$ respectively and integrating along the length of the shell the following ordinary differential equations can be obtained:

$$a_{11} g_1 + a_{12} \overset{\circ\circ}{g_1} + b_{11} \overset{\circ}{g_2} + c_{11} g_3 = a_{13} g_1$$

$$a_{21} g_1 + b_{21} \overset{\circ\circ}{g_2} + b_{22} g_2 + c_{21} \overset{\circ}{g_3} = b_{23} g_2 \tag{4}$$

$$a_{31} g_1 + b_{31} \overset{\circ}{g_2} + c_{31} g_3 + c_{32} \overset{\circ\circ}{g_3} + c_{33} \overset{\circ\circ\circ\circ}{g_3} = c_{34} g_3$$

where: $\overset{\circ}{g_1} = \partial g_1 / \partial y$

$a_{11} = -\lambda_m{}^2$ $\quad a_{12} = (1-\upsilon) r^2 I_m / 2$ $\quad a_{31} = -\upsilon \ell \lambda_m I_m / a$

$a_{21} = -(1+\upsilon) r \lambda_m I_m / 2$ $\quad a_{13} = \lambda I_m$ $\quad b_{11} = -(1+\upsilon) r \lambda_m I_m / 2$

$b_{21} = r^2$ $\quad b_{23} = \lambda$ $\quad b_{22} = -(1-\upsilon)\lambda_m{}^2 I_m / 2$

$b_{31} = r\ell / a$ $\quad c_{11} = -\upsilon \ell \lambda_m I_m / a$ $\quad c_{21} = -r\ell / a$

$c_{33} = -r^4 h^2 / 12\ell^2$ $\quad c_{34} = \lambda$ $\quad c_{32} = 2r^2 h^2 \lambda_m I_m / 12\ell^2$

$c_{31} = -\ell^2 / a^2 - h^2 \lambda_m^4 / 12\ell^2$ $\quad I_m = \sigma_m - 2\sigma_m^2 / \lambda m$

7. When the curved edges of the shell are free the same procedure can be used. Thus the eigen function of the straight beam can be written in the following form (ref. 3):

$F_1(x)=1 \quad \lambda_1=0, \quad F_2(x)=1-2x \quad \lambda_2=1$

$F_m(x) = \sinh(\lambda_m x) + \sin(\lambda_m x) - \sigma_m(\cosh(\lambda_m x) + \cos(\lambda_m x)),\ 0 \le x \le 1$

$\lambda_m = (4.73,\ 7.8532,\ 10.996, \ldots,\ (2m-3)\pi/2,\ m = 3,4,\ldots,\infty)$

$\sigma_m = (\sin\lambda_m - \sinh\lambda_m)/(\cos\lambda_m - \cosh\lambda_m)$

8. In order to solve the ordinary differential equation, Eq. 4, the spline integral method is used. The shell is divided by equidistance lines parallel to the shell co-ordinate x. The highest derivatives of g_1, g_2 and g_3 with respect to x are denoted by: (ref. 5)

$$\overset{\circ\circ}{g_1} = -1, \quad \overset{\circ\circ}{g_2} = -t, \quad \overset{\circ\circ\circ\circ}{g_3} = -f \tag{5a-c}$$

9. Eq. 5 with the associated boundary conditions can be transformed to a homogeneous Fredholm integral equation of the second kind , by approximating the unknown functions (l, t, f) as a cubic spline function, Eq. 5-c can be written in the following form for example:

$$g_3(x_m, y) = \int_0^1 Ky(x_m, y, \eta)\ S\Delta(x_m, \eta)\ d\eta \qquad (6)$$

where $Ky(x_m, y, \eta)$ and $S\Delta(x_m, \eta)$ are the Green's function and the cubic spline function .

10. After integrating Eq. 6 using the procedure given in ref. 5 the following equation is obtained, which can be written in matrix form as:

$$\underset{Mx\,1}{w_m} = \underset{MxM}{A}\ \underset{Mxl}{f^{oo}} + \underset{MxM}{B}\ \underset{Mxl}{f} \qquad (7)$$

where A and B are square matrices obtained from Eq. 6, M = N +1 and N equals the number of intervals. By applying the continuity conditions of the unknown functions for equal intervals, the following relation in matrix form is obtained: (ref. 1)

$$f^{oo} = C^{-1}\,D\,f = F\,f \qquad (8)$$

where: $C^{-1}\,D = F$

Substituting Eq. 8 into Eq. 7 gives:

$$w_m = A\,F\,f + B\,f$$
$$w_m = G\,f + B\,f$$
$$w_m = H\,f \qquad (9)$$

where: $A\,F = G$, $G + B = H$

11. By using derivatives of the Green's function, the following expressions can be obtained:

$$\overset{oooo}{g_3} = -f = -B^{-1}g_3$$
$$\overset{o}{g_3} = \overset{o}{B}f = \overset{o}{B}B^{-1}g_3 \qquad (10)$$
$$\overset{oo}{g_3} = \overset{oo}{B}f = \overset{o}{B}B^{-1}g_3$$

12. In the same way the derivatives of g_1 and g_2 with respect to y can be written as follows:

$$\overset{\circ\circ}{g}_1 = -H^{-1}g_1 \qquad \overset{\circ\circ}{g}_2 = -G^{-1}g_2$$
$$\overset{\circ}{g}_1 = \overset{\circ}{H}H^{-1}g_1 \qquad \overset{\circ}{g}_2 = \overset{\circ}{G}G^{-1}g_2 \tag{11}$$

13. In order to solve Eq. 4 substituting expressions Eqs 10 and 11 in Eq. 4 and arranging the equations in the matrix form as follows:

$$\underset{3L\times 3L}{\begin{bmatrix} L_{11} & L_{12} & L_{13} \\ L_{21} & L_{22} & L_{23} \\ L_{31} & L_{32} & L_{33} \end{bmatrix}} \cdot \underset{3L\times 1}{\begin{bmatrix} g_1 \\ g_2 \\ g_3 \end{bmatrix}} = \lambda \underset{3L\times 3L}{\begin{bmatrix} I & 0 & 0 \\ 0 & I & 0 \\ 0 & 0 & I \end{bmatrix}} \cdot \underset{3L\times 1}{\begin{bmatrix} g_1 \\ g_2 \\ g_3 \end{bmatrix}} \tag{12}$$

where: L = N-1 and N = number of intervals.

$$L_{11} = -\lambda_m^2 / I_m - (1-\nu)r^2H^{-1}/2$$

$$L_{12} = -(1-\nu)r\lambda_m \overset{\circ}{G}G^{-1}/2$$
$$L_{13} = -\nu l\lambda_m / a$$

$$L_{21} = -(1+\nu)r\lambda_m I_m \overset{\circ}{H}H^{-1}/2$$
$$L_{22} = -(1-\nu)\lambda_m^2 I_m / 2 - r^2G^{-1}$$

$$L_{23} = -rl\overset{\circ}{B}B^{-1}/a$$
$$L_{31} = -\nu l\lambda_m I_m / a$$

$$L_{32} = rl\overset{\circ}{G}G^{-1}/a$$

$$L_{33} = -l^2/a^2 - h^2/12l^2(\lambda_m^4 - 2r^2\lambda_m^2 I_m \overset{\circ}{B}B^{-1} - rB^{-1})$$

$L_{11}...L_{33}$ are square matrices of order (LxL)
I is a unitary matrix of order (LxL)

Eq. 12 can be reduced to eigenvalue problem as follows:

$$C\delta = \overset{\circ}{\lambda} I\delta$$

or $$\left[C - \overset{\circ}{\lambda} I\right]\delta = 0 \qquad (13)$$

14. For non-trivial solution of Eq. 13 the determinant must vanish.

$$\left|C - \overset{\circ}{\lambda} I\right| = 0$$

where C is the global matrix of order 3Lx3L given in Eq. 12, δ is the displacements vectors g_1, g_2 and g_3 of order 3Lx1.

15. If the curved edges of the shell are simply supported, the solution can be obtained by assuming a sine series function in the x-direction, therefore, the displacements functions can be written in the following form:

$$\begin{aligned} u &= \cos(m\pi x).g_1(y) \\ v &= \sin(m\pi x).g_2(y) \\ w &= \sin(m\pi x).g_3(y) \end{aligned} \qquad (14)$$

where: $0 \leq x \leq 1$

16. By substituting Eq. 14 into Eq. 1, ordinary differential equations will be obtained, then the spline integral technique is used to solve these equations which can be reduced to set of simultaneous equations, Eq. 12, where:.

$$L_{11} = -m^2\pi^2 - (1-\nu)r^2H^{-1}/2$$

$$L_{12} = (1+\nu)rm\pi \overset{\circ}{G} G^{-1}/2$$

$$L_{13} = -\nu l m\pi / a$$

$$L_{21} = -(1+\nu)rI_m m\pi \overset{\circ}{H} H^{-1}/2$$

$$L_{22} = -(1-\nu)m^2\pi^2/2 - r^2G^{-1}$$

$$L_{23} = -rl \overset{\circ}{B} B^{-1}/a$$

$$L_{31} = -\nu l m\pi / a$$

$$L_{32} = rl \overset{\circ}{G} G^{-1}/a$$

$$L_{33} = -l^2/a^2 - h^2/12l^2(m^4\pi^4 - 2r^2m^2\pi^2 \overset{\circ\circ}{B} B^{-1} - r^4B^{-1})$$

In the same way Eq. 12 can be reduced to the eigenvalue problem.

RESULTS AND DISCUSSIONS

17. The natural frequencies of the cylindrical shell panel clamped at all edges are presented in Table 1 . The results agree well with those obtained by Petyt and Deb Nath (ref. 8), Olson and Lindberg (ref. 7), Hasson (ref. 5), Deb Nath (ref. 4) and Webster (ref. 11). The natural frequencies of the cylindrical shell panel simply supported at all edges are presented in Table 2 . This table shows that the results are found to be in good agreement compared with those obtained by Hasson (ref. 5) and Mizisaws (ref. 6).

18. Table 3 contains the results of natural frequencies of the cylindrical shell panel with straight edges clamped and curved edges free with different (ℓ/a) values. Comparison of the results shows that good agreement was obtained with respect to the results of finite element and Rayleigh-Ritz methods (ref. 2).

19. Fig. 2 shows the effect of the (ℓ/a) value on the lowest natural frequency of a cylindrical shell panel with different boundary conditions . From this figure it is observed that the results agree with those obtained from finite element method (ref. 2) and Spline integral method (ref. 5).

CONCLUSIONS

20. Kantorovich-spline integral method which has been developed in the present work can be used as a general method for the analysis of different structural problems. It has been shown that the results agree well with those obtained by other theoretical methods . In the Kantorovich method the required of computer storage and time are small compared with other methods.

REFERENCES

1. AHLBER J. H. (ed). The Theory of Splines and Their Applications. Academic Press, New York, 1967.
2. BLEVINS R. D. Natural frequencies of shallow cylindrically curved panels. Journal of Sound and Vibration, 1981, Vol.75, No.1, 145-149.
3. CHEUNG Y. K. Finite strip method in structural analysis Pergamon Press, New York, First Edition, 1976.
4. DEB NATH J. M. Dynamic of rectangular curved plates. PhD Thesis, University of Southampton , 1969, 5.
5. HASSON D. A. Vibration and stability analysis of circular cylindrical thin shell panels . MSc Thesis, University of Mosul, 1991.
6. MIZUSAWA T. Application of spline strip method to analysis vibration of open cylindrical shells. International Journal for Numerical Methods in Engineering, Vol.26, 988, pp. 663-676.
7. OLSON M. D. and LINDBERG G. M. Dynamic analysis of shallow shells with a doubly curved triangular finite element. Journal of Sound and Vibration, 1971,Vol.19, No .3 ,299-318 .

8. PETYT M. and DEB NATH J. M. Vibration analysis of singly curved rectangular plates. Journal of Sound and Vibration, 1971, Vol.13 No.4, 485-497. 9. SEWALL J. L. Vibration analysis of cylindrically curved panels with simply supported or clamped edges and comparison with some experiments. NASA-TND-3791, 1967.
10. UGURAL A. C. Stresses in plates and shells. McGraw-Hill Book Co., Inc., New York, 1981.
11. WEBSTER J. J. Free vibrations of rectangular curved panels. International Journal of Mechanical Science, 1968, Vol.10, 571-582.

Table 1. Natural frequencies of a circular cylindrical shell clamped at all edges.
ℓ = 7.62cm, b = 10.16cm, a = 76.2cm, h = 0.033cm, ν = 0.33 .

Mode	m,n	frequency parameter λ					
		K-S	ref .4	ref.8	ref .5	ref.11	ref.7
$\lambda=\omega(D/m)^{0.5}/\ell^2$ (Hz)							
1	1,1	63.26	57.65	63.03	62.13	61.61	61.61
2	1,2	68.58	66.60	68.41	68.25	67.84	67.84
3	1,3	92.14	89.23	91.71	92.54	91.21	91.21
4	2,1	97.70	92.49	97.38	96.90	96.60	96.52
5	2,2	103.05	102.83	102.69	102.64	101.98	101.98
6	2,3	125.76	127.61	123.58	125.84	124.14	124.36

K-S present method.

Table 2. Natural frequencies of a circular cylindrical shell simply supported at all edges.
ℓ = 1.0m, b = 1.0m, a =1.91m, h = 0.0191m, v = 0.3.

Mode	m,n	frequency parameter λ		
		K-S	ref .5	ref .6
$\lambda=\omega(D/m)^{0.5}/\ell^2$ (Hz)				
1	1,1	63.12	64.36	62.39
2	1,2	80.46	81.14	80.16
3	1,3	97.98	97.41	96.85
4	2,1	101.19	101.39	100.73
5	2,2	108.90	110.73	107.80
6	2,3	137.57	131.49	130.90

K-S present method.

Table 3. Natural frequencies of a circular cylindrical shell clamped at straight edges and free at curved edges with different ℓ/a values. ℓ = 50.8cm, b=50.8cm, h = 1.905cm, v = 0.3.

Mode	m,n	frequency parameter λ $\lambda=\omega(D/m)^{0.5}/\ell^2$ (Hz) Kantorovich-Spline Integral (5x5)	(6x6)	ref.2 Rayleigh-Ritz	ref.2 Finite Element
(ℓ/a)=0.00					
1	1,1	22.39	22.38	22.33	22.17
2	2,1	24.52	24.52	26.55	26.40
3	1,2	62.07	61.84	61.51	61.51
4	2,2	67.99	67.75	67.51	67.51
(ℓ/a)=0.25					
1	1,1	22.39	22.38	22.33	22.17
2	2,1	33.07	32.99	32.79	32.50
3	1,2	62.07	61.84	61.51	63.06
4	2,2	69.51	69.11	69.23	69.40
(ℓ/a)=0.50					
1	1,1	22.39	22.38	22.33	——
2	2,1	48.18	47.98	46.75	44.07
3	1,2	62.07	61.84	61.51	61.75
4	2,2	67.86	67.05	74.15	69.17
(ℓ/a)=1.00					
1	1,1	22.39	22.38	22.33	——
2	1,2	62.11	61.84	61.51	60.86
3	2,1	82.88	80.01	81.39	80.67
4	2,2	99.71	98.61	91.29	98.23

K-S present method.

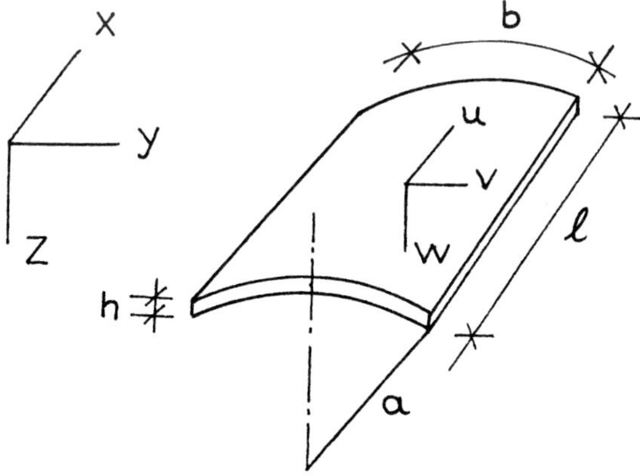

Fig. 1. The cylindrical shell panel.

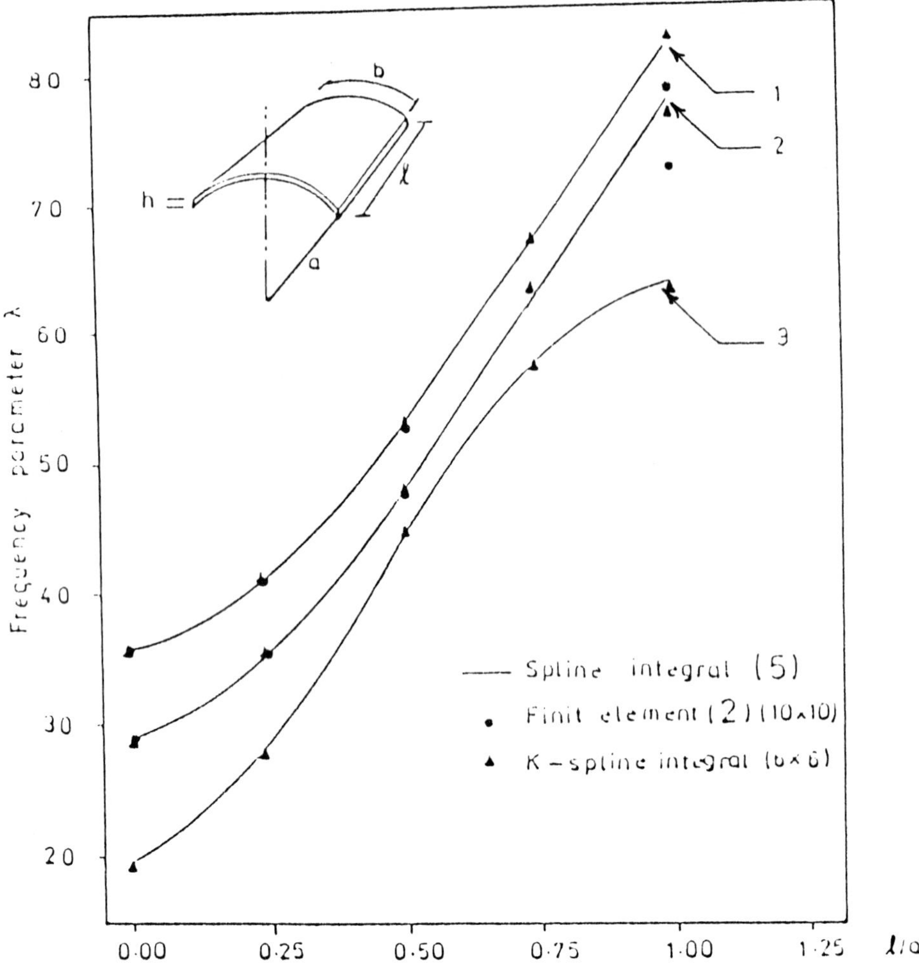

Fig. 2. Effect of (ℓ/a) ratio on the lowest value of frequency parameter λ for cylindrical hell panel. ℓ = 50.8cm, b = 50.8cm, h = 1.905cm, ν = 0.3

1. All edges are clamped.
2. Straight edges clamped, curved edges simply supported.
3. Simply supported at all edges.

58. Dynamic behaviour of cooling towers with initial imperfections

M. PIRNER, J. NAPRSTEK and T. HORYNA, Czech Republic Academy of Sciences

SYNOPSIS. The purpose of this paper is to express the dependence of dynamic characteristics of a hyperboloid of revolution on its initial imperfections. These imperfections are random functions of position on the middle surface of the shell and the service time of the structure. Influence of imperfections on natural frequencies and modes is investigated experimentally on a real structure and on its model. The theoretical solution is achieved using two methods. The first method uses a finite element solution which includes detailed layout of measured deviations from the ideal shape of the structure. The original semi-numerical solution by computational stochastic mechanic methods provides the second solution.

INITIAL IMPERFECTIONS AND ANALYSIS OF THEIR INFLUENCE

1. Some European cooling towers have meridional cracks, after 30 - 40 years of service. These are caused by the differential in the alternating thermal stresses between the inside and outside of the shell. For this reason, in the phase of design of the cooling towers for the Temelín nuclear power plant, the possibility of determining the serviceability and residual safety by means of the changes of dynamic response had been considered. The response can be obtained relatively easily by repeated tests, similarly to the procedure used to measure the time-dependence.

2. There are two types of initial imperfections:

(i) shell thickness deviations;

(ii) the deviations of the ideal shape of the structure, measured as diffences between the ideal distances of the middle surface from the axis of revolution of the structure.

Both these types are of technological origin.

3. Dimensions of a studied cooling tower are shown in Fig. 1. The tower is 155m high, its lower ring has a diameter of 112.8m, its top ring a diameter of 82m. The diameter of the throat is 77m. The wall thickness above the lower ring is 750mm, diminishing to 180mm in the throat and to 720mm below the top ring.

The lower ring is supported by 56 pairs of slanting columns. Total weight of the tower without machinery is 28450 tons.

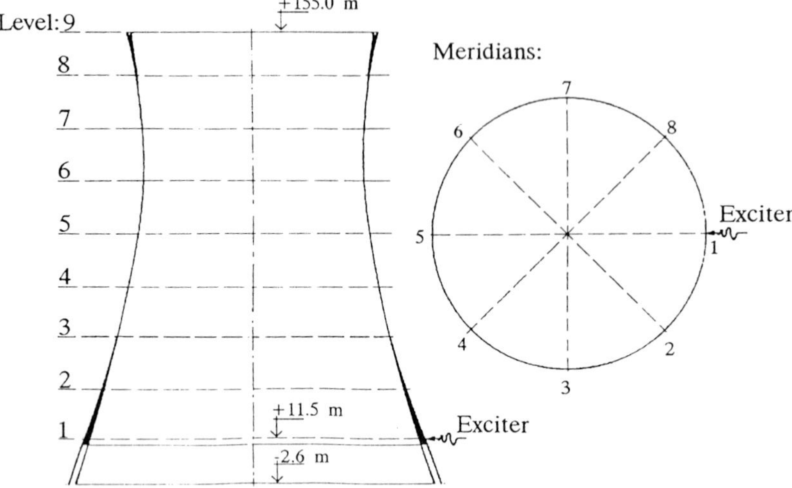

Fig. 1. Cross sections of the real structure and positions of measured points.

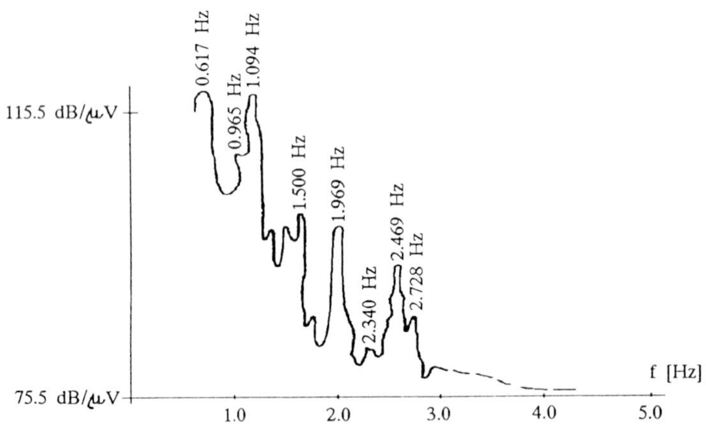

Fig. 2. Real structure response (at point 72) to environmental noise.

NATURAL MODES AND FREQUENCIES OF A REAL STRUCTURE AND ITS PHYSICAL MODEL

4. The cooling tower was excited by a controlled vector of force obtained by a rotary dynamic exciter fastened to the outer face of the shell above the lower ring. Five natural vibration modes were excited with frequencies given in Table 1. Natural frequencies were also verified by peaks of the response spectrum for environmental noise and wind loads (Fig. 20). The cooling tower response was measured at 72 points (Fig. 1) by type 8306 Brüel & Kjær accelerometers. The magnitude of the excited amplitude at the reference point (near the exciter) for the frequency of 1.969 Hz was 0.22mm; it was produced by the excitation force of 4270 N. The logarithmic decrement of tower vibration damping was obtained from frequency transfer functions.

5. The physical model on the length scale of 1: 100 has been made of reinforced epoxy resin. The physical model was created with imperfections in the shape of the external surface; therefore the thickness of the wall was different from theoretical value. The differences were nearly the same as the measured differences on the real cooling tower in power plant at Temelín. Radius imperfections of the cooling tower were not realised on the model. The imperfections were generated by wall thickness differences from the theoretical shape, and are given in Fig. 3, which shows the imperfections on the developed hyperboloid shell. Boundary conditions of the model approach reality. The length scale of the model was selected as to satisfy the technology of model production and to ensure that the damping of vibration of the model be comparable with the damping of vibration of the real structure. The damping of the model is about three times as high as the damping of the real structure. The model scale of frequencies was 1:40.

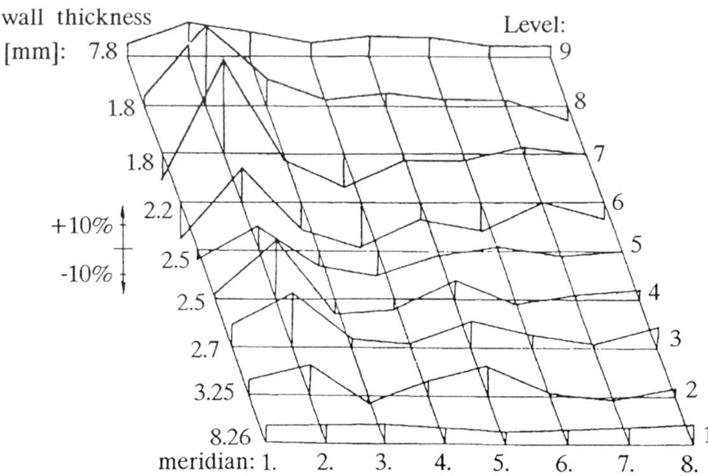

Fig. 3. The wall thickness imperfections of the physical model (%).

6. Natural frequencies of the model were determined from its response to the pseudo-random signal supplied by the type 2032 FFT Brüel & Kjær analyser to the RFT electrodynamic exciter. The frequency values given in Table 1 are mean values obtained from measurements at 72 points. The resonances were verified in the same way as in the case of the real structure. The response of the model was measured by a VPI-OMETRON laser sensor. Ten natural modes were excited. No more than 5 waves along the circle were excited on the model.

NUMERICAL SOLUTION USING THE FINITE ELEMENT METHOD AND COMPARISON WITH EXPERIMENTAL RESULTS

7. For the analysis of natural vibrations four theoretical models of the structure were derived from the experimental model. The difference between these models was differing shell thickness of the structure in horizontal section. The purpose was to create as many theoretical models as would enable monitoring of the influence of the wall thickness imperfections on natural frequencies and natural modes. For this reason the first theoretical model was provided with shell thicknesses measured on the experimental model (Fig. 3) that will be called the model with 100% imperfections. From this model another was derived - model with 0% imperfections - with constant wall thickness in horizontal section, determined by the mean value of measured wall thicknesses in horizontal section of the physical model. Between these two extreme models there are theoretical models with 50% and 25% imperfections.

8. The finite element analysis of natural vibration took into account linear geometrical and material relations (ref. 1). At the discretization of the structure the shell element was used for the modelling of the cooling tower skin while the slanting columns were modelled by beam elements. For each of the four theoretical models 100 natural frequencies and modes were computed.

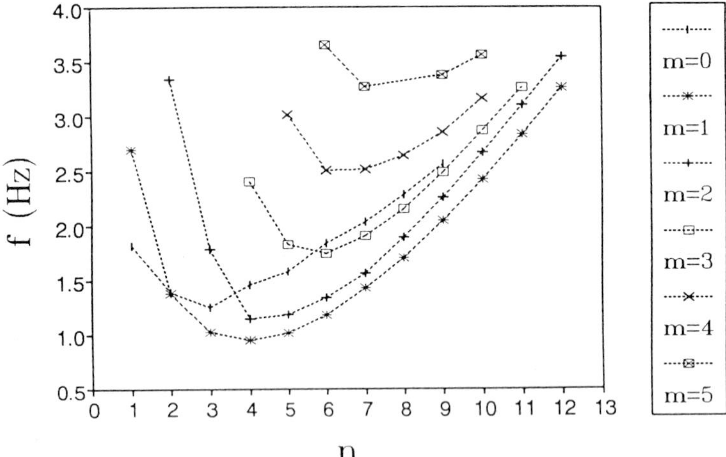

Fig. 4. Spectrum of natural frequencies of the theoretical model with 0% imperfections.

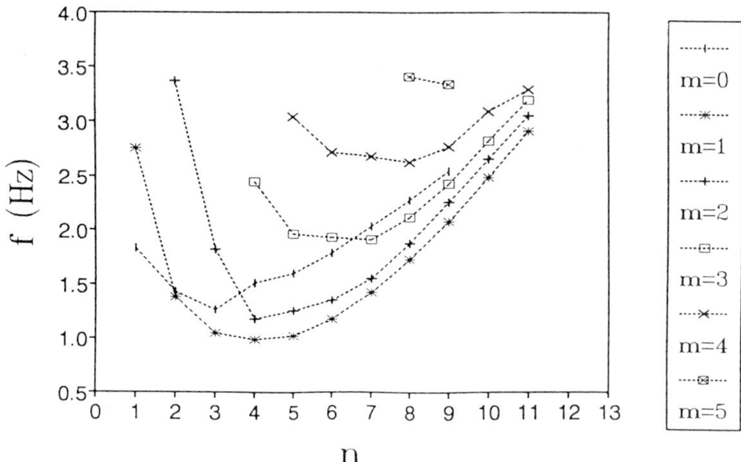

Fig. 5. Spectrum of natural frequencies of the theoretical model with 100% imperfections.

The natural modes and their respective frequencies were classified according to the number of waves in a horizontal section of the tower (*n*) and according to the number of nodes along a meridian (*m*).

9. Natural frequencies for the theoretical model with 0% imperfections are shown in Fig. 4 and the results for the model with 100% imperfections are shown in Fig. 5. In both figures the number of waves (*n*) in horizontal section is represented on the horizontal axis and the frequency (*f*) on the vertical axis. The frequency values pertaining to the modes with the same number of nodes along the meridian (*m*) are connected. It follows from the comparison of the natural frequency spectra for the model with 0% imperfections and that with 100% imperfections that considered variations of the cooling tower shell thickness do not exert any significant influence on the values of natural frequencies; the upward or downward shifts of frequency values represent maximally 3%, and the character of the spectrum is preserved.

10. Theoretical results of analyses of natural vibration have revealed natural modes that have different number of waves (*n*) in the horizontal sections of the structure. Similar phenomenon occurs also in the number of nodal points in meridians (*m*). This phenomenon increases with the increasing imperfections and may give rise to complications, not only in theoretical analysis, but also in experiments, where the model cannot be always excited with such stability resulting in the fluctuation of the excitation frequency difference between two adjacent natural frequencies. The measured "natural mode" is then a synthesis of two adjacent natural modes and may have different number of waves in horizontal sections.

11. Table. 1 reveals the differences between values obtained from experiments and theoretical values. The differences are expressed as a percentage of the

theoretical value of natural frequency of the theoretical model with 100% imperfections.

12. No vibration modes with more than 5 waves in the transverse horizontal section on the real structure, or its physical model, have been excited. This is due to energy requirements of existence of the natural modes with a higher number of waves. Some of the natural vibration modes require alternate shortening and elongating of the middle area of the hyperboloid consequently, they are also energetically exacting and their possible existence cannot be proved by measurement.

SEMI-NUMERICAL SOLUTION USING COMPUTATIONAL STOCHASTIC MECHANICS

13. A dominant part of hyperbolic shell imperfection in comparison with ideal shape of the structure is given by deviations of the wall thickness and of the radius of rotary middle surface. Based on the structure of the differential operator in the system of equations describing linear natural vibration of a thin-walled shell according to the bending theory (ref. 2):

$$\sum_{j=1}^{3}(\frac{1}{3}h^2 N_{ij} + L_{ij})u_j + \frac{(1-\nu^2)\rho}{E}(ib\omega - \omega^2)u_j = 0 \ ; \quad i,j = 1,2,3 \ ; \tag{1}$$

(N_{ij} and L_{ij} - are "bending" and "membrane" parts of the stiffness respectively), and considering a linear approximation of the influence of perturbations, it is possible to express the global stiffness matrix in finite element formulation (ref. 4):

$$G_{IJ} = G^0_{IJ} + \sum_{ij} G^r_{IJ,ij}\ r^\epsilon_{ij} + \sum_{kl} G^{bh}_{IJ,kl}\ h^\epsilon_{kl} \ ; \quad I,J = 1,...,M \ ; \tag{2}$$

$G^0_{IJ} = G^{m0}_{IJ} + G^{b0}_{IJ}$ is the stiffness matrix of the ideal structure composed of the "bending" and "membrane" parts.

$G^r_{IJ,ij} = G^m_{IJ,ij} + \Sigma_{kl}\, G^{br}_{IJ,kl,ij}\left(h^0_{kl}\right)^2$ is the stiffness matrix increment due to a "unit defect" of the shape at a node *i, j*; the "unit defect" of the shape is defined as the increment of the middle surface radius at node *i, j* equal to *one*, all the other nodes are in their positions in ideal shape of the structure. r^e_{ij}- is the defect magnitude of the radius at node *i, j*.

$G_{IJ,kl,ij}$ is the stiffness matrix increment of the element *k, l* due to a "unit defect" of the radius at node *i, j*.

Table 1. Natural frequencies and modes.

In the mode column the numerator means the number of waves about the periphery (n); the denominator means the number of nodes along the meridian (m); two values of n or m are given in case n or m alternates. Roman numerals in the Real structure column indicate the purity of excited mode (I. is the highest purity).

Real structure		Physical model		Theory with imperfections 100%		Theory with imperfections 50%		Theory with imperfections 25%		Theory with imperfections 0%		Coincidence with theory with imperfections 100%	
f (Hz)	Mode	f (Hz)	Mode	f (Hz)	Mode	f (Hz)	Mode	f (Hz)	Mode	f (Hz)	Mode	Real str.	Model
0.617	unkn.	0.798	4/1	0.983	4/1 1st	0.965	4/1 1st	0.957	4/1 1st	0.950	4/1 1st	-38%	-20%
0.965	unkn.	0.931	3/1	1.043	3/1 5th	1.032	3/1 5th	1.026	3/1 5th	1.022	3/1 5th	-7%	-10%
1.094 V.	3-4/2	1.355	4/2	1.171	4/2 7th	1.154	4/2 7th	1.148	4/2 7th	1.143	4/2 7th	-7%	+16%
1.500 II.	4/0-1	1.670	4/0-1	1.504	4/0 23rd	1.476	4/0 23rd	1.465	4/0 23rd	1.456	4/0 23rd	-0.3%	+11%
1.969 III.	3/2-3	2.081	3/2	1.815	3/2 34th	1.796	3/2 34th	1.788	3/2 34th	1.781	3/2 34th	+8%	+15%
2.340	unkn.	2.325	5/3	1.960	5/3 44th	1.881	5/3 40th	1.847	5/3 38th	1.827	5/3 38th	+19%	+22%
2.469 I.	4/3	2.638	4/3	2.432	4/3 59th	2.415	4/3 55th	2.408	4/3 55th	2.401	4/3 55th	+2%	+9%
2.728 IV.	2-3/3	2.915	1-2/2	2.770	1/1 75th	2.735	1/1 72th	2.713	1/1 72th	2.698	1/1 72th	+2%	+5%

$G^{bh}_{IJ,kl} = G^{b0}_{IJ,kl} . 2h^{0}_{kl}$ is the increment of the bending part of stiffness matrix due to a unit defect of the thickness of element *k, l*.

$G^{b0}_{IJ,kl}$ is the element of the bending part of stiffness matrix of element *k,l* for the ideal shape of the shell, added to elements *I, J* of the global stiffness matrix according to incidental table codes.

h^{0}_{kl}, h^{e}_{kl} is the ideal thickness of the element *k,l,* and its increment respectively.

$$\sum_{ij} \equiv \sum_{i=1}^{\mu} \sum_{j=1}^{\vartheta} ; \quad \sum_{kl} \equiv \sum_{k=1}^{m} \sum_{l=1}^{n} ;$$

μ and v are the number of nodes in the direction of the axis of rotation, and in the direction of the parallel of the surface of revolution respectively;
m, and *n* are the number of elements in these directions.
$I, J = 1, ..., M; \; M = N\mu v;$ N is the number of nodal degrees of freedom.

14. The natural vibration problem of the generally shaped shell structure solved by the finite element method leads to the following system of homogenous algebraic equations:

$$\sum_{J} G_{IJ}\, u_{J(s)} + (\mathrm{i}\omega_{(s)} B_{IJ} - \omega^{2}_{(s)} M_{IJ})\, u_{J(s)} = 0 \; ; \quad \sum_{J} \equiv \sum_{J=1}^{M} \; ; \tag{3}$$

G_{IJ} are elements of the global stiffness matrix, Eq. 2.
B_{IJ}, M_{IJ} are elements of the global damping matrix and mass matrix respectively.
$\omega_{(s)}$ is the s-th natural circular frequency (complex number).
Nodal point displacements in the s-th natural mode $u_{J(s)}$ may be expressed in the form of canonical decompositions (ref. 3):

$$u_{J(s)} = u^{0}_{J(s)} + \sum_{ij} u^{r}_{J,ij(s)}\, r^{\epsilon}_{ij} + \sum_{kl} u^{bh}_{J,kl(s)}\, h^{\epsilon}_{kl} \; ; \tag{4}$$

$u^{r}_{J,ij(s)}, u^{bh}_{J,kl(s)}$ are canonical coordinates corresponding to particular imperfections; they are the solution of the system, Eq. 8.
$u^{0}_{J(s)}$ is the response (natural mode) of the shell of ideal shape and wall thickness.
Similarly it is possible to write $\omega_{(s)}$:

$$\omega_{(s)} = \omega^{0}_{(s)} + \sum_{ij} \omega^{r}_{ij(s)} r^{\epsilon}_{ij} + \sum_{kl} \omega^{bh}_{kl(s)} h^{\epsilon}_{kl} \; . \tag{5}$$

Supposing r_{ij}, h_{kl} as independent and centered processes it can be proved that (ref. 4):

$$\sum_{J} [G^{0}_{IJ} + \mathrm{i}\omega^{0}_{(s)} B_{IJ} - (\omega^{0}_{(s)})^{2} M_{IJ}]\; u^{0}_{J(s)} = 0 \; ; \tag{6}$$

that is a problem of natural frequencies and modes of the ideal structure. Further it can be proved that the corrections (canonical coordinates) of natural frequencies may be evaluated from:

$$\omega^{\lambda}_{pq(s)} = \frac{-\sum_I \sum_J u^0_{I(s)}\ G^{\lambda}_{IJ,pq}\ u^0_{J(s)}}{\sum_I \sum_J\ u^0_{I(s)}(\mathrm{i}B_{IJ} - 2\omega^0_{(s)} M_{IJ})\ u^0_{J(s)}} \tag{7}$$

where either $l = r;\ \ p, q = i, j\ \ or\ \ l = bh;\ \ p, q = k, l.$
Canonical coordinates in summations of the formula in Eq. 4 are given by the systems of linear algebraic equations ($s \neq t$):

$$\sum_I \sum_J u^0_{I(t)}\ (\frac{B_{IJ}}{\mathrm{i}(\omega^0_{(s)} + \omega^0_{(t)})} + M_{IJ})\ u^{\lambda}_{J,pq(s)} =$$

$$= \sum_I \sum_J \frac{u^0_{I(t)} u^0_{J(s)}}{(\omega^0_{(s)})^2 - (\omega^0_{(t)})^2}\ [G^{\lambda}_{IJ,pq} + (\mathrm{i}B_{IJ} - 2\omega^0_{(s)} M_{IJ})\ \omega^{\lambda}_{pq(s)}]\ . \tag{8}$$

These systems always give M-1 equations for M-1 unknowns with the assumption of a negligible influence of damping; $s = t$ unknowns may be chosen:

$$u^r_{s,ij(s)} = u^{bh}_{s,kl(s)} = 0\ . \tag{9}$$

The matrix of the system in Eq. 8 is common for both series of canonical coordinates independent of s. This is a good feature from the conditionality point of view.

15. From Eqs 4 and 5 it is possible to derive (ref. 4) the mathematical mean value of natural frequencies and natural modes which is equal to values corresponding to the ideal structure. From these expressions and from Eqs 7-9 we can obtain the formulae for correlation of natural mode deviations:

$$R^{uu}_{IJ(s)(t)} = \sum_{ij} u^r_{I,ij(s)} u^r_{J,ij(t)} R^{rr}_{ij,ij} + \sum_{kl} u^{bh}_{I,kl(s)} u^{bh}_{J,kl(t)} R^{hh}_{kl,kl}\ ; \tag{10}$$

and for correlation of natural frequency deviation:

$$R^{\omega\omega}_{(s)(t)} = \sum_{ij} \omega^r_{ij(s)} \omega^r_{ij(t)} R^{rr}_{ij,ij} + \sum_{kl} \omega^{bh}_{kl(s)} \omega^{bh}_{kl(t)} R^{hh}_{kl,kl}\ . \tag{11}$$

In Eqs 10 and 11 $R^{rr}_{ij,ij}$, $R^{hh}_{kl,kl}$ are considered as known input data, measured on real structures or their models following evaluation by methods of mathematical statistics.

16. Considering the linear problem and input data of a Gauss character, the response is again given by Gauss processes. These processes are fully characterised by the mean value (as is given by solution of the problem in ideal shape of the shell, given by Eq. 6), and by correlations given by Eqs 10 and 11. This answers the basic question of the influence of random deviations of the shell shape and thickness on its dynamic characteristics.

CONCLUSIONS

17. Comparison of natural modes and frequencies of a cooling tower obtained from both experimental and theoretical methods shows the possibility of identifying particular types of defects in the shape of this structure. It is necessary to use a number of experimental and theoretical methods in order to obtain reliable results. The combination of the finite element calculations, based on detailed input data of a real structure, together with *in-situ* measurements and with analytical-numerical procedures, based on methods of stochastic mechanics, are presented in this paper. By means of all these methods it is possible to obtain global information regarding the influence of particular types of defects on the dynamic characteristics of the structure. It is obvious that using repeated measurements which can be simplified in the next stages, presents a possibility of determining the onset and evolution of defects.

REFERENCES

1. BATHE H. J., WILSON E. L. and PETERSON F. E.: SAP-IV; A structural analyses program for static and dynamic response of linear systems. Berkely, College of Engineering 1973.

2. GOLDENVEISEŘ A. G. *et al*: Free vibrations of thin-walled structures (in Russian). Nauka, Moscow 1979.

3. PUGACHEV V. S.: Introduction into Theory of Probability (in Russian). Nauka, Moscow 1968.

4. PIRNER M., NÁPRSTEK J. and HORYNA T. Influence of random and intended imperfections on dynamical properties of cooling towers. W. I. T. - Proc. of the conference "Comp. Methods and Exper. Measurements", to be held: University of Siena (Italy), May 1993.

59. Lack of fit as a design parameter of double-layer grids

U. R. MADI, University of Jordan

SYNOPSIS. The idea of incorporating lack of fit in selected members of double-layer grids as a design parameter of these grids with the aims of improving their behaviour under load and achieving more economy in their design is investigated. A simple procedure for determining to which members lack of fit is best applied and to what magnitude, is developed and presented. The whole presentation is supplemented by an example and aims to provide a useful means of improving practical designs of double-layer grids.

INTRODUCTION

1. The possibility of introducing favourable prestress in the critical members of double-layer grids by prestraining some selected members was observed some time ago (refs 1, 2).

2. It is well known that random imperfections represented by random lack of fit of the members of double-layer grids is responsible for a significant portion of the forces that exist in practical structures of this type. This is largely due to the high statical indeterminacy or redundancy characteristic of this structural form.

3. Therefore, manipulating this lack of fit to eliminate the possible unfavourable effects of its unavoidable random nature or even use it as a design tool that can achieve improved response under load and more economy in design is an appealing proposition. This idea has been probed with reported success in some previous studies (refs 3-5).

4. Furthermore, the idea of increasing or decreasing the intended lengths of some selected members of double-layer grids to improve on their qualities is not new. It has been used in many practical structures as a means of enhancing their drainage (ref. 6).

5. It is the aim of the present work to investigate the validity of this idea and attempt to develop easy to use design techniques that would make it handy for application by designers of practical double-layer grids.

LACK OF FIT AS A DESIGN PARAMETER

Double-Layer Grids Without LOF:

6. The statics and kinematics of space trusses can be presented in the alternative mesh and nodal descriptions (ref. 7) as follows:

STATICS KINEMATICS

$$P = [\,B : B_o\,]\begin{bmatrix} R \\ \lambda \end{bmatrix} \qquad \begin{bmatrix} V = o \\ \delta \end{bmatrix} \begin{bmatrix} B^T \\ B_o^T \end{bmatrix} U \qquad (1)$$

Mesh Description

$$Q = o = \begin{bmatrix} A^T & : & A_o^T \end{bmatrix} \begin{bmatrix} P \\ \lambda \end{bmatrix} \qquad \begin{bmatrix} U \\ \delta \end{bmatrix} = \begin{bmatrix} A \\ A_o \end{bmatrix} q \qquad (2)$$

Nodal Description

7. Where P represents the member forces, U the axial member deformations, λ the independent load sets and δ the deflections associated with the applied loads.

8. In the mesh description, R are the indeterminate or hyper-static forces and V represent the dual discontinuities at releases corresponding to R. In the nodal description, q are the hyper-kinematic displacements and Q are the dual constraint forces at these displacements. B, B_o, A and A_o are the mesh and nodal description matrices representing the equilibrium and compatibility relations. The elastic constitutive relations can be presented in the alternative stiffness and flexibility formulations as:

Flexibility

$U = F P$

Stiffness

$P = K U$

Assuming loads are applied only at the nodes with no prestressing present in the members and combining the constitutive relations with the static and kinematic descriptions results in the four alternative analysis formulations: the nodal stiffness, the nodal flexibility, the mesh stiffness and the mesh flexibility.

9. It is easy to show (ref. 8) that out of these four formulation, the nodal stiffness and the mesh flexibility always emerge as favourite in terms of ease of automation and least computational effort. These two formulations can be presented as:

$$(A^T K A) g = A_o^T \lambda \qquad (4)$$

Nodal Stiffness

or

$$(\underset{\sim}{B}^T \underset{\sim}{F} \underset{\sim}{B}) \underset{\sim}{R} + (\underset{\sim}{B}^T \underset{\sim}{F} \underset{\sim}{B}_o) \underset{\sim}{\lambda} = \underset{\sim}{o} \tag{5}$$

Mesh Flexibility

10. Double-Layer Grids with LOF: Lack of fit does not affect the behaviour of statically determinate structures. Therefore, lack of fit can only be applied meaningfully to a number of members of a space truss equal to its statical indeterminacy number α being defined as

$$\alpha = m + c - 3\,n \tag{6}$$

where n in the number of nodes, m is the number of members and c is the number of constrained degrees of freedom. However, it must be noted that these members must be chosen in such a manner that their removal from the structure does not affect its stability. i.e. the reduced structure must have a positive definite static matrix.

11. Introducing lack of fit into α appropriately selected members results in the following static and kinematic relations using the mesh and nodal descriptions:

STATICS KINEMATICS

$$\underset{\sim}{P} = \begin{bmatrix} \underset{\sim}{B} & : & \underset{\sim}{B}_o \end{bmatrix} \begin{bmatrix} \underset{\sim}{R} \\ \underset{\sim}{\lambda} \end{bmatrix} \quad \begin{bmatrix} \underset{\sim}{V} = -\underset{\sim}{U}_o \\ \underset{\sim}{\delta} \end{bmatrix} = \begin{bmatrix} \underset{\sim}{B}^T \\ \underset{\sim}{B}_o^T \end{bmatrix} \underset{\sim}{U} \ldots (7)$$

Mesh Description

$$\underset{\sim}{Q} = \underset{\sim}{o} = \begin{bmatrix} \underset{\sim}{A}^T & : & \underset{\sim}{A}_o^T \end{bmatrix} \begin{bmatrix} \underset{\sim}{P} \\ \underset{\sim}{\lambda} \end{bmatrix} \quad \begin{bmatrix} \underset{\sim}{U} \\ \underset{\sim}{\delta} \end{bmatrix} = \begin{bmatrix} \underset{\sim}{Q} \\ \underset{\sim}{A}_o \end{bmatrix} \underset{\sim}{q} \ldots (8)$$

Nodal Description

where $\underset{\sim}{U}_o$ are the imposed lack of fit in the α redundant members.

12. The elastic constitutive material relationships now become:

Flexibility Stiffness

$$\underset{\sim}{U} = \underset{\sim}{F} \underset{\sim}{P} + \underset{\sim}{\overline{U}}_o \qquad \underset{\sim}{F} = \underset{\sim}{K}(-\underset{\sim}{\overline{U}}_o + \underset{\sim}{U}) \tag{9}$$

where

$$\underset{\sim}{\overline{U}}_o = \begin{matrix} m-\alpha \\ \alpha \end{matrix} \begin{bmatrix} \underset{\sim}{o} \\ \hline \underset{\sim}{U}_o \end{bmatrix}_{1} \tag{10}$$

as such $\underset{\sim}{U}_o$ represents the lack of fit in all members ($\underset{\sim}{o}$ in members of reduced structure and $\underset{\sim}{U}_o$ in redundant members).
As such the analysis formulations become,

$$(\underset{\sim}{A}^T \underset{\sim}{K} \underset{\sim}{A}) \; \underset{\sim}{q} = \underset{\sim}{A}_o^T \underset{\sim}{\lambda} + \underset{\sim}{A}^T \underset{\sim}{K} \underset{\sim}{U}_o \quad \text{Nodal Stiffness} \quad \dots\dots (11)$$

$$(\underset{\sim}{B}^T \underset{\sim}{F} \underset{\sim}{B}) \underset{\sim}{R} + (\underset{\sim}{B}^T \underset{\sim}{F} \underset{\sim}{B}_o) \underset{\sim}{\lambda} + \underset{\sim}{B}^T \underset{\sim}{\bar{U}}_o + \underset{\sim}{U}_o = \underset{\sim}{o} \quad \text{Mesh Flexibility} \quad \dots\dots (12)$$

13. <u>Formulation Suitable for LOF</u>: Since LOF is applied to statically redundant members, the mesh flexibility formulation seems a more suitable formulation for representing space trusses with LOF. However, for the purposes of design, this formulation needs to be adapted so that member forces have two components: one due to applied loading and the other to LOF so that:

$$\underset{\sim}{P} = \underset{\sim}{B}_1 \underset{\sim}{\lambda} + \underset{\sim}{B}_2 \underset{\sim}{U}_0 \qquad (13)$$

needless to say $\underset{\sim}{B}_1 \underset{\sim}{\lambda}$ in this case replaces $\underset{\sim}{B}\underset{\sim}{R} + \underset{\sim}{B}_o \underset{\sim}{\lambda}$ in the usual formulation.

14. It can further be argued that normally space trusses are designed for a proportional loading system of variable intensity. For example dead load and/or live load is applied as uniformly distributed on the structure as such leading to point loads of fixed ratios depending on corresponding loaded area being applied at selected nodes of the structure. This suggests the following formulation:

$$\underset{\sim}{P} = \underset{\sim}{b}_1 \lambda + \underset{\sim}{B}_2 \underset{\sim}{U}_0 \qquad (14)$$

where λ is a scalar representing the load intensity of the proportional loading system.

15. Matrices $\underset{\sim}{b}_1$ and $\underset{\sim}{B}_2$ can be developed from first principles based on the geometry, topology and member properties of the structure and using software especially developed for this purpose. Or alternatively, any standard linear elastic analysis software package can be used for this purpose. As such the structure will first have to be analysed for a unit proportional loading system thus producing $\underset{\sim}{b}_1$ as the member forces under this loading. $\underset{\sim}{B}_2$ has then to be produced either by a direct solution for member forces under simultaneous application of end forces equivalent to unit LOF at the redundant members or on a column by column basis for unit LOF applied at one redundant member at a time.

16. <u>Design with LOF</u>: In designing DLG's with LOF, two questions have to be answered: To which members is LOF best applied and at what magnitude.

17. As to which members, a large number of designs with LOF applied to different members each time has shown that the choice of members to apply LOF

to, has little effect on the desired improvement. It seems that the additional load carrying capacity of the structure achieved by LOF is little affected by the choice of members. Almost identical improvement in load carrying capacity was achieved regardless of the chosen members to apply LOF to. However, it is vital that the choice of these members is done in such a manner to preclude inducing any instability. The reduced structure must remain stable and free from mechanisms. The choice of members can be done automatically if so desired by a node-mesh transform (ref. 9) that ensures the stability of the reduced structure through the non-singularity of the resulting mesh matrix. It has also been observed that locating some of the members with LOF in the vicinity of the most heavily stressed members in the structure tends to reduce the magnitudes of the necessary LOF.

18. As to the required amounts of LOF, they are to be calculated using a linear mathematical program. In this linear program, the load capacity of the structure is maximised subject to member forces not exceeding their allowable limits under two load combinations:

-Own Weight plus LOF, and

-Total load plus LOF.

The resulting linear program can thus be presented as:

$$
\begin{array}{cc}
 & \{\, 1 \quad \underset{\sim}{o}^{T} \,\} \begin{bmatrix} \lambda \\ \underset{\sim}{U}_{o} \end{bmatrix} \leq \text{Max } Z \\
\begin{array}{c} m \\ m \\ m \\ m \end{array} &
\begin{bmatrix} \underset{\sim}{b}_{1} & \underset{\sim}{B}_{2} \\ -\underset{\sim}{b}_{1} & \underset{\sim}{B}_{2} \\ \underset{\sim}{o} & \underset{\sim}{B}_{2} \\ \underset{\sim}{o} & -\underset{\sim}{B}_{2} \end{bmatrix}
\begin{array}{c} \leq \underset{\sim}{P}^{+}_{all} \\ \leq \underset{\sim}{P}^{-}_{all} \\ \leq \underset{\sim}{P}^{+}_{all,m} \\ \leq \underset{\sim}{P}^{-}_{all,m} \end{array} \\
 & \quad 1 \qquad \alpha \\
 & \lambda \gtrless o \;, \quad \underset{\sim}{U}_{o} \gtrless \underset{\sim}{o}
\end{array}
\qquad (15)
$$

Where $\underset{\sim}{P}^{+}_{all}$, $\underset{\sim}{P}^{-}_{all}$ are the allowable member forces in tension and compression respectively and $\underset{\sim}{P}^{+}_{all,m}$ and $\underset{\sim}{P}^{-}_{all,m}$ are the allowable member forces in tension and compression modified by the resulting member forces under own weight.

19. Specially developed software based on the Simplex Algorithm (ref. 9) can be used to solve this program or a variant from a standard spread sheet program can be used for this purpose. Obviously, not all the constraints need to be kept. Members with compressive forces without LOF need not be checked for allowable tensile forces and vice versa. The load parameter which is unrestricted

in sign can be maintained in this state as it will obviously acquire a positive value when it is being maximised. This does not apply to the LOF vector $\underset{\sim}{U}_o$ that needs to be replaced by two vectors one for elongation and another for shortening ($\underset{\sim}{U}_o = \underset{\sim}{U}_o^+ - \underset{\sim}{U}_o^-$) following the well-known simplex algorithm procedure.

20. The result will be the maximum load capacity λ together with $\alpha + 1$ basic variables. These basic variables will give the amounts of LOF to be applied at the selected members plus the complement to the most heavily stressed member which will signal that it has reached its allowable force value. If there are more than one member that reach their allowable forces, simultaneously, then one of these is selected automatically by the program.

21. The allowable member forces can be determined by a selection of member sections through a design of the DLG without LOF for total load (dead plus live). In this case, the aim of introducing LOF would be an improved structural response under load together with an increase in allowable load carrying capacity. If more economy is desired, then the DLG can be designed for a fraction of the total load without LOF and then it should be checked that the allowable load capacity with LOF is greater than the total load level. If this approach is followed, then substantial economy can result as it will be shown that increases of up to 71% in total allowable load capacity have been observed when introducing LOF.

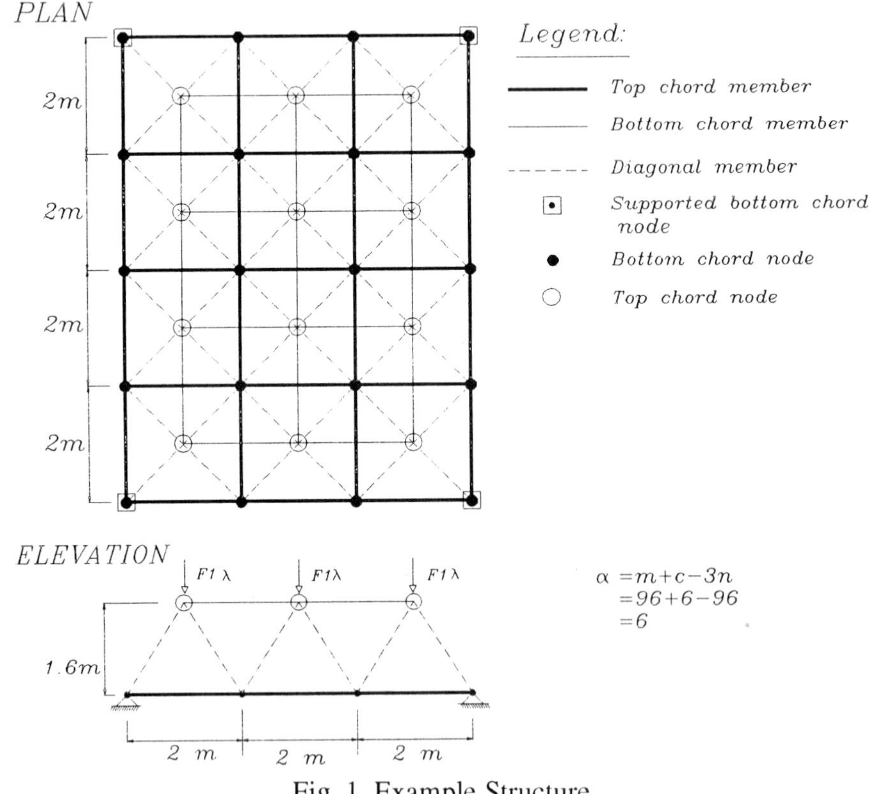

Fig. 1. Example Structure.

22. Example Structure: To demonstrate the workings of the previous discussions, the structure shown in Fig. 1, is selected to serve as an example.

23. This structure was designed for LOF applied to various selections of member groups. All groups of members share the common characteristic that their simultaneous removal from the structure results in a stable reduced structure. The resulting improvement to the allowable load capacity of the structure for all these designs was in the range of 71%. This demonstrates the arguments stated earlier that substantial improvement occurs with LOF and that this improvement is not sensitive to the choice of members to which LOF is applied.

24. Symmetry: It is to be noted that although the DLG may by symmetric, LOF need not be applied in a symmetric manner. It is also to be noted that the statical indeterminacy number α may change if parts of the structure are to be analysed and/or designed by virtue of symmetry and its associated additional constraints. This does not imply that in this case additional LOF can be applied as any LOF applied to members more than the real statical indeterminacy number will not have any effect on the force distribution in the DLG.

25. REMARKS AND CONCLUSIONS

The following remarks are deemed appropriate:

(a) LOF has been shown to have the potential of being used as a design parameter of practical double-layer grids.

(b) When used as a design parameters, LOF is not only capable of offsetting any undesirable effects that may accrue from its unfavourable random nature; but is capable of enhancing the behaviour of DLG's under load by reducing the possibility of any undesirable member instability under large compressive forces and increasing substantially the load carrying capacity.

(c) Substantial economy in the design of DLG's can result if LOF is applied effectively and in an efficient manner.

(d) Standard computer packages for the linear elastic analysis of structures together with standard spread sheet packages can be used to design DLG's with LOF. Alternatively specially developed software can be used for this purpose.

REFERENCES

1. HOFMEISTER, L.D and FELTON, L.P. "Prestressing in Structural Synthesis", AIAAJ., 8(2), Feb. 1970, pp 663-6.

2. HOLNICKI-SZULC, J., "Prestressing of Truss and Frame Structures", ASCE. J, Structural Division, 105 (ST3), March, 1979, pp 601-16.

3. SPILLERS, W.R and LEVY, R. "Truss Design: Two Loading Conditions with Prestress", ASCE J., Structural Engineering, 110 (4), April 1984, pp 667-87.

4. HANAOR, A., "Imposed Lack of Fit as a Means of Enhancing Space Truss Design", Space Structures, 1 (3), 1985, pp 147-154.

5. HANAOR, A., MARSH, C. and PARKE, G. "Modification of Behaviour of Double-Layer Grids: Overview", ASCE J., Structural Engineering, 115 (5), May 1989, pp 1021-1037.

6. MAKOWSKI, Z.S (Ed), "Analysis, Design and Construction of Double-Layer Grids", Elsevier, 1981.
7. MADI, U.R., "The Elastoplastic Analysis of Pin Jointed Spatial Structures", Space Structures, 1(4), 1985, pp 209 - 221.
8. MADI, U.R, "The Wolfe-Markowitz Algorithm Technique for the Nonholonomic Elastoplastic Analysis of Pin-Jointed Spatial Structures, Dirasat, University of Jordan, Volume XIII, No. 10, 1986, pp 91-107.
9. SMITH, D. L. "Plastic Limit Analysis and Synthesis of Structures by Linear Programming" PhD thesis, University of London, 1974.
10. DANTIZ, G, "Linear Programming and Extensions", Princeton, 1963.

60. Double-layer space structures of laminated timber tubular members

J. ESTEVEZ, J. PABLOS, S. MUNIZ, M. FREIRE, R. VAZQUEZ
and J. ALVAREZ, Universidad de la Coruna, Spain

SYNOPSIS. A new structural solution is described which combines the features of double-layer space trusses with the new use of laminated timber tubular members. This paper describe the spatial geometry of bar systems based on modularisation using repetitive elements which lead to a generalisation of complex geometrical sets. Shown as well, are those hollow sections considered and the typology of the nodes. The results available from testing clearly show the possibilities of this new structural concept, not only from the structural point of view, but also from an architectural perspective, which will open up a whole field for future structures.

INTRODUCTION

1. The present paper deals with the proposal of a new structural system to provide large span roofs, combining the performance of double-layer spatial truss typology with the use of hollow section bars of glued laminated timber. Solid sections, whether of laminated or solid wood, have already been used in spatial structures of different typology but rarely so in space trusses, the construction of which is largely monopolised by the use of metallic elements.

2. The use of solid timber sections in space structures presents serious disadvantages.

a) Firstly, the use of transverse sections of reduced dimensions considerably limits the length of the bars because their slenderness must be kept within reasonable limits. Care must also be taken so as not to penalise the structural performance of the compression members, which may cause the inefficient use of an increase in the number of nodes required for the structure, thus increasing both costs and complexity of construction.

b) Secondly, if longer members are to be used, the requirements of slenderness will imply a cross-section of considerable size and, therefore, require the use of laminated timber. In this case, the change in humidity levels within the section itself will certainly pose problems of dimensional variations.

3. Consequently, the use of laminated hollow sections will give the advantages inherent to laminated timber and provide members with a considerable cross-

section, thus minimising the disadvantages of this material. On the other hand, sections with a large radius of gyration can be obtained using small amounts of material with small thickness, therefore achieving a high performance circumventing problems due to changing humidity levels. At the same time, treatment of the timber against biological decay becomes considerably easier to implement.

GEOMETRY OF THE STRUCTURE

4. The complexity of the geometry of these structures, with a large number of nodes and members as well as the need for high precision in the dimensional control of the components of the truss, required the development of software which would allow the automatic deployment of the structures and the use of interchanging files for their manipulation within a CAD environment. The authors are at present working on the integral connection of a CAD graphic design environment with an analysis program, which is now in an advanced stage of development.

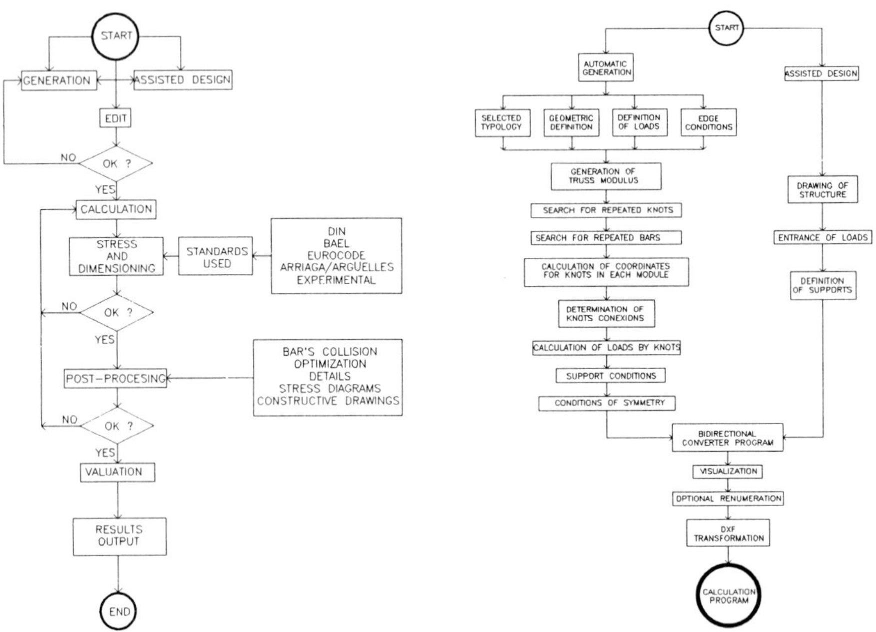

Fig. 1. General flow diagram for the software.

Fig. 2. Flow diagram of the generation program.

5. In the flow diagram shown in Fig. 1 the organisational hierarchy of the developed software is given which consists of three fundamental phases:

a) Generation and printing of the spatial truss. This allows for automatic generation and its manipulation in a CAD environment.
b) Calculation and dimensioning of the structure.
c) Post-processor requirements.

6. In this paper reference to the first phase, generation and printing are described. These generators allow the structure be flat, cylindrical or spherical, and include support conditions, conditions of symmetry and the automatic calculation of nodal loads.

7. In addition to the previous process shown in the flow diagram, the procedure can be used to design either the total structure or used for the modification of a structure already generated. This allows complete freedom in the choice of typology, permits openings in the structure, fitting to irregular borders, allows different supports conditions, special loading, and members constructed from different materials etc. To be certain of dealing in an environment which can generate design data in a totally graphic way, the basic modules considered in the automatic generation are as follows:

1. Semioctahedron, square-on-square space grid, mansard border.
2. Semioctahedron, square-on-square space grid, cornice border.
3. Semioctahedron, square-on-square space grid, upper layer set diagonally.
4. Semioctahedron, square-on-square space grid, lower layer set diagonally.
5. Prismatic modules, upper layer set diagonally.
6. Prismatic modules, lower layer set diagonally.
7. Tetrahedron, three-way space grid, mansard border.
8. Tetrahedron, three-way space grid, cornice border.

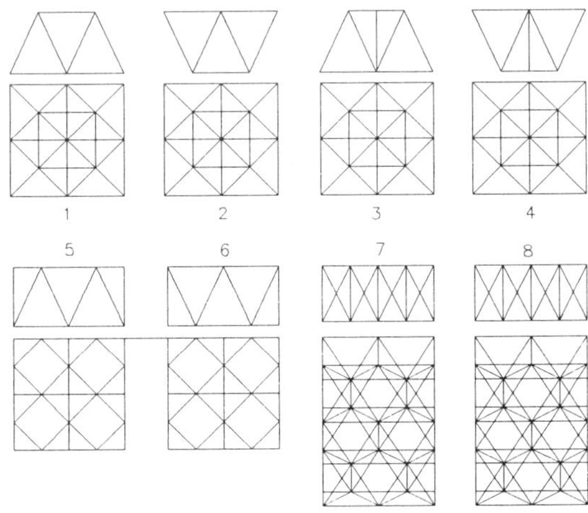

Fig. 3. Topologies for automatically generated trusses.

9. Before starting the analysis, the program examines the structure in order to eliminate repeated bars and nodes. It then undergoes renumbering of the nodes and members using the Cuthill-McKee method, searching to minimise the band width, thus increasing the speed of the remaining process.

10. A graphic visualiser is also available which shows in plan, elevation and perspective, the generated truss as well as enlargements of specific areas together with the numbering scheme. To exemplify this, two types of trusses generated by the program are shown in Figs 4 and 5.

11. After generating the structure topology the stiffness matrix is assembled, the equilibrium equations solved and the structure preliminary designed. This process is developed in an iterative form until the pre-set conditions of strength, stress and deformation are met. The types of members available leads to strict dimensioning of the structure with maximum utilisation of material and a minimum number of different member types used.

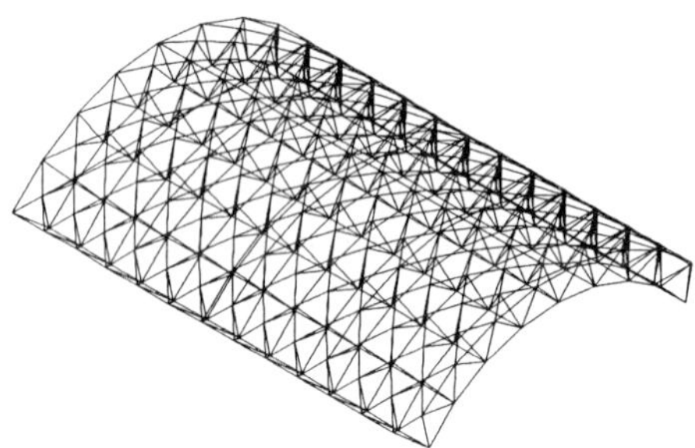

Fig. 4. Cylindrical truss, prismatic modulus.

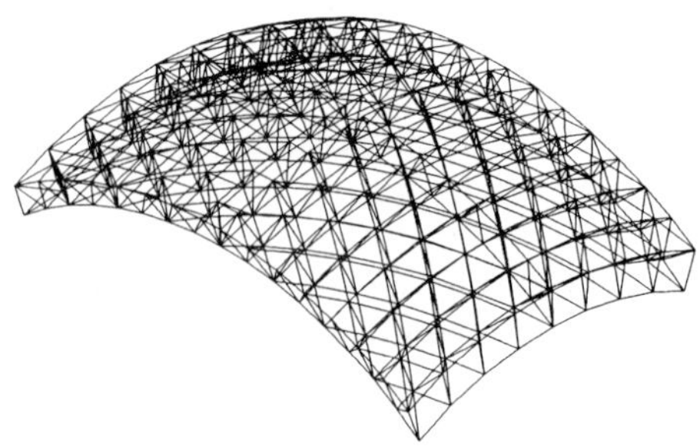

Fig. 5. Spherical tetrahedron truss.

DESCRIPTION OF COMPONENTS

Members

12. One of the aims considered important in the design of the members was to permit these to be easily made in the workshop at a reasonable cost and with dependable and adequate mechanical characteristics. These reasons led to the rejection of early solutions such as laminating hollow tubular sections obtained by rolling around a cylindrical matrix, opting finally for the proposal described in Fig. 6. Nevertheless, at the present time other variations are being studied which use boards as well as plywood.

13. The process of making the members is simplified enormously because the material used has a width four times the thickness of the wall of the hollow section. Before gluing the laminated timber, diecast anchoring plates are fixed into the member ends, and once the unit is glued together, the members are joined to the node with a simple fixed key.

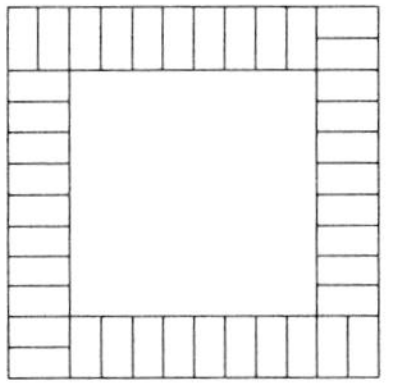
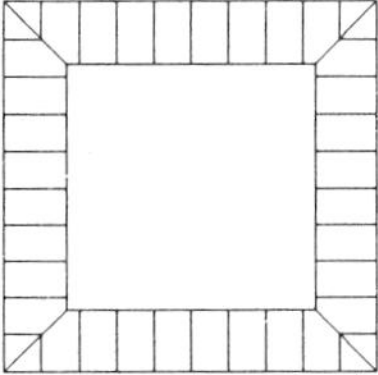

Fig. 6. Transverse sections through members.

Nodes

14. It is common knowledge that one of the greatest problems in timber construction is to guarantee the efficient transmission of forces, especially when dealing with tension forces. Therefore the design of the nodes where the truss members meet becomes very important. Having resolved this, a series of essential design requirements should be satisfied simultaneously. These are as follows:

a) To allow the efficient transmission of forces, both compression and tension.

b) To permit any member configuration, allowing the construction of single and double-layer trusses with single or double curvature.

c) To permit the construction of a simple structure easily integrated into an industrial process, even within a CAD-CAM system.

d) To allow simple erection and dismantling of the structure which would avoid welding on site.

e) To provide the possibility of adjustments which would absorb any imperfections in construction which may be significant in timber structures.

f) To allow for the ventilation inside the timber members.

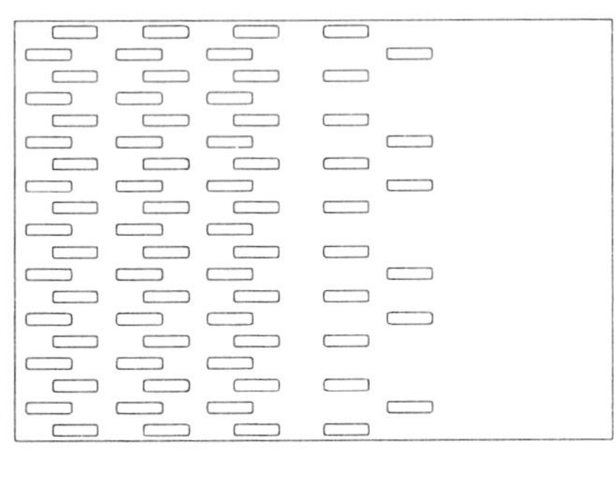
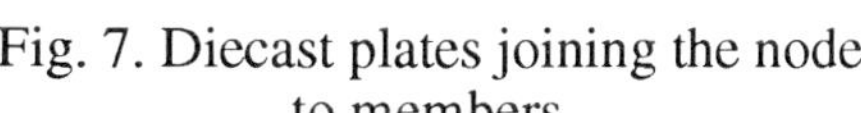

Fig. 7. Diecast plates joining the node to members.

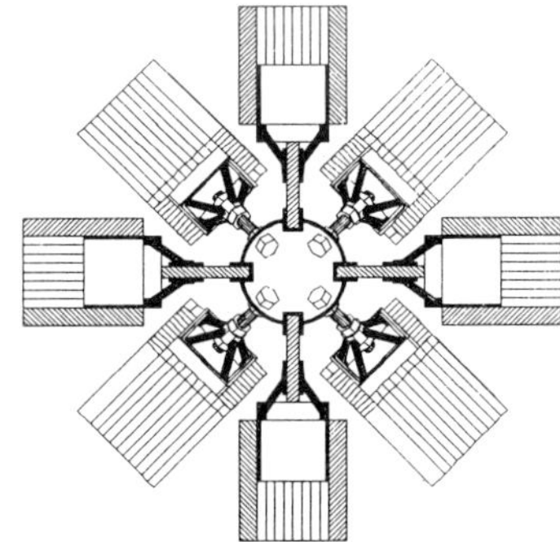

Fig. 8. A typical node.

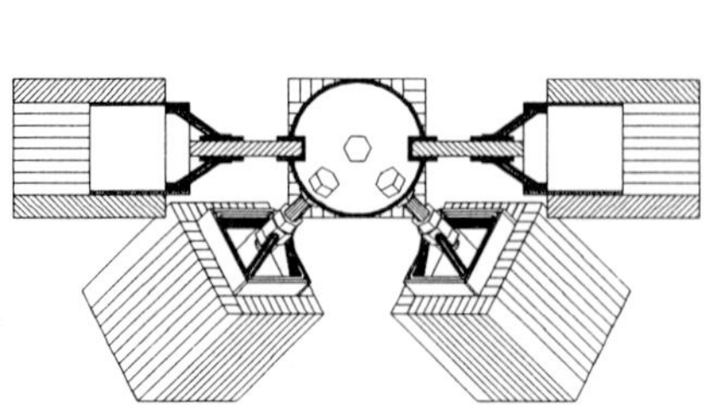

Fig. 9. Vertical section through the nodes

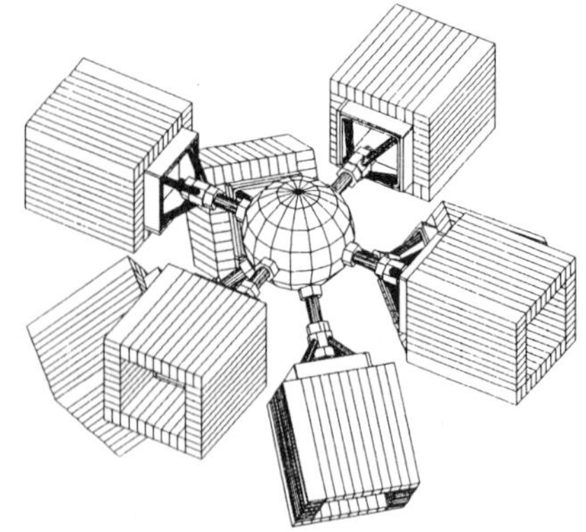

Fig. 10. A perspective view of the node.

Fig. 11. Photograph of the prototype.

15. Studies undertaken have resulted in the node which is shown in Figs 8 to 10 and in the photograph of the prototype (Figs 11-12). The node is at present in the process of being patented. It can be seen that the node consists of a hollow sphere with a screw-on top that allows access to the inside. This sphere is manufactured by machining and it incorporates non-threaded drilled holes in order to let the threaded rods pass through, which are then fixed into the inside of the sphere with screws. On the opposite end, the threaded bar is screwed into an octagonal nut onto which four solid round bars, which follow the edges of an octahedron, are welded. In order to joint the node to the wood, diecast plates with a special design based on Mitek Industries S.A., are fixed onto the inside of the timber member (Fig. 7). The proposed node, as shown, complies with all the requirements previously listed and the results of the analysis undertaken at present, demonstrate very satisfactory behaviour.

EXPERIMENTAL ANALYSIS

16. The proposals devised here have demonstrated that it was necessary to have an experimental basis in order to compare the designs carried out and direct the following research direction. For this purpose a preliminary test program was conducted, with the aim of determining the general performance of the most appropriate parts for future examination.

17. For this experimental investigation both solid and hollow sections tubes were devised. The dimensions of tubes established by using the Spanish regulation UNE 56-535-77 was not considered adequate for the type of sections designed, resulting in reference to the Eurocode to assess the requirements for the experimental investigation.

18. The solid tubes had a square cross-section, 6 x 6cms and were 36cm long, made from European White Wood (Piece Abies). This length is the maximum which can be obtained without requiring lateral restraints to prevent buckling, which would substantially complicate the experimental programme. At the same time, hollow bars of 6 x 6cm and 1cm thick walls were used in order to compare the performance of the two member types in compression. Gluing the corners of the hollow tubes was achieved by alternating the sides of the section and for some cases by dovetailing the edges at 45° to increase the area to be glued (Fig. 6).

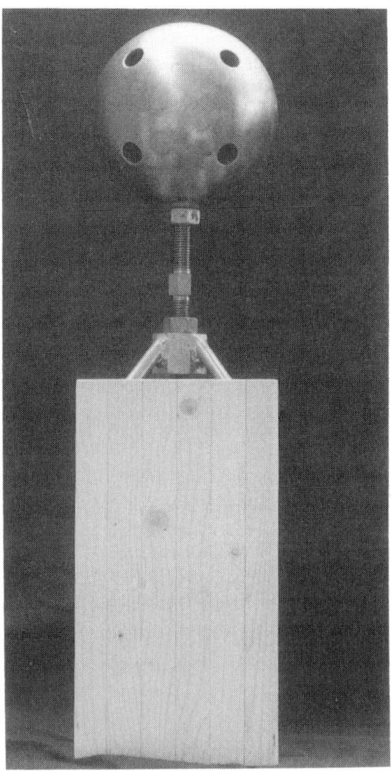

Fig. 12. Photograph of the prototype.

19. With the intention of studying members with dimensions as close as possible to members in a full size structure, six hollow trial tubes were constructed with the same cross-sectional dimension as those given previously but with a length of 140cm which corresponds to the maximum available clearance in the test machine.

Conditions for the experiments

20. The trials have been carried out using a model M-4 Suzpecar machine. The loading speeds have been established according to the regulation CEN-TC124 with a value of 5×10^{-5} L; L being the length of the trial tube in mm, which leads

to a load rate of 100kg/s. The failure sensitivity was established for all the test members at a laboratory temperature of 20°C.

21. The members were loaded as pin-ended tubes using a spherical ball and socket joint, especially designed for this purpose.

Characteristics of the test members

22. The following are the characteristics corresponding to the hollow test members used in the experimental investigation. Twelve members were constructed, six dovetailed (type A) and the other six alternating sides (type B). Of the twelve test members six were 140mm long, with a slenderness of 66 and the remaining six members were 362mm long, which is the maximum length that can achieved without lateral restraints. Table 1, summarises the member test data.

Table 1. Member test data.

	Piece 1A	Piece 2A	Piece 3A	Piece 4B	Piece 5B	Piece 6B
n°pieces	2	2	2	2	2	2.
h	1400 & 362	1400 & 362	1400 & 362	1400 & 362	1400 & 362	1400 & 362
b*d	61.2*60.6	61.8*61.8	61.2*61.1	60.6*60.2	60.8*59.4	60.9*60.9
e	10.20	10.10	10.05	9.35	9.70	9.32
A	2068.6	2088.7	2054.2	1909.3	1955.5	1922.9
H	13.8	13.9	13.9	14.7	14.4	14.8

h = length of the trial tube in mm.
e = average thickness of the wall in mm.

b x d is the dimension of the section in mm.
A = area of the section in mm^2.
H = Average humidity of the trial tube in %.

Results of tests

23. As an example the results corresponding to two test members, one dovetailed and the other with alternating sides on the edges are presented in Figs 13 and 14.

CONCLUSIONS

24. The studies undertaken concerning the structural proposal put forward in this paper, point towards the great architectural and structural possibilities for the use of hollow sections in the construction of space trusses. On one hand, the tubular members present a feasible solution from a technical point of view, easily applicable to an industrial process, with durable performance, together with the advantages which the use of reduced thickness offers towards the integrated use of wood protected against biological attack. Similarly, the proposed nodes permit the efficient transmission of forces between members of the truss, at the same

time allowing for ease in erecting the structure and absorbing tolerances on the site. Finally, the software developed, permits generation of the structure and the subsequent analysis and dimensioning, in order to optimise the structural proposal. Currently, the experimental work is progressing to obtain the constitutive equation and behaviour characteristics of the member to node connection.

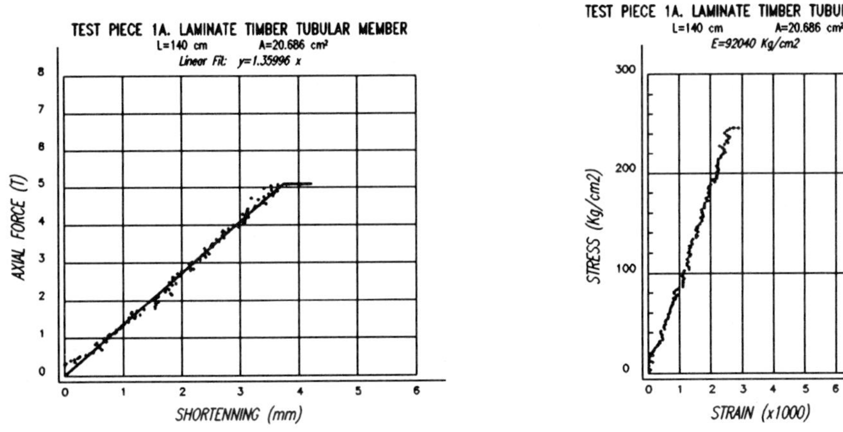

Fig. 13. Experimental behaviour for trial tube 1A.

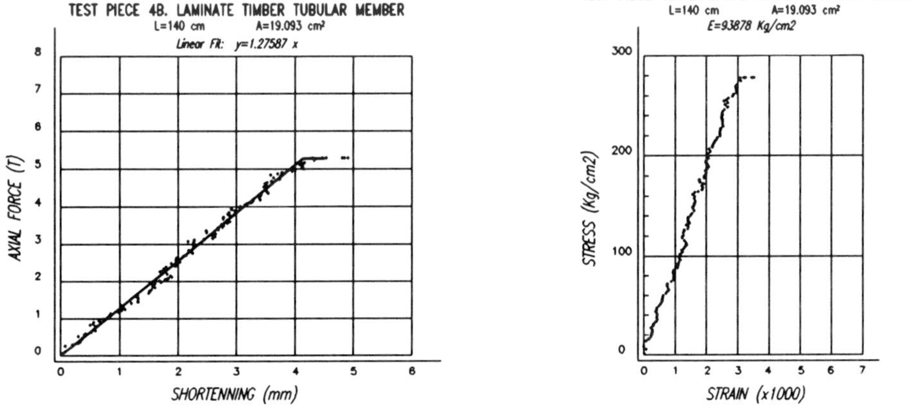

Fig. 14. Experimental behaviour for trial tube 4B.

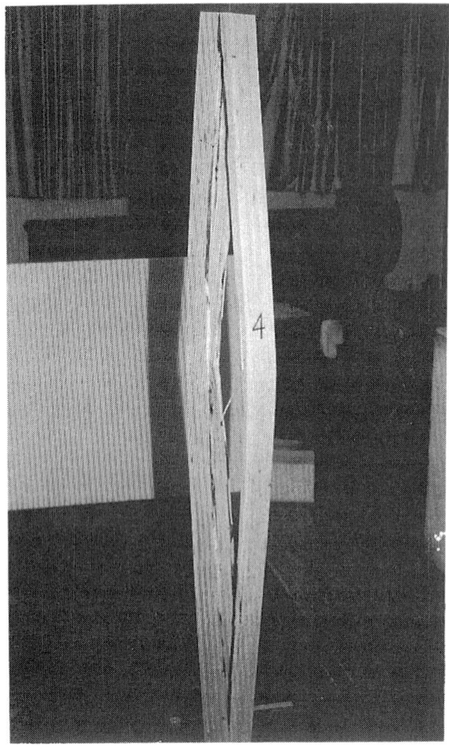

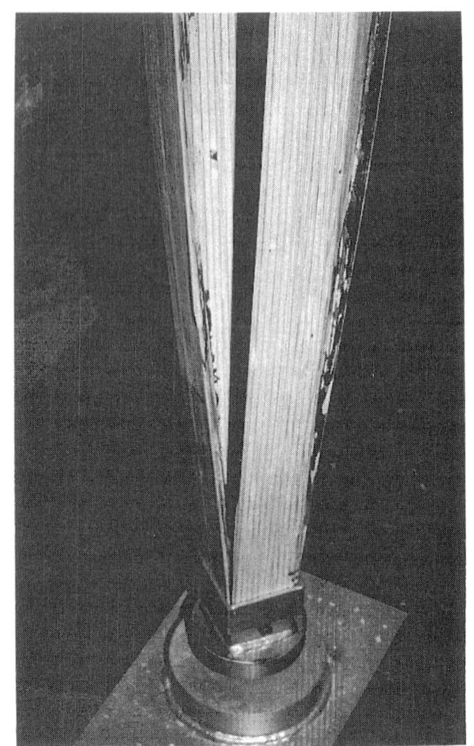

Fig. 15. Photograph of the test.

ACKNOWLEDGEMENTS

25 The authors would like to thank scholarship holder D. F. Caramés for his laboratory work.

BIBLIOGRAPHY

1. HUYBERS P. Timber pole space frames. Space Structures, 1986, Vol. 2, 77-86.

2. HUYBERS P. Wire binding technique for building structures of round-wood. Int. Conf. on the Design and Const. of Non-conventional Structures. London, 1987.

61. Double-layer bamboo space structures

K. GHAVAMI and L. E. MOREIRA, Pontificia University, Brazil

SYNOPSIS. In the assembly of a double layer bamboo space structure, the jointing is a major problem. Different types of joints were developed and their behaviour was studied. In this paper the detailed study of the connection using steel plates and pin joints in bamboo poles is presented. The optimum distance between the holes which has been established experimentally in relation to the shear behaviour of bamboo pole along the fibres, is discussed. Based on this data a method for calculating the strength of the joint was developed, and the results were verified in the laboratory on a 4m by 4m span bamboo space structure.

INTRODUCTION

1. Inventions and technical development since the 2nd World War have lead to an increase in the use of mineral resources, thus reducing the life-time of mines. Beside this, renewable resources such as timber and sea-plankton become extinct very rapidly because of deforestation and air and sea pollution. With this in mind, conscious of the necessity of managing natural wealth in a better and organised way, a research program on the usage of bamboo and natural fibres was started at PUC-Rio in 1979 (refs 1-6). In this paper the development of a joint for the construction of Double Layer Bamboo Space Structure (DLBSS) is described (refs 5-6).

2. Bamboo is a natural composite material with fibres and cellulose-lignin polymers, which give high mechanical strength and low specific weight to bamboo poles. Beside this, bamboo is a renewable and superproductive resource, which is abundant in tropical and subtropical regions of the earth, with low rates of energy of production in comparison with steel, aluminium, fibreglass, concrete and timber (refs 7-9). There are bamboos which grow 120cm in 24 hours and reach maximum strength in 3 years.

3. Double layer space structures are generally used to cover large areas without having to use intermediate columns. They are usually fabricated from steel or aluminium, but not bamboo. Up to the present time, whole bamboo poles have been applied in the production of plane trusses, but not DLBSS (refs 5-7).

4. The developed joints, with pins through the bamboo culms, have been used for the construction of a 4m by 4m space structure made of Dendrocalamus giganteus bamboo, for the verification of the theoretical results. The prototype model was instrumented in order to measure the deformations in the bamboo elements and also the local and global displacements. The prototype was subjected to a point load at the central joint and was loaded to failure. The test results were found to be satisfactory (refs 5-6).

DEVELOPMENT OF PIN CONNECTION

5. Differing from the metallic tubes commonly used in the construction of space structures, the connections between bamboo poles require the design of a joint assembly which takes account of the geometry and orthotropic properties of bamboo. In the developed node, rectangular plates welded to an octagonal base plate, were connected to the bamboo poles by two steel pins, as shown in Fig. 1.

6. To produce a safe connection of this type a preliminary parametric study based on an infinite plate subjected to a concentrated load was carried out. The parameters considered were the establishment of the optimum diameter, the number of pins, distance between the pin holes and the end of the bamboo member, and the minimum size of the plates. The results of the findings were also verified in an experimental study.

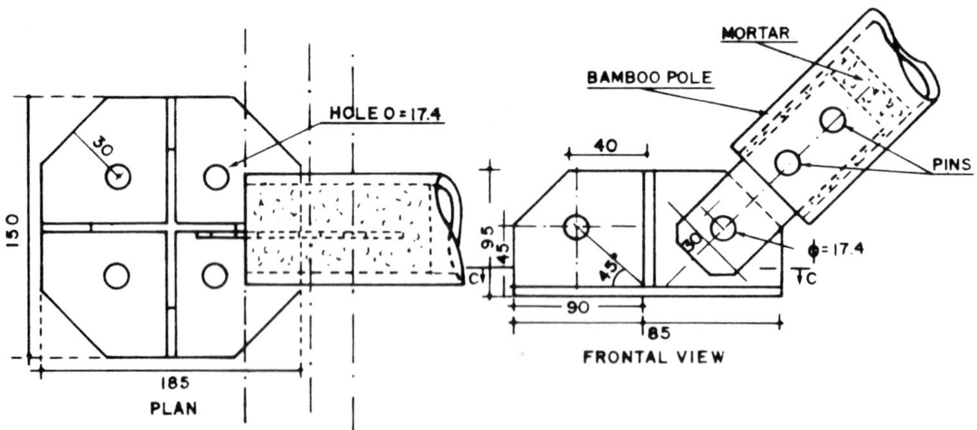

Fig. 1. Bamboo-pin connection.

PURE SHEAR TEST IN FIBRE DIRECTION

7. In previous work, the tensile, compression, bending and transverse shear strength were studied (ref. 5). In this investigation the shear strength in the fibre

direction of two types of bamboo, used in the development of DLBSS, is established, using a specially designed test specimen

8. The dimensions of the test specimen is shown in Fig. 2a. In this test the recommended width of the specimen was to be the same as the bamboo thickness. To establish the average shear strength in fibre direction, two cases of cross-section were studied. As the fibre distribution in the cross section of bamboo is not uniform, this was found to be an important factor. The microstructural study showed that the density of fibre distribution increases from the inner to outer part of the bamboo cross-section.

9. In case A the shearing occurred through an area with uniform distribution of fibres at the middle of the cross-section. However, in case B the shearing occurred through the cross section with non-uniform distribution of fibres, this represents the behaviour of the bamboo pole subjected to contact pressure as in the case of the developed joint. The tests were carried out in an Instron universal testing machine, as shown in Fig. 2.

10. The test results for both cases, for Bambusa vulgaris and Dendrocalamus giganteus species, are given in Table 1. In all the tests the width of the specimens was 8 mm, i.e. the thickness of the bamboo shell. The natural humidity of the 3 year matured bamboo was 11%, except for bamboo of column 3 and 6 which showed humidity of 15.7%. The average value of shear strength is given in column 7. In general, based on these limited results, the shear strength of bamboo which failed according to case B was higher.

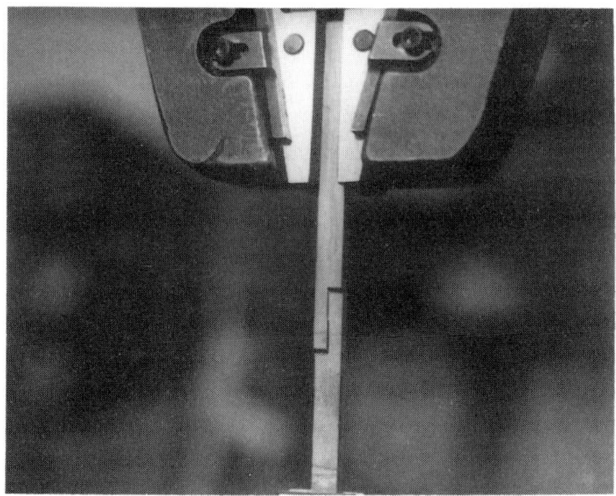

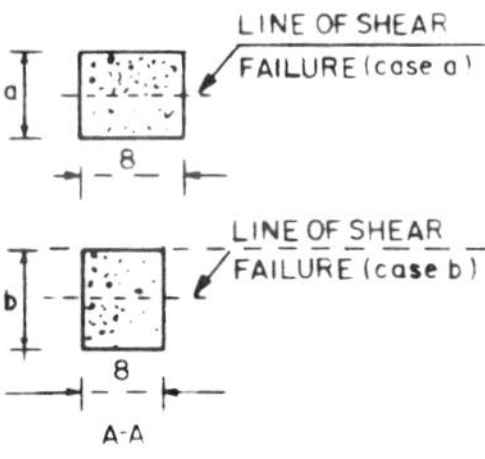

Fig. 2. Pure shear test in fibre direction

Table 1. Pure shear test results in fibre direction (MPa)

CASE	Bambusa vulgaris			Dendrocalamus giganteus			Mean
	1	2	3	4	5	6	7
A	6.1	6.2	5.8	7.1	7.4	6.5	6.5
B	7.8	7.8	6.3	7.4	7.0	5.6	7.0

TESTS ON BAMBOO PIN-CONNECTION

11. To establish the optimum diameter of the pin and pin holes, two series of tensile tests were carried out, as shown in Fig. 3. In the first series of tests the bamboo specimens had only one pin and in the second series two pins were employed. The best behaviour was observed for the 19 mm diameter pin at the distance of 75 mm to 80 mm.

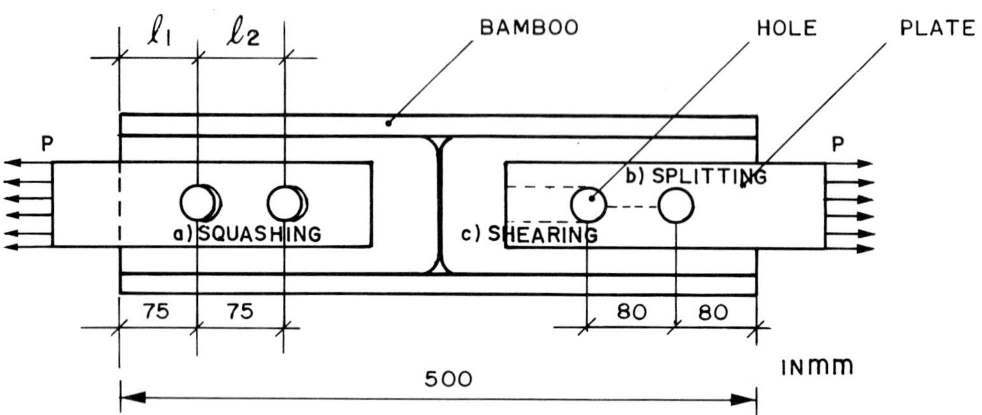

Fig. 3. Bamboo-pin tensile tests and their failure mode.

Table 2. Test results and theoretical shear stress for series 2 (MPa).

				τ_{yz}	τ_{yz}	σ_z	σ_y
CP	D(cm)	a(mm)	P(kN)	Eq. (4)	Eq. (7)	Eq. (1)	Eq. (2)
1	2	3	4	5	6	7	8
D1T	9.7	7.7	21.00	6.5	6.1	18.3	4.6
D2T	10.0	8.0	23.40	7.0	6.5	19.6	4.9
D3T	10.2	8.3	28.30	8.1	7.6	22.8	5.7
D4T	9.3	7.4	18.30	5.9	5.5	16.6	4.1
D5T	9.2	7.0	20.00	6.8	6.4	19.1	4.8
D6T	9.0	6.8	19.20	6.7	6.3	18.9	4.7
V1T	8.0	6.2	17.80	6.8	6.4	19.2	4.8
V2T	8.5	7.0	20.20	6.9	6.4	19.3	4.8
V3T	7.8	6.0	20.10	8.0	7.5	22.4	5.6
V4T	7.8	7.7	20.50	6.3	5.9	17.8	4.5
V5T	7.8	8.8	21.80	5.9	5.5	16.6	4.2
V6T	9.8	10.0	28.30	6.7	6.3	19.0	4.7

12. In general three types of failure in the bamboo-pin connections were observed:
(a) Progressive squashing of bamboo shell. In this type of failure with an increase of load the bamboo shell in contact with the pin started to squash and changed the form of the hole at failure from circular to elliptic, as shown in Fig. 3a.
(b) Splitting of bamboo. In this type of failure the bamboo shell, at the pin position, split along the centre line of the pin, as shown in Fig 3b.
(c) Shearing of bamboo shell. In this type of failure a segment of bamboo shell sheared out at a width smaller than the diameter of the pin, as shown in Fig. 3c and Fig. 4.

13. The two pin bamboo connection proved adequate for the construction of the prototype bamboo space structure. The test results for test series 2 for Bambusa vulgaris and Dendrocalamus giganteus denoted as V and D, respectively, are presented in Table 2. The external diameter, D, and the thickness of bamboo, a, are given in columns 2 and 3, respectively, and the maximum load at failures, given in column 4.

MATHEMATICAL MODELLING OF BAMBOO PIN CONNECTION

14. The ultimate strength of bamboo-pin joints, as in pin connections in steel structures, depends on the type of failure. However, the bamboo pole is an anisotropic material and has relatively low shear strength along in the fibre direction. For this reason the contact stress theory had been used to find the state of stress in bamboo pole subjected to contact stress. Although this method considers all the parameters, the stresses calculated were found to be much higher than the those found from the test results. Therefore, a second approach for the dimensioning of the joints, which produced better results, was adopted.

15. In this method the load P applied through the centre of pin was considered as a point load subjected to an infinite plate under the pin, as shown in Fig. 4. It is a simple method compared with the contact stress method. The state of stress in the bamboo shell was found using Eqs 1-3.

$$\sigma_z = \frac{P\cos\theta}{4\pi ra}\left[-(3+\nu_1)+2(1+\nu_1)\sin^2\theta\right] \tag{1}$$

$$\sigma_y = \frac{P\cos\theta}{4\pi ra}\left[1-\nu_1-2(1+\nu_1)\sin^2\theta\right] \tag{2}$$

$$\tau_{yz} = \frac{P\sin\theta}{4\pi ra}\left[1-\nu_1+2(1+\nu_1)\cos^2\theta\right] \tag{3}$$

In these equations the σ_y is responsible for splitting, σ_z for crushing and maximum τ_{yz} causes the shearing of bamboo along the shearing failure surface. The type of shearing, shown in Fig. 4, was the main failure mode in the experimental investigation.

16. The shear stress at a distance of $R_2 + d$, along the Y axis is found by substitution of $r = (R_2 + d)/\cos(\theta)$, into Eq. 3 and by differentiation of $\delta\tau_{yz}/\delta\theta = 0$ we find θ, for maximum shear stress parallel Y axis:

$$\theta = \pm \arccos\left[\frac{2\nu_1 + 1 + \left(2\nu_1^2 + 4\nu_1 + 3\right)^{1/2}}{4(1+\nu_1)}\right]^{1/2} \quad (4)$$

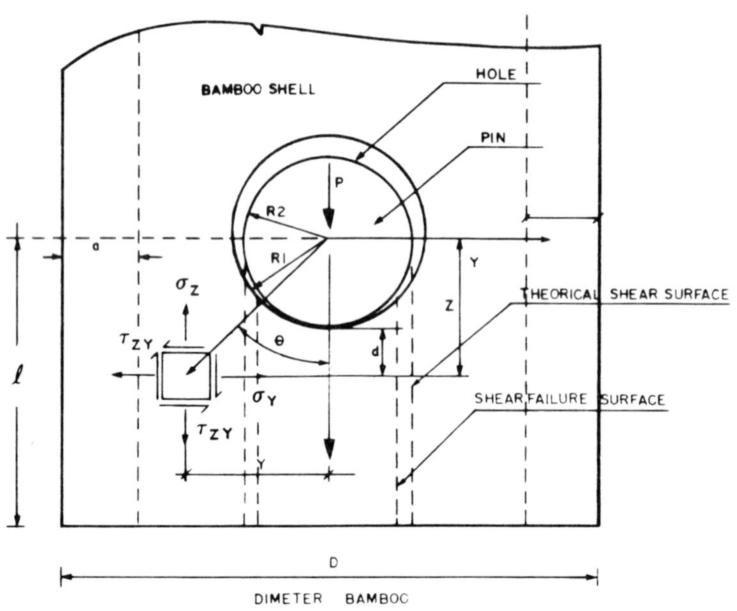

Fig. 4. Theoretical modelling of bamboo pin connection.

17. In Table 2, column 5, the calculated results for the case when $d = 0$ is also given. It can be seen that the result is compatible with the shear strength of the test result, as given in Table 1. Theoretical shear failure occurs at $y = 6.2$mm for the pin radius of 9.5mm. This value is close to the experimental shear failure which occurred at approximately $y = 7.0$mm.

18. The shear stress parallel to the Z axis at a distance $Y = R_2$ can be calculated using Eq. 6.

$$\tau yz_{R2} = P\sin^2(\theta)\left[1 - \nu_1 + 2(1+\nu_1)\cos^2\theta\right] / \left(4\pi a R_2\right) \quad (5)$$

Following the same procedure as before, we find the expression for θ and hence find τyz_m as given by Eq. 7.

$$\tau yz_m = \frac{P(3+\nu_1)}{16\pi a R_2(1+\nu_1)}\left[1-\nu_1+\frac{(1-3\nu_1)}{2}\right] \tag{6}$$

19. The ruptured element under the pin is considered to be an infinite plate subjected to a point load. Therefore the shear stress at the extreme end should be very small. The magnitude of shear stress is assumed to be a small fraction, f, of τyz, given by Eq. 7:

$$\tau yz_{R2} = f.\tau yz_m \tag{7}$$

20. By substitution of Eq. 5 and 6 into eq. 7 and resolving the expression for H one finds:

$$tg(\theta) = \left\{\frac{(3+\nu_1)\left[1-(1-f)^{1/2}\right]}{4(1+\nu_1)-(3+\nu_1)\left[1-(1-f)^{1/2}\right]}\right\}^{1/2} \tag{8}$$

from where the minimum distance between pin and end of bamboo and between two pins can be calculated by Eq. 9:

$$Lmin = R_2/tg(q) \tag{9}$$

From the test results presented in Table 2, the value of f is considered to be 5% of τyz_m . Based on this assumption the minimum value of the plate depth is found to be about 75mm, which corresponds to the experimental result. For this depth the value of τyz_m is given in column 6 of Table 2.

TEST ON FULL SCALE MODEL

21. To observe the behaviour of the connection in full scale structure, a DLBSS simply supported prototype with a 4m by 4m free span consisting of 32 elements, 13 nodes and 64 two pin end connections was tested, as shown in Fig. 5 . The structure was loaded through the centre joint by a hydraulic jack of maximum 12 tons capacity with the increment of 0.5 kN. The detail of the test is given in (refs 5-6).

22. The first failure of the prototype occurred in a pin connection of a tension element at a total load of 32 kN, resulting in a force of 11.24 kN in the element. The shear failure mode caused the collapse. The calculated shearing strength at failure corresponded very well with the shear resistance of bamboo used in the fabrication of this structure.

CONCLUDING REMARKS

23. The test results on the DLBSS prototype showed the soundness of the pin bamboo connection for the construction of such structures. The shear failure mode caused the collapse of the joint, as was predicted. Based on the experimental results a simplified formulation, considering the shear area under the pin as an infinite plate, was developed to calculate the shear strength of such joints. The contact stress theory, which considers most parameters involved in the design, overestimated the state of the stress in the connection. At present a study is underway to improve the mechanical properties and durability of bamboo through treatment in to enable the design of DLBSS of spans larger than 12 m. A modification of contact stress theory for the design of DLBSS connection is also being studied.

REFERENCES

1. GHAVAMI K. and HOMBEEK R. V., Application of bamboo as a construction material, Part I - Mechanical properties and water-repellent treatment of bamboo, Part II - Bamboo reinforced concrete beams. Proceedings of Latin American Symposium on Rational Organisation of Building Applied to Low Cost Housing, 1981, IPT CIB, São Paulo - Brazil, 49 - 66.

Fig. 5. Test of 4 m by 4 m bamboo space structure.

2. GHAVAMI K. and FANG H. Y., (ed.). Low cost and energy saving construction materials, 624 pp, Volume 1, ENVO Publishing Company,Inc. Lehigh, USA, 1984.
3. GHAVAMI K. (ed.). Low cost and energy saving construction materials, 300 pp, Vol. 2, EXPED - Expressão e Cultura, Rio de Janeiro, Brazil, 1986.
4. GHAVAMI K. Application of bamboo as a low cost construction material. Proceedings of the International Bamboo Workshop, publ. the Kerala Forest Research Inst. India, and Inst. IDRC, Canada, 1990, 270 -279.
5. MOREIRA L. E. Desenvolvimento de Estruturas Treliçadas Espaciais de Bambu, M.Sc. thesis Civil Eng. Dept., PUC-Rio, Brazil, 1991.
6. GHAVAMI K., and MOREIRA L. E., "Bamboo Space Structures". Procedings of the Int. Symposium on Industrial use of Bamboo, Beijing, China, December, 1992, 199 - 209.
7. JANSEN J. J. A. Bamboo in building structures, Ph.D. thesis, Eindehoven University of Technology, Holland, May, 1981.

62. Cylindrical bending of square-on-square triple-layer grids

C. ALDEA and M. V. SOARE, Civil Engineering Institute of Bucharest, Romania

SYNOPSIS. The behaviour of square-on-square triple-layer-grids is analysed first by considering the discrete structure and then the equivalent continuum. Explicit formulae are established assuming that the layers are identical and constant normal loads are applied to the middle layer nodes. The analogy of the behaviour with that of a Timoshenko beam is emphasised.

INTRODUCTION

1. The space frame can be considered as square-on-square triple-layer-grid, which consists of three parallel layers having identical square meshes, the bars being parallel with the sides of a rectangular boundary. The top and bottom layers are superimposed, in the vertical projection, while the middle layer is shifted by half the length of the module in both directions. The bracing members join the nodes of the middle layer to the neighbouring nodes in the top and bottom layers; it is easily seen that the bracing members in the upper an lower zones are in pairs. Although the construction of such space frames is mentioned in the literature beginning in 1971, only a very restricted number of papers deals with them (refs 1-4). Due to the existence of specialised computer-programmes for the analysis of reticulated space structures, the analysis of triple-layer-grids presents only routine difficulties.

2. When one side is considerably larger than the other, the support conditions and loading does not vary with the longer sides, the triple-layer works in cylindrical bending. The purpose of our paper is to elucidate the behaviour in this case. The analysis is concerned with the discrete structure and the equivalent continuum, the last one giving an overall view of the structure behaviour.

EQUILIBRIUM EQUATIONS

3. For the typical nodes A, B, and C in the three layers, the equilibrium equations are written (see Fig. 1):

Space structures 4. Thomas Telford, London, 1993

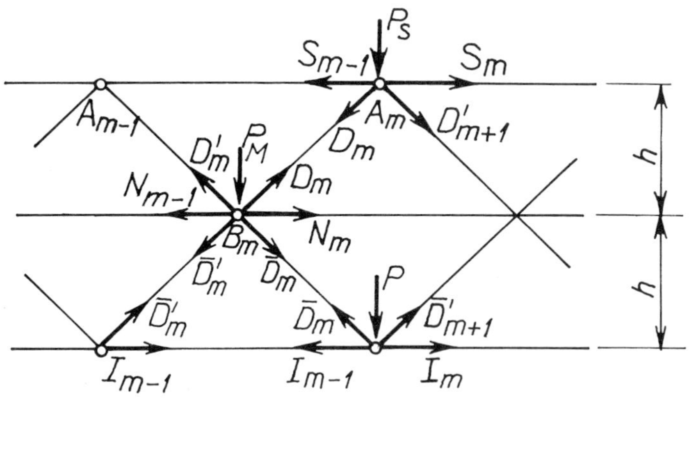

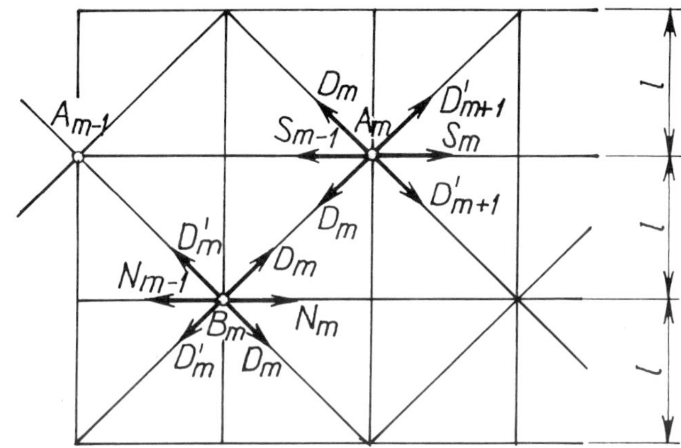

Fig. 1a. Vertical section. Fig. 1b. Top plane and middle plane.

$$S_m - S_{m-1} + \sqrt{2}\cos\gamma\left(D'_{m+1} - D_m\right) = 0 \quad (1a)$$

$$D_m + D'_{m+1} + P_s / 2\sin\gamma = 0 \quad (1b)$$

$$N_m - N_{m-1} + \sqrt{2}\cos\gamma\left(D_m + \overline{D}_m - D'_m - \overline{D}'_m\right) = 0 \quad (1c)$$

$$\overline{D}_m + \overline{D}'_m - D_m - D'_m + P_M / 2\sin\gamma = 0 \quad (1d)$$

$$l_m - l_{m-1} + \sqrt{2}\cos g\left(\overline{D}'_{m+1} - D_m\right) = 0 \quad (1e)$$

$$\overline{D}_m + \overline{D}'_{m+1} - P_l / 2\sin\gamma = 0 \quad (1f)$$

where only components of the nodal loading normal to the three layers have been considered and $\cot\gamma = l / h\sqrt{2}$.

ELASTICITY RELATIONSHIPS

4. With $k_s = EA_s/l$, $k_M = EA_M/l$, $k_l = EA_l/l$, $k_d = EA_d/l_d$ the axial rigidities of the bars, the bar forces are related to the displacements u_s, u_M, u_l (in the planes of the three layers) and W_s, W_M, W_l (normal to the same layers) by the following relationships:

$$S_m = k_S\,(u_{S,m+1} - u_{S,m})$$
$$N_m = k_M\,(u_{M,m+1} - u_{M,m})$$
$$I_m = k_I(u_{I,m+1} - u_{I,m})$$
$$D_m = (k_d \cos\gamma/\sqrt{2})[u_{S,m} - u_{M,m} + \sqrt{2}\tan\gamma\,(w_{S,m} - w_{M,m})]$$
$$D'_m = (k_d \cos\gamma/\sqrt{2})[u_{M,m} - u_{S,m-1} + \sqrt{2}\tan\gamma\,(w_{M,m} - w_{S,m-1})]$$
$$\overline{D}_m = (k_d \cos\gamma/\sqrt{2})[u_{I,m} - u_{M,m} + \sqrt{2}\tan\gamma\,(w_{I,m} - w_{M,m})]$$
$$\overline{D}'_m = (k_d \cos\gamma/\sqrt{2})[u_{M,m} - u_{I,m-1} - \sqrt{2}\tan\gamma\,(w_{M,m} - w_{I,m-1})]$$

Introducing the shifting operator **E** (defined by $\mathbf{E}S_m = S_{m+1}$, $\mathbf{E}^{-1}S_m = S_{m-1}$ ....) and denoting:

$$W = \sqrt{2}\tan\gamma . w \tag{2}$$

the previous system is written:

$$S_m = k_S\,(\mathbf{E}-1)u_S \tag{3a}$$
$$N_m = k_M\,(\mathbf{E}-1)u_M \tag{3b}$$
$$l_m = k_l\,(\mathbf{E}-1)u_l \tag{3c}$$
$$D_m = \left(k_d \cos\gamma\sqrt{2}\right)\left(u_S - u_M - W_S + W_M\right) \tag{3d}$$
$$D'_m = \left(k_d \cos\gamma/\sqrt{2}\right)\left(u_M - \mathbf{E}^{-1}u_S + W_M - \mathbf{E}^{-1}W_S\right) \tag{3e}$$
$$\overline{D}_m = \left(k_d \cos\gamma/\sqrt{2}\right)\left(u_l - u_M + W_l - W_M\right) \tag{3f}$$
$$\overline{D}'_m = \left(k_d \cos\gamma/\sqrt{2}\right)\left(u_M - \mathbf{E}^{-1}u_l - W_M + \mathbf{E}^{-1}W_l\right) \tag{3g}$$

APPLICATION OF THE DISPLACEMENT METHOD

5. Denoting the dimensionless coefficients

$$\psi_S = k_S/\left(k_d\cos^2\gamma\right);\quad \psi_M = k_M/\left(k_d\cos^2\gamma\right);\quad \psi_l = k_l/\left(k_d\cos^2\gamma\right) \tag{4}$$

introducing the bar forces as given by Eqs 3a-3g into the equilibrium Eq. 1 and conveniently arranging the terms, results in the difference equation system given in Table 1.

Table 1.

u_S	u_M	u_I	W_S	W_M	W_I	= Free term
$\psi_S(E-1)(1-E^{-1})-2$	$E+1$	•	•	$E-1$	•	0
$1+E^{-1}$	$\psi_M(E-1)(1-E^{-1})-4$	$1+E^{-1}$	$(1-E^{-1})$	•	$1-E^{-1}$	0
•	$E+1$	$\psi_I(E-1)(1-E^{-1})\ 2$	•	$-(E-1)$	•	0
•	$E-1$	•	-2	$E+1$	•	$-\frac{P_S}{\sqrt{2}\,k_d \sin\gamma \cos\gamma}$
$-(1-E^{-1})$	•	$1-E^{-1}$	$1+E^{-1}$	-4	$1+E^{-1}$	$-\frac{P_M}{\sqrt{2}\,k_d \sin\gamma \cos\gamma}$
•	$E-1$	•	•	$-(E+1)$	$+2$	$+\frac{P_I}{\sqrt{2}\,k_d \sin\gamma \cos\gamma}$

6. The general solution of the system given in Table 1 will be discussed elsewhere. We shall limit ourselves to the case when the top and bottom layers are identical; the analogy with the Timoshenko beam will be emphasised.
Considering $k_S = k_I = k$ and $\psi_S = \psi_I = \psi$ respectively, the system in Table 1 is considerably simplified. Carrying out the following change of variables:

$$u_S + u_I = U_1; \quad (u_S - u_I)/2 = u; \quad W_S - W_I = W_1; \quad (W_S + W_I)/2 = W \tag{5}$$

summing up and subtracting conveniently the equation, we get the difference equation system given in Table 2, where the normal loading was restricted to that of the middle layer.

Table 2.

$u_S + u_I$	u_M	$W_S - W_I$	$\frac{u_S - u_I}{2}$	W_M	$\frac{W_S + W_I}{2}$	= Free term
$\psi(E-1)(1-E^{-1})-2$	$2(E+1)$	∘	∘	∘	∘	0
$1+E^{-1}$	$\psi_M(E-1)(1-E^{-1})-4$	$-(1-E^{-1})$	∘	∘	∘	0
∘	$E-1$	-1	∘	∘	∘	0
∘	∘	∘	$\psi(E-1)(1-E^{-1})-2$	$E-1$	∘	0
∘	∘	∘	∘	$E+1$	∘	0
∘	∘	∘	$1-E^{-1}$	$+2$	$-(1+E^{-1})$	$\frac{P_M}{2\sqrt{2}\,k_d \sin\gamma \cos\gamma}$

7. It is noted that the initial system is divided into two systems of three difference equations each. The first system is homogeneous and the main determinant is non zero; therefore the trivial solution is valid. This results in:

$u_S = -u_I = u$ i.e. the in-plane displacements in the two extreme layers are equal and of opposite sense,

$u_M = 0$ i.e. the in-plane displacements in the middle layer are identically null,

$W_S = W_I = W$ i.e. the normal displacements in the extreme layers are equal and have the same sense.

Keeping in mind previous results, the second system is written:

$$\begin{bmatrix} \psi(\mathbf{E}-1)(1-\mathbf{E}^{-1})-2 & \mathbf{E}-1 & 0 \\ 0 & \mathbf{E}+1 & -2 \\ 1-\mathbf{E}^{-1} & +2 & -(1+\mathbf{E}^{-1}) \end{bmatrix} \begin{bmatrix} u \\ w_M \\ w \end{bmatrix} = \begin{bmatrix} 0 \\ 0 \\ P_M/2\sqrt{2}k_d \sin\gamma \cos\gamma \end{bmatrix} \quad (6)$$

SOLUTION OF SYSTEM

8. The simplest way of obtaining the unknown displacements is to introduce a displacement function Φ, using the procedure described in ref. 5. The potential Φ satisfies the difference equation:

$$\begin{vmatrix} \psi(\mathbf{E}-1)(1-\mathbf{E}^{-1})-2 & \mathbf{E}-1 & 0 \\ 0 & \mathbf{E}+1 & -2 \\ 1-\mathbf{E}^{-1} & +2 & -(1+\mathbf{E}^{-1}) \end{vmatrix} \cdot \Phi = P_M/2\sqrt{2}\, k_d \sin\gamma \cos\gamma \quad (7)$$

or developing the determinant:

$$(\mathbf{E}-1)^2(1-\mathbf{E}^{-1})^2\Phi = -P_M / 2\sqrt{2}k_d \sin\gamma\cos\gamma = -P_M \cot\gamma / 2\sqrt{2}k \quad (8)$$

The three unknown displacements may be expressed by means of difference operators applied to the potential Φ; each of these operators results by replacing the column corresponding to the unknown in Eq. 7, by a column where the only non-zero term equal to unity, is that corresponding to the loading line. Thus:

$$u = -2(\mathbf{E}-1)\Phi \quad (9a)$$

$$W_M = 2[\psi(\mathbf{E}-1)(1-\mathbf{E}^{-1})-2]\Phi \quad (9b)$$

$$W = (\mathbf{E}+1)[\psi(\mathbf{E}-1)(1-\mathbf{E}^{-1})-2]\Phi \quad (9c)$$

Replacing the results from Eq. 9 into Eq. 3 gives:

$$S_m = -I_m = -2k(\mathbf{E}-1)^2\Phi$$
$$D_m = -\overline{D}_m = -k_d \cos\gamma\ (1-\mathbf{E}^{-1})(\mathbf{E}-1)^2\Phi/\sqrt{2}$$
$$D'_m = -\overline{D}'_m = +k_d \cos\gamma\ (\mathbf{E}-1)(1-\mathbf{E}^{-1})^2\Phi/\sqrt{2}$$

In order to get the difference equation satisfied by each unknown, Φ is eliminated successively between Eq. 8 and each relationship Eq. 9 and 10:

$$(\mathbf{E}-1)(1-\mathbf{E}^{-1})^2 u = P_M \cot\gamma / \sqrt{2}\, k \quad (10a)$$
$$[(\mathbf{E}-1)^2(1-\mathbf{E}^{-1})^2] W_M = \sqrt{2} \cot\gamma\ [1-\psi(\mathbf{E}-1)(1-\mathbf{E}^{-1})/2] P_M / k \quad (10b)$$
$$[(\mathbf{E}-1)^2(1-\mathbf{E}^{-1})^2] W = \cot\gamma\ [1-\psi(\mathbf{E}-1)(1-\mathbf{E}^{-1})/2] P_M / k\sqrt{2} \quad (10c)$$
$$(1-\mathbf{E}^{-1})^2 S = P_M \cot\gamma / \sqrt{2} \quad (10d)$$
$$(1-\mathbf{E}^{-1})\, D = P_M / 4\sin\gamma \quad (10e)$$
$$(\mathbf{E}-1)\, D' = -P_M / 4\sin\gamma \quad (10f)$$

GENERAL SOLUTION FOR THE CASE $P_M = P$(= constant.)

9. The developed Eq. 8 is written as:

$$(\mathbf{E}^2-4\mathbf{E}+6-4\mathbf{E}^{-1}+\mathbf{E}^{-2})\Phi = -P\cot\gamma / 2\sqrt{2}\, k \quad (11)$$
$$\Phi_{m+2} - 4\Phi_{m+1} + 6\Phi_m - 4\Phi_{m-1} + \Phi_{m-2} = -P \cot\gamma / 2\sqrt{2}\, k \quad (12)$$

The solution of Eq. 12 rendered homogeneous is sought under the form $\overline{\Phi} = \beta^m$, where β satisfies the characteristic equation:

$$\beta^4 - 4\beta^3 + 6\beta^2 - 4\beta + 1 = (\beta-1)^4 = 0\ ;$$

it has the multiple root $\beta = 1$.
The general solution is therefore:

$$\Phi_m = -P\cot\gamma\ (m^4/4! + C_1 m^3/3! + C_2 m^2/2! + C_3 m + C_4)/2\sqrt{2}\, k \quad (13)$$

where C_1, C_2, C_3, C_4 play the part of constants of integration. From Eq. 13 we deduce Φ_{m+1}, Φ_{m+2}, Φ_{m-1}, Φ_{m-2} by replacing m by m+1, m+2, m-1, and m-2 respectively. Bar forces and displacements are given by Eq. 11; they are still written as:

$$S_m = -I_m = -2k(\Phi_{m+2} - 2\Phi_{m+1} + \Phi_m) \quad (14a)$$
$$D_m = -\overline{D}_m = -k(\Phi_{m+2} - 3\Phi_{m+1} + 3\Phi_m - \Phi_{m-1}) / \sqrt{2} \cos\gamma \quad (14b)$$
$$D'_m = -\overline{D}'_m = k(\Phi_{m+1} - 3\Phi_m + 3\Phi_{m-1} - \Phi_{m-2}) / \sqrt{2} \cos\gamma \quad (14c)$$

$$u_m = -2(\Phi_{m+1} - \Phi_m) \tag{14d}$$

$$w_M = -2\sqrt{2}\cot\gamma\,[\Phi_m - (\psi/2)(\Phi_{m+1} - 2\Phi_m + \Phi_{m-1})] \tag{14e}$$

$$w_S = w_I = -\sqrt{2}\cot\gamma[\Phi_{m+1} + \Phi_m - (\psi/2)(\Phi_{m+2} - \Phi_{m+1} - \Phi_m + \Phi_{m-1})] \tag{14f}$$

Replacing expressions corresponding to Φ_{m+2}, Φ_{m+1}, Φ_m, Φ_{m-1}, Φ_{m-2} into Eq. 14 is a tedious operation. Finally we get:

$$S_m = -I_m = P\cot\gamma\,[m^2/2 + (C_1+1)m + C_1 + C_2 + 7/12]/\sqrt{2} \tag{15a}$$

$$D_m = -\overline{D}_m = (P/4\sin\gamma)(m + C_1 + 1/2) \tag{15b}$$

$$D'_m = \overline{D}'_m = -(P/4\sin\gamma)(m + C_1 - 1/2) \tag{15c}$$

$$u_{S,m} = -u_{I,m} = (P\cot\gamma/\sqrt{2}k)[(m^3/6) + (C_1+1/2)m^2/2 + (C_1/2 + C_2 + 1/6)m + C_1/6 + C_2/2 + C_3 + 1/24] \tag{15d}$$

$$w_{S,m} = w_{I,m} = (P\cot^2\gamma/2k)[m^4/12 + (C_1+1/2)m^3/3 + (C_1/2 + C_2 + 1/4)m^2 + (C_1/2 + C_2 + 2C_3 + 1/6)m + C_1/6 + C_2/2 + C_3 + 2C_4 + 1/24] - (P/k_d\sin^2\gamma)[m^2 + (2C_1+1)m + C_1 + 2C_2 + 2/3] \tag{15e}$$

$$w_{M,m} = (P\cot^2\gamma/2k)(m^4/12 + C_1m^3/3 + C_2m^2 + 2C_3m + 2C_4) - (P/4k_d\sin^2\gamma)(m^2 + 2C_1m + 2C_2 + 1/6). \tag{15f}$$

Note that expressions of deflections consist of two terms; the denominator of the first one includes the axial rigidity k of the top and bottom layers, while that of the second one includes the axial rigidity k_d of the bracing members.

BOUNDARY CONDITIONS

10. Consider the triple-layer grid shown in Fig. 2 supported in the middle plane. The four constants of integration are determined using the following boundary conditions:

- for $m = -1 \Rightarrow S_{-1} = 0$, - for $m = p_{-1} \Rightarrow S_{p-1} = 0$,
- for $m = 0 \Rightarrow w_{M,0} = 0$, - for $m = p \Rightarrow w_{M,p} = 0$,

where p is the extreme node in the middle layer. There results:

$C_1 = -p/2$, $C_2 = -1/12$, $C_3 = (p^3 + p)/24$, $C_4 = 0$.

Replacing these values into Eq. 15, there results

$$S_m = -I_m = -(P\cot\gamma/2\sqrt{2})(m+1)(p-1-m) \tag{16a}$$

$$D_m = -\overline{D}_m = (P/4\sin\gamma)[m-(p-1)/2], \tag{16b}$$

$$D'_m = -\overline{D}'_m = -(P/4\sin\gamma)[m-(p+1)/2], \tag{16c}$$

$$u_{S,m} = -u_{I,m} = (P\cot\gamma/6\sqrt{2}k)[m-(p-1)/2][m^2-(p-1)m-(p^2+p)/2], \tag{16d}$$

$$w_{S,m} = w_{I,m} = (P\cot^2\gamma/24k)[m^4-2(p-1)m^3-(3p-2)m^2+(p^3-2p+1)m+ +(p^3-p)/2]+(P/k_d\sin^2\gamma)[(p-1)+(p-1)m-m^2], \tag{16e}$$

$$w_{M,m} = (P\cot^2\gamma/24k)m(p-m)[m(p-m)+p^2+1]+(P/4k_d\sin^2\gamma)m(p-m). \tag{16f}$$

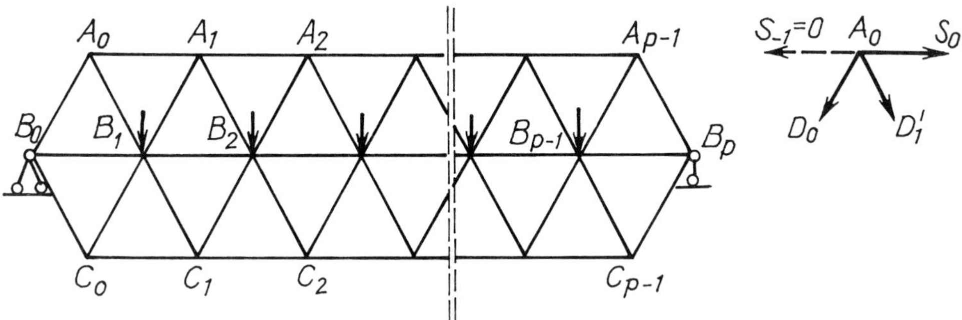

Fig. 2. Boundary conditions.

NUMERICAL EXAMPLE

11. Consider p=11 intervals, l=4m mesh grid, 2h = 5m structure height, P = 1000daN nodal force, A = 50cm^2 area of the cross-section of the top and bottom layer members, A = 20cm^2 area of the cross-section of the bracing members. With k = EA/l = 50E/400 = 0.125E,
k_d = (20/377.49172)E = 0.0529812E,
$\cot\gamma = \left(400/250\sqrt{2}\right) = 0.8\sqrt{2} = 1.1313708,\ \cot^2\gamma = 1.28,\ \sin^2\gamma = 0.4385964$
the diagrams of variation of the previous forces and displacements are plotted in Fig. 3.

EQUIVALENT CONTINUUM

12. Starting from the Taylor expansion
$\mathbf{E} = 1+(d/dx)+(l^2/2)(d^2/dx^2)+(l^3/6)(d^3/dx^3)+(l^4/24)(d^4/dx^4)+...$
and similar expression for $\mathbf{E}^2$, $\mathbf{E}^{-1}$, $\mathbf{E}^{-2}$, replacing in Eq. 13 results the equation:

$$d^4\Phi/dx^4=-(P\cot\gamma/2\sqrt{2}kl^4) \tag{17}$$

The approximation arises because terms proportional to l^6 and higher powers are neglected as l is small, with $P_M = P$ = constant. Using the same Eq. 16 we get:

$$S = -I = -2kl^2(d^2\Phi/dx^2), \tag{18a}$$

$$D = -D' = -\bar{D} = \bar{D}' = -(kl^3/\sqrt{2}\cos\gamma)(d^3\Phi/dx^3), \tag{18b}$$

$$u_S = -u_I = u = -2l(d\Phi/dx), \tag{18c}$$

$$w_S = w_M = w_I = w = -2\sqrt{2}\cot\gamma[\Phi-(\psi/2)l^2(d^2\Phi/dx^2). \tag{18d}$$

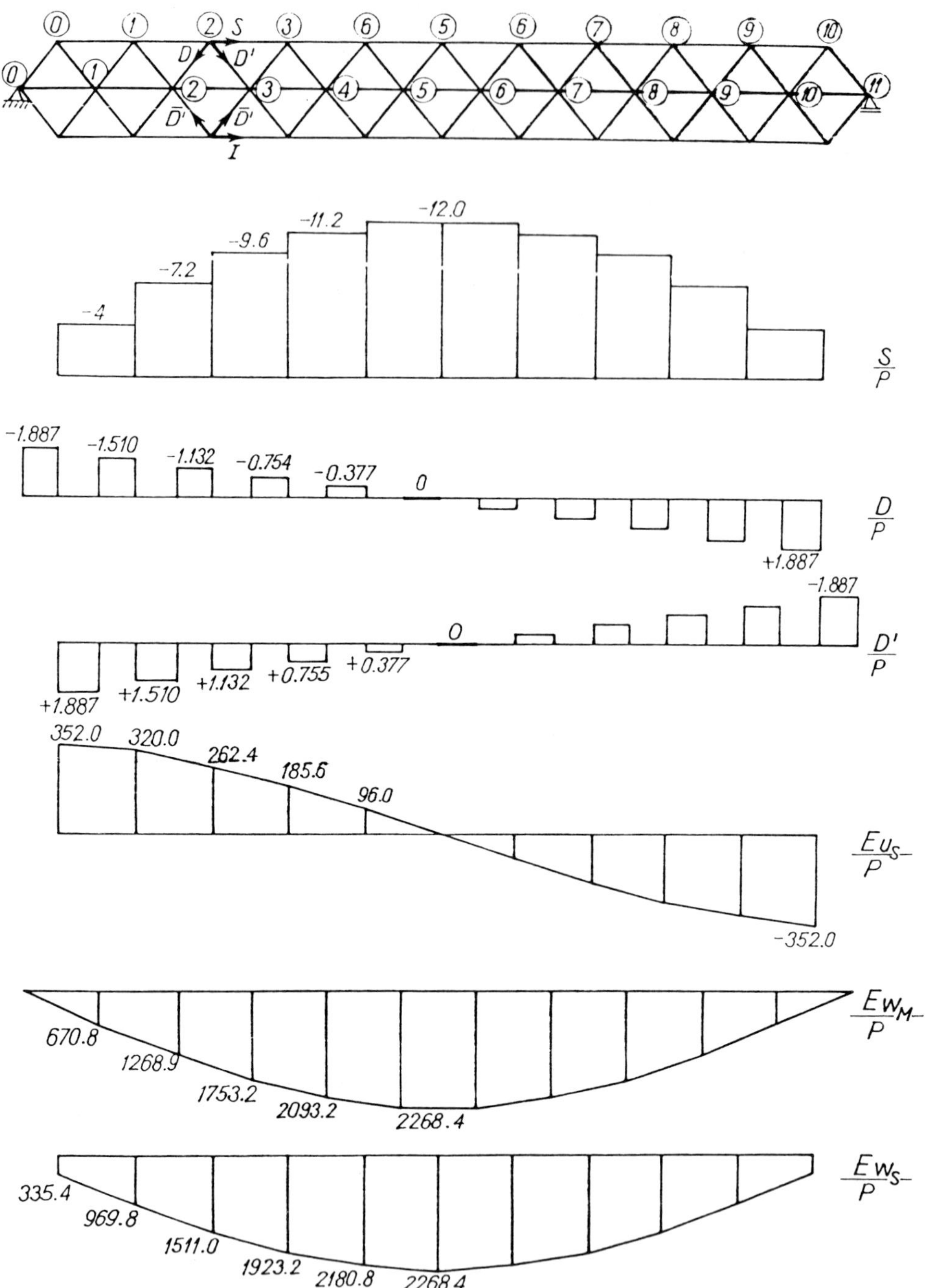

Fig. 3. Diagrams of variation of bar forces and displacements.

Keeping in mind Eq. 17 the general solution is:

and similar expression in Eq. 18 for the successive derivatives of F. Replacing these expressions in Eq. 18 there results:

$$S = -I = (P\cot\gamma/\sqrt{2}l^2)[(x^2/2)+C_1x+C_2], \quad (19a)$$
$$D = -D' = -\overline{D} = \overline{D}' = (P/4\sin\gamma)(x+C_1)/l, \quad (19b)$$
$$u_S = -u_I = (P\cot\gamma/\sqrt{2}kl^3)[(x^3/6)+(C_1x^2/2)+C_3x+C_4), \quad (19c)$$
$$w = w_S = w_M = w_I = (P\cot^2\gamma/kl^4)[(x^4/24)+(C_1x^3/6)+(C_2x^2/2)+C_3x+C_4]- -(P/2k_d\sin^2\gamma l^2)[(x^2/2)+C_1x+C_2]. \quad (19d)$$

The constants of integration are determined by use of boundary conditions corresponding to simple end support as considered at §10:
for x = 0 and x = a=> S = 0 and w = 0 (with a = pl).
The integration constants are: $C_1 = -a/2$, $C_2 = 0$, $C_3 = a^3/24$, $C_4 = 0$.
The final expressions of bar forces and displacements are:

$$S = -I = -(P\cot\gamma/2\sqrt{2}k)(x/l)(a/l-x/l), \quad (20a)$$
$$D = -D' = -\overline{D} = \overline{D}' = (P/4\sin\gamma)(x/l-a/2l), \quad (20b)$$
$$u_S = -u_I = (P\cot\gamma/\sqrt{2}k)(a/2l-x/l)[(a^2/2l^2)+(a/l)(x/l)-(x^2/l^2), \quad (20c)$$
$$w = (P\cot^2\gamma/24k)(x/l)(a/l-x/l)[(x/l)(a/l-x/l)+a^2/l^2]+(P/4k_d\sin^2\gamma)(x/l)(a/l-x/l). \quad (20d)$$

In the case where the equivalent continuum is considered the following abscissa x must be used for the computation of bar forces and displacements:

- for $S_m = -I_m$	$x=(m+1)l$	(m=0,1,2,...,p-2)
- for $D_m = -\overline{D}_m$	$x=[m+1/2]l$	(m=0,1,2,...,p-1)
- for $D'_m = -\overline{D}'_m$	$x=[m-(1/2)]l$	(m=1,2,3,...,p)
- for $u_S = -u_I$	$x=[m+(1/2)]l$	(m=0,1,2,...p-1)
- for $w_S = w_I$	$x=[m+(1/2)]l$	(m=0,1,2,...p-1)
- for w_M	$x=ml$	(m=0,1,2,...,p)

For the bar forces the formulae established by the two methods (discrete structure and equivalent structure) coincide. For the in-plane and normal displacements the differences are minor, the maxima differences amounting to 0.38% and 0.64% respectively.

FINAL REMARKS AND CONCLUSIONS

13. The fact that deflections of the three layers of the structure, replaced by its equivalent continuum are identical, denote that the considered strip subject to cylindrical bending, behaves like a beam.

Consider again Eq. 11b: with $w = W_M / \sqrt{2}\tan\gamma$, it may be written that:
$(\mathbf{E}-1)^2(1-\mathbf{E}^{-1})^2 w = (\cot^2\gamma/k)P-(1/2k_d \sin^2\gamma)(\mathbf{E}-1)(1-\mathbf{E}^{-1})P$
With $w = w_{(1)} - w_{(2)}$, deflections $w_{(1)}$ and $w_{(2)}$ will be governed by the difference equations:
$(\mathbf{E}-1)^2(1-\mathbf{E}^{-1})^2 w_{(1)} = (\cot^2\gamma/k)P$, $(\mathbf{E}-1)(1-\mathbf{E}^{-1})w_{(2)}=(1/2k_d \sin^2\gamma)P$
By a limiting process, the two equations become:

$$d^4w_{(1)}/dx^4 = (P\cot^2\gamma/kl^4) \tag{21a}$$

$$d^2w_{(2)}/dx^2 = (P/2k_d\sin^2\gamma l^2) \tag{21b}$$

14. Note that Eq. 21a represents the differential equation of the deformed axis in bending while Eq. 21b governs the deflection curve due to longitudinal shearing. Thus there results in a Timoshenko beam. Although in the numerical example the ratio $2h/pl = 5/44 = 1/8.8<1/5$ and we would expect reduced influence of longitudinal shearing, it is found that this has an effect, leading to larger deflections of about 15%. Considering again Eq. 21a it may be compared with the differential equation of the deformed middle surface of a plate working in cylindrical bending.

$$d^4w/dx^4=p/K$$

where K is the flexural rigidity of the corresponding plate.
Replacing $\cot^2\gamma = l^2/2h^2$ and $\lim_{l\to 0} P/l^2 = p$, there results in:
$K = 2kh^2 = (2Ah^2/l)E$
where $2Ah^2/l$ represents the moment of inertia of the two extreme layers, distributed over the interval l. However, the analogy with plates is less useful in the case of triple-layer parallel mesh grids due to the relative importance of deformations given by longitudinal shearing.

REFERENCES

1. BUNNI U.K. and MAKOWSKI Z. S. The structural behaviour of triple-layer grids. Proceedings of the 2nd International Conference on Space Structures, University of Surrey, 1975, 1-6.
2. MAKOWSKI Z. S. Trends and development in space structures. Bul. stiintific al Inst. Constructii Bucuresti, 1978, Vol 21, 1-2, 3-28.
3. BUNNI U. K., DISNEY P. and MAKOWSKI Z. S. Multi-layer space frames, Constrado, London, 1980.
4. SOARE V. M. and TOADER I. H. Contributions to the study of the state of stress and deformation of triple-layer rids. Rev. Roum. Sci. Techn.-Mec. Appl. 1983, Vol 28, no. 6, 651-659.
5. SOARE V. M. Discrete structures and equivalent continua in the mechanics of solids (in Romanian). Publishing House of the Romanian Academy, Bucharest, 1986.

63. Design against member collapse in double-layer space grids

R. I. PETERSEN and N. ROSENBERG, Technical University of Denmark

SYNOPSIS. Designing against collapse of an arbitrary member in a double-layer space grid is a time consuming process when using normal finite element programs, because the stiffness matrix has to be changed and re-factorised every time a member is removed from the static system. This paper presents an approximate method to calculate the redistribution of forces due to the failure of one single pin-connected member in a double-layer space grid by adding a load case to the existing load cases, for every member to be investigated.

INTRODUCTION

1. When designing space grids it is often specified in building codes that the grid should be investigated for the failure of a member. Some codes prescribe this investigation for structures with high safety requirements. Other codes prescribe the investigation when single bolt connections are used in the structure. In both cases, the redistribution of forces caused by the failure of a single member must not lead to collapse of other members and eventually a progressive collapse of the whole structure.

2. Most space grids are designed with the use of finite element method (FEM) programs. When a member in the static system is removed, then the stiffness matrix has to be rebuilt and re-factorised to calculate the redistribution of member forces. This is a time consuming process, when the space grid has several hundreds or thousands of members. A complete check demands, that the space grid is designed against the failure of an arbitrary member, so the total computing time will be extensive.

3. This paper presents an approximate method to calculate the redistribution of member forces, which does not need changing and re-factorisation of the stiffness matrix.

4. The investigation of the method is a part of a master thesis (ref. 1) undertaken in the Department of Structural Engineering of the Technical University of Denmark with Assistant Professor Henning Agerskov and Associate Professor Jesper Gath as advisors.

5. Nine different double-layer space grids were investigated including four commonly used grid configurations, and the results obtained by using the approximate method are compared with exact FEM-calculations.

6. A literature study, covering proceedings of the International Conference on Space Structures. (refs 2 and 3) and international periodicals (refs 4 and 5) was carried out in advance of these investigations. Only very sparse material on the subject was found. No investigations have put emphasis on designing a space grid against the consequences of the failure of an arbitrary member. Instead, most investigations on double-layer space grids have concentrated on determining the ultimate load, and also on the non-linear behaviour that leads to collapse of the space grid.

THEORY

7. The proposed method is called the Differential Force Method, because the basic principle of the method is to calculate a force determined as a difference between two member forces corresponding to two load cases on the member under investigation. Hereafter the method will be referred to as the DFM.

8. The DFM calculates the redistributed member forces with a member removed from the grid by considering two load cases. Load case no. 1, is the load that is prescribed by codes to be applied when examining the consequences of a failed member in the structure. Often codes use reduced partial safety factors on the load and on the materials, and only wind or snow loads have to be considered. Load case no. 2, is a member load applied to the member under investigation.

9. The differential force is defined as the difference between the member load in load case 2 and the member force in the member under investigation as a result of the external load; load case 1. The differential force thereby expresses how much of the ultimate member load the member carries, and thus the effectiveness of the grid to carry the member load.

10. All member forces are hereafter found by adding a ratio of the forces from load case 2 to the member forces in load case 1. To find the member forces in the space grid when member k is removed, then member k is loaded by the member load P_2 which results in the member force $N_{k,2}$ in the member k. The ratio K is now found as:

$$K = \frac{N_{k,1}}{P_2 - N_{k,2}} \tag{1}$$

The member forces N_i with the failure of member k are found as:

$N_i = N_{i,1} + K \bullet N_{i,2}$, i = 1, ... *number of members in the grid.*

11. The construction is not changed during the calculations, i.e. the stiffness matrix is unchanged. Thus, to investigate the failure of a member, requires only the addition of a load case in the calculations. In usual matrix-vector formulation

of the finite element method, a vector of node forces is added for every member investigated:

$$\underline{\underline{K}}\,\underline{v} = \underline{R} \quad , \quad \underline{R} = \begin{array}{c|c} \textit{Design load node forces} & \textit{Member loads for investigation of member outfall} \\ \underline{R}_{load\ case\ 1} & \underline{R}_1\ \underline{R}_2\ \underline{R}_3\ \cdot\ \underline{R}_j \end{array} \left[\begin{array}{c|c} r_{loadcase\ 1,1} & \cdot\ \cdot\ \cdot\ \cdot\ \cdot\ \cdot\ \cdot\ \cdot \\ r_{loadcase\ 1,2} & \cdot\ \cdot\ \cdot\ \cdot\ \cdot\ \cdot\ \cdot\ \cdot \\ \cdot & \cdot\ \cdot\ \cdot\ \cdot\ \cdot\ \cdot\ \cdot\ \cdot \\ r_{loadcase\ 1,i} & \cdot\ \cdot\ \cdot\ \cdot\ \cdot\ \cdot\ \cdot\ \cdot \end{array} \right]$$

i number of nodes
j number of investigations

12. This means, that the calculation of a member failure is a question of adding load cases to the existing design load cases, and calculating the displacements corresponding to these load cases simultaneously. No changing and re-factorisation of K is needed.

13. Example. For a given double-layer grid the distribution of member forces is assumed to be as shown in Fig. 1, with the code prescribed design load on the structure:

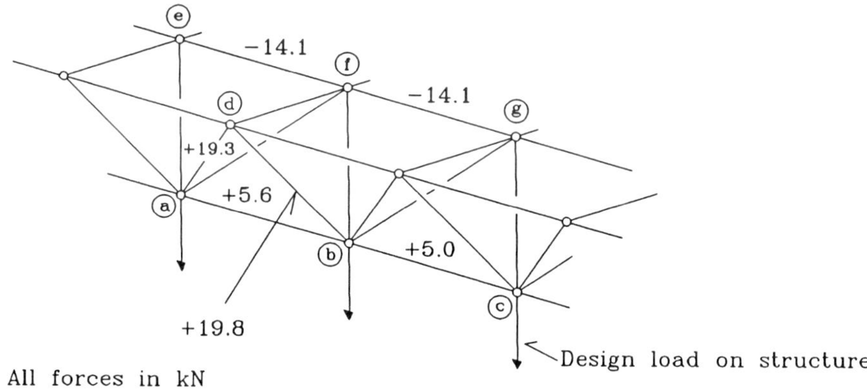

Fig. 1. Design load on the structure and corresponding member forces: Load case 1.

14. As the aim is to determine the redistribution of member forces when member b-d is removed, this member is loaded with P = 100kN. This load is applied in the following way:

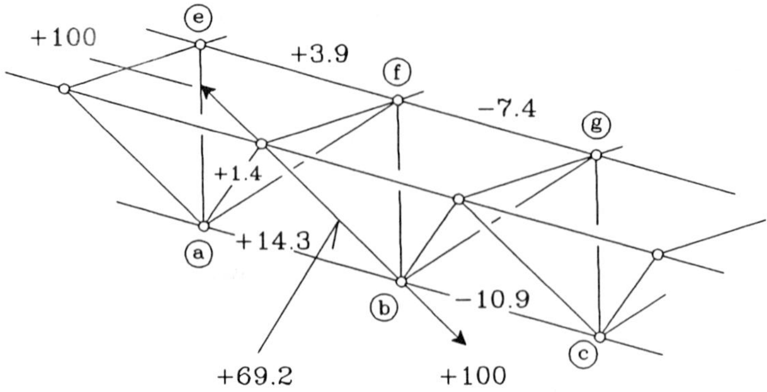

All forces in kN

Fig. 2. Member load on the member under investigation: Load case 2.

The ratio K is now found as:

$$K = \frac{19.8\ \text{kN}}{(100 - 69.2)\ \text{kN}} = 0.643$$

15. The member forces when member b-d is removed from the static system can now be calculated. Examples of comparison with exact FEM-calculations are shown below:

Table 1. Member forces calculated by the approximate method. Forces in kN.

Member	Member force with load case 1	Member force with load case 2	Member force with outfall of b-d	Exact calc.
b-c	+5.0	-10.9	-2.0	-2.0
a-b	+5.6	+14.3	+14.8	+14.8
a-d	+19.3	+1.4	+20.2	+20.2
f-g	-14.1	-7.4	-18.9	-18.8
e-f	-14.1	+3.9	-11.6	-11.6

Instability

16. An additional effect of the DFM is, that it reveals if the failure of a member results in instability of the grid. If the member considered is needed to carry the force, then removal of this member makes the grid a mechanism. If such a member is loaded with a member force of say 100kN, then the force in the member will be 100 kN, and K will be infinite. A short example illustrates this:

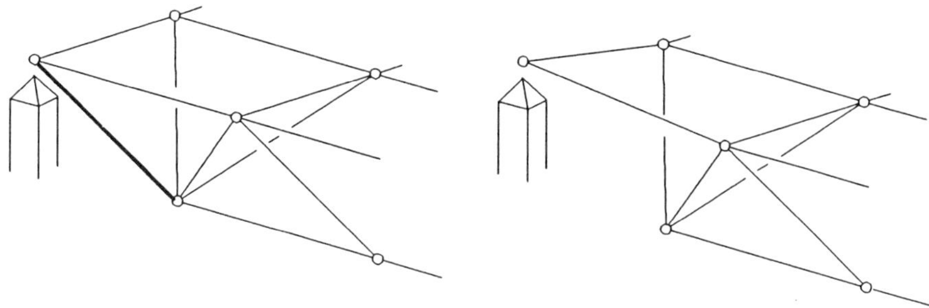

Fig. 3. Space grid where member removal will result in instability.

17. When the accentuated member at the support is removed, then the grid will be a mechanism. A member load applied to this member will only be carried by this member, thus making the differential force zero, and the K-ratio infinite.

Calculation time

18. The approximate method will only give significant time savings when many failures have to be investigated. If for instance the effect of 100 member failures has to be calculated, then the following comparison can be made:

Table 2. Comparison of calculation time.

Approximate FEM-calculaitons	Exact FEM-calculations
Re-arranging of stiffness matrix 1 time Re-factorization of stiffness matrix 1 time Calculation of displacements 100 times	Re-arranging of stiffness matrix 100 times Re-factorization of stiffness matrix 100 times Calculation of displacements 100 times

19. When the stiffness matrix is very large, the main part of the calculation time will be spent on re-arranging and re-factorisation of the matrix before the displacements are calculated. It is seen that the method gives a significant saving in the calculation time.

Accuracy of the DFM

20. A theoretical evaluation of the accuracy of the method has not yet been carried out. This will involve an evaluation of the influence of the removed member on the stiffness matrix, and therefore the sensitivity of the stiffness matrix to element changes and the influence of the changed matrix on the calculation of the displacement vector $\underline{v}$.

EVALUATION OF THE DIFFERENTIAL FORCE METHOD

21. Nine double-layer space grids were investigated to examine the precision of the method. All grids were first designed to carry an appropriate load combination of snow and self weight. This load was applied to all lower chord nodes. Seven tubes and six bolt dimensions were used in the design, corresponding to the dimensions often used in design of double-layer space grids in Denmark giving some well-optimised grids. All grids were based on the same module length of a = 3m. The characteristics of the grids are listed in Table 3.

Table 3. Grids used to investigate accuracy of the DFM.

Grid no.	Base area (a×a)	Number of members	Number of nodes	Hight of the structure	Support conditions
1	2×9	144	48	$\frac{1}{2}\sqrt{2}\cdot a$	At ends
2	3×9	216	67	$\frac{1}{2}\sqrt{2}\cdot a$	At ends
3	4×9	288	86	$\frac{1}{2}\sqrt{2}\cdot a$	At ends
4	5×9	360	105	$\frac{1}{2}\sqrt{2}\cdot a$	At ends
5	6×9	432	124	$\frac{1}{2}\sqrt{2}\cdot a$	At all sides
6	7×9	398	129	$\frac{1}{2}\sqrt{2}\cdot a$	At all sides
7	7×9	408	131	$\frac{1}{2}\sqrt{2}\cdot a$	At all sides
8	8×9	635	162	$\sqrt{2/3}\cdot a$	At all sides
9	4×4	128+1	41	$\frac{1}{2}\sqrt{2}\cdot a$	At corners

22. All member forces were calculated for the failure of several important members, including chord members at mid-span and diagonals at the supports. These members are shown in bold in Fig. 4. This figure also shows the support conditions.

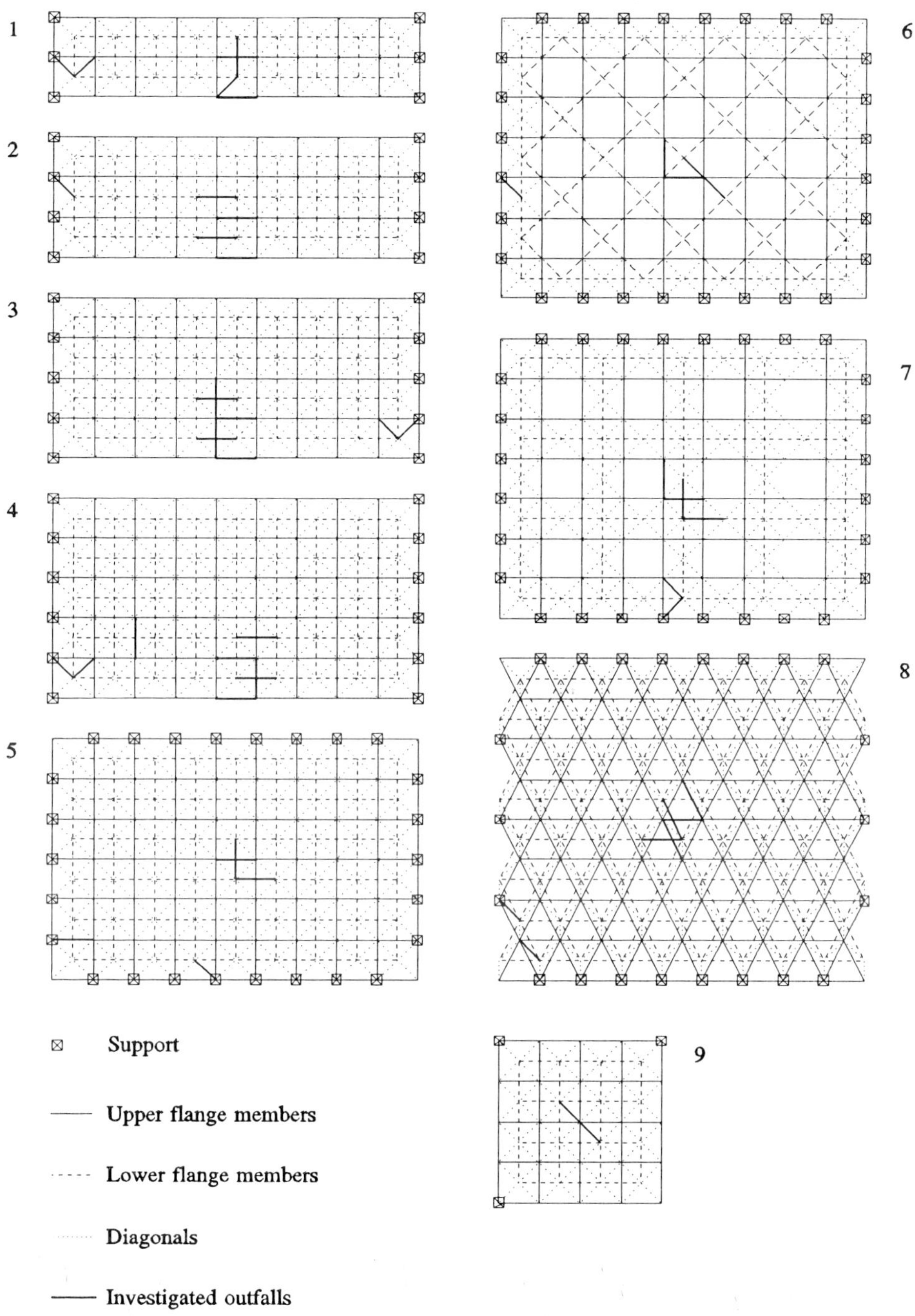

Fig. 4. Double -layer space grids used to investigate the accuracy of the DFM.

23. Grids 1 through to 4 were examined to decide if the precision of the calculation was vulnerable to the stiffness of the grid, as the torsional stiffness will grow as the grid is made broader.

24. Grids 5 through to 8 are normal type grids often used for construction. The grids are supported on all four sides to ensure uniform support conditions and make the results mutually comparable.

25. Grid 9 was designed to test the approximate method in the extreme. For this grid only chord member set an angle of 45° to the rest of the chord members in the bottom layer would contribute significantly to the torsional stiffness of the grid, and big differences in the member forces using the two methods could therefore be expected when investigating the failure of this member.

26. Evaluation of the results required a simplified approach. If the member force was less than a force of 0.1kN when calculated by both methods, then it was considered uninteresting to examine the difference further. The assumption was made, because one method might give for example 10^{-14} kN and the other 10^{-16} kN. This would give a large difference in percentages, but results of no consequences in the design practice. Furthermore some numerical errors in the equation solving could be expected, and these may give member forces of the above mentioned size (10^{-14} - 10^{-16}).

27. If either one or both methods gave a member force larger than 0.1kN, then the difference between the results was compared with the capacity of the member. If the difference was less than 0.1% of the member capacity, the approximation was considered to be sufficiently exact for practical design use. As the largest capacity was 895.3kN, the largest allowable difference using this measure was 0.001 x 895.3 = 0.8953kN. This measure of the precision was introduced to filter out a large percentage of insignificant results.

28. The accuracy of the DFM depends on the stiffness of the structure under investigation. The method mostly yields results that diverge less than 0.1% of the member capacity from exact calculations. For grids 1 to 4 and grid 9 there was a difference of 0.1-2% between the two results, but the member forces in these members were very small, and the difference was typically less than 0.5kN. The member forces calculated by the DFM were mostly larger than those calculated using the finite element method, so the DFM yields conservative results. For grids 4 to 8 there was practically no difference between the results obtained by the two methods.

CONSEQUENCES OF MEMBER FAILURE IN DOUBLE-LAYER SPACE GRIDS

29. As these investigations showed that the DFM yields accurate results, grids 5 through to 7 were investigated for a failure of an arbitrary member. These investigations showed, that the reduced partial safety factors on the load and on the materials gave an increased member capacity, which often compensated for the increase in member force due to a failed member. Failure of a flange member increases member forces in adjoining members in a narrow band surrounding the failed member including members in the opposite chord.

30. The most serious consequence however, was the redistribution of member forces which gave compression in lower chord members usually loaded in tension. The main rule for the design of double-layer space grids against member failure should therefore be to design all members to have a minimum compression strength. A compression strength of a few kilo Newton was sufficient.

31. None of these three grids showed any vulnerability to member failure, when all members were designed to have a minimum compression strength. However, it was not possible to conclude any further detailed rules for design practice, as the consequences of member failure is strongly dependent on the optimisation of the space grid.

32. The calculated deformations of grids 5 through to 7 were sufficiently accurate for common design use.

FURTHER INVESTIGATIONS

33. Because of the limited amount of time some topics were not investigated. First of all, a mathematical evaluation of the accuracy of the method is needed. Secondly, it should be examined if the method is suitable for FEM-calculations with other types of node connections. If for instance a lattice shell was to be investigated, it would be necessary to use the DFM assuming rigid node connections.

CONCLUSIONS

34. The approximate method yields results, which deviate less than 2% from exact FEM calculations, and is therefore sufficiently accurate for normal design use. The method also reveals if a failure member makes a structure unstable. If all members are designed to have a minimum compression strength, none of the investigated space grids are found to be vulnerable to the failure of an arbitrary member. The reduction in computing time is dependent on the number of member failures calculated, but is extensive for a complete check of large structures.

REFERENCES

1. PETERSEN R. I., ROSENBERG N., Strength of Double-Layer Space Grids, MSc Thesis submitted to the Department of Structural Engineering, The Technical University of Denmark, 1992.
2. Proceedings of the 2nd International Conference on Space Structures, 1975.
3. Proceedings of the 3rd International Conference on Space Structures, 1984.
4. Space Structures, Periodical, 1986-91. 5. IASS Bulletin, Periodical, 1980-91.

64. The aerodynamic advantages of a double-effect large-span suspension bridge under wind loading

C. BORRI and P. SPINNELI, University of Florence, Italy, and
M. MAJOWIECKI, University of Bologna, Italy

SYNOPSIS. In the paper, some recent advances in the design of large span suspension bridges are reviewed and a solution based on a design of S. Musmeci is finally investigated. This solution, based on a double effect cable system, has many advantages, outlined here, mainly in providing stiffness and, possibly, damping "reserve", that can face effectively many design uncertainties.

INTRODUCTION

1. In the design of large suspension bridges there is today a clear dichotomy in design choices between aerodynamically transparent decking on the one hand and very rigid deck structures providing a high degree of resistance to wind action on the other. These solutions have been dictated by the need to provide decking with a high degree of torsional rigidity and at the same time minimise the effect of drag, a crucial factor on large bridges.

2. The complete contrast between these two design approaches was clearly evident at the ISLAB 92 Symposium on the Aerodynamics of Large Bridges (Copenhagen 1992) (ref. 1), widely considered to reflect the state of the art and current tendencies in large bridge aerodynamics, and confirmed at the recent IASS congress (Toronto, July 1992) (ref. 2); the differences can clearly be seen in the contrasting design concepts of two impressive large span bridges currently under construction:

- the Stoere Baelt suspension bridge in Denmark with a 1624 m central span and highly efficient aerodynamic decking, and
- the Akashi Kaikyo suspension bridge in Japan with a 1990 m central span and a very rigid deck framework.

3. We appear therefore to have arrived at a true parting of the ways as regards design criteria (Fig. 1): on the one hand there is a tendency to reduce the pressures acting on a structure and on the other an attempt to increase structural resistance.

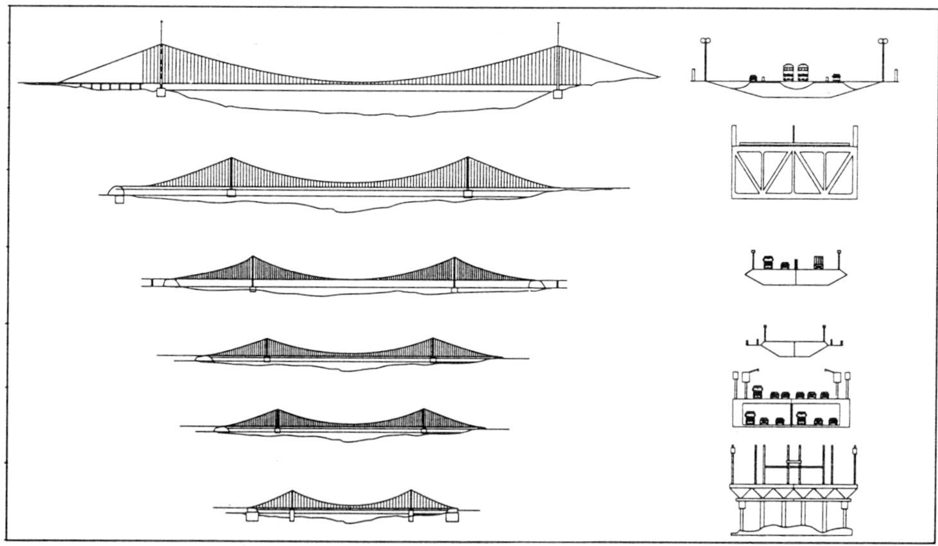

Fig. 1. Evolution of large span bridges.

4. It seems reasonable to ask therefore: can we continue to use the classical suspension bridge design with twin load-bearing cables and vertical hangers for very large spans without introducing structural innovations? N.J. Gimsing (ref. 3) has shown, in a series of interesting proposals, how considerable modifications to the classical vertical layout of cables and stays enables control of static and dynamic behaviour over spans far exceeding those currently possible (up to 5 kilometres using today's steels and theoretically up to 20 kilometres using carbon-fibre reinforced material). An equally stimulating contribution in this debate has been made by T.Y. Lin and P. Chow (ref. 4) in a proposal for bridging the Straits of Gibraltar.

THE EVOLUTION OF DESIGN ASPECTS OF LONG SPAN SUSPENSION BRIDGES

5. The two large suspension bridges mentioned earlier and currently under construction, both share the same main structural system, i.e. a classical suspension bridge design with decking on vertically parallel planes suspended on vertical hangers of variable length from two load-bearing cables. Given the well-known problems of aerodynamic stability in this type of bridge, recent theoretical and experimental research has been directed mainly towards the definition of aerodynamically efficient decking.

6. After the historic failure of the Tacoma Narrows bridge in 1940, caused by aerodynamic instability, most research analyses up until the 1960s concentrated on deck design .

7. An initial tendency was to provide suspension bridges with a high degree of torsional rigidity (see for example the heavy, stiff deck structures of the Verrazzano Narrows, Tago and Firth of Forth Road bridges). Rigid decking provides satisfactory aerodynamic resistance and stability but increases decking weight, construction costs, wind drag and lateral deflection (up to 14 meters on the Akashi Kaikyo (ref. 5). Not to be forgotten are the maintenance and inspection problems of this type of construction.

8. The end of the 1960s saw the introduction of a second approach to the problem: the basic strategy employed was to reduce the effects of wind action by concerning exclusively on the aerodynamics of deck profile. Because of their shape and behaviour under wind action, decks designed according to such criteria are defined as streamlined or aerodynamically transparent. Streamlined decks usually have a single- or multi-box girder construction with single or multiple cells and with an orthotropic decking plate to support and transmit loads. This type of construction reduces wind action, especially drag (either pseudo static or turbulent), and significantly contributes to overall torsional rigidity.

9. The single box girder solution does not, however, provide sufficient reliability on very large spans. A comparative study of some streamlined decks during preliminary design work for the Akashi Kaikyo bridge showed that the maximum span for a single box girder deck is 1700m. For this reason the height of the deck section, on two recent bridges the Stoere Baelt (1624 m) and the Humber (1100 m), has been increased in order to provide the torsional rigidity needed to counteract aeroelastic instability.

10. Another question currently debated is about the deck type that should be used for spans of over 2000m. An initial reply has been given by the Japanese designers of the Akashi Kaikyo bridge who ensure aerodynamic stability by means of a rigid deck framework despite the high price to be paid in terms of drag resistance and turbulence (buffeting and vortex shedding).

11. At 1624m the Stoere Baelt bridge was the longest of its type until the arrival of the Akashi Kaikyo at 1990m. The adoption of a rigid deck on the Akashi Kaikyo at the same time as this marked "jump" in scale from 1624m to 1990m, represents a historical departure in the development of suspension bridge design and raises a considerable number of questions for future applications. Even greater "jumps" in length are foreseen for the future: 3300m for the Straits of Messina, 5000m for a multi-span bridge over the Straits of Gibraltar.

12. The solution put forward by the designers of the Messina bridge (ref. 6) involves the use of a ventilated deck, i.e. a deck of three box sections separated by grill sections which permit the vertical passage of air. This drastically reduces the quasi-static coefficients of aerodynamic resistance to lifting and torque (moment).

13. Innovations in overall structural design can also be found in the large scale of the feasibility studies for bridging the Straits of Gibraltar (Fig. 3) (ref. 4). In this case attention is no longer exclusively concentrated on the aerodynamic performance of the deck but rather is mainly directed towards the use of purely

structural methods for increasing the resistance of a classic suspension bridge design so as to enable the large increase involved in this project.

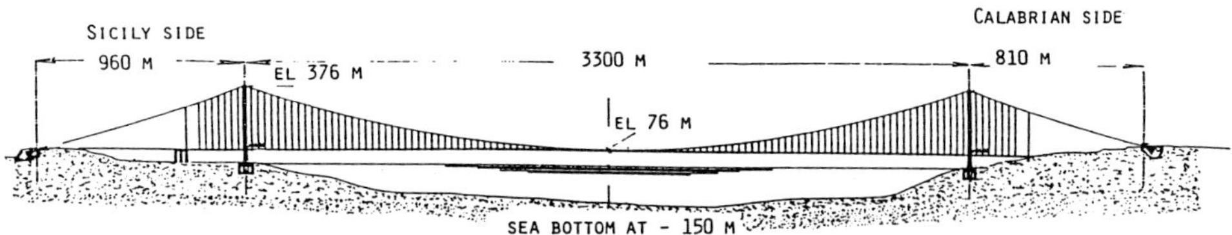

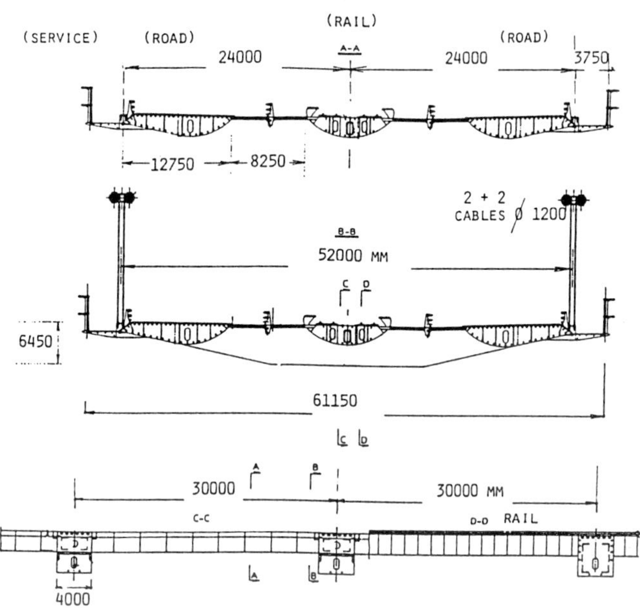

Fig. 2. Design of the Messina Bridge (ref. 6).

14. The discussion above will have made it clear that bridge designers are faced with two choices: a rigid deck or an aerodynamically transparent deck. Rigid decks can provide the aerodynamic characteristics needed to ensure efficiency and stability but entail increases in weight and therefore increases in the costs of structural works. Transparent decks reduce transverse pressures but are limited to use on spans of less than 1700m (as shown by Japanese studies) in those cases where considerable increases in length are not matched by corresponding increases width.

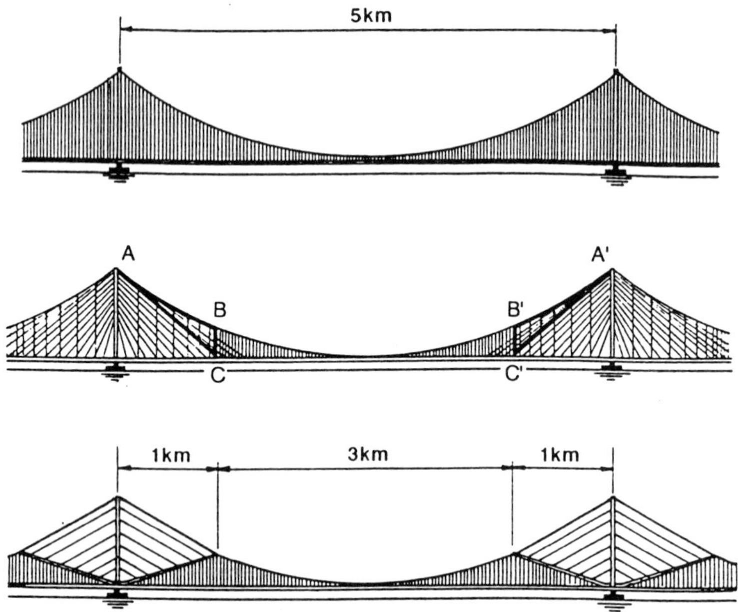

Fig. 3. Design of the Gibraltar bridge (ref. 4).

15. The lack of a synthetic solution combining these two design approaches is due mainly to the uncertainties in physical system modelling and more especially in the modelling of actions. Physical modelling does not yet make it possible to evaluate the sensitivity and degree of reliability of the overall design process, especially in the analysis of large bridges where wind action is undoubtedly a dominant factor.

16. Uncertainties arise from our incomplete knowledge of the stochastic properties of actions, the modelling and behaviour of materials, mathematic modelling and the modelling of structural properties such as mass, damping and rigidity. All these uncertainties mean that the whole design and construction process must be subject to general reliability testing (ref. 7). In a study of the Humber bridge (ref. 8), the uncertainties of the mathematical aeroelastic model were checked on the basis of structural parameter and aerodynamic uncertainties. For bridges of a similar type this study provides general indications concerning analytical model uncertainties through the use of critical flutter speed definitions. Maximum variation in this case is ± 10%.

17. An interesting debate during the ISALB 92 symposium gave the impression that from an engineering point of view the overall design process and structural system testing is reliable only on simple, classical suspension bridges of spans up to 2000 m. In order to remove some of these uncertainties, special wind tunnels have been constructed which are wide enough to take 1:100 aeroelastic bridge models complete with access viaducts. The wind tunnel for the Japanese project, for example, has a test width of 43 m. This was the only way of decreasing the degree of uncertainty when studying structural characteristics and wind-structure interaction (frequency, damping and inertial scaling) and the only way of faithfully modelling towers, decks and cable systems.

18. A second method for solving the problem is to deal with the "action" side of the equation. That is, by designing a deck section that is aerodynamically transparent or ventilated and which, as a consequence, drastically reduces the "actions" operating on the structure. To be applicable, however, this method presupposes that the methods for extrapolating from models to the finished project are reliable, i.e. that the uncertainties have been estimated on the basis of comparisons with existing constructions. In the case of large bridges, however, such a comparison is not practicable and extrapolation is not immediately possible. It should also be said that theoretical numeric methods for extrapolating from models to reality are also affected by numerous uncertainties which cannot be ignored.

19. A first series of uncertainties concerns the significance of the wind tunnel testing of models. According to Scanlan (ref. 9),"...the Reynolds number of the test will generally speaking be two to three orders of magnitude less... The common argument that the effects of the Reynolds number are negligible for structural elements with sharp edges still has to be fully demonstrated...".

20. Scanlan himself draws attention to some of the limits of his theory of aeroelastic instability: "... in the discussion reported above it is supposed that flutter and buffeting forces can be separated and do not interact... But given that physical phenomena and their interaction around bluff bodies are not linear effects, then this analytical separation, made for engineering analysis, is convenient but does not, in principle, include certain complexities of the fluid-structure interaction...".

21. Here it should also be emphasised that, even though model analyses and treatment might be exact, numeric extrapolation using algorithms to stimulate wind storms over time can itself lead to further uncertainties.

22. Firstly, a simulation must comprise at least 20 or 30 case histories of suitable duration (10-20 minutes) if it is to be statistically significant.

23. Secondly, all the algorithms used to generate histories in time do not take into account the quadrature components of the cross spectrum. This means that the phases cannot be simulated correctly. The same applies to the lee eddies caused by the shape of the object hit by wind currents or by the presence of transversal elements and roughnesses. The same is also true of cross winds and the eddy shedding running the entire length of a bridge which can trigger oscillations. These phenomenon cannot be correctly simulated given the current stage of development of generation methods.

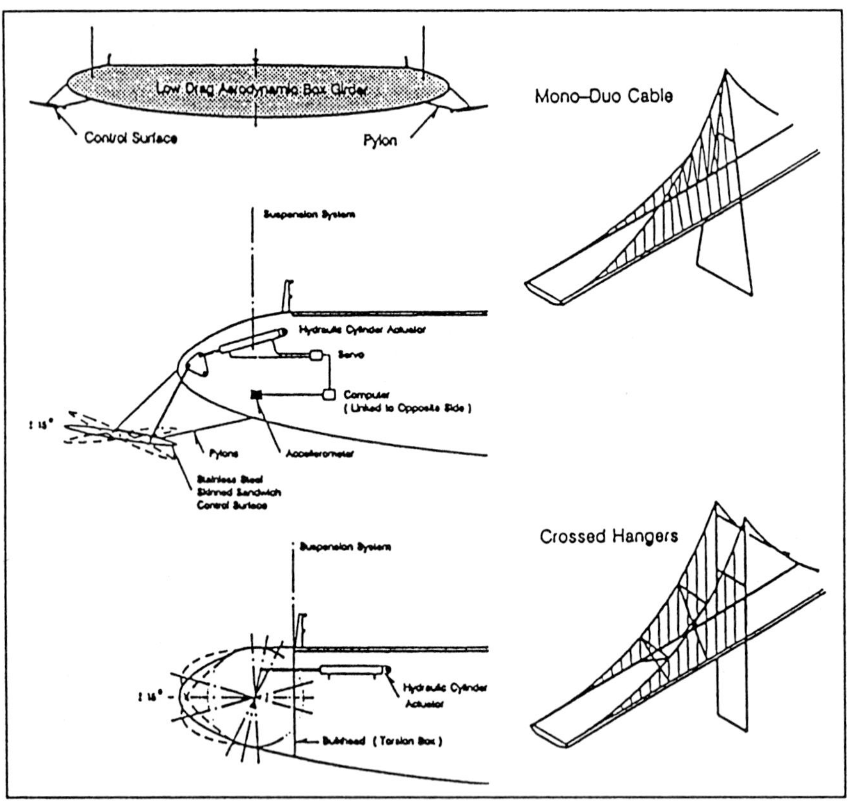

Fig. 4. (ref. 10).

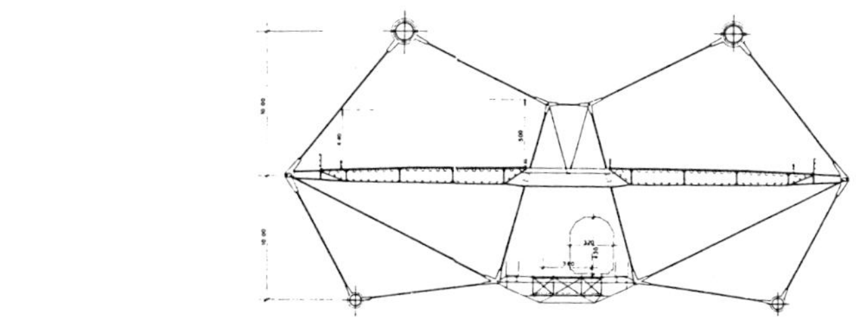

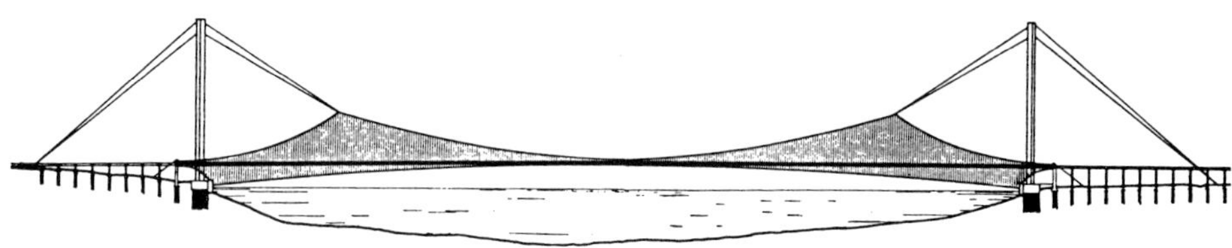

Fig. 5.

24. The considerations raised in the discussion above make it reasonable to ask the question: can we continue to propose the classical layout of a simple suspension bridge for large spans without introducing some structural innovations? In other words, can we hope to dominate all the problems which might arise through the zealous application of analyses to what is, in effect, an "old" layout without at the same time implementing an overall synthesis of structural concepts? Working along these lines, Lin and Chow (ref. 4), suggest increasing the deflection/ span ratio from 1/10 to 1/5 so as to reduce the stresses on the load-bearing cables. According to these authors, the increase in cost due to the higher towers required in their hypothesis would be offset by the reduction in cable diameter.

25. N.J. Gimsing (ref. 3) has shown, in a series of interesting proposals, how considerable modifications to the classical vertical arrangement of cables and stays enables control of static and dynamic behaviour over spans far exceeding those currently possible. K. Ostenfeld and A. Larsen (ref. 10) have proposed the introduction of a system of horizontal cables in order to control and limit lateral movement of the deck.

26. We are firmly convinced that theoretical and experimental analyses, the production, construction and maintenance of free spans greater than 2000m must

all be based on clear design concepts which remove, or at least substantially reduce, the uncertainties indicated by many leading authorities on the subject.

SOME REMARKS ABOUT A PROPOSED DOUBLE EFFECT TENSILE STRUCTURE

27. The aim of this paper is, therefore, to study the behaviour of a double-effect suspension bridge under the effect of wind action. To do this we will use a model inspired by S. Musmeci's design for the Straits of Messina bridge and we will be studying the reduction in aeroelastic instability provided by this model.

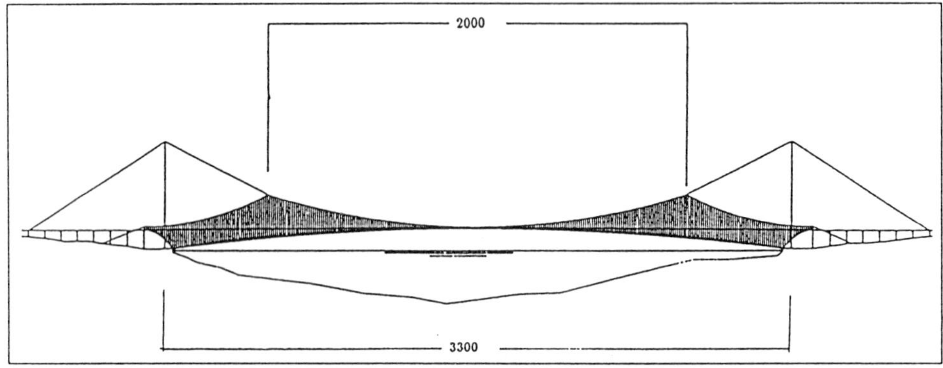

Fig. 6.

28. The model is based on four guidelines:

a) Reduction of bridge weight by means of a streamlined deck with high flutter stability.

b) Considerable increase in torsional rigidity (even though the deck is not torsionally rigid) by means of the double-effect produced by pull (or stabilising) cables with counter opposing curvature.

c) Increase in drag resistance through the adoption of multiple decks and the effect of second order rigidity produced by a spatial system of load-bearing and pull cables.

d) Increase in resistance to lift forces due to the double-effect in c above.

29. The aim of returning to the Musmeci project was to prove that it is possible to construct a large span suspension bridge through the introduction of certain structural innovations rather than by simply attempting to reduce the effects of actions to a minimum. Using the Musmeci structural layout means, in effect, giving a bridge design "structural reserves" in the form of pull cable rigidity, a rigidity which can be regulated. Structural reserves can also be provided in this layout by means of mechanical (but not aerodynamic) damping whereby the anchor points of the pull cables are fitted with viscous or hysteretic dampers. Put

in other terms, the structure in this case is no longer unprotected against external actions and we are no longer obliged to predict actions precisely (something which is in any case not possible). Rather, the structure is equipped with a rigidity and resistance which constitute the structural reserves needed to cope with the large increases in scale involved in this type of construction.

30. Therefore, the main advantages can be summarised as follows.

31. Increase of torsional and lateral stiffness: the increment of torsional stiffness leads, for bridges of about 3000 m span, to a reduction of the rotation at the mid-span to 80% of that computed for a classical suspension system (see Table 1).

Table 1. Input data of the example.

Central span of the bridge	3300m	
Main cable's sag	300m	A_f (main) 5.6m^2
Stabilising cable's sag	80m	A_f (stabilising) 0.8m^2
Initial slope of the lower cable	30°	
Vertical rigidity	1.40m^4	
Lateral flexural rigidity	276.1m^4	
$\lvert C_M \rvert \; \alpha = 0$	0.007	
$dC_M/d\alpha \lvert \alpha = 0$	0.1 rad ÷ 1 rad	
Mean wind velocity at the deck of the bridge	60 m/s	

32. This amount becomes significantly higher if one considers the variation of C_M (aerodynamic moment coefficient). Table 2 shows the percentile differences for some values of the initial C_M derivative.

Table 2. Results.

$dC_M/d\alpha \lvert \alpha = 0$	0.1	0.5	1
$\Delta\alpha$ (%)	20.36	25.78	36.21
Δd(%)		27.37	

$\Delta\alpha = (R_c - R_d)/R_c \times 100$

R_c quasi static rotation at mid-span (classical layout)
R_d quasi static rotation at mid-span (double effect layout)
$\Delta d = (D_c - D_d)/D_c \times 100$

D_c quasi static displacement at mid span (classical layout)
D_d quasi static displacement at mid span (double effect layout)

33. The stiffness increase has also an advantageous influence on the dynamic behaviour; in fact, the higher eigen frequency interacts with the lower spectral intensity of the wind turbulence. This will lead to a reduced dynamic response. Numerical tests on a 3D model are ongoing, whose results confirm substantially this effect.

34. Improvement of structural resource: the double effect suspension system allows the attribution of an additional design parameter for controlling the structural behaviour. This is given by the control one can activate on the pretension of the lower cables. This force is, primarily, an independent parameter, which can be easily changed. On the contrary, in the classical suspension layout the force in the cables is given by the equilibrium condition for given geometry and loads.

35. One should also stress the further advantage given by dampers, which can be easily installed at the extremities of the tower, stabilising cables, giving therefore a decisive contribution to the structural and aerodynamic damping.

36. Removing of the hangers at the extremities of the mid-span: as confirmed by the preliminary design of Gibraltar Strait Bridge by Lin and Chow, Musmeci's proposal allows the increase in the rigidity of the vertical hangers, which would become extremely low in the classical solution.

37. The factors listed above appear therefore to provide the hope for synthesis of the two approaches in suspension bridge construction. The solution proposed seems to satisfy all the primary requirements of large suspension bridges thus reconciling the two, previously opposing, design approaches. We can draw some comfort from the opinion expressed by R.H. Scanlan who believes that with the current design tendency towards high stability profiles, "instability due to pure and simple flutter is less and less probable... while the action of buffeting is attracting increasing attention as the main problem caused by wind on large bridges".

CONCLUDING REMARKS

38. Reproposing the concepts of the design of S. Musmeci, the authors tried to show an alternative way to face the design problems of long span suspension bridges, through the introduction of appropriate improvements in the structural layout. This concept is innovative with respect to the "classical" strategy of only reducing the loads due to the wind.

39. The double effect suspension bridge can provide some additional structural resources to the designer, consisting in the control of the stabilising cable rigidity as well as in the increased structural damping.

40. Comparative analyses into aeroelastic behaviour are still in progress, via numerical simulations on a 3D model. However, any worsening of the characteristic studied can be ruled out given that the proposal made does not influence deck aerodynamics.

REFERENCES

1. LARSEN, A. (ed.), A.A. Balkema, Rotterdam/Brookfield, 1992. Proceedings of the International Symposium on Aerodynamics of Large Bridges, Copenhagen, February 19-21, 1992.

2. IASS-CSCE International Congress. Innovative Large Span Structures: Concept, Design, Construction. Toronto, 1992.

3. GIMSING, N.J. Large bridges of the future. Proceedings of the International Symposium on Aerodynamics of large bridges, Copenhagen, 1992.

4. LIN, T.Y., CHOW, P. Gibraltar Strait Crossing - a Challenge to Bridge and Structural Engineering International, vol. 2, 1991.

5. MIYATA, T. et al. Aerodynamics of wind effects on the Akashi Kaikyo bridge. Atti del Convegno ANIV IN VENTO 1992, Capri.

6. FINZI, L., CASTELLANI, A. Innovative structural system for a 2 miles span suspension bridge. IASS-CSCE International Congress, Innovative large span structures: concept, design, construction. Toronto, 1992.

7. MADSEN, H.O., OSTENFELD-RONSETHAL, P. Wind criteria for long span bridges. Proceedings of the International Symposium on Aerodynamics of large bridges. Copenhagen, 1992.

8. DIANA, G., FALCO. Indagine analatico-sperimentale su un ponte sospeso di grande luce soggetto all'azione del vento. Atti del Convegno ANIV IN VENTO 1990.

9. SCANLAN, R.H. Wind dynamics of long-span bridges. Proceedings of the International Symposium on Aerodynamics of large bridges. Copenhagen, 1992.

10. OSTENFELD, K., LARSEN A. Bridge engineering and aerodynamics. Proceedings of the International Symposium on Aerodynamics of large bridge. Copenhagen, 1992.

65. On the performance of double-layer square-on-square grids

S. A. RIZWAN and A. A. BHATTI, Lahore University of Engineering and Technology, Pakistan

SYNOPSIS. A study is reported on the static linear elastic behaviour of an 8x8 square-on-square double-layer model grid having hollow circular skeletal members of 1/2" x 1/16" cross-section with a span/depth ratio of 10.66 for various support, subdivision and stiffening characteristics. The structure has been analysed using SAP90, a computer program run on a 386 machine. The theoretical and experimental behaviour of the grids have been compared in terms of nodal displacements and member forces. The results suggest that the structural behaviour of this type of double-layer grid depends heavily on both the support and stiffening conditions.

INTRODUCTION

1. Double-layer skeletal grids have been used very successfully by many engineers and architects the world over, due mainly to their superior efficiency in providing large unobstructed areas, associated with faster erection speeds and low maintenance cost, etc. An excellent account of various analysis and design aspects of double-layer grids is found in the literature (ref. 1) along with complete global information on some important parameters such as spans, design loads, self weights, member forces and maximum deflections, etc. (ref. 2).

2. It is generally believed that diagonal grids are more efficient than latticed grids under given loads, where efficiency is decided on the basis of self-weight, maximum member forces and deflections. It has also been reported that for latticed grids edge or stiffening beams are not usually a part of the most efficient grid structures (ref. 3). Although the response of various types of double-layer grids under a variety of loads is known, only a few attempts have been made to address areas like the effects of support constraints and localised stiffening of these structures (refs 3 and 4) wherein these restraints were considered as point-type columns or tree-type columns for a 10x10 square-on-square double-layer grid.

3. To the best of our knowledge, the effect of various types of support restraints and localised stiffening for square-on-square latticed grids, which are simple to design and erect, has not been studied in depth so far. Therefore, an

effort has been made to study these parameters and their effects on structural response in terms of maximum member forces and maximum displacements etc. This has been undertaken by assuming the structures to be pin-jointed trusses and by removing the inherent instability by using appropriate horizontal restraints which may be replaced by imaginative bracing of the lower chord. The theoretical response has also been compared with the experimental one after designing, fabricating and testing a similar model.

4. Theoretical analysis has been undertaken for two square-on-square grids, one 8x8 and the other 6x6, having the same span and span/depth ratio, with the latter structure being also the experimental model.

EFFECTS OF SUPPORT RESTRAINTS AND LOCALISED STIFFENING

5. To study these effects the behaviour of a 8x8 square-on-square latticed grid, having l/2"xl/16" hollow circular members with web members inclined at an angle of 56. 3° to the horizontal, loaded under a uniform pressure of 1 psi (144 lbs/ft^2) was investigated. A standard SAP90 static linear elastic program has been used and the loads corresponding to the above pressure have been assumed to act at all nodes in the bottom chord. The general layout of the structure along with the node numbering scheme is shown in Fig. 1.

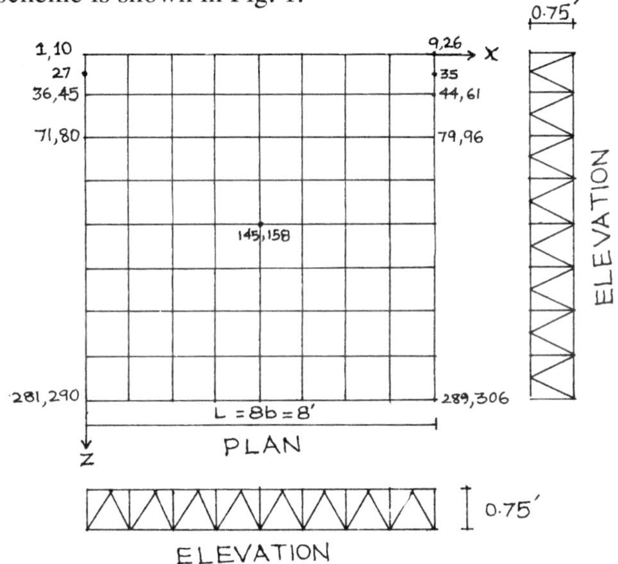

Fig. 1. General layout and node numbering of structure.

6. In the analysis, the structure has been supported using five different arrangements namely: S-1 to S-5 as shown in Fig. 2.

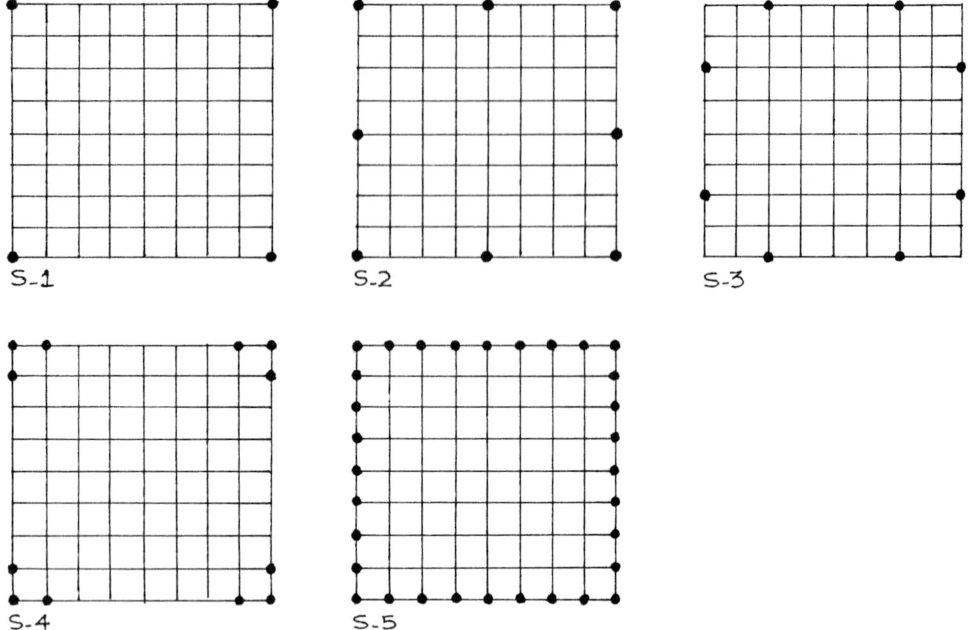

Fig. 2. Five different pin support conditions S-1 to S-5 for unstiffened system. (Note: S-1a to S-5a would mean no horizontal restraints at supports).

7. The trusses along the bold lines in Fig. 3 have been stiffened by 50%.

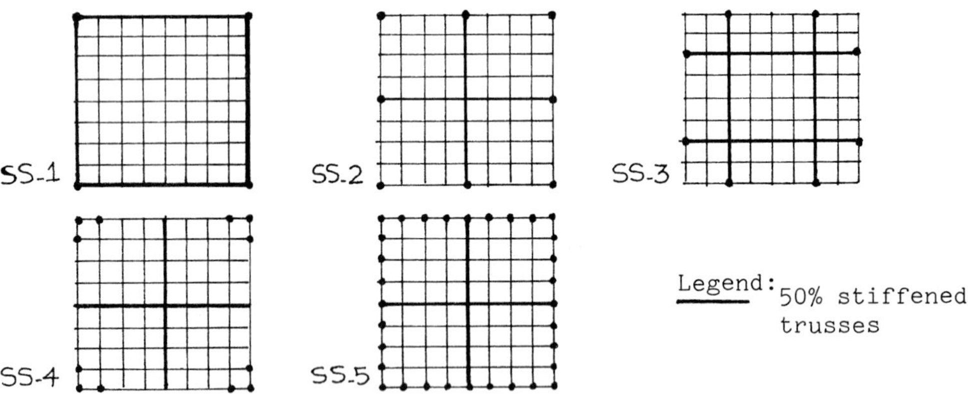

Fig. 3. Five different pin support conditions S-1 to S-5 for a 50% stiffened system as shown.
(Note: S-1a to S-5a would mean no horizontal restraints on supports).

8. The following types of support restraints and localised stiffening have been considered in the analysis.

Case 1: Support conditions S-1 to S-5 shown in Fig. 2 correspond to complete translational restraints in 3 directions without stiffening.

Case 2: Support conditions S-1a to S-5a corresponding to Fig. 2 are configured in a manner so that no horizontal reactions are generated due to vertical loads.

Case 3: 50% stiffened trusses along broken lines shown in Fig. 3 having only translational restraints at supports (SS-1 to SS-5).

Case 4: 50% stiffened trusses along bold lines as shown in Fig. 3 having no horizontal restraints at supports (SS-1a to SS-5a).

9. Table 1, gives the results of the analysis of the unstiffened system shown in Fig. 2 (Cases 1 and 2).

Table 1. Effect of edge restraint in a 8x8 square-on-square latticed grid.

Support conditions	Maximum chord forces (lbs.)	Max. Reactions		Maximum deflection central (in.)	Diff. $\frac{\text{Sia-Si}}{\text{Si}}$
		Horizontal lbs	Vertical lbs		
S1	-2948,+1340	±2045	2268	0.22912	28.5%
S1a	-2642,+2597	-	2268	0.294625	
S2	-1360, +332	±1004	1688	0.085591	16.7%
S2a	- 996, +747	-	1703	0.099968	
S3	-1002, +719	± 725	1152	0.08251	16.6%
S3a	- 837, 813	-	1152	0.096273	
S4	-1159, 667	± 764	1630	0.104593	16.7%
S4a	- 960, 866	-	1800	0.122148	
S5	- 938, +278	± 636	410	0.057722	55.3%
S5a	- 931, 907	-	406	0.089650	

10. The maximum central deflection corresponding to Case S5 is predicted to be 0.09 inches using the slab analogy method assuming pin connections.

11. Table 2, gives the analysis results of the 50% stiffened system shown in Fig. 3 (Cases 3 and 4) above.

Table 2. Effect of local stiffening of trusses in a 8x8 square-on-square latticed grid.

Support conditions	Maximum chord forces (lbs.)	Max. Reactions		Maximum Deflection central (in)	Difference in deflections	
		Horizontal lbs.	Vertical lbs.		SSia-SSi / SSi	Si-SSi / SSi
SS1	-3178, 1069	±2147	2268	0.188280	26%	18%
SS1a	-2940, 2872	-	2268	0.237338		19%
SS2	-1717, 432	±1261	1729	0.072255	21.8%	16%
SS2a	-1313, 1289	-	1736	0.088023		12%
SS3	-1282, 616	± 896	1152	0.070018	21%	15%
SS3a	- 980, 957	-	1152	0.084707		12%
SS4	-1136, 870	± 782	1642	0.102835	11.9%	2%
SS4a	-1012, 907	--	1813	0.115116		6%
SS5	-1229, 362	-	518	0.050794	56.5%	12%
SS5a	-1224, 1200	-	516	0.079357		11%

EXPERIMENTAL INVESTIGATION

12. A 6x6 square-on-square experimental model of a double-layer latticed grid having the same span (8') and span/depth ratio (10.66) has been designed, fabricated and tested in the laboratory of the Civil Engineering Department, at the University of Engineering and Technology, Lahore, Pakistan. This structure is shown in Fig. 4.

Fig. 4. The experimental model of a 6x6 double-layer latticed grid with a span of 8' and depth = 0.75'. Web member angle of inclination = 48.33° with the horizontal. Hollow circular Members(1/2"x1/16").

13. A special connector has been used in the structure and members were connected by running them full length through the holes in one direction and welding to connect cut members in the other direction. A slight spot-welding was also done on the nodes. The specially designed supports were assumed to have complete translational restraint at the supports. The model was first analysed by using a computer and was then tested in the laboratory by applying loads at the lower chord nodes by means of weight supported by hangers attached to the nodes. The following load cases have been considered.

A- 70 lbs load at 9 points adjacent to the centre of the model.

B- A central load of 324 lbs.

14. The results of the experimental investigation are given in Table 3.

Table 3. Results of a 6x6 square-on-square experimental model having a span = 8' and a span/depth ratio of 10.66.

Load case	Theoretical Maximum		Experimental Maximum	
	Central deflection (in)	Chord forces (typical) (lbs)	Central deflection (in)	Chord forces (typical) (lbs)
A	0.028	-163 (Member 124) +147 (Member 352)	0.031	-174 (Member 124) -132 (Member 352)
B	0.0175	-200 (Member 124) +128 (Member 352)	0.019	-175 (Member 124) +125 (Member 352)

15. After careful application of load and the instrumentation of the grid, which consists of deflection gauges and electrical resistance strain gauges at suitable pre-selected points, it was revealed that the originally designed ideal pin-supports for the grid actually behaved as roller supports due to some very minor play in the ball and socket arrangement used. Therefore, the theoretical results have been compared accordingly.

CONCLUSIONS

a) By eliminating horizontal restraints at supports, the resulting central grid deflections are increased. The greatest difference took place in support types numbers 5 and 1 being 55.3% and 28.5% respectively. In remaining cases the increase in deflection was constant at 16.7%.

b) By eliminating horizontal restraints the maximum compressive chord forces were reduced by a maximum of 26.76% in support type S-2 while there was no difference in case of support type S-5.

c) By eliminating horizontal restraints, maximum tensile chord forces increased in general for all types of supports, this increase being the greatest (70%) for the support type S-5.

d). By eliminating horizontal restraint in a continuously supported grid (S-5) there is no reduction in maximum compressive chord forces while tensile forces are increased by 70%. This was probably due to the fact that support restraints were applied to bottom chord joints only.

e). The magnitude of maximum horizontal reactions depends upon the number of supports and their position in plan. In general, by increasing the number of supports the magnitude of the horizontal reactions is decreased. The same applies to vertical reactions.

f). The magnitude of vertical reactions remains almost constant for a particular support system even if the horizontal restraint is removed.

g). In general by localised stiffening of trusses, achieved by increasing member cross-sectional areas and moment of inertia, the grid deflections are reduced when other parameters are kept constant. The local stiffening effect is more pronounced when the structure is made flexible by removing restraints. This confirms established results (ref. 5).

h). Stiffened trusses are useful only when they are supported adequately. If they are not supported at all (SS-4 type) the reduction in central deflection due to the additional stiffness is negligible.

i). In general, localised stiffening of trusses increases maximum compressive and tensile chord forces for the same support restraint condition.

j). The localised stiffening of trusses increases the magnitude of horizontal reactions while vertical reactions are not increased.

k). Localised stiffening effects become more pronounced when the number of supports is reduced. This means that there is no real advantage in stiffening continuously supported grids.

ℓ). For a constant span, loading and span/depth ratio, an increase in number of subdivisions and increases in the angle of inclination of the web members which reduces deflections and member forces.

ACKNOWLEDGEMENTS

The authors wish to extend their heartiest gratitude to Prof. Dr. Shaheen Akhtar, Pro-Vice-Chancellor of University of Engineering & Technology, Lahore, Pakistan for taking personal interest in this research work and for providing funds under the faculty research programme for the year 1992-93.

REFERENCES

1. MAKOWSKI, Z. S. (ed) Analysis, design and construction of double-layer grids, Applied Science publishers, Ltd. London, U.K, 1982.

2. MAKOWSKI, Z. S A world wide review of space structures in sports buildings. Proc. International Colloquium on space structures, Beijing, China, Elsevier Applied Science London, 1987, pp 1-4.

3. WEST, F.E.S. A study of efficiency of double layer grid structures, M.Phil Thesis, University of Surrey, U.K., 1967.

4. SALAJEGHEH, E. An analytical study of a double layer square-on-square grid. M.Sc. Thesis, University of Surrey, 1977.

5. LEE, H.G. & MAKOWSKI, Z. S. Study of factors affecting stress distribution in double layer grids of square and diagonal type, Architectural Science Review, Vol.-20, No. 4, Dec. 1977, pp 90-102.

CONVERSIONS: 1 Foot = 304.8mm; 1 psi = 145 N/mm^2; 1 lb = 4.448 N.

66. The concept of basic modules and its applications to lateral analysis of high-rise buildings

M. GRIGORIAN, MGA, USA

SYNOPSIS. This paper introduces a simple approach for the study of structural phenomena associated with lateral response of regular highrise frames. The concept is based upon the assumption that the structure is composed of imaginary modules that fit within the actual bays of the framework, and that all imaginary modules deform through similar shapes by formation of points of inflexion at similar locations. The imaginary modules of a story are combined to form an equivalent "Basic module" that possesses the same structural characteristics as the corresponding story level subframe. The use of the concept is further utilised to obtain a number of closed form formulae for the initial design of multistory structures. These include expressions for the determination of the natural frequencies, the P-delta effects, as well as the critical buckling loads of regular multistory frames.

INTRODUCTION

1. Despite the abundance of highly specialized software developed for the analysis and design of multistory buildings it is still deemed essential to have access to simple analytic methods that yield accurate design values as well as providing insight into the physical behavior of such structures. In this paper an attempt is made to introduce a simple theoretical technique that uses basic structural principles to develop practical design formulae for the lateral response of regular multistory structures subjected to transverse loading. The concept is based upon the assumption that the structure is composed of imaginary modules that fit within the actual bays of the framework, and that all imaginary modules deform through similar shapes by formation of points of inflection at similar locations. The imaginary modules of a story are combined to form an equivalent "Basic module" that possesses the same structural characteristics as the corresponding story level subframe. The concept is utilized to obtain a number of useful, closed form formulae for the initial design of multistory structures. Numerical examples have been provided to demonstrate the simplicity of the proposed solutions. In the forthcoming sections after introducing the concept of the "Basic module", the lateral stiffness of a story level subframe is derived. The theory developed in this section is then used to study the P-delta effect and module stability.

DEVELOPMENT OF THE BASIC MODULE

2. Under ideal conditions the total relative displacement δ_T of an imaginary module, as depicted in Fig. 1b, may take place due to the action of any one or any combination of the following major causes of displacements, (a) Racking or bent action $\delta_r = \delta_c + \delta_b$, which contains the components of bending displacements δ_c due to column bending and δ_b due to beam rotation, (b) Shear displacements δ_{sh} which also contains two components, δ_{shc} as the contribution of the shearing deformation of the columns and δ_{shb} as the shearing deformations of the beams. And (c), the cantilever displacements δ_a which are caused by the axial deformations of the columns. Therefore:

$$\delta_T = \delta_a + \delta_b + \delta_c + \delta_{shc} + \delta_{shb} \qquad (1)$$

caused by the external shear force V and the overturning moment Mo. Fig. 2 a, depicts the deformations of one such imaginary module composed of four rigidly connected prismatic members and simply supported at its lower nodes. To study the nature of any one component of the total drift δ_T it is convenient to separate the effects of that type of deformation from the others, then by the use of simple theory of structures;

$$\delta_b = \frac{Vh^2 l}{12E\bar{J}}, \; \delta_c = \frac{Vh^3}{12E\bar{I}}, \; \delta_{shc} = \frac{Vh^2}{G.h.\bar{A}_{shc}} \text{ and } \delta_{shb} = \frac{Vh^2}{G.l.\bar{A}_{shb}} \qquad (2)$$

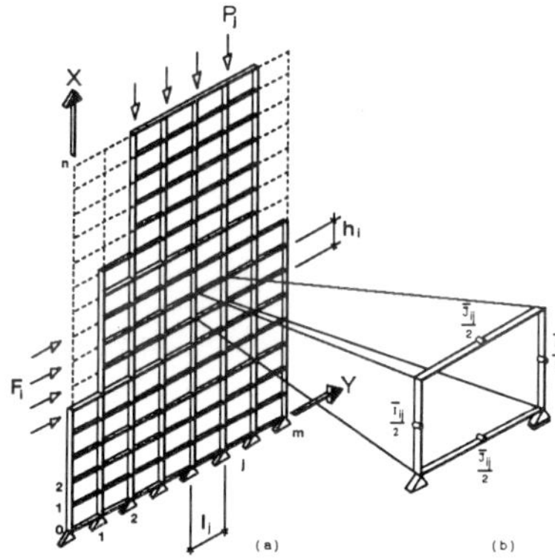

Fig. 1. Highrise moment frame, geometry and loading with enlarged imaginary module

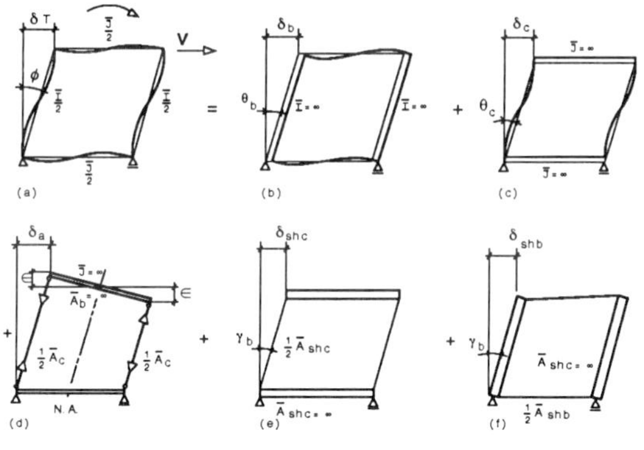

Fig. 2 . Deformations of a basic module.
a) Total elastic drift.
b) Drift due to bending of beams.
c) Drift due to bending of columns.
d) Drift due to axial strains.
e) Drift due to column shear
f) Drift due to beam shear.

where $\bar{I}$ and $\bar{J}$ are the sum of moments of inertia of the two columns and the two beams respectively. $\bar{A}_{shc}$ and $\bar{A}_{shb}$ are the total shear areas of the two columns and the two beams respectively. G is the shearing modulus of the material of the module. Assuming that the structure bends as an upright cantilever with a moment of inertia equal to the second moment of the column areas about their common neutral axis, then the axial deformations of the columns of the module would contribute their own share of the subframe drift caused by the fraction of the external moment acting upon it. i.e., $\delta_a = (M_o h^2/EI_o)$ where $I_o = \bar{A}_c.l^2/4$, and $\bar{A}_c/2$ is the assigned cross sectional area of each column of the module.

3. Therefore a complete presentation of module displacement may be summarized as;

$$\delta_T = \frac{Vh^2}{12E}\left(\frac{h}{\bar{I}} + \frac{l}{\bar{J}}\right) + \frac{Vh^2}{G}\left(\frac{1}{h.\bar{A}_{shc}} + \frac{1}{l.\bar{A}_{shb}}\right) + \frac{M_o h^2}{EI_o} \quad (3)$$

Since the terms inside the first and second parenthesis describe the sums of the flexural and shearing stiffnesses respectively of both beams and both columns of the basic module, then the merging of common members of adjacent modules would result in the regeneration of the original members at the same location. For instance if the module "ij" is merged in series with its neighbouring modules in the horizontal direction to form a new subframe containing all the bays in the same story as in Fig. (3) then Eq. (3) may intuitively be expanded to estimate the total lateral displacement of the newly formed story level subframe, where;

$$\frac{\bar{I}}{h_i} = \sum_{j=o}^{m} \frac{I_{ij}}{h_i}, \quad \frac{\bar{J}}{l} = \sum_{j=o}^{m} \frac{J_{ij}}{l_j}, \quad h_i.\bar{A}_{shc} = \sum_{j=o}^{m} h_i.A_{shcj}, \quad l.\bar{A}_{shb} = \sum_{j=o}^{m} l_j.A_{shbj} \quad (4)$$

l is arbitrarily selected as any convenient distance.

4. The general form of Eq. (3) suggests that it could be further condensed to define the deformations of an equivalent single bay elastic system possessing two none interactive stiffnesses K_{ri} and K_{ai} respectively at level i, i.e.,

$$\delta_i = \frac{V_i}{K_{ri}} + \frac{M_{oi}}{K_{ai}.h_i} \quad (5)$$

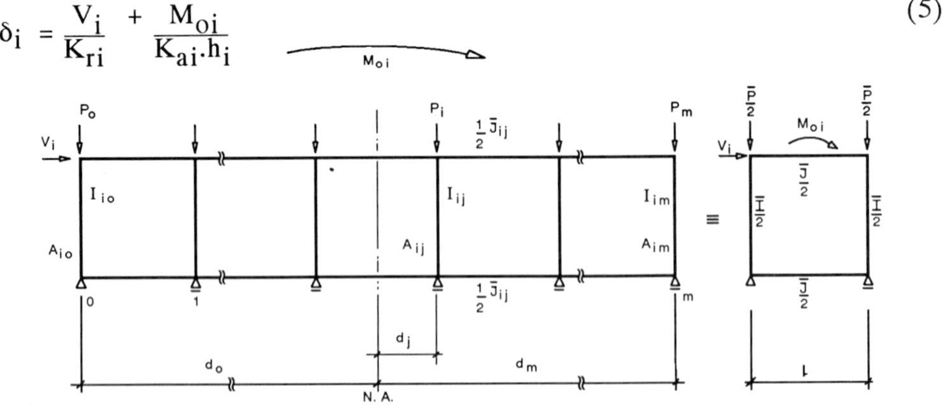

Fig. 3 . Horizontal subframe and equivalent module.

THE P-δ EFFECT AND MODULE STABILITY

5. The accumulation of gravity loads in the columns of high rise buildings tends to reduce the resistance of the structure to lateral displacements. Under extreme conditions, loss of stiffness may lead to structural collapse through three distinct, or combination of three distinct modes of elastic instability. The most commonly known mode of failure is that of strut buckling and is associated with overburdening of any specific column. Subframe or shear type instability is associated with racking stiffness and excessive loading of any story level subframe. Flexural or cantilever type instability is associated with beam-column type buckling of the entire structure under total gravity forces acting upon it. The effective racking moment of the subframe including the effects of the total axial load $\bar{P}_i$ may be expressed as:

$$M_{ri} = V_i.h_i + \bar{P}_i.\delta_i \qquad (6)$$

6. The product $V_i.h_i$ represents the story racking moment M_{ri} in the absence of axial loads. Rewriting Eq. (5) and replacing $V_i.h_i$ with M_{ri} then the story level drift becomes:

$$\delta_i = \frac{M_{ri}}{K_{ri}.h_i} + \frac{M_{oi}}{K_{ai}.h_i} \qquad (7)$$

Substituting for M_{ri} from Eq. (6) in Eq. (7) and rearranging it gives :

$$\delta_i = \frac{V_i}{(1-\bar{P}_i/P_{cr,r})K_{ri}} + \frac{Moi}{(1-\bar{P}_i/P_{cr,r})\,K_{ai}.h_i} \qquad (8)$$

where $P_{cr,r}=K_{ri}.h_i$ is the total racking critical load of story level i. It is instructive to note that the flexural buckling of the structure is governed by the total lateral displacements while shear type buckling is only affected by the increments of the lateral deflections. Now ignoring the terms including the quantities G and I_o, Eq. (8) and (3) reduce to:

$$\delta_i = \frac{Vh^3}{12\,E\bar{I}\left[\frac{\mu}{1+\mu} - \frac{\rho\pi^2}{12}\right]} \qquad (9)$$

which is a generalization of Eq. (5). $\mu = (\bar{J}h/\bar{I}l)$ and $\rho = (\bar{P}/P_{e,r})$ have been introduced for convenience, $P_{e,r} = (\pi^2 E\bar{I}/h^2)$ is the nominal racking Euler load of the basic module. The phenomenon of column softening may easily be included by using modified stiffness coefficients that describe the effect of axial forces on the rotations and deflections of uniform beam-columns. Therefore considering the equilibrium of a basic module under the action of both lateral and axial forces as in Fig. (4) then the governing matrix equation in terms of Livesly's stability functions may be written as:

$$\frac{2E\bar{I}}{h}\begin{bmatrix} f(\rho)+6 & -f(\rho) \\ -f(\rho) & +f(\rho) \end{bmatrix}\begin{bmatrix} \theta \\ \delta/h \end{bmatrix} = \begin{bmatrix} 0 \\ Vh+\bar{P}\delta \end{bmatrix} \quad (10)$$

where by definition the stability function $f(\rho) = s(1+c)$,

$$s = \frac{(1-2\alpha \cot 2\alpha)\,\alpha}{\tan \alpha - \alpha}, \quad c = \frac{2\alpha - \sin 2\alpha}{\sin 2\alpha - \alpha \cos 2\alpha}, \quad 2\alpha = \pi\sqrt{\rho}. \quad (11)$$

Elimination of θ from Eq. (10) results in:

$$\delta_i = \frac{Vh^3}{12\,E\bar{I}\left[\dfrac{\mu f(\rho)}{f(\rho)+6\mu} - \dfrac{\rho\pi^2}{12}\right]} = \frac{V}{K_{ri}} \quad (12)$$

7. The racking stiffness function K_{ri} now contains the effects of both column softening as well as shear type instability. Fig. (4) presents a number of "softening curves" for a wide range of stiffness ratios "μ"and force levels "r" . The numerical values of these curves are readily applicable to practical problems involving moment frames. It may be seen that for relatively flexible beams i.e., $0 \le \mu \le 1$ the two equations are in close agreement, whereas for stiffer beams i.e., $\mu > 1$ the differences could be significant. The modified stiffness solutions (12) derived for a "Basic Module" has been generalized to describe the softening of a story level subframe. However, in doing so a further assumption has been made that it is sufficiently accurate to use a mean value of ($\bar{P}/P_{e,r}$) at any story given by ($\bar{P}/P_{e,r}$) mean =($\sum P$)/($\sum P_{e,r}$) for all the columns instead of the values for each individual column. Eq. (12) and (5) indicate that the general softening of the structure is associated with both the absolute lateral displacement and the relative total story drift δ_i. Since the racking and cantilever components of the total lateral displacement and their respective mode shapes are independent of each other then the total critical load P_{cr} of the entire structure may be expressed by a Rankine type formula, i.e.,

$$\frac{1}{P_{cr}} = \frac{1}{P_{cr,r}} + \frac{1}{P_{cr,a}} = \frac{1}{h.K_{ri}} + \frac{1}{P_{e,a}} \quad (13)$$

$P_{cr,a}$ is equal to the Euler load of an equivalent solid beam-column. ie.

$$P_{cr,a} = P_{e,a} = \pi^2 EI_o / 4H^2 \quad (14)$$

where H = nh is the total height of the frame. $P_{cr,r}$ may either be calculated by trial and error from Eq. (12) or directly obtained from Fig. (4). However, the curves presented in this figure indicate an almost linear reduction of stiffness with increasing axial compression. This observation leads to the following important approximation.

$$f(\rho) = s(1+c) = 6-1.0652\rho \quad (15)$$

which when substituted in Eq. (12) results in the readily workable generalized displacement equation.

$$\delta_i = \frac{V}{K_{ri}} = \frac{Vh^3}{12E\bar{I}\left[\dfrac{1}{\dfrac{1}{\mu}+\dfrac{6}{f(\rho)}} - \dfrac{\rho}{\dfrac{12}{\pi^2}}\right]} \tag{16}$$

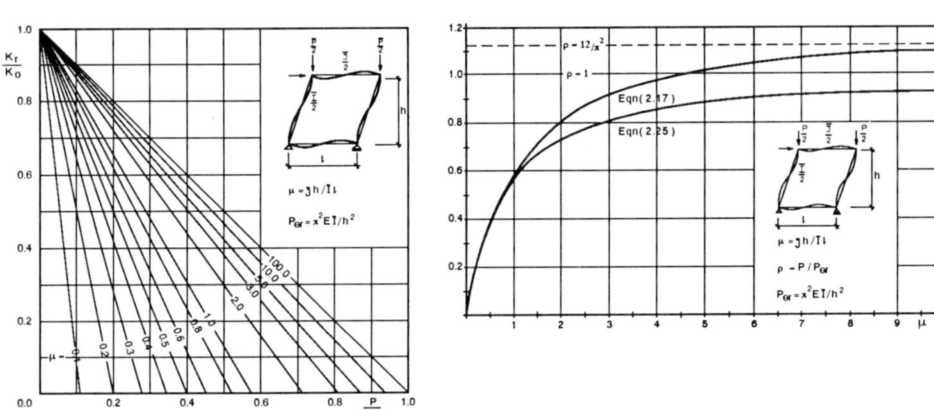

Fig. 4 . Softening curves for axially loaded modules.

Fig. 5 . Subframe critical load as function of beam to column stiffness ratio.

The sidesway or shear instability of the module is associated with total loss of racking stiffness, i.e., $K_{ri} = 0$, whence Eq. (16) yields:

$$P_{cr,r} = \frac{6E\bar{I}}{h^2\left(\dfrac{12}{\pi^2}-1\right)}\left[\left(1+\frac{12}{\pi^2}\mu\right) - \sqrt{\left(1+\frac{12}{\pi^2}\mu\right)^2 - 4\left(\frac{12}{\pi^2}-1\right)\mu}\right] \tag{17}$$

8. For a uniform frame, where $K_{ri} = K_r$ and $K_{ai} = K_a$, the results of this formula i.e., values of "ρ" corresponding to $(K_r / K_o) = 0$, Fig. (4), practically coincide with Bleich's and Merchant's solutions for antisymmetric buckling of regular rectangular multistory moment frames with fixed ends at the base and n>3, neither of which include the effects of cantilever buckling. An important conclusion drawn at this stage is that for multistory moment frames, i.e., n>3 the racking critical load is practically independent of the number of stories.

9. Substituting for P_{cr}, a from expression (14) and for K_{ri} from Eq. (16) in Eq. (13), results in the critical carrying capacity of the entire system as:

$$P_{cr} = \frac{P_{e,a}}{1 + \dfrac{P_{e,a}}{P_e}\left[\dfrac{(1+\mu)-(1-\pi^2/12)}{[1-(1-\pi^2/12)\rho](12/\pi^2-\rho) - \mu\rho}\right]} \tag{18}$$

10. Since "ρ" depends on P_{cr} then this equation can best be solved by trial and error. It may be of interest to note that for $\rho = 0$ and $I = 2I$ Eq. (18) coincides with Timoshenko's solution for a regular column with batten plates.

FRAME VIBRATIONS

11. Since the total lateral displacements of highrise structures are composed of two distinct components as in Eq. (5), it would seem reasonable to seek a Dunkereley type solution for the natural frequencies of such systems, i.e.,

$$\frac{1}{\omega^2_{\xi}} = \frac{1}{\omega^2_{\xi a}} + \frac{1}{\omega^2_{\xi r}} \tag{19}$$

Where ω_{ξ} is the natural frequency of the composite system corresponding to mode number "ξ". $\omega_{\xi a}$ and $\omega_{\xi r}$ are the natural frequencies of the flexural and racking components of the frame respectively. As far as the author can ascertain there are no known solutions for Eq. (19). However he has shown that the exact, undamped racking natural frequencies of a discretized cantilever, in the absence of axial compression, may be expressed as:

$$\omega^2_{\xi r} = \frac{4K_r . g}{W} \sin \frac{\eta\pi}{(4n+2)} \tag{20}$$

where "n" is the number of stories, $(z^2 = g/W)$ and K_r are the story level mass and racking stiffness respectively. $\eta = 1,3.....$ 2n-1 are odd numbers associated with the ξth mode of vibrations of the structure defined by $\eta=2\xi - 1$. The use of the concept of imaginary modules allows Eq. (19) to be generalized to include the effects of axial compression. This may be achieved simply by employing Eq. (9) or (12) for K_r in lieu of Eq. (5) and (7). Further in the absence of exact closed form solutions for the vibrations of uniform cantilevers with equal masses at regular intervals the author would like to suggest the following approximate solution based on Rayleigh's method of analysis, i.e.,

$$\omega^2_{\xi r} = \frac{16g . K_a}{W} \frac{A^2_{\xi}}{(2n+1)^4} \tag{21}$$

where for $n=1$, $\xi=1$ $A_1 = 3.8971$
for $n\geq 2$, $\xi \leq 2$ $A_1 = 3.6213$, and $A_2 = 24.2810$
for $n \geq 3$, $\xi \geq 3$ $A_{\xi} = (2\xi-1)^2.(\pi/2)^2$

12. Substitution of expressions (20) and (21) in Eq. (19) results in the following generalized frequency equation for uniform highrise structures:

$$\frac{1}{\omega^2_{\xi}} = \frac{W}{4gK_r}\left[\operatorname{cosec}^2 \frac{(2\xi-1)}{2(2n+1)} + \frac{K_r}{4K_a} \frac{(2n+1)^4}{A^2_{\xi}}\right] \tag{22}$$

As an indication of the accuracy of Eq. (21) it has been found that the error involved in calculating the fundamental frequency of a three mass system is less than 0.3% .

APPLICATIONS

13. To demonstrate the applications of the proposed concept and the corresponding solutions a number of practical examples discussing the statics, dynamics and the stability of a cantilevered moment frame are presented. The mathematical formulation presented above is not only applicable to highrise structures but also to any other structural form that may be shown to be composed of a series of regular basic modules, e.g. battened columns, vierendeel frames, etc.

Example I - Statics

14. Obtain the generalized deformation functions of a regular "n" story cantilevered structure carrying a uniform distribution of lateral nodal forces F(x)=F, and calculate the corresponding numerical values for the specific case of n=10, m=1, l=240 in, h=120 in, $I_o=I_1=400\,in^4$, $\bar{J}=2000\,in^4$, $A_o=A_1=25\,in^2$, and E=29,000 ksi.

15. The complete solution for this load case which is generally not found in the literature is contained in the double summation; (see Appendix).

$$Y(x)= \nabla^{-1}\Delta^{-1}\left[\frac{M(x)h^2}{EI_o}+\frac{F(x)}{K_r}\right] \tag{23}$$

which; with the origin of coordinates at the lower end gives:

$$Y(x) = \frac{Fh^3}{24EI_o}\left[(x-1)^2-4nx+6n(n+1)\right]x^2+\frac{F}{2K_r}\left[(2n+1)x-x^2\right] \tag{24}$$

$$\theta(x) = \frac{Fh^2}{12EI_o}\left[2x^3-3(1+2n)x^2+(6n^2+6n+1)x\right]+\frac{F}{2K_rh}(n-x) \tag{25}$$

16. These closed form solutions are exact since they satisfy both the governing difference Eq. (23) as well as the corresponding boundary conditions at both ends. However, in order to relate the flexural components of these deformation functions to mid-height of the modules the integer (x) must be replaced with (x-1/2) in the flexural terms of both Eq. (24) and (25).

Example II - Stability

17. Evaluate the magnitude of the critical uniform compression load associated with the structure of example I, and calculate the least beam stiffness required to stabilize a story level subframe.

18. Assuming the buckling of the structure is given by the form $Y(x) = B_\xi.(1-\cos\alpha x)$, where $\alpha = \xi\pi/(2n+1)$, and B_ξ is the amplitude of the deformed shape then the buckling of the frame would generate a bending moment $M(x) = P_{cr}.B_\xi - P_{cr}.Y(x)$ and a shear force $F(x)= \frac{P_{cr}}{h}\nabla Y(x)$, whence the governing equation becomes:

$$\nabla Y(x) = \frac{P_{cr}[B_\xi - Y(x)]h^2}{EI_o} + \frac{P_{cr}}{K_r.h} \nabla Y(x) \quad (26)$$

Substituting for the buckled shape in Eq. (26) and observing that

$$\nabla(1-\cos\alpha x) = 4\sin^2\frac{\alpha}{2}\cos\alpha x \quad (27)$$

then the desired solution becomes:

$$\frac{1}{P_{cr}} = \frac{1}{K_{r.}h} + \frac{1}{4K_{a.}h\sin^2\frac{\alpha}{2}} \quad (28)$$

19. Now letting (2n+1)h=2H and noting that as sin $\alpha/2$ diminishes towards ($\alpha/2$), the second term of Eq. (28) approaches the limit of ($4/P_{e,a}$), which is the inverse of the buckling load of the corresponding solid cantilever. In fact for n=10 the actual value is ($4.007/P_{e,a}$). For the present numerical example, μ=1.25, gives $P_{cr,r}=0.6310\ P_e$, $P_{e,a}=2.2525\ P_e$ and formula (18) which is a solution of Eq. (28) gives $P_{cr}=0.305P_e$ as the overall buckling strength of the structure. Theoretically racking stability is achieved when $K_r>0$, Eq. (16) then yields, after some rearrangement, for the smallest ratio insuring stability.

$$\mu_{min} = (\bar{J}h/\bar{I}l)_{min} > \frac{\rho\pi^2}{12(1-\rho)}\ [1-(1-\pi^2/12)\ \rho] \quad (29)$$

Example III - DYNAMICS

Calculate the first three natural frequencies of the composite structure of example I.

From Eq. (20) $\omega_{\xi r} = 18.9216\ z \sin\frac{(2\xi - 1)x180}{42}$, giving

$$\omega_{1r} = 1.4140\ z,\ \omega_{2r} = 4.2104\ z \text{ and } \omega_{3r} = 6.9128\ z \quad (30)$$

From Eq. (22) $\omega_{\xi a} = 0.9976\ z\ A_\xi$, giving

$$\omega_{1a} = 3.6126\ z,\ \omega_{2a} = 24.2227\ z \text{ and } \omega_{3a} = 61.5370\ z \quad (31)$$

Finally employing Eq. (20) it gives

$$\omega_1 = 1.3167\ z,\ \omega_2 = 4.1482\ z \text{ and } \omega_3 = 6.8696\ z \quad (32)$$

which compare favourably with results of a computer program developed for dynamic analysis of plane frames i.e.

$$\omega_1 = 1.3338\ z,\ \omega_2 = 4.0426\ z \text{ and } \omega_3 = 7.0028\ z \quad (33)$$

APPENDIX

$\nabla = \Delta\nabla = \Delta - \nabla = (E^{+1} - 1) - (1 - E^{-1})$, E is the finite difference operator defined by $E^{\pm1} f(x) = f(x\pm1)$.

67. Collapse patterns for double-layer grids

U. R. MADI and A. EL-TAYEM, University of Jordan

SYNOPSIS. The collapse patterns of double-layer grids (DLGs) are investigated by means of specially developed software that traces the elastoplastic response up to complete failure. A parametric study of DLGs is conducted in which four possible layout configurations, namely: square-on-square, square-on-diagonal, diagonal-on-square and diagonal-on-diagonal are used with varying side dimensions, depth, module, supports and number of member types and their response under vertical loading is traced to complete collapse. The main collapse patterns of DLG's are then identified and classified with the results of the parametric study presented in graphical forms. These forms help predict the failure pattern of a DLG once its parameters are specified. Also, based on these graphs, ways of enhancing the economy and load carrying capacity of DLGs can be identified.

INTRODUCTION:

1. The present study aims at investigating the collapse patterns of double-layer grids under various parameters that influence their occurrence. For this purpose, a computer program was developed that traces the structural response under a proportional loading system form the onset of loading and up to complete collapse. This program is based on the modified stiffness technique for the elastoplastic analysis of pin-jointed spatial structures (ref. 1). It employs the member constitutive relationship shown in Fig. 1 and conducts a series of scaled analyses to identify sequential member failure or change of state, i. e:

$$\underbrace{\sum_{i=1}^{n} \tilde{A}^T \tilde{K}_i \tilde{A}}_{\text{Stiffness matrix for stage i}} \; \underbrace{q_i}_{\text{nodal displacement}} = \sum_{i=1}^{n} \underbrace{\tilde{A}_o^T}_{\text{proportional loading system}} \; \underbrace{\lambda_i}_{\text{load factor for stage i}} \qquad (1)$$

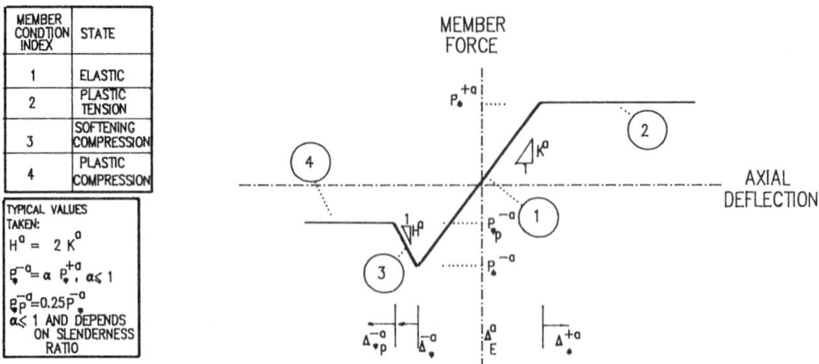

Fig. 1. Idealised member axial force-axial deformation constitutive relationship.

2. To enhance its efficiency and eliminate the need for laborious data preparation, the program was supplemented by an automatic data generation routine that produces all the required member topology and node co-ordinates once the grid configuration, dimensions, module and depth are specified.

PARAMETRIC STUDY

3. A parametric study was conducted with the help of example structures. This parametric study was based on rectangular grids of the four possible layout configuration shown in Fig. 2 namely; square-on-square, square-on-diagonal, diagonal-on-square and diagonal-on-diagonal (ref. 2). The decision to limit the study to square and rectangular plans supported on the outside boundary is justified because most double-layer grids in practice are of this type.

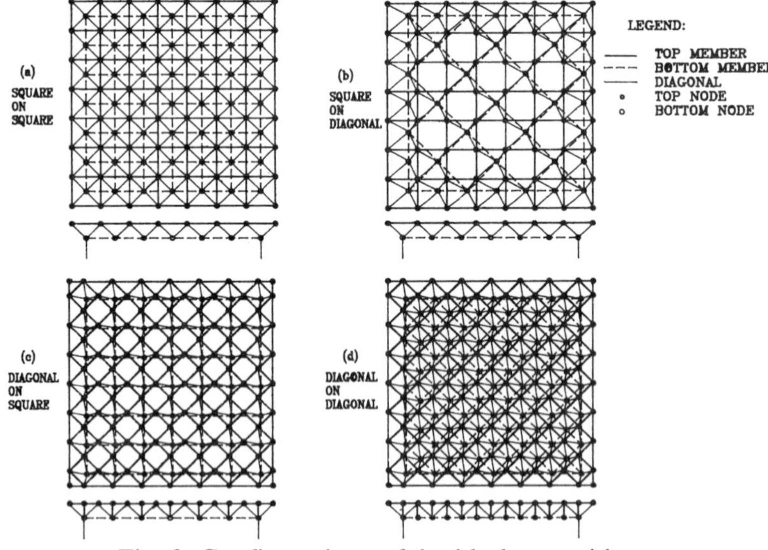

Fig. 2. Configurations of double-layer grids.

4. It covered various ratios of side dimensions, the support arrangements, the grid depth, the grid module and the number of member types. For each investigated case, the structure was loaded with a uniformly distributed load applied at the top grid nodes with the support constraints on the specified bottom chord nodes, and the response of the structure traced under increased load intensity up to complete failure with the collapse mechanism and pattern being identified and recorded. The results of the parametric study conducted can be presented as follows:

The Supports Arrangement and Length to Width Ratio:

5. To investigate the effect of length to width ratio, grids with various side dimensions were used in the investigation.

6. One side was fixed at 20m while the other side was varied according to the following dimensions; 10, 20, 30, 40 and 50m successively.

7. Also, the support system was varied. Different types of support systems used in previous parametric studies (refs 2-4) were used. These support systems are presented in graphical form in Fig. 3 and can be presented as:

(a) All sides are fully supported.
(b) Longer sides are fully supported.
(c) Shorter sides are fully supported.
(d) Five supports on each side.
(e) Three supports on each side.
(f) One mid support on each side.
(g) One corner support on each corner.

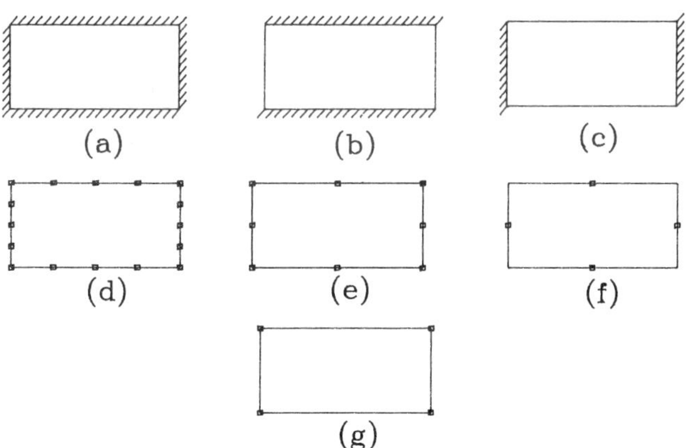

Fig. 3. Various support systems.

8. The collapse patterns of grids having the indicated side dimensions sets, each with all the support systems, one at a time, were determined for every selected layout configuration. The same grid module and depth were kept unchanged. The same member strength was used for all the top chord members, bottom chord

members and diagonal members. This strength was, however, adjusted to remain constant under the various configurations. i.e. same member section in square-on-square resulting in equal strength for all members; but in square-on-diagonal, the same strength was maintained in the top chord and diagonals with equivalent larger member strength reflecting the larger module to be used for the bottom chord members and so on for the other two configurations.

9. The effects on the collapse patterns due to the changes in the parameters under investigation can be presented as follows:

The Length to Width Ratio

10. Two main types of failure patterns were observed; a flexural type and a punching shear type. The flexural type is resembled by a 'yield line' like type of failure in which a failure line develops in either one or two directions starting from the vicinity of the middle of the grid and propagating to the sides causing complete failure. This failure line or plane passes through the failed members (yielding or buckling) in either one or both the top and bottom grids. The other main type of failure pattern is the punching type in which failure occurs in the diagonal members directly over the supports, as such the failure model resembles that of punching of the support columns through a flat slab. These two main failure patterns are demonstrated in Figs 4 and 5.

11. There also appears to be a less frequent type of a failure pattern in cases of cantilevers, as in the case of support type (f) in which a flexural yield line develops along the cantilevered part; this is demonstrated in Fig. 6.

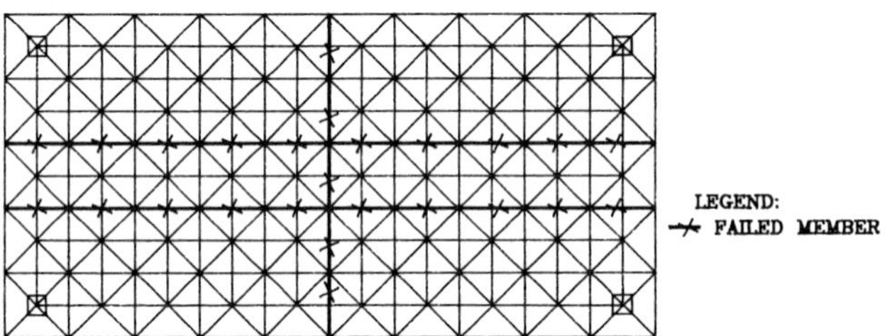

Fig. 4. Flexural yield line failure pattern.

12. Also based on the results obtained, it has been observed that the length to width ratio has little effect in determining the collapse pattern of being either of the flexural type or punching type. If other parameters dictate a flexural type of failure, then the length to width ratio will exhibit its effects indicating the number of failure lines, their location and direction. Fig. 7 shows the changes brought to the flexural failure lines of a typical square-on-square grid supported on all sides for various side dimension ratios. As can be observed, changing the length to

width ratio has marked effect on shifting the location of failure lines, their inclination and numbers.

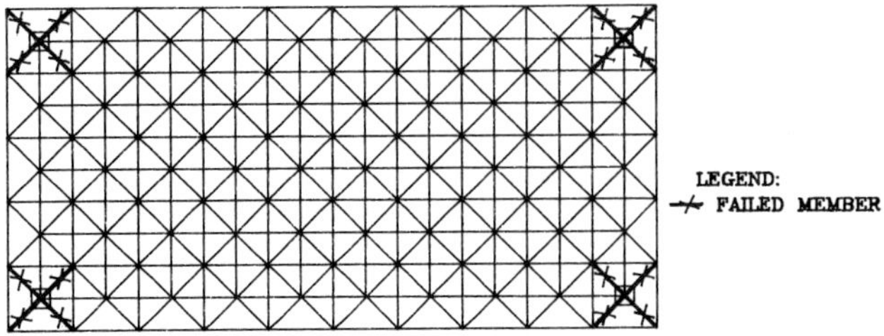

Fig. 5. Punching shear failure pattern.

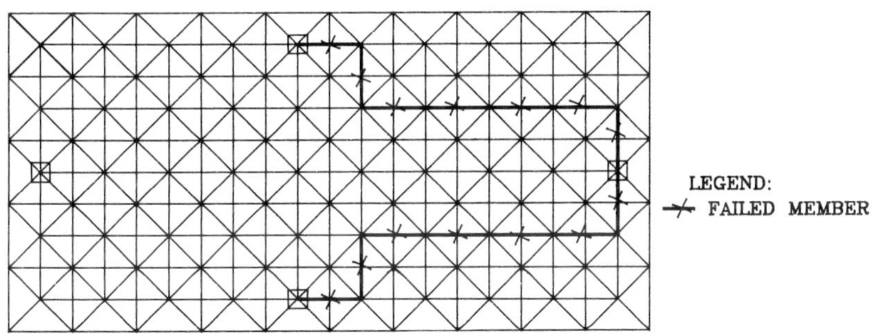

Fig. 6. Failure pattern in cantilevered parts.

13. The length to width ratio also appears to have no effect on the punching type collapse pattern if other parameters cause this pattern to occur.

14. As for the cantilevered type flexural collapse pattern, it also appears to be little affected by the length to width ratio. More important in dictating the occurrence of this pattern is the ratio of the cantilevered dimensions to the inside dimensions of the truss.

15. In the case of support system (f), the flexural failure pattern seems always to be of the cantilevered type as this system gives more forces in the cantilevered part than the middle of the grid.

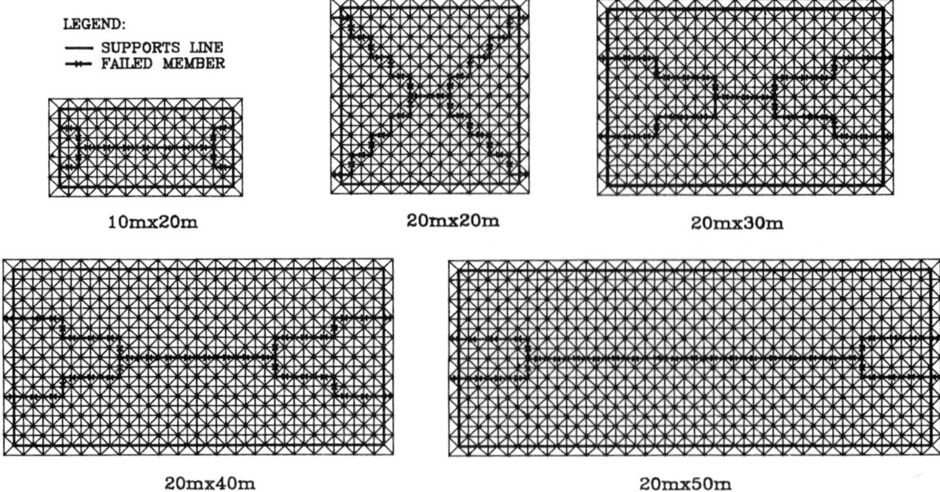

Fig. 7: Failure patterns of the flexural yield line type for various length to width ratios.

The Support System

16. The support system appears to affect the collapse pattern in two ways. The number of available supports dictate whether the failure pattern is of the flexural or punching type. Having more supports reduces the possibility of a punching type of failure as it reduces the forces in the diagonal members over the supports. Therefore, punching patterns of failures were only observed with support systems (f) and (g) as the number of available supports in both of these cases was limited only to four.

17. The other effect of the support system is to influence the location and direction of the failure line(s) if a flexural failure pattern is to occur. This is demonstrated in Fig. 8 that shows flexural failure patterns for a 20 x 30m square-on-square grid under the various support systems of the study.

The Depth And Module:

18. The depth of the grid appears to have a marked effort on dictating the type of failure pattern. Increasing the depth results in reduced forces in the top and bottom grid members, and in increasing the length of the diagonal members thus reducing their compressive load capacity. As such, increasing the depth of the grid moves the structure away from a flexural collapse pattern towards a punching type. Fig. 9 shows that after a certain depth, the collapse pattern shifts from the flexural to the punching type.

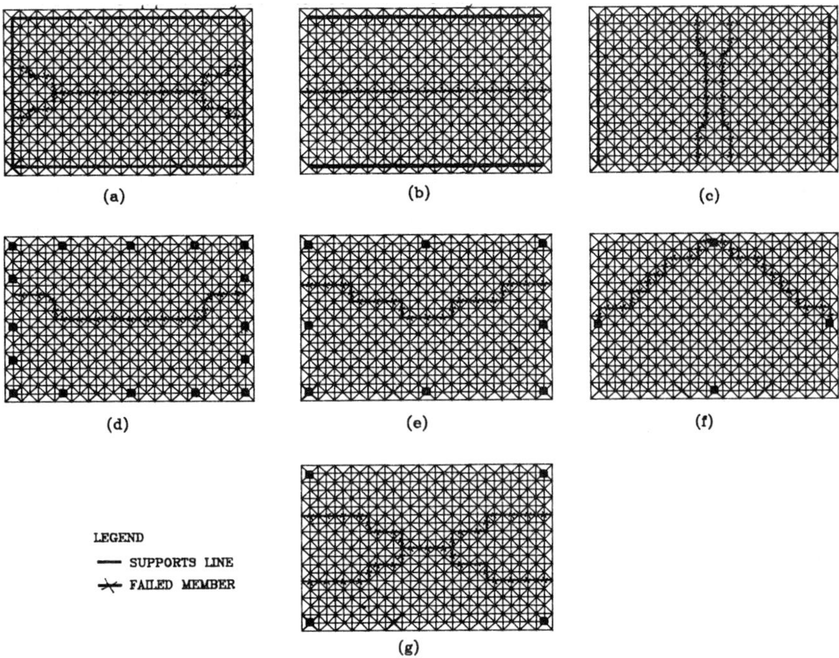

Fig. 8: Flexural failure patterns under the various support systems (Grid is 20 x 30m of the square-on-square type.)

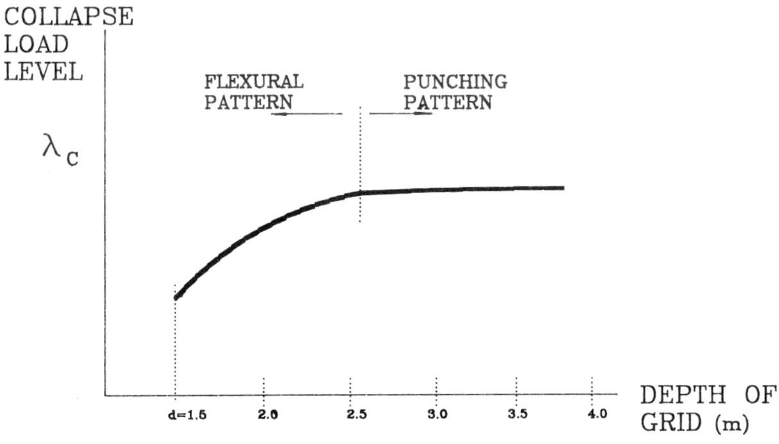

Fig. 9: Collapse pattern of a 15 x 20m grid of the square-on-diagonal type supported on all sides for varying depth.

19. The module seems to have no effect on the failure pattern if the strength of members is adjusted to reflect the change in module. It has been observed that the same failure pattern is maintained for the grid under different modules when the strength of members takes into account the change in module.

The Grid Configuration:

20. The grid configuration seems not to affect the collapse load level or the failure type (either flexural or punching). Its effect seems to be limited to dictating the location and direction of the flexural failure pattern if other effects rule in its favour. Nearly equal collapse load levels were observed for the different flexural collapse patterns of grids of equal dimensions and same support systems. Fig. 10 shows the flexural collapse patterns of 10 x 20m grid of each different configuration with four point corner supports. The observed collapse loads were observed to be nearly equal for all the four configurations.

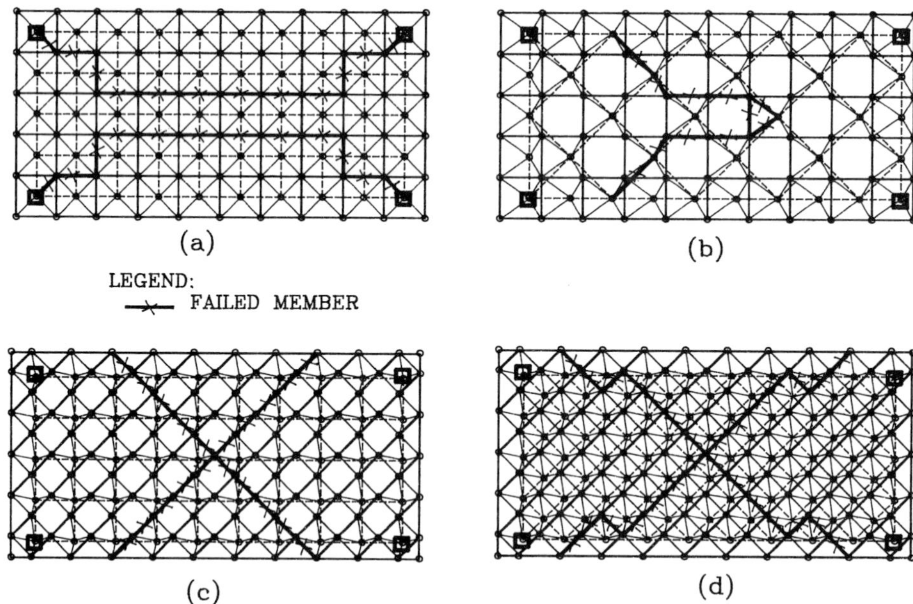

Fig. 10. Flexural collapse patterns of 10 x 20m grid under the various layout configurations.

The Number of Member Types

21. The investigation so far has been based on one member type for each of the top grid members, bottom grid members and diagonal members. Obviously, more economy can be achieved in the design of grids if more than one member type is selected as this will allow using smaller sections for the lightly stressed members. If this is done, it becomes difficult to identify collapse patterns and group them, as the possibility becomes open for local or global failures of unlimited number of

patterns depending on the number of member types and their relative strength. For this reason, this possibility has not been explored in this investigation.

REMARKS AND CONCLUSIONS

22. Based on the investigation carried out, the following remarks and conclusions are deemed appropriate.

- Double layer grids are very similar to concrete flat slabs in their failure patterns that fall into two main types: flexural yield line like and punching shear like.
- The number of supports, the grid depth and the relative strength of members are the main factors that dictate the collapse pattern. The grid module or configuration have no or little effect.
- Increasing the number of supports enhances the economy of double-layer grids as it reduces member forces and precludes the possibility of a punching shear type of failure.
- Having more member types enhances the economy of double-layer grids by reducing the member sections selected for lightly stressed members. However, having many member types increases the difficulty of predicting the collapse pattern as the number of possible patterns increases considerably.

REFERENCES:

1. MADI, Usama R., "The Elastoplastic Analysis of Pin-Jointed Spatial Structures", Space Structures, 1(4), pp 209 - 221, Elsevier, 1985.
2. MADI, Usama R., "An Investigation into the Design Parameters of Double-Layer Space Frame Grids", Space Structures, 2(4), pp 215-222, Elsevier, 1986-1987.
3. LEE, H.G and MAKOWSKI, Z.S. , "Study of Factors Affecting Stress Distribution in Double-Layer Grids of the Square and Diagonal Type", Architecture Science Review, 20, pp 90-102, 1977.
4. MAKOWSKI, Z. S, "Analysis, Design and Construction of Double-Layer Grids", Applied Science Publishers, London, 1981.

68. Space truss structural integrity

E. MURTHA-SMITH and S. F. LEARY, University of Connecticut, USA

SYNOPSIS. Structural integrity is a term used to indicate resistance of a structure to progressive collapse following localised damage. Double layer grid space trusses often experience rapid propagation of failure in the ultimate load condition and thus are potentially vulnerable to progressive collapse at loads less than the normal design load values. A study investigating the effect on the structural integrity of the parameters of support locations, span-module size ratio, span-depth ratio and long span to short span ratio is reported. The results indicate that vulnerability to progressive collapse is relatively insensitive to the parameters except that of the number and placement of supports. Space trusses with supports placed at the four corners have the greatest vulnerability to progressive collapse.

STRUCTURAL INTEGRITY AND PROGRESSIVE COLLAPSE

1. Structural integrity is a term used to indicate the quality of a structure to function as a viable stable system following localised damage. When a structure is locally damaged due to abnormal loads while the general loading is at or below design levels, there is a possibility that this localised collapse can propagate and lead to progressive collapse. Hence, structural integrity can be defined as the satisfactory resistance to progressive collapse.

2. Progressive collapse as a mode of failure received considerable attention following the collapse of the Ronan Point building following a gas explosion in one of the flats in the building (ref. 1). The debris overloaded the floor below which then collapsed and that accumulated debris caused the collapse to propagate down to the ground. Meanwhile, support was lost to the levels above the explosion and the collapse also propagated upwards to the roof. Changes were immediately made to the UK building regulations (ref. 2). In the years following, other countries have amended their building regulations (refs 3, 4 & 5).

FAILURE MECHANICS OF DOUBLE LAYER GRID SPACE TRUSSES

3. A typical double layer grid space truss is shown in Fig. 1. Changes in any one component behaviour from linear to non-linear behaviour will lead to non-linear changes in the forces in other components and the system behaviour

 Space structures 4. Thomas Telford, London, 1993

becomes increasingly complex. Failure can occur (ref. 6), for components in tension, by the yield and rupture of members, connectors and joints, and, for components in compression, by yielding, crushing or buckling. Pin-connected members can buckle individually. However, space truss systems which have continuity through the joints can function as space frames and thus "buckling" requires the bending of several members connected to each joint as each joint rotates in an unstable manner (ref. 7). Failure of a component can cause failure to propagate and can lead to geometric instabilities independent of buckling instabilities. For example, the rupture of a tension member can cause major redistributions required of the adjacent members. Similarly, compression members entering their post-buckling regime can require large redistributions. If these redistributions cause adjacent members to rupture or buckle then collapse can easily propagate.

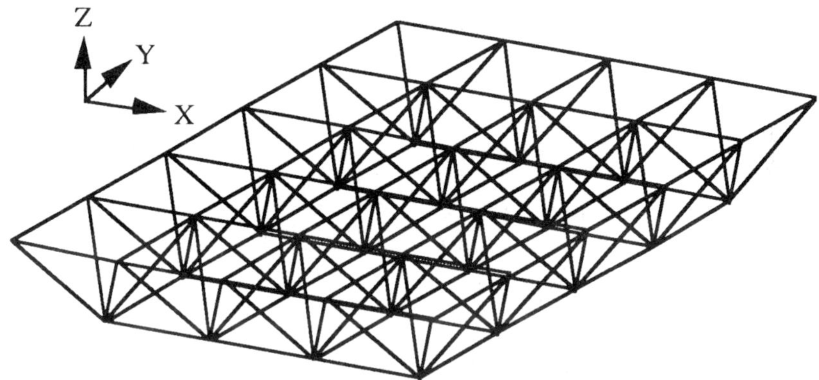

Fig. 1. Double layer grid space truss/frame.

SPACE TRUSS PROGRESSIVE COLLAPSE

4. In normal design, the usual loading environment considered is that of gradually increasing applied load until failure occurs. Certainly, the structure must be designed for this load condition. However, as the Ronan Point collapse demonstrated, structures with inadequate integrity are vulnerable to total collapse due to local damage even at load levels below the full normal design load. Double layer grid space trusses often experience rapid propagation of failure in the ultimate load condition and thus are potentially vulnerable to progressive collapse following local damage at loads less than the normal design load values.

RECENT STUDIES

5. It is only recently that studies of the structural integrity of double layer grid space trusses have been published (refs 8 & 9). Murtha-Smith, in 1988, published a study of alternate path analyses of space trusses (ref. 8). He considered the effect on the structure of the loss of a single member on the safety of the space truss. Linear and non-linear analyses were used to respectively

determine the resulting factors of safety in the remaining members and the factor of safety of the system due to the loss of a single member. The configuration of the space truss is shown in Fig. 2, in which diagonal members are shown only in one quadrant. The 12 module by 12 module space truss was supported near the corners to give 9 module spans and 1.5 module cantilever overhang spans. It was conventionally designed for uniformly distributed loading and members were selected from a range of twenty different member cross-sections. This space truss was then repeatedly analysed for different single members being removed.

6. The structures were analysed at a load level of the dead load plus 45% of the design live load. This load level has been suggested by Ellingwood and Leyendecker (ref. 10) that, together with an abnormal event causing local damage, has a probability of occurrence consistent with normal design. The space trusses were also analysed at the dead load plus the full live load.

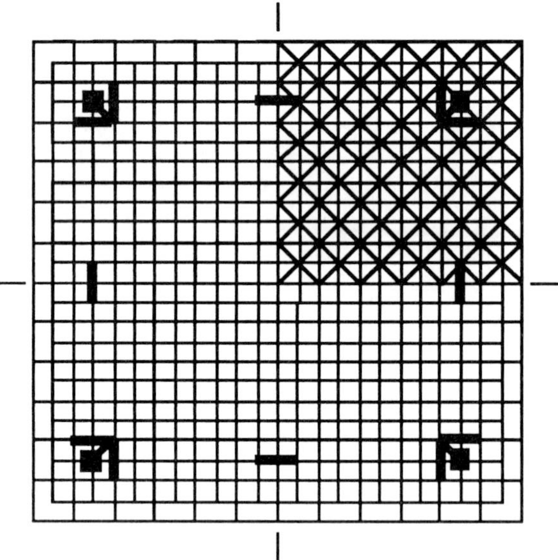

Fig. 2. Space truss configuration and vulnerable members. (ref. 8).

7. Linear analysis of the damaged structures gave the resulting factors of safety in each member. The linear analyses showed that when one of the more highly loaded members is removed, that is those shown in heavy lines in Fig. 2, a significant number of remaining members become unsafe. Thus, diagonal members in the region of any one of the four supports and chord members at mid-span along the support to support centreline cause the largest number of remaining members to become unsafe. The third worst case that resulted from the linear analyses was non-linearly analysed and had a very small (6%) margin of safety. Thus, it is likely that all the worst cases considered (as shown with heavy lines in Fig. 2) would collapse under the loading of dead plus 45% of the design live load.

8. Although a full parametric study was not undertaken the study suggested that a vulnerability to progressive collapse can occur for double layer grid space trusses. However, Murtha-Smith suggested further study to examine different support conditions, member types, span-depth ratios, etc.

9. Also in 1988, Hanaor and Ong (ref. 9) published a study investigating the structural redundancy of space trusses. They investigated the effect of the loss of one or more members on the load bearing capacity and deflections of double layer grid space trusses. The space trusses had members all of the same length and all of the same cross-section and thus did not represent a practical design condition. The trusses were all simply supported around their entire perimeter and loaded with a uniformly distributed load applied to the top chord joints.

10. The parameters varied were the number of modules per span, the number of members lost and the stress state (tension or compression) of the member removed. An increasing number of parallel chord members were removed from the centre of the space trusses. Hanaor and Ong measured the response of the structures using a "load factor" by comparing the maximum force in the damaged structure with the maximum force in the undamaged structure, and a "displacement factor" by comparing the corresponding maximum displacements. Their load factor appeared to be relatively insensitive to the number of modules per span while the displacement factor decreased, tending toward unity as the number of modules increased. Their load factor decreased and the displacement factor increased as the number of members removed increased.

CURRENT STUDY

11. In a study conducted by the authors (ref. 11), some of the parameters suggested by Murtha-Smith (ref. 8) were investigated and are reported herein. Specifically, the parameters investigated were:

• support locations: (a) along the entire perimeter, and (b) at the corners, and for both (i) on exterior joints and (ii) on interior joints (one half module in from the exterior joints);
• span-module size ratio: 8-1, 12-1, 16-1;
• span-depth ratio: 32-1, 19.2-1, 16-1, 13.7-1 and 10.7-1; and
• long span to short span ratio: 1-1 and 1.33-1.

A baseline space truss was designed using conventional design methods. The space truss had exterior perimeter support, 12-1 span-module size ratio, a 16-1 span-depth ratio and a long span to short span ratio of 1-1. Members for all the designs investigated were selected from a library of tubular members which offered 43 different members. The dead-live load ratio was 1-1.

12. It can be shown that the forces in the undamaged space trusses are proportional to the total load on the truss, proportional to the span-depth ratio and inversely proportional to the span-module size ratio. Thus changing any of these three variables will cause members to be selected from different parts of the available spectrum of members. It should be noted, of course, that, in addition, the force in each member will be determined by the specific location of the member in the space truss and the support conditions. In order to investigate the

effect of a variable, it is important to hold the other variables constant. For almost all of the trusses investigated, the load and spans were held constant.

13. Typically, most of the members carry small forces and thus are sized at the lower part of the available members while only a few members use the larger size members. Further, since the available members are discrete, then the overall efficiency of the design will change as the total load is changed since different members will be used. Thus a specific space truss might be very efficient for the undamaged state but be vulnerable to total collapse following localised damage, whereas just the opposite might be true for a space truss designed for a slightly different load.

14. Figs 3 and 4 show typical plots of members becoming unsafe following the removal of the member indicated at a load level of the dead load plus 45% of the live load. In those figures, only the safe top (compression) chord members are shown in addition to the removed and the unsafe members. Generally, the unsafe members have factors of safety that increase with distance from the removed member. Thus, for slightly different design loading, slightly different members might be used with slightly different reserves of capacity and hence, when the critical member is removed, the unsafe member pattern could be different particularly in members remote from the removed member.

15. Figs 3 and 4 present plots for 8x8 module space trusses, supported at the exterior joints, but plots not shown for 12x 12 and 16x 16 are similar to plots for the 8x8. This statement is particularly true for the perimeter supported cases, both for exterior and interior support joints. Indeed, for the perimeter supported cases, only diagonal members become unsafe and almost all of those diagonals are confined to a region 2 modules by 5 modules about the removed members. In the case of a removed tension chord member, as in Fig. 3(a), the unsafe diagonals "point towards" the removed member; in the case of a removed compression chord member, as in Fig. 3(b), the unsafe diagonals "point away" from the removed member.

16. Patterns for the corner supported cases as shown, for example, in Figs 4 (a) and (b) appear more complex. However, by examining Fig. 4(a) in which the maximum tension chord member is removed, it can be seen that there are patterns superimposed on each other. There is a set of unsafe diagonals in the vicinity of the removed member "pointing towards" the member. In addition, there is a set of unsafe tension chord members parallel to the removed member that experience increased tension; further, there is a set of unsafe tension chord members perpendicular to the removed member that go into compression safely; finally, there is a set of unsafe diagonal members, remote from the removed member, that experience compression although designed for small tension forces in the undamaged state. Fig. 4(b) shows the unsafe members caused by the removal of the largest compression chord member. Similar groupings of the members can be made as was done for Fig. 4(a). Plots not shown for the 12x12 and 16x16 cases are similar to Figs 4(a) and 4(b). As with the 8x8 case, the most unsafe members are in the 2 module by 5 module region adjacent to the removed member.

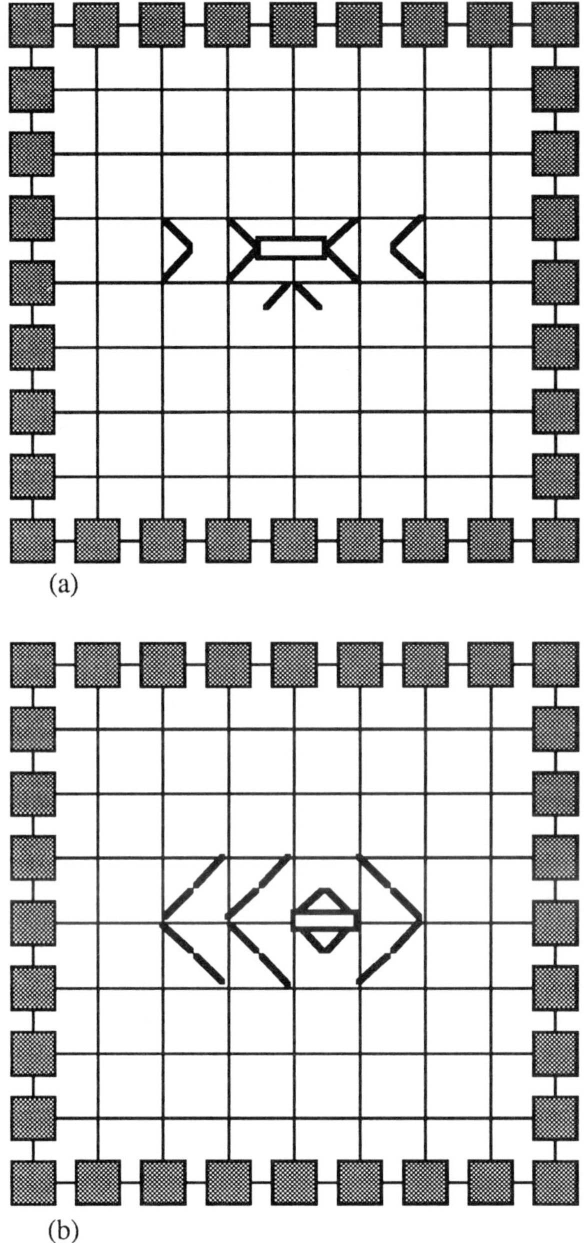

Fig. 3. Unsafe members: 8x8 bay space truss perimeter support;
(a) tension chord member removed, worst case;
(b) compression chord member removed, worst case.

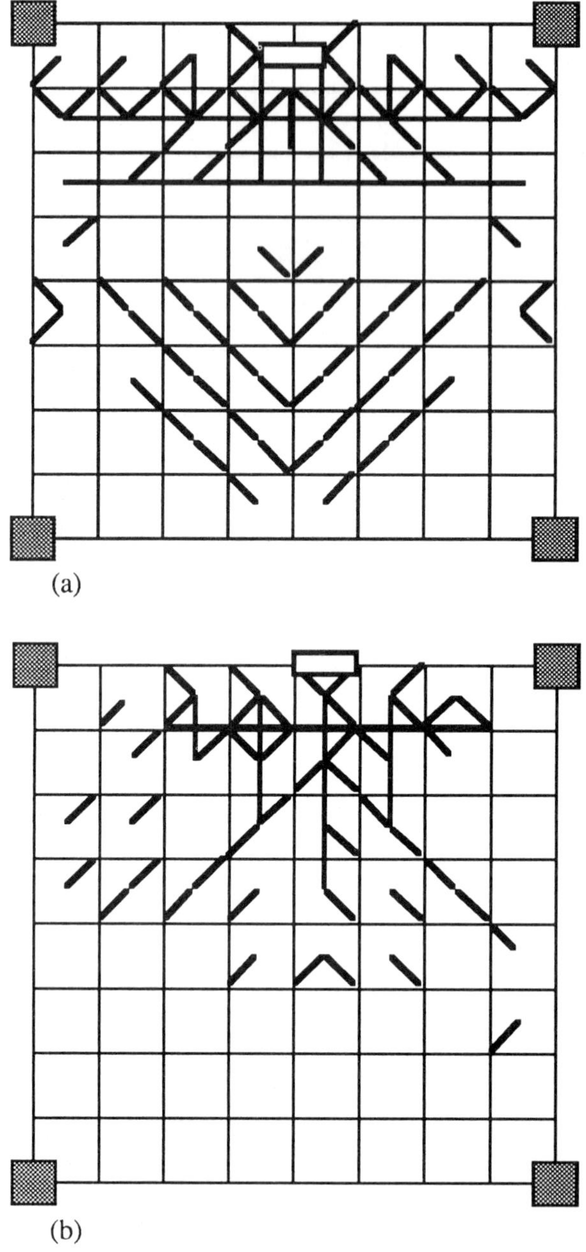

Fig. 4. Unsafe members: 8x8 bay space truss corner supports;
(a) tension chord member removed, worst case;
(b) compression chord member removed, worst case.

EFFECT OF THE PARAMETERS

17. For the exterior joint, perimeter supported space trusses, the maximum number of unsafe members for the worst case is relatively insensitive to both the span-module size ratio and the span-depth ratio, considering variation from the baseline case. Meanwhile, the next four worst cases show a slight increase in the number of unsafe members as the span-module size ratio increases. However, considering these data as percentages of the total number of members, the percentage of unsafe members decreases as the span-module size ratio increases. The number of members becoming unsafe is relatively insensitive to the long span to short span ratio. Similar trends exist for the interior joints perimeter supported space trusses as for the exterior joints case in varying the span-module size ratio. However, the specific values are smaller.

18. For the exterior joint, corner supported space trusses, considering variation from the baseline case with corner exterior supports, the worst case shows large variation in the number of unsafe members. The worst case for this condition is when the diagonal member at any one support is removed. This removal causes the support to be ineffective and the structure becomes geometrically unstable because of an inherent internal degree of freedom in the geometry. However, the numbers of unsafe members for the next four worst cases are insensitive to the span-module size ratio. The worst case is sensitive to span-depth ratio but this is due to the geometric instability introduced. The next four worst cases show some non-linear variation between the number of unsafe members and the span-depth ratio.

19. For the interior joint, corner supported space trusses, the worst case of losing a diagonal at the support still leaves three diagonals remaining at that support and so the geometry remains stable. Thus, the worst case is insensitive to span-module size ratio as are the next four worst cases.

20. Reviewing the summarised results above indicates that vulnerability to progressive collapse is very sensitive to the location of the supports but is relatively insensitive to the span-module size ratio and is relatively insensitive to the span-depth ratio.

FUTURE RESEARCH NEEDS

21. This current study has held the total load constant and observed the changes in the structural integrity of the designs due to changes in geometric parameters. Only linear analysis has been used. The study needs to be extended to different load levels since different designs will be developed that will use different parts of the available section spectrum with differing reserves of capacity. In addition, critical cases need to be non-linearly analysed. Further, the non-linear analysis should consider the dynamic effects that occur both during the loss of the critical member and in the subsequent behaviour particularly during "chordal displacement snap through" (ref. 6). Finally, although the equation of Leyendecker and Ellingwood (ref. 10) was developed considering probability and structural reliability, it was developed in the context of reinforced conrete panel

type structures. Thus a study is required applying reliability theory to this class of structure and the corresponding initiating causes for progressive collapse.

ACKNOWLEDGEMENTS

22. The authors gratefully acknowledge the sponsorship of the National Science Foundation (USA). The opinions expressed in this paper are those of the authors and do not necessarily reflect those of the sponsoring organisation.

REFERENCES

1. GRIFFITHS H, PUGSLEY A, and SAUNDERS O. Report of the inquiry into the collapse of flats at Ronan Point, Canning Town. HMSO, London, 1968.
2. Statutory instrument 1970, No. 109, building and buildings. The building fifth amendment regulations 1970. HMSO, London, 1970.
3. BOCA: basic building code, 7th edition. Building Official and Code Arninistrators International, Inc. Homewood, Illinois, USA, 1978.
4. Building code requirements for minimum design loads in buildings and other structures, ANSI A58.1-1982. American National Standards Institute, New York, NY, USA, 1982.
5. BREEN J E. Summary report of research workshop on progressive collapse of building structures. University of Texas at Austin, Austin, Texas, USA, 1975.
6. SMITH E. A. Space truss nonlinear analysis. Journal of structural engineering, ASCE, vol.110, no.4, April, 1984.
7. MURTHA-SMITH E. and BEAN J E. Double layer grid space frame buckling. International J. of Space Structures, vol. 4, no. 3, 1989.
8. MURTHA-SMITH E. Alternate path analysis of space trusses for progressive collapse. Journal of structural engineering, ASCE, vol.ll 4, no.9, September, 1988.
9. HANAOR A and ONG A-F. On structural redundancy in space trusses. International J. of Space Structures, vol.3, no.4, 1988.
10. ELLINGWOOD B and LEYENDECKER E. V. Approaches for design against progressive collapse. Journal of the structural division, ASCE, vol.104, no.ST3, March, 1978.
11. LEARY S F. A parametric study of the susceptibility of space truss structures to progressive collapse. MS thesis, Department of Civil Engineering, University of Connecticut, Storrs, CT, USA, 1992.

69. Geometric and material non-linear analysis of latticed shells of negative Gaussian curvature

NA WANG, XIN CHEN and SHI-ZHAO SHEN, Harbin Architectural and Civil Engineering Institute, P. R. China

SYNOPSIS. In order to investigate the elastic-plastic stability behaviour of hyperbolic paraboloid latticed shells, a finite element computer program was compiled using the Von Mises material, incremental stress-strain relationship and finite segment finite element method. Two model structures are at first analysed to prove the validity and efficiency of the program. Then an experimental model of a single-layer hyperbolic paraboloid latticed shell is studied in detail. Conclusions are drawn for the behaviour of this type of structure.

INTRODUCTION

1. From the consideration of economy and convenience of construction, single-layer latticed shells are superior to the double-layer forms. But designers have been insisting on using double-layer rather then single-layer systems for latticed saddle shells, because of the uncertainty regarding the stability of the structure in general. The authors have made study on the geometric non-linear stability behaviour of this type of structure (refs 1,2), and have obtained some valuable results. It is revealed that the problem of elastic instability of hyperbolic paraboloid latticed shells does exist, and the buckling of the "bifurcation snap-through" type is characteristic of this structure, the buckling mode is unsymmetric and possesses several waves in the compressional direction of the surface; and because of the effect of the tensional elements, the loading capacity can still increase in the post-buckled region. Based on this work, the effect of material non-linear behaviour on the stability of the structure is discussed in this paper, and some further results are reported.

THEORY AND CALCULATION EXAMPLES

Basic theory

2. The finite segment finite element method is used for the elastic-plastic analysis (ref. 3). Fig. 1 shows the finite number of segments for three kinds of cross-section, that is, a tube element, round bar element and rectangle bar element, which will be used in the analysis later. The assumption of the Von Mises material and incremental stress-strain relationship is adopted. Newton-

Raphson iterative method and arc-length method are used to trace the equilibrium path, and the displacement of joint rotation is modified by the Oran orientation matrix.

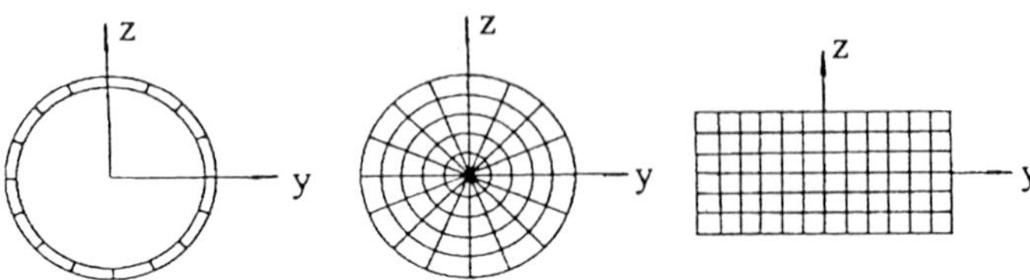

Fig. 1. Finite number of segment.

Cantilever beam

3. The geometry and parameters of the structure analysed are the same as those of Chan (ref. 4). The load - deflection curves obtained are shown in Fig. 2. It can be seen that the results presented by the authors agree well with those of Chan.

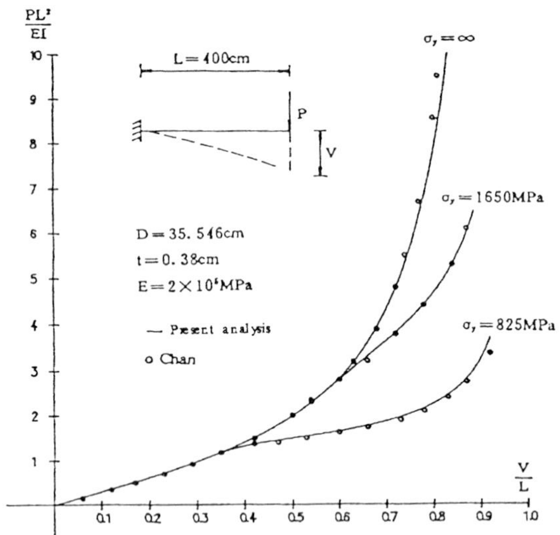

Fig. 2. Load-deflection curves of cantilever beam.

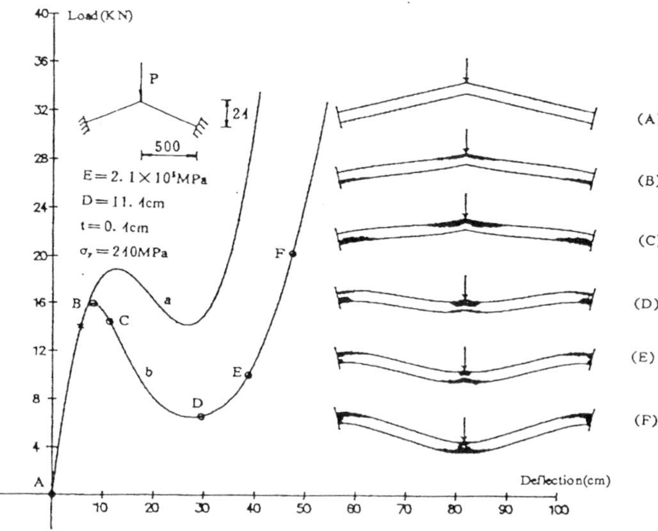

Fig. 3. Load-deflection curves of a toggle frame.

Toggle frame

4. A toggle frame structure is analysed to show the complete process of the development of plasticity in the structure. The geometry and parameters of this structure are shown in Fig. 3. Fig. 3 also shows the complete load-deflection curves, obtained in the case of elastic analysis and of elastic-plastic analysis, and the deformed shapes at several points of the elastic-plastic curve. The regions of developing plasticity are indicated by the dark areas. It can be seen clearly that the plasticity develops at first in the compressive regions, that is, the upper part of the crown joint and the lower part of the two end supports, but under large deflections these regions withdraw from compressive plasticity, and plasticity gradually develops in tension regions, that is, the lower part of the crown joint and the upper part of end supports.

EXPERIMENTAL STUDY ON HYPERBOLIC PARABOLOID LATTICED SHELL

5. Geometry of the structure is shown in Fig. 4. Uniform nodal loads are exerted on 16 interior nodes. In order to trace the post-buckling equilibrium path, a leverage system is used (Fig. 5). Load is exerted by turning a nut, which is fixed under a frame (Fig. 6). Fig. 7 shows the overview of the experiment model. Fig. 8 shows the finite element model, parameters of all the elements are shown in Table 1 (the serial numbers in which refer to Figs 4 and 8). The arrangement of the instruments is shown in Fig. 9, in which HP and LP indicate two high points and two low points, respectively. The complete load-deflection curves for the 16

interior nodes and 4 corner nodes, as well as the strains of some members were measured.

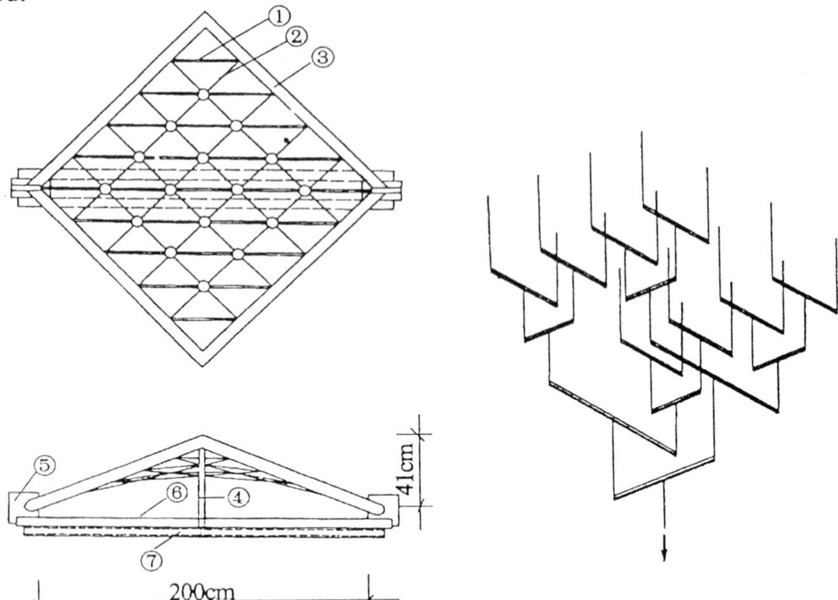

Fig. 4. Geometry of the structure.

Fig. 5. Leverage system.

Fig. 6. Supporting framework.

Fig. 7. Overview of the experimental model.

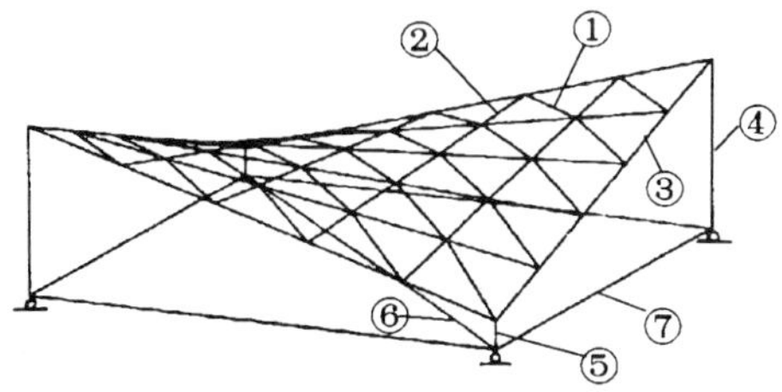

Fig. 8. Calculation model.

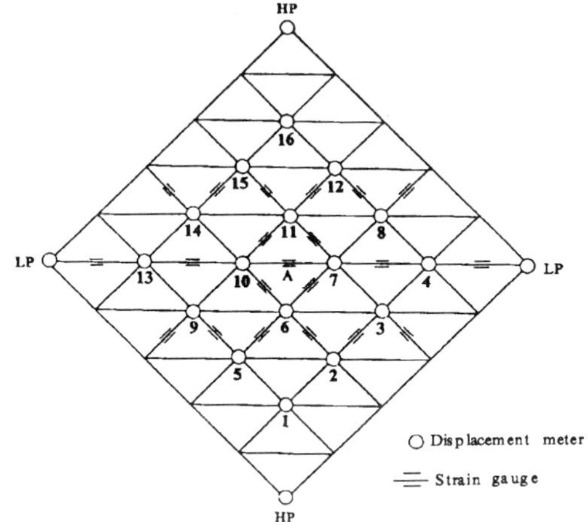

Fig. 9. Arrangement of the instrument.

Table. 1. Parameters of the structural elements.

Serial number	section	A (cm^2)	I_x(cm^4)	I_y(cm^4)	I_x(cm^4)	E(MPa)	G(MPa)	σ_y(MPa)
①		1.52	0.218	0.0732	0.507	1.98×10^5	8.11×10^4	400
②		0.196	0.0061	0.003	0.003	2.02×10^5	8.18×10^4	240
③		17.76	161.2	80.59	80..59	2.1×10^5	8.5×10^4	—
④		1.571	2.47	1.235	1.235	2.1×10^5	8.5×10^4	—
⑤		39.6	71.0	855.4	15.97	2.1×10^5	8.5×10^4	—
⑥		66.37	500.0	175.2	5584.0	2.1×10^5	8.5×10^4	—
⑦		8.4	30.0	51.0	11.9	2.1×10^5	8.5×10^4	—

6. As examples the load-deflection curves of nodes 7 and 9 are shown in Figs 10 and 11 respectively, in which the symbol * represents experimental results; curves a and curves b are the theoretical results of elastic and elastic-plastic analyses, respectively. It shows that when the effect of plasticity is considered both theoretical and experimental results agree well. The final theoretical deformed shape of elastic analysis and that of elastic-plastic analysis are similar to one another (Fig. 12) and agree with the experimental results (Fig. 13).

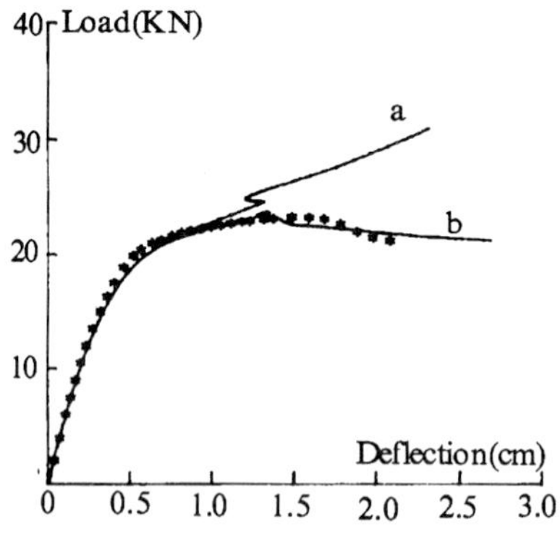

Fig. 10. Load-Deflection curves (node 7).

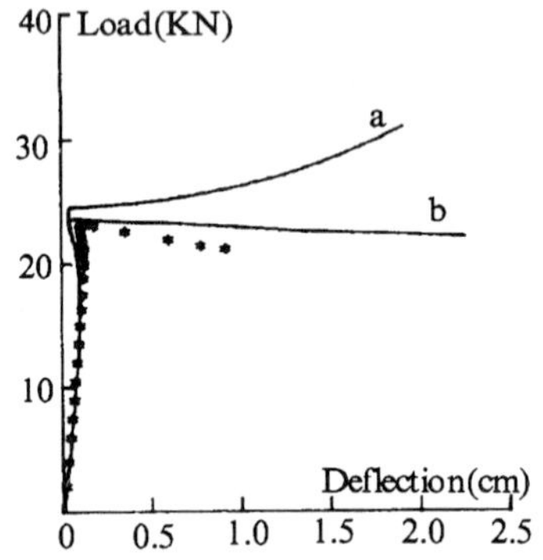

Fig. 11. Load-Deflection curves (node 9).

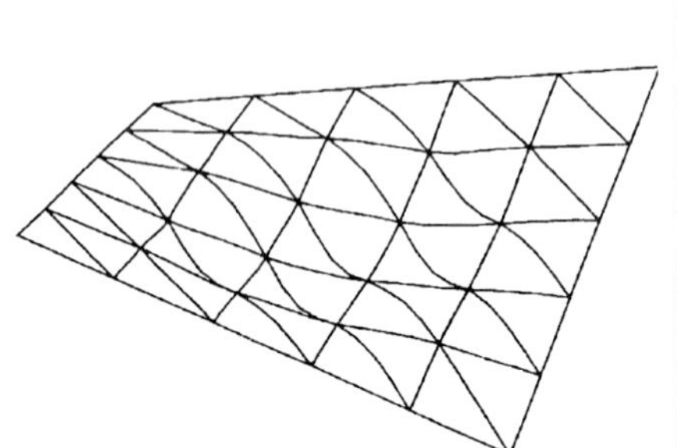

Fig. 12. Theoretical result of the final deformed shape of the tested model.

Fig. 13. Experimental result of the final deformed shape of the tested model.

CONCLUSION

7. Both the results of the experiment and elastic-plastic analysis indicate when plasticity develops in some parts of the elements, the rigidity and the loading capacity of the structure will be reduced, which must be considered in the structural design.

8. In the elastic analysis of hyperbolic paraboloid latticed shell, owing to the effect of the tensional elements, the loading capacity can usually increase in the post-buckled region. But when the plasticity is considered, both experimental and theoretical results, presented in this paper, show that the load will decrease with the development of plasticity. It also shows that the final deformed shape of the structure, from the elastic and elastic-plastic analysis, are similar.

REFERENCES

1. CHEN, X. and SHEN, S.Z. A probative study on non-linear behaviours of latticed shells of negative Gaussian curvature. Proceedings of the International IASS Conference - 10 Years of Progress in Shell and Spatial Structures. Vol. 1, Madrid, 1989.

2. CHEN, X. and SHEN, S.Z. Analytical and experimental study on stability of single-layer latticed shells of negative Gaussian curvature. Proceedings of the International LASS Symposium. Vol. 3, Denmark 1991,pp117-122.

3. MEEK, J.L. and LOGANATHAN, S. Theoretical and experimental investigation of a shallow geodesic dome. International Journal of Space Structures. Vol. 4, No. 2, 1989, pp89- 105.

4. CHAN, S.L. Geometric and material non-linear analysis of beam-columns and frames using the minimum residual displacement method. International Journal of Numerical Method in Engineering. Vol. 26, 1988. pp2657-2669.

70. Non-linear analysis and experimental study of bamboo roofing sheets

Q. YANG, H. L. ZHAO, Q. G. SONG and S. J. HWANG, Southeast University, P. R. China

SYNOPSIS. This paper presents a nonlinear finite element analysis, and an experimental study, of the behaviour of bamboo roofing sheets under transverse and in-plane forces. In the numerical analysis, a degenerate isoparametric laminate element and a total Lagrangian formulation are employed. The element stiffness matrix is at first evaluated for each layer by Gauss quadrature, and then assembled. The incremental Newton-Raphson method combined with the arc-length method is used in solving the nonlinear equations. The numerical results are in good agreement with that of the experiment.

INTRODUCTION

1. The bamboo roofing sheets, made of several bamboo laminae in a similar manner to that for plywood, play an important role in our new grid-sheet structure system. Unlike the roofing sheets in conventional grid structures which are considered as only an enclosure, bamboo sheets form one of the two classes of load bearing members in the new system. The special characteristic of bamboo make our new system a light, strong and economical space structure system.

2. In the bamboo roofing sheets, the bamboo trips are first laid in one direction side by side to form one layer, then each of the layers is laminated and cemented orthogonally (see Fig. 1). Because the bamboo consists of fibrin and wood, and can be treated as composite material (ref. 1), we can treat the bamboo roofing sheets as composite laminates. In the numerical analysis of this kind of roofing sheets, the degenerate isoparametric laminate element can be used, which takes into account the transverse shear correction, coupling between in-plane forces and out-off-plane displacements, coupling between out-off-plane forces and in-plane displacements, and the difference of material between each of the layers (ref. 2). In order to take into account the geometric nonlinearity, a total Lagranging formulation is adopted. The element stiffness matrix is at first evaluated for each layer by Gauss quadrature, and then assembled. The choice of the techniques to solve the nonlinear governing equations depends on the working states of the sheets. The incremental Newton-Raphson method serves well if the sheets are under direct transverse loading only; if, on the other hand, the sheets are also

Space structures 4. Thomas Telford, London, 1993

under in-plane forces, the incremental Newton-Raphson method has to be combined with the arc-length method, because the buckling of the sheets leads to an ill condition of the tangent stiffness matrix around the critical points.

3. The experimental model is a 0.9 x 0.9 x 0.01m bamboo laminate plate, with a simply supported boundary. The external load is a transverse pressure uniformly distributed over the top surface of the plate. The test not only gives knowledge about the whole load bearing process of the plates, but also enables us to judge the accuracy of a numerical method. Indeed, it shows good agreement with the method proposed in this paper.

METHOD OF NUMERICAL ANALYSIS

4. Taking into account the shear deformations and coupling, and neglecting the torsion deformations, the displacement interpolation functions can be chosen as follows:

$$\begin{Bmatrix} u \\ v \\ w \end{Bmatrix} = \sum_{i=1}^{n} N_i(\xi,\eta) \begin{Bmatrix} u_i \\ v_i \\ w_i \end{Bmatrix} + \frac{t}{2} \sum_{i=1}^{n} N_i(\xi,\eta)\zeta \begin{bmatrix} 1 & 0 \\ 0 & -1 \\ 0 & 0 \end{bmatrix} \begin{Bmatrix} \alpha_i \\ \beta_i \end{Bmatrix} \tag{1}$$

in which u_i, v_i, w_i are the nodal displacements in global co-ordinate system, α_i, β_i are rotations of the normal at the node along local y_l and x_l co-ordinates (see Fig. 2), $N_i(\xi,\eta)$ $(i = 1,n)$ are the element shape functions corresponding to the surface ζ = constant, t is the thickness of the plate, the global co-ordinate interpolation formulations are as follow:

$$\begin{cases} x = \sum_{i=1}^{n} N_i(\xi,\eta) x_i \\ y = \sum_{i=1}^{n} N_i(\xi,\eta) y_i \\ z = \frac{t}{2} \sum_{i=1}^{n} N_i(\xi,\eta)\zeta \end{cases} \tag{2}$$

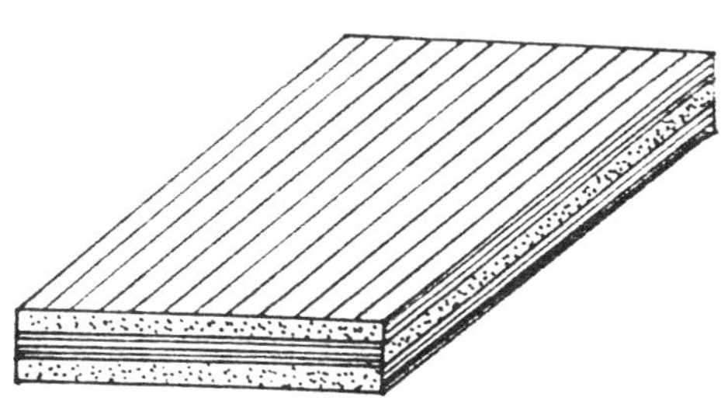

Fig. 1. Bamboo roofing sheet.

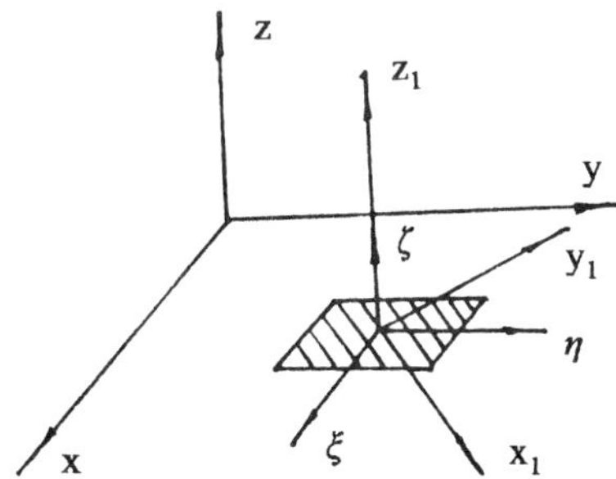

Fig. 2. Co-ordinate system.

5. In the local o_1-$x_1y_1z_1$ co-ordinate system (in which the x_1 axis is parallel to the fibre direction of this layer), the stress component $\sigma_{z1}= 0$. Introducing the Von karman assumptions, the Green-Lagrange strains may be expressed in local co-ordinates as:

$$\{\varepsilon\}_1 = \begin{Bmatrix} \varepsilon_{x_1} \\ \varepsilon_{y_1} \\ \gamma_{x_1y_1} \\ \gamma_{x_1z_1} \\ \gamma_{y_1z_1} \end{Bmatrix} = \begin{Bmatrix} \partial u_1/\partial x_1 \\ \partial v_1/\partial y_1 \\ \partial u_1/\partial y_1 + \partial v_1/\partial x_1 \\ \partial u_1/\partial z_1 + \partial w_1/\partial x_1 \\ \partial v_1/\partial z_1 + \partial w_1/\partial y_1 \end{Bmatrix} + \begin{Bmatrix} 1/2(\partial w_1/\partial x_1)^2 \\ 1/2(\partial w_1/\partial y_1)^2 \\ \partial w_1/\partial x_1 + \partial w_1/\partial y_1 \\ 0 \\ 0 \end{Bmatrix} \tag{3}$$

Thus, for the kth layer, the elastic geometrically nonlinear stress strain relations are obtained:

$$\{\sigma\}_1^k = [D]_1^k \quad \{\varepsilon\}_1^k \tag{4}$$

where $\{\sigma\}_1^k = \left\{\sigma_{x_1}, \sigma_{y_1}, \tau_{x_1y_1}, \tau_{y_1z_1}\right\}^T$, is the stress vector and $\{\varepsilon\}_1^k$ is the strain vector, $[D]_1^k$ is the elasticity matrix:

$$[D]_1^k = \begin{bmatrix} D_1 & D_{12} & 0 & 0 & 0 \\ D_{12} & D_2 & 0 & 0 & 0 \\ 0 & 0 & D_3 & 0 & 0 \\ 0 & 0 & 0 & D_4 & 0 \\ 0 & 0 & 0 & 0 & D_5 \end{bmatrix}_1^k$$

and $D_1 = E_1/(1-\mu_{12}\mu_{21})$, $D_2 = E_2/(1-\mu_{12}\mu_{21})$
$D_{12} = E_2\mu_{12}/(1-\mu_{12}\mu_{21})$, $D_3 = G_{12}$, $D_4 = G_{13}$, $D_5 = G_{23}$.

6. Refer to the original geometric configuration and displacement field, give the current configuration of the system in relation to its initial position, by use of the variation method or the virtual work method, the element tangent stiffness matrix of the composite laminate plate element can be achieved (refs 3 and 4).

$$[K]^e\{\triangle\delta\}^e = \{\triangle R\}^e \tag{5}$$

in which:

$$[K]^e = \sum_{k=1}^{m}\int_{-1}^{1}\int_{-1}^{1}\int_{-1}^{1}[B]^T [D]^k [B] \, |J| \, \frac{t_k}{t} d\xi d\eta d\zeta_k$$

$$\{\triangle\delta\}^e = \{\cdots\cdots\triangle u_i^e, \triangle v_i^e, \triangle w_i^e, \triangle\alpha_i^e, \triangle\beta_i^e \cdots\cdots\}^T$$

$$\{\triangle R\}^e = \{\cdots\cdots\triangle R_{xi}^e, \triangle R_{yi}^e, \triangle R_{xyi}^e, \triangle M_{xzi}^e, \triangle M_{yzi}^e\}^T$$

$$[D]^k = [T][D]_1^k [T]^T$$

and: $\zeta = -1 + \frac{1}{t}[-t_k(1-\zeta_k) + 2\sum_{j=1}^{k} t_j]$

t_k is the thickness of kth layer.
[T] is the co-ordinate transformation matrix from the local o_1-$x_1y_1z_1$ co-ordinate system to the global o-xyz co-ordinate system.

7. In solving the nonlinear equations, the incremental Newton-Raphson method has proved an efficient method, but breakdown at the first onset of instability occurs due to the ill-condition of the tangent stiffness matrix. In this paper, we use the incremental modified Newton-Raphson method combined with the arc-length method (ref. 5).

EXPERIMENTAL STUDY

8. The model is a 0.9m x 0.9m square bamboo laminated plate, the total thickness is of 0.01m, the total number of layers is three, each layer has the same thickness. The fibre direction of the top and bottom layers is parallel to the global x direction, and that of the middle is parallel to the global y direction, the four boundaries are all simply supported and can move freely in the o-xy plane (see Fig. 3). The external load is an uniformly distributed pressure on the top surface of the plate, supplied by water in a rubber bag which is enclosed by a box.

9. Displacement transducers and electric resistance foil strain gauges were used to measure the vertical displacement of selected points and strains of the top and bottom surface of selected points. These points are mainly distributed at the quarter points of the plate, but certain symmetric points were gauged to enable a study to be made of imperfections in the system. The value of the displacements and strains of any gauged point were automatically printed out by the 7V07 gauge at every load stage.

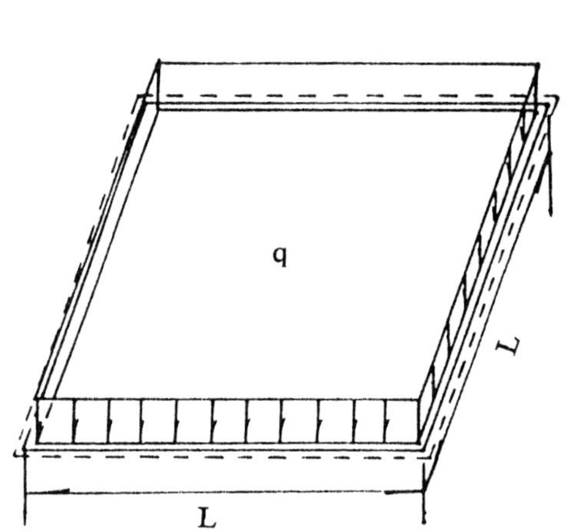

Fig. 3a. Idealised model.

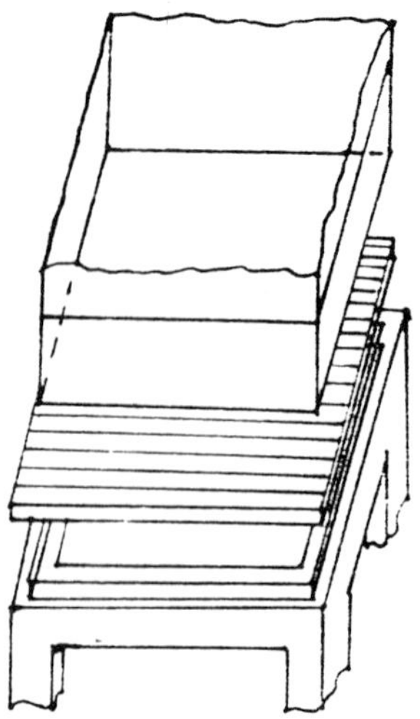

Fig. 3b. Test arrangement.

10. The model was first tested under its linear elastic limit load, in order to adjust the supports and check the instrumentation of the system. It was then loaded gradually from q = 0 to q = 11.76kN/m^2 with an incremental load of Δq = 980N/m^2 at every stage. Three plates were tested using the same procedure.

THE RESULTS AND COMPARATION

11. Because of symmetry, the discretisation is only taken for a quarter of the plate. The 4 x 4 grid quadratic laminate element used is shown in Fig. 4. The material characteristics used are as follows:

$E_1 = 0.76 \times 10^{10}$ N/m^2
$E_2 = 0.38 \times 10^{10}$ N/m^2
$G_{12} = G_{13} = G_{23} = 0.3 \times 10^{10}$ N/m^2
$\mu_{12} = \mu_{21} = 0.25$
the reference load is 9.8kN/m^2

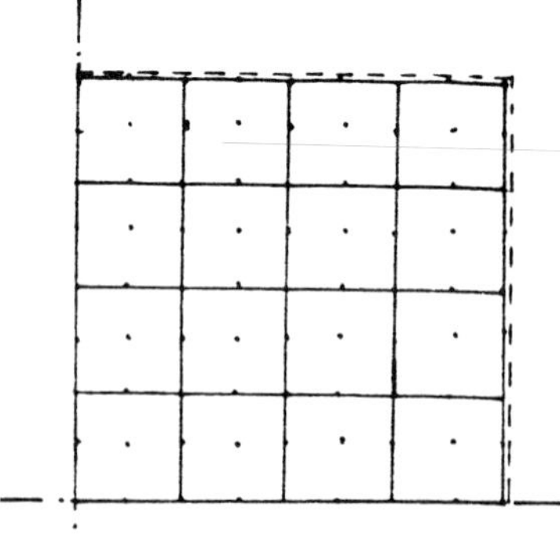

Fig. 4. Element grid.

12. The experimental and numerical load-deflection curves for the central point are shown in Fig. 5, the load-strain (ε_x bottom) curves of the central point are shown in Fig. 6. The experimental and analytical curves are in good agreement during the small displacement stage, but the results diverge when the load is increased. The maximum error occurs at a load factor of 1.2, that is

q=11.76kN/m². We think this is because the experimental plate cannot move freely in its plane as it is assumed to do in the analysis; it is partially restrained at the boundary by the water box (see Fig. 3). The computational results show that when in-plane displacements are restrained at the boundary, the load deflection curve is in better agreement at the large displacement stage, the plate is in some measure a membrane at this stage.

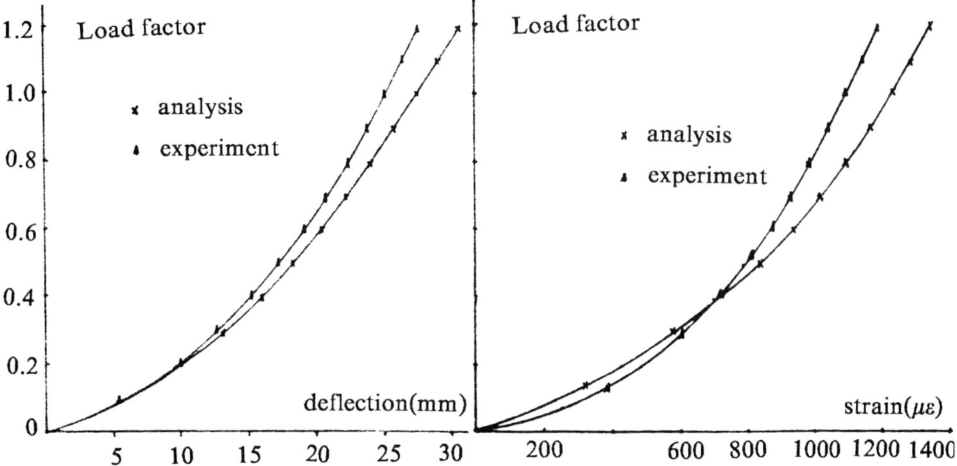

Fig. 5. Load-deflection curves of the central point.

Fig. 6. Load-strain (ε_x) curves of the central point (bottom).

13. The distribution of deflection w and strain ε_x (bottom) along the x axis at the largest load stage (load factor 1.2, load intensity q = 11.7kN/m²) are shown in Figs 7 and 8; only half of the curves are plotted because of symmetry. The numerical results are also in good agreement with that of the experiment.

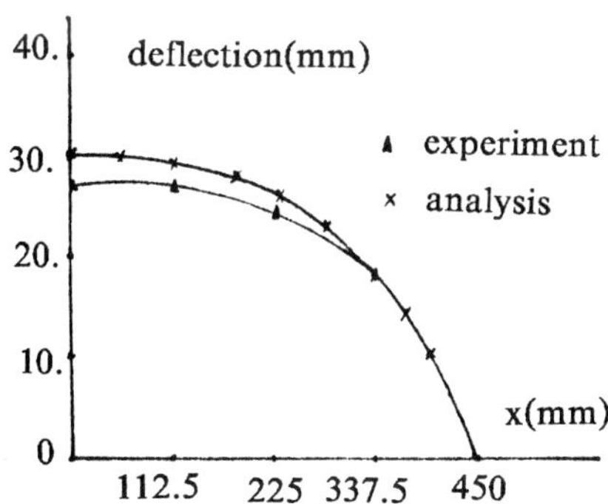

Fig. 7. Distribution of deflection plateunder along x axis.

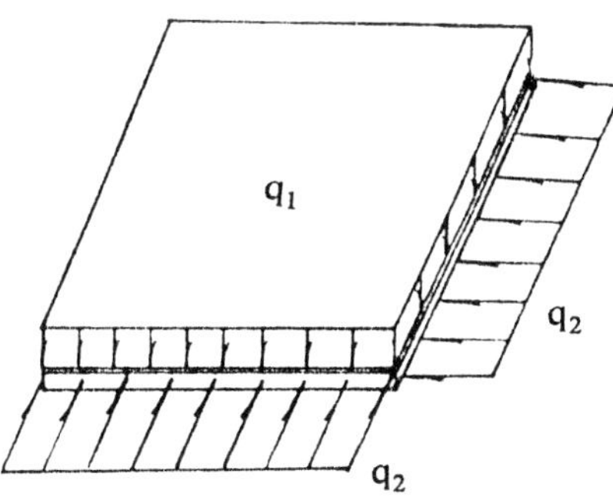

Fig. 9. Laminate bamboo under combine load.

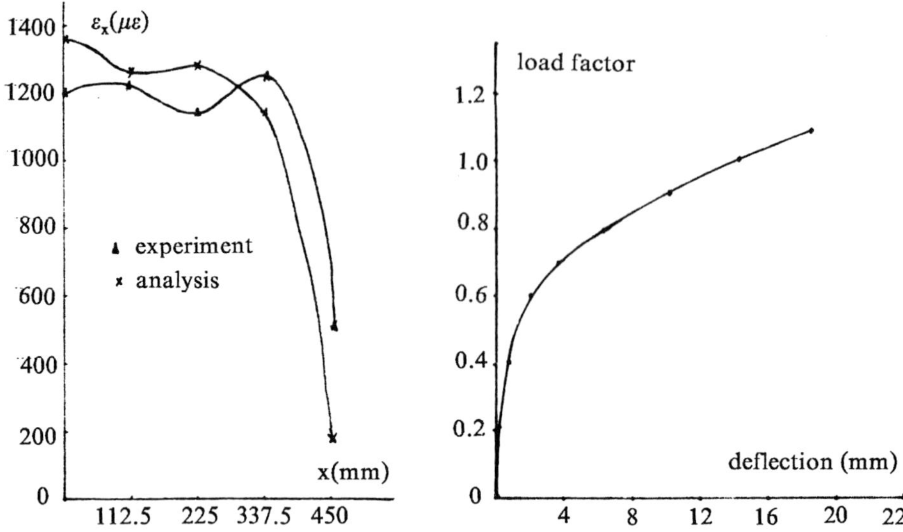

Fig. 8. Distribution of ε_x along x axis.

Fig. 10. Load-deflection curve of the central point.

LAMINATED BAMBOO PLATE UNDER IN-PLANE AND TRANSVERSE LOADS

14. when used as a roof plate, the laminated bamboo plate always carries a combination of in-plane and transverse load, as is shown in Fig. 9. The reference load used in the computational model is q_1 = 80N/m^2, q_2 = 8000N/m, the geometric and material properties are the same as before (see §8 and §11). The load-deflection curve of the centre point is shown in Fig. 10, contrary to that observed for the plate under transverse load, when forced by the in-plane load, the deflection will become large after a critical load. The load-strain (bottom) curves of the central point are shown in Fig. 11. Because of the repeated addition of bending strain (positive) and process strain (negative), the strain of the central point (bottom) is small and oscillates at the lower load stage, but after the critical load load is reached, bending is the main deformation and the strain increases. In the model the critical load factor is about 0.6~0.8. This value is important for engineering designers.

15. The distribution of deflection w along the x axis is shown in Fig. 12, it is of a similar character to that under transverse load (see §13 and Fig. 7). The distribution of strain ε_x(bottom surface) along the x axis is shown in Fig. 13, its maximum value occurs at the central point, but decreases sharply towards the edge, and at the edge point it decreases to a maximum negative value (compression).

CONCLUSION

16. The good agreement of experimental results with that of the numerical analysis shows the accuracy of the method and program. This program can be used in the design of the roof plates for our new space structure systems.

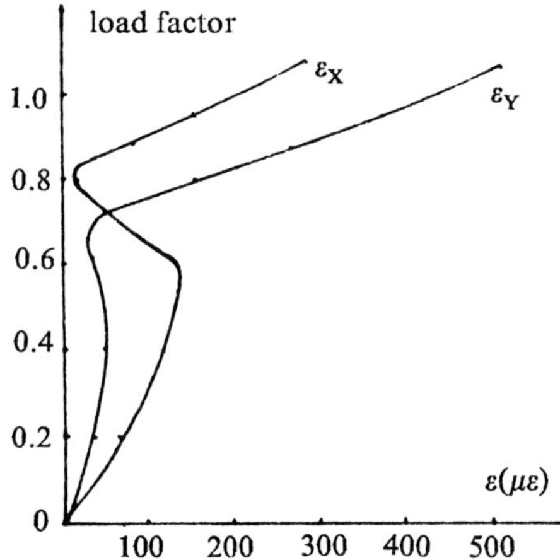

Fig. 11. Load-strain curves for the central point (bottom).

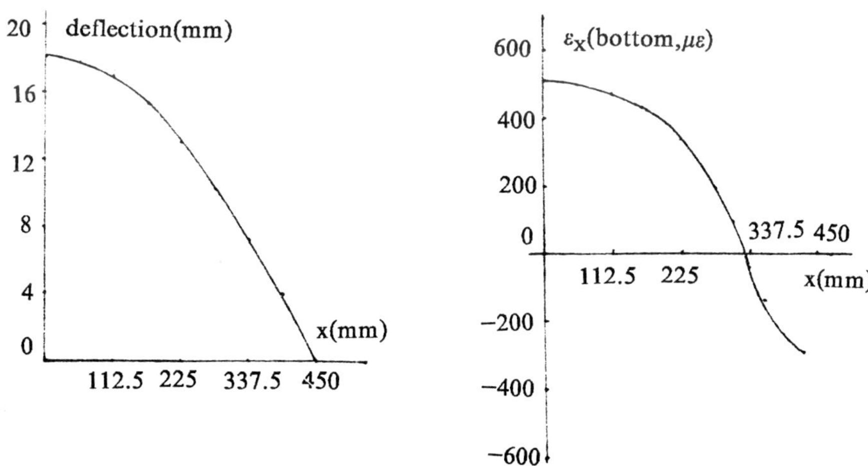

Fig. 12. Distribution of deflection w along x axis.

Fig. 13. Distribution of strain ε_X along x axis.

REFERENCES

1. GUAN X. H. Bamboo-a natural composite, Journal of Composite Material (in chinese), 1987, vol. 4, No.4, 79-83.
2. HONG Z. Q. Review of the finite element method in laminate composite plates and shells, Journal of Composite Material (in chinese), 1985, vol. 2, No.2, 80-88.
3. PANDA S. C. Finite element analysis of laminated composite plates, International Journal of Numerical Methods Engineering, 1979, vol. 14, 69-79.
4. CHANG T. Y. Large deformation analysis of laminate shells by finite element method, Comp. Struct., 1981, vol. 13, 331-340.
5. CRISFIELD M. A. A faster incremental / iterative solution procedure that handles snap-through, Comp. Struct., 1981, vol. 13, 55-62.

71. The non-linear behaviour of large space bar and beam structures

M. VASEK, Czech Technical University, Czech Republic

SYNOPSIS. Many space bar and beam steel structures have been designed during the last few years throughout the world. In future the spans of structures are going to increase. It will be necessary to conduct geometrically non-linear analyses in such cases for the solution of internal forces and deflections. The approach which uses the so called incremental stiffness matrix (refs 2 and 3) seems to be insufficiently accurate for large systems. Some members in the energy expression for the space member are omitted by this method. The error in the solution can be significant for a large structural system. The trigonometric series are used by other authors (ref. 4). The relation between the joint displacement and the displacement of any point of the element is expressed by polynoms and satisfies differential equations of engineering beam theory. To be able to prepare data for a structural system, the philosophy of formex algebra (refs 6 and 7) has been used as a basis for the computer program (ref. 8).

THE BASIC ASSUMPTIONS

1. Whenever we design a large space bar or beam structure, we are faced with the problem of assessing the true behaviour of our structural model. The geometrically non-linear solution gives us more accurate results. The incremental methods are often used for a calculation. The derivation of non-linear solution usually deals with the energy of a structural member due to the axial deflection. The deflection of the space bar (Fig. 1) with six degree of freedom is expressed in Eq. 1.

$$\varepsilon xx = \frac{\partial u_x}{\partial x} + \frac{1}{2}\left(\frac{\partial u_y}{\partial x}\right)^2 + \frac{1}{2}\left(\frac{\partial u_z}{\partial x}\right)^2 \qquad (1)$$

2. The deflection of a general member Fig. 2 with 12 degrees of freedom is written in Eq. 2 where ε_{xx} is the axial strain of the member, u_x, u_y and u_z are displacements of a point of a bar or a beam, E is Young's modulus, A is the cross sectional area, and L is the length of a member.

$$\varepsilon xx = \frac{\partial u_x}{\partial x} + \frac{1}{2}\left(\frac{\partial u_y}{\partial x}\right)^2 + \frac{1}{2}\left(\frac{\partial u_z}{\partial x}\right)^2 - y\frac{\partial^2 u_y}{\partial x^2} - z\frac{\partial^2 u_z}{\partial x^2} \tag{2}$$

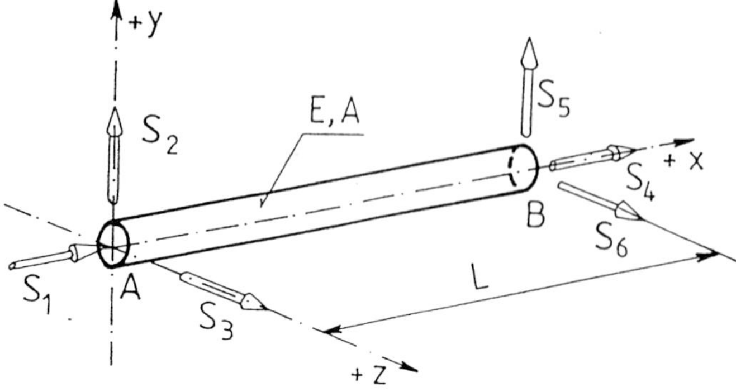

Fig. 1. Axially loaded bar in the space.

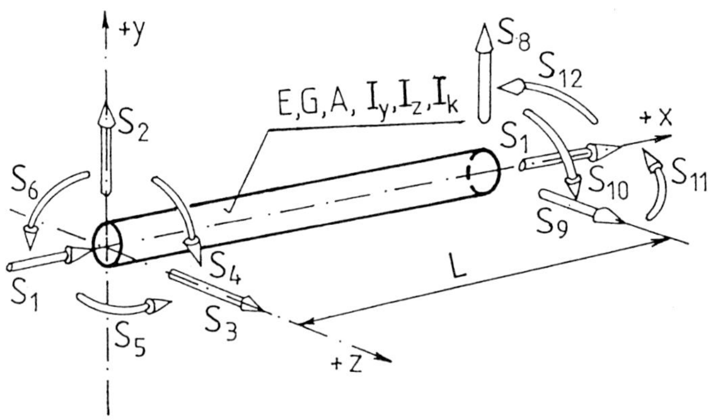

Fig. 2. Beam element in the space.

3. The curvature of a beam is supposed to be expressed as the second derivative of the corresponding deflection as is written in Eq. 3.

$$\frac{1}{r}=\frac{\partial^2 u_j}{\partial x^2} \tag{3}$$

where u_j is u_x or u_y deflection.

4. If we use Eq. 1 and/or Eq. 2 to express the potential energy U of a bar and/or a beam, together with Eq. 4 and Eq. 5 the energy stored in the bar may be found.

$$U=\frac{EA}{2}\int_0^L\left\{\left(\frac{\partial u_x}{\partial x}\right)^2\frac{1}{4}\left(\frac{\partial u_y}{\partial x}\right)^4+\frac{1}{4}\left(\frac{\partial u_z}{\partial x}\right)^4+\frac{1}{2}\left(\frac{\partial u_y}{\partial x}\right)^2\left(\frac{\partial u_z}{\partial x}\right)^2\right.$$
$$\left.+\frac{\partial u_x}{\partial x}\left(\frac{\partial u_y}{\partial x}\right)^2+\frac{\partial u_x}{\partial x}\left(\frac{\partial u_z}{\partial x}\right)^2\right\}dx \tag{4}$$

for the beam

$$U=\frac{E}{2}\int_0^L\left\{A\left(\frac{\partial u_x}{\partial x}\right)^2+Iz\left(\frac{\partial u_y}{\partial x}\right)^2+Iy\left(\frac{\partial u_z}{\partial x}\right)^2\right\}dx+\frac{E}{2}\int_0^L\oint_A\left\{\frac{\partial u_x}{\partial x}\left(\frac{\partial u_y}{\partial x}\right)^2+\frac{\partial u_x}{\partial x}\left(\frac{\partial u_z}{\partial x}\right)^2\right.$$
$$\left.-\left(\frac{\partial u_y}{\partial x}\right)^2\frac{\partial^2 u_y}{\partial x^2}y-\left(\frac{\partial u_y}{\partial x}\right)^2\frac{\partial^2 u_z}{\partial x^2}z-\left(\frac{\partial u_z}{\partial x}\right)^2\frac{\partial^2 u_y}{\partial x^2}y-\left(\frac{\partial u_z}{\partial x}\right)^2\frac{\partial^2 u_z}{\partial x^2}z\right\}dAdx$$
$$+\frac{E}{2}\int_0^L\oint_A\left\{\frac{1}{4}\left(\frac{\partial u_y}{\partial x}\right)^4+\frac{1}{4}\left(\frac{\partial u_y}{\partial x}\right)^4+\frac{1}{2}\left(\frac{\partial u_y}{\partial x}\right)^2\left(\frac{\partial u_z}{\partial x}\right)^2\right\}dAdx \tag{5}$$

5. If we apply Castigliano's theorem (part I), we are able to determine, from Eq. 4 and Eq. 5, the relations for joint forces S_1 to S_6 of a bar and similarly S_1 to S_{12} of a beam (refs 2 and 1). The usual assumption which is applied at this procedure is the constant change of the length of an element during the load step. This means that $\delta = (u_4 - u_1)$ for the bar, or $\delta = (u_7 - u_1)$ for the beam element is constant and so as an initialisation force $H = EA\delta/L$ is constant during a loading step. This assumption leads to the usually used formula for geometrical non-linearity Eq. 6.

$$\mathbf{S} = (\mathbf{K}_e + \mathbf{K}_g)\,\mathbf{U} \tag{6}$$

where **S** is the matrix of joint forces, $\mathbf{K}_e$ is the well known elastic stiffness matrix, $\mathbf{K}_g$ is geometrical stiffness matrix, which express the non-linear behaviour of a member and **U** is the matrix of deflections of joints of a structure.

$\mathbf{K}_g$ for the beam in space is shown in Eq. 7, where a=6/(5L), b=1/10, c=$2(I_z+I_y)/L$, d=2/(15L) and e=L/30.

$$
K_g = \left|\begin{array}{cccccccccccc}
0 & 0 & 0 & 0 & 0 & 0 & 0 & 0 & 0 & 0 & 0 & 0 \\
. & a & 0 & 0 & 0 & b & 0 & -a & 0 & 0 & 0 & b \\
. & . & a & 0 & -b & 0 & 0 & 0 & -a & 0 & -b & 0 \\
. & . & . & c & 0 & 0 & 0 & 0 & 0 & -c & 0 & 0 \\
. & . & . & . & d & 0 & 0 & 0 & -b & 0 & -e & 0 \\
. & . & . & . & . & d & 0 & -b & 0 & 0 & 0 & -e \\
\hdashline
. & . & . & . & . & . & 0 & 0 & 0 & 0 & 0 & 0 \\
. & . & & \text{SYMMETR.} & & . & . & a & 0 & 0 & 0 & -b \\
. & . & . & . & . & . & . & . & a & 0 & b & 0 \\
. & . & . & . & . & . & . & . & . & c & 0 & 0 \\
. & . & . & . & . & . & . & . & . & . & d & 0 \\
. & . & . & . & . & . & . & . & . & . & . & d
\end{array}\right| \quad (7)
$$

INTRODUCTION OF THE METHOD

6. The simplifying assumption regarding the constant elongation of the member during the loading step, is not observed in this work. All elements of the energy expression are evaluated. The degree of accuracy of the solution is controlled by members which are taken into account in the energy expression. The deflections at any point of the structural member are transformed to the joint deflection of the element through polynoms, which are derived from geometrical relations of the deformed element (ref. 2). After the substitution of polynoms to the deflection formulas, and after the integration, we can apply Castigliano's theorem.

Bar element

7. Let us proceed with the derivation for the bar element (Fig. 1). We will receive the matrix Eq. 8 for the joint forces **S**.

$$
\begin{aligned}
S &= EA/L.k_e.U + V_0{}^T.EA/2L^2.h.V + W.V_0{}^T.EA/2L^3.q.V \\
&\quad - [W_B.W_A]\ V_0{}^T.EA/2L^3.g.V
\end{aligned} \quad (8)
$$

Where **h**, **q**, **g** are 36 x 36 square matrices. The terms in the matrices are mainly zero, 1 or -1 values. The matrix **V** is of the size 36 x 1 and matrix $V_0{}^T$ has dimensions 6 x 36. Matrix **W** is the square matrix of dimensions 6 x 6. $\mathbf{W}_B$, $\mathbf{W}_A$ are diagonal submatrices of joint deflections which corresponds to the far end B and to the near end A of the bar element. All these matrices are composed from the vector **U**, in which all joint deflections u_1 to u_6 are included. It is beyond the scope of this paper to define all the matrices, but we can show some of the substantial matrices. Let us write in Eq. 9 the general form of matrix **h** and matrix $\mathbf{h}_1$ as an example. The matrices **q** and **g** are built in the same manner as **h**, where a=1/2.

$$h=\begin{vmatrix} h_1 & 0 & 0 & 0 & 0 & 0 \\ 0 & h_2 & 0 & 0 & 0 & 0 \\ 0 & 0 & h_3 & 0 & 0 & 0 \\ 0 & 0 & 0 & h_4 & 0 & 0 \\ 0 & 0 & 0 & 0 & h_5 & 0 \\ 0 & 0 & 0 & 0 & 0 & h_6 \end{vmatrix}_{(36x36)} \qquad h_1=\begin{vmatrix} 0 & 0 & 0 & 0 & 0 & 0 \\ 0 & -a & 0 & 0 & a & 0 \\ 0 & 0 & -a & 0 & 0 & a \\ 0 & 0 & 0 & 0 & 0 & 0 \\ 0 & a & 0 & 0 & -a & 0 \\ 0 & 0 & a & 0 & 0 & -a \end{vmatrix} \qquad (9)$$

8. The Eq. 8 can be written after the transformation to the global co-ordinate Cartesian system as follows:

$$S = K_E.U + T^T.V_0{}^T.EA/2L^2.H.V + \mathbf{V}\ .V_0{}^T.EA/L^3.Q.V - [\mathbf{V}_B.\mathbf{V}_A]\ V_0{}^T.\ EA/L^3.G.V \qquad (10)$$

9. This equation needs an iteration calculation to receive final solution. We can write symbolically:

$$\mathbf{S} = \mathbf{S}_E + \mathbf{S}_G + \mathbf{S}_Q + \mathbf{S}_H \qquad (11)$$

where $\mathbf{S}_E$ corresponds to the linear elastic solution, $\mathbf{S}_G$ corresponds to the obvious incremental matrix $\mathbf{K}_g$, and $\mathbf{S}_Q$ and $\mathbf{S}_H$ express the influence of higher order terms, which are important for the nonlinear deflection of large type of structures. Eq. 11 may be solved by an iteration procedure. The final solution of forces **S** has to correspond to the resultant vector of deflections **U** resp. **V**. We don't know the subdivision of joint forces S to the parts $\mathbf{S}_E$, $\mathbf{S}_G$, $\mathbf{S}_Q$ and $\mathbf{S}_H$. This iteration could be introduced as follows: let us take $\mathbf{S}_E$ as a fragment of the loading force **P** and solve the vector of deflections **U** for the elastic part of Eq. 10 only. After the substitution of the vector **U** to the all members in Eq. 10 we can calculate forces $\mathbf{S}_j$ and compare the result and starting approximation of $\mathbf{S}_j$ (the j-index corresponds to the indexes G,Q,H). The procedure may be repeated until the prescribed accuracy is achieved. The accuracy could be prescribed as the difference between estimated and calculated value of $\mathbf{S}_j$. This procedure has a good convergency after only a few steps (ref. 1).

Beam element

10. We can proceed in the similar way with the derivation for the beam (Fig. 2). Let us use the first two integrals from Eq. 5 which correspond to the elastic solution and to the geometric stiffness matrix. The last integral corresponds to the higher order terms, which are usually not included in the nonlinear solution. The sizes of matrices are rapidly increased, but nonzero values are sparse and concentrated along the main diagonal. We can write the matrix Eq. 12 which corresponds to the Eq. 6.

$$S = k_E.U + V_0{}^T(EA.{}^7h).V + V_0{}^T(EA.{}^1h).V \qquad (12)$$

where ${}^{7}\mathbf{h}$ and ${}^{l}\mathbf{h}$ are square matrices of the sizes 144 x 144 with the main diagonal composed from matrices ${}^{7}\mathbf{h}_j$ and ${}^{l}\mathbf{h}_j$ (j =1 to 12) which are of the size 12x12.

$$
h = \left|\begin{array}{cccccccccccc}
h_1 & 0 & 0 & 0 & 0 & 0 & 0 & 0 & 0 & 0 & 0 & 0 \\
. & h_2 & 0 & 0 & 0 & 0 & 0 & 0 & 0 & 0 & 0 & 0 \\
. & . & h_3 & 0 & 0 & 0 & 0 & 0 & 0 & 0 & 0 & 0 \\
. & . & . & h_4 & 0 & 0 & 0 & 0 & 0 & 0 & 0 & 0 \\
. & . & . & . & h_5 & 0 & 0 & 0 & 0 & 0 & 0 & 0 \\
. & . & . & . & . & h_6 & 0 & 0 & 0 & 0 & 0 & 0 \\
\hline
. & . & . & . & . & . & h_7 & 0 & 0 & 0 & 0 & 0 \\
. & . & . & \multicolumn{3}{c}{\text{SYMMETRY}} & . & h_8 & 0 & 0 & 0 & 0 \\
. & . & . & . & . & . & . & . & h_9 & 0 & 0 & 0 \\
. & . & . & . & . & . & . & . & . & h_{10} & 0 & 0 \\
. & . & . & . & . & . & . & . & . & . & h_{11} & 0 \\
. & . & . & . & . & . & . & . & . & . & . & h_{12}
\end{array}\right| \quad (144\text{x}144) \tag{13}
$$

$$
{}^{i}h_l = \left|\begin{array}{cccccccccccc}
0 & 0 & 0 & 0 & 0 & 0 & 0 & 0 & 0 & 0 & 0 & 0 \\
. & 0 & 0 & 0 & 0 & 0 & a & 0 & 0 & 0 & 0 & 0 \\
. & . & 0 & 0 & 0 & 0 & 0 & 0 & 0 & 0 & 0 & 0 \\
. & . & . & 0 & 0 & 0 & 0 & 0 & 0 & 0 & 0 & 0 \\
. & . & . & . & 0 & 0 & 0 & 0 & 0 & 0 & 0 & 0 \\
. & . & . & . & . & 0 & b & 0 & 0 & 0 & 0 & 0 \\
\hline
. & . & . & . & . & . & 0 & -a & 0 & 0 & 0 & b \\
. & . & \multicolumn{3}{c}{\text{SYMMETRY}} & . & . & 0 & 0 & 0 & 0 & 0 \\
. & . & . & . & . & . & . & . & 0 & 0 & 0 & 0 \\
. & . & . & . & . & . & . & . & . & 0 & 0 & 0 \\
. & . & . & . & . & . & . & . & . & . & 0 & 0 \\
. & . & . & . & . & . & . & . & . & . & . & 0
\end{array}\right| \quad (12\text{x}12) \tag{14}
$$

11. Matrix ${}^{7}\mathbf{h}$ corresponds to the terms which are multiplied by the deflection $\mathbf{u}_7$, and ${}^{l}\mathbf{h}$ corresponds to the terms which are multiplied by the deflection $\mathbf{u}_l$. Vector **U** has dimensions 12xl, $\mathbf{V_0}^T$ has dimensions 12x144 and **V** has dimensions 144xl. We can show matrices **h** and ${}^{l}\mathbf{h}_l$ in Eq. 13 and Eq. 14, where the upper index i is either 1 or 7, $a=3/(5L^2)$ and $b=1/(20L)$. After the transformation to the global co-ordinate system we have:

$$
S = K_E.U + T^T\{V_0(EA.{}^{7}H)V + V_0{}^{T}(EA.{}^{1}H)V\} \tag{15}
$$

This equation could be symbolically written as:

$$S = S_E + S_G \tag{16}$$

12. The forces S_G correspond to the forces calculated by the mean of geometrically stiffness matrix.

13. If we take into the derivation the last integral from Eq. 5, we have to introduce some terms which we call "cross-sectional properties of the second order". These are:

$$K_y = \int_V z^4 dA, \quad K_z = \int_V y^4 . dA, \quad K_{yz} = \int_V z^2 . y^2 \, dA \tag{17}$$

14. These expressions are actually "moments of moments of inertia about principal axes". After the procedure as was described for the bar element we can derive a similar equation to Eq. 11 for the beam element. We can apply the same iteration procedure as used for the bar element.

INFLUENCE OF SHEAR ENERGY

15. We were concerned about the energy due to the axial deformation only. The influence of the shear energy due to torsion should also be included for the space system. For the expression of the shear deflection we use the total strain tensor with respect to second order terms. The energy stored in the element due to the shear deformation is given by Eq. 18.

$$U = \frac{1}{2}\int_V G\gamma^2 dV = \frac{1}{2} G \left(\int_V \gamma^2 xy dV + \int_V \gamma^2 xz dV + \int_V \gamma^2 yz dV \right) \tag{18}$$

where **V** is the volume of an element, γ is the general shear deflection and **G** is the shear modulus. We assume further the validity of Navier's hypothesis and that the cross-section is non-deformable , so we can write for $z_0 = 0$ and $y_0 = 0$:

$$\frac{\partial u_x}{\partial y_0} = 0 \qquad \frac{\partial u_x}{\partial z_0} = 0 \qquad \gamma zy = 0 \qquad \frac{\partial u_y}{\partial y} = 0 \qquad \frac{\partial u_z}{\partial z} = 0$$

we can write finally:

$$\gamma xy = \frac{\partial u_y}{\partial x} + \frac{\partial u_z}{\partial x}\frac{\partial u_z}{\partial z} \qquad \gamma xy = \frac{\partial u_z}{\partial x} + \frac{\partial u_y}{\partial x}\frac{\partial u_y}{\partial z} \tag{19}$$

16. The procedure for the derivation of joint forces is the same as above. We apply Castigliano's theorem (part I) after the integration of energy expression. The result is written in matrix form in the local co-ordinate system in Eq. 20.

$$S = GA/2 \{e.U\ (I + U^T.c_1.U) + m.e.U.U^T.n.c_1.U\} \tag{20}$$

where **U** is the vector (12x1) of joint deflections u_1 to u_{12}. The matrices **m**, n and $\mathbf{c_1}$ are square transformation matrices (12x12), **I** is the unit matrix and **e** is a matrix (12x12), which we call the shear stiffness matrix. After the transformation by transformation matrices **T** to the global co-ordinate system we have relations Eq. 21 and Eq. 22. The matrix **e** is written in Eq. 23, where a=12/(SL), b=1/5, c=2(I_y+I_z)/(LA), d=4/(lSL) and e=1/(lSL). We can see, that values of terms in the shear stiffness matrix **e** are of a comparable value with the values of the geometrical matrix.

$$E = T^T.e.T \quad C = T^T.c_1.T \quad N = T^T.n.c_1.T \quad M = m.e.T \tag{21}$$

$$S = GA/2 \{E.U\ (I + U^T.C.U) + M.U.U^T.N.U\} \tag{22}$$

$$e = \left|\begin{array}{cccccccccccc}
0 & 0 & 0 & 0 & 0 & 0 & 0 & 0 & 0 & 0 & 0 & 0 \\
. & a & 0 & 0 & 0 & b & 0 & a & 0 & 0 & 0 & b \\
. & . & a & 0 & -b & 0 & 0 & 0 & a & 0 & -b & 0 \\
. & . & . & c & 0 & 0 & 0 & 0 & 0 & -c & 0 & 0 \\
. & . & . & . & d & 0 & 0 & 0 & -b & 0 & -e & 0 \\
. & . & . & . & . & d & 0 & -b & 0 & 0 & 0 & -e \\
\hline
. & . & . & . & . & . & 0 & 0 & 0 & 0 & 0 & 0 \\
. & . & \multicolumn{4}{c}{.\text{SYMMETR}..} & . & a & 0 & 0 & 0 & -b \\
. & . & . & . & . & . & . & . & a & 0 & b & 0 \\
. & . & . & . & . & . & . & . & . & c & 0 & 0 \\
. & . & . & . & . & . & . & . & . & . & d & 0 \\
. & . & . & . & . & . & . & . & . & . & . & d
\end{array}\right| \tag{23}$$

17. The first part of Eq. 23, which is **S** = GA/2 **E U**, corresponds to the elastic forces due to shear torsional energy. We can apply again the same iteration procedure for the final nonlinear solution as above.

DATA PREPARATION FOR LARGE SPACE BAR OR BEAM SYSTEMS

18. The formex algebra concept is used as a basis for the computer program (refs 6 and 8). The unlimited growth of the structural data needs allocatable variables during a calculation. This should be introduced by pointer variables in Pascal and by allocatable variables in the Fortran language. The seed program for a topology generation was developed, where the general formex is introduced as the pointer variable. The basic topology functions are introduced for a translation (TRAN), a composition (COMP), copying (RIN), a vertition (VER), a rosette(ROS), and a libra notation (NOT). The transformation functions to the basic co-ordinate systems are prepared.

REFERENCES

1. VASEK M. Solution of bars or beams with respect to geometrical nonlinearity, Stavebnicky casopis, vol.24, 1976, May, 415-427.
2. PRZEMIENIECKI J. S. Theory of Matrix Structural Analysis, New York, McGraw Hill Company, 1968.
3. MAJID K. L. Non-Linear Structures, Butterworths Ltd, London, 1972.
4. GOTO Z. and CHEN W.F. Second Order Elastic Analysis for Frame Design, Journal of the Structural Division, vol.113, No 7,1978.
5. MAKOWSKI Z. S. Space Structures of Today and Tomorrow, 3rd Int.Conf. on Space Structures, Elsevier publishers Ltd, 1984.
6. NOOSHIN H. Formex Configuration Processing in Structural Engineering, Elsevier publishers Ltd 1984.
7. VASEK M. The Automatisation of Data Preparation for Structural Systems, Rep. RVT-NVP-R12-371/082, Prague 1988, 1989 (in Czech).
8. VASEK M. Program for Space Frame Amalysis, Prague 1972, PC version 1991.

72. Using cyclic symmetry to analyse space structures

W. S. DOYLE, Computer Assisted Analysis, UK, and G. A. R. PARKE,
University of Surrey, UK

SYNOPSIS. Three different types of cyclic symmetric structures are analysed by means of the ADINA finite element program utilising cyclic symmetry to reduce data preparation, computer storage and computation time. These structures are:

(a) a skeletal tower, using truss finite elements;
(b) a solid pentagon using tetrahedron elements and;
(c) a cooling tower with a combination of different types of elements.

In the latter, a dynamic analysis is demonstrated where mode shapes and natural frequencies are computed. Geometry and boundary conditions are only described for the fundamental part of the structure while the loading, is defined for each cyclic part individually.

INTRODUCTION

1. Many structures can be classified as cyclic symmetric, there are examples in skeletal structures, plate structures, shell structures and three dimensional solid structures. Typical of these are:

Skeletal Structures.

2. Tower structures with non-axisymmetrical loads and with 4 and 6 cycles of symmetry respectively are shown in Figs 1a and 1b. Also dome configurations with 4 and 12 cycles of symmetry respectively are shown in Figs 2a, 2b and 2c. Grid structures with 4 and 6 cycles of symmetry are also shown in Figs 3a and 3b.

Plate Structures

3. Structures which require plate finite element solutions with either quadrilateral elements or triangular elements or a combination of both are also often cyclically symmetrical; two examples are shown in Fig. 4. In the flywheel, Fig. 4b, plate elements and 3-D elements are combined.

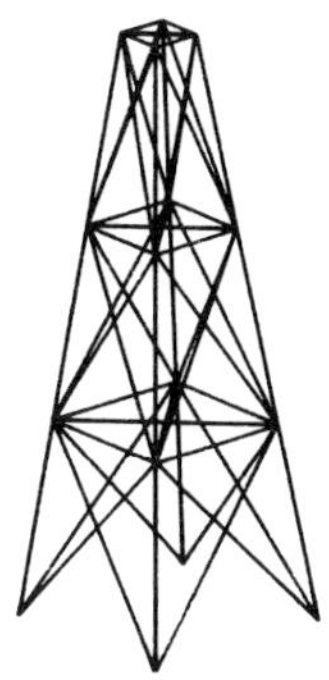

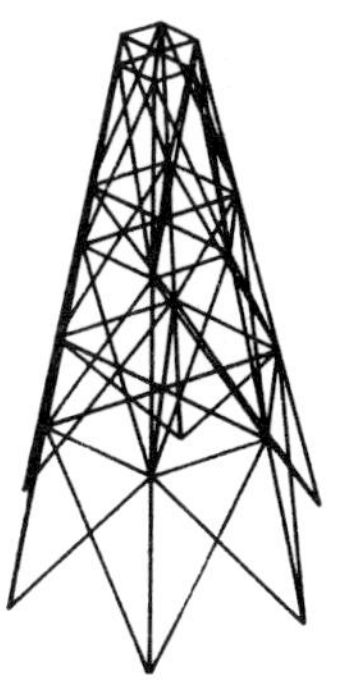

(a) 4-sided tower (4 cycles). (b) 6-sided tower (6-Cycles).
Fig. 1. Cyclic Symmetrical Tower Structures.

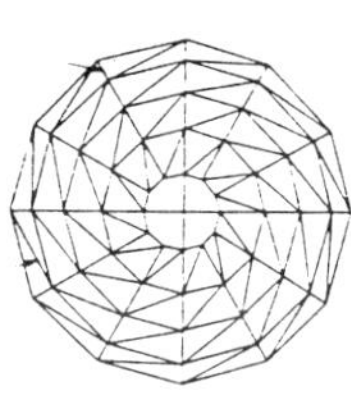

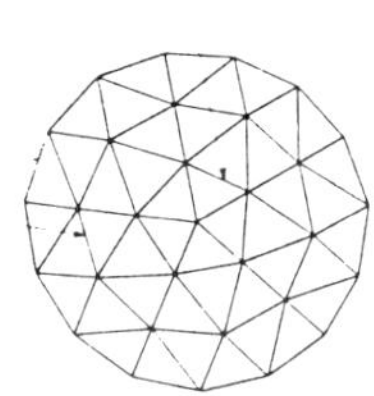

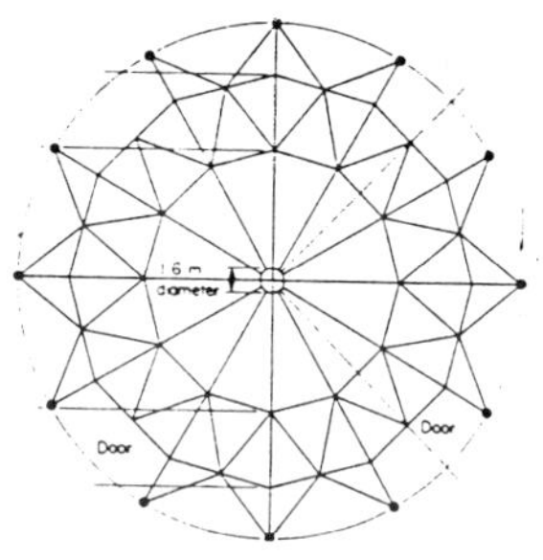

(a) Schwedler (12-Cycles). (b) Network (5-Cycles). (c) Lamella (12-Cycles).
Fig. 2. Cyclically symmetrical dome structures.

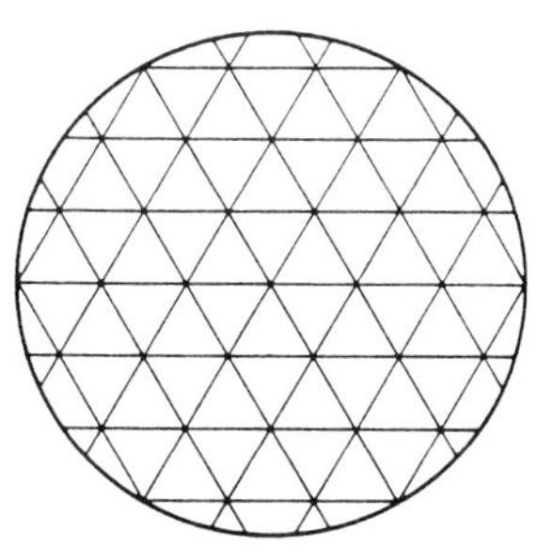

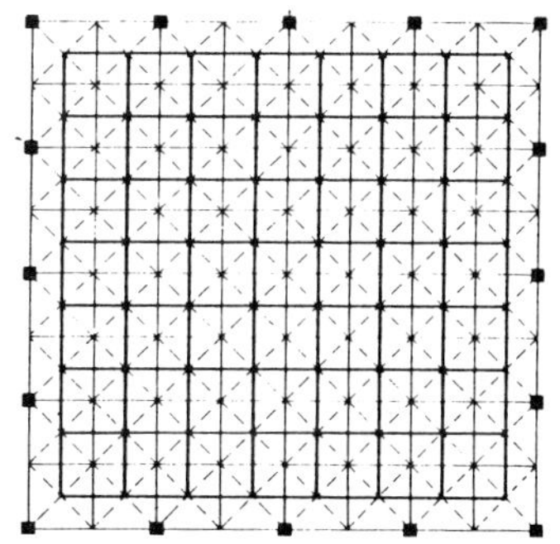

(a) Three-way grid (6-Cycles). (b) Space grid (4-Cycles).
Fig. 3. Cyclically symmetrical grid structures.

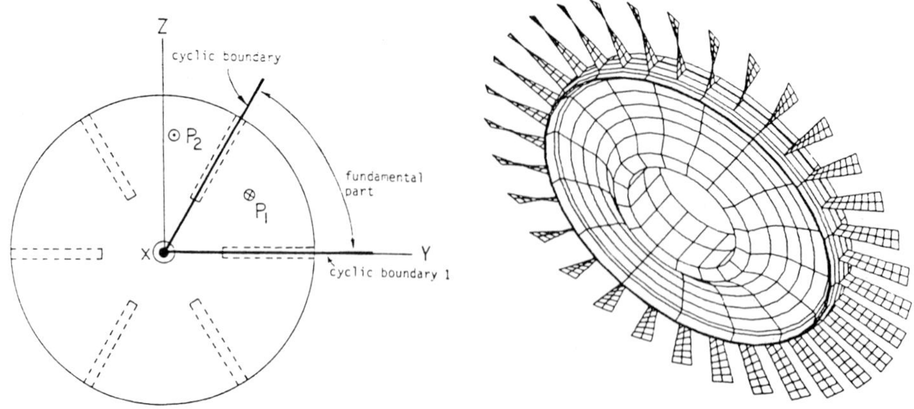

(a) Circular plate (6-Cycles). (b)Flywheel and turbine (32-Cycles).
Fig. 4. Cyclically symmetrical plate structures.

Shell Structures.

4. Numerous axisymmetric as well as non-axisymmetric shell structures are cyclically symmetric. Fig. 5a shows a chimney structure with wind forces and Fig. 5b shows a water tower also with applied wind forces.

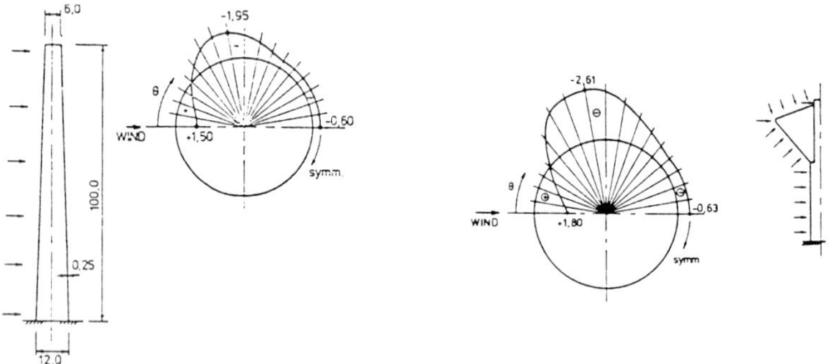

(a) A chimney (30-Cycles). (b) A water tower (30-Cycles).
Fig. 5. Cyclically symmetrical shell structures with wind loading.

Solid Structures

5. A solid pentagon where tetrahedral finite elements are used for its stress analysis is shown, Fig. 6.

Fig. 6. A solid pentagon structure (5-cycles).

ANALYSIS

6. The computer program used for these analyses is the ADINA general purpose finite element system (refs 1-2). In this system, cyclic symmetric structures can be analysed with any combination of truss, plate, beam, pipe, plane stress, shell or solid elements. It is necessary, however, that the complete structure is geometrically cyclically symmetric. Furthermore, the material properties and the displacement boundary conditions must be cyclically symmetrical. Hence, the complete structure must consist of N parts that are geometrically identical to the fundamental part such that N-1 non-fundamental parts are generated from the fundamental part about an axis of cyclic symmetry. Fig. 7 shows a schematic example.

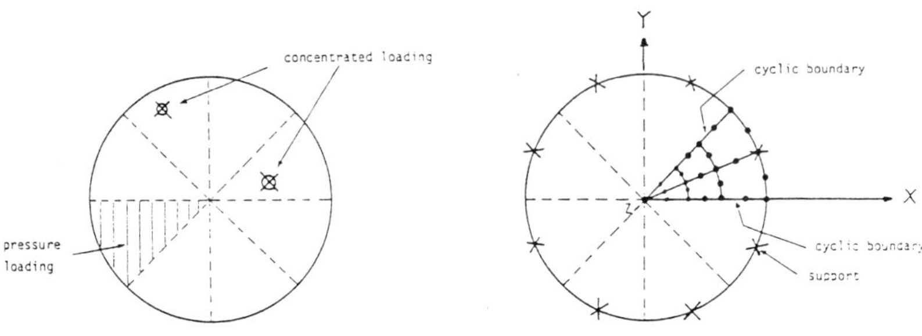

Fig. 7. Analysis of a circular plate using cyclic symmetry.

7. The externally applied loading need not be axisymmetric and can consist of arbitrarily chosen pressure loads, concentrated loads, centrifugal loads or mass loading. The loads in each of the N cyclic parts are defined independently in the input data, (see later). Provision is also made for prescribed displacements in a cyclic symmetric analysis. It is also important that the actual physical displacement boundary conditions at all points within the interior and on the boundary of the fundamental part are defined. The displacements across the boundary between the fundamental part and the adjacent parts are left free to displace, except where supports exist along the cyclic boundary.

8. The analysis is performed by using the theory of cyclic groups (ref. 2), where N analyses of the fundamental part are performed with changing, (mathematically evaluated) boundary conditions along the cyclic boundaries corresponding to the different load cases. The solutions of the analyses are superimposed in order to obtain the final solution. It is important to note, however that there is no approximation in the analysis results when using this cyclic symmetry technique, i. e. the same results are obtained when the complete structure is modelled.

9. This technique has also be used for frequency solutions and linear elastic dynamic analyses, (using mode superposition or direct integration with the Newmark method).

10. The mesh generation facilities available in the program enables the fundamental part of the structure to be modelled rapidly and conveniently. This, together with the fact that data for one segment only needs to be input, reduces the overall amount of data for a particular problem dramatically.

ANALYSIS OF A SKELETAL TOWER.

11. The input data for a simple 4 sided 6m high tower, tapering from 2m square at the base to 0.5m square at the top, is analysed. A non-axisymmetric load is applied horizontally and vertically on one side, Fig. 8.

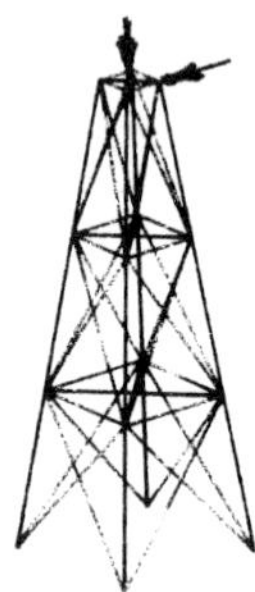

Fig. 8. Skeletal tower with non-symmetrical loading.

12. The data input for this simple tower is as follows:

```
DATABASE CREATE
*
HEAD STRING='SIMPLE RECTANGULAR TOWER USING CYCLIC SYMMETRY'
*
MASTER IDOF=000111 REACTIONS=YES  CY=4
*
COORDINATES
1  0.00 1.4142 0.00 STEP 2 TO 7  6.00 0.3536 0.00
9  2.00 0.0000 0.00        TO 11 6.00 0.0000 0.00
SYSTEM 1 PHI=90.0/COO/ENT NODE XL YL ZL
2  0.00 1.4142 0.00 STEP 2 TO 8  6.00 0.3536 0.00
*
EGROUP 1 TRUSS/ENODES/ENT EL N1 N2
1 1 3 TO 3 5 7
4 2 4 TO 6 6 8
7 1 4 STEP 2 TO 11 5 8
8 2 3 STEP 2 TO 12 6 7
13 3 4 TO 15 7 8
16 3 9 STEP 2 TO 20 7 11
17  4  9 STEP 2 TO 21 8 11
*
EDATA/ENTRIES  EL AREA
 1 853.0E-6  STEP 1 TO  6 853.0E-6
 7 235.0E-6  STEP 1 TO 12 235.0E-6
13 142.0E-6  STEP 1 TO 21 142.0E-6
*
MATERIAL 1 ELASTIC E=200.0E6
*
CYCLICLOAD 1/EGROUP 1/LOAD CONCENTRATED TYPE=NODES
7 1  -15  1/8 2    7.071/8 3   -7.071
*
CYCLICBOUNDARIES OPTION=1/2 1/4 3/6 5/8 7/9/10/11
BOUNDARIES 111111/1 2/BOUNDARIES 100111/9 10 11
*
FRAME O=U/VIEW 1 3.0 2.0  2.0
MESH VIEW=1 NODES=11 ELEMENT=1 BCODE=ALL
*
ADINA
```

13. The actual data required for a tower of this type is minimal. Apart from the standard data required for every problem there are only two lines of additional relevant data. The plotting data is simply to verify the correctness of the co-ordinates, connectivities and loading.

14. It is also shown that in order to change from a 4 sided tower to say a 6 sided tower only one item of data needs to be changed.

15. For the post-processing data we have asked for the member forces and the reactions to be printed out and a displaced shape to be plotted, Fig. 9.

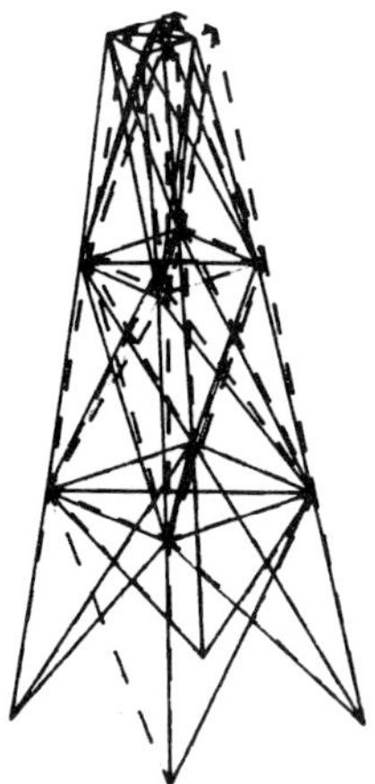

Fig. 9. Displaced shape superimposed on original configuration.

SOLID THREE DIMENSIONAL PENTAGON

16. A solid pentagonal structure is analysed by using 32 hexahedral elements and 32 tetrahedral elements per fundamental segment. The loading consists of a uniform pressure load of 10kN/m^2 applied to segments 1 and 3 respectively. Prescribed displacements varying from lmm to 9mm are also applied to all 5 segments. In addition, the structure is subjected to mass proportional loading and a centrifugal force is also applied.

17. The relevant input data is shown to be 52 lines of data apart from the standard data used generally including plotting data, see Fig. 10. Very little data is therefore required for this problem. In the post processing we have asked for plots of, (a) displaced shape, (b) stress vectors, (c) stress isobands and (d) displacement isobands respectively, see Fig. 11.

18. These results were compared with a full model analysis and the results were identical.

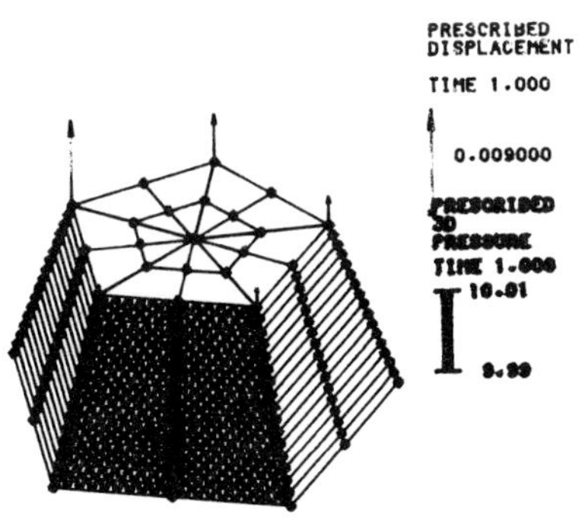

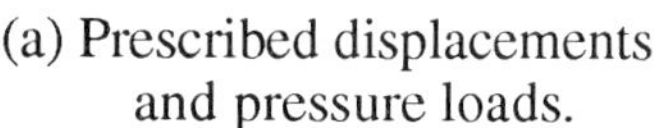

(a) Prescribed displacements
and pressure loads.

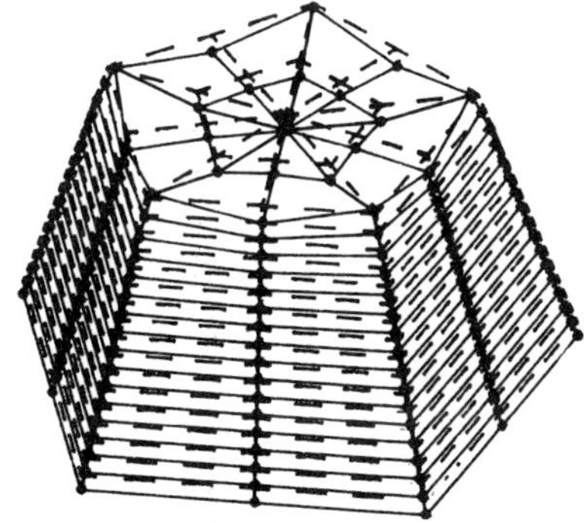

(b) Displaced shape on
original configuration.

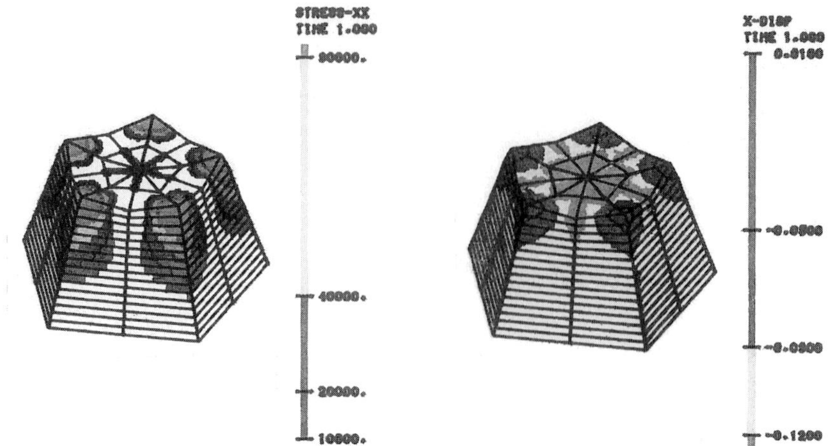

(c) Displacement isobands. (d) Stress isobands.

Fig. 11. Solid 3-D pentagon structure.

FREQUENCY ANALYSIS OF A COOLING TOWER.

19. Mode shapes and fundamental frequencies are computed for a cooling tower, where the fundamental part is modelled with; ten shell elements for the tower, one iso-beam element for a stiffener at the base of the tower, two beam elements for the supporting feet, two beam elements for the supports to the service platform, two 3-dimensional elements for the foundation, eight plate elements for the service platform, two truss elements to support the service platform, one pipe element for the supply system and two 3-dimensional plane-stress elements for the cover to the foundations (Fig. 12). This fundamental segment is one of 32 segments required to model the structure.

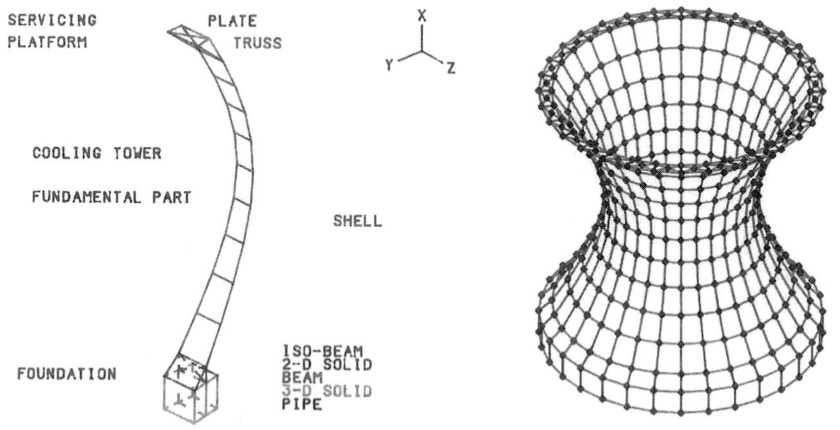

Fig. 12a. Fundamental part. Fig. 12b. Finite element model.

20. The results were compared with those obtained by modelling the whole structure and were found to compare exactly. The first four frequencies obtained from both analyses are shown in Table 1. The use of cyclic symmetry drastically reduces the amount of input data as well as the solution time compared with a full model, (Table 2).

21. The first 4 mode shapes are shown in Figs 13a and 13b.

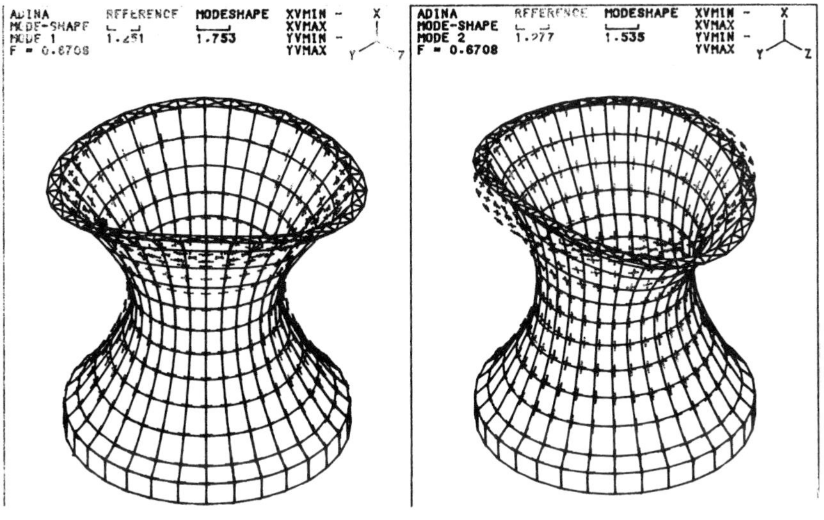

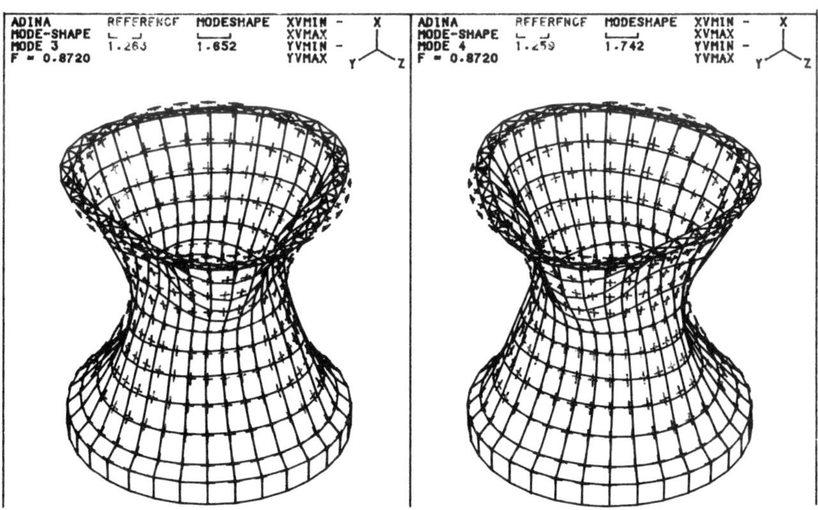

Fig. 13. Cooling tower mode shapes.

Table 1. Natural Frequencies.

Mode Shape	Frequency (Hz)
1	0.6708
2	0.6708
3	0.8702
4	0.8702

Table 2. Computation time.

	With cyclic symmetry	Without cyclic symmetry
No. of equations	198	3168
No. of non-zero entries in K matrix	11745	647721
Solution time normalised	1	20.5

CONCLUSIONS

22. This cyclic symmetry analysis technique has three advantages:

i) Only a relatively small part of the structure needs to be modelled, hence a considerable saving in data is achieved.

ii) Solution times can also be significantly less than when a complete structure is analysed. As an estimate, if there are N cyclic parts, a cyclic symmetric analysis can reduce computational time by a factor of up to N/2; the larger N is, the more time reduction is achieved.

iii) The computer storage requirements are reduced drastically, reductions up to a factor of N/2 have been achieved.

REFERENCES

1. ADINA, Finite element analysis programs developed by ADINA R D. Inc, 71 Elton Avenue, Watertown, MA 02172, USA.

2. BATHE, K.J. Finite Element Procedures in Engineering Analysis Prentice Hall, 1982.

3. ZHONG, W. AND QUI, C. Analysis of Symmetric or Partially Symmetric Structures", Computer Methods in Applied Mechanics and Engineering, 38, 1 - 18, 1983.

73. An introduction to the fire analysis of double-layer grids

A. YARZA and P. CAVIA, ORONA Space Structures, Spain, and
G. A. R. PARKE, University of Surrey, UK

SYNOPSIS. The fire analysis of double-layer grids requires a knowledge of the fire behaviour in large compartments and a study of the collapse behaviour of the grid under increasing temperatures. In order to perform practical analyses, structures exposed to fire need to be modelled numerically. Fire models provide the time-dependant temperature distribution within the compartment, based on a study of the heat flow. The Natural fire model has been used in the fire analysis of double-layer grids. The behaviour of the grid depends on the heat transfer from the hot gases to the structural members, the change of steel properties with temperature and the idealisation of the member behaviour. Collapse of double-layer grids is commonly characterised by the sequential failure of individual members. As fire develops, temperatures increase and the progressive weakness of steel causes failure of members. A step by step analytical method is performed in order to establish the fire resistance of the grid.

INTRODUCTION

1. The fire phenomenon has received a large amount of attention during the last decade, and it is found to be a complex field, mainly due to the high number of parameters involved such as, fire load density, geometry of the compartment, thermal characteristics of the compartment boundaries and ventilation rate etc.

2. The effect which elevated temperatures produced in fires has on structures, depends undoubtedly on the changes in the properties of materials from which the structures are constructed. In particular, structural steel becomes progressively weaker with increasing temperatures and consequently, the overall structural response is deteriorated.

3. Codes of Practice for fire design reduce the fire analysis of structures to the study of individual members. However, for highly redundant structures, such as double-layer grids, the failure of a member due to the effects of a fire does not necessarily produce the overall failure of the complete structure. Consequently, double-layer grids require a specific analysis to be carried out.

4. Structural response in a fire is studied as a fire limit state. The fire design of double-layer grids requires the study of the collapse behaviour of the whole

structure when subjected to increasing temperatures (ref. 1). The loss of member load carrying capacity due to both, decaying steel properties and possible plastic buckling, will result in a redistribution of internal forces within the structure. At a certain temperature, the ability of the structure to cope with the redistribution may no longer be possible and the grid may become a mechanism.

5. The complex nature of both fire behaviour and structural response leads to the necessity of mathematical models in order to perform practical analyses. These models present different degrees of sophistication and their accuracy is essential in obtaining designs close to the real behaviour.

FIRE ANALYSIS PROCEDURE

6. The fire analysis process for double-layer grids consists of four main steps as shown in Fig. 1.

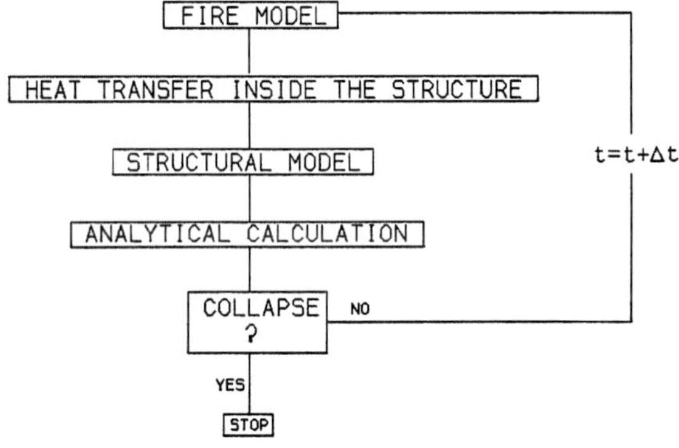

Fig. 1. General fire analysis procedure.

7. The **fire model** provides the gas temperature distribution in the compartment with time, after flashover has occurred. The transient temperature conditions lead to a heat transfer from the hot gases to the surface of the structural elements. A **heat flow** is then transmitted by conduction inside the structural elements, which is governed by Fourier's law of heat transfer. The temperature across the section depends on the shape factor and the amount of insulation that the section is provided with.

8. The **structural model** computes the changes in the material properties as a consequence of the elevated section temperatures and provides the basic data for the stability analysis of the structure. A **structural analysis** is carried out at each time step, based on the fire and member models, in order to study the stability of the overall structure at increasing temperatures. The analysis is carried out up to collapse. If the structure does not comply with the corresponding fire requirements, the structure needs to be enhanced either by increasing the cross-section of elements or by providing higher levels of protection.

FIRE MODELS

9. Obtaining data in real fire experiments tends to be very difficult due to the random nature of this phenomenon. Investigations carried out by various Fire Research Centres (ref. 2) classify the parameters involved in the behaviour of a fire as shown in Fig. 2.

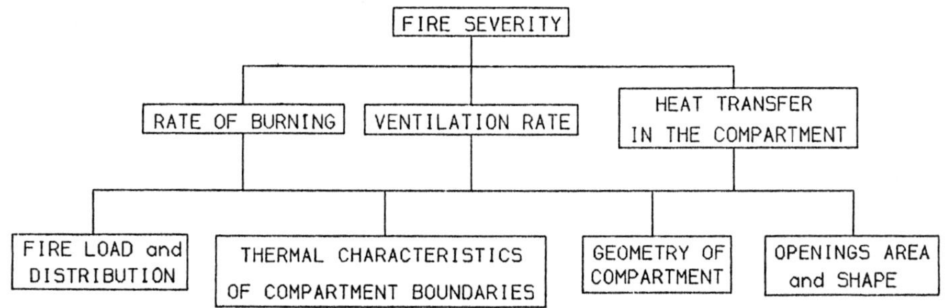

Fig. 2 Fire severity components and parameters.

10. In order to simplify the complex behaviour of fire, mathematical models are defined. Three fire models are available distinguished by their different levels of refinement, rather than by different scientific concepts. They are as follows:

1. ISO 834 Standard Fire Model.
2. T-Equivalent Standard Fire Model.
3. Natural Fire Model.

11. ISO 834, 1975 (ref. 3) describes the **Standard Fire Model** which is characterised by a monotonically increasing temperature-time curve. This is considered to be the traditional fire model and it is widely applied by most of the International Codes. The analytical expression to this curve is given by:

$$T-T_0 = 345 \log_{10}(8t+1) \tag{1}$$

where: T = Temperature at time $t\,^{\circ}C$.
T_0 = Ambient temperature ($20^{\circ}C$).
t = time (min).

12. The T-Equivalent Standard Fire model is a hybrid between the ISO 834 model and Natural fire model. It is mainly based on the former but it also introduces approximate fire severity components (ref. 4).

13. These two models do not closely represent a real fire, but the fact of being standardised enables different international tests to be compared and a relatively simple grading assessment method to be applied. The fire resistance that an

element is required to present is usually graded in National Codes according to these curves and it is defined as 15, 30, 60. etc. minutes depending on the type of building.

14. The Natural Fire Model is presented as the most sophisticated method of defining structural fire resistance. A detailed fire behaviour knowledge and modern computer-aided calculation methods are required to compute the heat flow developed within the compartment. The main variables to be taken into account are:

- the amount of combustible material and distribution;
- the combustion rate;
- the ventilation rate;
- the thermal characteristics of compartment boundaries;
- the geometry of the compartment.

In most cases, uniform temperature is considered throughout the compartment and therefore the whole structure is affected by the same temperature.

15. Double-layer grids are usually designed for covering large areas such as stadiums, sport-halls, warehouses, etc. The assumption that a large space frame structure will be subjected to a uniform temperature throughout the structure, during a fire, is not applicable; thus, local fires need to be studied. The following list of parameters need to be considered:

- location of the fire bed;
- growth of fire;
- location of the structural members.

16. Sophisticated 2D or 3D finite element programs are required to calculate the temperature contours in the compartment (ref. 4). The structural elements will be affected by a different temperature, depending on their location.

17. This model provides the only method to assess rationally the effect of local fires in large compartments. It may even allow the stability of bare steel structures during a fire to be proven. However, this model is not yet widely accepted by International Codes and it is still under development. The application will depend on the ability of the model builders to simplify the input data without losing accuracy and to prove the reliability of their fire modelling.

HEAT TRANSFER INSIDE THE STRUCTURE

18. The exposure of a structure to transient heating conditions leads to a heat transfer from the hot gases to the surface of the structural elements. The heat is then transmitted by conduction inside the elements, which is governed by Fourier's 3D law of heat transfer.

19. The temperature distribution across the member section depends on the degree of insulation and massivity of the section. The massivity is inversely

proportional to the shape factor (P/A = perimeter/area) and it is the dominant factor in determining the temperature reached in bare steel sections.

20. Numerical methods have been devised to deal with 1D, 2D or 3D heat flow inside steel element. The ECCS Recommendations (ref. 5) reduce the heat transfer to a 1D problem. The temperature across the section is considered to be uniform and thermal properties of steel to be independent of temperature. This simplification is based on the high thermal diffusivity of steel (=k/ρc; where k is the conductivity, ρ the density and c the specific heat). It is known that a uniform temperature is a conservative simplification (ref. 2). Guidance is given in the ECCS document to determine the temperature across the steel section for both unprotected and protected sections.

21. Double-layer grids are commonly composed of tubular steel sections. These sections present a high shape factor and for P/A>300 m^{-1} (e.g. 90/4 circular tubular section), the steel temperature can be considered to be equal to the temperature of the hot gases. In this case, it would not be necessary to take into consideration the heat transfer problem (ref. 5).

STRUCTURAL RESPONSE

22. The structural response to fire conditions depends on the changes in the properties of the materials from which the structure is constructed. Structural steels retain their strength up to about 250°C and thereafter there is a sharp decline in strength (ref. 5). Fig. 3 shows stress-strain curves for Fe430 steel for a range of temperatures from 20° to 600°C.

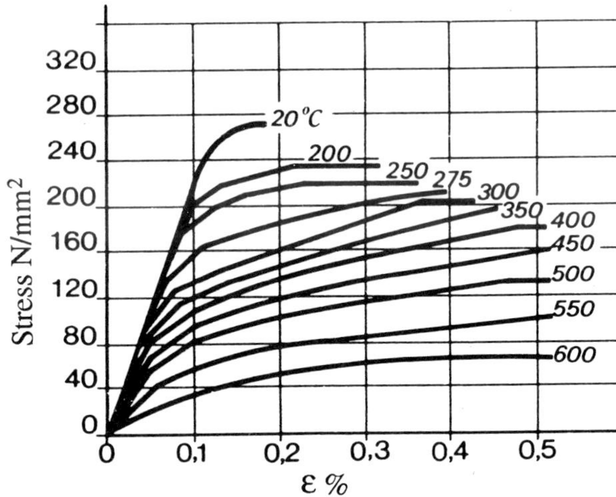

Fig. 3 Stress-strain curves for Fe430 ranging from 20° to 600°C.

23. An important effect in the behaviour of steel subject to a temperature rise, is the gradual disappearance of a well-defined yield limit. Both the yield limit stress and the modulus of elasticity are temperature dependant parameters. For temperatures above 600°C the effect of creep has to be analysed explicitly.

24. Table 1 shows the temperature dependence of the mechanical and thermal properties of constructional steel. These expressions are established by the ECCS Recommendations (ref. 5) and are accepted by a large number of authors. These recommendations also include approximate buckling curves at elevated temperatures.

25. The progressive reduction in the load-carrying capacity of the structure under increasing temperatures may produce the collapse of the structure. The temperature at which the structure collapses is known as the critical temperature.

26. Double-layer grids are normally highly statically indeterminate and consequently the overall structural collapse is, in general, due to a sequential failure of members. In pin-jointed double-layer grids, members are axially loaded and failure may be due to yielding of tension members or buckling of compression members. Most practical double-layer grid systems avoid the use of very stocky or very slender sections and the buckling encountered will be plastic in nature. Member instability due to plastic buckling causes a sudden loss of member load carrying capacity.

Table 1. Thermal and mechanical properties of steel at elevated temperatures.

THERMAL PROPERTIES	
COEFFICIENT OF THERMAL EXPANSION	$\frac{\Delta L}{L} = 0.4\times10^{-8}\,T_s^2 + 1.2\times10^{-5}\,T_s - 3\times10^{-4}$ [m/m] Approximated: $\frac{\Delta L}{L} = 1.4\times10^{-5}\,T_s$ [m/m]
SPECIFIC HEAT	$C_s = 520$ J/Kg°C
THERMAL CONDUCTIVITY	$K_s = 34.9$ W/m°C
MECHANICAL PROPERTIES	
ELASTICITY MODULUS	$E = (1 + 15.9\times10^{-5}T_s - 34.5\times10^{-7}T_s^2 + 11.8\times10^{-9}T_s^3 - 17.2\times10^{-12}T_s)E_o$ where, T_s = steel temperature $0<T_s<600$ °C $E_o = 2.05\times10^5$ N/mm²
COEFFICIENT OF REDUCTION OF EFFECTIVE YIELD STRESS	$\Psi = \frac{\sigma_{y,t}}{\sigma_{y,20}} = 1 + \frac{T_s}{767\ \ln(T_s/1750)}$ $0<T_s<600$ °C $\Psi = \frac{\sigma_{y,t}}{\sigma_{y,20}} = \frac{108(1-T_s/1000)}{T_s - 440}$ $600<T_s<1000$ °C

27. The general loss of strength of structural members due to an increase in temperature and the plastic buckling effect results in a redistribution of internal forces within the structure. If the structure is capable of withstanding this redistribution it may be able to allow an increase in the temperature and it will provide longer fire resistance. However, if the redistribution causes other members to fail, then the failure may be progressive and the structure will finally

become a mechanism. A structure can also be considered to be unserviceable, if excessive deflections are involved.

28. The behaviour of the sequential collapse phase depends critically upon the failure characteristics of the individual members. The use of a correct representation of the member is important in order to obtain an accurate structural response. There have been several different mathematical models used in performing incremental load-collapse analyses. These models are generally linear representations of the members stress-strain curves (ref. 6).

29. The yield limit, elasticity modulus and critical buckling stress depend on the temperature and the post-failure behaviour is also temperature-dependent. Consequently, the models mentioned above have been modified and a different idealisation will be required for each temperature.

ANALYTICAL METHOD

30. Several analytical methods for the collapse of double-layer grids have been studied (ref. 7). The fire collapse analysis is a particular application of these methods, the important difference being that the external load is maintained constant while the temperature is increased.

31. A general scheme of the analytical procedure in fire design is shown in Fig. 4. A sequence of linear analyses is performed for determining the minimum change in temperature that causes the stress in any member to reach a critical value. As members fail, modifications are made to the structure to simulate the action of these failures. This process is carried out until a mechanism is formed or a predetermined deflection is reached. Finally, the fire resistance of the structure is checked. If the fire resistance does not comply with the Fire Safety Codes, the structure is enhanced and the general procedure repeated.

DESIGN CRITERIA

32. The stability of a grid in a fire depends basically on the load level, fire severity, location of the structure in relation to the fire bed, support conditions, degree of insulation and the size of the tubular sections. Fire is considered to be an accidental load case, thus, in checking the stability of the structure unfactorised loads are taken.

33. It has been noted that in fire load analysis there might be two sources of loss of strength typically, progressive weakening of steel and plastic buckling. Temperatures above 250°C produce a sharp loss in the strength of structural steel. The members that remain elastic will not be able to withstand large load redistribution from failing members. In general, any attempt to improve the structure against fire should be directed to postponing the first failure.

34. Failure is expected to start from the most highly stressed members and thereafter the failure sequence depends on the stress redistribution characteristics. Preliminary simple analyses (ref. 8) can be carried out to determine the collapse patterns of the grid. The members can be fire protected according to these collapse patterns, so that the group of members prone to fail first would be insulated,

especially if they are in compression. This solution avoids general insulation of the grid which would be a expensive solution.

35. Intumescent paints are practically the unique protection measure that can be considered for pin-jointed double-layer grids, as this guarantees the aesthetical aspect that these structures are required to present. Intumescent paints swell up at temperatures around 150°C and give a thick viscous foam with good insulation characteristics; but they are expensive.

36. Boundary conditions are very important in fire design due to the large displacements that thermal expansion involves. Sliding supports are required to allow for the horizontal displacements expected. Unrestrained expansion of the grid is desirable, so that the temperature rise does not introduce directly any increase in internal stress distribution. This can be achieved by suitable support conditions, however if local fires are considered, free expansion is difficult to achieve.

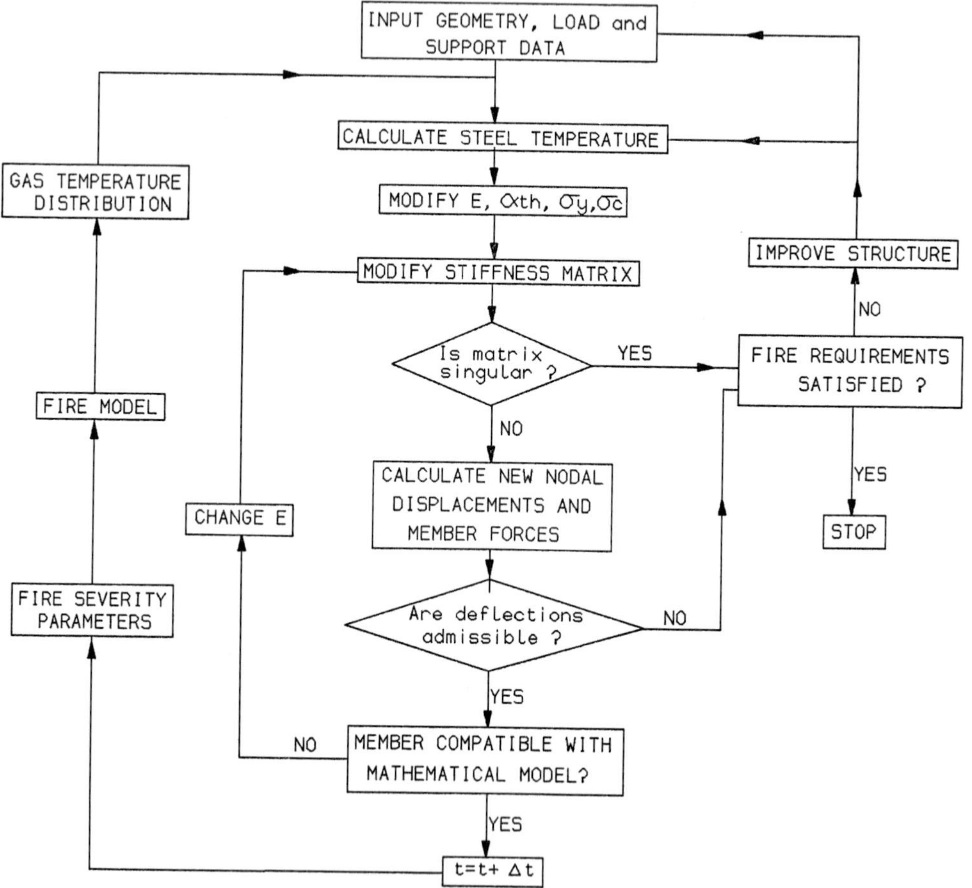

Fig. 4. General analytical procedure.

37. If the grid is supported only at the corners, the failure of the diagonal elements connected at the supports may produce the overall collapse of the grid. Consequently, it is recommended to support the grid around the boundary rather than adopt four corner supports. The diagonals members at the supports are very important for the fire resistance of the structure. It is preferable to over design these members with a larger cross-section than the one determined by elastic analysis and to support the structure at the top layer nodes, so that these diagonal members are in tension.

38. The assumption of uniform temperature distribution is not applicable in large compartments. This is usually the case for double-layer grids, which commonly cover large areas. Local fires need to be studied and their effect may be limited to a certain zone of the grid. Members located far from the fire bed should not be directly affected by the rise in temperatures and a higher stress reserve could be expected. However, an internal stress increment may also be expected as cold members restrain the expansion of hot members. Tests have shown that members subject to non-uniform temperature distribution attain significantly higher level of fire resistance than uniformly heated ones (ref. 4).

39. Although the fire bed can have a random location, it is necessary to reduce the analysis to the locations where a higher fire severity will be reached or where the structure is more vulnerable. Another important factor is the height at which the structure is located in relation to the probable fire bed. The higher the structure is located the lower the steel temperatures that will be reached.

CONCLUSIONS

39. A general review of the fire analysis of double-layer grids has been presented. It is emphasised that fire behaviour is a complex matter requiring extensive numerical modelling. Most International Codes still apply Standard Fire Models, which lead to high insulation requirements in steel construction. However, Codes are progressively accepting Natural Fire Models, as they provide the only reliable representation of fire behaviour.

40. Double-layer grids need a specific fire analysis method. A general step by step procedure has been described. It is necessary to study the behaviour of the whole structure and for large compartments, a local fire concept is introduced.

REFERENCES

1. KLIMKE H., POSCH J. Application of the modified fictitious force method for the fire resistance analysis of space trusses. Proceedings of the Third International Conference on Space Structures. Ed. H.Nooshin. University of Surrey, 1984. Pp 589-593.

2. MALHOTRA H.L. Design of fire resistant structures. Surrey University Press, 1982.

3. ISO Fire Resistance Tests. Elements of building construction. International Standard 834, 1975.

4. Steel and Fire Safety: A Global Approach. Steel Promotion Committee of Eurofer. Brussels, 1990.

5. European Recommendations for the Fire Safety of Steel Structures. European Convention for Constructional Steel. Amsterdam, 1983.

6. COLLINS, I.M. Collapse analysis of double-layer grids. PhD Thesis. University of Surrey, 1981.

7. PARKE G.A.R., Collapse analysis of double-layer grids. Studies in Space Structures. Ed. H.Nooshin. Multi-Science Publishing Co Ltd, 1991, pp. 153-179.

8. YARZA A. Fire analysis of double-layer grids. MSc Thesis. University of Surrey, 1992.

74. Qualification analysis of a new structural system

R. LANDOLFO and F. MAZZOLANI, University of Naples, Italy, and
E. VENTRELLA, Centro Acciai of Bitonto, Italy

SYNOPSIS. A new space structure system, called VESTRUT, is used in the design of the roofing of the new airport in Milan (Malpensa 2000). The system is comprises a quasi-spherical joint and tubular members. The joint is a combination of two semi-spheres separated by a central disk, which are connected together by an axial bolt. The ends of the tubular members are connected to the joint by means of high strength steel terminal bars. In order to fulfil the structural requirements of the design, the system has been submitted to a series of qualification tests and numerical analyses.

DESIGN REQUIREMENTS

1. The space structure of the roofing of the new airport in Milan (Malpensa 2000) is a double-layer grid which is simply supported on the longitudinal sides of each 42x75m rectangular unit. Nine units constitute the total roof. Under the most severe loading condition (dead plus live loads, in total about 5800 Nm^{-2}) the maximum member force is approximately 1150kN; this means that the system must provide a strength of at least 2000 kN in order to provide a suitable safety factor (1.70 minimum).

2. The former VESTRUT system was able to provide a maximum strength of approximately 250 kN (ref. 1),. It was, therefore, necessary to modify the system, by revising the dimensions and materials, in order to increase the load carrying capacity by at least 8 times. The new system was submitted to very severe qualification tests, which were conceived to demonstrate the system reliability in relation to the above design requirements.

STRUCTURAL TYPOLOGY

3. The space structure is composed of pyramidal units with a rectangular base of 2.40x3.60m and depth of 2.80m. The joint has a diameter of 300mm and is composed of two spherical bowls separated by a central disk and connected together by means of high strength steel bolts with diameter D=36−39 mm (Fig. 1). The members have tubular sections with diameter D=139−219 mm and thickness t=4−9 mm. Their ends have a conical shape which is connected to the

joint by means of high strength steel terminal bars with diameter D=60 mm. The ends of terminal bars have the shape of a hammer which is locked inside of the spherical joint.

4. The materials are the following:
   - the tubular members are in Fe 510 steel grade;
   - the terminal bars are in high strength steel (type 38 NiCrMo4);
   - the bolts are 10.9 type.

QUALIFICATION PROCEDURE

5. The qualification of the new space structure, called VESTRUT, has been carried out by means of the following tests:

a) Tensile tests on the joint

Tests were performed on the joint-terminal connection by applying a uni-axial tensile force, in order to establish behaviour under ultimate conditions.

b) Tensile tests on the member

Tests were performed on a full length tubular member, connected at the ends to joints, by applying a uni-axial tensile force, in order to evaluate the actual behaviour, in terms of strength and deformability, both in service and under ultimate load conditions (i.e. measurement of the equivalent modulus of elasticity).

c) Flexural and shear test on a prototype

Tests were performed on a specimen composed of two full scale units, in order to check the correct behaviour of the support devices and to submit the diagonals to the worst working conditions.

d) Complete bending test on a full span prototype

Tests were performed on an element of 42m span, transversally composed of two units with the same dimensions and with the same supporting condition as the actual structure. Preliminary tests on the thermal effect and dynamic performance were undertaken. Elastic cyclic tests under serviceability conditions were undertaken before the load was increased to achieve collapse.

6. The tests detailed in (a) and(b) were performed at the Centro Sviluppo Materiali Laboratory (CSM) in Rome. The tests detailed in (c) and (d) were carried out at the Centro Acciai factory in Bitonto. The measurements for tests (c) and (d) were undertaken by means of the equipment and the staff of Centro Sperimentale di Ingegneria (CSI) of Naples. The certification of the experimental results was assured by the Engineering Faculty of the University of Naples.

7. The numerical analyses were undertaken by means of finite element programs. The results of the experimental activity have been elaborated and compared with the numerical results, giving useful information on the reliability of this structural system in accordance with the design requirements and allowing the use of simplified models in the design calculations.

Fig. 1.

Fig. 2.

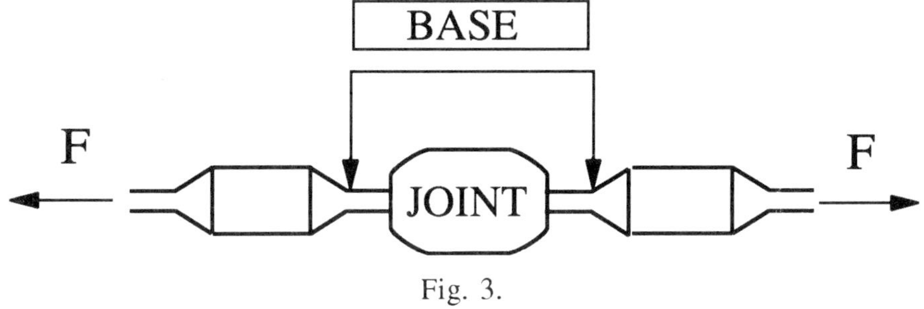

Fig. 3.

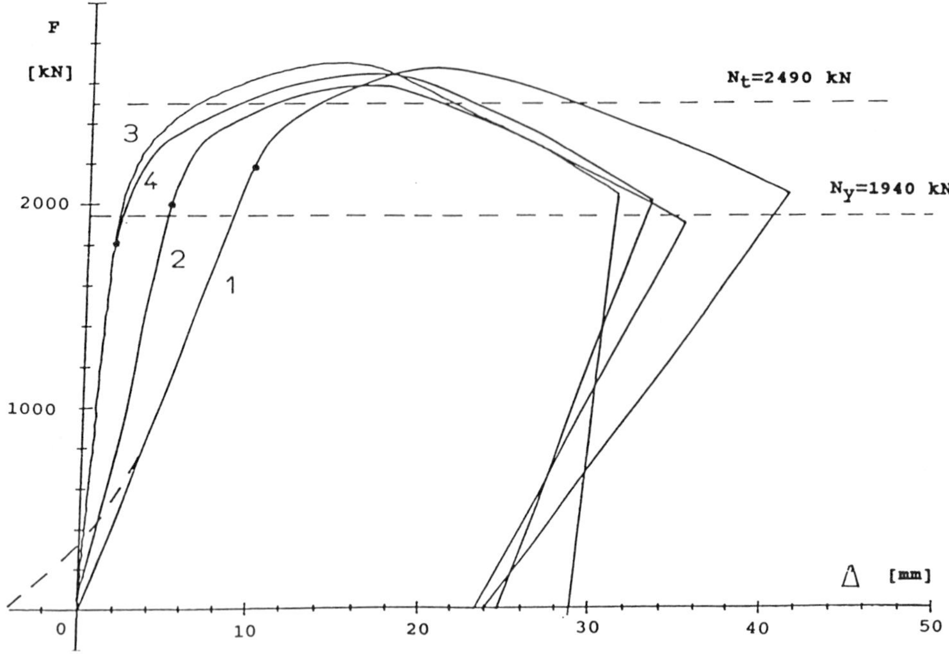
F
[kN]
N_t=2490 kN
N_y=1940 kN
3
4
2
1
2000
1000
Δ [mm]
0
10
20
30
40
50

Fig. 4.

TEST N.1
N_{max} = 267 t

Fig. 5.

TESTS ON JOINTS

8. Several specimens were submitted to axial force which was applied to the terminal bars (see Figs 2 and 3). The force versus elongation curves (F-Δ) are plotted in Fig. 4, referring to the four more significant cases, which give in practice the same value of the load bearing capacity.

9. From these results, it can be concluded that:

a) the behaviour of the joint can be considered as a full strength type, because the rupture occurred in the terminal bars, not in the joints (Fig. 5);

b) the experimental collapse load was 7% greater than that predicted on the basis of the nominal value of ultimate strength;

c) the local yielding in the contact zones between the hammer end of terminals and the internal housing slot, inside the spherical joint, was negligible.

TESTS ON MEMBERS

10. The specimens were composed of a tubular member (D=219mm, t=10mm) of 3600 mm length, connected at the ends to the joints (Figs 6 and 7). The typical F-Δ curves are plotted in Fig. 8, where curve A corresponds to the total elongation of specimen measured on the base A of Fig. 6 and curve B corresponds to the elongation of member only measured on the base B. The different slopes of the elastic part of both curves are due to the influence of local assessments in the joints.

11. From these results, it can concluded that:

a) the tests confirmed that the system is characterised by a full strength joint, as the rupture was always located in the terminal bars (as in fig. 5);

b) the experimental values for the elastic limit and of the collapse loads were in good agreement with those predictions, based on the nominal material properties;

c) from diagram A of Fig. 8 the equivalent modulus of elasticity was found to be 135.000 N/mm^2.

FLEXURAL AND SHEAR TEST

12. The prototype shown in Figs 9, 10 and 11 was submitted to two concentrated loads in the central interior joints (nodes 5 and 6). After three cycles of loading in the elastic range up to a force of 300kN, the specimen was loaded up to collapse, which occurred at a force of 820kN. The collapse condition was produced due to the rupture of the shank bolt in the joint (node 7), due to the combined effect of tensile and shear forces, which was apparent by the typical rupture surface inclined at 45°.

13. From these results, it can be concluded that:

a) the agreement between the experimental evidence and the numerical analysis was very satisfactory, which validates the reliability of the mechanical model, in which the above value of the equivalent elastic modulus was used;

b) the behaviour of the support devices was in agreement with the design requirements;

c) the safety factor at collapse was approximately 2.7.

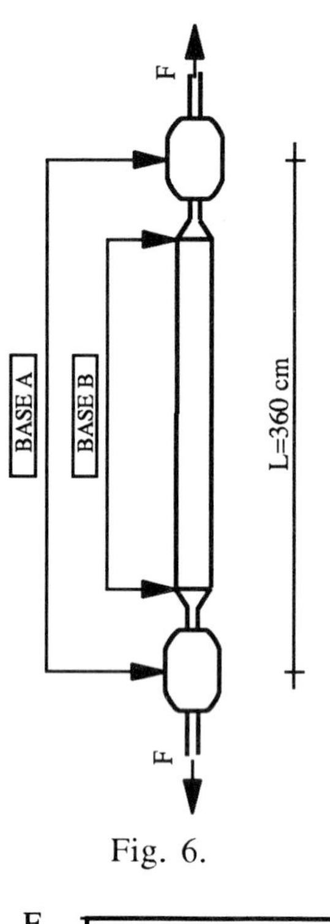

Fig. 6.

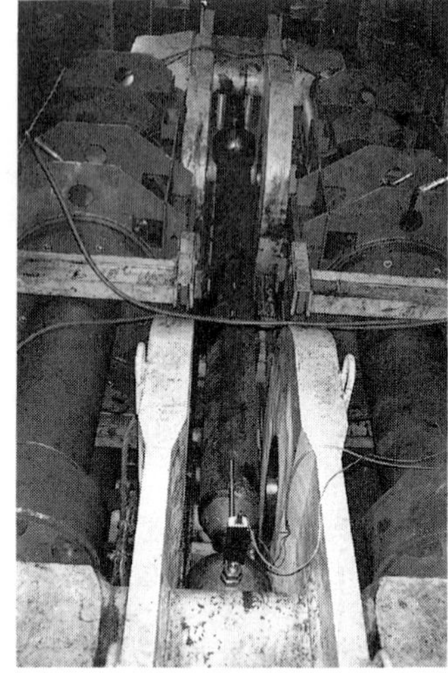

Fig. 7.

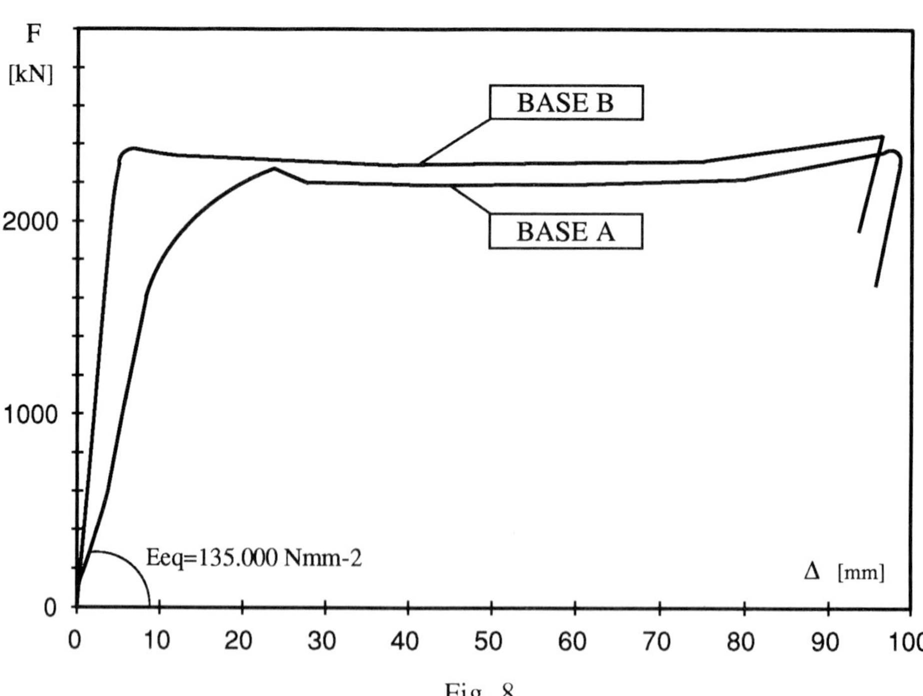

Fig. 8.

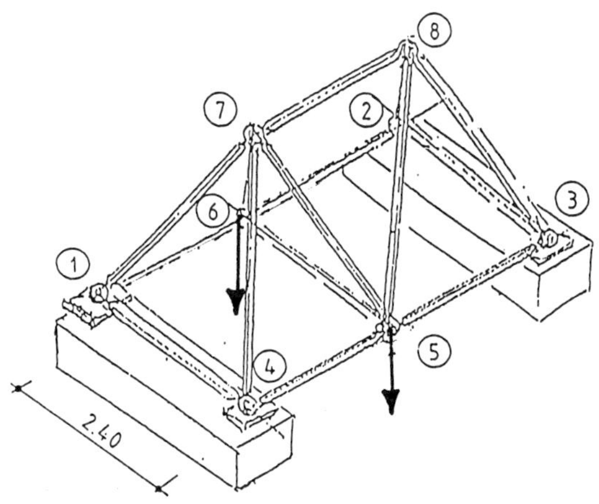

Fig. 9.

Fig. 10.

Fig. 11.

COMPLETE BENDING TESTS

14. The prototype (see Figs 12 and 13) is shown in plan in Fig. 14. It was designed with a initial camber of 66cm, in order to compensate for both the elastic and non-elastic deformations in serviceability conditions. The loading positions are shown in Fig. 15; they correspond to concentrated forces on six pairs of upper joints.

The dynamic test

15. The scope of this test was to evaluate the natural period of vibration, which has been obtained by exciting the prototype in the vertical plan. A frequency of about 3 Hz has been measured, which corresponds to T=0.33 sec. This result was confirmed by the numerical analysis. It is useful to observe that the value of the natural period under serviceability conditions and the maximum values of masses tends to decrease up to a predicted value of 0.7 sec.

The thermal test

16. Measurements were taken to establish the influence of temperature on the prototype over a period of one day. The difference between the environmental temperature and that measured on members of the structure is plotted in Fig. 16. The intensity of the thermal variation on the structure was $\Delta T = +4$ to $-14^{o}C$. The effects in terms of stress and deflection were negligible, according to the numerical analysis.

Cyclic test in serviceability conditions

17. Four cyclic loads have been applied to the prototype up to a maximum force of 60kN on each loaded joint (see Fig. 15). After the erection, a non-elastic deflection of about 7cm at the mid-span was measured, due to the clearance effect between joints and members. The elastic mid-span deflections are plotted in Fig. 17, where the comparison with the theoretical prediction shows that the equivalent elastic modulus includes the clearance effect, whereas the use of the nominal value of $E=210kN/mm^2$ should be used in the prediction of the elastic values under serviceability conditions.

Collapse test 5.

18. Each force has been increased up to a value of 130 kN, which produced the collapse of the prototype. The ultimate conditions have been obtained without a brittle rupture, a ductile failure occurred due to the plastic rotation in horizontal plan of the joint (node 24), which reacted to the sudden buckling of the compression member. The real safety factor was approximately 2.16, greater than that required in the design (1.70).

Fig. 12. Fig. 13.

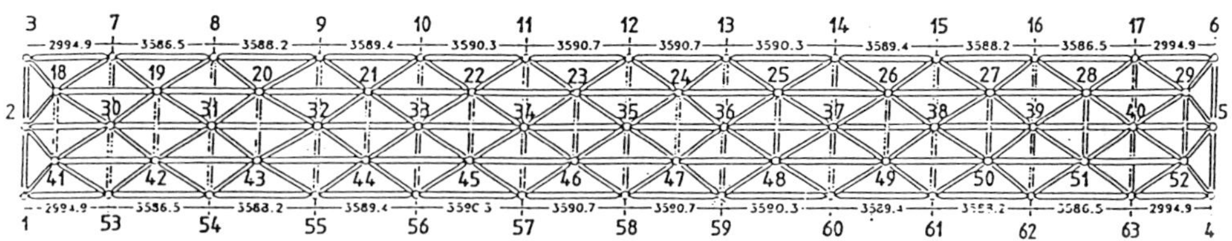

Fig. 14.

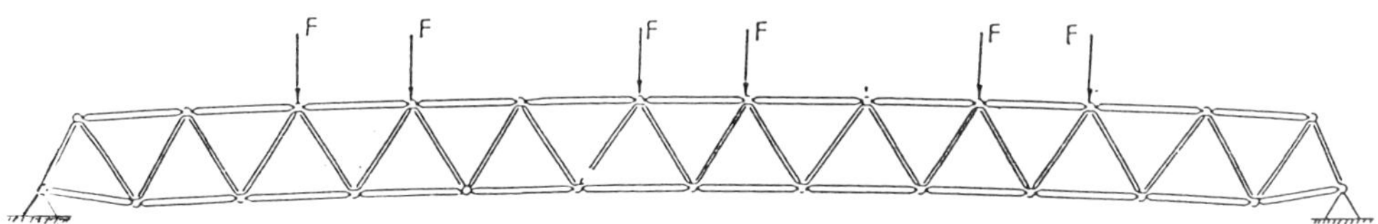

Fig. 15.

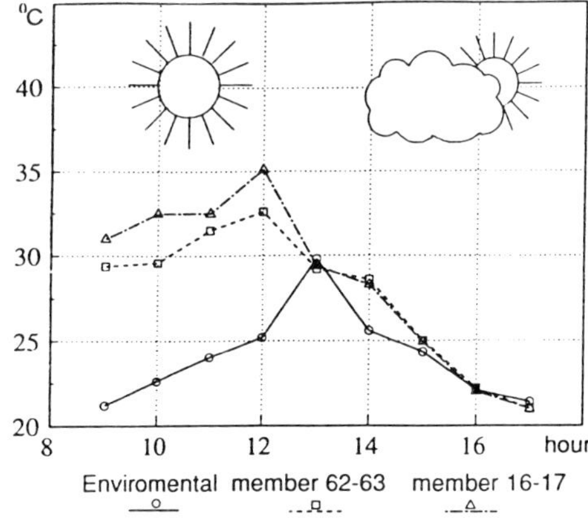

Fig. 16.

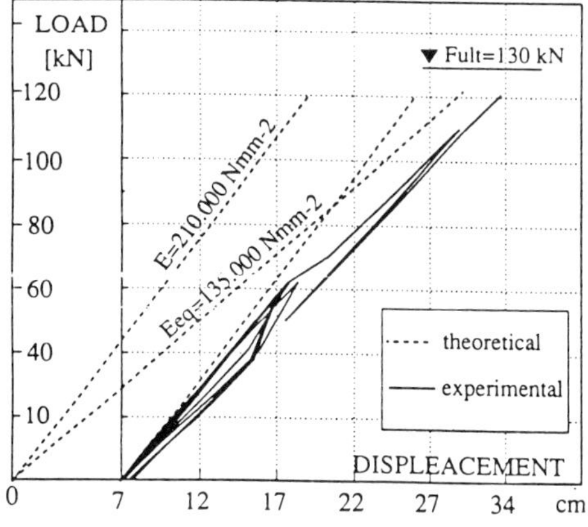

Fig. 17.

CONCLUDING REMARKS

19. The VESTRUT system has been conveniently modified in dimensions and materials in order to fulfil the design requirements, which arise from the structural scheme of the roof units of the new airport of Malpensa. The severe set of tests on specimens and prototypes provided suitable results in terms of strength, deformability and ductility. The new system is, therefore, qualified not only in view of this specific application, but also in general way. Due to its geometrical and mechanical features, it can be considered one of the most reliable space structures, mainly due the joint, which is of a full strength type.

REFERENCES

1. LANDOLFO, R.and MAZZOLANI, F.M. Indagine sperimentale sul sistema VESTRUT. Proceedings of the XIII Congresso C.T.A., 1991, October, 225-236.

2. BALLIO, G. and MAZZOLANI, F.M. Theory and Design of Steel Structures. Chapman and Hall, London, 1983.

75. Dynamic behaviour of slender tubular members of space structures under wind loading

I. ARANBURU, IKERLAN, Spain, and P. CAVIA, ORONA Space Structures, Spain

SYNOPSIS. Space structures often use slender tubular members. In many cases, structures stand outdoors for long periods while members remain under direct wind loading. Actually, slender members under the direct action of the wind are subjected to dynamic excitation with alternating loading resulting in some transversal oscillations. Several pin jointed systems for space structures offer very weak sections under transversal bending moments. In addition, the alternating nature of the loads can produce unexpected failure due to the fatigue of the connecting bolt. Furthermore, when dynamic resonance occurs, the resulting amplified oscillations drastically reduce the life of the bolt. This paper deals with the forecasting of this problem and some of the solutions which can be applied.

INTRODUCTION

1. The current trend of erecting more and more slender structures brings about a higher occurrence of failures and collapses related to the dynamic response of structures under wind loading. These events have been observed on cooling towers, chimneys, long span suspension bridges and circular or square shaped slender structures.

2. Skeletal space structures are built with slender members. Circular hollow sections are efficient when carrying, tensile as well as compressive loads having equal slenderness in any direction. In addition, they permit simple welding together with ease of manufacturing, offering one of the best choices for the section of the members.

3. In many cases, assembly and erection of parts of the structure is carried out with members remaining under direct wind loading. In other cases, the space structure is designed for outdoors. As a result of this, tubular members will be prone to the local effects of dynamic wind loading.

4. Von Karman (ref. 1) showed that slender circular members under direct wind action are subjected to a dynamic excitation with alternating loading in a direction perpendicular to the wind flow. These forces are due to the turbulence generated by the alternating vortices described by Von Karman. This turbulence appears in

the wake of the obstacle and result in transverse oscillations of the member which yield alternating bending moments and shear forces.

5. Several pin jointed systems for space structures offer a tapered section at the connecting ends which, although able to withstand large tensile and compressive forces, is inherently very weak under the action of bending moments and shear forces.

6. Even worse is when one of the natural frequencies of the bar happens to be close to the frequency of the alternating wind loading, resulting in some amplified transverse oscillations. In addition, it must be pointed out that due to the alternating nature of the loading, the connecting ends of the member will suffer fatigue damage which may eventually result in unexpected failures of the pin joints.

7. Therefore, it is necessary to forecast which of the members in a structure could be seriously damaged by this effect under wind velocities feasible during the assembly process, or for as long as the structure will last if it remains outdoors. This paper describes the above wind loading effect, considers how to forecast its occurrence and presents some of the solutions which can be applied.

DYNAMIC WIND LOADING OF SLENDER STRUCTURES

8. The dynamic response of any structure under wind is strongly dependent on its aerodynamics. Loading in the wind direction can be significantly reduced by suitable aerodynamic shapes. In addition to this advantage, cylindrical shapes offer the same characteristics in any transverse direction. As a result of this structures designed for relevant wind loading usually present an aerodynamic profile and they are often cylindrically shaped. Unfortunately, under some conditions an air stream going over this kind of shape can be distorted yielding some vortices (Fig. 1) which alternate leaving the external surface, resulting in some transverse oscillations.

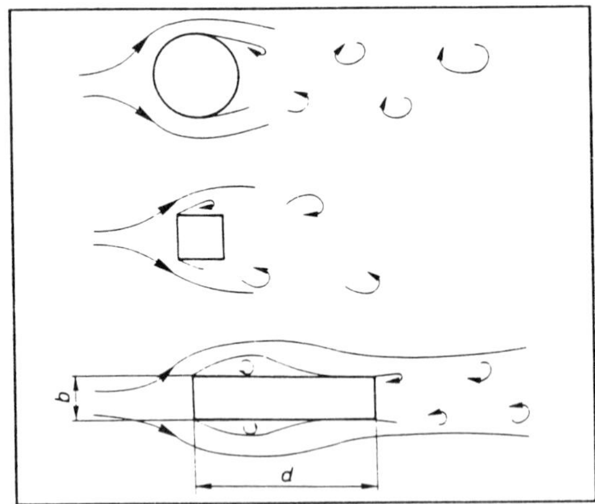

Fig. 1. Von Karman vortices.

9. The alternating nature of these vortices, highlighted by Von Karman, produces a pressure which is periodically followed by a suction. Usually this loading can be neglected but it has to be considered when the pulsation of these forces is close to one of the natural frequencies of the structure (ref. 2). For the usual wind velocities this resonance is more likely to occur in light or slender structures. The resulting resonance will appear as amplified transverse oscillations yielding bending or torsional strains and even ovalization deflections for thin shells.

10. The pulsation of the vortices is related to stream velocity. Several wind critical velocities can appear when the stream speed is raised since the pulsation increases with the wind velocity. When the transverse oscillations are relevant, they can influence the frequency of vortices, tuning them with the natural frequency of the structure, even when the wind velocity is slightly different to the critical value. This frequency tuning phenomenon is known as "lock in" (ref. 2).

11. The dynamic loading is due to the alternating Von Karman vortices. The transverse forces L and L,' resulting when the air vortex is leaving the outside surface are cyclically changing from $L > L'$ to $L < L'$ therefore, the resultant force (L-L') shows successively positive and negative sign, giving alternating oscillations in a perpendicular plan to the air stream (Fig. 2).

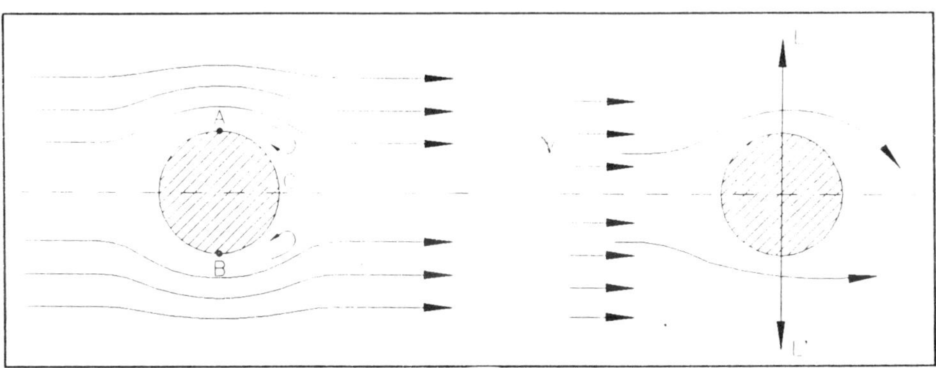

Fig. 2. Dynamic wind loading for circular sections.

12. Von Karman, neglecting fluid viscosity, showed that some relations remain constant. For smooth cylindrical surfaces it can be said that:

$$S_t = \frac{f_s D}{V} \tag{1}$$

f_s is the vortex shedding frequency (Hz); D is the outside diameter (m); V is the fluid stream velocity (m/s); (St) is defined as the Strouhal number, which takes a constant value of 0.2.

13. It can be shown that the phenomenon remains steady for Reynolds numbers (Re = 0.7E5 · V · D) between 1000 to 3.5E5 since for lower velocities, viscosity has to be taken into account. For higher speeds, the vortices are distorted becoming more unstable up to values of Re about 4E6 when vortices lose contact with the surface. For intermediate Re values, lateral forces become random instead of harmonic and as a result of this, the structure can oscillate with lower amplitudes at one or several natural frequencies. For very high Re values (Re > 6E6) the previous expression cannot be applied because fluid compressibility has not been taken into account, but it can be seen that the stream becomes extremely turbulent.

14. As the Strouhal number remains constant (within some limits of Re) for a given wind velocity and outside diameter, lateral force pulsation will remain steady becoming harmonic. If this pulsation approaches one of the natural frequencies of the cylinder, the response will be amplified. As the wind velocity can vary within a wide range, there are many specific stream velocities (critical velocities) which cause strongly amplified oscillations due to the coincidence of one of the natural frequencies of the cylinder with the pulsation of the vortices.

15. Steiman derived theoretically, the harmonic lateral force due to the Von Karman vortices as:

$$L = q\, C_l\, D \cos 2\pi f_s t \qquad (2)$$

Where:
L is the lateral force per unit length in (N/m);
D is the outside diameter in (m);
q is the dynamic wind pressure in (N/m^2);
f_s is the vortex shedding frequency(in (Hz);
C_l is the aerodynamic coefficient;
t is the time in (s).

16. Some tests on cylindrical shapes give for C_l the values indicated in Fig. 3 while Strouhal number follows the curve of Fig. 4.

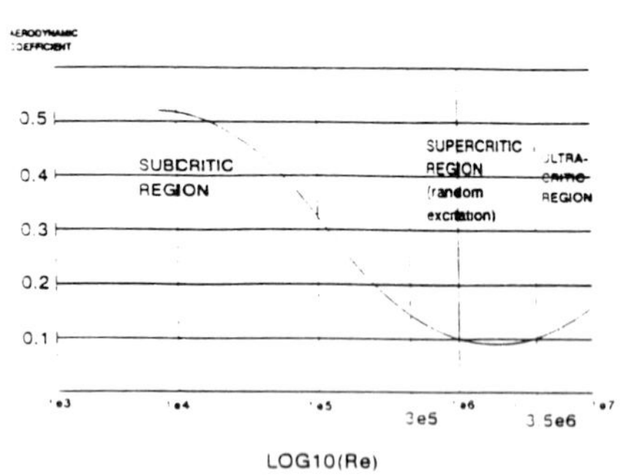

Fig. 3. Aerodynamic coefficient for cylindrical shapes.

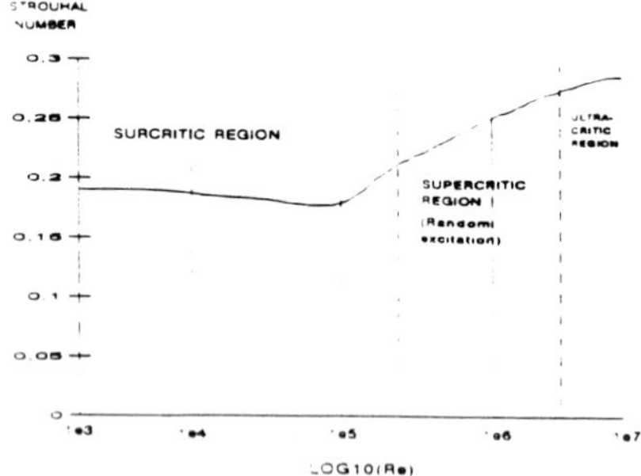

Fig. 4. Strouhal number for cylindrical shapes.

17. Other sections also exhibit this phenomenon but its effects are considerably lowered. The E.C.C.S. Standard (ref. 3) provides an easy way of evaluating Strouhal number for different sections and Re numbers (Table 1).

Table 1. Strouhal Number - E.C.C.S.

SECTION TYPES		S_t
V → □ — ⊔ ⊢⊣ ⌋ ⊥		0.15
V → ○	$10^3 < Re < 10^5$	0.2
	$10^5 < Re < 2.10^6$	0.2 to 0.3
$R = 0.7 \times 10^5 \times Vm \times b$	$Re > 2.10^6$	0.3

DYNAMIC BEHAVIOUR OF TUBULAR MEMBERS

18. It has been explained in the introduction why circular hollow section members are often used to construct skeletal space structures. As a result of standing outdoors, tubular members can withstand direct wind loading and due to the slenderness ratios usually employed, they will be prone to the local effects of the dynamic wind loading described in the previous section (Fig. 3).

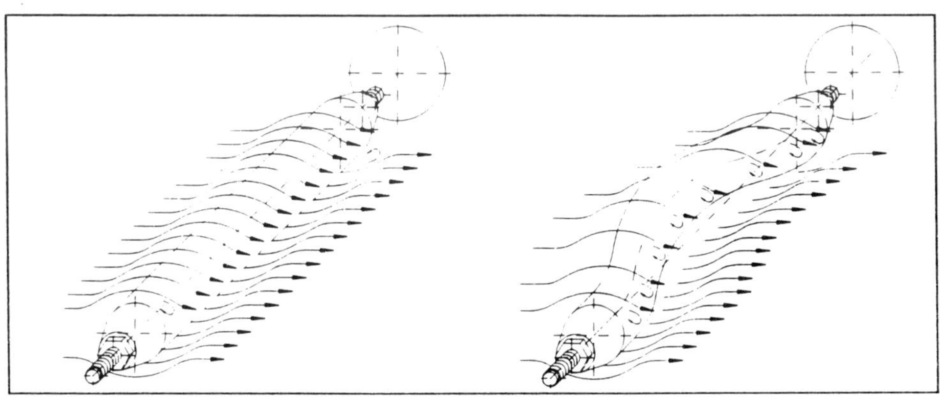

Fig. 5. Transversal oscillations of space structure members.

19. Natural frequencies for simply supported thin walled (thickness < < diameter) tubular members can be estimated by the following expression (ref. 4):

$$f_n = \frac{a_n}{4\sqrt{2\pi}} \frac{\sqrt{E}}{\sqrt{\rho}} \frac{D}{l^2} \tag{3}$$

where:

f_n = natural frequency of mode n (Hz). ρ = material density (Kg/m^3).
a_n = constant value dependent on the n mode.
D = outside diameter (m). E = Young modulus (N/m^2).
l = total length of the member (m).

20. The related critical wind velocities can be obtained from this expression:

$$v_{crn} = \frac{3.6 a_n}{\pi S_t} \frac{\sqrt{2E}}{\sqrt{\rho}} \frac{1}{\lambda^2} \tag{4}$$

where:
v_{crn} = Critical wind velocity for mode n (Km/h)
S_t = Strouhal number for a given Re.
λ = Slenderness ratio for the member $\left(2\sqrt{2}\, l/D\right)$.

For a given material with $1000 < Re < 3.5E5$ (S_t=0.2), β_n becomes a constant; with:

$$v_{crn} = \frac{\beta_n}{\lambda^2} \quad \text{and} \quad \beta_n = \frac{3.6 a_n}{\pi S_t} \frac{\sqrt{2E}}{\sqrt{\rho}} = \text{constant} \tag{5}$$

It is commonly accepted to consider only the two lowest (n=1&2) natural frequencies (ref. 2).

21. In addition, it must be remembered that natural frequencies are modified by the axial load level in the member due to the stress (un)stiffening effect (ref. 5). Tensile axial loads raise the natural frequency while compressive axial loads tend to lower the first natural frequency which becomes zero just as the member reaches the compressive critical load. Expressions used to obtain the modified natural frequencies can be obtained from many references (ref. 6).

FORECASTING THE LIFE OF TUBULAR MEMBERS

22. Several pin jointed systems for skeletal space structures use single bolted ends. The transverse sections at the connecting ends are able to withstand large tensile and compressive forces, but they are inherently very weak carrying bending moments and shear forces generated by the dynamically amplified wind loading. The alternating nature of these forces and moments can result in strong

fluctuating stresses which must be added to the mean tensile or compressive stress in the bolt. If alternating stresses are above the fatigue limit the occurrence of a failure, possibly resulting in the collapse the structure, may occur within a finite number of load cycles.

23. Guarantee against fatigue failure can only be certified by Goodman's criterion which also ensures that the alternating stresses will remain below the fatigue limit. For high tensile bolts the fatigue limit can be estimated as about 10% of the ultimate stress.

24. A simple forecast of the behaviour of the member will be obtained by considering the life of the bolt subjected to winds at the critical velocities associated with the member. The alternating stress can be estimated by considering the resulting end forces for a clamped-clamped beam under the Steiman (Eq. 2) lateral force per unit length amplified by about 20 times. This amplification is related to the damping of the member, which some references (ref. 5) generally place between 1 and 7% of the critical damping. A severe value of 2% is usually taken resulting in 25 times lateral loading amplification.

25. One example of such a simple forecast for a real structure is presented in Table 2, for critical wind velocities below 110 Km/h and for bolts with an ultimate stress of 100 kg/mm^2.

Table 2. Member life simple forecast application.

D th (mm)	Lengt (m)	M (mm)	fnat1 (Hz)	Re1 (*1E5)	S_t1	vcrit1 (Km/h)	fnat2 (Hz)	Re2 (*1E5)	S_t2	vcrit2 (Km/h)	σ_1 (Kg/mm^2)	σ_2 (Kg/mm^2)	Life (h)
50 2	2.45	14	22.8	.2	.20	20.5	91.1	.8	.20	82.0	2.1	26.3	.15
50 2	2.18	14	28.8	.3	.20	25.9	115.3	1.0	.20	103.7	2.7	23.8	.16
50 2	1.80	14	42.2	.4	.20	38.0	168.7	1.5	.20	-	3.9	-	-
50 2	1.23	14	91.1	.8	.20	82.0	364.3	3.2	.21	-	6.6	-	-
60 2	2.45	14	27.5	.3	.20	29.7	110.0	1.4	.20	-	5.3	-	-
60 2	2.18	14	34.8	.4	.20	37.6	139.3	1.8	.20	-	6.8	-	-
75 2	2.45	14	34.6	.7	.20	46.7	138.5	2.7	.21	-	12.8	-	3.7
75 2	2.18	14	43.8	.9	.20	59.2	175.3	3.5	.21	-	16.2	-	1.4
75 3	2.45	14	34.2	.7	.20	46.1	136.6	2.7	.21	-	12.5	-	4.0
75 3	2.45	20	34.2	.7	.20	46.1	136.6	2.7	.21	-	4.3	-	-
75 3	2.18	14	43.2	.9	.20	58.4	173.0	3.4	.21	-	15.8	-	1.5
75 3	2.18	20	43.2	.9	.20	58.4	173.0	3.4	.21	-	5.4	-	-
90 3	2.45	20	41.3	1.2	.20	66.5	165.0	4.7	.22	-	7.7	-	-
90 3	2.18	20	52.2	1.5	.20	83.6	208.9	5.9	.23	-	9.5	-	-
90 3	1.23	20	165.0	4.7	.22	-	660.2	18.7	.29	-	-	-	-
90 4	2.45	20	40.8	1.2	.20	65.8	163.2	4.6	.22	-	7.5	-	-
90 4	2.18	20	51.7	1.5	.20	82.7	206.6	5.9	.23	-	9.3	-	-
90 4	2.18	27	51.7	1.5	.20	82.7	206.6	5.9	.23	-	3.8	-	-
100 4	2.45	20	45.5	1.6	.20	80.7	182.2	6.4	.23	-	12.5	-	3.0
100 4	2.45	27	45.5	1.6	.20	80.7	182.2	6.4	.23	-	5.1	-	-

26. Fig. 6 shows the approach which has been implemented for ORONA Space Structures. The block diagram also includes an approach which is much more useful, but also more complex.

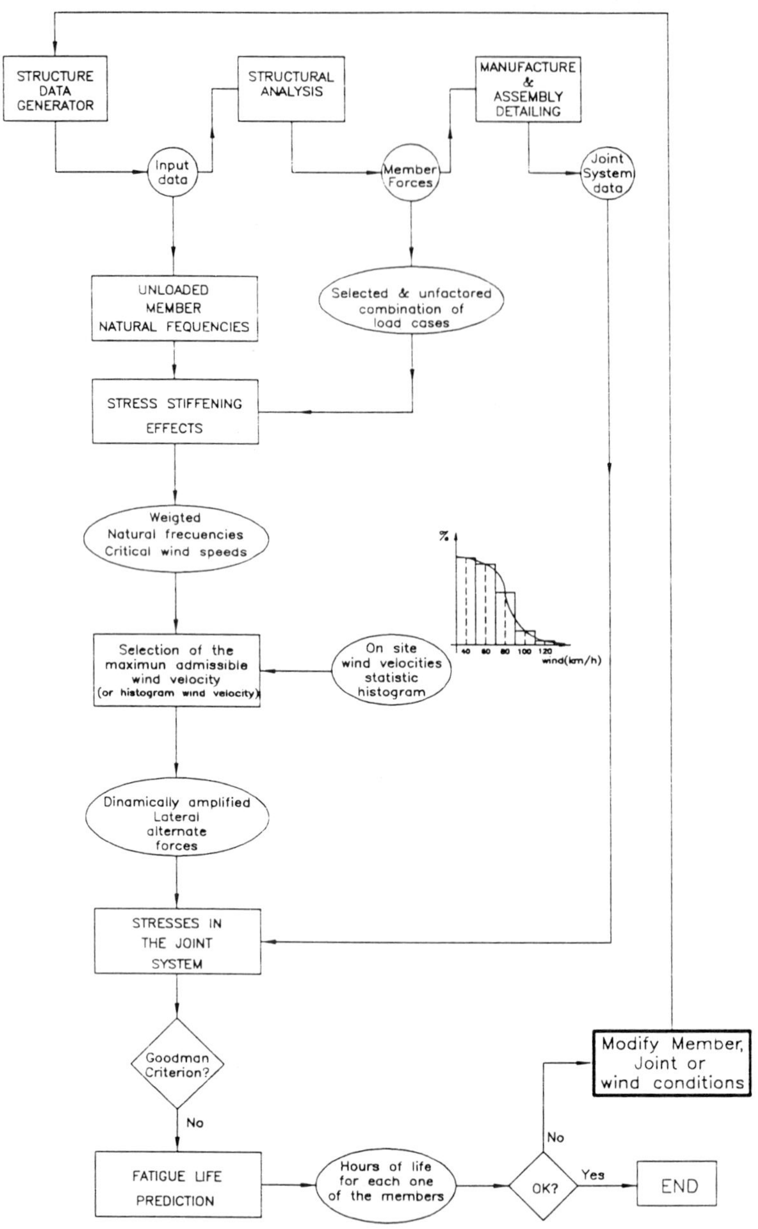

Fig. 6. Member life forecast approaches used by ORONA.

27. This method consists of using a real wind velocity histogram for the site where the structure is to be erected. A dynamic complex transfer function is

derived for each one of the members to be considered in the analysis. The vortex pulsation rate is calculated for each characteristic wind velocity and it is input into the complex transfer function. This results in an amplification factor for the lateral Steiman distributed load. Alternating stresses and the number of cycles of life estimated for each one of the characteristic wind velocities are then obtained. Finally Goodman's criterion and any accumulative fatigue life approach, like Palgrem-Miner are applied, producing an estimate for the fatigue life of the bolts exposed to winds defined by the wind velocities histogram.

SOLUTIONS TO BE APPLIED

28. It has been shown how Von Karman local effects affect members under the direct action of the wind. Therefore, the first solution to be considered is the protection of members against wind. During the assembly and erection phases as well as for uncovered structures, wind detectors can be used. For pin jointed skeleton space structures, it must be pointed out that after covering the structure, wind loads are distributed to the joints avoiding any direct wind loading of members. As a result of this, this solution only applies during the assembly and erection phases because once the structure is covered, no local alternating lateral oscillations will take place.

29. In many cases, it is not feasible to protect members and some preventive actions should be considered at the design or even at the assembly stage.

30. Looking through the combinations of outside diameters and lengths usually encountered in skeletal space structures some of the following conclusions can be drawn. First, the smallest diameter members (40, 50, 60, 75, 90 and 100 mm) appear to be most prone to suffer from Von Karman local effects. Depending on their lengths, one or two of the lowest modes can be excited. With respect to the wind, velocities below 40 or 50 Km/h don't seem to have enough energy to produce relevant lateral oscillations. On the other hand, the aforementioned effects are mainly produced by nearly steady wind conditions which are unlikely to be present at high wind velocities above 100 or 110 Km/h.

31. As a result of this and using expression (5) for the unloaded members critical velocities, it can be shown that in order to avoid resonance to be produced within wind velocities of a lower and an upper limit, only some member slenderness ratios are permissible. Table 3 shows the allowed slenderness values (less than 64, between 90 and 128 and greater than 181) for wind velocities within 50 and 100 Km/h and the corresponding allowable bar lengths for different outside diameters.

32. It can be shown from Table 3 that members with lengths between L2 and L3 will have their first critical velocity within 50 and 100 Km/h while in the case of lengths between L4 and L5, the second critical velocity will be found within these limits. The latter of this cases appears much more damaging, probably due to its higher energy content and usually forces a change of the outside diameters of the members or even the topology of the structure (lengths). For the first case, the appearance of the resonance can, in most of the cases, be allowed provided

that enough bolt section is supplied to the joints (recommended minimum bolt size = Met).

Table 3. Allowable bar lengths to avoid member critical wind velocities between 50-100Km/h.

		λ1 0		λ2 64	λ3 90		λ4 128	λ5 181		λ6 200
Dext (mm)	Met (mm)	L1 (m)	<L<	L2 (m)	L3 (m)	<L<	L4 (m)	L5 (m)	<L<	L6 (m)
40	14	0.00		0.90	1.28		1.81	2.56		2.83
50	14	0.00		1.13	1.60		2.26	3.20		3.54
60	20	0.00		1.36	1.92		2.71	3.84		4.24
75	20	0.00		1.70	2.40		3.39	4.80		5.30
90	20	0.00		2.04	2.88		4.07	5.76		6.36
100	27	0.00		2.26	3.20		4.52	6.40		7.07

33. These allowable lengths can be used at the design stage bearing in mind that they are approximately obtained, that they don't take into consideration any axial loading in the member, and that only two modes have been taken into account.

34. In the event that resonance cannot be avoided, other kinds of solution may be analysed. First of all changing the aerodynamic profile of the member can be considered. This means that the member should be provided with a rough outside surface or should have many holes in order to allow the wind to pass through it. Another way can consist of supplying an external perforated cover to each member or installing straight or helicoidal wings around the outside diameter of the tubes in order to stabilise the member.

35. Probably the best solution can be achieved when the jointing system provides not only a bolt, but also a tightening ring, as is the case with MERO or the new ORONA system. This offers a bigger radius from the member axis and a larger section to withstand bending and shear forces, provided that system preload is never lost under any condition. Even better is the case when the tightening ring stiffness is much greater than that of the bolt. In this case any alternating loading of the connecting system is shared proportionally between the stiffness of its components. This drastically reduces the alternating loading on the bolt which is more sensitive to fatigue than the tightening ring, resulting in a more suitable system not only for lateral wind loading but also for any kind of loading causing fatigue damage.

36. Fortunately smaller bolts which are more prone to be damaged under alternating lateral loading, can easily be preloaded. Different devices and systems

are being designed and tested in order to supply, under assembly and erection working conditions, the huge preload levels required for bigger bolts.

CONCLUSIONS

37. The paper has presented the effects of direct wind loading on simply supported slender members found in skeletal space structures. It has been shown how Von Karman vortices yield lateral alternating oscillations of the member whose actual effects have to be analysed taking into account the dynamic amplification due to the local natural frequencies of the member, and the fatigue strength of the joining system.

38. Two local dynamic behaviour forecasting approaches currently in use in ORONA Space Structures have been introduced. Simple forecasting of the lateral alternating oscillations is carried out on the basis of considering the fatigue life of the member under each one of its critical wind velocities. A second approach, related to the real wind conditions on the site, which offers a better estimate of the fatigue life of the member, has also been outlined.

42. It has been shown that the first solution dealing with the Von Karman local effects, is always related to protecting members against wind. However, this is not feasible in many cases and some actions must be considered at the design or even at the assembly stages.

43. For slightly axially loaded members (no stiffening effects), it has been recommend that some slenderness ratios should be avoided in order to move the critical wind velocities of the members outside the dangerous wind interval (40 to 110 Km/h).

44. In the event that resonance cannot be avoided, other kinds of solutions at the assembly stage may be considered. Primarily changing the aerodynamic profile of the member. However, better solutions can be achieved by obtaining a suitable joining system which will include not only a bolt, but also a tightening ring, whose bending and shear stiffness is much greater than that of the bolt.

45. This is the case with the new ORONA system which will work properly provided that system preload is never lost under any condition. Fortunately smaller bolts which are more prone to be damaged under alternating lateral loading, can easily be preloaded. Different devices and systems are being designed and tested in order to supply, under assembly and erection working conditions, the huge preload levels required for bigger bolts.

REFERENCES

1. DAVENPORT A.G. The response of structures to gusty wind, Von Karman Institute for Fluid Dynamics, Lecture series 89: Wind Effects on Buildings and Structures, Rhode St-Genèse, 1976.
2. National Building Code of Belgium: NBN B 03-002, Actions du vent sur les constructions, Part 2, Institute Belge de Normalisation, Bruxelles, 1988.
3. EUROCODE No 3, Calcul des effects du vent sur les constructions, ECCS, 1979.

4. DAWE D.J. Matrix and finite element displacement analysis of structures, Chapter 12, Clarendon Press, Oxford, 1984.
5. COOK R.D., MALKUS D.S., PLESHA M.E. Concepts and applications of finite element analysis, 3th edition, Chapter 12, John Wiley & Sons, Inc., New York, 1989.
6. ARANBURU I. Lateral vibration of axially loaded beams, IKERLAN report 65.31, Arrasate, 1992.

76. General theory and numerical analysis of single- and double-layer pin-jointed space grids

J. SZABO and T. TARNAI, Technical University of Budapest, Hungary

SYNOPSIS. The paper focuses on three problems related to singl- and double-layer pin-jointed space grids, namely:

(a) automated data generation;
(b) determination of rigidity properties;
(c) investigation of small and large displacements, stresses and change of state.

A quasi-inverse corresponding to the equilibrium matrix is introduced. In the investigation, rotation of the discrete vector field described by the equilibrium matrix plays an important role.

TOPOLOGY

1. Automated generation of basic geometric data (numbering of nodes and bars, co-ordinates and unit vectors of bars) of a single or double (or multi) layer grid is closely connected with the generation method of the coefficient matrix of the equilibrium equation denoted by G, which is given in § 6.

2. As shown by examples in ref. 10, topological relationships of a grid are determined by three factors: (a) a mathematically defined surface on which the nodes lie (e.g. elliptic paraboloid, hyperbolic paraboloid, parabolic cylinder, spherical cap, and so on), (b) formation of the network of the grid in the plane z = 0 (this network is projected parallel or centrally onto the given surface), (c) location of the points of the grid, lying on a rigid boundary (on the grid surface or not), that is, the sort of support of the grid.

STATIC AND KINEMATIC PROPERTIES

3. Due to the investigations by Szabó and Rozsa (ref. 5), Szabó and Roller (ref. 8), Pellegrino and Calladine (ref. 2), which cover similar points to each other, the equilibrium matrix G of the grid is built up of unit vectors determined by nodal co-ordinates whose number is equal to the number of bars μ; and its mathematical investigation, that is, the rank of G, gives unique information about the static and kinematic determinacy (indeterminacy and/or over determinacy) of the grid. Denoting the total number of joints (nodes) in the grid by ν_0, the number of

foundation joints by ν_1, the number of internal (non-foundation) joints by ν ($\nu = \nu_o - \nu_1$), and the number of bars by μ, the necessary condition of static and kinematic determinacy of a pin-jointed space grid is given by: $\mu = 3\nu$. A stricter formulation of this condition is used in ref. 10 for the generation of a grid and simultaneously of its matrix G. With each internal joint there are associated unit vectors of three bars joining it (in agreement with the requirements in paragraph 2) such that each of the μ unit vectors can occur only once, and each vector triplet (so-called trihedron) associated with a joint cannot be coplanar. If there are no more bars than in the ν trihedra, that is, there is no redundant bar and unit vector ($\mu - 3\nu = 0$), then the grid can be statically and kinematically determinate. If there are redundant bars, that is, $n = \mu - 3\nu > 0$, then the grid is at least n times statically indeterminate (it can have n independent states of self-stress); and if $m = 3\nu - \mu > 0$, then the grid is at least m times kinematically indeterminate (it can have m independent inextensional infinitesimal mechanisms).

4. Static and kinematic properties of a grid can be uniquely established (that is, with necessary and sufficient conditions) by the rank of the equilibrium matrix G. The oldest known procedure for determination of the rank of matrix G denoted by $\rho = \rho(G)$ and its variants published, for instance, by Bodewig (ref. 1) can be derived from the Gauss elimination algorithm of sets of linear equations, and the rank can be defined as the minimum number of dyads necessary for composition of the given matrix. During the procedure, in ρ steps, a dyad of decreasing size, that is, the product of a column matrix and a row matrix, is subtracted from G in each step; and in the ρth step G becomes the zero matrix, while there arise a lower trapezoidal matrix denoted by L and an upper trapezoidal matrix denoted by U (in special case two triangular matrices) such that $L \cdot U = G$.

INVESTIGATION OF GRIDS

5. In the procedure developed for the investigation of grids (especially for the analysis necessary in the neighbourhood of critical states and change of post critical state), the quasi-inverse of matrix G introduced by ref. 9 and denoted later in this paper by $Q = Q(G)$ plays an important role, and so does the dual quasi inverse of matrix G^T (that is, the transpose of G) of the kinematic (compatibility) equation introduced by (ref. 11) and denoted by $Q^D = Q(G^T) = Q^D(G)$. For these matrices, independent of their decomposition algorithm, the following relationships are valid:

Equilibrium equation	Kinematic equation	
$G \cdot s + q = 0$	$G^T \cdot dr + dt = 0$	(1)

where the notation below is applied:

- s vector of forces in bars;
- q vector of nodal loads;
- dr vector of change of nodal co-ordinates;
- dt vector of elongation of bars.

(Note that vector t is composed of two parts: $t = t_s + t_k$, the first of them, $t_s = F \cdot s$, depends on the forces in the bars; the second, t_k, is independent of the forces in the bars, it is the so-called kinematic load, for instance, change in temperature, manufacturing imperfection and so on. Here F is the flexibility matrix.) The matrix G and G^T, during their decomposition, are reorganised and can be written as follows:

$$\begin{bmatrix} G_{11} & G_{12} \\ G_{21} & G_{22} \end{bmatrix} \bullet \begin{bmatrix} s_1 \\ s_2 \end{bmatrix} + \begin{bmatrix} q_1 \\ q_2 \end{bmatrix} = 0 \qquad \begin{bmatrix} G_{11}^T & G_{21}^T \\ G_{12}^T & G_{22}^T \end{bmatrix} \bullet \begin{bmatrix} dr_1 \\ dr_2 \end{bmatrix} + \begin{bmatrix} dt_1 \\ dt_2 \end{bmatrix} = 0 \tag{2}$$

Emphasising that G_{11} is non-singular, that is, $det(G_{11}) \neq 0$ gives the generalised solution of Eqs. 2 by means of the quasi inverses Q and Q^D in the form:

$$\begin{bmatrix} Q_{11} & Q_{12} \\ Q_{21} & 0 \end{bmatrix} \bullet \begin{bmatrix} q_1 \\ s_2 \end{bmatrix} + \begin{bmatrix} s_1 \\ q_2 \end{bmatrix} = 0 \qquad \begin{bmatrix} Q_{11}^D & Q_{12}^D \\ Q_{21}^D & 0 \end{bmatrix} \bullet \begin{bmatrix} dt_1 \\ dr_2 \end{bmatrix} + \begin{bmatrix} dr_1 \\ dt_2 \end{bmatrix} = 0 \tag{3}$$

where:

$$Q_{11} = G_{11}^{-1} \qquad Q_{11}^D = Q_{11}^T$$

$$Q_{12} = G_{11}^{-1} \bullet G_{12} \qquad Q_{12}^D = -Q_{21}^T \tag{4}$$

$$Q_{21} = -G_{21} \bullet G_{11}^{-1} \qquad Q_{21}^D = -Q_{12}^T$$

Eqs 3 in more compact form are given by:

$$Q \bullet x + y = 0 \qquad Q^D \bullet x + y = 0 \tag{5}$$

where x denotes the input vector and y denotes the output vector.
It can be seen that Q^D is the skew transpose of Q:

$$Q^D = \begin{bmatrix} Q_{11}^T & -Q_{21}^T \\ -Q_{12}^T & 0 \end{bmatrix} = \begin{bmatrix} Q_{11}^D & Q_{12}^D \\ Q_{21}^D & 0 \end{bmatrix} \tag{6}$$

6. In the procedure developed for the investigation of large grids the authors have chosen a way different from that outlined in § 3 and 4, consisting of the following steps.

- Generation of the network of the grid on the plane z = 0.
- Assign the notation and x and y co-ordinates of its nodes, and calculate the z co-ordinates corresponding to the prescribed grid surface.

- Taking § 2 and 4 into consideration determine the trihedra associated with the internal nodes (trihedron vectors associated with the nodes are indicated by arrows in Fig. 1), and write the trihedron vectors one after the other in a μx3 matrix G_0.

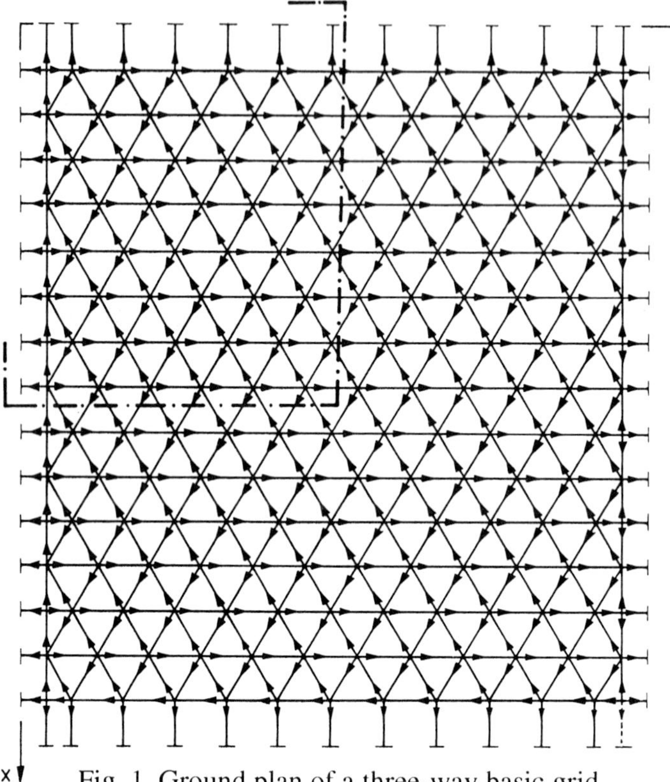

Fig. 1. Ground plan of a three-way basic grid.

- Determine the orthogonal trihedra associated with the trihedra (two orthogonal vector triplets are in question); this is achieved by checking whether a trihedron is not coplanar.
- Determine directly the quasi inverse Q of matrix G, as presented in § 5, in the following way:

  (a) writing the 3x3 submatrices of the ν orthogonal trihedra in the hyperdiagonal of Q_{11} then in μ-ν_1 steps generate the required quasi-inverse matrix essentially by the algorithm suggested by Sherman and Morrison (ref. 4) and extended by Rózsa (ref. 3):

$$\left(Q_{11}^{-1} + f \bullet h^T\right)^{-1} = Q_{11} - Q_{11} \bullet f \bullet h^T \bullet Q_{11} \big/ D \qquad (7)$$

where

$$D = 1 + h^T \bullet Q_{11} \bullet f$$

h^T is a row vector, in which in the place corresponding to the serial number of the unit vector in question, there is a 1 and all the other elements are 0;

f is a column vector, in which in the place corresponding to the mark of the unit vector in question, the three elements of the unit vector considered with a negative sign are positioned and the other elements are 0.

- If in some step of the above algorithm D = 0, then the next unit vector cannot be built in Q_{11} (that is, it gets into Q_{12}) and in all such cases the rank (starting with the value $\rho = 3\nu$) decreases by 1.

CHANGE OF STATE

7. Investigating the change of state of a grid with large displacements using the differential equation described in refs 6-9 namely:.

$$\begin{bmatrix} C & G \\ G^T & F \end{bmatrix} \bullet \begin{bmatrix} dr \\ ds \end{bmatrix} + \begin{bmatrix} dq \\ dt \end{bmatrix} = 0$$

Here C·dr shows that the scalar values s_{jk} of forces in the bars remain unchanged while a change in the direction (rotation) of the bar of length l_{jk} changes the equilibrium of nodes j and k; in other words, load increments dq_j and dq_k are needed to maintain equilibrium. An obvious answer could be given by the rotation coming from the vector product of the unit vector e_{jk} of the bar denoted by jk and the vector of differences of displacements of the bar ends $dr_{jk} = dr_k - dr_j$. However, using a simpler solution instead, which allows the built up of the equations. The component of displacement vector dr_{jk} parallel with unit vector e_{jk} is $dr_{jk}^{\parallel} = e_{jk}(e_{jk}^T \bullet dr_{jk})$ and its component perpendicular to e_{jk}, which results in the rotation, is $dr_{jk}^{\perp} = dr_{jk} - dr_{jk}^{\parallel} = (E - e_{jk} \bullet e_{jk}^T) \bullet dr_{jk}$. Thus, the effect to be equilibrated by the force increment coming from the rotation is $(E - e_{jk} \bullet e_{jk}^T) \bullet dr_{jk} \bullet s_{jk}/l_{jk}$ where E is a 3x3 unit matrix whose elements are defined by the Kronecker symbol δ_{ih}. The symmetric matrix C is composed of ν^2 submatrices of size 3x3:

- if j = k then $C_{jj} = -\sum_{\sigma}(E - e_{j\sigma} \bullet e_{j\sigma}^T)s_{j\sigma} / l_{j\sigma}$ and the summation is extended to all bars joining node j,
- if j ≠ k then (a) $C_{jk} = C_{kj} = (E - e_{jk}\, e_{jk}^T)s_{jk} / l_{jk}$ if j and k are internal nodes, (b) $C_{jk} = 0$ if j or k is not an internal node.

GENERATION OF GRIDS

8. The procedure concerning the generation of a single-layer grid (basic grid) for many sorts of grid surface, bar layout and support type has been developed. For the generation of double-layer grids this is considered as a base. The second layer is constructed under (or above) the first one adopting the following proceedure:

(a) Construct trihedra fitted to node triplets of the basic grid, with proper height and footpoints meeting the design requirements; connection of these trihedra to the basic grid does not change the static and kinematic properties (degrees of indeterminacy or over determinacy) of the grid.

(b) Connecting the new nodes by bars generated by the design requirements; the degrees of the static indeterminacy and kinematic over determinacy of the grid increase by the number of bars belonging to the second layer.

9. The procedure for the generation of double-layer grids will be presented in a simple example. The basic grid is constructed in the plane z = 0 as a network composed from equal equilateral triangles (Fig. 1); the network is projected parallel with the axis z to an elliptic paraboloid surface. The supporting points lie on this surface, too (Fig. 2). The steps in paragraphs 8(a) and 8(b) are illustrated in Figs 3 and 4. The detail in Fig. 1 is shown in Figs 2 to 4 from below. The double-layer grid presented here has altogether 538 nodes (56 of these are supporting points), 1687 bars, and the number of its degrees of static indeterminacy is 241 (the basic grid is statically once indeterminate, the redundant bar is indicated by the dashed line in Fig. 1).

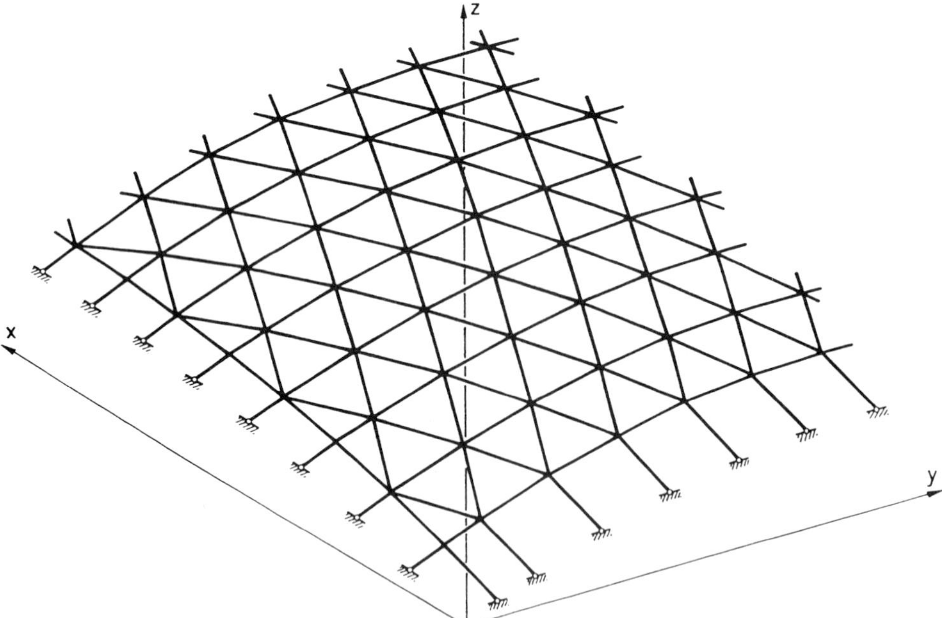

Fig. 2. Basic grid.

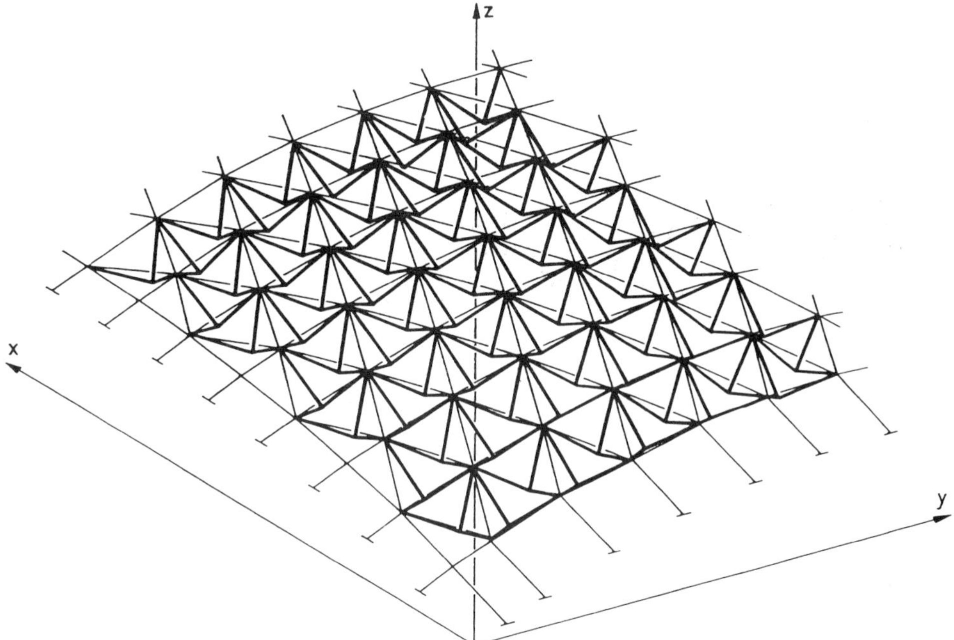

Fig. 3. Trihedra on the basic grid.

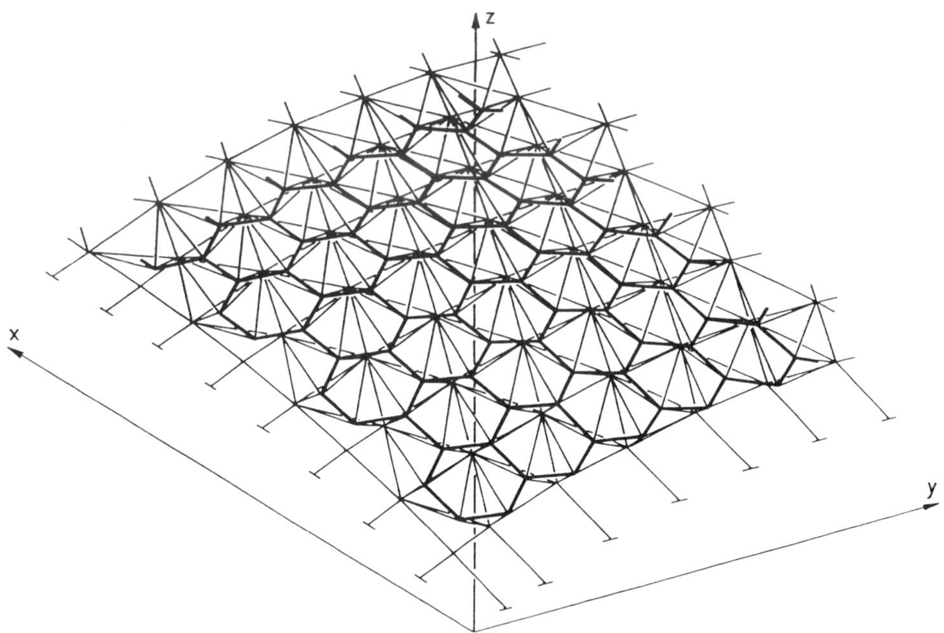

Fig. 4. Second layer of the double-layer grid.

The research reported here was supported by OTKA I/3 Grant No 41 awarded by the Hungarian Scientific Research Foundation.

REFERENCES

1. BODEWIG E. Matrix Calculus. North-Holland Pub. Comp., Amsterdam, 1956.
2. PELLEGRINO S. and CALLADINE C. R. Matrix analysis of statically and kinematically indeterminate frameworks. Int. J. of Solids and Structures, 1986, Vol 22, 409-428.
3. RÓZSA P. Linear algebra and its applications (in Hungarian). Müszaki Könyvkiadó, Budapest, 1974.
4. SHERMAN J. and MORRISON W. J. Adjustment of an inverse matrix corresponding to changes in a given column or a given row of the original matrix. The Annals of Mathematical Statistics, 1949, Vol 21, 124.
5. SZABÓ J. and RÓZSA P. Die Matrizengleichung von Stabkonstruktionen (Im Falle kleiner Verschiebungen). Acta Technica Acad. Sci. Hung., 1971, vol.71, 133-148.
6. SZABÓ J. and RÓZSA P. Grosse Verschiebungen von Stabkonstruktionen. Acta Technica Acad. Sci. Hung., 1972, vol.73, 53-60.
7. SZABÓ J. and ROLLER B. The general matrix differential equation of the skeletal frames. Proc. Int. Symp. on CAD, Coventry, 1972.
8. SZABÓ J. und ROLLER B. Anwendung der Matrizenrechnung auf Stabwerke. Akadémiai Kiadó, Budapest, 1978. (In Hungarian: Muszaki Könyvkiadó, Budapest, 1971.)
9. SZABÓ J. and KOLLÁR L. Structural Design of Cable-Suspended Roofs. Akadémiai Kiadó, Budapest and Ellis Horwood, Chichester, 1984. (In Hungarian: Müszaki Könyvkiadó, Budapest, 1974.)
10. SZABÓ J. and TARNAI T. Rigidity of the truss model of membrane shells. IASS-CSCE Int. Cong. on Innovative Large Span Structures, Toronto, 1992, vol.2, 438-449.
11. TARNAI T. and SZABÓ J. Numerical analysis of inextensional, kinematically indeterminate assemblies. Manuscript for publication, 1992.

77. Dynamic behaviour of a non-symmetrical space structure TV tower

M. PIRNER and O. FISCHER, Academy of Sciences of the Czech Republic

SYNOPSIS. Television towers are in general one dimensional structures of the beam type and also their response can be considered as deflection in one plane. Nevertheless some masts, especially those guyed in three directions or the nonsymmetrical ones, can be excited by wind not only in longitudinal and transversal directions, but also in torsion. This occurs especially if the mast support has been displaced by static wind and the forces in the guys are different. The measurements of the torsional movements of one real mast and of a wind tunnel model are described. The second part of the paper deals with the behaviour of the new TV tower in Prague, which is of nonsymmetrical construction.

A GUYED MAST AS A SPACE STRUCTURE

1. Slender structures like masts and towers have in general the advantage of simple mechanical behaviour (beam bending action, deflection in one plane). But this simple approach fails with guy-supports, if they have been analysed in detail, even in the linearised case of small displacements. The first difficulty lies in the fact that the guy support has very small rigidity in torsion and in the case of eccentric loading (masts with large antennae on horizontal arms) special arrangements of the anchors are to be used. Nevertheless provided that the structure remains symmetrical, the bending and torsion of the mast are independent and can be analysed separately.

2. Another problem occurs when the mast has been loaded by static wind and the mast-supports have been displaced so that the forces in windward and leeward guys of the support are different ($S_W \neq S_L$). Respecting this difference it can be easily shown that the translatory displacement of the guy support (perpendicular to the wind-direction) evokes at the same time the torsion of the support and vice versa. Under such conditions the flexural and torsional vibrations cannot be separated, the extent of their interaction depends on the spring constant D_{12} (see Fig. 1). Such a coupling may act as one of the conditions for the occurrence of flutter-type vibrations of the mast; it could become significant with masts guyed in three directions in strong wind, when the stresses in leeward cables and also the

support rigidity corresponding to the transverse displacement D_L decrease substantially.

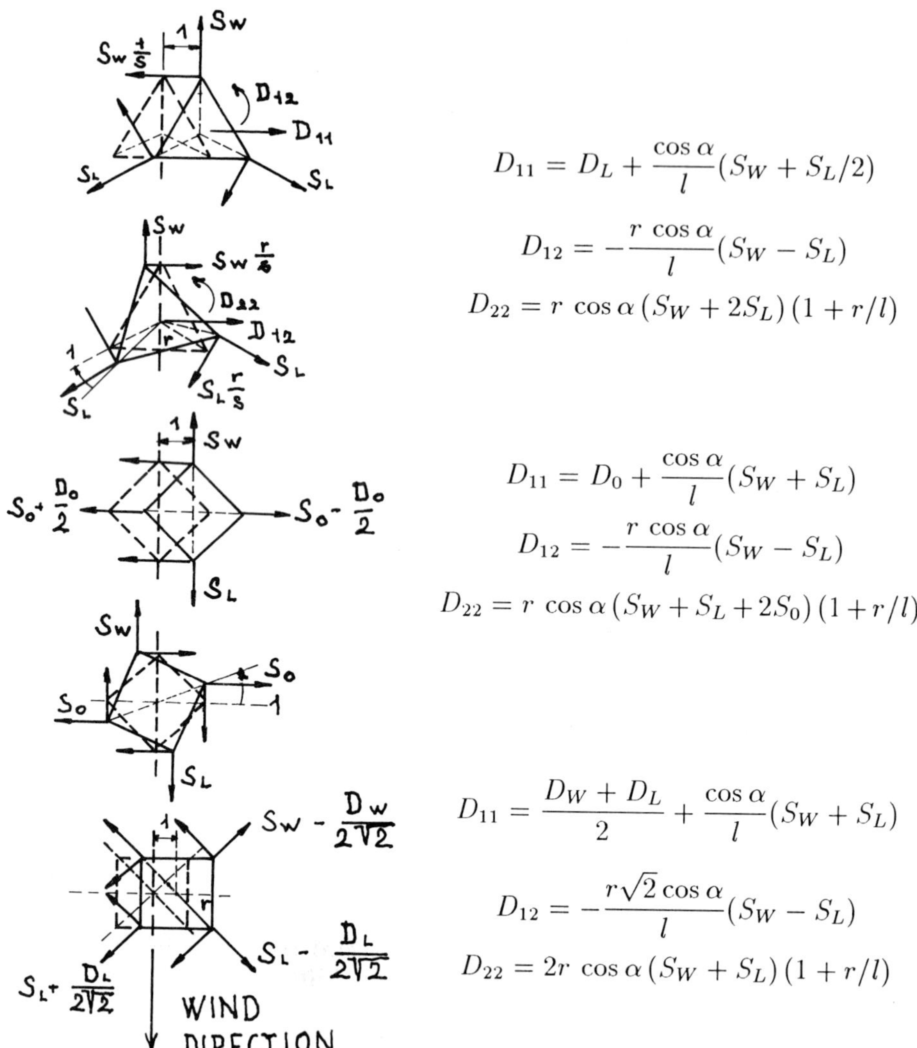

$$D_{11} = D_L + \frac{\cos\alpha}{l}(S_W + S_L/2)$$

$$D_{12} = -\frac{r\,\cos\alpha}{l}(S_W - S_L)$$

$$D_{22} = r\,\cos\alpha\,(S_W + 2S_L)\,(1 + r/l)$$

$$D_{11} = D_0 + \frac{\cos\alpha}{l}(S_W + S_L)$$

$$D_{12} = -\frac{r\,\cos\alpha}{l}(S_W - S_L)$$

$$D_{22} = r\,\cos\alpha\,(S_W + S_L + 2S_0)\,(1 + r/l)$$

$$D_{11} = \frac{D_W + D_L}{2} + \frac{\cos\alpha}{l}(S_W + S_L)$$

$$D_{12} = -\frac{r\sqrt{2}\cos\alpha}{l}(S_W - S_L)$$

$$D_{22} = 2r\,\cos\alpha\,(S_W + S_L)\,(1 + r/l)$$

Fig. 1. Spring constants for guy supports in different arrangements (s - length of the chord of the guy, ℓ = s cos α, α - inclination of the guy, S_W, S_L - forces in windward, leeward guys, S_0 - guy-force in equilibrium position).

OBSERVATIONS OF TORSIONAL MOVEMENTS ON MASTS

3. *In situ* measurements on a steel tubular mast were made several years ago. The mast is 178.94m high, with a diameter of 3.0m and 2.0m, anchored in two levels and fixed at the base. There were 3 seismic transducers on the level of 125m, in the positions given in Fig. 2. The power spectral densities of the response Y and (X_2-X_1) confirms the presence of the torsion of the shaft; its magnitude was evaluated as α_{RMS} = 2.5776E-4 rad, the displacement being v_{RMS} = 3.3mm. The wind velocity and direction are given in Fig. 2.

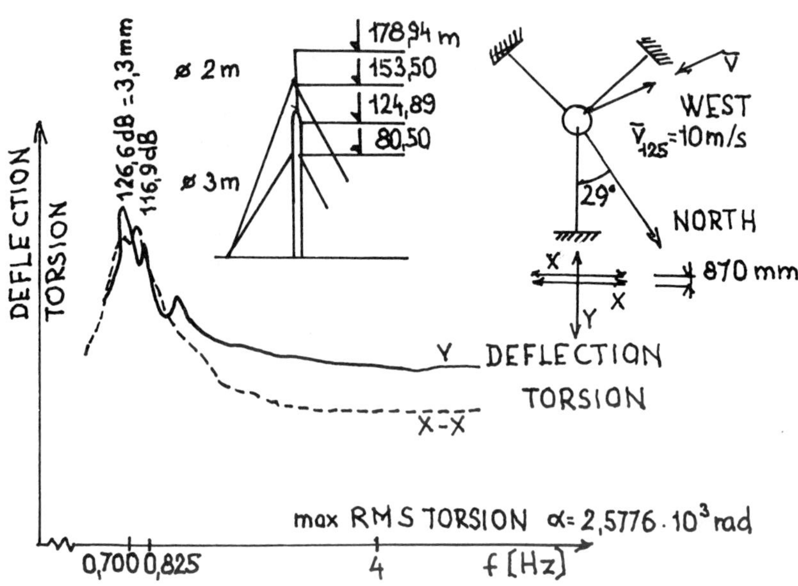

Fig. 2. Measured response (power spectral density of deflection and torsion) of a mast.

THE VERIFICATION OF THE PHENOMENON ON A MODEL

4. The rotation of a rigid mast-shaft was examined in stationary conditions in a wind tunnel. The model was a rigid tube of the length 1.99m and 62mm diameter, hinged at the base and anchored with 3 wires, fixed to the tunnel bottom by means of springs simulating the elasticity of real guys. Instead of the bending of the mast-shaft the rocking of the rigid model was considered, and its rotation instead of the torsion. The upper end of the model overlapped the tunnel air stream; it was furnished by a horizontal arm, on which the accelerometers were fixed. The length of this arm determined the sensitivity of the measurements of the rotation. The response of the top shows evidently the orbit of elliptical shape,

the longer axis of which is perpendicular to the wind direction (ref. 1). The frequency of the rotation of the shaft (both, positive and negative) f_t is an integral multiple of the frequency of bending (rocking) f_b. The examples of the orbits (f_t/f_b equal to 1 and 5) are given on Fig. 3. The magnitude of the rotation is represented by the ordinate, superimposed to the ellipse of the orbit.

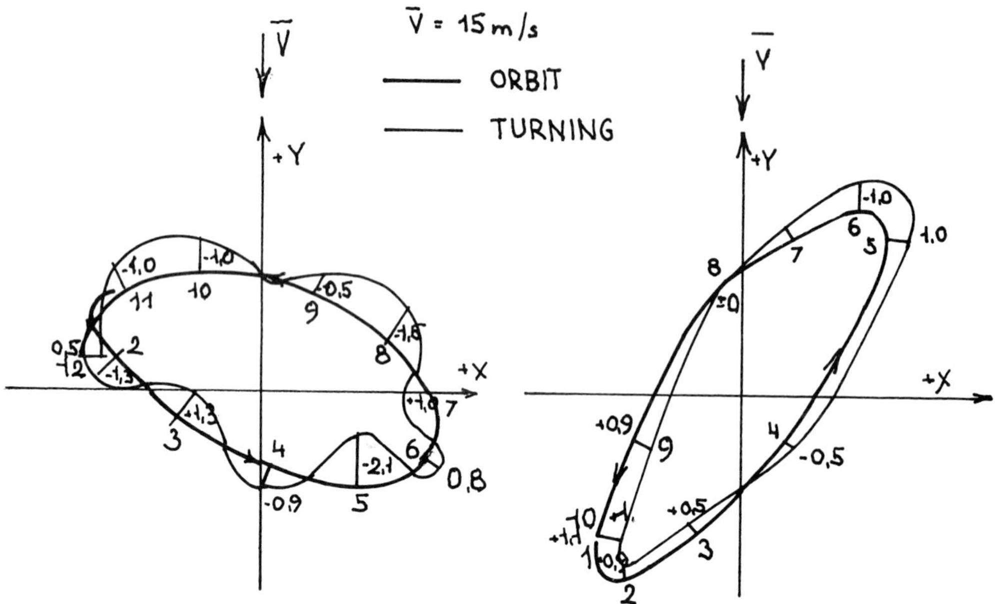

Fig. 3. Two orbits of the top of the model (displacement and rotation).

THE NONSYMMETRICAL TV TOWER OF PRAGUE

5. The new (1991) television tower of Prague is self-supporting consisting of three tubes (2 x 4.80m, 1 x 6.40m diameter), axial distance about 12m, up to the height of 120m, bearing some cabins which connect the tubes together. Above this level only the largest tube continues up to the height of 216m bearing the antennae for TV, FM radio and telecommunications. The material of the shafts are steel tubes combined with concrete, the cantilever is a steel tube with a plastic cylinder on the top. The pendulum vibration absorber has been hung on the steel lid of this cylinder (see Fig. 4).

6. Due to the structural system the response of the tower to wind loading is evidently three dimensional, as results also from the numerical solution made by the use of FEM. From the resulting great number of natural frequencies those of

practical importance were selected according to the experience of the authors and are given in axonometric view on Fig. 5.

7. When measuring the response of the tower two couples of seismic displacement transducers were used, each of them being placed on the levels 115 and 198m; from the distance between the transducers of one couple the centre of the rotational displacement of the floor could be determined. One typical example of the tower response, measured on the level of 138m, is given in Fig. 6. Above the diagram there are the calculated values of those natural frequencies which correspond best to the measured ones.

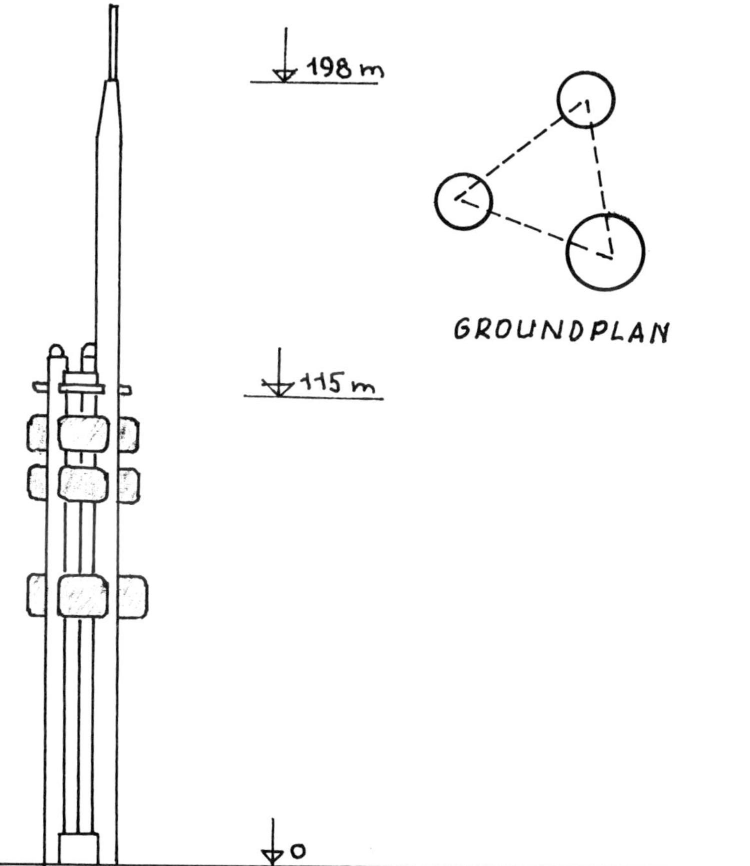

Fig. 4. The scheme of the TV tower Prague, the position of the vibration transducers.

REFERENCES

1. Kolousek V. et al. Wind effects on civil engineering structures. Elsevier, Amsterdam - Academia, Praha, 1983.

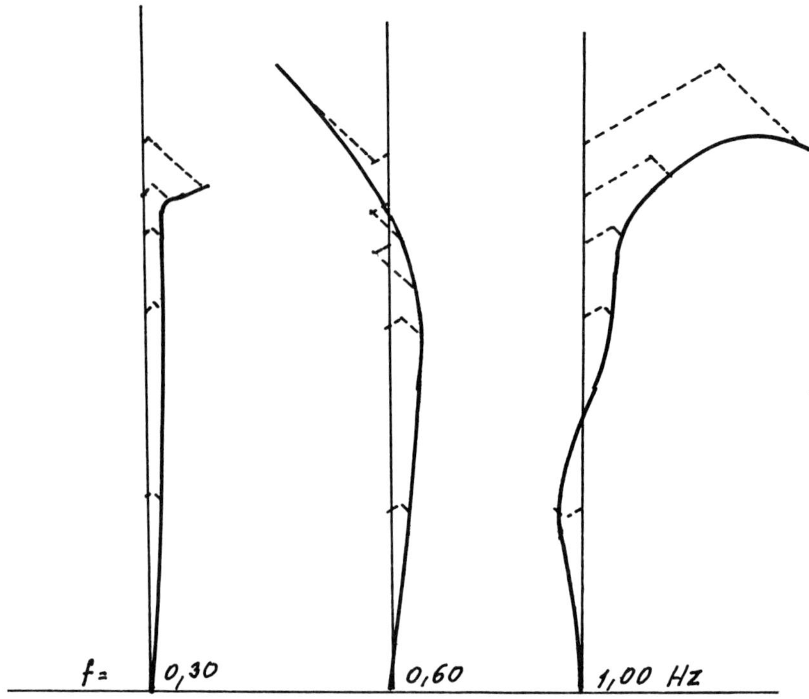

Fig. 5. The calculated first natural modes of the tower.

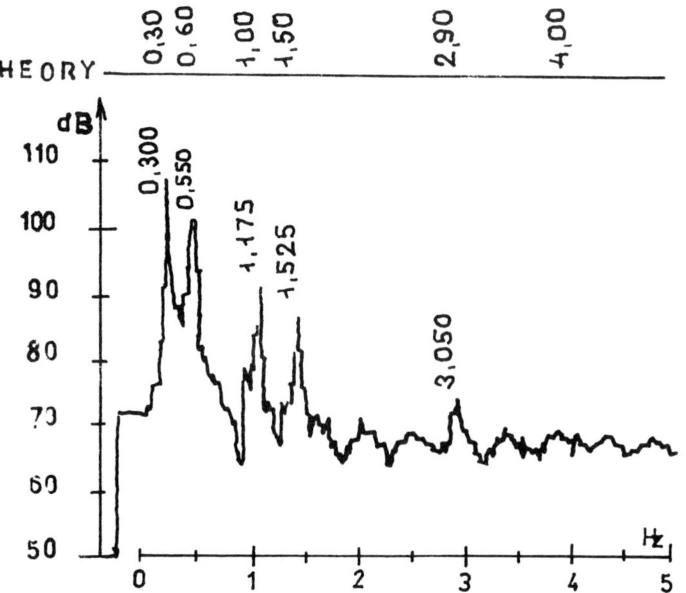

Fig. 6. The dynamic response of the tower on the level 198m.

78. Investigations on a 220 kV cold-formed latticed transmission tower

N. SUBRAMANIAN, Computer Design Consultants, India

SYNOPSIS. An attempt was made to analyse and design a 220kV transmission line tower using cold-formed lipped angle sections. Since 60 degree angles are not presently rolled in India, the tower was designed as a tower with a square base. This introduced some difficulties in fabrication and detailing. To verify the assumptions made in the design and to study the behaviour of the joints, full scale testing, up to failure, of this 36m tower was undertaken. From this investigation, it was found that cold-formed towers can be considered as a viable alternative to hot-rolled towers even for square based towers.

INTRODUCTION

1. Transmission line towers represent about 35-40 percent of the total cost of the transmission line project. Hence greater emphasis is now given to develop designs that are more efficient and cost less and give a better aesthetic look. Also, even a small percentage of the saving in the weight of one tower will result in substantial overall savings of costly material such as steel. Transmission line towers were traditionally designed using hot-rolled steel sections. The main difficulty of using hot-rolled steel angles is the non-availability of thinner sections and also the code restriction on the allowable thickness. This has resulted in the use of some minimum angle which lead to more weight. Hence, several countries have tried to substitute these hot-rolled angles with cold-formed shapes. Cold-formed shapes are readily available in thinner sections and provide a viable alternative for more economical structures due to savings in weight (ref. 1).

2. The added advantage is that any section can be rolled according to specification even in small quantities. (Most of the major cold-rolling companies in India accept a minimum of 5t for any section to be rolled). Hence, long unsupported lengths may be used for the leg members, which will in turn reduce the number of bracing members and the number of connecting bolts. Cold-formed members with thickness of 3mm - 6mm are used for transmission towers.

3. However in India, the transmission line tower code does not include the use of cold-formed steel sections in these towers. Hence an attempt was made to analyse and design a 220kV double circuit tangent tower using cold-formed

Space structures 4. Thomas Telford, London, 1993

lipped angle sections. Since 60 degree angles are not presently rolled in India, the tower was designed as a square based tower. This introduced some difficulties in fabrication and detailing which are discussed in this paper. In order to verify the assumptions made in the design and to study the behaviour of the joints of the tower, a full scale test of this 36m high tower was conducted. The tower was tested up to failure. From the test results, it was found that the analytical model predicts the behaviour within reasonable limits and that cold-rolled sections offer a viable alternative even for square based towers, resulting in 10 percent savings in cost.

EXAMPLE TOWER

4. In order to demonstrate the economy that can be achieved by using cold-formed sections, a 220kV double circuit tangent tower having the configuration shown in Fig. 1 was analysed . The loads acting on this tower based on sag-tension calculations are shown Fig. 2.

5. The wind load acting at different nodes of the tower were calculated by using a module of the TRANSTOWER program, developed by the author. These wind loads were assumed to act at each and every node of the tower. Similarly, the dead load of the tower was also distributed to act at every node of the tower. The analysis of the tower was made by using the analysis and design module of the TRANSTOWER program for the various loading conditions as specified in the IS code (ref. 2).

6. This program is based on the stiffness method of structural analysis and assumes the tower as a three-dimensional truss. The details of this program are discussed elsewhere (ref. 3). In the analysis, the secondary bracings were omitted, though their effect is considered in the design. Note that while designing cold-rolled members, the extra strength available due to cold rolling at corners was also taken into account (clause 6.1.1.1 of IS 801).

7. From an earlier analysis and design of a 400kV transmission line tower based on Indian specifications for cold rolled members (refs 4 and 5) it was found that there is a weight saving of about 30 percent. However, since the cost of cold-formed sections are higher than hot rolled sections, the savings in cost will only be of the order of 10 percent. Even this small saving will result in large savings in the overall project because several identical towers will be erected in the same project.

8. It is to be noted that completely new configurations such as triangular based tower will result in more economy, since the number of members will be 29 percent lower (refs 1 and 6). Triangular based tower results in about 27 percent reduction in the cost of foundation (ref. 6) .

DRAWBACKS OF INDIAN CODE PROVISIONS

9. It has been found in Italy that towers made of cold-formed sections having a triangular base (which is quite stable) are more economical than square base towers (ref. 6). However, Indian code does not specify the use of towers with a triangular base. (However, a mention about triangular based towers is made in the

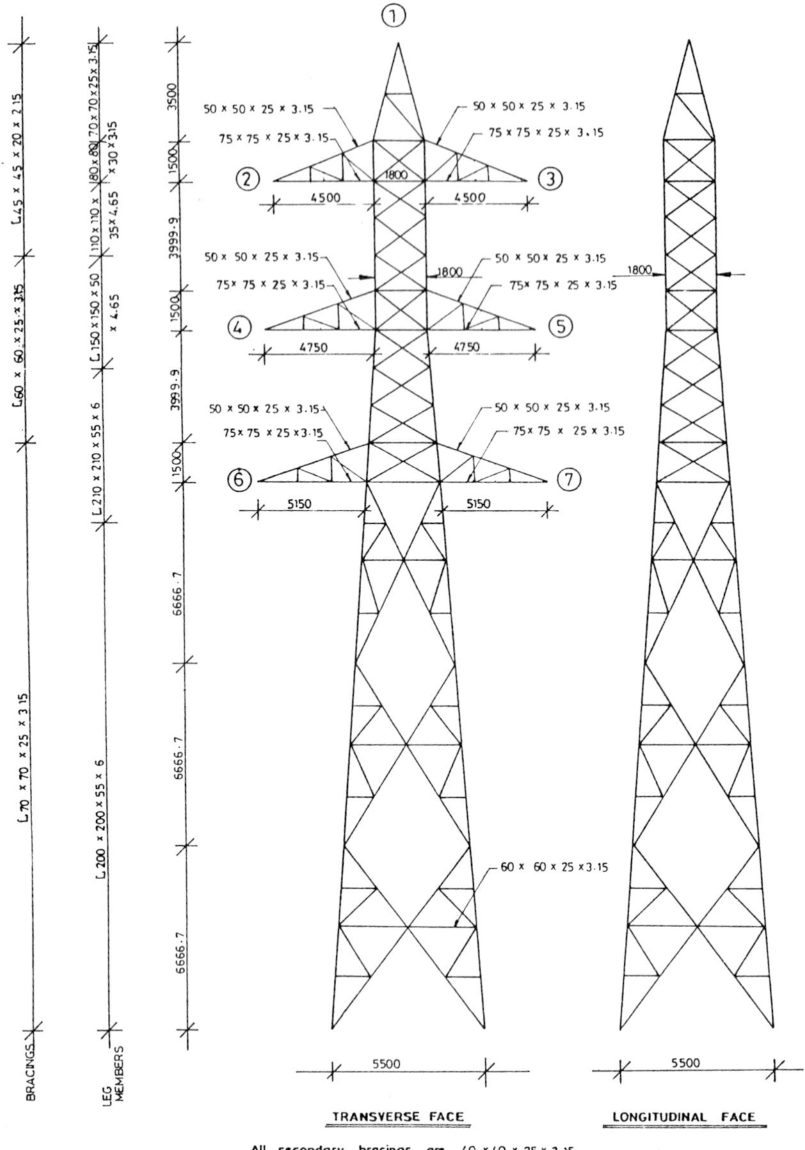

Fig. 1. Configuration of 220kV double circuit tangent tower.

draft Indian code on antenna towers). Moreover, the Indian code does not give any methods to calculate the warping constant C_W. Further, explicit equations are not given to find the allowable stress for singly symmetric and non-symmetric members subjected to lateral torsional buckling and for which the form factor Q is less than 1. It may be of interest to note that in the recent Indian Specification for cold-formed light gauge sections, the column factor Q has been omitted and the value of C_W has been included.

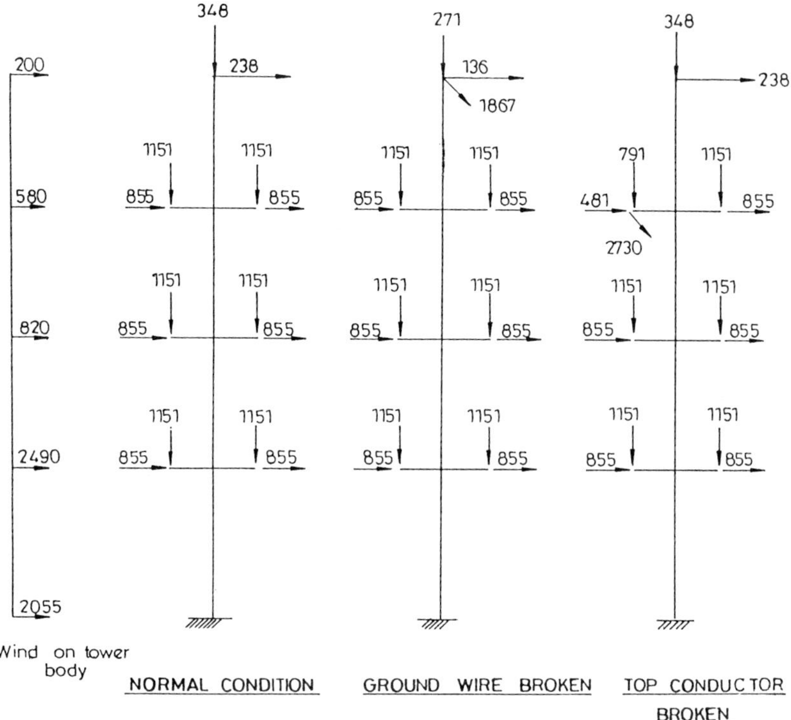

Fig. 2. Loading diagram.

CONNECTIONS

10. For cold-formed members, the same connection design philosophy used for hot-rolled angles is valid. Members meeting at a joint should be arranged with the centroidal axis meeting at one point. Eccentricity should be minimised. However, in cold-formed sections, other than angles, bolted connections on one leg or one plane may create eccentricities higher than those occurring in hot- rolled angles. In these cases, the flexural component of the stress caused by the eccentricity must be considered in the design. Moreover, the effect of local distortion in cold-formed steel connections should also be studied.

11. Further difficulties of connecting cold-formed lipped channels to the bracings are explained in Fig. 3. In normal hot-rolled towers, the bracings and leg members are angle sections and hence the bracings can be directly connected to the leg members on both faces of the leg angle, thus eliminating the use of extra gusset plates. It also, to some extent, reduces the bolt weight. As seen in Fig. 3, such a method cannot be adopted for cold-formed sections.

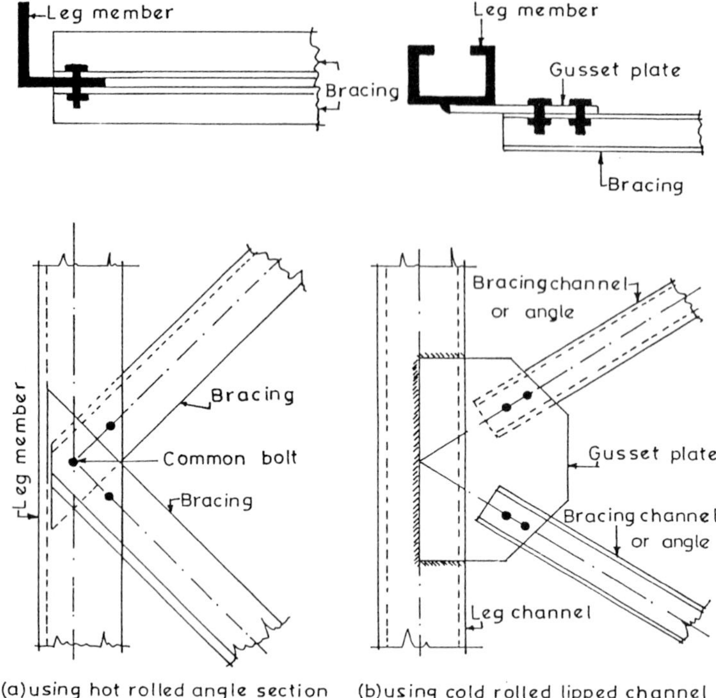

Fig. 3. Connection for hot- and cold-rolled members.

MODEL TOWER AND DETAILS OF TESTING PROCEDURE

12. In order to check the validity of the assumptions made in the design and to study the behaviour of the tower connections, a prototype tower with the configuration and member sizes as shown in Fig. 1 was tested in the Tower Testing and Research Station of Structural Engineering Research Centre, Madras. It is to be noted that this is the first time a prototype cold-formed steel tower has been tested in India. The tower at the test rig is shown in Fig. 4a. Since it was a preliminary test to ascertain the validity of using cold-formed sections in towers, the tower was tested only for the normal condition, though facilities are available to test it for broken wire conditions also.

13. In the test set-up, the transverse and vertical load components are combined into a resultant force, the longitudinal component being applied in the horizontal plane. Fig. 5 schematically shows the wire rope arrangement for the application of the resultant load at the required angle at a typical load point on the tower. In the rope applying the resultant load, a load cell to measure the load and an angle sensor to monitor the angle of load application are attached. The resultant load is controlled by the hydraulic ram at the top and the angle of application is controlled by the vertical ram at the lower level.

a) Before testing.

b) After testing.

Fig. 4. Tower at the testing station.

14. The double acting hydraulic rams are remotely controlled from the control room loading channels. This control is automatic by means of a closed loop servo system explained in the figure. The closed loop system adjusts the resultant transverse, and longitudinal components simultaneously at all load points and ensures that the loads are proportional from zero to maximum. Limit switches help to see that the load applied never crosses the ultimate value. The magnitude of load and the deviation in the angle of application are automatically recorded on charts. To identify the transducers range, each is given its own channel number e.g. 7(1), 7-(2), 7-(3), 7-(4). Accuracy of the measuring system during a test is determined by calibrating the transducer/recorder combination before and after testing. A calibration rig compares the transducer output against a N.P.L. calibrated proving ring.

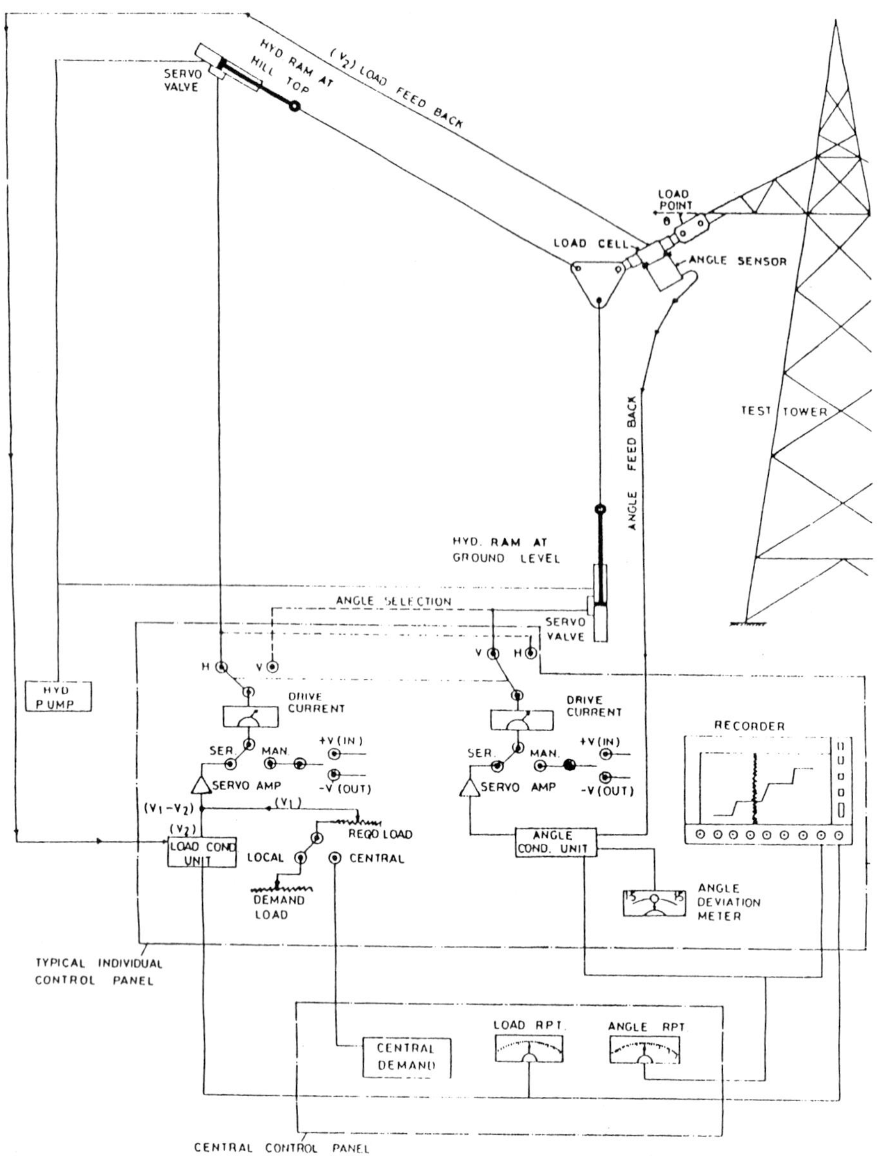

Fig. 5. Control system.

15. Depending on the minimum values of the ultimate test loads required to be applied at any particular load point during a series of tests, a load cell for that load point is chosen, and the load cell-recorder combination is calibrated for the

required ranges of the load cell, allotting one control channel for each load cell. The details of the tabulation for the tested tower is given in Table 1.

Table 1. Test schedule table for normal condition FS-2.0.

Load Pt	Dir & Type of Load	Horz. Load (KN)	Vert Load (KN)	Rigg Load (N)	Resul Load (KN)	Resul Angle (deg min)	Load Cell (KN)	Control Channel No. & (Range)	Test Load as % of Range
1	Resl.	8.76	6.96	250	10.82	37-27	50	1 (3)	86.56
2	Resl.	22.9	23.02	500	31.50	44-31	300	2 (4)	84.00
3	Resl.	22.9	23.02	500	31.50	44-31	500	3 (4)	50.40
4	Resl.	25.3	23.02	500	33.22	41-40	300	4 (4)	88.59
5	Resl.	25.3	23.02	500	33.22	41-40	500	5 (4)	53.15
6	Resl.	42.0	23.02	500	46.74	28-12	300	6 (3)	62.32
7	Resl.	42.0	23.02	500	46.74	28-12	500	7 (4)	74.78

16. After completing the rigging of the test tower, a trial test was conducted with around 10% of loading in all load points, to check the rigging pattern, hydraulic loading and control system functioning. During the actual test, a pause in loading was made after every 5% or 10% of ultimate test load has been applied. This is done to allow deflection readings to be taken and to assess the ability of the tower to accept a further increase in load. In addition to visual inspection of the tower throughout the test for probable distress in any member, at selected stages of loading, theodolite readings of the movement of graduated scales fitted on to certain parts of the tower were also taken for assessing the tower deflection. Table 2 gives the deflection readings taken during the test.

17. After reaching 84% of the ultimate test load, the tower failed. Subsequent inspection of the tower showed that both the compression bracings in the transverse face, immediately below the bottom cross arm level, had buckled. Also the compression legs below the bottom cross arm had buckled (see Fig. 4b).

Table 2. Structural deflections (mm).

Direction &	Percentage of Ultimate Test Load				
Location	Datum	25	50	75	at failure
Transverse At					
Top Conductor Level	0	115	250(63)*	360	390
Bottom Conductor Level	0	55	120(33)*	180	-

* Computed values

DISCUSSIONS AND CONCLUSIONS

18. Most of the cold-rolled section manufacturers in India are rolling only a range of standard sections. Hence it was not possible to adopt the actual designed sections in the tower. The sections shown in Fig. 1 are the sections used in the

testing and not the sections designed by using the computer program. Hence in order to compare the computed results with those of the test results, the sections and loads as adopted in the test were substituted in the computer model. Comparison of the computed and tested results show that the tower failed at a lower load which is 16 percent less than the calculated load. This may be attributed to the large deformations (which are four times more than the calculated values) noticed during the test. The test also showed that the initial imperfections in the tower members will have greater influence in the case of cold-rolled sections. It was also found that there was no appreciable deformation at the gusseted joints. A series of tests on parts of the towers are being planned in order to determine the effect of eccentricity of the gusseted joints.

19. However, this preliminary test ensured that cold-formed sections could be successfully adopted in four legged transmission line towers, which will result in economy of material. It was also seen that the gusset plates, though marginally increasing the cost of the tower, does not change the structural performance of the tower. More tests, including those on triangular based towers will throw-more light on the behaviour of such towers.

ACKNOWLEDGEMENTS

20. The tower was fabricated and tested on behalf of M/s.T. I . Metal Sections. The tower was tested at the Tower Testing and Research Station of the Structural Engineering Research Centre, Madras. The author was a consultant to M/s.T.I. Metal Sections in this project.

REFERENCES

1. GAYLORD, E.H. and WILHOITE, G.M., Transmission Towers Design of Cold-formed Angles, Journal of Structural Engineering Proc. ASCE, 1987, Vol., No., Aug., 1810-1825.

2. IS 802 (part 1) - 1977, Code of Practice for use of Structural Steel in Overhead Transmission Line Towers, Indian Standards Institution, New Delhi, 1977.

3. SUBRAMANIAN, N., and ANNAPURANI, C.R., Design of Transmission Line Towers using a Micro Computer, The Bridge and Structural Engineer, 1986, Vol.16, No.1, March, 21-31.

4. IS 801 - 1975, Indian Standard Code of Practice for use of Cold-formed Light gauge steel structural members in general building constructions (First Revision), Indian Standards Institution, New Delhi, Jan. 1976.

5. SUBRAMANIAN, N., Design of Transmission Towers using Cold-formed Members, Civil Engineering and Construction Review, 1990, Vol. 3, No. 5, May, 63-67.

6. ZAVELAND, M.A., and FAGGIANO, P., Design of Cold-formed Latticed Transmission Towers, Journal of Structural Div., Proc. ASCE, 1985, Vol. 111, Nov., 2427 - 2445.

7. Test Report on 220kV Double Circuit Tangent Tower, Report No.18, Tower Testing and Research Station, SERC, Madras, Dec. 1991.

79. Design of hybrid microwave towers

P. R. NATARAJAN, K. MURALIDHARAN and S. J. MOHAN,
CSIR, India

SYNOPSIS. With the increased telecommunication network in India, a large number of self-supporting microwave towers are being built. The microwave towers, which are space structures in steel, carry mainly communication antennae. The primary load on towers is due to wind gusts only. With the development of methods for the realistic estimation of wind loads, and probabilistic methods of design, the concept in microwave tower design has undergone a significant change. Hybrid microwave towers, triangular in shape, with tubular members for legs and angle sections for bracings are used replacing the conventional square towers with angle sections. These hybrid towers work out to be economical for all aspects of tower design.

INTRODUCTION

1 Microwave towers are steel space structures and they carry, at the top and other levels, electronic communication antennae. The early eighties saw a rapid growth in the telecommunication network in India for which tall towers were designed and built, ranging from 80m to 175m in height. These microwave towers were square in plan and used mostly standard angle sections as members of the tower, bolted and connected together. In the early nineties, schemes were started for linking the entire country with a telephone network system. This paved the way for the development of designs for small sized microwave towers of 15m height and medium sized towers of 40m to 50m height. Since the requirement of these towers are very large, economy aspects would be the prime objective in the design development.

2. With the development of probabilistic based design methodologies introduced in all facets of structural engineering, the design of microwave towers has undergone a further change. The risk of computed design loads being exceeded is set to an acceptably low level in this approach. This provides a consistent and logical way of relating loads and strengths resulting in economical and safe microwave towers, wherever the statistical environmental data is available.

3. Triangular towers provide better aerodynamic characteristics compared with square towers and hence are being increasingly used in towers structures. Tubular

members which offer less resistance to wind flow, and provide a high strength to weight ratio, are being used in place of angle sections in the design of members of microwave towers. But the connection between leg and bracings poses some problems, if both are tubes. Hence to achieve an optimum solution from all considerations, the design of a triangular tower with a combination of tubular members for the legs and angle sections for the bracings is attempted and is discussed in this paper.

OPTIMISATION OF TOWER DESIGN

4. The price of steel in India is very high and a good proportion of steel consumed by the structural steel industry could be saved if a more efficient procedure is adopted in the analysis and design methodology or by developing new configuration for towers. Triangular towers with tubular members can lead to an optimum design, for the following reasons:

- triangular towers are economical compared with square towers as they require less expensive foundations (three against four);
- erection can be undertaken faster as the number of members in a triangular tower will be less than the numbers required for a square tower;
- the overall force coefficient, for a given solidity ratio, is less, and hence triangular towers experience less wind loads, which would in turn result in less forces compared with square towers. If the members are of tubular cross-sections the force coefficient is even lower compared to flat sided members or angles;
- triangular towers occupy less area on the ground and therefore the cost of land needed to be acquired is less compared with square towers.

5. However tubes are expensive compared with angle sections in India, and unless a 30% reduction in terms of weight are achieved in the design, a cost effective design may not be possible. Another problem associated with towers using tubular members occurs during galvanising, when the tube ends are sealed for connections. To overcome these difficulties angle sections provided back to back in T-shape or in diamond form (Fig. 1) for bracings, and tubes for legs leads to a reasonably cost effective design.

6. The most important factor in the design of a microwave tower is the estimation of loadings on the tower. With powerful analytical tools available for the three dimensional analysis of space truss or frame structures, the analysis of structures has been simplified. But, an over estimation or underestimation of the loading, may result in a too conservative or unsafe design. The loads on microwave towers are mainly due to wind gusts acting on the body of tower, platforms, ladder, and antennae. In addition, gravity loads due to self weight of members, platforms, ladder, and antennae are also considered.

Wind load computation

7. The wind load on any panel of a tower (Fig. 1) is given by the formula :

$$F = Ae \times Pz \times CF \qquad (1)$$

where, Ae is the area of the members exposed in the panel; Pz is the wind pressure at height Hz; and CF is the force coefficient for the panel.

Wind pressure (Pz)

8. The wind velocity is not a constant. It varies with time and space. A power law is used to describe the mean velocity profile. The design wind speed is obtained using the relation:

$$Vz = Kl.K2.K3.Vb \qquad (2)$$

where, Kl = risk coefficient, K2 = terrain roughness coefficient, K3 = local topography coefficient and Vb = basic wind speed in m/sec. After evaluating the design wind speed, the design wind pressure is obtained using the relation.

$$Pz = 0.6\ Vz.Vz \qquad (3)$$

where, Pz is the design wind pressure in N/m^2.

9. The risk factor Kl is associated with the design life period of the structure. This approach incorporates different reliability levels depending upon the importance of the microwave tower. This means that a small size tower (up to 15m) which presents a low degree of hazard to life and property in the event of failure, can be designed for a 25 year return period (Kl = 0.94 to 0.85). The medium sized towers shall be designed based on a 50 year return period (Kl = 1.0) and the tall towers, which are exceptionally important shall be designed based on a 100 year return period (Kl = 1.05 to 1.08). This factor is very important as otherwise the wind load on a 15m high tower, when the design is based on a 100 year return period, would be about 47% greater than required and hence the weight of the tower would increase.

Force coefficient (CF)

10. The value of the force coefficient applied to a structure as a whole and when multiplied by the effective frontal area Ae by the design wind pressure Pz gives the total wind load on the structure. The value of the force coefficient differs for the wind acting on different faces of a structure. For microwave towers as the structure is a tapered one, the tower is divided into a number of panels and the load on each strip is calculated individually taking the area of each strip. The above coefficient for towers composed of flat sided members and for towers composed of rounded members are given for different solidity ratios in IS:875 (ref. 1)

11. The code gives the following method for computing the force coefficient for a single frame consisting of flat sided members and tubular members.

12. The wind flow can be sub-critical or super-critical depending upon the diameter of tubular members (d) and wind velocity (Vz).

$d.Vz < 6$ for sub-critical flow.
$d.Vz > 6$ for super-critical flow.

Accordingly, in a panel, some of the members may be in sub-critical flow. Angle members or flat sided members are always in sub-critical flow. For a specific solidity ratio, CFt(sup), CFt(sub), and CFf are the force coefficient for tubes in super-critical, sub-critical and for angle sections respectively, these are obtained from IS:875. If, At(sup), At(sub), and Af, are the area of tubes in supercritical, sub-critical and the area of angle sections respectively, then Ae, A(sub) and g are computed as:

Ae = At(sup) + At(sub) + Af
A(sub) = At(sub) + Af
g = At(sup)/Ae.

CF = {g.CFt(sup) + (1-g).[At(sub).CFt(sub) + Af.CFf] } / A(sub) (4)

13. Based on the above method, an example is given for the computation of CF values for a tower panel with four different bracing schemes (Fig. 1) and is given in Table 1.

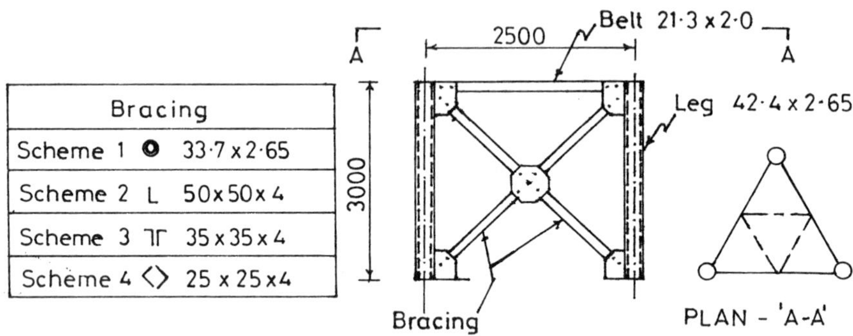

Fig. 1. Typical tower panel with different bracings.

<u>Loads on ladders and antennae</u>

14. For the analysis of the tower due to wind, two main directions of wind are needed to be considered (Fig. 2). The direction of wind in the first case acts perpendicular to any one face, which would give the maximum force in the legs for the design. The second case, in which, the wind acts parallel to any one face, also referred as the 30 degree wind, would give maximum forces in some of the bracing members of the tower. In the same figure, the location of the ladder in a triangular tower is given, specifically to emphasis the calculation of loadings on them. For the first case only 50% of the area is considered exposed, as against the full area in the second case. A force coefficient of 1.2 is used for tubular members for the ladder, treating them as infinitely long members.

Table 1. CF computation and comparison.

Heading	Scheme 1	Scheme 2	Scheme 3	Scheme 4
Leg	Tube	Tube	Tube	Tube
Size	42.4x2.65	42.4x2.65	42.4x2.65	42.4x2.65
Exposed	3x2x.0424=			
area	0.2544	0.2544	0.2544	0.2544
Belt	Tube	Tube	Tube	Tube
Size	21.3x2.0	21.3x2.0	21.3x2.0	21.3x2.0
Exposed	2.5x1x.0213=			
area	0.0533	0.0533	0.0533	0.0533
Bracing	Tube	Angle L	Angles T	Angles <>
Size	33.7x2.65	50x50x4	35x35x4	25x25x4
Exposed	3.905x2x	3.905x2x	3.905x2x	3.905x2x
area	.0337	.050	.035	.025x1.414
	= 0.2632	=0.3905	=0.2734	=0.2761
At	0.5709	0.3077	0.3077	0.3077
Af	0.0	0.3905	0.2734	0.2761
Ae	0.5709	0.6982	0.5811	0.5838
Gross area(Ag) = (2.5+0.0424)x3.00 = 7.626sq.m				
Ø=Ae/Ag	0.075	0.092	0.076	0.077
At/Ae	1.0	0.441	0.530	0.527
Af/Ae	-	0.559	0.470	0.473
CFt	1.750	1.716	1.748	1.746
CFf	-	3.1	3.1	3.1
CF	1.750	2.490	2.380	2.390

Ø = Solidity ratio

15. The loads on the antennae due to wind gust is normally given by the manufacturer. For any other wind velocity, the load can be computed using the formula.

WL = Ws x (Vr / Vm) 2 (5)

Where, Ws is the wind load supplied by the manufacturer, Vm is the corresponding maximum wind speed, Vr is the wind speed at the antenna location and WL is the wind load on the antenna. Depending upon the number of antennae at any particular level, a number of loading conditions are required to be analysed for the microwave tower; each considering the two main directions of wind. From these number of loading conditions, the maximum forces in the members for the design are evaluated.

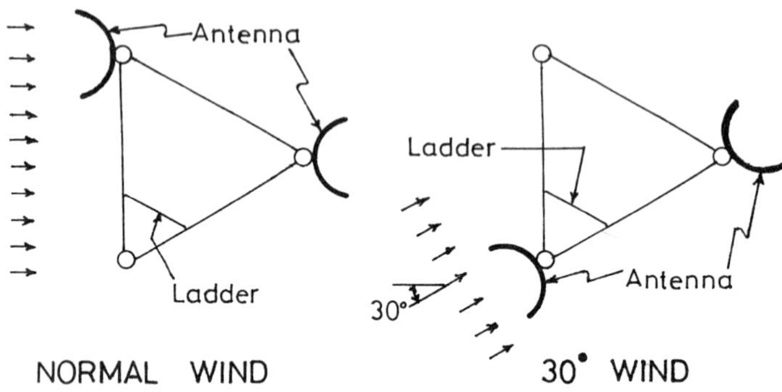

Fig. 2. Wind directions considered for analysis.

DESIGN OF MEMBERS

16. The design for the maximum wind loads may be considered as the design for deformations as well as strength. The tower design involves the design of tension members, compression members and connections. The design of members and connections are not much different from the conventional design of steel structural members as covered in IS:806 (ref. 2) and IS:800 (ref. 3). The most significant part of the design of the microwave towers is the design of the compression members. In order to reach minimum weight design, different bracings systems are followed to give maximum stability to the compression members. For a given bracing system, the capacity of the bracing members for the four different schemes discussed earlier is given in Table 2. The table also gives the loading on the panel and the weight of the panel.

COST EFFECTIVE SOLUTION

17. It can be seen from Table 2 that a triangular tower with all tubular members will give the minimum weight design. The load that the panel experiences is also the least. But, the connection between leg members and bracings becomes rather complex; joining the members together without gussets. Also in the case of bracings, additional gusset plates are required at the intersection points. Therefore, the cost of connections would be rather high compared to the hybrid tower. From a number of microwave tower designs undertaken, it is estimated that at least 20 to 25% by weight, is required to be added to the member weight to

allow for the connections. In addition, when the tube ends are squeezed or sealed by plates for the purpose of connections, many problems occur during galvanisation. In the case of hybrid towers, as the bracing members are made up of angle sections, connection to the leg is made through gusset plates which are welded to legs, prior to galvanising. Eccentricities in connection is avoided in the case of angles arranged back to back as bracing members. The exposed area to wind also reduces.

Table 2. Design of bracings and comparison of schemes.

CF	1.750	2.490	2.380	2.390
Ae(m½2)	0.5709	0.6982	0.5811	0.5838
Load (kN)	2.0	3.48	2.77	2.79
Bracing Size r(mm)	Tube 33.7x2.65 11.0	Angle L 50x50x4 9.7	Angles T 35x35x4 10.5	Angles <> 25x25x4 9.3
(kl) = 0.85x0.5x3905 = 1659.6mm				
kl/r	151	171	158	178
Capacity	12.82kN	13.93kN	21.81kN	12.80kN
Weight	0.34kN	0.42kN	0.51kN	0.40kN

pz = 2kN/sq.m, & Yst 25 Steel grade for members used

18. The cost of the tower includes material cost, cost of fabrication, cost involved in the connections, the galvanising cost, the cost involved in assembly and erection etc. The galvanising cost per tonne of steel is more in the case of tubes compared with angle sections. The cost of tubes are about 30% greater than the cost of standard angle sections. Hence a hybrid tower would lead to a cost effective solution.

19. Therefore, hybrid microwave towers with tubes as leg members, and angle sections provided in diamond form for the bracing members would be the optimum system for microwave towers.

CONCLUSION

20. A design with a combination of tubular members for the legs and angle sections for the bracings offers a good solution for microwave towers. Angle sections provided in diamond shape for the bracings would lead to a cost effective solution, both from the strength aspects as well as in terms of the exposed area for wind action.

ACKNOWLEDGEMENT

21. The paper is being presented with the kind permission of the Director, Structural Engineering Research Centre, Madras, India.

REFERENCES

1. IS:875 (Part 3) - 1987. Indian Standard Code of practice for design loads (other than earthquake) for buildings and structures, BIS, New Delhi, 1989.
2. IS:806 -1968. Indian Standard Code of practice for use steel tubes in general building construction, BIS, New Delhi, 1968.
3. IS:800 - 1984. Indian Standard Code of practice for use of Structural Steel in general building construction, BIS, New Delhi, 1984.

80. Influence of secondary bracings on the stability of towers

K. MURALIDHARAN, P. R. NATARAJAN and S. J. MOHAN,
CSIR, India

SYNOPSIS. Three types of members are used in transmission line towers, namely leg members, main bracings, and secondary bracings or redundants. Many interesting patterns of secondary bracings are used in towers with the aim of increasing the buckling strength of the main members to which these secondary members are connected. Methods to assess the buckling strength of main members supported by secondary members are not adequately covered in the literature. Results from analytical and full-scale tower tests are used to study the influence of the secondary bracings on the buckling strength of the main tower members.

INTRODUCTION

1. Transmission line towers are probably the most commonly used space structures. Their design is generally based on a minimum weight criteria as they are mass produced. The towers are generally of lattice type consisting of leg members or chord members, and bracing members, and they are designed for several load conditions. Most of the structural members of towers are subject to compression in at least one load condition. In order to optimise the design of the main leg and bracing members in compression, secondary bracing members are introduced in a selected pattern between the main members of the tower. These secondary bracing members (also known as redundants) are meant to support the main members at their intermediate points to give them better stability and thus increase their buckling strength. The redundants are not taken into account in the structural analysis of towers and their design is generally based on slenderness consideration.

2. The design of transmission line towers is often checked by conducting full-scale tests. The role of the redundants on the stability of the main members of towers has been studied using the data obtained from some of the full-scale tests and other studies conducted at the Structural Engineering Research Centre (SERC), Madras and reported in this paper.

DESIGN PRACTICE

3. The tower members are generally made of steel equal angle sections. Fig. 1 shows the cross-section of an equal angle member and the main axes. Buckling strength of these members about the VV-axis and the XX-axis are to be considered in the design. The VV-axis is the weakest axis from the buckling point of view. The radius of gyration about the XX-axis is about 50% more than that about the VV-axis. Therefore, if buckling about the VV-axis of a given length of equal angle section is prevented using redundant members, the slenderness ratios decreased by 50% as the member has to buckle about the XX-axis for the same length. This principle is made use of in the general design practice of transmission line towers. The assumptions made on the behaviour of main members with reference to redundants are as explained below.

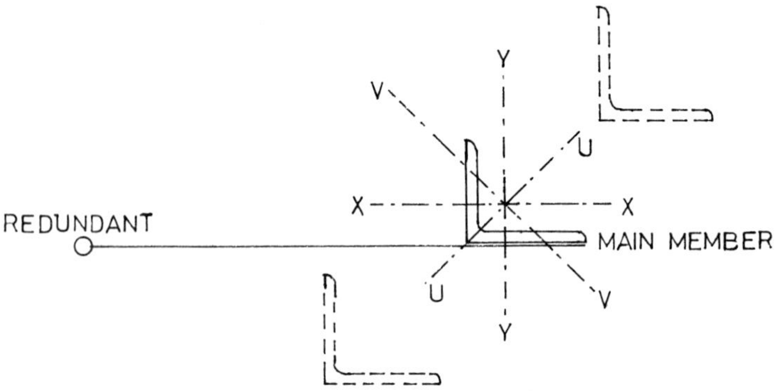

Fig. 1. Redundant preventing VV-axis buckling.

4. Between two main nodes in a tower an angle strut buckles at its critical load about its weakest axis, the VV-axis. Therefore, L/r_{VV} would govern the buckling strength (Fig. 2a). If a redundant member is provided at the middle of the strut as shown in Fig. 2b, buckling about the VV-axis of the full member length (L) is prevented. Therefore, the member as a whole can buckle about XX-axis. A corner stay at the middle (Fig. 2c) would also prevent XX-axis buckling, if the whole member length (L) is considered. The buckling would, therefore, be in one-half (L/2) of the member about the VV-axis.

5. The previous assumptions are acceptable to some extent. A redundant in the plane of the bracing connected at an intermediate point can prevent VV-axis buckling of the full member and thus increase its buckling strength. This can be clearly seen in Fig. 1. If a diagonal member has to buckle about its VV-axis, the original position of the cross-section of the member has to take one of the displaced positions shown by the dotted lines, after buckling. To allow this displacement, the redundant member has to undergo axial deformation as well as a lateral movement. But, these redundants are axially very stiff and they are also connected at the other end to a relatively rigid member. Therefore, the

displacement component of the main member along the axis of the redundant member is prevented. Since the redundant has very little bending rigidity, the displacement in the lateral direction or out of plane cannot be prevented. Therefore, the main member of length "L" can buckle about its XX-axis only. This has been experimentally verified (ref. 2).

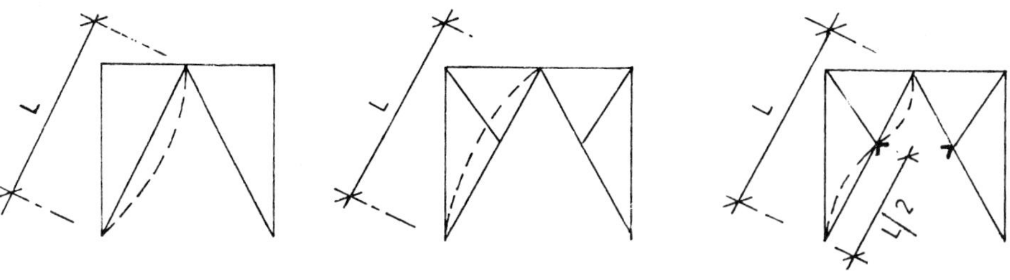

(a) Without redundant. (b) A redundant at centre. (c) A redundant and a hip bracing at the centre.

Fig. 2. K-braced tower panel.

6. In the case of the corner stay as in Fig. 2c, if it has adequate stiffness, it can give support to the main member against XX-axis buckling. Therefore, one-half of the main member alone can buckle about its weakest axis (VV) and thus the buckling strength is further increased. But, how effective is the corner stay? Even though the axial stiffness of the corner stay is good enough, the other end of it is connected to the main diagonal bracing in the adjacent perpendicular plane and, therefore, the overall stiffness of the corner stay depends on the out of plane bending stiffness of this diagonal bracing. These aspects have to be considered while designing the corner stay.

STIFFNESS REQUIREMENT FOR REDUNDANTS

7. If a compression member of symmetric cross-section of length 2L is supported laterally at the centre by a linear spring of stiffness K, then the stiffness of the spring to prevent buckling of the strut in first mode is given by, (ref. 1):

$$K_{req} = 2 \times 2\, P_{cr} / L \qquad (1)$$

Where, K_{req} = required stiffness of the intermediate support.

P_{cr} is the theoretical buckling load for the column of length L.

8. As a part of the research work carried out at SERC, a K-braced panel of a tower (Fig. 3) was tested to study the buckling of the compression bracing. The

stiffness of the corner stay was found analytically and experimentally (ref. 2). The analytical value based on the bending stiffness of the interconnected bracings, considered as a simply supported beam and the redundant as a cantilever beam, worked out to 484N/mm. The experimental value was found by measuring the deflection perpendicular to the XX-axis of the main bracing member for an applied load in the same direction. The measured value was 578N/mm which is greater than the required stiffness of the corner stay viz. 372N/mm as given by Eq. 1, obtained by substituting a value for P_{cr} = 58kN and L = 625mm. The bracing member failed by buckling about VV-axis in the bottom half portion of the bracing, as predicted.

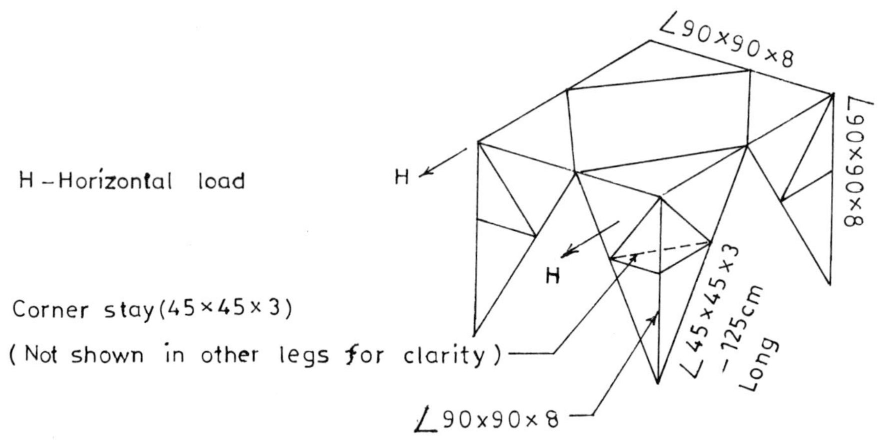

Fig. 3. A test panel with K-bracing and corner stay.

RESULTS FROM FULL-SCALE TEST

Leg member in compression

9. The bottom most panel of a parallel leg tower with X-bracing and horizontal redundants on adjacent faces at the mid-point of the leg member is shown in Fig. 4. The tower was tested to destruction. Strain values obtained during the test from electrical resistance strain gauges fixed to the leg member are plotted in Fig. 5. Assuming that the redundants would provide adequate support, the slenderness value was taken as 0.5 x l / r_{vv} to calculate the capacity of the leg member. The buckling stress of the member was found to be 235N/mm^2. During the destruction test, the member (which was the most critical one) failed by buckling in the same mode as predicted, and the stress in the member reached a value of 250N/mm^2, demonstrating the role of the redundants in the stability of the main leg member.

Fig. 4. X-Braced tower panel.

CORNER STAY

10. Fig. 6 shows the lowest panel of a 220kV tower. The main bracing members and the leg members are supported by a system of redundants. The leg member is divided into six equal lengths and the bracing into three. All these main members and redundants lie in the same plane on each side of the tower. In addition, to prevent the out of plane buckling (about the XX-axis), the main bracing member is supported at the third points using corner stays.

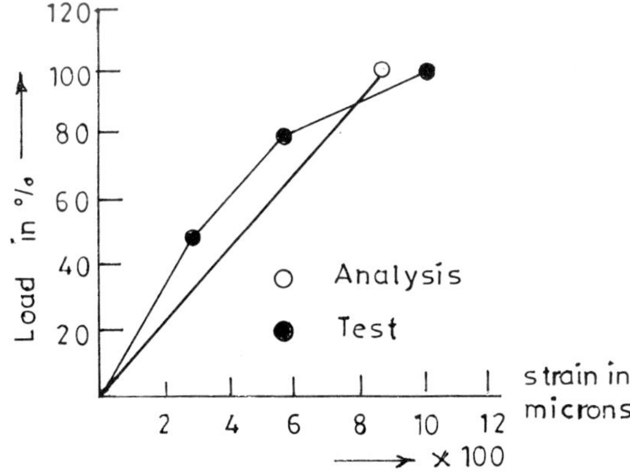

Fig. 5. Load v/s strain in tower leg.

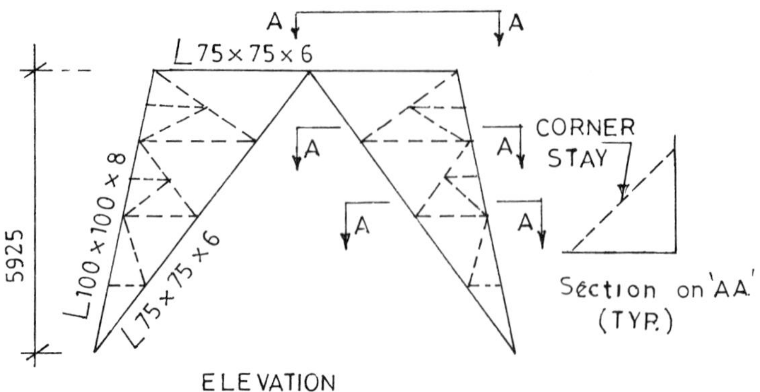

Fig. 6. Tower panel under test.

11. This corner stay arrangement was not found to be effective. During the tower test, the bracing member buckled about the XX-axis as shown in Fig. 7. It can be seen in the figure that the corner stays could not give adequate support. The ideal stiffness K_i required (ref. 1) at the one-third points of the main bracing is:

$$K_i = 2 \times 3\, P_{cr} / (l/3) = 385{\cdot}0\text{N/mm} \quad (2)$$

Where, $P_{cr} = \pi^2\, EI / (l/3)^2$

The other end of the corner stay is connected to the corresponding K-bracing member on the perpendicular plane. Therefore, the bending stiffness of this bracing member at its one-third point is approximately the stiffness of the corner stay and it is equal to $56.47\, EI / L^3 = 55$N/mm which is very low when compared to the required stiffness of 385N/mm.

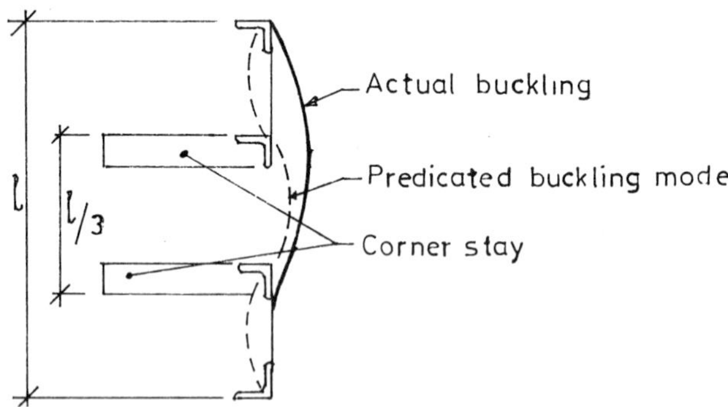

Fig. 7. Buckling of bracing member about the XX-axis.

12. A well designed secondary bracing system can control the buckling of the main members as required and an example is given in Fig. 8 showing a buckled main bracing member after a tower test.

13. It has been observed from full-scale tests that: (i) the force in the redundants is less than 2.5% of that in the main connected member, and (ii) well triangulated pattern for redundants are generally very effective.

Fig. 8. Well designed secondary bracing system.

CONCLUSION

14. The following conclusion can be drawn :

- secondary bracings or redundants have an important role in providing better stability to the main members of a tower;
- it is necessary to assess the stiffness requirement of the redundants and to design the redundants to meet the requirement;
- it is not always sufficient to design the redundants from the point of view of a maximum limit for the slenderness ratio.

ACKNOWLEDGEMENT

15. The paper is being presented with the kind permission of the Director, Structural Engineering Research Centre, Madras, India.

REFERENCES

1. SALMON.C.G and JOHNSON J.E, Steel Structures - Design & Behaviour, p.500, second edition, Harper & Row, New York, 1980.

2. NATARAJAN P.R, MURALIDHARAN K. and MOHAN S.J, Buckling of K-bracing, Report RD/30/2, Structural Engineering Research Centre, Madras, Aug. 1991.

81. A method for the calculation of lateral deflections of lattice towers

J. J. CAO and A. J. BELL, UMIST, UK

SYNOPSIS. The paper deals with the calculation of lateral deflections of lattice towers under transverse loading including concentrated loads, uniformly distributed loads and applied moments. Towers with four parallel legs or with four sloping legs and a variety of bracing types are considered. Formulae, which are simple to use and are similar to formulae for beams, are developed and presented for the determination of lateral deflections of the towers. Results obtained using the formulae are compared with the results of finite element (FE) analyses and with test results and very good agreement is demonstrated.

INTRODUCTION

1. Lattice towers are widely used to support telecommunications, aerials, transmission lines and in applications in cranes and offshore structures. Whilst finite element computer programs are relatively readily available for the analysis of these structures, there is in practice still a need for simple, accurate hand methods of analysis particularly perhaps in the preliminary stage of design or in areas where computers are not readily available. The accurate determination of the forces in the members of lattice structures is relatively straightforward and quick particularly since the forces in all members are seldom required in design. However the accurate determination of deflections, which are frequently important design criteria, is more lengthy and complicated. Two frequently used methods of determining deflections are based on:

(i) considering a tower as an "equivalent beam" (bracing members ignored);

(ii) using energy methods.

2. In Method (i) the second moment of area of an "equivalent beam" is approximated using only the area of the chords or legs of the lattice structure. This second moment of area is then used in ordinary cantilever beam deflection calculations. Method (ii) uses energy methods, and though accurate, is time consuming and tedious.

3. An accurate method is presented in this paper in which energy principles are used to obtain formulae which are similar to those used in beam calculations. The

formulae, for both deflection f and rotation θ as shown in Fig. 2, are derived for common bracing configurations and take full account of sloping leg members.

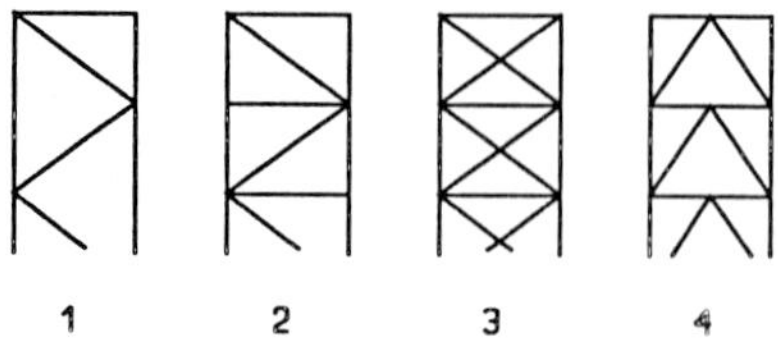

Fig. 1. Bracing types.

TOWERS WITH PARALLEL LEGS

4. The four types of bracing shown in Fig. 1. are considered. The analysis is developed first for Type 1 bracing (Fig. 1) and then extended to include the other types of bracing.

5. The tower shown in Fig. 2 carries a concentrated horizontal load P and an applied moment M at the top together with a uniformly distributed lateral load of q. The loads P, M and q are treated separately and the results are then superimposed.

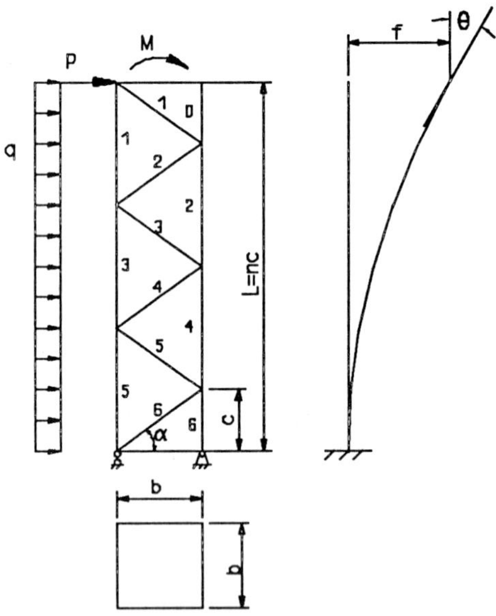

Fig. 2. Tower with parallel legs.

6. Using the member numbering system in Fig. 2. the axial forces in leg member i and bracing member i, due to the action of the concentrated load P are:

$$N_i = \frac{P}{2b} i c \qquad T_i = \frac{P}{2\cos\alpha} \tag{1}$$

When P equals unit load, these forces are $\overline{N}_i$ and $\overline{T}_i$ respectively.

7. Using the energy method (ref. 1), the displacement, f_P, of the tower due to P can be obtained as:

$$\begin{aligned} f_p &= \sum_{i=1}^{n} \frac{N_i \overline{N}_i}{EA_L} l_{Li} \cdot 2 + \sum_{i=1}^{n} \frac{T_i \overline{T}_i}{EA_B} l_{Bi} \cdot 2 \\ &= \sum_{i=1}^{n} \frac{P i^2 c^2}{4 b^2 E A_L} l_{Li} \cdot 2 + \sum_{i=1}^{n} \frac{P}{4\cos^2\alpha \, E A_B} l_{Bi} \cdot 2 \\ &= \frac{PL^3}{3EA_L b^2}\left(1 + \frac{1}{2n^2}\right) + \frac{PL}{2EA_B \cos^2\alpha \sin\alpha} \end{aligned} \tag{2}$$

l_{Li}, l_{Bi} - the lengths of leg member i and bracing member i.
A_L A_B - the cross-section areas of leg member and bracing member.
n - the number of diagonal bracing members.

Generally the term $1/2n^2$ in Eq. 2 is small compared to 1 and so can be neglected giving:

$$f_P \approx \frac{PL^3}{3EA_L b^2}\left(1 + 1.5 \frac{A_L}{A_B} \frac{1}{n^2 \sin^3\alpha}\right) \tag{3}$$

8. The first and second terms within the brackets of Eq. 3 indicate the deflections caused by the deformation of the leg and the bracing members respectively. The contribution to the deflection represented by the first part of Eq. 3 is same as that which would be obtained using beam theory with the equivalent second moment of area based on leg members only (i.e. Method (i)) and it can be seen that this leads to an underestimate of deflection.

9. The deflections caused by M and q can be obtained in a similar way and values added to f_P give the total deflection of the tower under the action of P, M and q as:

$$\begin{aligned} f = {} & \frac{PL^3}{3EA_L b^2}\left(1 + 1.5 \frac{A_L}{A_B} \frac{1}{n^2 \sin^3\alpha}\right) + \frac{ML^2}{2EA_L b^2} \\ & + \frac{qL^4}{8EA_L b^2}\left[1 + \frac{2}{n^2 \sin^3\alpha} \frac{A_L}{A_B}\left(1 + \frac{1}{n}\right)\right] \end{aligned} \tag{4}$$

10. This approach has been applied to the other three types of bracing systems in Fig. 1 and has also been extended to determine rotation θ. General expressions for the deflection and rotation of any of the tower forms shown can be expressed as:

$$f = k_p \frac{PL^3}{3EA_L b^2} + \frac{ML^2}{2EA_L b^2} + k_q \frac{qL^4}{8EA_L b^2}$$
$$\theta = \frac{PL^2}{2EA_L b^2} + \frac{ML}{EA_L b^2} + \frac{qL^3}{6EA_L b^2} \qquad (5)$$

where k_p, k_q are coefficients which take account of the different configurations of bracing members and may be expressed as:

$$k_p = 1 + \eta\lambda_p \qquad k_q = 1 + \eta\lambda_q \qquad (6)$$

where $$\lambda_p = \frac{0.75}{n^2 \sin^3\alpha} \quad \text{and} \quad \lambda_q = \frac{1}{n^2 \sin^3\alpha}(1 + \frac{1}{n})$$

and η is obtained from Table 1 in which A_H is the area of horizontal bracing members.

11. Examination of Eq. 5 shows that it is the same general form as would be obtained using beam theory together with the equivalent second moment of area, based on leg members alone, being modified by k_p and k_q to take account of bracing members.

Table 1. Value of η for towers with parallel legs.

Bracing Type	1, 2	3	4
η	$2\frac{A_L}{A_B}$	$\frac{A_L}{A_B}$	$4\frac{A_L}{A_B}(1 + \frac{A_B}{A_H}\cos^3\alpha)$

TOWERS WITH SLOPING LEGS

12. The method which has been developed for towers with parallel legs may be extended to towers with sloping legs by considering the latter as "equivalent beams" of variable section.

13. The tower shown in Fig. 3 carries a concentrated horizontal load, an applied moment and a uniformly distributed lateral load. As is usually the case in practice the slope of the legs is small. At any section, the equivalent second moment of area, J_x, based on legs only, is given by:

$$J_x = A_L \left(\frac{x}{a}\right)^2 b_0^2 \qquad (7)$$

At the same section the bending moment due to the applied loading is given by:

$$M_x = P(x-a) + M + \frac{q}{2}(x-a)^2 \qquad \textbf{(8)}$$

and the moment M_X due to a unit load for P is given by:

$$\overline{M_x} = x - a \qquad \textbf{(9)}$$

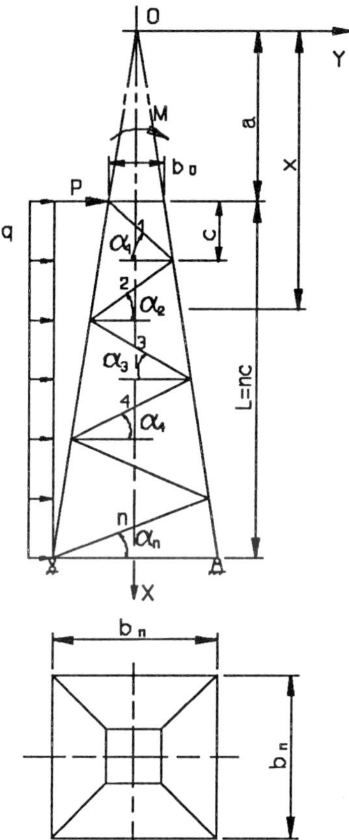

Fig. 3. Tower with sloping legs.

14. Using the energy method (ref. 2) the deflection, f_L, due to leg deformation can be obtained as:

$$f_L = \int_a^{a+L} \frac{\overline{M}_x M_x}{EJ_x} dx$$

$$= \frac{a^2}{EA_L b_0^2} \int_a^{a+L} \left[P(\frac{x-a}{x})^2 + M(\frac{x-a}{x^2}) + \frac{q}{2}\frac{(x-a)^3}{x^2} \right] dx \qquad \textbf{(10)}$$

$$= \frac{PL^3}{3EA_L b_0^2} k_1 + \frac{ML^2}{2EA_L b_0^2} k_2 + \frac{qL^4}{8EA_L b_0^2} k_3$$

and the rotation, θ, at top of the tower due to leg deformation is:

$$\theta = \frac{PL^2}{2EA_L b_0^2} k_2 + \frac{ML}{EA_L b_0^2} k_4 + \frac{qL^3}{6EA_L b_0^2} k_1 \qquad \textbf{(11)}$$

where

$$k_1 = 3k^2 [\frac{2k+1}{k+1} - 2k \ln(1 + \frac{1}{k})]$$

$$k_2 = 2k^2 [\ln(1 + \frac{1}{k}) - \frac{1}{k+1}]$$

$$k_3 = 4k^2 [\frac{1}{2} - 2k + 3k^2 \ln(1+\frac{1}{k}) - \frac{k^2}{1+k}] \qquad \textbf{(12)}$$

$$k_4 = \frac{k}{k+1}$$

and

$$k = \frac{a}{L} = \frac{b_0}{b_n - b_0}$$

15. The common types of bracing used in towers with sloping legs are the same as those used in towers with parallel legs as shown in Fig. 1. The effects of bracing members depend on their configurations. As an example of how they are included, Type 1 bracing as shown in Fig. 1 and Fig. 3 is considered.

16. For the case of the concentrated load P, the axial force in bracing member i, T_i, can be obtained as:

$$T_i = \frac{P}{\cos\alpha_i} \left[\frac{1}{2} - (\frac{i}{b_i} + \frac{i-1}{b_i - 1}) \frac{c \cdot \tan\theta}{2} \right] \qquad \textbf{(13)}$$

When θ is small:

$$\frac{i}{b_i} \approx \frac{i-1}{b_{i-1}} \qquad \frac{b_0}{b_{i-1}} \approx \frac{\tan\alpha_i}{\tan\alpha_1} \qquad \text{and} \quad \tan\theta = \frac{b_{i-1} - b_0}{2(i-1)c} \qquad \textbf{(14)}$$

Using these approximations in (13) gives:

$$T_i \approx \frac{P}{2\cos\alpha_i}\frac{b_0}{b_{i-1}} \approx \frac{P}{2\cos\alpha_i}\frac{\tan\alpha_i}{\tan\alpha_1} \tag{15}$$

When P equals unit load $T_i = \overline{T}_i$ and the tower deflection, f_{BP}, caused by the deformation of bracing members under P can be calculated as:

$$\begin{aligned} f_{BP} &= \sum_{i=1}^{n} \frac{T_i \overline{T}_i}{EA_B} l_{Bi} \cdot 2 \approx \frac{2}{EA_B}\sum_{i=1}^{n} \frac{P}{4\cos^2\alpha_i}\frac{\tan^2\alpha_i}{\tan^2\alpha_1}\frac{c}{\sin\alpha_i} \\ &= \frac{P\,c}{2EA_B\tan^2\alpha_1}\sum_{i=1}^{n}\frac{\sin\alpha_i}{\cos^4\alpha_i} \end{aligned} \tag{16}$$

17. Eq. 16 may be simplified and made more convenient, with little error, if the summation is replaced by the following integration:

$$\sum_{i=1}^{n} \frac{\sin\alpha_i}{\cos^4\alpha_i} \approx \frac{n}{\alpha_n - \alpha_1}\int_{\alpha_1}^{\alpha_n} \frac{\sin\alpha}{\cos^4\alpha} d\alpha \tag{17}$$

which leads to:

$$\begin{aligned} f_{BP} &\approx \frac{PL^3}{3EA_B b_0^2}\frac{1}{2n^2}\frac{1}{\tan^4\alpha_1}\frac{1}{\alpha_n-\alpha_1}\left[\frac{1}{\cos^3\alpha_n} - \frac{1}{\cos^3\alpha_1}\right] \\ &= \frac{PL^3}{3EA_L b_0^2}\, 2 \cdot \frac{A_L}{A_B}\frac{1}{n^2}\lambda_P \end{aligned} \tag{18}$$

where
$$\lambda_P = \frac{\cot an^4\alpha_1}{4(\alpha_n - \alpha_1)}\left[\frac{1}{\cos^3\alpha_n} - \frac{1}{\cos^3\alpha_1}\right] \tag{19}$$

18. The deflections, f_{BM} and f_{Bq}, caused by bracing deformations due to M and q can be obtained by a similar manner.

19. The same general analysis can be applied to towers with bracing Types 2, 3 and 4. A general expression for deflection similar to Eq. 5 for towers with parallel legs may then be written as:

$$f = k_P\frac{PL^3}{3EA_L b_0^2} + k_M\frac{ML^2}{2EA_L b_0^2} + k_q\frac{qL^4}{8EA_L b_0^2} \tag{20}$$

where
$$\begin{aligned} k_P &= k_1 + \eta_P\lambda_P \\ k_M &= k_2 - \eta_M\lambda_M \\ k_q &= k_3 + \eta_q\lambda_q \end{aligned} \tag{21}$$

Eq. 11 can be used to calculate the slope angle. Values of η_p and η_q in Eq. 21 are given in Table 3 and $\eta_M = \eta_p$.

Table 3. Values of η_p and η_q for towers with sloping legs.

Bracing Type	1,2	3	4
η_p	$2\frac{A_L}{A_B}\frac{1}{n^2}$	$\frac{A_L}{A_B}\frac{1}{n^2}$	$4\frac{A_L}{A_B}\frac{1}{n^2}(1+\frac{A_B}{A_H}\omega_p)$
η_q	$2\frac{A_L}{A_B}\frac{1}{n^2}(1-\frac{1}{n})$	$\frac{A_L}{A_B}\frac{1}{n^2}(1-\frac{1}{n})$	$4\frac{A_L}{A_B}\frac{1}{n^2}(1+\frac{A_B}{A_H}\omega_q)(1-\frac{1}{n})$

λ_M and λ_q are given by the following expressions:

$$\lambda_M = \frac{\cot an\alpha_n - \cot an\alpha_1}{6\tan^3\alpha_1}\left[\frac{1}{\cos^3\alpha_n} - \frac{1}{\cos^3\alpha_1}\right]\frac{1}{\alpha_n - \alpha_1} \quad \textbf{(22)}$$

$$\lambda_q = \frac{1}{\tan^2\alpha_1}\frac{\tan\alpha_n}{\tan\alpha_1 - \tan\alpha_n}\frac{1}{\alpha_n - \alpha_1}\left[\frac{1}{\cos\alpha_n} - \frac{1}{\cos\alpha_1} + \ln\frac{\tan\frac{\alpha_n}{2}}{\tan\frac{\alpha_1}{2}} - \frac{1}{3\tan^2\alpha_1}\left(\frac{1}{\cos^3\alpha_n} - \frac{1}{\cos^3\alpha_1}\right)\right] \quad \textbf{(23)}$$

20. In Table 3 the coefficients ω_p and ω_q take account of the effects of deformation due to horizontal bracing members and can be calculated (ref. 2) approximately from the following expressions:

$$\omega_p \approx 1.2141 - 0.01294\alpha_1 - 0.0056\alpha_n$$
$$\omega_q \approx 1.2386 - 0.00796\alpha_1 - 0.01146\alpha_n \quad (24)$$

TOWER WITH DIFFERENT TYPES OF BRACING OR BOTH PARALLEL AND SLOPING LEGS

21. In practice a tower may include different types of bracing configurations or include both parallel and sloping legs. Such towers can be analysed by dividing the whole tower into several tower sections, as is shown for the tower in Fig. 4.

Each section can be treated separately using Eq. 5 for the part with parallel legs or using Eqs 20 and 11 for the part with sloping legs. After the deflection f_i and rotation θ_i for each section have been determined, the total deflection of the tower, f_T, can be obtained as:

$$f_T = f_1 + f_2 + L_2 \cdot \theta_1 \qquad \textbf{(25)}$$

In general for a tower with n sections,

$$f_T = \sum_{i=1}^{n} f_i + \sum_{i=1}^{n-1} \left(\theta_i \sum_{j=i+1}^{n} L_j \right) \qquad \textbf{(26)}$$

where L_j is the height of jth tower section.

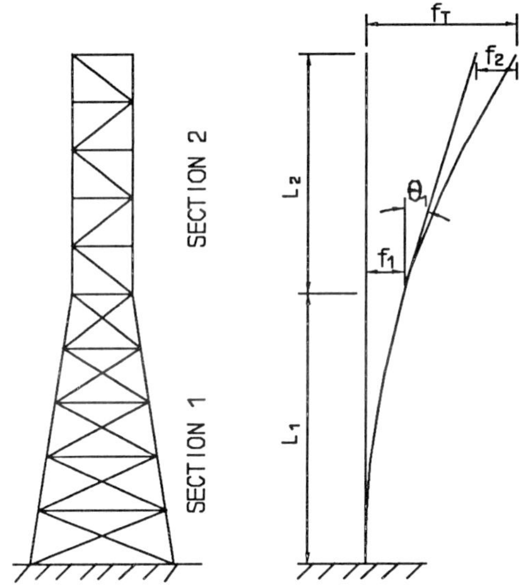

Fig. 4. Two section tower.

TESTS AND FEM ANALYSES OF MODEL TOWERS

22. Results obtained using the formulae which have been developed have been compared with the results of FE analyses and tests on model towers. Details of the towers used are outlined in Fig. 5 and Table 4.

23. Two loading cases were considered for each tower and comparisons of the results obtained are given in Table 5.

24. It can be seen from Table 5 that the formulae presented in this paper produced results which compared very favourably with those obtained from finite element analyses and from the tests. Results obtained using the "equivalent beam" (ignoring the bracing members) were found to underestimate deflections by up to 10%.

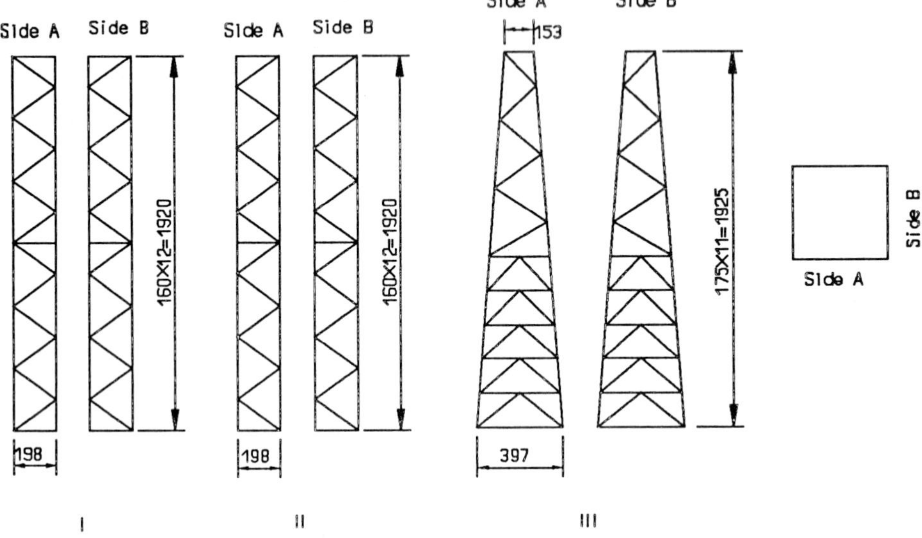

Fig. 5. Model towers.

Table 4. Dimensions of tower models.

Model	n	b (mm)	A_L (mm^2)	A_B (mm^2)
I	12	198	143	50
II	12	198	143	50
III	11	153-397	126	90

Table 5. Comparison of displacements of models

Model	Load		Calculation and Test (mm)				Comparison %		
	P (N)	M (N-m)	(1) Method (i)	(2) FEM	(3) Author's Formulae	(4) Test	(2)-(1) / (2)	(2)-(3) / (2)	(4)-(3) / (4)
I	3000	270	6.58	7.30	7.32	7.12	10.0	-0.3	-2.8
	0	5000	8.01	7.94	8.01	7.67	-0.9	-0.9	-4.4
II	3000	270	6.58	7.30	7.32	7.28	10.0	-0.3	-0.5
	0	5000	8.01	7.94	8.01	7.97	-0.9	-0.9	-0.5
III	6000	540	6.14	5.89	5.97	5.80	-4.2	-1.4	-2.9
	0	6000	5.12	4.65	4.54	4.60	-10.1	2.4	1.3

CONCLUSIONS

25. Formulae which are quick and straightforward to use have been produced for determining the lateral deflection and rotation of lattice towers with parallel legs Eq. 5 or sloping legs Eqs 20 and 11. It has also been shown how the formulae may be used for towers with sections with both sloping and parallel legs. The formulae have been found to produce accurate results.

Acknowledgement - All tests were carried out in Taiyuan Heavy Machinery Institute, P. R. of China and special thanks are given to Professor Kejin Xu of the Institute for his help.

REFERENCES

1. COATES R. C., COUTIE M. G. and KONG F. K. Structural analysis. William Clowes & Sons, London, 1975.
2. CAO J. J. Static analyses of lattice towers. MSc thesis, Taiyuan Heavy Machinery Institute, Taiyuan, P. R. of China, 1984.

82. Transient dynamic analysis of hyperbolic cooling towers

P. RERICHA, Czech Technical University in Prague, Czech Republic

SYNOPSIS. Three case studies of transient dynamic loading of a reinforced concrete hyperbolic cooling tower are described. The main goal is to sort out appropriate analytical models for the respective loading types - blast wave, local impact and earthquake. The models should enable the assessment of the ultimate load carrying capacity of the structure and therefore geometric and material non-linear behaviour up to failure has to be considered. A proprietary finite element code is utilised in the research.

DYNAMIC SOLUTION CONCEPTS AND MATERIAL LAWS

1. The computer code, (ref. 1), employs explicit time integration and co-rotational concept of the geometric non-linearity, (ref. 2). Triangular shell finite elements are similar to those of ref. 3 but moderate constant curvature is taken into account. The curvature must comply with the shallow shell geometric assumptions within an element. The initial stress may affect geometric stability and stress-strain relations, all dynamic analyses reported herein therefore were carried out on the structure previously stressed by dead load.

2. Adequate stress-strain relations for non-linear reinforced concrete shells and plates are necessarily based on the layer model of the cross-section since available integral constitutive equations do not represent the coupled bending and membrane actions with sufficient generality. The layer model adopted in the code employs the Mindlin-Reissner kinematic hypothesis. However, the presence of the transverse shear strain is rather confusing in layer models because conventional layers assume biaxial stress and strain states. In the present model, the Mindlin transverse shear strain is associated with the stress pattern of the limit shear strength theory of reinforced concrete cross-sections to form a consistent bending shear deformation hypothesis. It works well up to ultimate failure of the cross-section, (ref. 4).

3. Triaxial stress-strain states arise in layers which impairs computational efficiency. Fortunately, substantial transverse shear is confined to relatively small areas around concentrated loads and supports. An appropriate threshold shear strain level is set to trigger the more demanding bending-shear model where

Space structures 4. Thomas Telford, London, 1993

necessary, while the conventional biaxial stress layers are used elsewhere. Layers degenerate to columns in beams but otherwise similar concepts may be utilised.

4. The material law of concrete need not be very accurate since reinforcement extension rather than concrete compression governs the deformation of the cross section after the concrete has cracked in the tension regions. Nevertheless, unloading and strain softening has to be accounted for if the field of application is dynamics. Unloading in concrete is somewhat confusing since the smeared cracks concept introduces principal differences between unloading in tension and compression domains. In tension, cracks close without stress change in unloading. The present material law employs a direct extension of the deformation theory of plasticity in tension whereas the flow theory is adopted in compression. Strain softening is isotropic and independent for tension and compression. The plasticity condition is that of the maximum principal stresses.

5. Reinforcing steel may be either elastic-plastic or elastic-viscoplastic. In this paper, the viscoplastic option is not used owing to lack of the material data.

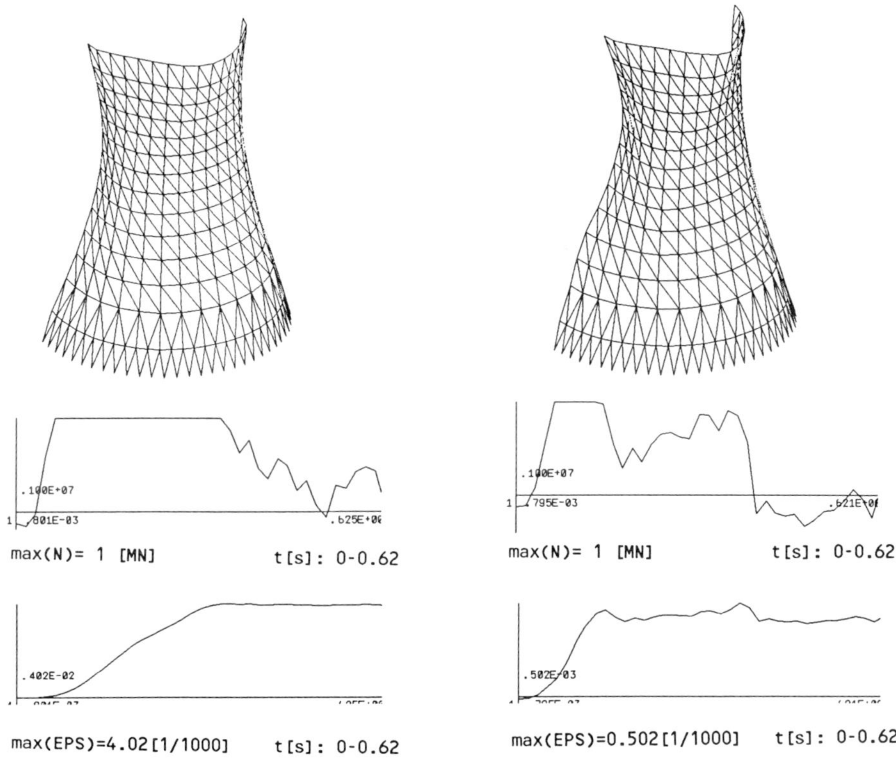

Fig. 1. Blast wave effect. Left - elastic shell, right - layer model. Deformed shapes at time 0.62s, max. displacement 0.34m. Time histories of the axial force (upper graphs) and strain in the leg facing the blast wave.

BLAST WAVE

6. The distribution of the air pressure on the tower surface is complex in time and space and not well known. Instead, a very simple substitute is used. The air pressure affects just the face surface of the tower. It is constant in the height direction and decreases with the square cosine of the angle between parallel and perpendicular radii. In time it is a triangular pulse with zero rise time and 0.33s downfall time. This idealisation may simulate the blast wave from an external gas cloud explosion, (ref. 4).

7. The legs obviously are apt to fail first. It is then a tempting idea to simplify the structure to an elastic or even rigid shell supported by layered elastic-plastic failing legs. The main goal of the present analysis is to validate or reject this proposal.

8. In Fig. 1 the elastic shell and layer shell solutions are compared. The peak air pressure was $10kN/m^2$. This load level is well below the ultimate bearing capacity. Cracking of concrete is widespread in the layer model but no substantial plastic yielding of the reinforcement has yet occurred. In spite of this, softening of the shell (compare the deformed shapes) implies significantly lower strains in legs even at this load level. At higher load levels the effect is more pronounced. We therefore conclude that the above simplification is not justified. Although the legs fail first their loading is greatly influenced by the actual shell stiffness and the shell cannot be deemed elastic or rigid. Axial strain and force are decisive for the failure of the legs since their bending stiffness is exhausted in the early phase of the deformation.

LOCAL IMPACT

9. The overall response to a local impact certainly is affected by the effective stiffness of the impact area. While an elastic shell obviously is not an adequate idealisation, the question may arise as to whether the conventional layer model is sufficient. Horizontal normal impact of the mass 1t and velocity 100m/s was simulated by both conventional and bending-shear layer models. Ideal plastic contact was assumed upon the area of $25m^2$ and the structure was loaded by the resulting initial velocity. The loading may represent a light aircraft crash or a similar event.

10. The deformed shapes at two times are shown in Fig. 2 for both shell models considered. The displacement scales are identical. The shapes differ significantly which indicates the insufficiency of the conventional layer model in these circumstances. The pronounced rebound of the impact point in the bending-shear model is due to the fact that concrete has been crushed in the early phase of the event. The compression crushing occurs in the immediate vicinity of the impact, where the transverse shear and bending strains combine to the negative principal strain, reaching beyond the softening threshold strain. Afterwards, the corresponding elements are nearly loose.

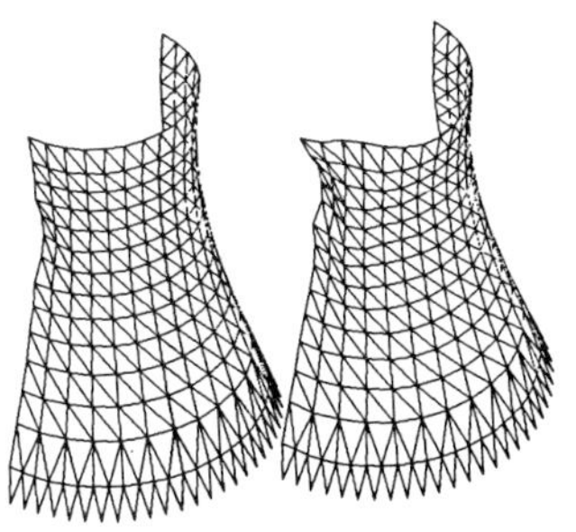

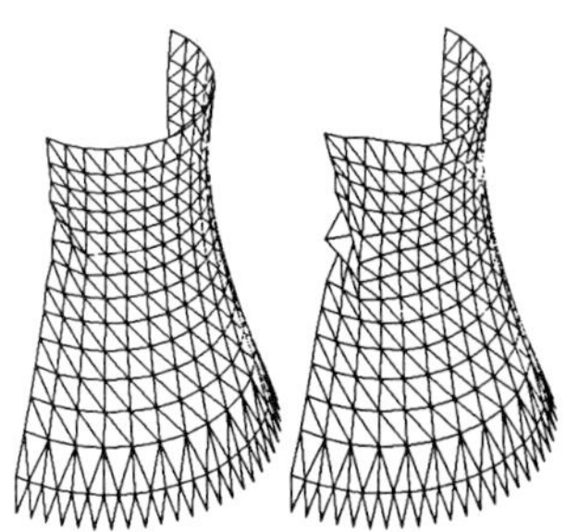

Fig. 2. Deformed shapes at 0.016s and 0.105s after impact. Left - conventional layer model, right - bending-shear layer model. Maximum displacement 0.090m occurs in the latter model at the node above the impact.

11. The displacement and damage spread wide in the conventional layer model whereas they remain contained around the impact area in the bending-shear model. Due to economic reasons, the mesh is regular in the impact area although local loading normally would require a finer mesh. Actual stresses and strains in this region certainly differ from those found analytically. Nevertheless, it is believed that the local stiffness is closely approximated and the strains and damage incurred outside the impact region are correct.

EARTHQUAKE

12. In the third case we are concerned with a computational model for earthquake loading. Usually real ground motion records are used as the kinematic excitation. Here an actual quantitative evaluation is not the goal, thus harmonic horizontal base motion is used instead. It is well defined and reproducible. The numerical experiments should also reveal to what extent the analytical model permits plastic shakedown, and explore the feasibility of the explicit time integration schemes to plastic shakedown analysis. Harmonic loading well suits these purposes. The frequency 3cps and amplitude 0.06m correspond to the maximum of the Czechoslovak design response spectrum.

13. Once again, elastic shell and layered shell models are compared. The legs assume always the same layer (column) model. A dramatic decrease of the axial strain in legs is seen in Fig. 3 in the layer shell model. Qualitatively different deformed shapes are observed. The elastic shell apparently fails to reflect the basic features of the structure. The deformed shapes indicate that the bottom edge of the shell is too stiff in the elastic shell model.

14. Both models seem to absorb the energy and shake down successfully. A serious assessment would require low cycle fatigue analysis in case of an actual design. Unrealistic fluctuations of the strain in legs nevertheless seem to disqualify the elastic shell model for this purpose.

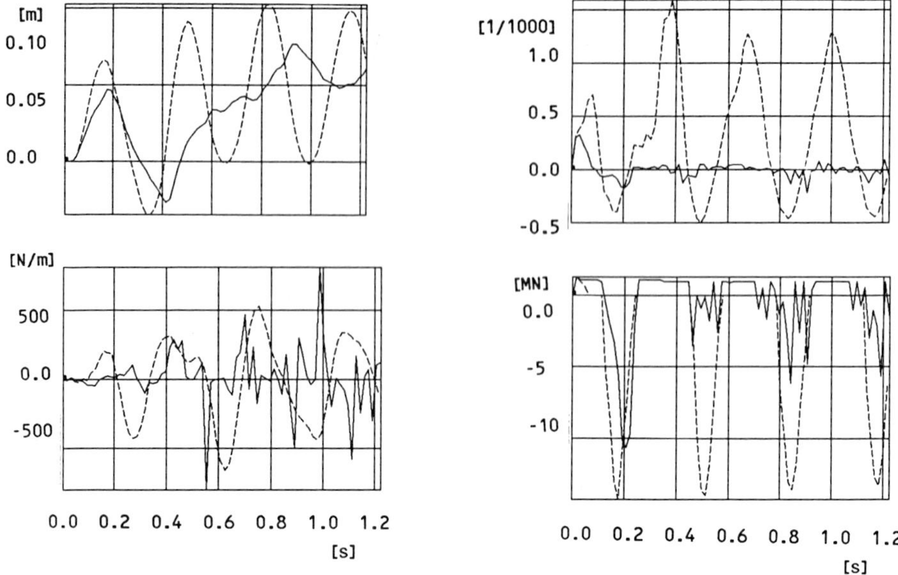

Fig. 3. Harmonic ground motion. Full lines denote the shell layer model, dotted lines the elastic shell model. Top left - horizontal deflection at the neck, right axial strain in the leg on the plane of symmetry. Bottom left - normal vertical membrane force at the shell neck, right - axial force in the leg.

15. The layer shell model took approximately 300s to run on a RS6000 model 320 workstation. This type of earthquake analysis obviously becomes acceptable even for the routine design computations, and similar conclusions can be drawn regarding the explicit time integration.

16. In one of the variant computations, integral elastic-plastic (flow theory) constitutive equations were used in legs. The conjugate pairs of the cross-section variables were considered independent. The results were rather amusing. The plastic extension in legs built up in the cyclic loading so that the tower steadily 'climbed up' on the extending legs. This curious behaviour testifies to the necessity of a proper unloading rule in earthquake analyses.

CONCLUSIONS

17. The elastic shell on elastic-plastic layered legs proves to be an insufficient model in transient dynamic loading conditions. Correct stiffness of the shell is necessary in order to get correct stresses in the legs. The layer shell model appears to be unavoidable in this context.

18. Transient dynamic analysis by explicit time integration schemes is a promising tool in plastic shakedown assessment and, consequently, in earthquake engineering.

19. The support of the IBM Academic Initiative CSFR is acknowledged.

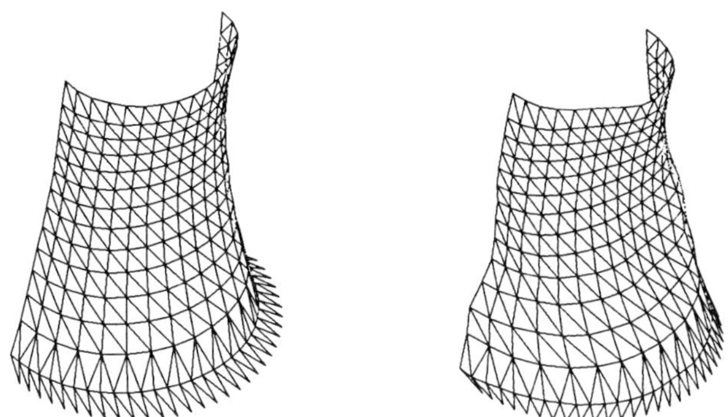

Fig. 4. Harmonic ground motion. The deformed shapes at time 1.1s. Left - elastic shell model, right - layer shell model.

REFERENCES

1. RERICHA, P. Fast transient analysis of non-linear structures. Acta polytechnica CTU Prague 32(1992), pp.41-47.

2. KENNEDY, J.M., BELYTSCHKO, T.and LIU J.I. Recent developments in explicit finite element techniques and their application to reactor structures. Nuclear Engng Design 97 (1986), pp.1-24.

3. BELYTSCHKO, T., STOLARSKI, H.and CARPENTER, N. A C0 triangular element with one point quadrature. Int. J. Numer. Meth. Engng 26(1984), pp.787-802.

4. RERICHA, P. Layer model of the bending-shear failure in RC beams and plates. J. Struct. Engng 117 (1991), pp.2865-2883.

83. Development of a self-deployable structural element for space truss applications

D. NEOGI and C. D. DOUGLAS, University of Massachusetts, USA

SYNOPSIS. This paper presents the progress and preliminary results of an on-going development program for a Self Deployable Structural Element (SDSE) for space truss applications. Deployable space structures can either be fabricated and assembled on earth and subsequently deployed in space or assembled in orbit with raw materials transported to the site by a launch vehicle. It is beneficial to pack and stow a collapsed structure in a minimum volume configuration. Ideally, this structure will deploy at the site without human intervention. Such a structure is flexible in its unheated state, allowing it to be folded and compactly stowed in the space vehicle. It is formed of a core of thermally-activated expanding foam or pressurising agent, an internal bladder, a load-carrying member of braided advanced composite material, and an outer retaining jacket. The core material, upon heating with a resistance wire, internally pressurises the structural element which leads to deployment. The same heat source also cures the advanced composite material.

BACKGROUND

1. Structures which acquire their final mechanical configuration after a chemical reaction can be used advantageously to support a wide range of functions. They are generally deployed from their stowed configuration by inflation, then cured by heating using sources like batteries, solar energy or exothermic chemical reactions.

2. Efforts are being made to minimise transport costs and human intervention during assembly of these structures. In the recent past, all efforts in these fields have had a concentrated goal to improve the strength and stiffness per unit deployed mass, small total mass and stowed volume, high interface adaptability, high deployment length, high number of possible deployment / retraction cycles, good pointing accuracy, and above all reliability.

Contemporary Space Structures

3. Currently, there are two types of deployable space structures. The first type of structures has components fabricated and assembled on earth and subsequently

deployed in space. The second type are advanced composite structures that are assembled in orbit with raw materials transported to the site by a launch vehicle. This approach is cumbersome and requires the transportation of the processing equipment for candidate space composite materials. Wilson and MacConochie (ref. 1), at NASA Langley Research Centre, propose on orbit post-fabrication of structures using pultrusion methods. The machine itself would require a space of 3m length and 1 m width to be packed in the space shuttle cargo bay which is only 4.5m in diameter.

4. Structures fabricated on earth employ two types of deployment systems. The first system relies on mechanical strain energy stored in the system. These structures deploy by release of boundary conditions or end constraints, allowing the structure to seek a minimum potential energy configuration. The advantage of this approach is that joint and hinge type movement is minimised since most of the deployment motion occurs within the structure itself. Here the ability to store potential energy in the structure is governed by the modulus and strength of the material and therefore is limited. Further, the stored configuration possesses high internal loads resulting in creep or stress relaxation which could cause a deviation from the desired deployed configuration. Examples are the development of the Astromast by the Astro Research Corporation, and the CTM biconvex mast developed by ESTEC, in the Netherlands (ref. 2).

5. The second deployment system includes telescoping booms, folding plate designs, articulated arms, etc. Recent efforts include development of an in-step inflatable antenna experiment by Freeland and Bilyeu (ref. 3). Another effort involves an extendible and retractable mast (ref. 4) by Dornier. The design is based on a telescoping carbon fibre structure with high stiffness, strength and accuracy. Conventional scissor designs have been investigated for use in heat radiator applications (ref. 5) by Roy L. Cox et al. of Vought Corporation. Such other structures are a telescopic axial boom (ref. 6), the development of an offset reflector structure (ref. 7), and an offset unfurlable tracking antenna (ref. 8).

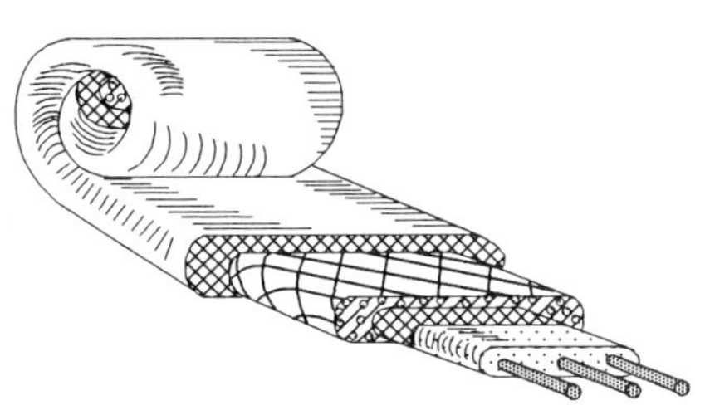

Fig. 1. Structural element in undeployed state.

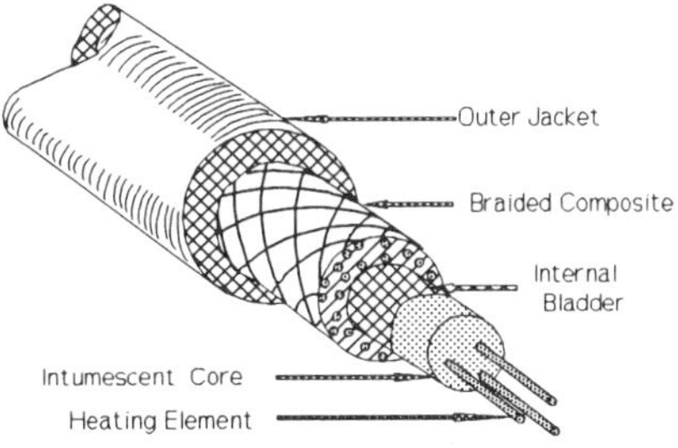

Fig. 2. Structural element in deployed state.

Summary of the Patent

6. This paper presents the results of the development program for a new US invention (Patent # 4,978,564 dated December 18,1990) (ref. 9). The self-deploying structural element is flexible in its unheated state, allowing it to be folded and compactly stowed in the space vehicle (Fig. 1). It is formed of a core of thermally-activated expanding foam or pressurising agent, an internal bladder, a load carrying member of braided advanced composite material, and an outer retaining jacket. Within the core material is a resistance heating wire. Passing current through the resistance heating wire provides the energy to internally pressurise the structural element which leads to the deployment of the structure, as shown in Fig. 2. The same heating element which deploys the structure also cures the pre-impregnated, braided, carbon/epoxy composite material. The cross-section of the structure in its stowed configuration is collapsed, or flat-oval and becomes circular after deployment.

COMPONENT MATERIAL SELECTION

7. Selection of the proper materials plays a key role in the functional performance of the SDSE. An experimental approach was used to select candidate materials for the SDSE. Theoretical parametric studies were performed to aid in the selection process, but will not be reported here. This section presents preliminary results of the investigations performed.

Braided Advanced Composite

8. NASA's space station design consists of a large rectangular truss structure to support the various modules and solar arrays. This structure will be constructed from tubular members. Use of carbon/epoxy composite tubes instead of aluminium tubes could save over 9000 kg in the weight of the structure (ref. 10). The lighter weight of the frame structure will require fewer space shuttle flights for earth-to-orbit transfer of the truss material. Carbon/epoxy tubes are also superior to aluminium tubes in strength, stiffness, thermal-fatigue cycling and dimensional stability. The coefficients of thermal expansion of carbon/epoxy composites are much lower than aluminium, and are suitable for space applications. A braided carbon/epoxy composite manufactured by Newport Composites (AS4/12K/NC-76/96 tow braided tube) was selected for its high longitudinal and torsional stiffness, and also high interlaminar strength. Braiding is superior to both tape-rolling and filament-winding in damage tolerance applications because the interweave in the braid limits the damage to a local area. A braided configuration also maintains the integrity of the structure in its uncured state. The composite tube is fabricated by fully impregnating and partially curing the carbon yarns, then braiding the tows to obtain the desired shape before the final cure.

Inner Bladder and Outer Jacket

9. The inner bladder is required to provide circularity to the composite as its cross-section changes from flat-oval to circular. The outer jacket acts as a

retaining member for the composite tube. The inner bladder and the outer jacket material have to withstand the pressures and temperatures at which the SDSE deploys and cures. After deployment, the internal pressure also consolidates the braided composite. One functional requirement is that the material will have to maintain flexibility and stability above 200 ^{o}C temperature and 0.14 MPa pressure. Very few materials were available off the shelf with the required characteristics. Silicone rubber tubes were selected due to the fact that they met all temperature, pressure, flexibility and weight requirements.

Intumescent Core

10. The core of the SDSE model needs to be flexible in its stowed configuration and upon heating to generate enough pressure to unfold the member before it cures. However, the material needs to be non-flammable. The materials that were considered include expandable foams, fire barrier caulks and blowing agents. The Uniroyal Chemical Company's azodicarbonamide blowing agent was selected for its lightweight and excellent decomposing properties to give the required pressure. Azodicarbonamide is a yellow crystalline solid which decomposes to produce a high yield of gas (220-240 cc per gram, STP). Unlike most other blowing agents, azodicarbonamide does not support combustion and is self extinguishing. A simple experiment was designed for pressure measurement, as shown in Fig. 3. The pressures measured were above the atmospheric pressure. The material to be tested is weighed and placed in the pressure vessel. The pressure vessel is an iron tube 0.3 m long and 0.05 m diameter with end caps and is attached to a manifold. The pressure vessel is then placed inside a batch oven and slowly heated. The temperatures are recorded by the three thermocouples. The pressure generation inside the system upon heating is measured by a pressure transducer and read by a multimeter in DC volts and later converted to MPa. A pressure relief valve is also provided in the event that the pressure generated inside the system is higher than the calibrated pressure for the pressure transducer.

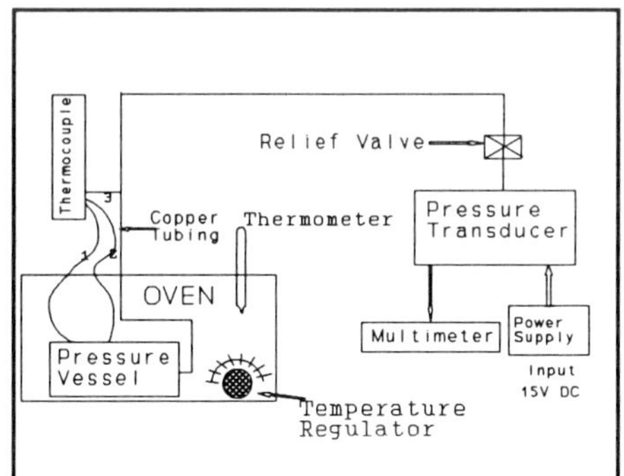

Fig. 3. Schematic diagram of the pressure measurement experiment set-up.

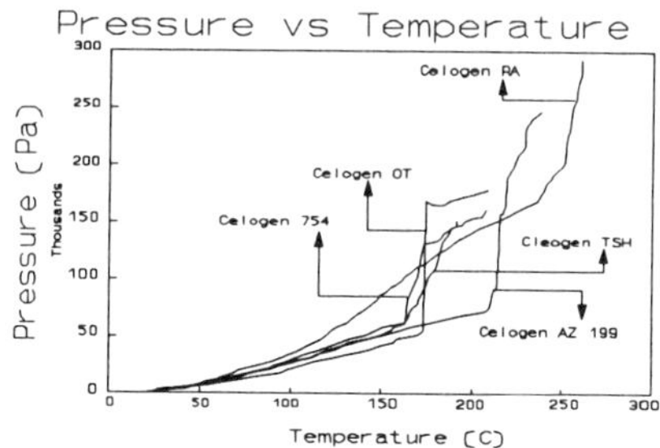

Fig. 4. Plot of temperature vs pressure for 2.03gm of different azodicarbonamides.

11. Fig. 4 presents the pressure generated by different azodicarbonamides as a function of temperature. For comparison purposes, all of the experiments were performed with 2.03 gm of the material. Of all the azodicarbonamides, Celogen OT (ref. 11) decomposed very quickly, causing a sudden increase in pressure. Other compounds resulted in higher pressure upon decomposition, but at much higher temperatures which could start premature curing of the carbon/epoxy composite. The behaviour of Celogen OT was appropriate for the SDSE and was selected for the prototype core material.

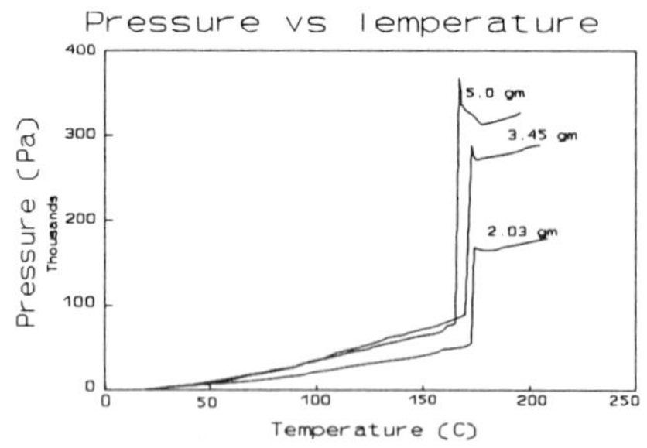

Fig. 5. Plot of pressure vs temperature for different masses of Celogen OT.

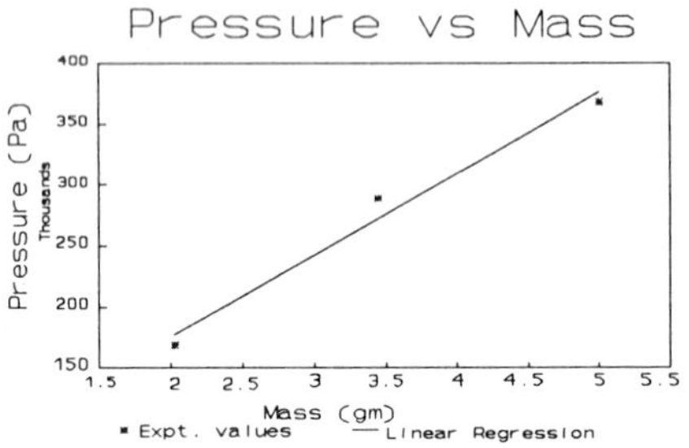

Fig. 6. Plot of pressure generated by different masses of Celogen OT.

12. The experimental results are presented in Table I. The different Celogens are classified according to the activator which governs the decomposition temperature. The approximate rate of increase of temperature maintained during the experiments was 6.5 °C/min. Table I presents the maximum pressures and maximum temperatures when the material decomposes.

Table I. Results of pressure measurement experiment on different azodicarbonamides.

Material	Mass (gm)	Maximum Pressure (KPa)	Maximum Temperature (°C)
Celogen OT	2.03	168.6	187
Celogen OT	3.45	288.8	173
Celogen OT	5.00	367.9	168
Celogen RA	2.03	292.3	260
Celogen TSH	2.03	154.9	201
Celogen 754	2.03	141.2	185
Celogen AZ 120	2.03	189.2	218
Celogen AZ 199	2.03	196.1	221

Resistance Heating Wire

13. The intumescent core needs to be thermally-activated by an internal heating element. A high resistance wire is an ideal choice. For the SDSE prototype model, nickel-chromium alloy wire was selected for its high resistance. Ni-Cr alloy composition is Ni 60 % and Cr 40 %, with resistance of 8.76 Ω/m. The wire is insulated using a flexible ceramic braid.

PROTOTYPE DEVELOPMENT

14. The SDSE prototype models were developed after a comprehensive study to select materials which would meet the specific requirements. Three experimental SDSE prototypes were developed. All prototypes deployed successfully.

Assembly of a SDSE Model

15. The composite tube has to be assembled layer by layer; 3.81mm thick silicone rubber is wrapped on a 27.0mm aluminium mandrel to form a tube with an internal diameter of 29.0mm. To form the silicone tube from a silicone sheet, the GE silicone rubber caulk (48 hour cure) is used, which gives a sealed bond at temperatures as high as 450 °C. The first layer forms the internal bladder of the composite tube. The approximate length of the tubes are shown in Table II. Over the internal bladder, the prepreg carbon/epoxy braided tube of internal diameter of 29.8mm is assembled. The last layer is the outer jacket, which is applied over the braided tube. The outer jacket is also fabricated from silicone rubber and assembled in a similar manner as the internal bladder, with internal diameter of 31.8mm. After all the layers have been assembled the mandrel is removed. A flexible lightweight composite tube is now ready for end caps.

A Novel Sealing Concept

16. After the tube is removed from the mandrel, it needs to be folded and capped or sealed. The cross-section of the end cap is shown in Fig. 7. The caps are moulded from room-temperature cured epoxy. The cap B has two electrodes, embedded during moulding, for the Ni-Cr internal resistance wire. The Ni-Cr wire enters and exits from the same end forming a closed loop circuit. The electrode has very low resistance compared to the Ni-Cr wire. The temperature of the Ni-Cr wire may be 400 °C but the electrode is so designed with a lower resistance that its temperature does not exceed 50 °C. This prevents the cap from decomposing. During the deployment experiment, cap B is fixed to a frame while cap A is on the end of the tube that unfolds. For assembly of the caps, the inner moulded cap 1 is first placed inside the composite tube. The inner moulded cap and the composite is then placed in a pot and epoxied, thus forming the moulded cap 2. The cap 2 integrally bonds the composite tube with the moulded cap 1, giving a pressure tight seal.

Maximising the Stowability

17. To obtain the minimum stowed volume, both the end caps are not sealed at the same time. Cap A is first sealed to one end of the composite tube. Next, the

insulated Ni-Cr wire is inserted into the composite tube. The tube is next flattened so that its cross-section is a flat-oval instead of its previous circular section. The flat-oval section makes it very easy to fold and achieve low-stowed volume of the composite tube. During the folding process of the tube, the blowing agent is added to the tube in measured quantities so that it is distributed uniformly throughout the length of the tube. After the entire tube is folded, cap B is assembled. Table II presents the measured characteristics of the tubes before deployment.

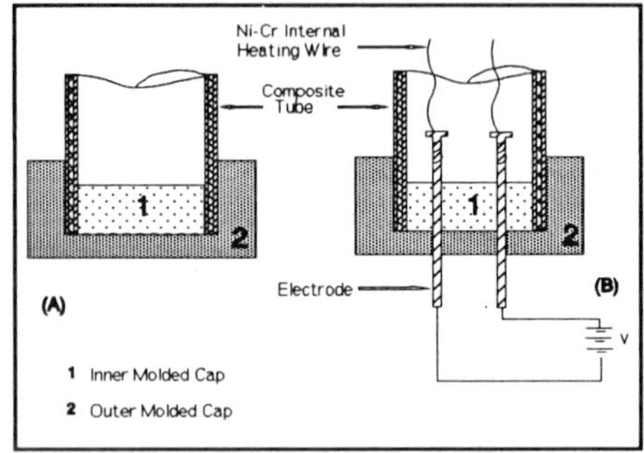

Fig. 7. Cross section of the end caps:
A is without electrodes, B is with electrodes.

Table II. Parameters of SDSE Models measured before deployment.

Model Number	Length (m)	Weight (kg)	Blowing Agent (gm)	System Resistance (Ω)
SD 004	0.70	0.252	4.00	11.7
SD 005	0.70	0.242	4.00	15.6
SD 006	0.70	0.255	4.25	12.4

EXPERIMENTAL RESULTS

18. All three SDSE prototypes deployed successfully. The following presents a summary of the deployment experiments.

SDSE Deployment Study

19. The experiments were recorded on video and then analysed frame by frame using an image analyser. Fig. 8 presents six frames acquired from the image analyser at various times during the deployment. As can be seen the SDSE is initially in a tightly folded configuration and then eventually deploys to the form of a straight load carrying member, once cured.

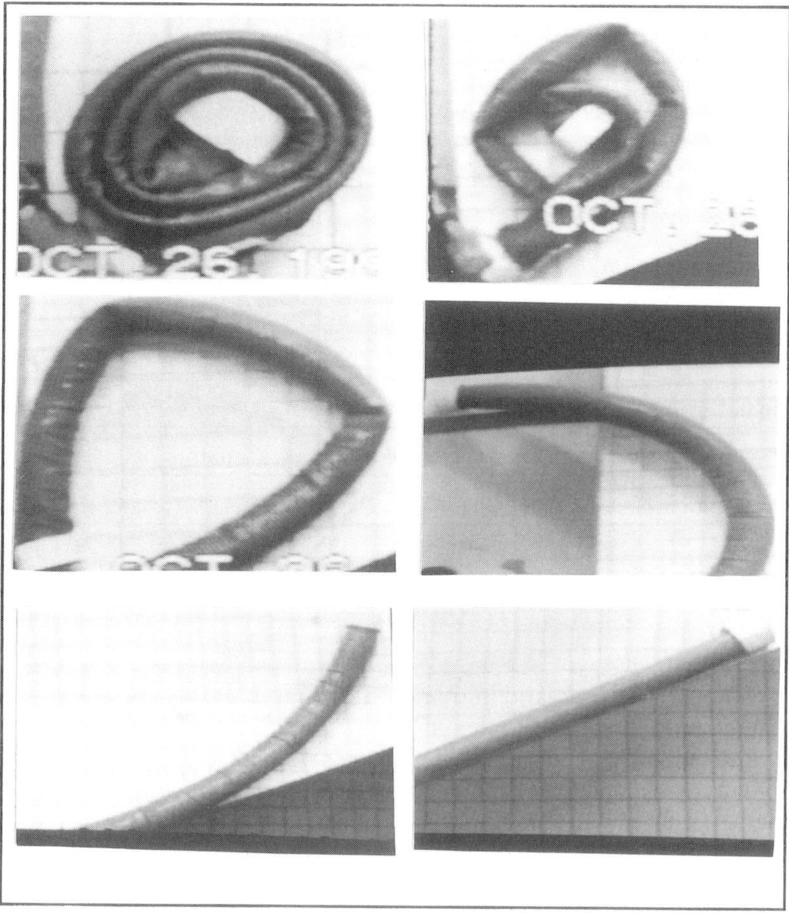

Fig. 8. Deployment process at different times for SD005.
Time to deploy was 25 seconds.

Physical Properties

20. The density was calculated and the apparent stiffness in the longitudinal direction of the cured braided composite was experimentally evaluated using an Instron testing machine for a SDSE model and compared with an aluminium tube of the same length, diameter and wall thickness.

$$\left(\frac{E_{xx}}{\rho}\right)_{composite} = 8.46 \text{ x } 10^6 \ m^2/s^2, \quad \left(\frac{E}{\rho}\right)_{aluminium} = 25.4 \text{ x } 10^6 m^2/s^2$$

Measured Parameters

21. Table III presents some measured experimental results before, during and after deployment.

Table III. Measured experimental results for three SDSE models.

	SD 004	SD 005	SD 006
Stowed Volume (m^3)	3.392 E-04	2.95 E-04	3.327 E-04
Deployed Volume (m^3)	5.745 E-04	6.73 E-04	6.274 E-04
Volume Increase	70 %	128 %	88 %
Time at Deployment (min.)	52	52	60
Time to Deploy (sec.)	Not recorded	25	20
Time to Deploy and Cure (min.)	390	400	390
Average Power (Watt)	43.33	44.76	50.06
Total Energy (Watt hr)	436.97	446.13	659.38
$\Delta V/\Delta t$ Before Cure (Volt/min)	0.264	0.377	0.332

22. In Table III, the time at deployment (min.), represents the actual time that had elapsed from the start of the experiment. The time to deploy (sec.), was the time that the folded SDSE prototype model took to unfold after the blowing agent had started to decompose and cause a sudden increase in the internal pressure.

23. The stowed volume was the approximate volume that the model occupied when folded up. This volume is different from the internal volume of the collapsed and folded tube. The cross-section of the folded tube is flat-oval and has an approximate cross-sectional area A_s = 1.532 E-04 m^2 and internal volume $V_s = A_s \times L$ = 1.07 E-04 m^3 (volume of the end caps have not been taken into account). Comparing this with the deployed circular cross-sectional area A_d = 7.92 E-04 m^2 and the deployed volume V_d = 5.542 E4 m^3, there is a 418% increase in the internal volume. The average power required for deployment and curing has been calculated to 65.75 watts/deployed meter.

CONCLUSION

24. The SDSE prototype developed is very flexible and can be compactly stowed in a limited space environment and can result in an 80 % volume saving. The specific stiffness measured on the SDSE models are lower than that of an aluminium tube, however the braided tubes used for the SDSE models did not have any unidirectional reinforcements, which can increase the stiffness substantially. In general, the use of carbon/epoxy braided composites will result in higher specific stiffness and lower CTE properties over aluminium, or other

metals, making it ideal for space applications. The success of the SDSE model will now lead to the development of more complex building modules as shown in Fig. 9, where more than one member will be linked to generate a triangle or a tetrahedron. The advantage of the SDSE proposed over the existing or ongoing projects is that it gives a much higher reliability at much lower cost. For example, the absence of any linkages eliminates the chance of any mechanical failure. Many other uses could be envisioned, such as emergency sailboat masts, deployable antennae, reflectors, tent frames, military communication applications, and any other field erectable structures.

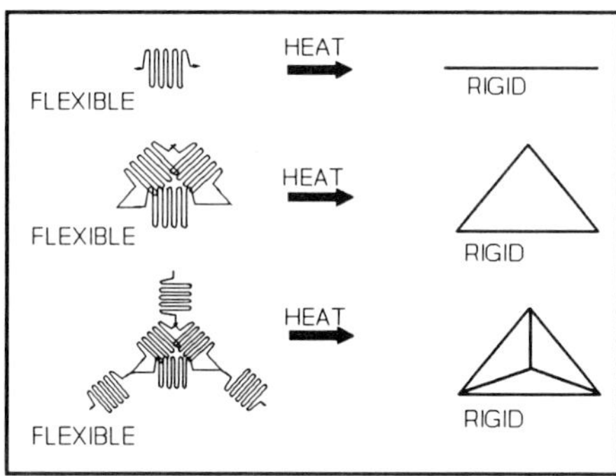

Fig. 9. The different building modules that can be assembled using SDSE members.

REFERENCES

1. WILSON, M.L., MacCONOCHIE, I.O. and JOHNSON, G.S.,"On-orbit fabrication of large space structures using pre-impregnated graphite tapes and a pultrusion process ", 43rd Annual Conference, Composites Institute, The Society of Plastics Industry, Feb. 1988

2. AGUIRRE, M., "Development in deployable masts to support flexible solar arrays", Proc. 4th European Symposium, Cannes, September 18-20, 1984

3. FREELAND, R.E., "In-step inflatable antenna experiment", 43rd Congress of the International Astronautical Federation, 1992

4. SCHMID, M. and AGUIRRE, M., "Extendible retractable telescopic mast for deployable structures", 20th Aerospace Mechanisms Symposium, May 1986

5. COX, R.L., et al., "Deployable radiators for ... payloads", Raumfahrtforschung Heft, May 1976

6. FELKAI, R., "Design and development of a telescopic axial boom", Proc. 20th Aerospace Mechanisms Symposium, May 1986

7. BERNASCONI, M.C., "Development of a 2.8m offset antenna reflector using inflatable, space-rigidised structures technology", Proc. 2nd ESA Workshop on Mech. Tech. for Antennas, 1986

8. LABRUYERE, G., PASSERON, L. and PASTORINO, M., "Mechanical and control study of a 7 m offset unfurlable tracking antenna ", 2nd ESA Workshop on Antennas, 1986
9. DOUGLAS, C.D., "Self deploying structural element", U.S. Patent Document No.4,978,564, dated December 18, 1990
10. MUNJAL, A.K., et al., "Design and fabrication of high quality graphite/epoxy braided composite tubes for space structures", 35th Int. SAMPE Symposium, 1990, p 1954-1967
11. "Technology of Celogen blowing agents", Uniroyal Chemical Company Inc., USA

84. Foldable ring structures

S. PELLEGRINO and Z. YOU, University of Cambridge, UK

SYNOPSIS. The general conditions for strain-free folding of symmetric ring structures are discussed and, after giving two simple examples, three ring-shaped pantographs are presented. For each pantograph the special geometries which permit strain-free folding are obtained, and the implications of finite joint size are considered.

INTRODUCTION

1. Foldable structures have many applications. In Space, typical uses include foldable solar arrays, booms, and antennas. This paper is concerned with foldable ring structures to support thin-membrane or wire-mesh reflectors, used in deployable antennas. Despite considerable developments over the past twenty years (refs 1-3), there is still a fairly urgent need for lightweight deployable antennas of about 8-10m diameter, particularly in Europe.

2. Figure 1 shows a pantographic deployable ring structure which we have developed as the edge beam of a deployable antenna. The figure is based on a model with outer and inner diameters, when deployed, of 3.2 and 1.9m respectively. The ring has twelve sides and consists of forty-eight pairs of rods. The rods in each pair are free to rotate relative to one another in their own plane and all other motions are prevented by an intermediate connector. The rods are connected by pin connectors to twenty-four identical inner joint assemblies and to twenty-four outer joint assemblies, also identical. Following an approach to foldable structures introduced by Kwan, You and Pellegrino (ref. 4), this deployable ring structure is completed by forty-two passive cables and two active cables. These elements are not shown in Fig. 1 and will not be discussed in this paper. The geometry of the rod assembly is such that it can be folded compactly, with all rods in a bundle, and then deployed synchronously until the rods in each pair become orthogonal. In fact, were it not for the passive cables, which become taut at this point, deployment could continue until the whole assembly becomes flat. The deployment process is entirely strain free; the structure is a large-displacement, highly redundant, single-degree-of-freedom mechanism. Only a small amount of power would be required for deployment in a zero-gravity environment, provided friction at the joints is low.

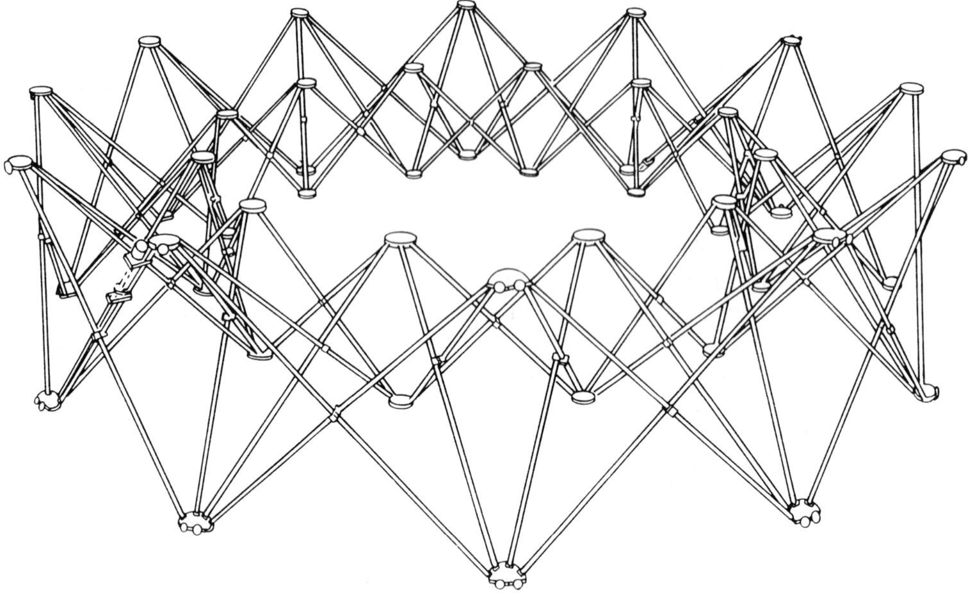

Fig. 1. Deployable ring structure based on Concept (b), with n=12. See paragraph 12.

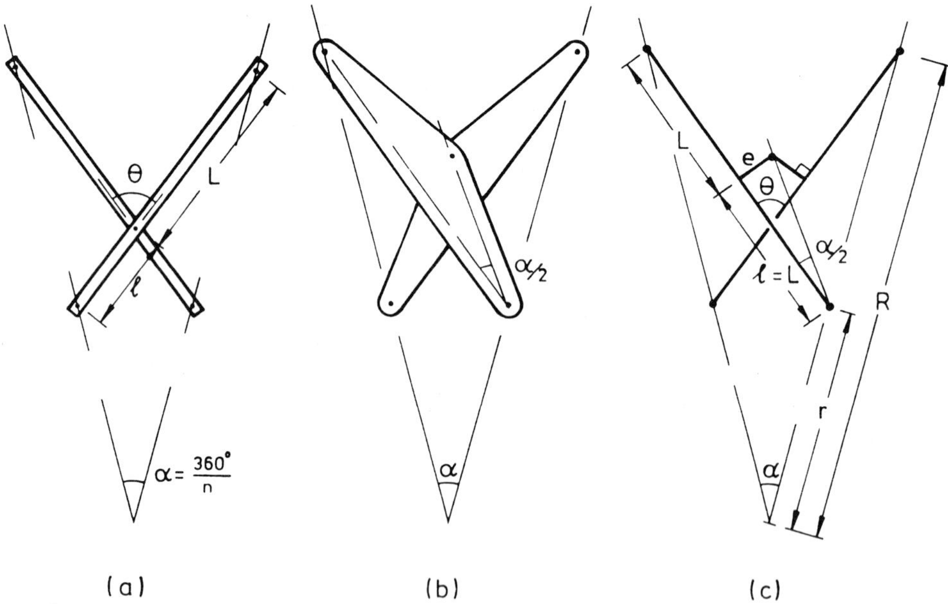

Fig. 2. Basic modules of two-dimensional pantographic rings which (a) cannot be folded and (b, c) can be folded without strain.

3. The layout of the paper is as follows. First, we discuss the general conditions for strain-free folding of symmetric ring structures and apply them to two simple examples. Then we present three pantographs which can form n-sided rings, where n can be any integer number. For each pantograph we obtain the special geometry which permits strain-free folding and also consider the implications of finite joint size. Finally, we compare the solutions obtained.

FOLDABLE RINGS: TWO SIMPLE EXAMPLES

4. A natural way of approaching three-dimensional foldable ring structures is through the two-dimensional solutions due to Hoberman (ref. 5). We begin by noting that it is always possible to make a flat, n-sided pantographic ring out of n identical pairs of rods joined by shear connectors (sometimes called scissor hinges). These connectors are not in the middle of the rods or a 'straight' pantograph would be obtained. Rings of this type, though, are always *rigid in the plane*. This follows from Zanardo (ref. 6), but can be shown also by noting that in the pair of rods of Fig. 2(a) there is a one-to-one correspondence between the deployment angle θ and the subtended angle α. For n-fold rotationally symmetric rings:

$$\alpha = \frac{360^{\circ}}{n} \tag{1}$$

hence θ is also fixed and therefore folding requires changes of length, or bending, of some rods.

5. This difficulty can be resolved (ref. 5) by making the rods in each pair a little wider and by offsetting the shear connector from the radial line between the two end connectors, Fig. 2(b). The schematic diagram in Fig. 2(c) defines e, the extra parameter introduced by this approach. It can be readily shown that:

$$\tan\frac{\alpha}{2} = \frac{L-l}{L+l}\tan\frac{\theta}{2} + \frac{2e}{L+l} \tag{2}$$

For $L = l$ the first term on the right hand side vanishes and hence α becomes a constant for all θ's. Therefore, strain-free folding of an n-sided ring based on the unit shown in Fig. 2(b) requires that the following two conditions be satisfied, in addition to Eq. 1.

$$L = l$$

$$\alpha = 2\arctan\frac{e}{l} \tag{3}$$

This idea can be extended to rings lying on conical surfaces from which deployable domes can be formed (ref. 5).

6. This two-dimensional foldable ring is unsuitable for the edge beam of a deployable antenna because it does not pack compactly and its out-of-plane stiffness is too small, therefore three-dimensional ring structures are considered next. The simplest one is the hexagonal pantograph shown in plan view in Fig. 3. Note that here n is always equal to six and hence $\alpha = 60^\circ$. The basic module is a triangular prism formed by three identical pairs of rods joined in the middle, as in Fig. 4(a). A triangular backbone for a deployable mast has been obtained by stacking these modules on top of each other (ref. 4). Here, though, they are connected side by side with the number of modules in each 60° sector given by 3 + 2i, where i is any integer number. All hexagonal rings of this type can be folded without strain because any number of rod pairs arranged as above fit together, provided θ is the same for all of them. If θ decreases, starting from 90°, all lengths in the plan view of Fig. 3 become shorter precisely by the same amount. Of course, there is a corresponding increase in the height of all rod pairs. In this three-dimensional structure, and also in the other ones presented later on, all pairs of rods lie in planes orthogonal to the mid-plane. The two-dimensional pantograph described in paragraph 5 is the only known foldable ring structure where all rods lie in this plane.

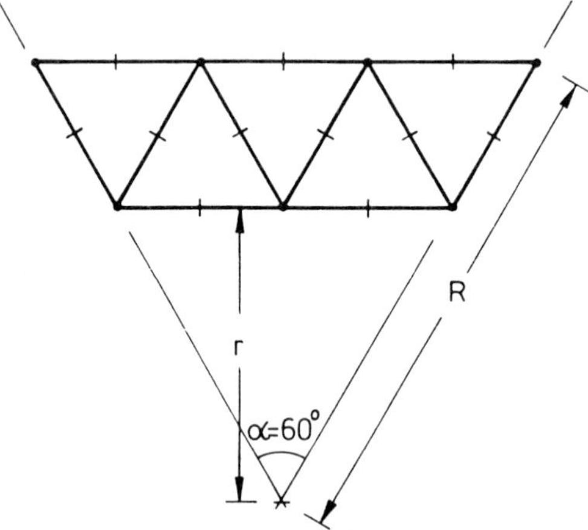

Fig. 3. 60° sector of three-dimensional hexagonal pantograph (plan view) consisting of identical pairs of rods joined in the middle. The projection of each rod pair is shown by a solid line and the shear connector by a short segment perpendicular to the rods.

7. The hexagonal pantograph provides an elegant and fairly intuitive example of a three-dimensional structure that satisfies all conditions stated in paragraph 5, but it has some potential drawbacks. First, it cannot approximate closely to a circle. Second, R/r is restricted to the range $4/\sqrt{3}$, with only six rod pairs on the inside face of the ring, to $2/\sqrt{3}$ in the limit for an infinite number of rod pairs. Third, it requires three different joint assemblies, at least, because the number of rods

connected to a joint varies from three to five. All ring structures described in the next section do not have these drawbacks.

THREE-DIMENSIONAL PANTOGRAPHIC STRUCTURES

8. Fig. 4 shows the three types of rod pairs which make up all pantographs presented in this section. Two pairs of rods, Fig. 4(a, c), have connecting pins mid-way between the end connectors and, in fact, the only difference between them is that the total rod lengths are $2l$ and 2L, respectively. In the third pair of rods, Fig. 4(b), the connector is at a distance l from one end, but at a distance L from the other end.

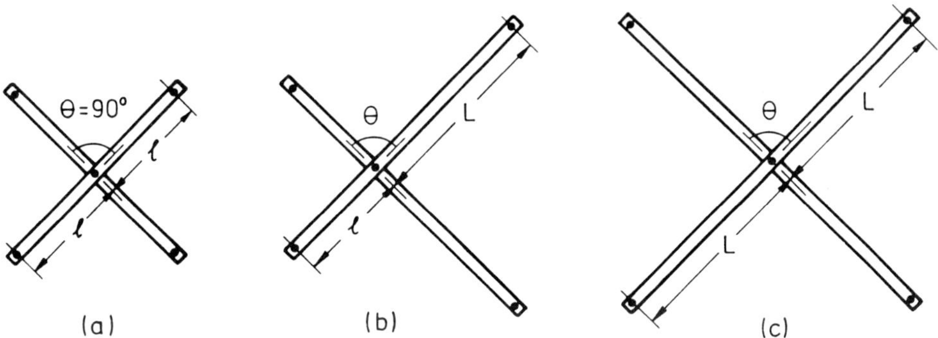

Fig. 4. Three types of rod pairs to make three-dimensional foldable pantographs.

9. A simple two-dimensional pantograph can be made by connecting these pairs of rods in the sequence (a), (b), (c). This pantograph can be folded without any strain. In particular, when $\theta = 0^{\circ}$ the right-hand-side end connectors of pair (a) are $2l$ apart and fit with the corresponding left-hand-side connectors of pair (b), also $2l$ apart; the kinematic condition imposed by the four-bar-chain is satisfied. Similarly, the right-hand-side connectors of pair (b) fit with the left-hand-side connectors of pair (c). These considerations are also valid if the three pairs of rods are not in the same plane, which is the case for three-dimensional pantographs. Also, they can be extended to pantographs with any number of rod pairs and also to ring-shaped pantographs, where the first and last rod pairs are connected. The key folding condition is that the semi-lengths of the rods included in any four-bar-chain must all be equal to either l or L. If this condition is satisfied, then (i) the whole pantograph can be folded without strain and (ii) the angle between rods in the same pair is θ everywhere in the pantograph. This folding condition is a special, almost trivial case of a general one derived by Escrig (ref. 7).

Concept (a)

10. This is the simplest of the three-dimensional pantographs because it consists of n (≥ 3) identical modules. Figure 5(a) is a plan view of the basic module; rod

pairs on faces common to adjacent modules are not repeated. AB is a pair of rods of length $2l$, Fig. 4(a); BC and AD are pairs of rods of length l + L, Fig. 4(b), with the semi-length l on the inside of the ring; CD is a pair of rods of length 2L, Fig. 4(c). This trapezoidal module satisfies the folding condition of paragraph 9 and hence it can be folded without strain. To put together a closed chain of n identical units there is a further geometric condition, to ensure that the angle α subtended by each unit satisfies Eq. 1. Therefore:

$$\overline{CD} = \overline{AB} + 2\,\overline{AD}\sin\frac{\alpha}{2} \tag{4}$$

and, expressing $\overline{CD}$, etc., in terms of rod lengths and deployment angle

$$2L\sin\frac{\theta}{2} = 2l\sin\frac{\theta}{2} + 2(l+L)\sin\frac{\theta}{2}\sin\frac{\alpha}{2} \tag{5}$$

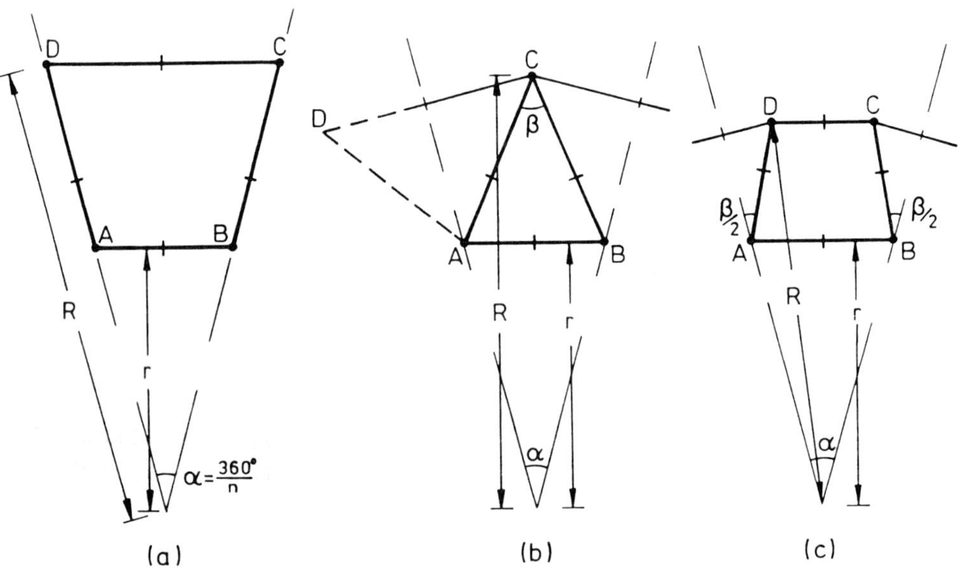

Fig. 5. Basic modules (plan view) of Concepts (a), (b) and (c).
See caption of Fig. 3.

From which:

$$\frac{L}{l} = \frac{1+\sin\frac{\alpha}{2}}{1-\sin\frac{\alpha}{2}} \tag{6}$$

The radii R and r respectively of the circumscribed and inscribed circles defined in Fig. 5(a) are related by:

$$\frac{R}{r} = \frac{\overline{CD}}{\overline{AB}\cos\frac{\alpha}{2}} = \frac{L}{l\cos\frac{\alpha}{2}} \tag{7}$$

Concept (b)

11. It consists of 2n triangular modules. In plan view, Fig. 5(b), the modules form a chain of isosceles triangles with alternate long bases, e.g. CD, corresponding to a pair of rods of length 2L, and short bases, e.g. AB, corresponding to a pair of rods of length $2l$. The angle β has a unique value, for a given n. All pairs of rods connecting the inner and outer faces of the ring are of length l + L. Both triangular modules satisfy the folding conditions of paragraph 9, hence all that remains to be done is to find the geometric condition to form a closed pantograph. Here:

$$\overline{AB} = 2\overline{AC}\sin\frac{\beta}{2} \quad \text{and} \quad \overline{CD} = 2\overline{AC}\sin\frac{\alpha+\beta}{2} \tag{8}$$

from which:

$$\frac{L}{l} = \frac{\cos\frac{\alpha}{2} + \sin\frac{\alpha}{2}\left(\sin\frac{\alpha}{2} + \sqrt{1+2\cos\frac{\alpha}{2}}\right)}{\cos^2\frac{\alpha}{2}} \tag{9}$$

and

$$\beta = 2\arcsin\frac{l}{l+L} \tag{10}$$

The ratio R/r is still given by Eq. 7.

Concept (c)

12. It consists of n trapezoidal modules and n triangular modules, arranged so that 2n pairs of rods of length 2L form the outside of the ring, while only n pairs of rods of length $2l$ form the inside face. In plan view, Fig. 5(c), this results in a

chain of alternate isosceles triangles and isosceles trapezia. Unlike the previous two concepts here in most cases, i.e. for $n > 6$, the shorter rods are on the outer face of the ring. For $n = 6$, $L = l$ and the trapezia become squares. It can be shown that:

$$\frac{L}{l} = \frac{\left(1+\sqrt{2}\sin\dfrac{\alpha}{4}\right)^2}{2\cos^2\dfrac{\alpha}{4}} \tag{11}$$

$$\beta = 2\arcsin\frac{L}{l+L} \tag{12}$$

and

$$\frac{R}{r} = 2\frac{\cos\dfrac{\alpha}{4}}{\cos\dfrac{\alpha}{2}}\frac{L}{l} \tag{13}$$

<u>Geometric design of joints</u>

13. The above derivations are based on the implicit assumption that the joints have no physical size. Of course, this is not realistic but our treatment would be perfectly valid if the axes of all connectors on the same joint assembly could be arranged to meet at a single point. Thus the behaviour of the real joint would be kinematically equivalent to a zero-sized joint located at this special point. This can be arranged at the outer joints of the three pantographs but is impossible at the inner joints.

14. The model structure shown in Fig. 1, based on Concept (b), has only two different types of joints: inner and outer joints. Top and bottom joints are identical because the structure is symmetric with respect to its mid-plane and also with respect to radial planes through the joints. In practice, each joint has been made as an assembly of two plates parallel to the mid-plane of the ring, with slots for the rods. The two plates are separated to insert the rod connectors and then fastened together.

15. For simplicity, in each joint type the axes of all connectors are at the same distance from the theoretical position of a zero-sized joint. Therefore all connectors on an external joint are offset by Λ from the theoretical joint, while all internal connectors are offset by λ. Note that the axis of each connector is, of course, still orthogonal to the rod axis. By extending the analysis in paragraphs 10-12 to pantographs with external joints of size Λ and internal joints of size λ it can be shown that, in addition to the geometric conditions already derived, strain-free folding requires:

$$\frac{\Lambda}{\lambda} = \frac{L}{l} \tag{14}$$

In Concept (a), for example, Eq. (5) becomes:

$$2\left(L\sin\frac{\theta}{2} + \Lambda\right) = 2\left(L\sin\frac{\theta}{2} + \lambda\right) + 2\left[(l + L)\sin\frac{\theta}{2} + \lambda + \Lambda\right]\sin\frac{\alpha}{2} \tag{15}$$

which is satisfied for any θ only if Eq. 6 and Eq. 14 are both satisfied.

16. More general joint designs, with two different rod eccentricities at each joint, have been considered by You (ref. 8).

DISCUSSION

17. The three concepts for pantographic rings presented above (note that further concepts could be generated in a similar way) lead to ring structures with rather different properties. For example, the most effective way of restraining the rods, which have to carry the compressive pre-stress induced by the active and passive cables against buckling, is to have $L/l \cong 1$. Figure 6(a), obtained by plotting Eq. (6, 9, 11), shows that Concepts (a) and (b) are almost equivalent in this respect, but (c) is much better for small valuse of n. The ring thickness, measured by R/r, also varies considerably, see Fig. 6(b).

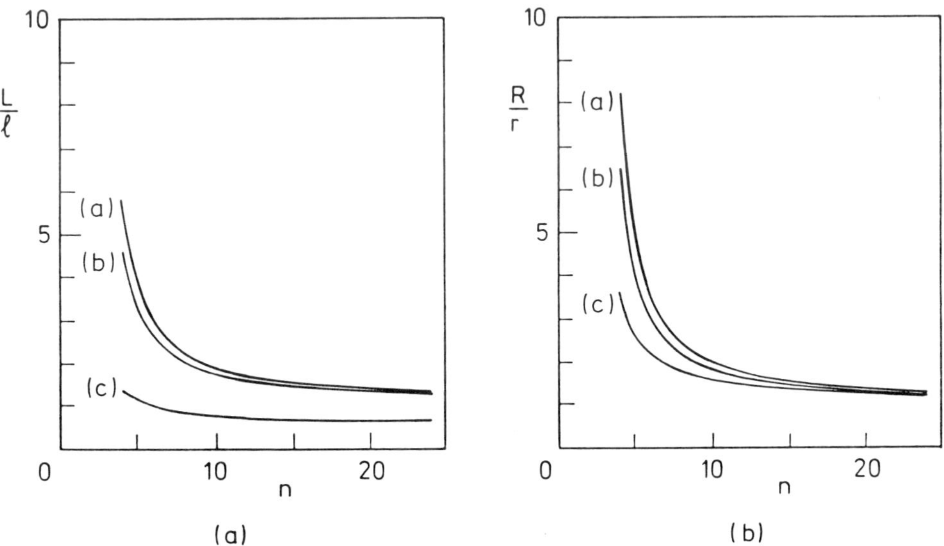

Fig. 6. Geometrical parameters for Concepts (a), (b) and (c).

18. For n = 12, the value chosen for the model of Fig. 1, Concept (c) has L/l = 0.71 which is the closest to one and also produces the most compact ring with R/r = 1.46. These values are respectively 1.58 and 1.64 for Concept (b), the second best. Also, the total length of rod for rings with equal inner radius is $194.3l$ for Concept (a) and almost the same, $198.9l$, for (c) but $247.9l$ for (b). Concept (c) would appear to be the best from these purely geometric considerations, however it requires twenty-four more joints than either (a) or (b) and leads to a softer structure than (b). Thus, the best compromise appears to be Concept (b), which has been selected for further investigation.

19. The model shown in Fig. 1 is part of this investigation, which will include extensive tests. Other small scale models of the concepts presented in this paper have been made during the course of this study and all have been found to behave precisely as expected. Also, it has been found that high geometric accuracy is crucial when making these models, because the gradual build-up of small errors during assembly can lead to large built-in stresses when the two ends of a ring are connected.

ACKNOWLEDGEMENTS

We thank Professors C.R. Calladine and K.H. Hunt for their advice. Financial support from the Cambridge Commonwealth Trust and SERC is gratefully acknowledged.

REFERENCES

1. FREELAND R. E. Survey of deployable antenna concepts. In: E. B. Lightner (ed.), Proc. Large Space Antenna Systems Technology-1982, NASA Langley Research Centre, 1982, 381-421.

2. ROEDERER A. G. and RAHMAT-SAMII Y. Unfurlable satellite antennas: a review. Annales des Telecommunications, 1989, vol. 44, 475-488.

3. ROGERS C. A., STUTZMAN W. L., CAMPBELL T. G. and HEDGEPETH J. M. Technology assessment and development of large deployable antennas. ASCE Journal of Aerospace Engineering, 1993, vol. 6, 34-54.

4. KWAN A. S. K., YOU Z. and PELLEGRINO S. Active and passive cable elements in deployable masts. International Journal of Space Structures, 1993, vol. 8.

5. HOBERMAN C. Art and science of folding structures. Sites, 1992, vol. 24, 34-53.

6. ZANARDO A. Two-dimensional articulated systems developable on a single or double curvature. Meccanica, 1986, vol. 21, 106-111.

7. ESCRIG F. and VALCARCEL J. P. Curved expandable space grids. In: B. H. V. Topping (ed.), Proc. International Conference Design Construction Non Conventional Structures, Civil-Comp Press, 1987, 157-166.

8. YOU Z. Deployable structures for reflector antennas in Space. First year report. Cambridge University Engineering Department, 1991.

85. Geometric constraints in assembling polygonal deployable units to form multi-unit structural systems

C. GANTES,Gantes Brothers and Co., Greece

SYNOPSIS. Deployable prefabricated space frames are reusable and relocatable, and offer the advantages of speed and ease of erection and dismantling which can be significant for certain types of applications, mainly in the aerospace industry and the temporary construction industry. A new type of such structures offer the additional advantage of being stable and stress-free in both the folded configuration and in a predetermined deployed configuration. A detailed geometric design is required in order to achieve this feature. A set of generic geometric constraints for flat and spherical structures of this type are proposed here.

INTRODUCTION

1. The deployable structures examined here are prefabricated space frames consisting of straight bars that are linked together in the factory into a compact bundle, and can be unfolded into large-span, load-bearing structures by simple articulation. Because of this feature they offer significant advantages in comparison to conventional, non-deployable structures for a wide spectrum of applications ranging from temporary structures to the aerospace industry. Such structures have been investigated by several researchers over the last 30 years (refs 1-5), and are growing in popularity as evidenced also by the recent publication of a textbook dealing with their analysis, synthesis, and design (ref. 6).

2. The present study deals with structures having as their basic module the so-called scissor-like-element (SLE), two bars connected to each other at intermediate points with a pivotal connection and hinged at their end points to end points of other SLEs. Such an element is illustrated in Fig. 1. Several SLEs are connected to each other to form units with plan view of regular polygons. Each side of the polygon is a symmetrical SLE, and each half-diagonal a non-symmetrical one.

3. A new type of deployable space frames that are based on the concept of scissor-like elements were introduced a few years ago in the Department of Architecture at MIT (refs 7-8). They have the additional advantage of being self-

standing and stress free in both their folded and their deployed configuration. This helps them overcome disadvantages of previous designs by having faster erection and dismantling times and higher load bearing capacity. The geometric and structural characteristics of such structures have been investigated in the Civil Engineering Department at MIT in recent years (refs 9-18), and analysis and design guidelines have been proposed.

4. As mentioned above, a fundamental design requirement of these structures is that each polygonal unit is self-standing and stress free when fully closed or fully deployed. However, at intermediate geometric configurations during the deployment process incompatibilities between the member lengths lead to the occurrence of second order strains and stresses resulting in a snap-through phenomenon that "locks" the structure in the deployed configuration. Hence, the objective of the design is to find a satisfactory compromise between desired stiffness under service loads, and desired flexibility during deployment in order to avoid residual stresses. Issues pertaining to numerical analysis in those two phases are addressed in refs 12-16.

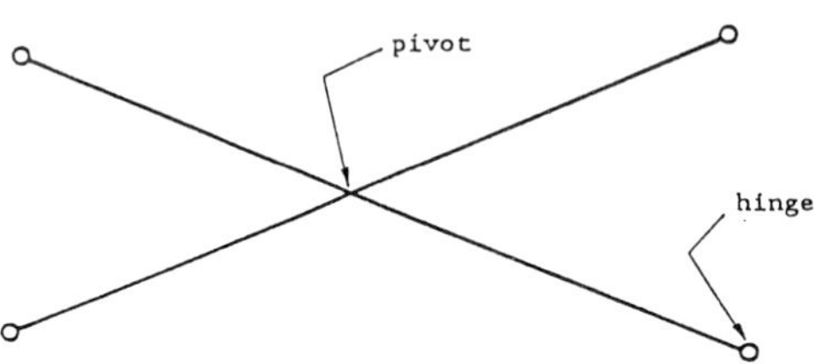

Fig. 1. A Typical Scissor-Like-Element (SLE).

5. But the present paper deals primarily with geometric design issues. A set of geometric constraints resulting from the requirement of zero stresses at the two extreme configurations are used to calculate the member lengths for given external dimensions of the unit. Such constraints have been derived in refs 17-18, and are briefly reviewed and expanded here in the next paragraphs.

6. Since single units cannot span large distances, it is desirable to assemble many of them to form larger structures (Fig. 2). A set of additional geometric constraints have to be satisfied by such assemblages in order to maintain the desired features of deployability and stress-free condition in both the folded and the deployed configuration. The present paper specifies these constraints for structures that are either flat or spherical in their deployed configuration.

GEOMETRIC CONSTRAINTS FOR SINGLE DEPLOYABLE UNITS

7. The geometric constraints for single units are derived on the basis of stress-free members in both the folded and the deployed configuration. The bars are

assumed to be perfectly straight before they are assembled to form a unit. Therefore, the required stress free condition is equivalent to straightness of the members in both extreme configurations once they have been assembled. A set of geometric constraints relating member lengths and angles between members in the deployed configuration are derived by looking at a development of all SLEs of the unit on a common plane, and applying basic geometric and trigonometric rules. A second set of geometric equations is obtained by applying the so called deployability constraint which requires that the sum of lengths between pivot and end nodes of the bars of SLEs that are connected to each other are equal, or referring to Fig. 3:

$$a + b = c + d \tag{1}$$

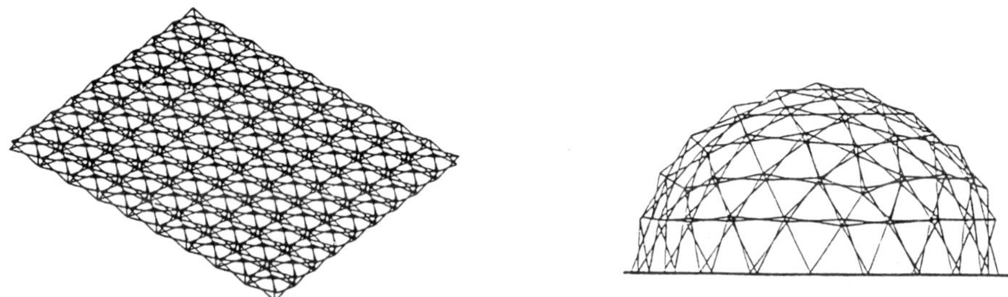

Fig. 2. Assemblages of Polygonal Units into Larger, Flat and Curved Deployable Structures.

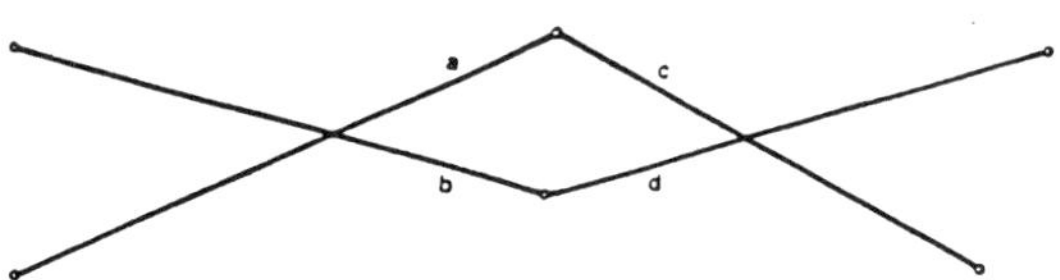

Fig. 3. The Basic Deployability Requirement.

8. For the example of polygonal units for flat structures (Fig. 4) the deployability constraint can be written as:

$$c + d = 2.e \tag{2}$$

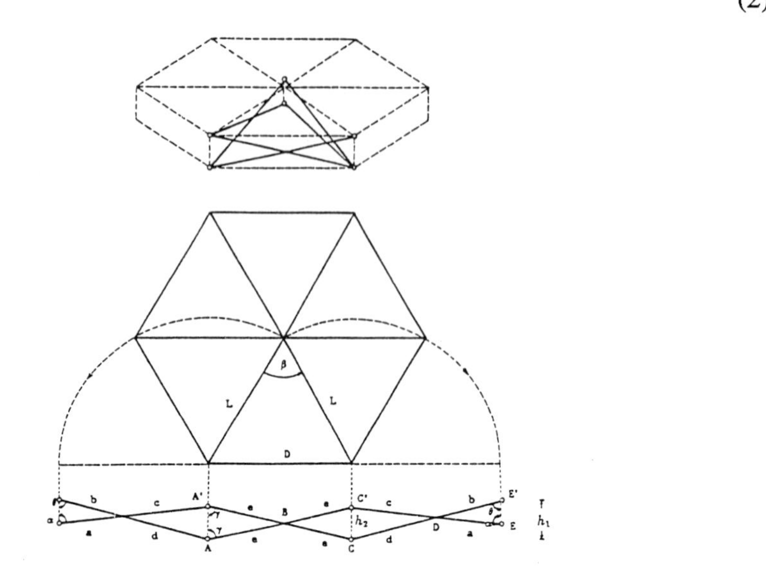

Fig. 4. Geometric Characteristics of a Flat Hexagonal Unit.

9. In all geometric design problems some of the variables have to be selected as known design parameters. In this case, the angle is known for any given type of regular polygon. In addition, it is meaningful to treat the external dimensions of the unit L, h_1, and h_2 as known quantities that are usually imposed by architectural requirements. Then, from the geometry in the deployed configuration one can observe:

$$\frac{a}{b} = \frac{c}{d} \tag{3}$$

$$\frac{b}{a} = \frac{\sin\alpha}{\sin\theta} \tag{4}$$

$$c.\cos\alpha + d.\cos\theta = h_2 \tag{5}$$

$$2.e.\cos\gamma = h_2 \tag{6}$$

$$L = \tan\theta.h_1 \tag{7}$$

$$(b + d).\cos\theta = h_1 \tag{8}$$

$$D = 2.L.\sin\frac{\beta}{2} = 2.e.\sin\gamma \qquad (9)$$

10. Eqs (2)-(9) are the geometric constraints for this unit. They are 8 independent, non-linear trigonometric equations for the 8 unknown quantities a, b, c, d, e, a, θ, γ. They can be simplified through algebraic manipulations, and yield the following analytical expressions for the unknowns:

$$\theta = \tan^{-1}\frac{D}{2.h_1.\sin\frac{\beta}{2}} \qquad (10)$$

$$\gamma = \tan^{-1}\frac{D}{h_2} \qquad (11)$$

$$e = \frac{D}{2.\sin\gamma} \qquad (12)$$

$$d = \frac{4.e^2 - h^2}{2.(2.e - h_2.\cos\theta)} \qquad (13)$$

$$c = 2.e - d \qquad (14)$$

$$b = \frac{h_1}{\cos\theta} - d \qquad (15)$$

$$a = \frac{b.c}{d} \qquad (16)$$

$$\alpha = \sin^{-1}\frac{b.\sin\theta}{a} \qquad (17)$$

11. The derivation of geometric constraints, and the geometric design process is similar for polygonal units for curved structures (Fig. 5), and is illustrated in ref. 17. In that case the design parameters are the three external unit dimensions (O_1O_2), (A_1A_2), and (A_2B_2). The process can be slightly modified to take the discrete joint size into account (ref. 18).

GEOMETRIC CONSTRAINTS FOR FLAT UNIT ASSEMBLAGES

12. The deployable units that are considered here can have a plan view of any regular polygon. Each side of the polygon is a symmetric SLE and each half-diagonal a non symmetric one. The lines connecting corresponding upper and

lower nodes are parallel to each other and perpendicular to the plane of the polygon. Hence, adjacent units form a flat structure in their deployed configuration.

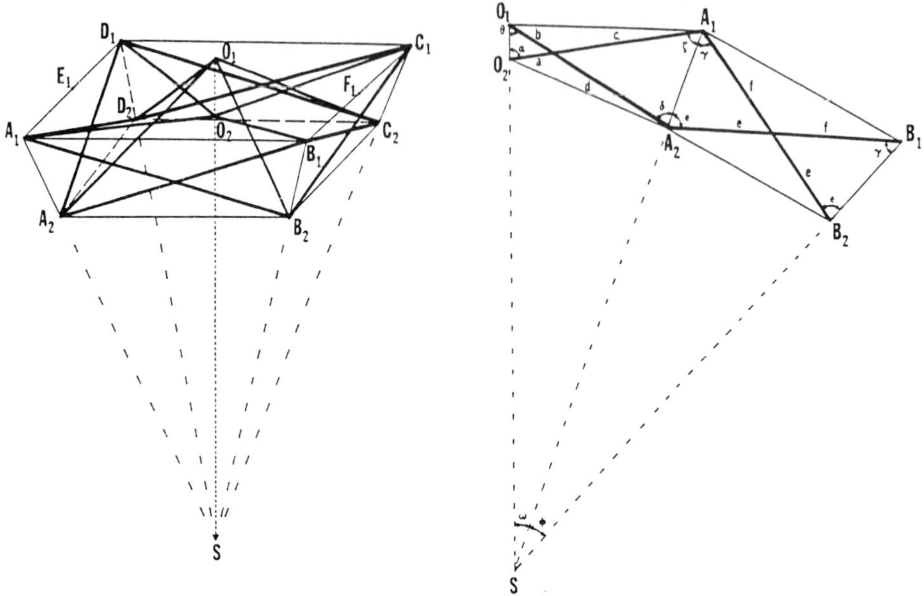

Fig. 5. Geometric Characteristics of a Curved Polygonal Unit.

13. The following requirements have to be satisfied in order for such an assemblage to behave in the desired way:

i) Each one of the units included in the assemblage must behave individually in the desired way. In other words they have to satisfy equations (2)-(9), and their geometric characteristics must be calculated from equations (10)-(17).
ii) The polygonal plan views of all units must fill the plane.
iii) Adjacent units must share the same nodes across their interfaces. This can be achieved if all units are designed for the same values of the side length D of the polygon, and the vertical distance h_2 between upper and lower corner nodes.

14. It is interesting to expand on point (ii), and list all possible tessellations of regular polygons that fill the plane. Let us first define a regular tessellation as one consisting of regular polygons of the same order. If p is the order of the polygon and q is the number of polygons that are brought together to fill the plane about a centre, it can be shown (ref. 19) that:

$$(p - 2)(q - 2) = 4 \tag{18}$$

Since p and q are integers we observe that the only regular tessellations that fill a plane are equilateral triangles, squares, and hexagons (Fig. 6).

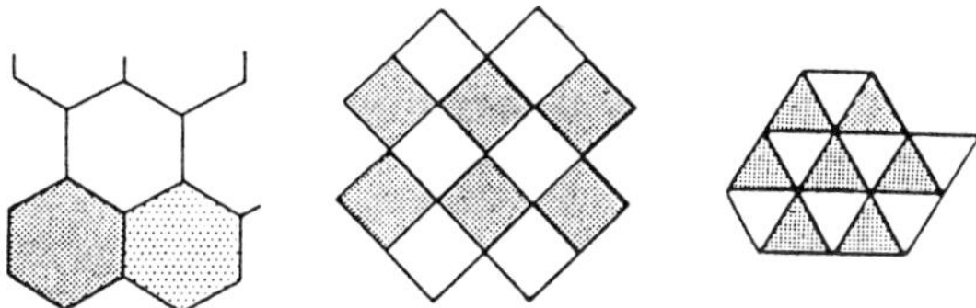

Fig. 6. The Three Regular Tessellations that Fill a Plane.

15. Let us now define as semi-regular tessellation one that consists of regular polygons of different order that are mixed in such a way that the number of edges and faces at all nodal junctions are the same. All semi-regular tessellations are characterised by three, four, or five way junctions depending on the number of edges meeting at each node. There is a total of eight possible mosaics of that type, illustrated in Fig. 7.

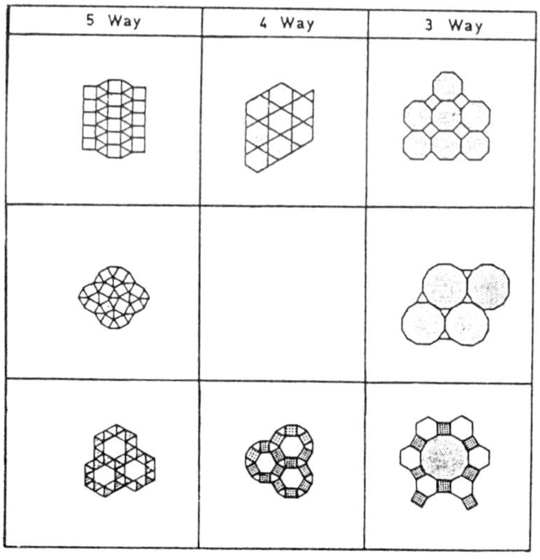

5 Way	4 Way	3 Way

Fig. 7. The Eight Semi-Regular Tessellations that Fill a Plane.

16. The three regular and the eight semi-regular mosaics that are shown in Figs 6 and 7 are the only possible ways in which flat polygonal deployable units can be assembled to each other to form larger deployable flat roofs.

GEOMETRIC CONSTRAINTS FOR SPHERICAL UNIT ASSEMBLAGES

17. The deployable units that are considered here can also have a plan view of any regular polygon. Each side of the polygon is an SLE forming the diagonals of a symmetric trapezoidal, and each half-diagonal is an arbitrary SLE. The lines connecting corresponding upper and lower nodes are concurrent, meeting at the same point S (Fig. 5). Hence, adjacent units form in their deployed configuration a spherical structure having S as the centre of the sphere.

18. The following requirements have to be satisfied in order for such an assemblage to behave in the desired way:

i) Each one of the units included in the assemblage must behave individually in the desired way. In other words they have to satisfy the constraint equations defined in ref. 17.
ii) The polygonal plan views of all units must form part of a convex polyhedron that is inscribed in the sphere.
iii) All external nodes have to lie on one half or a smaller part of the sphere so that the deployment process can be physically achieved.
iv) Adjacent units must share the same nodes across their interfaces. This can be achieved if all units are designed for the same values of the three external unit dimensions (O_1O_2), (A_1A_2), and (A_2B_2).

19. It is interesting to expand on point (ii), and list all possible convex polyhedra that have faces of regular polygons. Similarly to flat structures, there are the five so called regular polyhedra, also known as Platonic solids. All their faces are regular polygons of the same order, and they satisfy the Euler topological equation:

$$V - E + F = 2 \tag{19}$$

where V is the number of vertices, E the number of edges, and F the number of faces. Depending on the order of the polygonal faces there is the tetrahedron (4 triangles), the cube (6 squares), the octahedron (8 triangles), the dodecahedron (12 pentagons), and the icosahedron (20 triangles), as illustrated in Fig. 8. The tetrahedron and the cube cannot be considered for deployable structures because they do not satisfy criterion (iii). On the contrary, four faces of an octahedron can form a deployable pyramid, while six or less faces of a dodecahedron, and ten or less faces of an icosahedron make up deployable semi-spherical or shallow domes. In fact, all domes resulting from dodecahedra as well as domes resulting from icosahedra and having more than five faces violate slightly criterion (iii) because some of their bottom nodes are just beyond the circumference that separates the two semi-spheres. Therefore some additional secondary deployment stresses are introduced, hence, these configurations should, in general, be avoided. Consequently, the most interesting geometric configurations that can emanate from regular polyhedra are the 4-face pyramid resulting from half octahedron, and the 5-face pyramid resulting from one fourth of an icosahedron.

20. In addition, there are the so called semi-regular polyhedra (refs 19-22), shown also in Fig. 8. They are also inscribed into a sphere, and their faces are regular polygons, but of different order. Some additional geometric constraints have to be enforced in order for deployability to be maintained for such shapes. A characteristic example of a semi-regular polyhedron is a soccer ball consisting of pentagons and hexagons. The additional constraints have been derived in detail for this example (refs 11,17). They can be generalised for all semi-regular polyhedra, and are expressed here as follows:

i) All units must have the same curvature, and the lines connecting corresponding upper and lower nodes must be concurrent. This is achieved if the angle φ (Fig. 5b) is the same for all units.

ii) If ξ is the angle between a side and a radius of a regular polygon, and ξ' the projection of ξ on a plane tangential to the sphere, then the sum of all angles ξ' at any node of the polyhedron must equal 2π. The angle ξ' is given by:

$$\tan\xi' \frac{\sin\beta}{\cos\omega.(1-\cos\beta)} \tag{20}$$

where β equals $2\pi/n$, n is the order of the polygon, and ω is the angle shown in Fig. 5b.

iii) In addition, there are the angle relations:

$$\tan\frac{\phi}{2} = \sin\frac{\psi}{2}.\tan\frac{\beta}{2} \tag{21}$$

$$\tan\omega = \frac{\tan(0.5.\psi)}{\cos(0.5.\beta)} \tag{22}$$

where ψ is the angle E_1SF_1 in Fig 5a. Combining the above equations we can calculate the angles φ,ω,ψ and ξ' for all units of the structure.

SUMMARY AND CONCLUSIONS

21. Geometric design issues for flat and spherical deployable space frames that are self-standing and stress-free in the fully folded and the fully deployed configuration have been proposed. The flat structures emanate from tessellations of regular polygons while the spherical structures are based on the concept of regular and semi-regular convex polyhedra.

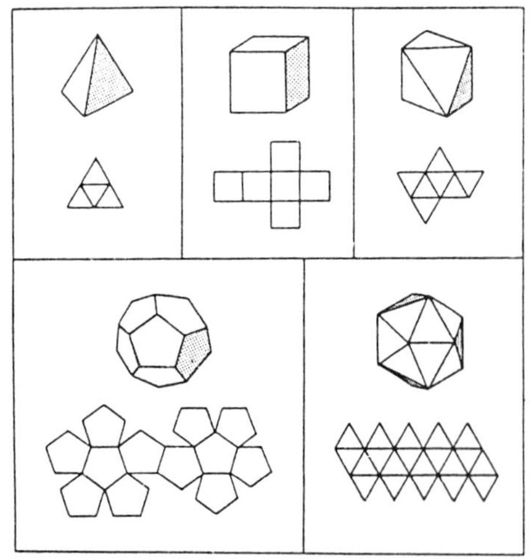

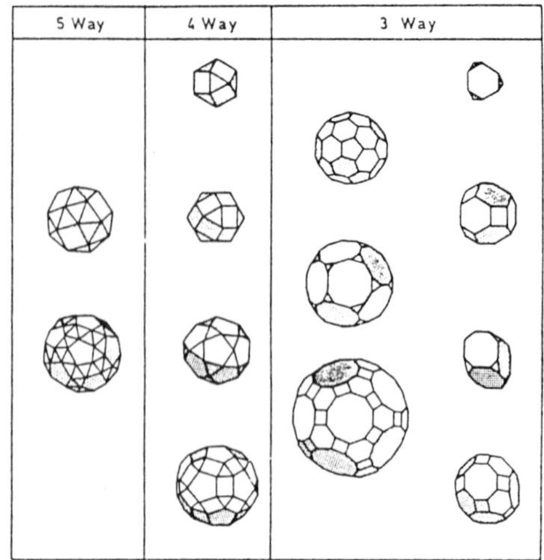

Fig. 8. The Five Regular and the Thirteen Semi-Regular Polyhedra.

REFERENCES

1. PINERO, E.P. Expandable Space Framing. Progressive Architecture, Vol. 43, No. 6, pp.154-155, June 1962.
2. ZEIGLER, T.R. Collapsible Self-Supporting Structures. U.S. Patent No. 4,437,275, 1984 (filed Aug. 7,1981).
3. RHODES, M.D. New Concepts in Deployable Beam Structures. NASA Conference Publication 2368, Part 1, pp. 331-348, Large Space Antenna Systems Technology Conference, Hampton, Virginia, Dec. 4-6,1984.
4. MIURA, K., and FURUYA, H. Adaptive Structure Concept for Future Space Applications. AIAA Journal, Vol. 26, No. 8, pp. 995-1002, Aug. 1988.
5. ESCRIG, F, VALCARCEL, J.P., and GIL DELGADO, O. Design of Expandable Spherical Grids. Proceedings of the XXX IASS Congress, Madrid, Spain, 1989.
6. KUZNETSOV, E.N. Underconstrained Structural Systems. Springer Verlag, New York, 1991.
7. KRISHNAPILLAI, S., AND ZALEWSKI, W.P. The Design of Deployable Structures. Unpublished Research Report, Department of Architecture, M.I.T., Cambridge, Massachusetts, Oct. 1985.
8. MERCHAN, C.H.H. Deployable Structures. Master's Thesis, M.I.T. Department of Architecture, June 1987.
9. GANTES, C. Developing New Concepts and Design Procedures for Deployable Structures. Master's Thesis, available as Research Report No. R88-13, Department of Civil Engineering, MIT, Cambridge, Massachusetts, September 1988.
10. GANTES, C., LOGCHER, R.D., CONNOR, J.J., and ROSENFELD, Y. Developing New Concepts and Design Procedures for Deployable Structures.

Research Report No. R90-07, Department of Civil Engineering, MIT, Cambridge, Massachusetts, May 1990.
11. GANTES, C. A Design Methodology for Deployable Structures. PhD Thesis, available as Research Report No. R91-11, Department of Civil Engineering, MIT, Cambridge, Massachusetts, May 1991.
12. GANTES, C., CONNOR, J.J., and LOGCHER, R.D. Geometric and Structural Design Considerations for Deployable Space Frames. MARAS '91, International Conference on Mobile and Rapidly Assembled Structures, Southampton, U.K., April 9-11,1991.
13. GANTES, C., CONNOR, J.J., and LOGCHER, R.D. Combining Numerical Analysis and Engineering Judgement to Design Deployable Structures. Computers & Structures, Vol. 40, No. 2, pp. 431-440, 1991, also presented at the 8th Conference on Non-linear Finite Element Analysis and ADINA, M.I.T., Cambridge, Massachusetts, July 17-19, 1991.
14. GANTES, C., CONNOR, J.J., and LOGCHER, R.D. Finite Element Analysis of Movable, Deployable Roofs and Bridges. Third Biennial Symposium on Heavy Movable Structures and Movable Bridges, St. Petersburg, Florida, November 12-15, 1990.
15. GANTES, C., CONNOR, J.J., and LOGCHER, R.D. A Simple Friction Model for Scissor-Type Mobile Structures. Journal of Engineering Mechanics, A.S.C.E., March 1993, (accepted for publication).
16. GANTES, C., CONNOR, J.J., and LOGCHER, R.D. An Equivalent Continuum Model for the Analysis of Flat Deployable Slabs. Journal of Aerospace Engineering, A.S.C.E., Jan. 1994, (accepted for publication).
17. GANTES, C., LOGCHER, R.D., CONNOR, J.J., and ROSENFELD, Y. Deployability Conditions for Curved and Flat, Polygonal and Trapezoidal Deployable Structures. International Journal of Space Structures, (accepted for publication in issue 2-3 of 1993).
18. GANTES, C., LOGCHER, R.D., CONNOR, J.J., and ROSENFELD, Y. Geometric Design of Deployable Structures with Discrete Joint Size. International Journal of Space Structures, (accepted for publication in issue 2-3 of 1993).
19. GASSON, P.C. Geometry of Spatial Forms. Ellis Horwood Limited Publishers, Chichester, U.K., 1983.
20. WOOD, G.D. Space Enclosure Systems: Identification and Documentation of Cell Geometries. Bulletin 203, Engineering Experiment Station, The Ohio State University, Columbus: University Publications, 1967.
21. WOOD, G.D. Space Enclosure Systems: The Variables of Packing Cell Design. Bulletin 205, Engineering Experiment Station, The Ohio State University, Columbus: University Publications, 1968
22. ZALGALLER, V.A. Convex Polyhedra with Regular Faces, 1969.

86. Expandable domes with incorporated roofing elements

J. B. VALCARCEL, F. ESCRIG and E. MARTIN, ETS Arquitectura, Spain

SYNOPSIS. In this paper the problems of roofing non-textile elements in expandable structures are studied. Two solutions have been proposed. These are systems made by folded plates that can be expanded with the whole structure and roofing plate elements, which may be assembled on the ground and later adapted in the final stage of unfolding. It is possible for these elements to add to the final strength of the structure, improving its performance. Design aspects of such structures are studied as well as, the problems related to their unfolding and the effect of these roofing elements on the strength of the structures. At the same time, specific examples of the possibilities of this system are proposed.

INTRODUCTION.

1. Unfolding bar structures offer an interesting solution in almost all situations that require easy transport and a quick means of providing roofing. Over the past two years our team has studied various typologies of groups of bars, which also offer in addition the ability to cover a large area when unfolded. Figs 1 and 2 show an example of a dome of 6 m diameter built as a model to 1/10 scale in the two stages of entirely folded up and unfolded.

2. The main difficulty posed with this type of structure lies in covering the unfolded structure. Designing effective covering mechanisms that can be folded with the complete package of bars and which cover the structure when unfolded, is a challenge. In actual fact, only textile covers offer effective solutions. If one requires a covering consisting of rigid elements, these must be placed in the structure after complete unfolding, which often results in a dangerous task of working at a great height.

3. Textile covers can be designed in such a way that the structure itself causes its own folding and unfolding together. Certain facts are offered relating to this problem (see ref. 5) and those solutions developed by our team are given. However, the aim of this project is to consider the possibility of covers formed from rigid elements, with which greater performance and durability can be achieved.

Fig. 1. Fig. 2.

ROOFING SYSTEMS WITH RIGID PANELS

4. One solution could be unfolding the structure and attaching to it a set of rigid panels joined to its nodes. As shall be seen later, fixing these roofing elements considerably improves the resistance of the unfolding structure and also has the advantage that the panels can be adjusted to achieve a nearly air tight space. One disadvantage, already mentioned, is that it is necessary to work at a certain height and for prolonged periods of time, negating one of the main advantage of unfolding structures namely their ease of erection.

5. As far back as 1971, the Spanish architect Emilio Pérez Piñero proposed for the Dali Museum in Figueras, Spain, an unfolding glass roof. This was formed by a truss of square modules consisting of bundles which held glass sheets by means of cross-pieces. These sheets were to be painted by Dali himself. The glass work was not constructed due to Pérez Piñero's death, but a model of the structure was built. Fig. 3 shows two phases of unfolding for this model.

6. Similar systems to this can be used as well for unfolding structures consisting of square cross-pieces. Fig. 4 shows the functioning of one of these modulus in two successive stages of unfolding. It is worth noting that in the unfolding process each of the rigid panels fits into the previous ones, thus producing a compact unit.

7. This system is more convenient the more regular the chosen module is. For this reason, the best advisable applications would be those trusses using consistent modules such as flat trusses and cylindrical domes.

8. Flat trusses (Fig. 5) are very simple structures but they allow for rather small spans, because they are not terribly efficient from a structural point of view. They are, however, very uncomplicated to build and a system such as that described could effectively solve the problem. In the figure, it can be seen that is possible to design the panels in such a way as to permit the overlapping of these and therefore increase the efficiency of the covering process.

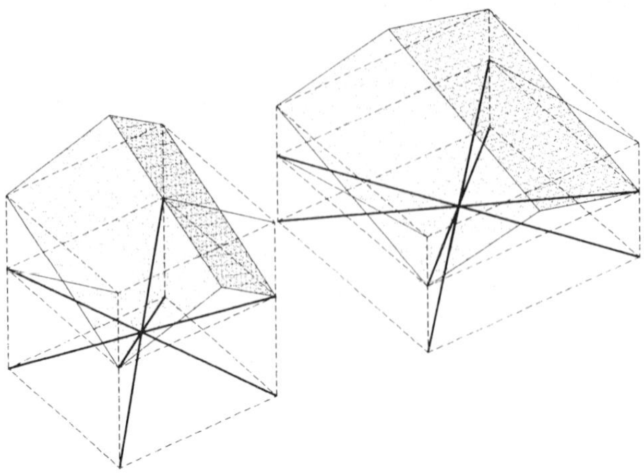

Fig. 3.

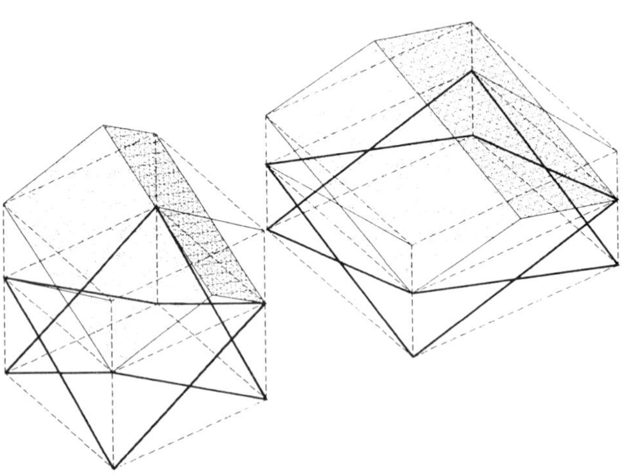

Fig. 4.

9. Cylindrical domes (Fig. 6) are the most useful structures. If they are constructed from square modules, the panels can even be made to unfold in two directions by which the problem of the top joint can be solved through an appropriate sealer.

10. When designing suitable roofing elements using triangular modules a serious problem is encountered. A geometric incompatibility exists which prevents the complete folding up of the structure (Fig. 7). It can be easily shown that, on closing, the panels trip one another up, thus preventing complete closure. Nevertheless, the authors believe that this structural solution offers such clear advantages that the system should not be rejected for this reason. It is decidedly possible to use this structure in a similar way as other foldable structures. This can be carried out by partially unfolding the structure on the ground, assembling the roofing elements in very favourable construction conditions, and subsequently continue the unfolding process until total unfolding is reached. This strategy opens up an enormous range of possibilities, far greater than those of systems based on the complete folding of the truss and roofing elements.

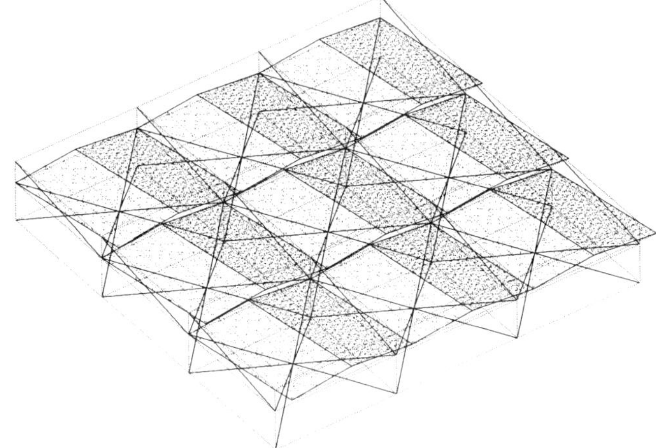

Fig. 5.

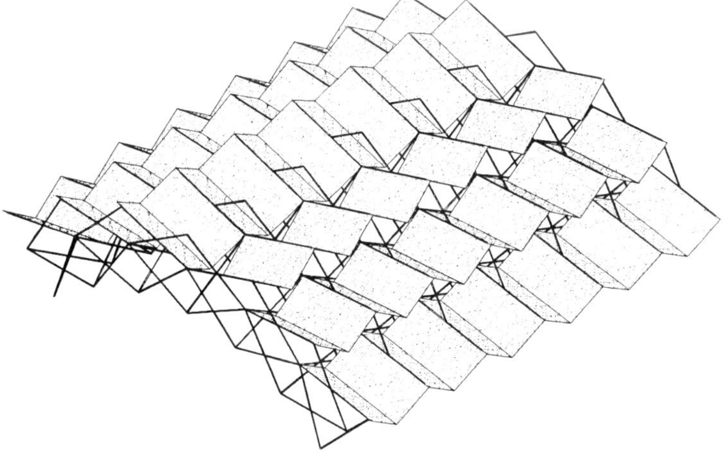

Fig. 6.

11. These type of elements can be used in numerous structures based on triangular modules either of cross-pieces or bundles. Fig. 8 shows a design for a hexagonal sun-shade with the successive unfolding stages. At this point in time the authors are working on several projects dealing with unfoldable domes with large spans. The authors have been able to ascertain that the way in which they unfold is not very conventional, because the structure is kept flat until almost the final stages of unfolding. Only at the end, does the structure curve up on its own, unfolding and establishing itself in its final position. This process is described in detail in the references stated and can be observed in Fig. 11.

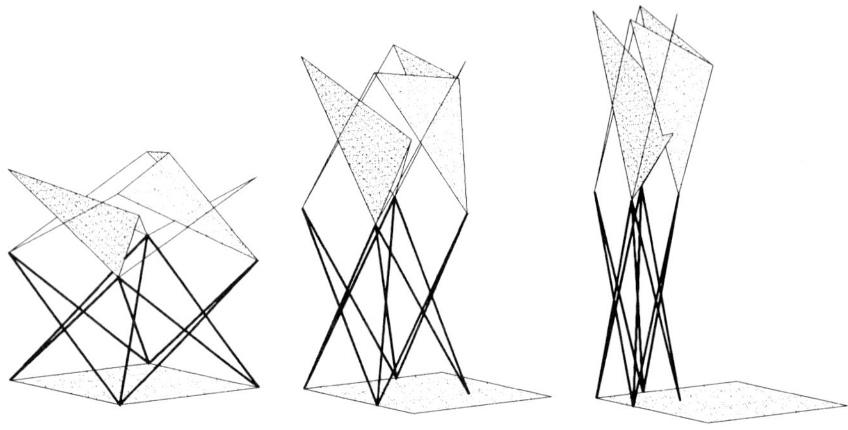

Fig. 7.

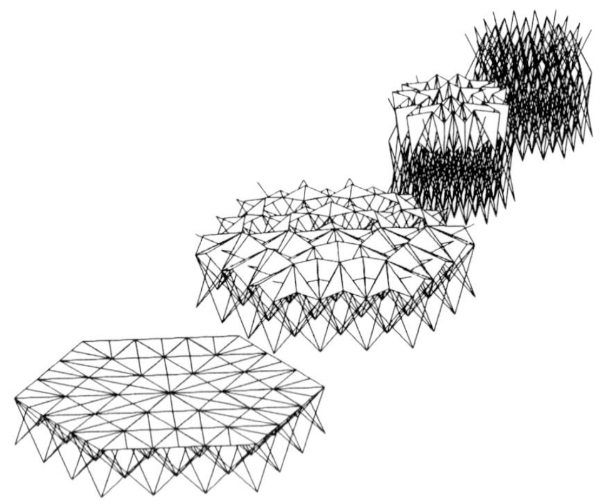

Fig. 8.

12. This special manner of unfolding permits a new roofing strategy; that is, the use of rhomboid rigid panels with a hinged joint. These panels could be placed over the structure in the fourth stage of unfolding, overlapping like the scales of a fish. This process can be carried out almost entirely on the ground, so the actual construction on site is simplified and later the unfolding may be completed. If then the free ends of the panels are fixed onto the corresponding nodes, excellent structural performance can be achieved, as will be seen later.

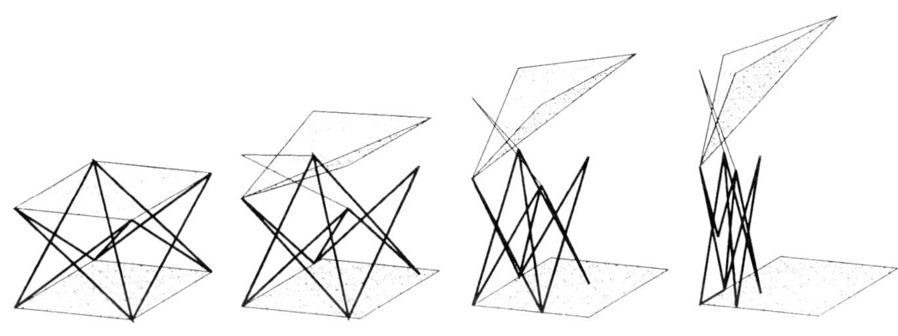

Fig. 9.

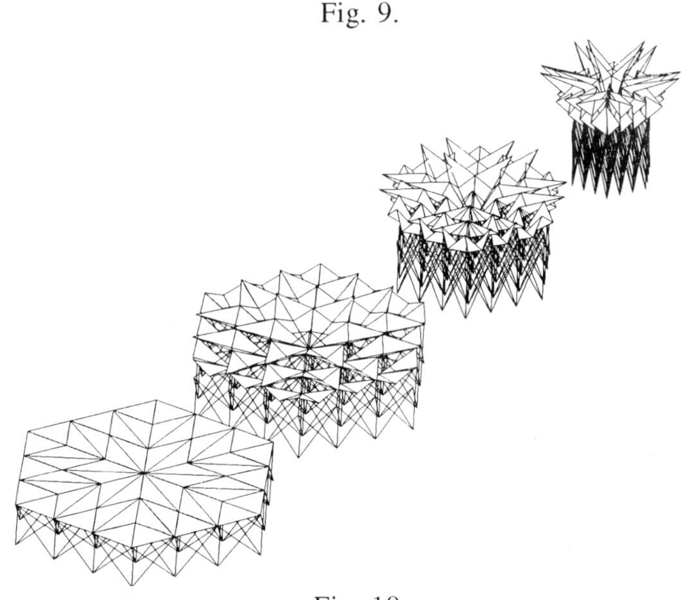

Fig. 10.

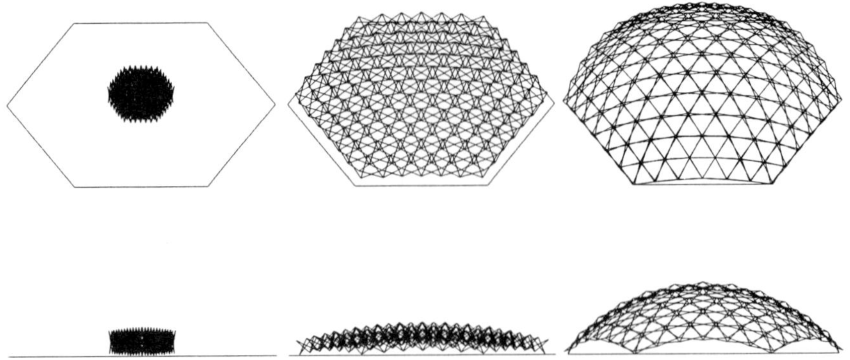

Fig. 11.

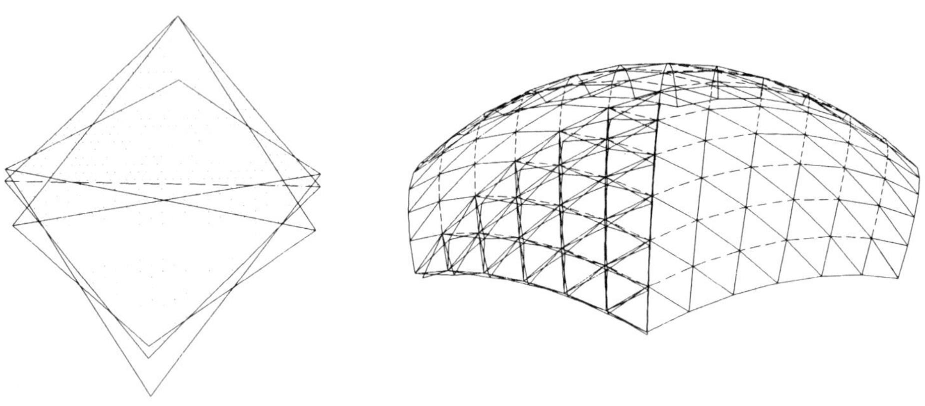

Fig. 12.

In Fig. 12, the final state of the dome with its roofing is shown.

ROOFING STRENGTH

13. If the roofing elements are correctly fixed, it is possible to count on their strength adding to the rigidity of the structural unit, which is of paramount importance, because these structures tend to have high deflections, requiring control.

14. As indicated in refs 3,4 the method of calculus developed is based on the use of matrix methods for the deployment of bars with inside joints. In these references the stiffness matrix employed is given and this differs considerably from the stiffness matrix for conventional applications.

15. It is clear that the study of these elements must be undertaken using the finite element method. The interaction of finite elements with a matrix program already developed, resulted in several problems. The solution adopted consisted of calculating the stiffness of each one of the roofing panels and condensing the degrees of freedom corresponding to the restraints at the nodes. This allows calculation of the stiffness matrix corresponding to the elements which can then be assembled to form the stiffness matrix of the whole structure.

16. The element used was a six node triangular type, because it allows greater precision with a less close-woven truss.

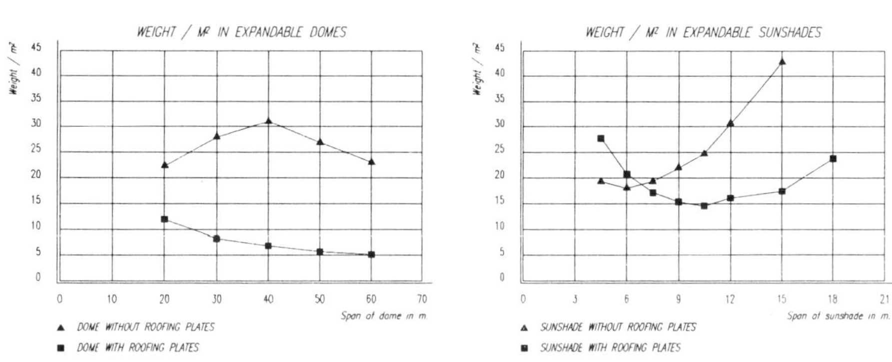

Fig. 13.

17. To demonstrate the efficiency of the strength factor of the roofing elements, two types of structures have been considered. Firstly, a hexagonal sun-shade was studied, and secondly a spherical dome. In both cases, dimensions of the bars were adjusted strictly to carry the forces exerted upon them. Hollow tubes of rectangular section were used, complying with the Spanish Standard MV-108. The design load was 60 kg/m^2, corresponding to the simultaneous loading of self weight plus snow for the area.

18. The drawings show the variation in weight per m^2 as the span of the structure increases, both when rigid elements in the roofing collaborate with the skeletal structure and also when they are inoperative.

CONCLUSIONS

19. The above explanation indicates that several roofing solutions with unfolding structures of rigid elements are possible. In some cases the covering can be incorporated in the bundle of bars forming the structure, while in others a

slight initial unfolding will be necessary, first with the subsequent incorporation of the roofing elements and later completion of the unfolding process. It is also possible in certain cases to unfold the covering near completion of the structure in order to fit the roofing elements and subsequently finish unfolding.

20. The resistant efficiency of these elements depends almost entirely on whether they are fixed in their final position following unfolding. If this is done, considerable material saving and greater efficiency in resistance and deformation of the structure will result.

ACKNOWLEDGEMENTS

21. The authors are grateful to scholarship receivers José Antonio Vázquez for his collaboration in C.A.D. graphics and José M. Asorey Brandariz for his help with the photographs.

REFERENCES .

1. PÉREZ PIÑERO, E., CANDELA, F., DALI, S. "La obra de Pérez Piñero". Arquitectura. Madrid n° 163-164. 1972.pp 1-28.
2. ESCRIG, F., P.VALCÁRCEL, J.B. "Curved Expandable Space grids". Non-conventional Structures '87. London 1987.
3. P.VALCÁRCEL, J.B., ESCRIG, F. "Analysis of curved expandable space bar structures" Int. Symposium on 10 Years of Progress in Shell and Spatial Structures. IASS. Madrid 1989.
4. ESCRIG, F., P.VALCÁRCEL, J.B. "To cover a Swimming Pool with an Expandable Structure". International Conference on Mobile and Rapidly Assembled Structures. MARAS'91. Southampton 1991.
5. P.VALCÁRCEL, J.B.; ESCRIG, F. "Expandable Structures with Self-folding Textile Cover". International Conference on Mobile and Rapidly Assembled Structures. MARAS'91. Southampton 1991.
6. P.VALCÁRCEL, J.B. ESCRIG, F.ESTÉVEZ, J. MARTIN, E. "Large Span Expandable Domes" Int. Symposium; on Large Span Structures. IASS. Toronto 1992.

87. Numerical analysis for folding of space structures

K. KAWAGUCHI and Y. HANGAI, University of Tokyo, Japan, and
K. NABANA, Shibaura Institute of Technology, Japan

SYNOPSIS. The main interest of the paper is to find the optimum process to fold structures into desired shapes. The numerical procedure is carried out until the geometrical measurer of folding takes the minimum value. This procedure generates not only an intelligent procedure for folding, but also the quickest sequence for building up or expanding the structures. The analytical procedure is based on the generalised incremental method in which generalised inverse is effectively used. Several illustrative examples are also presented.

INTRODUCTION

1. There are very few studies which address methods for the folding of structures. Most of the studies which are related to the folding of structures can be found in the field of so called deployable or expandable structures. In this area, many special structural systems or mechanisms are proposed and analysed (refs 1-5). Some geometrical rules for regular polygons to expand them were also investigated (refs 6 and 7). However, they are not applicable for complex general structural assemblies.

2. In this paper, we propose the numerical procedure to find the shortest and quickest sequence to fold general space structures from the given initial state to the predetermined folded configuration.

3. Firstly, the initial frames are assumed to consist of rigid links which are freely hinged at their ends. Then, we calculate the shortest path along which the geometrical measurer is minimised until the folding is accomplished.

4. After the calculation, one can eliminate the hinges which are not positively necessary and can specify the hinges which are necessary for the main process of the folding.

5. Such a path will also give an optimum process to build up the structure when the path is trailed back in the inverse sequence.

GEOMETRICAL CONSTRAINTS

6. Geometrical constraint conditions which represents topological configuration of the structure can be given as:

$$g_i(x_1,\cdots,x_n) = 0 \qquad (i = 1,\cdots,m) \tag{1}$$

in which $x_i\,(i = 1,\cdots,n)$ are the variables in the generalised coordinate system, n is the number of degrees of freedom and m is the number of the constraints.

7. *Maclarin* expansion of Eq. 1 as the function of the generalised incremental parameter s gives:

$$\frac{\partial g_i}{\partial x_k} x'_k s + \frac{1}{2}\left(\frac{\partial^2 g_i}{\partial x_k \partial x_l} x'_k x'_l + \frac{\partial g_i}{\partial x_k} x''_k\right) s^2 + \cdots = 0 \tag{2}$$

where ' indicates a derivative as a function of s. The repeated index represents a sum over that index. Neglecting higher order terms and regarding every coefficient as zero, we obtain:

$$\frac{\partial g_i}{\partial x_k} x'_k = 0 \tag{3}$$

$$\frac{\partial g_i}{\partial x_k} x''_k + \frac{\partial^2 g_i}{\partial x_k \partial x_j} x'_k x'_j = 0 \tag{4}$$

8. These last two equations can be written in matrix forms as:

$$\boldsymbol{Ax}' = \boldsymbol{o} \tag{5}$$

$$\boldsymbol{Ax}'' + \boldsymbol{A}'\boldsymbol{x}' = \boldsymbol{o} \tag{6}$$

wherein:

$$\boldsymbol{A} = \left[\frac{\partial g_i}{\partial x_k}\right] \tag{7}$$

$$\boldsymbol{A}' = \left[\frac{\partial^2 g_i}{\partial x_k \partial x_j} x'_k\right] \tag{8}$$

$$\boldsymbol{x}' = \left\{\frac{dx_i}{ds}\right\} \tag{9}$$

$$\boldsymbol{x}'' = \left\{\frac{d^2 x_i}{ds^2}\right\} \tag{10}$$

Both $\boldsymbol{A}$ and $\boldsymbol{A}'$ are $m \times n$ rectangular matrices.

9. Arranging a set of orthonormal bases, $\boldsymbol{h}_1,...\boldsymbol{h}_p$, of the null space of $\boldsymbol{A}$, an orthogonal complement matrix $\boldsymbol{H}$ can be written as:

$$\boldsymbol{H} = [\boldsymbol{h}_1,...,\boldsymbol{h}_p] \tag{12}$$

in which p is the dimension of the null space, that is, the remaining degree of freedom.

$$p = n - rank\,(\boldsymbol{A})$$

10. Using the *Moore-Penrose generalised inverse* of $\boldsymbol{A}$, $\boldsymbol{A}^{\dagger}$, the solution of Eqs 5 and 6 becomes:

$$\boldsymbol{x}' = \boldsymbol{H}\alpha' \tag{13}$$

$$\boldsymbol{x}'' = -\boldsymbol{A}^{\dagger}\boldsymbol{A}'\boldsymbol{x}' + \boldsymbol{H}\alpha'' \tag{14}$$

where both α' and α'' are any arbitrary p-dimensional column array.

GEOMETRICAL MEASURER

11. Now we define the geometrical measurer r, which indicates the geometrical accomplishment level of the folding of the structure, as:

$$r = \Re\,(x_1, \cdots, x_n) \tag{15}$$

where r should be set to take a less positive value during the progress of the folding and to take the least value, zero, only in the case of the complete folding. For example, the sum of the heights of all the nodes from the ground level, or the total volume which is enclosed inside between the structure and ground level, can be employed as the definition of r. One can choose the definition for r according to the motivation and aims of the folding.

12. The first and second derivatives of r as the function of the parameter s can be written as:

$$r' = \boldsymbol{R}'^T\boldsymbol{x}' \tag{16}$$

$$r'' = x'^T\boldsymbol{R}''x' + \boldsymbol{R}'^T\boldsymbol{x}'' \tag{17}$$

where ^{T} means the transpose of a matrix and

$$\boldsymbol{R}' = \{R'_i\} = \left\{\frac{\partial\Re\,(x_1, \cdots, x_n)}{\partial x_i}\right\} \tag{18}$$

$$\boldsymbol{R}'' = \left[R''_{ij}\right] = \left[\frac{\partial^2 \Re(x_1, \cdots, x_n)}{\partial x_i \partial x_j}\right] \tag{19}$$

$\boldsymbol{R}'$ is an n-dimensional column array, while $\boldsymbol{R}''$ is an n x n matrix.

13. Substituting Eqs 13-14 and Eqs 16-19 into Eq. 15 and neglecting higher order terms, the *Maclaurin* expansion of Eq. 15 as a function of s becomes:

$$r = \boldsymbol{R}'^T \boldsymbol{H}\boldsymbol{\alpha}' s + \frac{1}{2}[\boldsymbol{\alpha}'^T \boldsymbol{H}^T \boldsymbol{R}'' \boldsymbol{H}\boldsymbol{\alpha}' - \boldsymbol{R}'^T \boldsymbol{A}^\dagger \boldsymbol{A}'\boldsymbol{x}']s^2 \tag{20}$$

14. Let the element of m-dimensional vector $\boldsymbol{R}'^T \boldsymbol{A}^\dagger$ be represented as follows:

$$\boldsymbol{R}'^T \boldsymbol{A}^\dagger = (\gamma_1, \cdots, \gamma_m) \tag{21}$$

15. Then we can replace the second term of Eq. 15 with:

$$\boldsymbol{R}'^T \boldsymbol{A}^\dagger \boldsymbol{A}'\boldsymbol{x}' \rightarrow \boldsymbol{x}'^T \boldsymbol{B}'\boldsymbol{x}' \tag{22}$$

in which

$$\boldsymbol{B}' = [B'_{ij}] = \left[\frac{\partial^2 g_k}{\partial x_i \partial x_j}\gamma_k\right] \tag{23}$$

16. We also replace other terms as follows:

$$\boldsymbol{a}^T = \boldsymbol{R}'^T \boldsymbol{H} \tag{24}$$

$$\boldsymbol{T} = \boldsymbol{H}^T (\boldsymbol{R}'' - \boldsymbol{B}')\boldsymbol{H} \tag{25}$$

17. Substituting Eqs 22, 24 and 25 into Eq. 20, we obtain:

$$r = \boldsymbol{a}^T(\boldsymbol{\alpha}' s) + \frac{1}{2}(\boldsymbol{\alpha}' s)^T \boldsymbol{T}(\boldsymbol{\alpha}' s) \tag{26}$$

FORWARDING TO THE FOLDING

18. The aim of the calculation is to find the shortest path along which the geometrical measurer is minimised until it reaches one of the least values. To find the optimum path to reach the final shape which gives the minimum value of r, we apply the generalised incremental procedure employing Eq. 26 in each incremental step.

19. Although a conventional method such as Newton-Raphson procedure might be employed to obtain the numerical solution of Eq. 26, the generalised incremental method is used in this paper. The method does not require any numerical iteration scheme, and therefore will not suffer from numerical divergence during the calculating process.

20. Roughly speaking, the procedure consists of the following two steps:

- Firstly, the solution for the first term of Eq. 26 is successively derived.
- When the first term becomes zero, the second term is considered and the proceeding is judged.

21. The procedure gives $\alpha'\Delta s$ which satisfies Eq. 26 for the given increment value r and requires the minimum norm, $\|\alpha'\Delta s\|$, as a solution in each step because the *Moore Penrose generalised inverse* is utilised to get the solution of a set of simultaneous equations.

22. In the proposed procedure, the higher order terms can be employed to improve the accuracy of the solution so as to make it accurate enough to grasp the general features of the path.

ANALYTICAL SEQUENCE

23. In the present procedure, we give a certain appropriate value for Δr as the incremental parameter and carry out following numerical calculations.

(a) Solve the first term of Eq. 26 using:

$$\alpha\Delta s = (\alpha^T)^\dagger \Delta r \tag{27}$$

where the *Moore-Penrose generalised Inverse* of a row vector $\boldsymbol{a}^T$,which gives the least square and minimum norm solution, is given as:

$$\left(\boldsymbol{a}^T\right)^\dagger = \frac{\boldsymbol{a}}{\|\boldsymbol{a}\|^2} \tag{28}$$

(b) Using the value obtained from Eq. 27, the displacement vector $\Delta \boldsymbol{x}$ is derived as:

$$\Delta \boldsymbol{x} = \boldsymbol{H}(\alpha'\Delta s) \tag{29}$$

we repeat these two processes, (a) and (b), one after the other until the absolute value of $\boldsymbol{a}$ is estimated as almost zero. Instead of the last equation, the following equation is also available:

$$\Delta \boldsymbol{x} = \boldsymbol{H}\left(\boldsymbol{\alpha}'\Delta s\right) + \frac{1}{2}\left(-\boldsymbol{A}^{\dagger}\boldsymbol{A}'\boldsymbol{H}\boldsymbol{\alpha}'\right)\left(\Delta s\right)^2 \tag{30}$$

where the higher order terms improve the accuracy of the calculation, as is mentioned above.

(c) If $\boldsymbol{a} = \boldsymbol{H}^T\boldsymbol{R}' = \boldsymbol{0}$, we introduce the next two criteria.

(i) When $\boldsymbol{R}' = \boldsymbol{0}$, we consider that r takes the critical value and the calculation is terminated.

(ii) When $R' \neq \boldsymbol{0}$ and $\boldsymbol{R}'^T\boldsymbol{H} = \boldsymbol{0}$, we consider that r is not the critical value and consider the least eigenvalue of real Hermitian matrix $\boldsymbol{T}$ in Eq. 26. If the least eigenvalue is not positive, we continue the calculation along the direction of its corresponding vector.

24. In case of (ii), the restricted direction of the path and the steepest descent direction of r are perpendicular, while the configuration of the structure has not reached the final completed collapsed state.

NUMERICAL EXAMPLES

25. In this section, some illustrative results of numerical examples are shown. The structures of the examples are arched rather than simple space structures. We assume many hinges so that the structures form assemblies of rigid links.

26. The geometrical constraint is the rigidity of the members, which can be written as:

$$\left[\left(\boldsymbol{x}_k - \boldsymbol{x}_j\right)^T\left(\boldsymbol{x}_k - \boldsymbol{x}_j\right)\right]_i^{1/2} - l_i = 0 \qquad (i = 1, \cdots, m) \tag{31}$$

where li is the length of the i-th member and m is the number of members. As the geometrical measurer of the folding, the summation of the squares of each node height from the x - y plane is used. The measurer can be written as:

$$r = \boldsymbol{x}^T\boldsymbol{Z}^T\boldsymbol{Z}\boldsymbol{x} \tag{32}$$

where:

$$\boldsymbol{Z} = \begin{bmatrix} 0 & 0 & 1 & 0 & 0 & 0 & \cdots & 0 \\ 0 & 0 & 0 & 0 & 0 & 1 & \cdots & 0 \\ \vdots & \vdots & \vdots & \vdots & \vdots & \vdots & \ddots & \vdots \\ 0 & 0 & 0 & 0 & 0 & 0 & \cdots & 1 \end{bmatrix} \tag{33}$$

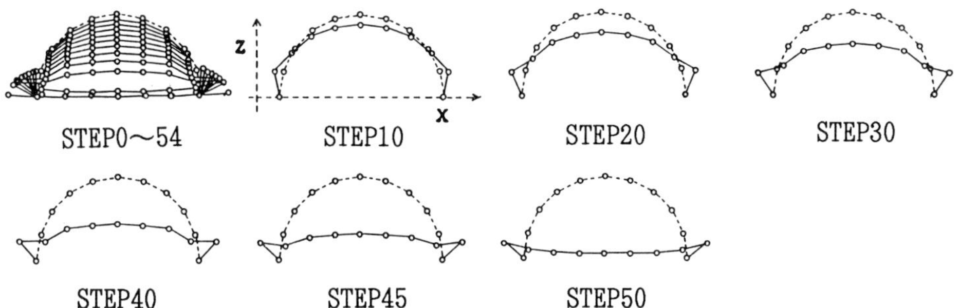

Fig. 1. The folding of the upper half of a regular icosagon frame.

The upper half of an icosagon

27. The upper half of a regular icosagon frame is folded. The lowest two foot nodes are fixed on the x-axis. The folding is limited in the x - z plane. The incremental value Δr is -0.1. The results are shown in Fig. 1. The step-r curve and step-$\|\boldsymbol{a}\|$ curve are shown in Figs 2(a,) and (b), respectively.

28. Since the initial value of r is 5.00, almost 50 steps are required to accomplish the folding. Observing the general displacement sequence of the folding, it is obvious that the four hinges at first and second lowest level above the foot are playing more important roles than the others during the folding process.

Dodecahedron

29. A regular dodecahedron frame, Fig. 3, is folded being fixed at the lowest pentagon. The incremental interval value Δr is initially set to be -0.1 and its absolute magnitude is gradually reduced as the value of the geometrical measurer decreases.

30. The calculation is firstly paused at step 231 because of the condition, $\|\boldsymbol{a}\| = \mathbf{o}$ and the first three least eigenvalues of $\boldsymbol{T}$ at this step are shown in Table 1. They are negative, which means the folding is not completed and the structure can be folded still further.

31. Since the first two eigenvalues are the same, one typical eigenvector is selected to represent the vectors which is associated with these two eigenvalues, and both the positive and the negative directions are examined to proceed with the calculation .

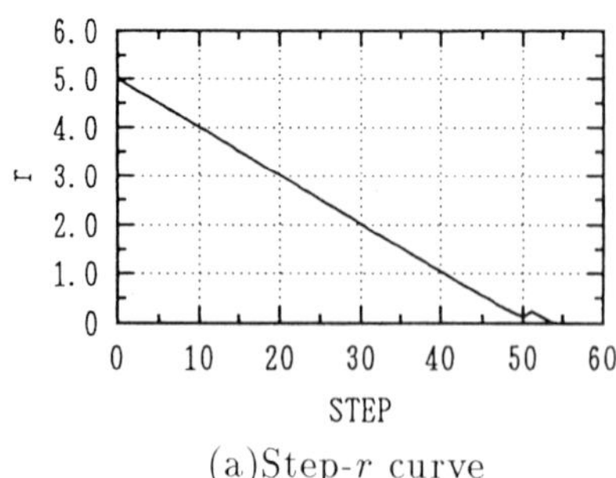

(a)Step-r curve

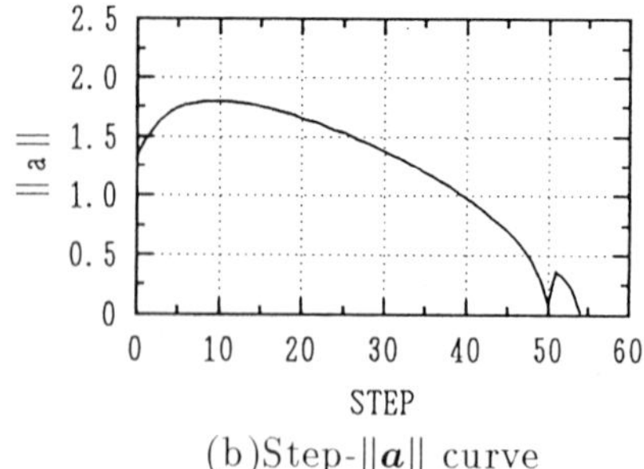

(b)Step-$||\boldsymbol{a}||$ curve

Fig. 2. Step-r and step-$||\boldsymbol{a}||$ curves for the upper half of a regular icosagon.

Table 1. The first three least eigenvalues of **T** at step 231.

	Eigen Value of **T**
1st Least	-3.714835
2nd Least	-3.714835
3rd Least	-3.071208

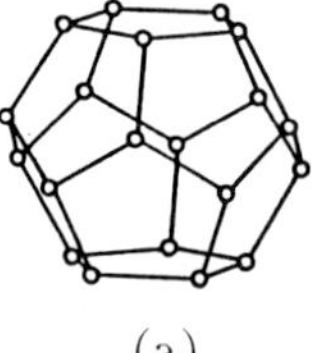

(a)

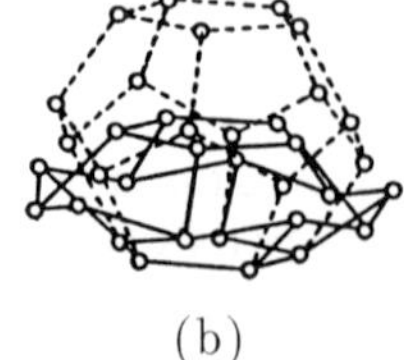

(b)

Fig. 3. The configuration of a regular dodecahedron:
(a) The initial configuration;
(b) The configuration at step 231.

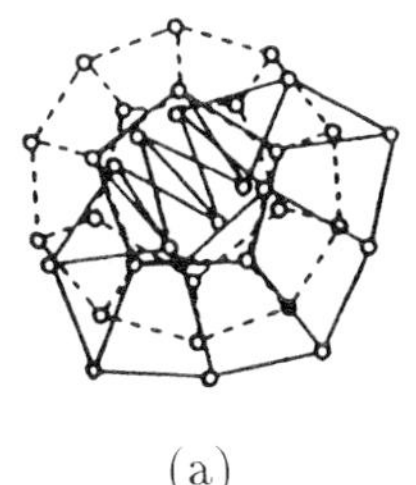

(a)

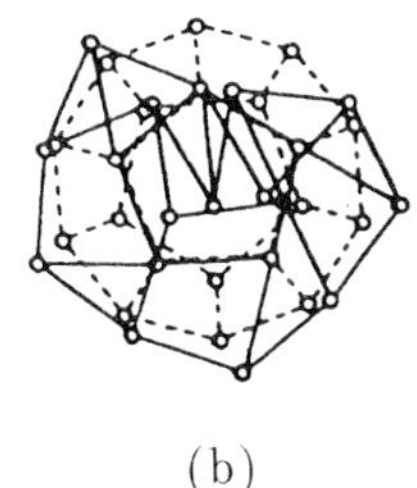

(b)

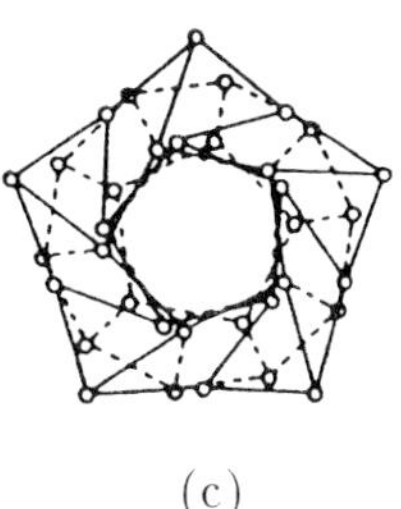

(c)

Fig. 4. The plan view of the folded configurations of a regular dodecahedron:
(a) The positive sign direction for the 1st least eigenvalues (step 513);
(b) The negative sign direction for the 1st least eigenvalues (step 457);
(c) For the 3rd least eigenvalue (step 427).

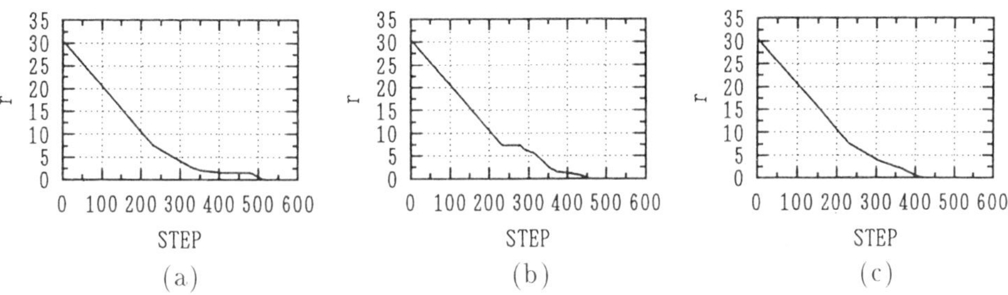

Fig. 5. Step-r curves for a regular dodecahedron frame:
(a) The positive sign direction for the 1st least eigenvalue;
(b) The negative sign direction for the 1st least eigenvalue;
(c) For the 3rd least eigenvalue (step 427).

32. The vector which corresponds to the third eigenvalue shows a twisted mode and is also used to continue the calculation.

33. The results are shown in Figs 4-5. The final shape for the third eigenvalue shows the most symmetrical regular pattern and the process requires the smallest numbers of steps among the results. On the other hand the other two show more complicated and irregular patterns and processes. In both cases, the structure experiences other stages where the process should be paused, $\|\boldsymbol{a}\| = \boldsymbol{o}$, and $\boldsymbol{T}$ has negative eigenvalues, several times.

A lamella dome without braces

34. A lamella-pattern dome without braces which has 36 nodes and 84 members is calculated. The initial value of Δr is -0.1 and it is controllable owing to the magnitude of *r*.

35. From these results we can easily distinguish the importance among the nodes. The rotation of the members connected to the nodes in the two lowest levels are extreme, while the members connected to the highest level rotate very little. Therefore, we can find that the hinged nodes in the two lowest levels are more important than those are in the highest levels. In the real project, such nodes need not be hinged.

CONCLUSION

36. A numerical method to establish the process of folding space structures has been presented.

37. One of the features of the proposed procedure is the introduction of the geometrical measurer, which is formulated according to the aim of the folding, case by case.

38. The method gives, not only the final folded shape, but also the quickest sequence of the folding. For all examples, it is obvious that inverted paths suggest the quickest process for building up or unfolding of the space structures.

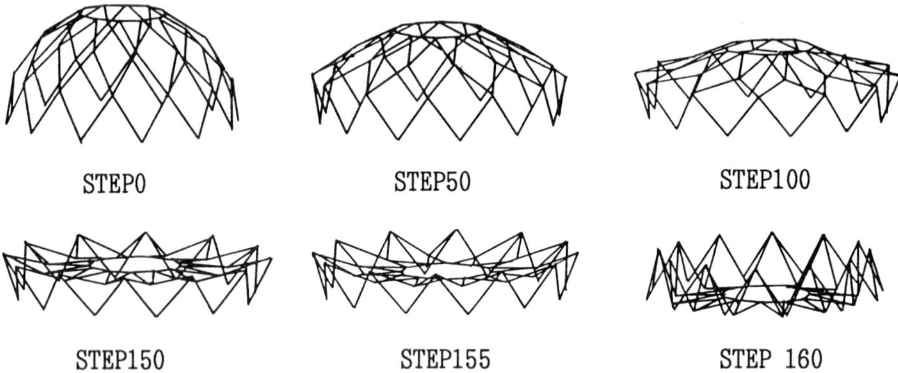

Fig. 6. The folding of a lamella-pattern dome without braces.

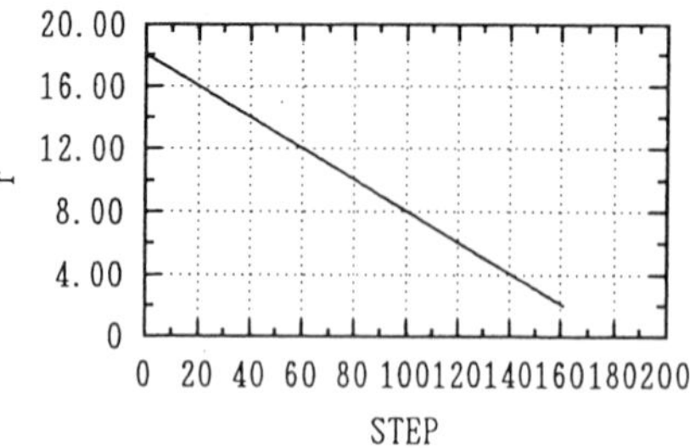

Fig. 7. Step-*r* curve for a lamella-pattern dome without braces.

REFERENCES

1. CRAWFORD R., F. Strength and efficiency of deployable booms for space applications, AIAA Paper No.7, 1971 Apr.
2. PINERO E. Three dimensional reticular structures, 1965, U.S.Pat. 3,185,164
3. ZEIGLER Collapsible self-supporting structures, 1984, U.S.Pat 4,437,275.
4.NATOR.I M., MIURA K. and FURUYA H. Deployable truss concepts in space technology, Proceedings of IASS Symposium, Osaka, Vol.3, 198, 261268.
5. SHAN W. Computer analysis of foldable structures, Computer and Structures, Vol.42, No.6, 1992, 903-912.
6. JOSEPH D. CLINTON Advanced structural geometry studies, Part II-Geometric transformation concept for expanding rigid structures. NASA Contractor Report, NASA CR-1735, 1971, Sept.
7. HUGO F. VERHEYEN Expandable polyhedral structures based on dipolygonids, Proceedings of Third International Conference on Space Structures, 1984, 88-93.
8. PELLEGRINO S. and CALLADINE C. R. Structural Computation of an assembly of rigid links, frictionless joints, and elastic springs, ASME Journal of Applied Mechanics, Vol.58, 1991 Sept., 749-753.
9. HANGAI Y. and KAWAGUCHI K. Analysis for shape-finding process of unstable link structures, Bulletin of IASS n.100, Vol.30, 1990, A11-128.

88. Configuration studies of foldable structures

W. SHAN, Taiyo Kogyo Corporation, Japan

SYNOPSIS. For the computer analysis and design of a foldable structure, one needs to create numerical models to represent such aspect of the structure as element connectivity, nodal co-ordinates, support arrangement and loading points. Formex algebra provides an ideal medium for creation of these numerical models in an efficient and convenient manner. The objective of the present paper is to demonstrate the application of the formex approach, in relation to foldable structures, in terms of a number of examples.

INTRODUCTION

1. A foldable structure is a structure which is capable of being folded into a bundle. The umbrella is the most familiar example. Foldable structures may be used as temporary exhibition structures, sales display systems, travelling theatres and shelters for emergency use. They are also applicable for use as extraterrestrial structures, since they may be sent into the outer space in the folded state and deployed rapidly by simple unfolding.

2. The foldable systems considered in the present paper are assemblies consisting of straight bars connected through pin joints and pivots, similar to those discussed in (refs 1-2 and 5-8). The folding mechanism of such structures is based on that of a pantograph. Consequently, they are called pantograph structures or p-structures, (ref. 7). The main issue to be dealt with here is configuration processing for p-structures by formex algebra. The reader is assumed to be familiar with formex algebra and Formian, (refs 3 and 4).

DUPLETS

3. The folding unit of a p-structure is a duplet, see Fig. 1a, consisting of two uniplets connected together with a hinge as an intermediate pivot. Each uniplet is capable of rotating freely with respect to the pivot and the joints connected to its ends. Duplets are divided into regular and irregular ones. The regular duplets include rectangular, trapezoidal and rhombic duplets as shown in Figs 1b to 1d.

Space structures 4. Thomas Telford, London, 1993

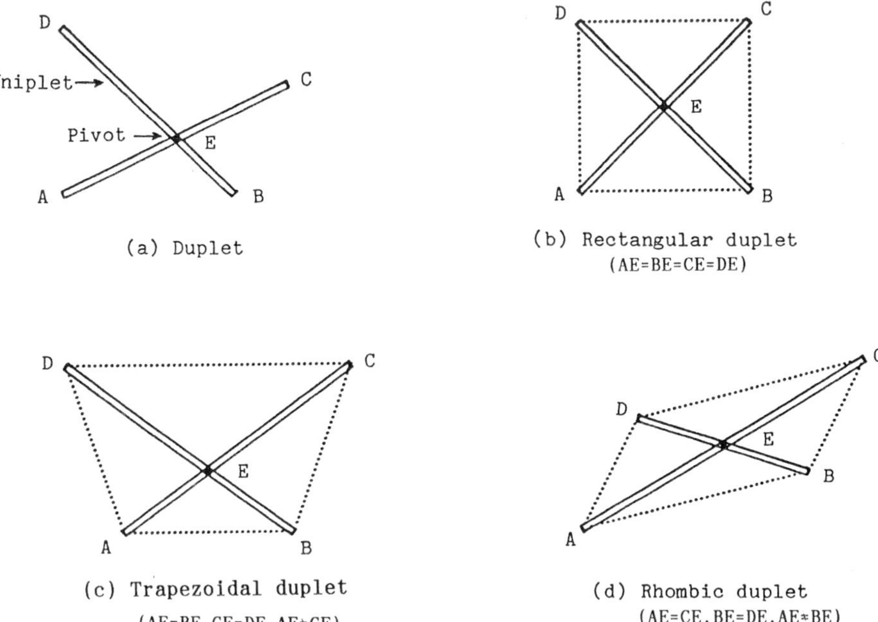

Fig. 1. Types of duplets.

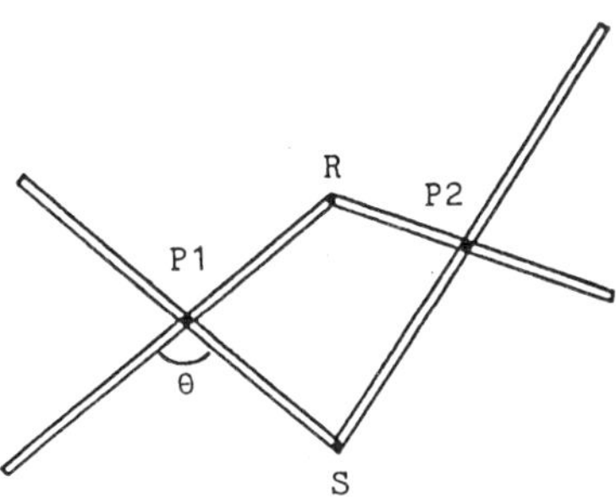

Fig. 2. Connection of a pair of neighbouring duplets.

4. In this study, every pair of neighbouring duplets, see Fig. 2, in a p-structure satisfy the following relation:

$$P1R + P1S = P2R + P2S \quad (1)$$

where P1R, P1S, P2R and P2S are respectively the original lengths of the segments of the uniplets as shown.

FOMEX FORMULATION FOR COMPATIBLE P-STRUCTURES

5. P-structures may be classified into two groups, depending on the manner in which the duplets are deformed in the folding process. A foldable p-structure is said to be 'compatible' provided that no internal deformation in the uniplets exists at any folded state and is said to be 'incompatible', otherwise. In this section, the basic approach in formex formulation of compatible p-structures is illustrated in terms of examples.

6. As the first example, consider the foldable assembly of Fig. 3 which consists of identical rectangular duplets. A formex formulation of this configuration may be written as:

CONNECTIVITY = '::

```
E1={[0,0,0;1,0,1;2,0,2],[2,0,0;1,0,1;0,0,2]}
E2=pex¦rinid(6,6,2,2)¦ros(1,2,1,1)¦E1
N ={[6,6,0],[6,6,2]}
F =lux(N)¦E2'
```

where formex variable F represents the configuration of the assembly with respect to the normat of Fig. 3a. The expression E1 represents the duplet at the bottom left corner in the first direction. A uniplet is represented as a cantle of the 3rd plexitude. The reason is that to carry out the stress analysis of a p-structure by the standard stiffness method, a convenient approach is to use the element stiffness matrices based on a 3-noded uniplet, (ref. 7). The formex represented by F may be readily transformed into element connectivity information in terms of a suitable node numbering pattern and used as data for structural analysis.

7. Attention is now turned to the determination of the geometric particulars of the p-structure at different folding stages. The assembly of Fig 3a may be freely folded to assume a state shown in Fig. 3b, which in turn may be freely deployed to assume the configurations in Figs 3c to 3e. Suppose that the structure does not experience diagonal distortion during the folding process. Any folded state is then specified by a parameter, for instance, the angle θ between the uniplets of a typical duplets as shown in Fig 3a and 3f. In terms of θ, the geometric particulars of the structure is obtained from the formex F by using the following co-ordinate transformation.

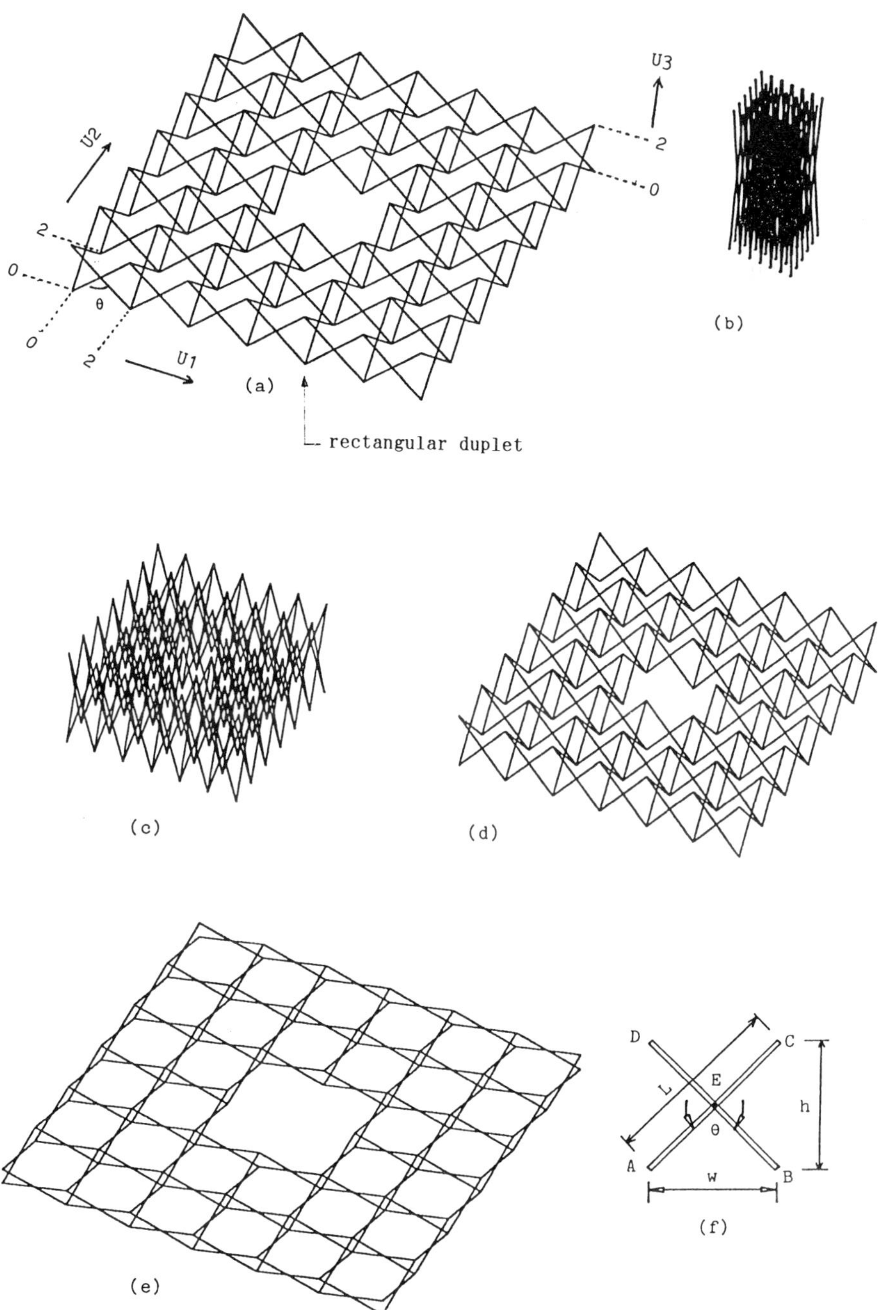

Fig. 3. Different folded states of a plate shaped p-structure.

```
x = L * sin ¦ (θ/2) * U1/2
y = L * sin ¦ (θ/2) * U2/2
z = L * cos ¦ (θ/2) * U3/2
```

where L is the length of a uniplet as shown in Fig. 3f. The configurations of Figs 3b to 3e are plots of F for which θ are taken as 3, 30, 75 and 150 degrees respectively for the above retronorm.

8. As another example, consider the foldable p-structure of Fig. 4a which assumes the shape of a barrel vault. It consists of two types of identical duplets, namely, trapezoidal duplets in the circumferential direction and rectangular duplets in the longitudinal direction. The assembly is compatible, thus capable of being freely folded to assume the configurations shown in Figs 4b to 4c. A formex formulation for this assembly with respect to the normat shown is written as:

```
F=pex¦rinit(6,6,2,2)¦ros(2,3,1,1)¦
   {[0,0,0;1,1,0;2,2,0],[0,2,0;1,1,0;2,0,0]}
```

The retronorm specifying different folded states of the assembly may be written with respect to the cylindrical co-ordinate of Fig. 4a as:

```
   | Ra  if U1=0
   | Rb  if U1=2
r= | Re  if U3 is even and U1=1
   | Rh  if U3 is odd  and U1=1
θ = α/2 U2
z = w/2 U3
```

where Ra, Rb, Re and Rh are respectively the radii of points A, B, E and H and where α and w are parameters associated with angular and linear co-ordinates as shown in Figs 4a and 4d to 4e. Given the dimensions of the trapezoidal duplets, one may calculate the lengths of the rectangular duplets by using Eq. 1. Once the dimensions of the duplets are determined, one may expressed all the parameters in the above retronorm in terms of the angle α, see Figs 4a to 4d and the configurations of Figs 4a to c may be obtained as plots of formex F by choosing appropriate values of α.

9. The formex formulation illustrated in these examples uses formices of the integer type only. This approach involves selection of a normat consisting of integer co-ordinates. The formices are formulated with respect to a partially folded configuration. A main task consists of setting up an appropriate retronorm for determination of geometric particulars of the assembly at various folded states. In some cases, however, one may find that it is natural to employ a normat consisting of co-ordinates that are real numbers. Consequently, the formices obtained are of the floatal type and application of it is described in the sequel.

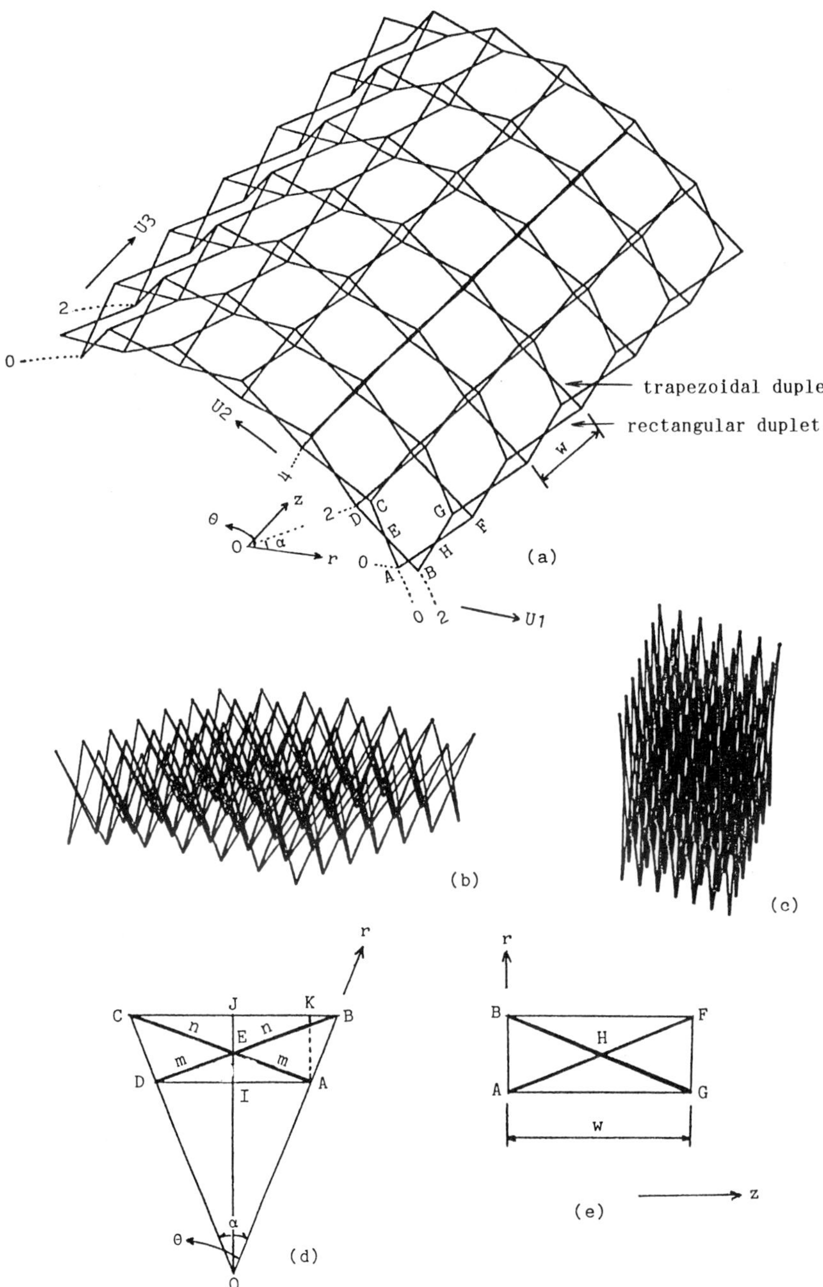

Fig. 4. Different folded states of a barrel vault shaped p-structure.

FORMEX FORMULATION FOR INCOMPATIBLE P-STRUCTURES

10. Formex formulations for incompatible p-structures is rather similar to that for compatible ones. For example, consider the fully deployed assembly of Fig. 5a. It is based on a hexagonal module consisting of six identical irregular duplets and six rectangular duplets, the dimensions of which are shown in Figs 5c and 5d. The duplets remain in their undeformed and stress-free state at the fully deployed and fully folded configurations, whereas internal deformations occur at other states.

11. The formex formulation of this p-structure is established with respect to the fully deployed state. The normat used here is coincident with the rectangular coordinate system xyz, with U1, U2 and U3 axes of the normat being identical with x, y and z axes, respectively. Thus, one may write:

```
E1={[0,0,50;217.5,-125.6,75.1;433,-250,100]
    [433,-250,0;217.5,-125.6,75.1;0,0,151]}
E2={[433,-250,0;433,0,50;433,250,100],
    [433,250,0; 433,0,50; 433,-250,100]}
F=rosad(0,0,6,60)¦(E1#E2)
F1=rinid(5,3,866,1500)¦F
F2=rinid(6,2,866,1500)¦tranid(-433,750)¦F
GRID=pex¦(F1#F2)
```

where E1 and E2 represent the irregular duplet AB and rectangular duplet BC, see Fig. 5b. GRID represents the configuration of the whole assembly. To obtain the geometric particulars for the fully deployed configuration, one may simply use the paritrifect retronorm,

BT(1,1,1).

12. Since the uniplets are deformed during the folding process, the partially folded configurations do not have simple retronormic relations with each other as in the case of compatible structures. To trace the deformation pattern for an incompatible structure during the folding process, one needs to make use of a suitable structural analysis program.

13. In using formices of the floatal type, one should bear in mind that any process that involves comparisons of the uniples of the formex variables can be meaningfully carried out only if an appropriate 'tolerance' is used, (ref. 4).

BEHAVIOUR OF P-STRUCTURES

14. The p-structures described in the previous sections consists of duplets. In many cases additional bars and cables are incorporated to stabilise or improve the capacity of the structure. The basic procedure of formex formulation, however, is the same as described in the previous sections. An analytical approach for p-structures is described in (ref. 7). In general, the following points apply to the p-structures.

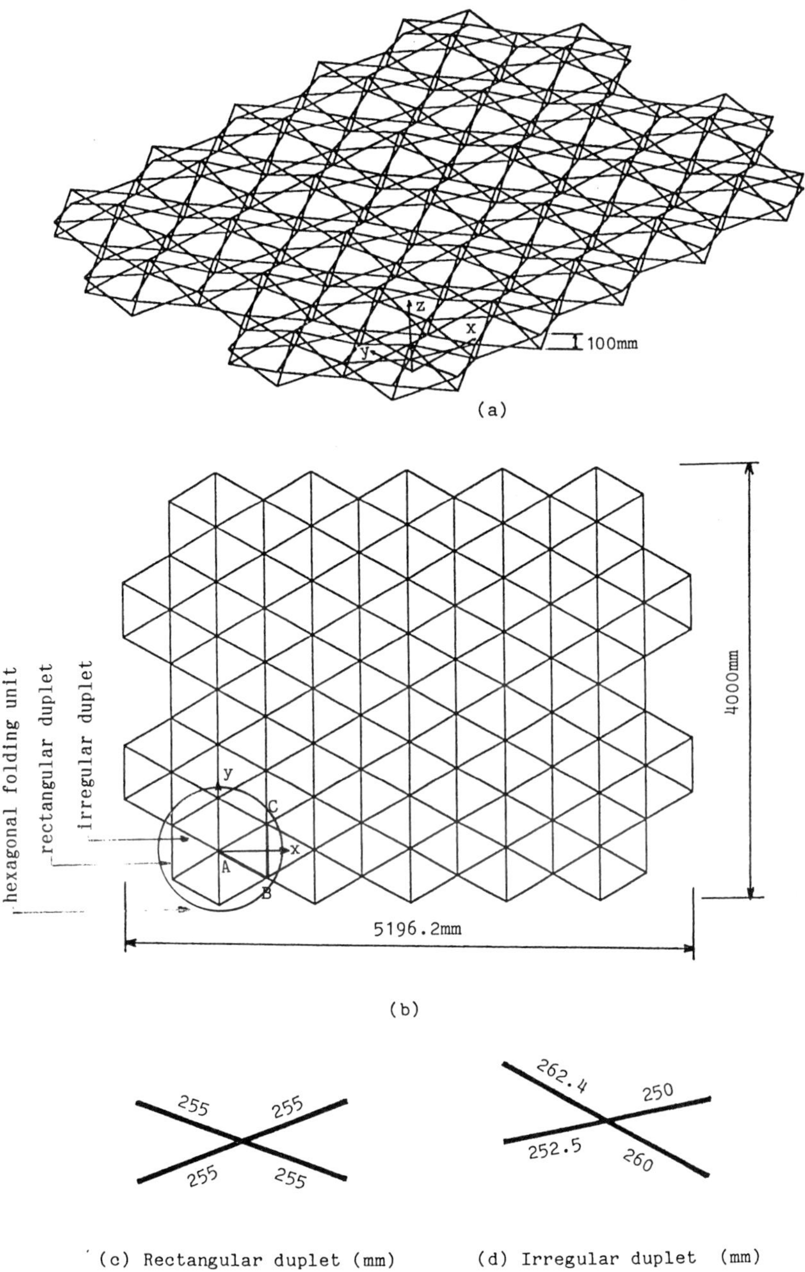

Fig. 5. Fully deployed state of an incompatible p-structure.

- The internal stresses in a p-structure are primarily bending stresses. The bending stresses in the uniplets are predominant, while the axial and shear stresses are small. Uniplets may be regarded as bending members subjected to axial forces.
- Incorporation of suitably arranged cables may result in significant reduction in the bending stresses and joint displacements.
- The effects due to the interaction between the axial forces and bending moments in a p-structure may be significant.

REFERENCES

1. ESCRIG F. Expandable space structures, Int. J. Space Structures, Vol. 1, No. 2 (1985).
2. ESCRIG F. and VALCARCEL P. J. Curved expandable space grids, Proceedings of the International Conference on the Design and Construction of Non-Conventional Structures, Edited by B. H. V. Topping, Civil-Comp Press, London, (1987).
3. NOOSHIN H. Formex Configuration Processing in Structural Engineering, Elsevier Applied Science Publishers (1984).
4. NOOSHIN H. and DISNEY P. Elements of Formian, Proceedings of the Fourth International Conference on Civil and Structural Engineering Computing, Edited by B. H. V. Topping, Civil-Comp Press, London(1989).
5. PINERO E. P. Three dimensional reticular structures, United States Patent 3185164(1965).
6. SHAN W. Foldable space structures, Ph.D Thesis, University of Surrey (1990).
7. SHAN W. Computer analysis of foldable structures, Computers and Structures, Vol. 42, No. 6(1992).
8. ZEIGLER T. R. Collapsible self-supporting structures, United States patent 3968808(1976).

89. The use of shape memory alloys in space structures, either as an actuator or as a passive damping element

W. VAN MOORLEGHEM, AMT, Belgium

SYNOPSIS. Shape memory alloys (SMA) have already been known for several years. However, we can only observe the first indications of industrial applications of SMA, especially for spatial structures. In future, it is expected that the use of shape memory actuators will increase, due to their advantages such as: reliability, volume savings, weight savings, noiseless operation and the suitability for micro G experiments. SMA actuators have also been recognised as possible actuators for use in smart structures (smart materials).

SHAPE MEMORY ALLOYS

Fig. 1. Difference in shape of a Cu-Zn-Al-shape memory coil at low (left) and high (right) temperature.

1. The basic mechanism governing the properties of shape memory alloys (SMA) is a change in crystal structure: a martensitic structure transforms, at a predefined temperature, into an austenitic structure during heating and reverts to the martensitic structure when the temperature is lowered again.

SHAPE MEMORY EFFECT

2. Through careful processing, it is possible to associate a specific shape with each of the structures: one shape in the martensitic state, and another one in the austenitic state. Since each structure is linked to a temperature range, this simply means that a switch from low to high temperature (or vice versa) will result in a shape change or, if this shape change is prevented, force change. The temperature range at which transformation occurs, is relatively small and can be chosen anywhere below 200 °C.

SUPER ELASTICITY

3. The thermoelastic process of martensite formation is both temperature (SME, shape memory effect) and stress (super-elasticity) dependent. Due to the reversibility of the martensite transformation, SIM (stress induced martensite) transforms back to the parent phase if the stress is removed and causes the macroscopic super elasticity. Superelastic strains up to 15% are possible in single crystal copper base alloys. Fig. 2 shows a typical superelastic stress-strain curve of a commercial polycrystalline Ni-Ti SMA (shape memory alloy). The plateau in the curve is temperature dependent (or composition dependent for a given working temperature).

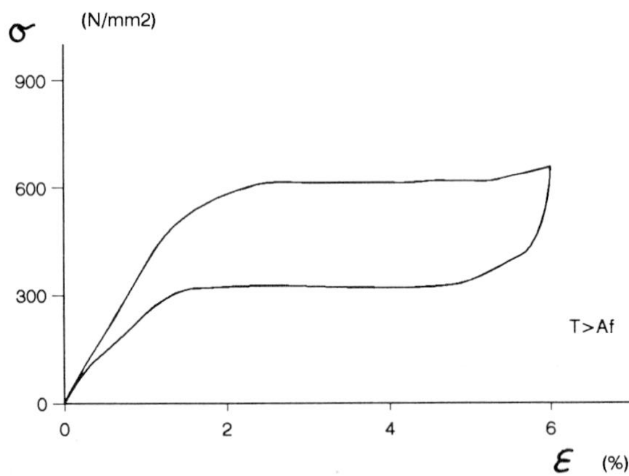

Fig. 2. Stress-strain curve of a superelastic Ni-Ti alloy.

Using SMA in the highly deformed martensitic condition, it is also possible to achieve a superelastic behaviour which is totally temperature independent. In this case, the curve doesn't show a plateau and the strains are limited up to about 2.5%.

DAMPING

4. Internal friction in thermoelastic martensitic alloys generally can be attributed to matrix-martensite and martensite-martensite interface movements. The former cause a damping peak at the transformation temperature and the latter occur in alloys below the transformation temperature. Other mechanisms may contribute significantly to the amount of damping, such as the movement of twin boundaries in Cu-Al-Ni and in Ni-Ti alloys.

5. Based on a comparative evaluation of different high-damping alloys, PROTEUS SMA-alloys stand out as having the highest damping effect of all. This occurs over a wide temperature range and is due to martensite-martensite interface movements. These PROTEUS high-damping-alloys can be used in two ways: as free unconstrained layers or in damped vibration absorbers.

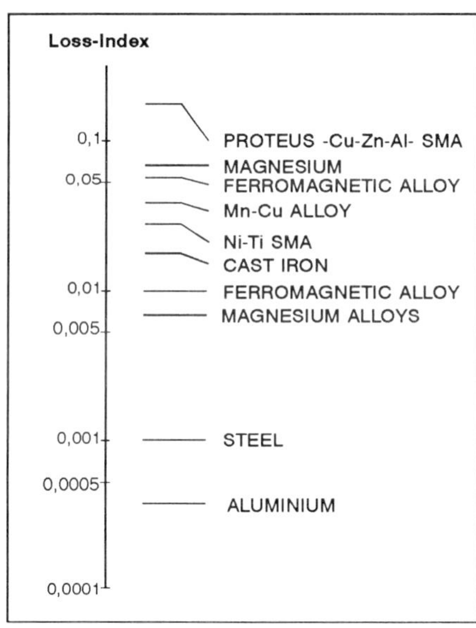

Fig. 3. Comparison of loss-index for different alloys.

Fig. 4. Space craft using PROTEUS-high-damping-alloy to protect a spectrometer against vibrations during launching.

SHAPE MEMORY ACTUATORS (ref. 1)

6. Comparison with alternative active elements. The main advantages of using shape memory alloys, as active elements in actuators are: (i) volume and weight savings in compact powerful actuators, (ii) noiseless operation, as mechanical and electrical of actuators, (iii) reliability: a SMA active element "has to" transform to its original parent phase/shape, even when the actuator has not been in use for some decades.

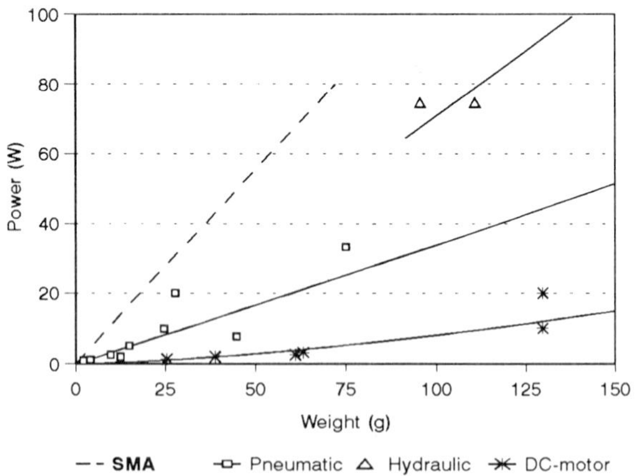

Fig. 5. Power/weight ratio.

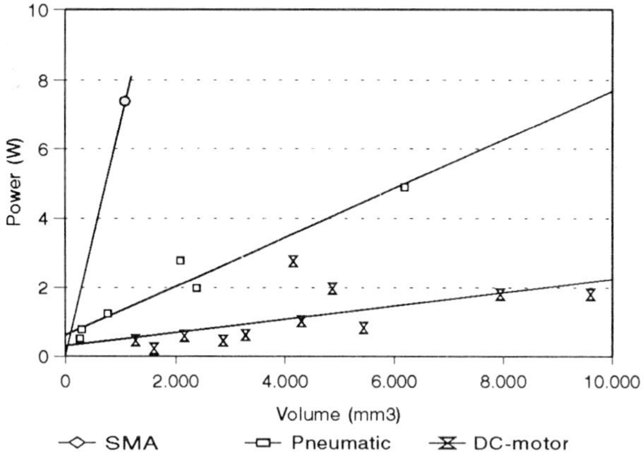

Fig. 6. Power/volume ratio.

7. The design of a SMA actuator starts from the knowledge of the temperature-stress-strain SMA behaviour (Fig. 7) (ref. 2), and of the limitations of SMA such as: overheating, lifetime, maximum transformation speed, stabilisation, hysteresis, ageing, etc.

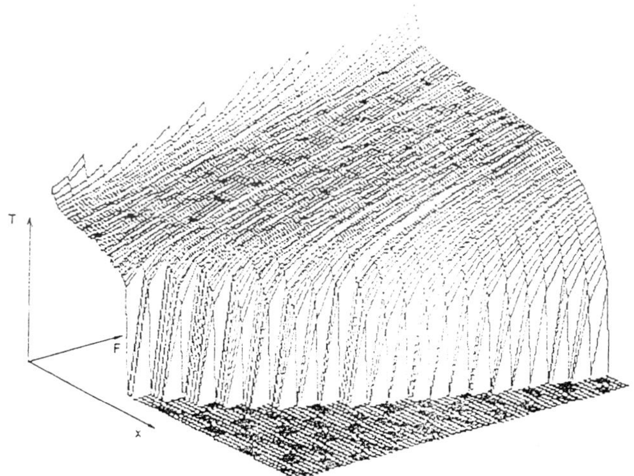

Fig. 7. Stress-strain-temperature behaviour of a Ni-Ti wire during heating.

SHAPE OF THE ACTIVE ELEMENT

8. Most SMA actuator designs use SMA wires or SMA coil springs as the active element. The main advantage of a coil is that it generates a large perceptible displacement out of a relatively small microscopic strain. As the stress distribution over the cross-section of the coil is not constant, a greater material volume is needed for generating the same work. When using tensile wires, a uniform stress distribution or optimum use of the material can be obtained. To generate sufficient

displacement there are two alternatives: first, to use a serial system, consisting of a long wire to generate the desired displacement and force; secondly, to use a parallel system of multiple wires in combination with a lever mechanism to generate a large force and a small displacement. Figs 8, 9 and 10 show some user data on a straight Ni-Ti SMA wire.

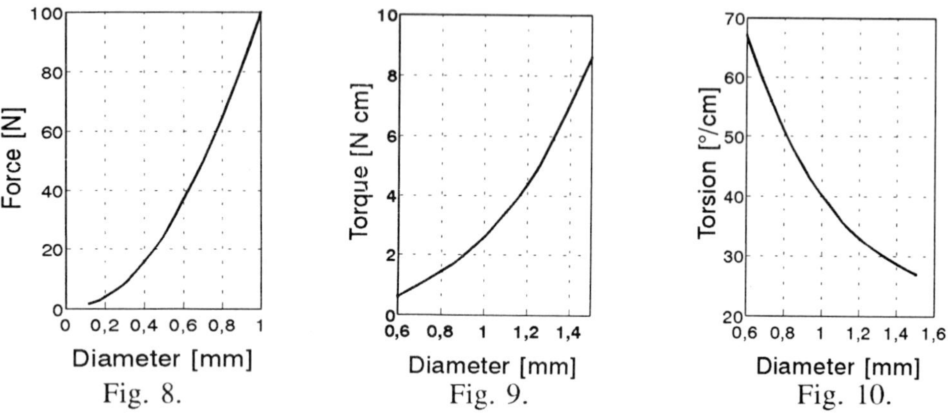

Fig. 8. Fig. 9. Fig. 10.

SOLAR PANEL ACTUATOR

9. In this case a torsion action of a SMA wire opens a solar panel. The actuator is heated by means of an electrical current via a resistive wire.

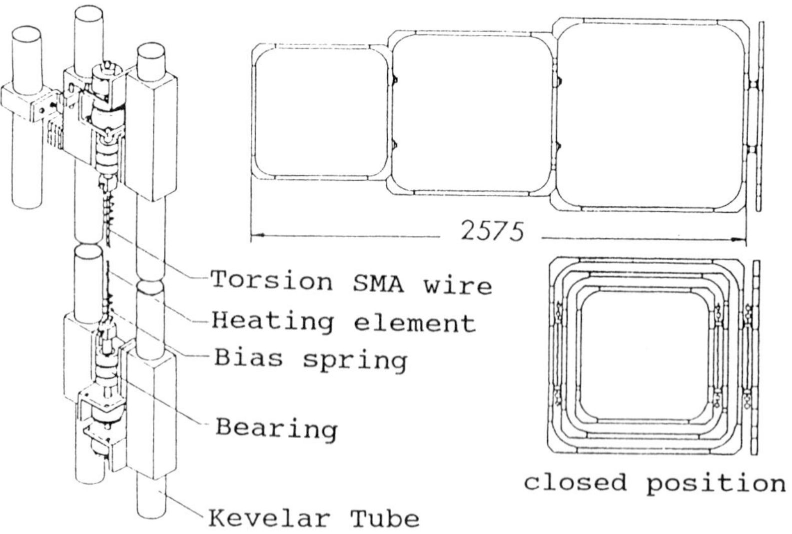

Fig. 11 . Solar actuator.

CUSTOM DESIGN SHAPE MEMORY MICRO ACTUATORS

10. A typical design of a micro-actuator is shown in Fig. 12. To maximise the efficiency and especially the bandwidth of the actuator, a straight wire (diameter 150 μm) is used as the active element.

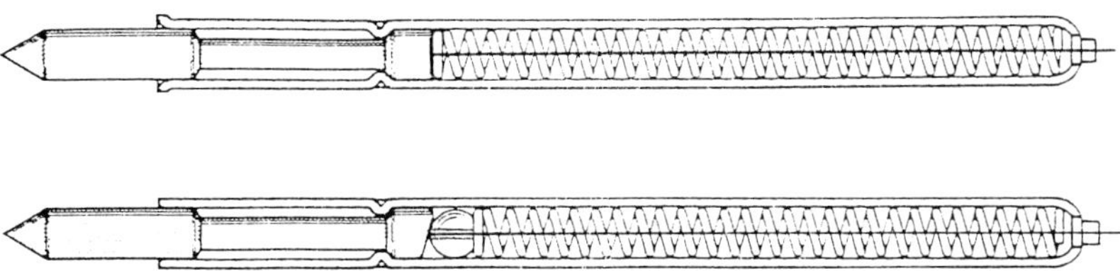

Fig. 12. Micro-actuator with a SMA wire as the active element.

When the SMA wire is activated through direct resistive heating, it pulls the micro piston against a bias spring, which is situated around the SMA wire. The housing of the actuator can be kept very small. Typical diameters of about 1.8 mm makes them attractive for aeronautical, medical, electronic and micro-robotics use. Other advantages of the configuration are its noiseless operation and short activation time. Typical overall characteristics of a micro-actuator are: diameter 1.8mm, length 40mm, weight 500 μgr, load 1N, activation time 0.1sec, voltage 0.5V - 5V, power 0.5W.

SPACE GRIPPER ACTUATOR

11. A first version for actuating a space gripper has been constructed, Fig. 13. It drives a light-weight gripper for manipulating scientific micro G experiments in a space station.

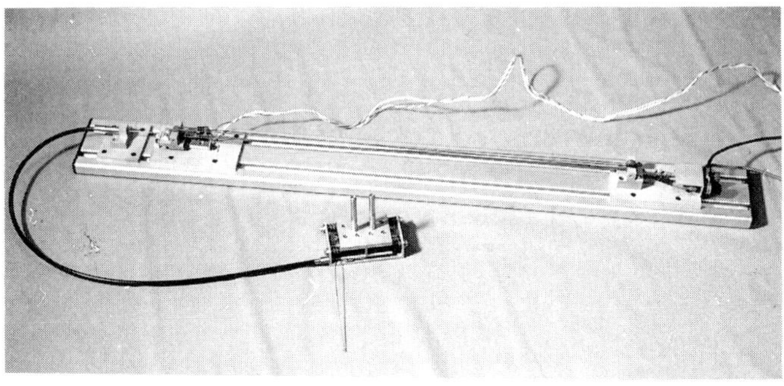

Fig. 13. Space Gripper Actuator.

The main design parameters for the gripper are: space qualification, high precision, gripping force 20N, opening/closing time < 3min and lifetime of 10 years (1000 opening/closing cycles). When not powered, the gripper is closed. The gripping force is provided by 4 parallel springs. The SMA actuator is used to open the gripper. To deliver the required force, it uses 4 parallel Ni-Ti 1.5 x 0.18 mm wires. One end of these wires is fixed while the other end is mounted on a linear guide. In this case, a levering pulley system is provided to obtain an equal stress distribution over the 4 wires. The system acts as a piston-like linear actuator (Max. 130 N). The total length is about 300 mm. A shielded steel cable is used to interconnect the actuator and the gripper.

Driving electronics

12. To obtain a higher electrical resistance, a serial (electrical) connection of the 4 driving wires is used. Using power MOSFETs, the driving electronics were further developed towards a pulse width modulated (PWM) system. This system offers the advantage of a much higher bandwidth compared to the thyristor amplifier.

Control

13. As the actuator only needs to keep the gripper open, a very simple position control is used. A limit switch, detecting the open position, switches the current amplifier on/off so that the "hot" position is maintained. A second command (power off) is used to close the gripper. The system is designed for atmospheric (mainly convective cooling) as well as for vacuum (only radiation) operation. For industrial use, an oil cooled version of this actuator, provided with a closed loop position control, has been developed.

SMA RELEASE MECHANISM

14. Hubble. A small light SMA release mechanism has been developed for actuating solar panels to their working position. Two bending elements were used.

SMA POSITIONING SYSTEMS

15. In space structures there is often a need for accurate positioning. Large space platforms need some parts to be positioned to micron accuracies. Satellites quite often use precision rotation equipment for pointing antennas into the desired direction. Shape memory actuators offer in those cases interesting solutions from the point of view of the smoothness of action and the freedom or flexibility during the design stage.

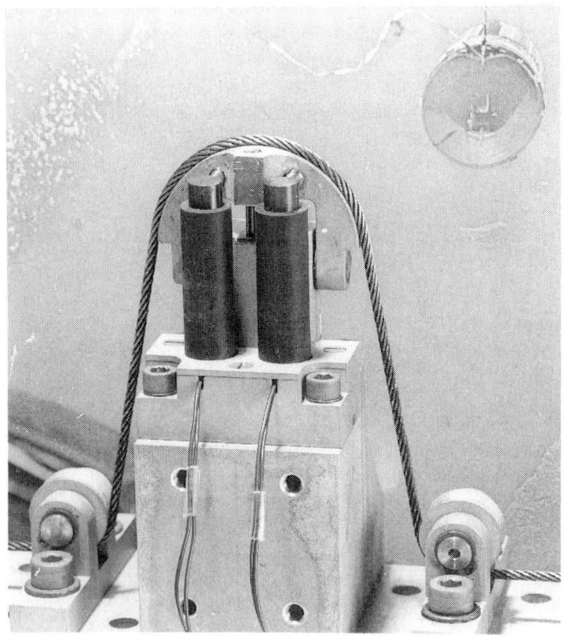

Fig. 14. "Hubble" actuator.

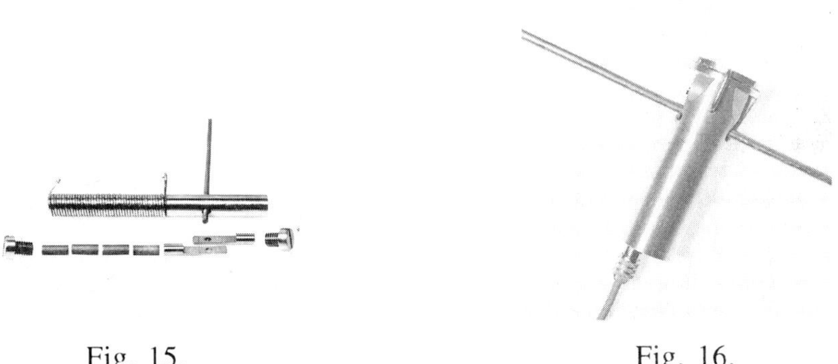

Fig. 15. Fig. 16.

Figs 15. and 16. show another release mechanism, the SMA-rod elements are used to activate a wire cutter.

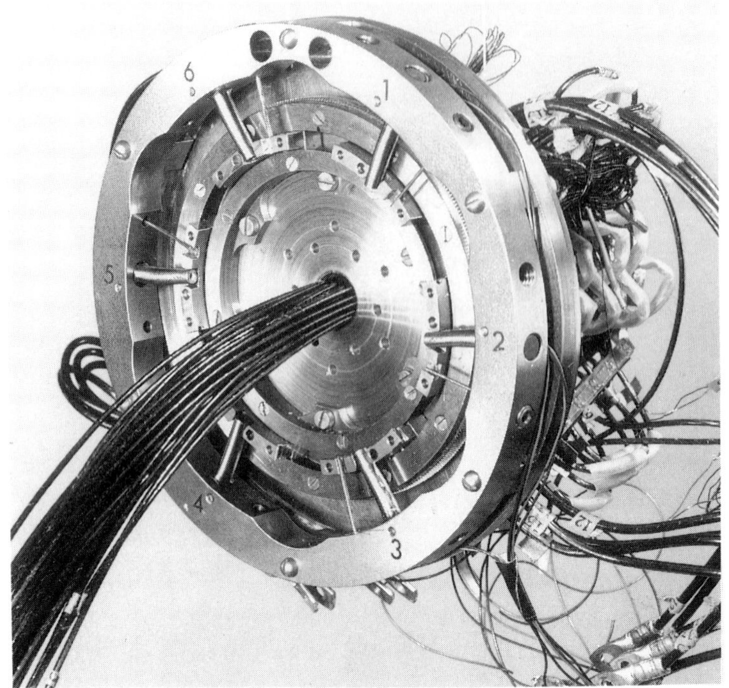

Fig. 17. SMA actuated rotor in a satellite mechanism.

REFERENCES

1. VAN MOORLEGHEM, W. et al. General discussion of the use of Shape Memory Actuators. Conference proceedings Actuator 92, 1992, June, 225-227.
2. FRATEUR, W. et al. Toepasbaarheid van vormgeheugenlegeringen als actief element in een robotactuator, Eindwerk, Katholieke Universiteit Leuven, Leuven, 1990.

90. Design and performance of the octahedral deployable mast

A. S. K. KWAN and S. PELLEGRINO, University of Cambridge, UK

SYNOPSIS. This paper presents the recently developed Octahedral Deployable Mast (ODM) which consists of a deployable backbone of prismatic rods and a set of cable segments, and is deployed and pre-stressed by a multi-segment cable. A full description of the concept behind the ODM is given, including the deployment mechanism. The general computational techniques for analysing the ODM and also for setting up optimal states of pre-stress in the fully-deployed configuration are briefly discussed. Finally, the manufacture and testing of a 1m high, two module prototype which can be collapsed to a 10cm high stack are described. The testing program included stiffness measurements and verification of deployment accuracy. The tests have shown that the imposition of suitable states of pre-stress can eliminate free play in the joints of the structure.

INTRODUCTION

1. Deployable retractable structures feature regularly in spacecraft due to the large size of satellite components combined with limited size of launcher envelopes. For example, the Hubble Space Telescope has two deployable solar arrays, measuring 12m x 2m each, which were launched in a compact, rolled-up configuration. New structural concepts are still needed and the deployable mast presented in this paper is a by-product of an ongoing project on deployable mesh antennae.

2. The Octahedral Deployable Mast (ODM) is a new modular deployable mast, which can be deployed from a fixed support. Fully folded, the ODM is a compact stack of square frames and connecting rods. Its deployment is essentially two dimensional, with the deployment envelope limited to the width of the ODM. In the fully deployed configuration, the mast consists of a stack of geometrically identical octahedra made of struts and cables, see Fig. 1. Because of its modular nature, ODMs with any number of modules can be made without any conceptual changes. Next, we shall describe the bottom module (two octahedra) of the ODM starting from the octahedron at the bottom of Fig. 1.

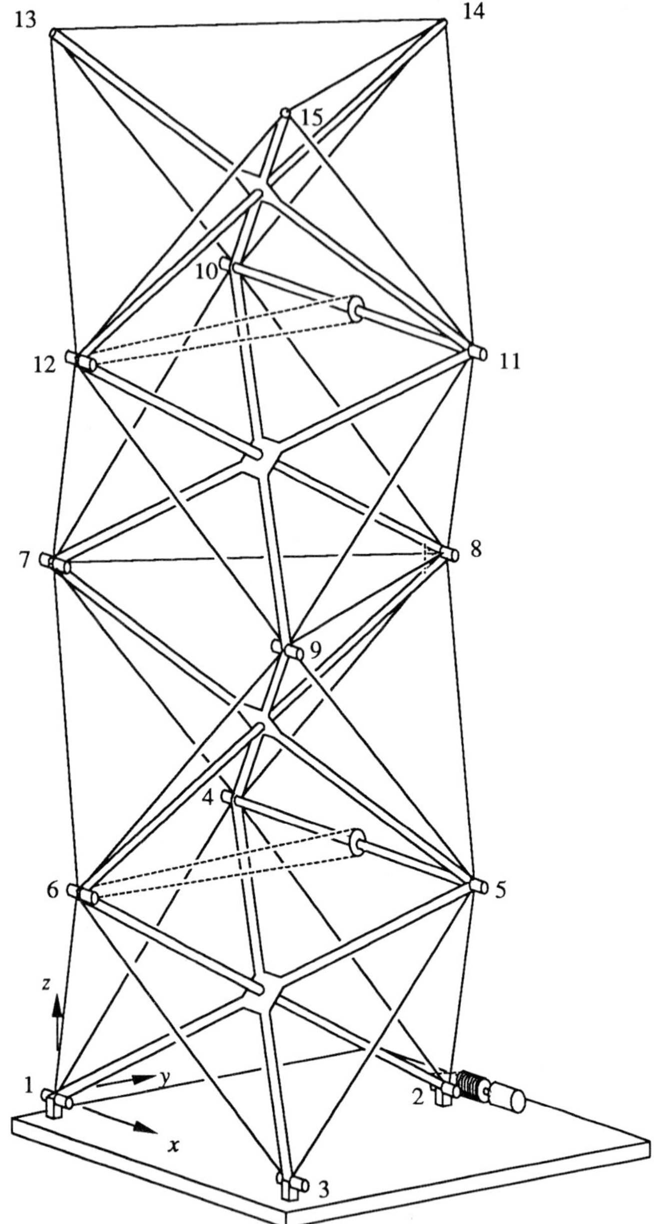

Fig. 1. A two-module ODM. Passive cables are drawn by solid lines and the active cable by a dotted line. Some parts of the active cable are hidden inside rods 2-6, 6-8, etc. All edge lengths are equal.

Deployable backbone

3. The deployable backbone of the ODM is mostly arranged along the space diagonals of the octahedra, see Fig. 1. In the octahedron at the bottom of Fig. 1 two rods, 1-5 & 3-4, form a cross which is stiffened by two pre-stressed cable elements, 1-4 and 3-5 while the third rod, 2-6, is connected to the centre of the cross with a pivot-and-slider joint which allows rod 2-6 to slide along its own axis, to rotate about an axis parallel to the x-axis and about its own axis. The bottom face (1-2-3) of the octahedron is fixed in size and therefore, as the pivot-and-slider joint moves up and down rod 2-6, the height of the octahedron will increase or decrease, respectively. This is the deployment mechanism. Actually, an ODM module is formed when a second octahedron is placed on top of the first, so that the two octahedra have a common face. The mobility of an ODM module is illustrated by Fig. 2. Only two of the three degrees of freedom allowed by the pivot-and-slider joint are required for deployment; the remaining degree of freedom is removed by connecting rods 2-6 and 6-8 to each other and to the base by revolute joints.

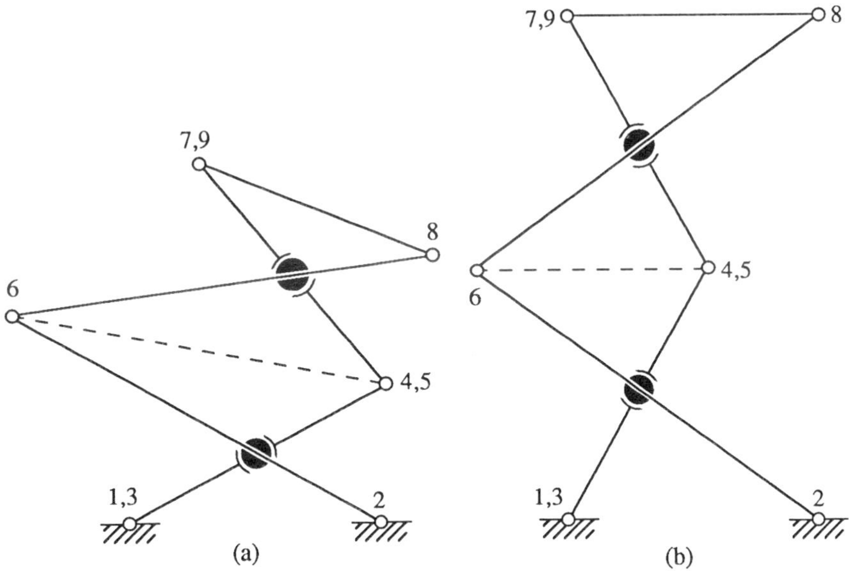

Fig. 2. Side view of bottom module of the ODM during deployment. As the active cable 6-4, 5 is shortened, the mast gains height, until the passive cables (not shown) become taut. Note that the top of the module, as well as the base, remain of constant size.

Active and passive cables

4. Passive cables connecting joints of the ODM whose distances increase during deployment (ref. 1) are arranged along 4 edges of each octahedron, e.g. 1-6 and 3-6. In addition, in our prototype there are two more cables, 7-8 and 8-9, connecting joints 7, 8 and 9 at the interface between the two octahedra. Folding does not require that the relative distance of these joints should change and, indeed, Fig. 2 assumes a constant size of triangle 7-8-9. Cables have been used, instead of triangular battens, to reduce the weight of the model, although at the expense of some extra mobility during deployment. All passive cables become taut when the ODM approaches its fully-deployed state.

5. The deployment of the ODM and, at the end of deployment, the prestressing of its passive cables are controlled by the active cable. At one end, this cable is firmly attached to joint 12; at the other end it is attached to a drum driven by an electric motor. The route of the active cable is mostly inside the backbone, the only visible parts are the loops between points of the backbone that have to come closer during deployment, e.g. the loop joining joint 12 to the pulley half way between joints 10 and 11.

6. The ODM deploys when the active cable is wound on to the drum. However, because the backbone consists of a series of modules each with its own independent mechanism, see Fig. 2, a single overall length constraint on the active cable cannot couple the deployment of many separate modules. Our model does not have further deployment constraints and hence its deployment can follow many different paths within a pre-determined envelope. Due to the tightening up of the passive cables, though, the fully-deployed configuration is unique.

Pre-stressing

7. A simple pin-jointed space frame consisting of 12 edge bars lying on the edges of an octahedron and 3 internal space-diagonals has 3 independent states of self-stress (each associated with one space-diagonal). If the 3 space-diagonals carry unit compressive forces, all 12 edge bars have tensile forces of $1/2\sqrt{2}$. Therefore, provided the correct state of pre-stress is imposed to it, such a frame can be made with 12 cables of unit length, say, and three bars of length $\sqrt{2}$ (the only practical difficulty is that the three bars have to go through the same point in the middle). Clearly, if two such pre-stressed space frames are joined together, the three cables at the interface will have tensile forces of $1/\sqrt{2}$. A fully deployed ODM module is in many aspects similar to this space frame, but in the middle there is an active cable instead of three fixed length cables. Naturally, the task of predicting the actual state of pre-stress of an ODM is made more complicated by the complex connections between different elements, but the simple analysis above turns out to be a reasonable first estimate of pre-stress levels, see paragraph 8.

COMPUTATIONS

8. We have developed numerical techniques for pre-stress analysis, and also for predicting force resultants and displacements due to any loads, of deployable

backbones connected to active and passive cables. First, we set up the equilibrium matrix of the complete assembly (refs 2, 3) from which we can compute a full set of independent states of self-stress. Then we identify suitable locations for a minimal set of turnbuckles by which any statically admissible pre-stress can be obtained (ref. 4). Within the same computational framework we can compute also the stress resultants and nodal displacements due to any loading.

9. For the prototype mast (see next section for details) this analysis produced 64 states of self-stresses in the fully deployed configuration. Some of these states of self-stress involve only members of the backbone and have nothing to do with pre-stressing the passive cables. We estimated that a nearly-uniform pre-tension (47-50N) of all passive cables could be achieved by shortening all passive cables by suitable amounts (in the range 0.8-2.9mm) and the active cable by 4.2mm from their nominal lengths.

DESIGN AND MANUFACTURE OF MAST

10. A two module mast has been built and tested (i) to validate the structural concept of the ODM; (ii) to highlight practical difficulties not shown in the theoretical models; and (iii) to compare the response of the physical model with computer predictions. The collapsed mast has a height of only 10cm with active cable length of 2.97m, while the fully deployed model is 1m high and the active cable is 2.31m long. The octahedra have nominal edge lengths of 300mm.

Backbone

11. The deployable backbone has been made out of Al-alloy tubes (16mm OD, 1mm thickness, EA=1.9×10^6 N, EI=17.1×10^6 Nmm^2 and GJ=12.7×10^6 Nmm^2) connected together with specially manufactured joints. The most complex component of the mast is the pivot-and-slider joint, shown in Fig. 3, which allows the three diagonals of each octahedron to intersect at a single point. An X-piece holds the two bars forming the cross. A short tubular element which can rotate with respect to the X-piece forms the pivot-and-slider support for the third bar. There are two other types of joints. One connects, e.g. rods 2-6 and 6-8 together by a hinge with axis parallel to the x-axis so that these rods are constrained to rotate in a plane parallel to y-z plane. The second joint type (e.g. joint 5) connects the remote ends of two crosses by a hinge parallel to the x-axis. The axis of rotation is therefore always at 45° to the axes of the rods making each cross. These hinge assemblies also support members 4-5 and 10-11. All connections between tubes and joints allow small length adjustments. In our model, short lengths of active cable are visible when the cable is taken out of a member and into the next one, over two small pulleys, to avoid the complication of building hinge assemblies with a pulley in the middle, see Fig. 4.

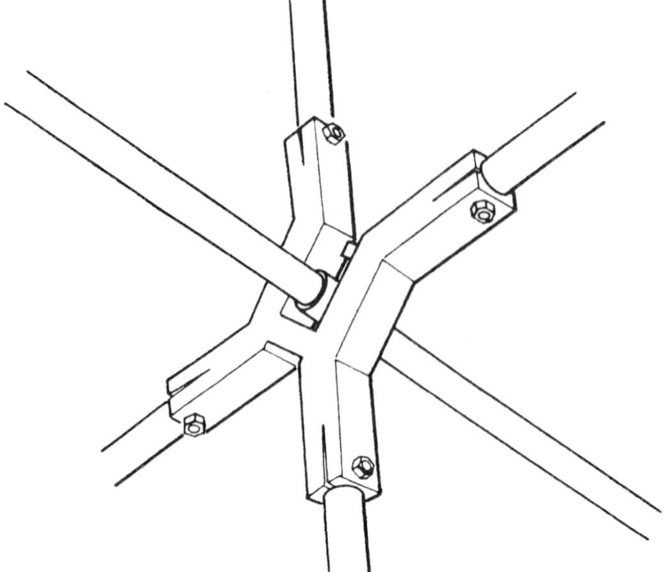

Fig. 3. Pivot and slider joint.

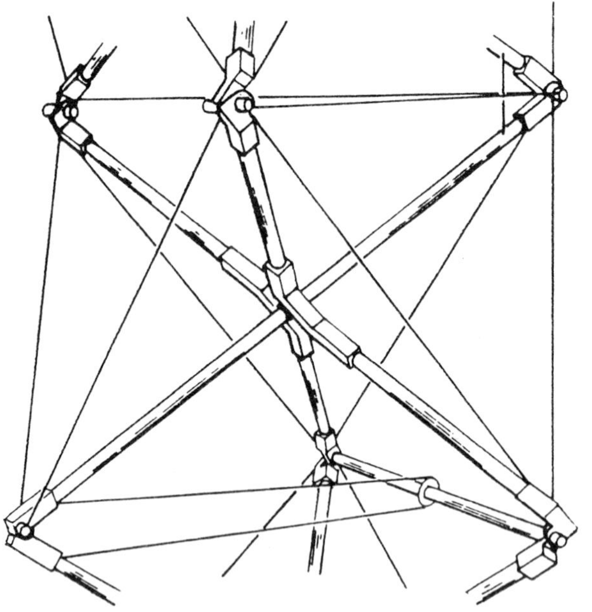

Fig. 4. Close-up view of the second octahedron from bottom.

Cables

12. Kevlar cables (1mm diameter, EA=25.6kN) have been used for the passive cables. These cables have been terminated simply by passing their ends through a hole in the joint assembly and tying a knot. Each cable contains a turnbuckle for accurate length control (down to 0.1mm). Strain gauges are glued on each turnbuckle to measure cable forces.

13. A steel cable (1mm diameter, EA=41.3kN) has been used for the active cable because the breaking load of the Kevlar cable is too low. All pulleys are mounted on ball bearings; the diameters of the pulleys are 25mm (large pulleys) and 10mm (small pulleys). A small strain gauged plate has been connected to the active cable, near joint 12.

Support and deployment mechanism

14. The backbone is connected to Al-alloy blocks fastened to the supporting base plate by stainless steel pins. A brushed DC motor (RS type PM10) is also connected to the base plate. The active cable runs over small pulleys as it comes out of rod 2 (see Fig. 1) onto the drum.

STIFFNESS TESTS

15. In this section we report on the behaviour of the fully deployed and prestressed ODM under 3 static load cases. Equal loads have been applied to the 3 top joints of the mast respectively in the x-, y- and negative z-directions. The displacement components of joint 13 in the direction of the load are plotted in Fig. 5 for the 3 load cases, together with the theoretical response (except for the negative z-direction loading, where experiment and theory practically coincide). For both the x- and y-loading cases, there are significant residual displacements although, after the first cycle, subsequent cycles tend to follow almost identical load paths. Table 1 compares the initial stiffnesses of the ODM in the 3 directions: clearly the lateral stiffness is much lower than the axial stiffness.

Table 1. Stiffness of ODM.

Loading	Measured	Predicted
x	0.64	0.50
y	0.80	0.86
-ve z	20.0	19.3

16. The axial forces in the active cable and two other cables in the lower module of the ODM are plotted in Fig. 6 for the 3 load cases above. Both experimental values and theoretical predictions are shown. Only the theoretical prediction for the active cable is not shown because it is practically identical to the measured response. The measured response is (i) practically linear, (ii) close to the theoretical prediction, and (iii) 'elastic' except for cable 2-4 for y-loading. Note the non-linearity in the response of cable 2-5 for x-loading as the axial force

becomes small. Theoretically, its response should be linear until the tension becomes zero, but in practice non-linear behaviour begins at about 10N.

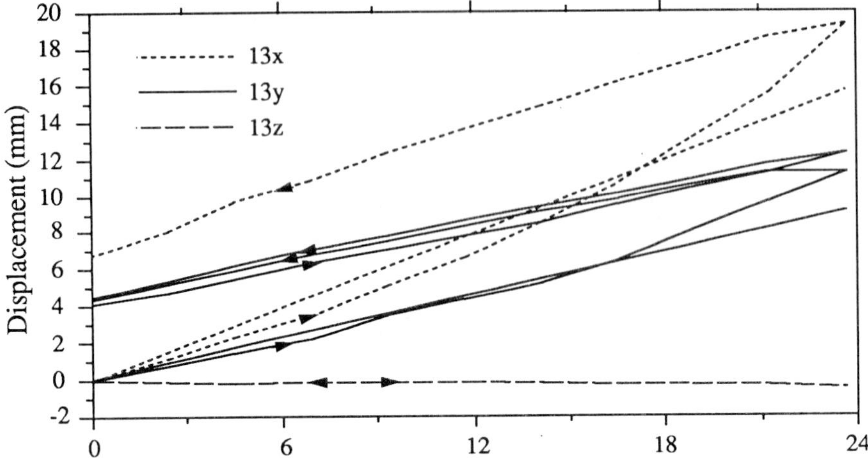

Fig. 5. Displacement components of joint 13 in the direction of the load, for 3 load cases.

DEPLOYMENT ACCURACY TESTS

17. Shape accuracy is important in most deployable structures for space applications, especially antennae. In these tests we have deployed and folded the prototype ODM twenty times and measured the position of joint 14 at the top of the mast with an Industrial Measurement System (ref. 5) using 3 electronic theodolites linked to a personal computer. After each deployment the motor was switched off when a set length of active cable had been wound onto the drum, thus producing the same level of pre-stress. Fig. 7(a) shows the position of joint 14 in the x-y plane after each deployment.

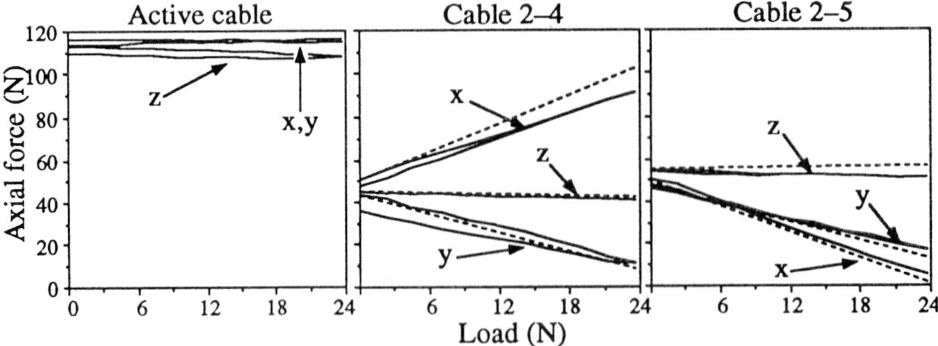

Fig. 6. Axial forces in selected cables for the three load cases. Theoretical predictions (except for the active cable) are shown in broken lines.

It can be seen that, while there is no consistent pattern of movement in the x-direction, there is a tendency for the mast to 'move' in the y-direction. The height of the ODM after successive deployments is plotted in Fig. 7(b): the mast increases in height with deployment. These two observations are in fact consistent, and the most probable cause is a general tightening up of the knots in the passive cables.

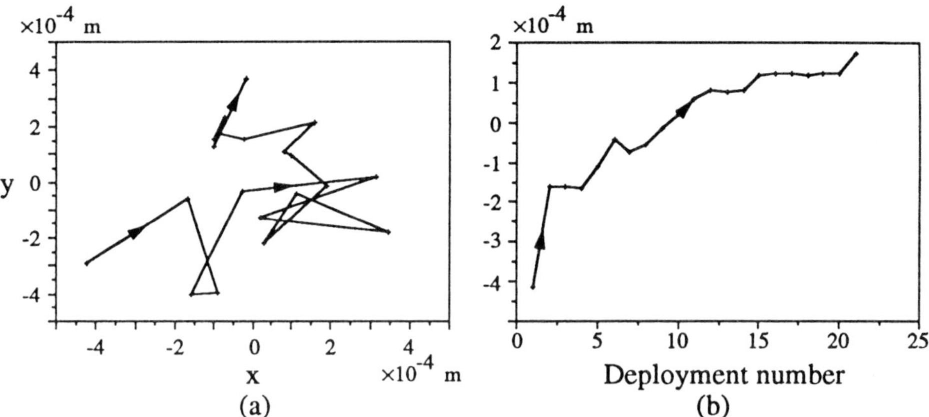

Fig. 7. Positions of joint 14 after successive deployments (a) x-, y-co-ordinates and (b) z-co-ordinate.

CONCLUSIONS

18. Overall, we have demonstrated that the concept of the ODM is a successful one. Our model showed a lack of free-play, or backlash, in the joints, leading to almost linear response to load (provided that the cables remain taut). Furthermore, when the pre-stress level is controlled, the ODM has good shape accuracy. These are two advantages of cable pre-stressed backbones over more conventional deployable structures. Many lessons have been learnt from the manufacture and testing of the prototype which would lead to improvements in future designs of the ODM. An obvious improvement is in the cable termination. We believe that successive tightening and slackening of the passive cables causes slip in the knots, leading to the progressive drift in the y-direction observed in the deployment tests (Fig. 7). In our current work cables are terminated with Al-alloy balls which are then connected to the rest of the structure, thus eliminating any possibility of slip.

19. Since each module of the ODM has its own independent mechanism, the ODM is particularly suitable for sequential deployment. The model described in this paper has no coupling between the two independent mechanisms of the two modules and it can be easily deployed in a sequential fashion, simply by preventing deployment of the bottom module until the top one is fully opened. No automatic mechanism, though, has yet been made. A mechanism to achieve this automatically may be a worthwhile improvement.

20. The key element in the ODM is the active cable. Although we have only used one active cable in our prototype for the sake of simplicity, it is possible to have more than one active cable so that failure of an active cable would lead, at most, to loss of some pre-stress but not to a total failure of the mast.

ACKNOWLEDGEMENTS

We would like to thank Professor K Miura for helpful comments. Financial support from SERC and DRA (Farnborough) is gratefully acknowledged.

REFERENCES

1. KWAN A.S.K., YOU Z. and PELLEGRINO S. Active and passive cable elements in deployable / retractable masts, International Journal of Space Structures, Special Issue on Deployable Structures, 1993, vol 8.

2. KWAN A.S.K. and PELLEGRINO S. Matrix formulation of three macro-elements for deployable structures, submitted for publication.

3. PELLEGRINO S., KWAN A.S.K. and VAN HEERDEN T.F. Reduction of equilibrium, compatibility and flexibility matrices, in the force method, International Journal for Numerical Methods in Engineering, 1992, vol 35, 1219-1236

4. KWAN A.S.K. and PELLEGRINO S. Pre-stressing a space structure, to appear in AIAA Journal.

5. KWAN A.S.K. and YOU Z. User guide to Theo3, an industrial measurement system using three Zeiss ETh2 theodolites, 1993, CUED/D-STRUCT/TR142, Cambridge University.

91. New spatial structures by innovative application of existing materials

R. NIJSSE, ABT Consulting Engineers, The Netherlands

SYNOPSIS. Responsible engineering implies a constant search for a better use of existing, or even introducing new, materials in our daily practice. Two methods are available to reach this goal: technology transfer and stretching of existing technology. In technology transfer we use experience assembled in a completely different domain. Stretching technology describes the constant push against the frontiers of the possible; gaining progress from a constant flow of small improvements.
The properties for two materials, glass and poly-aramide, were studied. In collaboration with architects we are designing structures using glass or poly-aramide as structural elements to create surprising concepts.

GLASS AS A STRUCTURAL COMPONENT

1. Glass is not an obvious structural material. However, it has an enormous structural potential. But, as in the world of Art where genius lays close to insanity, glass is equally dualistic. Take for instance its transparency next to its water imperviousness. Also, it couples an impressive potential in strength to a very brittle behaviour. Fig. 1 shows the potential for both the tensile and compressive strength but its brittle behaviour is evident from everyday practice.

2. The key to both the great strength and the brittleness lies in the amorph organisation of the molecular structure in the glass. Fig. 1 shows the basic element of glass.

3. This tetraëder causes both the firmness of glass and the brittleness, for if we note that the exterior has a negative electrical loading the brittleness will be understood. A crack started somewhere in the material will not be stopped but may grow continuously; the tetraëders are parted easily.

4. The high strength also results from the tetraëder-wise build-up since this creates a beautifully filled-up molecular structure, held firmly together by other forces working at the molecular level.

5. But does high strength automatically mean a impressive structural material? The answer must be no. In general a structure has to have three properties: strength, stiffness and stability.

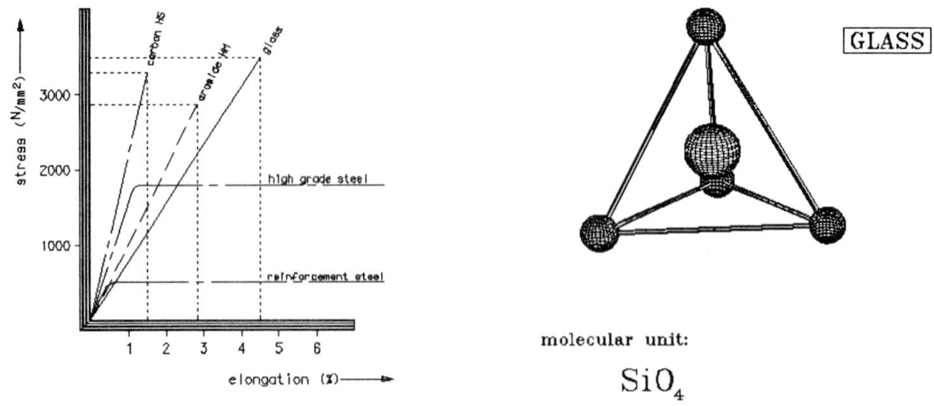

Fig. 1.

But I would like to add another criterion.

6. Each structure should give warning of overloading. The best warning is that of excessive deformation, which is clearly evident, and is an early warning of possible collapse giving enough time for escape.

7. Being a brittle material glass, unlike steel or reinforced concrete, fails without excessive plastic deformation. It is with this necessity for warning (plasticity) in mind that we started to design our glass-structures. Technology transfer implies a search beyond the borders of building technology. A study of products produced by the glass industry include standard plates, solid bars and cylinders. These elements we plan to transfer to building technology. But even with the commonly known plates we can stretch technology. The following projects both in study and under construction will show this.

8. European Patent Office. The Netherlands. An international architectural competition for $150m^2$ of high quality offices was won in 1990 by Neutelings & Roodbeen Arch. The original plan was to intersect office wings in an irregular grid. The spaces between the grid were to be roofed by glass (see Fig. 2). For this roof we developed a grid of steel beams spanning the approximately 25m wide atrium opening. Upon this beam-grid glass boxes were placed (see Figs 3 and 4).

9. Unfortunately this project was stopped in the final design (1991) due to a dramatic decrease in patent-applications from industry involved in military work.

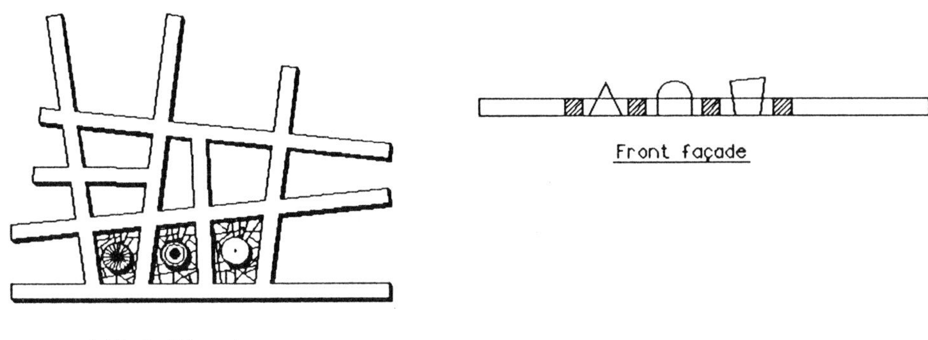
Plan European Patent Office
Front façade
Grid of office wings

Fig. 2.

Fig. 3.

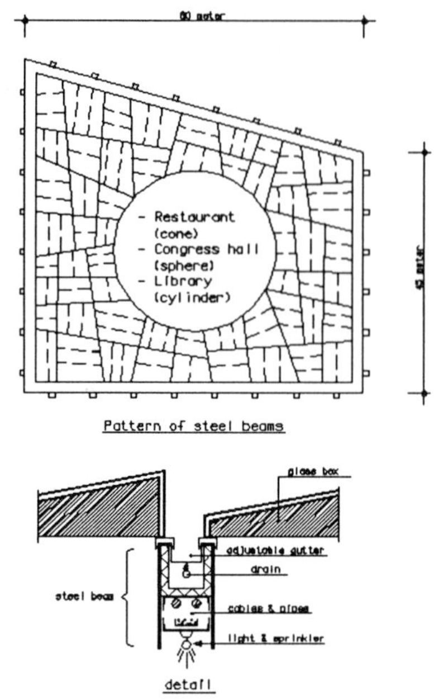
- Restaurant
(cone)
- Congress hall
(sphere)
- Library
(cylinder)
Pattern of steel beams
glass box
adjustable gutter
drain
cables & pipes
light & sprinkler
detail

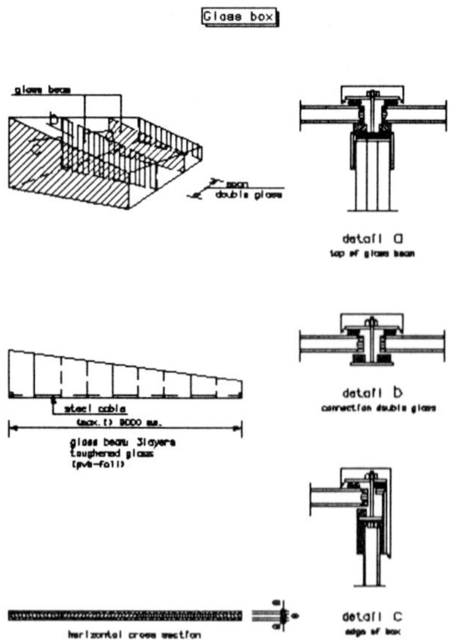
Glass box
double glass
detail a
detail b
steel cable
glass beam 3 layers
toughened glass
detail c
horizontal cross section

Fig. 4.

Glass roof, Budapest

10. With an open eye for the potentials of the Eastern European Market, a Dutch insurance company bought a stately mansion in the centre of Budapest. The Dutch architects group, Mecanoo, transformed this old building into a modern office, crowned with a big "egg", housing meeting rooms hanging in a glass roof. For this roof we developed glass beams balancing on glass columns (see Figs 5 and 6). By the time of the Conference the offices and the roof will be in use.

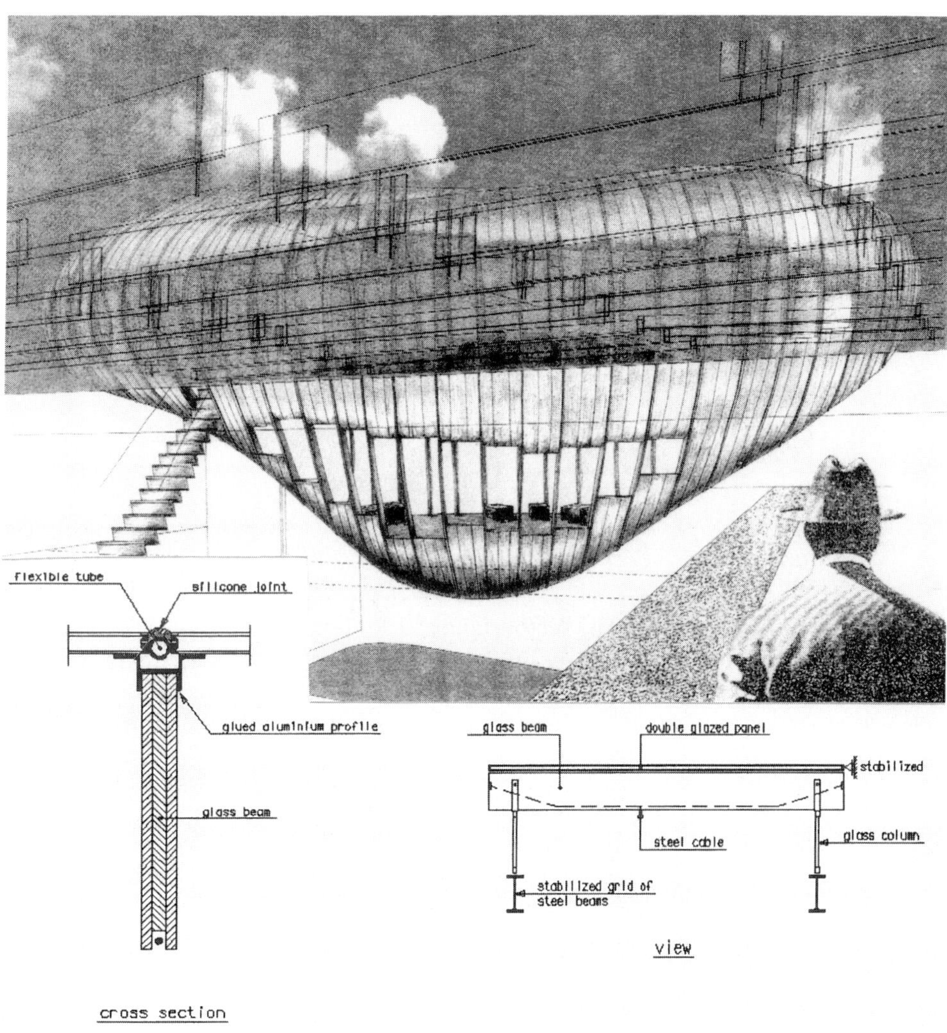

Figs 5 and 6.

Two study projects

11. In the search for new structural possibilities we created two special elements: the glass column (see Fig. 7) and the glass truss (see Fig. 8), in which glass is used for compression-loaded elements, and poly-aramide cables for the tensile-loaded elements. These structures will at least be good for a moment of surprise and will react in a special way to both natural and artificial light.

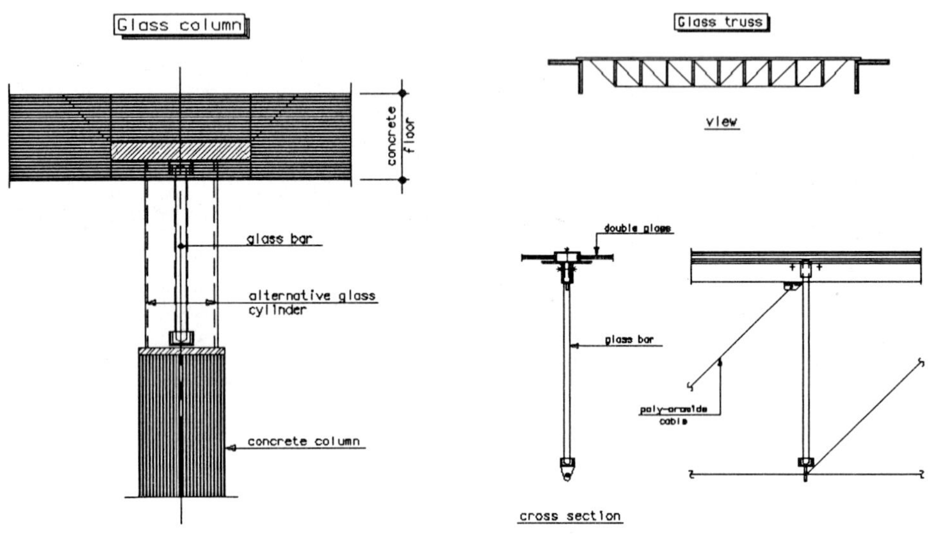

Fig. 7. Fig. 8.

Sonsbeek pavilion, Arnhem, The Netherlands

12. For the Sonsbeek '86 art exhibition a glass pavilion was designed by Benthem & Crouwel Arch. This temporary building serves as a glass covered walkway filled with vulnerable statues. Glass plates were used for walls, columns and the roof. Also structural silicone-joints were used (see Fig. 9). This project represented the start of the interest of ABT in structural glass.

POLY-ARAMIDE AS A STRUCTURAL COMPONENT

13. In the 1980s the chemical industry succeeded in making high strength fibres, poly-aramides on a large, commercial scale.

14. Despite the very high strength the connections proved to be problematic and expensive. We present three projects demonstrating the appealing properties of poly-aramides.

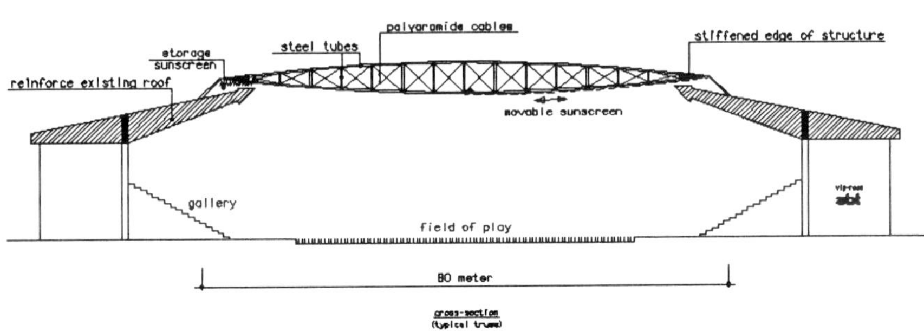
polyaramide cables
stiffened edge of structure
steel tubes
storage
sunscreen
reinforce existing roof
movable sunscreen
gallery
field of play
80 meter
cross-section
(typical truss)

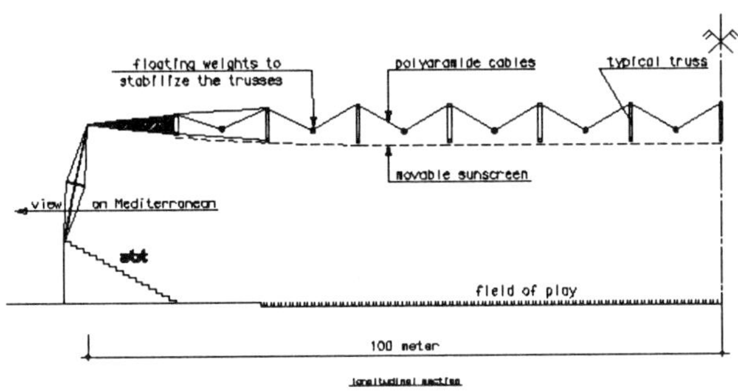
floating weights to
stabilize the trusses
polyaramide cables
typical truss
movable sunscreen
view on Mediterranean
field of play
100 meter
longitudinal section

Fig. 9.

Shadow roof-Monaco (Fr.), trusses stabilised by floating weights

15. To avoid sunburn on the grass of the football-field of AS Monaco we developed, in co-operation with a Dutch agricultural firm, a movable sun shade roof.

16. The sunscreens can move along the lower part of the trusses. The trusses are built up of small steel tubes and crossing diagonals made of poly-aramide cables. A special feature is the fact that trusses are stabilised against buckling by floating weights. (see Fig. 9). Unfortunately this project was cancelled.

Educatorium, Utrecht. The Netherlands, Mega-Truss

17. The famous Dutch architect Rem Koolhaas (O.M.A.) has been asked to design a lecture room-complex with a big examination hall for the university of Utrecht.

18. Due to the fact that no columns are allowed in the two large lecture rooms (400 and 500 persons) we were forced to think about a method to make a large free span within the building. Finally the marriage of a bridge and a building resulted in the shape of the building now under construction. It was the wish of the architect to clearly express the internal forces in the choice of the elements of the Mega-Truss.

19. Steel hollow sections filled with concrete were selected for compression-loaded bars, and poly-aramide cables for the tensile elements. (see Figs 10 and 11)

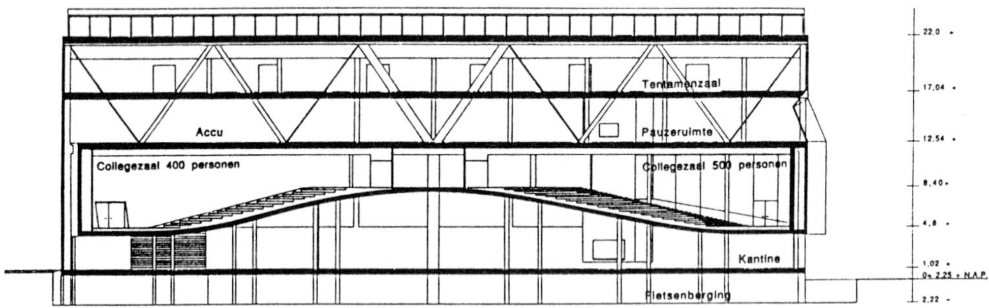

Fig. 10.

Fig. 11.

Foil-greenhouse. Wageningen. The Netherlands

20. A constant search for improvement of Dutch agriculture led to this project. A green house with foil (100 m PVDF), instead of glass, cable stayed by poly-aramide cables, was developed and constructed in 1991. Large areas of the roof are spatially curved by push-up points to avoid wind induced vibrations and to guarantee the the run-off of rainwater flow (see Fig. 12).

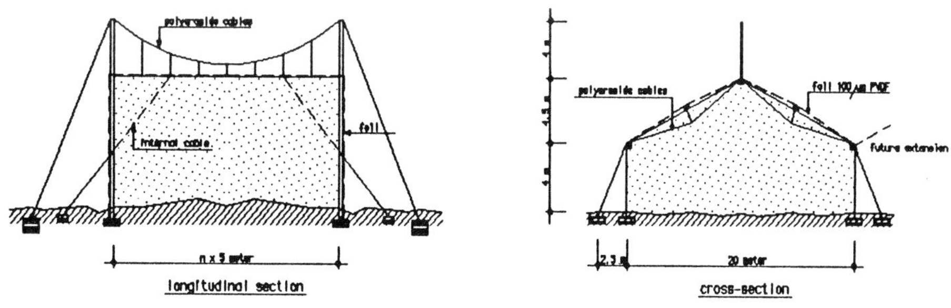

Fig. 12.

92. Wide-span structures using fire- and corrosion-resistant steel

T. ONO, Nagoya Institute of Technology, Japan, M. IWATA, E. NAGAI and T. OKAGAWA, Nippon Steel Corporation, Japan, and K. NISHIO, Kozo Keikaku Engineering, Japan

SYNOPSIS. This paper describes a large, spacious building that the authors recently designed. In this structural design, a large interior space was created by integrating an external wall structure (truss-in-wall) composed of Vierendeel trusses, and a wide-span roof structure (NS Space Truss), in a three-dimensional manner. A new fire and corrosion-resistant steel was employed for the truss-in-wall elements. The members of the truss-in-wall are rectangular hollow sections rotated 45 degrees around their axis and made into a diagonal grid. The NS Space Truss not only supports the roof dead load but also transmits the wind pressure and horizontal seismic force from external glass sashes to the truss-in-wall elements on both sides.

CONCEPT OF DESIGN

Outline of design

1. The characteristic of the present project is its structural design which positively incorporates structural members in the architectural design (Photo 1 and Fig. 1). In this structural design, a large interior space was created by integrating an external wall structure (truss-in-wall) composed of Vierendeel trusses, and a wide-span roof structure (NS Space Truss) composed of system trusses, in a three-dimensional manner. This structural design permits the standardisation of structural members and connections as well as automation of the fabrication of structural members, thereby contributing to an improvement in labour productivity and the cutting of construction costs.

2. Another major characteristic of the present design is that a new fire- and corrosion-resistant steel was employed for the truss-in-wall elements. This new material achieves the intended design effect and significantly reduces the maintenance cost of the building.

Structural design

3. The NS Space Truss roof spanned over the enclosed area and was supported by the truss-in-wall Vierendeel elements to form a large steel frame. This steel frame, consisting of the NS Space Truss and truss-in-wall elements, is supported

by 12 inverted-V SRC columns in such a way that it appears to be floating in the air.

Photo 1. External appearance.

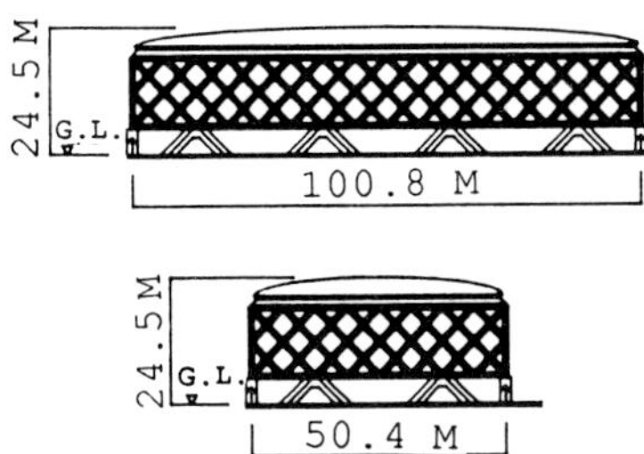

Fig. 1. Elevation.

Thus, the present building was designed as a three dimensionally-integrated structure in which the truss-in-wall elements, NS Space Truss and reinforced concrete frame are linked together to carry the load (Fig. 2).

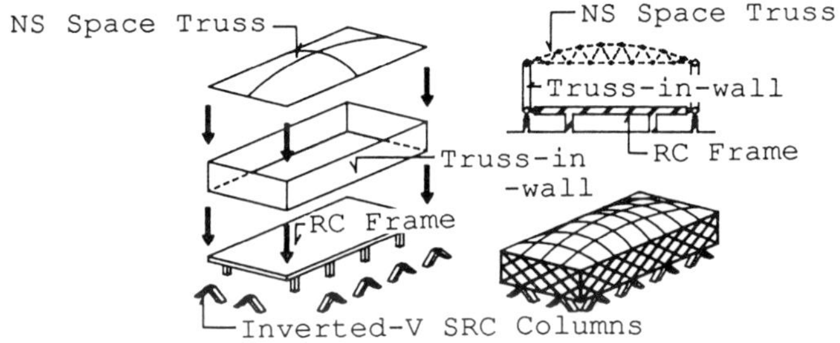

Fig. 2. Structural model.

4. The truss-in-wall is a Vierendeel truss consisting of 550 x 550mm rectangular hollow sections rotated 45 degrees around its vertical axis and made into a diagonal grid. This truss is the main structural member which not only supports the vertical load from the roof and part of the reinforced concrete frame, but also withstands seismic force and wind pressure. Since the truss-in-wall elements act as a Vierendeel truss resisting vertical loads and as a braced structure to resist horizontal load, it is expected to offer exceptional strength and stiffness.

5. The NS Space Truss is a system truss, made from circular hollow section members bolted to spherical nodes. This NS Space Truss not only supports the roof dead load but also transmits the wind pressure and horizontal seismic force from external glass sashes to the truss-in-wall elements on both sides.

6. The slab, except for a part of its periphery, is supported by reinforced concrete frame, set at 45 degrees against the external wall line. Since the girder

runs across the slab at an inclination of 45 degrees, this slab is expected to offer exceptional horizontal stiffness.

7. The 12 inverted-V SRC columns which are arranged at the periphery transmit the vertical and the horizontal force from the truss-in-wall elements to the foundation.

REALISATION OF THE CONCEPT

Design of the truss-in-wall structure.

8. Use of fire- and corrosion-resistant steel. The material used for the truss-in-wall elements is required to have the following three properties.

i) The material shall ensure the required fire resistance without any fire-proof sheathing.

ii) The material shall offer the required corrosion resistance without any coating.

iii) The material shall have good properties for fabrication.

9. A material which meets the above requirements, a new fire and corrosion-resistant steel, (FCR steel), was employed for the main structural members for the first time in the present project.

10. Fire resistance (high temperature characteristic). The FCR steel maintains a yield point of two-thirds or more of the standard value at a normal temperature (long-term allowable stress) even when the steel temperature reaches around 600°C. In addition, the decrease in the gradient in its stress-strain curve is small at temperatures up to 400°C to 500°C. Thus, compared with ordinary structural steel, the FCR steel has excellent high temperature characteristics. In the present design, therefore, based on a temperature analysis which assumes a fire, the proper space between the partitioning glass sashes and truss-in-wall elements was determined. As a result, the steel temperature in a fire could be kept below 600°C and the required fire resistance could be obtained without any fire protection.

11. Corrosion resistance. When the FCR steel is exposed to the air, it initially rusts as does ordinary steel. However, a dense, stable film of iron oxide (stable rust) is gradually formed on it, preventing the development of corrosion. The construction site for the present building is a hillside facing the Pacific Ocean, a severe natural environment. There was concern that non-uniform rust or rust propagation might occur in the process of formation of stable rust, consequently the steel was subjected to rust stabilising treatment. With the rust stabilising treatment, the need to protect the oxide film on the steel surface is eliminated since the coated film is gradually oxidised and replaced with stable rust, which covers the steel surface. This contributes significantly to decreasing the maintenance costs of the building.

12. Design of truss-in-wall connection (rectangular hollow section). Because the truss-in-wall members are subject to a large axial force and bending moment at the same time, the truss connection is required to have sufficient strength and stiffness to transmit these stresses smoothly. In the present design, in view of the design effect and fabrication of the truss, a coupling type reinforcement using a ribbed cross plate as shown in Fig. 3 was devised and adopted for the truss connection. The details of the truss connection are described below.

13. Design effect of connection. To impart a clear-cut image to the building, the truss members (rectangular hollow section) are rotated 45 degrees around its vertical axis and connected together. At the connection, the edge lines of members cross at right angles. Therefore, the coupling type reinforcement has a cross-sectional shape adapted to the valley formed at the connection, so that it does not impair the clear-cut image of member edges at the connection.

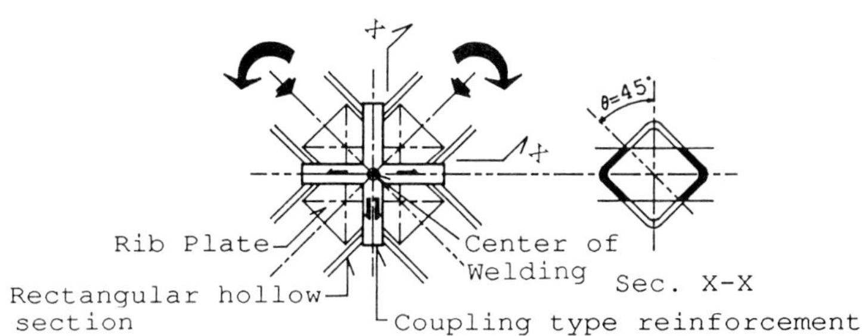

Fig. 3. Connection.

14. Structural performance of the connection. Because the cross-sections of the truss members (rectangular hollow section) concentrate at the centre of the weld of the connection as shown in Fig. 3, the axial force can be transmitted efficiently. In addition, since the coupling plate end is subject to out-of-plane bending stress, the plate is reinforced with a rib to increase its out-of-plane stiffness and to obtain a connection which efficiently transmits bending moment and axial force.

15. Confirmation of the structural safety of the connection. Structural safety of the connection was confirmed by the experiment described below. A test specimen made of rectangular hollow sections having the same length-to-diameter ratio as the members used in the structure was used to make the connection and was then subjected to compressive stress. The connection test specimen was provided with a coupling type reinforcement plate at the joint. The shape and mechanical properties of material are shown in Fig. 4 and Table 1. Fig. 5 shows the stress-strain curve obtained by subjecting the test specimen to a compression test. In the experiment, an axial force of 500 tons was introduced into the vertical member. The compressive force was made dimensionless by dividing the value by the yield axial force of the member, and the axial deformation was made dimensionless by dividing the value by the corresponding yield deformation. The maximum strength reached the yield axial force. At approximately 67% of the yield axial force, deformation increased and stiffness decreased.

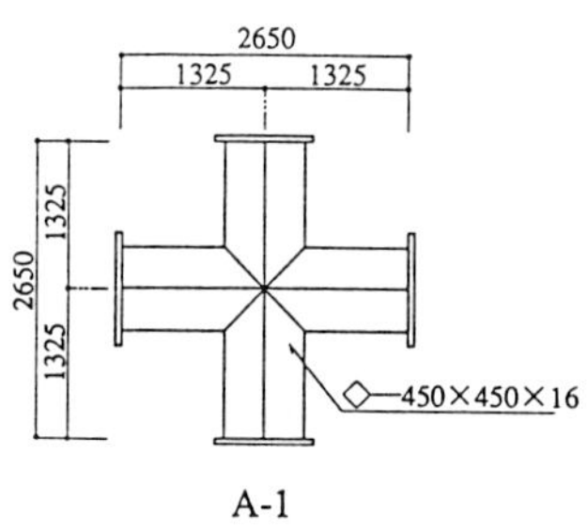

Fig. 4. Test Specimen.

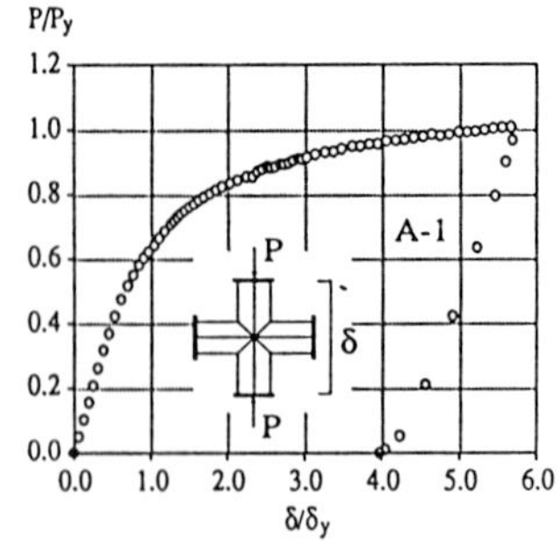

Fig. 5. Load-Deformation Relation.

Table 1. Material Properties.

Young's modulus (tf/cm^2)	Yield point (tf/cm^2)	Strength (tf/cm^2)	Elongation
2090	3.7	5.5	28%

The reason for this is probably that at yield axial force, the unloaded member was subjected to cross-sectional deformation via the coupling type reinforcement plate because a 500 ton axial force had been introduced into the other member. Based on the above result, in the present design, a rib plate constraining the coupling plate deformation was inserted in the cross section.

16. Fabrication of truss-in-wall connection (for rectangular hollow section). Because the truss-in-wall is a diagonal grid formed by rectangular hollow sections rotated 45 degrees around their axis, joints become extremely complex in shape. Although the maximum member thickness is 32mm, the thickness at the joint becomes as large as 70mm because the members are connected diagonally. This requires high dimensional integration, consequently, the members are required to be fabricated with extremely high accuracy.

17. Practical application of a cutting robot. In view of the above problems, in the present design, the external dimensions of the truss members were unified to standardise the connections and a cutting robot was employed to improve the efficiency of the member connection and secure the required accuracy of fabrication. In addition, to secure the required cutting accuracy, a work bed permitting the levelling and centring of work pieces to be adjusted within an accuracy of ±l mm was designed.

18. Improvement of weldability. The truss-in-wall connection requires the welding of extremely thick plate. Therefore, cracks can easily occur in the welds. In view of this, a new welding solvent was developed to improve the weldability. The new solvent is a conventional solvent to which accurate amounts of Mo, Cr, Ni, etc. are added.

Design of the NS Space Truss roof structure

19. To impart a buoyant impression to the building, the shape of the roof was designed so as to give the image of a bubble. So, the roof members were required to be simple and slender. However, structurally, the roof as a dimensionally integrated structure was required to have sufficient strength and rigidity. To meet all of these requirements, an NS Space Truss was adopted as the wide-span roof structure in the present design. Characteristics of the NS Space Truss are described below.

20. Structural advantage of the NS Space Truss. As shown in Fig. 6, the NS Space Truss consists of spherical nodes and circular hollow section members bolted together in a three dimensional manner. The axial centre of the circular hollow section members passes through the centre point of the node member to avoid secondary bending due to eccentricity of connection. Therefore, it is unnecessary to consider the deterioration of strength and secondary stress due to local buckling of the connections. Thus, design can be implemented very economically.

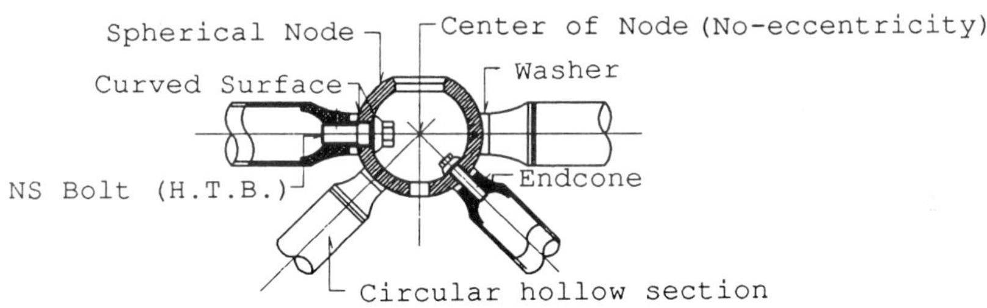

Fig. 6. NS Space Truss.

21. Adaptability to complicated shapes. The NS Space Truss is a system truss which was developed with emphasis on its adaptability to the complicated shape of roof structures. It can easily be adapted to any shape merely by adjusting the angle of node-bolt-holes and the length of the circular tubular members. Because the NS Space Truss components are fabricated by a completely established industrial system employing such numerically controlled machines as NC cutter, boring machine and automatic welder, the effect of the complexity of the roof structure shapes on production cost is minimal.

Securing strength as a dimensionally integrated structure

22. Horizontal stiffness of building. Used as a partitioning wall in the building, a glass sash is arranged inside the truss-in-wall structures. The top end of this sash is supported by the NS Space Truss via secondary beams and the bottom end is supported by the RC frame. Therefore, in order to secure the horizontal stiffness required of the building, the allowable deformation angle of the sash (h/120) must be met. To increase the horizontal stiffness of the present building, it is most effective, from the viewpoint of structural design, to improve the horizontal stiffness of the NS Space Truss. Therefore, in the present design, the NS Space Truss lower chord member was set in the horizontal position and diagonals having a large cross-sectional area were arranged effectively to improve the horizontal stiffness of the NS Space Truss and secure the buildings horizontal stiffness.

23. Strength of the RC Frame. Because of its plan, sufficient earthquake-resisting walls cannot be constructed. Hence, the horizontal force occurring during an earthquake is partly transferred to the truss-in-wall structure. In the present design, therefore, consideration was given to ensuring smooth transfer of the horizontal stress during an earthquake. Consequently, the RC frame was arranged at an angle of 45 degrees to the truss-in-wall elements. In addition, the end of the slab girder was connected to the truss-in-wall lower chord member to secure the joint strength necessary for the transfer of stress.

24. Safety against temperature stress. Because the present building is a large structure of 50.4m by 100.8m, the effects of temperature change on the building cannot be ignored. In view of this, a three dimensional analysis was conducted under suitable conditions to study the effects of temperature stress. As a result, it was confirmed that a stress concentration due to a temperature change occurred at the four corners of the building. Nevertheless, the effect of the temperature change on the truss-in-wall structure and NS Space Truss was negligibly small, although the amount of reinforcement of the girder connecting the truss-in-wall elements and RC frame was increased slightly.

Construction method for the present building

25. In the construction of the present building, four points can be cited.

i) The present building is unstable until its structural components are completely connected together.

ii) The construction work requires extremely high accuracy.

iii) Because the size of the truss-in-wall blocks that can be fabricated in the fabricating plant is limited by Traffic Regulations, nearly half of the welding has to be done on site.

iv) To shorten the construction period, the different structures have to be constructed concurrently. Taking the preceding points into consideration, proper temporary structures and construction methods were designed to carry out the construction of the present building, as described below.

26. Temporary scaffolding. The temporary scaffolding employed to construct the present building is shown in Fig. 7. Temporary scaffolding shown in Fig. 7; (1) and (2) are for the erection of the truss-in-wall elements. The horizontal force (wind pressure) is supported by structure (1). The structure vertical load is

supported by structure (2). Temporary scaffolding structures (3) and (4) are for the erection of the NS Space Truss; structure (3) supports the vertical load and gives level adjustment. Temporary scaffolding (5) supports the vertical load of the cantilevered RC frame. The temporary scaffolding (1) was installed outside the truss-in-wall structure so that it cold be dismantled easily and the work could be executed independently to the RC frame construction.

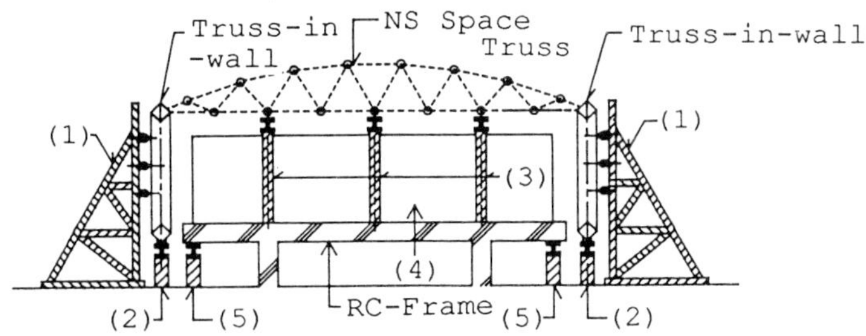

Fig. 7. Temporary scaffolding.

27. Erection of truss-in-wall elements. The truss-in-wall elements were erected by stacking factory-made blocks of the members as shown in Fig. 8. The welding work was performed stack by stack from below, except for the bottom members, and the welding quality was checked throughout using ultrasonic test methods. The bottom members were welded ahead of the other members because they served as the base for erection of the upper members and hence required sufficient stiffness. At the same time, they were connected to the RC frame.

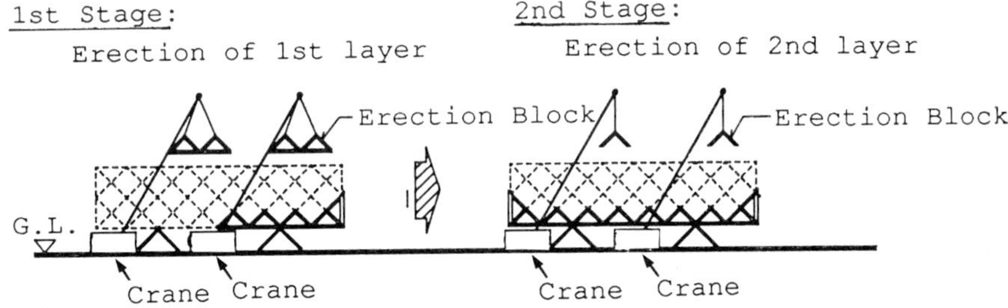

Fig. 8. Erection.

28. Erection of NS Space Truss. The connection between the truss-in-wall structure and the NS Space Truss is the point where erection errors in the

individual structures are accumulated. Therefore, the work on the connection is extremely difficult. In the present work, a span adjustment mechanism was provided for the connection in order to properly process the cumulative erection errors and secure the soundness of the connection. Span adjustment is performed by fabricating the boundary members 25mm short of the design length and inserting a filler plate of appropriate thickness between the endplates. In addition, the brackets were site-welded to the truss-in-wall structure to improve the erection accuracy.

CONCLUSION

29. The authors consider that the present project is representative of the architecture of the future in terms of design effect, productivity and economy. Also, we are confident that the successful expression of an innovative new design in the form of the dimensionally integrated structure obtained by combining the truss-in-wall structure and the NS Space Truss is of great significance and should be able to be effectively applied to other similar projects.

93. The manufacture of steel structures for large-span structures

E. P. LUKIANENKO and G. G. BONDARENKO, Metal Structures Works, Ukraine

SYNOPSIS. Large span steel structures used for building construction have often been made at the Dniepropetrovsk Plant of Metal Structures. Along with the assurance of reliability in the fabrication process, the principle aim of reducing labour costs during manufacturing and assembly works has also been achieved. During the last few years several types of steel structures have been fabricated at the plant. The structures fabricated include the aviation plant in Kiev, the so-called "Singers' Field"; the enclosing "sarcophagus" at the Chernobyl Atomic Power Station and the bridge over the river Dnieper. All of these structures were designed and built to comply with the requirements of reliability in operation, minimum labour intensity during manufacturing and assembly, together with sufficiently high quality at a competitive cost.

INTRODUCTION

1. Steel structures used for roofing arches at the aviation plant in Kiev had spans of 96m and 144m (Figs 1 and 2). Construction joints were made using high-strength friction-grip bolts (Fig. 3). Due to the complex shape of these structures it was necessary, for their transportation to choose technology capable of providing complete coincidence of the bolt holes, minimising the amount of the manual labour required on site. The design suggested welded H-sections for the main bearing sections of the arches, which had sufficient stiffness in operation and during erection, with the units having dimensions of 48 by 9m and 144 by 48m, with a mass of 1500t. For this case the consumption of steel per m^2, remained within specified limits.

2. When studying the manufacture of these designs special technological measures were developed to maximise the use of mechanised labour in welding, drilling and other fabrication operations. As a result, automatic welding was used providing minimial distortion, ensuring ease in the formation of the complex bolted joints (Fig. 3).

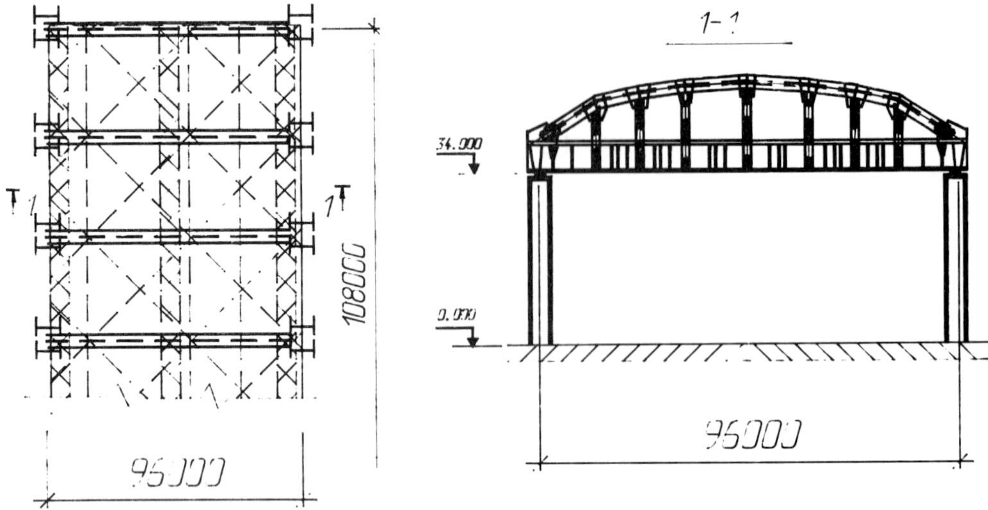

Fig. 1. Construction details.

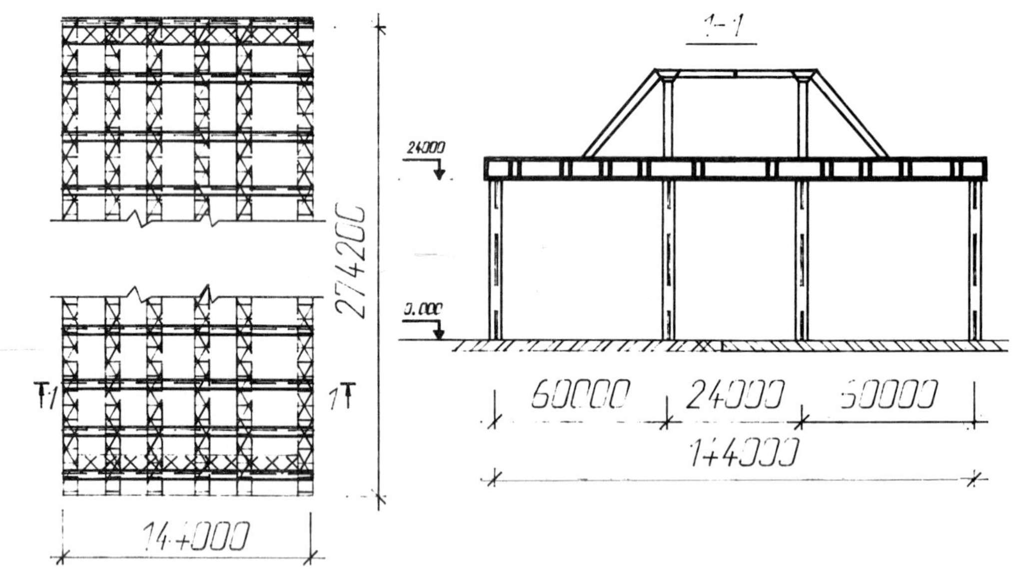

Fig. 2. Plan and cross-section details.

3. The proposed solution which improved the final assembly during the installation (Fig. 4), saved thousands of man-hours during the erection of the large structure.

4. A very beautiful building is now being erected for the so-called "Singers' Field" in the town of Khmelnitsky (Fig. 5). It is intended for large performances and exhibitions and with the help of extensible partitions can be divided into two halls. The steel girders for this building were made in 1992 (Fig. 5).

5. According to customer requirements, this building had to be erected in the shortest possible time. Thus, after studying the design possibilities simple girder sections were proposed, made from rolled angle sections. The cantilever part of the construction required complex bolted joints to ensure continuity with the main girders. The use of special templates provided increased accuracy, and ensured continuity of these parts during assembly, resulting in labour savings. Also the need for overall assembly at the plant was eliminated. During a two month period over 950t of steel was manufactured complying with the customers requirement.

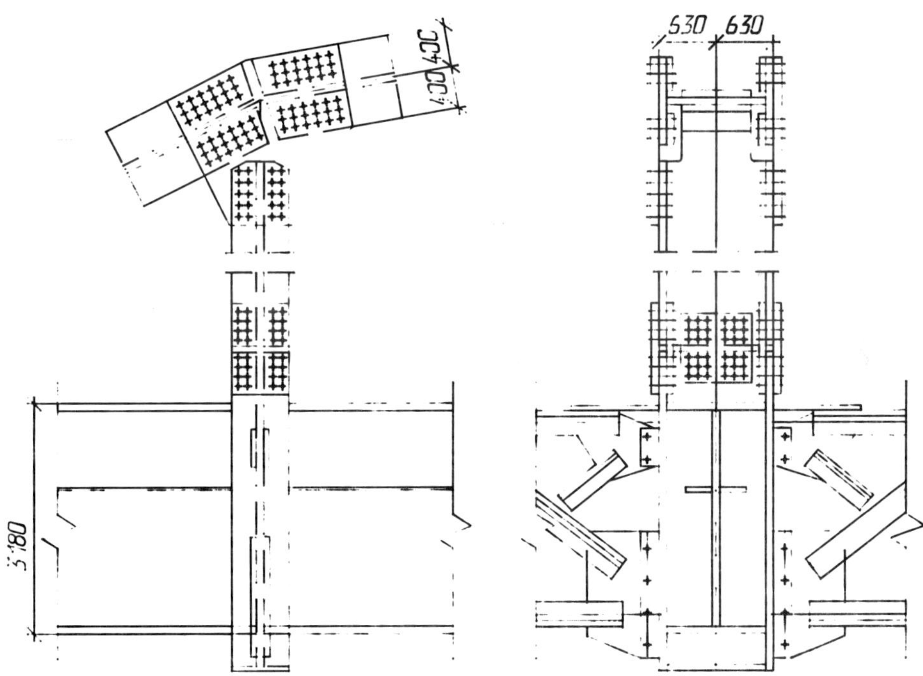

Fig. 3. Arch units.

6. Under extraordinary circumstances and within two months after the tragedy, the block beams for enclosing the "sarcophagus" over the 4th energetic unit at the Chernobyl Atomic Power Station were also made. Due to the severe radiation conditions, they had to have a sufficiently small mass to be lifted by crane situated a long distance from the place of erection. The beams were designed to be lifted into position and assembled without people working in the dangerous zone. This became the decisive factor in choosing the design shape. To meet the requirements, two types of spatial beams were used with lengths of 40m, 54m and 70m and corresponding masses of 54t, 68t, and 118t, respectively which were installed in pairs near the walls (Fig. 6). Assembly joints were welded, to reduce the manufacturing and assembly period. The steel structures were made in a week and erected over a period of several days. This allowed covering of the 4th energetic unit to begin immediately.

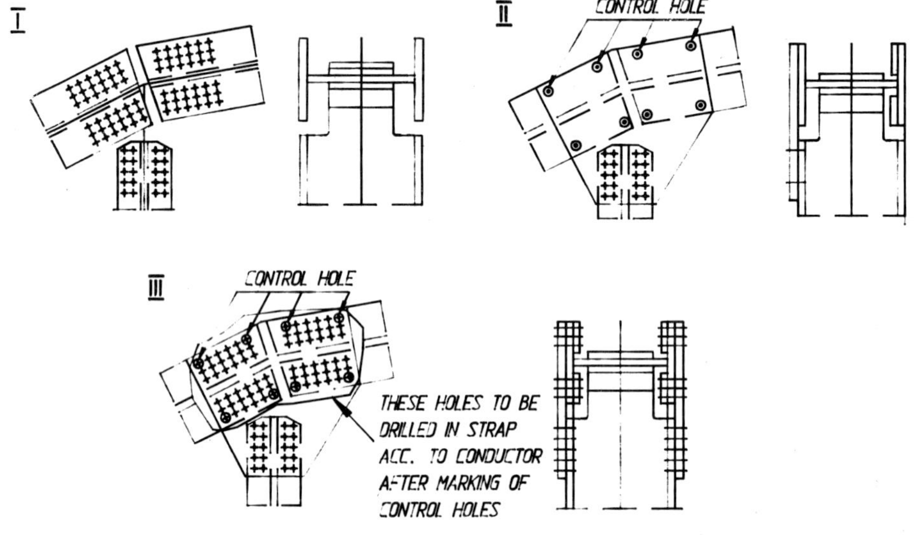

Fig. 4. Typical connections.

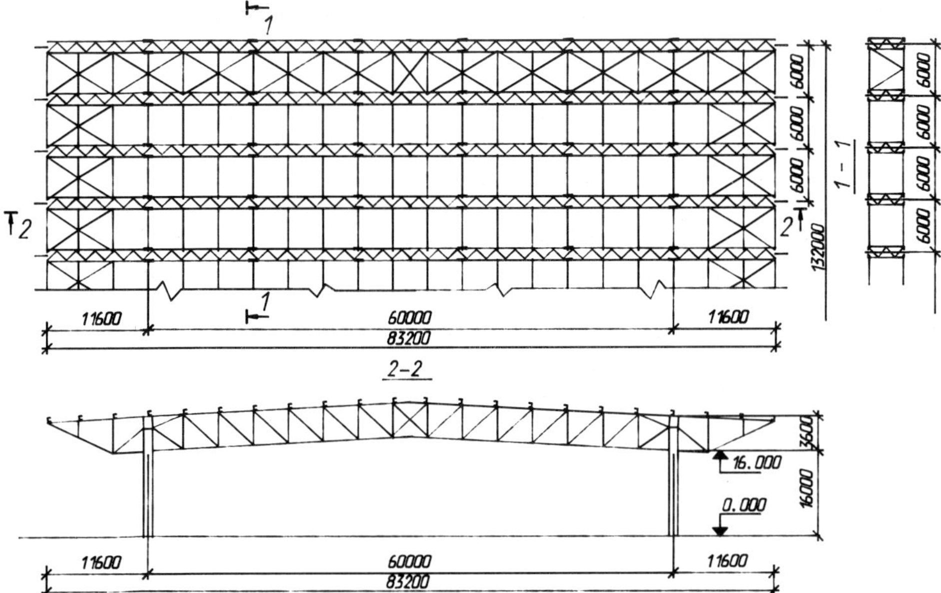

Fig. 5. Roof details.

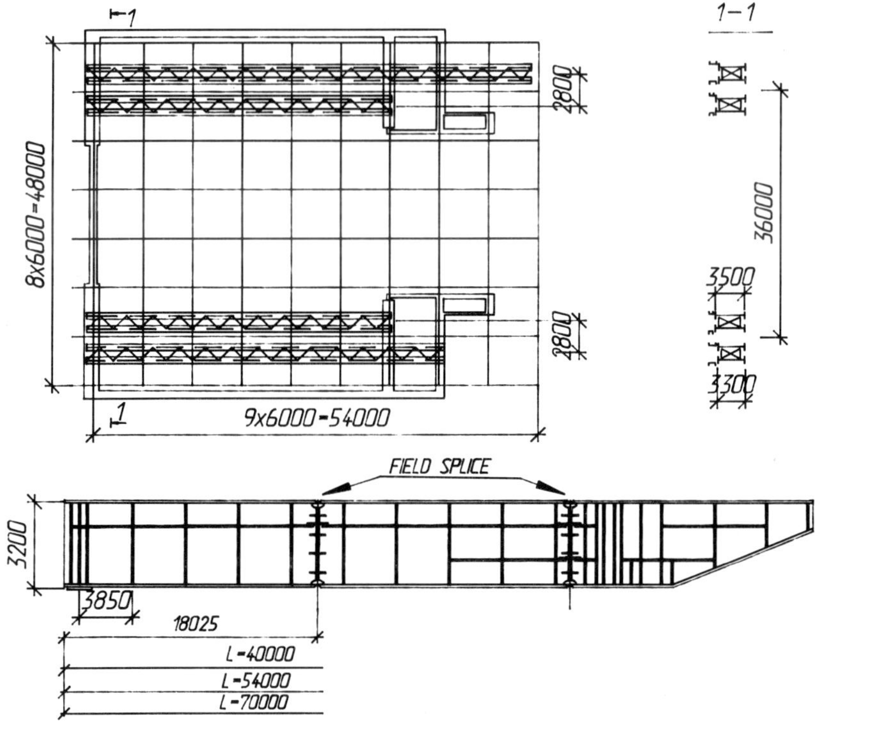

Fig. 6. Beam details.

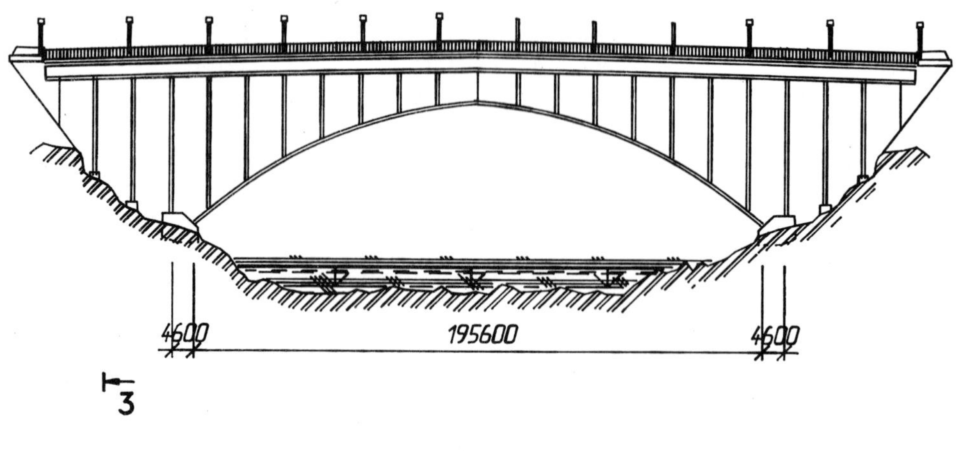
4600
195600
4600
3

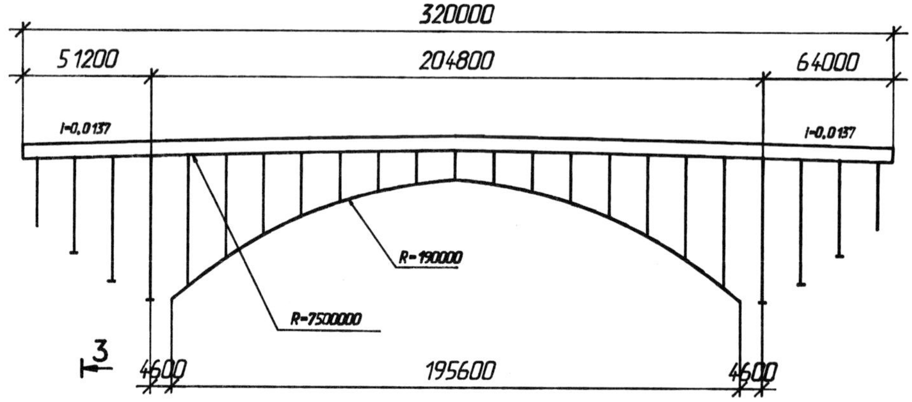
320000
51200
204800
64000
I=0,0137
I=0,0137
R=190000
R=7500000
3
4600
195600
4600

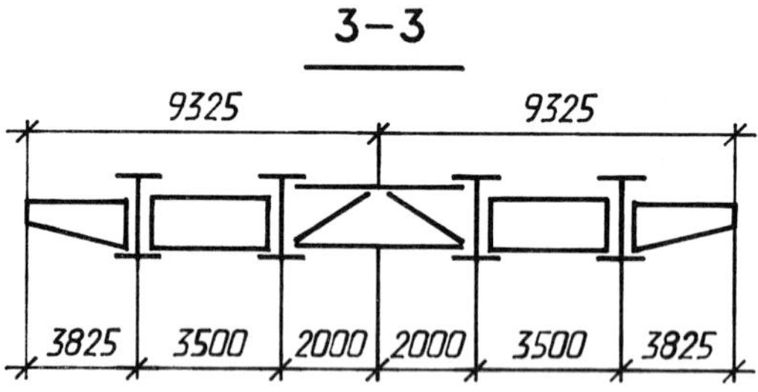
3–3
9325
9325
3825
3500
2000
2000
3500
3825

Fig. 7.

7. The road bridge across the river Stary Dnieper (Dnieper's branch) which spans from the rocky right-bank side to Khortitsa Island in Zaporozhye (the Ukraine), is shown in Fig. 7. The steelwork for the bridge consists of four, flanged beams in a grillage, each pair of the beams being supported by a flange-cross-sectioned arch. Stresses from the longitudinal beams pass on to the transverse beams an then pass via the stays to the arch. The horizontal rigidity of the span structure is ensured by the arches which are tied together. All the field joints are made using high-strength bolts, class 10.9 with an ultimate strength 500MPa. A distinctive feature of the material used, for the arch elements was a steel 16Г2АФ having a higher strength, radically differing from bridge steels in use in the USSR both in its composition: C-0·16%; Mn-1·3 to 1·7%: Si-0·3.to.0·6%; Cr-0·40%; Ni-0·30%; Cu-0·30%; V-0·08.to.0·14%; N-0·015.to.0·025%, and yield point 392MPa. The use of this steel required the development of suitable welding technology to exclude cold and hot cracking in the welds.

8. Several problems had to be solved in manufacturing the steelwork, principally to put into practice the erection and welding of a new steel material 16Г2АФ. Also to provide accurate location of the holes in the construction joints made to the design diameter and to reduce the number of elements to ensure that manufacturing and erection can be achieved in the shortest possible time. During the fabrication, the DneproGES electric power station was under repair, and there was an urgent need to transfer vehicle traffic onto the bridge.

9. Therefore, the specialists in the fabrication plant quickly developed and tested, using their previous experience of handling similar steels to optimise production conditions, metalwork erection and welding. The welding materials and conditions were selected to exclude cold, hot and especially hydrogen cracking. Techniques were developed to exclude the necessity of extra tack welds in weld zones. Jigs were made to drill holes to the full diameter and the erection sequence was carefully planned for the main beams and arch elements. Drilling, using manual pneumatic drillers, was practically excluded.

10. The production procedure allowed all the steelwork to be manufactured within a two months period. The 2mm difference between the mounting holes and the bolts in the main beams and the 3mm clearance in the ties had ensured matching at assembly. Note should be taken that the climatic conditions at the construction site (bright sunshine in the day-time and torrential rain at night) greatly affected the final stage of the bridge construction. In order to speed erection, the bridge was constructed from both banks where upon the final construction joint showed a miss-match in the bolt holes. On observation it was revealed that in day-time, with ambient temperatures over 35°C, the steelwork elongated, and at night, contracted. Therefore, the time for making the connection was selected between 8 and 9 a.m. Everything was suitably prepared, personnel and material and this short period was sufficient for the bolts to be inserted in the construction joints.

94. Cable-supported space frame roof structures

P. KNEEN, University of New South Wales, Australia

SYNOPSIS. This paper describes some analysis and design behavioural studies performed on a variety of cable-supported, space frame roof structures. The basic ideas of cable-stayed roof behaviour are discussed such as the use of the funicular load path concept to improve the capacity of the structure to carry uneven load distributions due to wind or snow. A cable-stayed system will impose inclined concentrated loads onto a space frame which suggests that the preferred analysis method is a discrete stiffness approach. Modelling the cables requires care. The behaviour is very dependent on initial prestress levels in the cables. This may be achieved by defining the manufactured lengths correctly.

INTRODUCTION

1. This paper studies a number of variations of a cable-stayed, two-way, space frame roof which is rectangular in plan. The variations are in the manner and details of the cable supporting system. It is not a complete study but should highlight the importance of a good layout of the supporting system.

2. For comparative purposes, the space frame component was kept constant in topology. The simple flat two way space truss (square on offset square) was used with 54 bays of 2.4m length in one direction and 44 bays in the other giving overall dimensions of 129.6m x 105.6m. The structural depth was assumed to be 2m. A range of ten suitable tube sizes as follows were chosen for the members, and the 19008 member sizes were to be selected by the computer during various stages of the large displacement non-linear stiffness analysis.

Section	Comp-capacity	Tens-capacity	Section
1	15000.000	15000.000	60x2.3-M12
2	30300.000	30300.000	76x2.3-M12
3	53300.000	53300.000	88x2.6-M16
4	94000.000	94000.000	102x32-M20
5	143000.000	143000.000	114x36-M20
6	225000.000	225000.000	114x6-1in
7	205000.000	205000.000	127x4-1in
8	328000.000	328000.000	140x5-M30
9	431000.000	431000.000	168x48-M36
10	570000.000	570000.000	168x64-M42

Space structures 4. Thomas Telford, London, 1993

3. It was assumed that four internal mast positions were possible and that these could be varied within reason. The outer perimeter of the roof was also assumed to be supported. For locations where main supporting cables reached the perimeter, suitable means of carrying the reactive forces could be provided. A possible application could well be a rectangular area with inclined seating along all four sides. Strong frames could be incorporated into this seating, or a horizontal ring beam consisting of continuous floor slabs may be used. These details are beyond the scope of this paper.

LARGE DISPLACEMENT ANALYSIS

4. When large, or significant displacements are involved, the normal equations of force equilibrium at each nodal degree of freedom should be formulated in the deformed configuration of the structure.

5. The normal linear stiffness equation P = KD between applied loads and displacements must take into account changes in geometry and the effects of strain-displacements on the stiffness matrices of the elements. In essence, the member stiffness matrix, **k**, is replaced by: $k = k_E + k_G$ where the first term is the normal elastic stiffness matrix which is a function of the elastic properties of the material and the geometric cross section properties. The second term is a function of the geometry of the member (its length), and the internal forces present in the member, and in particular the axial force. It is termed the geometric stiffness matrix.

6. For two dimensional cases, the combined member force-displacement equation for a pin-ended member such as a cable segment or truss element is given as:

$$\begin{bmatrix} S_1 \\ S_2 \\ S_3 \\ S_4 \end{bmatrix} = \frac{EA}{L}\begin{bmatrix} 1 & 0 & -1 & 0 \\ 0 & 0 & 0 & 0 \\ -1 & 0 & 1 & 0 \\ 0 & 0 & 0 & 0 \end{bmatrix}\begin{bmatrix} u_1 \\ u_2 \\ u_3 \\ u_4 \end{bmatrix} + \frac{F}{L}\begin{bmatrix} 0 & 0 & 0 & 0 \\ 0 & 1 & 0 & -1 \\ 0 & 0 & 0 & 0 \\ 0 & -1 & 0 & 1 \end{bmatrix}\begin{bmatrix} u_1 \\ u_2 \\ u_3 \\ u_4 \end{bmatrix}$$

7. Since the geometric stiffness matrix depends on the internal forces present, it will require that we go about solving the equations in an iterative manner. For the first step, we could assume the internal forces to be as prescribed, or as zero and perform an analysis. For zero internal initial forces, the geometric stiffness matrix is zero and the first step would correspond to a normal linear stiffness analysis. This first step will provide values for node displacements as well as the member forces.

8. If we were to take all the calculated internal forces in the deformed position, we would find that equilibrium was not satisfied at the nodes - there would be some "out of balance" nodal forces present. However, by performing another step, and incorporating the geometric stiffness matrices for members, based on the force estimates obtained, and calculated at the displaced position, we would

expect to reduce these out of balance forces to a smaller amount. In fact, a normal non-linear analysis of a double layered space frame may take from 2 to 5 steps to converge to an acceptable state. This state is described as being when the maximum absolute value of any out of balance force component is less than a prescribed value, and/or the maximum absolute value of any incremental displacement component is less than a prescribed value.

STRUCTURAL MODEL

9. All members were considered pin-ended. The main loading condition was taken as the equivalent of 1kPa all over the roof in addition to the self weight of the space frame tubes. Any supporting cables were expected to remain taut under these conditions. In terms of cable actions, each cable would have very little sag over its length and was therefore modelled as a single truss member with an effective modulus of 150000MPa for these preliminary studies. The cables were taken to be coincident with a joint of the space frame and all masts were pinned top and bottom. Mast heights extended to 15m above and below the top chords.

10. The nodes of the space frame were numbered systematically first going across the top nodes in the 44 bay direction and returning to the same side before numbering the next set of bottom chords. Hence, excluding the cables and masts which were numbered afterwards, the number of nodes was 4851 giving 14553 equations with a bandwidth of (45+44)x3 = 267. By numbering the masts last, and not reordering the numbers, a very large bandwidth resulted in portions of the stiffness matrix.

11. The program used had an in-built generation facility for several forms of space frames and it was more convenient to adopt the above approach. The program would carry out a large displacement analysis which incorporated the geometric stiffness matrices for the members. Aside from some cable elements which had either initial strains or forces specified, the geometric matrix is ignored on the first cycle. After each cycle, the out of balance nodal forces were calculated and applied to subsequent cycles until both the maximum out of balance force was acceptably small (<100 Newtons) or the incremental displacements were very small (< 0.2mm). Generally only three iterations were required in order to meet both criteria.

12. After convergence, the member capacities were checked. If not adequate, a new size was chosen and a record kept of the changes made and the numbers of each size. There was no attempt made to pre-design the tube sizes at the start of the analysis. Once the revised sizes were determined, the non-linear procedure was repeated until convergence was again achieved with the new frame. The number of times this was done was under the control of the user, and for this study typically three such design loops were called for, after which the number of changes was less than about 50 members out of the 19008.

COMPUTATIONAL FACTORS

13. The major drawback with performing a non-linear analysis is the fact that only one loading case at a time can be solved, as the structure stiffness matrix

incorporates the geometric stiffness terms which vary depending on the member forces, which of course depend on the loading. Consequently, the computational effort is not trivial and the automatic member selection scheme may end up with different tube sizes for different loading cases. To overcome this, the selected sizes could be recorded for a subsequent analysis (different loads maybe), and the selection process invoked only if an increase in member capacity was called for.

14. Traditionally, the use of structural and loading symmetry has reduced the numerical problem size enormously - typically if one quarter of the structure is analysed then the number of equations might be approximately 1/4 and the bandwidth reduced by about 1/2 making storage requirements about 1/8th and solution times about 1/16th. Symmetry in this non-linear application is a little more difficult. Members on planes of symmetry should be "halved", meaning extra coding to detect this for member selection. Those crossing a plane of symmetry should introduce an extra node, which is tricky for pin ended members as this node should move with each iteration, otherwise these members will be kinked and the inclined forces help stabilise the structure. Linear analyses generally fix these extra nodes entirely as only the original geometry is used. To permit uneven loadings, the entire structure was modelled for this work.

15. The work was carried out on a Sun Sparc2 workstation with 32Mb of memory. An in-core skyline solver was used which then required more virtual memory than the physical memory available. A typical complete run with 4 design cycles and a total of 9 iterations took about 2.5 hours of which about 42% was CPU time. Interestingly, the total time for equation solution took 21.3 minutes, of which the back substitution phase took 3.8 minutes. During the forward elimination phase, the first 14000 equations took 7.3 minutes, the last 565 equations 10.2 minutes, due to the localised very large skyline where the support cables came into play. The disk I/O and paging due to the use of virtual memory was evident for the last portions and the back substitution phases.

PRELIMINARY ANALYSES - Runs 0, 1 and 2

16. Perimeter springs were added to all edge top chord nodes and to nodes at locations where cable support connections were contemplated. This run (Run 0) was used to verify the operation of the program and to gain some feel for the expected results. It was used during the writing of the member selection code and for the CAD interface. Run 1 used what might be a reasonable starting point with the cable stays and masts initially unstressed. The central deflection after automatic member selection was 622mm with the tube sizes indicated below.

material	#cables	ave.force	max.force	min.force
1	6336	-66.7	20604.3	-19301.2
2	4160	-2111.7	36786.9	-43003.4
3	2954	-3466.5	61053.0	-63907.4
4	2344	-16539.5	102319.5	-110690.5
5	1528	-34456.1	163642.3	-163476.8
6	1252	-48325.3	245038.5	-237407.9
8	338	29715.0	333242.2	-324042.5
9	72	93528.3	428217.6	-416546.2
10	24	221791.0	645013.1	-557441.8

17. In reality, whilst a drawing of the flat roof may indicate all nodes in the desired horizontal position, the frame would really need to have the loads, or at least the self weight component, applied as a starting point under loads, rather than start from the horizontal plane and then apply the loads.

18. Rather than adopt the notion of a deliberate pre-cambering by perhaps shortening all bottom chord members, our second analysis introduced an initial strain in the supporting stays of 0.002 corresponding to a stress level of 300MPa based on the effective modulus of 150000MPa. In addition, the masts were made an extra 20mm long, and the resulting force fitting the masts to the given end co-ordinates. Thus after loading, the mast top is expected to be near the mathematical position. This, together with the shortened inclined cables, produced significant improvements in performance.

19. The deflected form of Run 2 shown in Figure 2 indicated that there was a large area of one way action in the central span, as also evidenced by the presence of the heavier tubes in the top and bottom chords. One contributing factor was that there were only one set of back-stays from each mast to the stiff perimeter points as shown in Figure 1. The central downward deflection was 152mm. A summary of the required space frame tube sizes and quantities is given below. Note the maximum forces for the space frame members are 556kN and -825kN, this being beyond the capacity of section 10.

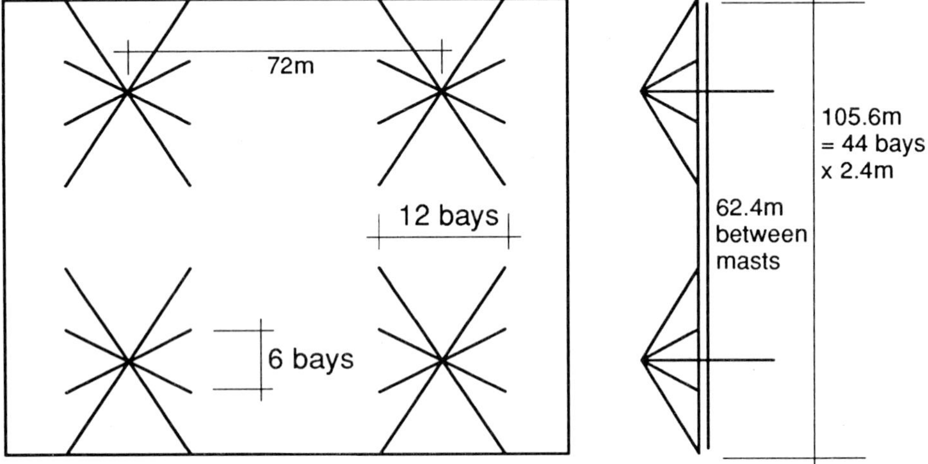

Fig. 1 Basic space frame roof dimensions for runs 1,2.

material	#cables	ave.force	max.force	min.force
1	6680	-1093.0	17756.2	-18761.3
2	4870	-5845.3	32892.4	-31188.1
3	3014	-11037.6	55885.0	-56908.7
4	2242	-23350.8	93053.6	-104730.7
5	1284	-46471.6	141955.2	-160142.3
6	650	-50965.8	226605.1	-237391.7
8	172	-128186.7	305839.5	-334456.2
9	44	-98188.9	409878.2	-431557.4
10	52	148350.0	556093.6	-825552.6

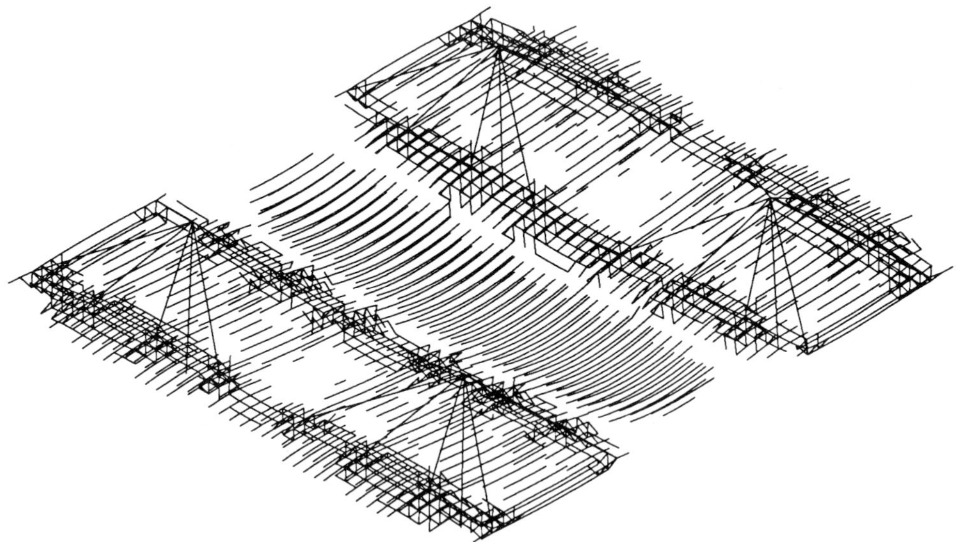

Fig. 2. Form of deflected shape showing heavier tubes only and deflections scaled by 40. Note presence of one way action.

RUN 3 - ALTERATION TO THE NUMBER OF CABLE STAYS

20. Two additional back-stays were provided to each mast in the second direction. Together with these being prestressed gave considerable improvements in behaviour. The maximum central deflections were reduced from 152mm in Run 2 to 80mm. In addition, the maximum member force in the space frame proper, was reduced from -825 kN to -665kN.

material	#cables	ave.force	max.force	min.force
1	7176	-1163.1	16465.7	-17215.4
2	4558	-4478.6	31174.5	-34321.6
3	3164	-8786.0	54968.1	-59785.5
4	2442	-28610.1	100433.4	-96565.3
5	924	-74104.7	138185.8	-153039.0
6	408	-106988.7	225476.7	-243821.0
8	272	-188123.1	311348.8	-342152.1
9	44	173780.6	432454.1	-412116.8
10	20	-526859.0	-472306.4	-665324.1

21. There were still very large forces around the points where the more centrally located cables attached to the frame. It was concluded that it would be much better to eliminate the effects of the inclined reactions by joining up the opposite cables thus allowing the space frame pick-up forces to become almost vertical and spread the reactions into all incident web members rather than be concentrated into a smaller number. This was the basis of Run 4.

RUN 4 - REDUCING THE INCLINED REACTIVE CABLE FORCES

22. This run also changed the mast locations in the long span direction so as to be more central. The clear span in this direction was decreased from 72m to

62.4m, the short span direction being unchanged at 62.4 metres. In addition, the spread of the cable stays was reduced from 12 bays to 6 bays. The result was that the deflections were reduced from 80mm to 65mm, and the maximum space frame member forces from -665kN to -611 kN. Figure 3 clearly indicates two way action being developed.

material	#cables	ave.force	max.force	min.force
1	7302	-648.3	16240.7	-18498.7
2	4668	-5361.6	31779.0	-31466.5
3	3148	-14258.4	53091.2	-63760.2
4	2812	-23296.3	91716.0	-97223.1
5	730	-79350.5	138779.1	-162082.6
6	236	-118447.2	232567.2	-238993.0
8	76	-113592.8	276795.6	-341764.2
9	20	-395325.1	-355330.9	-438758.8
10	16	-516155.8	-478508.4	-610896.9

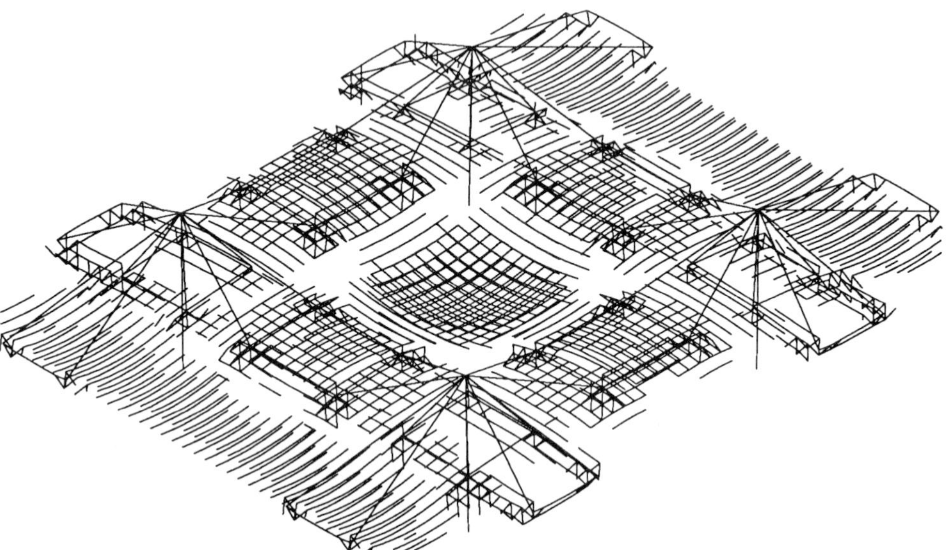

Fig. 3. Scaled deflected shape for Run 4 highlighting the heavier member sizes and the development of two way action in the central spans.

RUN 5 - INTRODUCTION OF STRUTS BETWEEN CABLES AND STIFFER SIDE SUPPORTS

23. To further reduce the inclination of the pickup reactive loads, struts between the spreading cables were arranged as shown in Figure 4. The large space frame forces in Run 4 were all located at the perimeter back-stay locations. For this run, the spring stiffness values used were increased so as to throw more load into the

supports rather than into the space frame. The deflections were reduced from 65mm downwards to 75mm upwards, but more importantly, the maximum space frame member force decreased from -611kN to -269kN, with the result that a sensible tube size could be used rather than have a special region of fully welded members.

material	#cables	ave.force	max.force	min.force
1	6546	188.4	15452.9	-15999.9
2	4960	1596.0	30615.0	-28865.3
3	3404	3048.2	55792.8	-51183.3
4	2948	-3455.5	98181.3	-97118.8
5	846	-27063.6	143242.1	-137198.3
6	248	-59176.9	210950.3	-219764.0
8	56	125277.1	295946.6	-268938.1

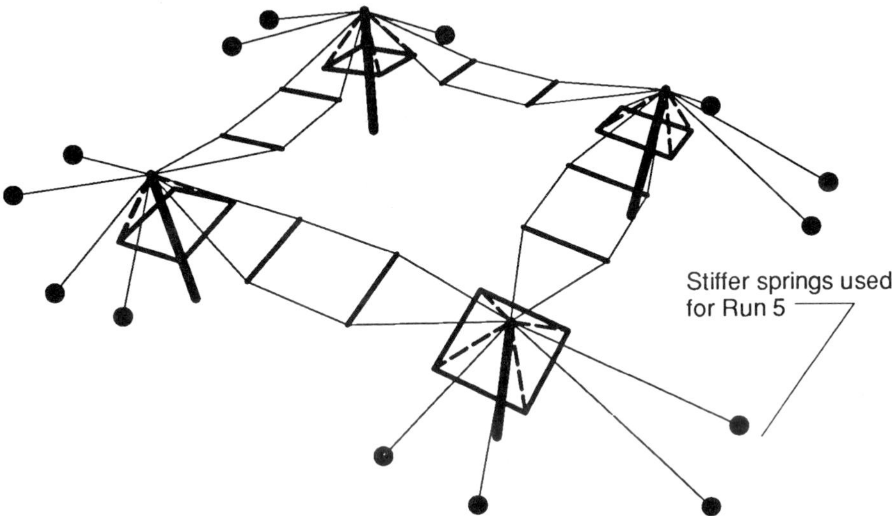

Fig. 4 Additional struts between cables to reduce inclined loads into the space frame members-Run 5.

CONCLUDING REMARKS

24. This paper has discussed the importance of realistic starting conditions for the cable stay lengths and has outlined the effects of cable stay configuration on the performance of the structure. In particular, it seems entirely feasible that large span, cable-stayed, space frame, roof structures can be constructed of lightweight members. It is advantageous to anchor any back-stays independently from the space frame roof. It would seem prudent to use the in plane stiffness of the space frame to resist any horizontal components of wind loads taking these forces out through the perimeter support system.

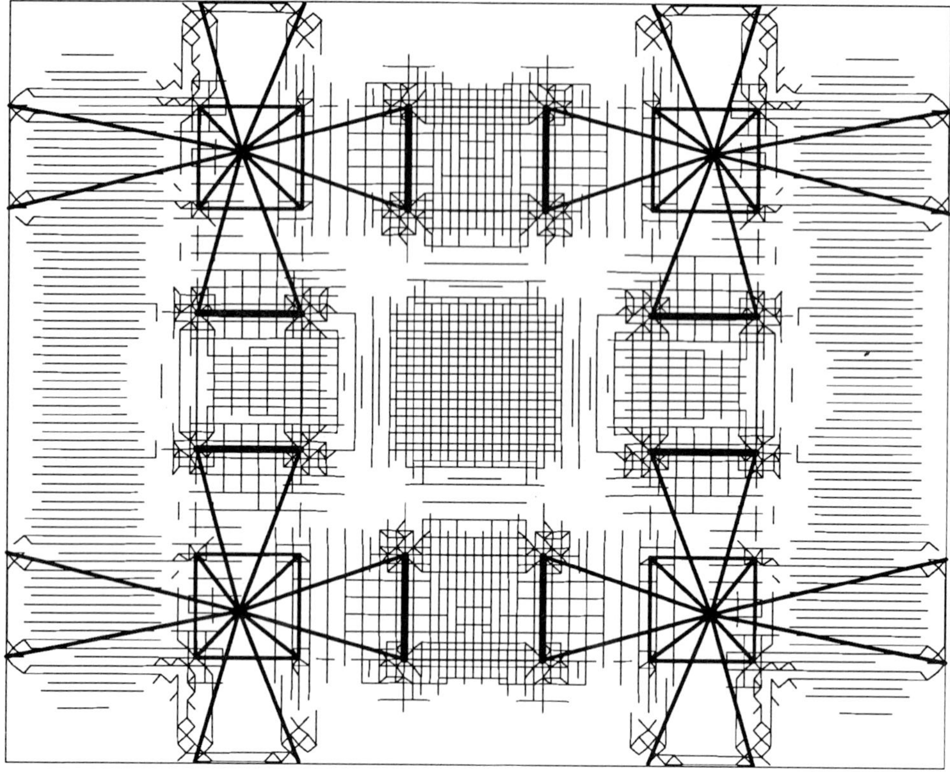

Fig. 5 Distribution of heavier tube sizes and location of cable stays.

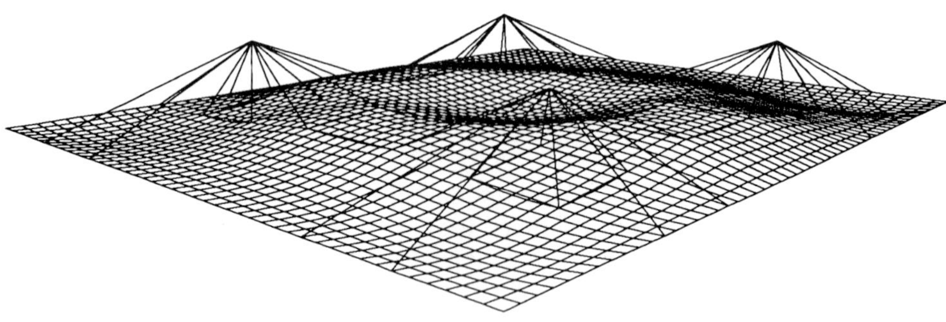

Fig. 6 Deflections under UDL for run 6 (top chords only shown).
Deflections scaled by 40.

95. Studies of the evolution of prestressing forces in members of large-span suspended structures

E. CRAILA, Building Research Institute, Romania

SYNOPSIS. As suspended roof structures consist of members with no compression stiffness (cables) it is necessary, during the assembly phase, to induce a tensile stressed state in the structure to be able to annul the compression stress generated under various hypothetical loading cases. In time, the cables undergo elongation which modify the initial geometry and consequently the initial prestressing force. In order to determine losses in the prestressing force, over time, theoretical and experimental studies were undertaken on an existing suspended structure presently used as the roof of a multi-purpose hall.

STRUCTURE DESCRIPTION

1. The suspended structure used as the mult-purpose hall roof consists of 16 spatial cable trusses, each with a triangular cross-section having the following characteristics:

- a truss upper boom (bearing boom) consisting of two groups, each of three 27mm diameter cables, placed at 1.8m from each other;
- a truss bottom boom (stretching), made from two 27mm diameter cables;
- struts and braces made from 10mm diameter cables;
- the span of the upper booms is 67.5m and that of the bottom booms is 63.3m;
- the upper boom deflections vary between 4.6 and 5.59m and the bottom boom deflections vary between 3.7 and 2.7m.

EXPERIMENTAL INVESTIGATIONS

2. In order to estimate the stresses within the structure, experimental investigations were performed in two stages:

- laboratory investigations;
- and *in situ* investigations.

3. The aim of the laboratory experiments is to calibrate the MEC device used for measurements on the cable structure. Calibration is performed on a horizontal

tension test stand made from two reinforced concrete abutments supporting the cable and device allowing loading of the tension cable samples.

4. The MEC device consists of an 80cm long beam with a high bending stiffness, supported on the cable by two clamp shaped small posts. In the middle of the span the device is supplied with a tensometric gauge and a screw used in applying a constant deflection on the stressed cable, in various stressing stages. Cable stressing is performed using an 120tf ram and a hydraulic pump.

5. The tensometric gauge in the middle of the device is subjected to compression between the screw force inducing deflection "z" in the cable. The device is equipped with resistive electro-tensometric transducers which turn physical values (compression micro-deformations) into electric values (potential differences) read at the tensometric bridge.

6. The service diagram of the device shows the relation between the stress in the cable and the effect on the gauge of the cable deformation caused by the lateral deflection "z".

7. The same static scheme is used on site, applying a lateral deflection "z" to the cable, using the MEC device connected to the tensometric bridge (TEPIC).

8. The resulting readings are used in the laboratory to determine the corresponding stress directly, representing the cable stress.

9. Calibration experiments, were performed on the MEC device using 27mm diameter cable sample. Three tension tests were undertaken in seven stages, using 3tf increments at each step.

ESTIMATION OF PRESTRESSING FORCES WITHIN STRUCTURE

10. The *in situ* measurements and their transformations from micro-deformations into stresses using the calibration diagram, provide the stress state in the cable structure, for hypothetical prestressing and self-weight loading cases.

11. In order to obtain the prestressing force values the self-weight stress values have to be known. By subtracting the stress values corresponding to the self-weight from the values corresponding to the experimental data, the stress values corresponding to the prestressing forces are obtained.

12. Table 1 presents the stress values in the bottom booms for the 16 roof trusses, under various loading conditions.

13. The first column provides the average stress values in the bottom booms generated by the prestressing and self-weight loading cases, which are estimated using the experimental tests performed in 1990.

14. The second column introduces the stresses in the bottom boom generated by permanent weight loading.

15. The third column gives the prestressing force values in the structure in 1990 at 20° C obtained by adding the modulus of values in column 2 to column 1.

16. Column 4 presents the temperature correction values of -10°C related to the assembly year, 1974. Consequently, the stress excess accounted for by a 10° C temperature decrease, was subtracted and the prestressing forces values in the structure were obtained, for 1990 at a 30° C construction temperature.

Table 1. Stress state in the bottom booms for various hypothetical loading cases, in 1990 (tons).

Truss no.	Stress values provided by measurements, hypothesis: Prestressing + specific weight (tons)	Stress values provided by hypothesis: Specific weight (tons)	Stress values provided by overlapping column (1+/2/) (tons)	Calculated stress values provided by temperature difference (-10^{0}C) compared to assembly (tons)	Stress values provided by correcting of values in column 3 by effect t for comparison with stresses in 1974 (tons)
1.	20.43	9	29.4	2.86	26.54
2.	20.57	12	32.57	2.68	29.89
3.	20.54	11	31.45	2.64	28.81
4.	18.97	11.5	30.47	2.6	27.87
5.	20.98	15.5	36.48	2.5	33.98
6.	14.95	15.2	30.15	2.4	27.75
7.	15.65	16.8	32.45	2.3	30.15
8.	11.15	17.1	28.3	2.3	26.0
9.	7,1	17.1	24.2	2.3	21.9
10.	12.3	16.8	29.1	2.4	26.7
11.	14.05	15.2	29.25	2.5	26.75
12.	16.0	15.5	31.5	2.6	28.9
13.	19.0	11.5	30.5	2.64	27.86
14.	22.4	11	33.4	2.68	30.72
15.	27.65	12	39.65	2.86	36.79
16.	28.55	9	38.55	2.86	35.69

NOTE: Environment average temperature during measurements in 1990 was t=20^{0}C

COMPARATIVE ANALYSES BETWEEN THE STRESS STATES IN THE STRUCTURE ASSESSED IN 1974 AND 1990

17. Table 2 gives the values by which the prestressing forces decreased from the assembly phase in 1974 to the time of analysis in 1990.

18. Columns 1 and 2 give the prestressing force values in 1974 and 1990 respectively, in the bottom booms and columns 3 and 4 gives the decreased values in tons and percentages respectively.

DETERMINATION OF RESTRESSING FORCE VALUES AND ANALYSIS OF STRUCTURE STABILITY

19. The safety of cable structures depends on two parameters:

- prestressing forces that increase structure stiffening and thus eliminate instability under unsymetrical loading and dynamic wind actions;
- cables safety factors which should be observed reducing the risk of cable failure under loading.

20. Both parameters have a mutual effect on each other as an increase in tensile prestressing force increases the stiffness of the structure improving the dynamic response under wind loading but reduces the overall safety factor for individual cables.

Table 2. A comparison of prestressing force variation in the years 1974 and 1990 (bottom booms).

turns no.	Prestressing force values		Variation of prestressing force values between 1974-1990	
	Year 1974 (tons)	Year 1990 (tons)	tons	%
1.	4.8	26.54	- 21.64	-44.7
2.	46.6	29.89	- 17.7	-37.98
3.	45.2	28.81	- 16.4	-36.28
4.	43.8	27.87	- 15.93	-36.36
5.	42.4	33.98	- 8.42	-19.85
6.	41.0	27.75	- 13.75	-33.53
7.	39.6	30.15	- 9.45	-23.86
8.	38.2	26.0	- 12.2	-31.93
9.	38.2	21.9	- 16.3	-42.6
10.	39.6	26.7	- 12.9	-32.5
11.	41.0	26.75	- 14.25	-34.75
12.	42.4	28.9	- 13.5	-31.83
13.	43.8	27.86	- 15.94	-36.4
14.	45.2	30.72	- 14.48	-32
15.	46.6	36.79	- 9.81	-21
16.	48	35.69	- 12.31	-25.6

21. For concave trusses, such as those in the multi-purpose hall structure under investigation, the upper boom supports the whole surface of the roof. Negative stresses occurring in the upper boom under wind loading may cause variable deformation of the roof and therefore fluttering.

22. The analyses that were performed proved that no negative stresses occur in the truss upper boom under any of the loading cases. If the stresses caused by both wind loading are added in turn to the stresses caused by permanent loading and the prestressing existing within the structure, an excess of tensile force remains in each upper boom, which accounts for the satisfactory behaviour of the roof under wind loading.

23. An analysis of the stress state at the bottom boom level proves that compression forces occur within them, caused by snow loading. But in the same situation, the upper booms stay in tension. In order to ensure that no compression stress occurs in any cable elements, the prestressing force should be increased by a sufficient ammount to annul compression forces occurring due to one snow loading condition.

ANALYSIS OF THE STRUCTURAL CAPACITY UNDER THE PRESENT CONDITIONS

24. The structure carries at present a higher load than that given by the design, because the roof has been refurbished. At the same time, the codes (STAS) on wind and snow loading have been modified, causing an increase in climatic loading requirements. The analysis of the structural capacity compared to the maximum stress in the structure gives a corresponding safety factor of two.

25. The appraisal of the safety factor corresponding to the structures maximum stress proved that in the existing situation it is greater than two. If restressing stresses are added to the present stresses generated by the present prestressing force and self-weight, in order to counteract the compression force occurring in the bottom boom due to snow loading, the safety factor decrease below two. Prior to restressing the structure, the load carrying capacity is insufficient and overloading is inadvisable.

96. The theory of suspended cable roofs

J. KADLCAK, Brno Technical University, Czech Republic

SYNOPSIS. The basic load-bearing part of economically progressive suspended cable roofs, suitable for roofing large spaces, is a system of cables of one or more spans and, most frequently, the space cable net. Within the frame of our research the authors have developed and applied, theories of the static analysis of the basic load-bearing elements and systems for some important types of suspended cable roofs. The theories derived have been fully automated through computer programs. The paper presents a brief outline of these methods.

SINGLE-SPAN CABLE

1. The author's papers present precise theories (refs 1,4-7) for the static analysis of a plane elastic single-span cable (Fig. 1) considered to be pinned to yielding, or rigid-immovable, or to elastically restrained supports in both the horizontal and vertical directions. The cable may also be stretched by a weight G (a cable with a counterweight) fed through a pulley fixed on to an immovable or elastic support.

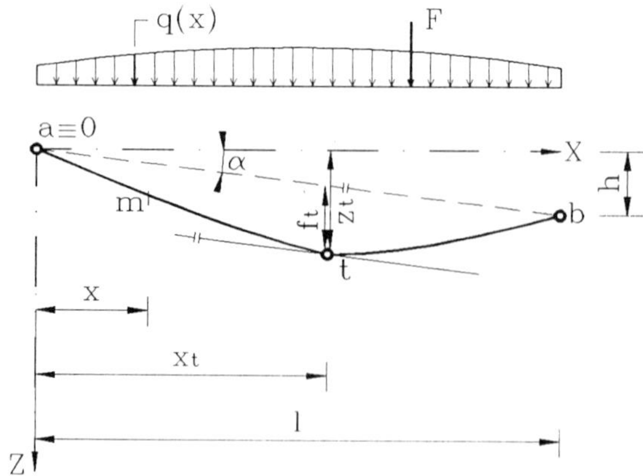

Fig. 1. Single-span cable under load.

The cable is considered to be vertically loaded with the load distributed along the horizontal projection of the cable. Thus, considering the loading of small sag cables when:

Space structures 4. Thomas Telford, London, 1993

$$f_{max} = f_t = 0.1\ \ell \tag{1}$$

where ℓ is the span and f_{max} is the largest sag of the cable. The effects of changes in temperature and of changes in the elastic and permanent deformation of the cable is also taken into account.

2. The problem investigated a simple cable of a given span ℓ, the height difference h of suspension points a, b and the load q(x) and is indeterminate to the first degree. Consequently to determine this relationship, another so-called supplementary condition must be used. Relationships have been derived for the horizontal force H_1 of the cable at the initial (assembly) state; for nine shapes and force conditions occurring in suspended cable roofs. The dependence between the horizontal forces H_1 and H_2 of the cable under an arbitrary initial loading state at temperature t_1 and the calculation of the loading state at temperature t_2 is expressed by a cubic equation in the unknown H_2. Any static or geometrical quantity of a cable under load, for example; a cable length L_2, may easily be determined by using the horizontal force H_2 from the relation:

$$L_2 = \frac{\ell}{\cos\alpha} + \frac{\cos^3\alpha}{2\,H_2^2} \int_0^{\ell} Q_2^2\,(x)\,dx \tag{2}$$

here Q_2 (x) is the shearing force in the cross-section x of a simple horizontal beam $\overline{ab}$ of the same span and load as that of the cable (Fig. 1).

CONTINUOUS CABLE

3. A new method for the static analysis of a plane continuous cable (Fig. 2) of an arbitrary number of n spans, suspended on hinged supports, has been derived in refs 4 and 7 and used to analyse continuous cylindrical suspended cable roofs (Fig. 4).

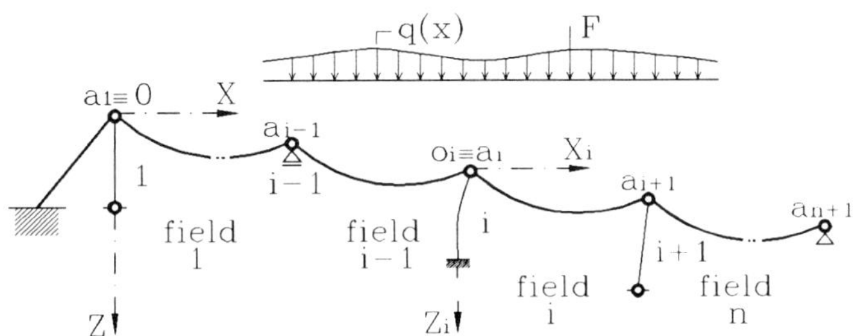

Fig. 2. The diagram of an n span continuous cable under load. Fixed supports, n+1; movable supports, i-1, i+1; elastic supports 1, i .

The support conditions considered are as follows:

i) Rigidly fixed;
ii) Movable;
iii) Elastic yielding both the horizontal and vertical directions;
iv) Attached to other cables.

For small deflections of the suspension system, the support conditions considered in the static analysis of the continuous cable may be represented by an elastic restraint in the horizontal direction.

4. The cable may also be tensioned by a weight G fed through a pulley fixed on to a rigid or elastic end support. The cable load is arbitrary vertical distributed along the plan projection of the cable.

5. Thirty three different cases of supporting cable spans have been analysed obtained by the combination of four basic constraints namely; fixed, movable, elastic and the counterweight on the end support. Several pertinent forms of the equations of state have been derived in terms of the unknown horizontal forces H_i; for the spans i=1,2,...,n for a continuous cable.

6. Having solved the system for a continuous cable of n spans, the remaining static and geometric properties of the cable will be determined.

SPACE CABLE NET

7. Most suspended cable roofs are provided with a basic load-bearing system represented by a space cable net (Fig. 3).

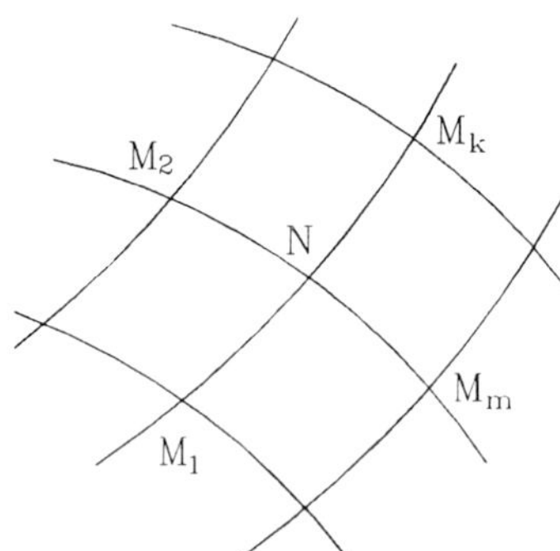

Fig. 3. Pre-stressed space cable net.

8. An original method for the static analysis of geometrically non-linear and also linear cable nets of a completely arbitrary geometric shape, and nodal loading; suspended on rigid or elastically yielding, supports, are derived in refs 3-5. The effects of the changes in nodal loading and temperature, on the elastic strain in the cables and of the initial pre-stressing of the net are considered in the theory. Elastic cable nets are analysed by means of the deformation method described in greater detail in ref. 3. Larger geometrically non-linear cable nets are solved, allowing for both linear and non-linear material behaviour, by using a deformation variant of the finite element method (ref. 5).

9. In the monograph (ref. 3) attention is also given to the problem of the modulus of elasticity of the steel cables. This work deals with the real, conventional, computed and computational modulus of elasticity of the steel cables. The work distinguish between new un-stretched cables, partially stretched cables and stretched cables.

SUSPENDED CABLE ROOFS

10. The theories derived for the static analysis of single-span and multi-span cables and the space cable nets have been successfully applied to compute the most important types of single-chord and double-chord suspended cable roofs.

The suspended cylindrical cable roof

11. The theory of the static analysis of a simple cable (Fig. 1) or of a continuous cable (Fig. 2) has been used to determine the force distribution in single-span or multi-span cylindrical cable roofs (Fig. 4).

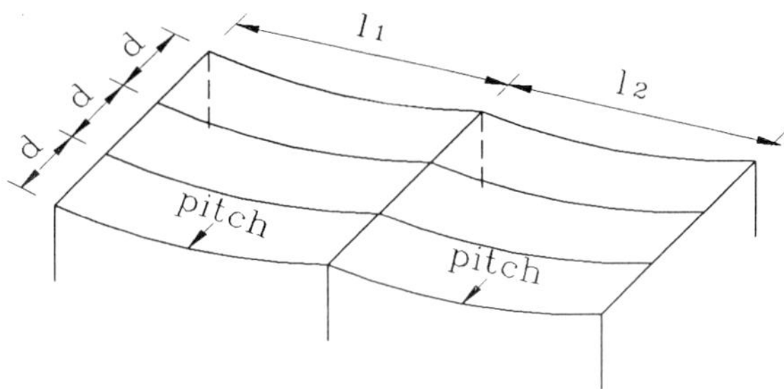

Fig. 4. Cylindrical two-span cable roof.

The suspended circular cable roof with a central column

12. The structural cables of the roof (storehouses) diverge radially from the top of the central hinged column or from the unilaterally fixed column towards the peripheral circular ring (Fig. 5a).The method used to undertake the static analysis of the roof (ref. 7) is based upon a discrete calculation model represented by a system of m cables each of which is being placed on two supports: a (the top of the column), b (the peripheral circular ring).

The suspended annular cable roof without the inner ring

13. The load-bearing cables of the roof are fixed to the peripheral circular ring of radius R and arranged to form the central polygonal opening suitable for the illumination of the roof space (Fig. 5b). The radius r of the circle inscribed in the central polygon is dependent upon the number of cable sockets on the peripheral circular ring. This economically very advantageous type of the cable roof, has been patented by J. Postulka (Czechoslovakia) and it has been used for the

construction of many exhibition and sports halls both in Czechoslovakia and abroad.

14. The static analysis of the cable roof is based upon a discrete calculation model represented by a space cable net with nodes at the intersection of the cables (refs 4,5,7).

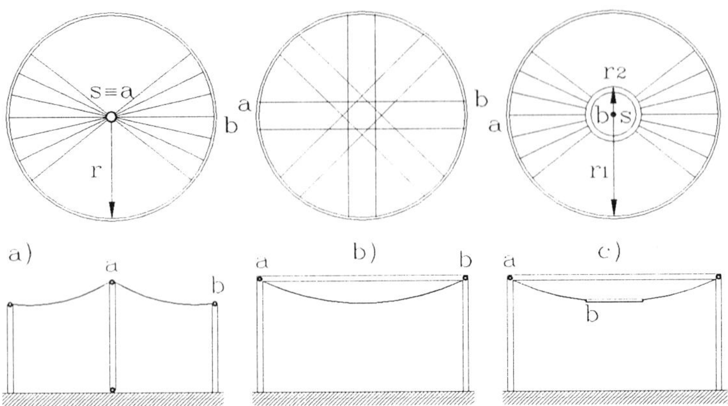

Fig. 5. Single-chord suspended cable roofs of the circular and annular type.

<u>The suspended annular cable roof with the inner ring</u>

15. The load-bearing cables of the roof diverge radially from the inner ring, stressed in tension, to the peripheral circular ring, stressed in compression (Fig. 5c). The static method of analysis presented (refs 2,5,7) is based upon a discrete calculation model represented by a system of n cables, each of them considered to be hinged and supported by, a (peripheral ring) and b (inner ring).

16. The horizontal displacement of the centre s of the inner ring does not occur under symmetrical load on the roof. All the bearing cables are loaded and stressed equally, and it is possible to investigate statically one single cable with a point of suspension b movable in the vertical direction according to the theory in ref 7. In calculating the cable forces the effect of temperature change is taken into consideration together with both the elastic and permanent strain of the cable and the strain in both rings.

17. Both the horizontal and vertical displacement and the inclination of the inner ring occur under an arbitrary load on the roof, and the system of n load-bearing cables has to be solved as a whole. The intensity and the direction of this displacement and inclination are given by the iteration as outlined in ref. 6.

18. The annular suspended cable roof, with an inner ring (Fig. 5c) of radius r_2 -> O, may be statically investigated as a suspended circular cable roof with a central mass point m.

REALISATION

19. Using the derived theories and computer programs, several annular suspended cable roofs, without inner rings (Fig. 5b), have been designed and used for halls both in Czechoslovakia and abroad; for example:

a) The 65m diameter circular exhibition halls for the sale of cars in Augsburg, Germany (ref. 7).

b) The multi-purpose circular halls at Bekescaba, Hungary (ref. 7) which had a diameter D_1 = 65m, on the roof of which another concentric circular hall is constructed for the air-conditioning equipment; the diameter of this smaller hall D_2 = 22m, and it is also provided with a cable roof.

c) The 60m diameter circular pools at Duslo Sala.

These theories have been applied not only to calculate many cylindrical suspended single-span cable roofs (Nove Zamky 55x75m, Dunajska Streda 36x45m,) and two-span cable roofs (a swimming pool in Brno), but also to evaluate the load-bearing cables of a three-span footbridge in Prague-Troja, etc.

REFERENCES

1. KADLČAK J. The static analysis of a perfectly elastic and flexible cable under vertical load. International Civil Engineering Monthly, 1971/72, vol. 2, No 4, 172-188.

2. KADLČÁK J.Obliczenie statyczne liny nośnej kołowego dachu wiszacego z pierścieniem środkowym. Inżyniernie i Budownictwo, 1972, No 4, 129-134, (in Polish).

3. KADLČAK J.; MATERNA A. Statics of elastic cable network, p. 88. Academia Prague, Rozpravy ČSAV, 1978.

4. KADLČAK J. Teorie lanových soustav. (The theory of cable systems.) Doctoral thesis, p. 241. The Faculty of Civil Engineering, Technical University of Brno, 1982.

5. KADLČAK J. at al. Statika visutych lanových střech (Static of suspended cable roofs.) Scientific and research report III-3-4/10, p.250.Faculty of Civil Engineering, Technical University of Brno, 1985.

6. KADLČAK J.; KYTÝR J. et al. Teorie vypočtu visutych lanových střech. (The theory of calculation of suspended cable roofs.) Scientific and research report, p. 290. Faculty of Civil Engineering, Technical University of Brno, 1990.

7. KADLČAK J. Statika nosných lan visutych střech. (Statics of bearing cables of suspended roofs.), p. 157. Academia Prague, Studie ČSAV, 1990.

97. Statics of tent structures

D. DE TOMMASI, S. MARZANO and M. D. PICCIONI, Politecnico di Bari, Italy

SYNOPSIS. The static problem for tent structures in which the sheet does not support compressive stresses is considered. The unilateral constraint on the stress is analysed and the consequent condition on the external loads compatible with equilibrium is obtained. Further, as an application of the general results, the prestressed state is examined. Finally, an example concerning a membrane bounded by four cables, is discussed.

INTRODUCTION

1. In this paper we investigate the restrictions on the external loads for the static problem of tent structures. As is well known, these structures are composed of a sheet of synthetic material with high mechanical strength, supported by steel cables arranged along the edges (cf., e.g., ref. 1). Since the sheet is very thin it supports only tensile membranal stresses. For this reason, the sheet is subjected to prestressing stresses, easily transmitted by the sustaining elements.

2. The constraint on the stress in the sheet induces restrictions both on the systems of external loads and on the geometrical configurations of the structure. In particular, the assumption of null surface forces leads to prove that only prestressed surfaces with negative gaussian curvature are possible (cf., ref. 1). Further, as shown in our previous paper (ref. 2), equilibrium imposes that edge cable configurations must coincide with the characteristic lines associated with the differential equation of the membrane equilibrium.

3. Here, we are mainly concerned with the characterisation of the external loads which can be carried by tent structures at equilibrium. Following some recent ideas on the study of unilateral constraint on the stress suggested for a different context and other purposes (cf., ref. 3), the authors obtain a general compatibility condition for the external loads. Then, in the case of membranes bounded by cables, we derive the corresponding restriction on the external forces. We also apply the results of our analysis to tent structures in the absence of surface loads (prestressed state). Finally, in a special case of technical interest, the authors show that compatibility condition on the loads implies a geometrical restriction on the prestressed surface.

Space structures 4. Thomas Telford, London, 1993

GEOMETRY OF A MEMBRANE

4. The authors begin by recalling some geometrical results from differential geometry of a surface which are essential in the development (for greater details, see refs 4-6).

5. For our purposes, a membrane is a regular surface $\mathcal{S}$ embedded in the three-dimensional Euclidean space. As usual, suppose that there exists a regular bijective map between the points (z^1,z^2) of the regular region S of R^2 and the points of the surface $\mathcal{S}$ (Fig. 1):

$$x:(z^1,z^2)\mapsto x(z^1,z^2)\in\mathcal{S} \qquad (1)$$

At every point x of $\mathcal{S}$, the set $\mathcal{T}_x$ of the vectors tangent to $\mathcal{S}$ is defined by [1]:

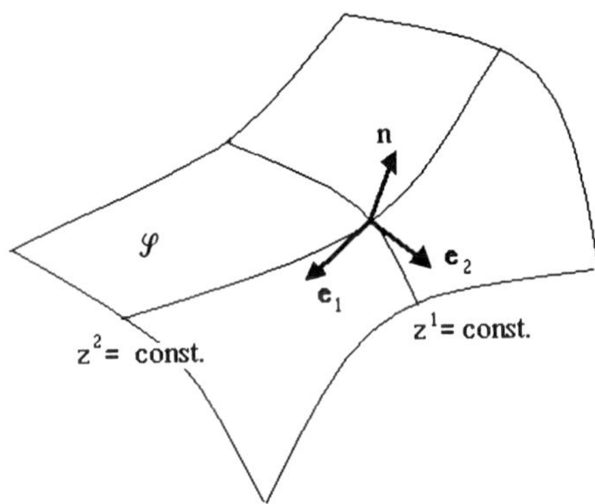

Fig. 1.

$$e_\alpha(z^1,z^2)=x_{,\alpha}(z^1,z^2) \qquad (\alpha=1,2) \qquad (2)$$

Moreover, e_1, e_2, together with the unit normal vector n

$$n(x)=|e_1(x)\times e_2(x)|^{-1}(e_1(x)\times e_2(x)) \; , \qquad (3)$$

denote the vectors of the covariant basis. We shall also consider the contravariant basis $\{e^1(x), e^2(x), n(x)\}$, where

$$e^\alpha=z^\alpha{}_{,x}(x), \quad \alpha=1,2 \qquad (4)$$

$(x\mapsto z^\alpha=z^\alpha(x)$ with α = 1, 2, is the inverse map of Eq. 1). For v and T

---

(1) A subscript preceded by a comma designates differentiation with respect to the corresponding co-ordinate.

smooth fields on $\mathcal{S}$, with v vector valued and T tensor valued, we can write:

$$v = v^i e_i = v_i e^i, \qquad T = T^{ij} e_i \otimes e_j = T_{ij} e^i \otimes e^j. \quad ^2 \tag{5}$$

We write ${}^s\nabla v$ for the surface gradient of v:

$${}^s\nabla v = v_{,\alpha} \otimes e^\alpha. \tag{6}$$

In particular, the negative of the surface gradient of the normal field n gives the Weingarten tensor field ${}^sL$:

$${}^sL := -\,{}^s\nabla n. \tag{7}$$

At each point $x \in \mathcal{S}$, ${}^sL$ is symmetric and its eigen-values κ_1 and κ_2 represent the principal curvatures of $\mathcal{S}$; further, the determinant of ${}^sL$ is the gaussian curvature κ_g of $\mathcal{S}$ at x.

6. Consider curves lying on the surface $\mathcal{S}$. As is well known, if $\mathcal{C}$ is a (regular) curve on $\mathcal{S}$ parametrised by arc length ℓ, the following formulae hold:

$$\frac{dx}{d\ell} = t, \quad \frac{dt}{d\ell} = \frac{1}{\rho} r, \tag{8}$$

$$\frac{1}{\rho} r.n = {}^sL.(t \otimes t), \tag{9}$$

where $\rho(\ell)$, $t(\ell)$ and $r(\ell)$ denote, for the typical point $x(\ell)$, the radius of curvature of $\mathcal{C}$, the tangent and normal unit vectors, respectively.

Let T be a tangential tensor field on $\mathcal{S}$, i.e.,

$$Tn = 0 \text{ on } S; \tag{10}$$

the surface divergence of such a field is defined by

$${}^s\mathrm{Div}\ T := T_{,\alpha} e^\alpha. \tag{11}$$

---

(2) We employ the summation convention. The range of Latin indices is $\{1,2,3\}$, and of Greek indices $\{1,2\}$. Notice that, for convenience, we set $e_3(x)=e^3(x)=n(x)$.

7. Then, for every regular vector field v on $\mathscr{S}$, the surface divergence theorem implies:

$$\int_{\partial\mathscr{S}} \mathrm{Tm.v}\, d\ell = \int_{\mathscr{S}} {}^{s}\mathrm{DivT.v}\ \ da + \int_{\mathscr{S}} \mathrm{T.}^{s}\nabla v\ \ da\ , \tag{12}$$

where $m(x) \in \mathcal{T}_X$ is the normal (unit vector) to the edge $\partial\mathscr{S}$, and

$$da = |e_1 \times e_2|\ \ dz^1 dz^2 \tag{13}$$

is the surface element on $\mathscr{S}$.

EQUILIBRIUM OF TENT STRUCTURES

8. Let S be the stress tensor field on $\mathscr{S}$. As $\mathscr{S}$ is a membrane, we shall assume S symmetric and tangential at each point $x \in \mathscr{S}$ (cf., e.g., refs 5,6). For each $m \in \mathcal{T}_X$, the vector

$$s = Sm\ , \tag{14}$$

lying in the tangent plane $\mathcal{T}_X$, gives the force (per unit length) acting on the line elements of normal m. If b denotes the field of the external forces (per unit area) on $\mathscr{S}$, the equilibrium equation takes the form:

$${}^{s}\mathrm{DivS} + b = 0 \ \ \text{in}\ \mathscr{S}\ , \tag{15}$$

or rather, in components,

$$S^{\alpha\beta}\big|_{\beta} + b^{\alpha} = 0,\ \alpha = 1,2\ ,$$

$$S^{\gamma\delta}\ \ {}^{s}L_{\gamma\delta} + b^{3} = 0.\ \ ^{3} \tag{16}$$

9. The stress field S must also satisfy the boundary condition. Thus, in view of Eq. 14, we require that

$$Sm = \hat{s}\ \ \text{on}\ \partial\mathscr{S}\ , \tag{17}$$

---

(3) $S^{\alpha\beta}\big|_{\beta}$ denotes the covariant derivative of $S^{\alpha\beta}$. We recall that

$S^{\alpha\beta}\big|_{\beta} = T^{\alpha\beta}{}_{,\beta} + T^{\beta\gamma}\Gamma^{\alpha}_{\beta\gamma} + T^{\alpha\beta}\Gamma^{\gamma}_{\beta\gamma}$ where $\Gamma^{\alpha}_{\beta\gamma} = e_{\beta,\gamma}.e^{\alpha}$ are the Christoffel symbols.

where m is the normal to $\partial\mathcal{S}$ and $\hat{s}(x) \in \mathcal{T}_x$ is the traction prescribed on the edge of $\mathcal{S}$. As remarked in the introduction, for tent structures the forces $\hat{s}$ are generally transmitted by cables (Fig. 2).

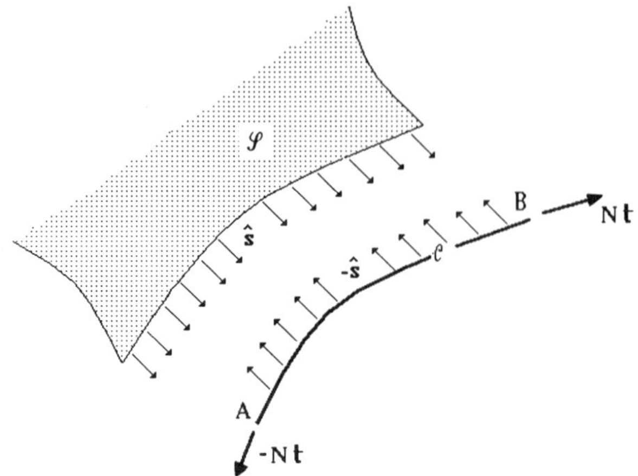

Fig. 2. Contact forces between cable and membrane.

10. In this case, if $\mathcal{C}$ ($\subset \partial\mathcal{S}$) is the curve occupied by the cable, equilibrium considerations for $\mathcal{C}$ yield:

$$\hat{s} = \frac{dN}{d\ell}t + N\frac{1}{\rho}r \quad , \tag{18}$$

where N (with N $\geq$ 0) denotes the normal force in the cable, and t, r and ρ are defined by Eq. 8. In particular, $\hat{s}(x) \in \mathcal{T}_x$ implies that:

$$N\frac{1}{\rho}r \cdot n = 0 \ , \tag{19}$$

or, in view of Eq. 9,

$$N({}^{s}L \cdot t \otimes t) = 0 \quad \text{at each point } x \in \mathcal{C}. \tag{20}$$

11. Hence, excluding the trivial case N = 0, it follows that $\mathcal{C}$ is an asymptotic curve of the surface $\mathcal{S}$.

Remark. It is well known, that if the gaussian curvature $k_g > 0$, there are no asymptotic directions on $\mathcal{S}$. Thus, when N $\neq$ 0, it is impossible to have edge cables at the points of elliptic type.

12. Our main concern here is the unilateral constraint on the stress. In fact, stability considerations for $\mathcal{S}$ imply that the stress tensor S is positive-semi-definite, i.e.,

$$s(x)m \cdot m \geq 0 \quad \forall m \in \mathcal{T}_x, \text{ at any point } x \in \mathcal{S} \tag{21}$$

13. In other words, the normal stress across every line element is either positive or zero (cf., e.g., ref. 7). Clearly, the condition of Eq. 21 imposes restrictions on the external forces (b, $\hat{s}$) which can be undergone by the membrane $\mathcal{S}$ at equilibrium.

14. In order to consider this problem we introduce the following definitions. We say that the stress field S on $\mathcal{S}$ is equilibrated with the system of external forces (b, $\hat{s}$) if S satisfies equilibrium Eq. 15 and boundary condition of Eq. 17. Further, we say that S is admissible if it satisfies the condition Eq. 21. Finally, a system of forces (b, $\hat{s}$) for $\mathcal{S}$ is said to be compatible (with equilibrium and condition Eq. 21) if there exists an admissible stress field S which is equilibrated with it.

15. Firstly, we observe that, in view of identity Eq. 12, S is equilibrated with (b, $\hat{s}$) if and only if the virtual work equation:

$$\int_{\mathcal{S}} S \cdot {}^s\nabla v \, da = \int_{\mathcal{S}} b \cdot v \, da + \int_{\partial\mathcal{S}} \hat{s} \cdot v \, d\ell \tag{22}$$

is satisfied for any virtual displacement field v on $\mathcal{S}$. On the other hand, introducing the strain tensor E corresponding to the displacement v;

$$E_{\alpha\beta} = E_{\beta\alpha} = \frac{1}{2}\left(v_{,\alpha} \cdot e_\beta + v_{,\beta} \cdot e_\alpha\right), \tag{23}$$

the first integrand in Eq. 22 can be written in the form;

$$S \cdot {}^s\nabla v = S \cdot \frac{1}{2}\left({}^s\nabla v + {}^s\nabla v^T\right) = S^{\alpha\beta} E_{\alpha\beta}, \tag{24}$$

since S is symmetric.

16. Now, consider the virtual displacements which satisfy:

$$(E_{\alpha\beta} e^\alpha \otimes e^\beta) m \cdot m \geq 0 \quad \forall m \in \mathcal{T}_x, \text{ at any point } x \in \mathcal{S}, \tag{25}$$

i.e., those displacements which produce (at each point $x \in \mathcal{S}$) an extension along every direction of the tangent plane. Writing $\mathcal{U}$ for the collection of such displacements. In view of Eqs 21 and 25, we obtain by Eq. 24 the inequality:

$$S\cdot{}^{s}\nabla v \geq 0 \ , \tag{26}$$

for any admissible stress field S and for any displacement $v \in \mathcal{U}$. Further, we assume that the admissible stress field S is equilibrated with (b, $\hat{s}$). Thus, Eq. 22 and Eq. 26 together imply that:

$$\int_{\mathcal{S}} b \cdot v \, da + \int_{\partial\mathcal{S}} \hat{s} \cdot v \, d\ell \geq 0 \ , \tag{27}$$

for any virtual displacement $v \in \mathcal{U}$. Hence, (b, $\hat{s}$) are restricted to those which satisfy the inequality Eq. 27, which may be interpreted as a compatibility condition on the systems of forces (b, $\hat{s}$) possible for the membrane $\mathcal{S}$.

17. For the case in which boundary forces are transmitted by cables, another compatibility condition can be easily derived from Eq. 27. Let $\mathfrak{C}$ be the part of $\partial\mathcal{S}$ occupied by the edge cable, and A, B its extreme points (see Fig. 2). In view of equation Eq. 18, an integration by parts shows that:

$$\int_{\mathfrak{C}} \hat{s} \cdot v \, d\ell = [Nt \cdot v]_A^B - \int_{\mathfrak{C}} Nt \cdot \frac{dv}{d\ell} d\ell \ . \tag{28}$$

Using the definition Eq. 23, we have:

$$t \cdot \frac{dv}{d\ell} = \left(E_{\alpha\beta} e^{\alpha} \otimes e^{\beta}\right) t \cdot t \ . \tag{29}$$

18. So, if one assumes $N \geq 0$ for the normal force in the cable, it follows that the integral on the right hand of Eq. 28 is non-negative for any virtual displacement $v \in \mathcal{U}$. Therefore, Eq. 27 implies the desired condition:

$$\int_{\mathcal{S}} b \cdot v \, da + \int_{\partial\mathcal{S}-\mathfrak{C}} \hat{s} \cdot v \, dl + [Nt \cdot v]_A^B \geq 0 \ , \ \ \forall v \in \mathcal{U}. \tag{30}$$

19. The specialisation of Eq. 30 for a membrane bounded by two or more cables is obvious.

PRESTRESSED STATE

20. Now, let $\mathcal{S}$ be a prestressing configuration for a tent structure. For such extremely light structures it is reasonable to neglect forces due to self weight, so that:

$$b = 0 \tag{31}$$

in the prestressed state. Thus, the only forces acting on the membrane $\mathcal{S}$ are the prestressing forces $\hat{s}$ which are generally transmitted by edge cables.

21. In this situation, equilibrium equation Eq. 16 $_3$ takes the form:

$$s^{\gamma\delta}\,{}^{s}L_{\gamma\delta} = 0 \ . \tag{32}$$

22. If the stress field S is admissible, Eq. 32 imposes some restrictions on the surface $\mathcal{S}$. It is easily seen that, at any point $x \in \mathcal{S}$ where S is non-vanishing, Eq. 32 and Eq. 21 are satisfied at the same time only if x is not an elliptic point, i.e.,

$$\kappa_g \leq 0 \tag{33}$$

23. In particular, if S(x) is positive-definite

$$S(x)m \cdot m > 0 \qquad \forall m \in \mathcal{T}_x - \{0\} \tag{34}$$

(i.e., the normal stress is positive for any direction), x is either a planar point ($\kappa_1 = \kappa_2 = 0$) or a hyperbolic point ($\kappa_g < 0$).

24. It is interesting to apply to the present situation the condition of compatibility for the external loads expressed by Eq. 27. Indeed, in view of Eq. 31, condition Eq. 27 reduces to:

$$\int_{\partial \mathcal{S}} \hat{s} \cdot v \, dl \geq 0 \qquad \forall v \in \mathcal{U} \ , \tag{35}$$

which is a restriction on the prestressing forces $\hat{s}$.

25. As an example of the last inequality consider a membrane $\mathcal{S}$ bounded by four cables. Let $\mathcal{S}$ be symmetric with respect to the planes π and σ, with π orthogonal to σ (Fig. 3). Assuming that the normal force is constant along each cable (i.e., the contact between cable and membrane is without friction) and equal to N ($N > 0$) for all the cables. In view of Eq. 18, we have:

$$\hat{s} = \frac{N}{\rho} r \ \ \text{at each} \ \ x \in \partial \mathcal{S} \ . \tag{36}$$

Then, by Eq. 35 it follows that:

$$\int_{\partial \mathcal{S}} \frac{1}{\rho} r \cdot v \, d\ell \geq 0 \qquad \forall v \in \mathcal{U} \ , \tag{37}$$

a geometrical restriction on the prestressed surface $\mathcal{S}$.

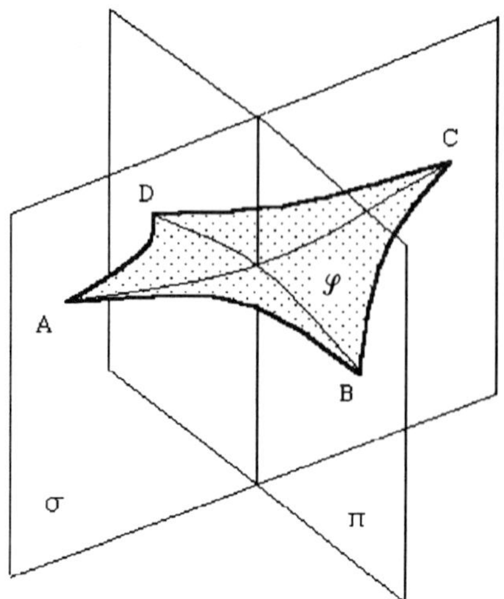

Fig. 3. (A, C $\in \sigma$; B, D $\in \pi$).

Acknowledgements

26. This research was supported by Ministero dell'Università e della Ricerca Scientifica e Tecnologica.

REFERENCES

1. OTTO F., SCHLEYER F.K. Tensile structures, Vol. II, The M.I.T. Press, Cambridge, 1969.

2. DE TOMMASI D., MARZANO S. Statics of edge cables in tent structures. International Journal of Space Structures, 1988, vol. 3, 15-19.

3. DEL PIERO G. Recent developments in the mechanics of materials which do not support tension. Proceedings of International Colloquium on free Boundary Problems, Irsee, Bavaria, 1987.

4. DO CARMO M. Differential geometry of curves and surfaces, Prentice-Hall, New Jersey, 1976.

5. NAGHDI P.M. The theory of shells and plates, Handbuch der Physik VIa/2, Springer-Verlag, Berlin, Heidelberg, New York, 1972.

6. PODIO-GUIDUGLI P. Lezioni sulla teoria lineare dei gusci elastici sottili, Masson, Milano, 1991.

7. ZAK M. Dynamics of film. Journal of Elasticity, 1979, vol. 9, 171-185.

98. Mathematical formulae as a medium for shaping membrane forms

W. J. LEWIS, University of Warwick, UK

SYNOPSIS. The complexity of the behaviour of lightweight membrane structures can be modelled in a variety of ways. One of the common approaches adopted in architecture relies on understanding the relationship between geometry and shape, observed in physical models. It is suggested here that the most universal medium for shaping membrane structures is offered by mathematical formulae which describe the form as a function of prestress and boundary conditions, and which are capable of predicting the deformations under a variety of load conditions. However, the accuracy of such predictions depends on the degree of understanding of this medium. In this study, attention is focused on the advantages (and limitations) of matrix algebra and finite difference methods for the form-finding and analysis of surface-stressed membranes.

INTRODUCTION

1. A definition of the surface geometry of structural membranes can be achieved by means of experimental techniques involving physical models, or numerical methods in which mathematical formulae serve to preserve the essential features of a physical entity.

2. A membrane is described in mathematical terms as a 'geometrically non-linear system', because the load-displacement relationship is not linear, due to significant changes of surface geometry under the applied loads. Consequently, an iterative solution is required when defining the equilibrium state.

3. A variety of numerical algorithms has been proposed to model the response of geometrically non-linear systems. Initially, attention was directed towards discrete structures such as beams, frames, and cable networks (refs 1-3). The analysis of membranes is complicated by the need to replace a continuum with an equivalent discrete system. In this process approximations are made. In general, two approaches can be adopted in surface discretisation: the Finite Element, or the Finite Difference method. The two approaches have developed independently of each other; nevertheless, a recent attempt has been made to combine the best features of each method into an efficient numerical solution for geometrically non-linear problems.

4. The Finite Element Method initially involves a discretisation of the surface into a number of finite elements. A stiffness matrix for each element has to be formulated, in order to compile a global stiffness matrix which enters the equation of equilibrium between the internal and external loads:

$$K\delta = P \qquad (1)$$

where K is a global stiffness matrix, δ is a vector of nodal displacements (at points of intersection of elements), and P is a vector of external loads.

5. The stresses, consistent with the geometry of the structure and applied loads, are translated into forces at the nodes of the elements, as given by the left hand side of Eq. 1. The global stiffness matrix K, formulated on the basis of current geometry remains unchanged for linear systems, allowing the nodal displacements to be found from the flexibility matrix. The solution is performed on the complete system of equations simultaneously; the size of the system depending on the total number of nodes in the structure, or more precisely, the total number of the degrees of freedom in the discretised surface. For non-linear systems, K becomes affected by the magnitude of displacements and hence Eq. 1 cannot be satisfied in one iteration. Various iterative techniques may be employed for the solution of the simultaneous equations, such as a method of simple iteration, or a Newton-Raphson method (ref. 2).

6. The Finite Difference method makes use of partial differential equations describing the variation of forces and displacements within the surface. A finite difference mesh is placed upon the surface to enable centred finite difference approximations of the first and second order derivatives to be calculated. Because of the geometrically non-linear nature of the problem, the variables change after each iteration, and hence another dimension is required; namely time. The basic relationship of static equilibrium remains the same as given by Eq. 1, but the method of solution is quite different. Using lumped stiffnesses at the points of a finite difference mesh, the system of equations is solved by a successive relaxation of the out-of balance forces at those points. The number of arithmetic operations performed at any one time is limited, as the computations are performed at each point in turn, rather than at all points in the surface simultaneously. Such an approach minimises round-off errors, and contributes to stability and accuracy of the numerical solution.

7. In current design practice, a method of Dynamic Relaxation with kinetic damping is used for the form-finding and static analysis of prestressed membranes. The method combines certain features of the finite difference and finite element approaches. The merging of the two methods has been possible, because, ultimately, each solution relies on defining equilibrium at the points of the discretised surface. The main objective of this study is to perform a close examination of some of the assumptions inherent in the above approach, which affect the accuracy of the final solution.

DYNAMIC RELAXATION ALGORITHM

8. The method makes use of the D'Alembert principle in which forces are related to the nodal displacements via the equation of dynamic equilibrium:

$$P_{ji} - K_{ji}\delta_{ji} = M_{ji}\,\ddot{\delta}_{ji} + C\dot{\delta}_{ji} \tag{2}$$

where the subscripts ji refer to the jth node in the ith direction, $\ddot{\delta}_{ji}$, $\dot{\delta}_{ji}$ are the acceleration and nodal velocities respectively, K_{ji} and M_{ji} are stiffness and mass respectively (assumed to be lumped at the nodes), P_{ji} is the external load vector, and C is viscous damping coefficient.

9. The left hand side of eq. 2 represents the residuals, or out-of balance forces at the nodes. The viscous damping coefficient is assumed to be zero, in which case Eq. 2 is reduced to:

$$R_{ji} = M_{ji}\,\ddot{\delta}_{ji} \tag{3}$$

and kinetic damping is used to control the solution, by monitoring the total kinetic energy of the system and stopping the iteration whenever a peak in kinetic energy is detected. The iterations are then restarted from the current geometry, but with the initial velocities set to zero. The process is continued, with accumulating displacements, until the residuals are reduced to a minimum.

10. The acceleration term in Eq. 2 is represented as the variation of velocity over the time increment Δt, using centred finite difference approximation:

$$\ddot{\delta}_{ji} = \frac{\dot{\delta}_{ji}^{\,t+\Delta t/2} - \dot{\delta}_{ji}^{\,t-\Delta t/2}}{\Delta t} \tag{4}$$

Combining Eq. 3 and 4 gives the following recurrent equation for velocities:

$$\dot{\delta}_{ji}^{\,t+\Delta t/2} = \dot{\delta}_{ji}^{\,t-\Delta t/2} + R_{ji}^{t}\,\frac{\Delta t}{M_{ji}} \tag{5}$$

and displacements:

$$\delta_{ji}^{\,t+Dt/2} = \delta_{ji}^{\,t-Dt/2} + \dot{\delta}_{ji}^{\,t+Dt/2}\,\Delta t \tag{6}$$

11. The following relationship has been suggested to ensure numerical stability (ref 3):

$$\Delta t \leq \sqrt{2\frac{M_{ji}}{K_{ji}}} \tag{7}$$

Substituting Eq. 7 into Eq. 5 shows that the size of the iterative step depends on the value of the residual and the ratio of mass to stiffness. The implication of this for the form-finding of surface stressed membranes, is that an ill-chosen initial surface, or incorrect value of stiffness would produce too large displacements, which in turn would contribute to numerical instability of the solution. Study of the efficiency of the Dynamic Relaxation algorithm showed that the use of fictitious masses, proportional to stiffness components in x, y, and z directions speeds up the convergence of iterations (refs 4 and 5).

ELEMENT FORMULATION

12. The continuum is discretised using triangular elements with 3 degrees of freedom (ref. 3). The choice of the element for the form-finding and static analysis of membranes is dictated by the simplicity of element formulation, which does not require the compilation and integration of element shape functions. The formulation relies mainly on the solution of a series of three simultaneous equations and lends itself very easily to the application of the Dynamic Relaxation algorithm.

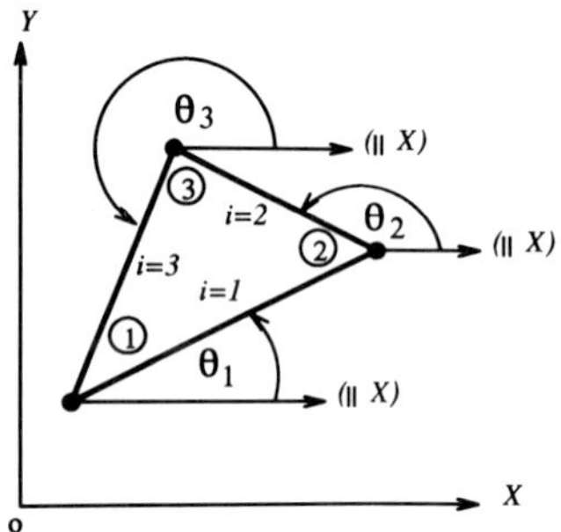

Fig. 1. Triangular element with three degrees of freedom.

13. The element, shown in Fig. 1 is represented by a triplet of pseudo-cables simulating the geometry and mechanical properties of a continuum element. The direct strain ε_i in the element side i is given as

$$\varepsilon_i = \varepsilon_x \cos^2\theta_i + \varepsilon_y \sin^2\theta_i + \gamma_{xy} \sin\theta_i \cos\theta_i \tag{8}$$

where ε_x and ε_y constitute strains in x and y directions, and γ_{xy} is the shear strain. In this formula, an important assumption is made, concerning the magnitude of strains which are considered to be small. The vector of strains in each side of the element can be expressed as:

$$\begin{bmatrix} a_1 & b_1 & c_1 \\ a_2 & b_2 & c_2 \\ a_3 & b_3 & c_3 \end{bmatrix} \begin{Bmatrix} \varepsilon_x \\ \varepsilon_y \\ \gamma_{xy} \end{Bmatrix} = \begin{Bmatrix} \varepsilon_1 \\ \varepsilon_2 \\ \varepsilon_3 \end{Bmatrix} \tag{9}$$

where the coefficients of the matrix [A] are:

$$a_i = \cos^2\theta_i\,,\quad b_i = \sin^2\theta_i\,,\quad c_i = \cos\theta_i \sin\theta_i$$

and the right hand side of Eq. 9 is given by:
$\varepsilon_i = \sigma_i / l_i$, l_i representing the initial lengths of the pseudo-cables

16. Solving for ε_x, ε_y, and γ_{xy} gives:

$$\begin{Bmatrix} \varepsilon_x \\ \varepsilon_y \\ \gamma_{xy} \end{Bmatrix} = [A]^{-1} \begin{bmatrix} l_1^{-1} & 0 & 0 \\ 0 & l_2^{-1} & 0 \\ 0 & 0 & l_3^{-1} \end{bmatrix} \begin{Bmatrix} \delta_1 \\ \delta_2 \\ \delta_3 \end{Bmatrix} = [B]\{\delta\} \tag{10}$$

where [B] is the transformation matrix, relating the element side extensions $\{\delta\}$ to the vector of local element strains.

14. The element stresses can be calculated as:

$$\{\sigma\} = \begin{Bmatrix} \sigma_x \\ \sigma_y \\ \tau_{xy} \end{Bmatrix} = [E][B]\{\delta\} \tag{11}$$

Assuming isotropic plane stress, the coefficients of the elasticity matrix, relating the element strains to the element stresses are given as:

$$d_{11} = d_{22} = \frac{E}{1-\upsilon^2}\,;\quad d_{12} = d_{21} = \upsilon\, d_{11}\,;\quad d_{33} = \frac{E}{2(1+\upsilon)}\,;\quad d_{31} = d_{13} = d_{23} = d_{32} = 0$$

in which E is the Young's modulus and v is the Poison's ratio.

15. The relationship between the elastic straining of the element and the vector of element side forces can be obtained from:

$$\begin{Bmatrix} T_1 \\ T_2 \\ T_3 \end{Bmatrix} = \left[[B]^T[E][B]V\right]\{\delta\} \tag{12}$$

where V is the volume of the element.
The element elastic stiffness matrix [K] is therefore:

$$[K_E] = \left[[B]^T[E][B]\right]V \tag{13}$$

16. During the form-finding, it is desirable to keep stresses within the element constant, (denoted here by σ_c), hence:

$$\{T_i\} = [B]^T \{\sigma_c\} V \tag{14}$$

in which case the forces in the pseudo-cables (element side forces) are dependent only on the geometry of the element, given by the matrix [B].

20. Once the forces in the pseudo-cables are found, they can be resolved into the x,y, and z components at the nodes of the element, allowing for the calculation of the nodal residuals. The nodes of the elements are assumed to coincide with the intersection points of the finite difference mesh. Subsequently, the Dynamic Relaxation algorithm can be employed to provide the numerical solution for the unknown displacements, hence defining the geometry of the surface. In the form-finding of stressed membranes, the solution may proceed with, or without inclusion of elasticity, resulting in the use of either Eqs 12, or 14 respectively. However, as will be pointed out later, the exclusion of elasticity may have undesirable consequencies, such as the numerical instability and divergence of the solution.

EFFECTS OF THE ASSUMPTION OF SMALL STRAINS

18. It has been suggested earlier, that the expression relating strains within the element to direct strains (Eq. 8) is based on the assumption of small displacements and small strains. This assumption is bound to be violated, particularly in the initial stages of the form-finding, when gross changes in the element geometry take place. As a result, forces in the pseudo-cables are related to element stresses via an incorrect transformation matrix [B]. Subsequently, numerical instability may occur, unless the chosen initial surface is fairly close to the final form-found shape.

19. Numerical instability manifests itself by element inversion; the problem which, it has been suggested, can be cured by applying a load to the offending apex node and hence correcting the topology of the discretised surface (refs 3 and 4). It is recommended here that rather than attempting to correct the topology of an inverted element, an onset of numerical instability should be controlled by inclusion of elasticity, which would control the surface strains. The numerical procedure, illustrating this approach is given below.

20. At time $t = i\Delta t$, the vector of element side forces is given as:

$$\{T_i\}^{t=i\Delta t} = [B]^T \{\sigma_c\} V\Big|^{t=0} + \sum_{i=1}^{i=n} \Big[[B]^T [E][B]V\Big]\Big|^{t=i\Delta t} \Big[\{\delta\}\Big|^{t=i\Delta t} - \{\delta\}\Big|^{t=(i-1)\Delta t}\Big] \tag{15}$$

where i is the current iterative loop, and n is the number of pre-defined loops. The first term on the right hand side of Eq. 15 shows the element side forces describing the initial constant stress. The second term gives the contribution of the elastic straining of the element to the vector of the element side forces.

21. The element stresses, initially taken as constant, can be revised accordingly.

$$\{\sigma\}^{t} = \{\sigma_c\}\Big|^{t=0} + \sum_{i=1}^{i=n} [[E][B]]\Big|^{t=i\Delta t}\left[\{\delta\}\Big|^{t=i\Delta t} - \{\delta\}\Big|^{t=(i-1)\Delta t}\right] \tag{16}$$

If it is found that at the time t the element stresses are negative, it is assumed that the element has buckled, in which case the element direct stresses are set to zero and the shear stresses calculated accordingly. The revised vector of element stresses , $\{\sigma_r\}$ is then used to recalculate the forces in the pseudo-cables $\{T_i^r\}$

$$\{T_i^r\}^t = V[B]^T\Big|^{t=i\Delta t}\{\sigma_r\}^t \tag{17}$$

22. The inclusion of elasticity in the numerical procedure leads to a violation of the condition of constant stress at equilibrium. By re-imposing the constant stress vector periodically, the condition of uniform stress may be represented more closely, as illustrated by the example below.

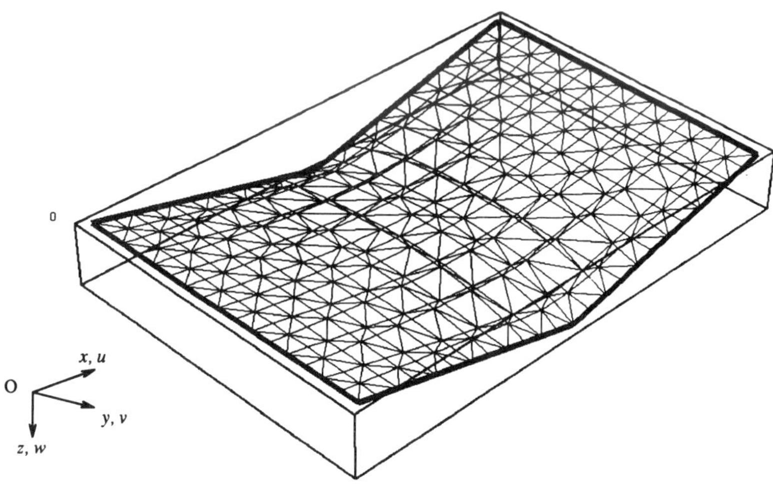

Fig. 2. Equilibrated three dimensional surface.

23. The surface illustrated in Fig. 2 has overall dimensions of 24.0m and 16.0m in the global x and y axes respectively, and is discretised by 768 elements. The element aspect ratio (of width to height of the triangle)) is initially equal to unity (approximately). The boundaries along the longer sides of the surface drop down to 3m at the mid-sides. The initial surface is described using Hermitian polynomials (ref. 6). The example is analysed using a range of coefficients of the polynomial affecting the initial slopes of the surface. These may reduce, or increase the proximity of the initial surface description to the final, equilibrated form.

24. It is found that for a well chosen initial surface the solution converges easily, even without the inclusion of elasticity. However, for an ill-chosen initial surface, the solution either diverges, or produces equilibrium at the nodes while grossly violating the condition of uniform surface stress. The latter case is illustrated in Fig. 2, where the residual is reduced to 0.025% of maximum pseudo-cable force, but the value of the maximum element stress deviation is 420.5%. The poor representation of the condition of uniform stress, leading to errors in the surface topology can be attributed to the assumption of small strains inherent in the element formulation. To negate these effects, the procedure outlined in paragraphs 19 to 21 is implemented, resulting in a smooth topology of the surface and maximum element stress deviations at equilibrium equal to 0.4%, as shown in Table 1.

25. Table 1 presents the maximum stress deviations, from the prescribed uniform stress in the surface. The results are presented for the first five re-impositions (N_{GI} = 1 - 5) of the constant stresss vector, $\{\sigma_c\}$.The numbers of iterations (N_{IT}) to achieve equilibrium for each re-imposition of constant stress are presented in the the second column.

Table1. Variation of element stresses in 3D surface (Fig. 2)

N_{GI}	N_{IT}	$\Delta\sigma_x^{max}$	$\Delta\sigma_y^{max}$	$\Delta\tau_{xy}^{max}$
1	1392	18.35 %	17.76 %	12.99 %
2	1208	2.31 %	2.23 %	1.65 %
3	899	-0.31 %	-0.31 %	0.23 %
4	627	-0.38 %	-0.37 %	-0.17 %
5	505	-0.39 %	-0.38 %	-0.18 %

CONCLUSIONS

26. The Dynamic Relaxation method incorporating triangular elements with three degrees of freedom used, is a well established numerical tool for the form-finding and static response of surface-stressed membranes. However, the element formulation includes the relationship between the direct strains (aligned with each side of the element) and the element strains along the x, y and z axes, which is based on the assumption of small displacements and small strains. This assumption cannot be satisfied during the solution procedure, particularly in the

initial stages of the form-finding, when gross changes in the surface geometry take place. As a result, numerical instability may occur, manifesting itself by element collapse, or inversion.

27. It is shown here that the errors induced into the numerical solution, due to the presence of large displacements and large strains, can be negated through the inclusion of elasticity in the numerical model. Such a procedure represents a self regulating mechanism which brings stability into the numerical algorithm and improves the accuracy of the surface description.

REFERENCES

1. TURNER M. J., *et al.* Large deflections of structures subjected to heating and external loads. Journal of Aero-Space Sciences, 1960, vol.27, February, 97-106.
2. ARGYRIS H. A., and SCHARPF, D. W. Large deflection analysis of prestressed networks. Journal of Structural Division, ASCE, Number St3, 1972, vol.98, March, 633-654.
3. BARNES M. R. Form-finding of minimum surface membranes. Proceedings of IASS World Conference on Space Enclosures. Building Research Centre, Concordia University, Montreal, 1976, July, 15-124.
4. BARNES M. R., and WAKEFIELD D. S. Form-fonding analysis and patterning of surface stressed structures. 1st International Oleg Kerensky Conference on Tension Structures, 1988, vol.1, 8/4-15/4.
5. SHAN J. Application of the sub-region mixed energy principle to numerical modelling of prestressed clad cable nets. PhD Thesis, University of Warwick, Department of Engineering, 1990.
6. LEWIS W. J., and GOSLING P. D. Computer modelling of stable minimal surfaces. Research Report CE32, University of Warwick, department of Engineering.

99. Design construction and maintenance of air-supported membrane structures in the UK

I. P. T. FIRTH, Flint & Neill Partnership, UK

SYNOPSIS. In this brief review of air-supported structures in the UK, the author considers some of the factors affecting their design, construction and maintenance, and comments on some common problems and their solutions.

INTRODUCTION

1. Air-supported structures provide a low cost and efficient means for covering a large space without any visible means of internal support. They therefore find many particular applications for indoor sports and other facilities where uninterrupted playing and viewing spaces are called for.

2. In the UK, the majority of air-supported structures tend to be ground-mounted airhalls enclosing two or three tennis courts or an equivalent plan area of about 2000m^2. In the USA and Japan there are several giant elevated air-supported stadium roofs, such as the Silverdome in Pontiac, Michigan which has a plan area of about 39,000m^2, and these are very impressive structures indeed.

3. In all cases the membrane envelope is supported by the internal overpressure of air which is constantly maintained by an inflation system. The membrane perimeter is held down either to a ground foundation or to an elevated anchorage structure, and entry is through air-lock doors designed to minimise losses. Cables are sometimes used to add strength and stiffness, such as in the case of the large low profile stadium roofs.

4. Air-supported structures are gaining popularity again in the UK following earlier uncertainties after the failures during the severe storms of 1987 and 1990. The author undertook an investigation into those failures which led to the publication of new design guidance for these structures (§. 7). One of the fundamental problems which is common with typical low cost airhalls is the fact that they have frequently been designed, constructed and maintained without adequate engineering input and often with little or no control by local authority engineers. The lessons of recent years and the new design guidance have led to a better understanding of the need for these structures to be properly engineered and to an increased level of control in most cases.

5. The reported failures were for various reasons but two predominant underlying causes were identified:

• The inflation pressure was too low to provide stability at the time of failure, either because of an inadequate air supply or due to excess leakage, or both.

• The membrane and cable anchorages were poorly designed and detailed and were constructed without due care and attention to detail, such that local stress concentrations and notches occurred in such a way as to initiate failure.

6. A more rigorous engineering approach to the design, construction and maintenance of these structures is called for, with a greater understanding of their peculiar and particular nature. For typical small airhalls it is appropriate to adopt properly engineered standardised details wherever possible in order to minimise overall costs.

DESIGN

Standards

7. The current UK standard governing airhall design is BS6661 (ref. 1) but following the recent failures it was recognised that this did not provide adequate guidance and certain enhancements have been proposed. These are embodied in the Performance Specification for Airhalls (ref. 2) which is now widely adopted and used in many indoor sport applications. Most UK suppliers and manufacturers now use this as a benchmark for setting the new industry standard.

8. The Standing Committee on Structural Safety (SCOSS) in their recent report (ref. 3), recommended that BS6661 should be withdrawn and that design guidance should be developed for all types of air-supported structures. This is to be welcomed and strongly supported, particularly in view of the increasing popularity of these structures.

9. The American Society of Civil Engineers have recently developed a new standard for the design of air-supported structures, which embodies a change from the traditional working stress approach using simple overall factors of safety to a more rigorous partial factor approach. This and other aspects of the ASCE standard should be considered in any new British or European initiative.

10. There is currently a good deal of confusion about the application of the Building Regulations to some of these structures, but there should be no doubt that the structural design and construction should comply with the Regulations. Local authorities sometimes concern themselves only with aspects such as fire safety, means of escape and planning issues, and ignore the structural design issues altogether. A more consistent and rigorous approach is necessary in order to provide the required controls, and this is also recommended in the SCOSS report.

Loading

11. Airhalls are subject to wind and snow loading which are the main environmental conditions affecting their design, and the inflation pressure necessary to support these loads contributes significantly to the forces in the structural elements.

12. Dead loads. Dead loads are small and are often ignored in practice, but they should be included in snow load combinations and in the treatment of collapsed elevated air-supported roofs hanging freely between supports. The dead load affects the collapse time of an air-supported structure, and it is therefore advisable to avoid suspending any equipment such as lights or services from the membrane wherever possible.

13. Imposed live loads. Imposed live loads apart from those due to environmental effects are usually small, and are related to the need for occasional access onto the membrane for maintenance purposes. These are commonly ignored in the design since the internal pressures are normally sufficient to support the weight of one or more persons by virtue of the local deformations of the membrane that occur. BS6399 (ref. 4) specifies a minimum uniformly distributed loading of 0.60 kN/m^2 over the entire roof where the slope is less than 30°, ignoring snow loading. This is unrealistic for most air-supported roofs but could be sustained by an inflation pressure greater than 600 N/m^2. It is difficult to envisage such high intensity loading arising from anything except snow loading, and this is dealt with separately by the Standard.

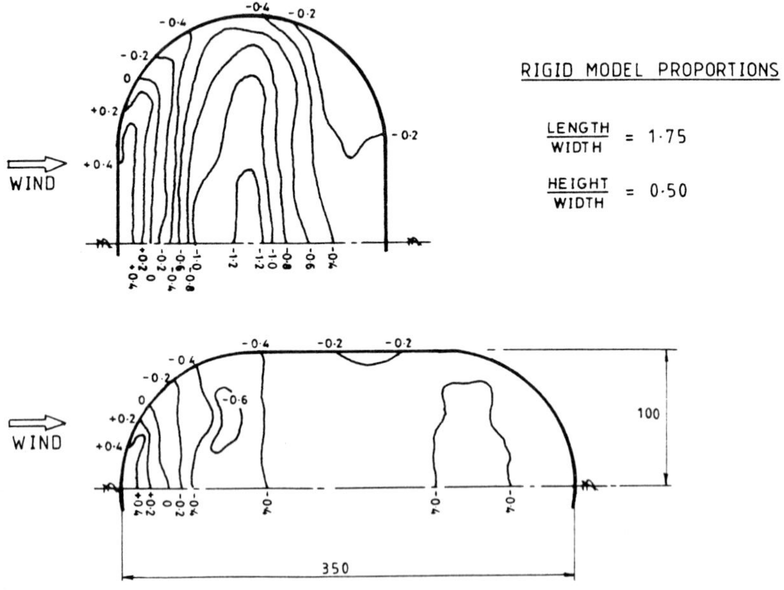

Fig. 1. Pressure Coefficients on a Semi-Cylindrical Model.

14. Wind loading. Wind loading is currently calculated for design purposes from CP3 Chapter V Part 2 (ref. 5) but this is soon to be replaced by BS6399 Part 2. However neither of these standards were prepared with air-supported or fabric structures in mind and special measures need to be taken to deal with their rapid response time and dynamic sensitivity. In using CP3, Class A should be

used for the derivation of the S_2 factor in determining the design windspeed for the membrane, appropriate to a short duration gust, instead of Class B or C. The proposed rules in BS6399 will permit a similar approach but may require a full dynamic analysis in order to determine the appropriate design wind speed. Any new UK standard for airhalls should give particular guidance in this respect to assist designers.

15. The distribution of wind pressures over the outside surface of the air-supported structure are determined by wind tunnel tests on representative models. Results are available for certain typical shapes of air-supported structure and an example is reproduced in Fig. 1 for a rigid semi-cylindrical model with semi-circular ends. The behaviour of a flexible membrane differs from that of a rigid surface and this should be considered, preferably by modelling the membrane flexibility and internal pressure correctly in the wind tunnel. As a general rule, the maximum positive pressure on the upwind side is lower for a flexible membrane than for a rigid surface, and the maximum negative pressure which occurs near the top of the structure is greater for the flexible membrane. Fig. 2, shows a typical distribution for a long semi-cylindrical model.

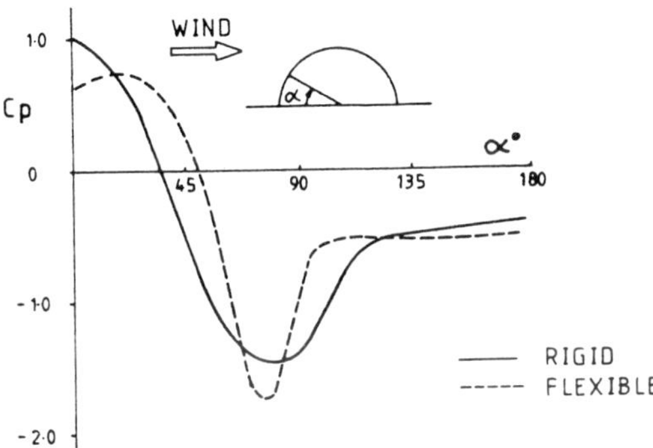

Fig. 2. Pressure Variation Around a Rigid and a Flexible Semi-Cylinder.

16. Model testing has to be undertaken with care and the results used in a suitable analysis to determine the behaviour of the structure. This is not usually appropriate or cost-effective for the majority of small standard typical airhalls and there is a need for codified guidance in such instances. BS6661 (ref. 1) provides such guidance by specifying the membrane tensions and internal pressures to be used in the design, in the form of simple formulae:

$P_{i(min)} = K_p\, q$
$N_\phi = K_\phi\, q\, R + P_i\, R$
$N_q = K_q\, q\, R + 0.5\, P_i\, R$

where q is the dynamic wind pressure, R is the radius of curvature of the membrane, P_i is the inflation pressure and $N_ø$ and N_q are the membrane tensions in the meridional and longitudinal direction respectively. The coefficients K depend upon the proportions of the standard typical airhall shapes covered by the Code.

17. The inflation pressure can be affected by the wind, particularly if one or more of the emergency exits are opened during a storm, and this has to be considered in the design. It is normally sufficient to allow for an internal pressure variation of about $\pm$ 0.2q in accordance with the provisions of CP3. In a typical airhall K_p may be about 0.5, so this could mean a change of between 40 and 140 percent in the inflation pressure. This aspect is too frequently ignored and is not addressed at all in BS6661.

18. Snow loading. Snow loading may be dealt with either by the snow melt method in which accumulation is prevented by maintaining an elevated internal temperature, or by the snow support method in which the weight of snow is supported by the internal inflation pressure.

19. The snow melt method depends on a reliable system for maintaining the required internal temperatures, and is really only practicable with relatively small enclosed volumes. Some degree of low level heating is often required in indoor sport facilities, usually for the benefit of spectators, and this may be sufficient in most cases. The required temperature depends on a number of factors including the snow fall rate, but is commonly of the order of 10 - 12°C. Higher temperatures are not economic because of the high heat loss through the membrane. Physical removal of snow using drag ropes and other mechanical means can also be effective but should not be relied upon.

20. The snow support method requires the inflation pressure to exceed the intensity of snow loading determined from BS6399 (ref. 4). The governing case is usually the asymmetric load case where drifting has caused a build up of snow on the downwind side, and at high altitudes in Scotland this can lead to peak intensities greater than 1.0 kN/m^2 However, intensities of between 0.2 and 0.5 kN/m^2 are more typical for most likely airhall sites in the UK, allowing for the short design life, requiring inflation pressures which are not very different from those commonly required for wind loading.

21. Inflation effects. Inflation effects are unique for air-supported structures because they are both a load and a resistance simultaneously. Increasing the inflation pressure increases both the forces in the structure and the ability of the structure to sustain load. Special care is therefore needed to ensure proper application in the design. BS6661 is a traditional working stress document, specifying overall minimum factors of safety, and gives little guidance on this. The proposed new ASCE standard adopts a more logical partial factor approach which enables the designer to select appropriate load factors for inflation pressure depending on the combination of loads being considered.

22. In the case of airhalls with a relatively small volume, the pressure can vary widely under extreme conditions, particularly if emergency exits are opened and if the inflation systems have a slow response time, and this must be reflected by the

assumptions made in the design. This has an influence on the inflation system control settings which are critical to the integrity of the structure.

Analysis

23. The analysis of membrane structures is complex and requires the application of relatively sophisticated computing techniques, allowing for geometric and material non-linear effects. In most cases, however, it is impractical to require such sophistication for every common airhall design, and simplified rules are called for, such as those provided in BS6661.

24. Computer methods. The rigorous analysis of a fabric structure first involves a form finding process to determine the shapes and forms of the membrane envelope. This has traditionally been done by physical modelling for tensile membrane structures using soap film or a stretch fabric such as ladies tights, but this is practically impossible for an air-supported structure so analytical techniques are used instead. Form finding is linked with the determination of the fabric cutting pattern, and if this process is not carried out properly the end product may have bulges and wrinkles, especially in the corners, and undesirable local stress concentrations may occur in the fabric.

25. The procedure is an iterative one, usually using the dynamic relaxation method of analysis, and is carried out in three stages of form finding, analysis under applied load, and cutting pattern derivation. Much has been written about these techniques by M. Barnes, D. Wakefield and others, and the interactive design/analysis process can be undertaken relatively simply with the necessary computer software.

26. It is important to model the membrane boundaries carefully and to incorporate their stiffnesses and those of any cable elements into the modelling correctly, as it is frequently in these zones that the highest concentrations of stress occur.

27. Simplified methods. Most airhalls in the UK are of a common standardised form for which it is usually reasonable to use the simplified formulae in BS6661 (§. 16). Other similar formulae exist such as those in ref. 6 and 7, but all should be used with care and not without an understanding of the basic principles of the structural behaviour. Such formulae give approximations to the maximum tensions to be expected under normal conditions but do not generally describe where such maxima occur.

28. A standardised form lends itself to a standardised design for which the full analysis for forming finding, load effects and cutting patterns only needs to be done once for a range of similar sized airhalls. Thereafter the approximate formulae may be used as a simple check that the required margins of safety are achieved in each particular case. This is commonly the way most airhalls in the UK are treated.

29. Simplified formulae can be misleading and misused, particularly without proper engineering input, and it is interesting to note that no such guidance is given in the new ASCE Standard which requires instead that the tensions are determined explicitly.

Membrane Materials

30. Most airhalls in the UK use PVC coated polyester fabric membranes, usually of about 1050 g/m^2 or 900 g/m^2 grade. Heavier grade fabrics are used in particularly exposed sites. This is a readily available material which is simple to cut and weld, is easily handled and can be readily folded. These materials typically have tensile strengths when new of about 100 - 115 kN/m for the 1050 g material and 80 - 90 kN/m for the 900 g material. Typical load-extension curves are given in Fig. 3.

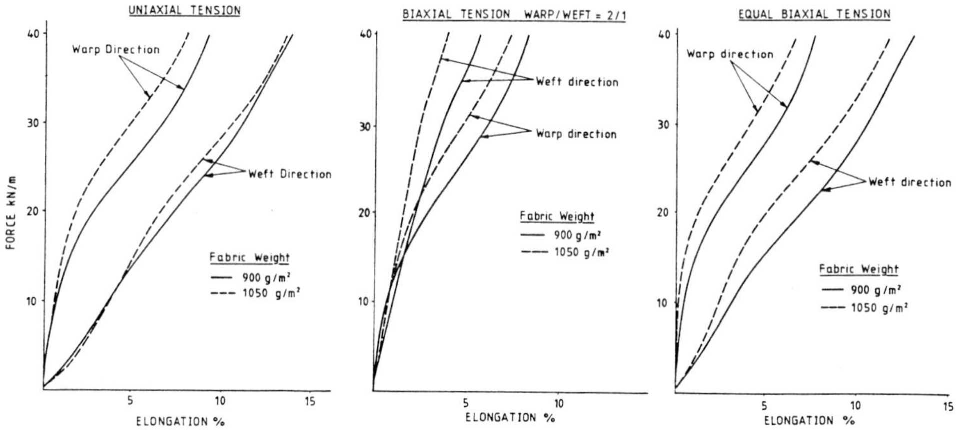

Fig. 3. Load-Extension Curves for Typical PVC Coated Polyester Fabric.

31. The more sophisticated designs, including low profile cable stiffened roofs are more usually constructed from higher strength more durable material such as the teflon or silicone coated glass fibres. These materials have a longer life and are generally superior to the polyester fabrics but are more difficult to weld, are much more expensive and require very careful handling since they are easily damaged by creasing.

32. Several small very low cost airhalls have been erected in the UK using a transparent visqueen material in conjunction with a system of stiffening cables. This is a low strength, low durability material which typically requires replacement every 4 or 5 years and these structures are only suitable for sheltered sites and then only with careful engineering design if the structure is to meet the necessary requirements.

Resistance

33. Membrane materials deteriorate with time depending on the nature of the fabric, the protective coating and the exposure, and this must be considered in the

design. BS6661 requires an overall factor of safety of at least 4.0 on membrane tension at the end of its design life, allowing for reductions in strength at welded seams, but this requires knowledge of the likely deterioration during that design life. Strength reductions of 10 - 20% over 10 years are commonly assumed for the polyester fabrics which is likely to be conservative, but the actual life of the fabric may be 20 years or more and there is insufficient data on the long-term behaviour of stressed fabric for reliable predictions to be made. Recent test evidence suggests that the strength reductions may not be as significant as previously thought, but there is still a need for more test results on fabrics before and after actual use in real structures. The main deleterious effect of ageing remains the embrittlement of the fabric material under the action of natural ultra-violet light. In a permanent well stressed fabric there is little bending and this should not be significant unless snatch loads or local stress concentrations are allowed to occur.

34. The resistance of the anchorages is determined by traditional methods using the relevant standards and criteria for such structural elements, taking care that the appropriate overall factors of safety required by BS6661 are achieved. A partial factor approach, would be more consistent with these standards and should be adopted in future.

35. Achieving the correct inflation pressure under all circumstances is of utmost importance in ensuring proper stability, and in preventing sudden snatch loads under the influence of severe gust loading, and this is an essential part of the overall structural resistance. The investigations and developments since the failures in January 1990 have concentrated on this as one of the main issues requiring more careful engineering input. This involves mechanical and electronic engineering more closely allied with structural engineering than perhaps in any other building type.

Inflation Systems

36. Fan units must be designed to supply a sufficient quantity of air to the interior to maintain the required inflation pressure under all circumstances, allowing for air losses through openings and around the perimeter, and for the permeability of the membrane itself. Standby systems are required to operate independently of the main fan in the event of power or fan failure, and the sophistication of these systems is constantly being developed to improve the overall reliability of the structure.

37. Most systems now in use in the UK are imported from overseas, and careful controls are required to avoid problems with adapting the electrical supplies to suit British Regulations.

38. Audible and visual alarms are also an essential part of the inflation system controls, and developments have been made in recent years to these in order to improve the security and safety of these structures and their occupants.

CONSTRUCTION

General

39. The construction and inflation of a simple ground mounted airhall is relatively straightforward. Because of this there has been a dangerous tendency for purchasers and contractors to do without proper engineering supervision, resulting in easily avoidable problems which are costly to rectify later. Proper contract co-ordination is also vital because it is common for the foundation to be constructed by someone other than the airhall supplier, and problems have occurred when the airhall designer and manufacturer have not had sufficient input to the foundation design and construction.

40. There is a perception among many that these are only temporary structures and that like a tent or marquee they do not need proper foundations. This notion must be dispelled so that at least the same level of care and control is given to these structures as to traditional buildings of the same size. The foundations and anchorages are normally designed for a 50 year design life even if the membrane may be replaced after 15 years or so, and these are certainly not temporary structures in the normal sense.

Foundations and Anchorages

41. Proper compaction around the ground beam is essential, particularly against the inside face where ground pressure usually provides some of the overturning resistance. Site joints between precast units require special attention. Owing to the speed with which it is possible to construct an airhall, care needs to be taken to ensure that the concrete has reached the required strength before loading the anchorages.

42. All anchorages must be carefully detailed and constructed to avoid stress concentrations in the membrane. It is important that all sharp corners are ground smooth, joints between adjacent steel sections are properly aligned, edge cables are correctly tensioned, holding down bolts are properly installed and so on. These things may seem obvious but it is surprising how often they are overlooked, and poorly constructed details such as these can lead to serious problems in the finished structure.

Inflation and Control Units

43. These are stand-alone prefabricated units which are simply lifted into place, but care is needed to position them correctly considering security, noise, access for maintenance, and other factors. The fan unit usually incorporates the anemometer which is a vital part of the control system, so it is no good placing it on the north or east side of the structure where it will be sheltered from prevailing winds.

44. Commissioning the system involves making adjustments of a series of controls to ensure that the internal pressures conform to the design requirements. Rigorous tests must be performed to the designer's satisfaction and a commissioning report should be included with the handover documentation detailing the final control settings.

Costs

45. Simple ground mounted airhalls such as those used for indoor tennis in the UK cost around £50 - 80 per square metre compared to around £150 - 180 per square metre for a traditional steel portal frame building, and even allowing for operation and maintenance costs they represent the lowest cost solution for covering a large space.

MAINTENANCE

46. The operation and maintenance of an air-supported structure must be addressed by the purchaser from the outset and not left until completion. There must be a proper maintenance policy, and a maintenance officer should be appointed with responsibility for routine inspections and all maintenance and emergency activities. He must be familiar with the principles of operation and behaviour of the airhall and with all of the controls for the inflation and alarm systems, and should ideally be present during commissioning. The supplier is required to provide a detailed maintenance manual, including a log book for recording the necessary daily, weekly and monthly inspections.

47. Proper security protection is vital since ground mounted airhalls seem to present an irresistible target for vandals armed with a pen knife. Localised damage to the membrane can be readily repaired by patching during calm conditions, but large tears can be difficult and particularly dangerous during strong winds, and money spent on a good perimeter fence and security system is money well spent.

48. Maintenance matters have too often been neglected in the past, but a proper systematic approach and a serious attitude is essential to the long-term integrity of these structures. Something as simple as a slack fan belt, a flat battery or an empty fuel tank can result in thousands of pounds worth of damage, and the importance of proper maintenance cannot be over emphasised. It is here that air-supported structures differ from most other types of building and it is often here that either they stand or fall.

CONCLUSIONS

49. Air-supported structures represent an economic way of covering a large area and are gaining popularity. A number of developments in their design, construction and maintenance have occurred in the UK in recent years, and further improvements are constantly being made. There is a need for a complete review of current UK criteria and of the controls exerted by local authorities, and with proper engineering input some of the exciting possibilities for constructing innovative air-supported structures in the UK might be realised.

REFERENCES

1. BRITISH STANDARDS INSTITUTION. BS 6661. Design, Construction and Maintenance of Single-Skin Air Supported Structures, 1986.
2. FLINT & NEILL PARTNERSHIP. Performance Specification for Airhalls. Lawn Tennis Association, January 1992. 3. STANDING COMMITTEE ON

STRUCTURAL SAFETY. Ninth Report, July 1989 to June 1992, SETO Limited, October 1992.
4. BRITISH STANDARDS INSTITUTION. BS 6399 Loading for Buildings, Part 3, Code of Practice for Imposed Roof Loads, 1988.
5. BRITISH STANDARDS INSTITUTION. CP3. Code of basic data for the design of buildings, Chapter V Loading, Part 2, Wind loads.
6. German Standard. DIN 4134, Air Supported Structures.
7. NATIONAL STANDARD OF CANADA. CAN3-S367 - M81, Air Supported Structures.

100. Anticlastic membrane correlation of form, prestress and behaviour

G. G. SCHIERLE and M. YIN, University of Southern California, USA

SYNOPSIS. The structural behaviour, or strength, stiffness, and stability, of anticlastic membranes are closely correlated with structural form and prestress. The objective of the presented research was to investigate, by computer analysis, correlation of form, prestress and stability for materials of different elasticity. The term *membrane* in this study includes fabric membranes as well as cable nets. Previous studies have demonstrated that structural stability of anticlastic membranes is critically reduced when load conditions reduce prestress to zero in some areas of the membrane. This research investigated the amount of prestress needed to prevent or minimise zero stress and thus instability. It will also be shown that, under certain conditions, prestress can reduce load induced deformation to half.

INTRODUCTION

1. Anticlastic membranes have strong correlation between form, prestress, and structural behaviour. For a given span and load, a small radius of curvature yields smaller membrane forces than a large radius. Further, the anticlastic curvature, combined with prestress, provides stability. While a chain or cable hanging between two points will adjust its form for any load condition, an anticlastic membrane resists such deformations by layers of mutually opposing curvature; just like two cables that cross each other stabilise their mutual cross point (Fig. 1A). Similarly, two layers of cables of mutually opposing, anticlastic curvature (Fig. 1B) stabilise a series of cross points and thus an entire surface.

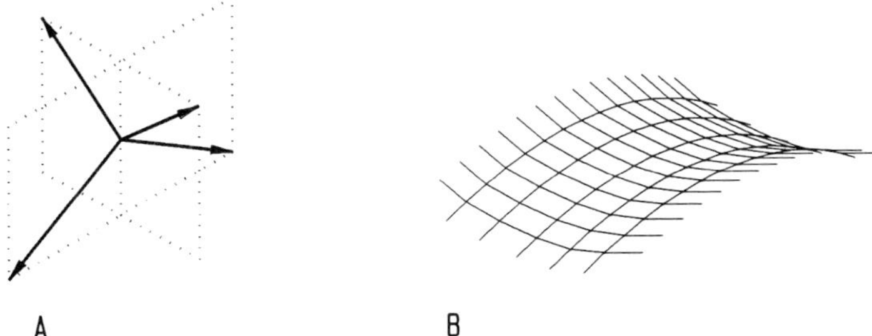

Fig. 1. Effect of curvature to stabilise anticlastic membranes.

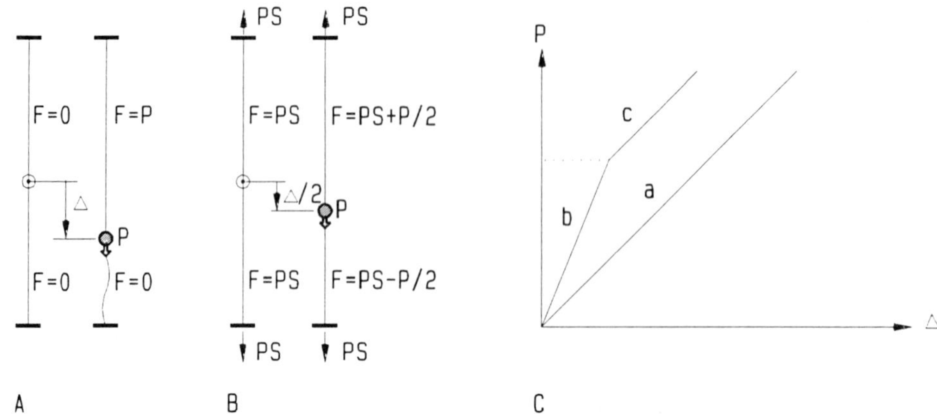

Fig. 2. Effect of prestress to provide stability and reduce deformation.

2. The effect of prestress to reduce deformation and give stability to a membrane can be illustrated with idealised models (Fig. 2) of an elastic string without prestress (A) and with prestress (B); where F = force, PS = prestress, and P = applied load. The load applied at mid-height of the string causes a deformation Δ in model (A); the upper link carries the entire load while the lower link carries no load and becomes slack. The string in model (B) is subject to an initial prestress. When load P is applied, the upper and lower link will each carry half; the latter by reducing its prestress. Since each link carries only half the load, the resulting deflection is Δ/2. Once prestress in the lower link is reduced to zero, any additional load will cause the same deflection as in the un-prestressed model (A), as illustrated in the P/Δ diagram (C) where a is the P/Δ curve of model (A); b is the P/Δ curve of model (B) with prestress; and c is the P/Δ curve of model (B) after the lower link prestress is reduced to zero. When the lower link in either model becomes slack, it will be unstable. These observations can be applied to anticlastic membranes, such as the model of Fig. 1B. If the model is without prestress, an applied gravity load would be carried by the concave layer only and the convex layer would become slack and thus unstable. But if the model has the proper amount of prestress, the concave and convex layers would each carry half the load, the latter through reduction of prestress; resulting in only half as much deflection as in the un-prestressed case. Model tests of anticlastic cable nets have demonstrated that when gravity load reduces prestress to zero such structures are highly unstable (ref. 1). Similar observations were made by Schierle during prototype testing for the membrane structures of Expo'64, Lausanne, Switzerland (Fig. 3). A first test prototype of natural canvas with edge cables deformed excessively, due to creep which resulted in slack fibres and thus instability. The addition of a supporting cable net resolved the problem. Therefore prestress in anticlastic membranes and cable nets should always be at a level to prevent zero

stress and slack members under any expected load condition. At least the majority of cables or membrane fibres should retain positive prestress. Therefore, the objective of this research was to determine, by computer analysis, the required amount of prestress for anticlastic membranes and cable nets with rigid beam edges and flexible cable edges.

Fig. 3. Anticlastic membrane structures for Expo'64 Lausanne, Switzerland. Architect: M.J. Saugey and G.G. Schierle; Frei Otto, consultant.

ASSUMPTIONS

3. The computer analysis to determine correlation of form, prestress, and structural behaviour was based on the configurations of Fig. 4 and the following assumptions. The quadrilateral anticlastic cable nets and membranes had side lengths of 50ft. (15.24m); with two low and two high points each; with both, rigid edge beams, and flexible edge cables. The high points were positioned to yield sag/span ratios at the lines of principal curvatures of 1:5, 1:10, and 1:15. The cable nets have mesh sizes of 3.5ft (1.07m) along the edges, reducing toward the centre according to lines of least length; providing one of several possible mesh configurations (ref. 2). The cable orientation is in the direction of the principal curvature for greatest stiffness. Tests have demonstrated that cables arranged parallel to the edges in the direction of generating lines and, thus, of least curvature, will have deformations about six times greater than those running

in the direction of principal curvature (refs 3 and 4). The net cables were of 1in (25mm) diameter strands, with metallic areas of 0.6in^2 (387mm^2) and elastic modules of 24,000ksi (165,480MPa). Edge cables were of 3in (76.2mm) diameter strands with metallic areas of 5 4in^2 (3,484mm^2) and elastic modules of 22,000ksi (151,690MPa).

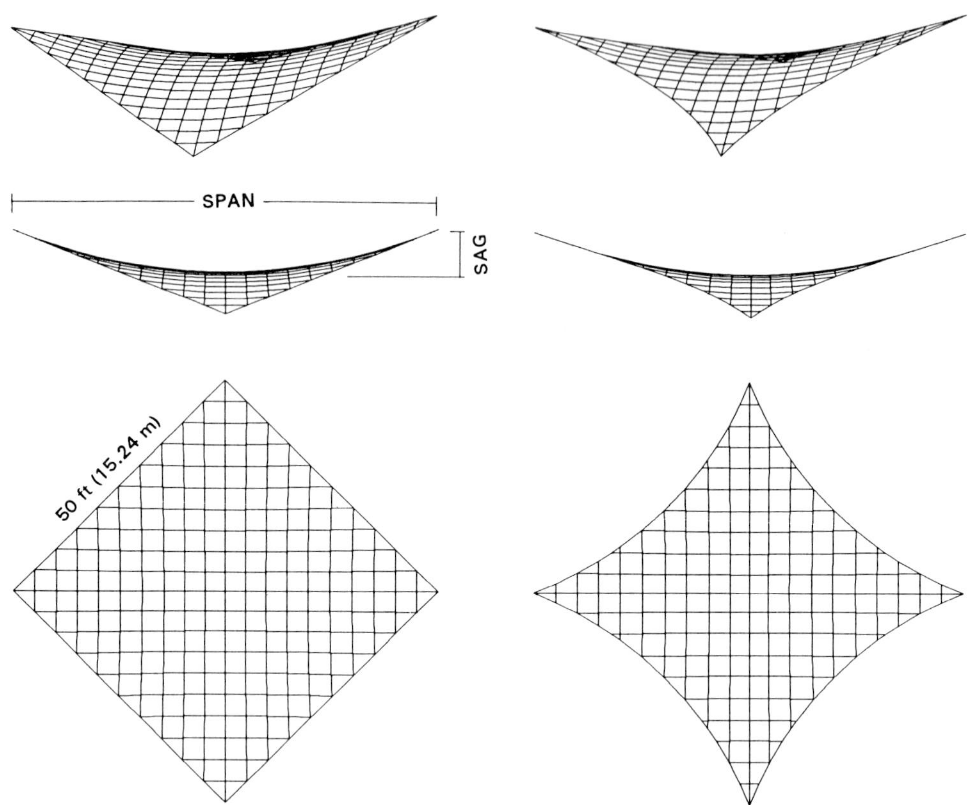

Fig. 4. Analysed prototypes with rigid edge beam and flexible edge cable.

4. Membranes were assumed to be fabric bands of the same width as the cable mesh, with elastic moduli of 9000/6750 pli (10943/8207kN/m) in warp and fill directions respectively.

5. Prestress was applied equally in concave and convex directions, to approximate minimal surface areas. Five levels of prestress were simulated, namely 30%, 40%, 50%, 60%, and 70% of the maximum stress under load. This was achieved by iterative runs, using a trial and error approach for each load case.

6. Gravity load was assumed 0.5k (2.2kN) applied at each joint of the cable nets and equivalent intervals of the membranes. Load was pro-rated for smaller tributary areas along the edges. The load was applied downward and normal to the ground.

METHODOLOGY

7. The analysis computer program *TRITRS* (ref. 5) was used for form finding and structural analysis. This finite element analysis program is based on a modified direct stiffness method for structural analysis of truss type structures with geometrically non-linear behaviour. The program includes a special prestress element suitable for form finding. Starting with a user defined approximate geometry, defined by X, Y, Z joint co-ordinates, connecting elements, and bi-axial prestress, the program finds the exact form for equilibrium, based on an iterative algorithm that converges in usually about 5 to 20 iteration steps. After the initial form is defined, subsequent runs compute member stresses, joint displacements, and support reactions under applied load, also in iterative analysis steps. The program treats cable elements such that if compressive forces reduce prestress to zero or less, the stiffness of such elements will be set to zero, to simulate the actual behaviour of slack cables which have no effective stiffness. However, prestressed cables can absorb compressive forces through reduction of prestress, as described for the idealised model of Fig. 2. But when the prestress is reduced to zero under compressive forces, the stiffness of the member will also be zero.

RESULTS

8. Analysis results are illustrated in Figs 5 and 6 for structures with rigid edge beams and flexible edge cables respectively; both for cable nets and membranes, and for 3 sag/span ratios. Our discussion refers to prestress force as a percentage of maximum force under load, as presented on the graphs. Comparing the edge beam structures of Fig. 5 with the edge cable structures of Fig. 6 reveals only minor differences. While the edge beam yields slightly higher forces, the differences are not significant, especially for sag/span ratios of 1:10. For sag/span ratios of 1:5 and 1:15 the edge beam structures require somewhat higher prestress to prevent slack members and instability. Hence edge conditions have only minor impact on required prestress in the case for the investigated saddle shapes. The following discussion is based on observations of the edge beam cases (Fig. 5).

9. For cable nets, the required prestress to prevent any cable from getting slack is 50% for 1:5 sag/span ratio and 40% for sag/span ratios of 1:10 and 1:15. As expected, a small sag yields higher maximum forces than a larger sag.

10. Slack members caused by zero prestress occur initially only in one or two cables and would probably not critically impair the structure's stability. The average force in convex cables is considered to be a more realistic measure of stability. The graphs show average convex cable forces still above zero at 30% prestress.

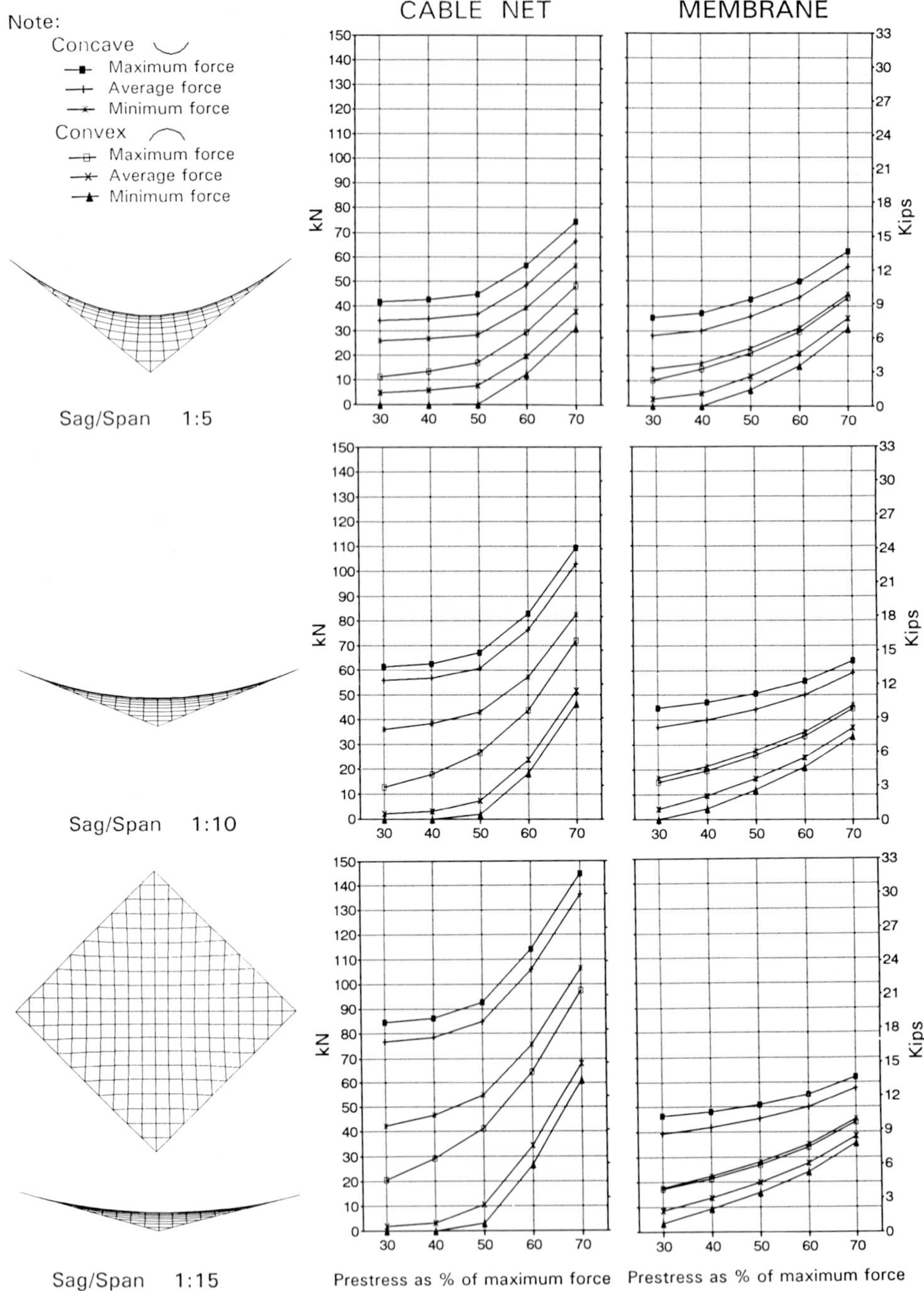

Fig. 5. Graphs correlating form and prestress as percentage of maximum force for anticlastic cable nets and membranes with rigid edge beam.

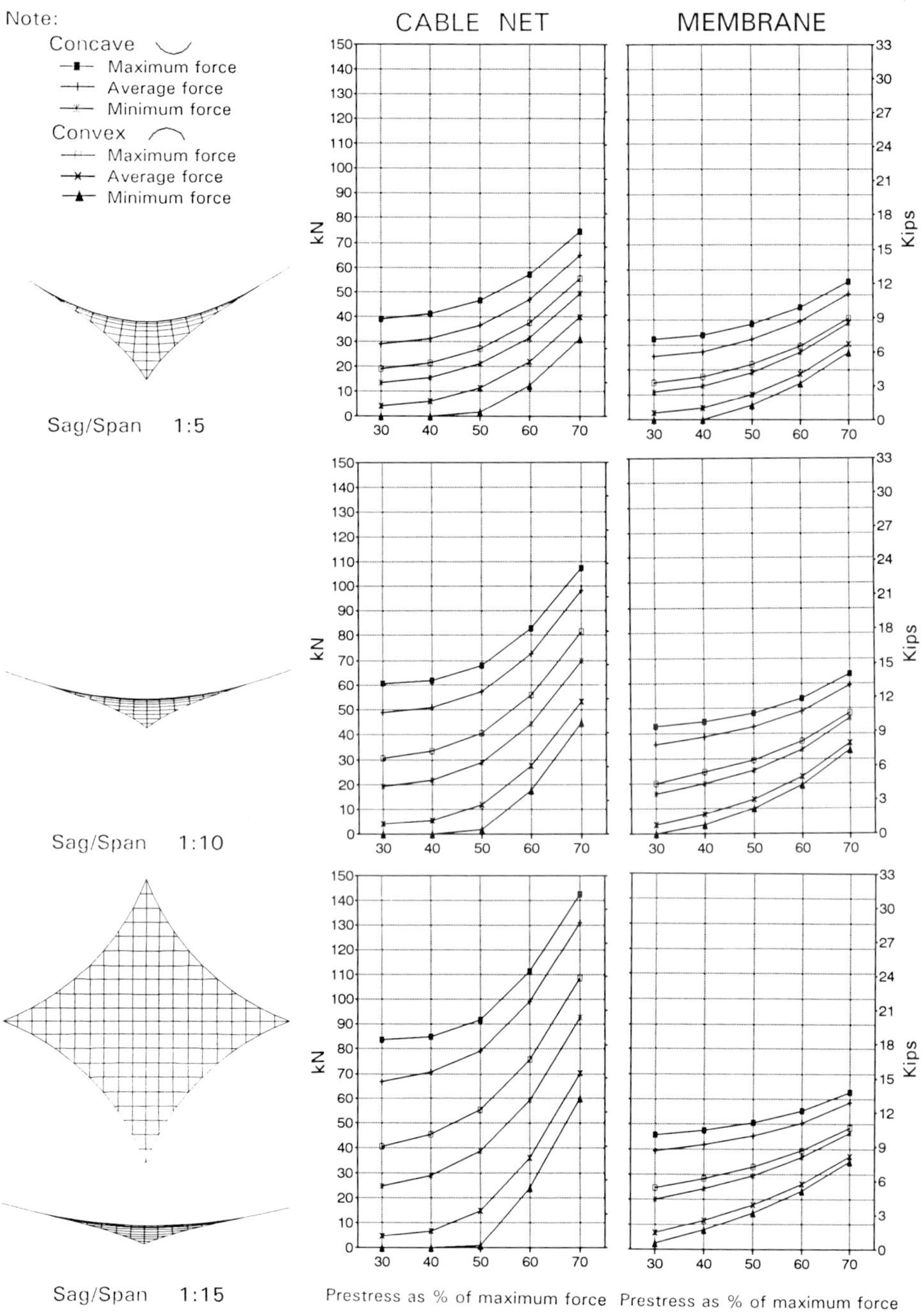

Fig. 6. Graphs correlating form and prestress as percentage of maximum force for anticlastic cable nets and membranes with flexible edge cable.

11. The initial level of prestressed should be adjusted for expected temperature variations, time dependent creep (ref. 6) and possible support displacements. An increase in temperature would reduce the prestress and so would creep and support displacements.

12. For membranes the required prestress to prevent any fibres from getting slack is less than in the cable net: 40% for 1:5 sag/span ratio, 30% for 1:10 sag span ratio, and less than 30% for 1:15 sag/span ratio. The average stress in convex membrane layers is well above zero even at 30% prestress for all sag/span ratios. Variation of maximum forces in membranes is minimal due to greater deflection than cable nets.

CONCLUSIONS

13. The graphs of Figs 5 and 6 and the above discussion of results demonstrate significant correlation between form, prestress, and structural behaviour; more so for membranes than for cable nets, due to their higher elastic modulus. More research is needed to:

- determine by dynamic analysis the real impact on structural stability when only one or two cables or a small membrane area become slack;
- determine the quantitative effect of temperature variation and creep on prestress,
- determine the non-linear elastic and plastic behaviour of some membrane materials, to provide better data for more accurate analysis.
- Determine the effect of asymmetrical and free-form shapes on prestress.

Data provided by such research would substantially help to make the design process for membrane and cable net structures more reliable and predictable. This is an essential factor for their wider use. It is hoped that the research presented here will make a contribution toward that end.

REFERENCES

1. SCHIERLE G. G. Lightweight tension structures, UC Berkeley, 1968.

2. BACH K. Nets in nature and technics, Institute fuer Leichte Flaechentragwerke, Universitaet Stuttgart, Stuttgart, Germany, 1975.

3. SCHIERLE G. G. Design considerations for lightweight structures, Proceedings, First International Conference on Lightweight Structures in Architecture, Sydney, Australia, 1986.

4. SCHIERLE G. G. Computer aided design and optimisation of surface structures, Work Documents, International Symposium, Weitgespannte Flaechentragwerke, Stuttgart, Germany, 1976.

5. HAUG E. and POWELL G. H. Analytical Shape finding of cable nets, Proceedings, IASS Pacifc Symposium, Tokyo, 1971.

6. BLUM R. Zeltbaumaterialien, Leicht und weit (Brinkman, ed.), Deutsche Forschungsgemeinschaft, Bonn, Germany, 1990.

101. Experimental studies on the flutter behaviour of membranes in a wind tunnel

H. MINAMI, Y. OKUDA and S. KAWAMURA, Osaka City University, Japan

SYNOPSIS. The flutter behaviour of PTFE-glass fabric has been investigated. Wind tunnel tests were conducted on fabric with both ends supported and one end supported. On the former condition the flutter critical wind speeds were explained by mass ratio, sag ratio and Froude number. The flutter did not occur when the interval between the supports of the membrane was below a critical value, or when the sag ratio of the membrane was below 0.06. The bending rigidity of the membrane was related to the frequency of the membrane on the flutter state. On the latter condition, the flutter appeared when the Froude number was more than 1.8 regardless of the length of membrane. When the length of the membrane was increased, flexure of the specimen became more severe near the trailing end of the membrane and the tensile breaking strength of the specimen was reduced.

INTRODUCTION

1. Generally, membrane structures are designed such that flutter of the membrane does not occur even under a strong wind condition. The large dynamic forces resulting from flutter may cause damage to the supporting edge and possibly cause tensile failure of the membrane. There is a possibility of flutter in the following membrane states:

i) when the initial stress in the membrane structure is insufficient;
ii) when the membrane is damaged and hangs down from the roof;
iii) when the full initial stress has not been introduced to the membrane under construction;
iv) when the membrane is supported at one side only during construction.

It is of interest to obtain basic knowledge concerning the flutter characteristics of a membrane in the above-mentioned states. How high is the flow speed to cause flutter to occur in a membrane? What effect would the frequency of the flutter and the magnitude of the dynamic forces on the part of supporting boundary have on the flutter state?

2. This experimental study was carried to obtain this basic knowledge. The flutter behaviour of an actual membrane material, PTFE-coated plain weave fabric, which has been used in practice for the long-term use membrane

structures, has been investigated in this study. The experiments were conducted for the membrane set horizontally in wind tunnel, and the upper and lower surfaces of the membrane exposed to wind.

3. On the first supported condition the membrane supported at the leading and trailing ends. We measured the flow speed, the frequency of the flutter, the maximum horizontal component of the bearing force at the leading end and the velocity of the progressive wave against each parameter; the interval between windward and leeward supports, the sag of the membrane was also measured. The membrane materials were two types of polyester film and a rubber fabric as well as PTFE-glass fabric. The critical flow speed, oscillation mode and the bearing force at the leading end on the flutter state were discussed with respect to dimensionless numbers. These numbers were sag ratio, Froude number, mass ratio, dimensionless frequency, kinematic bending rigidity, maximum horizontal bearing force coefficient and dimensionless velocity of the progressive wave.

4. On the second support condition the membrane was supported at the leading edge only, the flow speed, the frequency, and the maximum horizontal and vertical components of the bearing force on the support were measured for three different lengths of membrane on the flutter state. These results were illustrated with respect to dimensionless numbers as given above.

5. In the concluding remarks, a brief discussion of how the tensile breaking strength of the PTFE-glass fabric was reduced by the flutter for the two support conditions is given. The tensile breaking strength of the specimen was expected to be reduced when the breaking of the glass fibres in the base fabric occurred and increased owing to the continuous flutter oscillations.

6. There have been some studies in which the oscillations of a membrane roof in a membrane structure were measured in a wind tunnel experiment (refs. 2-4). In these studies the behaviour of the membrane roof, which was expected to be similar to that on actual membrane structures, were investigated with small scaled membrane models. The outer surface of the model was exposed to wind. The displacements of a one-way hanging membrane made of PVC-coated polyester fabric were measured in natural wind conditions (ref.5). The measured displacements were very small, less than 30cm, compared with the interval between the supports, 18.0m long, or sag, 1.7m deep. The flutter of a sail was analysed (ref.6), in which the experimental results on the flutter oscillation of membrane were shown. Taneda's study (ref.7) may serve as a reference when the behaviour of a membrane on the one-end-supported condition is discussed. In his study the specimens were very thin and flexible fabrics but they were very different from the ones used in actual membrane structures.

EXPERIMENTAL APPARATUS AND SPECIMENS

7. Fig. 1 shows the experimental apparatus for the both-end-supported condition in the wind tunnel. On the one-end-supported condition, the leeward clamp bar and the supporting columns were removed from the wind tunnel. The horizontal component of the force acting on the windward clamp bar was measured with the two force-detectors using a differential transformer system.

The vertical component of the force was measured with the torsional moment detector on the one end of the clamp bar. The quotient of each force component divided by the width B in the y-direction of the specimen was estimated as the horizontal or vertical component per unit width of the bearing force on the windward clamp bar. The section of the windward clamp bar was 3.2cm deep in the z-direction and 4.5cm long in the x-direction, which was similar to a streamline. The windward and leeward bars were set horizontally at the height of 48.4cm from the floor in the wind tunnel. The working-section of the wind tunnel was lm wide and lm deep. The membrane was subjected to air flow on the upper and lower surfaces, which exerted aerodynamic force on both surfaces of the membrane.

Table 1. Membrane materials.

	PTFE-coated glass fiber fabric	Rubber coated fabric	PET(polyester) film	
			110μm	44μm
Tensile strength(KN/m)	177	18.6	14.6	5.1
Mass per unit surface area, m(N·sec^2/m^3)	1.27	0.684	0.140	0.052
Bending stiffness per unit width, D (N·m)*	331x10^{-5}	14.1x10^{-5}	39.0x10^{-5}	2.1x10^{-5}
Thickness(mm)	0.78	0.61	0.110	0.044

* Measured with the device KES-FB2[1)]

8. The displacements of the membrane were measured using a laser beam system on the floor at the location of 37.8cm leeward from the windward support. The configurations of the membrane in the wind tunnel were recorded on video tape. The flow in the wind tunnel was uniform and the turbulent intensity of the flow was less than 1%. There was a clearance of 3.9cm between the membrane and side plates so that the flow pattern around the membrane became as similar to a 2-dimensional flow as possible.

9. The main specimen was PTFE(polytetraflouroethylene)-coated glass fibre fabric and was 40.0cm wide. PET(polyester) film, 110μm or 44μm in thickness, or rubber-coated fabric was added to the experiment for the both-end supported condition. Table 1 shows tensile breaking strength, mass per unit surface area, bending rigidity per unit width and thickness of each specimen.

EXPERIMENT USING THE BOTH-END-SUPPORTED CONDITION

Behaviour of the membrane in wind

10. When the flow speed U was 0m/sec, the configuration of the membrane was in the freely deformed shape (see Fig. 2(a)). As U increased, the windward part of the membrane was lifted up slightly and the freely deformed shape was warped (see Fig. 2(b)). When U reached the critical value, the membrane suddenly began to oscillate with progressive waves in most cases (see Fig. 2(c)). This behaviour of oscillation of the membrane is called 'flutter' in this paper. This critical flow speed was defined as the lower flutter critical flow speed U_{FL}. As U

approaches U_{FL}, it was occassionally observed that the membrane showed the wave form of the flutter but did not show any progressive waves, or the membrane did not show the progressive waves continuously. It was difficult to determine U_{FL} in these cases. When the progressive waves appeared continually, U_{FL} was estimated. As U exceeded U_{FL}, the oscillation of membrane appeared continuously. When U reached the another critical value, the membrane lifted upwards and shifted to the stationary state (see Fig. 2(d)). This behaviour is called 'diverged flutter' in this paper. When the diverged flutter occurred, U was defined as the upper flutter critical flow speed U_{FU}. The region in flow speed between U_{FL} and U_{FU} is called the 'flutter region in flow speed'. As the sag ratio, sag divided by the interval l, decreased, the flutter region in flow speed became gradually narrow, and finally U_{FL} had a tendency to be equal to U_{FU}.

11. Reynolds number Re, estimated with l, was 1.2×10^6 to 2.3×10^6 for PTFE glass fabric or rubber fabric. Re was also 0.3×10^6 to 1.0×10^6 for PET films.

Lower and upper flutter critical flow speed

12. The characteristics of U_{FL} and U_{FU} were examined using the co-ordinates of the Froude number Fn, mass ratio Mr, and sag ratio Sr.

$$Fn = U / \sqrt{gl}, \quad (1)$$
$$Mr = m/\rho l, \quad (2)$$
$$Sr = sag/l. \quad (3)$$

g = gravitational acceleration, ρ = density of air

U_{FL} and U_{FU} were plotted on the Mr-Fn or Sr-Fn plane in Fig. 3. As the membrane became lighter, U_{FL} and U_{FU} became lower. The flutter region in flow speed tends to be narrowed with the decrease of Sr as well as l. It was observed from the results of the experiment using PTFE-glass fabric or PET films that U_{FL} was very close to U_{FU} in the case of $Sr<0.06$. The behaviour of the membrane was shifted from the flutter state to the diverged flutter state owing to a slight change in flow speed that was less than 0.5m/sec. The Mr-Fn curve on the case of $Sr=0.06$, drawn as a broken line in Fig. 3(a), follows;

$$Mr = 0.0235 \; Fn^{2.09}. \quad (4)$$

13. U_{FL} or U_{FU} may be plotted on the following curved surface against each material in the Mr-Sr-Fn co-ordinates.

$$fL(Mr, Sr, Fn(U_{FL})) = 0, \qquad fu(Mr, Sr, Fn(U_{FU}) = O. \quad (5)$$

The experimental results for the PTFE-glass fabric in Fig. 3 are as follows; the longer l is, the lower U_{FL} became, and the flutter region in flow speed had a tendency to be expanded, while the shorter l is, the narrower the flutter region in flow speed became, and the flutter phenomena disappeared even in the cases of

Sr>0.06 on the critical value of 1. The consideration of R instead of Fn, was given to the above-mentioned procedures. The results estimated with R were less effective than that with Fn.

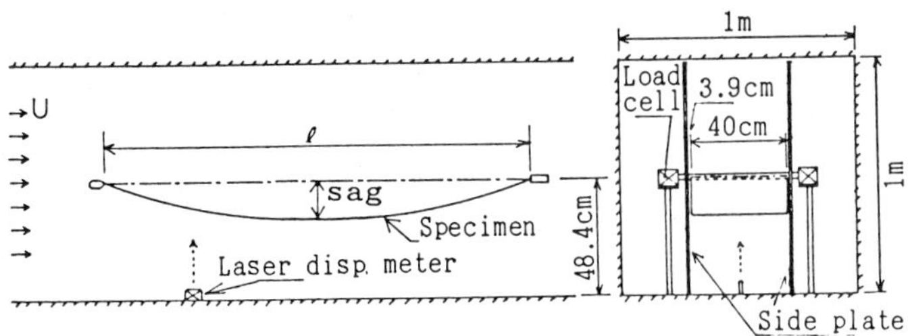

Fig. 1. Experimental apparatus.

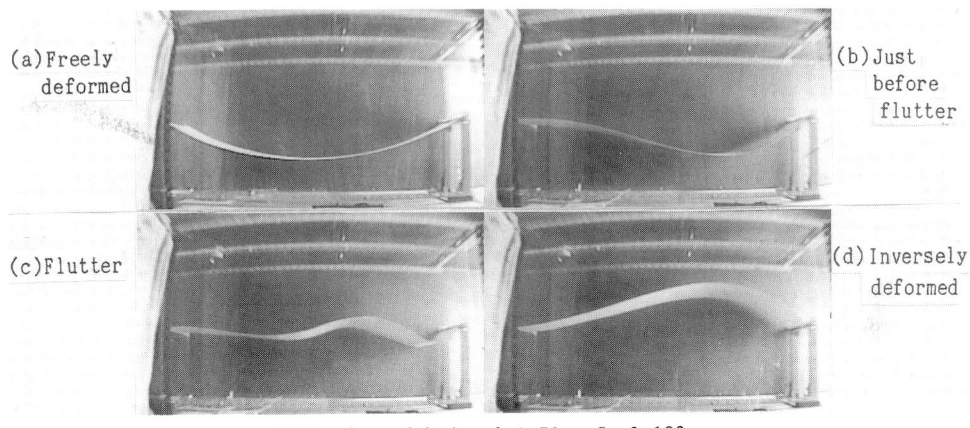

PTFE-glass fabric: l=1.78m, Sr=0.128.

Fig. 2. Configurations of both-end-supported membrane.

Flutter frequency

14. Figs 4 and 5 show the characteristics in the frequency f of the membrane on the flutter state. Dimensionless frequency Sn and kinematic bending rigidity K_B are as follows;

$$Sn = fl/U, \tag{6}$$

$$K_B = D/mf^2l^4 \tag{7}$$

KB is the ratio of terms related to bending moment and inertia force in the dominant dimensionless equation of oscillation.

15. Fig. 4 shows Fn-Sn curves on the flutter state. The frequency f tended to increase in proportion to U. There were some cases when f decreased in accordance with U, which depended on the material or the interval *l*. The maximum value of f was 4Hz for the case, *l*=1.78m, Sr=0.071, for the PTFE-glass fabric. It is meaningful to compare the results of two kinds of material, PTFE-glass fabric and rubber fabric at the case of *l*=1.78m and Sr=0.127 to 0.128. The mass m of rubber fabric was roughly half that of the PTFE-glass fabric. The bending rigidity D was, however, less than 1/23.

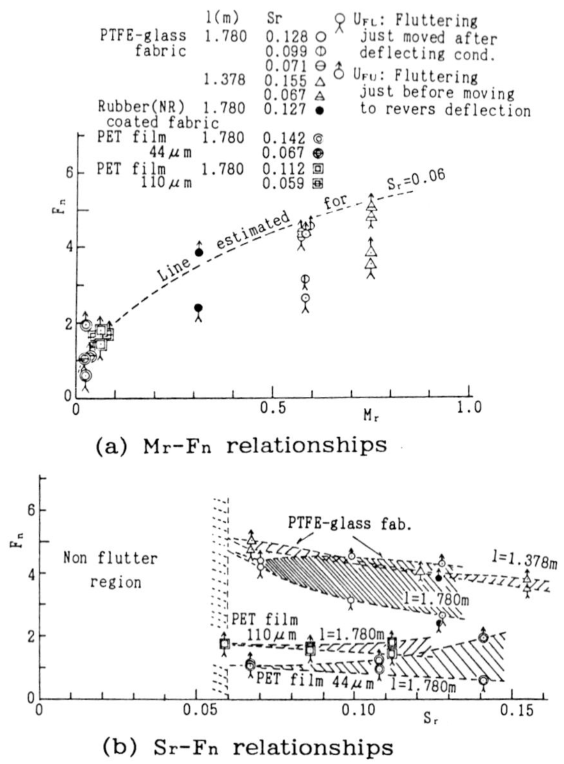

Fig. 3. Mr-Sn-Fn relationships on or off the flutter state.

The values of UFL for the both specimens were almost equal, while there was a remarkable difference between the frequencies at UFL for both specimens. The frequency f at UFL for the rubber fabric was near 4 times of that of the PTFE-glass fabric, which mainly was due to the large difference of the bending rigidity D.

16. Fig. 5 shows Fn-KB curves on the flutter state. There was a large difference between the kinematic bending stiffnesses KB of the PTFE-glass fabric and rubber fabric at UFL. In consideration of the above stated facts, KB was regarded as an effective parameter to examine the relationship between the states of the flutter oscillation and the characteristics of the materials.

Maximum horizontal bearing force

17. The maximum horizontal bearing force TH was the quotient, divided by the width B of the specimen, of the maximum value of the horizontal component of the bearing force on the windward support on the flutter state. Dimensionless value C_{TH} was estimated as follows:

$$C_{TH} = T_H/(0.5\,\rho\,U^2 l) \qquad (8)$$

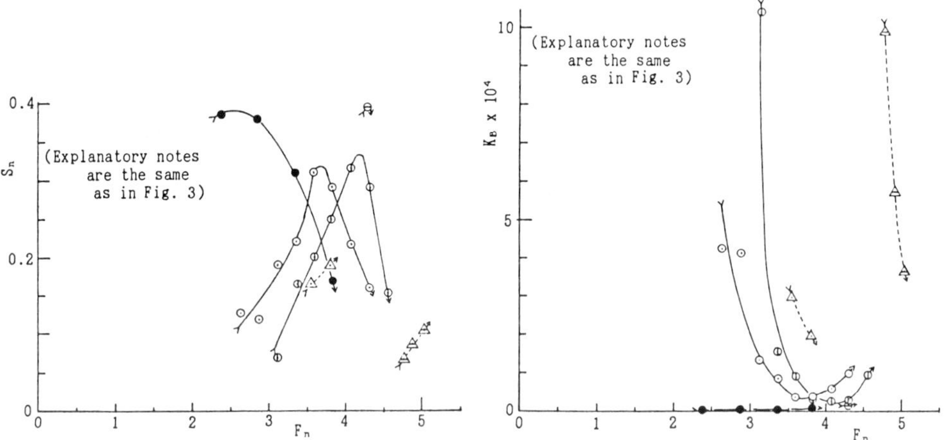

Fig. 4. Fn-Sn relationships on flutter state.

Fig. 5. Fn-KB relationships on flutter state.

Fig. 6 shows the Fn-CTH curves. The symbols unconnected with a solid line in the figure show the measurement values on the stationary state before or after flutter for reference. CTH on the flutter state increased and decreased with the increase of U corresponding to materials or Sr. U for the maximum value of CTH was different from U for the maximum value of the frequency f of the flutter. The maximum value of C_{TH} for the PTFE-glass fabric was 1.2 at U=13m/sec in the range of Sr<0.13 in this experiment. The membrane stress corresponding to this value was 220N/m and about 0.1% of the initial breaking strength of the specimen. Further the authors should like to add that the frictional resistance of the membrane and the aerodynamic resistance of the clamp bar were too small to be estimated in this experiment.

Velocity of flutter oscillation wave

18. The wave length V_w of the flutter oscillation was measured in the vicinity of the centre of the membrane using VTR. Fig. 7 shows the relationship between V_w/U and F_n. The maximum value of V_w/U was about 0.2 and V_w was apparently smaller than U. U for the maximum value of f was equal to U for the maximum value of V_w.

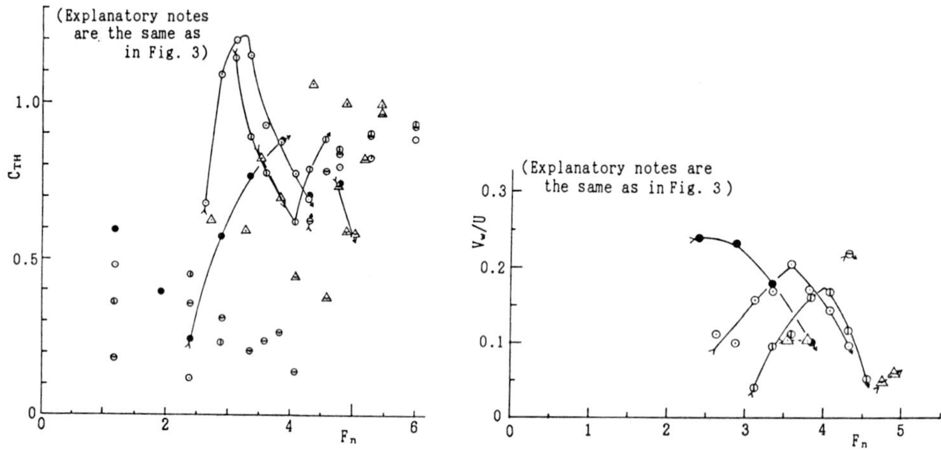

Fig. 6. F_n-C_{TH} relationships on flutter state. Fig.7. F_n-V_w/U relationships on flutter states.

Effect of clearances between side plates and edges of membrane

19. The thickness δ_s of the turbulent boundary layer on the side plate was about 8cm. It was found that the clearances d_{sc} between the side edges of the membrane and the side plates had a remarkable influence on U_{FL} and U_{FU} from the experimental results on the varying widths of membrane for the constant interval between the side plates. U_{FL} was the minimum value for $d_{sc}=0.5\delta_s$. Each clearance d_{sc} was about $0.5\delta_s$ in the other experiment of this study.

EXPERIMENT USING THE ONE-END-SUPPORTED CONDITION

Behaviour of the membrane in wind

20. In the case of U=0, the membrane was hanging from the windward clamp bar and resting on the floor in wind tunnel. The windward part of the membrane was lifted up with the increase in U. When U reached the critical value, the membrane began to oscillate with progressive waves. This behaviour of oscillations was also called 'flutter' in this paper. The flow speed for the occurrence of the flutter was defined as the flutter critical flow speed U_F. As U further exceeded U_F, the oscillation with the progressive waves occurred continuously. The membrane near the trailing edge had severe flexures and was

greatly deformed. Re estimated with the length L of the membrane was varied from 0.1×10^6 to 1.3×10^6. Fig. 8 shows the shapes per 1/15 sec of the membrane.

21. The frequency f of the membrane on the flutter state was in proportion to U in a narrow range of U. As L increased, the rate of increase of f for U decreased. In the case of L=lm, f was regarded as almost constant value for U>15m/sec. This f was 4Hz. As L increased, the membrane near the trailing end began to flex severely and to undergo large deflections.

22. Fig. 9 shows the Fn-Sn curves. Fn was calculated with Eq. 1 using L instead of *l*. For the case of L=0.5m, Sn remained an almost constant value. As L bacame longer than this value, Sn tended to have a maximum value and decrease with the increase of Fn. This tendency was considered to be related to the very large deformation of the membrane near the trailing edge. When the wave form of the flutter was a simple shape, Sn was regarded as almost constant to be independent of Fn. This figure shows that the flutter was observed to begin for almost constant values of Fn, 1.8, against each L.

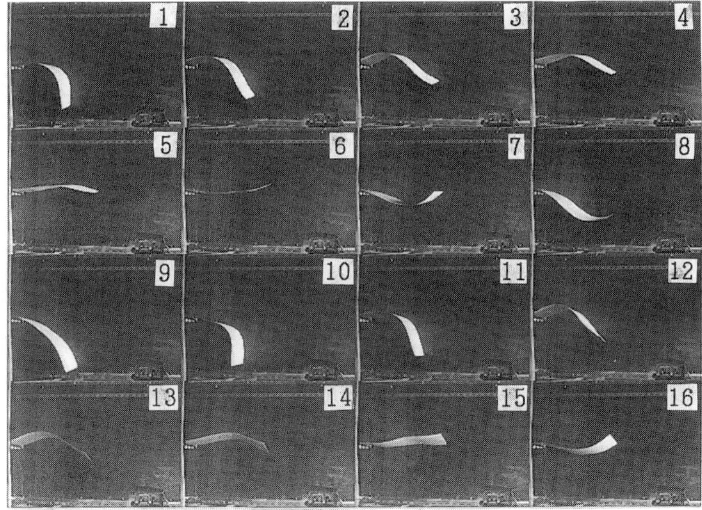

PTFE-glass fabric: L=0.7m, U=5m/sec.

Fig. 8. Configurations of flutter state of one-end supported membrane.

23. Fig. 10 shows Fn-C_{TH} and Fn-C_{TV} curves. C_{TH} and C_{TV} are as follows,

$$C_{TH} = T_H/(0.5\ \rho\ U^2 L), \qquad C_{TV} = T_V/(0.5\ \rho\ U^2 L) \tag{9}$$

T_V is the maximum value of the vertical component of the bearing force corresponding to T_H. C_{TH} and C_{TV} increased, as L became shorter for most experimental conditions. As Fn increased, C_{TH} and C_{TV} had a tendency to converge for each value regardless of L. It can be supposed from these experimental results that the convergent value of C_{TH} or C_{TV} was 1.5 or 0.3 respectively. The membrane stress was estimated to be about 370N/m in the condition, C_{TH}=1.5, L=1m, U=20m/sec, which was about 0.2% of the tensile breaking strength.

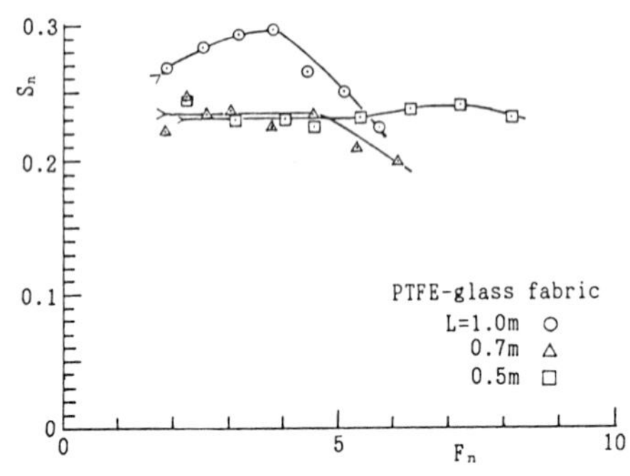

Fig. 9. F_n-S_n relationships on flutter state.

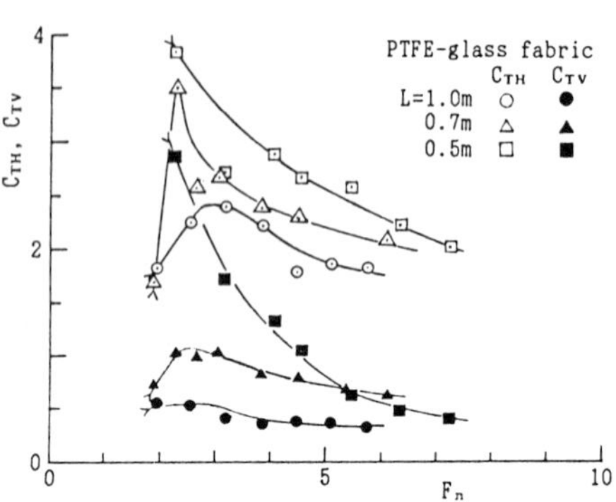

Fig.10. F_n-C_{TH} and F_n-C_{TV} relationships on flutter state.

CONCLUSION AND DISCUSSION

24. The flutter behaviour of the membrane was investigated on the both-end-supported and one-end-supported conditions. The conclusions for the design or construction of membrane structures can be arranged as follows.

25. Both-end-supported membrane. When the flow speed reached U_{FL}, the flutter occurred. The flutter region at the flow speed became narrow with the smallness of Sr or *l*. When Sr was smaller than 0.06, for any material, the flutter did not occur in practice. The flutter region in flow speed was able to be estimated effectively with three dimensionless numbers Mr, Sr and Fn, which formed a curved surface of the flutter region shown in the Eq. 5 for each material. The relationship between Mr and Fn at Sr=0.06 in these formulae was assumed to be shown in Eq. 4 in practice. The bending rigidity of membrane was supposed to be related to the frequency of the flutter. The maximum value of C_{TH} was about 1.2. The progressive velocity of the flutter waves was less than 20% of the flow speed U.

26. One-end-supported membrane. The long membrane underwent severe flexure near the trailing end on the flutter state. S_n was independent of F_n for the shorter membrane. S_n was, however, depended upon F_n under the severe flexing for the longer membrane. F_n at the beginning of the flutter was a constant value of 1.8, which proved to be independent of L. C_{TH} or C_{TV} converged to 1.5 or 0.3 respectively, independent of L, with an increase in F_n.

27. Brief discussion on reduction of tensile breaking strength. The reduction of the tensile breaking strength of PTFE-glass fabric, exposed to the flutter for three hours on the both-end-supported condition, was 3% of the initial strength. The authors also measured the reduction of the tensile breaking strength of the membrane near the trailing end. This membrane was used for the experiment on the one-end-supported condition. The reduction was 16% of the initial strength. We should pay more attention to this reduction in tensile strength, caused by the breaking of glass fibres in the specimen owing to severe flexing of the membrane near the trailing end.

ACKNOWLEDGEMENTS

28. The authors wish to express their sincere thanks to technical officer, Mr. S. Nakai in Osaka City University for his important role of these experiments, to Mr. H. Nose in Taiyo Kogyo Corporation for his kind supply of specimens, and to Professor S. Kawabata in Kyoto Univ. for his kind help in the measurement of the bending rigidity of the specimens.

REFERENCES

1. KAWABATA S. The standardization and analysis of hand evaluation. The Hand Evaluation and Standardization Committee, The textile Machinery Society Japan, 1980.
2. SIEV A. Experimental study of flutter in supported roofs. Bull. IASS, 23, 1965, 3-10.
3. KAWAMURA S. and KIMOTO E. Aerodynamic stability criteria of one-way type of hanging roofs in smooth uniform flow. Proc. 5th Int. Conf. Wind Eng., 1981, 939-948.
4. MATSUMOTO T. Self-excited oscillation of a pretensioned cable roof with single curvature in smooth flow. J. Wind Eng. and Ind. Aerodynamics, 34, 1990, 303-318.
5. KIMOTO E. and KAWAMURA S. Aerodynamic behavior of one-way type hanging roof, J. Wind Eng. and Ind. Aerodynamics, 13, 1983, 395-405.
6. NEWMAN B.G. and PAIDOUSSIS M.P. The stability of two-dimensional membranes in streaming flow. J. Fluids and Structures, 5, 1991, 443-454.
7. TANEDA S. Waving motions of flags. J. Physical Society of Japan, 24, No.2, 1968, 392-401.

102. The overlapping membrane structures of Ephesos and Carnuntum

D. S. HOPPE, Vienna University of Technology, Austria

SYNOPSIS. A brief description is given outlining the concepts behind the architectural design of the self-supporting, overlapping membrane roof structures used to cover the ruins of Ephesos and the excavation site for the Roman ruins at Carnuntum.

EPHESOS

1. The ruins of Ephesos is the largest archaeological open-air museum within the eastern part of the area round the Mediterranean seas. It is visited by more than one million tourists every year. Since 1893 the Austrian Archaeological Institute has been working there, uncovering the remains (ref. 1). From the time of the early Roman emperors, Ephesos was the most important city of the eastern part of the empire. Up to 300,000 inhabitants lived in differently equipped residential quarters. The archaeologists called these residential quarters "Hanghäuser" (insulae), which are terraced housing-units. The rooms of these units are grouped around a peristyl-court and exist partially as ruins as far as the second floor (ref. 1).

Protective Structures

2. The proposed self supporting membrane-construction creates a quasi suspended shelter above the antique landscape. Its curving and its bodiless and flimsy fabric structure, does not compete with the powerful forms of the ruins nor does it show any kind of style, or formalistic or fashionable modernism.

3. Tent-structures are visually timeless as the shape of the reversibly curved membrane merely results from the conformity with the natural laws of the statical lines of forces, determined by construction and material. These elemental flows of forces, translated into pure form, create a concept of form and space which is valid in all periods of time, and therefore was used in ancient times; a concept escaping any kind of arbitrary architecture.

4. The basic structure of the membrane-construction, consists of elements like steel latticed or tubular masts, compressed tubular bars, wire ropes (15 - 20 mm diameter), hinges, corner plates, screwed joints, tension bars and so on, using

foundation anchorage (tension) or pad foundations (compression). The membrane used as a roof skin is part of the construction and therefore effectively, reversibly curved, and consequently stabilised. The material used is 1·5 - 2mm thick, light-admitting, light-tight or clear. The range of possible materials goes from natural fibre tissue up to Teflon-coated fibreglass that is translucent and self-cleaning, and offers an extremely high durability, span condition and UV-resistance. This material is virtually weather resistant.

5. For the Ephesos construction this kind of Teflon-coated-fibreglass membrane has been proposed, the same material which has been in use for more than 6 years now in downtown Vancouver, B.C. on top of the EXPO-pavilion, still shining and showing no sign of any damage.

6. The colour chart is theoretically endless in variety, though for the proposed site, colours like natural, fallow, beige or light grey are intended to be used. The intended horizontal cover in the inner court would be made of snow-white fibreglass tissue, which would reflect light into the courts because of the translucent effect of the exterior skin, possibly intensified by small clear bays, like spotlights integrated in the roof membrane. In this way a white shining ceiling would simulate the inside lighting of the court.

7. The construction is made up of three stages, dividing the membrane into three layers due to visual, structural and construction-schedule reasons. Each layer or stage can be built and completed separately, e.g. stage 1 and 2 can be added to the currently covered stage 3, if the existing stage 3 cover is to remain.

8. When constructing the roof above stage 3, large parts of the existing internal and structural ranges remain untouched because of the entire cover of this stage only needs three masts. These masts are situated on existing foundations, and therefore do not interfere with any interior floor or wall structures. The three stages can be anchored separately, in each case to walls or columns. Stages 1 and 2 for example, are anchored to the existing wall-structure of the side of the present housing.

9. The proposed overall fastening is in fact the more comprehensive solution and offers many technical advantages. Besides, the pattern of three layers enhances the optimal ventilation of the roofing. The overlapping of the membrane layers and the cantilevering parts above the exterior walls are a protection against extremely heavy rain. As an additional protection, rimless clear lexanpanes (scratch-proof perspex) including distance panes are directly attached to the stone walls. The upper edge could show the same curving as the membrane or end in a straight line. In this way the silhouette of the walls is not concealed and still a bodiless shelter is achieved, as the ruins always stay visible.

10. The circa 15cm distance from the wall causes an additional ventilation. No change of climate or humidity will take place underneath the covered area of the Insula and the membrane also optimises ventilation. This seems very important taking into consideration the humidity caused by large tourist groups, which has destroyed the wall-paintings.

Fig. 1. General view.

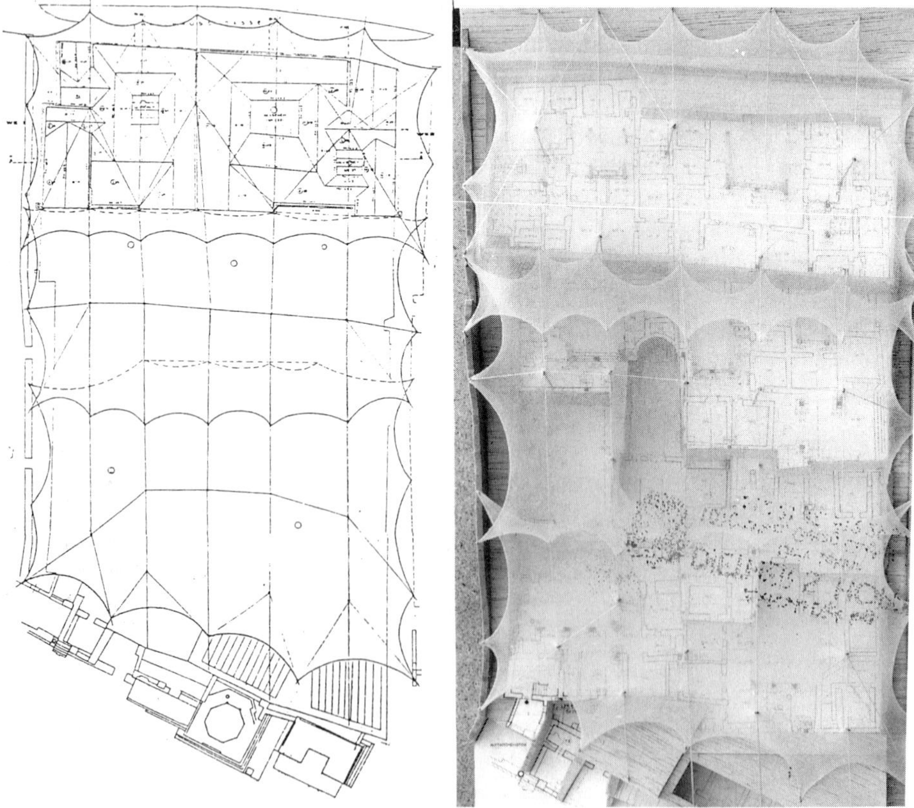

Figs 2 and 3. Site plan and model.

Fig. 4. Elevation view.

Fig. 5. Sectional view.

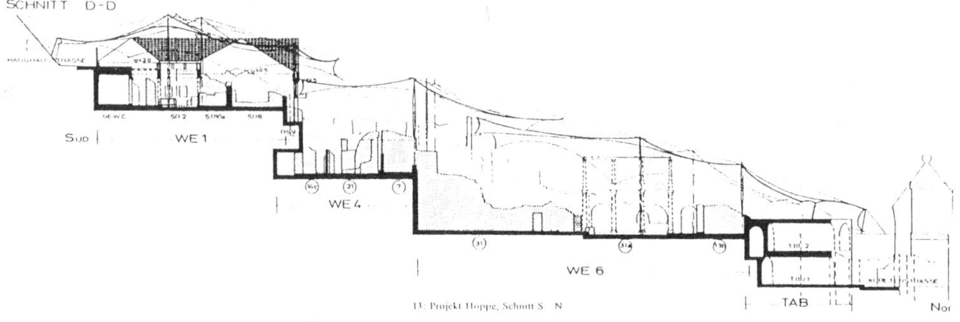

Fig. 6. Section.

11. The construction time is reduced enormously through prefabrication of membrane and supporting structure. The three layers forming the roof dematerialise the large surface and integrate it into the landscape. The construction is not submitting but integrating.

Jury's Opinion:

12. "In this project, a covering of the whole Insula by means of a self supporting membrane-construction in three overlapping layers is proposed. From the architectural point of view this roof system surely is an optimum, because it largely adapts to the landscape and impresses through its lightness and elegance of forms. However it has to be said that the large light surface would create a new dominant in the landscape. Especially positively mentioned is the fact that the architect succeeded in leaving the antique ruins absolutely untouched. On part of the Turkish representatives and the archaeologists doubts concerning the effectivity as for a weather and pollution protection are expressed." (ref. 1).

CARNUNTUM

13. Situated approximately 30km east of Vienna, the temple district of the Roman camp-town of the emperor Septimius Severus covered 150ha of land during its most flourishing times. Today it is the largest Roman excavation site within Middle-Europe, north of the Alps.

14. Started in the second half of the first century (about 50 AD), the main part dates from the second century AD. Before the invitation to the competition, the Archaeological Institute of the University of Vienna was looking for a roof shelter for the excavations. Consequently the author was invited to design a wide span textile construction to keep the excavation area free from supports.

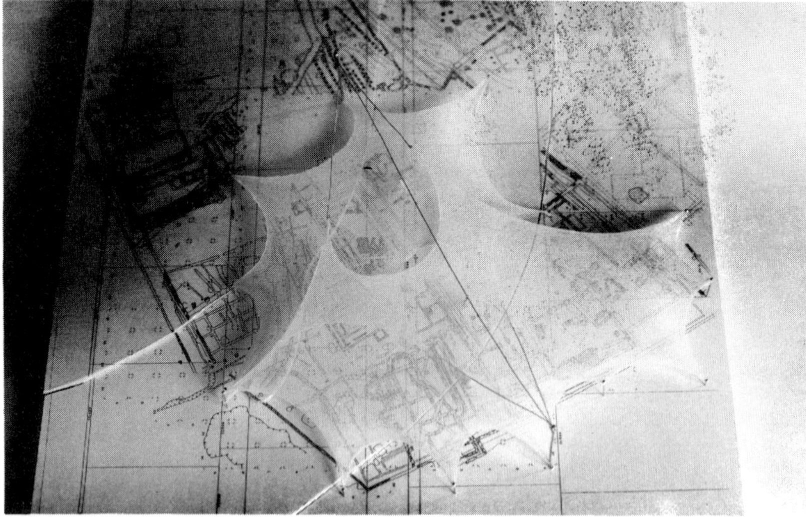

Fig. 7. Variant 1, plan view of the overlapping membrane construction

15 Three variants have been proposed for the roof structure. One with two overlapping textile membranes, one with a suspended construction and the third with poles standing outside and supporting the high-points of the large surface. These principal proposals have been presented and discussed within the project of an archaeological park for the district of Lower Austria. Membrane materials and the support system are quite similar to the overlapping structures of Ephesos.

16. Tension membrane constructions are found to provide ideal sheltering roofs for antique excavation sites and therefore are likely to be used more and more.

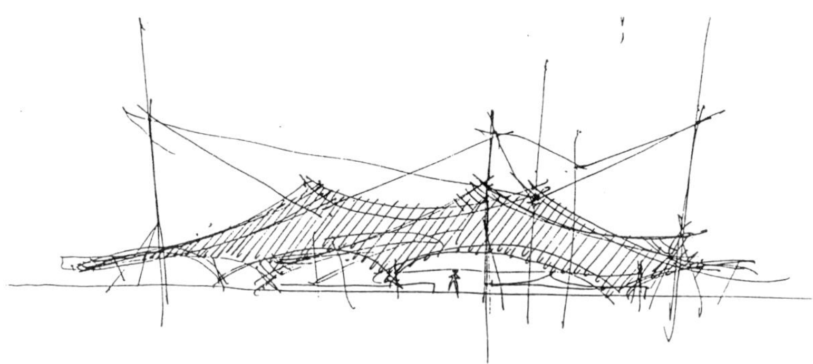

Fig. 8. Elevation study.

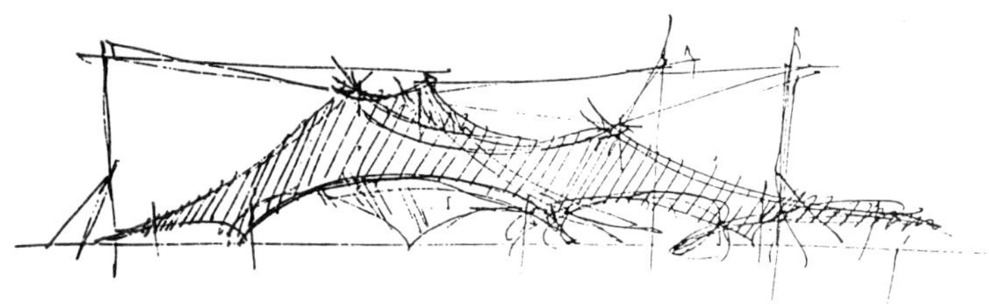

Fig. 9. Elevation study.

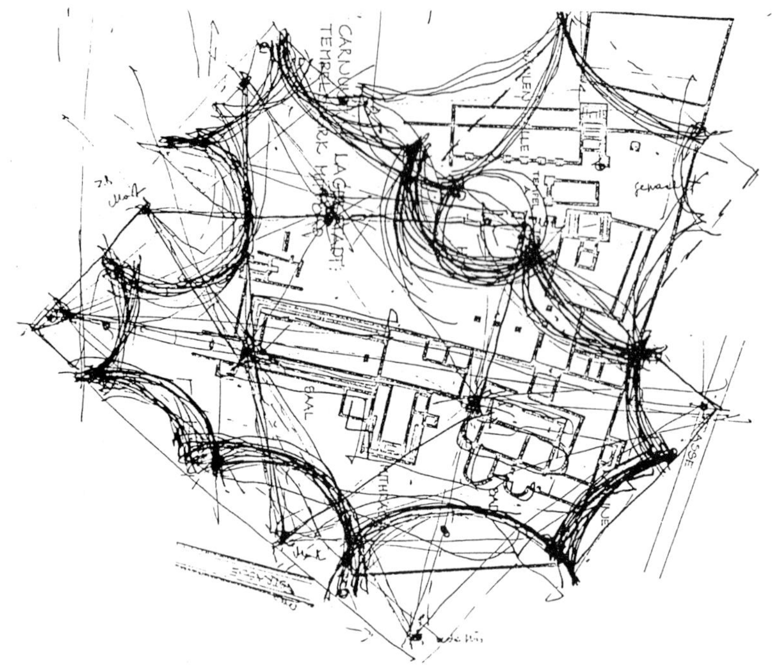

Fig. 10. Site study.

REFERENCES
1 . LANGMANN G. (ed.), Österreichisches Archäologisches Institut Grabungen 1989, Wien 1989.

103. A numerical method of optimum design analysis for cable assemblies

Y. NAMITA, Setsunan University, Japan, and H. NAKANISHI, Kobe Steel Ltd, Japan

SYNOPSIS. The optimum design and analysis of structural cable networks is shown, where, by using an optimisation technique, the authors obtain solutions which satisfy not only statical equilibrium conditions under stationary loadings, but also equilibrium conditions for member forces and deformation under superimposed loadings, making the cross-sectional areas of the chosen members a minimum. The objective function in the minimising process is formulated in terms of "desirability" regarding shape, deformation and member forces. Various solutions are possible in accordance with design requirement for shape, member forces or stiffness of structures by adjusting the value of a weighting attached to each term of the objective function.

INTRODUCTION

1. It is often the case that the equilibrium shape of long-span structural cable networks, such as cable net and cable truss, cannot comply with to the preset target shape owing to the geometrical non-linearity of the cable members. In such a case, attempts are made to bring the equilibrium shape as close to the required shape as possible, by controlling member forces and lengths.

2. The authors named the numerical process aimed at this purpose, the shape determination analysis of cable structures, and have presented several practical methods (refs 1-5). Among them, in refs 4 and 5 the authors showed a numerical analysis utilising the method of optimum design.

3. These analyses addressed the problem under fixed loads only (such as dead weight). In practical design calculations, however, as the fixed loads include the dead weight of members, sectional areas of members have to be assumed prior to shape determination. On the other hand, member sections are to be determined so that stresses and deformation satisfy design conditions (allowable stresses and displacements, etc.) under superimposed (additional) loads. The shape determination, therefore, should be finished before these additional loads are applied.

4. Hence, in practical designs, it is not rational to undertake shape determination and section calculation separately and it becomes indispensable to establish the

method which enables us to perform simultaneously both of these analyses in an automatic process.

5. Until recently the authors could find only a few references on this topic. Nishino, et al (ref. 6) presented a new method of cable design analysis which copes with the problem under multiple loading conditions. Nishino, et al claim that a variety of solutions are possible owing to setting appropriate design criteria regarding shape, stresses, dead weight, etc.

6. Recently the authors have also presented a method of analysis coping with such multiple loading conditions (ref. 7) utilising the optimisation technique shown in refs 4 and 5. This gives a solution to the problem described in paragraph 4 and covers totally the shape determination analyses by the authors in the past.

7. This paper shows the theory of optimum design analysis based on ref. 7 and gives some numerical examples, which will show the scope of application of the theory.

STATICAL AND DESIGN CONDITIONS

8. In the following the authors discuss two kinds of equilibrium states, "completed state (**C**-state)" and "deformed state (**D**-state)". In the former the structure is in equilibrium under stationary loadings (fixed loads) and in the latter the structure is deformed due to superimposed (additional) loadings.

9. As to these two states, the authors prescribe statical conditions, S-1 and S-2, and design conditions, D-1 through to D-6, i. e.,

S-1: Internal and external forces shall be in stable equilibrium (**C**- and **D**-state).

S-2: No compressive force shall act on any members (**C**- and **D**-state).

D-1: The equilibrium shape shall be as close to the preset target shape as possible (C-state) .

D-2: Tension of preset value shall act on a specially assigned member (**C**-state).

D-3: Tension of an arbitrarily selected member shall be as close to the preset target value as possible (**C**-state).

D-4: Member forces shall not exceed the prescribed allowable values (**C**- and **D**-state).

D-5: Nodal displacements due to additional loads shall be as small as possible (**D**-state).

D-6: Total weight of structural members shall be as small as possible.

10. A word may be in order to explain the meaning of conditions D-1 through D-6. Condition D-1 is the design requirement of shape prescribed from the structural scheme or functional viewpoint. D-2 is to assume a situation in which the design of a cable anchorage can be made easy by, for example, setting the tensions of boundary members beforehand. D-3 is to cope with the design requirements for achieving a nearly uniform distribution of the member forces over the whole structure by preventing the member forces from becoming too large or too small locally, or for keeping the tensions of prestressing members at prescribed values wherever possible. D-4 is an ordinary design condition for

calculating the cross-sectional area of a member. D-5 controls the extent of deformation when additional loads are applied to a completed cable assembly. D-6 is self-evident with regard to the cross-sectional design of members .

FORMULATION OF THE PROBLEM

Condition S-1

11. The equilibrium equation at an arbitrary point i in **C**-state (Fig. 1) is:

$$\sum_{m=1}^{Ni}(X_i - X_j)\varphi_m = F_{iX} \tag{1}$$

in the X-direction, where m is the number of members connected to point i; N_i is total number of members joined at point i; j is the number of the other end of member ij; X_i and X_j are nodal co-ordinates of member m; $\varphi_m = P_m / L_m$ is the tension coefficient (P_m and L_m are the tension and length of member m, respectively); F_{iX} is an applied force in X-direction to point i. Similar equations are obtained in the Y- and Z- directions at every free nodal point.

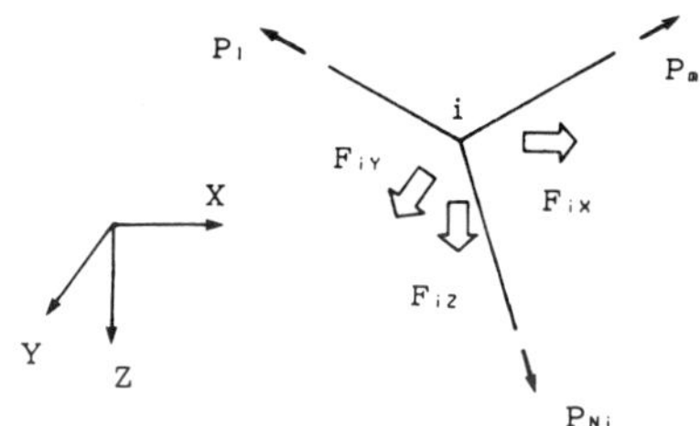

Fig. 1. Equilibrium at a joint.

12. Eq. 1 is rewritten in matrix form as:

$$\Phi_T X_T = F \tag{2}$$

where Φ_T is an f × 3n matrix consisting of φ_m (n: total number of joints); $\mathbf{X}_T$ and $\mathbf{F}$ are 3n × 1 nodal co-ordinates vector and f × 1 external force vector, respectively; f is the degrees of freedom.

13. Here, we classify the nodal points into two kinds of boundary and non-boundary points. Writing the nodal co-ordinates of the former $\mathbf{X}_C$ and the latter $\mathbf{X}$, rewriting Eq. 2 in the form:

$$\begin{bmatrix}\Phi & \Phi_C\end{bmatrix}\begin{bmatrix}X \\ X_C\end{bmatrix} = F \tag{3}$$

where Φ and Φ_C are f × f and f × (3n - f) matrices respectively. Therefore,

$$X = \Phi^{-1}(F - \Phi_C X_C) \tag{4}$$

14. The fixed load $\mathbf{F}$ is rewritten in the form:

$$\mathbf{F} = \mathbf{F}_C + \mathbf{F}_R \tag{5}$$

where $\mathbf{F}_R$ is a vector of member dead weight and $\mathbf{F}_C$ is a vector of fixed loads except member dead weight.

15. Rewriting $\mathbf{F}_R$ in the following form

$$\mathbf{F}_R = \mathbf{C}\gamma\,[L_m]\,\mathbf{A} \quad \text{or} \quad \mathbf{F}_R = \mathbf{C}\gamma\,[A_m]\,\mathbf{L} \tag{6}$$

where $[L_m]$ and $[A_m]$ are diagonal matrices of M_T-order consisting of member lengths and cross-sectional areas, respectively (M_T is the total number of members). $\mathbf{L}$ and $\mathbf{A}$ are vectors of M_T-order having member lengths and sectional areas, respectively, as their elements. $[L_m]\,\mathbf{A}$ and $[A_m]\,\mathbf{L}$, therefore, mean member volume vectors. $\mathbf{C}_\gamma$ is an $f \times M_T$ matrix which connects the member volume vectors with member dead weight vector $\mathbf{F}_R$. We assume here that the weight of a member m is distributed equally to both ends of the member of which weight per unit volume is γ_m.

16. Meanwhile, a cable network is composed of many ropes, each of which is erected between two anchoring points and is considered to be divided into several structural members spanning two adjacent nodal points. This means that members within a single rope have certain uniform sectional value. Hence, the sectional area vector A is rewritten in the form:

$$\mathbf{A}=\mathbf{C}_P\mathbf{A}_P \tag{7}$$

where $\mathbf{A}_p$ is a vector of N_p-order with uniform sectional areas and $\mathbf{C}_p$ is an $M_T \times N_p$ matrix, of which m-th row has unity as k-th element and otherwise zero elements when the cross-sectional area of the m-th member is in common with that of the k-th member. (N_P means the total number of uniform sectional values.) Hence Eq. 6 becomes:

$$\mathbf{F}_R=\mathbf{C}_\gamma[L_m]\mathbf{C}_P\mathbf{A}_P \tag{8}$$

17. From Eqs 4 and 5, nodal co-ordinates in $\mathbf{C}$-state are:

$$\mathbf{X} = \Phi^{-1}\,(\mathbf{F}_C + \mathbf{F}_R - \Phi_C\mathbf{X}_C) \tag{9}$$

and in $\mathbf{D}$-state (under additional loads), the equilibrium equations are written in the form:

$$\mathbf{X}'=\Phi_L^{-1}(\mathbf{F}_C+\mathbf{F}_R+\mathbf{F}_L-\Phi_{LC}\mathbf{X}_C) \tag{10}$$

where $\mathbf{X}'$ and $\mathbf{F}_L$ are the joint co-ordinates vector, after deformation, and an additional load vector of f × 1, respectively. Φ_L and Φ_{LC} are matrices consisting of tension coefficients in **D**-state, i.e.

$$\Phi_L = \Phi\Big|_{\varphi_T=\varphi_L}, \qquad \Phi_{LC} = \Phi_C\Big|_{\varphi_T=\varphi_L}$$

where φ_T and φ_L mean tension coefficient vectors in **C**- and **D**-state, respectively.

<u>Condition S-2</u>

18. The condition of incompressibility of a member is:

$$\varphi_k > 0 \quad \text{and} \quad \varphi_{Lm} \geq 0 \tag{11}$$

for **C**- and **D**-state, respectively, where k = 1,2,...,M; m = 1,2,...,M_T and M is the total number of members excluding those which are under the control of the following condition D-2.

<u>Condition D-2</u>

19. Rewriting the tension coefficient vector φ_T in **C**-state in the form:

$$\varphi_T = [\varphi \;\; \varphi_C]^T$$

where φ is an M × 1 unknown tension coefficient vector and φ_C is an (M_T - M) × 1 vector of tension coefficients in members which are under control of D-2. The elements of φ_C are no longer design variables.

<u>Condition D-4</u>

20. Expressing this condition in the form:

$$P_m \leq \frac{P_{cm}}{n_c}, \quad P_m + \frac{E_m A_m}{L_m} e_m - \alpha E_m A_m \Delta T \leq \frac{P_{cm}}{n_s} \tag{12}$$

for **C**- and **D**-state, respectively, where P_{cm} is the ultimate strength of the m-th member, and n_c and n_s are safety factors for **C**- and **D**-states, respectively. Assuming $P_{cm} = kA_m$ (k: proportional constant), Eq 12 become:

$$L_m \varphi_m \leq C_c A_m, \quad L_m \varphi_m + \frac{E_m A_m}{L_m} e_m - \alpha E_m A_m \Delta T \leq C_s A_m \tag{13}$$

where $C_c = k/n_c$, $C_s = k/n_s$ and m=1,2,...., M_T.

<u>Conditions D-1, D-3, D-5 and D-6</u>

21. Defining the quantity:

$$W = \sum_{i=1}^{f}\left(\frac{R_i}{qR_i}\right)^2 + \sum_{k=1}^{M}\left(\frac{\varphi_k - \varphi_{0k}}{q_{\varphi k}}\right)^2 + \sum_{i=1}^{f}\left(\frac{x_i}{q_{xi}}\right)^2 + \sum_{m=1}^{M_T}\left(\frac{L_m A_m}{q_{Am}}\right)^2$$

$$= \|Q_R R\|^2 + \|Q_\varphi(\varphi - \varphi_0)\|^2 + \|Q_X\, x\|^2 + \|Q_A(L_m)A\|^2 \qquad (14)$$

where R_i is the difference between the joint co-ordinates of the target shape and those of the completed shape, φ_k is an unknown tension coefficient, φ_{0k} is a target tension coefficient (target tension divided by member length in target shape), x_i is the joint displacement and $\mathbf{R}$, φ, φ_0 and x are vector expressions of these quantities.

22. In Eq. 14 the reciprocals of q_{Ri}, $q_{\varphi k}$, q_{xi} and q_{Am} mean the weights given to variables R_i, φ_k-φ_{0k}, x_i and L_mA_m, respectively, in compliance with design requirement; $\mathbf{Q}_R$, $\mathbf{Q}_\varphi$, $\mathbf{Q}_x$ and $\mathbf{Q}_A$ are diagonal matrices of which elements are (1/q)-values mentioned above. It is shown also in refs 4 and 5 that q-values mean desirable domain of variables in Eq. 14.

23. Vector $\mathbf{R}$ in Eq. 14 is rewritten in the form

$$\mathbf{R} = \Phi^{-1}(\mathbf{F}_C + \mathbf{F}_R - \Phi_C\mathbf{X}_C) - \mathbf{X}_0 \qquad (15)$$

by using Eq. 9, where $\mathbf{X}_0$ is the co-ordinates vector of the target shape. The joint displacement vector x has the form:

$$x = \Phi_L^{-1}\,(\mathbf{F}_C + \mathbf{F}_R + \mathbf{F}_L - \Phi_{LC}\mathbf{X}_C) - \mathbf{X} \qquad (16)$$

24. Calling the reciprocal of W from Eq. 14 "desirability" in design and consider attaining the design optimum when enhancing the desirability to a maximum.

25. Consequently, the prime subject of the analysis is to solve the optimisation problem: "Obtain the values of φ and $\mathbf{A}_p$ which minimise W of Eq. 14 under the conditions (11) and (13), where m tension coefficients and N_p uniform sectional areas of cable members are taken as independent variables."

NUMERICAL PROCESS OF SOLUTION

26. The numerical method for minimising W is based on the one presented in refs 4 and 5 and the authors will refrain from explaining it in detail and show only an outline of the method.

27. Using Gauss's method to get correction vectors. Let $\varphi^{(i)}$ and $A_p^{(i)}$ be approximate solutions at the i-th iteration, and $\Delta\varphi^{(i)}$ and $\Delta A_p^{(i)}$ be correction vectors for each of them. In order to express the vectors $\mathbf{R}$, x and $[L_m]\mathbf{A}$ in Eq. 14 in terms of correction vectors, we expand them in a Taylor series with respect to $\Delta\varphi^{(i)}$ and $\Delta A_p^{(i)}$, and rewrite them $\mathbf{R}_L$, x_L and $\{[L_m]\mathbf{A}\}_L$ respectively, i. e.,

$$\left.\begin{aligned} R_L &= R^{(i)} + M_1{}^{(i)}\Delta\varphi^{(i)} + M_2{}^{(i)}\Delta A_P{}^{(i)} \\ x_L &= x^{(i)} + M_3{}^{(i)}\Delta\varphi^{(i)} + M_4{}^{(i)}\Delta A_P{}^{(i)} \\ \{[L_m]A\}_L &= \{[L_m]A\}^{(i)} + M_5{}^{(i)}\Delta\varphi^{(i)} + M_6{}^{(i)}\Delta A_P{}^{(i)} \end{aligned}\right\} \quad (17)$$

where terms of higher order are neglected. Superscript i means the value at the i-th iteration and $M_1{}^{(i)}$ through $M_6{}^{(i)}$ are partial derivative matrices of **R**, x and $[L_m]$**A** with regard to independent variables.

28. The value of the objective function at the (i + 1)-st iteration, therefore, becomes:

$$\begin{aligned} W^{(1+1)} = &\left\| Q_R\left(R^{(i)} + M_1{}^{(i)}\Delta\varphi^{(i)} + M_2{}^{(i)}\Delta A_P{}^{(i)}\right)\right\|^2 + \left\| Q_\varphi\left(\varphi^{(i)} + \Delta_\varphi{}^{(i)}\right) - \varphi_0\right\|^2 \\ &+ \left\| Q_x\left(x^{(i)} + M_3{}^{(i)}\Delta\varphi^{(i)} + M_4{}^{(i)}\Delta A_P{}^{(i)}\right)\right\|^2 \\ &+ \left\| Q_A\left(\{[L_m]A\}^{(i)} + M_5{}^{(i)}\Delta\varphi^{(i)} + M_6{}^{(i)}\Delta A_P{}^{(i)}\right)\right\|^2 \end{aligned} \quad (18)$$

29. The correction vectors $\Delta\varphi^{(i)}$ and $\Delta A_p{}^{(i)}$ are determined by

$$\frac{\partial W^{(i+1)}}{\partial \Delta\varphi_k{}^{(i)}} = 0, \qquad \frac{\partial W^{(i+1)}}{\partial \Delta A_{P,n}{}^{(i)}} = 0 \quad (19)$$

where k = 1, 2, . . ., M, n = 1, 2, . . ., N_P. Eqs 18 and 19 give the simultaneous equations as to correction vectors.

30. For the sake of rapid convergence the authors developed a method of improving the correction vectors which decides the optimum step-size of numerical the process. Its detailed discussion is made in refs 4 and 5.

NUMERICAL EXAMPLE

31. The authors determine the shape of the cable truss shown in Fig. 2 under fixed loads and then obtain the optimum member sectional areas taking into consideration the restraint condition on deformation due to superimposed loadings.

32. In the second and the third columns of Table 1 are shown the fixed loads F_C (excluding member dead weight) and superimposed loads F_L respectively. The fourth and fifth column preset the target shape of the structure. The second and third columns of Table 2 give the target tension coefficients and target member forces, respectively. (Young's modulus: E = 2 x 10^7t/m^2, member weight per unit volume: $\gamma = 8.32$t/m^3, proportional constant in Eq. 13: k = 1.32 x 10^5t/m^2, safety factors in Eq. 12: $n_c = 3$ and $n_s = 2.7$)

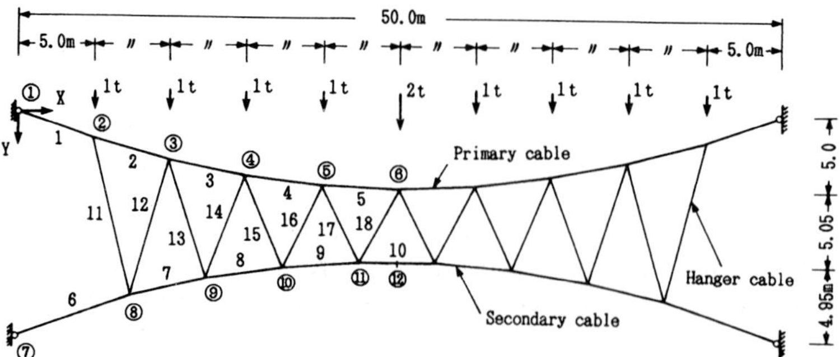

Fig. 2. Cable truss

Table 1. Loads, target shape and completed shape

Node No.	F_C (t)	F_L (t)	Target shape X	Target shape Y	Case 1 X	Case 1 Y	Case 2 X	Case 2 Y
1	0	0	0	0	0	0	0	0
2	1.0	1.0	5.0	1.8	5.084	1.692	5.012	1.758
3	1.0	1.0	10.0	3.2	10.109	3.090	9.995	3.176
4	1.0	1.0	15.0	4.2	15.081	4.070	14.984	4.210
5	1.0	1.0	20.0	4.8	20.028	4.672	19.981	4.874
6	1.0	1.0	25.0	5.0	25.0	4.881	25.0	5.182
7	0	0	0	15.0	0	15.0	0	15.0
8	0	1.0	7.5	12.45	7.674	12.975	7.308	12.202
9	0	1.0	12.5	11.25	12.652	12.007	12.350	10.850
10	0	1.0	17.5	10.45	17.623	11.352	17.379	9.997
11	0	1.0	22.5	10.05	22.623	11.038	22.436	9.595
12	0	1.0	25.0	10.05	25.0	11.039	25.0	9.596

(1ton=9.8kN) (unit: m)

Table 2. Member force and tension coefficient (**C**-state)

Member No.	Target value Ten. coef.	Target value Tension	Case 1 Ten. coef.	Case 1 Tension	Case 2 Ten. coef.	Case 2 Tension
1	3.258	17.313	10.550	56.527	3.478	18.484
2	3.219	16.716	10.583	55.200	3.493	18.101
3	3.173	16.178	10.647	53.952	3.490	17.783
4	3.147	15.850	10.616	52.914	3.478	17.536
5	3.119	15.606	10.504	52.271	3.453	17.362
6	0.546	4.326	6.017	47.750	0.328	2.563
7	0.865	4.449	9.304	47.171	0.471	2.457
8	0.866	4.386	9.370	46.981	0.473	2.412
9	0.877	4.400	9.366	46.919	0.478	2.425
10	1.793	4.482	19.828	47.131	0.965	2.474
11	0.009	0.1	0.175	2.024	0.012	0.129
12	0.010	0.1	0.131	1.336	0.018	0.170
13	0.012	0.1	0.223	2.063	0.018	0.147
14	0.013	0.1	0.120	0.997	0.015	0.107
15	0.015	0.1	0.275	2.122	0.028	0.176
16	0.016	0.1	0.188	1.332	0.011	0.061
17	0.017	0.1	0.291	1.999	0.033	0.175
18	0.018	0.1	0.189	1.245	0.010	0.050
unit	ton/m	ton	ton/m	ton	ton/m	ton

(1ton=9.8kN)

Table 3. Member force (**D**-state)

Member	Case 1	Case 2
1	76.439	47.848
2	74.546	46.551
3	72.505	45.370
4	70.559	44.232
5	68.778	42.723
6	35.279	2.983
7	35.179	3.169
8	35.671	3.509
9	36.522	4.033
10	37.480	5.133
11	25.638	1.278
12	0.791	0
13	3.158	1.341
14	0.120	0
15	3.466	1.406
16	0.191	0
17	3.639	2.364
18	1.007	0.242

(ton: 1ton=9.8kN)

33. As to setting q-values in Eq. 14 consider the following two cases.

Case 1: $q_{Ri} = 0.5m$, $q_{\varphi k} = \varphi_{0k}$ (φ_{0k}: target tension coefficient), $q_{xi} = 0.01m$, $q_{Am} = 10^{-3}m^3$ (i = 1,...,18; k = 1,...,18; m = 1,...,18).

Case 2: $q_{xi} = 10000m$ (Other q's shall be of the same value as above.)

34. These conditions mean that the desirable domains, in relation to shape and member forces, are common to both cases, while the domain of displacement, due to superimposed loads, is restricted severely in case 1 compared to case 2.

35. In the right half portions of Table 1 and Table 2 are shown the numerical results as to nodal co-ordinates and member forces in **C**-state. Table 3 and Fig. 3 show the member forces and displacement, respectively, in **D**-state.

36. The optimum solutions for member sectional areas are 18.2cm^2, 15.2cm^2 and 0.863cm^2 for upper chord, lower chord and web members respectively, in case 1. In case 2 these values become 9.84cm^2, 1.04cm^2 and 0.494cm^2 respectively.

37. The following conclusions are drawn from these numerical results:

(a) The shape of case 2 in **C**-state has conformed better to the target than that of case 1;

(b) While member forces in case 1 considerably exceed the target values, those in case 2 are close to the targets;

(c) Case 1 gives a much stiffer structure than case 2.

38. In summary, when preset tight restraints to displacement due to superimposed loads (case 1) are imposed, although the difference between completed shape and target shape become large in **C**-state, member tensions and sectional areas are greater than those of case 2, thus yielding a stiffer structure.

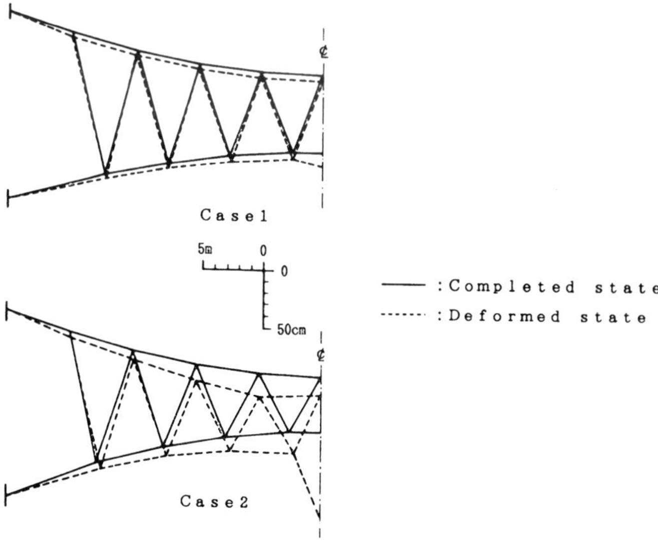

Fig. 3. Equilibrium shape (**C**- and **D**-state).

CONCLUSION

39. The authors have presented a method of design and analysis for cable network undertaken by utilising the optimisation technique, where not only the design conditions for shape and member forces in the completed state but also the constraint conditions for stresses and displacements in the deformed state were taken into account, to determine the design shape, design member forces and sectional areas of members in an automatic process.

40. By the present method, various kind of solutions are obtained in compliance with the design requirements. By increasing the weight of constraint for deformation, for instance, it is possible to obtain a cable network with a high stiffness or reduce total dead weight by setting large weight to dead weight term in the objective function.

REFERENCES

1. NAMITA Y. and NAKANISHI H. A consideration regarding the computation for shape determination of tension structures. Transactions of Architectural Institute of Japan, No. 212, Oct. 1973 (in Japanese).
2. NAMITA Y. and NAKANISHI H. A method of computation for determining the shape of cable structures. International Conference on Tension Roof Structures (IABSE London), Paper 2.1, Apr. 1974.
3. NAKANISHI H. and NAMITA Y. On a computation technique regarding the shape determination problem of tension structures. Transactions of Architectural Institute of Japan, No. 224, Oct. 1974 (in Japanese).
4. NAKANISHI H. and NAMITA Y. Shape determination analysis of cable structures by means of the methods for optimisation problem. Transactions of Architectural Institute of Japan, No. 238, Dec. 1975 (in Japanese).
5. NAKANISHI H. and NAMITA Y. Shape determination analysis of cable structures by means of the methods for optimisation problem. International Symposium on Wide-Span Surface Structures, Sonderforschungsbereich 64, University of Stuttgart, Apr. 1976.
6. NISHINO F., DUGGAL R. and LOGANATHAN S. Design analysis of cable networks. Journal of Structural Engineering, ASCE, Vol. 115, No. 12, Dec. 1989.
7. NAMITA Y. and NAKANISHI H. Optimum design analysis of structural cable networks. Proceedings of Japan Society of Civil Engineers, No. 432 / I-16, July 1991.

104. Investigation on the non-linear static and dynamic response of cable-stayed and suspension bridges

L. GAO, China Academy of Railway Sciences, P. R. China

SYNOPSIS. In this paper, the important earlier investigations on the nonlinear analysis of cable-stayed and suspension bridges are firstly presented. Secondly, the authors put forward two new geometrical nonlinear finite element formulations for a 3D beam element. The validity of the formulations have also been demonstrated by several simple examples. In the last part of the paper, a synthetical computer program for nonlinear static and dynamic analysis of cable-stayed and suspension bridges has been introduced. A case study for a cable-stayed railway bridge has also been presented.

INTRODUCTION

1. With the development of transportation, many long-span cable-stayed and suspension bridges have been constructed or are under construction. About thirty highway and one railway cable-stayed bridges have been built in China. A research project on super-long-span railway cable-stayed and suspension bridges is ongoing at the China Academy of Railway Sciences, which is financially supported by the Ministry of Railway of China.

2. Cable-stayed and suspension bridges are composite structures composed of main girder, bridge tower and pier as well as the cable system. Under the service load and in the construction stage, the bridges usually produce large displacement and the structural configuration is considerably changed. Because of this, the structural behaviour is geometrically non-linear. In order to ensure the safety and reliability of the bridges, nonlinear static and dynamic analysis for the structures must be done in structural design.

3. In general, the major factors causing geometric nonlinearity of the structures include three aspects: cable sag, the beam-column effect and structural configuration change due to large displacement. How to adequately model these nonlinear effects is one of the most significant questions in the analysis and design of cable-stayed and suspension bridges.

BRIEF REMARKS ON NONLINEAR MODELLING

Cable Nonlinearity

4. Cable nonlinearity becomes greater as linear density and horizontal projection become greater and the pre-tension force in the cable becomes less. There are many methods to model cable nonlinearity. The Effective Modulus Method, developed by Pippard and Chitty in 1944, is used frequently and has been improved by Ernst, et al. The investigation on the method has concluded that the precision of the method will decrease as the displacement of nodes becomes greater. Another method applied to model cables of the bridges is the multiple link method developed by Pugsley in 1952. However, this method necessitates the use of more computer time.

Beam-Column Effect

5. In 1961, Schreier developed the deflection theory to analyse nonlinearity of cable-stayed bridge through considering the so called beam-column effect caused by large displacements. Deflection theory is based on the differential equilibrium equation on the deformed configuration of the structure, with a continuous form and an analytical integral result. However for cable-stayed and suspension bridges with complicated form, deflection theory is not easy to use.

6. The Stable Function Method proposed by Saafan, et al is one of the conventional methods based on the discrete viewpoint. A series of stable functions are used to describe the coupling of axial force and bending moment in structural elements.

Further Research Topics on Nonlinearity

7. From previous research investigations on nonlinearity of cable-stayed bridges, it has been found that the elastic modulus of the cables may decrease to 40% of initial modulus, and the beam-column effect may reach 10%. Nonlinearity due to large displacements is in the region of 6-12%. Hence, these nonlinear effects can be significant and it is therefore important that in-depth investigations are undertaken on the nonlinear behaviour of cable-stayed and suspension bridges.

8. In previous investigations, the modelling methods used for whole structure can be classified into planar, space and simplified space (fish born model) models. The finite element solution approach includes the total and the tangent method. Discrete and mechanical modelling methods directly affect the solution precision. Therefore, the following topics must be further investigated.

(a). The nonlinear element model and its precision.
(b). The nonlinear formulation and solution approach.
(c). The space analysis method.
(d). The discrete method and its sensitivity.

9. Considering the significance of nonlinear effects for cable-stayed and suspension bridge design, a lot of investigations have been done in the China Academy of Railway Sciences on theoretical and experiental, as well as, practical

aspects. In the next section, two kinds of finite element formulation for a beam element will be presented.

GEOMETRIC NONLINEAR BEAM ELEMENT FORMULATION

10. According to variation theorem developed by R Hill, the virtual work equation on reference configuration θ is:

$$\int_{v\theta}(\tilde{S}_{ij}+\triangle\tilde{S}_{ij})\delta\left(\frac{\partial\triangle u_j}{\partial\theta_i}\right)dv$$
$$=\int_{v\theta}(\tilde{F}_i+\triangle\tilde{F}_i)\delta\triangle u_i dv+\int_{s\theta}(\tilde{T}_i+\triangle\tilde{T}_i)\delta\triangle u_i ds$$

where $\tilde{S}_{ij}$ is the P-K II stress tensor at time t refered to θ,
$\Delta\tilde{S}_{ij}$ is the nominal incremental tensor form t to t + Δ t,
$\tilde{F}_i$ is the body force in time t,
$\tilde{T}_i$ is the surface force in time t,
$\Delta\tilde{F}_i$ is the body force increment from t to t + Δt,
$\Delta\tilde{T}_i$ is the surface force increment from t to t + Δt

Total Lagrangian formulation

11. Using the initial configuration as the reference configuration, the Green strain tensor is:

$$E_{ij}(t)=\frac{1}{2}\left[\frac{\partial u_i}{\partial x_j}+\frac{\partial u_j}{\partial x_i}+\frac{\partial u_k}{\partial x_i}\frac{\partial u_k}{\partial x_j}\right]$$

hence
$$\triangle E_{ij}=E_{ij}(t+\triangle t)-E_{ij}(t)$$
$$=\triangle E_{ij}^L+\triangle E_{ij}^N$$

where
$$\triangle E_{ij}^L=\frac{1}{2}\left[\frac{\partial\triangle u_l}{\partial x_j}+\frac{\partial\triangle u_j}{\partial x_l}+\frac{\partial u_k}{\partial x_l}\frac{\partial\triangle u_k}{\partial x_j}+\frac{\partial u_k}{\partial x_j}\frac{\partial\triangle u_k}{\partial x_l}\right]$$

$$\triangle E_{ij}^N=\frac{1}{2}\frac{\partial\triangle u_k}{\partial x_l}\frac{\partial\triangle u_k}{\partial x_j}$$

The Piola-Kirchhoff II stress incremental tensor is

$$\triangle S_{ij}=C_{ijkl}\triangle E_{kl}$$

so

$$\int_{v0}C_{ijkl}\triangle E_{kl}^L\delta\triangle E_{ij}^L dv+\int_{v0}S_{ij}\delta\triangle E_{ij}^N dv$$
$$=R(t+\triangle t)-\int_{v0}S_{ij}\delta\triangle E_{ij}^L dv$$

Using the cubic displacement interpolation function,

$$\begin{cases} u = a_0 + a_1 x \\ v = b_0 + b_1 x + b_2 x^2 + b_3 x^3 \\ w = c_0 + c_1 + c_2 x^2 + c_3 x^3 \end{cases}$$

The total Lagrangian formulation is obtained

$$([K_L]+[K_D]+[K_\sigma])\{\triangle\delta\}=\{F_{ex}\}+\{\triangle F_{ad}\}$$

12. Where $[K_L]$ is infinite displacement linear stiffness matrix, $[K_D]$ is nonlinear stiffness matrix due to initial displacement, $[K_\sigma]$ is nonlinear stiffness matrix due to initial stress, and {R} is external force vector, {F} is nonequilibrium force.

$$[K_\sigma] = N \begin{bmatrix} \frac{1}{l} & 0 & 0 & 0 & 0 & -\frac{1}{l} & 0 & 0 & 0 & 0 \\ & \frac{6}{5l} & 0 & \frac{1}{10} & 0 & 0 & -\frac{6}{5l} & 0 & \frac{1}{10} & 0 \\ & & \frac{6}{5l} & 0 & \frac{1}{10} & 0 & 0 & -\frac{6}{5l} & 0 & \frac{1}{10} \\ & & & \frac{2}{15}l & 0 & 0 & -\frac{1}{10} & 0 & -\frac{l}{30} & 0 \\ & & & & \frac{2}{15}l & 0 & 0 & -\frac{1}{10} & 0 & -\frac{l}{30} \\ & & & \text{对} & & \frac{1}{l} & 0 & 0 & 0 & 0 \\ & & & & & & \frac{6}{5l} & 0 & -\frac{1}{10} & 0 \\ & & & & & \text{称} & & \frac{6}{5l} & 0 & -\frac{1}{10} \\ & & & & & & & & \frac{2}{15}l & 0 \\ & & & & & & & & & \frac{2}{15}l \end{bmatrix}$$

13. The matrix element in $[K_D]$ has a complicated form, which is associated with the nodal displacement, such that:

$$k_{11} = -\frac{2EA}{l^2}(u_i - u_j) + \frac{EA}{l^3}(u_i - u_j)^2$$

$$\begin{aligned} k_{12} = & -\frac{EA}{l}\left[\frac{6}{5l}(v_i - v_j) + \frac{1}{10}(\theta_{iy} + \theta_{jy})\right] \\ & + \frac{EA}{l^2}(u_i - u_j)\left[\frac{6}{5l}(v_i - v_j) + \frac{1}{10}(\theta_{iy} + \theta_{jy})\right] \end{aligned}$$

Updated Lagrangian formulation

14. Using the deformed configuration at time t as the reference configuration:

$$\triangle \tilde{S}_{ij} = C_{ijkl} D_{kl}$$

where $\Delta\tilde{S}_{ij}$ is the Jauman rate of the Kirchhoff stress tensor.

The deformation rate tensor is given by:

$$D_{ij} = \frac{1}{2}\left(\frac{\triangle \partial u_I}{\partial x_j} + \frac{\triangle \partial u_j}{\partial x_I}\right)$$

15. Also using a cubic interpolation function, the updated Lagrangian formulation can be obtained from virtual work equation,

$$([K_L] + [K_N])\{\triangle \delta\} = \{R\} + \{F_{ad}\}$$

where: $[K_L]$ is the linear stiffness matrix,
$[K_N]$ is the nonlinear stiffness matrix,
$\{R\}$ is the external load vector,
$\{F\}$ is the nonequlibrium force vector.

$$[K_N] = [K_N^1] + [K_N^2]$$

$$[K_N^2] = \begin{bmatrix} \frac{N}{l} & -\frac{1}{l^2}(M_{zi} + M_{zj}) & -\frac{1}{l^2}(M_{yi} + M_{yj}) & -\frac{1}{l}M_{zi} & -\frac{1}{l}M_{yi} & \cdots\cdots & \\ & \frac{12NI_z}{Al^3} & 0 & \frac{6NI_3}{Al^2} & 0 & \cdots\cdots & \\ \text{对} & & \frac{12NI_y}{Al^3} & 0 & \cdots\cdots & & \\ & \text{称} & & \frac{4NI_z}{Al} & 0 & \frac{1}{l}M_{zi} & \cdots\cdots \\ & & & & & \cdots\cdots & \cdots\cdots \end{bmatrix}$$

16. In the formulations, because of the introduction of the high order deformation, the methods have higher precision than the conventional beam-column models.

ANALYSIS APPROACH AND FINITE ELEMENT PROGRAM SNAP

17. The nonlinear finite element incremental equilibrium equation may be expressed as:

$$[M]\{\ddot{u}(t+\Delta t)\}+[C]\{\dot{u}(t+\Delta t)\}+[K]\{u(t+\Delta t)\} = \{R(t+\Delta t\}-\{F(t)\}$$

where: [M] is the mass matrix,
[C] is the damping matrix,
[K] is the tangent stiffness matrix at time t,
$\{R(t+\Delta t)\}$ is the external load vector at time $t + \Delta t$,
$\{F(t)\}$ is the non-equilibrium force at time t,
$\{\ddot{u}\}$, $\{\dot{u}\}$ and $\{U\}$ are the nodal acceleration, velocity and displacement vector respectively

18. For the above equation, a step-by-step method is used. The Wilson and Newmark methods have been included in the program SNAP. A nonlinear finite element program called SNAP developed by the China Academy of Railway Sciences for cable-stayed and suspension bridge analysis has the following main functions:

(a) Linear and nonlinear static analysis
(b) Linear and nonlinear dynamic analysis including free vibration and time-domain analysis
(c) Linear and nonlinear highway loading analysis
(d) Linear and nonlinear railway loading analysis

19. The program has four types of element: a truss element, beam element, 2D element and 3D element. For every type of element, the program includes various material and geometric nonlinear models.

20. The program needs a computer with at least 640K memory and a EGA/VGA graphics adapter.

EXAMPLE ANALYSIS AND DISCUSSION

Railway Cable-Stayed Bridge across Hongshui River

21. The bridge was constructed in 1983, with spans of 48, 96 and 48m. The main girder and cable tower of the bridge were constructed from high strength concrete. The structure of the bridge is shown in Fig. 1. Table 1 illustrates some mechanical and geometrical details of the bridge.

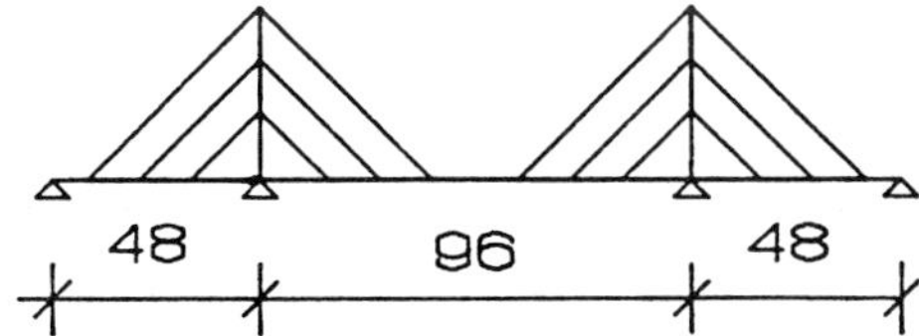

Fig. 1. Railway bridge across Hongshui River.

Table 1. Mechanical and geometrical parameters.

	Main Girder	Cable	Tower
Elastic Modulus (MPa)	0.51E5	0.19E6	0.51E5
Area (m^2) (average)	5.0	0.01716	7.0
Bending Inertia (m^4)(av.)	8.0	----	4.6

22. The bridge carries single-line railway traffic. The bending moment envelope under the load is shown in Fig. 2. Fig. 3 shows the deformed configuration of the bridge structure under the load of maximum bending moment in the mid-span.

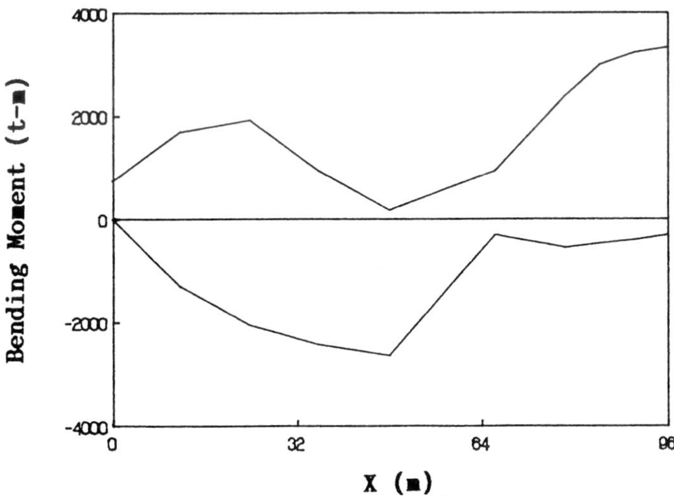

Fig. 2. Bending moment envelope.

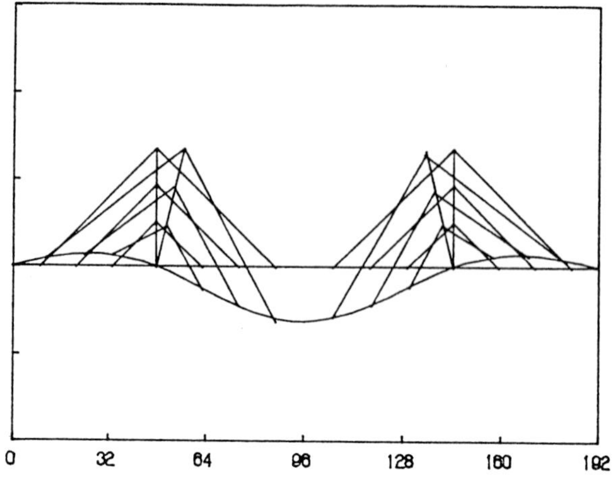

Fig. 3. Deformed configuration.

Concrete Suspension Bridge

23. Fig. 4. illustrates a concrete suspension bridge with the span of 400 meters. Mechanical and geometric details of the bridge are listed in Table 2.

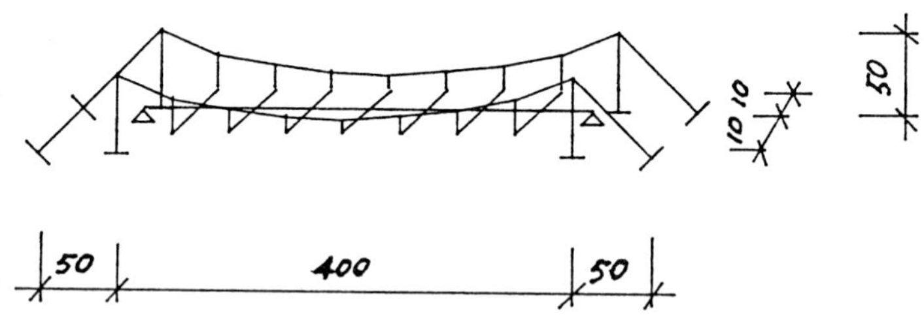

Fig. 4. Suspension bridge.

Table 2. Mechanical and geometrical parameters.

	Main Girder	Cable	Tower
Elastic Modulus (MPa)	0.51E5	0.19E6	0.51E5
Area (m^2) (average)	2.718	0.01716	2.718
Bending Inertia (m^4) (av.)	1.93	----	1.93
Density (Kg/m^3)	0.38E3	0.793E3	0.38E3

Fig. 5 shows the effect of cable tension on the first 5 frequencies of the suspension bridge.

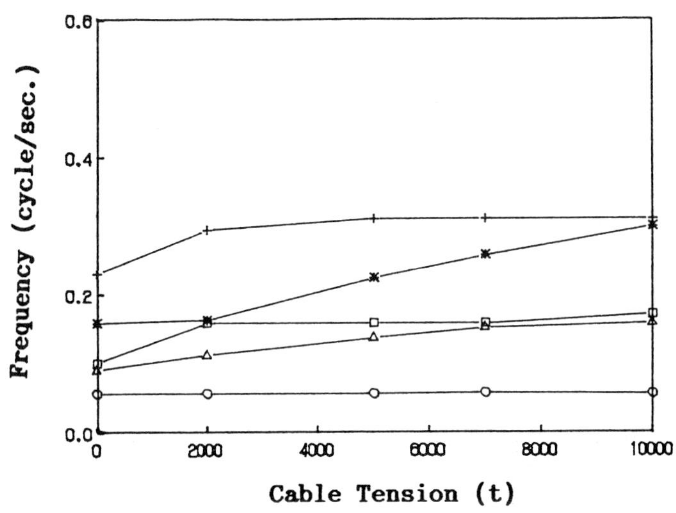

Fig. 5. Effect of cable tension on frequencies.

CLOSING REMARKS

24. Long-span cable-stayed and suspension bridges behave nonlinearly under service loads. The investigation on the nonlinearity is directly associated with safety and reliability of design and construction of these types of bridges. It is necessary to undertake further research on the methods of nonlinear analysis for cable supported bridges.

ACKNOWLEDGEMENT

25. The author would like to give his thanks to director of the research project, Prof. Qingguo Cheng, Senior Engineer Jiaying Pan, and my colleague Xuezhong Xin for their assistance and discussion. The financial support from the Ministry of Railway of China is also acknowledged.

REFERENCES

1. KANOK-NUKULCHAI, W. etc. Mathematical modelling of cable-stayed bridges, J. IABSE, SEI V.2, No.2, 1992.
2. TANG, M. C., Nonlinear behaviour of cable-stayed bridges, Sino-American Symposium on Bridge and Structural Engineering, Beijing, 1982.
3. PAN, J.Y., Study of long span cable-stayed railway bridges, Proc. 2nd EASEC, Vol. 1, Thailand, 1989.
4. GAO, Lubin, On nonlinearity of long span railway cable-stayed bridges, Proc. of 4th Conference on Bridge Engineering, 1992.

105. A simulation of cable-dome construction

P. LABOSSIERE and S. LEBLANC, Universite de Sherbrooke, Canada

SYNOPSIS. Tensile structures are often used to cover areas that must be left free from columns, such as sport buildings. Some of the most original roof structures designed in recent years use the cable-dome configuration. This lightweight system consists of a combination of cables, membrane and posts. A typical circular cable-dome is composed of radial trusses intersecting at a central tension ring. The purpose of this work is to demonstrate the behaviour of a cable-dome structure during its construction. The conventional method of analysis for a cable structural network was modified to include the posts.

STRUCTURAL CONFIGURATIONS OF CABLE-DOMES

1. Cable-domes are concentric structures in which radial trusses intersect through a central tension ring. This structural form evolved from the tensegrity principles developed by R.B. Fuller (ref. 1), and the first applications were conceived by D. Geiger (ref. 2). A typical truss part of a circular roof structure is shown in Fig. 1. The upper chord of the truss is made from fixed-length cables (Fl, F2, F3). These cables are supported in their final position by posts. These are in turn supported by diagonal cables (D1, D2, D3). The diagonal cables transfer the loads from the bottom of a post to the top of the next outside post along the truss. In order to satisfy equilibrium conditions, the bottom node of all posts at the same distance from the centre of the cable-dome are attached to a circular tension ring. On the figure, these rings shown in dotted lines, are replaced by an "equivalent" cable. The entire circular structure is supported by a circular compression ring. Except for this compression ring and the posts, all elements inside the structure are in tension. Actual roof structures built upon this principle have been constructed in Korea and the USA.

2. Since cables are the main element of the structure, a non-linear load deformation behaviour is expected. Also, since cables must remain in tension for all potential loading conditions, stability of the structure depends on the prestressing of the system. The analysis of cable-domes is performed by following conventional principles for cable networks. However, the system of equations and the identification of failure modes must be modified to include the

posts. The purpose of this work is to demonstrate the behaviour of a cable-dome structure during construction.

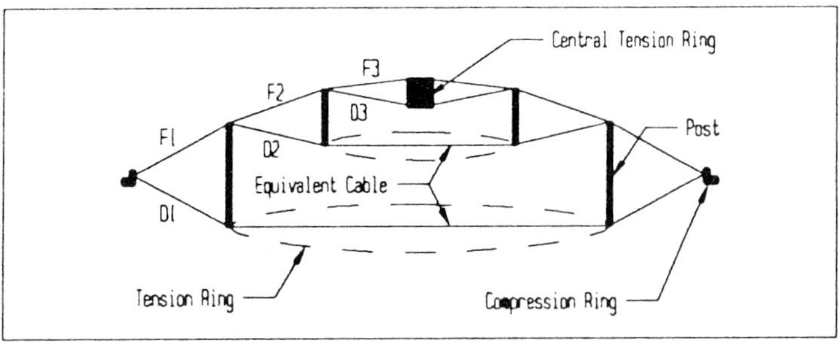

Fig. 1. Radial truss.

CABLE NETWORK ANALYSIS

Single cable in pre-tension

3. The effect of pre-stressing is usually dominant for all members of the cable-dome, and the geometry of the cables can be approximated by a straight line. Tension T in a cable of cross section A and length S produces an elongation of the cable, given by:

$$\Delta S_e = \frac{TS}{EA} \tag{1}$$

Cable network

4. The location of a typical node q in the structural system is defined by a set of co-ordinates (x_q, y_q, z_q). This node is attached to adjacent nodes by n cables of length S_i. The co-ordinates of these nodes are (x_{pi}, y_{pi}, z_{pi}). The axial force in each element attached to node q is denoted Fj. The equilibrium equations at the node, when the system is submitted to internal pre-stressing only, are given by:

$$\begin{aligned} \sum_{i=1}^{n} \frac{F_i}{S_i}(x_{pi} - x_q) = 0 \\ \sum_{i=1}^{n} \frac{F_i}{S_i}(y_{pi} - y_q) = 0 \\ \sum_{i=1}^{n} \frac{F_i}{S_i}(z_{pi} - z_q) = 0 \end{aligned} \tag{2}$$

The length S_i of each element is calculated as:

$$S_i = \sqrt{(x_{pi} - x_q)^2 + (y_{pi} - y_q)^2 + (z_{pi} - z_q)^2} \quad (3)$$

5. The application of loads (P_{xq}, P_{yq}, P_{zq}) to node q triggers modifications in the network geometry thus leading to a new equilibrium position. In the process, the element length changes from S_i to S_i', and the internal force becomes F_i'. The displacements of points p_i and q are identified as (u_{pi}, v_{pi}, w_{pi}) and (u_q, v_q, w_q), respectively, in the x - y - z system of co-ordinates. The new equilibrium equations for the cable can be written in the following form:

$$\begin{aligned} \sum_{i=1}^{n} \left[\frac{F_i'}{S_i'} \left((x_{pi} + u_{pi}) - (x_q + u_q) \right) \right] + P_{xq} = 0 \\ \sum_{i=1}^{n} \left[\frac{F_i'}{S_i'} \left((y_{pi} + v_{pi}) - (y_q + v_q) \right) \right] + P_{yq} = 0 \\ \sum_{i=1}^{n} \left[\frac{F_i'}{S_i'} \left((z_{pi} + w_{pi}) - (z_q + w_q) \right) \right] + P_{zq} = 0 \end{aligned} \quad (4)$$

and the new cable length is measured as:

$$S_i' = \sqrt{\begin{aligned} &((x_{pi} + u_{pi}) - (x_q + u_q))^2 \\ &\quad + ((y_{pi} + v_{pi}) - (y_q + v_q))^2 \\ &\quad\quad + ((z_{pi} + w_{pi}) - (z_q + w_q))^2 \end{aligned}} \quad (5)$$

6. It can be shown (ref. 3) that these equations can be combined in such a way that the relationship between the loads and displacements be expressed as:

$$\sum_{i=1}^{n}\left[\frac{F_i}{S_i}\Delta u_i+\left(\frac{E_iA_i-F_i}{S_i}\right)\Delta x_i a_i\right]=-P_{xq}+R_{xq}$$
$$\sum_{i=1}^{n}\left[\frac{F_i}{S_i}\Delta v_i+\left(\frac{E_iA_i-F_i}{S_i}\right)\Delta y_i a_i\right]=-P_{yq}+R_{yq} \tag{6}$$
$$\sum_{i=1}^{n}\left[\frac{F_i}{S_i}\Delta w_i+\left(\frac{E_iA_i-F_i}{S_i}\right)\Delta z_i a_i\right]=-P_{zq}+R_{zq}$$

where

$$R_{xq}=-\sum_{i=1}^{n}\left(\frac{E_iA_i-F_i}{S_i}\right)(\Delta u_i c_i+\Delta x_i d_i)$$
$$R_{yq}=-\sum_{i=1}^{n}\left(\frac{E_iA_i-F_i}{S_i}\right)(\Delta v_i c_i+\Delta y_i d_i)$$
$$R_{zq}=-\sum_{i=1}^{n}\left(\frac{E_iA_i-F_i}{S_i}\right)(\Delta w_i c_i+\Delta z_i d_i)$$

where a_i, b_i, c_i and d_i are values calculated from the structural configuration. The values for R_{xq}, R_{yq}, and R_{zq} correspond to the difference between a linear and a non-linear load-deformation relationship.

7. The equations of equilibrium for the whole structure are obtained by combining relationships such as the above, for all elements of the structure. The resulting system can be written in the following matrix form:

$$\mathbf{K}\cdot\mathbf{U}=-\mathbf{P}+\mathbf{R} \tag{7}$$

where **K** is a rigidity matrix containing first-order coefficients of the node displacements vector **U**, **P** is the applied loads vector, and **R** is a residual vector filled with high-order coefficients of **U**.

SOLUTION OF THE EQUATIONS

8. The system of equations is solved with a Newton-Raphson numerical method. A BASIC programme was written specifically to analyse the cable-dome on a micro-computer. In the current configuration, the programme can solve problems with a maximum of 125 nodes. In order to reduce the length of the

calculations, the analysis of a typical symmetric three-dimensional dome is reduced to that of a comprised radial truss. This involves transforming the tension rings into equivalent radial cables, as shown in Fig. 1. The program was validated by comparing results with those published in ref. 4.

APPLICATION OF PRE-TENSION IN THE CABLES

9. The program was used to simulate the overall behaviour of a typical cable-dome and its elements during construction. The simulation refers to a model presented in ref. 4. The roof structure has a diameter of 30 meters, and includes eight radial trusses and two tension rings. Dead loads are applied to both ends of each post. These loads are calculated from the weight of the cable net, plus other non-structural elements attached to the roof. The dimensions and material properties of all elements are given in Table 1. In the intended final configuration, the node above the central ring should be 4550 mm above the outer compression ring.

Table 1. Element properties.

ELEMENT	A (mm^2)	E (kN/mm^2)
D1	1734	186.3
D2	867	186.3
D3	434	186.3
F1	2361	127.5
F2	1494	127.5
F3	1060	127.5
Inner tension ring	1186	127.5
Outer tension ring	1719	127.5
Middle posts	3087	205.9
Outer posts	7021	205.9
Central ring (equiv.post)	10000	205.9

10. Two alternative erection sequences were considered; in both cases, the procedure was based on the gradual application of tension into the diagonal cables D1, D2 and D3, until the cable-dome reached its final position. The structure was analysed under a quasi-static hypothesis at the end of each step.

11. The posts were specifically identified to allow compression in the analysis. The stresses caused by the loads in the structural elements were compared to material strength. In the case of the posts, compressive strength included the possibility of buckling.

Procedure #1

12. In the first case considered, the cable net was first hanging from the perimeter compression ring. Tension was applied successively into the diagonal cables, starting with the ones directly attached to the perimeter. Over 50 intermediate positions were considered during the elevation of the roof, and the stress condition in all members was monitored.

Procedure #2

13. In the second construction method considered, a temporary tower was installed at the centre of the structure, from which the central tension ring was hung. Tension was then applied to the diagonal cables, and the structure lifted to the same final position as with the first method. The tower was removed as soon as it was no longer necessary to the lifting procedure.

STRESSES IN THE CABLE NETWORK

14. The following part of the article will focus on the effect of the loading sequence on the various elements of the cable-dome. The behaviour of individual cables, posts, and tension rings during the two alternative sequences will be compared in order to comment on the construction technique.

Loads in the diagonal cables

15. The loading history of the diagonal cables, for the first erection sequence, is shown in bold lines in Fig. 2. The results are expressed in terms of the tensile load applied in diagonal Dl.

16. Tensile stress in the other diagonal cables diminishes slightly when the first tensile loads are applied in any cable. At the end of the erection sequence, additional tension is applied in diagonal Dl, both to improve the rigidity of the system and to attain the proposed configuration. Tension in the other diagonal cables then increases proportionately to the load in diagonal cables Dl.

17. The loading history of the diagonal cables, for the alternative erection sequence, is shown in dotted lines in Fig. 2. Tension in the diagonals D2 and D3 appears earlier than for the first loading sequence. The load increase is more constant, and no slackening effect of the cables is observed when a new set of diagonals is tensioned.

Loads in the form cables

18. The form cables are initially in tension since they support the dead load of the net at the beginning of the loading sequence, in either procedure #1 or #2. The loading history of the form cable is shown in Fig. 3; the results are expressed in terms of the tensile load applied in diagonal Dl.

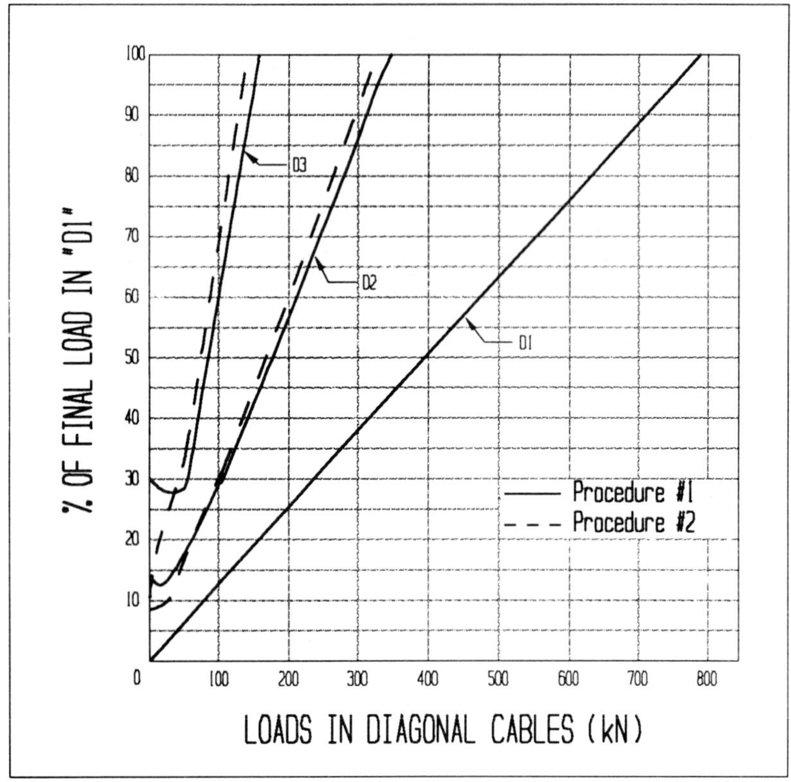

Fig. 2. Loading history of the diagonal cables.

For the first loading sequence, the results are shown in bold lines. As prestressing is gradually imposed to the system through the diagonals, tension in form cables first reduces. It remains low until the structure reaches an equilibrium position, and then increases gradually. There is a slackening effect and the tension in the cables Dl drops again when prestressing is introduced into diagonals D2 and D3.

19. For procedure #2, the general form of the curves is very similar to the first one. There is one "peak" in the curve when tension is introduced in diagonal D2. The slope of the curve changes when tension in Dl is approximately 35% of the final value. This change corresponds to the removal of the central supporting tower. When it is removed, most points of the structure have already reached their final position.

Loads in the tension rings

20. The behaviour of the tension rings is similar to that of the diagonal cables. A slight reduction of tensile stresses is observed in both tension rings when the diagonals D2 and D3 are first submitted to pretension. For the configuration chosen here, tension in the larger tension ring remains approximately double that of the smaller one during all the loading process.

Loads in the posts

21. The behaviour of the posts is similar to that of the diagonal cables. The compression gradually increases with the pretension, except when pretension is introduced in a new set of diagonals. There is then a small reduction of the loads in the posts. During the loading process, compression in posts B2 remains approximately one quarter of that in Bl; compression in elements B3 is about one half of that in posts B2.

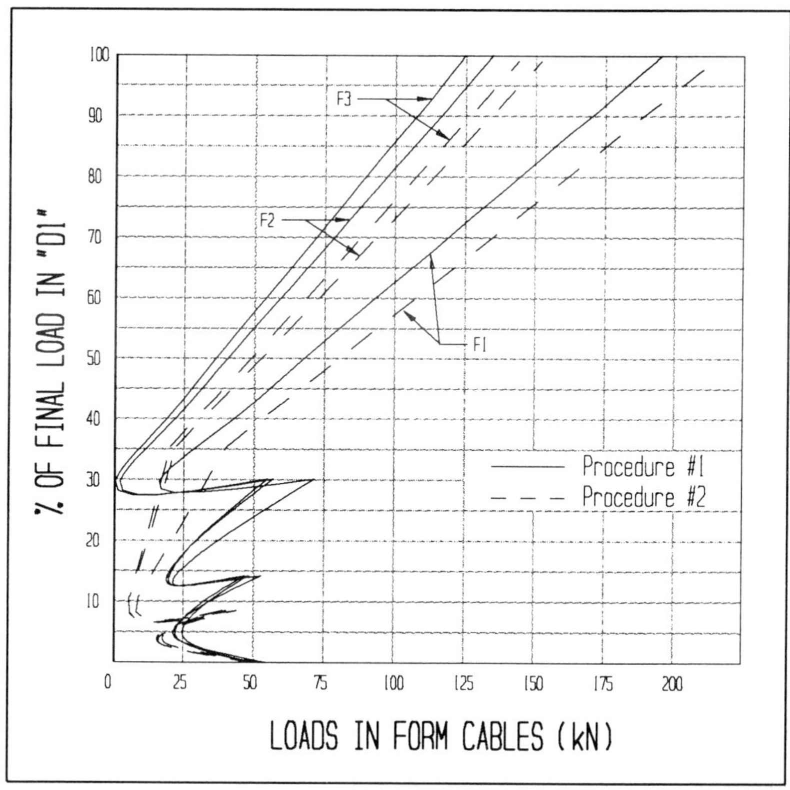

Fig. 3. Loading history of the form cables.

Position of the nodes

22. The position of each node and the tensile loads applied to the diagonal cables are closely linked together. The relation between both was carefully monitored during the entire simulation process, and the value of the stresses in the elements can be printed with respect to the location of any node of the structure.

CONCLUSION

23. The equilibrium equations used in the above calculations are those for a classical network of pre-stressed cables, slightly modified to account for the compression elements inserted throughout.

24. The results for two different loading sequences were presented in this article, but many other ways of lifting the cable net can be imagined. The two loading sequences simulated were selected in order to attain a predetermined position for all nodes, and the corresponding level of prestress.

25. Since the stability of a structure of this type is assured by the level of prestressing and by the continuity between elements, sudden variations of the internal loads should be avoided. By carefully selecting the order and level of tensioning of the cables in the network, it is possible to minimise the sudden load variations in the elements. In this respect, the second procedure simulated here appears more stable than the first one.

26. It should be obvious that the failure of a single element in the cable dome system would lead to the collapse of the entire structure. For this reason, parallel cables were installed in all actual cable-domes. During the simulation of the lifting process, the roof membrane was not considered. It is obvious, however, that a complete study of the structure in its working condition should account for the rigidity and stabilising effect of that membrane.

REFERENCES

1. FULLER, R.B. Inventions: The Patented Work of R.B. Fuller, 1983.

2. GEIGER, D. The Design and Construction of Two Cable Domes for the Korean Olympics. Proceedings of the IASS Symposium, Osaka, 1986, 265-272.

3. KRISHNA, P. Cable-Suspended Roofs, 1978.

4. YAMAGUCHI, I., OKADA, K., KIMURA, M., MAGARA, H., OKAMURA, K., OHTA, H., OKADA, A., and OKUNO, N., A Study on the Mechanism and Structural Behaviours of Cable Dome. Proceedings of the Inter. Colloquium on Space Structures for Sports Buildings, Beijing, 1987, 534-549.

ACKNOWLEDGEMENTS

This work was supported by a grant from the National Research Council of Canada.

106. Non-linear seismic response of saddle form cable-suspended roofs

F. JIASHEN and F. WEISHUI, Yunnan Institute of Technology, P. R. China

SYNOPSIS. This paper gives the asymptotic solutions of the non-linear governing equations of the saddle form cable-suspended roof in continuous theory, under vertical seismic force with Rayleigh damping.

INTRODUCTION

1. Cable-suspended roofs are used for large span buildings such as aircraft hangers, sport stadiums and pavilions. Such a saddle form roof as shown in Fig. 1 has two layers of cables with the upper one carrying the load and the lower one used to maintain stability. Under vertical seismic force together with Rayleigh damping, the governing equations in continuous theory are from refs 1, 2 and are given in Eqs 1', 2' and 3'.

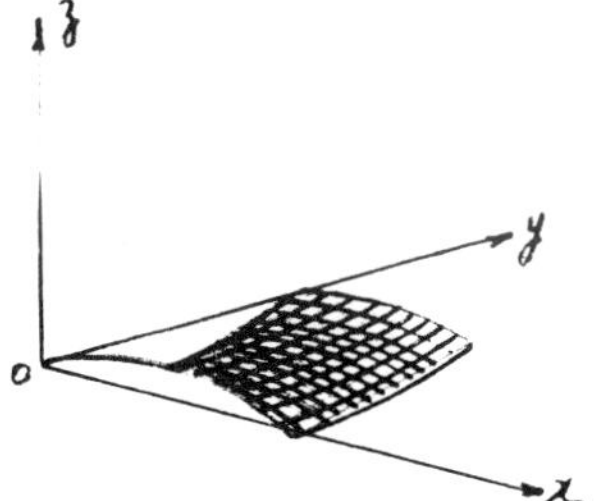

Fig. 1. The cable-suspended roof.

$$H_x \frac{\partial^2 w}{\partial x^2} + H_y \frac{\partial^2 w}{\partial y^2} + \triangle H_x \frac{\partial^2}{\partial x^2} (z_0 + w) + \triangle H_y \frac{\partial^2}{\partial y^2} (z_0 + w)$$

$$+ \left[- A + B\left(\frac{\partial w}{\partial t} \right)^2 \right] \frac{\partial w}{\partial t} + \rho \frac{\partial^2 w}{\partial t^2} = - \rho \ddot{w}_g \qquad (1)'$$

$$\triangle H_x \frac{a}{EA_x} = \int_0^a \left[\frac{\partial z_0}{\partial x} \frac{\partial w}{\partial x} + \frac{1}{2} \left(\frac{\partial w}{\partial x} \right)^2 \right] \mathrm{d}x \qquad (2)'$$

$$\triangle H_y \frac{b}{EA_y} = \int_0^b \left[\frac{\partial z_0}{\partial y} \frac{\partial w}{\partial y} + \frac{1}{2} \left(\frac{\partial w}{\partial y} \right)^2 \right] \mathrm{d}y \qquad (3)'$$

Where in Eqs 1',2,'3,' there are three unknown functions w, ΔH_x and ΔH_y. The dimensions of symbols in these equations are: x, y, z, w, a and b (m); t(s), A_x and $A_x(m^2)$, $A(kg/s\bullet m)$, $B(kg\bullet s/m^3)$, ρ (kg / m), $\ddot{w}_g(m/s^2)$, H_x, H_y, ΔH_x and $\Delta H_y(N)$, $E(N/m^2)$.

The boundary and initial conditions are:

$$w(0, y, t) = w(a, y, t) = w(x, 0, t) = w(x, b, t) = w(x, y, 0) = \dot{w}|_{t=0} = 0 \tag{4}$$

ASYMPTOTIC GOVERNING EQUATIONS AND THEIR SOLUTIONS

2. Let T be the longest period of free vibration of the roof, and f is the height of the roof, transforming all the symbols into non-dimensional parameters as:

$$\left.\begin{aligned}
&\zeta=\frac{x}{a},\ \eta=\frac{y}{b},\ \overline{w}=\frac{w}{f},\ \overline{z}_0=\frac{z_0}{f},\ \overline{H}_\xi=\frac{H_x}{Ea^2},\ \overline{H}_\eta=\frac{H_y}{Eb^2},\\
&\Delta\overline{H}_\xi=\frac{\Delta H_x}{Ea^2},\ \Delta\overline{H}_\eta=\frac{\Delta H_y}{Eb^2},\ \overline{A}=\frac{A}{ET}\\
&\overline{B}=\frac{Bf^2}{ET^3},\ \overline{\rho}=\frac{\rho}{ET^2},\ \lambda=\frac{a^2}{A_x},\ \mu=\frac{b^2}{A_y},\ \psi=\frac{f^2}{a^2},\ \phi=\frac{f^2}{b^2},\\
&\tau=\frac{t}{T},\ \ddot{\overline{w}}_g=\frac{\partial^2\overline{w}_g}{\partial\tau^2},\ \overline{w}_g=\frac{w_g}{f}
\end{aligned}\right\} \tag{5)'}$$

The non-dimensional governing equations and the B-I conditions are:

$$\overline{H}_\xi\frac{\partial^2\overline{w}}{\partial\xi^2}+\overline{H}_\eta\frac{\partial^2\overline{w}}{\partial\eta^2}+\Delta\overline{H}_\xi\frac{\partial^2}{\partial\xi^2}(\overline{z}_0+\overline{w})+\Delta\overline{H}_\eta\frac{\partial^2}{\partial\eta^2}(\overline{z}_0+\overline{w})+\left[-\overline{A}+\overline{B}\left(\frac{\partial\overline{w}}{\partial\tau}\right)^2\right]\frac{\partial\overline{w}}{\partial\tau}+\overline{\rho}\frac{\partial^2\overline{w}}{\partial\tau^2}=-\overline{\rho}\ddot{\overline{w}}_g \tag{1}$$

$$\Delta\overline{H}_\xi=\frac{\psi}{\lambda}\int_0^1\left[\frac{\partial\overline{z}_0}{\partial\xi}\frac{\partial\overline{w}}{\partial\xi}+\frac{1}{2}\left(\frac{\partial\overline{w}}{\partial\xi}\right)^2\right]d\xi \tag{2}$$

$$\Delta\overline{H}_\eta=\frac{\phi}{\mu}\int_0^1\left[\frac{\partial\overline{z}_0}{\partial\eta}\frac{\partial\overline{w}}{\partial\eta}+\frac{1}{2}\left(\frac{\partial\overline{w}}{\partial\eta}\right)^2\right]d\eta \tag{3}$$

$$\overline{w}(0,\eta,\tau)=\overline{w}(1,\eta,\tau)=\overline{w}(\xi,0,\tau)=\overline{w}(\xi,1,\tau)=\overline{w}(\xi,\eta,0)=\dot{\overline{w}}|_{\tau=0}=0 \tag{4}$$

Where in Eqs 1', 2' and 3', z = f (x,y) is the geometrical function of the saddle form roof.

3. Now, the parameter $\overline{B}$ <1 is a small parameter (ref. 3) and we may expand:

$$\left.\begin{aligned}
\overline{w} &= \overline{w}_0 + \overline{B}\overline{w}_1 + \overline{B}^2\overline{w}_2 + \cdots\cdots + \overline{B}^n\overline{w}_n + \cdots\cdots \\
\triangle\overline{H}_\xi &= \triangle\overline{H}_{\xi 0} + \overline{B}\triangle\overline{H}_{\xi 1} + \overline{B}^2\triangle\overline{H}_{\xi 2} + \cdots + \overline{B}^n\triangle\overline{H}_{\xi n} + \cdots \\
\triangle\overline{H}_\eta &= \triangle\overline{H}_{\eta 0} + \overline{B}\triangle\overline{H}_{\eta 1} + \overline{B}^2\triangle\overline{H}_{\eta 2} + \cdots + \overline{B}^n\triangle\overline{H}_{\eta n} + \cdots
\end{aligned}\right\} \tag{5}$$

The asymptotic governing equations of order zero are:

$$\overline{H}_\xi\frac{\partial^2\overline{w}_0}{\partial\xi^2} + \overline{H}_\eta\frac{\partial^2\overline{w}_0}{\partial\eta^2} + \triangle\overline{H}_{\xi 0}\frac{\partial^2}{\partial\xi^2}(\overline{z}_0 + \overline{w}_0) + \triangle\overline{H}_{\eta 0}\frac{\partial^2}{\partial\eta^2}(\overline{z}_0 + \overline{w}_0) - \overline{A}\frac{\partial\overline{w}_0}{\partial\tau} + \overline{\rho}\frac{\partial^2\overline{w}_0}{\partial\tau^2} = -\overline{\rho}\ddot{\overline{w}}_g \tag{6}$$

$$\triangle\overline{H}_{\xi 0} = \frac{\psi}{\lambda}\int_0^1\left[\frac{\partial\overline{z}_0}{\partial\xi}\frac{\partial\overline{w}_0}{\partial\xi} + \frac{1}{2}\left(\frac{\partial\overline{w}_0}{\partial\xi}\right)^2\right]\mathrm{d}\xi \tag{7}$$

$$\triangle\overline{H}_{\eta 0} = \frac{\Phi}{\mu}\int_0^1\left[\frac{\partial\overline{z}_0}{\partial\eta}\frac{\partial\overline{w}_0}{\partial\eta} + \frac{1}{2}\left(\frac{\partial\overline{w}_0}{\partial\eta}\right)^2\right]\mathrm{d}\eta \tag{8}$$

The asymptotic governing equations of order one are:

$$\overline{H}_\xi\frac{\partial^2\overline{w}_1}{\partial\xi^2} + \overline{H}_\eta\frac{\partial^2\overline{w}_1}{\partial\eta^2} + \triangle\overline{H}_{\xi 0}\frac{\partial^2\overline{w}_1}{\partial\xi^2} + \triangle\overline{H}_{\xi 1}\frac{\partial^2\overline{w}_0}{\partial\xi^2} + \triangle\overline{H}_{\eta 0}\frac{\partial^2\overline{w}_1}{\partial\eta^2} + \triangle\overline{H}_{\eta 1}\frac{\partial^2\overline{w}_0}{\partial\eta^2} - \overline{A}\frac{\partial\overline{w}_1}{\partial\tau} + \overline{\rho}\frac{\partial^2\overline{w}_0}{\partial\tau^2} = -\left(\frac{\partial\overline{w}_0}{\partial\tau}\right)^3 \tag{9}$$

$$\triangle\overline{H}_{\xi 1} = \frac{\psi}{\lambda}\int_0^1\left(\frac{\partial\overline{z}_0}{\partial\xi}\frac{\partial\overline{w}_1}{\partial\xi} + \frac{\partial\overline{w}_0}{\partial\xi}\frac{\partial\overline{w}_1}{\partial\xi}\right)\mathrm{d}\xi \tag{10}$$

$$\triangle\overline{H}_{\eta 1} = \frac{\Phi}{\mu}\int_0^1\left(\frac{\partial\overline{z}_0}{\partial\eta}\frac{\partial\overline{w}_1}{\partial\eta} + \frac{\partial\overline{w}_0}{\partial\eta}\frac{\partial\overline{w}_1}{\partial\eta}\right)\mathrm{d}\eta \tag{11}$$

The dimensionless B-I conditions are:

$$\overline{w}_i(0,\eta,\tau) = \overline{w}_i(1,\eta,\tau) = \overline{w}(\xi,0,\tau) = \overline{w}_i(\xi,1,\tau) = \overline{w}_i(\xi,\tau,0) = \dot{\overline{w}}_i|_{\tau=0} = 0, \quad (i = 0, 1, 2, \cdots) \tag{12}$$

4. For the solutions of zero order asymptotic governing equations, assuming:

$$\left.\begin{aligned} \overline{w}_0\ (\xi,\ \eta,\ \tau) &= \sum_{i,\ j=1,\ 3,\ \cdots}^{\infty} \sum^{\infty} \psi_{ij}\ (\tau)\ \sin i\pi\xi \sin j\pi\eta \\ \overline{z}_0 &= 4\left[\left(\xi-\frac{1}{2}\right)^2-\left(\eta-\frac{1}{2}\right)^2\right]+1 \end{aligned}\right\} \tag{13}$$

Substituting the values in Eq. 13 into Eqs 2 and 3, gives:

$$\Delta\overline{H}_{\xi 0}=\frac{\psi}{\lambda}\left[-\frac{16}{\pi^2}\sum_i\sum_j\frac{1}{i^2}\psi_{ij}\ (\tau)\ \sin j\pi\eta\right.$$

$$\left.+\frac{\pi^2}{2}\sum_i\sum_j\sum_{j'} i^2\psi_{ij}\ (\tau)\ \psi_{ij'}\ (\tau)\ \sin j\pi\eta \sin j'\pi\eta\right] \tag{A}$$

$$\Delta\overline{H}_{\eta 0}=\frac{\Phi}{\mu}\left[\frac{16}{\pi^2}\sum_i\sum_j\frac{1}{j^2}\psi_{ij}\ (\tau)\ \sin i\pi\xi\right.$$

$$\left.+\frac{\pi^2}{2}\sum_i\sum_{i'}\sum_j j^2\psi_{ij}\ (\tau)\ \psi_{i'j}\ (\tau)\ \sin i\pi\xi \sin i'\pi\xi\right] \tag{B}$$

Substitute the values in Eqs A and B into Eq. 1, there is (ref. 1):

$$\frac{\pi}{2}\psi_{mn}\ (\tau)\ \ (-m^2\overline{H}_\xi+n^2\overline{H}_\eta)\ +\frac{\psi}{\lambda}\left[\frac{128}{\pi^3}\sum_i\frac{1}{i^3}\psi_{in}\ (\tau)\ +\frac{2}{m}\sum_i\sum_j\sum_{j'}\right.$$

$$\left(\frac{1}{j+j'-n}-\frac{1}{j+j'+n}+\frac{1}{j-j'+n}+\frac{1}{j-j'-n}\right)\psi_{ij}\ (\tau)\ \psi_{ij'}\ (\tau)$$

$$+4n^2\times\sum_i\sum_{i'}\sum_j\frac{1}{j^2}\left(\frac{1}{i+i'-m}-\frac{1}{i+i'+m}+\frac{1}{i-i'+m}\right.$$

$$\left.+\frac{1}{i-i'-m}\right)\psi_{ij}\ (\tau)\ \psi_{i'n}\ (\tau)\ -$$

$$-\frac{\pi^3}{16}\sum_i\sum_j\sum_{j'}\sum_s s^2\ \ (\delta^{\eta}_{j+j'-s}-\delta^{n}_{-j-j'+s}-\delta^{\eta}_{j+j'+s}-\delta^{n}_{-j+j'-s}-\delta^{\eta}_{j-j'-s}$$

$$\left.+\delta^{n}_{-j+j'+s}+\delta^{\eta}_{j-j'+s})\ \psi_{ij}\ (\tau)\ \psi_{ij'}\ (\tau)\ \psi_{ms}\ (\tau)\ \right]+\frac{\Phi}{\mu}\left[-\frac{128}{\pi^3}\sum_j\frac{1}{j^3}\psi_{jm}\ (\tau)\right.$$

$$\left.+\text{terms in first}\ [\quad]\ \text{by replacing}\ i\ \text{to}\ j\ \text{and}\ m\ \text{to}\ n\right]-\frac{\overline{A}}{4}\dot{\psi}_{mn}\ (\tau)\ +$$

$$+\frac{\overline{\rho}}{4}\ddot{\psi}\ (\tau)\ =-\frac{4\overline{\rho}}{mn\pi^2}\ddot{w}_K\ (\tau) \tag{13$'$}$$

5. In obtaining the above equation, use has been made of the orthogonality of the Fourier series. This equation can be solved by successive iteration. First, taking i = j = 1 only, gives :

$$\ddot{a}_1^{(1)}(\tau)+\omega_1^2 a_1^{(1)}(\tau)-\gamma\left[a_1^{(1)}(\tau)\right]^3=\epsilon\left[\alpha\cos\beta\tau+F-2H\dot{a}_1^{(1)}(\tau)\right],$$

$$a_1^{(1)}(\tau)=\psi_{11}(\tau)-64\left(\frac{\psi}{\lambda}-\frac{\Phi}{\mu}\right)/9\pi^3\left(\frac{\psi}{\lambda}+\frac{\Phi}{\mu}\right)$$

$$\omega_1^2=\frac{4}{\bar{\rho}}\left[\frac{128}{\pi^2}\left(\frac{\psi}{\lambda}-\frac{\Phi}{\mu}\right)-\frac{\pi^2}{2}\left(\bar{H}_\xi+\bar{H}_\eta\right)\right]$$

$$-1024\left(\frac{\psi}{\lambda}-\frac{\Phi}{\mu}\right)^2/3\pi^3\bar{\rho}\left(\frac{\psi}{\lambda}+\frac{\Phi}{\mu}\right),$$

$$\gamma=\frac{3\pi^3}{4\bar{\rho}}\left(\frac{\Phi}{\mu}-\frac{\psi}{\lambda}\right),\quad \epsilon\alpha\cos\beta\tau=\frac{16}{\bar{\rho}}\ddot{w}_g(\tau),\quad \epsilon F=\frac{16\left(\mu\psi-\lambda\Phi\right)}{9\pi^3\bar{\rho}\left(\mu\psi+\lambda\Phi\right)}\times$$

$$\times\left[\frac{128}{\pi^2}\left(\frac{\psi}{\lambda}-\frac{\Phi}{\mu}\right)-\frac{\pi^2}{2}\left(\bar{H}_\xi+\bar{H}_\eta\right)-24\left(\mu\psi-\lambda\Phi\right)\right],\quad 2\epsilon H=-\frac{\bar{A}}{\bar{\rho}} \qquad (14)$$

6. Now, Eq. (14) is the Duffing equation with excitation, its solutions is (refs 3,4); (where β is the predominant frequency of the site):

$$a_1^{(1)}(\tau)=a_1^{(\beta)}(\tau)=a_{1\beta}^{(0)}(\bar{\tau}+\delta)+\epsilon a_{1\beta}^{(1)}(\bar{\tau}+\delta)+\cdots\cdots$$

$$a_{1\beta}^{(0)}(\bar{\tau}+\delta)=c_\beta\cos\bar{\tau}+c_\beta^3 a_{1\beta}^{(3)}(\bar{\tau})+c_\beta^5 a_{1\beta}^{(5)}(\bar{\tau})+\cdots\cdots$$

$$a_{1\beta}^{(3)}=-\frac{\gamma}{32\omega_1^2}\left(\cos\bar{\tau}-\cos 3\bar{\tau}\right),\quad \tau=\bar{\tau}+\delta$$

$$a_{1\beta}^{(5)}=\frac{23}{1204}\left(\frac{\gamma}{\omega_1^2}\right)^2\cos\bar{\tau}-\frac{3}{128}\left(\frac{\gamma}{\omega_1^2}\right)^2\cos 3\bar{\tau}+\frac{1}{1204}\left(\frac{\gamma}{\omega_1^2}\right)^2\cos 5\bar{\tau}$$

$$-\frac{3\gamma}{8\omega_1^2}c_\beta^2+\frac{57\gamma^2}{256\omega_1^4}c_\beta^4+\cdots=\frac{1}{\beta}\left(\omega_1-\beta\right)$$

$$a_{1\beta}^{(1)}(\bar{\tau}+\delta)=\frac{\alpha\cos l\beta}{2MN\omega_1}\cos\beta(\bar{\tau}+\delta)+\left(\frac{Nc_\beta^3}{8\omega_1\beta\bar{\rho}M}-sc_\beta-\frac{c_\beta}{2\omega_1\bar{\rho}M}\right)\sin\beta(\bar{\tau}$$

$$+\delta)+\left(\frac{\alpha c_\beta}{4\beta\bar{\rho}\omega_1 M}-\frac{2Nc_\beta^3}{16\omega_1 M\bar{\rho}^2}\right)\cos\beta\bar{\tau}+\frac{\alpha Nc_\beta^3}{16\omega_1 M\bar{\rho}^2}\cos\left(3\beta\bar{\tau}+2\beta\delta\right)$$

$$+\frac{FNc_\beta^3}{4\omega_1 M\beta^2}\cos 2\beta(\bar{\tau}+\delta)+\frac{FNc_\beta^3}{8\omega_1\beta M}\sin 3\beta(\bar{\tau}+\delta)$$

$$\sin l\beta=-\frac{4\beta c_\beta}{A\alpha},\quad S=\frac{2\omega_1\bar{\rho}}{3\gamma c_\beta^3},\quad M=\frac{\left(1+c_\beta\right)^2}{\beta}\omega_1 c_\beta,$$

$$N=\frac{\beta^2}{\omega_1}\left(-\frac{3}{43}c_\beta+\frac{57\gamma^2}{64\omega_1^2}c_\beta^3+\cdots\cdots\right)$$

If c_β^2 must be real roots, there should be:

$$\omega_1 - \beta \geqslant 0$$

Next, take i = 1 and j = 2, gives:

$$\ddot{\psi}_{12}(\tau) - \bar{c}_0\dot{\psi}_{12}(\tau) + \bar{c}_1\psi_{12}(\tau) + \bar{c}_2\psi_{12}^2(\tau) + \bar{c}_3\psi_{12}^3(\tau) = G(\tau)$$

$$\psi_{12}(\tau) = a_1^{(2)}(\tau) - \bar{c}_2/3\bar{c}_3$$

$$\ddot{a}_1^{(2)}(\tau) + \bar{c}_3[a_1^{(2)}(\tau)]^3 + (\bar{c}_1 - \bar{c}_2^2/3\bar{c}_3)\,a_1^{(2)}(\tau)$$
$$= G(\tau) + \bar{c}_0\dot{a}_1^{(2)}(\tau) + \bar{c}_2/27\bar{c}_3\,(9\bar{c}_1 - 2\bar{c}_2^2/\bar{c}_3)$$

7. Because the first order and higher asymptotic governing equations are linear, they are much easier to solve, and consequently are not discuss in this paper.

REFERENCES

1. FAN JIASHEN, et al., Proc. 4th Int'l Conf. on Steel Structures and Space Frames, Singapore, 1990, 73-76.
2. Harbin Archit. Eng Instit., Steel Structures for Large Span Buildings (in Chinese), Beijing Archit. Eng Press, 1985, 115-119.
3. A H NAYFEH, Introduction to Perturbation Techniques, New York, John Wiley and Sons, 1981, 147-188.
4. E G MALKIN, Methods of Liapunov and Poincare in Non-linear Vibration Theory (Chinese translation), Beijing, Science Press, 1959, 154-163.

107. Developments in tensegrity systems: an overview

A. HANAOR, National Building Research Institute, Israel

SYNOPSIS. Tensegrity structures are free-standing, self-stressed, pin-jointed networks. Whilst the exact definition is a matter of controversy, the term was coined by R. B. Fuller and developments in the field can be traced back to the late forties-early fifties and to the collaboration of Fuller with K. Snelson. Since then there have been major developments in generating a large range of morphologies and in understanding the mechanical properties and structural response of these structures; yet, despite several patents issued, no functional building of significant dimensions has been built to date. Practical implementation of the concept awaits further technological developments, and would probably be limited to special applications .

INTRODUCTION - BRIEF HISTORICAL NOTE

1. Since the definition of tensegrity structures is controversial, it is impossible to trace its origins with any degree of certainty. A widely accepted definition is: free-standing, self-stressed, pin- jointed network, consisting of compression members, bars or struts and tension members, tendons or cables. Under this definition, the first tensegrity structure was probably a kite. However, since the term Tensegrity was coined by R. Buckminster Fuller, and resulted, in all probability, from Fuller's collaboration with the artist Kenneth Snelson, the origin of modern tensegrity structures can be assigned to the late forties-early fifties, when this collaboration took place.

2. The first documented reference to tensegrity structures appears to be Fuller's patent Tensile-Integrity Structures of 1962 (filed August 1959 - ref. 1), followed in 1964 by Emerich's patent Construction de Rèseaux Autotendants (filed April 1963 - ref. 2), and Snelson's Patent Continuous Tension, Discontinuous Compression of 1965 (ref. 3). Further patents appear regularly, though in lower frequency over the following years (refs 4-8). These patents deal mostly with morphological aspects and technological means of construction and deployment. The first attempt to utilise the deployability potential of these structures appears in Vasilev 's patent of 1978 (ref. 6).

3. The first scholarly works published on tensegrity structures in readily accessible forms are Fuller's own Synergetics: Exploration in the Geometry of Thinking (ref. 10) in 1975, and the works by Pugh and Kenner which relate to Fuller's work, (refs 11-13). These works are all concerned with morphological aspects. Emerich has carried out extensive studies on the morphology of tensegrity networks over the years but much of his work is not readily available to the public at large (refs 27-29,43). The first work dealing with the mechanics of these structures appears to be Vilnay's paper of 1973 (ref. 14). Vilnay also introduced some new basic morphological concepts (see below). Since the early 80s, there appears to be an increasing interest in the field, as reflected in the published literature (refs 18-46). To date, however, no known functional structures of significant dimensions have been constructed, and very few full scale designs and feasibility studies have been conducted in order to demonstrate the feasibility and constructibility of the concept (refs 37,45).

MORPHOLOGICAL STUDIES

4. As mentioned above, the majority of works deal with the morphology of tensegrity structures. Most of the morphologies studied have the feature that bars are not in contact with one another. Only a few examples, representing milestones in the development of tensegrity structures morphology are highlighted here.

5. Snelson's work (ref. 9) is of artistic nature (Fig. 1). Many of his "sculptures" are of more or less free form or of low degree of regularity. From the structural point of view they are all characterised by having a single state of pre-stress, that is, the tensioning of one cable pre-stresses the whole structure.

6. Fuller's tensegrity dome (Fig. 2) is a multifaceted, single-layer tensegrity polyhedron (it has one layer of cables).

7. Emmerich (ref. 2) was the first to conceive double-layer tensegrity networks. In this configuration, bars are confined between two layers of cables, or tendons (Fig. 3a). This configuration is obtained by joining together tensegrity prisms (T prisms) or truncated pyramids (T pyramids), or simplexes as Emmerich terms these objects. Flat as well as curved surfaces are possible. Emmerich also investigated extensively and systematically tensegrity polyhedra (refs 28,29). He identified 17 basic single-layer polyhedra (Fig. 3b) and 18 double-layer or Hyper polyhedra (Fig. 3c), which can produce double-layer domes.

8. Vilnay introduced the concept of infinite tensegrity networks (ref. 15). His networks are plane-filling networks of cables, with bars connecting non-adjacent nodes in varying patterns (Fig. 4a). Although the network is essentially planar, actual structures have to be singly or doubly curved as it is a single-layer configuration (Fig. 4b).

9. Motro produced double-layer tensegrity grids by joining together tensegrity prisms at their nodes (Fig. 5). As a result his grids contain bars joined together at nodes, unlike other configurations studied, but bars are still confined between two parallel layers of tendons.

10. Hanaor carried out an extensive investigation of DLTGs with non-contacting bars, especially their mechanics and engineering properties (see

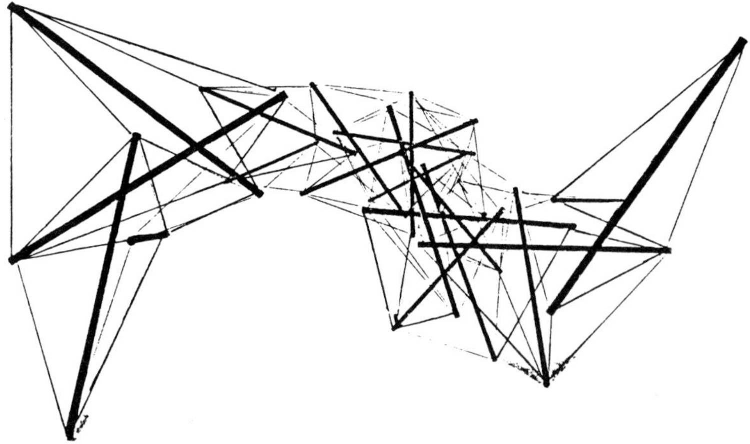

Fig. 1. Snelson's sculpture (ref. 9).

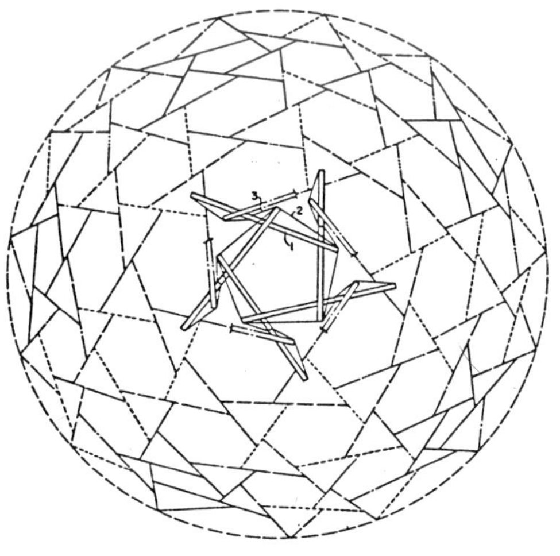

Fig. 2. Fuller's dome (ref. 1).

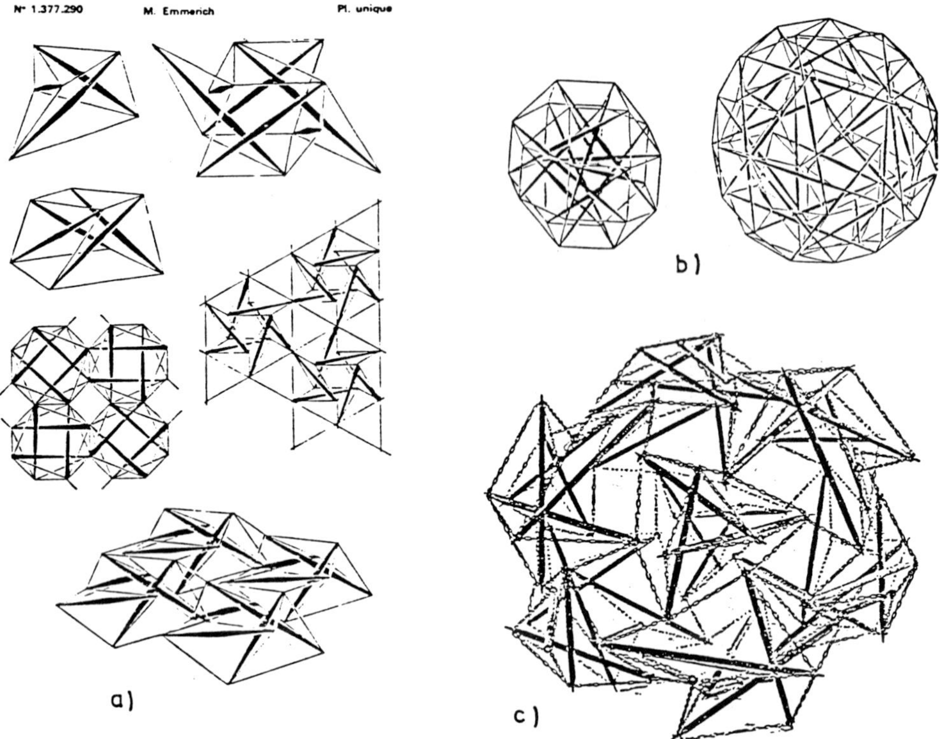

Fig. 3. Emmerich's tensegrity structures: a) DLTG (ref. 2), b) polyhedron, c) composite polyhedron (double-layer) (ref. 29).

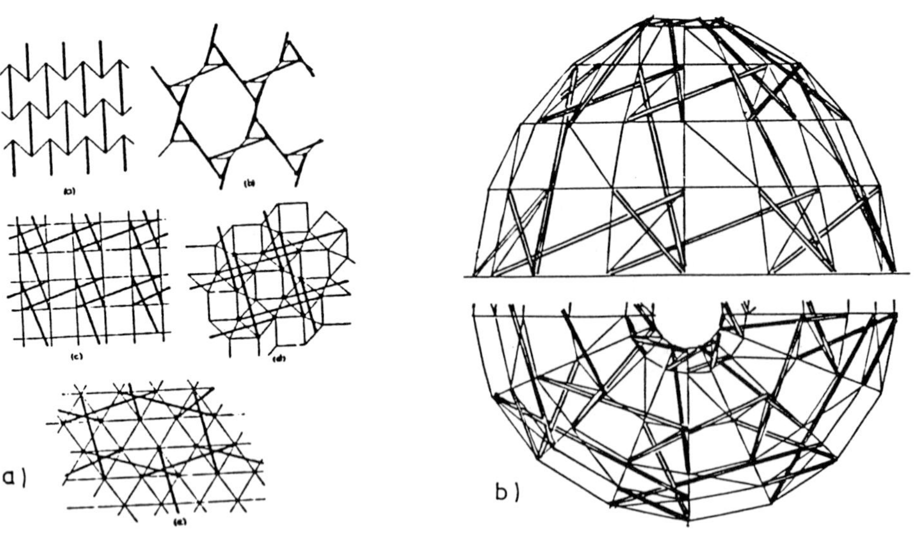

Fig. 4. Vilnay's structures: a) infinite net, b) dome (refs 15,20).

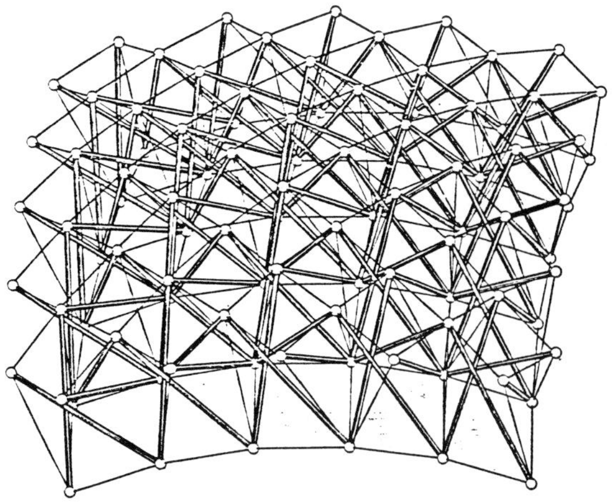

Fig. 5. Motro's double-layer tensegrity grid (refs 23,30).

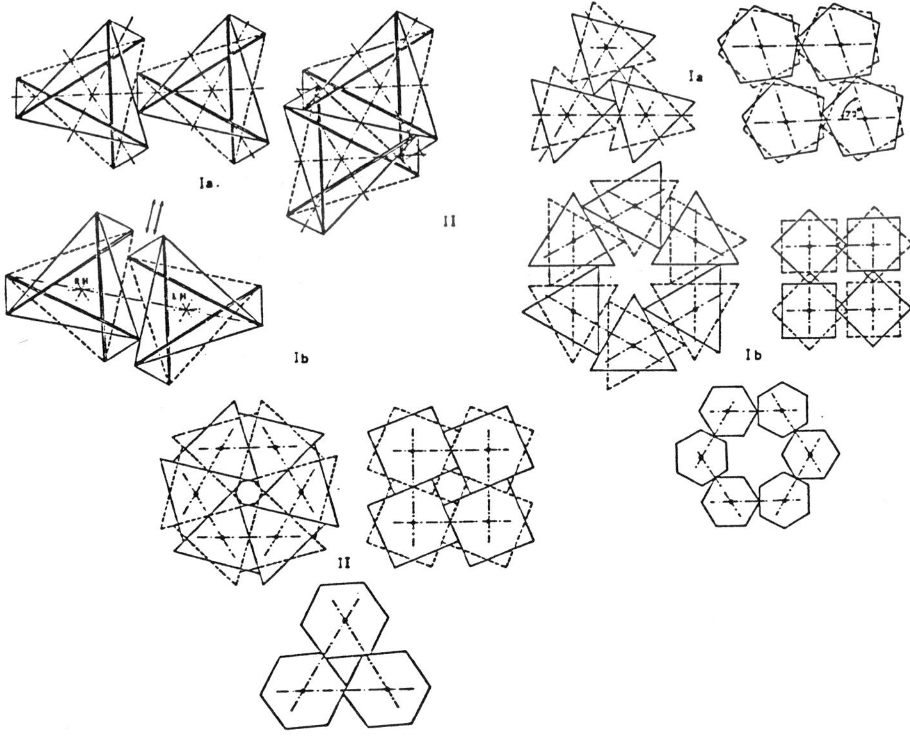

Fig. 6. Hanaor's DLTGs (ref. 34)

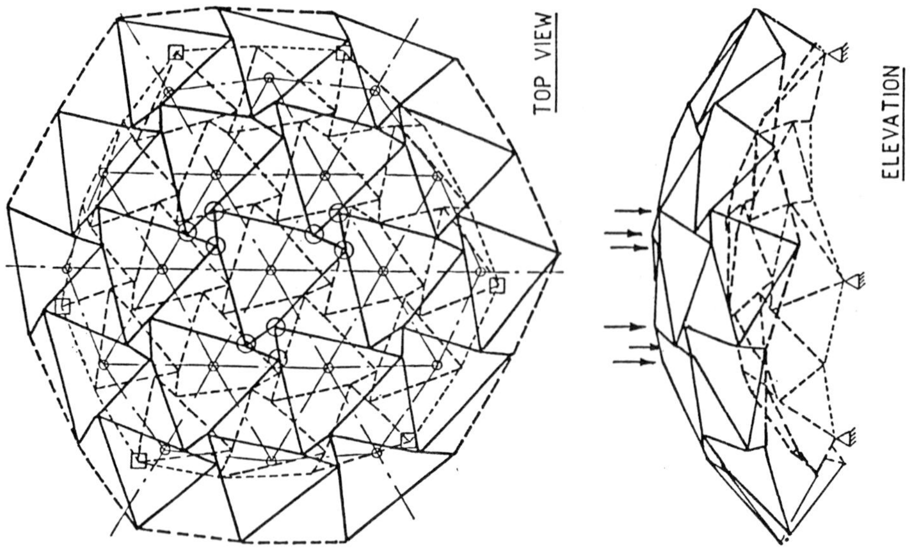

Fig. 7. Hanaor's double-layer dome (ref. 44), only top and bottom layers shown.

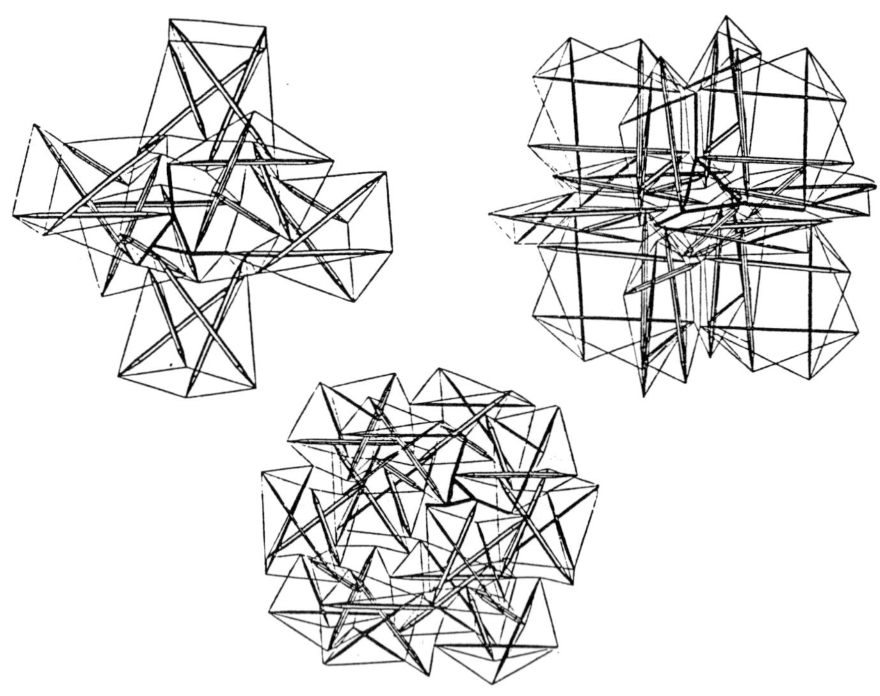

Fig. 8. Grip's plyhedral networks (ref. 39).

below). Three different methods of connecting T prisms were identified, producing a large variety of grids (Fig. 6). He also investigated the geometry of double layer tensegrity domes (DLTDs) by a geodesic subdivision of a pyramid, employing DLTGs made of T pyramids (Fig. 7).

11. Grip (ref. 39) extended the inventory of tensegrity polyhedra and hyper polyhedra, through the concepts of duality and space filling polyhedral networks, to produce a virtually unlimited range of very intricate and complex networks (Fig. 8).

12. It appears that the morphological study of tensegrity networks has reached a degree of saturation, whereby the range of conceivable patterns exceeds by far the likely range of applications, and is comparable to the range of possible bar networks. It remains to be seen whether the range of practical applications in any way approaches that of bar structures.

STRUCTURAL ANALYSIS AND DESIGN, EXPERIMENTAL RESULTS

13. Tensegrity structures are, from the structural analysis point of view, pre-stressed pin-jointed networks similar to cable networks. Except in some special configurations, they are geometrically flexible - they contain internal mechanisms, and therefore significant geometrical non-linearities. As in the case of cable networks, the analysis consists typically of two major phases: a shape-finding phase, aimed at obtaining the equilibrium pre-stressed geometry, and the static (or dynamic) analysis which determines member forces and nodal displacements under the applied loads. An intermediate phase of investigation of stability, mechanisms and states of pre-stress is sometimes employed for research purposes.

14. Analytical models for the non-linear analysis of cable networks are ubiquitous in the technical literature. Following, are some references to analytical procedures, specifically applied to tensegrity structures. Motro and his co-investigators (refs 21-23,31) applied the technique of dynamic relaxation to the shape-finding problem. They also performed dynamic analysis of T prisms and simple double-layer networks. Vilnay, in his book (ref. 32) outlines in a very succinct way the principles of the investigation of mechanisms and states of pre-stress and the shape-finding, static and dynamic analysis phases, but his procedure is not very suitable for computer implementation. Hanaor (ref. 26) presented a flexibility based model for investigating mechanisms and states of pre-stress and for static analysis, but employed a stiffness based procedure, which is more computationally efficient, for the actual analysis (shape-finding and static) of sizeable networks.

15. Hanaor (refs 33,37,45) carried out parametric analytical investigations of double layer tensegrity grids of varying geometries, including domes and geometrically rigid configurations (statically indeterminate, with no internal mechanisms). The main findings are that domes are stiffer than plane grids over a given span and geometrically rigid configurations are stiffer than geometrically flexible ones. There is a direct correlation between stiffness and load carrying capacity for a given span and topology.

16. Hanaor (refs 37,45) carried out a full design of various configurations of DLTGs covering a space of roughly circular plan, some 27 m in span. The DLTGs were compared with the design of a double-layer grid (DLG) bar structure of similar plan. The result indicate that, so long as the principle of non-contacting bars is maintained, all DLTG configurations were heavier than the DLG, due to the long bars. Although the design was carried out on a specific layout, the conclusion appears to be general for all tensegrity networks with non-contacting bars, as the feature of long bars (compared with tendon lengths) is inherent in all these configurations (see all figures). The conclusion that some bar to bar connections are essential to achieve light weight appears to be inescapable. This conclusion is of major impact on potential applications of tensegrity structures

17. Experimental tests were carried out by Hanaor (ref. 36), and by Motro et.al. (refs 22,31) on small models. Hanaor performed static tests on small DLTG models of various configurations. Motro et. al. performed dynamic tests on single triangular prism (simplex) models. All tests indicate a generally good agreement with theoretical predictions. The DLTG tests indicate a high degree of structural redundancy and the ability to maintain load carrying capacity with the loss of a small percentage of members.

18. Limited experience with small deployable models (refs 37,46) indicate that deployment by means of telescoping bars and air pressure is technically feasible and potentially very promising.

PROSPECTS

19. Suitable morphologies and the analytical tools exist to realise a large variety of tensegrity networks. For the actual construction of functional tensegrity structures several technical problems are yet to be resolved, some of which may put some heavy constraints on the range of potential applications.

20. The relative structural inefficiency of tensegrity networks with non-contacting bars (see para.16) restricts the potential applications for these structures. The feature of non-contact of bars is central for the implementation of tensegrity structures. It simplifies the connections between members (as compared with 3D bar structures) and makes them inherently collapsible and deployable. It is also the main architectural asset, giving these structures their peculiar appearance. For application as deployable structures, however, structural weight is of cardinal importance.

21. In the near future, it seems the main motivation for constructing tensegrity buildings is the architectural feature mentioned above. It can be anticipated that the first structures to be built will be light structures with high architectural content, such as exhibition halls, conservatories, arboreta, aquaria and the like.

22. One of the main technical aspects in the implementation of tensegrity structures is the roof covering. Most configurations are geometrically flexible. It is therefore almost certain that the roof covering has to be a flexible membrane, yet none of the studies carried out to date, consider the surface membrane. It is important that the membrane forms an integral part of the design, as the problem

of ensuring a crease-free, flutter-free surface, is by no means trivial. For deployable applications it is highly desirable that the membrane serves a structural, as well as functional role, reducing the size of cables or even eliminating them.

REFERENCES

1. FULLER R.B. Tensile-integrity structures. US patent 3,063,521, Nov. 1962.
2. EMMERICH D.G. Construction de rèseau autotendants. Brevet d'ivention (French patent) 1. 377. 290, Sep. 1964.
3 . SNELSON K . D. Continuous tension, discontinuous compression structures. US patent 3,169,611, Feb. 1965.
4. GUYON J.F.M. Combinaison de solides à éléments rigides non jointifs pour la réalisation de structurres tridimensionelles. Brevet d'invention (French patent) 1.519.240, Feb. 1968.
5. FULLER R.B. Non-symmetrical tension-integrity structures. US patent 3,866,366, Feb. 1975.
6. VASILEV V.V. Three dimensional roofing element. Soviet patent SU 586247, Jan. 1978.
7. KITRICK C.J. Tensegrity module structure and method of inter-connecting the modules. US patent 4,207,715, Jun. 1980.
8 . GWILLIAM T. S , Ohu R . Octet structures using tension and compression . US patent 4,711,062, Dec. 1987.
9. KENNETH SNELSON. Exhibition catalogue and essay by H.N. Fox, Albright-Knox art gallery, Buffalo, New York, 1981.
10. FULLER R.B. Synergetics: Explorations in the geometry of thinking. Macmillan, New York, 1975 .
11. PUGH A. An introduction to tensegrity. University of California Press, Berkeley, 1975.
12. PUGH A. A brier introduction to tensegrity. Supple W.J. ed. Proc. 2nd International conference on space structures, Univ. of Surrey, 1975, 707-711.
13. KENNER, H. Geodesic maths and how to use it. University of California Press, Berkeley, 1976.
14. VILNAY O. and Staierman B. Tensegric shells. Proc. IASS Symp. on Prefabricated Shells, Haifa, 1972, V2, Paper E2, 348-472.
15. VILNAY O. Structures made of infinite regular tensegric nets. IASS Bulletin, V18 (63), April 1977, 51-57.
16. CALLADINE C.R. Buckminster Fuller's "Tensegrity structures and Clerk Maxwell's rule for the construction of stiff frames. Int. J. of Solids and Structures, V14 1978, 161-172.
17. VILNAY O. Determinate tensegric shells. Architectural Science Revue, V22 N2 June 1979, 39-43.
18. VILNAY O. Determinate tensegric shells. ASCE J. of Struct. Div. V107 N ST10, Oct. 1981, 2029 -2033.
19. ROTH B. and Whiteley W. Tensegrity frameworks. Transactions of the American Mathematical Society, V265, N2, June 1981, 419-446.

20. VILNAY O. and Soh S.S. Tensegric shell behaviour. ASCE J. Struct. Div. V108, ST8, Aug. 1982, 1831-1845.
21. MOTRO R., BROCHIERO E. and JOUANNA P. Etud quantitaative de systèmes autotendants. World congress on nontypical structures, IASS, Mar del Plata, Argentina, 1982.
22. MOTRO R. Formes and Forces in tensegrity systems. Nooshin H. ed. Proc. 3rd. Int. Conf. on Space Structures, Univ. of Surrey, 1984,283-288.
23. MOTRO R. Tensegrity systems for double- layer space structures. Topping B.H.V. ed. Proc. Int. Conf. on the design and construction on non-conventional structures, Civil Comp Press, London 1987, V2, 43-52.
24. HANAOR A. Preliminary investigation of double-layer tensegrities. Topping B.H.V. ed. Proc. Int. Conf. on non-conventional structs., Civil Comp, London 1987, 35-42.
25. HANAOR A. and KING B. Tensegrity structures: aspects of design and analysis. Sherman D.R. ed. Building structures. Proc. Structures Congress 87, ASCE Struct. Div. 1987, 514-525.
26. HANAOR A. Pre-stressed pin-jointed structures - flexibility analysis and pre-stress design. Computers and Structures V28 N6 1988, 757-769.
27. EMMERICH D. G . structures tendues et autotendants. Monographies de Géométrie Constructive, Ecole d'Architecture de Paris la Villette 1988.
28. EMMERICH D.G. Simplex stables dans les systèmes constructifs. Monographies de Géométrie Constructive, Ecole d'Architecture de Paris la Villette 1988.
29. EMMERICH D.G. Self-tensioning spherical structures: single and double layer spheroids. Int. J. of Space Structures V5,N3,4 1990, 355-374.
30. MOTRO R. Tensegrity systems and geodesic domes. Int. J. of Space Structs. V5,N3,4 1990, 343 -354 .
31. MOHRI F. Analyse Théoretique et expérimentale des systèmes réticulés spaciaux autocontraints - incidence du niveau d'autocontrainte. Thèse de doctorat, Université Montpellier 11, 1990.
32. VILNAY O. Cable netas and tensegric shells: analysis and design applications. Ellis Horwood, Chichester 1990.
33. HANAOR A. Design, analysis and response of double-layer tensegrity grids. Topping B.H.V. ed. Developments in structural engineering, Proc. Forth Rail Bridge centenary conf. Edinburgh 1990 V1 579-590.
34. HANAOR A. Double- layer tensegrity grids. Nooshin H . ed . Studies in space structures, Multiscience 1991, 245-264.
35. HANAOR A. and LIAO M.K. Double-layer tensegrity grids: static load response. I - analytical study. J. of Struct. Eng. ASCE V117 N6 June 1991, 1660-1674.
36. HANAOR A. Double-layer tensegrity grids: static load response. II - experimental study. J. of Struct. Eng. ASCE V 117 N6 June 1991, 1675-1684.
37. HANAOR A. Engineering properties of double-layer tensegrity grids. Spatial structures at the turn of the millennium, Proc. IASS Symp. Copenhagen 1991, V.II, 195-200.

38. MOTRO R. Tensegrity systems: the state of the art. Int. J. Space Structs.V7,N2 - Special issue on tensegrity systems, Motro R. Ed., 1992, 75-84.
39. GRIP R. The Correspondence between convex polyhedra and tensegrity systems: a classification system. Int. J. Space Structs. V7,N2,1992, 115-126.
40. FUJURA H. Concept of deployable tensegrity structures in space application. Int. J. Space Structs. V7 N2, 143-152.
41. CONNELY R. and WHITELEY W. The stability of tensegrity frameworks . Int. J. Space Structs. V7 N2 1992, 153-164.
42. CHASSAGNOUX A., CHOMARAT A. and SAVEL J. A study of morphological characteristics of tensegrity structures. Int. J. Space Structs. V7 N2 1992, 165-172.
43. EMMERICH D.G. Stable simplex introduction to the resistance of forms. Proc. seminar on structural morphology, IASS, Montpellier, 1992, 43-52.
44. HANAOR A. Aspects of design of double-layer tensegrity domes. Int. J. Space Structs. V7,N2, 1992, 101-114.
45. HANAOR A. Generation and properties of some double-layer tensegrity grid geometries. Proc. seminar on structural morphology. IASS, Montpellier 1992, 113-124.
46. HANAOR A. Double-layer tensegrity grids as deployable structures. Int. J. Space Structs. V7,N4. 1992.

108. Absolute minimal self-tensioning configurations

D. G. EMMERICH, L'Ecole Nationale Superieure des Beaux-Arts, France

SYNOPSIS. Further structural lightness can be obtained in self tensioning configurations having dissociated compression members, by the reduction of the length of rigid struts relative to the size of tension elements. In an all space filling equi-partition, or packing, this relation is approximately two and in any case the size of compression members is more than $\sqrt{2}$ times greater than that of the tension elements. This limit can be reached only in a non-compact space-packing or hypertorus, having neighbouring prismatic subsets, where by the application of a torsional effect, the de-rotation of the originally anti-prismatic self tensioning simplex can be redressed. Three such configurations are presented here.

MINIMISATION OF STRUCTURAL WEIGHT

1. Normal space-frames are based on simple and regular bodies; pyramids, prisms, anti-prisms, regular polyhedra etc; forming as many stable simplex, kind of constructive molecules, with possibilities to be assembled in linear, planar or 3D arrangements; making masts, double or multiple layer space-packing (ref. 1).

2. Beyond their morphological characteristics, these structures can be classified according to the statical behaviour of their components, allowing evaluation of the efficiency of each structure. There are two main categories for these structures. The undifferentiated structures composed only of rigid elements working as efficiently in tension as in compression; even if some elements are never stressed in compression. The differentiated structures are composed not only from rigid struts but also by flexible tension members, thus achieving a greater economy in self-weight.

3. These differentiated structures use tension members in variable proportions. The most primitive frameworks have only some bracing ties, an overwhelming number of the elements remaining in compression. But in the most developed self-tensioning simplex, having an anti-prismatic shape, the proportion of rigid components to flexible ones can be decreased to one to three; a quarter of all structural members. In an all space filling space-packing the proportion can be reduced even to one to four as shown in Fig. 1.

4. On first sight, there is no way to achieve improvements beyond this limit. However, without being able to change their relative number, in some condition, there is a possibility to redress the shape of the anti-prism in a prism and reduce the length of the compression members relative to the length of the tension members. This providing an additional step towards minimisation.

HYPER-POLYHEDRA AND HYPERTORUS

5. In any self tensioning structure, this ratio comes close to two, or even more, the rigid strut being the diagonal of a non-planar rhombus; the lowest boundary of this ratio can be reached when the strut becomes the diagonal of a square.

6. The initial idea to obtain this limit is based on the utilisation of self-tensioning hyper-polyhedra built up by anti-prismatic simplex around an empty central polyhedron (ref. 2). Among the seven, or together with the allomorphs, eighteen hyper-polyhedra; at least three can form the composition of a non-compact periodic space-packing. This is also called in the morphological literature, loose-packing, infinite polyhedron, space labyrinth or sponge-space (refs 3-4). Topologically, having a set of holes, they form a multitorus. Therefore, these self-tensioning structures will be called "hypertorus".

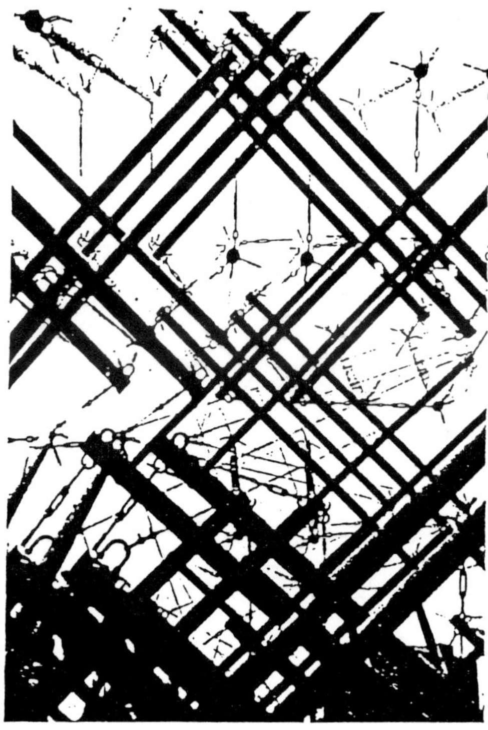

Fig. 1. The most regular self-tensioning space packing.

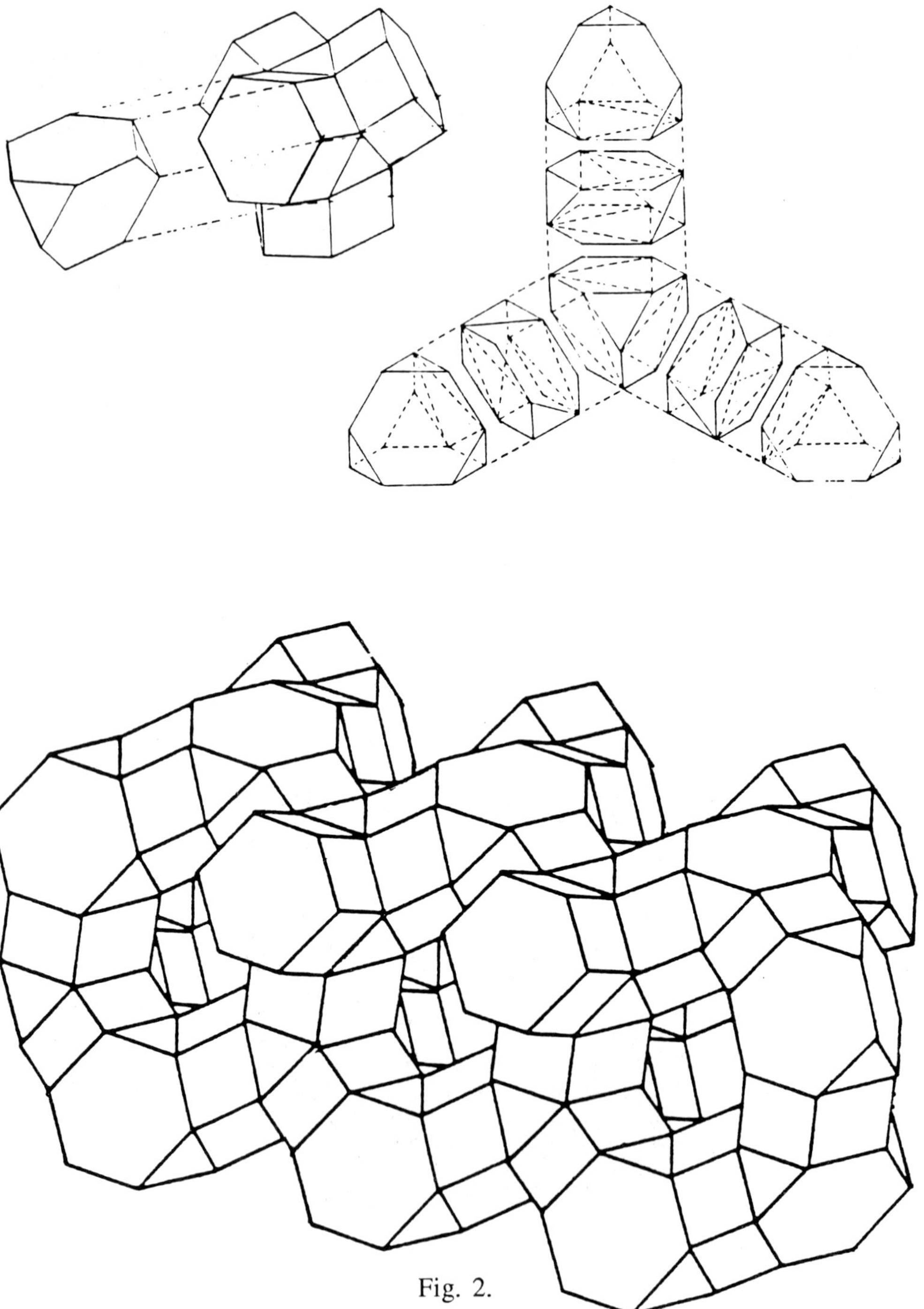

Fig. 2.

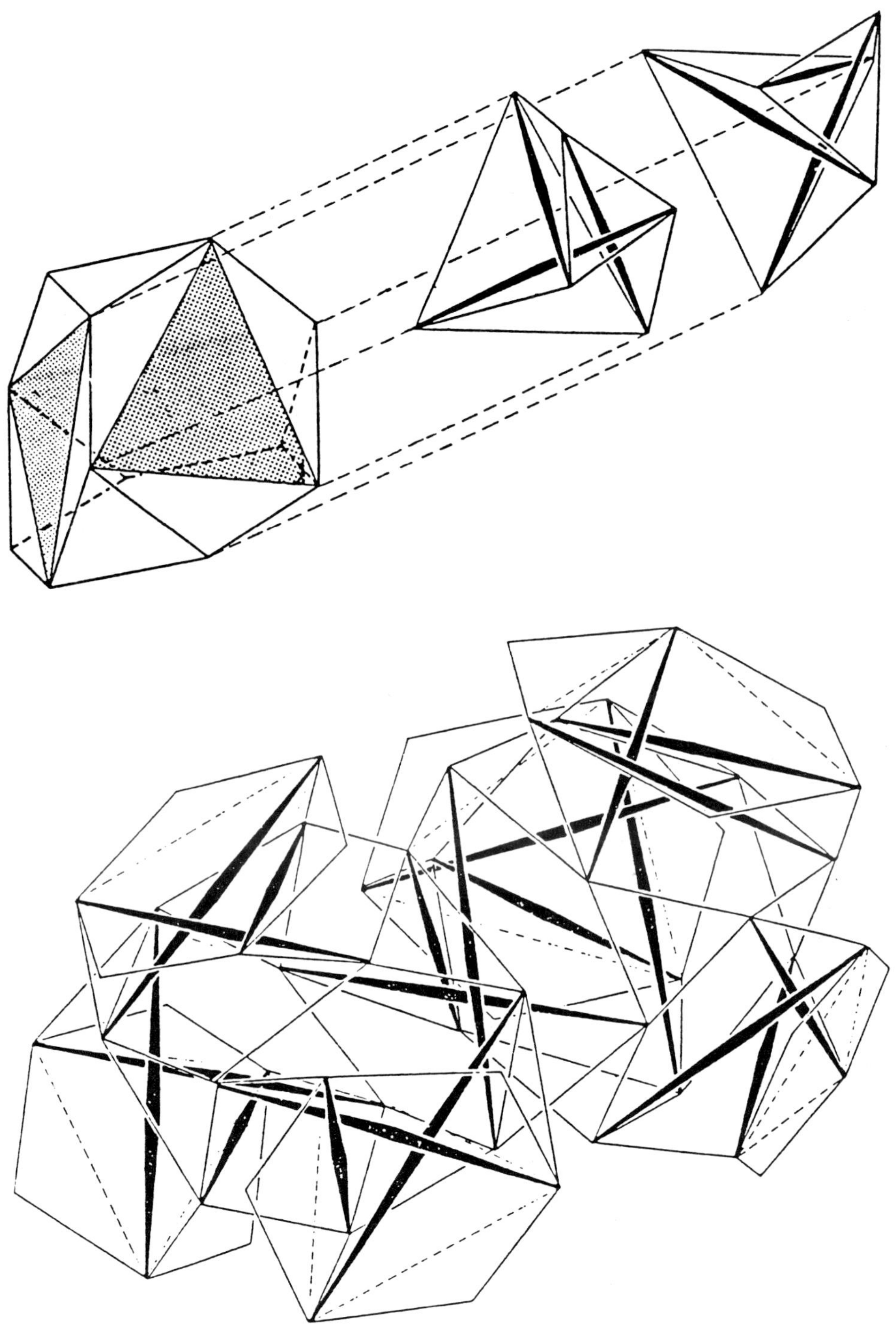

Fig. 2. (continued)

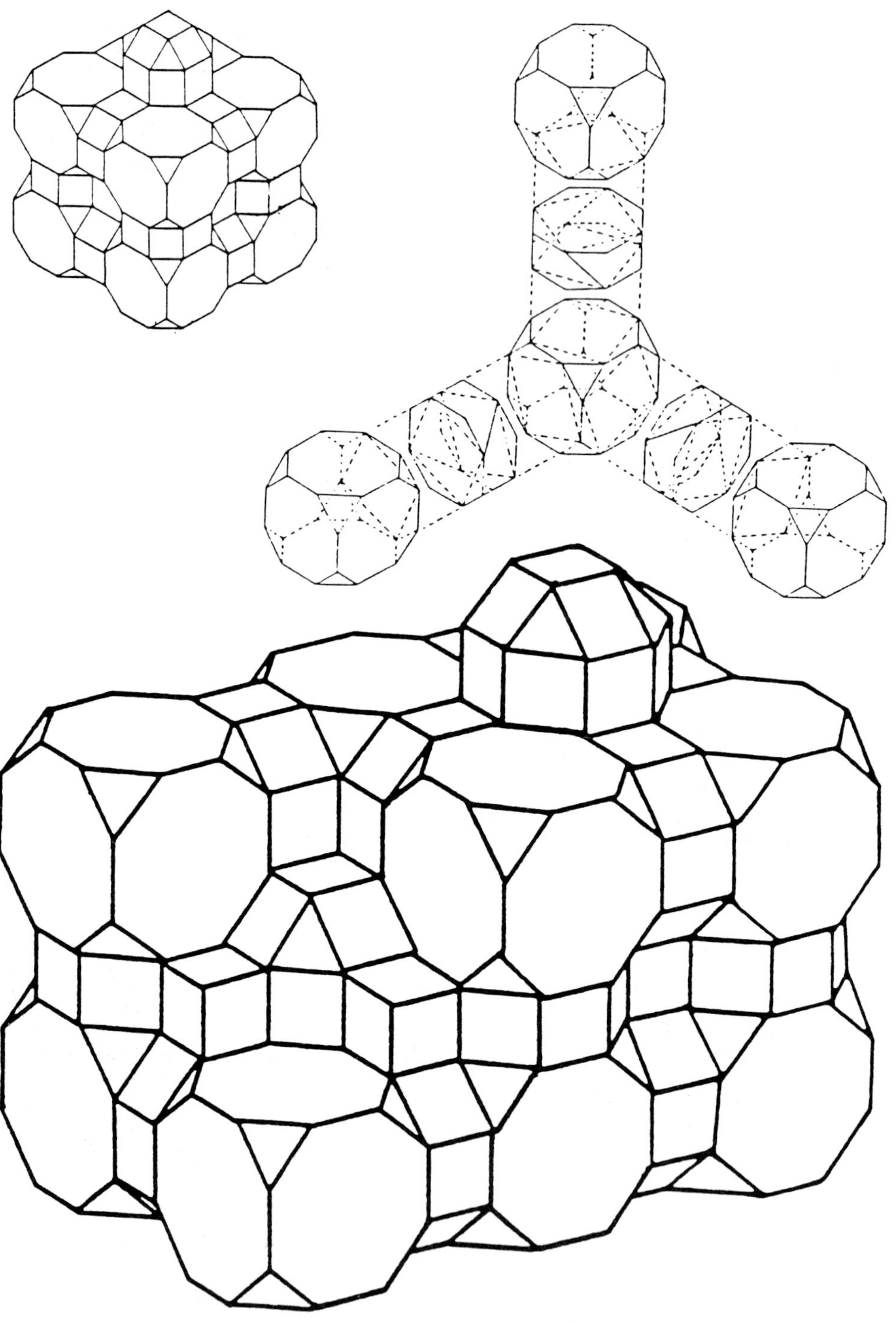

Fig. 3.

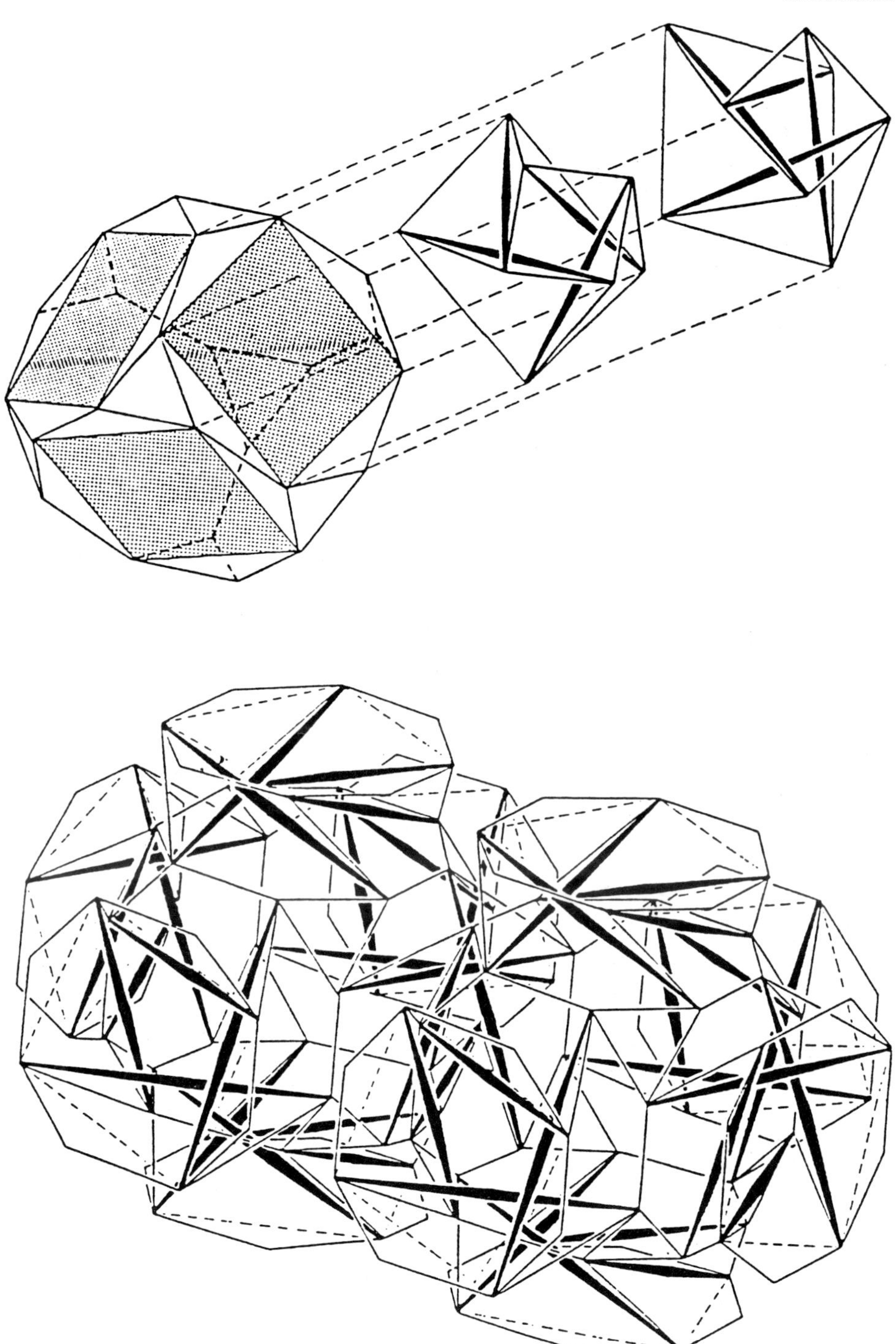

Fig. 3. (continued)

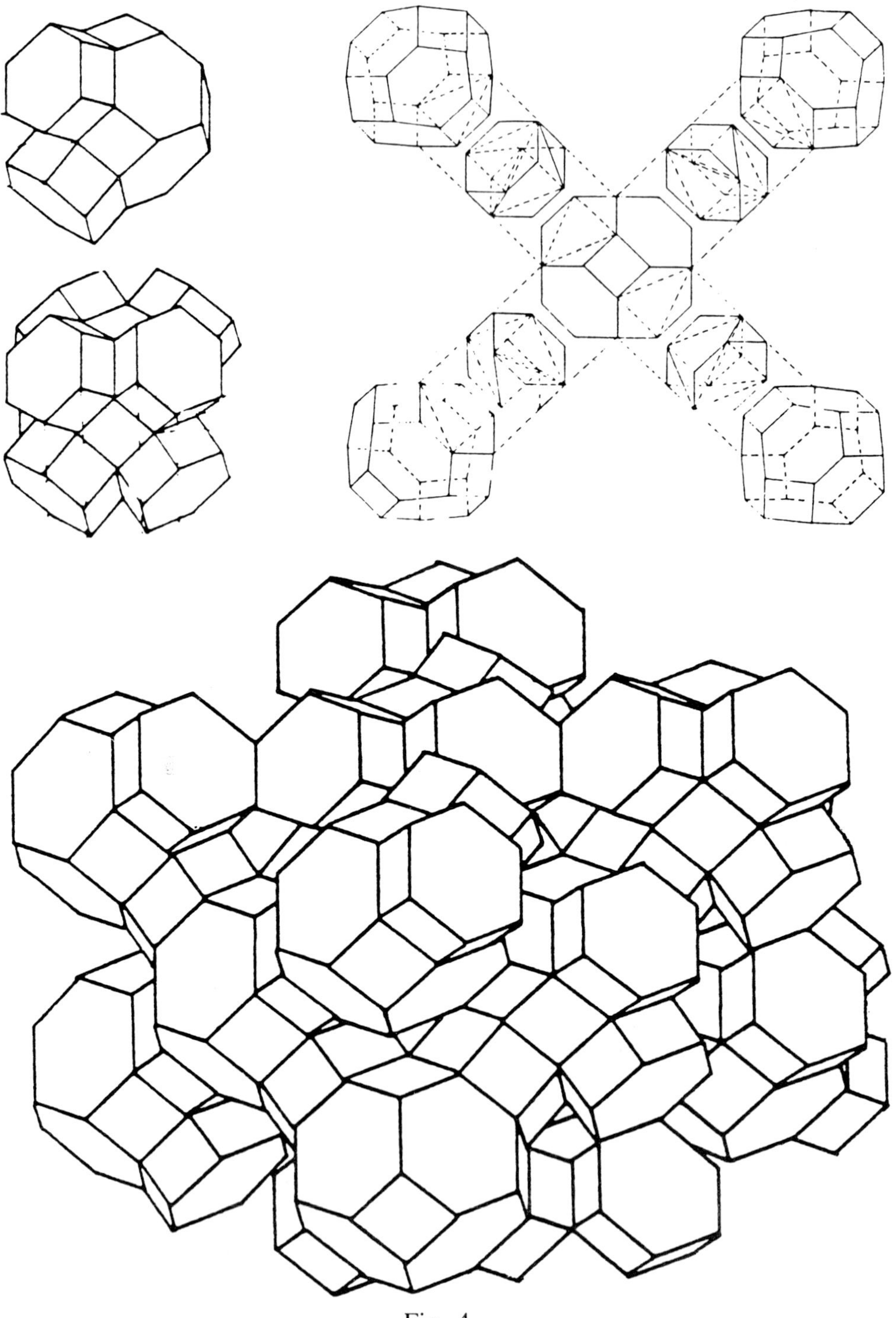

Fig. 4.

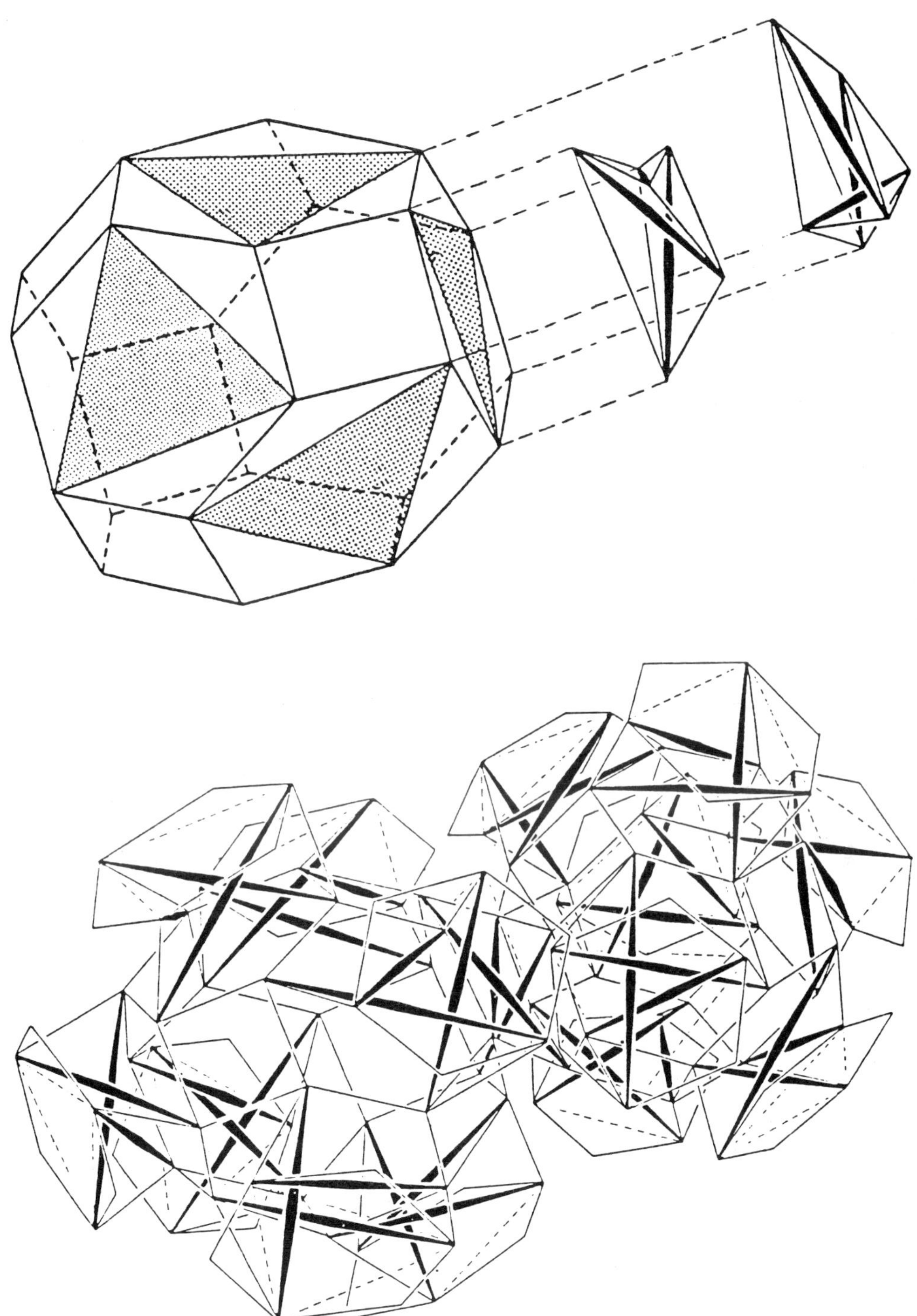

Fig. 4. (continued)

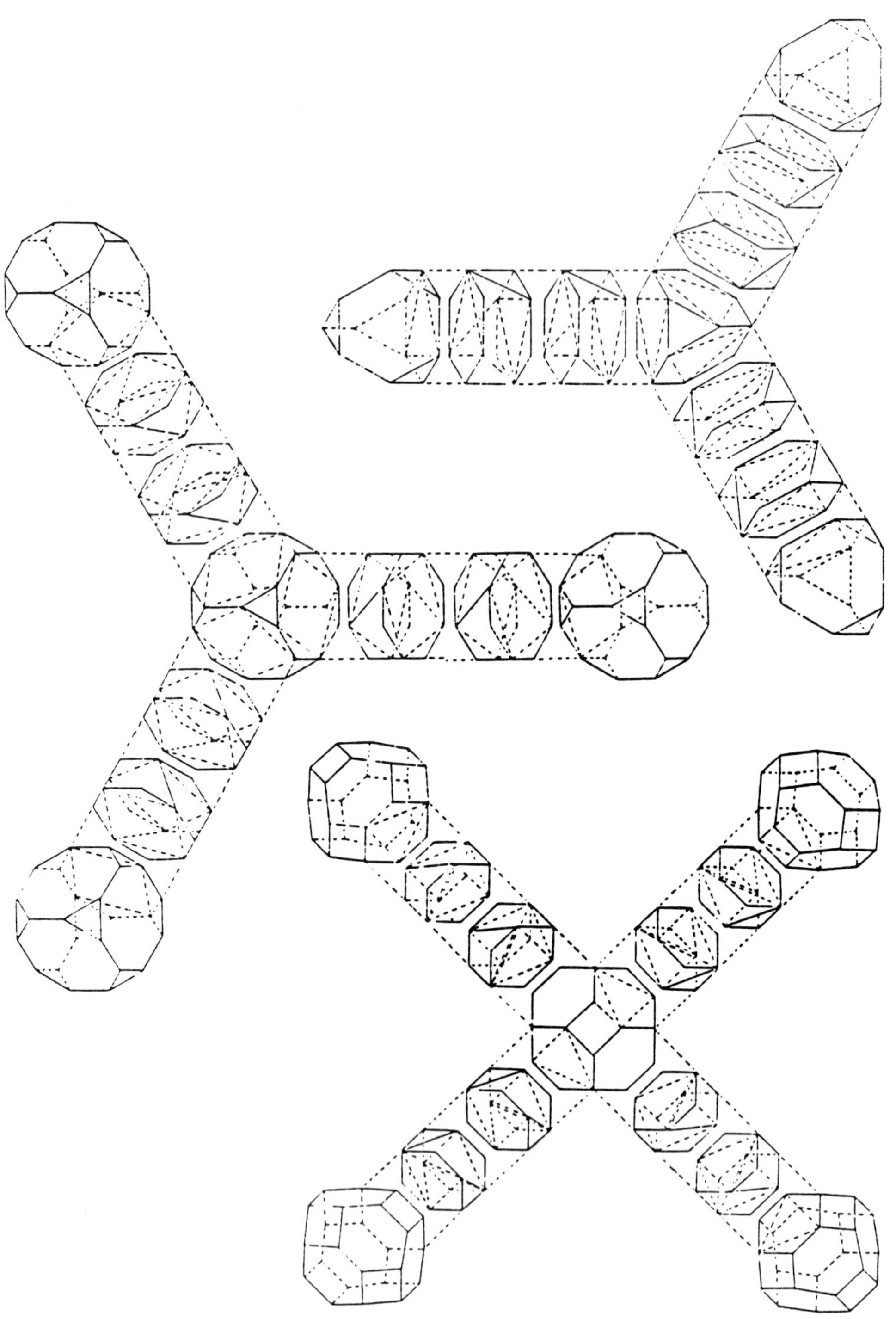

Fig. 5.

7. Effectively, the redressing operation, transforming the shape of the anti-prismatic simplex to a prismatic one, and their rhombus side loop to a square, it is possible only in the context of a non-convex structural arrangement, in a torus or in a hypertorus, where a specific moebiusian torsion is applicable. This torsion will balance the internal rotation of the anti-prismatic self-tensioning simplex, and transform it into a prismatic one, which on its own would not be stable. This redressed state is the extreme limit having both the advantage to reduce the length of the struts to $\sqrt{2}$ relative to the length of tension bracing, and at the same time creating a completely empty structural system, where any components are situated on the periphery.

8. Practically, there exists only three homogeneously contractible self-tensioning hypertorus, each composed continuously by empty prisms having square facets and diagonal $\sqrt{2}$ struts as a modular length. These hypertorus are defined by their connecting hyper-polyhedral core and the related symmetry order of the derived space lattice. The corresponding self-tensioning hypertoroidal structures are built up as follows:

Hyper-truncated tetrahedron core with hexagonal prisms; having four way connections in a diamond lattice symmetry (Fig. 2).
Hyper-truncated cubus core with octagonal prisms; having six way connections in a cubic lattice symmetry (Fig. 3).
Hyper-tetrakaïdecahedron core with hexagonal prisms; having eight way connections in a rhombododecahedral lattice symmetry (Fig. 4).

9. Figs 2 to 4, show the morphological data: the axonometrical representation of the basic polyhedra, the schema of their assemblage and the spatial arrangement of the packing; and also the structural composition of the self-tensioning hyper-polyhedron and the schema of their connection to form a hypertorus.

10. Without changing the basic morphology, the effect of minimisation can be somehow exalted by the superposition of two or several prismatic simplex, creating a stretched out oblong connecting mast, giving on a higher level a 3D system, looking like an indifferentiated structure at first sight, but, however, a self-tensioning one (Fig. 5). By this asymptotic process, structural lightness, which in the case of the three hypertorus seems to reach an absolute limit, can approach zero .

REFERENCES

1. EMMERICH D.G., Structures Tendues et Autotendantes. Edition: Paris-la Villette. Paris.1988.
2. EMMERICH D.G., Self-tensioning Spherical Structures. International Journal of Space Structures. Vol. 5/3-4. 353 - 374. Multi-Science.1990.
3. WACHMAN A., BURT.M., and KLEINMANN M., Infinite Polyhedra. Technion-Israel Institut of Technology. Haifa.1974.
4. LALVANI H., Morphological Aspects of Space Structures. Studies in Space Structures. H. Nooshin. Multi-science. 1991.

109. Analysis of the properties of integrally tensioned tensegrity domes

M. NESTOROVIC, University of Belgrade, Yugoslavia

SYNOPSIS. This paper details the analysis of a tensegrity structure with inclined bars under pressure between cables of upper and lower levels tensioned together with diagonals (Fig. 1). The cables of the tensegrity dome are linked with the contour framework on a square base. Counterbraces of the frame are incorporated in the tensegrity structure ensuring their stability. Results of the analysis show how tensegrity structures behave under prestressing and other loads, and serve to explain the phenomenology of these structures in general.

1. The intensity of forces of the primary (initial) prestressing, introduced into the structure with its own weight only, are determined in order to achieve the prescribed deformability, expressed by the maximum permitted deflection, by analysing all possible load combinations. In the first chosen geometrical configuration of the structure the horizontal components of forces in cables between levels were too large. This led to a correction of the geometry of the section between the levels by replacing h by hl (Fig. 2), which produced much smaller forces in elements caused by the primary (initial) prestressing and weight. The structural deformations under these loads is not high (0.10m - 0.17m), which is acceptable in regard to the spans 34.8 x 34.8 m, since this amounts to L/200.

2. The initial (reference) state of the prestressed tensegrity structure is defined by the initial position of the nodal configuration of the structure under the action of prestressing and weight. By applying a set of chosen prestressing forces in combination with the weight of the structure, a deformed state of the structure is obtained. The reference position of the nodes in the configuration was adopted after repeating the analysis, nine times, which brought into accord forces and deformations.

3. Six load cases were analysed (Fig. 3), as well as possible optimisation of the structure in regard to the stiffness. The analysis has shown that the dominant load cases are (1) self-weight, (2) snow loading, (4) positive temperature change and (7) non-symmetrical snow loading with optimisation of the stiffness of the contour.

 Space structures 4. Thomas Telford, London, 1993

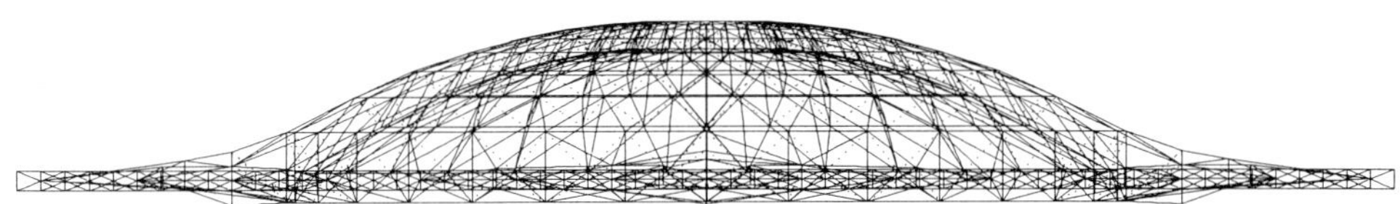

Fig. 1. Plan and diagonal elevation of the analysed structure.

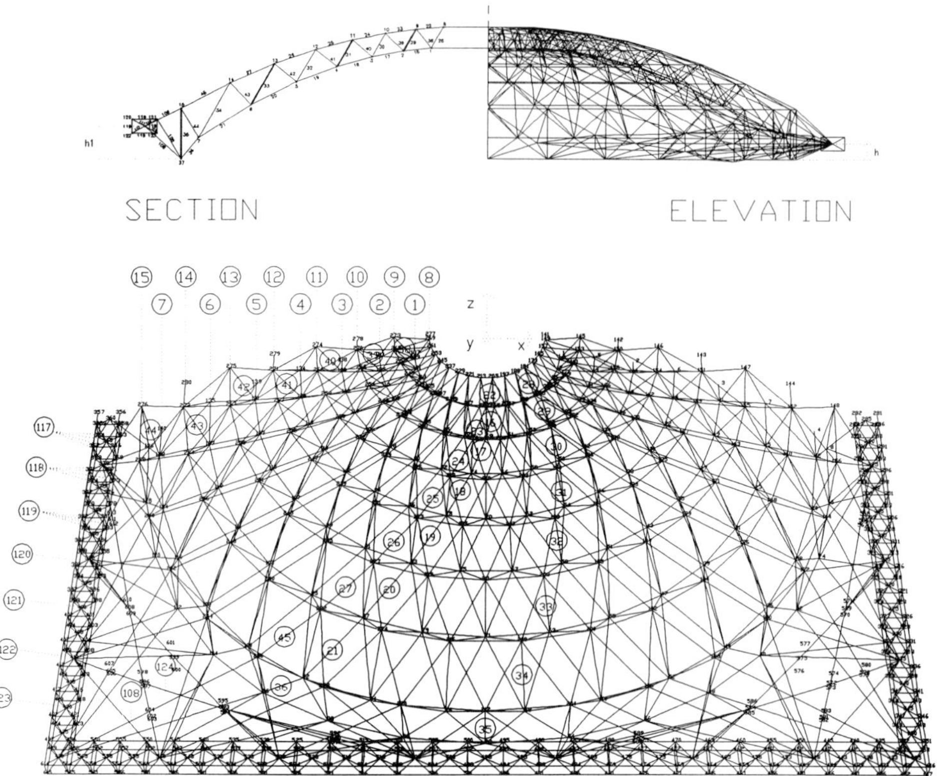

Fig. 2. Perspective drawing of the analysed structure with element group numbers.

4. The results of the analyses confirm that the chosen prestressing forces were suitable for all the load cases. Comparison of forces of the primary case (prestressing and weight) with forces of the other five load cases shows the measure of their influence on changes of forces and deformations in the total analysis. The primary case is dominant in regard to forces, and the nodal positions under these loads were adopted as reference positions. Sensitivity of the system is defined by the ratio of the extreme of the forces (or deformations) of the considered load case and the extreme of the forces of the primary case. The values of extreme forces for element groups of the ten analysed groups (from 54 appearing in the structure) show that they differ very little, except near the contour, where differences are considerable (Fig. 4.).

5. The diagram of maximum displacements for the same load groups Fig. 5. shows similarity between deformations of the tensegrity structure and the contour

node group	Fi(32) x10^6 (mm^2)	Qi(S.T.) (KN)	Qi(S.) (KN)	Qi(S.T.+S.) (KN)	Qi(W.) (KN)	Qi(S.T.−W.) (KN)	Qi(S.T.+N.S.) (KN)
8	1.14	−0.3	−1.2	−1.5	0.5	0.2	combination loads
9	1.67	−0.4	−1.6	−2.0	0.7	0.3	
10	2.69	−0.7	−2.8	−3.5	1.1	0.4	
11	3.99	−1.0	−4.0	−5.0	1.3	0.3	
12	5.63	−1.4	−5.6	−7.0	1.4	0.0	
13	7.70	−1.9	−7.6	−9.5	1.9	0.0	
14	10.29	−2.5	−10.0	−12.5	1.6	−0.9	
15	12.30	−3.1	−12.4	−15.5	1.0	−2.1	

Qi(S.T.+S.)=0.125*10^4Fi=......N =...........kN

Qw=0.8 KN/m^2*Fi=

Qs=1.0 KN/m^2*Fi=

αT = temperature coefficient (0.000012)

ΔT = temperature increment (relative −ΔT = −46.2 c°
+ΔT = +21.8 c°)

Table of chosen groups for the analysis of the phenomenology of the structure

gr. type	group name	groups in the structure
1	lower ring	1,2,3,4,5,6,7,37
2	upper ring	8,9,10,11,12,13,14,15
3	lower belt	16,17,18,19,20,21,36
4	uper belt	22,23,24,25,26,27,45
5	diagonals	38,39,40,41,42,43,44
6	counterbraces	28,29,30,31,32,33,34,35
7	elements betwenn contours	108
8	diagonals of the contour	117,118
9	external, internal and connecting elements of the contour	119,120,121,122,123
10	counterbraces of the contour	124

TOTAL: 10 TYPES TOTAL: 54 GROUPS

Types of external loads

i = A(S.T)
II = B(S.T.+S)
III = C(S.T.+W)
IV = D(S.T.+ T)
V = E(S.T.− T)
VI = F(S.T.+N.S)
VII = G(S.T.+N.S.O)

TOTAL: 7 LOADS

Fig. 3. Tabular survey of loads at the nodes of the upper level of the dome (weight, snow, wind and their combinations).

structure. In non-symmetrical cases of snow load and of the same load after the accomplished stiffness optimisation, there were considerable differences in displacements of the tensegrity structure.

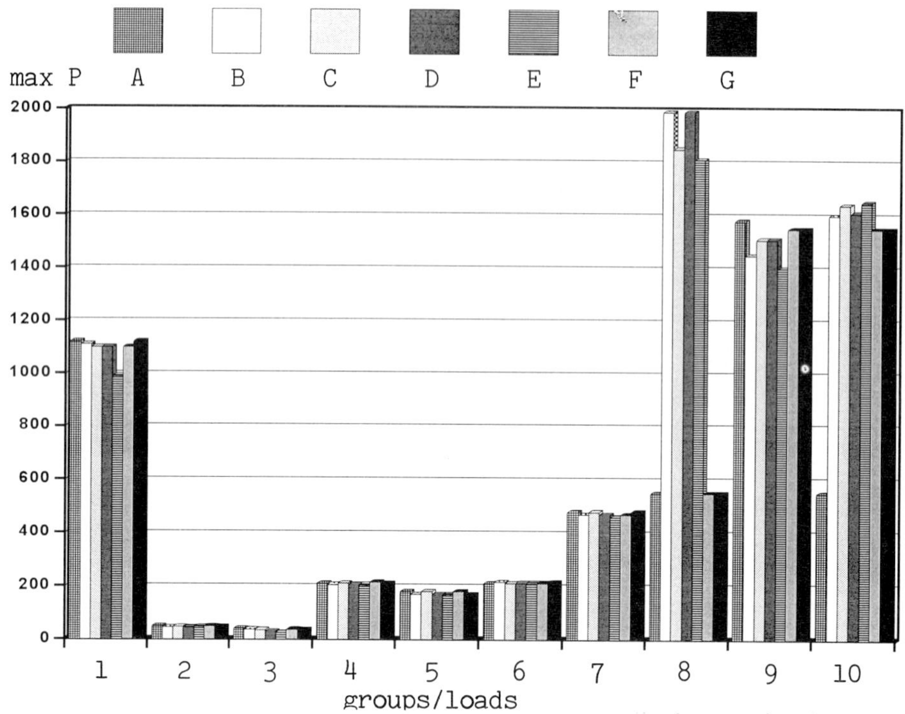

Fig. 4. Diagram of extreme influences of forces according to analysed groups in relation to the considered load.

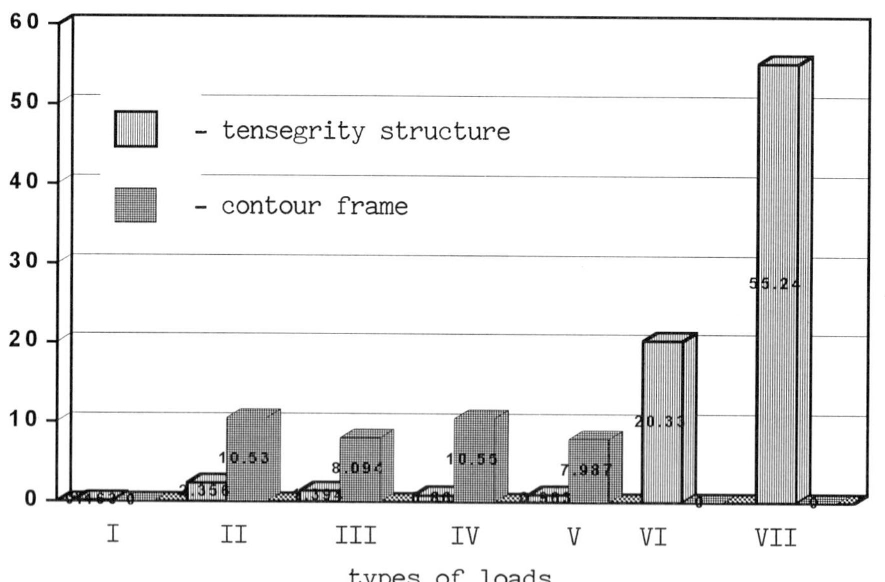

Fig. 5. Diagram of maximum displacement of given loads.

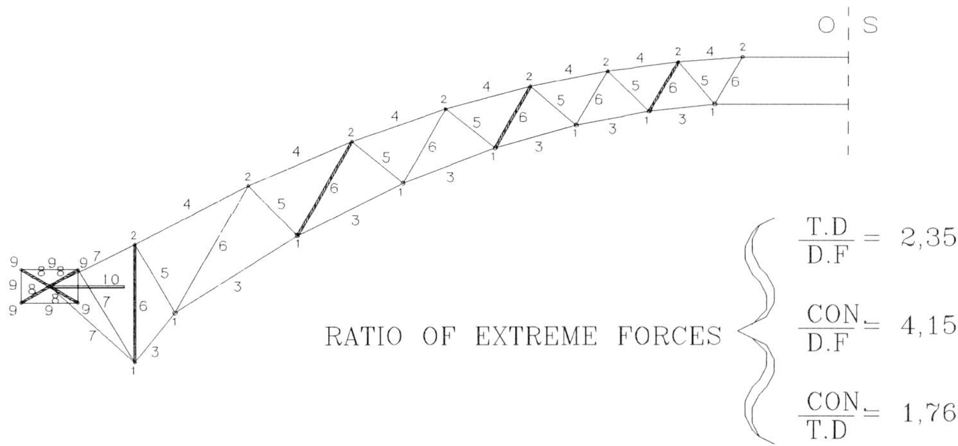

T.D–tensegrity dome

D.F–part between dome and contour frame

CON–contour

C– cable

T– tube

TYPE GROUP	MIN.	MAX.	TYPE	Ø mm	%	Σ GROUP	MAX. DISP.	
1	V	I, VII	0	3-20	3,82	7	IV, VII	T.D
2	IV	I	0	3-20	35,61	2		
3	V	I	0	3-20	115,55	1		
4	II	I, VI	0	3-20	5,82	5		
5	III	I, II	0	3-20	10,82	3		
6	III	I, II	1	38-76,1	6,45	4		
7	V	I, VII	0	20	7,18	6		D.F
8	VII	II, IV	1	193,7	265,08	10	II, IV	CON.
9	V	I	1	193,7	11,97	8		
10	I	V	1	193,7	198,0	9		

Fig. 6. Tabular survey of the sensitivity of the system.

6. The load cases that produce extreme forces (minimum and maximum) are shown in the tabular survey of the sensitivity of the structure for the ten chosen element groups (Fig. 6). In the tensegrity structure the largest forces appear in group 1 (forces in lower rings for load cases (1) prestressing and self-weight and (7) non-symmetrical snow load with accomplished stiffness optimisation). In the contour the largest forces appear in group 8 (forces in diagonals for load cases (2) symmetrical snow loading and (4) positive temperature change). In the table the types of elements are designated by O (cable) and 1 (tube), as well as their sections. The column denoted by % at the top of the table shows the percentage difference between the largest and the smallest extreme forces (all load cases taken into account). The next column gives the sequence of chosen groups according to the values of previously determined differences. The last column shows that maximum displacement of the tensegrity structure occurs for load cases (4) positive temperature change, and (7) non-symmetrical snow loading with accomplished stiffness optimisation, and of the contour in load cases (2) symmetrical snow load and (4) positive temperature change.

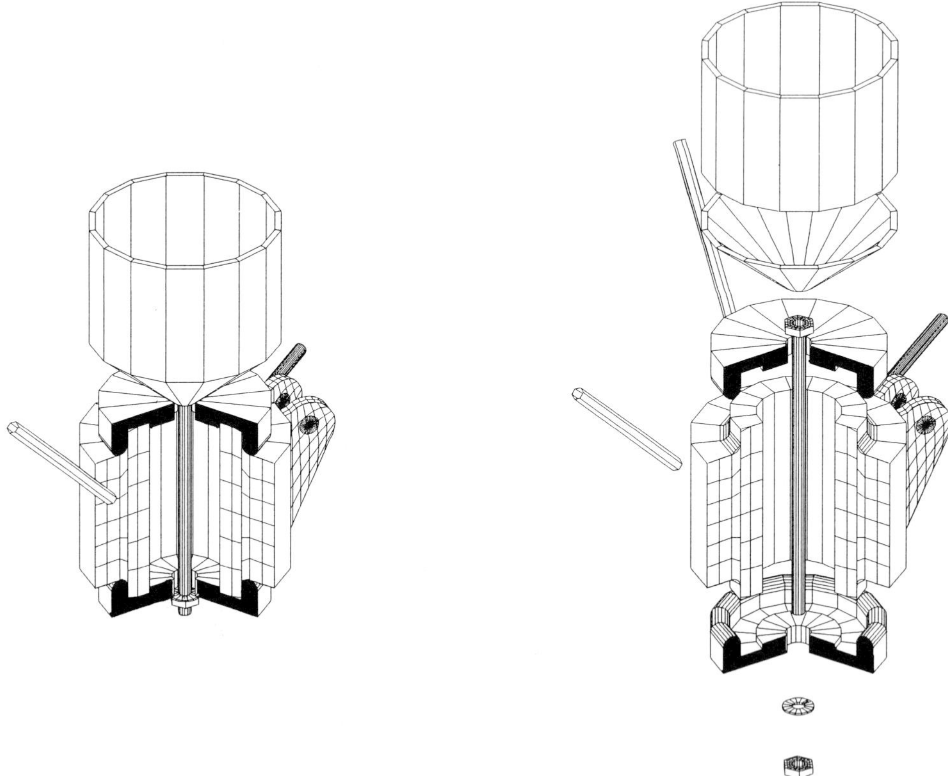

Fig. 7. Assembly (a) and disassembled state, (b) of the analysed typical node of the tensegrity structure (1- plate, 2- segment and 3- nose).

Fig. 8. Survey of the phases of mounting of the tenesgrity structure.

7. The typical node, developed for the considered tensegrity structure, may be regarded as a combination of the well-known IFI and BBRV nodes (Fig. 7). The geometry and structure of the node makes possible in-plane rotations of tensioned elements, while all forces remain centred in the node. The finite element analysis for each of its parts of the node was undertaken (1- plate, 2-segment and 3-nose).

8. Mounting and prestressing of the structure can be performed gradually - belt after belt, thus obtaining autonomous and stable structures after the mounting of each belt (Fig. 8).

9. On the basis of results of the analysis of the tensegrity structure it may be concluded that the influence of the deformation of the pneumatic roofing on the bearing structure is not significant (Fig. 9). The stiffness of pneumatic cushions is small and the contribution to the overall stiffness of the tensegrity structure may be disregarded.

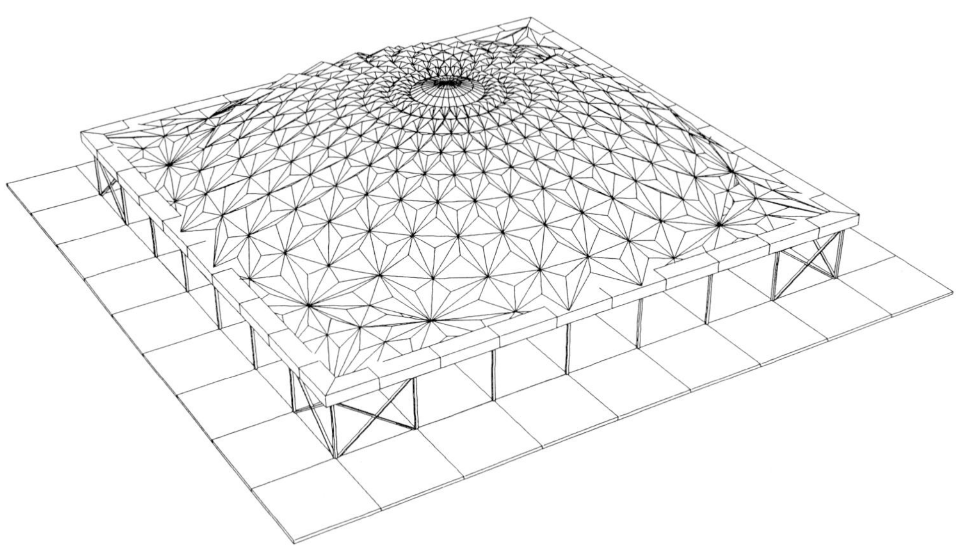

Fig. 9. Perspective drawing of the building with pneumatic roofing.

REFERENCE

1. NESTOROVIC M., Integrally tensioned (tensegrity) structure systems with momentless contours. Doctoral dissertation, Faculty of Architecture University of Belgrade, 1990.

110. Shape of tensegrity frames with an optimum rigidity

PEI SHAN CHEN, MASARU ABE and M. KAWAGUCHI, Hosei University, Japan

SYNOPSIS. The tensegrity dome is usually very deformable, especially when it is subjected to an unsymmetrical loading, and it is known that the deformation is sensitive to its erection (initial) shape. Based on this, the present paper gives a method to find the shape of minimum deformation and optimum distribution of prestresses in a tensegrity frame.

INTRODUCTION

1. To make a rigid tensegrity frame, a high prestressing force is required. But, this is not economical, and will make the construction more difficult. For a tensegrity frame with a certain configuration and subjected to a certain loading system, the nodal displacements and the rigidity is sensitive to its erection shape and prestressing. It is therefore important to find the erection shape and prestresses to give optimum rigidity.

2. The 'optimum shape' denotes a shape after prestressing and the 'deformation' is that due to loading only. A non-linear programming, gradient projection method, is adopted in the paper to minimise the nodal displacements.

GENERALISING THE PROBLEM

3. Generally, the displacement vector D is a function of the nodal co-ordinate vector X, the vector of cross-sectional areas of members A, and the vector of prestressed internal forces P. In a non-linear analysis, the nodal displacements are usually expressed in implicit functions. For a tensegrity frame with m members and $\tilde{n} = n/3$ nodes, via the composite method (ref. 1) the implicit functions are the equilibrium of the structure.

$$\sum_{k=1}^{m} \frac{\partial \delta_k}{\partial D_i} \left(\frac{E_k A_k}{L_k} \delta_k + P_k \right) - Q_i = 0 \ , \qquad (i = 1, 2, \cdots, n) \qquad (1)$$

where:

$D = \{D_i\}$	: displacement vector;
$A = \{A_k\}$	: cross-sectional area of members;
E_k	: Young's modulus for members;

$P = \{P_k\}$	: internal forces induced by prestressing;
Q_i	: nodal loads;
L_k	: length of members before loading;
δ_k	: change in length of the members due to loading.

and:

$$N_k = \frac{E_k A_k}{L_k} \delta_k + P_k \qquad (k = 1, 2, \cdots, m) \tag{2}$$

indicates the internal forces introduced by prestressing and external loading.

4. To minimise a vector, means to minimise the module of the vector, and minimising the nodal displacement vector is equivalent to minimising the module of the displacement vector. So, taking the module of:

$$F = \sum_{i=1}^{n} D_i^2 \tag{3}$$

as the objective function to be minimised, and taking X, A and P as unknown vectors to be solved by optimisation. Hence the problem can generally be expressed as:

Minimize $F = \sum_{i=1}^{n} D_i^2$

Domain Equation:

$$\sum_{k=1}^{m} \frac{\partial \delta_k}{\partial D_i} \left(\frac{E_k A_k}{L_k} \delta_k + P_k \right) - Q_i = 0 \qquad (i = 1, 2, \cdots, n) \tag{1}$$

Subject to

Geometrical Constraint:

$$\phi_j (X, A, P) = 0 \qquad (j = 1, 2, \cdots, s) \tag{4}$$

$$h_l (X, A, P) \leqq 0 \qquad (l = 1, 2, \cdots, r) \tag{5}$$

Mechanical Constraint:

$$\phi_j (X, A, P) = 0 \qquad (j = 1, 2, \cdots, t) \tag{6}$$

$$g_l (X, A, P) \leqq 0 \qquad (l = 1, 2, \cdots, q) \tag{7}$$

ABOUT THE CONSTRAINTS

5. Generally, the geometrical and mechanical constraints can be divided into active constraints (equalities) and inactive constraints (inequalities).

6. Some of the geometrical constraints may be given by architectural requirements or conditions for some specific purposes. In fact, tensegrity frames are often geometrically symmetrical structures. In the optimisation, however the nodal co-ordinates and member sections change, the values should be

symmetrical, if one wants to get a symmetric structure. It is necessary to control all of the joints between two surfaces which are defined beforehand, such that the joints cannot become too high or too low.

7. The mechanical constraints are always given according to the mechanical principles. Firstly, for a tensegrity frame which is, as is well known, constituted by cables and struts, the internal forces in the cables should be taken as inactive constraints. However, the shape may change, the internal forces introduced by prestressing must be in equilibrium without loads, which are the active constraints.

THE DESCENT DIRECTION OF THE OBJECT FUNCTION

8. We use the gradient projection method of Rosen (1960) (refs 2 and 3). First inducing the descent direction of the objective function.

9. Differentiating the domain equation with respect to the nodal co-ordinates, gives:

$$\frac{\partial}{\partial X_j}\left\{\sum_{k=1}^{m}\frac{\partial \delta_k}{\partial D_i}\left(\frac{E_k A_k}{L_k}\delta_k + P_k\right) - Q_i\right\} = 0 \quad (i, j = 1, 2, \cdots, n) \quad . \tag{8}$$

Hence

$$\sum_{k=1}^{m}\left\{\frac{E_k A_k}{L_k}\frac{\partial \delta_k}{\partial D_i}\left(\frac{\partial \delta_k}{\partial X_j} + \sum_{q=1}^{n}\frac{\partial \delta_k}{\partial D_q}\frac{\partial D_q}{\partial X_j} - \frac{\delta_k}{L_k}\frac{\partial L_k}{\partial X_j}\right)\right.$$

$$\left. + \left(\frac{E_k A_k}{L_k}\delta_k + P_k\right)\left(\frac{\partial^2 \delta_k}{\partial D_i \partial X_j} + \sum_{q=1}^{n}\frac{\partial^2 \delta_k}{\partial D_i \partial D_q}\frac{\partial D_q}{\partial X_j}\right)\right\} - \frac{\partial Q_i}{\partial X_j} = 0 \tag{9}$$

Eq. 9 may be rewritten as:

$$\sum_{q=1}^{n} T_q^i\left(\frac{\partial D_q}{\partial X_j}\right) + C^{ij} = 0 \qquad (i = 1, 2, \cdots, n) \qquad , \tag{10}$$

where

$$T_q^i = \sum_{k=1}^{m}\left[\frac{E_k A_k}{L_k}\frac{\partial \delta_k}{\partial D_i}\frac{\partial \delta_k}{\partial D_q} + N_k\frac{\partial^2 \delta_k}{\partial D_i \partial D_q}\right] \tag{11}$$

$$C^{ij} = \sum_{k=1}^{m}\left[\frac{E_k A_k}{L_k}\frac{\partial \delta_k}{\partial D_i}\left(\frac{\partial \delta_k}{\partial X_j} - \frac{\delta_k}{L_k}\frac{\partial L_k}{\partial X_j}\right) + N_k\frac{\partial^2 \delta_k}{\partial D_i \partial X_j}\right] - \frac{\partial Q_i}{\partial X_j} \quad . \tag{12}$$

We can obtain $\left\{\frac{\partial D_q}{\partial X_j}\right\}$ (j = 1, 2, ..., n) by solving Eq.10.

10. By differentiating the domain equation, we obtain

$$\frac{\partial}{\partial A_t}\left\{\sum_{k=1}^{m}\frac{\partial \delta_k}{\partial D_i}\left(\frac{E_k A_k}{L_k}\delta_k + P_k\right) - Q_i\right\} = 0 \qquad (i = 1, 2, \cdots, n) \tag{13}$$

$$\sum_{k=1}^{m}\left(N_k \sum_{q=1}^{n}\frac{\partial^2 \delta_k}{\partial D_i \partial D_q}\frac{\partial D_q}{\partial A_t} + \frac{E_k A_k}{L_k}\frac{\partial \delta_k}{\partial D_i}\sum_{q=1}^{n}\frac{\partial \delta_k}{\partial D_q}\frac{\partial D_q}{\partial A_t}\right) + \frac{E_t}{L_t}\frac{\partial \delta_t}{\partial D_i} = 0 \tag{14}$$

Eq. 14 takes the form:

$$\sum_{j=1}^{n}{}'T_j^i\left(\frac{\partial D_j}{\partial A_t}\right) + H^{it} = 0 \qquad (i = 1, 2, \cdots, n) \tag{15}$$

where

$$H^{it} = \frac{\partial \delta_t}{\partial D_i}\frac{E_t}{L_t}\delta_t \tag{16}$$

$\left\{\frac{\partial D_j}{\partial A_t}\right\}$ (t = 1, 2, ..., m) can be obtained by solving Eq.15.

11. Also by differentiating the domain equation,

$$\frac{\partial}{\partial P_t}\left\{\sum_{k=1}^{m}\frac{\partial \delta_k}{\partial D_i}\left(\frac{E_k A_k}{L_k}\delta_k + P_k\right) - Q_i\right\} = 0 \qquad (i = 1, 2, \cdots, n) \tag{17}$$

Hence

$$\sum_{k=1}^{m}\left(N_k \sum_{q=1}^{n}\frac{\partial^2 \delta_k}{\partial D_i \partial D_q}\frac{\partial D_q}{\partial A_t} + \frac{E_k A_k}{L_k}\frac{\partial \delta_k}{\partial D_i}\sum_{q=1}^{n}\frac{\partial \delta_k}{\partial D_q}\frac{\partial D_q}{\partial A_t}\right) + \frac{\partial \delta_t}{\partial D_i} = 0 \tag{18}$$

Eq. 18 takes the form:

$$\sum_{j=1}^{n}{}'T_j^i\left(\frac{\partial D_j}{\partial P_t}\right) + R^{it} = 0 \qquad (i = 1, 2, \cdots, n) \tag{19}$$

Where

$$R^{it} = \frac{\partial \delta_t}{\partial D_i} \tag{20}$$

$\left\{\frac{\partial D_j}{\partial P_t}\right\}$ (t = 1, 2, ..., m) can be obtained by solving Eq. 19.

12. Take

$$\nabla = \left(\frac{\partial}{\partial X_1}, \cdots, \frac{\partial}{\partial X_n}, \frac{\partial}{\partial A_1}, \cdots, \frac{\partial}{\partial A_m}, \frac{\partial}{\partial P_1}, \cdots, \frac{\partial}{\partial P_m}\right)$$

as the operator, then the descent direction of the object function takes the form:

$$\mathbf{d} = -\nabla^T F = -2 \begin{pmatrix} \partial D_1/\partial X_1 & \cdots & \cdots & \partial D_n/\partial X_1 \\ \cdots & \cdots & \cdots & \cdots \\ \partial D_1/\partial X_n & \cdots & \cdots & \partial D_n/\partial X_n \\ \partial D_1/\partial A_1 & \cdots & \cdots & \partial D_n/\partial A_1 \\ \cdots & \cdots & \cdots & \cdots \\ \partial D_1/\partial A_m & \cdots & \cdots & \partial D_n/\partial A_m \\ \partial D_1/\partial P_1 & \cdots & \cdots & \partial D_n/\partial P_1 \\ \cdots & \cdots & \cdots & \cdots \\ \partial D_1/\partial P_m & \cdots & \cdots & \partial D_n/\partial P_m \end{pmatrix} \begin{pmatrix} D_1 \\ D_2 \\ \cdot \\ \cdot \\ \cdot \\ \cdot \\ \cdot \\ \cdot \\ D_n \end{pmatrix} \tag{21}$$

LINEARISING THE CONSTRAINT AND THE PROJECTION OPERATOR

13. The non-linear constraints can be linearised by differentiating them with respect to the variables X, A and P. This is to find the Jacobian matrix of the constraints of Eqs 4 and 6. At a feasible point k $(X, A, p)^{Tk}$ we obtain:

$$\nabla\phi_i \left\{ \begin{pmatrix} X \\ A \\ P \end{pmatrix} - \begin{pmatrix} X \\ A \\ P \end{pmatrix}^k \right\} = 0 \qquad (i = 1, 2, \cdots, s) \tag{22}$$

for the active geometrical constraints, and

$$\nabla\psi_j \left\{ \begin{pmatrix} X \\ A \\ P \end{pmatrix} - \begin{pmatrix} X \\ A \\ P \end{pmatrix}^k \right\} = 0 \qquad (j = 1, 2, \cdots, t) \tag{23}$$

for the]mechanical constraints. $B = \begin{pmatrix} \nabla\phi_1 \\ \vdots \\ \nabla\phi_s \\ \nabla\psi_1 \\ \vdots \\ \nabla\psi_t \end{pmatrix}$ is the Jacobian matrix of

Eqs 4 and 6, and the linearised Eqs 22 and 23 can be rewritten in matrix form:

$$B \left\{ \begin{pmatrix} X \\ A \\ P \end{pmatrix} - \begin{pmatrix} X \\ A \\ P \end{pmatrix}^k \right\} = 0 \tag{24}$$

14. The constraint space defined by Eqs 4 and 6 is indicated by S .While the space defined by Eq. 24 is the tangential space at point k, we indicate this as $\overline{S}$.

$$P = I - B^- B \tag{25}$$

is the projection operator for an arbitrary vector to be projected onto space $\overline{S}$, where $B^{-1} = B^T(BB^T)^{-1}B$ is the generalised inverse of matrix B (ref. 4). Thus the projection of the descent direction onto the space is given by:

$$\widetilde{d} = - P \nabla^T F \tag{26}$$

15. The next point in space $\overline{S}$ is given by:

$$\begin{pmatrix} \tilde{X} \\ \tilde{\Lambda} \\ \tilde{P} \end{pmatrix} = \begin{pmatrix} X \\ \Lambda \\ P \end{pmatrix}^k + \alpha \widetilde{d} \quad ; \qquad \begin{pmatrix} \tilde{X} \\ \tilde{\Lambda} \\ \tilde{P} \end{pmatrix} \in \overline{S} \tag{27}$$

where α is the so called step length,

$$0 < \alpha \leq \min \left\{ - \frac{\phi_i}{\nabla \phi_i d} \middle| \nabla \phi_i d \; ; \; - \frac{\psi_j}{\nabla \psi_j d} \middle| \nabla \psi_j d \right\} \tag{28}$$

16. However, this point is out of the constraint space, in other words it is not a feasible point. So a special procedure must be devised to make it return to space S. There are many methods that may be chosen for this procedure, we have chosen Newton's method in this paper (ref. 5). It is assumed that the active constraints are non-degeneracy. Matrix B can be partitioned into two parts:

$$B = ({}^r\overset{r}{G}, C) \qquad \text{Rank}(G) = r \tag{29}$$

In accordance with this, the unknown variable vector can also be partitioned into two parts $\begin{pmatrix} \tilde{X} \\ \tilde{\Lambda} \\ \tilde{P} \end{pmatrix} = \begin{pmatrix} \tilde{\chi}_G \\ \\ \tilde{\chi}_c \end{pmatrix}$ Now, take $\tilde{\chi}_c$ as invariable, $\tilde{\chi}_G$ can be calculated by Newton's method,

$$\tilde{\chi}_C' = \tilde{\chi}_C - G^{-1}\{\phi_1, \cdots, \phi_s, \psi_1, \cdots, \psi_t\}^T \tag{30}$$

After this computation, we can obtain the next point in real constraint space. Of course, the value of the objective function at this point should be less than that at the former point $F^{k+1} < F^k$

THE ALGORITHM

17. step 0 : The initial feasible point $(X, A, P)^0$

step 1 : Descent direction of the objective function $d = - \nabla^T F$

step 2 : Linearising the constraints, and calculating B ,B^{-1}, and the projection operator P.

step 3 : The projection $\tilde{d} = Pd$

If $\tilde{d} = 0$, stop the calculation.

step 4 : Define the step length using Eq. 28.

$$\text{Then} \begin{pmatrix} \tilde{X} \\ \tilde{A} \\ \tilde{P} \end{pmatrix} = \begin{pmatrix} X \\ A \\ P \end{pmatrix}^k + \alpha\tilde{d}$$

But, if $\alpha = 0$, stop the calculation.

step 5 : To return to the constraint space by Newton's method, and $k = k + 1$, go to step 1.

AN EXAMPLE

18. The simple tensegrity frame, shown in Fig. 1, is taken as an example for the optimisation. But, in this example, the cross-sectional area of members is taken as constant, and variables to be solved by the optimisation are nodal co-ordinates and the internal forces introduced by prestressing.

19. The geometrical constraints are a symmetrical condition; and the inequality constraints are induced to control the structure to be a dome-like one. The active mechanical constraints are given by considering equilibrium under prestressing, without loads. It is taken as the inequality constraints, that the internal forces in the cables remain positive.

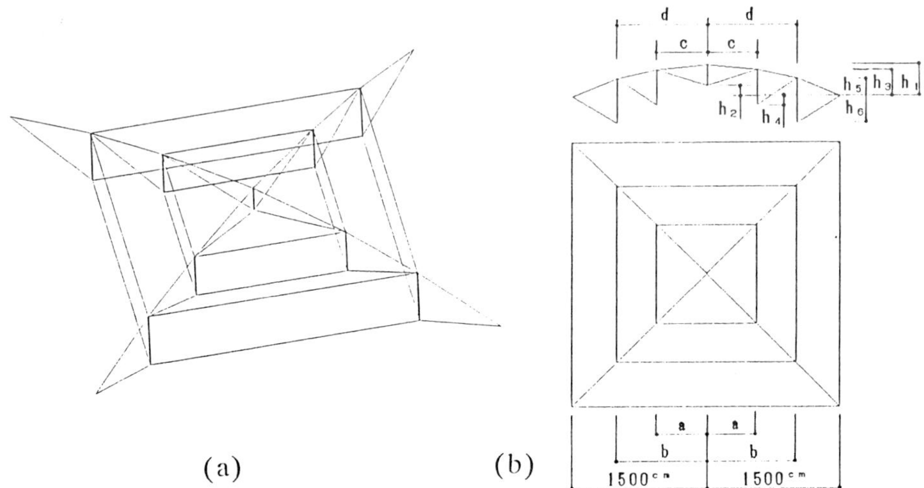

Fig. 1. A numerical example with a square plan.

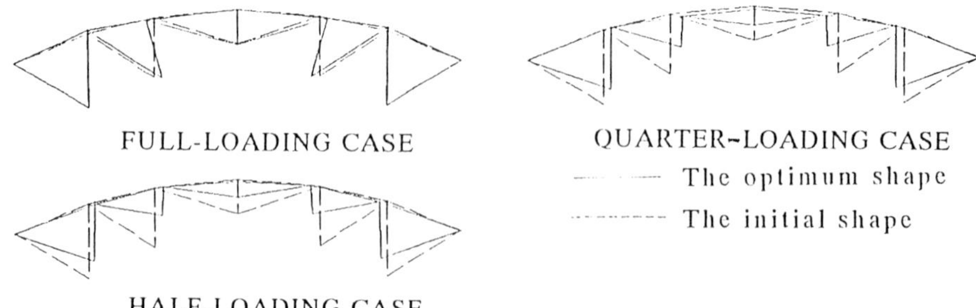

Fig. 2. The section of the optimum shape.

Table 1. The dimensions (cm).

	a	b	c	d	h_1	h_2	h_3	h_4	h_5	h_6
Initial	560.00	1000.00	560.00	1000.00	350.00	120.00	300.00	100.00	200.00	300.00
full-loading	506.62	993.53	613.67	1005.66	348.19	110.99	282.60	102.56	217.89	300.85
half-loading	511.80	962.14	500.52	954.01	331.01	231.41	305.36	113.71	195.45	182.52
quarter-loading	483.86	951.74	481.04	949.27	324.44	221.01	284.19	81.03	212.80	197.77

Table 2. The internal forces introduced by prestressing (kN).

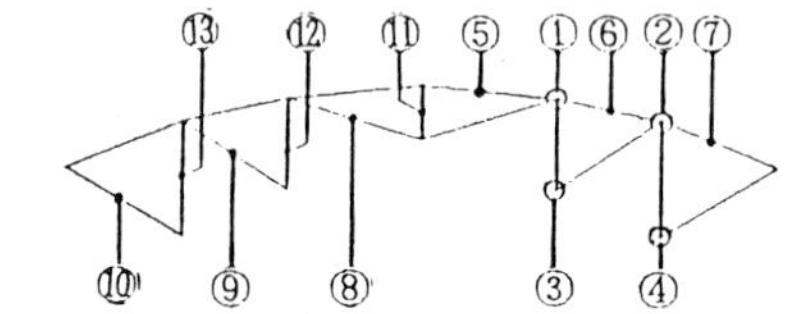

	initial	full-loading	half-loading	quarter-loading
①	18.422	26.806	3.123	34.749
②	33.167	21.194	22.166	9.292
③	7.333	6.049	27.703	19.758
④	41.667	41.426	75.468	70.499
⑤	3.968	3.233	18.367	1.488
⑥	31.512	40.782	29.991	51.984
⑦	91.856	83.866	103.605	96.278
⑧	1.128	1.256	6.401	0.951
⑨	11.513	11.436	39.087	28.385
⑩	64.010	64.439	108.962	102.671
⑪	-1.000	-0.975	-2.660	-0.352
⑫	-5.000	-5.080	-5.083	-5.573
⑬	-25.000	-24.968	-25.435	-25.376

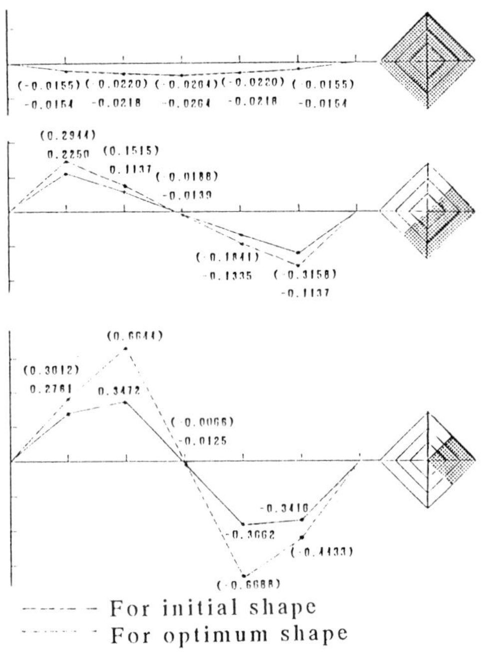

Fig. 3. The vertical displacements.

CONCLUSION

20. Our method is effective in finding the shape of a tensegrity frame having an optimum rigidity, especially for half-loading and quarter-loading. By this method, some of the nodal displacements can be reduced by as much as 30%, or 50% of their initial values.

21. Our analysis showed that the rigidity of a tensegrity frame can be increased by choosing the shape and prestressed internal forces, which are economical and very convenient. It is hoped that such a method is useful in finding the basis for economic designs and for the implementation of automation in design.

REFERENCES

1. EIKI NAGARA, JIRO KUNITA and SHIGEYA KAWAMATA, Analysis of cable nets in mixed formulation, part II, Rigerous solution of the geometrically Non-linear problems, Transactions of AIJ, No.22U Jun 1974.

2. MINOUX M. Mathematical Programming, Theory and Algorithms, John Wiley and Sons, 1986, 167-177.

3. ROSEN J. B. The gradient projection method for non-linear programming, Part 2, Non-linear Constraints, Journal S.J.A.M.,9, 1961, 514-532.

4. PENROSE.R. A generalised inverse for matrices, Por. Cambridge Philos, Soc.51, 1955, 406-413 .
5. ABADIE J. and CARPENTIER J. Generalisation of the Wolfe Reduced Gradient Method to the case of non-linear constraints, in Optimisation, R.(ed), H., Academic Press, 1969.

Index

Space structures 4

Volume 2

Space structures 4

Edited by G. A. R. Parke
and C. M. Howard

Volume 2

This volume contains the papers presented at the Fourth International Conference on Space Structures, held at the University of Surrey, Guildford, UK, on 5 - 10 September 1993

Organising committee: Dr G. A. R. Parke, Mr C. M. Howard, Dr P. Disney, Dr H. Nooshin, Professor J. E. Harding and Mr H. Wickens

Sponsors: Institution of Civil Engineers, UK; Institution of Structural Engineers, UK; Royal Institute of British Architects; IASS; Orona Space Structures, Spain; Tomoe Corporation, Japan; Kawatetsu Steel Products, Japan; Taiyo Kogyo Corporation, Japan; Mero, Germany

Published by Thomas Telford Services Ltd, 1 Heron Quay, London E14 4JD

First published 1993

Distributors for Thomas Telford books are
USA: American Society of Civil Engineers, Publications Sales Department, 345 East 47th Street, New York, NY 10017-2398
Japan: Maruzen Co Ltd, Book Department, 3-10 Nihonbashi 2-chome, Chuo-ku, Tokyo 103
Australia: DA Books and Journals, 11 Station Street, Mitcham 2131, Victoria

A catalogue record for this book is available from the British Library

Classification
Availability: Unrestricted
Content: Collected papers
Status: Invited authors' opinion
User: Structural engineers

ISBN: 0 7277 1968 8

Printed and bound in Great Britain

Contents

111. Exploring configuration processing

H. NOOSHIN and D. HADKER,University of Surrey, UK

SYNOPSIS. The objective of the present paper is to examine and classify different processes involved in generating configurations. It is shown that every activity in configuration processing involves a sequence of one or more of the following basic processes:
(1) the process through which configurations with regular patterns are generated,
(2) the process through which a configuration is modified by removing parts from it or adding parts to it,
(3) the process through which a configuration is deformed or displaced and
(4) the process through which the components of a configuration are ordered in a manner to comply with specified requirements.
The reader is assumed to be familiar with the concepts of formex algebra.

INTRODUCTION

1. The term "configuration" is used to mean an "arrangement of parts". The definition, at first glance, may sound too general. But this is intentional since there are good reasons for not making the definition of a configuration too specific. Examples of configurations in a structural context are: (i) the collection of all the nodes of a structure or any subset of these, (ii) the collection of all the elements of a structure or any subset of these and (iii) the collection of all the points of a structure that are under a particular kind of load or any subset of these.

2. Configurations are often represented graphically. However, a configuration may also be represented by a numerical model. In particular, when a computer is involved in representation of a configuration then the information stored in the computer about the configuration is bound to be in a numerical form.

3. The term "configuration processing" is used to mean creation and manipulation of numerical models for representation of configurations. When dealing with space structures, the concepts of formex algebra provide a suitable medium for configuration processing and the term "formex configuration processing" is used to mean configuration processing with the aid of formex algebra.

4. Formex configuration processing is a young discipline. However, there has been enough accumulation of experience in the field to allow emergence of various strategies, techniques and principles. In the present work, the focus of attention is on the basic processes that are involved in formex configuration processing and the study begins by considering a typical data generation problem.

AN EXAMPLE

5. Consider the configuration shown in Fig. 1. This is a view of a compound single layer braced barrel vault consisting of three connected sections denoted by A, B and C. The structure consists of 968 elements connected together at 350 nodes. The dimensions of a cross-section of the barrel vault are shown in Fig. 2.

6. Suppose that it is required to produce a formex formulation describing the element connectivity of the structure in terms of a suitable node numbering scheme together with the node coordinates relative to the global x-y-z coordinate system.

7. To begin with, it would be convenient to concentrate on generation of information about the "constitution" of the structure. This involves listing the elements of the structure and specifying the manner in which they are interrelated. The collection of all such items of information is referred to as the "compret" of the structure. The term may also be used in relation to a part of a structure.

8. As the first step in generating the required data for the configuration of Fig. 1, the comprets of sections A, B and C are formulated separately relative to a cylindrical normat. These formulations are shown in Figs 3, 4 and 5 where A1, B1 and C1 are formices relating to sections A, B and C, respectively. The formulations in Figs 3, 4 and 5 are intended to represent the basic patterns without any regard to minor deviations from the actual configurations, such as the missing elements that create the "windows".

9. In the next stage, one can deal with the necessary modifications to obtain formices that represent the comprets of sections A, B and C in every detail. The formulations are shown in Figs 6, 7 and 8. A2 in Fig. 6 is a formex representing the compret of section A and is obtained from A1 through operations that effect: (i) removing the unwanted elements inside the windows (through a luxum function), (ii) removing the superfluous overlapping elements (through the pexum function) and (iii) adding six new elements to create the windows.

10. The formex B2 in Fig. 7 represents the compret of section B and is obtained from B1 through operations that effect: (i) removing the unwanted elements inside the top windows (through a luxum function), (ii) removing the superfluous overlapping elements (through the pexum function) and (iii) removing the elements along the edges (through a conexum function). The last operation is performed to allow sections A, B and C to fit together without any overlapping of the elements along the edges where the sections are to be joined.

11. The formex C2 in Fig. 8 represents the compret of section C and is obtained from C1 by removing the parts of C1 that represent overlapping elements (through the pexum function). It may be noticed that the graphical representations of formices C1 and C2 in Figs 5 and 8 appear to be identical.

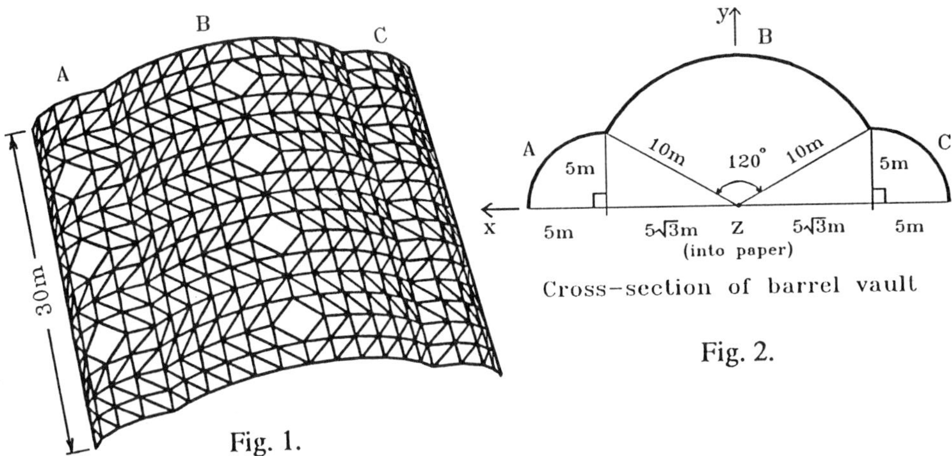

Fig. 1.

Fig. 2.

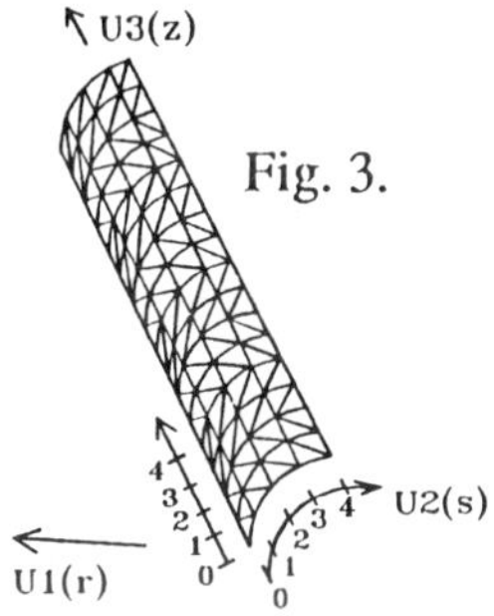

Fig. 3.

A1=RIN(3,4,4)¦LAM(3,2)¦
RINIT(4,2,1,1)¦([1,1,0;1,0,1]#
ROSAT(0.5,0.5)¦[1,0,0;1,1,0])

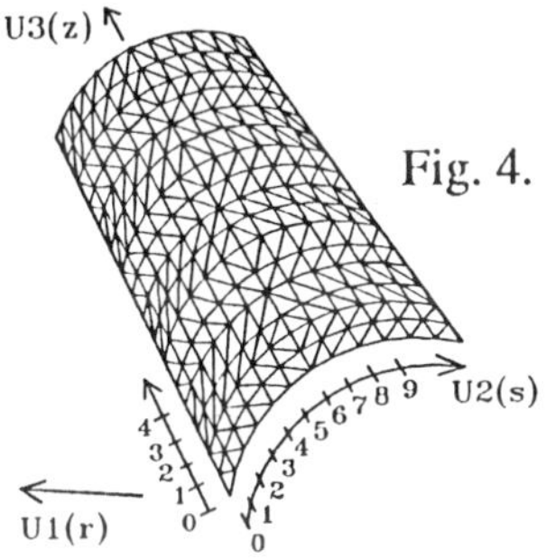

Fig. 4.

B1=LAM(2,6)¦RIN(3,4,4)¦LAM(3,2)¦
RINIT(6,2,1,1)¦([1,1,0;1,0,1]#
ROSAT(0.5,0.5)¦[1,0,0;1,1,0])

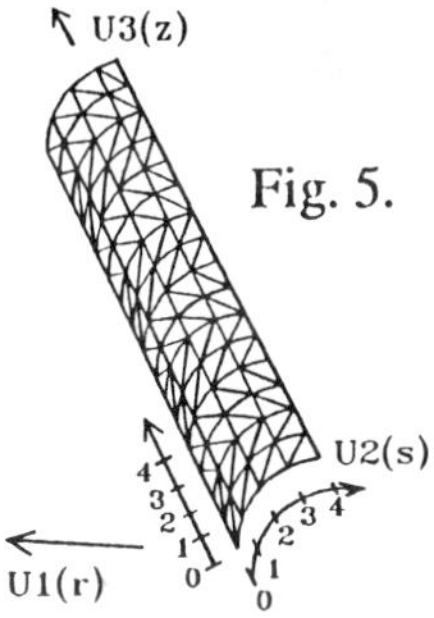

Fig. 5.

C1=REF(2,2)¦A1

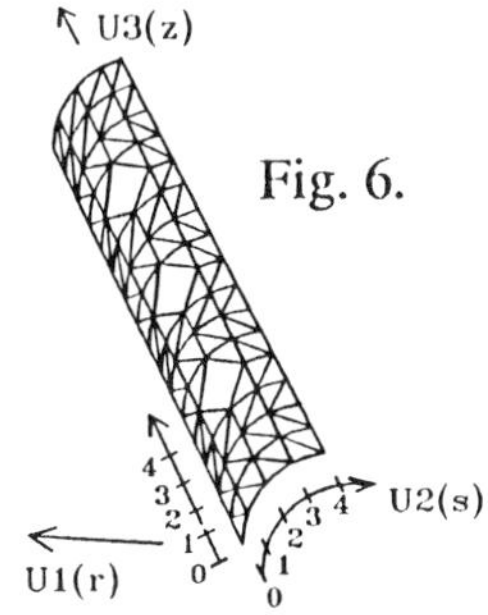

Fig. 6.

A2=PEX¦LUX({[1,2,4],[1,2,8],
[1,2,12]})¦A1#RIN(3,3,4)¦
{[1,2,3;1,1,4],[1,1,4;1,2,5]}

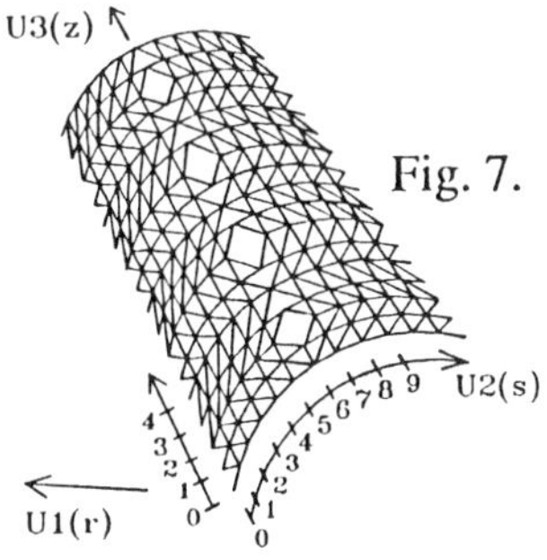

Fig. 7.

B2=PEX¦LUX(RIN(3,4,4)¦
[1,6,2])¦CON(RINIT(2,17,12,1)¦
[1,0,0])¦B1

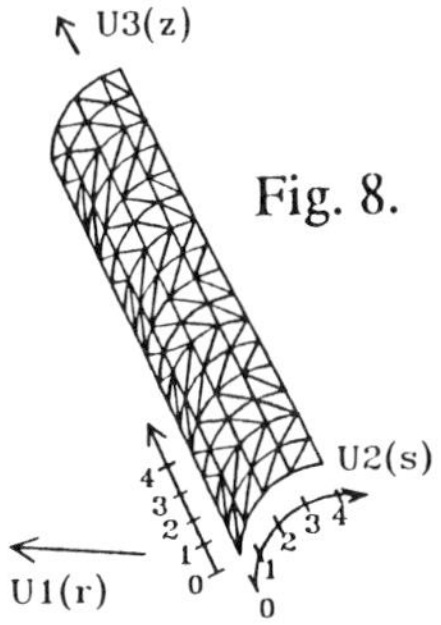

Fig. 8.

C2=PEX¦C1

This is due to the fact that the plotting style employed here does not distinguish between single and overlapped elements.

12. The next stage is concerned with the normic properties of the structure. The term "normic" is used to refer to all items of information that determine the actual "positions" and "measures" of objects, as contrasted with compretic properties that are only concerned with the "presence" of objects and their "interrelations". In dealing with the normic aspects, one may begin by first obtaining formices that represent the comprets of sections A, B and C in terms of the global x-y-z coordinate system. These are obtained as A3, B3 and C3 from A2, B2 and C2 through retronormic functions, as shown in Fig. 9. To elaborate, if one tries to plot formex A1, it would give rise to the configuration of Fig. 3 only if the uniples are regarded as cylindrical coordinates where every linear unit in the r-direction is 5 units of length (since the radius of section A is 5m, Fig. 2), every angular unit in the s-direction is 90/4 degrees (since 4 divisions in the s-direction correspond to a central angle of 90 degrees, Figs 2 and 3) and every linear unit in the z-direction is 30/16 units of length (since 16 divisions in the z-direction correspond to 30m, Fig. 1). The same applies to the plot of formex A2 in relation to Fig. 6. On the other hand, if one wants to produce the configuration of Fig. 6 in terms of the Cartesian global x-y-z coordinate system, then the uniples of A2 must be appropriately transformed and this is achieved through the first equation in Fig. 9. The other two equations in Fig. 9 may be interpreted analogously. The functions used in the equations of Fig. 9 are examples of retronormic functions. The role of a retronormic function can be described as follows: If F is a formex representing a configuration relative to a reference system other than the Cartesian global x-y-z coordinate system, then a retronormic function may be used to transform F into a formex that represents the configuration relative to the x-y-z coordinate system.

13. Formex A3 represents the compretic as well as the normic properties of section A of the barrel vault of Fig. 1 as a separate piece. Similarly, formices B3 and C3 represent the compretic and normic properties of sections B and C as separate pieces. A problem that is to be addressed at this point is that sections A, B and C of the structure of Fig. 1, as represented by A3, B3 and C3 are not in their correct relative positions. This may be seen from Fig. 9, where views in the z-direction of the plots of A3, B3 and C3 are shown and are denoted by A3', B3' and C3', respectively. Therefore, it is necessary to transform A3, B3 and C3 into formices that represent sections A, B and C in their correct relative positions. The required transformations are shown in Fig. 10. In this figure, AA is a formex representing section A in its correct position. AA is obtained from A3 by a translation in the x-direction (indicated graphically by TA). A view in the z-direction of the plot of AA is shown in Fig. 10 and is denoted by AA'. The formex BB is obtained from B3 by a rotation (indicated by RB) and the plot of BB is denoted by BB'. Also, in Fig. 10, CC' denotes the plot of CC where CC is obtained from C3 by a translation (indicated by TC) followed by a rotation (indicated by RC).

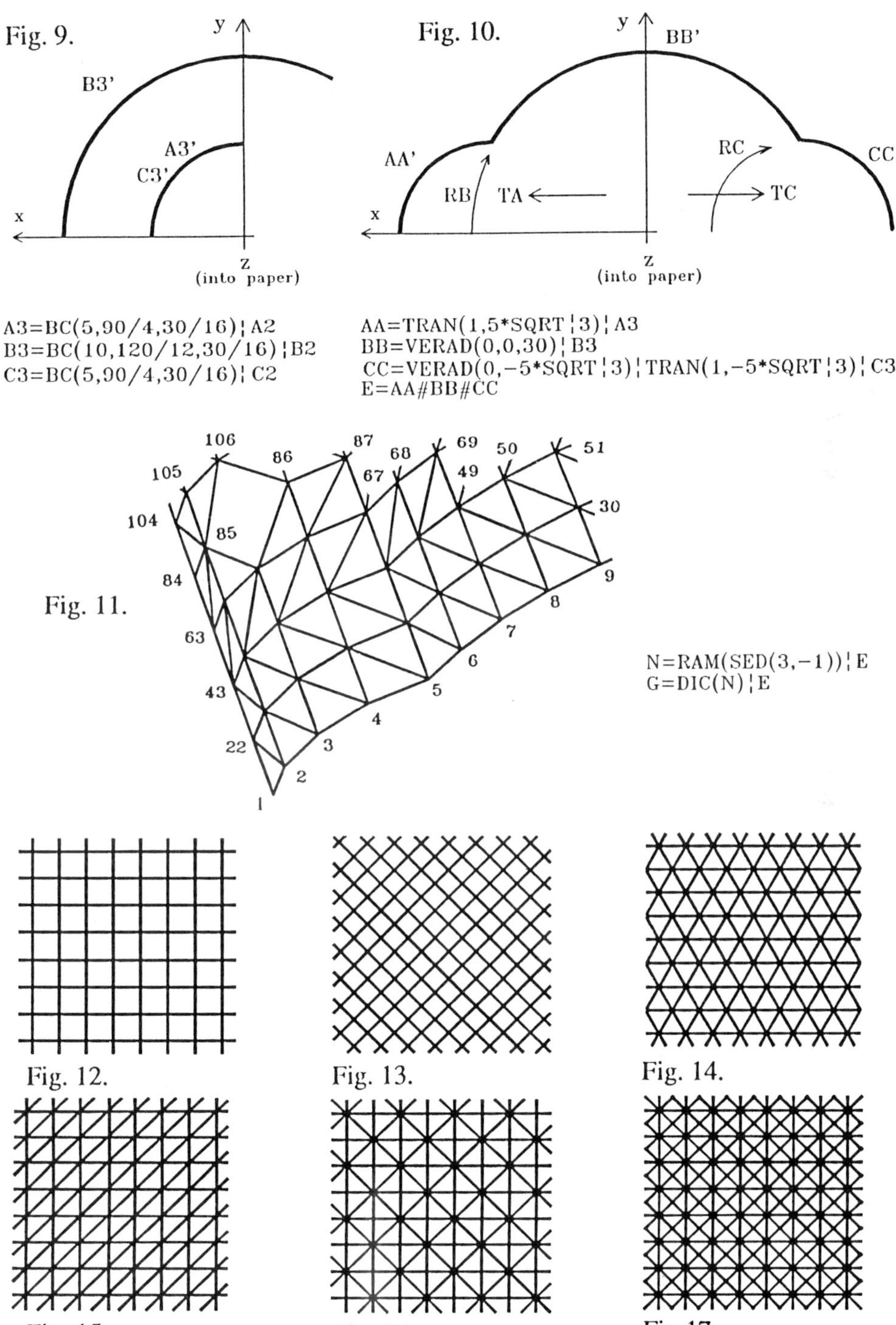
Fig. 9.
y
B3'
A3'
C3'
x
z
(into paper)
A3=BC(5,90/4,30/16)¦A2
B3=BC(10,120/12,30/16)¦B2
C3=BC(5,90/4,30/16)¦C2
Fig. 10.
y
BB'
AA'
RB
TA
RC
TC
CC'
x
z
(into paper)
AA=TRAN(1,5*SQRT¦3)¦A3
BB=VERAD(0,0,30)¦B3
CC=VERAD(0,-5*SQRT¦3)¦TRAN(1,-5*SQRT¦3)¦C3
E=AA#BB#CC
Fig. 11.
N=RAM(SED(3,-1))¦E
G=DIC(N)¦E
Fig. 12.
Fig. 13.
Fig. 14.
Fig. 15.
Fig. 16.
Fig.17.

14. Formices AA, BB and CC provide complete information about the compretic as well as normic properties of all the elements of the structure of Fig. 1 relative to the global x-y-z coordinate system. If one finds it convenient, one may also create a single formex to represent the configuration of Fig. 1 by composing AA, BB and CC. This composition is carried out in Fig. 10 with the resulting formex being E.

15. One may now proceed to create the list of node coordinates and establish a node numbering scheme. Both of these objectives are achieved in a single step by generation of a "numerant" for the structure. Formex N in Fig. 11, represents a numerant for the structure of Fig. 1. It lists all the joints of the structure in terms of their x-y-z coordinates and the order in which the nodes are listed indicates the order of node numbering (N is obtained from E through a rapported medulla function with the ordering regime being dictated by a sedilate perdicant). The node numbering pattern is (partially) shown in Fig. 11.

16. The final step in data generation for the barrel vault of Fig. 1 is to obtain the compret of the structure in terms of the node numbering pattern of Fig. 11. The formulation for a fomex G that represents this compret is shown in Fig. 11 (G is obtained through a dictum function).

GENERALISATION

17. The above example brings out a number of basic processes that are recurring themes in all configuration processing activities. These basic processes are discussed in the sequel.

18. The first basic process consists of all the operations that are performed to formulate one or more "regular" configurations which provide a suitable basis for creation of a "target configuration". This process is referred to as the process of "propagation". During the propagation stage one normally ignores the "irregularities", leaving them to be dealt with subsequently. Also, in this stage, one normally focuses the attention on the compretic aspects and does not worry about the normic properties. In the case of the example of Fig. 1, the propagation stage consists of all the operations that have given rise to the creation of formices A1, B1 and C1 (Figs 3 to 5). The term "target configuration" as used above implies the configuration that is required to be created. The terms "regular" and "irregular" as used above are subjective and relative. It is not really possible to define regularity or irregularity in a general sense. However, in spite of the difficulty in providing precise definitions for regularity and irregularity, there is a large measure of consensus on what is regular and what is irregular in practical situations. It should also be mentioned at this point that in the context of the propagation process, regularity or irregularity is only considered from a compretic point of view (rather than a general geometric point of view).

19. The second basic process consists of all the operations that are performed to modify the propagated regular configuration(s) by removal or addition of parts to take care of irregularities. This process is referred to as the "modification" process. In the case of the example of Fig. 1, the modification stage consists of all the operations that have been carried out to obtain formices A2, B2 and C2

from A1, B1 and C1 (Figs 6 to 8). Like the propagation stage, the modification stage is normally only concerned with compretic properties.

20. The third basic process consists of all the operations that are necessary for generation of information about the normic properties of the target configuration. This process is referred to as the "adaptation" process. The adaptation process involves "translocations" and/or retronormic transformations. In the case of the example of Fig. 1, the adaptation stage consists of all the operations that have been carried out to obtain formices AA, BB and CC from A2, B2 and C2 (Figs 9 and 10). The operations that have resulted in obtaining A3, B3 and C3 from A2, B2 and C2 involve retronormic transformations and those that have resulted in obtaining AA, BB and CC from A3, B3 and C3 involve translocations.

21. The fourth basic process involves all the operations that are performed to "order" specified parts of a configuration to satisfy required arrangement patterns. This process is referred to as the "organisation" process. In the case of the example of Fig. 1, the organisation stage consists of the creation of formex N that represents the numerant of the structure, Fig. 11.

22. Every configuration processing activity is bound to involve one or more of the above mentioned four basic processes. However, the order in which these processes follow each other need not necessarily be the same as for the example of Fig. 1. Also, a configuration processing activity may consist of cycles involving repeated applications of the basic processes in various combinations.

PROTOMORPHS

23. In the example of Fig. 1, the target configuration was completely specified at the beginning. However, in many other instances, one may be required to investigate various possibilities for a suitable configuration. A major aspect in such an investigation relates to the choice of patterns.

24. The idea of "protomorphs" provides a valuable aid in the search for patterns. A protomorph is a basic pattern that recurs frequently in a particular field of application. For instance, experience has shown that the patterns of Figs 12 to 17 are the bases of configurations that are commonly used in single layer grids, domes and barrel vaults. Thus, in the context of these types of structures, the patterns shown in Figs 12 to 17 may serve as protomorphs.

25. Protomorphs may be used in their pure forms or in combinations or in various modified forms. Configurations shown in Figs 19 to 22 show patterns that may be obtained by modifying the protomorph in Fig. 18.

26. A set of protomorphs may be regarded as an "alphabet" for pattern generation. As another example, Figs 23, 26, 29 and 32 show some suitable protomorphs for double layer grids. In these figures, the top layer elements are drawn in thick lines and the bottom layer elements as well as the web elements are drawn in thin lines. Two possible modifications of the protomorph of Fig. 23 are shown in Figs 24 and 25. Similarly, each of the protomorphs of Figs 26, 29 and 32 are accompanied by two modifications as shown in Figs 27, 28, 30, 31, 33 and 34.

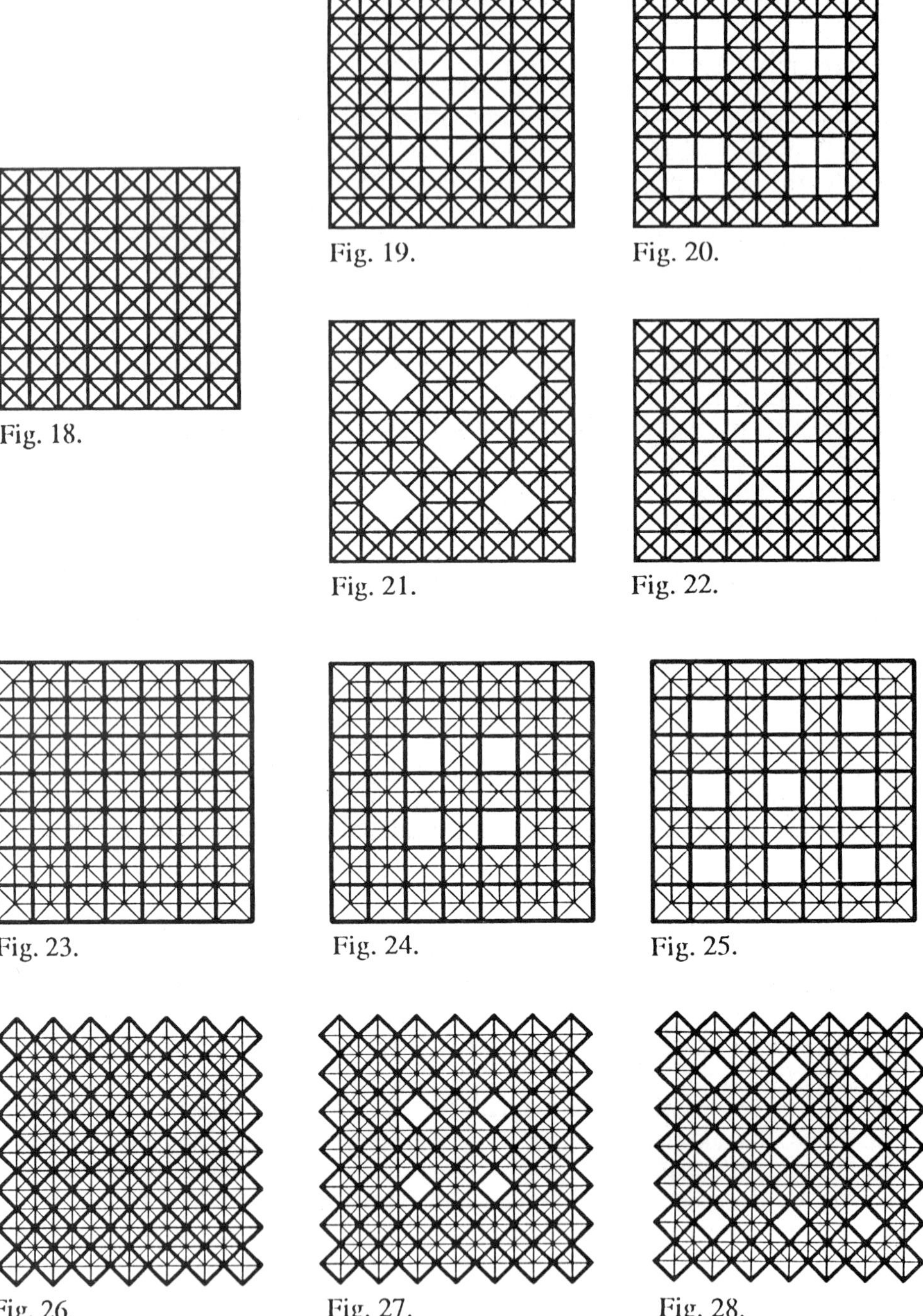

Fig. 18.

Fig. 19.

Fig. 20.

Fig. 21.

Fig. 22.

Fig. 23.

Fig. 24.

Fig. 25.

Fig. 26.

Fig. 27.

Fig. 28.

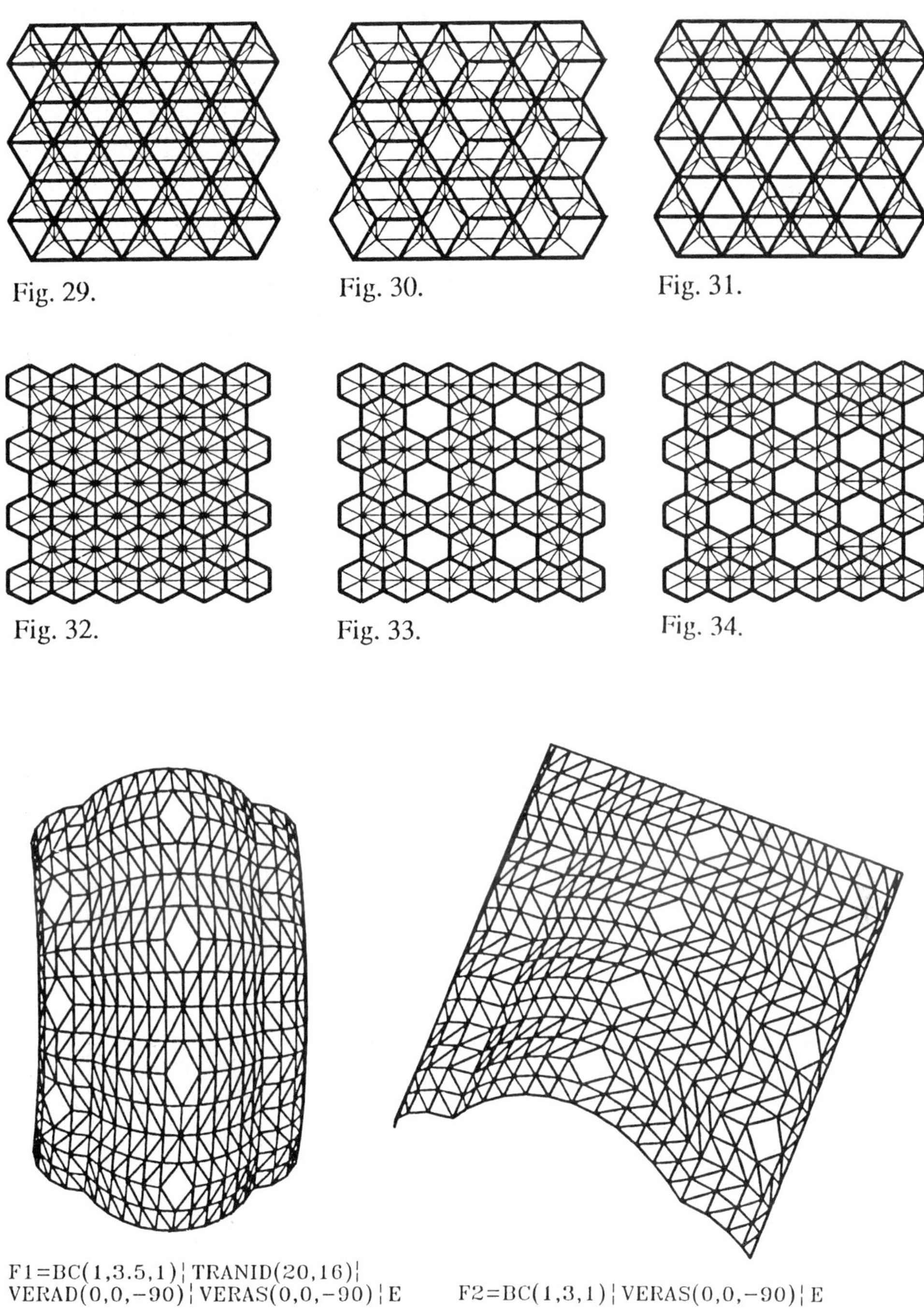

Fig. 29.

Fig. 30.

Fig. 31.

Fig. 32.

Fig. 33.

Fig. 34.

Fig. 35.

Fig. 36.

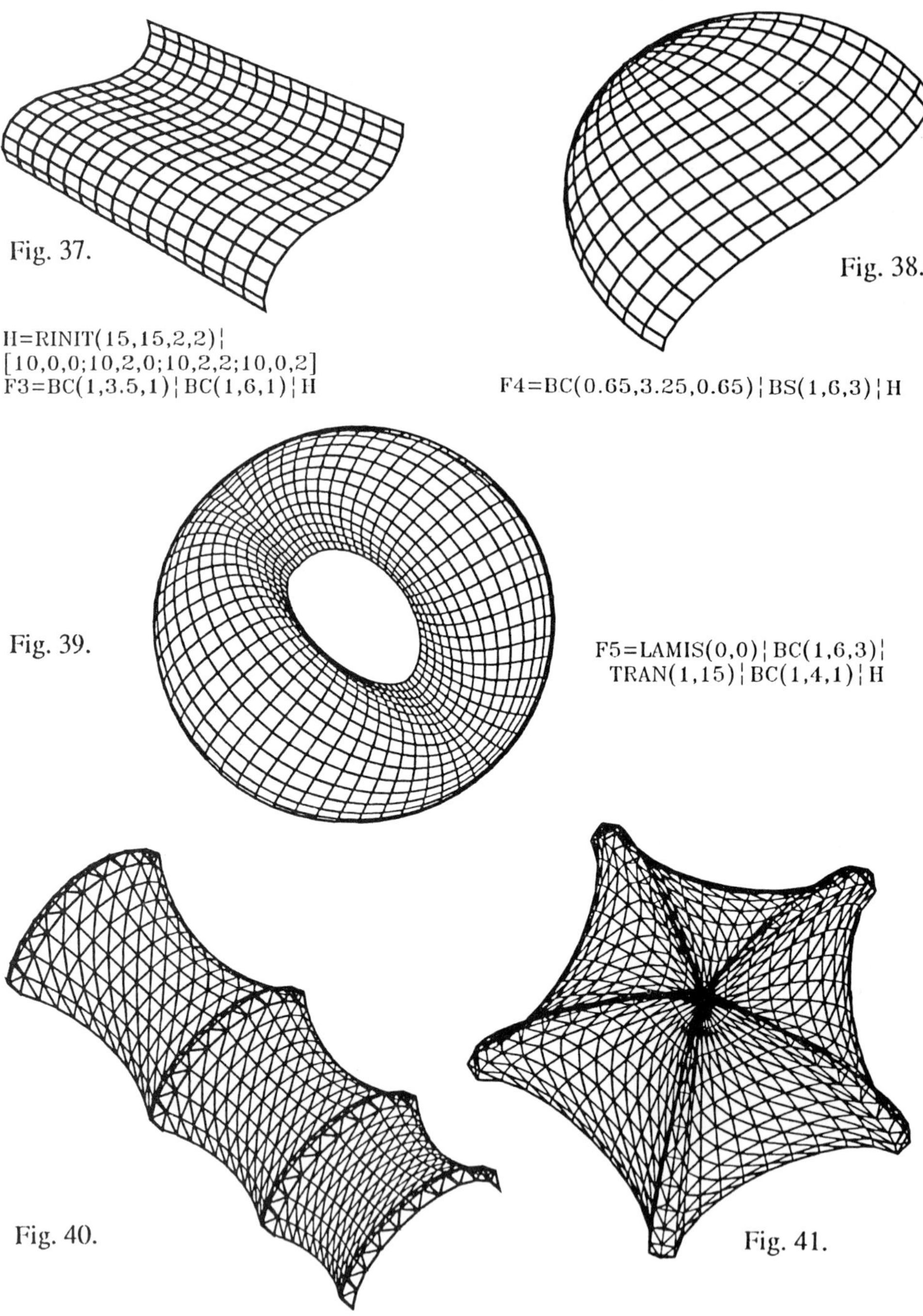

Fig. 37.

H=RINIT(15,15,2,2)¦
[10,0,0;10,2,0;10,2,2;10,0,2]
F3=BC(1,3.5,1)¦BC(1,6,1)¦H

Fig. 38.

F4=BC(0.65,3.25,0.65)¦BS(1,6,3)¦H

Fig. 39.

F5=LAMIS(0,0)¦BC(1,6,3)¦
TRAN(1,15)¦BC(1,4,1)¦H

Fig. 40.

Fig. 41.

27. In the context of structural configurations, protomorphs may represent patterns involving cable, bar or beam elements or finite elements for modelling of plate, shell or membrane structures.

28. Conceptually, protomorphs are patterns of infinite extent. In any particular case of application, however, one would deal with finite pieces of the protomorphs.

29. In working with protomorphs, the focus of attention is on their compretic aspects and these aspects are the main concern during the propagation and modification stages in a configuration processing problem.

30. Awareness of the idea of protomorphs encourages a systematic approach in pattern finding. Thus, one begins by concentrating on the overall compretic aspects and finds a simple pattern or a combination of simple patterns that can provide a suitable basis for a target configuration. The pattern(s) may then be subjected to "modification", "adaptation" and "organisation" processes as appropriate. Catalogues of protomorphs for particular applications are of value since they act as helpful aide-memoires.

PARAGENIC TRANSFORMATIONS

31. As regards the adaptation stage, the normic properties of the target configuration are sometimes fully specified at the outset. This, for instance, was the case for the example of Fig. 1. In many situations, however, it is required to search for suitable shapes. There are a number of well established retronormic transformations, including cylindrical and spherical ones that can be employed in this context. In addition, there is a new development that provides a rich repertoire of interesting and useful surfaces. These surfaces are referred to as "paragenic surfaces" and are defined through composite functions involving multiple usage of cylindrical and spherical transformations. A composite function of this kind is referred to as a "paragenic function", "paragenic transformation" or "paragenic retronorm". Two instances of the use of paragenic retronorms are illustrated in Figs 35 and 36. The configurations in these figures are obtained through paragenic retronorms involving formex E that represents the compret of Fig. 1 as formulated in Fig. 10.

32. Further examples of paragenic configurations are shown in Figs 37, 38 and 39 where the same formex H has been used with three different paragenic retronorms. Also, configurations shown in Figs 40 and 41 are two examples of compound shapes obtained by combining paragenic surfaces.

112. An application of formex configuration processing in space structures

T. OSHIRO, University of Ryukyus, Japan, and C. YAMAMOTO and Y. KISHIMOTO, Taiyo Kogyo Corporation, Japan

SYNOPSIS. Formian is an interactive programing language that may be employed for data generation using the concepts of formex algebra. This paper presents an application of Formian for generating a complex structural configuration composed of five spherical segments. Formex formulations for the five segments are derived on a plane co-ordinate system. The plotting of the formulations presents plane figures as well as the connectivity of all nodes. The formulation of the nodes to be connected is achieved using the Formian graphics or proper medium. After completing the formex formulations, including the connectivity list, a supplementary retronorm is applied to describe the spatial configuration.

ABSTRACT

1. Formex algebra, developed by H.Nooshin at the University of Surrey, is a powerful tool for the generation of data used in the design and analysis of space structures (ref.1). The algebraic nature of the formex approach makes it suitable for implementation in a programming language. Formian is one such a language, embodying the constructs of formex algebra together with many of the facilities found in a scientific computer language, which has been developed by H.Nooshin and P.Disney at the University of Surrey (refs 2 and 3). The objective of this paper is to show an application of Formian for generating a complex space structure composed of five spherical segments, as shown in Fig. 1 (ref. 4). Also, the concepts of plenix are introduced in configuration processing, where the concepts are used in linkage between Formian and the software package which provides graphic facilities. This technique enables one to use a work station which improves the processing significantly (ref. 5).

2. The idea of the configuration used as an example is that of the space frame covering Palau Sant Jordi, a sports stadium for 1992 Barcelona Olympic Games. (Architect: Arata Isozaki & Associates, Structural Engineer: Mamoru Kawaguchi & Engineers).

3. To proceed with the formex approach, formex formulations for each main segment are to be derived on a plane co-ordinate system, where each one is

Space structures 4. Thomas Telford, London, 1993

composed of a double layer grid. The plot of the formulations presents plane figures of the segment as well as connectivity of all nodes. The next step is to formulate a formex representing the nodes to be connected, where Formian graphics, or other proper medium such as a work station, may be used. When a work station has to be used, the concepts of plenix could be introduced in configuration processing. The application of the concepts is presented by SUBMIT statement, which includes the connectivity list in a plane co-ordinate system and the supplementary retronorm describing the spatial configuration.

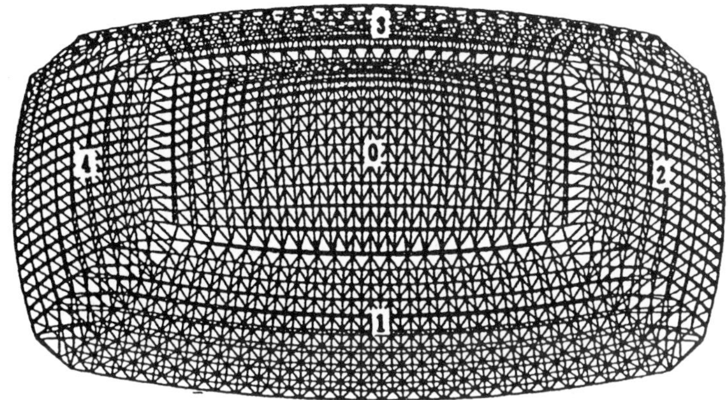

Fig. 1. General View of Stadium.

FORMEX FORMULATION FOR THE MAIN SEGMENT

4. As an example of this technique, the process of forming the configuration of the stadium shown in Fig. 1 is illustrated, where the stadium is composed of main five spherical segments. The numbers referring to each segment are noted in the figure.

Plane Figure of Centre Segment

5. Using a Cartesian co-ordinate system (I1,I2,I3), the plane figure of the centre segment, noted as 0 in Fig. 1, is to be formulated . The basic element representing the upper and diagonal members is shown in Fig. 2, and the formex formulations for the element are written as follows:

$$A1=ROS(1,2,1,1) \mid [(0,0,1;2,0,1),(0,0,1;1,1,0)] \qquad (1)$$

$$A2=RINID(23,13,2,2) \mid A1 \qquad (2)$$

where ROS(1.2,1,1) is a cardinal function referring to rotational replication in plane I_1-I_2 at (1,1) and RINID(23,13,2,2) is a tendid function referring to translational replication, and A2 represents all upper and diagonal members. Similarly, lower members are formulated by the following statements, and the total system of the members, represented by formex AB, is obtained by formex composition as follows:

B1=ROS(1,2,1,1)¦[1,1,0;3,1,0] (3)

B2=RINID(22,12,2,2)¦B1 (4)

AB=A2 # B2 (5)

The plot of formex AB represents the plane figure of a double layer grid. As shown in Fig. 2, the upper members at four corners are removed by using a luxum function (LUX) as follows:

A3=RINID(2,2,30,18)¦[0,0,1] (6)

B=LUX(A3)¦AB (7)

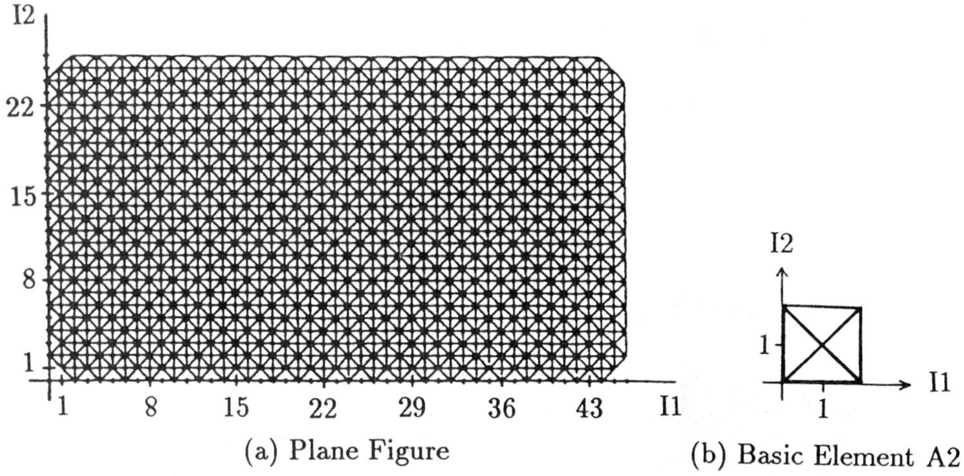

(a) Plane Figure (b) Basic Element A2

Fig. 2. Plane Figure of Centre Component.

Plane Figure of Side Segments

6. The plane figure of the side segments, 1 and 3, in Fig. 1 represents a double layer grid forming the trapezoidal shape shown in Fig. 3. The basic element constructing the side segments can be found by a close examination, which is shown in the same figure. Formex formulations for the basic element CC are represented as follows:

C1=ROS(1,2,3,2)¦[2,1,0:4,1,0] (8)

C2=LAM(2,3)¦C1 (9)

C3=TRANID(-1,-1)¦TRAN(3,1)¦C1 (10)

C4=ROS(1,2,1,1)¦[0,1,0:1,0,1] (11)

C5=ROS(1,2,2,2)¦[1,2,1:2,1,0] (12)

C6=C3 # C4 # C5 (13)

C7=TRANID(2,2)¦C6 (14)

CC=C2 # C6 # C7 (15)

where LAM(2,3) is a lambda function referring to reflectional replication with respect to $I_2 = 3$, and TRAN(3,1) is a translation function representing 1 unit of translation in the I3 direction, and TRANID(-1,-1) is a tendial translation function representing -1 unit of translation in the I1 and I2 directions.

7. Using the basic element CC, one obtains the double layer grid formulated by the following statements:

C8=LIB(J=0,3)¦LIB(I=J,14)¦TRANID(2*I,4*J)¦CC (16)

C=LAM(1,30)¦C8 (17)

where LIB(J=0,3) represents a libra composition and the variable J is referred to as a libra variable.

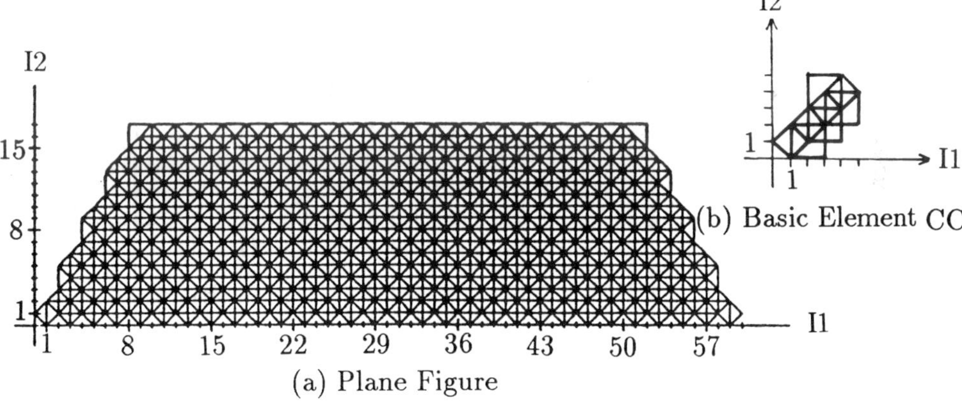

Fig. 3. Plane Figure of Side Segment.

Similarly, by using formex CC, the side segments, 2 and 4, in Fig. 1 are formulated. Formex formulations for the segments are obtained in a similar way except for the last two equations. Corresponding to size of the side segments, Eqs 16 and 17 are changed as follows:

D8=LIB(J=0,3)¦ LIB(I=J,6) TRANID(2*I,4*J)¦CC (18)

D=LAM(1,14)¦D8 (19)

COMPOSITION OF FIVE SEGMENTS

8. Formex formulations for each segment have been derived and the next step is to compose all of the formices representing the five segments. This composing process can be performed successfully by introducing a pansion function (PAN) as follows:

SPO0=PAN(4,0)¦B (20)

SPO1=PAN(4,1)¦C (21)

SPO2=PAN(4,2)¦D (22)

SPO3=PAN(4,3)¦C (23)

SPO4=PAN(4,4)¦D (24)

SPORT= SPO0 # SPOl ,# SPO2 # SPO3 # SPO4 (25)

NODE=MED¦SPORT (26)

E1=DIC(NODE)¦SPORT (27)

S=SPORT_R(0)¦SPORT (28)

DRAW S (29)

where a pansion function represented by PAN(4,0)¦B enables one to locate the segment B at point 0 in the I4 co-ordinate. Similarly, the segments, C and D, are located at corresponding locations in the I4 co-ordinate. Eq. 25 represents the total system, where formex composition is applied. To prepare a node numbering scheme, a medulla function (MED) takes an important role in generating the formex (NODE) representing all of the nodal points in series. Eq. 27 represents the node numbering, where a dictum function (DIC) enables one to represent the connectivity list of the total system. SPORT_R(0) in Eq. 28 is the supplementary retronorm, which has a rule for transformation of the plane figure into a spatial configuration. This supplementary retronorm is shown in the next section. As the connectivity list of all nodes for the five segments has been obtained using Eq. 28, the next important step is to connect corresponding nodes in the adjacent segments. For finding node numbers, Formian graphics should be applied properly for small scale configurations. However, it is very difficult for PC Version Formian to cope with such a large scale configuration. Graphic power on a work station is extraordinary compared with that for a PC. To eliminate this difficulty, the configuration generated by Formian on PC is sent to a work station through the network and the graphic software which have been developed on a work station. This system becomes powerful for visualisation of the configuration and finding node numbers. To realise this method, it becomes necessary to translate the information into the format that the particular computer package requires . This may be achieved through the SUBMIT statement (ref. 5).

9. The SUBMIT statement for the configuration processing could be of the form:

SUBMIT{10,"[SPO.DATA]"},{NODE,SPORT_R(0) NODE},{SPORT}}

where SUBMIT is a keyword and the construct that follows SUBMIT is an example of a data structure referred to as a plenix. SPORT_R(0) in the second panel is the supplementary retronorm representing co-ordinate transformation, which will be shown in the next section in detail.

10. By applying the technique mentioned above, one finds the node numbers to be connected, and the connectivity list is written by the formex as follows,

E2=[1;514],[1;522],...............,[165;1095] (30)

To represent the total system, the complete node connectivity list is obtained by formex composition as follows:

E =E1 # E2 (31)

SPORT_NEW = RED(NODE)¦E (32)

S=SPORT_R(0)¦SPORT_NEW (33)

where E represents the complete node connectivity list. All members are represented by Eq. 32, where a redictum function (RED) is introduced. This function enables the formex representing a configuration to be reconstituted from formices representing the member list and nodal data. The complete configuration could be shown by the last statement, Eq. 33.

RETRONORM

11. A rule for the transformation of a plane figure into a geometric point is referred to as a retronorm. The particular retronorm for this purpose is referred to as the supplementary retronorm. This is defined by program segments that are added to Formian interpreter. The co-ordinate system (I1,I2,I3), used for formex formulations, is transformed into a new co-ordinate system (X,Y,Z).

12. For the centre segment, the spherical surface in Fig. 4, two different circular curves with radii, r_1 and r_2, are considered in the two directions, a-a' and b-b'. All nodes on the section a-a' are assumed to be located on the circular segment with radius r_1. Consider a typical node at I_1=m in Fig. 5. Since the values of h and r_1 are given, the angles, $\theta_{m'}$ and θ_m, are to be determined by using trigonometric relations, from which the values of h' and I are determined. The next step is the transformation in the Y-Z plane, where the view from the

pointed direction in Fig. 5 gives the Y-Z plane, as shown in Fig. 6. Assuming that a typical point at $I_2 = n$ is located on the circular segment with radius r_2, the angles, $\theta_{n'}$ and θ_n, are determined by the values of r_2 and I. Further, the value of d is determined by using the angle θ_n. Assuming that the centre of the surface coincides with the centre of radius r_1, the values of X and Z are determined by radii, r_1 and r_2 and the angle θ_m as follows,

$$X = ((r_1 - r_2) + d) * \cos(\theta_m) \tag{34}$$

$$Z = ((r_1 - r_2) + d) * \sin(\theta_m) \tag{35}$$

13. Referring to the value of Y, the transformation mentioned above yields the configuration represented by full line in Fig. 7 which is unable to represent the desired transformation shown by the dotted line. Therefore, a modification to the configuration becomes necessary, which gives the following relation:

$$Y=(b'-a')/L^2*(I_1-I_m/2)^2+b' \tag{36}$$

14. For side segments, three steps of transformation are performed. The first transformation is in the Y-Z plane, which represents the transformation to the cylindrical surface as shown in Fig. 8. The transformation yields the following relations:

$$Y = r_1 * \cos(\theta_1)-(r_1 - I_4) * \cos(\theta_1 + d\theta_1 * I_2) \tag{37}$$

$$Z = (r_1 - I_4) * \sin(\theta_1 + d\theta_1 * I_2) - r_1 * \sin(\theta_1) \tag{38}$$

where the angle θ_1 is determined from the given values of d and h.

15. The second transformation is in the X-Y plane, which yields the circular segment as shown in Fig. 9 and the following relations are obtained:

$$X = L/2 - (r_2 - y) * \cos(\theta_2 + d\theta_2 * I_1) \tag{39}$$

$$Y = r_2 - (r_2 - y) * \sin(\theta_2 + d\theta_2 * I_1) \tag{40}$$

16. The third transformation is to assign the four side segments to the proper position, which can be performed by rotating and translating the corresponding segment.

CONCLUSION

17. Formian is a computer language that allows problems of data generation to be handled with ease and effectively. By using Formian, one may generate information of space structures effectively, including element connectivity, nodal co-ordinates, nodal number and element number, which are valuable for graphic visualisation and input data for structural analysis.

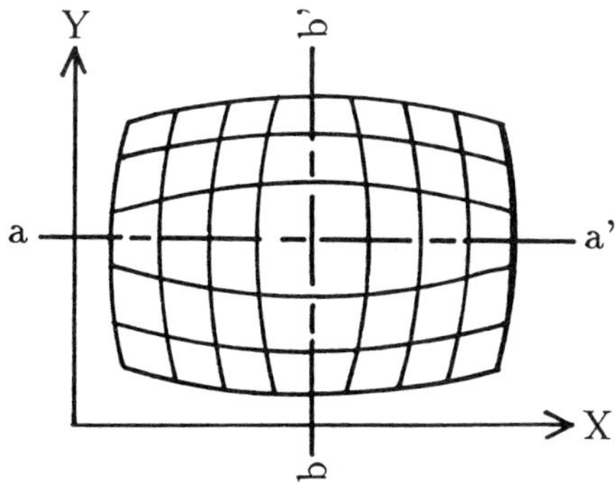

Fig.4 Surface Configuration of Center Segment (X-Y plane)

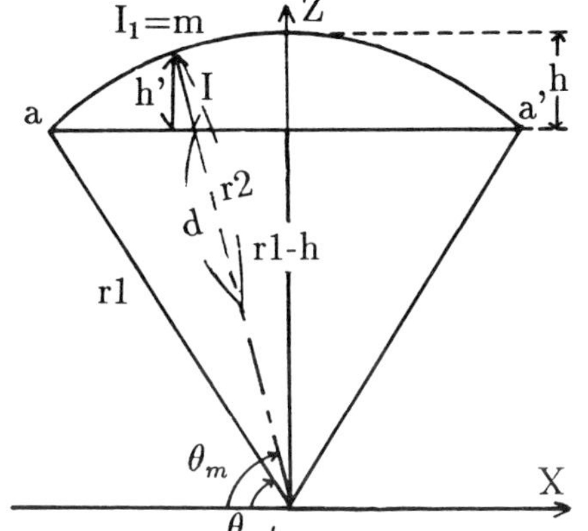

Fig.5 Transformation in X-Z plane

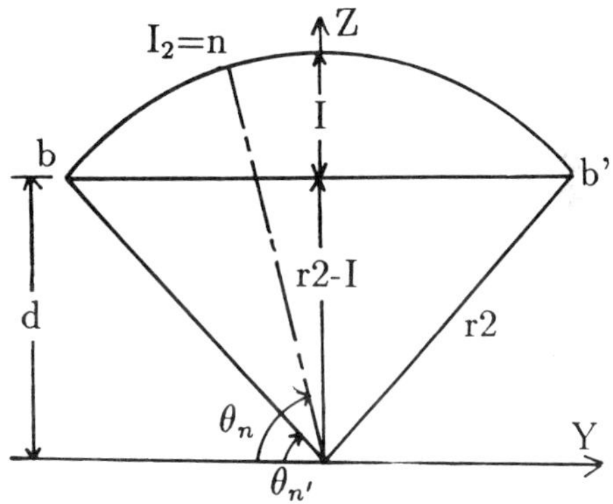

Fig.6 Transformation in Y-Z plane

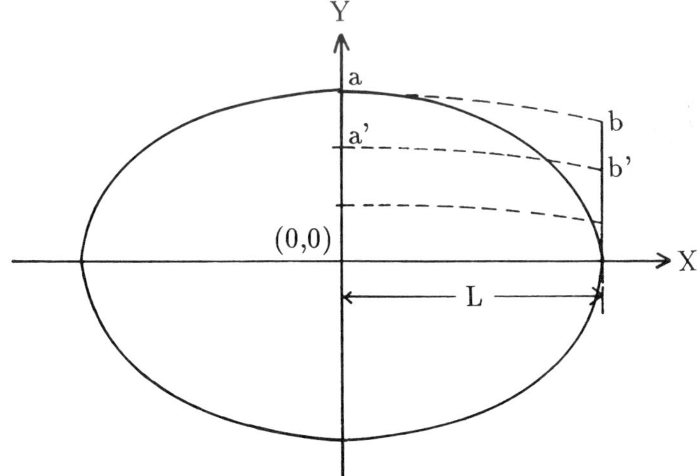

Fig.7 Modification in X-Y plane

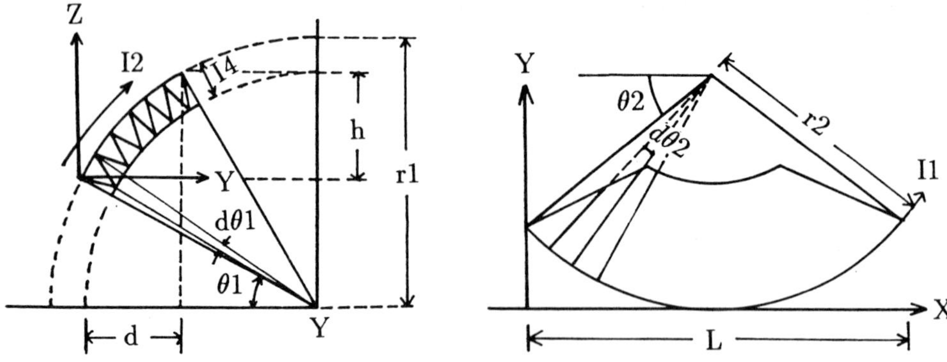

Fig. 8. Transformation in Y-Z plane. Fig. 9. Transformation in X-Y plane.

18. The proposed technique enables one to compose different segments and is powerful for configuration processing, particularly for complex surface configurations. The data generated by Formian are submitted to other software using the concepts of plenix.

REFERENCES

1. NOOSHIN H. Formex Configuration Processing in Structural Engineering, Elsevier Applied Science Publishers, 1984.
2. NOOSHIN H and DISNEY P. Elements of Formian, Proceedings of the Fourth International Conference on Civil and Structural Engineering Computing, Edited by BHV Topping, Civil-Comp Press, September 1989.
3. DISNEY P. The Programming Language Formian, Ph.D. Thesis, University of Surrey, June 1989.
4. OSHIRO T and ODA K. Application of Formex Configuration Processing in Space Structures, The Fourth International Conference on Computing in Civil and Building Engineering, July 1991.
5. HARISTCHIAN M. Formex and Plenix Structural Analysis, Ph.D.Thesis, University of Surrey, September 1980.

113. Resitition functions in Formian

S. YAMADA, Taiyo Kogyo Corporation, Japan

SYNOPSIS. The new formex functions called "restition functions" are developed by the author. These functions may be suitable for creating data for a hole or a boundary part of a configuration. The objective of the present paper is to introduce these functions and several examples. The reader is assumed to be familiar with the concept of formex algebra and Formian.

INTRODUCTION

1. A space truss may be regarded as an array of many units of the same shape such as squares or triangles. However, if there is a hole in a space truss, or a truss is be built to match a boundary of another structure, the shape of the units in these parts must be modified to suit the boundary. It may take much time in generating this configuration data.

2. The author is developing the restition functions as supplementary formex functions in Formian, under Dr.H.Nooshin in the Space Structures Research Centre of the Department of Civil Engineering at the University of Surrey. These functions may be suitable for problems dealing with holes or curved boundaries.

APPROACH OF RESTITION FUNCTIONS

3. Restition functions belong to a family of formex functions, which consist of seven functions called "juncture", "cojuncture", "vecture", "covecture", "septure", "cosepture" and "s-septure" functions. By using these functions, one may specify elements in a specific region of a configuration and these elements may be curtailed and modified. The concept of the region is firstly described, followed by the description of the restition functions.

REGION

4. In restition functions, there are two types of regions called the "ambit region" and the "circular region". Shapes of ambit regions are convex polygons as shown in Fig. 1. Ambit regions are generated from vertex points of the convex body. These points are referred to as "ambit points". Also, shapes of circular regions are circles or spheres as shown in Fig. 2. This region is generated from a centre point

and a radius. In the restition functions, these regions are given by a canonic parameter, whose constructs may be written as:

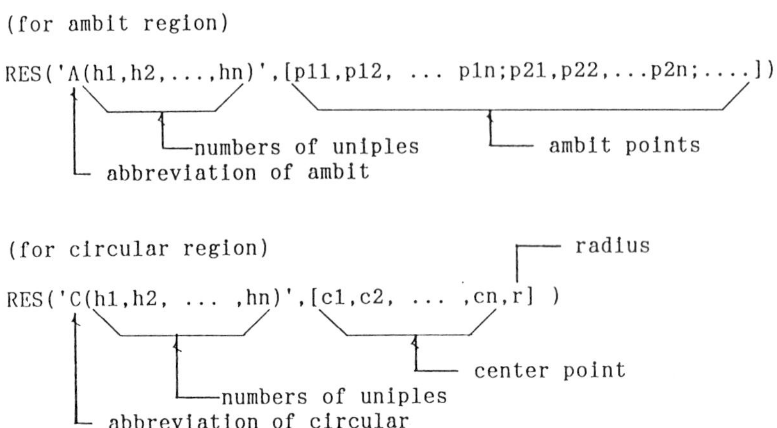

where RES implies abbreviation for restition functions, which are "JUN", "COJ", "VEC", and so on.

JUNCTURE, COJUNCTURE, VECTURE AND COVECTURE FUNCTIONS.

5. Consider the configuration shown in Fig. 3, where the plot of formex F is represented, and a circular region is represented by a circle.

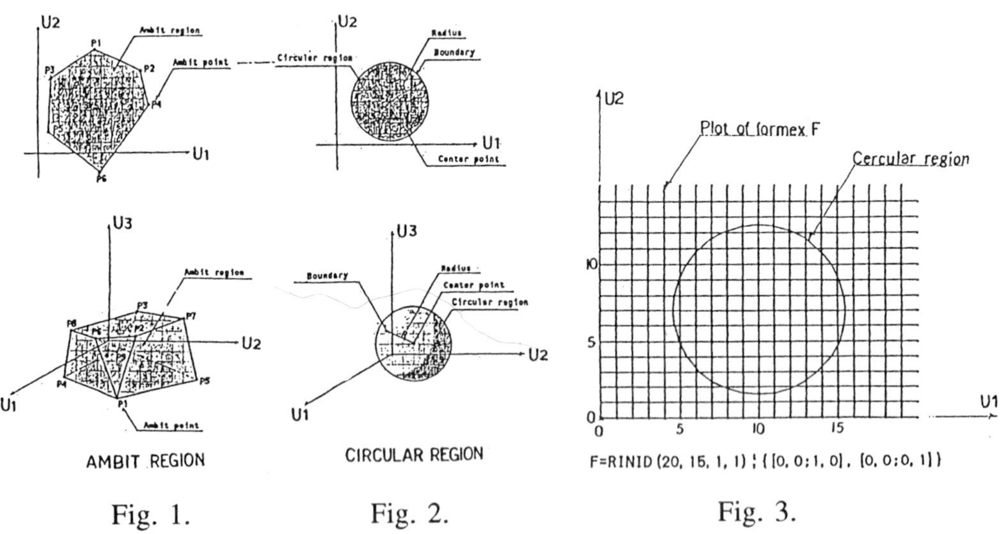

Fig. 1. Fig. 2. Fig. 3.

This formex F is obtained by the following formulation.

```
F=RINID(20,15,1,1)¦{[0,0;1,0],[0,0;0,1]}
```

Also, this region is specified by the centre point (10,7) with radius of 5.5 units. Suppose that it is required to represent all the elements of the plot of F, that are completely inside of the region, or outside of the region, as shown in Fig. 4 and 5, respectively. Formices representing these element may be written as:

```
JUN('C(1,2)',[10,7,5.5])¦F

COJ('C(1,2)',[10,7,5.5])¦F
```

where, JUN and COJ are abbreviations for "juncture", and "cojuncture" functions.

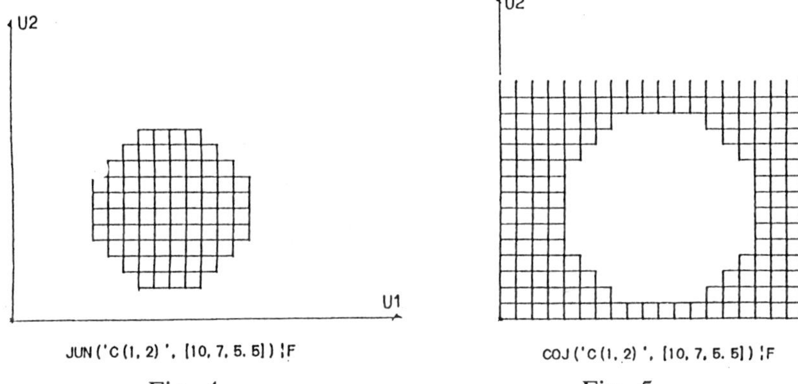

Fig. 4. Fig. 5.

6. Suppose that it is required to represent every element of the plot of F that is inside of the region, including the new elements which are created by cutting the plot of F on the boundary of the region, as shown in Fig. 6. Also, suppose that it is required to represent every element of the plot of F that is outside of the region, including the new elements which are created by cutting the plot of F on the boundary of the region, as shown in Fig. 7. These formices may be written as:

```
VEC('C(1,2)',[10,7,5.5])¦F

COV('C(1,2)',[10,7,5.5])¦F
```

where, VEC and COV are abbreviations for "vecture", and "covecture", respectively.

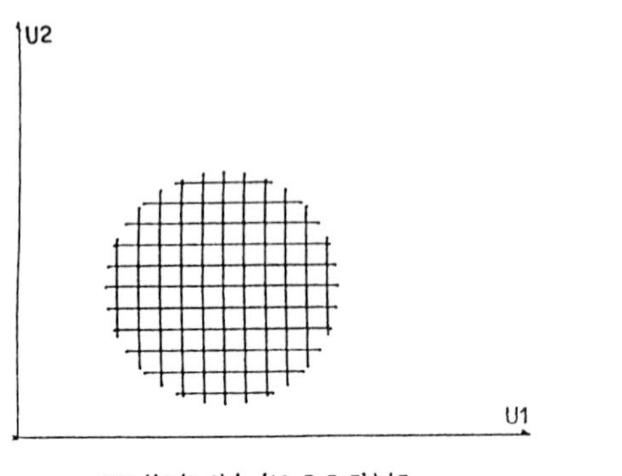

VEC('C(1,2)',[10,7,5.5])¦F

Fig. 6.

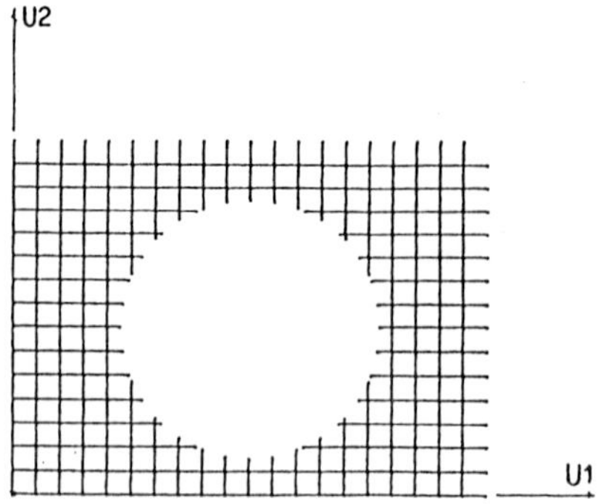

U2
U1

COV('C(1,2)',[10,7,5.5])¦F

Fig. 7.

SEPTURE, COSEPTURE, AND S-SEPTURE FUNCTIONS

7. Informally, the novation function may be used to translate several signets of a formex. The information to translate them is given by a canonic parameter of the novation function, which is also a formex. Septure, cosepture and s-septure functions generate the formex as a canonic parameter of the novation function. By these functions with novation function, the plots of several signets are moved to on the boundary of the region. Here, let Nl, N2, and N3 be formices obtained by the following formulations.

```
N1=SEP('C(1,2)',[10,7,5.5])¦F

N2=COSEP('C(1,2)',[10,7,5.5])¦F

N3=SSEP('C(1,2)',[10,7,5.5])¦F
```

where, SEP, COSEP and SSEP are abbreviations for "septure", "cosepture", and "s-septure". The plots of N1, N2 and N3 are represented in Figs 8, 9 and 10, respectively, where each arrow indicates directions to move the plot of signets, whose destinations are always on the boundary of the region. By septure function, the new elements are created from formex F. Plots of these elements are the same as the new elements created by the vecture function. By the cosepture function, new elements are created from formex F. Plots of these elements are the same as the new elements created by the covecture function. To describe the new elements created by s-septure function, consider the combination of elements from septure, and cosepture functions, and pick up the shortest elements for each plot of signet. These elements are obtained by the s-septure function.

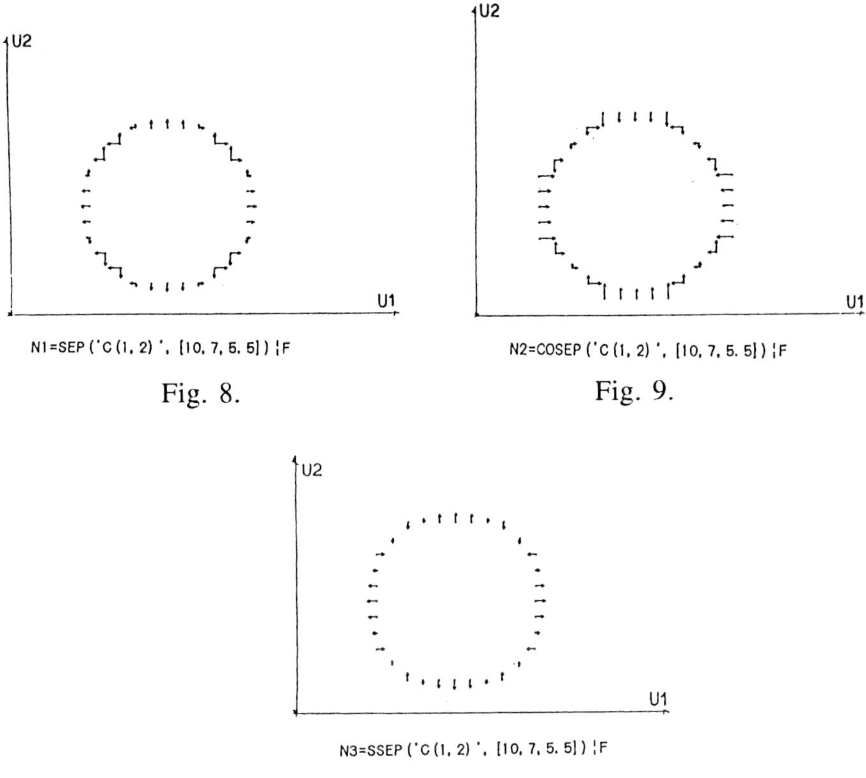

Fig. 8.

Fig. 9.

Fig. 10.

As a example of how restition functions are used, let E1 and E2 be a formex obtained from the following formulations:

```
E1=NOV(N3)¦F
E2=COJ('C(1,2)',[10,7,5.5])¦F
```

where formex N3 is obtained by s-septure function represented in Fig. 10. The plots of El and E2 are represented in Figs 11 and 12, respectively.

EXAMPLES

8. Consider a grid representing the shape of a pyramid as shown in Fig. 13, where in the pyramid, there are two circular holes, whose radius is 10m. The formex of the grid may be given by the following formulation.

```
F1={[0,0,12;-1,-1,11],[-1,-1,11;1,-1,11],[1,-1,11;0,0,12]}
F2=LIB(I=1,12)¦RIN(1,I,2)¦TRANIX(1-I,1-I,1-I)¦F1
F3=PEX¦ROS(1,2,0,0)¦F2
F =BT(35/24,35/24,20/12)¦F3
```

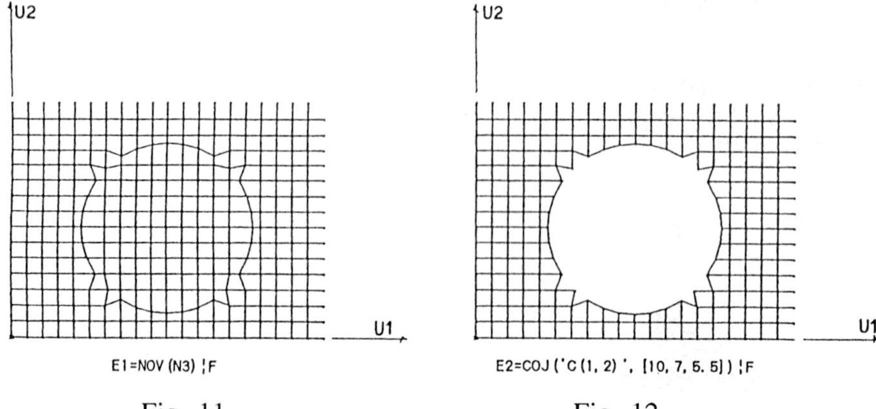

Fig. 11.   Fig. 12.

The plot of formex F is represented in Fig. 14. To create the holes, use the s-septure, and cojuncture functions which may be written as:

```
N=SSEP('C(1,3)',[0,0,10])¦F
P=COJ('C(1,3)',[0,0,10])¦NOV(N)¦F
```

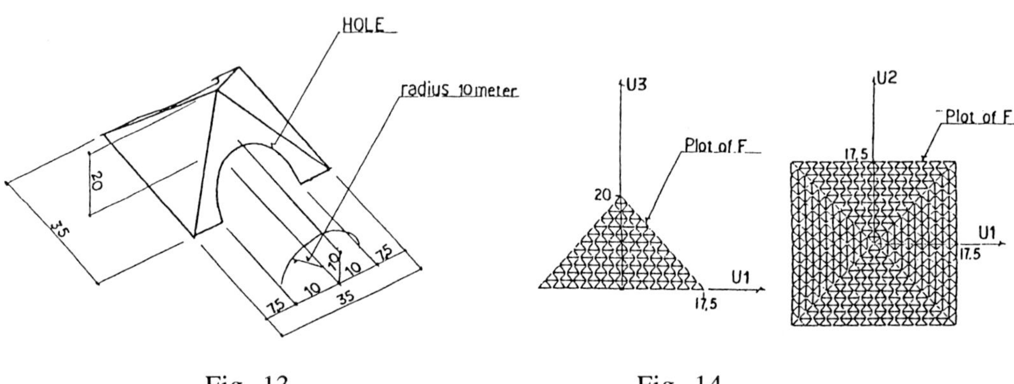

Fig. 13.   Fig. 14.

The required data may be obtained as the formex P, the plot of P is represented in Fig. 15. As an interesting example, let E be a formex, the configuration of whose plot is an arch shown in Fig. 16. Also, let A and AP be formices obtained by the following formulation:

```
S='A(1,2,3)'
R=[0,0,20.5;18,18,0;18,-18,0;-18,18,0;-18,-18,0]
N=SSEP(S,R)¦F
A=COJ(S,R)¦NOV(N)¦F
AP=A#P
```

where, S and R specify the ambit region. Here, R represents a set of ambit points which are slightly displaced from the vertex points of the pyramid. Formices A and AP are plotted in Figs 17 and 18, respectively.

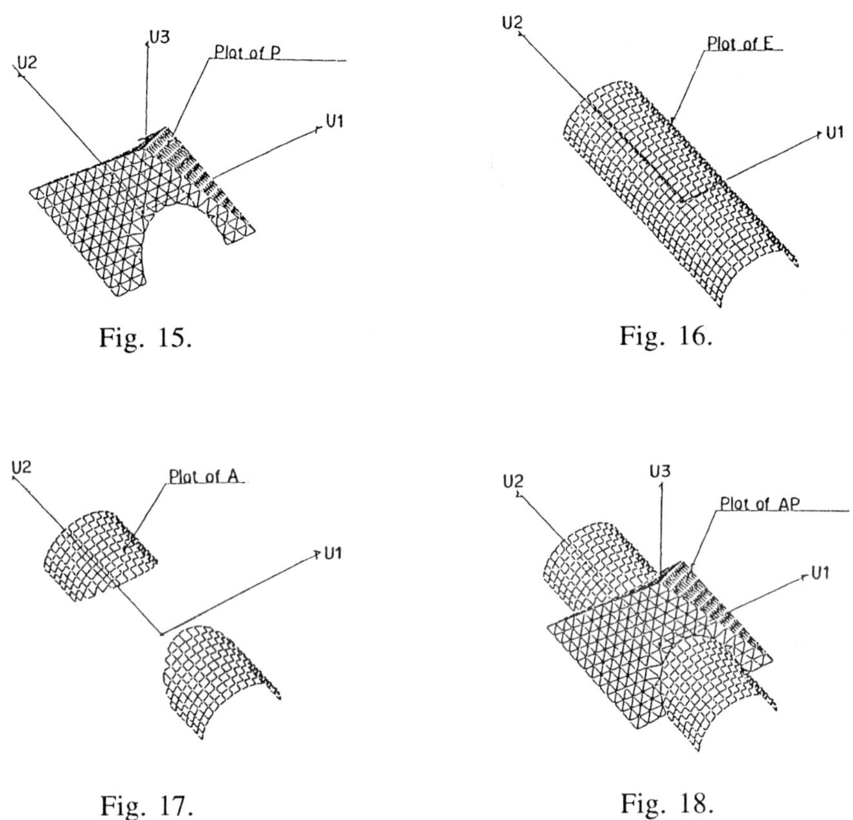

Fig. 15.

Fig. 16.

Fig. 17.

Fig. 18.

Similarly, the configurations shown in Figs 19 to 20 may be easily generated by using restition functions.

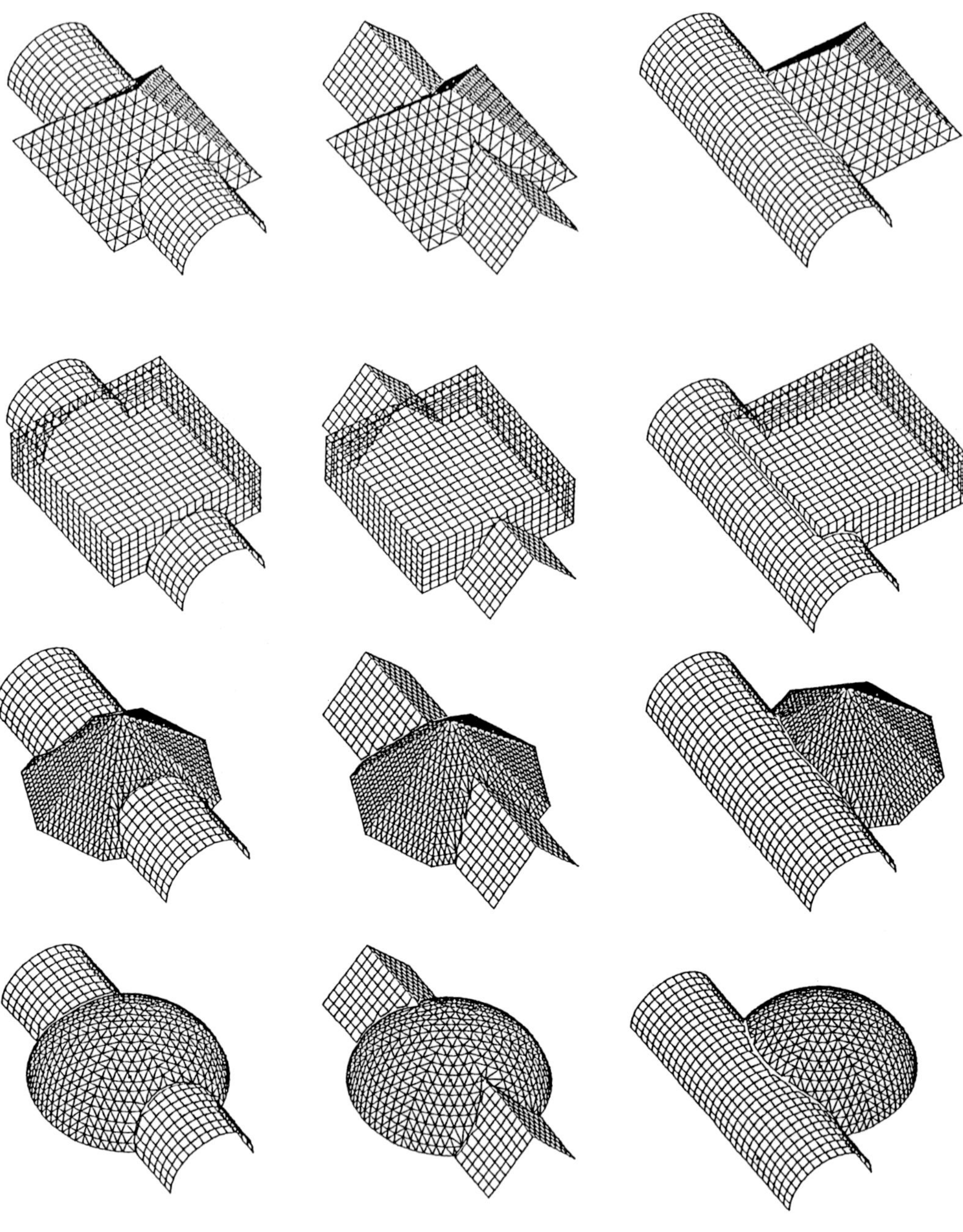

Fig. 19.

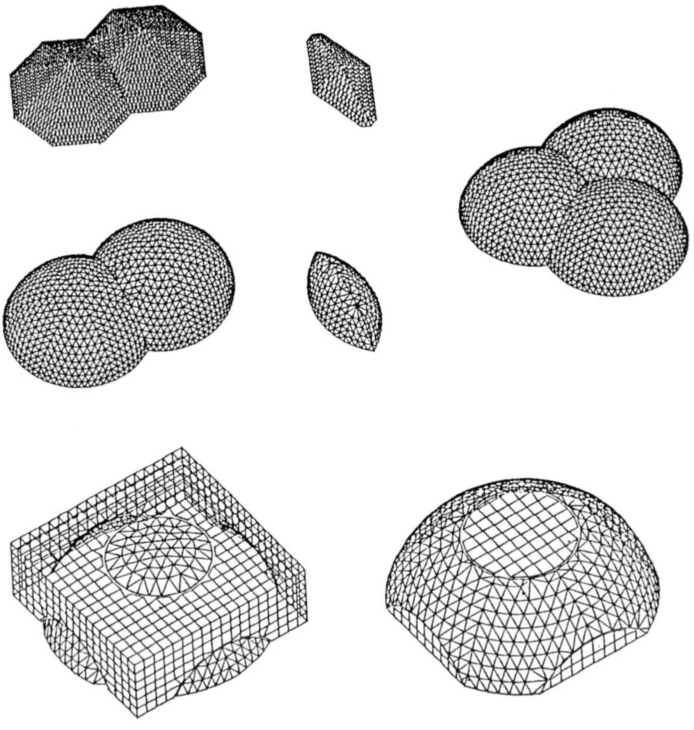

Fig. 20.

REFERENCES
1. Nooshin H., Formex Configuration Processing in Structural Engineering, Elsevier Applied Science Publishers, 1984.
2. Nooshin H. and Disney P., Elements of Formian, Proceedings of the Fourth International Conference on Civil and Structural Engineering Computing, Edited by B H V Topping, Civil-Comp press, London, September 1989.

114. A comparative study of space structures produced using Kenner retronorms

G. M. HAMMOUD, M. R. KHALIL and O. EL-LABBAR, University of Garyounis, Libya

SYNOPSIS. This paper presents a study of steel skeletal single-layer domes covering a large circular area. A single equation, which is one of the family of Kenner retronorms, is used to generate a large number of different dome shapes. Computer software is developed and utilised in the data generation, graphics, analysis and design of the different skeletal systems. Material requirements, joint displacements and support reactions are compared for the different cases. A number of shapes were found to consume less material and have less deformation. The effects of some other factors are included.

DOME SHAPES AND EQUATIONS

1. A single equation is used to produce a large variety of dome shapes. The mathematical basis of the equation and its use in the literature are reviewed in the sequel.

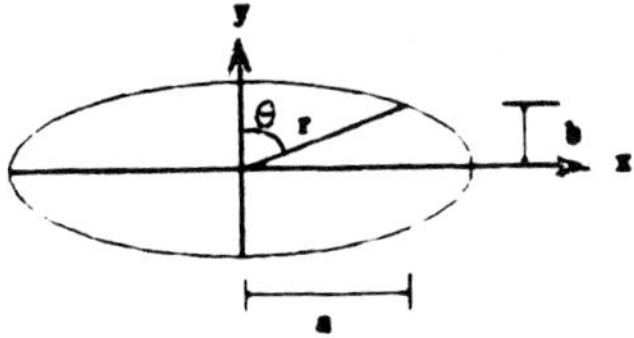

Fig. 1. Ellipse.

Mathematical basis

2. The equation representing the ellipse shown in Fig. 1. may be written using a two dimensional cartesian coordinate system as:

$$\left(x^2 / a^2\right) + \left(y^2 / b^2\right) = 1$$

Using a polar coordinate system, where $x = r \sin\theta$ and $y = r \cos\theta$, this equation becomes:

$$\left[\left(r^2 \sin^2 \theta\right) / a^2\right] + \left[\left(r^2 \cos^2 \theta\right) / b^2\right] = 1$$

Space structures 4. Thomas Telford, London, 1993

Let E = b/a hence: $E^2 r^2 \sin^2\theta + r^2\cos^2\theta = E^2 a^2$

Which means that:

$$r = E\,a / \left[\left(E^2\sin^2\theta\right) + \cos^2\theta\right]^{1/2}$$

Now, if n is used instead of 2, then this equation becomes

$$r = E\,a / \left[\left(E^n\sin^n\theta\right) + \cos^n\theta\right]^{1/n}$$

when n may take any positive value (between 0 and ∞). Fig. 2 shows a quarter of an ellipse having different n values.

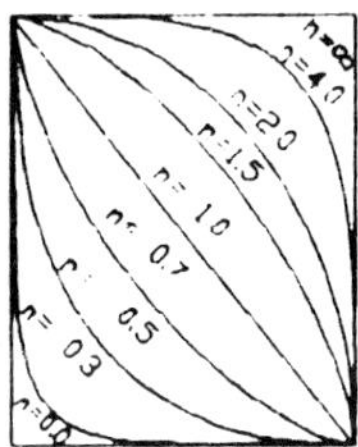

Fig. 2. Quarter of an ellipse having different values of n.

3. This equation can be extended to three dimensions to represent a double elliptic hemisphere or a superellipsoid. If the maximum x, y and z coordinates of any point at the produced surface are a,b and c respectively, and if $E_1 = b/a$, $E_2 = c/a$, θ is the horizontal angle and φ is the vertical angle, then:

$$R_1 = a\,E_1 / \left[\left(E_1^{n1}\sin^{n1}\theta\right) + \cos^{n1}\theta\right]^{1/n1}$$

and

$$R_2 = R_1 E_2 / \left[\left(E_2^{n2}\sin^{n2}\phi\right) + \cos^{n2}\phi\right]^{1/n2}$$

where: nl = power of the horizontal curve,
n2 = power of the vertical curve and
both nl and n2 are positive numbers.

Fig. 3 shows a case with nl = n2 = 2. Every combination of nl and n2 with values from 0 to ∞ gives a superellipsoid of a different shape.

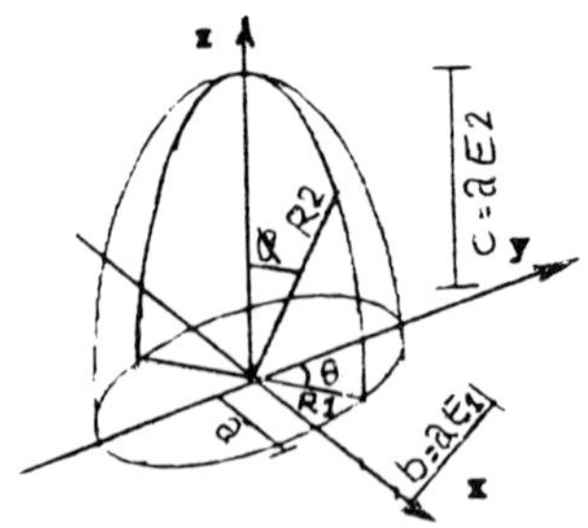

Fig. 3. Schematic sketch of a superellipsoid.

Kenner spherical retronorms

4. The suggestion of Kenner (ref. 1) to use the equation of nth order to represent the shape of a superellipsoid and the applications of this use by Huybers and Arends (ref. 2) and by El-labbar (ref. 3) are the basis of this study. Formex algebra and the computer language based on it (Formian) were utilised in the later reference to produce various shapes. In this study a computer program has been developed in BASIC language and used along with the drafting and design package (AutoCAD release 10) (ref. 4) to produce three dimensional views of the domes and prepare the required data for structural analysis.

5. The Kenner spherical retronorm is defined in ref. 4 and may take the form KS(a, θ, ϕ, E_1, E_2, n1, n2) where these symbols have the same meaning as given in § 2 & 3.

Graphical presentation

6. El-labbar (ref. 3) used Kenner spherical retronorms and Formian to plot a number of different dome shapes. The software developed by the authors attempted producing these shapes using the same variables. Two sets of shapes are shown in Fig. 4 where a, b and c may be produced using the retronorms KS(1, 36, 9, 1, 2.1, 100, 0.7), KS(1, 36, 9, 1, 1, 2, 1) and KS(1, 36, 9, 1, 1, 2, 2), respectively.

7. In this study, the number of meridians and rings of the domes were used instead of the respective horizontal and vertical angles. Also the horizontal and vertical angles that correspond to the first and last meridian and ring, respectively, were given as variables. Thus, a part of or a complete dome can be dealt with.

RESEARCH PROGRAM

8. A large number of steel skeletal domes are analysed and designed. Only vertical loads are considered. A number of computer codes are developed and employed in the study.

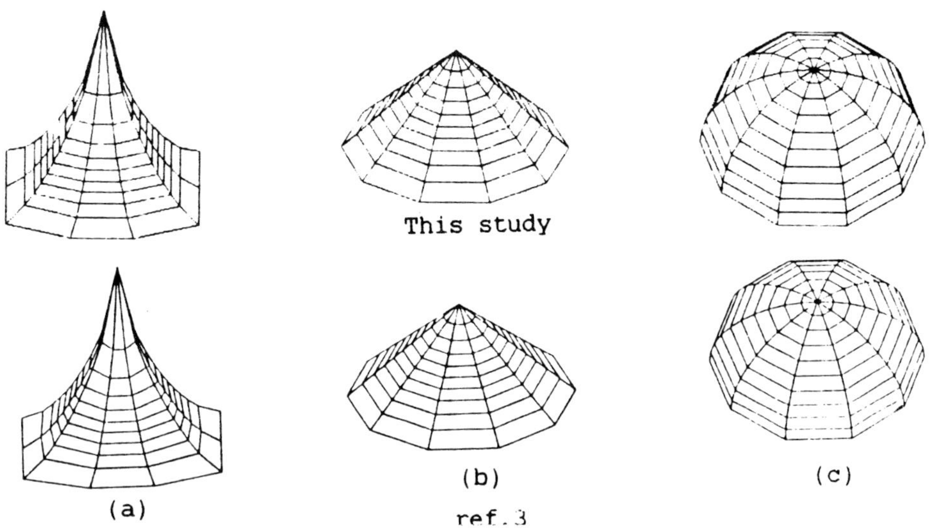

Fig. 4. Dome shapes compared to those of (ref. 3).

Space structures considered

9. A circular area of 50m diameter is chosen to be covered by steel skeletal domes of different shapes. This means that nl and E_1 values are 2 and 1, respectively, for all the cases. The dome shapes considered are covering a range of n2 from 0.5 to 5.0. Eleven shapes are considered with n2 of 0.5, 0.75, 1.0, 1.5, 2.0, 2.5, 3.0, 3.5, 4.0, 4.5, 5.0. It is assumed that the top part of the dome from the location of 5m horizontal diameter and above is to be manufactured separately and placed in position after erecting the lower part. The lower part which is between the levels of 50m and 5m diameters are referred to as the structural part of the dome. The upper part which is from 5m diameter level and above is referred to as the topping part. Thus the topping part is considered in loading but not in analysis nor in design. The height of the dome is recommended to be about one fifth of the diameter of the covered area (i.e. l0m) (ref. 5). This height is given the values of 8, 10, 12 and 15m. E_2 values is adjusted such that the horizontal diameter of 5m is at the specified height. A sample of the domes considered is shown in Fig. 5.

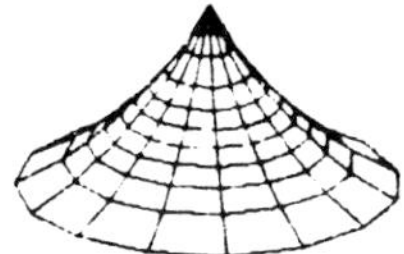

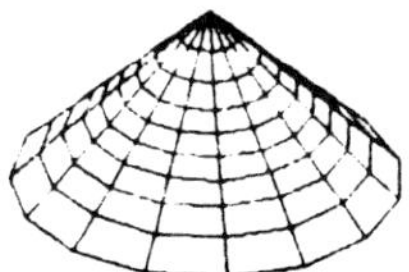

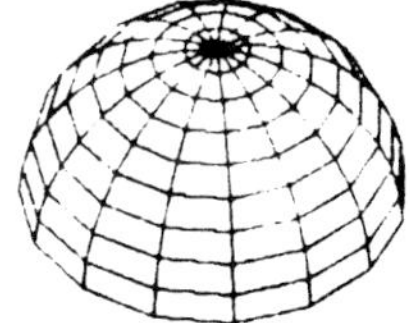

Fig. 5. Dome shapes.

10. The dome skeleton members are rigidly jointed to each other forming a single layer dome. All the considered domes are of the popular Scwedler type (ref. 6). The parts between the skeleton members may have any of the three shapes given in Fig. 6. These shapes are denoted 2-Way, 3-Way and 4-Way grids, referring to the number of directions taken by the members at any joint.

Fig. 6. Grid types.

11. In order to reduce the buckling effects in the compression members, a member length of about 3m has been found to be reasonable. 52 meridians and 13 rings gave member lengths in this range. However, the number of meridians is varied to 44 and 60 to study its effect.

12. Various supporting arrangements for the domes are considered. A support at every, every other and every third joint are tried. These provide 52, 26 and 18 supports at the lowest ring of the dome, respectively.

13. It is too complicated to represent the wind load on such shapes. The idea of increasing the vertical live load to account, somehow, for the wind effect (ref. 6) is adopted in this study. A value of live load of 1.5 kN/m^2 of horizontal projection is used. A dead load of 0.9 kN/m^2 is used to represent dome cover, skeleton own weight and other roof fixtures. The dead load value is an estimate based on similar space structures in the locality (Libya).

Data generation

14. Using a computer is a must, specially if the number of nodes is in hundreds. Making use of symmetry, a quarter of the dome is considered and a personal computer is employed in the whole study.

15. A computer program was developed in the BASIC language to handle the problem of data generation. The input of this program is the raw data of the dome which include a, E_1, E_2, nl, n2, θ_1, θ_2, ϕ_1, ϕ_2 and the grid type. Where θ_1, θ_2, ϕ_1, ϕ_2 are the start and ending angles of meridians and rings respectively. The remaining variables are as defined in § 2 and 3. Also the joint constraints and the loading are given as data. One of the output files of this program contains the required data for AutoCAD (ref. 4) in order to produce three-dimensional views of the considered configurations. A second output file gives the total surface area of the structural part of the dome and that of the topping part. The third output file gives the nodal coordinates, the member connectivities, the nodal forces and the

joint constraints. This last file is arranged in a manner that reduces both computing time and memory requirements.

Analysis and design

16. The direct stiffness method was employed in the structural analysis of the domes. Linear elastic behaviour was assumed. The analysis program was written in the FORTRAN 77 language to benefit from the capabilities of the Lahey compiler (ref. 7).

17. Both analysis and design are for the structural part of the dome. The topping part is included in load calculations and its surface area is considered in the comparison.

18. The analysis was performed with dome members of the same cross-sectional area. The resulting internal forces were used to design more realistic cross-sections which were used in a second analysis. This process was repeated until the changes in the cross-sections became insignificant. Tubular (CHS) steel cross-sections were assumed for all the skeleton members. The requirements of BS 5950:Part 1:1990 were followed in the design of all the skeleton members. Cross-sections were selected from the Structural Steel Work Handbook (ref. 8). The member cross-sections produced by the last design stage along with the member length were used to determine the total weight of the skeleton members. Also the joint displacements and the support reactions were determined for each considered case.

Criteria for comparison

19. In order to compare the different domes considered and study the effects of the variables, the following criteria were used:

(i) Total weight of material for members of the structural part.
(ii) Maximum joint displacement.
(iii) Total surface area of the dome structural and topping parts (a measure of the material required for covering the dome).
(iv) Reaction components at the ground supports.
(v) General considerations, like:
(a). The aesthetics. (b). Ease of construction.
(c). Total number of joints. (d). Total number of supports.

20. The comparison of the dome cases was qualitative in many instances. However, some quantitative comparisons are also included.

DISCUSSION AND ANALYSIS OF RESULTS

21. The total weight of the members, maximum joint displacement, surface areas (of structural and topping parts) and support reactions for the different domes considered are shown in Fig. 7(a-g) and Fig. 8(a-e). The first figure shows the effects of changing the dome height while the second shows the effects of

changing the grid type. All the domes in Fig. 7 are of the 2-way grid type and all the domes in Fig. 8 are of 10m height.

22. Only the effects of changing the number of meridians on maximum joint displacement and vertical support reaction are shown in Fig. 9(a,b). The effect of changing the support arrangements on the maximum joint displacement is shown in Fig. 10.

Maximum joint displacement

23. It is clear from the displacements shown in Fig. 7(b) and Fig. 8(b) that linear analysis is not valid for a large number of the domes considered. Although it may still need geometric non-linearity to be considered, the authors limited their discussions and conclusions to the range of dome powers between 0.75 and 3.0.

24. The maximum joint displacement is of a minimum value for domes of powers between 1.0 and 1.5. The larger the height the smaller the maximum deformation under load. It is clear from Fig. 8(b) that using 4-way grid reduces the joint displacements specially for domes with a power of 1.0 to 2.5. The effect of changing the number of meridians on the maximum displacement is not clear as shown in Fig. 9(a). On the other hand Fig. 10 indicates that only supporting all the ground joints produces reasonable displacements. Hence, the other support arrangements are disregarded.

Members weight

25. Minimum weight of members is found to be that for domes of vertical curve power between 1.5 and 2.0. The increase in weight becomes significant for powers away from the range 1.0 to 3.0. Increasing the dome's power will produce a flattened dome with high bending moments and this seems to be responsible for the increase in weight. The increase in dome height reduces the total weight slightly. The saving in weight by using 2-way grid instead of the 3-way or 4-way grids is more significant. Changing the number of meridians was found to have a negligible effect on member weight.

Surface area

26. From Fig. 7(c,d) it is clear that the surface area of the topping part is minor compared to that of the structural part of all domes. Also the higher the dome power the larger the total surface area. Furthermore, the increase in dome height increases the total surface area for all the dome cases. The topping part surface area increases with height only if the dome's power is less than 1.0.

Vertical reaction component

27. The vertical reaction component seems to be insensitive to the change of the dome power. However, it increases with the increase in dome height or with the use of 3-way grid. As expected, it increases with the reduction in the number of meridians due to the consequential reduction in number of supports.

Horizontal reaction component

28. The horizontal component of the support reaction decreases rapidly with the increase of dome power and slightly decreases with the increase in dome height. The above two results are due to the greater verticality of the members at the supports. The 2-way grid is found to produce a value less than that of the 3-way grid and more than that of the 4-way grid. Changing the number of meridians has a negligible effect on this kind of reaction component.

Support moment

29. The support moment is oscillating with dome power in the cases of large heights (12 and 15m). However, it tends to be minimum at a dome power of 1.5. The increase in dome height decreases the support moments for domes having power more than 1.0 and increases moment reaction in cases of domes having powers less than 1.0. The use of the 2-way grid increases the support moments in most of the dome cases. The change in the number of meridians does not give a noticeable effect on the support moments.

Other considerations

30. It should be noted here that the reduction of the number of meridians gives an important saving by reducing both the number of joints and the number of supports.

31. The aesthetic consideration is mainly a matter of taste. However, the authors believe that the domes of powers less than 1.0 or between 1.5 and 2.0 are more appealing than those of powers outside these ranges. Furthermore, the closer the dome power is to 1.0, the easier the erection of the skeleton and the placing of dome cover becomes.

CONCLUSIONS

32. For domes covering a circular area of 50m diameter, excessive joint displacement makes it unfeasible to use dome powers outside the range 0.75 to 3.0. A double-layer dome may be a better solution. Dome powers between 1.5 and 2.0 have less weight, less displacement and less vertical reaction than domes of other powers. Furthermore, the results confirm the suitability of having a dome height 0.15 to 0.2 of the dome diameter. A 2-way grid provides a large saving in the members weight without causing a substantial increase in joint displacements. Reducing the number of meridians increases the vertical reaction component, decreases displacement and has a minor effect on other reaction components and weight of members. Also it gives some saving by reducing the number of joints and supports.

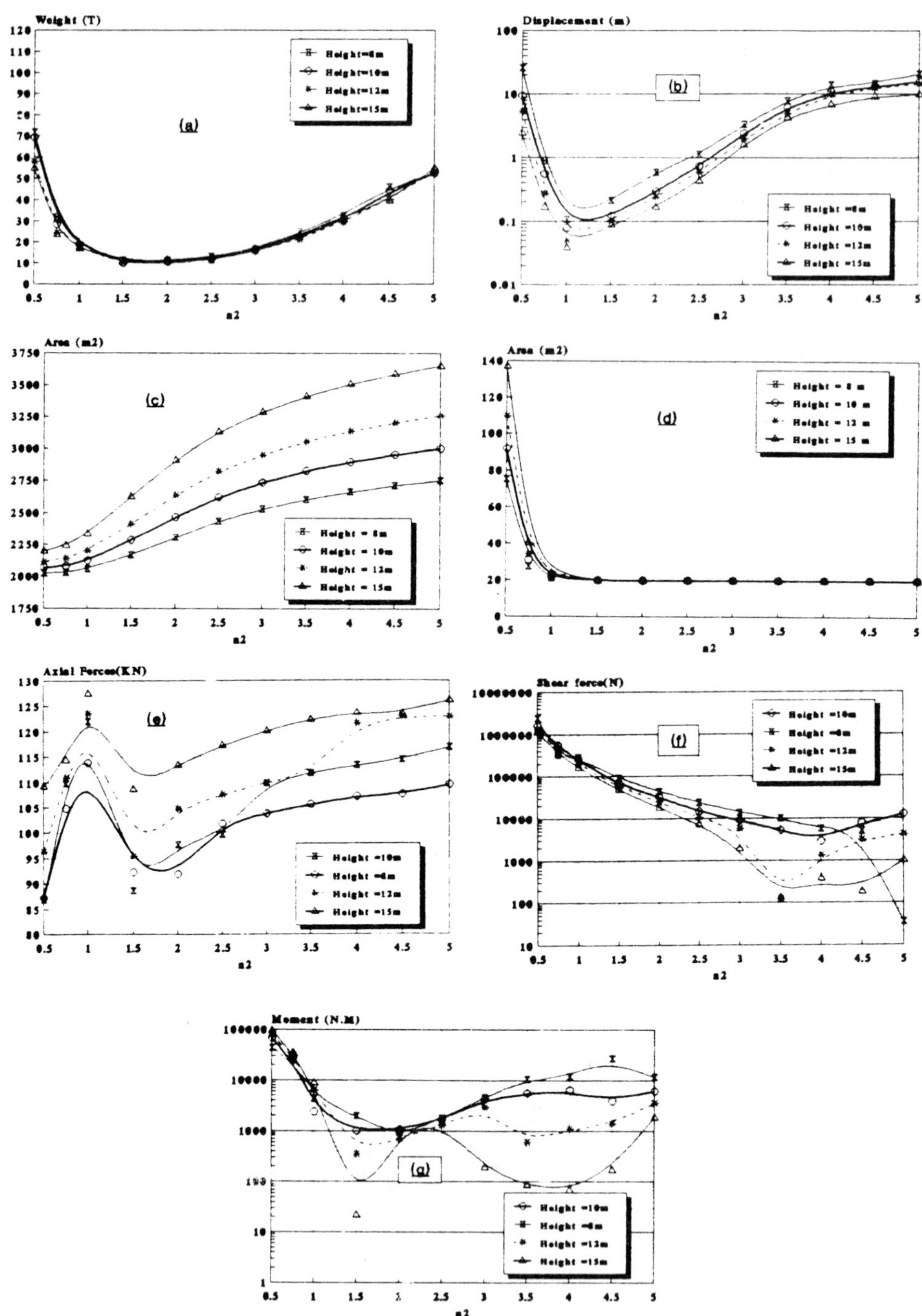

Fig. 7. Dome height effects.

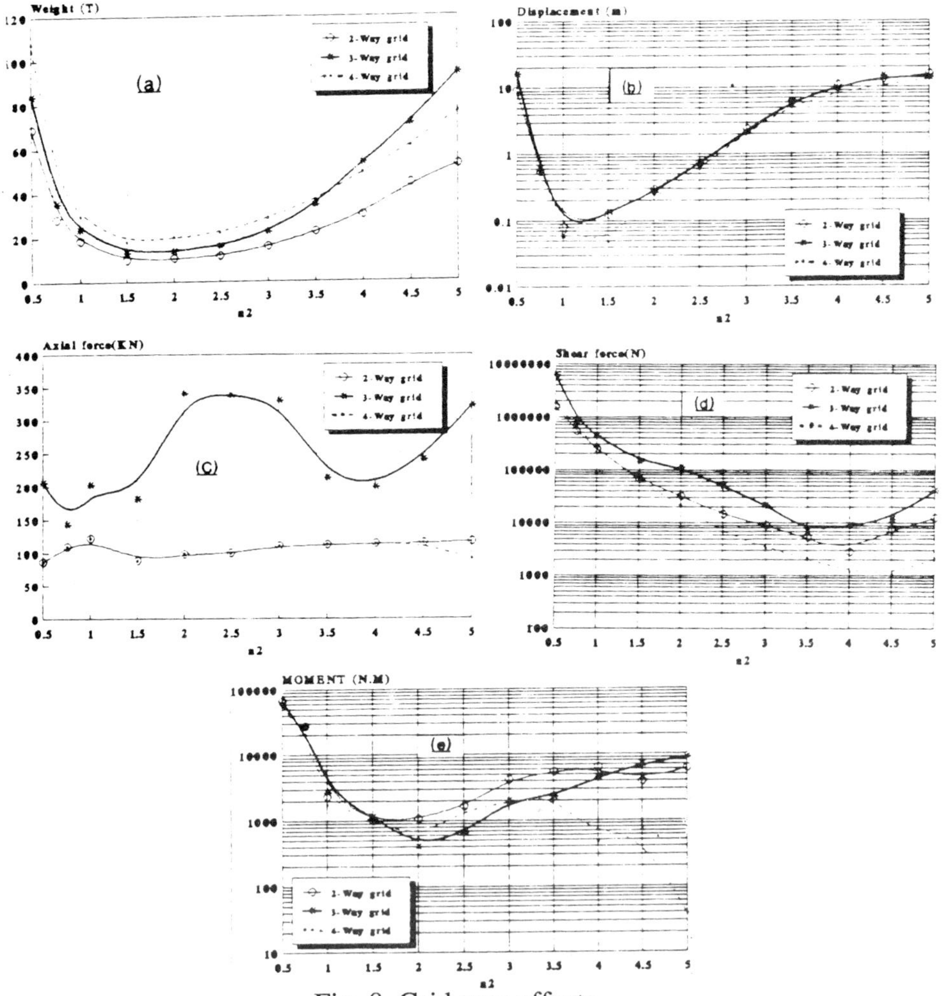

Fig. 8. Grid type effects.

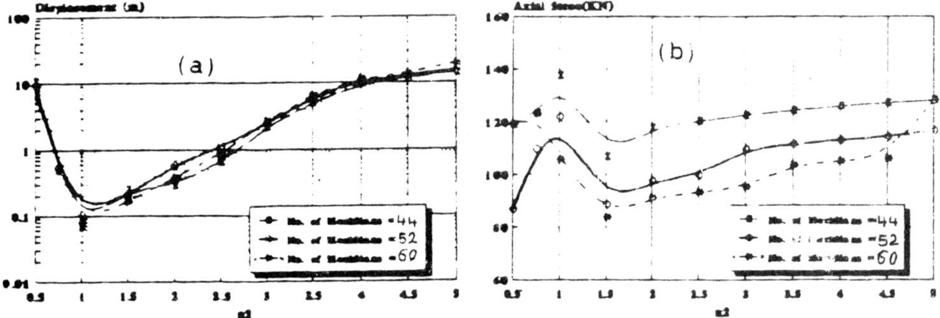

Fig. 9. Effect of number of meridians.

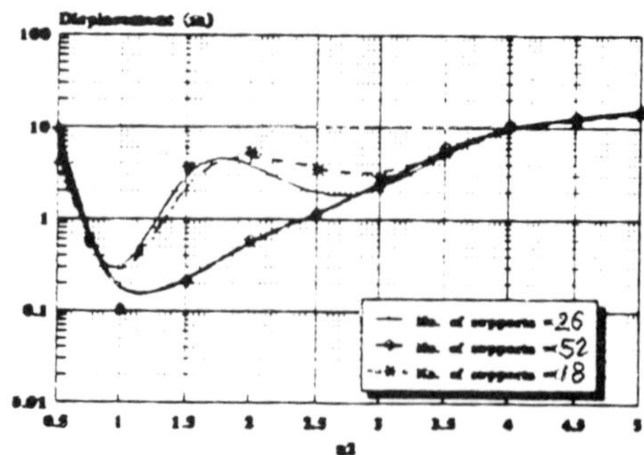

Fig. 10. Support arrangement effects.

REFERENCES

1. KENNER H. Geodesic math and how to use it, chapter 13, University of California press, Berkeley, 1976

2. HUYBERS P. ARENDS G. J., Double-elliptic vaults, analysis, design and construction of braced barrel vaults, edited by MAKOWSKI Z. S., Elsevier Applied Science publishers, 1985.

3. EL-LABBAR 0. Kenner retronorms, Proceedings of the International Symposium on Innovative Applications of Shells and Spatial Forms, Bangalore, India, 1988, 97-107.

4. NELSON & JOHNSON AutoCAD the complete reference, second edition, McGraw-Hill, 1991

5. Baco Contracts Ltd, Triodetic Structures, A British Alcan Company, Publication No. M5004/5M/7.85, London, September 1985.

6. MAKOWSKI Z. S. Part A, General concepts, analysis, design and construction of braced domes, organised by Space Structures Research Centre, University of Surrey, England, 1980.

7 Lahey Computer Systems, Lahey fortran language reference manual, Incline Village, NV 89450-6091, 1990.

9 British Steel Corporation, Structural Steelwork Handbook Sections to BS 4: Part 1: 1984.

115. A formex approach to substructuring

S. MAALEEK, University of Tehran, Iran

SYNOPSIS. A formex technique has been presented for multi-level sub-structuring. Basic sub-structures are formed as suitable data packages in terms of appropriate formex formulae in generic forms. By introducing a consistent formex approach, the process of multi-level assembly of sub-structures can be made fully automated and indeed, extremely easy. The method serves to organise a computer environment with an overall efficiency, not only in the generation of multi-level topological information of partitioned configurations, but also, in the actual sub-structure analysis process. Where necessary, new formex notions have been introduced to provide for the required capabilities.

INTRODUCTION

1. Multi-level sub-structuring of complex structures involves complications in handling the topological properties of different sub-structures at different analysis levels (refs 1,2) and the aim of the present work is to reduce these complications. Formex algebra (ref. 3) plays a paramount part in the approach presented here.

A SUPER-ELEMENT / SUPER-NODE APPROACH

2. Figs 1 and 2 show two types of spatial structural forms (taken from refs 4 and 5) which may be considered to be purely skeletal, purely continuum or combined continuum-skeletal systems. Let these structures be partitioned into a number of bounded sub-regions as shown in Figs 1(a) and 2(b). Each sub-region, Ω_m, may be assumed to correspond to a sub-structure S_m. Without introducing any further approximation, the d.o.f. associated with the internal nodes, (i.e. the nodes situated within Ω_m and also on the part of the boundaries of Ω_m on which natural boundary conditions are defined), can be eliminated in favour of interface boundary nodes (i.e. the nodes that lie on the part of the boundaries of Ω_m on which essential boundary conditions are prescribed) by static condensation. The resulting condensed matrices of sub-structure S_m now represent a super-element $\tilde{S}_m$. The set of internal nodes $\tilde{N}_{im}$ of the sub-structure S_m is defined as:

$$\tilde{N}_{im} = \{n_{im}^{(1)}, n_{im}^{(2)}, \ldots, n_{im}^{(ni)}\} \tag{1}$$

where $n_{im}^{(j)}, 1 \le j \le ni$, is an internal node and *ni* is the total number of internal nodes of S_m. Similarly, the set of boundary nodes $\tilde{N}_{bm}$ of S_m is given by:

$$\tilde{N}_{bm} = \{n_{bm}^{(1)}, n_{bm}^{(2)}, \ldots, n_{bm}^{(nb)}\} \tag{2}$$

in which $n_{bm}^{(k)}, 1 \le k \le nb$, denotes the kth boundary interface node of S_m, where *nb* is the total number of boundary interface nodes in S_m. Also, the set of

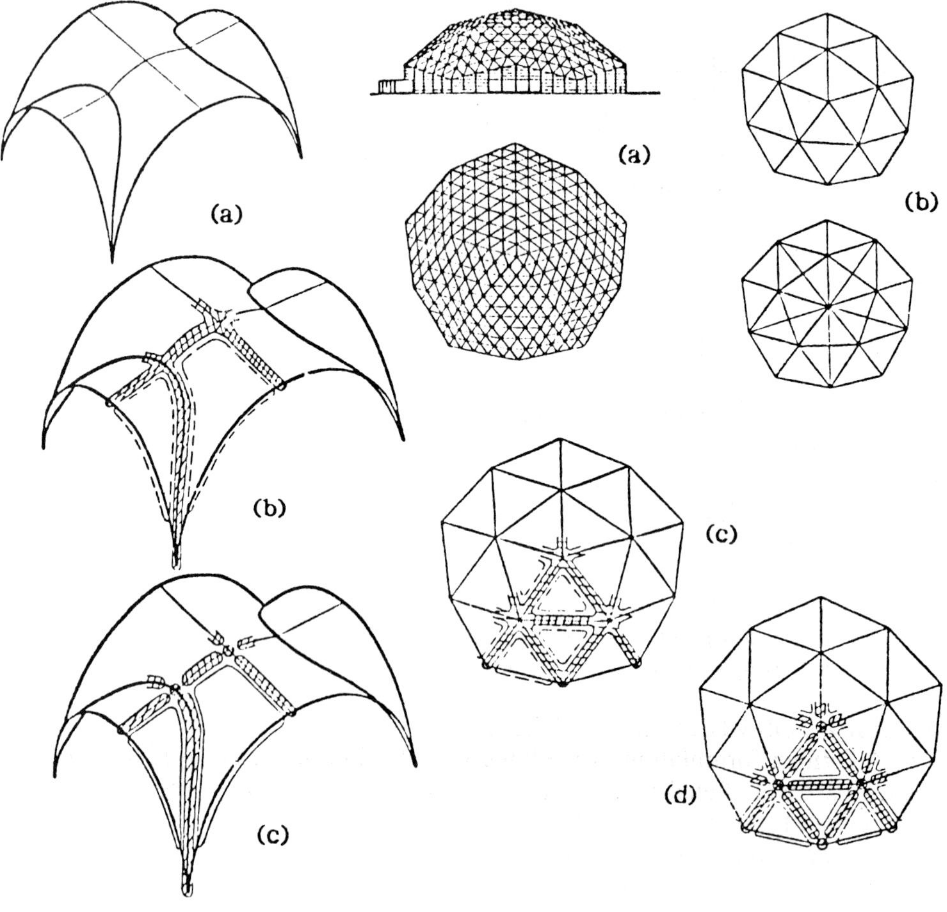

Fig. 1. Fig. 2.

total internal nodes of the whole discretised model which are to be eliminated in this level of analysis is denoted by $\tilde{N}_i$ and that of the total nodes to be retained is denoted by $\tilde{N}_b$ where:

$$\tilde{N}_i = \bigcup_{m=1}^{q} \tilde{N}_{im} \text{ and } \tilde{N}_b = \bigcup_{m=1}^{q} \tilde{N}_{bm} \tag{3}$$

where q is the number of sub-structures. Now, $\tilde{N} = \tilde{N}_i \cup \tilde{N}_b$; where $\tilde{N}$ denotes the total number of nodes in the structure.

3. The notion of super-nodes. The super-element assembly process will be greatly simplified by introducing the concept of super-nodes. A super-node is defined as a finite set of distinct nodes. A detailed treatment has been given in ref. 1.

4. Grouping procedure. An example of a suitable grouping of nodes with the aid of super-nodes is given in Figs 1 and 2. In Figs 1(b) and 2(c), for each structure, the set of total nodes $\tilde{N}$ is divided into two main groups $\tilde{N}_i$ and $\tilde{N}_b$; while Figs 1(c) and 2(d) show further subdivisions of the interface node set $\tilde{N}_b$ into smaller groups to facilitate super-element assembly process. Let super-nodes $\tilde{n}_i^{(j)} \varepsilon \tilde{N}_i$ and $\tilde{n}_b^{(k)} \varepsilon \tilde{N}_b$ represent the *j*th super-node of $\tilde{N}_i$ and the kth super-node of $\tilde{N}_b$ respectively. The node partitioning of a sub-structure may be represented by:

$$\tilde{N}_{im}^{(p)} = \{\tilde{n}_{im}\} \text{ and } N_{bm}^{(p)} = \{\tilde{n}_{bm}^{(1)}, \tilde{n}_{bm}^{(2)}, \ldots, \tilde{n}_{bm}^{(nb)}\} \tag{4}$$

where nb is the number of partitions (super-nodes) in the boundary node sets of S_m. The subscripts im and bm are used to denote internal and boundary (interface) super-nodes pertaining to the sub-structure S_m.

5. Summary of the procedure. The structure is divided into a number of (preferably similar) simple basic sub-structures/super-elements in the form of quadrilateral or trilateral sub-domains. Exploiting repetitions, the analysis can be carried out efficiently. The internal and boundary interface nodes are arranged in two separate main groups/super-nodes. The boundary nodes are further divided into smaller groups to match the appropriate interfaces.

FORMEX FORMULATION APPROACH

6. A formex formulation of the lowest level sub-structure is used to carry out the lowest level of analysis. The assembly of the resulting super-elements is then fulfilled with the aid of suitable application of appropriate compositions and transflections on super-element formices. A super-element formex should be expressed in terms of the signets representing the interface nodes. Hence, in the condensation process, not only is it needed to condense the stiffness matrices of the sub-structures, but also analogous condensation is beneficial when applied on the sub-structure formices.

7. Partitioned forms. Let F1, F2,.. ,Fn be n formices of the same grade, where $n \geq 1$ and consider a formex F such that:

$$F= F1 \# F2 \# .... \# Fn\text{-}1 \# Fn= \text{lib}(i=1,n) \mid Fi \qquad (5)$$

A plenix (ref. 6) of the form:

$$\{F1, F2, .., Fn) = \{Fi\}_{i=1,n} \qquad (6)$$

is referred to as a partitioned form of F (ref. 2). Partitioned forms may be used for the purpose of grouping.

8. Formex condensation. Formex condensation (ref. 2) is a formex procedure which performs the task of derivation of a super-element formex from a sub-structure formex.

SUB-STRUCTURE ANALYSIS OF MODULAR STRUCTURES

9. Consider the simple open web grids shown in Fig. 3 (ref. 2). A typical unit of a castellated beam is identified in Fig. 4. For illustration purposes, a simple mesh is considered.

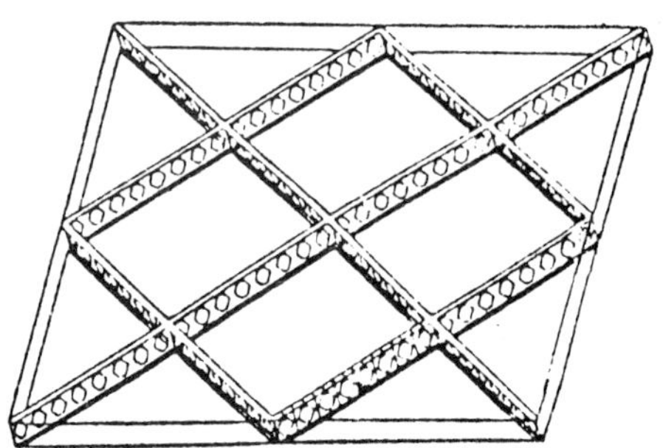
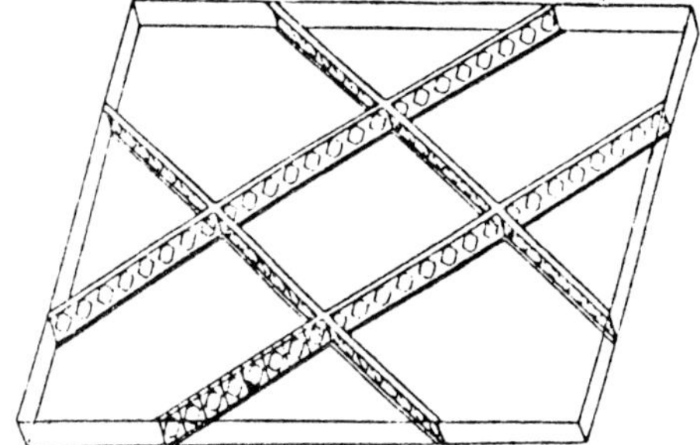

Fig. 3.

10. Formex formulation of basic sub-structures. Consider the idealisation of the web of a typical unit shown in fig. 5(a). A basic sub-structure is initially (roughly) defined by a formex of the first order, whose signets correspond to the corner nodes of the basic sub-structure. For instance, such formices (maniples) Fl and F2 for basic sub-structures No1 and 2 (denoted by S_1 and S_2) of Figs 5(b) and (c) are:

F_1= [O.0; 4,0; 4,12; 0,12] , F_2= [1,1; 7,13; 1,13]

The uniples of F_1 and F_2 are given in their local normal co-ordinate systems (U'_1,U'_2) and (U''_1,U''_2). The local co-ordinates chosen are such that for S_1, $U_1=U'_1$ and $U_2=U'_2$ and for S_2, $U_1=U''_1+7$ and $U_2=U''_2-1$. Formices F_1 and F_2 are referred to as the *father formices* for S_1 and S_2. The generation of the mesh configuration for each basic sub-structure is carried out with the aid of appropriate formex formulation (which may be available as a library of generic formulae). Distortion functions (ref. 2) have been used for the generation of the type of mesh layout for the sub-structure S_2 (ref. 2).

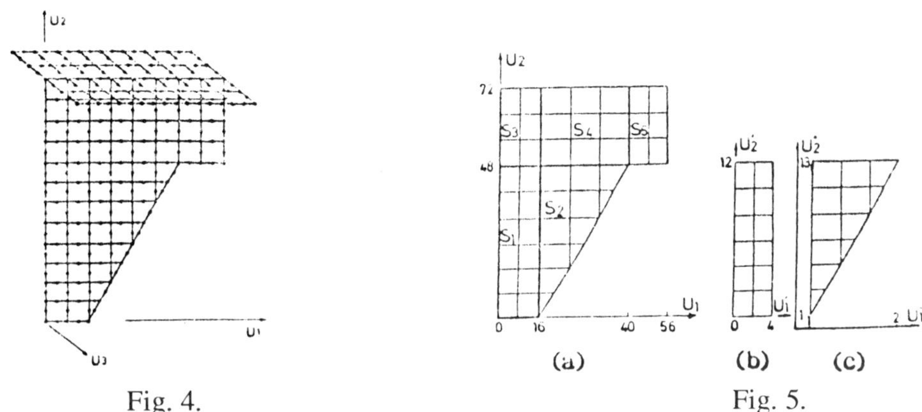

Fig. 4.

Fig. 5.

11. Super-node grouping and labelling. Fig. 6 shows some examples of such grouping for a quadrilateral basic sub-structure. Super-nodes of a basic sub-structure are automatically labelled as follows: all the nodes of a sub-structure that are to be eliminated are gathered in one group referred to as the internal super-node, and is labelled 1; the boundary interface super-node labelling is carried out in the ascending order in the direction of the local normal co-ordinates, beginning with the number 2. Let a formex F_m represent the mesh layout of sub-structure S_m. Also, let an ingot g_m be an exclusive catena of F_m. That is, g_m represents the set of total nodes of the sub-structure. In order to comply with the above super-node grouping, g_m is generated accordingly. Now, the appropriate grouping for g_m can be represented by a suitable partitioned form of g_m, denoted by $g_m^{(p,nm)}$, where *nm* is the number of the partitions (super-nodes), as follows:

$$g_m^{(p,nm)} = \{g_m^{(1)}, g_m^{(2)}, \ldots, g_m^{(nm)}\} = \{g_m^{(j)}\}_{j=1,nm} \tag{7}$$

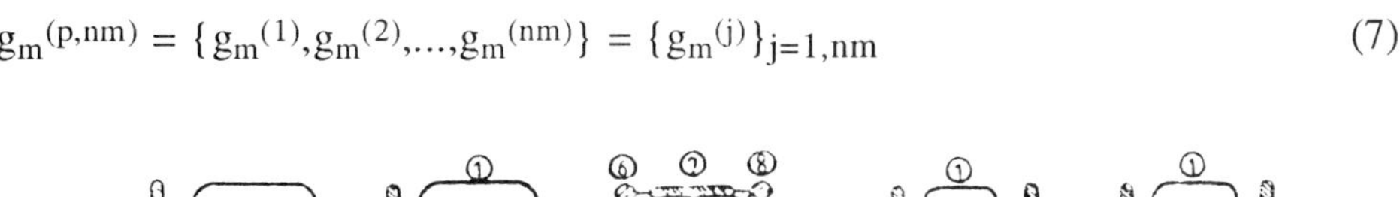

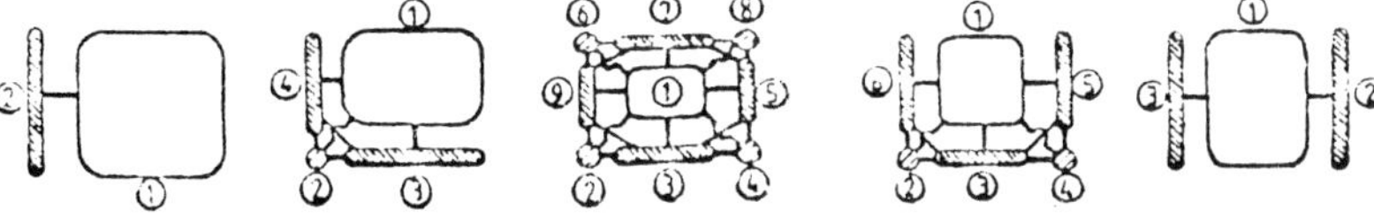

Fig. 6.

where each ingot $g_m^{(j)}$ is a nonprolate ingot representing the jth super-node of S_m denoted by $\tilde{N}_m^{(j)}$. Moreover, the set of total nodes of S_m, denoted by $\tilde{N}_m$ may also be partitioned into two main groups, $\tilde{N}_{im}$ and $\tilde{N}_{bm}$ as defined in 2. In an analogous manner, the ingots corresponding to such a partitioning can be represented by:

$$g_m{}^{(p,2)}=(g_{im}.g_{bm}) \tag{8}$$

where g_m and g_{bm} are nonprolate ingots representing the set of internal and boundary (interface) nodes respectively. The above super-node grouping also indicates the super-node labelling scheme at the super-element level given by the orderate (j) of the partitions in the partitioned form $g_m^{(p,nm)}$. Furthermore, it implies that:

$$g_{im} = g_m^{(1)}$$

and: $$g_{bm} = g_m^{(2)} \# g_m^{(3)} \# .... \# g_m^{(nm)} = \mathrm{lib}(j = 2,nm) \mid g_m^{(j)} \tag{9}$$

The ingot g_m is generated in the order of the desired super-node labelling. After elimination of the internal super-node, the resulting super-element contains the boundary nodes labelled from 2 to nm.

12. Node numbering at the basic sub-structure level is carried out using dictum functions. An example of node and super-node numbering for basic sub-structures of a typical unit is shown in Fig. 7(a).

<u>A consistent formex approach to sub-structuring</u>

13. When the condensed matrices of super-elements are obtained, the super-elements are ready for assembly. At this stage, the information regarding the connectedness of the super-elements is required. One way is to supply such information as input. This is the common way in which sub-structuring is treated in commercial packages. However, this is not to be considered as desirable in an automated multi-level analysis process. This is also inconsistent with ease of modification of the partitions and can be error prone and time consuming. It is therefore of importance to improve the current methods and provide the means for the automation of the procedure and the elimination of the (unnecessary) additional input data at this stage. From a formex standpoint, in cases where, topologically equivalent super-elements are repeatedly used, such super-elements may be considered as super-generants for the spread of information. In the present work, an automated super-element assembly process is described in which the information regarding connectedness of super-elements is created automatically from the basic sub-structure formices. The procedure is as follows: Appropriate transflections are applied on the basic sub-structure formices so that all such formices be expressed in terms of a global normal co-ordinate system. These formices are used in a searching process in which matching super-nodes are identified and the assembly of super-matrices is performed automatically

through an automatic formation of the formices representing the connectivity or super-nodes in the assembly. That is, the adjacency relations between super-nodes are automatically established; super-nodes are then grouped at the assembly of super-element level. Each super-node represents a block of nodal variables and the assembly of super-elements is carried out by positioning these blocks in their appropriate locations in the block adjacency super-matrix that indicates the block structure of the assembled (uncondensed) stiffness matrix for a higher level of analysis.

14. With regard to the definition of a partitioned form $F^{(P)}$ of a formex F, it may be considered natural for a partitioned form of a formex to be viewed as a *super-formex* and its partitions as *sub-formices*. From this point of view, we may allow the application of formex functions on partitioned forms in a natural way. For instance, let:

$$F=F_1\#F_2\#........\#F_n \quad \text{and} \quad F^{(p)}=\{F_1, F_2,..........,Fn\} \tag{10}$$

Also let θ be a formex function for which the distributive law applies; that is,

thus:
$$\begin{aligned} &\theta\,\vert\, F= \theta\vert F_1 \# \theta\vert F_2 \#......\# \theta\vert F_n \\ &(\theta\vert F)^{(p)}= \{\theta\vert F_1, \theta\vert F_2,......,\theta\vert F_n\} \end{aligned} \tag{11}$$

The direct application of θ on $F^{(p)}$ is defined as follows:

$$\theta\,\vert\, (F^{(p)})= (\theta\vert F)^{(p)}= \{\theta\vert F_1, \theta\vert F_2,......,\theta\vert F_n\} \tag{12}$$

Now, let F_m, g_m, $g_m{}^{(p,2)}$, g_{im}, g_{bm} and nm be as defined in 11. That is:

$$\begin{aligned} &g_{bm} = lux(g_{im}) \,\vert\, g_m, \\ &g_{bm}{}^{(p)} = \{g_{bm}{}^{(1)}, g_{bm}{}^{(2)},...., g_{bm}{}^{(nm-1)}\} \end{aligned} \tag{13}$$

Moreover, let:

$$g'_{bm}{}^{(p)} = \{g'_{bm}{}^{(1)},g'_{bm}{}^{(2)},...., g'_{bm}{}^{(nm-1)}\} \tag{14}$$

be a partitioned form whose jth partition $g'_{bm}{}^{(j)}$ is given by:

$$g'_{bm}{}^{(j)} = agg \,\vert\, g_{bm}{}^{(j)} \tag{15}$$

where the symbol 'agg' denotes aggregation function (ref. 2). Now, the following notation is used:

$$\left[\tilde{s}_{bm}{}^{(j)}\right] = g'_{bm}{}^{(j)} = [s_1{}^{(j)};s_2{}^{(j)};.....;s_{1j}{}^{(nm-1)}] \tag{16}$$

where 1j is the order of $g_{bm}{}^{(j)}$.

Noting that the cantles of an ingot are of the first plexitude, $s_i^{(j)}$ represents the single signet of the ith cantle of the jth partition, $g'_{bm}{}^{(j)}$, of the partitioned form $g'_{bm}{}^{(p)}$. On the other hand, $\tilde{S}_{bm}{}^{(j)}$ may be considered as an ordered collection of connected signets of the corresponding partition. That is, it is the aggregation of these signets and includes all the signets of $g_{bm}{}^{(j)}$ in connected form; while these signets in $g_{bm}{}^{(j)}$ are in an isolated form. In other words, the plexitude of the cantles of the homogeneous formex $g_{bm}{}^{(j)}$ is one, while the plexitude of the maniple $g'_{bm}{}^{(j)}$ is equal to the order of $g_{bm}{}^{(j)}$.

15. $g_{bm}{}^{(j)}$ is said to be the split image of $\tilde{s}_{bm}{}^{(j)} = g'_{bm}{}^{(j)}$. On the other hand, $\tilde{s}_{bm}{}^{(j)}$ is referred to as the complete image of $g_{bm}{}^{(j)}$. As a matter of convenience, one may write:

$$F''m = \text{agg} \mid g_{bm}{}^{(p)} = [\tilde{s}_{bm}{}^{(1)}; \tilde{s}_{bm}{}^{(2)}; .....; \tilde{s}_{bm}{}^{(nm-1)}] \quad (17)$$

where nm is the total number of super-nodes in sub-structure S_m when the internal nodes are gathered in a single super-node. Alternatively, the following notation may be used:

$$M_{bm} = \Psi(gim, gm^{(p)}) \mid F_m = [\tilde{s}_{bm}{}^{(1)}; \tilde{s}_{bm}{}^{(2)}; .....; \tilde{s}_{bm}{}^{(nm-1)}] \quad (18)$$

to represent indirect condensation of F_m through the partitioned form $g_m{}^{(P)}$ with respect to g_{im}. The construct on the right hand side implies a partitioned representation of the maniple that corresponds to the super-element s_m, resulting from the condensation of F_m. The objects $\tilde{s}_{bm}{}^{(j)}$, $1 \leq j \leq nm-1$, that are employed to represent a collection of connected signets corresponding to a complete super-node (i.e. the aggregation of these signets) are referred to as super-signets. A maniple, $M_{bm}{}^{(p)}$ of the form given above is referred to as a partitioned maniple and the number of super-signets in the partitioned maniple, i.e. nm-1 in the above, is referred to as the plexitude of the partitioned maniple which is the same as the number of super-nodes in the condensed super-element $\tilde{S}_m$. Based on the above concepts, a formex procedure for super-node labelling at a higher level of analysis has been given in ref. 1.

16. A formex approach to the assembly process. Using the concepts introduced above, let g_m, g_{im}, g_{bm}, $g_{bm}{}^{(p)}$, $g'_{bm}{}^{(p)}$, $M_{bm}{}^{(p)}$ and F_m be as defined in paragraphs 11, 14 and 15. Furthermore, let:

$$F = \text{lib}(m=1,q) \mid F''_m \quad (19)$$

where q is the number of basic sub-structures, so that:

$$F^{(p)} = \{M_{b1}{}^{(p)}, M_{b2}{}^{(p)}, ..., M_{bq}{}^{(p)}\} \quad (20)$$

and let: $g = \text{lib}(m=1,q) \mid g_{bm}$

that is $g^{(p)} = \{g_{b1}{}^{(p)}, g_{b2}{}^{(p)}, ..., g_{bq}{}^{(p)}\}$ (21)

where a principal panel $g_{bm}^{(p)}$ of $g^{(p)}$ is itself a partitioned form. Also let:

$$h^{(p)} = pex \mid g^{(p)} = (pex \mid g)^{(p)} \tag{22}$$

The order of this partitioned form equals the number of super-nodes in the assembly. Now a partitioned form $h'^{(p)}$ is constructed by rearranging the principal panels of $h^{(p)}$ so that all the partitions corresponding to *ni* internal super-nodes in the assembly precede those corresponding to nb interface super-nodes. It is also convenient to form partitioned forms $h'_i{}^{(p)}$ and $h'_b{}^{(p)}$ (as given in 18) where $h'_i{}^{(p)}$ contains those ingots of $h^{(p)}$ that represent the internal super-nodes at a higher level assembly and $h'_b{}^{(p)}$ contains those partitions of $h^{(p)}$ representing boundary super-nodes.

17. Super-node labelling. By the application of a transformation analogous to the dictum function on the resulting partitioned form $h'^{(p)}$, the super-nodes can be labelled. Let a formex F' be obtained from F by replacing every super-signets $\tilde{S}_{bm}^{(j)}$ of every partitioned maniple $M_{bm}^{(p)}$ of F by the orderate Oi of its split image $h'^{(Oi)}$ in $h'^{(p)}$. The relationship between F' and F may be written as:

$$F' = dic(h'^{(P)}) \mid F \tag{23}$$

F' contains the super-element/super-node list at the higher level assembly.

18. Remarks. To the partitioned forms:

$$h'_i{}^{(p)}=\{h'_i{}^{(1)},h'i^{(2)},....,h'_i{}^{(\hat{n}i)}\}$$

and $$h'_b{}^{(P)}=\{h'_b{}^{(1)},h'_b{}^{(2)},...., h'_b{}^{(\hat{n}b)}\} \tag{24}$$

there corresponds plenices l_i, with the same order as $h'_i{}^{(p)}$, and 1_b, with the same order as $h'_b{}^{(p)}$, such that:

$$li=\{l_i^{(1)}, l_i^{(2)}, ..., l_i^{(\hat{n}i)}\} \tag{25}$$

where the jth panel $l_i^{(j)}$ of li, $1 \leq j \leq ni$ is an integer equal to the order of the jth ingot $h'i^{(j)}$ of $h_i'^{(p)}$; and:

$$1_b = \{1_b^{(1)}, 1_b^{(2)}, ... 1_b^{(\hat{n}b)}\} \tag{26}$$

where the kth panel $1_b^{(k)}$ of 1_b, $l \leq k \leq \hat{n}b$, is an integer equal to the order of the kth ingot $h'_b{}^{(k)}$ of $h'_b{}^{(p)}$.

19. Condensation process at a higher level of analysis. Consider the case study of the castellated beam and let subscript u be used to indicate that the associated symbols are concerned with a typical unit. Consideration of Figs 7(a) and 7(b), with $\hat{n}i = 7$ and $\hat{n}b = 7$ leads to:

$h'_{iu}{}^{(p)} = \{h'_{iu}{}^{(1)}, h'_{iu}{}^{(2)}, \ldots\ldots, h'_{iu}{}^{(7)}\} = \{h'_{iu}{}^{1}, h'_{iu}{}^{2}, \ldots, h'_{iu}{}^{7}\}$ (27)
where the notation $h'_{iu}{}^{j}$ is used to represent the ingot that corresponds to the internal super-node j of Fig. 7(b). Thus:

$$l_{iu} = \{11,3,1,5,1,6,6\} \tag{28}$$

gives the number of the nodes in the internal super-nodes; so that there is a one to one correspondence between the iso-images of $h'_{iu}{}^{(p)}$ and 1_{iu}. Similarly:

$$h'_{bu}{}^{(p)} = \{h'_{bu}{}^{(1)}, h'_{bu}{}^{(2)}, \ldots., h'_{bu}{}^{(7)}\} = \{h'_{bu}{}^{8}, h'_{bu}{}^{9}, \ldots., h'_{bu}{}^{14}\} \tag{29}$$

where $h'_{bu}{}^{k}$ represents the ingot that corresponds to the boundary super-node k of Fig. 7(b). Thus:

$$l_{bu} = \{1,3,1,1,11,6,7\} \tag{30}$$

The condensed partitioned maniple F''_{u} (also denoted by M_{bu}), that represents the condensed typical unit, can be obtained from:

$$M_{bu} = F''_{u} = \Psi(h'_{iu}{}^{(p)}, h'_{u}{}^{(p)}) \mid F_{u}{}^{(p)} = [\tilde{S}_{bu}{}^{(1)}; \tilde{S}_{bu}{}^{(2)}; \ldots.; \tilde{S}_{bu}{}^{(\hat{n}b)}] \tag{31}$$

where $h'_{u}{}^{(p)} = \{h'_{iu}{}^{(p)}, h'_{bu}{}^{(p)}\}$ and $F_{u}{}^{(p)}$ is given by:

$$F_{u}{}^{(p)} = \{M_{bi}{}^{(p)}, M_{b2}{}^{(p)}, \ldots\ldots, M_{bq}{}^{(p)}\} \tag{32}$$

Furthermore, let $\tilde{S}_{bu}{}^{k}$ be the super-signets that represents the boundary super-node k of Fig. 7(b); thus:

$$M_{bu} = F''_{u} = [\tilde{S}_{bu}{}^{8}; \tilde{S}_{bu}{}^{9}; \ldots.; \tilde{S}_{bu}{}^{14}] \tag{33}$$

It is to be noted that the super-node labelling is implied by the order of appearance of the corresponding ingots in the partitioned form:

$$h'_{u}{}^{(p)} = \{h'_{iu}{}^{(p)}\ h'_{bu}{}^{(p)}\} \tag{34}$$

and connectivity of the assembly is represented by:

$$F'_{u} = \mathrm{dic}(h'_{u}{}^{(p)})\ F_{u} \tag{35}$$

Now let the set of internal super-nodes of Fig. 7(b) be gathered into a single partition $h'_{iu}{}^{(p,1)}$ as shown in Fig. 7(c); so that:

$$h'_{u}{}^{(p,8)} = \{h'_{iu}{}^{(p,1)}, h'_{bu}{}^{(p)}\}$$
$$h'_{u}{}^{(p,8)} = \{h'_{iu}{}^{(p,1)}, h'_{bu}{}^{8}, h'_{bu}{}^{9}, \ldots., h'_{bu}{}^{14}\} \tag{36}$$

Moreover, let a super-node labelling be assigned to the condensed typical unit in accordance with the partitioned form $h'_u{}^{(p,8)}$. The resulting super-node numbers and the appropriate partitioning are given in Fig. 7(c). Consequently:

$$F'_u = dic(h'_{iu}{}^{(p,8)'} \,\vdots\, F''_u \tag{37}$$

is a maniple whose signets are of the first grade and represents the typical unit in terms of the super-node labelling of Fig. 7(c) assigned after condensation. That is, F'_u = [2,3,4,5,6,7,8]. A similar procedure can be employed at the typical module level (ref. 1).

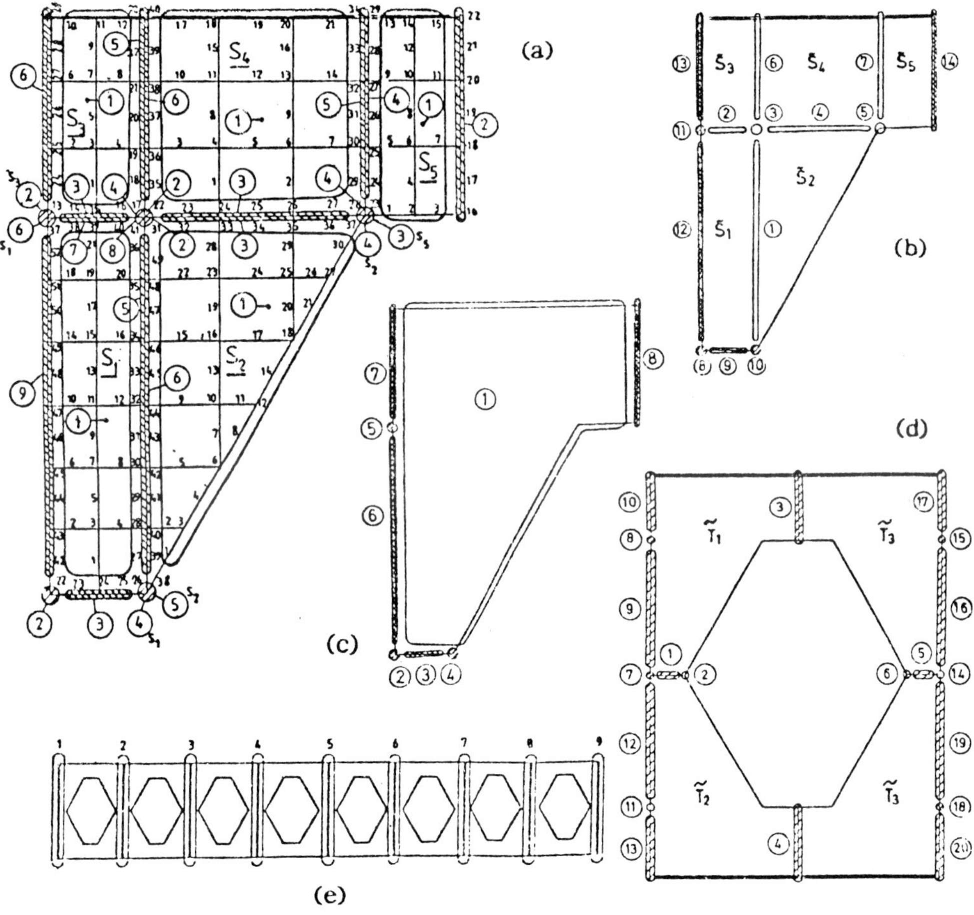

Fig. 7.

20. The case of modular structures with translational symmetry. These structures (e.g. a castellated beam) are composed of a chain of typical modules as shown in Figs 7(d) and 7(e). The properties of the whole assembly in such systems are characterised by the repeating typical module. These repeating geometric and physical properties are in fact reflected in the structure of the corresponding coefficient matrix. The case of cyclically symmetric structures has also been described briefly in ref. 1.

21. The castellated beam super-element. When a castellated beam itself is a component member of a more complex structure, the condensation of the block tri-diagonal system results in a reduced system that represents the castellated beam super-element. By the application of appropriate transflections on the formices representing the castellated beam super-element, grids of the type shown in Fig. 3 can be treated (ref. 2).

CONCLUDING REMARKS

22. The above formex procedure can be employed to establish the super-element / super-node adjacency relations in block forms which govern the block structure of the resulting matrices. The ideas introduced in this work have been used in the implementation of a model computer program with an automated multi-level sub-structure analysis capability.

REFERENCES

1. MAALEK S. Finite element analysis of modular structures and the behaviour of castellated beams. PhD thesis, University of Manchester Institute of Science and Technology, U.K., 1990, pp 54-150.

2. MAALEK S. A formex formulation for sub-structure analysis of open web grids. International Journal of Space Structures, 1989, vol.4, No. 1, pp 43-64.

3. NOOSHIN H. Formex configuration processing in structural engineering, Elsevier applied science publishers, London, 1984.

4. GHEROGHIU A. and DRAGOMIR V. Geometry of structural forms, p.69. Applied science publishers, London, 1978.

5. EBERLEIN H. Single and double-layer MERO domes, p.554, in: Analysis, design and construction of braced domes, edited by Z.S. Makowski. Granada, 1984.

6. HARISTCHIAN M. Formex and plenix structural analysis. PhD thesis, University of Surrey, U.K., 1980.

116. An approach for generation of geodesic forms

H. NOOSHIN and D. TZOURMAKLIOTOU, University of Surrey, UK

SYNOPSIS. The generation of geodesic forms has presented a challenge for engineers and architects for decades. The objective of the present paper is to show how the concepts of a mathematical tool called formex algebra can be used to represent and generate geodesic configurations in a convenient manner. Generation of geodesic forms is solved in two stages. Firstly, a function called the "polyhedron function" is used to generate a configuration modelled on a polyhedron. The resulting configuration is referred to as a "polyhedric configuration". In the next stage, a transformation referred to as the "tractation retronorm" is employed to obtain the projection of the polyhedric configuration on one or more specified surfaces. This transformation allows choice of different types of surfaces such as sphere, ellipsoid and paraboloid.

INTRODUCTION

1. Geodesic forms constitute an important family of structural systems. They are efficient and appealing and are employed frequently for dome structures. Data generation for geodesic forms has always been a difficult problem, from the days of hand data generation using tables of coefficients to today's specially written computer programs with many limitations and shortcomings. However, the approach presented in this paper provides a methodology that allows data generation for geodesic forms of all kind to be handled with ease and elegance. The method is based on the concepts of formex algebra and its programming language Formian. In actually using the method one has to be familiar with the formex concepts. However, the present paper is written in such a way that allows a reader to follow the basic ideas without any knowledge of formex algebra.

POLYHEDRIC AND GEODESIC FORMS

2. Consider the configuration shown in Fig. 1. This is obtained by first placing the triangular pattern of Fig. 2 on the top five faces of an icosahedron, as in Fig. 3, and then projecting the resulting configuration on a sphere which is concentric with the icosahedron, using the centre of the sphere as the centre of projection.

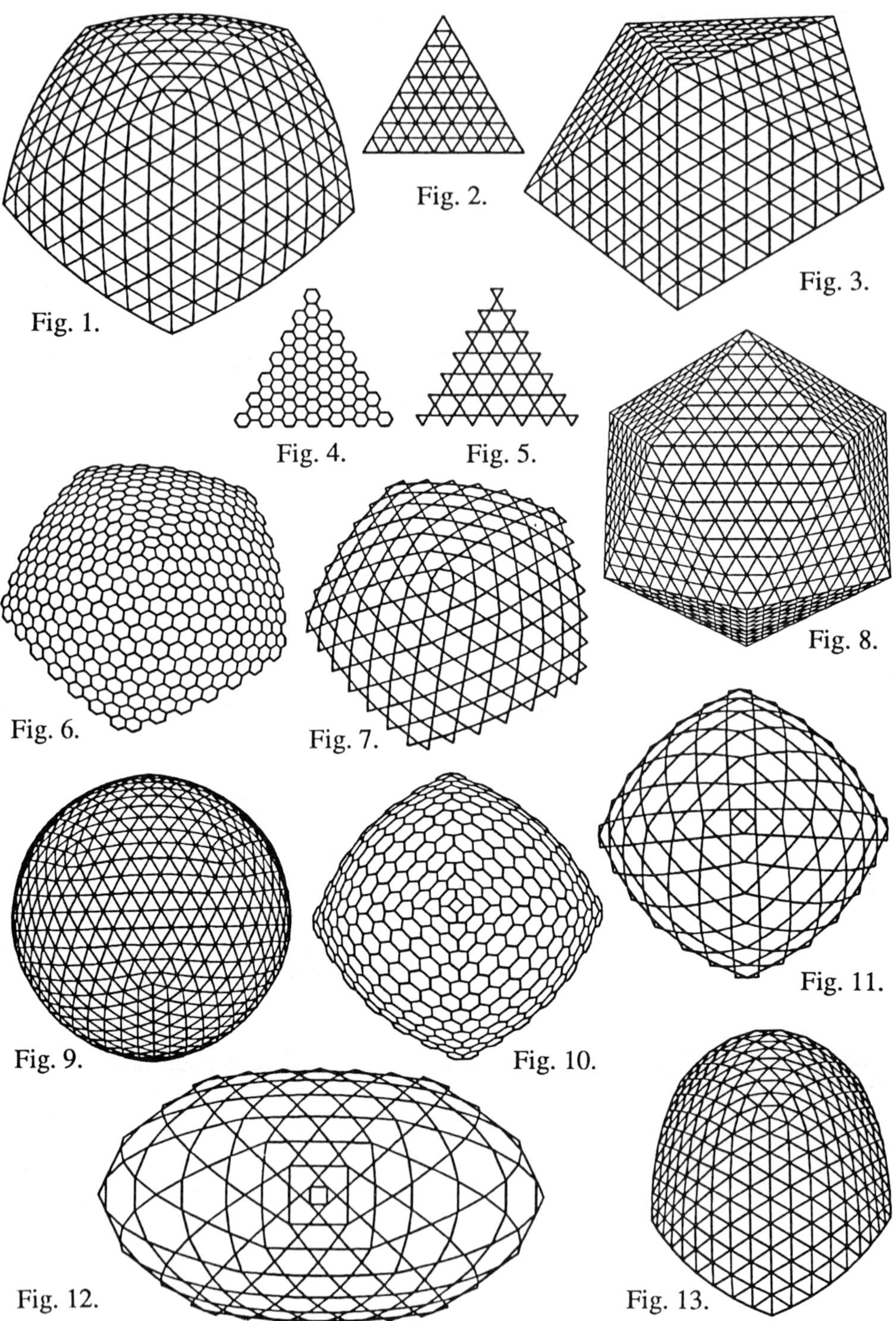
Fig. 1.
Fig. 2.
Fig. 3.
Fig. 4.
Fig. 5.
Fig. 6.
Fig. 7.
Fig. 8.
Fig. 9.
Fig. 10.
Fig. 11.
Fig. 12.
Fig. 13.

3. The configuration of Fig. 1 is an example of a "geodesic form" or "geodesic configuration". Also, the configuration of Fig. 3 is an example of a "polyhedric form" or "polyhedric configuration". The last two terms, in general, are used to refer to a configuration that is obtained by placing one or more patterns on one or more faces of a polyhedron. A polyhedron which is used as the basis for creation of a polyhedric or geodesic form is referred to as the "base polyhedron" of the polyhedric or geodesic form.

4. The above described procedure for obtaining a geodesic form involves a number of changeable features: Firstly, the pattern that has been mapped onto the faces of the polyhedron may be changed. For instance, the patterns of Figs 4 and 5, used instead of that of Fig. 2, will give rise to the geodesic forms shown in Figs 6 and 7. Secondly, the part of the polyhedron on which the pattern is mapped may be varied. For instance, the pattern of Fig. 2 mapped onto the faces on one side of an icosahedron will give rise to the polyhedric form shown in Fig. 8 and the geodesic form shown in Fig. 9. Thirdly, the type of polyhedron may be changed. For instance, the patterns of Figs 4 and 5 mapped onto the top four faces of an octahedron will give rise to the geodesic forms of Figs 10 and 11. Fourthly, the surface on which projection is obtained may be changed. Thus, instead of a sphere concentric with the polyhedron, one may use a nonconcentric sphere or a different type of surface altogether. For instance, the geodesic form of Fig. 11 will turn into that of Fig. 12 if an ellipsoid is used instead of a sphere. Also, Fig. 13 shows the result of projection of the polyhedric configuration of Fig. 3 onto a paraboloid.

5. Finally, the type of projection may be altered. To elaborate, the projection of every polyhedric configuration in the examples so far has been achieved using the centre of the polyhedron as the "centre of projection". However, there are other ways of producing the projected forms. Four different types of projection are considered in the present work. Namely, (i) "central projection", where the projection Q of a point P on a surface S is obtained as the intersection with S of a straight line that connects P to a given "centre of projection", as shown in Fig. 14, (ii) "parallel projection", where the projection Q of a point P on a surface S is obtained as the intersection with S of a straight line that passes through P and is parallel to a given "direction of projection", as shown in Fig. 15, (iii) "axial projection", where the projection Q of a point P on a surface S is obtained as the intersection with S of a straight line that passes through P and intersects a given "axis of projection", perpendicularly, as shown in Fig. 16 and, finally, (iv) "radial projection" which is a special case of central projection and is only applicable in relation to a surface S that has a centre (such as a sphere or an ellipsoid) and where the projection Q of a point P on S is obtained as the intersection with S of a straight line that connects P with the centre of the surface, as shown in Fig. 17.

6. In a Formian environment, a formex representing a geodesic form may be generated in two stages. In the first stage, a function is used to create a formex representing a polyhedric configuration. This formex is then transformed in the second stage into another formex representing the projection of the polyhedric

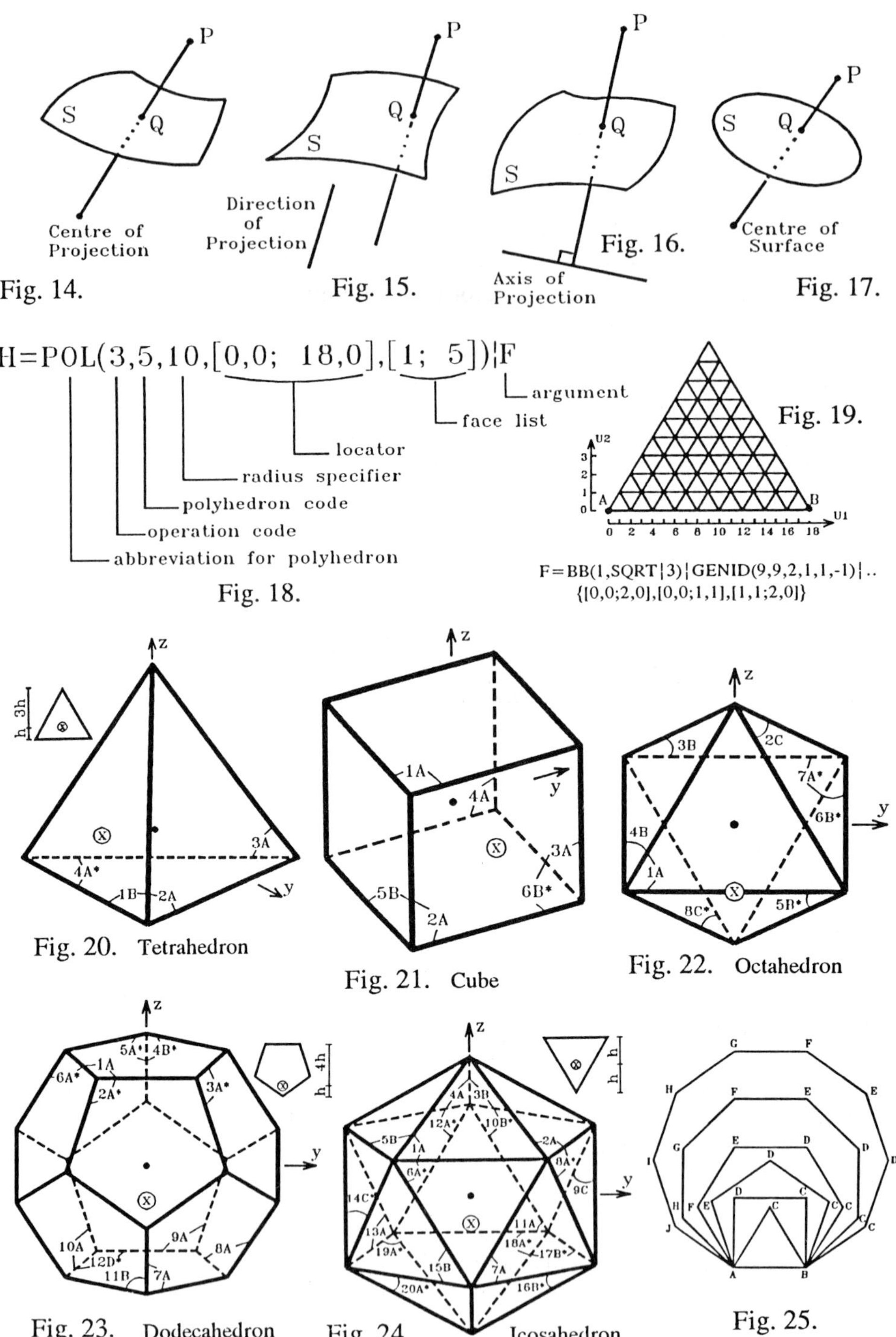

Fig. 14.

Fig. 15.

Fig. 16.

Fig. 17.

Fig. 18.

Fig. 19.

Fig. 20. Tetrahedron

Fig. 21. Cube

Fig. 22. Octahedron

Fig. 23. Dodecahedron

Fig. 24. Icosahedron

Fig. 25.

configuration on one or more specified surfaces. These two stages are discussed in the sequel.

POLYHEDRON FUNCTION

7. A "polyhedron function" may be used to generate a formex representing a polyhedric configuration. For instance, the polyhedric configuration of Fig. 3 can be represented by the "formex variable" H, as shown in Fig. 18.

8. In actually using Formian for generation of polyhedric or geodesic configurations, one must be familiar with "formex algebra". However, reference to formex algebra in this paper is kept to a minimum and a reader who is not familiar with formex algebra can follow the material without any problem. Such a reader may simply assume that a "formex" is a mathematical entity that can represent a configuration and ignore the details of the formulations.

9. Returning to the discussion of the arrangement shown in Fig. 18, the constituent parts of the formulation are as follows: POL is an abbreviation for "polyhedron" and is followed by a sequence of parameters enclosed in parentheses. The "operation code" specifies the type of action which is required to be performed. The integer 3 given as the operation code in Fig. 18 specifies the operation of mapping a pattern onto the faces of a polyhedron. There are a number of other operations that can be performed through a polyhedron function and these operations may be specified by suitable code numbers, as will be briefly discussed later. The "polyhedron code" specifies the type of polyhedron which is to be used as the basis for the operation. Table 1 lists the code numbers for the Platonic and Archimedean polyhedra. The integer 5 given as the polyhedron code in Fig. 18 specifies an icosahedron. The "radius specifier" determines the size of the polyhedron by specifying the radius of the circumsphere of the polyhedron, that is, the sphere that contains all the vertices of the polyhedron. This parameter is given as 10 (units of length) in Fig. 18. The "locator" specifies the manner in which a given pattern is to be mapped onto a face of the polyhedron. To elaborate, consider the pattern shown in Fig. 19. This pattern is the same as that of Fig. 2 but is shown here together with a reference coordinate system. A formex variable F representing the pattern relative to the reference system U1-U2 is also given in Fig. 19. Two corners of the pattern are denoted by the letters A and B. The pattern is intended to be placed on a face of the polyhedron in such a way that AB fits an edge of the polyhedron. This intention is conveyed by including the U1-U2 coordinates of A and B in the locator of Fig. 18.

10. The role of the "face list" in Fig. 18 is to specify those faces of the polyhedron onto which the pattern is to be mapped. The specified faces are 1 to 5. The "face numbers" for the Platonic polyhedra are shown in Figs 20 to 24. In these figures, each polyhedron is shown together with the global Cartesian x-y-z coordinate system. The origin of the coordinate system is at the centre of the polyhedron and is indicated with a large dot (The centre of a polyhedron is the same as the centre of its circumsphere). The point where the positive side of the x-axis intersects the polyhedron is indicated by a little circle with an enclosed x. This point is referred to as the "x-point". The positions of the positive directions

of y- and z-axes are indicated by arrows, with z being always vertical. Each face of a polyhedron has an associated "face code" which is given at one corner of the face. A face code consists of a number followed by a letter and possibly followed by an asterisk. The number in a face code is the "identification number" of the face. The letter in a face code determines the point A, B, C,.., etc of the pattern that is to be placed on the indicated corner of the face. These letters for patterns corresponding to different shapes of polyhedral faces are shown in Fig. 25. As far as the Platonic polyhedra are concerned, a face is an equilateral triangle, a square or a pentagon. However, Fig. 25 also contains information about the corner letters for a hexagon, an octagon and a decagon. This information relates to Archimedean polyhedra, as discussed latter. If a face code has an asterisk, it implies that the pattern placed on the face is the reflection, with respect to a horizontal plane, of the given pattern.

11. The "argument" in Fig. 18 is the formex variable that represents the pattern of Fig. 19. The symbol preceding the argument in Fig. 18 is used to separate the polyhedron function from its argument.

12. As the next example, consider the configuration shown in Fig. 26. This is a view of a polyhedric configuration obtained by mapping the diagonal pattern of Fig. 27 onto three faces of a cube. The formulation of the formex variable C1 that represents the polyhedric form of Fig. 26 is given in the same figure where the argument P1 represents the diagonal pattern as given in Fig. 27.

13. In creating a polyhedric form, it is possible to use different patterns for different faces of the base polyhedron. For instance, in the case of the polyhedric configuration of Fig. 28, the diagonal pattern of Fig. 27 is placed onto the 2nd and 3rd faces of a cube and the pattern of Fig. 29 is placed onto the 1st face. An important point to be noticed here is that the pattern of Fig. 29 consists of "tile" elements whereas the pattern of Fig. 27 consists of "line" elements. The difference between the element types is implied through the formex formulations in Figs 27 and 29. Namely, in the formulation of Fig. 27 an element is specified by the coordinates of its two end points, whereas in the formulation of Fig. 29 an element is specified by the coordinates of its four corner points. In general, there are no restrictions regarding the types of elements that may be involved in patterns for creation of polyhedric configurations.

14. A pattern that is mapped onto a face of a polyhedron need not necessarily "fill" the face or "match" the boundaries of the face, as demonstrated by the polyhedric configuration shown in Fig. 30. Here, the pattern of Fig. 31 which has a rectangular shape of boundary is mapped onto the 1st face of a cube and the pattern of Fig. 32 which has a triangular shape of boundary is mapped onto the 2nd and 3rd faces of the cube. The pattern that is placed onto the 1st face extends beyond the edges of the face and the pattern that is placed onto the 2nd and 3rd faces cover the faces only partially.

15. A polyhedric configuration may have more than one layer. For example, consider the configuration shown in Fig. 33. This is a view of a double layer polyhedric configuration which is based on the top five faces of two concentric icosahedra. The double layer pattern used for mapping is shown in Fig. 34. In

Table 1

Polyhedron	Code	Polyhedron	Code
Tetrahedron	1	Small Rhombicuboctahedron	10
Cube (Hexahedron)	2	Great Rhombicuboctahedron	11
Octahedron	3	Icosidodecahedron	12
Dodecahedron	4	Truncated Dodecahedron	13
Icosahedron	5	Truncated Icosahedron	14
Truncated Tetrahedron	6	Snub Cube	15
Cuboctahedron	7	Small Rhombicosidodecahedron	16
Truncated Cube	8	Great Rhombicosidodecahedron	17
Truncated Octahedron	9	Snub Dodecahedron	18

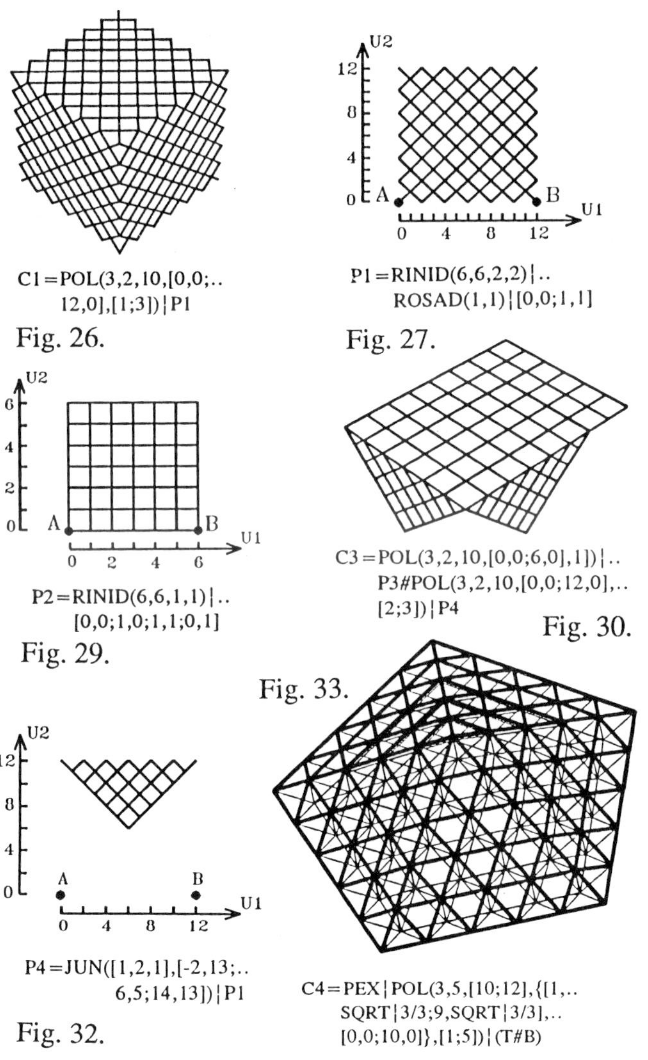

C1=POL(3,2,10,[0,0;.. 12,0],[1;3])|P1

Fig. 26.

P1=RINID(6,6,2,2)|.. ROSAD(1,1)|[0,0;1,1]

Fig. 27.

P2=RINID(6,6,1,1)|.. [0,0;1,0;1,1;0,1]

Fig. 29.

C3=POL(3,2,10,[0,0;6,0],1])|.. P3#POL(3,2,10,[0,0;12,0],.. [2;3])|P4

Fig. 30.

P4=JUN([1,2,1],[-2,13;.. 6,5;14,13])|P1

Fig. 32.

C4=PEX|POL(3,5,[10;12],{[1,.. SQRT|3/3;9,SQRT|3/3],.. [0,0;10,0]},[1;5])|(T#B)

Fig. 33.

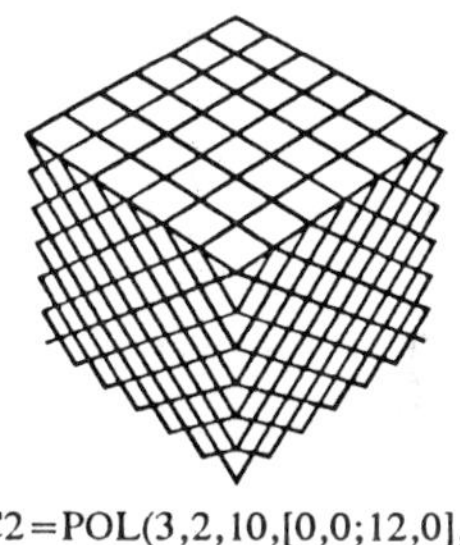

C2=POL(3,2,10,[0,0;12,0],.. [2;3])|P1#POL(3,2,10,.. [0,0;6,0],1)|P2

Fig. 28.

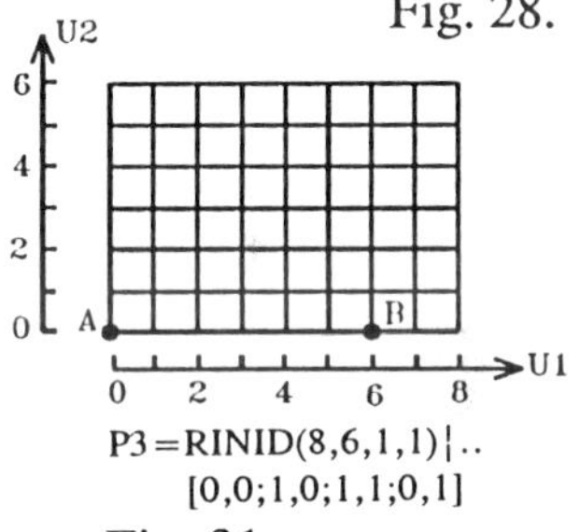

P3=RINID(8,6,1,1)|.. [0,0;1,0;1,1;0,1]

Fig. 31.

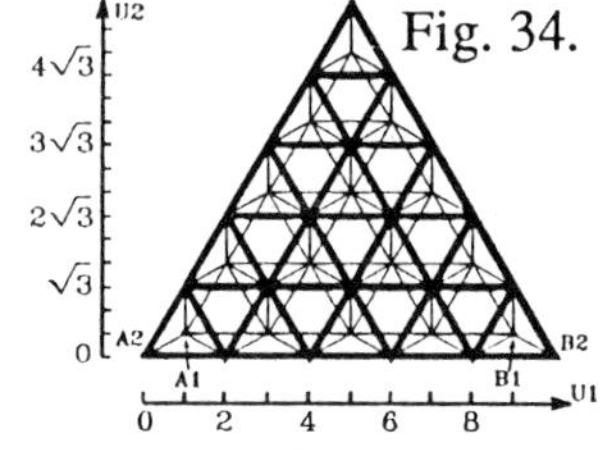

Fig. 34.

T=GENID(5,5,2,SQRT|3,1,-1)|.. ROSAD(1,SQRT|3/3,3,120)|{[0,0,2;.. 2,0,2],[0,0,2;1,SQRT|3/3,1]}
B=GENID(4,4,2,SQRT|3,1,-1)|.. ROSAD(2,2*SQRT|3/3,3,120)|.. [1,SQRT|3/3,1;3,SQRT|3/3,1]

Figs 33 and 34, the top layer elements are drawn in thick lines and the bottom layer elements as well as the web elements are drawn in thin lines. The new points to be noticed in this example are: (i) In formulating the pattern of Fig. 34, each layer has an identification number. The identification number of the bottom layer is 1 and that of the top layer is 2. These identification numbers appear as "third direction coordinates" in the formex formulations. (ii) The radius specifier in Fig. 33 contains two entries. The first entry, that is, 10, gives the radius of the circumsphere of the base polyhedron for the first layer (bottom layer) and the second entry, that is, 12, gives the radius of he circumsphere of the base polyhedron for the second layer (top layer). (iii) The locator in Fig. 33 contains two parts. The first part of the locator specifies the coordinates of points A1 and B1 (in Fig. 34) that should fit the end vertices of an edge of the base polyhedron of the first layer (bottom layer). The second part of the locator specifies the coordinates of points A2 and B2 (in Fig. 34) that should fit the end vertices of an edge of the base polyhedron of the second layer (top layer). Situations when polyhedric configurations involve more than two layers are dealt with in an analogous manner.

16. In the case of Archimedean polyhedra the faces are not all of the same shape. In spite of this, one may create a polyhedric configuration by mapping a single pattern on all or some of the faces of an Archimedean polyhedron. However, in creating polyhedric configurations that are based on Archimedean polyhedra, normally, one would use different patterns for different face shapes. For instance, the polyhedric configuration of Fig. 35 is based on an icosidodecahedron and is obtained by mapping the patterns of Figs 36 and 37 onto the triangular and pentagonal faces, respectively. A selection of Archimedean polyhedra together with their face codes and axes are shown in Figs 38 to 46.

17. The applications of polyhedron function as described so far relate to a particular mode of operation that can generate suitable polyhedric forms as the first stage in creation of geodesic forms. However, the polyhedron function may also be employed for other types of applications. For instance, the function may be used to place configurations along the edges or on vertices of a polyhedron. Figure 47 provides an example of these other applications. The configuration shown in Fig. 47 is obtained by placing the configuration of Fig. 48 on the edges of a tetrahedron. The value of the operation code in a polyhedron function specifies the type of operation required. However, the discussion of the details is beyond the scope of the present work.

TRACTATION RETRONORM

18. A geodesic configuration may be represented by a formex. For example, the geodesic configuration of Fig. 1 may be represented by the formex variable G whose formulation is shown in Fig 49. The constituent parts of this formulation are as follows: TRAC is an abbreviation for "tractation" and is followed by a sequence of parameters enclosed in parentheses. The construct that consists of TRAC together with the ensuing parameter list is referred to as a "tractation retronorm". The "projection code" specifies the type of projection to be used.

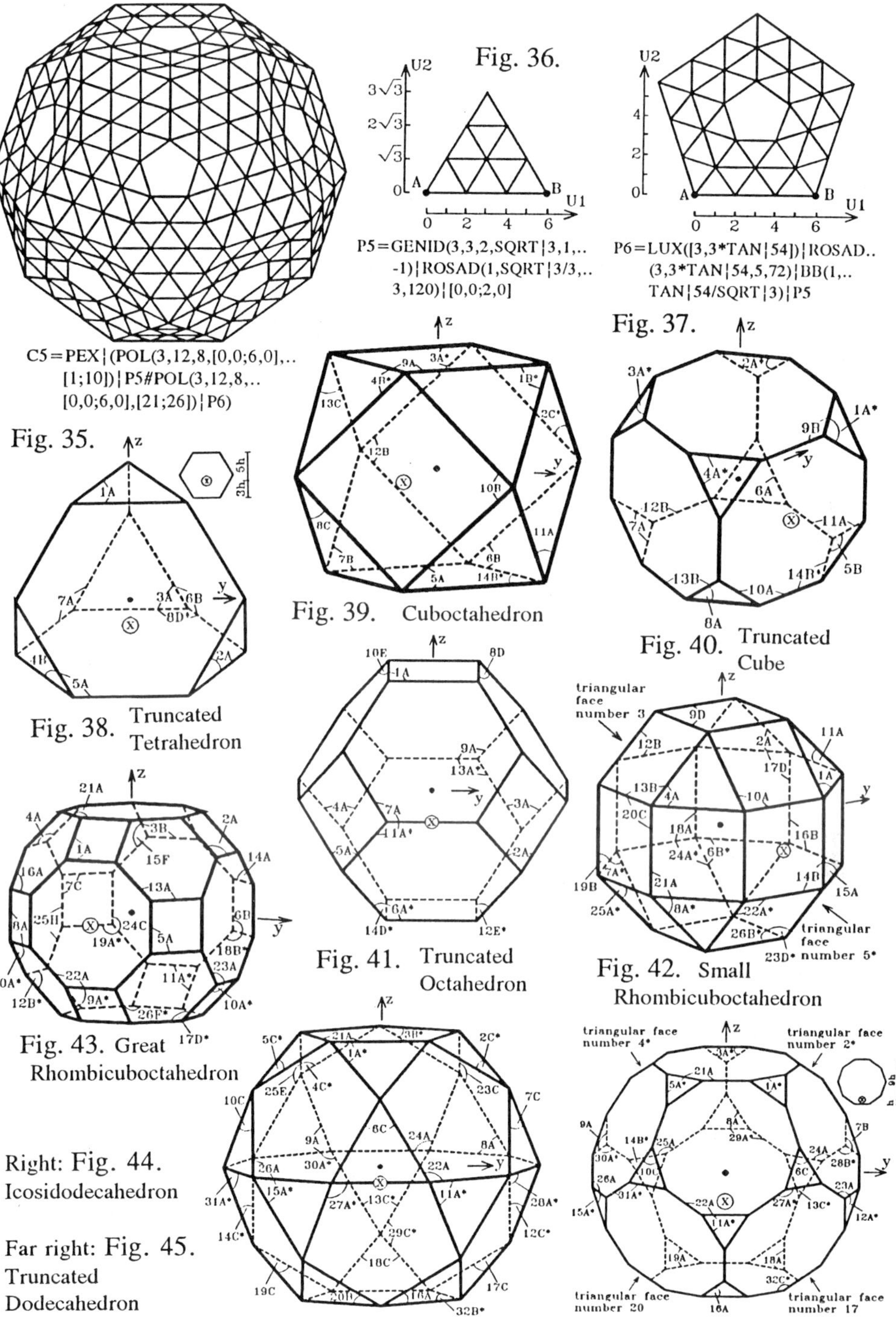

C5=PEX|(POL(3,12,8,[0,0;6,0],..
[1;10])|P5#POL(3,12,8,..
[0,0;6,0],[21;26])|P6)

Fig. 35.

Fig. 36.

P5=GENID(3,3,2,SQRT|3,1,..
-1)|ROSAD(1,SQRT|3/3,..
3,120)|[0,0;2,0]

P6=LUX([3,3*TAN|54])|ROSAD..
(3,3*TAN|54,5,72)|BB(1,..
TAN|54/SQRT|3)|P5

Fig. 37.

Fig. 38. Truncated Tetrahedron

Fig. 39. Cuboctahedron

Fig. 40. Truncated Cube

Fig. 41. Truncated Octahedron

Fig. 42. Small Rhombicuboctahedron

Fig. 43. Great Rhombicuboctahedron

Right: Fig. 44. Icosidodecahedron

Far right: Fig. 45. Truncated Dodecahedron

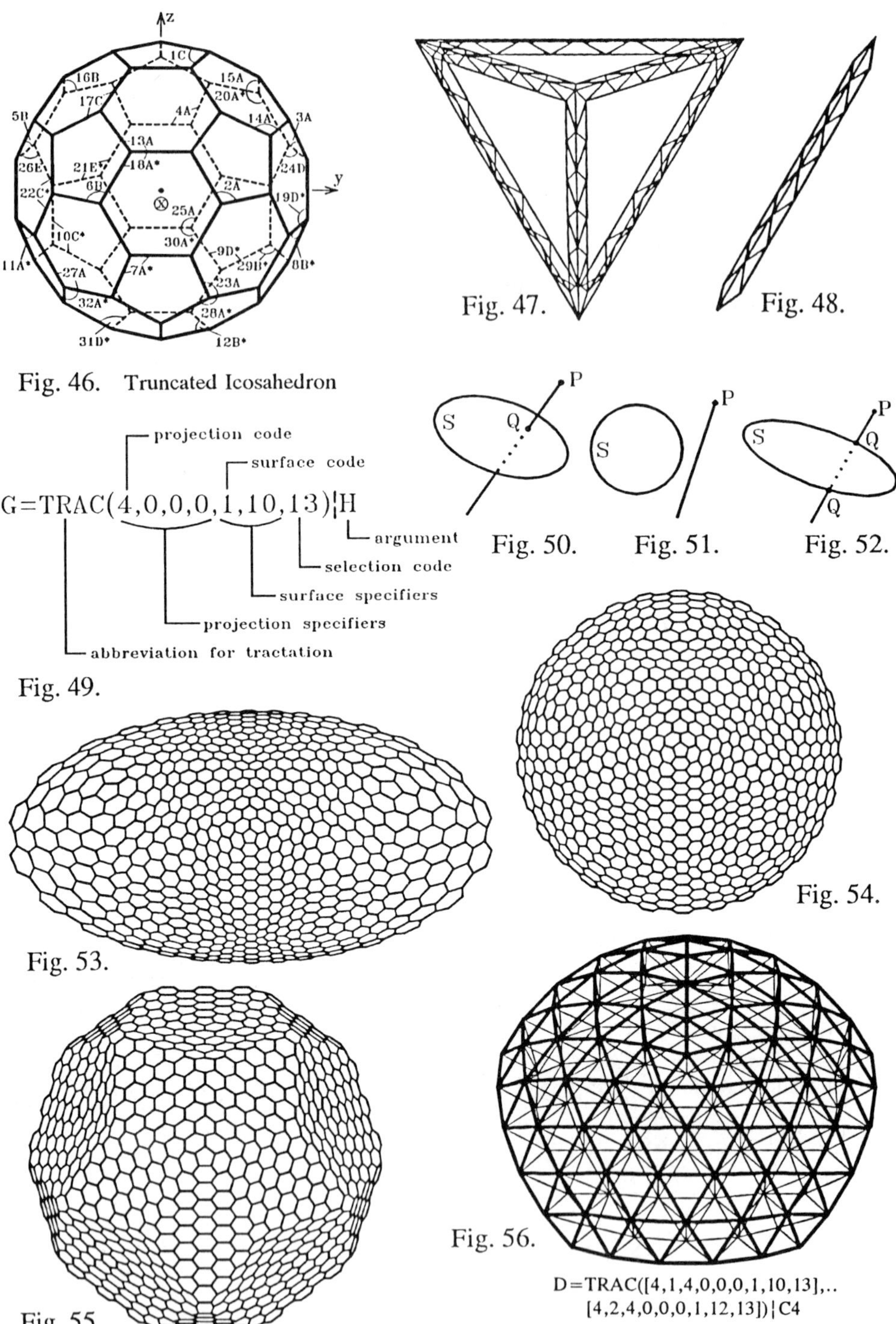

Fig. 46. Truncated Icosahedron

Fig. 47.

Fig. 48.

Fig. 49.

Fig. 50.

Fig. 51.

Fig. 52.

Fig. 53.

Fig. 54.

Fig. 55.

Fig. 56.

Table 2

Surface	Code	Items That Should Follow the Surface Code
Sphere	1	x, y and z coordinates of the centre of the sphere followed by the radius of the sphere *
Ellipsoid	2	x, y and z coordinates of the centre of the ellipsoid followed by three semiaxes of the ellipsoid in the x, y and z directions *
Paraboloid	3	x, y and z coordinates of the centre of the paraboloid followed by three semiaxes of the paraboloid in the x, y and z directions *

* In the case of radial projection the coordinates of the centre of the surface are not given since these coordinates are the same as those of the centre of projection.

Table 3

Selection Code	Indicated Solution		Selection Code	Indicated Solution	
	In the case of multiple solutions for Q choose the one which has the	In the case of nonexisting solution		In the case of multiple solutions for Q choose the one which has the	In the case of nonexisting solution
11	greatest x component	ignore the cantle (a part of) which represents point P	21	greatest x component	accept point P as the solution
-11	least x component		-21	least x component	
12	greatest y component		22	greatest y component	
-12	least y component		-22	least y component	
13	greatest z component		23	greatest z component	
-13	least z component		-23	least z component	
14	greatest distance from the origin		24	greatest distance from the origin	
-14	least distance from the origin		-24	least distance from the origin	
15	greatest distance from P		25	greatest distance from P	
-15	least distance from P		-25	least distance from P	

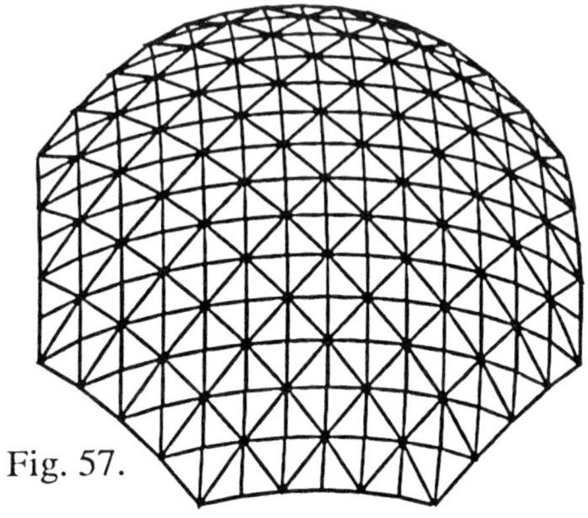

Fig. 57.

F=TRAC(2,10,10,0,10,10,..
1,1,10,10,-5,12,13)|E

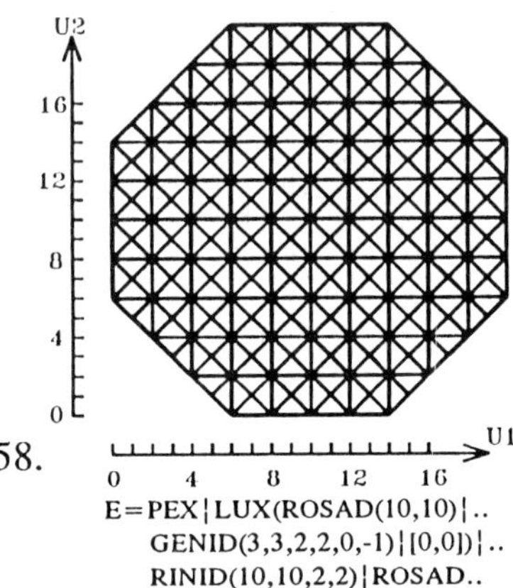

Fig. 58.

E=PEX|LUX(ROSAD(10,10)|..
GENID(3,3,2,2,0,-1)|[0,0])|..
RINID(10,10,2,2)|ROSAD..
(1,1)|{[0,0;2,0],[0,0;1,1]}

The projection code may have the value 1, 2, 3 or 4 indicating central, parallel, axial or radial projection, correspondingly (See Figs 14 to 17). The projection code in Fig. 49 is given as 4, implying radial projection, and this is followed by three zeros. These zeros are the x-y-z coordinates of the centre of projection. In the case of parallel or axial projection, the projection code must be followed by the coordinates of two points, specifying the direction or axis of projection (See Figs 15 and 16). The "surface code" specifies the type of surface on which projection is to be made. Table 2 lists three surfaces together with their corresponding codes. The surface code in Fig. 49 is given as 1, implying a sphere, and this is followed by the number 10 that specifies the radius of the sphere. The items that should follow the surface code in different cases are listed in Table 2. The "selection code" specifies the course of action to be taken when the projection of a point cannot be determined uniquely. To elaborate, in obtaining the projection Q of a point P on a surface S, the following situations may arise: (i) Q is determinable uniquely, Fig. 50, (ii) Q is nonexistent, Fig. 51, (iii) there is more than one solution for Q, Fig. 52. Various possible courses of action in the case of nonexistent or multiple solutions for Q, together with the corresponding selection codes are listed in Table 3. The "argument" in Fig. 49 is a formex variable representing the polyhedric configuration of Fig. 3. The formulation of this formex variable is shown in Fig. 18.

19. Application of the tractation retronorm is further exemplified through Figs 53 and 54 that show central projections of the polyhedric form of Fig. 55 on an ellipsoid and a sphere, respectively.

20. In using the tractation retronorm, it is possible to have more than one surface on which projection is made. For instance, the double layer polyhedric form of Fig. 33 is projected on two concentric spherical surfaces, using radial projection, and the result is shown in Fig. 56. The formulation for this double layer geodesic form is also shown in the figure. Two new points are to be noticed in this formulation. Firstly, the particulars for each surface are given separately, enclosed in square brackets. Secondly, when there are more than one surface for projection, it would be necessary to identify the intended surface for each point to be projected. This identification is achieved through two new items that are added at the beginning of the set of particulars for each surface. The new items in the particulars of the first surface in Fig. 56 are 4 and 1. These indicate that the first surface should be used for the projection of every point whose fourth coordinate is equal to 1. The new items in the particulars of the second surface in Fig. 56 should be interpreted analogously. Insertion of “layer identification numbers” in fourth coordinate positions is effected through the polyhedric function.

21. The applications of the tractation retronorm as described so far concern the usage in relation to geodesic forms. However, the tractation retronorm may be employed in any situation when it is required to obtain the projection of a single layer or multi-layer configuration on one or more surfaces. An important application of the tractation retronorm is in generation of "grid domes". For example, the grid dome of Fig. 57 is obtained by parallel projection of the pattern of Fig. 58 on a spherical surface. The formulations are included in the figures.

117. The formation of polyhedra by the rotation of polygons

P. HUYBERS, Delft University of Technology, The Netherlands

SYNOPSIS. The regular and the semi-regular polyhedra can be considered as portions of space that are completely surrounded by one or more kinds of regular polygons. The numbers and positions in space of these polygons are strictly ruled by universal criteria. It is therefore possible to form these polyhedra by placing polygons around the centre in distinct numbers, at certain distances and under certain angles in accordance with these rules. This paper describes a method where this is done for the regular and semi-regular polyhedra. Related figures that are found by derivation from the polyhedra, may be formed similarly by rotating planar figures that are not regular polygons. This method thus also allows the rotation of arbitrary figures - spatial ones not excluded and the rotation procedure can even be used repeatedly, so that very complex configurations can be described. A computer programme, called CORELLI, is being developed by the author in close co-operation with G van der Ende and it is meant mainly for study purposes. The outcome can be viewed in graphical and in numerical form. The latter data can be used as the input for further elaboration, such as in external 3D CAD or other computation programmes.

AN AUTOMATIC GENERATION ROUTINE FOR POLYHEDRA

1. The surface of the so-called 'uniform' polyhedra is formed by a closed pattern of regular polygons. All regular polyhedra consist of a set of polygons of one kind, whereas the semi-regular figures are formed either by two or by three different kinds of polygons. These polygons are organised in specific arrangements and numbers (ref. 2).

2. Such a polyhedron may be generated by firstly placing a polygon in the centre of the system and in the XY-plane, then translating it over the distance z_n along the Z-axis and at last rotating it around the Z-, X- and Y-axes respectively (angles a, γ and β in Fig. 2.). This has to be done for each of the polygons occurring in the polyhedron and also for each kind of polygon, that compose the polyhedron (Fig. 1).

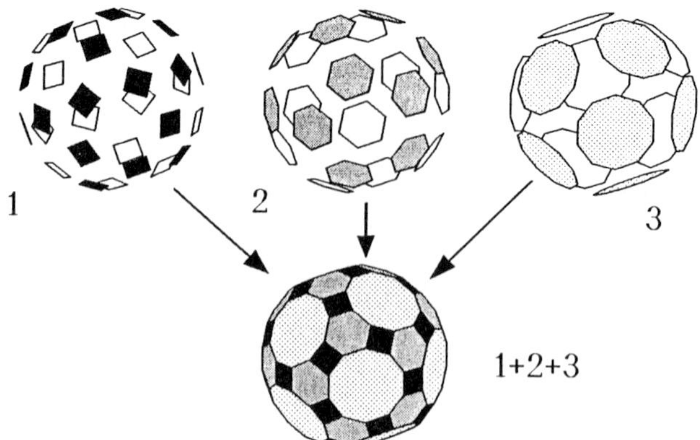

Fig. 1. Example of a polyhedron, consisting of 3 kinds of polygons and each kind having its own specific number and arrangement.

3. According to (ref. 2) the following relation exists between the rotated co-ordinates (x', y' and z') and the original co-ordinates (x, y and z):

$$\begin{vmatrix} x' \\ y' \\ z' \end{vmatrix} = (T_{\alpha\gamma\beta}) \begin{vmatrix} x \\ y \\ z \end{vmatrix} = (T_{\alpha})(T_{\gamma})(T_{\beta}) \begin{vmatrix} x \\ y \\ z \end{vmatrix}$$

where

$$T_{\alpha} = \begin{vmatrix} \cos\alpha & \sin\alpha & 0 \\ -\sin\alpha & \cos\alpha & 0 \\ 0 & 0 & 1 \end{vmatrix}, \quad T_{\gamma} = \begin{vmatrix} 1 & 0 & 0 \\ 0 & \cos\gamma & \sin\gamma \\ 0 & -\sin\gamma & \cos\gamma \end{vmatrix} \text{ and } T_{\beta} = \begin{vmatrix} \cos\beta & 0 & -\sin\beta \\ 0 & 1 & 0 \\ \sin\beta & 0 & \cos\beta \end{vmatrix}$$

The new coordinates are found as the inverse product of these matrices.

$$\begin{vmatrix} x \\ y \\ z \end{vmatrix} = (T_{\alpha\gamma\beta})^{-1} \begin{vmatrix} x' \\ y' \\ z' \end{vmatrix}$$

The coordinates after rotations are:

$P_X = \cos\alpha\ \cos\beta + \sin\alpha\ \sin\beta \sin\gamma - \sin\alpha\ \cos\beta + \cos\alpha \sin\beta \sin\gamma + \sin\beta \cos\gamma$

$P_Y = \sin\alpha\ \cos\gamma + \cos\alpha\ \cos\gamma - \sin\gamma$

$Pz = -\cos\alpha\ \sin\beta + \sin\alpha\ \cos\beta \sin\gamma + \sin\alpha\ \sin\beta + \cos\alpha\ \cos\beta \sin\gamma + \cos\beta \cos\gamma$

4. All vertices in a particular polyhedron lie on a circumscribed sphere, which is unique for this polyhedron. Each polygon has a distance to the centre C of the co-ordinate system of $CM = z_n = \sqrt{R_1^2 - R_2^2}$, (Fig. 3) with $CQ = R_1$ = radius of the circumscribed sphere and $MQ = R_2$ = radius of the circumscribed circle of the polygon = $^1/_2 \sin(\pi/n)$, where n is the number of edges of the considered polygon (ref. 1).

5. The value of z_n can thus be calculated for each kind of polygon occurring in a polyhedron, if R_1 and n are known. It is necessary to know also the numbers in which each of the different polygon kinds occurs and the angles according to which they have to be rotated. The answer to this question is quite obvious for the cube: it counts 6 squares that first are translated along the Z-axis over a distance of $z_n = 0.5$ (times the unit edge length) and then are successively rotated around the Z-, X- and Y-axes through the angles: 0°-0°-0°, 0°-0°-180°, 90°-0°-90°, 90°-0°-270°, 90°-270°-0° and 90°-90°-0°. With 7 different rotation cases, that each represent a specific combination of rotation angles and number of faces, all known polyhedra can be formed.

6. The required rotation angles for all uniform polyhedra were determined in ref. 3 and incorporated in the computer program CORELLI. The rotations take place according to:

1. The 4 triangles in the tetrahedron.
2. The 6 squares in the cube.
3. The 8 triangles in the octahedron.
4. The 12 pentagons in the dodecahedron.
5. The 20 triangles in the icosahedron.
6. The 12 squares in the truncated cuboctahedron.
7. The 30 squares in the truncated icosidodecahedron.

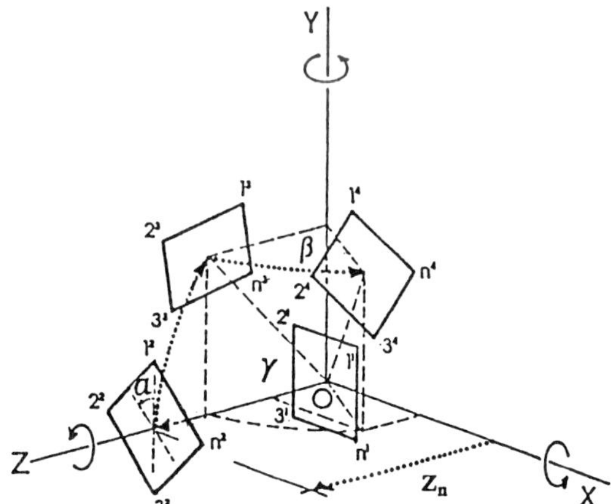

Fig. 2. Principle of the translation and rotation procedure of polygons for the formation of polyhedra.

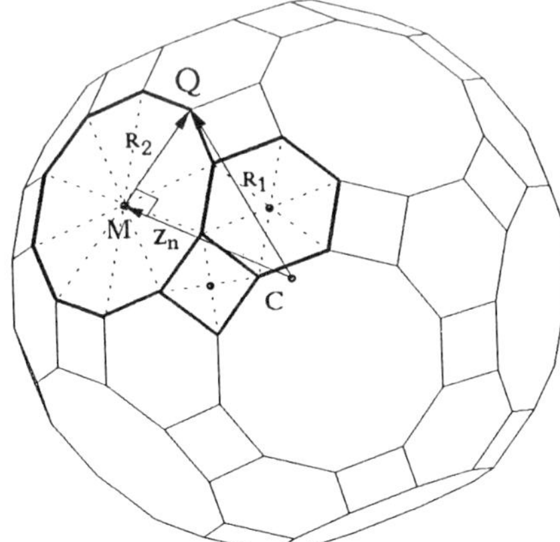

Fig. 3. Some characteristics of a particular polyhedron.

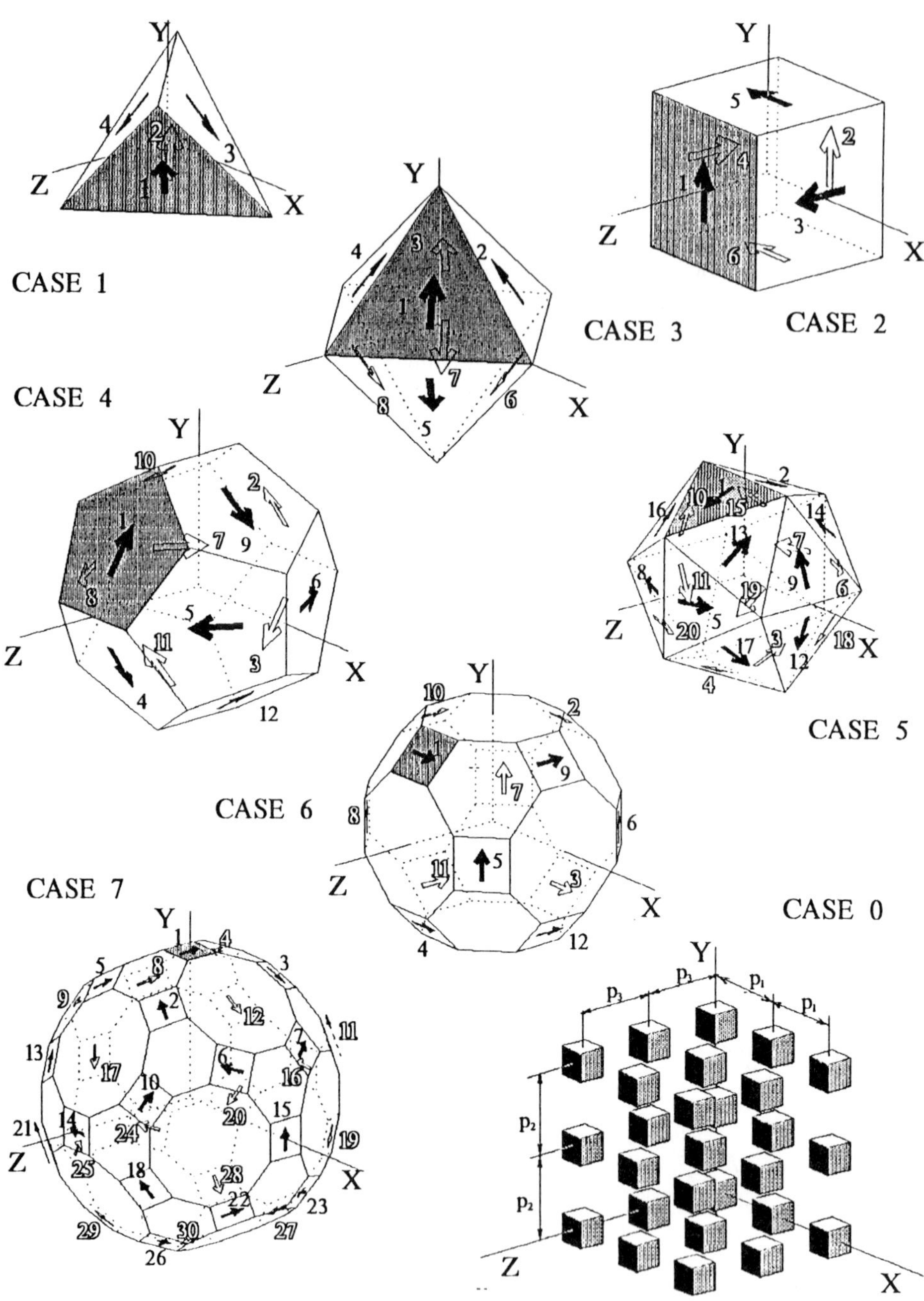

Fig. 4. The different rotation cases that were discerned.

FORMATION OF POLYGONS

7. An additional rotation case has been introduced, with which a matrix can be formed, where the required input is: the number of successive elements in the X-, Y- and Z-directions and their mutual distances. All can be chosen independently (See Case 0 in Fig. 4.). But a matrix can have one item in all directions. This means, that one specific object (plane or solid) can be put in the origin and observed or manipulated. This option is often of great use.

8. Instead of the mentioned polygons, that form the original polyhedra, any other suitable polygon can be rotated in the same way, using such a case. With one rotation case, and one kind of polygon, various solids can be described, just by varying distance z_n and the initial rotation a of the polygon. For the regular polyhedra the rotation of one kind of polygon suffices, whereas the semi-regular solids usually require more than one run. The polyhedron shown in Fig. 1 is composed of 30 squares rotated according case 7, 20 hexagons according case 5, and 12 decagons according case 4.

9. The co-ordinates of the corners in a polygon are computed as is shown in Fig. 5. The corner with the sequence number v has as its co-ordinates in the XY-plane:

$$y_v = R_2 \cos(v.2\varphi) = R_2 \cos(v.2\pi / n)$$

$$x_v = R_2 \sin(v.2\pi / n)$$

$$z_v = 0$$

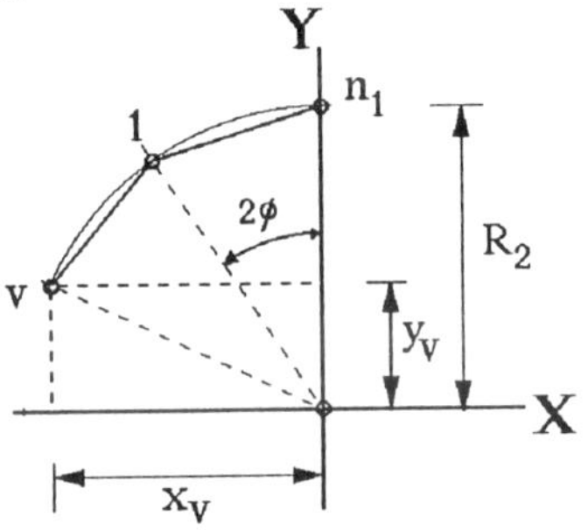

Fig. 5. The formation of a polygon in the XY-plane.

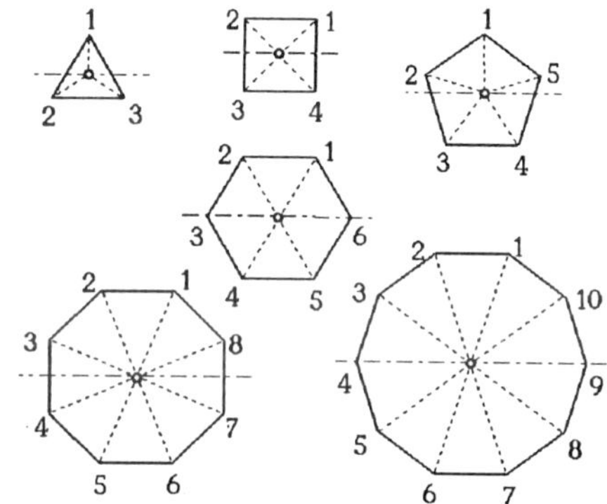

Fig. 6. The different polygons that constitute the uniform polyhedra (polygon gets extra rotation - φ if n_1 is even).

ROTATION DATA OF THE UNIFORM POLYHDRA

10. The various solids can be generated, using the following data as input:

n_1 = number of edges of polygon
n_3 = number of rotation cases
a, γ, β = initial rotation of polygon about Z-, X- or Y-axis
z_n = initial translation of polygon along the Z-axis

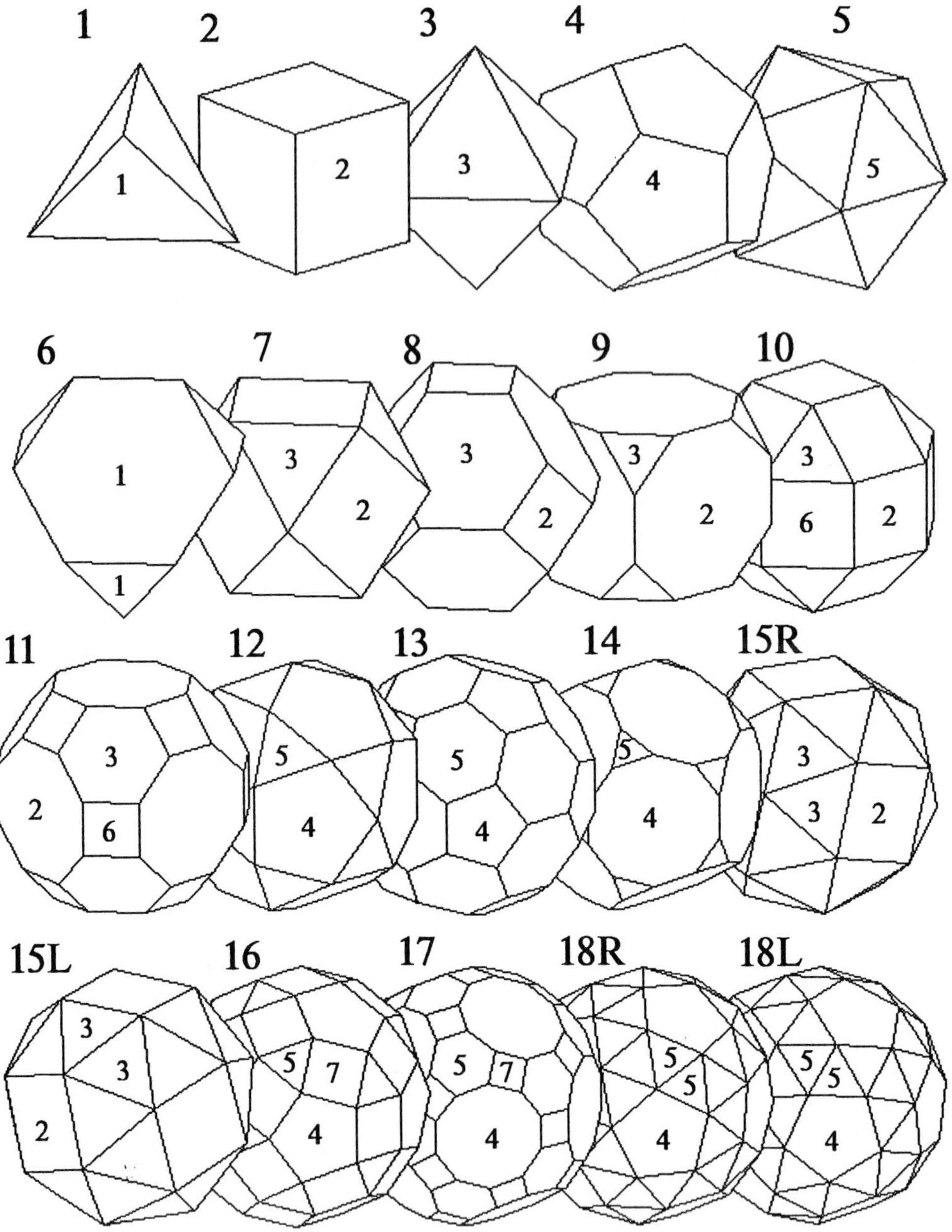

Fig. 7. Review of the 20 uniform polyhedra, 4 of which are left- and right-handed in pairs (P15 and P18).

11. In the next caption the relevant data of the familiar solids are summarised. The polyhedron is indicated with a sequence number P.

P	n_1	n_3	a	z_n	R_1
1	3	1	180	0.20412415	0.61237244
2	4	2	0	0.5	0.86602540
3	3	3	0	0.40824829	0.70710678
4	5	4	0	1.11351636	1.40125854
5	3	5	0	0.75576131	0.95105652
6.1	3	1	0	1.02062073	1.17260394
6.2	6	1	0	0.61237244	
7.1	4	2	-45	0.70710678	1.00000000
7.2	3	3	180	0.81649658	
8.1	4	2	-45	1.41421356	1.58113883
8.2	6	3	0	1.22474487	
9.1	3	3	180	1.68252198	1.77882365
9.2	8	2	0	1.20710678	
10.1	3	3	0	1.27427369	1.39896633
10.2	4	2	0	1.20710678	
10.3	4	6	0	1.20710678	
11.1	4	6	0	2.20710678	2.31761091
11.2	6	3	0	2.09077028	
11.3	8	2	0	1.91421356	
12.1	3	5	180	1.51152263	1.61803399
12.2	5	4	180	1.37638192	
13.1	5	4	180	2.32743844	2.47801866
13.2	6	5	0	2.26728394	
14.1	3	5	180	2.91278117	2.96944902
14.2	10	4	0	2.48989828	
15.1	cap	3	-/+20.31501434	0	1.34371337
15.2	4	2	+/-16.46756040	1.14261351	
16.1	3	5	0	2.15701985	2.23295051
16.2	4	7	0	2.11803399	
16.3	5	4	0	2.06457288	
17.1	4	7	0	3.73606798	3.80239450
17.2	6	5	0	3.66854248	
17.3	10	4	0	3.44095480	
18.1	cap	5	-/+19.51792257	0	2.15583738
18.2	5	4	+/-13.10640338	1.98091595	

12. The snub solids with the numbers 15 and 18 have a left- and a right-handed version. The signs of the value a refer to these. They both have two kinds of triangles: 8 (resp. 20) which touch only the corners of the n-gons and 24 (resp.

60) which have one common edge with one of the n-gons. These solids can not be formed in one operation.

13. Therefore triangular caps have to be made in advance, that are composed of four equilateral triangles. Such a cap is shown in Fig. 8. For this purpose use can been made of rotation case 0.

14. At first a triangle has to be generated and to this can be given the translation z_n and the successive rotation γ (around the X-axis). Both are different for the two snubs P15 and P18 (ref. 5).

P	γ	z_n
15	26.65741228	1.21335580
18	15.82463394	2.07708966

A cap is finally formed by the composition of three of these shifted triangles (ABC, BDE and EFC) with respective additional rotations about the Z-axis of a = 60°, 120° or 180°. To these three surrounding triangles is then added a fourth triangle BCE in the middle at a distance z_n and placed in an upside down position by the rotation a = 180°. The caps and the n-gons are organised in the way that is shown in Fig. 9.

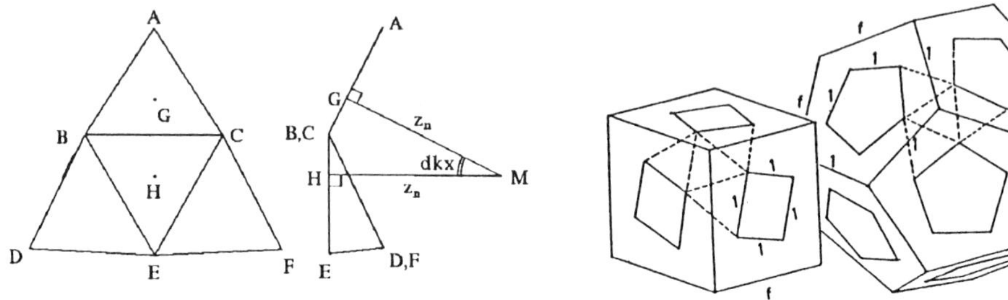

Fig. 8. A typical cap of a snub polyhedron.

Fig. 9. Position of the triangular caps in the snub solids P15 and P18.

THE ROTATION OF OBJECTS

15. As already anticipated in the previous paragraph, instead of starting the rotation procedure with a polygon, objects of any shape may be rotated. The program has the built-in facility to rotate not just regular polygons, but also star polygons (polygrams) and even arbitrary figures, as long as these are constituted by flat faces. With the help of this feature the faces of the reciprocal solids may be described and rotated in space in a similar way as their originals. This is also

applicable to the compounds of both types of figures, but in this case the object to be rotated usually has a 3-dimensional form (ref. 6).

16. These data may even be saved in a library, so that they remain available for later use. This is useful as in this way individual figures are directly available for read-out.

17. The rotation of objects can be used also to generate subdivision patterns of dome structures. The triangular regular polyhedra (tetrahedron, octahedron and icosahedron) are very suitable in this respect. The original polyhedron has to be subdivided and the resulting pattern is projected upon a spherical surface. This method is well-known, but the procedure may be executed with any of the polyhedra, even the semi-regular ones.

18. Such a geodesic shape may eventually again be treated as an object and rotated according one of the applicable possibilities (ref. 4).

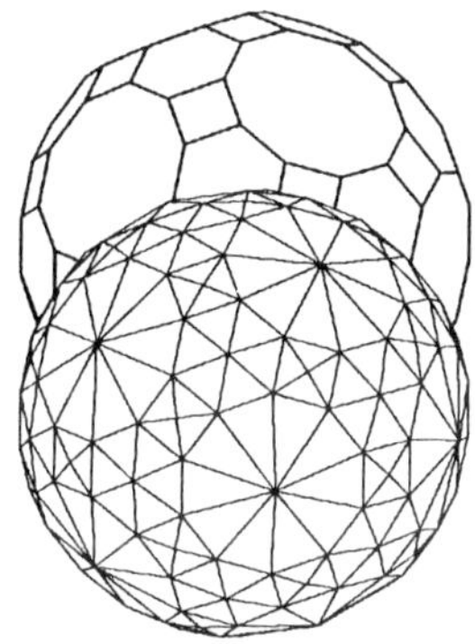

Fig. 10. 'Pyramidized' version of P17.

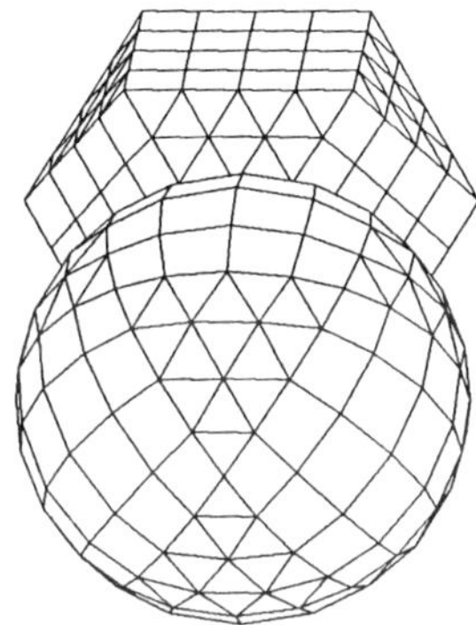

Fig. 11. Subdivision of the sphere from any of the semi-regular solids.

REFERENCES

1. HUYBERS P., The geometry of uniform polyhedra. Architectural Science Review. Vol. 3, No. 2, June 1980, p. 36-50

2. HUYBERS P., De coordinaten van uniforme polyeders (The coordinates of uniform polyhedra), TU-Report 10-76-08, Delft, 1979.

3. HUYBERS P., Rotatiemogelijkheden voor de Platonische en Archimedische veelvlakken (Rotation possibilities of the Platonic and the Archimedean polyhedra), TU-Report 10-78-03,1980.

4. HUYBERS P., Excursions in polyhedral geometry. Proceedings of the Int. IASS-Symposium on Spatial Structures at the Turn of the Milennium, Copenhagen, 2-6 september 1991, Vol. 2, p. 129-136

5. HUYBERS P., The snub polyhedra on closer examination, Proc. IASSCSCE Int. Congress on Innovative Large Span Structures, Toronto, 12-17 July, 1992.

6. HUYBERS P., The geometry of the dual uniform solids, 1st Int. Seminar on Structural Morphology, 7-11 September 1992, Vol 2, p. 1-11.

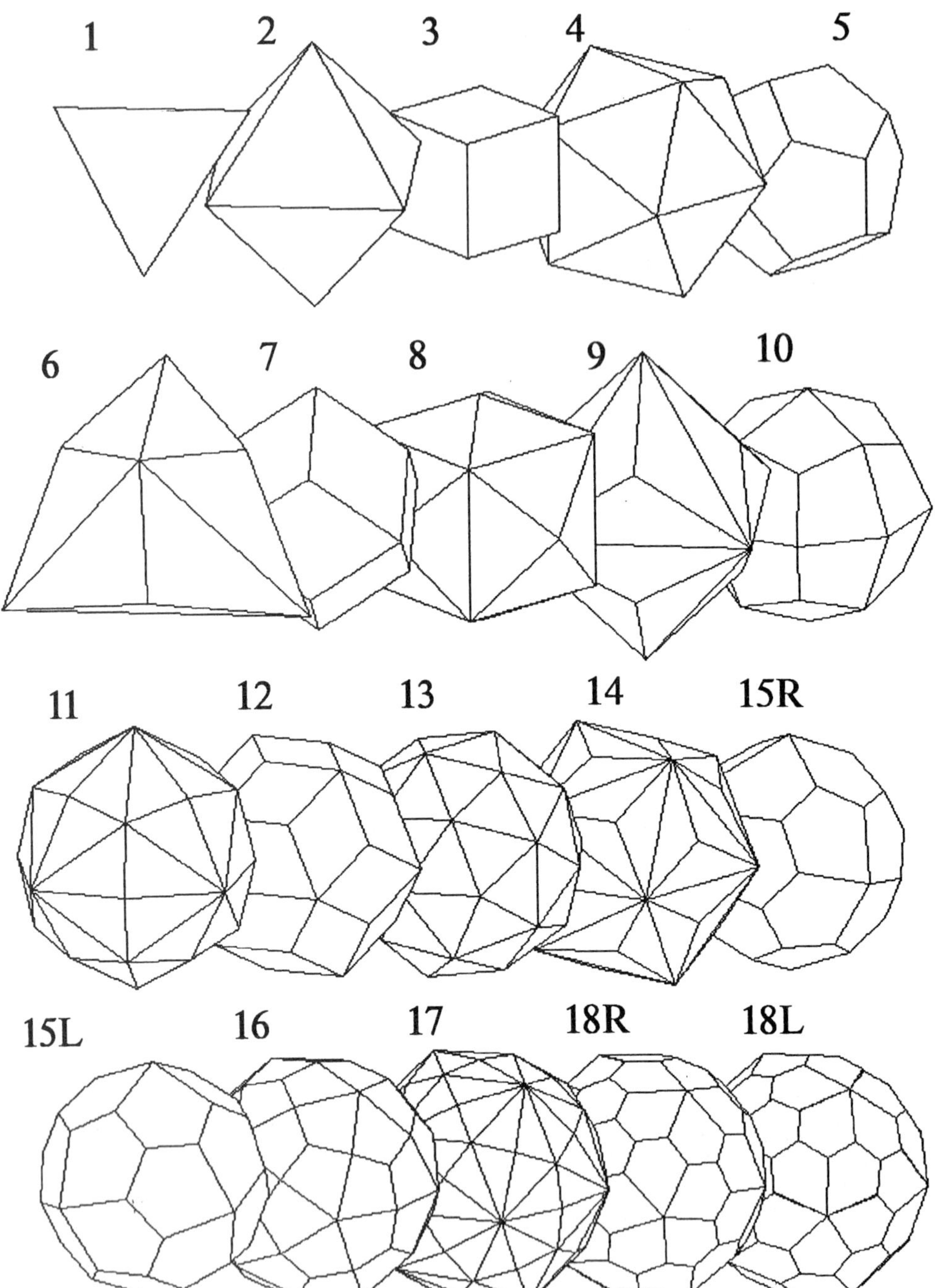

Fig. 12. The reciprocal figures of the uniform polyhedra.

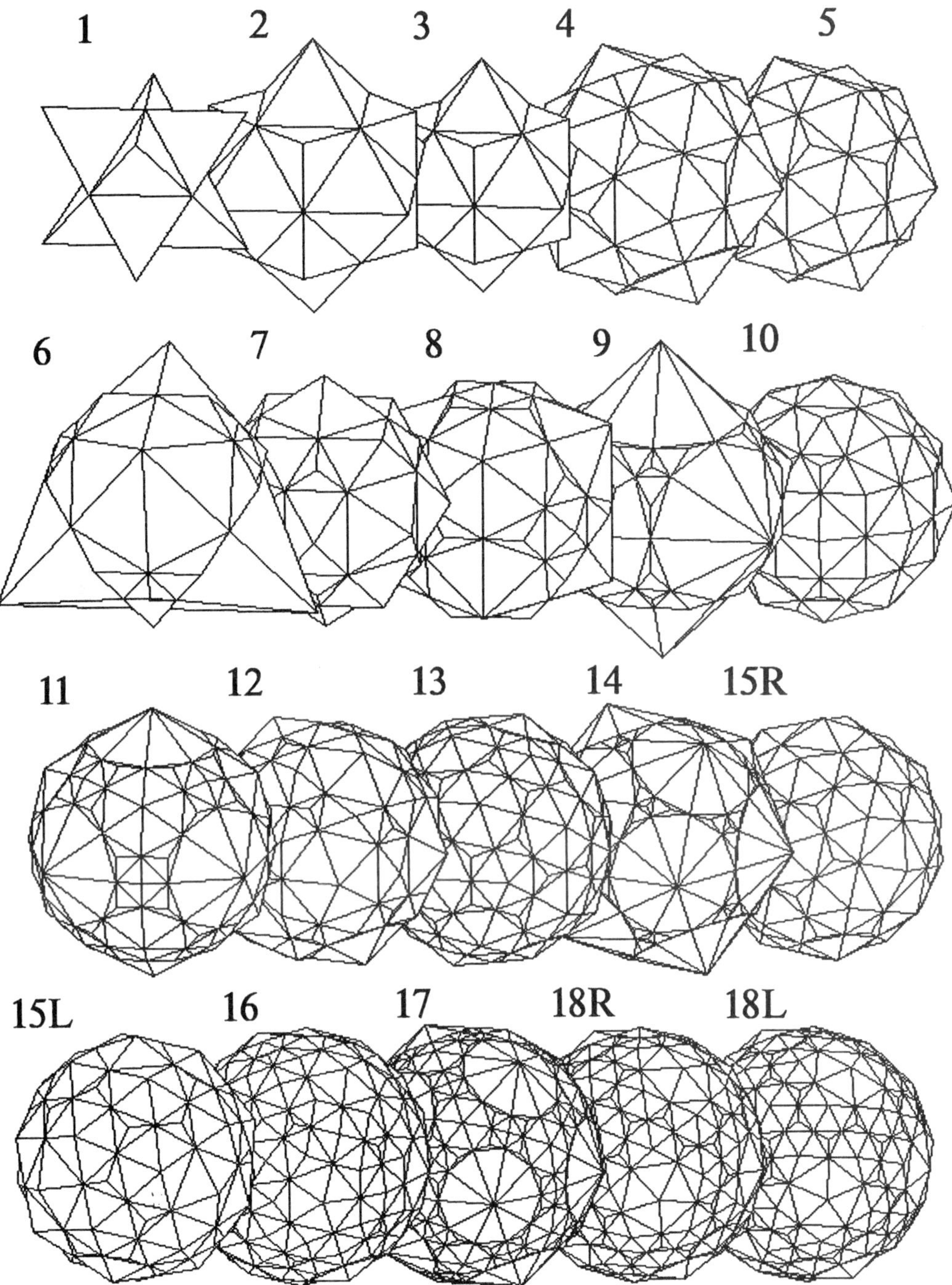

Fig. 13. Compounds of the uniform solids and their reciprocal figures (see also Figs 7 and 12).

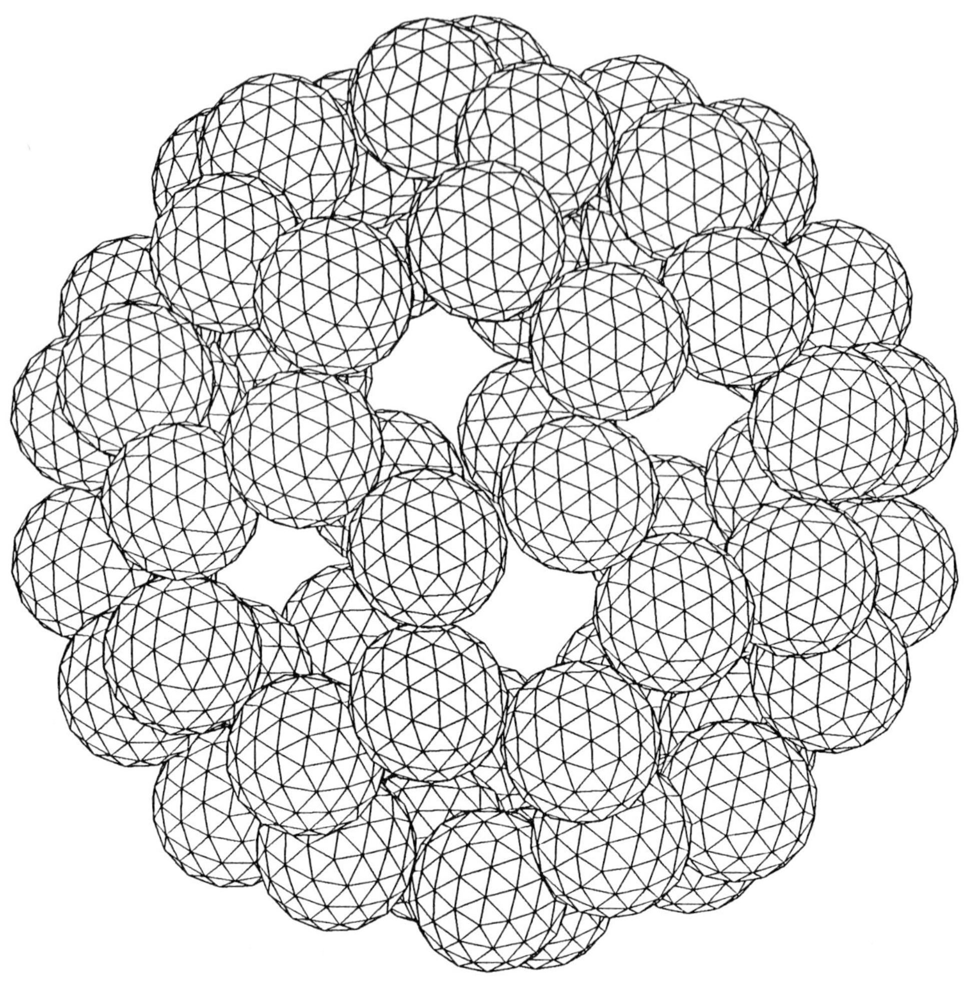

Fig. 14. Rotation in space of 60 geodesic spheres according the vertices of a truncated icosahedron (P13) in order to form the so-called 'Bucky-ball' (C_{60} - molecule).

118. Structural performance of clustered rhombic dodecahedral buildings

B. T. ROSSON, A. MALLER and R. K. FALLER, University of Nebraska-Lincoln, USA

SYNOPSIS. A three-dimensional finite element method of analysis has been used to determine the stress resultants and displacements of various rhombic dodecahedral (r-d) clusters, subject to static and dynamic loading conditions. The work focuses on the structural response of r-d clustered buildings of one, two, three and ten stories. The packing patterns of the clusters include square and hexagonal plan projections. The loading conditions include static gravity loads and simple dynamic ground excitations; in addition, the elastic buckling loads and natural frequencies of the various building types are determined. The results of the FEM analyses are compared to those of conventional rectangular building systems under similar loading conditions.

INTRODUCTION

1. Spatial clusters of non-orthogonal polyhedra offer a relatively large unexplored area of application to the architectural and structural design of buildings. Historically, architects and engineers have concentrated their efforts on the use of only two polyhedra — orthogonal and tetrahedral polyhedra. The use of non-orthogonal polyhedra in the architectural design of various building systems has been inhibited in the past due to an inability to sufficiently address the visual representation of the spatial disposition of the polyhedral clusters. Also, the structural design of non-orthogonal polyhedra inherently requires more sophisticated three-dimensional structural analysis techniques.

2. The objectives of this study are to evaluate the response of clustered r-d buildings to static and dynamic loading conditions, as well as to determine the first three elastic buckling loads and the first five natural frequencies. Numerous plots are made for response comparisons between the two r-d cluster types and with those of conventional rectangular buildings. The study is restricted to investigation of buildings of one, two, three and ten stories. The loading conditions are idealised for the static loads as concentrated vertical loads and for the dynamic loads as simple harmonic ground motions. The loads considered are not meant to be inclusive, or even representative of typical loading patterns or magnitudes, but simply those that readily provide for a means of comparison and

replication for the various clusters. The material and cross-sectional properties have remained constant for all members in each cluster.

3. The structural response of the clusters subject to the aforementioned loading conditions has been determined through a comprehensive finite element (FEM) investigation by implementation of the ADINA (Automatic Dynamic Incremental Non-linear Analysis) computer programs (refs 1-2). ADINA is particularly well suited to model the complex three-dimensional geometry through implementation of the 3-D beam element. The programs were developed for the analysis of structures with geometric and material non-linearities, subject to static and dynamic loads. Thus, this program was chosen for the analyses because no further modification was needed for the prescribed study parameters. A separate computer program was written to calculate the nodal co-ordinates and element connectivity for any r-d cluster.

4. Response comparison plots and tables are presented to indicate the relative behaviour of the various clusters. They have been non-dimensionalised where possible so that reasonable comparisons can be made.

FINITE ELEMENT MODELS

5. The use of rhombic dodecahedral clusters, from an architectural viewpoint, has been previously studied (ref. 3). However, the structural investigation of these clusters has received less attention. This was due primarily to the lack of sophistication of existing computer programs, as well as the required computer speed and memory needed to analyse the complex three-dimensional shapes. The finite element method of analysis has been shown to be effective in predicting the response of structures such as those presented in this study.

6. The rhombic dodecahedron can be used as a structure with a hexagonal or square plan project. It has unique geometric qualities that provide for convenient replication; all the members have equal length, with all sides having the same panel dimensions. With all the panel dimensions the same, clustering many rhombic dodecahedron together is relatively easy. The co-ordinates for the geometry of one r-d have been input directly into the computer program code; the program then generates additional r-d in the x, y and z directions based upon user input, while continuously checking to see if a nodal co-ordinate has been previously defined at that location. The element connectivity has been accomplished by checking for all possible nodal connections of a particular length, since all member lengths are identical. If a particular connection has been previously established, it is bypassed. The search then continues until all elements have been defined. No attention has been given to the ordering of the nodes because ADINA performs a bandwidth optimisation routine.

7. The bottom nodes are fully restrained, with all nodes having full moment continuity. The static loads are applied as a concentrated load applied downward at all top level peaks. The elastic buckling loads are applied at the same locations. The ground motion is defined as a sinusoidal wave with a period of 0.1 seconds and amplitude of 0.15 g's. The members are modelled as steel circular shafts with the following geometric and material properties: Diameter, D = 5.08 cm; Length,

L = 3.0 metres; Modulus of Elasticity, E = 200 GN/m^2; Poisson's Ratio, $\nu = 0.3$; and Mass Density, ρ = 7,800 kg/m^3.

R-D clusters, hexagonal plan projection

8. The hexagonal plan projection allows for six vertical columns, with a three faceted "natural" roof and support structure on the bottom. A single hexagonal plan r-d has been studied, as well as a single level of seven hexagonal plan r-d in a cluster. Levels two through ten have six r-d on each level as shown in Fig. 1.

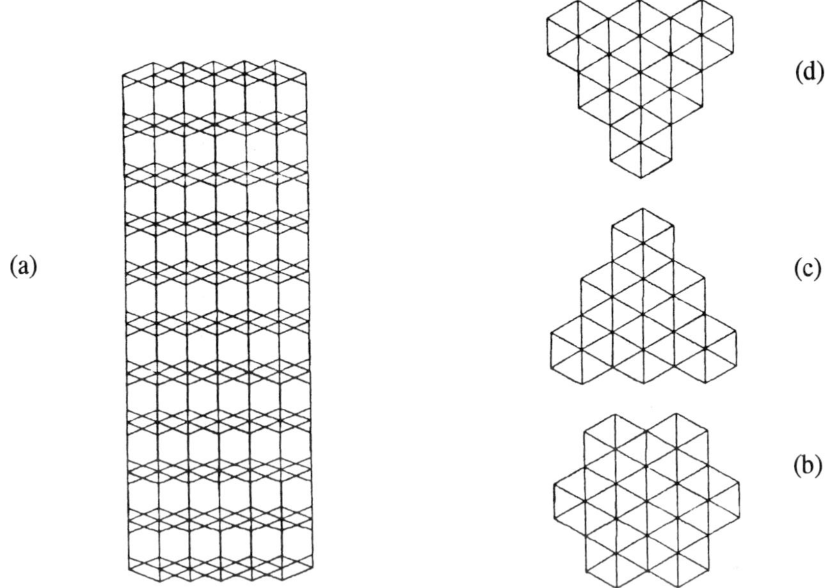

Fig. 1. Hexagonal plan projection r-d cluster, a) ten-storey elevation; b) plan section of first level; c) plan section of second, fourth, sixth, eighth, and tenth levels; and d) plan section of third, fifth, seventh and ninth levels.

R-D clusters square plan projection

9. The square plan projection allows for four symmetric vertical faces, with no vertical columns. The "natural" roof and bottom support structure have four facets. A single square plan r-d, and a cluster of nine r-d on a single level, have been investigated. The two storey square plan consists of nine r-d on the first level and four interior r-d on the second level. The three- and ten-storey buildings are created in a similar manner with alternating floors having three or nine r-d on a level as shown in Fig. 2.

Conventional rectangular buildings

10. Conventional building frame structures also have been studied, with the beams and columns having the same cross-sectional and length dimensions. Although this is not common, it is necessary for equitable comparisons with the

r-d clusters. A single square frame, with four columns and beams has been studied, as well as a single square floor of 16 columns and 24 beams, all equally spaced. The two-, three- and ten-storey buildings have the same pattern as the one-storey building.

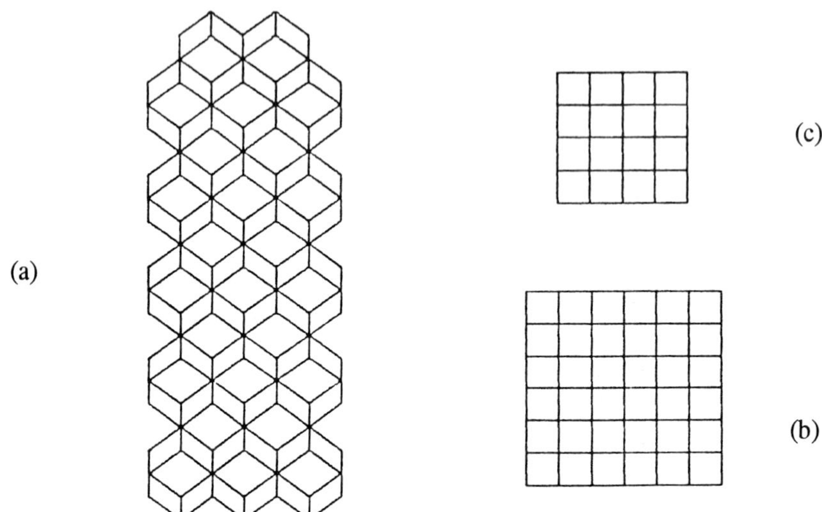

Fig. 2. Square plan projection r-d cluster, a) ten-storey elevation; b) plan section of first, third, fifth, sevength and ninth levels; c) plan section of second, fourth, sixth, eighth and tenth levels.

STATIC LOADING RESULTS

11. The responses of the hexagonal and square plan r-d clustered buildings to static concentrated loads are presented in Figs 3-6. Axial force comparisons have been made for members on the bottom floor level only; the vertical members for the hexagonal r-d clusters and the bottom diagonal members for the square r-d clusters. The figures indicate the percentage of the total axial load carried by each member vs. the percentage of the members with a particular axial stress resultant. This gives an indication of how the loads are distributed through the cluster for each r-d cluster type (hexagonal or square) and the effect of increased stories on load distribution. The conventional building results are not presented because this particular loading condition of loads applied at the peaks of the structure is not readily reproducible for this building type. For concentrated loads applied to top level columns, all of the members would resist the same percentage of the total load (the actual percentage would depend upon the total number of columns).

12. The figures indicate that the square r-d clusters exhibit a more uniform distribution of total load than do the hexagonal r-d clusters, and that the distribution of the loads does not vary appreciably as additional floors are added. The axial loads rarely exceed five percent of the total axial load for any of the

square r-d clusters. For the hexagonal r-d clusters, a relatively small percentage of members consistently have member loads of ten to fifteen percent of the total axial load. It is also noticed that some members actually experience an axial tension stress resultant, although this is true for a relatively small number of members and primarily for those with square r-d clusters.

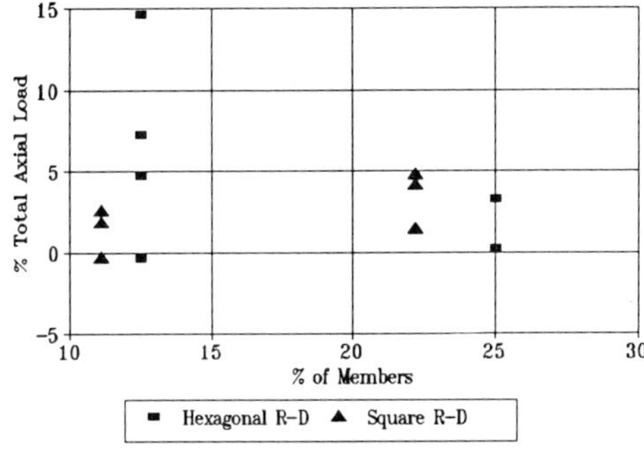

Fig. 3. Axial load distribution comparisons for one-storey configuration.

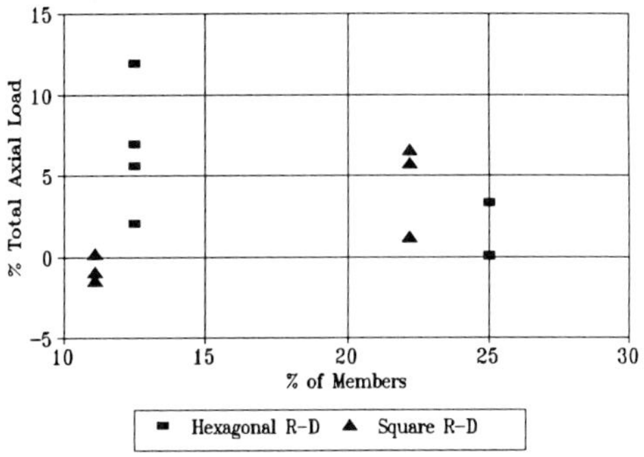

Fig. 4. Axial load distribution comparisons for two-storey configuration.

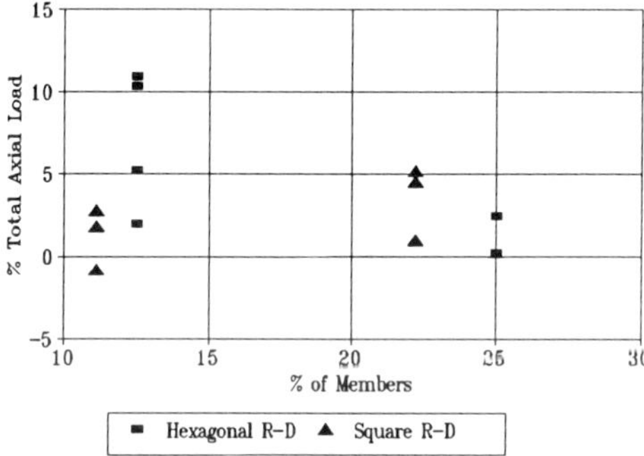

Fig. 5. Axial load distribution comparisons for three-storey configuration.

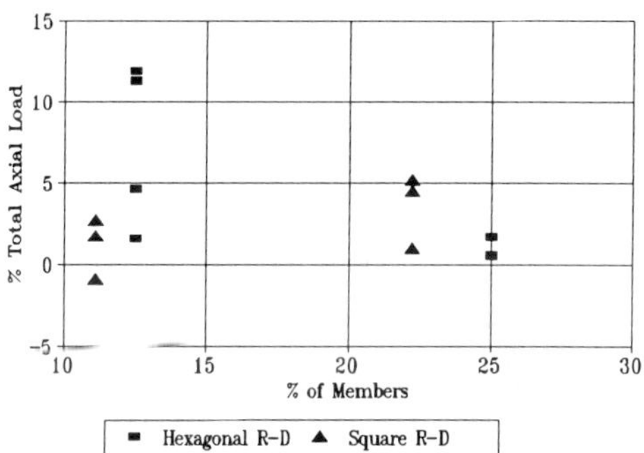

Fig. 6. Axial load distribution comparisons for ten-storey configuration.

ELASTIC BUCKLING RESULTS

13. The first three buckling loads have been determined for each r-d cluster type, as well as for each conventional building. The results are presented in Table 1. It is evident from the table that the buckling loads for all r-d clusters are significantly lower than for their conventional building counterparts. This is primarily due to the location of the concentrated loads; the r-d clusters are loaded at their peaks, producing significant bending prior to buckling, while the conventional buildings are loaded at the top of each column, thus avoiding this bending prior to buckling.

14. It is noticed that for both r-d cluster types, the buckling loads increase from the single unit to the one-storey cluster, and from the one-storey to the two-storey cluster, but then decrease as additional levels are added. This is in contrast to the conventional building; as the floor levels are increased the buckling load consistently decreases and at a much faster rate than that for the r-d clusters. It is also noticed that the buckling loads for each r-d cluster type are essentially the same for all levels.

Table 1. Elastic buckling loads, kN.

General Configuration	Mode	Buckling Load Per Load Location (Modes 1 through 3)				
		Single	1-Story	2-Story	3-Story	10-Story
Hexagonal R-D	1,2	4.85	6.32	6.72	6.41	5.52
	3	8.10	6.67	7.47	7.25	8.14
Square R-D	1,2	5.03	5.78	8.50	5.52	5.20
	3	8.63	5.96	9.74	6.58	8.59
Conventional	1,2	54.2	57.6	43.3	39.2	35.2
	3	66.3	60.1	46.3	42.3	36.8

FUNDAMENTAL FREQUENCY ANALYSIS RESULTS

15. The first five natural frequencies have also been determined for each building type. The results are presented in Table 2. The largest natural frequencies for both r-d clusters occur in the one-storey building; as the floor levels are increased, the frequencies decrease, but at a slightly faster rate for the hexagonal r-d clusters. The frequencies for each cluster type are relatively similar, with the hexagonal r-d cluster consistently having smaller values. The conventional building has the largest frequencies occurring in the single unit; the frequencies consistently decrease with increasing floor levels.

GROUND MOTION RESULTS

16. The responses of the r-d clusters and conventional buildings to simple harmonic ground excitation are presented in Figs 7-11. Only the displacement time histories for the uppermost node in the middle of the plan projection are presented for brevity. The displacement response, in the direction of the applied ground motion, has been divided by the height to that node from ground level; this is necessary because the height to the node for each floor level is not the same for any of the three building types. The member lengths were held constant at 3.0 metres, but due to the different packing configurations, the individual storey heights do not have the same elevation.

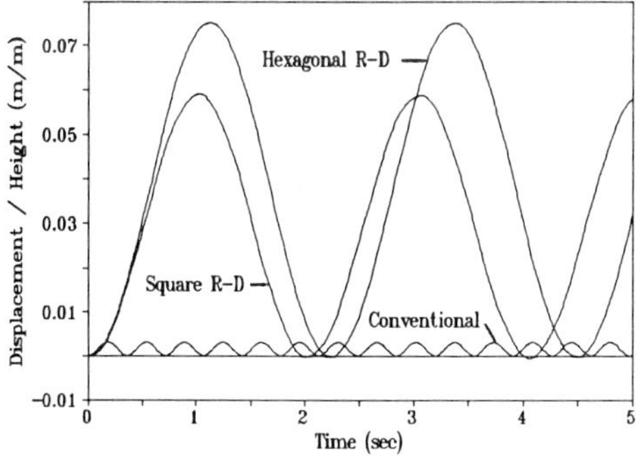

Fig. 7. Ground motion displacement responses for single configuration.

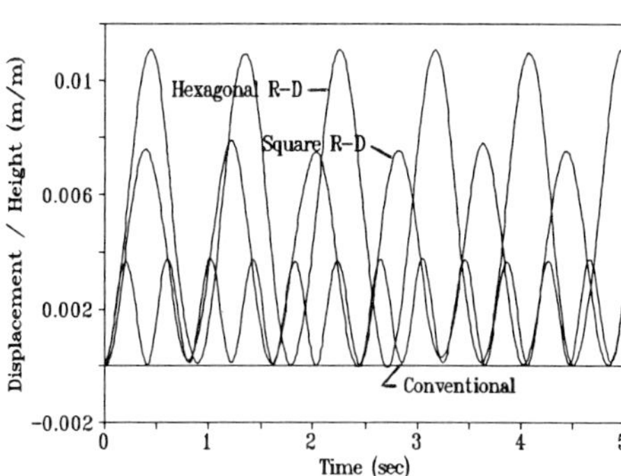

Fig. 8. Ground motion displacement responses for one-storey configuration.

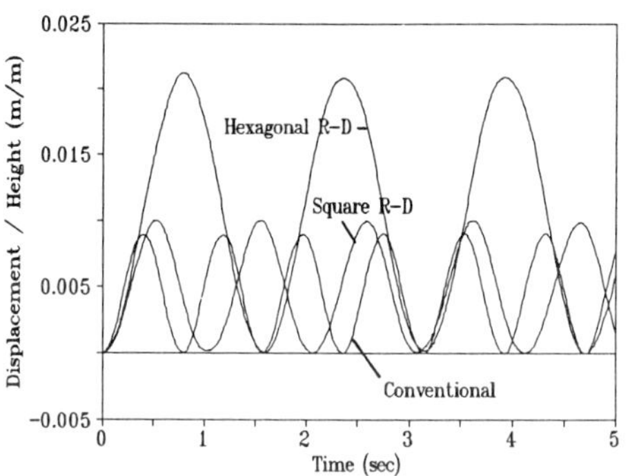

Fig. 9. Ground motion displacement responses for two-storey configuration.

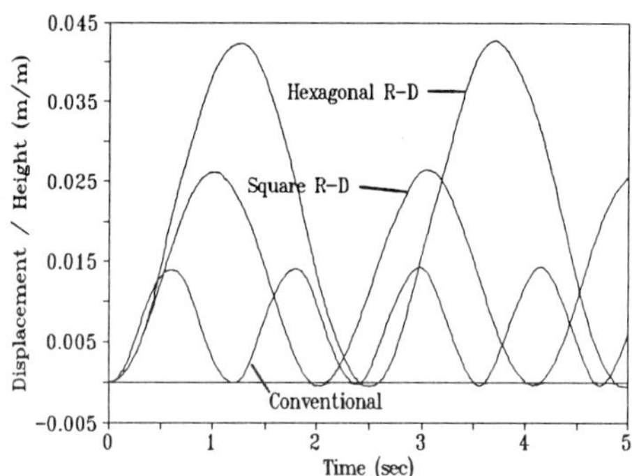

Fig. 10. Ground motion displacement responses for three-storey configuration.

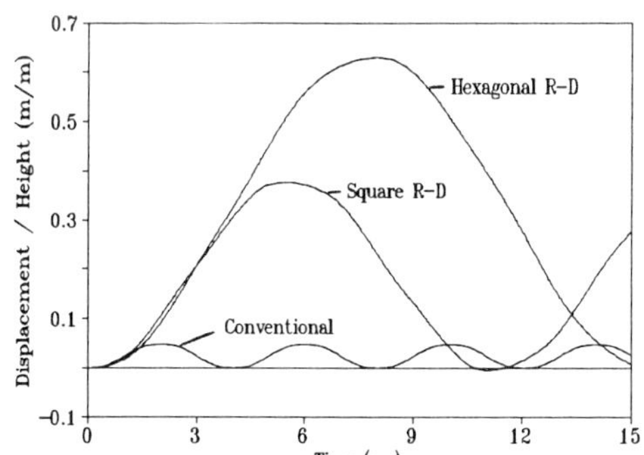

Fig. 11. Ground motion displacement responses for ten-storey configuration.

Table 2. Fundamental frequencies, Hz.

General Configura-tion	Mode	Frequency Analysis Results (Modes 1 through 5)				
		Single	1-Story	2-Story	3-Story	10-Story
Hexagonal R-D	1,2	0.431	1.08	0.630	0.401	0.0641
	3	0.781	1.43	0.919	0.620	0.173
	4	1.56	1.88	1.55	1.24	0.340
	5	2.40	1.88	1.70	1.35	0.343
Square R-D	1,2	0.479	1.22	0.957	0.489	0.0885
	3	0.838	1.38	1.36	0.783	0.315
	4	1.68	1.44	1.48	0.976	0.353
	5	2.58	1.63	1.56	1.10	0.421
Conventional	1,2	2.64	2.36	1.26	0.841	0.249
	3	2.93	2.55	1.37	0.914	0.271
	4	5.23	3.21	2.38	2.12	0.753
	5	219	3.94	3.24	2.56	0.753

17. As shown in the figures the hexagonal r-d clusters consistently have the largest response when compared with the other two types. The single unit configuration response, as depicted in Fig. 7, shows the largest difference in the peak responses between the conventional building and the two single rhombic dodecahedron. In Fig. 8, it is noticed that the conventional building peak response for the one-storey building is approximately one-half that of the square r-d cluster, and approximately one-third that of the hexagonal r-d cluster. As the floor level is increased to the two-storey configuration as depicted in Fig. 9, the conventional building and square r-d cluster peak responses are essentially the same, and approximately one-half that of the hexagonal r-d cluster. As shown in Fig. 10, the three-storey building has essentially the same percent differences in peak response between the three building types as that for the one-storey building. Fig. 11 indicates that the r-d clusters show a significant increase in the peak responses as compared to that of the conventional building; this is anticipated from the frequency analysis results (ref. 4). A considerable decrease in the natural frequencies is shown in Table 2 as the clusters are increased from three to ten floors.

CONCLUSIONS

18. In this study, a comprehensive finite element method analysis has been conducted to evaluate the response of square and hexagonal plan r-d buildings to static and dynamic loads. In addition, the first three buckling loads and first five natural frequencies have been determined for the single r-d, as well as the one-, two-, three and ten-storey clusters. Response comparisons have been made between the two cluster types and with that of comparable conventional buildings.

19. From the static loading results, it has been shown that the square r-d clusters exhibit a more uniform distribution of total load than do the hexagonal r-d clusters. It is suggested that possibly—since all possible loading conditions have not been studied —the square r-d clusters can be more readily designed with all members on a level having the same geometric properties than the hexagonal r-d clusters. It has also been shown that some members actually exhibit an axial tension stress resultant. From the elastic buckling load results, it has been concluded that all the r-d clusters exhibit a significantly lower buckling load than that of a conventional building, due to bending of the members prior to buckling. However, the buckling loads do increase for the r-d clusters as the storey levels are increased to the second floor, but then begin to decrease as additional floors are added. From the fundamental frequency analysis, it has been shown that the fundamental frequencies are considerably lower than those of the conventional building, and that the largest frequencies for the r-d clusters occur in the one-storey building. Also, the two cluster types have relatively similar natural frequencies, with the hexagonal r-d cluster consistently having smaller values. The ground motion results indicate that the hexagonal r-d clusters consistently have the largest peak response, and that the responses of the two r-d clusters were consistently higher than those of the conventional buildings.

20. As a result of this study it is suggested that further modelling of the clusters be conducted with additional static and dynamic load cases to determine conditions under which the r-d clusters might more favourably be compared with conventional buildings. In addition, further modelling is needed of the r-d clusters with support constraints directly below each lower level column; it is anticipated that this condition of restraint will produce higher buckling loads and natural frequencies, and reduce the displacement responses due to ground motion.

REFERENCES

1. ADINA, A finite element program for Automatic Dynamic Incremental Non-linear Analysis, Report ARD 87-1, ADINA R&D Inc., Watertown, MA, 1987.
2. ADINA-PLOT, A program for display and post-processing of ADINA results, Report 87-7, ADINA R&D Inc., Watertown, MA, 1987.
3. MALLER A. An analytical visual representation of non-rectangular deep space structures. International Journal of Space Structures, 1991, vol. 6, no. 4, 315-324.
4. CLOUGH R.W. and PENZIEN J. Dynamics of structures, McGraw-Hill, New York, 1975.

119. Dimpled single-layer geodesic shells

C. J. KITRICK, S-MOS Systems, USA

SYNOPSIS. As dimensions for geodesic shells increase local regions of the spherical surface become near planar resulting in highly non-linear response to radial loads. Design alternatives to offset this effect include the use of rigid-joints and spherical trusses. This study examines an alternative approach, a simple technique referred to as dimpling, to increase the stiffness of joints. Analysis indicates that the typical non-linear response is effectively modified toward a near linear response. Dimpling therefore, can be used to improve structural performance while at the same time retaining the simplicity of pin-connections for single-layer geodesic shells.

STATEMENT OF PROBLEM

1. As geodesic domes become larger the subdivision of the spherical surface also increases so that the dimensions of individual components can be kept within some acceptable range. As a result the dihedral angle between adjacent faces sharing a command edge increases towards 180 degrees as each bar subtends a smaller angle of the overall sphere. Thus local regions of the dome's surface approach a near planar condition, which induces a geometrically non-linear response to radially applied loads. This local dome instability results in a phenomenon known simply as dimpling, the result of a loaded joint inverting or turning inside out.

2. If the engineer feels that the dome is locally unstable, there are several options; one, to provide for a fixed joint rather than a pin connected one, or two, to create a spherical truss. Either choice will help resolve the local instability problem but both introduce another dimension in complexity to the dome's design and construction. A truss in particular increases the number of joints and members by about a factor of three.

3. The purpose of this study was to explore an alternative design approach. The approach proposed here is to, in effect, pre-invert or dimple, selected joints. This approach involves no more than a slight modification of the dome's geometry and does not increase the dome's complexity; the number of bars and joints remains the same.

Space structures 4. Thomas Telford, London, 1993

JOINT INVERSION

4. Dimpling vertices involves the re-positioning of a vertex along its radial path until its distance below the surrounding vertices is equal to what its previous height was originally above the surrounding vertices. Since the bars that surround the dimpled vertex are typically not equal in length we select the largest to determine the amount of radial contraction. In this way the shorter bars will elongate but the longest will not change. The new contracted radius can be computed by multiplying the original radius by the nodal inversion factor (NIF) given by Eq. 1.

5. An entire dome surface requires only a portion of the vertices to be dimpled. The resultant effect is a tiling of shallow pentagonal and hexagonal regions. The method of inversion outlined in §4 has little or no effect on the overall range of bar lengths. As shell frequency increases the dimpled dimension (2h) decreases accordingly.

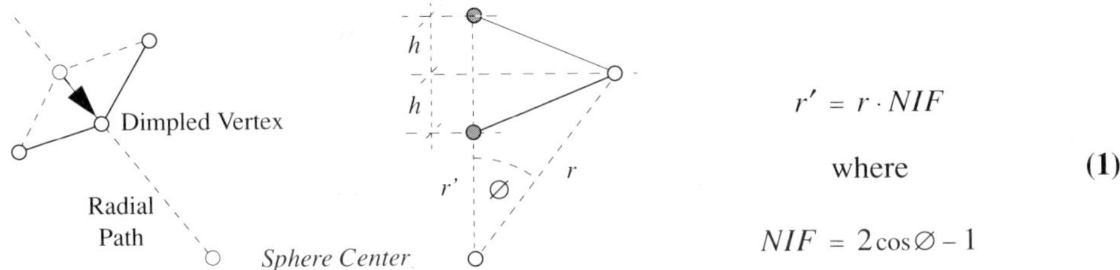

Fig. 1. Vertex contraction.

TEST GEOMETRY

6. The test geometry was composed of five pairs of single-layer frames, all based on the subdivision of the icosahedron. Their overall floor areas were the same, with only a range of height to baseline ratios (Table 1). The actual icosahedral subdivisions were of the class II variety; $\{3,5\}_{bc}$ where b = c (refs 1, 2).

Table 1. Geometric configurations.

Configuration	Radius	Height to Baseline Ratio
$\{3,5\}_{4,4}$	60'	0.1688
$\{3,5\}_{5,5}$	70.713'	0.1386
$\{3,5\}_{6,6}$	81.930'	0.1167
$\{3,5\}_{7,7}$	93.500'	0.1016
$\{3,5\}_{8,8}$	105.224'	0.0892

7. All configurations consisted of 61 vertices, 156 edges, and 64 faces. Each pair contained both a normal and dimpled configuration with the same radial

curvature. Three control vertices (1, 35, 57) were fixed to keep the overall floor dimension of each configuration approximately equal. Dimpled vertices are the centre of the hexagonal areas outlined in Fig. 2. The five curvatures are illustrated to scale in Fig. 3.

8. All boundary joints in the peripheral ring of the dome were constrained in the x, y, and z directions. Because of this the peripheral bars are zero stress bars.

LOADING

9. Radial loading was tested on multiple joints. Due to the symmetry of the shells certain joints were selected due to their placement. Joint numbering is indicated in Fig. 2. On the normal shells radial loads were applied to joint 30. On the dimpled shells joints 22 and 30 were loaded radially independently. It was assumed for comparative purposes that for the normal frame response at joint 22 would be equivalent to joint 30.

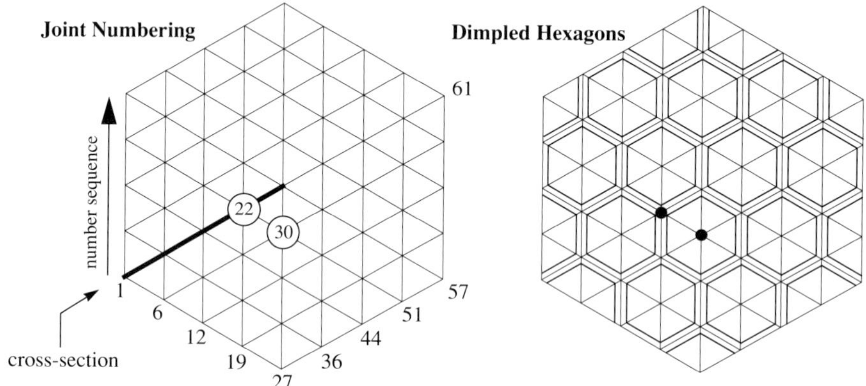

Fig. 2. Geometric configuration.

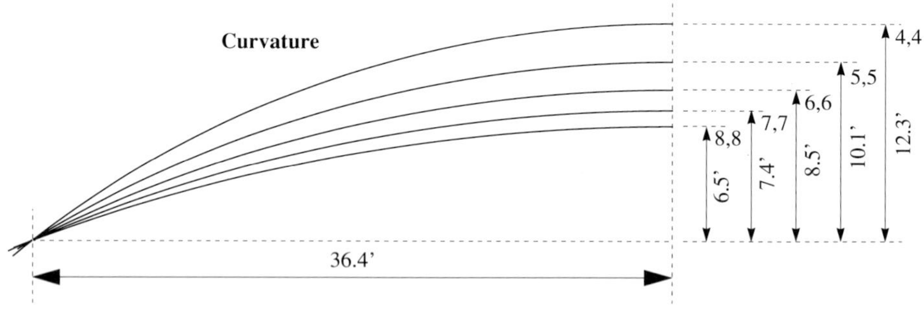

Fig. 3. Shell curvature cross-section.

ANALYSIS

10. All analysis was performed using the stiffness method in matrix form for space trusses (ref. 3). Assumptions included elastic material response, frictionless

bar/joint connections, constant cross-section of bars, and non-eccentrically (axially) loaded bars. Bar members and properties were taken from the AISC Manual of Steel Construction (ref. 4). All bars dimensions are given in Table 2.

Table 2. Member properties.

Nominal Diameter	3 inches
Cross Sectional Area	2.3 square inches
Radius of Gyration	1.16
Length Range	85-131 inches

11. Failure was defined to be either buckling of bars or the dimpling of a joint. Buckling column action was based on AISC formula 1.5-1 and 1.5-2 with exclusion of the safety factor. Since the slenderness ratios were only about a third of C_c, AISC equation 1.5-2 was not critical and equation 1.5-1 reduced F_a by a maximum of 5%.

$$F_a = \frac{\left[1-\left(\frac{Kl}{r}\right)^2\right]}{2C_c^2} \quad \text{where} \quad C_c = \sqrt{\left[\frac{(2\pi^2 E)}{F_y}\right]} = 384.76 \quad \text{and} \quad \begin{aligned} K &= 1 \\ E &= 30000ksi \\ F_y &= 40ksi \end{aligned} \qquad \textbf{(2)}$$

12. Dimpling of a joint was defined to be the zero volume of the tetrahedron that was defined by the surrounding bar members of the loaded joint (Fig. 4). As the apex joint of the tetrahedron was loaded it would displace toward the base of the polyhedron, eventually becoming coplanar with the three other base joints. The volume of the tetrahedron is simply defined as one sixth of the determinate of the four co-ordinates (Eq. 3). Obviously it would be unlikely that the volume would ever be exactly zero under any particular load. Fortunately if the determinate is initially arranged to yield a positive volume then, it will yield a negative value when the apex has inverted through the base.

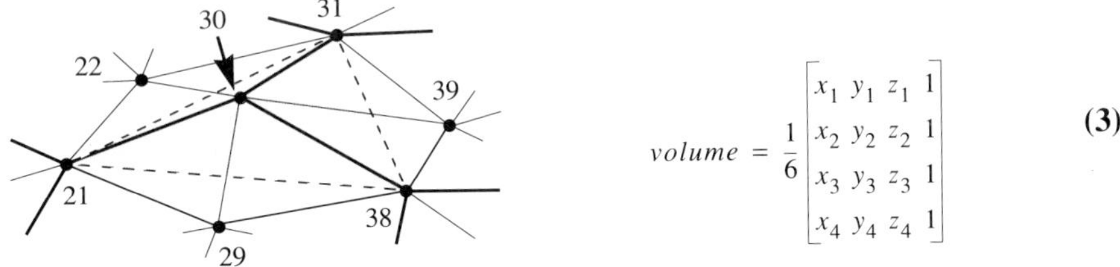

$$volume = \frac{1}{6}\begin{vmatrix} x_1 & y_1 & z_1 & 1 \\ x_2 & y_2 & z_2 & 1 \\ x_3 & y_3 & z_3 & 1 \\ x_4 & y_4 & z_4 & 1 \end{vmatrix} \qquad \textbf{(3)}$$

Fig. 4. Tetrahedral volume.

13. Non-linear behaviour was evaluated with a simple iterative loading technique. An initial radial load value was determined and joint displacements evaluated. Failure conditions were tested. If failure had not occurred the geometry would be modified by the joint displacements and the resultant geometry would be re-loaded with the incremental load value. This would continue until a failure

condition was reached. Then the original load value would be halved and the entire failure sequence repeated to determine whether the geometry was still too sensitive to the size of the load and yielded a significantly reduced failure load.

14. The initial load was set at 1kip for the 4,4 frequency configurations (normal and dimpled). The failure load differed very little from the next test at 1/2 kip. As frequency increased the sensitivity increased and the initial load values were reduced until they produced a stabilised total failure load. The normal configurations were much more sensitive to the loading values and were the limit cases in determining the final incremental load values. The final incremental loading values were 1 kip for the 4,4 frequency, 1/2 kip for frequencies (5,5), (6,6) and (7,7), and 1/4 kip for the 8,8 frequency.

RESULTS

15. For the normal frames joint sensitivity at the higher frequencies was a direct function of the geometry. On the 4,4 normal frame joint 30 lies approximately 12 inches above the six surrounding joints while only about 6 inches on the 8,8 normal frame. This greatly increases the effect of the radial loads on bar members. Figs 6, 7 and 8 illustrate the typical displacement response for the loaded joint and some surrounding joints using the 6,6 frames (normal and dimpled) as an example.

16. Table 3 summarises the failure load for each frame and the percent increase in load carrying capacity of the inverted frames over the normal frames at the same frequency. As the shells become increasingly shallow the increase in failure load capacity of the dimpled versus normal frames is dramatic. Fig. 5 is a graph of the tetrahedral volume used to measure the geometric inversion of joint 30 and clearly illustrates the non-linear response of the normal frames to geometric deformation.

Table 3. Non-linear failure loads.

Freq	Failure Loads[a]						%Change @30	%Change @22
	N^{30}	FM	D^{30}	FM	D^{22}	FM		
4,4	40.0	C	66.0	C	41.0	C	65.0	2.5
5,5	28.0	C	53.0	C	34.5	C	89.3	23.2
6,6	18.0	D	44.0	C	30.0	C	144.4	66.7
7,7	12.0	D	37.5	C	27.0	C	212.5	125.0
8,8	7.8	D	32.5	C	24.5	C	319.4	216.1

a. Symbols are $\mathbf{N^i}$ - normal frame at joint **i**, $\mathbf{D^i}$ - dimpled frame at joint **i**, **FM** - failure mode, **C** - compression, **D** - dimpling. All load values are in kips.

17. To verify the near linear response of the dimpled frames, all frames, both normal and dimpled, were loaded with the failure loads found from the non-linear analysis, and analysed linearly.

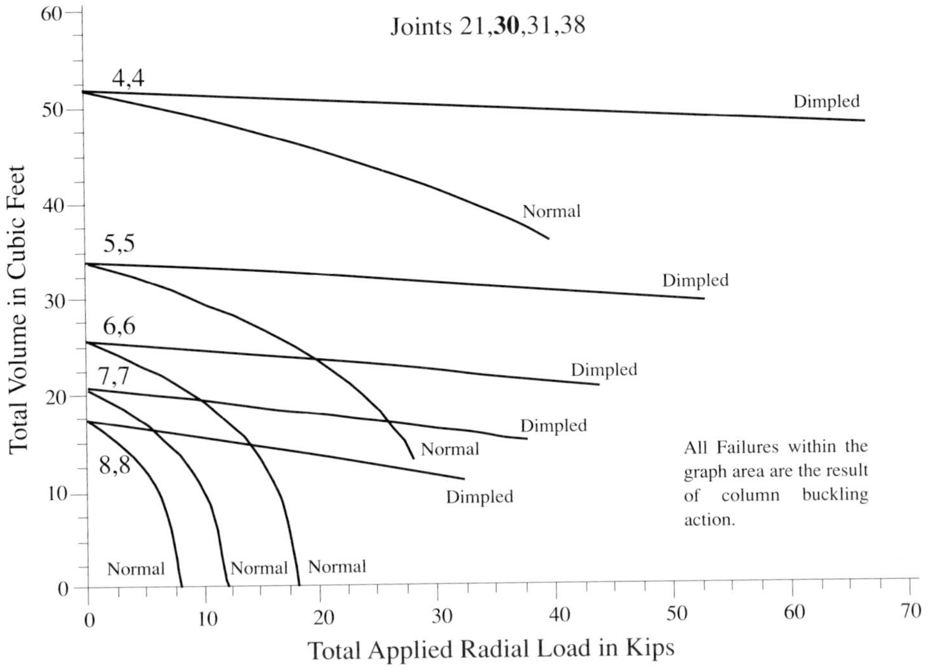

Fig. 5. Applied radial load versus tetrahedral volume.

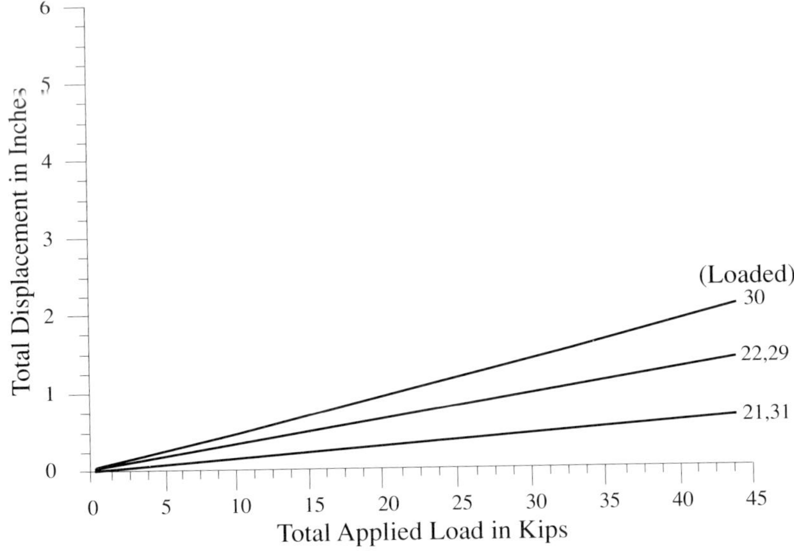

Fig. 6. Inverted frame (6,6).

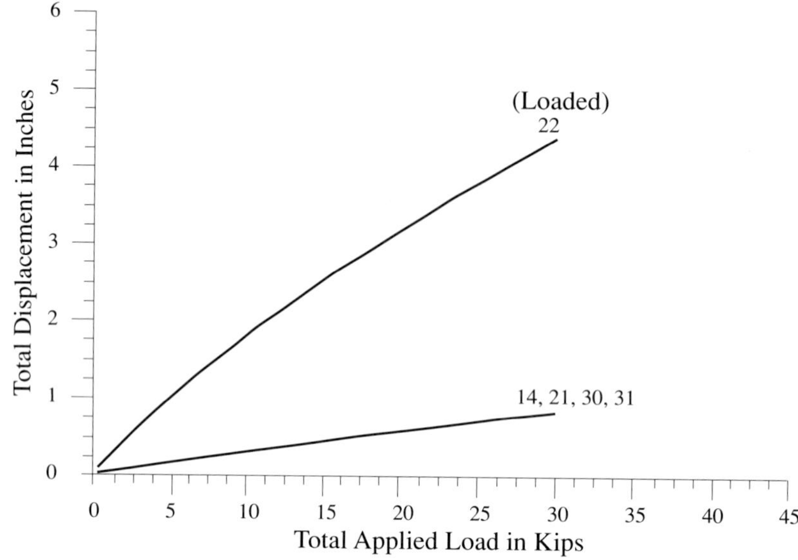

Fig. 7. Inverted frame (6,6).

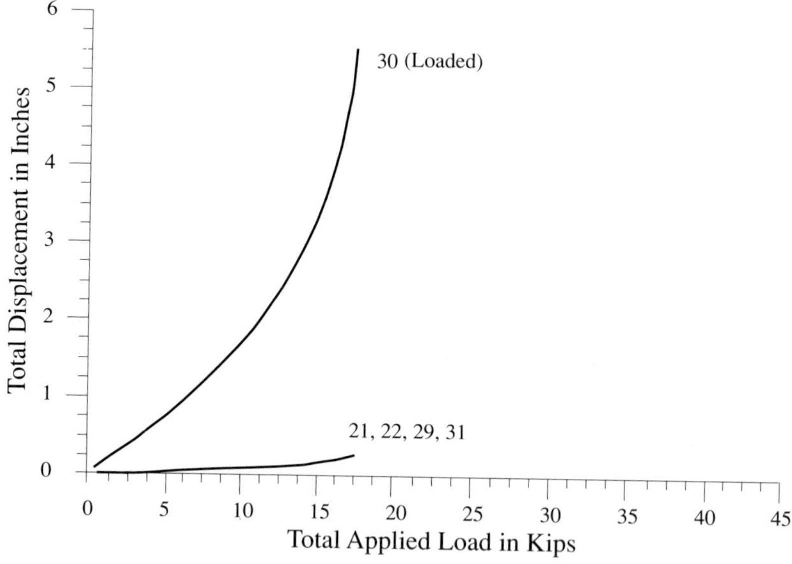

Fig. 8. Normal frame (6,6).

Table 4 summarises the response of the frames using both types of analysis. A check of the normal frames shows that linear analysis at joint 30 gave displacements 28-60% less than those from non-linear analysis. Clearly, as expected unmodified linear analysis gives the impression of the frame being much stiffer than it actually is and inadequately measures the frame's ability to withstand geometric deformation.

18. Looking at the response of the dimpled frames loaded at joint 30 it can be seen that linear analysis gave displacements 2-10% less than those from non-linear analysis while bar forces were within 4%.

19. The response of the dimpled frames loaded at joint 22 showed that linear analysis gave displacements 20-70% greater than those from non-linear analysis. Linear analysis also indicated bar forces in excess of the elastic limit.

20. The response of joint 30 indicates that use of linear analysis would yield reasonably correct load capacities, while the response of joint 22, a dimpled joint, indicates that linear analysis would yield very conservative load capacities.

21. As indicated by this analysis the dimpled frames displayed a much higher stiffness and ability to sustain concentrated radial loads, effectively preventing geometric deformation before material failure. Displacement of neighbouring joints was also much greater for the dimpled frames indicating that loads were able to spread over a wider area of the shell surface. In contrast the majority of the normal frames failed from geometric deformation.

22. If we take the lower load capacity from the two joints tested on the dimpled frames and compare them with the normal frame we find that the 6,6 dimpled frame has approximately double the failure load capacity, and the 8,8 dimpled frame almost triple. This increase in capacity makes high frequency and shallow height frames more feasible without the necessity of joint design modifications.

23. One implication of this study is that dimpled frames can be adequately analysed with linear techniques. Just as importantly it must be realised the normal geodesic frames do not respond linearly and analysis must take this into account. This was clearly the implication of previous studies on normal frames (refs 5-10).

24. The concept of pre-dimpling joints is not a technique applicable to all domes. Obviously domes of large enough dimensions and/or high frequency will require a double-layer (truss) configuration of some kind, and small to medium sized domes in the lower frequency range will probably not gain any benefit from this technique. Domes that can take significant advantage of the joint inversion technique are medium to large domes in the 80-200 foot diameter range or domes with a very small height to baseline ratio. Since many domes have skins attached directly to the frame the joint dimpling technique could induce ponding problems in the apex region of the dome. Domes with suspended skins would be easier to adapt.

Table 4. Linear versus non-linear analysis.

Freq	Type[a]	Load[b]	Linear[c] Δ	Non-Linear[c] Δ	% Change	Linear Force[d]	Non-Linea Force[d]	% Change
4,4	N	40.0[30]	2.569	3.581	-28.3	73.24	85.39	-14.2
4,4	D	66.0[30]	1.517	1.552	-2.3	84.64	85.25	-0.7
4,4	D	41.0[22]	4.224	3.546	+19.1	94.15	85.83	+9.7
5,5	N	28.0[30]	2.654	5.211	-49.1	64.25	85.85	-25.2
5,5	D	53.0[30]	1.769	1.837	-3.7	85.08	86.13	-1.2
5,5	D	34.5[22]	5.102	3.938	+29.6	97.44	85.53	+13.9
6,6	N	18.0[30]	2.367	5.498	-56.9	49.49	68.49	-27.7
6,6	D	44.0[30]	2.028	2.146	-5.5	84.65	86.21	-1.8
6,6	D	30.0[22]	6.253	4.404	+42.0	102.18	85.89	+19.0
7,7	N	12.0[30]	2.098	5.252	-60.1	38.46	53.88	-28.6
7,7	D	37.5[30]	2.291	2.480	-7.6	84.08	86.27	-2.5
7,7	D	27.0[22]	7.555	4.853	+55.7	107.61	86.54	+24.3
8,8	N	7.8[30]	1.742	4.128	-57.8	28.37	39.36	-26.0
8,8	D	32.5[30]	2.548	2.843	-10.4	83.26	86.27	-3.5
8,8	D	24.5[22]	8.867	5.212	+70.1	111.82	86.15	+29.8

a. Abbreviations are **N** - Normal, **D** - Dimpled frame

b. Loads are in kips. Superscripted numeral indicates load joint. (ex. 40.[30] is joint 30)

c. Linear and non-linear displacements are given in inches.

d. Force is given in

Table 5. S1 Unit conversions.

Symbol	US units	Conversion Factor	SI units
kips	1000 pounds	4.448	N - newtons
ksi	kips per square inch	6.895	kPa - kilopascals
"	inches	2.54	cm - centimetre
'	feet	0.305	m - metre
	cubic feet	28.317×10^{-3}	m^3 - cubic metre
	square inches	0.645×10^3	mm^2 - square millimetre

REFERENCES

1. COXETER H. S. M., Virus Macromolecules and Geodesic Domes. A Spectrum of Mathematics, 1972, pp. 98-107.
2. WENNINGER M. J., Spherical Models, Cambridge University Press, 1972.
3. WEAVER W. and Gere J. M., Matrix Analysis of Framed Structures Models, D. Van Nostrand, 1980.
4. AISC Manual of Steel Construction, 8th Ed., Part 5, Section 1.5, Chicago.
5. AGUILAR R. J., A Study of the Stability of Framed, Triangulated Geodesic Domes Under the Action of Concentrated Loads. Thesis presented to the North Carolina State University, 1964.
6. AGUILAR R. J., Snap-Through Buckling of Framed Triangulated Domes. American Society of Civil Engineers Proceedings, 93, ST 2, 1967, pp. 301-317.
7. AGUILAR R. J. and Huang T. A., Snap-Through Buckling of Triangulated Lattice Domes under the Simultaneous Application of Concentrated Loads. Space Structures, Ed. R. M. Davies, 1966, pp. 459-470, John Wiley & Sons, New York.
8. BUCHERT K. P., Buckling Considerations in the Design and Construction of Doubly Curved Space Structures. Space Structures, Ed. R. M. Davies, 1966, pp. 485490, John Wiley & Sons, New York.
9. SMITH GLENN ROOSEVELT, JR., A Non-Linear Static Analysis for Spherical Truss Domes. Thesis presented to the North Carolina State University, 1964.
10. WRIGHT D. T., Membrane Forces and Buckling in Reticulated Shells. American Society of Civil Engineers Proceedings, 91, ST 1, 1965, pp. 173-201.

120. Calculation of geometrical parameters and projection of geodesic domes and shells based on the network subdivision of the system D

G. N. PAVLOV, Institute of Architecture and Civil Engineering, Russia

SYNOPSIS. This paper describes the method for carrying out practical work on the subject "Projection of Geodesic Domes and Shells" for students studying the speciality "Architecture" and "Civil Engineering". General data on geodesic shell geometry and principles of the sub-division of a sphere according to the system indicated by the index "D", together with directions on the use of computer programs used for obtaining values of shell parameters, are presented in this work.

THE TASK

- To derive tables of geometrical parameters and computer drawings for various network sub-divisions of system "D" employing the programs from the Institute Computer Centre library.
- To obtain one variant of a dome making use of computer drafting.
- To investigate the assembly of the dome and to define general parameters by using tables of parameters and a scheme of the networks.
- To make up a scheme of typical details and to define their dimensions.
- To assemble the model of the dome taken as a whole or in components.

GENERAL INFORMATION ON THE GEOMETRY OF THE D" SYSTEM

1. Domes and shells up to the middle of the twentieth century were designed only on the basis of meridian-and-circular subdivision of the rotation surfaces. At the end of the nineteen forties there appeared new systems of sub-divisions based on the geometry of regular polyhedrons such as an icosahedron, an octahedron and tetrahedron.

2. The first step in the sub-division of a sphere which encircles a polyhedron, is the drawing of geodesic lines between the vertices of the polyhedron lying on the sphere surface. The resultant equilateral spherical triangles are called the Mobius triangles. The second step in the construction of the sub-division is drawing lines connecting the vertices of the Mobius triangles. The last triangle being divided into six similar rectangular spherical triangles, called Schwarz triangles.

Space structures 4. Thomas Telford, London, 1993

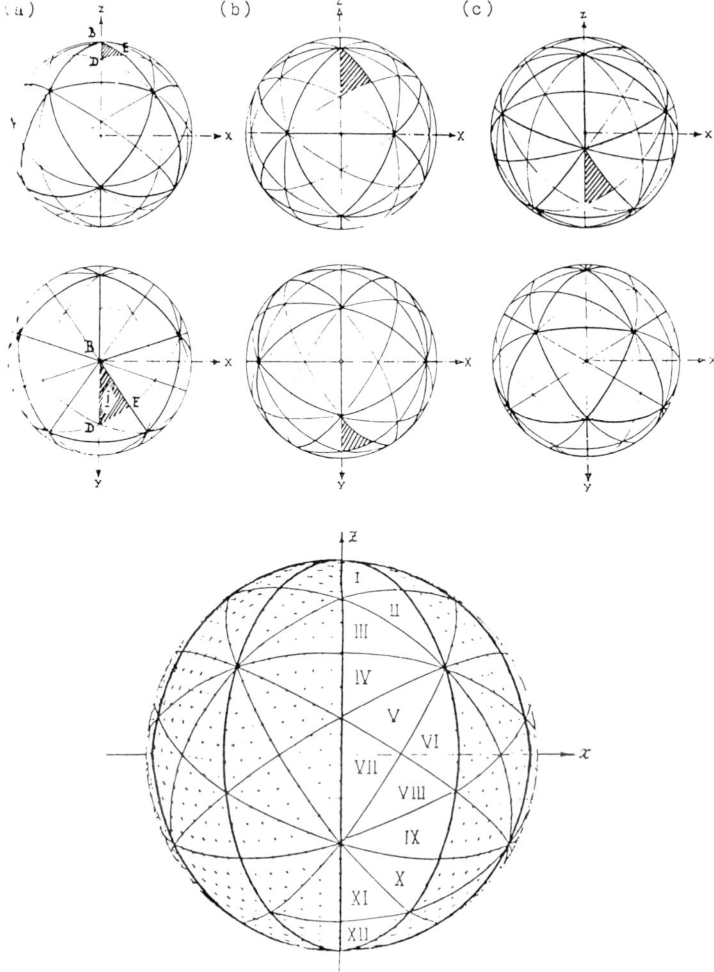

Fig. 1. Three kinds of space orientation of network division lines on the surface of a sphere: (a) orientation No. 1; (b) orientation No. 2; (c) orientation No. 3. Meridional sectors (heavy lines) and notation of Schwarz triangles from I to XII in icosahedral sub-divisions.

3. Fig. 1 shows the arrangements of icosahedral Schwartz triangles DBE on the surface of the full sphere with their numeration on the meridional sector. At the

zenith point of the dome all the three points of the Schwarz triangle; the B, D or E points are found. In accordance with this, Fig. 1 shows three possible orientations in the space of network surface of No 1, 2 and 3 orientations. The choice of network sub-division orientation greatly influences the total composition of the shells.

4. For the real projection of shells it is necessary to breakdown the surface into much smaller triangular cells. The latter can be obtained by the sub-division of typical sections into triangular cells, when the sphere surface assumes a pattern of a small-celled network. The choice of the mode of a typical section breakdown will be dictated by the peculiarities of the projected shell construction, where the main requirement (criterion of optimality) is the minimum number of types of the shell precast elements panels of corresponding forms, pivots, rods, nodal details, etc. Investigations showed that in a geodesic system there is a direct connection between a constructive solution of a shell and its geometrical basis. The geometrical network being optimal for a given type of dome construction. Our methodical directions discuss the mode of geometrical sub-divisions allowing the assembly of domes and shells from the least possible number of types of hexahedral pyramidal elements. The mode of sub-division is indicated by an index "D" and the whole plurality of the network types obtained as a result of its employment is united by the words - "network breakdown of the D system".

PRINCIPLE OF THE SPHERE SUB-DIVISION ACCORDING TO THE "D" SYSTEM

5. Fig. 2. shows a principle scheme of the geometrical Pavlov sub-division of a sphere according to the "D" system. The main criterion of the "D" system sub-division is the minimum number of hexagonal pyramidal detail types. It is achieved by means of arranging rows of similar symmetrical hexagonal figures in the direction of the line BE of a Schwartz triangle (Fig. 2a). It is clearly seen that the arrangement of similar hexagonal figures alongside the lines DE and DB gives much more types than along the line BE.

6. The main breakdown scheme of the "D" system is being formed in the following way: the line BE is divided into an equal number of sections and right-angled spherical triangles with an acute angle, equal to a B angle, are placed on these sections (see Fig. 2, hatched triangles). The lines of the first stage of the geometrical construction are formed by drawing large circles as a continuation of the hypotenuses and small cathetus of initial triangles, while the centre of the hexagonal figures of all subsequent rows are always on the continuations of the hypotenuses lines of the initial triangles. Then the centre of the hexagonal figures are connected by geodesic lines and large circles; perpendiculars to the line BE are then drawn through the vertices of the acute angles of the initial triangles. While dividing the line BE into more sections, the spherical Schwartz triangle will be divided into rows of hexagonal cells (Fig. 2b; thick lines). These cells are arranged towards the line BE, in all rows only one hexagonal detail does not repeat the configuration of typical details of the row; its axes of symmetry is placed on the line DB (Fig. 2a; detail 4).

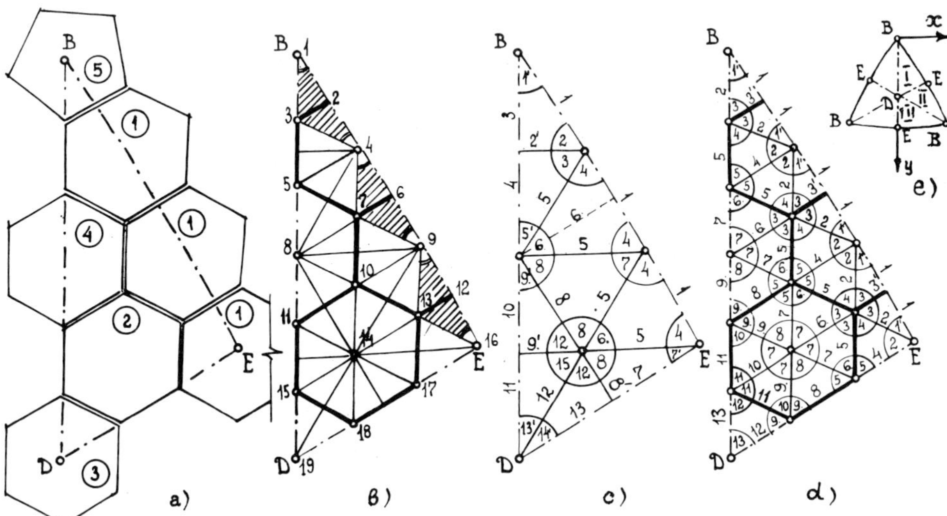

Fig. 2. Calculating scheme for the "D" system of sub-division and representation of the parameters of the two dual schemes: (a) positioning of the standard members within a Schwartz triangle, (b) the principle scheme of the sub-division with numeration of the nodal points, (c) marking of the parameters for "D1" sub-system schemes, (d) marking of the parameters for "D2" sub-system schemes, (e) numeration of the Schwarz triangle sectors within a Mobius triangle.

The main scheme of the "D" system network (Fig. 2b), except the initial hexagonal triangles and some triangles adjoining the lines DE and DB, consist of oblique-angled triangular facets. In spite of this, the sub-divisions possess rather favourable properties for constructing domes and shells. They are clearly obvious in dividing the main scheme into two dual networks of the "D1" and "D2" sub-systems (Fig. 2c and d).

7. The network of the "D1" sub-system consists of isosceles triangles, their inter-location being very convenient for constructing dome and shell details. In the limits of a Mobius triangle, equal details are located in concentric rows and all nodal elements are symmetrical. Breakdown of the "D2" sub-system consists of oblique-angled triangular cells but they are located symmetrically and by considering their unification it is possible to form a network on the sphere surface, not consisting of symmetrical hexagonal cells. The number of their types will be a minimum in comparison with all other systems of geodesic sub-divisions. The next feature of the "D2" sub-system is that the network sub-division in the limit of the Schwarz triangles consists of inter-crossing lines of large circles.

8. Table 1 presents the classification of "D" network types where the basis is the number of triangular cells on the full sphere. The icosahedronal sub-divisions

have index "D", and their corresponding dual sub-systems of "D1" and "D2". Octahedral networks are indicated by index "D0" ("D01" + "D02"); indexes of tetrahedral networks are "DT" ("DT1" + "DT2"). If designing a shell both networks are used by overlaying one on the other, the sub-division is described as follows: "17280-D" ("2880-D1"+"17280-D1"). It is the twelfth variant of sub-division.

Table 1.
Classification of types of sub-division in the "D" system.

Breakdowns in subsystems "D1", "DO1" and "DT1"									Breakdowns in subsystems "D2", "DO2" and "DT2"										
Number of divisions on sides of a Möbius triangle	Types of breakdowns			compositional mark	Number of types of elements				Number of divisions on sides of a Möbius triangle	Types of breakdowns			compositional mark	Number of types of elements					
							nodes								facets				
	"D_1"	"DO_1"	"DT_1"		sides	facets	total number	including mirror images		"D_2"	"DO_2"	"DT_2"		sides	total number	including mirror images	nodes	pyramidal units
1	20	8	4	△	1	1	1	-	2	60	24	12	◁	2	1	-	2	1
2	80	32	16	▽	2	2	2	-	4	240	96	48	▷	4	3	1	4	2
3	180	72	36	⊛	3	3	3	-	6	540	216	108	⊛	6	5	2	6	3
4	320	128	64	△	4	5	4	-	8	960	384	192	◁	8	7	3	8	3
5	500	200	100	▽	5	6	5	-	10	1500	600	300	▷	10	9	4	10	4
6	720	288	144	⊛	6	7	7	1	12	2160	864	432	⊛	12	11	5	12	5
7	980	392	196	△	7	9	7	1	14	2940	1176	588	◁	14	13	6	14	5
8	1280	512	256	▽	8	10	8	1	16	3840	1536	768	▷	16	15	7	16	6
9	1620	648	324	⊛	9	11	11	2	18	4860	1944	972	⊛	18	17	8	18	7
10	2000	800	400	△	10	13	10	2	20	6000	2400	1200	◁	20	19	9	20	7
11	2420	968	484	▽	11	14	11	2	22	7260	2904	1452	▷	22	21	10	22	8
12	2880	1152	576	⊛	12	15	15	3	24	8640	3456	1728	⊛	24	23	11	24	9
13	3380	1352	676	△	13	17	13	3	26	10140	4056	2028	◁	26	25	12	26	9
14	3920	1568	784	▽	14	18	14	3	28	11760	4704	4352	▷	28	27	13	28	10
15	4500	1800	900	⊛	15	19	19	4	30	13500	5400	2700	⊛	30	29	14	30	11
16	5120	2048	1024	△	16	21	16	4	32	15360	6144	3072	◁	32	31	15	32	11
17	5780	2312	1156	▽	17	22	17	4	34	17340	6936	8468	▷	34	33	16	34	12
18	6480	2592	1296	⊛	18	23	23	5	36	19440	7776	3338	⊛	36	35	17	36	13
19	7220	2888	1444	△	19	24	19	5	38	21660	8664	4332	◁	38	37	18	38	13
20	8000	3200	1600	▽	20	26	20	5	40	24000	9600	4800	▷	40	39	19	40	14
..	...	...	...	..	..	..	..	..	..	...	...	...	..	..	..	..	..	..

EMPLOYING THE PROGRAMME FOR CALCULATING GEOMETRICAL PARAMETERS

9. In the Institute Computer Centre (CC) personal computers have programs on them located under the name of SYSTD. A student must come to the CC with prepared values of the initial parameters to work with the computer on the chosen type of sub-division. On the display screen after pressing "Enter" there appears the following legends: "introduce the number of the system"; "indicate the number of sectors needed (Schwartz triangles)"; "introduce the dimensions of the page", etc. By entering various combinations of the parameters the student may obtain drawings at any scale. Fig. 3. shows an example of typical computer drawings.

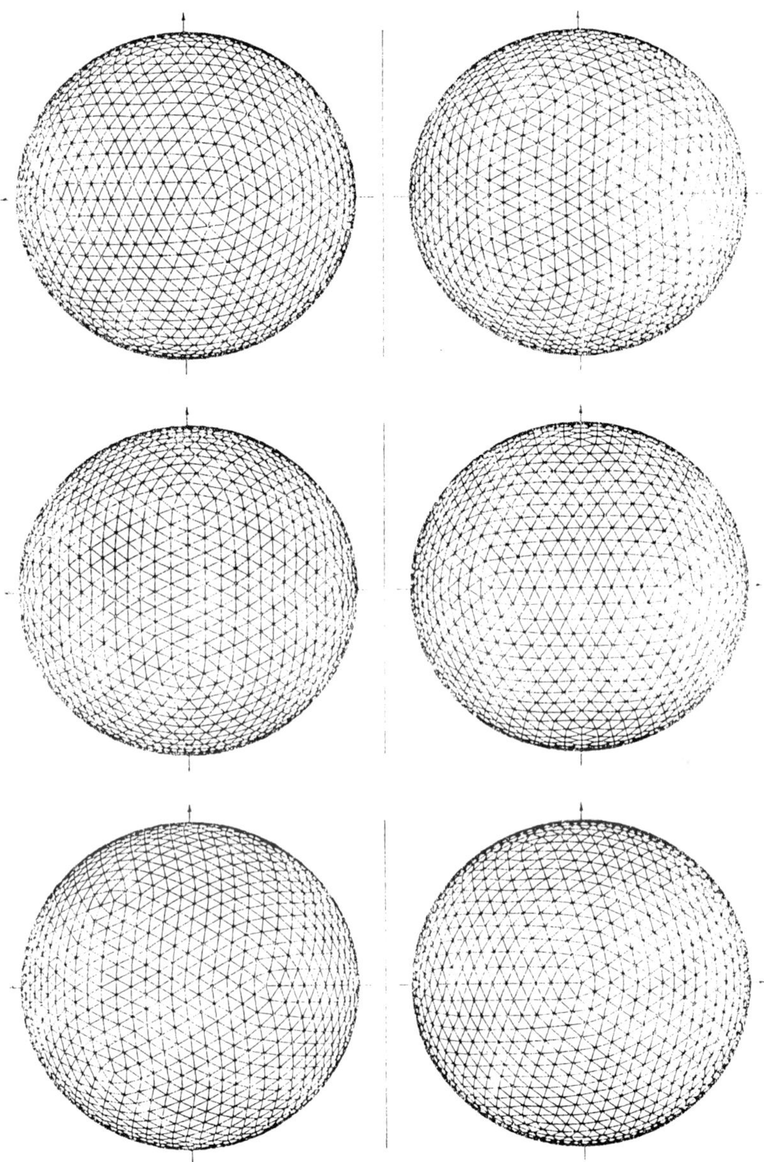

Fig. 3. Grid sub-division type "2880-D1".

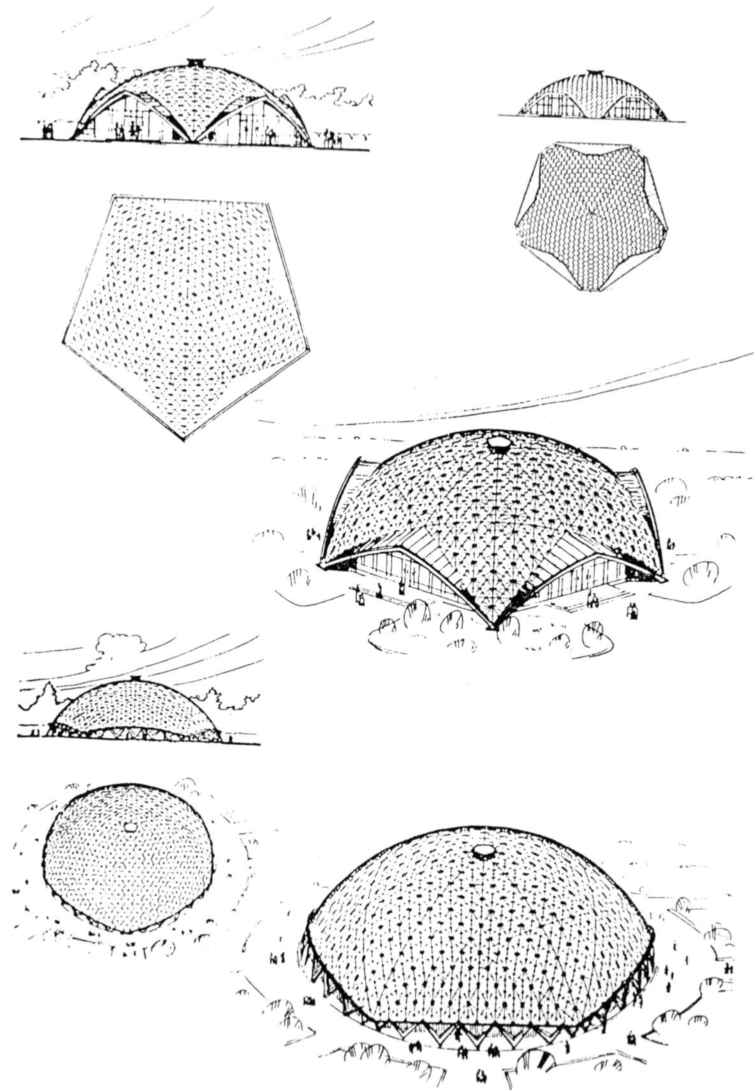

Fig. 4. Development sketches for compositional variants of domes using computer-produced drawings of grid sub-divisions (type "17280-D").

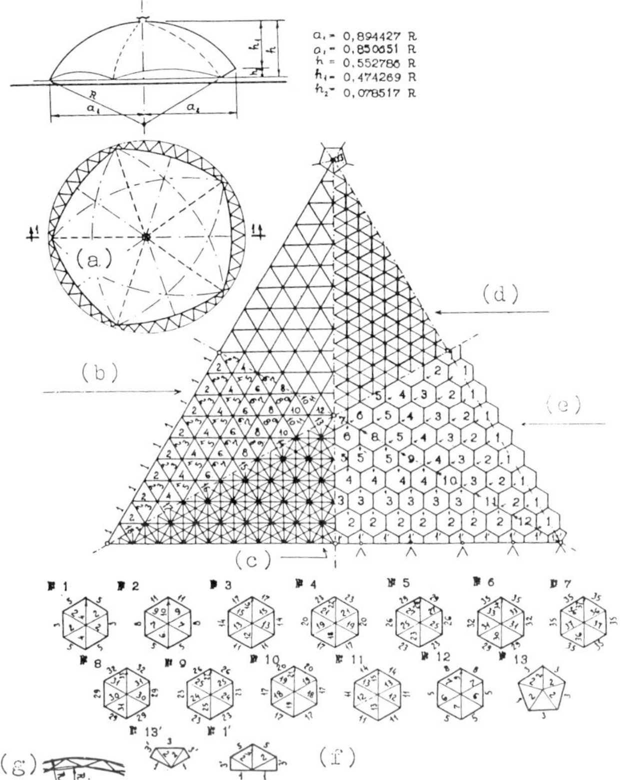

Fig. 5. Composition of the assembly drawing and schemes for standard members of the dome based on sub-division "6480-Dl" + "19440-D2": (a) plan; (b) the assembly drawing of the upper layer members. The length of the members is determined by the tables of scheme "6480-Dl"; (c) general view of the dome surface; (d) general view of the surface without the upper layer members; (e) the assembly drawing of standard pyramidal members of the dome; (f) compositional schemes of standard pyramidal members. Dimensions of ribs are determined using the table of scheme "19440-D2"; (g) cross-section of the dome shell.

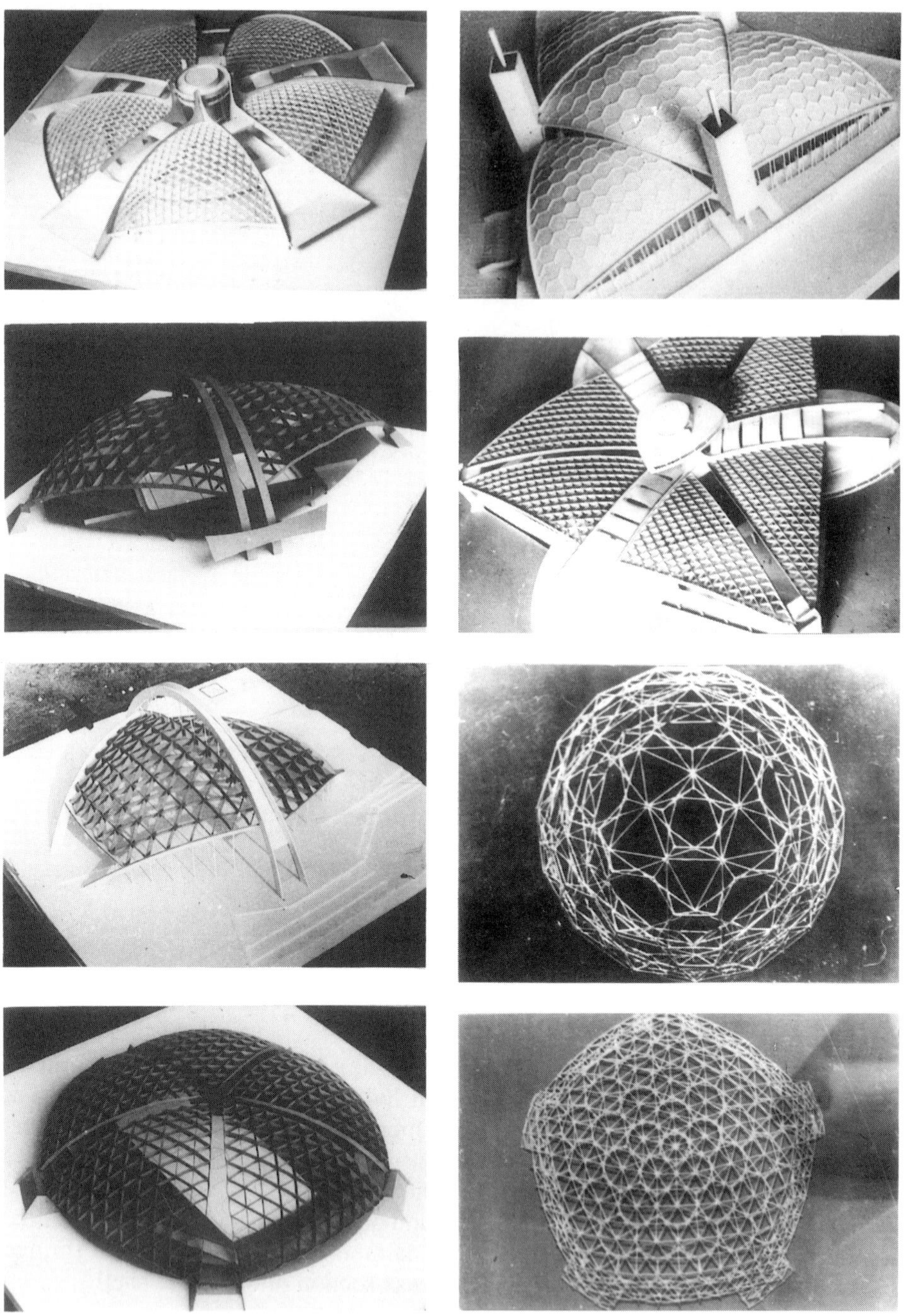

Fig. 6. Models of the diploma projections.

THE EMPLOYMENT OF SYSTEM "D" NETWORKS IN ARCHITECTURAL DESIGN

10. The first emergence of Geodesic domes is associated with the name of two researchers; M.S. Tupolev, professor of the Moskov Architectural Institute and R.B. Fuller, an American engineer, and were designed on the basis of grid surface division of various types. Nowadays there are more than 20 systems of spherical surface divisions available but only 6 of them are preferable as optimal for the existing structural schemes for shells. All the geodesic domes of large spans are calculated on the basis of the "D" system of network division. The type of division, the feature of which is the sub-division of a typical Mobius triangle into equal parts, was found to be the most convenient way of projecting steel shells composed of hexagonal pyramidal units. The "D" system provides the least number of types of elements in network divisions compared to all others.

11. The largest domes constructed on the basis of the "D" system are those of the car repair plants in Baton Rouge and Wood River (1958,1960), which span 114·4m. The first was designed using the type "19440-D2" division; the latter, by using two types of division: "6480-D1" and "19440-D2". The best known domes of the "D" system are the Climatron Dome in St.Louis (1959), spanning 53·4m, type "8640-D2" division; the dome of the main building of the Metal Corporation in Cleveland (1959), spanning 76,3m, type "8640-D2" division; the dome of the Exhibition Pavilion in Montreal (1967), 75m in diameter with sub-division types "5120-D1" and "15360-D2" for its upper part.

12. Examples of various domes shown by using computer draughting are given in Fig. 4. The diagram shown in Fig. 5 illustrates a method of working out an assembly scheme for the domes as well as their typical elements. Fig. 6 gives examples of students' project and diploma work.

REFERENCES

1. PAVLOV G.N. Compositional form-shaping of crystal domes and shells (in Russian), Architectura SSSR, No.2, 1977, 30-41.

2. Spherical grid structures (edited by T. Tarnai). Hungarian Institute for Building Science, Budapest, 1987.

121. The Tetraktys of Polyhedra

R. MEURANT, Institute of Traditional Studies, New Zealand

SYNOPSIS. In a previous paper "A New Order in Space" I advanced a three-fold order for the regular and semi-regular polyhedra. I now revise that order in the light of certain traditional schema of Sacred Geometry, particularly the Vesica Pisces and the Pythagorean Tetraktys. This situates Polyhedral Geometry within a traditional perspective, and allows the dimension of human meaning to he incorporated. The polyhedra exhibit structural patterns which in a fundamental sense embody the qualities of space, and reveal the structure of the human psyche. Polyhedra, space and consciousness are perhaps different aspects of the same underlying reality, namely Natural Harmony.

INTRODUCTION

1. Previously I have advanced a three-fold order for the regular and semi-regular polyhedra (ref. 1) with each class containing polyhedra which exhibit the same pattern of symmetry. Each pattern is characterised by neutral, male and female axes of symmetry with characteristic orientations and frequencies. The three classes are Class I: {2,3,3}; Class II: {2,3,4}; and Class III: {2,3,5}. Faces are neutral, male or female as they occur perpendicular to axes, except for paired neutral faces of the snub polyhedra. Within each class, polyhedral elements are disposed relative to one another according to a structure, and situated along sequences of truncation, rotational displacement, and transcendence.

2. Certain polyhedra reoccur according to the different symmetry patterns they exhibit, having regard to their colouring of faces. Individual polyhedral elements correspond precisely from class to class according to their position within the structure, and the sequences within each class correspond precisely from class to class. The order of polyhedra is also extended to two further classes to account for various colourings of all but one of the regular and semi-regular tilings of the plane (the exception ($3^3.4^2$) being a degenerate case). The extra tiling classes are Class IV: {2,3,6} and Class V: {2,4,4}.

3. This paper revisions the interrelationship of polyhedra into a more integral order, according to ancient schema of the traditional science of Sacred Geometry, in particular the Vesica Pisces and the Pythagorean Tetraktys. This situates the

discipline of Polyhedral Geometry within a traditional perspective, which facilitates further research whilst allowing the important dimension of human meaning to be incorporated.

4. Polyhedra exhibit structural patterns of interrelationship which embody fundamental qualities of space and reveal a deep structure of the human psyche which entertains them. This deep structure is not imposed, but rather discovered. Polyhedra, space, and human consciousness have embedded within them the potential for these orders of harmony. They are different aspects of the same underlying reality, I term Natural Harmony (ref.2).

SCHEMA OF SACRED GEOMETRY

5. The schema of Sacred Geometry which are relevant are the Circle and Sphere; the Vesica Pisces; the Trinity; and the Pythagorean Tetraktys. The Tetraktys is then developed into triadic arrays of other frequencies, with the Circle, Vesica Pisces, and a further schema the Star of David being considered as subsets or extensions of these.

6. Firstly the Circle is the most fundamental form of Sacred Geometry, and traditionally denotes the sacred centre, the cycle of creation, maintenance and completion, and the perfection of totality. It exhibits symbolism adequate to the monad (identity), duad (sacred temenos and profane without) and triad (spiritual centre, psychic domain, and physical boundary). The regular and semi-regular polyhedra are all bound by regular polygonal faces, each inscribable on a circle. In traditional symbolism these represent different qualities of manifestation. These faces are restricted to just the triangle, square, pentagon, hexagon, octagon, and decagon; and if the tilings are also included the dodecagon. Each face is disposed according to the frequency of axis of symmetry on which it lies, being either of single or double frequency. They are each constructed with compass and straight-edge, and embody $\sqrt{2}$, $\sqrt{3}$ or $\emptyset = 1.618...$ (and thus $\sqrt{5}$) proportional harmonies, which in tradition accord with qualities of generation, formation and regeneration.

7. The traditional symbolism of the Sphere is similar to that of the Circle, particularly as an exemplar of perfect form: a finite embodiment of the infinite (ref.3). Critchlow shows each of the regular and semi-regular polyhedra constructed on the Sphere (ref.4); the various polyhedra thus express different specific qualities of universal manifestation.

8. Secondly the Vesica Pisces, the most important schema of the Sacred Geometry of mediaeval Christianity, is regarded as the "Mother of Form". It represents an archetypal division of unity into duality, which generates a higher trinity. It also represents a quaternity, as the initial duality of centres generates a second duality of intersections; it thus organises the world into an above and below, and a left and right. I show later how this symbolism is particularly rich and adequate to its use with the polyhedra. Its proportional harmonies are $\sqrt{3}$; but it is also used to form $\sqrt{2}$ and $\emptyset$ harmonies.

9. Thirdly, I develop a symbol of the Trinity from three overlapping circles of equal radius, the centre of each circle lying at the overlap of the other two circles to yield three interpenetrating vesica pisces. The three-in-one symbolises the

integral interwoven three fold order of polyhedra, and hence represents the potential of three-dimensional space.

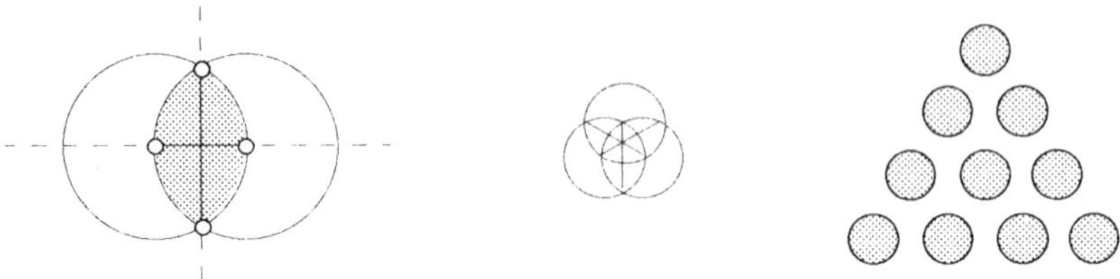

Fig. 1. The Vesica Pisces (Mother of Form), Trinity logo and the Pythagorean Tetraktys.

10. Fourthly, the Pythagorean Tetraktys, a four-frequency triadic array, exhibits symbolism of the central monad, the mirror-image duad, the triangular triad, the four level tetrad, and finally the decad as 1 + 2 + 3 + 4 = 10. The integration of three- and four-fold symbolism is particularly apt, three symbolising the heavenly realm and four the earthly (also, 3 x 4 = 12, the Sun, and 3 + 4 = 7, the Moon). The tetraktys was an essential part of the Quadrivium which, according to cabalistic tradition, was preserved by inscriptions on the pillars of Seth when God destroyed the world by flood and flame. Proclus says:

-Sacred Number springs
From th'uncorrupted Monad, and proceeds / To the Divine Tetraktys, she who breeds All; and assigns the proper bounds to all, / Whom we the pure immortal Decad call.

11. The triadic array is generalised to obtain a cosmogonic sequence of monaktys, duaktys, triaktys, tetraktys and pentaktys. The heptaktys is used with a subset yielding a further schema of traditional geometry the Star of David; or it is developed as an extension of the tetraktys. The Star of David usually represents the duality of complementary triads, but here the symbolism is of a three-fold interwoven order of quaternal vesica pisces.

COSMOGONIC SEQUENCE OF THREE INTERWOVEN CLASSES OF POLYHEDRA

12. Firstly I commence with the monaktys as the three-in-one symbol of the Trinity, which symbolises the interwoven harmony of the three-fold order of the polyhedra. This also represents what Lawlor describes as the Hindu concentric cosmic model of polyhedra inscribed one inside another (ref.5), one octave comprising Icosahedron, Dodecahedron, Cube, Positive and Negative Tetrahedra, and Octahedron through to inner Icosahedron.

13. Secondly the duaktys comprises three quasi-regulars, each the key solid of its class: the Class I Octahedron, the Class II Cuboctahedron, and the Class III Icosidodecahedron.

14. Thirdly, two forms of the triaktys are derived; firstly, each quasi-regular solid is located opposite its dual. Secondly the same dualities locate the respective male and female polar solids of each class, with male elements without and female within.

15. Fourthly the tetraktys locates three triplets at this level of manifestation, one for each class, of outward quasi-regular and withdrawn male and female poles to either side, the three triplets being symmetrically arranged about the central three-in-one. Quasi-regulars form an outer triangle; a hexagon consists of alternating male and female poles, which comprise two male and female interpenetrating triangles, with the polar opposites of each class lying opposite each other. (Here the tetraktys is shown within the Star of David).

THE STAR OF DAVID GIVES RISE TO THREE CLASSES OF VESICA PISCES

16. The tetraktys is developed into the Star of David by the addition of peripheral dualrhombic solids, being the duals of the quasi-regulars. Three interpenetrating Vesica Pisces one for each class surround a central three-in-one symbol. There are two outer triangles, one of the quasi-regular solids, the other of their duals, with each quasi-regular lying opposite its dual. There are two interpenetrating inner triangles, one of male poles and the other of female, with each male pole lying opposite its dual female pole.

17. This structure is teased apart into three separate Vesica Pisces, one for each Class. Each Vesica Pisces is reoriented into quasi-regular solid above and its dual below, with the male pole on the left and its dual the female pole on the right. The perfect quasi-regular above separates into the two dual polar regulars, and recombines in the dual solid below.

THE GRADUAL UNFOLDING OF INNER ORDER WITHIN EACH CLASS

18. A cosmogonic sequence for each Class represents the staged unfolding of inner order, from the single quasi-regular solid as source through to the full development into the complete class of regular and semi-regular elements. The duals of the quasi-regulars (which are not in general semi-regular) are at times included and at others excluded.

19. The sequence commences with the seed monaktys of the quasi-regular solid, which contains in potential the entire class. It proceeds to the quaternal Vesica Pisces of orthogonal dualities, with vertical axis of quasi-regular and its dual, and horizontal axis of dual polar extremes of the regular solids. I elsewhere conclude that the regular solids though perfect, are really extremes polar to their mediating and harmonising perfect quasi-regular solid, which is thus a more perfect form. Its shadow below is its dual (ref.6).

20. The triaktys for each class locates the quasi-regular at the centre of the base between its regular polar extremes on either side. The dual rhombic solid of the

quasi-regular lies at the apex, so that quasi-regular and dual lie on the central vertical axis. The great and small rhombic element lie on either side on the second level, so that the neutral elements of first, second and third degree form the inner triangle.

21. This is developed into the tetraktys which comprises all of the elements of that class. In accord with its central nature, the quasi-regular element is located at the centre, with its dual at the apex. Regular polar elements are located at two opposite vertices of the outer triangle. Their truncates are situated between them, along the base line. Great and small rhombic elements are located as before relative to the central quasi-regular element, forming a triangle which allows for the alternate sequences possible connecting the three. Enantiomorphs of the snub polyhedra lie either side of the central quasi-regular element.

22. Finally the subset of the pentaktys structures the completely unfolded order for each class (leaving aside the dual of the quasi-regular). The base line comprises the truncation sequence of regular male pole, truncated male pole, quasi-regular neutral centre, female truncated pole and regular female pole. The rotation-displacement sequences arrange an inner Vesica Pisces of small rhombic Second Order element and quasi-regular First Order neutral centre above and below, and intermediary left- and right-handed enantiomorphs of the snub polyhedra on the left and right. The central vertical axis erects a transcendent sequence of quasi-regular First Order element below, small rhombic Second Order element in between, and culminates in the great rhombic Third Order element on high.

23. Contemplation of this polyhedral order according to Vesica Pisces and Pythagorean Tetraktys, recalls Plato's dictum that they mediate a world of goodness, beauty and truth.

REFERENCES

1. MEURANT, R.C. A New Order in Space - Platonic and Archimedian Polyhedra and Tilings. International Journal of Space Structures, Vol. 6, No. 1, 1991.

2. MEURANT, R.C. Natural Harmony ~ A Structural Morphology of Consciousness? 1° International Seminar on Structural Morphology, Montpellier France, Sept. 1992.

3. CRITCHLOW, K. The Soul as Sphere and Androgyne. Golgonooza Press, Ipswich.

4. CRITCHLOW, K. Order in Space . Thames and Hudson, London, 1969.

5. LAWLOR R. Sacred Geometry - Philosophy and Practice, pp. 98-103. Thames and Hudson, London, 1982.

6. MEURANT, R.C. The Myth of Perfection of the Platonic Solids, Proceedings of the PAPER Conference on Myth Architecture History Writing, Auckland, 1991.

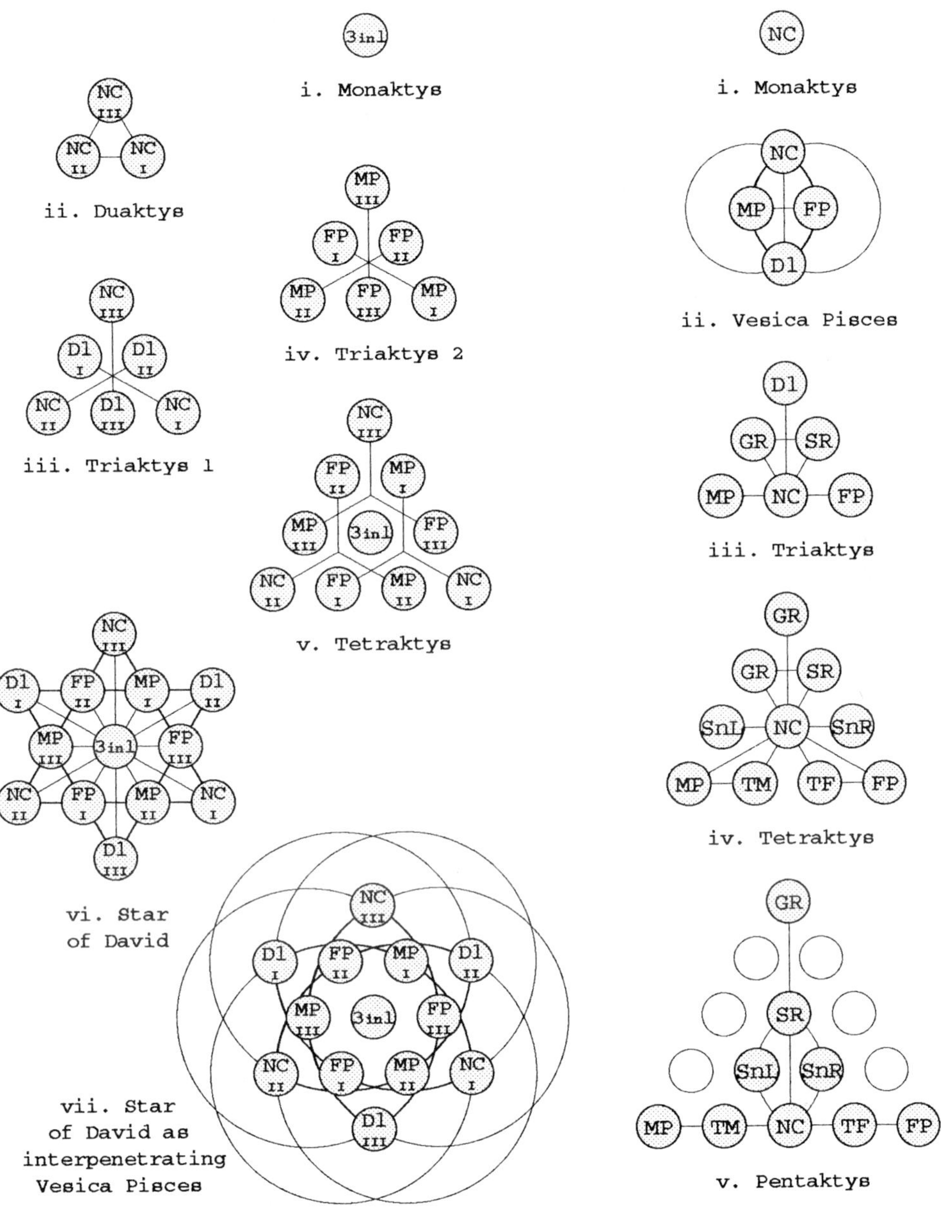

Fig. 2. Key to Schema of All Three Classes of Polyhedra.

Fig. 3. Key to Schema of Each Individual Polyhedral Class.

NC: Quasi-regular Neutral Centre, Dl: Dual Rhombic Solid, GR: Great Rhombic Solid, SR: Small Rhombic Solid,
SnL: Left-handed Snub Solid, SnR: Right-handed Snub Solid,
MP: Regular Male Pole, TM: Truncated Male Pole,
TF: Truncated Female Pole, FP: Regular Female Pole.

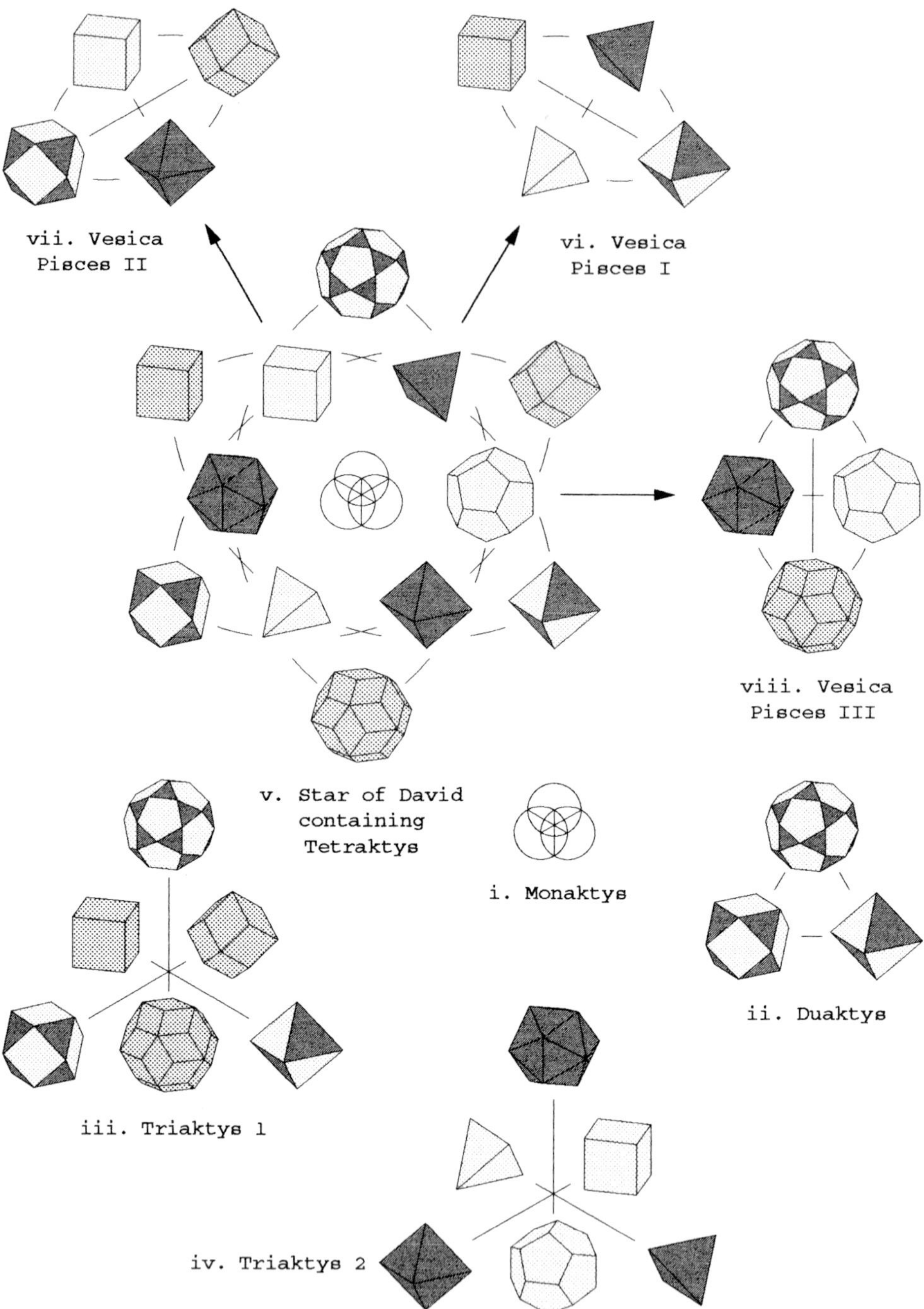

Fig. 4. Cosmogonic Sequence of All Three Classes of Polyhedra; Star of David (above) gives rise to three separate Vesica Pisces.

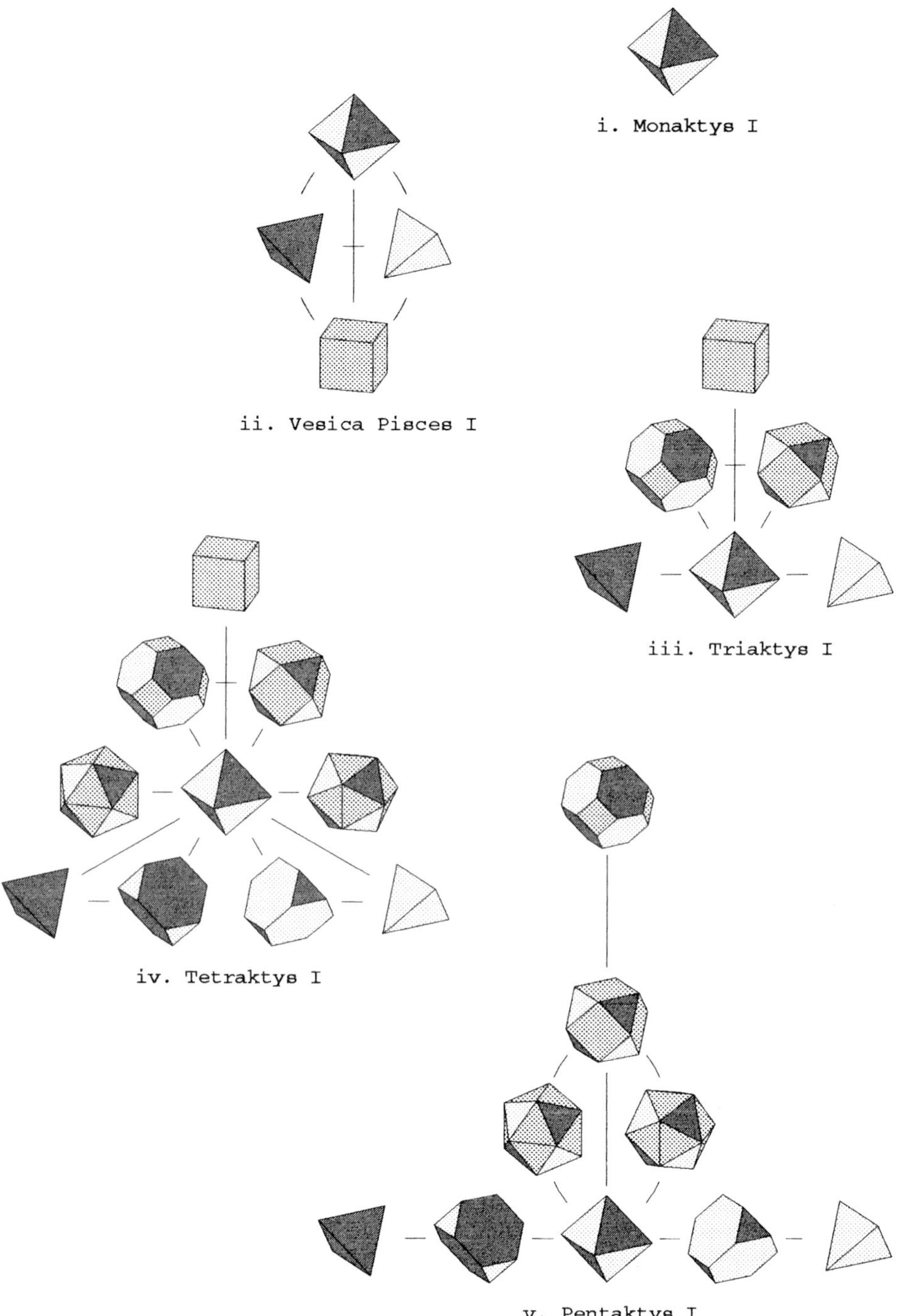

Fig. 5. The Unfolding of Inner Order - Class I : {2, 3 ,3}.

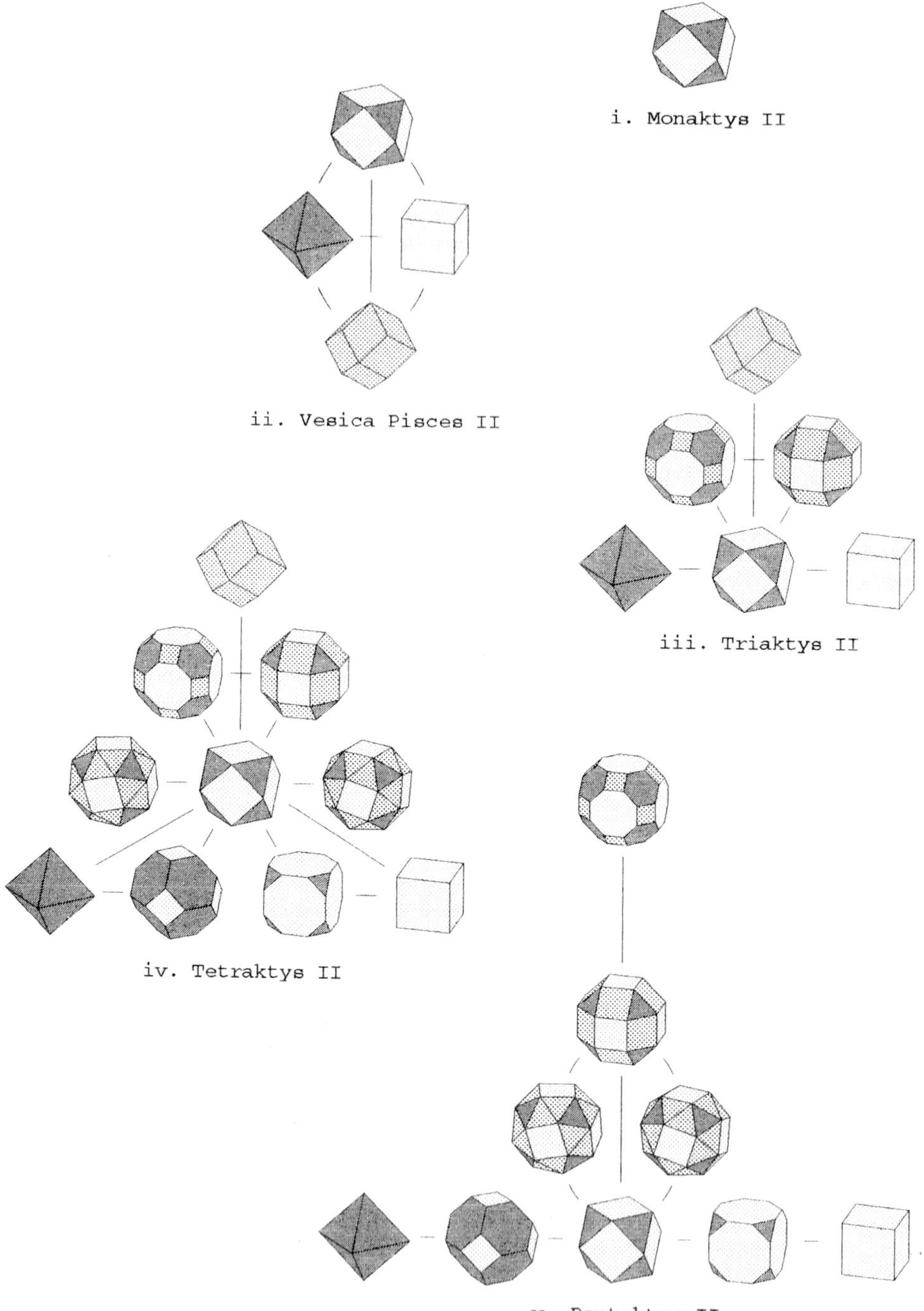

Fig. 6. The Unfolding of Inner Order - Class II : {2, 3, 4}.

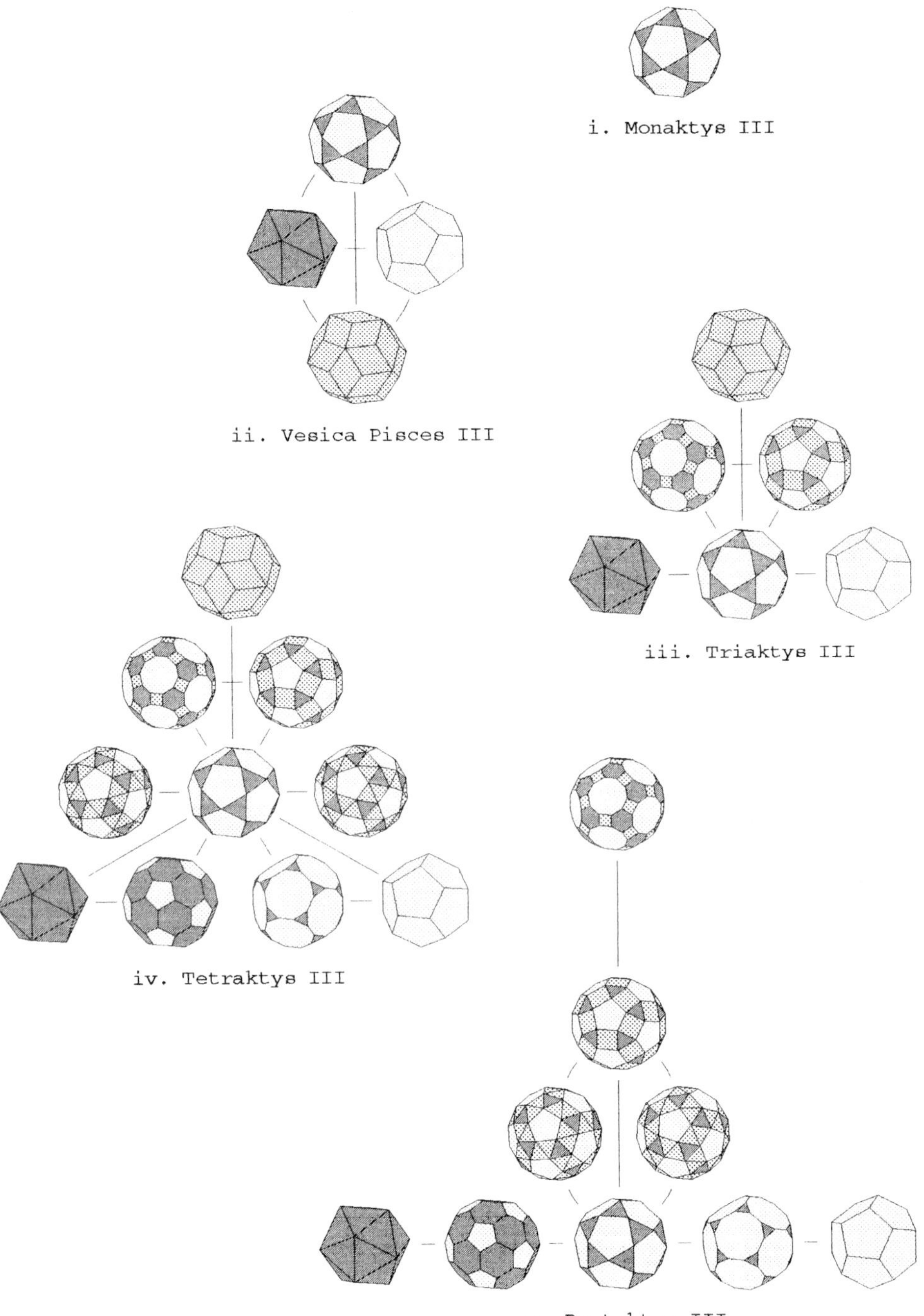

Fig. 7. The Unfolding of Inner Order - Class III : {2, 3, 5}.

122. Morphological aspects of spatial structures requiring formfinding processes

R. MOTRO, Université de Montpellier, France

SYNOPSIS. New space structures have been developed during the last years namely; cablenets, membranes, tensegrity systems. They all require "form finding" processes in order to stabilise initial mechanisms, but several classes can be identified according to the rigidification mode and the stability under external actions. In the first part of the paper the nature of formfinding and its place in the mechanical behaviour equations are underlined. Then three kinds of systems are distinguished: selfstressing systems, prestressed systems and selfstressed systems. In each case reference to usual numerical form finding methods is given and morphological aspects are presented in relation with basic examples.

FORM FINDING

1. Studies on space structures include the so-called "formfinding" calculations, which are needed for kinematically undetermined systems, for which one or several stable forms, under some determined conditions, are found. Causes of kinematic indeterminacy, and the way to eliminate them, need a re-reading of equations governing reticulated space structures mechanical behaviour.

2. Equilibrium equations may be written in terms of internal forces in members, or of nodal displacements. If we consider a pin-jointed structure, or a structure modelled in the form of a pin-jointed structure, with "b" members and "n" nodes, it can be written:

Equilibrium equation $$[A].\{t\} = \{W\} \quad (1)$$

with [A], "equilibrium" matrix (dimensions 3n x b), {t} internal forces vector (dimension b) and {W}, external actions vector (dimension 3n);

Compatibility equation $$[B].\{d\} = \{e\} \quad (2)$$

with [B], "compatibility" matrix (dimensions b x 3n), {d} node displacements vector (dimension 3n) and {e}, member relative deformation vector (dimension b);

Elastic material behaviour equation $$[H] . \{ e \} = \{ t \} \quad (3)$$

 Space structures 4. Thomas Telford, London, 1993

with [H], "elastic" matrix (dimensions b x b).

3. Combining these three equations, and taking into account the relationship between the equilibrium matrix [A] and compatibility matrix [B] (one is the transposed of the other as it can be demonstrated by applying the virtual work principle) leads to the equilibrium equation in terms of displacements as follows:

$$[A].[H].[A]^t.\{d\} = \{W\} \qquad (4)$$

which underscores the elastic rigidity matrix [Ke]:

$$[Ke] = [A].[H].[A]^t \qquad (5)$$

4. To be more precise, in the form finding problem it is also useful to split up the equilibrium matrix [A], which can be set in a product of two matrices:

$$[A] = [C]^t.[F] \qquad (6)$$

Matrix [C], the transpose of which appears in Eq. 6, is characteristic of the relational structure of the studied system (ref. 1). It is built, line by line, by introducing values -1 and 1 in the columns corresponding respectively to the origin node and the final node of each member "b" belonging to the space structure; all the other terms in this matrix are equal to 0. Matrix [F] is a square diagonal matrix of dimension "b": its terms are the direction cosines of each member, it depends upon the co-ordinates of the "3n" nodes . This matrix might be called "form matrix" and Eq. 4 can be written in the following form:

$$[C]^t.[F].[H].[F]^t\ [C].\{d\} = \{W\} \qquad (7)$$

5. Apart from the wrong choices for support conditions and the related rigid body displacements, three causes can be found for kinematic indeterminacies. They are related to:

- relational structure of the system, matrix [C];
- member rigidities which are included in the matrix [H] (with geometrical terms and terms depending on relational structure);
- co-ordinates of the "n" nodes, matrix [F].

6. Assuming that member rigidities and the relational structure of a system are fixed, "form-finding" problems are equivalent to co-ordinates calculations. Form, found in this way, is always conditionally stable. The problem is generally solved using iterative procedures.

7. It is useful to classify these systems, which require form-finding processes into three categories, in order to avoid some confusion and to underscore their morphological potentialities. Three classes may be distinguished and are named as follows:

- funicular systems, also called selfstressing systems;

- prestressed systems;
- selfstressed systems.

The presentation included in the this paper, is illustrated by examples, which are very simple but consistent with the complexity of the systems under examination.

FUNICULAR OR SELFSTRESSING SYSTEMS

8. These systems are well known, and many studies have been carried out, specially in the case of systems composed of members with only tension or compression rigidity; these rigidities are dual one to the other and experimental form finding processes developed by A. Gaudi are well known ("reverse hanging" method). The Roman root of the word "funicular" is related to the notion of rope, which is the fundamental member used for transmission of tension. The form of these systems is directly connected to the applied external actions. To be comprehensive funicular denomination has to associated with the corresponding loads. In this case, it can be claimed that for a cable every form characterised by $[F]_i$ is associated with an applied external load system $\{W\}_i$. The stability of form is directly and strongly dependent on $\{W\}_i$.
Three examples may be quoted.

Cables and cable-nets

9. Tri-dimensional action funiculars, applied to an unidirectional system such as a rope, may be considered as the generalised spatial cable problem (ref. 2). Form is the result of an iterative procedure. In the specific case of coplanar external actions, a graphic iterative form finding process can be used (ref. 3). Cable nets and inflated membranes (which could be considered as the limit of cable nets when decreasing the cables spacing) belong also to this class. Experimental three dimensional tests and the results of the algorithms have been largely developed at the Institute for Lightweight Structures by F. Otto and his team.

Mixed systems with bars and cables

10. The case of reticulated systems with rigid members in compression and tension has been recently investigated by Y. Hangai (ref. 4) with a method based on a generalised inverse matrix; we dealt also with this problem using an incremental method (ref. 5). The method used is to follow the evolution of the form between an initial situation $[F]_0$, and the stable equilibrated one, namely $[F]_e$., each of them being compatible with the initial manufactured lengths of members. Accuracy of numerical results has been found to be satisfactory. A simple example is shown in Fig. 1. Table 1 gives the numerical results compared with the theoretical ones.

Table 1. Initial and final co-ordinates.

Coordinates	X1	Y1	X2	Y2
Initial	0,707	0,707	1,414	0
Theoretical	0,707	0,707	1,707	0,707
Calculated	0,706	0,708	1,706	0,706

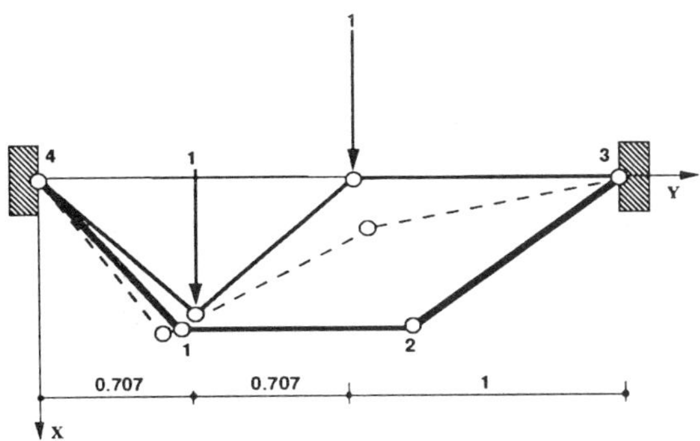

Fig. 1. Geometrical evolution under external loads.

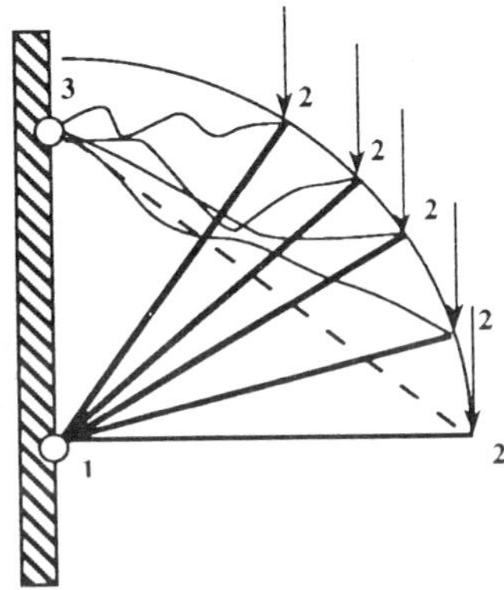

Fig. 2. Kinematics of unfolding.

11. It is thus possible to describe incrementally several geometrical configurations in the case of a basic system composed of one cable and one strut; this has been used to validate our numerical model (Fig. 2). The initial geometry is an unstable arbitrary geometry. The final geometry is stable under the applied loads: the cable between nodes 3 and 2 is tightened. It is possible to use this method for describing folding and unfolding of space structures. It is not sufficient to examine only initial and final stages, intermediate geometries have to be investigated. This method eventually coupled with a second order matrix analysis, necessary for determining form instabilities, may be useful for the description of the kinematics of foldable space structures. Concerning morphological aspects, it seems that the number of different geometries is not so high as in the cable and cable nets class. Several external loads result in the same equilibrated geometry.

Degenerated tensegrity systems

12. Lastly, the case of systems which may be called "degenerated tensegrity systems" must be quoted: these forms only reach a stable state under the action of external loads, such as selfweight. It may be said in opposition to cable nets for instance, is that the corresponding range of forms is significantly reduced. The

example of a square prismatic tensegrity module is illustrated in Fig. 3. This module is composed of 4 struts of identical length "g" and 12 cables of identical length "l". For a given relational structure and a fixed value of the ratio g/l, two distinct geometries, parametered by θ1 and θ2 may be found and can be regarded as funicular forms resulting from either "upward" or "downward" loading (ref. 5). θ is the characteristic geometrical parameter and it is the value of relative rotation of top square and bottom square of the modulus. Geometries corresponding to the limit values of θ1 and θ2 are given by Figs 4a and 4b. The definition given by Emmerich (ref. 6), "systèmes autotendants" ("selfstressing systems") is understandable by reference to downward loading ($\theta 2 \leq 90^o$). Yet rigidification in this case is of a different nature from rigidification developed in selfstressed systems; this is studied further in this paper. This specific class corresponds to the only double solution on the curve of Fig. 3.

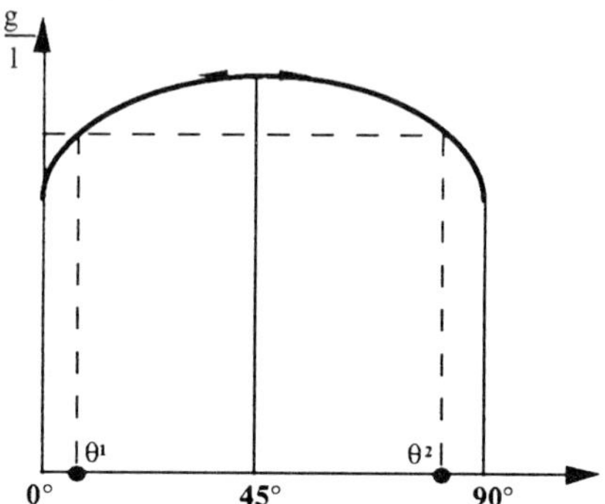

Fig. 3. Range of geometries of the 4-prismatic tensegrity module.

13. In this first class "funicular or selfstressing systems", the range of potential forms, even if large, is not very useful for the designer, except in very specific applications such as Colonia Guëll from Gaudi. Furthermore stability of funicular forms is very closely related with the associated loading system: a simple reversal of actions re-opens the problem. Nevertheless, developments in correlation with the study of foldable space structures appears to be of high interest.

PRESTRESSED SYSTEMS

Prestress

14. For space structures, as for other building systems, it may be useful to introduce an initial prestress. The effects of the initial prestress are calculated in combination with those of environmental and natural loadings in order to be adapted to admissible capacities of the structure. According to the building requirements, this principle corresponds to an adjustment of the manufactured system geometry to boundary conditions which do not fit it exactly; this may be understood as specific support conditions. If we refer to equilibrium Eq. 7,

$$[C]^t.[F].[H].[F]^t\,[C].\{d\} = \{W\} \qquad (7)$$

the prestress is equivalent to the fixity of some nodal co-ordinates in a calculation mesh; these nodes are then considered as "fixed nodes" and only co-ordinates of "free nodes" are searched. It corresponds to a partition of vector {d}. Calculation is then carried out for any value of {W} equal to zero or not. The force density method used for membrane form finding is a good illustration of this principle. The prestress is used for limiting deformations of flexible systems and to avoid flutter.

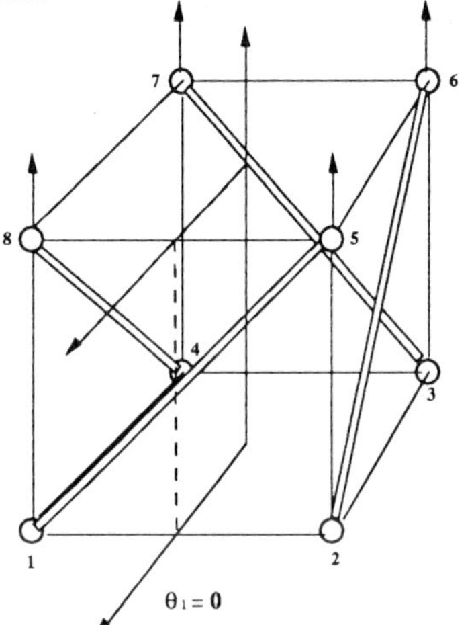

Fig. 4a. Upward loading funicular degenerated tensegrity system, $\theta = 0^o$.

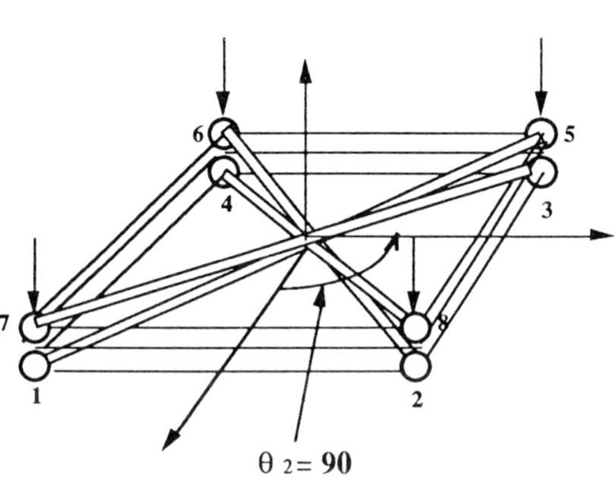

Fig. 4b. Downward loading funicular degenerated tensegrity system, $\theta = 90^o$.

The bi-cable

15. Illustration of the methods required for the analysis of a prestressed system is provided using a simple example. Fig. 5 shows a cubic frame with two cables "5-7" and "1-3", lengths of which are defined such as it is necessary to strain them to join the two cables at a common point M. Equilibrium is reached for a specific value of co-ordinate "z" of point M. This elementary form finding problem may be restricted to only one direction because of symmetry considerations. Required equilibrium is then restricted to the simple equality between the forces Rl and R2, respectively, exerted by bottom cable "1-3" and top cable "5-7" on the middle point M. This example allows the schematisation by means of co-ordinate "z" of:

- prestressed state characterised by z_1 which is equivalent to the equilibrium equation, Eq. 8.
- equilibrium under external load F, characterised by z_2; corresponding equilibrium is given by:

$$R'1 + F = R'2 < R2+F \qquad (9)$$

tension in the upper cable is increasing (correlated with R'2), tension R'1 in the bottom cable is decreasing.

- fluttering of bottom cable for a value of z greater than z_{max}.

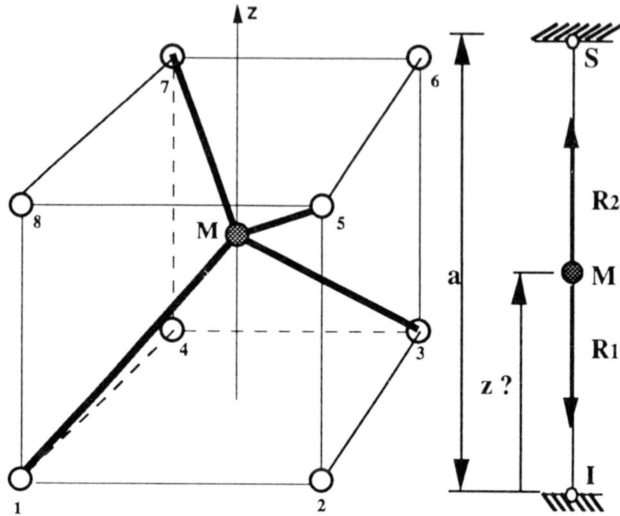

Fig. 5. Bi-cable example.

With this simple example two main properties of prestressing are underlined: it limits the system deformations by increasing its, and its effects are combined with those of external loads and not simply added. A study of internal energy shows there is progressive decreasing of prestress when external loads are applied, which is not true for funicular systems which are the limit of the above bi-cable, when z is greater than z_{max}.

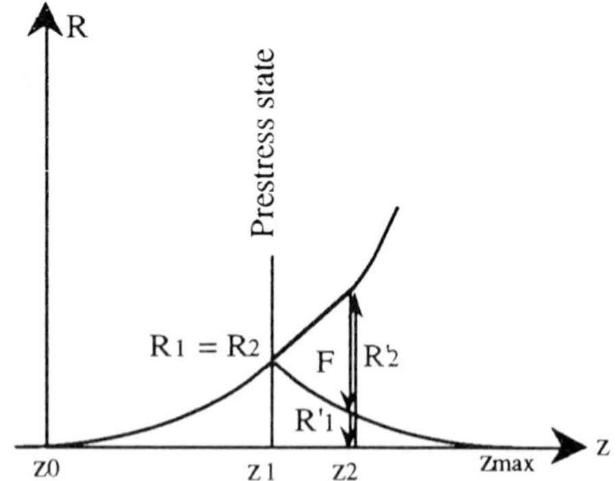

Fig. 6. Functioning points of bi-cable.

Force density method

16. The force density method developed by Sheck generalises the form finding solution for any spatial surface in equilibrium with fixed boundary nodes (ref. 7). The square diagonal matrix of force densities [Q] is the shape parameter. Its terms are calculated as the ratio between the internal force in a member and its length. To each matrix [Q] is associated a different surface shape. Surfaces of minimum area are obtained with a set of identical values of force densities; specific constraints may be implemented to respect morphological needs (solutions result from a linear programming procedure). Inside our own computer aided design tool (called Archimedes, (ref. 9)), it is possible to regulate co-ordinates of fixed points, to modify curvatures place to place; the designer is able to model the membrane shape to his convenience as shown in Fig.7 a-f.

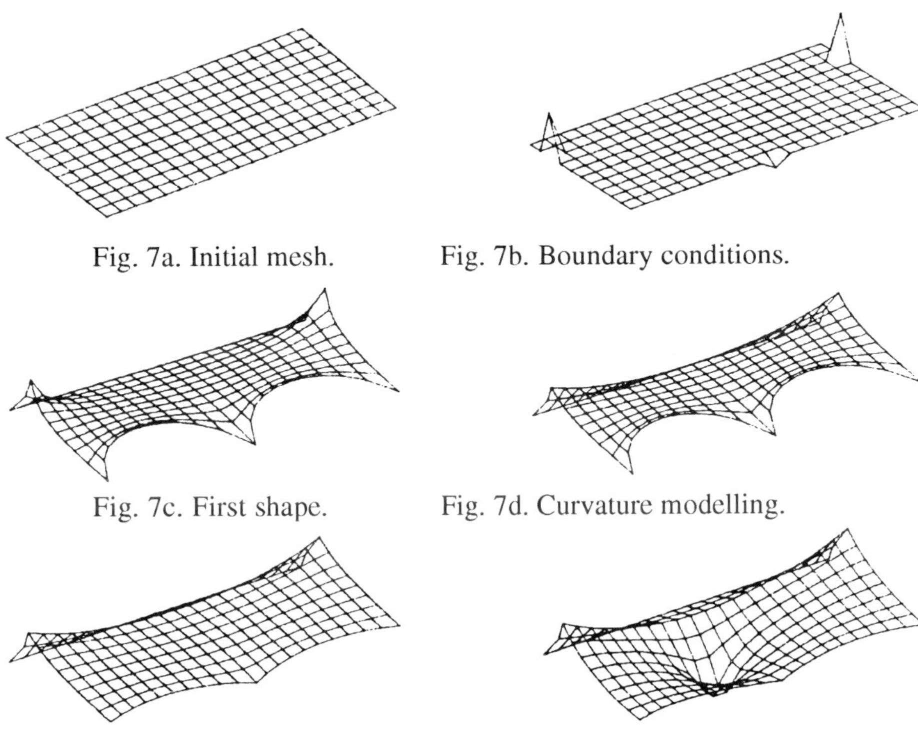

Fig. 7a. Initial mesh. Fig. 7b. Boundary conditions.

Fig. 7c. First shape. Fig. 7d. Curvature modelling.

Fig. 7e. Bolt ropes. Fig. 7f. Impluvium insertion.

SELFSTRESSED SYSTEMS

Definition

17. If we study general reticulated space structures composed of cables and struts, the most interesting class, in many respects, seems to be that of selfstressed systems, the so-called tensegrity systems (ref. 1). Concerning the mechanical approach, the form and the relational structure of these systems are

such that there exists one or more internal member force vectors {ti} which satisfy the homogeneous system of equilibrium equations:

$$[A].\{ti\} = \{0\} \tag{10}$$

or using Eq. 6:

$$[C]^t.[Fl.\{ti\} = \{0\} \tag{11}$$

These equations lead to selfstressed states if:

- the solutions vectors {ti} are compatible with the member rigidities (a selfstress state can not be achieved if cables are compressed);
- the relational structure satisfies qualitative equilibrium conditions (one strut and three cables at least, incident to each node);
- the geometry is such that the infinitesimal mechanism may be eliminated by selfstressing and this is only true for specific geometries which require form-finding processes.

Form finding

18. For some systems, the specific geometry of selfstress may result from an analytical process, like a search for the maximum value for the ratio g/l as can be seen in the Fig. 3 (paragraph 12.). The solution is reached with a relative rotation θ of top and bottom squares equal to 45°. The statical approach is also successful (Fig. 8).

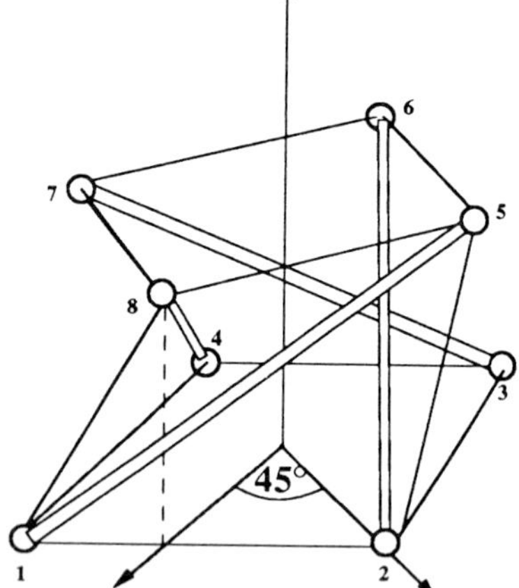

Fig. 8. Selfstress geometry for 4-prismatic tensegrity system.

A general method for this form finding problem has been developed by Belkacem (ref. 9). It is based on an adaptation of dynamic relaxation, which is also used for form finding membranes. Using the kinetic energy damping Belkacem found some interesting results and underscored the discrepancy between regular polyedric geometries and selfstressed geometries as may have been anticipated by those who have conducted testing on experimental models (particularly for the truncated tetrahedron).

Selfstress compatibility. Mechanisms elimination

19. Dealing with a method proposed by Calladine and Pellegrino for bar systems (ref. 10), Mohri established an algorithm to check the compatibility of the selfstress states when tensegrity systems are concerned (ref. 11). The number of selfstress states "s" and the number of mechanisms "m" are determined by the calculation of the rank r_A of the equilibrium matrix [A]:

$$s = b - r_A \qquad \text{and} \qquad m = N - r_A \tag{12}$$

(N being the number of degrees of freedom)
A dynamic study of the eigen-values of the systems shows the stiffness of the system and the elimination of some mechanisms (ref. 14).

Morphological considerations

20. In the author's opinion a great morphological variety is available with selfstressed systems. Realisation of free form selfstressed systems may be viewed. A first step was achieved with the design of double-layer plane selfstressed system which results from adjonction of elementary selfstressed modules (Fig. 9). Specific nodes were designed and an industrial prototype has been built. The case of double-layer single-curvature vaults has been studied according to geometrical and mechanical requirements. A model test has been undertaken for a double-layer single-curvature self-stressing system but the mechanical requirement calculation, has yet to be completed (ref. 13).

21. These are only sketches of the morphological potentialities of selfstressed systems. Selfstress allows the extention of the variety of forms and the creation of free forms without the need of extra devices for equilibrating stresses like in the funicular and the prestressed systems. But it is necessary to develop mathematical form finding models which allow a great number of design parameters to be studied. The difficulty lies in the complexity of these new space structures, and in the necessity to eliminate the mechanisms, and the need to ensure the stability of the selfstress state under the action of a wide range of multi-directional external loads.

CONCLUSION

22. New kinds of space structures have been mentioned; they all need form finding processes but they are also characterised by an internal initial energy which can react against external loadings. This research field is rich with potentials. Is it realistic to hope more from lightness ? Actual selfweights of space structures are now very low and reaching the asymptote given in Le Ricolais' joke "zero weight, infinite span". Solutions are elsewhere, perhaps in new structural principles like monitored selfstressed space structures which include transducers and actuators.

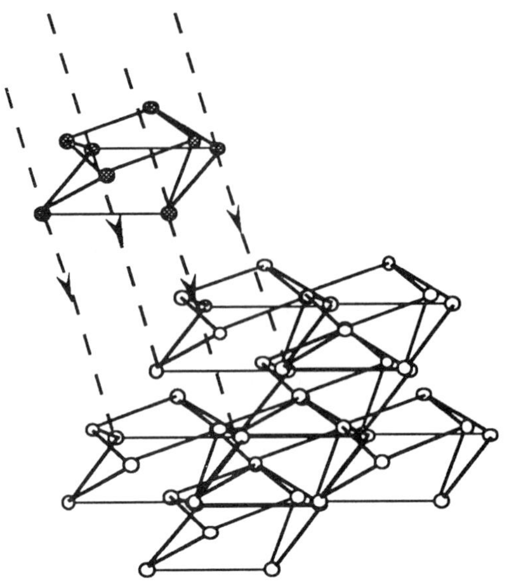

Fig. 9. Double-layer plane tensegrity grid.

REFERENCES

1. MOTRO R. Formes et Forces dans les Systèmes Constructifs. Cas des Systèmes Réticulés Spatiaux Autocontraints. Thèse d'Etat, Montpellier, 1983.
2. MOTRO R. and MALINOWSKY M., Formes Forces et Matériaux en Architecture Textile. Symposium International: Matériaux composites à renforts textiles appliquás à la construction. Forum International des Technologies Appliquées du Textile. Lyon, Juillet 1990.
3. MOTRO R., GROUPE ARCORA Un élément oublié de l'Architecture: le textile. Ouvrage sur la conception, le calcul et la réalisation de l'Architecture Textile, 1987. Not published.
4. HANGAI Y. Theoretical analysis of structures in the unstable state and shape analysis of unstable structures. Institute of Industrial Science. University of Tokyo, Sept. 1991.
5. MOTRO R. and YU-XIN L. Energy and Morphology for Structures. 1992. (in press)
6. EMMERICH D.G., Réseaux. Proceedings of the First International Conference on Space Structures. Z.S. Makowski ed., London, 1966.
7. SHECK H.J., The force density method for formfinding and computation of general networks. Computer Methods in Applied Mechanics and Engineering 3, pp 115-134., 1974.
8. PAULI N., MARTY A., PIERRE D. and MOTRO R., Two CAD systems for architectural membranes. Proceedings of the first International Seminar on Structural Morphology, pp239-249, Montpellier,1992.

9. BELKACEM S. Recherché de Forme par Relaxation Dynamique des Structures Réticulées Spatiales Autocontraintes. Thèse de Docteur Ingénieur INSA Toulouse 1987.
10. PELLEGRINO S. and CALLADINE C.R., Matrix analysis of statically and kinematically indeterminate frameworks. Int. J our. of Solids and Structures, Vol. 22 No 4, pp. 409-428,1986.
11. MOHRI F. Analyse Théorique et Expérimentale de Systèmes Réticulés Spatiaux Autocontraints Multimodules. Thèse de Doctorat. Université Montpellier II, decembre 1990.
12. MOHRI F. and MOTRO R. Static and kinematic determination of generalised Space Reticulated Systems. To be published in Structural Engineering Review.
13. MOTRO R., Tensegrity Systems and Geodesic Domes. Special Issue of the International Journal of Space Structures. "Geodesic Space Structures". Vol 5, No 3, 1990, pp343-354.

123. A non-conventional description of complicated space structures

J. B. OBREBSKI, Warsaw University of Technology, Poland

SYNOPSIS. In the present work new aspects of the difference description of complicated space structures are discussed. The method was first adapted for use on a computer in about 1970, and has been successively developed since by the author. The method refers to different types of structures. It is most convenient to describe regular structures with nodes placed at selected points of intersection of regular families of curved parametric lines. But may also be applied to non-regular structures. It seems to be very easy to use and competitive with such methods as Formex Algebra, Auto-CAD and other data generators.

INTRODUCTION

1. The method, which was first adapted for use on large old generation computers, offers the possibility to describe structures using small files of input data, much smaller than for other methods. It has now been adapted for use on PC computers and is still attractive for the same reasons.

2. Some details regarding the principles of the method are presented in refs 1-7 and particularly in ref. 3. This paper describes some features which are important to the application of the method on the IBM PC computer.

3. When introducing the input data, especially for large structures, two primary problems exist:

(a) how to determine the nodal co-ordinates, the member lengths of the structure and its position is 3D space - especially when the structure is curved;

(b) how to check co-ordinates from the input data.

4. The above tasks can be achieved by using the PC computer application and may be assisted by graphic display on the screen. Here the difference description for mentained space structures, gives the following possibilities:

- easy, general and exact description of any large structure, including geometry, loading and boundary conditions;
- graphical interpretation of generated nets of points, configuration of structure elements etc.;
- graphical interpretation of obtained results;

- easy modification of input data of the global shape of the structure , e.g. from plane through conical or membranal to double-curvature;
- easy modification and extension of program procedures responsible for the input data of N new types of structures.

GENERAL ASSUMPTIONS OF PROGRAM ARCHITECTURE

5. (a) The method should be oriented on selected types of structures and be efficient and fast in action.
(b) New users may start the program trained only in the use of thesystem of input data, and only after a short explanation displayed on the screen, he should be led only by program.
(c) Program should exactly describe structures and display their general view, or any part view. Moreover this view of a required part should be presented from any direction.
(d) It should be possible to change the scale of drawings - to focus on selected parts of structures.

ELEMENTS OF GEOMETRICAL DESCRIPTION

6. The structures considered are mostly regular with nodes at points of intersection of three families in general of curved parametric lines given in Euclidean space. As a standard 4 basic nets of points have been proposed (refs 3,4 and 7) namely:

(a) orthogonal (Fig. 1a);
(b) double-curvature net of revolution with two centres of curvature (Fig. 1b);
(c) translational net of points (Fig. 1c);
(d) multiple net of revolution (Fig. 1d).

In all the above cases we can define the relation:

$$\zeta_i = f_i(x_i) , \qquad i=1,2,3. \tag{1}$$

where ζ_i are global co-ordinates of the structural nodes in Euclidean space, x_i are integer numbers called the parametrical co-ordinates of the nodes, f_i are functions built into the program. The definitions of functions f_i should be easy to change in the program by the user, as contents of small procedures or by input data, thus changing the general shape of the structure, curvatures, etc.

7. A general view of the structure depends on the orientation of co-ordinates ζ_i (Eq. 1), defined by the position of 4 points on the screen. If we only change the positions of these 4 points, (Fig. 1e): O, A_1, A_2, A_3, we obtain the effect of a structure rotation. These points have co-ordinates: a_{oj}, a_{ij}, i,j = 1, 2, 3.

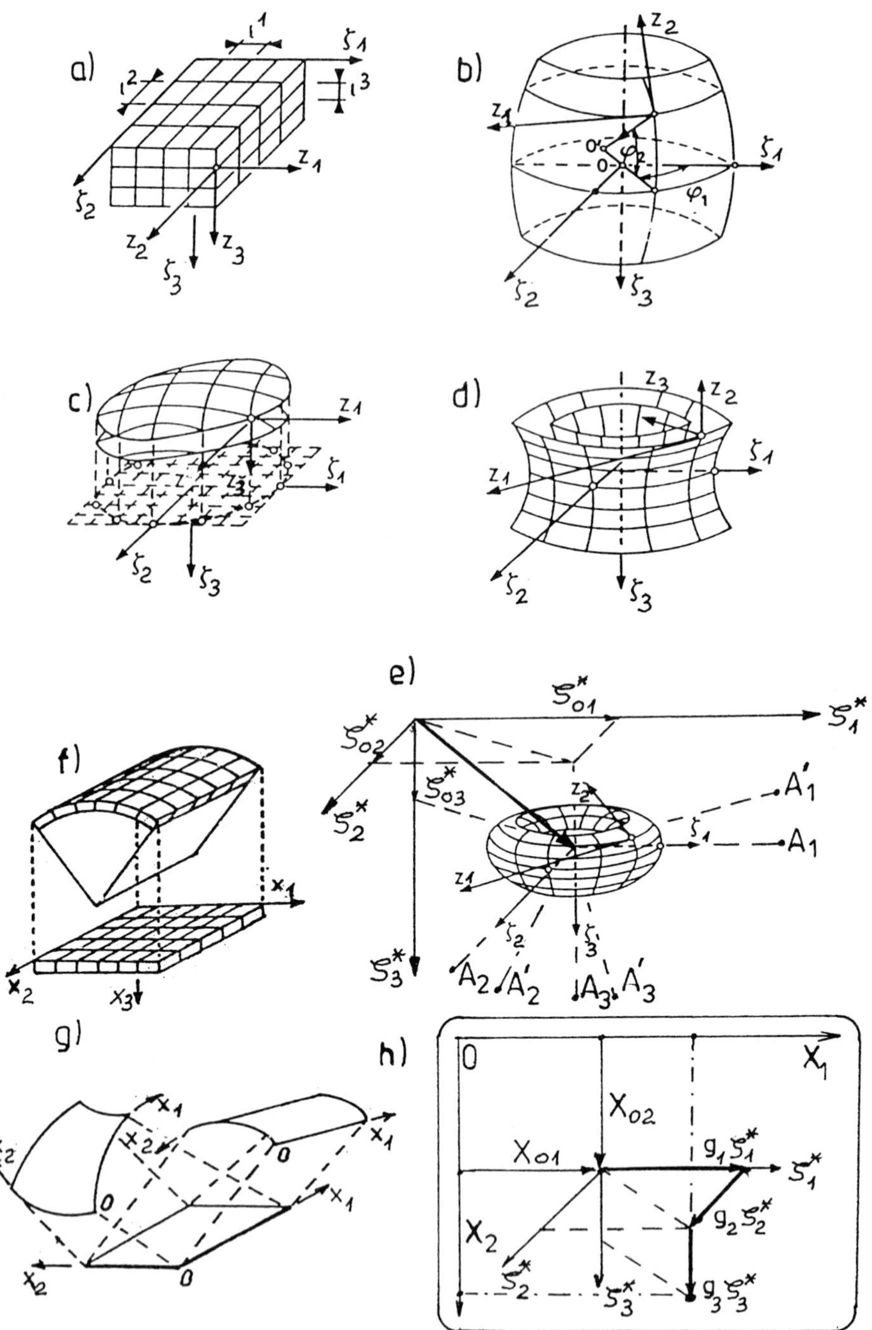

Fig. 1.

8. To obtain the effects of moving and scaling the structure, with regard to the active screen window, it is sufficient to undertake the following transformation to the new co-ordinates (Fig. 1e):

$$\zeta_i^* = \zeta_{0i}^* + s \cdot f_i(x_i) \;, \qquad i=1,2,3. \tag{2}$$

where ζ_{oi}^* is a vector of movement, and s is a scale parameter for focusing the part of structure.

9. Finally, it should be noted that the picture should be displayed on a two dimensional screen (x,y), with dimensions I_{x1} and I_{x2} only. A second transformation is needed, (Fig. 1h):

$$X_\alpha(\zeta_i^*) = X_{0\alpha} + g_{i\alpha}\zeta_i^* \;, \qquad , \; \alpha=1,2, \quad i=1,2,3. \tag{3}$$

where $\mathbf{g}_i$ are basis vectors of co-ordinates.

10. It should be explained, that the user declares only the positions of points: O, A_i, parametrical co-ordinates x_i, and the multiple coefficient s. The rest of the data are generated automatically.

11. By introducing a list of parametrical co-ordinates x_i, (integer numbers), all the data may be input from the keyboard, where decisive parameters can be introduced and displayed in an active text window. The picture of a structure can be observed simultaneously in another window.

Element arrangement description

12. The members of structure are co-linear with certain vectors appointed by Boolean displacement operators defined as follows, (Fig. 2a):

$$E_\Lambda\phi = E_1^{a_{\Lambda 1}} E_2^{a_{\Lambda 2}} E_3^{a_{\Lambda 3}} \phi(x_1,x_2,x_3) = \phi(x_1+a_{\Lambda 1}, x_2+a_{\Lambda 2}, x_3+a_{\Lambda 3}) \tag{4}$$

13. If we know the net of points (Eq. 1), and powers a_i of the above displacement operator, we can compute the lengths of all members, and their orientation defined with regard to the local nodal co-ordinates z_i. These local co-ordinates are defined (in the program), for each class of function f_i. The set of integer parameters $a_{\Lambda i}$ is usually very small for repeatable structures. These local co-ordinates are declared in such a way that, in all nodes of identical type, a member configuration can be regarded (observed) as identical. This permits, by writing nodal equilibrium equations, the application of a certain modification of Lagrange's mode of structural description.

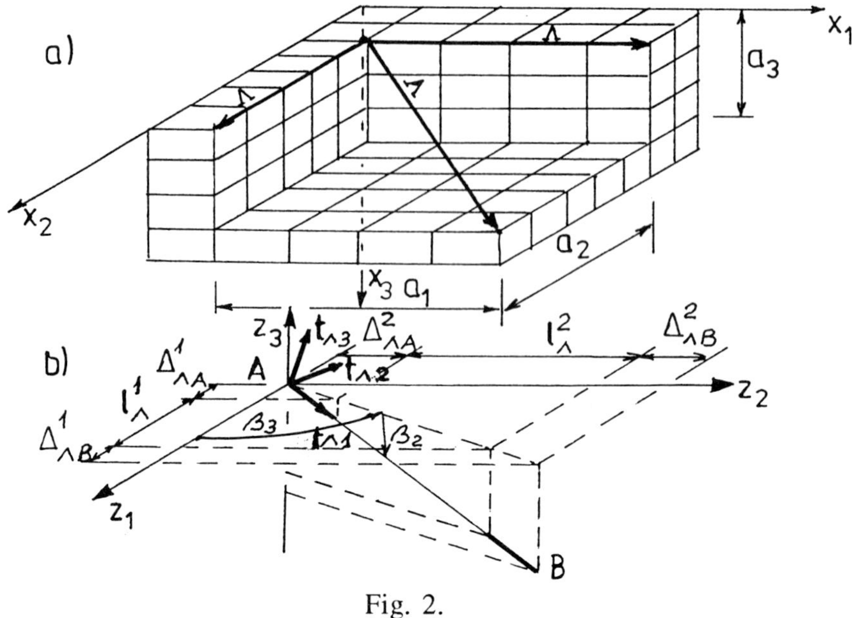

Fig. 2.

14. Additionally, for full definition of the member geometry we should add the angle of rotation of the principal axis of the member cross-section. These data permit the program to compute certain unit vectors of direction $t_{\wedge i}$, (Fig. 2b). Their components are obtained as the elements of the following matrix, (ref. 3):

$$D_o = B_o^T = [\{t_{\wedge 1i}\}, \{t_{\wedge 2i}\}, \{t_{\wedge 3i}\}] \,, \tag{5}$$

where

$$B_o = \begin{bmatrix} c_2c_3 & c_2s_3 & -s_2 \\ s_1s_2c_3 - c_1s_3 & s_1s_2s_3 & s_1c_2 \\ c_1s_2c_3 + s_1s_3 & c_1s_2s_3 - s_1c_3 & c_1c_2 \end{bmatrix} \tag{6}$$

and $s_i = \sin \beta_i$, $c_i = \cos \beta_i$. The angle β_1 is a torsion angle of the member in local co-ordinates y_i about axis y_1. Moreover:

$$\sin\beta_2 = -\frac{l_\wedge^3}{l_\wedge} \,,\ \cos\beta_2 = \frac{l_\wedge^*}{l_\wedge} \,,\ \sin\beta_3 = \frac{l_\wedge^2}{l_\wedge^*} \,,\ \cos\beta_3 = \frac{l_\wedge^1}{l_\wedge^*} \,,\ l_\wedge^* = \sqrt{(l_\wedge^1)^2 + (l_\wedge^2)^2} \tag{7}$$

and $l_{\wedge i}$ are the projections of the effective member length in the local co-ordinates z_i, (Fig. 2b).

15. It is necessary to mention, that if we substitute the values $s_i = \sin a_i$, $c_i = \cos a_i$, and angles a_i, into Eq. 6, defined in Fig. 3a and introduced from the keyboard, we can describe an orientation i.e. of given support co-ordinates defining given planes of movement of support nodes. Here is a valid transformation $\mathbf{z}=\mathbf{Bv}$, between the local and support co-ordinate systems.

16. Worthy of mention is also the possibility of describing a similar method for positioning of a typical finite element with number Γ in the structure, and its connection with some nodes, (eg. Fig. 3b), (ref. 3). Here, Boolean displacement operations $\mathbf{E}_{\Gamma i}$ similar to operators $\mathbf{E}_{\wedge}$, are defined. Determined are a modified set of unit vectors $\mathbf{t}_{\wedge i}$, (compare Eqs 4-6), which help in the composition of the stiffness matrix for the structure, etc. This aspect is beyond the scope of this paper and is covered in ref. 3.

17. Moreover, vectors $\mathbf{t}_{\wedge i}$ (and $\mathbf{t}_{\Gamma i}$) play an important part in the composition of the equilibrium equations for particular nodes, and computations of internal forces, (refs 1-7).

The member properties description

18. The description of the member cross-sectional properties does not involve large amounts of data input and may be achieved from the keyboard or prepared data file.

19. This information is connected with number $\wedge$ of typical operation $\mathbf{E}_{\wedge}$ (Eq. 4) - this means with a typical member (or Γ and $\mathbf{E}_{\Gamma i}$ for element). The following information belongs to this group:

- geometrical properties of cross-section e.g. A, I_i I_ω, etc.,
- material properties, E, G, ν, etc.,
- member boundary conditions.

It is possible to declare only type and number of, for example, rolled cross-sections, and the above data should be read from the appropriate files.

The loading description

20. The loading can be declared in global, local or local member co-ordinates systems. For a complicated space structure e.g. a domes, all these descriptions may be applied together. Of course, the dead weight of members is applied via the nodes of the structure. Then the loading should be transformed to local nodal co-ordinates. The later are used by forming equilibrium equations.

21. If we declare, for example, the self weight of the members, a problem of transformation arises, from member local principal co-ordinates y_i to the special fictitious additional nodal forces defined with respect to local nodal co-ordinates z_i. These fictitious forces are equivalent to continuous loadings. This transformation has the form:

$$y = B_o z \;, \tag{8}$$

ON THE COMPOSITION OF EQUILIBRIUM EQUATIONS

22. It is necessary to explain, that in this approach the stiffness matrix of the whole structure is written with respect to the nodal local co-ordinates systems of all particular nodes (Lagrange's description), (refs 1-7). All transformations should be performed by the program with the aid of the above described information, including Eqs 1-8 avoiding tedious input of data.

23. It is clear, that by assuming the truss model of, for example a dome structure, most enumerated problems are not important and can be neglected. But in the case when it is necessary to compute such a structure as a curved space frame, all the above described mathematical operations are important and probably should be applied.

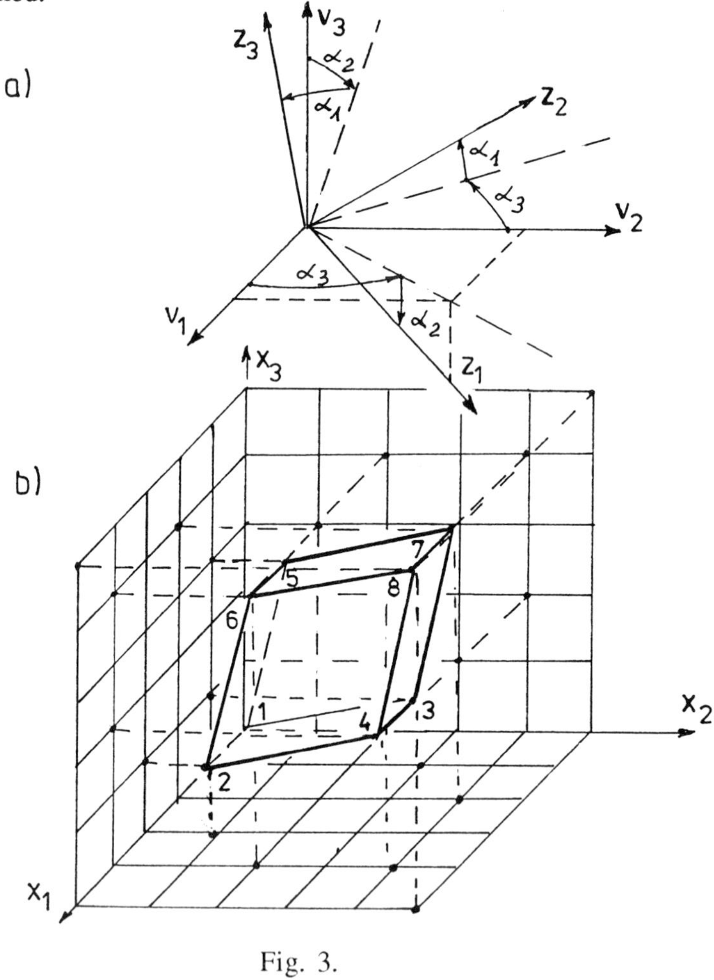

Fig. 3.

24. Additionally, composing nodal equilibrium equations, or computing internal forces of double- or triple-curvature structures, there is a need to add the vectors

of two displacements defined in two other local co-ordinate sets z_i, and z_i^*. Here certain geometrical objects (sets of $\cos(z_i, z_i^*)$ between each above axis), $\mathbf{A}_\wedge$, are needed. They can be computed by means of the following formula (ref. 3):

$$A_\wedge = g \cdot (E_\wedge g)^T = \beta g^* \cdot g^* E_\wedge \beta^T = \beta(E_\wedge \beta^T) \quad (9)$$

where g_i and g_i^* are basis vectors of two co-ordinates systems, and is a valid relation:

$$g_i = \beta_i^k \cdot g_k^* \qquad \vee \qquad g = \beta \cdot g^* \quad (10)$$

25. This is the relationship between basis vectors of local nodal z_i, and the global whole structure ζ_i co-ordinates systems. The definitions of elements of matrix β should be given in the program, or replaced by the user, for each of declared type of nets of points (Eq. 1).

OUTPUT INFORMATION

26. Taking into account the destination of described method for large sets of elements, the graphical interpretation of internal forces for the whole structure is here not foreseen. The particular diagrams for separate members (elements) are displayed on a screen in conversational form. A general view of the structure can be only shown in colours, highlighting the nodes which are connected with members experiencing the largest stresses.

FINAL REMARKS

27. The method presented relates to the description of structural input data, is here slightly extended in comparison with similar procedure applied by the author for use on old generation computers. The main problem here concerns only the graphical procedures, which help the user to describe the input data, as well as to display the resultant form. The graphical display of the results of an analysis, to enable interpretation, causes problems due to the large volume of the files saving these data.

28. The input data description was successfully applied by the author for programs based on DMEM (Difference-Matrix Equation Method, ref. 1-7), FEM and the Finite Differences Method. The most useful PC languages for writing input data procedures are TURBO PASCAL 6, and C++.

REFERENCES

1. OBREBSKI J. B. Application of digital computers to the solution of regular hexagonal gridworks. (in Polish). Mechanika Teoretyczna i Stosowana, Warsaw, 1, 10, 1972, pp, 117- 126.

2. OBREBSKI J. B. Statics of hexagonal member nets. (in Polish), IBTP Reports, Warsaw, 36,1972, pp.217.

3. OBREBSKI J. B. Numerical analysis and synthesis of large structural sets. (in Polish), IBTP Reports, Warsaw, 24,1979, pp.196.

4. OBREBSKI J. B. Application of the WDKM program system to analysis of space structures, Space Structures, 1, 1985, pp 93-98.
5. GUTKOWSKI W., OBREBSKI J. B. *et al*, Statical computations of structural roofs. (in Polish), ARKADY, 1980, and (in German), Werner/Arkady, Warsaw, 1985.
6. OBREBSKI J. B. New approach to geometrically non-linear problems in statical and dynamical analysis of large elastic thin-walled member structures. Proceedings of Int. Coll on Space Structures for Sport Buildings. Science Press Beijing & Elsevier Applied Science (London), Oct 1987.
7. OBREBSKI J. B. Difference equations in analysis of repeatable structures. Proceedings of Int. IASS Cong., Madrid, Sept., 1989.

124. Initial shape generation for membrane structures

N. YAMADA, University of Surrey, UK

SYNOPSIS. Membrane structures have been given much attention as rational structural systems to enclose large open spaces in recent years. An important initial step in the design of a membrane structure is the generation of an initial shape. This is a first suggestion for the shape of the structure which will be modified through an analysis process to find the initial equilibrium form and, finally, the actual form of the structure. The main purpose of this work is to introduce a method for the creation of the initial shape of a membrane configuration in the design of a membrane structure by using the formex approach (ref. 1). Also, the relationship between the Formian method (ref. 2) and the following analysis (prestress and stress analyses) process is introduced.

INTRODUCTION

1. In the last few years the rapidly increasing use of membrane structures has resulted in the development of a research field relating to the analysis, design and construction of membrane structures. One aspect of this field is concerned with the evolution of suitable techniques for determination of initial shapes for membrane structures. There are a number of computer methods which are considered ideal for initial shape determination and each method has an area of application in which the method is superior to other alternatives.

2. The formex approach provides a convenient means for initial shape determination in many situations. The applicability extends to a whole structural system that consists of combinations of space frames, membrane parts and cable elements.

3. In generating data for initial shapes, Formian is supposed to be used in conjunction with an interactive package for design of membrane structures. Thus, when required, Formian is employed to generate data for an initial shape and the data are then transmitted, through a SUBMIT statement bringing suitable functions, to the package for further processing.

BASIC DATA GENERATION

4. Consider the configuration of Fig. 1 that represents a quadrilateral finite element mesh on a cylindrical surface. A formex formulation for the configuration may be given by

```
MESH  =  '::
A1  =  LAMIT(6,4)¦RINIT(6,4,1,1)¦[1,0,0;1,1,0
                                      ;1,1,1;1,0,1]
A  =  BC(8,120/12,12/6)¦A1'                          (1)
```

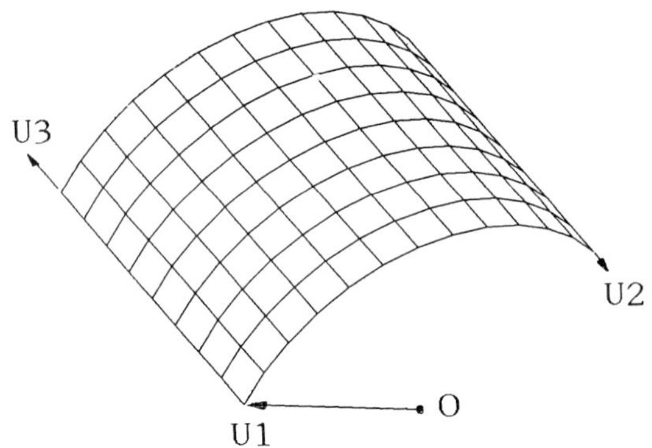

Fig. 1. Quadrilateral finite element mesh.

5. The formulation for the configuration can be written in a generic manner as a Formian scheme. For instance, one can write:

```
MESH  =  ':M,N,L,F,B:
A1  =  LAMIT(M/2,N/2)¦RINIT(TIC¦(M/2),TIC¦(N/2),1,1)¦
                         [1,0,0;1,1,0;1,1,1;1,0,1]
R  =  (F^2+L^2)/(2*F)
P  =  ASIN¦((R-F)/R)
A2  =  BC(R,(180-2*P)/M,B/N)¦A1
A  =  VERAT(0,0)¦VERAD(0,0,P)¦A2                     (2)
```

6. Here, M denotes the number of divisions for an arch, N denotes the number of divisions for a straight edge, L is the width of cylinder, B is the length of the cylinder and F is the height of the cylinder. The retronorm 'BC' in the above scheme transforms the signets of A2 into Cartesian co-ordinates of nodes relative to the x-y-z co-ordinate system of Fig. 2.

7. The scheme may give rise to an infinite number of configurations, specified by different values of the parameters, M, N, L, F and B.

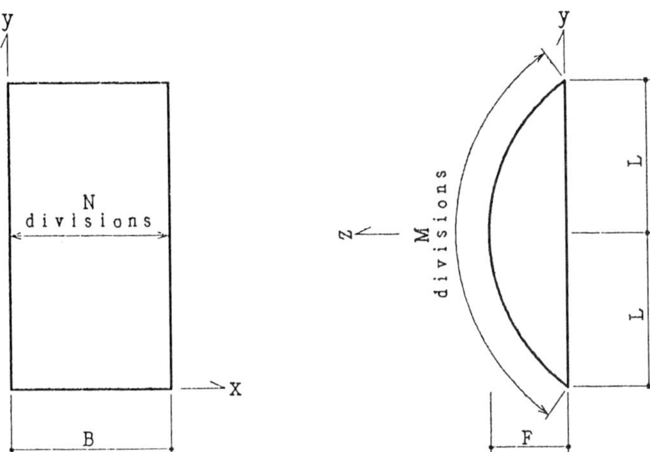

Fig. 2. Parameters of a cylinder in Cartesian co-ordinates.

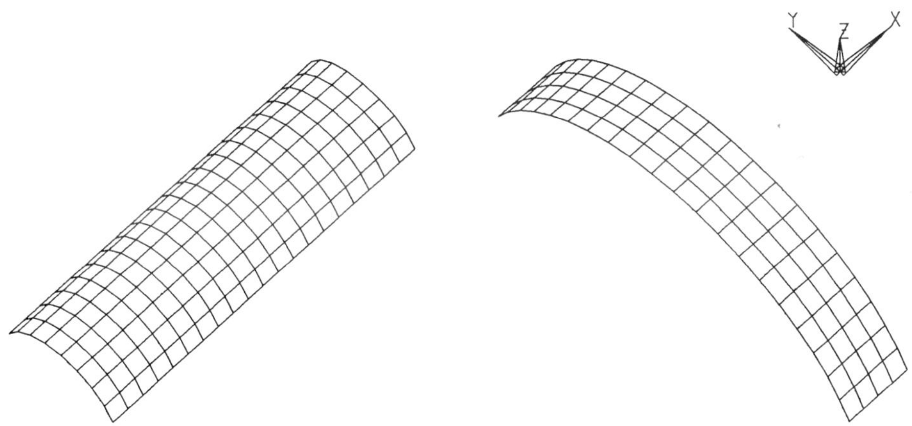

Fig. 3. Fig. 4.
Cylindrical shapes from different parameters.

8. The scheme is executed through induction statements listing the required values of the parameters. For example, the induction statement

MESH(8, 20, 3, 2, 18) (3)

corresponds to the configuration shown in Fig. 3, and the induction statement

MESH(20, 4, 9, 6, 3) (4)

corresponds to the configuration shown in Fig. 4.

9. The use of the above scheme for the representation of the configuration has the following advantages:

(a) The scheme describes the configuration by parametric rules and therefore is applicable in similar situations when cylindrical surfaces are required. The scheme may be stored in the computer memory and used whenever the need arises.

(b) The scheme may be easily modified to take account of new requirements. The manner in which a scheme is modified is similar to the way in which a text is edited using a word processor.

(c) The scheme is stored as a character string in the computer and therefore its storage requirement is quite small (ref. 3)

DATA GENERATION FOR AN INITIAL MEMBRANE SHAPE

10. As explained previously, Formian is a convenient tool for generating a finite element mesh. However, to generate the initial shape of a membrane structure, more information is required. Membrane structures often include cables. Thus, the cable elements should also be included in the data. Other information required are the boundary conditions, element orientation and principal axes (warp, fill) for each membrane element and properties for the cables.

11. The formex formulation of the membrane structure of Fig. 6 for our shape finding program SHAPE (ref. 4) is given in Fig. 5.

12. The five parameters in part (a) correspond to those in Fig. 2. Part (b) of the scheme consists of two expressions. R and P are linear and angular factors for the cylindrical retronorm. The formex variable Q is used to specify the orientation of the membrane surface as will be explained in the YARN function in part (h). The two variables Ml and C1 generate the 3-noded membrane elements and 2-noded cable elements, respectively. Information about the boundary conditions is generated in part (e).

13. Through the statements in part (f), the formices are obtained representing the nodal co-ordinates N, connectivity patterns of the membrane and cable elements M and C, and the boundary nodes BB.

14. The statement in part (g) gives rise to the actual node co-ordinates from the normat co-ordinates. The YARN function in part (h) specifies the orientation and principal axes of the membrane elements. The effect of the SUBMIT statement in part (i) is to prepare the data in a suitable form for input to the shape finding program. Here, T2 is an identifier to be a formex variable which contains the co-ordinates by shape finding analysis. As a result of the shape finding analysis, the equilibrium shape shown in Fig. 7 is obtained.

15. Using the same procedure, we may write formex formulations for the initial shapes shown in Figs 8a and 9a, which result in the equilibrium shapes of Figs 8b and 9b, respectively.

```
C1  = TRAN(3,TIC¦(J/2))¦RINIT(TIC¦I,3,1,J)¦..
                        [1,0,0;1,1,0]      —— (d)
B1  = RINIT(TIC¦(I+1),4,1,J)¦[1,0,0]      —— (e)
N  = MED¦M1                                ┐
M  = DIC(N)¦M1                             |
C  = DIC(N)¦C1                             |
BB  = DIC(N)¦B1                            ┴—— (f)
T1  = BC(R,(180-2*P)/I,B/J)¦N             —— (g)
               ┌──────────u3 is warp
MM  = YARN(3,2,Q)¦M                        —— (h)
               └──────────u2 is fill

SUBMIT {{100,'[SHAPE]',1,10},{N,T1},..
        {{MM,{200,200}},{C,{0.0,1.0,0.001}}},..
         {BB,[1,1,1]},{T2}}'               —— (i)
```

Fig. 5. The formulation of data for shape finding analysis program.

DATA GENERATION FOR STRESS ANALYSIS

16. The results of SHAPE (shape finding analysis) are used as input data for the stress analysis. The following SUBMIT statement (Fig. 10) is used to specify information about the properties for membrane elements, cable elements and wind loading coefficients.

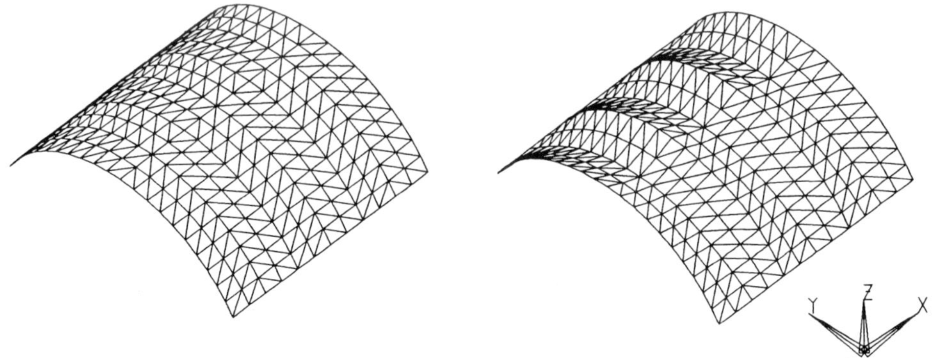

Fig. 6. Initial shape.

Fig. 7. Equilibrium shape.

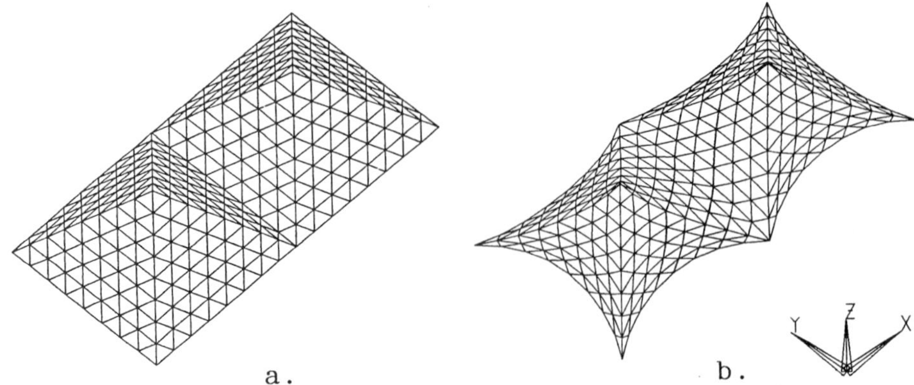

Fig. 8. Initial shape and equilibrium shape for a suspension structure.

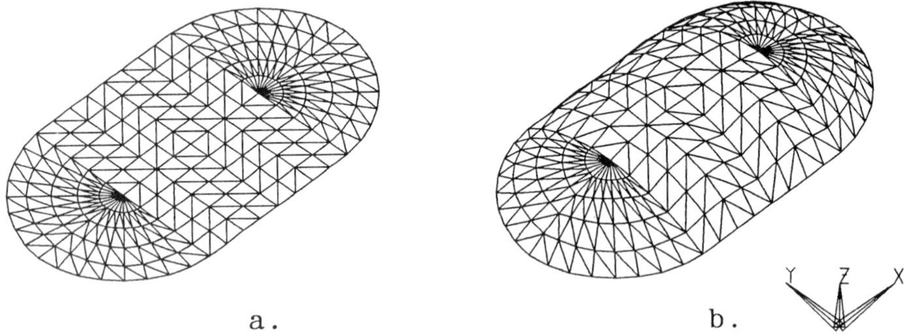

Fig. 9. Initial shape and equilibrium shape for a pneumatic structure.

```
SUBMIT {{101,'[STRESS]'},{N,T2},..
         {{MM,{W,[E_x,E_y],[V_yx,V_xy],G,[T_x,T_y]}},..
          {C,{W,E,A}}},..
          {P,{C_pe}}}
```

Fig. 10. SUBMIT statement for stress analysis.

CONCLUSIONS

17. It has been shown that the Formian process may be used to generate data for initial shapes of membrane structures. The procedure allows creation of general formulations, which is a feature of great value in dealing with shapes that may be defined by specification of suitable sets of parameters.

18. This paper mainly deals with initial shapes of membrane structures for shape finding analysis. Formian may also be used to generate data for stress analysis.

REFERENCES

1. NOOSHIN, H. Formex configuration processing in structural engineering, Elsevier Applied Science Publishers, 1984.
2. NOOSHIN, H. and DISNEY, P. Elements of Formian. Proceedings of the Fourth International Conference on Civil and Structural Engineering Computing, Edited by B H. V. Topping, Civil-Comp Press, London, September 1989.
3. NOOSHIN, H. and YAMADA, N. Formex pre-processing for membrane structures. Proceedings First U.S. National Congress on Computational Mechanics, Chicago, Illinois. July, 1991.
4. TODA, I., SUZUKI, T., KIKUSHIMA, M. and SAITO, Y. The influence of directional property of the membrane materials on the membrane analysis, p.41,42. Proceedings IASS Symposium, Osaka, 1986, Vol. 2 Edited by K. Heki.

125. Geodesic domes and golf balls

T. TARNAI, Technical University of Budapest, Hungary

SYNOPSIS. In this paper the analogy between different golf ball designs and networks of geodesic domes is investigated. Dimples of golf balls correspond to nodes of geodesic domes. Several examples of golf ball designs are presented, and cases are shown where dimple patterns can be used for geodesic domes.

INTRODUCTION

1. In North Yorkshire people call radomes "giant golf balls". The name comes from the fact that the arrangement of nodes on the radome sphere resembles the dimple pattern on golf balls. Although this similarity in many cases is only a semblance, in many cases it is well established.

2. In this paper the analogy between different golf ball designs and networks of geodesic domes (radomes in particular) will be investigated. The geometrical problem in golf ball design is to construct a dense packing of equal or different circles on the spherical surface under subsidiary conditions. Many of the modern dimple patterns have octahedral, icosahedral or quasi-icosahedral symmetry. In general, a triangular network can be associated with circular patterns on the sphere which are suitable also for construction of geodesic domes. In the paper several examples of actual contemporary golf ball designs will be presented where dimple patterns can be used, or have already been used, for geodesic domes (radomes).

GOLF BALLS

3. The history of the evolution of the golf ball has a vast literature (refs 1, 5, 7, 14). In the early period (about in the fifteenth century) feather-filled balls were used whose cover was constructed from strips of leather. About 1860, rubber-like balls were made from gutta percha which were more resistant but did not provide better performance than the feathery balls. These golf balls still had a smooth surface. Some years later, however, it was discovered that the performance of a gutta percha ball could be improved if an even pattern was made on the surface of the ball with a sharp hammer. At the end of the last century a bramble-patterned cover was introduced for the gutta percha ball, where raised

round bumps were situated along parallel small circles of the sphere. This ball conveyed an impression similar to that of the ball in Fig. 6. Later raised bumps were replaced by dimples, and by 1930 the use of dimples became dominant.

4. The role of dimples is to provide better performance by improving the aerodynamic properties of the ball. When a golf ball flies through the air, small pools of turbulence form in its dimples. The turbulence pulls the air rushing by closer to the ball which results in a smaller wake. That means less drag and longer flight. Dimples also help the ball soar higher. The dimples on a ball with a backspin grip the air on top of the ball, and propel it at greater speed, thus lowering the air pressure above the ball, which results in a rising flight.

5. An important requirement of dimple pattern design is that the ball has to have a parting line, that is, a great circle which is not intersected by any of the dimples. This requirement is due to the fact that the mould consists of two hemispheres which come together during manufacturing.

6. Over the years several different dimple patterns were designed and applied. There is a tendency to use a high order of symmetry in the patterns in order to produce balance by addition of "false" parting lines to the pattern. The number of dimples on a ball used by golfers is in general between around 300 and around 500.

OCTAHEDRAL SYMMETRY

7. If a geodesic dome forming a complete sphere has 300 to 500 nodes then the dome in general is of medium size, and the dome is designed with axial or icosahedral symmetry. However, a modular dome (Fig. l(a)) suggested by Spunt (ref. 9) contains two planes of symmetry, so the framework of the dome can be considered as a part of packing of equal circles on the sphere with octahedral symmetry (in the neighbourhood of a vertex of the octahedron). This structural system is very similar to the golf ball having 336 equal dimples in an octahedral layout (Fig. l(b)). Its dimple pattern (Fig. l(c)) has three great circles which do not intersect dimples. Although the modular dome in Fig. l(a) has no such great circles it is easy to construct modular domes having two parting lines.

8. The dimple pattern in Figs 1(b) and 1(c) is not only octahedral but it has the properties of the small rhombicuboctahedron (ref. 4). There are packed 16 dimples into a square face and 6 dimples in a triangle face. Geodesic domes of the shape of the small rhombicuboctahedron were suggested recently by Obrebski (ref. 6).

9. Dimple patterns are also designed with the properties of the cuboctahedron (ref. 4) which have 4 great circles which do not intersect dimples (Fig. 5(d)) (ref. 14).

ICOSAHEDRAL SYMMETRY

10. Geodesic domes are constructed mainly with icosahedral symmetry as are Fuller's fly's eye domes whose circular openings convey an impression of a golf ball (ref. 10).

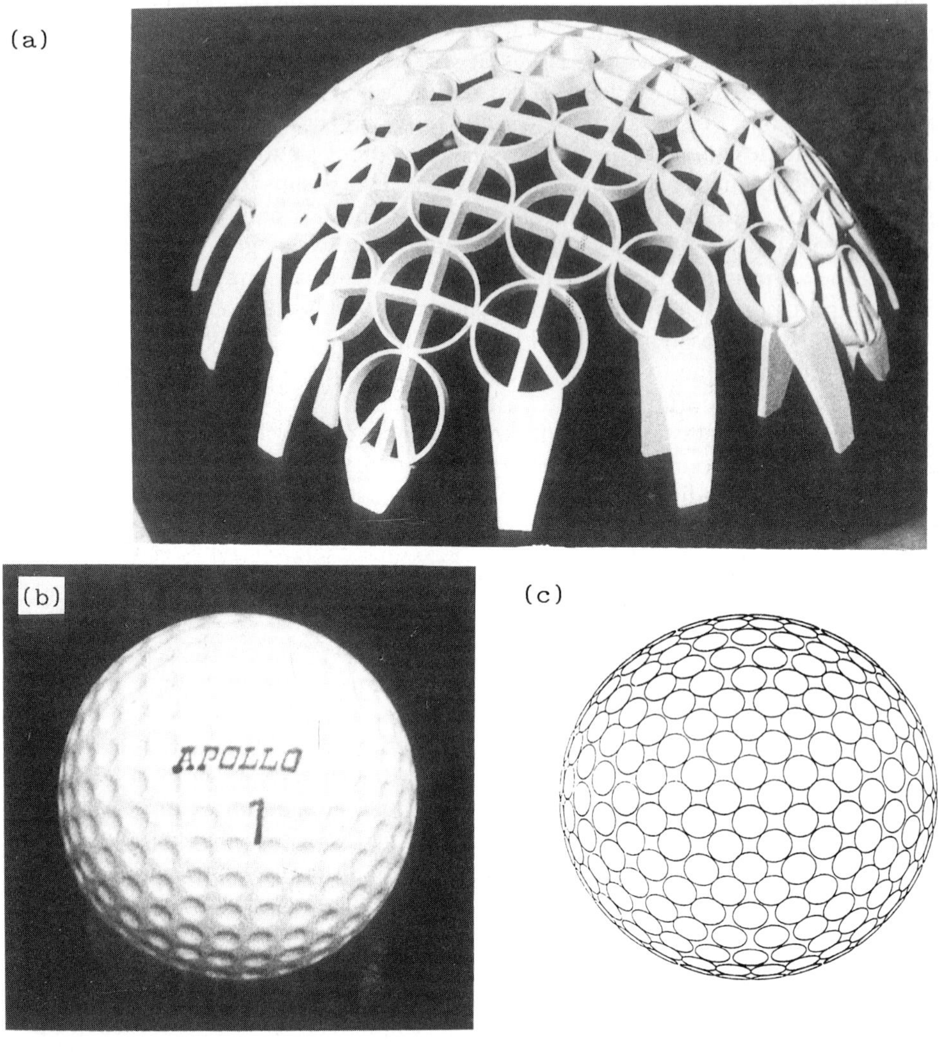

(a)

(b)

(c)

Fig. 1. (a) Solid rib model of Sput's modular dome (Courtesy of Prof. L. Spunt.) (b) A 336 dimple octahedral golf ball. (c) Dimple pattern design of a 336 dimple ball. (Courtesy of G.W. Slater.)

To our knowledge, the first icosahedral geodesic dome was constructed in the 1920s and the fly's eye domes were designed and modelled in the late 1960s, but the first icosahedral dimple pattern was introduced only in 1973 by Titleist (refs 1, 14). This pattern (Fig. 2) having 392 dimples and two dimple sizes, however, shows only quasi-icosahedral symmetry. Here, for the triangulation of the spherical surface, an edge of the icosahedron is subdivided into six parts which does not provide a parting line. Therefore, by an additional row of dimples at the equator the icosahedral arrangement is modified. In this way a parting line is obtained and the patterns on the two hemispheres are equal (Fig. 2(b)). By a rotation of one hemisphere relative to the other, similar to that in ref. 4, a version of the pattern can be obtained where the parting line provides a plane of symmetry of the arrangement of dimples. The centres of dimples lying in two rows adjacent to the equator determine a prism instead of an antiprism. This version is also intensively applied by golf ball manufacturers.

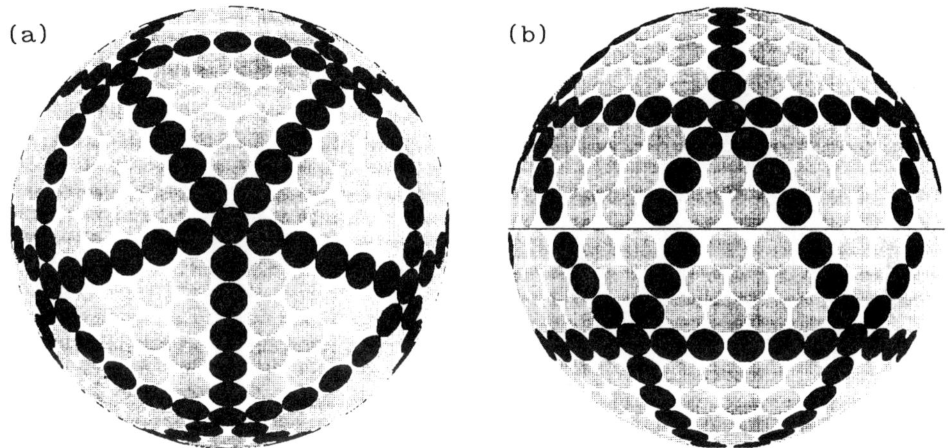

Fig. 2. Titleist's 392 icosahedron pattern, (a) pole view, (b) equator view. (Courtesy of S. Aoyama.)

11. If we want to apply true icosahedral symmetry, one of the ways can be the use of the icosidodecahedron (ref. 4) as a base. In this case we pack the dimples in triangular and pentagonal faces of the icosidodecahedron (IDH), and we have 6 great circles not intersecting dimples. It is interesting to see that a radome in Fig. 3(a) designed by Schonbach (ref. 8) and the dimple pattern of Maxfli MD 432 ball in Fig. 3(b) have almost the same construction. The 432 nodes of the radome considered as a complete sphere and the 432 dimples on the ball are arranged almost identically in an IDH system. The triangulation of the icosahedral surface in this case is not a regular triangular tessellation $\{3,5+\}_{b,c}$ defined by Coxeter (ref. 2), since we have not only pentavalent and hexavalent vertices but also heptavalent vertices. It can be seen in Fig. 3(b) that only one of the parting lines is real, the five others do not really separate the two halves of the configuration. This is why dimples positioned adjacent to the equator (seam line) are different from that adjacent to the "false" parting lines, and five dimple sizes are applied.

12. The other application of true icosahedral symmetry is the use of the dodecadodecahedron (ref. 4) as a base. The edges of the dodecadodecahedron (DDH) define ten great circles, and the dimples are packed in star-pentagonal and hexagonal faces of the spherical polyhedron. We do not know any geodesic dome

in this system. However, in golf ball design it is successfully employed. Using three dimple sizes Dunlop arranged 360 dimples in the DDH system (Fig. 4(a)). Pattern designs try to incorporate more dimple coverage in order to optimise the flight characteristics of the ball. Configuration of 360 equal dimples in Fig. 4(b) results in larger dimple coverage than that in Fig. 4(a), but it cannot be used as a dimple pattern of a golf ball since it has no parting line. Packing of equal circles on the sphere with icosahedral symmetry can result in high packing density if the circles are arranged due to a skewed regular triangular tessellation (having no plane of symmetry) (ref. 13), and at the same time it defines an advantageous network for geodesic domes (refs 11, 12).

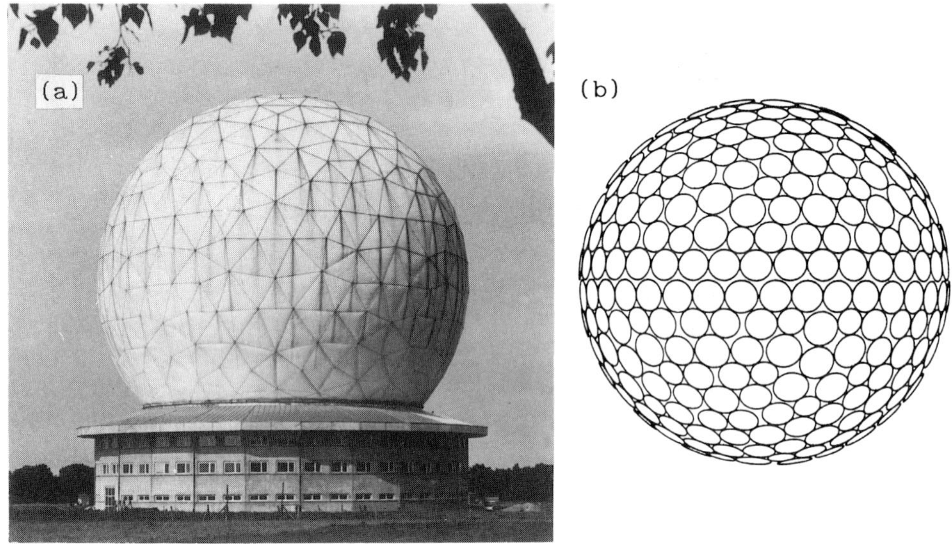

Fig. 3. (a) Schonbach's radome with a diameter of 49m. (Courtesy of Dr.W. Schonbach.) (b) Dimple pattern design of the Maxfli MD 432. (Courtesy of G.W. Slater.)

13. Because of symmetry the number of dimples on a ball and the number of nodes in a geodesic dome cannot be arbitrary. There are "magic numbers" which can occur as dimple numbers or node numbers. Such a magic number is 480. 480 dimples can be arranged in IDH system as shown in Fig. 5(a), using 4 dimple sizes. The dimples determine a regular triangle tessellation on the sphere having a plane of symmetry such that the triangulation number (the number of triangles on a face of the icosahedron) is 49. As Goldberg (ref. 3) has shown, this is the smallest triangulation number for which the triangulation is not unique. Indeed, we can determine another regular triangulation having no plane of symmetry. On the basis of this, triangle tessellation packing of 480 equal circles has been constructed (refs 12, 13) as can be seen in Fig. 5(b). 480 dimples can be

arranged also in the DDH system. Fig. 5(c) shows such a design where three dimple sizes were applied. A cuboctahedral arrangement is also available to accommodate 480 dimples as seen in Fig. 5(d), where four dimple sizes were used.

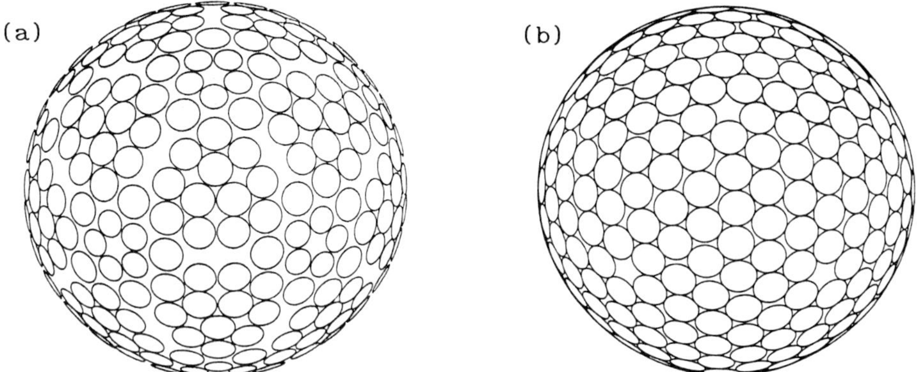

Fig. 4. (a) Dimple pattern design of the DDH 360. (Courtesy of G.W. Slater.) (b) Packing of 360 equal circles on a sphere with icosahedral symmetry. (Drawn by G. Koller.)

14. As mentioned in paragraph 6 dimple patterns having more then around 500 dimples are not in use. Once, to demonstrate the possibility, a golf ball manufacturer produced an IDH ball with 1212 dimples. The dimple configuration topologically was exactly the same as the configuration of small spheres in the sculpture in Fig. 6 which defines a perfect geodesic dome. (At sculpture, the parting line is not exactly a great circle but is a facewise waving curve.)

CONCLUSIONS

15. It is not difficult to imagine that in the 1970s geodesic domes influenced golf ball dimple pattern designs. During the years many new ideas arose in golf ball design which could also be useful for dome designers. Here, however, one can observe two opposite tendencies: (1) In golf ball design there is an intention to produce dimple patterns with high order of symmetry; (2) In radome design there is an intention to reduce the symmetry of the structure in order to decrease the disturbance of the radio signals caused by the structure. Sometimes radomes are constructed without symmetry since in this case the structure has no dominant spherical harmonics.

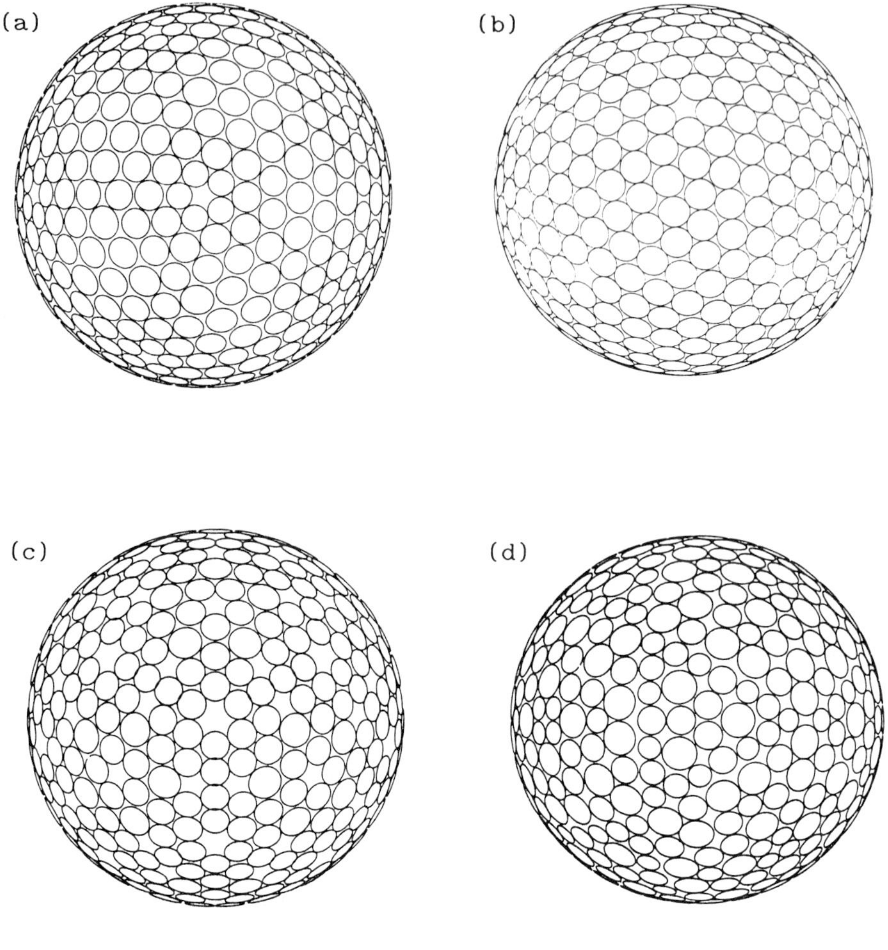

Fig. 5. (a) Dimple pattern design of the Slazenger Interlok ICOSDDH 480. (Courtesy of G.W. Slater.) (b) Packing of 480 equal circles on a sphere with icosahedral symmetry. (Drawn by G. Koller.) (c) Dimple pattern design of the Slazenger Interlok DDH 480, (d) CUBOCT 480. (Courtesy of G.W. Slater.)

Fig. 6. Spherical sculpture in the Congress Hall in Prague.
(Courtesy of Prof. A. Mackay.)

ACKNOWLEDGEMENTS
The author thanks G.W. Slater for repeated help and fruitful discussions. Research reported here was supported by OTKA IX3 Grant No.41 awarded by the Hungarian Scientific Research Foundation.

REFERENCES

1. CAMPBELL M. The Encyclopedia of Golf. Prentice-Hall Canada Inc., Scarborough, Ontario, 1991.
2. COXETER H.S.M. Virus Macromolecules and Geodesic Domes. A Spectrum of Mathematics (ed. J.C. Butcher). Auckland University Press and Oxford University Press, 1972, 98-107.
3. GOLDBERG M. A class of multi-symmetric polyhedra. The Tohoku Mathematical Journal, 1937, vol.43, 104-108.
4. HOLDEN A. Shapes, Space, and Symmetry. Dover, New York, 1991.
5. LUMB N., HOBBS M. and PINNER J. The Complete Book of Golf. Quintet Pub. Ltd., London, 1990.
6. OBREBSKI J.B. On the conception of the family of spherical domes with two-levels of repeatability. Proc. of the IASS-CSCE Congress, Toronto, 1992, vol.2, 596-607.
7. PEPER G., MCMILLAN R. and FRANK J.A. Golf in America. The First One Hundred Years. H.N. Abrams, Inc., Pub., New York, 1988.
8. SCHÖNBACH W. Als Netzkuppel ausgebildetes Radom mit 49m Durchmesser. Der Stahlbau, 1971, vol.40, 45-55.

9. SPUNT L. Modular dome structure. IASS World Congress on Space Enclosures. Build. Res. Centre Concordia University, Montreal, 1976, vol.1, 235-240.
10. TARNAI T. Spherical circle-packing in nature, practice and theory. Structural Topology, 1984, vol.9, 39-58.
11. TARNAI T. Optimization of spherical networks for geodesic domes.Proc. of the Third Int. Conference on Space Structures, Guildford (ed. H. Nooshin). Elsevier Appl. Sci. Publ., London, 1984, 100-104.
12. TARNAI T. Dense sphere packing on a sphere. Proc. of the First Int. Seminar on Structural orphology (ed. R. Motro and T. Wester). Montpellier - Le Grand Motte, France, 1992, 53-62.
13. TARNAI T. and GÁSPÁR ZS. Multi-symmetric close packings of equal spheres on the spherical surface. Acta Crystallographica, ser.A, vol.43, 612-616.
14. The Golf Ball. Titleist, Acushnet Comp., New Bedford, 1992.

126. Form generation and patterning procedures for tensioned membrane structures

P. KNEEN, University of New South Wales, Australia

SYNOPSIS. This paper describes the initial form finding procedures developed for tensioned fabric structures. The load carrying ability is predominantly determined in a global sense by the shape of the surface. In a local sense, the quality of the cutting patterns is vital. For tensioned fabric structures, some forms such as cones and hypars are very common and specific original procedures are described which lead to a very realistic starting point for the more detailed design. A simple but powerful smoothing procedure has been developed and applied to architectural structures. The paper also describes several computer procedures for the determination of fabric cutting patterns for membrane type structures. A starting point is to simply designate a sequence of node points describing the edges of a strip. Such patterns may not fit onto a supplied roll of fabric and then it is a problem of either redefining the layout of nodes, or somehow pulling these nodes into line, or ignoring the precise location and adopt a more general procedure for defining the proposed edges of the strips.

INTRODUCTION

1. Membrane or fabric structures are exciting three dimensional surface forms and it is important that the structural engineer can generate a sensible geometric form of the surfaces with a sensible layout of nodes and elements as a first step in the computer aided design process.

2. This paper describes portions of a computer program **fabdes** being used for this form finding process. The load carrying ability of surface structures is very dependent on the three dimensional form of the surface and the initial prestress forces. The first problem is to arrive at a satisfactory shape which is in equilibrium under the action of the initial tensions and self weight.

3. Generally, the surface is defined in regions or patches which are joined together to form the overall surface. Patches exist for simple triangles, hypars, bubbles, conical surfaces, surfaces of revolution, circular arches and many more. These preliminary shapes are initially generated without regard to the forces present. The patches are defined by a small number of control points. Some of the

available patches are described along with a powerful iterative scheme for smoothing surfaces in a plausible manner for anticlastic surfaces.

4. Having arrived at a suitable prestressed shape taking the prestressing forces into account using a large displacement, non-linear analysis package which satisfies equilibrium, this becomes the starting point for the structural analysis under applied loads as well as the form from which the cutting patterns are derived for membrane structures. These stages are beyond the scope of this paper. The specialised form finding software can be combined with readily available three dimensional CAD systems to provide a reasonably well integrated system for the design of many forms of fabric structures.

SHAPE DEFINITION OF CONICAL SURFACES USING FABDES

5. Cones are one of the most popular forms for tension membrane structures. Many have a four sided perimeter base with a central mast supporting a circular tension ring. The approximate surface is defined by the edges and the central ring location and orientation. The plane of the ring is taken to be normal to a theoretical mast defined as a vector between the base and its top. This top of mast is an imaginary (variable) point where the tangents to the surface at the central ring would meet the theoretical mast. Note that these concepts may not correspond exactly to the true physical situation but are useful in generating an initial surface of the approximate form.

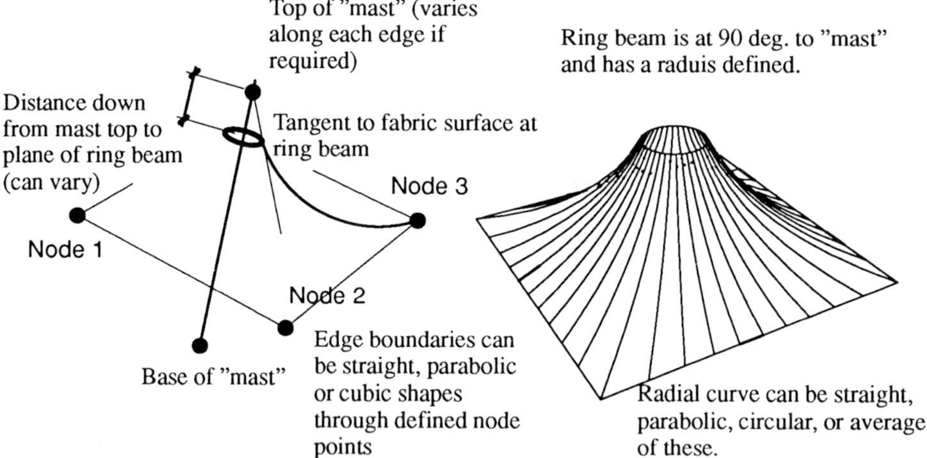

Features of Conical Surfaces in **Fabdes**.

6. The perimeter edges or "base" to the conical surface are defined by isoparametric shape functions through 2, 3 or 4 control points describing a straight line, parabola or cubic curve. Other curves are also supplied. Other features are:

- the number of edges is not restricted to four;
- the type of each edge can vary; some may be straight, others parabolic, circular arcs, etc;

- the number of divisions along different edges can vary to allow for edges of different lengths;
- there can be a variation of the mathematical point at the "top" of the mast - the conical surface need not be complete.

7. A simple example is a "ribbed" structure. Adopting the following assumptions that the "base" is a 20m square at height z = 0. There is a central ring beam at a height of z = 8000mm of diameter 2000mm. All base edges are straight and thus require only the co-ordinates of the end points. Dividing the mesh into 10 intervals along each edge and 8 in the radial direction.

8. The main feature used to simulate the ribs and dishing of the fabric between them is to supply the 3 values for "**h**" where **h** is a notional measure of how far *above* the plane of the ring beam a radial line would meet the *mathematical mast.* Note *above* is determined by the relative positions of the mast base and point in the plane of the ring beam. The mathematical mast need not be the same orientation as a real structural mast - its modelling function is to define the plane of the ring beam and the height **h** (above the ring beam) of an imaginary intersection point of the surface tangents at the ring beam.

9. The 3 values of **h** are for the start, middle and end of the particular edge under consideration. Here, the start and end will correspond to the ribs whilst the middle will be where the fabric is dished in. Recalling that the radius of the central ring beam is 1000mm, using **h** at the start and end of each edge as 100mm and 900mm in the central area. Given the three values, the program interpolates in a parabolic fashion to obtain intermediate values as it progresses along the given edge.

10. Each edge is treated somewhat independently but the node and element numbering assumes that the edges are adjacent and that they will be specified in a continuous fashion. The cone need not be fully closed but what edges are defined, should be joined.

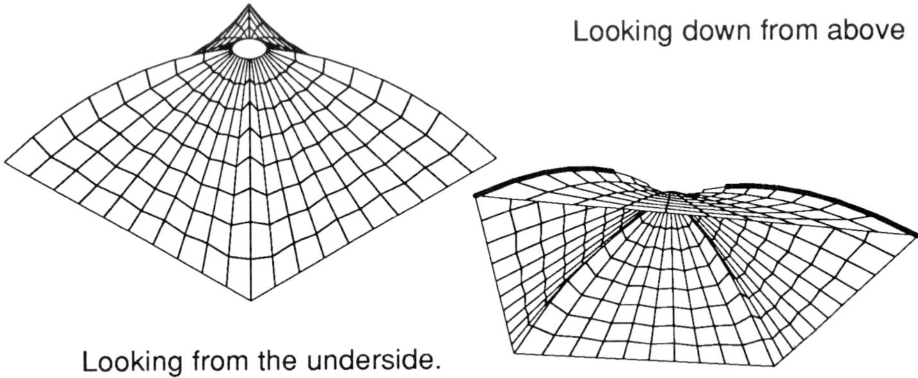

SURFACE SMOOTHING FEATURES

11. A more complex iterative routine based loosely on the Laplace equation in a finite difference form, has an initial starting point as elements in a region defined topologically together with the known co-ordinates of a subset of the nodes (e.g.

along fixed lines of support, edge cables etc.). The procedure then adjusts the co-ordinates of the remaining nodes. Quite complex but smooth surface shapes can be generated. For fabric structures, if the mesh layout is equal in both directions then the surface shape has equal curvatures in the two orthogonal directions.

12. Typically, nodes along boundaries or at isolated points may be known but interior nodal positions are not. Isolated points (e.g. mast supports, internal tie downs for fabric structures) can be read in as well. The next stage is to define node fixity conditions or specifically those nodes which are not permitted to move in x, y or z directions or a combination of these. It should be noted that this surface fitting procedure is purely based on geometry and is not related to physical stresses in fabrics, edge cables or masts.

13. The iterative procedure employs the finite difference form of the Laplace equation $\partial^2 z/\partial x^2 + \partial^2 z/\partial y^2 = 0$. It treats all adjacent nodes as fixed temporarily during an iteration. Thus, for the z co-ordinate, this can be reduced, in simple cases, to the average of the four adjacent nodes along orthogonal mesh lines. Similarly for the x and y co-ordinates.

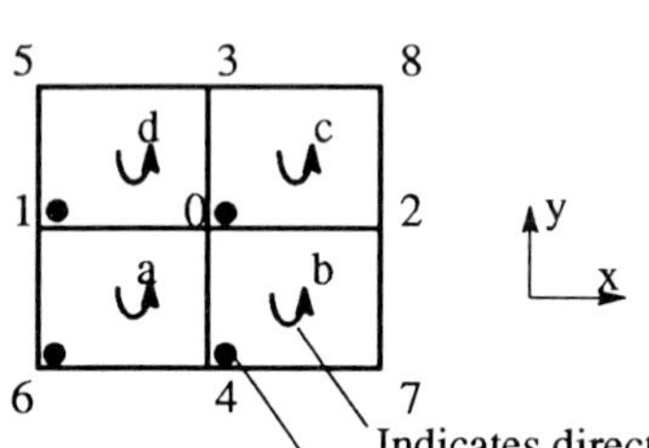

Indicates direction of incidences

Indicates first element incidence

For standard case of equal weights in x and y, an internal node 0, a revised z coordinate is:

$$Z_0 = \frac{(z_1 + z_2 + z_3 + z_4)}{4}$$

For alternate case of different weights in x and y – but constant for all elements

$$Z_0 = \frac{W_x(z_1 + z_2) + W_y(z_3 + z_4)}{2(W_x + W_y)}$$

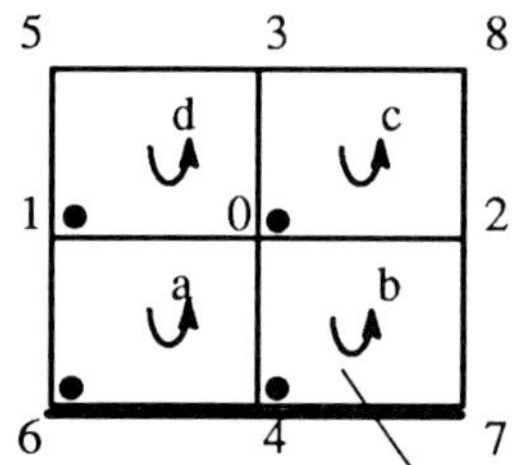

For a sample edge node 4 which also has two cable segments incident (with equal weightings W_c)

$$Z_4 = \frac{W_x(z_6 + z_7) + 2W_y z_0 + W_c(z_6 + z_7)}{2(W_x + W_y) + 2W_c}$$

Edge cable Element with weighting W_c

Basis of Surface Smoothing Operation

SMOOTHING APPLIED TO MULTIPLE EDGED REGIONS

14. The edges are defined as curves and a series of triangular patches is generated each with one side being one of the edges to the region. The third vertex is taken to be the average co-ordinate of all edge control points. The common nodes between adjoining triangles are eliminated. All nodes assigned to the perimeter are optionally fixed in position and the interior surface is smoothed as before. Alternatively, edge cables are generated and connecting links to known points used to simulate the panel attachment to a mast with a rigging screw or

shackle. The following diagram shows a series of hexagonal fabric panels formed by this procedure.

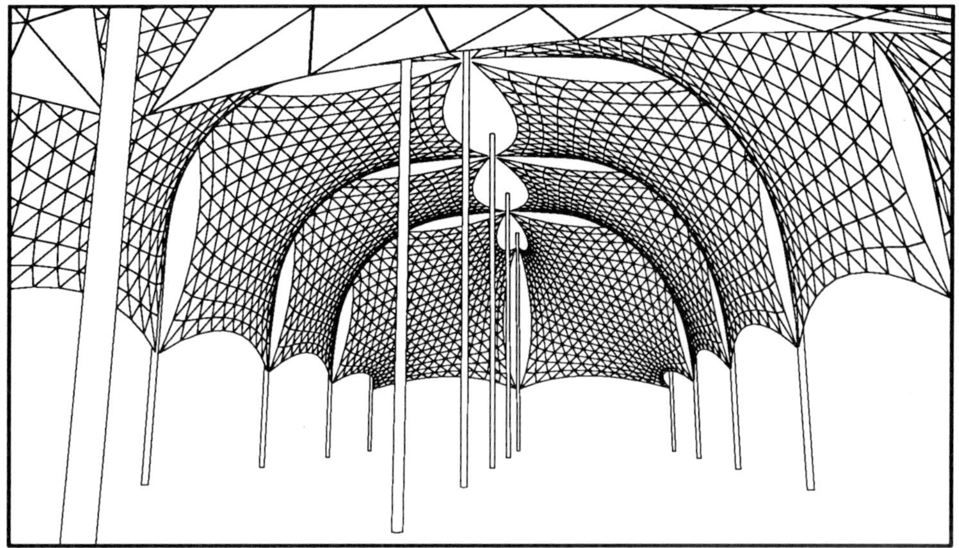

DETERMINATION OF CUTTING PATTERNS

15. For tensioned fabric structures, of paramount interest is the ability to create the desired surface shapes from strips of planar fabric. This fabric comes in rolls of standard width and is reasonably expensive. There is pressure on the designer to ensure a minimum degree of waste. Of course the fabric itself does not have any knowledge of the mesh used to represent the surfaces. The first stage in the production of the cutting patterns is to define the boundaries of the strips. There are several ways in which this can be done but, within **fabdes** they fall into two main methods as follows.

- By direct definition of node numbers which lie on the left and right sides of the strip.
- By defining the plan (i.e. the X-Y positions) layout of the four corner points of the strip and having suitable points along the edges calculated with a local surface fitting procedure. This procedure can be semi-automatic or under user control.

16. The first method then requires that care was taken in the first instance to have a suitable layout of nodes to fit the strips whereas the second is more independent from the node positions. The earlier form-generating procedures are without regard to potential patterning and hence the second is the preferred method.

17. Several designers will use two meshes. The first is done for basic form finding, presentation drawings and the structural analysis. Should the design proposal be converted into an actual project, then a more detailed (and time consuming) mesh might be generated for patterning. Several packages adopt the

first general method and to produce straight strips, the nodes are pulled into line by being attached to geodesic strings. **Fabdes** has these capabilities as well but they are generally not warranted.

18. The procedure consists of the following steps.

- To define a number of (x, y, z,) points along both sides of the proposed strip using either of the basic approaches outlined above.
- Form a series of triangles between the points and "flatten" these onto a plane.
- To apply "compensations" to allow for expected stretch in warp and weft.
- To form a smooth curve passing through the points along each flattened edge and to use the interpolated curves to produce the "offsets" at regular intervals to left and right of a reference line (either down the middle of the strip, or along the right hand side).
- To produce a scaled plot of the strips with the offsets and "running lengths" dimensioned.

19. The second stage "triangulates" these points and flattens the strip out into a two dimensional object. The planar strip has a local x axis going from the start of the strip to the end and the local y axis is positive to the left (looking along the strip from the start). The origin can be taken as at the centreline, or on the right hand edge of the roll.

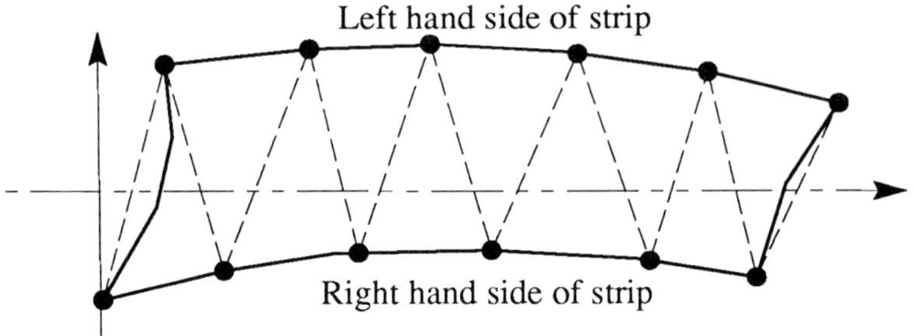

FACTORS AFFECTING PRACTICAL IMPLEMENTATION

20. The definition of an apparently straight strip (in plan) will normally produce a curved strip when flattened out. It is possible to define "bows" for both sides in plan, which will cause the basic plan shape to be curved, (perhaps in the opposing direction) such that when flattened out, it will adopt a straighter shape. This is not feasible when the strips must pass through nodes. This feature is easy to apply and serves the same purpose as the geodesic string concept.

21. It is necessary that the left hand side of one strip be the same three dimensional shape as the right hand side of an adjacent strip and so means of ensuring this are to be provided. Should the flattened strip be too wide for the roll of fabric, a chance to redefine the strip may be in order before proceeding to the next strip. The flattening out process assumes the fabric has no "give". It is known that some degree of shearing is possible and this may well be a suitable option to producing straighter strips.

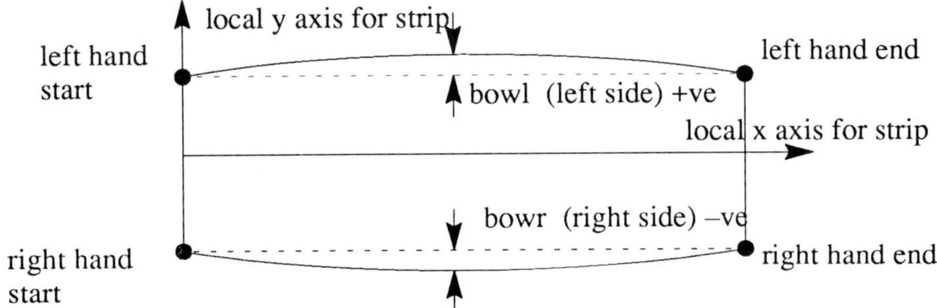

22. When the strip is not completely defined by the user supplying a list of node numbers then the computer will expect the four corners to be defined in a specific order. There are several methods available to enable the corner points to be defined and these are described as follows, noting that any one of these options can be selected for each of the four corner points.

- A specific plan (x, y) co-ordinates could be defined and a local surface fit given if there was no node at the point.
- A specific node number could be supplied if the corner coincided.
- The point could be defined as the intersection point of two lines.
- The point could be defined as having a given x (or y) value and lying on a defined curve such as an edge cable.

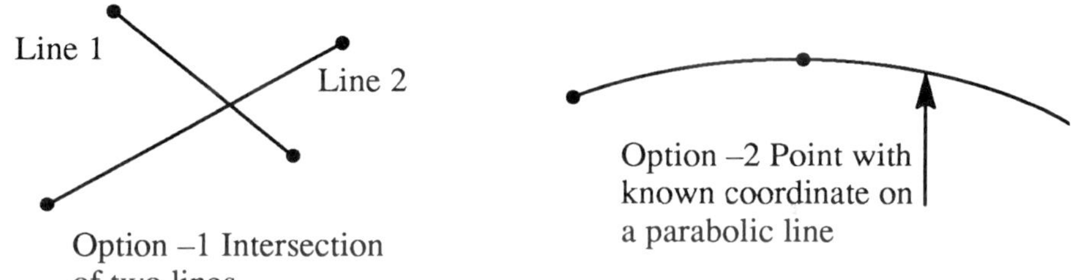

23. Thirdly, the compensation factors for the width of the roll at the start, middle and end of the strip are supplied. The flattened co-ordinates (in width) are *multiplied* by these factors. Typical values are 0.98 for a 2% stretch in the weft and a factor of 0.99 in the warp or lengthways direction. The weld or seam width and distances where the offsets are required, complete the physical factors in this production stage.

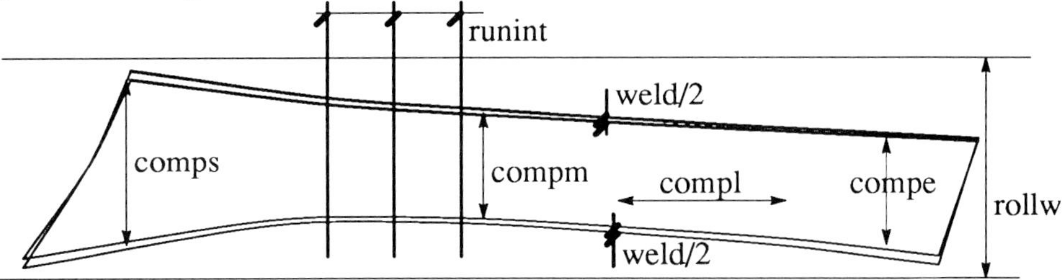

24. If the flattened strip does not fit on the roll, then there are two options available. The first allows for small rotations of the strip on the roll and the second attempts to apply a correcting shear so that the strip is straightened as illustrated below.

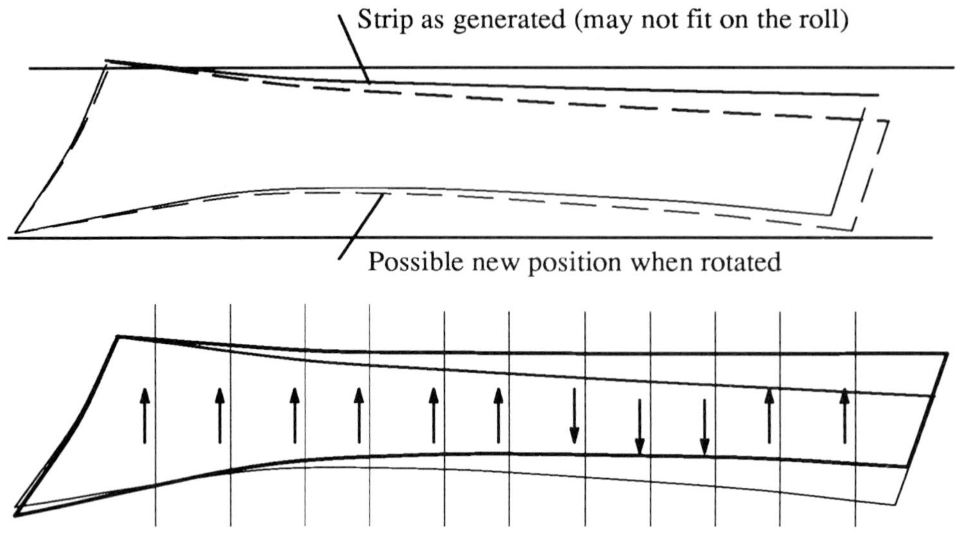

SAMPLE OUTPUT OF PATTERNS

25. A printer listing is generated as well as the chance to plot the individual strips directly or to transfer the patterns to a CAD system. Two types of data can be transferred - the detailed plots of each strip along with running lengths and offsets. The second is the basic patterns in their original three dimensional form. This latter case may be manipulated to produce a reasonable picture of the final structure.

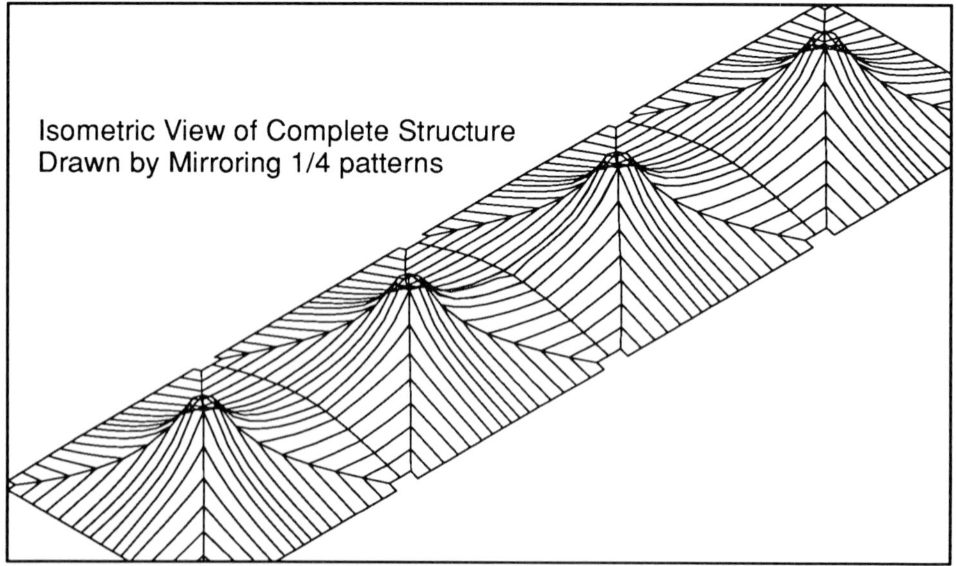

CONCLUDING REMARKS

26. This paper has described the work completed to date which recognises the specialised form finding procedures required for surface structures. It has concentrated on the initial shape generation aspects and has described a powerful smoothing procedure.

APPENDIX. PRINTER OUTPUT OF PATTERNS (SAMPLE)

```
Output of Strips     1 to     1 on file
strip   1 label   301 - interpolated points
  right offset   run length   left offset
      +         :                *      rhs            start              lhs

       865.6       0.0         0.0   !+                  :                    !
       803.2     300.0         0.0   ! +                 :                    !
       779.8     412.5       865.6   !  +                :                  * !
       740.8     600.0       864.9   !   +               :                  * !
       678.4     900.0       863.8   !    +              :                  * !
       616.1    1200.0       862.6   !     +             :                  * !
       553.7    1500.0       861.5   !       +           :                  * !
       491.3    1800.0       860.4   !        +          :                  * !
       448.1    2100.0       859.2   !         +         :                  * !
       417.4    2400.0       858.5   !          +        :                  * !
       386.7    2700.0       857.8   !           +       :                  * !
       356.0    3000.0       857.1   !           +       :                  * !
       325.3    3300.0       856.4   !            +      :                  * !
       294.6    3600.0       855.7   !             +     :                  * !
       269.7    3900.0       855.0   !             +     :                  * !
       269.8    4200.0       854.8   !             +     :                  * !
       269.9    4500.0       854.7   !             +     :                  * !
       270.0    4800.0       854.6   !             +     :                  * !
       270.1    5100.0       854.5   !             +     :                  * !
       270.2    5400.0       854.4   !             +     :                  * !
       270.2    5700.0       854.3   !             +     :                  * !
       293.9    6000.0       855.0   !             +     :                  * !
       324.7    6300.0       855.6   !            +      :                  * !
       355.4    6600.0       856.2   !           +       :                  * !
       386.2    6900.0       856.8   !           +       :                  * !
       416.9    7200.0       857.5   !          +        :                  * !
       447.7    7500.0       858.2   !         +         :                  * !
       489.4    7800.0       859.5   !        +          :                  * !
       551.5    8100.0       860.7   !       +           :                  * !
       613.7    8400.0       862.0   !      +            :                  * !
       675.8    8700.0       863.3   !    +              :                  * !
       737.9    9000.0       864.5   !   +               :                  * !
       778.6    9196.6       865.3   !  +                :                  * !
       800.0    9300.0         0.0   ! +                 :                    !
       862.2    9600.0         0.0   !+                  :                    !
       865.3    9615.4         0.0   !+                  :                    !
                                                       end

left hand side start/finish                       412.53     9196.59
right hand side start/finish at                     0.00     9615.40
max offsets to left and right are                 865.60      865.60
max width of roll required                         1771.
including an allowance for a weld of                 40.

width compensations at start, middle/end were       0.9900    0.9900    0.9900
and for lengths along roll                             0.9900

setting out check dimensions for strip 301
diagonal lhs start to rhs end =            9364.245
diagonal rhs start to lhs end =            9358.065
```

127. Forms and structures generated by identical elements

A. LOBEL, Architect DPLG, France

SYNOPSIS. This research deals with generation of double curvature forms with a unique modulus, an equilateral triangle.

General remarks

1. In February 1986, I published in the review L'Architecture d'Aujourd'hui the first results of the research which are presented here. Since that time important developments have been undertaken. Presented here is the present state of the research. The aim is to generate double curvature surfaces composed of identical elements. Double curvature implying rigidity of form, identical elements implying industrialisation, industrialisation implying economy. The element in question is the equilateral triangle composed of three identical sides. There is thus constituted an equilateral screen articulated to the joints. It can then assume an infinite number of positions, including the most improbable and the least practical.

First Stage

2. The aim (and the difficulty!) is now to identify positions which present some interest, for example arch forms to provide shelters. In creating these forms, it is interesting that the edges may be situated on a vertical or end plane whose position may be varied (alpha, beta). Thus by making the form symmetrical in relation to the plane, it can be continued in space (see Fig. 1). These symmetric forms have a fold where the symmetrical planes pass. It is possible to eliminate the fold by modifying the curvature of the arch (see Fig. 2).There are fewer joints but with more potential as will be seen later. These forms are only part of the first stage of the research. As may have been noticed, they rest on points. It is of course possible to undertake their construction, but they have some disadvantages such as the concentration of stress on a small number of points and the need to make the edges of the form rigid in space in order to make the whole form rigid. This is possible of course, but nevertheless complicated.

Second stage - creation of parent forms

3. The second phase of the research consists of using the forms arising from the first stage, in particular those which have no folds, in order to create totally closed volumes which will not lose their geometric shape. As has been seen in the first

stage, it is possible to create surfaces inscribed in sectors delimited by planes where the angle is variable (beta) and where the inclination of the opening is also variable (alpha).

4. By the interplay of symmetries we then obtain envelopes resting on 2,3,4 and 5 points and even more (in the case of envelopes with folds.) It is now necessary to take one of these envelopes (without folds and five supports) and place it on each of the 12 pentagons creating a dodecahedron. By adjusting the angle of inclination of the openings between each point, all these surfaces join exactly and constitute a closed volume inscribed in a sphere by 12 points of contact. I have called this result a parent form (see Fig. 3). By taking a section along one of its symmetrical planes, we obtain a form resting on a plane, which, when its points of contact in this plane are rigid, itself becomes rigid. Moreover, the section gives rise at its base to several semi-triangles (see Fig. 4).

Creation of derived forms

5. I have indicated frequencies on some of these diagrams (f/fl/f2/f3). I have used them to name each of these forms, for example: 3-3-3-4 csm (csm refers to the family). Finally, and this is the essential point, it is possible to vary these frequencies and to obtain forms of different proportions. At present, in this family, I have catalogued 600 forms (see Fig. 5).

6. By calculation it is possible to ensure that all the points of the base are situated in the same plane. These calculations were carried out with great precision. All the points are defined by their Cartesian co-ordinates. By taking a section of the parent form through other planes, a new family of forms with other characteristics is created.

Continuation of this research (in progress)

7. Finally, and this is the research in progress, it is possible to create other parent forms. Beginning with envelopes resting on 4 points and covering the 6 sides of a cube, we again obtain a parent form inscribed in a sphere by 6 points of contact (see Fig. 6), then by the interplay of frequencies, we change the proportions of the volume as before.

8. Similarly, one may begin with 3 point envelopes and cover the four sides of a tetrahedron (see Fig. 7). There also exist envelopes resting on 2 points which can be assembled round an axis and which in turn constitute parent forms for which the same process begins. (They are not illustrated here).

The joint

9. After the study of the morphology of forms, the most important point is the assembly joint of the bars. My research has therefore simultaneously concentrated on this important question, and an answer has been found. But...as Kipling would say...."that is another story"...!

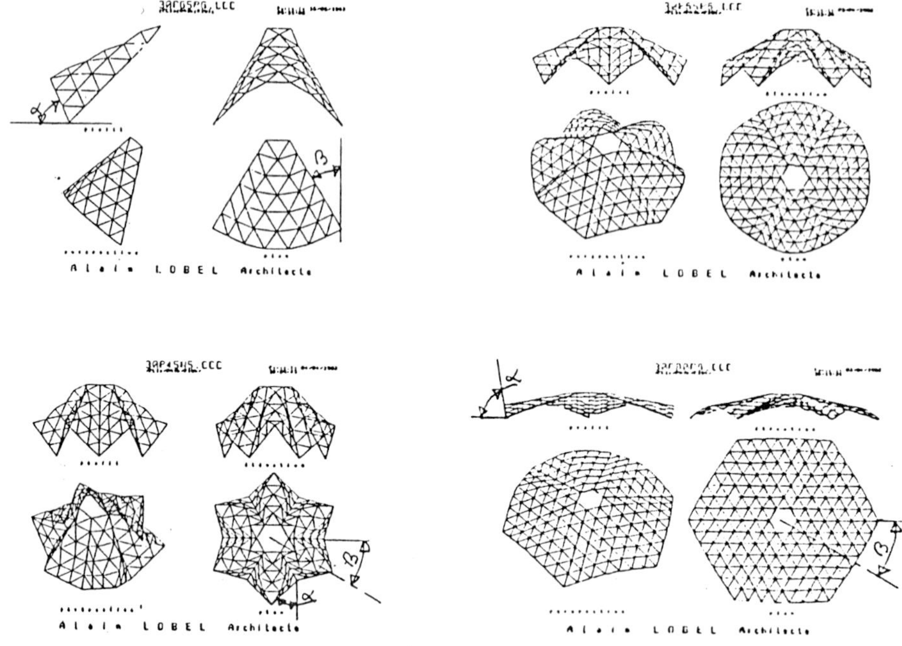
Alain LOBEL Architecte
Alain LOBEL Architecte
Alain LOBEL Architecte
Alain LOBEL Architecte

Fig. 1.

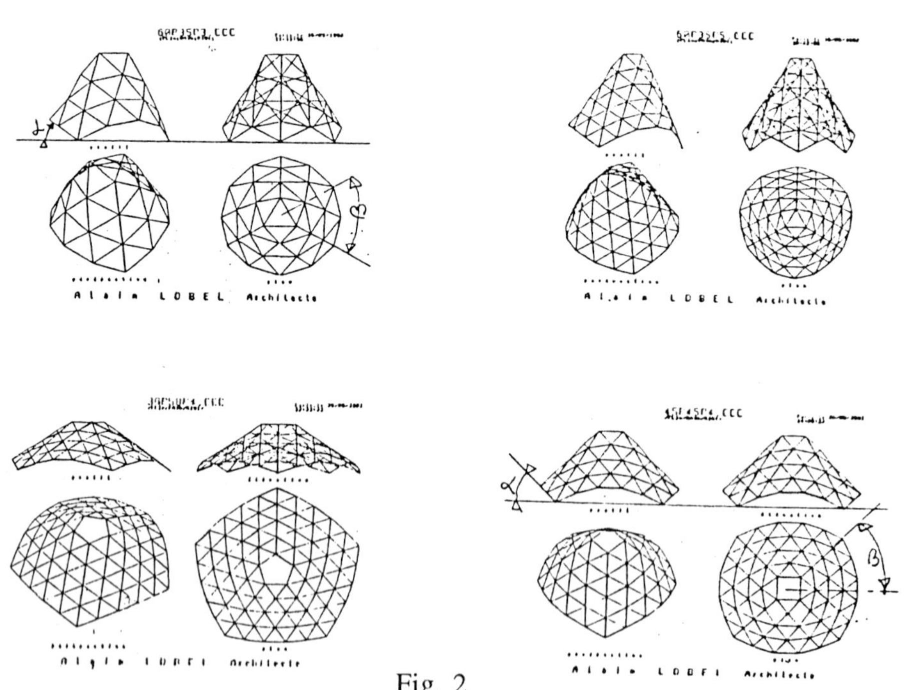
Alain LOBEL Architecte
Alain LOBEL Architecte
Alain LOBEL Architecte
Alain LOBEL Architecte

Fig. 2.

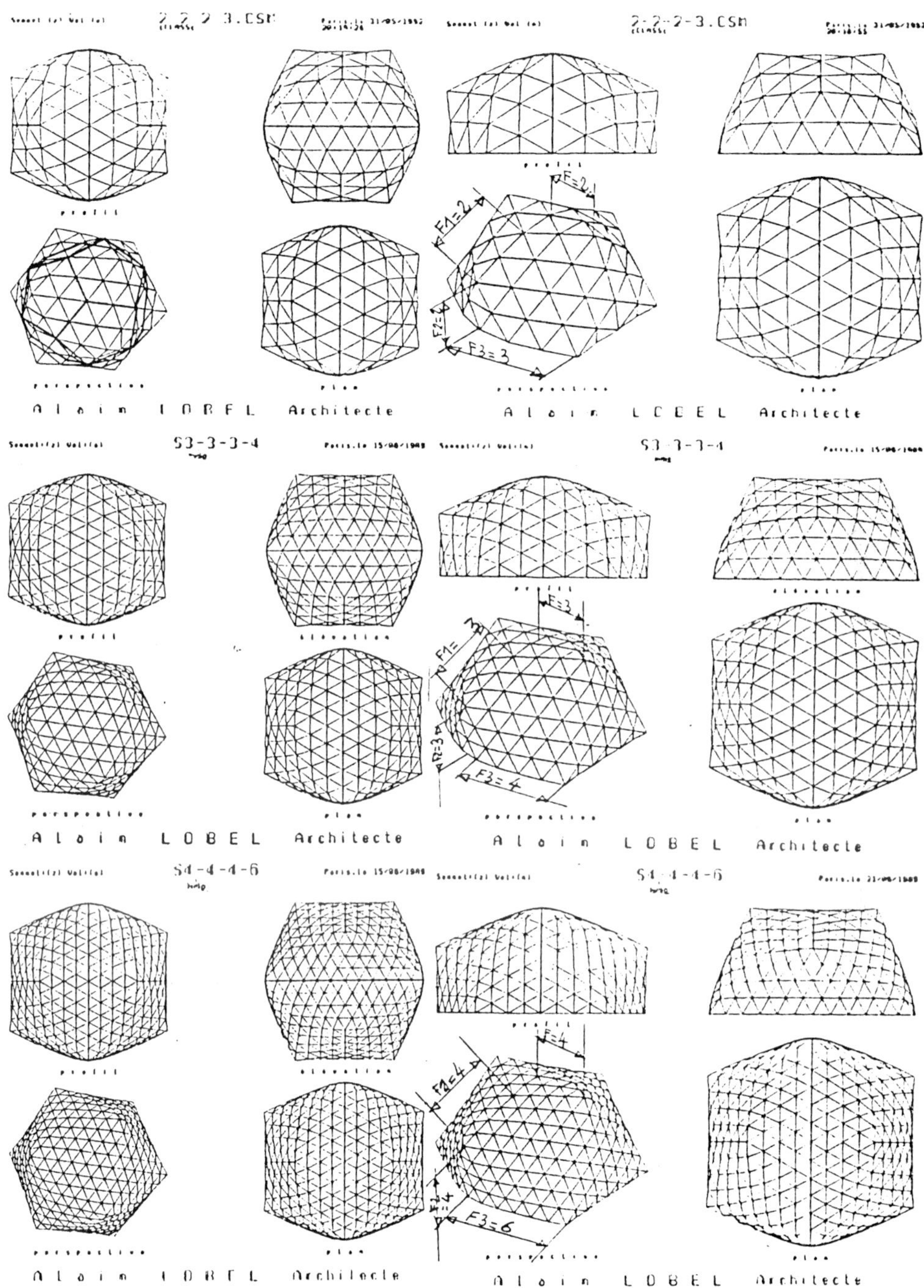
2-2-2-3.CSM
F=2
F1=2
F2=2
F3=3
Alain LOBEL Architecte
S3-3-3-4
F=3
F1=3
F2=3
F3=4
Alain LOBEL Architecte
S4-4-4-6
F=4
F1=4
F2=4
F3=6
Alain LOBEL Architecte

Fig. 3.

Fig. 4.

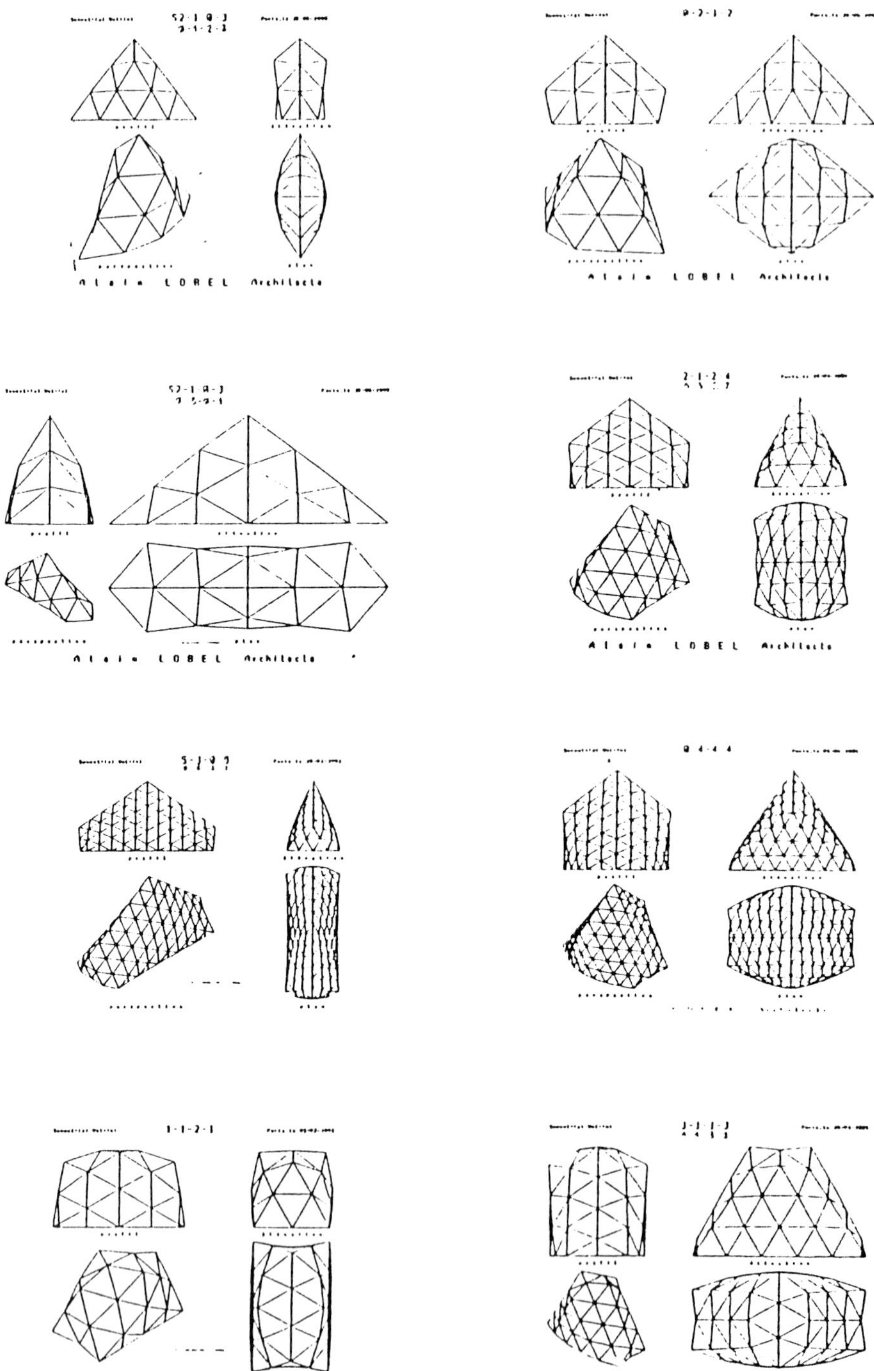

Fig. 5.

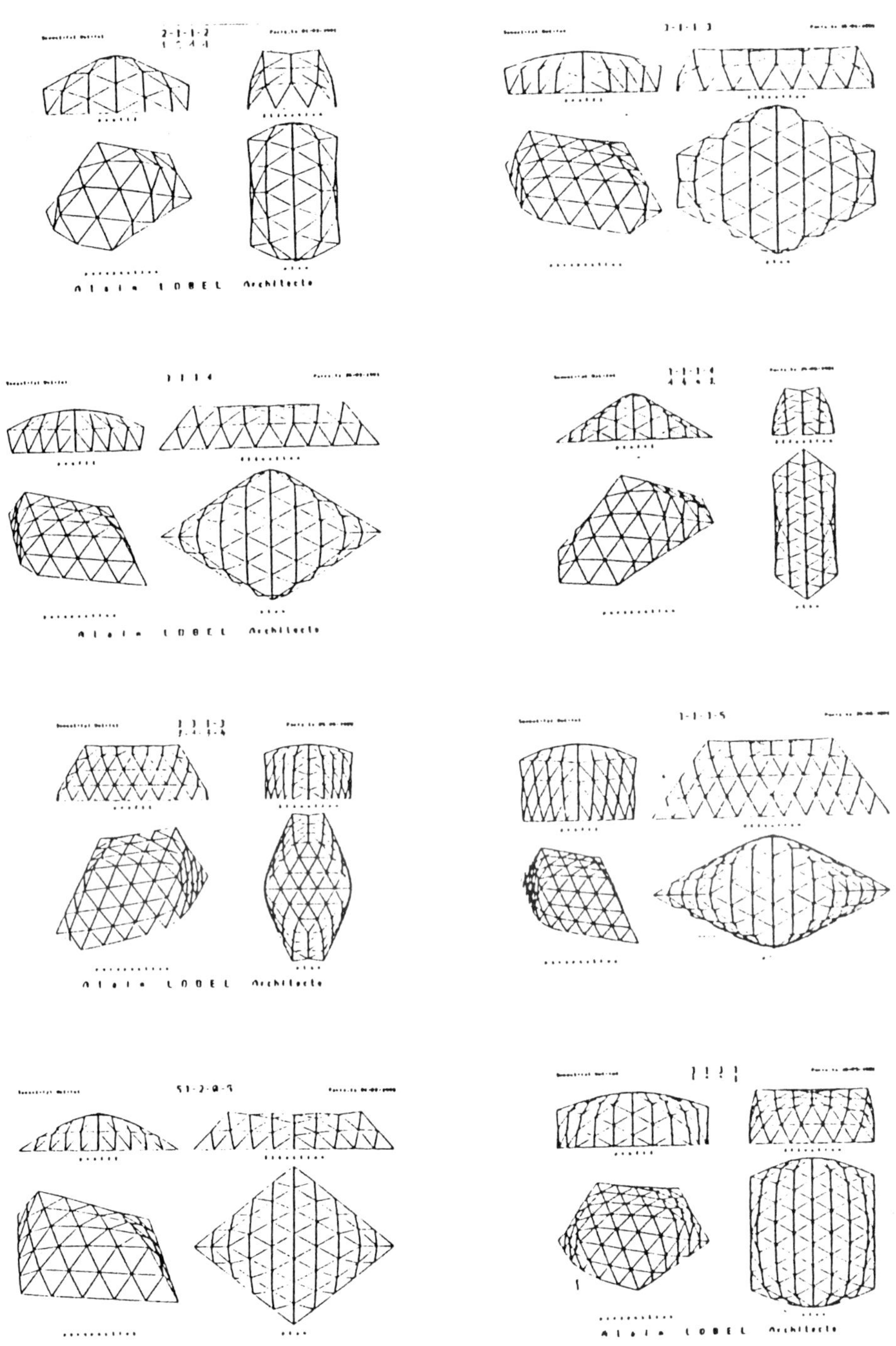

Fig. 5 continued.

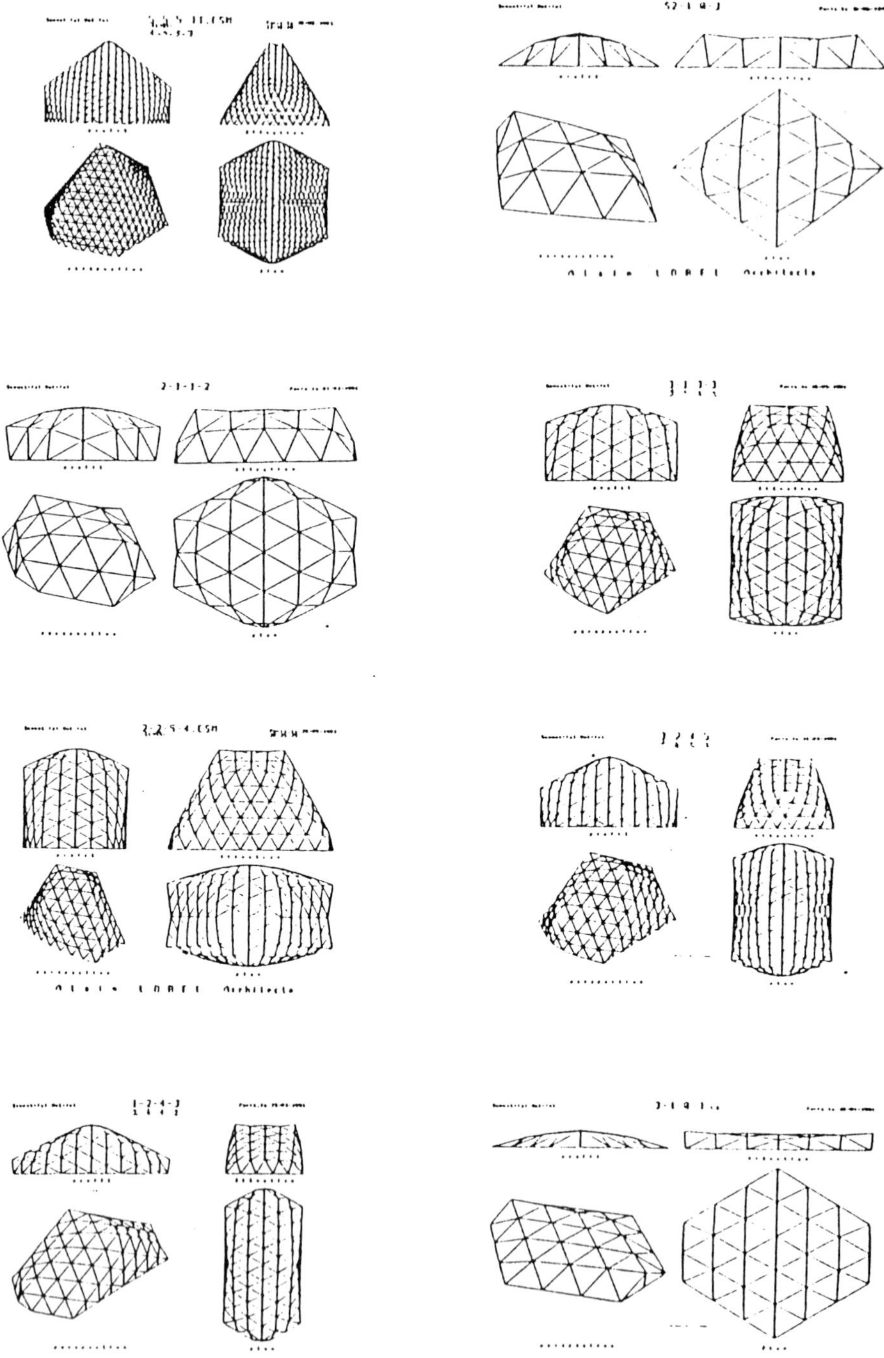

Fig. 5 continued.

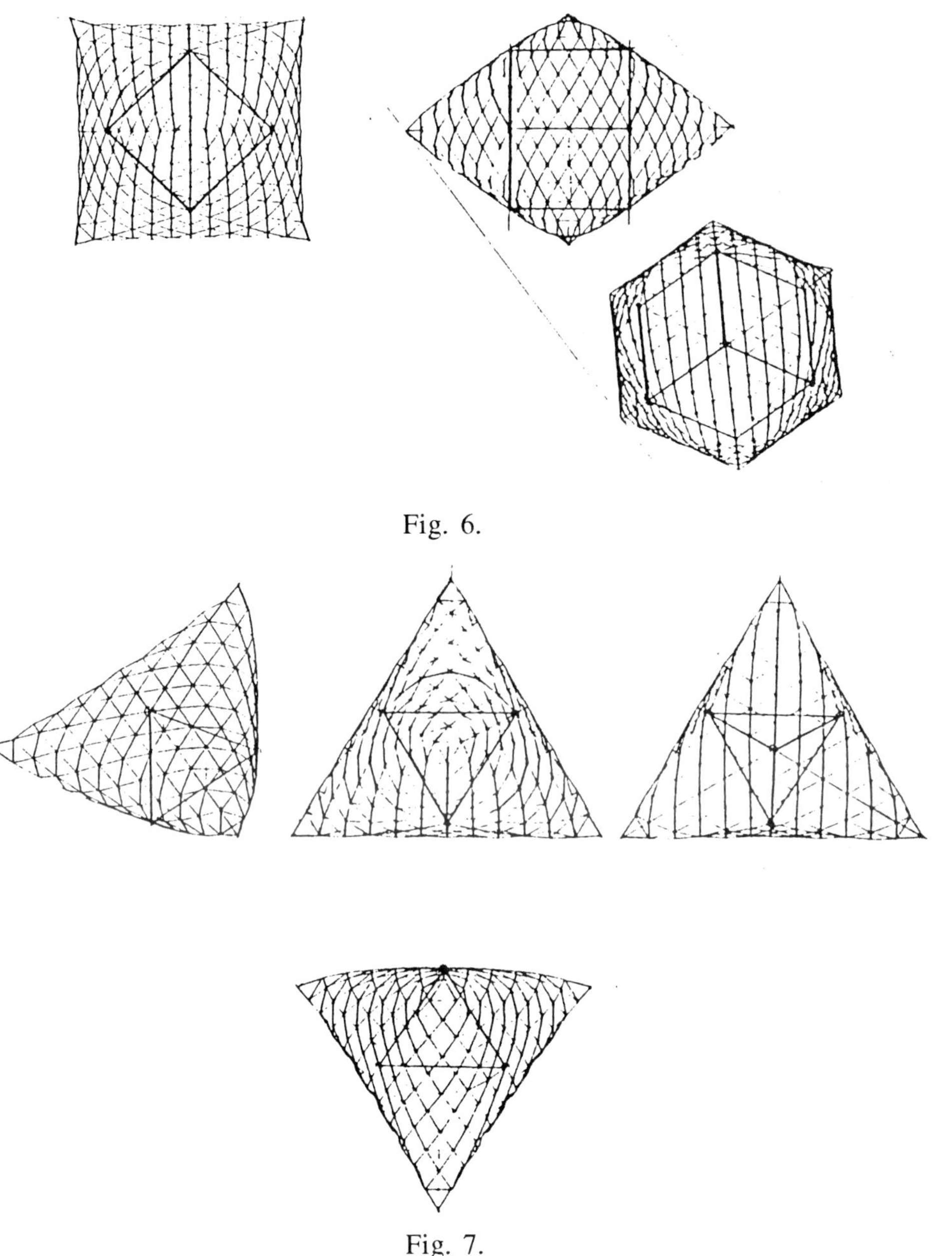

Fig. 6.

Fig. 7.

128. Computer-generated transformations of geodesic spheres

H. LALVANI, Pratt Institute, USA, and N. KATZ, Skidmore, Owings and Merrill, USA

SYNOPSIS. This paper presents computer-generated images of continuous transformations of geodesic surfaces within one 4-dimensional cubic framework. The morphologic concepts and the underlying hyper-scheme organisation follow earlier work (refs 1 and 2) where geodesic surfaces were characterised by symmetry, frequency of subdivision, and type of subdivision. For the purposes of demonstration, transformation within one family of icosahedral spheres with skewed subdivisions is illustrated in detail. The intermediate geodesic spheres produced in the transformation process, are published here for the first time.

1. Geodesic surfaces (spheres, cylinders, toroids, and planes in Euclidean and hyperbolic spaces) can be characterised by:

(i) symmetry P,Q, where P and Q are any positive integers representing the axes of symmetry (alternatively the Schlafli symbol {p,q} can be used for a structure composed of polygons {p} and vertex-figures {q});

(ii) the frequency of subdivision of the surface designated by fl, and f2, also a pair of positive integers; and

(iii) four types of subdivisions designated as p, q, r, and $\underline{r}$.

These make up a total of eight variables, which can be mapped in 8-dimensional space. The n-dimensional meta-morphological approach and the four types of subdivisions were first described in ref. 1, extend to the general symmetry P,Q and general frequency fl, f2 (ref. 2) and to n-dimensional structures.

2. This paper selects one example of a family of icosahedral spheres of symmetry (3,5), where P=3 and Q=5, and frequency (2,1), where fl =2 and f2= 1, and shows computer images of continuous transformations between the 16 types of skewed subdivisions within this family (for details of the 16 types of subdivisions, see refs 1 and 2). The continuous transformations between structures leads to many intermediate structures which have interesting geometries, thus broadening the repertory of geodesic structures. Each intermediate structure is a candidate "new" architectural space structure.

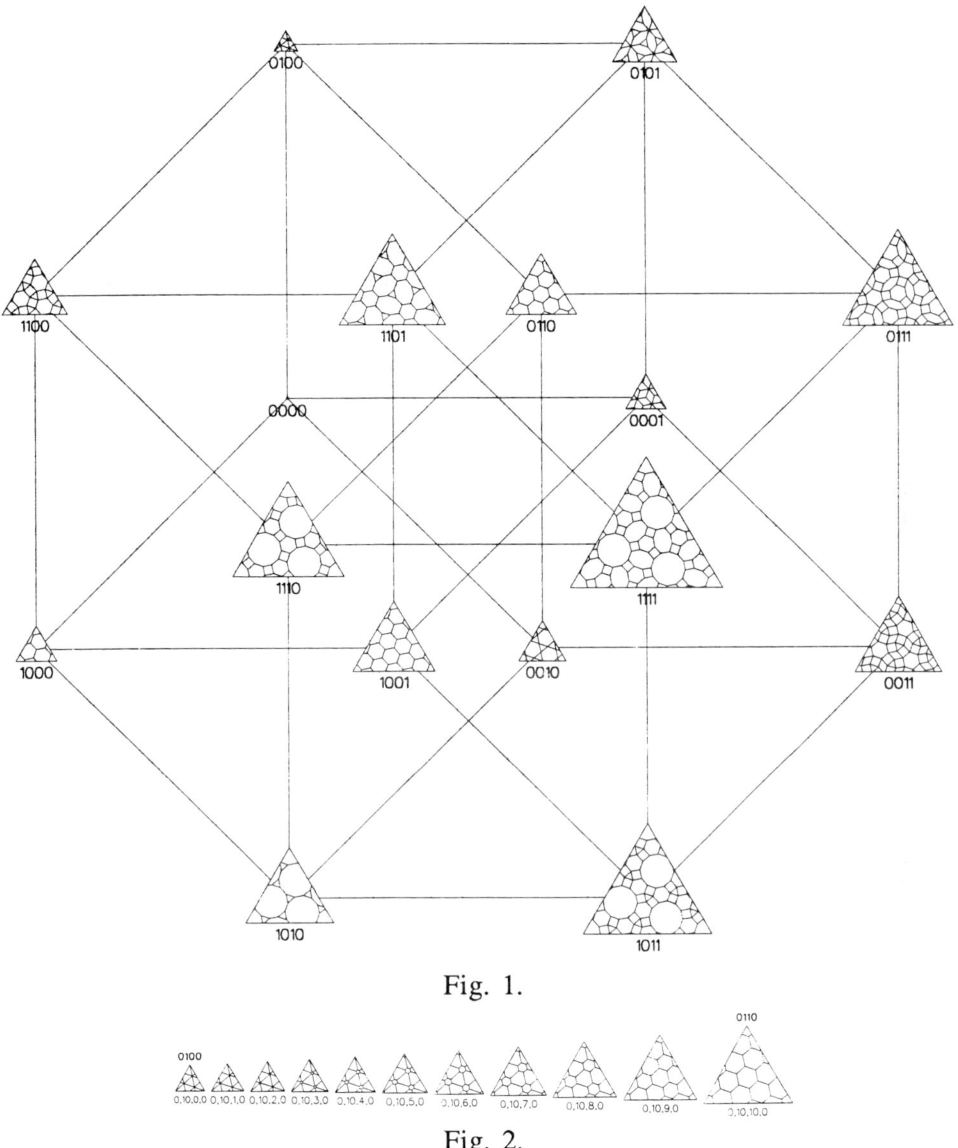

Fig. 1.

Fig. 2.

3. As before, the 16 subdivisions are mapped on the vertices of a 4-dimensional cube (Fig. 1) and are shown here with one Euclidean triangular face (P=3) which corresponds to the faces of geodesic structures (3,Q), where Q=3, 4, 5, or 6; the same subdivisions apply to hyperbolic or spherical (ellipsoidal) triangles. The 16 subdivisions are derived from combinations of four basic subdivisions, p (0100),

q (1000), r (0010), and r̲ (0001), and are binary-coded as before, but are shown here with equal-edged polygons. As a consequence of the equal-edge restriction, the sizes of the triangles vary in accordance with the algebraic laws of unions, intersections and complements (ref. 1).

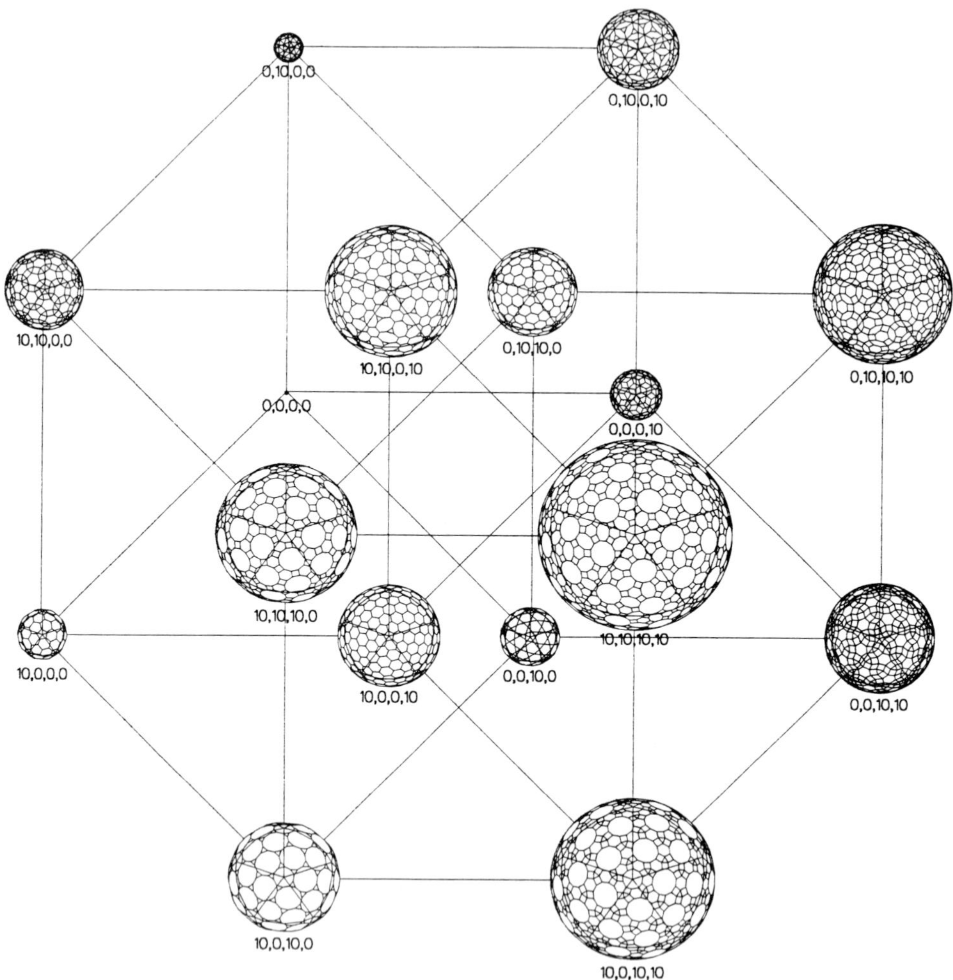

Fig. 3.

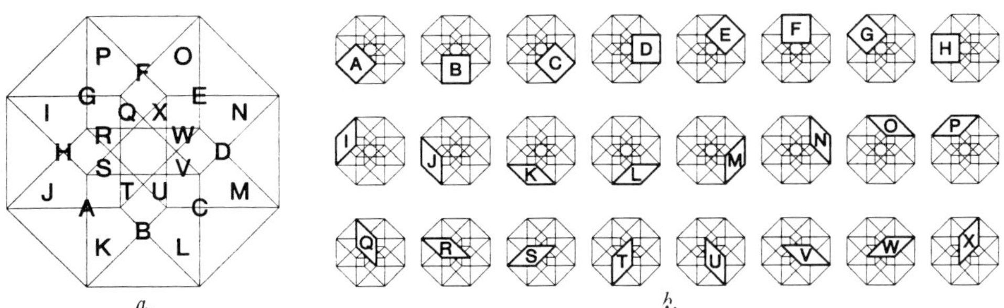

Fig. 4.

4. In Fig. 2, one example of a continuous transformation between the subdivision p (0100) and pr (0110) is shown in ten steps. Since the transformation is continuous, any number of steps or infinitely many steps can be introduced as intermediates. For the ten steps shown, the structures can be coded using 4-dimensional Cartesian co-ordinates, with the general intermediate having the co-ordinate (0,l0,x,0) where x is any integer between 0 and 10. In the illustration, the co-ordinates for all the intermediates are shown alongside.

5. In Fig. 3, all the 16 skewed subdivisions of triangles in Fig. 1 are used as faces of corresponding icosahedra and projected onto a sphere. The 16 spherical icosahedra are shown along their 5-fold views and their 4-dimensional co-ordinates are given assuming ten transformation steps between two adjacent spheres. The binary 1 is converted to the decimal 10. The remaining figures in this paper show intermediates lying on the edges and faces of the hyper-cube, and can be extended to include intermediates lying within the interior space of 3-cubic cells. Due to space restrictions, only one intermediate is shown along each edge and on each face. Note that each edge (of a total of 32 edges) of the hyper-cube is a l-dimensional transformation space, each face (of a total of 24 faces) of the hyper-cube is a 2-dimensional matrix of continuous transformations, and each cell (of a total of eight cells) is a 3-dimensional matrix of continuous transformations. Once again, due to space restrictions, only face matrices are shown.

6. In Fig. 4a, all 24 faces of the 4-cube are identified by the letters A, B, C, D,. ... X which are placed at the mid-points of each face. These 24 comprise three groups, each having eight faces as shown in Fig. 4b. The top row shows faces A through H, the middle row shows faces I through P, and the bottom row shows faces Q through X. The faces within each group are lettered sequentially in a counter-clockwise manner. Note that each face has a complementary face located at a complimentary location across the centre of the 4-cube. For example, faces A and E are complements, faces I and M are complements, and so on.

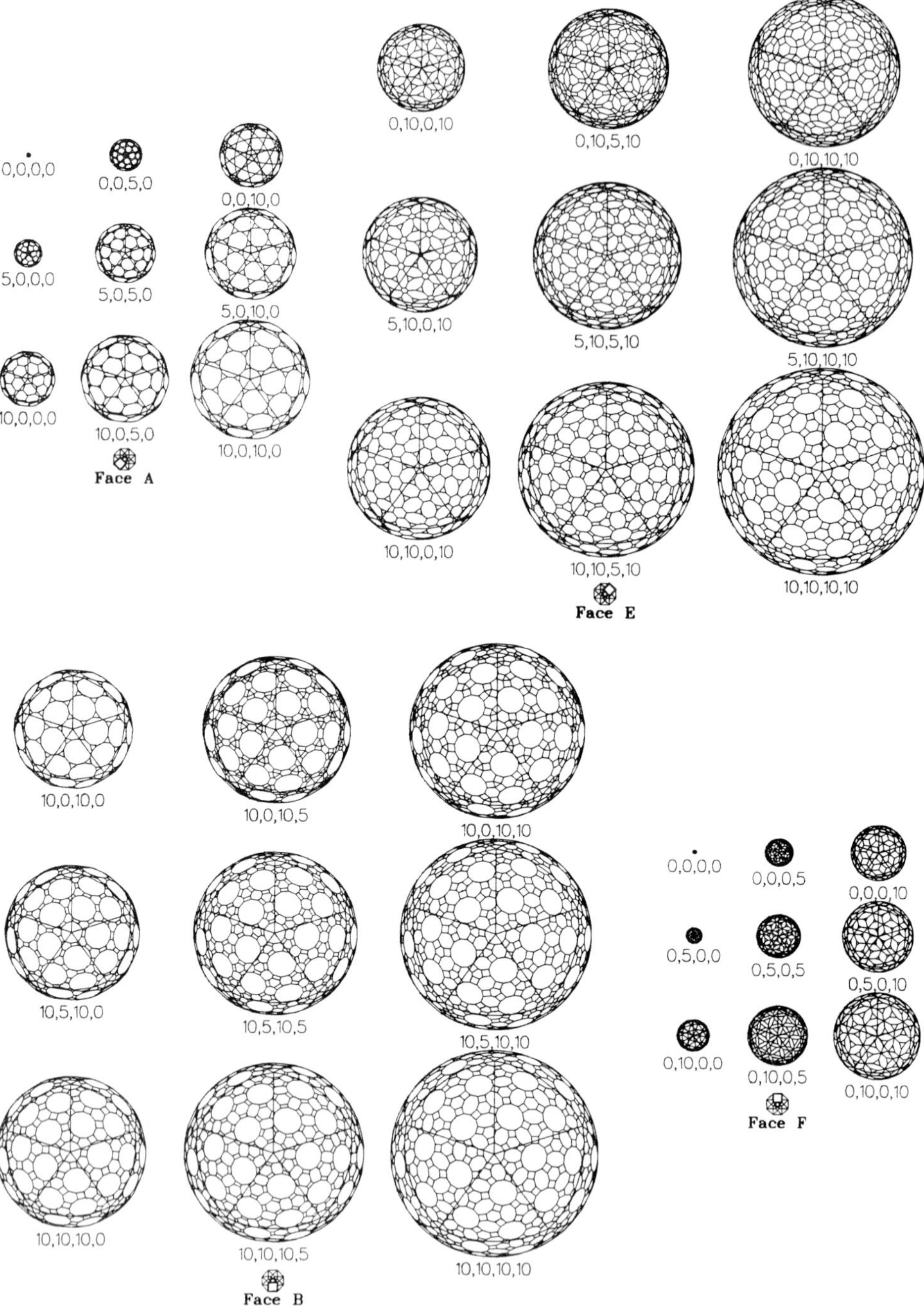
0,0,0,0
0,0,5,0
0,0,10,0
5,0,0,0
5,0,5,0
5,0,10,0
10,0,0,0
10,0,5,0
10,0,10,0
Face A
0,10,0,10
0,10,5,10
0,10,10,10
5,10,0,10
5,10,5,10
5,10,10,10
10,10,0,10
10,10,5,10
10,10,10,10
Face E
10,0,10,0
10,0,10,5
10,0,10,10
10,5,10,0
10,5,10,5
10,5,10,10
10,10,10,0
10,10,10,5
10,10,10,10
Face B
0,0,0,0
0,0,0,5
0,0,0,10
0,5,0,0
0,5,0,5
0,5,0,10
0,10,0,0
0,10,0,5
0,10,0,10
Face F

Fig. 5.

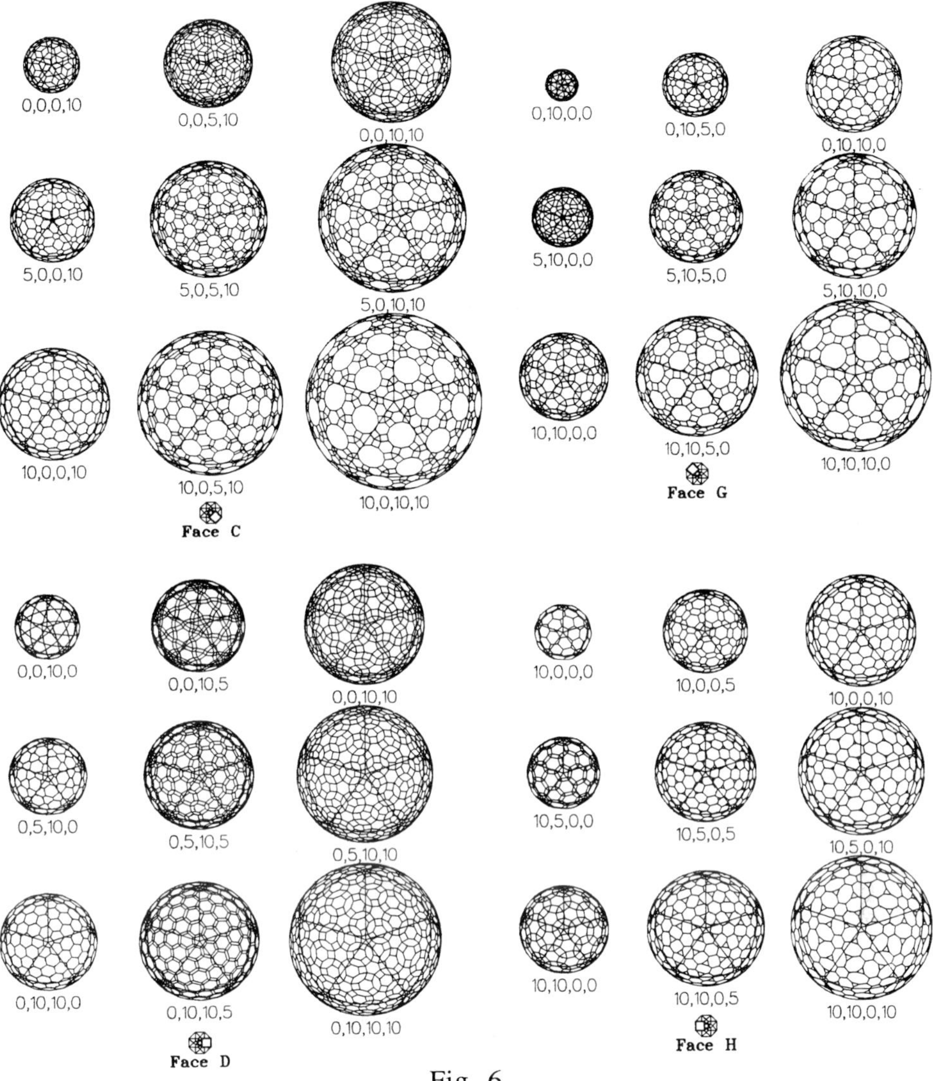

Fig. 6.

7. The remaining figures, Figs 5-10, show the 24 face matrices separately. Each matrix illustrates nine structures, four on the vertices, four on the mid-edges, and one lying on the face. For each individual matrix, a small key diagram identifies the face location within the hyper-cube. Figs 3 and 6 show the eight face matrices of the first group, Figs 7 and 8 show the face matrices of the second group, and Figs 9 and 10 show the eight matrices of the third group. In each figure, the complementary face matrices are shown adjacent to one another. For example, in Fig. 5 the matrix Face A (top left) is shown besides the complementary matrix Face E (top right). Each sphere is identified by its 4-dimensional co-ordinates

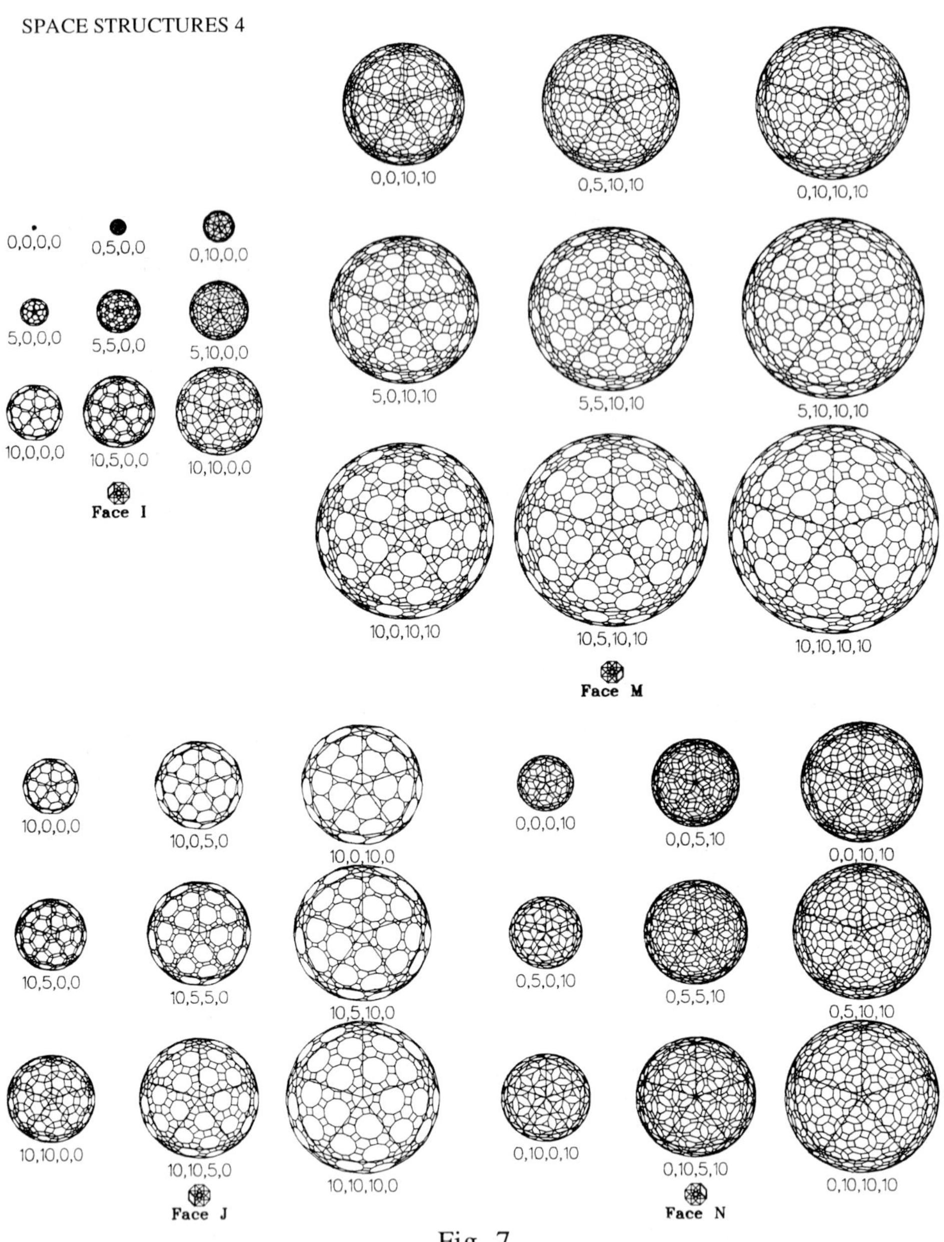

Fig. 7.

8. Note the following complementarities in Fig. 5: the co-ordinates (0,0,0,0) of top left position in Face A is the complement of (10,10,10,10) located at the bottom right of Face E; the mid-face locations (5,0,5,0) of Face A and (5,10,5,10) of Face E are complements; the top mid-edge (0,0,5,0) of Face A is a complement of the bottom mid-edge (10,10,5,10) of Face E, and vice versa; the

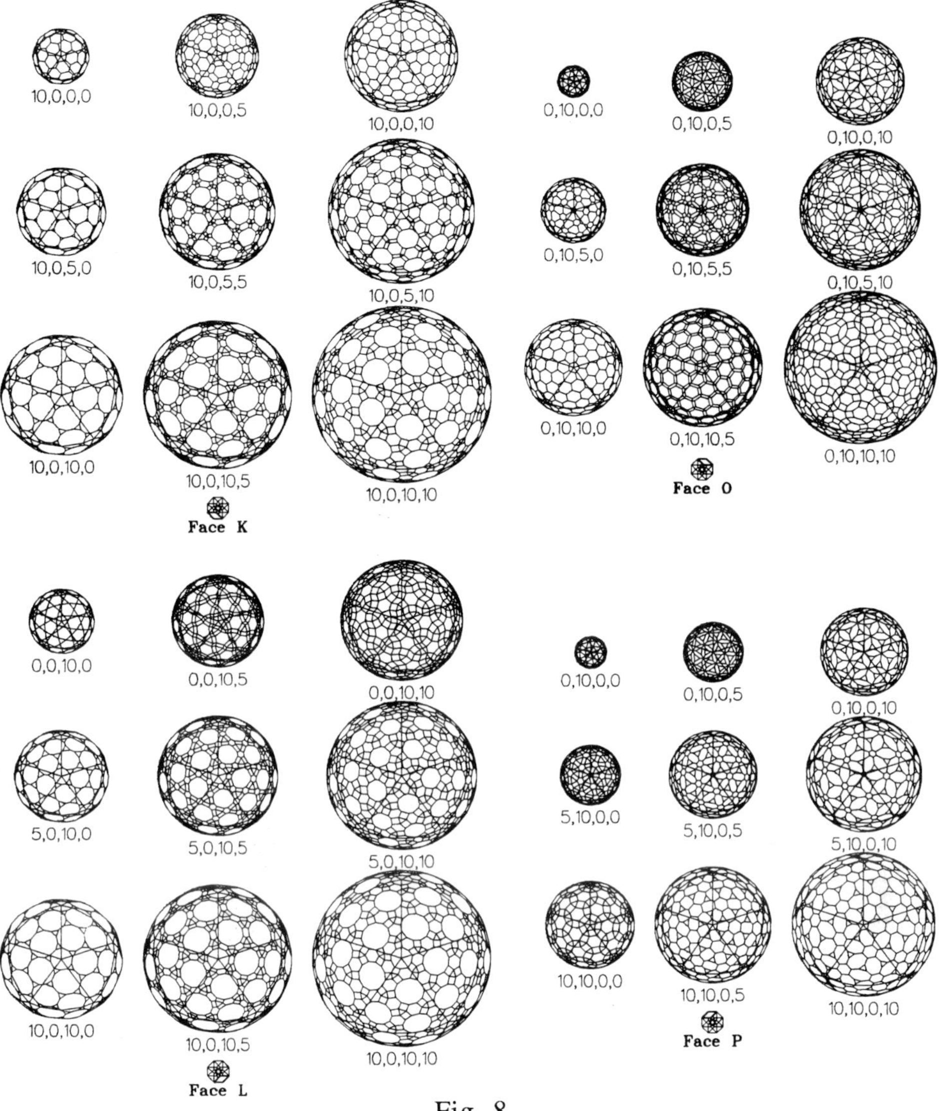

Fig. 8.

right mid-edge (5,0,10,0) of Face A is a complement of the left mid-edge (5,10,0,10) of Face E, and vice versa. These relationships hold throughout the 24 face matrices The relationships extend to arbitrary locations within the hypercube, for example, the sphere located at (3,7,9,5) is a complement of the sphere located at (7,3,1,5).

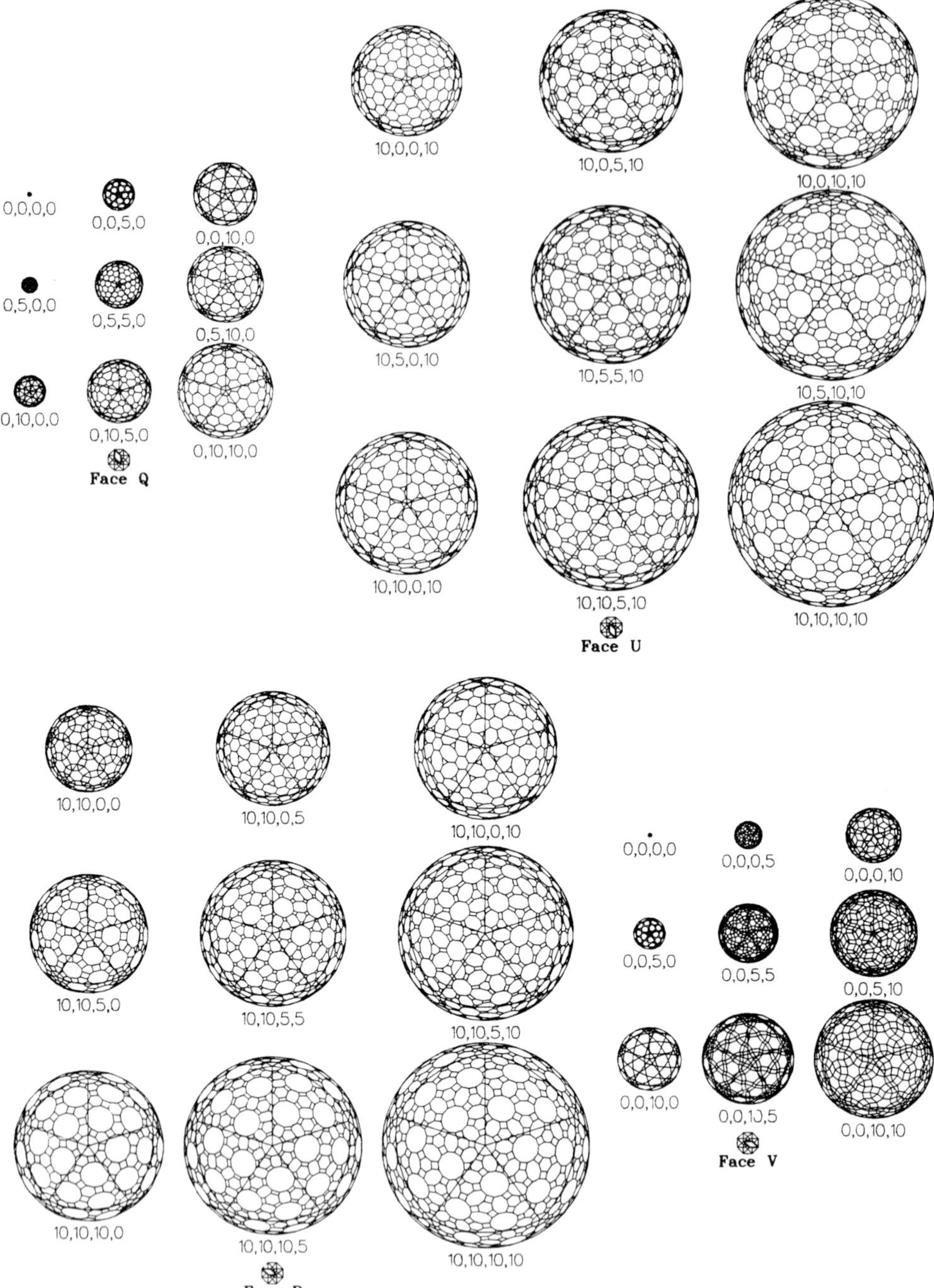

0,0,0,0
0,0,5,0
0,0,10,0
0,5,0,0
0,5,5,0
0,5,10,0
0,10,0,0
0,10,5,0
0,10,10,0
Face Q
10,0,0,10
10,0,5,10
10,0,10,10
10,5,0,10
10,5,5,10
10,5,10,10
10,10,0,10
10,10,5,10
10,10,10,10
Face U
10,10,0,0
10,10,0,5
10,10,0,10
10,10,5,0
10,10,5,5
10,10,5,10
10,10,10,0
10,10,10,5
10,10,10,10
Face R
0,0,0,0
0,0,0,5
0,0,0,10
0,0,5,0
0,0,5,5
0,0,5,10
0,0,10,0
0,0,10,5
0,0,10,10
Face V

Fig. 9.

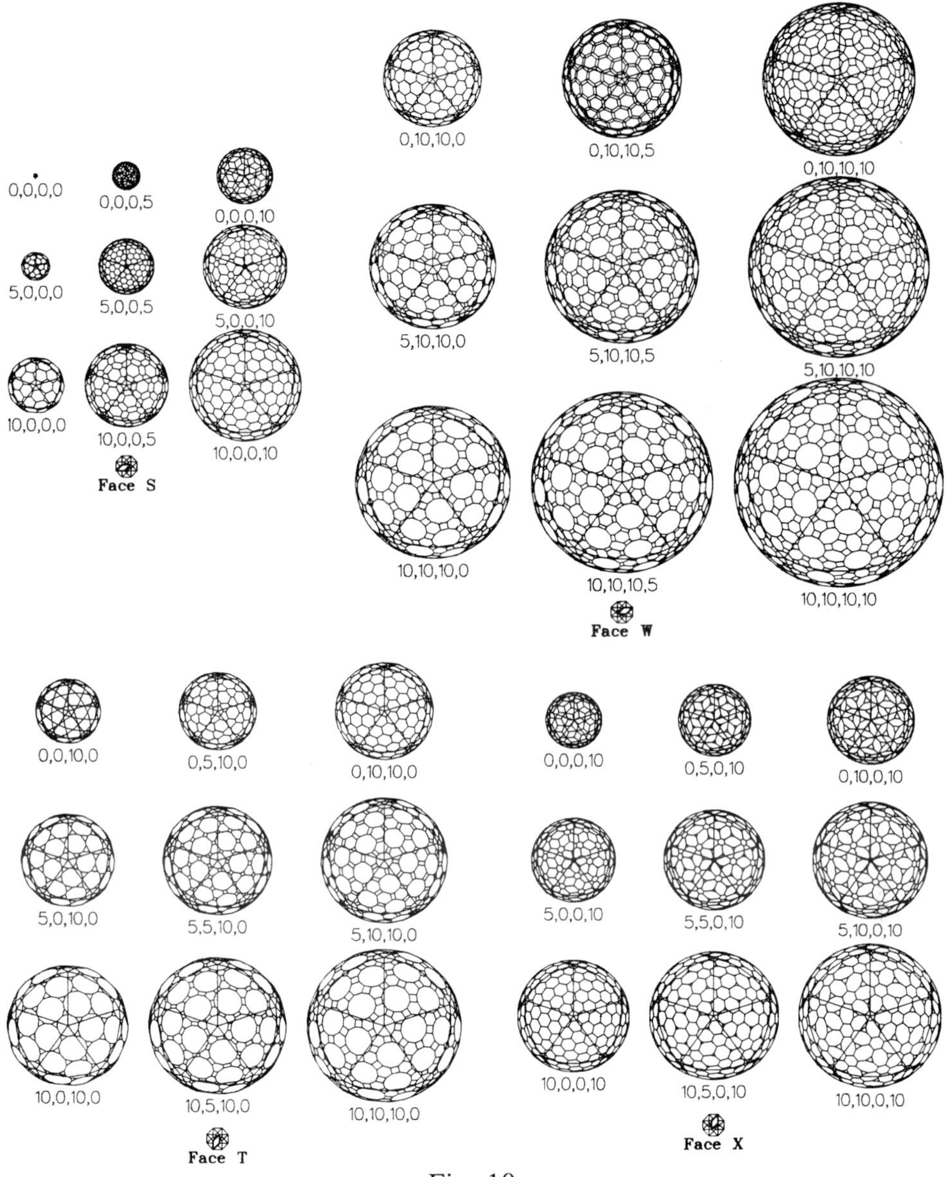

Fig. 10.

9. The readers are invited to scan through the catalogue of remaining images and discover for themselves the beauty of the order-underlying-order we term *meta-order*.

REMARKS: This project is part of our ongoing effort at establishing a unified morphologic framework for the *meta-morphology* of space structures. We have been able to extend this scheme to a large variety of space structures which include the present geodesic surfaces as a small portion of a much wider inclusive repertory of space structures, with each space structure occupying a specific location within an increasingly complex, open-ended hyper-cubic *meta-structural* space.

REFERENCES.

1. LALVANI, H. Structures on Hyper-Structures, Lalvani, New York, 1982; based on Multi-dimensional Periodic Arrangements of Transforming Space Structures, Ph.D. Thesis, Univ. of Pennsylvania (1981), published by University Microfilms, Ann Arbor, Michigan (1982).
2. LALVANI, H. Continuous transformations of subdivided periodic surfaces, space structures, Vol. 5, Nos. 3 and 4, 1990, pp. 255-279.

129. A modified method of automatic data generation

M. H. OSMAN, Malaysian University of Technology

SYNOPSIS. Many commercially developed software packages for structural analysis provide methods, in various forms, for the automatic generation of the input data. This paper presents a new scheme, which is a modification of two established methods. The concept of the new scheme is illustrated with the aid of some case studies.

INTRODUCTION

1. The stiffness method and the finite element method, require large amounts of data input for the analysis of large skeletal structures The principal data are those of the joint co-ordinates and the member incidences.

2. A large structure may requires hundreds or even thousands of rows of data input. To input the data line by line is obviously very tedious. Most of the commercial software packages for structural analysis provide methods, in their own styles, for the automatic generation of the data.

3. This paper presents a method which is in fact a combination of two established methods, *Formex* (ref. 2) and the automatic generation command menu found in *GTSTRUDL*, a structural analysis package (ref. 3). In principal, Formex is based on an algebraical formulation to generate data, whereas the data generation menu in GTSTRUDL is based on conversational commands. Both methods use prescribed abbreviated symbols. The method used in GTSTRUDL is the same as that used in the method of macro language input in the microSTRAN structural analysis software.

4. From experience using both methods in structural configuration processing, the author has found that for certain cases, each of them has advantages, as well as disadvantages, over the another. Therefore, the features in these two different methods are combined and modified to produce a new method. A number of program subroutines for the setting up and transformation of objects has been written, initially in BASIC language and then translated into FORTRAN 77.

THE MODIFIED METHOD OF AUTOMATIC GENERATION.

5. From the author's brief experience using both methods, the Formex method is most suitable when used for a single space structure which has a uniform modular unit, for example, a double-layer grid, a dome, or barrel vault roofing structure. The generation commands in GTSTRUDL cover all types of structural topology, but for multi-layer grids, lengthy commands are required. This led the author to develop a new modified method which combines the features of both methods.

6. The modified method uses abbreviated symbols in English terms, which resembles the generation commands in GTSTRUDL, rather than the ancient Greek terms used for algebraic formulation in Formex. For the initial stage, it deals only with a topologically uniform structural grid.

7. In this method, joints can be at any point in the plane or space, can be integer or real. Nodal numbers are not specified in the commands, the program will automatically give the nodal numbers, starting with number 1, and increased by 1. The numbering pattern depends on the way the commands are written, which is useful if bandwidth minimisation is needed by the structural analysis program.

8. Members or meshes are generated by specifying the end node co-ordinates of the elements. The numbering is not required to be specified in the commands, it will be given automatically by the program.

9. The structures are divided into three types of geometry, namely, Cartesian (two-dimension and three-dimensional), cylindrical, and spherical. Unless for the two-dimension cartesian, they have three co-ordinate directions in the space i.e. x, y and z for three-dimension Cartesian; r, ϕ and z for Cylindrical; and r, ϕ and θ for Spherical (see Fig. 1.). The type of the geometrical shape is specified right at the beginning of the joint co-ordinates and member co-ordinates generating commands. The general form is:

JOI COOR MEM COOR	TYPE	CAR CYL SPH

where CAR, CYL, and SPH are abbreviations of Cartesian, Cylindrical and Spherical respectively.

10. Joints or groups of joints and members or groups of members are called objects and are denoted by the letters OBJ 1, OBJ 2, .., and MEM 1, MEM 2, etc. If the numbering is required in alphabet names, for example, BEAM1, BEAM2, etc., for the member identification, the notation 'MEM' shall be replaced with MEMBEAM.

11. Joints can be generated by the command:

GEN n JOI @ (x_0, y_0, z_0) INC [X dx Y dy Z dz]

where, x_o, y_o, z_o are the co-ordinates of the first joint, dx, dy, dz are the co-ordinate increment for the generated joints (integer or real). At least one in the bracket [ ] to be specified.

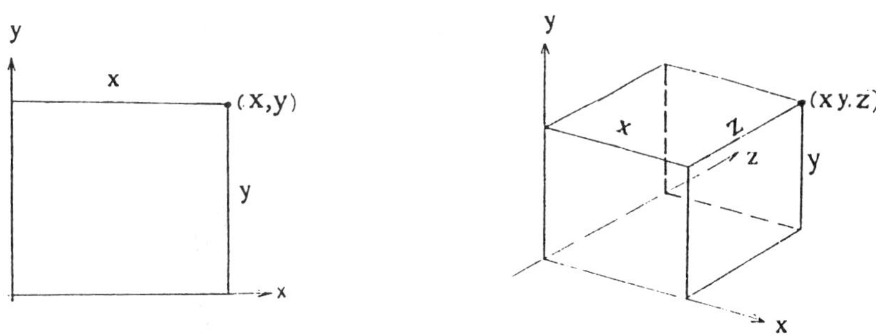

Two-dimension Cartesian. Three-dimension Cartesian.

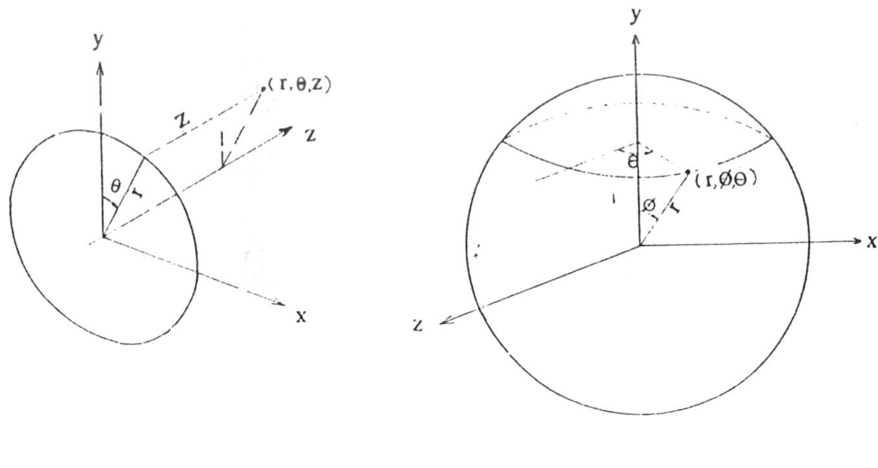

Cylindrical. Spherical.

Fig. 1. Types of co-ordinates systems.

12. For the member object, the command is:

GEN n MEM @ (x_{o1}, y_{o1}, z_{o1}; x_{o2}, y_{o2}, z_{o2})
INC [X dx Y dy Z dz]

where:

x_{o1}, y_{o1}, z_{o1} are the co-ordinates of end 1 of the first member,
x_{o2}, y_{o2}, z_{o2} are the co-ordinates of end 2 of the first member,
dx, dy, dz are the increments of the co-ordinates of both end 1 and end 2

13. Transformations of these objects are achieved by a number of commands. The most common is the Repeat command which has the general form of:

$$\text{REP} \quad \left| \begin{matrix} \text{JOI} & i \\ \text{MEM} & i \end{matrix} \right| \quad n \quad \text{TIM INC} \quad \begin{bmatrix} x & d'x \\ y & d'y \\ z & d'z \end{bmatrix}$$

where:

i is the name of the object
n is the number of repetitions

d'x, d'y and d'z are the increments (real or integer) between successive joints in the directions x, y, and z respectively.

14. Other commands which are quite rarely used are COMBine, COPY, ROTate, MOVE and REFlect, some of which will be used in the examples to explain their function. All of these commands work through different subroutines.

CASE STUDY

15. The application of the commands are explained here using examples of the generation commands for joint co-ordinates and member incidences . The '$' sign at the left end of any line signifies a remark, therefore words or characters to the right of the sign are ignored.

16. Case 1- Two-dimension Cartesian. These commands are for the generation of the plane grid interconnection pattern as shown in Fig. 2.

```
JOINT COOR TYPE CAR
JOI 1 , GEN 4 JOI @ (0,0) INC X 3.0
JOI 2 , REP JOI 1 4 TIM INC Y 3.0
JOI 3 , GEN 3 JOI @ (1.5,1.5) INC X 3.0
```

```
JOI 4 , REP JOI 2 3 TIM INC Y 3.0
MEM COOR TYPE CAR
MEM 1 , GEN 4 MEM @ (0,0;1.5,1.5)
                ROT ABOUT (1.5,1.5)
MEM 2 , REP MEM 1 2 TIM INC X 3.0
MEM 3 , REP MEM 2 3 TIM INC Y 3.0
MEM 4 , GEN 3 MEM @ (0,0;3,0) INC X 3.0
MEM 5 , REP MEM 4 1 TIM INC Y 12.0
MEM 6 , GEN 4 MEM @ (0,0;0,3.0) INC Y 3.0
MEM 7 , REP MEM 6 1 TIM INC X 9.0
MEM 8 , COMB MEM 3 5 7
$
MEM INC
MJ 1 MEM 8 ON JOI 4
LIST MEM INC
$ finish
```

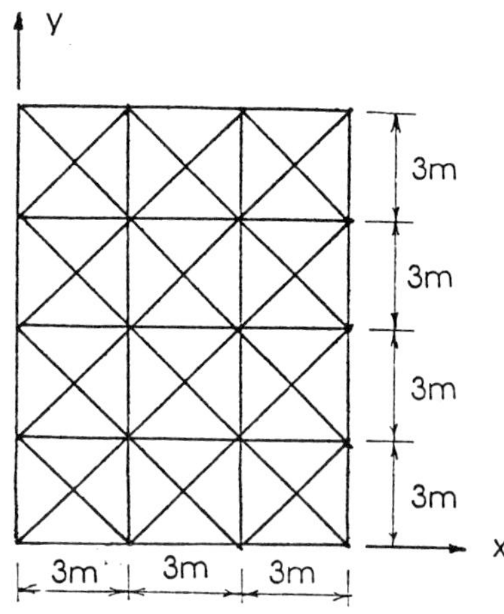

Fig. 2. Two-dimensional Cartesian.

Case 2 - Three-dimension Cartesian

17. The commands below are for the generation of joint co-ordinates and member incidences for a two-storey steel framed building which is 24m in length, 10m wide and 7m high, as shown in Fig. 3, ie an example of a three dimensional structure in cartesian co-ordinates.

```
JOI COOR TYPE CAR
JOI 1 , GEN 5 JOI @ (0,0,0) INC X 6.0
JOI 2, REP JOI 1 2 TIM INC Y 5.0
JOI 3 , REP JOI 2 2 TIM INC Z 3.5
LIST JOI COOR
$
MEM INC TYPE CAR
$ column
MEM 1 , GEN 5 MEM @ (0,0,0;0,0,3.5) INC X 6.0
MEM 2 , REP MEM 1 2 TIM INC Y 5.0
MEM 3 , REP MEM 2 1 TIM INC Z 3.5
$ beam
MEM 4 , GEN 4 MEM @ (0,0,3.5;6,0,3.5) INC X 6.0
MEM 5 , REP MEM 4 1 TIM INC Z 3.5
MEM 6 , REP MEM 5 2 TIM INC Y 5.0
MEM 7 , GEN 2 MEM @ (0,0,3.5;0,5,3.5) INC Y 5.0
MEM 8 , REP MEM 7 4 TIM INC X 6.0
MEM 9 , REP MEM 8 1 TIM INC Z 3.5
MEM 10 , COMB MEM 3 6 9
$
MEM INC
MJ 1 , MEM 10 ON JOI 3
$ see Fig.3
LIST MEM INC
```

18. The command LIST will output the listing of the joint co-ordinates or member incidences.

Case 3 - Cylindrical.

19. The single-layer barrel vault shown in Fig. 4 is an example of a cylindrical configuration. 'PZ' symbolises the ϕ co-ordinates, measured around the longitudinal axis of the cylinder i.e. z axis. The data generating commands are:

```
JOI COOR TYPE CYL
JOI 1 , GEN 13 JOI @ (10,0,0) INC Z 2.5
JOI 2 , REP JOI 1 8 TIM INC PZ 12.0
$
MEM COOR TYPE CYL
MEM 1 , GEN 12 MEM @ (10,0,0;10,0,2.5) INC Z 2.5
MEM 2 , REP MEM 1 8 TIM INC PZ 12.0
MEM 3 , GEN 8 MEM @ (10,0,0;10,12,0) INC PZ 12.0
MEM 4 , REP MEM 3 12 TIM INC Z 2.5
MEM 5 , GEN 4 MEM @ (10,0,2.5;10,12,0) INC PZ 12
MEM 6 , REF MEM 5 ABOUT Z 2.5
MEM 7 , REP MEM 6 5 TIM INC Z 5.0
MEM 8 , REF MEM 7 ABOUT PZ 48.0
MEM 9 , COMB MEM 2 4 8
$
MEM INC
MJ 1 , MEM 9 ON JOI 2
LIST MEM INC MJ 1
$ see Fig. 4
```

Case 4 - Spherical

20. These commands are to generate the data for the spherical configuration shown in Fig. 5. 'PY' symbolises the ϕ co-ordinates measured around the principal axis of the sphere i.e. y axis. 'TY' is meant for θ co-ordinates measured from the y axis.

```
JOI COOR TYPE SPH
JOI 1 COOR (30,0,0)
JOI 2 , GEN 9 JOI @ (30,18,0) INC PY 9.0
JOI 3 , REP J1 17 TIM INC TY 20
JOI 4 , COMB JOI 1 3
$
MEM COOR TYPE SPH
$ main member
MEM 1 , GEN 18 MEM @ (30,0,0;30,18,0) INC TY 20
$
MEM 2 , GEN 9 MEM @ (30,18,0;30,18,20) INC PY 9
MEM 3 , REP MEM 2 17 TIM INC TY 20
$
MEM 4 , GEN 8 MEM @ (30,18,0;30,36,0) INC PY 9
MEM 5 , REP MEM 4 7 TIM INC TY 20
MEM 6 , COMB MEM 1 3 5
```

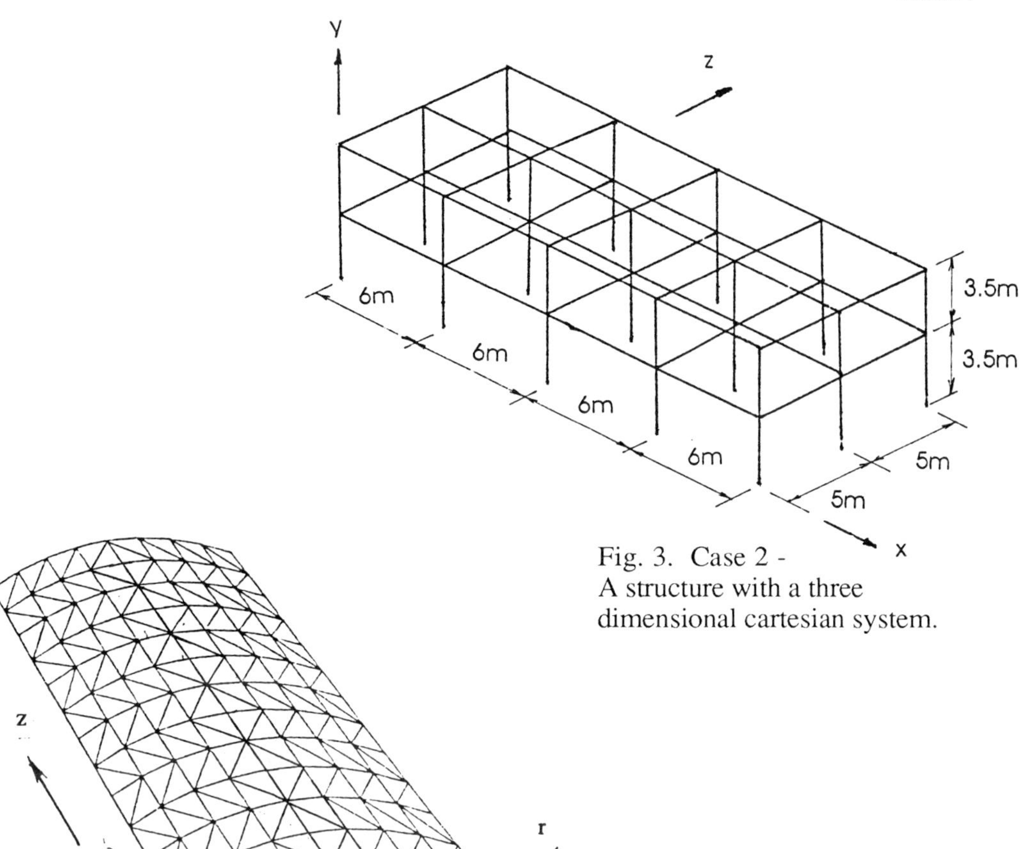

Fig. 3. Case 2 -
A structure with a three dimensional cartesian system.

Fig. 4. Case 3 -
A structure with a cylindrical co-ordinate system.

Fig. 5. Case 4 -
A dome structure with a spherical co-ordinate system.

```
$ bracing member
MEM 7 , GEN 4 MEM @ ((30,18,0;30,22.5,10) ROT ABOUT TY 22.5 PY 10
MEM 8 , REP MEM 7 7 TIM INC PY 9
MEM 9 , REP MEM 8 17 TIM INC TY 20
MEM 10 , COMB MEM 6 9
$
MEM INC
MJ 1 , MEM 10 ON JOI 3
$ see Fig. 5
```

Limitations

21. There are, of course, some limitations to this newly designed scheme. For uniform modular configurations, there are no problems, but in regard to structural configurations in general, few problems might occur. It cannot be applied to configurations with non-uniform modular units or to finite element meshes.

22. However, these limitations may be solved and even enhanced in the next stage, as it is the objective of the work to combine the features of other software into a single scheme.

CONCLUSION

23. The illustration of a new method of automatic data generation for cartesians, cylindrical and spherical structural configurations has been presented in this paper.

24. There are some limitations where subroutines for other features have yet to be worked out. However, the formulation for simple cartesian, cylindrical and spherical space structures with uniform modular units presented here, is expected to be the first step towards developing a new method that could be able to cover a large variety of geometrical and topological configurations of skeleton trusses and finite element meshes, including graphic facilities.

REFERENCES

1. NOOSHIN H., The Formex approach, Proceedings of the International Colloquium on Space Structures for Sports Buildings, Beijing, 1987.
2. NOOSHIN H., Formex Configuration Processing for Structural Engineering, Applied Science Publishers, London, 1985.
3. GTSTRUDL Users Manual Vols 1,2 and 3, GTICES System Laboratory, Revised 1985.
4. OSMAN M.H., Automatic Data Generation for Structural Analysis PreProcessing, Jurnal Kejuruteraan Awam UTM, Julai,1989.
5. SETHARES G.C., Bent R.J, Basic:An Introduction to Computer Programming, Brooks Publishing Co, 1985.
6. WEINMEN D.G., Fortran for Scientist and Engineers, PWS KENT Publishers, Boston, 1989.
7. ATSStructE User Manuals, ATS Software Pte Ltd, Singapore,1987.
8. MicroSTRAN Users Manual V4, ES Pty Ltd, Turramurra, NSW, Australia, 1990.

130. A natural generator of optimum topology of a space frame for specified fundamental frequency

MAKOTO OHSAKI and TSUNEYOSHI NAKAMURA, Kyoto University, Japan

SYNOPSIS. A computational method is presented for finding the optimal topology, as well as the optimal cross-sectional areas, of a space frame for specified fundamental frequency. The nodal location and the members which may exist are to be specified or tentatively assumed in advance. A sequence of optimum design problems is defined in terms of an artificial parameter for the lower bounds of the cross-sectional area of members. Then the sequence of optimal solutions is traced with respect to the parameter to find the solution with vanishing lower bounds. It is demonstrated in the examples that the frame with optimal topology is always stable, which is contrary to the case of pin-jointed trusses.

INTRODUCTION

1. Various computational methods have been presented for topology optimisation of trusses and frames (refs 1-5) such that the necessary members are selected from a prescribed set of candidate members. In most of the papers, stress and displacement constraints are considered. No efficient method has however been developed for this problem due to the possible instability and singularity of the solution (refs 3 and 5). Lin *et al.* (ref. 6) proposed a method for finding an optimal configuration for specified fundamental frequency. In that method, however, the nodal locations are taken as the design variables and topology has not been optimised.

2. Nakamura and Ohsaki (ref. 7) have presented a practical and powerful method for generating a sequence of optimal trusses parametrically with respect to fundamental frequency. Shin *et al.* (ref. 8) developed a similar method, independently, based on the theory of homotopy, and applied it to an optimisation problem with buckling constraints. Nakamura and Ohsaki (ref. 9) extended their method so as to generate the optimal topology of a truss for specified fundamental frequency. In their method, an optimisation problem is first formulated with a sequence of lower bounds for cross-sectional areas. A sequence of optimum designs is then considered as a function of the parameter which defines a common

level of the lower bound. Among the optimum designs traced with respect to the parameter, the optimum design for vanishing lower bound is found.

3. In this paper, the authors' method is extended further to find the optimal topology of space frames. It is shown that difficulty of approaching an unstable frame no longer arises in the process of decreasing the lower bound, cross-sectional area to zero, in order to generate the optimal topology.

Fig. 1. Local co-ordinate system.

TOPOLOGY OPTIMISATION PROBLEM OF SPACE FRAMES

4. Optimal topology is to be found for a frame with the candidate or probable joint locations and members which are to be specified in advance. The problem considered here is to find the optimal location of the joints and members as well as the optimal cross-sectional areas that minimise the total volume. The local co-ordinate system is defined as shown in Fig. 1. The lateral displacements in the y and z directions are approximated using cubic functions of x. The longitudinal displacement and the angle of twist are linear functions of x. The cross-sectional area of the *i*th member is denoted by A_i. The geometrical moment of inertia $I_i^y(A_i)$ and $I_i^z(A_i)$, respectively, with respect to y and z axis are regarded as functions of A_i only. The polar moment of inertia $I_i^x(A_i)$ is also regarded as a function of A_i

5. Let $\mathbf{K}(\mathbf{A})$ and $\mathbf{M}_D(\mathbf{A})$ denote the stiffness matrix and the mass matrix due to the member masses with respect to a global co-ordinate system. The mass matrix due to the non structural mass is denoted by $\mathbf{M}_J$. From the equation of motion of free vibration, the following equation is derived:

$$\boldsymbol{K}(\boldsymbol{A})\boldsymbol{\Phi}_r(\boldsymbol{A}) = \Omega_r(\boldsymbol{A})\{\boldsymbol{M}_D(\boldsymbol{A}) + \boldsymbol{M}_J\}\boldsymbol{\Phi}_r(\boldsymbol{A}) \qquad (r = 1, 2, \ldots, f) \qquad (1)$$

where $\Omega_r(\mathbf{A})$ and $\Phi_r(\mathbf{A})$ are *r*th eigenvalue and eigenvector, respectively, and *f* is the degree of freedom of nodal displacements. The *r*th eigenvector is normalised with respect to the mass matrix $\mathbf{M}_D(\mathbf{A}) + \mathbf{M}_J$.

6. The number of members and the length of the *i*th member are denoted by m and L_i, respectively. Then the Topology Optimisation problem of Frames (TOPF) to minimise the total volume $V(\mathbf{A})$ for the specified fundamental eigenvalue Ω_a is formulated as follows:

$$\text{[Problem TOPF]} \quad \text{Minimize} \quad V(\boldsymbol{A}) = \sum_{i=1}^{m} (A_i L_i) \tag{2}$$

$$\text{subject to} \quad \Omega_r(\boldsymbol{A}) \geq \Omega_a \quad (r = 1, 2, \ldots, f) \tag{3}$$

$$A_i \geq 0 \quad (i = 1, 2, \ldots, m) \tag{4}$$

In the optimal solution of TOPF, the cross-sectional areas of some members turn out to be equal to 0. The members with $A_i = 0$ may be removed and the frame with optimal topology may be found.

A SEQUENCE OF OPTIMAL SPACE FRAMES TOWARD OPTIMAL TOPOLOGY

7. There exist serious difficulties for solving TOPF using a mathematical programming approach or an optimality criteria method, because a lower bound for A_i is equal to 0 and the principal parameters such as f and m change if the set of the members with $A_i = 0$ and/or the number of joints change in the iterative process. In the following, the authors' method of finding the set of optimal solution parametrically with respect to a lower bound parameter of cross-sectional areas is briefly summarised.

8. An artificial lower bound $\overline{A}_i$ is introduced here for $\{A_i\}$. The vector $\overline{\mathbf{A}} = \{\overline{A}_i\}$ is defined using a scalar parameter ξ as $\overline{\mathbf{A}} = \xi \overline{\mathbf{A}}^b$ where $\overline{\mathbf{A}}^b$ is a prescribed vector. Then an Optimum Design Problem of Frames (ODPF) is formulated as follows:

$$\text{[Problem ODPF]} \quad \text{Minimize} \quad V(\boldsymbol{A}) = \sum_{i=1}^{m} (A_i L_i) \tag{5}$$

$$\text{subject to} \quad \Omega_r(\boldsymbol{A}) \geq \Omega_a \quad (r = 1, 2, \ldots, f) \tag{6}$$

$$A_i \geq \xi \bar{A}_i^b \quad (i = 1, 2, \ldots, m) \tag{7}$$

The solution of ODPF exists for each specified value of ξ. Therefore the optimal solution is considered to be a function of ξ (ref. 9). The optimality conditions for ODPF for the case of s multiplicity of the fundamental eigenvalue follows:

$$\frac{\partial V(\boldsymbol{A})}{\partial A_i} - \sum_{r=1}^{s} \left(\mu_r \frac{\partial \Omega_r(\boldsymbol{A})}{\partial A_i} \right) - \nu_i = 0 \quad (i = 1, 2, \ldots, m) \tag{8}$$

$$\nu_i(\bar{A}_i - A_i) = 0, \quad \nu_i \geq 0 \quad (i = 1, 2, \ldots, m) \tag{9a, b}$$

$$\mu_r\{\Omega_a - \Omega_r(\boldsymbol{A})\} = 0, \quad \mu_r \geq 0 \quad (r = 1, 2, \ldots, s) \tag{10a, b}$$

where μ_r and υ_i are Lagrange multipliers.

9. Consider a case where $I_i^y(A_i)$, $I_i^z(A_i)$ and $I_i^x(A_i)$ are linear functions of A_i. Then the conditions given in Eqs 8, 9a, 9b 10a and 10b are the necessary and sufficient condition for optimality (refs 10 and 11) and the solution of ODPF with $\overline{\mathbf{A}} = \mathbf{0}$ turns out to be the solution of TOPF.

10. The solution vector of ODPF considered to be a function of ξ is denoted by $\hat{\mathbf{A}}(\xi)$. Then the optimal solution with $\overline{\mathbf{A}} = \mathbf{0}$, i.e., $\xi = 0$ may be found using the Taylor series expansion method developed in ref. 9. Eqs 1, 8, 9, and 10 are differentiated with respect to ξ up to the desired order. The procedure is started with a trivial initial optimal solution $\hat{\mathbf{A}} = \xi^u \overline{\mathbf{A}}^b$. The initial value ξ^u for ξ is to be assigned such that $\Omega_1(\xi^u \overline{\mathbf{A}}^b) = \Omega_a$. Once such an optimal solution is available, the derivatives of all the functions ξ characterising the optimal solutions of ODPF which are evaluated at each initial optimal solution are found by solving a set of simultaneous linear equations. Then the optimal solution of ODPF with $\xi = \xi^s$ which is slightly smaller than ξ^u may be expressed in terms of the following Taylor series:

$$\hat{\boldsymbol{A}}(\xi^s) = \hat{\boldsymbol{A}}(\xi^u) + \hat{\boldsymbol{A}}_i'(\xi^u)(\xi^s - \xi^u) + \frac{1}{2}\hat{\boldsymbol{A}}_i''(\xi^u)(\xi^s - \xi^u)^2 + \cdots \tag{11}$$

where a prime denotes differentiation with respect to ξ. Series expansion is consecutively repeated whenever the number of active equality constraints in Eq. 6 and/or Eq. 7 changes. The parameter ξ is decreased from ξ^u to 0 to find the solution of TOPF which defines the optimal location of the members as well as the optimal cross-sectional areas.

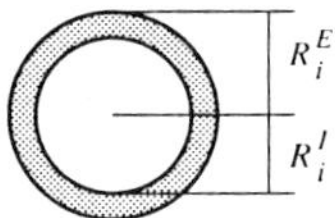

Fig. 2. Cross-section of the *i*th member.

11. It has been discussed in ref. 9 that the original procedure for a pin-jointed truss turns out to generate a truss of optimal topology which has the fundamental frequency with a great number of multiplicity. In ref. 9, the solution with $\xi = 0$ is extrapolated at a small value of $\xi = \xi^m$ using the differential coefficients of $\hat{\mathbf{A}}$ with respect to ξ. The truss of the solution to TOPF has the optimal topology if it is stable. Otherwise the truss should be stabilised in the process of $\xi \rightarrow 0$ even if the truss turns out to have an approximate optimal topology. It is demonstrated in the following examples that no such difficulty arises for rigidly-jointed space frames.

EXAMPLES

12. Optimal topologies are found for a space frame which consists of tubular members with hollow circular cross-sections as shown in Fig. 2. A_i, I_i^y, I_i^z and I_i^x may be written in terms of R_i^E and R_i^I defined in Fig. 2 as:

$$A_i = \pi \left\{ (R_i^E)^2 - (R_i^I)^2 \right\}, \qquad I_i^y = \frac{\pi}{4} \left\{ (R_i^E)^4 - (R_i^I)^4 \right\},$$

$$I_i^z = I_i^y, \qquad I_i^x = \frac{\pi}{2} \left\{ (R_i^E)^4 - (R_i^I)^4 \right\} \qquad (12a \sim d)$$

Consider a case where the ratio $c_i = I_i^y / A_i$ is fixed throughout the process of decreasing ξ to 0. Then $I_i^y, I_i^z, I_i^x, R_i^E$ and R_i^I are functions of A_i as:

$$I_i^y = c_i A_i, \qquad I_i^z = c_i A_i, \qquad I_i^x = 2c_i A_i,$$

$$R_i^E = \sqrt{\frac{1}{2}\left(4c_i + \frac{A_i}{\pi}\right)}, \qquad R_i^I = \sqrt{\frac{1}{2}\left(4c_i - \frac{A_i}{\pi}\right)} \qquad (13a \sim e)$$

It should be noted that I_i^y, I_i^z and I_i^x are linear functions of A_i and the optimality conditions derived in the foregoing section are necessary and sufficient conditions.

13. In the following examples, the elastic modulus and the Poisson's ratio are205.8GPa and 0.3, respectively. For all the members, c_i is equal to 24.5 cm^2.

<u>An axisymmetric space frame.</u>

14. Optimal topologies are first found for a space frame as shown in Fig. 3, where one of the eight equal parts is illustrated. The frame is supported rigidly at nodes 2 and 6. Since responses due to vertical excitations are critical for a space frame of this type, design requirements should be made for the lowest eigenvalue corresponding to the symmetric modes which are excited by a vertical motion. Therefore, condition of symmetry is assumed for the nodes in the planes 1-2-13-14 and 5-6-13-14. Each node lies either on a circular cylinder with a 2.0m, 4.0m or 6.0m in radius. The sets of nodes 1~5~13 and 2~6~14 are on a spherical surface, each of which has the radius 12m. Consider Case 1 where a non structural mass of 2000kg is located at node 13. The specified fundamental eigenvalue is 1576.03 rad^2/sec^2. The frame with optimal topology is illustrated in Fig. 4. Note that the members in the plane 5-6-13-14 are the same as those in the plane 1-2-13-14. A kind of dome has emerged and the inertia force due to the vertical motion of the non-structural mass at the node 13 is transmitted to the supports through the dome. It is interesting to note that two rings a and b should exist to support the axial force of diagonals c and d. The fundamental eigenvalue of this frame is 1575.29 rad^2/sec^2. Note that the truncation error due to Taylor series expansion has been made sufficiently small. For Case 2 with a non-structural mass at 14, the optimal topology is as shown in Fig. 5, where the

specified fundamental eigenvalue is 679.096 rad^2/sec^2. Here a dome-like frame partially double-layered in the central portion is enough for transmitting the inertia forces distributed in the lowest members.

15. Consider Case 3 where non-structural masses of 2000kg are located at all the lowest interior nodes in 2-6-14 surface. The masses at the nodes along the lines 2-4-6 are 1000kg. The specified fundamental eigenvalue is 954.423 rad^2/sec^2. The optimal topology is illustrated in Fig. 6. In this case, the second and third eigenvalues turned out to be equal to Ω_a and the three lowest eigenvalues of the frame shown in Fig. 6. are 954.331 rad^2/sec^2, 954.385 rad^2/sec^2 and 954.398 rad^2/sec^2.

A double-layer grid with rigid joints.

16. An optimal topology is found for a double-layer grid as shown in Fig. 7. The non structural mass at node 1 is 2000kg. The inclination angle of a diagonal member from the *xy* plane is 45°. The optimal topology is as shown in Fig. 8, where Ω_a is 1447.31 rad^2/sec^2. The inertia force due to the mass at node 1 is transmitted by the diagonals which are supported by the squares 2-3-4-5 and 6-7-8-9. The squares are finally supported by four parallelograms 10-11-12-6, 13-14-9-15, 18-7-16-17 and 8-21-19-20.

CONCLUSIONS

17. A practical and efficient method has been proposed for finding the optimal topology of a space frame for specified fundamental frequency. With the candidate nodal location and members, a sequence of optimum design problem is first formulated for a sequence of lower bounds of cross-sectional areas.

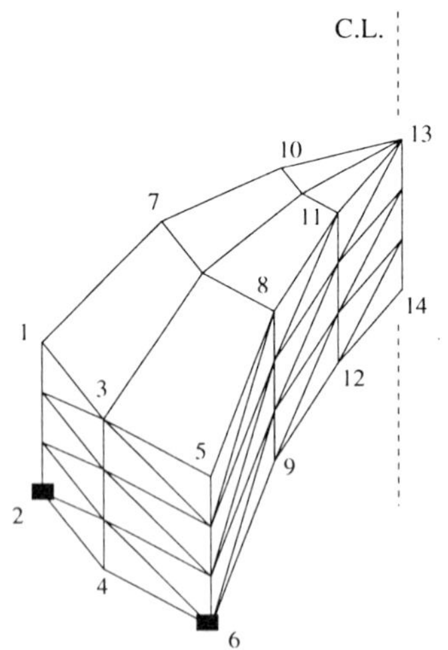

Fig. 3. An axisymmetric space frame.

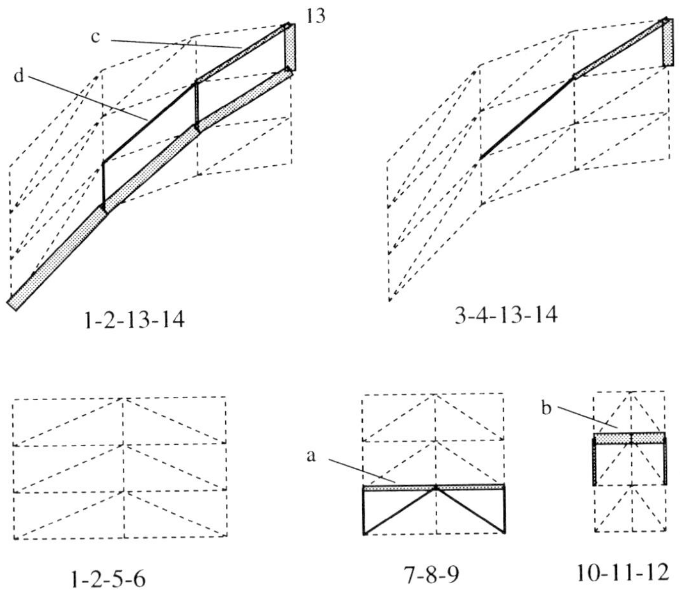

Fig. 4. Optimal topology of an axisymmetric space frame (Case 1).

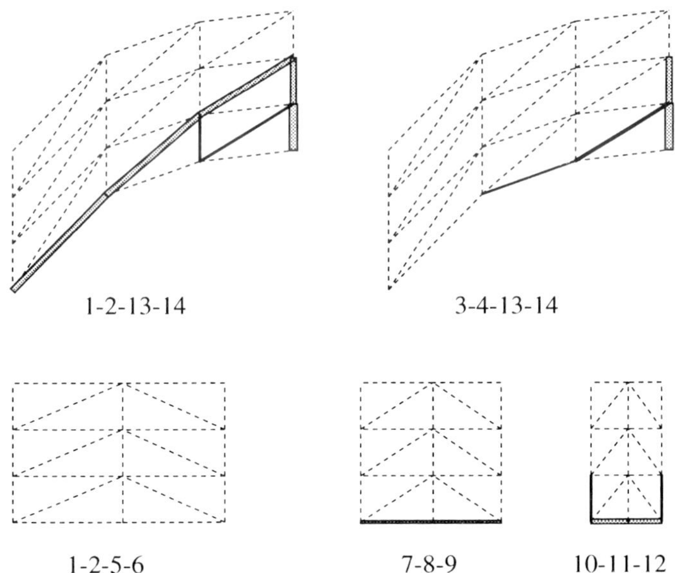

Fig. 5. Optimal topology of all axisymmetric space frame (Case 2).

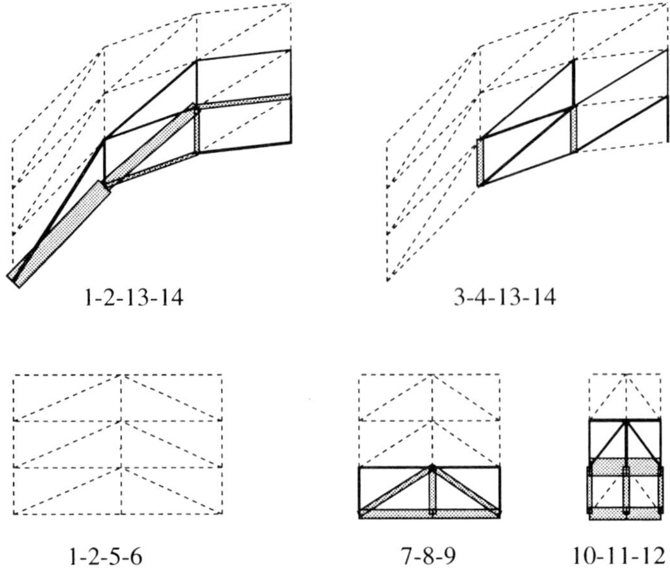

Fig. 6. Optimal topology of aIl axisymmetric space frame (Case 3).

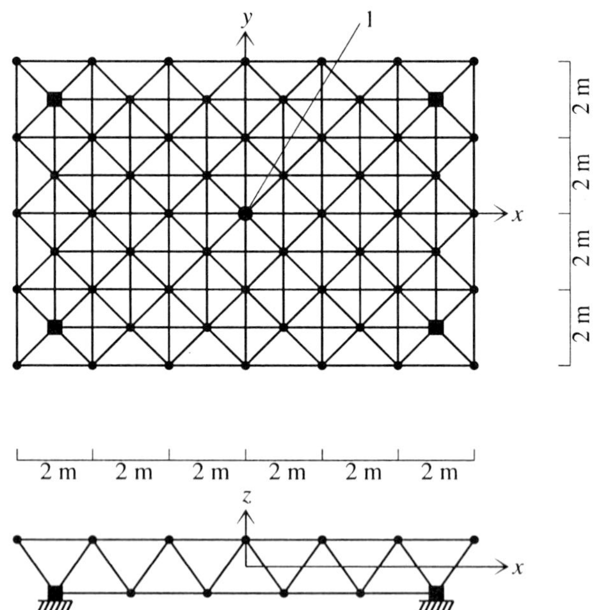

Fig. 7. A double layer grid with rigid joints.

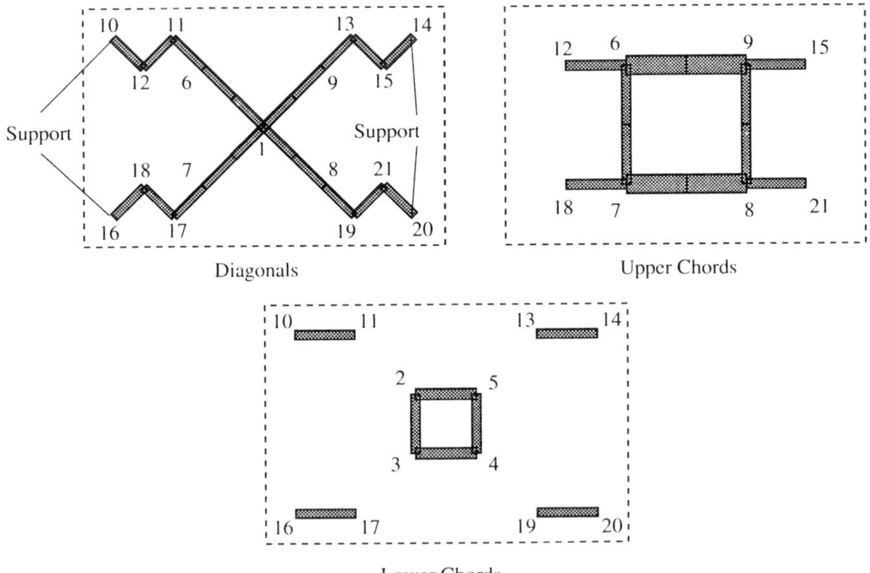

Fig. 8. Optimal topology of the double layer grid with rigid joints.

The sequence of optimal solutions is expressed in terms of a Taylor series expansion with respect to the parameter. The latter is then decreased to zero resulting in the optimal topology. It is well recognised that a pin-jointed truss with an optimal topology may be unstable. On the contrary, the results in the examples here demonstrate that a space frame with optimal topology is stable.

18. In the examples, optimal topologies are first found for axisymmetric frames with different locations of nodal non-structural masses. The optimisation results indicate that a dome-like frame is generated in the optimal topology and some tension or compression rings are necessary to support the inertia forces transmitted by radially generated arches. It is interesting to note that the optimal path of transmitting the inertia forces have been generated naturally not only in the axisymmetric space frame, but also in a double-layer grid with rigid-joints. The inertia forces in the latter are transmitted by the diagonals which are linked and stiffened by some squares.

19. It should be noted that an optimal topology found using the proposed method depends strongly on the candidate joints and members specified by the designers. Extensive numerical experiments need to be carried out to investigate the effect of the location of the candidate joints and members to the optimal topology.

REFERENCES

1. SVED G. and GINOS G. Structural optimisation under multiple loading, Int. J. Mech. Sci., Vol.10, pp 803-805, 1968.

2. KIRSCH U. Optimal topologies of truss structures, Comput. Meth. Appl. Mech. Engng., Vol.72, pp 15-28, 1989.

3. RINGERTZ U.T. On topology optimisation of trusses, Eng. Opt., Vol.9, pp 209-218, 1985.

4. TOPPING B.H.V. Shape optimisation of skeltal structures: A review, J. Struct. Div. ASCE, Vol.108(ST8), pp 1933-1951, 1983.

5. KIRSCH U. On the relationship between optimum structural topologies and geometries, Struct. Optimisation, Vol.2, pp 39-45, 1990.

6. LIN J.H., CHE W.Y. and YU Y.S. Structural optimisation on geometrical configuration and element sizing with statical and dynamical constraints, Int. J. Comp. & Struct., Vol.15(5), pp 507-515, 1982.

7. NAKAMURA TSUNEYOSHI and OHSAKI M. Sequential optimal truss generator for frequency ranges, Comp. Meth. Appl. Mech. Engng., Vol.67(2), pp 189-209, 1988.

8. SHIN Y.S., HAFTKA R.T., WATSON L.T. and PLAUT R.H. Tracing structural optima as a function of available resources by a homotopy method, Comp. Meth. Appl. Mech. Engng, Vol.70, pp 151-164, 1988.

9. NAKAMURA TSUNEYOSHI and OHSAKI M. A natural generator of optimum topology of plane trusses for specified fundamental frequency, Comput. Meth. Appl. Mech. Engng., Vol.94, pp 113-129, 1992.

10. SHEU C.Y. Elastic minimum-weight design for specified fundamental natural frequency, Int. J. Solids Structures, Vol.4, pp 953-958, 1968.

11. MASUR E.F. Optimal structural design under multiple eigenvalue constraints, Int. J. Solids Structures, Vol.20(3), pp 211-231, 1984.

131. Efficient faceted surface structures

T. WESTER, Royal Academy, Denmark

SYNOPSIS. The concept of *minimal structures* has always been a challenge to structural morphologists and very much effort has been put into this field. The meaning of *minimal* seems to be relative and not absolute as the word itself indicates. The numerous different interpretations have caused much confusion and misunderstanding. The paper will deal with yet another definition of *minimal structures* based on force and moment loaded plane faceted kinematic nets. The concept seems obviously relevant for plate structures while the relevancy for lattice structures will be discussed. The method implies that the *efficiency* of arbitrary plate structures can be measured (0-100%) and used interactively in the process of shaping and faceting structures. The determination of the *efficiency* is implemented in the author's computer program *CADual*.

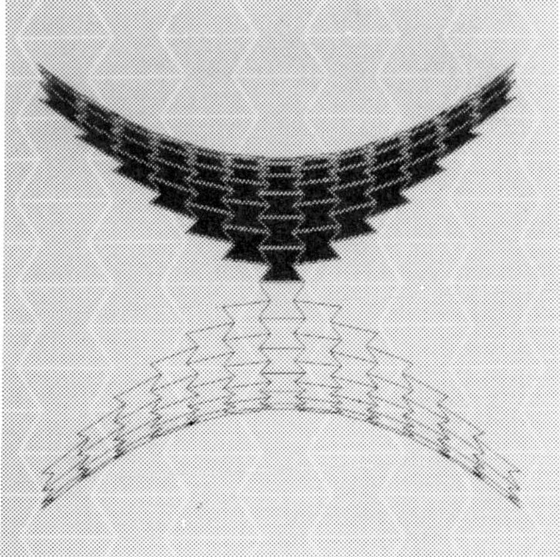

Fig. 1.

CHARACTERISTICS OF SINGLE LAYER PLATE STRUCTURES

1. As described (ref. 3), a topological characteristic of a single layer pure plate structure is that all vertices are 3-valent. The stabilisation by lattice (bar-and-node) action of a 3-valent structure (except the Platonic master polyhedron: the Tetrahedron) is impossible. The only possibility of stabilisation is by pure plate action .

LOAD INTRODUCTION

2. If a pure 3-valent plate structure is externally loaded, the load will be introduced to the structure by transferring the loads to the vertices, resolving them into forces in the directions of the three shear lines connecting the plates. These forces will act as the operational external forces for the structural analysis of the plate structure. All external loads on one plate are therefore to be transferred by bending (slab action) and plane membrane forces (wall action) to the corners of the plate polygon (= the vertices of the spatial shape) (Fig. 2). In this way any 3-valent vertex adds on load from the adjacent plates, and the sum of these loads is resolved unambiguously into the three directions of the adjacent shear lines. As one shear line always connects two vertices, the plate structure must be loaded with the resultant of these two forces (Fig. 3). This is the load which activates the plates and creates shear forces between the plates. If the resultant is zero, it is obvious that the plate structure will be 0-loaded along this shear line.

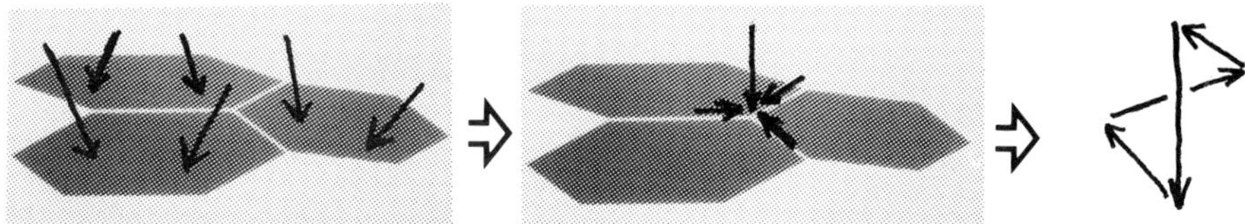

Fig. 2.

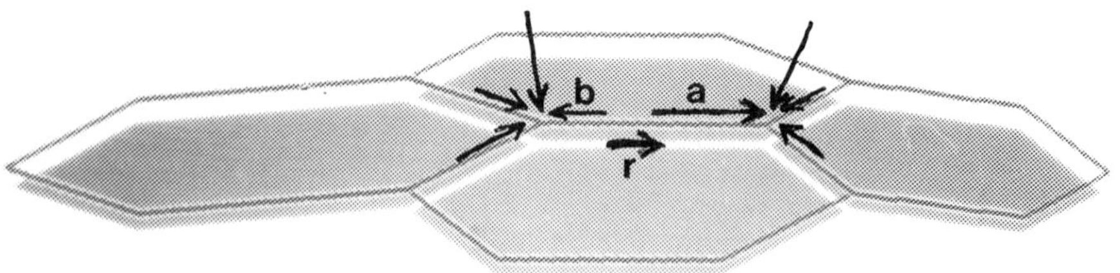

Fig. 3. Resultant $r = a + b$.

MINIMAL PLATE STRUCTURES

3. If all shear lines in a loaded pure plate structure are 0-loaded, no plates are forced and no shear forces are transferred across the shear lines. In this case all the plates can be removed from the structure, leaving the edges as the only (axially) forced element. This exceptional case, when a pure plate structure carries external loads without any plate action at all, is of course of the utmost interest. The situation where the shape of a pure plate structure is so neatly adjusted to the actual load, that all the plates and shear lines are zero-stressed, seems to be a kind of ultimate definition of a force-to-form minimal plate structure. This concept is obviously very important for plate structures, while its relevancy for pure lattice structures, i.e. the dual situation, is more ambiguous as explained later. If the valency is more than three, the external vertex load cannot be unambiguously resolved, hence the minimal situation cannot be uniquely determined.

4. It is significant and important that the considerations are local equilibrium investigations between shape and force and do not imply boundary or support conditions. Therefore edge conditions are not considered in Figs 1 and 9.

LOCAL FORCE NETS AND STRUCTURAL EFFICIENCY

5. The 3-valent loaded net is called the local force net, and in case of a minimal structure, this kinematic net carries the external load in perfect balance.

6. In the common situation, where the local force net is not in equilibrium with the external load, the difference between the force in either end of an edge will load the plate structure. Note that this force difference is closely related to the so-called *buffer force* (ref. 5) . If the difference is zero, the *efficiency* (e) along this edge is defined to be 1 (or 100%), and if the two forces have the same direction or one of them is zero, the *efficiency* is defined to be 0. In this way the *efficiencies* for all the n edges can be evaluated and a measure for the efficiency for the whole structure (E) can be calculated:

$$\text{If } \vec{a}.\vec{b} \leq 0 \text{ then } e = 1 - \frac{|\vec{a} + \vec{b}|}{|\vec{a} - \vec{b}|}$$

$$\text{If } \vec{a}.\vec{b} \succ 0 \text{ then } e = 0$$

$$E = \frac{\sum_{1}^{n} e_n}{n} \tag{1}$$

7. E is seen to be a rough measure and the precise value should therefore be taken with caution, particularly for comparing completely differently shaped and faceted plate structures.

8. As the *efficiency* is dependent on the facet configuration of the structure, a geometrical change of a facet will result in a change in the *efficiency,* i.e. the *efficiency* depends on both the shape and the faceting of the structure. This effect is used interactively in the computerized design process in *CADual* for both shaping and faceting a structure in order to control (increase) the structural *efficiency.*

9. Note that the forces in the *local force net* for a minimal structure do not need to be either in compression or in tension or of the same magnitude all over, as long as the forces in either end of an edge are equal and oppositely directed the *efficiency* is 100%. Also note that scaling of the geometry and/or the load does not affect the *efficiency*. The *efficiency* may be regarded as the portion of the load which is carried by the *local force net.*

10. In the following two highly significant loading cases, the hydrostatic and the uniform vertical load will be discussed in relation to the configurations which create minimal plate structures.

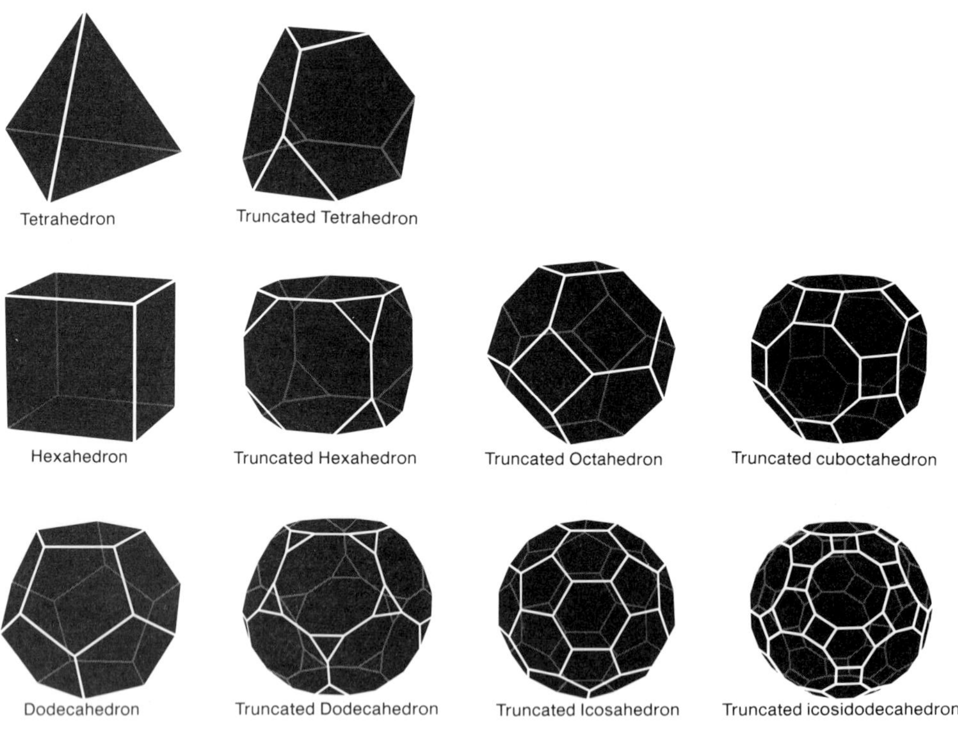

Fig. 4.

THE HYDROSTATIC UNIFORM AREA LOAD

11. That is, the same area load perpendicular to all plates. The free spherical soap bubble is the classic example on a minimal structure subjected to this load. All polyhedral 3-valent configurations where all vertices are congruent will also be minimal plate structures. The three regular and seven semi-regular 3-valent polyhedra (Fig. 4) and all the regular prisms (but not the antiprisms as they are 4-valent) fall into this category. Only the three regular of the polyhedra have equal forces in their local force nets. The fact, for example, that a shallow prism turns out to be a minimal plate structure shows a limitation of the concept, as the bending (slab action) perpendicular to the plates is not taken into consideration.

12. If the 3-valent regular polyhedra are hexagonated (ref. 3), i.e. dual to the triangulation (e.g. ref. 1) methods, the typical pattern will be as follows: The basic polyhedron is, as mentioned previously, a *minimal* plate structure, the first breakdown will cause a sharp decrease in *efficiency* and from there the *efficiency* will increase with the frequency and approach 100% as the polyhedron approach to the spherical shape. An example based on the cube is shown in Fig. 5.

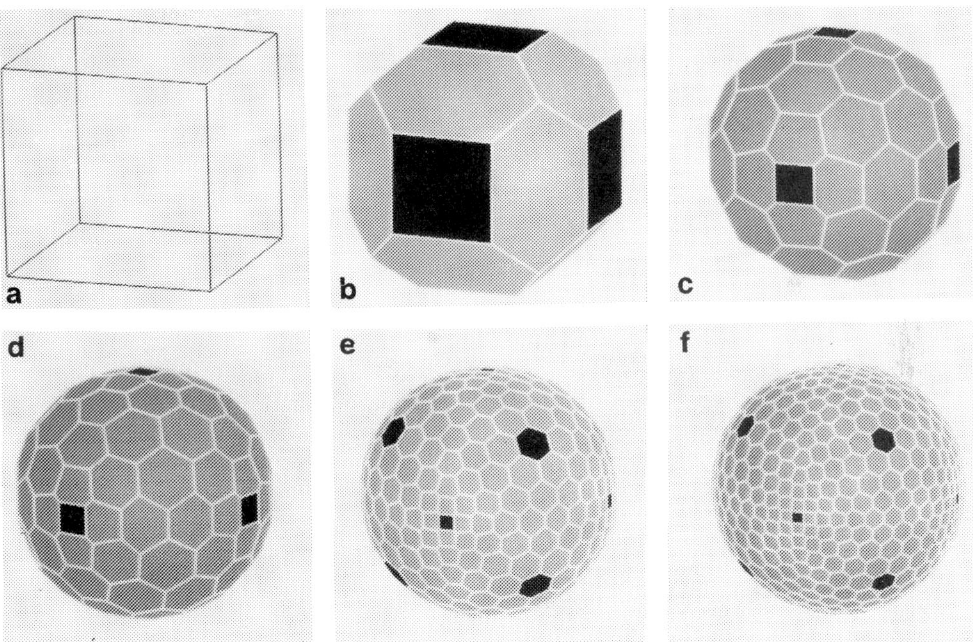

Figs 5(a)-5(f) show a cube which is subdivided to increasing frequency (γ). If all are hydrostatically loaded the global efficiency (E) will be: (a) the cube, E=100%; (b) 2γ, E=86%; (c) 4γ, E=95%; (d) 5γ, E=96%; (e) 9γ, E=99%; (f) 12γ, E=99%. The plates, that may be removed because of local efficiency e=100, are shown in black.

13. If the loading case is changed, for example, to idealized selfweight (all plates have the same weight per unit area), the efficiency of the cube is 0 and increasing with the frequency up to 67% for the 12 frequent cube, see Figs 5(a) and 5(f). This leads to the conclusion that a high frequency plate polyhedron is normally more efficient than a low frequency plate polyhedron, except for hydrostatical uniform loading.

THE VERTICAL PROJECTED UNIFORM AREA LOAD

14. That is, the total load on one plate is the horizontally projected plate area times the unit area load. This is the type of load that in the plane situation gives a parabolic moment curve for a beam, or a parabolic shape for a kinematic chain.

15. A paraboloid of revolution enveloped by tangent plates which vertical projection gives congruent regular hexagons (Figs l(a) and 9(a)) turns out to be a minimal plate structure for this particular loading case (all plates have equal total load). The local force net may either be in tension all over as a hanging net or in compression all over as a standing net. As in the plane kinematic chain, the internal forces in the net increase with the distance from the apex.

16. This shape was used in 1990 for a 2.4 m high and 2.2 m wide dome sculpture by the Danish ceramic artists Gunhild Rudjord and Esben Madsen in collaboration with the author (Fig. 6). The plates are made up of blue glazed 12-15 mm thick ceramic tiles where the thickness of the tiles was required due to the fabrication and not to the structural demands. The wet clay was cut directly from the computer drawn pattern and the joints are ordinary cement mortar with no reinforcement at all. As the selfweight is absolutely dominant and very close

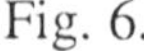

Fig. 6.

Fig. 7.

to the ideal load, the shape is very advantageous. The sculpture is now on display at a Danish art museum.

17. The concept was also used in 1991 in a final examination project by two Danish civil engineers (ref. 2) as a proposal for a 30m high palm house with plane 15mm hardened glass used as structural plates (Fig. 7). The project showed clearly that today's high quality glass is an absolute possibility for structural plates, especially as a structure with high structural efficiency. The project consists of three combined paraboloidal units. As the shape is close to the *minimal* plate structure for selfweight, the 15mm hardened glass-skin is selfcarrying. The plate structure will only be assisted by the underlying coarse meshed steel lattice in case of heavy wind load. The project pointed out that the shear resistant join between the glass plates has to be further developed as no existent standard joint solution seems to fulfil all the necessary requirements for forces, movements and temperatures.

18. The minimal solution has also a negatively curved configuration as shown in Figs l(b) and 9(b). The main shape is a hyperbolic paraboloid (HP) and the plates are tangent planes to this. The horizontal projection of the plates gives a congruent and symmetric hourglass shaped pattern where the length of the "top" and "bottom" is the double of the inclined "sides" and the angles are 60° and 240°. In this case some members are in tension and others are in compression. The hourglass shape is a necessary requirement in negatively curved single layer pure plate structures (ref. 3). Depending on the direction of the vertical load, the "top" and "bottom" of the hourglass will be in tension and the "sides" in compression - or opposite. The length and the width of the plates change with the distance from the centre point of the shapes, but in a different pattern as for the positively curved solution .

EFFICIENCY OF LATTICE STRUCTURES

19. The present concept of minimal structures is as shown above quite useful in the design of highly efficient single layer plate stabilized structures. A bar-and-node structure of the same main shape and a lattice pattern which is analog to the faceting of a plate structure with high *efficiency*, will probably also be an excellent configuration for the same loading. The concept for plate structures may therefore also be very useful in the design of efficient lattice structures.

20. The dual of the *force-to-form* concept for plate structures is a *moment-to-form* lattice structure, which means that the triangular facets of the lattice structure must be loaded with moments. A simple example, the hydrostatically loaded cube, a *minimal* plate structure, is dual transformed to an octahedron subjected to hydrostatic moment load (Fig. 8).

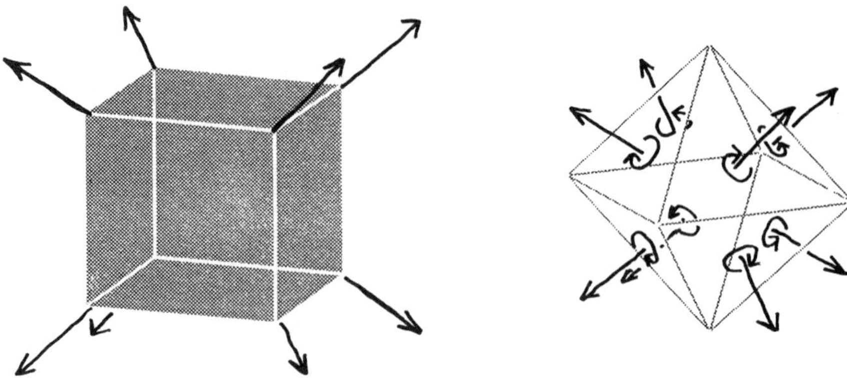

Fig. 8 Examples on dual minimal structures.

21. All moments create forces along the edges and as these forces will be equal and oppositely directed on each side of the edge, they will create equilibrium as pure shear, i.e. without any axial forces. The local force net will in this case consist of moment-loaded triangular plates in local equilibrium with its neighbours. This concept of *minimal* lattice structures does not seem very relevant, while its plate counterpart is directly operational .

CONCLUSION

22. The use of highly efficient plate structures seems particularly interesting when the selfweight is dominant to other loads (e.g. wind, snow, live load etc.), as the *efficiency* in this case differs less than if the selfweight is relatively small. The result of this effect is that a heavy selfweight-loaded minimal structure often has a higher *efficiency* for the different relevant load combinations and is therefore more "ideal" than a lightweight minimal structure.

23. The concept is implemented in the author's computer program *CADual*, and has turned out to be operational and fairly simple to manage for manipulation of the main shape (e.g. affine-transformations) and of single plates or groups of plates (e.g. tilting, translation, rotation etc.), and investigations for different loading cases and their combinations.

ACKNOWLEDGEMENTS

24. The research on which this paper is based was granted by The Danish Technical Research Council 1988-90 and later supported by the Royal Danish Academy of Fine Arts.

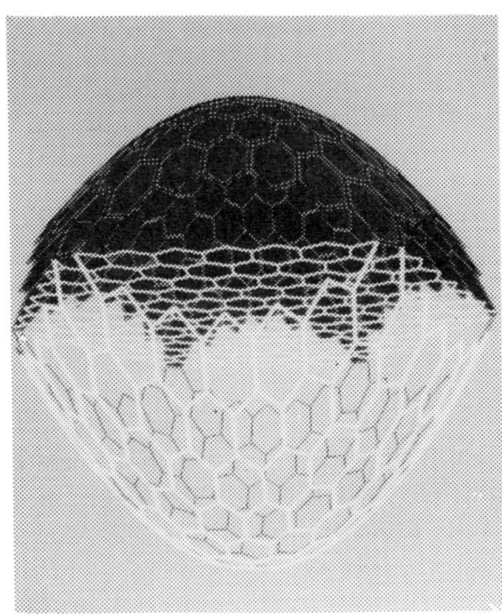
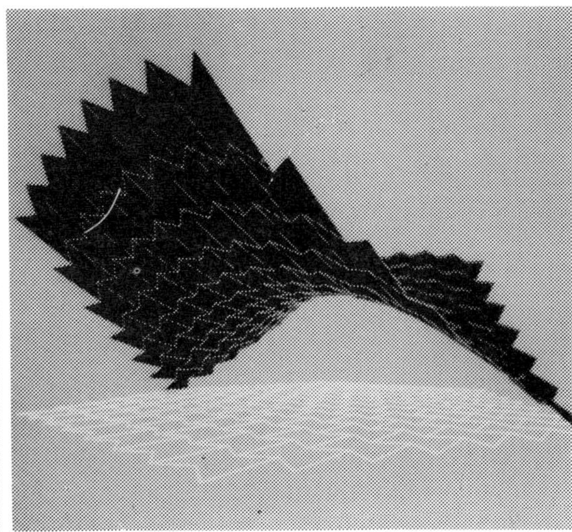

Fig. 9.

REFERENCES

1. KENNER, H. 1976, Geodesic Math., University of California Press, USA.
2. OHANNESSIAN, P. and GRUNNET, 1991, Projektering af en Glasskivekuppel (Design of a Glass Plate-Dome). Final Examination Project Report, (in Danish), The Engineering Academy of Denmark, Lyngby.
3. WESTER, T. 1984, Structural Order in Space, Royal Danish Academy of Fine Arts, School of Architecture, Copenhagen.
4. WESTER, T. 1987, Plate-Domes, Proc. Int. Conf. on the Design and Construction of Non-Conventional Structures, Vol.1, pp 241-251, London.
5. WESTER, T. 1991, The Structural Behaviour of Arbitrarily Plane-Facetted Spatial Nets. Proc. IASS Coll. on Spatial Structures at the Turn of the Millennium: Vol.II, pp119-124, Copenhagen.

132. Structural models as design and education tools for space structures

G. G. SCHIERLE, University of Southern California, USA

SYNOPSIS. The design of structures requires analytical abilities and, like any creative process, also requires strong intuition. The old master builders could develop their intuition by empirical experience with materials and processes in the field, much like a carpenter develops a feel for strength of wood by working with it. Today's designers are often far removed from direct empirical experiences, making it difficult to develop intuition. It will be argued that structural models help to develop structural intuition, besides serving for quantitative analysis, especially for unique structure concepts.

INTRODUCTION

1. Structural models have long been used to test unique structures (refs 1-2). While today's computer models, based on finite element analysis, are better in many ways than physical models, the latter have important advantages. However, computer models provide greater accuracy and flexibility to evaluate the impact of parameters, such as variable elasticity, for example. Physical models have important advantages too. Foremost they can provide intuitive understanding as well as analytical data. On the analytical side they yield stress, deformation and general cause/effect correlation of form and behaviour for unique structures, even when no mathematical model exists, as demonstrated by the author's test of various cable nets (ref. 3). Working with such models can be a highly educational experience, not only for students, but also for experienced professionals. The tactile feedback to touch, and observations regarding deformations to it, can be of invaluable benefit in developing informed intuition regarding correlation of form and structural behaviour. Such intuition is invaluable for creative design, where a cyclical process of analysis and synthesis is repeated for the final solution. Both the analytical and intuitive aspects of structural models will be discussed.

MODEL SCALES

2. Structural simulation models have 4 scale relationships to the original structure:

• Geometric scale $S_L = L_m/L_o$, model dimensions / original dimensions

Space structures 4. Thomas Telford, London, 1993

- Force scale S_p = P_m/P_o, model forces / original forces
- Strain scale S_ε = $\varepsilon_m/\varepsilon_o$, model strain / original strain
- Stress Scale S_f = f_m/f_o, model stress / original stress

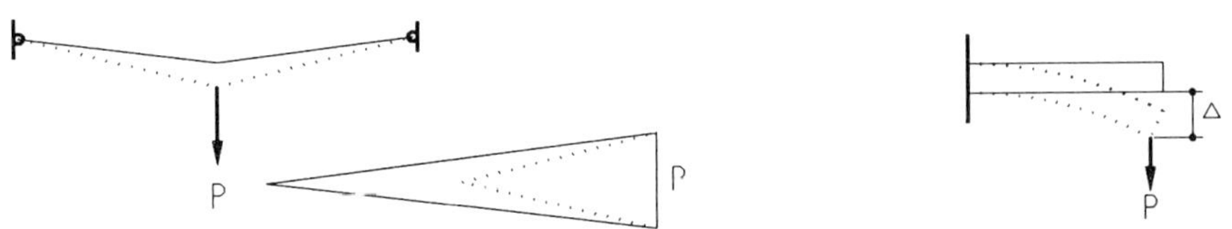

(a) Suspended Cable Deflection. (b) Beam Deflection.
Fig. 1. Suspended cable (a) and beam (b) under point load.

3. The geometric scale is frequently used for architectural and similar models to convey visually formal and spatial relationships in ratios, such as 1:100, or 1/4" = 1'. For simulation models this scale should be chosen as large as possible for accuracy. The force scale relates model loads and forces to corresponding loads and forces in the original structure as ratios, such as 1:100,000. If, for example, a given joint in a space truss carries a load of 12,000N and the force scale is 1 :10,000 then the load in the corresponding model joint would be 12,000/10,000, or 1.2N. The strain scale relates model strains to corresponding strains in the original structure as ratios, such as 5:1 . If the strain scale is chosen 1:1, then the model deformations relate to deformations in the original in the geometric scale. When the expected deformations are very small, then a strain scale greater than 1:1 may be chosen to exaggerate deformations for better visualisation. However, the strain scale should be 1:1 if a structure has geometrically non-linear behaviour. For example, in Fig. 1a, the load P generates a large force in the suspended cable, due to its shallow angle, as shown in the adjacent equilibrium force polygon. The large force causes large deformation. Under the deformed geometry the cable force will be smaller, which, in return causes a smaller deformation. Clearly this mechanism has geometrically non-linear behaviour. But in the beam of Fig. 1b the deformation Δ will increase linearly with the load P, as long the deformation remains within the linearly elastic range of the material, thus it has geometrically linear behaviour. A structure like Fig. 1b can be modelled with a strain scale greater than 1:1 for better visualisation, assuming linearly elastic material, while a structure like Fig. 1a must be modelled with a strain scale of 1:1 to avoid errors. The stress scale relates model stress to stress in the original structure. It is determined by the strain scale and the relationship of the E-module

in model and original, namely $S_f = [E_m\varepsilon_m]/[E_o\varepsilon_o]$ Since $\varepsilon_m/\varepsilon_o = S_\varepsilon$, the stress scale is:

$$S_f=S_\varepsilon E_m/E_o \quad (1)$$

If the E-module is equal in model and original and the strain scale is 1:1, then the stress scale is also 1:1. If the stress scale is greater than 1:1, then one must ensure that the model stress is within linearly proportional stress/strain limits and does not exceed the breaking strength of the model material.

Axial stress

4. For models with axial stress only, such as a space truss with pinned-joints, for example, the force scale may be derived as follows:
Unit strain $\varepsilon = \Delta L/L$ and strain $\Delta L=PL/AE$ (A = cross-section area, E = modulus of elasticity) hence the force $P=AE\ \Delta L/L$, or $P=AE\varepsilon$; and the force scale is $S_p = P_m/P_o = [A_mE_m/A_oE_o]\ [\varepsilon_m/\varepsilon_o]$; since $\varepsilon_m/\varepsilon_o = S_\varepsilon$, the strain scale is:

$$S_p = [A_mE_m/A_oE_o]\ S_\varepsilon \quad (2)$$

If the strain scale $S_\varepsilon = 1$, then

$$S_p = A_mE_m/A_oE_o \quad (3)$$

If $S_\varepsilon=1$ and $E_m/E_o = 1$ (model and original have the same E-module) then

$S_p = A_m/A_o = S_L^2$ (the force scale is equal to the square of the geometric scale), i.e:

$$S_p = S_L^2 \quad (4)$$

Bending Stress

5. For models with bending stress only, the force scale may be derived as follows:
Unit strain $\varepsilon = \Delta/L$ and strain $\Delta = kPL^3/EI$ (I = inertia moment). The integration constant k, being identical in model and original for like members, can be ignored. Thus force is $P = EI\Delta/L^3 = EI/L^2(\Delta/L)$ and force scale $S_p = P_m/P_o = [E_mI_m/E_oI_o]\ [L^2{}_o/L^2{}_m]\ [(\Delta/L)_m/(\Delta/L)_o]$. Since $L^2{}_o/L^2{}_m = 1/S_L^2$, i.e. the inverse of the geometric scale and $(\Delta/L)_m/(\Delta/L)_o = S_\varepsilon$, i.e. the strain scale; the force scale can be defined as:

$$S_P = [E_mI_m/E_oI_o]\ [1/S_L^2]\ [S_\varepsilon] \quad (5)$$

If the strain scale $S_\varepsilon = 1$, then:

$$S_P = [E_mI_m/E_oI_o]\ [1/S_L^2] \quad (6)$$

If $S_\varepsilon = 1$ and $E_m/E_o = 1$ (model and original have the same E-module) then

$S_P = [I_m/I_o]\ [1/S_L^2]$, or

$S_P = S_L^2$ (if all elements have geometric similitude) (7)

Eq. 7 is mathematically identical to Eq. 4; but the stated condition about geometric similitude must be satisfied when applied to models with bending stress or combined axial and bending stress.

Combined Stress

6. Models of combined axial and bending stress must be based on Eq. 7, or, if different E-modulus are used for model and original, must satisfy that the force scales are equal for axial and bending stress, namely $S_P = A_m/A_o = [I_m/I_o]\ [1/S_L^2]$. For example, if a square column under combined bending and axial stress is modelled with $E_m/E_o = 0.5$ and one column dimension is doubled to adjust for the smaller E, then A_m increases 2 times and I_m $2^3 = 8$ times; yielding the wrong results.

TESTING

7. Model testing must follow precautions, particularly for lightweight structures:

- Loads should be applied gradually to prevent rupture.
- Loads should be applied for short periods only to minimise creep.
- The support apparatus must be rigid to prevent distortions under load.

The first 2 points are achieved by a movable platform to temporarily hold loads. Lowering the platform by a pulley mechanism with a winch provides gradual loading. After displacement and stress measurements are taken the platform is raised and may be prepared for another load case.

8. The following brief test descriptions demonstrate that models are not only tools for analysis but can be instrumental in developing informed intuition. The space truss of Fig. 2 (based on ref. 4) revealed good stiffness in general, but large deflections at unsupported ends (left and right). It is composed of a series of one-way trusses, running in 3 directions, which stabilise each other. But the long trusses do not stabilise the others at unsupported ends, resulting in large deflections. The delta-truss roof (Fig. 3) had initially no bracing between top chords. The student designers found this acceptable based on computer analysis, but added bracing after model tests revealed rotational instability. The tests were more convincing and educational than prior arguments by the professor. The model shown in Fig. 4 behaved similarly. Tests of the model shown in Fig. 5 led to the design of end truss supports to ensure rotational stability. The cable stayed roof structures (Figs 6-9) revealed correlation of stay cable slope and system

stiffness, and allowed the study of the means for rotational stability in a tactile intuitive manner. The orthogonal grid shell model (Fig. 10) served both for testing and form-finding, by hanging it through a cut-out with weights superimposed; a process similar to one developed by the IL institute (ref. 5). Tests of various dome forms and loads provided information correlating form with stiffness and stability (ref. 6). A similar form-finding process was used for the anticlastic grid shell of Fig. 11. The arch supported grid shell (Fig. 12) was tested with and without stay cables to stabilise arches laterally and in-plane. The stay cables were effective in stabilising the arches in both directions under asymmetrical load. The synclastic timber vault model (Fig. 13) was initially unstable, until tests of various bracing schemes determined a configuration of good stability and stiffness. The anticlastic membranes (Figs 14-15) revealed a strong correlation of boundary condition and surface curvature. Flat surface areas could be detected visually and by feeling soft (low stress) areas. A grid drawn on the membranes allowed for the measurement of stress patterns under load; these correlated with charts from elasticity tests for both warp and fill directions. The prestressed cable truss model (Fig. 16) revealed much greater stiffness than the lentil-shaped cable girder of Fig. 17. It established the feasibility of concepts without curvature (ref. 7). The case study models of the Sears tower and Bank of China (Fig. 18 and 19 respectively) served to study the load-bearing behaviour of bundled tube and braced framing. The models of Figs 20-21 for suspended high-rise structures demonstrated the dual function of the suspender trusses as top and belt trusses to reduce lateral drift. The dynamic model (Fig. 22) revealed rotational tendencies under lateral seismic load - even without eccentricity - simulated for selected earthquakes by an electro-dynamic shaker table.

Fig. 2. Space truss roof.

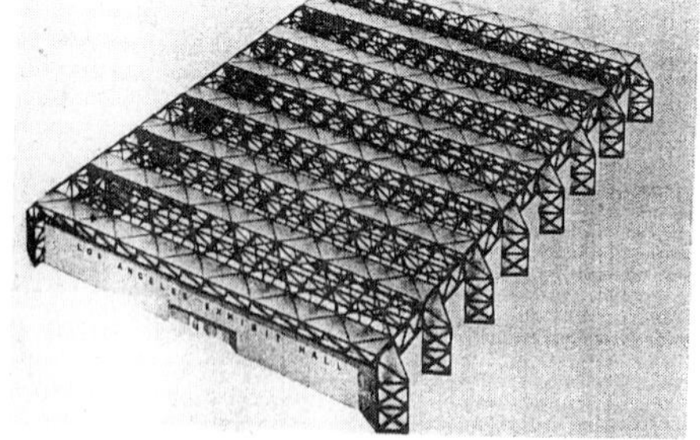

Fig. 3. One-way delta-truss roof.

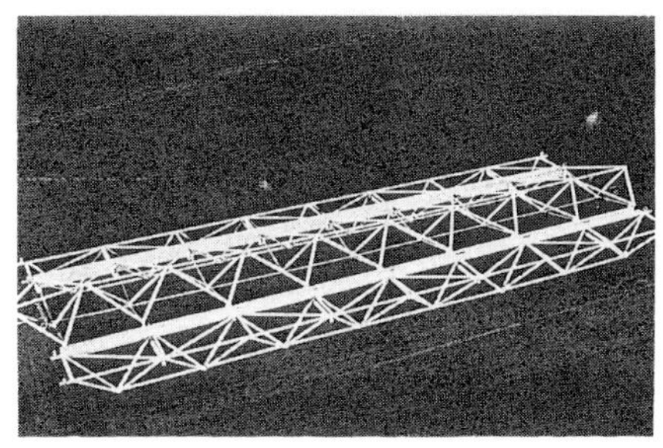
Fig. 4. One-way delta-truss roof.

Fig. 5. Folded plate truss roof.

Fig. 6. Cable stayed girder roof.

Fig. 7. Cable stayed cantilever roof.

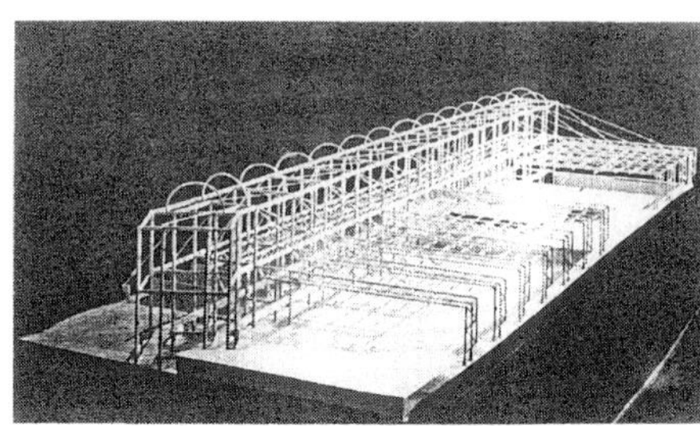
Fig. 8. Cable stayed delta-truss roof.

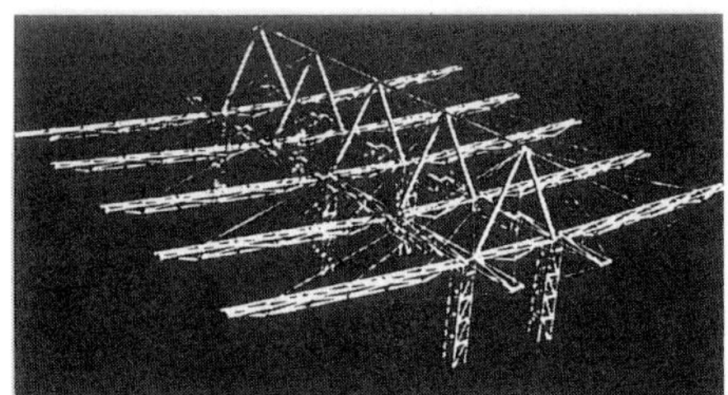
Fig. 9. Cable stayed delta-truss roof.

Fig. 10. Dome grid shell.

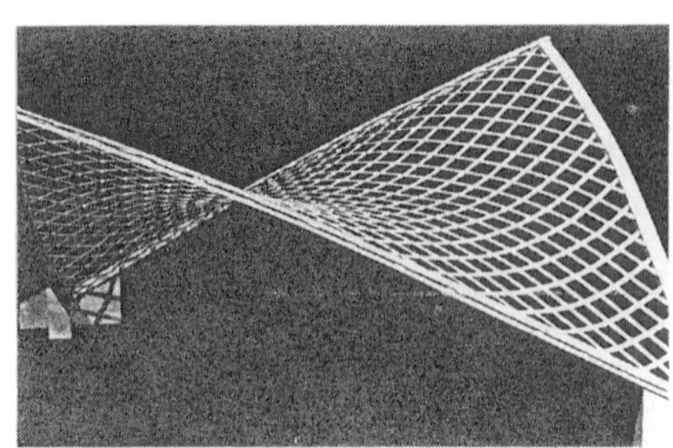

Fig. 11. HP grid shell.

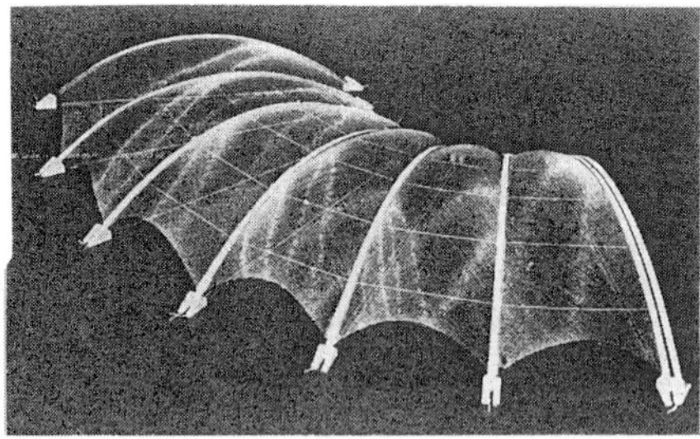

Fig. 12. Arch supported grid shell.

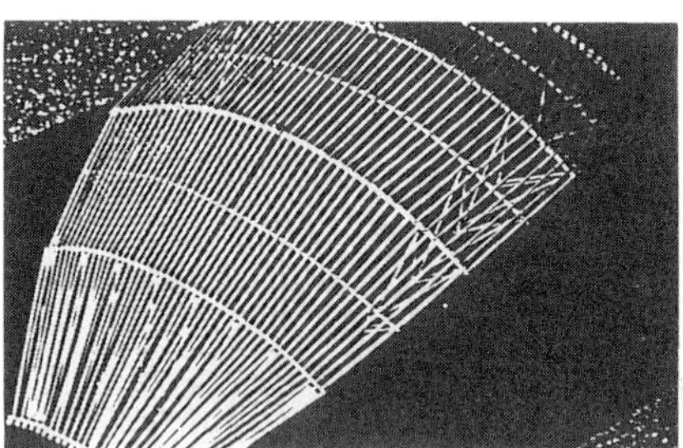

Fig. 13. Synclastic timber vault.

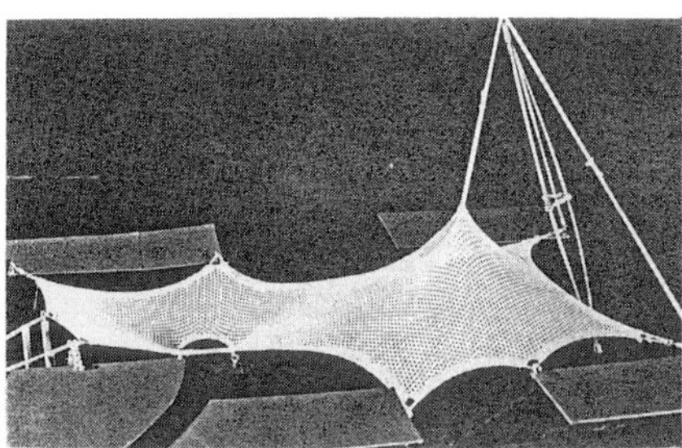

Fig. 14. Anticlastic membrane.

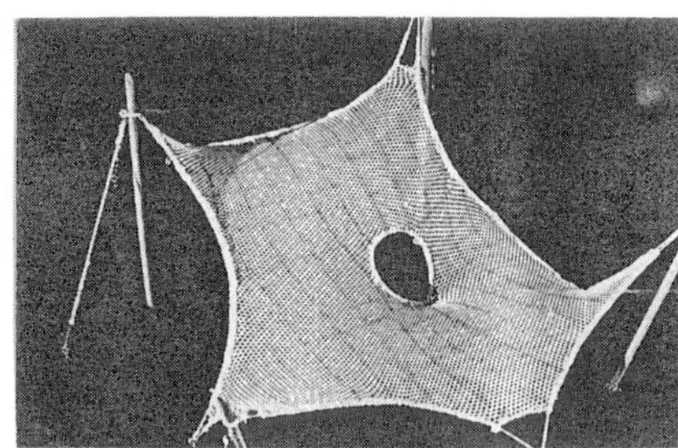

Fig. 15. Anticlastic membrane.

Fig. 16. Prestressed cable truss.

Fig. 17. Prestressed cable girder.

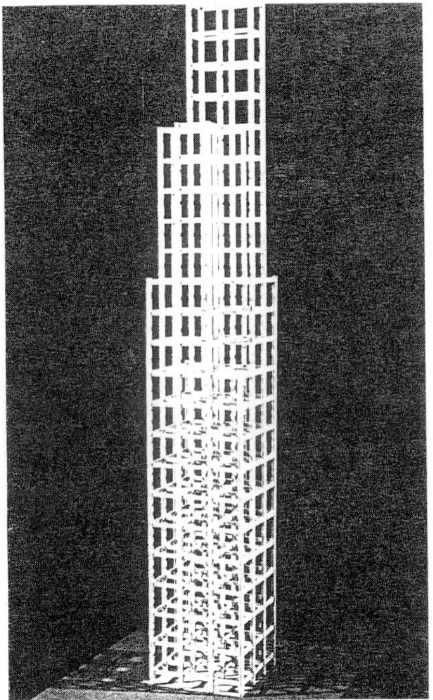

Fig. 18. Sears tower.

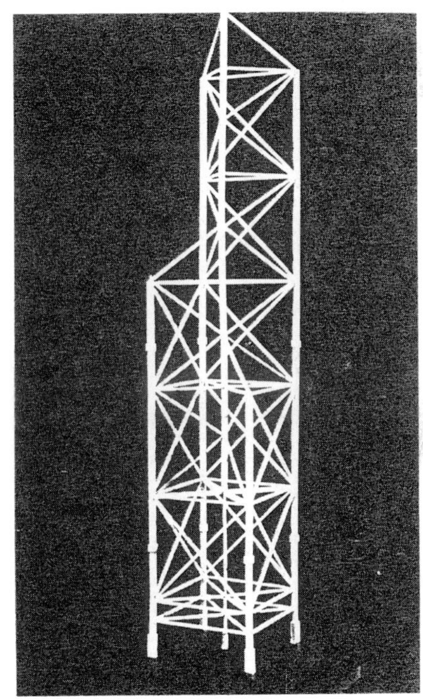

Fig. 19. Bank of China.

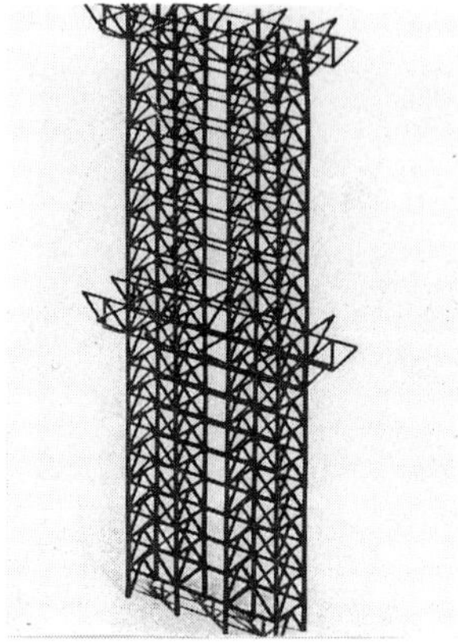

Fig. 20. Suspended high-rise.

Fig. 21. Suspended high-rise.

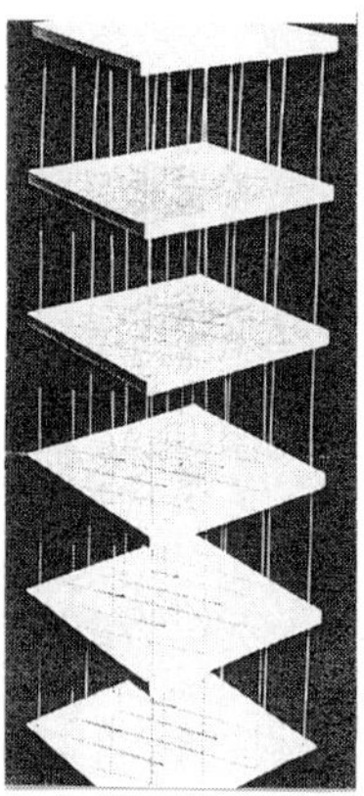

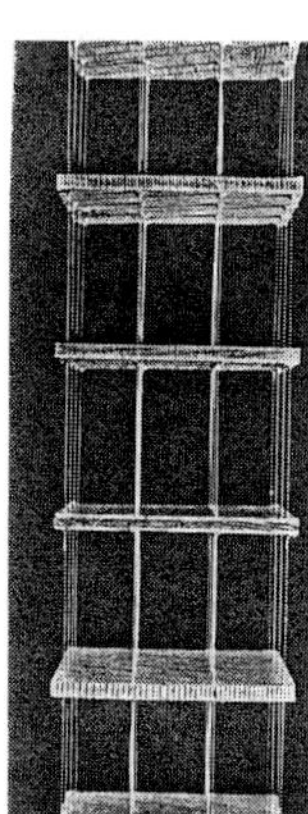

Fig. 22. Seismic dynamic high-rise model.

CONCLUSIONS

9. Instead of detailed test reports the purpose of the foregoing description is to give an overview of the possibilities and benefits of structural model testing. In architectural education the greatest benefit of such testing is to develop informed intuition, based on tactile feedback to touching, careful observations, and intuitive analysis of observations. While model testing is not a substitute to experience of actual construction, it provides some advantages, such as visual and tactile

feedback to various load conditions. Building of models can also provide insight regarding sequence and method of construction. While computer simulation has certainly the advantage of greater speed, accuracy and flexibility, it cannot fully replace physical models in developing informed intuition. Given the great advances in computer technology and programing techniques over the last 30 years, education in architecture and engineering should now focus on the creative aspects of design; on methods of using computers for creative exploration; but also continue do use and refine methods of testing physical models as design and education tools.

ACKNOWLEDGEMENTS
10. The author wishes to acknowledge the students who prepared the presented models: Bharali; Boyajian; Cruikshank; Chen; Chong; Dytoc; Fung; Gee; Guthertz; Jordan; Lau; Liang; Metha; Miller; Ogami; Suekama; Wudofsky; Yhe; and Yu. The Fig. 13 model is from a seminar the author conducted at the IBOI institute of Professor Natterer at EPFL Lausanne, Switzerland.

REFERENCES

1. MUELLER R. K. Handbuch der Modellstatik, Springer Verlag, Berlin, 1971.
2. MUELLER R. K., STIEMER, S. F. and WESSOLLY L. Modelltechnik fuer Seilnetze, Mitteilungen des Sonderforschungsbereichs 64, Nr 40, University Stutgart 1976.
3. SCHIERLE G. G. Design considerations for lightweight structures Proceedings, International Symposium, Lightweight Structures in Architecture, University of New South Wales, Sydney, Australia, 1986.
4. BOREGO J. Space Grid Structures, MIT Press, Cambridge, MA, 1986.
5. OTTO, F. et al, Grid Shells, IL Report No. 10, Institute for Lightweight Structures, University of Stuttgart, 1974.
6. KUO P. H., WU W. and SCHIERLE, G. G., Orthogonal grid shell: static model tests, comparing deformations to form, Developments in Structural Engineering, Vol. 1 (Topping, ed.) Chapman & Hall, London, Van Nostrand Reinhold, New York, 1990.
7. SCHIERLE G. G., Prestressed trusses: analysis behaviour and design, Developments in Structural Engineering. Vol 1 (Topping, ed.) Chapman & Hall, London, Van Nostrand Reinhold. New York, 1990.

133. Computer-based modelling of space structures with the program KONDAR

J. SANCHEZ ALVAREZ, MERO - Raumstruktur, Germany

SYNOPSIS. KONDAR is a program system for the computer supported geometric construction and representation of spatial objects. The system has been developed under the direction of Professor Dr. Helmut Emde in the specialised area of processing of geometric information at the Technical University in Darmstadt. The space structures company MERO in Wurzburg has intensively supported and collaborated on the development of the program, which has been strongly oriented to cope with the enormous variety of topological networks and geometric shapes that are encountered in the computer based practical design of space structures. Recent MERO structures are presented to illustrate the capabilities of KONDAR.

REQUIREMENTS TO A PROGRAM SYSTEM

1. Any design process in architecture and engineering aims at the realisation of spatial objects. The design of a spatial object comprehends the following aspects:

- geometric construction in space;
- graphic representation;
- logistic organisation.

Consequently, a program system for the design engineer or architect must enable these three aspects.

2. In the context of design of spatial objects, geometric construction involves the determination of form and structure. The form describes the visually perceptible and tangible extension of an object, while the structure describes the composition of the whole and the interrelations among its parts. Furthermore, any spatial object of the human environment can be seen either as a complex of simpler objects or as an element of even more complex objects. This approach yields the possibility of understanding reality as a hierarchically composed world of objects.

3. Representation. During the process of geometric construction, it must be possible to easily check and graphically document the current state of development and also other stages in the design of a spatial object.

Space structures 4. Thomas Telford, London, 1993

4. Organisation. As the results of the design process grow partly following a pre-conceived development scheme and partly spontaneously, it is normally necessary to be able to refer to and modify results of previous construction steps in a simple agile way at any time. Besides geometric characteristics, design objects possess many other non-geometric properties, which assume relevance at different stages of design or belong to other disciplines, such as structural analysis, technical production, aesthetics or economics. Consequently, it should be possible to relate non-geometric attributes to the geometric construction of spatial objects.

REALISATION THROUGH THE PROGRAM KONDAR

5. KONDAR can be used for the numerical, graphical and automated construction of complex spatial objects, their graphical representation for checking and documentation purposes and the organisation of the construction results. The name KONDAR, incidentally, comes from the German: "KONstruktion" and "DARstellung", that is, construction and representation, respectively.

6. Construction in KONDAR applies to structured "objects" and spatial "models". A KONDAR-object is a hierarchically ordered complex of simpler elements, where every element can in turn be a complex of still simpler elements. This organisation principle enables simple generation, modification and handling techniques. Spatial models are geometrically conceived and implemented as three dimensional bodies, which consist of points, lines and facets (or faces). KONDAR-models can be geometrically transformed, added, subtracted and further modified in a large variety of ways.

7. Representation in KONDAR includes quick check drawings of every construction stage, precise documentation drawings and elaborate presentation graphics of the results of the construction process.

8. Organisation in KONDAR refers to the handling and administration of the various stages in the construction process and the graphic representation of results. The relation and handling of non-geometric attributes and spatial objects is now in development.

9. The program system consists of various program parts or modules, each of which includes a family of functions. The application of the system does not require programming knowledge and all functions can be addressed through "commands" in an engineering oriented user language.

10. KONDAR provides a module with a comprehensive collection of optimised pre-programmed networks on a variety of analytical carrier surfaces, the main ones being the cylinder, the cone, the sphere and more general surfaces of revolution; the helicoid, and various paraboloids. A set of regular and semi-regular polyhedra, various spherical geodesic networks and a complementary "Grid" module having a dozen of frequently used topological networks, complete the choices of ready-geometries.

11. The construction module allows individual or collective definition and modification of points, lines and facets on the "current" model, that is, the model

being processed. A powerful geometry and vector calculator and the possibility of user-defined co-ordinate transformations complement the menu of geometric construction functions.

12. The combination module allows the convenient generation of complex "objects" out of one or a few simpler "object elements", which can be both simpler objects and models. Object elements are hierarchically organised in an inverted-tree-type structure, the ramification nodes of which are given by sub-objects and the terminal nodes are given by models. The highest level in this hierarchy, that is, level 0 or "root of the tree", is the object itself. The maximum number of hierarchy levels in this organisation has conventionally been set to ten. A KONDAR-object has been realized as a list containing the names of the constituent objects and/or models and their position within the object"s co-ordinate system.

13. The graphic representation module provides frequently-used orthogonal, isometric and other types of parallel projections and central perspectives, which can be graphically enhanced by functions like colouring, hidden-line and hidden-surface elimination and shading.

14. A large collection of standard functions and default "switches" allow graphical and numerical monitoring of the geometric construction process.

15. The recent implementation of a journal enables not only the recording of a KONDAR session, but also provides the user with the possibility of programming construction and representation procedures in the form of macro-commands having variable input parameters .

16. KONDAR has both alphanumeric and graphic input modi and its data interface with other CAD systems is achieved through binary, neutral-ASCII and DXF-files.

KONDAR IN THE DESIGN OF MERO SPACE STRUCTURES

17. The following examples of space structures are presented as summaries of actions and commands describing their geometric construction processes in the KONDAR environment. In order to make the text readable to anyone not familiar with KONDAR, the program's specialised terminology has been replaced with "words" of a more conventional language. The main construction steps are illustrated with the aid of schematic drawings also produced with KONDAR.

The roof space structure for the State Legislative Assembly Building in Sabah, Malaysia, (Architect: JHR, Sabah).

18. The double-roof space structure shown in Figs 1 and 2 is geometrically constructed using a number of doubly-curved, anticlastic surface portions, which span within given boundary lines. The basic carrier surface for the networks is a conoid, which can be generated by connecting with straight lines the corresponding subdivision points of a series of equally-segmented circular arches, as illustrated in Figs 1a and 1b.

19 The topological network of the structure corresponds to that of the commonly known square-on-square double-layer grid, from which every layer is mapped onto a different conoid surface, as shown in Fig. 1d.

20. In order to enable a handy element organisation, the element colour attribute is used to separate points and lines in sub-sets, which are conventionally called bottom and top-layers or chords and, by extension, inter-layer diagonals. To this aim, colour identification numbers are assigned in ascending order to the sequence: bottom layer, inter-layer diagonals and top layer. Furthermore, lines within a layer are distinctly coloured to distinguish the straight generatrix line segments from the circular arches lines. The construction strategy for one roof is as follows:

21. Create a "macro-command" to generate a basic arch, where the radius (r), the number of equal subdivisions (n), the height (h) and the width (w) are variable input parameters (Fig. la).

22. Generate a model having the terminal points of every arch at the straight edges of a top-layer conoid segment, as shown in Fig. 1 a.

23. Generate the various top-layer arches using the above mentioned basic-arch command and use a further macro-command for positioning and addition of the individual arches with the aid of the above mentioned model having the arches terminal points (Fig. la).

24. Connect the corresponding sub-division points of the arches with straight lines, as shown in Fig. 1b.

25. Let the four-sided surface facets be automatically generated, to complete the top-chord of the grid.

26. Generate a "provisional" bottom-chord and the inter-layer diagonals using the "secondary net" function, as shown in Fig. 1d. However, the resulting bottom-chord points are not in the correct locations, but the topological network is the required product here.

27. Generate a separate model for the correct bottom-chord arches, using the macro-commands created for the top-chord arches, as Fig. 1c shows.

28. Add to the model with the correct bottom-chord arches the provisional double-layer grid model.

29. "Unite" the model, so that the provisional bottom-chord points collapse into the correct bottom-chord points and the repeated elements are eliminated. The resulting grid corresponds to one half of the roof and it looks as shown in Fig. 1d.

30. Generate the ridge of the half-roof and add it to the correct two-layered conoid grid model, see Fig. 1e.

31. Solve special edge situations and store the current model.

32. Obtain the second half of the roof by reflecting the current model about the vertical plane defined by the y-z axes, add the reflection to the original half and "unite" the resulting model to eliminate repeated elements. The result is as shown in Fig. 1f.

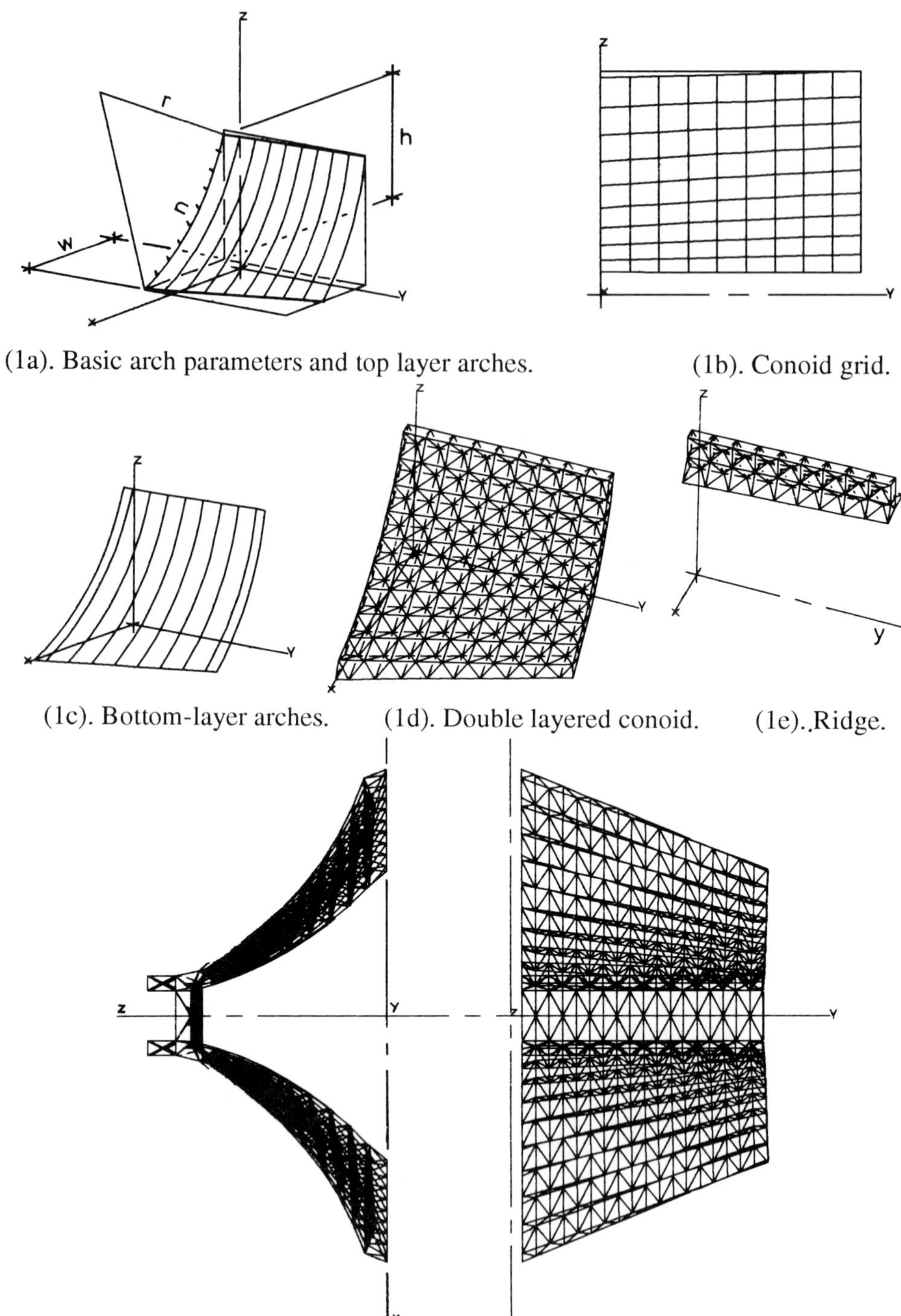

(1a). Basic arch parameters and top layer arches.

(1b). Conoid grid.

(1c). Bottom-layer arches.

(1d). Double layered conoid.

(1e). Ridge.

(1f). A reflection about the y-z plane completes roof-one.

Fig. 1. Sabah roof. Geometric construction summary of roof-one.

33. Re-order the model elements at this point to get easy to handle sub-sets of similar elements. Bottom-chord points are to precede top-chord points. The re-numbering is to start with the arch that is closer to the origin of the co-ordinate system and progress, arch by arch, towards the end facade, that is, in the positive direction of the y-axis. The points in every arch are to be numbered in the direction -x to +x. Lines are to be re-ordered according to their colour "index", that is, the colour attribute identification number.

34. Subject the model to the "Form" function to get the desired inclination of the arches, which have a maximum deviation from the vertical direction at the end facade and gradually turn vertical towards the origin of the co-ordinate system. Translate the model to its final position to get roof one, which looks as shown in Fig. 1f.

35. Construct the second roof following the above generation procedure for the first roof. Add the model of roof two to the model of roof one to produce the final double-roof model.

36. Fig. 2 presents a central perspective of the final model of the double-roof structure, where only the visible facet edges at the top-layer are drawn to improve visualisation. Also, a "floor grid" has been added to enhance the picture.

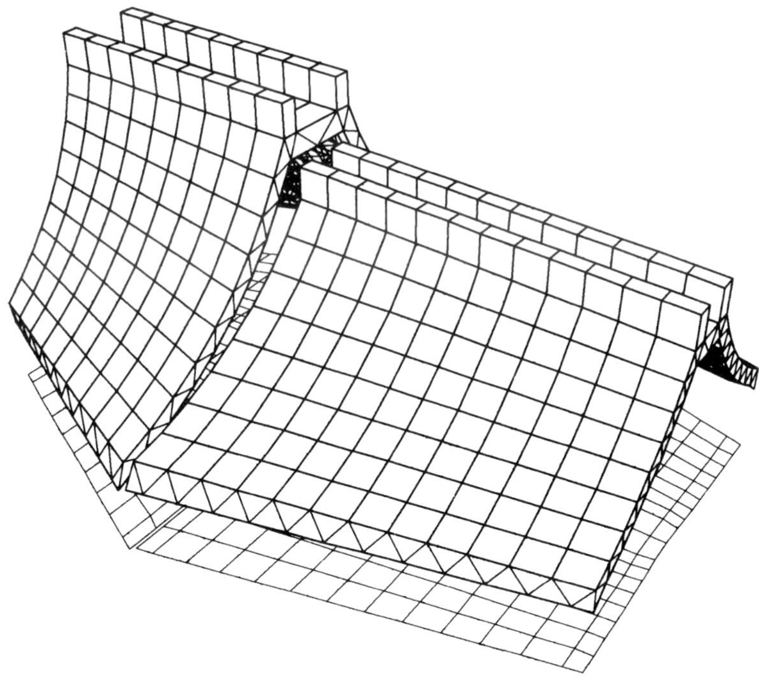

Fig. 2. Sabah roof; perspective view highlighting surface facets.

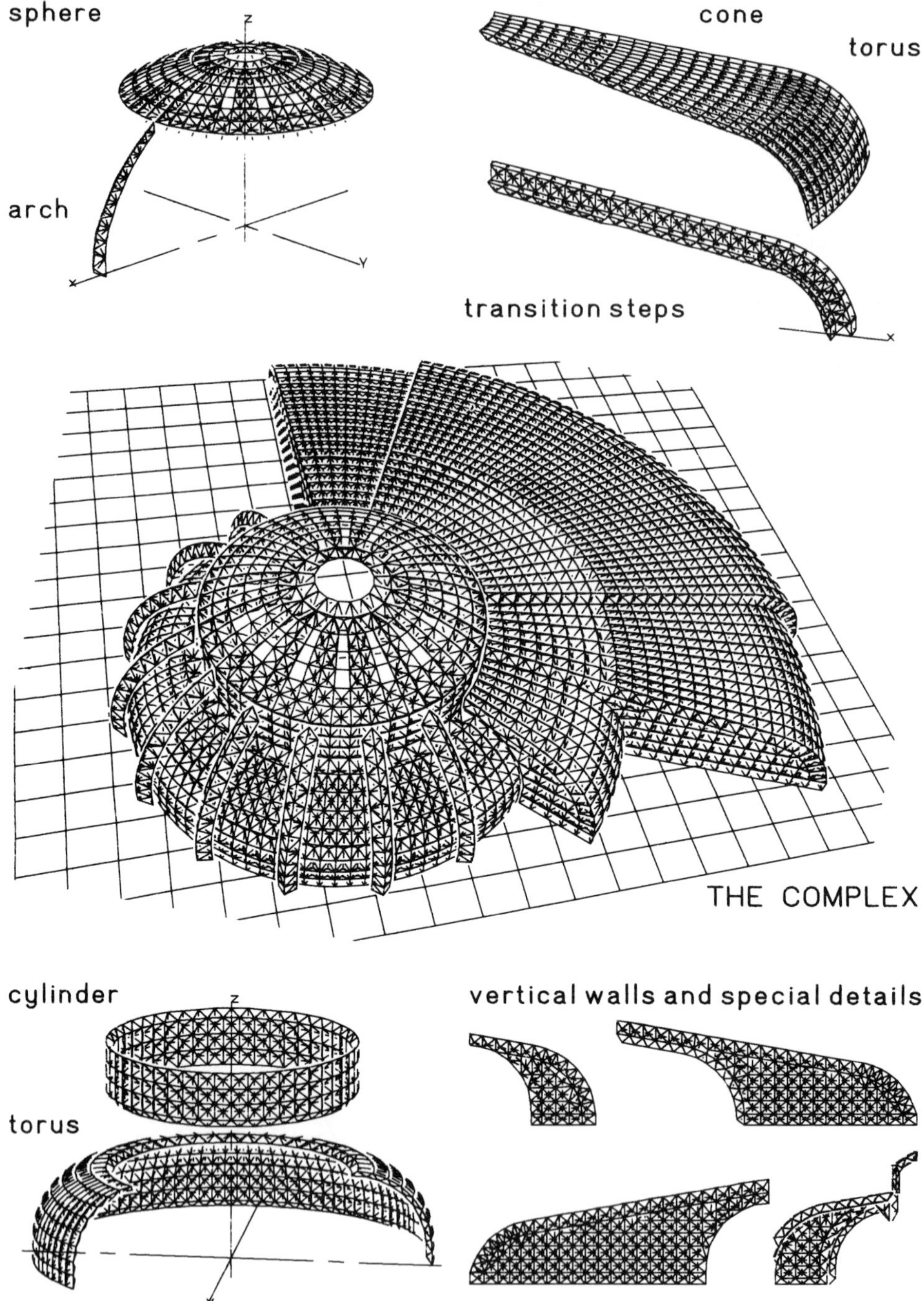

Fig. 3. Grand Slam Canyon structure, geometry complex summary.

The roof space structure for the glass envelope of the Grand Slam Canyon leisure resort at the Circus-Circus Hotel-Casino, Las Vegas, USA. (Architect: V. Simpson, USA).

37. The geometric shape of the space envelope is a complex combination of various double-layer flat plates and double-layer curved shells. The basic geometric forms used to shape the complex are portions or segments of spheres, cylinders, cones, tori and vertical planes, which are "seamless" inter-connected. The basic topological network of the space structure corresponds to that of the commonly known square-on-square double-layer grid, mapped onto the various surfaces. The geometric construction process of this complex structure is very elaborate and cannot be described here in a few paragraphs. Instead, Fig. 3 graphically summarises the complexity of the process by schematically presenting the various basic geometric sub-models around a perspective view of the resulting complex.

38. Further to the geometric complexity and the overall dimensions of the structure (some 28000 square metres of surface area extending approximately 160 by 160 metres in plan view and 40 metres in height), the amount of geometric elements (some 6000 points, 22000 lines and 2760 facets), the large set of non-geometric attributes and the volume of actual construction material required to fabricate the structure turned out to be so large that the modelling of the complex had to be organised in easier-to-handle sub-structures, following not only structural considerations, but also a logistic plan for the workshop production, the packing, the shipping and the installation sequence of the real structure.

39. In order to facilitate data handling in subsequent processes, such as structural analysis, tube and joint dimensioning, graphic representation of the results and computer-aided construction detailing, the various geometric elements of every sub-structure, that is, points, lines and facets, are organised in a manner similar to that of the previously described double-roof for Sabah.

REFERENCES

1. EMDE H. Modellversuch der Bund-Länder-Komission. CAD-Ausbildung in Bauwesen. Leaflet for the Hannover Fair CeBIT' 89, Technische Hochschule Darmstadt, 1989.

2. EMDE H.,SCHMIDT M. and SÖLLER C. KONDAR Benutzerhandbuch, KONDAR-C / V1.7, TH-Darmstadt, 1989.

134. SFCAD and its applications

F. CHU, Xinhong Spaceframe Co. Ltd, P. R. China

SYNOPSIS. A practical spaceframe computer aided design system SFCAD is introduced. The software runs on a Sun SPARC station in the windows environment with the graphic interface of menus, buttons and icons making it very convenient to use. Formex algebra concepts were adopted as the programming basis of configuration data processing, some source codes in C are listed. SFCAD also plots out spaceframe fabrication drawings which can be directly used in our workshops. The author has analysed and designed over 600 spaceframe projects within the past 5 years. Some special solutions recently completed are illustrated.

INTRODUCTION

1. In 1986, the author first introduced Dr. H. Nooshin's formex algebra theory to China and wrote some FORTRAN subroutines on a VAX-II computer to generate spaceframe configuration data for analysis. At that time, graphic monitors and pen plotters were not popular in China, a text menu driven data generating system was developed, configuration and analysis results were only shown on plots via a graphic printer. As the technical partner of Xinhong Spaceframe Co. Ltd, the author has analysed over 100 spaceframes each year during 1987 to 1989. After receiving the computer analysis results, the technicians in the workshops need to design manually the fabrication details of the bolted spherical joints, which was very labour-intensive and error-prone. Due to the rapid development of space structures in China, the text screen based spaceframe analysis program could not meet the requirements.

2. In early 1990, a Sun SPARC station with high resolution colour monitor and a HP 7585B plotter were chosen for developing a spaceframe computer aided design package. SPARC station 1 is a 32 bit RISC graphic workstation with 12.5 MIPS processing speed, and its price only a little higher than that of a high-end 386 PC. The author quickly transferred his own three-dimensional truss analysis program to the SunOS(Unix), and wrote a number of new programs in the C programing language based on Sun View windows environment. A year later, an integrated system called SFCAD was completed with graphic interface of menus,

 Space structures 4. Thomas Telford, London, 1993

buttons and icons. Most types of spaceframe can be easily generated, graphically modified, automatically analysed, and design drawings of the bolted spherical jointed spaceframe can also be conveniently plotted out, which is then be sent directly to the workshop for fabrication and carried to the site for erection. The total design time is thus greatly reduced. In the fiscal year of 1992, SFCAD has accomplished over 250 spaceframes designs with over 1,500 A1 size drawings.

FORMEX ALGEBRA

3. The author's personal opinion of formex algebra is that the concepts are very useful for high level programers of configuration data processing, but the theory is quite abstract for ordinary design engineers to understand. The Formex interactive interpreter cannot be easily popularised because one must be able to write formex formula first.

4. In SFCAD, the formex algebra is hidden behind a graphic interface. Users can click the mouse buttons on the screen menus and buttons, input a few parameters, and follow the screen prompts to generate most types of double layer grids, barrel vaults and some types of domes, then one can move, remove, add nodes or struts, rotate or mirror, etc., to customise the structure. Although most of these functions have been programed using formex concepts, the user feels that he is operating the graphic spaceframe, not the abstract algebra. For some curved space structures, simplified configuration data are generated and modified using SFCAD, then special programs are written to do transformations and to reduce the types of member length. In SFCAD, a standard programing interface is also provided to make programing easy. All the applications illustrated below were completed using this approach.

5. The following shows how the square-on-square offset double-layer grid shown in Fig. 1 was programed behind the menu; the formex representations of node coordinates and strut connectivity are as follows:

```
F=rinid(m+1,n+1,2,2):[1,1,0]#rinid(m,n,2,2):[2,2,-1]

G=vin(1.73,2):F
```

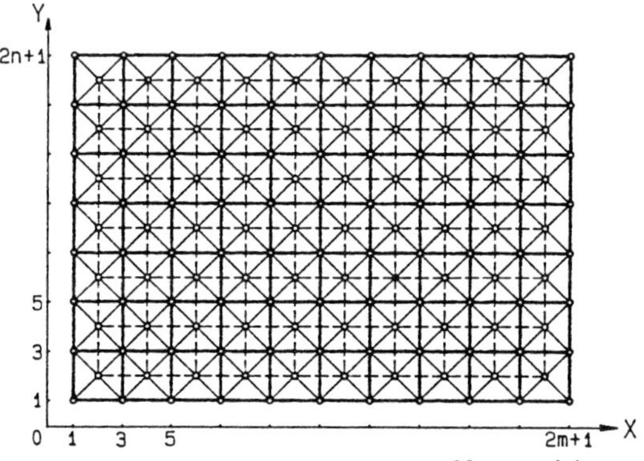

Fig. 1. Square-on-square offset grid.

The simplified C programing source codes are:

```
#include "formex.h"
void soso(m,n,sx,sy,sz) int m,n; float sx,sy,sz;
{ rinid(m+1,n+1,2,2,1,1,0);        /* Upper chord nodes    */
  rinid(m,n,2,2,2,2,-1);           /* Lower chord nodes    */
  vin(1.73,2);                     /* Strut connecttivity */
  for(i=0;i<node;i++) {            /* Transformations      */
    x[i]*=sx/2;
    y[i]*=sy/2;
    z[i]*=sz;
  }
}

void rinid(A,B,C,D,x0,y0,z0) int A,B,C,D,x0,y0,z0;
{ int i,j;
  for(i=0;i<A;i++)
    for(j=0;j<B;j++)
      tranid(C*i,D*j,x0,y0,z0);
}

void tranid(tx,ty,x0,y0,z0) int tx,ty,x0,y0,z0;
{ x[node]=x0+tx;
  y[node]=y0+ty;
  z[node++]=z0;
}

void vin(lower,upper) float lower,upper;
{ int i,j;
  for(i=0;i<node;i++)
    for(j=i+1;j<node;j++) {
      len=sqrt((x[j]-x[i])*(x[j]-x[i])+(y[j]-y[i])*
               (y[j]-y[i])+(z[j]-z[i])*(z[j]-z[i]));
      if(len>=lower&&len<=upper) {
        left[strut]=i;
        right[strut++]=j;
      }
    }
}
```

where the header file "formex.h" defines the node coordinate arrays x[], y[], z[], strut connectivity arrays left[], right[] and other required global variables.

SFCAD GRAPHIC INTERFACE

6. As a structural engineer, the author had to devote a great deal of effort on the computer graphics programing. The rapid development of workstation hardware and software makes this task quite difficult. In addition to the notifier based windows, SunView or OpenWindows there are the graphics libraries Pixrect, CGI, Core, GKS, PHIGS and XGL, etc. The ISO Graphic Kernel System (GKS) in canvas of SunView windows environment was finally selected for programing SFCAD.

7. Formex generating programs were built into icons and buttons (Fig. 2), in the preprocessing window, the generated spaceframe can be graphically modified as deleting, creating, shutting, folding, rotating, mirroring, and assigning node loadings or support conditions (Fig. 3). In GKS, first you need to understand Normalised Device Coordinate(NDC), World Coordinate, window and viewport, then the polyline, polygon, text, colour, primitive and segment, input event and segment transformation. In SunView, you need to know very well Window, Frame, Canvas, Panel, Text, TTY, Menu, Icon, Scrollbar and alert, etc.

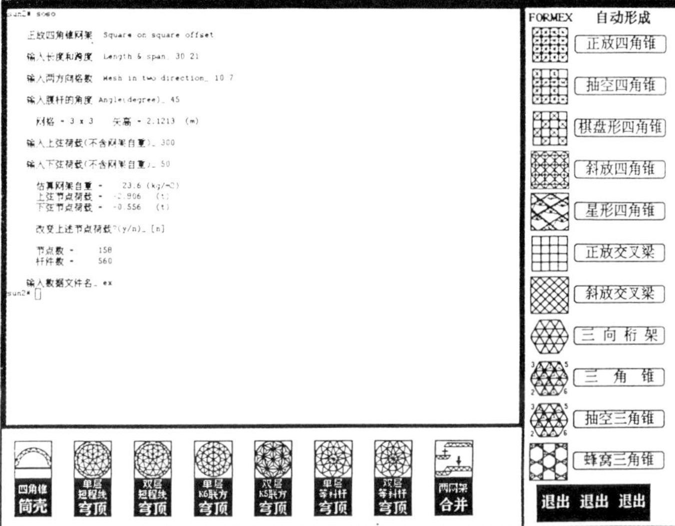

Fig. 2. Formex icons and buttons.

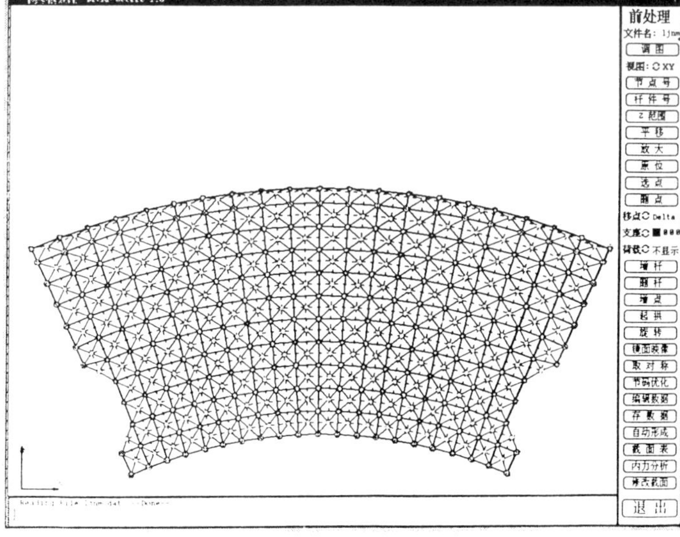

Fig. 3. Pre-processing window.

8. In China, Chinese characters are essential in engineering drawings. As SunGKS does not support them, our own vector set with over 8,000 characters was developed. All menus and prompts are in Chinese making it easier to use by other engineers.

9. Post-processing and detail design are strongly demanded by engineers and spaceframe manufacturers. Fig. 4 shows the fabrication table of struts, and Fig 5 shows the fabrication sketches of nodes, these can be plotted automatically in SFCAD. SFCAD has its own drawing file format and plotter drivers.

No.	ID	SECTION	CUT_L	WELD_L	QTY	BOLT	SLEEVE	CAP/CONE	UNIT_WT	TOTAL_WT
			(mm)	(mm)					(kg)	(kg)
1	1	Φ48x3.5	1924	1944	137	M20	32/35	48x14	7.39	1012.4
2	1A		2310	2330	210				8.87	1863.3
3	1B		2302	2322	160				8.84	1414.7
4	1C		2210	2230	23				8.49	195.2
5	1D		2190	2210	4				8.41	33.6
6	2	Φ60x3.5	2190	2210	4			60x14	10.68	42.7
7	2A		1906	1930	2	M27	45/42	60x16	9.30	18.6
8	2B		2294	2314	16	M20	32/35	60x14	11.19	179.0
9	3	Φ76x4	2210	2322	16			76x60/14	14.66	234.6
10	3A		1906	1930	17	M27	45/42	76x60/16	12.65	215.0
Sum					589					5209

Fig. 4. Fabrication table of struts.

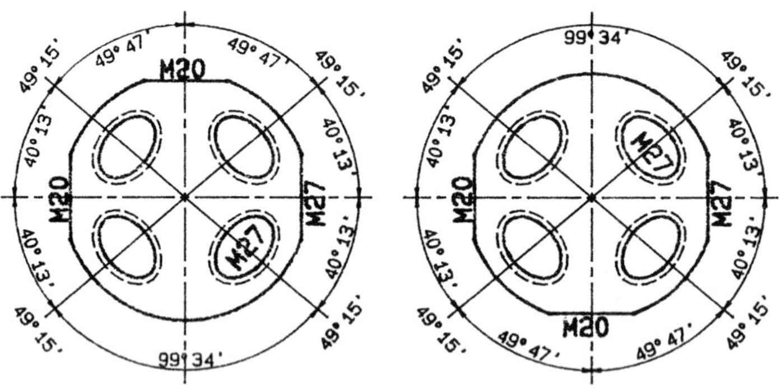

Fig. 5. Fabrication sketches of nodes.

APPLICATIONS

10. Various types of space structure have been accepted by Chinese architects and engineers in recent years. Using SFCAD, the author has completed some special solutions.

11. The cinesphere for the China Science and Technology Museum in Beijing is a double-layer geodesic dome with 35m outer diameter and a 25.5m rise and was completed in 1990 (Fig. 6).

12. A warehouse for the Zhongyuan Fertilizer Plant in Puyang (Fig. 7) is a double-layer barrel vault with a 58m span, 23m rise and a length of 135m and was completed in 1989.

13. An olive dome for the Jinan International Trading Center (Fig. 8) is a double-layer three-way grid which is 58x29m in plan and has a rise of 7m. The structure was completed in 1989.

Fig. 6. Cinesphere geodesic dome in Beijing.

Fig. 7. Puyang barrel vault in construction.

Fig. 8. Jinan Trade Centre olive dome.

14. A double-layer ring-net dome for the Jinan Long Distance Bus Station (Fig. 9) was completed in 1991. The dome is 45m in diameter in plan and has a rise of 12.6m. GRC pyramid cladding panels were used for better appearance.

15. A pair of butterfly spaceframes for the Huaqiaocheng Food Market in Shenzhen (Fig. 10) were completed in 1991.

Fig. 9. Jinan Bus Station ring-net dome.

Fig. 10. Butterfly spaceframe in Shenzhen.

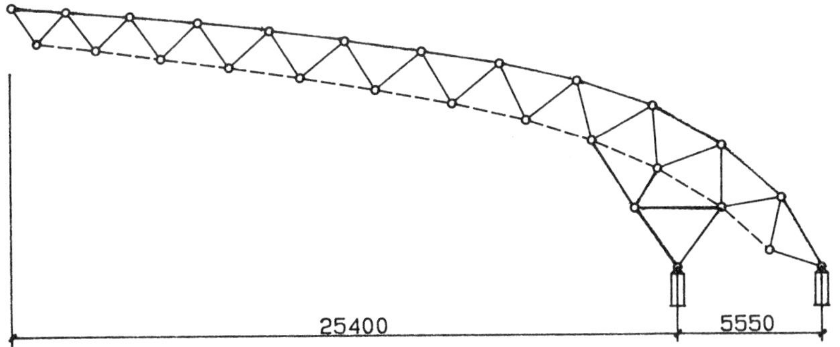

Fig. 11. Section of Shenzhen Stadium spaceframe.

16. A full ring spaceframe for the Shenzhen Stadium (Figs 11 and 12) was completed at the end of 1992. It covers a ground area of 20,000m^2, and the whole roof consists of 6,000 spherical nodes and 22,536 tubular struts with 395 and 640 different types respectively. It is a very good example of using SFCAD for configuration, post-processing and bolted spherical joint details design.

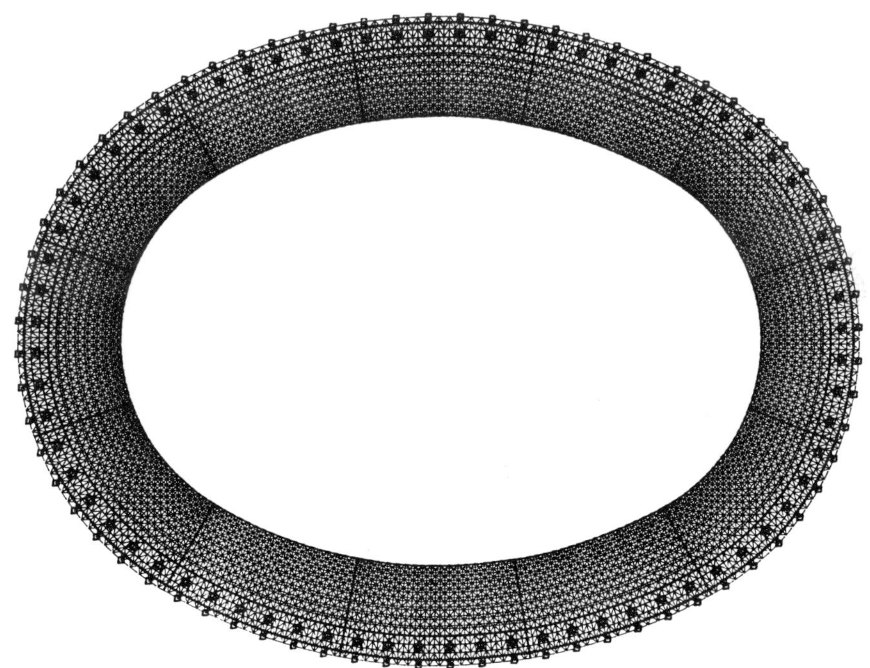

Fig. 12. Layout of Shenzhen Stadium spaceframe.

17. In 1992, the author finished the design of several special projects that will be constructed in 1993. Among them there are the Rongqi Gymnasium (88x76m folded plate, Fig. 13), the Shengyang Summer Palace (75m diameter high and low dome, Fig. 14), and the Zhuhai Gymnasium (84m diameter arch supported dome, Fig. 15).

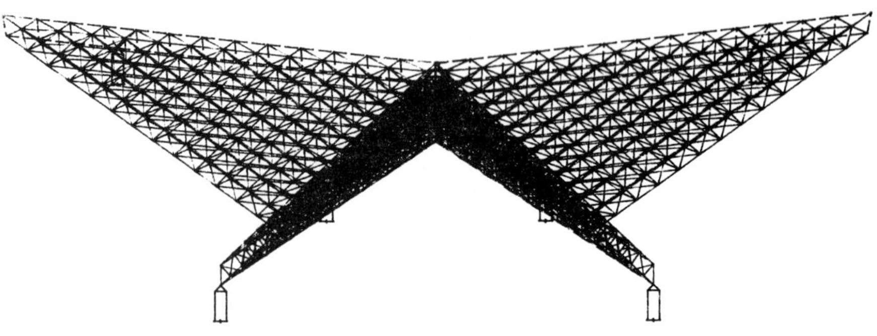

Fig. 13. Rongqi Gymnasium folded plate spaceframe.

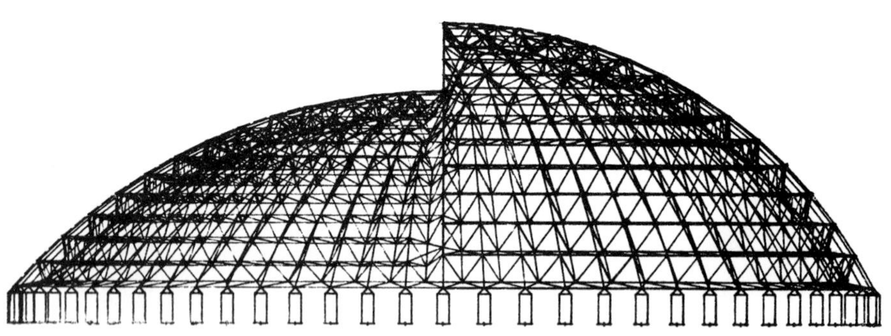

Fig. 14. Shengyang Summer Palace dome.

18. Founded in 1982, Xinhong Spaceframe has completed over 1,000 projects in China and abroad, and SFCAD was used in the design of most of them.

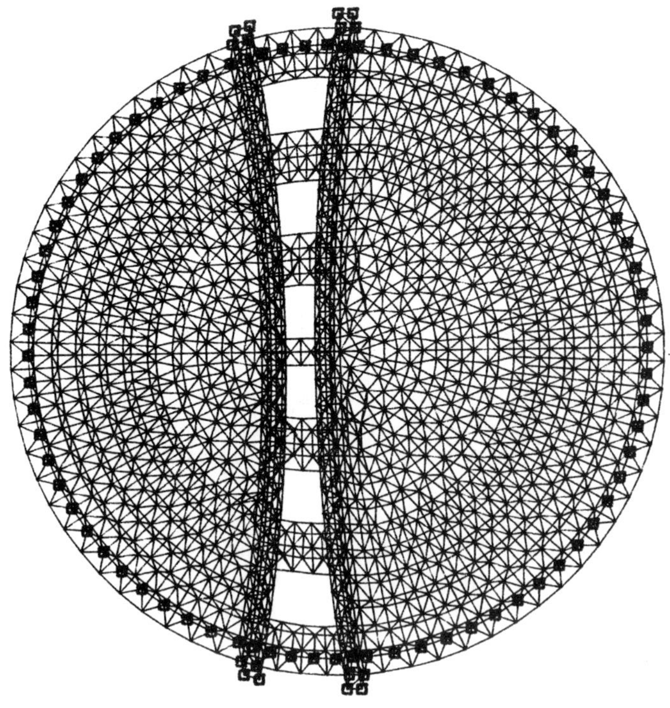

Fig. 15. Zhuhai Gymnasium dome.

REFERENCES

1. NOOSHIN H. Formex Configuration Processing in Structural Engineering, Elsevier Applied Science Publishers, London, 1984.

2. SANCHEZ ALVAREZ J.S. Formex processing of symmetric structural systems, Proceedings of the Third International Conference on Space Structures, Elsevier Applied Science Publishers, London, 1984.

135. Computerized shape generation of shells

I. VIZOTTO, V. ARCARO and F. IGUTI, State University of Campinas, Brazil

SYNOPSIS. This paper presents mathematical modelling for the shape definition of shell structures through the use of mathematical programming techniques combined with the finite element method. The model allows the shape definition of shell structures with any horizontal projection. The applied loads result in an almost pure state of compressive stress. The shapes are generated by loading plane flexible membranes. The model allows the use of concentrated forces, distributed forces, uniform pressure and also non-zero displacement prescriptions. Large displacements and large strains are considered without simplification.

NOTATION. A scalar is expressed by a Greek letter, a vector is always a column matrix and is expressed by a lower case letter and a matrix is expressed by an upper case letter.

FINITE ELEMENT DEFINITION

1. The triangular membrane element with constant strain is modified to allow displacements orthogonal to its original plane, which is established at the xy plane. The interpolating functions for the displacement field on the element are given by :

$$u_x(x, y) = \alpha_1 x + \beta_1 y + \gamma_1 \tag{1}$$

$$u_y(x, y) = \alpha_2 x + \beta_2 y + \gamma_2 \tag{2}$$

$$u_z(x, y) = \alpha_3 x + \beta_3 y + \gamma_3 \tag{3}$$

2. The nine nodal displacements are numbered according to the node number k (k = 1, 2, 3) such that :

u_{3k} : displacement along z axis

u_{3k-1} : displacement along y axis

u_{3k-2} : displacement along x axis

STRAIN COMPONENTS

3. Considering the quadratic strain definition, and also that the material is homogeneous and isotropic, the strain components can be written as :

$$e_{xx} = (c_{11}u_1 + c_{12}u_4 + c_{13}u_7) + \Big[(c_{11}u_1 + c_{12}u_4 + c_{13}u_7)^2 + (c_{11}u_2 + c_{12}u_5 + c_{13}u_8)^2 + (c_{11}u_3 + c_{12}u_6 + c_{13}u_9)^2 \Big] / 2 \tag{4}$$

$$e_{yy} = (c_{21}u_2 + c_{22}u_5 + c_{23}u_8) + \Big[(c_{21}u_1 + c_{22}u_4 + c_{23}u_7)^2 + (c_{21}u_2 + c_{22}u_5 + c_{23}u_8)^2 + (c_{21}u_3 + c_{22}u_6 + c_{23}u_9)^2 \Big] / 2 \tag{5}$$

$$e_{xy} = 0.5 \Big[c_{11}u_2 + c_{12}u_5 + c_{13}u_8 + c_{21}u_1 + c_{22}u_4 + c_{23}u_7 + (c_{11}u_1 + c_{12}u_4 + c_{13}u_7)(c_{21}u_1 + c_{22}u_4 + c_{23}u_7) + (c_{11}u_2 + c_{12}u_5 + c_{13}u_8)(c_{21}u_2 + c_{22}u_5 + c_{23}u_8) + (c_{11}u_3 + c_{12}u_6 + c_{13}u_9)(c_{21}u_3 + c_{22}u_6 + c_{23}u_9) \Big] \tag{6}$$

where c_{ij} is an element of the matrix C,

$$C = \begin{bmatrix} x_1 & y_1 & 1 \\ x_2 & y_2 & 1 \\ x_3 & y_3 & 1 \end{bmatrix}^{-1} \tag{7}$$

where x_k and y_k are initial co-ordinates of node k (k = 1, 2, 3).

CONSTITUTIVE RELATIONSHIP

4. A linear stress strain relationship is considered according to the following expression :

$$s = D\,e \tag{8}$$

with vector s, vector e and matrix D given by :

$$s^T = [\ s_{xx}\ \ s_{yy}\ \ s_{zz}\ ] \tag{9}$$

$$e^T = [\, e_{xx} \;\; e_{yy} \;\; e_{zz} \,] \tag{10}$$

$$D = \frac{\varepsilon}{(1 - \nu^2)} \begin{bmatrix} 1 & \nu & 0 \\ \nu & 1 & 0 \\ 0 & 0 & (1 - \nu) \end{bmatrix} \tag{11}$$

where,
ε : elasticity modulus
ν : Poisson coefficient

POTENTIAL STRAIN ENERGY

5. Defining ω as the undeformed volume of the element, it follows that the potential strain energy π can be written as :

$$\pi(u) = \omega\, e^T D\, e \,/\, 2 \tag{12}$$

GRADIENT AND HESSIAN OF THE POTENTIAL STRAIN ENERGY

6. The gradient of the potential strain energy, known as the internal forces vector, and the hessian of the potential strain energy, known as the stiffness matrix connected to the internal forces vector, can be computed with the assistance of the derivatives of the components of vector e with reference to the nodal displacements.

$$\partial\pi/\partial u_i = \omega\, s^T \partial e/\partial u_i \tag{13}$$

$$\partial^2\pi/(\partial u_i \partial u_j) = \omega \left[\partial e^T/\partial u_i D\, \partial e/\partial u_j + s^T \partial^2 e/(\partial u_i \partial u_j)\right] \tag{14}$$

UNIFORM PRESSURE

7. Considering the deformed structure configuration, defining ρ as the uniform pressure acting on the element, and defining w(u) as a vector orthogonal to the surface of the element, such that its Euclidean norm equals twice the area of the element, it follows that the pressure force p(u) acting on the element can be written as :

$$p(u) = \rho\, w(u) / 2 \tag{15}$$

8. The vector w(u) can be written as the vector product of two other vectors a and b, which can be defined with the assistance of the co-ordinates of the element nodal points on the deformed configuration, thus :

$$a = \begin{bmatrix} (x_2 + u_4) - (x_1 + u_1) \\ (y_2 + u_5) - (y_1 + u_2) \\ (0 + u_6) - (0 + u_3) \end{bmatrix} \tag{16}$$

$$b = \begin{bmatrix} (x_3 + u_7) - (x_1 + u_1) \\ (y_3 + u_8) - (y_1 + u_2) \\ (0 + u_9) - (0 + u_3) \end{bmatrix} \tag{17}$$

therefore,

$$p(u) = \rho/2 \; a \wedge b \tag{18}$$

9. Note that the definition of the vector w(u) requires that the elements incidence are established clockwise. It can be shown that the pressure equivalent nodal forces can be written as :

$$f^T(x) = 1/3\,[\,p_x\ p_y\ p_z\ p_x\ p_y\ p_z\ p_x\ p_y\ p_z\,] \tag{19}$$

where,

$$\begin{aligned} p_x = {} & (y_3 + u_8 - y_1 - u_2)(u_6 - u_3) - \\ & (y_2 + u_5 - y_1 - u_2)(u_9 - u_3) \end{aligned} \tag{20}$$

$$\begin{aligned} p_y = {} & (x_2 + u_4 - x_1 - u_1)(u_9 - u_3) - \\ & (x_3 + u_7 - x_1 - u_1)(u_6 - u_3) \end{aligned} \tag{21}$$

$$\begin{aligned} p_z = {} & (x_3 + u_7 - x_1 - u_1)(y_2 + u_5 - y_1 - u_2) - \\ & (x_2 + u_4 - x_1 - u_1)(y_3 + u_8 - y_1 - u_2) \end{aligned} \tag{22}$$

10. The stiffness matrix F(u) connected to the pressure forces can be written as:

$$F_{11} = F_{41} = F_{71} = 0 \quad (23)$$
$$F_{12} = F_{42} = F_{72} = u_9 - u_6 \quad (24)$$
$$F_{13} = F_{43} = F_{73} = y_2 + u_5 - y_3 - u_8 \quad (25)$$
$$F_{14} = F_{44} = F_{74} = 0 \quad (26)$$
$$F_{15} = F_{45} = F_{75} = u_3 - u_9 \quad (27)$$

$$F_{16} = F_{46} = F_{76} = y_3 + u_8 - y_1 - u_2 \quad (28)$$
$$F_{17} = F_{47} = F_{77} = 0 \quad (29)$$
$$F_{18} = F_{48} = F_{78} = u_6 - u_3 \quad (30)$$
$$F_{19} = F_{49} = F_{79} = y_1 + u_2 - y_2 - u_5 \quad (31)$$

$$F_{21} = F_{51} = F_{81} = u_6 - u_9 \quad (32)$$
$$F_{22} = F_{52} = F_{82} = 0 \quad (33)$$
$$F_{23} = F_{53} = F_{83} = x_3 + u_7 - x_2 - u_4 \quad (34)$$
$$F_{24} = F_{54} = F_{84} = u_9 - u_3 \quad (35)$$
$$F_{25} = F_{55} = F_{85} = 0 \quad (36)$$
$$F_{26} = F_{56} = F_{86} = x_1 + u_1 - x_3 - u_7 \quad (37)$$
$$F_{27} = F_{57} = F_{87} = u_3 - u_6 \quad (38)$$
$$F_{28} = F_{58} = F_{88} = 0 \quad (39)$$
$$F_{29} = F_{59} = F_{89} = x_2 + u_4 - x_1 - u_1 \quad (40)$$

$$F_{31} = F_{61} = F_{91} = y_3 + u_8 - y_2 - u_5 \quad (41)$$
$$F_{32} = F_{62} = F_{92} = x_2 + u_4 - x_3 - u_7 \quad (42)$$
$$F_{33} = F_{63} = F_{93} = 0 \quad (43)$$
$$F_{34} = F_{64} = F_{94} = y_1 + u_2 - y_3 - u_8 \quad (44)$$
$$F_{35} = F_{65} = F_{95} = x_3 + u_7 - x_1 - u_1 \quad (45)$$
$$F_{36} = F_{66} = F_{96} = 0 \quad (46)$$
$$F_{37} = F_{67} = F_{97} = y_2 + u_5 - y_1 - u_2 \quad (47)$$
$$F_{38} = F_{68} = F_{98} = x_1 + u_1 - x_2 - u_4 \quad (48)$$
$$F_{39} = F_{69} = F_{99} = 0 \quad (49)$$

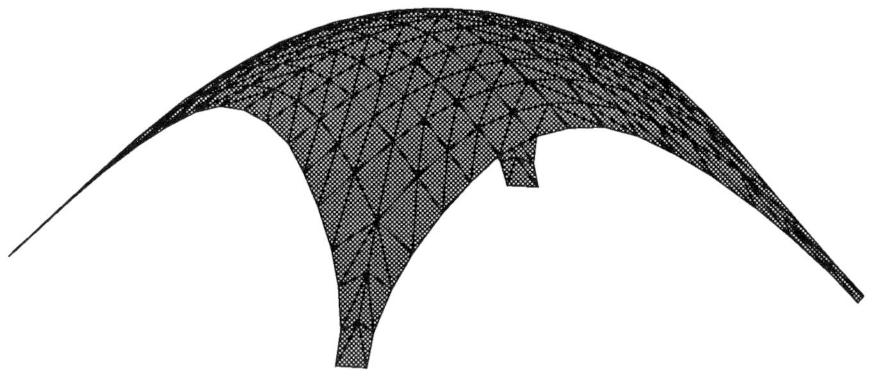

Fig. 1. Weight acting upward.

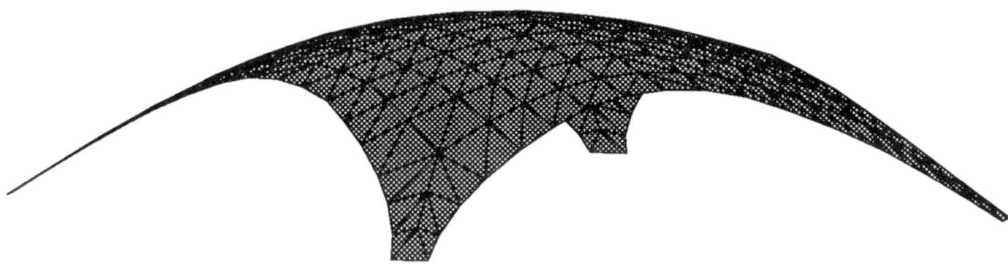

Fig. 2. Uniform pressure.

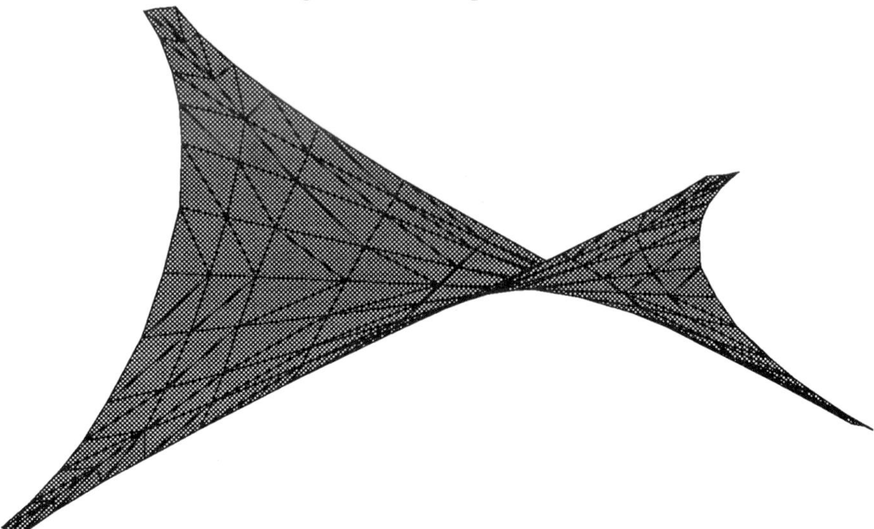

Fig. 3. Non-zero displacement prescription.

11. Note that F(u) is not symmetric and therefore the potential energy function connected to the pressure forces does not exist.

EQUILIBRIUM CONFIGURATIONS

12. When the total potential energy function exists, the stable equilibrium configurations correspond to local minimum points of this function. It is advisable to use a Quasi Newton type method to find these local minima because this does not require the evaluation of the stiffness matrix. Considering x as the vector of unknown displacements and considering f as the vector of nodal forces, the total potential energy function Π can be written as :

$$\Pi(x) = \sum_{elements} \pi(x) - f^T x + \Pi_0 \qquad (50)$$

and its gradient can be written as :

$$\nabla\Pi(x) = \sum_{elements} \nabla\pi(x) - f \qquad (51)$$

where,

Π_o : constant

13. When the total potential energy function does not exist, it is advisable to use an incremental Newton Raphson type method to find the solutions of the non-linear system of equations given by the equilibrium equations :

$$\sum_{elements} (\nabla\pi(x) - f(x)) = 0 \qquad (52)$$

14. Note that in this case the evaluation of the stiffness matrix is required, although a close estimate of the stiffness matrix can be found through finite differences. A comprehensive discussion concerning equilibrium configurations can be found in Arcaro (ref. 1).

EXAMPLES

15. The following three shape definitions of shell structures were generated by loading plane flexible membranes.

REFERENCES

1. ARCARO V. F. Conceitos Fundamentais e Metodos Computacionais Basicos para Encontrar Configuracoes de Equilibrio de Estruturas Nao Lineares, p30. Unicamp, Campinas, 1992.

136. A reduction method for the solution of large sets of equilibrium equations

F. SUTO, Tokyo Denki University, Japan

SYNOPSIS. This paper investigates ways of reducing, into a sub-structured form, the large stiffness matrix equation, emanating from the analysis of complex space structures, into a suitable form for solution on a very high speed super computer.

INTRODUCTION

1. In this paper a basic feasibility study is described into ways of reducing the stiffness matrix equation associated with the analysis of space frame, with the aim of applying the method to very high speed super computers or connection machines. Until recently the sub-structuring method after the Neumann-type single processing has been used for this purpose, whereas the method described here, differs as it belongs to the non-Neumann-type plural processing, suitable for systems of multi-linked computers. It is characterised by a mechanical and algebraic idea intending to divide the very large structural matrix into a proper number of independently blocked, sub-matrix equations, so as to be able to use the super computer or connection machines in their most efficient way.

2. The mechanical and algebraic process treated here consists of (a) the division into blocks of the whole original stiffness matrix equation by intersecting along boundaries so as not to loose the physical values of the stiffnesses and loads at selected co-existant nodal points, and (b) the formation of the rearranged matrix equation formed as independent blocks of sub-matrix equations obtained by the relocation and addition of rows and columns in the original matrix equation. Thus the paper develops the derivation of a generalised formula on the reduction method for a large scale stiffness matrix equation.

3. On one hand, from the general viewpoint of the mechanisms, the super computer or connection machines are at present characterised as an assemblage of many large interconnected computers or micro-processors. Therefore they can exert their utmost to reduce the original matrix equation size very effectively, when applied in parallel operations to structural problems needing very time-consuming processes, as is the case with the analysis of large frameworks or the non-linear analysis of space structures. In order to optimise recent multiprocessor

computer systems thought must be given as to how to partition the large structural stiffness matrix into proper, independent and sub-structured blocks and consequently solve the system of equations.

4. For any general space frame to be considered, the following points are relevant:

a. the space frame is constructed from tubular members;
b. the angle made between the z-axis of the member co-ordinate system (o-x_m y_m z_m) and the z axis of the global co-ordinate system (O-X Y Z) is set at right-angles for each member;
c. the original structural stiffness matrix equation is formed where the member stiffness are transformed from the local co-ordinate system into the global co-ordinate system.
d. each sub-structured block of the matrix, concerning the global co-ordinates O-XYZ is arranged so as to be processed independently of others when solved by a constituent computer accessed on a multiprocessor computer system;
e. discretization and the establishment of each block are carried out using completely regular, algebraic and mechanical means without any complicated exceptions.

5. A simple dome space frame is used to exemplify the second phase of this study in order to verify the method. In addition some remarks on comparing the results obtained by parallel processing with those obtained using conventional solution methods are given.

THE FLOW SEQUENCE FOR THE COMPUTING PROCESS

6. The entire flow sequence of the process described is shown in Fig. 1. Generally the procedure consists of the following items: i.e. (item 1) formation of the original structural stiffness matrix equation transformed into the global co-ordinate system from the local co-ordinate system; (item 2) relocation of rows and columns of the above matrix equation into blocks to be divided; (items 3 & 4) by intersecting boundary members along each block circumference and re-evaluating their stiffness values; (item 5) forming each sub-structured stiffness matrix in a completely independent block; (item 6) dividing each original nodal force relevant to block-boundaries into those to be used for the intersected nodes common to adjacent blocks; (items 7 & 8) after the above items are processed, the re-arranged structural stiffness matrix equation is reduced based on the parallel processing of respective blocks by the super computer or connection machines; (items 9 & 10) the finally reduced structural stiffness matrix equation is solved for the unknown displacement vector; (item 11) solving for all the unknown displacements related to the individual blocks, applying the inverse process of reduction for respective blocked equations; (item 12) transforming the above displacements into the global co-ordinate system; (item 13) resulting in the calculation of all the inner forces of the constituent members.

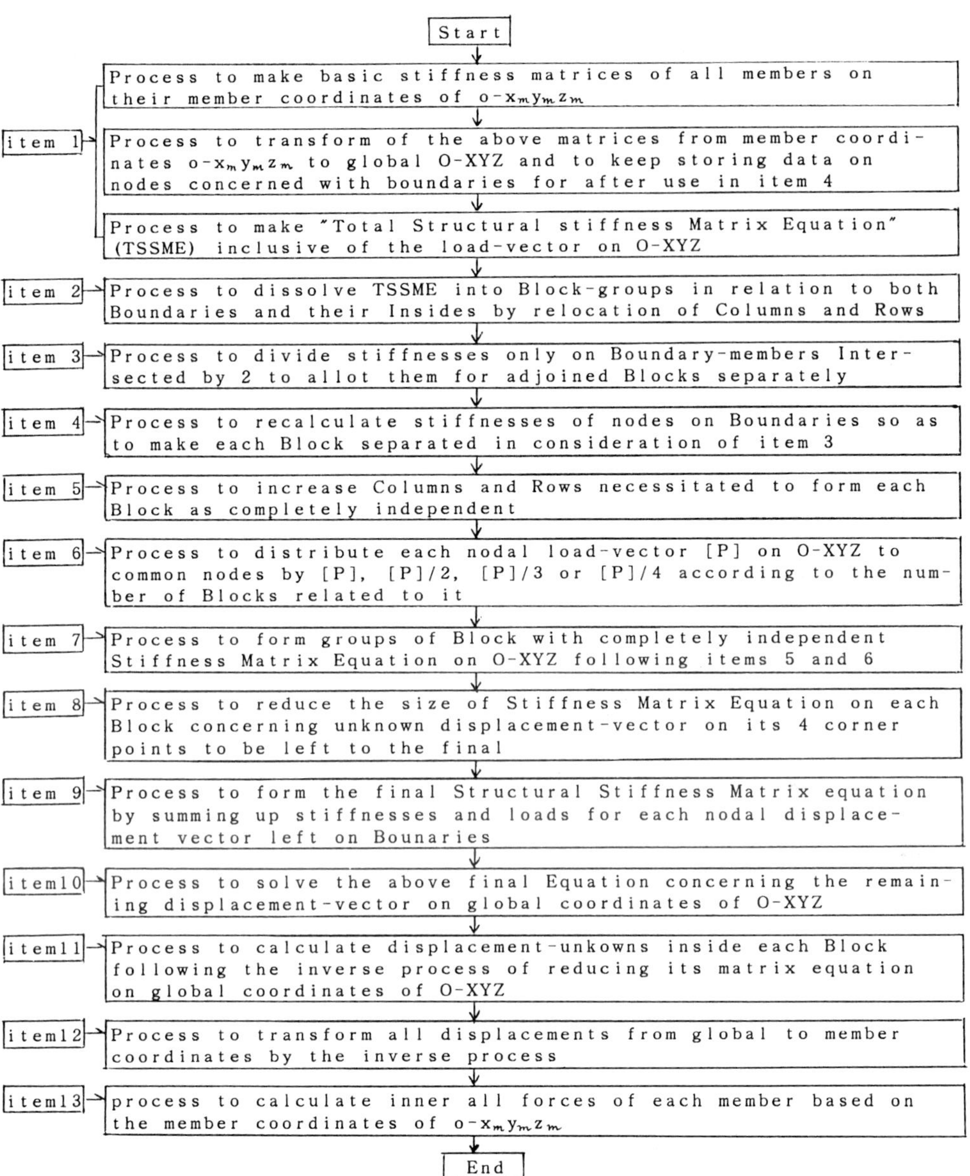

Fig. 1 Flow chart.

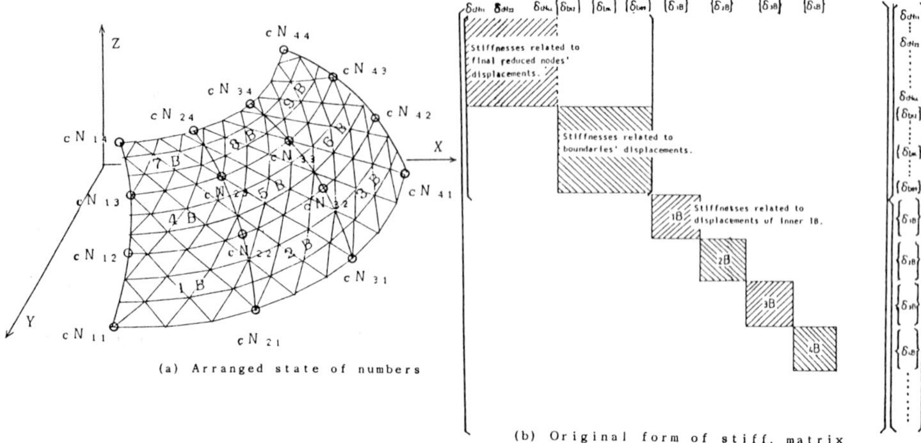

Fig. 2 Arrangement of node, boundary, and discretised block-numbers.

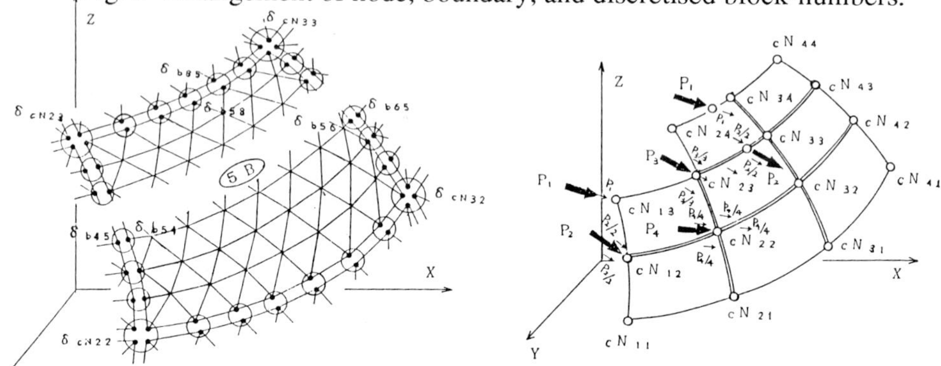

Fig. 3. Common nodes arround block Number 5.

Fig. 4. Distributed nodal forces corresponding to surrounding blocks.

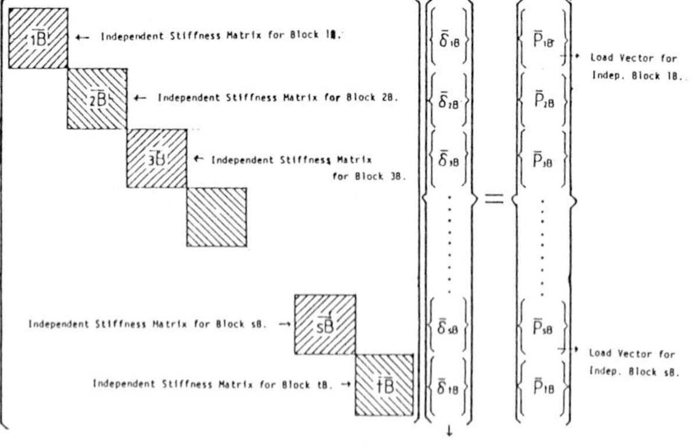

Fig. 5. Arranged form of the whole structural stiffness matrix sub-structured into independent blocks.

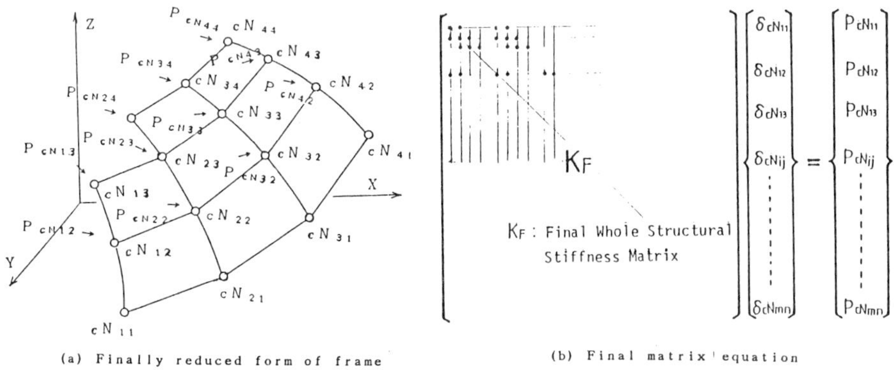

Fig. 6 Final reduced matrix equation.

PRINCIPAL PROCESS

7. The basic algebraic processing making the sub-structured matrix consisting of independent blocks is described showing how to divide an original structural stiffness matrix into blocks so as to make it amenable to parallel processing.

8. The process of reducing an extremely large structural stiffness matrix emanating from the analysis of a large space frame into a series of small matrices by this sub-structuring method is performed as follows:

a. first the original structural stiffness matrix equation related to the 3D global co-ordinate system is formed in the usual way;
b. the above matrix is then grouped into blocks in sub-structured form with consideration to the number of constituent processors available;
c. nodes on the boundaries of the block are numbered orderly related to surrounding blocks (Fig. 2);
d. stiffness values of each boundary member are divided by 1 or 2 respectively depending on whether there exists an adjacent block having the same member common (Fig. 3);
e. the load for each node on the block boundaries is divided by 2, 3 or 4 depending on whether there exist 2, 3 or 4 blocks having the same common node (Fig. 4);
f. in the original structural stiffness matrix, columns and rows are increased in relation to "d" and "e" above in order to divide the matrix into independent sub-structured equations;
g. the matrix transformed as outlined in "f" above is implemented by replacing columns and rows pertinently so that each sub-structured block can be formed into completely independent matrix equations (Fig. 5);
h. the reduction process is then carried out for each of the independent sub-structured matrix equations using the multiprocessor computer system;
i. finally the reduced structural stiffness matrix equations are obtained by re-arranging the matrix elements summed together to achieve the correct displacements (Fig. 6).

DETAILS OF THE MATRIX REFORMATION

9. The precise process for reforming the stiffness matrix is shown in Fig. 7. The figure expresses the situation around Block 1 in Fig. 2. As shown the Block 1 matrix is processed as follows:

i. the initial state of the stiffness matrix of the block is rearranged into three matrices, i.e. a sub-matrix related to the nodal displacement vector and sub-matrices related to those of boundary blocks;
ii. the columns and rows concerning the nodes cN12, cN21 and boundaries are increased referring to the former Section 8-f describing the division of member-stiffness and consistent allocation of stiffness-elements for adjacent block-nodes as shown in Fig. 8;
iii. after increasing the number of columns and rows related to nodes on cross-points and on boundaries, independent matrices of lB, 2B etc. are formed by relocating them to their respective defined places;
iv. corresponding with the above independent matrices, the nodal loads are divided, referring to the former Section 8-e, and set to form independent matrix equations.

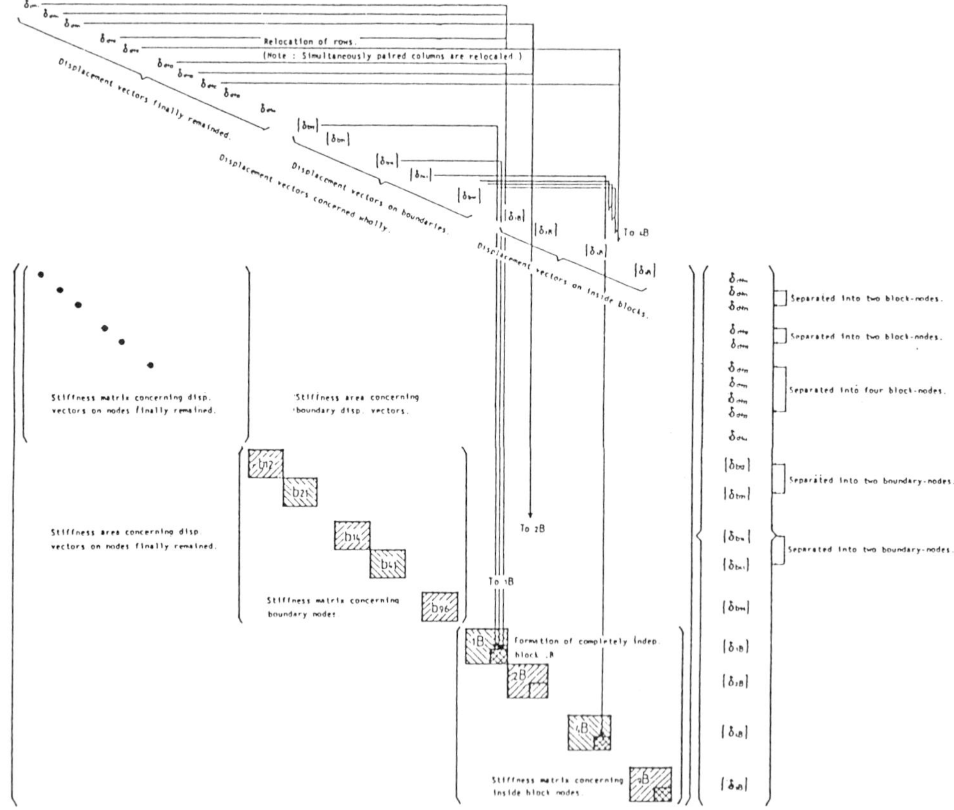

Fig. 7 Matrix reconstitution by the relocation of columns and rows.

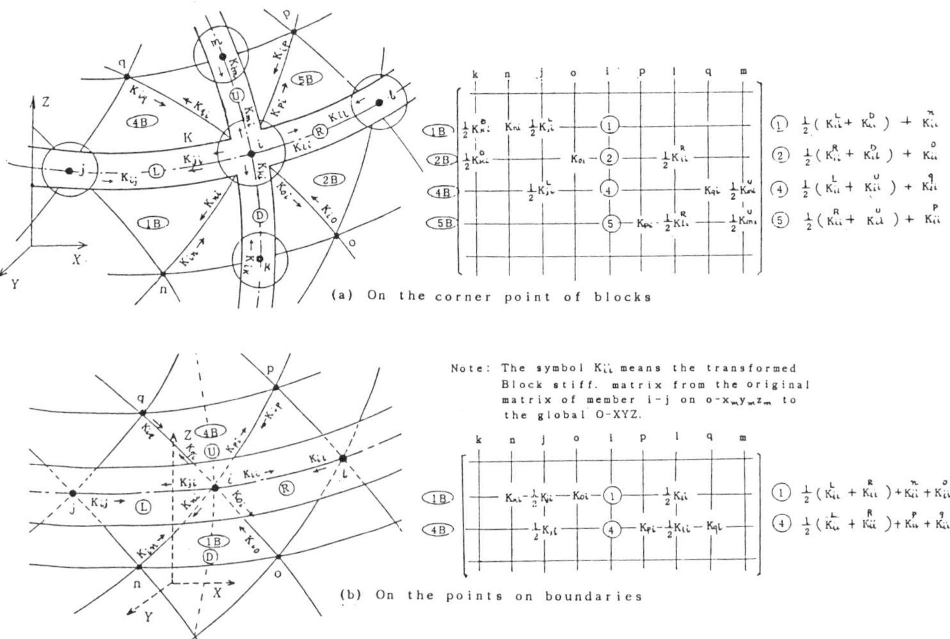

Fig. 8 Divided stiffness at nodes on intersected boundaries.

CONSIDERATION OF THE COMPARATIVE NUMERICAL ANALYSES

10. In order to verify this approach for independently sub-structured stiffness matrices, a test on a simple half dome, constructed from rigidly connected steel tubes with diameters from 30 to 60cm in the form of a single layer network, has been analysed for both vertically and horizontal loading of magnitude 1 ton.f applied to all nodal points as shown in Fig. 9. As the main aim of the investigation was to review the analytical results obtained by the two approaches, their nodal translations and rotations, u,v,w,θ_X,θ_Y,θ_Z, are shown in Table 1. Further when the values of comparative differences *y* described in Fig. 10 of the frequency distribution have been calculated, it is apparent that there appears to be larger mean values i.e. about 26% obtained using the parallel technique compared with the conventional approach. It is thought that this tendency suggests modification of the method and the following points require further consideration:

i. investigating how to divide the full stiffness matrix into an optimum number of blocks;
ii. imposing the correct boundary restrictions by incorporating more nodes on the boundaries to prevent a decrease in the final stiffness;
iii. modifying, by a reasonable method, both the stiffness and nodal loads to common boundary nodes using a more accurate physical way so as to be

able to apply this method extensively even for more general and irregular forms of space frames;

iv. evaluation of relative numerical errors occurring during the calculation process in both methods.

11. Although the above problems are to be pursued, the relative differences obtained from the two approaches will decrease if careful consideration is given to the preceding points.

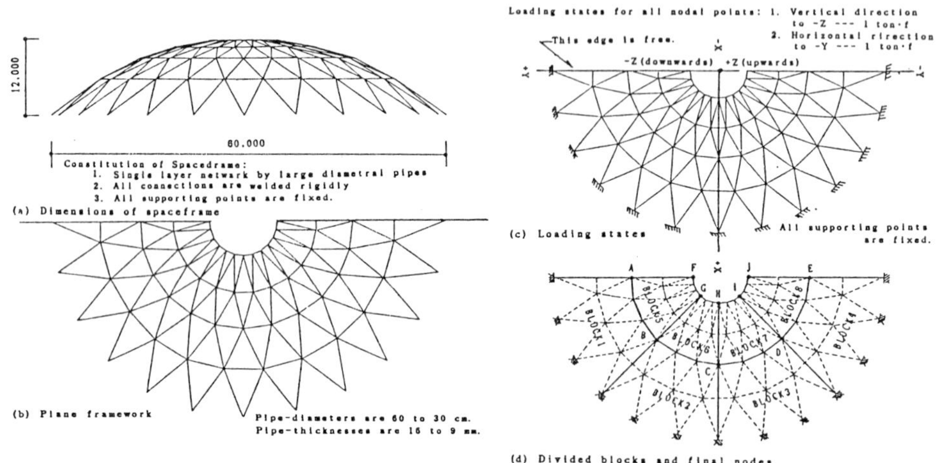

Fig. 9 Space frame model for test.

Table 1. Comparison of displacements on nodes finally left.

Points / Direct.		node A method (1)	node A method (2)	node B method (1)	node B method (2)	node C method (1)	node C method (2)	node D method (1)	node D method (2)	node E method (1)	node E method (2)
Vertical loading	u	-0.705036E+00	-0.954542E+00	-0.336282E-01	-0.345951E-01	0.249743E-05	0.295082E-04	0.336399E-01	0.347153E-01	0.705042E+00	0.954692E+00
	v	-0.552281E-01	-0.792634E-01	-0.252755E+00	-0.278813E+00	-0.339530E+00	-0.419950E+00	-0.252740E+00	-0.278780E+00	-0.552147E-01	-0.792155E-01
	w	0.521713E+00	0.855747E+00	-0.325181E+00	-0.429629E+00	-0.508615E+00	-0.628666E+00	-0.325151E+00	-0.429435E+00	0.521753E+00	0.856005E+00
	θ_x	-0.412535E-03	-0.174618E-02	0.341778E-03	0.943001E-03	0.518963E-03	0.117781E-02	0.341739E-03	0.943123E-03	-0.412566E-03	-0.174630E-02
	θ_y	0.203240E-02	0.226562E-02	0.832117E-03	0.109990E-02	-0.135702E-07	0.231020E-06	-0.832101E-03	-0.110032E-02	-0.203228E-02	-0.226562E-02
	θ_z	-0.746853E-03	-0.137440E-02	-0.280006E-03	-0.238321E-03	0.915418E-08	-0.111852E-06	0.280012E-03	0.238379E-03	0.746839E-03	0.137446E-02
Horizontal loading	u	0.645144E+00	0.883482E+00	0.365374E+00	0.471624E+00	0.212001E+00	0.231382E+00	0.365380E+00	0.471638E+00	0.645207E+00	0.883746E+00
	v	-0.370812E+00	-0.459688E+00	-0.288618E+00	-0.358948E+00	-0.876328E-05	-0.215986E-03	0.288612E+00	0.358610E+00	0.370824E+00	0.459548E+00
	w	-0.136660E+01	-0.179486E+01	-0.639710E+00	-0.715232E+00	-0.127619E-04	-0.845874E-04	0.639707E+00	0.715568E+00	0.136671E+01	0.179510E+01
	θ_x	0.116592E-02	0.164186E-02	0.610751E-03	0.808628E-03	0.151613E-07	-0.557070E-06	-0.610757E-03	-0.809465E-03	-0.116607E-02	-0.164131E-02
	θ_y	0.137051E-02	0.161176E-02	-0.146257E-05	-0.159389E-03	-0.496618E-03	-0.159175E-03	-0.150170E-05	-0.159011E-03	0.137050E-02	0.161137E-02
	θ_z	0.212533E-03	0.347200E-03	0.246180E-03	0.377766E-03	0.312397E-03	0.248300E-03	0.246201E-03	0.378059E-03	0.212613E-03	0.347306E-03

Note: 1. methods mean (1)-the Direct reduction and (2)-this Block one
2. units are taken as (1) cm for u, v, w and (2) radian for $\Theta_X, \Theta_Y, \Theta_Z$

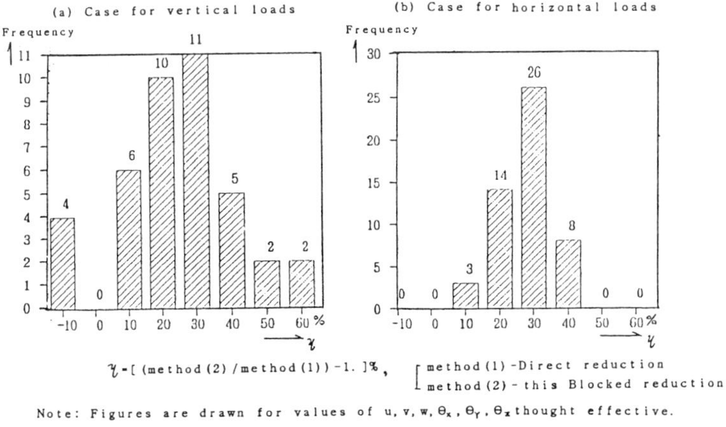

Fig. 10 Histogram of deviation.

CONCLUSION

12. The author concludes that his method aiming at non-Neumann parallel processing described in this paper can be appropriate for the general reduction of large structural stiffness matrix equations when applied to a super computer or connection machines. By its application the efficient speedy processing of large structural problems associated with static, dynamic, linear or non-linear analysis can be achieved.

ACKNOWLEDGEMENT

13. The author thanks Mrs H. Hagiwara, M. Okajima and N. Shimizu, and all students in his laboratory, for their earnest co-operation in computing the test example and for drawing the necessary figures .

REFERENCES

1. SUTO, F. and ASAYAMA, S., Matrix-algebraic process on large reduction of its rank on super computers, Extended Abstract Book, The 4th Int. Con. on Computing in Civil and Building Eng. Tokyo 1991 (IV-ICCCBE Tokyo 1991)
2. SUTO, F., ASAYAMA, S. and ASADA, M., A Large Reduction Method of Matrix Struc. Equation for Super-Computer Part 1. Theory based on Intersected Blocks set with Boundary Springs, summaries of techcl. papers of annual meeting 1991, AIJ
3. SUTO, F., ASAYAMA, S. and HAGIWARA, H., A Large Reduction Method of Matrix Struc. Equation for Super-computer Part 2. Considerations on Appropriate Spring-values inserted, summaries of techcl. papers of annual meeting 1991, AIJ
4. KAMAL, O.and ADELI, H., Automatic Partitioning of Frame Structures for Concurrent Processing, Microcomputers in Civil Engineering 5,269-283, 1990, Elsevier Science Publishing Co.,lnc.

5. SUTO, F., A Large Matrix Reduction Method for Multiprocessors, Microcomputers in Civil Engineering, 7, (1992), 191-199, Elsevier Science Publishing Co., Inc.
6. SUTO, F., A Large Reduction Method on Structural Stiffness Matrix by Substructuring with Virtual Springs for Microprocessors, Proceedings of The 2nd Canadian Conf. on Computing in Civil Engineering, Ottawa, 1992

137. Linked interactive graphic software addressed to static and dynamic analysis of 3D structural systems

M. MAJOWIECKI, Universita di Bologna, Italy, and R. TRAVISAN, Enexsys, Italy

SYNOPSIS. After a brief introduction on the state of the art, the main features of a new package dedicated to civil and structural modelling is illustrated. The package, named STRAND (STructural ANalysis & Design), written in the C language, is organised in accordance with CAD technology. With a WYSIWYG interface it is possible to manage data in order to minimise the errors in the design process and increase the reliability of the User-Software-Hardware system.

INTERACTIVE GRAPHIC LANGUAGE ADDRESSED TO THE STRUCTURAL DESIGNING PROCESS

1. As far as structural analysis by means of an electronic computer is concerned, we are presently witnessing the development of "specialised software" for applied fields (civil, mechanical, aeronautical) or even more specialised ones for construction types (buildings, skyscrapers, bridges, lightweight structures, etc.). This development is based on past experiences relative to "general purpose" programs, originating principally in the field of aeronautic construction (SAP, ADINA, NASTRAN, MARC, etc.). At the beginning of the 60's, the first "batch" modellings, aimed at civil buildings, were made by Clough et al. (ref. 10) with the FRMSTC and FRMDYN programs, TABS by Wilson and Dovery starting from 1972 (ref. 11) and ETABS by Wilson et al. (ref. 12), written in 1975.

2. A new generation of programs directed towards research into the pre-stressed membrane form and the analysis of 3-D buildings of the interactive graphic type were developed at the Computer Graphic Laboratory of the University of Cornell (USA) and introduced into the technical literature at the end of the 70's (refs 13-15).

THE STRAND PROGRAM (STRUCTURAL ANALYSIS AND DESIGN)

3. The program presented in this paper represents the most advanced state of the extensive research on the use of interactive graphic techniques (I.G.) in the field of structural engineering, carried out beginning in 1970 in the Department of

Structural Engineering of the University of Bologna, with the assistance of software houses specialised in CAD techniques (refs 16-21).

4. The research development was carried out principally in the field of lightweight structures (tensile structures, prestressed and pneumatic membranes) and three-dimensional framed systems in reinforced concrete and steel, subject to seismic loading.

5. The addressed interactive graphic language was obtained developing and implementing special hardware and software interfaces and a special command filter for the user-hardware communication (Fig. 1).

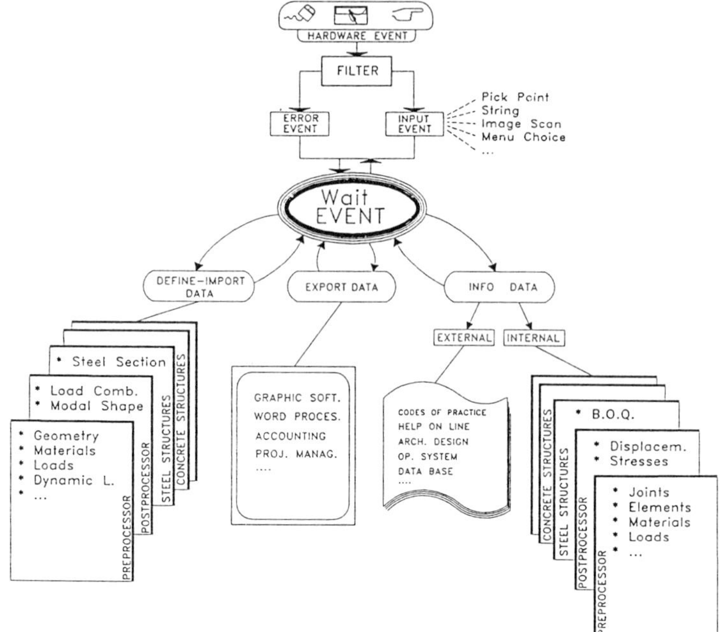

Fig. 1. Command filter.

6. The following illustrated program operates at the second level (CAD), while it is in a research and testing phase as far as the implementation of the semantic simulation of artificial intelligence is concerned (EXPERT SYSTEM) (Fig. 2) (ref. 22).

7. With respect to the first operative version written in FORTRAN (ref. 21) and in accordance with the profound transformation taking place in the technical-scientific hardware systems, the program was re-written in the C language and its interactive functions were completely restructured in order to optimise its use on the 32 bit UNIX work-situation, the PC-DOS, the 386 DOS EXTENDER (PHAR-LAP) and, more recently, the WINDOWS 3.1 version.

8. The hardware and software system is organised as an INTERFACED NETWORKED SYSTEM (Fig. 3). The flow chart of the general and addressed functions of the program is illustrated in Fig. 4.

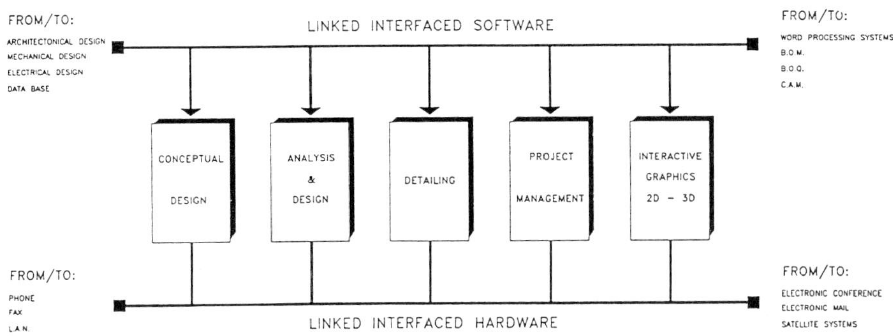

Fig. 2. User communication levels.

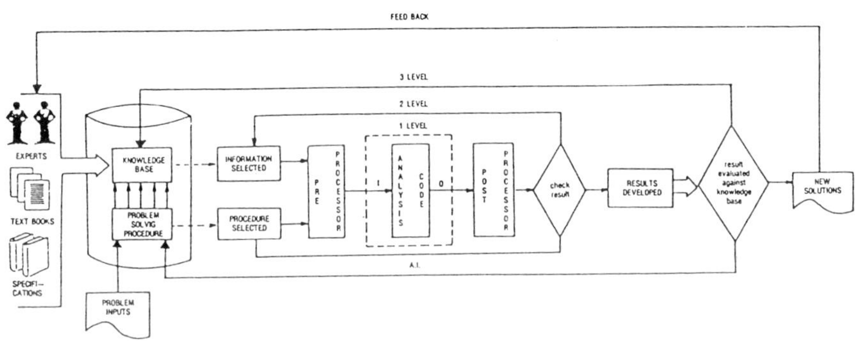

Fig. 3. Networked interfaced software-hardware system of modular interchangeable computational programs/system.

INTERACTIVE GRAPHIC PRE-PROCESSOR

The areas of the screen

9. The screen is divided into 7 areas, each with a different function.

10. Area 1 (Fig. 5) is dedicated to the clipping commands and therefore allows a zoom-in on the structure and the display on the screen of only a part, section or element of the same (windowing).

11. Area 2 allows the display of views of the structure. Graphic images in the xy, xz and yz planes, isometric images or perspective views can be obtained with the possibility of defining the viewpoint.

12. Area 3 is dedicated to the main program menu. Each line of the menu allows access to a dynamic (Fig. 6) sub-menu (which can be superimposed on the image already existing on the screen).

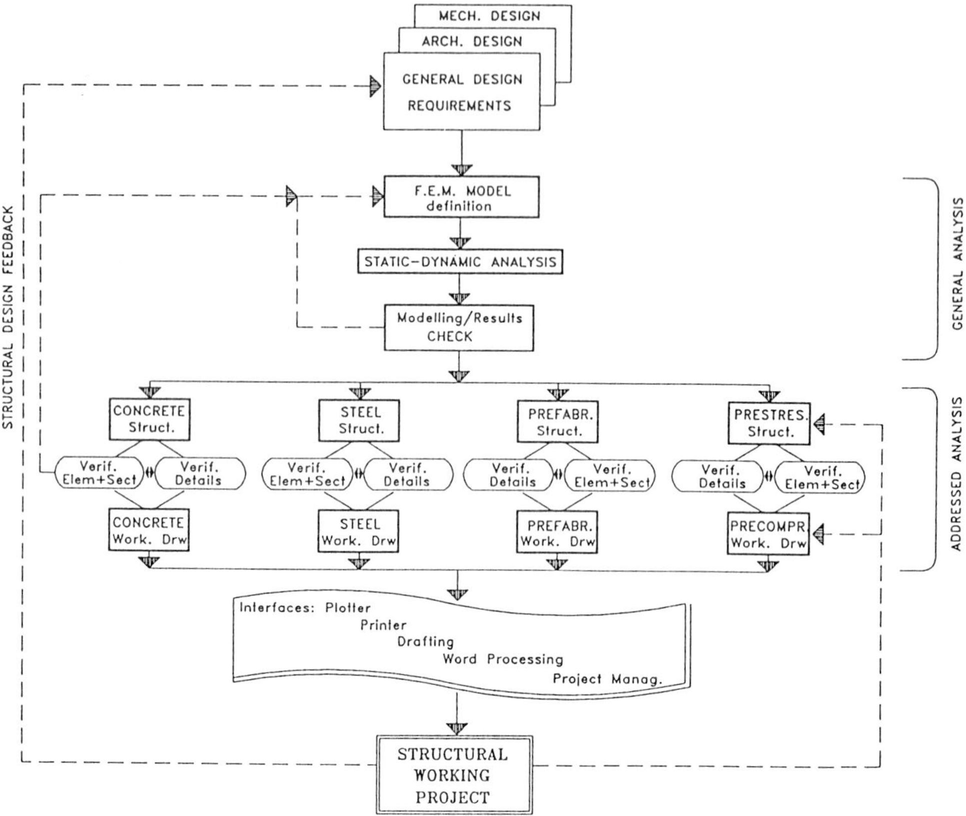

Fig. 4. Flow-chart of the general and addressed functions of the program.

BOX 1 (Clipping functions)	BOX 3 Dinamic menus
BOX 7 (DISPLAY AREA)	BOX 4 Reference system
	BOX 5 Memo & Help
BOX 2 (Projections)	BOX 6

Fig. 5. Topological organisation of the screen.

13. Area 4 contains 3 different types of information depending on the working phase: normally, either the orientation of the reference system appears or it contains the design of the section in the element size assignment phase, or the input map for assigning the external boundary conditions to the nodes.

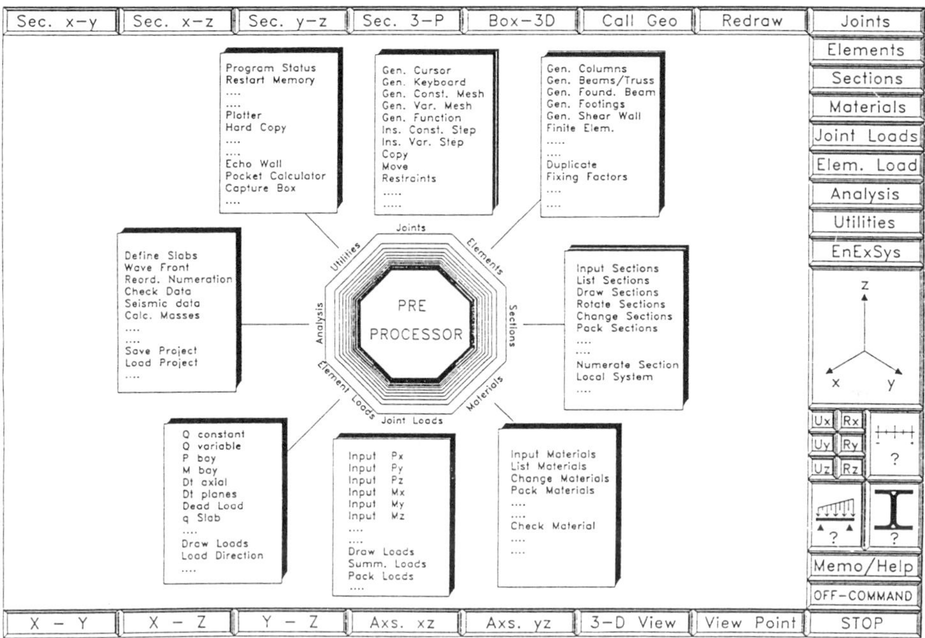

Fig. 6. Typical pre-processor page, main and secondary menu commands.

14. Area 5 simultaneously contains three kinds of information regarding the setting of some parameters and, more precisely, the current node boundaries, the type of current section, the loading condition and an area for change from the graphic to the alphanumerical page, containing numerical information related to the input data.

15. All the nodes or elements or loads generated after the assignment of one of these parameters will automatically have the respective assignment of that external boundary or that section or its assignment to that load condition.

16. Area 6 has 3 different functions: these make it possible to return to input procedure by positioning the cursor in the "OFF COMMAND" area; exit from the module by means of the "STOP" and; return to the module menu or read the type of command required (intermediate area). The intermediate box reminds us, instant by instant, of step we have reached in the input procedure.

17. Finally, area 7 is dedicated to the display of the structure.

CENTRAL PROCESSOR

18. The mathematical model of the STRAND code, programed entirely in the C language, allows the following types of analyses to be carried out according to the displacement method:

(i) general static analysis with 6 degrees of freedom (DOF) per node;

(ii) modal dynamic analysis by means of modal superposition with a response spectrum in accordance with Eurocode 8 and Italian regulations. In the case of multilevel buildings, a DOF floor condensation can be made (3 per node + 3 per floor);

(iii) dynamic analysis by integration of time histories, in the time-domain according to the Newmark method;

(iv) random dynamic analysis, in the frequency-domain through modal superposition with density power spectra description for seismic or wind input.

The library of available finite elements consists of:

(i) truss-type element;

(ii) beam type-element with shear deformability and fixing factors for internal flexional boundaries according to G.R. Monforton and T.S. Wu in "Matrix analysis of semi-rigid connected frames", ASCE, Dec. 1963;

(iii) isoparametric 8-node plate-shell elements;

(iv) brick-type solid element with 21 DOF;

(v) beam element on elastic soil in accordance with Bing Y. Ting and Eldon F. Hockry, "Beam on elastic foundation finite element", ASCE, 1985.

19. The STRAND program is also equipped with a non-linear processor in the geometric field for the analysis of flexible structures.

INTERACTIVE GRAPHIC POST-PROCESSOR

20. The post-processor runs the executive project-checking phases of the previously analysed structural elements.

21. The executive design and verification is based on the "criteria of design and/or verification" and is controlled by the regulations in force regarding the materials used:

- Eurocode 2 and Italian Rules for structures in reinforced concrete;
- Eurocode 3 and Italian Rules for structures in steel;
- ACI 318-89 and Canadian Concrete Code.

The areas of the screen

22. The screen is divided into areas, as in the case of the pre-processor.

23. The content of the two areas changes: area 3 contains the dynamic menus and area 5 holds the setting of some parameters.

24. Picking with the mouse in the main menu, area 3, it is possible to access 8 dynamic menus containing all the post-processing functions (Fig. 7).

25. The first is the File menu which contains, as well as the CALL GEOMETRY (the first to be selected to display the project required on the video), a series of useful support options for use during the various phases of the data

control. The second menu displays both the modal deformations and those in the various load combinations of the structure under examination.

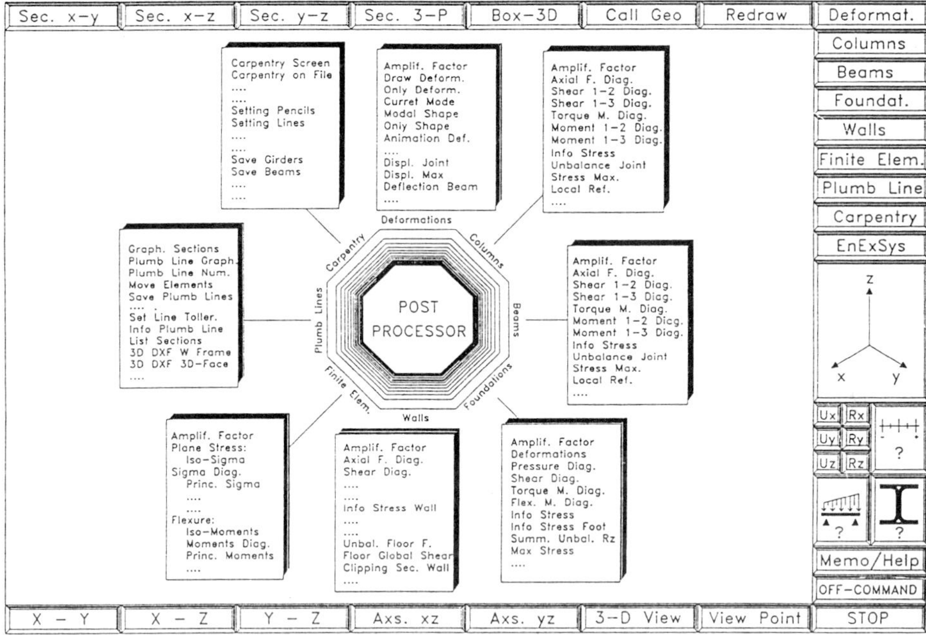

Fig. 7. Typical post-processor page, main and secondary menu commands.

26. The next 3 menus contain the display options of the stress diagrams in the various load combinations for beams or column type elements, foundations beams or plinths, and shear walls respectively.

27. The menu, "Finite Element", contains options used for the display of tension for the mesh-type finite elements with stress coloured distribution or principal stress arrows.

28. The final menu supplies a schematic design of steel floor structures: this design can later be completed with the graphic programs with which the system is integrated.

29. As far as structures in reinforced concrete are concerned, the post-processor manages:

(i) the reinforcement design with two possible alternatives:

- bent reinforcing bars and stirrups,
- straight running reinforcing bars and stirrups;

(ii) the executive drawing of the steel floor structure with a tabulated drawing of the reinforcement;
(iii) the executive tabulated drawings for columns;
(iv) metric computation for concrete and steel, distinct for each element;
(v) structural specification characteristics;
(vi) technical report.

30. As far as metallic structures are concerned, the post-processor manages are:
(i) processing of drawings as a whole;
(ii) design and control of standard details and parametrically defined type details;
(iii) shop drawings by means of a separate program (STEEL-DRAW);
(iv) list of materials and B.O.Q;
(v) structural technical-specification characteristics;
(vi) technical report.

31. Some output results are shown in Fig. 8, illustrating the synthetic graphic diagrams obtained on a HP 9000 UNIX-XWINDOWS minicomputer and a Compaq 486 DOS-EXTENDER PC, under WINDOWS 3.1.

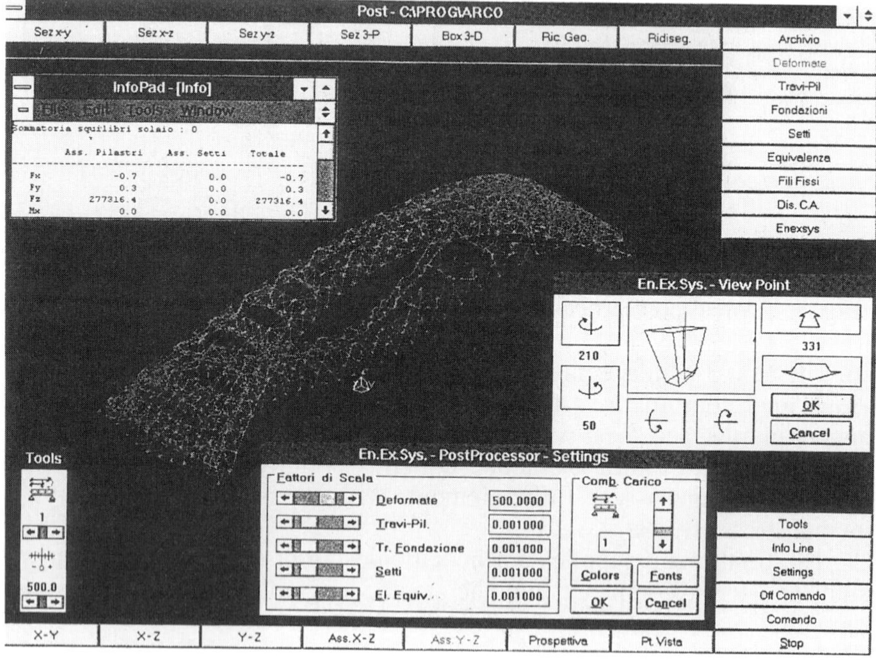

Fig. 8. Graphic post-processing of output results:
a) Modal vibration shape in a steel framed arch structure.

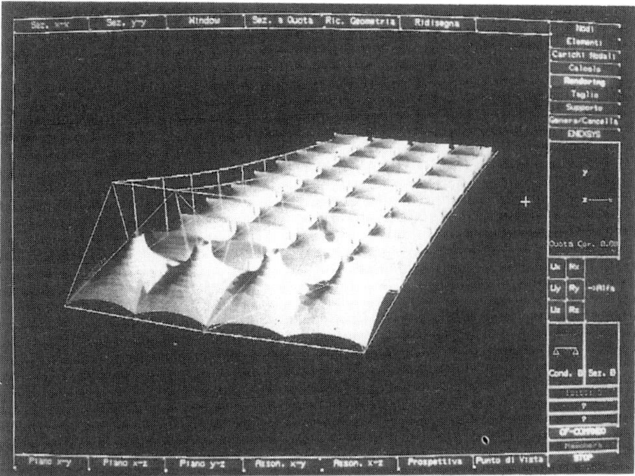

Fig. 8(b). Shape finding of a prestressed membrane structure (Olympic Stadium membrane covering, Rome).

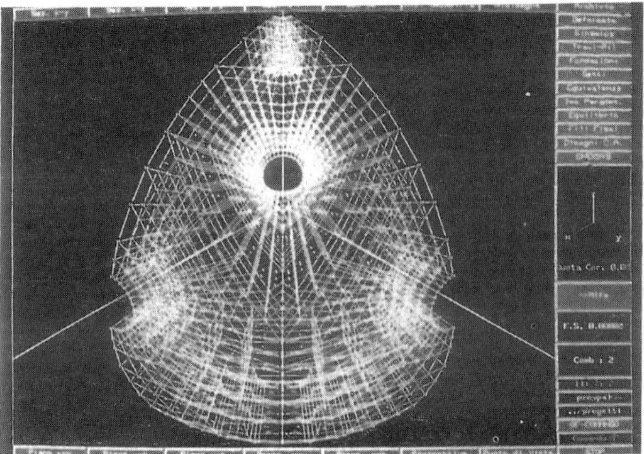

Fig. 8(c). Analysis of a structural joint type "Mero" with brick finite elements - 3D vectorial stress distribution.

REFERENCES

1. PEARSON, K.and TODHUNTER, I. History of the Theory of Elasticity, Cambridge Univ. Press, 1886.

2. TIMOSHENKO, S. History of Strength of Materials, McGraw, New York, 1953.

3. BENVENUTO, E. La scienza delle costruzioni e il suo sviluppo storico, Sansoni, 1981.

4. CROCI, C. La scienza delle costruzioni tra teoria e realtà. Nota III: L'industria delle costruzioni, 1979.

5. POZZATI, P. Teoria e tecnica delle strutture, UTET, 1977.
6. ODEN, J.T. and BATHE, K.J. A commentary of computational mechanics. Appl. Mech. Rev., 1978, vol. 61.
7. FANELLI, M. and GIUSEPPETTI, G. Mathematical analysis of structures: usefulness and risks. IABSE, 11th Congress, Vienna, 1980.
8. FANELLI, M., Informatics and the use of computers in structural engineering. IABSE, Periodica, vol. 1, 1983.
9. WORKING COMMISSION V. Using computers in the design of structures. IABSE Proceedings, P-80/84.
10. CLOUGH, R.W., KING, I.P. and WILSON, E.L. Structural analysis of multi-storey buildings. J. Structural Dic. ASCE, 1964, vol. 90 (ST3), 19-34.
11. WILSON, E.L. and DOVEY, H.H. Three-dimensional analysis of building systems-TABS. Earthquake Engineering Research Centre, Rep. EERC 72-8, University of California, Berkeley, Dec. 1972.
12. WILSON, E.L., DOVERY, H.G. and HABIBULLAH, A. Three-dimensional analysis of building systems-TABS 80, vol. 1: Theoretical Manual; vol. II: Users Manual. A report to the U.S. Engineer Waterways Experiment Station, Vicksburg MS, June 1980.
13. McGUIRE, W. Interactive computer graphics and the design of steel frames, presented at the First Symposium on Metal Structures, Mexican Society of Structural Engineers, Queretaro, Mexico, July 1978.
14. ABEL, J.F., GREENBERG, D.P., McGUIRE, W. and GALLAGHER, R.H. Interactive graphics for finite element analysis. Proceedings of Seventh Conference on Electron Computation, ASCE, St. Louis, Mo., 1979.
15. PESQUERA, C., McGUIRE, W. and ABEL, J. Interactive graphical pre-processing of three-dimensional framed structures", Computer & Structures, vol. 17, 1983.
16. MAJOWIECKI, M. Membrane coverings in steel and high strength textiles: aspects of interactive design and construction. Acier, Stahl, Steel, 1975, no. 12.
17. MAJOWIECKI, M. and TIRONI, G. Geometrical configuration of pneumatic and tent structures obtained with interactive computer aided design. IASS, World Congress on Space Enclosures (W-Code), Montreal, 1976.
18. MAJOWIECKI, G. and TIRONI, G. Alcune applicazioni di progettazione interattiva mediante calcolatore nello studio di strutture fortemente deformabili. INARCOS, Bologna, aprile 1978.
19. MAJOWIECKI, M. and MONTANARI, M. Sulla progettazione interattiva di telai piani metallici. Acciaio, 1980.
20. MAJOWIECKI, M. Tecniche interattive nella moderna metodologia della progettazione strutturale. Acciaio, 1980.
21. MAJOWIECKI, M. and ZUCCARELLO, F. Static and dynamic analysis for multilevel buildings: a computer aided building design methodology. Third International Conference on Space Structure, Guildford, U.K., 1984.
22. FENVES, S.J., MAHER, M.L. and SRIRAM, D. Knowledge-based expert system in civil engineering. IABSE Periodica, 1985, vol. 4.

138. Computer-aided design of reticular space structures with random behaviour elements

A. LA TEGOLA, L. OMBRES and C. PECORA, University of Calabria, Italy

SYNOPSIS. In this paper we describe the computer aided design of space reticular structures with random behaviour elements. The design system is based on a finite element solution, with reference to optimised structures by analytical model of minimum volume of bars. The method allows the analysis of the static behaviour of simulated structures, by the definition of statistic distribution and the evaluation of the collapse multiplier, and the analysis of the post-critical behaviour by definition of the equilibrium path, described by the load-displacement curve drawn by an incremental process using the "arc length method" taking into account also second order terms of deformation.

INTRODUCTION

1. There is a widespread use in civil engineering of reticular space structures particularly for long span roof coverings.

2. Such structures usually have the shape of a single or double-layered reticular shell, which enables the structures to absorb external actions. Their weights are moderate and they can be easily produced and assembled.

3. Space reticular structures are usually made of materials, such as steel and aluminium, which are endowed with high mechanical properties that allow the construction of structural elements of small thickness and low weight. Modular design and prefabrication techniques, which allow the mass-production of modules by defining assembling modes, are also used to produce space reticular structures.

4. The structural design of these constructs is very complex because of the high number of parameters involved. These parameters are all aleatory and depend on the variation of material per unit of volume (geometrical and mechanical imperfections, instability phenomena, etc.). Such parameters characterise the behaviour of the bar, which is the key element of the structure. It is important to take account of these parameters since they influence both the local behaviour of the bar and the overall behaviour of the structure.

5. From an analytical viewpoint the problem is very difficult and must be solved by numerical methods. The structural design must be based on a

methodology, that, besides being quick and efficient, can use calculus tools ordinary and reasonable in cost.

6. This paper describes a computer aided design system of reticular space structures, based on the use of low cost computers(personal computers) and takes into account of both the random behaviour of the bars (variation in mechanical and geometrical parameters) and the influence of instability phenomena.

7. The computer aided design system, based on a solution approach to finite elements, involves the following steps:

- the lst step is the problem data input;
- the 2nd step is the check or design of the structure;
- the 3rd step is the analysis of the critical and the post-critical behaviour of structure.

8. In the lst step, the space reticular structure is defined geometrically by means of formex algebra procedures. Constraint and load conditions are also defined. The functions typical of formex algebra suffice to define the geometrical parameters (maximum size in space, basic module) required to obtain the connection of all nodes and the space definition of all the bars of the structure.

9. In the 2nd step, different possible procedures are investigated depending on the needs. The structure is tested by means of the geometrical characteristics, which are in this way designed and assigned to the bars.

10. The problem of design is approached through the phase of structural optimisation seen as the search for the minimum weight solution following the procedure proposed by the authors in (ref. 3).

11. In the 3rd step, the analysis of the static behaviour of the space reticular structure, by now completely defined geometrically and from the viewpoint of loads and constraint conditions, is carried out.

12. The behaviour shown up to the ultimate limit state is analysed. The collapse multiplier of the loads is determined according to the different mechanical and geometrical parameters which characterise the bars. Hence, the appropriate laws of statistical distribution are determined for each bar.

13. Such an analysis is carried out by means of a numerical simulation procedure, based on the Monte Carlo method. We assume a constitutive law for the material which takes into account the instability phenomena which may occur with compressed bars. A distribution of the collapse multipliers is therefore obtained upon which basis it is possible to determine, for a given probability, the collapse multiplier of the structure.

14. The analysis is completed through the construction of an equilibrium path described by means of a load-displacement curve. Such a curve is drawn by way of an incremental process and represents both the fundamental and the split path. The latter is not interrupted from the collapse condition.

15. The analysis is also carried out by taking account of the second order terms of the deformation field and using the "arc length method".

16. The results obtained from the numerical applications of the technique to the analysis of single or double layered reticular structures, with annular section bars, make it is possible to examine the fundamental aspects of the static behaviour of

space reticular structures and the effectiveness of the proposed computer aided design approach.

GEOMETRICAL DEFINITION OF SPACE RETICULAR STRUCTURES

17. The geometrical definition and graphic representation of a reticular space structure, usually consisting of triangular grids, are extremely complex due to the high number of bars and connecting nodes located by an equally high number of co-ordinates in a Cartesian space reference system. The difficulties which may arise can be overcome only by means of an automated approach such as the one based on formex algebraic transformation functions (ref. 2).

18. Formex algebra is a mathematical support system used for planning in the CAD environment. Such as system is based on simple algebraic functions which allow the automated preparation of data and graphics for structural analysis. In fact, once the basic modules are defined, it is possible to generate any structure by copying a given module, that is pasting one module to another. In this way the principles of modular design are applied.

19. The use of formex procedures, now consolidated as far as the geometrical representation of space reticular structures, allows complicated encoding mechanism to be avoided. The process of geometrical definition becomes, consequently easier and, most important, interactive.

20. A reticular space structure of any shape and size can therefore be described with only a few data such as, its size on the plan and the function defining the shape and the basic module.

21. In the design system we propose, the phase of the geometrical representation of space reticular structures is completed by defining load and constraint conditions. These latter can be defined also considering current design codes.

STRUCTURAL DESIGN: BAR SIZING

22. Once the structure is geometrically defined, and the magnitude and position of external actions are known, it is necessary to size the bars cross-sections. This process is carried out through the solution of the optimisation problem, which is seen as the search for minimum weight solution. Usually, the criteria used to solve the optimisation problem imply costs in terms of computation and time, which makes their application to small structures too expensive. Such a process requires, in fact, long elaboration times and a check on the solution, which, technologically speaking, yields unreliable results, is not sufficient.

23. In this paper, an optimisation procedure, which was defined by the authors (ref. 3) is used. This procedure is based on analytical models, such as the Feasible Regulation Method and the Optimality Criteria Method, which are widely used for the solution of optimisation problems, and are adapted to the requirements. In such models, account is taken of the limits on tensional and deformation values as prescribed by the design codes, and the analysis is carried out according to the structural material currently on the market.

24. Analytically, the minimum weight problem is defined by the condition:

$$\min V = \sum_{i=1}^{n} A_i l_i \tag{1}$$

where V is the overall volume of the material used; A_i and l_i are the areas of the cross-section and the length of the i-th bar, respectively, and n is the total number of the bars used. The following limits are to be satisfied:

$$\sigma_c(A_i) \leq \sigma_i \leq \sigma_t; \qquad \delta_{ij} \leq \bar{\delta}_{ij}; \qquad A_i \leq \overline{A}_i \tag{2}$$

where σ_t is the tensile strength of the material, $\sigma_C(A_i)$ the compression strength of the material function of the area of the bar cross-section, $\bar{\delta}_{ij}$ the ultimate value of the displacement δ_{ij} of the i-th node in the j-th direction, $\bar{A}_i$ the value of the area of the bars cross-section, A_i is defined on the basis of both technological limits (minimum commercial section of area) and design codes (area of the cross-section corresponding to the slenderness ultimate value of compressed bars).

25. The solution of the problem is obtained by iteration through optimisation recursive formulae. In particular, the procedure used to redefine the areas consists in the evaluation of the scale factors which take the design point on to the constraint surface. Hence the formulae adopted are the following:

$$\delta_{ij} = \frac{\sigma_i}{\bar{\sigma}_i} \quad \text{for the tension} \tag{3}$$

$$\xi_i = \left[\left(\frac{V}{\bar{\delta}_{ij}}\right)\frac{\sigma_i \sigma_{ik}^{j}}{E}\right]^{\eta} (\text{F.R.M.})\,;\quad \xi_i = \left[\frac{\sum_{j=1}^{p} \lambda_{ij} \frac{F_i F_{ik}^{j}}{E}}{(A_i)_v}\right]^{\frac{1}{q}} (\text{O.C.M.}) \text{ for the deformation} \tag{4}$$

with $\quad (\lambda_{ij})_{v+1} = (\lambda_{ij})_v \left(\frac{\delta_{ij}}{\bar{\delta}_{ij}}\right)^{\frac{1}{c}}$

where p is the number of constraints imposed on nodal displacements, F_i the deformation of the i-th bar, F_{ij} the deformation of the i-th bar caused by a load in the j-th direction, q is the computation pass, υ the iteration index, σ_i the tension in the i-th bar, $\sigma_{ik}{}^{j}$ the tension in the i-th bar caused by a unitary load applied on the k-th node and in the j-th direction, η and c the parameters for the check on the procedure of convergence and stability, which are defined on the grounds of a numerical test that takes into account both problems linked to the automated procedure (overflow) and of their contribution to the final solution. We assume $\eta = 0.20$, $c = 0.75\text{-}0.85$ (ref. 3).

STATIC BEHAVIOUR ANALYSIS

26. The analysis of the static behaviour of space reticular structures is carried out by referring to the ultimate limit state, which corresponds to collapse, and by considering that these structures are made of real bars. Consequently, these bars are imperfect both geometrically and mechanically and their imperfections are randomly distributed.

27. The imperfections considered are the following:
-changes in the diameter and thickness of the cross section of the bar with respect to the mean values;
-the non linearity of the geometrical axis of the bars; -change in the yield stress of the material used, to which a distribution function (refs. 4,5) is associated.

28. The number of imperfect structures, for which the collapse multiplier is determined, is unlimited. This is due to the fact that imperfections are randomly distributed along the bars. As a consequence there are remarkable analytic difficulties which can be overcome only through numerical simulation procedures, i.e. the pseudo-random generation of a specimen of imperfect structures. For each structure the collapse multiplier is determined: through this method a distribution of collapse multipliers is obtained. On the basis of the above mentioned distribution the collapse multiplier of the structure is determined for a given probability.

29. The generation of simulation structures is carried out through the Monte Carlo method (ref. 4) which makes it possible to generate a specimen of random variables (imperfections) included in a certain interval in which a probability function is evaluated. By means of the Neuman procedure a random distribution of the imperfections considered is generated. Since these imperfections affect the mechanical behaviour, each bar and consequently the whole structure, is characterised by random behaviour.

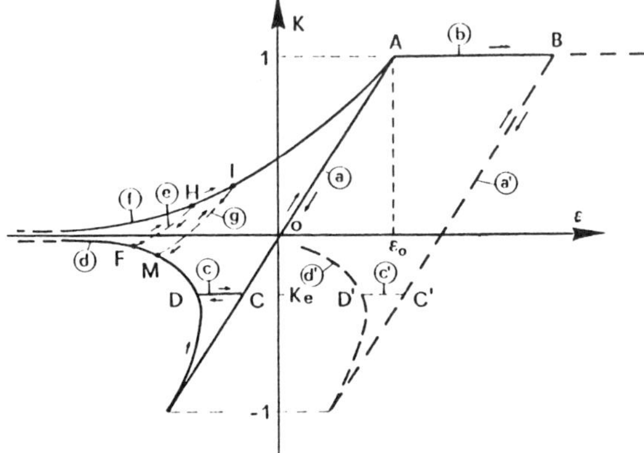

Fig. 1. Constitutive law of material.

CONSTITUTIVE LAW

30. The constitutive law of the bars of the imperfect structures analysed as shown in Fig. 1 and is characterised analytically by the following:

Curve a : $\dot{K}_{ij} = \dfrac{\dot{\varepsilon}_{ij}}{\varepsilon_0}$ for $K_{Eij} < K_{ij} < 1;\ K_{Eij} \geq -1$

Curve b : $\dot{K}_{ij} = 0$ for $K_{ij} = 1;\ \varepsilon_{0ij} \leq \varepsilon_{ij} \leq \varepsilon_{rti}$

Curve c : $K_{ij} = K_{E\,ij}$ for $\varepsilon_{Dij} \leq \varepsilon_{ij} \leq \varepsilon_{Ci}$

Curve d: $\varepsilon_{ij} = \varepsilon_0 K_{ij} - 2\left(\dfrac{d_{ij}}{l_{ij}}\right)^2 \left(1 + \dfrac{1}{K_{ij}}\right)^2$ for $\varepsilon_{ij} \leq \varepsilon_{Di}$

Curve e: $\dot{K}_{ij} = \dfrac{\dot{\varepsilon}_{ij}}{\varepsilon_0 + 4\left(\dfrac{d_{ij}}{l_{ij}}\right)^2 \left(1 + \dfrac{1}{K_{Fij}}\right)^2 \dfrac{(K_{Fij} - K_{Eij})^2}{(K_{ij} - K_{Eij})^3}}$ for $K_{ij} \leq K_{Hij}$

Curve f: $\dot{K}_{ij} = \dfrac{\dot{\varepsilon}_{ij}}{\varepsilon_0 + 4\left(\dfrac{d_{ij}}{l_{ij}}\right)^2 \dfrac{1 - K_{ij}}{K_{ij}^3}}$; $K_{ij} = K_{Hij}$ for $\dot{\varepsilon}_{ij} > 0;\ \dot{K}_{ij} > 0;\ K_{ij} < 1$

Curve g: $\dot{K}_{ij} = \dfrac{\dot{\varepsilon}_{ij}}{\varepsilon_0 + 4\left(\dfrac{d_{ij}}{l_{ij}}\right)^2 \left(1 + \dfrac{1}{K_{Iij}}\right)^2 \dfrac{(K_{Iij} - K_{Eij})^2}{(K_{ij} - K_{Eij})^3}}$ for $K_{ij} \geq K_{Mij};\ \dot{\varepsilon}_{ij} < 0$

Curve h: $\varepsilon_{ij} = \varepsilon_0 K_{ij} - 2\left(\dfrac{f_0}{l_{ij}}\right)^2 \left[\dfrac{1}{\left(\dfrac{K_{Eij}}{K_{ij}} - 1\right)^2} + \dfrac{2}{\left(\dfrac{K_{Eij}}{K_{ij}} - 1\right)}\right]$ for $\dot{\varepsilon}_{Dij} \leq \varepsilon_{ij} \leq 0$

with

$K_{ij} = \dfrac{N_{ij}}{N_{0ij}}$; $\dot{K}_{ij} = \dfrac{\dot{N}_{ij}}{N_{0ij}}$; $\varepsilon_{ij} = \dfrac{\Delta l_{ij}}{l_{ij}}$; $\dot{\varepsilon}_{ij} = \dfrac{\Delta \dot{l}_{ij}}{l_{ij}}$

where N_{ij} is the normal stress of the bar ij; N_{oij} is the tensile limit stress along the bar ij; l_{ij} the length of the bar ij; Δl_{ij} the length variation along the bar ij.

COLLAPSE MULTIPLIER DETERMINATION AND ANALYSIS OF POST-CRITICAL BEHAVIOUR

31. The collapse multiplier of the generic simulated structure is determined on the basis of the construction of the equilibrium path, that is the curve which represents a continuous path in an incremental load process of type λF (where λ is the load multiplier). Such a curve is generated from the undeformed configuration.

32. In particular, in the computer aided design system the solution proposed is the "arc length method", which was initially introduced by Riks (ref. 7). From a methodological point of view this procedure refers to an analytic formulation of the curve of the equilibrium path ($\lambda,\underline{u}$) as follows $\lambda = \lambda(\alpha)$, $\underline{u}=\underline{u}(\alpha)$ so that the incremental strategy affects the parameter a which is a curvilinear abscissa linked to the equilibrium curve.

33. From a computational point of view, the pairs (λk, $\underline{u}k$) are sought so that there is a balance between the external forces $\lambda\underline{p}$(where $\underline{p}$ is the loading vector) and the internal forces $s(\underline{u}_k)$ which are the reaction of the structural system to displacement $\underline{u}_k$. Once the total increase is defined ($\Delta\lambda,\Delta\underline{u}$) the current solution ($\lambda_k,\underline{u}_k$) is improved by an equilibrium equation and orthogonality, following iteration:

$$\lambda_{j+1}=\lambda_j-\frac{\omega\,\Delta\underline{u}^T\underline{K}^{-1}\underline{r}_j}{\Delta\lambda+\omega\Delta\underline{u}^T\hat{\underline{u}}} \quad ; \quad \underline{u}_{j+1}=\underline{u}_j+\omega\left[\underline{K}^{-1}\underline{r}_j+\left(\lambda_{j+1}-\lambda_j\right)\hat{\underline{u}}\right] \tag{5}$$

with $\hat{\underline{u}}=\underline{K}^{-1}\underline{p}$, where $\underline{K}$ is the stiffness matrix, $\underline{r}_j$ the nodal instability which corresponds to the increase considered, and ω the accelerator with a value between 1 and 2.

34. This procedure improves the convergence due to loads next to collapse, and considers geometrical non-linearity. It also allows the curve in the post-collapse phase to be described by considering the complex link between internal forces and displacements. In this way complex numerical operations are avoided. The collapse multiplier is determined as a value λ_{max} along the first ascending part of the equilibrium path, or as value λ corresponding to the formation of a mechanism in the structure.

NUMERICAL APPLICATION

35. The design system described in the previous sections has been codified in a computation program for small computers (personal computers). Some of the results obtained from the numerical applications point out its effectiveness and usefulness. They describe the fundamental aspects of the static behaviour of space reticular structures made of bars with random behaviour and instability phenomena.

36. In Fig. 2 the automated description of a space reticular structure is shown. This has the shape of a simple curvature vault whose size on the plan is 19 x 37 m and the sagitta is 6 m. According to these data and by choosing a triangular module, the structure presents 63 nodes and 158 bars. Vertical nodal loads are equal to 200 N/m^2.

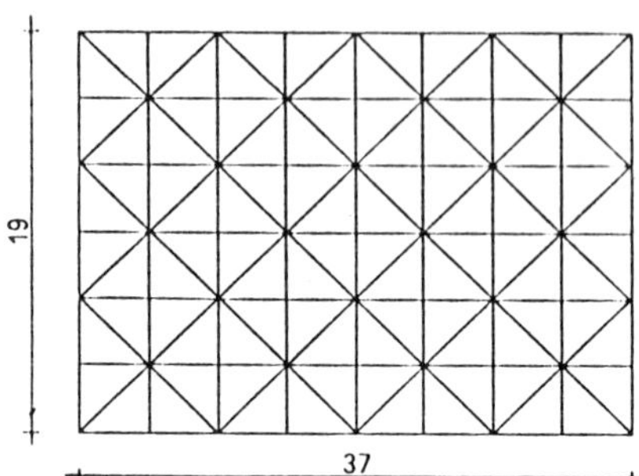

Fig. 2. Geometrical representation of the structure.

37. The results obtained in the planning stage through the optimisation process are shown in Fig. 3 . Annular section bars which can be found on the market are used in this method. The optimisation has been carried out by imposing a limit both on stress and on the value of the nodal displacement.

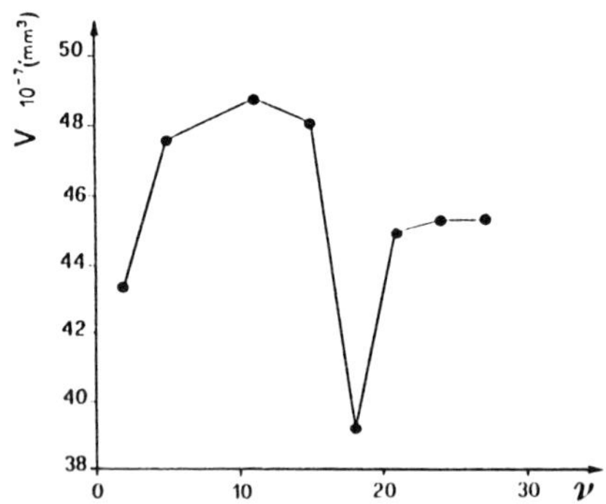

Fig. 3. Size optimisation: diagram volume versus iterations (constraints: w=35mm).

38. In Fig. 4 the results of the probabilistic analysis are shown. These results are related to the collapse multiplier variation on the basis of the trust interval θ and of probability p.

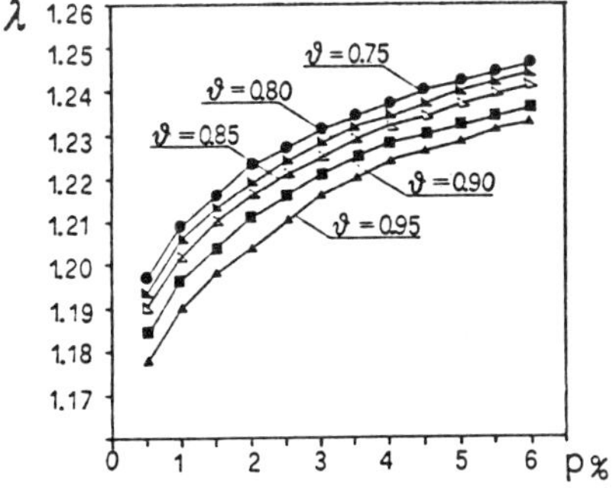

Fig. 4. Collapse multiplier of the structure versus probability.

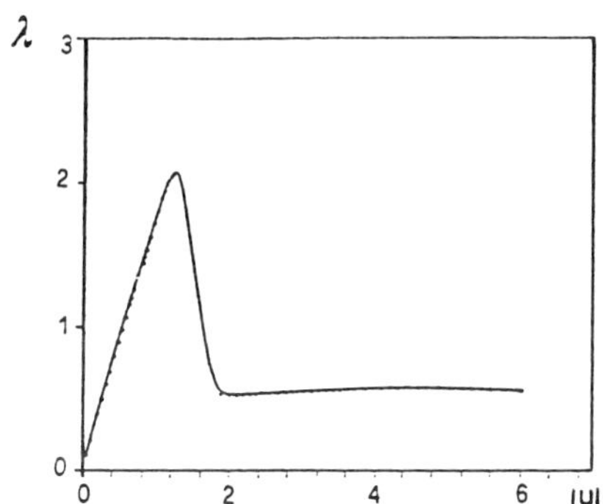

Fig. 5. Equilibrium path of the structure.

39. The equilibrium path of the structure is show in Fig. 5 . A drastic change in the slope immediately after the collapse is pointed out from the study of the curve. This proves the effectiveness of the optimisation process which has been performed. In fact, when a bar collapses the others undergo a stress next to the limit. For this reason a further increase in stress would sharply increase the nodal displacements leading to a total change in the shape of the structure.

CONCLUSIONS

40. The results obtained allow us to draw the following conclusions:

- the design of space reticular structures is to be based on automatic computation, since there are various parameters a number of complex problems need to be solved. The computer aided approach is very expensive, especially in the case of extended structures which would require the use of large and powerful computers;
- the structural optimisation phase turns out to be essential for good planning. This method involves problems related to the stability of equilibrium, which frequently occur with these structures. Such a phase is influenced by the

strength limits of the material and by the deformation limit of each element. The structural elements are almost always prescribed by the Codes;
- the analysis of the ultimate limit state carried out through the determination of the collapse multiplier in relation to the imperfect structures as a whole, points out the importance of the completion stage in order to minimise the influence of imperfections. Moreover, by properly changing the statistical parameters, it is possible to show the degree of danger and to arrange the most suitable means to cope with it;
- the post-critical behaviour of space reticular structures, which is described through the equilibrium path, is very useful. In fact, it makes it possible to point out the degree of danger of some aspects of the constitutive law of plastic phase under compression. Therefore, a proper and exact design must also take into account this aspect of the structural behaviour.

REFERENCES

1. MAKOWSKI Z.S. Strutture spaziali in acciaio, CISIA, 1978.
2. NOOSHIN H. Formex configuration processing in structural engineering, Elsevier Applied Science Publishers, London, 1984.
3. LA TEGOLA A., OMBRES L. and PECORA C. Minimum weight design of space reticular structures: a computer aided system, in press, International Journal of Space Structures, 1993.
4. LA TEGOLA A. Il carico di collasso delle strutture con imperfezioni random, Giornale del Genio Civile, n.7-8-9, 1979.
5. LA TEGOLA A. Probability analysis of braced barrel vaults composed of random strength elements, Analysis design and construction of braced barrel vaults, Elsevier Applied Science Publishers, London, 1985.
6. GIONCU V. Instability behaviour of triangulated barrel vaults, Proceedings of "Traian Vuja" Polytechnic Institute, Timisoara, 1990.
7. RIKS E. An incremental approach to the solution of snapping and buckling problems, International Journal of Solids and Structures, Vol. 15,1979.

139. DacsIIs network for supporting design, analysis, construction and manufacturing of space frames

Y. YAMAOKA, K. IMAI and T. MORITA, Kawasetsu Steel Products Corporation, Japan, and K. WAKIYAMA and Y. YUASA, Osaka University, Japan

SYNOPSIS. The paper presents a software network called DacsIIs (Design, Analysis and Construction of Space-frame Supporting Software). Functions of DacsIIs cover shape generation, analysis, presentation, construction simulation, estimation of material quantity and manufacturing of members and nodes, and the design of secondary materials. Some examples are demonstrated through the design of sophisticated space frames.

BACKGROUND AND OUTLINE OF THE NETWORK

1. It is inevitable in designing sophisticated space frames such as structures with curved surfaces and/or folded plate surfaces that an effective software network should be built for shape configuration, analysis, simulation of construction, manufacturing, estimation of material quantity, presentation and the design of secondary members and finishes.

2. The DacsIIs network can deal with the items mentioned above, by connecting various kinds of packages including graphic software. Basically, software in DacsIIs is accessed by the standardised common data file. DacsIIs organises software for personal computers, large analysis codes for main frames and NC packages and edits results automatically. Pre-, intermediate and post-processors connect the packages effectively in this network.

3. The DacsIIs network is shown in Fig. 1. In the network, the pre- and intermediate processors link packages and generate necessary data effectively. Formian is placed as a main part of the pre-processor. Several intermediate processors are applied between packages. These processors are *FormPstStr*, *Numbering*, *Const*, *PstToBklng*, *PstToSTRAP* and *PstToCG*. Post processors are *TrsChecker*, *PnlChecker*, *PostTrs* and *AutoEdit*.

SHAPE CONFIGURATION

4. At the first stage, Formian generates the shape and necessary data such as co-ordinates, connectivity and attributes of nodes (load points, supported points and boundary loaded points). When designing roof panels, Formian generates co-ordinates and element connectivity.

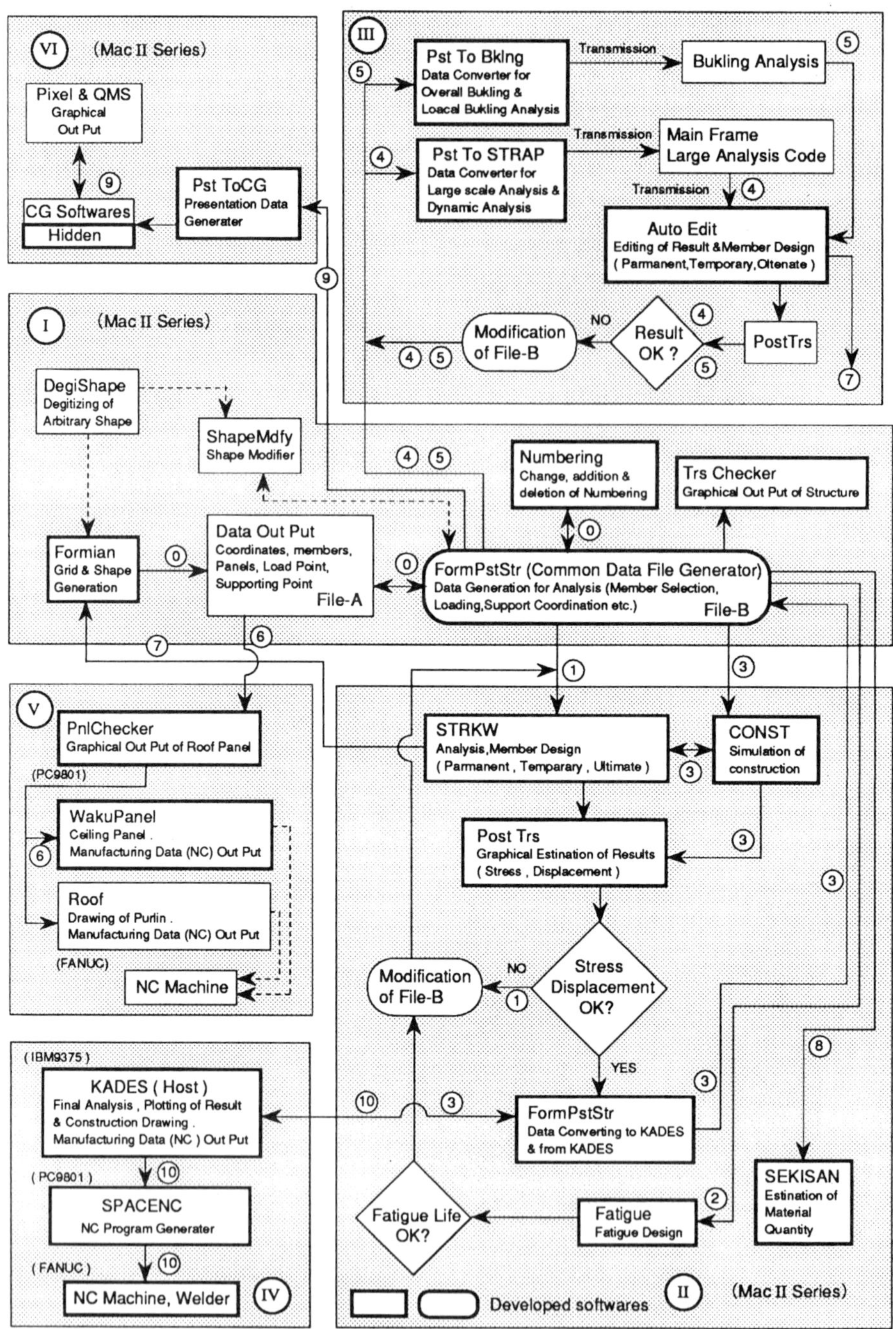

Fig. 1. Diagram of DacsIIs network.

In DacsIIs Formian does not send the load data and selection of members in order to ease user interaction. Formian sends the integer data such as load points and boundary loaded points for calculating the load in the succeeding processor. Moreover Formian can utilise the analysed results for shape generation.

COMMON DATA FILE GENERATION

5. Processor *FormPstStr* generates the standardised common data file. This common data file is a core element of the network and offers necessary data for various packages. *FormPstStr* changes the data format, suitable for succeeding packages and generates necessary data such as the selection of members, member attributes, member end condition, support condition and load.

6. Processor *Numbering* changes the numbering of a common data file when deletion or addition of nodes and members is required. *Numbering* can also join together several independent structures by using individually generated common data files (Fig. 2). By this method, if two independent structures with different numbering systems join together, a new unified structure can have a partially different numbering system, that will be convenient for the designer.

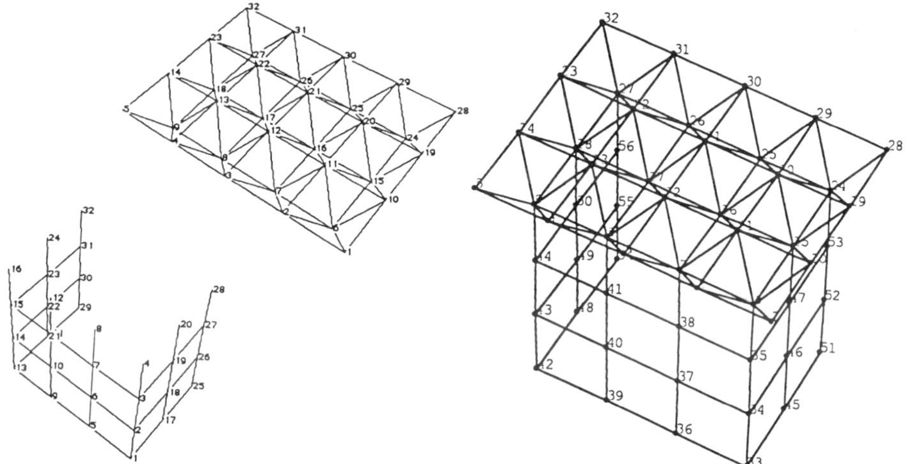

Fig. 2. Joining of structures with different numbering systems.

Load data generation

7. The generation of loads for the analysis of complicated structures such as curved surfaces requires considerable amounts of labour, especially for wind loads. One of the important functions of processor *FormPstStr* is load data generation. Load types obtained by *FormPstStr* are dead load, snow load, wind load and seismic load.

8. Definition of influence area. The definition of the area of influence and an outward normal vector of load point is inevitable for the calculation of load data.

9. Formian designates loaded nodes and boundary loaded nodes in integer form for *FormPstStr* (Fig. 3). *FormPstStr* calculates the influence areas of loaded

nodes by using these integer data. At first, *FormPstStr* defines triangular and quadrangular panels by referencing topological data of the connected loaded nodes and loaded node 'i' such as those shown as 1 to 4 in Fig. 4. For every triangle, the influence area of 'i' is one third of the triangle's area, and for a quadrangle, every influence area of the loaded node 'i' is half of the dummy triangle's area, consisting of 'self' and 'partner' nodes. If the loaded node (self) is on the boundary of the loaded plane, area '5' for example, should not be counted (Fig. 4). This judgement is made by using a boundary loaded node identifier, that is, the triangle consisting of three boundary loaded nodes is ignored in the calculation.

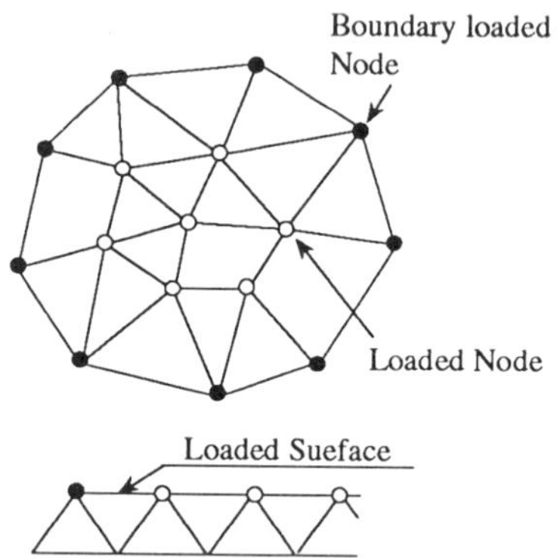

Fig. 3. Identification of load points.

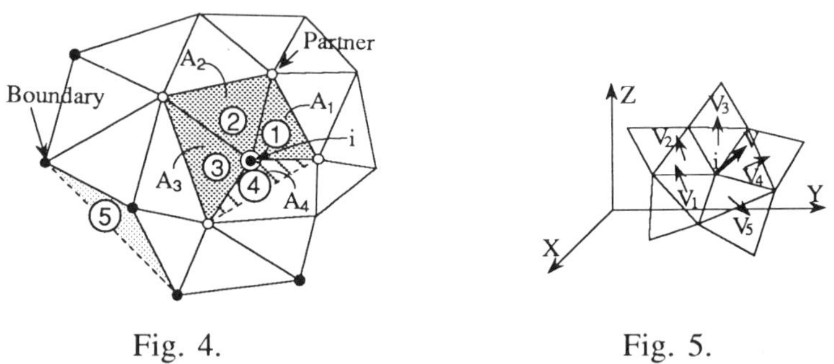

Fig. 4. Fig. 5.

10. Definition of an outward normal vector.
FormPstStr defines the average normal vector of a loaded node by normalising the accumulated direction cosines (λ, μ, ν) of the normal vectors of triangles '1' to '5 for example, as shown in Fig. 5.

11. Generation of load data.
Loads are calculated as follows:

(Dead load)

Piz=p * Ai	p: Distributed unit dead load. Ai: Influence area of loaded node.

(Seismic load)

Pi=p * Ai * Ci	Ci: Seismic coefficient at loaded node.

(Wind load)

Pix=pw * cw * Ai * λ	pw: Wind pressure at loaded node.
Piy=pw * cw * Ai * μ	cw: Wind force coefficient.
Piz=pw * cw * Ai * ν	λ,μ,ν: Direction cosines of normal vector.

(Snow load)

Piz=ps * Ai' * h	ps: Distributed unit snow load. Ai': Projected influence area to horizontal plane. h: Reduction factor based on the maximum inclination angle of the surface.

12. Other special cases. Special cases can be dealt with by adding some identifying data to Formian output data, and the same process outlined in paragraphs 8 to 11 is available. One case is shown in Fig. 6. In this case a loaded node is connected to more than two boundary load nodes (partners). Another case is shown in Fig. 7. Here, more than two surfaces are connected at the edge lines with sharp angles. The average normal vector of these surfaces on the edge lines has no meaning when generating wind and snow loads. For these two cases, *FormPstStr* generates correct load data, by adding surface defining data, to the Formian load point data.

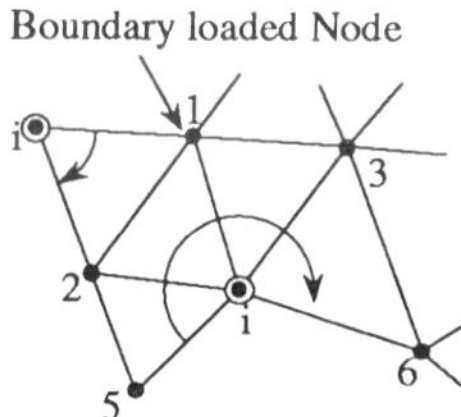

Fig. 6. Special boundary.

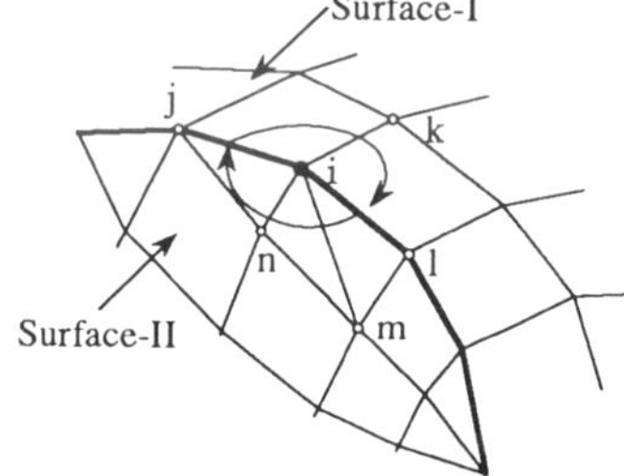

Fig. 7. Connected edge of surfaces.

13. Definition of ordering of partner points. When defining the influence area and normal vector, partners should be ordered in a definite direction around the loaded point, seen from the outside of the surface, for example. This is the definition of the 'head-and-tail' of the surface. The order is defined by geometrical sorting of partners in the local Cartesian co-ordinate system (X'-Y'-Z', in Fig. 8). Axis Z' is a vector from a base point that is inside the surface to the loaded point. Some menus of defining a base point are shown in Fig. 9. The menus are as follows: 1) fixed base point; 2) base point moves parallel to the x, y,

or z axis; and 3) the base point moves in the plane parallel to the x-y, y-x, or z-x planes.

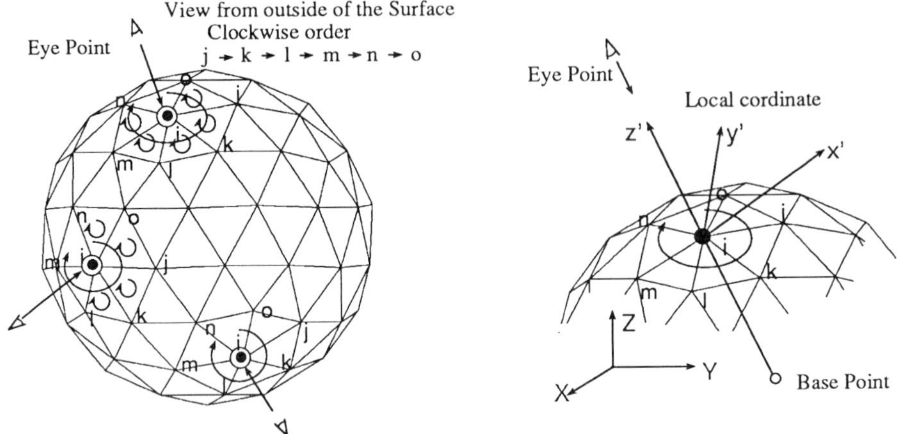

Fig. 8. Ordering of partners and definition of the 'head-and-tail' of the surface.

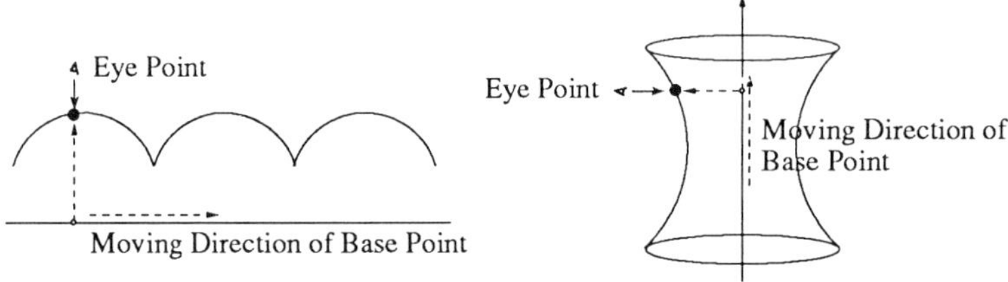

Fig. 9. Some examples of defining base points.

ANALYSIS AND ESTIMATION OF RESULTS

14. *STRKW* analyses the structures by using common data files and estimates the allowable strength of members. *PostTrs* is the post-processor used to show the results of *STRKW* by graphical output i.e. stresses of members are shown by a colour scale and displacements are shown by a magnified scale.

FATIGUE DESIGN

15. *FATIGUE* estimates the fatigue life of joints by utilising the results of *STRKW* under different load conditions. *FATIGUE* calculates the fatigue life based on the range of the bolt tension that is defined by the spring constant of the joint and member force. *FATIGUE* also defines the initial clamping force of the bolts in order to ensure contact between the member and node for every joint, for all load conditions.

SIMULATION OF CONSTRUCTION

16. *CONST* simulates the construction process of the structure by using a common data file, active node numbers and *STRKW*. Active node numbers define the nodes that have been already assembled. With the help of *CONST*, stresses and displacements of a structure and the reaction of temporary supports etc. under constructions are easily obtained.

USE OF LARGE ANALYSIS CODES

17. *PstToSTRAP* converts the common data file and adds necessary data for the large displacement analysis codes of a main frame. The generation of data for the large displacement analysis code and estimation of the results require a vast amount of work. With the help of *PstToSTRAP*, the large displacement analysis code can be easily and quickly accessed from the terminal. *AutoEdit* edits the results and estimates the allowable strength of members in the same format as *STRKW* results. Therefore the results are shown graphically by *PostTrs*. For eigenvalue analysis, each mode is also shown graphically by *PostTrs*. *PstToBklug* is used for the buckling analysis and has similar functions as *PstToSTRAP*.

ROOF PANEL DESIGN

18. *PnlChecker* shows the panel layout graphically. *WakuPanel* makes the assembly and component drawings for the roof and ceiling panels by using Formian output data. The *WakuPanel* can design triangular and quadrangle panels. Some example are show in Fig. 10.

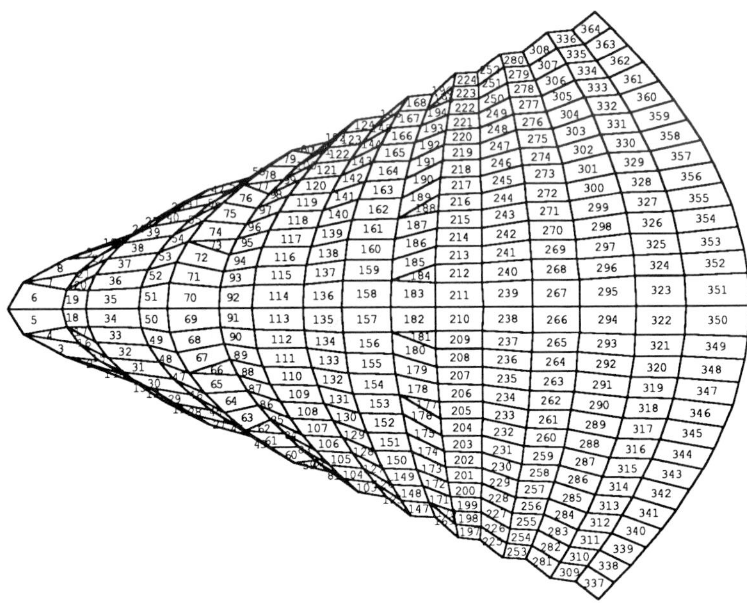

Fig. 10(b). Panel layout.

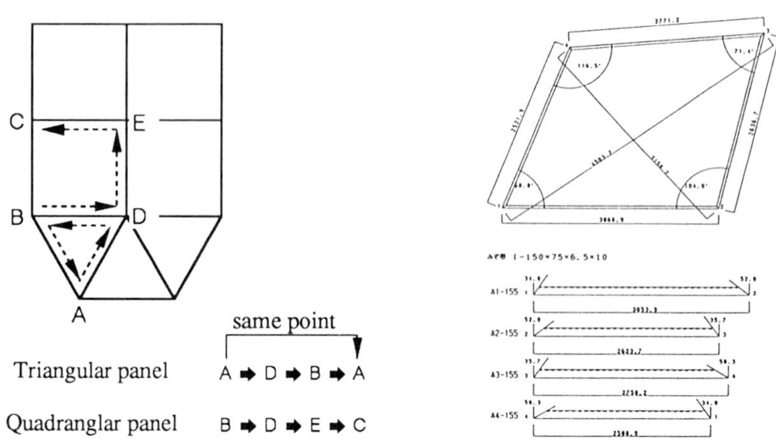

Fig. 10(a). Definition of connectivity. Fig. 10(c). Fabrication drawing of panels.

PRESENTATION

19. *PostToCG* converts the common data file and adds necessary data for the computer graphic (CG) software. Figs 11 and 12 show the output example of CG software.

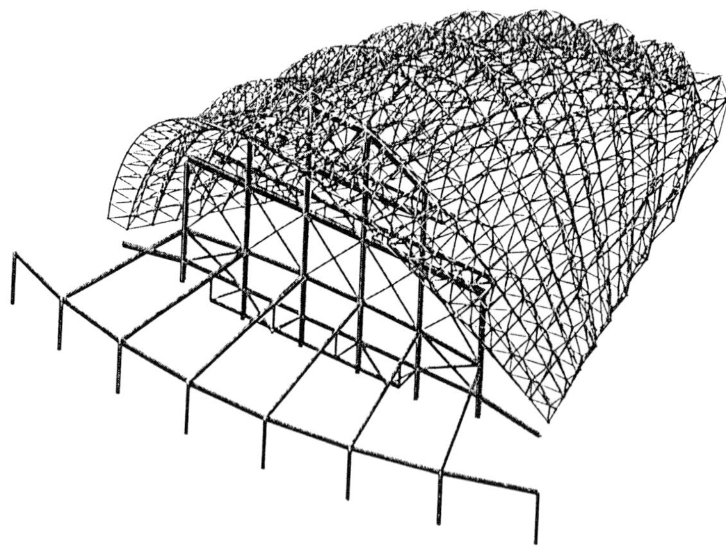

Fig. 11. Computer graphic output of framework.

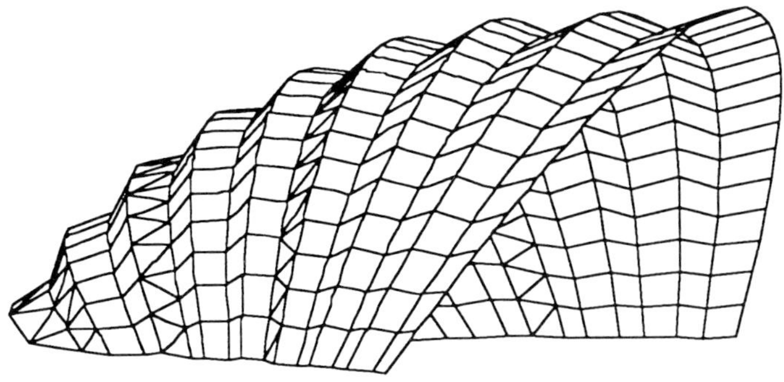

Fig. 12. Computer graphic output of roof panels.

FINAL DOCUMENTATION AND MANUFACTURING

20. *KADES* is used to make final documentation such as the plotting of results and construction drawings etc.. Also *KADES* generates the data for manufacturing the spaceframe. By using the manufacturing data from *KADES*, *SPACENC* generates the numerically controlled (NC) programmes for NC machines.

ACKNOWLEDGEMENTS The authors express their gratitude to Mr. F. Asayama, Mr. K. Yoshida and Mr. M. Yoshinaga of Kawatetsu Steel Products Corporation for making the drawings in this paper.

REFERENCES

1. YANO, K. Analytical Geography in Three Dimensions, p.80-135. Shokabo, Tokyo, 1990.
2. YAMAOKA, Y., WAKIYAMA, K., IMAI, K., MORITA, T., TAGAMI, K., and YUASA, Y. DacsIIs network for supporting design, analysis, construction and manufacturing of space frames(Part-l Framework of the network and it's function and Part-2 Generation of common data file). Summaries of Technical Papers of Annual Meeting Architectural Institute of Japan, 1992, vol. A, August, 1571-1574.
3. RUH, M. A. Computer aided construction system for the generation of geometry for reticulated space frames. Third International Conference on Space Structures, 1984, September, 110-116.
4. TACHIBANA, E. Formulation of topological concepts in discrete structural systems. Proceeding of the first symposium on the use of computer in building engineering, Architectural Institute of Japan, 1979 March, 253-258.

140. Numerical simulation of elastic-plastic buckling behaviour of a reticular dome

H. TAKASHIMA and S. KATO, Toyohashi University of Technology, Japan

SYNOPSIS. A procedure is developed to simulate the structural behaviour of the ball-joint system which is generally accepted in steel space frames. The structural features of the ball-joint system that are noticed by the present authors, are the looseness between the screw and socket, and an initial axial force at the connector. To simulate this behaviour, the section at the connection is divided into small finite elements subjected to material non-linearity, described as bi- or tri-linear function. To confirm the applicability of the present procedure, numerical analyses for the section are performed under several conditions and the results are compared with the experimental ones. Several calculations for single-layer reticular domes using the ball-joints are performed to investigate the effects of the joints on the buckling strength of the reticular domes.

INTRODUCTION

1. With developments in both the performances of computers and analytical procedures for framed structures, the detailed calculations, for example, with consideration of material non-linearity and semi-rigid connection like a real joint etc., have been performed relative easily. Some analytical studies especially for single-layer reticular domes are found in refs 1 to 7. They also give fruitful results to describing the buckling behaviours of the single-layer reticular domes.

2. In those studies, however, the connections have been assumed to be simply hinges, rigid or elastic spring. The ball-joint system generally used in steel space frames has some distinctive mechanical characteristics. From these features, the authors selected as the important parameters, the looseness between the socket and screw and an initial axial stress caused by inserting and tightening the screws. In the present study, the analytical procedure considering the mechanical features of the ball-joints is developed to investigate their effects on the buckling behaviour of space frame structures especially structures like reticular domes.

MODELLING OF BALL-JOINT

3. There are various ball-joint systems available for practical construction. From them, a relatively standard system was chosen to develop a simulation procedure. The typical system is illustrated in Fig. 1. The nose of the member is in contact

 Space structures 4. Thomas Telford, London, 1993

with a rigid ball. The noticeable feature is that an initial stress occurs between the faces by screwing the connector.

4. In a more realistic model, one has to consider yielding of the circumference of the connector, but such a detailed analysis is not necessarily needed to discuss the whole buckling behaviour of reticular domes. Consequently, a well-known and effective method in which the section of the connection is divided into small elements, subjected to material non-linearity, may be accepted (see Fig. 2).

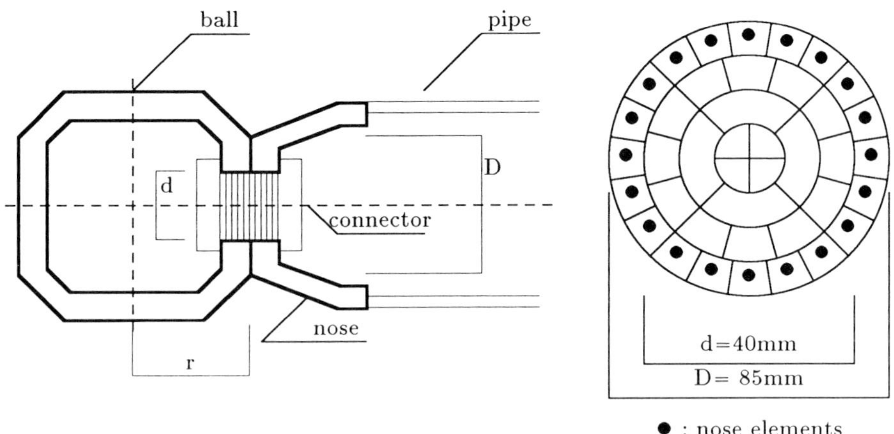

Fig. 1. Geometry of objective connection. Fig. 2. Fibre elements.

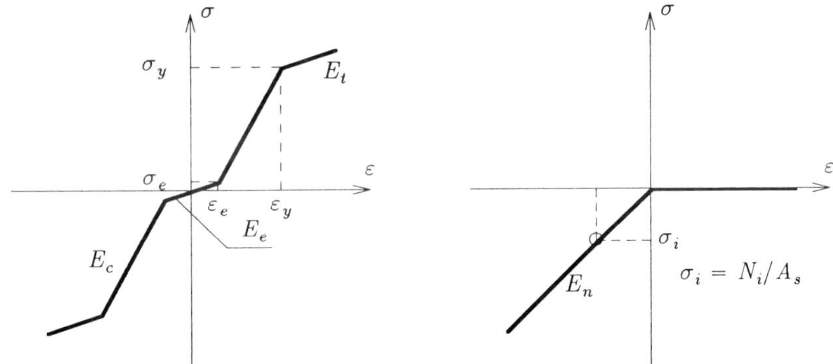

Fig. 3. Stress-strain relation at screw. Fig. 4. Stress-strain relation of the nose elements.

5. In Fig. 2, the elements marked by dots represent the nose section and the others constitute the connector section. Former elements become active only in compression and carry no tension. Figs 3 and 4 give the stress-strain relationship containing a relaxation between the connector. In these figures, E denotes the tangent modulus of the stiffness at each state. The nose part of the joint assembly is assumed not to yield. The yield strain ε_y of the connector is defined as:

$$\varepsilon_y = \left(1 - \frac{E_e}{E_c}\right)\varepsilon_e + \frac{\sigma_y}{E_c} \tag{1}$$

where ε_e is the strain expressing the slack between the connector and the socket, calculated by $\varepsilon_e = d_e / l_s$, where d_e and l_s are the amount of slack and the length of the connector respectively. In Fig. 2, d and D respectively denotes the diameter of the connector and the nose surface. In the present model, the nose and the connector is assumed to have a length of l_s.

BENDING BEHAVIOUR OF THE CONNECTION

Geometry of a simple beam

6. The bending behaviour of a simple beam is simulated by using the method described in a previous paper. The method was used to investigate the applicability of the present connection model to practical problems.

7. Fig. 5 and Table 1 respectively, explain the geometry and material properties of the simple beam. In Fig. 5, the upper part describes the practical beam and the lower is the analytical model composed of 11 finite elements. The central circle shows the ball, assumed to be rigid, and the developed model for the connection is connected at both sides of the ball. A concentrated load is applied to the centre of the ball in the vertical direction.

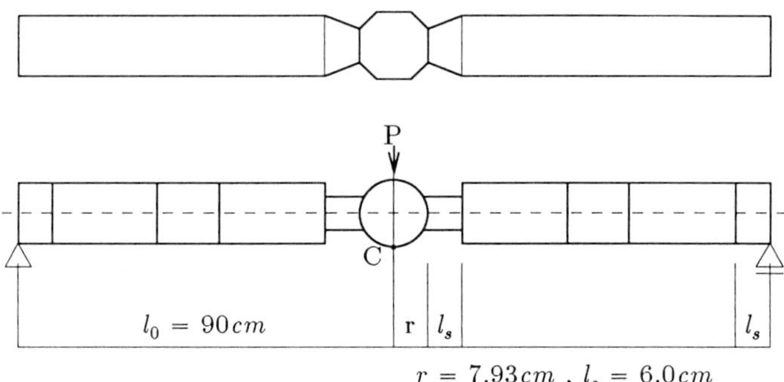

Fig. 5. Simple beam to be analysed.

Table 1. Material properties of the member.

properties	pipe $\phi140\times4$	connector $\phi40$
moment of inertia(cm^4)	395.47	11.46
sectional area(cm^2)	17.09	12.57
yield stress(tf/cm^2)	2.4	2.7
Young's modulus(tf/cm^2)	2100	
modulus of strain hardening(tf/cm^2)	21	

Table 2. Coefficients in material non-linear functions.
(units tf/cm²)

E_e	210	E_t	21
E_c	2100	E_n	2100×0.25

8.The experimental data to be considered are given in ref. 8. In Table 2, the coefficient E_n is assumed to be 2100x0.25tf/cm^2 because the present analytical model for the connector has a tendency to overestimate the bending stiffness of the nose compared with its physical bending stiffness.

Applicability of analytical model

9. In this section, the applicability of the present analytical model is verified by comparing both the analytical and experimental results.

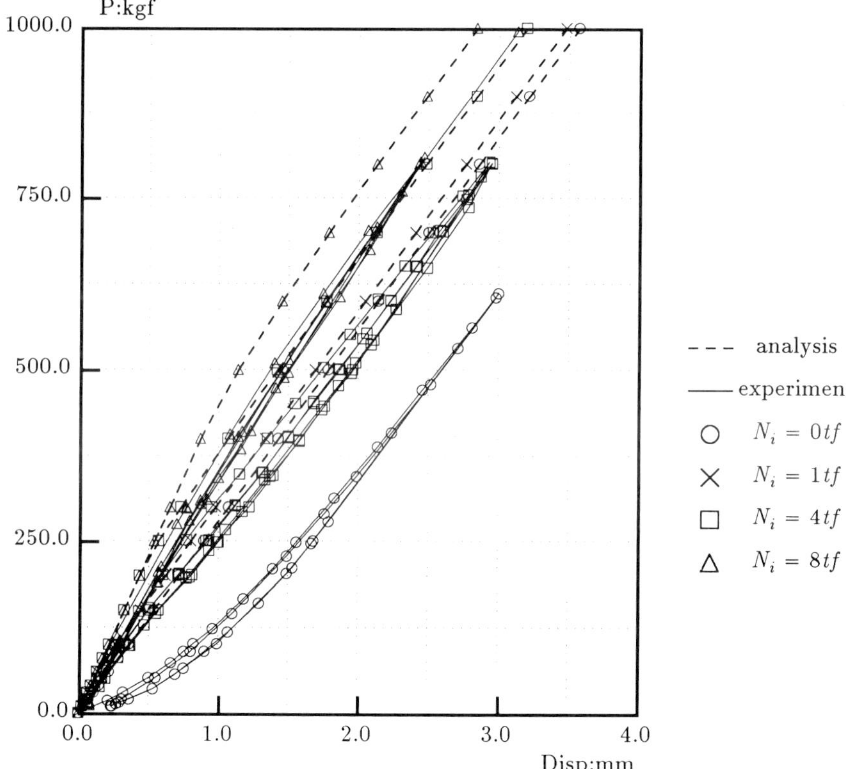

Fig. 6. Relationships between the load and displacement for the case of no relaxation at connectors.

10. Fig. 6 shows the relationships between the load and the displacement of the central nodal point, C, from both the experiments and the analyses. In the analyses, the relaxation at the connector is not considered. The analytical results include the case of $N_i = 1$ tf, while there is no experimental data for $N_i = 1$ tf.

11. It is confirmed that the analyses are judged not to be sufficient to follow closely the experimental results. Especially, in the case of $N_i = 0$ tf, the load-

displacement curve of the analysis is steeper than the experimental one where the initial slope of the load-displacement curve is relatively gentle.

12. Accordingly, the relaxation d_e = 0.0lcm is considered at the connector to express the initial behaviour in the case of N_i = 0 tf. The results of the analyses are drawn in Fig. 7. The slopes of the load-displacement relations are gentle in comparison with the previous case described in Fig. 6.

13. The gentle slope during the premiere loading can be followed, with some errors, by the analyses considering the relaxation at the connectors. So, it may be confirmed that the tendency of the bending behaviour of the simple beam is reasonably expressed by the present analytical model.

14. However, in the present analysis, the agreement between the analysis and experiment is insufficient to some extent. Adjustments of the parameters, d_e,E_e and E_n, are required in future studies.

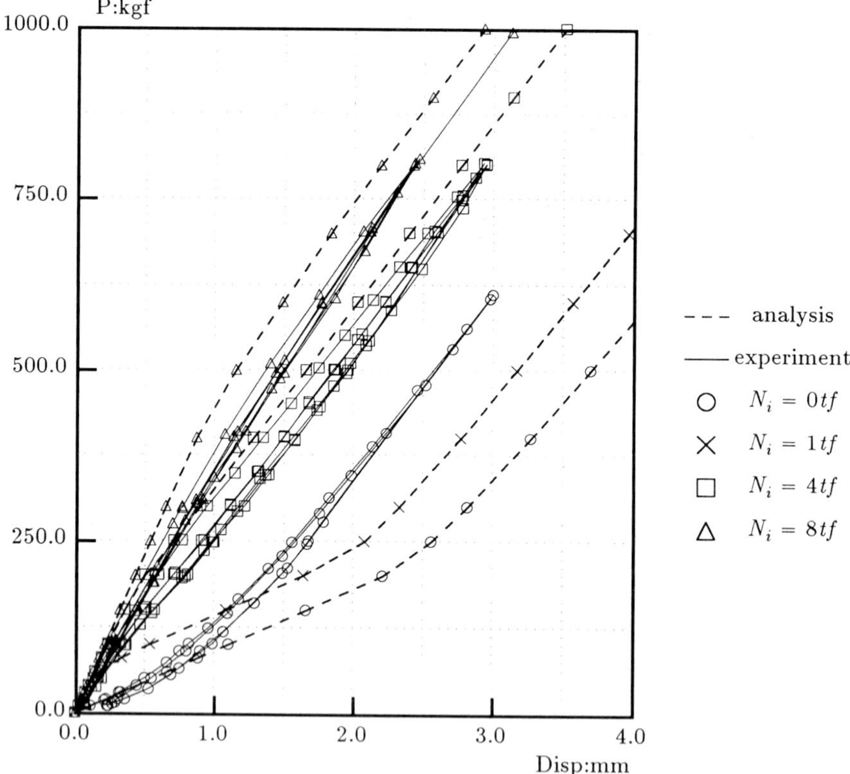

Fig. 7. Relationships between the load and displacement for the case with a relaxation $d_e = 0.01$ cm.

DESCRIPTION OF ANALYSED DOME

Geometry of the dome

15. The geometrical configuration of the dome was based on a hexagonal plan and is illustrated in Fig. 8. Each nodal point on the dome is placed on a spherical surface whose radius of curvature is R (see the table in Fig. 8). The positions of the nodal points are determined by dividing equally the principal diagonal line of the dome structure. The present reticular dome has a triangulate network pattern because of the uniformity in member density and rigidity distribution. The parameter θ_0 expresses the half angle subtending a member and is assumed to be 2 degrees in the present calculation.

16. The boundary conditions of the dome are assumed to be a pin supported which restricts the displacements of the peripheral nodes in all directions. The external loads are applied at the centre of the ball nodes, representing a uniformly distributed load in the vertical direction. For the elastic-plastic buckling analysis, a partial model of the dome was used, in order to reduce computational time.

Geometries and material properties of constitutive member

17. The material properties of the dome members are the same as those for the simple beam shown in Tables 1 and 2, but the length of members are different from the previous beam and assumed to be almost 300cm.

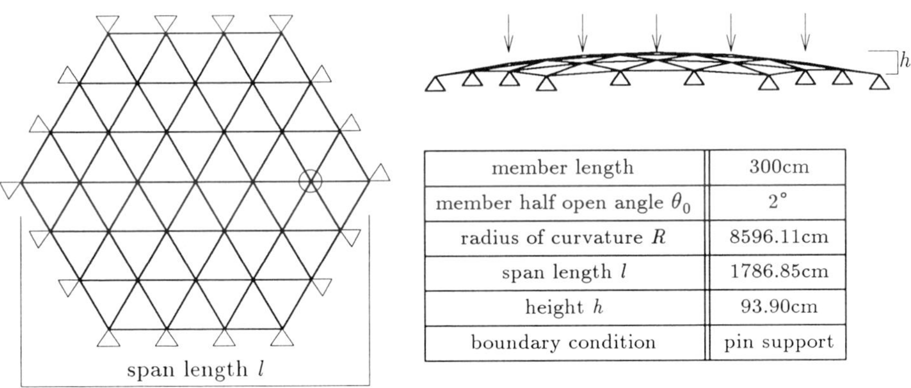

member length	300cm
member half open angle θ_0	2°
radius of curvature R	8596.11cm
span length l	1786.85cm
height h	93.90cm
boundary condition	pin support

Fig. 8. Geometry of the reticulated dome based on a hexagonal plan.

18. Each constitutive member is generated from a series of elements in the order of the rigid ball, a connection element, a tubular member, a connection element and a rigid ball. The tubular member includes the elastic-plastic element at the central part and the element length is equal to the diameter of the pipe. The material non-linearity of the pipe is assumed to be a bi-linear function.

RESULTS OF ELASTIC-PLASTIC ANALYSIS OF A RETICULAR DOME

19. Several calculations allowing for material plastic yielding of the connector elements are performed based on the modelling developed for the ball-joints.

20. Fig. 9 describes the results in the case for no relaxation, $d_e = 0.0$cm. These analyses are undertaken to investigate only the effects of the initial stress on the elastic-plastic buckling strength. In this figure, "Disp" denotes the vertical displacement at the nodal point marked by a circle in Fig. 5 and "Load" means the concentrated loads applied at each nodal point.

21. In the very first loading state, the initial stiffness depends on the magnitude of the initial stress, as confirmed in previous work, but the decrement of the maximum load with respect to the initial stress N_i , is not steep and the load-displacement curves are close to each other. From these results, it is observed that the initial stress can not give large changes to the elastic-plastic buckling strength of the reticular dome under consideration.

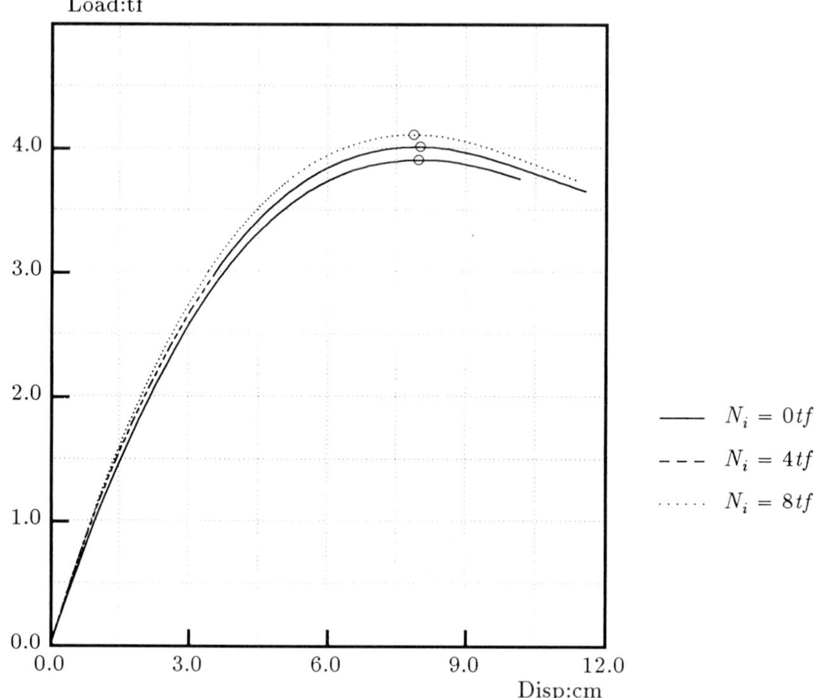

Fig. 9. Relationships between the load and displacement for the case of no relaxation at the connectors.

22. Fig. 10 shows the results of the analysis allowing for relaxation at the connectors in the dome. The magnitude of the relaxation is assumed as d_e = 0.01cm being the same as in the case of the simple beam analysis. The three equilibrium paths are obtained also from the elastic-plastic analyses of the reticular dome illustrated in Fig. 8.

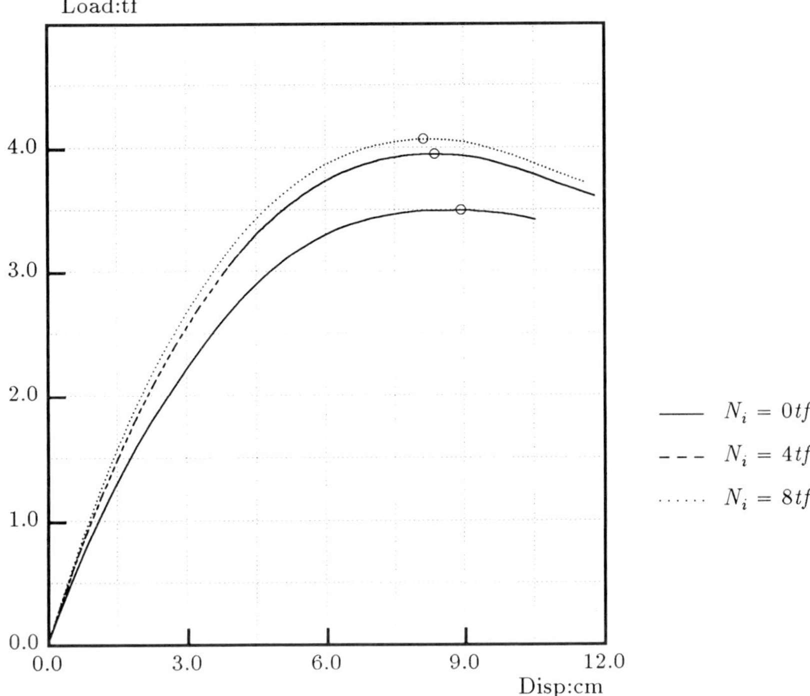

Fig. 10. Relationships between the load and displacement for the case of a relaxation $d_e = 0.01$ cm.

23. In the cases of N_i = 4 tf and 8 tf , both paths are similar to the previous ones considering no relaxation, respectively. However, in the case of N_i = 0 tf the difference is pronounced and the results deviate from those without any relaxation. The reason for the decrease in the maximum load for the case of N_i = 0 tf and d_e = 0.01cm is considered to be due to the initial deflection being amplified by the relaxation but without the initial stress by locating and tighteningf the screws.

24. From the above-mentioned observations, it is predicted that the initial tightening of the screw plays an important role when the relaxation at the connector is intensive.

DISCUSSIONS AND CONCLUSIONS

25. Through the present study, some discussions and conclusions can be drawn as follows. First of all, the conclusions are enumerated as:

(1) The applicability of the analytical model of the connection is verified by comparing the analytical results with the experimental results obtained from bending test on a simple beam;

(2) the elastic-plastic equilibrium paths of reticular domes may be followed by using the present connection model;

(3) the elastic-plastic buckling strength of the present dome is not greatly influenced by a change in the initial stress at the connectors and the relaxation at the connectors has more infulence on the strength than the initial stress;

(4) the ball-joint system used in steel space frames can be numerically simulated by using the proposed finite element method in which the influence of the structural features of the connection, is reflected in the load carrying capacity of steel reticular domes;

26. Meanwhile, the problems to be solved in future studies, are:

(1) the coefficients in the material non-linear functions of the analytical model of the connection have to be established with suitable values through iterative calculations;

(2) the effects of the structural features of the connection on the elastic-plastic buckling behaviours of reticular domes must be investigated to establish the design criteria.

ACKNOWLEDGEMENT

The present research was financially supported in part by The Naito Research Grant.

REFERENCES

1. WRIGHT, D.T., Membrane forces and buckling in reticulated shells, ASCE ST1, pp. 173-201, Feb., 1965.

2. LIND, N.C., Local instability analysis of triangulated dome framework, The Structural Engineering, pp. 317-324, Aug., 1969.

3. FORMAN, S.E. and HUTCHINSON, J.W., Buckling of reticulated shell structures, Int. Journal Solids Structures, vol. 6, pp. 909-932, 1970.

4. DEL POZO,F.F. and DEL POZO,V.F., Buckling of ribbed spherical shells, Proceedings of IASS congress, vol. 1, pp. 199-222, Sep.,1979.

5. MEEK, J.L. and LOGANATHAN, S., Geometric and material non-linear behaviour of structures comprising circular and rectangular hollow section, Proceedings of I.A.S.S. - 10 Years Of Progress In Shell And Spatial Structures -, vol. 4, Sep., 1989.

6. MEEK, J.L. and LOGANATHAN, S., Geometrically non-linear behaviour of space frame structures, Journal of Computers Structures, vol. 31, no. 1, pp. 35-45, 1989.

7. KATO, S., UEKI, T., and TAKASHIMA, H., Study on the buckling behaviour of semi-rigidly jointed latticed dome of single layer, Proceedings of I.A.S.S. 10 Years Of Progress In Shell And Spatial Structures, vol. 4, Sep., 1989.

8.SHIBATA,R., TAKASHIMA,H., YAMADA,S., and KATO,S., Bending test of a joint in constitutive members which are applied to system truss, Summaries Of Technical Papers Of Annual Meeting AIJ, vol. B, Structures I, pp. 1663-1664, Aug.,1992.

141. Tangent stiffness matrix for space frame members with both member and joint imperfections

F. A. FATHELBAB, Alexandria University, Egypt

SYNOPSIS. The conventional procedure for the study and analysis of space frame structures assumes that the members of the structure are geometrically perfect. Also the procedure assumes that the joints of the space structure are pinned or completely rigid. Space frame members may suffer from geometric imperfections, also most joints of these structures in reality are semi-rigid. A new space frame member tangent stiffness matrix which incorporates both the effects of member geometric imperfection and joint properties has been developed.

INTRODUCTION

1. Space frame structures are widely used in practice. These structures can be used to cover large areas without the need for intermediate supports, while at the same time attain a very small structural weight per unit area. The two main components of any space structure are the members and the joints. Many new prefabricated joints have been developed. Most of these joints are not perfectly pinned nor perfectly rigid but semi-rigid. Also space frame members may suffer from geometric imperfections along their length. These imperfections may be due to errors in fabrication, errors in erection and/or due to misuse of the structure after erection.

2. It has been realised that the pattern and the value of imperfections along the length of the compression member affect its behaviour significantly. Also the joint behaviour makes an important contribution to the behaviour of the whole structure, and the uncertain joint performance has been one of the critical factors in many structural collapses. The object of this research is to develop a tangent stiffness matrix for a space frame member. This matrix incorporates both member and joint imperfections.

MEMBER IMPERFECTIONS

3. For space frame members, imperfections along the member length are represented by initial curvature (bow or out of straightness) about the two principal axes of the member cross-section. Also initial twist may exist. Twist imperfections have only significant effect in case of open cross-section members.

For closed section members, which most space structures consist of, the geometric imperfection that needs to be considered is the initial bow of the members. Any arbitrary crooked shape of an imperfect member, as shown in Fig. 1, can be expressed as an appropriate summation of first, second and higher order sinusoidal curves (refs 1, 2), such as:

$$y_o = \sum_{i=1}^{\infty} e_i \sin\frac{i\pi x}{L} \tag{1}$$

where y_o and e_i are as shown in Fig. 1.

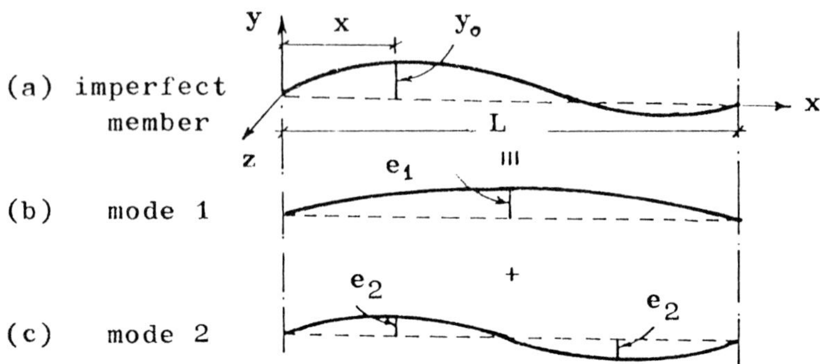

Fig. 1. Member geometric imperfection.

4. In practice members buckle in the shape of their first and/or second order modes. It is therefore considered sufficient to model the geometric imperfections that are likely to enhance the possible modes of failure in compression. So the member imperfections will be expressed using only the first two terms of Eq. 1 which gives:

$$y_o = e_1 \sin\frac{\pi x}{L} + e_2 \sin\frac{2\pi x}{L} \tag{2}$$

This approach has been considered by Hatzis (ref. 2).

5. Referring to Fig. 2 for a prismatic imperfect member, and from beam-column theory, the deformation y_1 of the member measured from its initial position is:

$$y_1 = \frac{M_1^c}{Q}\left(\frac{\sin\phi(L-x)}{\sin\phi L} - \frac{L-x}{L}\right) + \frac{M_2^c}{Q}\left(\frac{\sin\phi x}{\sin\phi L} - \frac{x}{L}\right) + q\left(\frac{e_1}{1-q}\sin\frac{\pi x}{L} + \frac{e_2}{4-q}\sin\frac{2\pi x}{L}\right) \tag{3}$$

where

M_1^c, M_2^c are the end moments

Q is the axial force in the member

q is non-dimensional axial force parameter = Q/P_E

e_1, e_2 are the amplitudes of the imperfections

$\phi = \sqrt{Q/EI}$

L is the member length

P_E is the member Euler load = $\pi^2 EI/L^2$

E is the modulus of elasticity

I is the moment of inertia of the member cross-section.

The total deformation y measured from the x-axis can be obtained by adding Eq. 2 to Eq. 3 to give:

$$y = y_1 + y_o = \frac{M_1^c}{Q}\left(\frac{\sin\phi(L-x)}{\sin\phi L} - \frac{L-x}{L}\right) + \frac{M_2^c}{Q}\left(\frac{\sin\phi x}{\sin\phi L} - \frac{x}{L}\right) + \frac{e_1}{1-q}\sin\frac{\pi x}{L} + \frac{e_2}{4-q}\sin\frac{2\pi x}{L} \tag{4}$$

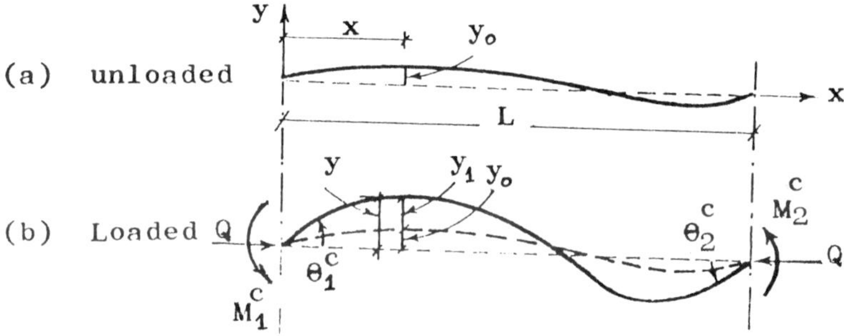

Fig. 2. Deformation of unloaded and loaded imperfect member.

The differential of Eq. 4 with respect to x evaluated at x=0 and x=L for θ_1 and θ_2 respectively gives:

$$\theta_1^c = \frac{L}{EI}(v_1 M_1^c + v_2 M_2^c) + \frac{\pi e_1}{L(1-q)} + \frac{8\pi e_2}{L(4-q)} \quad ,$$

$$\theta_2^c = \frac{L}{EI}(v_2 M_1^c + v_1 M_2^c) - \frac{\pi e_1}{L(1-q)} + \frac{8\pi e_2}{L(4-q)} \tag{5}$$

where v_1 and v_2 are given in Table 1. Solving Eqs 5 for M_1 and M_2 leads to:

$$M_1^c = \frac{EI}{L}(c_1\theta_1^c + c_2\theta_2^c - A_1e_1 - A_2e_2)\ , \quad M_2^c = \frac{EI}{L}(c_2\theta_1^c + c_1\theta_2^c + A_1e_1 - A_2e_2) \qquad (6)$$

where,

$$c_1 = \frac{v_1}{v_1^2 - v_2^2}\ , \quad c_2 = \frac{-v_2}{v_1^2 - v_2^2} \qquad (7)$$

$$A_1 = \frac{\pi(c_1 - c_2)}{L(1-q)}\ , \quad A_2 = \frac{8\pi(c_1 + c_2)}{L(4-q)} \qquad (8)$$

c_1 and c_2 are the ordinary stability, functions given in Oran (ref. 3) and are listed in Table 1.

Table 1. Stability functions of frame member.

Q	Compression	Zero	Tension
v_1	$\frac{1}{\varphi}(\frac{1}{\varphi} - \frac{1}{\tan\varphi})$	1/3	$\frac{1}{\psi}(\frac{1}{\tan\psi} - \frac{1}{\psi})$
v_2	$\frac{1}{\varphi}(\frac{1}{\varphi} - \frac{1}{\sin\varphi})$	- 1/6	$\frac{1}{\psi}(\frac{1}{\sin\psi} - \frac{1}{\psi})$
c_1	$\frac{\varphi(\sin\varphi - \varphi\cos\varphi)}{2(1-\cos\varphi) - \varphi\sin\varphi}$	4	$\frac{\psi(\psi\cosh\psi - \sinh\psi)}{2(1-\cosh\psi) + \psi\sinh\psi}$
c_2	$\frac{\varphi(\varphi - \sin\varphi)}{2(1-\cos\varphi) - \varphi\sin\varphi}$	2	$\frac{\psi(\sinh\psi - \psi)}{2(1-\cosh\psi) + \psi\sinh\psi}$

where $\varphi = \phi L = \pi\sqrt{q}\ , \quad \psi = \varphi\sqrt{-1}$ (9)

JOINT IMPERFECTIONS

6. Joint imperfections considered are joint stiffness (not pinned nor rigid but semi-rigid), and joint size. The joint stiffness is represented by the stiffness of a zero-length spring at the ends of the member. The joint size is represented by a rigid arm with its length equal the size of the joints.

Joint bending stiffness

7. Fig. 3 shows a geometrically imperfect member with springs at its ends. The slip rotations (the difference between the base member end rotations and the joint

rotations), due to the bending flexibility of the joints are ϕ_1 and ϕ_2 at ends 1 and 2 respectively. The total joint rotations are:

$$\theta_1^m = \theta_1^c + \phi_1 \ , \quad \theta_2^m = \theta_2^c + \phi_2 \qquad (10)$$

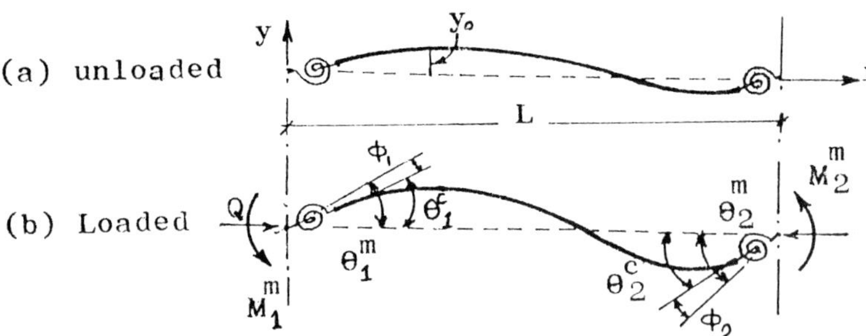

Fig. 3. Imperfect member with end springs.

The slip rotations ϕ_1 and ϕ_2 depend on the bending stiffness k_{1b} and k_{2b} of joints 1 and 2. These stiffnesses can be related to the bending stiffness of the base member by:

$$\phi_1 = \frac{M_1^m}{k_{1b}} = \frac{M_1^m L}{EI}\varepsilon_1 \ , \quad \phi_2 = \frac{M_2^m}{k_{2b}} = \frac{M_2^m L}{EI}\varepsilon_2 \qquad (11)$$

where $\varepsilon_1 = (\frac{EI}{L})/k_{1b}$, $\varepsilon_2 = (\frac{EI}{L})/k_{2b}$ (12)

Substituting the values of ϕ_1 and ϕ_2 from Eq. 11 and the expression for θ_1^c and θ_2^c from Eq. 5 into Eq. 10 and solving for M_1^m and M_2^m gives:

$$M_1^m = \frac{EI}{L}(c_{11}\theta_1^m + c_{12}\theta_2^m - A_{11}e_1 - A_{12}e_2) \ , \quad M_2^m = \frac{EI}{L}(c_{12}\theta_1^m + c_{22}\theta_2^m - A_{21}e_1 - A_{22}e_2) \qquad (13)$$

where

$$c_{11} = \frac{\varepsilon_2(c_1^2 - c_2^2) + c_1}{\sigma} \ , \quad c_{12} = \frac{c_2}{\sigma} \ , \quad c_{22} = \frac{\varepsilon_1(c_1^2 - c_2^2) + c_1}{\sigma}$$

$$A_{11} = \frac{\pi(c_{11} - c_{12})}{L(1-q)} \ , \quad A_{12} = \frac{8\pi(c_{11} + c_{12})}{L(4-q)} \ , \quad A_{21} = \frac{\pi(c_{22} - c_{12})}{L(1-q)} \ , \quad A_{22} = \frac{8\pi(c_{22} + c_{12})}{L(4-q)} \qquad (14)$$

and $\sigma = \varepsilon_1\varepsilon_2(c_1^2 - c_2^2) + (\varepsilon_1 + \varepsilon_2)c_1 + 1$ (15)

It is worth noting that any combination of end conditions (pinned, rigid, and flexible) can be obtained from Eq. 13 by substituting the appropriate value of ε_1 and ε_2 ($\varepsilon = 0.0$ for rigid joint and $\varepsilon=\infty$ for a pinned joint).

Joint stiffness and size

8. Fig. 4. shows an imperfect member with rigid parts at the ends, the length of which represents the size of the joint. The total length of the member between the centres of the joints is L, and the size of the joints is $\lambda_1 L$ and $\lambda_2 L$ for joints 1 and 2 respectively, and the length of the base member is λL.

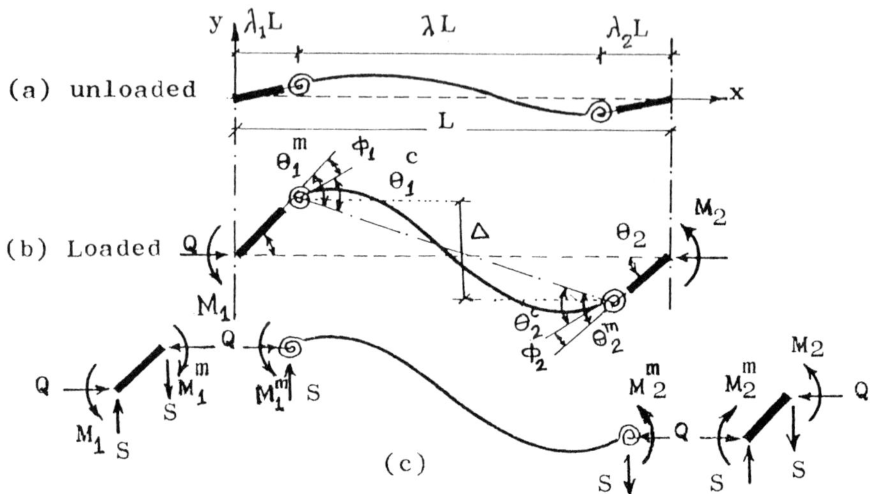

Fig. 4. Imperfect member with springs and rigid arms.

Consider the equilibrium between the forces at the ends of the base member with springs and the forces at the centres of the joints as shown in Fig. 4(c), then:

$$M_1 = M_1^m + \lambda_1 LS - Q\lambda_1 L\theta_1 \ , \quad M_2 = M_2^m + \lambda_2 LS - Q\lambda_2 L\theta_2 \qquad (16)$$

where M_1, M_2, θ_1, θ_2 and S are as shown in Fig. 4, and M_1^m,M_2^m are as given in Eq. 13 after replacing L by λL. The rotations at the ends of the springs θ_1^m and θ_2^m with respect to the chord of the base member are related to the rotations θ_1 and θ_2 at the centres of the joints 1 and 2 with respect to the chord of the whole member by:

$$\theta_1^m = \theta_1 - \frac{\Delta}{\lambda L} \quad , \quad \theta_2^m = \theta_2 - \frac{\Delta}{\lambda L} \qquad (17)$$

The shearing force S is related to M_1^m,M_2^m and Q as:

$$S = \frac{M_1^m + M_2^m + Q\Delta}{\lambda L} \tag{18}$$

The value of Δ can be obtained from the condition

$$\Delta + \lambda_1 L\theta_1 + \lambda_2 L\theta_2 = 0 \tag{19}$$

Substituting the values of M_1^m, M_2^m, S and Δ into Eq. 16, gives:

$$M_1 = \frac{EI}{L}(\gamma_{11}\theta_1 + \gamma_{12}\theta_2 - \beta_{11}e_1 - \beta_{12}e_2) \quad , \quad M_2 = \frac{EI}{L}(\gamma_{12}\theta_1 + \gamma_{22}\theta_2 + \beta_{21}e_1 - \beta_{22}e_2) \tag{20}$$

where $\gamma_{11} = \frac{1}{\lambda}[(h_1+1)^2 C_{11} + 2h_1(h_1+1)c_{12} + h_1^2 c_{22} - \phi^2 h_1(h_1+1)]$,

$$\gamma_{12} = \frac{1}{\lambda}[h_2(h_1+1)C_{11} + (2h_1h_2+h_1+h_2+1)c_{12} + h_1(h_2+1)c_{22} - \phi^2 h_1 h_2] \quad ,$$

$$\gamma_{22} = \frac{1}{\lambda}[h_2^2 C_{11} + 2h_2(h_2+1)c_{12} + (h_2+1)^2 c_{22} - \phi^2 h_2(h_2+1)] \quad ,$$

$$\beta_{11} = \frac{\pi}{\lambda L(1-q)}[h_1(c_{11} - c_{22}) + c_{11} - c_{12}] \ , \quad \beta_{12} = \frac{8\pi}{\lambda L(4-q)}[h_1(c_{11} + 2c_{12} + c_{22}) + c_{11} + c_{12}]$$

$$\beta_{21} = \frac{\pi}{\lambda L(1-q)}[h_2(c_{22} - c_{11}) + c_{22} - c_{12}] \ , \quad \beta_{22} = \frac{8\pi}{\lambda L(4-q)}[h_2(c_{11} + 2c_{12} + c_{22}) + c_{22} + c_{12}]$$

(21)

and $h_1=\lambda_1/\lambda$ and $h_2=\lambda_2/\lambda$. It can be seen that if the size of joint is neglected (i.e. $h_1 = h_2 = 0$), the expressions in Eqs 20 will lead to the same expressions in Eqs 13 for the case of an imperfect member with end springs. Also if the member is geometrically perfect (i.e. $e_1 = e_2 = 0$) Eqs 20 lead to the same expressions derived by Fathelbab (ref. 4) for a member with joint imperfections only.

EFFECT OF IMPERFECTIONS ON AXIAL FORCE

9. For the space frame member shown in Fig. 1, and considering imperfections about the two principal axes y and z, the initial relative axial strain, μ_o due to imperfection is:

$$\mu_o = \frac{\pi^2}{L^2} \sum_{n=y,z} (e_{1n}^2 + e_{2n}^2) \tag{22}$$

where e_l and e_2 are as shown in Fig. 1, and n refers to axes y and z. For the loaded member, shown in Fig. 2, the total relative axial strain will be:

$$\mu = \frac{QL}{EA} + \sum_{n=y,z} \mu_{bn} \tag{23}$$

where μ_{bn} is the bowing axial strain (including the initial strain due to the initial bowing) about the two principal axes, and A is the member cross-sectional area. The value of μ_{bn} is:

$$\mu_{bn} = b_{1n}(\theta^c_{1n} + \theta^c_{2n})^2 + b_{2n}(\theta^c_{1n} - \theta^c_{2n})^2 \tag{24}$$

b_{1n} and b_{2n} are the so called bowing functions given by Saafan (ref. 5). the value of these functions are:

$$b_{1n} = \frac{(c_{1n} + c_{2n})(c_{2n} - 2)}{8\pi^2 q} \quad , \quad b_{2n} = \frac{c_{2n}}{8(c_{1n} + c_{2n})} \tag{25}$$

and c_{1n} and c_{2n} are the stability functions given in Table 1.

10. If the joints at the ends of the member are not completely rigid under the applied axial force Q, but have axial stiffnesses k_{1a} and k_{2a} at ends 1 and 2 respectively, then the zero length spring representing the joint should have an axial stiffness equal the stiffness of the joint. The axial force in Fig. 3 is related to the net relative axial strain by:

$$\frac{QL}{EA}(1 + \varepsilon_{1a} + \varepsilon_{2a}) = \mu - \sum_{n=y,z} \mu_{bn} + \mu_o \tag{26}$$

$$\text{or} \quad Q = EA\varepsilon_a(\mu - \sum_{n=y,z} \mu_{bn} + \frac{\pi^2}{L^2} \sum_{n=y,z} (\frac{e^2_{1n}}{4} + e^2_{2n}) \tag{27}$$

where

$$\varepsilon_a = \frac{1}{(1 + \varepsilon_{1a} + \varepsilon_{2a})} \quad , \quad \varepsilon_{1a} = \frac{EA}{L}/k_{1a} \quad , \quad \varepsilon_{2a} = \frac{EA}{L}/k_{2a} \tag{28}$$

11. If the joint size is considered as for the member in Fig. 4, Eq. 27 should be modified to include the effect of joint size and L should be replaced by L. The modification of Eq. 27 was performed by Fathelbab (ref. 4) as follows:

$$Q = EA\varepsilon_a(\mu - \sum_{n=y,z} c_{bn} + \sum_{n=y,z} c_{on}) \tag{29}$$

where

$$c_{bn} = \frac{1}{2}(h_1\theta_{1n}^2 + \theta_{2n}^2) + \lambda[b_{1n}\{(1+2h_1)\theta_{1n} + (1+2h_2)\theta_{2n} - \phi_{1n} - \phi_{2n}\}^2$$
$$+ b_{2n}\{\theta_{1n} - \theta_{2n} - \phi_{1n} - \phi_{2n}\}^2 + \frac{(h_1\theta_{1n} + h_2\theta_{2n})}{2(1-B_n^{\ m})}] \tag{30}$$

and $B_n^{\ m} = b_{1n}(\theta_{1n}^m + \theta_{2n}^m)^2 + b_{2n}(\theta_{1n}^m - \theta_{1n}^m)^2$, $c_{on} = \frac{\pi^2}{(\lambda L)^2} \sum_{n=y,z} (\frac{e_{1n}^2}{4} + e_{2n}^2)$ (31)

12. From Eq. 29, the non-dimensional axial force parameter q_n can be written in the form:

$$q_n = \frac{Q}{P_{En}} = EA\varepsilon_a(\mu - c_{by} - c_{bz} + c_{oy} + c_{oz}) / \frac{\pi^2 EI_n}{(\lambda L)^2} = \frac{s_n^2 \varepsilon_a}{\pi^2}(\mu - c_{by} - c_{bz} + c_{oy} - c_{oz}) \tag{32}$$

where s_n is the slenderness ratio of the base member about n-axis = $\lambda L / \sqrt{I_n / A}$ and P_{En} is the Euler load about the same axis. If the value of q_n from Eq. 32 is substituted into Eq. 27, this may be re-written in the form:

$$QL = \frac{EI}{L}(\frac{\pi^2}{\lambda^2} q) \tag{33}$$

where I is a reference moment of inertia of the member cross-section, and q is the axial non-dimensional parameter with respect to this inertia.

EFFECT OF IMPERFECTION ON TORSIONAL MOMENT

13. As the twist imperfection effect is negligible for closed section members, only joint imperfection is considered. If the joints have torsional stiffnesses k_{1t} and k_{2t} at ends 1 and 2 respectively, then the relation between the torsional moment M_t and the relative twisting angle θ_t will be:

$$M_t = \frac{GJ}{L}\varepsilon_t\theta_t \tag{34}$$

where G is the shear modulus, and J is the polar moment of inertia of the member cross-section, and

$$\varepsilon_t = \frac{1}{\lambda(1+\varepsilon_{1t}+\varepsilon_{2t})}, \quad \varepsilon_{1t} = \frac{GJ}{L}/k_{1t}, \quad \varepsilon_{2t} = \frac{GJ}{L}/k_{2t} \tag{35}$$

MEMBER TANGENT STIFFNESS MATRIX

14. The expression in Eq. 20, after generalising for the two axes y and z, along with Eq. 34 and Eq. 33 relate the independent member end forces to its relative end deformation. These expressions can be written in the form:

$$\{S\} = \{f(u)\} \tag{36}$$

where $\{S\}$ is the vector of the independent member end forces, (u) are the relative end deformations, and f are a set of non-linear functions in (u). These functions include the effects of member and joint imperfections as well as the effect of axial force.

15. For any given displacement u, Eq. 36 can be differentiated to give:

$$\{\Delta S\} = [t]\{\Delta u\} \tag{37}$$

where

$$\{\Delta S\}^T = \{\Delta M_{1z}, \Delta M_{2z}, \Delta M_{1y}, \Delta M_{2y}, \Delta M_t, \Delta GL\} \quad ,$$

$$\{\Delta u\}^T = \{\Delta\theta_{1z}, \Delta\theta_{2z}, \Delta\theta_{1y}, \Delta\theta_{2y}, \Delta\theta_t, \Delta\mu\} \tag{38}$$

and [t] is the member tangent stiffness matrix with respect to its local current axes. The elements t_{ij} of [t] are given by:

$$t_{ij} = \frac{\partial S_i}{\partial u_j} + \frac{\partial S_i}{\partial q} \cdot \frac{\partial q}{\partial u_j} \qquad \text{for } i,j = 1,2,...,6 \tag{39}$$

Here only one element (t_{11}), as an example, will be evaluated and the rest follow in similar way.

$$t_{11} = \frac{\partial M_{1z}}{\partial \theta_{1z}} + \frac{\partial M_{1z}}{\partial q} \cdot \frac{\partial q}{\partial \theta_{1z}}$$
$$= \frac{EI_z}{L}[\gamma_{11z} + \gamma'_{11z}\theta_{1z} + \gamma'_{12z}\theta_{2z} - \beta'_{11z}e_{1y} - \beta'_{12z}e_{2y}]\frac{\partial q}{\partial \theta_{1z}} \tag{40}$$

' denotes differentiation with respect to q, and γ'_{11z}, γ'_{12z}, β'_{11z}, β'_{12z} can be evaluated from Eqs 21 in terms of c_1, c_2, c'_1 and c'_2. The value of c'_1 and c'_2 can be obtained in terms of b_1 and b_2 from the relation presented by Saafan (ref. 5), where:

$$b_1 = \frac{c'_1 + c'_2}{4\pi^2} \quad , \quad b_2 = \frac{c'_1 - c'_2}{4\pi^2} \tag{41}$$

The differentiation of q in Eq. 39 can be evaluated from Eq. 32.

15. After evaluating all elements of the tangent matrix [t], this matrix can be transferred from the member local current axes to the structure global axes. This transformation is done by considering geometry, equilibrium, and orientation of the member. Oran (ref. 3) presented a procedure for this transformation.

CONCLUSION

16. Expressions relate the independent space frame end forces to its end relative deformations were obtained. These expressions include the effects of member and joint imperfections as well as the effect of axial force and bowing functions. A procedure to obtain the member tangent stiffness matrix in its current local axes was presented.

REFERENCES

1. TIMOSHENKO, S. P. and GERE, T. M. Theory of Elastic Stability. McGraw-Hill, New York, 1961.
2. HATZIS, D. T. The influence of imperfections on the behaviour of shallow single layer lattice domes. Thesis submitted at the university of Cambridge for the degree of Doctor of Philosophy, March, 1987.
3. ORAN, C. Tangent stiffness in space frames. J. Struct. Div., ASCE, 1973, Vol.99, No.ST6, June, 987-1001.
4. FATHELBAB, F. A. The effect of joints on the stability of shallow single layer lattice domes. Thesis submitted at the university of Cambridge for the degree of Doctor of Philosophy, September, 1987.
5. SAAFAN, S. A. Nonlinear behaviour of structural plane frames. J. Struct. Div., ASCE, 1963, Vol.89, No.ST4, August, 557-579.

142. Application of group theory to the analysis of space frames

A. ZINGONI, M. N. PAVLOVIC and D. LLOYD-SMITH, Imperial College, UK, and G. ZLOKOVIC, Serbian Academy of Science and Arts, Yugoslavia

SYNOPSIS. Compared with conventional methods of exploiting symmetry in structural analysis, use of group theory in the description of the symmetry of the structure presents a more systematic approach to analysis, with the result that all the symmetry properties of the structural configuration are taken into account. Furthermore, the problem is decomposed into independent subspaces within which the number of unknowns is only a fraction of those of the original problem, with a consequent reduction in computational effort. In this paper, we illustrate how the technique may be used to simplify the static analysis of indeterminate space frames.

INTRODUCTION

1. Application of group theory to the solution of problems in physics and chemistry is well established (refs 1-8). In structural analysis, application of the theory to eigenvalue problems of the vibration and stability of beams, the vibration of taut strings, and the buckling of a plate, has been shown (ref. 9) to lead to polynomial equations of lower degree than would be obtained in a conventional analysis. As demonstrated by Zlokovic (ref. 9), the same approach applied to the static analysis of structures by the force method results in a block-diagonal structure flexibility matrix consisting of independent submatrices of the smallest possible dimensions, symmetry adapted redundants within the corresponding subspaces of the original vector space of the problem being independently obtained through the inversion of the associated submatrix of the block-diagonalised structure matrix. Conventional analysis, on the other hand, would in general require the inversion of an entire n-dimensional structure flexibility matrix, in arriving at the solution for n redundants. It must be noted, however, that application of group theory to a physical problem does not, in general, lead to the discovery of new results; its purpose, as pointed out by Leech and Newman (ref. 10), is to facilitate analysis or computation, enabling results to be obtained more quickly. Structural-engineering applications are not an exception to this.

 Space structures 4. Thomas Telford, London, 1993

2. Zlokovic (ref. 9) has considered a pin-jointed space frame with pentagonal symmetry. In the present paper, redundant pin-jointed space frames based on the square and the equilateral triangle are considered, and a more direct procedure for adapting arbitrary applied loads to the symmetry of the subspaces of the symmetry adapted redundants, and for obtaining final redundants corresponding to any arbitrary combination of applied loads, is illustrated in detail. While the redundant structure itself may have symmetry properties that are quite recognisable in the conventional sense (i.e. planes of reflection), quite often the only possible choice of releases, or at least the most obvious one to a structural analyst, may result in a configuration of the statically determinate primary system and set of redundants that no longer has symmetry in the conventional sense (i.e. no reflection planes), while still retaining less obvious symmetry properties (like rotational permutability) which only group theory can exploit. Such an instance will be illustrated in this paper.

3. The method consists of identifying the symmetry group of the configuration of the structure, and then choosing a system of releases that can be permuted by elements of the group, or its subgroups if the only possible system of releases has a configuration with less symmetry than the original structure. Using the character table of the symmetry group - such tables are widely available in group-theory literature (see, for example, refs 1, 8, 10 and 11) - for the configuration of the assumed arbitrary redundants, idempotents of the centre of the group algebra (refs 9, 12) are deduced. These correspond to the various possible irreducible representations (refs 9-14) of the symmetry group, such representations in turn being associated with particular independent subspaces into which the original vector space of the problem has to be decomposed. For each subspace, basis vectors (i.e. symmetry-adapted redundants) are isolated by applying the corresponding idempotent to each arbitrary redundant function ϕ_i (i=1, 2,..., n; n is the total number of arbitrary redundants); unit values of these symmetry-adapted redundants are then applied to the released structure, in accordance with the coefficients (+1 or -1) of the ϕ_i's making up the basis vector(s) of that subspace. Provided the arbitrary applied loads are permuted in the same way by elements of the symmetry group of the configuration of the assumed set of arbitrary redundants as those arbitrary redundants themselves, the same coefficients also give the manner in which the applied loads should be decomposed into configurations consistent with the symmetry of the subspace in question. Symmetry-adapted redundants for each subspace are solved for independently of those in other subspaces; superposition of these then gives the final redundants.

SPACE FRAME OF SYMMETRY GROUP C_{2v}

4. The first example is a pin-jointed space frame shown in plan and elevations in Fig. 1. Joints 1, 2, 3 and 4 form the corners of a square with sides of length $2a$, lying in a horizontal plane a distance a above the horizontal support plane containing the joints A, B, C and D, the latter forming the corners of a square with sides of length $4a$ and symmetry planes coinciding with those of square

1234. Points 1', 2', 3' and 4' lie in the support plane, vertically below joints 1, 2, 3 and 4, respectively, to which they are connected by vertical struts. The space frame is subject to arbitrary vertical loads P_1, P_2, P_3 and P_4, applied at joints 1, 2, 3 and 4, respectively.

5. With 16 members (m=16), 24 reaction components (r=24) and 12 joints (j = 12), the degree of static indeterminacy of the structure is given by the usual formula m+r-3j, and is therefore 4.

6. The configuration of the structure belongs to symmetry group $C_{2\upsilon}$, with elements E (identity), C_2 (rotation about the vertical axis 0 - see plan of Fig. 1 - through 360°/2), σ_x and σ_y (reflections in the vertical planes passing through 0 and orientated along x-x and y-y, respectively - see plan of Fig. 1). The four redundants are taken as the forces in the vertical members 11', 22', 33' and 44'. Releasing these members leaves a statically determinate structure that preserves the $C_{2\upsilon}$ symmetry of the original structure. The four redundants are also permutable by the elements of symmetry group $C_{2\upsilon}$.

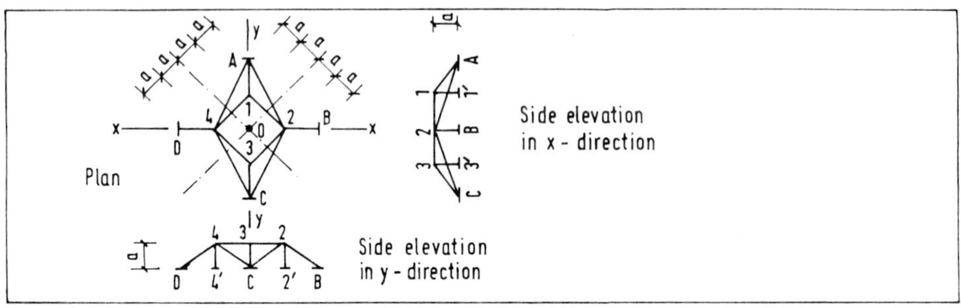

Fig. 1. Space frame of symmetry group $C_{2\upsilon}$.

7. The character table of group $C_{2\upsilon}$ is shown in Table 1. The rows, labelled A_1, A_2, B_1 and B_2 in accordance with standard notation (see, for example, refs 9-14), correspond to the various irreducible representations Γ_i (i = 1,..., 4) of the group, while the four columns correspond to each of the four classes K_j (j = 1,..., 4) of $C_{2\upsilon}$: $K_1 = \{E\}$, $K_2 = \{C_2\}$, $K_3 = \{\sigma_x\}$ and $K_4 = \{\sigma_y\}$. Classes of a group are the distinct sets of symmetry elements formed by collecting into a set, for each symmetry element α of the group, the distinct results of the conjugates $\beta^{-1}\alpha\beta$ for all elements β of the group. According to representation theory, the number of different possible irreducible representations of a group is equal to the number of classes, and all elements belonging to the same class of a given irreducible representation have the same character (see, for example, refs 9 and 10). Methods of generating character tables may be seen in (refs 1, 13 and 14), among others.

8. Orthogonal idempotents $\pi(\pi \neq 0)$ of the centre of the group algebra, which satisfy the relations $\{\pi_i^2 = \pi_i,\ \pi_i\pi_j = 0 \text{ if } i \neq j\}$, can be written down directly from the character table, using the expression (ref. 9):

$$\pi_i = (h_i/h) \sum_{\beta} \chi_i(\beta^{-1})\beta \tag{1}$$

Table 1. Character table of group $C_{2\upsilon}$.

C_{2v}	E	C_2	σ_x	σ_y
A_1	1	1	1	1
A_2	1	1	−1	−1
B_1	1	−1	1	−1
B_2	1	−1	−1	1

where h_i is the dimension of the i-th irreducible representation, given by $\chi_i(E)$, the first value of the i-th row of the character table; h is the order of the group (i.e. the number of elements in the group); χ_i is a character of the i-th irreducible representation; β is an element of the group, and β^{-1} its inverse. Noting that $E^{-1}= E$, $C_2^{-1} = C_2$, $\sigma_x^{-1} = \sigma_x$ and $\sigma_y^{-1} = \sigma_y$, the orthogonal idempotents of group $C_{2\upsilon}$ are:

$$\pi_1 = (1/4)\ (E + C_2 + \sigma_x + \sigma_y) \tag{2a}$$
$$\pi_2 = (1/4)\ (E + C_2 - \sigma_x - \sigma_y) \tag{2b}$$
$$\pi_3 = (1/4)\ (E - C_2 + \sigma_x - \sigma_y) \tag{2c}$$
$$\pi_4 = (1/4)\ (E - C_2 - \sigma_x + \sigma_y) \tag{2d}$$

9. Idempotents act as projection operators (refs 9, 10, 13, 14) by means of which the basis vectors of a subspace U of a particular symmetry type can be identified. Each idempotent, corresponding to a particular irreducible representation, also corresponds to a particular subspace of a specific symmetry type (i.e. π_i corresponds to Γ_i and U_i, where i = 1,..., k; k is the number of idempotents = number of irreducible representations = number of independent subspaces). Applying idempotents to arbitrary functions of a problem generates symmetry-adapted functions for the respective subspaces.

10. The four-dimensional space V of the present problem may thus be decomposed into four independent subspaces U_1, U_2, U_3 and U_4 of distinct symmetry types, whose basis vectors are the symmetry-adapted redundants obtained by applying the linear operators π_1, π_2, π_3 and π_4 in turn, to each arbitrary (redundant) function ϕ_1, ϕ_2, ϕ_3 and ϕ_4, as follows:

Subspace U_1

$$\pi_1\phi_1 = (1/4)(E+C_2+\sigma_x+\sigma_y)\phi_1=(1/4)(\phi_1+\phi_3+\phi_3+\phi_1)=(1/2)(\phi_1+\phi_3) = \pi_1\phi_3$$
$$\pi_1\phi_2 = (1/4)(E+C_2+\sigma_x+\sigma_y)\phi_2=(1/4)(\phi_2+\phi_4+\phi_2+\phi_4)=(1/2)(\phi_2+\phi_4) = \pi_1\phi_4$$

Basis vectors of U_1 may thus be taken as:

$$\bar{\phi}_1^{(1)} = \phi_1 + \phi_3; \qquad \bar{\phi}_2^{(1)} = \phi_2 + \phi_4 \tag{3a, b}$$

Subspace U_2

$$\pi_2\phi_1 = (1/4)(E+C_2-\sigma_x-\sigma_y)\phi_1=(1/4)(\phi_1+\phi_3-\phi_3-\phi_1)=\pi_2\phi_3 = 0$$
$$\pi_2\phi_2 = (1/4)(E+C_2-\sigma_x-\sigma_y)\phi_2=(1/4)(\phi_2+\phi_4-\phi_2-\phi_4)=\pi_2\phi_4 = 0$$

U_2 is a null subspace

Subspace U_3

$$\pi_3\phi_1 = (1/4)(E-C_2+\sigma_x-\sigma_y)\phi_1=(1/4)(\phi_1-\phi_3+\phi_3-\phi_1)=\pi_3\phi_3 = 0$$
$$\pi_3\phi_2 = (1/4)(E-C_2+\sigma_x-\sigma_y)\phi_2=(1/4)(\phi_2-\phi_4+\phi_2-\phi_4)=(1/2)(\phi_2-\phi_4)= -\pi_3\phi_4$$

A basis vector of U_3 may thus be taken as:

$$\bar{\phi}^{(3)} = \phi_2 - \phi_4 \tag{4}$$

Subspace U_4

$$\pi_4\phi_1 = (1/4)(E-C_2-\sigma_x+\sigma_y)\phi_1=(1/4)(\phi_1-\phi_3-\phi_3+\phi_1)=(1/2)(\phi_1-\phi_3)= -\pi_4\phi_3$$
$$\pi_4\phi_2 = (1/4)(E-C_2-\sigma_x+\sigma_y)\phi_2=(1/4)(\phi_2-\phi_4-\phi_2+\phi_4)=\pi_4\phi_4 = 0$$

A basis vector of U_4 may thus be taken as:

$$\bar{\phi}^{(4)} = \phi_1 - \phi_3 \tag{5}$$

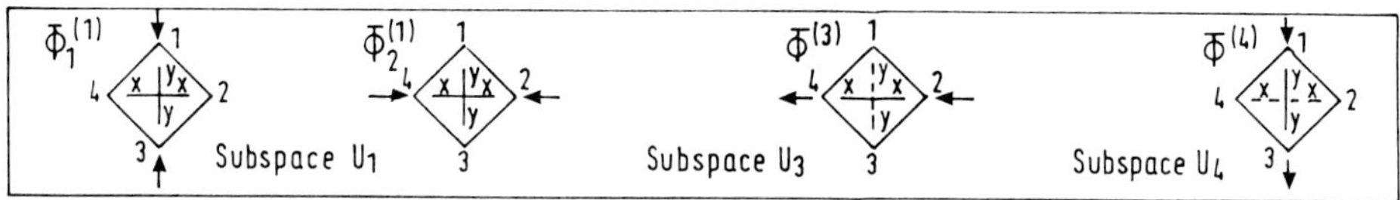

Fig. 2. Symmetry types of the subspaces of the C_{2v} problem.

11. The symmetry types of the various basis vectors, which must be the same for all basis vectors within the same subspace, are best visualised by plotting the components of the basis vectors onto the vertices of the square 1234 in the plan of Fig. 1. This has been done in Fig. 2, where inward-pointing arrows denote positive coefficients of components ϕ of the basis vectors $\bar{\phi}$, and outward-pointing arrows denote negative coefficients. The solid and the dashed lines at the centre of the square mark the orientation of symmetry and antisymmetry planes, respectively, of the $\bar{\phi}$ configurations. It is clear from Fig. 2. that subspace U_1

has symmetry about both the x- and y-axis, as well as cyclic C_2 symmetry; U_3 has symmetry about the x-axis and antisymmetry about the y-axis, as well as cyclic C_2 antisymmetry; U_4 has symmetry about the y-axis and antisymmetry about the x-axis, as well as cyclic C_2 antisymmetry.

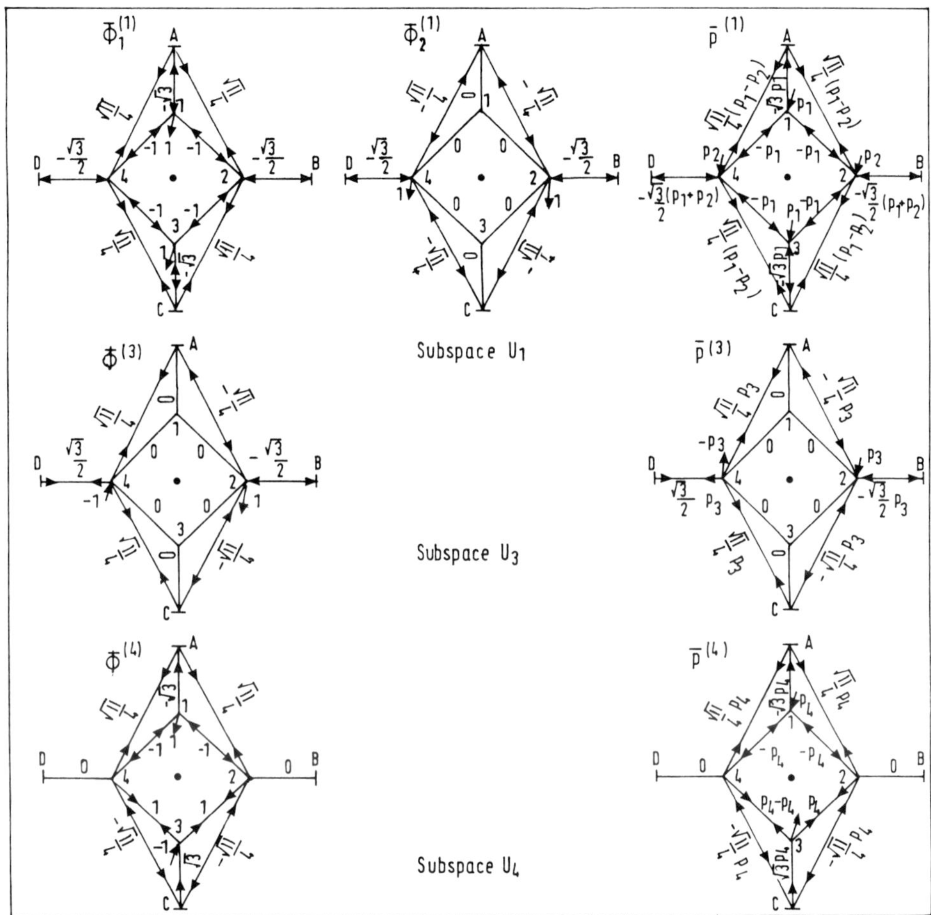

Fig. 3. Member forces due to symmetry-adapted redundants and loads.

12. The member forces, when unit values of symmetry-adapted redundants $\overline{\phi}_1^{(1)}$ and $\overline{\phi}_2^{(1)}$ (in subspace U_1), $\overline{\phi}^{(3)}$ (in subspace U_3) and $\overline{\phi}^{(4)}$ (in subspace U_4) are applied to the released structure in accordance with the co-ordinates of the ϕ's of a given basis vector $\overline{\phi}$, are shown in Fig. 3, where positive values denote tension and negative values compression. These forces have been calculated from ordinary statics using the dimensions of the space frame shown in Fig. 1. The co-ordinates of $\overline{\phi}_1^{(1)}, \overline{\phi}_2^{(1)}, \overline{\phi}^{(3)}$ and $\overline{\phi}^{(4)}$ also give the pattern of application of $\overline{p}_1^{(1)}$ and $\overline{p}_2^{(1)}$ (in subspace U_1), $\overline{p}^{(3)}$ (in subspace U_3) and $\overline{p}^{(4)}$ (in subspace

U_4). Denoting the actual values of the $\bar{p}_1^{(1)}, \bar{p}_2^{(1)}, \bar{p}^{(3)}$ and $\bar{p}^{(4)}$ components by p_1, p_2, p_3 and p_4, respectively, the member forces due to $\bar{p}_1^{(1)}, \bar{p}_2^{(1)}, \bar{p}^{(3)}$ and $\bar{p}^{(4)}$ are shown in Fig. 3 alongside their $\bar{\phi}$ counterparts, with those due to $\bar{p}_1^{(1)}$ and $\bar{p}_2^{(1)}$ combined as $\bar{p}^{(1)}$, since $\bar{p}_1^{(1)}$ and $\bar{p}_2^{(1)}$ belong to the same subspace and possess the same symmetry type.

13. From the given dimensions of the space frame, member sets {1A, 3C, 2B, 4D}, {A4, A2, C4, C2}, {14, 12, 34, 32} and {11', 22', 33', 44'} have element lengths (L) of values $a\sqrt{3}$, $a\sqrt{11}$, 2a and a, respectively. Denoting by $T_1^{(1)}, T_2^{(1)}, T^{(3)}, T^{(4)}, T_0^{(1)}, T_0^{(3)}$ and $T_0^{(4)}$ (the member forces due to systems $\bar{\phi}_1^{(1)}, \bar{\phi}_2^{(1)}, \bar{\phi}^{(3)}, \bar{\phi}^{(4)}, \bar{p}^{(1)}, \bar{p}^{(3)}$ and $\bar{p}^{(4)}$, respectively, application of virtual work gives the following deformation coefficients, with the extensional-rigidity parameter EA having been assumed, for simplicity, to be the same for all members:

Subspace U_1

$$f_{11}^{(1)} = (1/EA)\Sigma(T_1^{(1)})^2L = 32.1111(a/EA);\ f_{22}^{(1)} = (1/EA)\Sigma(T_2^{(1)})^2L = 13.7188(a/EA)$$

$$f_{12}^{(1)} = f_{21}^{(1)} = (1/EA)\Sigma T_1^{(1)}T_2^{(1)}L = -6.5226\ (a/EA)$$

$$\delta_{10}^{(1)} = (1/EA)\Sigma T_1^{(1)}T_0^{(1)}L = (30.1111p_1 - 6.5226p_2)(a/EA)$$

$$\delta_{20}^{(1)} = (1/EA)\Sigma T_2^{(1)}T_0^{(1)}L = (-6.5226p_1 + 11.7188p_2)(a/EA)$$

Subspace U_3

$$f^{(3)} = (1/EA)\Sigma(T^{(3)})^2L = 13.7188(a/EA);\ \delta_0^{(3)} = (1/EA)\Sigma T^{(3)}T_0^{(3)}L = (11.7188p_3)(a/EA)$$

Subspace U_4

$$f^{(4)} = (1/EA)\Sigma(T^{(4)})^2L = 29.5130(a/EA);\ \delta_0^{(4)} = (1/EA)\Sigma T^{(4)}T_0^{(4)}L = (27.5130p_4)(a/EA)$$

14. The general solutions for the symmetry-adapted redundants, based on the usual compatibility considerations of the force method, are as follows, for subspaces U_1, U_3 and U_4, respectively:

$$\begin{bmatrix} \bar{\phi}_1^{(1)} \\ \bar{\phi}_2^{(1)} \end{bmatrix} = - \begin{bmatrix} f_{11}^{(1)} & f_{12}^{(1)} \\ f_{21}^{(1)} & f_{22}^{(1)} \end{bmatrix}^{-1} \begin{bmatrix} \delta_{10}^{(1)} \\ \delta_{20}^{(1)} \end{bmatrix} \quad (6)$$

$$\bar{\phi}^{(3)} = -\left\{f^{(3)}\right\}^{-1}\delta_0^{(3)};\quad \bar{\phi}^{(4)} = -\left\{f^{(4)}\right\}^{-1}\delta_0^{(4)} \quad (7, 8)$$

Substituting the coefficients given in §13 into the above expressions, one obtains:

$$\bar{\phi}_1^{(1)} = -0.9310p_1 + 0.03276p_2; \quad \bar{\phi}_2^{(1)} = 0.03276p_1 - 0.8386p_2 \tag{9a, b}$$

for subspace U_1, and

$$\bar{\phi}^{(3)} = -0.8542p_3; \quad \bar{\phi}^{(4)} = -0.9322p_4 \tag{10, 11}$$

for subspaces U_3 and U_4, respectively.

15. Notice how, through a group-vector decomposition of the original four dimensional space of the problem, symmetry-adapted redundants of the two-dimensional symmetrical subspace U_1 have been obtained through the solution of only two simultaneous equations, while those of the one-dimensional antisymmetrical subspaces U_3 and U_4 have been obtained by solving simple equations in only one unknown. A conventional analysis would have required the solution of four simultaneous equations for the four unknown redundants, for the situation of the space frame subjected to completely arbitrary vertical joint loads P_1, P_2, P_3 and P_4.

16. By reference to Fig. 3, the algebraic sum of the subspace components of loading at a given joint must equal the applied vertical force, as follows:

$$p_1 + p_4 = P_1; \quad p_2 + p_3 = P_2; \quad p_1 - p_4 = P_3; \quad p_2 - p_3 = P_4 \tag{12a-d}$$

This gives the symmetry-adapted components p_1, p_2, p_3 and p_4 in terms of the actually applied joint loads P_1, P_2, P_3 and P_4, as follows:

$$p_1 = \tfrac{1}{2}(P_1 + P_3); \; p_2 = \tfrac{1}{2}(P_2 + P_4); \; p_3 = \tfrac{1}{2}(P_2 - P_4); \; p_4 = \tfrac{1}{2}(P_1 - P_3) \tag{13a-d}$$

Instead of deducing the above relationships from Eq.12, they could also have been written down straightaway from the basis-vector expressions Eqs 3a, 3b, 4 and 5, with the common multiplier 1/2 (earlier dropped out in arriving at these) now restored. Substituting into results Eqs 9a, 9b, 10 and 11, expressions Eq. 13 for symmetry-adapted loads enable the symmetry-adapted redundants $\bar{\phi}_1^{(1)}, \bar{\phi}_2^{(1)}, \bar{\phi}^{(3)}$ and $\bar{\phi}^{(4)}$ to be evaluated for any arbitrary combination of applied vertical loads at the platform level of the space frame.

17. Finally, superposition of the symmetry-adapted redundants occurring at each of the joints 1, 2, 3 and 4 gives the final redundants X_1, X_2, X_3 and X_4 in members 11', 22', 33' and 44' of the space frame. Denoting the actual values of the $\bar{\phi}_1^{(1)}, \bar{\phi}_2^{(1)}, \bar{\phi}^{(3)}$ and $\bar{\phi}^{(4)}$ components by $\phi_1^{(1)}, \phi_2^{(1)}, \phi^{(3)}$ and $\phi^{(4)}$, respectively, and using Fig. 3 to visualise the superposition of the components, we obtain:

$$X_1 = \phi_1^{(1)} + \phi^{(4)}; \; X_2 = \phi_2^{(1)} + \phi^{(3)}; \; X_3 = \phi_1^{(1)} - \phi^{(4)}; \; X_4 = \phi_2^{(1)} - \phi^{(3)} \tag{14a-d}$$

18. As a numerical example, consider a space frame with vertical applied loads of $P_1 = 20\text{kN}, P_2 = 15\text{kN}$ and $P_4 = 30\text{kN}$. For the subspace loads, Eqs 13 give:

$p_1=(1/2)(20+5)=12.5\text{kN}$; $p_2=(1/2)(15+30)=22.5\text{kN}$
$p_3=(1/2)(15-30)= -7.5\text{kN}$; $p_4=(1/2)(20-5)= 7.5\text{kN}$

which, upon substitution into Eqs 9-11, yield the symmetry-adapted redundants as:

$$\bar{\phi}_1^{(1)}=-0.9310(12.5)+0.03276(22.5)=-10.9\text{kN};\quad \bar{\phi}^{(3)}=-0.8542(-7.5)=6.4\text{kN}$$
$$\bar{\phi}_2^{(1)}=0.03276(12.5)-0.8386(22.5)=-18.5\text{kN};\quad \bar{\phi}^{(4)}=-0.9322(7.5)=-7.0\text{kN}$$

final redundants then follow from Eqs 14 as:

$$X_1 = (-10.9)+(-7.0)=-17.9\text{kN};\quad X_2 = (-18.5)+(6.4)=-12.1\text{kN}$$
$$X_3 = (-10.9)-(-7.0)=-3.9\text{kN};\quad X_4 = (-18.5)-(6.4)=-24.9\text{kN}$$

SPACE FRAME OF SYMMETRY GROUP $C_{4\upsilon}$

19. If the vertical members 11', 22', 33', and 44' of Fig. 1 are removed, and inclined members introduced between joints 1 and B, 3 and B, 1 and D, and 3 and D, the configuration shown in plan in Fig. 4(a) results. This configuration belongs to the symmetry group $C_{4\upsilon}$ of order 8, with elements E (identity), C_4 and C_4^{-1} (rotations about the vertical axis 0 - see Fig. 4(a) - through 360°/4 and -360°/4, respectively), C_2 (rotation about the vertical axis 0 through 360°/2), σ_x, σ_y, σ_1 and σ_2 (reflections in vertical planes passing through 0, as shown in Fig. 4(a)).

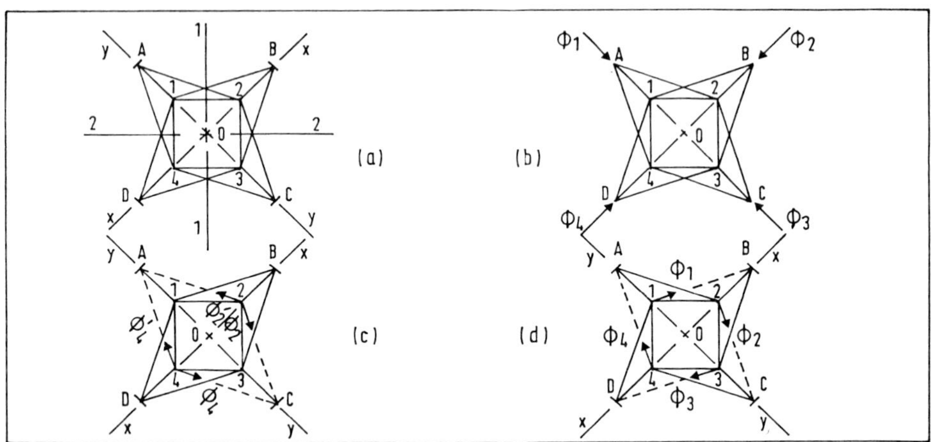

Fig. 4. Space frame of symmetry group $C_{4\upsilon}$: (a) complete frame; (b), (c) and (d) show release systems of $C_{4\upsilon}$, $C_{2\upsilon}$ and C_4 configurations, respectively.

20. The space truss has a degree of static indeterminacy of 4. Releasing the four radial reactions at the supported joints A, B, C and D (and adopting a system of external redundants directed perpendicularly to the vertical axis of rotation 0, as shown in Fig. 4(b)), while preserving the full $C_{4\upsilon}$ symmetry of the original

structure, unfortunately results in an inadequately restrained, and hence geometrically unstable, structure. (Such behaviour is typical of space trusses based on n-sided polygons, where n is even (ref. 15), a point which will be noted again in connection with the third and final example.) Releasing the inclined members 2A, 2C, 4A and 4C results in the statically determinate and stable configuration of the first example, but with redundants now as shown in Fig. 4(c). Although the released structure and its set of four redundants $\{\phi_2, \phi'_2, \phi_4, \phi'_4\}$ still retain the symmetry of the subgroup $C_{2\upsilon}$ of $C_{4\upsilon}$ this choice of releases is not suitable for a group-theory analysis because the arbitrary vertical loads applied on the structure at joints 1, 2, 3 and 4 cannot be adapted to match the symmetry of the subspaces of the redundants; these vertical loads P_1, P_2, P_3 and P_4 at the platform joints, with lines of action lying in the x and y reflection planes, are permuted by elements of group $C_{2\upsilon}$, in a different manner from that of the redundants $\phi_2, \phi'_2, \phi_4, \phi'_4$ whose lines of action do not lie in the x and y planes of reflection.

21. In conventional analysis, releasing the frame in the cyclic manner shown in Fig. 4(d) does not result in any additional computational effort to that which would be incurred were the system of Fig. 4(c) adopted, as in both cases four simultaneous equations would still require to be solved eventually, assuming complete arbitrariness of the applied loads P_1, P_2, P_3 and P_4. However, the advantage of the release system of Fig. 4(d) over that of Fig. 4(c) lies in the former's amenability to group-theory analysis, which makes use of the residual cyclic symmetry about O of that configuration; conventional analysis cannot make use of such rotational symmetry in order to reduce the dimensions of the eventual structure flexibility matrix.

22. The configuration in Fig. 4(d) belongs to the symmetry group C_4, a subgroup of $C_{4\upsilon}$, with elements E, C_4, C_4^{-1} and C_2. The character table of group C_4 is given in Table 2.

Table 2. Character table of group C_4.

C_4	E	$2C_4$	C_2
A	1	1	1
B	1	−1	1
E	2	0	−2

The idempotents corresponding to the representations A, B and E, respectively, are, from Eq. 1, as follows:

$$\pi_1=(1/4)(E+C_4+C_4^{-1}+C_2) \quad (15a)$$
$$\pi_2=(1/4)(E-C_4-C_4^{-1}+C_2) \quad (15b)$$
$$\pi_3=(2/4)(2E-2C_2)=E-C_2 \quad (15c)$$

Applying these, in turn, to the arbitrary variables ϕ_1, ϕ_2, ϕ_3 and ϕ_4 gives the basis vectors (in the subspaces indicated by the superscripts) as follows:

$$\bar{\phi}^{(1)} = \phi_1+\phi_2+\phi_3+\phi_4; \qquad \bar{\phi}^{(2)} = \phi_1-\phi_2+\phi_3-\phi_4 \qquad (16, 17)$$

$$\bar{\phi}_1^{(3)} = \phi_1-\phi_3; \qquad \bar{\phi}_2^{(3)} = \phi_2-\phi_4 \qquad (18a, b)$$

The cyclic C_4 symmetry (in subspace U_1) of $\bar{\phi}^{(1)}$, cyclic C_4 antisymmetry (in subspace U_2) of $\bar{\phi}^{(2)}$, and cyclic C_2 antisymmetry (in subspace U_3) of $\bar{\phi}_1^{(3)}$ and $\bar{\phi}_2^{(3)}$ are illustrated in Fig. 5; corresponding sets of symmetry-adapted loads are also shown. Once $\bar{\phi}^{(1)}, \bar{\phi}^{(2)}, \bar{\phi}_1^{(3)}$ and $\bar{\phi}_2^{(3)}$ have been solved for, redundants X_1, X_2, X_3 and X_4 at joints 1, 2, 3 and 4 follow by superposition:

$$X_1=\bar{\phi}^{(1)}+\bar{\phi}^{(2)}+\bar{\phi}_1^{(3)}; \qquad X_2=\bar{\phi}^{(1)}-\bar{\phi}^{(2)}+\bar{\phi}_2^{(3)}$$

$$X_3=\bar{\phi}^{(1)}+\bar{\phi}^{(2)}-\bar{\phi}_1^{(3)}; \qquad X_4=\bar{\phi}^{(1)}-\bar{\phi}^{(2)}-\bar{\phi}_2^{(3)} \qquad (19a\text{-}d)$$

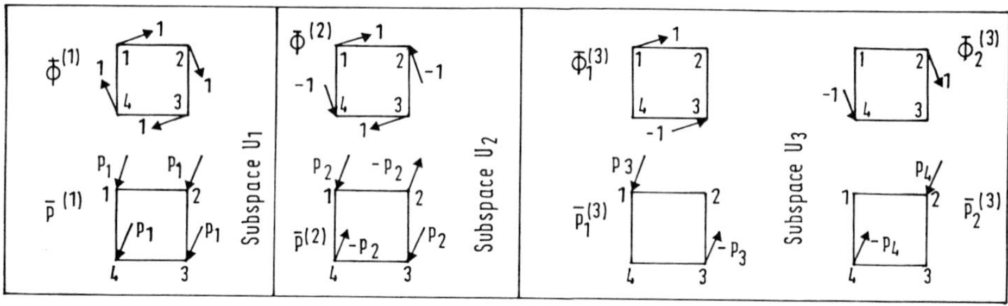

Fig. 5. Symmetry-adapted redundants and applied loads for the C_4 configuration.

SPACE FRAME OF SYMMETRY GROUP $C_{3\upsilon}$

23. Fig. 6. shows the final example, a pin-jointed space frame based on the equilateral triangle. Joints 1, 2 and 3 lie in a horizontal plane a certain distance above the support plane containing joints 1', 2' and 3'. The three symmetry planes of the equilateral triangles 123 and 1'2'3' (in the plan) coincide. The frame has m=12, j=6 and r=9, so that i=m+r-3j=3. Releasing members 11', 22' and 33' results in a statically determinate structure that has the same symmetry as the original indeterminate frame, and redundants that can be permuted by symmetry elements of the group $C_{3\upsilon}$, as shown in Fig. 7. (Radial horizontal reactions at the supports 1', 2' and 3' could equally have been chosen as releases, since the ensuing configuration also preserves the full symmetry of the original structure.) Incidentally, space trusses of the type shown in Fig. 7 have a critical form (ref. 15) when the regular polygon upon which they are based has an even number of sides; a little rotational distortion, such as is shown in Fig. 8 for the square-based frame, will remove such instability, but at the expense of reflectional symmetry,

again leaving group theory as the only means by which the residual (rotational) symmetry can be effectively utilised.

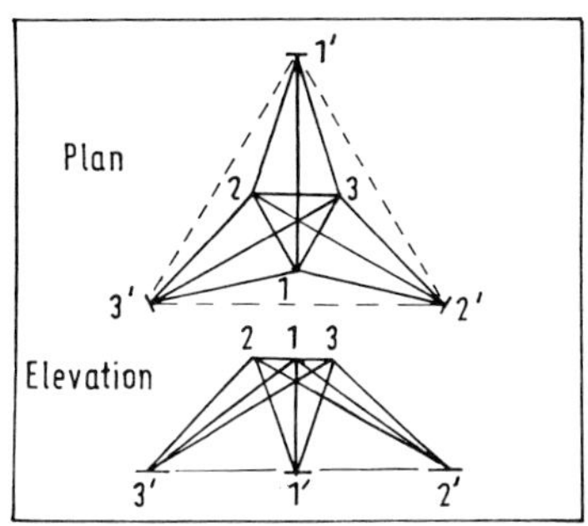

Fig. 6. Space frame of symmetry group C_{3v}.

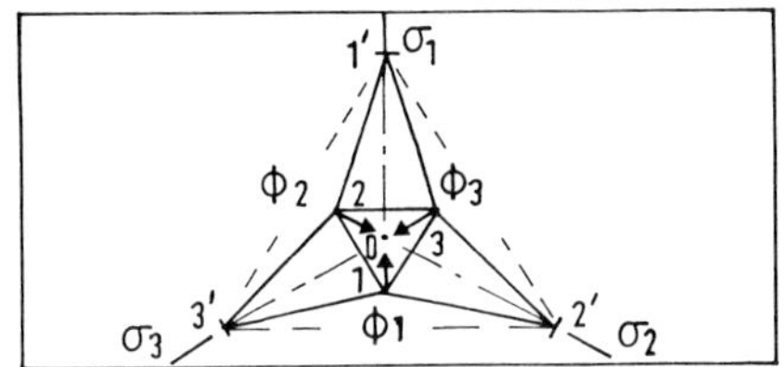

Fig. 7. C_{3v}, configuration of redundants.

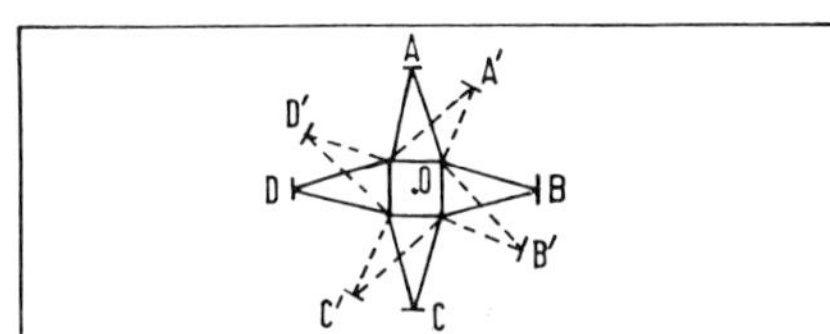

Fig. 8. Rotational distortion of C_{4v} space frame to C_4 form.

24. With reference to Fig. 7, group C_{3v} has six symmetry elements, namely E (identity), C_3 and C_3^{-1} (rotations about the vertical axis of symmetry O through 360°/3 and -360°/3 respectively), and σ_1, σ_2 and σ_3 (vertical reflection planes as shown in Fig. 7). The character table of group C_{3v} is given in Table 3 below.

Table 3. Character table of group C_{3v}.

C_{3v}	E	$2C_3$	$3\sigma_v$
A_1	1	1	1
A_2	1	1	-1
E	2	-1	0

Using Eq. 1 and the values in the above table, idempotents of the centre of the group algebra are obtained as follows:

$$\pi_1 = (1/6)\ (E+C_3+C_3^{-1}+\sigma_1+\sigma_2+\sigma_3) \tag{20a}$$
$$\pi_2 = (1/6)\ (E+C_3+C_3^{-1}-\sigma_1-\sigma_2-\sigma_3) \tag{20b}$$
$$\pi_3 = (2/6)\ (2E-C_3-C_3^{-1}) \tag{20c}$$

Applying these to ϕ_1, ϕ_2 and ϕ_3 gives the basis vectors:

$$\bar{\phi}^{(1)}=\phi_1+\phi_2+\phi_3 \tag{21}$$

$$\bar{\phi}_1^{(3)}=2\phi_1-\phi_2-\phi_3;\quad \bar{\phi}_2^{(3)}=2\phi_2-\phi_3-\phi_1 \tag{22a, b}$$

in the subspaces of U_1 and U_3 (U_2 being null), as depicted in Fig. 9.

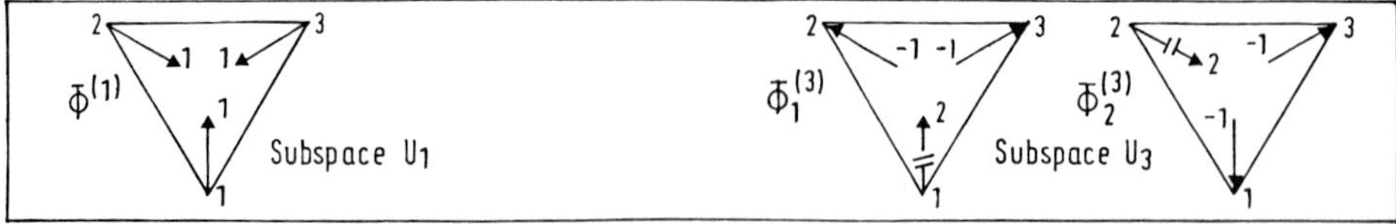

Fig. 9. Symmetry-adapted redundants for the C_{3v} configuration.

Thus, the original problem simplifies to solving, separately, one equation in one unknown $\bar{\phi}^{(1)}$ (for subspace U_1), and two equations in two unknowns $\bar{\phi}_1^{(3)}$ and $\bar{\phi}_2^{(3)}$ (for subspace U_3), instead of solving for three redundants as three equations in three unknowns. The final redundants would then be given by superposition, as follows:

$$X_1=\bar{\phi}^{(1)}+2\bar{\phi}_1^{(3)}-\bar{\phi}_2^{(3)};\ X_2=\bar{\phi}^{(1)}-\bar{\phi}_1^{(3)}+2\bar{\phi}_2^{(3)};\ X_3=\bar{\phi}^{(1)}-\bar{\phi}_1^{(3)}-\bar{\phi}_2^{(3)} \qquad (23a\text{-}c)$$

CONCLUSIONS

25. In this paper, the application of group theory to the analysis of various indeterminate space frames has been illustrated on the basis of the force method. Not only is such an approach more systematic than conventional use of symmetry, but also it affords considerably greater reductions in computational effort, and, indeed, may be the only effective way of making use of certain types of symmetry in the structure. Research is ongoing on the assimilation of the outlined techniques into numerical schemes for the generalised force and direct stiffness methods, as well as the finite element method.

ACKNOWLEDGEMENT

26. The ongoing research on the application of group theory to structural analysis is supported by the Royal Commission for the Exhibition of 1851, through a Research Fellowship held by the first author.

REFERENCES

1. HAMERMESH M. Group theory and its application to physical problems. Pergamon Press, 1962.
2. WIGNER E.P. Group theory and its applications to the quantum mechanics of atomic spectra. Academic Press, 1959.
3. TINKHAM M. Group theory and quantum mechanics. McGraw-Hill, New York, 1975.
4. FALICOV L.M. Group theory and its physical applications. University of Chicago Press, Chicago, 1966.
5. BHAGAVANTAM S. and VENKATARAYUDU T. Theory of groups and its application to physical problems. Andhra University Press, Waltair, 1951.

6. COTTON F.A. Chemical applications of group theory. Wiley-Interscience, New York, 1971.
7. HALL L.H. Group theory and symmetry in chemistry. McGraw-Hill, New York, 1969.
8. SCHONLAND D. Molecular symmetry. Van Nostrand, 1965.
9. ZLOKOVIC G.M. Group theory and G-vector spaces in structural analysis. Ellis Horwood, Chichester, 1989.
10. LEECH J.W. and NEWMAN D.J. How to use groups. Methuen & Co., London, 1969.
11. CRACKNELL A.P. Applied group theory. Pergamon Press, 1968.
12. ZLOKOVIC G.M. Group supermatrices in finite element analysis. Ellis Horwood, Chichester, 1992.
13. SWAMY N.V.V.J. and SAMUEL M.A. Group theory made easy for scientists and engineers. Wiley-Interscience, New York, 1979.
14. HALL G.G. Applied group theory. Longman, London, 1967.
15. TIMOSHENKO S.P. and YOUNG D.H. Theory of structures. McGraw-Hill, 1965.

143. Reliability analysis of space structural systems

R. A. AL-GEROUSHI and G. M. HAMMOUD, University of Garyounis, Libya

SYNOPSIS. Reliability analysis may be used to evaluate structural safety and hence improve the structural design process. The exact evaluation of the reliability, or the probability of survival, of structural systems having several inter-dependent failure modes, requires very lengthy numerical integration. This paper outlines the formulation of the probability of failure and the reliability index for the space structural systems in a form suitable for practical applications. A computer program has been developed to check the reliability of different space structural systems. This reliability analysis allows the safety to be included in the design process for a new structural system or in the use of a new material, in the absence of code regulations.

THE STATEMENT OF THE CONCEPT

1. The concept is to develop a computer program which is capable of the design of space structural systems using a suitable probability of failure or reliability index. In the absence of code regulations, this design can be calibrated by the existing level of safety in the design codes for different materials, such as the steel codes.

METHOD OF ANALYSIS

2. Monte Carlo probabilistic structural analysis is used for the complete analysis of the space structure. In the following, each element of the analysis will be discussed. However, theories supporting the analysis are not given, since they are well established.

Linear analysis

3. Linear elastic analysis is considered in the whole study and the direct stiffness method has been implemented and programmed using the FORTRAN language, with the new lahey compiler, which provides for large capacity and low calculation time on personal computers.

Space structures 4. Thomas Telford, London, 1993

Probability distribution

4. Loads. The uncertainties in the loads are considered by treating them as random variables. In the studies on the reliability of loading, it is concluded that the dead load has a normal distribution and the live load has a gamma distribution.

5. Section properties. All the sectional properties are considered to be random variables and having a normal distribution. However, the computer program has the capability to consider any probability distribution.

6. Monte Carlo analysis. The Monte Carlo analysis is used for load data generation from the mean and standard deviation. It is also for generating the outcome list of the random variables from functional relationships, such as the bending stress from the combined stress functional relationship. The Monte Carlo program, firstly, generates the uniform random variables, secondly, the required distribution can be generated, by using the transformation techniques.

7. Probability of failure. The case of failure is defined as when the load effect is greater than the section capacity. Three behaviours are considered: bending stress, shear stress and torsional stress. Therefore, the load effects are as follows:

$$\sigma = P/A + M_y/Z_y + M_z/Z_z$$
$$v = F/A$$
$$\tau = Tc/J \qquad (1)$$

and the section capacity is fy for bending stress, for shear stress and for torsional stress.

Let the load effect be represented by R and the section capacity by S. Hence:

when R-S > 0 the section has failed
and R-S < 0 the section is safe.

Therefore, when the randomness is considered for the load and the section capacity, the probability of failure P_f can be written as follows (ref. 1):

$$P_f = P.(R - S > 0) \qquad (2)$$

Using the Monte Carlo analysis method, the probability of failure can be calculated. The probability of failure is obtained by dividing the total number of Monte Carlo sample analyses, which experienced failure by the total number of Monte Carlo samples. The dependency between the random variable is neglected in the program, but in the undergoing research program this is considered.

Sample analysis

8. Reliability index. A linear safety margin is considered in this study and the reliability index has been defined by Cornell (1969), on the basis that it is the ratio between the mean of the safety margin (F) and the standard deviation of that margin σ_F.

$$\beta = u_F / \sigma_F = (u_R - u_S) / (\sigma_R^2 + \sigma_S^2)^{1/2} \quad (3)$$

The reliability index equation shows that the load effect R and the section capacity have normal distribution as shown in Fig. 1.

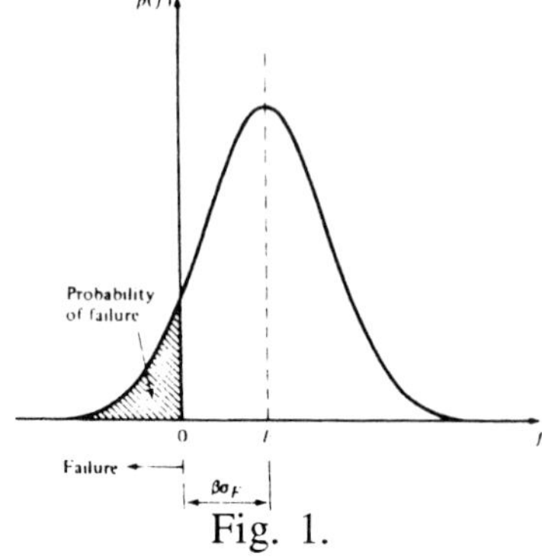

Fig. 1.

APPLICATION ON A DOME STRUCTURAL SYSTEM

9. A steel dome with a radius of 25m and a height of 10m has been generated. The dome has 13 rings and 52 meridians. The dome has been analysed and designed initially by using the BS5950. The resulting sectional properties were placed in a special file for the Monte Carlo use. Table 1 shows the mean and coefficient of variation for the load and property data, which were assumed to be close to the real data. Fifty dead load and live load values were, generated using the normal and gamma distribution respectively. In the same way fifty values were generated for the member properties, using the mean and the coefficient of variation of the material yield strength and sectional properties, assuming normal distribution. Hence, 2500 Monte Carlo analyses were done to determine the probability of failure and the reliability index of each ring and meridian. Fig. 2 shows a flow-chart which gives the complete analysis and design with the probabilistic analysis and reliability calculations.

10. Ring member removal effect. The purpose of ring member removal is to investigate the system redundancy of the space structure. After studying the probability of failure of each member from the whole dome analysis the effect of the removal of critical members, on the values of the new probability of failure and reliability index, is studied. Fig. 3 shows the relationship between the sector number and the probability of failure with member removal, for the three behaviours considered (bending, shear and torsional stresses). Fig. 4 shows the relationship between the sector number and the reliability index for the bending stress. The reliability indices for the torsion and shear did not change, and have constant value throughout. It can be noted that the first and second vertical member sectors, which are connected to the first and second rings, were affected more than other members.

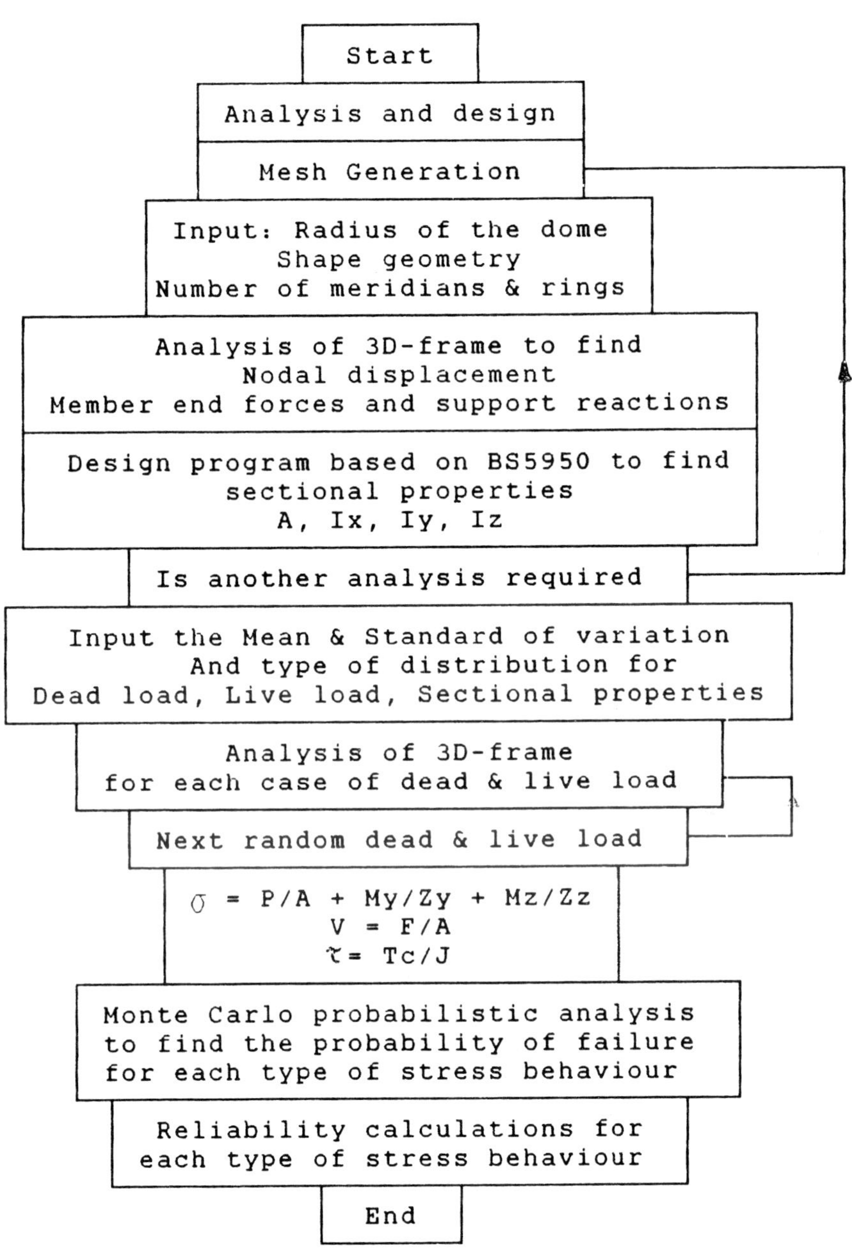
Start
Analysis and design
Mesh Generation
Input: Radius of the dome
Shape geometry
Number of meridians & rings
Analysis of 3D-frame to find
Nodal displacement
Member end forces and support reactions
Design program based on BS5950 to find
sectional properties
A, Ix, Iy, Iz
Is another analysis required
Input the Mean & Standard of variation
And type of distribution for
Dead load, Live load, Sectional properties
Analysis of 3D-frame
for each case of dead & live load
Next random dead & live load
σ = P/A + My/Zy + Mz/Zz
V = F/A
τ = Tc/J
Monte Carlo probabilistic analysis
to find the probability of failure
for each type of stress behaviour
Reliability calculations for
each type of stress behaviour
End

Fig. 2.

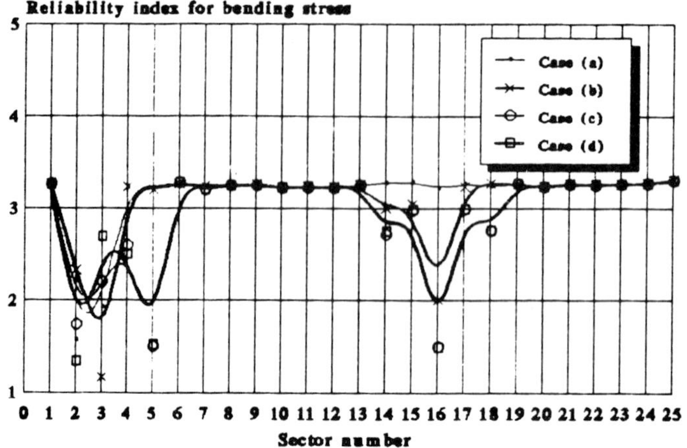

Fig. 3.

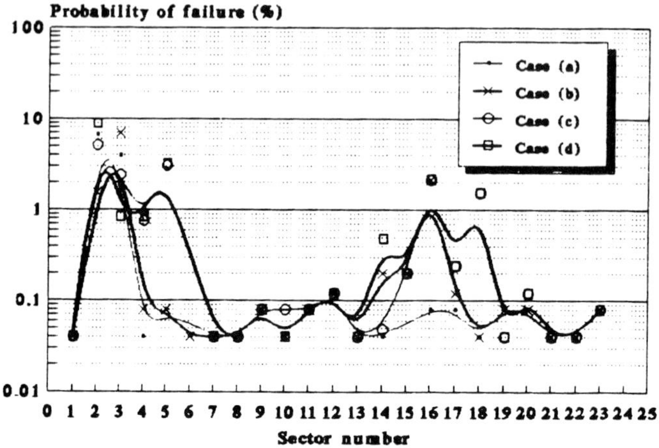

Fig. 4.

Case (a) Analysis of dome without removing any member.
Case (b) Half of the members of the second ring are removed.
Case (c) Half of the members of the third ring are removed.
Case (d) All the members of the second ring and half of the members of the third ring are removed.

CONCLUSION

11. The following points summarise the conclusions:

(a) The safe reliability index β is a very useful measure for safety level of the members and the space frame system.

Table 1. Load, resistance and material property statistics.

Type	Mean value	Coefficient of variation	Distribution
Dead load	1.0 D	0.1	Normal
Live load	0.37 L	0.6	Gamma
Static yield stress	1.05 F_y	0.3	Normal
Static yield stress in shear	1.05 F_y	0.3	Normal
Static yield stress in torsion	1.05 F_y	0.3	Normal
Cross sectional area	1.05 A_g	0.3	Normal
Sectional modulus	1.05 Z	0.3	Normal

Where:
D is the dead load
L is the live load
F_y is the material yield strength
A_g is the gross cross-sectional area
Z is the sectional modulus

(b) When the dependency between the random variables is included, and wind load effect is also included, the computer program, which is developed here, can be used for the analysis and design of space structural systems with materials other than steel.

(c) From the values of the reliability indices for bending stress, torsional stress and shear stress it is concluded that the member removal has a noticeable affect on the bending stress and a negligible affect on both torsion and shear stresses.

REFERENCES

1. BAKER, M.C. Structural Reliability Theory and its Applications. London 1982.
2. BENJAMIN, J.R. and Cornell C.A. Probability, Statistics, and Decision for Civil Engineers. McGraw-Hill Book Company, New York, 1970.
3. HART, G.C. Uncertainty Analysis, Loads, and Safety in Structural Engineering. Prentice-Hall, Inc., Englewood Cliffs, New Jersey, 1982.
4. ANG A.H. and TANG W. H. Probability Concepts in Engineering Planning and Design. Volume II. John Wiley & Sons, Inc. New York 1975.

144. An alternative finite-element formulation for spatial beams and truss structures

F. KASZA, University of Agriculture, Hungary

SYNOPSIS. Betti's law of reciprocal work is introduced to general cases. Making use of a spatial cantilever, with general end unit loading as a virtual system of forces and displacements, the element end forces are expressed via element properties as nodal loads. Nodal equilibrium equations can be constructed by adding them together. The technique also holds for thermal loads, dynamic and stability cases. Thus a new finite-element-like method is introduced for rod structures. Only spatial rod structures are considered in this paper.

APPLICATION OF BETTI'S LAW

1. Finite element formulations are usually based on some form of energy method. One of them is **Betti's law** of reciprocal work (ref. 1) which is in close relation to the Somigliana identity (ref. 2). The latter serves as a basic equation to boundary element solutions as the exact solution to Laplace's equation (refs 3,4). The equations are widely used in the solution of complicated 3D volumes and structures in solid mechanics. The principle can be easily adapted to the solution of structure mechanics problems. In this paper a wide interpretation of Betti's law is used to obtain a system of equations for space structures, either frame or truss or mixed type structures, fabricated from linearly elastic, homogenous and isotropic materials. Other cases are excluded even though the terms could hold.

2. The method offers a direct technique to construct a system of nodal equations with unknown variables of displacements. Some forms of statement of Betti's law are as follows:

$$W_{12}=W_{21}, \quad i.e. \quad \int_V \vec{f}\vec{u}^* dV=\int_V \vec{f}^* \vec{u} dV, \quad or$$
$$U_{12}=U_{21}, \quad i.e. \quad \int_V \vec{\sigma}\vec{\varepsilon}^* dV=\int_V \vec{\sigma}^* \vec{\varepsilon} dV \tag{1}$$

The equations express that the two external, or inner, system of forces mutually do the same work when the displacements due to the other are formed. One system represents real loads and displacements on the structure while the other is

Space structures 4. Thomas Telford, London, 1993

the virtual one marked with stars. All of them depend on the space vector **r**. The integral statement relates to the volume taken and should be understood in a general sense, i.e. (a) it holds not only for the whole structure but any separated part of a linearly elastic structure, provided that the inner forces are applied at the separation interfaces, (b) the load vector **f** represents all the external forces and moments while the displacement vector **u** stands for all translational and angular displacements.

3. The virtual system of loads and displacements can be chosen freely, but they should satisfy the equations of elastic equilibrium and compatibility. In spite of this either forces or displacements can be chosen as primary virtual values and the others as resultant (secondary) ones, forces are used as primary values throughout this paper.

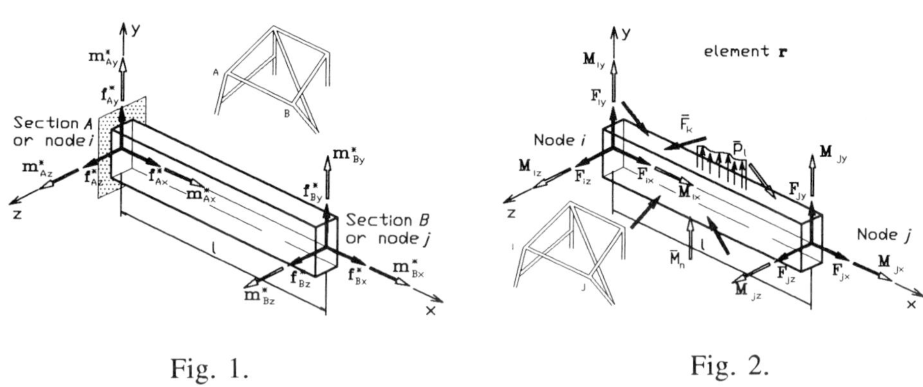

Fig. 1. Fig. 2.

4. Many combinations of virtual loads and the resultant displacements can be chosen to apply the principle to a beam or truss element, but only a few will simplify the solution. One version per component is shown herein with the aim of transforming the effect of the element geometry, load and material properties to the two nodal ends of the rod element of structure.

5. A cantilever with general load at its end is the simplest case of a space beam that may be used in forming the necessary virtual system (Fig. 1). It is not important if the clamped rod is a single beam or it is a part of a more complex structure, the applied virtual system may be the same. The elastic solution to the cantilever is used to produce the contribution of the structure elements to the nodal equations.

FINITE ELEMENT FORMULATION

6. The steps of a computer program for finite element solution of a structural problem are: (a) data input describing geometry, material properties and boundary conditions; (b) calculation of element stiffness matrices and assembling the system matrix; (c) application of boundary conditions to the system equations; (d) solution of the system equations; (e) computing of secondary unknowns; (f)

editing and producing final outputs. The alternative formulation differs in steps (b) (c) and (e) from the usual one.

7. In this alternative formulation, nodal contributions are produced as nodal loads i.e. partly as actual nodal loads, partly as functions of the element end unknowns of displacements and rotations. Nodal inner forces are expressed directly if a larger system matrix is assembled in a similar way.

8. Fig. 2 shows an element (numbered r) connecting two nodes (i and j) of the structure. The figure shows the inner nodal forces and moments and the external concentrated and distributed loads, forces and moments. Table 1 summarises the virtual system of unit forces and the resultant virtual displacements.

Table 1.

		Applied system of virtual loads & displacements resulted		
		Cross section		
		A (clamped)	*B (free)*	*any (at x)*
Loads	force	$f^{\bullet}_{Ax}=-f^{\bullet}_{Bx}$	$f^{\bullet}_{Bx}$	$f^{\bullet}_{x}=f^{\bullet}_{Bx}$
		$f^{\bullet}_{Ay}=-f^{\bullet}_{By}$	$f^{\bullet}_{By}$	$f^{\bullet}_{y}=-f^{\bullet}_{By}$
		$f^{\bullet}_{Az}=-f^{\bullet}_{Bz}$	$f^{\bullet}_{Bz}$	$f^{\bullet}_{z}=-f^{\bullet}_{Bz}$
	moment	$m^{\bullet}_{Ax}=-m^{\bullet}_{Bx}$	$m^{\bullet}_{Bx}$	$m^{\bullet}_{x}=-m^{\bullet}_{Bx}$
		$m^{\bullet}_{Ay}=-m^{\bullet}_{By}+f^{\bullet}_{Bz}l$	$m^{\bullet}_{By}$	$m^{\bullet}_{y}=-m^{\bullet}_{By}+f^{\bullet}_{Bz}(l-x)$
		$m^{\bullet}_{Az}=-m^{\bullet}_{Bz}-f^{\bullet}_{By}l$	$m^{\bullet}_{Bz}$	$m^{\bullet}_{z}=-m^{\bullet}_{Bz}-f^{\bullet}_{By}(l-x)$
Displacements	translational	$u^{\bullet}_{Ax}=0$	$u^{\bullet}_{Bx}=\frac{f^{\bullet}_{Bx}l}{AE}$	$u^{\bullet}_{x}=\frac{f^{\bullet}_{Bx}x}{AE}$
		$u^{\bullet}_{Ay}=0$	$u^{\bullet}_{By}=\frac{1}{I_zE}\left(m^{\bullet}_{Bz}\frac{l^2}{2}+f^{\bullet}_{By}\frac{l^3}{6}\right)$	$u^{\bullet}_{y}=\frac{1}{I_zE}\left(m^{\bullet}_{Bz}\frac{x^2}{2}+f^{\bullet}_{By}l\frac{x^2}{2}-f^{\bullet}_{By}\frac{x^3}{6}\right)$
		$u^{\bullet}_{Az}=0$	$u^{\bullet}_{Bz}=\frac{1}{I_yE}\left(f^{\bullet}_{Bz}\frac{l^3}{6}-m^{\bullet}_{By}\frac{l^2}{2}\right)$	$u^{\bullet}_{z}=\frac{1}{I_yE}\left(-m^{\bullet}_{By}\frac{x^2}{2}+f^{\bullet}_{Bz}l\frac{x^2}{2}-f^{\bullet}_{Bz}\frac{x^3}{6}\right)$
	rotational	$\varphi^{\bullet}_{Ax}=0$	$\varphi^{\bullet}_{Bx}=\frac{m^{\bullet}_{Bx}l}{I_xE}$	$\varphi^{\bullet}_{x}=\frac{m^{\bullet}_{Bx}x}{I_xE}$
		$\varphi^{\bullet}_{Ay}=0$	$-\varphi^{\bullet}_{By}=\frac{1}{I_yE}\left(-m^{\bullet}_{By}l+f^{\bullet}_{Bz}\frac{l^2}{2}\right)$	$-\varphi^{\bullet}_{y}=\frac{1}{I_yE}\left(-m^{\bullet}_{By}x+f^{\bullet}_{Bz}lx-f^{\bullet}_{Bz}\frac{x^2}{2}\right)$
		$\varphi^{\bullet}_{Az}=0$	$\varphi^{\bullet}_{Bz}=\frac{1}{I_zE}\left(m^{\bullet}_{Bz}l+f^{\bullet}_{By}\frac{l^2}{2}\right)$	$\varphi^{\bullet}_{z}=\frac{1}{I_zE}\left(m^{\bullet}_{Bz}x+f^{\bullet}_{By}lx-f^{\bullet}_{By}\frac{x^2}{2}\right)$

VIRTUAL LOADS FOR DIFFERENT DISPLACEMENTS

9. In Figs 1 and 2 a straight clamped (at A) prismatic bar in compliance with an element taken from a structure is shown. End forces and moments are applied as the nodal effects. Expressions are formulated in the local system of co-ordinates

and transformed into the global one. This procedure leads to a system of nodal equations. They are formally equilibrium equations of forces and moments. In reality, however, they contain the compatibility conditions implicitly, as well. It is not important how the beam or truss element is attached to the nodes because the method introduced here poses no restriction on it.

10. The external loads of the rod element *i-j* are $\mathbf{f_k}$, $\mathbf{m_l}$ (the vector components are external forces $\mathbf{f_k}$ and external moments $\mathbf{m_l}$). As an effect of the loading, an internal system of forces arises to satisfy the equilibrium of the element such as:

$$\vec{F}_i^T=\langle F_{ix}, F_{iy}, F_{iz}, M_{ix}, M_{iy}, M_{iz}\rangle\ , \vec{F}_j^T=\langle F_{jx}, F_{jy}, F_{jz}, M_{jx}, M_{jy}, M_{jz}\rangle \tag{2}$$

The element has a state of deformation depending on the load, material constants and geometry. The deformation can be characterised by its nodal values in the points *i* and *j* with vectors:

$$\vec{u}_i^T=\langle u_{ix}, u_{iy}, u_{iz}, \varphi_{ix}, \varphi_{iy}, \varphi_{iz}\rangle, \qquad \vec{u}_j^T=\langle u_{jx}, u_{jy}, u_{jz}, \varphi_{jx}, \varphi_{jy}, \varphi_{jz}\rangle \tag{3}$$

11. The method is illustrated by applying 6 different virtual loadings to the element at node *j* (section B). At the same time node *i* (section B) is clamped so that its displacements and rotations are zero in both direction. Thus the beam has been clamped using the load as follows:

$$\vec{f}_B^{*T}=\langle f_{Bx}^*, f_{By}, f_{Bz}^*, m_{Bx}^*, m_{By}^*, m_{Bz}^*\rangle \tag{4}$$

The well known general solution to this is given in Table 1. The vector reactions in the clamped end are:

$$\vec{f}_A^*=-\vec{f}_B^*, \qquad \vec{m}_A^*=-\vec{m}_B^*-\vec{r}_{AB}\times\vec{f}_B^* \tag{5}$$

The functions of axial force, shear forces, twisting moment and bending moments for any arbitrary section are:

$$\begin{aligned} &f_x^*=f_{Bx}^*, \quad f_y^*=-f_{By}^*, \quad f_z^*=-f_{Bz}^* \\ &m_x^*=-m_{Bx}^*, \quad m_y^*=-m_{By}^*+f_{Bz}^*(l-x), \quad m_z^*=-m_{Bz}^*-f_{By}^*(l-x) \end{aligned} \tag{6}$$

On the basis of the elementary terms of the strength of materials (ref. 5) the virtual angle displacements are given by:

$$\begin{aligned} &\varphi_x^*=\frac{m_{Bx}^*x}{I_xG}, \quad -\varphi_y^*=\frac{1}{I_yE}(-m_{By}^*x+f_{Bz}^*lx-f_{Bz}^*\frac{x^2}{2}), \\ &\varphi_z^*=\frac{1}{I_zE}(m_{Bz}^*x+f_{By}^*lx-f_{By}^*\frac{x^2}{2}) \end{aligned} \tag{7}$$

and the translational displacements are given by:

$$\begin{aligned} &u_x^*=\frac{f_{Bx}^*x}{AE}, \quad u_y^*=\frac{1}{I_zE}(m_{Bz}^*\frac{x^2}{2}+f_{By}^*l\frac{x^2}{2}-f_{By}^*\frac{x^3}{6}), \\ &u_z^*=\frac{1}{I_yE}(-m_{By}^*\frac{x^2}{2}+f_{Bz}^*l\frac{x^2}{2}-f_{Bz}^*\frac{x^3}{6}) \end{aligned} \tag{8}$$

These local system virtual equations are adapted to establish the nodal contribution values in the following.

FORMULATION FOR STATICS

Determining nodal effects

12. In order to establish the x directional nodal contributions at node j one can apply an x directional virtual system characterised by the force and displacement values:

$$f^*_{Bx}=1, \quad f^*_{Ax}=-1, \quad f^*_x=1, \quad u^*_{Ax}=0, \quad u^*_{Bx}=\frac{l}{AE}, \quad u^*_x=\frac{x}{AE} \tag{9}$$

which is the solution of a simple cantilever beam of length l subjected to a vertical unit force at end B. When the force is put on the element (clamped at A) the real and virtual systems of loads undergo (reciprocal, or mutual) work, while the displacements belonging mutually to the other system take shape. The relevant terms for the work are:

$$W_{21}=f^*_{Bx}u^*_{jx}+f^*_{Ax}u^*_{ix}=W_{21}=F_{Bx}u^*_{Bx}+W_{fx} \tag{10}$$

So the equation for the (local) x directional nodal force contribution is:

$$u_{jx}-u_{ix}=F_{Bx}\frac{l}{AE}+W_{fx} \quad \text{and} \quad W_{fx}=\int_i^j f_x u^*_x dx \tag{11}$$

where W_{fx} is the work of the external loading with x directional element deformation. These types of expressions are to be formulated later.

13. With an entirely analogous deduction the equation of x directional moment contribution is expressed by:

$$\varphi_{jx}-\varphi_{ix}=M_{Bx}\frac{l}{I_xG}+W_{mx} \quad \text{and} \quad W_{mx}=\int_i^j m_x\varphi^*_x dx \tag{12}$$

when $m^*_{BX}=1$ used as external virtual load.

14. Next the bending components, deflections and rotations of the section are considered with the provision that shear effects, and consequently the shear work, are negligible. There are many different possibilities in choosing virtual loads. For instance all the four virtual force and moment components in directions y and z may have unit value. In this case the equations may need to be rearranged to contain only one component of element to node j interaction. Herein another procedure is shown.

15. Let the virtual force be $f^*_{By}=1$ and m^*_{Bz} chosen to result in zero rotation at B, so that:

$$\varphi^*_{Bz}=0=\frac{1}{I_zE}\left(m^*_{Bz}l+f^*_{By}\frac{l^2}{2}\right) \rightarrow m^*_{Bz}=-\frac{f^*_{By}l}{2}=-\frac{l}{2} \tag{13}$$

16. For determining the moment about z, let $m^*_{Bz}=1$ and f^*_{By} chosen to result in $u^*_{By} = 0$, the formula is:

$$\varphi^*_{Bz}=0=\frac{1}{I_zE}(m^*_{Bz}\frac{l^2}{2}+f^*_{By}\frac{l^3}{3}) \rightarrow f^*_{By}=-\frac{3m^*_{Bz}}{2l}=-\frac{3}{2l} \quad (14)$$

17. For virtual bending about y the expressions are:

$$f^*_{Bz}=1 \text{ and } m^*_{By}=\frac{l}{2} \text{ for } \varphi^*_{By}=0 \quad (15)$$

expressing the nodal force contribution in the z directional, and to determine nodal contributing bending about y:

$$m^*_{By}=1 \text{ and } f^*_{Bz}=\frac{3}{2l} \text{ for } u^*_{Bz}=0 \quad (16)$$

Although there are other ways to express nodal contributions, this approach has been found to be useful.

Force contribution in the y direction

18. In this case the virtual load is selected according to paragraph 15 and Eq. 13 and by using terms of Table 1. The non-zero force and displacement values are:

$$f^*_{jy}=1,\quad m^*_{jz}=-\frac{l}{2},\quad f_{iy}=-1,\quad m_{iz}=\frac{3l}{2}$$
$$u^*_{jy}=\frac{l^3}{12I_zE},\quad u^*_y=\frac{3lx^2-2x^3}{12I_{zE}},\quad \varphi^*_z=\frac{l^2-lx}{2I_zE} \quad (17)$$

Hence the final formulas after substitution are expressed by:

$$W_{21}=f^*_{jy}u_{jy}+m^*_{jz}\varphi_{jz}+f^*_{iy}u_{jy}+m^*_{iz}\varphi_{iz}=W_{12}=F_{jy}u^*_{jx}+W_{fy},$$
$$u_{jy}-u_{iy}-\frac{l}{2}\varphi_{jz}+\frac{3l}{2}\varphi_{iz}=\frac{l^3}{12I_zE}F_{jy}+W_{fy}, \quad (18)$$

and $W_{fy}=\int_i^j f_y u^*_y dx+\int_i^j m_y\varphi^*_z dx$

Moment contribution in the z direction

19. Paragraph 16 gives the necessary loads at node j. On this basis the suitable formulas are given by the following equations:

$$f^*_{By}=-\frac{3}{2l},\quad m^*_{Bz}=1,\quad f_{Ay}=\frac{3}{2l},\quad m_{Az}=\frac{1}{2}$$
$$\varphi^*_{Bz}=\frac{5l}{4I_zE},\quad u^*_y=\frac{1}{I_{zE}}(\frac{3x^3}{12l}-\frac{x^2}{4}),\quad \varphi^*_z=\frac{1}{I_zE}(\frac{x}{2}+\frac{3x^2}{4l}) \quad (19)$$

and

$$W_{21}=f^*_{jy}u_{jy}+m^*_{jz}\varphi_{jz}+f^*_{iy}u_{jy}+m^*_{iz}\varphi_{iz}=W_{12}=M_{jz}\varphi^*_{jz}+W_{mz},$$
$$-\frac{3}{2l}u_{jy}+\frac{3}{2l}u_{iy}\varphi_{jz}+\frac{1}{2}\varphi_{iz}=\frac{l}{4I_zE}M_{jz}+W_{mz}, \quad (20)$$

and $$W_{mz}=\int_i^j f_y u_y^* dx+\int_i^j m_z \varphi_z^* dx$$

Its important to note that although the expression of the work done by external load is the same as before the calculation is different due to the different virtual load.

Force in the z directional and moment contribution in the y directional

20. It is clear from paragraphs 15 to 17 that these two contributing components are formulated in exactly the same way as the latter two. Writing only the results, they are:

$$u_{jz}-u_{iz}+\frac{l}{2}\varphi_{jy}+\frac{l}{2}\varphi_{iy}=-\frac{5l^3}{12I_zE}F_{jz}+W_{fz}, \quad and \quad W_{fz}=\int_i^j f_z u_z^* dx+\int_i^j m_y \varphi_y^* dx \tag{21}$$

$$\frac{3}{2l}u_{jz}-\frac{3}{2l}u_{iz}+\varphi_{jy}-\frac{5}{2}\varphi_{iy}=\frac{l}{4I_yE}M_{jy}+W_{my}, \quad and \quad W_{my}=\int_i^j f_z u_z^* dx+\int_i^j m_y \varphi_y^* dx \tag{22}$$

Consideration of virtual work of external load

21. All the necessary formulas have already been developed herein to introduce virtual work terms of external forces and moments. General expressions can easily be formulated for concentrated or any type of distributed loads. For further information the reader is referred to the works of the author currently being developed . Due to the limited length of this paper only examples on dynamic and geometric non-linearity loads are presented.

Transfer to global system

22. The developed expressions are valid in the local system (Fig. 1). Two ways of assembling the system matrix are mentioned here: (a) the force and moment, as well as translational and rotational displacement components expressed with global co-ordinates, should be substituted in the place of those given in local co-ordinates or (b) only nodal displacements are put into local equations then expressions of local load co-ordinates are transformed to global ones and used as a force or moment in the nodal equilibrium equations. In any case the steps should be repeated for both ends of all elements. The operation (a) is a well known simple rotational matrix transformation (refs 6, 7) in case of which a much larger system matrix is gained but nodal forces and moments are directly calculated.

23. When transferring to the global system one has to account for the element connection to the node. In this method, a wide range of joints may be chosen by using additional equation(s). For instance, in the case of a coplanar hinge joint instead of applying a nodal contribution, an additional equation should be used for a zero moment contribution. Truss and frame structures with typical joints are

examples of this. In the truss there is only one (local x) directional contribution. The frame has rigid-joints.

Constructing and solving system equations

24. The element contributions and nodal loads also have to be substituted into the global system nodal equations. In the case of transformation: (a) nodal loads are considered as acting on one of the elements joining at the given node; (b) nodal loads should be summed using the relevant nodal equilibrium equation. When the system equations has been built up, the solution of the linear system of equations gives the nodal unknowns. Finally the original local system terms are used to compute moment and shear values for the beam elements.

FORMULATION FOR THERMAL LOADS

25. Thermally induced load is taken into account through an equivalent statical load (refs 5,8). An element is asumed to have a linear temperature distribution along each co-ordinate axis. Let the normal temperature T_0 and temperature distribution between two sections C and D be expressed by:

$$T=\vartheta_0+\vartheta_x x+\vartheta_y y+\vartheta_z z \tag{23}$$

Which may approximate any temperature distribution complying with the applied theory. In this case the stress and strain are given by:

$$\varepsilon_x=\alpha(T-T_0) \quad \text{and} \quad \sigma_x=E\varepsilon_x=E\alpha(\vartheta_0-T_0+\vartheta_x x+\vartheta_y y+\vartheta_z z) \tag{24}$$

In general the axial stress in a beam is given by:

$$\sigma_x=\frac{f_x(x)}{A}+\frac{m_y(x)}{I_y}z+\frac{m_z(x)}{I_z}y \tag{25}$$

Combining Eqs 24 and 25 gives:

$$f_x=E\alpha A(\vartheta_0-T_0+\vartheta_x x), \quad m_y=E\alpha I_y\vartheta_z, \quad \text{and} \quad m_z=E\alpha I_z\vartheta_y \tag{26}$$

which correspond to the inner forces. According to Eq. 1 and the virtual energy expressions (refs 5, 8) the additional terms, due to thermal effects to the virtual work of loads, are:

$$W_{fx}^T=\int_C^D\frac{f_x f_x^*}{AE}dx, \quad W_{fy}^T=\int_C^D\frac{m_z m_z^*}{I_z E}dx, \quad W_{mz}^T=\int_C^D\frac{m_z m_z^*}{I_z E}dx \tag{27}$$

$$W_{fz}^T=\int_C^D\frac{m_y m_y^*}{I_y E}dx \quad \text{and} \quad W_{my}^T=\int_C^D\frac{m_y m_y^*}{i_y E}dx$$

The right hand side expressions have different meaning as stated previously, due to the fact that the virtual moment functions are different.

FORMULATION TO DYNAMICS

26. Apart from nodal excitation the simplest part of the dynamical formulation is the transformation of time-dependent transient force or moment from element to nodal excitation or impact. The same equations can be used as those for the static case. Free vibration examination of a structure starts from the equations of a beam element of finite length (refs 9, 10).

27. The longitudinal (tension-compression), torsional and bending vibration of a rod is described by the equations:

$$\frac{\partial^2 u_x}{\partial x^2}-\frac{\varrho}{E}\frac{\partial^2 u_x}{\partial t^2}=0,\qquad \frac{\partial^2 \varphi_x}{\partial x^2}-\frac{q_0}{I_xG}\frac{\partial^2 \varphi_x}{\partial t^2}=0,\qquad and$$
$$\frac{\partial^4 u_y}{\partial x^4}-\frac{\varrho}{E}\frac{\partial^2}{\partial x^2}\frac{\partial^2 u_y}{\partial t^2}+\frac{A\varrho}{I_zE}\frac{\partial^2 u_y}{\partial t^2)}=0,\qquad \frac{\partial^4 u_z}{\partial x^4}-\frac{\varrho}{E}\frac{\partial^2}{\partial x^2}\frac{\partial^2 u_z}{\partial t^2}+\frac{A\varrho}{I_yE}\frac{\partial^2 u_z}{\partial t^2}=0 \tag{28}$$

Without regarding details it can be stated that each the solution to each equation has a similar form which may be presented as:

$$u_x=Ce^{i\omega t}e^{i\lambda x}\quad and\quad u_z=Ce^{i\omega t}e^{i\lambda x} \tag{29}$$

where the multipliers and constants in the exponents should be distinct for the two components when the formulation is really made, not only indicated. In the equations i denotes the imaginary unit. For the axial load and longitudinal vibration the elastic reciprocal virtual work is calculated as follows:

$$f_x=AE\frac{\partial u_x}{\partial x}=CAEi\lambda e^{i\omega t}e^{i\lambda x},$$
$$W^D_{fx}=\int_i^j\frac{f_xf^*_x}{AE}dx=Ci\lambda\int_i^j e^{i\omega t}e^{i\lambda x}dx=Ce^{i\omega t}\left[e^{i\lambda x}\right]_0^1 \tag{30}$$

For torsion vibration the same form of expression is obtained with exactly the same steps:

$$W^D_{mx}=Ce^{i\omega t}\left[e^{i\lambda x}\right]_0^1 \tag{31}$$

It is emphasised that the same notations are used for different variables in order to show the possibility of identical handling of different types of internal resultants.

28. As for bending, the moment function and nodal contribution formula of the y component are given by the following equations:

$$m_y=I_yE\frac{\partial u_z}{\partial x^2}=-CI_yE\lambda^2e^{i\omega t}e^{i\lambda x},$$
$$W^D_{my}=\int_i^j\frac{m_ym^*_y}{I_yE}dx=-C\lambda^2\int_i^j m^*_ye^{i\omega t}e^{i\lambda x}dx \tag{32}$$

In the same way we obtain for the nodal contribution in the z directional:

$$m_z=I_zE\frac{\partial u_y}{\partial x^2}=-CI_zE\lambda^2e^{i\omega t}e^{i\lambda x},$$
$$W^D_{mz}=\int_i^j\frac{m_zm^*_z}{I_zE}dx=-C\lambda^2\int_i^j m^*_ze^{i\omega t}e^{i\lambda x}dx \tag{33}$$

The exponential form should be changed to trigonometric and hyperbolic form to get the proper description.

29. In every case the result is in the form:

$$W^{D}_{...}=Ce^{i\omega t}, \qquad ...=C_D\omega^2 e^{i\omega t} \tag{34}$$

which leads to an eigenvalue, eigenvector problem through nodal equilibrium assemblage. The brief description of the dynamic presentation is discussed in more detail in refs 9-11. However, it can be clearly seen that the formulation results in an eigenvalue, eigenvector task as was expected, the emerging forms raise more interesting aspects and consequences, which are beyond the scope of this paper.

GEOMETRICAL NONLINEARITY FORMULATIONS

30. An element of a structure considered here undergoes a deformation state. The original geometry is assumed not to change. One effect, however, will be examined, namely that of the bending due to the deformational eccentricity of the centroid of rod sections. According to ref. 5 (p. 477) it holds for a strut that:

$$\frac{d^2m_y}{dx^2}+\frac{f_x}{I_yE}m_y=\frac{f_z}{I_yE} \quad \text{and} \quad \frac{d^2u_z}{dx^2}=-\frac{m_y}{I_yE} \tag{35}$$

The form of the solution is:

$$m_y=Ce^{i\lambda t}+g(x)=C_S\lambda^2e^{i\lambda t}+g(x) \tag{36}$$

where function g(x) is a particular solution which needs to be produced for each given transverse load case; g(x) is easily computed for known concentrated or linearly distributed loads and there is no difficulty in expressing nodal contribution. The method used is exactly the same as previously described in the dynamic case, apart from the additional integral product of g(x) with m_y*. The procedure will be the same for bending about the *z* axis. The equations are of the same types as those obtained for dynamics. The solution will produce critical loads and deformed shapes through an eigenvalue and eigenvector calculation.

31. Once the system matrix has been formed with loads of linear and geometrically non-linear cases, the critical values of stability and the deformed shape can be calculated. This non-linear part of formulation gives the possibility of taking into account the effect on bending stiffness which decreases in compression and increases in tension.

CONCLUSIONS

32. Static, dynamic and stability formulations are presented here to solve truss, beam and mixed type structures. The method exposes a wide range of utilisation of Betti's law of reciprocal work and gives a general concept of building up the linear equation system describing any type of rod structures. Static formulation is shown to follow the procedure. In the other cases only a very short introduction

is given to show the wide range possibilities. Nevertheless the author is aware of many other aspects of the topic both in the possibilities of extension and concerning applications.

REFERENCES

1. BETTI, E. Teoria dell Elasticita, Il Nuovo Ciemento, 1872, tome 7-10.
2. SOMIGLIANA, C. Sopra l'equilibrio di us corpo elastico isotropo, Il Nuovo Ciemento, 1885, tome 17-19.
3. CRUSE, T.A. Boundary element analysis in computational fracture mechanics, 31-33. Kluwer Academic Publishers, Dordrecht, 1988.
4. BREBBIA, C.A., TELLES, J.C.F. and WROBEL, L.C.. Boundary element techniques. Theory and applications in engineering, 183-190. Springer-Verlag, Berlin, 1984.
5. HEARN, E.J. Mechanics of materials, Second edition, Pergamon Press, Oxford, 1985.
6. BREBBIA, C.A. and FERRANTE, A J. Computational Methods for the Solution of Engineering Problems, 3rd edition, Pentech Press, London, 1986.
7. RAO, S.S. The Finite Element Method in Engineering, 2nd edition, Pergamon Press, Oxford, 1988.
8. HIBBELER, R.C. Structure Analysis, 2nd edition, Macmillan Publishing Company, New York, 1990.
9. MEIROVITCH, L. Analytical Methods in Vibrations, Macmillan Publishing Company, New York, 1967.
10. TIMOSHENKO, S. Vibration Problems in Engineering, Van Nostrand, New York, 1937.
11. BREBBIA, C.A., et al. Vibrations of engineering structures, Computational Mechanics Ltd., Southampton, 1976.

APPENDIX FOR NOTATIONS

W and **U** stand for virtual work and energy with the first index for the system doing the virtual work while the second index serves the displacements.

f, **u**, σ, ε are external or virtual force, displacement, and stress, strain, respectively. **F** and **M** are nodal forces and moments. **m** and φ are external or virtual moments and angular rotations.

A is the cross section area. $\mathbf{I_x}$, $\mathbf{I_y}$, $\mathbf{I_z}$, are moments of inertia about the indexed axis. **E** and **G** are Young's and shear moduli.

An example of the virtual work nodal contribution of the external load is $\mathbf{W_{my}}$, which is the virtual work of the external loads when the y directional unit moment is applied.

α is the thermal expansion coefficient. ϑ**-s** are constants of assumed temperature distribution. ω and λ are the mass density, natural frequency and shape parameter of vibrating rod element. All the other notation can be found from the text and figures.

145. Simplified analysis of a space grid roof system

N. G. BHAGAVAN and N. GOPALAKRISHNAN, Structural Engineering Research Centre, India

SYNOPSIS. The plate analogy to the double layer grid roof system is adopted to estimate the member forces for a grid roof system. The ratio of brace to main member area is varied widely to establish the effect on the stress resultants and deflections. The analysis is carried out for three different types of boundary conditions. The grid is forced to behave in an orthotropic fashion by changing the area of main members along the two orthogonal directions. A comparison is made with the equivalent isotropic/orthotropic plates.

INTRODUCTION

1. The equivalent continuum method for the analysis of double layer space trusses, has been dealt with by Kollar (ref. 1). Kollar has formulated continuum differential equations for double layer space trusses. Abdul Karim et. al. (ref. 2) have given design charts for the analysis and fabrication of the Indian patented SPACE-SERC grid roof systems for spans up to 30m.

OBJECTIVE OF THE STUDY

2. The present work compares the behaviour of the space grid with that of the equivalent analogous plate with similar loading and support conditions. Thus, this methodology leads to the establishment of the member sections as preliminary design data, before proceeding with the detailed analysis and design of the space grid roofing system. The paper details a limited parametric study on this procedure. An extensive parametric study may be carried out, based on this methodology, to produce a handbook for preliminary design of space grid roofing systems.

3. The following parameters are considered in the study of the stress resultants and deflections for the square grid system:
a) The ratios of brace to main member area considered are: a_b/a_m = 0.2, 0.4, 0.6, 0.8, 1.0, where, a_b = area of brace members and a_m = area of main members in the top and bottom chords.

b) The boundary conditions for the grid roof system are:
hinge-supported on four corners (case 1);
hinge-supported on opposite edges (case 2);
hinge-supported on two corners and one opposite edge (case 3).
c) The ratios of the main member areas in the two orthogonal directions are:
$a_{mx} / a_{my} = 0.25, 0.50, 0.75, 1.0$, where, a_{mx} = area of main members in the X-direction and a_{my} = area of main members in the Y-direction.

For each of the above cases the analysis is carried out for the equivalent plate and for the grid; the stress resultants are compared.

ANALYTICAL MODEL

4. The basic structure taken for the analysis is a 10m x 10m square grid (Fig. 1). The spacing of the members is 1.0m for both top and bottom chords and they are parallel to each other in the X- and Y-directions. The bracing pattern is of tetrahedral shape. The joints are assumed to be pinned with only translational degrees of freedom for each joint. There are, in total, 221 joints and 800 members in the structure. The loading is assumed to be uniformally distributed with a magnitude of 0.1 tonne/m (1kN/m). The boundary conditions are shown in Fig. 2. The plate thickness is obtained using Eq. 1 for the isotropic case: Bending rigidity of the truss/metre = Bending rigidity of the plate/metre.

$$a_m . \frac{d^2}{2} = \frac{t^3}{12(1-\mu^2)} \qquad (1)$$

where d = depth of the truss; t = thickness of the plate; μ = poisson ratio of steel = 0.3

Similarly for the orthotropic case:

$$E.a_{mx} . \frac{d^2}{2} = \frac{E_x . t^3}{12(1-\mu_x . \mu_y)} \qquad (2)$$

$$E.a_{my} . \frac{d^2}{2} = \frac{E_y . t^3}{12(1-\mu_x . \mu_y)} \qquad (3)$$

$$\frac{E_x}{E_y} = \frac{\mu_y}{\mu_x} \qquad (4)$$

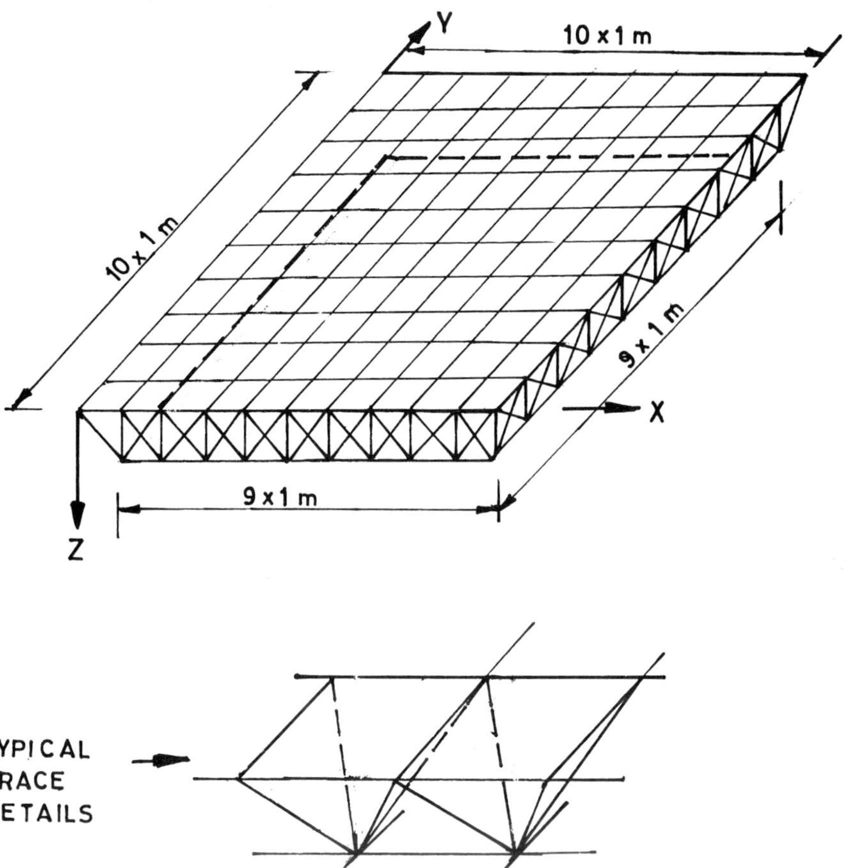

Fig. 1. Grid roof system taken for the analysis.

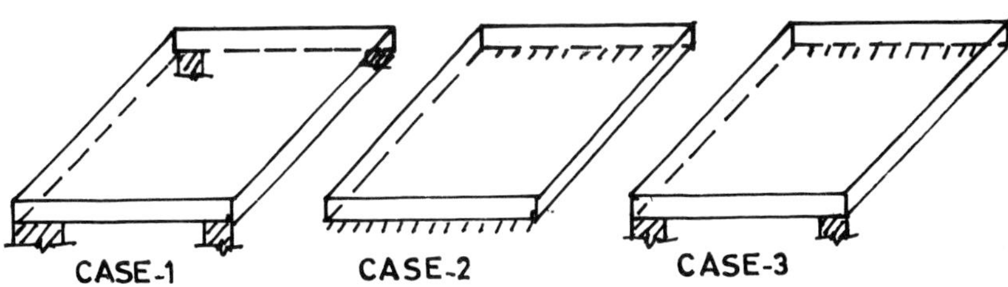

Fig. 2. Different boundary conditions of the grid roof and its equivalent plate.

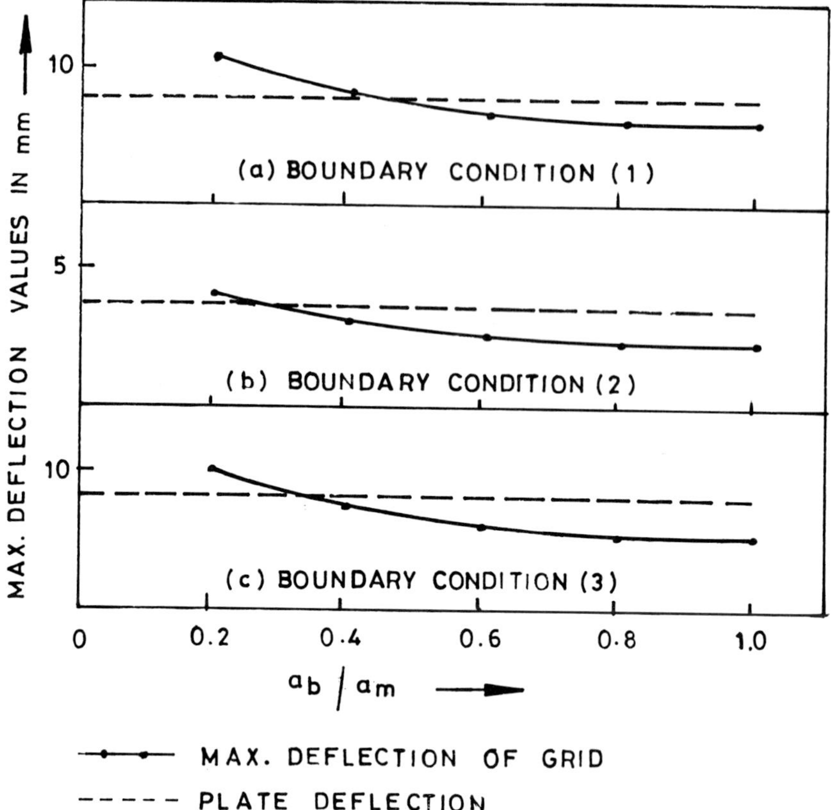

Fig. 3. Variation of deflection of the grid for the isotropic case.

DISCUSSION OF RESULTS

5. Deflections. The deflections of the grid at the critical points are compared with the plate deflections. The deflection of the grid corresponding to the case of $a_b/a_m \geq 0.6$, show proximity to the plate deflection. Since plate analysis is based on simple plate theory, neglecting shear deformation, it is likely to give a lesser deflection (Fig. 3).

Stress Resultants

6. Case 1 (grid hinge-supported at four corners): The loading pattern, the boundary condition and the structure are symmetric about the centre lines parallel to the X- and Y-axes and hence the stress resultants and deflections are also symmetric. The points of maximum bending moment for the plate are the mid-points of the edges connecting the supports. For the grid this is also the region of maximum stress. Fig. 4a shows the variation of the force in the top chord members at x = 0m. Fig. 4b and 4c show the variation of F_y in the bottom and top chord members at x = 0.5m and x = 1.0m respectively. $F_{y,max}$ occurs along the edge x = 1.0m for $a_b/a_m = 0.2$. For $a_b/a_m = 1.0$, $F_{y,max}$, occurs along the edge at x = 0. Fig. 4e shows the variation of F_y along the section y = 5.0m. The stress resultant at the centreline of the plate is more than that for the grid.

However, at the edges the grid suffers more force than is predicted by the plate analysis. The reaction thrust for the bottom chord can be estimated by applying the forces developed in the top chord members and finding the reaction at the end; this will cause contraction at the end equal to the elongation produced by the discrete forces.

7. Case 2 (grid hinge-supported at the opposite edges): This is the simplest of all the cases. The structure exhibits a beam like behaviour with M_y nearly uniform along the section y = 0.5m, with a value of $M_y = 0.123wl^2$ at the centre (x = 5.0m) and increasing slightly to $M_y = 0.132wl^2$ at the edges (x = 0m). The grid also behaves in a similar fashion.

8. Case 3 (grid hinge-supported at two corners and opposite edge): The deflections and the stress resultants are symmetric about the centreline parallel to the Y-axis. The points of maximum moment, as predicted by the plate analysis, are at the mid points of the unsupported edges. The M_y value along the mid-section, at y = 5.0m is nearly uniform, varying between 1.4 tonne.m along the edges to 1.2 tonne.m near to the centre of the plate. The points of maximum stress for the grid are also the same as for the plate (Fig. 5). The stress resultants deduced from the equivalent plate are nearly the same along the edges x = 0 and x = 10.0m. However, the plate analogy underestimates the forces in the X-direction to an extent of nearly 25%, for $a_b/a_m = 0.2$, and 12%, for $a_b/a_m = 1.0$. The value of F_y at the centre of the plate is over-estimated by the continuum results. In brief, the plate smooths out the distribution of moment along the central section, whereas the grid shows more moment at the edges and less moment at the centre.

ORTHOTROPIC GRID ANALYSIS

9. The orthotropic grid, supported on four corners is an assymmetric structure with F_x,(y = 0) $\neq$ F_y,(x = 0) and F_x (y = 5.0m) $\neq$ F_y (x = 5.0m). Because the area of the main members in the X-direction is less than the area of the main members in the Y-direction, the structure attracts more force in the Y-direction than in the X-direction. For example, $F_{x,max}$ (y =0) = 80% of $F_{y,max}$ (x =0). However, the equivalent orthotropic plate does not show this much variation and $M_{x,max}$ (y = 0) = 95% of $M_{y,max}$ (x=0). This causes an underestimation of the forces (F_y) by the plate to the tune of 15% (Fig. 6). The orthotropic analysis, for the Case 1 boundary condition, shows an interesting picture. The F_y forces along edges when x = 0 and x = 10m are estimated correctly by the plate analysis, but the F_x force variation along the edge when y = 0 is badly underestimated by the plate analysis. The underestimation is as much as 50% of the actual forces experienced by the grid.

SHEAR FORCE DISTRIBUTION

10. Fig. 4d shows the shear distribution at x = 1.0m for the grid and for the plate, for the Case 1 boundary condition. The heavily loaded braces are those above the supports. The maximum force is carried by the brace which points towards the centre of the grid and is 2 to 2.5 times the force in the adjacent two braces.

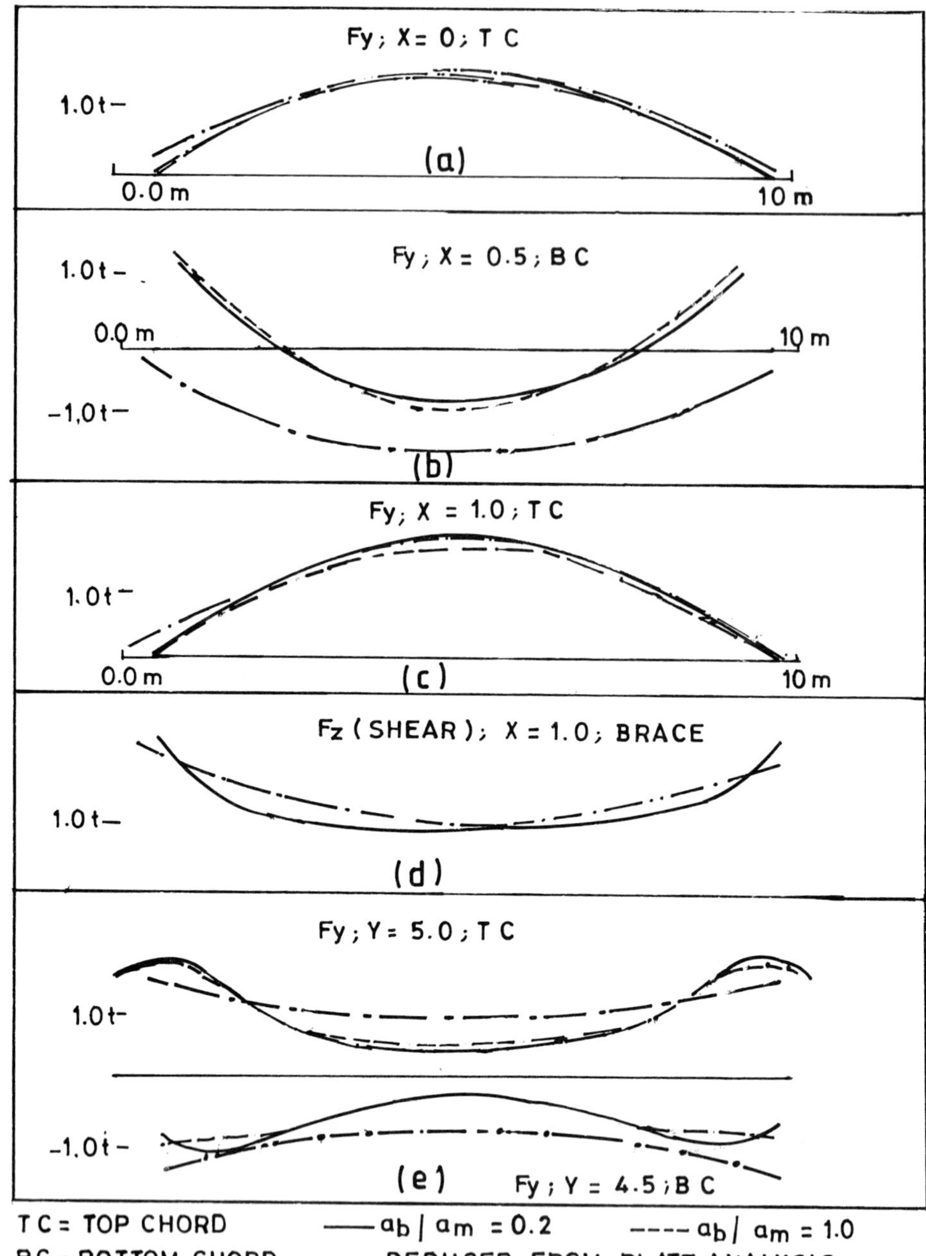

Fig. 4. Stress resultants for the isotropic grid supported on four corners.

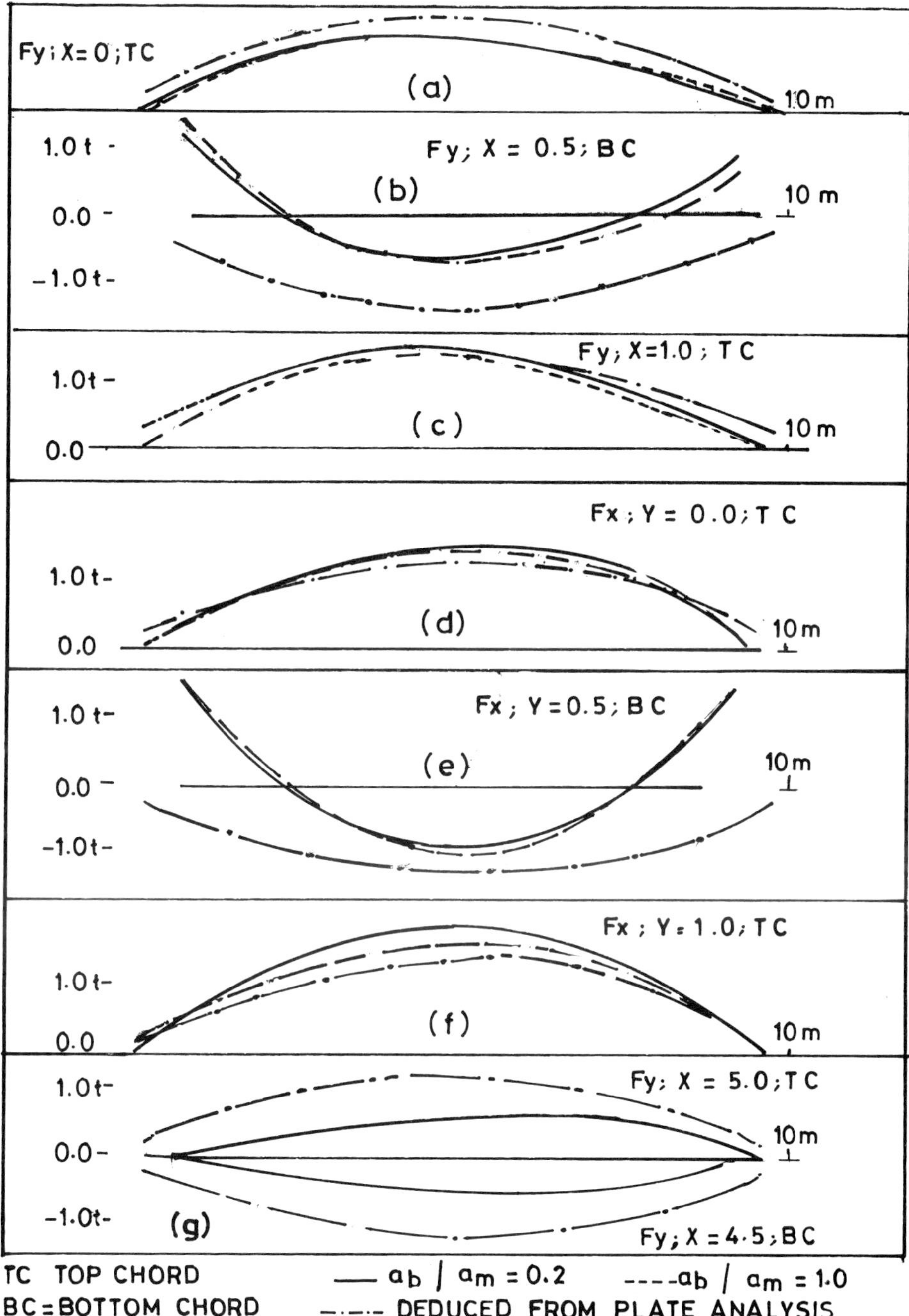

Fig. 5. Stress resultants for the isotropic grid supported on two corners and the opposite edge.

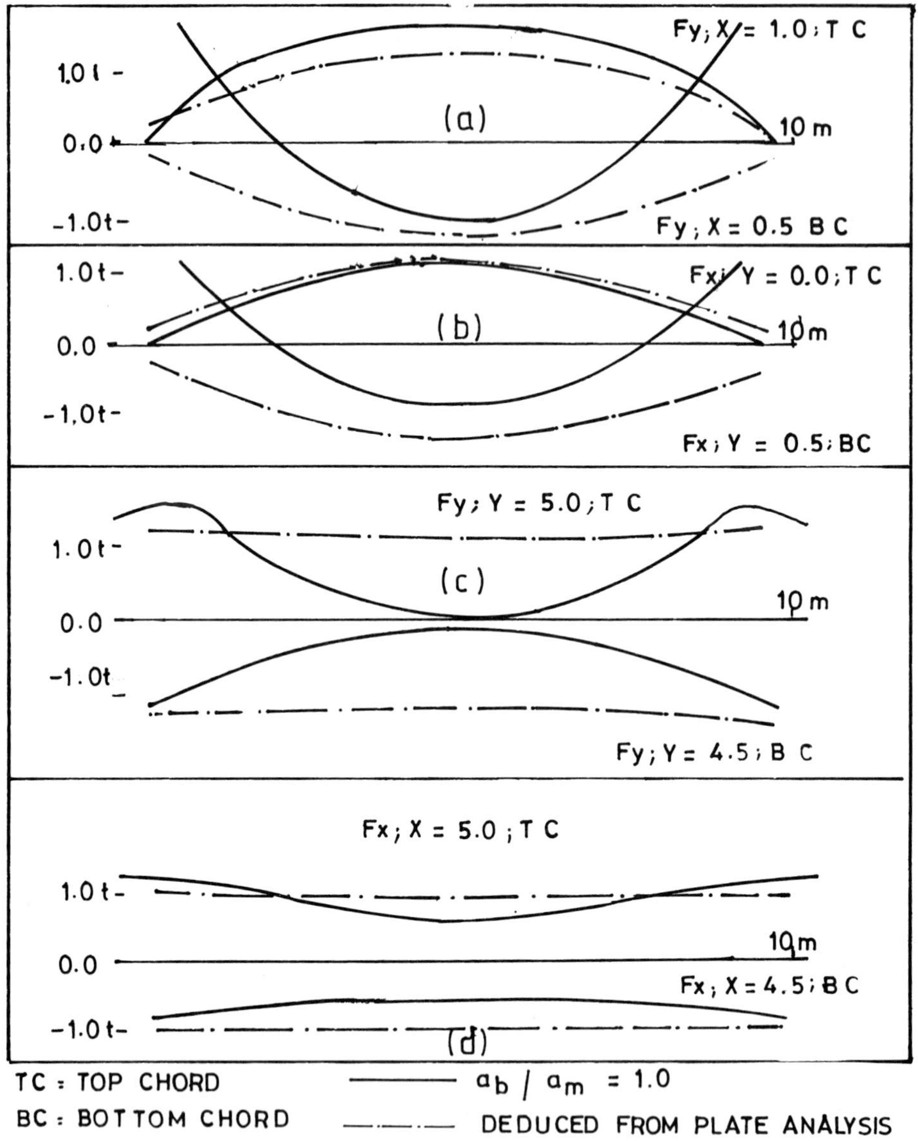

Fig. 6. Stress resultants for the orthotropic grid supported on four corners ($a_{mx}/a_{my} = 0.25$).

CONCLUSIONS

11. The variation of the brace area has very little affect on the distribution of forces in the members. For the case of an isotropic plate supported at the four corners, the plate predicts the internal stress resultants to a high degree of accuracy. In the case of an isotropic plate supported at the two corners and the opposite edge, the F_y force values can be safely assumed from the plate moments. The F_x force values are underestimated and they need to be increased by 25%. The equivalent plate analogy for orthotropic grids should be used with caution. The discrepancy is the worst in the case of case-3 boundary condition wherein the F_x force is to be increased by a factor of two on the value extrapolated from the equivalent plate.

ACKNOWLEDGEMENTS

12. The paper is being presented with the kind permission of the Director, SERC, Madras.

REFERENCES

1. KOLLAR L. Analysis of Double Layer Space Trusses by the Equivalent Continuum Method. Second International Conference on Space Structures, University of Surrey, U.K., 1975.
2. ABDUL KARIM E., SETHU NARAYANAN S. and NATARAJAN P.R. Handbook on Space Grid Roof System. Structural Engineering Research Centre, Madras, 1986.

146. The KT space truss system

K. IMAI, T. MORITA and Y. YAMAOKA, Kawatetsu Steel Products Corporation, Japan, K. WAKIYAMA, Osaka University, Japan, and S. TSUJIOKA, Fukui Institute of Technology, Japan

SYNOPSIS. The KT space truss system was officially approved by the Japanese Ministry of Construction in 1989 and has been applied to over 90 structures in Japan and patented in many countries of the world. The system is designed to endure severe loads and/or the detrimental effects of environmental conditions. The KT truss system consists of solid spherical nodes, hollow tubular members, solid section members, double tube members and specially designed joint assemblies.

THE JOINING MECHANISM AND SPECIAL FEATURES

1. The system consists of four series, that is KT-I, KT-II, KT-III and KT-FLD. All these series have threaded spherical nodes. Joining bolts are M10-75, tubes are 27.0-355.6mm in diameter, spherical nodes are 60-400mm in diameter. These four series have complete compatibility in assembling.

2. The range of combinations of bolt diameter with section diameter for the system is shown in Table 1.

KT-I Series

3. KT-I is the most basic in KT truss system and can be applied economically for small, medium and relatively large structures. Bars and nodes are elegantly powder coated for general member sizes up to 165.2mm diameter. The powder coating is durable and should be retained for more than 15 years even in an outdoors environment.

4. KT-I consists of the following components (see Fig. 1):

- i) A threaded spherical node (60-300 mm in diameter) .
- ii) High strength bolt with hexagonal boss in the middle of the shank (M10x1.5-M48x3 / the allowable tensile load for permanent use is 23 - 591kN). The hexagonal boss transmits the turning torque from the hexagonal sleeve to the bolt.
- iii) Hollow sleeves (sleeve nuts) are inserted into left-hand-threaded holes of stub cones at tubular bar ends. The threaded end of each bolt is inserted into the sleeve nut and then the anchor nut is attached to the threaded end of the bolt.

iv) Anchor nuts have left-hand threads.

v) Hexagonal hollow sleeves are installed between the nodes and the end of the tubular bars. The sleeves absorb the compressive stress between the nodes and the tubular bars and transmit the applied torque to bolts through the inner hexagonal sections that engage with the boss of the bolts.

vi) Spring or pushing thread on bolt shank (pushing device) to push bolts towards the node during the assembly of the space frame.

vii) Hollow tubular bars are 27.2 - 216.3mm in diameter with a length of up to 8 meters (the allowable tensile load for permanent use is 22 - 591kN and the allowable compressive load is 24 - 815kN).

Table 1. The range of the KT truss system.

CHS \ Bolt	M10	M12	M16	M20	M24	M30	M36	M42	M48	M56	M65	M75
ϕ 27.2	○											
ϕ 34.0	○	○								KT-I & I(rev)		
ϕ 42.7	○	○										
ϕ 48.6		○	○									
ϕ 60.5		○	○	○								
ϕ 76.3			○	○	○							
ϕ 89.1			○	○	○							
ϕ 101.6				○	○	○						
ϕ 114.3				○	○	○						
ϕ 139.8					○	○	○					
ϕ 165.2						○	○	○			KT-II	
ϕ 216.3							○	○	○			
ϕ 267.4								○	○	○		
ϕ 318.5									○	○	○	
ϕ 355.6										○	○	○

KT-III (M48 – M75)

5. The joint assembly (bolt, sleeve nut, anchor nut, hexagonal sleeve) is shown in Fig. 2. The node and the tubular bar are produced separately. After completion of the surface treatment process, the joint assembly is installed into the tubular bar. In the powder coating process, the bar element has openings at both ends. During the process of installing the joint assembly into the tubular bar, the threaded hole of the bar is glued in order to maintain the sleeve nut in the correct position.

6. During the assembly of the space frame, the threaded end of the bolt is pushed or pulled into the hexagonal sleeve and then the hexagonal sleeve is turned at the threaded hole of the node (see Fig. 4). In this manner, assisted by the pushing device which provides a pushing action, the bolt enters the node hole.

7. The assembly of a space frame is, therefore, simplified assuming that the distance between adjacent nodes accurately reflects the correct measurement of the final assemblage. This mechanism, which eliminates the necessity to consider the order in which bars are assembled, drastically simplifies dismounting in the event that disassembled bars occur.

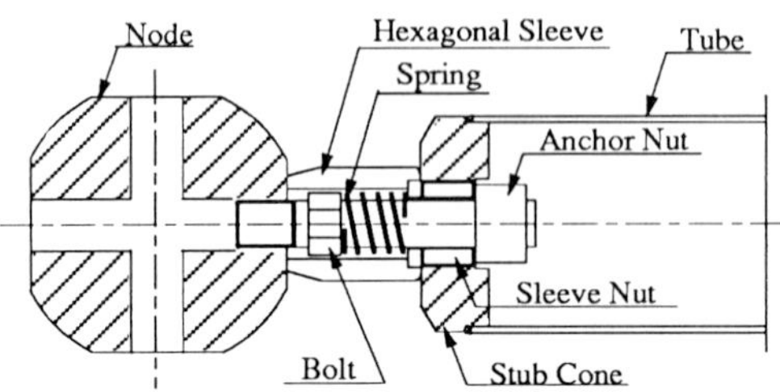

Fig. 1(a). Composition of KT-I Series.

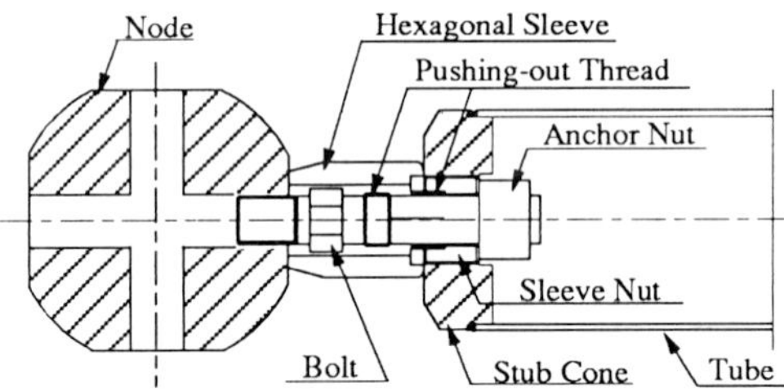

Fig. 1(b). Composition of KT-I (rev.) Series.

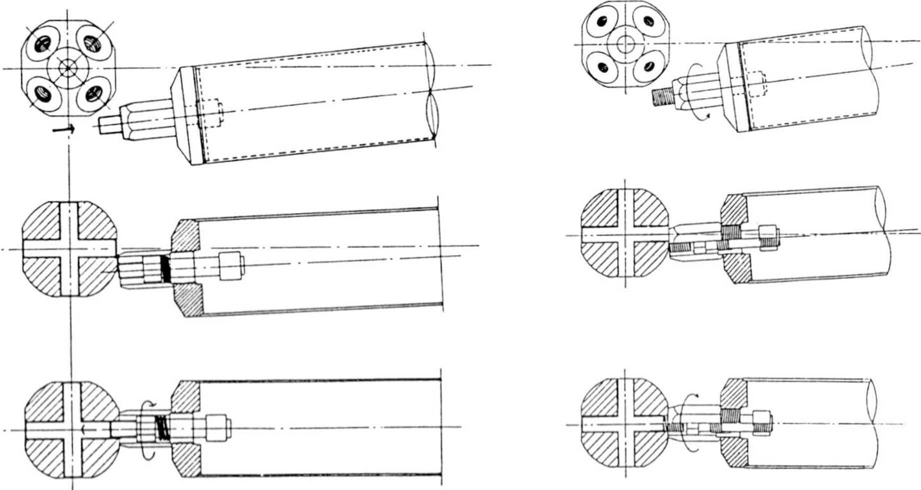

Fig. 2. Assemblage of KT-I (Left:KT-I, Right:KT-Irev).

8. Traditional systems cannot be assembled in this manner due to the fact that the bolt must project slightly from the hollow sleeve in order to be screwed into the node hole.

Fig. 3. Inside view of the gymnasium (60x100 m hipped roof).

Fig. 4. Stadium roof structure (Fatigue design was applied).

<u>Special features of KT-I.</u>

9. Some of the special features are:

i) High bolt reliability both under static and cyclic loading is realised because a significantly high torque can be applied. As a result, high clamping force, which significantly improves the fatigue characteristics of the joint, is effectively introduced. Another advantage of the high clamping force is that its use effectively tests all bolts at the construction site; if bolts have any serious defects, such as micro-cracks caused by the heat treatment process, the bolt will fracture at very low tensile force.

ii) Bar length tolerance is absolutely minimal; it becomes less than 0.1-0.2 mm regardless of the length of the bar element. Because of this severe tolerance, the stress distribution of the structures fits well with calculated values. As a result, fatigue specifications can be applied and the exact buckling strength of the bar element can be estimated accurately.

iii) Exceptional resistance to corrosion is achieved because there are no gaps or openings into the inside of the joint and the tubular bar. In addition, the high clamping force is effective in preventing even minor openings in the surface of the joint.

iv) All bar elements are easily assembled and changed with the aid of a pushing device. A fine thread is used for larger bolts (M30-48) to reduce the final tightening torque required.

10. Some typical examples of construction are shown in Figs 3-6.

KT-II Series

11. KT-II is normally suitable for the large structures with large members that exceed the range of application of the powder coating process. KT-II has a much simpler joining mechanism than KT-I.

Fig. 5. Single layer dome. Fig. 6. Entrance gate of the university.

12. KT-II consists of the following components (see Fig. 7):

i) The threaded spherical node (155-400mm in diameter).

ii) High strength bolt with hexagonal boss and pushing thread in the middle of the shank (M36x3-M75x4 / the allowable tensile load for permanent use is 336-1395kN.). The hexagonal boss transmits the turning torque from the hexagonal sleeve to the bolt. The pushing thread that engages with the thread of the stub cone pushes the bolt toward node hole.

iii) Anchor nut is the same to KT-I

iv) Hexagonal sleeve is the same to KT-I

v) Hollow tubular bars, 139.8-355.6 mm in diameter with a length of up to 10 meters (the allowable tensile load for permanent use is 528-1591kN and the allowable compressive load is 576-1735kN.

13. Concerning the joint assembly (bolt, anchor nut and hexagonal sleeve), the tube is welded after installing bolts, anchor nuts and hexagonal sleeves to the stub cones. After completion of fabrication, the member is painted.

14. The space frame is assembled in the same manner to KT-I(rev) with the aid of pushing device for the bolt.

Special features of KT-II

15. KT-II has the same special features as KT-I. Additionally KT-II is more economical for large members because it is not necessary to use powder coating and it has a much simpler joining mechanism than that of KT-I.

KT-III Series

16. KT-III is designed to joint elegantly compact sections such as solid sections and thick walled seamless tubes that have large sectional areas with high member stress. Bolts are screwed into the section directly. Normally KT-III is assembled

in combination with the KT-I joint (see Fig. 9) that is applied for the member with relatively low stress.

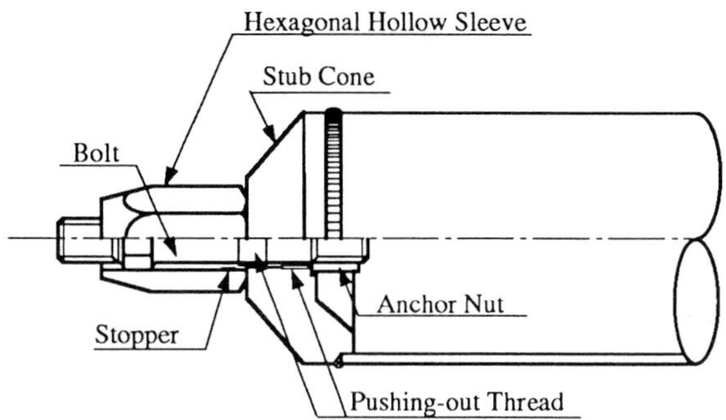

Fig. 7. KT-II Joint.

17. KT-III consists of the following components (see Fig. 8):
   i) The threaded spherical node (155-400mm in diameter).
   ii) High strength bolt with hexagonal boss between the two threads (M48x3-M75x4 / the allowable tensile load for permanent use is 591-1395kN). The bolt has right-hand thread at the node side and left-hand thread at the member side. The bolt acts as a turnbuckle. The left-hand thread is screwed into the member directly. By this mean the loss of sectional area of the member becomes minimal at the end of the member. In order to absolutely minimise the loss of sectional area, a type-2 bolt is applied. The bolt has two different diameter threads on the member side. The critical section moves from line(a) to line(b), accordingly a larger bolt can be used than type-1.
   iii) Hexagonal sleeve is the same for KT-I and II.
   iv) Bars are square or circular solid section or thick walled seamless tubes.

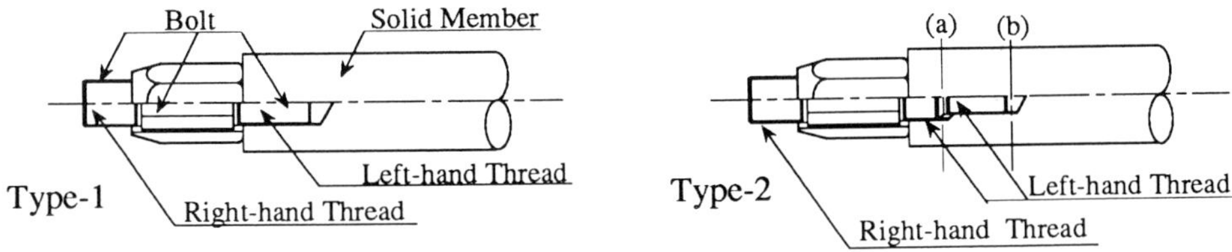

Fig. 8. Composition of KT-III.

18. An example of the process of assembling is shown in Fig. 10. The bolt of KT-III protrudes from the end of the hexagonal sleeve with the length of two times of a thread engaging length to the node. Therefore the node move towards the member with the distance of the thread engaging length when the bolt turns. Because of this protrusion, the whole structure cannot be assembled using only KT-III members in general. This problem is resolved by mixing the modified KT-I members which enable the assembly in the correct measurement of the final position of nodes. Fig. 10 shows the example of the assembling order by using KT-I and III members.

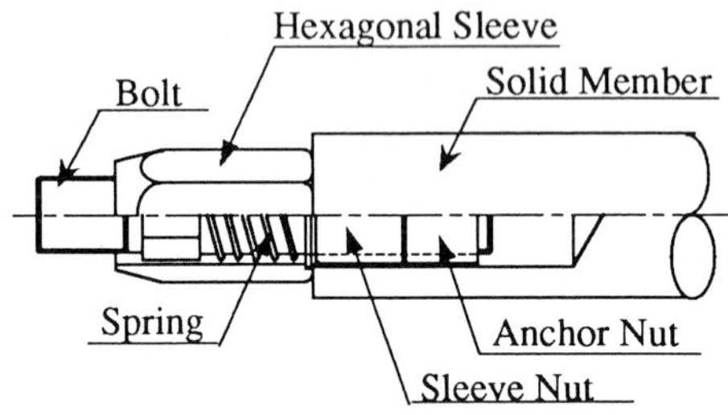

Fig. 9. Modified KT-I in assembling with KT-III.

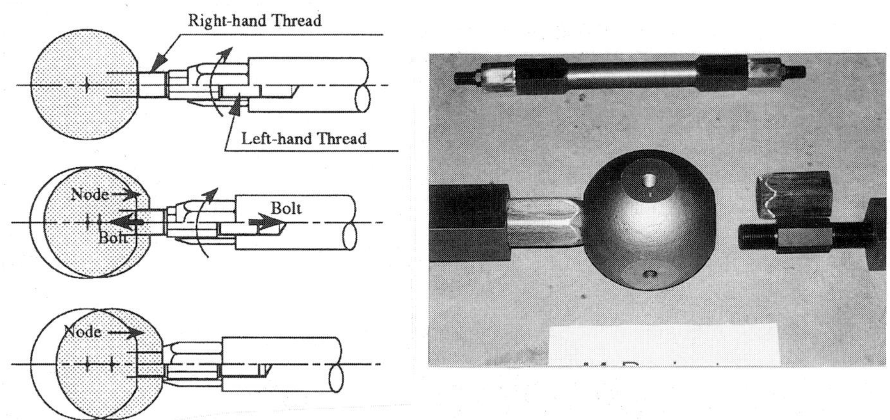

Fig. 10. Assemblage of KT-III.

Special features of KT-III

19. The significant feature is that KT-III connects the compact section with minimum loss of cross section. And KT-III has the same special features KT-II.

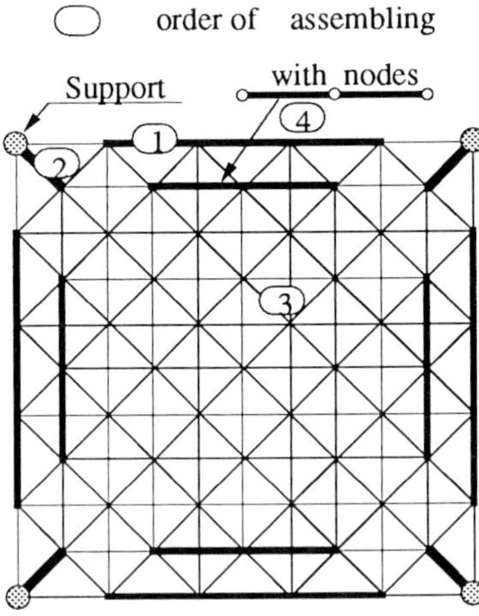

Fig. 11. Scaled model of KT-III.

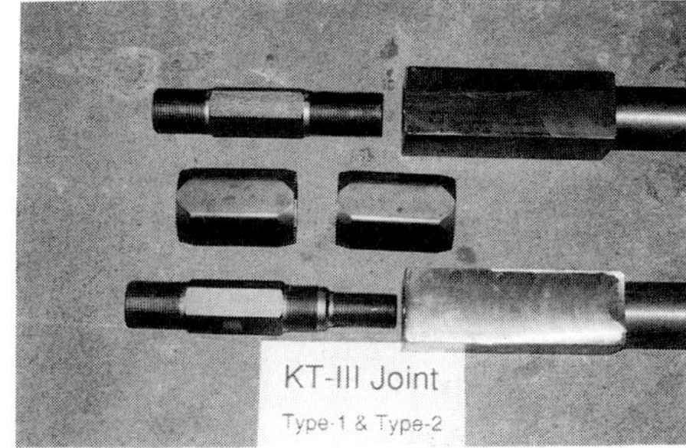

Fig. 12. KT-III joint.

KT-FLD Series

20. KT-FLD is designed to absorb energy beyond the yielding of members. Double tube type FLD is applied for this purpose. The joint is the same as for KT-I or II. The FLD member consists of the following components:

1) The outer tube withstands the axial stress and has reinforcing tubes at both ends. The reinforcing tube restricts the rotation at the member ends in co-operating with the inner tube when the outer tube becomes plastic.
2) The inner tube restricts the lateral deflection of the outer tube. A gap (g) that allows the axial movement of the inner tube caused by the plastic deformation of the outer tube is arranged between the end of the inner tube and the inner surface of the stub cone. The inner and the outer tube are connected by the positioning weld at the centre of the member. The gap (g) is defined in order to realise the required axial deformation of the member.

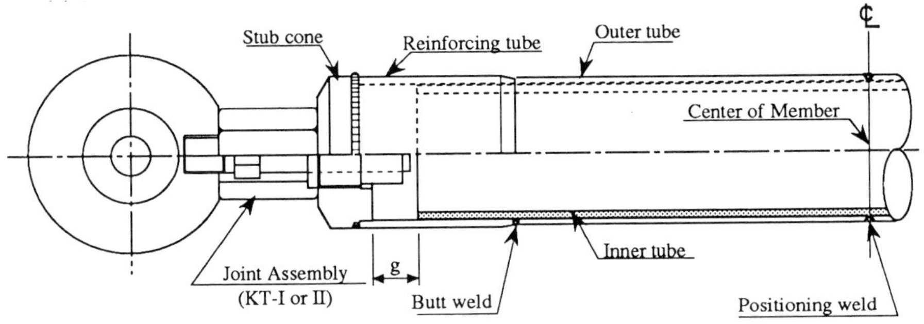

Fig. 13. Composition of KT-FLD.

21. Special features of KT-FLD. The significant feature is that KT-FLD member can endure axial force beyond the elastic limit without buckling, as a result an extraordinary large axial plastic deformation can be obtained without any decrease of axial stress.

ACKNOWLEDGEMENTS
The authors express their gratitude to Mr. F. Asayama, Mr. K. Yoshida and Mr. M. Yoshinaga of Kawatetsu Steel Products Corporation for producing the drawings for this paper.

REFERENCES

1. IMAI K., WAKIYAMA K., TSUJIOKA.S, YAMADA Y. Proposing a new joint system (KT-system) of space frame with threaded spherical nodes and it's fatigue characteristics. IASS MADRID, 1989.
2. MENGERINGHAUSEN M. Raumfachwerke aus Staben und Knoten. BAUVERLAG GMBH, Wiesbaden und Berlin, 1975.

147. The first 10 years of the CUBIC space frame

L. A. KUBIK, ASW-Cubic Structures, UK

SYNOPSIS. After briefly explaining the CUBIC Space Frame concept, this paper charts its development by describing a selection of the applications during the ten year period since it was first applied. In the process a number of special features are identified. The paper concludes with a review of some current developments.

INTRODUCTION

1. Following the first application of the CUBIC Space Frame in 1979, the innovative and versatile modular constructional system without diagonal bracing members has been used in a wide variety of building types with many different performance requirements. Its success has been due not only to its ability to economically frame buildings with widely differing spans, shapes and loadings whilst retaining maximum usable space within the structure depth, but has also been due to a continuing research and development programme linked to feedback from projects both during and after construction (refs. 1-5). A selection of these projects is described in this paper.

THE CONCEPT

2. The CUBIC Space Frame concept arose from the need to provide economical building structures which could easily accommodate services and which were also adaptable to suit changing building requirements without requiring expensive alterations (ref. 1).

3. It uses three basic module types (Fig. 1), the size of which can be chosen to suit individual project requirements. Each module consists of a vertical post with up to four chord members welded to each end. They are fully welded in controlled conditions in the fabrication shop using special jigs for setting out, and they then nest together for transport (Fig. 2). On site the modules simply bolt together to create the desired structure, using bolts in clearance holes. The location of the site joints away from the nodal joints between the posts and chords, and the absence of bracing members, means that only two members have to be connected at each point on site. It also permits the more complex nodal joints to be fully welded in controlled conditions in the fabrication shop, avoiding the need for special connectors.

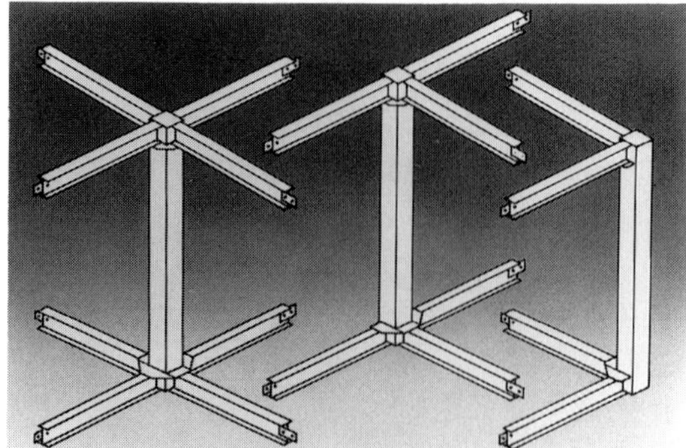

Fig. 1. CUBIC Space Frame modules.

4. All section sizes and connection details can be chosen to suit the incident member forces and moments. A variety of section types can be used for the members, although generally SHS has been used for the posts and I sections have been used for the chords. The connections are made from standard steel plates and flats.

5. Gentle cambers for roof drainage or to minimise the effects of frame deflections in service can be introduced into the structure by simply extending the length of the top chords and reducing the length of the bottom chords. This camber can be provided in one or two directions. The chords can also be angled relative to the post to achieve roof slope. All of these geometrical adjustments are accomplished during manufacture of the modules in the jigs, so that on site it is only necessary to bolt the modules together.

Fig. 2. Modules stacked for transport.

6. The absence of diagonal bracing members maximises the usable space within the structure depth, which can therefore be used not only for location of services and machinery, but also for walkways, offices or other accommodation. These features, and others, are considered further in the next section as a selection of actual projects is used to illustrate the development of the CUBIC Space Frame.

A SELECTION OF PROJECTS FROM THE FIRST 10 YEARS

Rehearsal Theatre, Nottingham

7. The first commercial application of the CUBIC Space Frame was for a roof over the Waverley Rehearsal Theatre at Trent Polytechnic (now Nottingham Trent University). The roof spanned 12m x 20m between load bearing walls, was 1m deep and had a grid size of 2m in each direction. In addition to supporting woodwool slabs and weatherproofing on the top chords, the structure was also required to support variable service and equipment loads.

8. Of great importance to the selection of the CUBIC Space Frame for this project was the difficult access to the site. The relatively small size and stable structural form of the modules meant that they could be transported through the corridors of the existing building and erected one by one mounted on temporary scaffolding. This method of erection enabled the roof to be constructed quickly and efficiently within a weather protecting envelope. The resulting benefits and cost savings offset the cost of the load testing, which was required to verify the design assumptions. Fig. 3 shows the frame being set up for load testing.

Fig. 3. Waverley roof prior to load test.

9. At the time of this application little information was available regarding the effect that the decision to use clearance holes at the splice joints would have on frame camber and deflection. Therefore, it was decided to aim to provide a flat roof under dead loading, with the required slope for roof drainage being achieved by variable depth timber firrings fixed to the top chords. The feedback from this project enabled cambers to be used with confidence in subsequent projects.

East Midlands Airport

10. The first use of camber for drainage was for the roof over a baggage handling bay at East Midlands Airport. A one way camber was provided across the short span of this 28.8m x 14.4m structure, and the roof cladding was fixed directly to the top chords. Gutters were accommodated within the structure depth by providing a step in the top chord as can be seen in Fig. 4 , the effect of which was to actually reduce the chord moment at the face of the perimeter posts.

11. During erection it was necessary to minimise the disruption to the operation of the baggage handling area. This meant that, in contrast to the module by module assembly in the previous example, the roof structure was assembled in three panels on the ground adjacent to the baggage handling area. These panels were then quickly lifted into place at times to suit the client and connected to the irregularly spaced supporting columns (Fig. 4).

Fig. 4. Large panel erection.

Food processing factories, Nottingham

12. The application of the CUBIC Space Frame to the roof structure of two food processing factories in Nottingham clearly demonstrated the benefits of the usable space within the structure depth.

Fig. 5. Plant installation in roof.

13. The structural form was originally chosen to minimise the number of columns due to the poor ground conditions on site, whilst maximising the available space for service installation within the structure depth. By increasing the structure depth slightly it was found that ground floor area could be saved by installing the main plant, services and walkways within the roof structure (Fig. 5). Not only did this reduce the plan area of the building, it also permitted the

building services to be more efficiently planned, maintained and modified in a completely separate compartment to the ground floor production area. The inherent bending strength of the space frame chords provided considerable freedom in locating even the heavier items of plant, which could be supported at various positions along the chords and not just at the nodes.

Multi-purpose theatre, Skegness

14. The roof structure over the octagonal theatre shown in Fig. 6 represents the first non-rectangular application of the CUBIC Space Frame. It was also the first use of significant two-way camber, with a requirement to achieve a minimum residual camber for drainage of 250mm over the 34m span. This was achieved using the simple geometric adjustments referred to in paragraph 5 in both grid directions.

15. The 1.5m deep roof structure is exposed to view inside the theatre and accommodates large ventilation ducts. It was later found to be feasible to use the load distributing properties of the space frame together with accurate determination of service load over the plan area to permit several tonnes of additional load to be suspended from the structure over the stage during concerts, despite this not having been allowed for in the original design brief.

Fig. 6. Octagonal theatre roof.

Demountable standard structure

16. The requirement for the demountable standard structure was to have an easily transportable temporary roof structure, adaptable to suit variable spans and loading conditions. It also needed to be amenable to rapid assembly and erection without craneage on a variety of sites, typically for pop concerts and music festivals.

17. The structure uses 1m deep modules with a 1.5m grid size. The modules are fully interchangeable and can be easily manhandled and assembled on a temporary stage to form a variety of roof shapes based on the 1.5m module. Once assembled the roof structure can be lifted into position using simple hoists mounted on scaffold towers which form the side of the stage (see Fig. 7). The structure can be

covered with a tarpaulin and offers many convenient points from which to hang lights or special effects above the stage.

Fig. 7. Demountable standard roof.

Stansted hangar

18. The roof over the award winning diamond shaped FFV Aerotech hangar at Stansted Airport is the largest application of the CUBIC Space Frame to date. Spanning 98m x 170m the roof is only 4m deep, significantly shallower than the alternative latticed girder solution (ref. 6). It was assembled from 1201 modules, each weighing between 0·5 and 1 tonne, and was fully supported on columns along two sides and on approximately 6m deep lattice girders, along the other two sides. The roof was cambered in two directions using a standard adjustment to the geometry of each interior module. Very close correlation was recorded between the predicted cambered shape and that measured towards the end of construction.

19. The use of the deeper latticed girders over the doors provided increased stiffness to limit deflections to the levels specified by the door supplier, whereas the shallower structure depth within the hangar provided the greater clearance required for maintenance of the aircraft. Fascia frames were bolted to the perimeter modules to create the desired architectural appearance and accommodate the sizeable gutters for roof drainage.

Fig. 8. Stansted hangar erection.

20. Erection commenced with the construction of a central bridge of modules across the 98m span, after which modules were added line by line until the structure had been completed (Fig. 8). In this way erection was able to proceed simultaneously on two faces.

Superstores

21. The benefits of reduced numbers of columns, together with flexibility concerning their position have been beneficial in the application of the CUBIC superstores (Fig. 9), as has the ability to provide and support a variety of perimeter details for both wall and roof, and the good accessibility for services within the structure depth.

Fig. 9. Superstore application.

Birmingham International Convention Centre

22. The final project to be described is the 51m span octagonal roof structure over Hall 3 at the Birmingham International Convention Centre (Fig. 10). In form, the structure was similar to the multipurpose theatre mentioned in paragraph 15, although both the applied loading and the span were significantly larger, the former to allow for walkways within the roof structure and a 30 tonne, exhibition load which could be applied to a small area anywhere over the span. In addition there were linking structures which applied large loads around the perimeter.

Fig. 10. Erection of the Birmingham roof.

23. The roof was supported on reinforced concrete columns at the 8 corners of the octagon. To avoid damaging the columns Glacier bearings were required at the interface and tight tolerances were specified for the point of load application to the column under all loading and environmental conditions. In addition, each column also had a positional tolerance. The positional tolerance of the columns was accommodated by the base plate details, and the point of load application was controlled by precise manufacture of the modules in the jigs mentioned in paragraph 3 above.

CURRENT DEVELOPMENTS

24. As can be seen above, the CUBIC Space Frame is an established and versatile framing system which offers many exciting opportunities to building designers and users. Nevertheless there are always opportunities for further improvement and development. A number of such opportunities are now being explored in conjunction with ASW-CUBIC Structures Ltd, a joint venture company between Kubik Enterprises Ltd and ASW Ltd, which was formed in 1989 to offer a comprehensive design, manufacture and erection service for the CUBIC Space Frame.

25. Two areas of particular interest are described in separate papers in this conference. The first is an investigation into the effect of geometry on the weight and cost of the CUBIC Space Frame using optimisation techniques. The ultimate aim is to establish a range of the most cost-effective standard modules which can be used for the widest range of spans and loading conditions (ref. 7). The second relates to the development of the CUBIC Composite Floor, taking advantage of the enhanced strength available by embedding the top chord of the CUBIC Space Frame in concrete (ref. 8). These investigations, and others, will no doubt open up new horizons for the CUBIC Space Frame in the years ahead.

ACKNOWLEDGEMENT

26. The author gratefully acknowledges the significant contribution which his late father, Mr M.L. Kubik, made to the advancement of the CUBIC Space Frame until his sudden death in 1991. He has left his mark on all of the structures described in this paper, and on many others as well.

REFERENCES

1. KUBIK M.L. and KUBIK L.A. An introduction to the CUBIC Space Frame. Int. J. Space Structs, 1991, vol.6, no.1, 41-46.

2. KUBIK L.A. The CUBIC Space Frame in theory and in practice. Steel Construction Today, 1991, vol.5, no.2, March, 59-63.

3. CHILTON J.C. An investigation of the elastic behaviour of the CUBIC Space Frame. PhD thesis, Trent Polytechnic (now Nottingham Trent University), UK, 1988.

4. KUBIK L.A. and McCONNEL R.E. The behaviour of CUBIC Space Frame joints. Proceedings of 3rd International Symposium on Tubular Structures, Lappenranta, Finland, Elsevier Applied Science, London, 1989, 410-417.

5. BUKY C. An elastic analysis of CUBIC Space Frames. The Structural Engineer, 1980, vol.58A, August, 249-250.
6. KUBIK L.A. and CHILTON J.C. Design and construction of the CUBIC space frame roof, maintenance hangar, stansted airport. Paper presented at the International Symposium on Space Structures, Copenhagen, Denmark, 1991.
7. SHANMUGANATHAN S. and KUBIK L.A. Optimisation of CUBIC Space Frame structures. Paper to be presented at the 4th International Conference on Space Structures, University of Surrey, UK, September 1993.
8. THOMPSON K.W. and KUBIK L.A. Development and testing of the CUBIC Composite Floor. Paper to be presented at the 4th International Conference on Space Structures, University of Surrey, UK, September 1993.

148. Space deck barrel vault to the Bentall Centre, Kingston upon Thames

S. C. BAIRD, N. L. JAMES and J. E. SHOTTON, Space Decks Ltd, UK

SYNOPSIS. This paper will discuss the analysis, design and construction of the 120m long glazed, barrel vaulted atrium to The Bentall Centre, Kingston Upon Thames. The concept for the steel space frame structure was based on the well proven Space Deck system of welded pyramidal units.

CLIENT BRIEF

1. In March 1991 Mowlem Management awarded a contract for the design and construction of the space frame and associated glazing on the Bentall Centre to Space Decks Limited.

2. The clients architect and consulting engineers, Building Design Partnership, had visited the U.S.A. to examine similar buildings and to assist in defining their design requirements for the barrel vault roof. Fig. 1 shows the comparative sections of various famous buildings which assisted in illustrating the problems of both environmental design and construction.

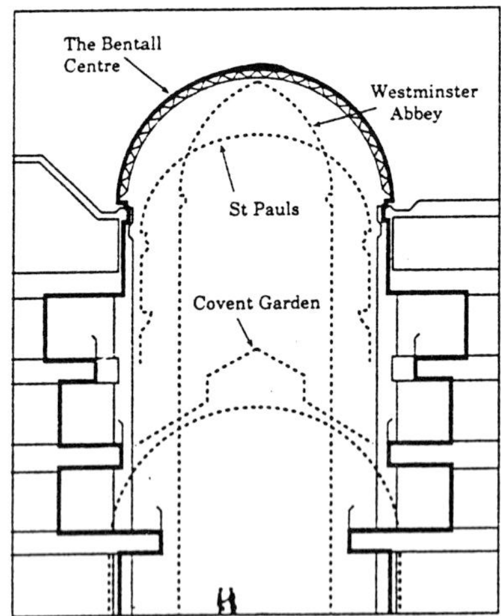

Fig. 1. Comparative sections.

Their basic requirements were:

(i) To construct an economical but reliable structure and glazing system.

(ii) To allow real, not reflected or filtered sunlight into the atrium. This was not to be reduced by tinted glazing.

(iii) To create an environment after dusk as pleasant as that during daylight hours.

The final design took several months to complete but these three basic requirements were behind all advances in the design development.

FRAME GEOMETRY

3. The large 15m semi-circular arch was formed using slightly rectangular modules 937.5mm wide along the length of the roof and 1010mm long around the curve of the arch, with an overall depth of 495mm. Twenty four modules around the curve of the arch meant that each modular unit was translated by 7.5 degrees from its neighbours. To achieve this two sides of the unit had their top chord angles specially manufactured to have the downstand leg set by 3.75 degrees. In order to accommodate a central hole of 2mm diameter for the installation of the fibre optic light, the apex node of the pyramidal unit had to be redesigned. Bottom or inner chord tie bars are angled at 3.7 degrees to the axis of the apex boss, this also affected the redesign of the node.

4. A flat disc 120mm diameter and 50mm thick was eventually decided upon.

5. The 10m wide arch again used slightly rectangular modules for the curved part of the arch. Maintaining the sight lines along the whole of the length of the atrium meant that smaller module sizes were required. Using the same number of modular units to form the arch meant that the angled sides of the units were set to the same 3.75 degrees.

6. Modular sizes around the curve were 698mm and along the roof 625mm. The vertical skirt to each side of the curve used modular units 625mm square. Depth of construction was maintained at 495mm and a similar apex boss used.

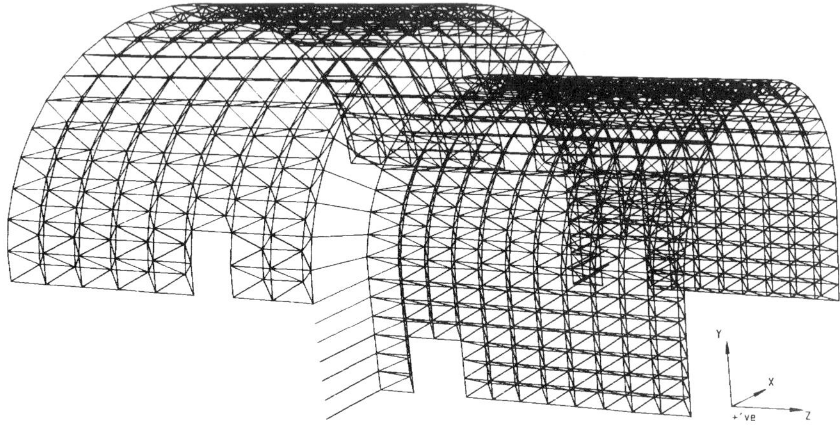

Fig. 2. Junction of 15m arch and 10m arch.

7. At the end of the 13m arch a decorative apse was installed. This took the form of one quarter of a sphere, 10m in diameter using a single plane of 33.7mm diameter CHS members interconnected by disc bosses similar to the main arches (Fig. 3). Decorative panels were fixed to the underside of this apse.

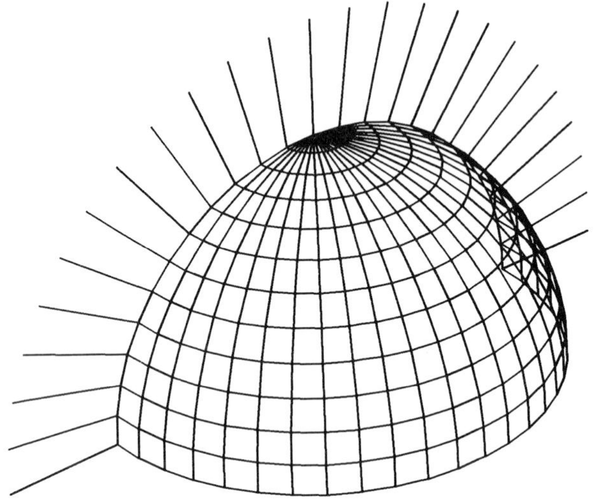

Fig. 3. Apse.

ANALYSIS

8. The springing points of the arch provided very stiff vertical and horizontal support. Consideration had to be given to an expansion joint in the main structure which traversed the 15m vault at approximately halfway along its length and an expansion joint which ran parallel to the longitudinal axis of the atrium, severing the continuity of the main structure between the springing points of the arch. Expansion joints across the arch were simply catered for in the normal manner by leaving a gap between the panels of the arched structure. However, the longitudinal expansion joint presented the problem of how to deal with differential movement between the two main blocks of shops each side of the arch. This could be accommodated by pinning the arches on one side and allowing sliding bearings at the other side, pinning both springing points of the arch and catering for stresses induced in the frame by movements of supports, or by using a three-pinned arch.

9. Initial computer analyses were carried out and the three deflected shapes drawn. It was decided at this stage to adopt the three-pinned arch approach which generated lesser deflections between the crown and springing point pins and hence reduced the effects on the roof glazing and internal coffer panels.

10. There were four main computer models established to determine the member forces and deflected form of the barrel vault. The models produced were,

the apse, the 15m barrel vault panel supporting the apse, an internal panel of the 15m vault and the junction of the two vaults. Openings in the sides of the vaults at the springings to allow for doorways had to be allowed for in this latter model. Doorways were necessary to gain access to both vaults from the surrounding flat roof to the internal maintenance gantry system (Fig. 4).

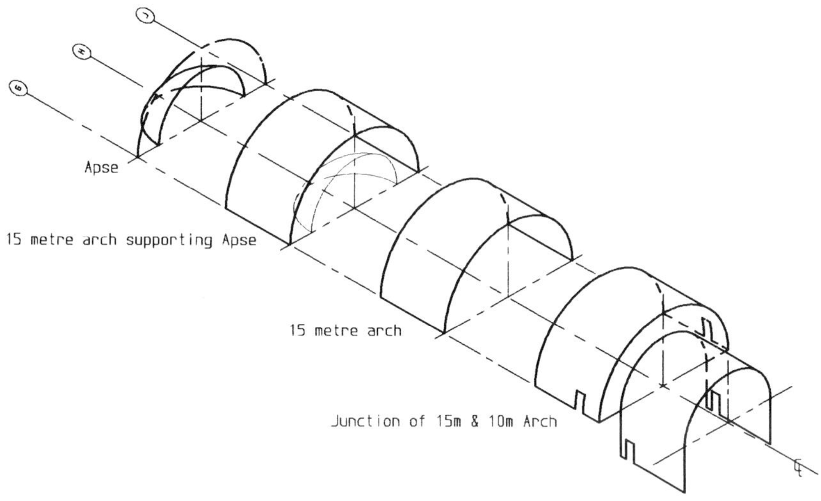

Fig. 4. Computer models.

11. The model was subjected to 12 primary load cases, 5 serviceability limit state load combinations and 5 ultimate limit state load combinations. Design loadings for the atrium were determined from BS 6399; Part 3; 1988 - Loading for buildings. CP3; Chapter V; Part 2: 1977 - Code of basic data for the design of buildings. Snow drifting effects at the junction between the 10m and 15m wide vaults were fully considered. Wind loading distribution was agreed in consultation with the Building Design Partnership as there is no obvious standard case in the code of practice. Dead load of the glazing system amounted to 0.3kN/m^2 for the superficial area. An allowance of 0.25kN/m^2 was specified for the perforated coffer panels, node fixings, fibre optics and other services, this also being applied to the superficial area. Self weight of the space frame was greater than normally expected for a space deck structure due to the small modular grid. The 15m wide vault amounted to 0.3N/m^2 and the 10m wide vault 0.56kN/m^2.

Fig. 5. Internal view of 15m barrel vault looking towards the apse.

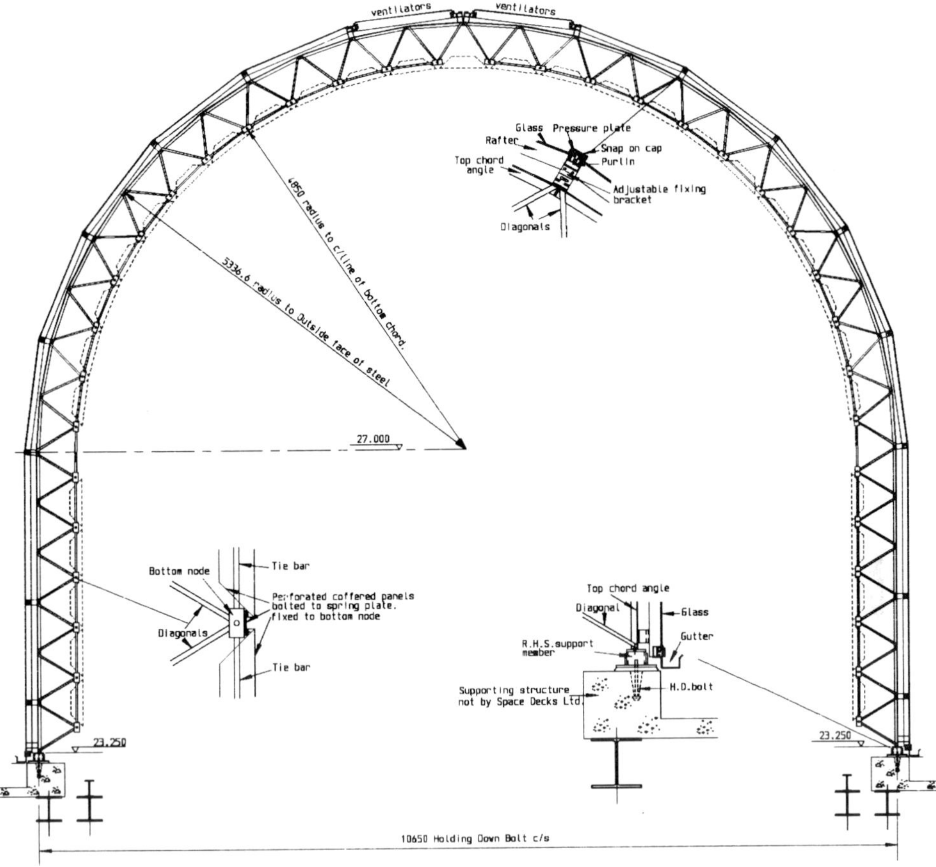

Fig. 6. Cross-section through the 10m vault.

TEMPORARY WORKS

12. To facilitate the erection of the space frame arches which are 30m above the ground level, it was decided to build a flat deck just below the springing of the arched roof. This deck was constructed from standard 1212 Space Deck units spanning the width of the opening. Metal decking and 20mm thick plywood formed the working deck. This not only facilitated the erection of the arched space frame and glazing, but also allowed work on the shops below to continue unhindered. Following trades fixing the coffered ceiling panels and fibre optic lights were also able to take advantage of this working platform running the full length of the atrium. The Space Deck platform proved to be a cheaper solution than tradition scaffolding.

13. After completion of the atrium, the temporary working platform was split into sections and lowered to the ground using the atrium space frame, and removed from site.

ERECTION

14. The maximum size of assembled space frame was dictated by the available tower cranes capacity and positions. In general assembled panels of space frame were half an arch from springing to crown by 7.00m long. See Fig. 7.

<u>Stage 1</u>

Assembly of the two halves of the three-pinned arch in 7.5m long panels on the space deck temporary platform.

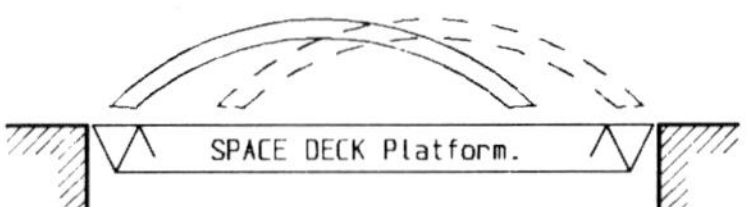

<u>Stage 2</u>

Lifting one half of the arch frame into position. The first half of the arch was held in position by the tower crane pending Stage 3, except for part of the 10m arch which was propped at the free edge of the panel.

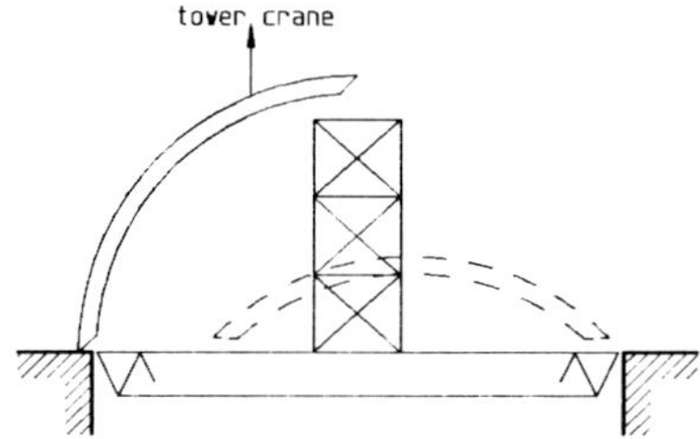

<u>Stage 3</u>

Lift second half of the arch frame into position, make ridge connection and connect to adjacent bay.

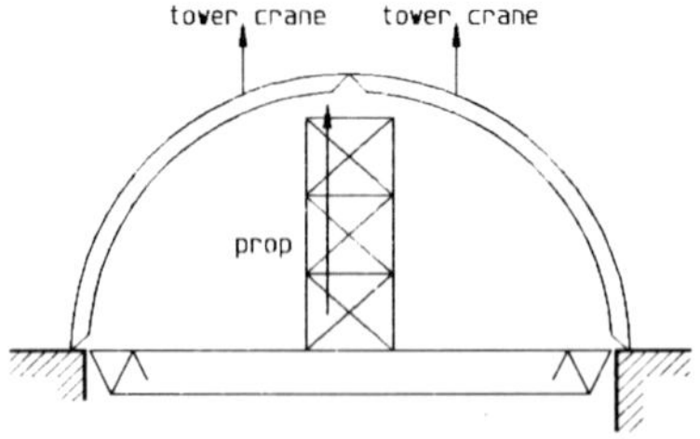

Fig. 7.

Fig. 8. 15m Vault complete.

In the foreground, a panel of the 10m vault is laid out ready for assembly on top of the Space Deck temporary working platform. Note the Internal Cleaning gantry was installed prior to the erection of the barrel vaults.

149. PolyFrame: the system solution space structures

D. L. RICHTER, Temcor, USA

SYNOPSIS. Richter PolyFrame is a system of construction, invented by the author, that structurally integrates a space truss with its closure panels into a totally interactive system for optimum efficiency. Originally developed at Temcor for aluminium domes, PolyFrame is now being expanded to include other three dimensional structural forms. The various design and engineering aspects of the system are reviewed in this paper.

BACKGROUND

1. This paper will not review the basic research preceding the invention of PolyFrame. My earlier work, beginning in 1949 has been well covered in papers presented at previous International Conferences on Space Structures and reviewed in Professor Makowski's excellent book on Braced Domes.

2. Suffice to say that the goal of this early research work was the development of high strength surfaces of convolution. Perhaps the best example of the culmination of this work is the Temcor Convention Centre dome shown in Fig. 1, erected in Denver, Colorado in 1973.

Fig. 1.

It is a frameless dome that spans 24m with a rise of 8m and is made entirely of formed 2mm thick aluminium sheet diamond panels. The dome only weighs 7kg/m^2 of surface or about 6% of the design live load.

3. The famous Kaiser Aluminium / Temcor Geodesic Dome erected in Hawaii in 1957 used a similar diamond panel configuration with one strut added to the exterior of each panel to form a 0.3m deep space truss. This dome design has a practical span limit of about 70m.

4. Fig. 2 is of a typical 70m Temcor Geodesic dome erected in Beaver, PA. Larger domes of this configuration are not practical since further increases in the diamond panel thickness for additional strength would be too costly.

Fig. 2.

POLYFRAME DOMES

5. Temcor's all aluminium, PolyFrame dome was invented to fill the need for circular domes up to 130m in diameter. Our 126m dome in Long Beach, California, designed to cover Howard Hughes "Spruce Goose" aeroplane, is close to the economic size limit for unstiffened PolyFrame domes. The completed dome only weighs 12kg/m^2 of surface area. Since the cost of the aluminium can represent 30% to 40% of the total cost, the unit weight of an erected dome enclosure is a good measure of its structural efficiency. Fig. 3 shows the Spruce Goose aircraft being removed through a temporary opening made in the dome in 1992. The seaplane was subsequently shipped up the Pacific coast on a barge to Oregon for display.

6. One of the unique feature of this kind of dome construction is the ease with which temporary openings can be made in the structure. In fact, the dome could be totally disassembled and reassembled on another site in the future.

7. Unquestionably the most important engineering aspect of Temcor Geodesic and PolyFrame domes is the system integration used in their design.

8. The more conventional engineering approach in space structure design is to build a self-supporting structural frame, to which a grid of beams or purlins are

added that span between nodes. Roof panels or sheeting are then applied over the beams followed by a weather resistant roofing material.

Fig. 3.

POLYFRAME PANELS

9. While it may not be readily apparent, the aluminium panels that enclose the fully triangulated geometry of Temcor PolyFrame domes are a vital and necessary component of the structure. They do far more than just close out the weather, they increase the dome strength by up to 600% by stabilising the struts about their minor axis. Since the panels are thus used to strengthen the dome, all panel discontinuities, such as at the dome base edge and around all openings through the dome surface must be locally reinforced.

10. In addition to restraining the struts, these 1.3mm thick triangular closure panels are designed to carry snow loads in excess of 800kg/m^2. This minimum panel design load was substantially exceeded by deep snow drifts on our 50m diameter PolyFrame dome erected 13 years ago for the Scientific Station at the South Pole.

11. This proven panel strength is an important design feature because a single panel thickness and closure detail can satisfy virtually all of the structural and weather tightness requirements, regardless of the size of the dome.

12. The thin aluminium panels function like tension membranes to span the four to five square meter triangular openings in the frame, therefore they must be fully supported on all three sides and continuously secured to the surrounding dome struts. To meet this requirement, Temcor PolyFrame domes use the patented interlocking batten system shown in Fig. 4. This batten joint design is strong enough to fully develop the yield strength of the aluminium alloy sheet, 3003-H16.

13. The Temcor interlocking batten joint has the required strength to make a water tight seal without slippage or permanent distortion. This kind of strength is not achievable with the use of sheet metal screws through the thin aluminium panel and struts, or with battens that rely on friction clamping to hold the panel edges on the surrounding struts. If the panels are not pulled taut and clamped securely across the triangular openings the struts will not be properly stabilised

and the light weight panels could flutter due to wind forces and cause metal fatigue.

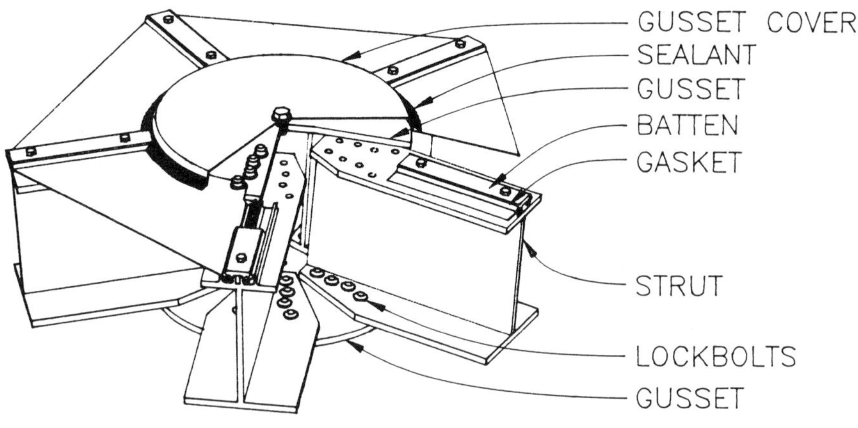

Fig. 4.

14. To function properly as a PolyFrame, the manufacturing tolerances must be tightly controlled for proper fit and a single, corrosion resistant, material such as aluminium must also be used to avoid differential expansion and joint slipping.

15. Many different physical tests were conducted to verify the Temcor edge joint and panel strength and its effect on the overall structure. Pressure loads in excess of 2000kg/m^2, or 2.5 times the minimum panel design load were applied in our most recent tests without a failure or critical distortion of the panels. That test pressure loading is equal to two meters of water or about fifteen meters of deep snow drifts.

Fig. 5.

16. To safely test our panels with sufficiently large loads required the construction of a special vacuum chamber. Pressure forces were applied on the panel by removing air from the clear space between the test specimen and the floor of the chamber. Nothing special was done to the dome panels or joints to make them air tight for our the tests, in fact, PolyFrame domes are frequently used with mild internal pressures in industrial applications.

17. Since a vacuum pressure loading does not make an impressive photograph, 15 of our engineers stood on the panel for the photo snap shot in Fig. 5. If our engineers were a bit more athletic, we could have added six more layers of 15 people standing on each others shoulders without causing a failure of the aluminium panel.

18. The construction of thousands of Temcor domes throughout the world without a failure, over the past 25 years stand witness to the viability of the PolyFrame system.

PANEL TENSION

19. These recent panel tests were conducted primarily for the purpose of refining the formulas used in our structural analysis programs that calculate the interaction between the panels and the dome frame. The factors affecting the membrane tension along the strut edge are the panel size, proportions, initial tautness and thickness, as well as the edge restraint, strut stiffness and the applied load. A reasonably good estimate of the maximum edge tension for 1.3mm thick aluminium, equilateral triangle panel can be found with the following formula:

$$P = 0.21\, t\ \sqrt[3]{E.(W.L/t)^2} \qquad (1)$$

Where P is the maximum panel tension, per unit length, near the mid-span of the strut. L is the length of a strut framing the panel. W is the design load, or perpendicular force per unit panel area and t is the panel thickness.

INTEGRATED DESIGN

20. An essential advantage of panel restraint along opposite sides of each strut is the lateral column stability it provides. This feature allowed us to design the dome struts and frame joints for maximum beam strength and dome snap-through stiffness. Full lateral buckling support of the strut is provided even when the panel sizes or loads on opposite sides of a strut are unequal. The opposing membrane forces are automatically balanced by slight lateral, but restrained deflections of the strut. While the tension forces establish their own equilibrium across the struts, the membrane tension must also be counterbalanced by axial compression within the strut.

21. The strut compression, induced by panel tension, must be added to the primary strut compression in PolyFrame structures. It can increase the primary strut compression by as much as 30% in small radius domes under heavy snow loads.

DOME SNAP-THROUGH

22. The aluminium alloy 6061-T6 used in the dome frame has about the same yield strength as common steel, but its Modulus of Elasticity is only one third that of steel. Therefore, all aspects of elastic instability are especially important when designing aluminium structures.

23. My paper presented at the Third International Conference on Space Structures included results of our snap-through tests on a very shallow, aluminium framed dome. The Geodesic test dome was 12.2m diameter and included 37 internal nodes and 24 node, support points, equally spaced around a steel base ring.

24. A total of 32 tests were conducted with various symmetric, non-symmetric and concentrated loadings. With the exception of the concentrated load test in which fasteners sheared, all of the snap through failures were elastic in behaviour with 100% recovery. The non-linear computer programs available today allow us to analyse the snap-through strength of that test structure with good and conservative results.

25. Larger and more complex structures like the aluminium PolyFrame dome required further refinement of the analytical process. It is extremely important that we recognise that the panel tension and joint stiffness must both be considered in any proper analysis of the dome structure snap-through because the panel tension amplifies the strut compression and the joint stiffness affects the dome stability.

JOINT STIFFNESS

26. Shallow domes with a single layer of struts joined with single gussets or small spherical fittings at each node are quite prone to snap-through failures. To avoid this weakness, PolyFrame domes are made with thick top and bottom gussets attached to the strut flanges with tight fitting lock-bolts. The degree to which the structure behaves as a single- or double-layer grid dome is determined by the joint stiffness.

27. PolyFrame joint stiffness is controlled by the gusset alloy, contour, diameter and thickness, as well as the quantity and tightness of the fasteners and most importantly, the tolerance and quality of the bolt holes. Unlike steel construction, the use of friction connection with high strength bolts are not approved for use in aluminium structures. The joints in PolyFrame domes can develop up to the full strength and stiffness of the dome struts while a poorly made joint will develop only a small fraction of the strut stiffness. The importance of joint stiffness cannot be over emphasised in dome design.

28. Fig. 6 is a photo of one of the 140 tests of different sizes and kinds of strut and gusset tests that have been conducted by Temcor to determine the proper stiffness to be used in our analysis.

29. Regardless of the procedure used to analyse the dome, we would recommend a minimum Safety Factor of 1.95 against snap-through failures in any occupied dome building. The Safety Factor might be reduced to 1.65 for any unoccupied industrial uses of the structure.

Fig. 6.

30. The 80m PolyFrame dome, shown in Fig. 7, was engineered, manufactured and erected by Temcor for the University of Alabama. It was designed to support up to 190,000kg of speakers, catwalks, light grids and curtains suspended from the dome, in addition to the normal live loads. These requirements are typical for College and University general activity buildings today.

Fig. 7.

POLYFRAME SPACE TRUSS

31. The stability of the PolyFrame structure can be improved by increasing the strut depth, or by integrating dome frame with a space truss. The potential configurations for the space truss that can be used are inherently limited by the PolyFrame design that includes inner and outer gussets.

32. Fig. 8, illustrates the ideal space truss configuration for PolyFrame because the truss lacing lie in the planes of the strut webs and the inner chords are only half the length of the outer dome struts, for improved column stability. Certain features of this system are now being patented.

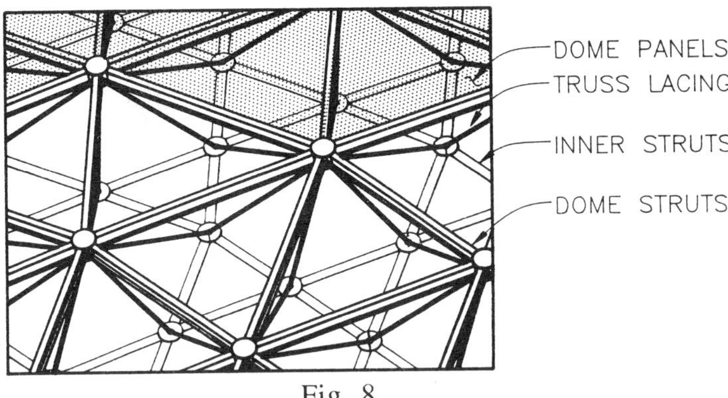

Fig. 8.

33.This new space truss is being incorporated in the top half of a 98m dome being erected for the University of Hawaii to stiffen the dome for the suspended speakers, lights and catwalks.

34. The project was competitively bid by international teams of "design build" Architects and General Contractors. The team that used the Temcor dome in their design were successful against all competition including those that proposed buildings using conventional steel structures.

THE RICHTER DOME

35. Recent developments in the Richter Dome system now make it possible to effectively build an aluminium dome over 300m in diameter. Improvements in the design and construction procedures make such super large, rigid domes practical.

36. Fig. 9 illustrates the general configuration of this new system. It has primary grid of 20m to 23m long trusses that define the primary geodesic dome shape. The large triangular openings between the primary trusses are closed in with PolyFrame subassemblies, made in the form of shallow "bubbles" for added strength.

37. In as much as the radii of the bubbles would be smaller than the radius of the primary Geodesic truss dome, the strut compression within the bubbles would be less than in the surrounding trusses. Under a uniform load, the trusses would have the greatest compression strain or contraction, thus causing the PolyFrame bubbles with a lower stress level to rise up relative to the surrounding trusses. By rising up against the load, they remain resistant to local snap-through.

38. The height of the bubbles in the drawing shown in Fig. 9 are greatly exaggerated for clarity. Very little change in the bubble curvature would be required to make them resistant to snap-through, so they will still contribute significantly to the overall strength of the dome as a shell structure.

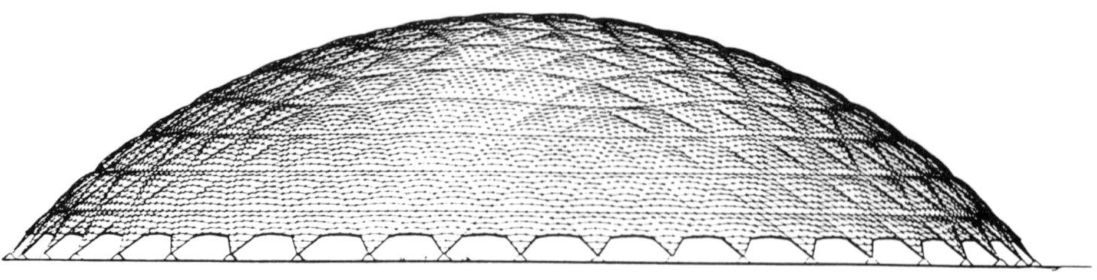

Fig. 9.

39. Proposals have been made for a Richter Dome stadium in Florida and for a 220m diameter Sports Dome for Basildon, UK. The 220m diameter dome proposed for the Basildon Arena had a projected total weight of 25kg/m^2 of surface area including the insulated panels.

40. In addition to the structural efficiency of the system, there is a significant advantage in the assembly procedures in that it allows these super large domes to be erected without expensive shoring. The primary trusses would be erected from the base up, with each successive row of trusses cantilevered up and out, like crane booms extending up from the lower, completed portion of the dome. The dome enclosure could then be completed by placing the large sub-assemblies of panelled, sealed and insulated PolyFrame bubbles over the primary trusses.

41. These are only a few examples of the new kind of structures being designed, developed and manufactured in aluminium by Temcor. There is still a great deal more that can be achieved by using the systems approach in the design of lightweight space structures.

150. MERO glass envelopes

H. EBERLEIN and W. KEMMER, MERO-Raumstruktur, Germany

SYNOPSIS. One of the most admired glazed structures using prefabricated steel elements was the glass palace built for the World Fair in 1851 in London by Joseph Paxton. This building has influenced the development of steel building systems even in the 20th century. Based on the early works of Max Mengeringhausen from 1940 the design of space frames became extremely popular in the second part of this century. Although space frames are appropriate for glazed structures by its lightness, it took three more decades before the technology for space frames and glazing became advanced enough to be successfully combined in skylight structures.

SPACE FRAMES

1. The growth of the space frame market has been accompanied by continuous research into geometrical configurations and nodal connections.

2. Both areas of development have created a tremendous amount of ideas and, last but not least, paper. But two simple concepts have proven to be still valid and are rising like a mountain over the dust; i.e.

- to use the topological and metrical relations of regular polyhedrons for the generation of space grids and
- to use spherical nodes to join members in space by single bolt connectors.

3. MER0 has used these concepts not only for the design of conventional envelopes for a variety of purposes, as e.g. hangers, sport halls etc., but also for skylight structures, as e.g. the greenhouse in Düsseldorf, using a geodetic single-layer configuration, and the roof of the Stadium in Split/Yugoslavia.

4. The topological layout of the latter was a square double-layer grid which was then transformed onto a cylindrical surface and cut by a series of ellipses. The magnitude of different structural parts could be managed by means of computer programs and automated fabrication.

5. Both structures have been covered by acrylic skylights and barrel vaults, for the greenhouse and stadium respectively.

Fig. 1. Greenhouse pyramid.

SKYLIGHTS

6. The architectural trends in the 80's created a growing market for skylight structures. The logical consequence was the search for lighter structures and new glazing techniques.

7. MERO's development was focused on the following items:

- decrease of **grid density**;
- **integration** of glazing and structure;
- **dry sealing** system.

All items were pursued and solved simultaneously by approximately 1985.

8. The decrease of **grid density** has been achieved by an extension of the traditional concepts. Spar grids as the square on larger diagonal square configuration have been used for a variety of pyramidal skylights, as e.g. for a greenhouse complex at the Gruga Park in Essen (Fig. l). The basic square grid was transformed to a diamond shape to fit into the triangular faces of the pyramid.

9. A square layout of the triangular faces was chosen for the skylight pyramid covering a hospital atrium in Bad Neustadt, Lower Franconia, Germany. To decrease the density of the space frame even more, secondary members in the upper layer have been applied to support the glass panels.

10. The **integration** of glass and structure was achieved by the development of the MER0-PLUS system. The system provides four different node types (NK, ZK, TK, BK).

11. The nodes are produced by means of NC-machinery with extremely narrow tolerances, providing an overall structural accuracy to allow the support of EPDM-gutter and glass panels directly on top of the structure. This technology avoids the often unattractive or at least uneconomical mullions, especially if the structures are double-curved, as e.g. skylight domes.

12. The skylight dome for another hospital in Bad Neustadt was one of the first attractive applications of the new technology by MER0. The system is a single-layer geodesic space truss and consequently uses a disk-type node (TK) together with insulated glazing.

13. Single-layer frames, such as the 60m skylight dome for the Mirage Hotel in Las Vegas, U.S.A. have to resist bending and therefore need stiff connections provided by cylindrical nodes (ZK) and two bolt connections.

14. A block type node (BK) is appropriate to solve the complex geometrical problems that arises at the corner points of a polyhedron as it appeared with another atrium skylight in Bad Neustadt (Fig. 2). To support the plane faces of the structure, single members have been used which are connected to the square members in the faces by means of bowl-type nodes (NK).

15. The MERO-PLUS technology provides advantages for glazed structures, high accuracy by NC fabrication, high stiffness by double-layer and/or double-curvature layout and increased safety by structural redundancy together with infinite variability of shapes by the modular concept.

16. A **dry sealing** system appeared to be one of the essential prepositions for a safe application of MERO-PLUS systems in an international scale under all climatic and weather conditions.

Fig. 2. Polyhedrical glass atrium.

Fig. 3. Montreal museum garden roof with hold-down cables.

Fig. 4. Garden roof under construction.

Fig. 5. Suspended hold-down of cables.

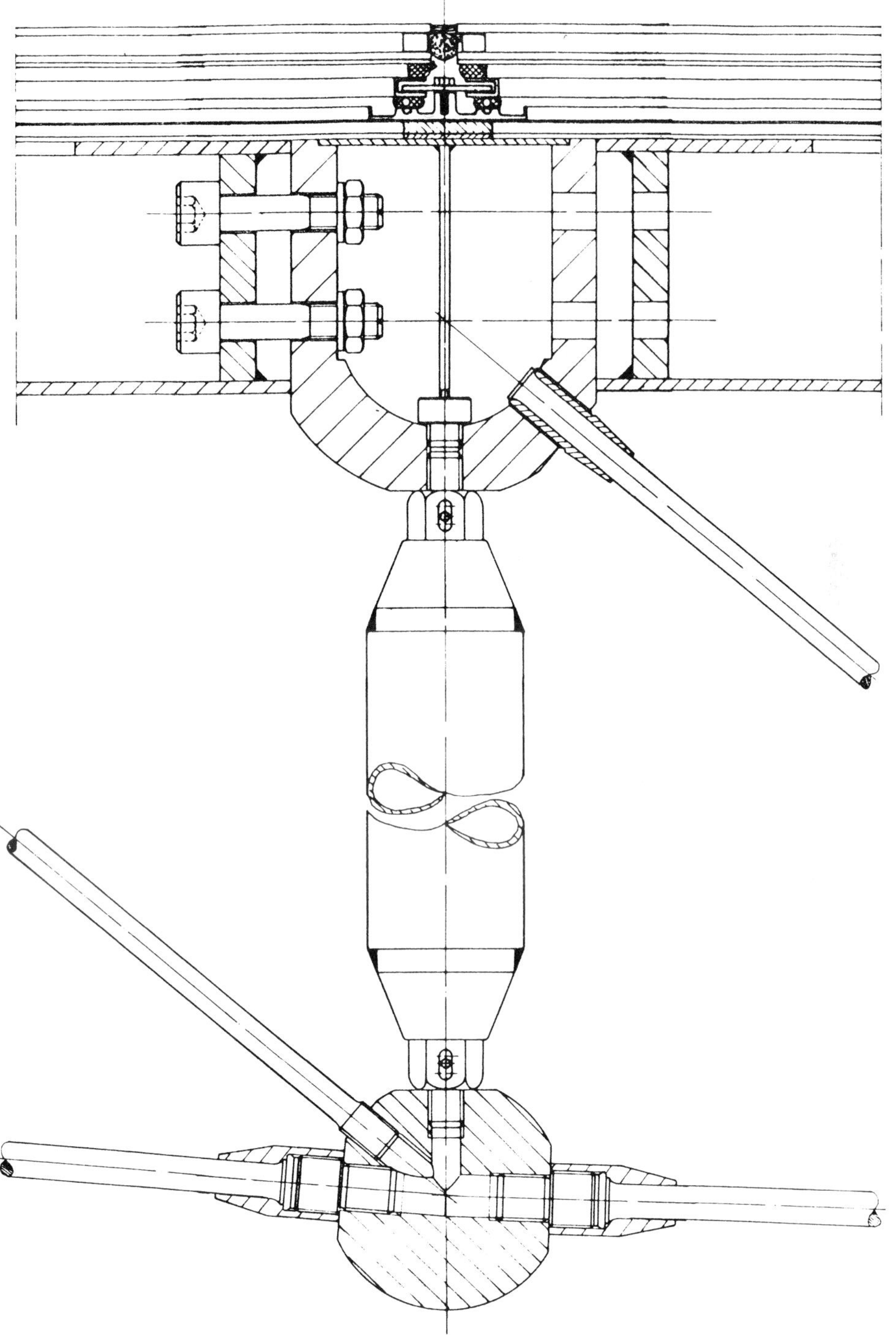

Fig. 6. Structural layout of top and bottom chord.

17. The VARIO-CLIP patented system provides all advantages necessary to meet functional and aesthetical demands as well as easy design and handling. The application can be extended to seal sandwich panel joints also in combination with glass, as is illustrated by an atrium facade and roof in Antalya, Turkey.

RECENT DEVELOPMENTS

18. When Fuller conceived the idea of tensegrity, he could not know that it would influence the design of lightweight structures to the extent as it does today. I.M. Pei's Louvre Glass Pyramid and Peter Rice's Parc de la Vilette Science Museum structural glass atriums represent "integrity by tension".

19. In 1989, when Moshe Safdi searched for an appropriate structural layout for the glazed lobby and garden roofs of the Montreal Museum of Fine Arts, MERO offered him a structural design, based on the tensegrity principle.

20. The structure comprises an orthogonal truss system, providing direct support for the structural glazing by a RHS top chord and CHS verticals. All other members are high tensile stainless steel rods. Connectors are bowls (NK) in the top chord and spheres (KK) in the bottom chord. To hold the structure down and prestress the tension rods, cables have been used (Fig. 3).

21. The roofs are sloped plates of dimensions 17.5 x 23m and 13 x 33m for the lobby and garden roof respectively. The top chord sparse layout (Fig. 4) includes secondary members. They are made of C-sections to be closed by clamped metal sheets to cover the sprinkler equipment.

22. Only three pairs of reversed stay cables are applied to secure the lobby roof against wind uplift. This implies some bending in the lateral top chord members. The connections to the bowl nodes in the top chord have been designed appropriately with two high tensile bolts. They are prestressed to avoid gaps in the joints and subsequent deflections of the structure (Fig. 6).

23. The cables are supported by simple rollers that are fixed to the vertical compression members or hung from the bottom chord nodes (Fig. 5). Only the relatively big bends at the upper cable supports in the garden roof required more elaborate saddle type bearings to provide the necessary radii for the cables.

24. The components of this structure represent half a century of MERO development.

QUO VADIS?

25. The design of lightweight glass structures may not have reached the climax with the use of structural glass, although it is still the next step for MERO. The first attempts have been made in Germany by the IKG (Institut für Konstruktion Glasbau) to substitute the steel supporting structure altogether by structural glass. Obviously, some ductility has to be provided by the connectors to secure the glass against brittle failure.

26. Beside these engineering problems, the design of glazed structures is related to environmental concepts. The extended use of solar energy and the increasing need for natural, instead of mechanical, air conditioning are challenges for the future design of glazed structures.

151. Development and projects of the NS space truss system

M. IWATA and K. KAMIYAMA, Nippon Steel Corporation, Japan

SYNOPSIS. This paper describes the NS Space Truss System focusing on development of an ideal systemized space truss. Here, the development means not only creating the concept of systemized space truss, but also realization of the concept and utilization such as, standardization, how to fabricate, erect, and control quality. Project examples are described in the latter part, for making sure that the development has been done sufficiently, and the system has enough quality for actual projects.

CONCEPT OF IDEAL SYSTEM TRUSS

1. We call the systemized space truss "the system truss". The system truss should ideally have extensive applications as a large-space structural system. It needs to be capable of very large spans and various shapes. To achieve this ideal, the following should be required.
(1) A great number of members can be joined to the node at any desired angle.
(2) The structural members must be capable of being any length.
(3) The components of the system truss must be highly resistant to axial forces.
(4) These components are also required to possess a given degree of dimensional accuracy.

2. One approach to meeting these requirements is to divide a system truss into components, steel pipe members (chords and diagonals) and spherical nodes which are joined together with a single bolt (Fig. 1).

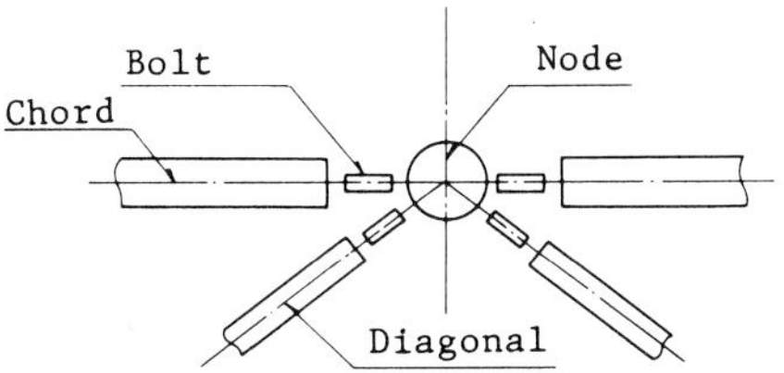

Fig. 1. Components of system truss.

The viability of this approach is supported by three facts.
(1) The steel pipe is an axially symmetrical member with no directionality in the section and having a high torsional rigidity, and is therefore suitable as an axial force member of a three-dimensional system truss.
(2) The spherical nodes, on the other hand, are symmetrical with respect to a point. This means that steel pipe members in a three-dimensional space can be jointed to the node without eccentricity as far as the axis of each pipe member is oriented to the centre of the sphere.
(3) The bolt joining the pipe members and nodes is axially symmetrical, like the pipe members.

3. Providing a system truss with an angle adjusting mechanism (Fig. 2) and thereby improving the tolerance of the length of each member is one way to attain a viable system. In fact, it is possible to obtain a system truss assembly, even without a member-length adjusting mechanism, by finishing each member to length accurate to within +1.0 mm, provided that such an angle-adjusting mechanism is incorporated in the system, as already verified by a stochastic method (ref. 1). This is an important factor and has an effect especially on the specification of fabrication and assembly.

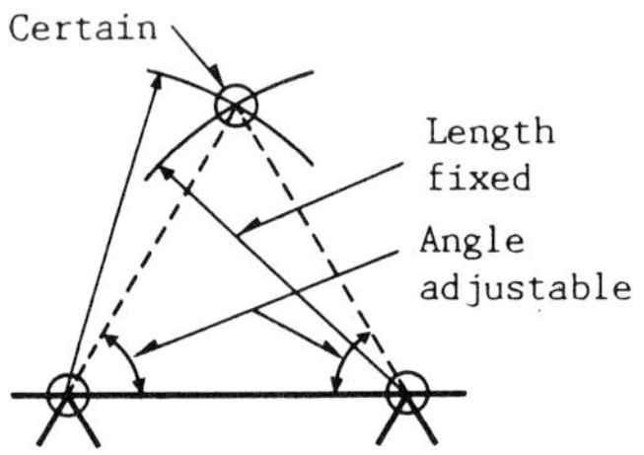

Fig. 2. Angle adjusting mechanism.

REALIZATION OF CONCEPT

4. The foregoing concept of the ideal system truss is hereunder developed into a specific joint system. First, a system truss way of joining was worked out, in which steel pipe members are joined to a spherical node by means of a bolt. The joint system was then closely analyzed and classified further. Specific joint systems were introduced from these preliminary studies.

Classification of Joint Systems

5. Joint mechanism. Situated between steel pipe members and a spherical node, the joint mechanism having an angle-adjusting mechanism serves to transmit tensile and compressive forces to and from them. It can be classified into four types, as illustrated in Fig. 3 (ref. 1).

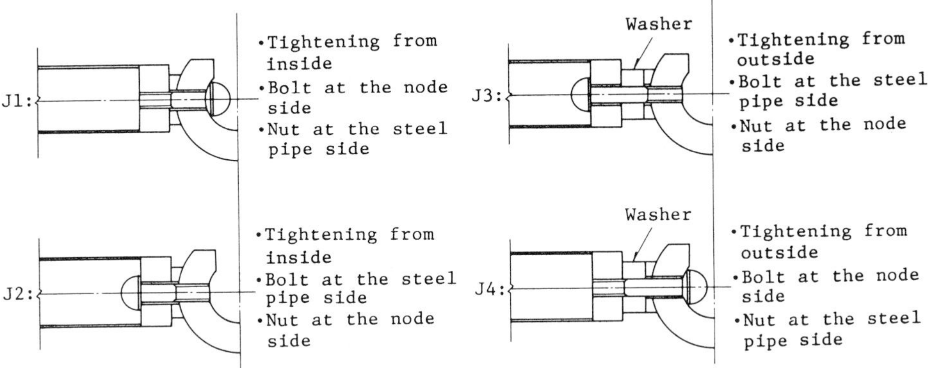

Fig. 3. Joint mechanism.

6. Node. The node is a joint provided to receive axial forces transmitted from steel pipe members and to stabilize the nodal point. The node has tapped holes or non-tapping holes for bolts. The kind of holes depends on the joint mechanism: tapped holes for J2 and J3, and no-tapping holes for Jl and J4.

7. Steel pipe member. To transmit the axial force on the steel pipe to the bolt, a conical part is welded to the head end of the steel pipe. This conical part is called the end cone. The transmission of axial forces can be accomplished by use of either the underhead of the bolt or the nut. The former is employed if the end cone is to have the angle-adjusting function (in the case of J2 and J3) and the latter if not. In the latter case (Jl and J4), however, the angle-adjusting function needs to be imparted to the node.

Realization of Joint System

8. No idea can be translated into reality with theoretical studies alone. It must be supported by experience including trial and error in the manufacture and assembly. "Know-how", gained by realization of one idea helps in the formulation of another. Several realistic ideas gained in the development of the NS Space Truss System are summarized in Table 1, and are shown in Fig. 4.

UTILIZATION OF NS SPACE TRUSS SYSTEM

9. For the commercial utilization of the NS Space Truss System it is necessary to standardize its components and establish the processes for designing, manufacturing and assembling the standardized components.

Table 1. Features of joint systems of the NS Space Truss System.

Type of NS Space Truss	Joint mechanism	Advantages	Disadvantages
A	J1	· Less costly because of fewer and simpler components	· Member latitude angle restricted
C	J1	· Member latitude angle less limited with the result of a widened range of applicable shapes	· Only a single opening having to be made on node, a negative factor making manufacture and assembly less efficient
UI	J2	· Lightweight components applicable to ultra-large spans · Eliminates the need for on-site selective use of bolts	· Lacks simplicity of rotating mechanism
UE	J3	· External bolting can be employed where internal bolting is restricted	· Increase in man-hours needed to make components
UX	J3	· Restriction on member set angle eliminated, thus making it possible to apply to all shapes	· Node opening eliminated, thus making fabrication a little more difficult

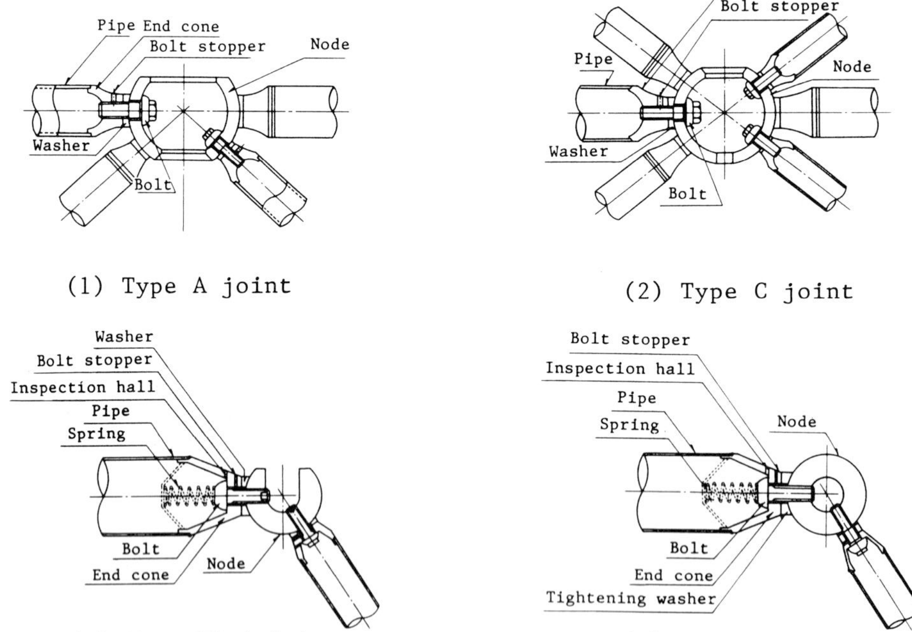

Fig. 4. Joint systems of the NS Space Truss System.

Standardization of Components

10. For each component of the NS space truss system, the shape, size and material must be standardized. Standardization is made firstly of the bolt, followed by the end cone and node.

11. Standard bolt. The bolt sizes range from M12 to M200 according to the ISO and JIS standards.

12. Standard end cone. The end cone sizes are standardized and systemized to match the bolt and steel pipe sizes. More precisely, the end cone size is determined first, based on the outside diameter of the steel pipe. Then, the maximum bolt sizes corresponding to the respective end cones, and the maximum pipe plate thicknesses corresponding to the strength of the bolts are determined.

13. Standard node. The most important considerations must be given in selecting the shape and size of the node. This includes securing sufficient bolt strength to withstand the axial forces transmitted from the steel pipe members. It is also necessary to keep the allowance for the interference or contact between end cones, with respect to the angles of pipe members, to a maximum.

14. Materials. The most suitable materials are selected for the components in compliance with the JIS standards or their modifications as appropriate (Table 2).

Table 2. Materials of components.

Component	Applied standard	Specification
Node	JIS G 3106	Equivalent to SM490A
End cone	JIS G 3106	Equivalent to SM490A
Pipe	JIS G 3444	STK400
Bolt	JIS B 1186	Bolt of special shape equivalent to F9T

Design Method

15. The design of the components means selecting the best components. To select optimum components, integration of sizes of individual components and verification of their geometric compatibility are needed in addition to the general calculation of structural strength. Design strength of each component was determined by experiments.

Manufacturing method

16. Major requirements in the production of the components are the abilities to manufacture them (i) with high precision, (ii) for a wide range of sizes each in small quantities, (iii) at a high efficiency and (iv) with stable high quality. To meet these requirements, a computer-assisted automatic manufacturing system has been established as described below.

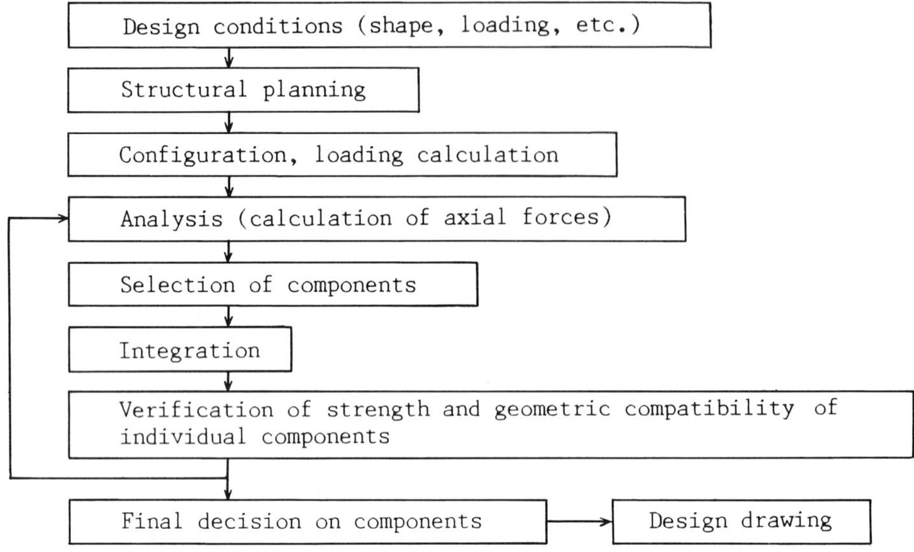

Fig. 5 Design flow.

17. Manufacture of nodes.

(1) Forging process: the node is shaped by forging after being heated.

(2) Machining: a numerical-controlled lathe is used to shape the outer surface of the node. The workpiece is then sent to a machining centre, where the bolt hole and other portions are machined with an accuracy of one thousandth of a radian based on the process parameters preloaded (Fig. 6).

18. Manufacture of pipe members. Pipe members are welded automatically based on the welding conditions pre-set to ensure high reliability and accuracy in length. In this welding process, the length of each member is controlled to an accuracy within ± 0.5 mm. Fig. 7 illustrates the frequency distribution of errors in member length.

Fig. 6. Drilling of node by machining centre.

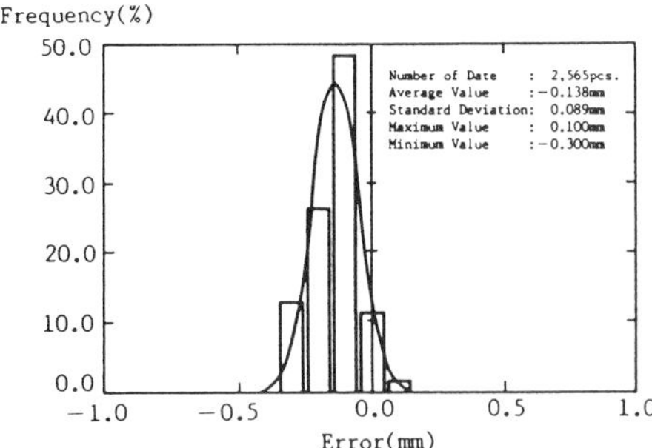

Fig. 7. Error frequency of member length.

Assembly

19. Assembly control. The assembly control is subdivided into the control of joints and the control of the shape of the whole structure. As regards joints, the bolt head must be kept in close contact with the inner surface of the node or end cone. Also, care must be taken not to produce any clearance between the node and washer and between the washer and end cone. The bolts shall be tightened to "snug tight". The shape control of the whole space truss structure, on the other hand, is based on the concept that components manufactured with a high accuracy and well-controlled joints assembly will lead to a high-accuracy shape of the whole structure. (See section 3 under CONCEPT OF IDEAL SYSTEM TRUSS.)

20. Erection methods. The lift-up method (Fig. 8), the sliding method, the movable scaffold method, and the block method have been developed as efficient erection methods.

Fig. 8. Ground assembly before lift-up.

PROJECTS

21. Singapore Indoor Stadium. This is a giant indoor stadium having a diamond structure of 219 m by 126 m. The diagonal lines are constructed of giant welded pipe trusses, the main trusses of which support the four faces constructed of the NS Space Truss System. This is an example of dynamic space structure created by combining system truss and other structures (Fig. 9).

Owner : Singapore Sports Council
Architect : Kenzo Tange
Structural consultant : M. Kawaguchi

Fig. 9. Singapore Indoor Stadium.

22. Dalian TV Tower. This 192m high TV tower was constructed on the top of a 170m high hill in Dalian. This is the first tower of the Space Truss System ever built in the world (Fig. 10).

Owner : Dalian Broadcast and Television Bureau
Architect : Nikken Sekkei

Fig. 10. Dalian TV Tower.

23. Jakarta GMF No. 3 Hangar. A large-scale hangar was constructed for Garuda Indonesia Airways in the Sukarno Hatta Airport in Jakarta. It is an ultra-large span structure measuring 312 m wide by 85 m deep and one of the biggest of its kind the world with a maximum front column spacing of 105 m (Fig. 11).
Owner : Garuda Indonesia Airways
Consultant : ENCONA Engineering, Inc.

Fig. 11. Jakarta GMF No. 3 Hangar.

Fig. 12. Canopy.

24. New Chitose Airport Terminal. In the newly opened large airport terminal building, the NS Space Truss System is adapted at three important places, because of its beauty (Figs 12-14).
Owner : Hokkaido Airport
Architect : New Chitose Airport Terminal Design Jointed Venture

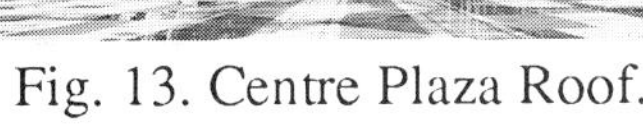

Fig. 13. Centre Plaza Roof.

Fig. 14. Airline Counter.

CONCLUSION

25. It took much time and effort to develop and utilize the NS Space Truss System. The authors are proud of the ideal system truss and many projects thereof have been achieved.

ACKNOWLEDGEMENTS

26. The authors would like to express their deep gratitude to all architects, consultants, engineers, site engineers and many other people concerned with the development and construction of the NS Space Truss System.

REFERENCE

1. IWATA, M., SHIRAI, T. and KAWAI, H. The System Truss as a Building System. International Journal of Space Structures, Volume 6, No. 3, 1991.

152. Space frame roof and pylons of the Olympic swimming centre and sports hall in Athens

A. KLIEM, MERO-Raumstruktur, Germany

SYNOPSIS. For the Olympic Games in 1996, for which the city of Athens had hoped to provide the sports facilities, among other facilities required, were a swimming hall, a sports hall and a training hall. This paper discusses these interesting support structures.

ARCHITECT'S IDEA

1. Already at an early stage, in 1979 on the occasion of the construction of the Olympic Stadium, WEIDLEPLAN/Stuttgart was also invited to undertake the master plan of the Sports Hall and the Swimming Complex (Fig. 1). The grandstands should provide seats for 15000 to 18000 spectators for the Sports Halls and seating for 4500 people for the Swimming Hall.

2. The construction of these two halls started at the end of 1988.

GEOMETRY

3. Apart from minor modifications, the geometry which was originally proposed by the architect was chosen for the final design. For the centre area of the hall the module is 7.2 x 7.2m with a structural depth of 4.86m. However, for the 15 sloping roof structures at the sides, a rectangular module of 3.6 x 3.723m with a structural depth of 2,70 m was chosen (Fig. 2).

4. Due to the slope of the side roofing and in connection with the uniformly directed perimeter diagonals of the centre part, there resulted unfavourable connection angles at the connecting nodes and also at the supports.

STRUCTURAL SYSTEM UNDER EARTHQUAKE LOADS

5. Greece which is situated at the periphery of the continental massivs, is a critical earthquake area, as has been proven several times in the past. The most recent activity being the disastrous earthquake in Kalamata in 1986. Therefore earthquake loading needed to be considered.

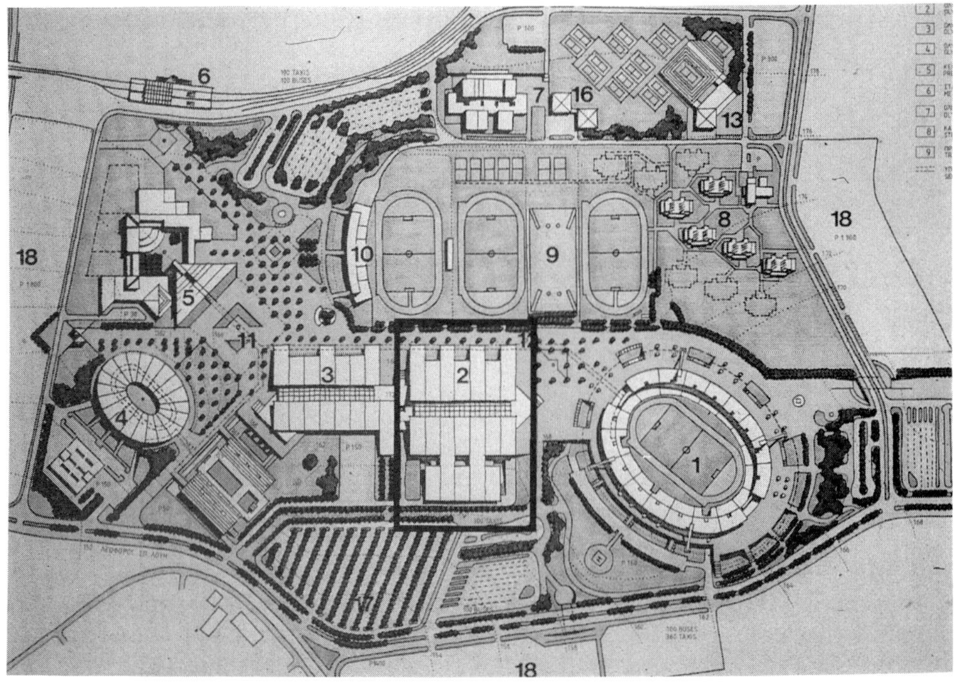

Fig. 1. Masterplan of Athens Olympic Sports Centre.

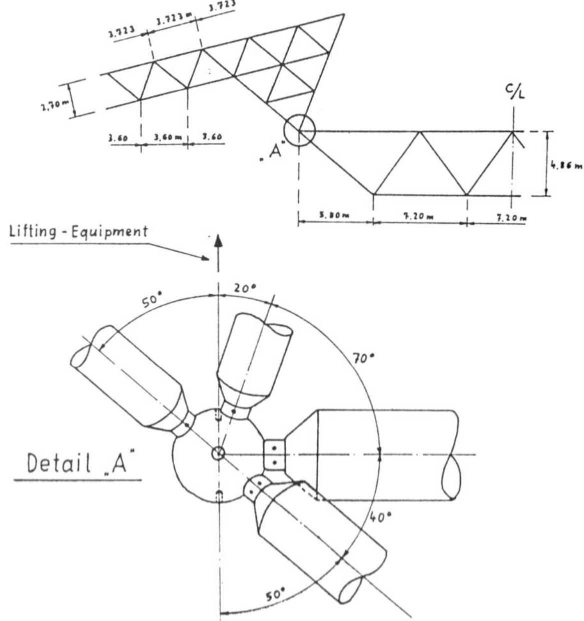

Fig. 2. Cross-section of space structures.

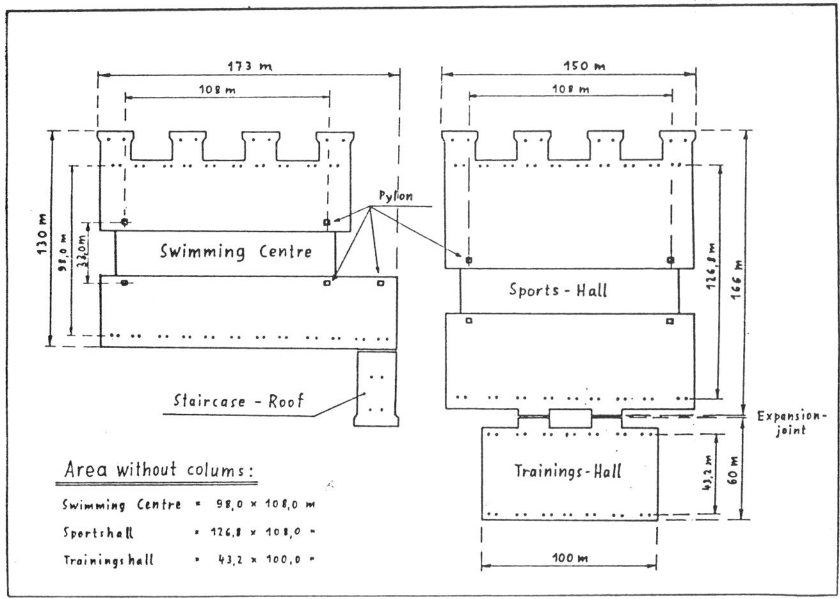

Fig. 3. Plan view of Swimming Hall, Sports and Training Hall.

6. For this reason, the complete concrete sub-construction was divided into 10 separate blocks, the high pylons having been considered as individual structures. The owner insisted on the outer bearings remaining movable in all directions.

7. Loads from wind and earthquake should only be transferred into the 20 - 30 m high pylons which are installed within the grid area of 33.2 x 108m (Fig. 3).

8. Due to the relatively large movements of the outer bearings caused by temperature differences, snow, wind and earthquakes the originally planned two large frameworks had to be separated into main and secondary frames with separate stabilisation.

9. Thus, the advantages from the framing effect of the structures in connection with the mushroom head supports were no longer given. Furthermore, the weight of the construction was increased.

LOAD ASSUMPTIONS AND STRUCTURAL ANALYSIS

10. For the sloping roof, additional to the self weight of the roof structure and the double-layer roof skin, service loads of 0.40kN/sqm from catwalks, lighting as well as ventilation ducts had to be considered. The snow load was taken as 0.75kN/m^2. The trench-like centre area of approximately 30m had to be treated in a different manner. Here, 10cm gravel filling as a top roof covering as well as different snow loads had to be considered.

11. This, resulted in 3 separate snow load cases, i.e. snow on the sloping roof areas at the sides and 2 snow load cases each acting on the centre area, i.e. a maximum snow load of 2.25kN/sqm and a minimum snow load of 0.60kN/sqm, alternatively considered both on the left as well as on the right side.

12. Apart from relatively small horizontal forces, which had to be absorbed only by the pylons, wind loads resulted in considerable uplift forces to the outer bearings. More detailed considerations were necessary with regard to the earthquake loads.

13. Because of the outstanding dimensions and the unusual column positions of these buildings, the owner required that a dynamic earthquake calculation of the steel reinforced concrete pylons was carried out, together with the space frame roof constructions.

14. This calculation was carried out in co-operation with the University of Aachen (Prof. Sedlacek) and by use of their programs.

15. Distorted deformations in this calculation showed that only 2 pylons in the longitudinal direction of the hall were used to absorb the horizontal earthquake loads. However, in the cross direction all 5 pylons of the Swimming Hall and all 4 pylons of the Sports Hall are involved (Fig. 4).

16. The combination of the individual earthquake load cases in the three main directions is in accordance with the draft of Eurocode 8.

17. According to the dynamic earthquake calculation, only minor reinforcements for the space frame were necessary. The tensile and pressure forces of up to 8000kN in the space frame tubes resulted from permanent load, live load and temperature differences between the outdoor and indoor areas of the roofings as well as from permanent load and wind uplift forces and according to Theory II.

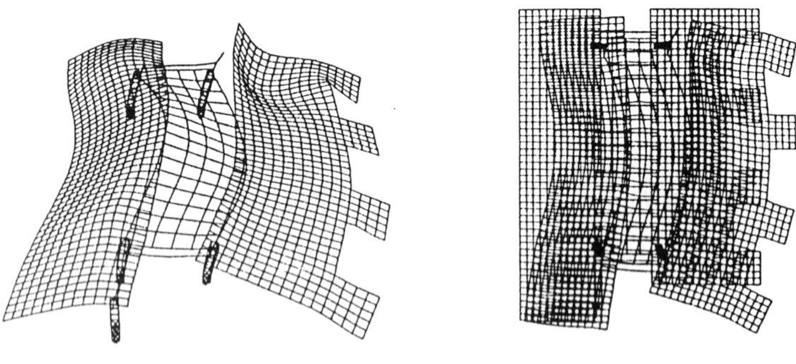

Fig. 4. A sample dynamic earthquake calculation.

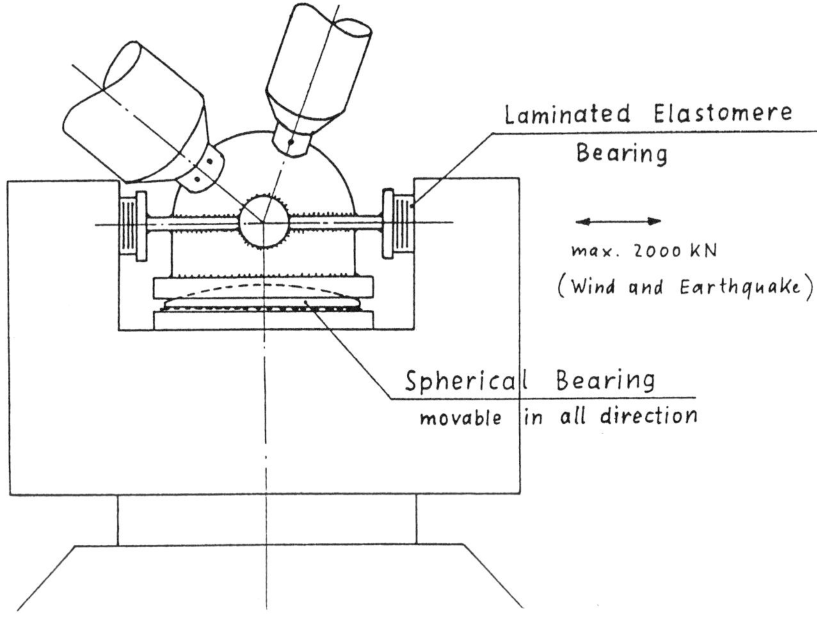

Fig. 5. Pylon-head-support for space frame.

18. Influenced by the failure of the highway bridge in Oakland (California), the main contractor of these buildings decided to use an interconnected solution for the pylons. Inside and outside of the steel box frames, shear connectors were used for the connection to the reinforced concrete. This solution is certainly the best.

19. The pylon received vertical loadings of up to 11000kN and horizontal loadings of up to 2000kN at heights of up to 30m (Fig. 5). Sliding bearings were used for the bridge bearings.

DIMENSIONING AND EXECUTION

20. The four independent frameworks with a total ground plan of approximately 50,000m^2 consisted of approximately 30,000 tubes with a total length of 116,000m and approximately 7,700 node connectors. From a total of 18 different load cases the most unfavourable forces in each tube are electronically determined and added to 5 different load combinations, as follows:

I permanent load, live load and temperature differences;
II permanent load, live load, temperature differences and wind loads;
III Theory II;
IV earthquake loads from permanent load and 50% live load;
V assembly loads.

Quite obviously, the member sizes and respective connectors could only be determined by comprehensive calculation.

21. Approximately 50 different tube sizes were used from tube diameters of 76.1 to 355.6mm and wall thicknesses of 3.0 to 3.6mm. Here, connecting bolts M20 to M90 of quality 8.8 and 10.9 were provided.

22. The extreme tensile forces of up to 800 tons in the area of the centre part of the bottom level were partly absorbed by continuous members of 355mm outer diameter and round steels of up to 150mm diameter.

23. These tubes arrived on site in lengths of up to 13m and were then welded together and finally ultrasonic tests were performed.

24. At the diagonal connections of these continuous chords the standard MERO connecting technique could not be applied. According to the German standard for tubular constructions DIN 18 808 tube sections with bolted flanges were required, in order to reduce site welding to a minimum.

25. Due to the 10 different types of joints, it was necessary to provide a variable tack-welding (stitcher). This should provide for the correct spatial direction and the precise distance from the theoretical node centre to the outside flange, taking into consideration of the weld shrinkage.

26. Apart from three cases, where spacers had to be used for tolerance balance, no difficulties occurred during the assembly. Considering the very large dimensions of the components, the workshop staff completed work of the highest quality.

27. The majority of the 10 different spherical node connections, from 110 mm to 40 mm diameter, were calculated by using a dimensioning program. Depending on the type of the connecting member, for which the permissible tensile and pressure connection forces were already determined, the computer checks whether the diameter of a spherical node is attributed in any section level to the corresponding forces, and also whether the spatial forces occurring at the right and left side of a section, remain in a pre-determined ratio.

28. The computer program also checks that the angle between two members to determine whether the threaded bolts will interfere inside of the node and check the connection of parts at the outside of the node.

29. Lastly, the program supplies all data for the computer aided node drilling machine, such as spatial angle, thread diameter and lengths of the thread as well as the size of the node flats. Furthermore, the program produces a schedule of parts according to their size.

30. For such a complicated tasks, a modern computer is essential. At least 1300 different MERO tubes as well as the minimum 3000 different node pieces were either hot dip galvanised or sand-blasted and galvanised in a plasma coating plant.

Fig. 6. Lifting process of the centre part of the space frame (approximately 60 x 120m and 560t).

Fig. 7. Lifting process of the centre part of the space frame (approximately 60 x 120m and 560t).

The connecting parts such as bolts and nodes are electro-galvanised and chromatised. In combination with an adhesive primer and a site paint finish, this corrosion protection provides an optimum solution.

Fig. 8. Hydraulic jack above the auxiliary cantilever beam and lifting cable beside the pylon.

ERECTION

31. The original assembly procedure using movable scaffolding, additional auxiliary columns in the centre, with high scaffolding at the sides, could not be used as it was difficult to achieve horizontal stability of the scaffolding against wind and earthquake forces, as stability had also to be guaranteed during installation. Also, the ground assembly of the 60 x 120m centre part, weighing 560t, was not possible as there were no appropriate cranes available.

32. In the end, assembly on the ground on 6m high scaffolding was possible, however, lifting was carried out via cantilevered beams and hydraulic jacks which were installed on top of the 4 pylons. (Figs 6, 7 and 8)

33. Thus it was possible, simultaneously with the space frame assembly, to complete the foundations and install the steel pylons . The centre part of the space frame was lifted approximately 15m using 4 lifting cables, to be subsequently placed on 10 auxiliary supports (Fig. 9).

34. After removal of the jacks, the bearings together with the necessary connecting members were installed and elastomeric material was applied to the space frame part at the level of the final bearing positions and to the final horizontal bearings.

35. At the same time, all temporary horizontal auxiliary construction could be removed.

36. Subsequently, the space frames of the side halls were assembled up to the mushroom like perimeter bearings (Fig. 10). As soon as the full size of the Swimming Hall with approximately 130m width and approximately 173m length was reached, the auxiliary columns could be removed and it was possible to proceed with the roof covering and other installation works.

Fig. 9. Centre part of space frame above the 10 auxiliary support towers.

Fig. 10. During assembly of the sloped space frame area.

Fig. 11. Architect design drawings.

153. A circular hanging roof in Sala

J. POSTULKA, Interexperta Ltd, Czech Federal Republic

SYNOPSIS. The paper describes the 60m diameter circular hanging roof used to cover swimming pools, 50 x 21m and 9·5 x 16·6m. The structure was constructed using precast reinforced concrete columns and an outside ring structure was used for anchoring the roof cables. The roof structure was clad using lightweight panels and supported at the centre an air-conditioning structure.

INTRODUCTION

1. A multi-purpose water storage reservoir was built in the town of ŠALA in 1986-87. The building was constructed to cover two swimming pools, with plan dimensions of 50 x 21m and 9·5 x 16·6m and also used to provide a cover to a reservoir used for storing water for extinguishing fires. A hanging roof structure with a circular ground-plan of 60m was built in order to cover the pools as shown in Figs 1 and 2.

Fig. 1. Circular swimming hall in Šala, ČSFR.

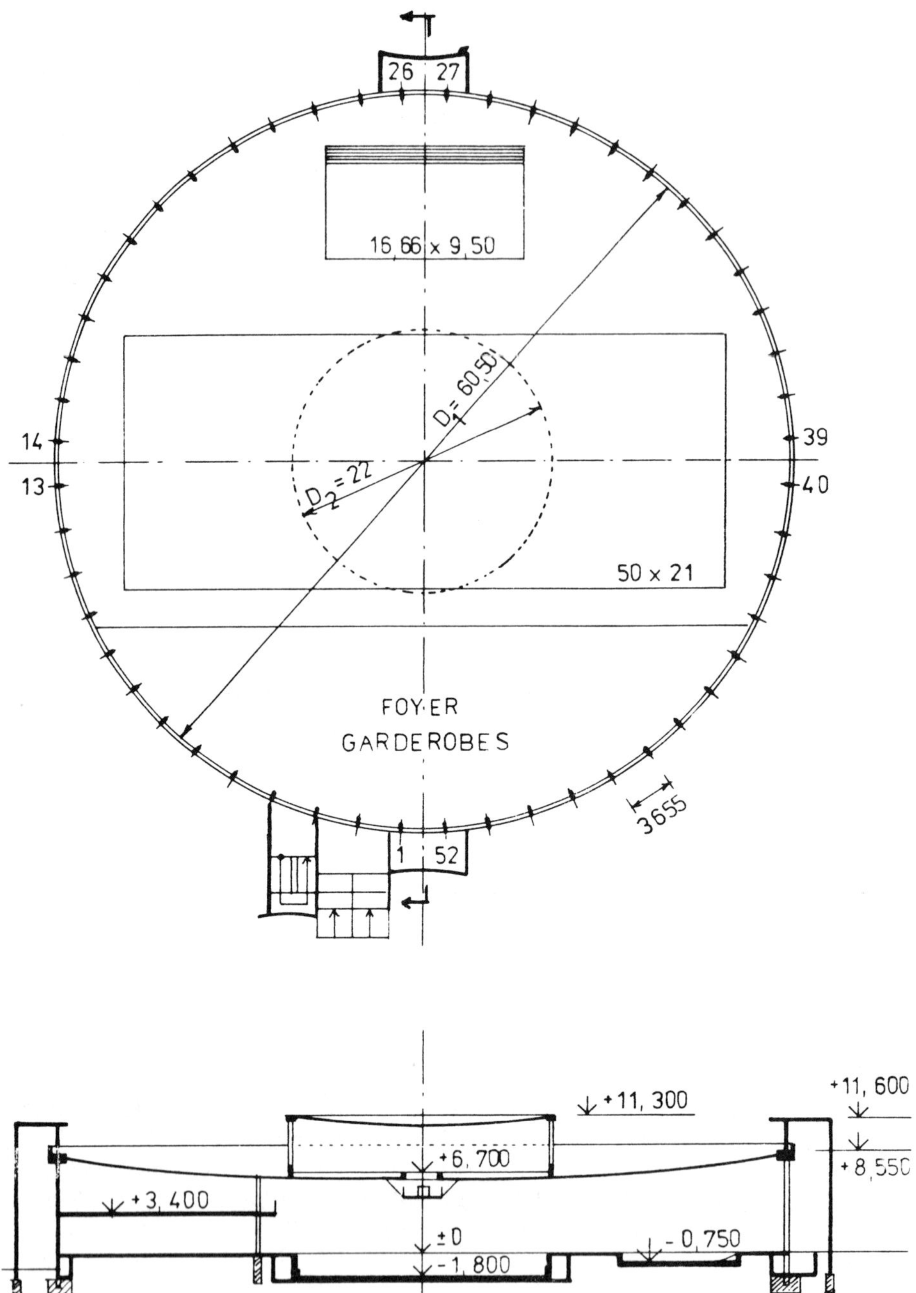

Fig. 2. Ground plan and a cross-section through the hall.

THE STRUCTURE

2. The outside walls of the circular hall are formed using precast reinforced concrete columns of 300 x 500mm cross section and at 3655mm centres. These columns support a precast reinforced concrete ring beam, formed from straight elements creating a polygon. This ring beam serves as a means of anchoring the 30mm diameter roof cables (Fig. 3).

Fig. 3. Precast hall structure.

3. The cable roof net is constructed with a 3m diameter circular opening in the roof centre. The roof cables are protected against corrosion in the wet swimming hall atmosphere by a coating of PVC.

4. The roof membrane supported by the cable net, is a stainless steel wire net with a mesh size of 20 x 20mm and a wire diameter of 2mm. The fine mesh is covered with a glass cloth / geotextile membrane, mineral-wool and wood-cement tiles, using rubber mastic for water proofing.

5. The main 60m diameter hanging roof supports, in its middle, a circular 22m diameter hall structure used to house air-conditioning plant. This hall structure has steel tubular columns on a reinforced concrete parapet supporting a steel-sheet outside ring structure used for anchoring the cable-net with 18mm diameter

cables. The steel ring is formed from a channel filled with concrete. The roof cladding for the air conditioning structure consists of the same layers as used for the main roof. The air-conditioning house floor is a horizontal reinforced concrete slab, 100mm thick on a pearlit-mats filling. The outside walls of this small structure are formed from Profilit-glass (Fig. 4).

Fig. 4. The circular 22m diameter hall structure supported by the main hanging roof.

6. There is also a lighting bridge in the main hall hanging from the cable roof, Fig. 5.

7. The lighting bridge supports in turn, the air-conditioning and rain-water drain pipes to and from the air-conditioning room on the roof. The warm air from the air-conditioning structure is conducted through several pipelines placed in two vertical masonry shafts located at the front and back of the main hall. It is then pumped into an underground air-duct around the hall. From this air-duct warm air heats up the windows inside the hall.

8. The water treatment plant for both swimming pools is located in a neighbouring building together with a water pumping station, drinking water reservoir and an emergency-power supply.

9. The outside walls of the hall are constructed partially in glass and partially using 250mm thick reinforced concrete sandwich panels. Construction details are shown in Figs 1, 3, and 6.

10. At the front of the hall is an annex to the main entrance with a staircase to the upper floor and roof, combined within the vertical shaft also used for the air-conditioning ducts as shown in Fig. 6.

Fig. 5. The interior of the swimming hall showing the lighting bridge supported by the cable roof.

Fig. 6. Exterior of the hall, showing walls and windows.

CONCLUSION

11. The very low costs of this structure, resulting from the economical use of material, combined with the pleasing appearance of the structure shows both the structural and architectural potential of such structures.

154. The laboratory for Schlumberger Cambridge Research Ltd: first floor and roof grillage design and construction

M. DICKSON and S. J. GREGSON, Buro Happold, UK

SYNOPSIS. This paper outlines the design, analysis and construction of the reinforced concrete grillage ribbed slab and the steelwork grillage roof for the Schlumberger Cambridge Research - Phase II building on the western outskirts of Cambridge. An architectural quality exposed soffit was formed to the first floor reinforced concrete slab using white ferrocement permanent formwork units to form a complex grillage of ribs. The roof consists of a grillage of 356mm deep universal beam members on a 3.6m module, connected at their junctions by spheroidal graphitic (ductile) iron cast nodes, which also provide the connections to the external columns.

INTRODUCTION

1. Schlumberger Cambridge Research Ltd specialise in high technology research primarily for oil companies. Their Phase 1 laboratory in Cambridge is on the City's western outskirts in the well-known 'tent' building.

2. Expansion of the company drove the need for a new laboratory and office building. The general concept of the laboratory evolved into a two-storey building with first floor cantilevered all round, full-height glazing, external first floor perimeter columns and a flat roof.

3. The building is separated into two square pavilions 21.6m x 21.6m at ground floor on a 3.6m architectural grid. The 1.8m cantilever gives first floors of 25.2m x 25.2m. The pavilions are joined by a two storey high atrium 10.8m wide at ground level, 7.2m wide at first floor, which is the reception area for both the new and existing buildings. The overall floor area is 2300 square metres.

STRUCTURAL ENGINEERING - GENERALLY

4. The structural engineering form developed into a reinforced concrete slab at first floor on steel tubular columns, with a steel-framed roof. Column positions were defined by the internal room layout: 3.6m square offices on the building's perimeter and much larger laboratories and computer suites in the centre.

Schlumberger Phase II - Front elevation

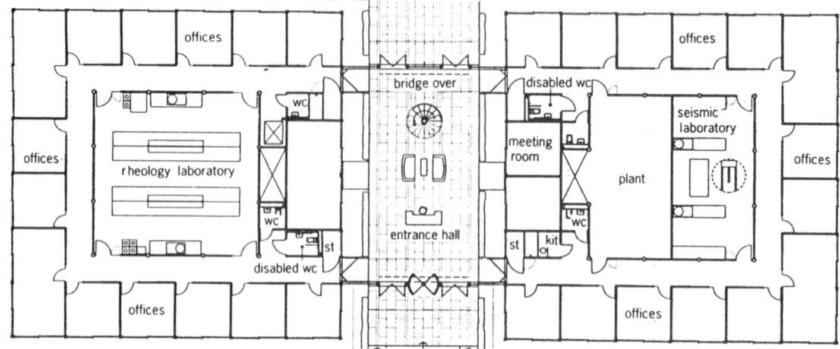

Architect's ground floor plan

First floor slab soffit and reinforcement

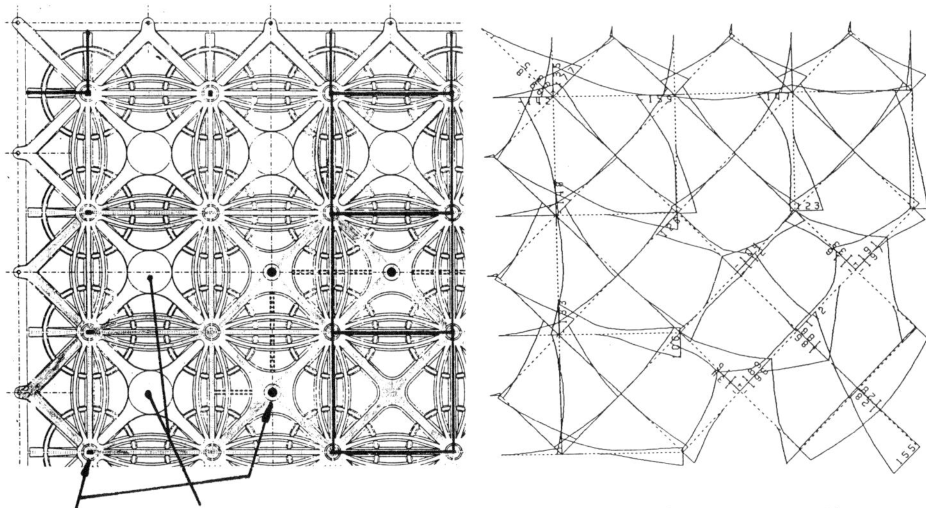

columns ribs arch upwards to 150 thick
ferrocement unit types outlined
primary structural ribs shaded

bending moment diagram
from the grillage analysis
(kNm)

Quarter of one of the two first floor slabs

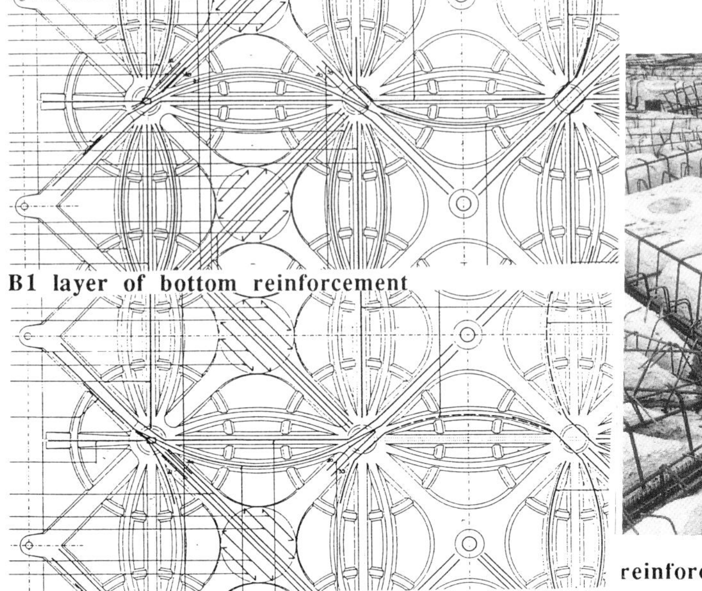

B1 layer of bottom reinforcement

B2 layer of bottom reinforcement

Details of bottom reinforcement showing the lap arrangement

reinforcement on site
showing the top of the
white ferrocement units

5. Fixed-base internal columns, with moment connections to the first floor slab and roof, were used to provide sway stability. Because bracing was not possible within the scheme, 244.5mm diameter steel circular internal columns at 3.6m centres were used in the walls around the central laboratories. This scheme effectively gave a line support around the first floor slabs' central 10.8m x 10.8m span, while fitting the partition layout and general 3.6m grid. Perimeter ground floor columns are 150 x 100mm steel rectangular hollow sections, also at 3.6m spacing.

FIRST FLOOR SLAB

6. The concept of the first floor slab called for an exposed soffit which reflected the room layout below, providing ribs or beams to take the partition heads. The first floor slab design was influenced by the work of the Italian architect/engineer, Nervi. In the 1950's he used ferrocement as a structural medium both for permanent formwork, and as an integral structural element of his buildings.

7. Ferrocement is a thin cement: sand mortar and light steel mesh composite material. The mesh is formed around a mould and the cement: sand mortar sprayed, or sometimes trowelled and forced around the mesh. The final ferrocement shell is usually of the order of 25 to 50mm thick. Moulds are normally glass reinforced plastic (GRP) or even ferrocement itself with a suitable debonding agent.

8. At Schlumberger, ferrocement was used to form what are effectively permanent formwork soffit units to the reinforced concrete first floor slab. This enabled the architect to have a fair-faced factory-produced "concrete" slab soffit in white ferrocement. Construction of the fairly complex system of primary and secondary ribs would have been extremely difficult to form in any other way. Furthermore, the on-site works of forming the grey concrete slab did not need to provide an architectural finish, with all the associated site quality assurance problems.

9. Coupled to the use of ferrocement permanent moulds was the concept that in the central 10.8m square zone a diagonal rib grillage within the slab would make great structural sense in terms of reducing its overall weight to depth ratio. These ribs could be formed from square moulds of ferrocement with primary diagonal elements. The overall slab depth was 400 mm including ferrocement.

10. At the offices around the building's perimeter, the diagonal ribs arch upwards from an overall depth of 400mm to become a 150mm thick slab. Each office thus has its own coffered ceiling, with the rib layout co-ordinating between the internal space and the external cantilever ribs. Structurally, the primary rib system is changed here from a diagonal to an orthogonal layout between the external columns and the line of internal columns. The system reverts to diagonal ribs on the external cantilever.

11. The moulds for the units were first formed as positives in timber and GRP; a quarter of each of the 3.6m square moulds was made. GRP negatives were then made from these; four quarters joined together to form the whole mould. Some ferrocement negative moulds were also made in order to increase the production

of units. One was even used to panel beat the mesh reinforcement into shape, although in hindsight a mould sized for the initial cover allowance to the mesh would have simplified the process. Five types of units were made: the main units are 3.6m square, with the edge cantilever units 3.6 x 1.8m and 1.8m square at the corners.

12. The layout of the rib system was brought about by very detailed and close liaison between Buro Happold and Michael Hopkins & Partners, to ensure that all the architectural, planning, services and structural requirements were fulfilled. All the ribs are reinforced, and act structurally. Strict co-ordination was required with the services engineer to ensure that all openings in the first floor for services from the raised floor above, fitted the rib module and provided the correct lighting levels and air flow. The main diagonal ribs are 260mm wide at the bottom, the narrow straight ribs 150mm, and the curved ribs 100mm, all including ferrocement.

13. The slab was analysed using 'STRAP' grillage analysis software. The primary ribs were modelled, using one beam element to model the 'triple' ribs of two curved and one central straight rib. One quarter only of the slab was analysed, due to its symmetry, with suitable boundary conditions. Load cases were run for full ultimate and pattern loading. A further, detailed computer model of just a 'triple' curved and straight rib unit was then run, using faceted, straight elements to model the curved ribs.

14. The rib layout called for a radical approach to the reinforcement design. The reinforcement was carefully organised keeping overlapping bottom bars at rib junctions to a minimum so that a multiplicity of layers was avoided. The design was for a maximum of three bottom layers, although in practice on site this became four in some areas, with secondary lap bars in the upper two layers. Top reinforcement was on a diagonal grid, mirroring the main diagonal ribs.

15. Reinforcement drawings were produced at 1:25 scale with separate drawings for each layer. This produced a considerable number of reinforcement drawings, twenty nine in all. For such an unusual slab the steel fixers had to completely understand the bar layering system, requiring comprehensive and clear drawings; hence the need for such an unusually large scale. Indeed the specification purposely called for a test panel to be cast, incorporating all the design conditions into a fully-blown workmanship trial.

16. Before steelfixing began, meetings were held on site with the steelfixers to ensure that they fully understood the design. The trial panel was further used to educate everyone to the practicalities and necessities of organised and rigorous work practices. While there were the inevitable on-site problems, steelfixing for the complex bottom reinforcement took approximately four days for the West Pavilion and two days for the East Pavilion, with much of the beam and rib steel previously pre-fabricated.

17. In order to ensure that the concrete could completely fill the narrow ribs and surround the reinforcement without voids, a 10mm maximum aggregate size was used with a plasticiser, and a design slump of 100mm.

Roof steelwork during erection

Internal and edge cast nodes - during construction

Internal view of the finished steelwork and roof panels

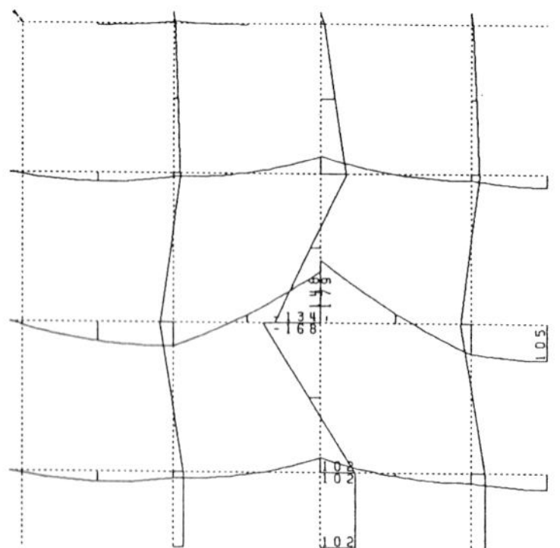

Typical bending moment diagram for one quarter of the roof grillage (kNm)

18. The test panel was also cast to check the practicality of the concrete mix. Areas of the soffit of the test panel ribs were broken-out to check the compaction around the bars in the narrow ribs. The first sector of the West Pavilion was cast in two horizontal layers with the top reinforcement fixed after the initial pour. There was concern that trying to pour concrete into the bottom of the narrow ribs through the top steel would be very difficult and also could lead to segregation of the mix. However, the concrete pour in fact went so well that subsequently concrete was poured full-depth, with each pavilion cast in two halves.

19. Few problems were in fact found on site in pouring the concrete using this mix, and very careful monitoring was carried out by both our Resident Engineer and the Management Contractor.

ROOF STRUCTURE

20. The external first floor columns are 88.9mm diameter circular steel tubes at 3.6m spacing, supported on cantilever nibs on the end of the first floor cantilever slab. The internal columns are 244.5mm diameter steel tubes, four per pavilion at 10.8m spacing, which bolt onto the tops of the ground floor internal corner columns.

21. The architectural concept for the roof evolved into a grillage of exposed I-beams, all the same 356mm depth, on a 3.6m module. The constant depth of the grillage enabled constant height partitions to be fixed below.

22. Various other schemes were investigated including space frames, glued-laminated timber frames, and composite all-welded double cross beam systems. The final solution was a grillage of 356 x 171 UB sections 3.6m long connected by moment connections at node points. Both fabricated steel and cast node connectors were investigated. Very quickly it became clear that cast node connections bolted to the steel beams via simple end-plate connections would provide the simplicity and clarity of form which the architect sought. From the structural engineer's point of view this detail also avoided complex full-penetration welded joints at all the nodes where the grid members cross, which are areas with high bending forces.

23. The grillage was analysed, again using 'STRAP' grillage analysis software. Similarly to the first floor slab grillage, a quarter only of one pavilion roof was analysed for both full ultimate load and pattern loading.

24. Both steel and SG Iron (Spheroidal Graphite or Ductile Cast Iron) castings were considered for the grillage nodes. Steel has the advantage of weldability, but the disadvantage of increased cost and difficulty of casting. There was no need for weldability in this case, and after much research SG Iron was selected.

25. SG Iron has been used for some years and is a significantly stronger and more homogeneous material than grey cast iron, in particular having good strength in tension. SG Iron's strength is due to the homogeneous spread of very small nodules of graphite throughout the material, in contrast to the heterogeneous nature of grey cast-iron. SG Iron exhibits good compression and tension properties, and some of the higher grades have allowable working stresses higher

than those of mild steel. As a material it does not have a pronounced yield point like steel, and generally the 0.2% proof stress is used as an alternative.

26. Grade 450/10 SG Iron was used here, with an allowable working tension stress of 152 N/mm^2, 190 N/mm^2 in compression and a 0.2% proof stress of 305 N/mm^2. The modulus of elasticity, E, for SG Iron is typically 169 kN/mm^2. lower than the 210kN/mm^2 (approximately) for steel.

27. The node to beam moment connection is central to the structural integrity of the grillage, and therefore a beam-node connection was load-tested to verify the calculations. Two beams were tested bolted to a central node casting. These were loaded at third points to provide the required design bending moment at the beam-node connection. Loading was incremented up to the working bending moment (BM) and then continued to ultimate bending moment and above. The test was stopped for safety reasons at 1.25 x ultimate BM at which point the bolts were beginning to yield, as indicated by the load deflection curves.

28. The load test results showed that the bolted node test piece deflected a maximum of approximately 1.4 x the elastic deflections that would be expected for a beam with no central bolted node. Elastic lengthening of the bolts, coupled with distortions of the beam end plates, however slight at working loads, leads to rotation of the connections and a deflection increase over a purely elastic beam. This 1.4 factor was applied to the computer-predicted deflections from the grillage analysis as a first approximation upper-bound for the grillage deflections.

29. Tolerances in the roof construction were critical. In order to achieve the correct aesthetic and structural integrity of the casting-beam connection the amount of shimming at these connections had be kept to a minimum. In the final event the fabricator (Tubeworkers) elected to machine both the casting connection faces and the beam end plates (post welding) to avoid any shims at all. The machining enabled the beams to be made to a final theoretical tolerance of +0,-1mm.

30. Pre-cambering the central 10.8 x 10.8m spans was achieved by machining appropriate node faces at a very slight angle, thus slightly raising the centre of the 10.8m spans above the rest of the building. A 40mm pre-camber was used. Erection on site went quickly with no major technical problems.

31. A consequence of the roof grillage arrangement is that the corner columns carry very little load, there being no major load path to them. This benefits the corner RC cantilever ribs below, which support these columns, in that being corner ribs they actually cantilever 2.55m rather than the normal 1.8m.

32. To provide the roof fabric various types of proprietary panel were investigated. Eventually a simple 3.6m square factory-made timber plywood and joist composite panel 125mm thick was selected. The soffit plywood has a final, lower ply of Finnish birch which becomes the ceiling finish. These panels sit directly on the top flange of the steel grillage. Electrical conduit for the fire alarms and lighting were pre-fixed in the panels in the factory, the cables then pass through holes in the centre of the node castings and hence to the switches in the partitions.

ATRIUM

33. The atrium was envisaged as being a very light and open space. To achieve this, planar glazed walls were designed with an ETFE foil cushion roof. The foil cushions are air-inflated three membrane cushions formed from "Hostaflon" translucent ETFE polymer film. At Schlumberger, the cushions are 7.2m x 3.6m and supported on a lightweight T-Section steelwork frame which in turn sits on the edge castings of the main roofs. The small inflation and back-up fans for this installation were located in the plant room and inflation provided by a small network of tubes in the roof structure. To provide shading and cover to the entrances, a membrane canopy is located on the front and rear atrium elevations.

34. The design and construction of this building has used many innovative and unusual structural forms, not least the ferrocement permanent formwork for the first floor slab. This material may show an alternative future way forward for designing architecturally finished structural elements, but without the complex on-site problems associated with fair-faced insitu concrete. SG Iron castings also provide a method of producing more flexible designed elements in steelwork. It is important to consider the long lead-in period associated with such castings very early in the programme.

35. Given the level of innovation in the design of this building, the co-operation of the Building Control Officer at Cambridge City Council and their checking engineers, WS Atkins and Partners also needs acknowledgement.

36. The £3.3m project was constructed using a management contract in one year from September 1991 to August 1992.

Architect:	Michael Hopkins & Partners
Structural Engineers:	Buro Happold (Leeds)
Services Engineers:	Buro Happold (Leeds)
Management Contractor:	Team Management Ltd
Superstructure RC:	Intex
Main Roof Steelwork Fabricator:	Tubeworkers

REFERENCES

1. NERVI P.L. 'Structures' Mc Graw-Hill, 1956.
2. NERVI P.L. 'Ferro-cement ,: Its characteristics and potentialities', translated by the C&CA ,1951.
3. BRITISH CAST-IRON RESEARCH ASSOCIATION, 'Engineering data on nodular cast irons SI units',1986.

155. Space frame supported membrane structures

A. KLIEM, MERO-Raumstruktur, Germany

SYNOPSIS. The development and especially the improvement in the quality of flexible roof membrane material during the past few years is linked with the increased use for enclosing systems. These materials are not only used for shading roofs (in areas without snow loads), but also for permanent structures such as sports stadiums, multi-purpose halls and pavilions common in our area. Due to the limited loading capacity of these membranes, steel, wood or reinforced concrete constructions are always required to suspend or anchor the membrane. The variety of shapes leave a lot of freedom for individual fantasy. Temporary structures, such as circus tents, party tents and small exhibition halls offer a large range of applications for tent structures. A large temporary exhibition hall and a temporary maintenance hangar for planes will be discussed in this paper. Both structures have been built for a design life of several years.

TEMPORARY EXHIBITION HALL - DUSSELDORF

1. With the clear span area of 64.0 by 153.0m giving a covered area of 9,790m^2 and a clear height of 14.0m (below the roof membrane), this space frame construction is of extraordinary size (see Fig. 1). The rather short time for the use of the structure, about 2 years, with subsequent storage of the material, led to an element-type steel tube construction method. Fast erection and disassembly is to be guaranteed but it is also required that storage of the material be achieved in quite a small space. Therefore, the MERO space frame construction system of tubes and nodes was the perfect choice.

2. The following firms were part of the construction execution:

• Owner	Carl Nolte in Greven
• User:	NOVEA-Messegesellschaft in Dusseldorf
• Consulting Office:	IPL-Ingenieurplanung in Radolfzell
• Delivery and erection of roof membrane:	Carl Nolte in Greven
• Steel construction:	MERO Raumstruktur GmbH & Co. in Würzburg
• Duration of erection:	approximately 6 weeks, at the end of 1988.

Space structures 4. Thomas Telford, London, 1993

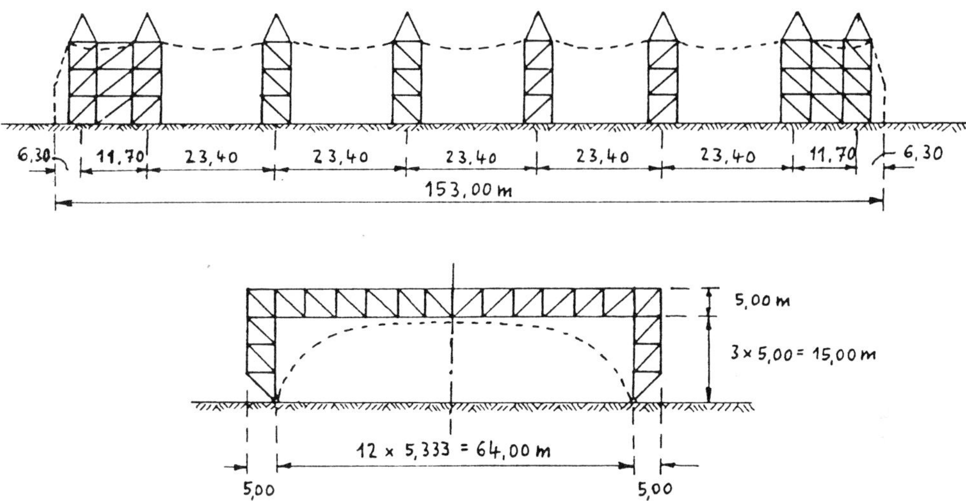

Fig. 1. Layout of structures. a) Elevation. b) Cross-section.

3. Eight double-hinged space frame portals of which only two end-portals each are connected by bracing, form the structure for the suspension of the flexible polyester-coated membrane. The spans to be covered by the membrane are quite unusual; 23.4m between centres and 11.7m between the end portals. The necessary stability of the entire space frame construction in the longitudinal direction of the hall was guaranteed by pre-stressed polyester edge cables together with the cable suspensions at distances of approximately 5.0m to the steel construction. Not only was it necessary to avoid fluttering and rupturing of this light skin due to wind suction loads, but also to accept safely the wind pressure and the snow loads. Reports by the weather bureau in Düsseldorf confirmed a measured maximum snow height of 30cm in the past 20 years. This corresponds to a snow load of approximately 0.25kN/m^2. The non-heat insulated, concave shaped roof membrane will limit the snow load, due to the melting, caused by heating the hall in winter. In case of a heavy snowfall and break-down of the heating system, the snow has to be removed manually. Winds in the cross direction to the hall are transferred through the space frame, into the low-level foundations. Fig. 2 shows the coefficients of the wind load distribution.

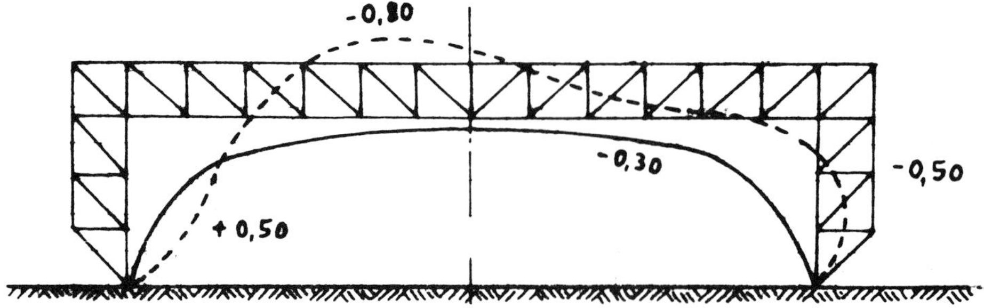

Fig. 2. Wind load coefficients.

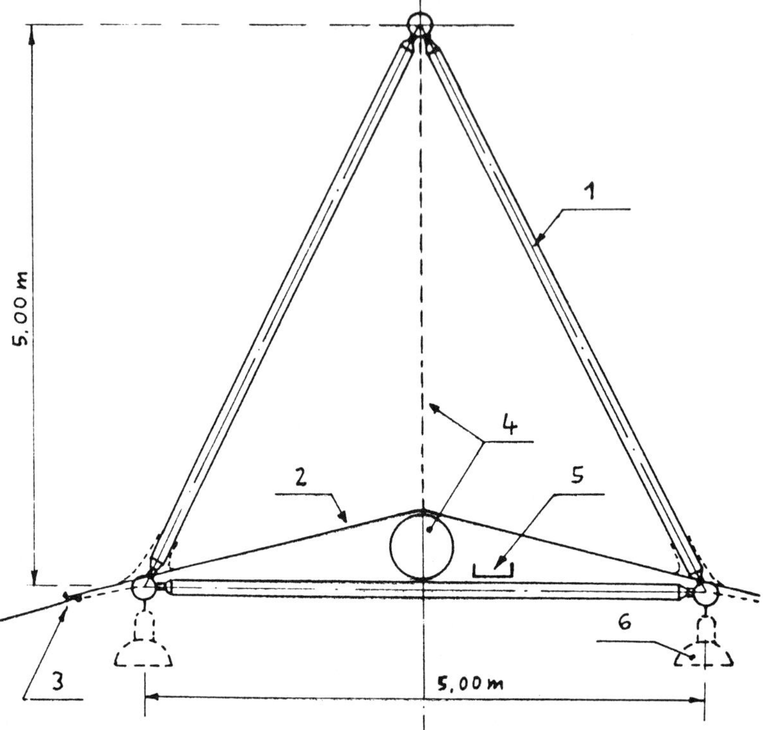

Fig. 3. Cross-section of one triangular girder.

where:
1. MERO-members
2. Membrane
3. Membrane-joint
4. Ventilation pipe (heating) and suspension rod
5. Cable tray
6. Lamps

Horizontal and vertical bracings between the end frame bars and posts, provide the transfer of the membrane prestressing forces, horizontal forces from both the permanent load and snow, as well as the wind loads in the longitudinal direction of the hall. Loads of 0.02kN/m^2 were considered adequate for the self weight of the roof membrane including the fixing components. The weight of the filigree steel construction itself, is only 0.20kN/m^2. Fig. 3 shows the arrangement of the ventilation system, the cable trays, the lamp placement as well as the membrane arrangement in the area of the space frame girder bottom chord. In this case it became necessary to secure, against the ingress of water, the penetrations of the space frame diagonals, using pre-shaped so-called hoses. (see Fig. 4).

ERECTION

4. The full length of horizontal frame was pre-assembled off-site on the ground, including the ventilation system and the roof membrane in the bottom chord area. (Figs 5 and 6).

Fig. 4. Membrane erection in the bottom chord area.

Fig. 5. Pre-assembled triangular girder (74m long).

Fig. 6. Ventilation duct within space frame girder.

The frame posts were erected simultaneously up to a height of 10m with one adjustable temporary support each. Two mobile cranes then lifted the frame bars to a height to allow installation of the frame post members of the upper section. The construction was pulled to its final position and the screw connections made between the frame post members, with the help of aerial work platforms (Fig. 7).

Fig. 7. Connection of frame parts with the help of aerial work platforms.

Fig. 8. Membrane assembled between the first erected portals.

Lifting of the end girders was undertaken together with the bracing aligned in between them, giving a total lift weight of 18t. The erection of the roof skin was executed with the help of cable hoists and also with scissors platforms see Figs 8 and 9. Figs 10 and 11 show photos of the completed hall.

Fig. 9. Lifting of a membrane element (18.4m wide).

Fig. 10. Completed hall from the outside.

Fig. 11. Completed hall from the inside.

TEMPORARY AEROPLANE MAINTENANCE HALL IN MUNICH

5. The membrane for the aeroplane maintenance hall in Munich, covers an area of 39.4m by 61.5m, see Fig. 12.

6. The following firms were part of the project:

• Owner and User:	German Wings Airline
• Consulting Office:	IPL-Ingenieurplanung in Radolfzell
• Delivery and Erection of the Roof Membrane:	Carl Nolte in Greven
• Steel Construction:	MERO-Raumstruktur GmbH & Co. in Wurzburg
• Erection Duration:	4 weeks

7. Five space frame portals of various heights, fixed to the foundation with two end-portals being connected by bracing, form the structure for the suspension of the flexible roof skin (Fig. 13).

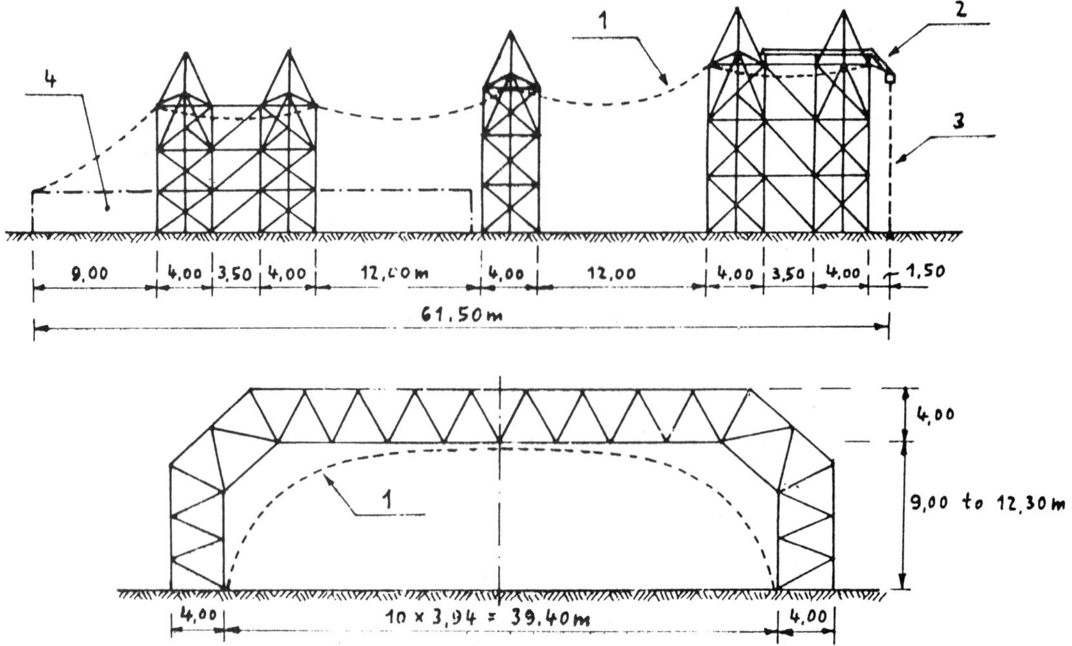

Fig. 12. Layout of structures

where:
1. Membrane
2. Door rail - substructure
3. Movable door
4. Spare parts storage made of reinforced concrete.

Fig. 13. Suspension point between membrane and steel structure.

Fig. 14. Completed hall from the outside including the door structures.

8. The various heights as well as the door height of approximately 12.0m resulted from the nose-in position of the aircraft with a tail unit of up to approximately 11m. (Figs 14 and 15).

9. The U-shape, approximately 5m high spare parts storage at the opposite site of the door, allowed the final fixation of the roof skin to this reinforced concrete construction. Additional front portal construction became necessary for the sliding door guides.

REFERENCES

1. Mengeringhausen, M., Raumfachwerke aus Staben und Knoten, Bauverlag, Wiesbaden, Berlin, 1975.

156. Styling dome for the Chrysler Technology Center

J. D. CLINTON, 3D Structures, USA

SYNOPSIS. The Chrysler Corporation broke ground October 30, 1986 for a Technology Centre to be completed in 1993. The central focal point is an acoustical styling dome with a spherical diameter of 43.28 m, a height of 10.67 m, and an acoustical lighting cove below the spring line of the dome. The cove has a height of 5.5 m and a spherical diameter of 38.40 m. The dome support frame is a geodesic icosahedron of the Class ll Method 3 type. The dome also supports the fire suppression system, a revolving platform and computer controlled lighting system.

THE CLIENT'S NEEDS

1. The $1 billion, 163.9 hectare facility in Auburn Hills, Michigan USA was designed to be one of the most advanced automotive design centres in the world. The Chrysler Corporation decided to include a styling dome designed after the dome located in the previous styling centre. However, the old hard plaster dome made it an acoustics echo chamber that was very undesirable. Chrysler insisted that the architects of CRSS, Inc. Houston, Texas duplicate the functions of the old dome with new state-of-the art acoustics and lighting.

2. The new styling dome was to include three turntables for model studies, a 24.38m wide projection screen that lowers into the floor, 85 computer controlled lighting fixtures with maintenance access from the domes exterior. In addition, a cove and floor parameter lighting system to simulate a horizon and control ambient daylight was to be provided. It was also to include a stage for the board of directors to meet and review new product designs, a place where car models could be viewed under controlled lighting and out of the view of competitors, and an acoustically controlled space where a whisper could not be heard across the 37.19 m span of the dome.

3. In the 5.5 m high cove one 14 m wide, 4 m high stage opening, two 7.3 m wide, 3.4 m high auto openings and two 3.4 m wide, 2.7 m high personnel openings were to be provided.

Space structures 4. Thomas Telford, London, 1993

CONTRACT AWARD

4. 3-D Structures, Inc. of Avondale, Pennsylvania and Star* Net Structures, Inc. of West Babylon, New York formed a joint venture to compete in a design award competition. The 3-D Structures design team has the experience of designing the largest spherical Omnimax theater screens in the entertainment field. They have worked out many of the difficult design details for providing accurate large compound curved surfaces having a monolithic appearance capable of accepting high resolution optically projected images yet providing an acoustically desirable volumetric space. 3-D Structures also designs and manufactures aerospace visual flight simulator screens and understands the need for high-tech applications of materials to provide the high quality surfaces demanded in the visual fields.

5. Star* Net Structures has the experience of designing large span space trusses capable of supporting large loads. They also have experience in design and manufacturing high-tech aerospace structures that demand the precision that would be required for the support frame of the Chrysler styling dome. Due to this unique design team, a novel and new system for a butt-seam spherical acoustical ceiling design was proposed. A full size section of the design was presented to CRSS and Chrysler and the decision to use the proposed design was made.

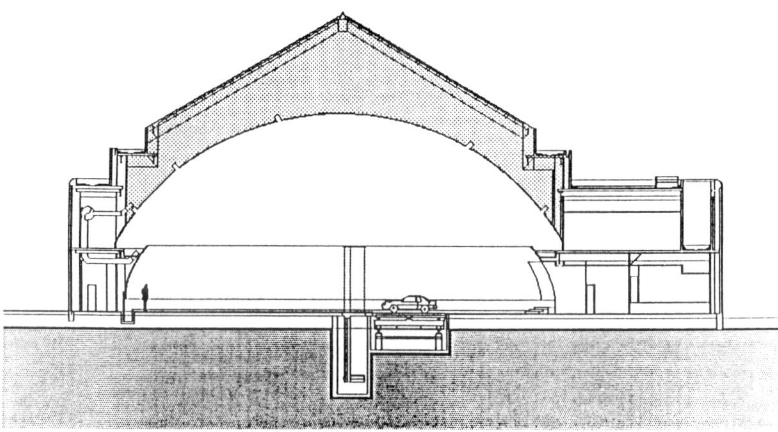

Fig. 1. Chrysler Technology Center styling dome.

DESIGN PARAMETERS

6. As part of the product design wing of the new Chrysler technology center a spherical ceiling was needed; and, the following criterion was to be met:

(a) A white spherical domed ceiling having a height of 10.67 m, a span of 37.19 m at the spring line, and a diameter of 43.28 m.

(b) The center of the dome was to be located 5.64 m below the floor, and the dome's structural support ring was to be located 5.5 m above the floor.

(c) The ceiling support frame was to be capable of supporting 85 computer controlled lighting fixtures and their necessary maintenance access structure.

(d) The ceiling frame was also to be designed to support the water sprinkler fire suppression system.
(e) Starting at the floor level a white 5.5 m high cove with a spherical diameter of 38.40 m and center located 4.88 m below the floor was desired. The cove was to have one 14 m wide, 4 m high stage opening, two 7.3 m wide, 3.4 m high auto openings and two 3.4 m wide, 2.7 m high personnel openings.
(f) Both the ceiling and the cove were to provide an acoustically desirable environment that would provide a show place where high quality lighting studies and presentations could be conducted by Chrysler's product design group

THE DESIGN SOLUTION

7. During the proposal stage a careful study was made of several spherical topologies in order to choose the best dome solution for the criterion presented. Having had experience in designing both Schwedler and triangulated Geodesic type Omnimax theater screens the design team of 3-D Structures was aware of the shortcomings and qualities of both systems. Most theater designs are of the Schwedler lap seam type, however, under ambient lighting conditions the lap seams become very visually pronounced and undesirable for the lighting conditions to be used in the Chrysler styling dome. The trapezoid shapes also present a visual problem in that the eye has a tendency to follow the continuous lines of the gores and zones present in the normal Schwedler type of cladding. The first butt-seam triangular Geodesic type screen located at the EPCOT center in Orlando Florida, USA solved many of the visual seam problems. However, the triangular nature of the panels presented new problems in controlling the accuracy of the seam widths and introducing many more panel types to be manufactured, which increased production costs.

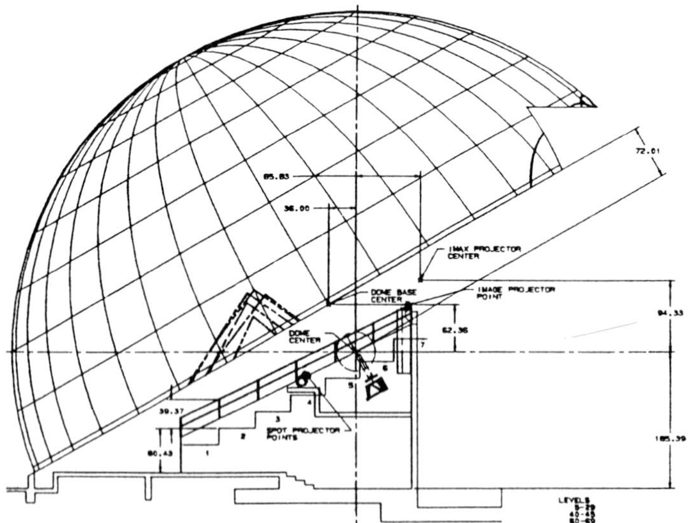

Fig. 2. An Omnimax screen using a Schwedler type dome.

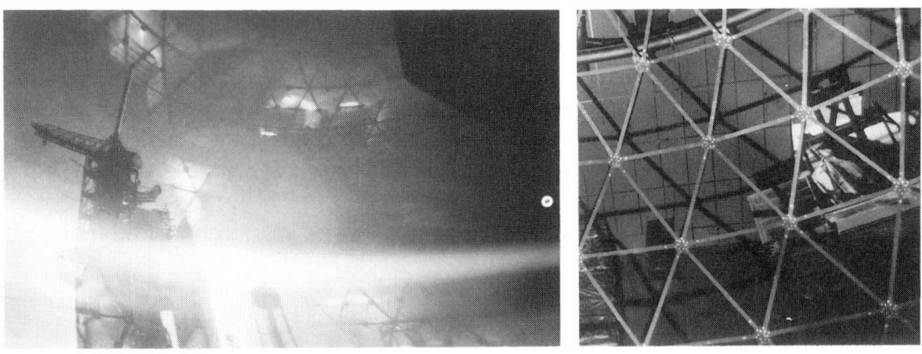

Fig. 3. An omnimax screen at EPCOT using a Geodesic type dome.

8. As a result of the study of panel types for several Schwedler and Geodesic systems it was learnt that a hexagonal panel supported on a triangulated geodesic frame would give the best visual appearance. It was also decided to use a butt-seam system to reduce the ability to see the seams. The Class II, method 3, 36 frequency, icosahedral geometry gave the best results in reducing the number of panel types and total seam length.

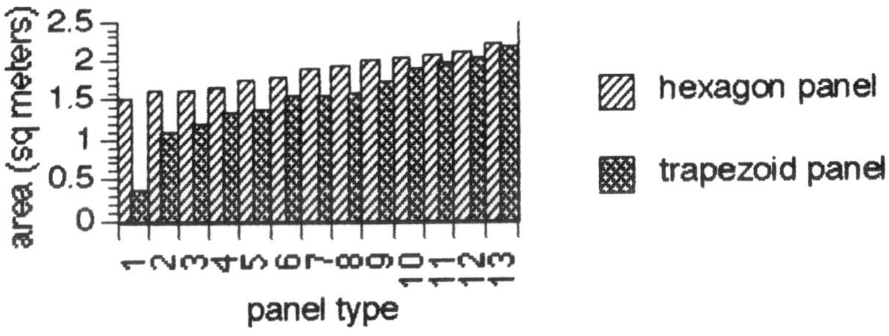

Fig. 4. Panel type comparison.

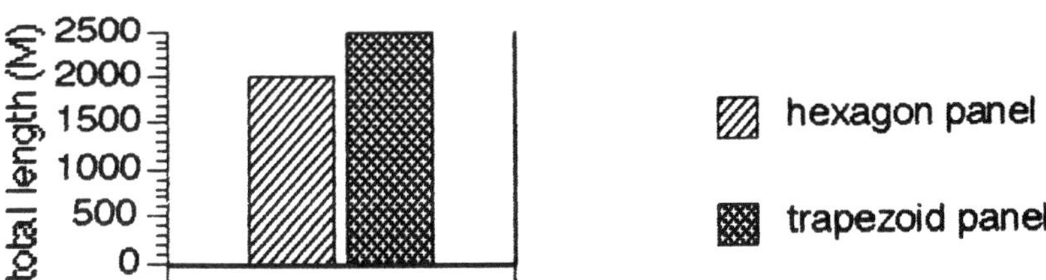

Fig. 5. Seam length Comparison.

9. From the same study it was determined that the support frame for the hexagonal panels would be a Class II, method 3, 12 frequency icosahedral geometry retaining the triangular nature so common to the geodesic type structures. The frame was also designed to support a revolving maintenance platform and ladder, a sprinkler system, and lighting system.

10. It was desirable for the cove to have the same appearance as the ceiling. However, a geodesic type geometry did not seem to be the best solution. A bi-polar geometry consisting of hexagonal panels within 180 gores and 18 zones was chosen. The frame was a triangulated system based on 60 gores and 6 zones.

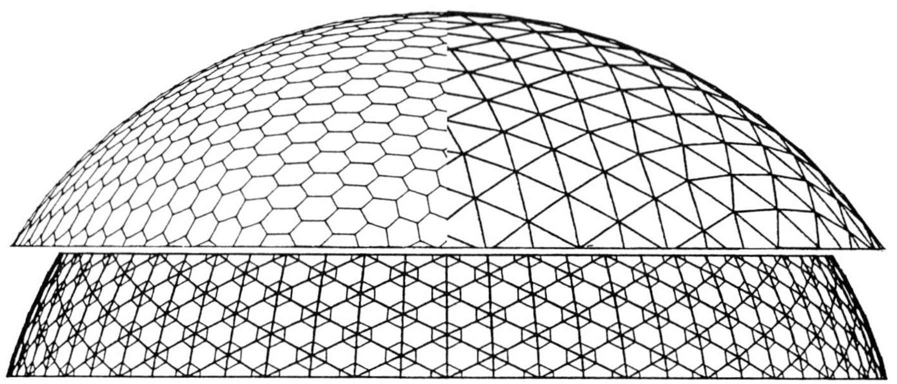

Fig. 6. Chrysler Technology Center styling dome.

11. The support frame consisted of straight members while the panels were to be formed into spherical shapes. From previous designs of butt-seam systems it was known of the importance of providing adjustability in the panel to frame connection. A unique solution was found that provided for the adjustability and at the same time remained hidden from view under the lighting conditions to be used within the structure.

12. To provide an acoustical surface for the dome and to prevent snap-through buckling of the aluminum panels a thickness of 2.36 mm with a staggered perforation pattern of IPA number 110 at 2.4 mm diameter perforation, and a 4.76 mm spacing providing 23% voids was chosen for both strength and acoustical properties.

13. All exterior surfaces were coated flat black to prevent reflected light from returning to the interior of the dome. And the interior surface was powdercoated with a white matt finish to provide an optically diffused screen.

14. The geometrical design and engineering analysis was jointly done by 3-D Structures and Star* Net Structures with a state-of-the art finite element modeler used to assist in the analysis of the structure. The FEA modeler was found to be

more than adequate in helping to determine the final selection of materials and sections used in its construction.

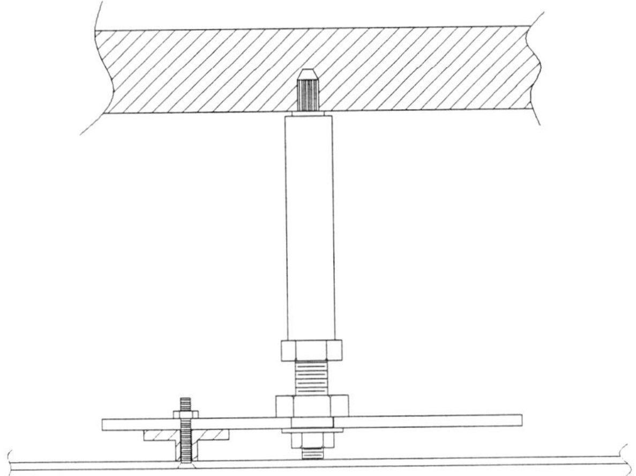

Fig. 7. Compliant frame to panel connection.

15. The detail design, fabrication and installation of the frame and revolving maintenance ladder was completed by Star* Net Structures. While the design, fabrication and installation of the panels was completed by 3-D Structures.

FABRICATION

16. Jig, fixture and precision machining techniques common to the aerospace field were used throughout the fabrication of the frame. This was necessary due to the precise installation techniques necessary for the installation of the butt-seam hexagonal panels. Stretching of the panels was done by using very accurate stretching dies and precise stretching equipment located in Canada. Fabrication of the panels was undertaken on specially designed cutting fixtures capable of holding .025 mm tolerances. The frame-to-panel connectors were designed with tight tolerances and machined on precision equipment.

17. A tight quality control was initiated on all parts to ensure the precision needed during the installation phase.

18. All parts were pre cut, inspected and finished with powdercoating prior to shipping to insure the best possible surface finish.

INSTALLATION

19. The installation of the frame was undertaken by supporting the structure at discrete points during construction and lifting the structure toward the building structure as a new row of struts were added. The final setting of the dome to its base ring was precisely placed by precision surveying techniques using a total engineering station so that the dome frame would be within 6 mm of uniform radius.

20. Installation and pressurising of the piping for the sprinkler system was completed before installing the panels to insure that no damage would occur to the finish on the panels. The revolving catwalk also was placed above and supported by the dome prior to installing the panels.

21. The most critical phase of installation was the setting of the frame-to-panel connectors and placement of the panels. Even though the location of the panel connectors on the frame were preset and their offset distances were factory set, final adjustments were made in the field. The installation crew consisted of people used to working with precision because each panel was aligned with precise optical equipment and specially designed distance measuring equipment originally developed for use in installing aerospace flight simulator screens. The panels were located to 5 arc second accuracy in azimuth and elevation and to 1.6 mm of dome radius, resulting in a panel separation at the seam of no more than 1 mm. Some separation is necessary for thermal expansion. Final edge adjustments were made to ensure spherical planarity along each edge so that shadowing from the cove and parameter lighting did not occur.

CONCLUSIONS

22. The new butt-seam acoustical perforated aluminum domed ceiling was successfully completed in 1992 and has advanced the state-of-the-art for structures of this type. The structure consisted of a fully triangulated self supporting geodesic frame with compound curved hexagonal panels used as interior cladding.

23. The hexagonal panels provide the optimum tessellation of a spherical surface, making a more visually pleasing pattern than the conventional trapezoid tessellation more commonly used. The uniquely designed compliant connector made possible a successful butt-seam dome with non-discernable features by its adjustable characteristics. The overall seam length was well below the seam length of the typical gore-zone based constructions seen in most installations.

24. By using a geodesic icosahedron geometry the structure provided optimum rigidity for the member size. The use of a completely triangulated structure of this type allows the geometry to be totally governed by the length of its members without the requirement for moment connections which are needed for the typical Schwedler type system.

25. The marriage between the hexagonal tessellation of the geodesic type domed ceiling and the hexagonal tessellation of the Schwedler type cove complemented each other well. The use of each system for its own optimum set of conditions provided the best choice for the total design of the styling dome for the Chrysler project.

157. New developments of space structures in China

T. T. LAN, Chinese Academy of Building Research, China

SYNOPSIS. This paper gives a state-of-the-art review of the recent developments of space structures in China. The space structure is an ideal solution for covering long-span buildings. For example, space trusses are popular to cover aeroplane hangars, and different types of hybrid structure have been applied to sports buildings. One of the most important fields of application of the space truss is in single-storey industrial buildings, this is illustrated by several recently built huge industrial plants. A type of composite space truss has been studied and found appropriate for floors of multi-storey buildings. Some of the general aspects regarding the executed projects on various applications are described.

INTRODUCTION

1. The theoretical research and engineering practice on space structures has been undertaken in China since the 1950s. Over forty years, the development of space structures has shown remarkable progress due to the great structural potential and visual beauty. New and imaginative applications of space structures are being demonstrated in the total range of building types, such as sports halls, exhibition pavilions, aeroplane hangars, workshops, storehouses, assembly halls and terminal buildings. There have been a variety of structural forms like space trusses, reticulated shells and cable-suspended structures, in addition, hybrid structures have also been developed. Examples of the widespread application may be found not only in short or medium span buildings, but also in long-span buildings, which varied from 10m or 20m to a maximum of 110m. The developments of space structures in China have been reviewed in refs 1-4.

2. With the advent of computers, analysis of the space truss by the matrix-displacement method is a routine procedure in the engineering office. Some of the more sophisticated computer programs can execute automated design and even CAD or CAM. Complicated calculations, such as the stability analysis of single-layer reticulated shells or the non-linear analysis of cable-suspended structures, have also put into practice using mainframe computers.

3. Many kinds of erection technologies were employed in the construction of space structures. The construction process falls into two main categories, the

structure can be either assembled in the air, or assembled at ground level and then hoisted into position. Some innovative methods have been developed with success, the unique feature being to erect a large structure using small non-specialised plant.

LONG-SPAN ROOFS

4. Covering of large areas by designing new types of structures has always presented a challenge to engineers. The emergence of larger aeroplanes and the need for greater operational efficiency has created a demand for hangars with greater spans. In the past few years, space trusses have been used to cover hangars in airports situated at Shanghai, Changsha and Xianyang with spans of around 50m. For Guangzhou Baiyun Airport, the maintenance hangar has a span of 80m (ref. 5). The roof design was complicated by the requirement to accommodate the high tail and low fuselage of the aeroplane and the suspended cranes with maximum point loads of 275kN. An integral space truss composed of high and low bays at different elevations was adopted. A triple-layer space truss was employed as an edge girder over the hangar door as well-as a stiffening box girder at the junction of high and low bays. The members for the main girders are shop fabricated H-shaped welded sections, while the members for the space trusses are double-angle sections. They are all connected by gusset plates by means of high-tensile bolts. All members were preassembled on the ground in blocks, then lifted by cranes and erected. Fig. 1 shows the erection of the roof structure.

Fig. 1. Hangar for Guangzhou Baiyun Airport.

5. The space truss has been adopted for even larger span hangars. A recent example is the maintenance hangar at Chendu airport with a span of 140m. The overall size of the roof structure is 140m x 87m with a clear height of 26m. Two-

way orthogonal lattice grids were used for the roof which has a module size of 7m and depth of 8m at the front decreasing to 6m at the back. Under the bottom chords, 3 travelling cranes of 10t capacity are suspended, and also fuselage maintenance docks. Thus the roof structure does not only sustain all the loads, but also takes care of the dynamic effects with a strict limitation on deflections due to the suspended equipment. The space trusses are supported on columns along three sides and on a hingeless portal frame spanning 140m over the door opening. The horizontal girder is composed of triple-layer lattice grids with a depth of 12m.

6. One of the noticeable trends of long-span buildings built in China is the variety of structural forms. Space structures can create a unique and aesthetically pleasing form and at the same time meet the functional requirements of the building. In this respect, the hybrid construction will allow designers to utilise a combination of different types of structures so as to enhance the performance of the system. It can fully make use of the advantages of certain type of structure and avoid or offset the disadvantages of other types of the combined structure. In many sports buildings, instead of the space truss used in the earlier stage, different types of hybrid construction have been developed which are shown in the following examples.

Cable and truss

7. A new type of transversely stiffened, single curvature, cable suspended structure can be treated as a kind of hybrid structural system combining lattice trusses with cables (ref. 6). Generally, the single curvature cables have low rigidity and vulnerable to mechanismic deformation under unsymmetrical loading. If trusses, as transverse stiffening elements, are set orthgonally on the cables and their end supports are then connected to the top of the column by forced displacement, pre-tensioning will be introduced in the cables. In such an integral system, the flexible cables can resist the main loading, utilising the high-strength steel, at the same time, trusses can improve the stiffness of the roof due to their inherent flexural rigidities, and also can share and transmit the unsymmetrical loading effectively. This type of construction has been used for the Anhui Gymnasium which is hexagonal in shape with an overall dimension of 72m x 69m, and also for the Yangpu Gymnasium in Shanghai, a rectangular building of 54m x 45m. Comparing with other roof structures of a similar size, they showed the distinct advantages of economy both in material consumption and construction cost. A recent example is the Chaozhou Gymnasium, a 56m square building. The roof structure comprises 24 cables and 14 trusses set orthogonally. Different from the previous buildings, the cables and trusses are laid parallel to the diagonals of the square plan so that they are skewed at 45 degrees to the boundary. The sagging cables and convex-shaped trusses form a saddle surface on the roof as shown in Fig. 2. The roof structure has a prominent architectural feature by the innovative application of the hybrid system.

Fig. 2. Exterior view of Chaozhou Gymnasium.

Rigid frame and cable

8. The Dandong Gymnasium is one of the few examples that employs a rigid frame as the main supporting structure in the middle of the competition hall. Two sets of singly curved cables span 40m across the hall, with the higher ends anchored at the rigid frame and the lower ends on the horizontal ring beams of the stand frames. The sag of each main cable is 2.36m giving sag/span ratio of 1/17. The girder and columns of the 48m span rigid frame were constructed in reinforced concrete box sections. The rigid frame took about 85% of the total roof load and consumed nearly half of the steel.

Arch and space truss

9. Arches, arising from the ground, were frequently used in combination with other types of structure. A pair of parabolic arches were used for the Jiangxi Gymnasium as the main support in order to reduce the span of the space truss covering the roof and to support monitors in the middle of the hall. The building is quasi-hexagonal in shape measuring 84.3m x 64.4m. The arches in reinforced concrete box sections with a span of 88m over the roof with a rise of 51m. The trapezoidal shaped space trusses, composed of triangular pyramids, were supported on a pair of lattice trusses suspended from the arches.

Arch and cable net

10. A pair of arches in combination with cable nets were used in two large gymnasiums. In Sichuan Provincial Gymnasium, the plan is in hexagonal shape which measures 72.4m x 79.4m. A pair of mutually inclined reinforced concrete parabolic arches span 102.5m across the middle of the building with a rise of 39m. On the roof, two trapezoidal cable nets of anticlastic curvature are supported on the arches and the surrounding rigid frames. In the Qingdao Gymnasium, the oval shape building is 72m by 87m in size. A similar pair of arches, intersecting with each other in the central portion, are used in combination with two semi-

elliptical cable nets. The arches have a span of 109m and a rise of 26m. The Qingdao Gymnasium is shown in Fig. 3.

Fig. 3. The Qingdao Gymnasium.

Cable-arch and cable net

11. A special type of cable-arch was employed as the supporting structure in the middle of the Chaoyang Gymnasium in Beijing built for the Asian Games (ref. 7). The cable-arch is composed of a pair of hanging cables and two parallel lattice steel arches connected by vertical and horizontal tie bars. The gymnasium is elliptical in plan with axes of 66m and 78m. Two saddle-shaped cable nets are supported on the cable-arch and peripheral ring beams, the maximum span for the load carrying cables being 31.5m.

Cable stayed hybrid structure

12. In the National Olympic Sports Centre in Beijing, two main buildings were built for the 1990 Asian Games. The Sports Hall and Swimming Hall have a similar exterior view but are of different construction. The size of the Sports Hall is 70m by 83m, while the size of the Swimming Hall is 70m by 117m, both in a quasi-rectangular plan. They both have two reinforced concrete towers at each end of the hall with cables suspended from them. For the Sports Hall, two cylindrical reticulated shells are connected to a spatial lattice truss along the ridge. For the Swimming Hall, 7 curved lattice trusses are connected to each side of a steel box-girder. The roof structure is suspended by the cables. As shown in Fig. 4, the buildings reflect an interesting combination of the Chinese traditional architectural style with the contemporary structural technique (ref. 8).

Fig. 4. The National Olympic Sports Centre.

LARGE AREA INDUSTRIAL BUILDINGS

13. One of the most important fields of application of the space truss is in single-storey industrial buildings. Large column grids of space trusses can provide an unobstructed area with minimum interference from internal supports creating maximum flexibility in production. The trend of using space trusses started in early 1980s and it has expanded to a very large scale. The space trusses are very competitive, both in construction time and cost, when compared with reinforced concrete girders or steel lattice trusses. It is estimated that there are at least a dozen space truss roofs which cover areas exceeding 10,000m^2. Applications have been widespread, and include automobile works, steel plants, rolling stock plants, chemical works, machinery factories and textile mills.

14. The processing factory of the Tianjin Steel Tubing Mill employs continuous space trusses supported on a column grid of 36m by 18m. In the transverse direction, it has 3 spans of 36m. In the longitudinal direction, the length is 564m, with a column spacing of 18m. The total area amounts to 60,000m^2. To accommodate the production of tube processing, triple-layer grids were used to form the transverse monitors on the roof and overhead cranes of 12.5t capacity were supported on crane girders. The members made of circular steel tubes are connected by bolted spherical nodes. The total steel weight was nearly 2000t, giving an average of 32kg/m^2. Compared with conventional steel trusses, the application of the space truss leads to a substantial saving of steel by 47%.

15. In the first automobile works in Changchun, a huge assembly plant for Golf sedans has recently been completed. The dimensions of the plant are 421.6m in length by 189.4m in width, covering a total area of 80,000m^2, which is believed to be one of the largest space truss roof structures in the world. The entire building is divided into 10 blocks by 4 transverse expansion joints and 1 longitudinal joint. The column grid is 21m by 12m. The space truss was designed

for heavy loading as required by the production technology, which included 2.5kN/m^2 for suspended conveying equipment and 0.5kN/m^2 for ventilating fixtures. Two-way orthogonal lattice grids were used so that the mechanical and electrical units could pass through the openings between the top and bottom layers. Corresponding to a rectangular column grid, the module size was taken as 3.5m x 3.0m. The depth varies from 3.0m at the mid-span to 2.8m at end supports. This will lead to a more rational distribution of forces in the members and also formation of slopes for drainage on the roof. More than 40,000 circular steel tubes were used for the members and connected by welded hollow spherical nodes. The total amount of steel is 2490t, corresponding to 31kg/m^2. Fig. 5 shows the construction of space trusses.

Fig. 5. Assembly plant for First Automobile Works.

MULTI-STOREY BUILDINGS

16. Extensive studies have been undertaken into a type of composite construction for the space truss. Since reinforced concrete slabs are frequently used as the cladding for space trusses, they can replace the top chord members and act integrally with the steel members. The upper portion of a space truss is generally in compression and lower portion in tension, such a composite structure combines steel and concrete materials using the major characteristic of each material acting together. Since 1980, more than 20 buildings have been built using composite space trusses. The largest one which measures 45.5m by 58m is a roof structure for the Fuzhou Gymnasium in Jiangxi province.

17. It was found that the composite space truss is especially suitable for floors of multi-storey buildings. A successful example is the extension of a department store in Xinxiang city, Henan province. Four storey floors of composite space trusses were constructed on top of the original two storey building without interference of the business (Fig. 6). The floor structure is 35m x 35m square with one clipped corner. Diagonal on square space grids were selected for the

space truss. The module size in the top layer is 3m x 3m, which is also the size of the precast reinforced concrete slab. In the bottom layer, the module is 4.25m x 4.25m while the depth of the space truss is 2.5m. The bottom chord and web members made from double angle sections were connected by welding. The detail of the top chord joint is shown in Fig. 7. A loading test has been performed, during construction. The excellent behaviour of the composite structure was verified by the integral action of reinforced concrete slabs and steel members.

18. In the two storey gymnasium of Zhibo Physical Training School in Shandong province, a composite space truss was also used for the floors. The floor, 36m x 51m in plan, was divided into two separate areas of 36m x 25.5m each by establishing intermediate columns. Two-way orthogonal lattice grids were used, the geometrical dimensions used for the module grid and depth were all 3m. Bolted spherical nodes were applied to connect the tube members. A special type of node for the top chord joint was developed as shown in Fig. 8. Since the node is made of high strength steel of grade C45, a circular plate of low carbon steel is attached to the node by high tensile bolts. During assembling, the lower anchor plates of the concrete slab were welded to the circular plate and then the joints were filled with in-situ concrete. This constructional detail will ensure the jointing of the steel members with the reinforced concrete slabs.

Fig. 6. Composite space truss for Xinxiang Department Store.

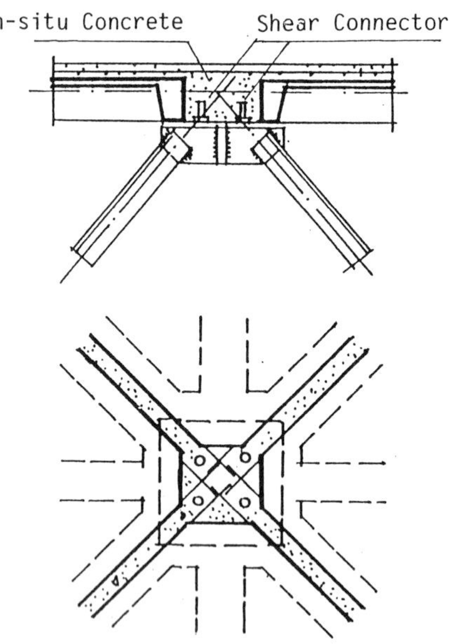

Fig. 7. Detail of top chord joint.

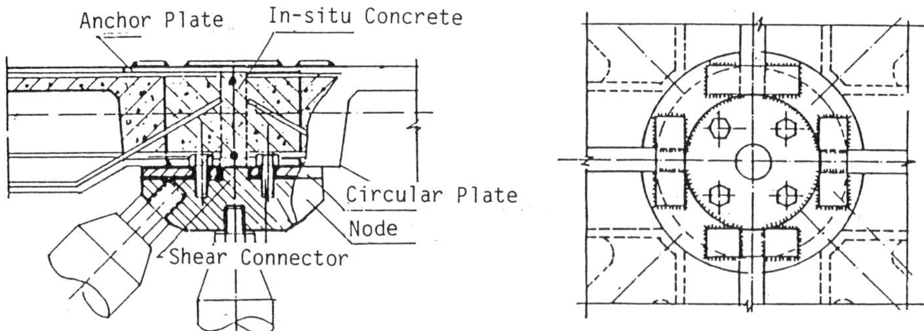

Fig. 8. Top chord joint for composite space truss.

REFERENCES

1. LAN Tien T. Recent trends and developments of space trusses in China. Proceedings of 3rd International Conference on Space Structures, 1984, 22-27.

2. LAN Tien T. Space structures in China, Space Structures, 1985, Vol. 1, No. 3, 155-160.

3. LAN T. T. A review of recent developments of spatial structures in China. Bulletin IASS, 1986, Vol. XXVII-2, No. 91, 51-56.

4. LAN T. T. Contemporary spatial structures for long-span roofs in China. Proceedings of IASS International Symposium, 1991, Vol. 1, 17-25.

5. LIU S. The 80m integral space truss with high and low bays used in Guangzhou Baiyuan Airport hangar. Journal of Building Structures, 1988, Vol. 9, No. 4, 54-61 (in Chinese).

6. LAN T. T. and ZHAO J. Theory and practice of transversely stiffened single curvature cable-suspended roof. Proceedings of IASS-CSCE International Congress, 1992, Vol. 2, 355-364.

7. JIANG Z. Application of large-span prestressed cable-arch and cable-grid structure to the roof of a gymnasium. Proceedings of International Symposium on Modern Applications of Prestressed Concrete, 1991, 407-414.

8. CIU Z. and ZHANG G. Roof structure design of the gymnasium in National Olympic Sports Centre. Journal of Building Structures, 1991. Vol. 12, No. 1, 24-37 (in Chinese).

158. Design extension of a multi-purpose hall

M. I. ROZVAN, Timisoara Design Institute, Romania, and
M. V. SOARE, Civil Engineering Institute of Bucharest, Romania

SYNOPSIS. A multi-purpose hall designed and erected in 1968-9 and having a seating capacity of 2500 is now under capacity for events such as international sporting events. Two alternative solutions were considered. Firstly, the design of a new multi-purpose hall having double the seating capacity of the existing hall. Secondly, doubling the capacity of the existing hall by an appropriate extension to the structure. The second variant was chosen on economical grounds. The existing roof consists of a double-layer grid supported by four reinforced concrete walls. The extension will involve the demolition of one wall, the extention of the neighbouring walls and the addition of a new closing wall. The new space will be covered by means of a new double-layer grid similar to the existing one but of opposite slope. The neighbouring sides of the two roofs are connected with bars, using *in-situ* welding. This new throat is suspended, by means of ties, from an external steel portal frame. The paper gives full details of the adopted solution.

INTRODUCTION.

1. The multi-purpose hall Olympia of Timisoara-Romania was designed in 1968 and erected in 1968-9 with a seating capacity of 2500. At the beginning it had a convenient capacity.

2. The new construction has lead to the development of the sporting activities and the multi-purpose hall became cramped for internal events and particularly for international sportive activities.

3. Two solutions to this problem were considered. The first consisted of the design of a new multi-purpose hall having double capacity. The second proposal was to double the capacity of the existing hall by means of additional construction.

4. The first solution, that of building a new hall to seat 5000, proved to be too expensive an option and the municipality could not offer a central site for the construction, as was the case for the existing hall.

5. The second solution was accepted and it is the purpose of the present paper to briefly describe the proposed solution.

6. The present multi-purpose hall to be extended consists of the arena and a single side stand with 2500 seats; doubling the number of seats of the hall would be possible by providing a second stand, identical to the existing one, on the other side of the arena. This fact supposes the demolition of a main bearing wall which supports one of the four sides of the reticulated roof.

7. The proposed solution ensures the extension of the actual hall by introducing an external steel portal frame supporting both the side of the existing roof (no longer supported by the demolished bearing wall) and the side of the new roof. Both sides are supported from the steel portal frame, while the other sides are supported on the existing bearing walls and the new bearing walls respectively.

8. Around the extended hall are provided locker rooms, protocol rooms, gymnasia, sauna, etc.

9. The sequence of execution is given as follows:

(i) building of the first group of locker rooms;
(ii) demolition of locker rooms on the extension side;
(iii) construction of the external steel portal frame;
(iv) construction of the new stand and of the three new bearing walls which enclose the new stand;
(v) erection of the new reticulated roof over the stand, which is supported upon the three new bearing walls and the wall which is to be demolished;
(vi) suspension of the existent reticulated roof (46 x 46m) and the new one (23 x 46m) on the external steel portal frame;
(vii) building of the locker rooms, protocol rooms, etc;
(viii) demolition of the intermediate bearing wall;
(ix) completion of finishing works in the extended hall and in the remainder of the complex.

DESCRIPTION OF THE ACTUAL MULTI-PURPOSE HALL

10. The structure of the multi-purpose hall consists of a steel double-layer grid supported on a skeleton in monolithic reinforced concrete, consisting of longitudinal and transverse frames.

11. The reticulated roof has a rectangular platform with the general dimensions 46 x 55m and a mesh size of 2m. The height of the structure is of 2.5m. The support of the reticulated structure is along a square contour with the sides 46 x 46m; thus, there results a central structure supported on the boundary and two cantilevers having a span of 4.5m, which lie along the sides parallel to the long side of the arena.

12. The stand is situated on one side of the arena, in the longitudinal direction; the framed structure consists of inclined girders over three unequal spans.

DESCRIPTION OF THE PROPOSED SOLUTION

13. The increase of the number of seats is achieved by the construction of a new stand opposite to the existing one.

14. The roofing of the new zone consists of a square-on-square double-layer grid having the same dimensions to the existing one (mesh size 2m, height 2.5m)

but of opposite slope. The general dimensions of the platform are 23 x 46m (Fig. 1).

15. This solution involves the demolition of the wall which also includes the longitudinal frame which supports the existing roof structure. Also the 4.5m overhanging roof structure on the same side has to be disassembled.

16. The support of the new double-layer grid will be on the monolithic reinforced concrete walls constituting the future three external sides and in the interface of the two reticulated roofs the previously mentioned steel frame will allows the suspension of the two neighbouring double-layer grids by means of ties, spaced at 6m.

17. The present solution has the advantage that it removes completely intermediate supports, the presence of which would reduce largely the view of the arena from stands.

18. The execution of the new construction does not impose any interruption on the running of the existing hall.

DESCRIPTION OF THE ADOPTED CONSTRUCTIONAL SOLUTION

19. The suspension frame is represented by a trussed space portal frame having the theoretical span of 62.0m and a total height of 17.40m. The girder height h = 5.40m has resulted from the limiting condition of the maximum deflection and from optimisation of the steel consumption. The limitation of deflections avoids the introduction of supplementary forces in the existing double-layer grid roof.

20. The space frame consists of two plane frames with bracing members at 4m centres, pinned together by means of transverse wind-bracings in every panel of the girder and the whole height of the columns.

21. The girders of the plane frames consist of a principal system of bars of length 6m and a secondary system of diagonals spaced at 2m in order to supply intermediate points at the bottom flange, corresponding to the points of the double-layer grids.

22. The frame columns consist of four members interconnected by crossed diagonals in both directions.

ANALYSIS DATA

23. The static analysis of the frame considered the reactions of the supporting sides of the two neighbouring double-layer grids; these reactions are induced by the following loads:

- self-weight of the two double-layer grids;
- self-weight of the roof cover;
- snow load (including snow drift loads due to the shape of the roof);
- horizontal seismic action.

24. For the determination of bar forces a computer program has been used based on the most unfavourable combination of loads.

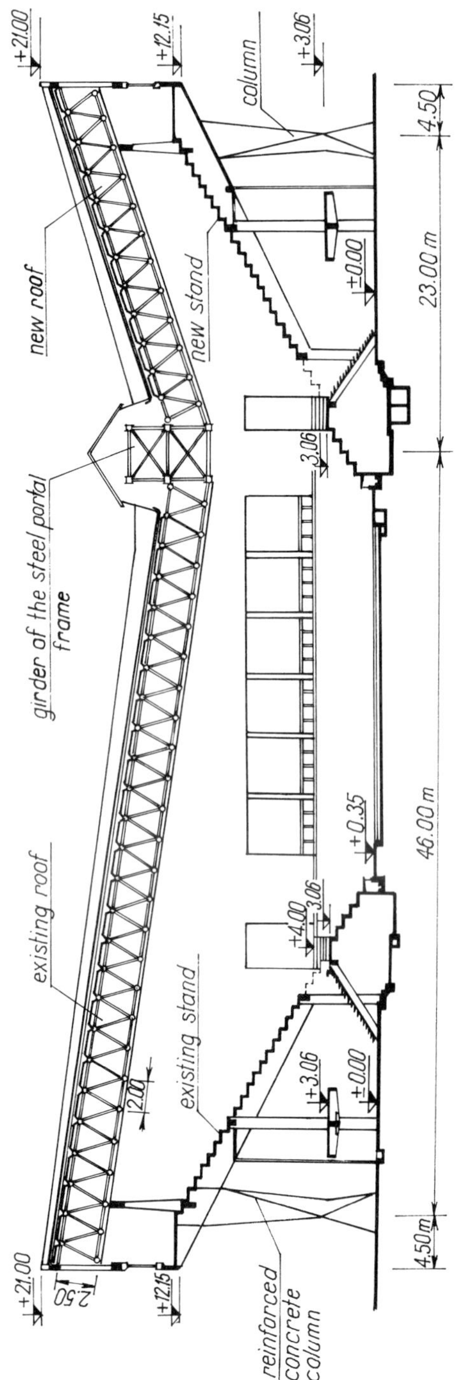

Fig. 1. Cross-section of the design extended multi-purpose hall.

25. The proportioning was based on the allowable stress method ($\sigma_a = 160$ N/mm^2) and using hot-rolled sections (according to Romanian standards).

26. The shape of cross-sections are as follows:
- π–shaped of cross-sections for the flanges of the girders and columns;
- double-tee-shaped cross-sections (consisting of welded bands) for the main diagonals of the girder and for the columns;
- two channel-shaped cross-sections (hot-rolled sections) for the secondary diagonals and the transverse wind bracings.

27. The frame consist of machined transoms assembled by means of high strength prestressed bolts.

28. For the assembling of these transoms in the final position a special scaffolding was provided along the whole length of the frame. The assembly of the roof structure will be done by means of transoms of 4.0 x 11.5m, assembled above ground two by two, on a special platform; the transoms obtained in this manner have the dimensions of 4.0 x 23.0m which will be erected in the final position. The connecting bars between transoms will be welded *in-situ* to the neighbouring points.

29. A detailed plan of manufacturing, pre-assembling, transport, erection and labour protection were detailed in the project, but they are not reported here.

159. Design and construction of a 1/30 scale model of the Palasport in Bologna

R. ZARRI, Instituto di Technica delle Costruzioni, and L. J. TASKOV, D. JURUKOVSKI and M. BOJADZIEV, University Sent Ciril and Metodius, Macedonia

SYNOPSIS. The dynamic behaviour of the specially shaped structure, such as the Palasport in Bologna, Italy, under the wind or earthquake exicitation, mostly depend on the dynamic characteristics of the structural system. On the other hand, the dynamic characteristics are closely related to the geometric shape of the structure, mechanical characteristics of the construction material and permanent loads (dead load, snow etc.). In order to investigate the stress distribution in double curvature, lamellar wood covering structure of the Palasport, a 1/30 scale model was designed and constructed at the Institute of Earthquake Engineering Seismology in Skopje, Republic of Macedonia. The model was made of almost the same material as the prototype, so the artificial mass simulation modelling technique has been applied. In this paper, the design of the model and some interesting details of the construction are presented. The testing will be undertaken in February 1993, from which the behaviour of the model under various static and dynamic conditions will be defined.

INTRODUCTION

1. The Palasport in Bologna is a very special structure, almost elliptical in plane with dimension of median axes of 80/120m, and a column height of about 15m. The cover has been constructed using lamellar wood beams linked by steel joints, forming one cylindrical and two half spherical parts consisting of 52 reinforced concrete columns with cross-section 60x60cm with 26 supported by two steel braces from the external side of the structure. The radius of the dome part in plan is 40m, while the radius of the cylinder is 63m, so the height of the cover is 14m. Connections between the wooden beams are made by steel rings (internal joints) while the external joints between the columns and beams are vertical steel plates and half rings (Fig. 2). The construction took place during 1992. In order to investigate the excessive loading of the cover by the snow (symmetric and nonsymmetric) as well as by the dynamic loads, a model in geometric scale 1/30 was constructed for testing on a shaking table at the Institute of Earthquake Engineering Seismology, University "St. Ciril and Metodius" in Skopje, Republic of Macedonia.

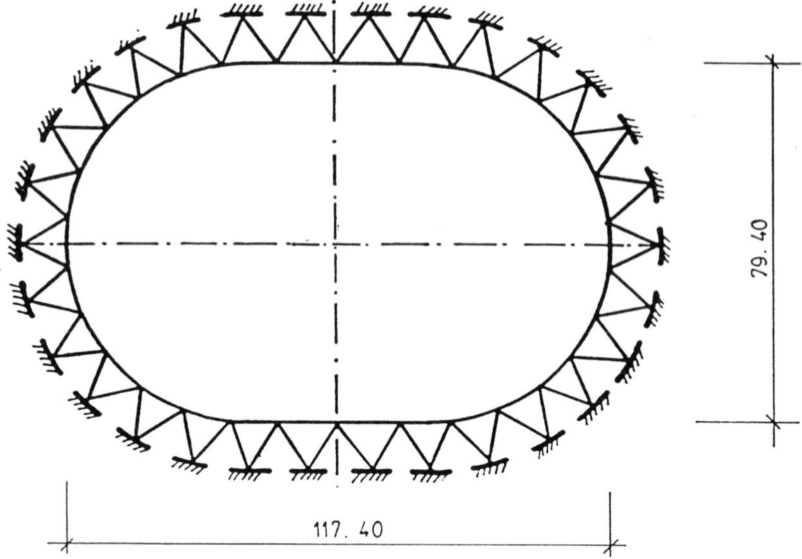

Fig. 1. Disposition in plan of the Palasport-Bologna.

DESIGN AND CONSTRUCTION OF THE MODEL

2. Considering the geometry of the Palasport in Bologna and the dimensions of the shaking table at IZIIS Dynamic Testing Laboratory (5/5), a 1/30 scale model was adopted as the most convenient one. Calculation of the model parameters have been performed following the principles of similitude between the model and prototype based on the Buckingham theorem. For proper simulation of the inertial as well as the gravity forces, an artificial mass simulation modelling technique has been applied. Table 1 shows the required proportions between the prototype and model parameters for the considered geometrical scale of 1/30.

3. Based on the scaling factors given in Table 1, numerical values of modelling parameters are given in Table 2. The model itself is built of the similar materials to the real structure. The columns are built of micro-concrete, the supporting braces are fabricated from steel, and the cover structure is of wooden beams. The Young's modulus of model materials is nearly the same as the prototype. This modelling technique offers reliable simulation of stress-strain relationships both in the linear and nonlinear state, if required additional mass is uniformly distributed on the roof of the model. Fig. 3 shows the model geometry. The plan dimensions of the model are 2.6x3.9m, the height of the columns is 0.5m, while the height of the roof is 0.47m, giving a total height of 0.97m. The dimension of the structural elements are adopted as follows: the cross-section of the concrete columns 2x2cm, of steel braces: d=1.8 cm, of the wooden beams: 0.5x4.0cm and of the steel rings: d=2.2cm. The steel supporting braces are supported on concrete foundations.

Fig. 2. Internal and external connections joints.

Table 1. Similitude requirements for 1/30 scale model of Palasport - Bologna.

Parameters of similitude	Required scaling factor	Achieved scaling factor
lenhgt (l_r)	l_r=1/30	1/30
time (t_r)	$l_r^{1/2}$ = (1/30)$^{1/2}$	(1/30)$^{1/2}$
frequency (ω_r)	$l_r^{-1/2}$ = (30)$^{1/2}$	(30)$^{1/2}$
gravity acceleration (g_r)	1	1
acceleration (a_r)	1	1
Joung's modulus (E_r)	E_r = 1	1
mass density (ρ_r)	1	1
added mass (m_r)	l_r^{-1} = 30	30
strain (ε_r)	1	1
stress (σ_r)	1	1
cross sections (A_r)	l_r^2 = (1/30)^2	(1/30)^2
moment of inertia (I_r)	l_r^4 = (1/30)^4	(1/30)^4
force (F_r)	E_r l_r^2 = (1/30)^2	(1/30)^2
displacement (δ_r)	l_r	1/30

All of the structure is constructed on the steel foundation plate which will be connected to the shaking table. All structural elements are assembled by welding the joints or by bolts with bolts of 3mm diameter. The welding is performed by the argon welding technique. Connection between the concrete and steel elements is made by built in the steel plates in the concrete elements.

4. In the first phase, the lower structure consisting of columns, braces and supporting ring is assembled (Fig. 4). In the second phase, the wooden roof structure is assembled using the wooden forms (Fig. 5).

THE PLANNED TESTING PROGRAMME

5. The model will be completed in January 1993. The test on the shaking table will be performed in February 1993. Before subjecting the model to dynamic excitation, a static test will be performed by application of symmetric and asymmetric load onto the roof to simulate snow loading conditions.

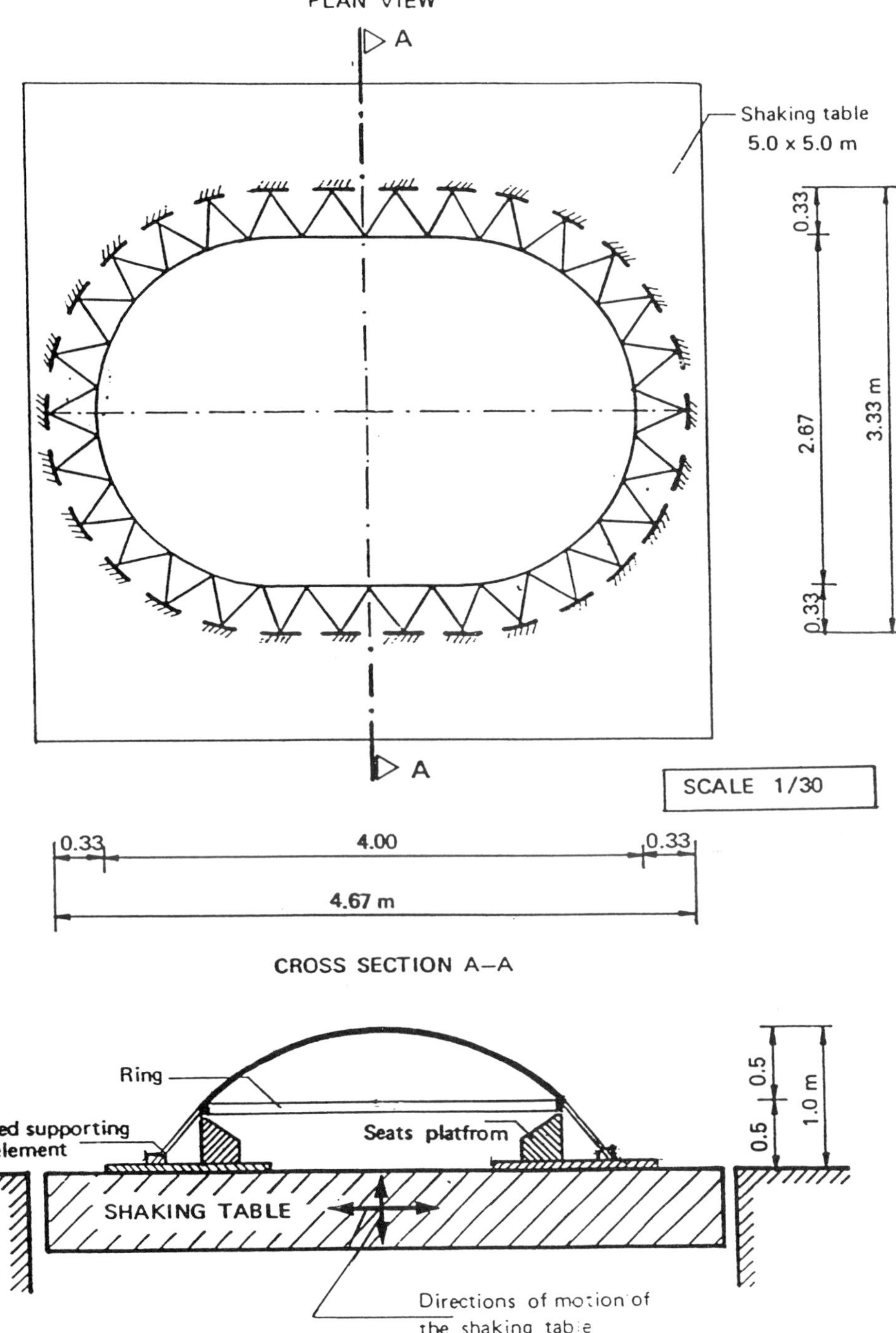

Fig. 3. Model of the Palasport in Bologna at a scale of 1/30 for the shaking table test.

Table 2. Prototype / Model parameters.

PARAMETERS		PROTOTYPE	MODEL
Dimensions at base (m)		80/20	2.60/3.90
Hight of the columns (m)		15.0	0.50
Hight of the cover structure (m)		14.0	0.47
Total hight of the structure (m)		29.0	0.97
Weight of the roof (kN)		9700	10.70
Compression strength (MPa)	concrete	$\sigma_c = 30$	$\sigma_c = 30$
	steel	$\sigma_s = 240$	$\sigma_s = 240$
	wood	$\sigma_w = 40$	$\sigma_w = 45$
Joung's modulus (MPa)	concrete	$E_c = 31000$	$E_c = 25000$
	steel	$E_s = 200000$	$E_s = 200000$
	wood	$E_w = 11000$	$E_w = 12500$

Fig. 4. Lower structure of the model of the Palasport in Bologna.

Fig. 5. Part of the wooden roof structure of the model.

Fig. 6. Overview of the Palasport - Bologna.

Fig. 6. continued.

The stress distribution in the covering structure will be measured in this case as well as the reaction forces at supporting columns and beams. During the dynamic test, the following testing programme should be undertaken: the dynamic properties test, harmonic excitation test for linear elastic state, low to moderate earthquake excitation test, high intensity earthquake excitation test to provoke the nonlinear response of the model in the critical joints.

CONCLUSIONS

6. The objective of the shaking table test on the 1/30 scale model is the verification of the seismic stability of the covering wooden shell structure as well as the supporting structure of the Palasport-Bologna, Italy. The most important problems that should be investigated are: behaviour of the structural elements made of three different materials (wood, concrete and steel) interconnected by a special connections, under various dynamic conditions; the weak points of the structural system; the non-linearities that are expected to occur under very strong earthquakes; the damage mechanism; the damping capacity; the limit capacity of structure in respect to strength and deformabilty; the safety factor etc. Analytical prediction of all these parameters by computer graphic simulation is possible but not sufficiently accurate because of the fact that some of these mechanisms are very complicated for analytical description. Experimental methods applied in laboratory or field conditions should provide the necessary information for proper modelling and analytical prediction for all other response parameters.

ACKNOWLEDGEMENTS

7. The authors would like to express their sincere appreciation for the technical support given by the following Italian companies to the laboratory test: COGEI SpA - Bologna; COOPCOSTRUZIONI - Bologna; C.F.M. - Modena; ERREVI PREFABBRICATI - Bologna; HOLZBAU SpA - Bressanone; and S.E.P. PREFABBRICATI - Bologna.

160. Structural solutions with cold-formed profiles for double and single layered grids

D. DUBINA, Technical University of Timisoara, Romania

SYNOPSIS. The paper presents three new structural systems for prefabricated double-layer grids formed from units using thin-walled cold formed steel members connected by bolded joints. Also a new bolted connection solution for single-layer grids is presented. Both the structural system and joint solution are covered by Romanian patents.

NEW PREFABRICATED DOUBLE-LAYER GRID SYSTEMS.

Plane Double-Layer Grid Structures - PDLG

1. This type of structure, Fig. la, (ref. 1) is built from standardised parallelipipedal sub-assemblies (modules) whose corners are joined by means of ordinary or prestressed high tensile bolts. Modules are made from angles, U or C sections or rectangular cross-section tubular members. Fig. 1b shows the connecting elements for a top-layer node.

2. Tables 1 and 2 present member cross-section dimensions obtained using angle sections for chords, diagonals and struts for a 30m by 30m span structure; the bay length and height are b = 3.0m and h = 1.8m respectively.

Rare Double-Layered Grid Structure - RDLG

3. This type of structure, Fig. 2a, is built from standardised pyramidal sub-assemblies (modules) joined together by means of square frames to form a top layer and square frames or independent bars to form the bottom layer. Fig. 2b shows the author's solution for a bottom-layer node. All joints are made using prestressed high tensile bolts. The jointing system must provide rigid connections, to prevent becoming a mechanism.

4. Top and bottom chord members can be made of U or C sections or rectangular cross-section members. Diagonal bars are made from tubular members. The jointing system uses a cross-shaped central connector similar to a jointing system patented by the author (ref. 2).

Space structures 4. Thomas Telford, London, 1993

5. Table 3 gives member cross-section dimensions for a 30m by 30m span structure; the upper bay length, the bottom bay length and the height are denoted respectively by b1, b2 and h.

Space Double-Layer Grid Structure - SDLG

6. This type of structure, shown in Fig. 3, is built from standardised pyramidal sub-assemblies and forms grids with two or three panels. The connection system and member cross-sections are basically identical to those in the two systems dealt with above. In this case it is not necessary for the joint to be rigid.

7. Table 4, presents cross-section dimensions for a 30m by 30m span structure.

NEW BOLTED JOINT SYSTEM FOR SINGLE-LAYER GRIDS

8. Basically, this jointing system (ref. 3) is similar to that described above for double-layer grids, where the members are made from two U or C sections placed side by side, or from square or rectangle cross-section tubular members interconnected at the nodes using ordinary or prestressed high tensile bolts passing through a central connector and corner plates. This bolted connection solution may be used for both single-curvature and double-curvature single-layer grids.

9. Fig. 4a, shows the structural solution and the main component parts of the joint, for a six members node of a spherical lattice structure.

10. Figs 4b, c and d present the experimental curves for both ordinary bolts and prestressed high tensile bolts.

CONCLUDING REMARKS

11. Both experimental research and numerical studies carried out by the Steel Department Laboratory from Timisoara, validate the new solution described in this paper. Comparisons made with other structural systems (refs 4 and 5) indicate good performances for these new structural solutions.

REFERENCES

1. DUBINA D., PACOSTE C., GRECEA D. Patent RO nr. 97611/ 1985.
2. DUBINA D., COLCEA E. Patent RO nr. 87969/ 1985.
3. DUBINA D., PACOSTE C. Patent RO nr. 95910/ 1988.
4. DUBINA D. Numerical and experimental research concerning two new systems for double-layered grids. Proceedings of the International IASS Symposium, Copenhagen, 1991 September, vol 3, 213-220.
5. DUBINA D. New structural systems for double layered grids with bolted joints. Proceedings of ICSAS' Singapore, 1991 May, vol. I.

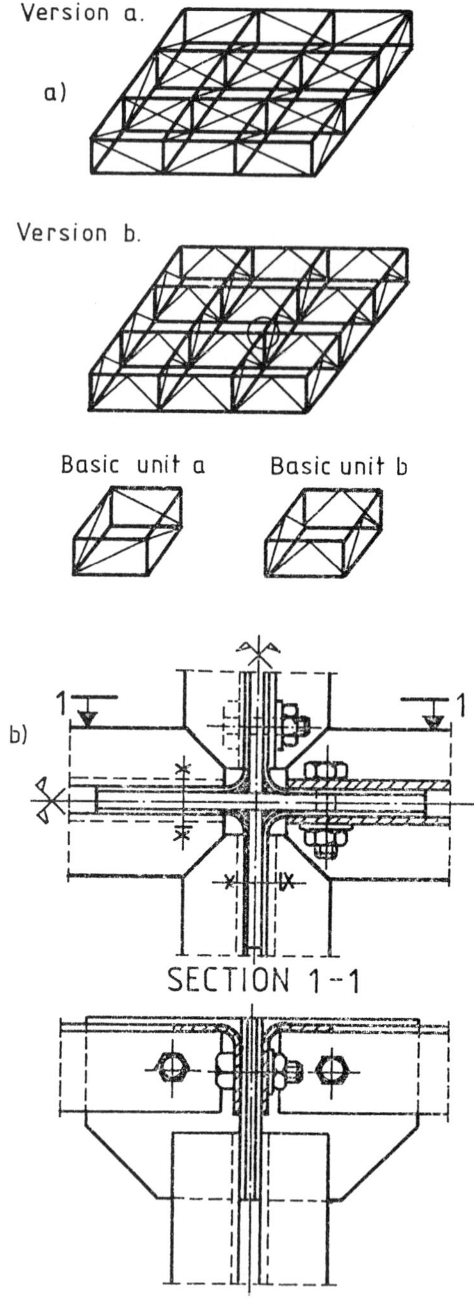

Fig.1. Plane Double-Layer Grid Structure.
a) Structure Geometry,
b) Upper Chord Connection.

Table 1. Cross-Section Dimensions for PDLG structure a version

Dimensions [m]	Load [daN]	Maximum axial forces [daN]			Deflection [cm]	Cross-Section Dimensions		
		Chords	Diagonals	Struts		Chords	Diagonals	Struts
L = 30.0 b = 3.0 h = 1.8	150.	− 15144.	− 4978.	−400.	5.60	60×60×3	50×50×3	40×40×3
	200.	− 20190.	− 6636.	−535.	7.46	60×60×4	50×50×4	40×40×3
	250.	− 25245.	− 8300.	− 670.	9.34	60×60×5	50×50×5	40×40×3

Table 2. Cross-Section Dimensions for PDLG structure b version

Dimensions [m]	Load [daN]	Maximum axial forces [daN]			Deflection [cm]	Cross-Section Dimensions		
		Chords	Diagonals	Struts		Chords	Diagonals	Struts
L = 300 b = 2.0 h = 1.8	150.	−12522.	− 4601.	− 4279.	5.40	60×60×3	50×40×3	40×40×3
	200.	−16692.	− 6133.	− 5704.	7.20	60×60×4	50×50×3	40×40×3
	250.	−20874.	−7670.	− 7133.	9.00	60×60×5	60×50×3	40×40×3

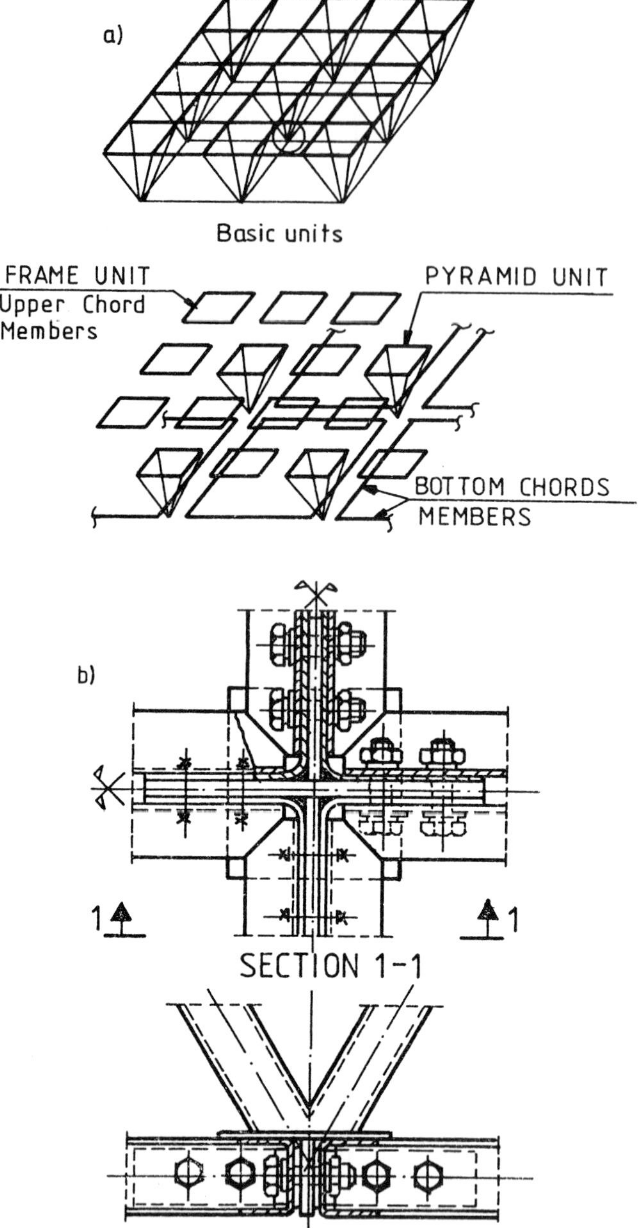

Fig. 2. Rare Double-Layer Grid Structure.
a) Structure Geometry,
b) Bottom Chord Connection.

Table 3 - Cross - Section Dimension for RDLG structure

Dimensions [m]	Load [daN]	Maximum Design Stresses: Stress	Upper Chord	Bottom Chord	Diagonals	Deflection [cm]	Cross-Section Dimensions: Upper Chord	Bottom Chord	Diagonals
	150.0	daN N	-4210.	-12019.	-7438.	10.45	][120×60×6	][160×80×6	Φ 76×10
		daNcm M_z	334670.	160972.	65462.		]C 120×60×20×6	]C 160×70×20×5	
		daNcm M_y	12030.	0.	2676.				
L = 30.0 b_1 = 1.875 b_2 = 3.750 h = 1.20	200.0	daN N	-5600.	-16025.	-9893.	13.89	][160×60×6	][180×80×6	Φ 83×10
		daNcm M_z	445111.	214093.	87065.		]C 160×60×20×5	]C 180×70×20×5	
		daN M_y	16000.	0.	3960.				
	250.0	daN N	-7017.	-20032.	-12397.	14.73	][180×60×6	][200×100×6	Φ 95×10
		daNcm M_z	557783.	268287.	109100.		]C 180×60×20×5	]C 200×80×20×5	
		daNcm M_y	20050.	0.	4460.				

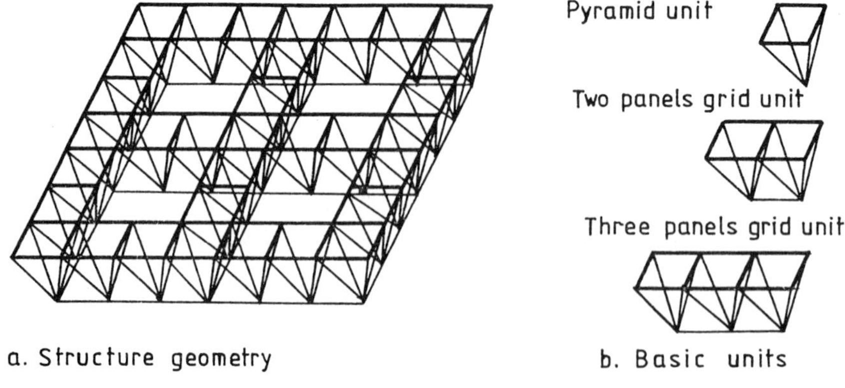

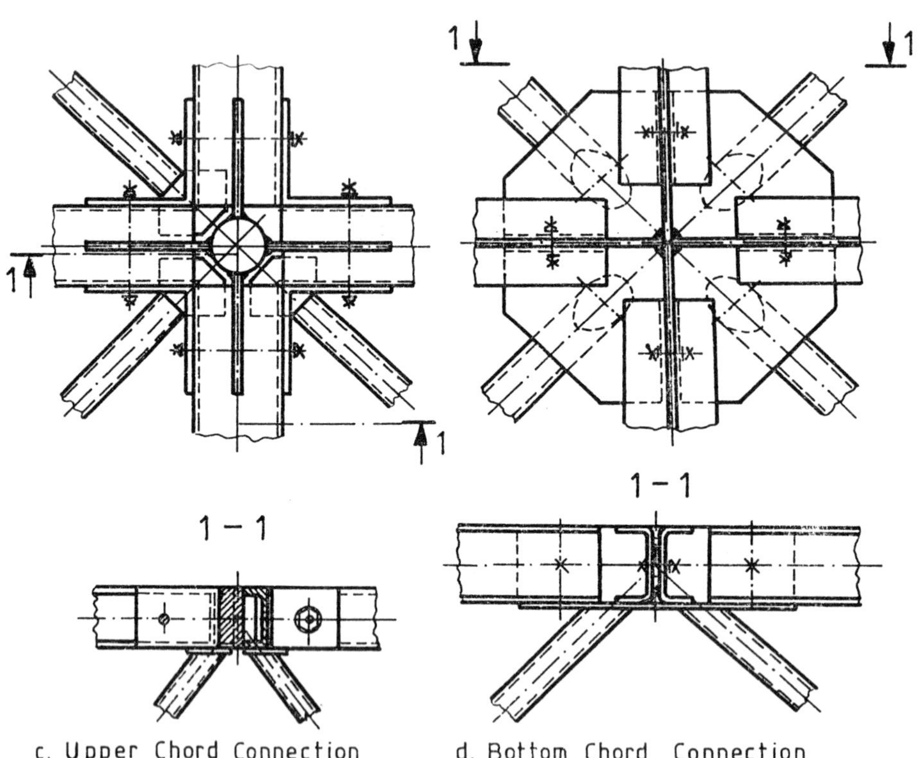

Fig. 3. Space Double-Layer Grid Structure.

Table 4. Cross-Section Dimension for SDLG structure

Dimensions [m]	Load [daN]	Maximum axial forces [daN]		Deflection [cm]	Cross-Section Dimentions	
		Chords	Diagonals		Chords	Diagonals
L = 30.0 b = 2.0 h = 2.0	150.	−7125.	−4846.	2.21	⊏⊐ 80×45×4 □ 80×4	ϕ 76×3,5
L = 30.0 b = 2.0 h = 2.0	200.	−8906.	−6058.	2.76	⊏⊐ 80×45×4 □ 80×4	ϕ 26×3,5
L = 30.0 b = 2.0 h = 2.0	250.	−11133.	−7572.	3.45	⊏⊐ 80×45×4 □ 80×4	ϕ 76×3,5

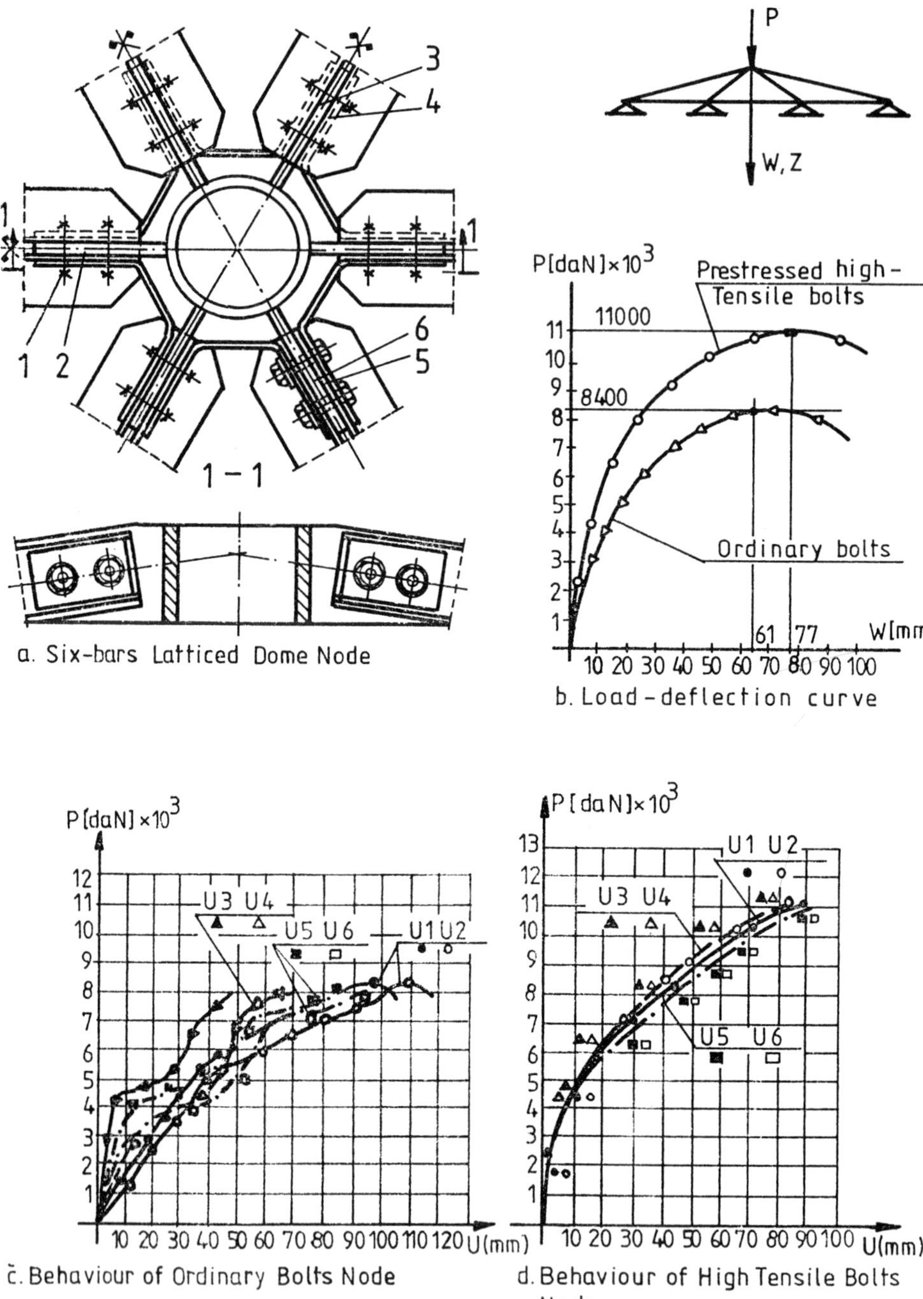

Fig. 4. Connection System for Single Layer Structure.

161. Imagination versus reality in space frame design

J. KLEPRLIK and I. COLLINS, MERO-Raumstruktur, Germany

SYNOPSIS. MERO is known worldwide for its commercial leadership in the design, supply and erection of extremely diverse spaceframe applications. Over 50 years of experience in the applied theory and commercial development of the general technology has made it possible to successfully execute the most creative and complex projects. The continued application of new design and fabrication technologies has made it possible to explore and form new philosophic, architectural and structural trends.

1966,1975,1984,1993?

1. The authors are pleased to present to the Fourth International Conference on Space Structures. This, perhaps simplistic, contribution is a reflection of real "daily life" design opportunities experienced by the MERO company. Our intent is to display a fragment of the ideas experienced in the past ten years and explore the applications made possible by continued changes in technology.

SPACE STRUCTURES AS AN ACCEPTED PHENOMENON

2. The latter half of this century has seen the acceptance of spaceframes by architects, engineers and building owners. What is a spaceframe, what is a space structure ? Many outstanding academics and practising designers have produced varied definitions, it doesn't require us to create new ones. Perhaps more importantly, what are the attributes of "spaceframes" ? The possible answers include: - endless versatility of form and shape, speed of design and production, optimal weight of elements, high strength to weight ratio, systemized construction, serial production, cost effectiveness, ideal medium for overcoming the separation between structure and architecture.

SPACEFRAME DESIGN - ART OR SCIENCE ?

3. Perhaps fortunately, the design of a "real" space structure is still more art than science. Despite the availability of high speed computing, the mathematical selection of a suitable system, form, geometry, etc., to meet all other prescribed conditions, is in general impractical. The designer must also consider other

factors such as aesthetics, functional requirements, site conditions, economics etc.. Frequently in the design planning process there are as many intangibles as knowns. Therefore, design and planning decisions are often based, out of necessity, on engineering judgement and experience. Issues adding to the complexity of decision making for the designer are in fact the new tools made available to him. For example, the introduction of numeric control (NC) machines enables the production of progressive spaceframe systems in complex designs, due to their extraordinary accuracy. These machines have produced the real change from civil spaceframes to mechanical parts and significantly alter the relationship between production labour and material costs. MERO has developed software to generate, analyse, estimate and revise complex spaceframes very quickly. This interactive system can even capture and update production costs as they occur. Human/computer interaction, however, is the key, as the steps in the "design chain " inevitable vary from project to project and, hence, all crucial decisions should be, and are, " man made ". Therefore, as more and more processable data is being collected, ever changing technology is providing more options to the designer, not less. Complexity inevitably increases with choice, hence the final project solution tends to become more a planned art than an optimal science.

DESIGN PARAMETERS AND EXAMPLES

4. Functional requirements. According to the prevalent MERO philosophy, an enclosure consists typically of two components, a skeleton (structural space structure) and a skin (non-structural cladding). The optimal interaction of these two elements, the "total system", should combine to meet all the functional requirements. The spaceframe structural function is to support catwalks, HVAC, sprinkler systems, hanging loads, glass cleaning devices etc., in addition to traditional live, wind, seismic and cladding loads.

Fig. 1.

Fig. 2.

The physical geometry must also provide sufficient support bearing for cladding materials. Fig. 1 shows the Al Ain dome interior which supports lighting and air ducts. As a double layer grid it is possible to utilise the space between the inner and outer chords for services. Fig. 2, in contrast, shows the Bad Neustadt single layer dome with hanging shade curtains. Functional requirements for the skin should be thought of as being similar to those of human clothing i.e. the cladding should provide adequate temperature protection from weather elements and be aesthetic. Moisture, thermal noise and light control are key physical properties of the cladding systems. Fire protection is both a skin and skeletal consideration. Cladding types vary but could be categorised as being "hard" or "soft". Hard cladding can be glass or composite panels, or traditional metal decking etc.. Soft claddings can be fibre reinforced fabrics or acrylic. Local structural deformations are a very important consideration for cladding systems. Softer claddings are more forgiving to larger local deformations. Therefore, different structural concepts are possible with different cladding types. Technological advances in glass production has vastly broadened the usage of glass in varied climates. The balance between daytime lighting and solar energy gain or loss is a good solution. Natural light has been shown to add to peoples' work performance and their satisfaction within an environment. Not surprisingly, as glass costs drop and the variety grows, we have seen a vast increase in glazed spaceframes and hence a need for "stiffer" structures.

5. Non-separation of structure and architecture. From the architectural viewpoint spaceframe structures can now provide a huge array of shapes and forms. The geometry "palate" enhances this configuration processing capability. Table 1 compiles spatial forms. Let us examine these groups.

Table 1.

1 planar	2 curved		3 composite
flat	2a single	2b double	partially developable
	developable	non developable	
horizontal vertical sloped folded stepped	quadrics surface of revolution surface of translation		polyhedra/ derivative folded composite free
	cylinder cone	sphere, conoid hypar, cylindroid	

6. Group 1. Planar. This application is generally well suited to universal connectors and typical assembly. Large projects in this group are increasingly rare as spans are more easily achieved in curved forms or via conventional structures. However, many interesting canopies and atriums have been executed. One such planar project was the triangular acrylic-clad entrance canopy at the Swiss Hotel in Illinois (Fig. 3).

Fig. 3.

Fig. 4.

The plate is suspended from the tower by three inclined cables. A fourth cable at the canopy tip is provided to resist uplift. A larger plate is the 60m x 60m Hyatt Hotel atrium in California (Fig. 4). This structure was subject to a major earthquake in 1989. Having been designed to "float" on neoprene bearings with very limited lateral support, the spaceframe actually moved in excess of 100 mm in all horizontal direction without sustaining damage. There is a major behaviour advantage for space structures, over say conventional trusses, in earthquake regions because of there inherent bracing systems.

7. Group 2. Curved. Curved structures can be conceived in single or double layer structural networks. Single layered structures can be achieved via triangular pin-jointed systems or quadrangular geometry with moment-resistant connections. Shell stability is a consideration in design. Many theoreticians and mathematicians have provided research on the elastic, inelastic and stability analysis of spaceframe shells. From a practical standpoint the following criteria influence the choice of system: shape, curvature, cladding type (can assist stability), loading, average member length, and method of assembly. Curved structures are easier to produce today as the advances in manufacturing technologies permit greater degree of freedom from the restrictions of producing many identical parts. However, advances in spaceframe production are not matched by all cladding materials. Therefore, caution must be exercised by the designer to achieve the economical "total system" as the economic structure may create a very expensive cladding.

8. Group 2A. Single curved. The barrel vault is the most common form in the group. Two examples are shown in Fig. 5. The form in Fig. 5a is not exactly cylindrical, the cross section curve is a composite circular arch to achieve more room inside without increasing the overall height.

Figs 5a and 5b. Fig. 6.

Fig. 5b shows a single layer triangulated network for a unique pedestrian bridge. Fig. 6 is a ribbed vault where two spaceframes' end walls are braced together by ribs which also serve to screen mechanical units on a high rise roof top. The other form in this group is conical. This, seldom used form, is a good solution for glass structures, as cladding elements are planar and repetitive. Presidential Circle in Florida (Fig. 7) is an example of a minimal structure supporting glass. The roof is a cone and the walls are vertical portions of a cylinder. In general, single curve solutions are both aesthetic and economical solutions for clad structures.

9. Group 2b. Double curved. The classical form of this group is the dome as part of a sphere. MERO has contributed to large scale dome construction and integrated cladding. The two domes referenced, Al Ain and Stockholm, were clad with aluminum sandwich panels. These triangular and trapezoidal (resp.) panels were in fact flat panels and, by use of a high frequency in the dome geometry, the surface becomes visually curved. However, Fig. 8 shows the Dome Imax project in Paris. Here the glass is also double curved over a classic double layer grid. The architecture is exciting and precise, the cost was less of a design parameter. An extraordinary futuristic project is shown in Fig. 9. The symbol of Islamic revolution has been "forged" into the outer surface of the dome. Architecturally, the pure form of the dome is being used less often. Fig. 10 shows some recent studies, which become easier to achieve production-wise and also take into greater consideration cladding capabilities.

10. Group 3. Composite and free forms. The oldest classical composite form is the pyramid. The common form being a half of a platonic polyhedra - octahedron. Every edge of the platonic polyhedron has the same length, each face is identical. MERO built a family of pyramids in Essen (Fig. 11) using a rotated geometry. Of increasing interest and application

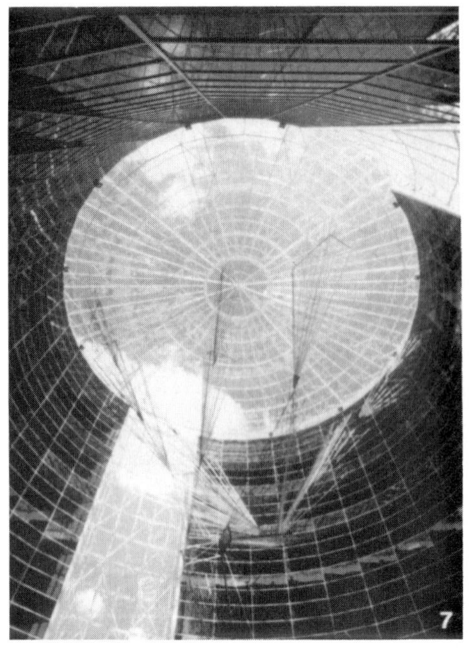

Fig. 7.

Fig. 8.

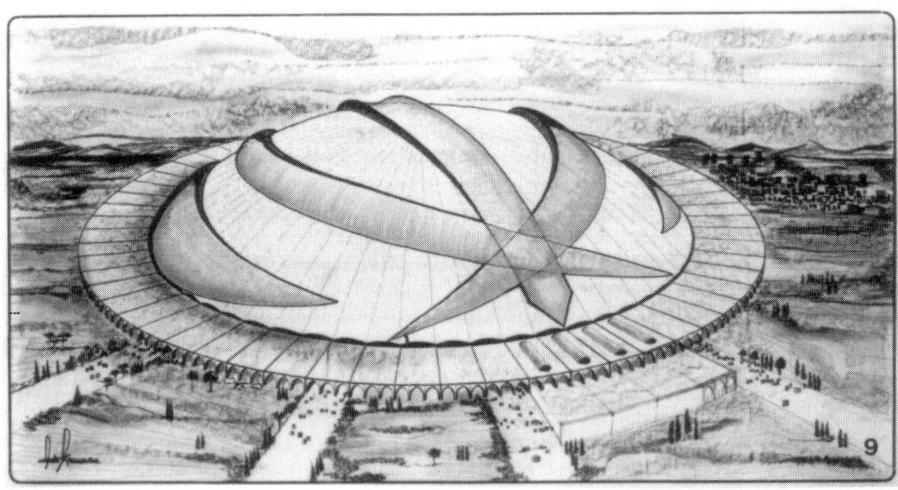

Fig. 9.

are the endless architectural opportunities made possible with composite forms. The combination of basic forms gives the freedom to create aesthetic-structural hybrids of outstanding properties to cover even the most irregular ground planes. The MERO range of structural systems permits the mixing of pinned and bending connectors (or no connectors) which allows an "anatomical" hierarchy, providing both unusual image plus effective structural behaviour. This capability of mixing forms, geometries, and systems creates huge design freedoms to the MERO designer and his customer. Fig. 12 shows the Mirage Hotel, Nevada. The dome-like structure is not part of a sphere. The single layer segments of stiff bending elements are connected by double layer pin-jointed ribs which help control lateral deflections. The combination of glazing and aluminum panels creates a spider-like expression with minimal glass supports. The Music Pavilion in Essen, Fig. 13 is an example of composite form.

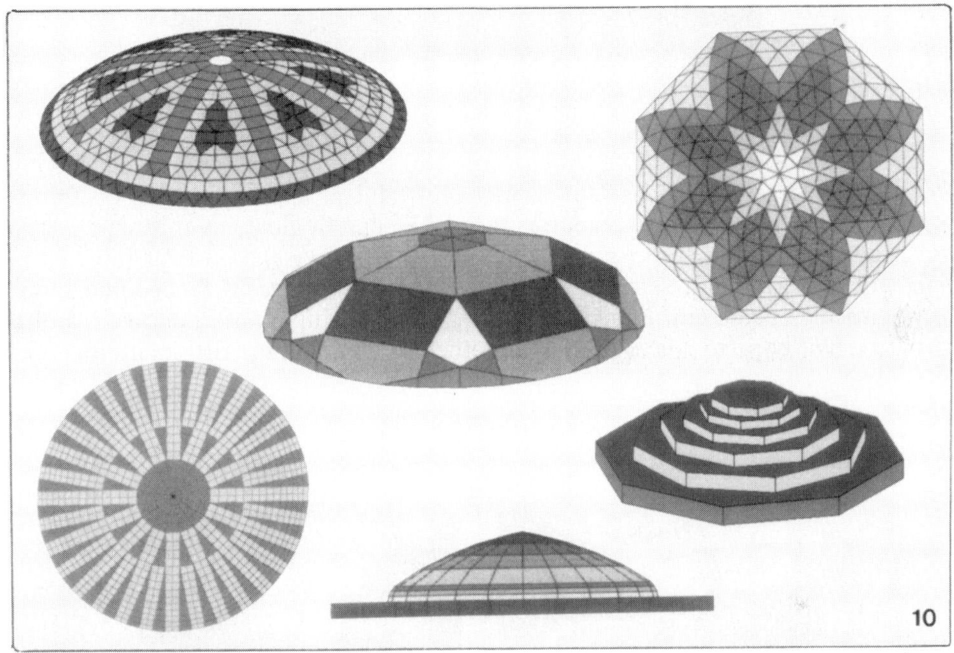

Fig. 10.

Five "spherical" single layer triangles of pin-ended elements are integrated with double layer torsion resistant ribs. The cladding fabric is suspended from the stiffened structure. Fig. 14 consists of cylindrical walls, circular rings, single layer spherical surface areas and a tower structure. Fabric, acrylic and timber are used in the cladding. This small project displays many of the new possibilities of combining form, materials and strength characteristics. City Place, Fig. 6, is an example of mixed systems. The vertical walls which support glass and stone cladding are comprised of bending resistant members. These walls are braced longitudinally by triangulated ribs comprised of axial only elements. The atrium in New Orleans, LA (Fig. 15) is a glass skin supported by a composition of non-regular polyhedron plates. The weight contrast between the skeletal frame and supporting beams is evident.

Fig. 11.

WHAT NEXT ?

11. The "Classic" space frame will have reduced applications as customer expectations, coupled with human nature and technological advancements, will push designers to exercise greater creativity. We envisage an ever increasing mixture of forms and systems to enable custom enclosures to become more economically achieved. Figs 16-18 show some examples of conceived future projects. We hope to convey these and other advances at the fifth conference in 2002 ?

Fig. 12.

Fig. 13.

ACKNOWLEDGEMENTS

12. The authors would like to thank their friends and colleagues at MERO, whose varied skills and talents have gone into the successful execution of such concepts. Thanks also to Y. Kleprlik for her efforts in preparing this paper.

Fig. 14.

Fig. 15.

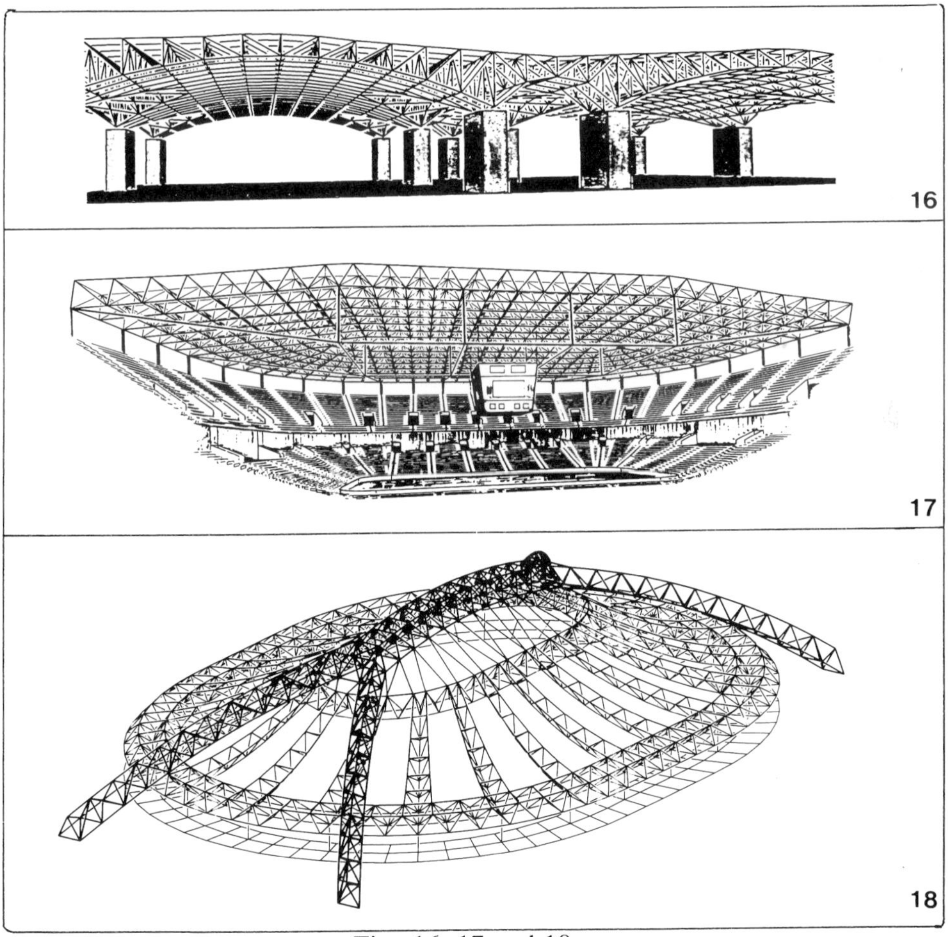

16

17

18

Figs 16, 17 and 18.

162. Development and testing of the CUBIC composite floor

K. W. THOMPSON and L. A. KUBIK, ASW-Cubic Structures, UK

SYNOPSIS. The benefits of composite action between concrete and steel for two dimensional structures are well known, but little work has been done to investigate the benefits of composite action in triactive structures. Little or no design guides are in existence. The CUBIC Composite Floor has the top chords embedded in the structural concrete slab and so eliminates the need for additional shear connectors. Extensive testing has been carried out both on an assembled frame and components to establish the structural action from which appropriate design guides are recommended. Both full and partial composite criteria are discussed. The departures from existing design guides for composite structures are identified.

COMPOSITE CONSTRUCTION - CONCEPT

1. A composite floor slab consists of profile steel decking, light or normal weight concrete and nominal reinforcing in the form of mesh. The overall thickness of the slab is usually controlled by fire insulation requirements and varies from 110mm to 130mm. A grillage of primary and secondary beams supports and acts compositely with the floor slab. The use of mechanical shear connectors welded through the decking and to the top flange of the supporting steel ensures good composite action.

2. Column grids vary from typically 6m x 7.5m in smaller buildings to 15m x 15m in prestigious commercial buildings. Many column grids are rectangular with an aspect ratio between 1.5 and 2.5 .This is done to optimise the relative size of the primary and secondary beams. Secondary beams at 2.4m, 3m or 3.6m centres generally span the greater distance onto the primary beams. Traditionally, column grids have been in the range of 6m to 9m. For these spans, primary and secondary beams are usually simply supported Universal Beam or Column sections acting compositely with the floor slab.

3. In conventional floor systems, services are placed under the steel beams i.e. interactive layout. This results in an unnecessarily large floor depth which increases the building height. Improvements have been made in order to facilitate

 Space structures 4. Thomas Telford, London, 1993

the services which have led to the introduction of new structural systems, namely Parallel Beam and Slimflor.

Parallel beam

4. This design makes effective use of continuous construction by displacing beams away from column centre lines and running non-composite primaries and composite secondaries in different horizontal planes. The composite action is effected by the use of shear connectors.

Slimflor beam

5. This consists of steel columns and beams, with floors of precast concrete elements. To allow a reduced total depth of floor, steel 'top hat' sections are integrated into the floor structure, exposing only the bottom flange and avoiding downstand beams. The effect is to provide a flat soffit so service runs are uninterrupted. Some composite action is achieved by the action of the structural grout embedding the steel beam. The degree of composite action is dependant upon the type of beam used and the addition of shear connectors.

CUBIC composite floor

6. The CUBIC Composite Floor is a three dimensional structure which uses the typical prefabricated steel modules (ref. 11). These modules are bolted together to form a steel structure which is capable of supporting all construction loads. A permanent soffit shutter is then positioned on the bottom flange of the top chords. A lightweight reinforcing mesh is laid over the top chords to provide crack resistance. Finally, concrete is cast embedding the top chords and providing a 30mm cover to the top flanges, (Fig. 1). No shear studs are necessary since all interaction between the steel and concrete occurs at the embedded nodes.

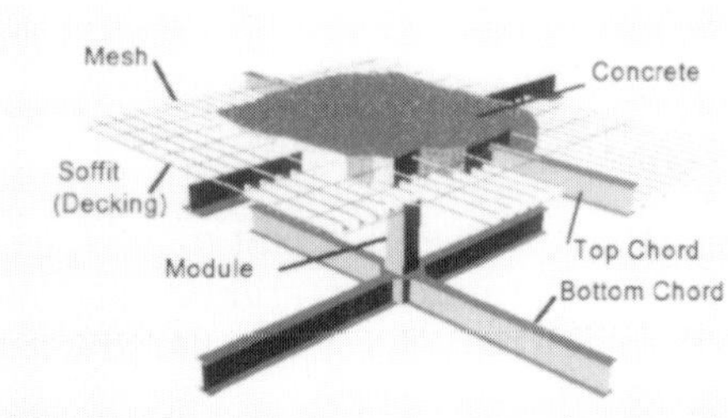

Fig. 1. Typical floor construction.

7. During construction the steel structure has to support the shuttering, reinforcement and wet concrete loads, in addition to the usual construction imposed loads. However once the concrete has hardened, all subsequent loads which are applied to the floor are shared between the steel and concrete. Since the concrete enhances the strength of the top chord, the greatest saving in steel weight can be achieved by adopting a smaller section size or by increasing the grid size of the module.

THEORY

8. The theory of conventional composite construction is based on the recommendations of BS5950 Part 3. These recommendations rely on the assumptions which have been found satisfactory for plane structural frames. The behaviour of three dimensional composite structural frames differs from that of planar frames in a number of ways, namely:

(i) Significant amount of bi-axial compression in the concrete slab.
(ii) Continuity of the slab over the nodal joints of the space frame.
(iii) Contribution of the concrete slab in resisting shear forces.
(iv) Distribution of member forces during the application of external loads due to the two way spanning effect of the space frame.

9. The present investigation deals with chords variable in two stages, namely:

(i) At the stage of casting the floor slab, the upper chords are less rigid than the lower chords.
(ii) Once the concrete has hydrated, the upper chords, reinforced by the enveloping concrete become stiffer than the bottom chords. The upper chord splice joints embedded in concrete behave differently from the splice joints of the lower chords.

10. Despite these differences, the fundamental concept of the analysis of the CUBIC Space Frame (i.e. the compatibility of deflections of chord lines at the nodes) remains true for the CUBIC Composite Floor.

Forces affecting deflection

11. Within the elastic range of materials used, the deflection of the CUBIC Space Frame is caused by three separate actions of the module members, namely:

(i) Bending due to (a) Local flexure of the chords under the action of the vertical shear forces transmitted through the chord joints and (b) Rotation of the vertical members.
(ii) Axial Loads causing shortening of the top chords under compressive action and lengthening of the bottom chords due to tension.
(iii) Shear Forces causing shear deformation of the webs of the chords.

12. The proportion of contribution by each factor to the total deflection at each node varies with the geometry of the structure (grid size and span-to-depth ratio), intensity of loading and the location of nodes within the structure.

ANALYSIS

13. The analysis has been carried out in a number of different ways to investigate the influence that the concrete has on the structural behaviour of the frame. Firstly, to provide control data for assessing the influence of the concrete, the test frame has been analysed assuming that the steel resists all the applied load without any contribution from the concrete. Secondly, with various scenarios of contribution of concrete, these are:

(a) the concrete is able to resist axial compressive forces only, so that all shear and bending in the top chord is resisted by the steel chords. The effective width of the concrete is based on the floor span, using the method described for simply supported beams in BS5950:Part 3:Section 3.1 (i.e. not greater than span/4). In fact the effective width is constrained by the grid size in this case, typically 1.5m;

(b) the contribution of the concrete to both the axial and the flexural strength of the top chord is included, again assuming an effective concrete width based on the floor span, determined in accordance with the provisions of BS5950:Part 3:Section 3.1 and;

(c) the contribution of the concrete to both axial and flexural strength of the top chord is also included. This time the effective width of the concrete slab is based on the module grid size and not the floor span (i.e. grid/4 and not span/4), to reflect the substantial influence of chord bending on structural behaviour.

14. Each analysis has been carried out using conventional linear-elastic theory, assuming linear elements for the top chords with section properties determined using the above assumptions.

15. The software that has been used for the calculations implicitly assumes that all nodal and splice connections are fully rigid. In practice this is not correct for the test frame, since the effect of bolt slippage and embedment in the clearance holes of the splice plates needs to be considered, as does the effect of nodal deformation on the perimeter girders due to the adopted stiffener detail. As found for the CUBIC Space Frame, the results of the floor test indicate that this primarily affects deflections on first loading, having little effect on deflections in subsequent loading cycles, or on strains. A separate allowance has been made for these effects in the analyses using the experience gained from CUBIC Space Frame applications (ref. 7), the results of which confirm this conclusion.

TEST FRAME

16. The test frame represents a full size structure of a typical commercial floor with a grid size of 1.5m, an overall depth of 890mm and a span of 6.75m x 6.75m between supports (Fig. 2). The unfactored design load is 3kN/m^2 at the casting stage of the concrete slab and 8.7kN/m^2 in service when acting compositely with the hydrated concrete slab.

17. The frame was supported at each corner on 200kN capacity load cells with a ball joint between the frame and load cell. After erection of the steelwork the lightweight concrete was poured to a total depth of 130mm, giving a 30mm cover to the top chord. The design mix was 30N/mm^2 with a dry density of 19kN/m^3, although the actual cube strength was 37N/mm^2.

18. The test frame was then instrumented with strain gauges and linear displacement transducers (LVDT). Optical markers were provided for precision optical surveys to be carried out during the test (Fig. 2).

TEST PROCEDURE

19. The test programme consisted of four stages which are described below. Stages 1, 2 and 4 were in accordance to BS5950: Part 1 Section 7 (ref. 1), and represent respectively a stiffness test, a strength test and a test to failure. Stage 3

was included to record the effects of including negative bending moments into the frame as could occur at internal column positions.

Fig. 2. View through test frame.

20. Stage 1. 1.0 Dead load + 1.25 Imposed load. The 35 tonne applied load was maintained for 24 hours and then removed, in accordance with the procedure in BS5950 (ref. 1).

21. Stage 2. 1.4 Dead load + 1.6 Imposed load. The load was applied in stages to obtain some load-deflection curves. After the maximum applied load of 48 tonnes had been maintained for 24 hours the frame was unloaded in a single step.

22. Stage 3. Reverse load application through upward jacking at node C3. For this test the frame was loaded as in Stage 2, with weights in the form of sand bags and cast steel billets (Fig. 3). An hydraulic jack positioned under node C3, was then used to apply an upward force to the frame until the deflection at C3 was zero. The upward load of 25 tonnes was then removed. Some residual deformation of the frame in the region of node C3 was evident from the deflection measurements.

23 . Stage 4. Imposed downward load to failure of the structure. An initial load of some 65 tonne was applied in stages using sand and billets. The additional load which was required to produce failure was applied using nine hydraulic jacks positioned above the frame (Fig. 4). The frame failed by failure in the bottom chord at grid location A/ 3-4 as predicted.

Fig. 3. Stage 2 loading.

Fig. 4. Stage 4 loading.

RESULTS

24. From the test results it is immediately clear that the test frame is significantly stiffer than would be expected from the steelwork alone, in other words the concrete has a marked effect on frame stiffness. Of the various methods of analysis, the deflection predicted by the analysis in paragraph 13(b), assuming that concrete resists both flexure and axial forces, using an effective width equal to grid/4 most closely predicts the measured frame deflection. Table 1 indicates typical results for the average ratio of predicted/measured deflection for all nodes.

	Table 1.	Table 2.
ANALYSIS	PREDICTED/MEASURED DEFLECTION	PREDICTED/MEASURED STRAIN
a) Steel frame:	1.76	1.52
b) Composite, concrete resists compression only, effective width = span/4:	1.37	1.44
c) Composite, concrete resists bending and compression, effective width = span/4:	0.88	0.70
d) Composite, concrete resists bending and compression, effective width = grid/4:	1.05	0.985

25. Similarly, the predicted strains, assuming that concrete resists both flexure and axial forces, using an effective width equal to grid/4, most closely predict the measured strains (see typical results for the bottom chord in Table 2).

26. From these tables it can be seen that the analysis assuming the concrete to resist axial compression only, leads to overestimates of both deflection and strain in the steel. However, since the effects of fully composite action between embedded nodes is difficult to justify without providing some form of mechanical shear connection, it is suggested below that this method of analysis is more prudent at present.

DESIGN RECOMMENDATIONS

27. The design of continuous and simply-supported steel/concrete composite beams is covered by BS5950:Part 3:Section 3.1 (ref. 2). This design code does not apply to three-dimensional structures. Nor does it apply to composite beams in which the steel is partially embedded in concrete and which have no added shear connectors. These are therefore the main issues which need to be addressed for the CUBIC Composite Floor.

28. It is generally accepted (ref. 4) that even partial encasement of a steel section increases its stiffness and strength. However, unless some form of mechanical shear connection is provided, or the beam is fully encased, the benefit cannot be reliably predicted since it relies upon bond strength, which can be reduced or even

totally eliminated by shrinkage. The benefit may eventually be found to be more predictable for a structure such as the CUBIC Composite Floor, which has restraint provided by the orthogonal members of the top chord grillage, but no tests have so far been carried out to investigate this possibility.

29. Since there is no significant mechanical shear connection between the embedded module nodes in the CUBIC Composite Floor it is prudent to base ultimate strength calculations on the assumption that the steel resists all shear and bending in the top chord without any assistance from the concrete. The concrete will clearly assist in restraining the top chord and preventing local buckling, as well as in carrying a proportion of the axial compression between the embedded nodes. As seen above, this approach was found to be conservative for the test frame.

30. It should be noted that the spacing of embedded nodes in the CUBIC Composite Floor does not satisfy the spacing rules for shear connectors in BS5950:Part 3:Section 3.1. It will therefore be desirable to include a buckling check on the slab between the nodes in the design process.

31. In practice, such a design approach means that the concrete effective breadth can be based on the floor span rather than the grid size. The force transfer between steel and concrete can also be checked for different top chord details using a modified simple push-off test similar to that described in BS5400:Part 5 (ref. 12), rather than a more complex arrangement to test different combinations of shear, moment and axial force.

32. It is worth noting that for the CUBIC Composite Floor the BS5400:Part 5 push-off test has been modified, since it was not possible to ensure adequate compaction of concrete around the nodal steelwork with the vertically cast specimens. This problem was overcome by casting the specimens horizontally (Fig. 5), and testing them in a specially designed horizontal test rig (Fig. 6). Results have so far been obtained for a number of nodal details, including assessments of different decking profiles and orientations.

Fig. 5. Mould for push-off test.

Fig. 6. View of test rig.

33. The push-off test identifies the load capacity of the tested node under uniaxial loading. In reality, each node will be subjected to biaxial loading. Until

biaxial loading tests have been carried out it would be prudent to assume that the following relationship is satisfied:

$$Fx/Px + Fy/Py \leq 1$$

where Fx = force in x direction Fy = force in y direction
Px = load capacity in x direction Py = load capacity in y direction

34. It was shown in paragraph 24 that the predicted deflections of the test floor were noticeably greater than the measured deflections for the analysis assuming concrete to resist only axial force. Checks are currently being made to see whether this is likely to be critical in practical designs, in which case allowance could be made for the stiffening effect of the concrete in deflection calculations in a similar manner to that made for semi-composite Slim Floor construction (ref. 4).

REFERENCES

1. BRITISH STANDARDS INSTITUTION, 1990. BS5950:Part 1, The limit state code of practice for design of steelwork in building using hot rolled sections.
2.BRITISH STANDARDS INSTITUTION, 1990. BS5950:Part 3:Section 3.1, Code of practice for design of simple and continuous composite beams.
3. BRITISH STANDARDS INSTITUTION, 1982. BS5950:Part 4, Code of practice for design of floors with profiled steel sheeting.
4. MULLETT D.L. Slim Floor Design and Construction. The Steel Construction Institute, 1992.
5. AMERICAN SOCIETY OF CIVIL ENGINEERS. Composite steel-concrete construction, proceedings of the American Society of Civil Engineers, vol. 100, no. ST5, May 1974,1085-1139.
6. KUBIK M.L and KUBIK L.A. (1991) "An introduction to the CUBIC Space Frame". International Journal of Space Structures Issue 6 (1).
7. KUBIK L.A. (1991). "The CUBIC Space Frame in the theory and in practice". Steel Construction Today. March 1991.
8. KUBIK L.A. and McCONNEL R.E. (1989). The behaviour of CUBIC Space Frame Joints." Proceedings of the Third International Symposium on Tubular Structures.
9.POLLARD R.R. (1988) "Composite construction with the CUBIC Space Frame". Final year project, Trent Polytechnic Nottingham.
10.THOMPSON K.W. (1990). "An investigation into the behaviour of the CUBIC Space Frame Composite Floor (a triactive structure)". Progress report, Nottingham Polytechnic, December 1990.
11. THOMPSON K.W and KUBIK L.A. (1991) "Prototype Testing of the CUBIC Space Frame Composite Floor". Proceedings of the Fourth International Symposium on Tubular Structures, Delft 1991.
12. BRITISH STANDARD INSTITUTION, 1979. BS5400:Part 5. Code of practice for design of composite bridges.

163. Composite double-layer bridge-grids

M. ASHRAF, M. A. JAVIED and Z. ALI, Lahore University of Engineering and Technology, Pakistan

SYNOPSIS. A double layer bridge-grid is composed of two interconnected layers of steel skeletal members and covered at its top by a steel plate or concrete slab called a continuum. This type of structure supported on two opposite short edges has proven more useful for covering large span highway bridges compared with conventional reinforced/prestressed concrete construction both in terms of economy in cost and time. The significance of the continuum is evident from the results obtained from testing of a model of a double layer bridge-grid without and with the top covering as an integral component of the structure. The results show that the continuum, not only increases the overall stiffness of the structure, but also decreases considerably critical stresses in the top layer skeletal members. The good agreement between the theoretical and experimental observations also helps in validation of assumptions employed in a new analytical girder analogy method, proposed by the author.

INTRODUCTION

1. Composite space structures are more rigid and have better stress distribution than the comparable skeletal ones, (refs 2-4). The top layer members of skeletal double layer grids are, generally, subjected to critical compressive stresses. However, the introduction of top sheeting or slab as an integral part of the structure results in a decrease in forces and deflections. Particularly, the top layer member forces decline at a much faster rate than enhancement in the overall stiffness of the resulting composite double layer bridge-grid with the increase in continuum thickness.

2. Makowski (ref. 6) in his extensive survey of the different types of space structures also pointed out various advantages of the stressed skin sheet space structures. He noted that the sheeting, covering the prefabricated skeletal units, and acting as the roof cover, can be used at the same time as an integral part of the structure relieving the stresses in the component pyramid units.

3. Castillo (ref. 5) was perhaps the first to extend practically the concept of space structures to composite space structures. To check the performance of composite space structures referred to as "Tridilosa system", Castillo undertook

 Space structures 4. Thomas Telford, London, 1993

tests on a double layer grid of 3.00m span, with and without a concrete slab as its top layer. It was observed that the load carrying capacity in the former case was 15 times greater than that of the latter case where failure occurred due to buckling of top layer skeletal members.

4. Al Bazzaz (ref. 1) investigated the behaviour of a composite double layer grid and concluded that a reinforced concrete slab can be used as a replacement of the top layer steel skeletal members. He extended the work of Castillo by conducting theoretical and experimental studies. Further, he suggested that the composite double layer grids could be used for comparatively large spans due to their potentially large load carrying capacity.

5. Makowski (ref. 7) presented a critical review of the recent development of space structures in Mexico. In addition to discussion on a modified use of Castillo's Tridilosa system for multi-storey buildings, he reported the application of composite triple layer grids for single or multi-bays of 50 m for highway bridges. These bridges composed of interconnected three layers of skeletal members and top concrete slab decks, are found to be economical in comparison with conventional reinforced/prestressed concrete bridges, both in terms of construction cost and time.

6. An extensive investigation was conducted into this type of bridges by Ashraf (ref. 2), who referred to them as "composite triple layer bridge-grids". The analysis, based on the stiffness method of these structures involves both the skeletal and thin plate rectangular shell elements having six degrees of nodal freedom. A general purpose program in FORTRAN77 was developed and results of several structures obtained using this program and a standard package LUSAS were found identical. Further, the validation of assumptions employed in the simulation of these structures by this program was carried out by making and testing of a reasonable size model composite triple layer bridge-grid. The nodal deflections, strains of skeletal members and of plate components recorded were found in good agreement with the predicted results for the two load cases used: uniformly distributed and concentrated loading.

7. The analysis based on the stiffness method requires a large amount of computer time and memory for the solution of a modest size composite triple layer bridge-grid. Therefore, the use of approximate methods of analysis, useful for hand calculations, was explored for the preliminary design of the structure, leaving final and detailed analysis to the stiffness method. This led to the introduction of a new method called "girder analogy", (ref. 2). The analysis of composite triple layer bridge-grids having different span/width and span/depth ratios by this method results which were in good agreement with those obtained from the general purpose program. Further, the girder analogy method was found to be applicable for bridge type space structures subjected to both uniformly distributed and concentrated loading. The girder analogy method was extended and tested for the analysis of composite double layer bridge-grids for a number of span/width ratios, with or without eccentricity between the top layer skeletal and continuum members, (refs 3-4). The efficiency of the girder analogy method was

obvious from its simplicity in use and reasonable accuracy for the solution of highly redundant skeletal and composite double/triple layer bridge grids.

8. In this paper, the role of continuum as an integral component of composite double layer bridge-grids has been investigated. For this purpose, a model of a double layer bridge-grid composed of hollow steel skeletal members has been tested, first without any top covering, under two types of loading: modified uniformly distributed loading and two point loading. This skeletal model was then converted into a composite one, by bolting a thin steel covering to the top layer joints as integral component of the structure. The composite model was also tested under the same two types of loading, for two different thicknesses of the continuum separately. The observations made on the models were found in good agreement with the theoretical solutions obtained from two methods: the standard stiffness and the girder analogy methods.

MODEL

9. The square on square offset configuration was selected, for the model of a double layer bridge-grid, due to its relatively high rigidity Fig. 1. The model was referred to as a skeletal model when it consisted of steel circular hollow pipes, with outer diameter of 12.87 mm and wall thickness of 2.06 mm. The model was converted into a composite one (Fig. 2) by bolting two steel sheets, of 1.08mm and 1.65mm on two separate occasions, at the top layer joints after carrying out testing, within the elastic limit, under the two types of loading. The steel sheets in the composite models were considered as top decks of a prototype bridge.

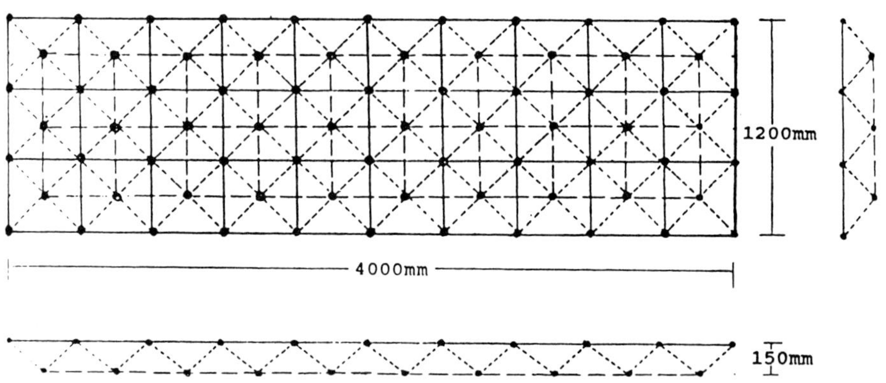

Fig. 1. Skeletal model of a double layer bridge-grid.
(Showing the square on square configuration)

10. For the model, a special joint was designed to produce a reasonable accuracy in the fabrication of the model. All the connection points of the model were precisely welded, without damaging the thin wall of the steel tubes. All the longitudinal members of the top and bottom layers were passed through the joints and were welded at each side of the joint. Then transverse and diagonal members were inserted into their respective slots in a joint and welded into position. Fig. 3

shows a typical joint of the model indicating bottom layer members and diagonal bracing members. The steel sheets were bolted to the top layer joints of the model. The steel sheets obtained from the market were in wrapped form. Therefore, in spite of all the efforts carried out to make them straight, there remained slight initial crookedness and imperfections in them.

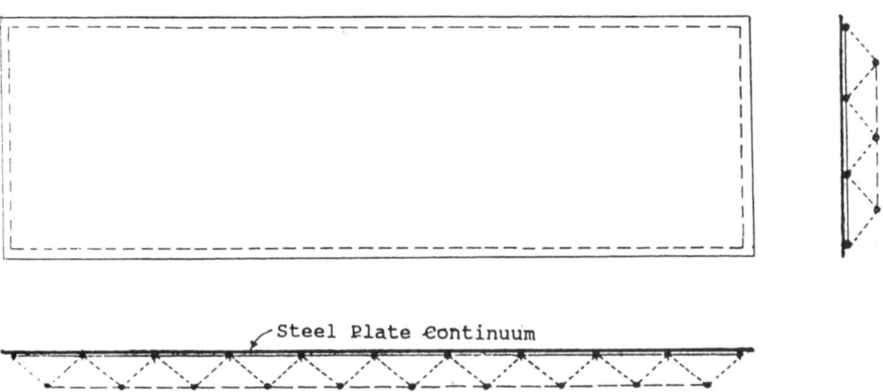

Fig. 2. Composite model of a double layer bridge-grid.

Fig. 3. Details of joints and the connection of the continuum with the top layer members.

11. Two loading cases A & B were used for both forms of the model, i.e. skeletal and composite. In loading case A, standard weights were attached to the hooks, hung from the top layer nodes of the model (Fig. 3). Care was taken to avoid eccentric loading. The load was applied at twenty eight points. A load of 267N (60 lb) in three equal increments of 89N was applied on each of the 14

interior top layer nodes, whereas half of this loading was applied on 14 exterior joints. The load in this case had been worked out by adding both UDL (uniformly distributed load) and KEL (knife edge load) type H.A. loading, in accordance with B.S. 5400 and then distributing this to each of the twenty eight top layer nodal points. The loading in this case has been referred to as the 'modified uniformly distributed loading'. In load case B, 2670N (600 lb) load was applied at each of the two points of the top layer at mid span of the model. This two point loading was carried out in five, approximately equal increments. This type of loading was meant to represent concentrated effects of a heavy vehicle on the bridge grid and is referred to as 'two point mid span loading'. The loading system comprised of a hydraulic jack, a load cell and a pressure gauge.

12. There were 240 members and 74 joints in the model and the available instrumentation to measure the stresses and deflections were insufficient. The preliminary analysis of the model revealed that axial forces in skeletal members were dominant and the other types of forces were relatively negligible. Therefore, the available instrumentation was connected to the nodes, considered to have large vertical deflections and to members which were to be highly strained. The positions of strain gauges and deflection gauges were kept the same for both the skeletal and composite models. On the top surface of the steel thin plate, in the case of the composite model, six 45° strain rosettes were used to determine the principal strains and stresses. Each strain rosette was composed of three filaments arranged in three directions in the horizontal plane and its middle filament makes an angle of 45° with the other two which were set at a right angle to one another. Normal strains in three directions were measured at each rosette point and then the maximum and minimum strains and stresses were computed independently.

DISCUSSION OF RESULTS

13. The experimental studies were conducted first, on the skeletal model and then, these tests were carried out on the same model by connecting two steel sheets, on two separate occasions, to its top layer as an integral part of the structure. Two loading cases (A & B) were used for both the skeletal as well as for the composite model. Applied loading was designed such that the member forces both in the skeletal and composite models remained well within the elastic limit of the material, and the same model was used for all the experiments. Skeletal and composite models with a continua thicknesses of 1.08mm and 1.65mm were tested individually. Then experimental results were compared with the theoretical results obtained by the standard stiffness and girder analogy methods. Experimental deflections and axial forces were found to be in good agreement with those obtained from the theoretical analyses. The important points of this study are as follows:

14. The continuum had a significant effect on the stiffness of a composite double layer bridge-grid. A decrease in deflections was noted with the increased continuum thickness. Experimental and theoretical mid-span deflections were reduced by 23% and 28% for 1.08 mm and 1.65 mm continua thicknesses respectively, as compared with zero thickness (skeletal model), Fig. 4.

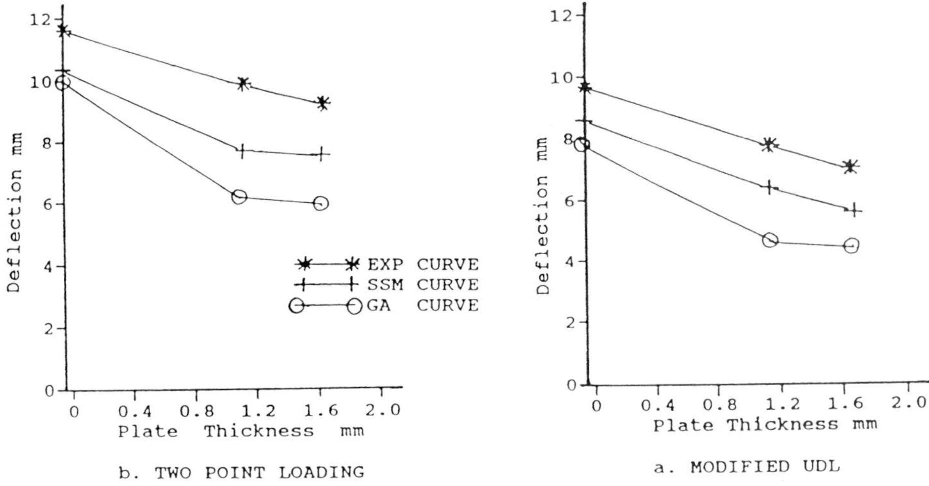

Fig. 4. Mid-span deflections in skeletal and composite models.

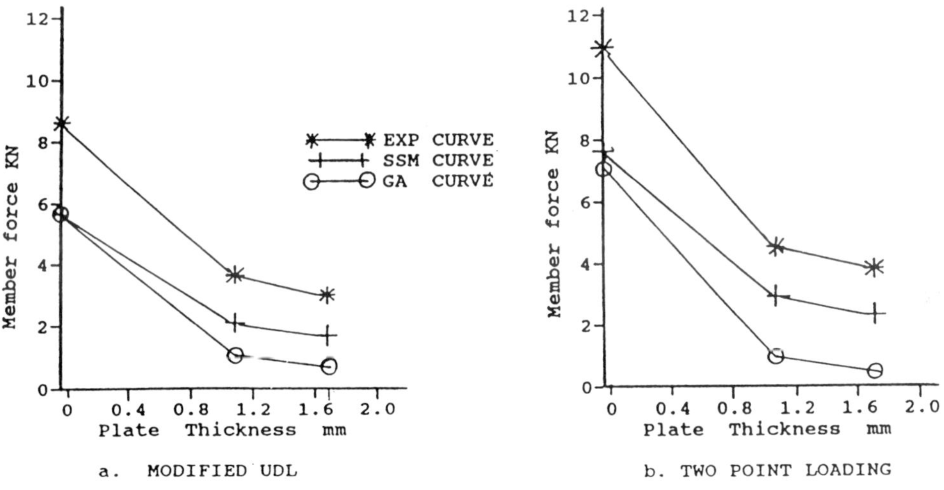

Fig. 5. Maximum top layer member forces on skeletal and composite models.

15. Axial forces in the top layer skeletal members, of the composite model were considerably reduced by providing a continuum as an integral part of the structure. The measured axial forces in the top layer decreased by 52% and 60% for 1.08mm and 1.65mm thicknesses of continuum, respectively when compared

with the skeletal non-composite structure. However, the decline in bottom layer member forces is small, Figs 5-6.

16. The smaller change of bottom layer forces and large decrease in top layer forces, can be attributed to the upward movement of neutral axis of the bridge-grid due to the integral effect of the continuum.

17. The principal stresses on top surface of the steel sheets are well within the elastic limit and decrease with the increases in continuum thickness Fig. 7.

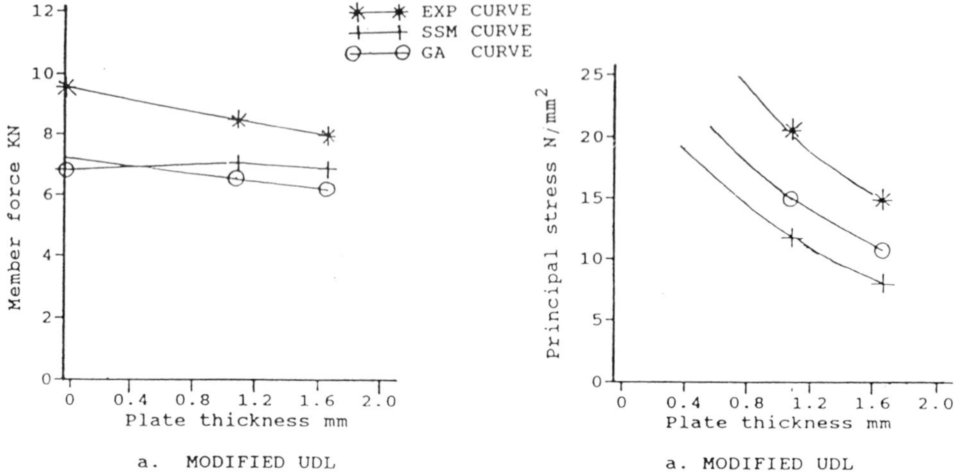

Fig. 6. Maximum bottom layer member forces in skeletal and composite models.

Fig. 7. Maximum principal stresses in composite models.

18. The decrease in deflections and member forces is gradual for loading case A, as compared with for loading case B, where these changes are abrupt. This can be attributed to the change of bending moments in two loading cases.

19. Experimental and theoretical results were, generally, in good agreement with each other, as they depicted patterns of decreases, both in deflections and member forces Figs 4-7. This agreement showed that the assumptions employed in theoretical analyses are reasonable.

20. The higher experimental values of nodal deflections and member forces compared with those obtained from the theoretical analysis can be attributed to:

(a) Continuum was not properly attached to the top layer members, but it was placed on the upward projected joints and was fastened there by bolts. This might cause uneven eccentricities of the continuum from the centre of the top layer members.

(b) The sheets/continua used had initial curvature and lack of straightness.

CONCLUSIONS

21. From the above discussion, it is concluded that the behaviour of a composite double layer bridge-grid improves significantly with the use of top sheeting as an integral component. It follows that the contribution of the top sheeting/continuum is small, though due to the increase in the stiffness of the structure it provides considerable decrease in critical stresses in top layer skeletal members.

REFERENCES

1. AL-BAZZAZ A. A. J. An investigation into composite double layer space grid structures. Ph.D. thesis, University of Strathclyde, UK. 1976.
2. ASHRAF M. Structural behaviour of composite triple layer ridge-grids. Ph.D. thesis, University of Surrey, UK. 1988.
3. ASHRAF M, ILYAS M. and AKHTAR M.N. Study of composite double layer bridge-grid using the girder analogy method of analysis. Journal of University of Engineering & Technology, Lahore, Pakistan, 1989.
4. ASHRAF M, CHAUDHRY M.A. and SIDDIQI Z.A. Behaviour of composite double layer bridge-grids having eccentricity. Proceeding of APSAC 89, Malaysia, 1988.
5. CASTILLO H. A space frame construction with steel and reinforced concrete - the "Tridilosa" spatial structure. In: Dvaies R.M. (ed.) space structures. Oxford: Blackwell 1966, 1089-1093.
6. MAKOWSKI Z.S. Steel Space Structures, London, Michael Joseph, 1965.
7. MAKOWSKI Z.S. Space structures in Mexico: A review of their recent developments. Building Specifications, October 1981.

164. Automation in construction of shells, space, multi-storey, super high-rise and extra-terrestrial structures

D. N. BINI, Binistar Inc., Italy

SYNOPSIS. Reinforced concrete shells are no longer fashionable. Space structures have taken their place "under the spot light" due to their multiple terrestrial applications and potential extra-terrestrial uses. Yet, with the advent of construction automation, both these forms of structures will become increasingly popular. This is due to their economy, simplicity, minimal use of material, and for their proven and lasting structural performance.The author has dedicated his professional life to the development of original automated construction technologies to render the construction of special structural forms such as concrete shells (refs 1 & 3) and space structures (refs 4 & 5) as efficient and economical as possible. This paper, presents the technologies that have been the fruit of twenty-five years of personal research in construction automation. It is divided in two sections.The first part describes and critiques the author's development and utilisation of "automated construction processes" applicable to reinforced concrete shells and space structures of different dimensions and shapes. The second part of the paper presents various joint research efforts that are now in their initial developing stages. Most of aforesaid research efforts are being closely developed together with Shimizu Corporation of Tokyo (Japan). They are applicable to a wide variety of "automated structures", from "semi-traditional" multi-storey buildings, to city infrastructures and from extra terrestrial space structure to self-shaping and self-sinking Shelters and Habitats for Lunar Outposts.

PART 1

1. On July 4th, 1965 in Crespellano, (Italy) at 4pm. the author made the world's first attempt at using an automatic self-shaping construction process to build a fully reinforced concrete structure on a previously built floor and footing system (Fig. 1). This experiment was successfully completed by 7pm. In only three hours, a dome, 12m in diameter and 6m in height had been pneumatically lifted from the ground and shaped by air into an hemispherical thin shell structure. This structure was soon proven sound by severe load-tests carried out by the University of Bologna, School of Engineering. During the pneumatic lifting

process, the auto-positioning of the required steel reinforcement was successful, but the control, and the final distribution and thickness of the wet concrete was inadequate. This was due to a lack of control of the "asymmetrical" lifting sequences lasting 60 minutes. For 20 years this building was utilised as an office and for commercial space.

Fig. 1. The worlds first experiment in Crespellano Bologna. Italy (1965). Sequence of the inflation and completed 12m diameter spherical dome.

2. A year later, on June 6th,1966, in Castelfranco Emilia, (Italy) after some spine tingling moments, another major experiment was undertaken. At 7am, from 25 ready-mix trucks, a special-blend, lightened-concrete, was poured at ground level on top of a pre-shaped, folded, 1mm. thick, semi-elliptical membrane anchored to a circular footing structure and covered with a different self-positioning steel reinforcement. Again, in spite of a severe lack of lifting control, at 6pm both steel and concrete had been lifted up and shaped by one man operating a small fan generating a pressure of 60cm of water.

3. In less than four hours, 115 cubic meters of wet concrete and flat steel was shaped into an elliptical reinforced concrete dome 36m in diameter and 12m in height. During the inflation process the previously built circular reinforced concrete footing structure, (0.70m x 0.60m x 114m) was almost lifted from its trench. Very hard work was also required to manually maintain uniform the thickness of the concrete during its suspenseful, uneven inflation. In spite of these difficulties during construction, a quarter of a century later this structure is still standing.

4. On May 16th, 1967 at 3:45pm an improved and definitive version of this automatic method of construction, provided with an unique control system necessary for its lifting process was presented in New York by the author, together with the Chairman of the Architectural Technology Division at Columbia University, Prof Mario Salvadori. The perfected lifting sequence was controlled by a powerful network of multi-directional steel springs (anchored to the footing and imbedded in the wet concrete). This innovation provided an efficient, symmetrical guiding force acting in opposition to the inflation process. The proper setting of the concrete was achieved through the use of an additional external membrane. After the inflation was completed, this membrane also allowed a number of vibrating carts to roll all over it and re-compact the still wet

concrete retained between the inner and outer membranes. For the first time in the United States, a fully controlled, compacted and sound, self-supporting shell structure was automatically shaped and built two hours after the concrete had been poured at ground level directly from two ready-mix concrete trucks. After all scientific tests were successfully carried out. The structure was demolished. This was in accordance with the request of the local authorities and this proved to be the only major problem related to this structure.

Fig. 2. The first 30m in diameter experiment in Castelfranco Emilia, Modena, Italy (1966). Detail of the footing system.

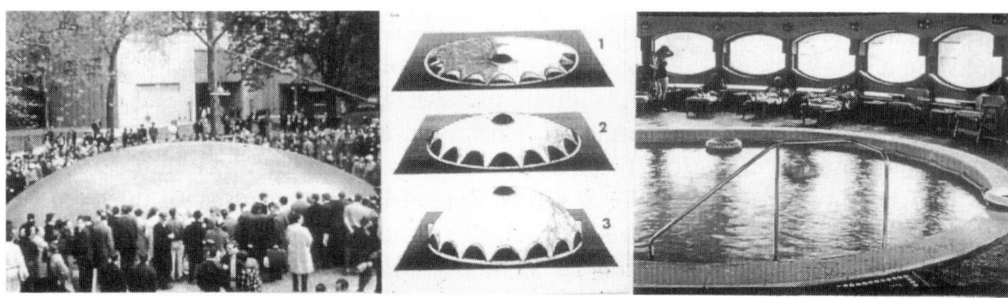

Fig. 3. Sequences of inflation of the first Binishells experiment successfully built in the U.S.A.(1967). Original design of the third Binishells built in Italy and photo of the interior (1967).

5. The above described method of construction is now known as Binishells. It requires some expensive and sophisticated equipment, but continues to be the cheapest and fastest method to produce circular-based, monolithic, reinforced concrete shell structures, with elliptical sections ranging in size from 12 to 36 meters in diameter. Yet even automation is affected by human error. It is important to report in this paper that between 1970 and 1990, 3 of over 1,600 Binishells successfully built in 23 countries of the world, have suffered severe failures and have been demolished or re-built (refs 2 & 3).

6. In the seventies the author, as a further improvement to the Binishells concept, devised two additional systems:
A) The first method of construction, called "Binix", was developed to reduce the amount of concrete necessary to build a ribbed dome structure while increasing its span. This method introduced an assembly of prefabricated elements, laid on top of a flat pneumatic system at ground level. These elements were positioned to provide the form-work for the reinforced concrete. The prefabricated steel was properly located between and within the forms and the concrete was poured in at ground level. The entire flat structure was then lifted up and shaped only by air pressure. This System may produce a large, hexagonal-based, monolithic, framed or ribbed dome structure which may span up to 100m. A small Binix prototype was successfully built in Australia in 1976. The only problem experienced during the construction of the prototype was related to the process of re-compacting and vibrating the concrete inside the raised form-network. (ref. 4)

Fig. 4. The Binix experiment produced in NSW Australia (1977).

B) The second system, called "Minishell", produces 8m x 8m or 10m x 10m square-based, monolithic, reinforced concrete shell structures, provided with four large reinforced openings which result from the automated construction process itself. Only 4.6m^3 of concrete reinforced with No. 400, 6 mm. diameter corrugated rods, combined No. 200, 36mm external diameter hard drawn 2mm spring wire are necessary to complete this structure.

7. The Minishell system is considered to be one of the fastest and cheapest systems for building permanent monolithic shelters and is therefore suitable for low-cost housing. After the floor, footing and reinforcing steel contained within the springs are prepared, the erection time for each structure is reduced to 30 minutes. The required man power results in 1.5 hours per man per sq.m. of covered area. Several Minishell Tourist villages are in use in Australia and Italy.

8. The only two problems experienced during the application of this method of construction were:

(a) Structural weakness in the "vertical" portion of the walls when the concrete was still green.

(b) Lack of precision and symmetry in the shape and size of the four automatic openings (refs 5 & 6).

Fig. 5. Sequences of construction of the Minishell Tourist Villa, in Cairns, Australia (1980).

9. The "Binistar" system was originally conceived and patented by the author in Australia in 1979. The first 40m prototype was successfully built in 1986 in Bari (Italy) by E.GE.CON. and F.lli Dioguardi S.p.A., with the direct assistance and supervision of Binistar Inc. California, U.S.A.

10. To our knowledge it is the only demountable, *automatic*, method of construction capable to shape large-span metal space-frame structures of different shapes and dimensions. Binistars are pneumatically-lifted from their footing-level to their final position by way of an air-tight, special fabric, pre-shaped membrane made of 1 mm. thick self-extinguishing polyester+PVC bands welded at high frequency with 4-6cm of overlap. At the end of the construction process, this membrane, remains in tension, suspended from the structure's nodal points as a leak-free final cover. The galvanised steel-space frame structure is composed of a series of telescopic pipes some (or all) of which are variable in length and mechanically fixed after the inflation is completed.

11. The weight of the metal portion of the structure is 10-19 kg/m^2. depending on the free span required. The set up and erection time fluctuates between 0.40-0.50 hours per man per m^2. of covered surface, depending on local conditions, shape and dimensions. (A 20m by 40m torroidal structure requires 400 working hours, a 40m by 42m barrel vault structure requires 1,000 working hours). The inflation process requires 3-4 blowers capable of producing a pressure of 70-100mm of water, each provided with a flow of 15-20,000m^3 per hour.

12. One of the unique advantages of a Binistar space structure, is it's potential for a quick and easy (partial or complete) *pneumatic disassembly*, with the possibility of immediate relocation. The extremely safe disassembly process is performed through sequences which are the inverse of those necessary for the assembly. A number of Binistar structures have been leased and re-utilised in different locations. This system, fully developed and engineered by the E.GE.CON - F.lli Dioguardi Group, has been designed for high winds (100 km/hr) and heavy snow loads (1m/m^2). It is presently used in Italy, France and Spain and will be soon utilised in U.S.A.

13. In the first prototype, the extension of each telescopic pipe was controlled by automatic locking devices. This additional level of automation was however subsequently abandoned due to cost and difficulties in manufacturing it. At

present each telescopic pipe is manually fixed and secured at it's final automatic extension. This was the case in all commercial Binistars produced and erected in Italy by the E.GE.CON.- F.lli Dioguardi Group for the '90 World Soccer Championship and in Spain for the '92 Olympic Games and for Seville World '92 EXPO (ref. 6).

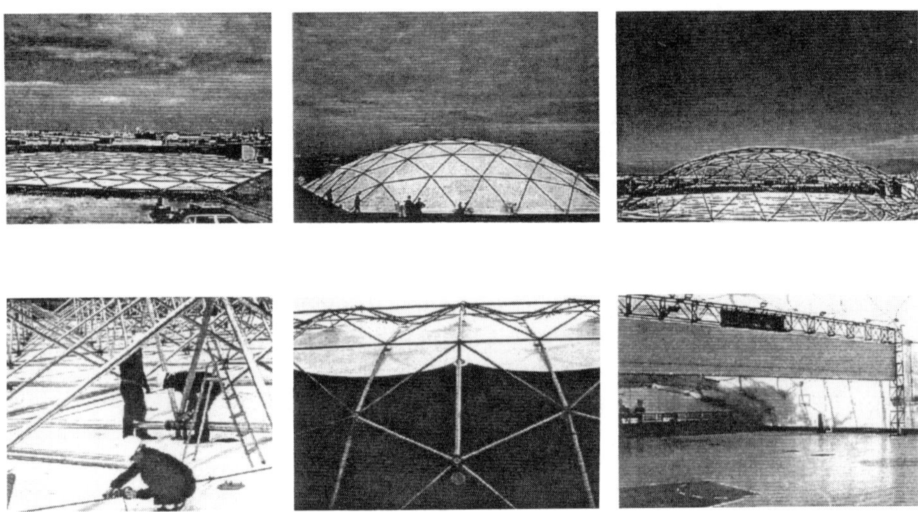

Fig. 6. Sequence of construction of Binistar structures in Italy and details (1968-86).

14. Unlike the Binishells, the Binix, the Minishell and the Binistar, the following systems have not as yet been proven commercially or tested through prototypes. In 1985 the author was requested by authorities of the former Soviet Union, to devise a pneumatic erection system for a prefabricated single family house, to help solve a severe lack of single low-cost housing in the former Soviet Union (ref. 5).

15. The result of such a study was a structural shelter made of 4 different basic prefabricated modular components which were designed to be mass-produced in a mobile or traditional factory. A total of 28 elements either completely finished, or unfinished (made from a combination of the same 4 basic components) are assembled at ground level . These elements coupled with a unique pneumatic footing system make a self-erecting, air-supported low-cost shelter. The working model presented in Moscow was nicknamed: "Pak-Home".

Fig. 7. Pak-Home model presented in Moscow (1986).

16. In the same period the author developed two other conceptual systems, the "Fold-A-Struct" and the "Autotent" (ref. 6). These systems respectively produce:
a) An instant, demountable, square-based, class-room unit, made up of four identical mass produced components provided with 'inner stored energy" which contribute to the self-assembly process:
b) A folded space frame which may be utilised as a mini-shelter. This shelter is designed as a civilian or military tent which is dropped by low-flying aircraft in case of emergencies. The autotent self-shapes during its fall and is provided with an identification light and with a survival kit.

17. The author also collaborated with his son Nicoló to design the "Nico-Space" (developed in the U.S.A. only as a working model). This object is a self-shaping, carbon-fibre and kapton-polycarbonate-graphite-composite-laminate icosahedron, developed for mini space-shelter or mini space-warehouse to store space debris.

Fig. 8. Fold-A-Struct self-assembling model (1987).

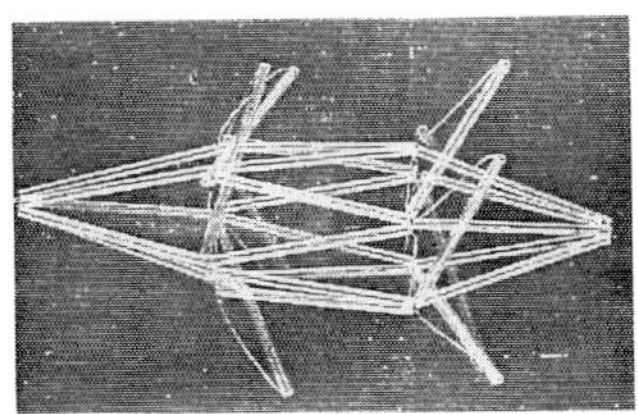
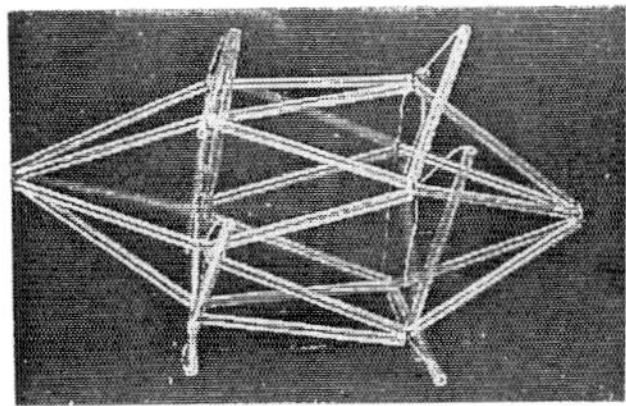

Fig. 9. Nico-Space self-shaping model (1988).

PART 2

18. The above mentioned, unique construction technologies, have relied on air pressure alone to shape, assemble, or lift building structures. The second part of this paper describes construction systems in which construction automation is complemented with the aid of robotics.

19. In 1989 a "joint-research" effort was developed by the author and Dr. Toshiaki Fujimori, General Manager of the planning department of Shimizu Corporation of Tokyo, Japan. This effort resulted in the conceptual development of a method to construct a mile high, seafloor-based, vertical, cable-braced, reinforced-concrete (or all-steel) hyperbolic megastructure, designed as the infrastructure for a city of 600,000 inhabitants. Such a megastructure, called "Tower City", is supposed to be constructed by two mobile, robotized factories.

20. The construction process also includes a new method for realising an underwater 100 meter thick continuous footing system. In essence this footing systems provides a circular "dam", one mile in diameter, which is able to resist external, dynamic hydraulic sea water pressure. The aforesaid dam allows for the recovery of new land from unused shallow sea-floors.

21. This concept was presented on January 25, at the M.I.T "1991 International Workshop on Innovative Structure, Materials, Designs and Construction for the 21st Century" (ref. 7).

22. In 1991, the "Lunit" (a self-shaping and self sinking, pressurised, mobile unit for a life-supporting Lunar Outpost) and the "Lunhab" (a self-shaping and self sinking pressurised Lunar structure designed as a life-supporting habitat) were jointly developed together with Dr. Shinji Matsumoto, General Manager of the Institute of Technology and Deputy General Manager of The Space Project Office and other experts of Shimizu Corporation. Both of these developments have been presented by Mr. H. Nakazawa in Washington DC at the 43rd Congress of the International Astronautical Federation (ref. 8).

23. The author, together with other experts of the Space Project Office of Shimizu Corporation, is presently working on the joint development of a

conceptual interior design for the "Lunit" system. A proposal has been put forward for the construction in 1993-94 of a working scale model of the "Lunhab" at the University of Florence (Italy) affiliated campus of the International Space University.

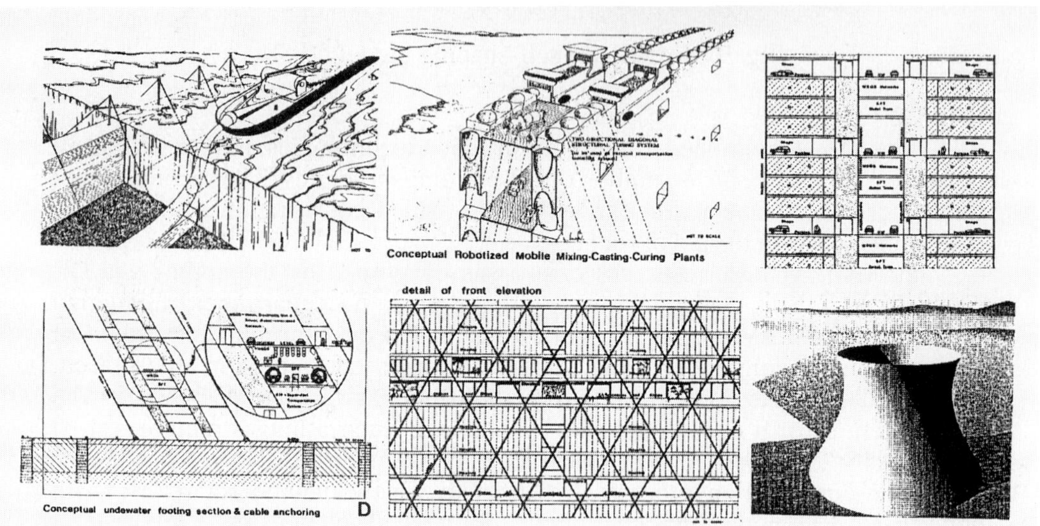

Fig. 10. The Tower City Concept. (1998-1990).

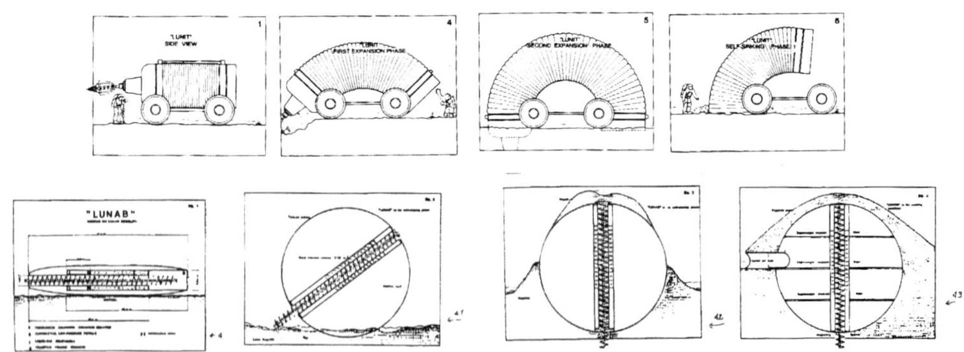

Fig. 11. Lunit and Lunhab concepts (1991-1992).

24. The author was also involved in a joint effort to develop a patented automated method of construction, owned by the Institute of Technology of Shimizu Corporation, called "Air-Lift-Up" system. This system will combine an advanced pneumatic construction process and robotics to construct semi-traditional multi-storey buildings. This system was presented in 1992 by Dr. Haruo Nakazawa, at the First International Symposium of CIB Working Commission W 82 (ref. 9).

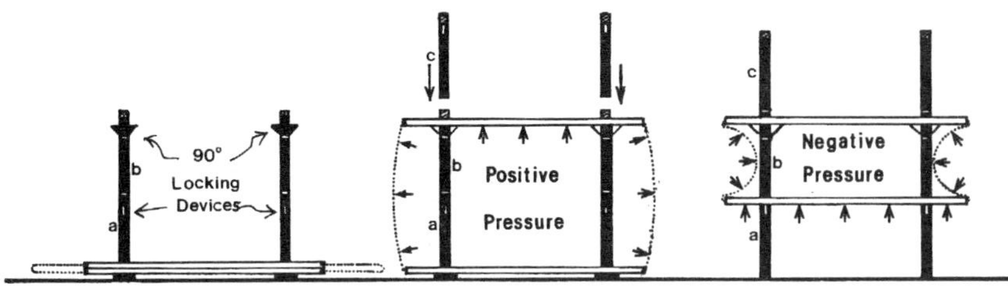

Fig. 12. The "Air-Lift-Up" system (1991-92).

25. The aim of this paper is to contribute to the promotion of construction automation and to encourage Universities, Institutes and Industries to invest more effort in this important subject.

CONCLUSION

26. Construction automation eliminates the need for workers to operate in uncomfortable and hazardous conditions, while concurrently saving time, reducing costs and improving the quality of the work place in the construction industry. It is well known that in spite of the employment of sophisticated machinery and equipment and some recent efforts to improve on-site working conditions, construction sites in general offer some of the worst working conditions existing today. As has already been experienced in other industries, the application of automation and robotics can cut costs and environmental pollution, and dramatically improve safety and quality control to the benefit of living standards in general.

REFERENCES

1. BINI, D. "A new pneumatic technique for the construction of thin shells" for the International Association for Shell Structures. Proceedings for the First IASS International Colloquium on Pneumatic Structures. University of Stuttgart, Germany, May 11 - 12, 1967.
2. BINI, D. "The Cause of the Collapse of the Pittwater Binishells" published by Engineers Australia, issue of May 3rd, 1991.
3. LEVY, M. & SALVADORI, M. "Why Building Fall Down". published by W.W. Norton & Co., New York and London ,1991.

4. BINI, D. "Place 1,000 tons of concrete at ground level, rise a 100 meter dome in one hour". Bini Consultants Australia, National Ready Mix Concrete Association Convention, Sydney N.S.W. Australia, September 1976.
5. BINI, D. and PIETROGRANDE, S. "Self shaping space structures" for the "International Journal of Space Structures", Volume 4, No 2, 1989, published by Multiscience Publishing Co. Ltd. U.K.
6. BINI, D. "Pak-Home and Minishell systems" '96 Expo Centre Seminar organised by The Chamber of Commerce of Moscow, USSR, June 9-10.
7. BINI, D. and BINI, N. "Tower City": A glance into the technologies of the 21 century. The International Workshop on innovative Structures, Materials, Designs and Construction for the 21st Century in MIT Cambridge, 11991.
8. MATSUMOTO, S., SASAKI, J., SAITO, T., NOZAKI, K., CANAMORI, H., KAY, Y., TAKAGI, K., UENO, H., NAKAZAWA, H., YAMAKAWA, Y. and BINI, D. "Inflatable lunar structure with reinforcing rings-self-shaping and self-sinking system for a lunar outposts". World Space Congress, Washington DC, 28 Aug.-Sept., 1992.
9. NAKAZAWA, H., MATSUMOTO, S., TAKADA, H. and BINI, D. "Air-Lift-Up Construction System" A case Study of future Building construction System. First International Symposium of CIB Working Commission W 82: Construction Beyond 2000, Future of Construction - Construction of the Future. June 15-18, 1992, Espoo, Finland.

165. Space structures in The Netherlands since 1984

J. M. GERRITS, Delft University of Technology, The Netherlands

SYNOPSIS. The paper contains two reviews of spatial structures in the Netherlands during the period 1983-1991. The first review reflects the development and research in the field of spatial structures, while the second deals with the architectural trends.

DEVELOPMENT AND RESEARCH

1. The development of space structures in the Netherlands has been marked by the development of several new space frame systems as well as research and development of polyhedra forms.

New space frame systems

2. Thin poles of roundwood are the main structural components of a frame system developed by the Delft University of Technology (P. Huybers, Fig. 1, ref. 1). A specific tool using a special connection technique has been developed for the steel wire bindings (Fig. 2). It is applicable for all kinds of structures. All dimensions and angles were calculated using two CAD programs. Since 1984 several structures have been built in the Netherlands and in England (dome structures, agricultural sheds and a watch-tower).

3. The D3BV bureau, consulting engineers and architects (H. Nieuwland and J.G. Kraus) developed a two-way single-layer framc system for the flat roof construction of a music hall in The Hague in 1987. The structural steel components are prefabricated welded plane girders, using hollow tubular sections and are bolted together in the 'square on square' frame arrangement. This system is applied for clear spans up to approximately 50m. Unfortunately, it has never been used.

4. The ABT bureau, consulting engineers and architects (M.C.M. Maarschalkerwaart) developed a flat two-way double-layer frame system on the 'square on square' arrangement, the so called Minode system. The main components are prefabricated welded half octahedra and high tensile tie bars within the node at the lower chord level (Fig. 3). The first application was the wall and floor construction of pavilions on the Waal front in Nijmegen (1990).

5. The Van Thiel system, developed by the firm Van Thiel Space Structures (L.G.A. Schrauwen), is suitable for all kinds of space frame structures. The system consists of solid spherical steel nodes and tubular members (Fig. 4). Since 1990 it has been used frequently in the Netherlands and in Belgium.

Polyhedra

6. At the Delft University of Technology, P. Huybers has carried out research in the field of the geometry of the dual uniform solids, including the reciprocal figures of semi-regular solids. He has developed an interactive computer program for the use of these solids (ref. 2) and has catalogued all the relevant data for these figures with a high degree of accuracy (ref. 3).

ARCHITECTURAL TRENDS

7. Space structures are accepted by occupants, architects and designers in the Netherlands especially because of their visual appearance. Unfortunately they are considered expensive. As a result most designers make too little use of the specific structural and architectural properties of these structures and therefore the cost is higher compared with other more traditional structures. This was particularly true in the eighties when the young designers had a great interest in space frames and membrane structures, as a consequence of their architectural composition. During the last years the field of interest of the architects and designers has changed from space frames and membranes to (single or spatial) guyed structures.

Space frames

8. The number of realised steel or aluminium space frames, mainly as flat roof constructions, is now stabilised. Most frames have been applied for projects with clear spans up to approximately 35m. In all cases modular prefabricated components are used, forming two-way double-layer frames with steel cladding on the outside. Except for projects with cladding beneath the frame. The usual frame arrangements are 'square on square', 'square on larger square' and 'square on diagonal'.

9. More and more applications of space frames were discovered by the Dutch architects and designers during the eighties. Then they have started using frames for:

- heavily loaded roof construction (permanent loading above 4.5 kN/m^2): concrete roof-plates on the frame of a theatre in Haarlem (design Wiek Rolling, gangway and hoisting machines between top and bottom layer of the Nodus space frame with suspended dress-circle, 13.5x13.5m, 1985,) and lightweight concrete roof on a theatre in Drachten (design Abe Bonnema, Tuball system, 17x25m, 1985),
- roof and wall construction: Sloterdijk railway station in Amsterdam (design H.C.H. Reijnders, Mero system, 50x60m, wall height 13 m, 1986),
- roof, wall and columns construction: the railway station in Almere (design P.A.M. Kilsdonk, Montal system, 45x153m, wall height 5.25m, 1987),

- floor construction: an experimental house in Almere (design Benthem & Crouwel, Octatube system, 8x10m, 1984) and two movable platforms in a cultural centre in Gouda (design Sliedrecht Zonneveld, Mero Deck system, each 5.6x8.6m, 1985).

10. Another trend is the application of glass systems and lighting equipment as integral parts of members, such as:

- combination of frame and structural glass system: a music pavilion in Haarlem , Fig. 6,(design Wiek Rölling, dome diameter 9 m, 1984) and the Glass Music Box in the former Exchange of Berlage in Amsterdam (design Pieter Zaanen, 13x23m, wall height 9m, Tuball system, 1989),
- combination of frame and artificial lighting: roof structures in a theatre in Drachten (design Bonnema, 17x25m, 1985) and a showroom in Deventer (design Postma, 600m^2, 1986).

11. The application of frame systems as integral parts of a completed construction is another trend, such as:

- combination of frame and membrane structures: a exhibition dome of the Autotron Car Museum, in Drunen (design Mick Eekhout, triangulated Octatube system with suspended single membrane underneath the frame, diameter 40m, 1985; two years later moved to Rosmalen) and a canopy in Almere (design architects office Op ten Noord/Blijdensteijn, flat octatube system with membrane elements suspended in the frame, 17.5x35m, 1988),
- combination of frame, membrane and tensigrity structures: an entrance canopy for the 'Nederlanden van 1870' office on Diemen, Fig. 5, (design Maarten Grasveld and Mick Eekhout, 5.4x12.6 m, height poles 14 m, 1987).

Domes

12. Several wooden geodetic domes were built by De Coene Bouwelementen, for example: two domes for a swimming pool in Nes, Fig. 7, (design Van der Zee and IJbema, each diameter 41 m, height 7,8 m, 1986) and the leisure dome in Goes (diameter 47m)

13. A new application for domes in the Netherlands is the covering of basins. At the end of the seventies the legislation no longer permited the odour nuisance of the open basins of waste-water purification plant. For this reason gastight domes were applied, such as aluminium geodetic domes with aluminium cladding for two oxidation basins in Hedel (design de Vries Zn, Altech system, diameter 52 m, height 8 m, 1991).

Membrane structures

14. A few shelters have been built in the Netherlands. Mostly they are applied as stands for temporary or mobile exhibitions or shows. Such as: an outside stage shelter for the AVRO broadcasting company in Kaatsheuvel (20x8 m, height 8 m, 1987). They have also been used for permanent buildings, for example: the liberation museum in Groesbeek, Fig. 8, (design A.G.M. Croonen and ABT, Φ20 m, clear height 11 m, 1987) and the 10 canopies for the Doe Mere-centre in

Almere (design Bakker en Verhoeff, each 10x20 or 10x10 m, height 20 m, 1988).

Tensigrity structures

15. Tensigrity structures are mostly applied as 'objects of art', sometimes in combination with membranes, such as a twist tensigrity in the innercourt of the National Maritime Museum in Amsterdam (design Krijn Giezen and Mick Eekhout, 12x12 m, height 14 m, 1984).

16. The only permanent tensigrity structure in the Netherlands as a building structure, is an exhibition green house in Arcen, Fig. 9. This structure utilises the first application of a plane tensigrity girder, a post stressed system was developed by Chiel Bartels (design Frans Royen, 22 units of 12x12 m, height 15 m, 1991).

CONCLUSION

17. Space frames have found a permanent place in the Netherlands next to the other more traditional structures. Since the eighties there is a trend to use space frames for more varied applications.

18. A few membranes and tensigrity structures have been built as a result of the limited knowledge of the designers and the climate in the Netherlands.

19. The last few years the field of interest of the architects and designers has changed from space frames and membranes to (single or spatial) guyed structures. As a consequence of these developments and the state of the economy there has been little research and development.

REFERENCES

1. HUYBERS P. Thin poles of roundwood for structural engineering applications in building. Structural Engineering Review, 1990, 2, 169-182.

2. HUYBERS P. Computer aided design of polyhedral building structures. Design Structures, 1, 1993.

3. HUYBERS P. The geometry of the dual unform solids. Proceedings of the First International Seminar on Structural Morphology. Montpellier, 1992, 179-190.

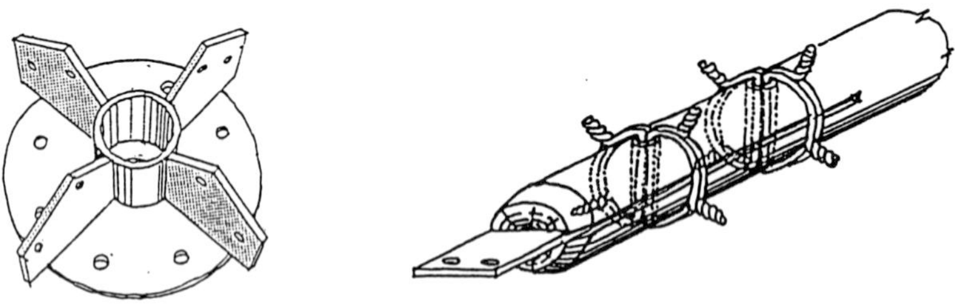

Fig. 1. The main structural components of the thin roundwood poles.

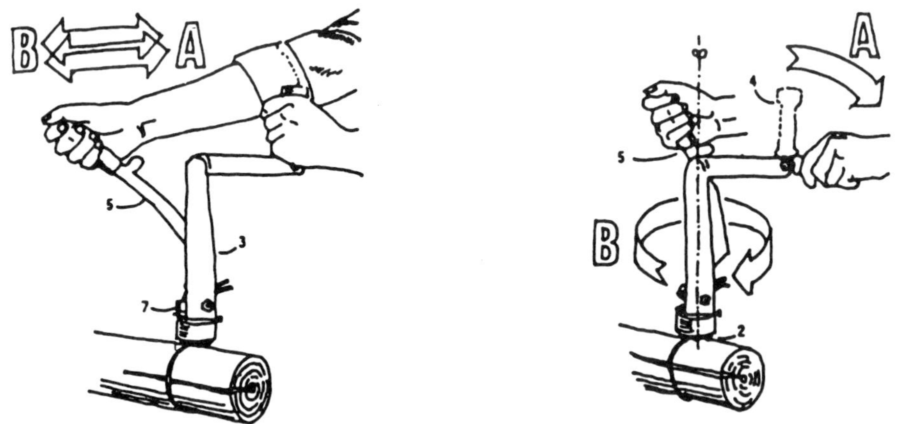

Fig. 2. The tool for the wire binding connection.

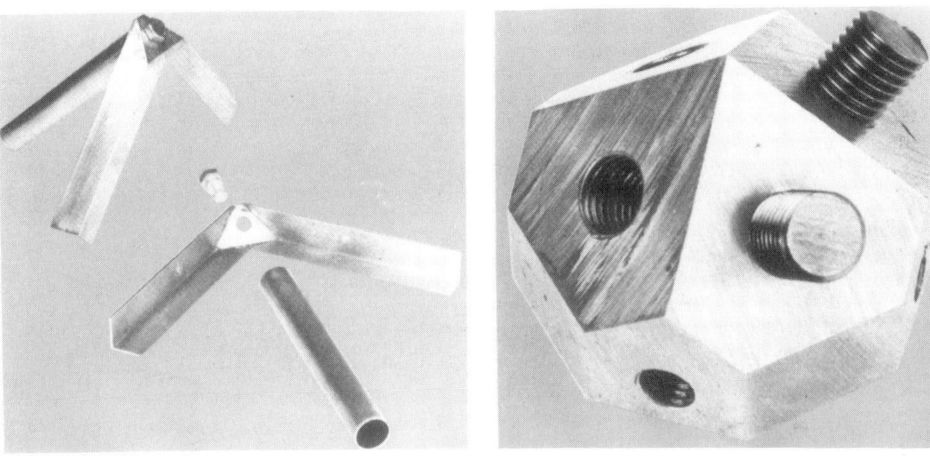

Fig. 3. The minode system (the top chord bracing connection and bottom chord node).

Fig. 4. The Van Thiel joint.

Fig. 5. A model of the entrance canopy for the 'Nederlanden van 1870' in Diemen.

Fig. 6. Music pavilion in Haarlem.

Fig. 7. A swimming pool in Nes.

Fig. 8. The liberation museum in Groesbeek.

Fig. 9. The exhibition greenhouse in Arcen.

Fig. 10. Exhibition dome of the Autotron Museum in Drunen/Rosmalen.

166. The space structures of CNRSM in Brindisi

C. DENTAMARO and N. DEVENUTO, Politecnico di Bari, Italy, and
M. MEZZINA,University of Cagliara, Italy

SYNOPSIS. An important Technological Park, where we find the National Centre for the Research and Development of Materials (C.N.R.S.M.) is being created near Brindisi (Southern Italy). The typologies utilised have provided important elements of integration to enhance the standardisation of the product. In this context, the designing procedure aims to achieve the link up of all the operators through careful choosing of the construction strategies. This objective has been realised in coupling those reinforced concrete elements cast-on-site with highly standardised steel components, characterised by high static efficiency and simplicity of assembly. This solution has been preferred by the designers and the contractor company because of the short delivery dates, the ease of material provision and the simple fabrication.

Fig. 1. The arcade in the Researcher Residence Building.

GENERAL DESCRIPTION

1. The C.N.R.S.M. is a vast complex which is to provide a centre for research activities, with noteworthy architectonic pavilions to be used as laboratories, a conference-hall, guest-residences and further facilities.

2. The overall surface of the lot amounts to more than 200,000 m^2, 25,000 m^2 of which is covered by the main buildings, whilst the remaining area is available for parking, internal roadways, landscaping, plant rooms, etc. Hereafter the structural and architectonic solutions adopted for the main pavilions of the technological park are described.

THE MATERIAL LABORATORY BUILDING

3. This building is characterised by a plan of rectangular form with a base area of 1,274 m^2 and is of 2 storeys with an underground level. At ground level are to be found the standard services: work-shops, laboratories, test-rooms and general services. At the second level, with a smaller area, are located studies and administrative services.

4. The supporting structures are fabricated of reinforced concrete cast on site; vaultings have been accomplished utilising lightened Bausta-type concrete floors for a depth of 25 cm, with ribs and deck cast on site, whilst steel reticulated plates have been used for the floor covering.

Fig. 2. The Material Laboratory Building.

5. The roof structure is a double-layered space grid, with tubular elements and special bolted joints. All the supports are sand blasted and painted and provided with special devices to facilitate assembling. The module of the steel truss mesh is $1.97m^2$. The members have hollow section of various diameter and were grouped on the basis of their state of stress, in order to reduce the structural weight and to avoid too high a number of different sections. They are connected to the nodes with special adjusting devices. The covering of the deck type is made of a pre-painted corrugated steel sheet, with an insulating layer and waterproofing.

Fig. 3. Double layered grid: a typical node (Uzay system).

6. The roof structure is divided into four separated units, to minimise thermal deformations and subsequent stresses. The roof loads are transferred to the upper r.c. frames through four bearings on each unit. One of the four supports is fixed, two allow a single directional displacement and the fourth allows for both the horizontal displacements. In this way the structure can permit free thermal deformations and restraint against wind effects.

7. The bearing allowing displacements is designed as follows: a base linked to the concrete structure; an intermediate steel disc with the lower face coated with a P.T.F.E. layer and provided with a guide bar to transfer the lateral force to the base; and an upper plate welded to the node of the steel truss. The fixed bearing has no intermediate sliding disc.

8. The erection work, for each unit, was carried out by pre-assembling on site groups of members, which were bolted to the spheres. The resulting elements were then lifted by crane and connected to the concrete structure. The erection of the purlins and the corrugated steel sheet of the covering followed.

THE CONFERENCE HALL BUILDING

9. The conference building spread over 800 m^2. It is characterised by a plan of polygonal form and is separated into the lecture area, at ground level, and the spectator seating, which rises to a level of +1.33 m from ground level, where the general services and technical plant spaces are located.

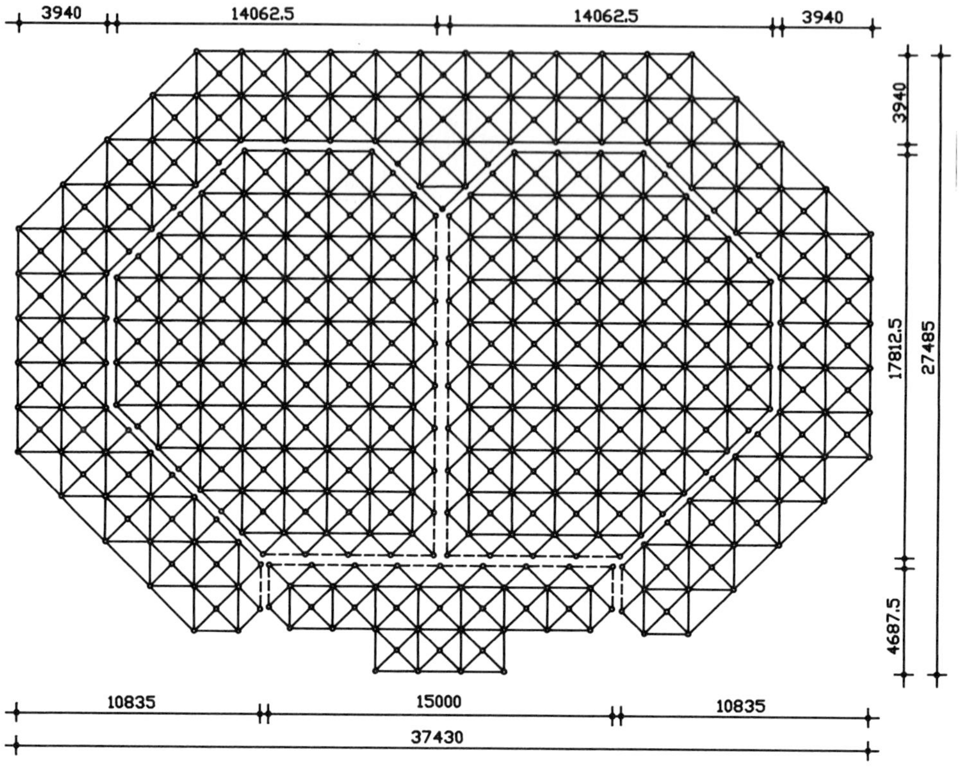

Fig. 4. Roofing of the Conference Hall.

10. The pavilion roofing is a double-layer reticulated steel plate supported by r.c. perimetral frames. The structural typology is the same as utilised in the laboratory building covering: the truss is articulated in 5 different units; the bearings have the same features.

RESEARCHER RESIDENCE BUILDING

11. The residence building has three storeys above ground-level and a basement. The internal spaces are well distributed, from the viewpoint of its floor plan and its elevation. The plan dimensions are 108 m x 28 m. The unit is composed of two main longitudinal bodies separated by an arcade which presents original architectural characteristics.

Fig. 5. The Conference Hall.

Fig. 6. Frontal view of the Researcher Residence Building.

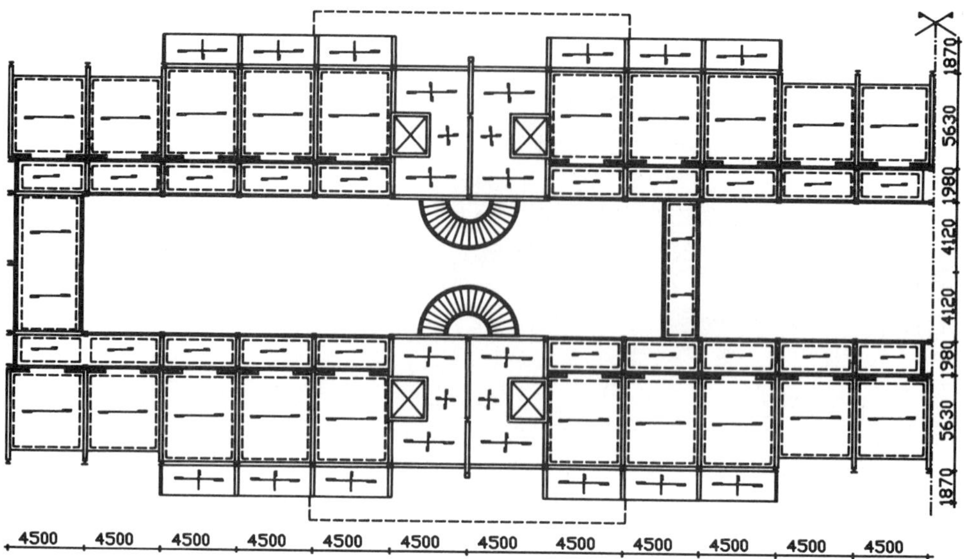

Fig. 7. Framing of the first floor.

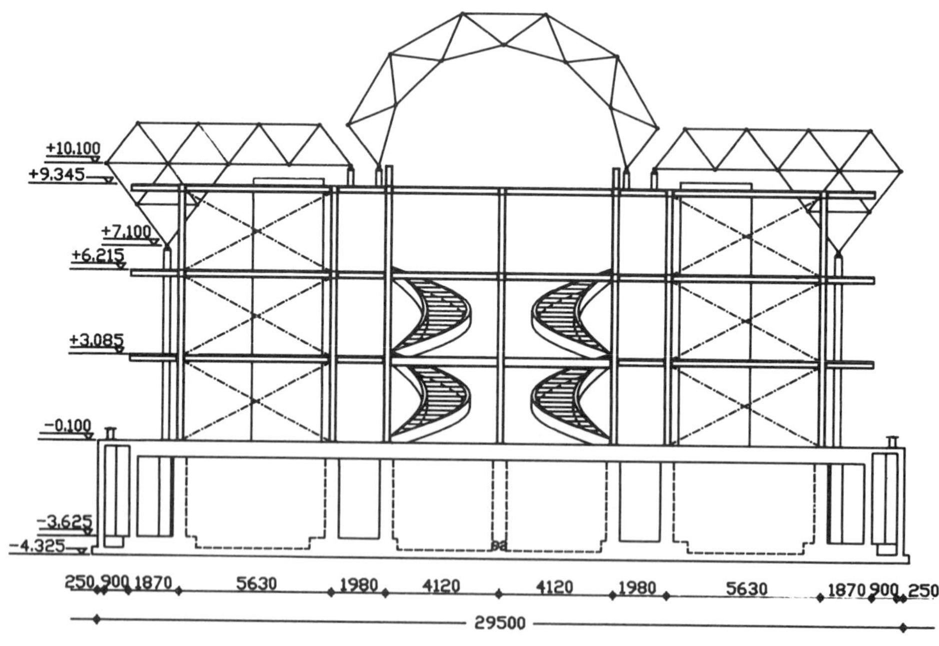

Fig. 8. Structural section.

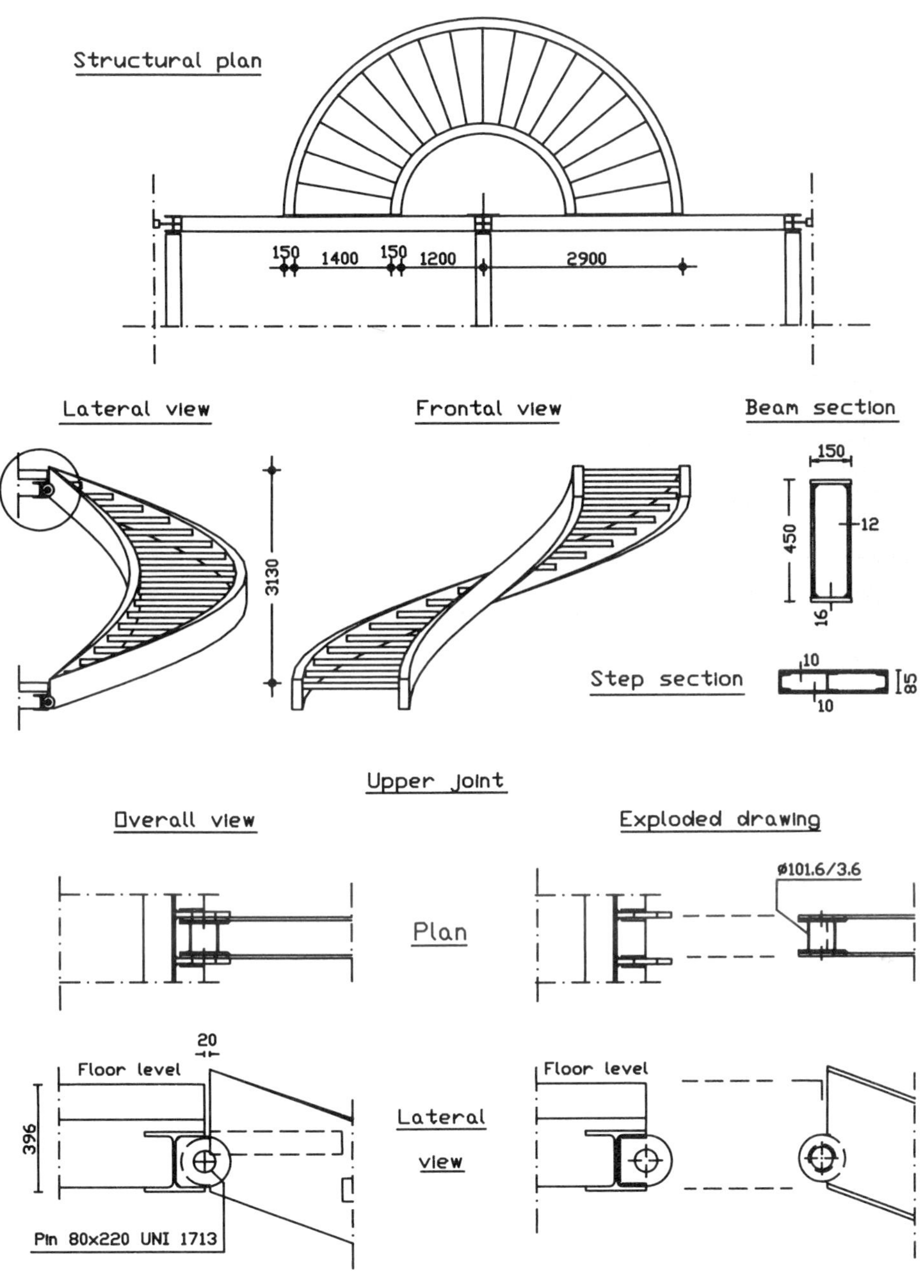

Fig. 9. Helicoidal stairway.

12. At ground level we find the entrances to the duplex apartments (subdivided into two levels, ground and first) and the central landscaped area. At the second level are located the remaining apartments.

13. The connection between the various levels and between the two wings is ensured by helicoidal stairways and by walk-ways, located in the arcade. The central area is covered by a double layer arched steel truss.

14. The supporting structures, below floors, and all lift-wells are fabricated of reinforced concrete cast on site. Steel is used for the vertical structure, all stairways and walkways and for the covering truss.

Fig. 10. Helicoidal stairways.

15. Notwithstanding the covering features, with careful study it has been possible to attain steel structures characterised by high-efficiency and simplicity of assembly.

16. The structural typology used envisages: HE type, hot-rolled profile steel pin-ended columns; vaultings formed by floor-beams also manufactured from HE type hot-rolled profile steel, serving solely as simple support jointed by means of absorption into the floor; lightened Bausta-type concrete floors, with ribs and deck cast-on-site for a total depth of 25 cm.

17. Stability of the structures during the erection phase has been assured by specific vertical and horizontal bracings, whilst the overall stability of the completed building is ensured also by the concrete lift-well, all able to absorb the wind forces.

18. All r.c. to steel connections are of the direct type attained by means of devices cast into the concrete elements.

Fig. 11. A lateral view of the Researcher Residence Building.

NORMS AND MATERIALS

19. The structural planning agrees with the "Technical Standard for reinforced normal or pre-stressed concrete and steel structures" D.M. 27-7-85 by the Ministry of Public Works and: D.M. 12-2-82; C.M.LL.PP. 24-5-82 n.22631; CNR 10011-85. The material provided for the steelworks are: for plates and sheets having thickness < 30mm Fe 360C UNI 7070; for section Fe 430B UNI 7070; for bolts 8.8 e 10.9 UNI 3740; and E 52 4B UNI 5152 type coated electrodes.

The materials for the reinforced concrete structures are: R_{ck} = 250 daN/cm^2 concrete available for all structures; and Fe B 44 k corrugated steel bars tested in workshop.

DESIGN AND CONSTRUCTION

Owner: C.N.R.S.M. - Brindisi;

Architectural design: arch.N.Barletti, arch.I.Ciracì, ing.L.Del Grosso;

Supervision of construction: prof.ing.N.De Venuto, arch.N. Barletti, arch. I. Ciracì, ing.L.Del Grosso;

Structural design: prof.ing.M.Mezzina, ing.C.Dentamaro, ing.D. Raffaele;

General contractor: "Cittadella" s.c.r.l.

167. A review of space structures in Greece

G. M. HADJISTERGIOU, Consultant Engineer, Greece

SYNOPSIS. The paper provides a brief review of the recent trends in the field of space structures in Greece.

INTRODUCTION

1. A large number of long span roof structures have been built in Greece during the last two decades. The majority of these space structures cover sports buildings, as there is a constantly growing demand for sports halls, gymnasia, swimming pools , tennis halls and large span sports stadia.

2. The emphasis on physical education is combined with the intention to utilise the sports centres not only for local community services but also for competition on a national and international scale. As a result, large scale, i.e. long span structures are required. Long span structures have also been built for other uses (i.e. helicopter hangars, warehouses, etc.).

3. The aim of this paper is to provide a brief review of the recent trends in the field of space structures in Greece.

LONG SPAN STRUCTURES OF THE LAST TWO DECADES

4. The list of the projects that follows is indicative, its aim being the tracing of the main trends in the development of Space Structures in Greece. Within the above mentioned perspective, this list summarises the projects that were presented during the Conference on Space Structures which was organised by the Technical Chamber of Greece (Athens, 2-4 October 1990). The list,being indicative, does not include the names of the structural engineers, architects and contractors of each project.

CABLE STRUCTURES

5. The roof structure of the new sports hall in Athens (1985)

Main Features include:

(i) The suspended roof: The roof structure consists of an orthogonal rope net in the shape of a hyperbolic paraboloid.

- Diameter : 113.96m

- Mesh dimensions : 4.0m x 4.0m

(ii) The peripheral ring: The border structure, to which the rope net is anchored consists of a perimetrical box-section ring in prestressed reinforced concrete which follows the movement of the border curve.

STEEL DOMES

6. National Swimming Pool at Mytilinl (1972)

Main features include:

The building has a circular plan and its structural framing consists of 30 reinforced concrete columns around the periphery, a circular prestressed beam connecting the columns near the top, and a steel framed dome supported by the circular beam.

- Outer diameter : 55.0m.
- Steel dome base diameter : 53.0m
- Rise : 7.15m

It consists of 30 main arch ribs, 6 parallel rings, purlins and wing bracings.

STEEL FRAMES

7. The following roof structures consist of a series of one-way truss systems which are supported by columns either of reinforced concrete or steel.

(a). Tanagra Hangar (1987)

- Main span : 56m
- Total floor area : $2400m^2$

(b). Warehouse for the Pireaus Harbour (1991)

- Main span : 60m

8. The indoor Swimming Centre and Sports Hall of Athens Olympic Sports Centre (1990)

Main features include:

(i) Total floor area of the

- Swimming Centre: $20{,}000m^2$
- Sports Hall: $30{,}000m^2$
- Maximum free span: 108m

(ii) Geometrical data of the double-layer grid

- Module of 7.2 x 7.2m with a structural depth of 4.86m , for the centre area.
- Module of 3.6 x 3.6m with a structural depth of 2.70m, for the sloping roof structures.

(iii) The roof is supported by 20-30m high pylons which are installed with the grid area of 33.2 x 108m.

9. Indoor sports halls in some Greek cities (Lamia, Peristeri, St. John Rentis, etc.)

The roof covers a total floor area of approximately $2{,}500m^2$ (50m x 50m or 40m x 40m). Approximate structural depth of the double layer grid is 2.70m.

MEMBRANE STRUCTURES

10. Covered Swimming Pool in Chios (1987)

The framework of the building is made of reinforced concrete, while the cover (surface area : 4.000m^2) is formed by a pair of attractable membranes of polyester fabric PVC. The exterior membrane is suspended at the centre by three parallel wire ropes bound on six metallic masts (height : 22m), while around the perimeter it is anchored on the framework of the building.

11. Coverage of the temple of Apollo Epikourios at Vassai.

Main features include:

(i) A temporary canopy shields the ancient temple in order to protect it from the bad weather conditions, one of the main causes of deterioration.

(ii) Total covered surface : 48m x 24m .

(iii) Height : 12.5m.

(iv) Material : Membranes of Polyester fabric PVC.

CONCRETE STRUCTURES

12. Coverage of the main theatre of the Military Academy at Vari nearby Athens.

Main features include:

(i) Elliptical shell covering a floor area of 40 x 31m^2.

The shell is supported by a peripheral ring-beam of 0.40m x 0.80m.

(ii) The shell structural depth at the apex : 12cm.

COMMENTARY

13. The proprietors of all the above mentioned projects are state owned organisations, mainly the General Secretary of Athletics.

14. It is worth mentioning that in most cases the supporting system of the roof is of concrete rather than steel. This is due to the fact that there is a considerable experience in the construction of concrete in Greece, however complex they are.

15. The Greek Architectural and Structural Engineering consultancies have demonstrated a significant ability in dealing with the problems of long span roof structures. In certain cases, the Greek offices have co-operated with consultancies from other European countries.

16. All these projects have been realised by Greek general contractors which employ a roof structure company as a subcontractor. Although a number of Greek firms have developed their own frame structure systems (i.e. joints, etc.) most of the double-layer grid roofs are constructed by firms based in other European countries i.e. Germany). Most of the membrane structures have been constructed by Italian firms. The steel domes, steel frames and the concrete roof structures have been constructed by Greek firms. Finally, we should underline that a significant effort is being conducted within the Technical Universities mainly at the theoretical level.

168. Lightweight structures in Argentina

P. D. REISSIG, Centre for lightweight structures, Argentina

SYNOPSIS. This study of lightweight structures in Argentina will focus on three areas. The first is a brief historical and contextual panorama of the built environment, with emphasis on advanced construction systems. The second area is devoted to an overview (inventory) of the last 30 years including professionals, industry and related research. The final part is concerned with reaching conclusions for future directions to pursue, both in terms of the appropriate technologies to use, as well as its applications in terms of function. This regional case study hopes to open up discourse regarding contextual factors that condition the developments in this field. One might say an "area studies" vantage point of lightweight structures.

HISTORICAL DEVELOPMENTS IN THE USE OF CONCRETE

1. The concept of structural performance in building began to develop in Argentina in the 1920s with the widespread use of reinforced concrete which replaced (mainly imported) metal skeletal structures, used until then. In 1932 the Kavanagh building was built measuring over 100m in height which was the tallest reinforced concrete building in the world for a number of years. Numerous grain silos, bridges and tank towers were built throughout the 1930s and 1940s. The teaching of shell structures began in the late 30s in the engineering schools based on the text of the engineer Julio Zuker. These shells, which began with cylindrical (Zeiss-Dywidag Patent) and dome types were built up until the 1960s (Fig. 1). At that point, hypar and ellipsoidal shells, as well as folded plates and reinforced ceramics, became more widely used.

Fig. 1.

Fig. 2.

2. The engineers Delpini and Gallo (ref. 1) are recognised as the two most eminent figures of that period regarding structural use of reinforced concrete, based on expressive and static performance criteria (structuralist). Significant works of these two engineers are the Mercado de Abasto in Buenos Aires 1928-37 designed by Delpini, (Fig. 2); and the Dome of Escuela Militar, Rio Santiago, La Plata design by Gallo. For industrial building and other large spans, metal truss structures were also economically competitive.

3. Beginning in the 1950s and up until the 70s pre-cast concrete was widely used. The major figures in this field were Luis Toschi and his follower, Arturo J. Bignoli (ref. 2) who executed a large and varied number of projects (Fig. 3). Also in the 1950s the use of pre-stressed concrete especially in bridges was instigated. (Rio Bermejo en Aguas Blancas 1953). An important pre-stressed shell structure was the F.A.T.E. Tire Factory, designed by Gallo and Dany. Folded plate structures were developed in the 1960s achieving spans of 30m. (Establecimiento Yamil by Massa).

4. It is worth noting that structural ceramics (ceramica armada) as developed by the Uruguayan master Eladio Dieste (ref. 3) has had important influence in Argentina, Uruguay and Brazil since the 1960s. This technique has received special media coverage due to its social-cultural roots in an era where credibility in design and technology is based on its "internationalness." Beyond Dieste's justification of regional context and structural wholeness, his work stands to show how much more there is to explore and learn from materials as old and heavy as brick (Fig. 4).

Fig. 3.

Fig. 4.

5. It is also important to mention that parallel to the above historical evolution of the construction techniques based on professional design and industrial applications, there has developed to an extremely widespread level, the use of 3-D truss sections in parabolic roofs, used to span garages/shops up to 30m wide (Fig. 5). These trusses, fabricated from steel bars, are also used for column supports. This construction technique considered "vernacular" by the profession and which falls short of meeting building code requirements for wind loads, never-the-less bares living testimony of the natural search for lightweight, economical building solutions for large spans. It is important to consider this

natural phenomena when evaluating a society's capacity and probabilities of adapting its built environment to meet their present needs and possibilities.

DEVELOPMENTS IN LIGHTWEIGHT SYSTEMS

6. Beginning in the late 1960's and coinciding with a world-wide spirit of change, alternatives and expression, Argentina started to employ space frames, geodesic domes and membrane structures for fairs, exhibitions and pavilions. Two important architectural studios produced projects of interest at that time, namely, Arquitectura Publicitaria run by J. Bruguera experimented sculpturally with space frames and polyhedral structures (Figs 6 and 7), also the company Tenso-Estructuras run by architects Bacciadune, Karlic and Lagomarsino began development in membrane structures. They produced pioneering work in the field by Argentine standards.

Fig. 5. Fig. 6. Fig. 7.

7. The trajectory of this company is noteworthy because it represents an example of commitment to design over profit, and the possibility of combining various professional functions in order to carry on work in such an unstable economy as in Argentina. This group of architects began their work in 1971 with a tunnelling, single-layer air-supported structure measuring 7m wide by 240m long with over a dozen pneumatic domes attached to it. By 1973 they reached production capacity of 4000m^2 per month, and the following years they exported to Brazil, Chile, Paraguay and Venezuela. Following is a list of mayor projects to date:

- External vault frame with stressed skin envelope for the River Plate Club swimming pool, 50m span covering 3000m^2, 1980.
- Internal vault frame with stressed skin envelope for the temporary offices, IBM Buenos Aires 16m by 40m; (this structure ended up as permanent offices and became the subject of interest in studying human reactions to such a work environment), 1985.

- Air-supported structure for grain storage sheds measuring 40m by 150m with an external steel cable net, 1990.

This team also developed flexible, storage, membrane structures for transporting 1.5 tons of cement (Figs 8-11).

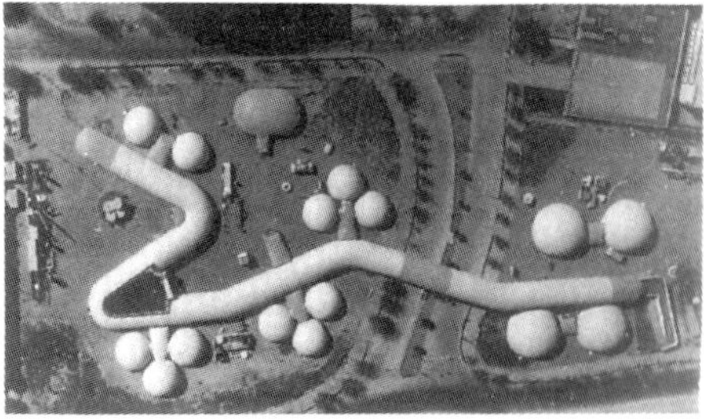

Fig. 8.

Fig. 9.

Fig. 10.

Fig. 11.

8. Tent type structures began to become popular promoted by the company APA. However, these structures consist mainly of metal frames with a membrane enclosure used non-structurally (Fig. 12). A company called Eurobra introduced glulams into the market in 1970 with moderate success. Another company which began producing lightweight structures in the 1960's, which is still going strong today, is Arpon run by Osvaldo Pons. These structures consist of parabolic surface space frames covered by a thin concrete shell. These structures span up to 35m. The company was founded in 1959 and has covered 350,000m^2 to date (Fig. 13).

Fig. 12.

Fig. 13.

9. Space frames became a common sights by the mid 70s constructed mainly by artisan means, employing flat plate nodes with bolted or welded connections. Various small companies sprouted up offering industrialised nodes consisting mainly of Unistrut replicas or other similar joints. Most of these examples used end-flattened round struts (Figs 14-17).

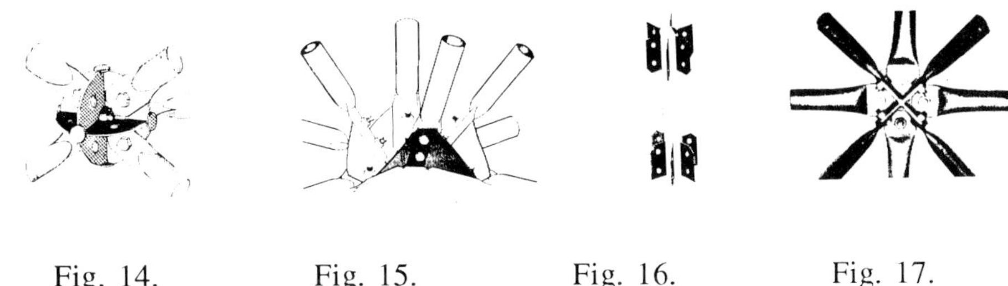

Fig. 14. Fig. 15. Fig. 16. Fig. 17.

10. Most recently a couple of metal structures companies have begun building geodesic dome prototypes for what they feel is an unfulfilled market demand. Apart from a few flat end strut and plate node domes, covered with fabric membranes, used as pavilions in the early 70s, the only other geodesic domes to be seen in Argentina were two Fuller domes imported from the U.S.A. for exhibition purposes.

PROFESSIONAL DESIGNERS AND LIGHTWEIGHT STRUCTURES

11. There have been a number of interesting projects built by important architectural firms over the past 25 years. The Cerro de la Gloria Stadium in Mendoza built by the team of Mateola, Sanchez-Gomez, Santos, Solsona and Vinoly in 1976 is partially covered by a doubly-curved space frame supported by a pre-stressed concrete beam (Fig. 18). The stadium for Mar del Plata designed and built by Bramante in 1978, uses a flat curved space frame measuring 40m by 150m and 1.6m deep supported on one edge by the tribune and on the other by tensile elements (Fig. 19). The architect Mario Roberto Alvarez employed a space frame as a supporting floor structure in the Ken Brown factory in Buenos Aires covering $5{,}300m^2$ (1969). The architect Mederico Faivre (ref. 4) has been working with lightweight structures since 1969 when he produced an innovative expanding aluminium dome with polyethylene strip nodes (Figs 20, 21). Since then he has completed various projects including a few churches which use lightweight systems adapted to each circumstance as opposed to applying a specific system *a priori*. Other important projects built over this last period, which also use space frame systems, are the Cordoba Bus Terminal built by C. Fontan-1970 and the Mendoza Bus Terminal built by G. Olguin in 1972.

Fig. 18.

Fig. 19.

Fig. 20.

Fig. 21.

RESEARCH AND LIGHTWEIGHT STRUCTURES

12. Regarding recent developments in research into lightweight structures, it is worth mentioning two theoretical works which have been published abroad (refs 5 and 6). The first belongs to Oscar Andres from the Universidad Nacional del Sur in Bahia Blanca. His work is concerned with the application of the Homeostatic Model Technique (HMT) relating to the conceptual design of shell roofs. Models are built starting from plastic sheets; simultaneous action of loads and heat develops their homeostatic evolution. Then the model compensates for the degradation of the materials mechanical properties, searching for new shapes to keep the forces in equilibrium. The model changes not only its geometry, but also its structural type, moving from a plate to a shell structure (Fig. 22). The architect Osvaldo Tonon (Buenos Aires) has analysed in depth the geometry of folding plate structures. He has generated by both graphic and analytical means, diverse types; vaults, domes (eccentric) mushrooms, cones, saddles, etc. All these forms which consist of straight edges and flat planes offer quite rigid structures adaptable for architectural applications (Fig. 23).

13. There are very few programs or curricula which are geared towards this field of study at the universities in Argentina. Yet it is important to recognise that the engineering schools nation-wide have traditionally been very strong and have produced top level work on an international level. There exists a Structural Engineers Association and the Instituto de Cemento Portland Argentino both of which concentrate their interest on concrete-based systems.

14. The Centro de Estructuras Livianas (Centre for Lightweight Structures) is a recently founded organisation dedicated to the promotion and development of lightweight systems. The Centre is of civil and independent status but maintains close ties with the universities, professional entities and industry. It is not a research centre, but hopes to serve as a medium for exchange and organisation of related activities. The C.E.L. recognises the importance of an interdisciplinary approach, incorporating both analytical as well as creative methods of design and experimentation. The Centre's interest is in the application of the systems developed and to fulfil this requirement it must recognise the social and economic context as well as the real industrial and commercial possibilities. There is only one other centre with similar objectives whose existence is known of in Latin America and that is the Laboratorio de Estructuras Laminares in Mexico City run by the late J. Mirafuentes. Their laboratory has produced very valuable work both in terms of research as well as built projects.

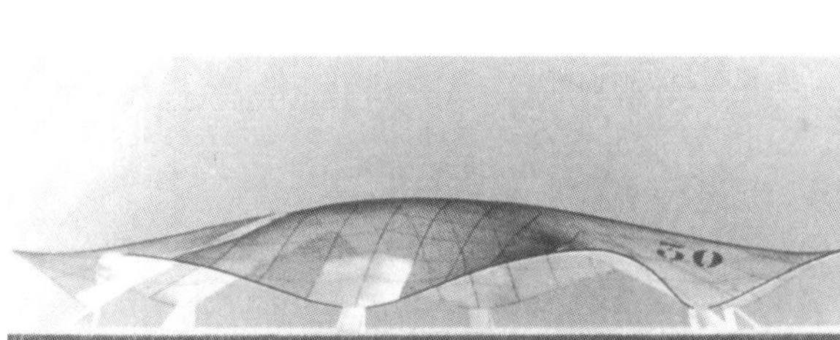

Fig. 22.

Fig. 23.

CONCLUSION

15. As this brief panorama of lightweight structures in Argentina shows, activity has been sporadic and diverse. The economic and political climate has been quite unstable and can by no means be ignored. There is only one significant design studio dedicated exclusively to lightweight systems (Tenso-Estructuras) and they work only with membrane structures. As far as the building industry is concerned, there are two significant companies devoted exclusively to lightweight systems; Arpon in Cordoba using combinations of space frame/concrete shell, and Fapym in Buenos Aires who manufacture industrialised space frames. One might wonder whether an economy of limited resources mixed with industrial potential (Argentina) can actually be more apt at producing lightweight solutions given the scarcity of materials and the flexibility of the construction industry (due to having to adapt quickly to changing market conditions)? This author believes that lightweight structures will not remain the domain of the more industrialised countries, but will eventually find their place in less developed regions due to necessity and accessible know-how. From an industrial point of view, the chances of small lightweight structure companies surviving are rather good, due to the lack of competition (including imports) and the large volume of construction which occurs whenever the economy is strong. This fact has raised interest of technology transfer agreements from large international companies

(Birdair and Starnet). From a professional perspective, Argentina has usually been up-to-date regarding design issues and trends. This has allowed for an openness on behalf of the "clients" towards innovative approaches. There has been a recent resurge of interest in membrane systems due to the Expo Sevilla, which was closely followed by the media. In general terms it seems that all the right ingredients are present to allow lightweight structural systems to grow; a large demand for building, many undeveloped areas and projects, few zoning and code restrictions, and receptivity for design and industrial criteria. On the negative side there is a conservative notion of what housing should be, and that limits severely any possibility for serious applications of these systems in that field. Nevertheless the ground is fertile, and the solutions might be innovative and the applications diverse.

REFERENCES

1. INST. DELL PORTLAND ARGENTINO. Bovedas, Cascaras. serie E, no. 21, Bs. As. 1943.
2. BIGNOLI A. J. Estructuras de hormigon premoldeados. Revista Construcciones, no. 229-231, Bs. As. 1971.
3. DIESTE F. La estructura ceramica. Editorial Escala, Bogota, 1987.
4. FAIVRE M. Estudio Faivre y Roman. Revista Arquitectura Sur, 5, Bs. As. 1991
5. ANDRES O. Homeostatic models for shell roof design. Proceedings from the I.A.S.S. Congress. Madrid, vol. 1, 1989.
6. TONON O. Geometry of spatial folded forms. Int. Journal of Space Structures, vol. 6, no. 3, 1991.

169. Design of hybrid space structures for large spans

H. KLIMKE and C. STUTZKI, MERO-Raumstuktur, Germany

SYNOPSIS. Extrapolating the design of space frames with small spans to those with large spans, leads to enormous dimensions for the structural elements, bolts, nodes and tube diameters. This is the reason for the development of a new branch in space frame design. This branch shows new qualities instead of pure growth of quantities. We can find these new qualities in the combinations of space frames with cables which together are more than the sum of its parts. They can form economic as well as aesthetic structures. This paper does not set up general principles or rules but aims to present some projects with their characteristic properties. These projects are taken from the everyday work on clients enquiries. They are - each with its typical structure - based on known existing examples and may serve as examples again, once the characteristic effects of space frame structures with cables have been understood.

THE ADVANTAGEOUS CHARACTERISTICS OF SPACE FRAME STRUCTURES:

1. The use of space frame structures is always advisable for area-like structures, as a continuous structure is better achieved by fine structural space frames contrary to the numerous hierarchical structures with principal girders, secondary girders, purlins etc. with single-axis load distribution. Space frames as support structures created by CAD design also apply for irregular and curved architectonic shapes with emphasis on filigree appearance at the same time. Domes are therefore an excellent area of application for classical space frame structures and their derivatives.

2. Triple chord girders are classical constructions for bending and pressure effected steel constructions. These triple chord girders may be assembled and erected easily by use of members and nodes of the space frame construction method. They are therefore a logical and effective way of construction for member-node systems. Lightness, filigree appearance, unlimited flexibility of the design, surface-like support, direct support of the cladding through its fine definition, easy erection are the advantages of space frame structures.

SENSIBLE USE OF TENSILE CONSTRUCTION

3. The common characteristics of all tensile construction is that the cables are arranged in such a manner that all unfavourable features of the individual parts are compensated. Tensile construction show less deformation under load effects compared with other structural forms. Thus, they are more stable and, above all, lighter as the high tensile members develop their full potential only by prestressing (ref. 1).

4. After extensive theoretical and technological research on tension structures initiated in the 60's and after assessment of the performance of a large number of such structures actually built all over the world, it is not possible to state that the pioneering period of theoretical and technical development of tension structures has ended (ref. 2)

5. The solutions, for example, the introduction of tensile forces and states of self stress, have been developed from simple, plane models to highly complicated systems (refs 2 and 3). In this method of construction, tensile forces are often not transferred to concrete foundations but to the compression elements of the construction which requires large space frame structures with substantial members.

COMBINATIONS OF SPACE FRAMES AND CABLES

6. Due to aesthetical reasons, cables can be used for structural elements that will be in tension under all load cases. Tensile cables which are placed in the bottom chord of the space frame structure may be of advantage in order to attract the tensile forces instead of having them transferred through the bolts of the members. Here, the cables act as reinforcement like in concrete, however, not as economical, as concrete is better suited to accept compression forces than the members of space frame structures.

7. The tensile forces of the cables are usually stabilised in the space frame structure which has to accept the corresponding compression forces. A saving in material by use of high tensile cables is therefore partly compensated by more material for the compression areas of the space frame structure.

8. Cables do not only play an important role in the acceptance of tensile forces, but additionally serve to stabilise structural areas which are in danger of buckling.

GARBAGE DUMP (OCHTRUP).

9. The roof structure consist of one arch and a tensile network. The roofing consists of a polyester membrane. The arch takes compression and bending and is designed as a triple chord girder using space frame components. The span of the arch is 125m.

10. The tensile network serves as the support construction for the roof and also as stabilisation for the arch, limiting deflections in the vertical direction of the arch as well as perpendicular to the plane of the arch. In such a way, the cable net considerably reduces the bending moments enabling a very light construction with a very slender arch.

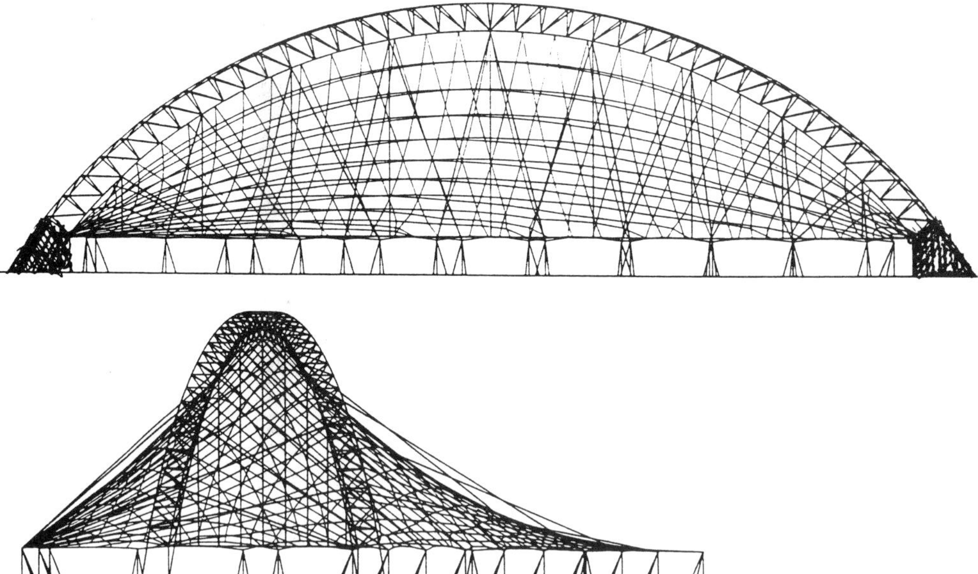

Fig. 1.

HANGAR (PEKING)

11. Building a hangar roof with a 150m clear span as a space frame structure, requires, according to common thinking, an edge girder above the door opening of up to 5 layers and bolts up to 90mm in diameter. The design presented here shows a roof consisting of a triple-layer space frame plate which is continuously supported at all four sides having dimensions of 150 by 85m. The total construction weight of the space frame is 30kg per square meter.

12. The space frame plate is the most economical solution with extremely low self-weight for such spans, provided it is supported at all four sides and the members are placed approximately along the main stress directions. This means that the door girder problem, of bridging of the 150m span with little deflection, has to be resolved separately. The space frame structure does not serve this purpose here.

13. A favourable solution is an arch together with tension rods supporting the edge of the space frame. The tension rods, the arch and the front side of the space frame as well as the connecting members between the arch and the space frame, form a system of compression and tension elements, introducing a state of self stress to the tension rods by pre-tensioning. This state of self stress can be adjusted in such a way that the tension rods will not be slack from extreme inside pressure caused by open doors plus wind suction on the roof surface.

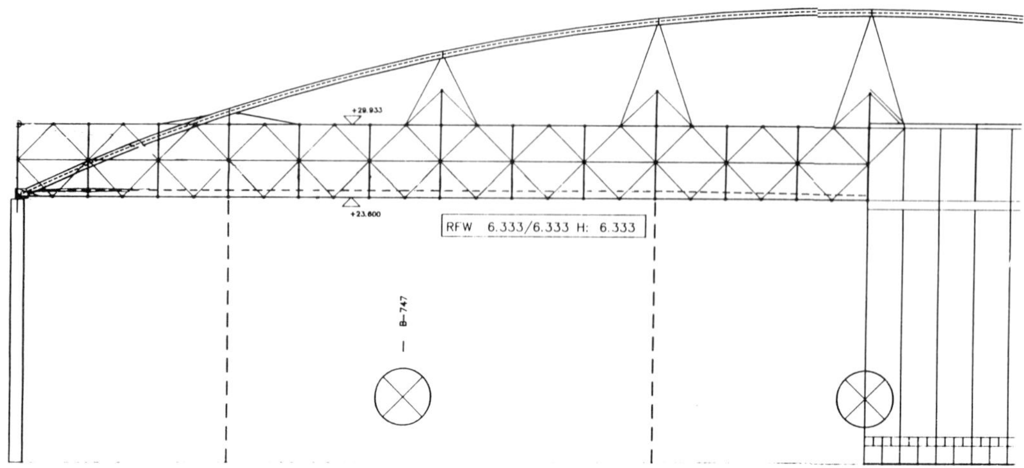

Fig. 2.

PEDESTRIAN BRIDGE (HAMBURG)

14. The 80m span pedestrian bridge is schemed as a triple chord girder of space frame components appearing as a pure beam in bending. In order to relieve the bolts in the bottom chord, a tension cable which is fixed at the ends will be provided in parallel direction to the bolts. Vibrations are avoided by passive tuned mass dampers beneath the walkway.

15. The installation of the tie rod is a post-tensioning process without shear connection between the space frame and pre-stressing element. The space between the tension strand and the enveloping tube will be filled with grout after tensioning. The saddles are curved according to the permissible shear compression. The end fixings consist of welded constructions which are bolted to the space frame structure. The position of the tension element leads to a constant bending moment overruling the single span moment of the girder. This considerably reduces the forces in the space frame .

HANGAR

16. The space frame plate supported at three sides is executed as a double-layer grid with two tensile systems which are tensioned against each other. Each one introduces additional compression and bending into the space frame plate. The suspensions in the short cross direction with the pylons bear the main load of self-weight and snow, and transfer these loads into the pylons. The tension forces

are transferred into the walls of adjoining buildings. These walls distribute those forces into the foundation.

17. A suspension cable is required above the door girder to minimise the deflections and stresses under wind suction on the roof, plus internal pressure present when the doors are open.

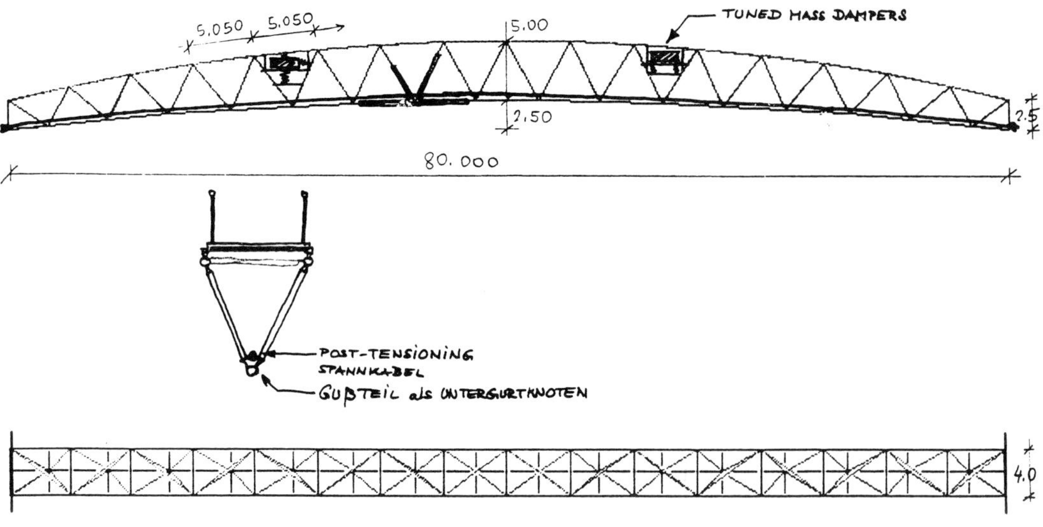

Fig. 3.

RIBBED STEEL DOME WITH TENSEGRITY STIFFENING

18. An architectural contest was held in preparation for the Olympic Games 2000. Based on the ideas of architects Gerkan, March and Partners, the structural layout of a lightweight glazed steel dome was prepared by MERO.

19. The design was based on the combination of a single-layer ribbed dome with the tensegrity concept, which was invented by Buckminster Fuller and successfully applied for a number of cable domes by David Geiger.

20. Tensegrity means that this type of structure gains its integrity by the resistance of pre-stressed tension elements, i.e. wire cables.

21. Nevertheless tensegrity structures undergo relatively high deformations under the action of live loads. This is the reason why this type of structure has only, in the past, been covered by flexible membranes.

22. A derivation of the basic principle allows both a tensegrity and cable dome to be used to ensure the integrity of a single-layer steel dome that would buckle without a stabilising substructure. The alternative would have been the use of a double-layer space frame.

23. The ribs and rings of the dome are stiffened against deformation from unsymmetrical loads by diagonal, continuous wire cables, clamped to the nodal points of the rib and ring structure after prestressing.

24. The steel members of the structure together with the secondary beam members provide direct support for the glass panels without an additional layer of aluminium rafters.

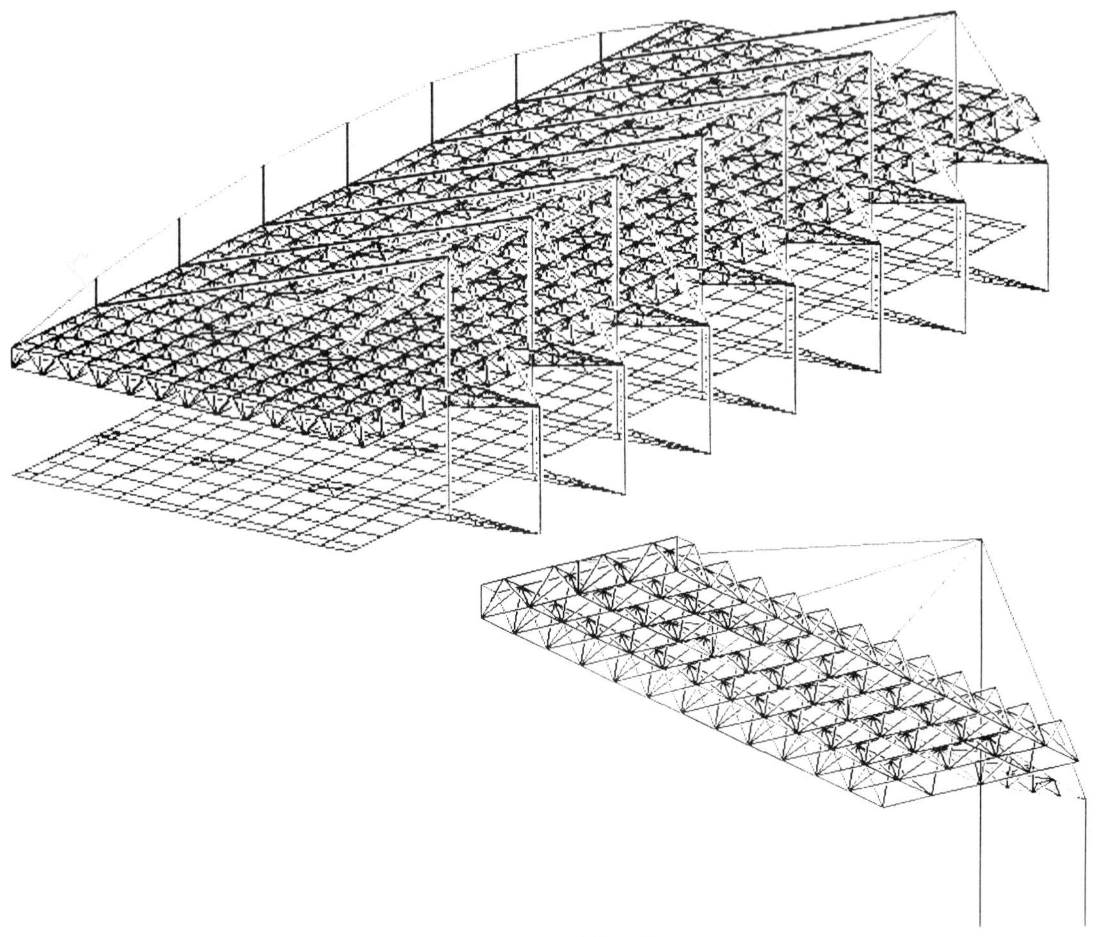

Fig. 4.

25. The stiffening tensegrity structure consists of vertical compression members, tension diagonals and circular tension chords. Diagonals could be high tensile rods, the circular chords could be wire cables, pre-stressed by means of shortening the diagonal rods. The vertical compression members have to be pin-connected to allow for the change of angles due to prestressing.

26. The combination of two known and proven concepts a single-layer ribbed steel dome and the tensegrity stiffening of cable domes, is creating a hybrid structure stiff enough for a glazed skylight, and light enough for maximum translucence.

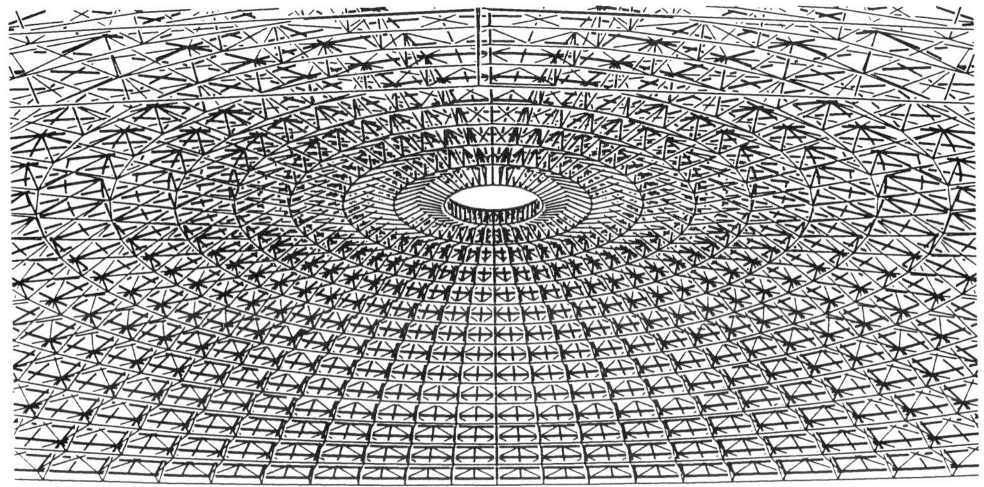

Fig. 5.

REFERENCES

1. SCHLAICH : About Prestressing
2. MAJOWIECKI and BOBROWSKI : Tension Space Structures
3. STAVOIDIS : Cable Structures with the Geiger System

170. Development and application of the composite space truss in China

DONG SHILIN, Zhejiang University, P. R. China, and XIA HENGXI, Agricultural University of Hebai, P. R. China

SYNOPSIS. During the last decade, a new structure combining a reinforced concrete slab and a steel space truss, forming the composite space truss, was developed. Sufficient stiffness, with lightness and, favourable technical economic characteristics are the features of this new kind of structure, which is suitable for both roof and floor construction. This paper summarises the history of the development of the composite space truss and its application for structures in China. The characteristic types, classification, calculation methods and key points; joint design, construction and installation of the structure, are also explained in this paper. Finally, further extensive applications of the composite space truss and required engineering research are proposed.

THE HISTORY OF THE COMPOSITE SPACE TRUSS AND ITS ENGINEERING APPLICATION

1. During the last decade, the composite space truss was developed. In this type of structure top steel chords are substituted by a reinforced concrete slab and the top joints are replaced by combined joints of steel and reinforced concrete resulting in a composite space structure.

2. In the early 1960's, in Mexico, a space plate structure was proposed. It was a plate of reinforced concrete in which the tension part is removed to reduce the weight of the structure. The span of this space plate, which is formed from 50mm thick reinforced concrete and a total depth of 300mm, is only three meters long. Its bottom chords and diagonal members were made from 10mm diameter circular steel members, from which the orthogonal square pyramid space truss was formed. This kind of space plate was used for the floor structure of a 15-storey building in Mexico, where the weight of every storey was reduced by 50% (ref. 1). A composite space truss proposed by MERO in 1981 is referred to as MERO - Massiv composite space truss system (ref. 2).

3. In Rumania, in 1981 a multiple - function gymnasium was designed with a diagonal square pyramid composite space truss. The space truss is 30m x 30m in area, 1.8m in height, and 3m x 3m grid for both the prefabricated top chord slab and the bottom chord grid. Detail analysis shows that the structure can save 20%

in steel weight, compared with an all steel grid, and the cost of production can also be reduced by 1-2% (ref. 3).

4. In China, in 1980 the first composite space truss was constructed in the form of a honeycomb triangular pyramid, which was used to roof two dining rooms (Jiahe Coal Mine, Xuzhou, Jiangsu Province), with areas of 21m x 54m and 9m x 18m, respectively (Fig. 1).

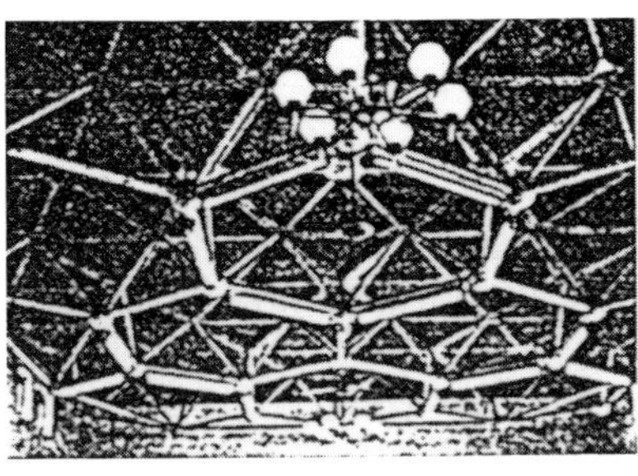

Fig. 1. Dining room of Jiahe Coal Mine.

Fig. 2. Construction of the composite space truss of Xinxiang Department Store.

Use of the composite structure saved 19.3% steel weight compared with an all steel grid. Subsequently, up until the end of 1992, a total of 20 composite space truss structures were built, not only for roofs, but also for multi-storey buildings, covering over 30,000m^2. A composite space truss with the longest span in China was used to the roof the gymnasium with an area of 45.5m x 58m, which was built in Fuzhou, Jiangsi Province in 1987. What is more, the added floor and extended construction of the Xinxiang Department Store, undertaken in 1988, which was built with an area of 34m x 34m, used four - floor diagonal-square composite space trusses covering two - floors (Fig. 2). This was the first use of a composite space truss for a multi-storey structure in China. An orthogonal square pyramid composite space truss with openings was used in the Changsha Great Textile Building, which had an area of 24m x 27m. In this structure two-stories are underground and eleven - storeys above ground. Columns are on a grid of 12m x lOm and 12m x 7m, respectively. A prestressed orthogonal square pyramid composite space truss was used in the Shanghai International Shopping Centre, which was finished in 1992, covering an area of 27m x 27m.

CHARACTERISTICS OF A COMPOSITE SPACE TRUSS

5. The reinforced concrete slab and the slab with ribs is used to replace the top chord of the normal steel space truss, so that a composite structure is formed. These are complicated structures similar to the plate-shell system.

6. In general, in the space truss, the top part of it is subjected to compression, and the bottom part subject to tensile forces therefore, in the composite space

truss, the stiffness advantages of the two kinds of material can be fully developed, that is, the reinforced concrete carries the compressive forces, and the steel takes the tension.

7. The composite space truss can support load and acts as an enclosure wall and covering in one function, so roof cladding is not required if the composite space truss is used for roofs.

8. The composite space truss has a high rigidity, good earthquake resistant behaviour, its vertical rigidity can increase by 30-50%, when compared with an all steel space truss of the same span.

9. The composite space truss can be successfully used in multi-storey structures. There are three advantages for this kind of structure:

a) Longer spans are possible for this kind of structure requiring fewer internal columns, giving large useful areas, available for multiple - floor markets, exhibition halls, workshops and other public or industrial buildings.

b) The total weight for the structure is less than that of conventional space trusses with possible reductions in weight in the order of 50% .

c) Due to the reduced weight of the structure, the effect of the horizontal earthquake force can be reduced by 20%. Also the internal force and the resistant force in the column and in the foundation can also be reduced by 15-20%, so that the weight of the whole building itself becomes less resulting in the economic use of building materials.

THE TYPES AND THE CLASSIFICATION OF COMPOSITE SPACE TRUSSES

10. The composite space truss can be classified according to the type and the direction of the top chord prefabricated slab. In general, there are four kinds of top chord prefabricated slabs; that is the orthogonal square slab (with the diagonal rib in the slab), the diagonal square slab (with the cross rib in the slab), the equilateral triangle slab and the slab alternative to the equilateral triangle shaped and hexagon - shaped (Fig. 3).

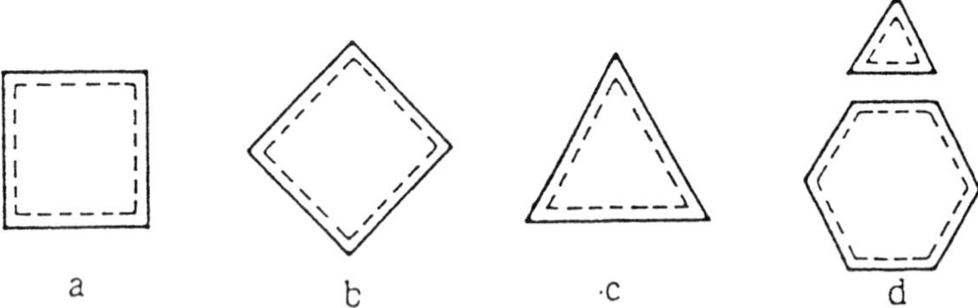

Fig. 3. Essential types of composite space trusses.

11. Therefore four kinds of structures can be obtained as follows:

a) Two-way orthogonal composite space truss type, including:

i) two-way orthogonal composite space truss;

ii) orthogonal square pyramid composite space truss;
iii) orthogonal square pyramid composite space truss with openings;
iv) chessboard-shaped square pyramid composite space truss.

b) Two-way diagonal composite space truss type, including:
i) two-way orthogonal diagonal composite space truss;
ii) diagonal square pyramid composite space truss;
iii) star-shaped square pyramid composite space truss.

c) Three-way composite space truss type, including:
i) three-way composite space truss;
ii) triangle pyramid composite space truss;
iii) triangle pyramid composite space truss with openings.

d) Honeycomb-shaped triangle composite space truss type. There is only one type of space truss in this category.

12. Above there are eleven kinds of composite space truss, and six of them having been built or designed in China.

CALCULATION OF THE COMPOSITE SPACE TRUSS

13. The bottom chord, the diagonal members and the bottom joint of the composite space truss have the same loading conditions as those for these elements in a general space truss structure. The top chord slab and the top joint have different loading condition compared to those in a conventional space truss. The vertical loads acting on the top chord slab of the composite space truss is transmitted into the top joint through the moment and the horizontal shear in the top chord slab and the ribs, so that local bending and deformation occur in the top chord slab and the ribs. The top chord slab and the ribs can be seen as the top surface layer of the composite space truss, so under the action of vertical loads, axial force can be found in the ribs and membrane forces can be found in the slab. After sorting out the loading acting on the composite space truss, a suitable analysis method would be chosen to determine the forces acting in the structure. At present, there are three kinds of useful approaches for analysis:

a) <u>Finite element method.</u>

A member element, beam element, and the plate and shell element are used to analyse the structure. In this method, the top chord slab of the composite space truss is dispersed into the beam element and the plate and shell element, the diagonal member and the bottom chord are still taken as truss elements acted on only by axial forces. The stiffness equations can be assembled according to the general finite element method, and the equilibrium equations solved to yield displacement and finally internal forces.

b) <u>Plate analogy method (ref. 4).</u>

The top chord slab of the composite space truss is considered as the top surface layer of a sandwich plate. The diagonal member and the bottom chord are considered as the filling and the bottom surface layer of the sandwich plate, respectively. Hence, this complicated composite space truss could be seen as the continued sandwich plate in a structure, the differential equations are used to describe its loading condition, the analytical method, or the effective approximate method, are applied to calculate the internal force and the displacement of the composite structure.

c) Simplified calculation method - equivalent space truss displacement method (ref. 5).

This method used a different approach to analyse the structure compared with the plate analogy method. Using the energy principle, the top chord slab can be equivalent to the four-group or three-group diagonal plane bar system in the top chord plane, so that this more complicated composite structure can be converted into a equivalent space pin-jointed bar system and a special computer program for the space truss displacement method can be used. Finally, the internal force in the top chord slab and its ribs can be obtained from the internal force in the equivalent diagonal plane bar system.

14. In the composite space trusses built in China, the simplified calculation method was used in most of the designs. For this reason, the two-way orthogonal composite space truss is used as an example to show the calculation steps.

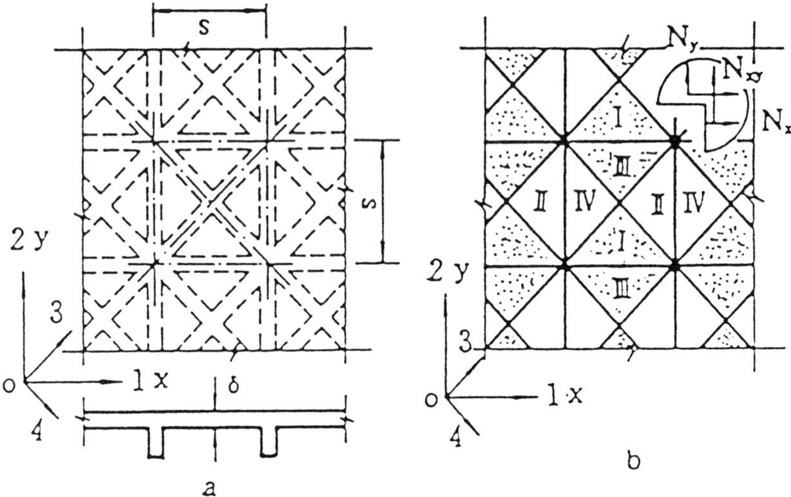

a) The slab with the ribs. b) Equivalent top chord bars.

Fig. 4. The internal force calculation for the top chord slab.

15. Fig. 4a shows the top chord slab with ribs of the two-way orthogonal composite space truss. The slab can be equivalent to four groups in the plane bar system (Fig. 4b), and the cross-sectional area can be determined by the following formulae:

$$A_1 = A_2 = 0.75\eta\delta s \qquad A_3 = A_4 = 0.375\sqrt{2}\eta\delta s \tag{1}$$

where: δ is the thickness of the slab,
s is the side length of the square slab with the ribs,
η is the correction coefficient related to the effect of the concrete's Poisson coefficient ν, when $\nu = 1/6$, $\eta = 0.825$,

16. Determining the stiffness $E\,\overline{A}_i$, of the equivalent top chord bar:

$$E\overline{A}_i = E\left(A_i + A_i^0\right) \qquad (i=1,2,3,4) \tag{2}$$

where E is the elastic modulus of the concrete and A_i^0 the section area of the rib.

17. Solving the equivalent space truss by the space truss displacement method, and obtaining the internal force of the equivalent top chord bar $\overline{N}_i$.

18. Obtaining the internal force in the equivalent plane bar system and the ribs according to the stiffener distribution:

$$N_i = A_i \overline{N}_i \sqrt{A_i} \qquad N_i^0 = A_i^0 \overline{N}_i \sqrt{A_i} \tag{3}$$

19. The internal force N_x N_y N_{xy} of I, III type triangle units and II, IV type triangle units in the slab can be calculated by the following formulae (Fig. 4):

$$\begin{Bmatrix} N_x \\ N_y \\ N_{xy} \end{Bmatrix} = \frac{1}{2S} \begin{bmatrix} 2 & 1 & 1 \\ -2 & 3 & 3 \\ 0 & 1 & -1 \end{bmatrix} \begin{Bmatrix} N_1 \\ \sqrt{2}N_3 \\ \sqrt{2}N_4 \end{Bmatrix} \tag{4a}$$

$$\begin{Bmatrix} N_x \\ N_y \\ N_{xy} \end{Bmatrix} = \frac{1}{2S} \begin{bmatrix} -2 & 3 & 3 \\ 2 & 1 & 1 \\ 0 & 1 & -1 \end{bmatrix} \begin{Bmatrix} N_2 \\ \sqrt{2}N_3 \\ \sqrt{2}N_4 \end{Bmatrix} \tag{4b}$$

20. Depending on the connection of the top slab to the top chords and nodes, the local bending internal force $\{M_x\ M_y\ M_{xy}\}^T$, $\{Q_x\ Q_y\}^T$ in the slab and the moment M^o, shear Q^o in the rib can be obtained depending on the support conditions. Following these conditions reinforcement calculation should be undertaken with known $\{N_x\ N_y\ N_{xy}\}^T$ and N^o the determined slab thickness and the section area of the rib should also be checked to see if they meet the design requirements. The design of the bottom chord and the diagonal members in the

composite space truss follow the same procedure as that used for their design in general space trusses.

THE JOINT STRUCTURE OF THE COMPOSITE SPACE TRUSS

21. Based on engineering practice and research results, there are several kinds of top joints suitable for composite space trusses. These are as follows:

a) Welding crossed plate joints.

This kind of joint is developed from crossed plate joints of the general steel truss system, and it is mainly used for angle iron composite space truss. Fig. 5 shows the structure of the joint. A similar joint has been used in Xinxiang Department Store and the Changsha Great Textile Building.

b) Semi- partial spherical joints.

This kind of joint is welded together from a partial sphere and steel cover plate. It has a large stiffness, is easy to make, and strong in the direction of the diagonal members connected. It is mainly used in tubular composite steel space trusses (Fig. 6).

c) Spiral ring joints.

This kind of joint is developed from a spiral ball joints (Fig. 7) and used for the composite steel tubular space truss. It was applied in the composite space truss of the seventh and eighth floors in the Shanghai International Shopping Centre. The other types of the joint are the punching plate joints, anchored face to face welding joints and so on.

THE MANUFACTURE AND THE INSTALLATION OF THE COMPOSITE SPACE TRUSS

22. The bottom chord, the diagonal member and the top chord prefabricated slab can be made in the prefabrication plant or prefabricated at site. The small unit can also be formed employing with restrictions on the weight lifting capacity and the transportation available.

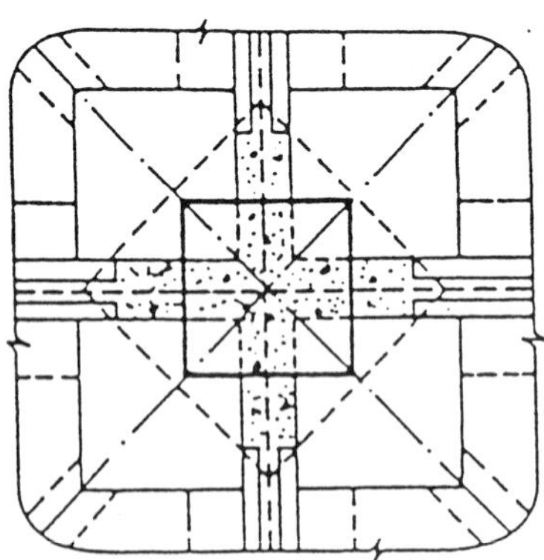

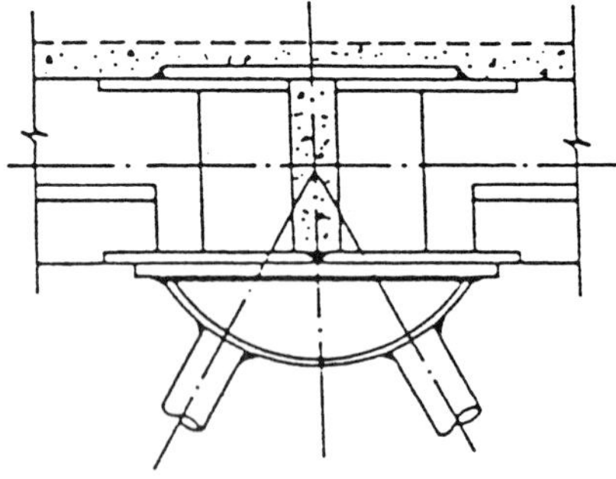

Fig. 6. The less partial spherical joint.

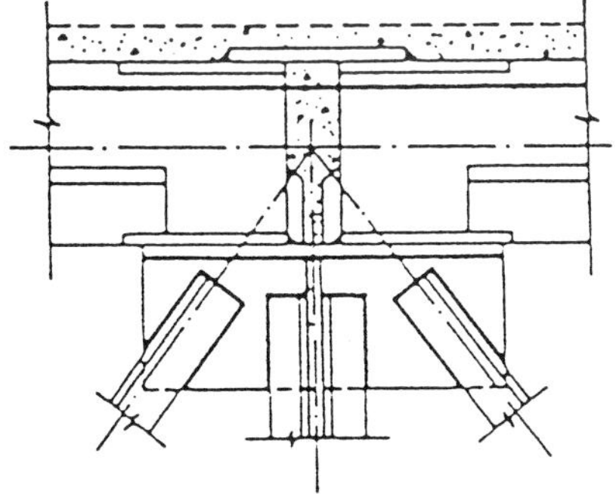

Fig. 5. The welding crossed plate joint.

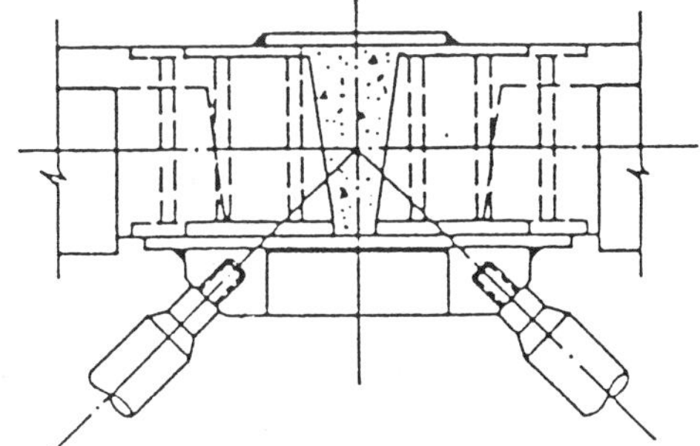

Fig. 7. The spiral ring joint.

23. For example, in the large and small dining rooms of Jiahe coal Mine in Xuzhou and in the gym of Fuzhou, the prefabricated composite triangle and square pyramids (shown in Fig. 8) were used in the composite structure.

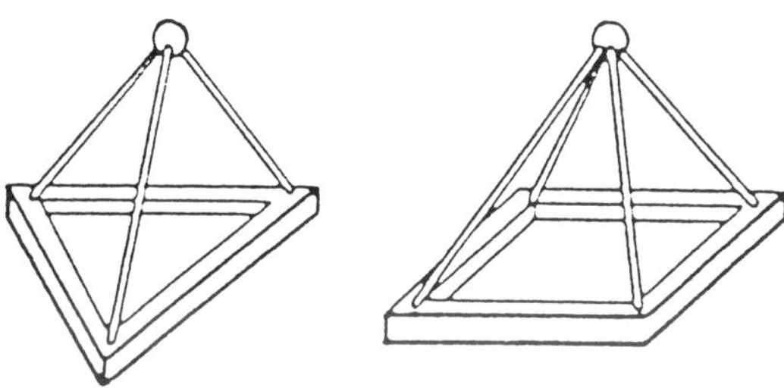

Fig. 8. The prefabricated composite pyramid.

24. There are primarily three kinds of construction methods suitable for composite space trusses, namely:

a) High-altitude bulk method.

In general, the full supports should be erected first and the single members or small units should be placed into position one by one.

b) High-altitude slide movement method.

The composite space truss can be installed on the ground, then lifted to a suitable height and slid into position. For example, this kind of method was used in the composite space truss roof of the storehouse in Shanghai Petroleum supply Station.

c) The whole body-lifting method.
This method is especially suited to the construction and the installation of the composite space truss in multi-storey and high-rise buildings. A small lifting machine or a hydraulic climbing jack can be used as the lifting equipment. This method was chosen to be used in Changsha Great Textile Building.

CONCLUSION

25. The development of composite space trusses and their engineering application is relatively new. In order that the composite space truss can be developed fully, some suggestions and technical problems which need to be studied, are described below:

a) The composite space truss used in the multi-storey structures has more future potential as roof structures. It also give a new way for transforming cities, adding floors and extending the use of existing buildings.
b) Calculation methods for analysis of composite space truss are available. Both accurate calculation methods and simplified calculation methods are also available, but the optimal design method still needs to be studied and refined.
c) The behaviour of the top joint of the composite space truss should be further investigated, developed and new ideas studied to reach its full potential.
d) With the construction of the composite space truss, the high - altitude bulk method and the high-altitude slide movement method are mainly applied, giving successful results. In order that the composite space truss can be extensively applied, the whole body-lifting method and other effective construction methods require investigation.

REFERENCES

1. CASTILLO, H. A Space Frame Construction with Steel and Reinforced Concrete - the "Tridilosa" Spatial Structure. Space Structures, Davies R. M. (ed.). Blackwell Scientific Publications, Oxford and Edinburgh, 1967.
2. KLIMKE and STICHT, "MERO—Massiv", MERO Development and Computation Centre, Report No. 9, 1981.
3. SOARE W.V., CRAINICESCU M. and TAROG D. Double - Layer Grid with Steel Member and Reinforced Concrete Slabs. Space Structures. Vol.1, No.1, 1985.
4. DONG SHILIN; GAO BOQING. Analysis of Composite Space Grids by Sandwich Plate Analogy Method. J. of Structural Engineering, No.1, 1989. (in Chinese)
5. DONG SHILIN; YANG YONGGE. Simplified Calculation for Analysing Space Truss and Plate Composite Structures. J. of Building Structures, No.4, 5, 1985. (in Chinese).

171. Space structures used in the construction of large-span roofs and tall buildings

J. REBIELAK, Technical University of Wroclaw, Poland

SYNOPSIS. Skeletal bar space structures, owing to their architectural and technological advantages, are often used to form of roof structures. Features of these structures such as the use of modular elements, the simple construction process, and their relatively small self weight per square metre are the reasons for their frequent application for covering large spans. Increases in span in turn cause a considerable increase in the problems connected with the design and building processes. This paper presents proposals for shaping large span space structures which can be made by using relatively short members. The manner of application of such structures in the construction of tall buildings has also been presented.

INTRODUCTION

1. Straight bars located in space in a uniform way and joined together by means of articulated nodes can create many different shapes of structures. They are often constructed in many forms forming roof covers (refs 1-3). Economic structures are created through use of a minimum number of bars of different lengths and the number of members necessary for every unit of area covered is small. An example of such a structure is shown in Fig. la. The space structure can be thought of as a solid system describing the arrangement of bars and nodes. The solid system of the example form is composed of a cubo-octahedron half and an octahedron half. The edge lengths (l_1 and l_2) of these solids, are the lengths of bars creating this form of structure.

2. The construction depth (h) of the flat form of the structure (see Fig. 1b) is defined as the distance between the external layers of horizontal bars. The economic span (L) of this structure is determined as being about 20h. Increasing the span of the structure usually causes an increase in member lengths. Because the load-carrying capacity of the strut depends, in an inversely proportional way, on the square of its reduced buckling length, in areas of compressive force the shortest bars should be used. One of the main goals of this paper is to present the proposal describing a way of shaping construction modules so as to allow the economic construction of large span building structures which meet the condition mentioned above.

DESCRIPTION OF THE STRUCTURES GEOMETRY

3. Structures with compound space arrangements, were shaped considering simple double-layer structures as solid systems, examples of which are shown in Figs 1c-e. In the proposed system the geometry description is made by writing in curly brackets, the symbols representing the chord topology used in the relevant structure. Triangular grids are denoted by the symbol T, orthogonal grids as O, diagonal grids as D, hexagonal grids as H and triangular-hexagonal grids as TH (refs. 4-6).

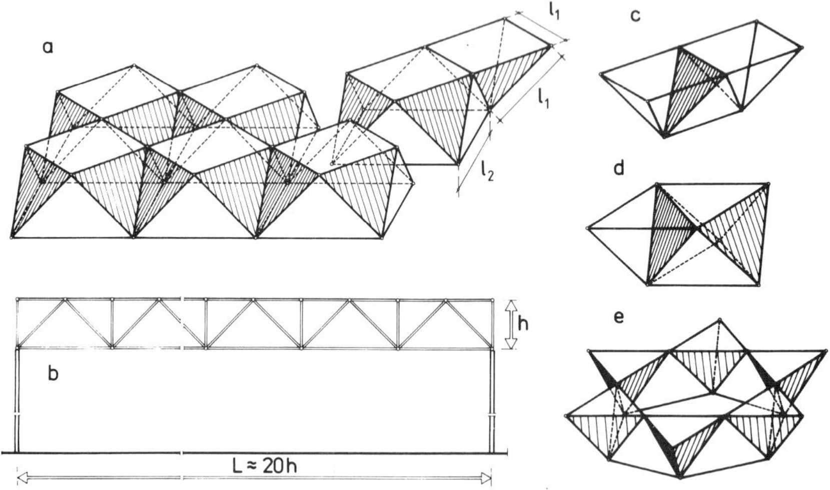

Fig. 1. Schemes of bar systems in simple form for space structures.

4. Symbols of these grids, in the curly brackets, are separated by means of dashes and the symbol representing the chord members is located at the beginning. Before and after the curly brackets additional denotation can be located functioning as ordinal symbols of a given variety within one group of structures. The shape of the structure denotation can inform about the type of basic structure and it can describe the way it is transform into its final shape. This kind of geometry description has the character of an open system which can be used to define names for a very large range of, sometimes complex, space structures. The space structure of the solid system shown in Fig. la is denoted as {D-O}. The structure type "square on square", the solid system of which is shown in Fig. 1c, is defined as {O-O}. The structure of the solid system shown in Fig. 1d has been denoted as {T-T} and the structure of the solid system presented in Fig. 1e is defined as {TH-H/2}. The proposed systems for large spans are presented in this form in refs 5 and 6.

EXEMPLARY SHAPES OF STRUCTURES

5. The simplest way of obtaining a space structure with a large construction depth and built of relatively short bars, consists of an appropriate composition of two double-layer structures. Figure 2a shows the solid system of a structure shaped by joining together two layers of structure {D-O}. This triple-layer form of space structure is denoted as 2x{D-O}A. The bar set, shown in Fig. 2b, creates a construction module of another type of the triple-layer structure which is defined as 2x{TH-H/2}A. The transformed shape of the construction module of this space structure, presented in Fig. 2c, enables the creation of a compound form of double-layer structure, which is denoted as 2x{TH-H/2}B. The bar arrangement of the structure 2x{D-O}A can be transformed in a similar way.

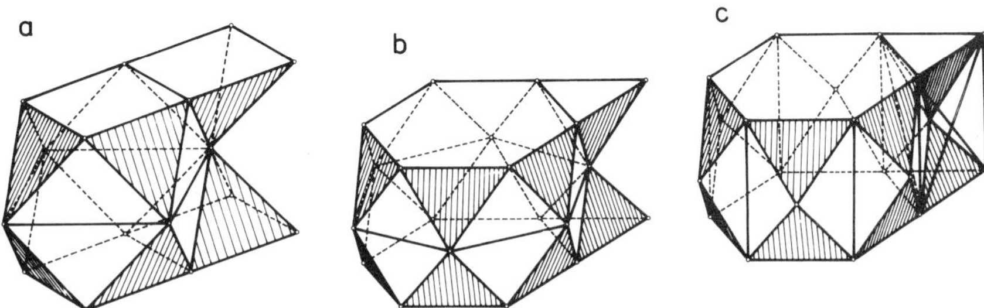

Fig. 2. Exemplary shapes of solid systems in structures of compound inner construction.

6. The construction module for structures should be of simple shape possessing considerably rigidity. Therefore, the regular tetrahedron is the best basic solid body for the modular form which, being built from short bars, could be applied in the construction of large span structures. This polyhedron is often a component part of systems in many types of simple forms of structures. The structure {T-T} is one of them, and its solid system is shown in Fig. 3a. The pre-specified shape construction module will be obtained through the division of all its edges into two equal parts and by placing three additional bars on each of its faces (see Fig. 3b). If this new module shape is made of bar lengths equal to the lengths of bars creating its basic shape (compare Fig. 3a) then the construction depth (h_o) of this module will be twice as big as the construction depth (h) of the basic module. Therefore it should be expected that the span of structures built by means of these "big" forms of modules can be about twice the span of classic forms of structures created using equal length members. Figure 3c shows schemes of the cross-section and the top view of the new form of structure {T-T} constructed by using "big" bar modules. This new "big" shape of the structure is denoted as B{T-T}A. The lower layer bars are twice as long as the bars in the other layers. If compressive forces act in the lower layer members then, in previously determined areas, the "big" module should be placed in an inverted position.

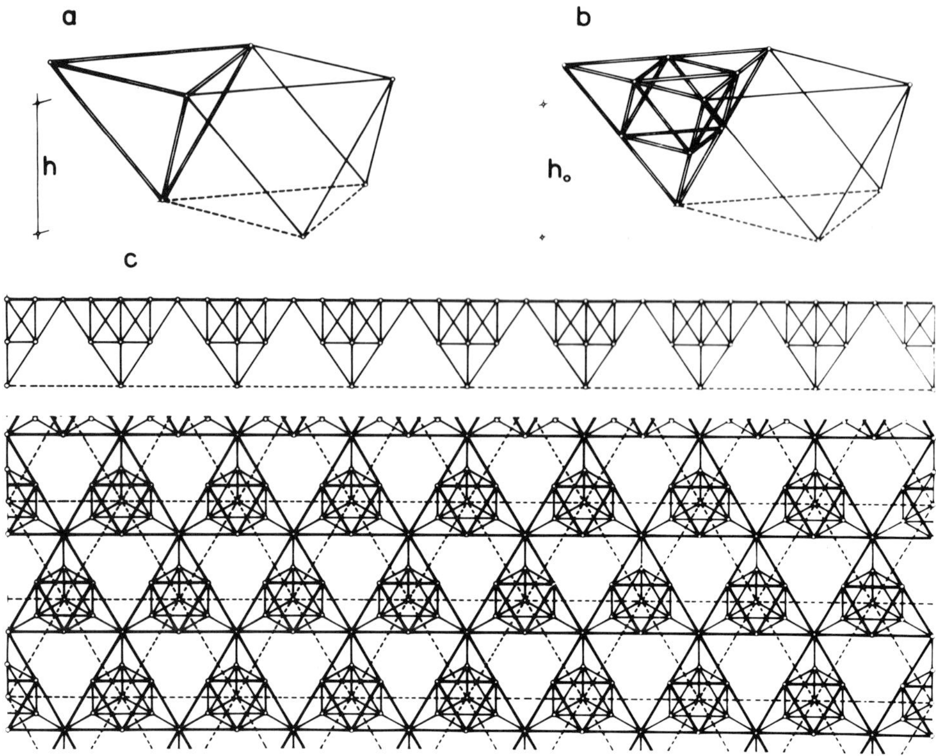

Fig. 3. Solid system of the structure {T-T} (a); and the shape of the tetrahedron bar module (b); applied in building of the structure B{T–T}A (c).

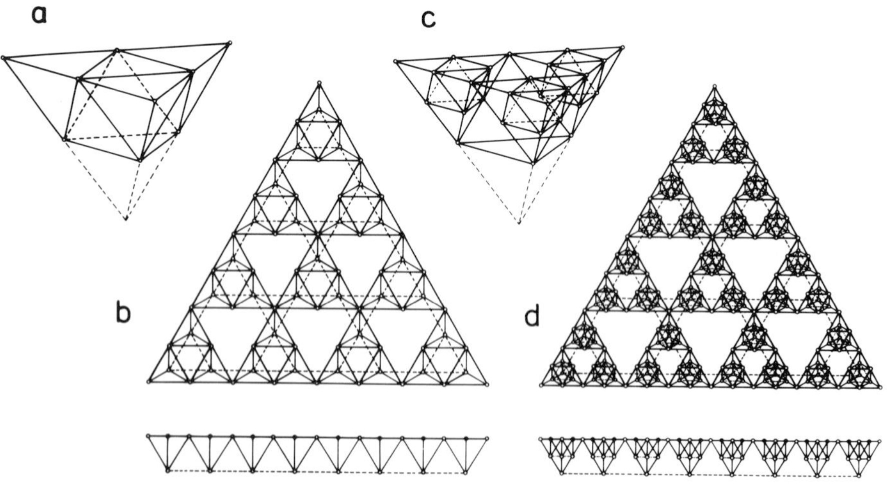

Fig. 4. Examples of forms of construction modules and schemes for structures created using them.

7. The proposed shape of the "big" tetrahedron module can be subject to many transformations. Figure 4 shows examples of structures shaped on the basis of the structure {T-T}. A reduced form of the "big" bar module (see Fig. 4a) makes possible the creation of a structure denoted as RB{T-T}A. Schemes of its front and top views are presented in Fig. 4b. When suitable spaces of the reduced form of construction module are refilled by the "big" forms of tetrahedron bar sets (see Fig. 4c) they will then create the complex shape of space structure named RDB{T-T}A. Figure 4d shows schemes of its front and top views. Member topology of this latter structure can also be the subject of further transformations consisting, for example, of an appropriate complement of bars in the upper layer. The construction module described as bars set in the shape of a regular tetrahedron can be applied in building various types of structures. Figure 5 shows exemplary forms of transformations and configurations of bar modules, obtained on the basis of its model shape, which enable shaping a wide range of structural forms of, sometimes, very complicated inner construction.

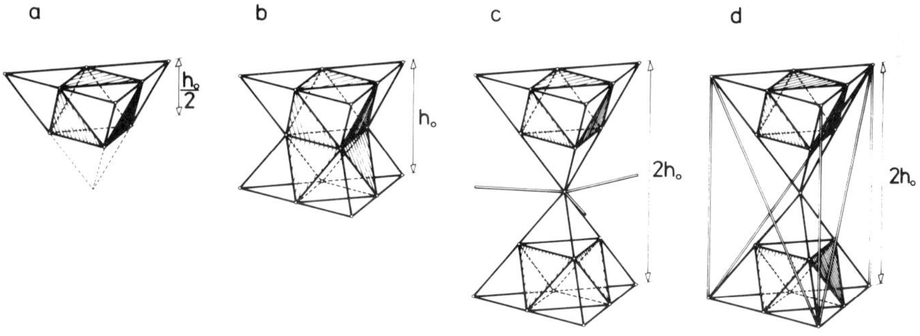

Fig. 5. Exemplary shapes of bar sets obtained as a result of tetrahedron module transformations.

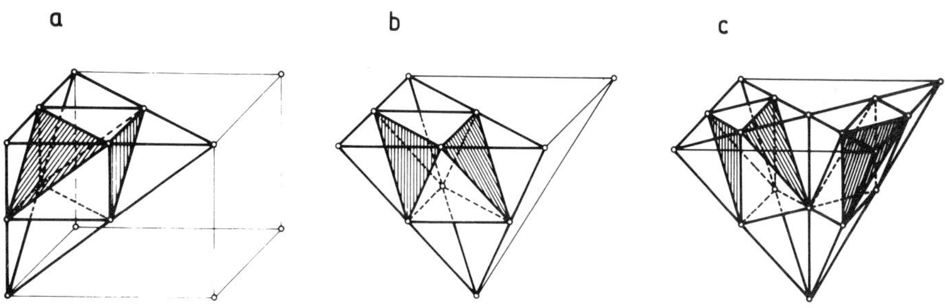

Fig. 6. Localisation of bar modules in spaces of chosen types of solids.

8. The presented form of the "big" construction module can take the shape of not only a regular tetrahedron. Examples of other shapes of these modules are shown in Fig. 6. If the tetrahedron bar set fills only a part of the space of the cube in the way shown in Fig. 6a, then it can be used for shaping many types of structures on the basis defined as {ODT-O}. Fig. 7 shows schemes of the top and front views of some structures of this group.

9. Halves of octahedrons are the component parts of solid systems in many types of space structures. A tetrahedron bar set can take up half of the space of this solid body (see Fig. 6b) or it can fill two of its four quarters (see Fig. 6c). The structure denoted as 3/6B{RT-T/30°}A (see Fig. 8a) was built by means of a suitably transformed form of module which is shown in Fig. 6a. The structure denoted as 2/4B{O-O}A, was shaped on the basis of a simple form of double-layer structure {O-O} by using the form of bar module shown in Fig. 6c. Schemes of its top and front views are presented in Fig. 8b.

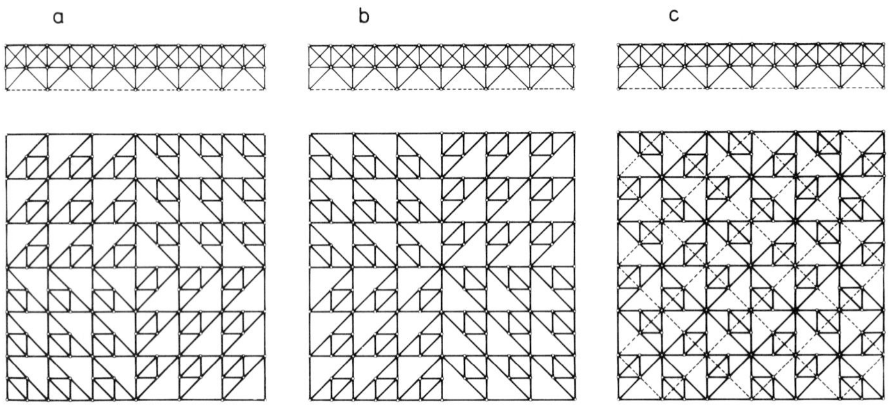

Fig. 7. Schemes of top and front views of the structures successively marked as: (a) structure B{ODT-O}D; (b) structure B{ODT-O}E; (c) structure B{ODT-O}F.

10. The endeavour to apply one of the presented forms of construction modules into the space structure {D-O} (compare Fig. la) does not give satisfactory results. In order to build the "big" shape of this structure it is necessary to divide skew triangular faces of cubo-octahedron halves in the same way as was done for the faces of the regular tetrahedron (compare Fig. 3b). The assumed way of sub-division requires that in every type of this structure group, it is necessary to apply diagonal bars in quadrangular, horizontal fields in the middle layer. These fields are located in the centre of the octahedron halves. The structure denoted as B2{D-O}A belongs to this group of structures. The modular bar set of this structure (see Fig. 9a) is completed by an additional bar system located in the boundary areas of the structure and its form is shown in Fig. 9b. Schemes of the top and front views of the structure B2{D-O}A are presented in Fig. 9c.

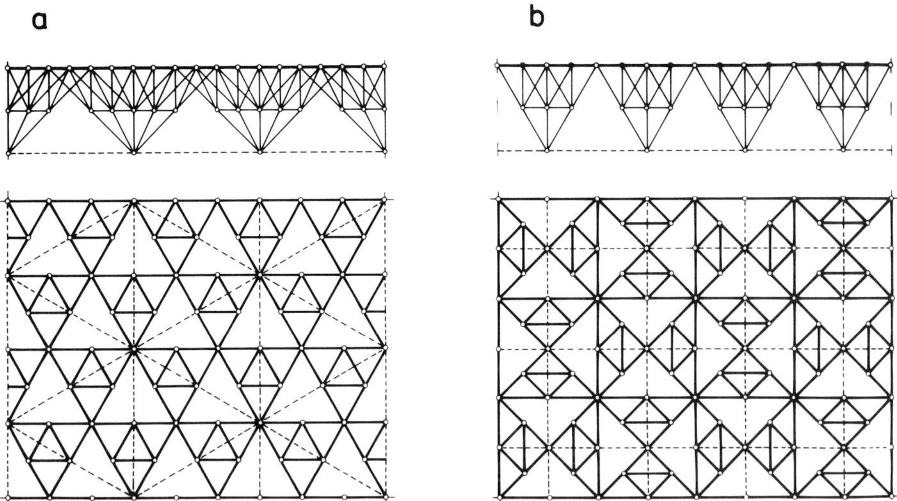

Fig. 8. Scheme of structures: (a) 3/6B{RT-T/30°}A; (b) 2/4B{O-O}A.

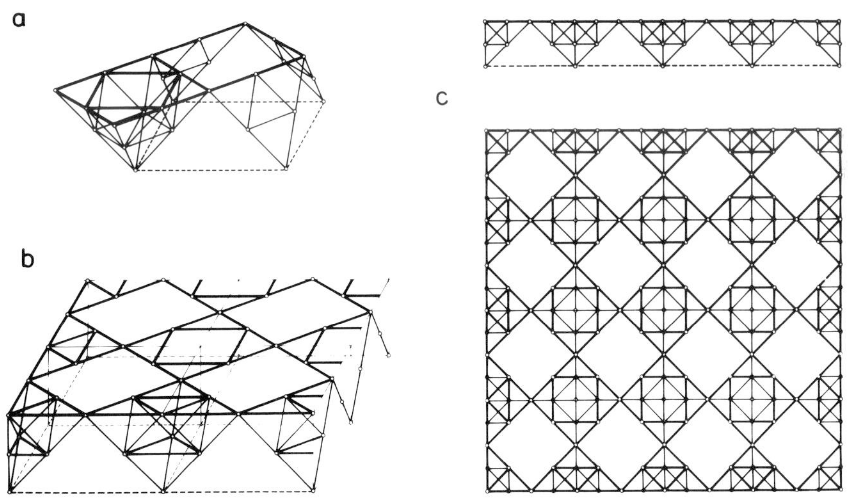

Fig. 9. Bar arrangements for a structure denoted as B2{D-O}A.

PROPOSED APPLICATION OF THE STRUCTURE {D-O} IN THE CONSTRUCTION OF TALL BUILDINGS.

11. The aim of applying some of the principles of member arrangement for space structures used for tall building construction can be observed. The process of shaping structures for large span roofs was concurrent with the search for advantageous forms suitable for tall structures possessing a high slenderness ratio (ref. 7). The construction of tall buildings should give an appropriate rigidity to the whole system of elements which are carrying forces caused by many types of loading.

12. The idea of the solution proposed is to apply the type of bar arrangement in the space structure {D-O} in shaping the steel construction of the external, circumferential load-bearing walls of a tall building. In the model shape it was assumed that the square grid of columns and spandrel beams of the external frame walls of a building, is the orthogonal bar grid of the vertically positioned {D-O} space structure (see Fig. 10a). Its mesh length 'M' corresponds with the main columns spacing. Between the main columns, intermediate columns can also be placed. Curtain walls are situated in the external layer of the vertically positioned space structure because all of the columns in the tall building are located in its inner space. In the vertical direction the module 'M' equals the height of one, or the height of a set, of several storeys. A scheme of the top view of a typical storey of this model shape is shown in Fig. 10b. The Fig. 10c presents the scheme of a vertical section of such a building in which the module 'M' corresponds to the height of two storeys. The whole building consists of a certain number of vertically positioned segments of the circumferential bearing structure. Boundaries of such segments are determined by the position of horizontal cross-sections of the type B-B. In levels of these cross-sections the external layer of the bearing structure is devoid of bars and only its cross-bracing runs to nodes joining the main columns with main edge beams. These beams are located in the middle layer of the structure 2x{D-O}A (compare Fig. 2a). Component parts of this multilayer space structure will fulfil the construction requirements as horizontal bracing for the whole building. Other types of space structures can also be applied in a similar manner to the one presented above with the purpose of shaping various forms of constructions for very tall buildings.

13. The construction system of the tall building, proposed in this paper, should still be the subject of further investigation. These studies should determine, among other things, the influence of thermal loading on the magnitude of the forces acting in the main columns of the circumferential structure. The optimum geometric proportion of a single segment of such a structure should be estimated, and the economic range of rigidity ratio between all elements in this type of construction should be defined. The proposed construction system can also be the subject of various transformations carried out in order to obtain many different forms in the design of tall buildings.

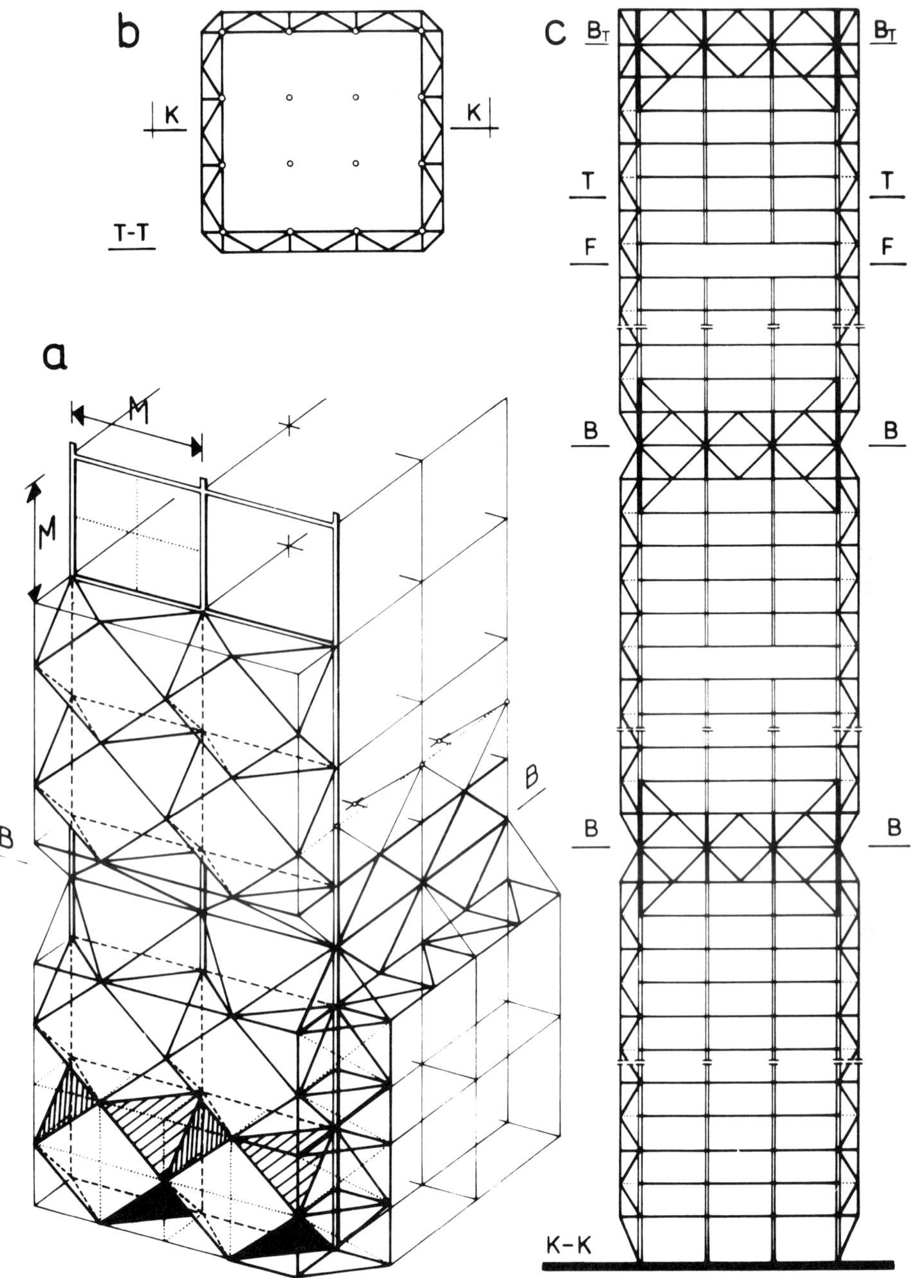

Fig. 10. Space structure of the type {D-O} applied in the construction of tall buildings.

CLOSING REMARKS.

1. Systems of structures proposed in the paper can be applied in the construction of large span roof covers and in structures for very tall buildings. The examples presented may testify to the great development potential of construction formula for bar space structures. This formula can be applied to shaping many types of construction having advantageous technical features and interesting architectural forms.

REFERENCES

1. MAKOWSKI Z.S. Space structures of today and tomorrow. A brief review of their future use. Proceedings of the Third International Conference on Space Structures, 1984, September, 1-8.
2. MENGERINGHAUSEN M. Raumfachwerke als Mittel der individuellen Baugestaltung. Proceedings of the Third International Conference on Space Structures, 1984, September, 73-80.
3. DU CHATEAU S. Structures spatiales. Phenomene d'ideation an architecture. Proceedings of the Third International Conference on Space Structures, 1984, September, 1058-1067.
4. BRÓDKA J. (ed.) Structural Covers (in Polish), p.24. Arkady, Warszawa, 1985.
5. REBIELAK J. Space structures of large span (in Polish), p.58, Scientific Papers of the Institute of Architecture and Town Planning of the Technical University of Wroclaw, No. 27, Monographs, No. 15, Wydawnictwo Politechniki Wroclawskiej.
6. REBIELAK J. Space structures-proposals for shaping, International Journal of Space Structures, 1992, Vol. 7, No. 3, 175-190.
7. REBIELAK J. Tall structures-proposal for shaping (in Polish), Report of the Institute of Architecture and Town Planning of the Technical University of Wroclaw, No. P-639, Wroclaw, 1991.

172. Large-span concrete shelters

M. ASHRAF, Lahore University of Engineering and Technology, Pakistan,
and A. MANSOOR, Corps of Engineers, Pakistan

SYNOPSIS. Requirements of large clear span shelters to house aircraft and factories and to provide indoor sports and recreational facilities has led structural engineers and architects to use space skeletal frames as well as thin walled closed structures. A box girder, is a typical thin walled closed concrete structure and due to a favourable pattern of flexural stresses and high torsional rigidity combined with low shear stresses is suited to large spans. Compressive stresses are found to dominate in a shell or arch. Therefore, the use of the box girder in the form of arch, herein referred to as "box section arch", has been studied for covering even large span structures which may also be employed as buried structures. For the preliminary analysis and design of a box section arch, an approximate method of analysis, based on simple theory of bending has been suggested. For the validity of assumptions employed in the method, a box section arch having a clear span of 500 ft has been analysed for the rise to span ratio varying from 0 to 0.5 by this method and the finite element method employing a standard package program-SAP80. The results obtained by the two methods are found to be in good agreement with each other. Further, the optimum rise to span ratio of a box section arch is found to be between 0.1 to 0.2.

INTRODUCTION

1. Analysis of box sections has been a focus of attention for many research workers in the recent past. A box section because of its slender proportions can be regarded as a thin walled section. Folded plate theory which was conceived by Goldberg and Leve (ref. 6) for simply supported prismatic shells has been developed as a direct stiffness approach by Scordelis et al (ref. 12), Jhonson and Mattock (ref. 7) and Chu and Dudnik (ref. 1) to deal with simply supported box girders. Richman (ref. 10) applied the approach to develop finite difference equations for the analysis.

2. The finite element method, which has been used extensively in structural mechanics, has also been used in the analysis of box sections. Davies et al (ref. 4) applied the finite element technique for the analysis of various types of bridges including box girders. Moffat (refs 8 and 9), however, undertook the application

of the finite element technique to box girder analysis exclusively and proposed some special finite elements for box section analysis. Fam and Turkstra (ref. 5) also proposed a finite element scheme for analysis of box sections. Several finite element program have been developed by Scordelis et al, however, they can prove expensive when fine meshes are required to obtain accurate results.

3. At the preliminary design stage when a designer is trying to optimise the proportions, a detailed analysis procedure, despite the increase in availability of relatively powerful computers, are considered costly. Therefore, some simplified approaches developed by Sawko and Cope (ref. 11) and Crisfield (refs 2 and 3) may be used which reduce the versatile finite element method to a limited usage by reducing computation time. Further, assuming perfect elastic behaviour, Shushkewich (refs 13 and 14) developed an approximate solution for box girder analysis employing simple strength of materials relationships forming an analogy between a box girder and a beam for different loading conditions.

4. Study of the box section indicates that it is suitable for large span applications and has considerable flexural stiffness. Further, by combining a box section with an arch/barrel form, it is possible to achieve even larger clear spans with reasonable economy due to the dominance of compressive membranal stresses with relatively small bending moments and shear forces.

5. The analysis and design of a box girder considered as a prismatic member has received ample attention. The possibility of using the box section as an arch has not obtained adequate consideration, probably due to the availability of other cheaper alternatives like braced barrel vaults, double layer grids and grid domes, etc. In a country like Pakistan, which has limited resources in steel the prospects of using concrete box sections as an arch may be good. Therefore, this paper attempts to suggest an approximate method of analysis for a box section arch which may be used at the preliminary design stage before more accurate analysis can he performed by the finite element method. This method is based on the analogy between a box section arch and a curved beam.

DERIVATION OF EQUATIONS

6. A box section arch of radius R, subtending an angle β at the centre is subjected to a uniformly distributed load w as shown in Fig. 1. Applying Castigliano's second theorem, support reactions for varying support conditions can be obtained. The reactions so determined are axial thrust in addition to bending moment and shear, Fig. 1. The bending moment and axial compression at any section at an angle of ϕ form the left support can be expressed as:

$$P = P_1 \cos\phi + V_1 \sin\phi - WR\,\{\sin\beta/2 - \sin(\beta/2-\phi)\}\,\sin(\beta/2-\phi) \quad (1)$$

$$M = M_1 + P_1 R\,(1-\cos\phi) - V_1 R \sin\phi - WR^2/2\,\{\sin\beta/2 - \sin(\beta/2-\phi)\}^2 \quad (2)$$

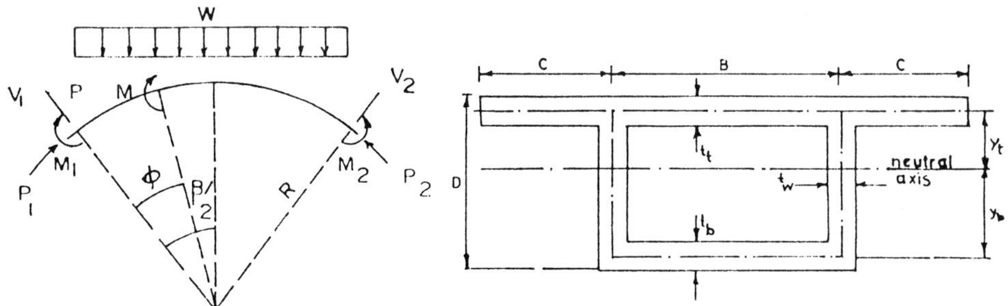

Fig. 1. Arch loading and Reaction. Fig. 2. Box Section.

7. The cross-section of a box section arch given in Fig. 2 is assumed to be subjected to a load w/2 at each web. The centroidal distance between the two webs is B while the width of the cantilevers is C and depth of section is D. The distances from neutral axis to the centres of top and bottom slabs are y_t and y_b respectively. The thicknesses of the top and bottom slabs are t_t and t_b, respectively whereas web thickness is denoted by t_w.

8. For the illustration of membrane forces and their sign convention, a segment of a box section arch is shown in Fig. 3 as a flat box. N_{xx}, N_{xy} and N_{yy} indicate longitudinal forces, membrane shear forces and transverse shear forces respectively.

9. Also shown in Figs 2 and 3 are the four distinct parts into which a box section may be divided. These are cantilevers, top slab (between webs), bottom slab and webs. Orientation and location of the local axes of each of these elements is also depicted in Fig. 3. Each of these elements is considered separately to determine the expression for membrane forces below:

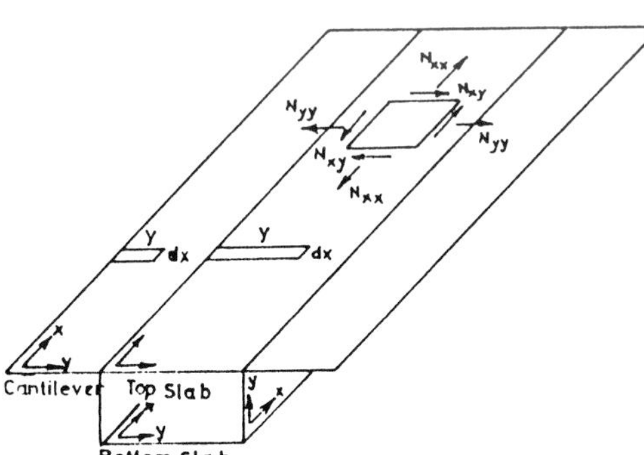

Fig. 3. Sign convention for element axis and membrane forces.

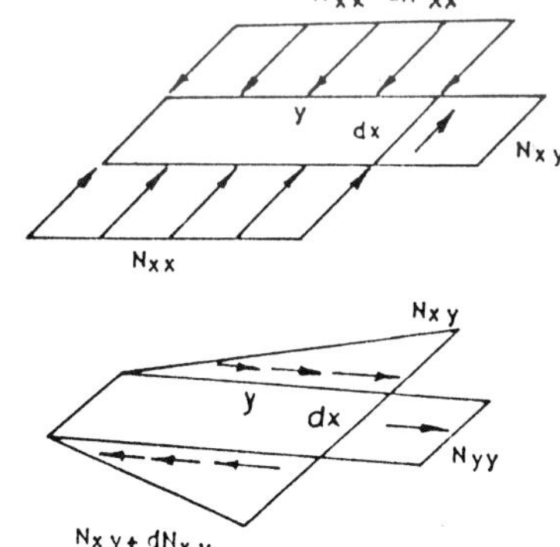

Fig. 4. Free body diagrams for cantilever.

Cantilevers.

10. If the effects of shear lag are not considered then the longitudinal stress at the centre of the top slab can be written as:

$$\sigma_{xx1} = - My_t/I - P/A \qquad (3)$$

where A and I are cross-sectional area and second moment of area, respectively. The longitudinal membrane force N_{xx1} for a unit length can be expressed as the product of σ_{xx1} and t_t. Hence:

$$N_{xx1} = \sigma_{xx1}\, t_t \qquad (4)$$

Free body diagrams for the cantilever are shown in Fig. 4. For equilibrium, summation of the forces in the X and Y directions (noting that dx = R dϕ) yields:

$$N_{xy1} = \frac{dN_{xx1}}{d\phi}\,\frac{y}{R} \qquad (5)$$

$$N_{yy1} = \frac{d^2N_{xx1}}{d\phi^2}\,\frac{y^2}{2R^2} \qquad (6)$$

Substitution of the values of Eqs 1 and 2 into Eqs 4 to 6 will give membrane forces in the cantilever slab as:

$$\begin{aligned} N_{xx1} = {} & - y_t t_t/I \{M_1 + P_1 R(1-\cos\phi) - V_1 R\sin\phi - wR^2/2\,[\sin\beta/2 - \sin(\beta/2-\phi)]^2\} \\ & - t_t/A \{P_1 \cos\phi + V_1 \sin\phi - wR\,[\sin\beta/2 - \sin(\beta/2-\phi)]\,\sin(\beta/2-\phi)\} \qquad (7) \\ N_{xy1} = {} & - y_t t_t y/I \{P_1 \sin\phi - V_1 \cos\phi - wR\,[\sin\beta/2\,\cos(\beta/2-\phi) - \sin(\beta/2-\phi)\cos(\beta/2-\phi)]\} \\ & - t_t y/A\{-P_1/R\,\sin\phi + V_1/R\,\cos\phi - w\,[\sin(\beta-2\phi) - \sin\beta/2\,\cos(\beta/2-\phi)]\} \quad . \qquad (8) \\ N_{yy1} = {} & - y_t t_t y^2/2I\,\{P_1/R\,\cos\phi + V_1/R\,\sin\phi - w[\sin\beta/2\,\sin(\beta/2-\phi) + \cos(\beta-2\phi)]\} \\ & - t_t y^2/2A\{-P_1/R^2\cos\phi - V_1/R^2\sin\phi + w/R[2\cos(\beta-2\phi) + \sin\beta/2\sin(\beta/2-\phi)]\} \qquad (9) \end{aligned}$$

Top Slab

11. The longitudinal membrane force in the top slab "N_{xx2}" will be same as that determined in cantilever slab as neither the thickness nor the centroidal distance of the slab from the neutral axis changes. The top slab is rigidly connected to the web and will therefore develop reactions R_x and R_y which can be determined by taking a strip from web to web and summing forces as:

$$R_x = \frac{dN_{xx2}}{R\,d\phi}\,\frac{B}{2} \qquad \& \qquad R_y = 0 \qquad (10)$$

Summing forces in the X-direction for the element as shown in free body diagrams of Fig. 5 gives:

$$N_{xy2} = \frac{dN_{xx2}}{Rd\phi}\left(y - \frac{B}{2}\right) \tag{11}$$

Carrying out the addition of the forces in Y-direction and taking account of the transverse membrane forces being transmitted from cantilever yields:

$$N_{yy2} = \frac{d^2N_{xx2}}{2R^2d\phi^2}\, y(B - y) + N_{yy1}(x,c) \tag{12}$$

By inputting values established earlier into Eqs 11 and 12, membrane forces in the top slab between the web can be written as:

$$\begin{aligned} N_{xy2} = {} & -y_t t_t (y-B/2)/I\,\{P_1 \sin\varnothing - V_1 \cos\varnothing - wR[\sin\beta/2 \cos(\beta/2-\varnothing) - \sin(\beta/2-\varnothing)\cos(\beta/2-\varnothing)]\} \\ & - t_t (y-B/2)/A\{-P_1/R \sin\varnothing + V_1/R \cos\varnothing - w[\sin(\beta-2\varnothing) - \sin\beta/2 \cos(\beta/2-\varnothing)]\} \quad (13) \\ N_{yy2} = {} & y_t t_t y\,(B-y)/2I\{P_1/R \cos\varnothing + V_1/R \sin\varnothing - w[\sin\beta/2 \sin(\beta/2-\varnothing) + \cos(\beta-2\varnothing)]\} \\ & + t_t y(B-y)/2A\{-P_1/R^2 \cos\varnothing - V_1/R^2 \sin\varnothing + w/R[2\cos(\beta-2\varnothing) \\ & + \sin\beta/2 \sin(\beta/2-\varnothing)]\} + N_{yy1}(x,c) \quad (14) \end{aligned}$$

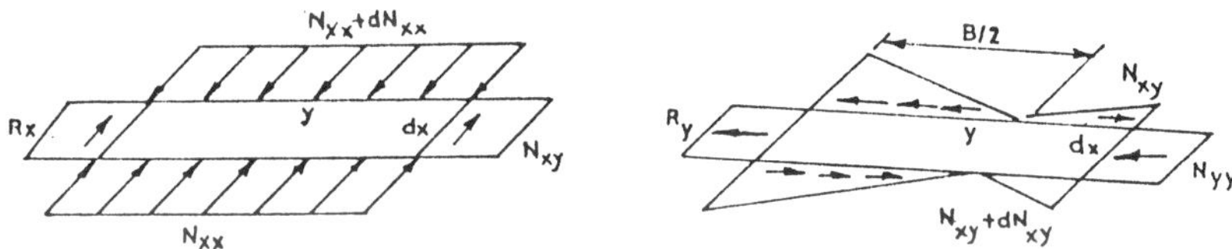

Fig. 5. Free body diagrams for top slab.

<u>Bottom Slab</u>

12. Stresses at centre of the bottom slab can be expressed as:

$$\sigma_{xx3} = My_b/I - P/A \tag{15}$$

Longitudinal membrane forces in the bottom slab can be expressed as product of σ_{xx3} and slab thickness t_b:

$$N_{xx3} = \sigma_{xx3}\, t_b$$
(16)

Recognising that bottom slab forces can be found from the expression for the top slab by replacing y_t with $-y_b$ and t_t with t_b, the equations for membrane forces in the bottom slab are found as:

$$N_{xy3} = \frac{dN_{xx3}}{Rd\phi}\left(y - \frac{B}{2}\right) \tag{17}$$

$$N_{yy3} = \frac{d^2N_{xx3}}{2R^2d\phi^2}\, y(B-y) \tag{18}$$

$$\begin{aligned} N_{xx3} = {} & y_b t_b/I\,\{M_1+P_1R(1-\cos\phi) - V_1R\sin\phi - wR^2/2\,[\sin\beta/2-\sin(\beta/2-\phi)]^2\} \\ & -t_b/A\,\{P_1\cos\phi + V_1\sin\phi - wR\,[\sin\beta/2-\sin(\beta/2-\phi)]\sin(\beta/2-\phi)\} \end{aligned} \tag{19}$$

$$\begin{aligned} N_{xy3} = {} & y_b t_b(y-B/2)/I\{P_1\sin\phi-V_1\cos\phi-wR[\sin\beta/2\cos(\beta/2-\phi)-\sin(\beta/2-\phi)\cos(\beta/2-\phi)]\} \\ & -t_b(y-B/2)/A\{-P_1\sin\phi/R+V_1\cos\phi/R-w[\sin(\beta-2\phi)-\sin\beta/2\cos(\beta/2-\phi)]\} \end{aligned} \tag{20}$$

$$\begin{aligned} N_{yy3} = {} & -y_b t_b y(B-y)/2I\,\{P_1/R\cos\phi+V_1/R\sin\phi - w[\sin\beta/2\sin(\beta/2-\phi) + \cos(\beta-2\phi)]\} \\ & +t_b y(B-y)/2A\,\{-P_1/R^2\cos\phi - V_1/R^2\sin\phi + w/R\,[2\cos(\beta-2\phi)+\sin\beta/2 \\ & \sin(\beta/2-\phi)]\} \end{aligned} \tag{21}$$

Web

13. Longitudinal stresses and forces at any section of the web can be expressed as:

$$\sigma_{xx4} = M(y_b - y)/I - P/A$$

$$N_{xx4} = \sigma_{xx4}\, t_w \tag{22}$$

Since the web is rigidly connected to the bottom slab, it would obtain a shear and transverse stress component from the bottom slab. The expression for membrane shear forces can be found by differentiating N_{xx4} with respect to X and integrating it with respect to Y, as is evident from the expression found in earlier sections. The expression for the transverse membrane force is found by further differentiation and integration of N_{xx4}. Substitution of values and taking account of forces transmitted by the bottom slab will yield equations for membrane forces in the web as:

$$\begin{aligned} N_{xx4} = {} & (y_b-y)t_w/I\,\{M_1+P_1R(1-\cos\phi) - V_1R\sin\phi-wR^2/2\,[\sin\beta/2-\sin(\beta/2-\phi)]^2\} \\ & -t_w/A\,\{P_1\cos\phi + V_1\sin\phi-wR\,[\sin\beta/2-\sin(\beta/2-\phi)]\sin(\beta/2-\phi)\} \end{aligned} \tag{23}$$

$$\begin{aligned} N_{xy4} = {} & (y_b-y/2)t_w y/I\{P_1\sin\phi-V_1\cos\phi-wR[\sin\beta/2\cos(\beta/2-\phi)-\sin(\beta/2-\phi)\cos(\beta/2-\phi)]\} \\ & -t_w y/A\{-P_1/R\sin\phi+V_1/R\cos\phi-w[\sin(\beta-2\phi)-\sin\beta/2\cos(\beta/2-\phi)]\}-N_{xy3}(x,0) \end{aligned} \tag{24}$$

$$\begin{aligned} N_{yy4} = {} & (y_b/2-y/6)t_w y^2/I\{P_1/R\cos\phi+V_1/R\sin\phi - w[\sin\beta/2\sin(\beta/2-\phi)+\cos(\beta-2\phi)]\} \\ & -t_w y^2/2A\,\{-P_1/R^2\cos\phi-V_1/R^2\sin\phi+w/R[2\cos(\beta-2\phi)+\sin\beta/2\sin(\beta/2-\phi)]\} \\ & -dN_{xy3}(x,o)/R\ d\phi \end{aligned} \tag{25}$$

DISCUSSION OF RESULTS

14. A flat box section along with five box section arches, having rise to span ratios between 0.1 and 0.5, covering a span of 500 ft, subjected to a uniform load of 15 kips/ft were analysed by the above proposed method (Eqs 1-24). Dimensions of the typical box section are shown in Fig. 6. To study the validity of the proposed method, six structures mentioned above were analysed by using the SAP80 package as well. Critical values of membrane forces obtained by the proposed method and SAP80 package are shown in Figs 7-9. The following important points are noted from the comparison of the two types of results.

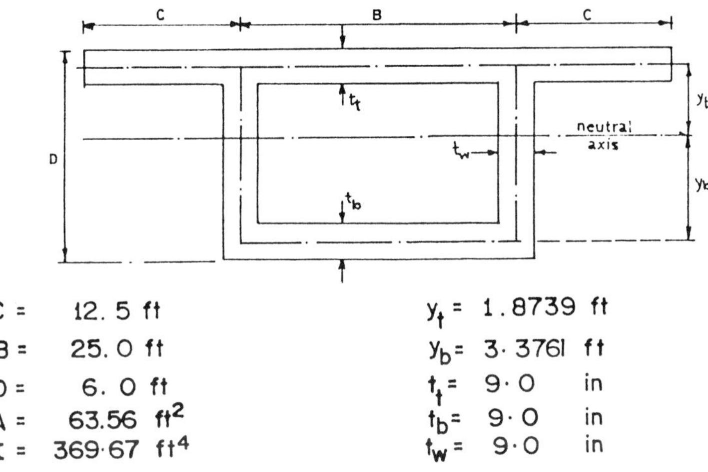

Fig. 6. Typical box section analysed.

15. There is an appreciable reduction in the membrane forces obtained in the box section arches in comparison with those of a flat box section. Maximum positive longitudinal membrane forces in the top and bottom slab in the case of box section arches are 25 to 40% of the values obtained in flat box section in adverse conditions. When the rise to span ratio is 0.1- 0.2, these forces even do not occur.

16. Maximum negative longitudinal membrane forces in the top and bottom slabs of box section arches are 30 to 50% of the values obtained in the flat box section. These forces are even smaller when the rise to span ratio is between 0.1 and 0.3. Membrane shear forces vary between 5 to 50% of those of flat box section ones. These results are indicative of the suitability of box section arches for large span coverage. Box section arches having a rise to span ratio of 0.1 to 0.2 would yield the most economical designs and rise to span ratios in this range can give optimum use of a box section arch.

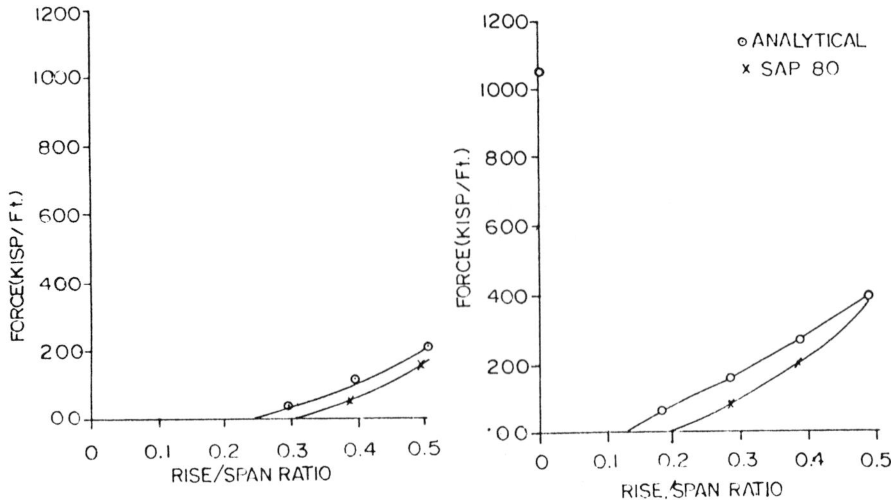

Fig. 7. Maximum positive longitudinal membrane forces.

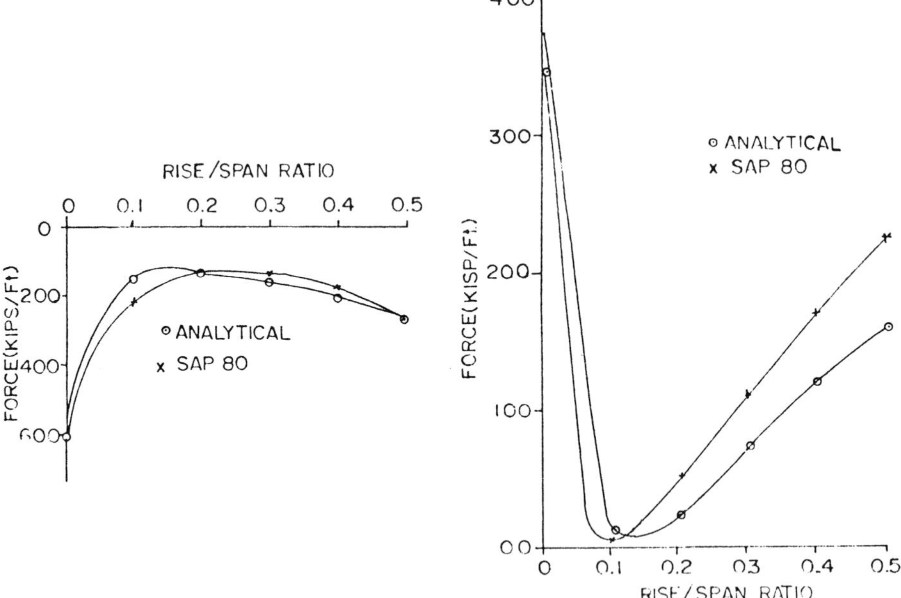

Fig. 8. Maximum negative longitudinal membrane forces in top slab.

Fig. 9. Maximum membrane shear forces in webs.

17. Analytical results follow trends indicated by the finite elements analysis. Negative longitudinal membrane forces obtained by both methods are in good agreement with each other. Positive longitudinal membrane forces also show close resemblance of the two types of results.

18. The analytical method, however, slightly overestimates the membranal forces which may be useful at the preliminary design stage. The analytical

method, generally underestimates the membrane shear forces to the order of 25 - 35%. The difference in the two sets of results of transverse membrane forces is large and can be attributed to incompatibility in simulation of the two types of model and also due to the fact of neglecting shear lag in the analytical method.

CONCLUSIONS

19. The analytical method developed is suitable for hand calculations and can be easily programmed as well. The results obtained are in good agreement with those obtained from finite element analysis. The suggested method provides a tool for the quick estimation of the dimensions of box section arches; especially at a preliminary design stage. Further, it is found that the optimum rise to span ratio of a box section arch lies between 0.1 to 0.2.

REFERENCES

1. CHU AND DUDNIK, Concrete box girder analysis as folded plate, A.C.I. Spec Pub SP-23, 1969.
2. CRISFIELD, Finite element method for the analysis of multi-cellular structures, Proc of ICE, March 1971.
3. CRISFIELD AND TWEMLOW, The equivalent plate approach for analysis of cellular structures, Civ Engg and Public works Review, Mar 1971.
4. DAVIES ET AL, Analysis of various type of bridges by finite element method, Dev in Bridge Design and Construction, Crosby Lockwood & Sons, 1971.
5. FAM AND TURKSTRA, A finite element scheme of box girder analysis, Int Jnl of Computer and Structures, Vol 5, 1975.
6. GOLDBERG AND LEVE, Theory of prismatic folded plate structures, IABSE Pub Vol 17.
7. JHONSON AND MATTOCK, Lateral distribution of loads in composite box girder bridges, Highway Research Board, Washington D.C.
8. MOFFAT AND LIN, Some finite elements having particular application to box girders, IABSE Proc P-1/77, Feb. 1977.
9. NATIONAL BUILDING STUDY SPECIAL REPORT NO. 24, HMSO, 1956.
10. RICHMAN, Matrix difference analysis of box girder, Proc of ICE, Aug 1969.
11. SAWKO AND COPE, Analysis of multi-cell bridges without diaphragms - a finite element approach, Structural Engineer, Nov 69, P 455-460.
12. SCORDELIS, DEFRIES AND SKENE, Direct stiffness solution for folded plate, ASCE, Jnl of Structural Div., Vol 112, No. ST8, 1986.
13. SHUSHKEWICH, Membrane forces acting on a box girder bridges, ASCE, Jnl of Structural Div., Vol 112, No ST8, 1986.
14. SHUSHKEWICH 1, Approximate analysis of concrete box girder bridges, ASCE, Jnl of Structural Div., Vol 114, No ST 7.

173. A multi-hypar roof in Ness-Tziona, Israel

M. PAZ-NER, Consulting Engineer, Israel

SYNOPSIS. The design and construction of a multi-hypar steel roof over the main hall of a small synagogue is described. The structure is composed of twenty nine small hypars and two flat elements. It was constructed on the site nearby the building and then hoisted to the top of the structure, where it was mounted to its walls.

THE DESIGN AND CONSTRUCTION OF THE STRUCTURE

1. The multi-hypar shape of the roof was determined by the architect. The roof covers the main hall of a small synagogue whose horizontal dimensions are 14.5m x 14.5m. These small dimensions do not justify, perhaps, a space structure and the task of the engineer was to design the most economically possible structure that could also be easily constructed. A reinforced concrete cast *in situ* structure was out of the question. A structure composed of precast elements was considered, but was found difficult to construct, mainly because of the problem of joining the elements into one cohesive whole and at the same time meeting the requirements of the architect; such as a smooth, white, water-retardant exterior surface, thermal insulation and that the structure be acoustically absorbent from the inside. Finally, a light steel structure was proposed and adopted.

2. The structure is actually a dome spanning over a square hall. It is composed of 29 identical hypars (Figs 1-3) and 2 flat elements that are mounted together in different positions to form the comprehensive structure. The hypar elements are composed of 10cm high "U" profiles as edge beams filled in with two layers of 19mm pipe struts that form the hypar surface. Between the two layers, and passing through the intersections of the pipe struts, flat steel bands of 30/3mm section are inserted, running in the direction of the two parabolas. The outside surface of the elements was designed to be covered with thin steel sheet strips and the inside ones with such strips, but perforated to meet the acoustical requirements. The inner space was designed to be filled with thermal-acoustical material. All connections between the struts forming the individual elements and those that connect them to form the whole structure, were welded ones. The edge

members of the individual elements, when mounted together, formed the skeleton of the comprehensive structure. The dimensions of the struts and edge beams follow mostly from practical considerations. As regards strength, they are rather exaggerated.

Fig. 1. Perspective view.

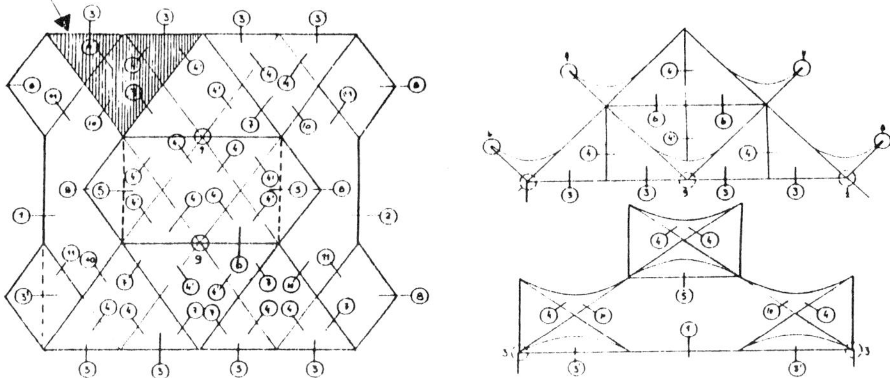

Fig. 2. Plan and elevation (schematic).

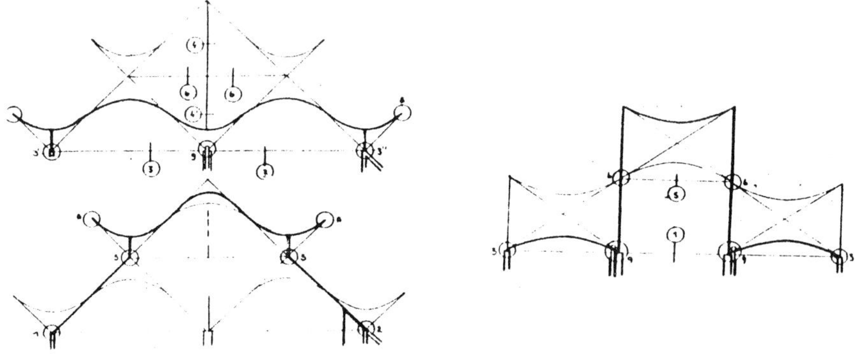

Fig. 3. Sections (schematic).

3) The construction of the roof was performed on a site nearby the building. The individual elements of the roof were constructed separately (Fig 4) and then joined together (Fig 5). The outer surface was covered with flat steel galvanised sheet strips, which had white latex and fabric-reinforced insulation applied. On the inner surface of the roof acoustical perforated sheets were attached, following the shape of the surface. During this procedure, glass wool sound absorbing and thermal insulating pads were inserted between the inside acoustical cladding and the outside one. After completion of the whole procedure, the completed multi-hypar dome was lifted by a light crane and placed in position upon the walls of the building and connected to the supports prepared in advance (Fig 6).

Fig. 4. Individual Hypar.

Fig. 5. Assembling the structure.

Fig. 6. Hoisting the completed roof structure atop the building.

ACKNOWLEDGEMENT: The author wishes to express his thanks to the building's architects, Avi Kaplan and Jehoshua Amit, Architects, Tel-Aviv.

174. Developments in cold-formed space structures

J. C. CHAPMAN, Chapman and Dowling Associates, UK, and D. BUHAGIER and P. J. DOWLING, Imperial College, UK

SYNOPSIS. The initial purpose of the investigation was to produce design methods for a space frame system consisting of back to back channel chords, with tubular diagonals, but many of the results have general application. The paper is mainly concerned with the local and overall buckling behaviour of thin walled channels. The design of connections is also discussed. The economy of the system has been demonstrated by structures built.

INTRODUCTION

1. A major advantage of reinforced concrete is that a wide range of structural types and forms can be produced from cement, aggregate, and a stock of about five bar diameters. Likewise cold formed steel sections and structures can be produced from a stock of about five thicknesses of pre-galvanised strip. Modern cold forming mills can produce purpose-designed sections with accurately located pre-programmed hole patterns, and with optional on-line painting, for a complete structure, in a few days.

2. A prime design objective, especially in space structures, must be to reduce the cost per connection. This can be achieved by using cold formed sections which are continuous across connections, which necessitates their being placed back-to-back. Tubular diagonals can be connected between the back-to-back chords. Simplicity in manufacture and assembly is not achieved without added complexity in design and analysis. The following paragraphs will outline some of the phenomena which were encountered and some of the design methods which were developed.

DESIGN

3. The simplest back-to-back connection, both for production and assembly, is made from channel sections, and the simplest method of connecting tubes is by flattening and punching the ends. An early system for truss and space frame systems incorporating these features was published in 1942 (ref. 1). More recently, a similar system, in which the diagonals consist of tubes with flattened ends which are bent on inclined crease lines through approximately 45° to enable

the ends (tabs) to be inserted between the channels, was developed by an Australian architect, Edwin Codd. Many such structures, using singly lipped channels have been built in Australia and the Far East (ref. 2). The connection is architecturally very neat but requires highly formable steel and special tooling. Also, the tabs are not wholly contained within the back-to-back channels, and this restricts the tube diameter and results in end eccentricities. For European application, an increased module size, with greater applied load intensity, was required, so deeper, doubly lipped, channels were adopted (Fig. 1, ref. 3). Alternatively, tooling and erection can be simplified, and dimensional variations more easily accommodated, at the cost of introducing connecting plates and additional bolts (ref. 4). The tubes are then loaded concentrically, the tabs are fully supported, and the tab hole end distance is not limited by the chord width.

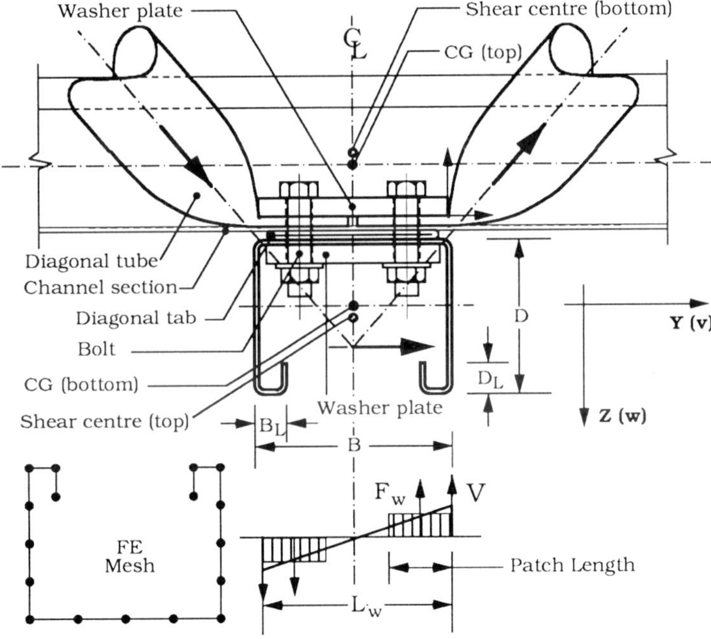

Fig. 1. Space frame connection - bottom chord diagonal on square.

4. The diagonal forces act at a point which is eccentric to the chord centroids, and the eccentricity moment becomes important in regions of high shear, especially at interior columns. It is therefore advantageous to orient the chords so that the greater moment causes lip tension. The moments are transmitted through the bolts and washer plates by vertical forces applied to the channel webs. The

section must be dimensioned to resist moment and axial force, and to accommodate the bolts in a position which permits entry of the wrench socket. The eccentricity moments are an unavoidable consequence of the joint design; the continuous chords, having the required moment capacity, do however provide a comforting degree of robustness. Chord section dimensions are constant for a structure; thicknesses currently being used vary between 1.5 and 4.0mm. Nodal distances are typically 3m.

LOCAL BUCKLING

5. Effectiveness in compression can be estimated on the assumption that all sides are simply supported. In fact, the cross-section has a single critical stress, and various approximations may be made to estimate the strength of the section. The assumption permitted in BS 5950 Part 5 (ref. 5), that the increased strength of the more slender side, resulting from restraint afforded by less slender sides, may be taken into account without any reduction in the strength of the less slender sides, was not adopted.

6. The eccentricity moment causes vertical forces in the channel webs. The lever arm of the couple is based on an assumed linear distribution; the intensity and length of equivalent uniform forces are then found such that the lever arm and the moment are unchanged (Fig. 1). BS 5950 Part 5 treats web crippling, but the equation was found to be non-conservative for the more slender sections, and a correction factor was therefore introduced. The load application in the tests on which the code method is based (ref. 6) was not the same as in the Conder Harley joint. The coexistent compression caused by axial force and bending is taken into account assuming linear interaction. Where the local strength of the channel section at a connection is insufficient, split channel inserts are provided. The inserts strengthen the table and webs of the channel, and transfer moment directly to the webs. If necessary they can be extended to reinforce the span between adjacent nodes.

LIP BUCKLING

7. When the lips of the section are subjected to compression, they may buckle as struts. The tendency to buckle is resisted by the deformational stiffness of the cross-section; the stiffness depends on the mode of deformation, which is symmetric about the z axis if the lip stresses are equal. The lips may then be treated as struts on an elastic foundation, for which the "foundation modulus" is given by the deformational stiffness. The stiffer the foundation the shorter the wavelength, and the greater the buckling stress. The lip buckling resistance can then be found from the Perry strut equation, for an assumed imperfection over the buckling wavelength (ref. 7). The lip buckling behaviour is found to be similar for lip compression caused by axial force or by bending (Fig. 2). Modal interaction between local buckling and lip buckling does not occur, because the buckling wavelengths are sufficiently different. The reduction in lip resistance caused by lip buckling can be taken into account by reducing the design stress at the lip.

TORSIONAL FLEXURAL BUCKLING

8. For singly symmetric sections, torsional flexural buckling can occur. For the section of Fig. 1, the critical torsional buckling stress is much smaller than for flexural buckling, as can be seen in Fig. 3. For shallower sections the difference diminishes, and the buckling stresses become equal when D/B = 0.45.

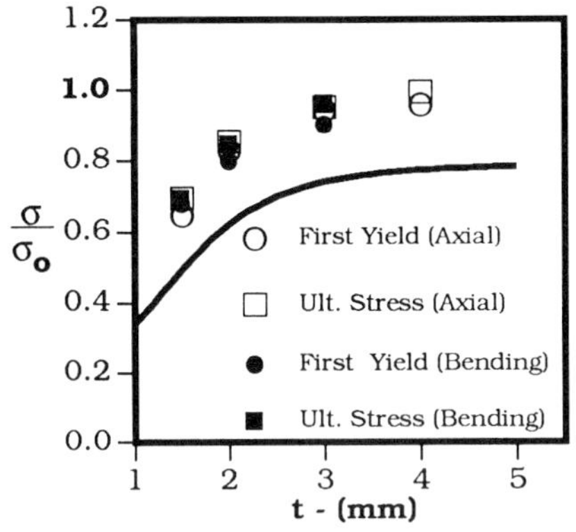

Fig. 2. Lip buckling strength-design curve and FE results, (L=1200mm, single half wave).

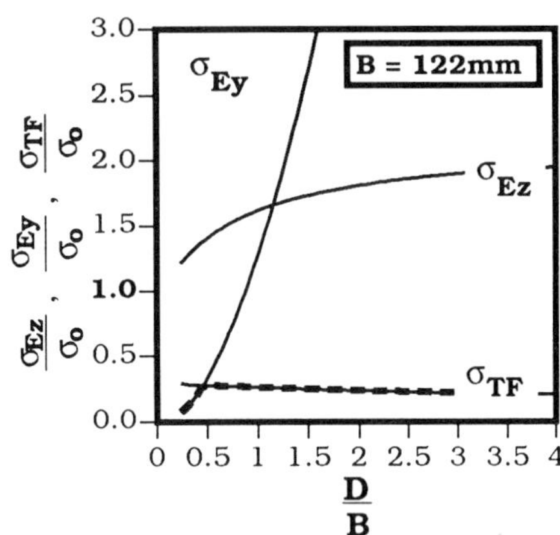

Fig. 3. Variation in critical buckling stress with depth.

9. Whereas for flexural buckling, compressive yielding occurs first at both lips, for torsional flexural buckling yielding occurs first at one of two diagonally opposite corners, whilst the stress at the other two corners is relatively small. This difference in behaviour is not recognised in current design codes, in which the design procedure implies that flexural buckling occurs, but with an effective length which is increased on the basis of the ratio of the flexural and torsional flexural buckling stresses. The governing equations for critical torsional flexural buckling enable the relative values of rotation and translation to be determined. These relative values can then be used to define an assumed imperfection which is affine to the critical torsional flexural buckling mode. Young's equation for buckling of an imperfect strut (ref. 8) can then be applied to each imperfection to find the rotation and translation of the shear centre at any given load level. Curvatures and stresses around the section caused by warping restraint and by translation of the shear centre can then be easily found (by addition of the axial stress) for a given load, and the load to cause first yield can also be found (Fig. 4, ref. 9). Elasto-plastic finite element studies confirm that the stress distribution and load to cause first yield are given with good accuracy by the above method, and that failure follows closely upon first yield. The average axial stress to cause first yield is 87 N/mm^2, compared with 89 N/mm^2 given by the above method. Ideally, a finer mesh would have been used (see Fig. 1).

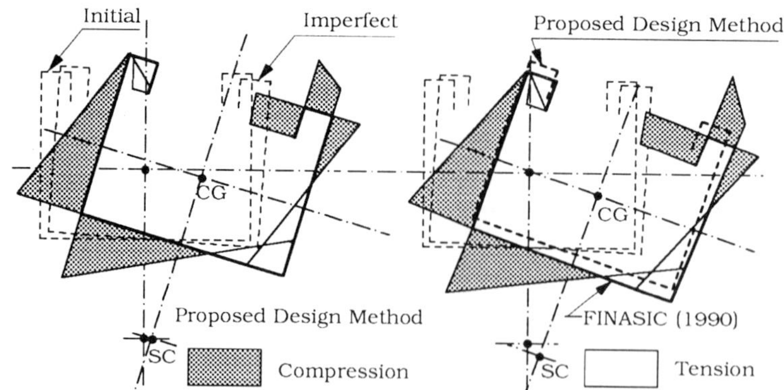

Fig. 4. Stress distribution and displacements at first yield given by proposed design method and FE analysis. Torsional flexural buckling ($B = 122mm$, $D = 97mm$, $B_L = 17mm$, $D_L = 18.5mm$, $t = 3.0mm$ $L = 3000mm$ $\sigma_o = 390\ N/mm^2$).

LATERAL TORSIONAL BUCKLING

10. According to BS 5950, lateral torsional buckling about the minor axis does not occur. This is correct for doubly symmetric beams because when $I_z = I_y$ the critical stress becomes infinite. For the singly symmetric section of Fig. 1, $I_y / I_z = 0.53$, but lateral torsional buckling does in fact occur; the critical stress $\sigma_{LT} = 336$ N/mm^2. The variation of σ_{LT} with B/D, according to refs 5, 10, 11 and 12 is shown in Fig. 5. The varying effect of considering the pre-buckling curvature can be seen. The variation of σ_{LT} with slenderness is shown in Fig. 6, which also shows finite element results.

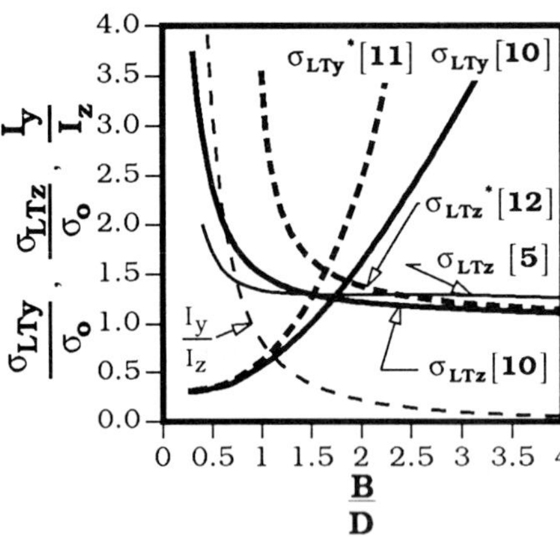

Fig. 5. Variation in critical buckling moment with breadth (D = 97mm).
* pre buckling curvatures included

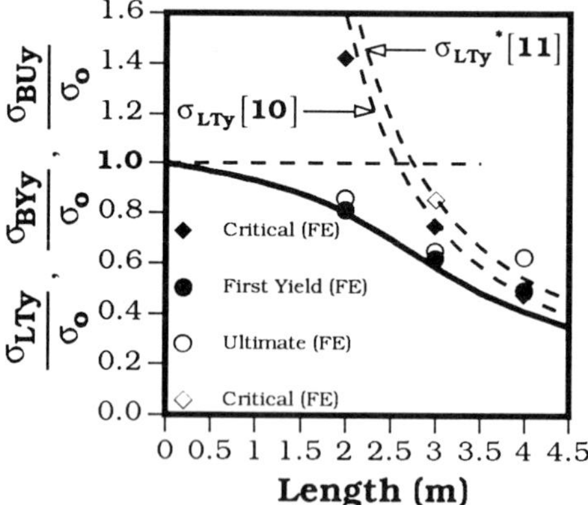

Fig. 6. Bending stress to cause first yield and failure according to proposed method (straight beam) and FE analysis.

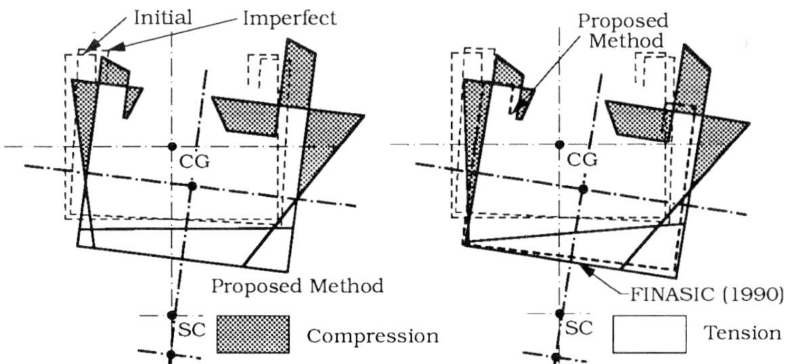

Fig. 7. Stress distribution and displacements at first yield given by proposed method (pre-buckling curvatures included) and FE analysis; (B = 122mm, D = 97mm, B_L = 17mm, D_L = 18.5mm, t = 3.0mm L = 3000mm σ_o = 390 N/mm^2).

11. The above method can also be applied to a member having imperfections affine to the critical mode, to give the stresses caused by warping restraint and by horizontal translation of the shear centre. The total stresses are then given by adding the stresses caused by vertical bending. The deformations so determined agree closely with those given by non-linear finite element analysis, but the stress distribution does not agree so well (Fig. 7, ref. 13). The calculated bending stress to cause first yield is 238 N/mm^2, compared with 247 N/mm^2 from FE analysis.

INTERACTION BETWEEN BUCKLING MODES

12. When the bending moment causes lip compression, there is a strong interaction between the two torsional buckling modes because both modes have a wavelength equal to the span. The modes do differ somewhat in respect of the relative magnitudes of rotation and translation, and the critical stress for bending is much higher than for compression. For the section of Fig. 1, and the nodal distances in the present application, modal interaction with lip buckling is a minor effect, because the lip buckling wavelength is much less than the span.

13. When the moment causes lip tension, there is no tendency for lateral torsional buckling to occur, and the moment reduces the rotation of the section. If the moment is large enough in relation to the axial force, torsional buckling is inhibited, and the interaction then approximates to that which would be expected between bending and flexural buckling. For smaller moments, the resistance varies little with moment. By arranging the chords so that the greater eccentricity moment causes lip tension, the strengths of intersecting members having equal compressive forces, are similar.

DESIGN OF DIAGONALS

14. Compression diagonals in the Conder Harley system are designed for axial force and for the eccentricity resulting from the unsupported part of the tab crease. The tensile strength is governed by the tab end distance, and where the tab ends are enclosed between the intersecting chords, the end distance is limited by the width of the chord. It is sometimes advantageous to provide sleeves for heavily loaded compression diagonals. This enables the strength to be increased without increasing the unsupported length of tab. Additional checks must then be made at the ends of the sleeve.

BEHAVIOUR OF CONNECTIONS

15. If moments are assumed to result in linearly varying forces between the washer plates and the chord table it can be shown that bolt forces given by vertical resolution of the diagonal forces are magnified by 25%, for the washer plates and bolt centres shown in Fig. 1. Bolt forces given by strain gauges bonded within small concentric holes confirmed this result. This magnification effect is not prying in the usual sense, but it happens to be the same as the allowance for prying which appears to have been made in the allowable bolt forces given in BS 5950. In designing the Conder Harley frame, the force magnification is explicitly taken into account, and the design bolt resistance has been increased accordingly.

16. Bending actions on the washer plates are extremely complex, because the forces depend not only on the diagonal forces but also on the thickness of the chord table, on the reserve of stability in the chord webs, and on whether node inserts are used. For a diagonal on square arrangement, the diagonal tabs in one direction enter between the washer plate and the table of a bottom chord (as in Fig. 1), in which case the bending moment on the plate is readily found. It may be conservatively assumed that the washer plate moments can be found in a similar manner where the tabs enter between the chords. Some allowance can however be made for chord thickness and for inserts. Shearing of the table must also be considered. Tests have shown the design assumptions to be conservative.

17. End bearing stresses according to BS 5950 Part 5 appeared to be non conservative for small end distances. This supposition was confirmed by tests on flat bars and on flattened tubes, and the design formula was adjusted accordingly. However, the reduction in bearing stress required by the code for thin plies was not applied, because the plies are confined by the washer plates.

EXPERIMENTS

18. Experiments were conducted on 9m x 9m frames consisting of full scale components (Fig. 8), to clarify the behaviour of compression and tension chords, diagonals, and connections, and to confirm the validity of design methods. Experiments were designed to cause failure of a particular component. The component was then replaced by a heavier component, or strengthened, so that failure of another component could then be achieved. Thus a succession of tests were conducted on each frame. In this way the resistances of chords, diagonals, bolts and washer plates were determined, for a range of sizes. Under-designed

bolts and washer plates were used to clarify behaviour and to establish maximum resistance. Experiments were also designed to study the strengthening effects of node inserts and of devices intended to inhibit torsional buckling. Component tests were also carried out to study web crippling, end bearing, torsional flexural buckling, and lip buckling.

Fig. 8. Diagonal on square test frame.

Fig. 9. Channel tunnel terminal canopy.

CONCLUSIONS

19. A reasonable basis for design has been established, but much scope remains for refinement to achieve further economy.

20. Considerable advances in understanding and in design methods which have general application to cold formed structures have been achieved.

21. The economy of the system has been successfully tested in the European market (see, for example, Fig. 9).

ACKNOWLEDGEMENTS

22. The authors would like to acknowledge the support and co-operation of Conder Harley Europe. The assistance received by the second author from the Commonwealth Scholarship Commission in the UK and from the University of Malta is also gratefully acknowledged.

REFERENCES

1. HARTMAN, R.B. US Patent 2284898, 2 June 1942.

2. CODD, B.E., WHITE, S.H. Eagle Centre Market refurnishment, Derby. The Structural Engineer. Volume 70, No.5, 1992.

3. CHAPMAN, J.C., BUHAGIAR, D, CODD, B.E., PRINGLE, R.J., DOWLING, P.J. Development of a space frame consisting of cold formed sections. Proc. First world conference on constructional steel design. Elsevier, December 1992.

4. PRINGLE, R. U K Patent applied for.
5. Code of Practice for design of cold formed sections. BS 5950 Part 5, 1987. BSI London.
6. YU, W.W. Cold formed steel design. John Wiley 1985.
7. BUHAGIAR, D., CHAPMAN, J.C., DOWLING, P.J. Design of C-sections against deformational lip buckling. Eleventh international conference on cold-formed steel structures. University of Missouri-Rolla, October 1992.
8. YOUNG, T. A course of lectures on natural philosophy and the mechanical arts. Royal Institution, London 1807.
9. CHAPMAN, J C., BUHAGIAR, D. The application of Young's buckling equation to design against torsional buckling. Submitted for publication ICE, November 1992.
10. TIMOSHENKO, S.P., GERE, J.M. Theory of elastic stability. McGraw Hill 1961.
11. ATTARD, M.M. General non-dimensional equations for lateral buckling. Thin-walled structures 9 (1990).
12. VACHARAJITTIPHAN, P., WOOLCOCK, S.T., TRAHAIR, N.S. Effect of in-plane deformation on lateral buckling. J. Struct. Mech 3(1) 1974.
13. BUHAGIAR, D., CHAPMAN, J.C, DOWLING, P.J. Lateral torsional buckling of channels about the minor axis under constant moment. Submitted for publication I.Struct.E., March 1993.

175. Preliminary tests on timber pole wire laced steel node space structures

D. R. GRIFFITHS, University of Surrey, UK, and T. R. N. WOOD, The John Farquharson Partnership, UK

SYNOPSIS With world-wide concern for green issues growing and timber being increasingly recognised as the most environmentally friendly of structural materials, this paper demonstrates a value enhancement use for forestry thinnings which would otherwise be put to low grade uses. The work on timber pole space frames conducted at the University of Surrey follows on from work by Huybers (refs 1,2 and 3) in the Netherlands and the Ministry of Agriculture, Fisheries and Food in the UK (refs 4 and 5). The objectives were to reduce the steel in the node, to reduce its visual intrusion and to minimise eccentricities. The work was successfully concluded by full scale tests on a small space frame.

INTRODUCTION

1. One of the main uses of timber in construction is in roofing. Due to the light loading situation, full advantage can be taken of the high strength to weight ratio of wood, the ease with which it can be shaped and jointed, its suitability for prefabrication and its lightness for transport and erection. Thus, structural timber can successfully compete with steel and concrete in the form of trussed rafters in small to medium span roofs. Longer spanning frames and arches can be made from glulam and timber space frames are often practical.

2. In some specialist roofing situations further advantages can be found from adopting a round-wood pole solution. Environmentally, debarked poles represent the most economic use of wood that is a by-product of the thinning process in the maintenance of the woodlands. The main advantages in the use of thin round-wood poles are seen to be:

(i) it is a cheap material gaining significant value enhancement if used structurally;

(ii) managed properly, forests are a renewable resource providing a continual supply of timber;

(iii) timber consumes far less energy during processing into a usable form; a sixth that of concrete and a hundredth that of profiled steel;

(iv) no energy is wasted in processing the timber into rectangular or square sections;

(v) used in space frames, timber has architectural merits over rival materials. This is especially true when careful attention is paid to making the connecting node area as compact and as neat as possible (ref. 6).

3. A further advantage to be gained from using timber poles is their strength superiority over sawn timber. Sawn timber is weaker because some of the microfibrils which give timber its tensile and compressive strength will be cut through when the section is shaped. Cutting the microfibrils also makes the section more prone to warping. A comparison of the permissible stresses for; (i) round de-barked, (ii) machine rounded and (iii) general sawn timber, is given in Table 1.

Table 1. Basic permissible stresses used in design for Scots Pine timber (N/mm^2) (After Paw and Rattray ref. 7).

	Round debarked	Machine rounded	Sawn BS 5268
Bending parallel to grain.	15.7	13.4	5.3
Tension parallel to grain.	12.6	16.7	3.2
Compression parallel to grain	9.7	8.9	6.3

4. In the UK, there has been a general resistance to the use of timber in the 20th century. However, it is in roof structures and barn frames, traditionally the province of timber, that innovations are beginning to be studied and adopted. Architecturally, round-wood structures can provide an aesthetic and natural solution.

5. These merits are improved by using the compact node and by tapering timber poles at their ends. Their development at Surrey is the subject of this paper.

6. In the United Kingdom, the advisory service of the Ministry of Agriculture, Fisheries and Food (ADAS), has undertaken projects on timber pole space frames (refs 4 and 5). Their work was based on ideas presented by Huybers in Holland, in the 1980s. Huybers (ref. 1) had demonstrated a new system for jointing round-wood poles using a tubular pin and wire lacing system. He then built a number of structures in Holland and made further developments to the steel nodes in his system (refs 2 and 3). ADAS built a small barn at their research farm near Winchester and attempted to provide a design method for UK use (ref. 4).

7. The University of Surrey has a world wide reputation in the field of space structures (refs 8 and 9) and has also carried out significant research in timber engineering (refs 10 and 11). In linking these backgrounds two students undertook projects investigating the use of parallel veneered timber (refs 12 and 13). Later Lomo and Tselingas (ref. 14) initiated the work on timber poles. Their

project culminated in the testing of a single pyramidal unit which indicated the problems in full scale structures caused by the eccentricities built up in their node.

8. The continuing research by Wood (ref. 15) as part of his Master of Engineering course firstly evaluated the effect of these eccentricities and explained the poor performance of the pyramid and secondly developed a node that minimised eccentricity and improved the effectiveness of the joint in assembly and use and benefited the appearance of the structure.

SELECTION OF A NODE SYSTEM

9. To connect the individual members of a space frame together specially developed nodes are required. These connections can often cost up to 50% of the total cost of the structure thus affecting the viability of the system. The node should be as simple and repetitive as possible to enable mass production and be easily modified for edge and corner connections.

10. An appraisal of the node systems currently available for structures in steel, aluminium and timber noted three systems for consideration in greater detail, these were the Nodus, Unistrut and Huybers systems. Huybers system for connecting the timber poles to the node was considered to be vastly superior to any alternative in terms of its strength, simplicity and visual appearance. Huybers node is in two parts. The first being the connection of the timber poles to steel plates using dowels and wire bindings, the latter holding the timber pole together and preventing it splitting. The second is the space frame connector to which the steel plates are bolted.

11. Surrey developed Huyber's work making only detail changes to the first part of his system but radically revising the second part by making the steel node a single unit. The Surrey node also tapers the ends of the poles to a set diameter enabling the poles to move closer to the hub of the joint. The steel tubular dowels and wire bindings are retained. The flat steel sections linking to the timber poles are then formed into a single unit by welding them to a central vertical tubular member. The tapered poles reduce the requirements for the outstand of the flat sections and improve the resistance of compression members to buckling.

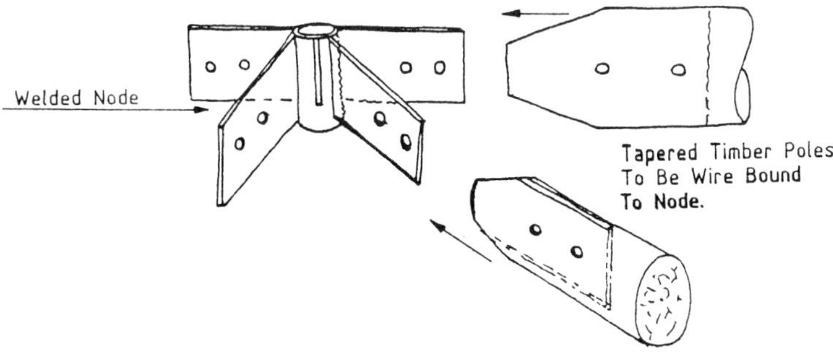

Fig. 1. Adopted node system.

12. The advantages of the new system include the following.

(i) No bolted connections are required.

(ii) The steel node can be shop welded in large numbers and transported easily for on-site erection of the space frame structure.

(iii) No eccentricity occurs in any of the members as the axial loads act through the centre of the node.

(iv) Fabrication of the truss on the ground is easy, there is sufficient flexibility in the system to insert all the members and the tubular pins before introducing the lacings which increase the rigidity of the structure.

(v) Checks on the joint are easy during assembly. If the presence of the dowel is checked when applying the bindings then the visual presence of the bindings checks the adequacy of the joint.

(vi) Edge and corner nodes are readily adapted from the standard node by omitting some of the radiating flat sections.

(vii) Shallow or deep web space frames can easily be adopted by altering the length and diameter of the poles and adjusting the thickness of steel plate. These alterations can also be incorporated to enable the frame to take higher or lower loads.

(viii) The hollow tubular node can be used to connect secondary roof elements or to support handrails for a footbridge if the space truss is inverted.

(ix) The node is architecturally suitable for a wider range of structures due to its compactness and low visual impact.

ANALYSIS AND TESTING OF MEMBER ELEMENTS

Materials

13. Due to its ready availability the timber used in the analysis and tests was Larch (*Larix decidua).* It is not indigenous to the UK, nonetheless it is now widely planted. In modern forests the trees are planted close together, (2.5 thousand plants per hectare) to ensure straightness of the tree with little or no growth of side branches. Upon reaching a height of 10m thinning occurs allowing the remaining trees to reach maturity. In the UK, thinnings production has increased in recent years and is set to increase further with farmers being persuaded by the Ministry of Agriculture to turn over productive fields to forestry. This is an attempt to try and decrease the food surpluses within the European Community. The poles used had a minimum diameter of 100mm with the taper at the ends having an initial diameter of 50mm tapering out at an angle of 26%.

14. The steel used in the nodes was grade 43 mild steel throughout. The legs were cut from 8mm plate, the node centres were made from 4.8mm thick 50mm diameter tube and the dowels used were 3.2mm thick 19mm diameter tube.

Analysis and Testing

15. Within the scope of the project the following elements of the structure were analysed and tested:

(i) compressive failure of the timber poles;
(ii) tensile failure of the timber poles;
(iii) failure of the steel dowels;
(iv) compressive / buckling failure of the steel plates;
(v) bearing failure of the steel plate.

16. Lomo and Tselingas analysed the individual elements of the structure and predicted the load carrying capacity for different failure modes. (See Table 2) They found the critical factor for design to be the buckling of the 8mm steel plate linking the timber to the node. They then completed five tensile and six compressive tests on individual members. The results correlated acceptably with the analysis and were incorporated into Wood's work.

Table 2. Analysed load capacities for the components of the space structure.

FAILURE MODE.	LOAD CAPACITY. (KN)
Buckling of the timber pole	55.9
Tensile failure of the timber	63.7
Failure in the dowel	68.9
Buckling of the steel plate (Lomo)	47.9
Buckling of the steel plate (Wood)	53.7
Bearing in the steel plate	64.4

17. Both the analysis and the individual member tests considered the application of the load without eccentricity. In building their pyramidal space frame, Lomo and Tselingas overlapped their members to form a node inducing eccentricities up to 7mm. This reduced the capacity of the frame such that the load in the critical compression members at failure was found to be 20kN, 12kN below that predicted. The back analysis of the results showed that the 12kN could be accounted for by the build up in eccentricity.

18. The back analysis concluded that the node be redesigned to eliminate eccentricity and reduce the length of the steel legs. Wood's solution achieved these requirements and significantly improved the buckling capacity of the steel plates (Table 2).

19. The tubular steel hub, fundamental to Wood's node, was too complex to analyse within the time scale of the project but simple tests were carried out to prove the adequacy of the joint in relation to other parts of the design. When loaded in tension through two opposing chord members the node carried 75kN before failing. When a similar test was undertaken in compression the failure load was reduced to 60kN.

ANALYSIS AND TESTING OF THE SPACE TRUSS UNIT

20. To confirm the predicted behaviour of the individual members and their interaction with the nodes, a full scale, three bay by one bay, space truss was constructed. This was analysed using the performance data for the individual members and the results checked by testing. The truss was based on a standard node spacing of 1.25m which gave a height of truss of 1.77m (measured between chord centrelines) and a slope of 45° in the web members.

21. The truss was modelled on a P.C. using a software program 'QSE Space'. Three loading regimes were considered in the model, these were:

(i) equal loading on each of the top three nodes;

(ii) full loading on the central top node and half loading on the outer top nodes;

(iii) a single load on the central top node.

22. The analysis predicted that the mode of failure for loading regime (ii) would be buckling of one of the steel plates, radiating from one of the top edge nodes, linked to the top horizontal members.

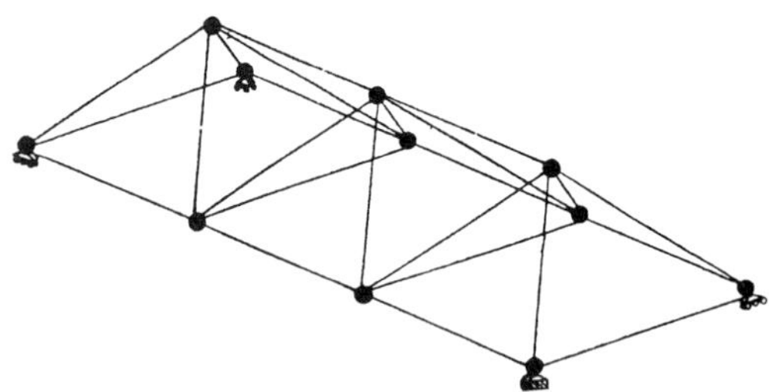

Fig. 2. Outline of modelled truss.

23. The construction of the space truss was used to check the practical adequacy of the node system. Assembly proved very easy and there were no difficulties in locating members on the rigid nodes. Once all the timber members had been positioned and fixed to the steel nodes using tubular pins, wire binding proceeded using the tool specially developed by the University of Delft (ref. 3).

24. The truss was mounted in a test rig such that one corner was held in a fixed position and the supports provided at the other corners allowed movement in the principal directions of the truss. Three loading jacks were positioned so that they could apply loads to the top three nodes independently. LVDT's were placed at strategic points on the truss and were linked to a data logger to record any truss movement during the tests.

25. Three stages of loading were applied to the truss namely:

(i) initial testing;

(ii) cyclic loading;

(iii) loading to failure.

26. Initial testing was to:

(i) bed in the truss;

(ii) establish the likely behaviour of the truss under the different loading regimes.

27. For each loading regime the truss was loaded to 20% of the predicted failure load at a steady rate, so that the full load was achieved at the end of three minutes. The load was held for short periods not exceeding three minutes then removed. The truss was allowed to recover between tests.

28. The maximum deflection at the centre of the truss occurred under loading regime (iii) and was approximately 4mm. After the first cycle, using loading regime (i), there was a 0.83mm set but after the third loading regime the set was negligible.

29. Cyclic loading was undertaken at 40% of the predicted failure load based on Eurocode test recommendations and representative of the working load. For each regime, the load was cycled three times. The loading and unloading rates were 15kN/minute. The load was held for five minutes and a recovery period of five minutes was allowed between cycles. From each cycle the truss was found to recover without any noticeable permanent deformation. Under the worst loading regime, (iii), the maximum deflection in the centre of the truss was found to be 7.22mm which represented 0.00133 of the span.

30. The truss was loaded to failure under regime (ii) in stages:

(i) load to 80% of the predicted failure and hold for twenty minutes monitoring the truss for creep deflection;

(ii) load to predicted failure load and hold for a further fifteen minutes;

(iii) load to failure.

31. During stage (i) linear elastic behaviour was noted up to 70% of the predicted failure load. Creep was measured during stages (i) and (ii) and the results, together with those at lower load levels are recorded in Fig. 3. Creep during stage (i) averaged 0.32mm/min and was slowly decreasing when stage (ii) begun. During stage (ii) the average creep had increased to 0.79mm/min and was also decreasing when stage (iii) commenced.

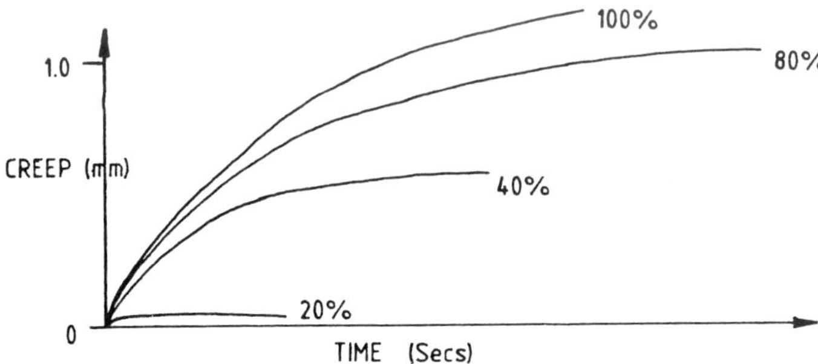

Fig. 3. Creep time behaviour of truss.

32. At stage (iii) the recorded failure load was 54.86kN with a central deflection of 19mm (see Fig. 4). This failure load was 5% greater than that predicted. The failure was initiated by buckling of the steel plate radiating from an end top node

and linked to a top horizontal member. This was as predicted by the individual member tests and model analysis (see Fig. 5).

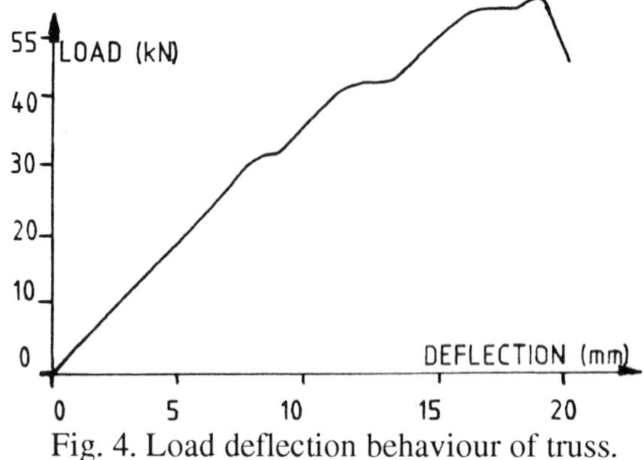

Fig. 4. Load deflection behaviour of truss.

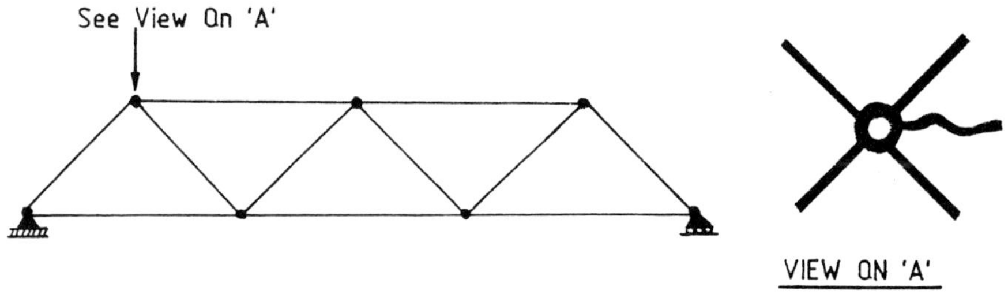

See View On 'A'

VIEW ON 'A'

Fig. 5. Failure shape of node after testing.

33. The analysis, based on the small scale tests, gave a good prediction of the failure load. However, it could not predict the deformations of the truss since it had not been possible to input load slip characteristics of the node connections. During the elastic phase, deformations were approximately twice those predicted by the analysis. Joint slip is thus very high, even compared with other timber structures, and it is suggested that an amplification factor of two be applied to the standard frame deflections at working loads.

34. Based on a deflection limit of 18mm and a factor of safety of 2.5 the safe working load for the frame would be 13.4kN/m^2 giving maximum member forces of 15kN in tension and 22kN in compression.

CONCLUSIONS

35. The conclusions that can be drawn from this paper are as follows:

36. Timber pole space frames are an excellent use of forestry thinnings, in any country. In the third world they may be considered as appropriate technology whereas in the developed world their usage would be more limited to rural or appropriate developments.

37. A node was successfully developed based on Huybers ideas for linking the timber poles to steel plates using timber dowels and wire bindings which firstly hold the dowels in place and secondly act against the splitting of the timber.

38. The node is structurally efficient with the adoption of the tubular hub eliminating eccentricities.

39. The node, due to its compactness and uniformity, is architecturally more acceptable and facilitates its use in structures with a higher architectural requirement than barns and farm outbuildings.

40. The node is easy to fabricate and only minor modifications are required for the manufacture of the edge and corner nodes.

41. The node offers the opportunity of being manufactured cheaply in bulk.

42. The nodes can also be used to support the fixings for, decking materials, roofing sheets or even hand rail posts for footbridges in rural areas.

43. With further tests and the improved preparation of the individual elements there is the possibility of an even greater efficiency and higher load capacity. For example, the timber poles could be placed closer to the node when an industrial procedure is adopted for sharpening the poles.

44. A basis for further work would be to give consideration to the hub of the node and to test a more practical structure.

REFERENCES

1. HUYBERS. P. Timber pole space frames. Space Structures 2, Elsevier Applied Science, England, 77-86, (1986).
2. HUYBERS. P. The use of round-wood poles for agricultural building structures. Paper to conference on Wooden Buildings in Agriculture, Sweden, 14-17.11, (1988).
3. HUYBERS. P. Building structures of timber poles connected with steel wire lacings. Paper to International Timber Engineering Conference, London, 2.511-2.518, (1991).
4. Ministry of Agriculture, Fisheries and Food. Pole buildings: design and construction. Technical note No TFS 815, Farm Buildings and Structures, March.(1984).
5. Ministry of Agriculture, Fisheries and Food. The use of timber in farm buildings. Technical note No TFS 812, Farm Buildings and Structures, October, (1984).
6. SADAKATA. K. Experimental study on the practical use of the space truss structure that utilised small diameter lumbers. Paper to International Timber Engineering Conference, London, 2.519-2.526, (1991).
7. PAW. C, RATTRAY. J. C., and ROBERTSON. A. M. Building with round-wood 1. Farm Building Progress (99), January, 7-11, (1990).
8. Studies in Space Structures. Edited by H.Nooshin, Multi Science Publishing Co, Brentwood, Essex, (1991).
9. Proceedings of the Third International Conference on Space Structures. Edited by H. Nooshin, Elsevier Applied Science, Barking, Essex, (1984).

10. GRIFFITHS, D.R. and ROBERTSON. B. A. Towards design stresses for orientated strand board. Proceedings of International Timber Engineering Conference Volume 3. TRADA, London, (1991).
11. GRIFFITHS. D. R. Determining the racking resistance of timber framed walls. Proceedings of Pacific Timber Engineering Conference Volume 2, IPEN2, Wellington, pp 504-512, (1992).
12. REDMOND. J. Microlam space frame. Bsc, University of Surrey, (1981).
13. KERRIDGE. N. V. Kertopou Space frame. Bsc, University of Surrey, (1982).
14. TSELINGAS. C, and LOMO. O. Roundwood space frames. BEng, Project Report, University of Surrey, (1991).
15. WOOD, T. R. N. Small diameter timber pole space frames. MEng, Dissertation, University of Surrey, (1991).

176. Conceptual design of space grids in industrial buildings

HAO ZOU, CERIS, P. R. China

SYNOPSIS. This paper presents the structural concepts and systems of space grids applied to industrial buildings. The interrelation between the basic concepts of mechanics in the structural design scheme and the architectural space-form is explained. Space grids can be used not only as sub-systems of horizontal structures for roof and floors of industrial buildings but also as sub-systems of vertical structures such as side walls or gable walls etc. All sub-systems can be considered as working together with columns in framed bents and other longitudinal structural sub-systems, thus forming an overall spatial structural system. This paper gives some design examples of space structures and also provides the principle of seismic design for industrial buildings.

THE CONCEPTS OF TOTAL-SYSTEM AND SUB-SYSTEMS

1. There is a close relation between the spatial concept of modern buildings and the principles of structural mechanics. The key point of structural design is the formation of an overall idea and the establishment of a total-system. Calculation is only the means of verification. In the course of shaping a general concept, it is very important to make inference and judgement by applying the basic concepts of structural mechanics. Designers usually aim at creating economic and advanced systems and forms of structures in accordance with the requirement of architectural functions and by making use of the mechanical properties of materials.

2. When designing a building form, the selection of appropriate basic sub-systems to substantiate the idea of a total-system will be of great importance. The major task of total-to-sub-system hierarchy is to visualise the major components of a space-form scheme, which can play the role of a basic structural sub-system. Thus a spatial component should first be thought of in terms of its structural role in the total scheme, and then its layout, size and connection details. Namely, the space-structure concept of the total system controls the requirements for details.

3. At the stage of total system design, the task is to form a structural scheme by overall relationship consideration. At the stage of sub-systems designing, the structural scheme was modified based on the local relationship. Some feedback

between total and sub-system designing is required for optimum design. Thus only after the refining of the key properties of total and sub-system schemes one's attention can turn to member design.

4. During the determination of the overall spatial idea and types of basic sub-systems, it is very important to achieve the most direct and shortest route for the transfer of force, so that a more efficient structure is obtained with less use of material.

5. Space grids are structural systems consisting of many stressed short-members and formed in certain orderly pattern. They are three dimensional structures, bearing forces in three directions. So space grids are good structural sub-systems having perfect structural forms for transfer of forces.

THE STRUCTURAL SYSTEM FOR TRANSFER OF FORCES IN SINGLE STORY INDUSTRIAL BUILDINGS

6. In the design of traditional single-story industrial buildings, the elements which carry forces only in one direction such as beams, columns, planar trusses and bar-systems etc. are generally considered as the supporting structures. A series of frames or bents are connected by various bracing to form an integrated system which can carry loads from various directions. This spatial structural system can be considered as a combination of some sub-systems.

7. One of the sub-systems is a horizontal sub-system, for instance, the roof structure which supports vertical and gravitational loads. This kind of sub-system consists of planar trusses to support the weight of roofing material, structural self-weight, live loads such as snow and ash, various weights hanging on the trusses, etc. All planar trusses are connected by horizontal top and bottom chord bracing as well as vertical bracing, to constitute what is called a horizontal sub-system. The planar trusses are also connected with columns to form framed bents. A column can also be considered as a sub-system of bars. A framed bent supports the vertical and gravitational loads transferred from the roof of a certain area, the vertical and lateral forces from cranes and other loading, and resists horizontal forces due to wind and earthquake. All of the above mentioned loads are transferred to the foundation through the columns of framed bents. Structures of bent and frames are another basic force-transferring vertical sub-system. In addition, longitudinal vertical sub-systems are formed by columns of a framed bent together with columns of side wall, cross bracing for columns, longitudinal structural systems of crane girder (including crane girders, auxiliary vertical trusses, vertical bracings and horizontal bracings at top and bottom flange of crane the girder, etc.), to transfer the longitudinal braking force, the wind force from gable walls, earthquake load and other loads in the longitudinal direction.

8. Vertical sub-systems are usually designed as longitudinal braced frames, which are very important for the transfer of forces in industrial buildings. Furthermore, gable walls, either solid or framed, can form another vertical sub-system to support lateral forces and other loads. In short, the external loads acting on an industrial building are transferred to foundations generally through horizontal and vertical sub-systems consisting of supporting structures and

bracing systems. Therefore, a building can be considered as a spatial overall system comprising of horizontal and vertical sub-systems (Fig. 1).

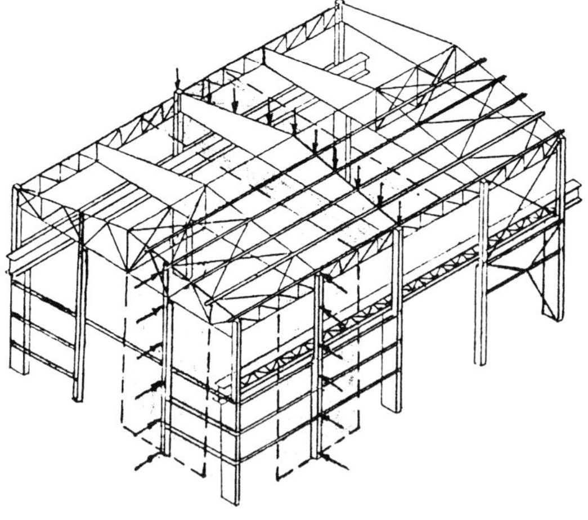

Fig. 1.

9. Because a space structure has the performance of sustaining forces from many directions, if it is considered as a combination of sub-systems, it combines an ordinary planar supporting structure and its bracing systems. For example, space grids can be used as the horizontal sub-system for a roof, instead of using ordinary roof trusses with various bracings. Likewise, vertical space grids can be used as lateral and longitudinal wall structures to replace usual wall column and bracing systems, for the transfer of wind load and other lateral forces (Fig. 2).

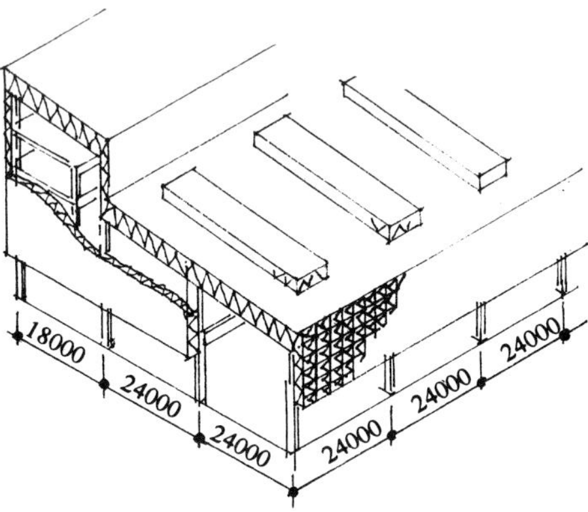

Fig. 2.

10. If the horizontal sub-systems of space grids works in co-ordination with the vertical one, the displacement of columns will be reduced enormously. Especially in higher buildings that are dominated by horizontal loads including earthquake loading and wind forces, the effect will be even more beneficial.

OVERALL DESIGN USING SPACE GRIDS AS HORIZONTAL AND VERTICAL SUB-SYSTEMS IN INDUSTRIAL BUILDINGS

Space grids as horizontal sub-systems

11. Usually the use of space grids in buildings is mostly for roofs and floors, to replace planar trusses and slab and beam structures.

12. Space grids for horizontal sub-system can be considered as two dimensional integrated structures. They carry the loading of roofs or floors in the vertical direction and, when connected with columns, they act as rigid diaphragms in the horizontal direction, bringing greater overall rigidity to the building. The joints at the top and bottom chords of a space grid are inter-connected by diagonal web-members both in the x and y directions. Systems of web-members were not only in the x direction but also in the y direction; so a spatial function can be played resulting in savings in materials.

13. In recent years, space grids are increasingly adopted for the design of industrial metallurgical plants in China. For instance, the author designed a double-layer space grid as a horizontal sub-system for the roof of a galvanised steel sheet shop in Guangdong Province. The roof covers an area of 21,600m^2. The shop has two 30m aisles. The longitudinal length of the shop is 360m, the basic longitudinal spacing of columns is 12m, while some columns are spaced at 18m.

14. The main roof of the structure was formed from a square-on-square offset grid which incorporated a slope for drainage (Fig. 3).

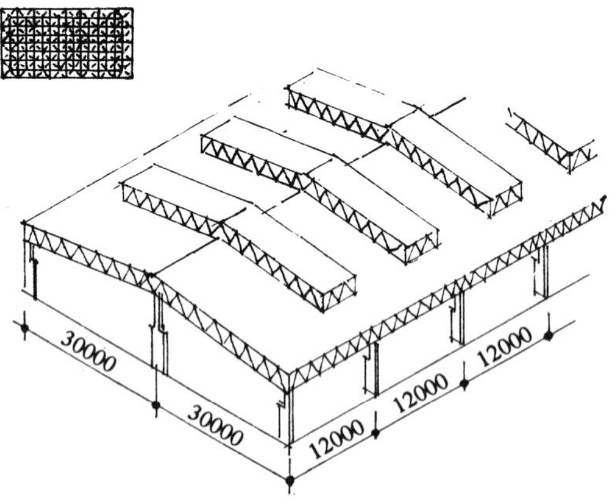

Fig. 3.

15. This factory roof has transverse skylights along the span direction. The skylights are arranged at the top of columns which support the load from the space grids. The skylights work in co-ordination with the roof to form an multi-layer space grid, and share the negative-moment effect at the top of the columns.

16. Another example is the use of braced barrel vaults as horizontal sub-systems for the roof of a stockhouse for a hot-rolled steel plate mill in the Northeast of China. The span of the mill building is 32.5m, the length is 240m and the main spacing of columns is 12m. The braced barrel vaults consist of three parts, i.e. latticed members on the shell surface, lateral edge trusses set up in the direction of the transverse span and bowstring trusses (as the stiffener ribs) at the ends of the shell in the longitudinal direction. Therefore, cylindrical braced shells are formed. The wavelength of the braced barrel vault was arranged to be in the direction of the longitudinal rows of columns. The span of shells is the same as the span of the mill building, so that the single-layer braced barrel vaults with 24 multi-waves work in conjunction with each other. The covered area of the roof was about 8000m^2 (Fig. 4). This project was designed by the author in the late seventies.

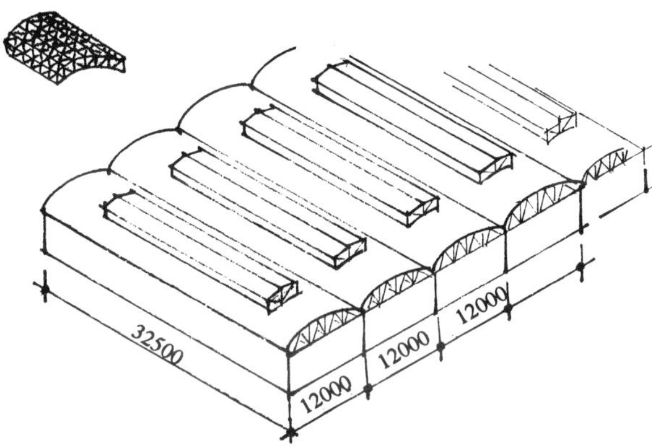

Fig. 4.

SPACE GRIDS AS VERTICAL SUB-SYSTEMS

17. Space grids can be placed vertically inserted between columns at the ends of a building to form the supporting gable walls, or inserted longitudinally between columns of bents to form the supporting structure of the side walls. The space grids so placed can be considered as working together with the framed bents of the building, so that the overall stiffness of the building will be greatly increased and significant economic savings will be achieved. Of course, in such a structural design, attention must be paid to the problems related to the connection details and opening holes for doors and windows.

SEISMIC DESIGN AND STRUCTURAL RESPONSE FOR INDUSTRIAL BUILDINGS

18. Since the violent earthquake in 1976 in Tangshan China, great attention has been paid to the seismic design of buildings. Some people consider that space grid structures have a favourable seismic response. Many space grid roofs were built in this city when the district was rebuilt. According to the Chinese Seismic Design Code for Building, the basic seismic design concept includes three levels of requirement. Firstly, buildings will not collapse when exposed to a big earthquake. Secondly, that buildings can be repaired after moderate earthquake loading; and thirdly, that buildings will not be damaged under a small earthquake loading. These are also the principal points in deciding earthquake loads. Earthquake resistance calls for energy absorption (or ductility) rather than strength only. If a building has good ductility, collapse of the building can be prevented even though the building is seriously damaged. Space grids and latticed shell structures certainly have better ductility than traditional structures. Of course, earthquake-resistant behaviour of a building is related to the stiffness, mass and damping property of the supporting structures.

19. Seismic design of industrial buildings consists of two parts; the seismic design of space grids as horizontal sub-system for roofs and the seismic design of the overall system of the building. The practical calculation of space grids as horizontal sub-systems under earthquake loads requires the consideration of the following points.

20 In the calculation of the horizontal action of an earthquake, the space grid (without skylights and other protuberances) is assumed to act as a stiffened slab, the mass of the roof is concentrated on the top of columns and the total horizontal earthquake load P acting on the space grids is determined according to the horizontal stiffness of the supporting structure. In the case of a single-mass system, the total horizontal earthquake load P is equal to the shear force Q_o at the bottom of structure. The horizontal earthquake load Pi at joint i of the space grid can be determined by the following formula:

$$Pi = \frac{Wi}{\sum_i Wi} P \qquad (1)$$

where, Wi is vertical load at joint i during the earthquake.

$\sum_i Wi$ is total vertical loads during the earthquake.

The internal forces and displacements of a space grid can be calculated by using the matrix displacement method.

21. When the vertical action of the earthquake is considered, a simplified approximate formula is used at present in China to determine the vertical earthquake loads at the joints of the space grid, this is given as:

$$Piz = \pm kWi \qquad (2)$$

where, Piz is the vertical force acting at joint i for both the up and down directions.
Wi is the vertical load at joint i when the earthquake occurs.
K is the coefficient of vertical load.
k = 0.1 for Chinese seismic intensity degree 8.
k = 0.2 for Chinese seismic intensity degree 9.
The internal forces and displacements of space girds under vertical earthquake loading can be calculated by using the matrix displacement method.

22. The seismic design of overall systems using space grids in industrial building includes:

23. Transverse calculation. The mass of the roof of the space grid is concentrated at the span centre of the framed bent. The effect of vertical earthquake loads are not considered in transverse seismic analysis of framed bents. The principle of mass concentration about the columns, walls and crane girder systems can be considered as the same as that for ordinary planar frames. Under the horizontal earthquake action, the seismic response of a structure can be analysed by earthquake response spectrum analysis where the seismic effect Sj for each mode of vibration is evaluated, and the combination of horizontal seismic effects can be determined by the following formula:

$$S = \sqrt{\sum Sj^2} \tag{3}$$

For ordinary cases, the number of modes is 3, but 5 - 7 can be adopted in multi-span buildings.

24. Longitudinal calculation. When the longitudinal framed bent of a mill building is analysed, it is usually assumed that the space grid acts as a shear slab connected with the columns in the longitudinal direction. In this case, transverse displacements of the bent are restrained and only longitudinal displacements are permitted. Longitudinal displacements of each column row at the bearing place are restrained by the corresponding shear slab of the roof. The mass of the space grid can mainly be concentrated at the centre of the roof span to verify longitudinal seismic effects. Alternatively the total mass of each roof can be concentrated at the two ends of the top chord to verify the seismic effect of the longitudinal sub-system.

CONCLUSIONS

25. The general design concept of industrial buildings is always formed by the combination of some horizontal or vertical sub-systems. In determining the overall spatial idea and basic types of sub-systems, the major subject is to achieve the best route for force transfer.

26. Space grids can be used not only as horizontal sub-systems of an industrial building, but also as its vertical sub-systems. Because a space structure has the combined function of a supporting structure and its bracing systems, so favourable economic effect can be achieved.

27. Space grids are mostly used as horizontal sub-systems for supporting structures of roofs and floors to replace planar trusses, slab and beam structures. They are used not only in various civil and public building but also in industrial buildings such as industrial metallurgical plants. Space structures can be placed in the vertical direction and inserted between columns to form vertical sub-systems; but the construction method and connection details need further study.

28. The seismic design of an industrial building includes the design of horizontal sub-systems of the roof structure, and the overall seismic design for the whole building. It is generally recognised that space grids have good ductility and are suitable for buildings in earthquake areas.

REFERENCES.

1. MAKOWSKI, Z.S. Analysis Design and Construction of Double-Layer Grids, Applied Science Publishers Ltd., London, 1981.
2. LIN, T.Y.and STOTESBURY, SIDNEY D. Structural Concepts and Systems for Architects and Engineers, John Wiley & Sons, Inc., New York, 1988.
3. SHEN, Z.Y., YIAN, H., MA K.J.and CHEN, Y.J. Structure of Space Grids, GuiZhou Renmin Publishing House, GuiZhou, 1987.

177. Some design aspects of 45 m diameter antennas for a giant metrewave radio telescope

B. JANARDAN and M. K. S. YOGI, Tata Consulting Engineers, India, and G. SWARUP and S. C. TAPDE, Tata Institute of Fundamental Research, India

SYNOPSIS. Design of large antennas is rather complex. A cost-effective design becomes possible if a satisfactory antenna configuration and the right structural shapes are selected. Proper values for aerodynamic parameters and coefficients for drag and lift must be chosen to obtain reasonably good estimates of wind loads to ensure that a safe design with respect to the dynamic and static the wind effects is obtained. Corroboration of assumed wind force parameters with satisfactory experimentation, wherever possible, enhances the confidence of the designer and the telescope user. This paper discusses these issues as well as other design aspects of the antennas for telescopes.

INTRODUCTION

1. The giant meterwave radio telescope (GMRT) is an array of thirty antennas arranged in a Y-pattern (Fig. 1). This configuration was selected by the Tata Institute of Fundamental Research (TIFR) from radio-astronomy research considerations. Fourteen antennas are located in a central square (1km x 1km) at the Y-arm junction. Six antennas are located on one arm of the Y, and five are located on each of the other two arms. Each arm is about 15 km long. The GMRT, presently under construction at the village of Khodad, about 80km from Pune, India, will be the largest steerable radio telescope of its type in the world. Each antenna is a 45m diameter paraboloid dish on an EL-AZ mount, having a near-hemispherical sky coverage. A stretched stainless steel wire mesh is used for the reflector surface. Antennas can receive signals of five different frequencies in the 38MHz to 1420MHz range.

STRUCTURAL SYSTEMS OF THE ANTENNA

2. Fig. 2 shows the structural systems of the antenna. The dish is supported on a yoke structure of steel plate construction. The conical reinforced concrete tower houses electronic equipment and supports the yoke structure, which is placed on an azimuth bearing. The circular raft foundation, which supports the tower, is designed for soil conditions at the antenna location.

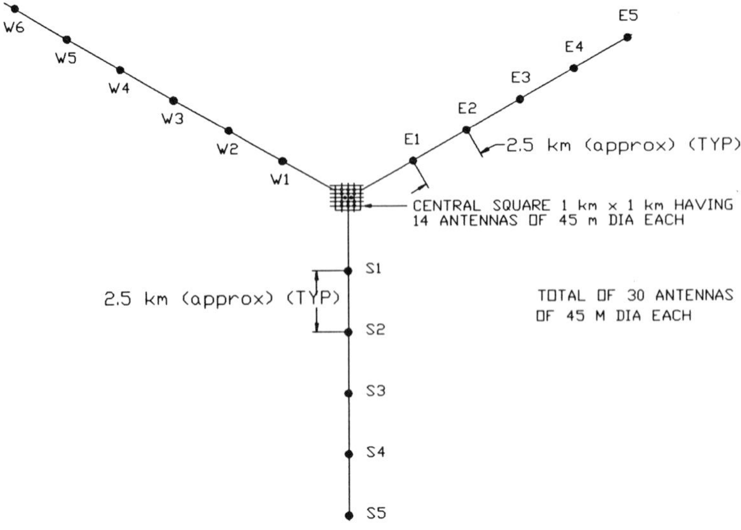

Fig. 1. Configuration of the GMRT.

3. The quadripod structure, of triangular frame tubular member (TFTM) construction, is supported at four points near to the rim of the dish. The feed cage structure, on which the antenna receivers are mounted, rotates about an axis parallel to the elevation axis of the dish.

4. The backstructure consists of sixteen TFTM parabolic radial frames (PRF) joined at their outer ends by a TFTM circular rim. Eight PRFs are terminated at their inner ends on a TFTM circular hub. The other eight PRFs are continued through the hub to an octagonal ring near the centre of the dish. Tension rods connect the bottom members of adjacent PRFs. The hub is supported at four points on the cradle structure.

5. The cradle structure is supported on two bearings on the elevation axis of the dish. A bull gear, integrally connected to the cradle structure, is moved about the elevation axis by the pinion of the elevation drive system, to point the dish to the desired direction. Elevation and azimuth drive systems are supported on the yoke structure.

6. A stretched stainless steel wire mesh reflector surface is stitched to the top wire ropes of the stainless steel wire rope truss system. A typical outer wire rope truss sector, having two radial wire ropes and twelve circumferential wire rope trusses (CWRT), is bounded by two adjacent PRFs, the rim and the hub. Radial wire ropes, connected at their outer ends to the rim and inner ends to the hub, span across the top wire ropes of the CWRTs. Top wire ropes of the CWRTs and the radial wire ropes are connected, through turnbuckles, to adjustment studs provided on PRFs, rim and hub. Link wire ropes of the CWRT system are connected, through turnbuckles, to the bottom nodes of the PRFs. Turnbuckles

are operated to obtain desired wire rope tension levels and the reflector surface profile. The inner rope truss sectors are similar to the outer ones, with radial wire ropes connected at their outer ends to the hub and inner ends to the octagonal ring.

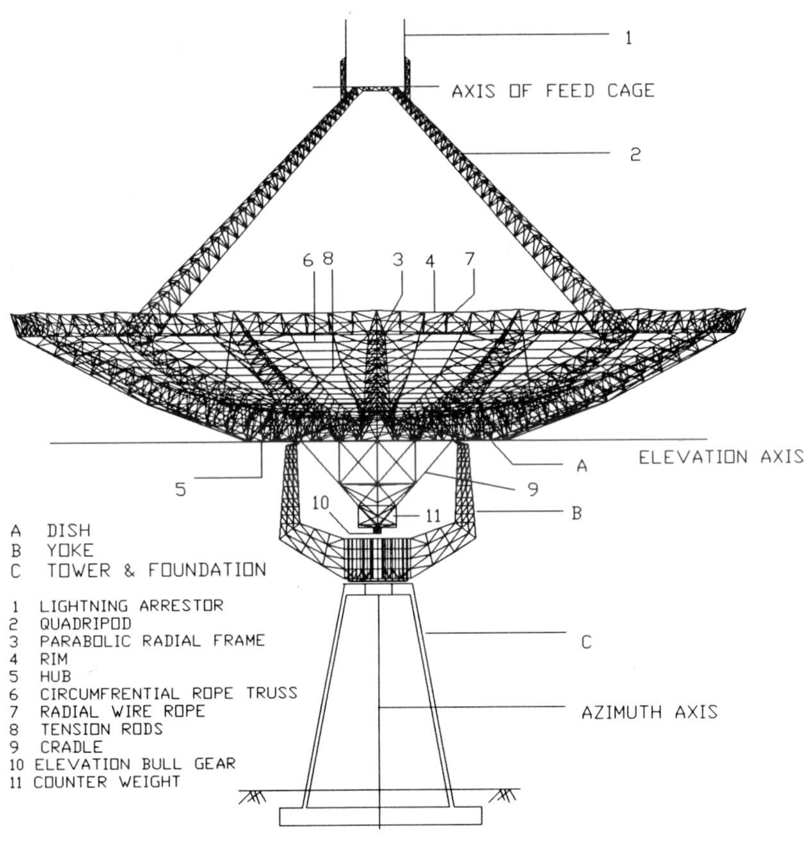

Fig. 2. Structural systems of the 45m antenna.

CHOICE OF STRUCTURAL SYSTEMS

7. The wire mesh reflector surface and its supporting wire rope truss system have a low self weight and attract a low wind load. The configuration in Fig. 3 was chosen to minimise the number of wire rope elements and to have a rope truss system with the least possibility of vibration under wind load. Wire rope elements are pretensioned to ensure that they do not slacken in winds of speeds up to 24 m/sec.

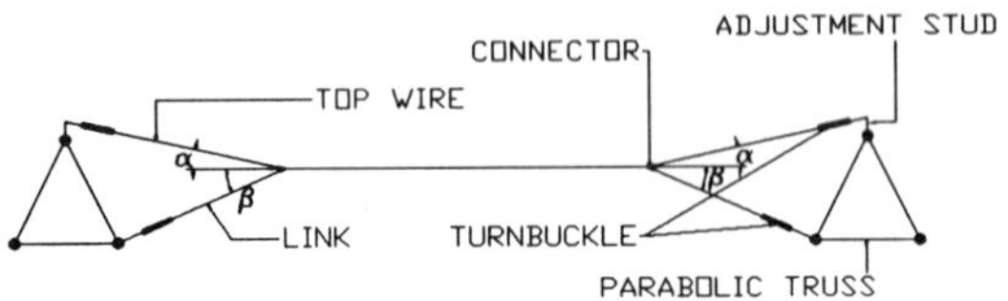

Fig. 3. Circumferential wire rope truss.

8. The members of the dish are the major contributors to its weight and wind loads. Steel tubes, with low drag compared to other structural shapes, were selected to ensure that the dish is subjected to the least wind load. The cost savings on mechanical systems (elevation and azimuth bearings, drive systems, etc.) were more than the cost increase for a tubular structure; this resulted in an antenna of lower total cost.

WIND LOAD ASPECTS

9. Wind speed and pressure. Design wind speed and pressure expressions are:

$$Vdz = K1*K2*K3*Vb \quad (1)$$
$$pz = 0.57*Vdz*Vdz \quad (2)$$

where:

Vb = 3-sec basic wind speed in m/sec
Vdz = 3-sec design wind speed in m/sec
K1 = Probability factor (risk coefficient)
K2 = Terrain, height and structure size factor for 3-sec wind speed
K3 = Topography factor
pz = Wind pressure in N/m^2

A basic wind speed of Vb = 39m/sec, for the Pune region (ref. 1), has been used. A probability factor of K1=0.95, for a thirty three year design life, and a value of K2 for terrain category-2 and class-A structure size, were used. A topography factor, of K3 = 1, for a fairly uniform terrain and an air density of 1.106 kg/m^3 (corresponding to the GMRT site elevation of 650 m above mean sea level) was used.

Drag coefficients for structural elements

10. Drag coefficients (Cd) used for estimating design wind loads on elements are based on ref. 1. Cd = 1.2 was used for the tubular members and wire ropes since the majority of members have Vdz*D < 6, and wire ropes are of the fine stranded type with Vdz*D<6 (where: D = tube or wire diameter in metres). For the RC tower a value of Cd = 0.7 was used. A value of Cd=2 was used for the yoke structure and other miscellaneous items.

Drag and lift coefficients for wire meshes

11. Curves 1and 2 of Fig. 4 show some data available for wire meshes of generally higher solidity ratios. It was felt that due to the low solidity ratios and the large ratios of wire spacing to wire diameter, values of Cd should be close to 0.6 for wind parallel to the mesh and 1.5 for wind perpendicular to the mesh. Details of the wire diameter, spacings and solidity ratios for the three GMRT meshes are given in Fig. 4. Experimental work was done at the National Aeronautical Laboratory (NAL), to obtain Cd values for these meshes. Two types of models, one for wind perpendicular to mesh and the one for wind parallel to mesh, were studied. The results of the studies (refs 4 and 5) are given in Fig. 4. Cd values of curve 3, which are generated based on the above values, were used in design.

12. Curves 1 and 2 of Fig. 5 show variation of lift coefficient with wind attack angle for wire meshes. Values of curve 2 were used in design.

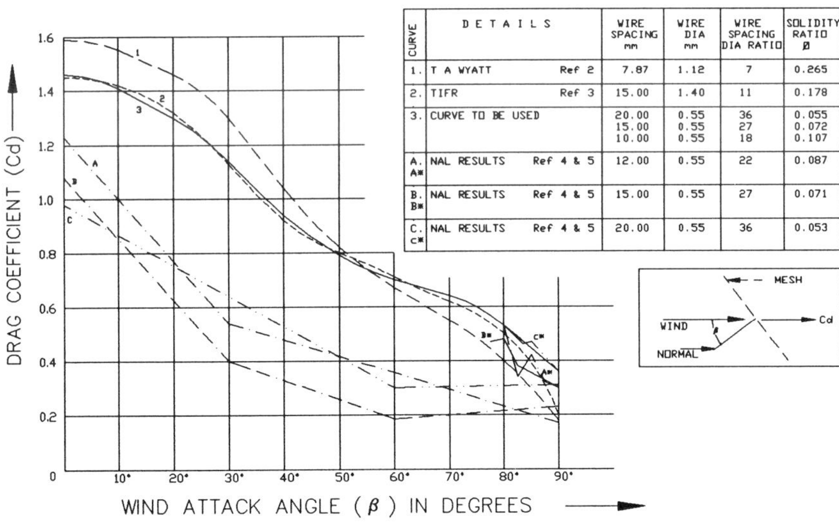

CURVE	DETAILS		WIRE SPACING mm	WIRE DIA mm	WIRE SPACING DIA RATIO	SOLIDITY RATIO Ø
1.	T A WYATT	Ref 2	7.87	1.12	7	0.265
2.	TIFR	Ref 3	15.00	1.40	11	0.178
3.	CURVE TO BE USED		20.00 15.00 10.00	0.55 0.55 0.55	36 27 18	0.055 0.072 0.107
A. A*	NAL RESULTS	Ref 4 & 5	12.00	0.55	22	0.087
B. B*	NAL RESULTS	Ref 4 & 5	15.00	0.55	27	0.071
C. C*	NAL RESULTS	Ref 4 & 5	20.00	0.55	36	0.053

Fig. 4. Drag coefficient curves, available data, NAL test results and design curves.

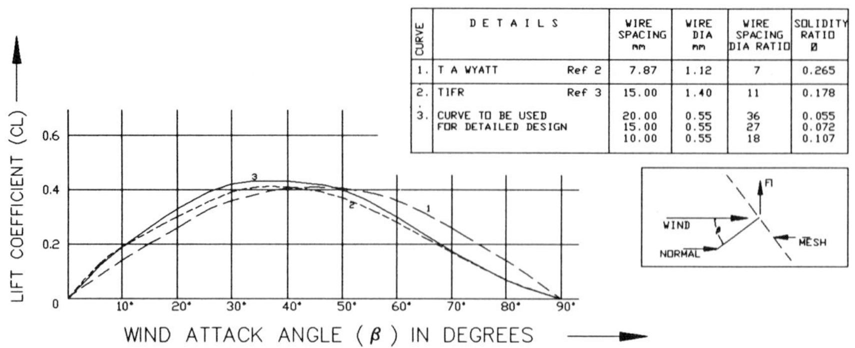

CURVE	DETAILS		WIRE SPACING mm	WIRE DIA mm	WIRE SPACING DIA RATIO	SOLIDITY RATIO Ø
1.	T A WYATT	Ref 2	7.87	1.12	7	0.265
2.	TIFR	Ref 3	15.00	1.40	11	0.178
3.	CURVE TO BE USED FOR DETAILED DESIGN		20.00 15.00 10.00	0.55 0.55 0.55	36 27 18	0.055 0.072 0.107

Fig. 5. Lift coefficient curves.

Shielding effects

13. Shielding effects reduce the wind load that can act on a structure. Two types of shielding, member to member and frame to frame, could be visualised. Ref. 1 gives the force coefficients for vertical equilateral triangular towers with round members for different solidity ratios. Tower to tower shielding effects are not available. Shielding effects for the dish cannot be considered precisely because PRFs and other triangular frames do not have an equilateral shape and are not vertical or parallel to each other. Moreover, they assume different orientations with respect to the wind direction for different antenna positions. Shielding effects were therefore considered approximately through force reduction factors.

Force reduction factors

14. Wind load on the dish is computed in a convenient manner by considering projected areas of members on a plane perpendicular to the wind direction. The force reduction factor accounting for member shielding, for front wind, defined as the ratio F1/F2, was estimated considering a PRF with the lowest solidity ratio. F1 is the force on the PRF, obtained by summing the individual panel forces, Fli, calculated with the force coefficients, given in ref. 1 for equilateral triangular towers. F2 is the force on the PRF, calculated with Cd=1.2 considering projected areas of all members of the PRF. The force reduction factor for front wind was calculated as 0.75. The force reduction factor for side wind, accounting for frame to frame shielding, is assumed as 0.68 (i.e. 90% of the force reduction factor for front wind). The force reduction factors for other conditions are assumed to be given by the following expression:

$$0.68 + 0.09 * \text{Cos A} \qquad (3)$$

where, A is the angle between the dish axis and the horizontal plane (A = 0° for the horizon position and 90° for the zenith position of the dish).

Calculated wind forces and results from the wind tunnel studies

15. Wind tunnel studies were conducted, at the University of Roorkee (UOR), to determine wind forces on the dish for various dish positions using a model (scale 1:50), fabricated by TIFR. The wind loads calculated with the force reduction factors were found to be higher (10% to 50%) than results of wind tunnel tests (ref. 6). It was decided that the force reduction factors, calculated as above, could be used for the computation of slightly conservative wind loads.

Buffeting effects

16. A few antennas in the central square are closely spaced (about 50 m apart) from radio astronomy considerations. It was felt that wind tunnel studies for these antennas would help to assess the severity of the buffeting effects. The results of wind tunnel studies at UOR indicated that no design changes are necessary for these closely spaced antennas.

DYNAMIC ALONG-WIND RESPONSE

17. Ref. 1 gives the gust factor method for computing dynamic along-wind response for structures. Two values of natural frequency, 1.5 Hz (obtained for the antenna) and 1.0 Hz (lowest antenna structure frequency limit from servo system design considerations), were used to estimate the maximum possible value for the gust factor. An overall structural damping ratio of 0.01 (ref. 1) was assumed for these calculations. The maximum gust factor was found to be sufficiently low to conclude that the dynamic along-wind effects need not be considered, if static wind loads for a 3-sec wind speed are used for the design .

DYNAMIC CROSS-WIND RESPONSE

18. Dynamic stresses could occur in a round tubular member due to cross-wind effects resulting in a narrow-band oscillatory response, a broad-band oscillatory response or response switching from narrow-band to broad-band and vice versa. Ref. 7 has been used to determine the peak cross-wind deflection response for the tubular members of the dish. Response switching effects have been ignored. The aerodynamic data and boundary conditions used for members have a large influence on the response. Suggested values for the damping ratio vary considerably from 0.003 (ref. 7) to 0.005 (ref. 8). Underestimation of the damping value could lead to an unsafe member/structure design. Based on measurements on the members of the prototype dish, a damping ratio of 0.0042 was selected. A surface roughness of 0.005mm, recommended for painted steel tubes (ref. 9), was used.

Boundary conditions and dynamic bending stresses

19. Members of the dish structure are welded at their ends. Since end conditions for members cannot be identified easily, it was decided to calculate peak responses for clamped-clamped and clamped-hinged boundary conditions to

ensure that the member is safe for both. Dynamic bending stresses were computed using the peak deflection responses and the corresponding mode shape characteristics.

Fatigue considerations

20. Fatigue effects were considered by reduced permissible bending stresses. A reduction factor for 2,000,000 cycles of stress reversals and for various ratios of maximum to minimum stress (for mild steel material conforming to IS:226), were taken from ref. 10. Stresses for a member were examined for the critical wind speed at the member natural frequency and the maximum site design wind speed for the member.

Observations and measurements on prototype dish

21. The backstructure, along with the rope truss system and the wire mesh reflector surface, was built and hoisted onto a frame of height 6m, in order to identify and sort out problems associated with the fabrication, assembly and erection of the tubular steel welded structure, as well as for assessing wind effects on the structure. The reflector surface response in high monsoon winds indicated that the rope truss system was quite stable. The damping for the steel tubular members was also measured.

OTHER DESIGN ASPECTS

22. Curvature effects were considered in the design of the PRF top chord members which were bent to a parabolic profile. All members on which a man is likely to stand during assembly, erection or maintenance stages, were designed for a concentrated mass of 75 kg at the centre.

23. The tubular structure was analysed as a pin-jointed space truss. Moments due to the rope tensions at the top of adjustment studs were considered when designing the members. Joint rigidity effects were considered through increased stress factors, ranging from 1.25 to 1.65, depending on member type and location. After the design was completed, the entire structure was analysed as a rigid-jointed frame and members checked for axial forces and moments.

Assembly and erection considerations

24. The dish was initially analysed to obtain the dead load stress pattern in members with the assumption that the entire dish is assembled from weightless members and gravity is then turned on. A margin was left in the design to cater for possible changes in the dead load stress pattern due to the assembly/erection scheme adopted. After obtaining the assembly/erection schemes from the two agencies (each entrusted with the task of completing fifteen antennas), and observing the actual implementation of these schemes, the entire dish was checked for its safety. One of the agencies preferred to assemble the dish at ground level, around the RC tower, and hoist it onto the cradle structure. The other agency preferred to assemble the entire dish on the already erected cradle

structure, using a localised tall scaffolding. However, the second agency has subsequently adopted the former scheme.

ACKNOWLEDGEMENTS

25. The authors thank Tata Consulting Engineers and the Tata Institute of Fundamental Research for their permission to publish this paper. The authors also thank the Indian Institute of Science, the National Aeronautical Laboratory, the University of Roorkee and the Indian Institute of Technology-Bombay for their valuable association, inputs and wind tunnel studies. The authors are indebted to Dr. Ben Hooghoudt, Consultant, and to the Netherlands Foundation for Radio Astronomy, for much valuable advice during the conceptual design of the antenna system.

REFERENCES

1. IS: 875 - 1987: Indian Standard Code of Practice for Design Loads (other than earthquake) for Buildings and Structures, Part - 3: Wind Loads.
2. T.A. WYATT, Annals of The New York Academy of Sciences, Vol. 116, 1964, pp222-238.
3. G. SWARUP, Drag and Lift Coefficients for Wire Mesh Screens, Internal Communication TFR(BNG): 162(11-1): TCE DES: :87, Dtd 1987-11-02.
4. Wind Tunnel Testing of Screen Elements used for a Radio Telescope Antenna (GMRT), Project Document FM 8723, National Aeronautical Laboratory, Bangalore, INDIA, October 1987.
5. Wind Loads on Sparse Meshes used in Antennae of Large Telescopes, Project Document FM 8815, National Aeronautical Laboratory, Bangalore, INDIA, September 1988.
6. Report on Wind Tunnel Studies on a 45m Diameter Antenna Dish, University of Roorkee, Roorkee, INDIA, February 1989.
7. ESDU - 85038. Circular Cylindrical Structures: Dynamic Response to Vortex Shedding, Part - 1: Calculation Procedures and Derivation, Engineering Sciences Data Unit, December 1985.
8. DANSK STANDARD DS 410.2. Dansk Ingeniorforenings Norm for Last Pa Bacrende Konstruktioner - 2 Vindlast, October 1977.
9. ESDU - 78006. Across-flow Response Due to Vortex Shedding: Isolated Circular Cylindrical Structures in Wind or Gas Flows, Engineering Sciences Data Unit, 251-259, Regent Street, London, WIR 7 AD, October 1978.
10. IS: 1915 - 1961, Indian Standard Code of Practice for Steel Bridges.

178. An investigation into the possibility of a composite space truss bridge

W. M. SEBASTIAN and R. E. MCCONNEL, University of Cambridge, UK, and P. R. HEAD, G. Maunsell & Partners, UK

SYNOPSIS. This paper introduces the ideas which lead to the conception of a composite space truss bridge. The structure, which is expected to have applications in the motorway widening situation, comprises a steel rectangular double layer space truss connected to an overhead concrete slab by headed shear studs. Resistance to corrosion is provided by a Glass Reinforced Plastic (GRP) plank enclosure around the sides and base of the truss. Enhanced structural action arising from adequate connection of the planks to the truss bottom chords is to be considered. Details of the analysis, design and testing of a proposed bridge specimen are presented, and a finite element computer program which will be used to predict the structural response to failure is discussed.

INTRODUCTION

Composite Bridges

1. Detailed views of a typical composite bridge are shown in Fig. 1. The structure comprises a reinforced concrete slab with two or more plate (I-Section) girders. Headed stud shear connectors welded onto the girder top flanges provide the slab to girder connection. K-type cross bracings between girders (not shown) transfer lateral wind forces on the deck to support bearings, and also protect the girders against lateral instability both at the bearings and at mid-span (in the latter case during deck concreting). Vertical and longitudinal stiffeners along the length of each girder (see Fig. l(b)) strengthen the webs against buckling.

2. Depth/span ratios are typically 1/20 and the slab thickness is generally 250mm and 400mm between and near girders respectively.

Deck Slab Thickness

3. The main design constraint on a conventional composite bridge's slab thickness is the magnitude of the transverse moments developed during the composite stage. Clearly, for transverse spans of less than the generally used 2.5m to 4.0m range, the slab thickness could be reduced.

Space structures 4. Thomas Telford, London, 1993

Motorway Widening

4. Wider motorways are now required in Britain in order that the ever increasing traffic flow can be accommodated. Providing extra structural supports for existing bridges which cross the carriageway will be impractical, so that demolition of existing structures will follow. Replacements will therefore have to be constructed, and significant benefits can be gained from making the replacement structures as light as possible.

5. The concrete slab accounts for a large proportion of a composite bridge's self-weight. Reducing the slab thickness would therefore be the most effective way of creating a lighter structure. From paragraph 3, this is possible only if the steel girders are placed closer together, hence more girders per unit width will be required. For a given span and construction depth, the required total flange area remains constant even for a reduced spacing, since design moments and the effective lever arm between flanges is unchanged. Moreover the slender webs used in normal design are already susceptible to shear-induced buckling. Reducing the web thickness is therefore not advisable, and the extra cost arising from additional girder web steel coupled with the increased number of girders would probably offset most of the savings from a thinner slab.

6. It would therefore appear that any realistic alternative to classical plate girder construction that might reduce slab weight is worthy of serious consideration for use in motorway widening situations.

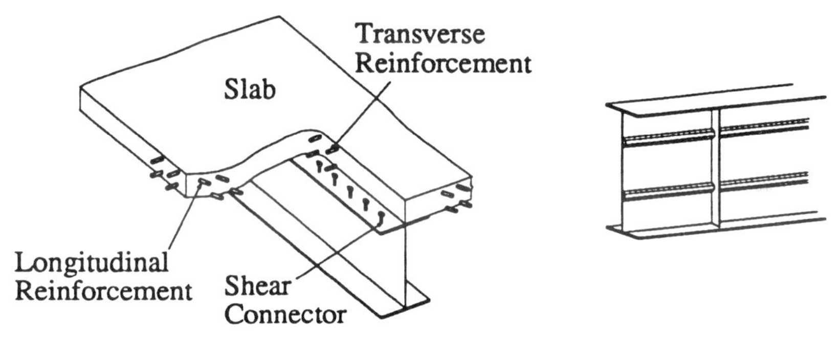

(a). Connection Details. (b). Girder Stiffener.

Fig. 1. Composite Bridge Slab and Girder Details.

Space Truss Bridges

7. Space truss bridges, to the best of the authors' knowledge, are very rare. This is primarily because bridges span in the longitudinal direction only, and therefore the two-way action of a conventional space truss would not be utilised in bridge applications. A further problem with corner-supported space trusses is that they are liable to sudden, progressive collapse, in spite of the fact that up to 25% of the members can be redundant. This was illustrated by the sudden

collapse of the Hartford Coliseum roof truss in 1978 at half its design load (ref. 1), which was initiated by the buckling of only a few highly compressed top chords, and involved only a small proportion of the redundant members.

8. Various means of protection against this type of pre-mature failure have been tested, particularly by Schmidt, refs 2,3. Probably the most successful method investigated to date has been that proposed by El Sheikh (ref. 4) which involved the use of a concrete slab acting compositely with the top chords of the truss. In his recent experimental programme, El Sheikh tested one composite and two non-composite square-on-square grids to total destruction. Design (factored) imposed floor loads were applied to all test pieces.

9. Generally, El Sheikh found that composite action enhanced the strength, ductility, and reliability of the structure. The tensile capacity of the lower chords was activated well before any discernible deterioration of the composite top chords. Failure of the composite truss was therefore gradual with adequate warning against collapse. The slab successfully restrained the steel top chords against buckling, resulting in an overall ductile behaviour.

10. Obviously, bridge load levels could be supported by suitably designed composite space trusses, and in particular, a composite form of this type of structure appears to be a possible way of significantly reducing self weight and improving peak load ductility. The amount of steel required is less than that for a conventional composite bridge and the top chords are naturally more closely spaced than the girders in a classical composite bridge, thereby allowing the use of a thinner slab. Calculations suggest that a 40% reduction in slab thickness from 250mm to 150mm could be possible.

11. The advantages of such a reduction in slab thickness would be:
(i) for moderate spans up to approximately 30m, the resulting lighter structure could be hoisted into position in a single lift. Closure of the motorway during the erection process is then kept to a minimum;
(ii) for longer spans, the steel sub-structure could be made strong enough to support the wet concrete. This would eliminate the need for props onto the motorway; and
(iii) the obvious reduction in cost of raw materials.

12. It was therefore decided that an experimental investigation of composite space truss bridges was justified, and the original intention was to test at full scale half a 20m span structure. However, reaction floor limitations dictated maximum imposed loads of approximately half those stipulated by BS5400, the British Code for bridge design. As a result, an 11.11m span bridge, loaded to half the maximum code load levels was eventually selected for construction and testing.

GRP Enclosure

13. Protection against corrosion of any structure with a significant proportion of exposed steel presents problems for the designer. This is particularly true of motorway bridges, and in the present study it is proposed to surround the truss in a Glass Reinforced Plastic enclosure. This enclosure will restrict the entry of water into, and the circulation of air around, the truss.

14. The main components of a typical GRP system are shown in Fig. 2. Direct connection between the planks is not possible, and special GRP connectors have been designed for this purpose. Each plank comprises two flanges connected together by eight thin, evenly spaced webs, all of GRP material. Specially shaped slots arc formed in the two outermost webs, and similarly shaped "ears" on the connector enable the planks to be slid into position on either side. In this way, several planks can be joined together to form a proper enclosure. This type of system has successfully been used as an enclosure around the A19 Tees Viaduct, and also for the construction of an all plastic footbridge in Aberfeldy, Scotland.

15. Investigations have shown that the GRP material modulus is roughly 10% that of steel, and the geometric distribution of material in the cross section means that the planks possess a large second moment of area about a horizontal axis through their centroid. These suggest that by proper connection to the truss bottom chords, the planks could act compositely with these members to enhance their tensile capacity. It is also anticipated that the shear stiffness of the planks will enable them to act as diagonal bracing elements between bottom chords, and so increase the torsional rigidity of the structure. The possibility of using the planks as structural elements in the present structure is therefore being considered.

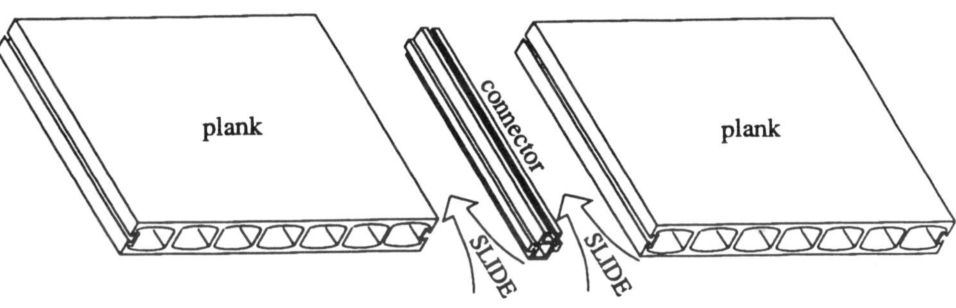

Fig. 2. Basic Components of GRP Enclosure.

DETAILS OF THE PROPOSED TEST SPECIMEN

16. An initial design was performed by G. Maunsell and Partners. This had twin back to back Tee Section chords, tubular diagonals, and joints using typical steel fabrication details. However, such a structure would not have been suited for construction using standard university machine shop practice, so alternative joint details will be employed in the actual test specimen.

17. The truss framing plan, typical section and elevation are shown in Fig. 3. Space limitations in the lab restricted plan dimensions to 11.11m span x 4.344m width.

18. Structural Tees and Circular Hollow Sections are used for the chords and webs respectively. Cold formed APHALOK floor decking, 1.2mm thick, spans

between the top chords and supports the wet concrete without any propping during the non-composite stage. Small tubes running laterally between the bottom chords complete the truss. A continuous B1131 mesh near the top of the slab ensures that design hogging moments can be developed. The longitudinal bars in the mesh are considered to be an adequate safeguard against shrinkage stresses, while 12mm re-bar acts with the permanent shuttering as bottom reinforcement. Headed shear studs, 19mm in diameter, are placed in every trough and welded through the decking plates directly on to the flanges of the top chords.

19. Chord to diagonal connection is achieved via a MERO type joint (Fig. 4). Cones are welded on to the ends of each diagonal tube, and M27 bolts pass through the cones into drilled and tapered joint blocks (described below) welded to the Tee Section chords. Each bolt has a perpendicular-to-axis hole through which a special pin is driven. The pin projects from the sides of the bolt and rests in slots cut into a sleeve which surrounds the bolt along part of its length. This allows tightening of the bolts without the need to rotate the tubular members.

20. The angles between the planes of the individual longitudinal trusses (see typical section, Fig. 3(c)) are not catered for in standard MERO construction, and consequently special joint blocks, with holes tapped to receive the M27 bolts (see Fig. 5), will be fabricated. Welds around the back and bottom edges of the blocks provide connection to the Tee Section chords. It should be noted that this detail holds for attachment of the diagonals to both top and bottom chords. The slightly unusual geometry of the end support detail, with tubular members in a vertical plane meeting the support diagonals, is taken from the original Maunsell design.

21. Details of the GRP plank to bottom chord connections are given in Fig. 6. The form of the GRP connectors is still under investigation, and the latest design comprises a Tee Section connector with the previously mentioned specially shaped "ears" to facilitate connection to the planks. The connectors are bolted to the bottom chords at regular intervals. A slightly loose fit at the plank/connector interface is inevitable, and a special sealant is inserted into this space. This sealant will prevent the entry of moisture through the gap, but, more importantly, will possess adequate bond strength so that some of the tension in the truss bottom chords can be transmitted through to the planks on either side. The manufacturing process for the planks allows them to be extruded in the lengths required for this project. Reduced structural action arising from any transverse discontinuities in the planks is therefore eliminated.

22. It should be emphasised here that the MERO type jointing system is not necessarily being proposed for use in a practical structure, and the joint details used for the test bridge are a compromise between conventional steel fabrication practice and what is feasible in a laboratory environment.

DESIGN

23. Straightforward linear elastic analyses were performed on a discrete member model of the structure for all appropriate load combinations. Design loads were based on those detailed in BS5400, reduced by a half. Grade 50D steel and 40 MPa concrete were used in the design.

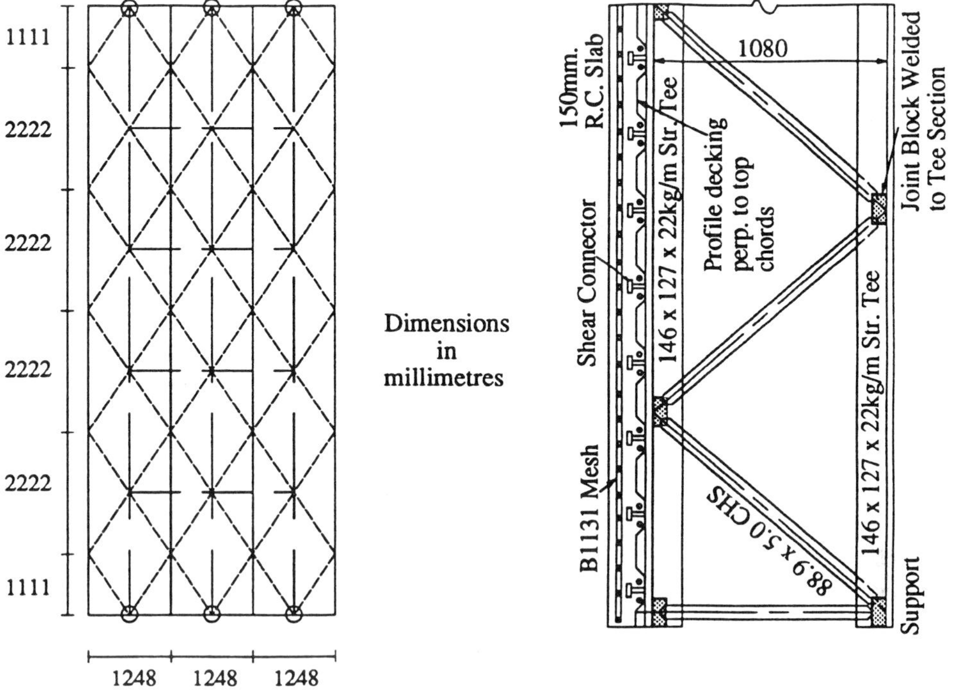

Fig. 3(a). Plan View. Fig. 3(b). Typical Part Elevation.

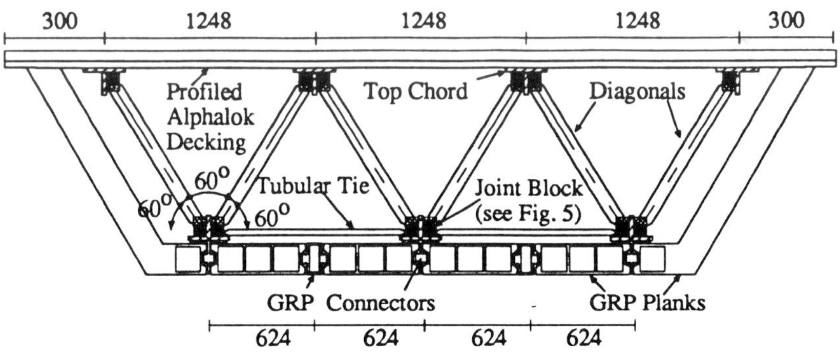

Fig. 3(c). Typical Section.

24. Superimposed loading consisted of a 50mm thick concrete surfacing, while HA loads, representing normal traffic in Great Britain, and sixteen concentrated wheel loads to model the abnormally heavy HB vehicle, were used as a basis for determining the live load levels. Individual analyses were done for each type of

live load in turn acting with the surfacing, and the HB vehicle was positioned to give maximum moment anywhere in the structure.

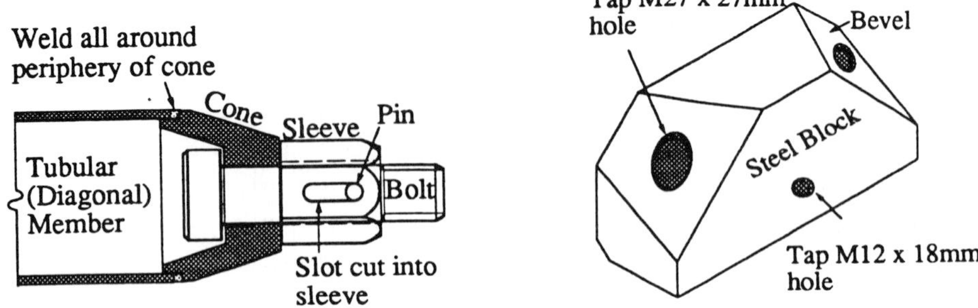

Fig. 4. Diagonal End Connection.

Fig. 5. Chord/Diagonal Joint Block.

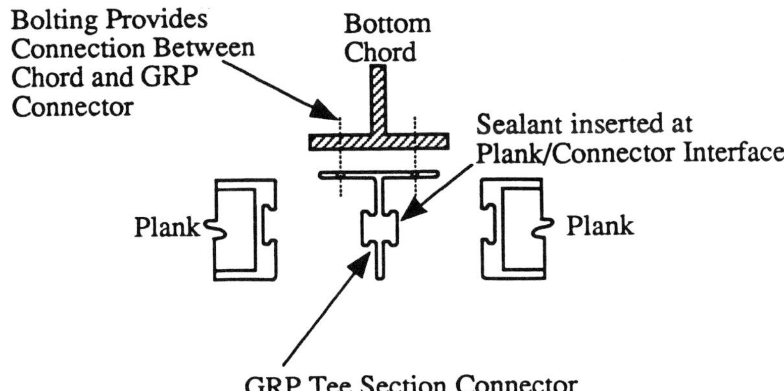

Fig. 6. Connection of Planks to Truss Bottom Chords.

25. The continuous chords were replaced in the model by line elements with both bending and axial stiffnesses, while pin-joints were assumed at the ends of the diagonals. Calculation of equivalent bottom chord properties assumed that the inverted Tee Sections acted compositely with a certain effective width of GRP plank. For the composite stage, predetermined widths of cracked slab (reduced to allow for shear lag effects) were considered to act with each top chord. Lateral elements between the top chords were proportioned to represent the cracked reinforced slab only.

26. All external loads were distributed to the top chords in the computer analysis. Output moments in these members were therefore due to both global and local effects. However only global lateral moments in the slab were directly accounted for, and the local moments had to be calculated separately by hand;

these moments being particularly significant for HB loading. Secondary moments arising from joint eccentricities were taken by the chords and slab.

27. BS 5950 Part 4 was used to check the profiled decking against yielding, web-buckling and excessive deflection at the wet concrete stage. Buckling considerations during the non-composite stage governed the design of the Tee Section top chords. Bottom chords were taken as the same size as top chords for simplicity. At the composite stage, maximum compression in support diagonals determined these member sizes, and lateral hogging moments from HB loading dictated the area of top reinforcement in the slab.

28. Punching effects in the presence of heavily localised HB loading require special consideration in a slab as thin as that being considered, and BS 5950 guidelines for punching failure in profiled decking indicated that additional tensile reinforcement near the slab soffit was required. However, tests will be conducted prior to construction of the bridge to ensure that the slab possesses adequate punching shear capacity .

29. Calculations at the serviceability limit states showed that over-stressing was not a problem anywhere in the structure.

30. Weld sizes were determined at the ultimate limit state for the maximum shear stresses due to a combined tangential force and torsional moment acting on the joint block; the torsion arising from eccentricity of the net force in the joint from the weld centroid.

31. High axial stresses in the bolts lead to the use of Grade 10.9 steel or equivalent for the bolts, cones and joint blocks, while EN16 material sufficed for the sleeves.

COMPUTER PROGRAM

32. A computer program written by El Sheikh (ref. 5) and developed for composite steel/concrete construction has been suitably modified and will be used to trace the non-linear collapse behaviour of the test specimen.

33. The Oran beam-column element (ref. 6) is used to model the steel truss behaviour. This element accounts for bi-axial bending and incorporates the destabilising effects due to high axial forces. The theory assumes no initial curvature in the truss members, and it does not model the behaviour in the post-buckling range. Experimentally measured buckling loads, and a rapidly descending post-buckling characteristic, are considered to be more representative of strut behaviour (see ref. 7), and are therefore being incorporated explicitly in the program.

34. A four-noded rectangular concrete element is used to model slab response. Both membrane and flexural effects are modelled, and the bending element is essentially a simplification of the nine-noded heterosis element developed by Hughes and Cohen (ref. 8). A smeared crack approach is adopted, and plane slab sections are assumed to remain plane but not necessarily perpendicular to the mid-surface after deformation. The latter effect is due to the (average) rotations induced by out of plane shear stresses. Full details of this element can be found in ref. 9.

35. Compression Field Theory (ref. 10) provides the constitutive equations for the concrete. Although this model assumes constant stresses on each element face, coupling between flexural and membrane effects in the slab will cause a variation of through-thickness stresses in practice. Slab elements are therefore divided into a number of thinner layers, and the above theory is applied to each layer. Provided the layers are thin enough, the actual variation of normal and shear stresses through the slab thickness can be closely followed. Shifting of the neutral axis as external loads are increased is also monitored.

36. All reinforcing bars are assumed to carry axial forces only, and these are replaced in the formulation by line elements with equivalent axial stiffness at the level of their centroids.

37. Eccentricities at the connections are modelled by short rigid links attached to the ends of all members connected to any particular joint.

38. Structural response near failure load levels is non-linear, and the Riks-Wempner method of iteration control is employed to obtain convergence on the exact solution at each increment of applied load.

39. In order to establish confidence in the use of the program, results obtained from its use were compared with published values, and good agreement was obtained.

CONSTRUCTION

40. Initial states of self-stress in the structure due to random inaccuracies in member lengths are clearly undesirable. Previous research (ref. 11) has shown that this inadvertent prestressing can induce compressive forces as high as 15% of the buckling loads in some members. Great attention is therefore being paid to the fabrication of the diagonals.

TESTING AND INSTRUMENTATION

41. All external loads will be simulated in the lab by an appropriate arrangement of jacks and spreader beams.

42. A punching shear test will be done on an individual slab specimen, with a simulated HB wheel load mid-way between slab supports. Shear connector performance will be assessed in separate push out tests.

43. For the main test piece, all loads will be first applied with the GRP planks disconnected. By comparison with the results of further tests done after the planks have been attached, the stiffening and strengthening effects of the planks can be assessed.

44. Strain gauges placed on truss elements, GRP planks, and slab reinforcement will follow the variation of stress in these members, provide data for comparison with the results of the computer program, and enable an appraisal to be made of fatigue effects near welded joints. Displacement transducers will monitor deflections at salient points along the structure.

45. The test specimen will be constructed and tested before July 1993. Processing of the data so obtained will be complete by early September 1993.

CONCLUSION

46. The use of a composite space truss bridge has been proposed for motorway widening situations. This type of structure has not been commonly used for bridges, and it is hoped that the results of the present experimental programme will confirm that, as well as being economic, it is also a viable structural alternative to conventional plate girder composite bridges.

REFERENCES

1. THORNTON, C. H. and LEW, I. P. Investigation of the causes of the Hartford Coliseum collapse. Third International Conference on Space Structures, September 1984, pp. 636-641.
2. SCHMIDT, L. C. and HANAOR, A. Force limiting devices in space trusses. Journal of the Structural Division, ASCE, 1979, vol. 105, No. ST5, pp. 939-951.
3. SCHMIDT, L. C., MORGAN, P. R., and HANAOR, A. Ultimate load testing of space trusses. Journal of the Structural Division, ASCE, 1982, vol. 108, ST6, June, pp. 1324-1355.
4. EL-SHEIKH, A. I. The effect of composite action on the behaviour of space structures. Ph. D. Thesis, Engineering Department, University of Cambridge, September 1991.
5. EL-SHEIKH, A. I. and McCONNEL, R. E. A finite element approach, Technical Report No. CUED / D-Struct. / TR. 128 (1989), University of Cambridge.
6. ORAN, C. Tangent stiffness in space frames. Journal of the Structural Division, ASCE, 1973, vol. 99, No. ST6, June, pp. 987-1001.
7. SCHMIDT, L. C., MORGAN, P. R., and STEVENS, L. K. The influence of imperfections on the behaviour of a space truss. University of Melbourne, Australia.
8. HUGHES, T. J. R. and COHEN, M. The heterosis finite element for plate bending. Computers and Structures, 1978, vol. 9, pp. 445-450.
9. COPE AND CLARK Concrete Slabs-analysis and design, pp 109-145. Elsevier Applied Science Publishers.
10. VECCHIO, F. J. and COLLINS, M. P. Predicting the response of reinforced concrete beams subjected to shear using compression field theory. ACI Structural Journal, May-June 1988, pp. 258-268.
11. AFFAN, A. Collapse of double layer space grid structures, Ph.D. Thesis, Engineering Department, University of Cambridge, September 1987.

179. Approximate expressions for the stresses in the boundary zone of spherical shells

K. KUNIEDA, Kyoto University, Japan, and Y. KOHNO, Taiyo Kogyo Corporation, Japan

SYNOPSIS. An approximate expression for the bending solution of spherical shells, in the vicinity of a boundary, is derived for axisymmetric and antisymmetric states. The expression provides a straightforward method for evaluating the stress disturbance due to boundary constraints in engineering practice and. makes clear the influence of conical angles, radius-thickness ratio and loading type. This paper also provides explicit expression of the positions of zero value and extremum value of stresses and displacement components. These positions manifest the area of the, so-called, edge zone, where stresses due to bending moments become dominant.

INTRODUCTION

1. The bending solution for spherical shells is obtained through superimposing the local solution in the vicinity of a boundary, which is referred to as the additional bending solution, on the membrane solution (ref. 1). Since the exact solution, of the additional bending solution, contains associated Legendre functions with complex order, several approximate solutions containing only elementary functions have been developed prior to the modern developments in computers. Geckeler's approximate solution (ref. 2) and Yokoo and Matsuoka's approximate solution (ref. 3) can be mentioned as representatives of approximate additional bending solutions. It has recently been confirmed, from the numerical examination, by the writers, that especially the latter almost coincides with the exact solution.

2. The additional bending solution of spherical shells is determined from Poisson's ratio, half-open angle, radius-thickness ratio, boundary conditions and loadings. It has been found, from numerical examination of the exact solution, that zero points and extremum points of the additional bending solution are almost fixed, regardless of changes in half-open angle and loadings, in the case of simple support. It is impossible to represent this phenomenon explicitly from the exact solution, and, as for the approximate solution, Geckeler's solution can be used only in the case of axisymmetric loadings. Yokoo and Matsuoka's solution is not suitable for representing the phenomenon since the expressions are somewhat

Space structures 4. Thomas Telford, London, 1993

complicated and constants in the expressions contain the Poisson's ratio, half-open angle and radius-thickness ratio as the parameters which cannot be separated. Therefore, in this paper, an approximate solution which is suitable for representing the above mentioned phenomenon is newly derived from approximate expressions of associated Legendre functions, when spherical shells are subjected to axisymmetric or antisymmetric loadings. Accordingly, it is derived not from the point of view of pursuing accuracy but from the point of view of clarifying the influences, mainly of half-open angle and radius-thickness ratio, on the additional bending solution.

3. The finite element method, rather than analytic solutions, is used as a tool for solving this kind of problem today. It is favourable that the distribution of boundary-zone stresses is easily grasped in making elements. The approximate solution, newly derived in this paper, also has this advantage.

EXACT SOLUTION

4. As a result of introducing a stress function ,φ, the governing equation of the additional bending solution can be written as (ref. 1):

$$(H_0 - \lambda_1)(H_0 - \lambda_2)\begin{Bmatrix}\psi\\ w\end{Bmatrix} = 0 \tag{1-a}$$

where

$$H_0(\ ) = (\ )_{,\phi\phi} + cot\phi\ (\ )_{,\phi} + \frac{1}{sin^2\phi}(\ )_{,\theta\theta}$$

$$\lambda_1 = -1 + i\sqrt{12(1-\nu^2)\eta^2-1}, \quad \lambda_2 = \overline{\lambda}_1$$

υ is Poisson's ratio and η is the radius-thickness ratio.

The solution of Eq. 1, the exact solution of the additional bending solution, is given below:

$$\psi = \sum_{i=1}^{2} A_i P^n{}_{\mu_i}(cos\phi)\ cosn\theta \tag{2}$$

$$w = \frac{r}{K}\sum_{i=1}^{2}\left\{A_i \frac{1}{2+\lambda_i} P^n{}_{\mu_i}(cos\phi)\right\} cosn\theta \tag{3}$$

where

$$\mu_1 = -\frac{1}{2} + \sqrt{\frac{1}{4}-\lambda_1} = k + l\,i, \quad \mu_2 = \overline{\mu_1}$$

A_i is an integral constant.

Another solution containing associated Legendre functions of the second kind, which appears when spherical shells have another boundary on the pole side, is omitted.

APPROXIMATE SOLUTION DERIVED FROM LEGENDRE FUNCTIONS

5. The approximations used are:

(a) the use of asymptotic expansion of associated Legendre functions of the first kind;

(b) the use of approximate expressions concerning the derivatives;

(c) the omission of the derivatives of lower order due to:

$$|P^n{}_{\mu_1}(cos\phi)| << |P^n{}_{\mu_1,\phi}(cos\phi)| << |P^n{}_{\mu_1,\phi\phi}(cos\phi)|$$

As for spherical shells, since $|\mu_1|$ is sufficiently large, the following expressions can be used:

$$P_{\mu_1}(cos\phi) = -\sqrt{\frac{2}{\mu_1 \pi \; sin\phi}}\; cos\left\{(\mu_1+\frac{1}{2})\phi + \frac{3}{4}\pi\right\} \tag{4}$$

$$P^1{}_{\mu_1}(cos\phi) = -\frac{\Gamma(\mu_1+2)}{\Gamma(\mu_1+\frac{3}{2})}\sqrt{\frac{2}{\pi \; sin\phi}}\; cos\left\{(\mu_1+\frac{1}{2})\phi + \frac{\pi}{4}\right\} \tag{5}$$

$$P_{\mu_1,\phi}(cos\phi) = -P^1{}_{\mu_1}(cos\phi) \tag{6}$$

$$P^1{}_{\mu_1,\phi}(cos\phi) = \mu_1(1+\mu_1)P_{\mu_1}(cos\phi) \tag{7}$$

$$P_{\mu_1,\phi\phi}(cos\phi) = -\mu_1(1+\mu_1)P_{\mu_1}(cos\phi) \tag{8}$$

$$P^1{}_{\mu_1,\phi\phi}(cos\phi) = -\mu_1(1+\mu_1)P^1{}_{\mu_1}(cos\phi) \tag{9}$$

The approximate solution of the additional bending solution, corresponding to axisymmetric or antisymmetric loadings, can be commonly written as:

$$X_i = \sqrt{\frac{2}{\pi}}\frac{e^{-l\phi}}{\sqrt{sin\phi}}\;\delta_A\;\delta_i cos\left\{(k+\frac{1}{2})\phi + \beta_A + \beta_i + s_i\right\} \tag{10}$$

where X_i represents the dimensionless stress resultants and displacements δ_i and β_i are constants determined from the Poisson's ratio and the radius thickness ratio. S_i is $\pi/4$ or $3\pi/4$. δ_A and β_A are integral constants. Sinθ or cosθ, which appears in the case of antisymmetric loadings and represents the distribution in the parallel direction, are omitted in this expression.

COMPARISON WITH EXISTING APPROXIMATE SOLUTION

6. Fig. 1 shows the deviation of Yokoo and Matsuoka's approximate solution and the writers' approximate solution from the exact solution. The ordinate represents the ratio of the maximum value of an approximate solution M_ϕ, to that of an exact solution M_ϕ, and the abscissa represents the half-open angle. The following conclusions about the error of the approximate solutions can be drawn:

(i) The error of Yokoo and Matsuoka's solution is extremely small.
(ii) The error of the writers' solution is small, around 90° with respect to half-open angle and in the domain of larger radius thickness ratios.
(iii) On the whole, the error of the writers' solution stays within 2% to 3%, and can be negligible in practical use.

The dimensionless displacement w corresponding to axisymmetric loadings is given below:

Yokoo and Matsuoka's solution

$$w = A\frac{e^{-\beta\bar{\phi}}}{\sqrt{\sin\phi}}\cos(\delta\bar{\phi}+B), \quad \bar{\phi} = \phi_0 - \phi \tag{11}$$

The writers' solution

$$w = \sqrt{\frac{2}{\pi}}\frac{e^{-l\phi}}{\sqrt{\sin\phi}}\delta_A\,\delta_7\cos\left\{(k+\frac{1}{2})\phi+\beta_A+\beta_7+\frac{3}{4}\pi\right\} \tag{12}$$

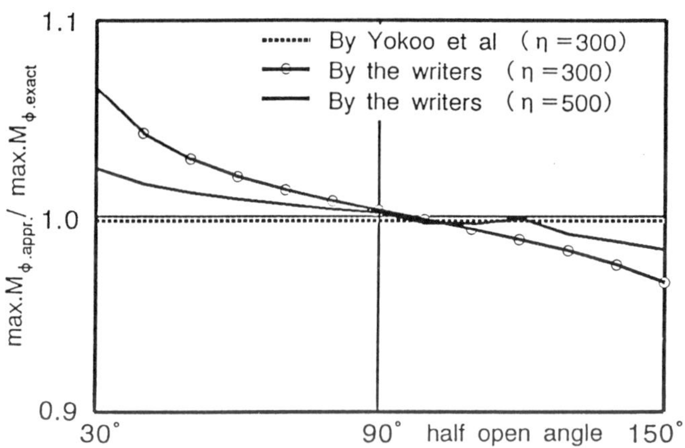

Fig. 1. The errors of the approximate solutions.

7. Although both solutions are represented as the combination of exponential functions and trigonometric functions, the constants in the expression of Yokoo and Matsuoka's solution, β and δ, contain Poisson's ratio, radius-thickness ratio and half-open angle as the parameters whereas those of the writers, k, l, δ_i and β_i, contain only Poisson's ratio and radius-thickness ratio. In addition, the other stress resultants and displacements of Yokoo and Matsuoka's solution are obtained through differentiating w or a stress function while all of stress resultants and displacements of the writers are represented as Eq. 10. Such advantages of the writers' solution represent explicitly the influences of half-open angle and radius-thickness ratio on the solution.

ZERO AND EXTREMUM POINTS OF ADDITIONAL BENDING SOLUTION

8. The boundary conditions used are given below:

Simple support

$$u = v = w = M_\phi = 0 \text{ at } \phi = \phi_0 \tag{13}$$

Rigid support

$$u = v = w = w,_\phi = 0 \text{ at } \phi = \phi_0 \tag{14}$$

where ϕ_0 is the half-open angle. As a result of applying the writers' solution to the boundary conditions, the integral constants, δ_A and β_A, can be written as:

Simple support

$$\beta_A = -(k+\frac{1}{2})\phi_0 - \gamma_n(\nu,\eta) \tag{15}$$

$$\delta_A = L_n(\nu,\eta,\phi_0)\ R_n(\nu,\phi_0;q) \tag{16}$$

Rigid support

$$\beta_A = -(k+\frac{1}{2})\phi_0 - \alpha_n(\nu,\eta,\phi_0;q) \tag{17}$$

$$\delta_A = \Lambda_n(\nu,\eta,\phi_0;q) R_n(\nu,\phi_0;q) \tag{18}$$

where q indicates that terms containing it are influenced by loadings.

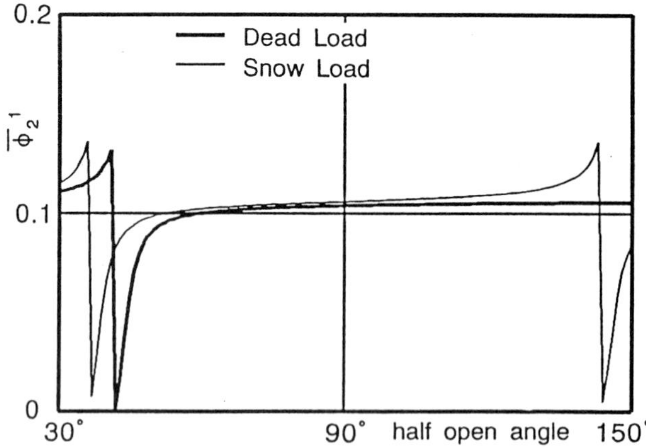

Fig. 2. The first point of $N_\theta = 0$ (fixed support).

From Eq. 10, the h zero point of the additional bending solution, which is measured from a boundary, is given below:

$$\overline{\phi}_i^{\,j} = \frac{\gamma_n - \beta_i - s_i + (\frac{1}{2}+m_i^{\,j})\pi}{-(k+\frac{1}{2})} \tag{19}$$

Rigid support

$$\overline{\phi}_i^{\,j} = \frac{\alpha_n - \beta_i - s_i + (\frac{1}{2}+m_i^{\,j})\pi}{-(k+\frac{1}{2})} \tag{20}$$

where m_i^j is an integer, and m_i^1 is an integer which makes $\overline{\phi}_i^{\,1}$ the smallest positive number.

9. The derivatives of the additional bending solution with respect to ϕ, can be obtained approximately not through directly differentiating Eq. 10 but through using Eqs 6-9 and omitting the derivatives of lower order. The jth extremum point of the additional bending solution can be given below:

Simple support

$$\overline{\phi}_i^{\,dj} = \frac{\gamma_n - \beta_{m(i)} - t_i + (\frac{1}{2}+m_i^{\,j})\pi}{-(k+\frac{1}{2})} \tag{21}$$

Rigid support

$$\overline{\phi}_i^{\,dj} = \frac{\alpha_n - \beta_{m(i)} - t_i + (\frac{1}{2}+m_i^{\,j})\pi}{-(k+\frac{1}{2})} \tag{22}$$

where m(i) represents a change in a subscript of β, and t_i is $\pi/4$ or $3\pi/4$.

10. Note that the zero points and the extremum points, in the case of simple support, are determined from only Poisson's ratio and radius-thickness ratio whereas those in the case of rigid support are determined from Poisson's ratio, radius-thickness ratio, half-open angle and loadings.

11. It is also found that the period of the zero points and the extremum points is $\pi/(k+0.5)$. Since k is determined from Poisson's ratio and radius-thickness ratio, the additional bending solution has the period of zero points and extremum points which is determined from Poisson's ratio and radius-thickness ratio.

12. The first zero point and extremum point, $\overline{\phi}_i^{\,1}$ and $\overline{\phi}_i^{\,d1}$, in the case of $\upsilon=0.3$, $\eta=300$ and simple support are given in Table 1. The following conclusions can be drawn from Eqs 19-22 and Table 1.

(i) The Eqs 19-22 give zero points and extremum points of the additional bending solution and can be used practically in comparison with the numerical examination of the exact solution.
(ii) When spherical shells are simply supported, the zero points and the extremum points of the additional bending solution are determined from only Poisson's ratio and radius-thickness ratio within the restriction of considering axisymmetric and antisymmetric loading. Half-open angle and loadings among axisymmetric and antisymmetric loadings play no part.
(iii) When spherical shells are rigidly supported, since α_n is determined from Poisson's ratio, radius-thickness ratio, half-open angle and loadings, the zero points and the extremum points are influenced by all of them.
Fig. 2 gives the first zero point of N_θ in the case of rigid support, and it shows that the zero point is influenced by half-open angle and loadings.

Table 1. The first zero points and extremum points in the case of simple support (degree).

	$\overline{\phi}_i^{\,1}$				$\overline{\phi}_i^{\,d1}$			
	$n = 0$		$n = 1$		$n = 0$		$n = 1$	
	Appr.	Exact	Appr.	Exact	Appr.	Exact	Appr.	Exact
N_ϕ	2.0	2.1	2.0	2.4	4.0	4.4	4.1	4.8
N_θ	4.0	4.1	4.0	4.1	6.1	6.2	6.1	6.2
$N_{\theta\phi}$	-	-	2.0	2.1	-	-	4.1	4.2
M_ϕ	0.0	0.0	0.0	0.0	2.0	2.1	2.0	2.1
M_θ	0.0	0.0	0.0	0.0	2.0	2.1	2.0	2.1
$M_{\theta\phi}$	-	-	6.1	6.2	-	-	0.0	0.1
Q_ϕ	2.0	2.1	2.0	2.1	4.0	4.1	4.1	4.2
Q_θ	-	-	0.0	0.1	-	-	2.0	2.2
u	-	-	0.0	0.1	-	-	2.0	2.2
v	2.0	2.1	2.0	2.1	4.0	4.1	4.1	4.2
w	4.0	4.1	4.0	4.1	6.1	6.2	6.1	6.3

MAXIMUM VALUES OF THE ADDITIONAL BENDING SOLUTION

13. Since the additional bending solution rapidly dies away in the region far from a boundary, it takes the maximum value at the boundary ($\phi = \phi_0$) or at the first extremum point ($\overline{\phi} = \overline{\phi}_i^{\,dl}$). As a result of substituting the equations giving the integral constants and the equations giving the first extremum point into Eq. 10, the absolute values of the additional bending solution at the boundary and at the first extremum point can be written as:

Simple support

$$|X_i|_{\phi=\phi_0}| = |L_n(\nu,\eta,\phi_0)||R_n(\nu,\phi_0;q)|\delta_i|\cos(-\gamma_n+\beta_i+s_i)| \qquad (23)$$

$$|X_i|_{\bar{\phi}=\bar{\phi}_i^{d1}}| = \frac{\sqrt{sin\phi_0}}{\sqrt{sin(\phi_0-\bar{\phi}_i^{d1})}} e^{l\bar{\phi}_i^{d1}} |L_n(\nu , \eta , \phi_0)|$$

$$\circ |R_n(\nu , \phi_0 ; q)| \delta_i |cos(\beta_i-\beta_{m(i)}+s_i-t_i+\frac{\pi}{2})| \quad (24)$$

Rigid support

$$| X_i|_{\phi=\phi_0} | = |\Lambda_n(\nu , \eta , \phi_0 ; q)||R_n(\nu,\phi_0;q)|\delta_i |cos(-\alpha_n+\beta_i+s_i)| \quad (25)$$

$$|X_i|_{\bar{\phi}=\bar{\phi}_i^{d1}}| = \frac{\sqrt{sin\phi_0}}{\sqrt{sin(\phi_0-\bar{\phi}_i^{d1})}} e^{l\bar{\phi}_i^{d1}} |\Lambda_n(\nu,\eta,\phi_0;q)|$$

$$\circ |R_n(\nu , \phi_0 ; q)| \delta_i |cos(\beta_i-\beta_{m(i)}+s_i-t_i+\frac{\pi}{2})| \quad (26)$$

THE MAXIMUM VALUES IN CASE OF SIMPLE SUPPORT

14. When spherical shells are simply supported, M_ϕ, M_θ, Q_θ and u take the maximum values at the first extremum points, and the other stress resultants and displacements are maximum at the boundary. When spherical shells are simply supported, $L_n(\upsilon,\eta,\phi_0)$ and $\sin(\phi_0 - \bar{\phi}_i^{dl})$ can be written as a product of a function with respect to radius-thickness ratio and a function with respect to the half-open angle, as a result of the approximation, that some functions with respect to Poisson's ratio and radius- thickness ratio which take small values compared to 1, are regarded as constants. Following this, the maximum values of the additional bending solution can also be written as a product of a function with respect to radius-thickness ratio and a function with respect to half-open angle as follows:

$$|X_i|_{\phi=\phi_0}| = H_i(\nu,\eta)\Phi(\nu,\phi_0;q) \quad (27)$$

$$|X_i|_{\bar{\phi}=\bar{\phi}_i^{d1}}| = H_i^d(\nu,\eta)\Phi^d(\nu,\phi_0;q) \quad (M_\phi ,M_\theta ,Q_\theta ,u) \quad (28)$$

where, in the case of axisymmetric loadings, H_i, Φ, H_i^d and Φ^d are given below:

$$H_i(\nu\ ,\ \eta)\ =\ \frac{\eta\ \delta_i\ |cos(-\beta_3+\beta_i+s_i-\frac{\pi}{4})|}{\delta_7\ |cos(-\beta_3+\beta_7+\frac{\pi}{2})|} \tag{29}$$

$$\Phi(\nu\ ,\ \phi_0\ ;\ q)\ =\ \frac{|R(\nu\ ,\ \phi_0\ ;\ q)|}{1\ -\ 0.035\,cot\phi_0} \tag{30}$$

$$H_i^{\,d}(\nu\ ,\ \eta)\ =\ \frac{\eta\ e^{l\bar{\phi}_i^{\,d1}}\ \delta_i\ |cos(\beta_i-\beta_{m(i)}+s_i-t_i)|}{\delta_7\ |cos(-\beta_3+\beta_7+\frac{\pi}{2})|\sqrt{cos\bar{\phi}_i^{\,d1}}} \tag{31}$$

$$\Phi^d(\nu\ ,\ \phi_0\ ;\ q)\ =\ \frac{|R(\nu\ ,\ \phi_0\ ;\ q)|}{(1\ -\ 0.035\,cot\phi_0)^{\frac{3}{2}}} \tag{32}$$

15. Fig. 3 shows the influence of half-open angle on the maximum value of an additional bending solution N_θ, in the case of simple support and own weight as the loading, and Eq. 30 and two results of the numerical examination of the exact solution, in the case of η=200 and η=400, are drawn. Since the three curves almost coincide, it is found that Eq. 30 has sufficient accuracy, and the influences of radius-thickness ratio and half-open angle on the maximum value can be separated.

16. Fig. 4 shows the influence of radius-thickness ratio on the maximum value of Q_ϕ, in the case of simple support, and Eq. 29 and two results of the numerical examination of the exact solution, in the case of ϕ_0=60 degrees, own weight and wind load as the loadings, are drawn. Since the three curves almost coincide, it is found that Eq. 29 has sufficient accuracy, and the influence of radius-thickness ratio on the maximum value is not disturbed by loadings.

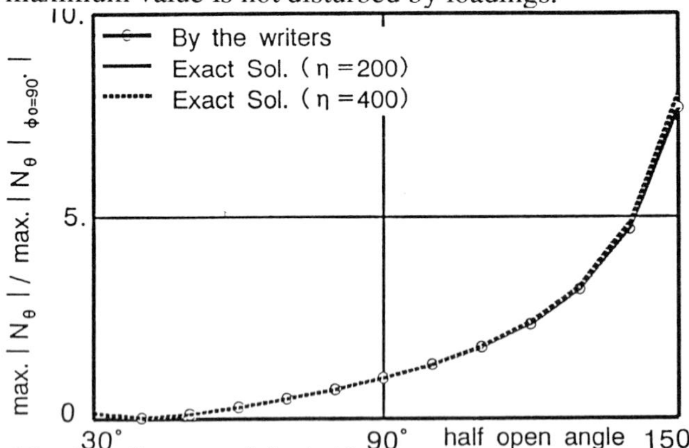

Fig. 3. Influence of the half open angle on maximum N_θ (ν = 0.3, dead load simple support).

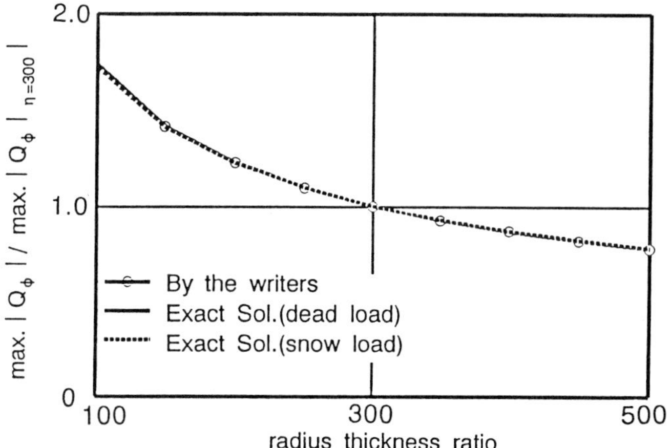

Fig. 4. Influence of radius-thickness ratio on maximum Q_{ϕ} ($\upsilon = 0.3, \phi_0 = 60^{\circ}$, simple support).

REFERENCES
1. YOKOO, Y, MATUOKA, O and KUNIEDA, H. General solutions of spherical shells in state of free vibration: Part 1 Exact solutions (written in Japanese). Trans. AIJ, Vol.83, 1963, pp.7-14
2. KRAUS, H. Thin elastic shells. John Wiley & Sons, 1967
3. MATUOKA, O. Research on shells (written in Japanese). Thesis, 1959

180. Influence of curvilinear non-homogenity and volumentary changes of the coupled problems of reinforced concrete shells

J. KAISER and J. SUMEC, Slovak Academy, Czech Federal Republic, and V. KAMENECKIJ, CHISI, The Ukraine

SYNOPSIS. In this paper the influence of curvilinear non-homogeneity and volumetric changes of the coupled problem of reinforced concrete shells is studied. Finite Element (FE) mathematical models are used for the analysis of the problem. The theoretical results obtained from the analysis of real structure are also used. From the solution of the problem some practical conclusions are formulated.

INTRODUCTION

1. During detailed analysis of the ruptured homogenisation silo it was found that combinations of temperature and dead-load were dominant factors influencing the destruction process in the cylindrical part of the structure. Temperature load along the height in the upper part of the structure was uniformly distributed with the significant quasi-linear gradient in the adjacent homogeneous material. In spite of the lower level, the temperature was relatively low due to sedimentation of homogenised materials in the jacket. The temperature variation through the thickness of the jacket is linear. From the different non-stationary temperature gradients, the highest is short-time dominant (Fig. lc, curve 6); for long time, see the stationary case of temperature loading (Fig. lc, curve 8).

2. The classical solution of the problem is based on the sudden rise in temperature at the homogenised media level, which divided the silo into two parts 1, 2 with different stiffnesses $EJ_2 << EJ_1$ (Fig. 2). The internal forces are rather different in comparison with the solution based on the assumption of isotropic properties of the material. In spite of the inclusion of the local deviations of stiffness near the media level or shear deplanations, the results made it possible to follow the other effects connected with 3D directions and moduli of material constants (especially for concrete, which depends on stress intensity).

3. Accurate results by using the FE method in a 3D formulation were obtained. The concrete structure of the silo will be divided on the tetrahedrons with curvilinear non-homogeneous orthotropic properties continuously distributed in the space occupied. Reinforcement will be modelled as one-dimensional compatibility elements.

Space structures 4. Thomas Telford, London, 1993

4. The fundamental treatise of the 3D formulation for non-homogeneous media is given in ref. 1, where the tetrahedrons T12 (3x4 parameters with the linear displacement approximations) are used.

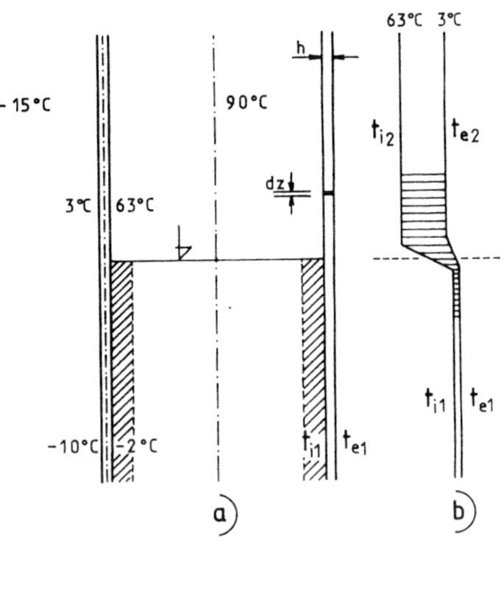

Fig. 1 The temperature changes in the structure of the jacket.

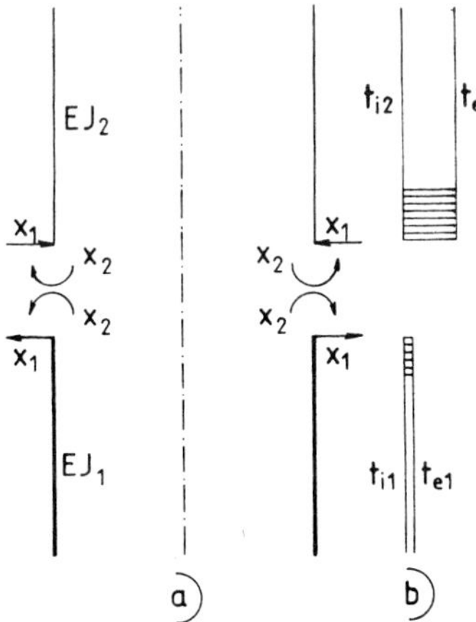

Fig. 2 The system of statically indeterminate forces.

The curvilinear orthotropy (or general curvilinear anisotropy) is defined in the centroids of the elements and linearised for each element medium. In the case of the T30, T60 and TEA8 elements (refs 2-4) the effects of continuous non-homogeneity as well as curvilinear anisotropy were approximated using the complete polynomial of the 2-degree, or 3-degree (i.e. 10 or 20 parameters respectively). Closed formulation of modified stiffness matrices (for the elements T12, T30, T48, T60 and TEA8) designed according to Argyris treatises for non-homogeneous media are detailed in ref. 8. In refs 6,7 and 9 the problems of temperature non-homogeneity or the refractory concrete media are discussed.

5. In the case of the reinforced concrete structures the effect of conditioned curvilinear non-homogeneity of concrete by using multi-parametric tetrahedrons complying with the FE method have been introduced. Reinforcement is modelled on the stiffness matrices of the individual 10 elements, which are simultaneously connected to some of the neighbouring nodal points of the concrete tetrahedrons media. The number of selected parameters is then equal to the maximum of the nodal parameters of the tetrahedron.

6. Let us suppose, that in the first step of the calculation (i.e. the linear step, where reinforced concrete is modelled as a homogeneous and isotropic material) we take the values and directions of the principal stresses in each nodal points. From the σ_1, after vertical projection of the Madryck' s parabola we obtain σ_1^* and a new corresponding modulus E_1, shown in Fig. 3a. If in the next principal direction (e.g. direction 2) the tensile principal stress is σ_2, then in $\sigma_2 \leq R_t$ the modulus of concrete E_o is unchanged. For $\sigma_2 > R_t$ we enter $E_2 = tg\eta_2$ or $E_2 \leq E_o/10$ (Fig. 3a). For reinforcement the linear branch of the stress-strain diagram up to the yield strength $\overline{R}_n$ and the plastic zones is assumed (Fig. 3b). Value $\overline{E}_o$ is changed if $\overline{\varepsilon}/\overline{\varepsilon}_n > 1$ or $\overline{\sigma}_1 > \overline{R}_n$, respectively. Then the modulus of steel material on the value $\overline{E}_1$ is reduced.

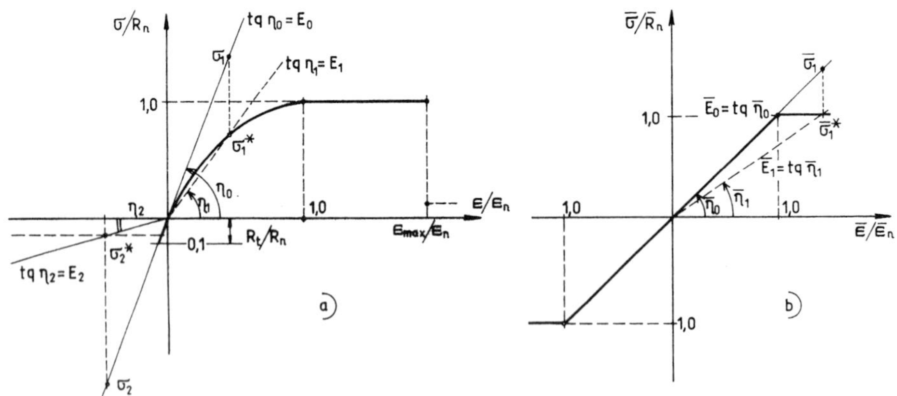

Fig. 3 The stress-strain relations.

7. At each nodal point of the concrete element (1 = 1,...,10; 1 = 1,...,20) three new moduli of elasticity in the principal directions of orthotropy are defined. At this stage of the solution, the Poisson ratio is set at ν = 0 and for shear moduli (ref. 5) we have:

$$1/G_{ij}^{l} = 1/E_i^{l} + 1/E_j^{l} \quad or \quad 1/G_{ij}^{l} = 2/\sqrt{E_i^{l}E_j^{l}} \tag{1}$$
$$i \neq j; \quad i,j = 1,2,3$$

8. From the tensor transformation we receive for each node one anisotropic characteristic in the global co-ordinate system (GCS). The curvilinear anisotropic continuum is given by the 10 (or 20) matrices of the elastic moduli. Each of the coefficients e_{ik} of the matrix E is a function of the domain of the element. The vector of base functions are defined in the form:

$$u^T = [L_1, L_2, L_3, L_4; L_1L_2, ..., L_3L_4; L_1^2L_2 - L_1L_2^2, ..., L_2L_3L_4] \equiv [\varphi_1, \varphi_2, ..., \varphi_{20}] \tag{2}$$

where the first ten basic functions will be used for the T30 element, and all twenty ones for the T60 or TEA8 elements, respectively. Each coefficient of matrix E can be write in the form:

$$e_{ij}(L_1, L_2, L_3) = \bar{e}_{ij}^{l}\varphi_l \tag{3}$$

9. In the case of T30 element we have (I = 1,...,10 nodes)

$$\bar{e}_{ij}^{n} = e_{ij}^{m} \qquad m = 1,...,4 \tag{4}$$
$$\bar{e}_{ij}^{4+k} = 4e_{ij}^{k} - 2(e_{ij}^{r} + e_{ij}^{q}) \qquad k = 1,...,6$$

and for T60 and TEA8 elements:

$$\bar{e}_{ij}^{4+k} = 9/4\,(e_{ij}^{4+k} + e_{ij}^{10+k} + e_{ij}^{r} + e_{ij}^{q}) \qquad k = 1,...,6$$
$$\bar{e}_{ij}^{10+k} = 9/4\,(3e_{ij}^{4+k} - 3e_{ij}^{10+k} - e_{ij}^{r} + e_{ij}^{q}) \qquad k = 1,...,6 \tag{5}$$
$$\bar{e}_{ij}^{16+m} = 27e_{ij}^{16+m} + 9/2\,(e_{ij}^{a} + e_{ij}^{b} + e_{ij}^{c}) - 27/4(e_{ij}^{4+a} + e_{ij}^{4+b} + e_{ij}^{4+c} + e_{ij}^{10+a} + e_{ij}^{10+b} + e_{ij}^{10+c})$$
$$m = 1,...,4$$

according to the index Tables 1,2:

Table. 1.

k	1	2	3	4	5	6
rq	12	13	14	23	24	34

Table. 2.

m	1	2	3	4
abc	123	124	134	234

10. Denoting:

$$\bar{e}_{ij} = [\bar{e}^1_{ij}, \bar{e}^2_{ij}, \ldots, \bar{e}^{20}_{ij}]^T \tag{6}$$

and partial derivatives at the corner nodal points m of the tetrahedron

$$\left(\frac{\partial e_{ij}}{\partial v}\right)_m = (e_{ij})^m_{,v} \qquad v = x, y, z; \quad (v = L_1, L_2, L_3) \qquad i, j = 1, \ldots, 6 \tag{7}$$

11. For TEA8 we define two vectors:

$$e^{AX}_{ij} = [e^1_{ij} \ldots e^4_{ij} \; (e_{ij})^1_{,x} \ldots (e_{ij})^4_{,x} \; (e_{ij})^1_{,y} \ldots (e_{ij})^4_{,z}]^T \tag{8}$$

and

$$e^{AL}_{ij} = [e^1_{ij} \ldots e^4_{ij} \; (e_{ij})^1_{,L_1} \ldots (e_{ij})^4_{,L_1} \; (e_{ij})^1_{,L_2} \ldots (e_{ij})^4_{,L_3}]^T \tag{9}$$

12. Let V_T be a co-ordinate matrix for T60 and V_A for TEA8, respectively. Denoting d_A as the transpose matrix for angular displacements; $A_T = V_T$, $B_T = V_T^{-1}$ (for T60) and $A_A = d_A V_A$, $B_A = A_A^{-1}$ (for TEA8), (refs 2, 4). The resulting polynomial coefficients are as follows, for T60 $e_{ij} = u^T \bar{e}_{ij}$ for TEA8 $e^{AL}_{ij} = u^T \bar{e}^{AL}_{ij}$, while:

$$\bar{e}_{ij} = V_T^{-1} e_{ij}, \quad \bar{e}^{AL}_{ij} = V_A \bar{e}_{ij} \tag{10}$$

13. The partial derivatives of the transposed vector of base functions appear on the well-known functional matrix P_N (3 x 10) or (3x20) in the form:

$$P_N = \begin{bmatrix} u^T_{,L_1} \\ u^T_{,L_2} \\ u^T_{,L_3} \end{bmatrix} = \begin{bmatrix} 1 & 0 & 0 & -1 & L_2 & L_3 & \cdots & \\ 0 & 1 & 0 & -1 & L_1 & 0 & \cdots & etc \\ 0 & 0 & 1 & -1 & 0 & L_1 & \cdots & \end{bmatrix} \tag{11}$$

and for the deformation vector in Cartesian co-ordinates:

$$e = [\varepsilon_x, \varepsilon_y, \varepsilon_z, \gamma_{xy}, \gamma_{yz}, \gamma_{zx}]^T = T[u_{x,x}, u_{x,y} \ldots]^T \tag{12}$$

or after introducing the matrix J^{-1}

$$e = T\, diag\left|J^{-1} J^{-1} J^{-1}\right|\, diag\left|P_N P_N P_N\right| [q_x\, q_y\, q_z]^T = TJ_3 P_3 q = P_q q \tag{13}$$

and for the basic part of the stiffness matrix k_q we have

$$k_q = \int_V P_q^T E P_q dV = \int_V P_3^T J_3^T T^T E T J_3 P_3 dV = \int_V P_3^T H' P_3 dV \tag{14}$$

Exact integrations given the closed formulation of the k_q for individual elements T30, T60 and TEA8, respectively (refs 6-8).

14. Introduction of volumetric changes, (which for T30 (T60) element are represented by the quadratic (cubic) function of the co-ordinates in the curvilinear non-homogeneous continuum) may be expressed in the form:

$$Q_0 = B_3^T P_0 = B_3^T \int_V P_3^T J_3^T T^T E e_0 dV \tag{15}$$

15. In this part (the first phase) the state of stress and strain distribution of the cylindrical homogenisation silo is analysed. Temperature loading, without including the rheological and long-time repeated thermal effects, are taken into account. The geometrical characteristics of the silo are as follows: thickness h = 0.25m and radius of the jacket a = 7.125m. The homogenisation mixture with an increased temperature 65°C with respect to the external environment temperature causing the temperature gradient of approximately 40°C in the upper part of the jacket. But in the lower part is only about 2°C. The possible short-time external temperatures (Fig. 1) induce the initial strength failure of the concrete part in the homogenisation mixture level. The critical pressure strain ($\varepsilon > \varepsilon_{max}$) was observed in some parts of the jacket.

16. The material characteristics: R_n = 15 MPa, R_t = 1.5 MPa, E_o = 27 GPa, ε_n = 0.002, ε_{max} = 0.0035, the vertical reinforcement (2x5ϕ V14mm/m[l]), the horizontal reinforcement (5ϕ V20 + 5ϕ V26mm/m[l]). The influences of two types of shear reinforcement were analysed (Fig. 4).

CONCLUSION

17. The analysed silo had a small cross-sectional area of vertical reinforcement and manufacturing tolerance was not kept. This caused the failure of the concrete elements by horizontal continuous crack propagation. After two years, 15cm deep holes approximately 6m wide were observed in the jacket. The decisive limiting load was the shear load (not adequately carried by the shear reinforcement).

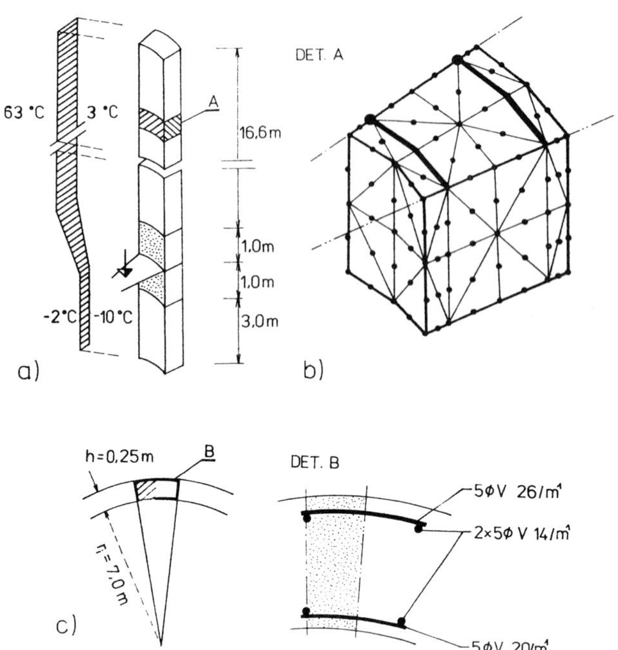

Fig. 4. Scheme of element discretization. Temperature distribution through the jacket.

REFERENCES

1. ARGYRIS, J.H. Continua and Discontinua. In: Proc. of the Matrix Methods in Structural Mechanics. Wright Patterson Air Force Base, Ohio, October 1965.

2. ARGYRIS, J.H., FRIED I. and SCHARPF, D.W. The TET 20 and TEA 8 Elements for the Matrix Displacement Method. The Aeronautical Journal of the Royal Aeronautical Society, 1968, p.618.

3. FJELD, S. The Finite Element Method in Three Dimensional Theory of Elasticity. Division of Structural Mechanics. The Technical University of Norway, 1968.

4. HOLAND, I. and BELL, K. Finite Element Methods in Stress Analysis. Tapir,Trondheim, 1969.

5. LEKHNICKLT, S. G. Theory of Elasticity of the Anisotropic Bodies. (In Russian).Gosizdat, Moscow, 1950.

6. KAISER, J. Stiffness matrix T30, modified for non-homogeneous curvilinear anisotropic continuum with non-linear volumetric changes. Part 1. (In Slovak). Stav. Cas., 1984, vol. 32, no. 3, March, 201-218.

7. KAISER, J. Stiffness matrix T30, modified for non-homogeneous curvilinear anisotropic continuum with non-linear volumetric changes. Part 2. (In Slovak). Stav.Cas., 1984, vol.32, no. 4, April, 315-328.
8. KAISER, J. Numerical analysis of space structures. (DSc Thesis).(In Slovak). Faculty of Civil Engineering, STU Bratislava, 1983, 89-263.
9. CESNAK, J. Mutual relation of components of steel and concrete bound in reinforced concrete structures. (In Slovak). Stav. Cas., 1979, vol. 27, no. 2, February, 119-138.

181. A modern design concept to increase durability and safety of natural draught cooling towers

W. B. KRATZIG and K. GRUBER, Ruhr University, Germany

SYNOPSIS. The improvement of safety and reliability of natural draught cooling towers (Fig. 1) requires highly sophisticated design tools for the analysis of the response behaviour and failure mechanisms. The evaluation of reliable statements concerning mechanical durability and safety prognoses demands the application of computer codes with realistic physically non-linear material models for both components of reinforced concrete: the concrete and the reinforcing steel. The simulation of the life-long behaviour of natural draught cooling towers requires the consideration of both the influences of time-dependent material properties and defects of concrete and steel, influencing serviceability and structural durability. Non-linear numerical analyses of crack evaluation processes up to the collapse, under the above mentioned influences, are necessary tools of modern design concepts in order to achieve the above mentioned goals for cooling tower shells.

DURABILITY OF REINFORCED CONCRETE

1. Reinforced concrete structures have to withstand the action for which they have been designed without deterioration over their life-time. Structures fulfilling this demand are called durable.

2. The absence of durability of concrete may be caused either by the environment to which the concrete is exposed or by internal effects within the concrete itself (refs 1-2). The external effects can be physical, chemical or mechanical: they may be due to high winds, weathering, occurrence of extreme temperatures, abrasion, electrolytic action and attack of natural or industrial liquids and gases. The extent of damage produced by these agents depends mainly on the quality of the concrete, although under extreme conditions any unprotected concrete will deteriorate.

3. Internal effects are for example alkali-aggregate reactions and volume changes due to the differences in thermal properties of aggregate and cement paste.

4. Under the influence of the above mentioned aggressive media the damage to concrete may lead to corrosion of steel, the second constituent of reinforced concrete. After only a few years the cross-sections and consequently the tensile

Space structures 4. Thomas Telford, London, 1993

strengths of the reinforcement bars can become greatly reduced. Fig. 2 shows the reinforced concrete shell of a cooling tower, erected in 1956, badly damaged by aggressive industrial gases.

Fig. 1. Cooling tower Herne IV (owner: STEAG AG, design: Balcke-Dürr, Ph. Holzmann AG, Krätzig & Partner, construction: Ph. Holzmann AG, 'Prüfingenieur': Prof. Dr. H. Duddeck), RGN-Photo.

Fig. 2. Damaged cooling tower shell.

CRACKING IN REINFORCED CONCRETE SHELLS

5. In reinforced concrete shells, hairline cracks are a normal phenomenon for this kind of construction. Since the introduction of limited prestressing in the first edition of DIN 4227 in 1953 (DIN: German Industrial Norm), hairline cracks have been accepted as unavoidable in reinforced as well as in prestressed concrete. New insight into this matters led to revisions of the guidelines for crack limitations (refs 3-4).

6. The ability of reinforced concrete structures to withstand all influences without serious damage occurring over their life-time depends to a high degree on the width and depth of tension cracks developed. Concrete surfaces, with wide and deep cracks, lose their protecting function for the embedded mesh of reinforcement bars.

7. The quantity, width and depth of cracks in reinforced concrete shells is influenced by the tensile strength of concrete, by the ratio of steel reinforcement and by the bond between steel and concrete. Higher tensile strengths prevent the formation of cracks, an increased ratio of reinforcement limits the deformation of the shell structure and the resulting crack extensions.

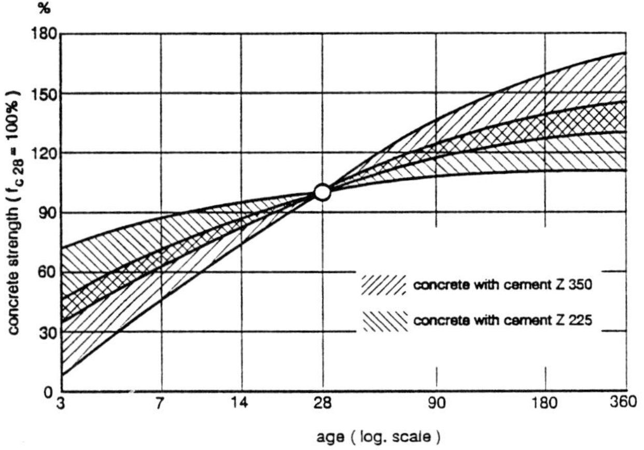

Fig. 3. Age - dependency of concrete strength according to ref. 5.

8. The tensile strength of concrete is assumed as a fraction of the compressive strength and therefore depending on the age of the structure (refs 5-6,12). After some years of service the compressive strength increases up to 140 % of the 28 day compressive strength (Fig. 3), generally assumed in design calculations. The modulus of elasticity, the tension strength and the bond between steel and concrete also change with time and lead to a more advantageous load-bearing behaviour of the shell structure.

THE NUMERICAL MODEL OF REINFORCED CONCRETE SHELLS

9. In order to simulate the real response behaviour of reinforced concrete shells by numerical analyses, a layered finite element model (Fig. 4) of high accuracy

has been developed at the Institute for Statics and Dynamics, Ruhr-University Bochum (ref. 7). Via geometrically and physically non-linear computations it admits the quantification of collapse and crack-levels hence developing into a design tool to determine safety margins and durability improvements for shell structures. Numerical supplements enable the user to consider effects of decay and damage of concrete, steel and bond during the analysis process.

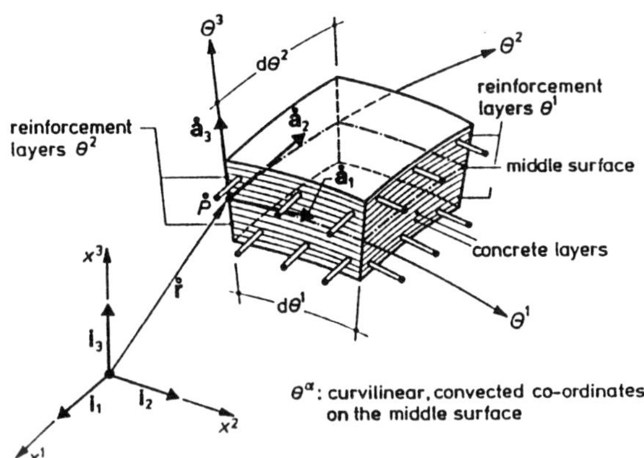

Fig. 4. Layered shell continuum.

Non-linear shell response

10. The reinforced concrete shell is modelled as a multi-layered continuum. Each layer is related to the reference surface, usually the middle surface. The individual layers are considered and described as membrane shells, fully bounded with their neighbours. The shell model is implemented into a finite element formulation (refs 8-9). To trace arbitrary load-deformation paths, incremental-iterative techniques are applied, e.g. the Riks-Wempner-Wessels technique (ref. 9). In order to avoid accuracy manipulations on the element level each strain increment will be divided into a number of sub-increments and the initial value problem solved at each material point with the required accuracy.

Modelling of the constituents

11. The non-linear behaviour of the concrete layers, particularly in the compression-compression-domain, is described by the elasto-plastic-fracturing model (EPFM) according to Bazant (ref. 10). This model considers concrete macroscopically as an isotropic material. A plastic yield surface of Drucker-Prager-type describes the plastification of the concrete in the stress space and dual to this, a fracturing surface indicates the micro-cracking in the strain space. The material constants are defined by inelastic material functions of stress and strain as well as by the strength of concrete.

12. The EPFM was originally formulated for stress states in the compression-compression-domain only. In the present work it is empirically expanded into the compression-tension-domain and to the tension-tension-domain, justified by the fact that plastification and micro-cracking occurs similarly (Fig. 5).

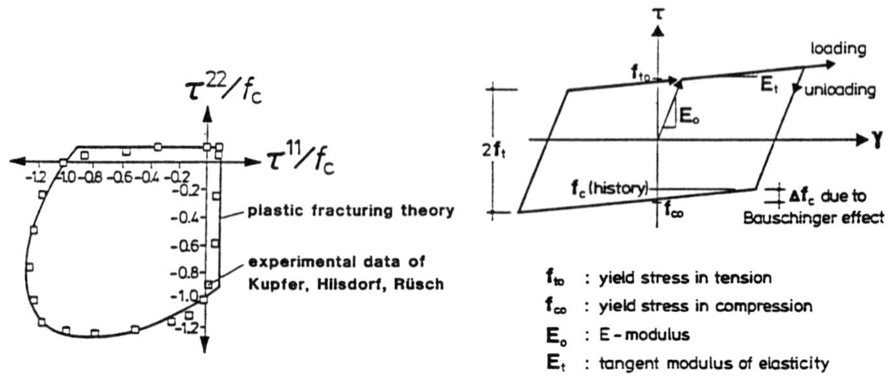

Fig. 5. Biaxial failure envelope according to KUPFER/HILSDORF/RÜSCH.

Fig. 6. Elasto-plastic material law for steel.

13. To determine the beginning of tensile cracking, a principal stress criterion is applied. After opening of a first crack, a secondary crack may develop perpendicularly to the first one. The closing of open cracks is described by a kinematic criterion (ref. 9). After the closing of cracks the concrete can again sustain compressive stresses but has irreversibly lost its tensile strength.

14. The individual bars of the reinforcement are considered as smeared into uniaxial steel layers. An elasto-plastic material law (Fig. 6) is applied to model the non-linear behaviour of the reinforcement.

The concept of tension-stiffening

15. The bond between reinforcement and surrounding concrete is assumed to be rigid. A bond slip which occurs after surpassing the maximum bond strength cannot be explicitly included in this macroscopic kind of model. It will be considered implicitly by the effect of "tension stiffening". In a cracked reinforced concrete shell, the intact concrete between each pair of adjacent tensile cracks assists partly the steel in carrying the internal tensile forces and thus contributes in parts to the stiffness of the structure (Fig. 7).

16. Several different techniques have been developed by numerous investigators (e.g. refs 11-12) for taking the tension stiffening effect into account. In the present study the method of Noakowski (ref. 13) is applied to model

properly the tension behaviour of the concrete between two cracks. This concept is based on phenomenological laws of the bond and slip-behaviour between concrete and steel. In contrast to other, simpler models, it depends not only on the concrete properties, but also on the quality, surface roughness and ratio of the steel bars. The internal tensile forces carried by the concrete over the cracks are represented by a fictitious additional reinforcement.

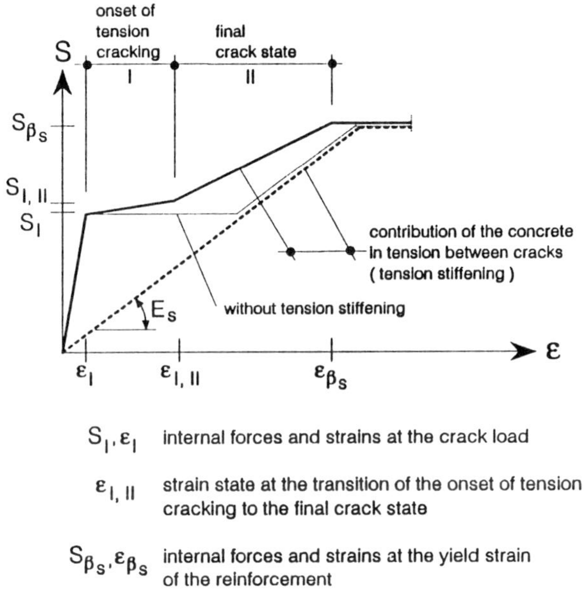

Fig. 7. Simplified representation of the principle of tension stiffening laws according to ref. 4.

17. The realistic estimation of distances and widths of tensile cracks represents a further important part of this model. Together with the calculation of crack depths it enables the designer to draw conclusions on the serviceability and structural durability of built or planned reinforced concrete shells.

DESIGN OF NATURAL DRAUGHT COOLING TOWER SHELLS

18. Safety and reliability of engineering structures, including natural draught cooling towers, depend on many factors. They are on one hand determined by the accuracy of the applied mechanical model. On the other hand they depend also on the intensity and distribution of external actions (dead-weight, wind, thermal loading etc.), the supposed strengths, the correspondence of the actual shape of the structure with the planned one (thickness and radial imperfections) and the quality of the actual construction on the site.

19. As already mentioned in §2 the mechanical durability and quality of reinforced concrete shells are influenced by environmental impacts to which the structures are exposed or by internal effects within the material itself. Shells of natural draught cooling towers are especially damaged by aggressive media like condensed vapour, dripping down the inner shell surface. Extensive sand formations, low surface strengths, spallings of concrete, and corrosion of the embedded reinforcement characterise the defects. In several areas high carbonation depths endangers the corrosion protection. Fig. 8 shows the geometry and wall thickness of a cooling tower shell, analysed in this study.

20. Since natural draught cooling towers are mostly built of reinforced concrete, in Germany the design guidelines of the individual structural members are found in the DIN 1045. Many of their elements resemble "classic" concrete members, such that their safety requirements are sufficiently formulated by the DIN. The cooling tower shells themselves, however, and to some extent the supporting columns and the foundations cannot be treated as standard elements. Their safety and reliability require a more advanced analysis (refs 14-15). The sufficient dimensioning of a cooling tower shell is assured by several rules and measures laid down in the design guidelines BTR of the VGB (German Technical Association of Large Power Plant Operators). The actual version of 1991 (ref. 16) constitutes the safety concept currently in use in Germany.

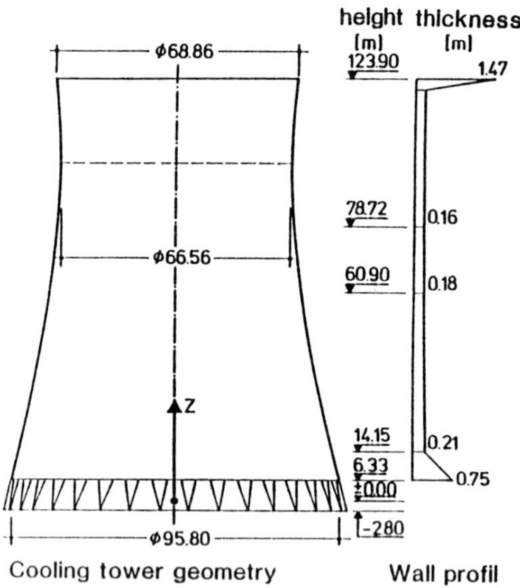

Fig. 8. Geometry of an analysed cooling tower.

A NEW CONCEPT TO INCREASE DURABILITY OF COOLING TOWER SHELLS

Linear part of the design concept

21. The first steps of the new design concept consist of the well known main rules laid down in the BTR-VGB. Input parameters are the shell geometry, loads and characteristics of the perfect materials. The required shell thickness of the natural draught cooling tower is determined by the requirements of the buckling criteria corresponding to § 1.3.4 of the BTR-VGB, using a safety factor against buckling of 5.0 according to DIN 1045; section 24.3 The dimensioning of the reinforcing mesh at both faces of the shell is based on a linear stress analysis. The quality and quantity of the inner and outer meshes should be equal, as far as possible, parts near the upper and lower shell edges being excepted. Minimum values of the shell thickness (16 - 18 cm) and ratios of reinforcement (0.30 %) are to be considered in accordance with the guidelines of the BTR, chapters 1.4.2 and 1.4.3 b.

Non-linear part of the design concept

22. Up to this point of the design concept no statements about the design quality and the mechanical durability of a certain cooling tower shell are possible. Following the linear dimensioning of the shell thickness and reinforcement bars, non-linear numerical concepts become necessary. With the help of these numerical calculations, the damaging influences during the time of operation of a natural draught cooling tower can be simulated by different load-histories.

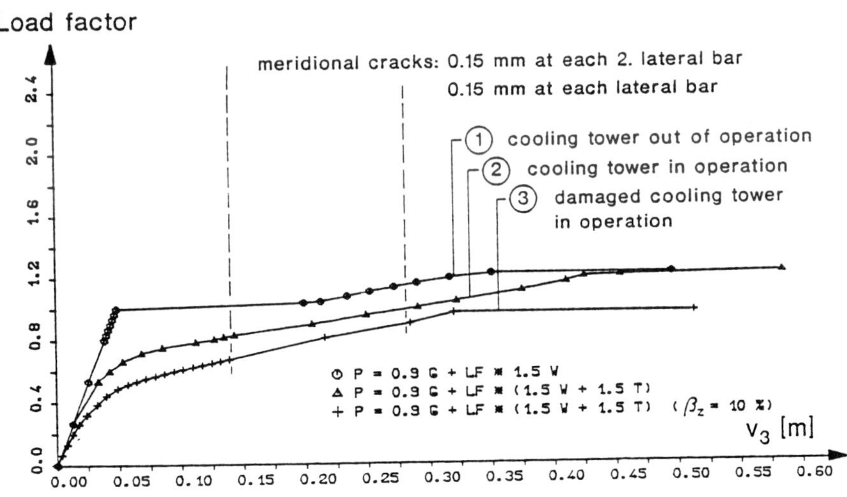

Fig. 9. Load-displacement curves for the cooling tower of Fig. 8 at $\Theta^1 = 67.5°$, z=107.7m.

23. Predamages, caused by returning strong wind loads, lead to a decrease of the load-bearing capacity of the structures. In the numerical analyses those effects

can be represented by a collective of loadings and unloadings. In addition to all important kinds of loads, quality reductions by deteriorations, like the above mentioned corrosion of concrete and steel, sand formations or spallings, can be integrated within the load-histories. Fig. 9 demonstrates some load displacement paths for the cooling tower shell of Fig. 8.

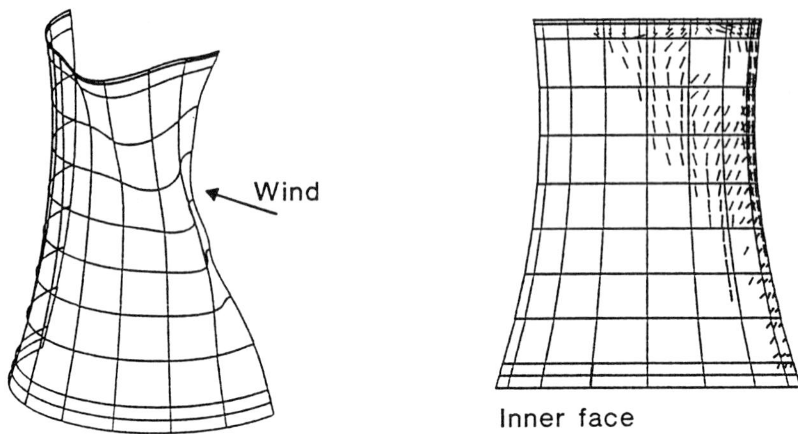

Fig. 10. Deformation and crack pattern of the cooling tower of Fig. 8 for the load-level 0.9g + 1.22·1.5 w.

24. At each point of the resulting load-displacement curves the deformations and damages of the reinforced concrete shell can be presented (Fig. 10). With the help of crack patterns in different layers of the shell structure, monitoring the resulting crack widths and depths at every load level, conclusions can be drawn concerning the reliability and mechanical durability of the structure. Local parts of the shell can be enclosed being critical with respect to tension cracks. In these areas it becomes necessary to increase the thickness of the concrete shell or the ratio of reinforcement bars.

25. This method has to be repeated, until crack widths and depths at defined load levels reach admissible values, laid down in several design guidelines. In this way all factors influencing the safety, quality and structural durability of cooling tower shells can be integrated in the non-linear design process.

CONCLUSIONS

26. To improve quality and structural durability of natural draught cooling towers, both the influences of time-dependent material properties and defects of concrete, steel and bond should be considered in the design process. A modern non-linear analysis concept, satisfying those requirements, has been developed to increase safety and reliability of these reinforced concrete shells. It enables the user to systematically design structures in order to withstand all planned conditions without deteriorations over their life-time.

REFERENCES

1. Durability of Concrete Structures, Comite Euro-International du Beton CEB, Paris, 1982.
2. NEVILLE, A.M. Properties of Concrete, 3rd edition, Pitman, London, 1981.
3. DIN 1045, Absatz 17.6, Beschränkung der Rißbreite unter Gebrauchslast (Limiting of the crack width under working loads),1988.
4. SCHIESSL, P. Grundlagen der Neuregelung zur Beschränkung der Rißbreite, Deutscher Ausschuß für Stahlbeton, Heft 400, Berlin, 1989.
5. SCHULZE, W. et al. Der Baustoff Beton, Band 1, VEB Verlag für Bauwesen, Berlin, 1972.
6. BACHMANN, H. Stahlbeton, Eidgenössische Technische Hochschule Zürich,1974.
7. BEEM, H. et al. FEMAS, - Finite Element Moduln Allgemeiner Strukturen, Benutzerhandbuch Version 90.1, Lehrstuhl für Statik und Dynamik, Ruhr-Universität Bochum, 1991.
8. BASAR, Y., KRÄTZIG, W.B. Mechanik der Flächentragwerke, Verlag Friedrich Vieweg & Sohn, Braunschweig,1985.
9. KRÄTZIG, W.B., ZAHLTEN, W. The application of plastic-fracturing theory to the finite element analysis of general reinforced shells, Proceedings SMIRT, Volume B, pp 311-322,1989.
10. BAZANT, Z.P., KIM S.-S. Plastic-fracturing theory for concrete, J. Eng. Mechanics Div., Proceedings ASCE,105, pp 407-428,1979.
11. GILBERT, R.I., WARNER, R.F. Tension stiffening in reinforced concrete slabs, ASCE, J. Struct. Div., Nr. ST12, Vol.104, pp 1885-1900, 1978.
12. GÜNTHER, G., MEHLHORN, G. Mitwirkung des Betons zwischen den Rissen, Forschungsbericht aus dem Fachgebiet Massivbau Nr. 3, Universität Kassel,1987.
13. NOAKOWSKI, P. Kontinuierliche Theorie der Mitwirkung des Betons auf Zug, Deutscher Ausschuß für Stahlbeton, Heft 394, Berlin,1988.
14. KRÄTZIG, W.B., ZHUANG, Y. Nonlinear physical and geometrical collapse analysis of cooling towers, Proceedings IASS,1, pp 135-144,1989.
15. KRÄTZIG, W.B., GRUBER, K., ZAHLTEN, W. Numerical collapse simulations of large cooling towers checking their safety and durability, Technical Report No. 92-3, Ruhr-University Bochum,1992.
16. Structural design of cooling towers (BTR-Bautechnik bei Kühltürmen), VGB Technische Vereinigung der Großkraftwerksbetreiber e.V., Essen,1991.
17. BASAR, Y., GRUBER, K., KRÄTZIG, W.B., MESKOURIS, K., and ZAHLTEN, W. Nichtlineare Analyseverfahren für Tragsicherheit, Schädigungsevolution und Restlebensdauer windbeanspruchter Naturzugkühltürme, Bauingenieur 67, pp 515-524,1992

182. Co-operation of structural plates and shells

U. A. TARNO,Tallinn Technical University, Estonia

SYNOPSIS. Structural plates are complex space frames which possess upper and lower chords and web elements. Unlike the ordinary structural plate the space between web elements is filled with a thin-walled surface consisting of hyperbolic paraboloid shells. The edge beams of the shells may be used as the web elements. In this study the number and the composition of the lower chords has been varied. According to the location of the lower chords the hyperbolic paraboloid shells receive part of the tensile and pressure forces produced by the whole structure. In this paper the analysis of the results of experiments, conclusions and recommendations are presented.

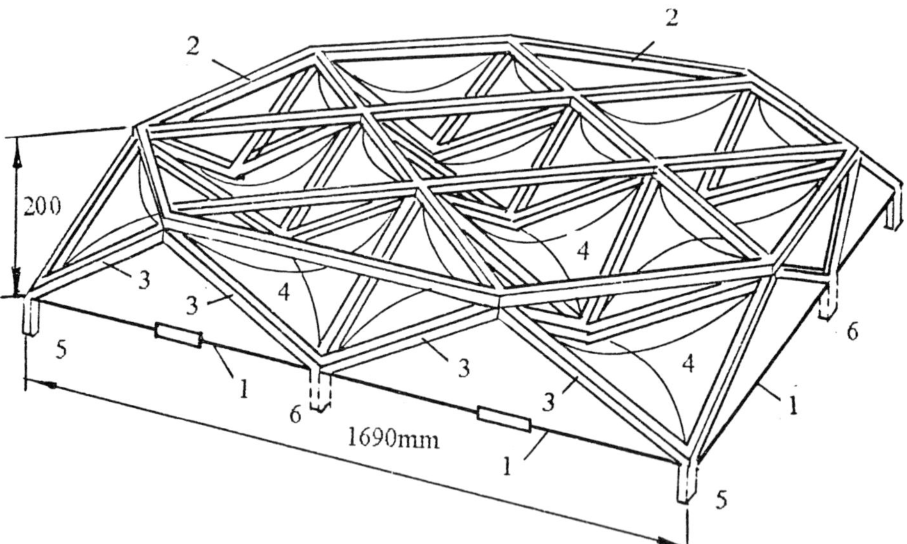

Fig. 1. General view of a structural plate model with thin-walled hyperbolic paraboloid shells (hypars). (1) - lower chords, (2) - upper chords, (3) - diagonals, (4) - hypars, (5) - main supports, (6) - additional supports.

Space structures 4. Thomas Telford, London, 1993

1. In this paper the analysis and results of an experimental study of a large-span structure is presented. The behaviour of structural plates is influenced by geometrical characteristics, the degree of statical indeterminacy and the manner in which the space frame is supported and loaded. The fundamental patterns of the double-layer space frame were obtained by combining the arrangement of the lower layer and web elements. Today the shells have become primary structural elements in combined structures in space and civil engineering. Unlike the ordinary space frame the space between web elements is filled with the thin walled surface consisting of the hyperbolic paraboloids (Fig. 1). The edge beams of thin walled double curvature shells may be used as the web elements.

2. In the course of investigation a 1:5 scale space frame model (1200xl200mm) was used. The model consisted of upper and lower chords and shells made of 4mm thick glass fibre reinforced plastic.

3. The distribution and quantity of the inner forces were examined experimentally: a theoretically examination would have been too complex. A series of tests were planned in order to facilitate the scientific investigation of the structures. The space structure examined consisted of 16 hyperbolic paraboloid shells and the chords.

4. The height-to-span ratio of the model was l:40 (Fig. 1). The upper chords were made of wood elements (30x30mm), the lower chords were made from of 5mm steel rods. The number and the composition of the lower chords was varied (variants 1 -12, Fig. 2). The minimum number of the supports on the perimeter of the structural plate is considered to be 4 (variants 1-4), the maximum number 8 (variants 5 -12, Fig. 2). The distributed static load was replaced by concentrated loads placed on the upper layer. The structure was investigated in the linear state under the 4 different steps of loading. Maximum load was 8.33 kPa. Deformations were measured by tensoresistors.

5. Distribution of the inner forces in the lower chords was independent of the fixing of the chords. In order to compare the inner forces in the lower chords the average values characterising every single type of chords are used. The total of the inner forces ΣN was calculated by summing the inner forces of different chords.

6. In the following paragraphs we will compare the inner forces of the lower chords under different supporting conditions and variants of chords to each other.

7. In the variant 1 the inner perimetrical chords (2) carry most load. Almost identical inner forces appear in the outer perimetrical chords (1) (95% of that of the inner perimetrical chords (2)). Chords (4) and (5) of the long diagonal have inner forces compared with rod (2) of 78% and 65% respectively. Chord (3) is relatively lightly loaded - its inner forces are of about 11% of the inner forces of chord (2).

8. Comparing variants 1 and 2 we can find that in variant 2 the medium chords (5) belonging to the long diagonals have been removed. The tension forces increase considerably in chords (2) (about 2.2 times) and in chords (1) (about 1.5 times).

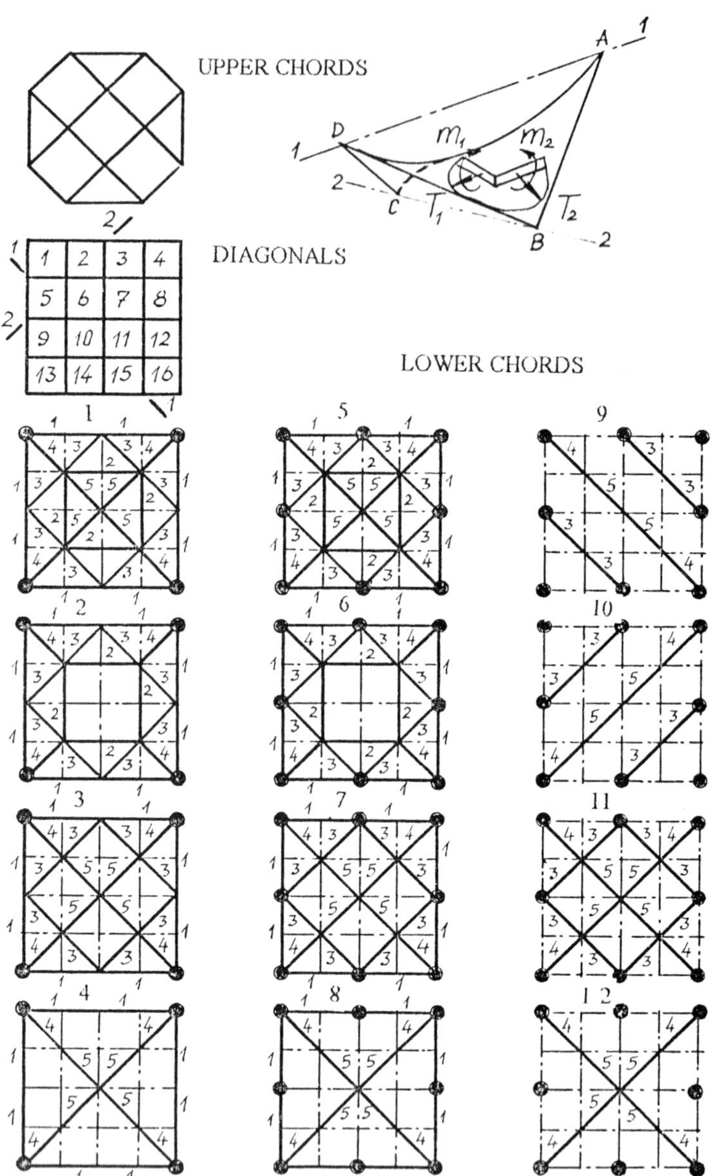

Fig. 2. Plan of the elements of the structural plate upper chords(l), lattice chords(2) and lower chords(3) (variants 1-12).

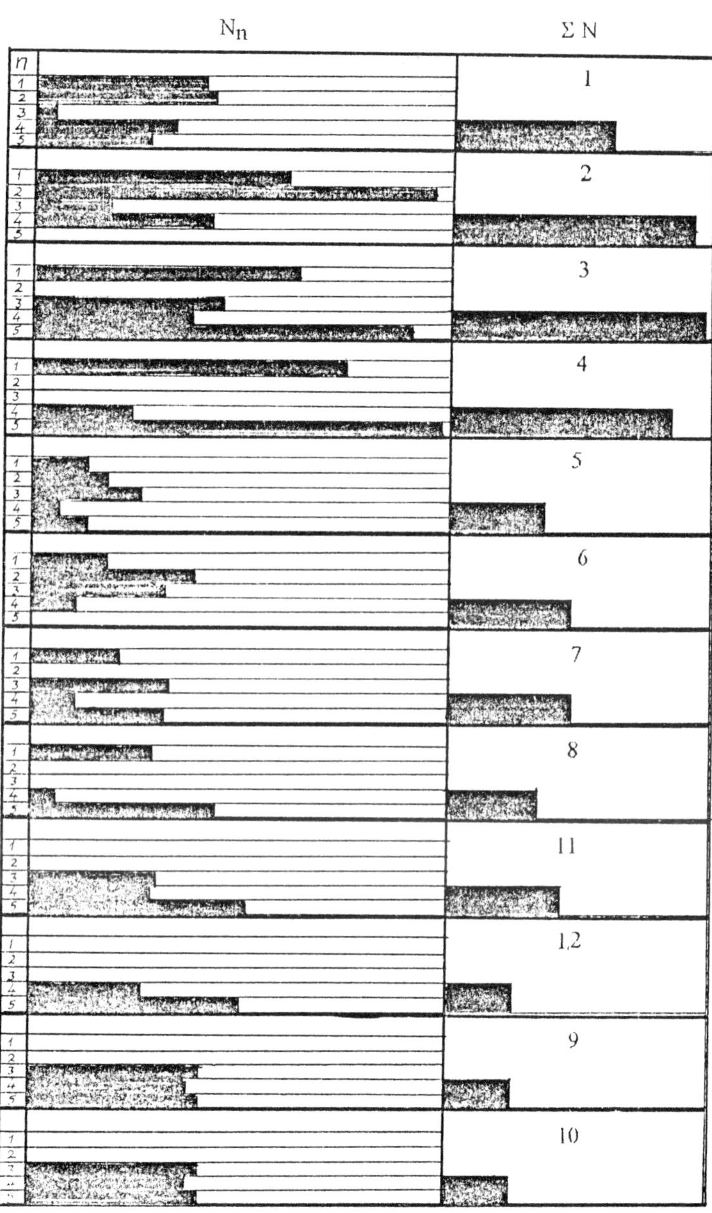

Fig. 3. Tension forces N_n in the lower chords n=1-5 and total forces $\sum N$ in variants 1-12.

The greatest inner forces appear in the inner perimetrical chords (2). The forces of the perimetrical chord (1) form about 63% of the inner forces of the chord (2).

9. The system of the inner perimetrical chords (2) is removed in the variant 3. The greatest inner forces appear in the chords (5) (comparing variants 1 and 3, the increase is about 3.25 times). The forces in the outer perimetrical chords (1) form about 70% of the forces in the chords (5).

10. If we compare variants 3 and 4 we can see that the chords (3) belonging to the short diagonals have been removed. Comparing variants 3 and 4 we say that the inner forces of the chords (5) have increased by 1.08 times. The greatest inner forces appear in the central element (5) of the long diagonal. The inner forces of the perimetrical chord (1) form 76% of the forces in the chord (5).

11. In the following variants the model has 8 supports on the perimeter. In the variant 5 the forces in the chords (2) form about 68% of the forces in the chords (3). The smallest load falls to this part of the diagonal chord that is near the corner support- its inner forces are about 22% of the forces in the chord (3).
In the variant 6 we have eliminated chord (5). Forces in the chord (2) increase by about 2 times and it starts functioning as chord (5). The total of the tension forces ΣN increases by about 1.25 times. Knowing this it may be concluded that it is unsuitable to remove the central chords (5) belonging to the long diagonals. Comparing variants 5 and 7 we notice that the chord (2) has been removed. The inner forces of the chord (1) increased by about 1.7 times. The forces in the chord (4) inceases by about 2 times and the forces in the chord (5) increases by about 2.5 times. Chords (3) and (5) are equally loaded . Chords (3) and (5) form the part of the structural plate that is resting on the intermediate supports.

12. In variant 8 all the lower chords except the outer perimetrical chords (1) and the chords (4) and (5) belonging to the long diagonals have been eliminated. The greatest load rests on the inner parts (3) of the diagonal chords. The inner forces of the perimetrical chords (1) form about 66% of the forces in the chord (5).

13. In the variant 11, additionally removed the chords are (1). The forces in the chords (3) remain the same the forces in the chord (4) increases by 2.8 times. The forces in the chord (5) increases by 1.5 times. The total ΣN of the inner forces in all the chords decreases. It forms about 92% of the total ΣN calculated in the variant 7.

14. Comparing variants 8 and 12 we notice that in the variant 12 only the chord (1) is missing. The forces of the chord (4) increase by 4.6 times, the forces in the chord (5) increase by 1.1 times. The total of the inner forces in the system ΣN decreases as the chord (4) is working in more optimum conditions than the chord (1). The total of the forces is about 60% of the total calculated in the previous variant. Variants 9 and 10 have only one-way diagonals (3), (4) and (5). The greatest load rests in the chords (3) and (5).

15. We examined the different variants of the lower chords to find out their influence to the inner forces of the hypar panels.

16. In Figs 4, 5 and 6 we can view the inner forces in case of different supporting variants and different sets of the lower chords (variants 1- 12). The inner forces and the bending moments in the diagonal lines of the hypars (5) and

(10) (see Fig. 2) are determined experimentally. The membrane forces T_1 take effect in the direction of the axes 1-1, that is connecting the upper corners of the hypars. The membrane forces T_2 take effect on the direction of the axis 2-2 connecting the lower corners of the hypars.

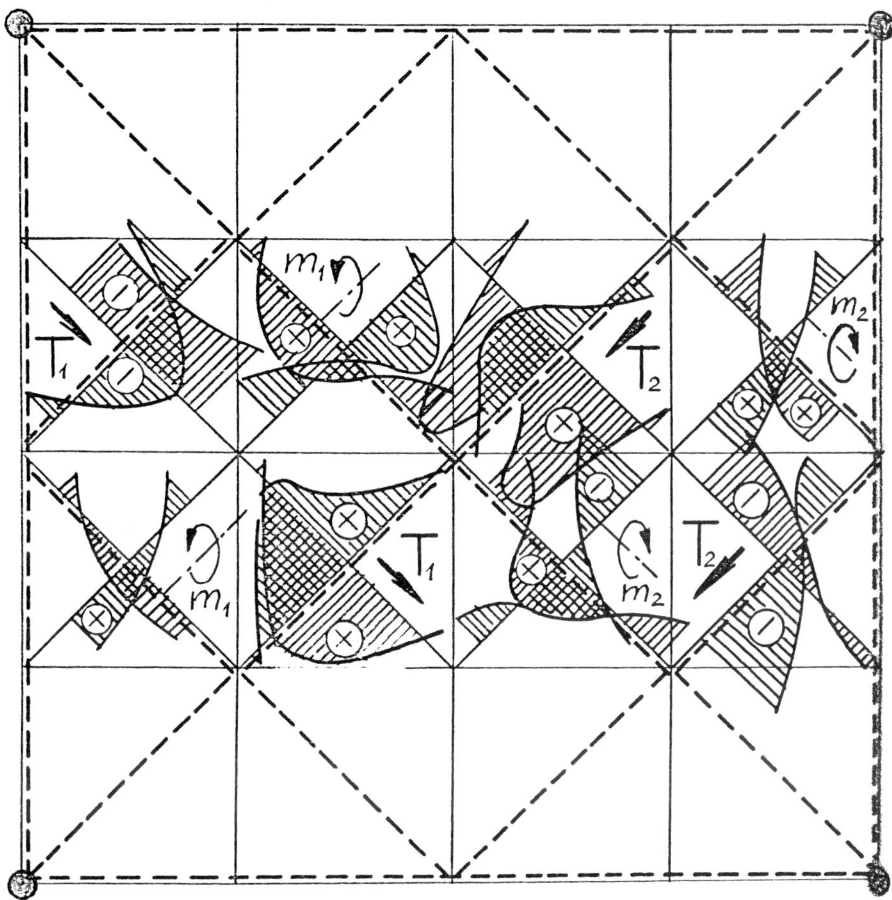

Fig. 4. Inner forces T_1, T_2, m_1 and m_2 in hypars in the variant 3.

17. The surface loading of the flat hyperbolic paraboloid shell is transferred to the edge beams by using the shear forces. This transference generates normal forces in the edge beams. These forces differ from each other in sign (+ or -) and value. The tension forces appear near the upper ends of the edge beams. The compression forces take effect near the lower ends of the beams. We can see an analogous set of forces in the thin-walled parts of the shell near the edge beams.

18. The tension forces appear in the upper zone of the hypars cross-section connecting the lower corners of the shell. Compression forces appear near the edge beams.

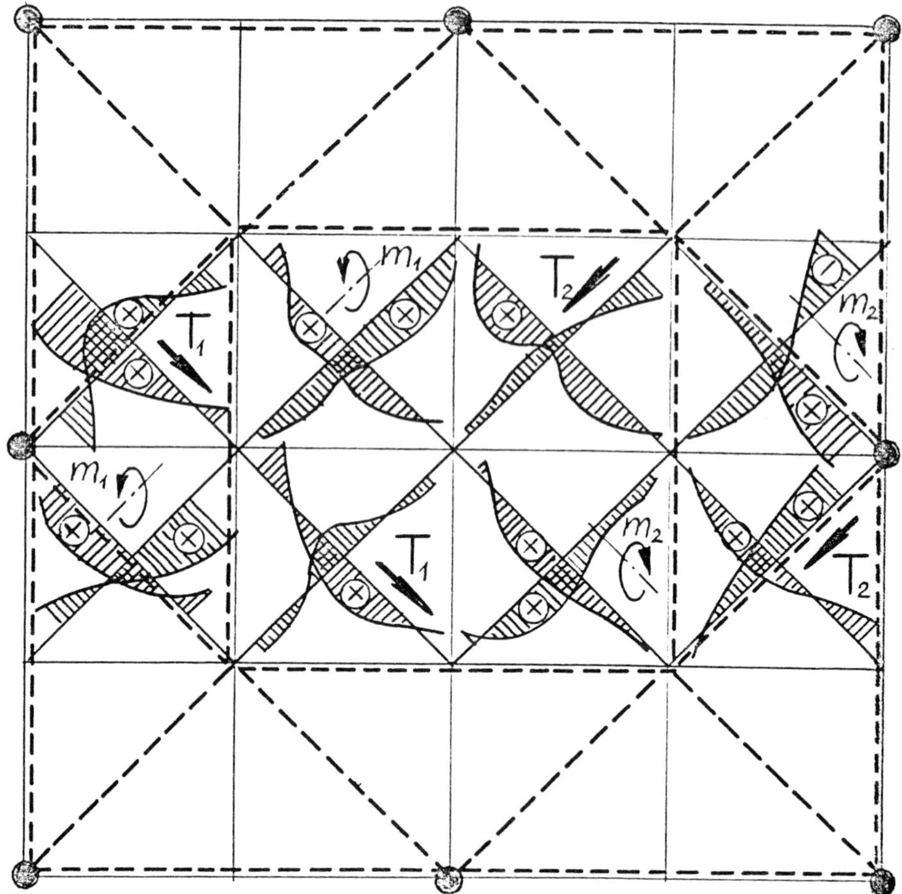

Fig. 5. Inner forces T_1, T_2, m_1 and m_2 in hypars in the variant 6.

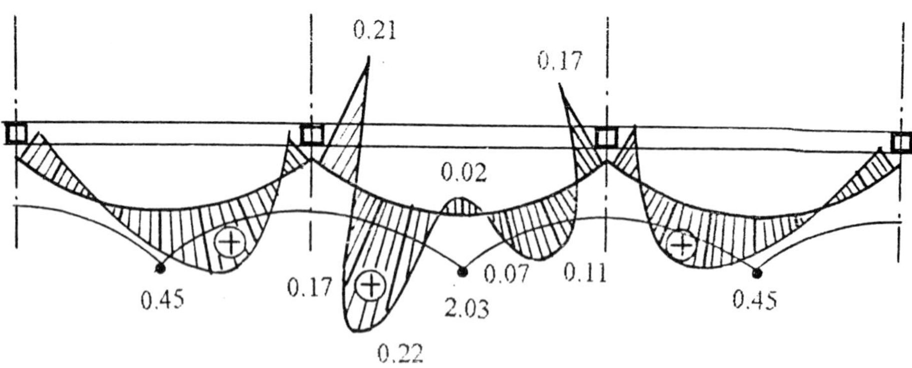

Fig. 6. Normal forces T_2 (kN/cm) in the cross section of the hypars and lower chords N (kN). Loading q = 8.33 kPa.

19. A part of the structural plate, the hypar, has in its edge beams additional inner tension and compression forces. These forces depend on the total statical work of the whole structural plate, and cause different influence on the hypar.

20. According to different sets of the lower chords the hyperbolic paraboloid shell receives different parts of the tension and compression forces of the whole structure. The distribution of the inner forces descend on the system of the lower chords and different locations of the supports. Additional space frames are formed in case of a certain composition of the upper and lower chords.

21. The direct surface loading has no effect on the surface of the hyperbolic paraboloid shell. The inner forces are transferred by the diagonals (edge beams) and upper and lower chords to the shell. Due to this, the forces cause considerable bending moments in the hypars. The majority of the surface of the hypar is influenced by positive bending moments. These bending moments cause the flattening of the cross-section in the direction of the axis 2-2 and increasing of the curvature in the direction of the axis 1-1. In both cases we notice tension stresses in lower fibres of the shell plate. It is important to pay special attention to the essential changes of the inner forces. These changes depend on the supporting conditions on the perimeter of the plate.

CONCLUSIONS

22. There are additional normal forces in different edge beams of the hypars. These forces depend on the location of the hypar in the composition of the structural plate and may be either tensile or compressive forces.

23. According to the set of the lower chords the hyperbolic paraboloid shell receives different parts of the tensile and compressive forces of the whole structure.

24. The direct surface loading has no effect on the hyperbolic paraboloid shell. Due to the transference of the inner forces by the edge beams considerable bending moments are caused in the shells.

25. Additional space frames are formed in case of a certain composition of the upper and lower chords. It is important to pay special attention to the essential changes of the inner forces.

26. The changes of inner forces depend on the supporting conditions in the perimeter of the plate. Working of a structural plate that has supports only in its corners is most intensive in the direction of the longest diagonal. Frequent perimetrical supports reduce the load resting on one single rod and reduce the bending moments that influence the whole frame. These supports cause a decrease in vertical deflections by 4-fold.

REFERENCES

Analysis, Design and Realisation of Space Frames. A State of the Art Report. Bulletin of the I.A.S.S., n.84/85, April-August 1984.

183. Solution porcedures for non-linear stability analysis of shallow shells

S. I. TRUSHIN and A. S. IVANOV, TsNIISK, Russia

SYNOPSIS. This paper describes the formulation of the problems and numerical algorithms for the non-linear analysis of shallow shells by the finite difference energy method. The development of the model is based upon the material and geometrically non-linear equations taking account of shear deformations and the action of multi-parameter loading. Solution procedures for discrete non-linear problems are founded on incremental/iterative techniques or initial value formulation. Any nodal displacement, load parameters or arc-length of load-deflection curve can be adopted as a control parameter. Some numerical examples of non-linear stability analysis of shallow shells with determination of equilibrium curves and surfaces are presented.

METHOD OF ANALYSIS

1. A specified engineering theory for orthotropic and isotropic shear-deformable shells and plates with geometrical and material non-linearity is used for the investigation of stability phenomena of thin-walled structures. The relationships of this theory allow: the satisfaction of the boundary conditions more strictly as compared to the theory based on Kirchoff-Love hypotheses; the building of more effective numerical algorithms and; the analysis of multi-layered composite constructions with low shear rigidity (ref. 1). Physical relations are formulated with respect to Hook's law or a theory of plastic yielding with isotropic hardening. To formulate a variational problem using displacements, the Lagrange principle of virtual displacements is employed. The functional of energy in elastic-plastic analysis can be written in following form:

$$L=\iiint_V \left(\frac{1}{2}\, e_m^T\, D_{m-1}^{ep}\, e_m + \left[\sigma_{m-1} - D_{m-1}^{ep}\, e_{m-1}\right]^T e_m \right) dV - \iiint_V q_m^T u_m\, dV \,, \quad (1)$$

where e_m is a vector of strains at a level of loading m; σ_{m-1} and e_{m-1} are stresses and strains at a previous level; D_{m-1}^{ep} is an elasto-plastic matrix; u_m is a vector of displacements and q_m is a vector of external load. In the case of an elastic problem the terms in square brackets may be omitted and the functional after integration through the thickness is given by:

$$L = \frac{1}{2} \iint_S \varepsilon_m^T D \varepsilon_m dS - \iint_S q_m^T u_m dS , \tag{2}$$

where ε is a vector of strains of the middle surface S.

2. Numerical realisation of the solution of this non-linear problem is based on the finite difference energy method of discretization. The region S occupied by the structure is covered by a grid, and the unknown displacement functions and derivatives, which come into the functional Eqs 1 or 2, are approximated, within the grid cell, by the nodal values and finite difference expressions. The integration over the region S are substituted by adding over the same region. As a result the function L transforms into a scalar function with vector arguments I(u,p), which is the discrete analogue (u is the nodal displacement vector, p is the vector of load parameters). The stationary conditions for this function are non-linear Eulerian equations:

$$F(u, p) = \nabla W(u) - (p_1 Q_1 + \ldots + p_N Q_N) = 0, \tag{3}$$

wherein $\nabla W(u)$ is the gradient of the potential energy of deformation; $Q_1,\ldots,Q_N$ are normalised vectors of nodal loads; $p_1,\ldots,p_N$ are parameters of external loads.

3. Non-linear problems, which are presented in the paper, are solved using the Newton-Raphson, modified Newton-Raphson, Euler and Runge-Kutta methods. According to these methods it is necessary to determine components of the tangent stiffness matrix $\nabla^2 W(u)$ and gradient vector $\nabla W(u)$. The ith element of the gradient vector and (i,j)th element of the stiffness matrix are expressed in the form:

$$\frac{\partial W(u)}{\partial u_i} = \iiint_V \left[D_{m-1}^{ep} (e_m - e_{m-1}) + \sigma_{m-1} \right]^T \frac{\partial e_m}{\partial u_i} dV, \tag{4}$$

$$\frac{\partial^2 W(u)}{\partial u_i \partial u_j} = \iiint_V \left[\frac{\partial e^T}{\partial u_i} D_{m-1}^{ep} \frac{\partial e}{\partial u_j} + (D_{m-1}^{ep} (e_m - e_{m-1}) + \sigma_{m-1})^T \frac{\partial^2 e}{\partial u_i \partial u_j} \right] dV$$

The first and a second derivatives of e are given by

$$\frac{\partial e}{\partial u_i} = \Delta e^L (r_i) , \tag{5}$$

$$\frac{\partial^2 e}{\partial u_i \partial u_j} = \Delta e^{NL} (r_i + r_j) - \Delta e^{NL} (r_i) - \Delta e^{NL} (r_j) ,$$

where Δe^L is the linear part of strain increment Δe; Δe^{NL} is the non-linear part of strain increment Δe; r_i and r_j are unit vectors.

NUMERICAL SOLUTION OF NON-LINEAR EQUATIONS

4. The non-linear system of algebraic equations $F(u,p) = 0$ define a N-dimensional manifold in space of variables $X = \{u,p\}^T$. In the case when N=1 a solution manifold presents an equilibrium path which can be determined by continuation methods. Riks, Wempner and Crisfield (refs 2-4) suggested the use of an arc-length as a control parameter to computationally trace an equilibrium path. The non-linear Eq. 3 is usually solved using of the Newton-Raphson (NR) and modified Newton-Raphson (mNR) methods. According to the algorithms of these methods, the unknown functions $u(s_m)$ and $p(s_m)$ are found by solution of iterative equations at each level of loading m, which corresponds to a prescribed value of control parameter s_m:

$$\nabla^2 W(\mathbf{u}^k(s_m))\Delta\mathbf{u}^{k+1} = Q\Delta p^{k+1} + Qp^k(s_m) - \nabla W(\mathbf{u}^k(s_m)), \tag{6}$$

where k is the number of the iteration, $u^k(s_m)$, $p^k(s_m)$ are values of the sought functions at the kth iteration; Δu^{k+1}, Δp^{k+1} are the increments of the functions at the next iteration. A system of Eq. 6 must be supplemented by a constrained equation in the following form:

$$\mathbf{a}^T\delta\mathbf{u}^{k+1} + a_{n+1}\delta p^{k+1} = \Delta s, \tag{7}$$

$$\text{or} \quad \|\delta\mathbf{u}^{k+1}\|^2 = \Delta s^2, \tag{8}$$

where the incremental functions δu^{k+1} and δp^{k+l} are given by:

$$\delta\mathbf{u}^{k+1} = \delta\mathbf{u}^k + \Delta\mathbf{u}^{k+1}; \quad \delta p^{k+1} = \delta p^k + \Delta p^{k+1}. \tag{9}$$

5. The algorithm suggested by Crisfield (ref. 4) leads to the solution of a square equation for determination Δp^{k+1}. The sign of Δp at the first iteration corresponds to the sign of determinant of the tangent stiffness matrix and the sign of Bergan's current stiffness parameter (ref. 5). In the elastic-plastic analysis authors use a double iteration loop, proposed by Bushnell (ref. 6). In the inner loop the set of algebraic equations, Eqs 6, 7 or 8, with given fixed material properties and plastic strains are solved. In the outer loop the strain-dependent material properties, the matrix D^{ep}_m and the plastic strains and stresses are calculated.

6. The second group of methods which are applicable for the numerical solution of the non-linear Eq. 3 jointly with a constraint Eqs 7 or 8, is based on the reduction to an initial value problem:

$$\frac{d}{ds}F(\mathbf{u}(s), p(s)) = 0, \tag{10}$$

$$\mathbf{a}^T du/ds + a_{n+1} dp/ds = 1, \tag{11}$$

$$\text{or} \quad \|du/ds\|^2 = 1 \tag{12}$$

with initial values $u(s_o)=u^o$; $p(s_o)=p^o$.

Stricklin, Haisler *et al* (ref. 7) proposed the consideration of a self-correcting form of these equations

$$\frac{d}{ds}\mathbf{F}(\mathbf{u}(s), p(s)) + z\mathbf{F}(\mathbf{u}(s), p(s)) = 0, \tag{13}$$

where z is a suitably chosen scalar parameter.

If we use Euler's method for the numerical solution of the initial value problem in Eqs 13, 11 or 12, the computational correlations at the mth step are defined in the following form (we assume $z=1/\Delta s$):

$$\nabla^2 W(\mathbf{u}(s_{m-1}))\Delta\mathbf{u} = Q\Delta p + Qp(s_{m-1}) - \nabla W(\mathbf{u}(s_{m-1})), \tag{14}$$

$$\mathbf{a}^T \Delta\mathbf{u} + a_{n+1}\Delta p = \Delta s, \tag{15}$$

$$\text{or} \quad \|\Delta\mathbf{u}\|^2 = \Delta s^2. \tag{16}$$

7. A more efficient algorithm, which possesses a higher order of accuracy and does not accumulate errors in the calculation, may be constructed by applying the Runge Kutta (RK) method to the solution of system of Eqs 13, 11, or 12.

$$\mathbf{u}(s_m) = \mathbf{u}(s_{m-1}) + \Delta\mathbf{u}; \quad p(s_m) = p(s_{m-1}) + \Delta p \tag{17}$$

where $\Delta\mathbf{u} = (\Delta\mathbf{u}^1 + 2\Delta\mathbf{u}^2 + 2\Delta\mathbf{u}^3 + \Delta\mathbf{u}^4)/6$; $\Delta p = (\Delta p^1 + 2\Delta p^2 + 2\Delta p^3 + \Delta p^4)/6$

Both incremental displacements Δu^k and loading parameters Δp^k (k = 1,2,3,4) are defined by the consequent solution of the following:

$$\nabla^2 W(\mathbf{u}^k)\Delta\mathbf{u}^k = Q\Delta p^k + Qp^k - \nabla W(\mathbf{u}^k), \tag{18}$$

$$\mathbf{a}^T \Delta\mathbf{u}^k + a_{n+1}\Delta p^k = \Delta s, \tag{19}$$

$$\text{or} \quad \|\Delta\mathbf{u}^k\|^2 = \Delta s^2. \tag{20}$$

The current values u^k, p^k are corrected at each sub-step with the assistance of the increments at the previous sub-step:

$$u^1 = u(s_{m-1}); \qquad p^1 = p(s_{m-1});$$

$$u^2 = u^1 + \Delta u^1/2; \quad p^2 = p^1 + \Delta p^1/2;$$

$$u^3 = u^1 + \Delta u^2/2; \quad p^3 = p^1 + \Delta p^2/2; \tag{21}$$

$$u^4 = u^1 + \Delta u^3; \qquad p^4 = p^1 + \Delta p^3.$$

The algorithm of calculations which is applied for the numerical solution of Eqs 18, 19, or 20 is the same as for the first iteration of the NR method.

8. Solution procedures which were developed for the numerical investigation of equilibrium surfaces of shells by the action of multi-parameter loading are founded on the application of the methods written above. In the two-dimensional case the n equilibrium equations involve n+2 unknowns u, p_1, p_2 and can be written in the form:

$$F(u, p_1, p_2) = \nabla W(u) - p_1 Q - p_2 Q = 0 \tag{22}$$

9. A surface determined by these equations can be constructed by tracing a grid of curves. A strategy of solution consists of four main stages:
1) initial values of loading parameters p_1, p_2 and nodal displacement vector u is stored;
2) Eq. 22 is solved with the assistance of continuation procedures for the increments Δp_1 and fixed value of p_2; the load-deflection path is determined.
3) execute one step increment solution respectively to Δp_2 and fixed value of p_1;
4) obtained values p_1,p_2 and the displacement vector u are stored and process of calculations continues at stage 2 etc. It is necessary to set the number of steps for increments of load parameters and an accuracy of sought solutions at each step. If we use, for instance, the NR method for tracing the equilibrium curve corresponded to the loading upon parameter p_1 with a fixed value of parameter p_2, the computation relations can be written in following form:

$$\nabla^2 W(u^k(s_m^1))\Delta u^{k+1} = Q_1 \Delta p_1^{k+1} + Q_1 p_1^k(s_m^1) + Q_2 p_2(s_n^2) - \tag{23}$$
$$-\nabla W(u^k(s_m^1)),$$

$$a^T \delta u^{k+1} + a_{n+1} \delta p^{k+1} = \Delta s^1, \tag{24}$$

$$\text{or } \|\delta u^{k+1}\|^2 = (\Delta s^1)^2, \tag{25}$$

where s_m^1, Δs^1 are a value and an increment of the arc-length of the curve upon a load parameter p_1; s_n^2 is the value of the arc-length according to p_2. An analogous relationship can be constructed for solutions obtained with the assistance of the initial value formulation.

NUMERICAL EXAMPLES

10. As a first example consider a shallow elastic spherical shell subjected to a uniform pressure loading q (Fig. 1). All edges are hinged and immovable. An non-dimensional curvatures of the shell, k_1^* and k_2^* and a parameter of loading q^* are given by:

The double symmetry of the problem was exploited. The quarter of the shell was divided into 16 elements which give 125 degrees of freedom. The load-deflection behaviour of this shell was analysed using the initial value formulation (RK method) and the incremental iterative technique (NR method). Both numerical solutions are in a close agreement. An equilibrium path for a central point is presented in Fig. 1.

11. A shallow cylindrical shell subjected to a uniformly distributed transverse pressure and compression pressure applied to the curvilinear edges provides an example for elastic-plastic analysis (Fig. 2). The shell dimensions and material properties are as follows: radius of curvature R=2.08m; shell thickness h=0.03m; dimensions in plane a = b = lm; elastic modulus $E=7*10^{10}N/m^2$; Poisson's ratio $\nu = 0.3$; yield stress $\sigma_y = 3.4*10^8 N/m^2$; plastic modulus $E_t=3*10^9\ N/m^2$. All edges are hinged and immovable along the axis of the edge. Loading was assumed dependent on one parameter and a transverse pressure q_2 connects with compression pressure q_1 by the relationship $q_2=0{,}0024q_1$. A five-by-five mesh was used to idealise one-quarter of the structure with seven nodes through the thickness. An equilibrium path for the central point is given in Fig. 2, where P denotes the appearance of plastic strains.

12. To illustrate the technique for numerical tracing of the equilibrium surface consider an elastic cylindrical shell subjected to loading of two types (Fig. 3). This test problem has been used by many researches in the past (loading has been assumed dependent on one parameter) (ref. 4). The shell is hinged along the straight edges and arches are free. The material and geometrical parameters of shell are as follows: R = 2.54m; L = B = 0.504m; $E=0.3105*10^{10}N/m^2$; $\nu = 0.3$; $h=0.635*10^{-2}m$. In the present analysis it is assumed that the first loading type is a vertical concentrated force at the central point and the second type is a unified horizontal tensile pressure applied to the arches. Fig. 3 illustrates the equilibrium surface for the shell which was determined using the computational algorithm presented above. For tracing the load-deflection path, at a fixed parameter of loading p_2 corresponded to tensile pressure, the NR method with constrained Eq. 25 was exploited.

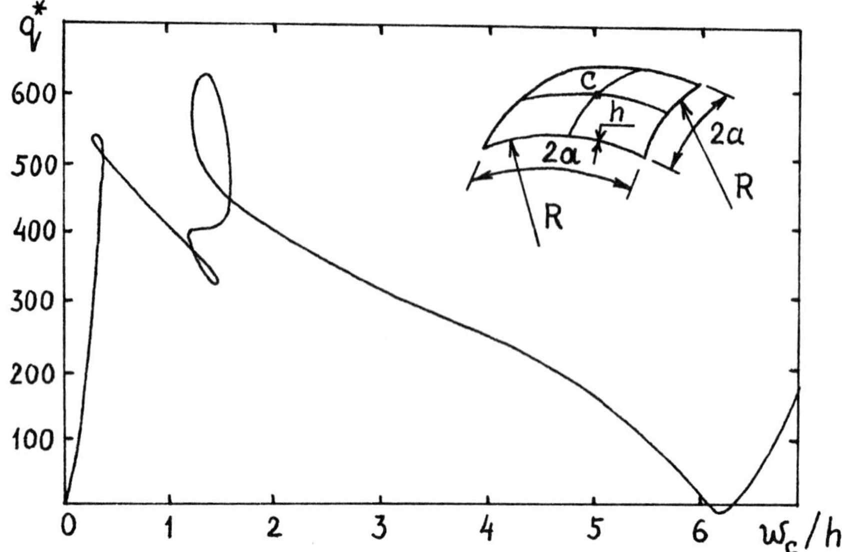

Fig. 1. Load-deflection curve for spherical shell.

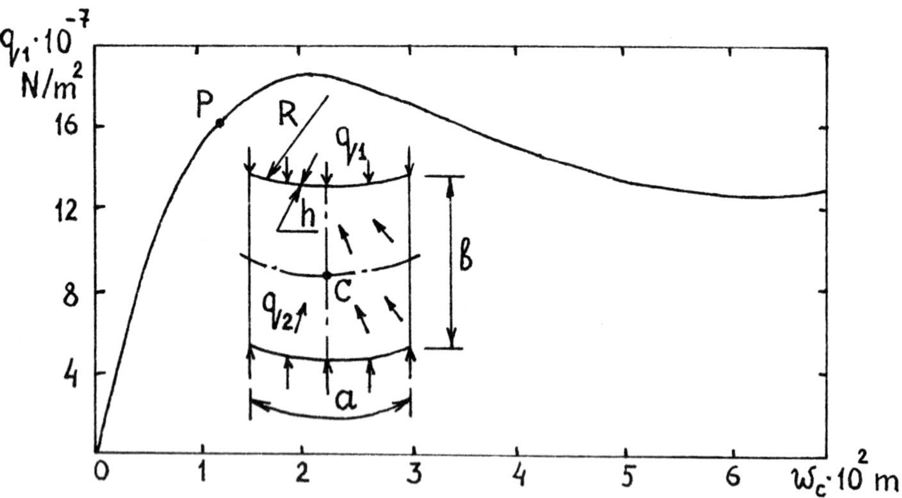

Fig. 2. Elastic-plastic analysis of cylindrical shell.

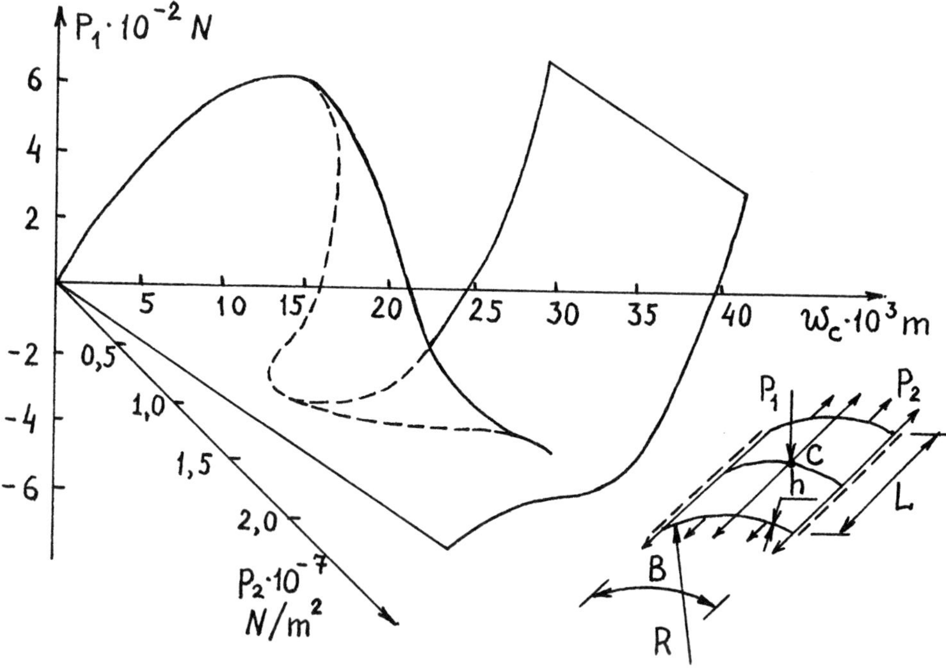

Fig. 3. Equilibrium surface for cylindrical shell.

REFERENCES

1. MILEYKOVSKY I. E. and TRUSHIN S. I. Analysis of thin-walled structures, pp.200. Stroyizdat, Moscow, 1989.
2. RIKS E. An incremental approach to the solution snapping and buckling problems, Int. J. Num. Meth. Eng., 1979, Vol. 15, pp.529-551.
3. WEMPNER G. A. Discrete approximations related to non-linear theories of solids, Int. J. Solids Structures, 19711 Vol. 7, pp.1581-1599.
4. CRISFIELD M. A. A fast incremental/iterative solution procedure that handles "snap-through", Comp. & Struct., 1981, Vol. 13, pp.55-62.
5. BERGAN P. G. Solution algorithms for non-linear structural problems, Comp. & Struct., 1980, Vol. 12, pp.497-509.
6. BUSHNELL D. Stress, buckling and vibration of hybrid bodies of revolution, Comp. & Struct., 1977, Vol. 7, pp.517-537.
7. STRICKLIN J. A., HAISLER W.E. and VON RIESEMANN W.A. Evaluation of solution procedures for material and/or geometrically non-linear structural analysis, AIAA Journal, 1973, vol. 11, pp.292-299.

184. The ultimate carrying capacity of imperfectly stiffened panels subjected to axial load

G. YANLIN and J. LINDNER,Institut fur Baukonstruktionen und Festigkeit, Germany

SYNOPSIS. This paper presents a theoretical study of the elastic plastic interaction buckling of imperfect longitudinally stiffened panels under axial loads. A material and geometric nonlinear spline finite strip method is developed to analyse such an interaction problem. The initial geometric imperfections and residual stresses due to welding are involved in the analysis. An iterative numerical procedure combining a modified arc-length method and Newton-Raphson iteration is employed in the solution of the nonlinear incremental equilibrium equations and a full range load deflection path is obtained. The numerical results obtained show that the authors' method is not only an efficient tool but also exhibits good accuracy in the analysis of large deflection elastic plastic behaviour of folded plate structures.

INTRODUCTION

1. Although a great deal of work done by Koiter and Pignataro (refs 1,2), van der Neut (ref. 3), Ali and Sridharan (ref. 4), and Sridharan and Peng (ref. 5), has been published on the elastic interaction between local and overall buckling modes in the stiffened panels under axial compression, relatively little has been done on the study of the elastic plastic interaction, which may be very important for moderately stocky structures or for welded structures. For the stiffened panel made from thin steel plates, its collapse is often caused by the interaction (Fig. lb) between the local buckling of the plate and stiffeners (Fig. la) and the overall buckling of the whole stiffened panel (Fig. lc). This interaction penetrates the full loading process for the stiffened panel with local (plate and stiffeners) and overall (whole stiffened panel) initial geometric imperfections or with local initial geometric imperfections and eccentric loads at the ends. This paper presents a theoretical study of the large deflection elastic plastic interaction behavior of a stiffened panel with local and overall initial geometric imperfections under in-plane load.

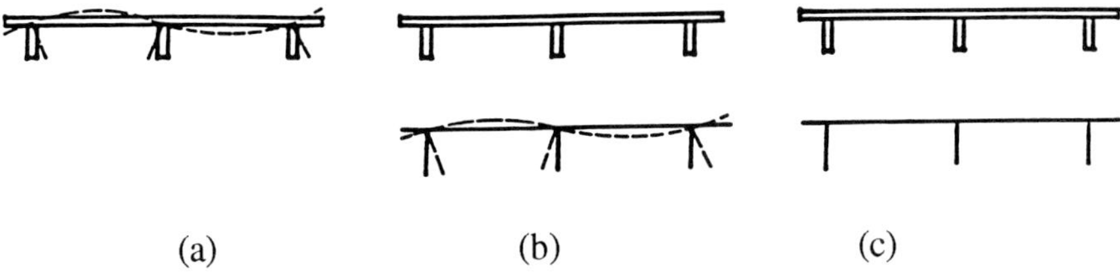

Fig. 1. Buckling modes of stiffened panels.
(a) Local Buckling.
(b) Interaction Buckling.
(c) Overall Buckling.

2. A complete interaction buckling analysis even for a simply supported thin-walled member, taking full account of the effects of plasticity and large deflections, involves great computation effort. Therefore, for a stiffened panel made of a large size plate and stiffeners, it is impossible to analyse accurately this interaction behaviour using a non-linear finite element method on micro-computers. To overcome this difficulty, a large deflection elastic plastic spline finite strip method, as a modification of the finite element method, is developed to analyse the interaction of the stiffened panels. This method, unlike the semi-analytic finite strip method (refs 6-9) which is only applicable to simply supported boundary conditions, is capable of dealing with structures involving abruptly changing material properties, concentrated and patch loading, internal supports, and any prescribed external boundary conditions. In addition the spline finite strip method significantly reduces the computation involved for folded plate structures compared with the finite element method.

PLASTICITY

3. In this study the von Mises yield criterion associated with a more rigorous multi-layer integration is used. Because of the inclusion of a volume integral, plate elements are divided into several layers through their thickness and the direct stress σ_z between layers is assumed to be zero. Thus, at any level z from the mid-plane of plate, the yield surface is given by:

$$f_{(z)} = (\sigma_x^2 + \sigma_y^2 - \sigma_x\sigma_y + 3\sigma_{xy}^2)/\sigma_Y^2 = \sigma_e^2/\sigma_Y \qquad (1)$$

where σ_Y is the uniaxial yield stress, σ_x and σ_y are direct stress components in x and y directions, and σ_{xy} is the shear stress in x,y plane.

4. In conjunction with the Prandtl-Reuss flow rule the relationship relating the stress increments in the elastic plastic range can be derived as:

$$\{\Delta\sigma\}_Z = [D_{ep}]_Z \{\Delta\varepsilon\}_Z \tag{2}$$

where D_{ep} is the tangential elastic plastic modular matrix (3x3) valid for the given layer z.

SPLINE FINITE STRIP FORMULATION

Displacement fields

5. A typical local B_3-spline function with equal section length is defined as follows:

$$\varphi_i = \frac{1}{6h^3}\begin{cases} 0 & x \le x_{i+1} \\ (x - x_{i-2})^2 & x_{i-2} \le x \le x_{i-1} \\ h^3 + 3h^2(x - x_{i-1}) + 3h(x - x_{i-1})^2 - 3(x - x_{i-1})^3 & x_{i-1} \le x \le x_i \\ h^3 + 3h^2(x_{i+1} - x) + 3h(x_{i+1} - x)^2 - 3(x_{i+1} - x)^3 & x_i \le x \le x_{i+1} \\ (x_{i+2} - x)^3 & x_{i+1} \le x_{i+2} \\ 0 & x_{i+2} \le x \end{cases} \tag{3}$$

6. A linear combination of (m+3) local B_3-splines is required to fully define all the displacement components in x direction (longitudinally) for the strip with m sections. Due to the localisation of the B_3-splines, only three boundary local B_3-splines need to be amended to satisfy any prescribed the in-plane and out-of-plane boundary conditions. Then, with a cubic polynomial interpolation for w and a lower order linear interpolation for u and v in y (transverse) direction , the displacement function for the strip can be obtained. The expressions of the displacement functions are given in ref. 10.

Strain-displacement relations

7. According to kirchhoff's hypothesis, the strain state at any level z from the mid-plane of a plate is defined by Eqs. (4) and (5). In Eqs.(4) and (5) v_o and w_o are the initial geometric imperfections, respectively, in y and z direction of the plate elements. The initial geometric imperfection v_o for the stiffeners is the same as overall flexural initial imperfection of the whole stiffened panel. In these strain expressions the underlined term in Eq. (5) is included to account for the influence of the finite membrane displacement (v) on the non-linear behaviour in the plate

and stiffeners. The other non-linear terms involving u and v are considered of secondary importance and have been ignored.

$$\varepsilon_x = \varepsilon_{xo} - z\frac{\partial^2(w-w_o)}{\partial x^2}$$
$$\varepsilon_y = \varepsilon_{yo} - z\frac{\partial^2(w-w_o)}{\partial y^2} \qquad (4)$$
$$\varepsilon_{xy} = \varepsilon_{xyo} - 2z\frac{\partial^2(w-w_o)}{\partial x \partial y}$$

$$\varepsilon_{xo} = \frac{\partial u}{\partial x} + \frac{1}{2}\left[\left(\frac{\partial w}{\partial x}\right)^2 - \left(\frac{\partial w_o}{\partial x}\right)^2\right] + \underline{\frac{1}{2}\left[\left(\frac{\partial v}{\partial x}\right)^2 - \left(\frac{\partial v_o}{\partial x}\right)^2\right]}$$
$$\varepsilon_{yo} = \left(\frac{\partial v}{\partial y} - \frac{\partial v_o}{\partial y}\right) + \frac{1}{2}\left[\left(\frac{\partial w}{\partial y}\right)^2 - \left(\frac{\partial w_o}{\partial y}\right)^2\right] \qquad (5)$$
$$\varepsilon_{xyo} = \frac{\partial u}{\partial y} + \left(\frac{\partial v}{\partial x} - \frac{\partial v_o}{\partial x}\right) + \left[\left(\frac{\partial w}{\partial x}\right)\left(\frac{\partial w}{\partial y}\right) - \left(\frac{\partial w_o}{\partial x}\right)\left(\frac{\partial w_o}{\partial y}\right)\right]$$

SOLUTION PROCEDURE

8. In the present study, a numerical procedure combining the modified arc-length method (refs 11,12) and Newton-Raphson iteration is presented. The solution procedure starts by using the Newton-Raphson iteration but as yielding occurs it switches to the modified arc-length method to surpass limit points of the load-deflection curves. The combined use of both method is found to be more efficient than if the modified arc-length method was used alone. Convergence is considered to be satisfactory when either the Euclidean norm of iterative deflections is within a specified percentage of the Euclidean norm of the total deflections or the maximum iterative deflection is within a specified percentage of the maximum total deflection.

9. To improve stress accuracy a sub-incremental technique (ref. 13), where the total strain increment is divided into a finite number of sub-strain increments and the corresponding stress for every sub-increment is determined, is adopted in the calculation of the stresses in the elastic-plastic range. This technique provides a good stress accuracy while relaxing the number of increments required to obtain the complete load deflection paths.

NUMERICAL RESULTS

Comparison with existing theoretical results

10. As an example, a large deflection elastic-plastic analysis of the plate with residual stresses and an initial imperfection magnitude of $w_0=0.005a$ in the local buckling mode is performed. The unloaded edges in the plate are free to pull in. The comparison with other analytical results reproduced by Bradfield in ref. 14 is shown in Fig. 2. The residual stress pattern assumed for the plate is also shown in the figure. Utilising symmetry, three strips, eight sections and three layers are chosen over a half-width of the plate. The results obtained are graphed as normalized load (P/P_Y) versus normalized axial shortening $\varepsilon/\varepsilon_Y$, where P_Y is the overall squash (yield) load and ε_Y is the yield strain defined by σ_Y/E. The comparison shows that the results obtained by the present spline finite strip method agree well with results given using other procedures.

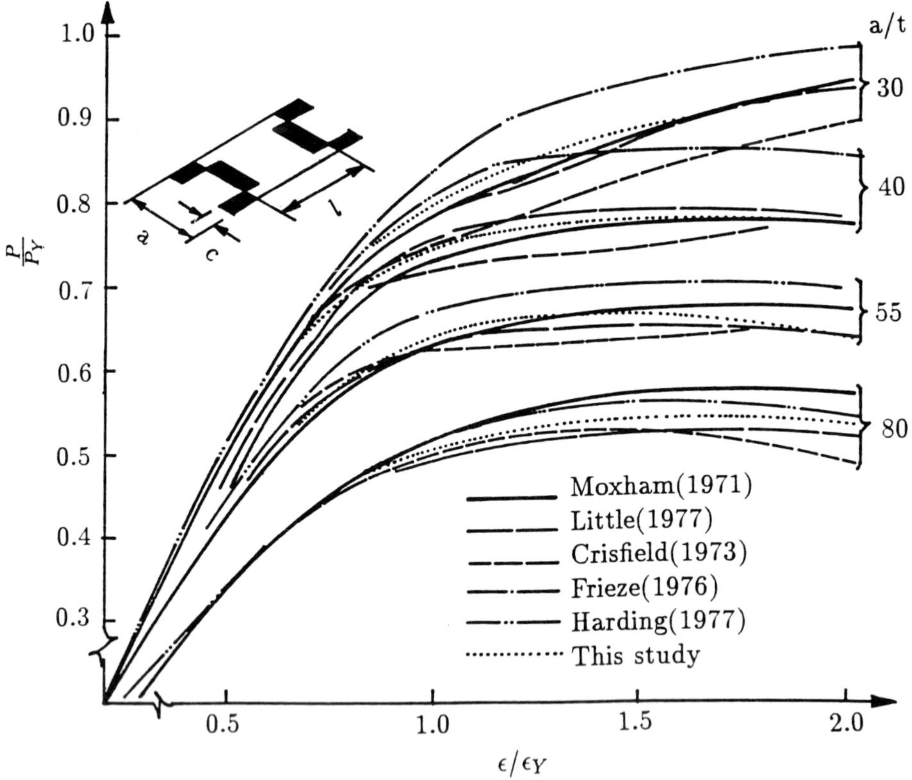

Fig. 2. Comparison with other analytical results.
($a/1=0.875$, $w_0/a=0.005$, $\sigma_Y=250$MPa, $c=3t$)

Idealised beam-column mode of a stiffened panel

11. In the analysis of the interaction buckling of stiffened panels there are usually two viable approaches. One is an approximate beam-column method where a wide stiffened panel with free edges is simplified as an idealised beam-column mode with T-section as shown in Fig. 3. The other is an accurate method where the stiffened panel is analysed as a whole rather than as the idealised beam-column mode. Little (ref. 15) introduced the first approach to the analysis of stiffened panels. But local plate buckling was allowed for by using an average compressive stress-strain curve of a single plate in forming M-P-Φ curves of plate buckle length. His analysis was restricted to the situation where the stiffeners did not buckle locally.

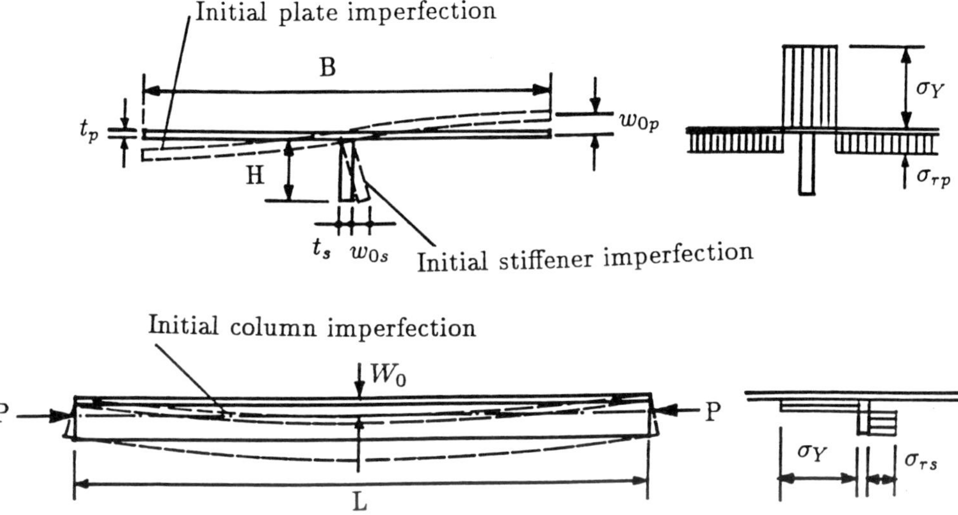

Fig. 3. Idealised beam-column mode of a stiffened panel.

12. In this study the first approach is used. However, the authors, unlike Little who treated the beam-column mode as an assembly of plate-buckle-length, short sections and predicted the strength of this mode by using a finite integration method with the aid of the approximate M-P-Φ curves of the short section, will use the spline finite strip analysis developed above to analyse accurately the interaction behaviour of this beam-column mode. The beam-column is assumed to be simply supported at the ends. Initial overall column imperfection (see Fig. 3) takes a sinusoidal curve longitudinally. The boundary conditions along the two edges of the plate in the T-section is taken so that they can keep the rotations about axis and the transverse membrane displacements as zero. The tip of the stiffener is free to deform. It is still assumed that the collapse of the idealised beam-column mode results from interaction between the local buckling of the T-

section and the overall buckling of the complete beam-column. The residual stress pattern recommended by Skaloud and Zornerova (ref. 16) is adopted for the stiffened panels and is shown in Fig.3 as well. In general, the residual stresses in the stiffener are usually significantly smaller than those in the plate (ref. 17) and are ignored. Initial stiffener imperfection in the out-of-plane direction of the stiffener is not considered. The initial plate imperfection is assumed to be sinusoidal across the plate outstand and takes the maximum value at the tip of the plate outstand. This imperfection pattern is approximate to one of the original stiffened panels. The initial plate imperfection is assumed to have the same wave shape as the local buckling mode.

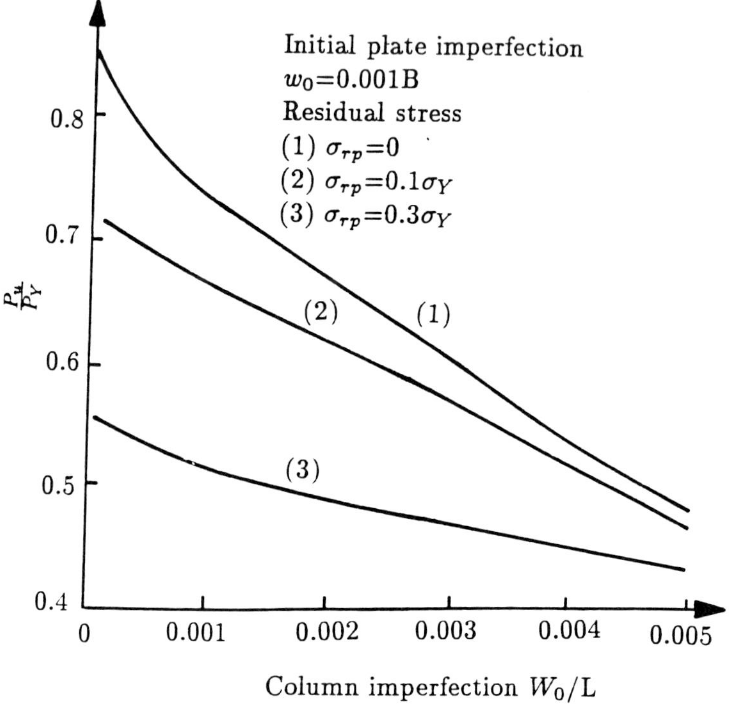

Fig. 4. Effect of column imperfections.

13. Figs 4 and 5 show the respective effects of initial column imperfections and residual stresses on the collapses load of the columns. The columns are assumed to be simply supported and centrally loaded at the ends. The material property and geometric sizes are: σ_Y=250MPa, B=550mm, t_p=lOmm (plate), t_s=lOmm (stiffener), H=80mm, L=1565mm. This column should therefore exhibit the most severe imperfection sensitivity because the Euler buckling load of the column and the local buckling load of the section are almost identical. But as the imperfections increase the sensitivity decrease.

SUMMARY AND CONCLUSIONS

14. A material and geometric non-linear spline finite strip method has been developed to analyse the interaction behaviour of the stiffened panel. A full range analysis is performed by employing the modified arc-length method and Newton-Raphson iteration. A lot of numerical results reveal that the present spline finite strip analysis for combined material and geometric non-linearities not only is an efficient tool but also has good accuracy.

15. A beam-column interaction mode from the stiffened panel is studied by using the large deflection elastic-plastic spline finite strip method. The respective effects of the column imperfections and the residual stresses on the collapse loads of the columns are investigated.

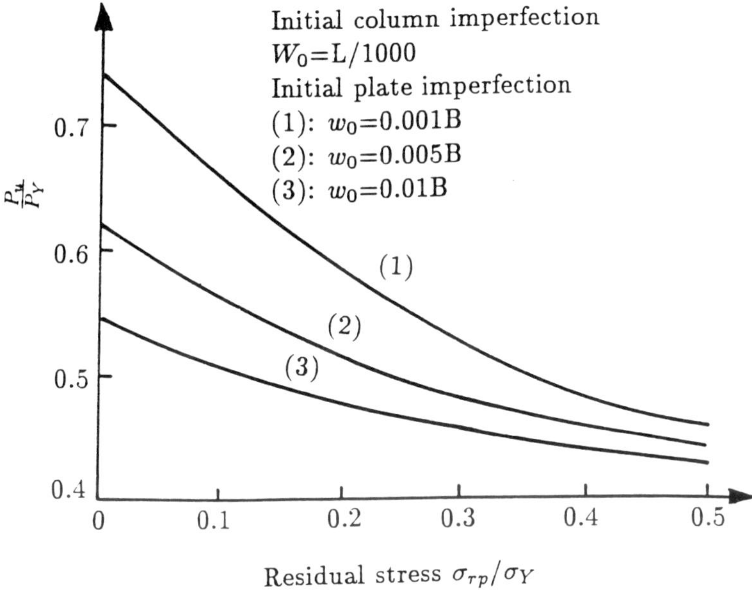

Fig. 5. Effect of residual stresses.

REFERENCES

1. KOITER,W.T. AND PIGNATARO, M. An alternative approach to the interaction between local and overall buckling in stiffened panels, IUTAM symposium, Cambridge, MA, Buckling of Structures, (Edited by B. Budiasky), Springer, Berlin, 1974, 133-148.

2. KOITER,W.T. AND PIGNATARO, M. A general theory for the interaction between local and overall buckling in stiffened panels, Delft University of Technology, Dept. of Mechanical Engineerng, Report WTHD-83, 1976.

3. VAN DER NEUT, A. Mode interaction in stiffened panels, IUTAM Spmposium, Cambridge, MA, Buckling of Structures, (Edited by B. Budiasky), Springer, Berlin, 1974, 117-132.
4. ALI, M.A. AND SRIDHARAN,S. A versatile model for interactive buckling of columns and beam-columns, Int. J. of Solids Structures, 1989, Vol.24, No.5, 481-496.
5. SRIDHARAN,S. AND PENG,M.H. Performance of axially compressed stiffened panels, Int. J. Solids Structures, 1989, Vol.25, No.8, 879-899.
6. YANLIN GUO AND SHAOFAN CHEN, Elasto-plastic analysis and experimental study of locally buckled cold-formed thin-walled channel sections, Chinese J. of Civil Engineering, 1990, Vol.23, No.3, 36-46.
7. YANLIN GUO ,Stability in the elastic-plastic range of cold-formed thin-walled box sections subjected to an eccentrical load, Engineering Mechanics, 1990, Vol.7, No.4, 1-9 (in chinese).
8. YANLIN GUO AND SHAOFAN CHEN, Post-buckling interaction analysis of cold-formed thin-walled channel sections by finite strip method, Thin-Walled Structures, 1991, Vol.ll, 277-289.
9. YANLIN GUO AND SHAOFAN CHEN, Collapse analysis of steel plate by finite strip method, Computer Methods in Applied Mechanics and Engineering, 1991, Vol.93, 319-333.
10. YANLIN GUO Nonlinear analysis of thin shells by spline finite strip method, J. of Xian Institute of Metallurgy and Construction Engineering, 1990, Vol.22, No,4, 366-373 (in chinese).
11. RAMM,E. Strategies for tracing the nonlinear respone near limits, Proceedings of the Europe-U.S., Workshop Ruhr-Universitat Bochum, West Germany, 1980, 153-163.
12. CRISFIELD, M.A. A fast incremental/iterative solution procedure that handles snap through, Computers and Structures,1981, Vol.13, No.l, 55-62.
13. NAYAK,G.C. AND ZIENKIEWICZ,O.C. Elastic-plastic stress analysis,a generalization for various constitutive relations including strain hardening, Int. J. Num. Meth. Eng., 1972, Vol.5, 113-135.
14. BRADFIELD,C.D. An evaluation of the elastic-plastic analyses of steel plates loaded by uniaxial in-plane compression, Int. J. of Mech. Sci., 1982, Vol.24, No.3, 127-146.
15. LITTLE,G.H. Stiffened steel compression panels--theoretical failure analysis, The Structural Engineer, 1976, Vol.54 , No.12, 489-500.
16. SKALOUD,M. AND ZORNEROVA,M. Effect of residual stresses on the stability problem of longitudinally stiffened compression flanges, Acta Technical, CSAV, 1984, Vol.20, No.2, 242-253.
17. OSTAPENKO,A. Ultimate strength design of wide stiffened plates loaded axially and normally, Symposium on Nonlinear Behaviour and Techniques in Structural Analysis, TRRL, Crowthorne, SR164UC, 1974.

185. Hyperbolic paraboloid reticulated shell roof structure for a gymnasium

T. T. LAN and J. ZHAO, Chinese Academy of Building Research, P. R. China

SYNOPSIS. This paper gives an outline of the design and construction of the roof structure for the Gymnasium of Beijing Institute of Physical Education. The structure is composed of four hyperbolic paraboloid double-layer reticulated shells with an overall dimension of 59.2m x 59.2m. The structural layout of the roof is discussed first. This is followed by some essential aspects of design and constructional details. The procedure of fabrication and erection of the roof system is also discussed.

THE STRUCTURAL LAYOUT

1. The Gymnasium of Beijing Institute of Physical Education is located near the scenic site of Yuanmingyuan in the north suburb of Beijing. The gymnasium is one of the sports facilities in the Institute for multi-purpose usage. During the 1990 Asian Games, boxing matches were staged here. The building is square in plan with a roof span of 52.2m, the overall dimension of the roof is 59.2m x 59.2m. The maximum height of the roof is limited to 18.6m which conforms to the limitations of building heights in Yuanmingyuan area. Fig. 1 shows the exterior view of the gymnasium.

Fig. 1. Exterior view of the gymnasium.

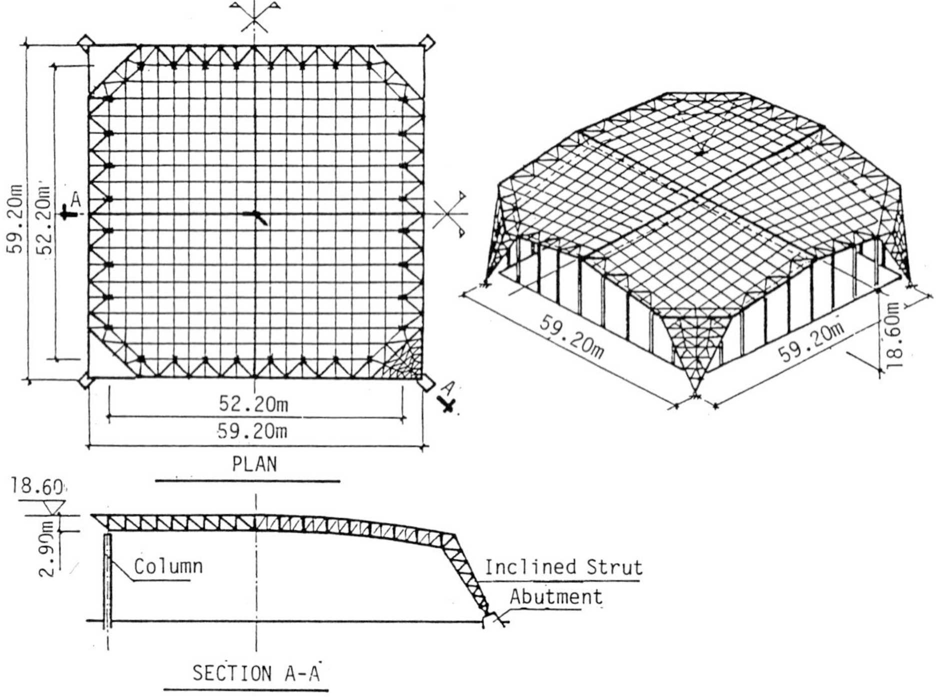

Fig. 2. Structural layout of the roof.

2. Considering comprehensively the architectural and structural requirements of the building, double-layer reticulated shells composed of four hyperbolic paraboloids were selected as the structure of the roof. Various grid arrangements of the space frame, such as two-way orthogonal lattice grids, two-way diagonal lattice grids and orthogonal square pyramid space grids were taken for comparison. Finally the orthogonal lattice grids were used in view of the ruled surface for a hyperbolic paraboloid shell. Thus each truss composing the reticulated shells is plane and straight, where all the top and bottom chords as well as diagonals are of equal length. This special feature will simplify considerably the fabrication and erection of the roof structure.

3. The decision of the module dimension and depth of the reticulated shell depends largely upon the span, column distance, loading, and the roofing. The limitation on the building height has restraint upon the depth. As the column distance along the perimeters is 5.8m, the grid module was taken as 2.9m x 2.9m. In accordance with the module dimension, a depth of 2.9m was used which took into account the arrangement of diagonal members and also space for ventilating ducts. The depth is only 1/18 of the roof span which is conducive to increase the interior height of the building.

4. The rise of the shell is limited by the following factors. Firstly, while observing the building height limitation, maximum space should be provided above the spectator seats. Secondly, the quantity and detailing of roof panels also require a smaller rise, since a large rise will cause more consumption of roofing material and difficulties in the connection of seams. Thus a comparatively flat roof was designed with a rise of 3.5m, resulting in a rise/span ratio of l/15.

5. The unique feature of the load-bearing system is that the reticulated shells are supported along perimeters and also on four inclined lattice struts at corners. The inclined struts are in the shape of an inverted triangle with upper side connected with the corners of the shell and lower points supported on the abutments at the outside ground level. The inclined struts act integrally with the shells to resist all kinds of loadings. The structural layout is shown in Fig. 2.

STRUCTURAL ANALYSIS AND DESIGN

6. The vertical loads acting on the reticulated shells are: (a) 1.6kN/m on the top chords, including roof panel and a live load of 0.5kN/m^2; (b) 0.4kN/m^2 on bottom chords, including the weight of catwalk, lighting and acoustic equipment etc. The self weight of the shell was estimated as 0.52 kN/m^2. As required by the boxing match, a lamp suspended at the centre of the roof will add a concentrated load of 30kN at mid-span. The loadings for calculating the vertical earthquake effect were taken as (dead load + $^1/_2$ live load). The thermal effect in the roof system was taken into account by checking the temperature change of ± 30°C. Circular hollow sections were used for the members. The layout of the structure is shown in Fig. 2. The structural layout of the roof used hollow spherical nodes of the Oktaplatte type. The steel used is grade 3, with a yield strength of 24 kN/cm^2.

7. The matrix displacement method was employed for the analysis of the reticulated shells with inclined struts as the integral part. The structure is assumed as a space truss with hinged joints and only one quarter is analysed by taking advantage of its symmetry. In the process of design, certain thoughts were given to the restraint of the supports. As the shell is rather flat, a strong restraint of the supporting columns to the shell will produce a considerable thermal stress in the shell members. Under the vertical load, the shell will exert a large horizontal thrust to the columns which will increase the column sections. For the benefit of both shells and columns, the horizontal restraints at the column top are released through constructional details so that the column will mainly resist vertical load. The horizontal thrust of the shell due to vertical load and also the horizontal earthquake effect are mainly taken care of by the inclined struts. Such construction shows the advantages of explicitness of force transferring and rationality in load resisting.

8. The following loading conditions were considered in the structural analysis:

(a) dead load + live load;
(b) dead load + 0.5 live load + vertical earthquake effect;
(c) dead load + live load + temperature increase 30°C;
(d) dead load + live load + temperature decrease 30°C.

In addition, the strength of the inclined struts was checked for horizontal earthquake effect and the strength of the sub-assembly units was checked for the erection procedure.

9. Under dead load plus live load, the maximum deflection in the mid-span is calculated as 16.01cm, which corresponds to 1/326 span. The maximum horizontal displacement at a column top is 1.70cm. The maximum compression in the top chord member is 672.6kN while tension in the bottom chord member is 613.2kN. The horizontal and vertical reactions at the base of the inclined strut are 262.0kN and 268.5kN respectively. The vertical load resisted by the inclined struts constitutes 11.6% of the total load.

10. The temperature change does not cause much influence on the member forces. When temperature variation is ± 30°C, the variation of the forces in top chord member is only 3.1% of the normal force. However, temperature effect has more influence on the horizontal displacement at supporting joints and reactions of the inclined struts. For example, on increasing the temperature by 30°C, the maximum horizontal displacement at the column top will increase to 2.61cm which is 53.5% more. The horizontal reaction of the inclined strut will increase to 558.5kN, more than twice the normal thrust.

11. The horizontal earthquake effect was checked by assuming the horizontal stiffness of the shell as infinite. As the columns have certain horizontal stiffness, they will act simultaneously with the inclined struts to resist horizontal earthquake displacements. However, the struts will take the main part, i.e. 85% of the total horizontal forces.

12. Seven types of circular hollow section are used with maximum sections of 159 x 19mm (diameter x wall thickness) and minimum of 57 x 4mm. The hollow spherical nodes have four types, the diameters are all 400mm, but with wall thickness varyies from 10mm to 16mm, most of the nodes are reinforced with diaphragms. The steel consumption of the structure is shown in Table 1. It is worthwhile to point out that the structure was designed to support a heavy type of roofing the load being 2.52kN/m^2. Yet the actual roofing used was lightweight decking with a load of 1.30kN/m^2. If the actual roof load is used for design, then the steel consumption is estimated to decrease further by 30%.

CONSTRUCTIONAL DETAILS

13. In order to release the horizontal restraint at the top of the column, laminated elastomeric pads were used for the support. Such a pad has the advantages of large horizontal shear deformation and good bearing capacity to resist vertical loads.

Table 1. Steel consumption of the structure.

Reticulated Shell tons	Inclined Strut tons	Purlin & Post tons	Total tons
Member 115.56 Node 39.31 Support 2.72	Member 14.38 Node 6.76	14.87	193.60
45.0 kg/m^2	6.0 kg/m^2	4.2 kg/m^2	55.2 kg/m^2

Fig. 3 shows the detail of the elastomeric bearing. The size of the pad is 300 x 300 x 78mm and the allowable horizontal deformation is 27.5mm.
14. A ball and socket type of supporting joint was used for the inclined strut. This will resist horizontal and vertical reactions effectively, but not transfer moments to the abutment. The supporting joint is composed of a semi-spherical node welded on a base plate and a ball hinge underneath. The detail of the support is shown in Fig. 4.

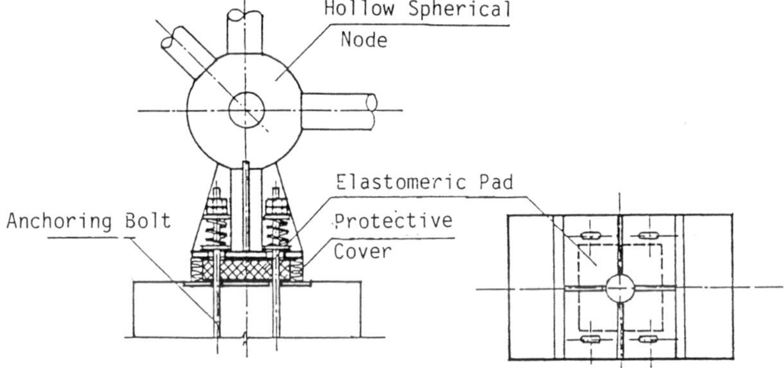

Fig. 3. Elastomeric bearing.

15. The roofing employs coloured sandwich panel supported on purlins of channel section. The purlins are resting on posts above the shell nodes and set diagonally with the grid, which will increase the plane stiffness of the shell. Thus the roof panels are laid in the direction so as to facilitate draining.

FABRICATION AND ERECTION

16. The reticulated shells were constructed by dividing them into individual blocks. These units were fabricated at ground level, then hoisted up into their final position and assembled. Such methods fully employ the advantages of the ruled surface of the shell. The construction proceeded as follows.

17. The primary unit of the reticulated shells is a plane lattice truss which was fabricated on an assembly platform by welding. Such a platform can be controlled

and levelled precisely. The welding of the trusses was carried out on the ground, so that a high working quality could be ensured.

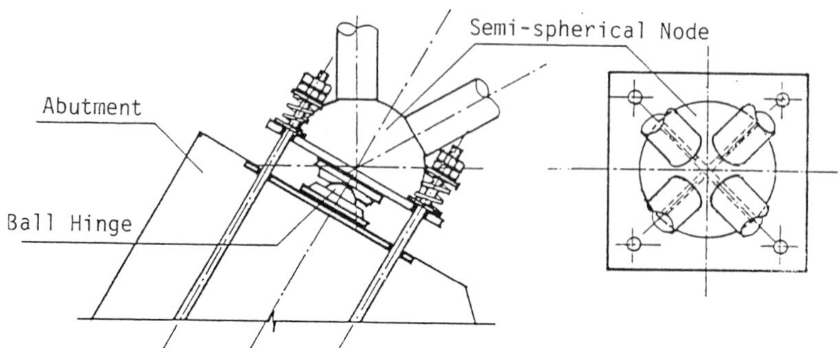

Fig. 4. Supporting joint for inclined strut.

18. After the fabrication of the lattice trusses had been completed, two or three pieces of truss were sub-assembled as a block. The twisting curvature was formed by the difference of elevation at both ends of the truss. Through the rigorous control of the elevation, the configuration of the shell conforms to the theoretical surface. The sub-assembly produced a spatial block with high rigidity which was hoisted up by a 40t crawler crane. Fig. 5 shows the erection.

Fig. 5. Erection of sub-assembly blocks.

19. On the construction site, a trestle was erected along the mid-span of the roof structure. The sub-assembly blocks were then erected symmetrically with one end resting on the trestle and other end on the ring beam. All the elevations of the blocks were adjusted by transit so that they would fit into their final positions accordingly. The shells were formed by welding the blocks together and then connected to the inclined lattice struts which were prefabricated on the ground. The entire roof structure was finished by removing the trestle.

20. The construction method employed proved to be very efficient in facilitating workmanship and speed of erection. The finished structure showed a high quality of welding as well as the precision of the dimension of the shells.

CONCLUSION

21. The roof structure of double-layer reticulated shells composed of four hyperbolic paraboloids is the first example application to sports buildings in China. It is believed to be one of the longest span roof structures of its type in the world. The completed building shows that it is not only visually pleasing but also economical in material consumption and construction cost. Comparing with other gymnasiums of similar size built for the Asian Games, the cost of the building either per square meter or per seat is among the lowest.

22. The roof system combines the reticulated shells with four inclined struts. The vertical loads are mainly supported by the peripheral columns while the horizontal loads are mainly resisted by the inclined struts. Such a system of load-resisting was made possible by using elastomeric pads on the column supports.

23. Due to the limitation on building height, the rise of the shell was taken as only 1/15 of the span which did not show a distinct characteristic of shell action. It is suggested that the rise could be increased to 1/6 - 1/8 of the span for better performance.

186. Expandable structure for the Venezuelan Pavilion at Expo '92

C. H. HERNANDEZ, Universidad Central de Venezuela, and
W. ZALEWSKI, MIT, USA

SYNOPSIS. Deployable structures are the structures of the future, offering tremendous potential for both engineers and architects. The paper describes the design and construction of the Venezuelan pavilion at the Exposition in Seville in 1992.

INTRODUCTION

1. The use of deployable structures began in ancient times when nomadic tribes moved from place to place. Their relatively light tents and tepees were made collapsible and transportable. The main advantages of deployable (transformable) structures are their re-usability, the small volume they occupy during storage and transportation, and the ease and speed of erection. The attractiveness of the idea of structural transformation and the expectations of its practical application to the new uses and demands in the contemporary technological and economical environment, brought recently new interest in the old idea; its revival was stimulated by the trend-setting works and concepts of E.P. Piñero and R. Buckminster Fuller.

2. Implicitly, the deployable structures of the future will have a far broader range of applications than ever before; in architecture, construction and aerospace industries.

3. They can serve either as once only deployed frameworks used as stationery objects or, more often, can be put to recurring uses, passing various cycles of deployment, collapsing and transportation.

4. The repetitive utilisations of deployable structures include emergency facilities (shelters, bridges, hangars etc.), funfair exposition pavilions, grandstands, sport facilities and various types of domed, vaulted and plane space trusses. In outer space the single, deployable only operations, are considered for the main framing support systems and manned stations and laboratories.

5. In these days, there exists already a multitude of contributions reflecting certain specialised interests in the theory of structural transformations. Unfortunately, in most cases these investigations are focused on the similar, relatively narrow studies, such as the mathematical formulation of geometrical and mechanical constraints and the kinematics of events associated with the evolution

Space structures 4. Thomas Telford, London, 1993

of structure from its compact to final form. Quite rare are however, realistic propositions and studies of the broader character, more directly related to the development, planning and erection of usable, easily buildable, economical structures, resisting the whole range of transportation and service loads.

6. In response to the professional and intellectual challenge of the idea of structural transformations and to involve the students of the Massachusetts Institute of Technology (MIT) in Cambridge -Mass and the researchers in the Instituto de Desarrollo Experimental de la Construccion (IDEC) in Caracas, in the new and very promising field of Architecture and Engineering; the investigations into deployable structures have been undertaken a few years ago with workshops and seminars conducted in both places. Their research and educatory efforts have been motivated by the firm belief in the rationality in the idea of structures adjusting their forms to the consecutive stages of construction and service, and also by the strong desire to see them built.

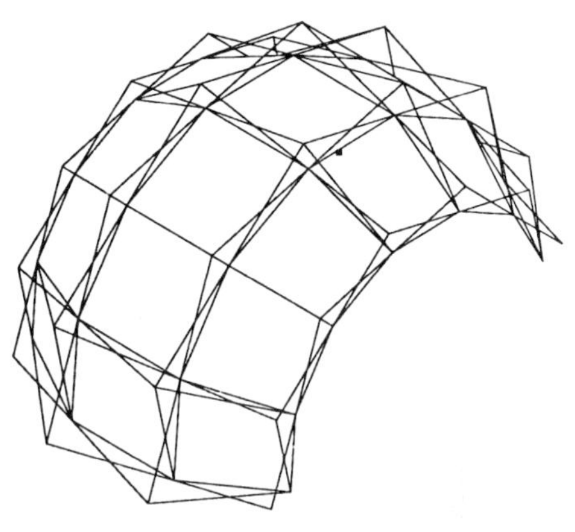

Fig. 1.

7. Until now the emphasis of our research was mainly related to under explored areas such as:

- Economical and functional feasibility of deployable structures.
- Classification and selection of their types and material.
- Design of joints and other hardware.
- Mechanics and management of the deployment and collapse operations.

The initial experience and understanding of technical problems pertinent to the implementation in practice of the concept of structural transfiguration have been developed through studies on models and large size prototypes as well as by the contacts with prospective clients, producers and contractors, resulting in a deeper involvement with various actual projects.

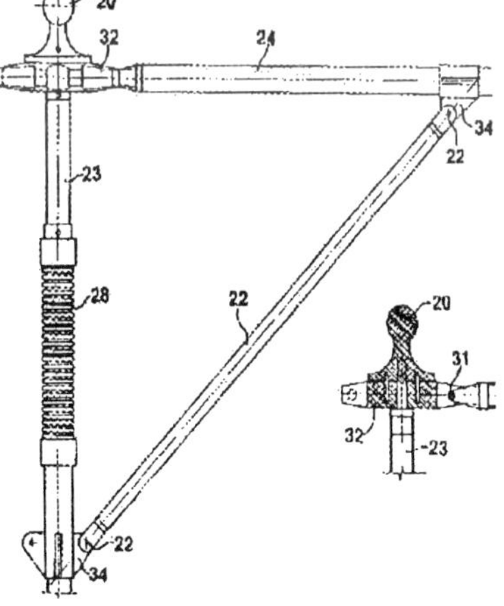

Fig. 2.

8. The largest of these projects, namely the pavilion of Venezuela for 1992 Exposition in Seville, has already passed the first stages of its realisation, offering a quite positive example of the team work in which the constituent, artistic, architectural and technical concerns and ideas were mutually interacting and incorporated into the final solution. It also inspired many conceptions for the broader uses of deployable structures in the future.

Fig. 3.

9. The opportunity for the project of the Venezuelan pavilion at the Expo'92 in Seville has been the first occasion to demonstrate, on a large scale, the application of the concept of deployable structures. The requirements of the project, the time limitations for its design and construction; the need of a temporary building, that had to be returned back to Venezuela and the high costs of building in Spain made the use of a deployable structures a convenient choice. On the other hand, to go to a universal exposition whose motto was THE ERA OF THE DISCOVERIES, presented itself as a unique opportunity to advance the development of deployable structures and provided a technological innovation in the Pavilion for Venezuela. Aluminium was selected as the building material due to its particular characteristics. Its low weight (density 2.71 kg/cm^3) allowed lighter pieces to simplify its transportation and handling. The use of this material was also favoured for its behaviour vis-a-vis the climatic agents, considering that its structure would be exposed to different conditions.

10. Its particular characteristics and production process made it possible to prepare the required alloy, thus obtaining the necessary resistance for each case

(Alloy N° 6261 with an ultimate strength of 3300kg/cm^2), thus rationalising its use. Also, the possible finishes attainable were in accordance with the image of the Pavilion; a symbol that represented the level of technological capabilities of the Venezuelan industry. In this way close team work developed between the researchers and professionals from Idec and the engineers of the aluminium industry Alcanven and Albarca. This led to the successful design, detail engineering, production engineering and construction of this structure, between July 1990 and May 1991, when the structure was shipped to Spain.

Fig. 4.

11. The raw material processing plants, VENALUM and ALCASA, donated the aluminium necessary for manufacturing the structure. The firm Aluminio de Venezuela, ALCANVEN, made the strength studies and tests, prepared the required alloy, manufactured the moulds and pieces and, finally, the firm Aluminio de Barquisimeto, ALBARCA, painted and assembled the components. All this process had the support and guidance of the foundation for the development of aluminium FUNDALUM, which had been participating in some of our investigation in the area of deployable structures. Their action was thus decisive for the development of this process.

12. For the manufacture of the components, the extrusion process instead of the injection process was used, because it permitted a greater flexibility in the design of the components and simplification in the manufacturing of the moulds; thus, considerably shortening construction time. The Pavilion building was conceived basically as the container of an audio-visual room, which, due to its geometric requirements, made it necessary to adopt a structure free from intermediate supports. The result was a building whose longitudinal section would form a triangular shape to orient the visual relations between spectator and screen (Fig. 5).

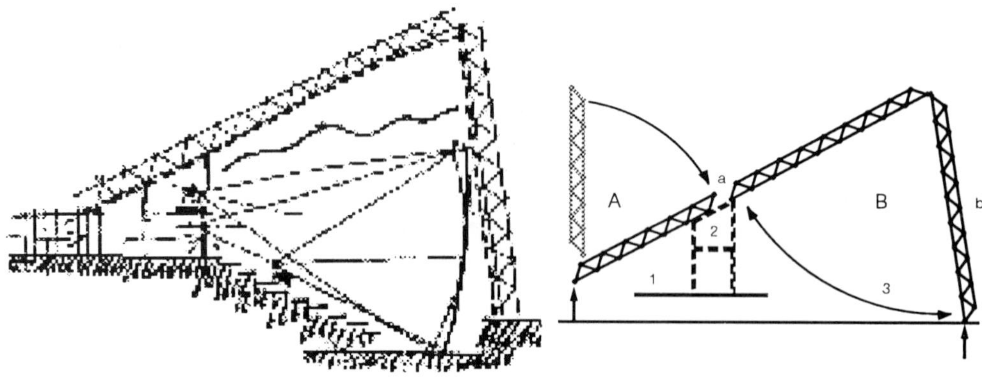

Fig. 5.

Fig. 6.

13. A deployable structure that folded in one direction, formed by parallel trusses hinged at its nodes to permit an accordion-like fold, was chosen due to its adaptation to the geometrical requirements of the Projection Room and because of the smaller number of pieces required together with simplification in the erection.

14. The structure, defined on two planes, (Fig. 6) the superior (A), which goes from the square to the top, and the rear (B), that goes from the top to the floor. The plane A has an intermediate support (the projection cabin) and it was divided into two groups of trusses, one with a span of 13m and the second with a span of 18m, plane B has trusses of 18m also. The side planes (C), that complete the ensemble and collaborate with the distribution of the side forces from the wind, are made up by variable size trusses, connected to the structural grid of the building.

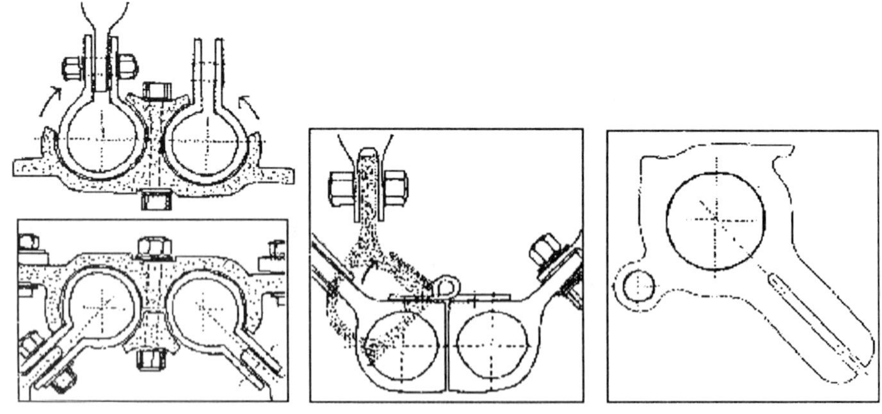

Fig. 7.

15. A system of low-weight panels is incorporated into the structure, designed to provide thermal and acoustical isolation, as well as the interior and exterior finish of the audio-visual room.

16. Because the connections are responsible for the accordion-like movement and for joining together the different components, the design work concentrated on their behaviour (Fig. 7). As a result, two separate pieces were designed, which formed the connection assembly, namely; the hinged-node and the staple.

17. The hinged-node makes up the trusses, allows the accordion like movement, and receives the transverse elements that give stability to the structure in its final state Fig. 8A). The staple maintains the hinge in the right place after unfolding of the structure and it is also the element that carries the cover (Fig. 8B).

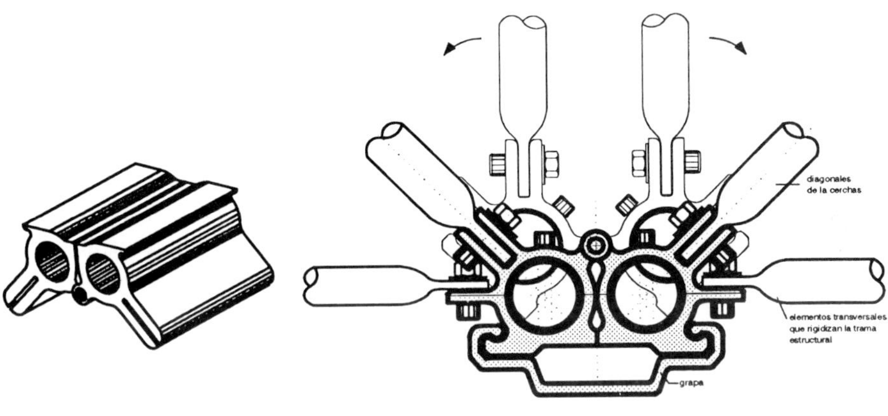

Fig. 8A. Fig. 8B.

18. For the rest of the components the use of standard production circular sections was considered, to simplify the production and lessen the construction time. Consequently, the trusses are made up of continuous tubular elements, on which the hinged nodes are fixed at 2meter intervals. The diagonals are tubular elements, flattened at the ends, that are bolted to the node leaves. The trusses are positioned parallel to each other and are joined by the nodes in an alternating way. The first truss is connected to the second at the inferior string, and the second with the third at the superior string, and so on, so that the trusses can rotate 45 degrees with respect to each other, allowing extension or folding as an accordion, forming a package to be shipped to Seville (Fig. 9). The process of the roof erection was undertaken according to the geometry of the audio-visual room and to the possibility for mobility on site. Because the superior plane required an intermediate support from the projection room, the erection process was divided into two stages. Stage A that covers the waiting area and the projection room, and stage B that corresponds to the audio-visual room, using the top of the projection room for the rotation axis.

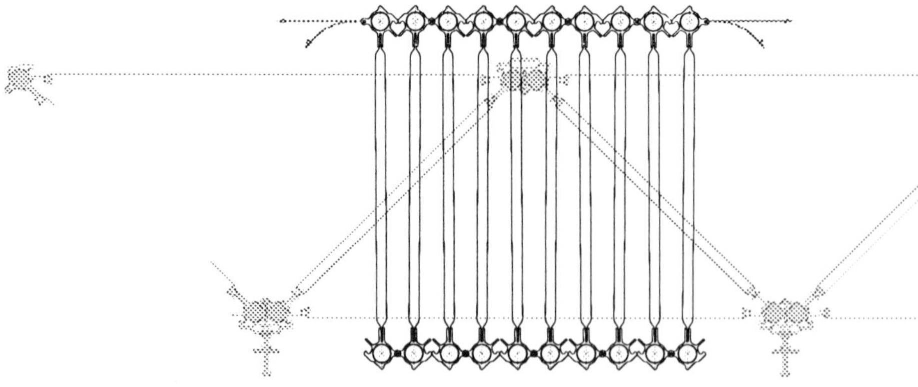

Fig. 9.

19. The first stage was assembled from a group of twenty two 13m long trusses that once deployed will cover an area of $308m^2$. The second stage was assembled from two groups each of twenty two 18m trusses, joined together at one of the extremes. In closed configuration this group was 3m high, 2.8m wide and 18.8m long (including packing) weighing 8000 kg. Two cranes were used to lift the packages with the help of a beam that worked as a rail for several trolleys joined by a close circuit of steel cable, to allow the symmetrical movement required during deployment. The structure was hung from trolleys and placed in the centre of the beam deploying in both opposite directions (right and left) in a movement similar to that of opening a vertical curtain held by its centre, thus changing from the initial close dimension of 2.8m (part A, Fig. 10) to the final developed size of 22m wide (part B, Fig. 10)

20. Once the structure was suspended and unfolded, it was fixed to avoid its refolding. This movement was restricted through the use of staples in the nodes. Some tubular perpendicular elements were used as well in the trusses to guarantee the side stability of the structure and the aperture size. All these operations were undertaken while the structure hung from the crane. Subsequently, the net was put into place, already set in its final configuration. In the case of the first section, this was put on top of the anchors on the square, inclining it until it reached the projection cabin (Fig. 11). Likewise, the second section was positioned the same way, which included, as mentioned before, two planes folded on themselves from the top, from which the structure was suspended and unfolded. The structure was made rigid, by using the staples and transverse elements. Subsequently, the structural net was moved vertically until the extreme of one of the two planes rested on temporary supports placed over the projection room, which fixes the bottom seam in place and permitted its rotation (Fig. 12). Then the beam was moved, inclining the structure to the rear wall. This movement permitted the two planes to separate, forming an inverted "V". Subsequently, the other plane took its

final position attached to the anchors on the rear wall (Fig. 9). Finally both nets were connected and the temporary supports from the projection room were taken away.

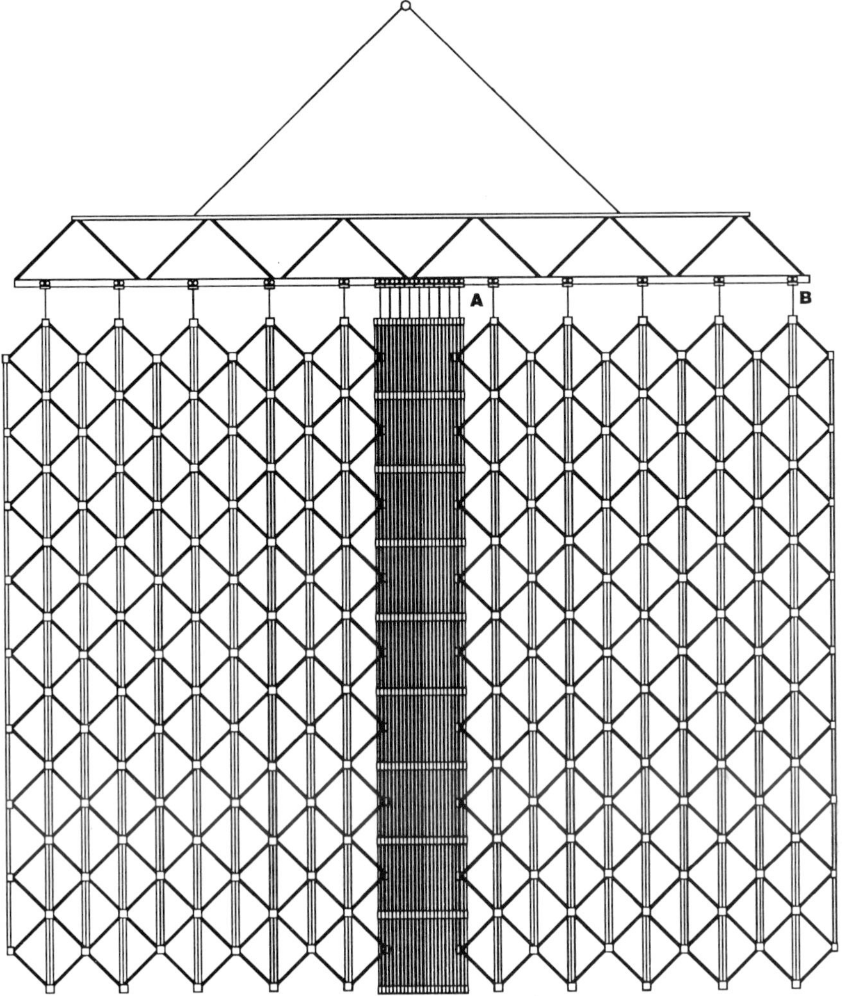

Fig. 10.

21. The system of light panels were then added to the structure. Each panel is composed of an exterior moulded surface, made of polyester resin reinforced with glass fibre 3mm thick, with a light grey gel-coat finish, and an interior surface made up from a galvanised steel sheet, 0.7mm thick, which is covered by a film of dark grey polyvinyl chloride. The interior is made from rigid polyurethane foam, see Fig. 14.

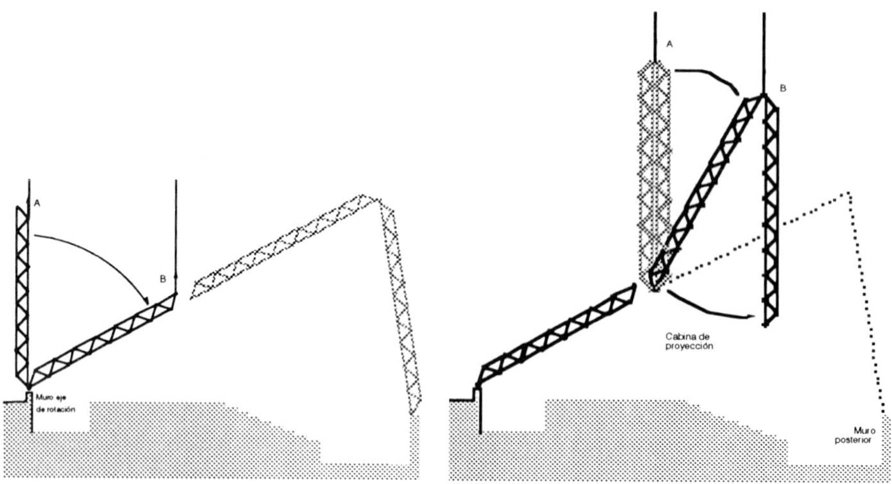

Fig. 11. Fig. 12.

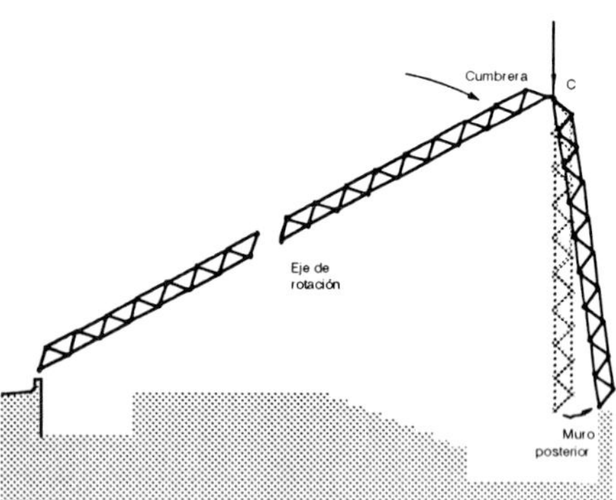

Fig. 13.

22. The panels are suspended from the structure's nodes by adjustable bolts. These bolts support tubular elements, on top of which the panels rest (Fig. 15).

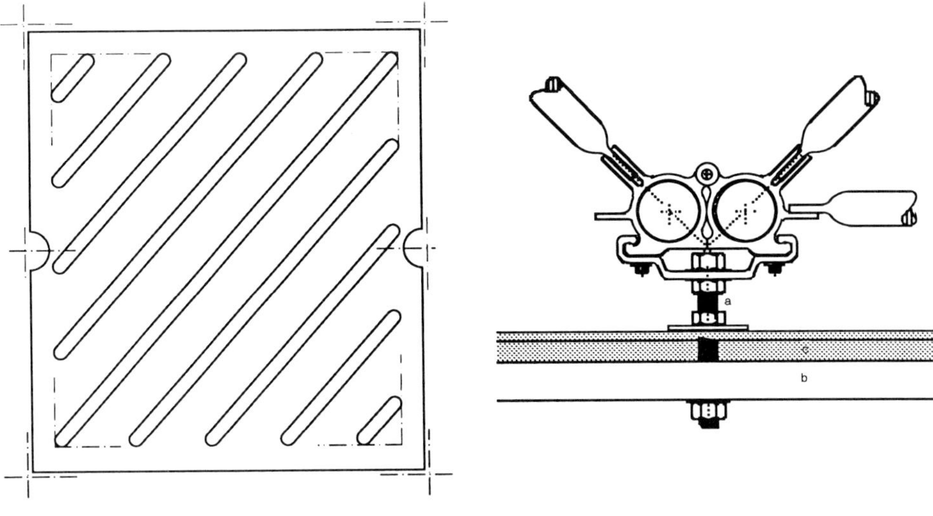

Fig. 14. Fig. 15.

23. The bolts permit adjustment to level the covered surface after the panels are hung. The panels also have an external moulding that eases rain water distribution. The joints between panels were finally filled with structural silicone to water-proof the building.

24. A complex aluminium structure formed from 6475 pieces, with a weight of 10,000kg providing 1242m^2 of structure area was built in a Venezuelan factory, with factory tolerances and quality, shipped 5235 miles and successfully erected in two days. This is in itself a record, but more importantly it shows what can be done using deployable structure technology, opening new possibilities for the use of these structures in architecture. Also, Venezuelan industry and research showed that it could respond to technological challenges, creating new ways of building using aluminium.

25. The technological contribution to construction that transformable structures bring to the present have transcendental importance.

26. The Pavilion of Venezuela in the Seville 1992 Expo is an example of this in the eyes of the world.

REFERENCES

1. INSTITUTO DE DESARROLLO EXPERIMENTAL DE LA CONSTRUCCION. Pabellón de Venezuela Expo '92. Editor IDEC. Ex Libris, Caracas. Venezuela. Primera edición 1992.

2. W. ZALEWSKI. Venezuelan Patent for ESTRAN 1, 1989.

3. CARLOS H. HERNANDEZ. Mobile and rapid assembly structure. Topic in Engineering vol. 8, p.237-248. Editor P.S. Bulson Computational Mechanics Publications, Southampton, UK and Boston USA 1991.

187. Defects in the fabrication and erection of 3D structures with spherical nodes

D. DAKOV and S. TCHUKANOV, Higher Institute of Architecture and Civil Engineering, Bulgaria

SYNOPSIS. Some typical defects and damage observed in 3-D structures such a double-layered grids of the type MArxI (modification of the Mero system) are discussed. The main reasons for the occurrence of such defects are considered. Special attention is paid to the influence of the gaps between the ends of the bars and the spherical nodes for the security of the entire space structure type. The redistribution of member forces due to the replacement of some bar members has been investigated. The results obtained by theoretical and experimental studies of these three dimensional structures are analysed.

TYPICAL DEFECTS IN THE FABRICATION AND ERECTION OF 3-D STRUCTURES OF THE MArxI TYPE.

1. Three dimensional steel space structures became popular in Bulgaria from 1977, after the opening of a specialised factory for their fabrication, which had a capacity of 10000 tonnes per year. The structures, which are a modification of the Mero system, are produced under a Russian licence. Because of their advantages, the use of 3-D structures quickly spread in many regions of the country, but simultaneously some specific defects of the structures became apparent. These defects led to doubts about their reliability.

2. After extensive participation in different damage inspections considering the faults and failures of 3-D structures, the authors have determined two primary reasons for these defects, namely:

(a) Omissions in the design of the structure, causing conditions for failures.
(b) Negligence and insufficient quality control in the production and erection processes.

3. The omissions in the design are relatively small in number, due to the high qualifications of the designers, specialists in this field and using reliable software. Sometimes the initial forces in the bars and nodes are under-rated and the influence of deformation of the structure with low rigidity is not accounted for.

4. The design of the node required to transmit member forces is also unsatisfactory. For often in practice it is very hard to determine whether the bolt's head is tight in the tube's sleeve. It is impossible to compensate for deviations in the bar's length. During the pre-tightening of the bolts, initial forces will be

Space structures 4. Thomas Telford, London, 1993

introduced into the members, tension in the tightened bars and compression in the adjacent bars.

5. The most frequent defects found in the 3-D MArxI type structures, obtained during fabrication and erection are as follows:

- gaps between the face of the nut and the sleeves, or between the face of the nut and the node;
- low quality welds, between the tube and the sleeve;
- lack of perpendicularity between the contacting surfaces in the node;
- tearing of the bolts;
- deflection of some low loaded bars with high slenderness.

6. The observations of erected 3-D structures proved that the dimensions of the gaps are from 0.1 mm up to 4.0 mm, but in single cases even bigger gaps can be found. The gaps between the node and the nut cause elimination of the corresponding compression member, and the gaps between the bolt's head and the sleeve cause disconnection of the corresponding tension members. However, with an increase in load, the smaller gaps will be closed. In these cases the relevant connections will become functional, but the real member forces will be considerably different from those calculated in the initial model.

7. According to some researchers (ref. 1), separate gaps with dimensions up to 3 mm, spread throughout the structure, do not necessarily affect its stress and strain condition. If the gaps are concentrated at a node, located in a heavily loaded part of the structure, a real danger of local geometrical instability appears, which may take the structure to a limit state.

8. Local geometrical instability may be prevented by considering the combined work of roof elements (purlins, corrugated steel sheets) and the structure, if reliable connections between the structure and cladding are made. To achieve this, continuous purlins have to be used and corrugated steel sheets with high rigidity are necessary.

9. The gaps in the nodes are due to inaccuracies in fabrication, as well as improper erection of the structure.

10. Theoretically, prescribed manufacturer's tolerances for the bar's length are $\pm$ 0.1 mm, but in practice there can be found deviations of up to $\pm$ 2.0 mm in 25% of the bars. These deviations from the designed dimensions, may cause a considerable change in the stress state of the 3-D structures, especially in larger structures.

11. Low quality welds, connecting the sleeves to tubes, form a considerable part of the total number of registered defects. This leads some of the designers to require 100% quality control for the complete member, which makes the structure very expensive. Obviously, effective quality control is of great importance, but 100% control is necessary only for the welds, as is recommended in ref. 2.

12. Lack of perpendicularity between the contact surfaces in the nodes and torn bolts are rather rare. Tearing of bolts or welds implies very often a pre-failure situation, which leads to immediate replacement of the corresponding structural elements. The deflections of some under stressed members, with small diameters, is usually due to a combination of defects in both the fabrication and erection. If

the deflections are small, it is recommended not to replace these bars, in order to avoid additional complications.

SOME LESSONS LEARNED FROM THE FAILURE OF THE 3-D STRUCTURE USED FOR STORAGE IN ROUSSE

13. The roof structure of the building consisted of four equal separate blocks, with dimensions of 42 by 70.5 m, in which there are strips free of bars (Fig. 1). The blocks have been assembled on site and were then lifted up to the prescribed level and placed on the top of reinforced concrete columns, using six cranes of two different types.

14. During the lifting of the first block three members detached from the node, due to failure of the weld connecting the sleeves to the tubes.

15. The fitters had replaced the defect members and then completed the erection of the first block. During the lifting of the second block similar failures started when the block was at 1.5 m above the site, resulting in considerable distortions of the whole block. After quickly lowering the structure to the ground additional damage appeared.

16. Inspection of the failure proved that serious faults in the fabrication of the structural elements had been made in the factory, coupled with unacceptable deviations from the designed dimensions of the members resulting from insufficient quality control. Several omissions have been made in assembling and erecting the 3-D structure as follows; namely a poorly prepared site, and no synchronised lifting of the structure, etc.

17. The stress analysis of the structure proved that the elimination of some members aimed at saving steel, is not the primary reason for the failure. After the occurrence of this damage designers avoid using 3-D structures with intermittent strips.

INVESTIGATION OF A 3-D STRUCTURE WITH REPLACED MEMBERS AFTER THE ERECTION

18. The structure investigated was the Technical University of Sofia. It consists of six equal blocks, having dimensions of 42 by 30 m (Fig. 2). After the erection of the whole structure, completion of the roof and partial installation of the equipment, a number of defects have been noticed, including; low quality welds, large gaps and torn bolts. The designers insisted, that the damaged elements should be replaced, but they gave no advice as to the consequence of the changes and have not prescribed ways to decrease the load in the neighbouring bars. The replacement of the bars has been chaotically undertaken, which resulted in increasing the number of damaged bars. Only after the deterioration of the structure was the aid of the laboratory for Testing Steel Structures-VIAS requested. The task was to establish the load distribution in the 3-D structure after the replacement of the damaged bars. A stress analysis of different loading states has been undertaken, as well as testing of the structure.

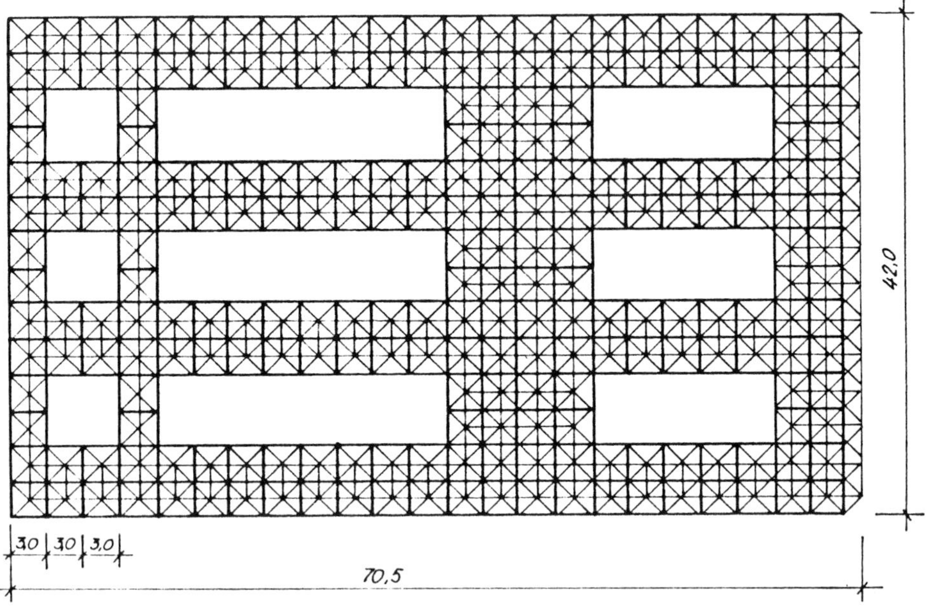

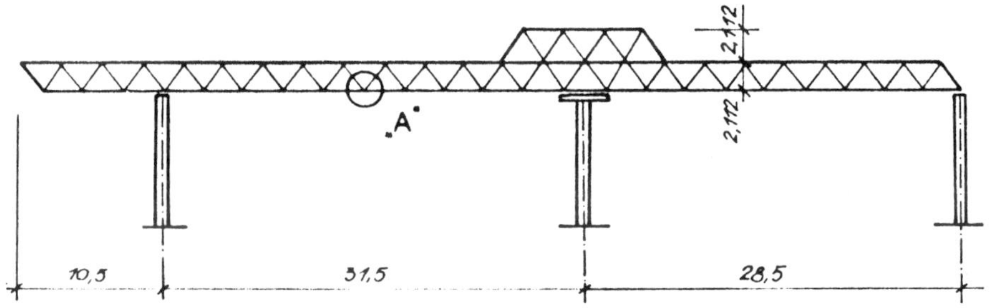

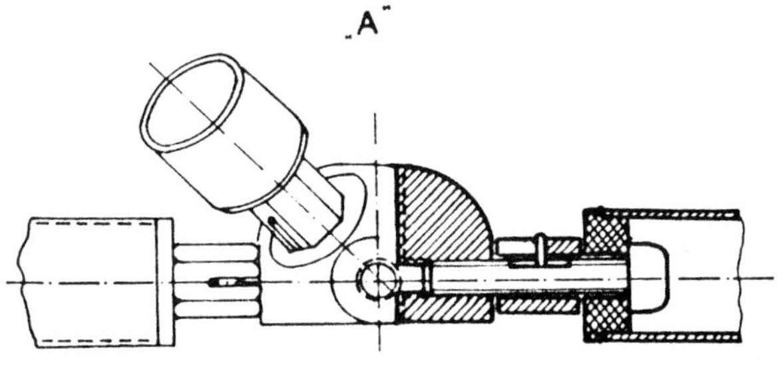

Fig. 1.

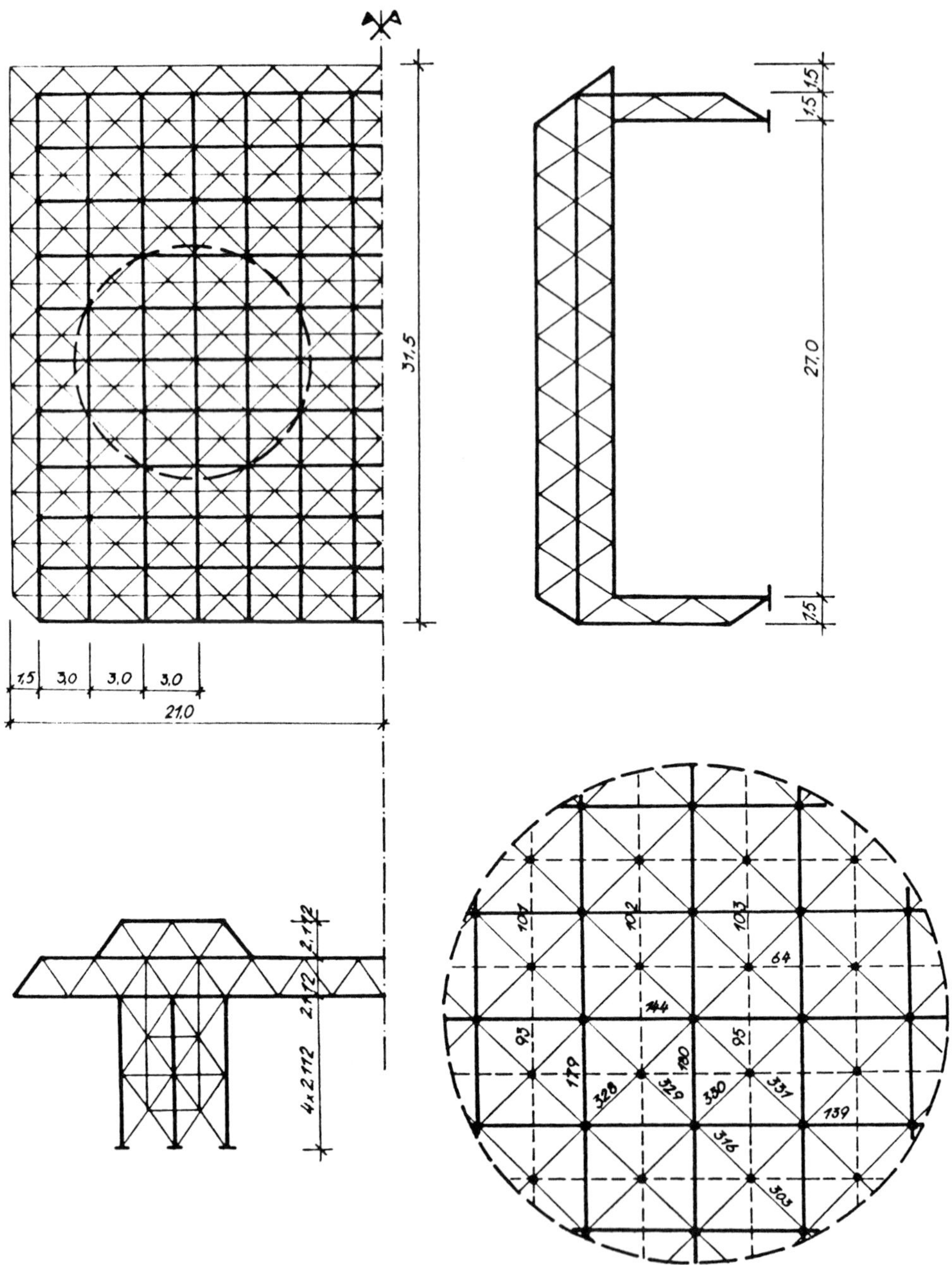
31,5
1,5
1,5
27,0
1,5
1,5
3,0
3,0
3,0
21,0
2,112
2,112
4 x 2,112
101
102
103
64
144
93
95
179
180
328
329
330
331
139
316
303

Fig. 2.

19. The static analysis proved that when a member is replaced, it is possible for the force in the diagonal members situated near the substituted members to change from tension to compression. Some of these diagonals are not able to carry the compression load and will stop working. There are some members with smaller cross sections in which the load direction does not change, but its magnitude may increase by up to six fold (Table 1).

Table 1. Characteristic loads in kN, for selected members.

Bars		Dead load $g = 1$ kN/m²				Snow s	Wind w
		Design state	Exper. result	Elim. bar 102	Elim. bar 103	0.7 kN/m²	
Lower grid	93	41.2	41.3	48.8	----	39.3	-15.6
	95	73.3	28.2	85.0	25.3	68.7	-26.8
	101	46.6	---	62.0	66.7	44.6	-17.7
	102	78.7	---	----	99.2	74.8	-29.4
	103	78.1	---	97.8	---	74.1	-28.8
Midle grid	144	0.8	---	-5.7	---	0.8	-0.3
	179	4.7	---	14.9	30.2	4.7	-2.0
	180	4.6	---	14.8	30.0	4.6	-1.9
Diago-nals	316	-6.3	---	-12.8	----	-5.9	2.5
	328	2.6	---	-14.0	---	1.9	-0.7
	330	5.5	---	11.4	-11.4	5.2	-1.9

20. The calculation undertaken to assess the stress in some of the bolts fitted with larger gaps on one side of the nodes, showed that the normal stresses, due to bending, may increase by a factor of up to 2.5 times. This is the main reason for the bolt failure. It has to be kept in mind, that in this case the members are also eccentrically loaded.

21. The analysis of the experimental investigation proved, that the load in some members of the real structure my differ by 2.5 times the calculated loads, obtained from the initial ideal static analysis. Therefore, the replacement of some individual members, without prior analysis of the state of the structure can be dangerous.

REFERENCES

1. DEMIDOV N. N., BRAZENAK G. U., Quantitative evaluation and assessment of structural ability of MArxI construction with defects, Buildings and Architecture, 1989, No 5, Novosibirsk.

2. SCHMIDEK, R., STUTZKI, CH. Quality control in a medium sized lightweight steel construction company. Proc. 1. Symposium spatial structures, 1990, Dresden.

188. Theoretical analysis of the sheet space structure system and engineering practices

H. L. ZHAO, X. M. CAO, M. Q. QIAN, Q. YANG, J. MA and C. Q. ZHAO, Southeast University, P. R. China

SYNOPSIS. This paper gives an introduction to the structural principles, advantages and construction of a sheet space structural system. The strength analysis and the method for calculation of the critical load of the combined structure are also investigated in detail. The mechanical performance and economic benefit of several engineering projects of various span are studied to give a suitable design range for these structures.

BRIEF INTRODUCTION

1 The merits of the sheet space structure system are represented by thin shells, grids, grid shells, suspended and gas filled structural forms. This is a new structural system adapted for Chinese conditions. The advantages are as follows:
(a) the structure has combined functions of load carrying, and enclosing, and hence it must have reasonable mechanical behaviour;
(b) the existing gas layer gives good thermal insulation even though the walls are thin;
(c) formed with sheets, the space structure is light and rigid;
(d) factory manufacturing guarantees the engineering quality;
(e) the structural configurations are nimble and unique. Consequently, these forms are greatly appreciated by architects and structural engineers.

2 Essentially, the structural system is an application of aero-structure principles in civil engineering; the main attributes being low-weight and high-strength. These characteristics may be obtained in two ways.

(a) Use of lightweight planking. There are two types of planking. One is the laminated sheet, its unit weight is about 10kN/m^3. The bending strength is 120N/mm^2. The axial tension strength is 80N/mm^2. It is produced by gluing bamboo at high temperature and under pressure with PRF structural glue (see Fig. 1). This results in good penetration resistance, fire performance and durability. The other is strengthened light composite cement sheet. Its unit weight is 10kN/m^3. The tensile (compressive) strength is 50N/mm^2. This type of planking is made from lightweight cement and steel wire mesh sheeting (see Fig. 2).

Fig. 1. Laminated bamboo sheet.

(b) Use of a lightweight thin-walled steel framework. The composite action of the sheets greatly reduces the forces within the structure. As a result, a strengthened light composite cement sheet framework can be made with thin-walled steel pipes (Fig. 2). However, aluminium penetration needs to be applied, which gives more than 50 corrosion free years.

3. The jointing arrangement between two of the skeletons and between the skeletons and the sheets is shown in Fig. 3.

THEORETICAL ANALYSIS AND MODEL EXPERIMENT

4. The theoretical analysis of the sheet space structure system includes strength analysis, vibration analysis and buckling analysis. Only the linear-elastic buckling analysis of the sheet space structure system is given in this paper.

The buckling analysis of the sheet space structure system

5. As the structures are composed of beam and shell members, it is difficult or impossible to give a theoretical solution. The finite element method has therefore been used. Basic assumptions made are as follows:

a). the beams are initially straight and the sheets are flat;

b). the material is linear-elastic;

c). the sheets are rigidly connected to the beams and the deformation of the beams and sheets is in accordance with the plane assumption;

d). the axial forces in the beams and the membrane stresses in the sheets are invariable while the structure is buckling.

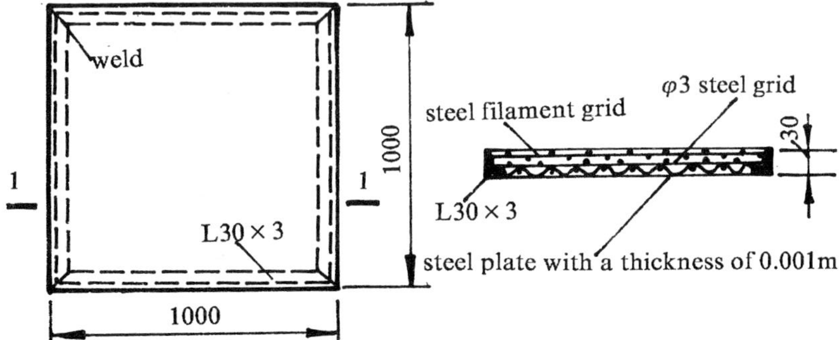

Fig. 2. Strengthened lightweight composite cement sheet.

The formulation of the FEM for linear-elastic buckling analysis

6. In the linear-elastic buckling analysis, the axial (membrane) forces of the elements and bending forces of the elements are connected, and the relation between node forces and node displacements is as below:

$$\begin{Bmatrix} F_{em} \\ F_{ef} \end{Bmatrix} = \begin{bmatrix} K_{em} & 0 \\ 0 & K_{ef} \end{bmatrix} \begin{Bmatrix} \triangle_{em} \\ \triangle_{ef} \end{Bmatrix} + \begin{bmatrix} 0 & 0 \\ 0 & N_e \end{bmatrix} \begin{Bmatrix} \triangle_{em} \\ \triangle_{ef} \end{Bmatrix} \quad (1)$$

where: $\{F_{em}\}$ are the axial (membrane) forces in the element;
$\{F_{ef}\}$ are the bending forces in the element;
$[K_{em}]$ is the axial (membrane) stiffness matrix of the element;
$[K_{ef}]$ is the bending stiffness matrix of the element;
$\{\Delta_{em}\}$ are the axial (membrane) displacements of the element nodes;
$\{\Delta_{ef}\}$ are the bending displacements of the element nodes;
$[Ne]$ is the geometric stiffness matrix of the element.

Transforming Eq. 1 to the global coordinates gives:

$$\begin{bmatrix} p \\ m \end{bmatrix} = \left(\begin{bmatrix} K_{pp} & K_{pm} \\ K_{mp} & K_{mm} \end{bmatrix} + \begin{bmatrix} N_{pp} & N_{pm} \\ N_{mp} & N_{mm} \end{bmatrix} \right) \begin{bmatrix} u \\ \theta \end{bmatrix} \quad (2)$$

or

$$\{p\} = ([K] + [N])\{u\} \quad (3)$$

7. Having given loads {p }, we can obtain the geometric stiffness matrix of the element [N], and according to the classical definition of stability, the minima eigenvalue of the structural eigenvalues (Eq. 4), found using the counter power method, gives the critical load of the structure.

$$([K] - \lambda [N])\{u\} = \{0\} \quad (4)$$

$$\{p_{cr}\} = \lambda\{p\} \quad (5)$$

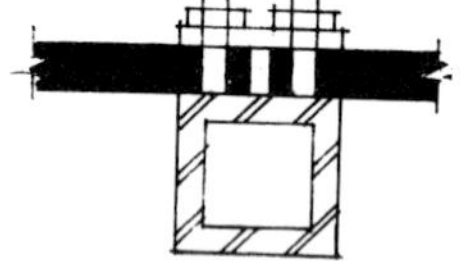

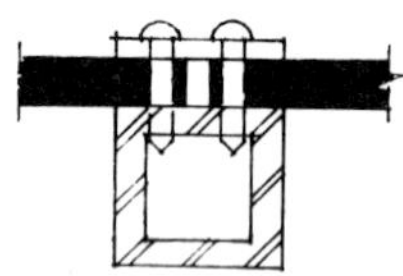

(a) Weld screw. (b) Self push bolts.

Fig. 3. Connections.

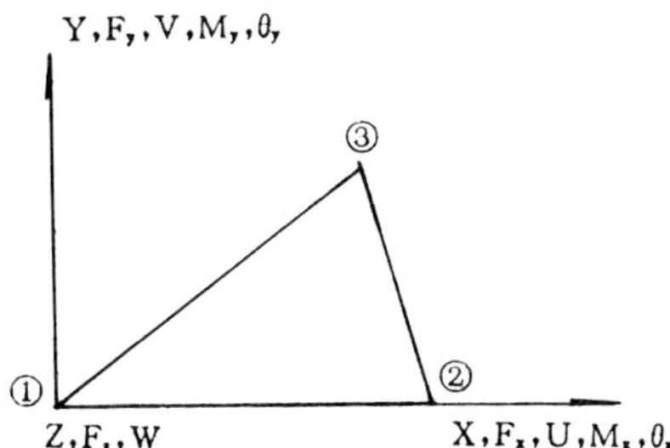

Fig. 4. Triangular plate and shell element.

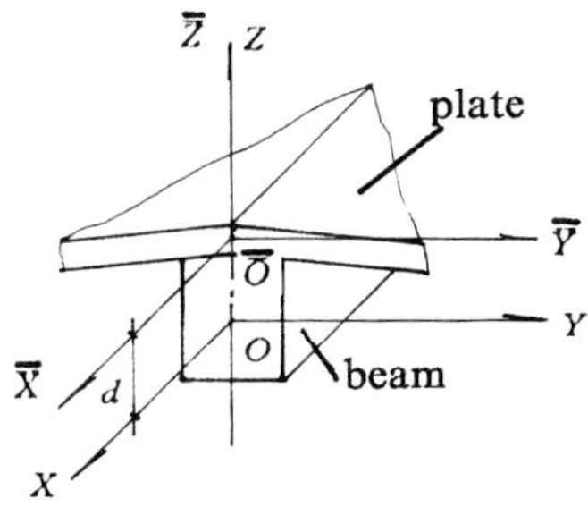

Fig. 5. Connection between beam and plate.

8. As beam members are fixed with the triangular members (Fig. 4), the relation between the nodal displacements $\{\delta\}^e$ of the beam and the node displacements $\{\bar{\delta}\}^e$ of the triangular shell (Fig. 5) is given by:

$$\{\delta\}^e = [c]\,\{\bar{\delta}\}^e$$

where [c] ——the conforming matrix.

$$[c] = \begin{bmatrix} \beta & 0 \\ 0 & \beta \end{bmatrix}$$

and

$$[\beta] = \begin{bmatrix} 1 & 0 & 0 & 0 & -d & 0 \\ 0 & 1 & 0 & d & 0 & 0 \\ 0 & 0 & 1 & 0 & 0 & 0 \\ 0 & 0 & 0 & 1 & 0 & 0 \\ 0 & 0 & 0 & 0 & 1 & 0 \\ 0 & 0 & 0 & 0 & 0 & 1 \end{bmatrix}$$

The nodal load $\{F\}^e$ of the beam elements is related with the nodal loads $\{F\}^e$ of the shell elements by:

$$\{F\}^e = [c]^T\{\bar{F}\}^e$$

The relation between the nodal loads and the displacements of beam elements in the local co-ordinate $\bar{x}\,\bar{y}\,\bar{z}$ (Fig. 5) is:

$$\{\bar{F}\}^e = [c]^T([K] + [N])[c]\{\bar{\delta}\}^e$$

where: [K] is the elastic stiffness matrix of the beam element in the local co-ordinate system and [N] is the geometric stiffness matrix of the beam element in the local co-ordinate system.

Fig. 6. The experimental model.

Fig. 7. Support beam.

Using the above formulae, a FEM structural stability analysis program in FORTRAN has been written, and a model analysed. The actual nodal coordinates of the experimental model were used in the analysis.

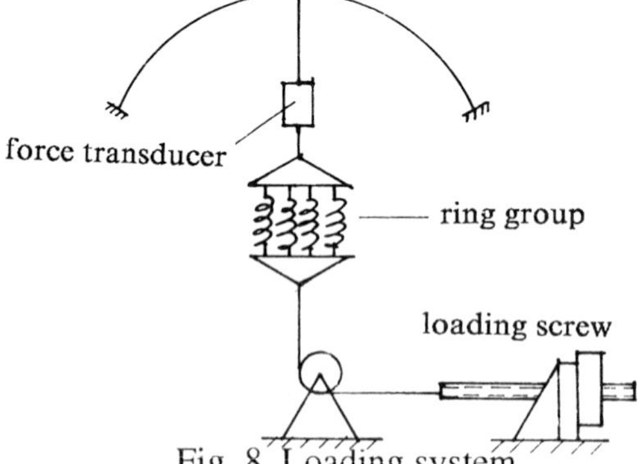

Fig. 8. Loading system.

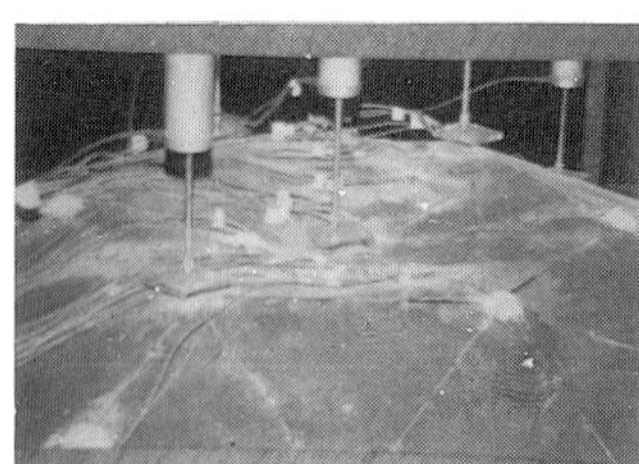

Fig. 9. Snap through of the grid shell.

The experimental model of the sheet structure

9. The model is shown in Fig. 6. It contains 37 steel nodes with a diameter of 30mm, 90 steel beams with a diameter of 8mm, and 54 steel triangular sheets with a thickness of 0.7mm. The nodes are on the surface of a sphere of diameter D = 1800mm.

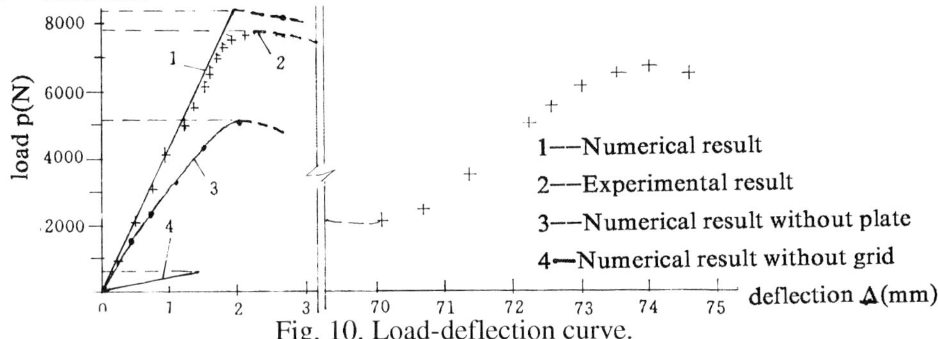

Fig. 10. Load-deflection curve.

The plan of the model is a hexagon with a side length of 75mm. The 6 boundary nodes of the model are connected with a strong hexagonal steel beam (Fig. 7). In order to trace the load deflection path at the post-buckling stage, a group of springs were used as the loading system; unloading automatically took place when snap-through occurred (see Fig. 8). The load is a single concentrated load acting at the apex, the test is controlled by the incremented deflection of the apex, the snap-through and the whole load deflection curve are shown in Figs 9 and 10.

10. It can be seen from Fig. 10 that the test critical load was 7960N and the computed critical load was 8390N; the analytical result is a little higher than that of the test. The autors think that this comes from the ideal assumptions and errors in the experimental system. It can also be seen that the experimental curve is basically linear and agrees well with the analytical curve before buckling, so the use of linear analysis is valid.

11. The model dome without sheets was also analysed taking into account geometric non-linearity, its load deflection curve is shown in Fig. 10. The response of the model was clearly non-linear and the critical load was less, that is to say, the sheets play an important part in load capacity.

ENGINEERING PRACTICE

12. The advanced features of the sheet system has raised great interest and concern from engineers. Table 1 gives economic comparison data on some constructed roofs. There is still about 24000m^2 of building under construction.

13. Because of the composite action of the grid and sheets, the sheet system is more economic than conventional space structure systems such as grid system. It is suitable for large and middle span roofs. But by elaborating the design, it can also be used for small span roofs, such as the Tongling dolphin raising house.

14. The Tongling dolphin raising house was commissioned in order to protect, and breed this endangered species. A sheet space structure system was chosen for the building. More details are shown in Fig. 11.

15. The span is not large and the design loads are small (self weight+live load = 0.8kN/m^2). The sheet space structure system was composed of bamboo glued rib board and a steel framework, which work together to carry the forces (Fig. 12).

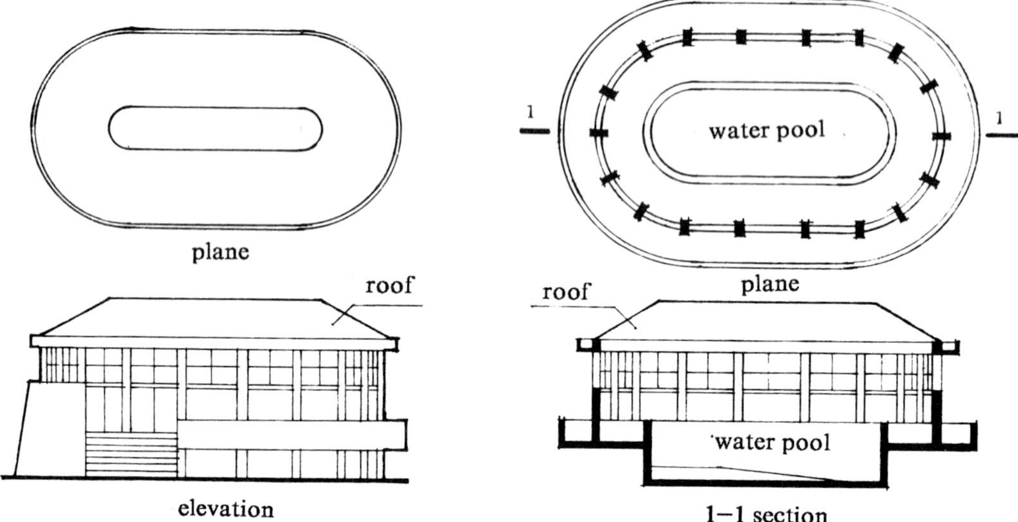

Fig. 11. The dolphin raising house.

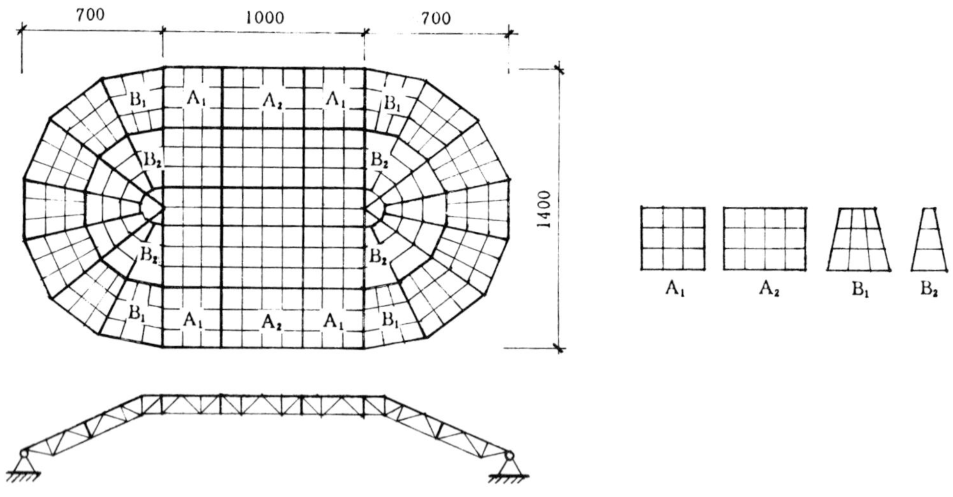

Fig. 12. Sheet system roof.

The Mero bolted joint with a diameter of 80mm, was used to joint the steel framework, while the self push bolts which can transmit forces homogeneously were used to join the sheets to the framework.

CONCLUSION

(a) The sheet space structure system has good structural performance and gives a superior economic index for large, middle and small span roofs, which indicates a good future for its application.

(b) The structural planking, jointing styles and anti-corrosion methods ensure the economy of the system, which means that it is well suitable for Chinese conditions.

(c) A counter power method is introduced to calculate the critical load of the composite structure. This approach is simple and meets the engineering designing needs; the method also has practical value.

Table 1. Economic comparison of some roofs.

project	span grade	plane size	load	steel use
The roof of store house of Nanjing new electric plant	large span	72m × 62m	1000N / m^2	200N/m^2 , 40% of thesame span plane arch system
The roof of the indoor sports field of Shanghai institute of physical education	large span	51m × 93.5m	1500N / m^2	300N/m^2 , 54% of thesame span grid shell
The roof of the new railway station in Hefei	large span	two 60m × 60m and one 40 × 40m	1000N / m^2	350N/m^2 40% of the same span grid plate
The roof of the mill building of Dayuko chemical fertilize plant	large span	30m × 150m	1500N / m^2	180N/m^2 , 40% of the same grid plate
The roof of the multifunction hall of Dayuko chemical fertilizeplant	large span	36m × 36m	1450N / m^2	211N/m^2 , 52% of the same span grid shell
The roof of the dolphin raising house in Tonglin	small span	14m × 24m	900N / m^2	107.6N/m^2 , 50% of the same span grid shell

REFERENCES

1. ZHAO H.L *et al*, Experimental study analysis and design of a sheet framed space structure, Proc. of third Int. Conf. on Space Structure, Sept. 1 984, U.K.

2. MAKOWSKI Z. S., Analysis, design and construction of braced domes, 1984.

3. GALLAGHER K.H., A discrete element procedure for thin shell instability analysis, AIAA, J. vol.5, No.l, 1966.

4. Southeast Institute of Water Conservancy: Finite element method in elastic mechanics, 1977, second edition, Water Conservancy Press.

5. GUE F.Y., Assemble of FORTRAN program, 1982, National defence industry press.

189. Efficient optimum design of structures

E. SALAJEGHEH, University of Kerman, Iran

SYNOPSIS. This study presents a method for structural optimisation, using high quality approximation concepts. The basic idea is to reduce the number of structural analyses required in the optimisation process, as well as reducing the overall computational cost of the optimisation. The structural response quantities such as forces, displacements, etc. are approximated as a function of design variables or some intermediate variables. By employing these approximate relations, the original design task is transformed into a sequence of explicit approximate problems, each of which can be solved by numerical optimisation techniques, very efficiently. The design variables can be considered as sizing and shape variables. These variables can also be treated as continuous or discrete. First a continuous variable optimisation is performed, the result of which is used as a starting point for the discrete variable optimisation. The discrete solution is obtained by branch and bound, or by a dual strategy method. Examples are offered to demonstrate the efficiency and reliability of the proposed method.

INTRODUCTION

1. Optimum design of structures is achieved by combining mathematical programming techniques and finite element analyses. The objective function and the constraint functions are first expressed as functions of the design variables. Then, by employing numerical optimisation methods, the objective function, which is normally taken as the weight of the structure, is minimised while the constraints are satisfied. The constraints include bounds on member stresses, deflections, frequencies, etc.

2. In the process of optimisation, because the constraint functions are implicit functions of the design variables, their precise numerical evaluation requires a complete finite element analysis. In addition, since the solution techniques are iterative, a large number of structural analysis is required to obtain the optimum solution.

3. In the past, to increase the efficiency of the method of optimisation, some approximate concepts have been introduced. The most attractive approach is to linearise the functions under consideration and solve a sequence of linearised optimisation problems. The solution of each approximate problem does not

require the analysis of the structure. Some other improvements such as design variable linking, constraint deletion, etc. have increased the robustness of the process (refs 1-3).

4. In the present work, a method will be introduced by which the highest quality approximation can be achieved. The implicit structural responses such as forces, displacements, frequencies, etc. appearing in the optimisation problem, are first approximated. By substituting these approximate functions into the original problem a non-linear explicit problem is created, the solution of which, often requires less than 10 analyses of the structure. This method is very robust and efficient for large structures where the computational cost of the analysis is high.

5. The same approximate technique is applied to the design of the structures where some or all of the design variables are chosen from a prescribed set of values (discrete variables). The discrete optimum design of the structures is achieved by combining the approximate techniques and integer programming. For practical design problems, where the design variables are linked and the number of independent design variables are chosen reasonably, the method can be used efficiently.

6. The same approximation concepts are used to achieve the optimum shape of structures. In addition to the sizing variables, the co-ordinates of joints are also considered as design variables. In this case, the numerical results also indicate that the optimum configuration of the structures with discrete variables can be achieved at little computational cost.

7. To further increase the efficiency of the technique for problems with a great number of design variables, a dual strategy is used. This approach first creates a separable approximation to the first level approximate problem and uses duality theory to search for the discrete solution. Thus two level approximations are used to solve the problem.

GENERAL OPTIMISATION PROBLEM

8. The general form of an optimisation problem can be stated as;

Minimise F(X) (1)

subject to:

$$g_j(X) \leq 0 \quad j = 1, m \tag{2}$$

$$X_i^l \leq X_i \leq X_i^u \tag{3}$$

$$X_i \in D_i \tag{4}$$

where F(X) and $g_j(X)$ are the objective function and constraints, respectively. The term X is the vector of design variables, including s sizing variables and (n-s) shape variables, where n is total number of variables, as;

$$X = [X_1, \ X_2, \ \ldots.., \ X_i \ \ldots.., \ X_s, \ X_{s+1}, \ \ldots.., \ X_n] \tag{5}$$

The total number of constraints is m, and X_i^l and X_i^u are lower and upper bounds on the design variable X_i, respectively. D_i is the set of discrete values, which may be different for each variable X_i.

9. The objective function F(X) is normally taken as the mass of the structure, which is a function of the design variables. The constraints provide bounds on various response quantities. Assuming the structure is modelled as an assemblage of finite elements, stresses can be calculated in the elements under each of the prescribed loading conditions. Then a typical stress limit may be stated as;

$$\sigma^l \le \sigma_{ijk} \le \sigma^u \tag{6}$$

where i=element number, j=stress component and k=load condition. The compression and tensile stress limits are σ^l and σ^u, respectively. Equation 6 is converted to the following two equations of the form of Equation 2.

$$g_1(X) = \left(\sigma^l - \sigma_{ijk}\right) / \left|\sigma^l\right| \le 0 \tag{7}$$

$$g_2(X) = \left(\sigma_{ijk} - \sigma^u\right) / \left|\sigma^u\right| \le 0 \tag{8}$$

Other typical constraints include displacement, frequency, local buckling, system buckling, dynamic response, etc.

10. The problem defined by Eqs 1-3 is a general non-linear programming task, and a variety of methods and software are available for its solution. In general, to reach an optimum design, the objective and constraint functions and their gradients with respect to the design variables must be evaluated dozens, and sometimes hundreds, of times. In structural synthesis, this implies a large number of detailed finite-element analyses, which would be prohibitive for practical design. Therefore, approximation techniques are resorted to, whereby the structural responses are approximated using some carefully chosen intermediate variables. From this, an approximate problem is created and solved using non-linear programming methods. This process is repeated until it converges to an acceptable optimum. The key issue then, is the creation of a very high-quality approximation so that the number of detailed finite element analyses can be as few as possible, and so the approximate optimisation phase is not strongly dependent on move limits.

APPROXIMATE PROBLEM PRESENTATION

11. Consider a simple stress constraint for a member of a pin-jointed structure, where the member i is under a tensile force P_i with cross-sectional area A_i. Then the stress $\sigma_i = P_i / A_i$ can be linearised as:

$$\overline{\sigma_i}(X) \cong \sigma_i(X^o) + \nabla\sigma_i(X^o).(X - X^o) \tag{9}$$

where ∇ is the gradient operator. In case of sizing optimisation, X can be a vector of cross-sectional areas and X^o is the initial design point. The equation $\sigma_i(X)=P_i(X)/A_i$ is quite non-linear in X and so the approximation given by Eq. 9 is valid for small changes in X.

12. If the reciprocal of the cross-sectional areas are considered as design variables,

$$\text{i.e. } X_i = 1/A_i \text{ (intermediate variable)} \tag{10}$$

then the stress $\sigma_i(X)=P_i(X).X_i$ is more linear in X and, in the special case of the statically determinate structure, P_i is independent of X, so the approximation is precisely linear in X. With reciprocal areas, the objective function will now be non-linear, but it can be evaluated easily and it is not necessary to approximate it.

13. In general, P_i is not constant and it is proposed that instead of linearising σ_i, the force P_i is linearised as:

$$\overline{P}_i \cong P_i(X^o) + \nabla P_i(X^o).(X - X^o) \tag{11}$$

Then the stress can be recovered as $\overline{\sigma_i}(X) = \overline{P}_i(X) / A_i$ This can be shown to be a higher quality approximation than found by using the reciprocal variable X and a much higher quality approximation than found by direct linearisation. With force approximation, both direct and reciprocal variables can be used, with this approximation, the need for move limits during the approximate optimisation is greatly reduced (refs 4,5)

14. It has also been shown that, force approximation is a very effective method, when the design variables are considered as shape variables (co-ordinates of joints and thickness of plates) (refs 6-8), as well as sizing variables.

15. In the case of rigid jointed structures, the linearisation of forces is created with respect to the member section properties (areas and moments of inertia), while the design variables can be the physical member section dimensions. The use of section properties as intermediate variables has also been used to create high quality approximate relations for frequency constraints in frame structures (ref. 9).

16. A higher quality of approximation for frequency constraints, can be achieved by using the Rayleigh quotient approximation (ref. 10);

$$\lambda_j = \frac{U}{T} = \frac{\phi_j^T K \phi_j}{\phi_j^T M \phi_j} \tag{12}$$

where **K** and **M** are the global stiffness and mass matrices, respectively and λ_j is the Jth eigenvalue and ϕ_j is the corresponding eigenvector. The approximation for λ_j can be achieved by linearising the numerator, U, and denominator, T, independently and use the result to estimate λ_j. The same intermediate variables can be used in this case.

17. Another important aspect of approximation concepts, is the idea of constraint deletion. To solve each approximate problem (design cycle), the gradients of the structural responses are required to construct the constraint functions. As in practical design problems, a great number of constraints are involved, the gradient calculation, is very costly. In addition, most constraints may be far from critical, it is reasonable to ignore many of the constraints for this design cycle. The simplest approach would be just to retain only those that are critical or potentially critical for the current design cycle. Typically, the number of retained constraints may be two to three times the number of independent design variables.

CONTINUOUS VARIABLE OPTIMISATION

18. By employing the approximate relations for forces, displacements, frequencies, etc. the original design problem can be converted into a sequence of approximate problems, the solution of which can be obtained easily by numerical optimisation methods. In this method, we avoid the large number of finite element analyses, normally needed for optimisation.

19. The basic steps to achieve an optimum design, with continuous variables can be summarised as:

1. Perform a detailed finite element analysis of the initial proposed design.
2. Evaluate all the constraints and retain those that are critical or potentially critical for further consideration during this design cycle.
3. Calculate gradients of the structural responses with respect to intermediate variables, included in the retained constraints. Using these gradients, construct approximations of the relevant responses.
4. Set up the high quality approximation to the original problem, by using the approximated responses.
5. Solve the approximate continuous variable problem, with move limits on design variables (and intermediate variables) to insure the reliability of the approximation.
6. Check for convergence to continuous optimum, if satisfied, proceed to discrete variable optimisation (if necessary). Otherwise update the problem and repeat from step 1.

DISCRETE VARIABLE OPTIMISATION

20. In most of the engineering design problems, some or all of the design variables are not free to assume arbitrary values, instead, the design variables are confined to a limited number of values determined by standard available sizes. In such a case, the discrete variable optimisation becomes essential. A recent review of the techniques dealing with discrete variables was reported in ref 11. The

literature shows that without a proper approximation technique, design of structures with discrete variables, is costly.

21. The same approximation concepts outlined in this paper, is used to solve the discrete problems effectively. Two methods are briefly introduced here.

Branch and bound method

22. The classical branch and bound method was originally developed for linear problems, but has been also applied to non-linear problems, with structural response approximation (refs 12,13).

23. First a continuous variable optimisation is performed. This provides a starting point, as well as a lower bound on the discrete solution. Now, a single variable increased to its next discrete value and the optimisation is performed with respect to the other variables. If the optimum is greater than before, the variable is set to its next lower value and the process is repeated. If an improvement is found, the search is continued in this direction until no improvement can be found. Then this variable is allowed to change subject to the bound currently imposed, and the process is repeated for the next discrete variable, until all discrete variables have been examined. All these operations are performed on the approximate problem, which dose not require the analysis of the structure. Thus for a reasonable number of independent design variables, the method is quite efficient. It is also to be mentioned that the first application of branch and bound is on the approximate problem, available from the last iteration of the continuous design problem. As the quality of approximation is controlled by move limits, further application of branch and bound may be necessary, each requires one analysis of the structure under consideration.

Dual method

24. After achieving a continuous solution, first all of the structural responses are approximated, thus a high quality approximate problem is available. Again the objective function and constraints in this approximate problem, are linearised with respect to design variables or their reciprocals, thus the first level approximate problem is converted into a second level approximate problem of separable form. Now, the duality theory can be applied to solve the problem, by constructing explicit dual function, which are maximised subject to non-negativity constraints on the dual variables, while a number of one-dimensional functions are minimised subject to side constrains. The minimisation of the one-dimensional function of primal variables are performed with discrete values. Because move limits are imposed on the problems, a sequence of first and second level approximation optimisation are performed to achieve the discrete variable optimisation. Further discussion of the method is presented in ref. 14.

NUMERICAL RESULTS

25. Here, two examples are offered to demonstrate the efficiency and reliability of structural optimisation process, including discrete variables. The first example is a double layer grid with 47 independent discrete sizing variables, solved by

branch and bound. The second example is a 25-bar space tower with eight independent discrete sizing variables and five independent discrete shape variables, solved by branch and bound and the results are compared with the dual approach. The DOT optimiser (ref. 15) is used to minimise or maximise the functions under consideration.

Example 1: Double-layer grid

26. A double-layer grid with a span of 21m and height of 1.5m shown in Fig. 1, is chosen from ref. 4. The grid is simply supported at every other boundary joint of the bottom layer. The structure supports a uniformly distributed load on the top-layer of 155.5kg/m^2 that is treated as a concentrated vertical load at the joints. The structure is treated as a truss, and the modulus of elasticity of the material is E=2.1 x 10^{6}kg/cm^2 with a material density of 0.008kg/cm^3. Member areas are linked to maintain symmetry about the four lines of symmetry in the plane of the grid. Thus, the problem has 47 independent design variables. The initial areas are taken as 20cm^2 with a lower bound of 0.1cm^2. Constraints include member stresses and displacements. The allowable stress is taken as $-1000 \leq \sigma_i \leq 1400$kg/cm^{2}and displacement limits of $\pm$2.5cm are imposed at each joint in each co-ordinate direction. The set of discrete values is considered as:

$$X_i \in \{0.1,\ 0.5,\ 1.0,\ 1.5,\ 2.0,\ 2.5,\ 3.0, \ldots\ldots\}(cm^2); \qquad i = 1,47$$

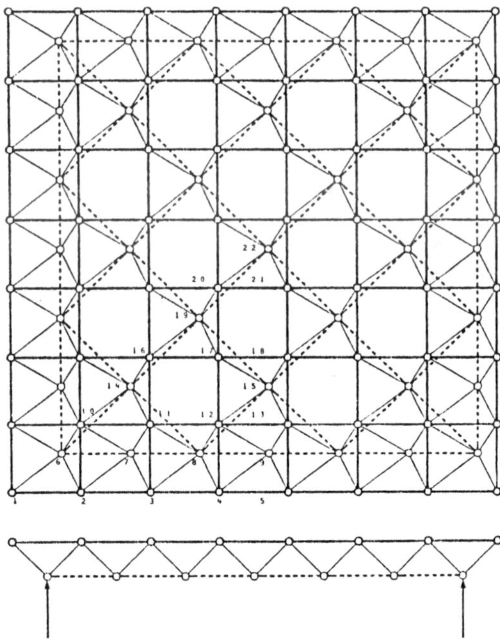

Fig. 1. Double-layer grid.

The continuous design problem is completed with 7 analyses of the structure and 2 more analyses are required for the discrete solution. The results are presented in Table 1. The structural weight corresponds to 1/4 of the grid.

Table 1. Results for double-layer grid.

DESIGN VAR.	MEMBERS	CONTINUOUS OPT.	DISCRETE OPT.	DESIGN VAR.	MEMBERS	CONTINUOUS OPT.	DISCRETE OPT.
1	1-2	0.35	0.5	26	6-2	0.75	1.0
2	2-3	0.12	0.1	27	10-7	0.22	0.5
3	3-4	0.71	1.0	28	7-3	0.16	0.5
4	4-5	0.11	0.1	29	14-11	1.71	2.0
5	10-11	0.10	0.5	30	11-8	3.16	3.5
6	11-12	2.96	3.0	31	8-4	0.86	1.0
7	12-13	4.09	4.0	32	12-9	0.10	0.1
8	16-17	5.20	4.5	33	19-17	3.25	3.5
9	17-18	0.10	0.1	34	17-15	5.39	5.5
10	20-21	20.17	19.5	35	2-7	0.13	0.5
11	2-10	0.70	1.0	36	7-11	0.30	0.5
12	3-11	0.70	1.0	37	3-8	0.97	1.0
13	11-16	2.45	2.5	38	8-12	6.90	7.5
14	4-12	0.70	1.0	39	4-9	0.10	0.1
15	12-17	6.71	6.5	40	12-15	5.38	5.0
16	17-20	12.06	11.0	41	1-6	0.43	0.5
17	6-7	0.53	1.0	42	6-10	1.15	1.5
18	7-8	0.10	0.1	43	10-14	0.38	0.1
19	8-9	3.45	4.0	44	14-16	3.24	4.5
20	14-8	4.65	5.0	45	16-19	0.22	0.5
21	19-15	10.82	11.5	46	19-20	7.03	7.5
22	8-15	2.47	2.5	47	20-22	0.10	0.1
23	6-14	0.10	0.1				
24	14-19	2.54	2.5	Weight *kg*.		545.82	572.29
25	19-22	9.14	9.5				

Example 2: 25-bar space tower

27. The 25-bar truss shown in Fig. 2, is designed to support two independent load conditions. The load conditions and material properties are given in Table 2. The member areas are linked in following groups: A1; A2=A3=A4=A5; A6=A7=A8=A9;AI0=A11;A12=A13;A14=A15=A16=A17; A18=A19=A20=A21 and A22=A23=A24=A25. The co-ordinate variables are also linked to maintain symmetry. The independent shape variables are the following co-ordinates X4, Y4, Z4, X8 and Y8. There are then a total of eight independent sizing variables and five independent shape variables. The constraints include stress and Euler buckling. Euler buckling compressive stress limit is taken as $-K_iE_iA_i/(L_i)^2$where K_i is a constant, determined from the cross-sectional geometry and E_i, A_i, L_i are the modulus of elasticity, area and length, respectively. The discrete values for the design variables are considered as:

$$X_i \in \{0.1,\ 0.2,\ 0.3,\ 0.4,\ 0.5,\ ......\}(in^2); \qquad i = 1,8$$

$$X_i \in \{1.0,\ 2.0,\ 3.0, 4.0,\ 5.0,\ ......\}(in); \qquad i = 9,13$$

Table 2. Load conditions and material properties for tower.

a) Load condition 1 (Lbs)

Joint	Fx	Fy	Fz
1	0.0	20,000.0	-5000.0
2	0.0	-20,000.0	-5000.0
3	0.0	0.0	0.0
6	0.0	0.0	0.0

b) Load condition 2 (lbs)

Joint	Fx	Fy	Fz
1	1000.0	10,000.0	-5000.0
2	0.0	10,000.0	-5000.0
3	500.0	0.0	0.0
6	500.0	0.0	0.0

c) Material properties:

Modulus of elasticty=10E7psi
Allowable stress=+-40,000psi
Dencity=0.1lb/in^3
Buckling coefficient Ki=39.274
Minimum area=0.1in^2

in = 2.54 cm
lb=0.45 kg
psi=0.07 kg/cm^2

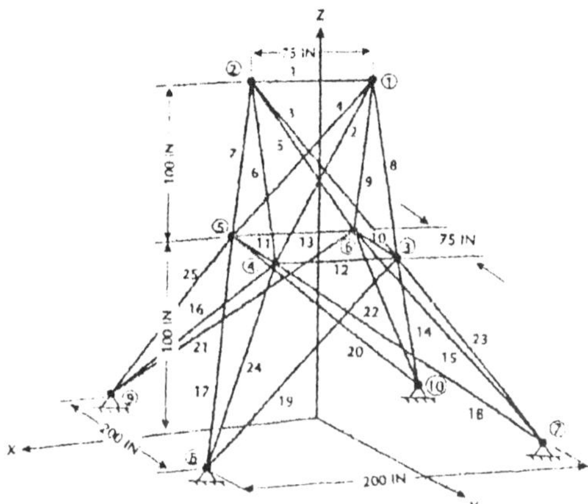

Fig. 2. 25-Bar Tower.

Table 3. Results for 25-Bar Tower.

Variable No.	Variable	Initial Value	Continous Opt.	Branch & Bound	Dual
1	A1	0.009	0.106	0.1^L	0.1^L
2	A2	0.782	0.420	0.5	0.5
3	A6	0.754	0.917	1.0	0.9
4	A10	0.001	0.109	0.1^L	0.1^L
5	A12	0.130	0.102	0.1^L	0.1^L
6	A14	0.558	0.153	0.2	0.2
7	A18	0.982	0.762	0.7	0.8
8	A22	0.801	0.551	0.6	0.6
9	X4	37.500	19.490	19.0	20.0
10	Y4	37.500	39.820	42.0	40.0
11	Z4	100.000	97.190	90.0	97.0
12	X8	100.000	15.380	10.0	15.0
13	Y8	100.000	77.500	76.0	77.0
Weight		229.410	130.980	140.20	140.5

L: lower bound

The number of structural analyses needed to perform the continuous and discrete optimisation was less than 10. The discrete variable optimisation is carried out by the branch and bound method and also by the dual method. The results for this problem are presented in Table 3.

CONCLUSIONS The high quality approximation of structural responses are used to obtain the optimum design of structures with continuous and discrete design variable. The variables are considered as sizing and shape variables. It is shown that the member forces can be approximated as functions of sizing and shape variables with high quality approximation. Often less than ten analyses of the structure are required to complete the optimisation process. Also, the method of branch and bound in conjunction with response approximation is an efficient approach to deal with problems involving discrete variables. As an alternative approach to the solution of problems with high number of continuous or discrete variables, dual methods can be used effectively.

ACKNOWLEDGEMENT The author would like to thank VMA Engineering for providing the facilities to carry out a part of the work and Dr. A. R. Ahmadi of Kerman University for reviewing the manuscript.

REFERENCES

1. SCHMIT, L.A. and MIURA, H., Approximation concepts for efficient structural synthesis, NASA CR 25.52. March 1976.
2. LUST, R.V. and SCHMIT, I.A.., Alternative approximation concepts for space frame synthesis, NASA CR, M
arch 1985.
3. SCHMIT, L.A. and FLEURY, C., Discrete-continuous variables structural synthesis using dual methods, AIAAJ., 1980, Vol.18, No.12, 1515-1524
4. SALAJEGHEH, E. and VANDERPLAATS, G.N., An efficient approximation method for structural synthesis with reference to space structures, Space Structures Journal, 1986/87, Vol. 2, No.3, 165- 175.
5. VANDERPLAATS, G.N and SALAJEGHEH, E., A new approximation--n method for stress constraints in structural synthesis, AIAA J., 1989, Vol.27, No.3, 352-358.
6. HANSEN, S.R. and VANDERPLAATS, G .N., An approximation method t or configuration optimisation of trusses, AIAA J., 1'390, Vol.28, 161-172.
7. VANDERPLAATS, G.N. and HAN, S.H., Arch shape optimisation using force approximation methods, Structural Optimisation, 1990, Vol.2, 193-201.
8. THOMAS, H.L. and VANDERPLAATS, G.N., An improved approximation for stress constraints in plate structures, (submitted for review to Structural Optimization).
9. VANDERPLAATS, G.N. and SALAJEGHEH, E., An efficient approximation technique for frequency constraints in frame optimization, International Journal for Numencal Methods in Engineering, 1988, Vol.26, 1057-1069.

10. CANFIELD, R.A., High quality approximation of eigenvalues in structural optimization, AIAA J, 1990, Vol.28, 1116-1122.
11. VANDERPLAATS, G.N. and THANEDAR, P.B., A survey of discrete variable optimisation for structural design, In: Proceedings of the Tenth AISCE Conference on Electronic Computation, ed. O. URAL and T.L. WANG, American Society of Civil Engineers, New York, 1991, 173-180.
12. SALAJEGHEH, E. and VANDERPLAATS, G.N., Efficient optimum design of structures with discrete design variables, (In press, Int. J. of Space Structures).
13. SALAJEGHEH, E. and VANDERPLAATS, G.N., Optimum design of trusses with discrete sizing and shape variables, (submitted for review to Structural Optimisation).
14. VANDERPLAATS, G.N. and SALAJEGHEH, E. An approximation method for structural synthesis with discrete sizing and shape variables, using duality theory (forthcoming).
15. DOT user's manual, VMA Engineering. GOLETA, CA, 1992.

190. Minimum cost configurations for steel space trusses

J. O. SURTEES, University of Leeds, UK, and P. S. FASHOLE-LUKE, L. G. Mouchel & Partners, UK

SYNOPSIS. Results are presented from a recent cost study on single span triangular section space trusses. Design is based on a practical scheme for rationalisation of material. Detailed costing of steel sections, fabrication, painting, transport and erection related to trusses and purlins, has been extended to include supply and fixing of roof cladding. Steel cost includes an accurate assessment of waste, while fabrication cost is based on operational times and unit labour rates with allowance for works overheads. The results confirm that minimum structure weight is by no means a reliable criterion for economic design and a strong case is provided for use of gap-joints rather than overlap-joints, notwithstanding any nodal eccentricity introduced in establishing the gap.

SPACE TRUSS ROOF SYSTEMS

1. Space truss roof structures have visual qualities comparable to those of space decks and are preferred increasingly for wide spans because of greater constructional economy. The cost of nodal connections in space decks is far in excess of actual material cost and the high degree of site assembly demands a firm, clean, level surface. In contrast, space trusses involve fewer and cheaper joints with minimal on-site fabrication.

2. The Vee girder is the most common form of space truss. This is a triangular cross-section truss consisting of two upper chords and one lower chord, with triangulated bracing in the plane of the inclined webs and cross bracing between the upper chords (not necessarily triangulated). Because of the high lateral stiffness of the Vee girder, it is not susceptible to lateral instability and does not need to be tied to a lateral bracing system. It is possible for roof decking to be fixed direct to the top chords, spanning between trusses rather than between purlins. Purlin systems are still commonly used, however. The form of bracing adopted for the inclined webs is clearly important if the trusses are visible. Options include Warren type bracing, Constant N-bracing and N-bracing with alternating diagonal direction.

3. Inverted Vee, or Delta, girders provide serrated roof surfaces with increased natural lighting, routes for service ducts etc. Trapezoidal cross-sections formed

from four main chords are also used. Purlin spans are reduced in this case but only at the expense of increased bracing material and number of joints.

SCOPE OF INVESTIGATION

4. There has been no reported cost study of space truss roof systems. Much has been published on space grids but optimisation studies have been concerned with minimum weight rather than minimum cost. Many have used simplified design models for the space grid merely to provide a vehicle for research into numerical optimisation procedures. The present investigation has rigorously applied design rules from current documents (refs 1-3), together with a high degree of refinement in detailing and costing methods (refs 4-6). The prime objective has been to provide design guidance towards efficient rather than minimum cost solutions.

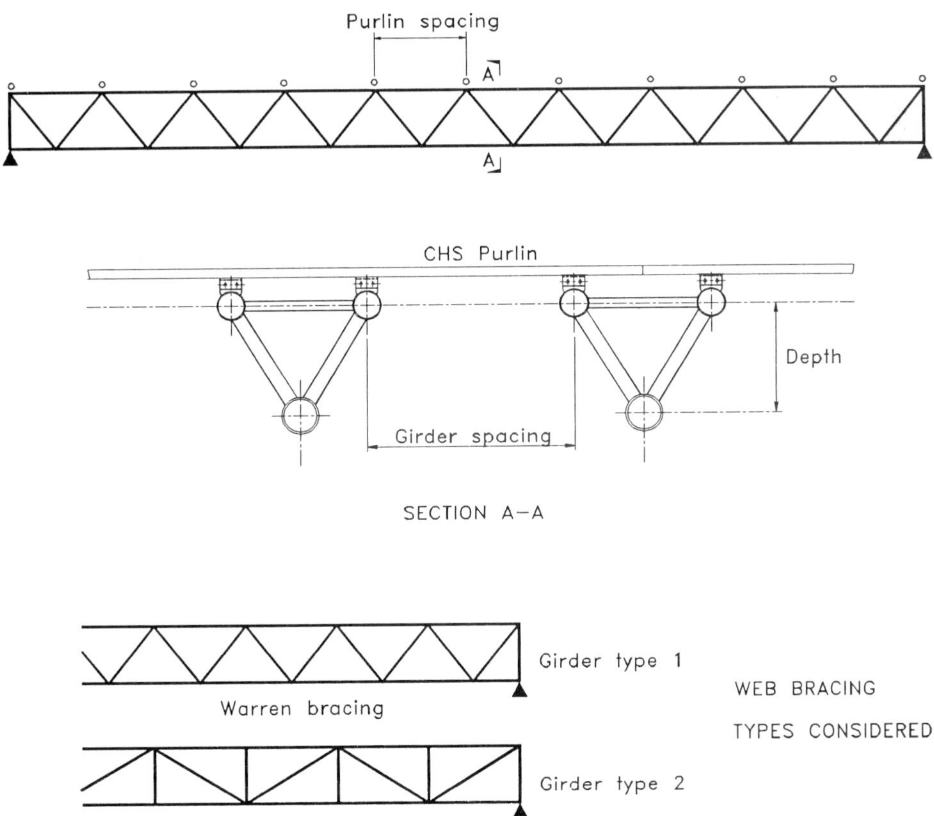

Fig. 1. General form of V-girder roof structure considered in this study.

5. The general form of roof structure investigated is shown in Fig. 1. Continuous CHS purlins were used to support roof cladding. Several forms of inclined web bracing were investigated, including Warren type, and N-types with either constant or alternating diagonal direction. The influence of the following factors on cost was investigated:

- steel grade;
- girder depth;
- girder spacing;
- purlin spacing;
- triangular versus trapezoidal cross-sections;
- form of bracing used in inclined webs;
- number of girders to be fabricated;
- type of joint detail.

Because of the potentially large permutation of these factors against girder span, the main part of the investigation was confined to a constant span of 40m. Selective designs for 60m and 80m spans have provided similar characteristics to those observed in the 40m case.

SCOPE OF COST MODEL

6. At the outset, it was assumed that the combined cost of steelwork, cladding and foundations would need to be used as a measure of structural efficiency. From preliminary studies, however, it was found that the variation in cost of stanchions, cladding and foundations as a result of changes in roof girder and purlin spacing was quite small in comparison with the variation in cost of the roof girders and purlins, only. This enabled roof girder and purlin design to be de-coupled from other complex issues related to ground bearing conditions, form of side cladding, eaves height etc.

7. Cost comparison is therefore based on the cost of one girder and its associated purlins and roof cladding, expressed as a cost per square metre of roof. The latter cost is strongly influenced by the number of girder and purlin joints required in the roof as a whole and is thus directly dependent on the number of girders used in a given roof. This investigation has therefore assumed a constant length/span ratio of 1.5 which, for 40m span girders at typical spacing of 4-6m, amounts to eight, nine or ten girders in a roof length of 61m. (Note that girder spacing refers to the distance between the nearer chord members of adjacent girders as shown in Fig. 1). To accommodate discrete values of number of girders (n) and spacing (s) in a fixed length of roof (L), the girder width (B) must satisfy: $n.B+(n-1)s=L$. Interpretation of results should therefore take account of the implied value for B (see Table 1).

Table 1.

Spacing s (m)	Girder width B (m)		
	n = 8	n = 9	n = 10
4.0	4.13	3.22	2.50
4.5	3.69	2.78	2.05
5.0	3.25	2.33	1.60
5.5	2.81	1.89	1.15
6.0	2.37	1.44	0.70

COSTING DETAILS

8. The cost per square metre of roof was based on the cost of one space girder and its associated purlins and roof cladding, broken down as follows.

- Material (steel plate and CHS, either grade 43c or grade 50c, bolts).
- Fabrication (marking, cutting, edge prep., drilling, welding, and assembly).
- Transport
- Erection
- Painting
- Cladding

Material cost

9. Circular hollow sections, available in the UK from British Tubes Stockholding (ref. 4), are supplied at standard lengths of either 7.5m, 10m or 12m in bundles weighing 3 tonnes and priced per 100m. For a given size of section, the amount of unusable waste in the roof as a whole largely determines which standard length to purchase. Automatic determination of the best option was carried out in the costing procedure. Unusable waste was allotted a redeemable value of 25% of the purchase value. It is not customary to rejoin small lengths of waste.

10. Plate material is priced per tonne. The main application of plate is in forming the purlin connections but fabrication cost is much more significant than material cost.

Fabrication cost

11. Overview. Actual fabrication costs are strongly influenced by the effectiveness of procedures for assembly and welding used by the fabricator but, for a given set-up, changes in length of preparation, length of welds etc. directly affect operation times and thereby modify fabrication cost. A change in the latter is usually much more influential than the corresponding change in material because large office and works overheads may be linked to workmanship. Thus, for minimum-cost design it is essential that time-based costing is applied consistently to all fabrication operations (refs 5, 6).

12. Handling. This involves fetching the steel from storage and placing ready for cutting, followed by subsequent movement to the assembly area.

13. Marking. This is for identification, cutting, drilling and welding.

14. Cutting. Flame cutting has been used throughout.

15. Edge preparation.

16. Setting out. Time is allocated for checking in, assembling tools and preparing the assembly area.

17. Welding.

18. Assembly. When all parts have been cut, edge-prepared and cleaned they are assembled and tacked joint by joint.

19. Inspection. This entails dimensional checks as well as weld inspections. Both may vary from cursory checks to intensive and lengthy tests and careful thought must be given to appropriate levels of inspection.

DESIGN PROCEDURE

20. Because of the costs involved in locating material and verifying section sizes and steel grades during fabrication, together with the relatively low price of raw steel, considerable rationalisation of section sizes is applied in practice when sizing truss members. The scheme adopted in this research was to use a constant section size throughout the top chord and likewise for the bottom chord. Internal bracings were also kept to a constant size. Thus, including the purlins, four CHS section sizes were selected in the course of the design process.

21. Concentric noding of member centre-lines often requires special overlap end preparation in order to accommodate geometric interference between members. An alternative approach is to provide a suitable gap between converging web members at the surface of the chord by displacing their centroidal intersection towards the latter. The bending moment resulting from this displacement can then be considered along with other forces. Often, the additional bending moment is small and may be neglected. Two approaches were therefore investigated. The first assumed a constant gap of 20mm and adjusted member sizes to satisfy all design constraints. The second allowed overlaps in the longitudinal direction in order to maintain concentric noding. Both strategies permitted overlaps in the transverse direction. Because of a stipulated minimum overlap rule, it was sometimes necessary to increase the size of diagonal members in order to comply.

RESULTS

22. A selection of results is presented in Figs 2-4. These refer to the two girder types defined in Fig. 1 with depths ranging between 2.2m and 2.9m, which, for a span of 40m, gives span/depth ratios ranging approximately between 14:1 and 18:1. All results are presented in the same format with side-by-side comparison of cost per square metre and total steel weight (directly proportional to weight per square metre). With the exception of those plotted in Figs 2(a) and 2(b), all results refer to minimum cost rather than minimum weight solutions. Likewise, with the exception of Figs 2(c) and 2(d) all results refer to grade 50 steel.

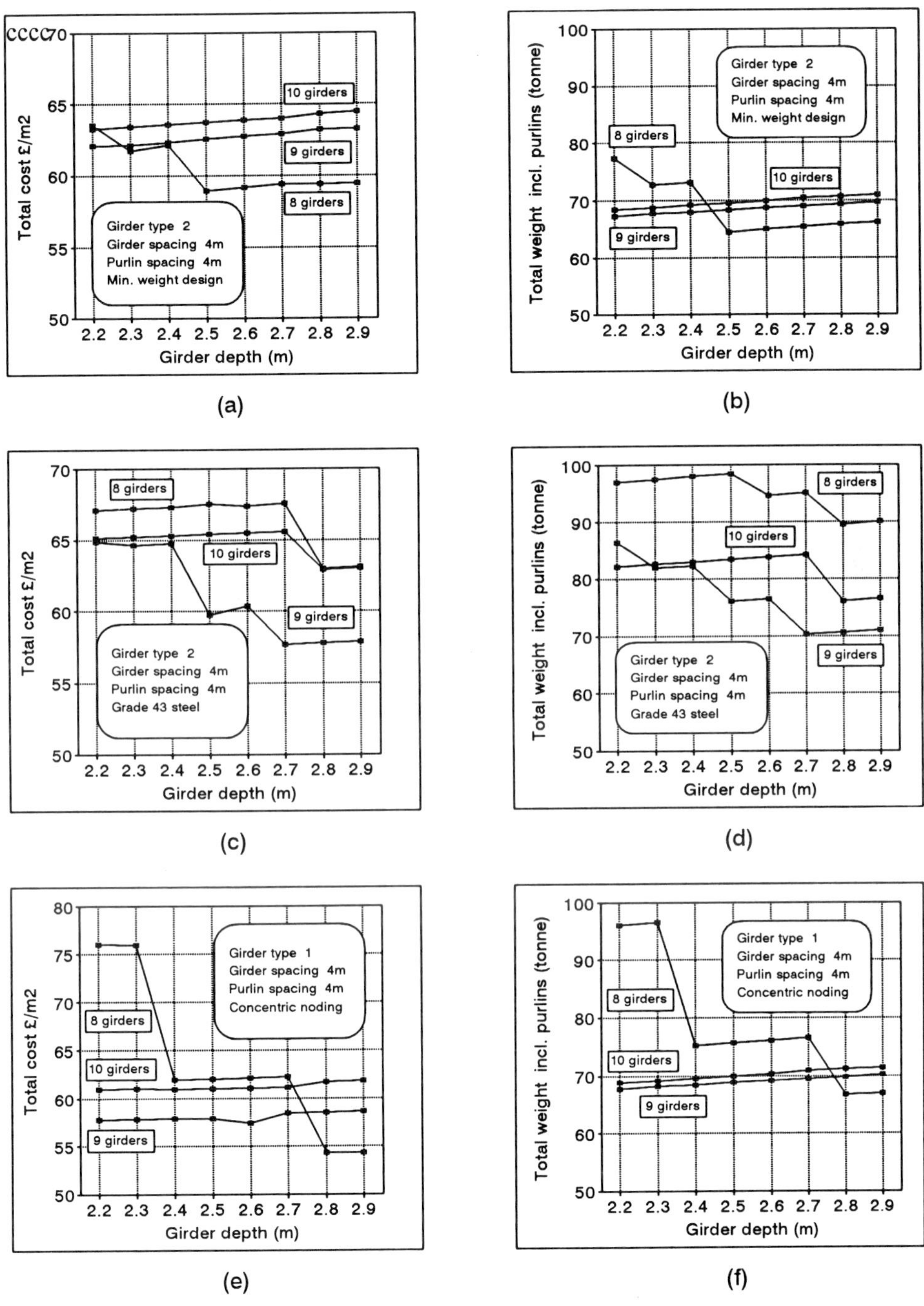

Fig. 2. Variation of cost and weight with girder depth and spacing for selected criteria (Grade 50 steel, 20mm gap joints, minimum cost design, except where stated).

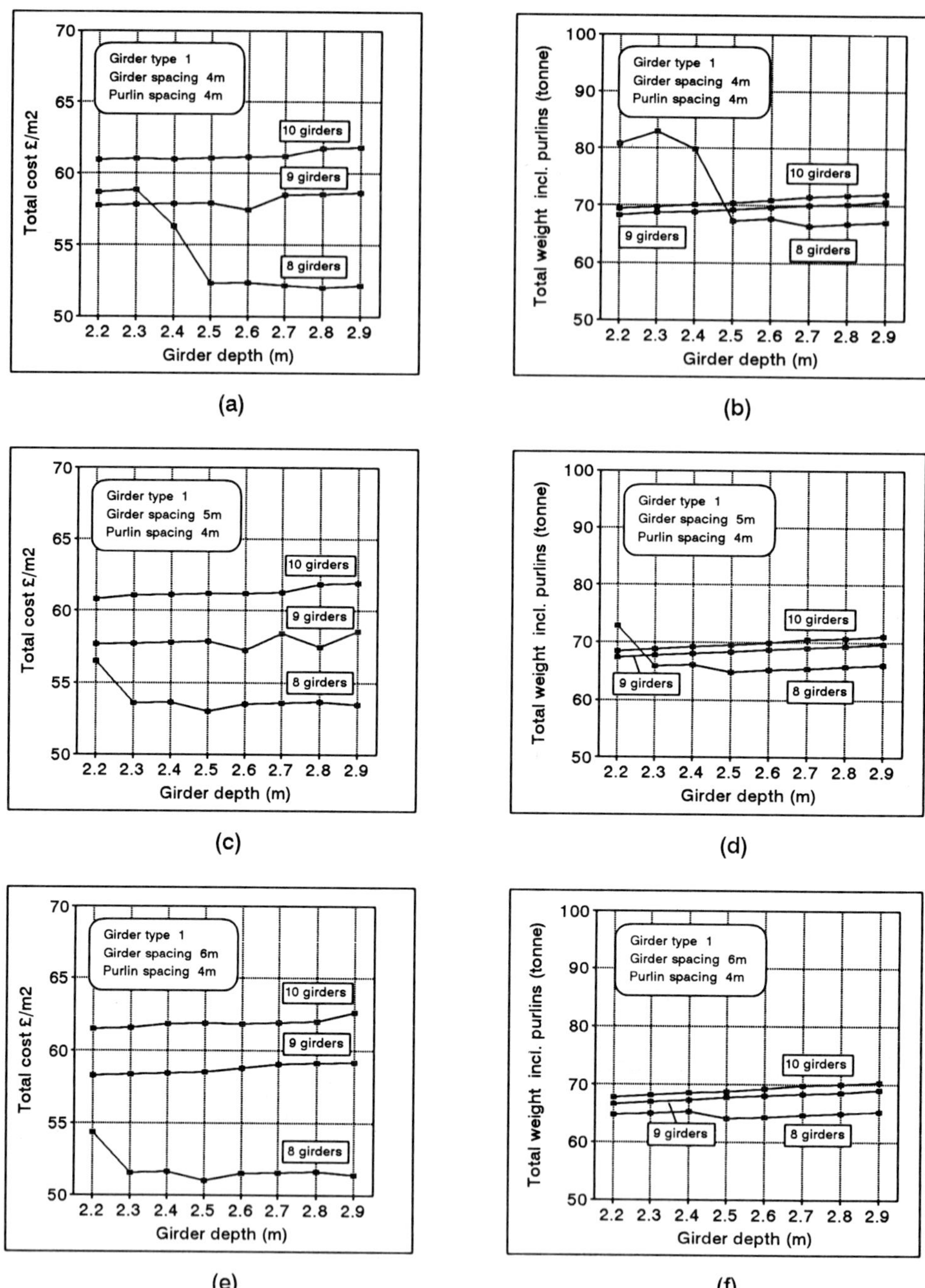

Fig. 3. Variation of cost and weight with girder depth and spacing for Type 1 girders (Least-cost member sizing; Grade 50 steel; 20mm gap joints; 4m purlin centres).

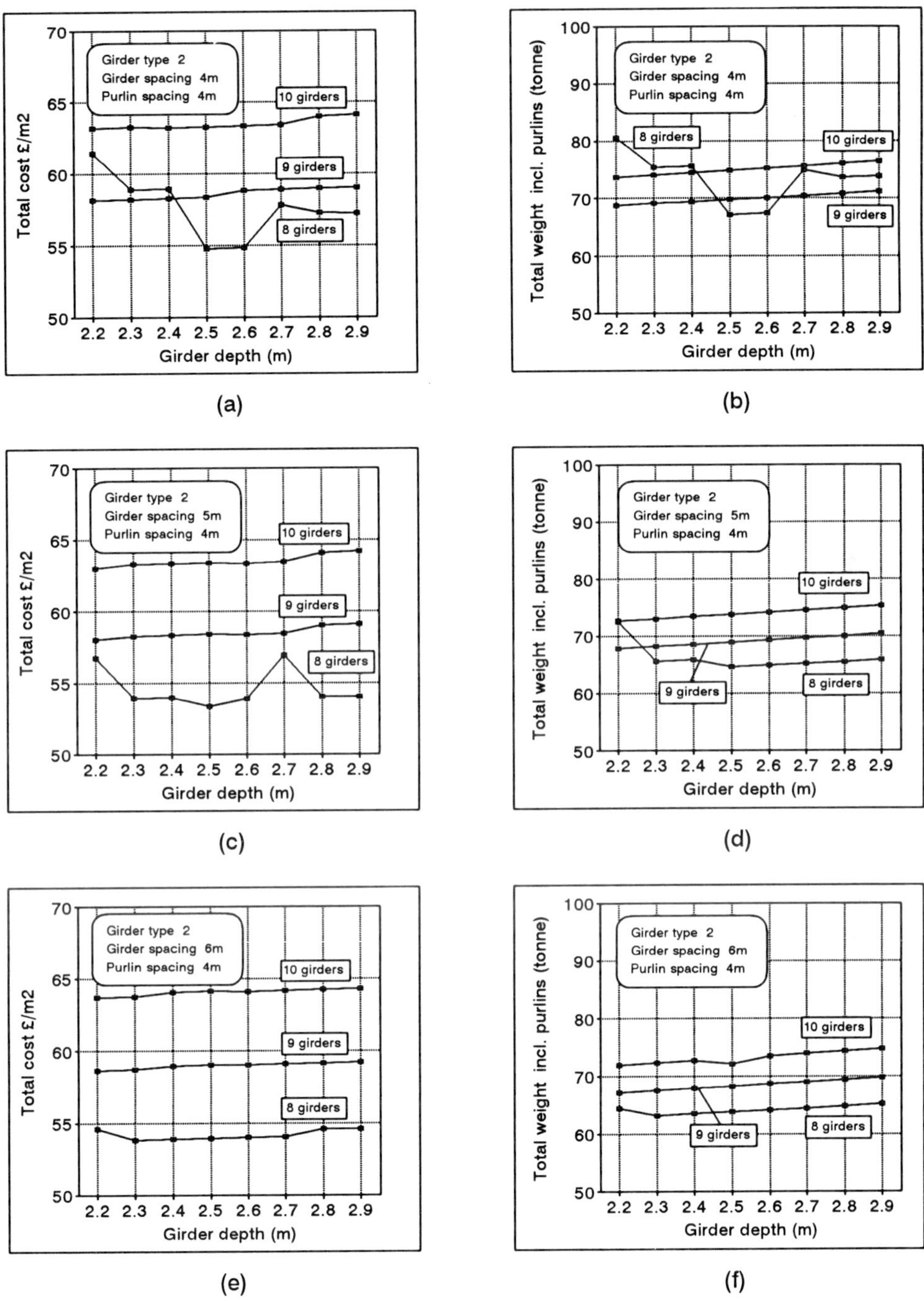

(a) (b) (c) (d) (e) (f)

Fig. 4. Variation of cost and weight with girder depth and spacing for Type 2 girders (Least-cost member sizing; Grade 50 steel; 20mm gap joints).

Minimum weight solutions

23. Figs 2(a) and 4(a) represent minimum weight and minimum cost solutions, respectively, for a typical roof configuration. Minimum cost solutions are cheaper than their minimum weight counterparts in all cases, as might be expected, but in some cases the difference is substantial. What tends to happen is that low weight solutions incur overlap joints in the transverse direction whereas low cost solutions permit gap joints to be used. The latter are cheaper to fabricate but require a larger bottom chord section to accommodate the joint. The cost of additional steel is always less than the saving in fabrication cost, however.

Grade 43 versus grade 50 steel

24. Figs 2(c)-(d) and 4(a)-(b) provide a typical comparison of minimum cost solutions for the respective grades of steel. Grade 50 steel is about 30% stronger than grade 43 steel but only 10% dearer. With few exceptions, the results confirm the advantage of using grade 50 steel. However, in the case of girders with low breadth/depth ratio (10 girders, 2.9m depth, for example), the balance swings marginally in favour of grade 43 steel because member size is determined more by joint geometry considerations than material strength. These girders are considerably heavier than their grade 50 equivalents, however.

Concentric versus eccentric member intersections

25. The effect of overlap joints as opposed to gap joints is typified in Figs 2(e)-(f) and Figs 3(a)-(b). The unit cost associated with fixed gap joints is always less than that for concentric joints though sometimes the margin is small. The higher cost of concentric joints is mainly caused by the increased workmanship in overlap joints. However, violation of the minimum overlap constraint may only be rectified by increasing the diameter of web members. This adds to the cost difference in some cases and also explains sudden rises in weight of steel. On the other hand, fixed gap joints frequently exceed the limits on joint eccentricity and require additional moments to be accommodated in joint and member design.

Influence of type of truss bracing

26. Two simple N-type bracing systems were evaluated in addition to the Warren and alternating N-types shown in Fig. 1. Because of its more suitable geometry for forming gap joints, Warren-type bracing (girder type 1 in Fig. 3) proved generally cheaper than the other forms, closely followed by N-type with alternating diagonals (girder type 2 in Fig. 4).

27. Similar characteristics may be observed in both sets of results. The cost per square metre increases with girder spacing, while weight of steel decreases slightly. The most noticeable characteristic of the results is the small change in cost and weight per square metre over the range of depth investigated (2.2m to 2.9m). The exceptions to this all relate to 8-girder systems. At low girder spacing, the latter comprise very wide girders (see Table 1) which, when associated with short panels (i.e. purlin spacing) and shallow depth, produce a

costly overlap in the longitudinal direction or, alternatively, increased chord sizes to permit gap joints to be used.

REFERENCES

1. BRITISH STANDARDS INSTITUTION BS5950: part 1: 1990. The structural use of steelwork in building.
2. BRITISH STEEL Welded Tubes design manual TD 338.
3. GIDDINGS T.W. and WARDENIER J. (ed.) Monograph no.6: The strength and behaviour of statically loaded welded connections in structural hollow sections, CIDECT, 1986.
4. BRITISH TUBES STOCKHOLDING. Price list for hot finished structural hollow sections, Jan. 1989.
5. CORKER J. Fabrication estimating, Welding and metal fab., Mar. 1966 Aug. 1968. passim.
6. THE WELDING INSTITUTE Standard data for arc welding, Cambridge, 1985.

191. An expert system for discrete optimization of space trusses

J. NICZYJ and W. M. PACZKOWSKI, Technical University of Szcezcin, Poland

SYNOPSIS. The rational design of space trusses requires the application of a properly organised optimisation analysis. A designer must have access to effective and easy-to-use software. The expert module allows the interactive co-operation with optimisation programs, facilitates introducing a task, allows the adoption of optimisation algorithm control, creates and uses the knowledge base while computing and also handles the results. An example of the ortho-diagonal methods utilisation in optimal analysis of a space truss roof for an industrial building is presented.

INTRODUCTION

1. For the last few years, an exceptionally rapid growth in the computational capacity of both workstations and personal computers has taken place. The common availability of this equipment offers an architect not only the ability to design a chosen variant or element of a structure automatically but also the capability to optimally shape entire objects, such as sports halls, exhibition pavilions, satellite aerials, etc. A non-trivial role is played here by the interaction between a designer and a computational program. Computational programs are more frequently equipped with specialised expert systems which assist the user at each of the stages of the design. The paper presents the problems associated with the discrete optimisation of halls covered with space trusses. Particular attention is paid to functions which, in the optimisation analysis, are performed by the expert system and to methods used in the system.

PROBLEMS OF FORMULATING AND SOLVING TRUSS OPTIMISATION TASKS

2. Optimum shaping of spatial trusses consists of the selection of geometrical parameters, strength features and architectonic, structural and assembling solutions for which analysed element or entire structure meets all requirements and, at the same time, is the best for the sake of some criteria. The essential difficulties for a designer to cope with during formulating and solving tasks involved with optimisation of spatial truss structures are (refs 4,5):

 Space structures 4. Thomas Telford, London, 1993

(a) the prolonged computation time for the strength analysis of the structure resulting from the necessity to use both decomposition and co-ordination methods,
(b) the selection of task-related, co-ordinational and superior optimisation criteria,
(c) the determination of both essential decision variables and feasible domain constraints as well as selection of an effective optimisation method.

3. The prolonged time of computation is influenced not only by the determination of objective function values but also by checking whether a given solution belongs to the feasible domain. The conservative conditions to be met by a structure can be checked after completion of the element design. An additional factor that extends the analysis time is the necessity to consider a number of load states and service phases of a structure. From the mathematical point of view, the objective functions of structure optimisation tasks are mostly non-linear (ref. 4) and determined on the discrete, quasi-concave feasible sets (Fig. 1).

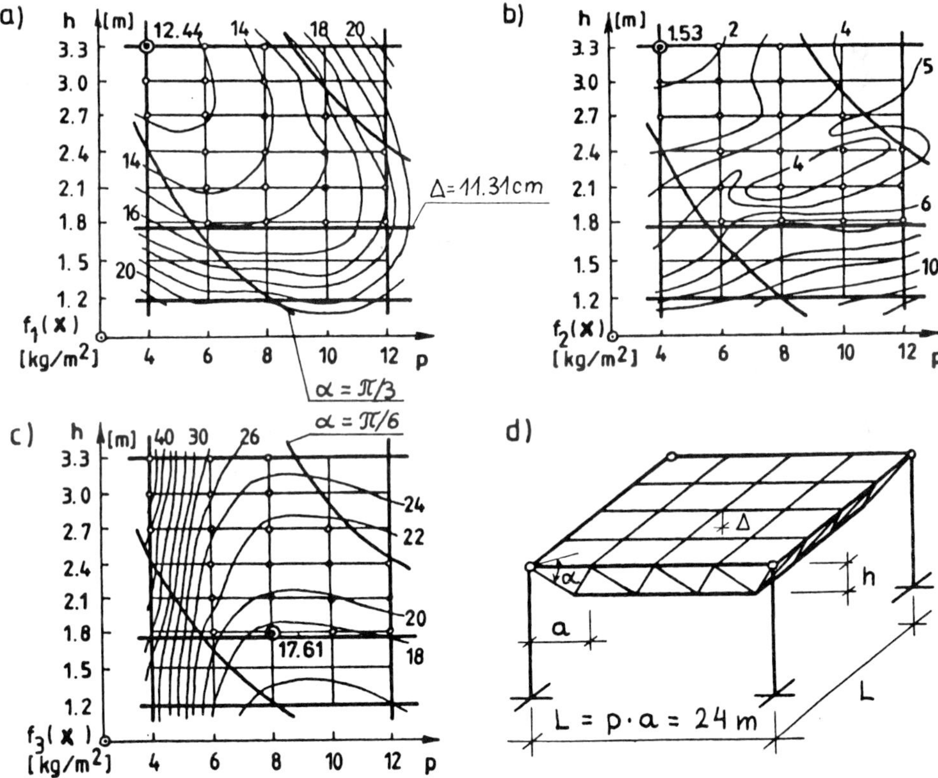

Fig. 1. The objective functions describing mass of: (a) truss bars, (b) truss joints (c) other steel elements of hall and, (d) the hall structure. (ref. 5).

4. The OPTIM program (ref. 8) used for the optimisation, was developed based on the algorithm of the ortho-diagonal (O-D) method (ref. 8) for seeking the minimum of the non-linear function within the discrete feasible domain. In the O-D method, two steps are distinguished: the test step and the working step. During the test step, the neighbourhood of given point $\mathbf{x}_{si}$ (especially that of starting point $\mathbf{x}_{so}$) is investigated in order to determine the direction of objective function value correction. In this step, five different methods to establish the point neighbourhood may be used. Every successive step utilises results from the preceding one, thereby checking the position more accurately, however this is time-consuming. During the working step, minimisation of the objective function along the direction of correction takes place in one of three possible ways. The vector $\mathbf{x}_{ri}$, that minimises the value of objective function in the correction direction, becomes the starting point $\mathbf{x}_{si+1}$ in the successive test step which determines a new direction of correction. This process proceeds up until the moment when, in the neighbourhood of point $\mathbf{x}_{si}$, the value of objective function cannot be corrected.

EXPERT SYSTEM OF STRUCTURE OPTIMISATION

5. An expert system is a group of problem-oriented programs which use both knowledge and analysis/inference procedures; the creation of which requires the participation of experts (ref. 1). In developing a computational system that assists an engineer's work, one should construct expert sub-systems in which the operation fields cover all the design stages (Fig. 2). An effective and economically justified design cannot be carried out without making use of the optimisation methods (refs 4,8,9).

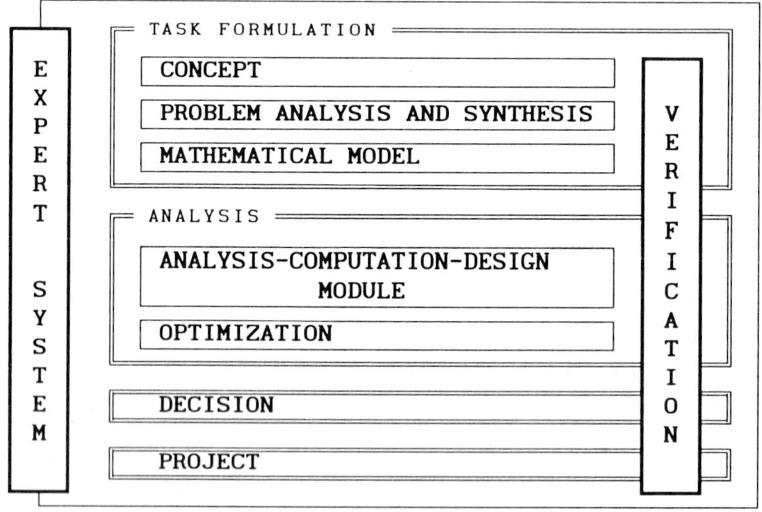

Fig. 2. Design process scheme.

6. In the presented expert system, the knowledge base that stores data on the optimisation process was formed, based on information data which were received from an expert and which resulted from his experience and knowledge (Fig. 3). The knowledge about the problem being solved is gained using the decision table-and-tree method (ref. 3). The knowledge representation is based on the principle of bases-rules (ref. 1) composed of both conditions and assumptions which present a computer model of control recognition. This, consequently, aims at determining the type of action to be taken by the program under given circumstances.

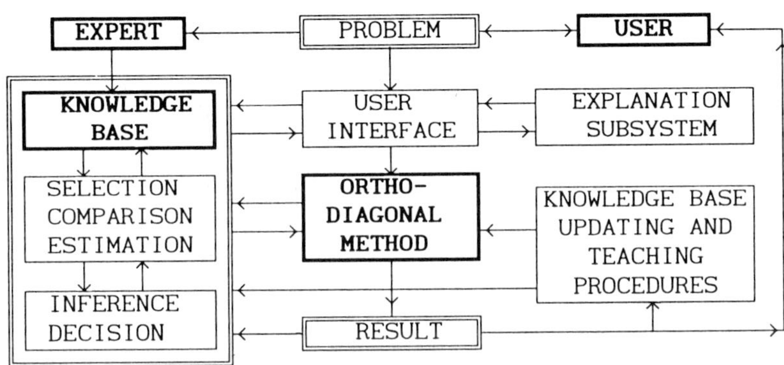

Fig. 3. Expert system components.

7. A dialogue with the user is organised using procedures for input data explanation, assisting the responses gaining process and for visualisation of system operation results. In the final phase, the system presents the user suggestions and remarks resulting from the task solving.

8. The interactive module allows, through posed questions, the following information to be obtained: information data which orientate the formulation and acceptance of a method for solving the given task, and information data that are necessary for the entire system to operate, with possible analysis variants being taken into account. An essential element of the optimisation process is to formulate a task. Suitable procedures of the system are set to carry out versatile tests which recognise the nature of task and the desired results of the analysis. For loaded decision variables, the expert system determines (through selective analysis and making use of information supplied by the user) their effect on given criteria. The aim of such an action is to reduce the size of the task or to decompose the task. The dialogue with the user leads to determination of all constraints for the given problem and relations between them. The recognition of character of the feasible domain (convexity, connectivity and continuity) has a direct impact on the selection and adequate controlling of the optimisation method algorithm. Also, the ability to indicate a correct starting point of the task may rest with the expert system. Having carried out a suitable initial analysis, the expert

system can make modifications to the feasible domain, which may consist, for example, in reducing it dimensionally or in narrowing the seek area.

9. The system modifies the vector of control parameters after termination of each optimisation analysis cycle. If the stop condition (determined by the O-D method) is satisfied ahead of declared computation time or declared number of iterations, the expert system can cause a restart of the task solving under other control. Of importance for the user is that the system processes results of the optimisation analysis and that there is a possibility of continuing a stopped task.

10. An important factor to be taken into account in the expert systems is the analysis time. Therefore, controlling the optimisation analysis should restrict the number of computations of time-consuming elements (e.g. values of objective function) in a given method. The expert determines the way of moving within the decision variables space with an option to omit steps that are needless at a given state of the analysis, according to his experience. The declared computation time, type of solution and the analysis of the initial course of the task are based on the knowledge base and they influence both the course and computation time of the optimisation task. The scope and quantity of the control variables offer a great number of possible tracks for the optimisation algorithm control. The expert system determines the method control (which is varied during the analysis) so that the solution is time-optimum and satisfactory for the user.

11. In continuous development of the knowledge base, an important element is the possibility of making use of completed optimisation analyses. The prepared procedures for the information selection and reduction, as well as for the inference and decision-making, allow the system knowledge base to self-teach and be updated. One of the tasks to be performed by the teaching procedures is to select, based on a limited initial course of the optimisation task, a further course of the task that is most probable for given class of solutions and, next, to accept it. A measure of division into control classes is the dimension of the decision space determined by number of decision variables.

12. The optimisation of controls selection using the O-D method is performed so that the averaging loss R is least. The loss R is associated with functional relationship between the erratic classification of initial control S and the control S that exists in the knowledge base. It is termed average risk and is expressed by the following formula:

$$\bar{R} = \underset{S_j, S^*}{\mathcal{M}} R(S_j, S_j^*) . \tag{1}$$

Averaging by means of the operator takes place within a set of all pairs of controls (Sj, S) that are feasible in a given class. The loss functions were assumed based on computed values of the objective functions f(**x**i):

$$R_{min} = \min_j R(S_j, S_j^*) = \min_j \left[\sum_{i=1}^{q} \beta_i \left(f(\mathbf{x}_i) - f^*(\mathbf{x}_i) \right)^2 \right]_{S_j} . \tag{2}$$

Introduced weight coefficients βi assume their maximum value in the final phase of the analysis by the O-D method, for the sake of their decisive significance to the final result.

13. The selection of that control to which a higher probability of correct classification $\mathcal{P}$ can be attributed was based on decision procedures that employ the Bayes' rules. On account of a finite teaching set, that results from finite quantity of calculations of the objective function values, the conditional probabilities in these rules were replaced with frequencies.

14. The quality of the classifier is considered on the assumption that there are two classes: S1 and S2, the *a priori* probabilities P(S1) and P(S2) of which are equal. The images f(**x**) are one-dimensional and can assume discrete values f(**x**i), i=1,2,.,q and, moreover, every conditional discrete distribution P(f(**x**i)|Sj) is equiprobable:

$$\mathcal{P}(q,S_1,S_2) = \sum_{i=1}^{q}\left[\max_{j=1,2}\left(P(f(\mathbf{x}_i)|S_j)\cdot P(S_j)\right)\right]$$

$$P\left(f(\mathbf{x}_i)|S_1\right)\simeq s_1/m_1,\quad P\left(f(\mathbf{x}_i)|S_2\right)\simeq s_2/m_2,\quad P(S_1)+P(S_2)=1 \tag{3}$$

where: m1 and m2 designate numbers of elements in classes S1 and S2,
S1, S2 are appearance numbers for values fi(**x**), in classes S1 and S2 respectively,
and i = 1, 2,..., q, q is the number of objective function computations.

15. Restarted analysis follows the track which, after the initial analysis, provides the highest probability of a correct classification, $\mathcal{P}$. As the entire task is completed, the average risk and the loss function are computed. In case of obtaining:
(a) a solution that is unsatisfactory for the user,
(b) great values of both loss function and average risk, and
(c) unsatisfactory score for obtained solution at assumed point score scale, the system renews the task, beginning from the last point of the initial analysis.

16. Acceptance of given control to the preferable controls set of the knowledge base takes place when this control has been classified, by the decision rules, as being new or belonging to a given class of controls. In such a case, the number of statistic elements of control models for various optimisation tasks is increased. The classifier employed here is a two-class model.

17. The complexity and scale of the computation may be considerably reduced through confining the considered problem to a sub-optimal subset of parameters by using the sequential selection procedure. In this problem, as a criterion function, the average risk and the loss function were assumed for the selection of the decision variables subset, while for the control vector parameters, the initial analysis of limited teaching set was assumed.

18. The problem of selection of decision variables is to transform N-dimensional space into other space of dimension Nl<N. This should be done in such a way that the quality of solution is affected as slightly as possible and that the obtained quality be sufficient enough for the purposes of the statistical or economical analysis of a structure. For setting the solution quality pursuant to the user's needs, the expert system with a suitably directed knowledge base is used. The space transformation may concern reduction and/or selection of the number of decision variables, the number of objective function computations, or the acceptance of control vector parameters. In case of multi-criterion tasks, the objective function vector, or the set of non-dominated estimations and solutions, can be subjected to the reduction and selection.

EXAMPLE

19. The example is of the analysis of the steel parts of a hall structure which is composed of a double-layer spatial truss which forms the roof, poles which support the truss at its corners, and a shell (ref. 5). The truss members are selected using a pipe catalogue comprising maximally 33 sizes arranged according to increasing cross-section area with monotonic growth of critical force being retained. Truss nodes are of the Okta-S system (ref. 2). The truss load is equal to 2.0 kN/m^2 resulting from the weight of the roof covering, snow load, dead weight of the structure and from technological loads.

20. The optimisation analysis was carried out according to two criteria:
(a) minimum mass of steel elements of both the structure and the hall shell (Fig. 1).

$$f_m(\mathbf{x}) = \sum_{j=1}^{3} f_j(\mathbf{x}) = \frac{\gamma}{L^2}\left[\sum_{k=1}^{Lp} l_k \cdot A_k + \sum_{k=1}^{Lw} V_k + \sum_{k=1}^{6} M_k\right], \tag{4}$$

(b) minimum conventional function of cost being dependent on the structure geometry

$$f_c(\mathbf{x}) = \frac{p^2 + (p+1)^2}{40} \cdot a^2 \cdot \left[1 + \left|\sin\left(\frac{\Pi}{4} - \alpha\right)\right|\right] \cdot \left(1 - 0.4 \cdot e^{-0.35t}\right), \tag{5}$$

where:
$\mathbf{x} = [p, h, t]^T$ - vector of decision variables,
$p = L/a$, a, h, α - geometrical variables of the space truss (Fig. 1d),
p - number of divisions in the upper layer of the truss, $4 \leq p \leq 12$,
h - spacing between the layers (depth of truss), $1.2 \leq h \leq 3.3$ m,
t - number of cross-sections in the truss members' series of types, $t \in \{1,2,3,5,9,17,33\}$,
γ - specific gravity of steel, $\gamma = 7850$ kg/m^3,
l_k - length of k-th member of truss,
Ak - cross-section area of k-th member of truss,
Lp - number of truss members, Lp ε <128, 1152>,

Lw - number of truss nodes, Lw ε <41, 313>,
Vk - material volume of k-th member of truss,
Mk - material volumes of purlins (k=1), trapezoidal sheet of covering (k=2), hall corner pylons (k=3), wall columns (k=4), wall beams (k=5) and of steel in wall layer panels PW8/B (k=6).

21. The feasible domain of the task was determined assuming constraints resulting from design-related, technological and computational conditions. A detailed description of both the decision variables and feasible domain for the presented structure, being solved making use of decomposition, can be found in ref. 5.

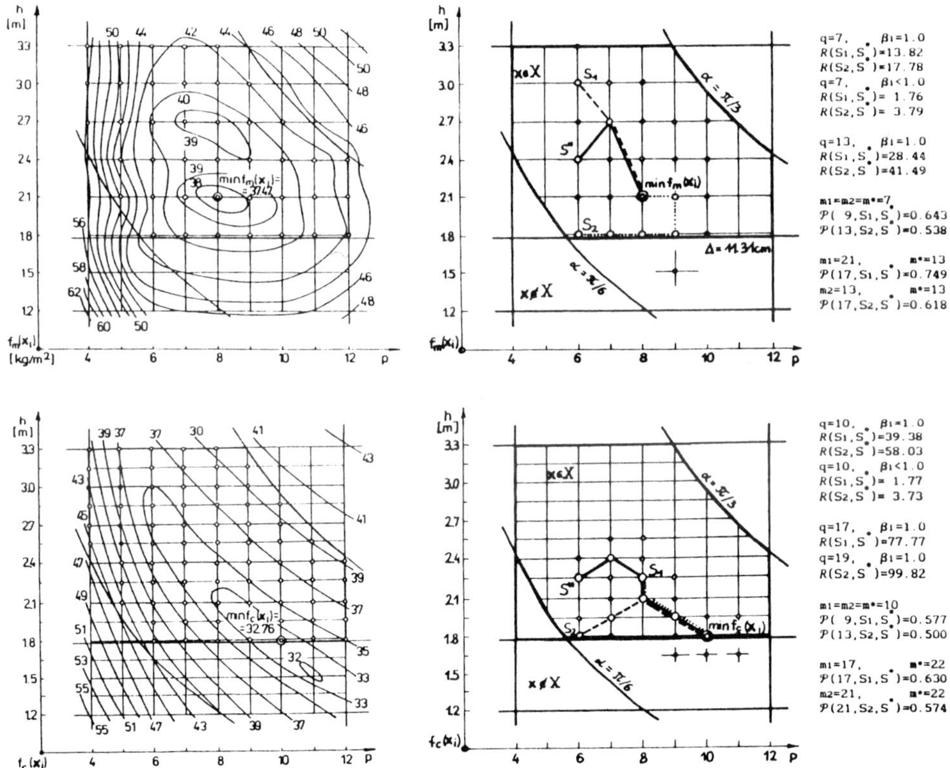

Fig. 4. The tracks Sl of moving of tasks fm(**xi**) and fc(**xi**).

22. Optimisation was performed independently for either of functions fm(**x**) and fc(**x**). Fig. 4 show the tracks Sj of moving within the feasible domain of the tasks with the use of the O-D method, at different starting points and for the variable

t=33. The selection of solution track carried out using the knowledge base selection-and-comparison procedures is presented for the initial control S*. The computed values of the loss function R(Sj, S*) for particular controls and the probability of correct classification $\mathcal{P}$ ascribed to them are shown in Fig. 4. For the objective functions fm(**x**) and fc(**x**), a further course of the analysis was assumed for the control S_1. After gaining the minimum objective function, the values of both the loss function and average risk were recalculated for the entire course of the task.

23. Fig. 5 presents the course of the solution track for a space determined by three decision variables. In case of the two-criteria optimisation (ref. 6), the preferable vector of decision variables was obtained as being equal to $\mathbf{x}_p$ = [8, 2.1, 3].

CONCLUSIONS

24. The computer program OPTIM based on the ortho-diagonal method of the discrete optimisation makes an effective instrument for structural design. The expert system controlling the process of the analysis enables satisfactory solution to be obtained in a time acceptable to the user.

25. The fact that the preferable solution is situated within the feasible domain, and that the determination of its co-ordinates is rather difficult without having well organised analysis, can be taken as characteristic property of the above problem.

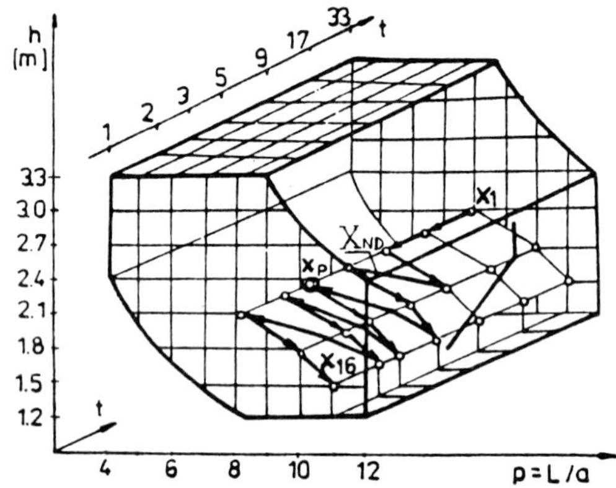

Fig. 5. The course of the solution track for a N=3 space.

REFERENCES

1. FROST, R.A. Introduction to knowledge base system. McGraw-Hill, New York, 1986.

2. HAUK, V. The mannesmann Okta-S joint for tubular space structures. Second Int. Conf. on Space Structures. Reports, University of Surrey, September 15-18, 1975.

3. HURLEY, R.B. Decision tables in software engineering. Van Nostrand Reinhold Company Inc., New York, 1983.

4. JENDO, S., PACZKOWSKI, W.M. Some problems of optimisation in civil engineering structural design, X Polish Conf. Computers Methods in Mechanics, Swinoujscie, Poland, 1991, Vol.1, 301-308.

5. JENDO, S.and PACZKOWSKI, W.M. Multi-criteria discrete optimisation of large scale truss systems. Structural Optimisation, Springer-Verlag (in preparing).

6. KARCZEWSKI, J.A., NICZYJ, J. and PACZKOWSKI, W.M. Discrete optimisation of the space truss being controlled by an expert system. IASS Symp. Spatial Structures at the Turn of the Millennium, Copenhagen, Denmark, 1991, Vol.III, 191-197.

7. NICZYJ, J. and PACZKOWSKI, W.M. An expert system of controlling a program for discrete structural optimisation. Int. Conf. on Optimisation Techniques and Applications (ICOTA'92), Singapore, 1992, Vol.2, 835-842.

8. PACZKOWSKI, W.M. Discrete optimisation program based on the ortho-diagonal method, Int. Conf. on Optimisation Techniques and Applications (ICOTA'92), Singapore, 1992, Vol.1, 248-255.

9. SHOUP, T.E. and MISTTREE, F. Optimisation methods with applications for personal computers, Prentice-Hall Inc., New Jersey, 1987.

192. Optimization of CUBIC space frame structures

S. SHANMUGANATHAN, Nottingham Trent University, UK, and
L. A. KUBIK, ASW-Cubic Structures, UK

SYNOPSIS. This paper describes a novel approach which is being applied to a complex cost optimisation problem for a 3-dimensional structure, the CUBIC Space Frame. By breaking the problem into several simple parts and looking at the effect of variables one by one, it is shown that the optimisation can be carried out easily and efficiently.

INTRODUCTION

1. The CUBIC Space Frame is a modular three-dimensional structure without diagonal bracing members (refs 1-3). The structure is assembled from modules such as that illustrated diagrammatically in Fig. 1. Many of the past applications have been one-off designs which used a variety of module types in terms of geometry, section size and connection details. However, recent experiences have suggested that by carefully selecting the grid size and depth, a small number of standard module types can be used for many structures without significantly increasing the weight of the structure. The result can be simpler and more cost-effective structures. With this in mind the research into optimisation of the CUBIC Space Frame was initiated at Nottingham Trent University in 1991.

2. Optimisation is the process by which the best solution is obtained for a given problem, taking into account constraints such as limits on structural depth or available materials. Although much early work on optimisation concentrated on minimum weight solutions (refs 4-6), the desirability for minimum cost solutions has been recognised by many researchers (refs 7-11).

3. The factors that determine the overall cost include material costs, fabrication costs, painting, transport and erection costs. Cost optimisation is therefore a complex process even for relatively simple structural forms. To simplify the process for the CUBIC Space Frame, the investigation has been considered in three stages. In the first stage attention is focused on finding minimum weight designs for a range of spans under prescribed loading, considering depth, grid size and section size as discrete variables.

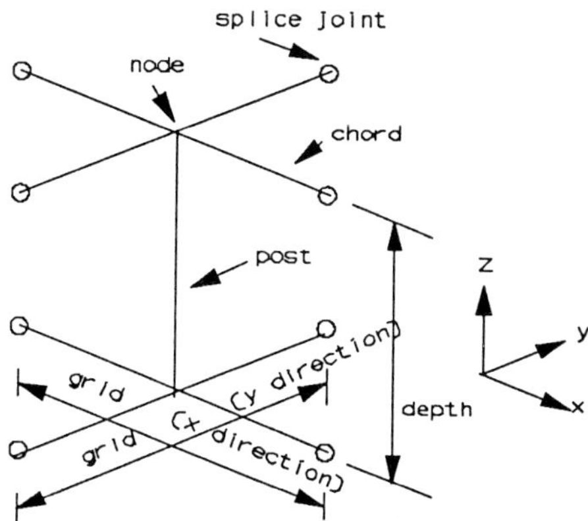

Fig. 1. Typical CUBIC Space Frame module indicating terminology.

The second stage introduces the repeatability of modules, or standardisation, and assesses the effect on the previously determined minimum weights. In the final stage costs are evaluated for a selection of the standard module types to identify the most cost-effective solutions. This paper deals mainly with the first two stages of the optimisation study.

PARAMETERS USED IN THE STUDY

4. Parameters that affect the design of the CUBIC Space Frame include the applied loading, span, support conditions, grid size, module depth and section size.

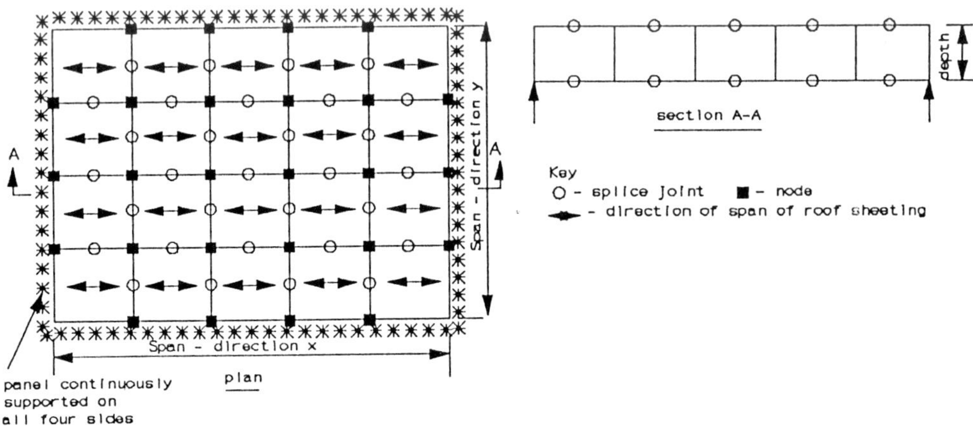

Fig. 2. Typical structural form used in study.

Loads, spans and support conditions are usually defined for a given structure, but structure depth, grid size and section size are not.

5. For the initial study, square and rectangular CUBIC Space Frame roof panels have been used, continuously supported around the perimeter (see Fig. 2). Spans between 15m and 60m were considered, being those most likely to be encountered in practice. Uniform grid size and constant depth have been assumed over the span. Grid sizes have been restricted to the range 1.5m to 7.5m and depths within the range span/40 to span/10. The space frame has been designed to support a uniformly distributed applied loading of $0.45 kNm^{-2}$, due to roof covering and services, and an imposed roof loading of $0.6 kNm^{-2}$, in addition to self weight. The roof decking has been assumed to span between the top chords as shown in Fig. 2.

6. By systematically changing the grid size, module depth and section size of the members, ideal combinations of these parameters can be found for a range of spans. The same section size has been used for both the upper and lower chords, assuming grade 50 steel throughout. The software which was developed to simplify this task is described in § 8 below.

7. As mentioned in §3, the study has been divided into three stages. The first stage focuses on finding minimum weight structures whilst the second stage examines the implications of using standard modules. Since the main concern in the first two stages is with producing safe overall member designs, connection details at nodes and splices are not considered. However, the connection details can have a significant effect on costs and are therefore considered in detail in the third stage of the programme.

THE SOFTWARE FOR MINIMUM WEIGHT DESIGNS

8. Fig. 3 outlines the algorithm used in the software to produce minimum weight designs for the CUBIC Space Frame roof structure. The software has been written in C++ to run on any IBM PC family of computers and most Unix platforms. The various stages are described below.

```
read Data( );
   for maxIterations do
      analysis( );
      design( );
      change Section( );
   done
```

Fig. 3. Algorithm of the software.

Analysis

9. The analysis routine calculates the forces and moments in the members of the space frame. The routine was previously developed by the second author to permit the rapid and simple analysis of CUBIC Space Frame structures by microcomputer (refs 1 and 2). The software is based upon linear elastic theory

and uses an iterative procedure to achieve compatibility of nodal deflection to the required accuracy. The first analysis is carried out using the smallest available section from the table described below, and subsequent analyses use revised sections, determined as outlined in §12. The self weight of the structure is calculated at each iteration to take into account the change of sections.

Design

10. Using the calculated member forces and moments and nodal deflections, the member sizes and frame deflection are checked for structural adequacy. The designs conform to BS5950 Part 1 (ref 12), both with regard to ultimate strength and serviceability.

11. The section size, a discrete variable, is chosen from a table of available rolled steel sections. Universal beams or columns are used for the chords and square hollow sections are used for the posts. The sections are tabulated in ascending order of mass.

Change of section size

12. A number of methods for changing section size were evaluated during the software development. The most efficient procedure involves starting from the smallest available section in the table. The section size is then progressively increased for any members found to be over-stressed during the design check until a safe design has been produced, the magnitude of the adjustment being dependent on the level of over-stress. Once a safe design has been found for the structure the section size of under-stressed members is reduced in proportion to the level of under-stress and the analysis, design and selection procedure is then repeated to produce several possible safe designs until the desired number of iterations has been completed.

13. During the preliminary trials it was noticed that after a few safe designs they started repeating. This was due to the discrete section sizes used and the standard method by which section sizes were increased or decreased. By incorporating a random perturbation into the section selection procedure, this difficulty was overcome and it was found that after 100 iterations several satisfactory safe minimum weight designs could be produced for all configurations. The perturbation technique picks a few members randomly and chooses the section size for them arbitrarily without obeying the rules for general section selection.

RESULTS OF MINIMUM WEIGHT STUDIES

14. Fig 4 shows some sample results for a typical roof panel, in this case a square panel of span 30m. These results show the effect of both grid size and depth on the minimum weight. As can be seen, there were several designs which had weights within 10% of the absolute minimum weight, including many with span/depth ratios of 30 or more. Similar results were obtained for other spans.

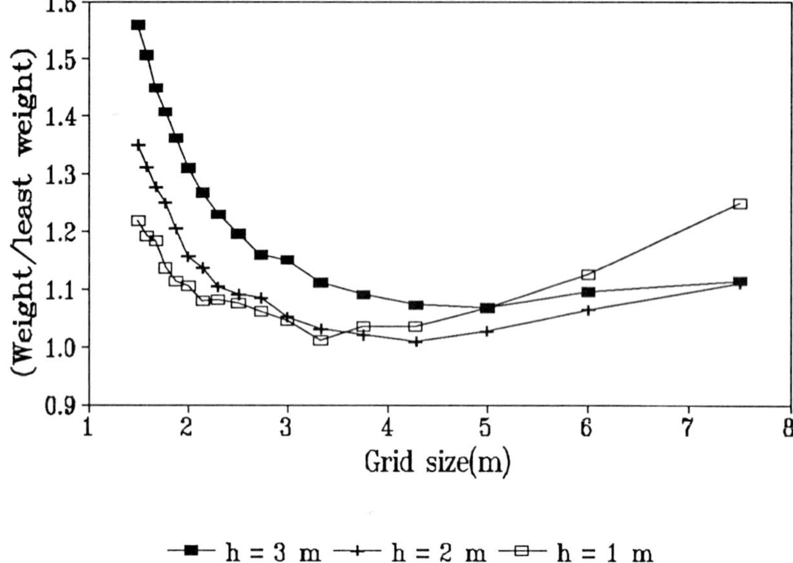

Fig. 4. Effect of grid size and depth on minimum weight for 30m x 30m panel.

15. Since there were no constraints on the variety of section sizes or on the number of different types of module in these schemes, they can be regarded as least weight solutions against which the schemes utilising standard modules in the second stage of the study can be compared.

THE EFFECT OF MODULE STANDARDISATION

16. In contrast to the minimum weight study, in which all member sizes were selected individually to suit the appropriate forces and moments, the introduction of standardisation means that large groups of members are considered together when selecting the section size. Three main methods of standardisation have been considered, namely:

- method 1, in which one section size is used for all chords and another section size is used for all posts.
- method 2, in which one section size is used for all chords in the x direction (Fig. 2), another is used for all chords in the y direction, and a third section is used for all posts.
- method 3, in which edge and middle strips of chords are considered in each direction, based on quarter spans as shown in Fig. 5. One section size is used for all chords in the edge strips and another for all chords in the middle strips. Four sections are used for the posts, one for each of the module types shown in Fig. 5.

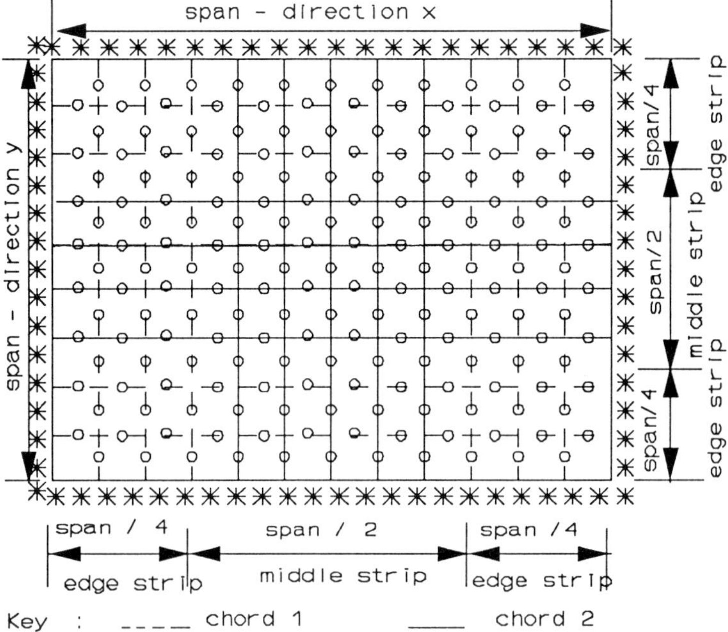

Fig. 5. Layout of the CUBIC Space Frame roof with three types of module.

17. The software which was used for the minimum weight designs was adapted to tackle the standard module designs. This resulted in a much simpler procedure for changing section size. In the case of the method 1 and method 2 standardisation it was simply necessary to systematically increase the size of the most over-stressed member (either post or chord) by one section until a safe design had been achieved. In method 3 standardisation a similar approach was used to produce the first safe design, although, due to the larger number of sections used, improved economy was found to be possible by then, systematically reducing the section size of the least stressed sections until the reduction produced an unsafe design. The design immediately prior to the last reduction was then adopted as the appropriate standard design.

18. Standard designs have so far been undertaken for a range of square panels with spans between 15m and 60m. The results indicate that by using a single standard module type (method 1) the CUBIC Space Frame roof structure is typically more than 30% heavier than the absolute minimum weight structure. No improvement results from using a different chord section in each direction (method 2). However, by using 3 types of module (method 3) the weight of the standard structure can be reduced further, so that it is typically only 20% heavier than the absolute minimum weight structure. In all cases a wide range of practical options exist for both grid size and depth without a significant additional weight penalty - see the typical results in Fig. 6 for the 30m span.

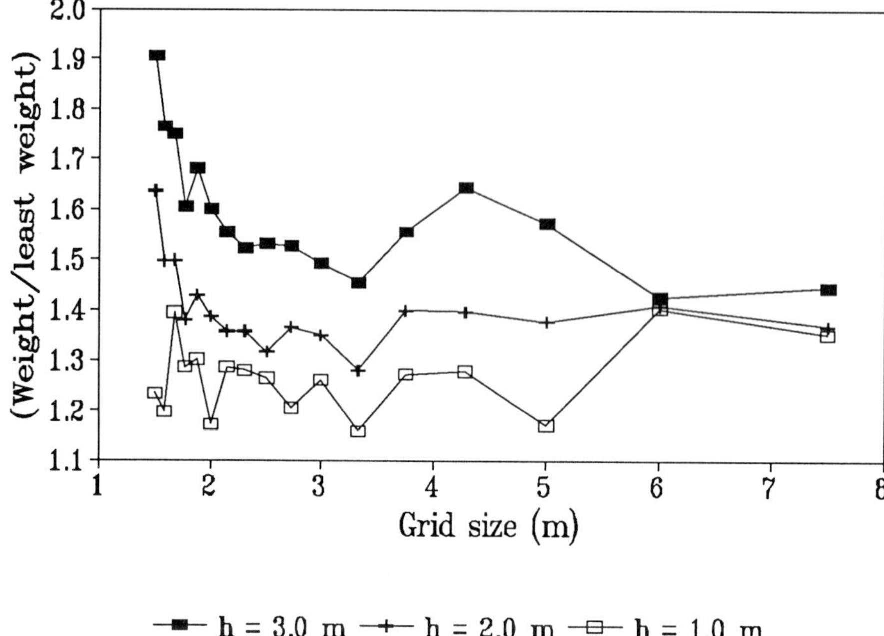

Fig. 6. Effect of standard modules using method 3 on minimum weight for 30m x 30m panel.

CONCLUDING REMARKS

19. In this paper, an effective three-stage process for optimisation of a three dimensional modular space frame structure has been described. Minimum weight solutions have been obtained for a range of spans and the effect of adopting standard module types has been investigated. A number of module types that have the least effect on the minimum structure weight have been identified. This range of standard module types is now being evaluated in the final stage of the investigation to establish costs and identify the most cost-effective solutions. As mentioned in §7, this will involve detailed consideration of the nodal and splice joint designs, for which separate optimisation studies are currently being undertaken (ref. 13). However, by limiting the cost investigation to a small number of standard module designs which are closest in weight to the absolute minimum weight solution, a rather complex problem has been reduced to one of manageable proportions.

ACKNOWLEDGEMENT

20. The research is jointly funded by ASW-CUBIC Structures Ltd. and Nottingham Trent University, to whom the authors are grateful.

REFERENCES

1. KUBIK L. A. The first 10 years of the CUBIC Space Frame. Paper submitted for the Fourth International Conference on Space Structures, September 1993.
2. KUBIK M. L. and KUBIK L. A. An introduction to the CUBIC space frame. Int. J. Space Structs, 1991, Issue 6, 1.
3. KUBIK L. A. The CUBIC space frame in theory and in practice. Steel Construction Today, March 1991.
4. ALLWOOD R. J. and CHUNG Y. S. Minimum-weight design of trusses by an optimality criteria method. International journal for numerical methods in engineering, Vol. 20, pp 697-713, 1984.
5. FLEURY C. and GERADIN M. Optimality criteria and mathematical programming in structural weight optimisation. Comp & Struct. Vol. 8, pp. 7-17, 1978.
6. SAKA M. P. Optimum design of space trusses with buckling constraints. The third int. conf. on space structures, edited by H Nooshin, pp 656-660, 11th-14th Sept. 1984.
7. SALAJEGHEH E. Optimum design of double-layer grids. The third int. conf. on space structures, edited by H Nooshin, pp. 661-668, 11th-14th Sept. 1984.
8. TEMPLEMAN A. B. Structural design for minimum cost using the method of geometric programming. Proc. Inst. Civil Eng., 46, pp 459-472, 1970.
9. LEE B.S. and KNAPTON J. Optimum cost design of a steel-framed building. Engineering Optimisation, Vol. 1, pp. 139-153, 1975.
10. CRAWFORD A. B. and JENKINS W. M. Optimum design of some steel roof structures. The Structural Engineer, Vol. 58A, No. 10, October 1980.
11. HORRIDGE J. F. and MORRIS L. J. Comparative costs of single-storey steel framed structures. The structural Engineer, Vol. 64A, No. 7, July 1986.
12. BRITISH STANDARDS INSTITUTION, 1990. BS5950: Part 1, the limit state Code of Practice for design of steelwork in building using hot rolled sections.
13. KUBIK L. A. and SHANMUGANATHAN S. Minimum cost nodal connections for the CUBIC space frame, paper submitted for the Fifth International Symposium on Tubular Structures, August 1993.

193. Weight minimization of non-linear three-dimensional frames with a buckling constraint

C. W. MARTIN, University of Nebraska-Lincoln, USA, and H. H. LIM, Valmont Industries, USA

SYNOPSIS. A procedure is presented for weight-minimisation of arbitrary three-dimensional frames made from tapered tubes. Deflections are presumed to be large enough to require non-linear analysis, and buckling is a possible failure mode, but the material remains elastic. The objective function and constraints are non-linear. The procedure accommodates multiple load conditions, imposing symmetry, and including "strings" of tapered members whose end diameters must match. The motivation for the work was the design of towers to support electrical transmission lines, e.g., at river crossings, and example calculations for these structures are presented, but the procedure is general and can be applied to any frame.

ANALYSIS REQUIREMENTS

1. Towers for electrical transmission lines require large-deflection analysis but the material remains elastic. A modified Newton-Raphson algorithm has been found to be most effective in iterating to equilibrium in the deformed geometry. Element types needed are truss, guy and tapered tubular frame. Cross sections of frame elements can be circular or equilateral polygons with 6, 8 or 12 sides. The effect of bowing on length should be included in the calculation of axial force in frame elements and the effect of axial force on bending stiffness is significant. The effects of bowing and distributed loads are also needed in the calculation of stress at sections within the frame elements.

2. Multiple load conditions are required and each load condition results in iteration to equilibrium in a different deformed geometry. The strategy adopted in the programs is to save the deformed nodal co-ordinates and the forces and moments at one end of each element for later calculation of stresses in the optimisation process. The analysis process has previously been described in (refs 1-3). One of the larger structures designed by the program of (ref. 3) is shown in Fig. 1. Note the man standing by the left leg.

DESIGN PROCESS

3. The objective function used here is total weight of the structure. The weight is a non-linear function of the design variables. Its gradient could be obtained

Space structures 4. Thomas Telford, London, 1993

without great difficulty but is not used in the algorithm. Other objective functions could be substituted with little effort.

Fig. 1. Large transmission line structure.

4. For unstable structures, the large-deflection analysis does not converge and considerable computing time may be wasted. Previous experience as related in ref. 1 has shown that a buckling constraint is really needed even if structural designs which have instability as a failure mode are not desired. If there is no buckling constraint, minimum values of diameters and thicknesses must be very carefully chosen to prevent the design algorithm from selecting a trial design for which instability is a failure mode. The focus of this paper is the addition of a buckling constraint to the design process described in (refs 1,3).

5. The design algorithm is a nested, two-stage process. The outer loop is called the design iteration. The inner loop is called the optimisation iteration. Complete analysis is performed only for each design iteration. One of the most important concepts in the programs implementing the algorithms is the definition of a design space and an analysis space. The design variable vector contains only the independent design variables, which are the minimum needed to uniquely define allowable variations in the structure. This may be written symbolically as;

$$a = T_I v \tag{1}$$

where the a is all of the radii, thicknesses, etc. required for finite element analysis, and v is the minimum set of independent design variables that are allowed to change in the design process. The transformation is, in these programs, non-linear. It is not actually performed as a matrix multiplication. This transformation process allows definition of strings of tapered finite elements, in which the radius at the terminal end of one element matches the radius at the initial end of the following element. It also allows prescribed discontinuities in radius, as at a slip joint. Tapers and thicknesses within a string of elements can be either independent, or constant throughout the string. Diameters of cross members can be made dependent on diameters of vertical members, etc.

6. A modified Box-complex algorithm is used in the search for minimum weight (refs 4,5). The optimisation iterations consist of moving the points in the complex. The best results have been obtained when the points in the complex are partly rational (to ensure that motion is possible in all design variable-co-ordinate directions) and partly random. When a trial point move violates a constraint, the backward move is based on the degree of constraint violation. One of the most important constraints is maximum effective (Von Mises) stress, which must be checked at several sections within each element, for each loading condition. Local buckling is avoided by limiting the width-to-thickness ratio, and the allowable ratio is stress-dependent. These constraints are non-linear, cannot be written explicitly in terms of design variables, and gradients cannot be obtained at any reasonable computing cost. During the optimisation iterations, the location of the stress constraint is estimated using the element moments and forces from the complete analysis performed in the previous design iteration.

7. The non-linear system stiffness equations may be written as:

$$K(X, a)X = P \tag{2}$$

in which K is the system stiffness matrix, X represents displacements, a represents the radii, thicknesses, etc. and P represents external loads. The displacements of an element in the local x-co-ordinates are given by:

$$x = T_2 X \tag{3}$$

and the element forces p are given by:

$$p = k(x, a)x \tag{4}$$

which is a non-linear relation, because the influence of axial force on bending is included. The stresses σ are obtained from the element forces as:

$$\sigma = H(x, a)p \tag{5}$$

The design variables, v are related to the analysis variables a, by:

$$a=T_1 v \tag{6}$$

In these programs, during the optimisation iteration, stresses are estimated by:

$$\sigma + \Delta\,\sigma \approx H(x,T_1(v + \Delta v))k(x,a)T_2 K^{-1}(X,a)P \tag{7}$$

This estimate of stress is exact for linear, small-deflection analysis of determinate structures, but significant errors may occur because of redistribution of load within an indeterminate structure or because of non-linearity. At the end of the optimisation cycle, a complete analysis is performed, which sometimes finds that the new design is over-stressed or that it would fail by buckling. In this case, the programs calculate a point on the hyperline in the design space between the last good design point and the new (over stressed) design point. Another analysis is then performed to confirm that the revised new design point does not violate any constraints.

BUCKLING CONSTRAINT

8. The gradient, V, to the buckling eigenvalue can be calculated by a "brute method of perturbation" of each of the design variables. The method of (ref. 6) uses much less computing time but requires program code for derivatives of stiffness matrices with respect to design variables. Using G to represent the "geometric stiffness" or "stress stiffening" matrix and $\overline{X}$ for the eigenvector:

$$[K - \lambda G]\,\overline{X} = 0 \tag{8}$$

and the gradient vector V can be calculated as in (ref. 6) by:

$$V = \nabla\lambda = \lambda,_{v_i} = \frac{\overline{X}^T K,_{v_i}\overline{X} - \lambda\overline{X}^T G,_{v_i}\overline{X}}{\overline{X}^T G\overline{X}} \tag{9}$$

Also, a unit vector, u, in the direction of the gradient is needed;

$$u = \frac{V}{|V|} \tag{10}$$

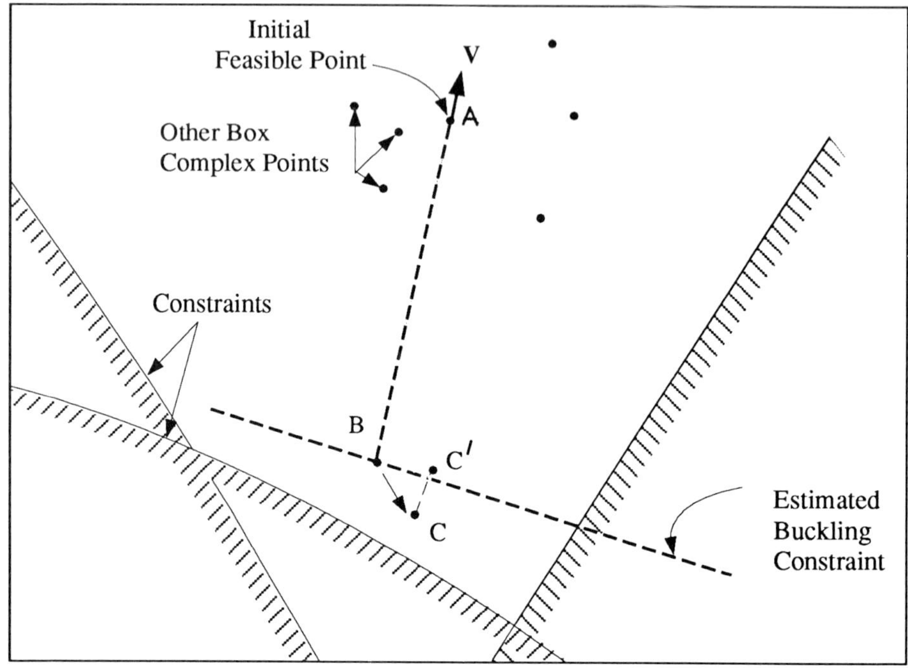

Fig. 2. Schematic sketch of Box complex and buckling constraint.

9. Incorporation of the eigenvalue constraint into the Box complex search algorithm is shown schematically in Fig. 2. The Box algorithm requires a feasible starting point. Let point A be the feasible starting point with design variable vector v_A and an eigenvalue λ_A. To implement the constraint, a point B with design variable vector v_B and eigenvalue λ_B must be found. The value of λ_B associated with point B should be equal to or slightly larger than one, say $\lambda_B = 1.05$. The hyperplane passing through point B and perpendicular to the gradient vector V is to be used as an approximation to the buckling constraint. The design variable vector v_B locating point B is found from:

$$v_A = v_B - \frac{[\lambda_A - \lambda_B]V}{V \bullet V} \tag{11}$$

10. The Box complex algorithm starts with an initial feasible point, such as point A in Fig. 2. A random number generator is used to create other points. If the initial random points violate constraints, they are moved toward point A until there is no constraint violation. The algorithm proceeds by moving the worst (heaviest, in our case) point through the centroid to the opposite (lighter) side. If a constraint violation is detected, the point is moved part way back toward the centroid. A trial point, such as point C in Fig. 2, is tested for violation of the estimated buckling constraint by computing:

$$\delta = [v_C - v_B] \bullet u \tag{12}$$

If δ is negative, the estimated buckling constraint is violated, and the trial point C is replaced by C', whose design variable vector is given by:

$$v_{C'}, = v_C - \delta u \tag{13}$$

Incorporation of the buckling constraint in the Box algorithm is thus easy to program. Very little additional computing time is required, except for calculation of the gradient vector, V.

11. After completion of the search in the design space, a complete analysis and constraint check is done on the new trial design. If a constraint violation is found, it is necessary to move backward toward the last feasible point and repeat the analysis and constraint checks to find a new feasible point.

NUMERICAL EXAMPLES

12. The examples presented here are taken from ref. 7. Small deflection analysis was used because of limitations on the amount of time available for a Master of Science research project. This does not make any difference in validity or functioning of the design algorithm, but a non-linear analysis probably would compute higher stresses and thus result in heavier final designs. The programs described in refs 1,3, which use the same design algorithm (but without the buckling constraint), have successfully designed many structures with multiple load conditions and large deflection analysis.

13. The first example is a vertical tapered pole with hexagonal cross-section. It is a typical support for lights at high-way intersections, 83 ft tall.* Only one load condition was specified, which is 5000 lb vertical and 200 lb horizontal, as shown in Fig. 3. There are three independent design variables: the top end radius, wall thickness, and taper. Constraints include stress, local buckling (width-to-thickness ratio) and general or system buckling. Analysis involved 9 elements, 10 nodes and 54 degrees of freedom. The initial and final designs are shown in

Table 1. The final design is on three constraints: system buckling, local buckling (width-to-thickness ratio), and (almost) minimum wall-thickness. The design history is shown in Fig. 4. It is believed that the final weight of 1181 lb is essentially at the minimum.

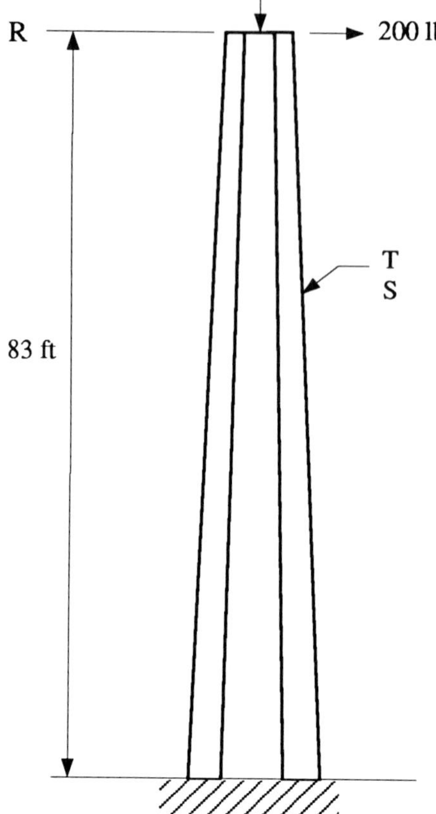

Fig. 3. Vertical pole design example.

Table 1. Initial and final design of the vertical tapered pole, example 1.

Variable	Minimum specified	Maximum specified	Initial value	Final value
Top-end radius, in.	3.0	15.0	12.0	5.447
Taper in/in	0.0	0.05	0.0	0.001
Wall thickness, in.	0.1	0.5	0.3	0.107
Weight, lb.			7233	1181

Note: The final design is on the system buckling constraint, on the local buckling constraint (width-to-thickness ratio) and very close to minimum thickness.

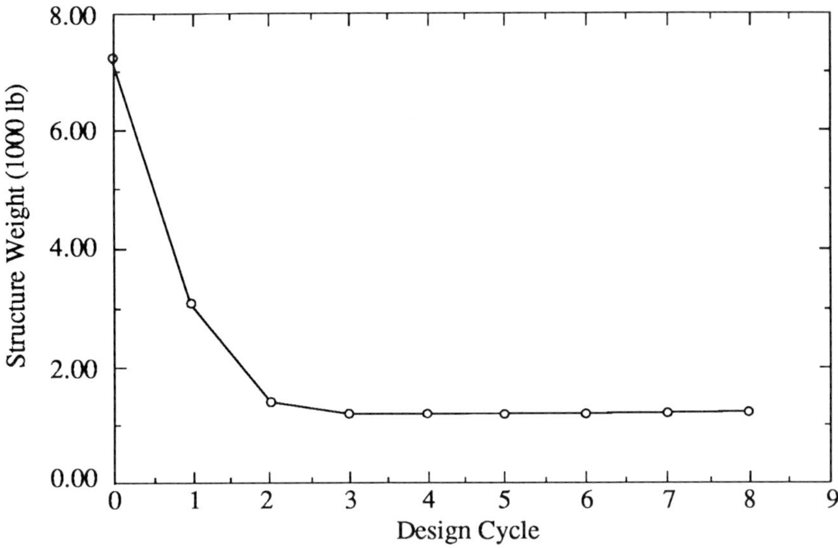

Fig. 4. Structure weight vs design cycle for vertical pole structure.

Table 2. Initial and final design of a river-crossing structure, example 2.

Variable	Minimum specified	Maximum specified	Initial value	Final value
Top radius, in.	5.0	12.5	8.0	7.1
Taper	0.0	0.0075	0.0025	0.0037
Column thickness 1, in.	0.1875	1.0	0.70	0.1876
Column thickness 2, in.	0.1875	1.0	0.70	0.1903
Column thickness 3, in.	0.1875	1.0	0.70	0.2417
Column thickness 4, in.	0.1875	1.0	0.70	0.2617
Cross arm thickness, in.	0.1875	1.0	0.90	0.1877
Weight, lb.			72982	33744

Note: The final design is on the system buckling constraint, on 3 local buckling constraints (width-to-thickness ratios at the bottoms of column segments T2, T3 and T4); on 2 thickness constraints and very close to a third thickness constraint.

14. The second example is a transmission line river-crossing structure, shown in Fig. 5. It is 164 ft tall, and in this example is designed for only one load condition as shown in the figure. Symmetry is imposed. Design variables for the vertical columns consist of the top-end radius, one taper and 4 thicknesses. The

upper two horizontal members are truss members with fixed radius but variable thickness. The lower three horizontal members were not allowed to change. There are 7 design variables, 19 elements, 14 nodes and 84 degrees-of-freedom in this example. The final design, as shown in Table 2, is on six constraints and very close to a seventh constraint. The design history is shown in Fig. 6. It is believed that the weight of 33744 lb for this design is essentially at the minimum.

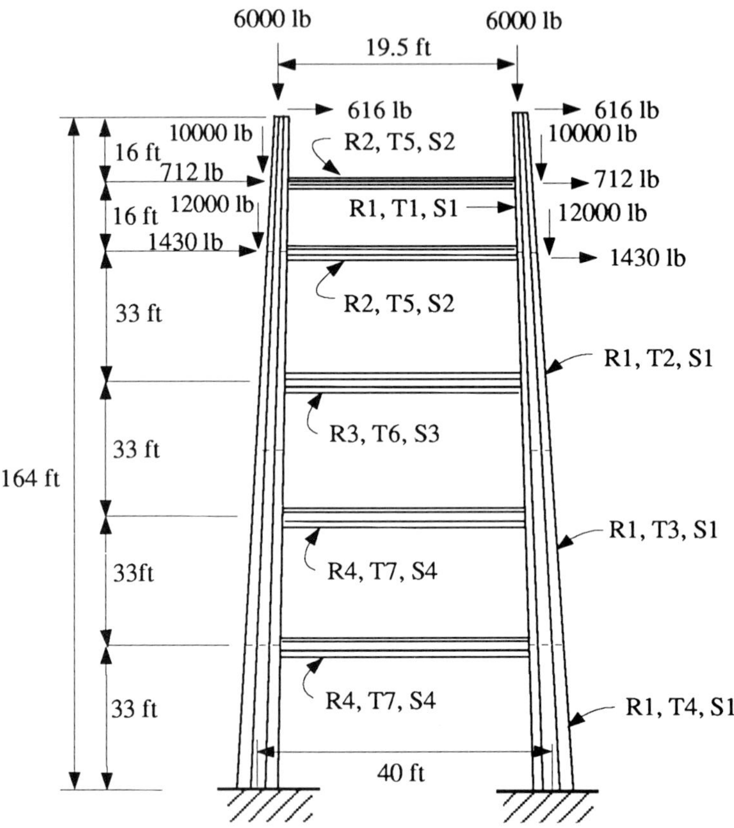

Fig. 5. River crossing design example.

CONCLUSION

15. A procedure has been developed for minimum-weight design of arbitrary three-dimensional frames made of tapered tubes which can fail by buckling. Application to two example problems has demonstrated that it is effective.

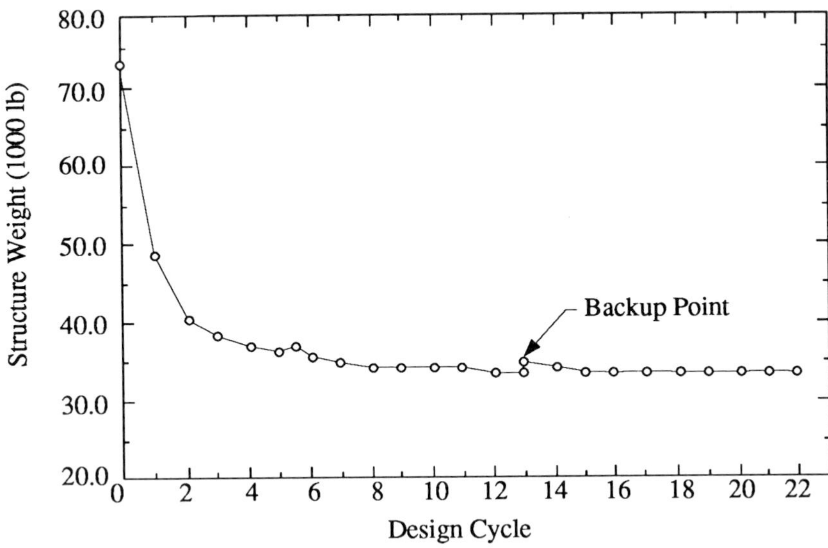

Fig. 6. Structure Weight vs Design Cycle for River Crossing Structure.

REFERENCES

1. MARTIN C.W., Automatic design of frames with tapered tubular members, p. 841-850 in Haug, E.J. and Cea, J. (ed.) Optimisation of distributed parameter structures, Sijthoff and Nordhoff, Rockville, MD, 1981.

2. MARTIN C.W. et. al., Non-linear analysis of space frames with tapered tubular members. Proceedings of the Sixth Canadian Congress of Applied Mechanics, the University of British Columbia, Vancouver, 1977, 161-162.

3. MARTIN C.W., DAVY D.T., ANDERSEN G. and LIN B., A computer program for optimum design of space frame transmission line structures: User Manual. Contract report to Valmont Industries, Valley, Nebraska, July, 1976.

4. BOX M.J., DAVIES D. and SWANN W.H., Non-linear optimisation techniques. p. 52 Imperial Chemical Industries Monograph No. 5, Oliver and Boyd, Edinburgh, 1969.

5. KUESTER J.L. and MIZE J.H., Optimisation techniques with FORTRAN, McGraw Hill, 1973.

6. PEDERSEN P., The integrated approach of FEM-SLP for solving problems of optimal design. pp. 757-780 in Haug, E.J. and Cea, J. (ed.) Optimisation of distributed parameter structures. Sijthoff and Nordhoff, Rockville, MD, 1981.

7. LIM H.H., Optimum design of frames with tapered tubular members and a buckling constraint. M.S. thesis, University of Nebraska, Lincoln, August 1991.

* UNIT CONVERSIONS

1 ft. = 0.3048 metre 1 in. = 0.0254 metre 1 lb. = 4.4482 Newtons

194. Optimization of shallow network domes

G. W. DUDA, Technical University of Wroclaw, Poland

SYNOPSIS. The paper deals with the optimisation of space trusses with fixed external dimensions under uniform snow and self-weight loading. Attempts are being made to find such a number of truss joints which minimise the material volume. The set of constraints imposed on a structure includes the effect on the stability of the loss of a bar. The statical problem of shallow domes, including geometrical non-linearity, was solved by using the Newton-Raphson iteration procedure.

INTRODUCTION

1. In the present paper network domes with joints which lie on a regular net of meridian and latitudinal circles of a spherical surface are considered. The bars connected at any joint of a truss make the configuration symmetrical with respect to the meridian plane and repeated many times in the structure Fig. 1. The number of truss joints depends on the number of meridians M and latitudinal circles N. In a dome with fixed external dimensions of its spherical surface with the base diameter D and height H, the numbers M and N will determine the length and direction cosines of its members. If the structure is subjected to a fixed axi-symmetrical load, the resultant forces acting on the joints, the axial forces, and the cross-sectional areas of the members, will depend on the M and N numbers. One can finally state that the material volume of the structure, considered as the objective function of the optimisation problem, depends on the value of M and N. In a feasible two-dimensional region of a permissible solution, a real structure will be represented by a point with the co-ordinates M and N. One looks for such values of the integer numbers $M = M^v$ and $N = N^v$, which minimise the theoretical material volume V of the structure ($V = V_{min}$). The optimisation mentioned above presents a non-linear programming problem. It was solved numerically by systematically searching for the points of the feasible region of solutions. In the calculation of bar forces in a shallow network dome, it is required to account for the effects of the geometrical non-linearities. This problem was solved by the stiffness method, using the Newton-Raphson iterative formula. The general problems of geometry, statical analysis and construction of bar

Space structures 4. Thomas Telford, London, 1993

domes are given in ref. 1. Problems of non-linear analysis of bar domes are discussed in many papers (e.g. in refs 2, 3 and 4). The foundations of optimum design in civil engineering are given in refs 5 and 6.

GEOMETRY AND LOADING OF A DOME

2. The net of N latitudinal circles on a sphere is created on the basis of a uniform angular division of the circle in the meridian plane. The basic geometrical elements of the dome are the radii of latitudinal circles r_i and the vertical distance between two adjacent latitudinal planes h_i, i = 1, 2,.., N. These elements are shown in Fig. 1. In a regular net of M meridian circles, the angular distance between two adjacent meridian planes-is constant, and it is equal to $\varphi = 2\pi / M$. Using r_i, h_i and φ it is possible to determine the length of bars $l\upsilon$ and their direction cosines according to the directions of unit vectors $\mathbf{b}\upsilon$ shown in Fig. 1.

$$l1_i = \sqrt{r_{i-1}^2 + r_i^2 + h_i^2 - 2\cos\varphi\, r_{i-1}\, r_i} \tag{1}$$

$$l2_i = 2\, r_i \sin\varphi \tag{2}$$

$$\mathbf{b1}_i = \left[r_{i-1}\cos\varphi - r_i, \quad r_{i-1}\sin\varphi, \quad -h_i \right]^T \frac{1}{l1_i} \tag{3}$$

$$\mathbf{b2}_i = \left[-\sin\varphi, \quad \cos\varphi, \quad 0 \right]^T = \text{constant} \tag{4}$$

In each joint of the truss a local system of cylindrical co-ordinates x, y, ,z was used (see Fig. 1). Any axi-symmetric load causes equal displacements of the truss joints lying on the same latitudinal circle i.

$$\mathbf{q}_i = \left[u_i, \quad v_i, \quad w_i \right]^T \quad \text{where } v_i = 0, \quad i = 1,2,\ldots N \tag{5}$$

In the apex joint $q_{N+1} = \left[0, 0, w_{N+1}\right]^T$. Thus the basic geometrical elements of the truss are:

$$r_i = r_{oi} + u_i; \quad h_i = h_{oi} + w_i - w_{i-1} \tag{6}$$

where r_{oi} and h_{oi} are the initial values of r_i and h_i (in the undeformed structure). The length and direction cosines of deformed members are computed according to the Eqs 1 to 4. The direction cosines presented in Eqs 3 and 4 are determined in the local co-ordinate system of a joint.

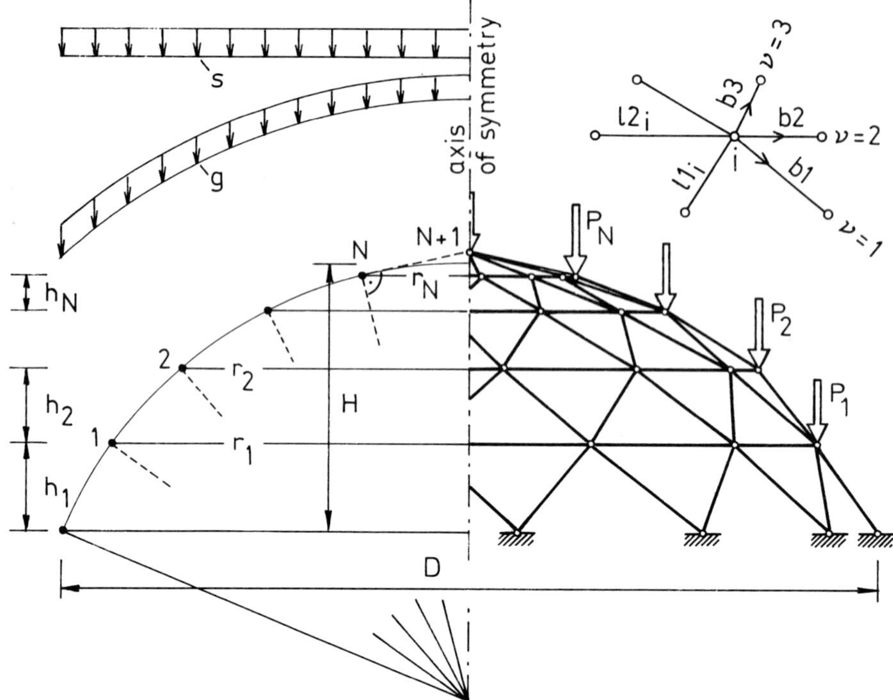

Fig. 1. Basic geometry elements and loads of the dome.

3. A uniformly distributed gravity load should be transformed into a system of resultant forces acting on the truss joint. In the present paper axi-symmetric snow load s, and self-weight g are considered. From this type of load the forces $\mathbf{P_i}$ in the joints which lie on the same latitudinal ring i are equal, and they are as follows :

$$\mathbf{P}_i = \left[P_{ix}, P_{iy}, P_{iz}\right]^T = \left[0, 0, P_{iz}\right]^T, \quad \text{where;} \tag{7}$$

$$P_{iz}(s) = s\,Fs(i), \quad P_{iz}(g) = g\,Fg(i), \quad \text{and} \tag{8}$$

$$Fs(i) = \frac{1}{3}\sin\varphi\left(r_{i-1} - r_{i+1}\right)\left[r_i + \cos\varphi\left(r_{i-1} - r_{i+1}\right)\right] \tag{9}$$

$$Fg(i) = \frac{1}{6}l2_i\sqrt{4.l1_i^2 - l2_i^2} + \frac{1}{6}l2_i\sqrt{4.l1_{i+1}^2 - l2_i^2} + \frac{1}{12}l2_{i-1}\sqrt{4.l1_i^2 - l2_{i-1}^2} + \frac{1}{12}l2_{i+1}\sqrt{4.l1_{i+1}^2 - l2_{i+1}^2} \tag{10}$$

STATICAL ANALYSIS

4. Determination of the axial forces in the members of a network dome, treated as a space truss, and subjected to a uniform axi-symmetric load is, according to the linear theory, a statically determinate problem. In this case it is possible to calculate on the basis of equilibrium conditions, the axial forces $N1_i$, in the diagonal members, and $N2_i$ in the latitudinal members, from the following formulae:

$$N1_i = l1_i \left(P_{iz} + P_{i+1,z} + \ldots + P_{Nz} + \frac{2}{M} P_{N+1,z} \right) \Big/ h_i \tag{11}$$

$$N2_i = \left[N1_i \left(r_{i-1} \cos\varphi - r_i \right) / l1_i + N1_{i+1} \left(r_{i+1} \cos\varphi - r_i \right) / l1_{i+1} \right] \Big/ \sin\varphi \tag{12}$$

where $\varphi = 2\pi / M$, and M is the number of meridians. The geometrically non-linear analysis of the structure was carried out by the stiffness method based on the equation:

$$\left(\mathbf{K}_E + \mathbf{K}_G \right) \mathbf{q} = \mathbf{P} \tag{13}$$

in which $\mathbf{K}_E$ is the standard elastic stiffness matrix; $\mathbf{K}_G$ the geometrical stiffness matrix which depends on the current value of axial forces in the truss members, $\mathbf{q}$ is the vector of joint displacements and $\mathbf{P}$ is the vector of joint loads. Matrices $\mathbf{K}_E$ and $\mathbf{K}_G$ are quadratic and symmetric. The respective formulas for the matrix elements are given in ref. 2. The stiffness matrix is most frequently built in one Cartesian co-ordinate system. In the case presented above, it is convenient to use a multi-global co-ordinate system, similarly to ref. 3. This leads to a considerable reduction of the number of degrees of freedom of the truss joints. Thus, the vector $\mathbf{q}$ consists of 2N+1 elements, and the matrices $\mathbf{K}_E$ and $\mathbf{K}_G$ has the dimension (2N+l) x (2N+1). Since the stiffness matrices depends on the displacement vector $\mathbf{q}$, the Eq. 13 is non-linear. It was solved by using the Newton-Raphson iteration procedure. The axial forces computed according to the Eqs 11 and 12 were used to build the matrix $\mathbf{K}_G$ at first approximation. Examples of the application of the Newton-Raphson method in the non-linear analysis of reticulated space truss are given, among others, in refs 3, 4.

DETERMINATION OF THE CROSS-SECTIONS OF THE BARS

5. The determination of the cross-sectional areas of the bars was made according to the limit state method design. The shape of the cross-section of a bar is described by the dimensionless factor $\psi = A^2/I$, where A is the cross-sectional area, and I the second moment of area. For a circular tube with dimensions r, t (r - radius of the centre line, t - wall thickness) obtained from $\psi = 16\pi\eta / (4 + \eta^2)$, where $\eta = t / r$. If the bar *i* is tensioned by the force N_i, its cross-sectional area is taken from the formula:

$$A_i \geq N_i/R \tag{14}$$

where R is the strength of the structural material. For the bar *j*, compressed with a force N_j, one should compute firstly its characteristic length:

$$l_j^c = \lambda_p \sqrt{2.N_j/\psi.R} \tag{15}$$

where $\lambda_p = \pi\sqrt{2E_0 / R}$, $E_0 = 2/3E$, E isYoung's modulus, and secondly its cross-sectional area, using one of the formulas:

$$A_j \geq \sqrt{2\psi N_j^{'}/R}\, l_j/\lambda_p, \qquad \text{If } l_j \geq l_j^c \tag{16}$$

$$A_j \geq \frac{N_j^{'}}{R} + \frac{\psi}{2}\left(l_j/\lambda_p\right)^2 \qquad \text{If } l_j < l_j^c \tag{17}$$

where $N_j^{'} = \text{abs}(N_j)$ and l_j is the real length of bar j. Eq. 16 is based on Euler's curvature. Eq. 17 is based on Ostenfeld's parabola and is valid in the range of elastic-plastic buckling.

FORMULATION OF THE OPTIMUM DESIGN PROBLEM

6. In the above described optimisation problem of the network dome the constant parameters are:

- external dimensions of the spherical surface D and H;
- the value of the loads s and g;
- the strength of the structure material R;
- the shape of the cross-section of the bars.

The integer numbers M and N are decision variables. The objective function V(M, N) expresses the theoretical volume V of the structure material. The criterion of the optimisation is formulated as:

$$V = V(M, N) = \min = V^v \tag{18}$$

7. The set of restrictive conditions imposed on a structure consists of :
   a) cross-sectional areas A_i of the tension members determined from Eq. 14;
   b) cross-sectional areas A_j of the compression members determined from Eqs 16 and 17;
   c) $A_k \geq A_{min}$, for all members;
   d) the bar length is limited to $l_{min} < l_k < l_{max}$.

Since the analytical formula of the objective function is unknown, its values are computed numerically from the formula $V = \sum l_k A_k$. The restriction d (above) creates a discrete region **Z** of permissible solutions, in which the optimal structure is to be found.

NUMERICAL EXAMPLE

8. In order to investigate the described optimisation problem, a computer program was written (in Turbo Basic). The following values of constant parameters have been accepted: D = 30m, H = 4m, g = 0.95 kN/m^2, s = 0.80 kN/m^2, R = 210 MPa (steel).

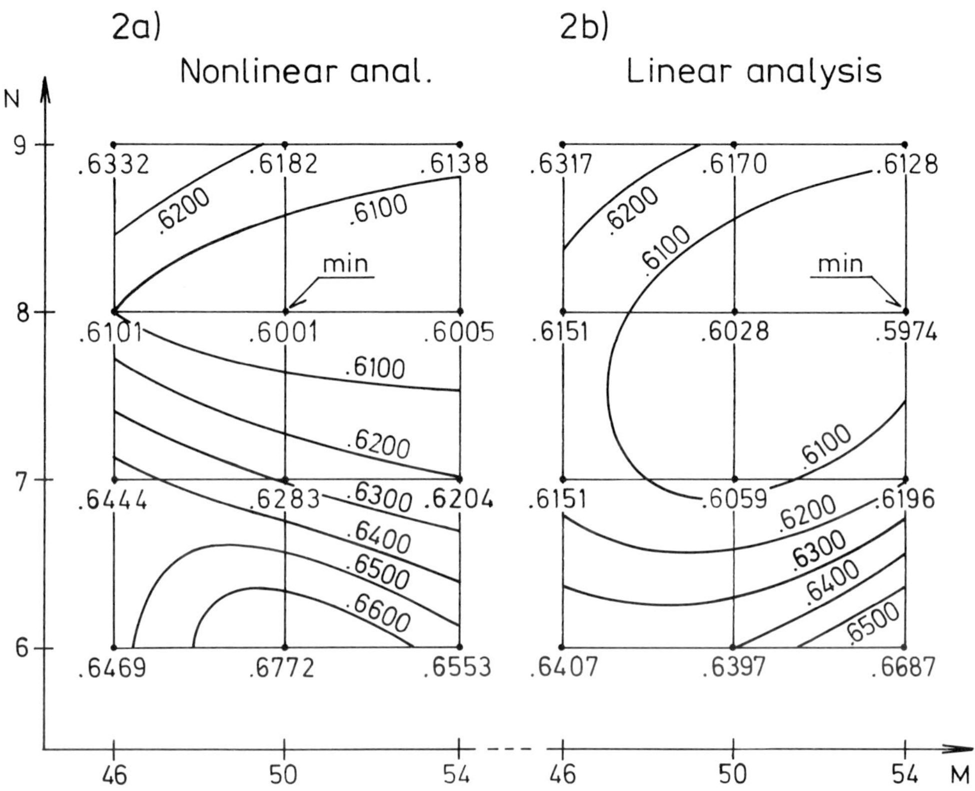

Fig. 2. Map of the objective function.

It is assumed that the structure's members are made of circular steel tubes described by the shape factor $\psi = 1.876$, and of which the smallest cross-section is A_{min} = max (A_o, $\psi\ l^2/200^2$), where A_o = 3.676cm^2 (tube Φ 42/3), l in centimetres. The bar length was limited to l_{min}= 0.4m and l_{max}= 4.0m. The condition $0.4 \leq l_k \leq 4.0$m creates the feasible region in which the variables M and N have the values M = 46, 50, 54 and N = 6, 7, 8, 9. Fig. 2 presents a map of the objective function (treated as a continuous function). The shape of the line V = constant is only probable. In Fig. 2a there are the results of the computations, with allowance made for geometrical non-linearity and in Fig. 2b the results obtained according to linear statics are given. In further analysis the optimisation of the dome with changing height H has been provided. Some results of the computation are shown in Fig. 3 .

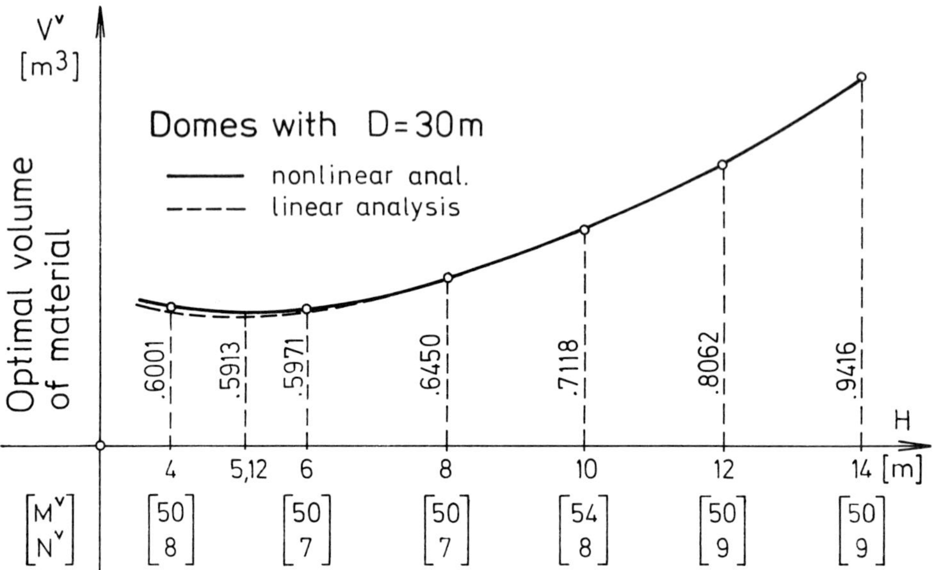

Fig. 3. Optimal material volume as a function of the dome height.

It has been found that the global minimum of the material volume reaches $V^* = 0.5913m^3$ of steel in the dome with $H^* = 5.12m^3$, ($M^* = 54$, $N^* = 7$). In region **Z** in Fig. 2, the maximal difference of material volume between domes with different M and N is equal to 12.8%. The optimal solution, according to non-linear statical analysis, has been found for (M^v, N^v) = (50, 8), but according to the linear analysis it has been found for (M^v, N^v) = (54, 8). The difference between the values V^v according to non-linear and linear statical analysis is small and equal to 0.45%. Greater differences can possibly be present in domes supported on an elastic ring.

REFERENCES

1. MAKOWSKI Z.S. Analysis, design and construction of braced domes. Nicholas Publishing Company, New York, 1984.

2. PRZEMIENIECKI J.S. Theory of matrix structural analysis. McGraw-Hill, New York, 1968.

3. JAGANNATHAN D.S., EPSTEIN H.J., and CHRISTIANO P. Non-linear analysis of reticulated space trusses. Journal of the Structural Division, ASCE, vol. 101, no. ST12, Dec. 1975, pp. 2641-2658.
4. ROTHERT H., DICKEL T., and RENNER D. Snap-through buckling of reticulated space trusses. Journal of the Structural Division, ASCE, no. ST1, Jan. 1981, pp. 129-143.
5. BRANDT A.M. Foundations of optimum design in civil engineering. PWN-Polish Scientific Publishers, Warszawa, Poland, Nijhoff Publishers, Dordrecht, Boston, London, 1989.
6. HAUG E.J., and ARORA J.S. Applied optimal design mechanical and structural system. A Wiley-Interscience Publication, John Wiley & Sons, New York, Chichester, Brisbane, Toronto, 1979.

195. Recent applications and development of steel space structures with welded hollow spherical joints in China

XI-LIANG LIU, Tianjin University, P. R. China

SYNOPSIS. This paper explains why the welded hollow spherical joint grids are so widely used in China. It also introduces the structure, calculation, manufacture and construction of the welded hollow spherical joint (abbreviated to WHSJ in the following text). Some typical projects of the WHSJ are given. Some personal views on this kind of joint are presented.

INTRODUCTION

1. The WHSJ, adopted earliest in China, has a history of about 27 years, and it is nowadays also one of the most popular joints all over the country. The first project built was the Tianjin Scientific Committee Hall in 1966. In this country, more than 2000 space grid structure projects have been built, over half of which are of the WHSJ grids. Particularly, of all the gymnasium roof structures which were built for the 11th Asian Games held in China, more than 90% are the WHSJ grids. This can be seen clearly from Table 1 .

2. The WHSJs are made of two hollow semi-spheres by butt welding (see Fig. l). The semi-sphere can be manufactured in two ways: cold pressing and hot pressing. It is easier to obtain a final shape by hot pressing, which doesn't need much pressure, so hot pressing technology is used more widely. While the cold pressing not only needs a higher pressure, but also wears and tears the machine to a large degree, thus, it is rarely used. In order to improve the load capacity of the WHSJs when the diameter of the sphere is fairly large, a stiffening plate can be added into the sphere at the butt welded seam and the three parts are welded into a whole body. The load capacity of the WHSJ can be improved by 15-40% by means of the stiffening plate.

3. The WHSJ grids have seven apparent advantages as follows: 1) Cheap in cost. As the WHSJ needs little machining, its manufactured cost is usually 25% cheaper compared to the bolt spherical joint grids, under Chinese conditions; 2) Simple in construction. It doesn't require precise machining and, provided the cutting face of the tube is perpendicular to the tube axis, the member will be naturally aligned on the sphere without any additional eccentricity; 3) It is not only suitable for mid-small span structures, but also large span structures.

Space structures 4. Thomas Telford, London, 1993

Table 1. Gymnasiums for the 11th Asian Games in Beijing (1990).

<table>
<tr><td rowspan="2">Steel Space Grid Structure</td><td>Welded Hollow Spherical Joints (WHSJ)</td><td>Beijing College Student Gymnasium (64m × 64m)
Haidian Gymnasium(48m × 52m)
Yuetan Gymnasium (Irregular Octagon 57m × 66m)
Tennis Gymnasium of BITC (60m × 60m)
Guang Cai Gymnasium (46.8m × 67.6m)
Gymnasium ofFengtai Sports Centre (54.6m × 76.6m)</td></tr>
<tr><td>Bolted Joints</td><td>Ditan Gymnasiums (Hexagon Side 30m)</td></tr>
<tr><td rowspan="2">Steel Latticed Shells</td><td>Welded Hollow Spherical Joints (WHSJ)</td><td>Gymnasium of the Olympic Sports Centre (70m × 83.2m)
Natatorium of the Olympic Sports Centre (70m × 100m)
Shijing Shan Gymnasium (3–△99.7m)
Gymnasium of Beijing Institute of Physical Education (53.2m × 53.2m)</td></tr>
<tr><td>Bolted Joints</td><td></td></tr>
<tr><td colspan="2">Cable Structure</td><td>Chaoyang Gymnasium (66m × 78m)</td></tr>
<tr><td colspan="2">Plane Truss</td><td>Capital Training Gymnasium for Speed Skating (88m × 180m)</td></tr>
</table>

While the bolt spherical joint grids are only suitable for mid-small span structures, due to the limit of the bolt diameters (ordinarily 56mm); 4) As both ends of the tube are sealed up, the anti-corrosion capacity of the WHSJ grids is higher than that of the bolt spherical joint grids which have some clearance; 5) It is most suitable for an integral hoisting because of its high rigidity; 6) With its joint rigidity, the joint can bear certain moments, so it is most fit for the latticed shell; 7) As it is simple in composition the joint needs only one kind of steel, such as low carbon steel. The only disadvantage of the WHSJ is that it requires a large amount of welding on site, although, this causes no problem in China.

STRUCTURAL DEMANDS OF THE WHSJ

4. The sphere has two factors: the outer diameter and the thickness of the sphere wall. Usually, the diameter is firstly designed according to the structural demands, then the thickness can be calculated. The diameter of the WHSJ should meet the requirements to arrange all the tubes welded onto the joint. In order to make it easier to weld and avoid the steel to be over-burned, the clearance between any two tubes connected to the same joint should not be less than 10mm. According to these requirements, the following formula to determine the joint diameter can be derived:

$$D = \frac{d_1 + 2a + d_2}{\theta} \tag{1}$$

where θ is the radian of any two tube lines meeting at the centre point of the sphere, a is a constant no less than 10mm, d_1, d_2 are the outer diameters of the two tubes respectively, see Fig. 2.

5. According to the design requirements, the ratio of the outer diameter to the thickness of the spherical joint can usually be chosen from the range 25 to 45. The ratio of the sphere wall thickness to the maximum tube wall thickness should be from 1.2 to 2. The thickness of the sphere wall generally should not be less than 4mm.

6. When the diameter of the sphere is more than 300mm, a stiffening plate can be added into the sphere in order to increase the load capacity of the joint. The thickness of the stiffening plate should not be less than the thickness of the sphere wall. The member with the largest internal force should be placed in the plane of the stiffening plate, special attention should be paid to this in construction.

7. To be connected to the sphere, the tube should be grooved at the end. A certain clearance should be left between the tube and the sphere in order to weld through. Thus, the weld seam can be regarded as having the same strength as the tube. Otherwise, the weld seam must be calculated as a fillet weld.

8. To simplify the fabrication and ensure the quality of the weld seam, at the end of the tube, a piece of casing tube can be used to weld with the sphere. The length of the tube is usually not less than 40mm. Of course, the casing tube will lead to some difficulty in the construction and probably add to the cost. If it is possible to ensure the quality of the weld seam without the casing tube, it is sensible to do away with it, see Fig. 3.

9. In order to make it easy to weld, the two semi-spheres with the stiffening plate, the plate can be made with flange. The height of the flange usually should not be more than 1mm a higher flange will make it difficult to achieve an acceptable weld.

THE CALCULATION FORMULAE OF THE WHSJ

10. On the basis of a vast amount of experiments, the following conclusions can be drawn: 1) In the experiment of collapse load, the uni-directional tension load is quite near the two-directional tension load. The uni-directional compression load is also quite near the two-directional compression load. In the case of different load combinations in two-directions, the load capacity, when the compression load is larger, is near that of the uni-directional compression load capacity. Also, the load capacity in tension is larger than that in compression. All these tests show that the collapse load can be gained based on the uni-directional load and it is just the uni-directional load capacity. So the formulae adopted in ref. 5 are derived on the basis of uni-directional load experiments; 2) the stiffening plate will improved the load capacity by a large amount; 3) Both thickening the sphere wall and enlarging the tube diameter will improve the load capacity of the WHSJ; 4) In the case of the tension load, the joint will usually fail with a broken tube. The load capacity is due to the strength of the weld seam and the tube. Only thickening the tube wall can lead to the sphere being broken first; caused by the stress concentration on the sphere surface; 5) In the case of the compression load

the joint will usually be destroyed when the shell plate buckles along the circumference of the tube.

11. Using the vast amount of experiment results, a distribution diagram of the collapse of the WHSJ for different load cases can be drawn. So the load capacity formula under compression or in tension can be regressed by using a statistic method, see Figs 4-6.

12. The formulae adopted by ref. 5 are:
For the compression case;

$$N_c = \eta_c(400td - 13.3\frac{t^2d^2}{D}) \tag{2}$$

where N_c is the design value of the WHSJ under axial compression, D is the outer diameter of the hollow sphere, t is the wall thickness of the hollow sphere, d is the outside diameter of the tube, η_c is the ratio that improved the load capacity of the hollow sphere when stiffened in compression, η_c = 1.0 without the stiffening plate, η_c = 1.4 when stiffened;
For tension;

$$N_t = \eta_t(0.6td\pi)f \tag{3}$$

where N_t is the design value of the WHSJ in axial tension, f is the design strength of steel in tension, η_t is the ratio that improve the load capacity of the WHSJ when stiffened in tension, η_t = 1.0 without the stiffening plate, η_t = 1.1 when stiffened.

OPTIMISATION OF THE WHSJ

13. In order to effectively improve the load capacity, the tube used should have its largest possible diameter and smallest wall thickness. However, this will correspondingly require a large diameter sphere, which is usually more expensive, which leads to an uneconomic design of the whole grid. Thus, the joint also bears the problem of optimisation, i.e., the problem of the diameter and the wall thickness of the sphere to best match that of the tube. To solve this problem, a rational curve and diagram of L/D can be used to obtain an optimum design. Designers can use them for reference, see Table 2. The specific steps are as follows: 1) to estimate the internal force of the member according to any simplified calculation method; 2) look up the rational L/D value in the table according to the known L and N; 3) calculate the outer diameter of the tube according to structural requirements or with D / d= 2.7, then use a type listed; 4) calculate the wall thickness of the sphere and the wall thickness of the tube.

Table 2. Rational Values of L/D.

L(m) \ N(t)	10	20	30	40	50	60	70	80	100
2.0	10.29	8. 4 4	8.32	8.33	8.16	8.16			
2.5	12.32	9.02	8.46	8.29	8.30	8.33	8.15		
3.0	12.89	10.75	9.00	8.43	8.29	8.20	8.23	8.14	
3.5	13.56	11.86	10.08	9.04	8.87	8.60	8.32	8.16	8.19
4.0	14.30	12.70	11.38	9.90	9.03	9.07	8.90	8.33	8.18
4.5	14.83	13.74	12.44	11.17	9.98	9.69	9.41	8.74	8.15
5.0	15.44	13.86	12.60	12.28	10.94	9.97	9.60	9.60	8.89
5.5	15.57	14.29	13.16	12.44	11.88	10.73	10.03	9.71	9.43
6.0	16.14	14.86	13.73	12.99	12.44	11.87	11.00	10.28	10.19

THE WHSJ SERIES

14. The most commonly used hollow spheres are listed in Table 3 and Table 4.

MEASURES TO IMPROVE THE LOAD CAPACITY

15. Except for stiffening plates (see Fig. 7a), some other measures can be taken to improve the load capacity of the WHSJ when thickening the wall of the sphere is not allowed. These include using double parallel stiffening plates; cross stiffening plate; stiffening steel ring (Fig. 7b); placing concrete into the sphere (Fig. 7c); the main loaded members passing through the sphere (Fig. 7d) and stiffening plate outside the sphere (Fig. 7e), etc.

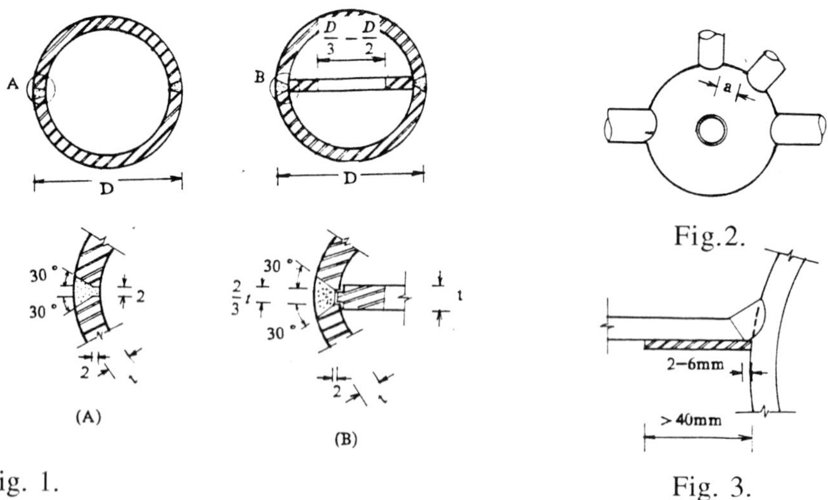

Fig. 1.

Fig.2.

Fig. 3.

Table 3. WHSJ Series.

No.	Designation	Size and Thickness (mm × mm)	Diameter of Tubes (mm)	Tensile Limit Capacity (kN)		Compressive Limit Capacity (kN)
				16Mn	A3	
1	WS1604	D160 × 4	60	212	145	148
2	WS1606	D160 × 6	60	318	217	216
3	WS1804	D180 × 4	63.5	224	153	156
4	WS1806	D180 × 6	63.5	336	230	228
5	WS2206	D200 × 6	76	403	275	273
6	WS2008	D200 × 8	76	537	366	355
7	WS2206	D220 × 6	89	471	322	319
8	WS2208	D220 × 8	89	629	429	412
9	WS2406	D240 × 6	102	540	369	363
10	WS2408	D240 × 8	102	720	492	470
11	WS2410	D240 × 10	102	900	615	568
12	WS2606	D260 × 6	114	604	412	405
13	WS2608	D260 × 8	114	805	550	523
14	WS2610	D260 × 10	114	1006	687	632
15	WS2808	D280 × 8	133	939	641	603
16	WS2810	D280 × 10	133	1174	801	727
17	WS2812	D280 × 12	133	1409	962	839
18	WS3010	D300 × 10	140	1236	844	767
19	WS3012	D300 × 12	140	1483	1012	887
20	WS3512	D350 × 12	146	1547	1056	948
21	WS3514	D350 × 14	146	1805	1232	1069
22	WS4014	D400 × 14	152	1879	1282	1136
23	WS4016	D400 × 16	152	2147	1465	1259
24	WS4018	D400 × 18	152	2415	1649	1371
25	WS4518	D450 × 18	152	2415	1649	1416
26	WS4520	D450 × 20	152	2684	1832	1529
27	WS5020	D500 × 20	152	2684	1832	1574
28	WS5025	D500 × 25	152	3355	2290	1842

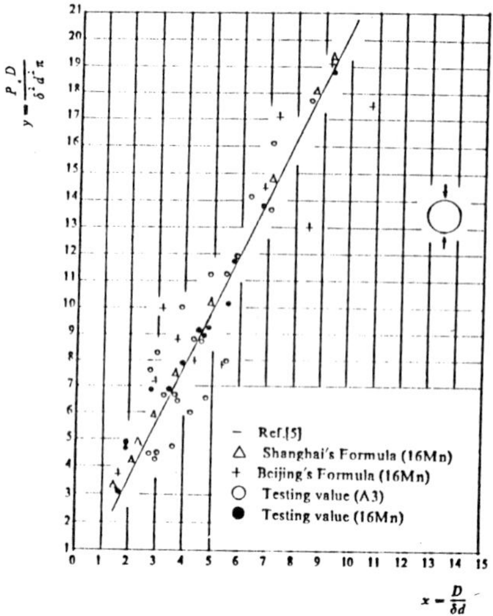

Fig. 4.

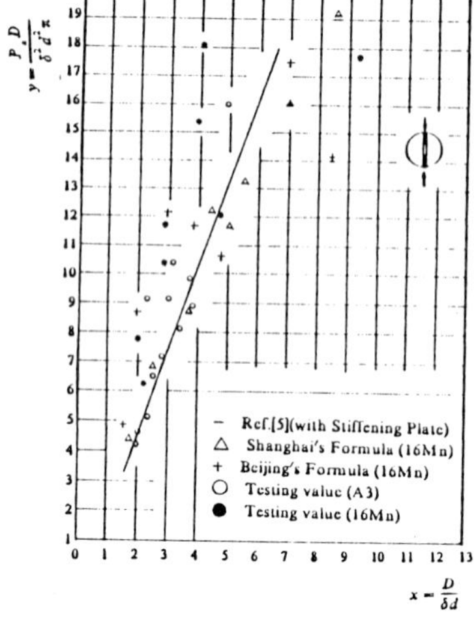

Fig. 5.

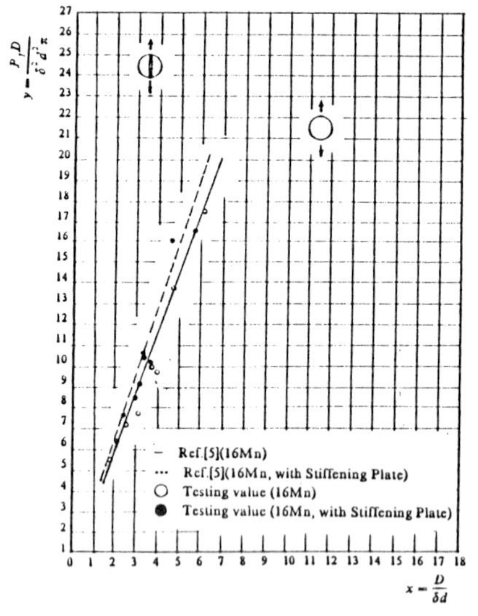

Fig. 6.

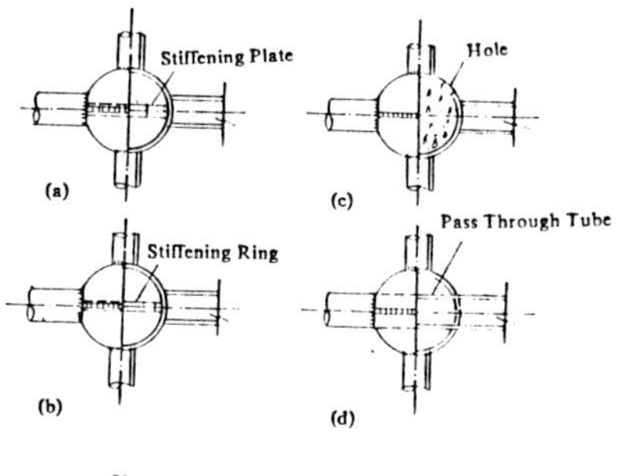

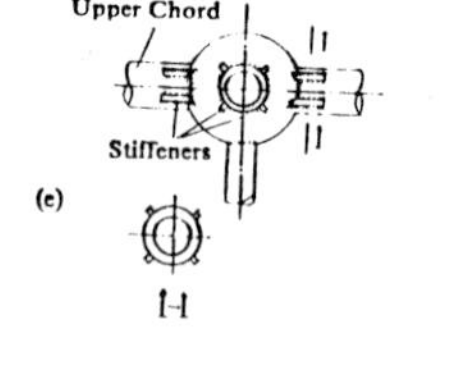

Fig. 7.

Table 4. WHSJ with stiffening plate series.

No.	Designation	Size and Thickness (mm × mm)	Diameter of Tubes (mm)	Tensile Limit Load Capacity (kN)		Compressive Limit Load Capacity(kN)
				16Mn	A3	
1	WSR3008	D300 × 8	140	1088	742	891
2	WSR3010	D300 × 10	140	1360	928	1074
3	WSR3012	D300 × 12	140	1631	1114	1242
4	WSR3510	D350 × 10	146	1418	968	1142
5	WSR3512	D350 × 12	146	1701	1161	1327
6	WSR3514	D350 × 14	146	1985	1355	1496
7	WSR4012	D400 × 12	152	1771	1209	1406
8	WSR4014	D400 × 14	152	2067	1410	1591
9	WSR4016	D400 × 16	152	2362	1612	1762
10	WSR4018	D400 × 18	152	2657	1813	1920
11	WSR4514	D450 × 14	159	2162	1475	1689
12	WSR4516	D450 × 16	159	2470	1686	1876
13	WSR4518	D450 × 18	159	2779	1897	2050
14	WSR4520	D450 × 20	159	3088	2108	2210
15	WSR4522	D450 × 22	159	3397	2319	2356
16	WSR5016	D500 × 16	168	2610	1782	2005
17	WSR5018	D500 × 18	168	2937	2004	2194
18	WSR5020	D500 × 20	168	3263	2227	2370
19	WSR5022	D500 × 22	168	3589	2450	2532
20	WSR5025	D500 × 25	168	4079	2784	2749

Fig. 8.

Fig. 9.

Fig. 10.

REFERENCES

1. LIU XI-LIANG, LIU YI-XUAN Design of double-layer grids, China Architectural Industry Press, Beijing, 1979.
2. TAN YU-CHENG, LAN TIAN Experimental investigation of the load-bearing behaviour of hollow spherical joint for space truss and its rational design, Page 147. Selected papers on space grid structures, Science Press, China, 1985.
3. LIU XI-LIANG, CHEN ZHI-HUA Analysis of collapse mechanism and experiment study on load capacity of welded hollow sphere joint. Proceedings of the Third East Asia Pacific Conference on Structural Engineering & Construction Shanghai 23-26 April, 1991, pp 138-144.
4. LIU XI-LIANG, CHEN ZHI-HUA Typical connections for welded hollow globe joints in space grids. Steel construction, Journal of the China steel construction society, 4th Quarter, 1991.
5. Code for the design and construction of double-layer grid structures, JGJ 7-91, China, 1991.

196. Greater flexibility in design with new rigid space frame connection

J. PARSONS, University of Southern California, USA

SYNOPSIS. Fabrication and erection are a major portion of the cost for space frames and double-layer grids. This paper introduces a new connector that permits such fabrication by utilising continuous cord members for top and bottom grids. Nodal connectors are shop welded, diagonal web members are bolted. The space frame or double-layer grid may be erected on the ground and lifted into its final position. Lack of fit and joint slippage can result in built-in member stress, resulting in premature member over-stress that can lead to buckling failure. Design of this node requires special attention because of eccentric forces.

OBJECTIVES

1. Space frame systems most commonly used to date generally utilise a node connection that is spherical in shape or composed of plate resisting elements for connecting cord and diagonal web members together. The force system is considered concurrent, intersecting at a common point. The stress flow path is through the members to the nodes. The nodal connection must be designed for these forces. This type of system generally requires a large number of elements that need careful fabrication and erected in such a manner as to produce the desired structure within certain tolerances. The node connection is a very cost sensitive item because of the expensive manufacturing or fabrication process.

2. In order to reduce fabrication, material and erection cost, a new space frame system utilising continuous members for the top and bottom cord grids, that are oriented at right angles to each other, completed together by bolted diagonal web members at the nodal points, appears practical . The connection of cord members to diagonal member requires an analysis with careful design considerations by the structural engineer since it requires an in-depth investigation of the stress flow path. Lack of fit should not become a problem (refs 1 and 2).

3. The grids may be square or rectangular. The top grid is offset to the bottom grid. Four diagonal web members connect all bottom grid layers nodes to top grid layer nodes. When assembly is complete, the space frame roof is lifted into place, either in segments or as one unit. This idea so intrigued Mr. Paul K. Mai, founder of PKM Steel Service, Inc. in Salina, Kansas, USA, that he built a small model

of a connector that he thought would work. He then contacted the author to work with him on the development of such a system. Mr. Mai's original idea with some modification is represented in Figs 1 and 2. This connector is known as the Mai Sky System (patent pending).

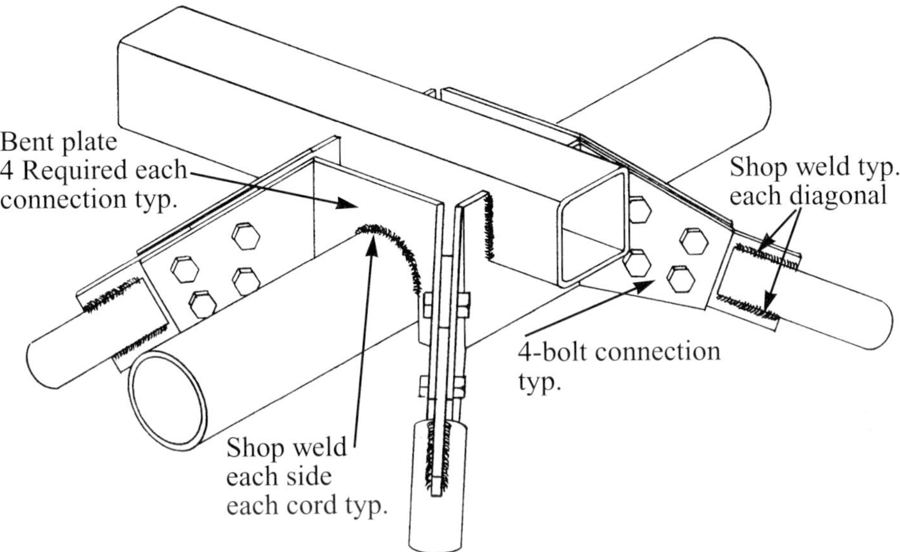

Fig. 1. Typical top node connection.

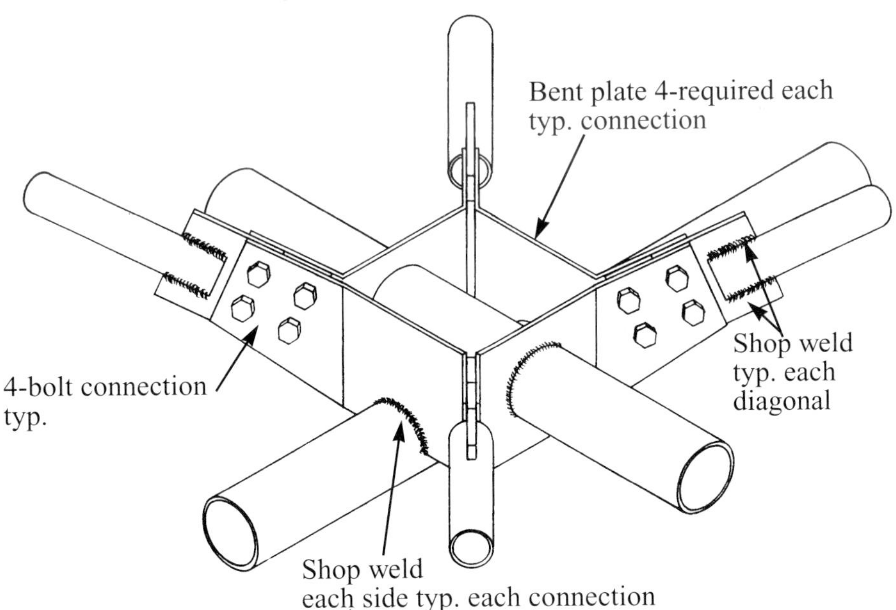

Fig. 2. Typical bottom node connection.

NODE AND MEMBER CONSIDERATION

4. When the preliminary nodal geometry was being calculated it was realised that eccentric forces would exist at the connection. The cord members are continuous but the axial forces, moments and torsion in the members vary each side of the node. Forces in the diagonal web members also vary in magnitude and direction. Since these members are bolted at the nodal connection, they are considered pinned but their axial forces do contribute to moments, shears and torsion in the nodal plates (ref. 1).

5. It is obvious that the plates comprising the node, need to be thoroughly investigated for all forces and checked for local buckling. This will be necessary for various nodes in a structure under different loading conditions. The structures designed to date by the author including those with concentrated loads, indicate generally three different nodal types. For ease of fabrication, placement and erection it might be advantageous to use only one or possibly two different types. Some load testing on a model structure have been undertaken to date and testing is now proceeding on a completed structure.

6. The top cord is continuously supported on steel beams around the perimeter of the structure. The perimeter beam is supported on columns spaced at a distance that will provide a stiff boundary support with minimum deflections. This exterior beam and column system also facilitate erection (refs 1 and 2). It is strongly recommended that the structural support design and space frame design, be one integral design with proper provision for temperature movement and other such considerations.

COMPUTER STUDIES

7. Computer studies have been made on numerous structures including a test on a model space frame, a 11.58m by 30.48m space frame roof structure now completed and a 33.33m by 62.28m space frame roof now under construction. The latter is an addition to a local steel warehouse. Numerous preliminary studies have been, or are now, in the process of being performed on various types of structures, including large aircraft hangers.

8. Since the ratio of length to width for the new structure for the steel warehouse is 2:1, approximately 75 % of the roof load spans in the 33.53m direction . The top longitudinal cord members are square tubes. Metal roof deck spans the 33.53m direction and a flat surface was desirable to support the roof decking. The transverse top cord members are circular tube sections. The bay or grid module is 4.95m longitudinally by 4.88m transversely. The depth of the centroid of the top grid layer to bottom grid layer is 2.44m at the high side and 1.73m at the low side. The top cord of the roof slopes 0.71m for roof drainage in the 33.53m direction while the bottom cord is level. The bottom cord members and diagonal web members are circular tube section (Figs 3 and 4).

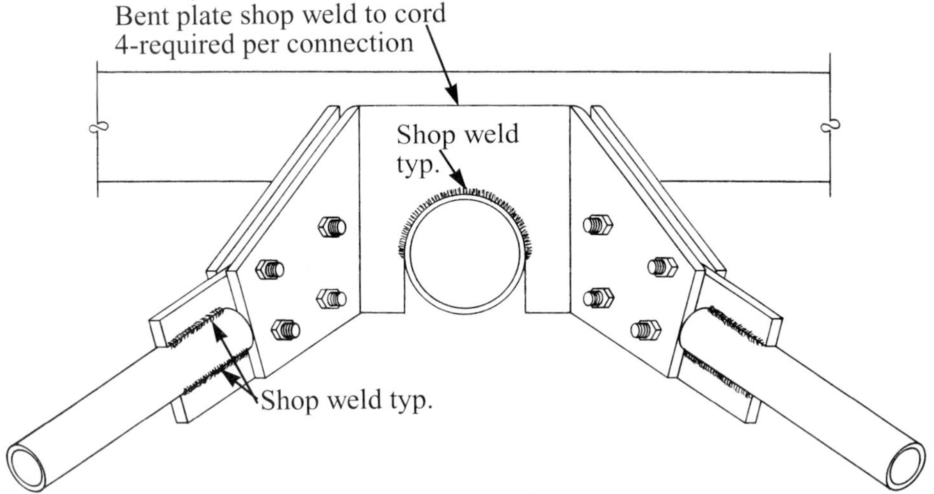

Fig. 3. Typical top node connector.

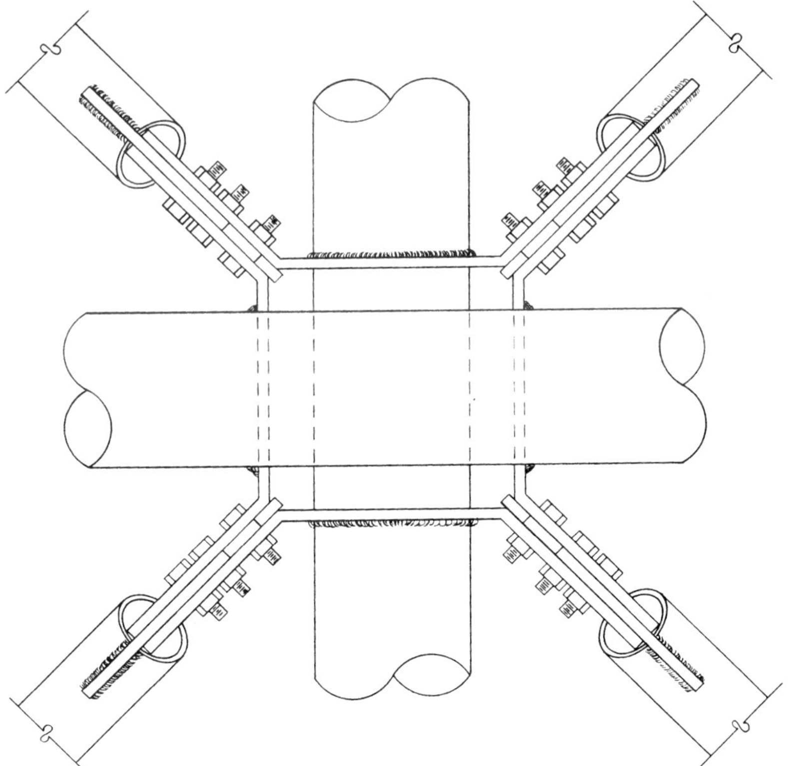

Fig. 4. Plan of a typical top or bottom node connector.

9. A number of computer models were investigated for the PKM Steel Service, Inc. In the first model the two layers forming the top grid were modelled in the same inclined plane, the two layers forming the bottom grid were modelled in the same horizontal plane. All cord members were input as continuous beams, pinned at their extreme end. Additional co-ordinates were entered into the model to orient the three axis of each cord member in the proper direction. The diagonal web members joining the top and bottom grids together. Member properties were assigned with the appropriate joint restraints and degrees of freedom. The roof dead load plus snow loads were applied as uniform line loads to the square tube, running in the 68.28m direction. Lateral loads were applied as horizontal joint loads, as required. The design loads were applied with required force combinations and the model was analysed with the SAP90 software using a finite element analysis on a 486 model, 50Mz computer.

10. A second model was prepared with the two top cord layers located in their actual position in space and joined together with an element representing the nodal connection. The length of this element was equal to the vertical distance between the centroids of the two members . The bottom grid was entered in the same manner (Fig. 5). Member properties assigned to this element were, cross-sectional area, modulus of elasticity, shear modulus, moments of inertia and shear areas about the major axis. Various degrees of freedom conditions were applied at each end of this element to study the resultant effects. The area and moments of inertia had little effect on the displacements and member forces. The connector is required to provide resistance to translation and rotational forces. This resistance is built into the connector by welding to the cord members and flange bolting.

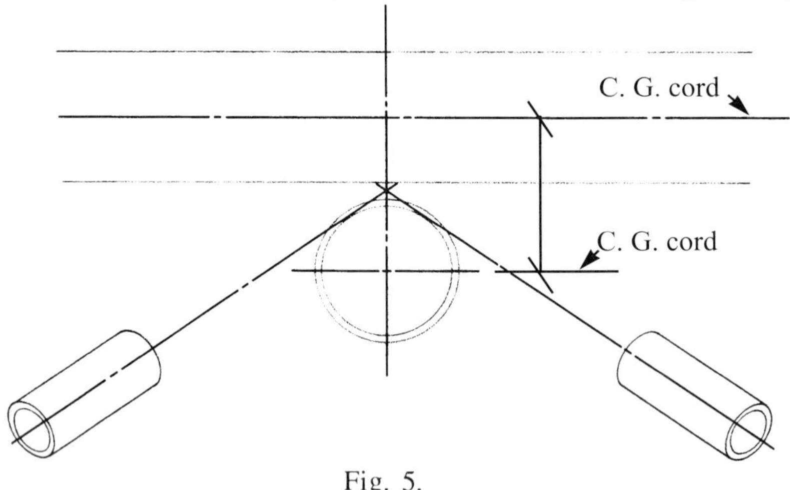

Fig. 5.

TESTS

11. A test on a model space frame was conducted in May, 1992. The model was constructed in the fabrication shop of PKM Steel Service, Inc., in Salina, Kansas, U.S.A. The model was four bays square at the base and three bays at the

top. Each bay was 1.48m giving a total base dimension of 5.89m. The depth was 0.38m measured from the top grid centre to the bottom grid centre. Top and bottom grid members were 50.8mm round tubes with 2.66mm wall thickness. All diagonal web members were 25.4mm round and standard pipe (Figs 6-8). The bolted connector used in the test was in single shear. This has now been modified.

1st Test: All exterior nodes, Top Cord supported.
2nd Test: Corner and mid-side nodes supported only.

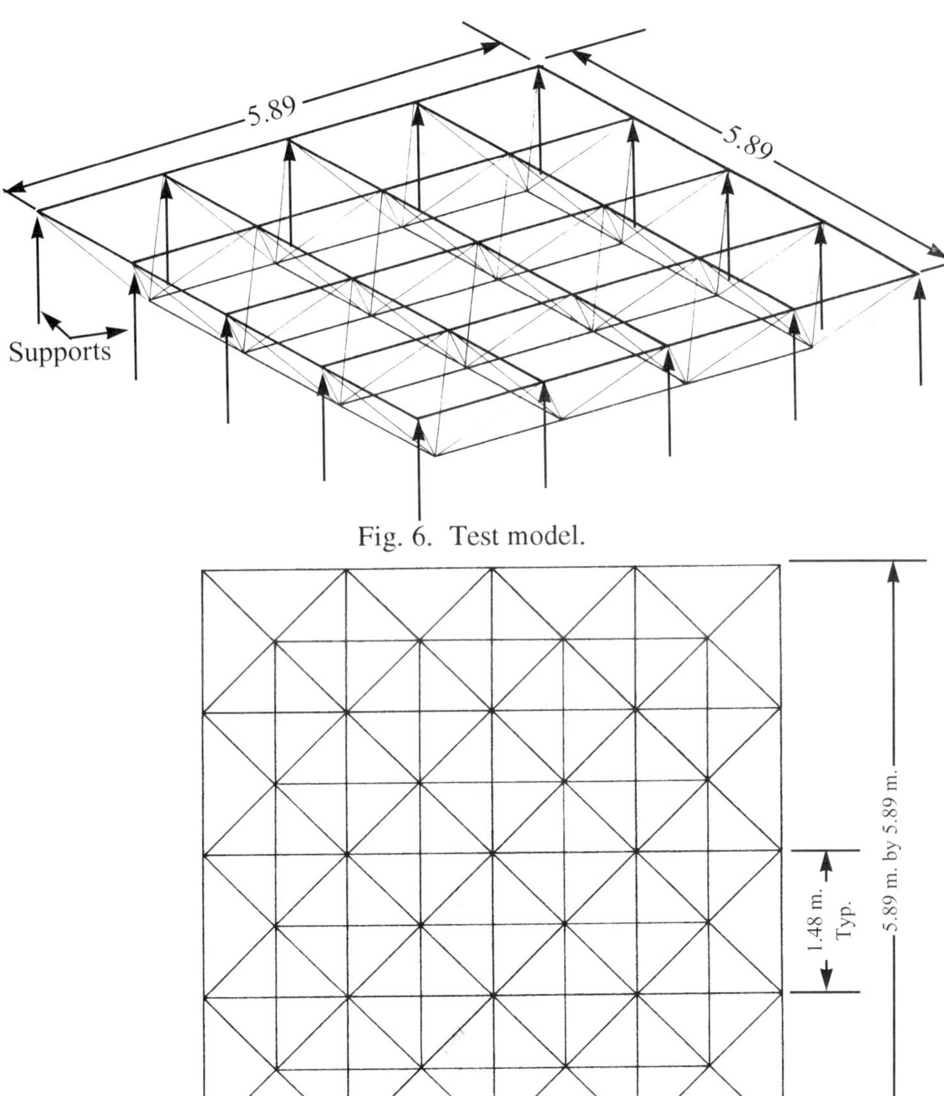

Fig. 6. Test model.

Fig. 7. Plan of test model.

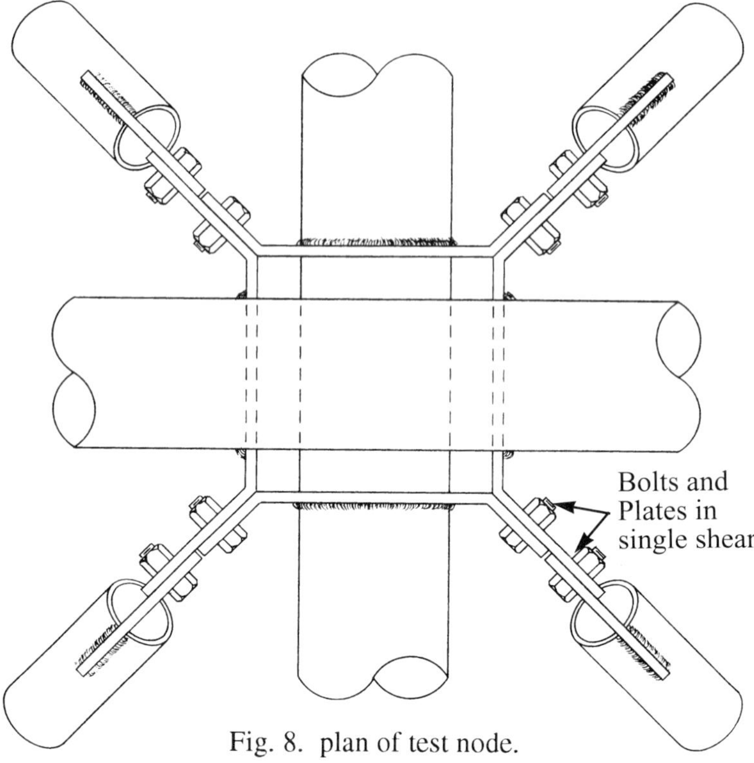

Fig. 8. plan of test node.

12. The frame was loaded with steel plate producing a uniform load of 3.06kPa. The normal design load is 1.44kPa. The deflection was 1/465. The load was left on for 24 hours, then removed giving an 87 percent recovery in the structure. The test was considered a success but it was decided to proceed with further testing until failure. The frame was re-tested with corner and mid-span nodes supported. The frame then failed by buckling of the single flange plate connection at supports. The connection was then revised, so that the connection bolts act in double shear.

ADVANTAGES

13. Since the Mai Sky System allows the cord members to be fabricated in longer pieces with connectors that are shop welded to the top and bottom cord members. Cord members may be field spliced as required in order to obtain the necessary member length. Generally pipe and rectangular steel tubing are used for cord members. These members are the most efficient sections for resisting moments, torsion and shear. The diagonal web members are pipe sections and are bolted to the top and bottom cord nodes. No moment resistance is considered to be present at the diagonal web to node connection. Because of the fixity provided by the nodal connection, the effective length of cord compression members may be reduced to 80 percent of the centre to centre nodal distance (ref. 3). This reduction could amount to a substantial savings in materials. The effective length

of the diagonal web members should be 90 percent of the centre to centre of node dimension, or equal to the centre to centre of bolt pattern length, at each end of the diagonal, whichever results in the least length. In any case the plates forming this connection should be checked for local bending and buckling as well as bolt capacity. Assembly may be undertaken at ground level with the space frame lifted into the final position. Lack of fit should not be a problem and therefore additional stress due to this condition need not be considered. The bottom cord may be an I-beam suitable for a travelling hoist (Fig. 9).

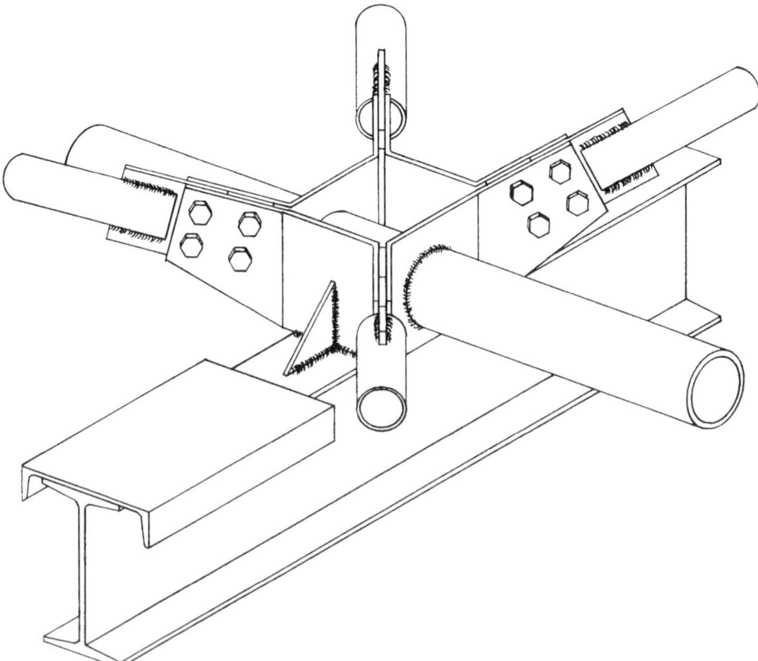

Fig. 9. Alternate bottom cord node with I-beam for hoist load.

14. Assembly may be completed in the field with unskilled workmen. Other building trades may also complete all or a major part of their work at ground level, prior to erection of the space frame. The structure may be lifted in one complete unit or in segments and joined together in its final position.

15. It should be noted that the ease of fabrication, placing of roofing, insulation, etc. along with speed in erection, without the use of scaffolding, will almost always result in a large cost saving.

FIELD ASSEMBLY

16. The bottom cord members are placed first, followed by the longitudinal layer and then the transverse layer on top, or as detailed. The top cord layers are placed next in the proper sequence followed by the diagonal members which are bolted to the top cord nodes. The flanges of these nodes have a rectangular piece of steel

bolted to them to act as positioning blocks; see Fig. 10. A crane or other method of lifting is attached to the top cord grid to lift up the grid. This allows the diagonal web members to drop down at the lower end into the proper position by bearing against the control block. The lower end of the diagonals are then bolted to the bottom cord nodes. Final adjustment and fitting of the space frame is made and the bolts are tightened before erection.

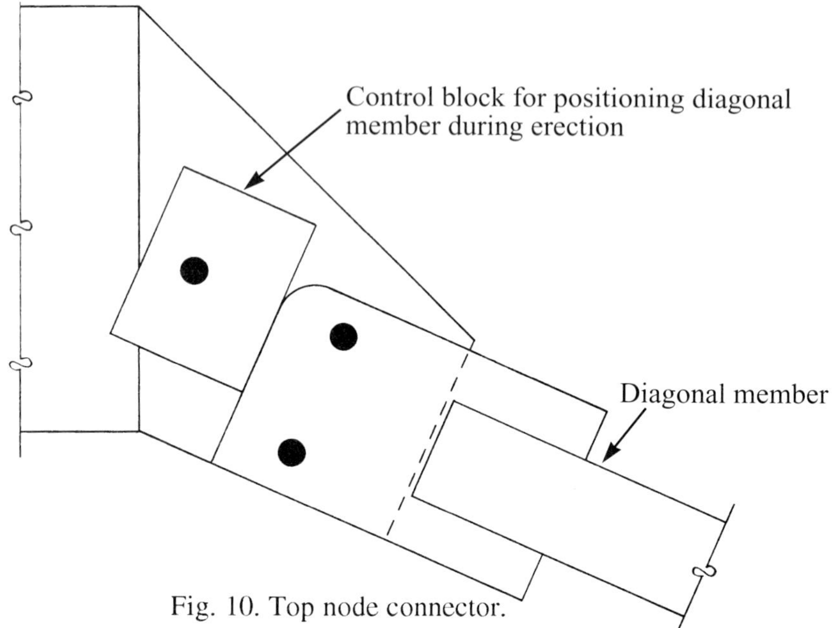

Fig. 10. Top node connector.

REFERENCES

1. SUTO F., WADA A. and SUGIHARA K. Some aspects of reliability and safety of space frames. International Journal of Space Structures, Vol. 3, No. 4, 1988.

2. MADI U. R. and EL-TAYERN A. A. On the effective length of compression members in double-layer grids. International Journal of Space Structures, Vol. 6, No. 1, 1991.

3. MAKOWSKI Z. S. Analysis, Design and Construction of Double Layer Grids. Applied Science Publishers London 1981

197. Joint stresses in steel frame space structures

J. E. GIBSON, City University, UK

SYNOPSIS. An investigation was pursued to determine the degree of correlation between experimental and theoretical computed values for the deflections and stresses at the joints in steel frame space structures. Two very different types of space structure namely a four bay six storey frame model and a skeletal hyper shell model were examined in which all members were rigidly welded at the joints. Deflections were measured by dial gauges and stresses due to bending moments and forces were determined by strain gauges at the joints. These experimental results were compared with computed values from a three dimensional program and the degree of correlation was established.

THEORETICAL INVESTIGATION AND COMPUTER PROGRAM

1. The theoretical forces, moments and deflections in the frame were determined using a three dimensional rigid frame program. The program was written in FORTRAN 77 for a micro computer with 4 Mbyte RAM.

2. Using the information for the connectivity for nodes and members and then satisfying displacement continuity and equilibrium at each joint of the structure, the overall stiffness matrix [K] for the structure was determined. Thus

$$[K][D] = [F] \tag{1}$$

in which [K] is the known overall stiffness matrix and, [F] is the load vector.

3. Inversion of Eq. 1, to determine the displacement vector [D] i.e.

$$[D] = [K]^{-1}[F] \tag{2}$$

was obtained by using a previously developed routine by the author for the analysis of multi-shell structures (ref. 1). A final routine outputs the value of moments and forces node by node.

4. In a large number of structures in practice many of the members have identical dimensions and advantage was taken of this to automatically generate the nodal co-ordinates.

5. Again in the case of the six storey model shown in Fig. 1, the basic dimensions required are, a, b, c, and d. From this data and the number of nodes, a sub routine to automatically determine the nodal co-ordinates was formed.

6. By suitably numbering the members, a subroutine was introduced to automatically generate the connectivity between nodal and member numbers.

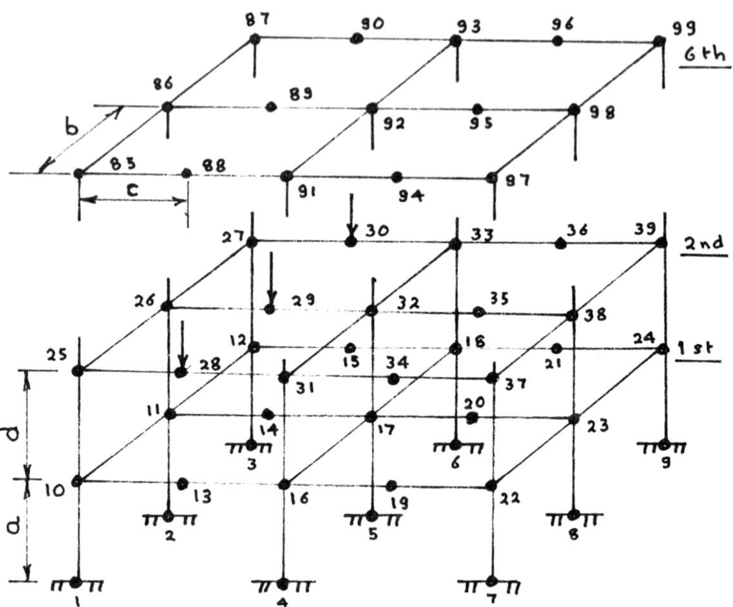

Fig. 1. Nodes in six storey steel frame.

Output from the program

7. The final output from the program yields the computed nodal deflections and rotations referred to fixed global axes namely u, v, w, Ox, Oy and Oz for each node in turn. Finally, though all the forces and moments at the node are computed, only dominant ones are printed out, these are the axial force Fx and the moments Mx, My and Mz.

INVESTIGATION OF MULTI-STOREY STEEL FRAME

Construction of model

8. The multi-storey model steel frame consisted of a four bay six storey all welded steel frame structure as shown in Fig. 1. Only the first two levels and the sixth level (the top level) are shown for convenience. As may be seen there were nine main columns extending over six storeys with twelve interconnecting beams at each level.

9. For convenience in manufacture all member were constructed from 19mm square steel cross section, accurately cut to the required size. The beams were welded to the columns, great care being taken to ensure that the welds were adequate and also that in the welding process little or no overall geometrical distortion of the structure as a whole occurred.

10. The bases of the nine column at nodes 1, 2, 3, 4, 5, 6, 7, 8, and 9 were each welded to base plates which in turn were firmly bolted to the testing frame in which the model was placed.

Analysis of model

11. Stresses and deflections were computed at the end points of both beams and columns to determine the structural behaviour of the joints. The stresses and deflections immediately in the vicinity of the applied point loads were computed by introducing mid point nodes as may be seen in Fig. 1. The introduction of such nodes at 13, 14, 15 and 28, 29, 30 etc., in the beams facilitated comparison between theoretical and experimental values.

Instrumentation and testing of the model

12. Deflections at specific nodes in the model were measured by dial gauges graduated in .01mm. For the particular series of tests now to be described, dial gauges were positioned at nodes 10, 13, 16 and 19 and also at 25, 28, 31 and 34, as shown in Fig. 1 and at 40, 43, 55, 58 and 61 (not shown).

13. Further dial gauges were located at nodes 11, 14, 17, 20, 26, 29, 32, 41, 44 and 47.

Strain gauging

14. To determine the stresses at selected points in chosen members, pair of foil gauges were fixed on opposite surfaces to allow both bending and axial stresses to be measured. Strains in each of these gauges were recorded by an Intercole data logging system.

15. Various types of data logging programs for processing the results were developed for the system, using FORTRAN. In this particular instance the program scanned the gauges under zero load and finally zero loading. From this information the bending and axial stresses in the member were finally obtained.

Loading mechanism

16. Point loads were applied to the model by means of three hydraulic loading cylinders. These were mounted on two screwed rods which allowed the loading ram heads to be positioned over selected loading points on the model.

17. The three loading cylinders had been previously calibrated to provide a load/pressure factor. The three cylinders were connected to a manifold and oil pressure was produced by a hydraulic hand pump. The valve on the manifold allowed any combination of the cylinders to be utilised. The hydraulic pressure in the cylinders themselves was monitored by a special highly sensitive Budenberg gauge.

Comparison of experimental and theoretical results

18. The overall investigation of this three dimensional steel frame involved the examination of the stress and deflection conditions that occurred in members due to various loading conditions. As there were some one hundred and twenty six members and thirty mid-points as possible loading positions, the number of analyses for every conceivable combination of loads proved insurmountable.

19. It was thus decided to limit the analyses to those in which sets of point loads were applied at the 1st, 2nd and 4th level only. To further reduce the amount of the output data only three point loads were applied at any one of these levels, any combination of these three loads could however be permitted.

20. From an experimental view point, the detailed examination of stresses in any one member required that a minimum of six strain gauges should be fixed onto the member. Namely two pair at either ends and one pair centrally placed. The strain gauges were limited to positions around the first, second and fourth levels.

21. After further consideration some ninety strain gauges were fixed to the beams and columns in the immediate vicinity of the joints and also at the mid-point of beams to provide sufficient experimental information about the structural behaviour of the model.

22. Consideration had also to be given to the placing of dial gauges recording deflection; obviously the smallest vertical deflection would occur at the beam column joint and the maximum values in the region of the point load at the centre. The dial gauges were also located at the first, second and fourth level and numbered twenty four in all.

23. These dial gauges were located in the vicinity of nodes 10, 13, 16 and 19; 11, 14, 17 and 20 at the first level and at the second level at nodes 25, 28, 31 and 34; 26, 29, 32 and 35. Information about deflections at the third level was recorded by some six dials, in all only three gauges were located at the fourth level.

24. Despite all the above restrictions in the number of strain and dial gauges employed, the actual experiments carried out provided adequate information about deflection and stresses at the joints. However, for the purpose of this paper, only typical information will be presented and this is for the test in which all three point loads were located at the second level.

25. Point loads of 200 lb (890 N) were applied simultaneously at nodes 28, 29 and 30 as shown in Fig. 1. The experimental deflection as measured by the dial gauges at node 25, 28, 29 and 34 are shown in Fig. 3(a) as crosses, the computed theoretical values are displayed as full lines. It should be observed that this full line is formed by joining the computed deflection points with a series of straight lines and thus varies slightly from the true theoretical curve.

26. As was to be expected, the maximum deflection occurred in the region of the point at node 28 giving a value of 60 (.01mm) units. The dial gauges at node 25 and 31 (offset from the columns by 50mm to avoid damaging the strain gauge) recorded values of 10 units.

27. The computed deflection at node 34 showed a reversal of direction giving a value of 10 divisions upwards which was confirmed by the experimental value as shown by the cross.

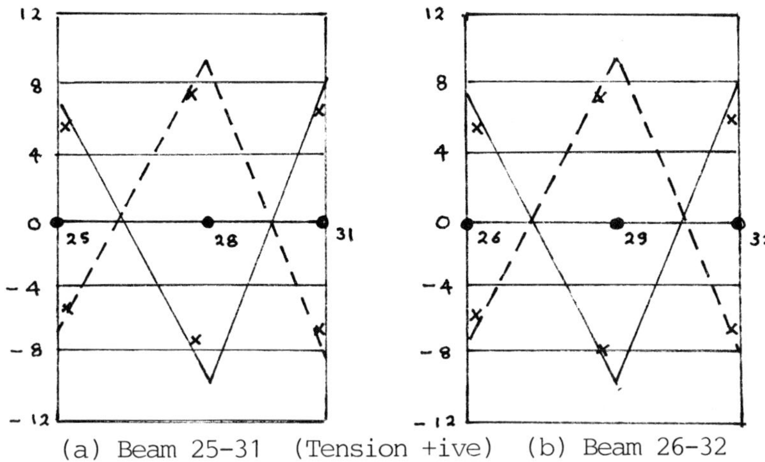

Fig. 2. Stresses in beams in 1000lb/in^2 (6.9N/mm^2).

28. The results for the nodes 26, 29, 32 and 35 are shown in Fig. 3(b), maximum deflection occurring in the region of the node 29, a reversal of direction at node 35 also occurred. Again the experimental values indicated by crosses compared favourably with the computed values shown as the full lines.

29. It was found convenient to reduce the computed moment and force to yield the actual stresses which then produced on the upper and lower surfaces of the beams. This allowed direct comparison with the experimental stresses as measured by the strain gauges.

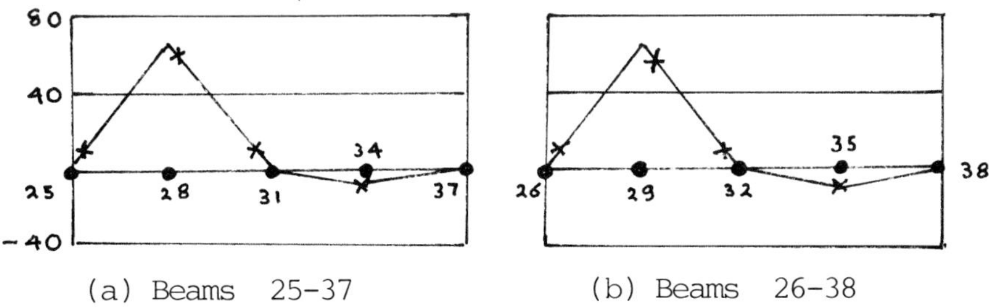

Fig. 3. Deflections in beams in .01 mm.

30. Again, only two typical sets of results which were representative of the structural behaviour of the frame will be presented. The first set are for the beams spanning between nodes 25 and 31 as shown in Fig. 2(a) .

31. In Fig. 2(a) the computed stresses on the upper surface for the sections 25 to 31 are shown as the full lines and those on the lower surface as the dashed lines. The corresponding experimental stresses are shown by crosses. It may be seen that the agreement is reasonably good.

32. The stress distribution in the section 31 to 37 was relatively very much smaller since it was far removed from the point load at node 28. Agreement between theory and experiment was good but has not been presented.

33. The second set of results are for the beams spanning between nodes 26 and 32, these are shown in Fig. 2(b). Again agreement between theoretical and experimental values is good.

34. It is to be observed that the theoretical stress profiles are almost identical for the beams 25, 28, 31 and 26, 29 and 32. This is of course to be expected, as the applied load condition of 200 lb (890 N) at the three nodes 28, 29 and 30 should by symmetry produce almost identical stress conditions in the three plane sections bounded by nodes 1, 7, 85, 97; 2, 8, 86, 98 and 3, 9, 87 and 99.

35. The above investigation has shown that the theoretical prediction of deflections and stresses as computed give reasonable agreement with experimental values, not only for the joints but also at the centres of the beams. Other combinations of point loads at different levels produced similar agreement between computed and experimental values. Thus it may be concluded that the program will certainly predict deflection and stress values for this type of multi storey welded steel structure in the region of the joint. An entirely different type of welded structure, namely a hyper skeletal structure will now be considered.

INVESTIGATION OF A HYPER SKELETAL STEEL FRAME

36. The second model to be investigated was a steel hyper skeletal structure formed by a series of steel bars welded together to form a hyperbolic paraboloid surface as shown in Fig. 4.

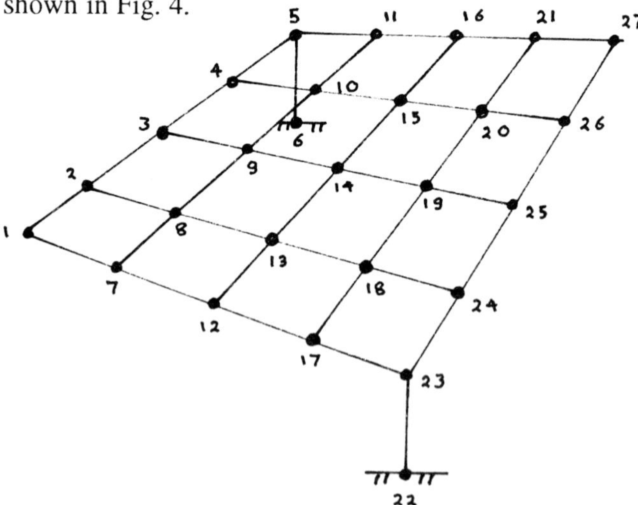

Fig. 4. Nodal points in hyper skeletal shell frame.

The outer edge beams connecting nodes 1, 5, 27 and 23 were of 25 by 25 mm cross-section. The internal members forming the grid of a hyperbolic paraboloid surface were all of 12 by 12 mm cross-section. The outer edge beams were supported by two columns at nodes 5 and 23 and the column bases at nodes 6 and 22 were rigidly attached to the base of the testing rig.

Analysis of the structure

37. The structure was analysed using the rigid frame program for a variety of loading conditions. Since the majority of members were located on a hyperbolic surface of the form:

$$z = kxy \tag{3}$$

advantage was taken of this to allow the automatic generation of the nodal co-ordinates. Again the numerical ordering of the nodes allowed member/nodal connectivity to be automatically developed. Several analyses were carried out with point loads applied to selected nodes and the resulting nodal deflections and forces and moments were printed out.

Instrumentation and testing of the model

38. It was decided to load the model by simply applying dead weight to selected nodal points. To this end, slotted weights placed on weight pans attached to the nodes provided a suitable loading system.

39. Deflections were measured by dial gauges which were graduated in .01 mm. The dial gauge were located at the following nodal points, namely 1, 8, 9, 13, 10, 14, 18, 15, 19, 20 and 27, thus giving a reasonable coverage of the overall deflections of the structure.

40. In order to determine the stresses due to bending moments and axial forces in selected member, pairs of foil gauges were located at the ends of the members close to the joints. The gauges were wired up to the data logging equipment and a permanent record of strain at selected gauge points was recorded by the printer.

41. For each loading condition the standard method of testing was employed. Thus, readings for deflection and strain were recorded for zero load, then for the loaded condition and finally on unloading. These values were then processed to determine the experimental stresses and deflections.

42. Owing to space limitations only the theoretical and experimental nodal deflections will be given for the following loading conditions. A single point load applied at node l, two equal loads were applied at nodes l and 27 and finally all internal nodes were loaded.

43. When only a single point load acts at node 1, the values for deflection are given in columns (1) and (2) of Table 1 and show reasonable agreement between theory and experiment. The greatest discrepancies occurring at nodes 1 and 27, the extreme high points of the structure, were of the order of some nine percent. However, when both extreme nodes 1 and 27 carry equal point loads, (the overall structure then being under a balanced load condition) agreement between theory and experiment as given in columns (3) and (4) is very much better.

44. Finally, when all nodes were loaded, agreement between experiment and theory given in columns (5) and (6) was very good. Many other loading conditions with varying types of column support were examined (ref. 2) and agreement between theory and experiment was found to be good.

Table 1. Nodal deflection due to various point loads.

Point Load Location	Load at Node [1] 10lb (44.5N)		Load at Nodes [1] and [27] 10lb		Loads at All Nodes	
	(1)	(2)	(3)	(4)	(5)	(6)
Node	Thtl in .01 mm	Exptl in .01 mm	Thtl in .01 mm	Exptl in .01 mm	Thtl in .01 mm	Exptl in .01 mm
(1)	143.2	151.0	93.9	90.5	34.5	31.7
(8)	57.7	64.5	33.6	35.3	56.6	49.2
(9)	25.9	29.0	15.1	18.0	66.1	58.0
(10)	3.7	2.0	7.3	9.0	44.1	41.5
(13)	25.9	29.0	15.1	14.5	66.8	52.3
(14)	4.7	3.5	9.3	10.1	90.8	89.6
(15)	-10.8	-10.0	15.1	19.0	67.2	66.5
(18)	3.7	3.5	7.3	5.3	44.5	38.9
(19)	-10.8	-14.0	15.1	21.0	67.4	70.8
(20)	-24.2	-18.0	33.5	36.3	57.9	65.3
(27)	-49.4	-58.0	93.8	98.0	36.4	29.8

General conclusion

45. Having examined the two very different types of space frame it can be concluded that the method of analysis would seem to give a fairly accurate prediction of the overall structural behaviour of these frames. Moreover the joint and deflections as predicted by the theoretical program (which assumed that all joints were rigid) gave reasonable agreement with experimental values. This implies that the types of weld used in the steel model were of adequate strength.

46. In the case of the hyper skeletal structure in which the members were placed one on top of the other at the joint, for welding purposes, no account of this eccentricity was considered in the analysis. Nevertheless, the computer analysis provided values that compared favourably with experimental information.

47. It would thus seem that the rigid frame program used in the investigation was reasonably capable of predicting the joint stresses and deflections in these two very different types of space structure.

REFERENCES

1. GIBSON J.E. Thin shells computing and theory, p.187. Pergamon Press, Oxford, 1981.
2. GIBSON J.E. An investigation of a steel framed hyper shell. Conference on 10 years of Progress in Shell and Spatial Structures. I.A.S.S., Madrid, 1989.

198. Diaphragm-stiffened multiplanar tubular joints for retractable roofs

Y. MAKINO and Y. KUROBANE, Kumamoto University, Japan, and
M. TAKAGI and A. HORI, Hazama Corporation, Japan

SYNOPSIS. Tubular members with a large diameter to thickness ratio are most efficient as compression members when used in three-dimensional space truss structures. However, profits in compression strength of members are earned at the expense of reducing the ultimate resistance of unstiffened tubular joints. In three-dimensional truss structures with various kinds of joints and under different loading conditions, it is difficult to design the joints assuming certain collapse mechanisms of trusses. In addition, to utilise the full strength of members, it seems better to have the joint resistance greater than the capacity of the members framing into the joints.

In this study, unstiffened joints and two types of diaphragm stiffened joints are tested. One type of stiffened joint has three stiffening diaphragms, a central through diaphragm and two diaphragms welded to the inside surfaces of the chord. The other type has only one central diaphragm. Two types of loading are investigated, the case of gravity loads acting on truss frames and that of horizontal loads perpendicular to the span direction of truss structure. The results obtained are promising with a suggestion that diaphragm stiffening of tubular structures could be advantageous as a design technique.

INTRODUCTION

1. The retractable roof under study is composed of four twin-fan-shaped double layered trusses as shown in Photo 1. The roof trusses are part of a sphere with a radius of 170m, the span length is about 200m and the truss depth is 3.25m as shown in Fig. 1(a). CHS 355.6x8.0 - 406.4x16.0 upper chords and CHS 406.4x9.0 - 508.0ϕ x 16.0 lower chords are used. All the members are of cold formed STK490 steel. The ultimate loads could be reached for the cases of both gravity loads and horizontal loads due to earthquakes. This is reported in detail in ref. 1.

2. To obtain a dome-like shape of roof trusses, the straight members used need to be welded together at an angle to the joints. Another way of fabricating the roof trusses would be by using curved members bent in the fabrication shop. However, for the latter case, combined compressive and bending stresses acting on the chord members are disadvantageous. Therefore, in terms of cost and

Space structures 4. Thomas Telford, London, 1993

fabricating technique, the first method was chosen due to its merits. Furthermore, as the actual distance between two joints is about 8 to 9m, considering members of standard length would enable the fabrication of only one unit using one member. In other words, there is at least one chord joint in one unit. In this respect, the choice of the position of a chord joint, whether it be at the end joint or the middle of the chord, becomes necessary. To overcome this problem, the stiffened joint shown in Fig. 1(b) is considered. The central diaphragm, as shown in Fig. 1(b), restrains the local deformation of the joint at which the members of standard length are connected keeping an angle to fit the curved shape of the roof trusses. The diaphragms inside restrain local deformation of chords caused by axial brace loads.

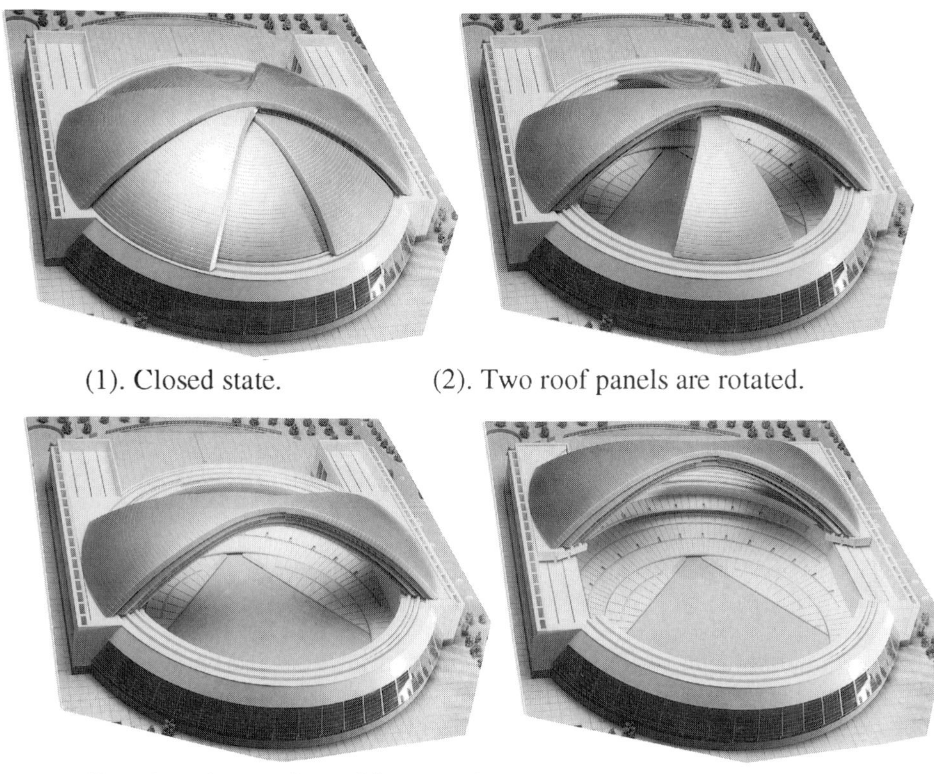

(1). Closed state. (2). Two roof panels are rotated.

(3). The other roof panel is rotated. (4).Opened state.

Photo 1. Opening and closing system of roof panel.

TEST

3. Specimens. Loading tests of the specimens were carried out by adopting 1/2 scale joint models. The specimens and the definition of symbols are shown in Fig. 2. The measured dimensions are given in Table 1. In the actual joints, the in-plane angle between the chord and the brace is 36.9° and the out-of-plane angle

between the planes of braces is 31.6°. In the specimens tested the corresponding angles are 45° and is 30° respectively because of the limitations imposed by the configuration of the test set-up. The angle between two chords at a joint resulting from the curvature of the roof trusses is 2.8°. In the specimens, the two chords were connected in a straight line, because the angle between the chords is very small.

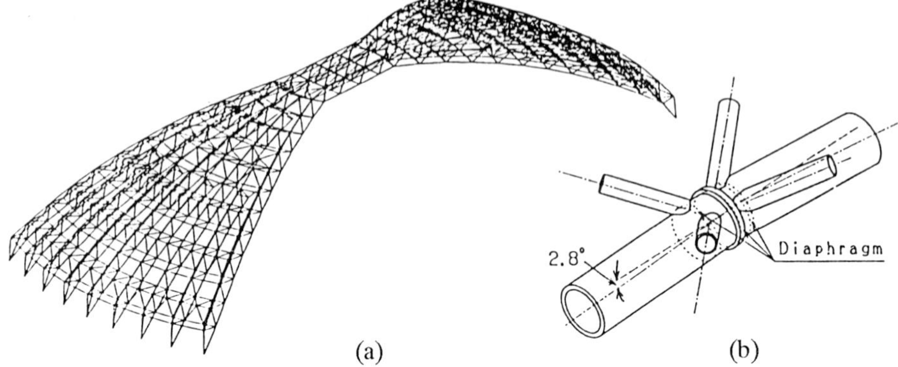

Fig. 1. Roof truss and diaphragm stiffened multi-planar tubular joint.

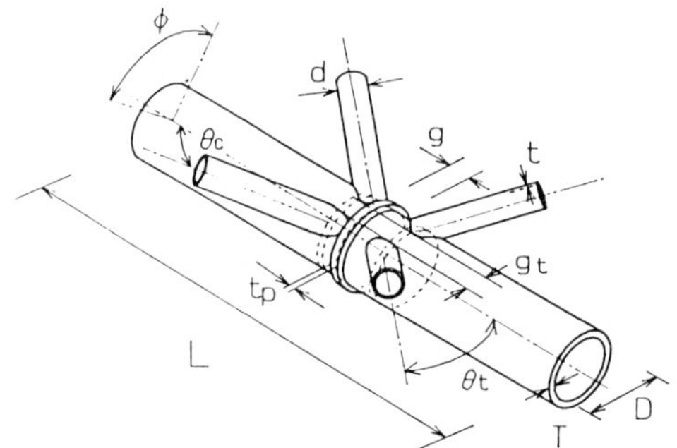

Fig. 2. Specimen and symbols.

4. Although the chord diameter to thickness ratios of actual lattice members are about 36, the value for the specimens is 19 so that the lattice members do not buckle before the joints fail under load. It is recognised that the thickness of lattice members does not effect the ultimate resistance of an unstiffened joint. Although the tube materials are STK490 in actual structures, STK400 tubes were used in this test. The effect of the difference in used material on the joint behaviour is negligibly small. Results of tensile coupon tests of the chord, lattice and stiffening plate materials are shown in Table 2.

5. In the specimen designation, the symbol DK means Double K-Joints, the symbols S and U denote loading type 1 (gravity loading: Symmetric deformation) and loading type 2 (horizontal loading: Unsymmetrical deformation) respectively. N, S1 and S2 denote unstiffened, stiffened type 1 (single diaphragm) and stiffened type 2 (three diaphragms) joints, respectively. NP denotes a trial test

carried out to check the test set-up performance (pilot test). The data obtained from this test was reliable and hence were used for the unstiffened joint data (loading type 1). The loading type 2 NP test was ignored and considered as unreliable because of errors related to support conditions.

Table 1. Dimensions of specimens (measured).

Specimen Designation	D (mm)	T (mm)	d (mm)	t (mm)	t_p (mm)	g (mm)	g_t (mm)	θ_c (deg.)	θ_t (deg.)	ϕ (deg.)
DK-S-N	215.9	4.4	101.7	5.2	—	52.8	7.9	49.1	49.1	60
DK-S-S1	215.9	4.4	101.7	5.2	5.6	56.1	7.6	49.1	49.1	60
DK-S-S2	215.9	4.4	101.7	5.2	5.6	56.3	7.4	49.1	49.1	60
DK-U-N	215.9	4.4	101.7	5.2	—	52.7	7.9	49.1	49.1	60
DK-U-S1	215.9	4.4	101.7	5.2	5.6	55.8	7.3	49.1	49.1	60
DK-U-S2	215.9	4.4	101.7	5.2	5.6	56.4	7.7	49.1	49.1	60

Table 2. Results of tensile coupon test.

	σ_y (N/mm^2)	σ_u (N/mm^2)	Elongation (%)	Grade of Steel
Chord(216.3ϕ×4.5)	388	470	35.2	STK400
Brace(101.5ϕ×5.7)	426	473	29.8	STK400
Plate(t=6mm)	321	453	38.5	SS400

Test set-up

6. Two test devices were used in the experiment. The case of gravity loads is denoted "loading case 1", while the case of horizontal loads applied perpendicular to the truss span is denoted "loading case 2". The two loading cases are shown in Fig. 4. and the test set-up for each case is given in Fig. 3. In both cases, the load is applied from the frame through a 100 ton hydraulic jack and a spreader beam to the tension lattices. Universal joints at the ends of the compression lattices allow rotation in all directions. In loading case 1, the chord is free at one end and has a pin at the other end allowing in-plane rotation of the truss. On the other hand, in loading case 2, the chord is braced at both ends to prevent the rigid body rotation of the chord about its centre after local deformation takes place. The support conditions are shown in Fig. 4.

Measuring system

7. Loads were measured by a load cell connected to the hydraulic jack and by load cells attached at the compression lattice ends. Strain gages were placed on the chord and lattices to measure the axial and the bending stresses. Stress measurements by strain gages seemed necessary to monitor those differences in loads between the two compression lattices which arise if a test specimen is not

set accurately into position. Because of difficulty in measuring local deformations of joints, displacements of the jack relative to the support points were measured instead. In other words, the overall deformation of a joint was obtained. Fig. 5 shows the deformation measuring devices and their corresponding positions.

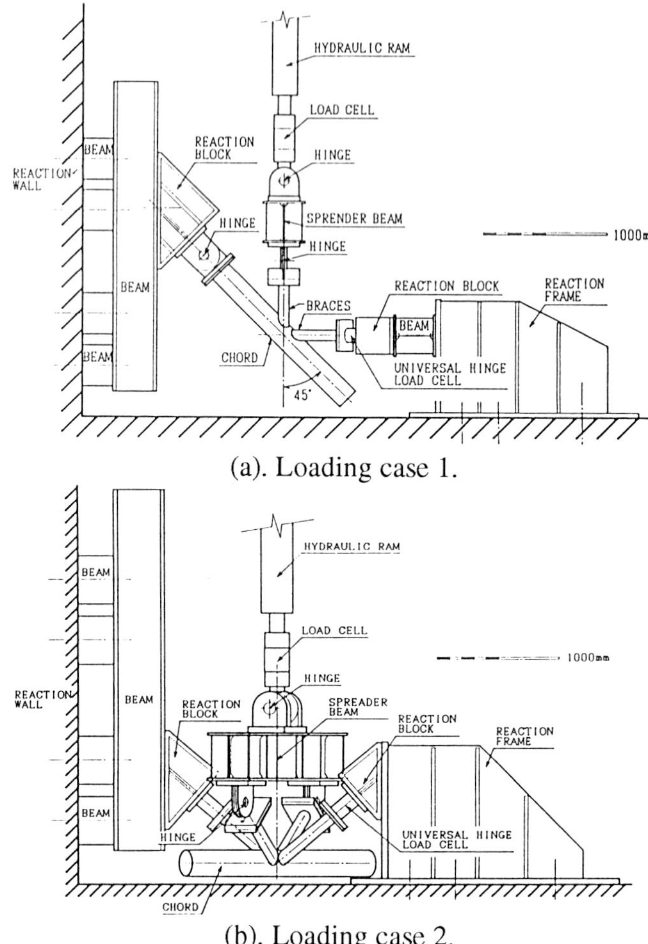

(a). Loading case 1.

(b). Loading case 2.

Fig. 3. Test set-up.

TEST RESULTS

Load deformation relationship

8. The test results are summarised in Table 3. The loads in the table are the values recorded by the load cells attached at the top of the hydraulic jack and at the ends of the compression lattice members. The load-deformation relationships are shown in Figs 6(a) and 6(b), where the load is that recorded by the load cell at the hydraulic jack and the deformation direction is that of the loading jack. The

ultimate capacities are denoted by the circles. In this test they are also the maximum loads.

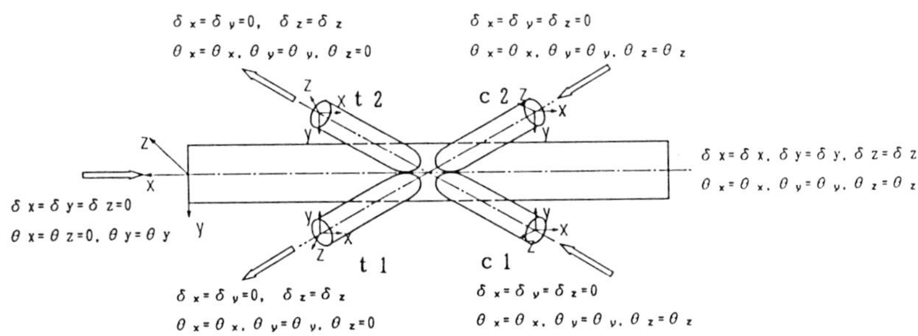

(a) Loading case 1.

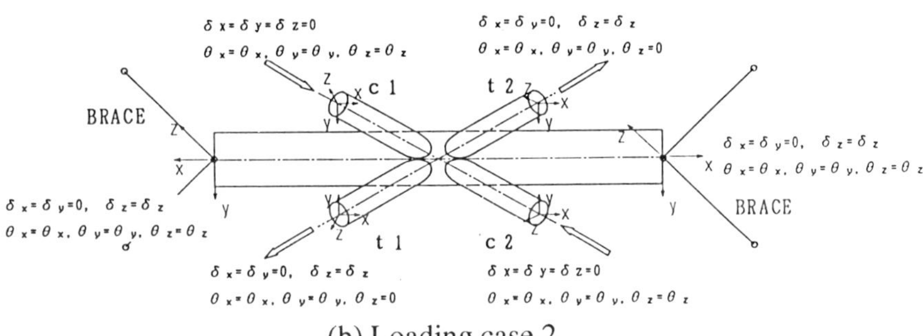

(b) Loading case 2.

Fig. 4. Loading and support conditions.

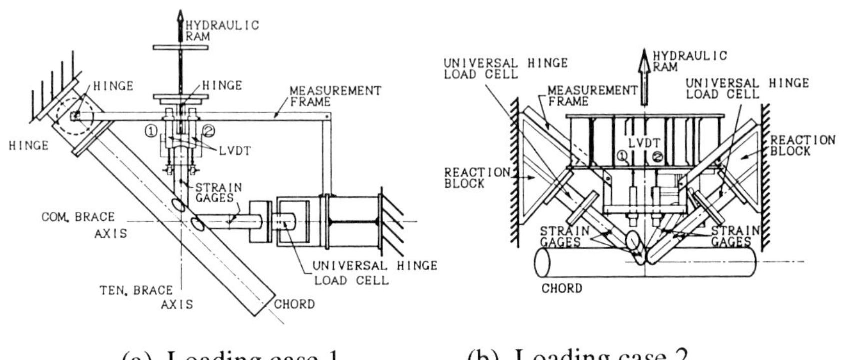

(a). Loading case 1. (b). Loading case 2.

Fig. 5. Measurements of deformations.

Loading case 1

9. The initial stiffness of unstiffened (DK-S-N) and stiffened type 1 (DK-S-S1) joints are nearly equal. The initial stiffness determined from pilot test is low because of lose connections between the measuring devices and test set-up. The diaphragm stiffened type 2 joint (DK-S-S2) have the highest stiffness. The ultimate capacities of type 1 and type 2 joints is about 1.5 and 1.9 times that of unstiffened joints. The deformation at ultimate load is larger for the stiffened joints as shown in Table 3 and Fig. 6(a). As for the post peak load-deformation relation, the unstiffened and stiffened type 1 joints exhibit a gradual decrease in the load, whereas sudden failure due to shear in the chord governed the capacity of stiffened type 2 joints. Cracks appeared at this time, in proximity of the maximum load.

Loading case 2

10. In the pilot specimen, the horizontal displacements of the chord ends were not restrained. This configuration introduced large joint deformation and a rotation of the whole joint in the horizontal plane. The test was stopped and for this reason the data obtained from this test was unreliable and hence was neglected. Therefore, the chord ends were braced to prevent the horizontal rotation of the joint and tests were carried out using the new configuration (see Fig. 4).

11. The initial stiffness becomes larger as the number of diaphragms increases. The ultimate capacities of stiffened type 1 and type 2 joints were about 1.3 and 2.2 times that of the unstiffened joint. The deformations at ultimate loads of stiffened joints, as shown in Table 3 and Fig. 6(b), are smaller than those for load case 1. In the post peak load-deformation relationships, cracks initiated at the weld toes or in the welds after reaching the ultimate loads and extended rapidly resulting in failure of the joints. This phenomenon was observed in all the specimens.

Table 3. Test results.

Specimen Designation	$_tP_u$ (kN)	$_cP_u$ (kN)	$_{chord}P_u$ (kN)	δ_u (mm)	K (kN/mm)
DK-S-N	200	179	498	8.2	101
DK-S-NP	201	189	511	5.9	143
DK-S-S1	288	284	749	15.4	130
DK-S-S2	371	361	956	19.5	168
DK-U-N	207	212	—	26.7	58
DK-U-S1	268	278	—	22.0	66
DK-U-S2	459	445	—	18.7	85

$_tP_u$: Axial Force in Tensile Lattice at Ultimate Load
$_cP_u$: Axial Force in Compressive Lattice at Ultimate Load
$_{chord}P_u$: Axial Force in Chord at Ultimate Load
δ_u : Deformation at Ultimate Capacity
K : Initial Stiffness

DISCUSSION

Ultimate strength of welds between chord and tension lattice members

12. The capacity of the DK-U-S2 specimen was governed by cracks which initiated in the welds between the chord and one of the tension lattices at the forked intersection and extended along the welds. Local deflections of the chord wall, sustained before cracks started, were smaller than those for the other specimens as observed from load-deformation curves and also from appearance of the joints after failure. Resulting from an examination of weld profiles after testing the throat thickness was found to be 5.2mm in leg intersection areas, which is considered as an appropriate size compared with the chord wall thickness of 4.4 mm and lattice wall thickness of 5.2 mm. The average stress over total throat area at the ultimate joint capacity was calculated on the assumption that the throat thickness is equal to 5.2 mm throughout the total weld length surrounding the tension lattice end. Its value was equal to 86% of maximum shear strength $(\sigma_U/\sqrt{3})$ of the chord material, which shows that the welds are not fully effective in carrying a tensile load in the lattice member. No doubt stresses concentrate at points of crack initiation because a diaphragm is welded at these locations from the inside of the chord. An effective weld length approach may be a feasible solution to prevent premature failure of joints due to failures in the welds, although this requires further study.

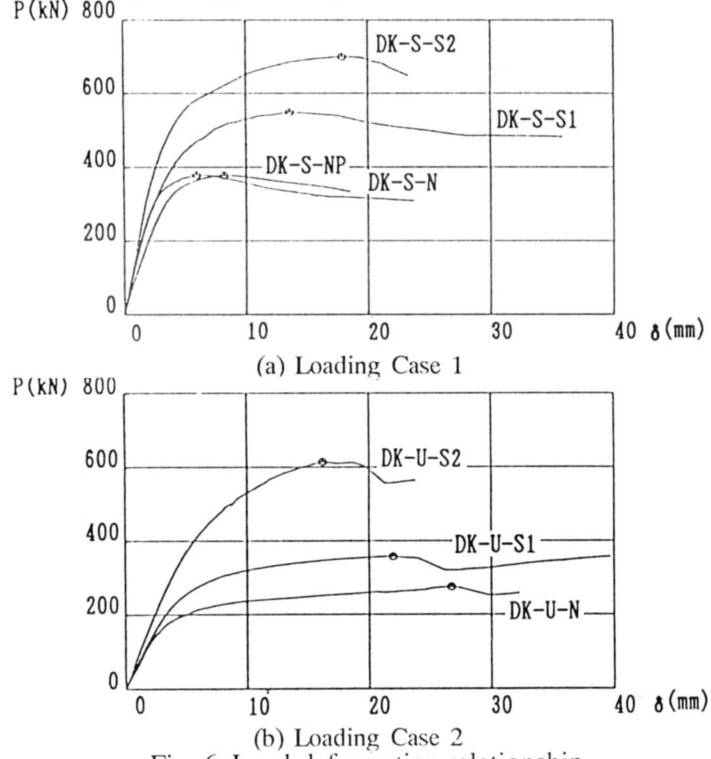

Fig. 6. Load-deformation relationship.

Joint efficiency

13. The ultimate capacity of joints is evaluated based on yield axial capacities of the lattice members. This is referred to as the "joint efficiency" for the sake of convenience. The joint efficiency is defined as the ratio of ultimate compressive axial loads on the lattice member to yield axial capacity of the same member. The results are given in Table 4.

$$\text{Joint Efficiency} = \frac{\text{Axial loads in lattices at joint ultimate capacity}}{\text{Yield axial load for lattice (measured)}}$$

Table 4. Joint efficiency.

Specimen Designation	d/t=19 (This Tests)			d/t=36	
	${}_cP_u$ (kN)	P_y (kN)	J.E.	P_y (kN)	J.E.
DK-S-N	189	622	0.28	346	0.55
DK-S-NP	179	622	0.27	346	0.52
DK-S-S1	284	622	0.42	346	0.82
DK-S-S2	361	622	0.54	346	1.04
DK-U-N	212	622	0.32	346	0.61
DK-U-S1	278	622	0.41	346	0.80
DK-U-S2	445	622	0.66	346	1.29

${}_cP_u$: Axial Load in Compressive Lattice at Joint Ultimate Capacity
P_y : Yield Axial Load for Lattice
J.E. : Joint Efficiency

14. As mention earlier, as the main objective of the test is to investigate the ultimate behaviour of joints, the tube thickness was increased to prevent buckling of lattice members before joints fail. In other words, the diameter to thickness ratio of the tested specimens was 19 while that in actual structures was 36. As the thickness of lattice members does not affect the joint ultimate capacity, the computed "joint efficiency" values are considered to apply to actual structures. Value of the joint efficiency calculated assuming that d/t = 36 are also summarised in Table 4. The joint efficiency of stiffened type 2 joints is more than 1.0.

15. The buckling strength of lattices in actual structures is estimated according to the buckling strength formula $\sigma_c = \pi^2 E_t/\lambda^2$, ref. 6. If the effective buckling length is taken to be 0.7 times the distance between joints (5.77m x 0.87), then λ = 47.2 for 216.3ϕ x 6.0 lattice members. Assuming that lattice members in the actual structures and the chords of the tested specimens are of the same material, the buckling strength σ_c computed from the stress-strain relationships observed in chord stub column test is 282 kN/mm^2. The buckling strength of lattice members is lower than the tensile yield strength of the material (σ_y = 3.88 kN/mm^2. In ref. 6, it is stated that the 95% confidence limit of buckling strength is σ_c = 3.73 kN/mm^2 which is still below the tensile yield stress. In other words, if stiffened type 2 joints are used in actual structures, buckling of the compression lattice

members takes place before tensile yielding in the tensile lattice members, which in turn occurs before the ultimate strength of the joint is reached.

16. The above consideration is based on the test results in which chords and lattices are made of STK400 material. The probability distribution of buckling strength for STK490 high strength steel used in actual structures is different from that of STK400. However, it is estimated that there is a little difference in coefficient of variation, unless members are of slender capacity, (ref. 6.). Therefore, in stiffened type 2 joints, the joint capacity is larger than the lattice member capacity.

CONCLUSIONS

17. In this paper, the improvement in the ultimate capacity of three dimensional tubular joints, which are fabricated by direct welding of lattice members to the chord and strengthened by appropriate stiffening diaphragms, is investigated. The following results may be concluded.

(1) Stiffening by diaphragms improves the ultimate capacity of joints and is quite appealing in terms of fabrication and cost. The respective ultimate capacities of stiffened type 1 and type 2 joints are 1.5 and 2.0 times that of unstiffened joints for load case 1, and 1.3 and 2.1 times for load case 2, respectively.

(2) The ultimate capacity of stiffened type 1 joints is nearly the same for both load case 1 and load case 2. The ultimate capacity of the other joints is greater for load case 2 than for load case 1. The joint design can be carried out based on the ultimate capacity for load case 1. This design is on the safe side and may be adopted for the actual design due to the fact that an equation for predicting the ultimate capacity of joints subjected to load case 2 has not been established yet.

(3) If stiffened type 2 joints are employed in the actual design, the ultimate strength of a joint is larger than the buckling strength of the corresponding lattice member, and hence larger than the tensile yield strength of the lattice members. In other words, the joint is stronger than the lattice members framing into it. Stiffened type 1 joints are also effective depending on the chord-lattice arrangement.

(4) According to the recommendations of AIJ for the design and fabrication of tubular structures in steel, the strength of welds conforming with specified size requirements is greater than the joint strength of unstiffened joints. However, in this research, there were cases where the strength of welds were lower than the joint capacity. To prevent the lattice member buckling before reaching the joint ultimate capacity, the thickness of the lattice member was taken to be 1.3 times that of the chord in the tested specimens.

REFERENCES

1. HORI A., et al. A structural design of a sports dome which has four ribbon-shaped retractable roofs. Annual Meeting, Arch. Inst. of Japan (AIJ), 1990, 1337-1342 (in Japanese).

2. MAKINO Y., KUROBANE Y.and OCHI K. Ultimate Capacity of Tubular Double K-Joints. Welding of Tubular Structures, Proc. Int. Conf., Boston, 1984,451-458.
3. PAUL J.C., UENO T., MAKINO Y.and KUROBANE Y. Ultimate Behaviour of Multiplanar Double K-Joints of Circular Hollow Section Members. Proc. 2nd Int. Offshore and Polar Engineering Conf., San Francisco, 1992, 377-383.
4. PAUL J.C., UENO T., MAKINO Y.and KUROBANE Y., The Ultimate Behaviour of Circular Multiplanar TT-Joints. Proc. 4th Int. Symposium on Tubular Structures, Delft, 1991, 448-460.
5. Architectural Institute of Japan, Recommendations for the Design and Fabrication of Tubular Structures in Steel. AIJ, Tokyo, 1990 (in Japanese).
6. OCHI K.and KUROBANE Y., Statistical Evaluation of Load-Carrying and Deformation Capacities of Cold-Formed Circular Tubular Members. Journal of Structural and Construction Engineering, Transaction of AIJ, No.391, 1988, 59-71 (in Japanese).

199. Space truss testing

E. MURTHA-SMITH, University of Durham, UK, and S. H. HWANG and Y. H. CHOU, University of Connecticut, USA

SYNOPSIS. The failure behaviour of double layer grid space trusses and frames is reviewed including some previous experimental studies. The design and results of the series of tests recently conducted by the authors is presented. The experimental programme conducted material, component, tetrahedral sub-assemblies and full space truss system tests. Supports and loading were designed to minimise the associated uncertainties so that measurement variations were due only to space truss and frame behaviour. Computer-assisted electronic data collection was used. Suggestions are made for the design of future space truss and frame testing.

FAILURE OF DOUBLE LAYER GRIDS

1. A typical double layer grid space truss is shown in Fig. 1. Changes in any one component behaviour from linear to nonlinear behaviour will lead to nonlinear changes in the forces in other components and the system behaviour becomes increasingly complex. Failure can occur (ref. 1), for components in tension, by the yield and rupture of members, connectors and joints, and for components in compression, by yielding, crushing or buckling. Pin-connected members can buckle individually. However, space truss systems which have continuity through the joints can function as space frames and thus "buckling" requires the bending of several members connected to each joint as each joint rotates in an unstable manner (ref. 2). Failure of a component can cause failure to propagate and can lead to geometric instabilities independent of buckling instabilities.

2. There is often significant divergence between experiment and analysis. For example, once nonlinear behaviour occurs in one member in a symmetrical double layer grid, subsequent failures are not necessarily symmetrical potentially leading to divergence between analysis and experiment. Chordal displacement snap-through (ref. 1) is another phenomenon that can occur without any change in external load but is often not modelled also leading to divergence between analysis and experiment. Indeed, the phenomenon can lead to rapidly propagating failure and even collapse for no change in external load. Another source of

divergence is due to differences between the assumed versus the actual loading environment: load controlled, displacement controlled or the equilibrium path. In addition, of course, the quality of the design or the measurements of an experiment can also lead to divergence.

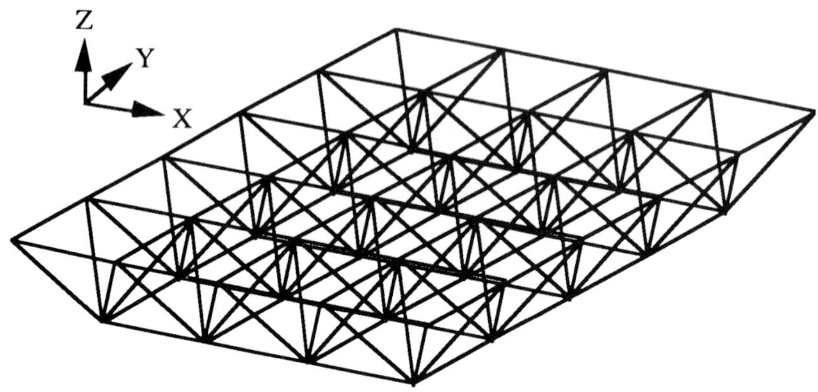

Fig. 1. Double layer grid space truss/frame.

PREVIOUS EXPERIMENTAL STUDIES

3. The following is not a comprehensive review of previous experimental studies. Rather, representative work is reviewed for its implication on the conduct of experimental studies.

Linear regime

4. An example of testing in the linear regime is given by Glogowczyka and Odrzanska (ref. 3). They reported an experimental study investigating the random variability in member forces in a double layer grid, as shown in Fig. 2. (Note, in Figs 2-6, support points are shown as solid black rectangles and load points are shown as white rectangles.). A 7 by 7 module double layer grid was tested in the linear regime and forces were measured and compared. The double layer grid was fabricated from components 150 mm in length. The space truss was equally loaded at 36 of the interior joints. The double layer grid was supported in three different manners: (i) around the 36 lower perimeter joints, (ii) along the 7 perimeter joints of each of two opposite edges, and (iii) at the 4 corner joints. These three cases had decreasing external indeterminacy. Thus fabrication variability would tend to cause higher lack of fit stresses in case (i) than in case (iii). Initial lack of fit stresses were not included in the measurements. Since the double layer grids were configured to have 4 axes of symmetry for cases (i) and (iii) and two axes of symmetry for case (ii), any one member would typically be

one of a group of eight symmetrical members for cases (i) and (iii) or of four for case (ii).

5. The authors reported the standard deviations, the means and the coefficient of variation (the ratio) of the measured forces for each group. For case (i), which has the highest external redundancy, the coefficient of variation was 15% for the highest stressed chord members whereas for both cases (ii) and (iii), the corresponding coefficient was 11%. However, examining all the forces reported gives that the average coefficient of variation was 55%, 13% and 24% for cases (i), (ii), and (iii) respectively, and the ratio of the average standard deviation to the average force was 18%, 10% and 18%. Finally, comparing experimental forces with theoretical forces based on linear space truss analysis gives ratios of 90% for all three cases and coefficients of variation of the averages of those ratios of 63%, 9% and 29% for cases (i), (ii), and (iii), respectively.

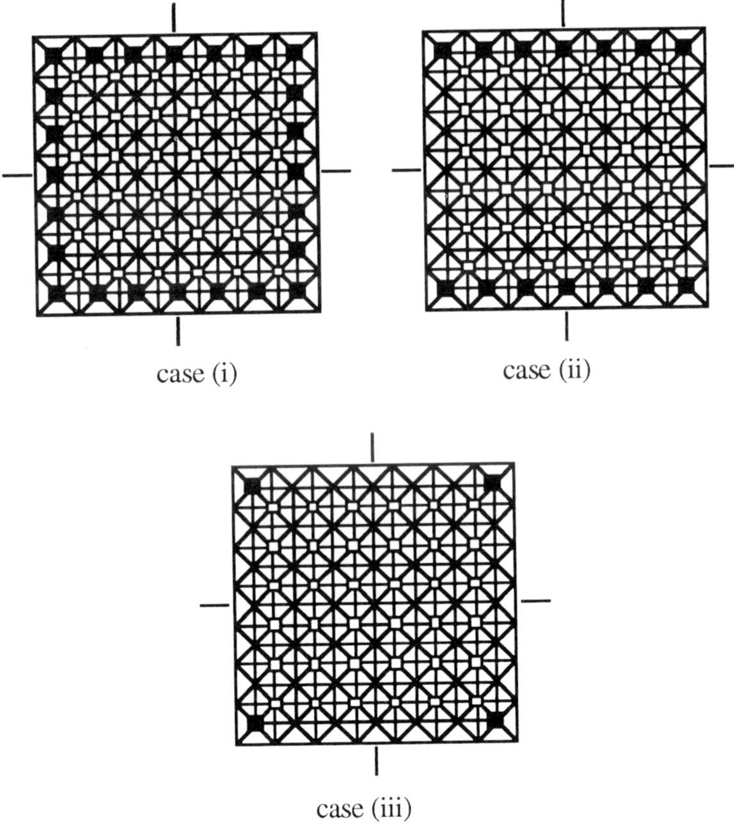

Fig. 2. Test configurations of Glogowczyka and Odrzanska (ref. 3).

6. Thus, one important conclusion that can be drawn from these results is that it could be expected that much better experimental results can be obtained by using the minimum number of supports required rather than supports around the entire perimeter as has often been used.

Nonlinear regime

7. Much of the pioneering work has been performed by Schmidt and his associated researchers (for example, refs 4-7) in a series of perimeter supported, square on plan, multiple load point tests.

8. In 1975, Schmidt *et al* reported (ref. 4) tests on single members, sub-assemblies and an 8 by 8 bay space truss, shown in Fig. 3. The system used had deliberate eccentricities in the configuration of the members. The chord members in orthogonal directions overlapped, were continuous through the joints and were welded together. Diagonal web members were also welded into the configuration eccentrically. Nominal slenderness ratios were 192 for the chord members and 250 for the diagonal members. For their analyses, Schmidt *et al* idealised their members as fix-ended because of the continuous and welded joints. Accordingly, they tested their single members in the fix-ended condition obtaining a range of peak loads between 3.90 to 5.62 kN. Nevertheless, in the sub-assemblage tests, the peak loads in the critical strut ranged from 1.90 to 2.17 kN. However, using an analysis method (ref. 2) that accounts for the continuity of the joints and the stiffness of adjacent members, the elastic buckling load in the critical strut in the sub-assemblage should range from a lower bound of 1.91 kN to an upper bound of 2.91 kN, consistent with the sub-assemblage tests. It appears that assuming full fixity of the members grossly overestimates their capacity.

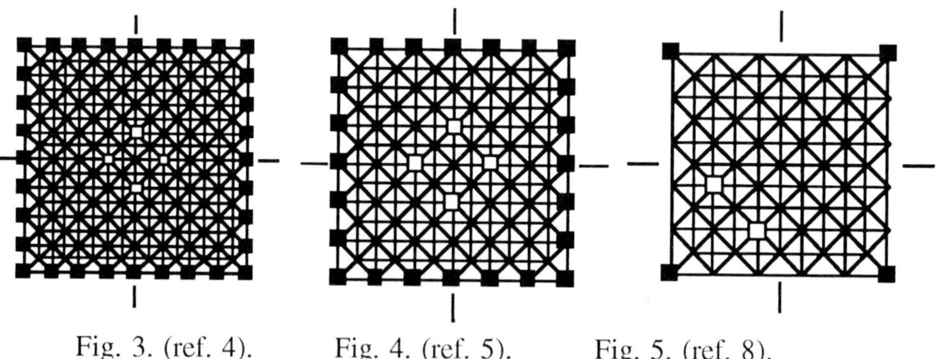

Fig. 3. (ref. 4). Fig. 4. (ref. 5). Fig. 5. (ref. 8).

9. The 8 by 8 bay space truss was supported around the entire perimeter and was symmetrically loaded at four points near the centre in a displacement controlled test. Schmidt *et al* reported linear behaviour up to an applied load of 8.50 kN, nonlinear behaviour thereafter until a peak load of 14.9 kN, after which the applied load reduced as the capacity reduced. Their analytical prediction of the peak load, based on fix-ended member capacities, was 17.00 kN. However, using the method of ref. 2, gives that first, second and third joint buckling should

occur at applied loads of 8.10, 11.1 and 13.1 kN, respectively, similar to the actual behaviour of the truss. Thereafter, the system is unstable. According to this analysis, when the first joint buckles, the equivalent effective length factor is 0.94 for the most highly stressed member. Schmidt *et al* pointed out the unsymmetrical behaviour of the nominally symmetric truss caused by the onset of buckling of one member of a symmetrical group.

10. In 1976, Schmidt *et al* reported (ref. 5) tests on a series of 6 by 6 bay space trusses, perimeter supported and loaded at 4 symmetric joints near the centre, shown in Fig. 4. The double layer grid was fabricated from components 305 mm in length. Member slenderness ratios were nominally 67. Such members have steep postbuckling curves. Despite the symmetry of the tests, Schmidt *et al* reported that the force distributions in the linear range were highly asymmetric and deviated noticeably from the theoretical predictions. The maximum force in the members was 15% above the average force in that group and the experimental value was approximately 10% above the theoretical value. Indeed, the results in the linear range have statistics similar to those reported by Glogowczyka and Odrzanska described above. The unsymmetrical force distributions in the linear range caused unsymmetrical buckling to occur at a much lower load level than predicted.

11. In 1984, Collins (ref. 8) referenced tests reported in his Ph.D. thesis of 1981 (ref. 9). He tested four double layer grid 5 by 5 bay space trusses, shown in Fig. 5. Members were 360 mm long. In contrast to Schmidt's earlier tests, Collins supported his trusses only at the four corner points thereby reducing the random errors introduced by lack of fit at the supports and reducing the number of load paths to the supports. He used only 2 load points and thus reduced the symmetry so that each member was only one of a pair of symmetrical members compared with Schmidt's configurations in which each member was one of a group of eight nominally symmetrical members. By reducing the symmetry, Collins was reducing the ambiguity of the structural failure sequence allowing him to better place instrumentation.

12. Nevertheless, natural variation would cause slightly different behaviour between a symmetrical pair, one would fail before the other and symmetry would then be lost potentially preventing the other member of the pair ever failing. Such behaviour did occur in Collins' tests. However, the analytical model predicted identical behaviour between symmetrical members and as soon as nonlinear behaviour began, there was divergence between the analysis and experiment.

13. An additional source of divergence was that the analysis developed an equilibrium path, while the experiment was displacement controlled. Thus when the equilibrium path called for a load and displacement reduction in order to maintain equilibrium, the test reduced only the load but not the displacement. Since these two load paths are most likely non-conservative, then it cannot be expected that when these two paths subsequently intersect, the state of the structure would be the same independent of the path taken. Collins reported (ref. 9) observing the phenomenon which the first author has labelled "chordal

displacement snap through" (ref. 1). However, Collins instrumentation did not allow measurement of the phenomenon.

RECENT TESTING BY THE AUTHORS

14. The design and results of the series of tests recently conducted by the authors is presented. The experimental programme conducted material, component, tetrahedral sub-assemblies and full space truss system tests. Supports and loading were designed to minimise the associated uncertainties so that measurement variations were solely due to space truss and frame behaviour. Computer-assisted electronic data collection was used.

Component testing

15. The authors conducted a series of tests using the Unistrut System 1, formerly known as "Moduspan". Materials, individual members, tetrahedral sub-assemblies and full space trusses were tested. Standard tension tests were conducted to obtain material parameters. Individual members were tested in tension (ref. 10) and in compression (ref. 11). In this space truss/frame system, load is transferred from the member to the joints through two lugs in the joint which engage holes in the member and through a bolt passing through the member and joint. In addition, there is eccentricity between the member and joint. Accordingly, an extensive investigation was made of the effect of each component along the load transfer path at the connection (ref. 12).

Tetrahedra testing

16. Many space truss systems are space frames because the jointing system provides continuity between members. Thus, while the primary action is one of a space truss, a secondary resisting system exists because of load transfer by frame action. Further, the buckling capacity of the members is increased due to the additional restraint provided to the critical members by the members adjacent to it (ref. 2). However, this additional restraint is not sufficient to provide the equivalent of full fixity. Thus, testing individual members either as pin ended or as fix ended does not represent nor correlate well with the conditions found in the space truss/frame.

17. As Murtha-Smith and Bean have shown (ref. 2), the degree of restraint provided is approximately equal to having between one and two of the diagonals at each end of the critical member. Accordingly, a series of tetrahedral sub-assemblies, shown in Fig. 6(a), were tested both in compression and in tension in the direction of the critical member (refs 10 and 13). The attached members completing the tetrahedron were not loaded. As expected, the additional diagonals had little effect on the behaviour of the critical member in tension. Also as expected, the diagonals did provide additional restraint to increase the capacity of the critical compression members.

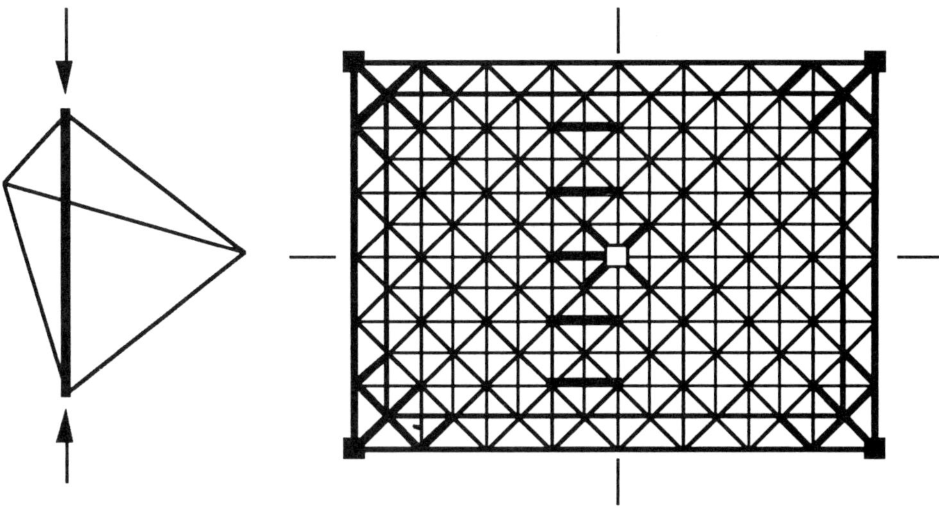

(a) Tetrahedral sub-assembly. (b) 6 by 8 space truss/frame.

Fig. 6. Murtha-Smith, Hwang and Chou.

Space frame testing

18. A series of space frames were tested (ref. 10). The preliminary frames used a 48 member, 2 by 3 module configuration. The frames were corner supported and centrally loaded by a single load. The series was designed to develop the testing technology and seek correlation between component and subassembly testing with the system test. Critical members were instrumented and gave good agreement with the tetrahedral subassembly behaviour.

19. The subsequent series used a 384 member, 6 by 8 module configuration to minimise symmetry, shown in Fig. 6(b). The frames were corner supported and centrally loaded by a single load. This configuration minimises errors due to support lack of fit and load indeterminacy. Members at the support and load points were stronger than the others to prevent localised failure from occurring. Further, to prevent the ambiguities associated with symmetry, the compression chord members adjacent to and to one side of the centreline were also formed using stronger members. These modifications allowed instrumentation to be concentrated in those seven compression chord members immediately adjacent to the other side of the centreline. A few other members were instrumented for check readings.

20. As can be seen in Fig. 7, excellent correlation was obtained for compression members in the tetrahedral tests and the space truss/frame tests (ref. 10).

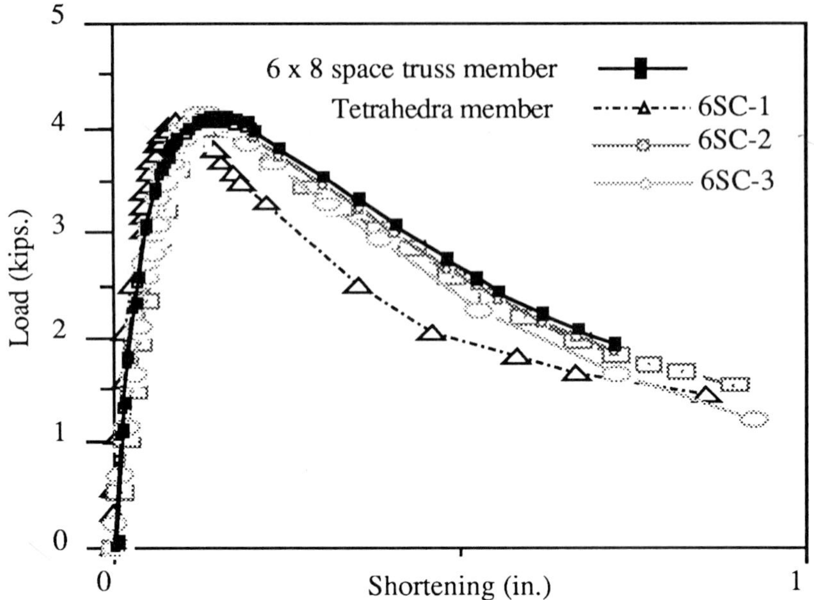

Fig. 7. Space truss and tetrahedra member behaviour comparison.
(1 kips = 4.448 kN, 1 inch = 25.4 mm)

PROPOSED TESTING PROTOCOL

21. It is recommended that testing programmes include testing materials, members and other components. In addition, the testing of sub-assemblies such as the tetrahedral assemblies used by the authors is recommended. In the testing of the space trusses, the number of supports and the number of loading points should be minimised. However, special care is required in the design of the support and load points to prevent localised failures from not allowing the test to demonstrate the phenomenon being sought.

22. Symmetry should also be minimised so that a unique, unambiguous and predictable failure sequence can be adequately instrumented, obtained and observed. Asymmetry can be obtained by placing the supports unsymmetrically, loading eccentrically and, if and where necessary, having different strength members on opposite sides of a potential symmetry axis.

FUTURE RESEARCH NEEDS

23. Both analytical methods and experimental procedures have improved significantly in the last two decades. Current analytical methods allow for complex modelling that was unavailable to earlier researchers. Further, current instrumentation and the lessons learned from the pioneers in this field allow for

significant improvement in the quality of results. Some unanswered questions still remain. Glogowczyka and Odrzanska's test in the linear regime demonstrated the degrading effect on the quality of the results of increasing external indeterminacy. Even so, there was still significant variation in their best case in the linear regime. Accordingly, it is desirable to obtain much better correlation in the linear regime. Larger scale components and modern instrumentation would probably give better results and smaller variation.

24. Experiments in the nonlinear regime have shown the improved quality of results by minimising the external redundancy. However, it has not been demonstrated that when continuity exists there is an increase in member capacity that matches theoretical predictions. For space trusses/frames with large slenderness ratio, the system buckling should be reversible, at least initially. When medium slenderness ratio members are used, the buckling is not reversible but is rapid and possibly dynamically unstable. While this phenomenon, chordal displacement snap through, has been qualitatively observed (ref. 9), it has yet to be properly demonstrated in a double layer grid space truss under properly controlled and instrumented conditions.

ACKNOWLEDGEMENTS

25. The authors gratefully acknowledge the sponsorship of the National Science Foundation (USA) and the Space Frame Division, Unistrut Corporation, Inc., (USA). The opinions expressed in this paper are those of the authors and do not necessarily reflect those of the sponsoring organisations.

REFERENCES

1. SMITH E. A. Space truss nonlinear analysis. Journal of structural engineering, ASCE, vol.110, no.4, April, 1984.

2. MURTHA-SMITH E. and BEAN J E. Double layer grid space frame buckling. International J. of Space Structures, vol. 4, no. 3, 1989.

3. GLOGOWCZYKA K. Z. and ODRZANSKA S. W. An attempt of measurement of random internal forces in bars of a regular space structure. Proceedings of the 2nd International Conference on Space Structures, University of Surrey, Guildford, Surrey, September, 1975.

4. SCHMIDT L. C., MORGAN P. R. and STEVENS L. K. The influence of imperfections on the behaviour of a space truss. Proceedings of the 2nd International Conference on Space Structures, University of Surrey, Guildford, Surrey, September, 1975.

5. SCHMIDT L. C., MORGAN P. R. and CLARKSON J. A. Space trusses with brittle-type strut buckling. Journal of the structural division, ASCE, vol.102, no.ST7, July, 1976.

6. SCHMIDT L. C., *et al* Ultimate load behaviour of full-scale space truss. Proceedings of Institute of Civil Engineers, part 2, vol.69, March, 1980.

7. HANAOR A. and SCHMIDT L. C. Space truss studies with force-limiting devices. Journal of the structural division, ASCE, vol.106, no.ST11, November, 1980.

8. COLLINS I. M. An investigation into the collapse behaviour of double layer grids. Proceedings of the 3rd International Conference On Space Structures, University of Surrey, Guilford, September, 1984.
9. COLLINS I. M. Collapse analysis of double layer grids. Ph.D. thesis, Department of Civil Engineering, University of Surrey, Guildford, February, 1981.
10. HWANG S-H. Experimental behaviour of space frames. Ph.D. dissertation, Department of Civil Engineering, University of Connecticut, Storrs, CT, USA, May 1989.
11. MURTHA-SMITH E and MAGYAR P. M. Cold-formed steel channel struts. Journal of structural engineering, ASCE, vol.117, no.4, April, 1991.
12. MURTHA-SMITH E, HWANG S-H, and BEAN J. E. Load transfer in a space frame connection. International J. of Space Structures, vol.7, no.3, 1992.
13. CHOU Y-H. Behaviour of steel tetrahedral frames. M.S. thesis, Department of Civil Engineering, University of Connecticut, Storrs, CT, USA, May 1989.

200. Ductile truss structures and scaled model experiments

H. MUKAI and A. WADA, Tokyo Institute of Technology, Japan, and
A. WATANABE and M. IWATA, Nippon Steel Corporation, Japan

SYNOPSIS. It is quite difficult to perform seismic design for steel truss structures by allowing plastic deformation of members, because of the brittle buckling behaviour of compressive members. In order to protect the member from buckling and enhance ductility, the authors insert a central tube into the truss tube member and implement a ductile truss structure using these non-buckling members. Experiments have been performed on three types of small scale truss structures, using an automated loading system with stepping motors which proved the effectiveness of the compound member in plastic design.

INTRODUCTION

1. The social needs of large span structures are increasing, for such purposes as atria attached to tall buildings or baseball stadium roofs. Such structures include shell structures using reinforced concrete and air-inflated membrane structures, but steel trusses are the most common. In Japan, a seismically active land, it is very important to ensure safety against earthquakes for steel truss structures as well as for general buildings.

2. In the seismic design of general buildings, plastic deformations are permitted for some structural members in order to absorb the large earthquake input energy. However, in steel truss structures, it is quite difficult to perform seismic design by allowing plastic deformation of members because of the brittle buckling behaviour of compressive members. If such plastic deformation could be allowed for in steel truss structures, a more rational seismic design method could be used.

3. To enable plastic deformation, members subjected to a large force or deformation are required to have high ductility. Actually, the slenderness ratio of many members is over 100 and not much ductility remains in the member after buckling due to compressive forces. Thus it is difficult to apply plastic design to truss structures.

4. In this study, the authors have insert a central tube into the tube of a truss member, which acts as a ductile compressive member. The elastic buckling force of this central tube is designed to be larger than the axial yield force of the outer tube. This combination protects the tube from buckling and enhances ductility and

consequently the whole structure should have a higher loading capacity and larger deformability. This tube will be referred to hereafter as a "non-buckling member". The non-buckling member is functionally equivalent to a plate steel brace inserted in a pre-cast concrete wall panel, or a steel brace covered by reinforced concrete to prevent buckling. Double tube truss members have also been investigated by Tada, et al.

5. In this study, an experiment using reduced scale models of steel truss structures, which included the non-buckling members as mentioned above, was performed to investigate the structural mechanical properties. As the model becomes smaller, the strength reduced and in the case of a distributed load, the load at each loading point becomes extremely small. Therefore it is difficult to control all the load increments simultaneously, and the applied load has been substituted by the load specified by the tournament method. Here, the authors developed an automatic loading system which controls the loads at multiple points concurrently.

SPECIMENS AND LOADING SYSTEM

Specimens

6. Experiments for the three types of specimen were carried out as follows:

(a) Normal Truss Structure (Fig. 1; Truss-1);
statically determinate plane truss structure, total span 120 cm, height 12.5 cm, grid length 20 cm.

(b) Truss structure with one non-buckling member (Fig. 1; Truss-2);
one non-buckling member installed in the central upper chord.

(c) Truss structure with five non-buckling members (Fig. 2; Truss-3);
non-buckling members installed in all upper chords.

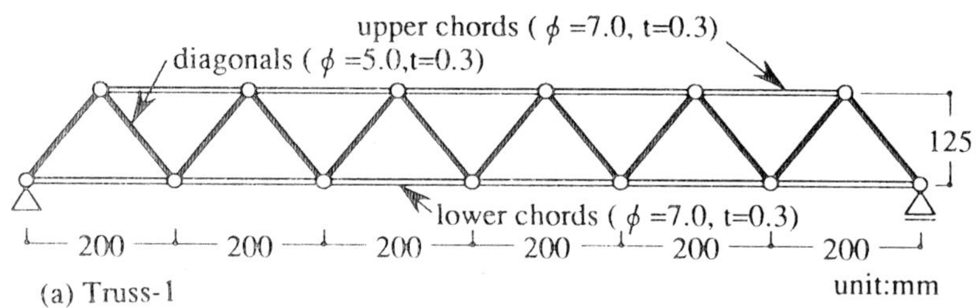

(a) Truss-1

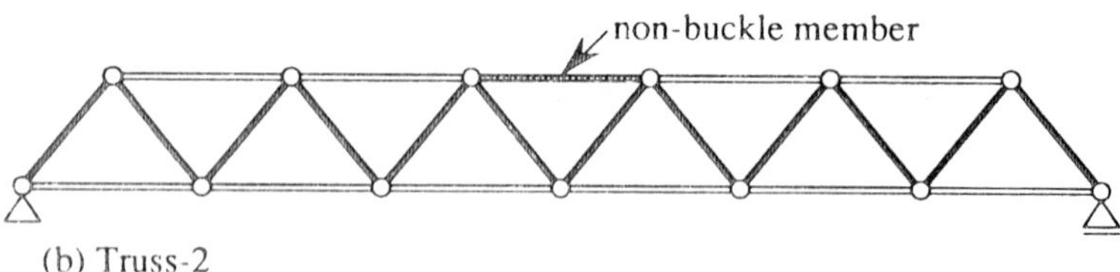

(b) Truss-2

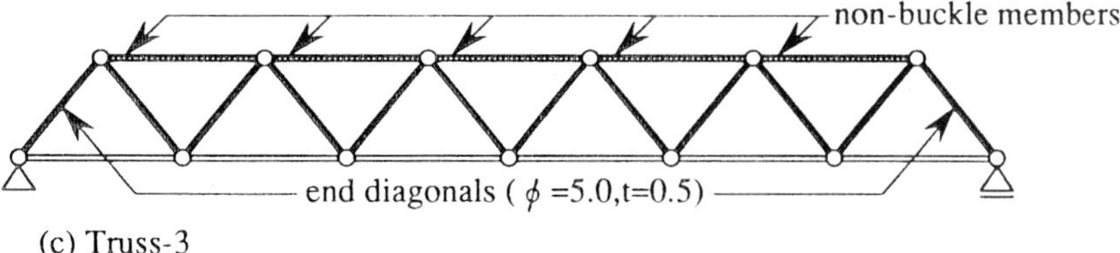

(c) Truss-3

Fig. 1. Three specimens.

7. For the truss members, stainless steel tubes were used, 7.0mm diameter with a wall thickness of 0.3mm, for chord members, and 5.0mm diameter and 0.3mm wall thickness for the diagonal members. Joints are connected to truss members by silver brazing (Photo 1). The details of the non-buckling member are shown in Fig. 2. The central tube has a 6.0mm diameter and 1.0mm wall thickness. By setting the central tube very slightly apart from the outer tube and both joints, the axial force will not be transmitted into the central tube. The sectional area is calculated so that the yield strength of the outer tube ($P_Y = 1523N$) will be lower than the Euler buckling load of the central tube ($P_E = \pi^2 EI/\ell^2 = 2428N$) and the safety factor (P_E/P_Y) is 1.59. The stainless tube is manufactured by cold rolling and the material test results indicate that the 0.2%-strength is 3.1 times higher than the strength of normal stainless steel. Accordingly the silver brazed joint is the weak point, and in order to compensate for this, the stainless steel tube was annealed by heating to 1050°C for 5 minutes and cooling in air at room temperature.

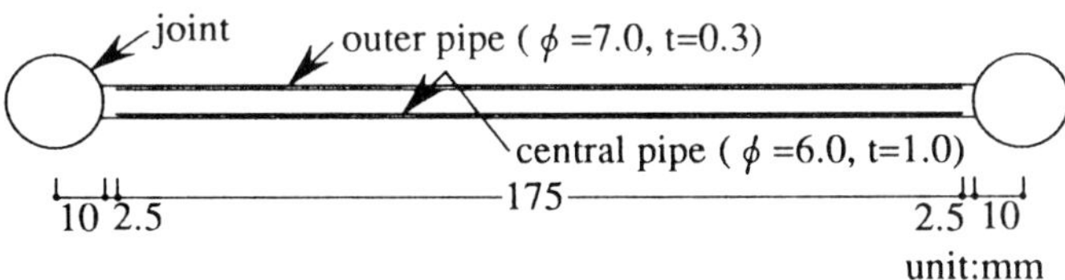

Fig. 2. Detail of non buckling member.

Loading System

8. Fig. 3 and Photo 2 show the loading apparatus. A female screw is placed in the centre of a worm-wheel and a worm-gear is rotated by a stepping motor controlled by a personal computer. This mechanism provides a tension force to the rod through the shaft. There is a groove in the shaft and a ball bearing is put

into this groove and supported by the top and bottom bearing cylinders to prevent co-rotating caused by the worm-gear.

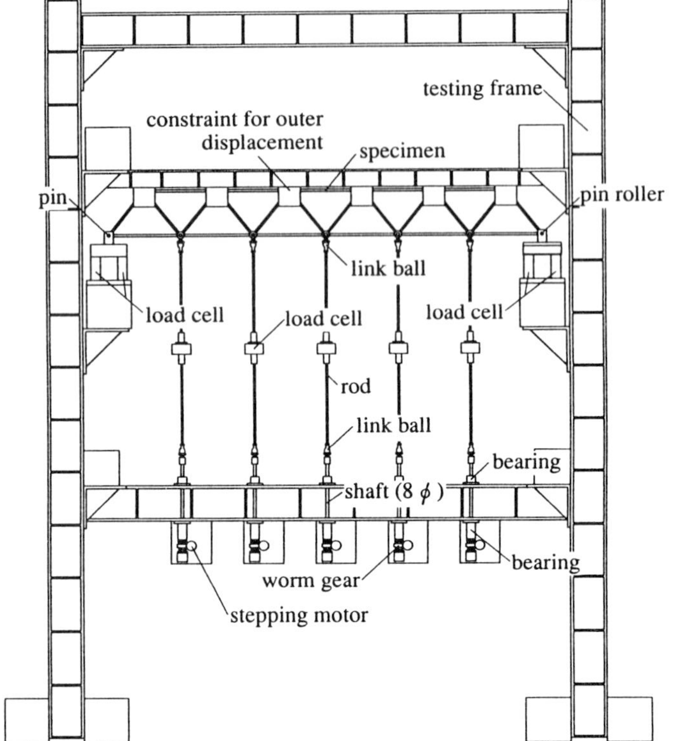

Figure 3. Loading system.

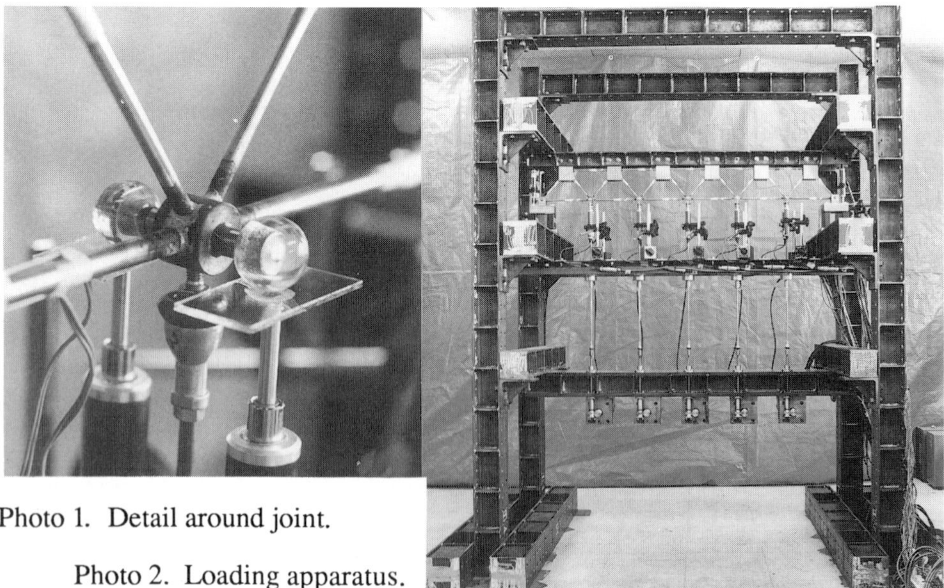

Photo 1. Detail around joint.

Photo 2. Loading apparatus.

9. Two link-balls are attached to each end of the rod so that the bending moment does not act on the central load-cell. Thus the loading point can move freely in accordance with the lateral and vertical deformation of each truss joint. Load is applied until the vertical displacement of the joint, of the lower chords, reaches 30mm; loading is controlled to be equal on all the load-cells attached to tensioned rods.

10. The deformations of the truss structure are measured by displacement transducers at each joint on the lower chords, the total applied load is checked by load-cells attached to the right and left supports and the strain is measured by strain gauges attached to the centre of each member.

LOADING CONTROL SYSTEM

Loading System

11. The personal computer is connected to the multi-axial controller through an RS232C and it is also connected to the measuring instrument through a GPIB. This system measures the load and displacement automatically and controls five stepping motors simultaneously. There is a displacement transducer and load cell for each stepping motor. The computer compares the measured load and displacement values with the base control value and controls the stepping motors accordingly. A stepping motor can control displacement more precisely than a hydraulic pump, and it has high stiffness as a loading mechanism. With these characteristics, the negative slope of load-displacement relationship after buckling can be controlled.

Motor Control System

12. The base control value is the load value of the load-cell corresponding to the base motor (P3), and other load values are set in proportion to this base value. In the elastic region, the motor is controlled by the load value of the base point. Around the maximum strength region or in the plastic region, it is controlled by the displacement value of the base point.

13. The on/off procedure of the motor is controlled as follows. The motor at the base point is set to be stopped when the value of corresponding load or displacement becomes greater than the base control value ± the allowance. Other motors are stopped when the measured load or displacement is greater than the load value of the base motor and they are activated when both values are less than it.

RESULTS

Motor Control

14. Unlike a hydraulic pump, a stepping motor can easily control the displacement and it has a stiffness as a loading system about 100 times higher than the truss specimens. Furthermore the negative slope of the specimens after buckling is of the same order as their elastic stiffness. Therefore the data in the negative slope region are easily measured even after the structure reaches an unstable condition.

Relationship between Load and Displacement

15. Fig. 4 shows the relationship between the total load of each truss structure and the vertical displacement of each joint of the lower chords. The nomenclature (a), (b) and (c) indicate Truss-1, Truss-2 and Truss-3. v_1, v_2, ~v_5 express the vertical displacement of each lower chord joint from left to right. Table 1 lists the maximum strength values of three truss structures. The value for Truss-3 is the strength at the end of the experiment, where v_3 = 21mm.

Table 1. Total maximum strength of each specimen.

Specimens	Truss.1	Truss.2	Truss.3
Total maximum strength(N)	812.80	912.44	1184.20
(Ratio)	(1.000)	(1.123)	(1.457)

16. In Truss-1, the strength is drastically decreased by the buckling of the central member of the upper chords. In Truss-2, the strength of the early stage is relatively higher than Truss-1 because the central member does not buckle. But later the strength is reduced due to the buckling of an upper chord member next to the central non-buckling member. As a result, the deformability of the truss structure itself was not improved. This reason is analysed as follows. The total load which makes the central member yield is quite close to the total load which makes the neighbouring member buckle. Thus the neighbouring member buckled before the non-buckling member reached a stable yield. Using one non-buckling member is not effective for the improvement of the deformability.

17. In Truss-3, the strength continued to increase even after the stiffness reduction around the point P = 941N and consequently large deformability was obtained.

Relationship between Total Applied Load and Axial Strain of Member

18. Figs 5 to 7 describe the relationship between the total applied load and the axial strain in upper chord members U2 and U3 (central member) for Trusses-1 to -3.

19. In Truss-1, shown in Fig. 5, the central upper chord (U3) buckled first. After buckling, its neighbouring members are subjected to a bending moment caused by the shortening of the central upper chord.

20. In Truss-2, shown in Fig. 6, plastic deformation without buckling of the central member starts at around the maximum structural strength point. The strength of the structure is reduced after the neighbouring upper chord member (U2) buckles. Truss-2 undergoes a similar change to Truss-1 after the buckling of member U2.

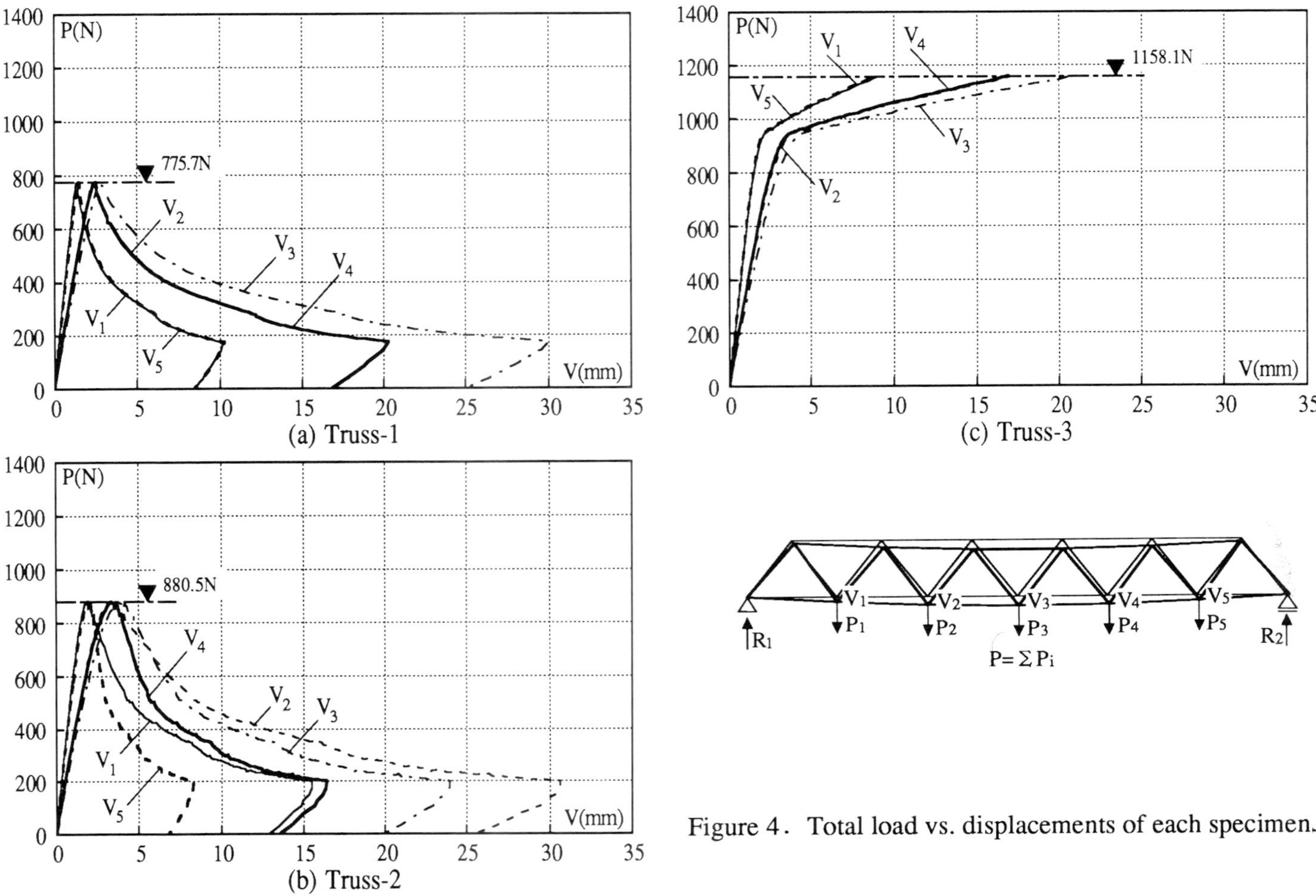

Figure 4. Total load vs. displacements of each specimen.

21. In Truss-3, shown in Fig. 7, the non-buckling members U2, U3 and U4 are subjected to a compressive force. The axial strains of U2, U3 and U4 are over 0.55%, 0.92% and 0.51%, respectively, when the plastic deformations take place. The non-buckling members are confirmed to work efficiently in this structure type. With this strength enhancement, the central two members of the lower chords yield in tension, forming a ductile truss structure.

22. From the experimental results of Truss-2 and Truss-3, the authors concluded that for the improvement of structural deformability, non-buckling members should be installed not only for the member with the largest compressive axial force but also for members with an axial force close to this value.

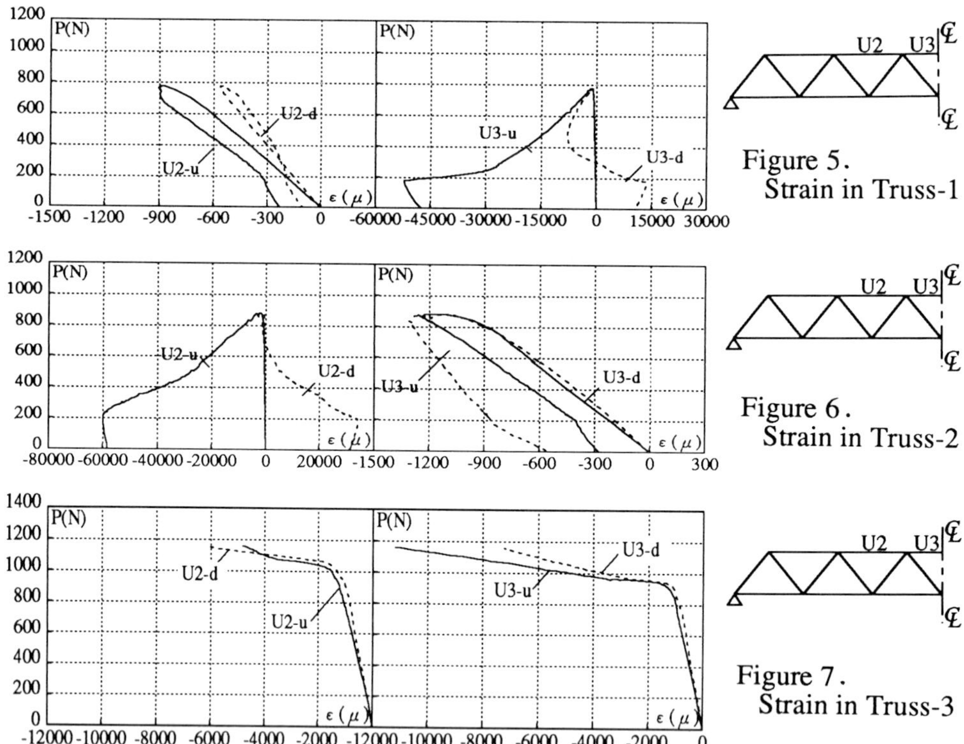

Figure 5.
Strain in Truss-1

Figure 6.
Strain in Truss-2

Figure 7.
Strain in Truss-3

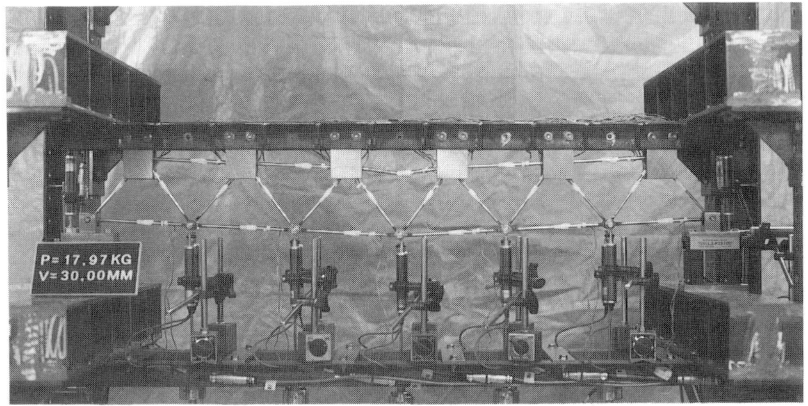

Photo 3 (a). Final Deformation of Truss-1

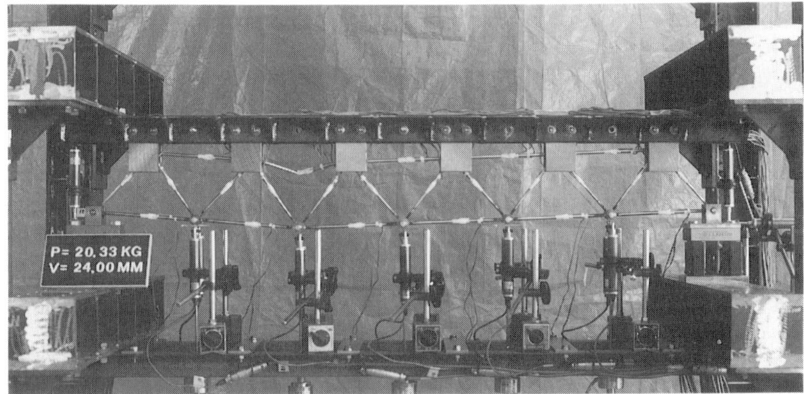

Photo 3 (b). Final Deformation of Truss-2

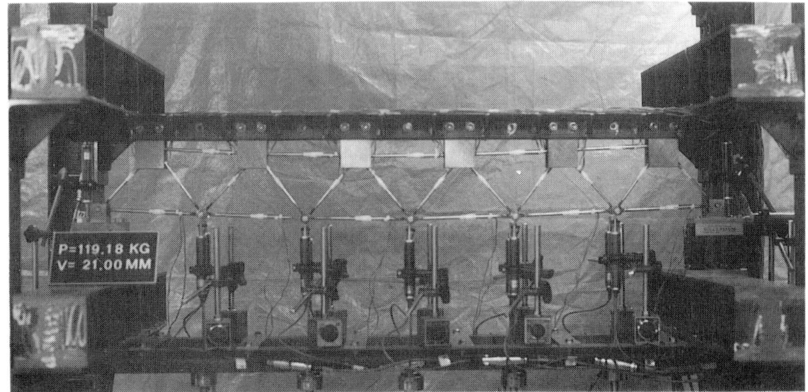

Photo 3 (c). Final Deformation of Truss-3

Deformation at the Experiment Completion

23. Photo 3 shows the final deformation state; (a) ,(b) and (c) are Truss-1, Truss-2 and Truss-3 respectively. In Truss-1 and Truss-2, a hinge in the structure was formed at the buckling point in the upper chord members, then the structure became a 3-hinged mechanism, and finally the whole structure collapsed. From Fig. 4, the proportion of the vertical displacements of the lower chord joints were obtained. In Truss-1, the proportion of v1: v2: v3: v4: v5 is 1.00: 1.99: 2.94: 1.99 :1.01. In Truss-2, this proportion is 1.00: 1.98 :1.55: 1.06 :0.54. The V line is determined by connecting the lower joint under the buckled upper member to both right and left supports. After buckling, it is found that this structure has transformed into this V shaped mode.

CONCLUSIONS

24. In this study the authors proposed a ductile truss structure and proved its effectiveness in plastic design by undertaking experiments. The authors performed experiments on three types of small scale truss structure using an automated loading system with stepping motors. The following conclusions were drawn.

a) Accurate load control is accomplished by controlling the stepping motors and worm-gears automatically using a personal computer. A high stiffness loading mechanism, which can catch negative slope data after the buckling of the chord member, was obtained.

b) In a statically determinate plane truss structure, our proposed non-buckling member has proved to work efficiently to protect buckling and enlarge the deformability. This non-buckling member is made by inserting a central tube into the truss tube, and is used for the members which are subjected to large compressive force.

c) For the improvement of the load capacity and deformability of the truss structure, it is necessary to install this non-buckling tube not only for the member with the largest compressive axial force but also for members with compressive axial force close to the largest one. But when the member with the largest compressive axial force is designed to be relatively small, it is considered that one non-buckling member placed in its position could be effective.

d) By using non-buckling members for compressive members, ductile truss structures could be created, where the plastic deformation were allowed during earthquakes.

ACKNOWLEDGEMENTS

Much help was received from Prof. S. Hayashi, Mr. S. Ichihashi, Dr. T. Yamada, Mr. A. Hori, Mr. Y. Huang, Mr. S. Onoe, Mr. H. Nishio, Mr. M. Gunji, Mrs. M. Yoneda and Mr. A. Sturt; all of whom were greatly appreciated.

201. Experimental analysis of stresses in the members of a sports hall roof structure

V. RADIANOV, Building Research Institute - Incerc, Romania

SYNOPSIS. The experimental analysis of the stresses in the members of a sports hall roof structure was undertaken. The novel structural solution could be applied to a large number of sports halls. The roof structure is a double-layer grid space structure with a 36 x 42m surface area. This analysis showed that the roof structure was adequate both under symmetrical and unsymmetrical loading, thus confirming the homogeneous design of the structure, the layout on supports, and the load distribution. The structure had an adequate behaviour, remained working in the elastic range under a load which exceeded the design load by 15%. No tendency towards failure due to local instability was noticed.

INTRODUCTION

1. In recent years, double-layer grid space structures have been continuously applied and improved. The double-layer grid roof structure that has been investigated brought about new elements in conception and execution, which are related to:

- structural geometry
- structural detailing
- support
- rain water discharge
- cladding

2. Taking into account that this type of roof structure was to be used and applied for a long period, the stresses in the members had to be analysed experimentally. The analysis of these buildings, designed for the use of large groups of people, was to confirm the performance of the system as well as of its innovating elements related to serviceability and in service safety.

CHARACTERISTIC FEATURES OF THE ANALYSED STRUCTURE

3. The roof structure is a double-layer grid space structure, with a 36 x 42 m surface area, has two slopes, and a 10m cornice height. The roof structure is supported at the boundaries on an RC beam, at the boundary nodes of the upper layer.

4. The roof structure is a diagonal-on-square type and has the following structural detailing:

- the bottom layer members and the bracing members are of steel circular hollow sections;
- the top layer members are of steel hot-rolled sections and are 80 x 80 x 10.2 angle profiles;
- the joints (nodes) are the "DISK" type;
- the rigid support of the roof cladding is of corrugated steel sheet bolted to the bottom layer members of the roof structure.

EXPERIMENTAL ANALYSIS

5. The experimental analysis was designed to evaluate the following structural characteristics:

(i) the structure ability to resist a loads exceeding the design load by 15%;
(ii) the general behaviour of the joints;
(iii) the manufacturing of the member and joint ends, which determine the members loading;
(iv) support wedging which determines the stress state in members;
(v) interaction between corrugated steel sheeting and the structure;
(vi) the structures bearing capacity under unsymmetrical loads;
(vii) the bearing capacity reserves of the structures.

6. The loading-unloading cycle testing method was used (ref. 1).

7. Taking into account the load assumed in the static design and the load existing on the roof structure when the experimental analysis was performed, the loading levels were established and are presented in Table 1.

THE LOADING-UNLOADING EQUIPMENT

8. In order to apply the testing method in loading-unloading cycles, adequate equipment was designed and manufactured consisting of the following elements:

(i) water supply tank used as a water source;
(ii) water pumping station;
(iii) water supply pipe from the pumping station to a general water distributor placed at the roof structure level;
(iv) pipe sections for water distribution to loading tanks;
(v) loading tanks.

A 20daN/m^2 dead load was applied when the experimental analysis was performed, this represented the gravel layer for the protection of the cladding.

9. The loading tanks were provided with openings and were calibrated by marking the water level corresponding to the loading levels with coloured stripes.

10. Loading was performed by pumping water into the tanks until it reached the level corresponding to the loading stage, while unloading was performed by pulling out the rubber stoppers from the tanks holes, the water in the tanks discharged through the rainwater discharge openings.

Table 1.

Experimental analysis loading levels	Notation	Loads	
		Total load daN/m^2	Test load daN/m^2
Zero level	N^0_{IP}	120	0
Substitution level	N^S_{IP}	140	20
Level I	N^1_{IP}	165	45
Level II (design)	N^2_{IP}	190	70
Level III	N^3_{IP}	210	90
Level IV	N^4_{IP}	220	100
Level V (maximum)	N^M_{IP}	225	105

EQUIPMENT FOR STRESSES AND DISPLACEMENTS MEASUREMENT

11. The evolution of the stress values in the structure members was measured by means of resistive strain gauges with a 10mm measuring base, taking into account the homogeneity of the steel structure along this length (ref. 2). The electrical resistive strain measurement method allowed centralisation of the measurements in automatic equipment for reading the strains in the roof structure members (ref. 3).

12. The strain measuring points were located as follows:
- on all boundary bracings;
- on all bottom layer members intersecting the symmetry axis;
- on the top layer members where maximum stresses occurred.

13. Ninty six measuring points equipped with 320 active strain gauges were established. In the case of the members made of steel circular hollow sections each measuring point consisted of 3 strain gauges located in the central zone of each member at 120° around the perimeter, and along the member (Fig. 1).

14. The top layer members were manufactured from 80 x 80 x 10.2 angle profiles and had measuring points made up of four transducers located in the central section of the members towards the ends of the angle profiles flanges.

15. The measurements were performed using an automatic AUTOLOG-PEEKEL INSTRUMENTS B.V. equipment together with the N. 23.100 equipment.

16. The displacement of the joints of the sports hall roof structures was measured with a U-200 HUGGENBERGER A.G. - type deflectometer with a 1/10 mm accuracy. The displacement measuring points were selected so that the deflected shape of the structure could be obtained and compared quantitatively and qualitatively with those resulting from the analytical calculation.

RESULTS AND CONCLUSIONS

17. The experimental analysis of stresses in the members of a sports hall roof structure was performed when the structure was subjected to a static loading in-situ test. The structure displayed highly symmetrical behaviour, the displacement measurements at symmetrical points were virtually identical (within 1/10ths of millimetres). The response of the structure confirmed the homogenous erection, the correct layout of supports and the uniformity of loading (Figs 1-2). The displacement values under symmetrical load were in accordance with the design values. The increases in nodal displacements were proportional to the increases of loads on the structure.

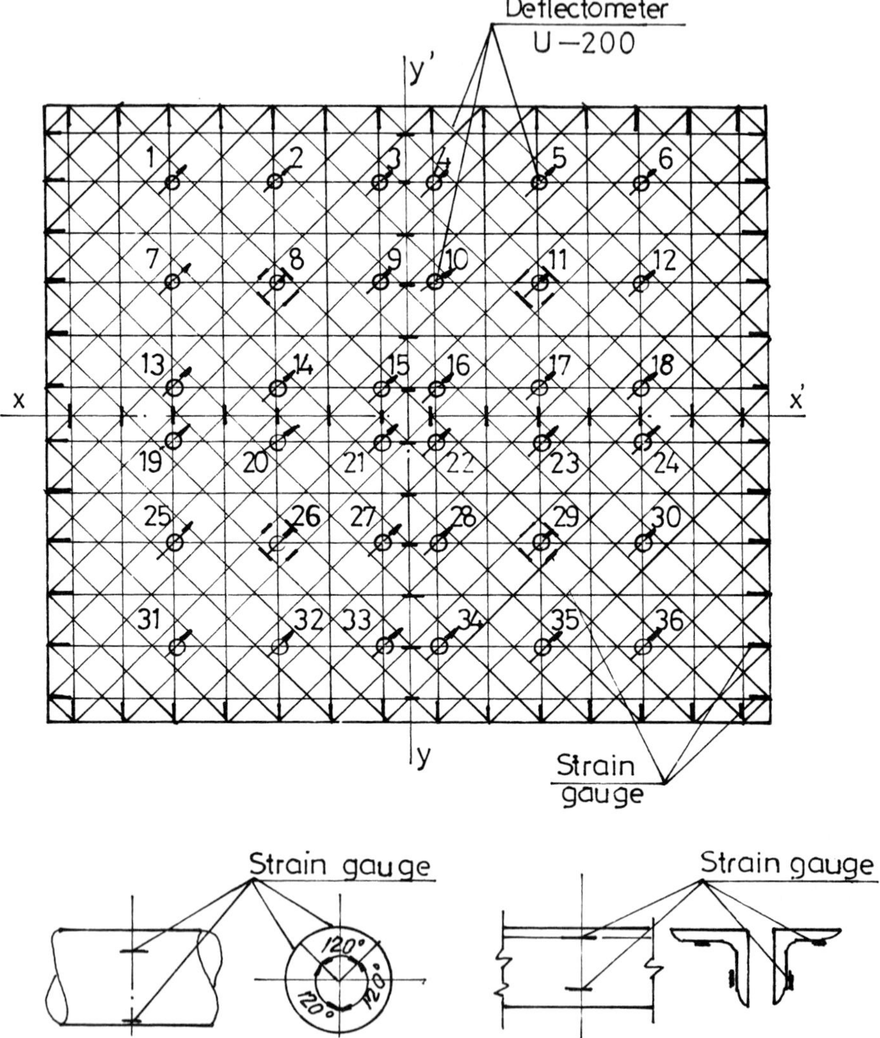

Fig. 1. Diagram showing location of points for measuring member strains and joint displacements.

18. The conclusions drawn from these measurements are as follows:

(i) the structure was homogeneous, this quality being provided during manufacturing and erection, and due to the uniform distribution of loading;

(ii) the joints are perfect spherical hinges, with no rigidity on any of the 3 axes, so that no bending or torsion moments occurred, thus corresponding to the considered design hypothesis;

(iii) the structure had an overall elastic behaviour.

19. In the case of unsymmetrical load, the nodal displacements at all points of the structure (Figs 3-4) were lower than those occurring under symmetrical loading. Therefore, the design loading hypothesis is that providing the loading on the whole roof area.

20. The stresses found in the structural members had values in agreement with those analytically calculated for the bottom layer members and for the boundary bracings, while for the top layer members, the stresses were 30% lower than those calculated.

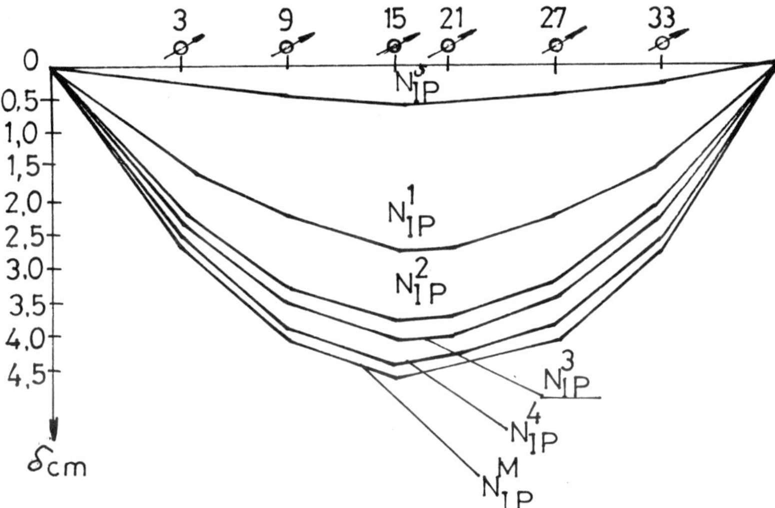

Fig. 2. Displacement of thre joints on the x-x' axis under symmetrical loading.

21. The large difference in the top layer member stresses can be explained by the interaction between the corrugated steel sheeting and the top layer members.

22. The values of the stresses under unsymmetrical loading were lower than those found under the symmetrical case, see Fig. 6.

23. The general conclusion is that the experimentally analysed structure had an adequate behaviour under symmetrical and unsymmetrical loads, remained within in the elastic range under a load exceeding 15% the design load, and showed no tendency towards failure due to local instability.

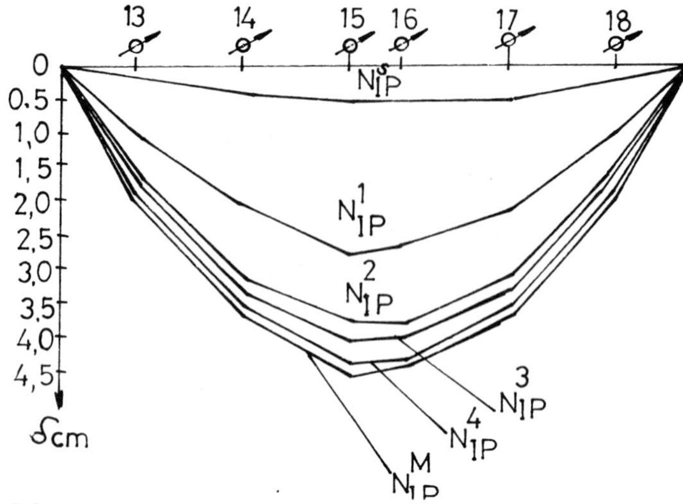

Fig. 3. Displacement of the joints on the y-Y' axis under symmetrical loading.

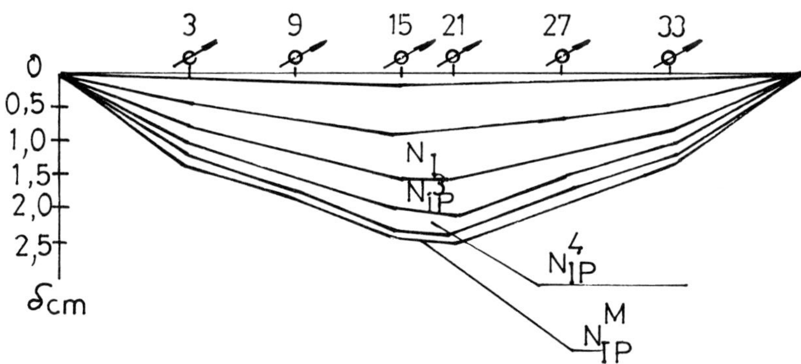

Fig. 4. Displacement of the joints on the the x-x' axis under unsymmetrical loading.

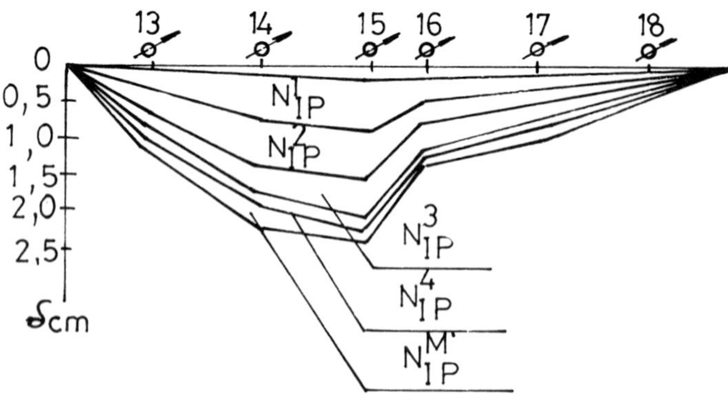

Fig. 5. Displacement of joints on the y-y' axis under unsymmetrical loading.

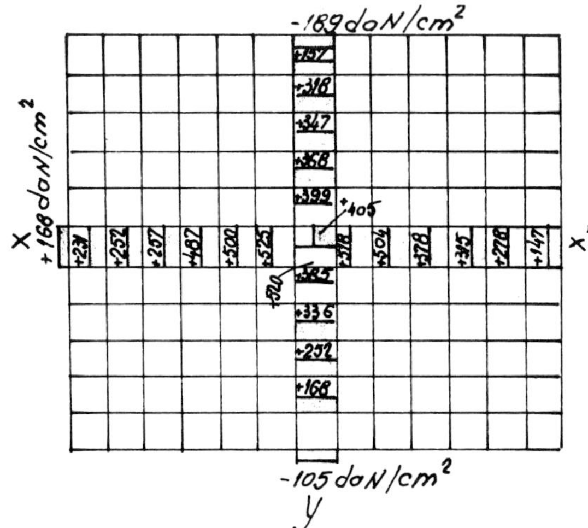

Fig. 6. Graph of stresses in the bottom layer members under symmetrical loading.

24. This experimental analysis of the stresses in the members of a sports hall roof structure contributed to the improvement in the design of this type of structure as well as of the testing methodology.

REFERENCES

1. Techical Standard STAS 1336-80 - Testing Buildings In Situ by Statical Loading, Romania, 1980.
2. BALAN, S. and ARCAN, M., Testing of Constructions. Technical Publishing House, Bucharest, 1965.
3. Service instructions for the AUTOLOG equipment manufactured by PEEKEL Instruments B V, Holland, 1979.

202. Theoretical and experimental study on the structural system of a long-span hangar

XI-LIANG LIU and JIANJUN YE,Tianjin University, P. R. China

SYNOPSIS. The structural plan and design of a long-span hangar has been proposed in this paper. Long-span hangar construction using structural systems such as triple-layer grids, grids with a door frame, cable-stayed grids and grids combined with an arch were analysed theoretically. The economical factors of these structural systems were contrasted. Further more, suggestions on how to deal with long-span hangars constructed as triple-layer grids is given and the feasibility of using concrete filled steel tubes in space structure is also discussed.

INTRODUCTION

1. A large aeroplane (such as a Boeing 747) needs a long-span hangar in which the plane can be maintained and overhauled. The basic function of a maintenance and overhaul hangar is to minimise aircraft down-time which affects the economics of fleet operation. The latter is also affected by the amount of hangar space and related equipment available. Essentially, aircraft down-time depends upon the accessibility of its components and systems to maintenance personal. The degree of accessibility is decisive also for aircraft positioning, tail-in or tail out, as well as for hangar equipment (work platforms, docks, cranes, etc.) and the arrangement of the workshops, spares, stores and offices. The structure surrounding the hangar space must be capable of meeting the hangar's internal functions. Another concept approach is to reduce the hangars internal "free-movement" space to an optimum which meets the requirements.

2. Since the aircraft can be docked only in the direction of its longitudinal axis, there are only two docking principles, tail-in and tail-out. Four basic hangar types can be designed, they are:

standard hangar, tail-in,
standard hangar, nose-in,
compact hangar, tail-in,
compact hangar, nose-in.

The standard hangar is also known as a "box" hangar. The compact hangars space is better adjusted to fit the aircraft and when compared with a conventional "box" hangar, reduces the internal space to approximately half that of the former. Compact hangars reduce building costs and maintenance, which is an important economic factor. But due to the requirements for different aircraft classes or types, it can be assumed that construction and maintenance costs of a hangar depends upon, but are not proportional to, the volume (ref. 1). So the standard hangar (or "box" hangar) is often used.

3. The structural system of a long-span hangar outlined in this paper is a standard hangar (or "box" hangar) which provides an uninterrupted span of 150m; enabling two Boeing 747 jets to be placed side by side. To meet the requirements for different aircraft classes and types, the normal height for long-span hangars is approximately 25m.

STRUCTURES AND MATERIALS

4. With space structures being widely used in the construction of long-span hangars, grid structures become the main structural pattern. According to the characteristic of a hangar, a standard hangar using a grid is always supported along three sides with one side free. In the free side of the grid, the forces and deflections are very large, which is not an ideal boundary condition. So the key to the question is to add support along the free edge and this leads to a more uniform stress distribution. Adding a girder along the free edge to strengthen the grid is a conventional solution. One should mention the two long span hangars, erected at Heathrow Airport, London, covered with prefabricated double-layer grids. In order to add support to the free edge of the double-layer grid, a fascia girder and a spine girder were used in the plan of the BA hangar 01 at Heathrow Airport (ref. 2). However, this kind of structural system has its drawbacks in that a great deal of structural steel was used in the girders, which is approximately forty to sixty percent of the total weight of structural steel. It is necessary to study and exploit structural systems for long-span hangars and improve their economical factors (refs 3-5).

Triple-layer grids

5. It is proposed in the paper to use a triple-layer grid structure for the long-span hangar, and to take full advantage of each material by making use of high tensile steel wire for the tension members and concrete filled steel tubes for the compression members.

6. Since the 1960's, triple-layer or multi-layer grids have been used to strengthen double-layer girds by adding layers at the support or at the free edge of double-layer grids. There are some examples:

(a) The hangar at Klaton airport, Switzerland, is a multi-layer grid which has an area of 128 x 128m, supported at four corners.

(b) Constructed in 1969, the Currigun exhibition pavilion in Colorado is a triple-layer grid which provides a support-free area of 205 x 72m. The pavilion was supported at multi-layer grids points.

7. Comparing the use of double-layer grids for long-span structures, triple-layer grids have the advantage of higher rigidity, less individual member force, fewer types of joint and member sizes resulting in a lower total steel weight, etc. (ref. 4).

Cable-stayed grids

8. Cable-stayed grids use the technique of cable-suspension bridges in pavilion construction. This structural system has the advantage of a novel type, flexible arrangement, using smaller amounts of structural steel and higher resistance to the vibration forces of wind, earthquake or crane, etc. It is proposed in this paper, to use a cable-stayed grid for the long-span hangar, adding supports to the grid by stayed cables.

Concrete filled steel tubes

9. Concrete filled steel tubular structures, exploited in recent years, have both the advantage of a steel and concrete structures in that it has the property of high strength, low weight, good plasticity, high resistance to fatigue and easily to construct. It is widely used for compression members that have a high axial force and small eccentricity (ref. 6). It is proposed in the paper to use concrete filled steel tubes as compression members in the space grids.

STRUCTURAL SYSTEMS AND ECONOMICAL FACTORS

Structural systems

10. In order to compare the economical factors of different structural systems, five layouts of long-span hangars were designed. As shown in Fig. 1. they are:

(a) double-layer square-on-square offset grids;
(b) triple-layer square-on-square offset grids;
(c) double-layer square-on-square offset grids with a door frame;
(d) double-layer square-on-square offset grids combined with a concrete filled steel tube arch;
(e) cable-stayed double-layer square-on-square offset grids.

In each case, the grid covers an area of 150 x 100m with a structural depth of 9m. It has been chosen to have a modular grid of 18 x 12 bays for the double layer grid and 25 x 17 bays for the triple-layer grid. Each of the structures was assumed to be simply supported along three sides on unyielding supports and the average intensity of the distributed loading acting upon the structure over the whole area was 2.5kN per square meter, corresponding to final constraint conditions for dead load.

11. For each case, instead of adding girders, a third or fourth layer of grids was added on the top layer of the double-layer or triple-layer grids at the free edge, which became a local triple-layer grid or four-layer grid near the free edge. From Fig. 1. it may be seen that:

- The roof structure of hangar (1) consists of: A double-layer grid and a local triple-layer grid near the free edge.
- The roof structure of hangar (2) consists of: A triple-layer grid and a local four-layer grid near the free edge.
- As well as a double-layer grid and a local triple-layer grid near the free edge, the roof structure of hangar (3) - (5) also consist of the following parts: Hangar (3): A door frame, supported by the two sides of the door,

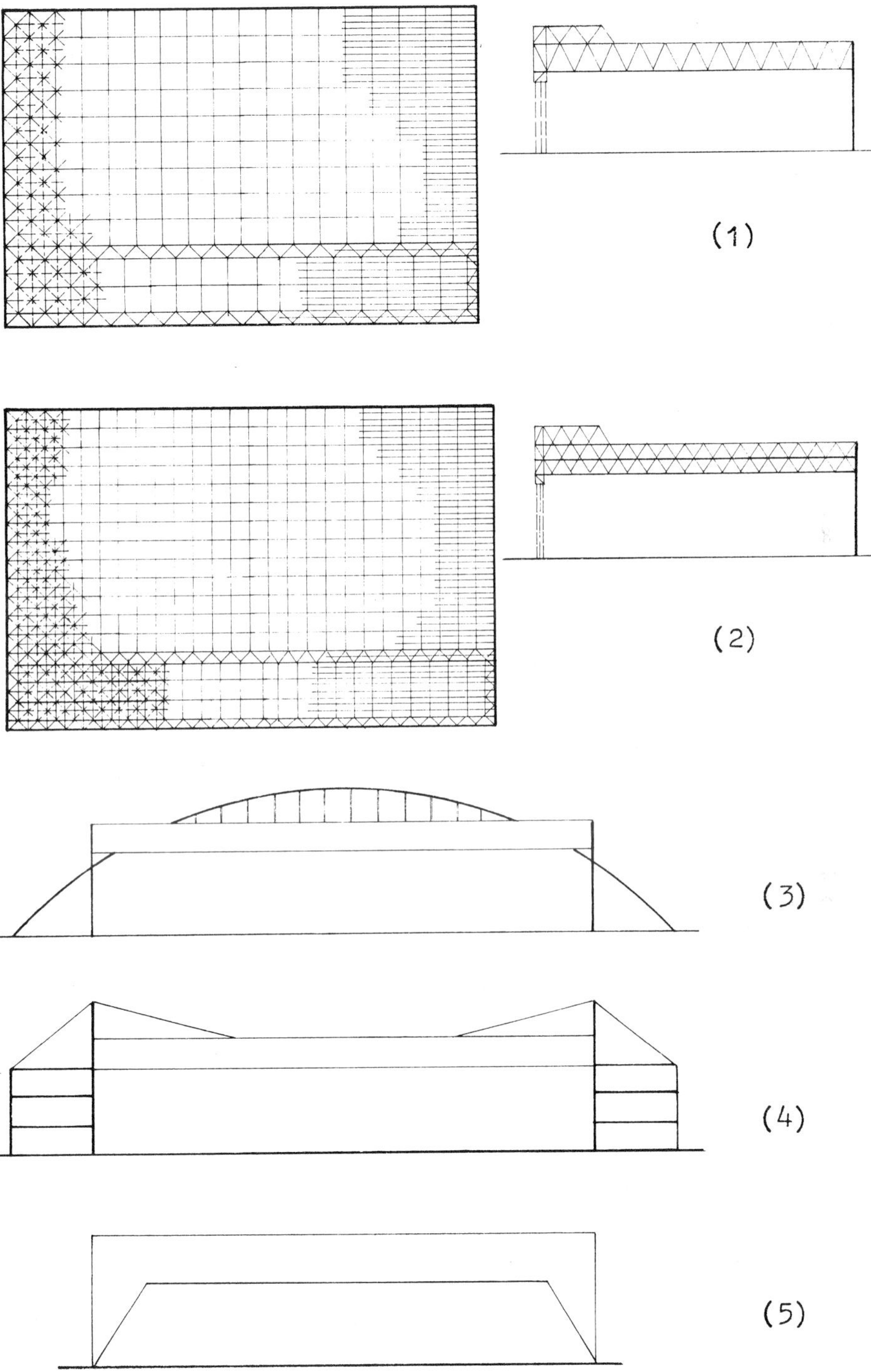

Fig. 1. Five layouts of structural systems.

Hangar (4): A concrete filled steel tube parabola arch, 45m high supported on two concrete foundations. The distance between the two foundations is 202.5m. Several steel bars connect the grid and the arch.
Hangar (5): A few stayed cables which lie between the grids and the 45m high stanchions.

Economical factors

12. The five structures were analysed using a computer and the stiffness method, together with an optimum design technique, assuming that all the bars of the grids are pin-connected, carrying axial forces only. Non-linearity of the stayed cable was taken into consideration (refs 7-9).

Table 1. Economical factors of the five structural systems.

Structural systems	(1)	(2)	(3)	(4)	(5)
Weight of strutural steel(kg/m^2)	175	160	181	138	152
Peak value of tensile force(kN)	5355	3812	2525	2356	4233
Peak value of compression force(kN)	-5674	-5168	-5046	-3268	-5131
Peak value of deflection(cm)	24.4	34.0	41.7	40.2	41.1

13. A careful comparison of the five alternatives leads to the conclusions (shown in table 1):

(a) The structural system of double-layer grids has the simplest structure plan, but it costs more due to the weight of structural steel.

(b) Comparing double-layer grids, with triple-layer grids shows that the latter has members with smaller axial forces giving a lower total weight of structural steel but the total number of members and joints increases.

(c) The structural system of the double-layer grid with the door frame has the greatest reduction in axial forces; but the total weight of structural steel increases slightly.

(d) The structural system of the double-layer grid combined with the concrete filled steel tubular arch has the lowest peak value of axial force, and consumes the least weight of structural steel. However it is difficult to construct the concrete filled steel tubular arch.

(e) The structural system of the cable-stayed double-layer grid consumes the lowest weight of structural steel; but the peak value of axial force is large and the structure of the stanchions are complex. In order to give full play to the stayed cable, the stanchions must be high enough. However the height of the hangar must also be within the height restriction relevant to the airport. Consequently, similar to the double-layer grid combining a concrete filled steel tubular arch, this structural system is also not economically viable.

14. Complying with the conclusions obtained above, in order to be more economical, one may design the following structural systems:

(a) triple-layer grid combined with a door frame;
(b) cable-stayed triple-layer grid;
(c) cable-stayed triple-layer grid combined with a door frame, etc.

Concrete filled steel tubes in space structures

15. In the layout mentioned above, if the compression members were replaced by solid concrete filled steel tubes, the calculation result revealed that the weight of structural steel could be decreased by twenty percent, but in addition the total weight of structure increased greatly. One may reason that the slenderness ratios of members in grids of large modular units are generally bigger than that of small modular units and this leads to stability controlling the design of compression members in grids based on a large module. Since a solid concrete filled tube has a small radius of gyration but an increased cross-sectional area compared with a conventional tube, a composite compression member under the same axial force as a hollow tube, will lead to a weight increase of the structure. It may be clear that if concrete filled steel tubes are used as compression members for grids, it would be most beneficial to use light weight concrete filled steel tubes or centrifuge concrete filled steel tubes and to use grids with small modular units. The use of concrete filled steel tube in space structures needs further consideration.

MODEL EXPERIMENT

16. Supported along three sides with one side free, the grid has a peak value of axial force and deflection in the free side. The value of axial force and deflection decreases gradually from the middle of the free edge to the three supported edges. It is important to analyse the free edge of the grid.

17. An experimental hangar model, which has the structural depth of 0.3m and plan dimensions of 3.6 x 2.4m, was produced using square-on-square offset triple-layer grids. The model was simply supported along three edges on unyielding columns. A door-frame and the fourth layer of the grid was added to the free edge. At the bottom of the door-frame, two tensile bars were used to resist the tensile force. A number of the same cross-section thin-wall steel tubes were used as structural members and welded to hollow balls, used as joints. It is virtually impossible to produce a model of a pin-ended triple-layer grid, but the rigidly connected joints may be manufactured more readily and more accurately. Axial force is the predominant action of members. Fig. 2. shows the layout of the experimental test model.

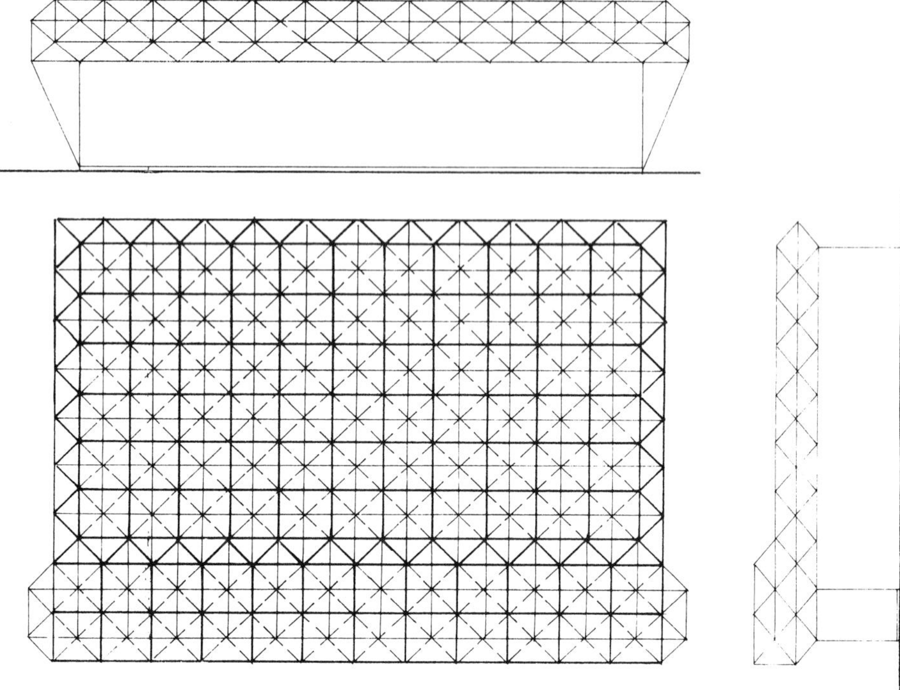

Fig. 2. Experimental model.

18. The model was loaded with equal vertical loads applied incrementally at each joint with an intensity of 1.111kN per square metre per increment. Finally, the intensity of 5.555kN per square metre at the fifth increment of distributed load was obtained. There are forty-four electrical resistance strain gauges used to measure member strains and eight displacements transducers recording joint deflections. Results were obtained experimentally and theoretically assuming pinned joints. Table 2 gives member strains under the fifth load increment and Table 3 gives the deflections of joints also under the fifth load increment. Fig. 3 shows the relative values of axial force and deflection, which decrease gradually from the middle of the free edge to the three supported edges.

19. Table 2 shows that there is reasonable agreement between the theoretical and experimental results. The value of axial force and deflection is slightly higher in the theoretical results than in the experimental results. The comparative error of axial force is smaller in the tensile members compared with the compression members and with the increase of axial force, the comparative error reduces. During the loading and unloading procedures, the difference in the experimental results for the same load increment varies within five percent.

Table 2. The strains (1.0 e-6) of members under the fifth load increment.

Members	Theoretical strains	Experimental strains	Members	Theoretical strains	Experimental strains
1	215	213	23	-280	-194
2	148	144	24	-101	-120
3	231	247	25	-130	-126
4	156	130	26	-10	-20
5	82	70	27	-71	-57
6	71	23	28	-93	-89
7	79	58	29	-108	-120
8	56	50	30	-162	-150
9	12	1	31	-89	-83
10	60	50	32	-261	-164
11	95	77	33	-270	-314
12	50	41	34	-310	-110
13	255	202	35	-217	-199
14	294	300	36	-119	-85
15	278	314	37	-204	-140
16	35	33	38	-94	-37
17	19	6	39	-114	-54
18	267	247	40	-148	-84
19	72	24	41	-282	-283
20	60	10	42	-311	-179
21	-285	-154	43	-280	-181
22	-204	-124	44	-161	-140

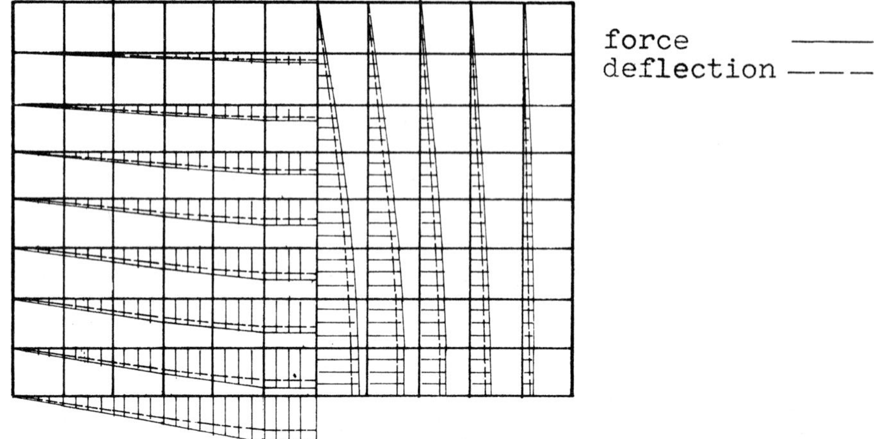

Fig. 3. Relative values of axial force and deflections.

Table 3. The deflection (cm) of joints at the fifth load increment.

Joints	1	2	3	4	5	6	7	8
Experimental deflections	0.76	1.08	1.45	1.69	1.97	1.96	1.53	1.65
Theoretical deflections	0.95	1.38	1.72	1.96	2.11	2.11	1.72	1.38

20. Summary. With the general agreement of the theoretical and experimental results, the theoretical values of axial forces an deflections are greater than the experimental results, indicating that the analysis procedure tends to be conservative.

REFERENCES

1. RUDALPH P., A three-dimensional Concept for multi-form Hangars, airport forum no.4, 1980.
2. MAKOWSKI Z. S., Analysis Design and Construction of Double Layer Grids, Applied Science Publisher Ltd, London, 1981.
3. Space Structure Committee, Bridge and Structure Engineering Institute, China Civil Engineering Institute, Proceeding of 4th Space Structure Academic Exchanging Conference, 31-35, 85-89, 576-580, 1988, Chengdu.
4. Space Structure Committee, Bridge and Structure Engineering Institute, China Civil Engineering Institute, Proceeding of 5th Space Structure Academic Exchanging Conference, 126-1356, 418-424, 527-532, 1990, Lanzhou.
5. Space Structure Committee, Bridge and Structure Engineering Institute, China Civil Engineering Institute, Proceeding of 6th Space Structure Academic Conference, 63-70, 257-261, 657-664, Earth-quake Publishing House, China 1992.
6. SHANTONG Z., Concrete Filled Steel Tube Structure, Heilongjiang People's Publishing House, 1985.
7. WEAVER W., PALD J. R., JOHNSTON R., Finite Elements for Structural Analysis, 1984 by Pretive-Hall, Inc. New Jersey . 07632.
8. XUCHENG JW, MING S., Basjcal Principle of Finite Element and Numerical methods, Qinghua University Publishing House, Beijing, 198.
9. WENLU J., Calculation of Suspended Structures, China Building Industry Publishing House, 1975.

203. Static model tests on long-span twin-fan-shaped space frames for retractable roofs

A. HORI, Hazama Corporation, Japan, A. WADA, Tokyo Institute of Technology, Japan, and Y. MAKINO, Kumamoto University, Japan

SYNOPSIS. Static loading tests on 1/100 scale models of a raised double-layered space truss of complicated-shape have been carried out. The truss has a twin-fan-shaped plan form, because it was designed as one of four retractable roofs to cover a circular space. Members were modelled as 5.0-2.0mm diameter copper pipes. Two specimens were tested under either asymmetrical vertical load or span-directional horizontal load. The main conclusions are that buckling length of a chord in a rigid jointed truss can be evaluated as that of two continuous chords, and that the macro-behaviour of these complicated trusses may be explained by a simple model.

INTRODUCTION

1. Many long-span structures have been erected in recent years, and the double-layered and raised space truss is a very effective form for spans of 100-200m. Experimental studies on double-layered space trusses have been reported by Schmidt (ref. 1), Collins (ref. 2) and many other researchers. However for the raised case, very few tests have been carried out. This paper reports on the elasto-plastic static loading tests on 1/100 scale models of a raised double-layered space truss of complex shape, which is designed for one of the retractable roofs in a proposed domed stadium (ref. 3). Specimens have a twin-fan-shaped plan form and one longitudinal general arch in each. The number of members and of translational freedoms in a specimen is about 2190 and 1890, respectively. Two specimens were examined under either asymmetrical vertical load or span-transverse horizontal uniform load.

SPECIMENS

2. Shape of specimens. Overall dimensions of the 1/100 scale specimens are shown in Fig. 1. They have a 2200mm length, about 850mm width and 450mm height, 32.5mm depth and a rise-span ratio of approximately 0.20. Fig. 2 shows the framed shape of the specimens. It consists of main-trusses (left half of Fig. 2 (a)), which contain two lower chords and one upper chord in each, and secondary members (right half of the figure). The number of main-trusses decreases toward

the top. Plans of upper chords, lower chords and diagonals are shown in the left half of Fig. 2(b)-(d), respectively. Diagonals between main-trusses are located basically at intervals of one row.

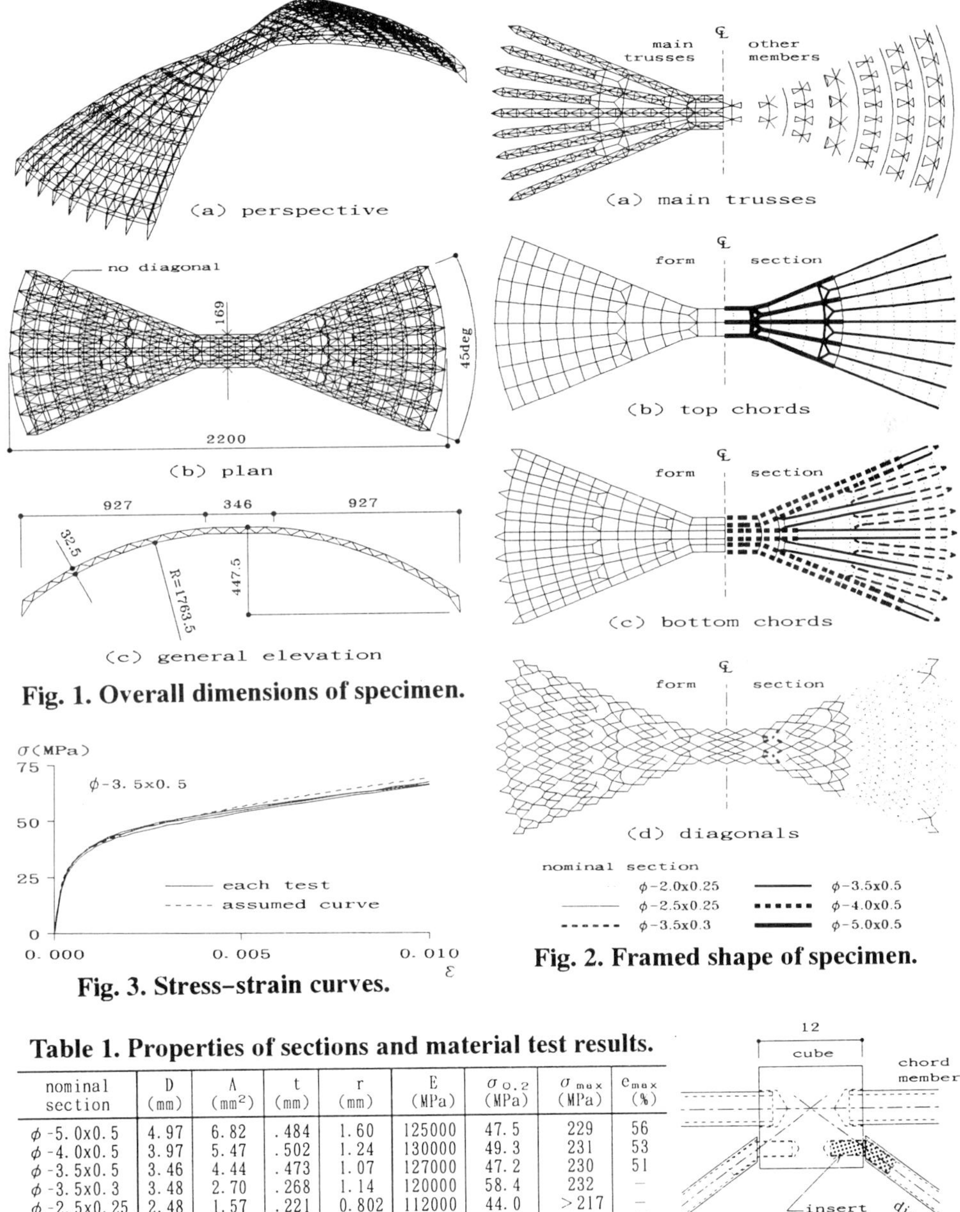

Fig. 1. Overall dimensions of specimen.

Fig. 2. Framed shape of specimen.

Fig. 3. Stress-strain curves.

Table 1. Properties of sections and material test results.

nominal section	D (mm)	A (mm²)	t (mm)	r (mm)	E (MPa)	$\sigma_{0.2}$ (MPa)	σ_{max} (MPa)	e_{max} (%)
φ-5.0x0.5	4.97	6.82	.484	1.60	125000	47.5	229	56
φ-4.0x0.5	3.97	5.47	.502	1.24	130000	49.3	231	53
φ-3.5x0.5	3.46	4.44	.473	1.07	127000	47.2	230	51
φ-3.5x0.3	3.48	2.70	.268	1.14	120000	58.4	232	–
φ-2.5x0.25	2.48	1.57	.221	0.802	112000	44.0	>217	–
φ-2.0x0.25	1.97	1.30	.238	0.617	117000	56.3	241	44

D: diameter A: area t: thickness (calculated from D and A)
r: radius of gyration E: young modulas $\sigma_{0.2}$: 0.2% offset strength
σ_{max}: maximum strength e_{max}: elongation

12
cube
chord member
insert
diagonal member

Fig. 4. Connections.

3. Members. Ready-made copper pipes were annealed and used for members. Sections were selected among the six sizes shown in Table 1, to maintain the original slenderness ratio between nodes as closely as possible. The size of each member is shown in the right half of Figs 2(b)-(d) for upper chords, lower chords and diagonals, respectively. Their sizes increase toward the top. Material test results of the annealed pipes are shown in the right hand columns of Table 1, and the stress-strain curve is shown in Fig. 3. Offset strength at 0.2% is 44-58MPa and hardening is greater than for structural steel.

4. Fabrications. Connections were replaced by 12mm regular cubes of copper. Most of the chord ends were brazed into machine-made holes in the cubes. Details are shown in Fig. 4. All diagonals and the rest of the chords, which join obliquely to cube-faces, are brazed through L-shaped-inserts as shown in Fig. 4. The support condition at the specimen base is fixed. The total weight of one specimen is about 12kg. The members were located and assembled on a twin-fan-shaped steel having high accuracy, bolting most of the cubes.

BUCKLING STRENGTH OF CHORDS

5. Chord tests. Axial force-displacement relations for single chords were examined for each section. Chord specimens have 86.5mm length for the standard lower chord and pinned ends (Fig. 5). Test results are shown by the solid line in Fig. 5. The strength of each section at first buckling exceeds the 0.2% offset value. The broken line in the figure shows the axial force calculated from experimental strains at the specimen centre, where damage is most-concentrated. For this calculation, a stress-strain model is extended from the Ramberg-Osgood Model as shown in Fig. 6. Furthermore, constant volume is assumed and stress is multiplied by 1/(1+strain). The calculated force (broken line in Fig. 5) agrees very well with the experimental results. Therefore, the axial force of the member is well-approximated from the elasto-plastic experimental strains under an appropriate stress-strain assumption.

6. Analysis. Results of a buckling analysis using fibres with the stress-strain model of Fig. 6 are shown by the dotted line in Fig. 5. It agrees well with the experimental results.

7. Buckling forms and strength. Ignoring rotational constraint by diagonals at chord ends, chord buckling can be modelled as 1:pinned-pinned, 2:tripled, 3:pinned-fixed or 4:fixed-fixed, as shown in Fig. 7(a). However, the difference between 2 and 3 above turns out to be small on the analytical axial force-displacement curves under monotonic compression for all section sizes. Therefore, by adding diagonals as rotational elasto-plastic spring at the chord ends, two further models were analysed, 5:springed-springed and 6:springed-fixed. Their axial force-displacement curves are shown by the solid and roughly-broken lines in Fig. 7(b), respectively. Buckling strength is significantly increased by rotational constraint of diagonals.

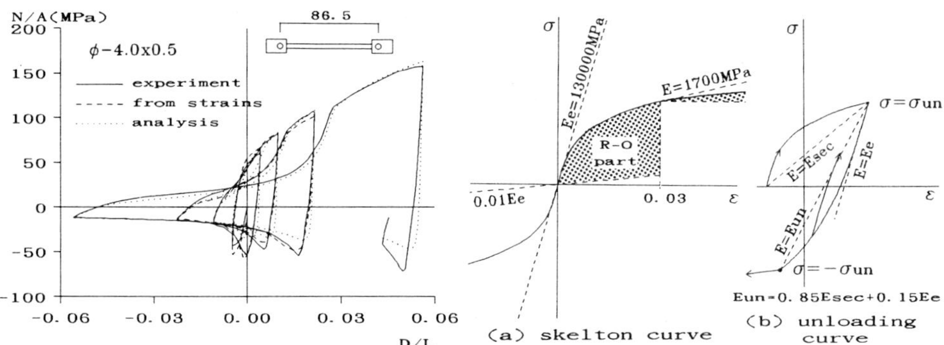

Fig. 5. Chord tests and analysis. **Fig. 6. Extended stress-strain model.**

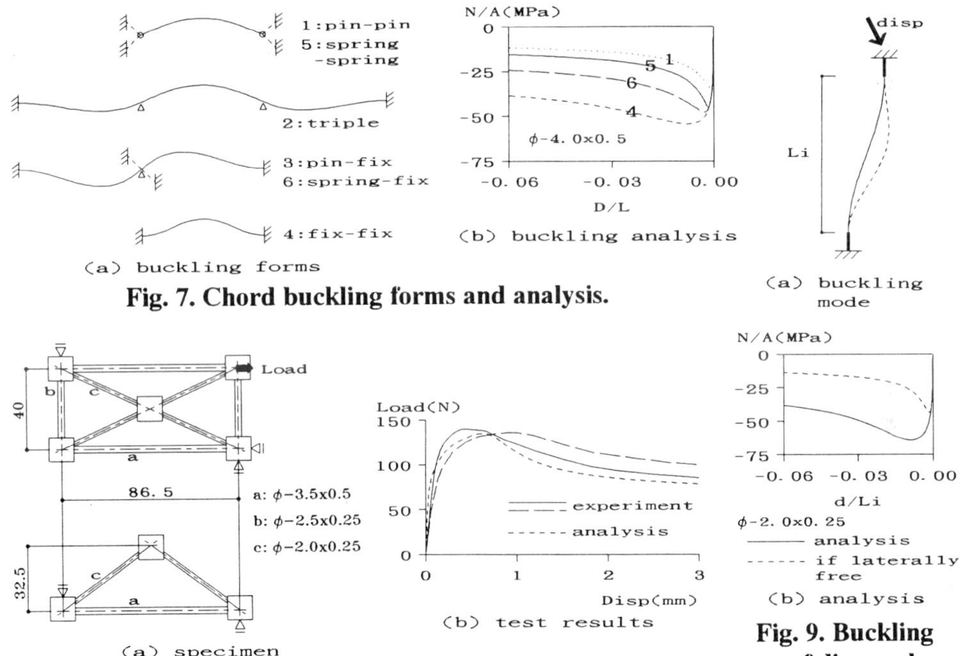

Fig. 7. Chord buckling forms and analysis.

Fig. 8. Pyramid tests.

Fig. 9. Buckling of diagonals.

BUCKLING STRENGTH OF DIAGONALS

8. Pyramid tests. The buckling strength of diagonals was examined by pyramid specimens (Fig. 8(a)) which correspond to one of the typical parts of the twin-fan-shaped specimen. Three specimens are loaded monotonically at one of the cubes in a pyramidal base, to put two opposite diagonals in compression and the others in tension. Load-displacement curves are shown in Fig. 8(b). The solid and broken lines correspond to typical and another type whose buckling waves are observed at one or two diagonals, respectively. Plastic hinges due to buckling formed at the inner ends of the L-shaped -inserts.

9. Forms and strength of diagonals. If both ends of the buckled diagonals are fixed rotationally and changes of other member lengths are small, buckling is modelled under relative translational displacement, which is oblique to the member axis as shown in Fig. 9(a). However, their analytical axial force-displacement curves are the same as the non-oblique results (one solid line in Fig. 9(b) shows both results). Therefore, diagonal buckling can be evaluated as if both ends are fully fixed. The load-displacement curve of the pyramid is calculated using the analytical curve of a diagonal, and shown by the dotted line in Fig. 8(b). It agrees well with typical experimental results.

ASYMMETRICAL VERTICAL LOAD TEST

10. Earthquake and wind loads are severe in some countries. Longitudinal earthquake excitation to an arch causes asymmetrical vertical inertia forces, which were severe in this design (ref. 3), because horizontal displacement at the arch top brings vertical displacement anti-symmetrically to the right and left halves.

11. Loading and measurements. A twin-fan-shaped specimen was tested under simplified asymmetrical load that is a downward load on half of its arch. Using the tournament method, 28 cubes on the lower chord surface of a half side were pulled proportionally to the dead load in design. Fig. 10(a) shows the tournament, which has a loading screw as well as a load-cell. Centripetal displacements of the spherical-shell-like specimen were measured at 22 cubes shown by blacked circles in Fig. 10(b). Horizontal values were measured at 14 cubes shown by arrows indicating their directions. Strains were measured with 3 strain-gauges for each section at the centres of 47 members, shown by bold lines in the figure.

12. Progress. Fig. 11(a) shows heavy or moderate damage after testing. Damage progress is as follows. (a) No damage is observed before the maximum strength. (b) After strength falls slightly, continued buckling of two or three chords is observed at mid-height in the loaded half. (c) Diagonals adjacent to the buckled chords are clearly bent. Chord buckling is newly observed at upper and middle height in the other half, and damage of chords and diagonals is observed just above both bases. This state means that four or more macroscopic hinges form along the overall arch, leading to macroscopic instability. Strength at this time is 90% of the maximum. (d) Buckling and damage becomes heavier and the test finishes after the maximum displacement reaches 5% of the span. The total number of damaged members was up to 12% of the total. Typical buckling of the

upper chord is out-of-plane as shown in Photo 1(a), though some buckled in-plane (Photo 1(b)).

13. Results. Relations of load and centripetal displacements are shown in Fig. 12. The maximum strength is 839N. These curves display very smooth shapes, slow strength reduction and asymmetrical deformation of the arch, which is convex downward at the loaded half and upward at the other half. In the arch-transverse direction, centripetal displacements are almost the same and show no remarkable asymmetrical deformation. Bending moment of the centre main-truss calculated from the chord strains is shown in Fig. 13. Bending strength at the most damaged part decreases in the final state.

14. Maximum strength by simple analysis. Modelling a specimen as a planer pinned truss as shown in Fig. 11(b), maximum strength is calculated using the springed-fixed strength of §7. Analytical buckled members are shown in the figure and the overall strength is 92% of the experimental value. These indicate good agreement with experimental results. Therefore, strength and macroscopic damage are also well estimated by a simple model which simulates the collapse mechanism well, as long as the member strength is properly evaluated.

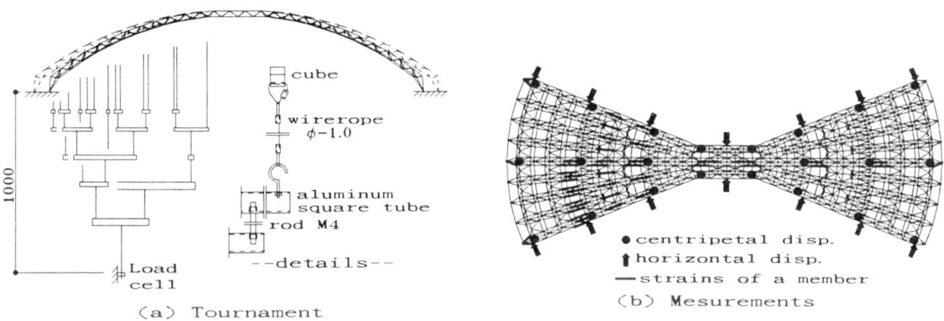

Fig. 10. Loading and measurements of asymmetrical vertical load test.

15. Considerations. The above results are summarised as follows for buckling length: (a) Buckling of continuous chords occurred first. (b) The number of buckling continuation was equal or less than 3, because of the gradient of the axial force. (c) The strength of three continuous chords is nearly equal to that of two. (d) The overall strength is explained by member strength of two chords with rotational constraint by diagonals. From (a)-(d) above, the buckling length of a chord in a rigidly-jointed truss can be evaluated as that of two continuous chords, if preceded by elasto-plastic chord buckling. Ignoring rotational constraint in the case of two equalised chords, it is simplified as 0.7 times their nodal length.

TRANSVERSE HORIZONTAL LOAD TEST

16. A specimen was tested cyclically under span-transverse horizontal load, which is the most severe in this design and makes the behaviour more complicated. Because distribution of the maximum inertia force of the earthquake response analysis in the design was approximately uniform span-directionally, the experimental distribution is also set to be uniform.

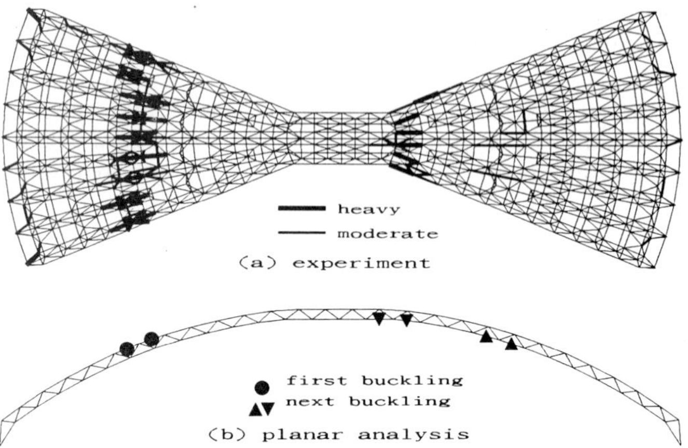

Fig. 11. Damage of asymmetrical vertical load test.

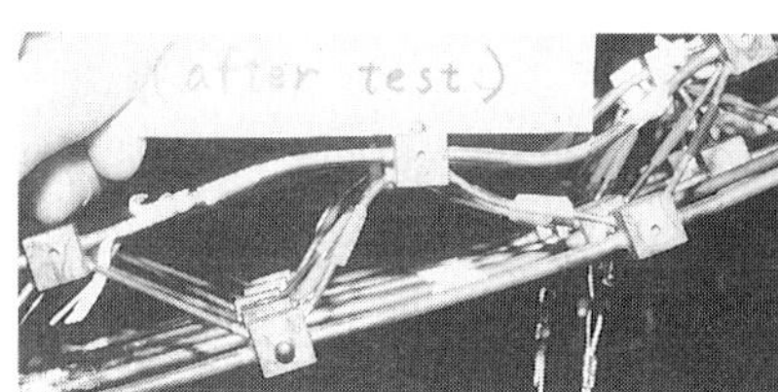

(a) typical one

(b) fewer type

Photo 1. Buckled upper chords.

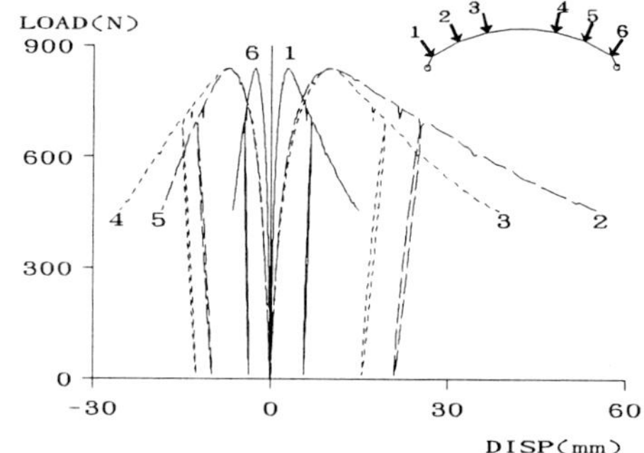

Fig. 12. Load–displacement curves.

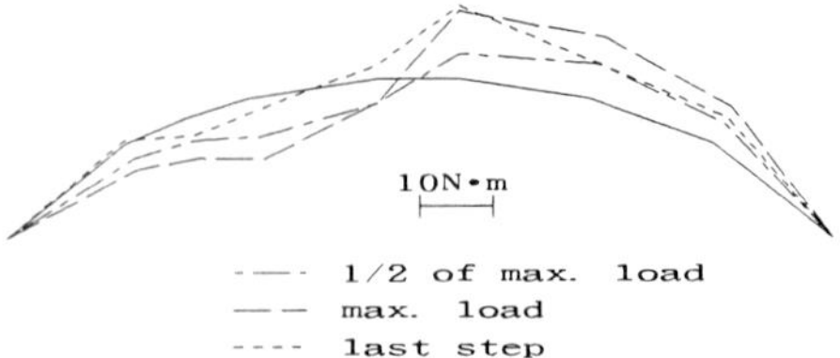

Fig. 13. Bending moment of *main–truss*.

17. Loading and measurements. Fig. 14(a) shows a plan of the tournaments. One tournament is used for loading and the other is used as a counter-weight. The total load is distributed to 16 cubes on the front surface of the specimen. A view of the test is shown in Photo 2. Cyclic loading is controlled by top displacement divided by span with a target of plus-minus 1/1000, 1/500, 1/200 or 1/100, for 1st-4th cycles, respectively. Final loading is carried positive. Centripetal displacements are measured at 22 cubes shown by blacked circles in Fig. 14(b) and horizontal displacements at 22 cubes shown by arrows. Strains are measured at 49 members shown by bold lines in the figure.

18. Progress. Fig. 15 shows major damage at positive cycles and that near the top at final loading. Damage progress is as follows: (a) No damage is observed in the first two cycles. (b) Buckling of diagonals is observed at 1/200 cycle in most of the 16 pyramidal portions, which are between main-trusses and just above the bases. They are considered as a result of the in-plane shear like pyramid-tests. Damage is also observed at lower chords on both bases in the front half of loading direction. (c) They spread to above the bases and to the other half at 1/100 cycles. Upper chords are damaged in the other half above the bases. Photo 3(a) shows a view around a base. (d) The above damage becomes heavier, and overall sinking is observed at final loading. Eventually, slight strength reduction is caused with upper and lower chord buckling near the top. Photo 3(b) shows an overall final view. The number of damaged members was up to 17% of the total.

19. Results. The load vs. top displacement curve is shown in Fig. 16. The maximum strength up to 1/200 cycle is plus 1160N and minus 1190N, values almost the same as the overall maximum. Strength reduction after diagonal buckling is not clear, and the shape of hysteresis loops is energy-dissipative. Overall deformed shape after test is shown by the broken line in Fig. 17, with the

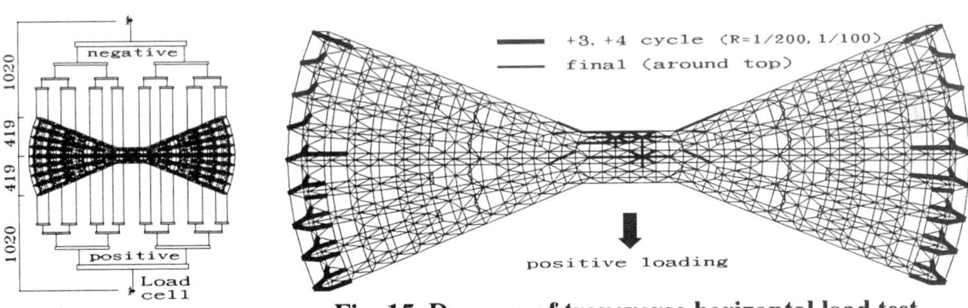

Fig. 15. Damage of transverse horizontal load test.

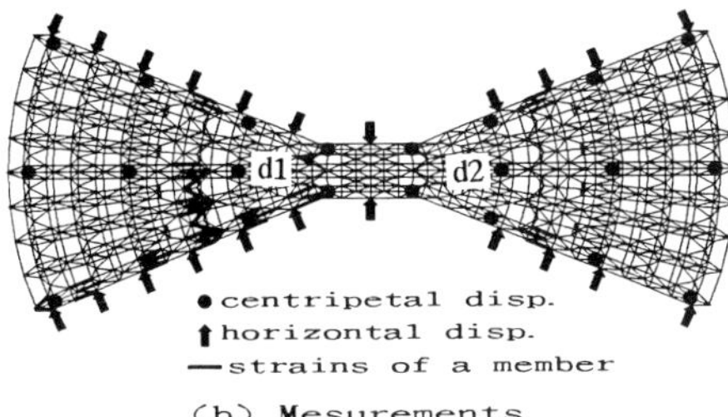

Fig. 14. Loading and measurements.

Photo 2. View of test.

solid line showing the initial shape. It is almost symmetric to the right and left. The sinking is visible and the lower parts of the front side project outward. Fig. 18 shows relations of centripetal displacements (dl and d2 in Fig. 14(b)) and horizontal displacements near the top. Sinking is accumulated by cyclic loading.

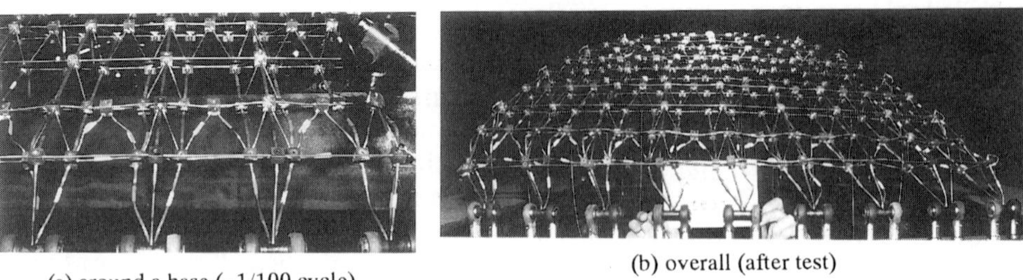

(a) around a base (−1/100 cycle)

(b) overall (after test)

Photo 3. View of damage.

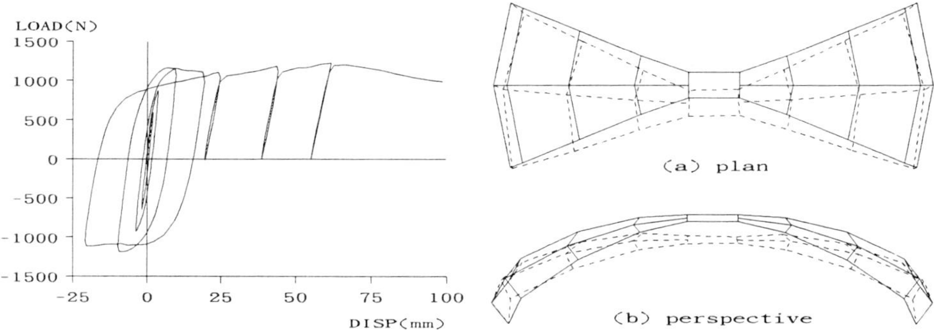

Fig. 16. Load–displacement curve.

Fig. 17. Deformed shape (scale=1).

20. Macroscopic partial model. The portion above a base is modelled macroscopically as follows. Out-of-plane curvature is ignored and the X-shaped diagonals and two adjacent lower chords are replaced by one beam initially. Thus, the portion is modelled as a planar open-frame as shown in Fig. 19(a). The frame is rectangulated next and forces are considered as shown in Fig. 19(b). The bending moment of the truss-beam is also shown in the figure. It explains the reversed shear of the strain-gauged diagonals in the trussbeam-trussbeam connections.

21. Macroscopic strength. Only the calculated results are reported herein. The overall collapse mechanism strength of the partial model is 1530N, which relates to strength of secondary diagonals. On the other hand, local mechanism strength just above a base does not effect diagonal strength and is 1220N, which is almost the same as the experimental values of 1160N and 1120N. Therefore, it is considered that observed diagonal buckling hardly affects the maximum strength. In observations, chord damage was heaviest at the lower chords on bases. In

addition, the overall strength reduction after diagonal buckling is not clear and is different from that of the pyramid tests. Consequently, the experimental mechanism is considered to be macroscopic bending-yield of the main-trusses on the bases, though the first major damage is to the secondary diagonals.

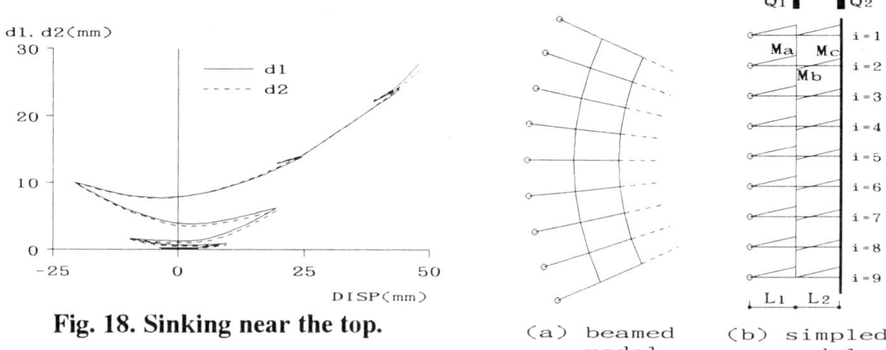

Fig. 18. Sinking near the top.

Fig. 19. Macroscopic partial model.

CONCLUSIONS

22. The following conclusions are obtained through this study. (a) A one hundredth scaled frame test is suitable and enables elasto-plastic tests for complicated complete structures. (b) Elasto-plastic member forces are calculated from experimental strains under an appropriate stress-strain model. (c) Buckling strength of chords is increased by rotational constraint from diagonals. (d) An asymmetrical vertical load test for a twin-fan-shaped specimen shows strength reduction under macroscopic instability. Strength and damaged positions in a simple planar analysis agree well with experimental results. (e) Buckling length of a chord in a rigid-jointed truss can be evaluated as that of two continuous chords, if preceded by elasto-plastic chord buckling. Ignoring rotational constraint in the case of two equal-sized chords, it is simplified as 0.7 times their nodal length. (f) A transverse horizontal load test shows diagonal buckling first and chord damage later, though the former is considered not to concern the overall collapse mechanism. Strength of the macroscopic model of the most damaged part is well matched with experimental results. (g) Macro-behaviour of a raised and double-layered space truss of complex shape may be explained by a simple model.

23. Direct comparison between design and experiment became difficult because of the copper-made specimens. This needs to be investigated analytically.

REFERENCES

1. SCHMIDT L.C., MORGAN P.R., et al. Space trusses with brittle-type strut buckling. Jour. of the Struc. Div., ASCE, 1976, vol. 102, July, 1479-1492.
2. COLLINS I.M., An investigation into the collapse behaviour of double layer grids. Proc. of the 3rd Int. Conf. on Space Struc., 1984, 400-405.
3. HORI A., et al. A structural design of a sports dome which has four retractable roofs. Annual Meeting, Arch. Inst. of Japan (AIJ), 1990, 1337-1342, (in Japanese).
4. HORI A., WADA A., et al. Static test of a model of a long-span double-layer space truss under unsymmetrical vertical load. Jour. of AIJ, 1992, June, (in Japanese).
5. HORI A., WADA A., et al. Static test of a model of a long-span double-layer space truss under transverse directional horizontal load. J. of AIJ, 1993, Feb., (in Japanese).

204. An experimental study of plane tubular trusses with two different welded joints

J. FILIPKOWSKI and A. KRYSZTOFORSKI, Technical University at Zozalin, Poland

SYNOPSIS. An experimental study is described to assess the ultimate strength of two kinds of plane rectangular hollow section trusses. The trusses are different only in the form of welded truss joints. In the first truss the bracing members are connected directly to the chord face. In the second truss the welded joints are strengthened by some inner construction. The complete truss tests indicate that the first truss is little serviceable because of large plastic deformation which takes place in the support joints.

INTRODUCTION

1. There has recently been considerable increase in the use of steel tubes for structural members. This is not surprising since the hollow sections - circular, square or rectangular - differ positively in many aspects in their physical properties from the orthodox rolled sections. Structures, particularly trusses, made of hollow section members are more pleasing and lighter than conventional ones and they have a higher corrosion resistance.

2. The main problem in the design and in the prefabrication of tubular trusses is the connections between tubular members. In the case of three-dimensional trusses this problem was solved by the development of several excellent connectors e.g. Mero and Nodus joints. As for two-dimensional trusses where fabrication of the whole structure would be carried out under factory conditions the joints are solved by direct welding, providing a cheap solution. Particularly thus made joints are not complicated when trusses are composed of square or rectangular hollow section tubes, since the cutting of the web members and welding is quite simple.

3. The behaviour of statically loaded typical rectangular hollow section welded truss joints is quite good. Several theoretical and experimental investigations are presented in the literature (refs 1-5). Analysing this type of connection one may conclude that all of them work poorly as the transmission of forces from one member to another is not very efficient. Besides, when web members are welded to one side of the chord members, as with the typical solution, they cause significant deflections in the connecting wall of the chord members and in

consequence the trusses are characterised by fairly large deformations. To prevent all these negative features occurring one has to reinforce the joints by applying a suitable inner construction.

4. The paper presents some inner reinforced joint constructions and experimental results of investigations of plane tubular trusses with different rectangular hollow section welded truss joints.

DESCRIPTION OF TRUSS MODELS AND MEASUREMENT TECHNIQUES

5. To get reliable experimental results the investigated plane tubular trusses were fairly large. The axial dimensions of these trusses are presented in Fig. 1.

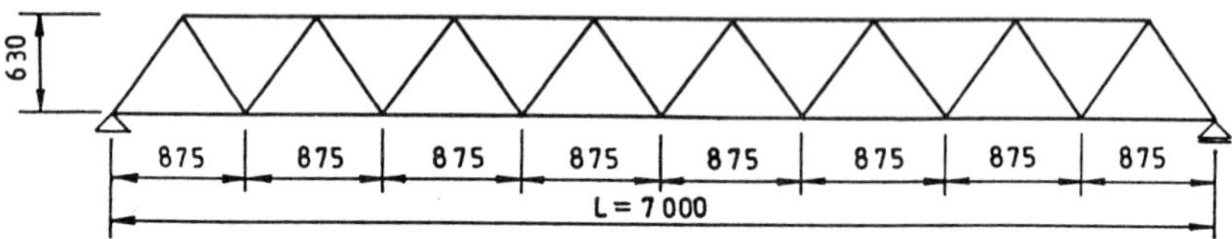

Fig. 1. Axial dimensions of investigated trusses.

The chords and bracing members were made from steel rectangular hollow section tubes and steel square hollow section tubes with cross-sectional dimensions of 80x60x3.5 and 40x40x3, respectively.

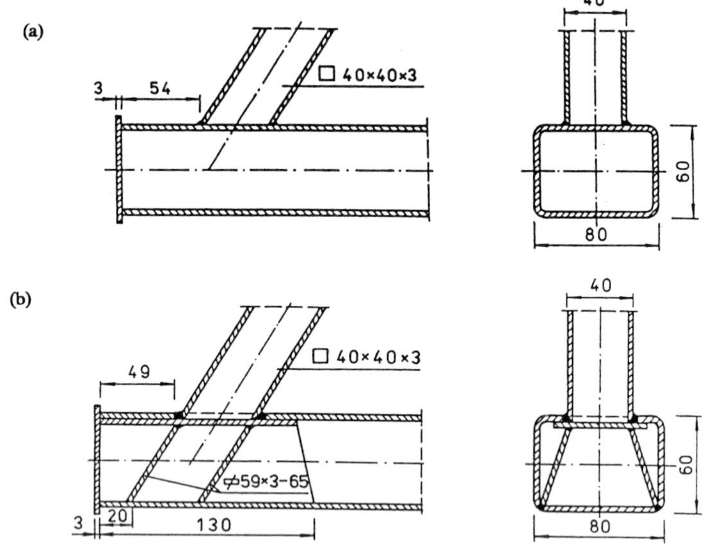

Fig. 2. Section and cross-section of supported joint; (a) joint without inner reinforced construction, (b) joint with inner reinforced construction.

One of the truss models has typical joints where the bracing members were welded directly to the faces of the chords. The second model has reinforced joints which differ only, from the first ones, by their inner construction (Fig. 2 and 3).

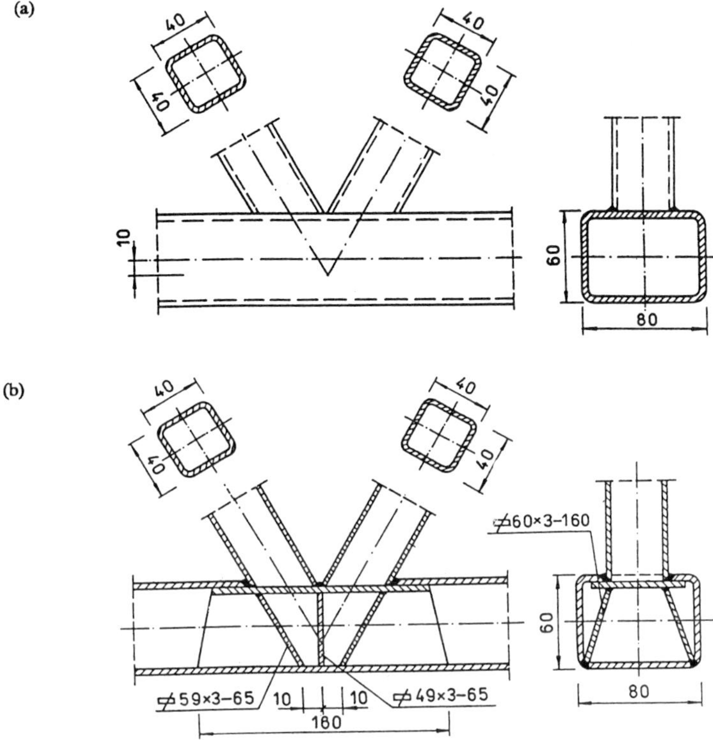

Fig. 3. Section and cross-sections of truss joints; (a) joint without inner reinforced construction, (b) joint with inner reinforced construction.

6. The sketch of the testing rig, with one truss installed thereon, is shown in Fig. 4. As one can see, the trusses were loaded by concentrated forces acting on truss joints. The forces were caused by a hydraulic cylinder through the traverse beams. To check independently the magnitudes of reactions dynamometers were installed. The top chord of the investigated truss had a horizontal flexible plane truss which prevented the chord from buckling.

7. During the investigation the vertical displacement of all truss joints and their deformations were measured with the help of dial gauges. The investigation was carried out for four forces acting on the truss joints, namely, for P = 5.5, 10.5, 15.5 and 20.5 kN. The extreme forces (P = 20.5 kN) caused normal stresses in the chords of 242 MPa.

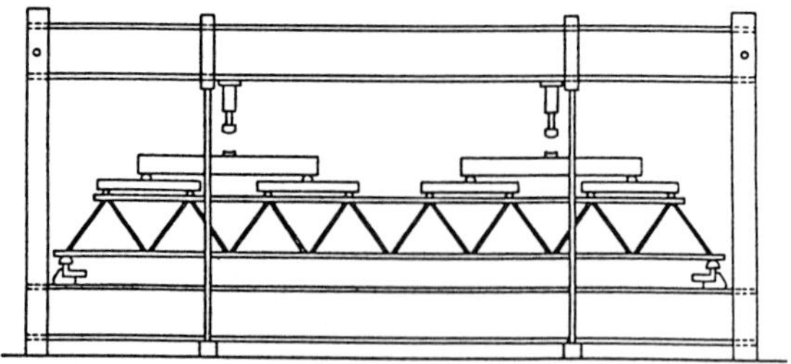

Fig. 4. Sketch of the testing rig.

RESULTS AND SOME MODIFICATION

8. The behaviour and load capacity of the investigated trusses are described, in the best way, by the vertical displacements of the central joints. According to the Polish Code the deflection of a truss should not be greater than fa=250^{-1} (ref. 1.). Fig. 5 presents these deflections for both trusses during the process of loading and unloading.

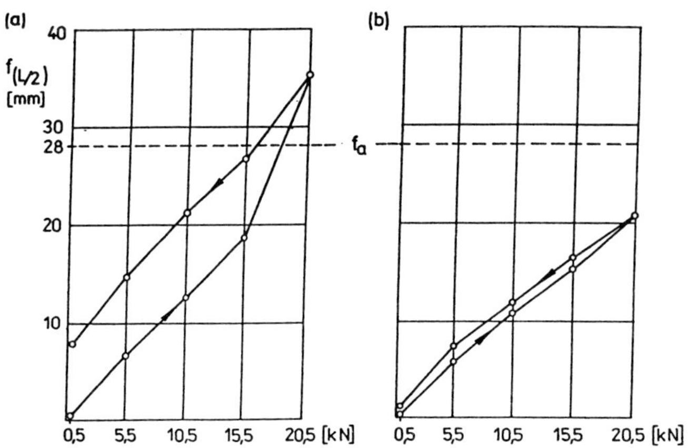

Fig. 5. Deflections of investigated trusses; (a) for the truss without inner reinforced construction, (b) for the truss with inner reinforced construction.

9. Fig. 5 clearly indicates the difference in the load capacity of the investigated trusses. The relatively large permanent deflection in the truss without inner reinforced construction is mainly caused by the deformation of the supported joints (Fig. 6).

10. Taking into account the difficulty in industrial prefabrication of trusses with the proposed inner reinforced construction of the joints, the third truss was investigated which had modified reinforced joints as shown in Fig. 7.

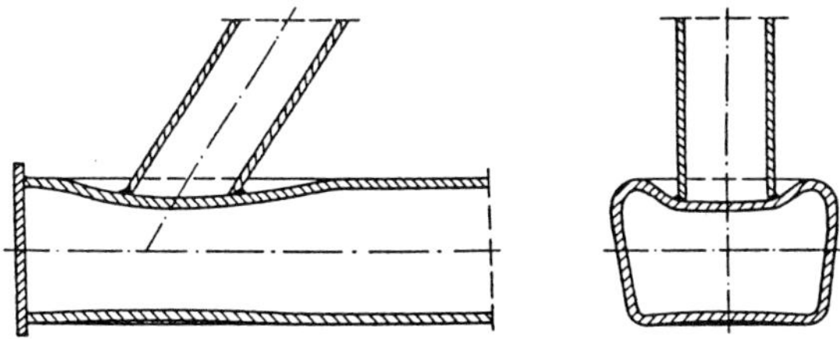

Fig. 6. Section and cross-section of the deformed supported joint.

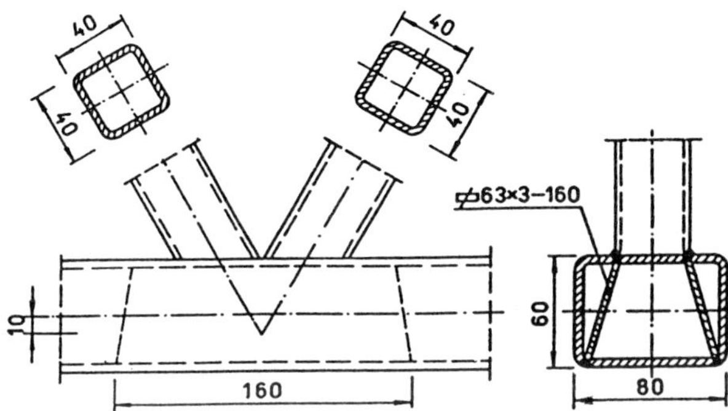

Fig. 7. Modified reinforced joints of the truss.

An analogical investigation carried out on the trusses described above, did not give any essential difference. This truss had nearly the same deflection in the range of the test loading as the trusses with reinforced joints shown in Fig. 2b and Fig. 3b.

CONCLUSION

11. The tests presented in the paper indicate that trusses made of rectangular and square hollow section tubes without inner reinforcement give serviceablility problems because of the large plastic deformation which takes place mainly in the support joints.

REFERENCES

1. CZECHOWSKI A. and BRÓDKA J. Etude de la resistance des assemblages soude's en croix deprofils creux rectangulaires. Construction Metallique, 1977, 3, 17-25.
2. DAVES G. and ROPER C. G. Gap joints with tubular members - a yield line approach. Building Science, 10, 1975, 199-205.
3. DAVIES G. and PACKER J. A. Analysis of wep crippling in a rectangular hollow section. Proc. Instn. Civ. Engrs, Part 2, 1987, 83, Dec., 785-798.

4. EASTWOOD W. and WOOD A. A. The static strength of welded joints in structural hollow sections. Construction Steelwork, January 1971, 6-20.
5. PACKER J. A. Developments in the desing of welded HSS truss joints with RHS chords. Can. J. Civ. Engng., 1983, 10, Mar. No. 1, 92-103.

205. The V-system braced barrel vault

N. BALUT, Building Research Institute, Romania

SYNOPSIS. The paper presents a braced barrel vault system which gives the possibility to attenuate the non-linear character of structural behaviour and to improve stability. This can be achieved by increasing the depth of the component trusses and consequently the angle between two adjoining trusses. In order to maintain the member lengths within reasonable limits, an adequate bracing system should be adopted. The V-system barrel vault meets the above mentioned requirements. Moreover, the I-section upright members and purlins provide an additional rigidity. On the other hand, the proposed construction system (mainly consisting of purlins and shop-welded triangles, with bolted site connections), allows for simple manufacturing and erection.

INTRODUCTION

1. A single-layer braced barrel vault is actually a folded (prismatic) shell-like reticulated structure, consisting of several plane trusses. The nodes lay on the surface of a circular cylinder. In the case of a roof, the vault shape is very different from the funicular polygon of the applied loads, especially when due to un-symmetrical snow loads. Therefore, it is necessary to have fully triangulated trusses if the structure is meant to carry the loads mainly by axial forces (refs 1 and 2) .

2. Causes and effects of non-linear behaviour. A simplified analysis could be carried out by decomposing the applied loads (Fig. la) and then considering each truss separately. However, this model is very often unacceptable. The very small angles between the adjoining trusses are the principal cause of non-linear behaviour. The effect of the radial load components P_r is strongly influenced by these angles (practically the f/a ratio, Fig. lb), and is increased by the radial deformation w, since equilibrium must take place in the deformed position. Any imperfection f_i (i.e. deviation from the theoretical rise f) directed inwards would intensify the non-linear behaviour of the structure.

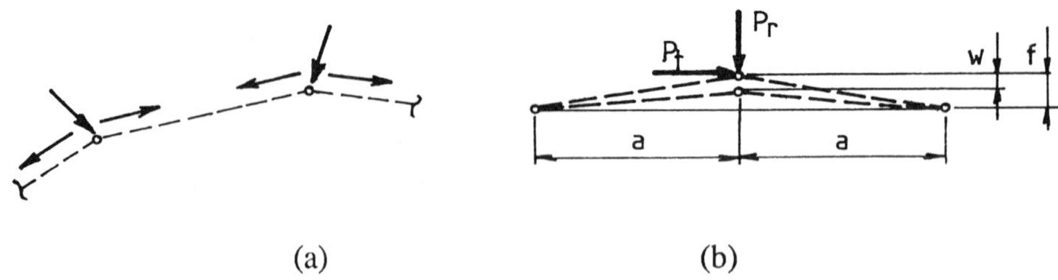

(a) (b)

Fig. l. The effect of loads on the component trusses.

3. The role of structure geometry. The effects of non-linear behaviour are: the forces in members and connections are greater than those predicted by a linear analysis; instability phenomena can occur. These aspects are reviewed in detail elsewhere (refs 3-5). Instability phenomena can be classified into two categories: those characterised by node displacements (node instability, line instability and general instability) and the instability characterised by the individual buckling of component members. Both categories of instabilities occur in practice by equilibrium limitation. In the case of the first category, the critical load has two components, one which is resisted by axial forces (corresponding to membrane stresses in continuous shells) and one which is resisted by bending of the members. Although the relationships derived in ref. 3 are based on a simplified model and are not meant to be a substitute for a proper non-linear analysis, it can be concluded that the critical load component resisted by axial forces depends essentially on the parameters $(f/a)^3$ and w/f (obviously, w depends on the initial geometry and on the axial rigidities of the members); moreover, it is substantially affected by the f_i/f ratio, which is a measure of geometrical imperfections. Increasing the rise f would increase f/a and diminish w/f and f_i/f. While f_i has a random character, reflecting the probable manufacturing and erection inaccuracies, f is entirely decided by the designer, who can reduce the imperfection sensitivity of the vault. The rise is:

$$f = R - \sqrt{R^2 - a^2} \cong a^2 / 2R \tag{1}$$

where R is the cylinder radius. Thus the angle between trusses is practically proportional to their depth:

$$f / a \cong a / 2R \tag{2}$$

But on the other hand, increasing the truss depth results in increased lengths of the web members, which favours their individual buckling. Besides, there is an interaction between individual buckling of the members and the other forms of instability (ref. 4). Based on the above considerations, it may be concluded that choosing a proper geometry is an essential option for the designer.

4. The role of flexural rigidity. A fully triangulated structure is able to carry loads of any distribution even if all node connections are hinged. The effect of providing semi-rigid or rigid connections is that a part of the applied load is resisted by bending moments. This part is not eroded by the deviations f_i of the nodes from the theoretical geometry although it is influenced by the imperfections of individual members) and is therefore a stable component of the critical load. It may be concluded that one can take advantage of the flexural rigidity of component members, especially those which are meant to resist primary bending moments e.g. longitudinal members that are designed as purlins). The most important effect of their flexural rigidity is to reduce geometrical non-linearity and imperfection sensitivity. However, one should bear in mind that a well-designed reticulated shell is supposed to resist the applied loads mainly by axial forces. In consequence, any member that is not subjected to primary bending should be just as rigid as necessary to attenuate non-linear behaviour. This is a matter of structural optimisation.

5. The effect of tangential load components. It should be noted that the effect of tangential load components (Fig. 1b), unlike that of radial components, is nearly the same as predicted by a linear analysis. The influence of f_i, w, and even f, is negligible.

THE V-SYSTEM VAULT

6. Five different bracing systems are shown in Fig. 2: the V (or K) system Fig. 2a, the rectangular (Fig. 2b), triangular (Fig. 2c), rhombic (Fig. 2d) and X (Fig. 2e) systems. Assuming that the length d of the diagonals is identical in all cases, and taking $\alpha = 45^0$ and $\beta = 60^0$, the truss depth in Figs 2a, 2d and 2e is:

$$b = 2b_1 \cong 1.63b_2 \tag{3}$$

The bracing systems in Figs 2a, 2d and 2e are best suited for increasing the truss depth without needing excessive member lengths. Transport and erection can be simplified to a great extent if the vault is made of small sub-assemblies, instead of prefabricated trusses. In the case of Fig. 2e, the sub-assemblies should be isolated members, while in Figs 2a and 2d they can be longitudinal members and triangles (Fig. 2a) or rhombs, in particular squares (Fig. 2d). Two successful examples of the structures shown in Figs 2d and 2e are reported in ref. 6. In comparison with the rhombic system, the V system has a greater number of members and joint connections, which is a disadvantage. In exchange, it requires a smaller size unit for transport in order to obtain the same truss depth b:

$$c \cong 0.71d \tag{4}$$

Moreover, the V system is more rigid, especially because the upright members can be given a sufficient flexural rigidity in the transverse direction to attenuate the non-linear behaviour. This can be an important advantage in the case of large span single-layer vaults.

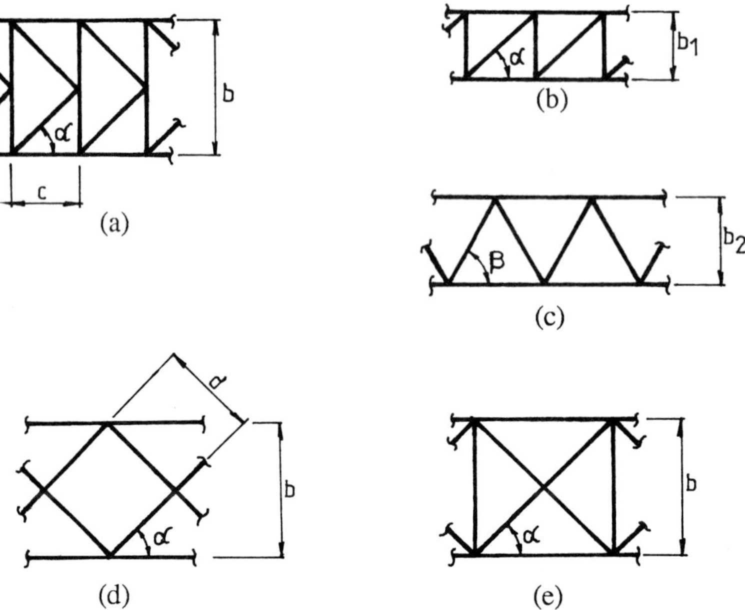

Fig. 2. Comparison of different bracing systems.

7. The geometrical scheme of the V-system vault is presented in Fig. 3a. It consists of triangles, a central upright member B, longitudinal members C (main purlins) and secondary purlins D. Fig. 3b shows an expanded view of the vault.

8. Each triangle (Fig. 3c) is composed of an I-section upright member 1 (which can be a hot rolled or welded I section, a castellated beam, or can be made of two cold-formed channels), and two hollow section diagonals 2. A plate 3 is welded at the connection between an upright member and a diagonal, and a stub hollow section 4 and two plates 5 and 6 are welded at the connection between two diagonals (Figs 3 and 4). An I-section central upright member B is placed at mid-span of each truss, since the arrangement of the triangles must be that shown in Figs 3a and 3b. This arrangement is necessary because the upright members of certain triangles are at the same time component beams of the gables, or chords of the stiffening arches or trusses of the vault.

9. The main longitudinal members C are I sections, disposed along the generatrices of the cylinder, with their webs in radial directions. Any of the solutions mentioned for the upright members can be chosen for these I sections. At the same time, they have the role of purlins.

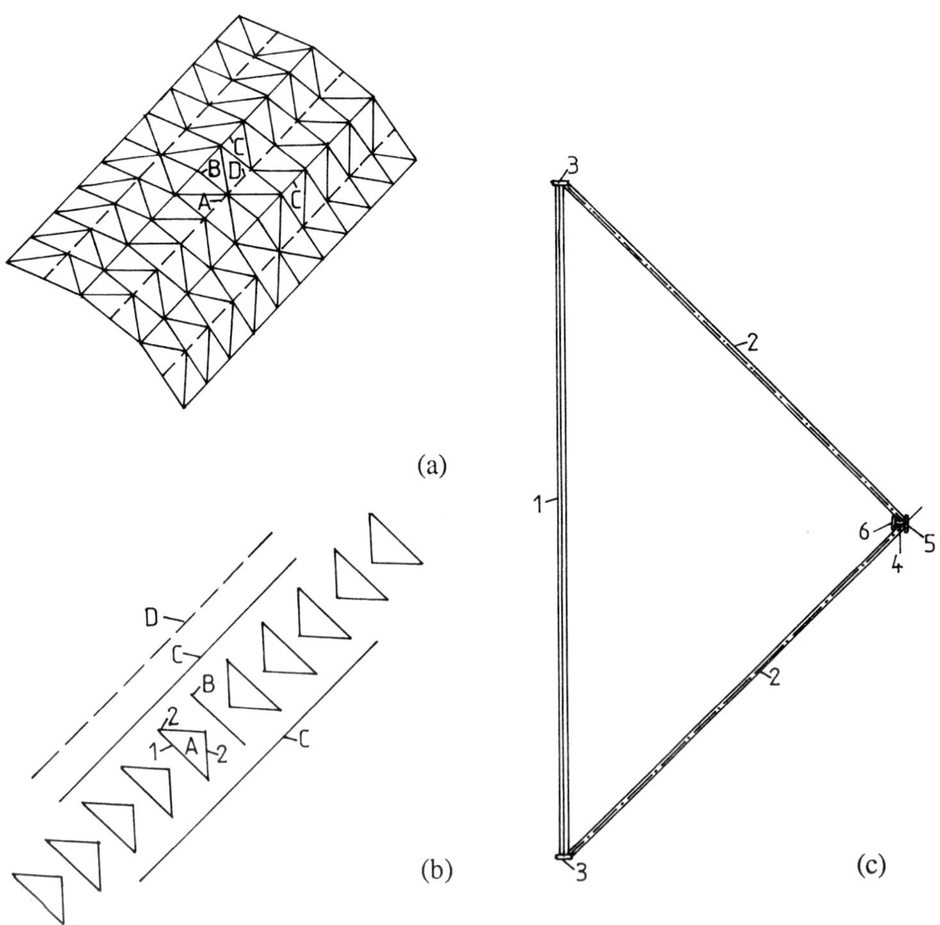

Fig. 3. The V-system vault.

10. Secondary purlins D (Figs 3a and 3b) can be added at mid-depth of each component truss, to comply with the maximum allowable span of the sheeting, or simply for aesthetic reasons (a better approximation of the cylindrical form).

11. Triangles are shop welded. Site connections are achieved by means of the bolts 7 and 8 (Fig. 4). Prestressed high strength bolts are preferable. The use of black bolts for erection, followed by site welding, is a possible but a technically poorer solution.

12. In order to reduce the number of splices of the main purlins C, these should be manufactured in lengths of 6-12 m. Splices must not be placed in the nodes.

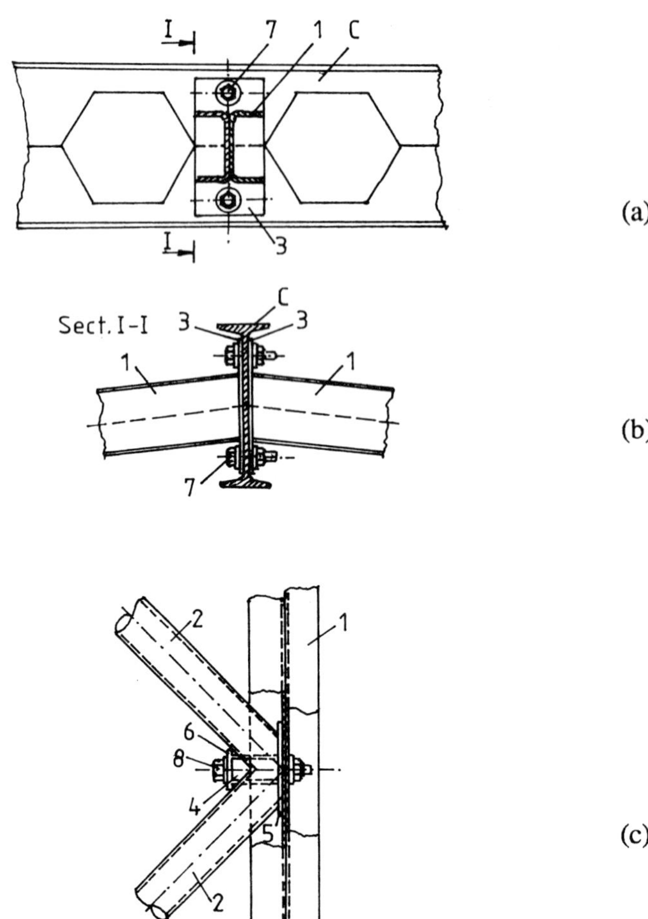

Fig. 4. Connection Details.

CONCLUSION

13. The advantages of the described system can be summarised as follows:

- by increasing the truss depth and consequently the angle between two adjoining trusses, the non-linear behaviour and imperfection sensitivity are attenuated, thus improving structural performance and especially stability;
- the I-section longitudinal members are normally used as purlins, and some of the upright members can be at the same time component beams of the gables, or chords of the stiffening arches or trusses;
- the flexural rigidity of these I-section members about the strong axis exerts a favourable influence on the structural behaviour (however, no member should be oversized without necessity);

- due to the flexibility of the members about the weak axis, the structure can accommodate small manufacturing accuracies without giving rise to significant residual stresses;
- the lengths of the diagonals and upright members are maintained within acceptable limits;
- in comparison with other possible alternatives (except for those where the whole structure is assembled from individual members), the V-system vault requires a smaller size unit for transport of the sub-assemblies to attain a given truss depth.

14. This system has been registered as a Romanian patent (RO 97613).

REFERENCES

1. MAKOWSKI, Z.S. History of development of various types of braced barrel vaults and review of recent achievements all over the world. In: MAKOWSKI, Z.S.(ed.) Analysis, design and construction of braced barrel vaults, p.1. Elsevier Applied Science Publishers, London and New York, 1985.
2. MULLORD, P. Introduction to the analysis of braced barrel vaults. Ibidem, p.36.
3. GIONCU, V., BALUT, N., POPUMB, D. and RENNON, N. Instability behaviour of triangulated barrel vaults, Ibidem, p.159.
4. GIONCU, V. and BALUT, N. Instability behaviour of single layer reticulated shells, International Journal of Space Structures (in press).
5. BALUT, N., PORUMB, D.and GIONCU, V. Some aspects concerning the behaviour and analysis of single-layer latticed roof structures, Stability of metal structures, Third International Colloquium, Paris, Nov. 1983, Preliminary report, pp133-140.
6. DU CHATEAU, S. Architectural aspects of braced barrel vaults, In: MAKOWSI, Z.S. (ed.) Analysis, design and construction of braced barrel vaults, p.223, Elsevier Applied Science Publishers, London and New York, 1985.

206. Structural design of a large-scale steel latticed vault

S. KONOMI and S. MAEHARA, NIHON SEKKEI Inc., Japan

SYNOPSIS. The roof structure of the Fukuoka Exhibition Hall Building is composed of large scale steel vaulted shells. Different analyses have been performed in order to design this beautiful large-scale steel-shell structure. One of the most important points in the design is that the maintenance routes, lighting and mechanical lines are distributed in the rib reinforcement for the thin shell. Prestressing is used to reduce vertical displacements.

INTRODUCTION

1. The architectural design of the roof was undertaken by imagining sea waves, because the structure is located in front of the sea in Fukuoka city (see Fig. 1). The Fukuoka Exhibition Hall Building consists of 4 stories (two basement floors) with a total architectural area of 40,581m^2, a maximum height of 31m and seating capacity for 15,000 people. This long span building will be mainly used as an exhibition hall; there is also a small gymnasium and council room. The availability of movable seats enables the area to be opened up for pop concerts, formal sport games, etc. The construction of this building was started in January 1993, and will be completed by the spring of 1995, ready for use in the "Universiad" games that will be held at Fukuoka city, Japan.

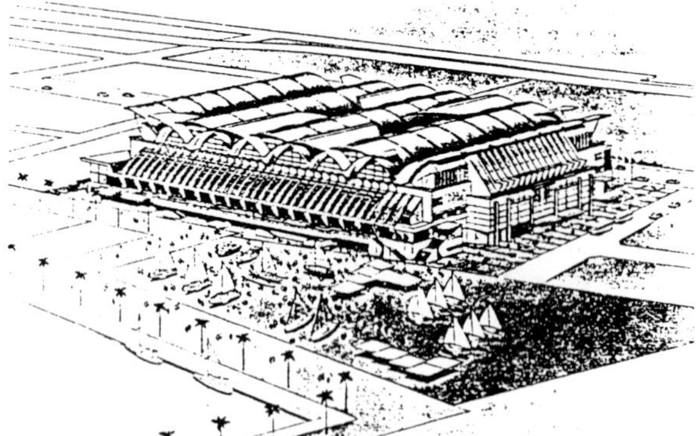

Fig. 1. Fukuoka Exhibition Hall.

Space structures 4. Thomas Telford, London, 1993

2. The dimensions of the roof structure are 144 by 120m and the height of the vault is 7.5m. This roof is composed of three continuous steel latticed vaults and two steel shell wings arranged at both sides of the vaults as shown in Fig. 2. The roof structure is supported by four big columns, whose spans are 36m and 100.8m, with other smaller columns at the shell wings.

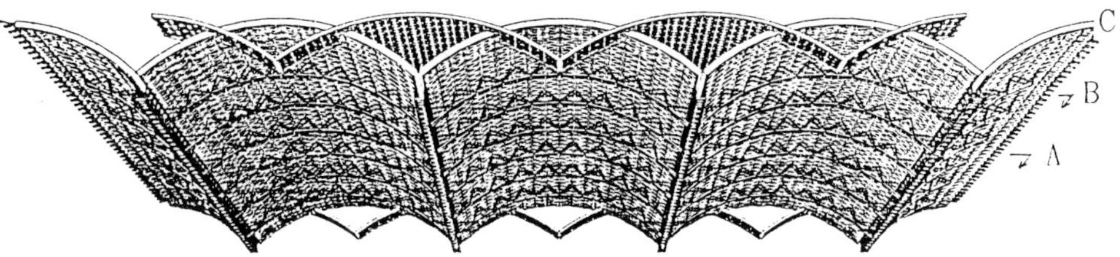

Fig. 2. Perspective drawing of the roof frame.

3. The steel latticed vaults and wings are composed of welded built-up H-steel sections (BH-850 by 200mm) arranged in the longitudinal direction, and steel trusses (H = 850mm) in the transverse direction as shown in Fig. 4. The steel latticed vault is reinforced by steel truss ribs arranged every 14.4m in the transverse direction in order to avoid buckling, general instability and to reinforce against non-symmetrical loads such as seismic and wind loads, as show in Fig. 3. Additionally, these ribs are utilised as maintenance routes and for lighting and mechanical uses.

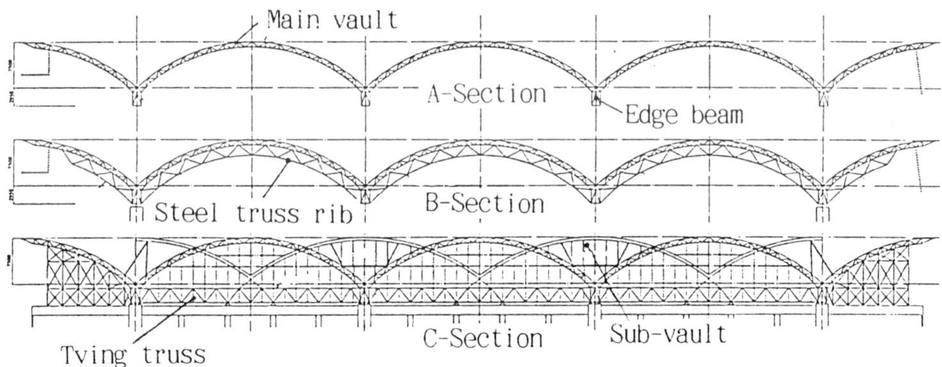

Fig. 3. Elevation of the roof framing.

4. Instability behaviour and the stiffening effect of ribs, together with the dynamic behaviour under vertical and horizontal earthquake vibrations, and prestressing effects on displacement and stress, have been analysed in order to design this beautiful shell structure. Simple connections are used and wind tunnel tests were performed to determine the wind force coefficient.

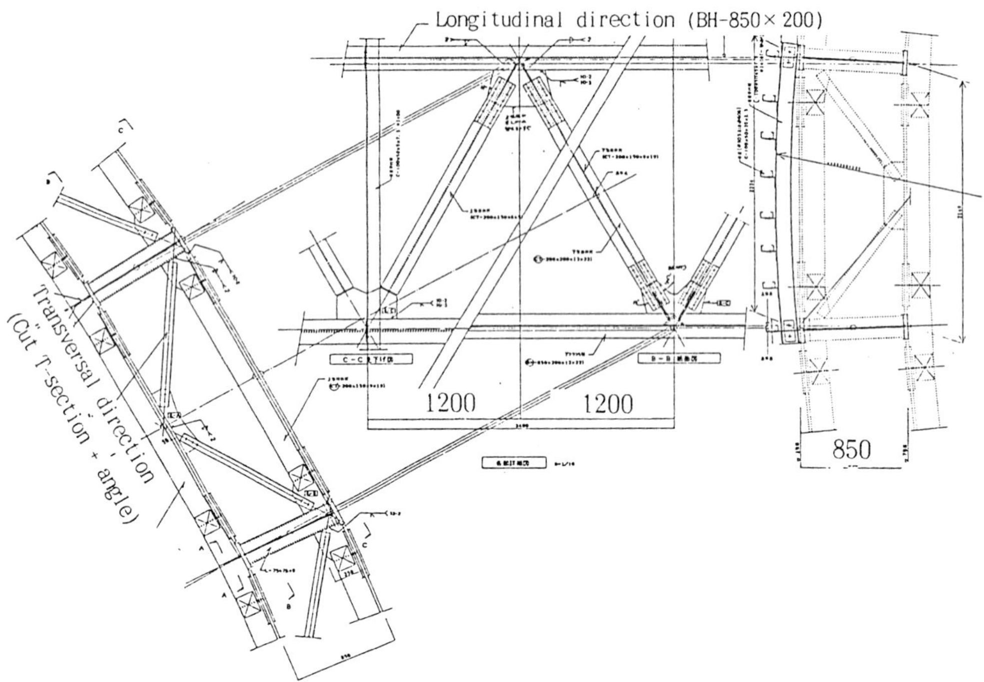

Fig. 4. Framing plan of the roof (Main shell vault).

STRUCTURAL ANALYSIS

Design Load

5. The design live load of the roof is 980 N/m^2 which allows for the suspension of various exhibits, and also the possibility of having concentrated loads of 20kN at the frame nodes.

6. Wind load is determined using the results of the wind tunnel test. The model scale is 1/300, and roughness of the surrounding condition of the site is considered. Finally, 0.9 to 5.5 kN/m^2 of wind load is adopted allowing for the effects of gust and a 150 years return period, using the Eq. 1 specified in ref. 1.

$$P = qCG_f,\ q = \frac{1}{2}\rho V_z^{\,2},\ V_z = 1.2(Z/10)^{0.1}R \qquad (1)$$

where:

P : Wind design load.
G_f : Gust effective factor = 2.1 (Ground roughness type I).
C : Pressure coefficient (wind tunnel test).
R : Statistical factor = 1.07.

7. As for the aseismic design, vertical and horizontal loads on the roof structure are considered. Vertical and horizontal seismic load coefficients are determined by numerical dynamic analysis in time domain, using three actual earthquake records, whose maximum velocities are scaled to 16cm/sec. Finally the horizontal seismic load coefficient for the roof design is 0.35 and the vertical seismic load coefficient for the roof design is 0.053 which correspond to the first step in the aseismic design, that is the allowable stress design for a moderate earthquake. In the second design step the elastic limit strength which corresponds to 0.75 of the seismic load coefficient is also confirmed against a severe earthquake. The first natural period in the vertical direction of the roof is 0.71 sec. and the first natural period in the horizontal direction, of the whole structure including the main frame that supports the roof structure, is 0.33 sec.

Stress Analysis

8. Seven cases of structural frame analysis have been undertaken to design the roof structure as shown in Table 1. There are two analytical models; one considers the whole model with some secondary members thinned out as shown in Fig. 5, and the second, considers a quarter of the structure taking into account all members and the symmetry conditions as shown in Fig. 6. Combinations of the stresses used to design the section are shown in Table 2.

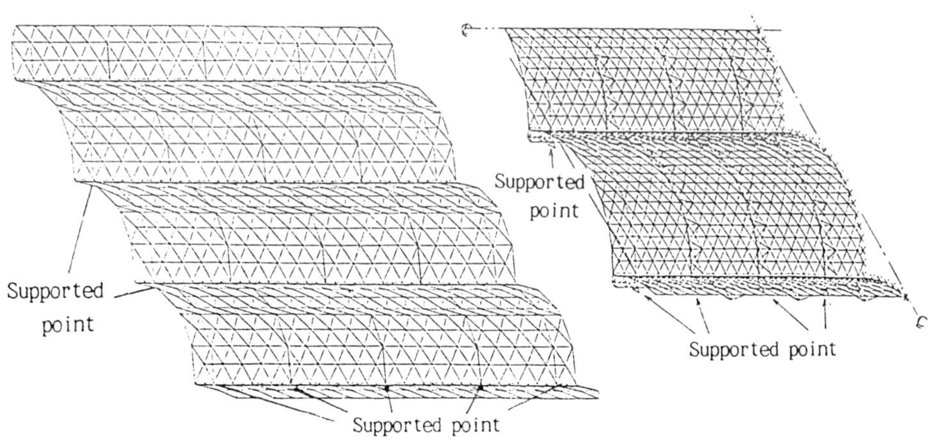

Fig. 5. Analytical model (whole model).

Fig. 6. Analytical model (a quarter of the model).

Table 1. External load cases.

No.	Load case
1	Vertical load (Steel weight only)
2	Vertical load (Live load etc.)
3	Thermal stress (± 15℃)
4	Thermal stress (± 30℃)
5	Earthquake
6	Wind forces
7	Prestressing

Table 2. Stress combinations for designing section.

No.	Combination	Allowable stress
1	1+2+3	long term
2	1+2+3+7	long term
3	1+2+4	short term
4	1+2+3+5	short term
5	1+2+3+5+7	short term

9. Boundary conditions (support conditions) are set considering the order of the construction. The support condition against self weight of the steel roof structure is initially considered as a roller (free) in the long span direction in order to avoid large internal forces in the supporting columns due to horizontal movements of the roof, when the temporary supports used for erection are removed. The support conditions against other loads such as live loads, seismic loads, etc. are considered as spring supports. The corresponding spring constants are determined using a three-dimensional stress analysis.

10. The results are shown in Table 3, where the vertical deflection at the centre of the roof and horizontal displacement at the supported point of the roof are presented. Fig. 8 shows one example result from the analysis of the structure supporting vertical loads. Axial forces are expressed separately for compression and tension. The design of the sections of the members is performed considering a combination of axial force and two directional bending moments.

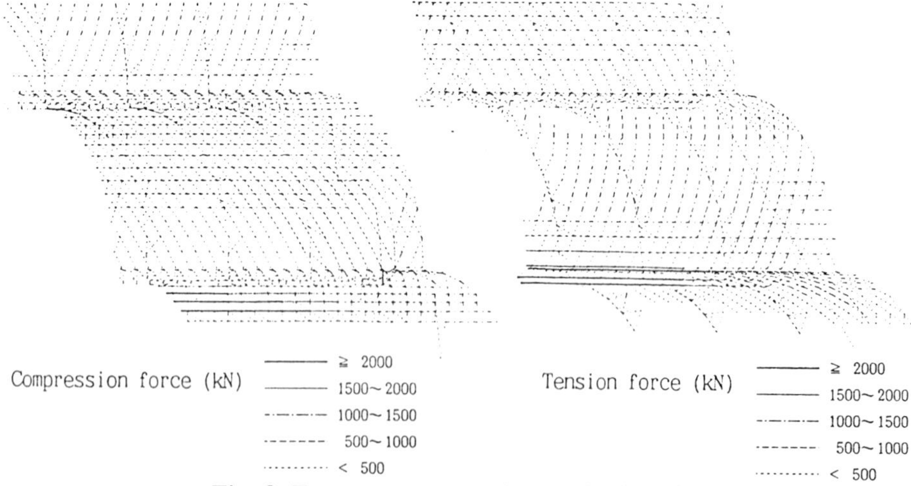

Fig. 8 Frame stresses under vertical load.

Table 3 Vertical deflection at the centre of the roof.

No.	Vertical deflection (cm)*1	Horizontal movement (cm)*2
1	− 8.6	
2	− 17.0	2.25
3	± 0.63	± 0.66
4	± 1.23	± 1.31
5	0	1.73
6	+ 8.74	
7	+ 7.2	

*1 − : downward, + : upward
*2 longitudinal direction (supported point)

11. As mentioned in §3, the steel vaulted shell is reinforced by steel truss ribs, which are used as maintenance routes, etc. after their effect in the structure is confirmed by a stress analysis. The preliminary study, investigating the effect of the truss ribs when using a single vault model, confirmed that the truss ribs are very effective in maintaining the initial shape of the vault and also in resisting non-symmetrical loads and horizontal loads.

Stability Analysis

12. Stability analysis is performed using the incremental load analysis which considers a geometrical non-linear behaviour (member behaviour is linear). Two analytical models are studied as shown in Fig. 9. The first is a single vaulted shell model, and the second is a continuously arranged shell model. Three types of shell height are considered to analyse the first model.

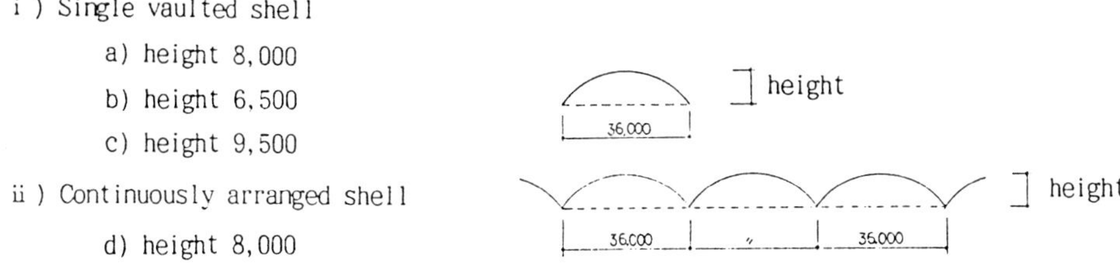

Fig. 9 Analytical model for stability analysis.

13. The results are shown in Fig. 10 and Table 4. For the first model, the taller the shell height, the more stable the structure is. The shell height of this roof structure is about 8,000mm, so the stability limit is thought to be 35 kN/m^2 (12 times the design load). The second model, a continuously arranged shell is thought to be 1.3 times more stable than the single vaulted shell model. The design load level seems to be stable in this analysis.

Table 4 The results of stability analysis.

	Stability limit		
Height	(a) Single vaulted	(b) Continuously arranged	(b)/(a)
6,500	25.0 kN/m^2	——	——
8,000	35.3 kN/m^2	45.6 kN/m^2	1.29
9,500	42.6 kN/m^2	——	——

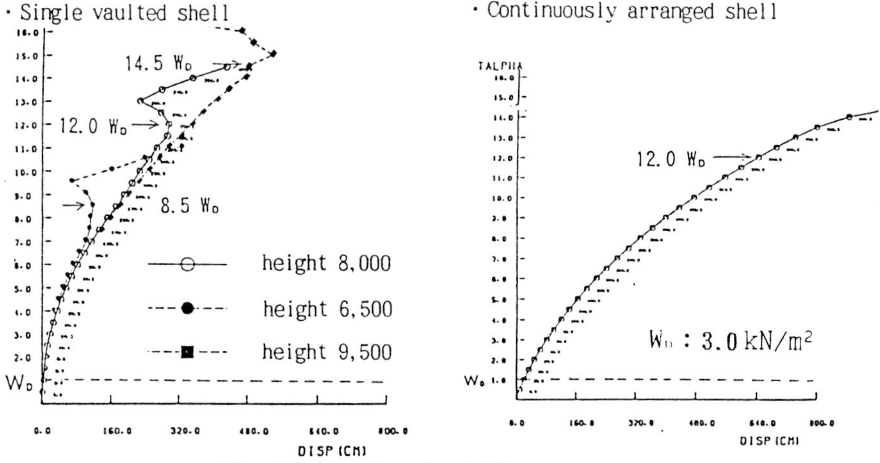

Fig. 10 Results of stability analysis
(Hysteresis load verses displacement).

Prestressing

14. It is too difficult pre-camber this roof structure, therefore, prestressing is used in order to get the least possible deflection. Details of the prestressing are shown in Fig. 11. Unbonded tendons, used to avoid rust, are installed in the steel pipe at the edge beam in the bottom of the steel vault. The tensile forces in these prestressing cables are restricted by the following equation for any load case (ref. 2).

$$\begin{aligned} \sigma_0 &\leq 0.6Pu \text{ and } 0.7Py \\ \sigma_{max} &\leq 0.7Pu \text{ and } 0.85Py \end{aligned} \tag{2}$$

where

σ_0 : Stress by prestressing

σ_{max} : Stress by $\sigma_0 + \sigma_\wedge$

$\sigma_\wedge$: Stress due to seismic load, thermal stress, live load, etc.

15. Stresses due to prestressing are shown in Fig. 12, and the deflection of the centre of the roof is +7.2 cm. The maximum tensile forces in the prestressing cables are obtained considering the worst load case condition and are equal to 0.67Pu and 0.79Py.

16. The anchoring plates for these prestressing cables are analysed using the Finite Element Method and plate elements. These anchoring plates are designed such that the maximum stress does not go beyond 0.75 Fy; where Fy is the yield strength of the anchoring plate. Thus the plate thickness for a prestressing force of 3000kN is 90mm, and for the case of 1500kN is 70mm.

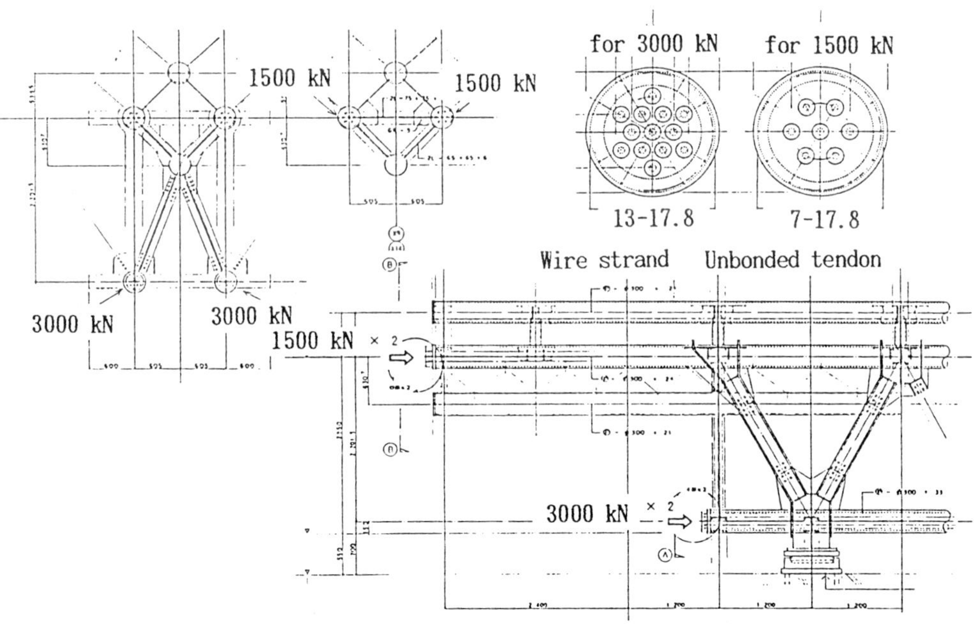

Fig. 11. Prestressing details.

CONCLUSIONS

17. Studies investigating the wind load, using wind tunnel tests; the seismic load through dynamic analysis; the effect of the rib reinforcement to increase the stability behaviour; and the effect of prestressing to reduce vertical deflections, were performed in order to design the large-span thin steel vault structure reported in this paper.

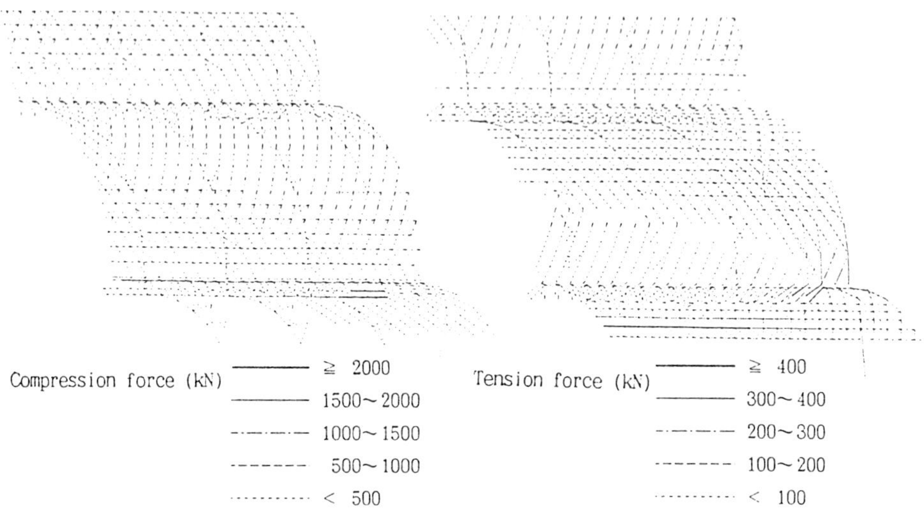

Fig. 12. Frame stresses against prestressing.

18. The construction of this building has started as a result of the co-operation between architectural and mechanical designers. Various reports regarding the construction will be presented in the near future.

REFERENCES

1. Architectural Institute of Japan, Recommendations for Buildings Design Load, Tokyo 1981.
2. BELENYA, E. Prestressed Load-Bearing Metal Structures, Moscow 1972.

207. Repair and rehabilitation of a historic masonry barrel roof structure

M. W. FANTOZZI, The Fantozzi Co. Inc., USA

SYNOPSIS. This paper presents a case study on the repair and rehabilitation of an early unreinforced masonry (URM) barrel roof structure. The rehabilitation included structurally upgrading the building to improve the performance of the structure in earthquakes. The reconstruction consisted of adding a concrete cylindrical shell and shear wall system. Substructure improvements were required to resist overturning and the potential for the top portion of the structure to "slide" on the rubble footing. The impact of the structural repair and original appearance is limited by retaining the historic fabric of the building.

INTRODUCTION

1. The Powder House building (Fig. 1) is one of the few existing structures still standing from the original quicksilver mining operation which began in around 1845 in the New Almaden district of Santa Clara County, California. The New Almaden mine was the first mercury mine claimed in North America and produced the largest total production in the world (ref. 1). The mine produced an average of about 500 tons of mercury each year. In 1926, mining production began to decline until mining ceased in 1976. The County of Santa Clara purchased the property in the early 1970s and established the New Almaden Quicksilver Park.

2. The County Historical Heritage Commission recognised the historical significance of the mine site and buildings as an important part of the County's history. The Commission designated the park a "historical place" and in 1992 undertook to restore the Powder House building.

3. Although the construction of this incidental building is typical for unreinforced masonry (URM) construction, it had a unique brick barrel roof. The unreinforced construction of this vaulted roof was a sample of time honoured construction techniques based solely on geometric stability. The Powder House is one of only two brick buildings in the western part of the United States with this unique vaulted roof. The 1989 Loma Prieta Earthquake severely damaged the building, collapsing the barrel roof.

4. The structural consolidation was designed to improve the condition and stability of the structure by the addition of new materials and construction methods. Structural lightweight concrete was used for the reconstruction. A series of structural analyses was conducted utilising a general purpose finite element computer program to evaluate the structural capacity and seismic load response of the retrofit. Geometry data collected during the site investigation was used in developing the computer model.

Fig. 1. New Almaden Quicksilver Mine Powder House.

INVESTIGATION

5. In order to restore the Powder House in a manner consistent with preserving its historical fabric, it was necessary to collect information from existing records and field inspections. This information was necessary to ascertain the extent of intervention required to satisfy criteria for structural stability and long-term preservation. Search of historical records produced little significant information regarding the original construction of the Powder House. This is likely due to the fact that the Powder House served as an ancillary building, similar to a garage, to store blasting powder and dynamite for use in the mining operations. The Powder House was probably built in 1871, coinciding with historical records of active mining operations in the area.

Field Investigation and Condition Survey

6. A site visit was performed to identify the overall spatial configuration, gravity and lateral load resisting systems, and to make a preliminary assessment of the present condition of the building. The Powder House was constructed with sand-molded clay bricks and had an overall plan dimension of 4m x 7m. The URM walls have four wythes of solid brick units above the cobble stone footing, up to the level of the wood floor. Above the wood floor level, the walls become three wythe cavity walls. The exterior part of the wall has two wythes of brick with the single interior wythe bonded to the interior wythe with header bricks spaced intermittently. The cavity walls are approximately 457 mm thick with a 146 mm wide cavity (Fig. 2).

7. The longitudinal walls continue with the three wythe cavity wall up to where the barrel roof begins. Here, the walls undergo transition to a solid pattern to support the barrel roof. The short end walls undergo transition to a solid four wythe wall above the door openings and then transition to a 15 m radius.

8. Steel bars, 25mm in diameter, were placed transverse to the building and used as tension ties for the barrel roof. Five bars were placed near the top of the longitudinal walls. Two bars were set within the end walls and secured to the posts supporting the wood roof. Three bars were equally spaced between the end walls and secured to the exterior wood beams running the length of the building on each side . In addition, two bars were set within the lower portion of the end walls near the floor level and secured to the exterior wood posts.

9. When the masonry barrel roof collapsed, it bent the exposed bars, pulling the longitudinal walls inward. The wood beams running the length of the building near the top of the walls prevented the walls from collapsing completely. Ironically, it was the rather incidental wood frame structure which prevented the complete collapse of the Powder House.

10. Diagonal cracking was observed along each wall with heavier cracking occurring in the north end of the building. The pattern of these cracks is typical of cracks associated with in-plane movement due to lateral loads. In addition, the longitudinal walls experienced severe cracking from the out-of-plane movement caused by the barrel roof collapsing in the earthquake. Conspicuously absent were cracks below the floor level where the brick wall becomes four wythes thick. Examination of the exposed top portion of the cobblestone footing at other locations around the building perimeter did not disclose evidence of cracking suggestive of long term settlement or earthquake induced displacement. Despite the extensive damage from the earthquake, no evidence of movement between the lower course of brick and the top of the cobblestone footing was apparent.

11. During our reconnaissance of the site, we examined the conditions exposed in an approximately 1 meter deep pit excavated along the outside face of the footing at the north-east corner of the building. The footing comprises angular, cobble size serpentine and silica-carbonate rocks bonded together by mortar. At the pit location, the top of the cobblestone footing extends approximately six inches above the adjacent ground surface and its outside face projects about four to six inches out from the base of the wall. Assuming the walls of the Powder

House are roughly centred on the cobblestone footing, it is likely that the width of footing is of the order of 0.75 to 1.0 meter and extends about 1 meter below the adjacent grade. The cobblestone footing was found to bear on stiff clay which appears to be a natural deposit rather than man-made fill. These materials appear to have provided good support for the cobblestone footing over the 122 year life of the Powder House.

12. The Powder House is situated on a ten percent slope, with the north end facing downhill. The taller north wall is more flexible than its companion south wall. The additional tributary weight shifted the centre of mass northward, whereas the stiffer south wall shifted the centre of rigidity slightly southward, resulting in a substantial eccentricity. The action of the earthquake ground motion induced a torsional moment, contributing to the significant damage to the north end of the building.

Non-destructive Testing

13. A test programme was conducted in support of the visual inspection for the purposes of assessing the general condition of the masonry and steel, and to better quantify construction techniques used in this period. The test programme consisted of non-destructive testing of mortar joints, spectrographic analysis of mortar and brick samples, chemical analyses of mortar samples, compressive strength tests on clay brick units and tensile test on a rod specimen.

14. In-place shear tests were not conducted for purposes of assessing the strength characteristics of the existing URM walls. Since the Powder House was severely damaged in the Loma Prieta earthquake it would be inconsistent with the goal of long-term preservation to rely on the strength of the existing URM to resist any portion of the design lateral force. Furthermore, it was felt that the earthquake may have caused micro-cracking to occur in the bed joints which would undermine the in-place shear test results and make it difficult to establish quantifiable data on the overall mortar quality and workmanship. For the purposes of restoration, it was determined that establishing the relative quality of the mortar and clay brick units would be sufficient to develop a restoration methodology. Destructive testing of the existing standing walls would not provide any useful information, and would require additional repair work. Samples of mortar and brick for laboratory analysis and testing were easily obtained from the collapsed roof without introducing any more damage to the building.

15. Since the individual masonry units are bound together by the mortar, the strength of the resulting assemblage is somewhat dependent upon the physical properties of the mortar. Although, the strength of masonry assemblages in shear and flexure depends primarily upon the mortar bond strength and not upon the compressive strength of the mortar itself; the compressive strength of the mortar does reflect the ability of the mortar to bind the assemblage together and, therefore, indirectly reflects the bond strength. Therefore, a measuring methodology which gives information about the compressive strength of the existing mortar was developed.

16. The Type PM Schmidt hammer was used to obtain rebound numbers on exterior and interior wall joints. The height of rebound of the impacting mass from a material is a function of the surface hardness of that material. The material characteristic "hardness" is derived for the elastic and strength properties of the material. Since the strength of the binder as a rule determines the bond strength of the mortar, the surface hardness gives an indication of the bond strength of the mortar.

17. The type PM Schmidt pendulum hammer was designed by the Netherlands Organisation for Applied Scientific Research and the Netherlands Institute for Building Materials and Structures (ref. 2) for testing the quality of light-weight building materials of low hardness and strength. The type PM version of the Schmidt hammer is actually a modified version of the Type P hammer. The PM hammer is equipped with an 9 mm diameter head with a shaft length of 10 mm, suitable for use in narrow pointing in masonry. The head has an impact energy equal to 0.883N-m. The instrument has a measuring head at the end of a pendulum which moves through a semicircular arc in the vertical plane. The driving force can be provided by either gravity or a spring. The latter makes it possible to test materials in the horizontal position. For testing the pointing in vertical masonry, gravity was used.

18. Rebound numbers were obtained from 35 points located along the four interior surfaces and 43 points located on the four exterior faces. The limited number of readings taken along the exterior of the west wall was due to the limited access allowed by the existing wood siding. The mean rebound reading at interior locations was 215 with a standard deviation of 4.6. At exterior locations, the mean rebound reading was 18.8 with a standard deviation of 43. Rebound readings less than 10 were obtained at locations at or near existing cracks, demonstrating the sensitivity of the procedure.

19. On average, rebound readings were about 12% higher at interior joints than at exterior locations, which is expected. The interior of the Powder House is not subjected to the same temperature and humidity extremes as the exterior of the building. Based on the rebound readings, the mortar was found to be of uniform quality, but inadequate for structural applications.

Laboratory Analysis

20. Several samples of brick and mortar were tested in the laboratory. The analyses included semi-quantitative emission spectrographic analyses, a chemical analysis on mortar samples, brick compression tests, and a tensile test on a steel rod specimen. For the spectrographic analyses, two brick samples and two mortar samples were taken from the Powder House from the front and rear of the building. The purpose of the spectrographic analysis was to determine whether the constituent materials used in the manufacture of the bricks came from the same source and whether the composition of the mortar was consistent.

21. The results of the spectrographic analysis on the brick samples showed no significant difference between the two samples. The analysis of the mortar samples also showed no significant differences in their constituent materials

except for silicon. Silicon is the basic element of sand and thus it is inferred from the analysis that one sample may have more sand in the mix than the other. However, we cannot be certain because silicon is also found in other compounds of nature in the form of silicates and the spectrographic analysis cannot distinguish between compounds of the basic elements.

22. As intimated above, the spectrographic analysis is limited in its ability to determine the compound from which the detected element was found. Thus, a spectrographic analysis will not be able to distinguish the difference between a cement-based mortar or a lime-based mortar since both materials are calcareous. Therefore, a chemical analysis of the mortar was performed to verify its base formulation. From these results, we identified the mortar as lime-based with mix proportions of approximately 1 part lime to 3 parts sand.

23. The durability of brick buildings is a function of their weatherability. A reasonably accurate index for measuring the durability of brick is its compressive strength. This stems from the fact that the durability of clay products results primarily from the firing or burning which fuses clay particles together by incipient vitrification. The fused outer skin of the brick is its main line of defence against water and thus the compressive strength of the brick is an appropriate indicator of its durability and absorption qualities. Therefore, in order to establish the quality of the brick units, compressive strength tests were performed on two bricks The compressive strength of the two samples tested were 6,570kPa and 7,950kPa.

24. A 12 inch long coupon of one of the steel rods was tested in tension for replacement in kind. Typically, wrought iron used in building construction between 1870 and 1900 had tensile strengths ranging from 205 MPa to 290MPa. Steel produced in the same period for use in building construction had tensile strengths between 345MPa and 415MPa (ref. 3). The tensile strength of the specimen was 402 MPa. From the tensile test, we may infer that the material is steel and not wrought iron.

Natural Hazards

25. Geologic hazard maps developed for the City of San Jose and for the County of Santa Clara (ref. 4) do not indicate the presence of any faults or landslides in the vicinity of the Powder House. The general area has been interpreted to have a relatively low susceptibility to earthquake induced landslides above or beneath the Powder House.

26. The nearest and most important source of large earthquakes that may cause damage in the project area is the San Andreas Fault, which is about 5 miles south-west of the site. The southern Santa Cruz Mountains segment of the fault was the source of the magnitude 7.1 Loma Prieta earthquake of 1989. Recent studies at this fault and others in the San Francisco Bay Area, have concluded that there is a 23 percent chance that a magnitude 7 earthquake will occur on the San Francisco Peninsula segment of the fault by the year 2020 (ref. 5). There is a lesser but quantified chance of a recurrence of an earthquake comparable to the magnitude 8.3 San Francisco earthquake of 1906. The expected peak horizontal ground

accelerations at the site associated with these two events could be on the order of 0.45g and 0.55g respectively.

STRUCTURAL ANALYSIS

27. When structural intervention is considered for historic buildings, special treatment to preserve those aspects of the building that have made it historic must be considered. The structural intervention must be designed to have a much longer life expectancy than typical upgrades of buildings. Therefore, the preservation of historic buildings requires some special procedures to achieve the dual goals of protecting life and conservation.

Design Criteria

28. The selection of the design method and criteria are, to a large extent, controlled by the expected performance criteria for the building. Life-safety in the event of ground shaking is the paramount consideration in the design criteria. However, preservation and authenticity are the principal goals of conservation. Therefore, the design criteria must balance the need for structural stability and maintenance with preservation goals.

29. In order to provide a basis for designing the lateral-force-resisting systems (LFRS) for historical URM buildings, it is necessary to develop a design methodology consistent with conservation goals. In keeping with this design philosophy, the design was based on an expected peak ground acceleration of 0.55g. By using 0.55g as the effective peak ground acceleration we have improved the life expectancy of the building by designing for an earthquake with only a 5 percent probability of being exceeded in 100 years.

30. The effectiveness of supplemental vertical resisting elements in reducing forces on overstressed components is dependent upon the stiffness, strength and ductility of the existing elements relative to the new elements. Any added element, to contribute effectively, must be stiff enough relative to the existing elements to attract sufficient load away from the existing system to prevent premature failure of weaker components. Therefore, the higher force level for structural consolidation is consistent with preservation goals. Under the proposed reconstruction plan, the long-term performance of the building is significantly improved.

Consolidation Method

31. The preferred solution from a conservation point-of-view would be a replacement-in-kind retrofit which would require reconstruction of the cavity wall masonry construction using current construction methods. Although, the replacement-in-kind solution would preserve the authenticity of materials and construction, it became apparent that the existing structural system was not suitable for the long-term preservation of the building and it was necessary to provide additional strength by structural intervention.

32. The reconstruction, shown in Fig. 2, consists of filling the existing wall cavity with structural lightweight concrete and constructing a 76mm thick

lightweight concrete cylindrical shell. Existing bricks were set on forms prefabricated to the 1.5m radius prior to pouring the shell. At the top of the wall, a reinforced concrete beam was added. The beam provides lateral support for the thin concrete shear wall and brick facing, and reduces the amount of longitudinal steel required near the edges of the shell. The brick walls were either repaired or reconstructed and lightweight concrete was placed in the cavity similar to high-lift grouted masonry construction. Due to the extensive damage to the interior wythe of bricks, this wythe was demolished and reconstructed, making placement of wall reinforcement relatively easily.

33. Lightweight concrete was used to minimise the added weight and because the expanded shale aggregate, having a maximum dry density of 1,600kg/m^3 with a design strength of 20.7MPa, has a coefficient of thermal expansion close to that of brick masonry. The coefficient of thermal expansion of normal weight concrete is approximately twice that of brick. It is desirable to have materials with similar physical properties to insure compatibility and minimise differential movement between the different materials. The long-term serviceability of the brick-lined concrete barrel roof will depend on the dimensional stability of the new concrete shell.

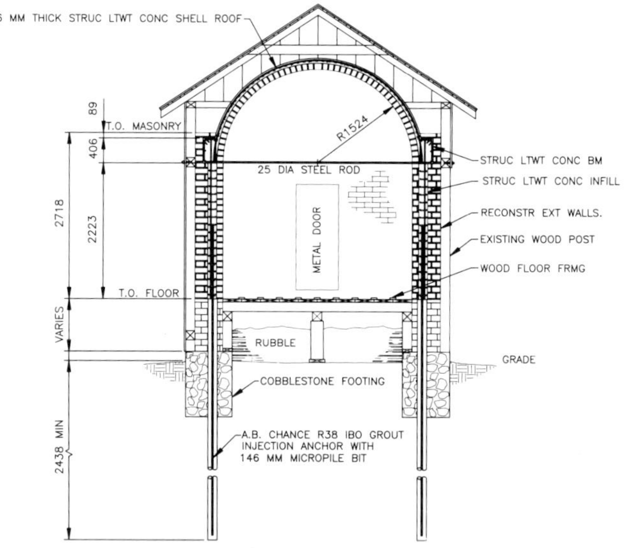

Fig. 2. Reconstruction of the Powder House.

34. The long-term effects of creep were also considered in design of the shell. In general, creep at early ages is largely unrecoverable, but much of that occurring in old or dry concrete is recoverable. The magnitude of this creep depends very much upon the period of the sustained load. Tests have shown that stress levels

greater than 30 to 40 percent of ultimate contribute significantly to creep. Therefore, design stresses in the concrete shell were limited to 30 percent of the ultimate stress.

35. The solid masonry wall below the floor level was core drilled to accommodate the placement of self-boring anchors. Twenty-two "IBO" grout injection/self-boring anchors were drilled to a depth of 2.4m below grade. The self-boring anchors acted as mini-piles resisting overturning and the potential for the top portion of the structure to "slide" on the cobblestone footing. The high strength drill rod is made of a heavy wall steel tube which has a continuous, cold-rolled drill steel rope thread over its entire length. The drill rod was fitted with an IBO micropile drill bit which has a dual-roller, cutter-wheel, rotary drill bit with central injection. Cement grout was continuously pumped during drilling providing a soil-cement column surrounding the rod in a one-pass installation. The drill rod was left in place acting as dowels, extending 1200mm into the lightweight concrete infill wall. Wall reinforcement was placed full height and transition steel set in place for the concrete shell. After the wall steel was in place, the interior wythe was reconstructed and the exterior wythes were repaired and repointed as required.

36. Reconstruction of the brick walls was accomplished by reusing the existing brick units, and mortar was designed to match the original pointing in colour, texture and detailing. A lime based mortar was used. This was necessary to prevent damage to the masonry units. It is a common error to assume that hardness or high strength is a measure of durability. Stresses within a wall or roof caused by expansion, contraction, moisture migration, or settlement must be accommodated in some manner. In a masonry wall, these stresses should be relieved by the mortar rather than by the bricks. A mortar that is stronger or harder than the bricks will not "give" thus causing the stresses to be relived through the bricks, resulting in cracking and spalling. High lime mortar is soft, porous, and changes little in volume during temperature fluctuations. In general, mortars should be softer (measured in compressive strength) than the masonry units.

CONCLUSIONS

37. The criteria governing the strengthening measures selected include occupant safety, extending the useful life of the building, and restoration. Special treatment to preserve those aspects of the building that have made it historic must be included in the design. Historic buildings require special attention because of the implicit objective of preserving these buildings for posterity. In addition to life safety, this objective dictates that strengthening criteria provide a higher level of structural protection than for regular buildings. The reconstruction offered the best opportunity to preserve the architectural look and "feel" of the original Powder House and at the same time enhance its structural performance for its conservation.

REFERENCES

1. BAILEY E. H. and EVERHART D. L. 1964. Geology and quicksilver deposits of the New Almaden District, Santa Clara County, California: U.S. Geological Survey Professional paper 360.

2. Netherlands Organisation for Applied Scientific Research, 'the IBBC-TNO Pointing Hardness Tester," Institute for Buildings, Materials and Structures; 1989.

3. American Institute of Steel Construction, Historical Record Dimensions and Properties of Rolled Shapes Steel and Wrought Iron Beams and Columns, Edited by Herbert W. Ferris, Chicago, IL; 1953.

4. Working Group, 1990. Probabilities of large earthquakes in the San Francisco Bay Region. California: U.S. Geological Survey Circular 1053.

5. SEED H. B. and IDRISS I. M. Ground motions and soil liquefaction during earthquakes. Engineering Monographs on Earthquake Criteria, Structural Design, and Strong Motion Records, Vol. 5. Earthquake Engineering Research Institute, 1982.

208. The Unistrut space frame for the main stadium roof of the Shah Alam sports complex, Selangor, Malaysia

S. C. HSIAO, Unistrut Space Frame Systems Inc., USA, and D. MAJKIC, Energoprojekt, Malaysia

SYNOPSIS. Two identical space frame roofs, crescent shaped facing each other over the grandstands of the Main Outdoor Stadium in the Shah Alam Sports Complex, Selangor, Malaysia, are currently under construction with a completion date of these structures in the spring, 1993 (Fig. 1). Because the scope of the work is complex and large, this paper will limit the discussions to the following items of interest. (1) Brief Overall Description. (2) Design considerations of the wind effect on the structure. (3) Allowable design load for cold-formed steel tubing with side wall bolt-access holes. (4) Assembly method and erection sequence of the space frame structure. (5) Project organisation to direct and monitor the project.

Fig. 1.

BRIEF DESCRIPTION

1. The Main Outdoor Stadium, with grandstands seat 72,000, is the centre piece of a large complex which will include a number of stadiums, a veladrome, swimming pools and shooting ranges which when completed will be a showcase for national and international sports competition.

2. The space frame geometry of the arched crescent form is generated from a cut section of a circular cylinder, with a radius 240m, tilted 10^0 and are cut 45^0 at the front and 23^0 at the back, as measured from a horizontal level to the ground. The maximum distance between cuts is 69m. The span of the arch is 284m, and the height measured from the playing field is 61m.

3. The crescent structure is made from Unistrut Space Frame System III and is constructed with a double-layer offset rectangular grid with a module size 3.8m x 3.6m and 3.0m depth. At the leading edge of the arch, a triple-layer truss in a diamond section is added both for structural and aesthetic functions. Structurally this stiffened edge truss contributes a more efficient distribution of loads within the framework and aesthetically it articulates the arch form. The surface area per frame is $17193m^2$. Each structure has 2778 nodes, 10775 members and 32 supports (Figs 2-3).

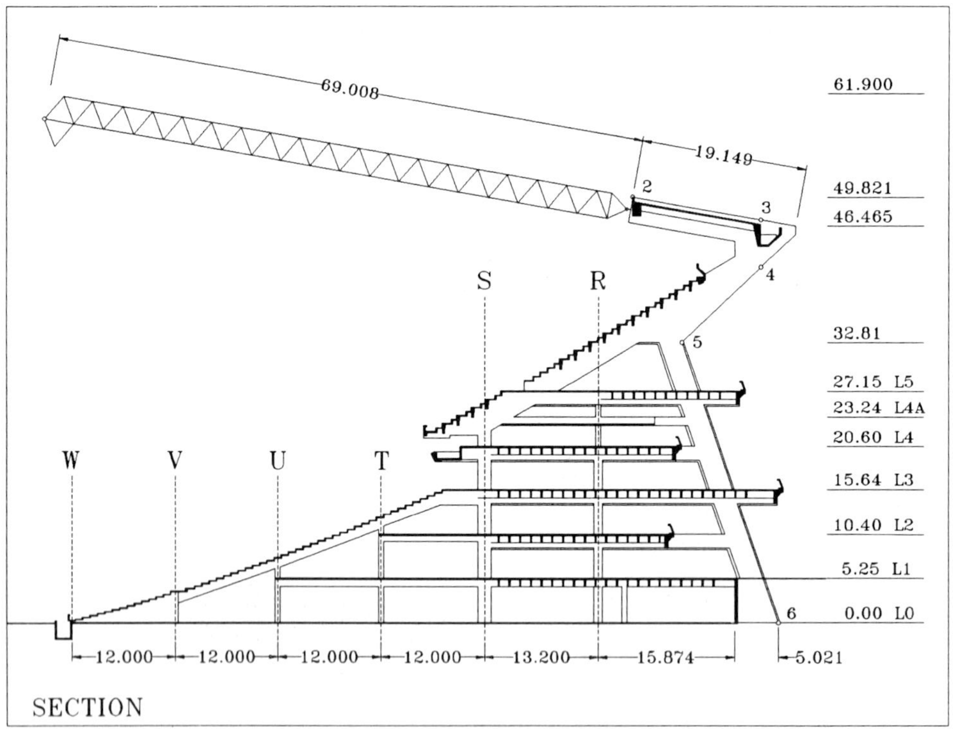

Fig. 2.

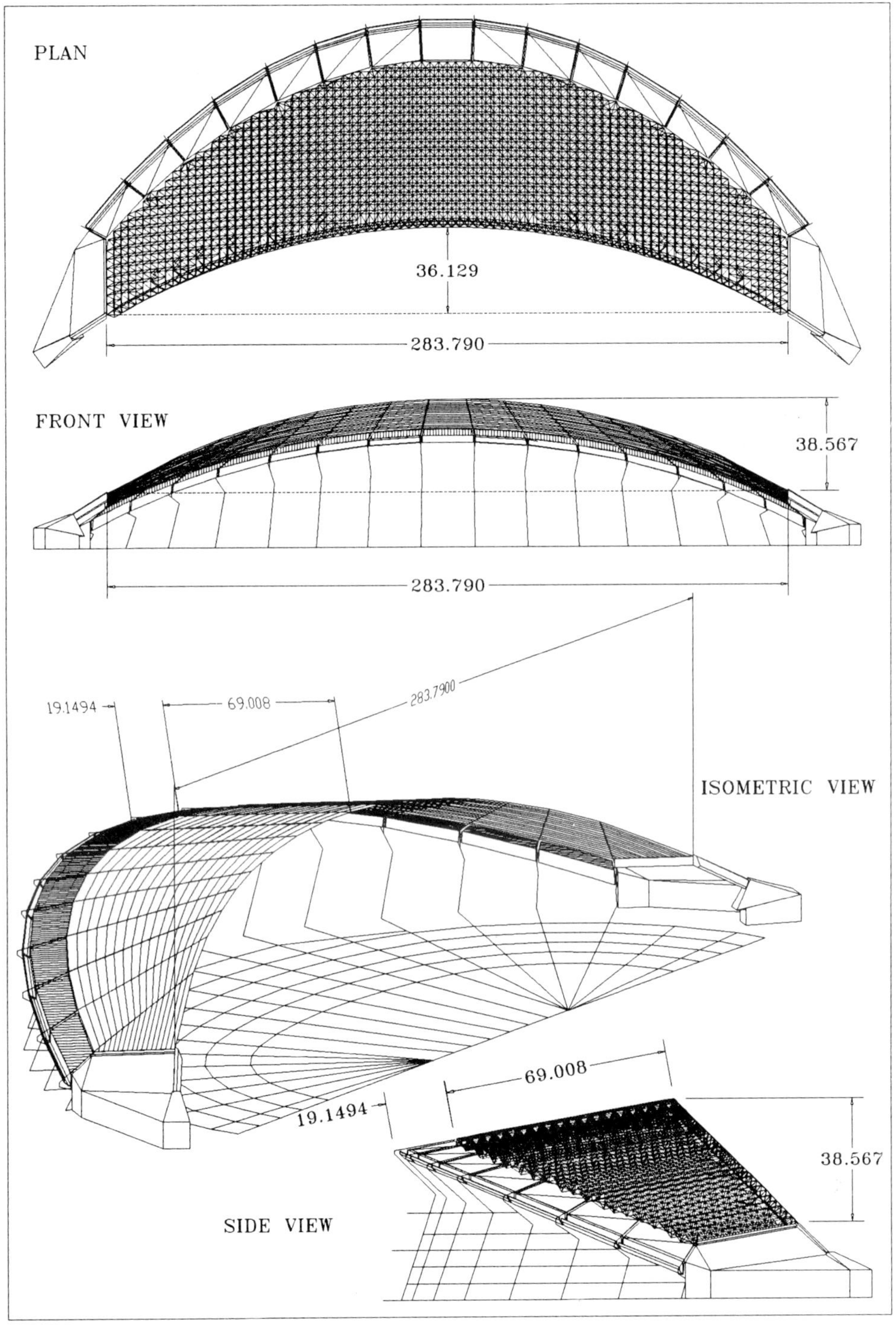
PLAN
36.129
283.790
FRONT VIEW
38.567
283.790
19.1494
69.008
283.7900
ISOMETRIC VIEW
69.008
19.1494
38.567
SIDE VIEW

Fig. 3.

4. The supports for the space frame consist of 16 concrete frames at the rear edge and two 8-point buttress supports at the end edge-beams leaving the front edge facing the playing field open without obstruction.

5. The specified loadings are Live Load 0.5 kN/m^2, Space Frame 0.75 kN/m^2 (vs actual 0.40 kN/m^2), Gutter and Glazing System 0.325 kN/m^2, Wind Speed 75 MPH with 50 year return period, Floodlight 15 kN/m, PA Speaker 30 kN/m at 15m c/c, Speaker Cluster 80 kN, Support Settlement 19mm, Temperature +22C -14C from a 30^{0}C average, Temporary Load at any nodal point 15 kN.

DESIGN CONSIDERATIONS OF THE WIND EFFECT ON THE STRUCTURE

6. For this long span lightweight structure, wind loading is one of the most critical loadings in design (ref. 1). The steps that were taken to arrive at a more accurate and realistic wind loading cases for design are as follows:

(a) Obtain local wind measurement data and perform a statistical analysis to obtain a 50 year return period one minute wind speed at a height of 19.2m above ground.

(b) Perform computer analysis to determine vibration modes of the structure.

(c) Conduct wind tunnel tests with a scale model of the structure and surrounding area for pressure model, pneumatic averaging and aeroelastic tests (Fig. 4).

(d) Determine wind loads on the structure for simplified load cases and for detailed load cases, taking into consideration the dynamic loadings from the vibration modes, as well as uniform fluctuating load components, superimposed on the mean wind loading.

(e) Loading combinations for 10 serviceability limit state requirements and for 63 ultimate limit state requirements are selected in the final structural analysis of the space frame roof. British Standards were used for this design.

Fig. 4.

ALLOWABLE DESIGN LOAD FOR COLD-FORMED STEEL TUBING STRUCTURE WITH SIDE WALL BOLT ACCESS HOLES

7. To use steel tubing material efficiently for this large project, a special member dataset was created utilising a number of thin wall tubing and bolt/sleeve assemblies that were designed, for a given tube diameter, to match a group of compression and/or tensile loads in the structure. This process was repeated after each structural analysis so the final dataset consisted of only 40 of the initial 100 member set. Considerations were also given to node sizing and the economics of material purchasing.

8. The specified finish requires an 80 micron minimum thickness, hot dipped galvanised finish for tubing material plus architectural top coats. Since the bolt/sleeve/pin are assembled after hot dipped galvanising, access holes in the tube are required for second stage bolt insertion.

9. The allowable load for a tube with thin wall material and with an access hole in a given diameter was determined by (1) a series of tests on tubing materials in a variety of wall thicknesses, tubing diameters and access hole diameters from which (2) a statistical model was constructed and (3) an empirical formula extracted.

10. The formula (1) that was extracted from test results, as shown below, also considered the local buckling of the tubing and code requirements of D/t to satisfy both American Institute of Steel Construction-Manual of Steel Construction and the British Standards (ref. 2).

$$P_{ye} = F_y \left[1.36275 - \frac{\frac{KL}{r}}{366.166} - \frac{\frac{d}{l}}{87.928} - 0.467296 \frac{F_y}{F_u} \right] \qquad (1)$$

Where:

p_{ye} = Experimental compressive design strength.
d = The diameter of the access hole.
F_y = Yield stress.
F_u = Ultimate stress.

The p_y used in design is the minimum of p_{ye}, p_y, stipulated by the code, F_y, or $0.84F_u$.

ASSEMBLY METHOD AND ERECTION SEQUENCE OF THE SPACE FRAME STRUCTURE

11. Five key phases are important for the space frame assembly. They are construction drawings review, erection method selection, material scheduling and sorting, shoring tower design and shoring removal procedure.

12. For this project, erection was accomplished by a built-in-place method. The entire space frame structure was divided into multiple sections, each was

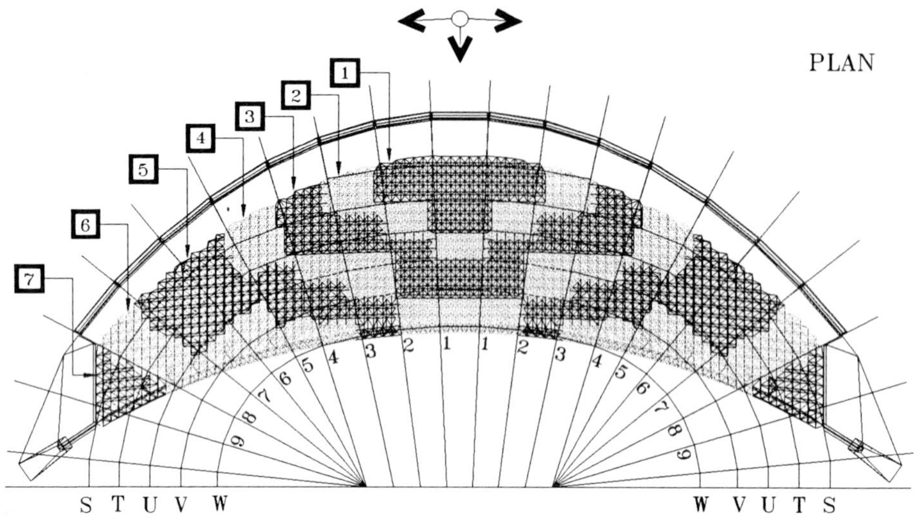
PLAN
1
2
3
4
5
6
7
1 2 3 4 5 6 7 8 9
S T U V W
W V U T S
STAGES OF SPACE FRAME CONSTRUCTION

Fig. 5.

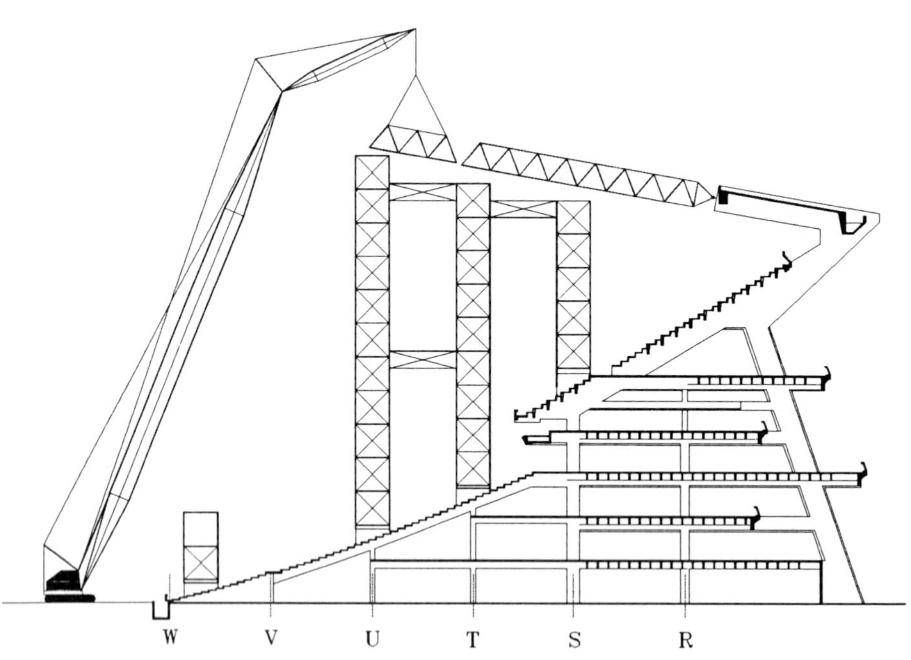
W V U T S R

Fig. 6.

supported by adjustable jacks mounted on the shoring towers. Assembly work began at the mid arch of the roof next to the concrete frames and continued both to the left and to the right and from support to support before moving out toward the playing field (Fig. 5).

13. Within a section, ground assembled sub-sections were lifted up by crane, positioned, and stitched together in the air (Fig. 6). In areas beyond the reach of a crane, individual members were lifted up by a hoist and the connection made in the air. A total of 70 shoring towers in heights from 7m to 60m and 3 mobile cranes with 150T capacity each were scheduled in this construction.

14. During erection, measurements were made periodically to check the overall frame dimensions. Variances, if any, are corrected before moving on to the next section.

15. The construction of the space frames for the Shah Alam Main Outdoor Stadium roofs has gone smoothly and there have been no surprises (Figs. 7-8).

Fig. 7.

PROJECT ORGANISATION THAT DIRECTS AND MONITORS WORK

16. Due to the scope and certain complexity of this project, Unistrut management set up certain agendas and organisations to assure that the project would proceed in an orderly fashion, with adequate resources and attention to detail. The project teams were organised, with assignments made to key personnel, and reporting and communication lines established so that "surprises" could be minimised and handled promptly. Highlights of these tasks are listed below.

(a) A management Steering Committee was formed consisting of the corporate chief executive officer, controller, legal representative, and the president of the space frame business unit, to meet regularly with the project manager, project engineer and others, as required. This organisation and its activities has proven to be extremely valuable in the conduct and success of this project.
(b) Manufacturing sourcing was an issue that required early decision. The objective was to source products from anywhere including in-house facilities that meet project requirements, quality standards and is cost effective. The decision was to source tube and node material from Hanmaek Industries Co., Ltd.
(c) Manufacturing Quality Control and Inspections were important to a successful operation. Manufacturer's quality assurance program must meet established standards and that it must be carried out prudently. In this project, Unistrut contracted AICO, an Asia Inspection Company to perform full inspections of all operations at the plant site as an additional quality check.
(d) Engineering issues, "Hot Topics", were raised early. All issues in engineering, from geometry to loading, from limit state design to frame stability, from material dataset selection to impact of bolt-access hole in tubing strength, etc. were reviewed. These were done in consultation with consultants early on so design could proceed with minimum back tracking.
(e) For erection, a construction specialist was provided by Unistrut. His function encompassed erection recommendations, trouble shooting, record keeping, etc.

17. In summary, this project is a success due to careful orchestration of project management and dedicated people who understood their functions and worked together as a team.

Fig. 8.

REFERENCE

1. GRIEVE J.B., STONE G.K. and IRWIN P.A., "Prediction of Structural Wind Loads on the Large Span Roofs", The IASS-CSCE International Congress, Toronto, Canada,July, 1992.

2. MHAIMEED K. and HSIAO S.C. "Shah Alam Sports Complex: "Design and Construction of Unistrut Space Frame Structure", The 11th International Speciality Conference on Cold-Formed Steel Structures, St. Louis, Missouri, USA, October, 1992.

PROJECT:	The Main Outdoor Stadium Shah Alam Sports Complex, Selangor, Malaysia
OWNER:	J.K.R. Public Works Ministry of Selangor
ARCHITECTS:	Hijjas Kasturi Associates SDN
STRUCTURAL ENGINEERS:	Tahir Wong SDN.BHD. & Jurutera Perunding Bersama
MAIN CONTRACTORS:	Energoprojekt (Malaysia) SDN.BHD. Kuala Lumpur, Malaysia
SPACE FRAME AND SKYLIGHT SUBCONTRACTOR:	Danzie SDN. BHD Unistrut Space Frame Systems, Inc. Binatama Akrindo/lniaga J.V.-Space Frame Erector

CONSULTANTS TO UNISTRUT SPACE FRAME SYSTEMS, INC.

(1) Professor Erling Murtha-Smith, Department of Civil Engineering, University of Connecticut, Storrs, Ct. USA.

(2) Mr. Curtis J. Hoff, Ph.D., Automated Analysis Corporation Ann Arbor, MI USA.

(3) Mr. Peter A. Irwin, Ph.D., P. Eng., Rowan, Williams, Davies & Irwin, Inc., Guelph, Ontario, Canada.

UNISTRUT PROJECT TEAM

Project Manager	James A. Moses, P.E.
Project Engineers	Francisco Castano, P.E., Sun Chien Hsiao, P.E., Khaldoun Mhaimeed, P.E., Todd Smith, P.E.
Construction Specialist	John Boyd

209. Experimental study of the behaviour of typical space frame joints

A. I. EL-SHEIKH, University of Dundee, UK, and R. E. MCCONNEL, University of Cambridge, UK

SYNOPSIS. The stiffness and performance of space (three dimensional) frames rely critically on the bending stiffness and strength of the vertical to chord member joints. This paper reports on a study of Cubic Space Frame (CSF) joints. In such space frames, a joint typically connects a single vertical member to four half length chord members. The general behaviour of such joints, and in particular their stiffness and ultimate strength, was studied experimentally. Amongst the parameters varied were the basic node dimensions and the ratio between the shearing and bending actions in the two orthogonal chord directions.

INTRODUCTION

1. Fig. 1, shows a view of a typical Cubic Space Frame structure. It is a modular system of construction consisting of two layers of rectangular grids connected by vertical members only. No diagonal bracing is used, and the

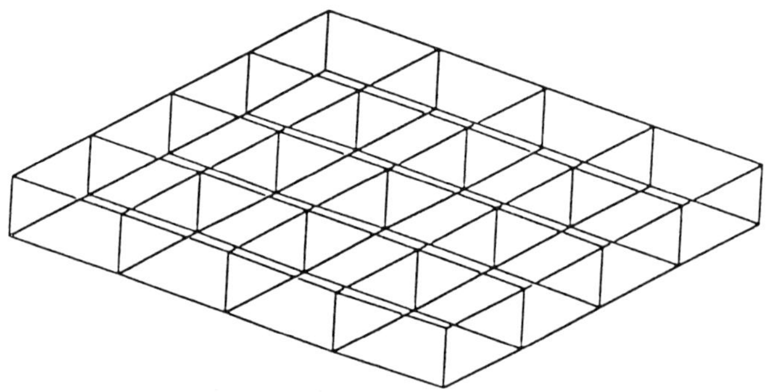

Fig. 1. View of a double-layer CSF.

modules are connected together by simple mid-chord bolted splices. Each module has a vertical member (usually a box section) and a number of half length chords connected to it at its upper and lower ends, see Fig. 2. While the mid-chord splices are required to transmit axial and shearing forces only, the joint connecting the chords to the vertical is required to transmit shearing forces and bending moments in two directions, as well as the global axial forces.

2. Unlike inherently light space trusses, space frames have received relatively little interest, and have not been popular for spanning large areas when only a few intermediate supports are allowed. This is due to their heavy self weight and the complexity experienced in making joints which are sufficiently rigid. Dividing the structure into small modules has also proven difficult. However, the Cubic Space Frame system has to a large extent overcome these difficulties by adopting shop-fabricated rigid joints in conjunction with site bolted mid-chord splices, so enabling the manufacture of the structure in small identical modules.

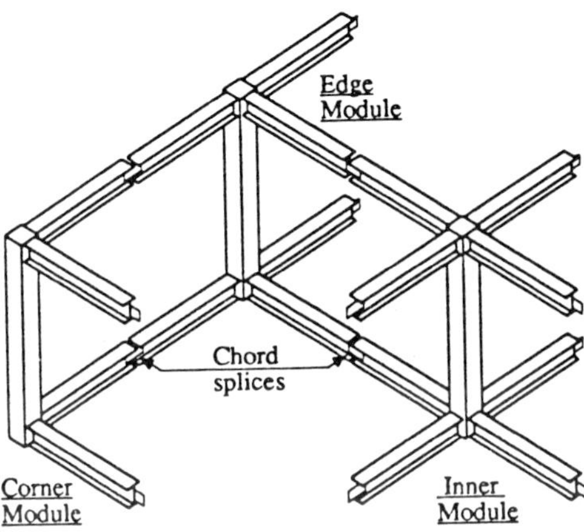

Fig. 2. Typical CSF modules.

Consequently, the Cubic Space Frame has provided a competitive alternative to the more usual space truss systems. The CSF has many of the advantages of space truss systems such as: manufacturing simplicity; easy packing; and rapid assembly. Compared to space trusses, its major disadvantage is its larger weight, but lower cost of joint manufacture can compensate for this. Refs 1,2 give a general description of the system and a discussion of its behaviour.

3. This paper discusses a particular aspect of the behaviour of Cubic Space Frames, namely the behaviour of CSF nodes under various combinations of loading in order to evaluate and improve existing models of the response of actual CSF nodes.

4. It is normal practice to consider the CSF node as a rigid joint under working loads. However, a full understanding of the behaviour of nodes is essential for any accurate analysis of the overall structure's performance.

CSF NODES

5. A typical CSF node connects a vertical member (usually a box section) to between 2 and 4 half length chord members as shown in Fig. 2. Often the node is formed by closing the vertical box member with end (outer) plates, and butt-welding the outer flanges of the chord members to these plates. The inner flanges of the chord members are usually fillet welded to collar (gusset) plates of the hexagonal shape shown in Fig. 3. This configuration was adopted in the manufacture of the CSF nodes tested in this work.

FAILURE MODES

6. From earlier work on CSF nodes, (refs 1,3 & 5) it was generally found that their ultimate strength could be predicted by an upper bound approach incorporating some yield line analysis. Four possible modes of failure were investigated in the aforementioned work, see Fig. 4. These modes are:
(a). The outer plate mode OP in which the chord member rotates by an angle β about the line connecting the collar plate to the box section. This mode also involves significant extension of the outer plate and distortion of the box sides. When the outer plate initially yields, its extension is likely to be insufficient to develop the full box-face yield-line mechanism. However, as the outer plate strains further, the box hinges begin to develop more resistance leading to the development of the full yield line pattern. This mode of failure may occur whether the chord web is welded to the box face or not.

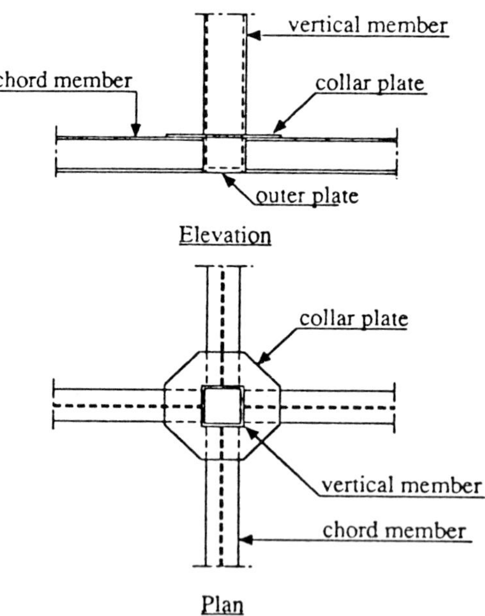

Fig. 3. Details of a typical CSF node.

(b). The collar plate mode CP which involves shearing of the collar plate, forming of box-face yield-lines, and rotation of the chord member by an angle α about the line connecting the outer plate to the box member. Angles β, α and θ represent the box surface rotations as shown in Fig. 4c.

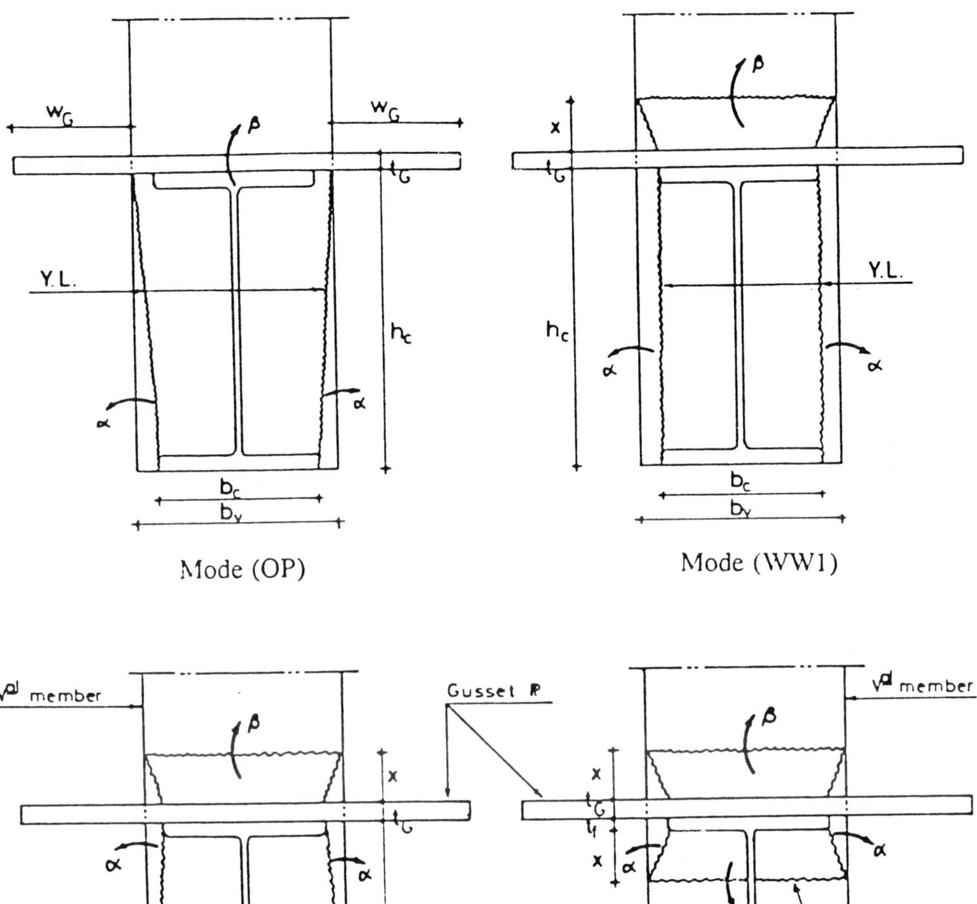

Fig. 4. Possible modes of failure of a CFS node.

(c). The welded-web mode WW1 which assumes shearing of the collar plate, and axial straining of the outer plate with no relative rotation between the chord member and the box section.

(d). The unwelded-web mode UWl which is applicable only to nodes with an unwelded chord-web to box-face connection. In this mode of failure, no chord rotation relative to the node is assumed to take place, Fig. 4d. As the outer plate strains axially, the collar plate shears accompanied by local distortions of the box section near the outer and collar plates.

EXPERIMENTAL TEST SERIES

7. In the work presented in this paper, further node testing was carried out to cover various variables which, it was believed, might have a significant influence on CSF node strength. Coupon tests were made before manufacture of the nodes to ensure that all the node components had consistent strength properties. Four nodes of the same dimensions were tested earlier by Saxton (ref. 3) but due to considerable variations in material strength between the various node components, the nodes suffered premature failures before a complete failure mechanism could develop. Three test series were considered in the present work.

Test Series 1

8. In Test series 1, two typical CSF nodes were tested to failure under uniaxial tension, see Table 1. Fig. 5, shows their details, and the loads applied to the test specimens. While the first specimen had a welded chord-web to box connection, the second was unwelded in this region. During the tests, deflection gauges were used to measure the displacements and internal deformations of the nodes under loading. Test series 1 was designed to investigate the significance of the web-to-box weld.

Test Series 2

9. In Test Series 2, biaxial tension tests were carried out on two CSF nodes, details of which are depicted in Fig. 5. These nodes were tested under biaxial tension forces with the ratio between the forces in the horizontal and vertical directions, T_2/T_1, chosen as 0.40 for the first test 2-1, and 0.80 for the second test 2-2, see Table 1. A ratio T_2/T_1 of 1.00 was not adopted because of concerns about possible damage to the horizontal loading frame if the node had failed in its horizontal direction. Deflection gauges were used to record the deformations of the nodes in both the horizontal and the vertical direction.

10. Since, in practice, CSF nodes are normally placed under biaxial action with varying ratios of the forces acting in each direction, this series of tests was intended to model actual cases of nodal loading. These tests also helped check the biaxial validity of the modes of failure proposed for uniaxial tension shown in Fig. 4.

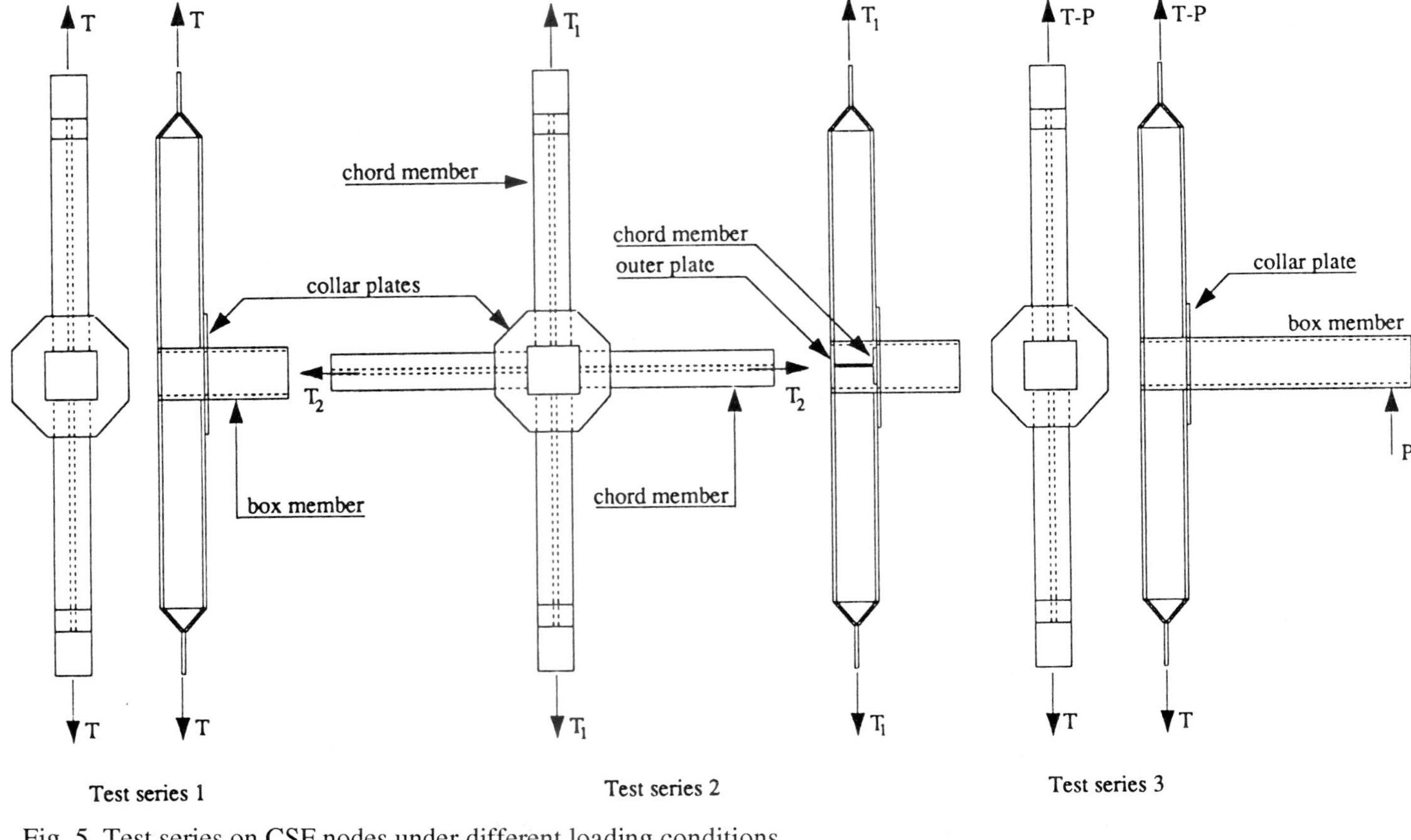

Fig. 5. Test series on CSF nodes under different loading conditions.

Test Series 3

11. In Test Series 3, three nodes were tested under combined uniaxial tension, T, and bending moment, M, see Table 1. The nodes were tested to failure with the M/T ratio set to 120 mm, 80 mm and 40 mm for the first, second and third tests respectively. The specimen details and the load applied are shown in Fig. 5. Deflection gauges were used to measure the development of the node deformations under loading. Test series 3, was designed to check the validity of the proposed failure models for nodes under such loading.

Test loading procedure

12. In all the tests, load was applied by a 5000kN load-control Amsler tensioning machine, using a constant loading rate as far as possible. Two cycles of loading and unloading were performed before finally loading to failure. The first loading cycle was up to approximately the design working load of the nodes. The second cycle was up to 1.6 times the working load. These loading values were similar to those used by Saxton (ref. 3). Finally, the nodes were loaded to failure.

Table l. Nodes included in the experimental programme.

Test series	Forces applied	Variables considered	Specimens		
1	uniaxial tension T	effect of chord-web to box weld	1-1 unwelded connection	1-2 welded connection	
2	biaxial tension T_1, T_2	ratio of forces T_2/T_1	2-1 $T_2/T_1=0.40$	2-2 $T_2/T_1=0.80$	
3	Uniaxial tension & bending T,M	ratio of forces M/T	3-1 $M/T=120$ *mm*	3-2 $M/T=80$ *mm*	3-3 $M/T=40$ *mm*

EXPERIMENTAL RESULTS

Test series 1: Uniaxial tension

13. The load deflection relationships recorded for nodes 1-1 and 1-2 showed that the nodes acted elastically up to load levels of 560kN and 680kN respectively. These values were 17% and 42% above the corresponding working loads respectively. Subsequently, the nodes gradually lost stiffness due to

yielding of the outer flange of the unwelded node 1-1, and of the outer plate of the welded node 1-2. Due to the lack of web welding in node 1-1, the box face distortion for this specimen was less than that of node 1-2.

14. From the observations of the behaviour of nodes 1-1 and 1-2 near failure, it could be deduced that failure modes UW1 and WW1 respectively had developed. Collapse eventually occurred at 732kN and 923kN for nodes 1-1 and 1-2 respectively, due to necking of the chord member outer flange and the outer plate. This in turn caused failure in the butt weld joining them together. The presence of the web-to-box weld clearly gave a significant enhancement in performance (26%).

Test series 2; Biaxial tension

15. From the extension in both directions of the chord flanges, opening of the collar plate and the box section, and the lateral node movement, nodes 2-1 and 2-2 could be seen to act as elastic joints up to load levels of 600kN and 560kN respectively. These values were 25% and 17% above the corresponding working load levels. Subsequently, the node stiffnesses degraded in the direction of the bigger (vertical) tension force with the gradual development of box hinge lines and yielding of the outer and collar plates. At the same time, the outer and collar plates were observed to undergo relatively smaller deformations in the (horizontal) direction of the smaller tension force.

16. From the observations of the behaviour of series 2 nodes near failure, it was noticed that the outer plates yielded first in the direction of the bigger force T_1, (vertical direction). The box sections suffered significant distortions due to web pulling. Observing the distorted faces of the box sections near failure indicated that failure mode WW1 was developing in both tests. This was accompanied by considerable deformations in the T_1, direction, while little increase in strength was developed between initial yielding and overall failure. After considerable elongation, collapse took place when the outer-plate to outer-flange connection in the T_1 direction failed, followed immediately by tearing of the box member. Collapse of nodes 2-1 and 2-2 took place at T_1 = 820kN and 760kN respectively.

Test series 3; Uniaxial tension and bending

17. Linear behaviour was observed up to load levels of 640kN, 720kN and 760kN for the three nodes respectively, these values being 33%, 50% and 58% above the corresponding working loads. Subsequently, the outer plate mode OP was clearly seen to be developing on the upper face of the box members in all three tests, i.e. the face connected to the upper chord member. This involved considerable outer plate extensions and rotations about the line connecting the inner flange of the upper chord member to the upper box face. These rotations reduced with the smaller M/T ratios. Failure was reached at total loads of 710kN, 800kN and 890kN for tests 3-1, 3-2 and 3-3 respectively. In all tests, collapse started at the outer-plate to outer-flange connection and spread by tearing of the box face.

18. The influence of bending moment in reducing the node ultimate axial tension strength was evident.

DISCUSSION

19. From the results of the test series carried out on CSF nodes, it could be seen that the nodes behaved linearly up to load levels well above the corresponding working load levels. The subsequent reduction in stiffness was gradual with the yield spreading through the outer and collar plates, and with the development of yield lines on box faces. Final collapse started in all tests with failure in the butt weld connecting the outer-flanges to the outer-plates. This was due to the high concentration of stresses close to the weld edges developed by excessive plastic deformations of the node components. See ref. 6. for full details of the tests carried out.

CONCLUSIONS

(a) The results of test series 1 showed that welding the chord-web to the box member resulted in a considerable increase in strength.
(b) Nodes under axial forces in two orthogonal directions suffered a slight reduction in strength as compared with similar uniaxially loaded nodes.
(c) Co-existent bending moment reduced the node axial strength significantly.

ACKNOWLEDGEMENTS

20. The authors wish to thank Dr L A Kubik for his agreement to the publication of the results of the experimental study carried out on the nodes of the patented Cubic Space Frame system.

REFERENCES

1. CHILTON, J. C. An investigation of the elastic behaviour of the cubic space frame, Ph.D. Thesis, Trent Polytechnic, Nottingham, U.K, 1988.
2. CHILTON, J. C., DAFFERN, C., KUBIK, M. L. AND McCONNEL, R. E. The cubic space frame - Its construction and methods of analysis using a micro-computer, Proceedings of IASS Symposium, Osaka, Elsevier Science Publishers, 1986, vol. 3, 301-308.
3. SAXTON, R.C. Cubic space frame node analysis, A final year report, Dept. of Engng., University of Cambridge, U.K, 1989.
4. CHILTON, J. C., DAFFERN, C., KUBIK, M. L. AND McCONNEL, R. E. Microcomputer analysis of the cubic space frame, Journal of Structural Engineers, 1986, vol. 64A(8), 203-205.
5. KUBIK, L. A. AND McCONNEL, R. E. The behaviour of cubic space frame joints, International Symposium on Tubular Structures, Finland, 1989.
6. EL-SHEIKH, A. I. AND McCONNEL, R. E. Testing of typical space frame nodes and splices, Technical Report No. CUED/D-Struct/TR.134, Cambridge University, Cambridge, U.K, 1991.

210. High-rise space frames: effect of configuration on lateral drift

CHENG-YU HO and G. G. SCHIERLE, University of Southern California, USA

SYNOPSIS. Lateral drift of high-rise space frames, caused by wind or seismic forces, is an important design consideration in areas with substantial wind or seismic activity. The research presented here compares various framing configurations regarding their capacity to minimise lateral drift caused by wind. The study was done using ELM, a structural analysis computer program based on the direct stiffness method for structures of geometric and material linearity. The framing concepts are based on identical beam, column, and bracing members for 10, 20 and 30 storeys respectively; to isolate and illustrate the effect of framing configurations on drift control. The results are presented in graphs to visualise the correlation between configuration and drift, for informed intuition. The graphs also include numeric data for quantitative comparing. The results show strong correlation between lateral drift framing configurations.

INTRODUCTION

1. Excessive lateral drift in high-rise frames can damage secondary systems, such as partitions or curtain walls; generate secondary column stress due to PΔ moments; and cause discomfort to building occupants under prolonged cyclical drift (ref. 1). Damage to secondary system can be controlled by reducing drift, or by uncoupling those systems from the primary structure. The PΔ effect is most severe in moment resisting frames; the Uniform Building Code (ref. 2) for example allows smaller seismic drift for moment resisting frames (.3% storey drift vs. .5 % for other systems). Dynamic wind forces are the greatest cause for human discomfort, due to more frequent occurrence and longer duration than seismic forces. Design for wind forces is, thus, usually based on objectives to minimise lateral drift. Our research focused on lateral drift caused by wind forces, based on the following assumptions.

ASSUMPTIONS

2. The bracing concepts were analysed for 10, 20, and 30 storey buildings of 3 bay frames (Fig. 1) with 30ft (9.14m) column spacing and 13ft (3.96m) storey height. The frames are assumed to be spaced 30ft (9.1 4m) apart.

3. The lateral wind load per floor was assumed as shown in Fig. 1, acting on wind and lee sides in proportion 8:5, as required by method / of the Uniform Building Code.

4. Gravity load was applied simultaneously with wind load, assuming combined live and dead loads of 100 and 50psf (pounds/square foot) for floors and roof respectively, yielding loads of 3 and 1.5kip/ft (4 and 2kN/m) on floor and roof beams respectively (Fig. 1).

5. Beam and column sizes are shown in Fig. 1. They are identical for all framing concepts of identical heights and vary every second floor. Member sizes are based on combined wind and gravity loads, with 1/3 stress increase for wind load per AISC Manual of Steel Construction, Allowable Stress Design (ref. 3). They were initially sized by the portal method and then adjusted, where necessary, based on moments and forces from the computer analysis. Beams were designed for bending stress and columns for combined bending and buckling stresses. Based on the above, the steel weight (in kip) per frame (for primary members only) is:

Type of member	10 storey frame	20 storey frame	30 storey frame
Columns and beams	131	421	817
Bracing	10	37	59
Total	141	458	876

6. Secondary stress, due to geometric non-linearity under deformed conditions, was ignored. While a moment resisting frame was used to compare the braced frames against, the main objective was relative stiffness of various bracing concepts. PΔ column moments are virtually zero in braced frames with pin-joints; and, due to small drift, are very small for moment resisting joints combined with bracing. In our case PΔ stresses are only about 3% of the allowable bending stress).

7. The analysis was done in two dimensions instead of three dimensions, to facilitate the large number of cases. The ELM analysis program (ref. 4) was run on 486-33 computers with SVGA display for visualisation.

BRACING CONFIGURATIONS

8. Fig. 2 shows the bracing configurations investigated for 10, 20, and 30 storey buildings, analysed with both, pin-joints and moment resistant joints. In all but one scheme the bracing is in the centre bay and includes MRF: moment resisting frame; DB: diagonal bracing; XB: X-bracing; KB: K-bracing; and X2B: double X-bracing. All bracing concepts were also investigated with cap truss and belt truss (the latter for 20 and 30 storey only). One scheme, TF: through frame, has the same number of diagonal braces as the DB scheme, but distributed through all 3 bays.

9. Fig. 3 shows drifts for the frames of Fig. 2 as a percentage of building height. White bars represent bracing concepts with all pin-joints for beam/column and brace/beam joints. Black bars represent concepts with all joints moment resistant.

Gravity Load : (kip/ft)
ROOF = 1.5, FLOOR = 3.0

Wind Load : (kip)

MEMBER SIZES (two stories per size)

Frame	Int-Column	Beam	Ext-Column	Bracing
10 Storey	W10x60	W18x71	W10x77	W6x20
	W12x 96	W18x76	W12x87	W6x20
	W14x132	W21x83	W14x99	W8x28
	W14x145	W24x84	W14x109	W8x31
	W14x211	W21x83	W14x132	W8x31
20 Storey	W10x68	W16x89	W10x68	W8x31
	W12x79	W18x97	W12x65	W8x31
	W14x99	W18x119	W14x82	W8x31
	W14x132	W21x111	W14x109	W8x31
	W14x159	W21x132	W14x132	W8x40
	W14x233	W24x131	W14x159	W8x48
	W14x257	W24x131	W14x193	W8x58
	W14x311	W24x146	W14x211	W8x67
	W14x370	W24x146	W14x257	W10x60
	W14x455	W24x146	W14x342	W8x67
30 Storey	W8x35	W18x65	W8x48	W8x31
	W12x65	W18x86	W10x68	W8x31
	W14x109	W18x106	W12x106	W8x35
	W14x145	W21x111	W14x109	W8x35
	W14x176	W21x132	W14x159	W8x40
	W14x233	W24x131	W14x176	W8x40
	W14x283	W24x146	W14x193	W8x40
	W14x311	W27x146	W14x233	W8x48
	W14x342	W27x161	W14x257	W8x58
	W14x398	W27x178	W14x283	W8x58
	W14x426	W27x178	W14x311	W10x54
	W14x455	W30x173	W14x342	W10x60
	W14x500	W30x173	W14x398	W10x68
	W14x605	W30x191	W14x426	W10x77
	W14x605	W27x178	W14x455	W10x68

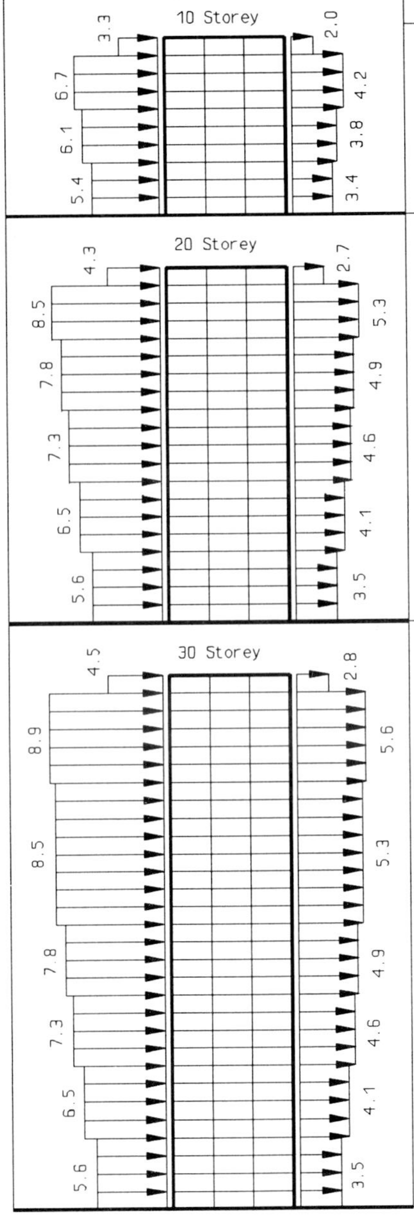

Fig. 1. Basic building frames with loads and member sizes.

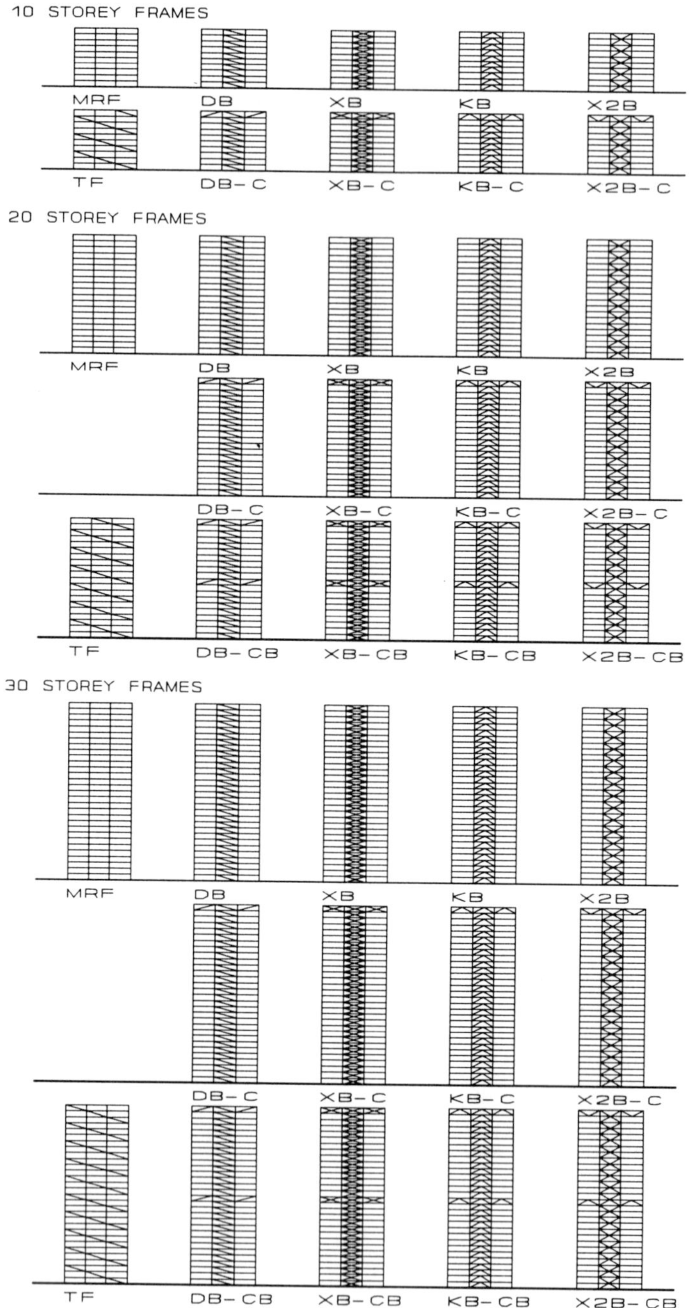

Fig. 2. Framing configurations.

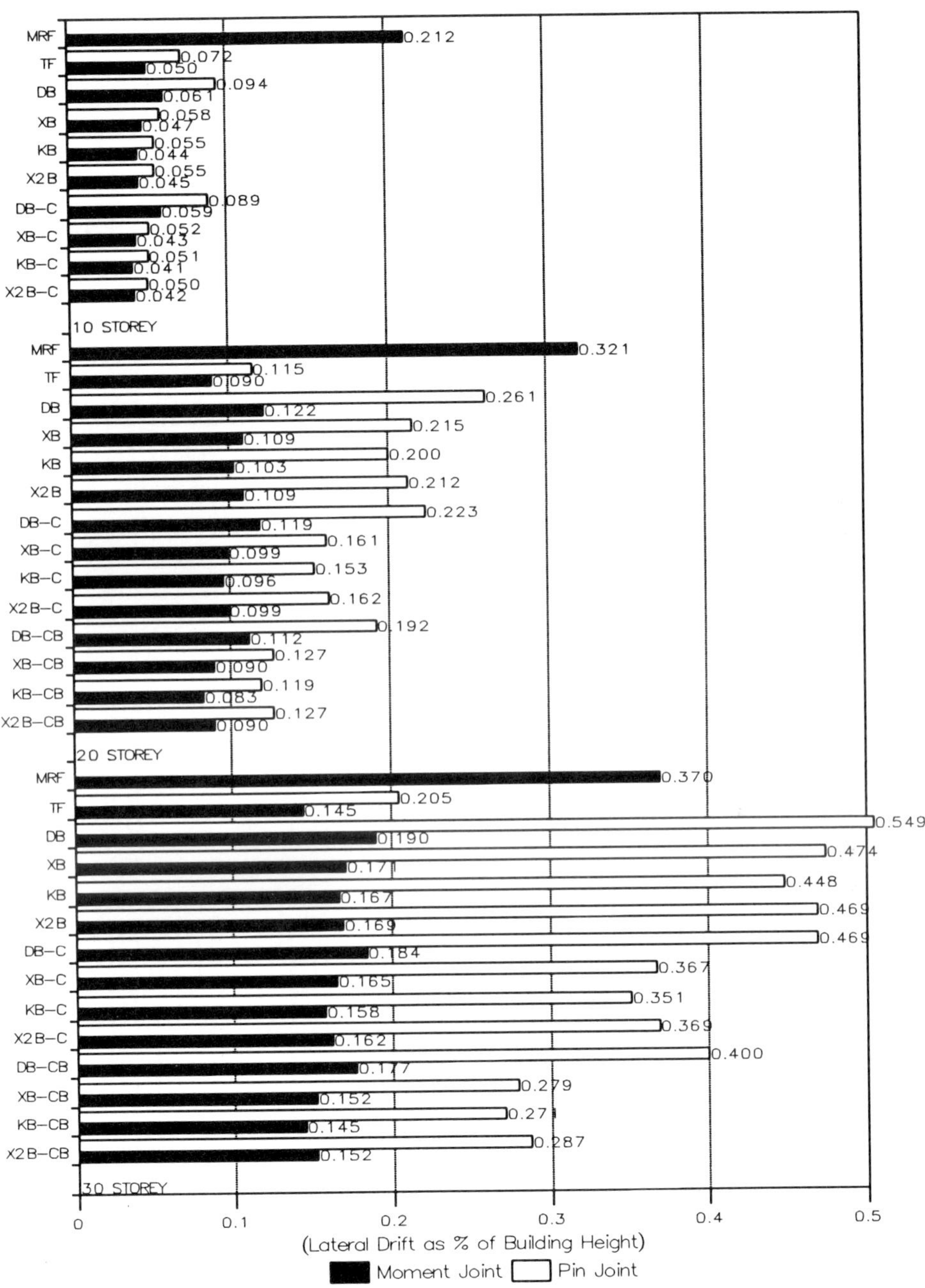

Fig. 3. Lateral drift of framing schemes with and without moment resistant joints.

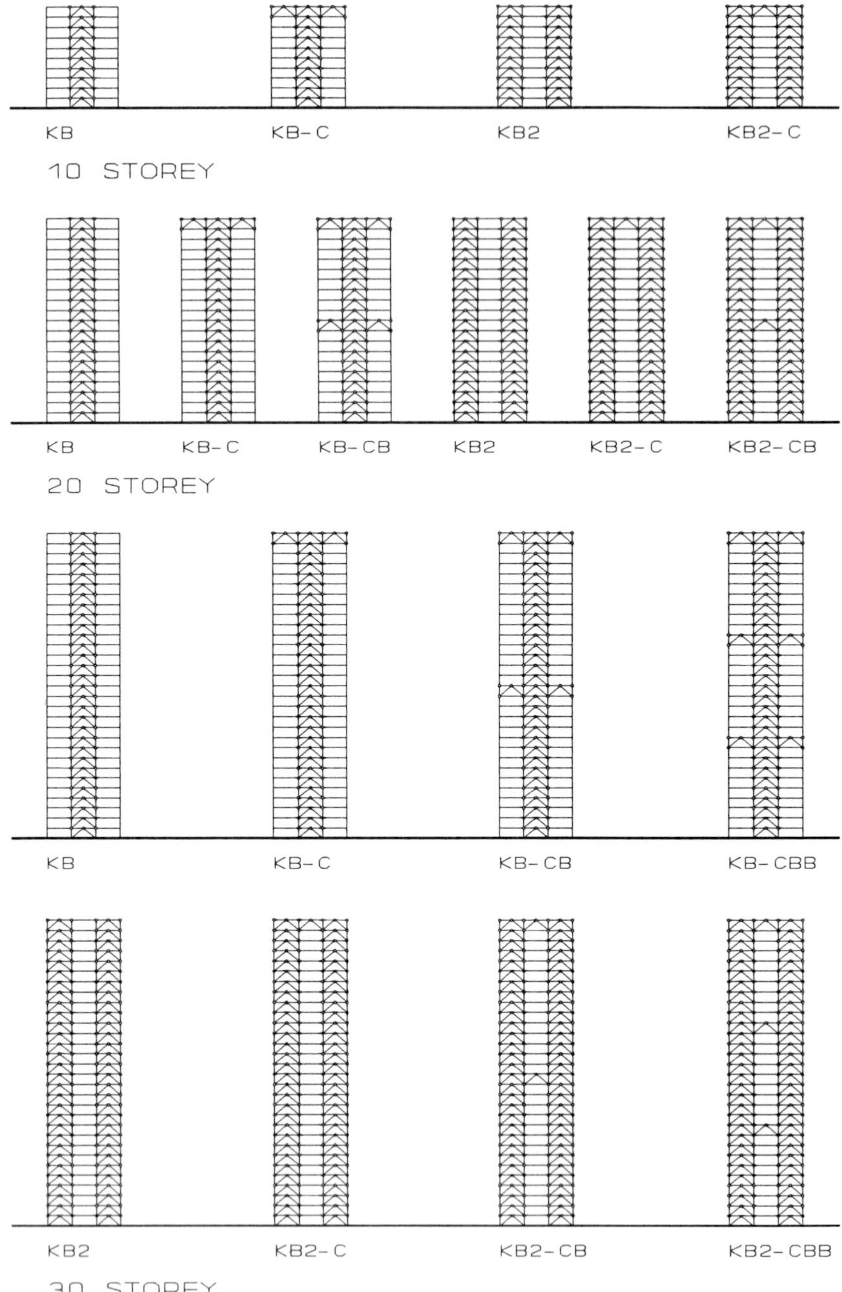

Fig. 4. K-bracing of 1 and 2 bays, with and without cap and belt trusses.

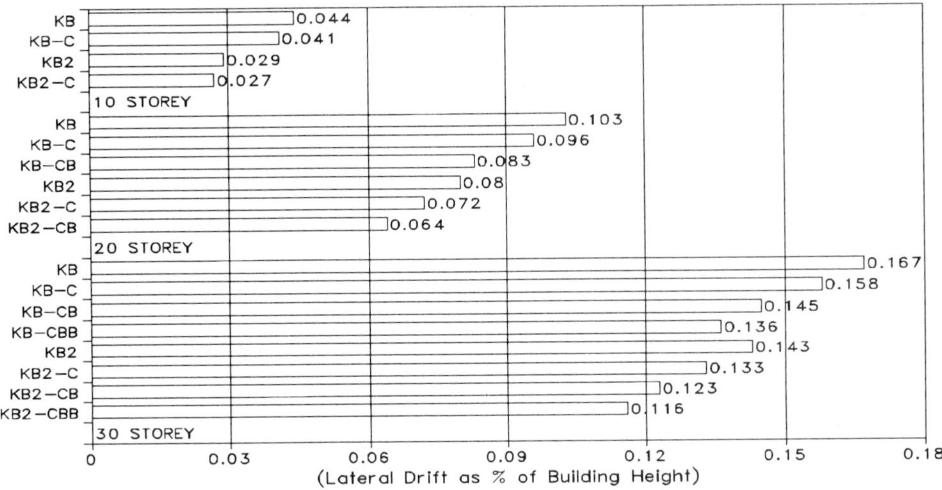

Fig. 5. Lateral drift of K-bracing with and without cap and belt trusses.

The performance of semi-rigid joints, sometimes used in practice (refs 5 and 6), can be approximately interpolated.

10. Fig. 4 compares the K-bracing (the most efficient configuration) in arrangements of 1 and 2 bays; and with a second belt truss for the 30 storey building.

11. Fig. 5 shows drifts for the frames of Fig. 4 as a percentage of building height; with all joints assumed moment resistant.

12. Fig. 6 shows drift distributions for the K-braced frames (centre bay only) comparing concepts with and without cap and belt trusses.

RESULTS

13. Comparing the various configurations and building heights of Figs 3, 5 and 6, we can make observations regarding general trends and specific configurations. There are definite differences in lateral drift between the various framing configurations and the moment resistant frame.

14. Fig. 3 shows that bracing of the centre bay is most effective for 10 storeys and least effective for 30 storeys because its height/width ratio is 4.3, 8.7, and 13 for the 10, 20, and 30 storeys respectively, i.e. the effectiveness of bracing decreases with increasing aspect ratio. Compared to the moment resisting frame, K-bracing reduces drift by 79%, 68%, 55% for 10, 20, 30 storeys of moment resistant joints respectively. Compared to the least effective diagonal bracing, the most effective K-bracing with cap and belt trusses shows drift reductions of 46%, 40%, 40% for 10, 20, 30 storeys with pin joints respectively.

15. Fig. 5 compares the schemes of Fig. 4 of moment resisting joints and single and double K-bracing with and without cap and belt trusses. Comparing single bay bracing with double bay bracing with cap and belt trusses shows drift reductions of 39%, 38%, 30% for 10, 20, 30 storeys respectively.

16. Fig. 6 shows drift distribution over the building height for single bay K-bracing with and without cap and belt trusses. The graphs show that cap and belt

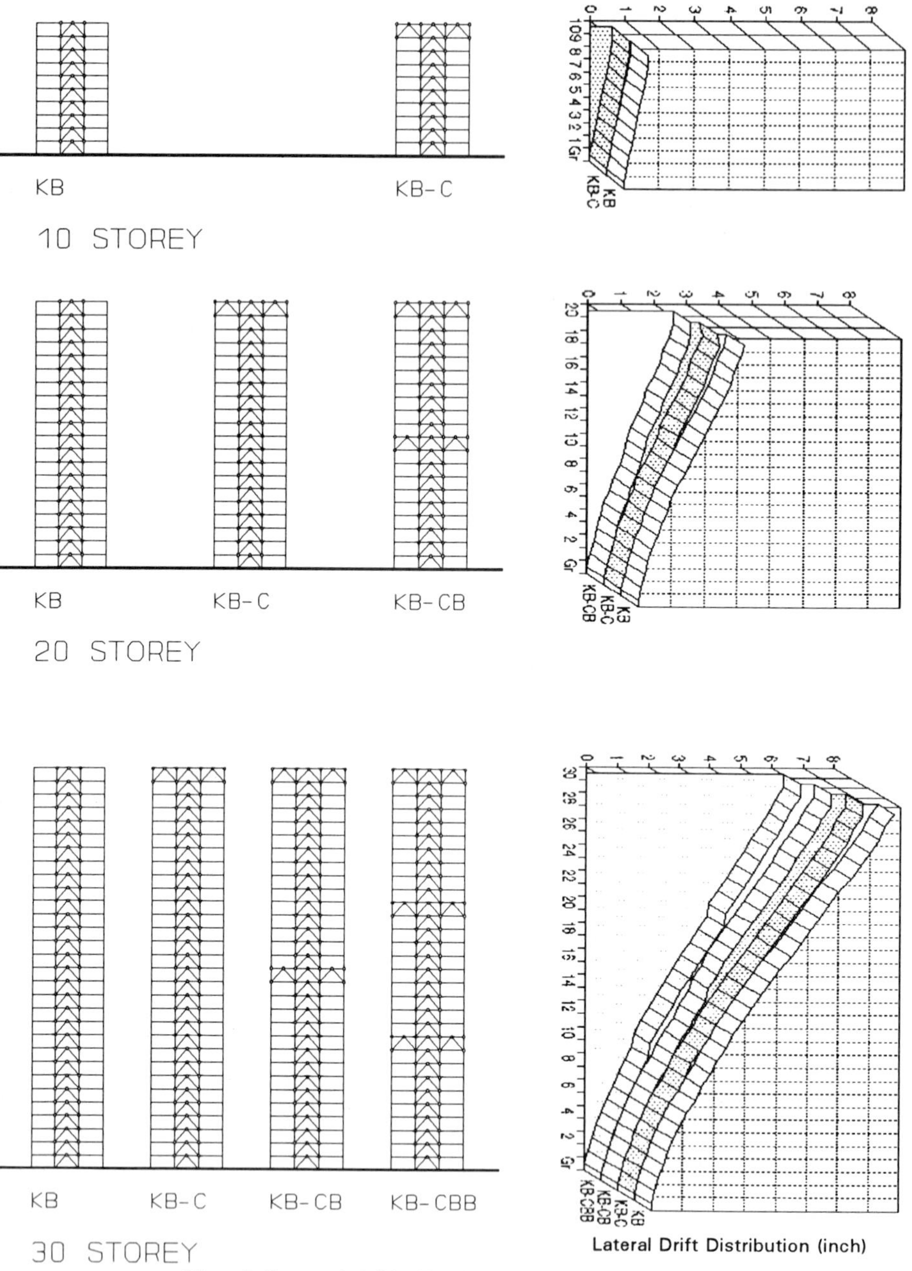

Fig. 6. Lateral drift distribution for K-bracing with and without cap and belt trusses.

CONCLUSIONS

17. To reduce lateral drift of high-rise buildings is an important design consideration in areas of high wind and/or seismic activity. The research presented here shows that selecting the most appropriate bracing system can substantially reduce drift with only minor cost differences (caused by variable number of bracing connections). A sensitivity analysis, doubling cross-section areas of cap and belt truss braces, revealed only very minor drift reductions. The reductions ranged from a minimum of 1% for the 20 storey K-braced frame with cap truss, to a maximum of 7.6% for the same frame combining cap and belt trusses. While the objectives of this study are for wind induced drift, they do not automatically apply to seismic situations. Seismic forces tend to increase with the stiffness of a building. For example in the equivalent static formula for seismic base shear $V=ZICW/R_w$, (see ref. 2) the R_w factor is 12 for Special Moment Resisting Frames but varies between 4 and 8 for various bracing schemes. This implies a lateral force increase of 50% to 300%. More research is needed to determine the effect of various framing configurations on lateral forces and drift in seismic situations.

ACKNOWLEDGEMENTS

The authors acknowledge a similar, unpublished, study of framing systems by P. Kuo, U.F. Yeh, and G.G. Schierle, which gave the impetus to this research.

UNIT CONVERSION

To convert from:	To:	Multiply by:
inch	centimetre	2.54
foot	meter	.3048
kip	kN	4.448

REFERENCES

1. NAEIM, F, Design for Drift and Lateral Stability, Seismic Design Handbook (Naeim, Editor), Van Nostrand Reinhold, New York, 1989.
2. Uniform Building Code, International Conference of Building Officials (ICBO), Whittier, 1991.
3. AISC Manual of Steel Construction Allowable Stress Design, American Institute of Steel Construction, Chicago, 1989.
4. ELM Analysis User Manual, Fujitsu Company, 1990.
5. NETHERCOT, D A, The influence of semi-rigid joint action on the behaviour of steel frame structures, Developments in Structural Engineering (Topping, Editor), Chapman and Hall, London, Van Nostrand Reinhold, New York, 1990.
6. ANDERSON, D, Analysis of semi-rigid steel frames and criteria for design, Proceedings Third International Symposium on Stability of Steel Structures, Budapest, 1990.

211. Domes of radial beam type for circular unheated buildings

I. V. MOLEV, Nizhny Novogrod, Russia

SYNOPSIS. Building systems without an outward thrust, generally known as domes of radial-beam type (ref. 1), are proposed and designed for roofs of circular unheated buildings, intended mostly for agricultural use. The radial-beam dome structures consist of radial ribs (three dimensional or plane) connected at the top by the upper ring, the lower ends of the ribs being joined together in pairs by diametrical collar beams. In present designs for such roofs, the purlins (ring member) can be included in the integrated spatial behaviour of the system. The objective of this paper is to improve the structural efficiency of circular unheated buildings constructed using spatial steel structures. Consideration is given to reducing material consumption and labour required for manufacture for buildings ranging in size from 9 m to 36 m in diameter.

RADIAL-BEAM DOMES WITH THREE-DIMENSIONAL RIBS

1. The bearing structure of the roof is made as a system of radially arranged three-dimensional truss units connected by a central member and forming a cone-shaped radial-beam dome (Fig. 1). The upper chords of the units (Fig. 2) are made of single channel sections. The lower chords and the brace are made from round bars. The triangular strut frame components are of angle section. The units incorporate roof elements of single angle section. The roof purlins of the same cross section with the truss elements are mounted between the bearing units. The roofing is of galvanised panels 0.8mm in thickness (Fig. 2). The roof panels are joined together by the stand-up welts and fastened to the purlins by cleats.

2. For ease of transportation and handling, the units are collapsible. The strut, the lower chord members and the brace may be put to transportation position between the chords when folded in their upper chord connecting joints (Fig. 2), thus forming a transportable panel. This can easily and quickly be erected in position by tightening three nuts only. The folded unit has no parts projecting beyond the size of the upper chord.

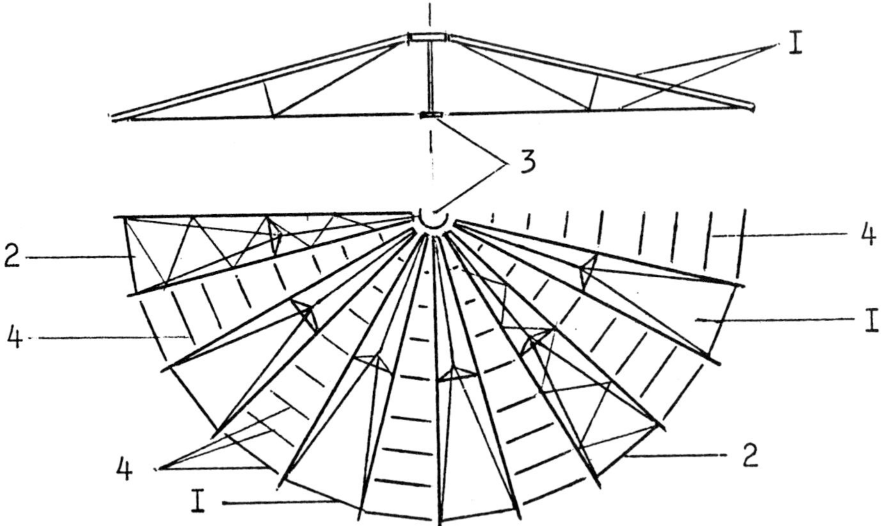

Fig. 1. Scheme of radial-beam dome structure with three-dimensional ribs:
1 - three-dimensional unit, 2 - bracing three-dimensional unit,
3 - central member, 4 - purlins.

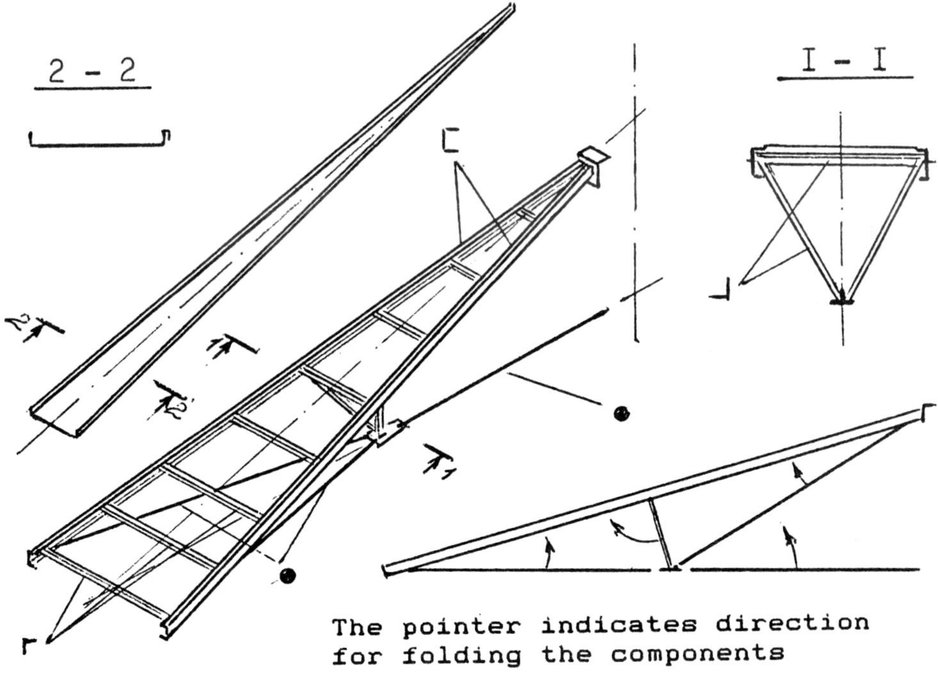

Fig. 2. Three dimensional unit and roof panel.

3. The structure can be erected without any assembling equipment or scaffolding. As each unit (radial rib of the dome) is an invariable system, no additional bracing is needed in the course of erection or operation.

4. The static calculation of the cone-shaped radial-beam dome was performed on a computer according to the spatial design schemes for two phases: operation and erection. The experimental study of the real behaviour of the roof structures, carried out on full-scale parts, has shown that the stress-strain condition of the dome constructions, under combined operational loads, corresponds to the assumptions made in the design and that the roof structure has the bearing capacity consistent with the calculated values.

5. The consumption of materials is an important characteristic of any structure. For the given 18 m diameter roof structure, the reduced steel consumption as a function of the covered area is equal to 23.75 kg/m^2 for the load-bearing constructions (including 4.85 kg/m^2 for the purlins supporting the roof covering), and 10.25 kg/m^2 for the roof panels (with allowance made for dead load of 0.4 kPa and snow load of 2.4 kPa). At lower snow loads the consumption of steel decreases. The regular variations of the mass index under changing live (snow) load for the 18 m diameter dome structure are shown on Fig. 3.

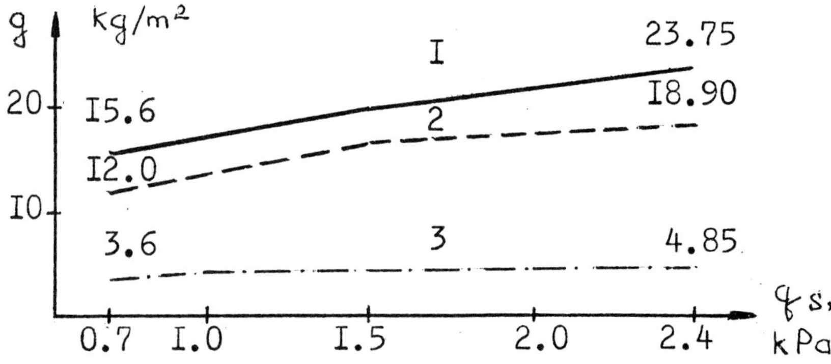

Fig. 3. Snow load variation (g) dependence of mass index (qs).

It is of interest to note the regular variation of the load-bearing construction mass under various spans, for the radial beam domes with three-dimensional ribs, which is presented in Fig. 4. The analysis of the data shows that such structural designs are most appropriate for diameters up to 24 m, the best performance being achieved by the 18 m domes. For this reason large scale construction of more than 1,000 buildings has been undertaken recently, using the roof structure design described to cover circular buildings of 18 m diameter.

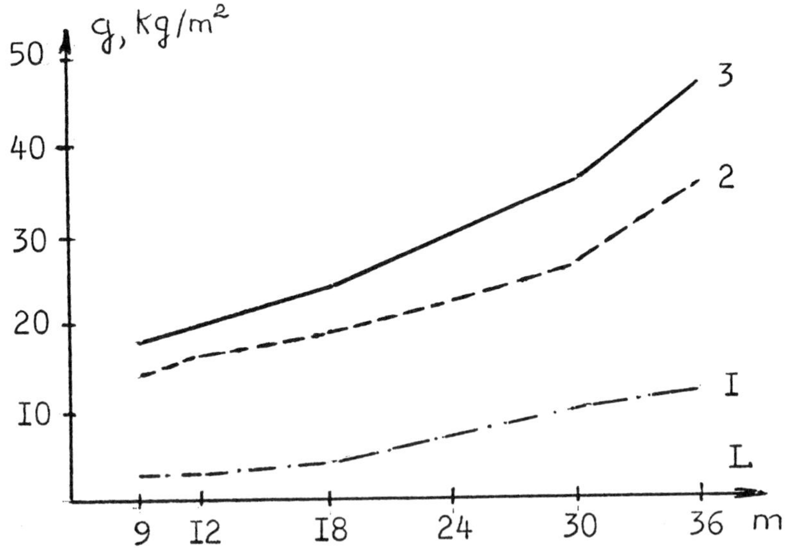

Fig. 4. Rule for variation of mass of the radial-beam domes with three dimensional ribs at various spans (L):
1 - steel consumption for purlins, 2 - steel consumption for bearing members of the dome, 3 - overall steel consumption.

RADIAL-BEAM WITH HANGING ROOFING

6. A cone-shaped radial-beam dome structure having 12 meridianal plane members and a hanging roofing has been designed for buildings 24m in diameter (Fig. 5). Each radial rib is made of channel sections in the form of a truss with few elements, in which all the joints but the lower one have no auxiliary inserts. This facilitates manufacture, and reduces the labour requirement and cost of construction (Fig. 5). In this case the radial ribs are joined together in the centre by 8 members having a compressed upper ring and a tensioned lower ring (at the level of the truss lower chord). The invariability of the roof structure and stability of the upper chords of the radial ribs are provided by the system of bracing. The braces are made of angle sections and assembled with single precision bolting. The weight of the bearing constructions of the roof and bracing (with allowance for dead load of 0.4 kPa intensity and live snow load of 2.4 kPa) is 9.976 tons, or 20.52 kg/1m^2 of the building floor area (the area covered by the dome structure is actually larger than that of the building floor).

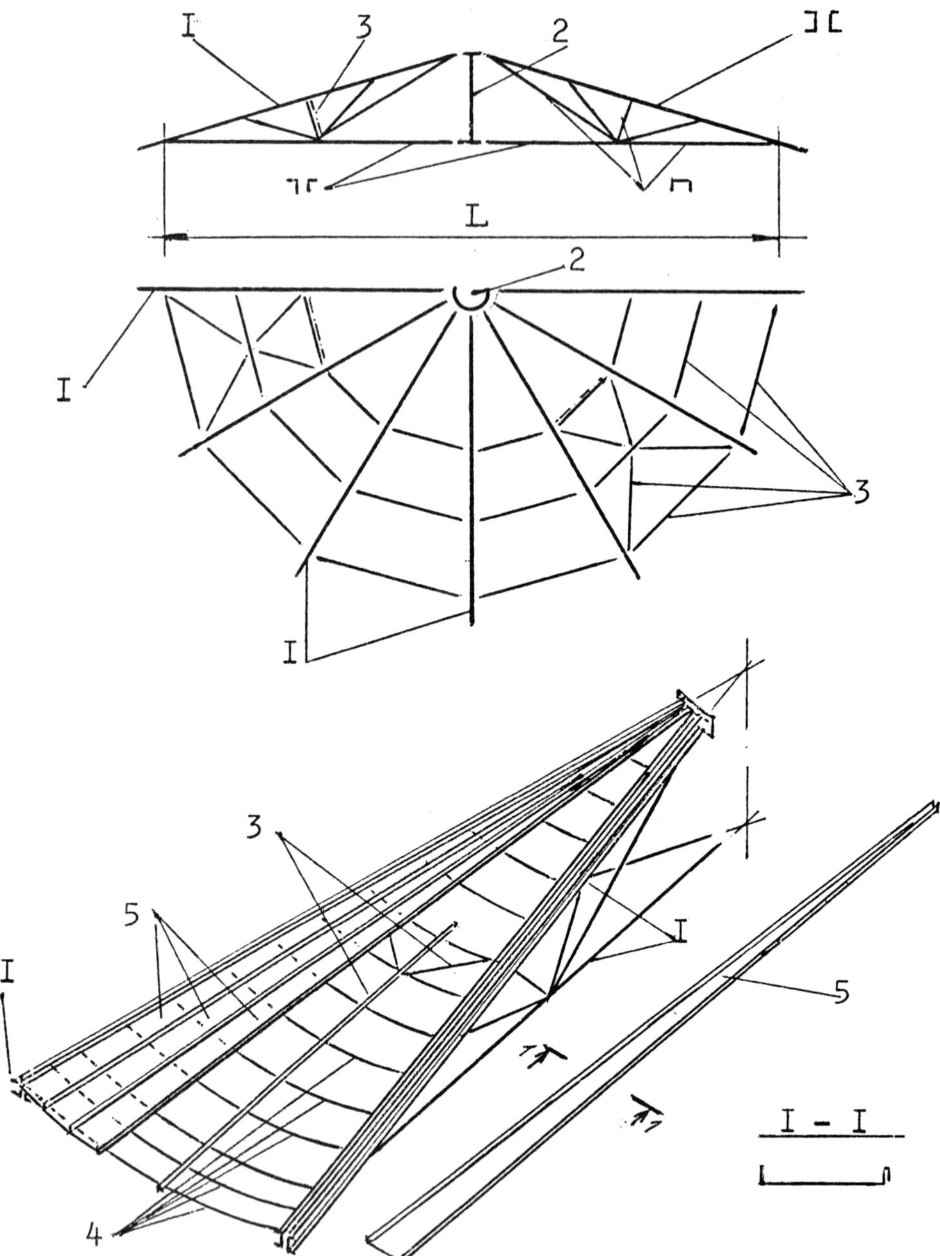

Fig. 5. Scheme of radial-beam dome structure with hanging roofing:
1 - radial rib, 2 - central member, 3 - bracing, 4 - cables (purlins).

7. The roof bearers mounted between the dome ribs are flexible cables made of steel rod with diameter of 12mm. The cables are doubled in places where they are more than 3m long. The cables are spaced 0.9 - 0.95m permitting the use of flat roof panels from sheet steel 0.8mm in thickness (like panels of the roof 18m in diameter). The assembled structure possesses adequate rigidity both under symmetrical and asymmetrical loads as shown by the practical performance of buildings having such roofs. The thrust developed in cables in case of asymmetrical loading is taken by the upper chords of the radial ribs in conjunction with the bracing. The weight of roof bearers (cables) is 1.213 tons (2.5 kg/m^2 on the floor area). The sheet steel consumption for the roofing panels, as a function of the building floor area, is 11.86 kg/m^2.

8. The structural design of the roof and framework of a polycylindrical circular building allows for erection by the lift-roof method. The roof structure is assembled at the ground level and then lifted into position and requires no temporary supports.

9. The study of rules governing the weight characteristics of such roofs has indicated that the best steel consumption indices are shown by the systems spanning 24m. Good results may be obtained with structures spanning 30m. Thus, the efficient span for radial-beam structures, with hanging roofing, ranges between 24 and 30m. A small scale construction project (up to 10 buildings), including buildings of 24 m in diameter with polycylindrical walls and a roof of a radial-beam dome structure with hanging roofing, lends support to its high reliability in operation.

RADIAL-BEAM DOMES WITH LATTICED UPPER CHORD

10. For dome structures more than 30 m in diameter the structural design adopted for the roof spanning 24 m proved to be inexpedient because of high material consumption. For this reason one more design of a radial-beam dome was proposed and developed spanning 36 m incorporating plane radial ribs with few elements and latticed upper chord (Fig. 6). This roof structure consists of 18 radial ribs joined together in the centre at the levels of the upper and lower chords by a special member. To provide for invariability of the system (under asymmetrical loads) and stability of the upper chord of the radial ribs there is a bracing system. The bearing members of this roof structure include circular and meridianal purlins incorporating ties which take the load component acting along the roof slope.

11. The radial ribs consist of a latticed upper chord, lattice elements (of a sub-diagonal type) and a lower chord. The upper chord of the rib is designed as a truss 1000mm deep having parallel chords made from channel sections which lie horizontally with their flanges in a vertical position. The lattice struts made from angle sections are attached to the channel flanges and oriented so that they are in tension under any possible loading. The lattice struts are of single angle sections. The radial rib lattice elements are constructed of thin-walled pipes and the lower chord members are made from hot-rolled equal angles. Prefabricated joints are welded and the erection joints are bolted. The roof covering elements are of two

types, either, wooden boards with continuous sheathing are laid on steel purlins and covered with roll roofing felt, or, the steel purlins are covered with wooden boards of rare sheathing underlying the steel sheet roofing.

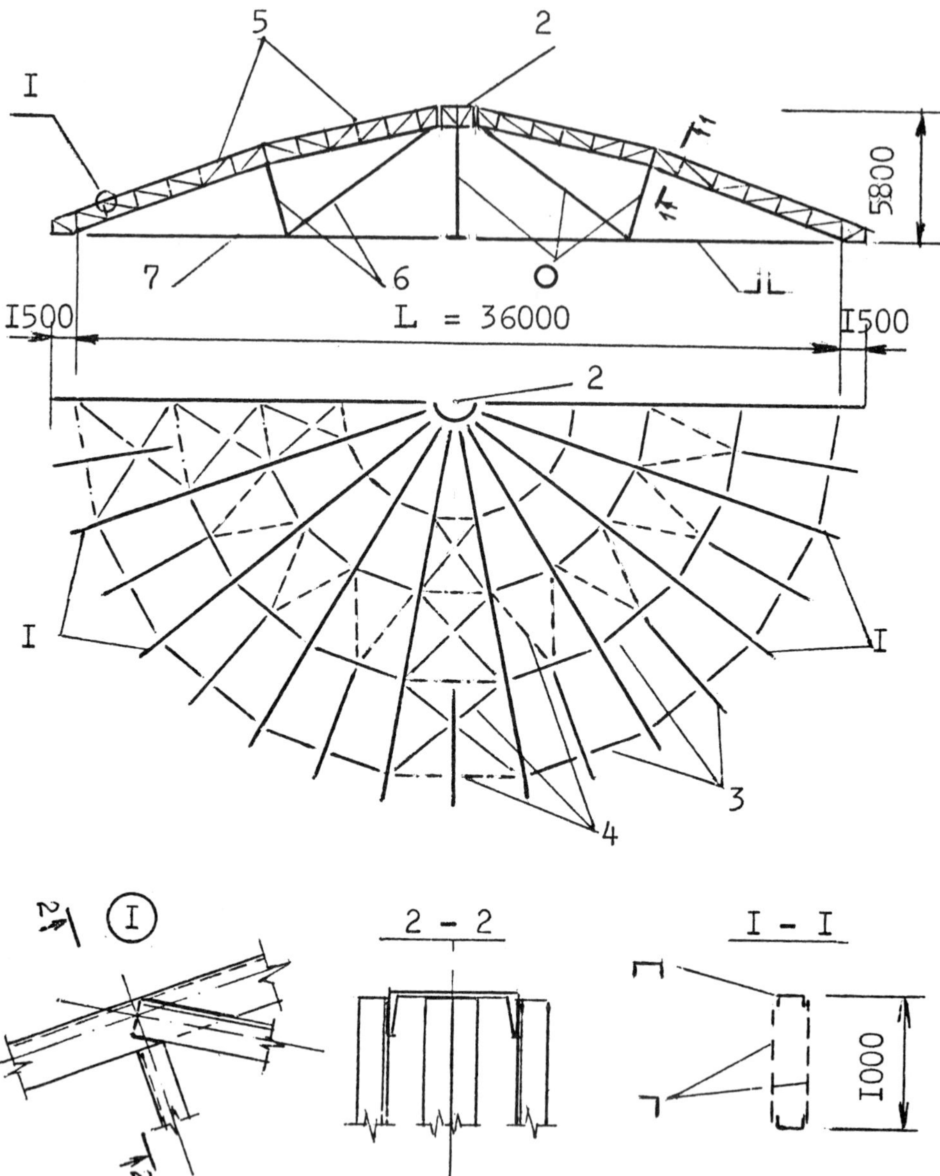

Fig. 6. Scheme of radial-beam structure with latticed upper chord:
1 - radial rib, 2 - central element, 3 - purlins, 4 - bracing, 5 - upper chord the radial rib, 6 - lattice, 7 - lower chord.

12. Roof constructions are designed for the following loads: dead load with intensity of 1.0 kPa; snow load of 2.4 kPa and a wind load of 0.42 kPa. The overall dimensions of the roof are: span of 36 m (between centre lines of supports); maximum diameter of 39 m (including the overhang); rise of 5.8 m and covered area of 1170 m^2. Steel consumption for the roof calculated as a function of the covered area is 27.77 kg/m^2. This includes: radial ribs at 21.33 kg/m^2; braces at 2.81 kg/m^2 and purlins at 3.63 kg/m^2. Consumption of wood is 36-51 m^3, depending on the type of roofing. In 1992 an experimental building 36 m in diameter with polycylindrical walls and a radial-beam dome structure of a latticed upper chord type was completed.

13. The study of the rules governing the weight characteristics of radial-beam domes, of various structural design depending upon the area covered (Fig. 7), has shown that for each design there is an optimal size range where the dome structure may be realised with the minimum material consumption. For the radial-beam domes with three dimensional ribs the optimal span is in the range of 9-18m, for those with hanging roofing the optimal range is 24-30m and for those with latticed upper chord the optimum size is 36m.

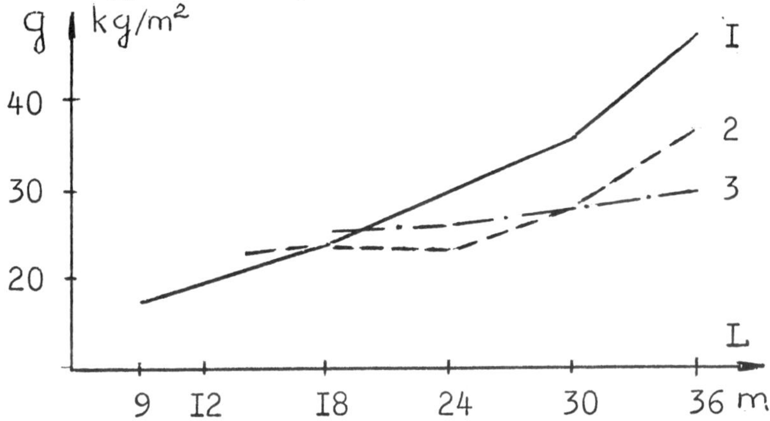

Fig. 7. Rule for variation of mass of the radial-beam domes at various spans: 1 - radial-beam domes with three-dimensional ribs, 2 - with hanging roofing, 3 - with latticed upper chord.

CONCLUSION

14. Radial-beam domes represent an efficient roof system to cover circular unheated buildings having no aerial conveyers.

15. The advantages of the structures described are the simplicity and clear-cut static scheme, production and erection effectiveness, and high reliability in operation. The positive technical and economic indices are beneficial for a wide application of the radial beam domes in agricultural construction.

REFERENCES

1. BELENYA Ye.I.(Ed). Steel Structure. p.432., Stroyizdat, Moscow, 1985.

212. A prefabricated long-span plywood/timber hyperbolic paraboloid in Dublin

D. TULLY, Daniel F. Tully Assoc. Inc., USA

SYNOPSIS. The Irish Basketball Association built its National Basketball Arena. It is the first prefabricated plywood/timber hyper structure of its type known in Europe. The hyper shell spans 36.9m x 65.85m (120' x 214') with a rise of 13.8m (45') to its highest peak. The components are plywood (CDX-STRUCT I) 15.4mm (1/2") laminated off-site into shells 46.2mm thick x 9.23m x 8.07m. Edge beams are laminated long leaf yellow southern pine beams.

1. In November of 1980, I met several members of the Board of Directors of the Irish Basketball Association (IBBA) at the Westchester County Airport in White Plains, New York. At this meeting there unfolded plans for a program which could provide the unemployed teenager a place to play. It was the intention of the Irish American and the Irish Basketball Association to fund several large multipurpose sports facilities and arenas throughout the country, beginning on a site somewhere in Dublin.

2. As the architectural plans developed over the coming two years, it showed a target budget of $8,500,000.00 for each building. The building was to contain an arena seating for 5,000 for basketball and other court activities together with a 50 meter pool and diving facility. Additionally provisions would be made for an indoor two hundred meter track, sports medicine facilities and other support spaces.

3. Raising the funds for this ambitious project became an impossible task. The Americans did not donate as expected and most of the funds came from the Irish government and Irish sponsors. Furthermore, the project size was by necessity, greatly scaled down, so that the current programme provides one building only which is approximately 37m x 66m and which contains a three court basketball facility, lockers, fitness spaces, lobby and a hospitality area. Its approximate cost was $2.5 million (U.S.).

4. Wood/plywood shell structures have been site built in many areas of the U.S. over the past twenty-five years. The system utilises a lamination of three layers of plywood and one layer of face boards placed over directrices or generators of the hyperbolic paraboloid (hyper). The first prefabricated shell was

built in Tilton, New Hampshire (see Photos 1 and 1A) and was for a square building of 110 ft. x 110 ft.

5. Prefabrication was developed to reduce the high cost of field fabrication and to increase the reliability of the fabricated product (quality control). The pieces to be shipped were cut from the square components of the structure 27.5' x 27.5' and were cut along lines parallel to the diagonal (see Fig. 1) cutting through the tensile plane thus requiring a reconstructed tensile joint (see Photo 2) which would carry the design load plus a factor of safety. Factors of safety along these lines of action were taken to be in excess of nine.

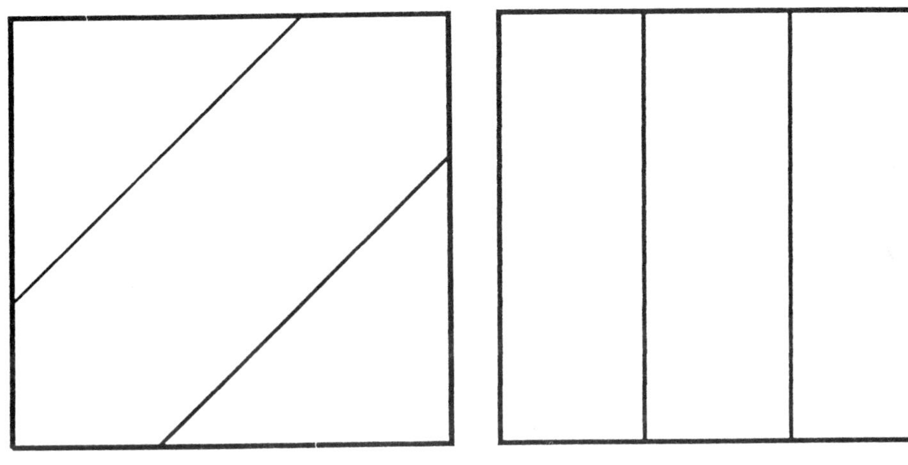

Fig. 1. Fig. 2.

6. Later as shell sizes and span requirements increased the shell sizes approached 40' x 40' yielding clear spans of more than 160' cutting the shells along the diagonals provided shipping problems which became formidable.

7. After the design and construction of many dozen athletic facilities utilising the larger size shells, effort was placed in preparing a design which would allow the shells to be cut along the generator lines (see Fig. 2). As can be seen, the diagonal cut was developed to allow the erection to proceed using the stiffness of the doubly curved centre sections and avoiding any supplementary stiffeners.

8. Cutting the shells along the generator lines would allow both the mass production of pieces and the bundled shipping. In the case of the Irish project, vastly reducing the bulk of the project shipped and thereby the cost of shipping. Container trailers are delivered to the factory site in upstate New York, loaded with the shell parts and shipped to the final destination in the states or overseas. The container trailers are delivered to the site and the building can be erected directly from these (see Photo 3). The penalty paid for the generator cut is the necessity for erection beams to provide the correct geometry for the individual

parts until one whole shell is assembled in place and has the stiffness of the entire geomertry.

Photo. 1.

Photo. 1A.

Photo. 2.

Photo. 3.

Photo. 4.

Photo. 4A.

Photo. 5.

The process which has helped the fabrication of the shells was the use of a tongue and groove edge in such a manner that the first piece in place has the lower edge of its groove extended slightly to receive the tongue of the mating part prior to its being correctly and finally in place (see Photo 4).

9. After years of performance of the joints with a factor of safety of nine or greater (see Photo 5), we determined that the same factors should be used on the generator joints. At least in so far as the calculated and accumulated safety factors were concerned.

10. At the joints of the Irish project, the membrane stress being approximately 900lbs per linear foot, both a tested adhesive (Dow 3780) and 24 - 8d nails per linear foot in double shear together with a cover plate of 1/2" plywood, 24 inches wide also nailed and glued in place with 24 nails per foot in single shear was used to provide a capacity of 200 lbs. per inch (2400 lbs per ft.). The total is approximately 8640lbs per linear foot. The principal reason for this abundance of caution is the lack of absolute knowledge as regards the true membrane stress at any given time as well as the considerable point of concentration of flexural stress in a plane of potential bending failure - the reassembly joints.

11. Shell to edge member connections remain 1/2" x 6" or 1/2" x 8" lag bolts and washers at 4" centres to supplement this detail.

12. A bead of 3/16" adhesive is placed along the top surface of the beam (Dow Corning 5230) which has the capacity to bond and cure whether the wood is wet

or not and which develops a glue line strength of about 120 psi on this basis. The lags take the full load of about 1100 lbs per foot and the glue line does the same.

13. Projects using this adhesive which we have disassembled twenty years after the installation still had substantial glue line strength. Although the glue is not used in the calculation of performance of the structure, it can be relied upon to assist in stiffening the structure after the occurrence of drying of the plywood around the lag bolts where the plywood may shrink away from the bolts either radially or in depth (tangentially).

14. In the United States of America, a surface of plywood which is more than 1-1/4" thick and which is uninterrupted by through joints or cracks is classified as "heavy timber" when used in conjunction with heavy timber beams or columns. This classification is for fire purposes and allows large primarily single storey athletic facilities, provided that the timber components are at least twenty feet above the floor, that proper egress exists and the access for fire trucks is on all sides of the structure. Only recently in some jurisdictions are sprinklers required and in fewer jurisdictions is fire preventive treated wood required.

15. The Republic of Ireland was in the midst of developing an updated set of building requirements as we began this project but during that time two fires of note took place, one in Britain and one in Ireland which involved a place of public assembly and loss of life. Understandably the trend in adopting fire regulations was conservative and authority given to the fire officer was virtually absolute. As a result of this, the regulations still not having been adopted, we were required to provide a Class "O" fire rating (impossible with wood) or use one approved fire protective treatment which was a pressure impregnation, Koppers Non Com-X.

16. Due to the California plywood incidents of structural degradation in plywood from that particular treatment, the company ceased manufacturing the product.

17. We were then faced with a dilemma. A contract having been signed for the production of the roof, we consulted the British testing laboratories who were certified to test for fire resistance and sought out those characteristics of the formerly approved product which made it acceptable. We then sought out an alternative product which was a two coat surface applied intumescent clear sealer that had acceptable performance.

18. The product was "Albert D.S." and it actually out performed the originally approved product in the tests made in Britain.

19. The licensed manufacturer of the hyper roof, Unadilla Laminated Products, had not dealt in international trade before with this client and the Irish Basketball Association had not dealt with the manufacturer. Therefore there were at first some concerns about the transfer of product, contract commitments and payment as there was some concern about the availability of a letter of credit for the client which has been recently organised. This was all resolved by escrow accounts. As the product was manufactured and inspected by me, the client would place progress payments in an escrow account in Ireland. The funds would be released on my authority which I would give starting with the loading of the roof elements

onto the ship in New Jersey and ending with the delivery of the product to the site in Dublin.

20. Roof elements were loaded onto trailers and containerised cargo carriers for shipment. The largest element of plywood was 7'6" x 37' and the longest piece of timber was approximately 70'. Plywood panels were stacked 24 deep and could have been stacked on edge or horizontally. Each piece was the exact replica of the adjacent one so that there was a sandwiching of all the components on five separate containers. Shipment took six weeks and all elements were unloaded on the same trailers which were used at the factory in the USA.

21. The contractor, Rohan Construction Limited of Dublin, had not erected such a structure before and the process had to begin with a thorough course in erection techniques for the contractor so that he could provide an intelligent tender and then so that he could safely and efficiently erect the structure. The erection process was smooth and uneventful. The sort every engineer hopes of his designs. After about six weeks interrupted by a period of intense rain, the roof was erected and temporary supports were removed.

22. The maximum deflections occurred at the intersections of the horizontal beams and were nominal, approximately ±3/8" to zero.

23. During the roof erection the sidewall masonry was begun but was not completed simultaneously with the roof decentering as the roof structure is functionally independent of the sidewalls, except that the roof does provide stability to the walls at the top. The sidewalls and roof were designed to resist 120mph sustained winds.

24. The design load assumptions were as follows:

Snow:	30lbs per ft^2
Wind:	120 mph
Dead:	14lbs per ft^2
Plywood:	1/2" nominal CD EXT Grade Douglas Fir 5 Ply
Laminated Beams:	Long leaf Southern Pine, architectural grade
Concrete:	4000 psi equivalent

25. The edge beam interaction was designed utilising the Hankinson analysis for merging elastic moduli of the combined assembly of timber and plywood taking into consideration the constantly variable angles of intersection of the plywood elements with the timber. The use of field applied adhesive, it is believed, enhances this interaction.

26. All factory adhesive was phenolic resin. Machine or roller applied. Fasteners between plywood layers were steel or brass staples. The function of the staples is to hold the plywood sheets in contact until the adhesive bonds. Nevertheless and in the event of incomplete curing of the adhesive or of an incomplete contact, the staples will carry the load. Plywood assemblies were randomly tested by sampling 2" cores and testing for shear. No tests of less than 120psi were uncovered and no glue line failures between sheets of plywood occurred.

27. Membrane stresses were transferred to the edge members by the lag bolt and field gluing technique. Tongue and grooved assembly of joint in the shell

field were nailed and glued together with the addition of a cover plate of 1/2" plywood which was nail-glued to the adjoining parts.

Architect/Engineer: Daniel F. Tully
Irish Architect: Michael Kenny - Dublin, Ireland

This article is dedicated to the memory of Alfred L. Parme without whose invaluable contribution to the design of shells, practitioners such as I would be found wanting.

213. Thin shells for space station and lunar station using gas-supported forms

A. WILSON and P. AALIPOUR, Brigham Young University, USA

SYNOPSIS. Thin-shell structures provide unique advantages for space construction. They utilise a minimum of structural materials, transfer loads efficiently, and provide a large factor of safety. Construction of thin shells in space has seemed impossible until the use of gas-supported forms made the process very economical. Spheres, cylinders and combinations are possible in space using gas-supported forms. Spheres can be created in space, one inside the other, providing several concentric spheres, which allows zoning of pressures for specific purposes. The smaller interior sphere with higher pressures could be used for living quarters while reduced pressures could be used for outer spheres where storage or other requirements are specified.

INTRODUCTION

1. The ability of man to enter space and return has created a desire to stay longer in space to experiment and learn about the universe from a different perspective. In order to stay longer in space an intermediate station, referred to as a "space station," is desired. This space station must be capable of providing living quarters for people, a docking area for space vehicles to come and go, life support facilities for humans to exist for an expanded length of time, protection from the elements and a friendly environment where experiments can safely be conducted for the benefit of mankind. The space station could be a necessary part of future space exploration which may involve space factories, space telescopes, manned flights to the planets, mining the asteroids, inhabiting the moon, building solar power satellites, and allowing the growth of space colonies.

2. Recent studies have suggested the possibility of using a unique structural system for creating a space station or moon station utilising thin-shell structures built with fibre-reinforced composites and using gas-supported forms as the construction system. Larger structures are possible when built at the final location rather than if pre-built on earth and transported into space.

3. The sprayable fibre composite materials can be applied using automated systems which limit the time required for people to be involved in the actual

construction process. Less human involvement means less risk and greater safety for construction personnel in space.

4. The unique environment in space is characterised by the presence of a vacuum, high temperature gradients, and micro-gravity. Structures are exposed to radiation generated from the sun's surface. The outside surface temperature of a structure in space will range from +300°F (149°C) on the hot side to -300°F (-149°C) on the cold side. Micrometeorite particles travelling at high speed will impact the space structures and must be considered in the design.

5. Shell structures that are fabricated in space will not be subject to the lift-off forces from earth but will be pressurised structures in space. The gas pressure used in each structure will vary depending on the use and application of the structure. Higher gas pressure will be used where astronauts spend most of their time.

6. Currently NASA has contracted with several aerospace companies to build a space station. The space station will consist of many parts. The living and laboratory parts are called modules. Each module has a cylindrical shape, 15ft (4.6m) in diameter by 60ft long (18.3m). The aluminum modules are designed for lift-off forces constructed on earth and will be shipped into space by shuttle. The size of the module is limited by the shuttle size and only one module will be transported for each shuttle.

7. Using the presently designed space station, or a portion of the space station, as the base of operation for thin-shell construction in space, much larger structures can be built having a great economical potential. For example, a 100ft (30.5m) diameter sphere would provide a space volume of 523,300ft^3 (14,810 m^3) compared with NASA's cylindrical module which provides 10,600ft^3 (300 m^3) of space volume, and each would take one shuttle flight for transportation into space.

8. Since the size of a thin-shell structure in space is far less restrictive than transporting a pre-built structure from earth, many sizes and shapes or combinations of different sizes and shapes can be constructed in space using inflatable fabric forms and sprayable fibre reinforced composite materials.

OBJECTIVE

9. The objective of this research was to develop an economic system for building a space structure capable of providing facilities in space for the space station as well as other applications. It is possible to develop and build larger structures in space and use less physical input by astronauts, creating a safer environment. Expandability, repair, replacement, and technological updating in the future are possible with less effort than other systems. A module can be used as an element in the space station system for laboratories, living areas, logistic transport, environmental control, and life support systems. This module will be built in space by using a composite structure created from liquid materials shipped into space. The advantages of composites used to create sandwich-type building components over conventional materials include superior strength to weight ratio, high stiffness, low thermal expansion characteristics, tailored mechanical

properties and the ability to combine several separate traditional parts into a single composite part (see Figs 1-3). Structures built from composites in space are not required to withstand the large forces created during lift-off but instead need only be designed for the much smaller loads that actually exist in space.

PROPOSED SPACE STRUCTURE

10. A sphere 100ft (30.5m) in diameter was arbitrarily selected for analysis and comparison purposes. Originally a structure consisting of floors and partitions rigidly connected to the shell was considered. This structure was statically indeterminate to a high degree making the analysis time consuming and expensive. Construction of the sphere with rigid partition-to-shell connections would be labour intensive and hence expensive. It was perceived that concentric spheres would have many advantages. The spheres become statically determinate under pressure loads, are much easier to build and easy to utilise. Several concentric spheres with different pressures could be held in a relative position with respect to each other simply by utilising a minimum of six shock-absorber-type flexible connections between any two adjacent spheres.

11. The concentric spheres provide an economical means of utilising different pressures based upon the space utilisation. The smallest sphere could be used for living areas using higher gas pressure for comfort and safety while larger spheres could use reduced pressures for such activities as research. Still lower pressures could be used in the larger spheres where storage was utilised. Gas locks between zones makes for ready access from one sphere to the next sphere. Larger spheres could be designed for smaller pressures, resulting in a more economical use of materials. There would be a greater degree of safety for astronauts with living quarters located within an inner sphere protected from space problems by several outer spheres.

12. The outside surface temperatures were calculated from radiation-conduction equations, using the principle of heat transfer (Aalipour 1990). The surface of the space station will be exposed to radiation from the sun. The actual temperatures on the structure surface can range from +300°F (149°C) on the sunny side to -300°F (-149°C) on the shady side. The actual temperatures can be reduced by using a reflective coating on the outside and by using an insulating layer directly beneath the gas-supported form. The insulating layer could be a low-density polyurethane foam layer.

13. Fabrics used for gas-supported forms must have good strength, be durable, resist impact forces and gas pressure, be resistant to extreme temperatures and have the desired surface characteristics. Gas-supported forms can be fabricated into many desired shapes such as spheres and cylinders and also have variable size. Fabric can be folded and transported in a small size before inflation. Kevlar is one fabric that could be used as a gas-supported form for the construction of space structures.

14. The amount of gas pressure needed to inflate the fabric varies with the size of the form. A smaller size sphere requires higher pressure, while a larger size sphere requires lower pressure. In order to remove wrinkles from the fabric a

minimum gas pressure of 0.05 to 0.10psi (345 to 690Pa) is required. Inflating a 100ft (30.5m) diameter sphere to 0.05psi (345Pa) gas pressure requires a cylinder 1.67ft (51m) in diameter and 6ft long (1.83 m) at 2000 psi (13,780 kpa) pressure.

15. The construction of a spherical structure 100ft (30.5m) in diameter would start with the inflation of the Kevlar gas-supported form. A 4 to 5inch (102 to 127mm) thick low density polyurethane foam (1 to 2lb/ft^3) (157 to 314N/m^3) would be sprayed on the inside of the slowly rotating form. Following the foam a 1/2-inch (12.7mm) thick structural layer of graphite/epoxy composite will be sprayed on the inside of the low density polyurethane foam.

16. The spray gun and chopper machine used to spray the graphite/epoxy composite plus the foam spray equipment are attached to a movable arm which in turn is attached to a telescoping central shaft within the gas-supported form. Hoses transferring the polyurethane and the epoxy resin transfer the liquids from outside the gas-supported form to the spray guns within the form. All the spraying of the polyurethane foam and the graphite/epoxy composite will be automated. Polyurethane foam was successfully sprayed in a vacuum under various conditions. (Penfold and Miller, 1990).

17. Other spheres as needed can be constructed inside the larger sphere. The inner spheres could be constructed using the same technique, except each sphere has one structural layer of graphite composite material.

18. The inside surface of the largest sphere can be utilised for attachment of equipment, machines, experiments, utilities, etc. The inside and outside surface of all interior spheres can be used as needed.

19. The flexible connections between the concentric spheres, used to hold each sphere in its proper relative position, consists of a shock absorber with a light spring attached. When forces produced by pressure and thermal loads try to transfer to other elements of the space structure, the flexible joints eliminate these forces and act like a weak spring. The loads do not transfer to other parts of the structure and local stress concentrations are minimised.

20. The cost of shipping materials into space by the space shuttle is around $4500 per pound (Gregg, 1990). If a structure could be built utilising much less material because of the shape or because of the design forces, the potential for decreasing costs becomes significant. Thin shells in space constructed using gas-supported forms and designed for much smaller forces because lift-off forces are minimised offer substantial savings. The size of the structure is not limited to the size of the space shuttle, thus very large space structures are possible.

21. It is feasible to construct a permanent space structure using gas supported forms and spraying liquid fibre reinforced composite materials.

22. Many sizes and shapes of structures, such as spherical, cylindrical, and elliptical, can be constructed in space using gas supported forms, composite materials, spraying techniques, zoning the pressures, and flexible connections. Various shapes and sizes can be combined together in space using this construction technique.

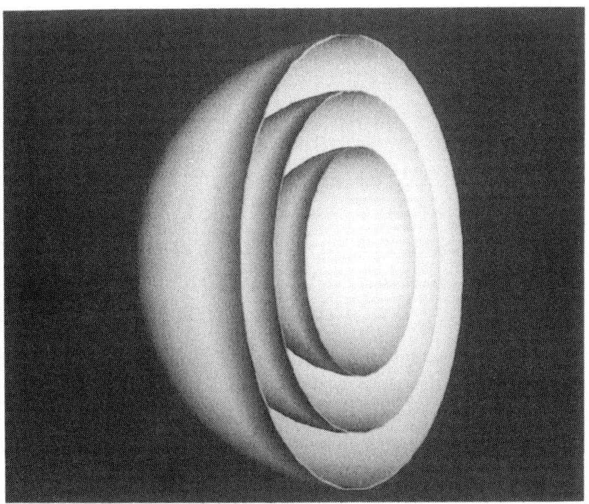

Fig. 1. Section through proposed space station.

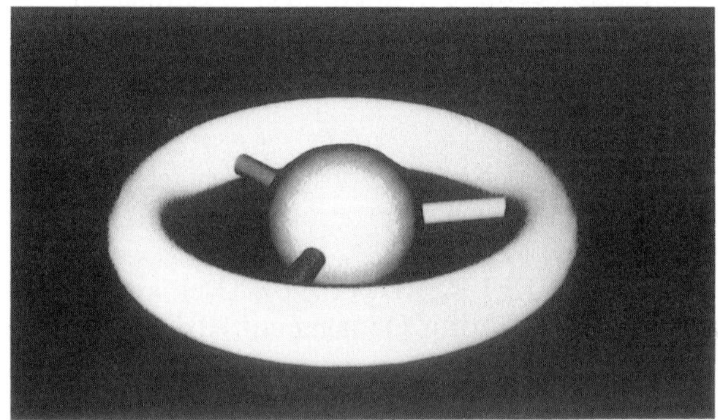

Fig. 2. Combination of inflatable structures.

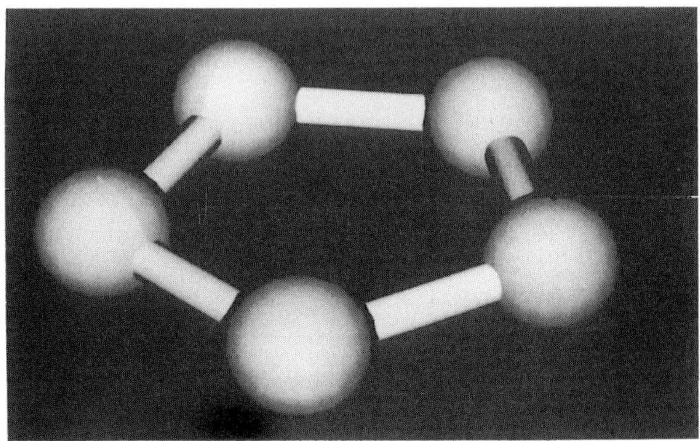

Fig. 3. Combination if inflatable structures.

23. It is possible to have several concentric spherical structures built inside a larger spherical space structure using gas-supported forms and flexible connections. Zoning the pressure within concentric spheres makes it possible to build larger structures more economically.

24. A single structural layer of fibre-reinforced composite material with flexible connections was more economical for a space structure than multi-layered structural composite material with rigid connections.

25. The estimated cost of constructing a 100ft (30.5m) diameter spherical space station with a 1/2inch (12.7mm) thick single structural layer of fibre-reinforced composite material, a 4-inch (102mm) thick layer of low-density polyurethane foam, with a Kevlar gas-supported form would be about 15 times more economical per unit of space volume than the current NASA cylindrical modules.

26. Structural forces at lift-off are eliminated because the structure is constructed in space using liquid materials.

27. Analysis for the inner spheres becomes statically determinate when pressure loads are applied to each sphere, no thermal loads are considered, and flexible connections are used.

28. When floors or partitions were rigidly connected to the space structure, large stress concentrations occurred at the rigid connections.

29. Flexible connections make it possible to increase and decrease the pressure in each sphere independently, when concentric spheres are used.

REFERENCES

1. AALIPOUR, P. (1990). Analysis and design of space station using inflatable forms and composite materials. Dissertation, Civil Engineering Department, Brigham Young University, Provo, UT.

2. PENFOLD, S.M.and MILLER, J.D. (1990). Urethane foams in a vacuum. Masters project, Civil Engineering Department, Brigham Young University, Provo, UT.

3. GREGG, C. (1990). "Personal letter". Boeing Company, Public Relations Department, Huntsville, AL.

214. Large-span steel structures constructed by prestraining methods

S. NAKASHIMA, Osaka Institute of Technology, Japan, S. IGARASHI, Osaka Sangyo University, Japan, K. INOUE, Osaka University, Japan, Y. HASHIMOTO, Mitsubishi Heavy Industries, Japan, and E. KATO, Toda Corporation, Japan

SYNOPSIS. Two new prestressing methods which can be effectively used for constructing large span roofs have been developed recently. These methods which are called the MPS Method and FPS Method are used mainly for frames composed of trussed beams and columns. This report is intended to verify, based on the scale model test results, the effectiveness of the methods which are used for constructing large span roofs by using hydraulic jacks and incorporating connectors for connecting the members.

INTRODUCTION

1. Recently, the social needs for buildings having large column free spaces have been increasing. Large span roofs may be formed by using space frames, suspension roofs, membrane structures, etc. Further, a number of methods in which major related members are prestressed during construction to minimise stresses and deflections due to their dead weight (generally known as prestressing methods) have been proposed lately.

2. The authors, too, have developed new methods which enable prestress appropriate for building forms and site conditions, to be undertaken without using special materials or providing special foundation work. This report describes the major characteristics of steel structures constructed using such methods.

BRIEF DESCRIPTION OF THE NEW METHOD

3. In these new methods, some specific members (called PS members) are tentatively detached from the roof trusses and then external restraining forces are applied in order to cause the structures to deform upwards so that inter-joint prestrains are introduced where PS members are located. (In this paper, the term "prestrain" is used concerning this phenomenon.) Next, in this condition, the detached members are replaced back into the deformed trusses and the restraining forces are released. When this procedure is followed, target residual strains are retained in the members. This means that prestresses are introduced into the whole frames by causing relative displacement to the specific joints. These methods are divided into two types by the modes of loading as shown in Fig. 1.

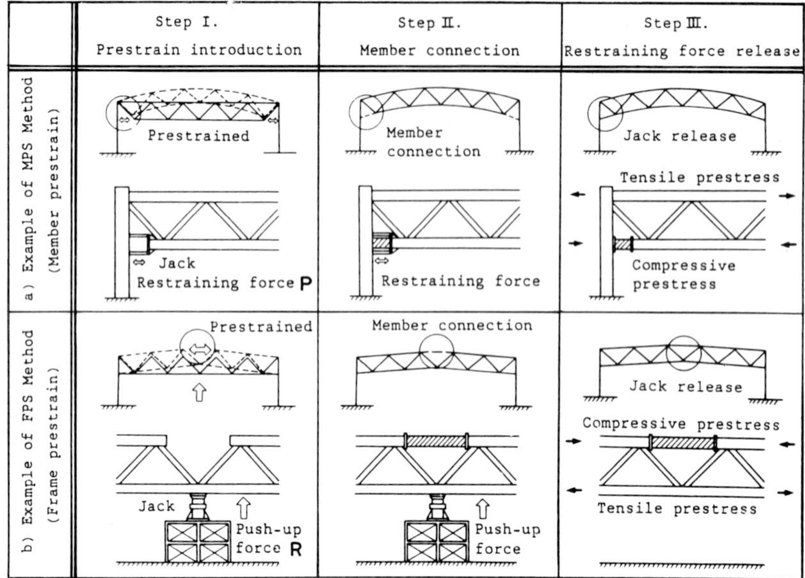

Fig. 1. Principle of prestressing methods.

4. MPS Method: In this method, hydraulic jacks are placed at the PS member locations to introduce prestrain into the beam directly. Fig. la shows a case where the lower chord end portions near the columns are selected as PS members whose joint intervals are lengthened directly by jacking; thus, they are prestrained and then connected. The beam prestressed in this way causes its lower chord member and upper chord member to be compressed and tensioned respectively, and this helps reduce the stress and the deflection at the midspan of the beam.

5. FPS Method: Fig. lb shows a case where the central portion of the upper chord member is selected as a PS member. In this case, the frame from which the PS member has been detached is pushed up and in this state the frame connection is made. This can help reduce not only the deflection at the midspan of the beam but stresses at the beam ends and in the columns.

6. Instead of selecting PS members as shown in Figs la and 1b, identical effects can be obtained by giving tensile prestrain to the both ends of the upper chord members or to the midspan of the lower chord members as shown in the next section.

EXPERIMENTS

7. The principle and the method of prestressing by the MPS Method and the FPS Method and the reasonableness of the analyses were examined by

undertaking tests using the same frame specimen but both with and without members. This section describes the results of such tests.

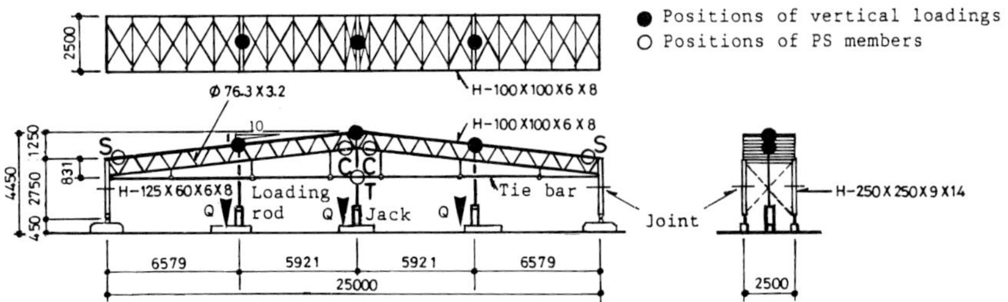

Fig. 2. Outline of test specimen and vertical loading apparatus.

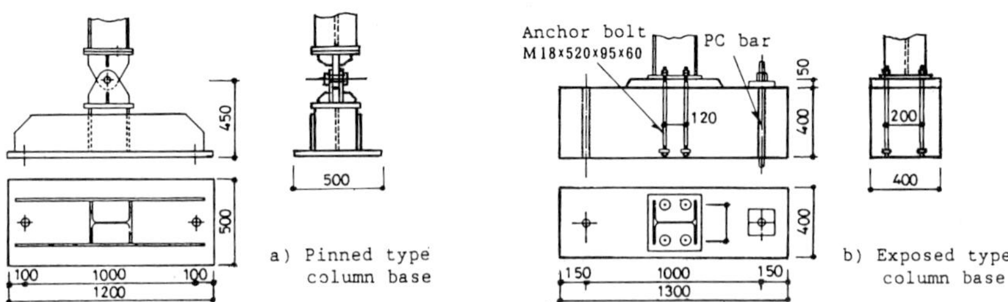

Fig. 3. Details of column bases.

Brief description of the experiments

8. Test specimens and experiment types. The test specimen is as outlined in Fig. 2. A 1/4 to 1/6 scale steel model was used to represent actual structures (100 to 150m in span). The specimen was composed of two frames connected by sub-beams and bracing.

9. As shown in Fig. 2, prestrain is introduced at the centre of the tie bars (T), to the centre of lower chord members (C), or to both ends of the upper chord members (S).

10. The types of column bases used are shown in Fig. 3. Two types of column bases, i.e., a pinned column base shown in Fig. 3a and an exposed column base connected to the concrete footing shown in Fig. 3b were used. The tests were conducted using these two different column support conditions.

11. As shown in Table 1, the tests were conducted for ten cases in which such factors as the methods and positions of prestraining and the column supporting conditions were changed in varying ways. Each of these test cases were named as indicated to the right side of Table 1.

12. Test Procedures and prestressing methods. For each test case, the PS members in tie bars were made discontinuous and a state in which the frame specimen was acted upon only by its dead weight was taken as the initial state. Then, prestrain was introduced and following this vertical loads (Q) were applied.

13. The prestraining apparatus and the push-up apparatus used in the tests are shown in Figs 4 and 5 respectively, and the loading positions are shown in Fig. 2.

14. The members were prestrained in the following manner. At the initial state, the frame was supported by push-up jacks located at the midspan in such a way that the support reaction would become zero.

Table 1. Test cases.

Test case	PS member position	Jack position	Positions of PS member and jack
MPS-T MPS-T-RC	T	T	P P
MPS-C	C	C	P P
MPS-S MPS-S-RC	S	S	P P
FPS-T FPS-T-RC	T	Midspan	R
FPS-C	C	Midspan	R
FPS-S FPS-S-RC	S	Midspan	R

Designations.

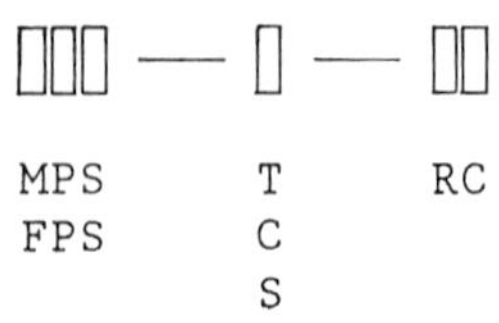

T: Center of tie bars
C: Center of lower chord members
S: Both ends of upper chord members

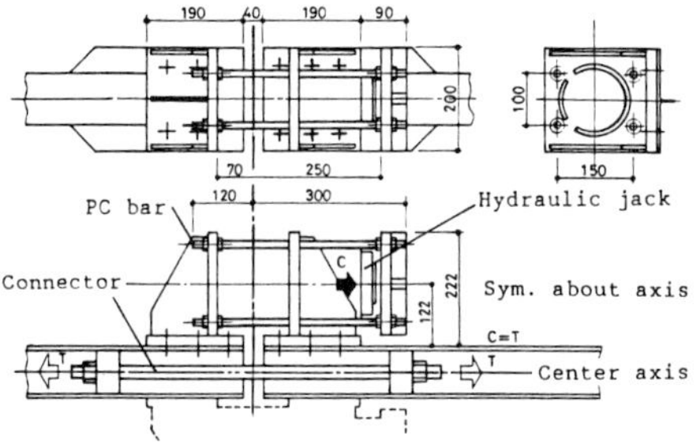

Fig. 4. Prestraining apparatus.

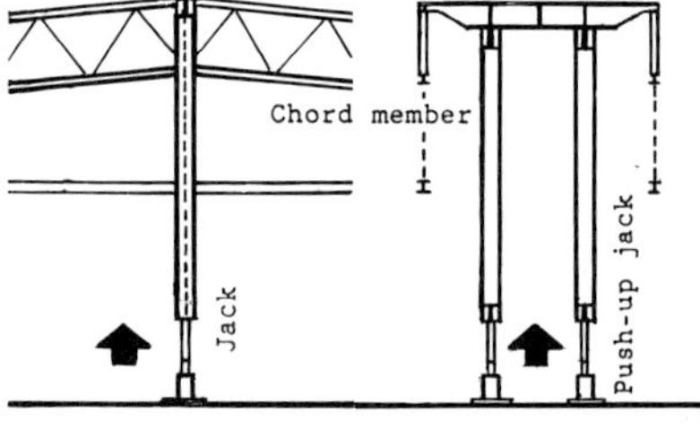

Fig. 5. Push-up apparatus.

Photo 1. PS member being prestrained by MPS method.

Photo 2. Joints not selected as PS member.

Then, the PS members were made discontinuous (by removing high strength bolts). The prestraining was made by the PS loading apparatuses located at the parts made discontinuous and the push-up apparatuses in the MPS Method and the FPS Method respectively. Then, the connectors were joined. Following this, the loads applied by jacks were released and thus the prestresses forces (Ps) were introduced.

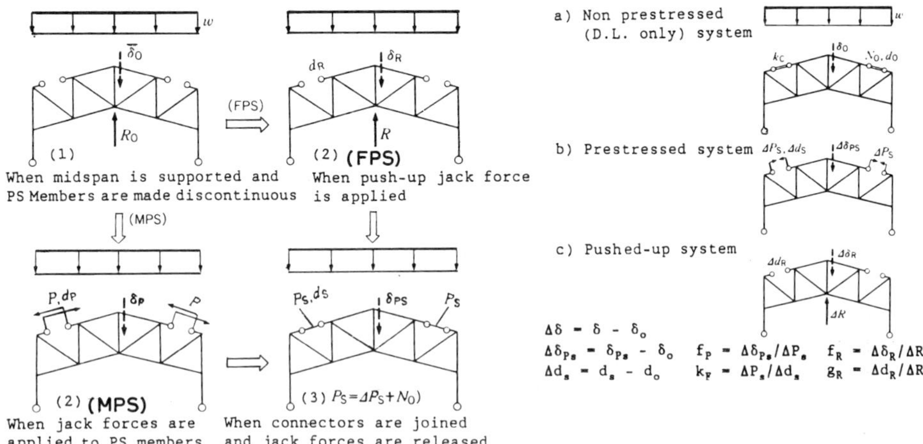

Fig. 6. Prestressing load steps.

Fig. 7. Basic structures for analysis of prestressed frames and their characteristic coefficients.

15. The prestressing was controlled in such a way that the difference between the strains measured by the strain gauges attached to the connectors and the target values would not exceed 5%.

16. Photos 1 and 2 show the PS member to which prestress is being introduced and the joints not selected as PS members.

Analyses

17. The loading steps for introducing prestresses in the cases of MPS-S and FPS-S are illustrated in Fig. 6.

18. Step (1): At this step, the displacement is $\overline{\delta}_0$ caused by the dead weight of the frame, and in that state the frame is supported in such a way that the supporting reaction of the jack becomes zero. After the PS members are made discontinuous, the frame is still supported by the push-up jack. (Supporting reaction = R_0)

19. Step (2): In the MPS Method, the loads (P) are applied directly to the PS members by jacks placed in the parts made discontinuous and in the FPS Method, the load (R) is applied to the midspan by a push-up jack. This creates the

prestrains d_P and d_R in the PS members and causes the roof to be displaced to δ_p and δ_R in the cases of MPS Method and FPS Method respectively.

20. Step (3): At this step, the frame is relieved of the loads after the connectors have been joined, and the roof shows a downward displacement to δ_{Ps}. Also, prestress (P_s) is introduced into the connectors, where P_s is the sum of the axial force (N_o) due to the dead weight and the genuine prestress (ΔP_s). The magnitude of ΔP_s or P_s can be related to the forced load (P or R) or to the initial camber ($\Delta\delta_{Mo}$ or $\Delta\delta_{Fo}$) at Step (2).

21. The relations between ΔP_s or $\Delta\delta_{ps}$ ($=\delta_{ps}$ - δ_o) which is to be finally introduced and the load (P or R) which is to be applied have a great infulence on carrying out the analyses in the prestrain method; hence, these relations need to be formulated. By using the characteristic coefficients (f_p, k_F, f_R and g_R) in the three basic structure systems shown in Fig. 7, the following relation formulas can be established.

In the MPS Method,

PS jacking force : $P = P_s/\eta_c$ (1)

Camber : $\Delta\delta_{Ps} = \delta_{Ps} - \delta_o = f_p\, \Delta P_s$ (2)

Initial camber : $\Delta\delta_{Mo} = \delta_p - \delta_o = f_p\, (P_s/\eta_c - N_o)$ (3)

In the FPS Method,

Jack-up force : $R = - P_s/(\eta_c\, k_F\, g_R)$ (4)

Camber : $\Delta\delta_{Ps} = f_p\, \Delta P_s$ (5)

Imitial camber : $\Delta\delta_{Fo} = \delta_R - \delta_o = - \delta_o = - f_R\, P_s\, (\eta_c\, k_F\, g_R) - f_p\, N_o$ (6)

At step (1), the reaction force to the jack-up apparatus is,

$$R_o = \left(\overline{\delta}_o - \delta_o + f_p.N_o\right)/f_R \tag{7}$$

If it is assumed here that a spring constant of a connector is k_c and the PS spring constant is k_F, a connector rigidity ratio (η_c) is then,

$$\eta_c = k_c / (k_F + k_c) \tag{8}$$

This amounts to a coefficient of force transfer to connectors at the time the loads (P) are released in the MPS Method.

22. Now, the analyses will be made for these experiments in order to predict the prestress characteristics in each case indicated in Table 2. The prestress characteristics of each case from this table are as follows. While there are three alternatives to choose the PS members from, as long as the camber efficiency (f_p) due to the prestress (ΔP_s) is concerned, Case S is favourable. However, since the camber efficiency (f_j) as related to the jack applied loads is influenced by the axial force (N_o) due to the roof's dead weight, there is not much difference between the three alternatives in case of the MPS method. As regards the FPS Method, Case

C is most favourable. Compared with the MPS Method, the FPS Method can give higher loading efficiencies (η_j and f_j); so, the prestress and the camber can be obtained more effectively with smaller loads if the latter method is used. On the other hand, however, the ratio (β) of the initial camber to the required camber is about 1.0 in the MPS Method whereas the value of this ratio is trebled or quadrupled in the FPS Method, and this means that special attention needs to be given to the influence caused by deformation.

Table 2. Characteristics of prestressing structures.

(ton, cm) 1Kgf=9.80665N

Test case	ΔP_s (=$P_s - N_o$) (ton)	α (= $\frac{N_o}{P_s}$)	f_P (= $\frac{\Delta\delta_{Ps}}{\Delta P_s}$)	e_j(= $\frac{P_s}{P_j \text{ or } R_j}$)	η_j(= $\frac{\Delta P_s}{P_j \text{ or } R_j}$)	f_j(= $\frac{\Delta\delta_{Ps}}{P_j \text{ or } R_j}$)	β(= $\frac{\Delta\delta_{Mo} \text{ or } \Delta\delta_{Fo}}{\Delta\delta_{Ps}}$)
MPS-T MPS-T-RC	10.0	0	−0.16	0.97	0.97	−0.16	1.0
MPS-C	8.5	0.24	−0.22	0.91	0.69	−0.15	1.1
MPS-S MPS-S-RC	5.0	0.40	−0.32	0.96	0.58	−0.19	1.1
FPS-T FPS-T-RC	10.0	0	−0.16	1.2	1.2	−0.20	4.0
FPS-C	8.5	0.24	−0.22	2.8	2.1	−0.47	2.8
FPS-S FPS-S-RC	5.0	0.40	−0.32	1.8	1.1	−0.35	3.2

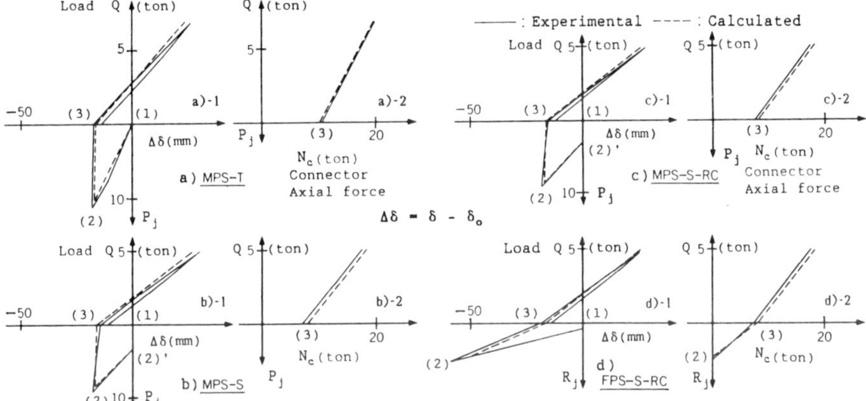

Fig. 8. Load-deformation relations and load-prestress relations.

Test results and consideration

23. Now, the test results will be compared with the analysis values and the aforesaid structural characteristics of the prestresses will be examined.

24. Load-deformation relations and load-prestress relations. Load (P_j, R_j or Q) vs deformation ($\Delta\delta$) relations and load (Q) vs connector axial force (N_c) relations for the typical four cases are shown in Figs 8a-8d. The load by jacking (P_j or R_j) and vertical load (Q) after prestressing are shown below and above the origin, respectively, of the ordinate. On the abscissa, the roof deflection ($\Delta\delta$), which does not include the initial deflection due to the dead weight, at the midspan is indicated. The values given a minus sign represent the cambers. Also, the connector axial force (N_c) is shown on the abscissa.

25. Points (1), (2) and (3) in the figure correspond to the load steps shown in Fig. 6. Namely, the loading and deformation at Point (2) are indicative of the maximum jacking force (P or R) and the initial camber ($\Delta\delta_{Mo}$ or $\Delta\delta_{Fo}$) while the deformation and the connector axial force at Point (3) represent the ultimate camber ($\Delta\delta_{ps}$) and the introduced prestress (P_s) respectively. Points (2) in Fig. 8 indicates the load at the time the central part of the roof was separated from the centre support point during the process of increasing the PS load and forming the cambers. P_j at this time is equal to the axial force (N_o) due to the dead weight.

(ton)

Test case		Prestress $\Delta P_s(=P_s-N_o)$			Jack force $\Delta P(=P-N_o)$, R		
		Target $\Delta\bar{P}_s$	Exp. ΔP_s		Exp.		Predicted
MPS	-T	10.0	10.5		11.2		10.3
	-C	8.5	8.3	8.9	10.4	10.4	9.6
	-S	5.0	4.8	4.7	6.0	6.0	5.3
	-S-RC		4.7	4.8	6.0	6.0	5.3
	-T-RC	10.0	10.5		11.0		10.3
FPS	-T	10.0	9.4		9.4		8.3
	-C	8.5	8.5	8.5	4.2		4.0
	-S	5.0	5.0	5.2	6.7		4.6
	-S-RC		4.8	4.5	5.2		4.6
	-T-RC	10.0	8.9		8.3		8.3

Table 3. Jack loads and prestresses.

26. The broken lines, which indicate the analysis values previously described, are in good agreement with the frame behaviours during prestressing.

27. Table 3 shows the applied jacking loads and the tensile forces actually introduced. At Step (2), the experimental values generally turned out larger than the computed values, and this is mainly accounted for by the unforeseen lateral slip of column bases which caused the apparent value of the PS spring constant to be smaller than was expected.

28. Fig. 9 shows the deflection modes of the trussed beam after prestressing (Step (3)) and under the maximum vertical loading. For FPS Method, the mode at Step (2) is also shown.

29. Stresses in members. As typical examples, the measured and calculated stresses in the upper and lower chord members and the lattices are shown in Fig. 10 for the case of the MPS-T.

30. Comparatively good agreement is seen between the two kinds of values. For other cases too, measured stresses were found to agree well with the calculated stresses.

31. Degree of fixity of exposed column bases. Comparison of b)-1 and c)-1 in Fig. 8 clearly shows that Q-$\Delta\delta$ relation in the MPS-S-RC, which has exposed column bases on concrete footings, is almost identical with that in the MPS-S which has pinned column bases.

32. The moment distributions in columns at the time of prestress introduction and vertical load application are shown in Figs 11a and 11b respectively. The rotational rigidity of the exposed column base can be obtained from ref. 1. Also, it is known from this figure that the exposed type column bases may be treated as pinned column bases.

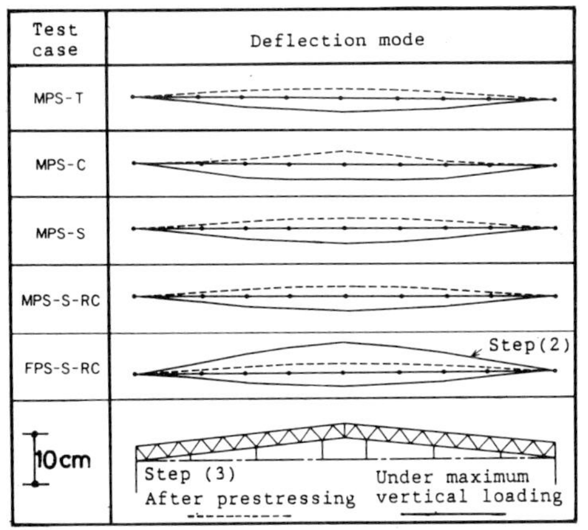

Fig. 9. Deflection modes.

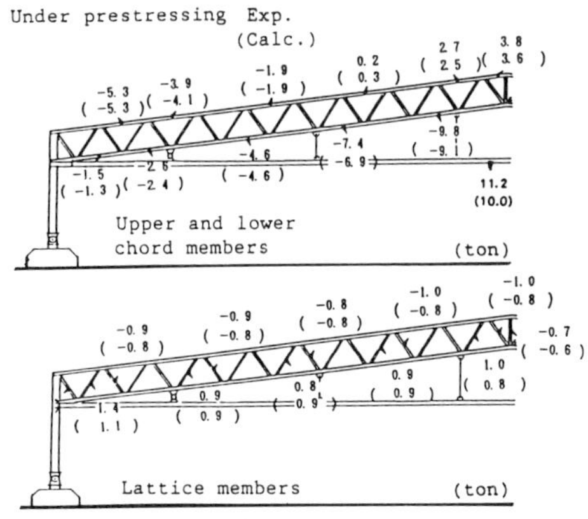

Fig. 10. Experimental and calculated values of axial forces in members. (MPS-T)

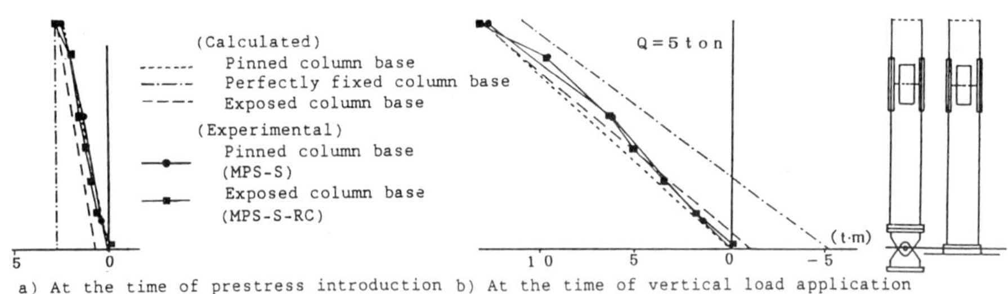

Fig. 11. Moment distributions in steel columns.

CONCLUSIONS

33. The methods to introduce prestresses by prestraining (the MPS Method and FPS Method) were developed to facilitate the construction of large span roofs. A number of model tests were conducted by the authors to verify the effectiveness of such methods and the following results were obtained.

34. These methods in which inter-joint prestrains are introduced where specific members are located were studied and their reasonableness in terms of both analysis models and design principles was confirmed.

35. Construction by these methods can be implemented by the use of jacks and connectors only which join PS members. Further, prestresses can be controlled easily by measurement of connector axial forces.

36. Pinned type exposed column bases placed on reinforced concrete footing beams as used in these tests may be treated as perfectly pin-supported models.

37. Both in MPS Method and in FPS Method, the conditions of cambers and stresses in members are determined by selected PS member positions. Further, decision on which of the aforesaid two methods is to be used should be made depending on the dead weight and the constructional conditions of a building under consideration.

REFERENCES

1. A. I. J. Recommendations for the design and fabrication of tubular structures in steel, p.128. Maruzen, Tokyo, 1990.
2. KINOMOTO, K., OZAKI, T., NOMURA Y., KIKUTA S., MURASE Y. et al. Experimental study on the long span structures, Abstracts. Annual meeting of A.I.J., 1991, pp 1231-1234.

215. Space frames and shopping malls

A. H. NOBLE, ABBA Space Structures Co., South Africa

SYNOPSIS. The paper discusses the architectural, engineering and commercial value of space structures for large shopping centres. Four of the salient features of such centres are considered: Signage Towers, Entrances, Concourses and Plazas. Each of these features is illustrated by recent buildings undertaken in South Africa by Abba Space Structures. Among the structures both the Southgate and East Rand Mall were winning entries in the South African Institute of Steel Construction Awards 1990 and 1991.

INTRODUCTION

1. The changing social conditions in South Africa, have resulted in a trend towards the construction of large shopping malls. Central Johannesburg has been ringed with a series of these centres: Northgate, Southgate, Eastgate, Westgate and Highgate malls. Such principal malls tend to be situated on the main arterial roads that straddle the city, and depend upon a community that has adequate means of transport.

2. The malls serve not only as emporiums but also as social centres, drawing large numbers of people into their secure environment; the market place atmosphere of the old cities being reinterpreted for the large modern urban communities. East Rand Mall, to the East of Johannesburg, for example, has a million people each month passing through its doors.

3. Successful centres require careful consideration to be given to their architectural, engineering as well as commercial aspects. It has been found that space structures are well suited to this type of building. They possess the necessary potential for striking architectural appeal as well as being an economic form of construction .

4. The key features of shopping malls have been developed to meet specific requirements. The signage towers are used to assist customers arriving by car in locating the centre. Entrances, as first points of contact with the mall are celebrated, yet at the same time, expressed on a human scale. Centres are usually planned around a system of interconnecting concourses; the areas formed at the

intersections of these concourses are expressed as plazas, providing " people-areas" for music, food and entertainment.

Fig. 1. Aerial view of East Rand Mall.

THE SIGNAGE TOWER

5. The signage tower is used as a marker and mood setter for a shopping centre. It tells the users even before entering a centre that it is a place of value, and informs them about the available stores.

6. Some towers simply mark the centre. The 42m high by 42m wide " H" tower at Southgate, on the borders of Soweto, bears only the centre's name. This large space frame tower makes a bold statement which is clearly visible from five highways adjacent to the mall. Some 10,000 high tensile bolts were used to secure the 50 tons of tubular steelwork in the tower. This grand architectural gesture transforms the facade of the shopping centre into something special.

7. The wind load was the principal design force that had to be considered in engineering the structure. The South African wind code required a regional basic wind speed of 40m/sec based on a 3second gust at 1 0m above ground, estimated to be exceeded once only in 50 years. The almost 2000m altitude of Johannesburg above sea level allowed a reduction in the air density of some 83%. The 6m by 4m legs were structured as hollow tubes of space framing. The square hollow section vertical members, of the legs, were turned at 45deg to the frontal wind thereby reducing their wind drag in the critical direction, while the horizontal and diagonal tubes were all circular. The final structural analysis, after preliminary hand calculations, was carried out using microStran, a commercially available programme based on the direct stiffness method of analysis, which can be run from a micro computer.

8. The whole structure was assembled piecemeal on site in a matter of four weeks and craned into position on prepared reinforced concrete footings. The accuracy of the manufacture and assembly was such that the 30m long by 10m deep cross-beam lifted in one piece, between the erected legs, with a mere 1 0mm clearance .

9. The structure cantilevers 42m from the reinforced concrete footings. No advantage was taken in the design by bracing the tower horizontally from the main two-storey structure of the shopping centre. Such an arrangement was proposed at preliminary design stage but the Architects, Bentel Abramson and Partners, insisted the structure be free standing. Six variants of the tower were considered before the design was finally accepted.

10. The space structure " H" erected in 1990, was surprisingly economic. Its entire cost, inclusive of design, manufacture, paint and erection, was approximately R450,000.00.

11. The Southgate monolith can be contrasted with the East Rand Mall's signage tower. It is only 23.5m high and does not form part of the architecture of the main shopping centre. It is a free standing structure adjacent to the highway on the boundary of the mall's car park. None-the-less the steelwork feature that occurs at the centre's main entrances (a half spoked wheel) is repeated on the tower.

Fig. 2. Southgate signage tower.

The tower is a delicate structure of 9.5 tons of light tubular steel: two 23.5m high by 1 5m wide archways, 4.2m apart, connected by a light bracing system. The legs of the arches are pairs of braced 219mm diameter tubes. Difficult soil conditions necessitated standing the tower on a piled foundation.

12. Due to the steel being of a more delicate nature and with manufacture taking place a year after the Southgate structure; increased labour and material costs and an 18% inflation rate, the steel in the East Rand tower cost was 50% more per ton. Yet it was still an economic structure, the cost of the tower being offset by the advertising which it carries.

THE ENTRANCE

13. The first intimate contact that a user has with a mall is the entrance. Each and every person visiting a centre passes an entrance; they may not experience the whole of the centre during their visit but they have to experience the entering of the building. This is a moment of expectation. Will my effort in coming to the centre be rewarded? Is this the kind of place that I should have come to? These are the type of questions we can assume the users are asking as they enter the mall. The architecture has to provide answers for this moment of expectation. Scale is important in this regard; grandeur yes, but never at the expense of the human being.

Fig. 3. East Rand Mall signage tower.

14. The l6m diameter entrance domes at the Northgate Mall are external to the main building. Their intimate size and visual appeal prepare the user:; it is as if they were saying, "Here you will enter a different world". South African architecture is at the moment increasingly expressing itself in "fantasy" and nostalgic elements. This is particularly so in the design of shopping centres.

Fig. 4. Northgate entrance dome.

15. The structural form of the seven Northgate domes is completely novel. Based on the work of Danish academic Ture Wester they were conceived as a series of open plates. As four plates meet at each " node", buffer-forces as well as plate actions are developed. The plates are bolted together in such a way that they appear to float, 45mm clear spaces being formed between them. Their kite shapes are formed from 50x50x5 angles. The effect of this arrangement is to give the structure a rich yet at the same time delicate quality. They have been left as exposed, uncovered steelwork.

Fig. 5. East Rand Mall Entrance.

16. On the East Rand Mall contract the entrances are a series of glazed space frame vaults. The East entrance is 13.75m long by 10m diameter, and the North and South entrances are 8.75m long by 15m. The barrels are supported along their sides on Vierendeel girders. The front end of each barrel is reduced down 1.66mm in diameter to terminate in a vertical glazed gable. A decorative wheel-shaped arch on two pairs of tubular columns completes each structure. Suspended from the underside of the Vierendeel girders and across the gable front is a 700mm wide steel box-gutter. The structure stands 4m to the underside of this gutter, rising to 12m at their crowns.

17. The barrelled space frame of these entrances mirrors the double layer, diagonal-on-diagonal, configuration used for the main concourse areas of the centre: an Abba Dek-Space of preformed pyramids arranged in a checker board pattern, connected on the inner chord by a series of cruciform members.

CONCOURSES

18. Large shopping malls will usually have in excess of 200 shopping outlets. Some are major supermarkets, but the majority will be small businesses. There will also be restaurants, fast-food outlets and cinemas. It becomes necessary therefore to design the centres around a system of shopping concourses. Where possible the concourses are limited in length to a maximum of 60m. The East Rand Mall has an interconnecting system of nine such concourses totalling 330m in length. By using short concourses with changes in direction and visual breaks an acceptable scaling can be achieved.

19. The elements of intrigue and surprise play an important role. The avoidance of the deadening effect of uniform lighting is also important. Good natural lighting sufficiently high above the concourse is essential. By providing a large enough volume in a concourse, the shop fronts can be changed in height and line to create a maximum of interest. Space structures are ideally suited to achieve these architectural requirements; high visual appeal, large column-free areas, height and flexibility of planning. The regular square grid that is usually favoured in the layout of a centre also plays into the economy of repetitive space frame units.

20. The concourses to the East Rand Mall are a series of double layer space frame barrel vaults (see paragraph 17). To achieve the maximum economy a checker board layout of pyramid units was chosen. The outer and inner grids running diagonally to the line of the concourse. This arrangement creates a pleasing spiral configuration along the vault's length. The outer grid is a net of back to back angles created by the bolting together of pyramids and tray units. The square tubular purlins for the sheeting are welded into the pyramids. Welding-in of purlins on the one hand speeded up the erection, and on the other provided additional rigidity to the structure. The resulting structure was an economic 15.8kg/m sq. on surface area. A total of 41 500 bolts were used.

21. Erection is a major part of this type of construction. The horizontal tie force, required for the structure's own mass, was provided by temporary ties during the erection sequence. The programme required that the space frame barrel vaults be erected after the supporting lattice girders (and adjoining trusses) were in position. The 15.1m wide arch sections had therefore to be craned up through the 14.9m gap left between the girders. To overcome this difficulty the outermost pyramids were manufactured with detachable corner elements.

22. The main Southgate concourse straddles a double storey volume that flows from the mall's main entrance; 15m in diameter, 67.5m long and rising 23.2m above the ground floor, the double layer space frame barrel vault is an impressive structure. The flush-glazed vault is supported on only eight columns, spanning 30m at one point. Using the diagonal-on-diagonal grid with the pyramids in checker board configuration gave the structure a powerful visual effect of two sets of enfolding spirals.

23. This pleasing arrangement of members (which was also used in the East Rand Mall) proved to be very economic. Only 65 tons of steel were used in the structure. The unglazed space frame vault cost R140/m sq. on surface area.

Ostensibly 2000 identical pyramids were manufactured. This was a saving of 32% in number, over the more normal diagonal-on-diagonal configuration, making a significant contribution to the economy of the structure. The connection of the pyramids at the inner grid was achieved by a simple cruciform connection.

Fig. 6. Southgate concourse.

24. The assembling of such large numbers of preformed units in a single structure presents its own special problems. Even the smallest discrepancy in the manufactured members would have caused intolerable assembly problems, not to speak of the distortion of the calculated stress patterns that would have occurred. Workmanship of the highest quality is demanded. Fortunately South Africa while being a developing country has a high level of structural steel skills.

PLAZAS

25. Plazas are an essential feature of a shopping mall: areas where the public can gather, be entertained, served with food as well as providing a place for product promotions. Even after normal shopping hours a good mall will be crowded with strollers and window shoppers. The human factor cannot be over-stressed. Although the mall is primarily an emporium it is most successful when the people using the centre are considered.

26. Space is perhaps one of the prime considerations of the architecture of a plaza. After walking along the confined route of a concourse, to emerge into the spacious area of a well planned plaza is attractive; add music, trees and perhaps the presence of a fountain and you have a great attraction.

Fig. 7. Dome to main plaza at Killarney Mall.

27. The plaza under the large dome at Killarney Mall is usually filled with music. A restaurant opens on to it on one side and interesting exhibitions stand on display. The plaza is covered by a 16.3m x 10m dome, elliptical in cross section and 1 8m rise, seven meters above ground. On plan the dome is rectangular with semi-circular ends. Glazed in fibre glass panels that simulate classical stained glass, it makes a striking plaza.

28. This dome is the largest of five feature domes erected as part of a refurbishment of the Killarney centre. The nature of the refurbishing meant that the dome had to be carried into the mall in small lengths, and site bolted. The main elliptical ribs are 150 x 40 RHS with rings of 90 diameter horizontals and secondary arches formed of 45mm "D" shaped tubes. The horizontals and "D" tubes are welded-up as panels and site bolted to the main arches with decorative filigree arches being welded in each of the segments. This complete structure has a mass of 3.6 tons.

29. The dome is supported on the steel ring girder that supports the decorative plaster surrounds, and is weather proofed by an upper roof sheeted in polycarbonate. Between the upper roof and the dome the air temperatures of the void on warm days is about 60° c. Ventilating the space would have required circulating enormous volumes of air, so the consultants simply designed for a bubble of 60° c. air to sit above the domes. The effect therefore on the dome, of having temperatures of 21° c. below and 60^{0} c above had to be planned for.

30. In contrast the eight East Rand Mall plazas are cross groins that rise 16m above the ground. They are supported on four columns with a 20m by 20m column grid. Their architectural shape is formed from two 20m diameter barrels intersecting at right angles. Whereas structurally the roof consists of two 28m elliptical arches intersecting at right angles that in turn support 20m diameter secondary arches braced by a light vault on which the sheeting rests. These

elliptical arches are constructed from 219mm diameter tubes with a 5mm web. The 20m diameter secondary arches are 165mm diameter and the light upper vault is a fine grid of 38.1 by 38.1mm SHS tubes. The upper vault is supported every 1.1 m from the secondary arches and the secondary arches occur every 2.0m. In order for the roof surface to clear the valley gutters, the upper vault had to be supported 250mm above the secondary arches. At the ends of the crossing the barrels decrease in diameter, sloping down 2.5m to meet the concourse vaults. The volume created in this way makes for a striking plaza.

Fig. 8. East Rand Mall plaza.

POSTSCRIPT

31. The building of successful shopping centres presents a unique set of problems, architectural, engineering and commercial. The various recurring features developed to meet this challenge are well suited to creative application of space structures. From the experience gained in South Africa it is clear space structures can be both aesthetically pleasing and economic.

216. A quasicrystal for Denmark's COAST

E. REITZEL, Danish Royal School of Art, Denmark, and T. ROBBIN, New York, USA

SYNOPSIS. The first true quasicrystal structure will be built in the summer of 1993 at COAST (The Centre for Art, Science, and Technology at the Danish Technical University). Seven meters high, with 190 nodes and 448 72cm stainless steel rods as well as coloured, mirrored and half-mirrored glass, it will be exquisitely and magically responsive to both changes in light and the viewer's movement.

1. Quasicrystals are non-repeating patterns and thus an apparent paradox: we usually think of patterns as composed of regularly repeating elements. In quasicrystals there is regularity: all the nodes, edge lengths, and two dimensional shapes are identical, all the joints are in rows, and only two similar three dimensional shapes make up the entire structure. Yet the pattern is non-repeating and consequently counter-intuitive and intriguing. Equally fascinating is the fact that quasicrystals have five-fold, three-fold, and two-fold symmetry at the same time. Experimental physicists, that have a sample of quasicrystal material, know that when they shoot x-rays through it they obtain a pentagonal scatter pattern, turning the sample results in a scatter pattern of isometric triangles and hexagons, and turning it again results in patterns with right angles. Knowing that photons and x-rays are the same thing, Robbin realised that such structures would cast shadow patterns that transmogrify as the sun passes overhead.

2. These visual effects can best be seen in the project study of a quasicrystal dome of 700 vertices (Fig. 1). From underneath, the dome appears to be a five-pointed star pattern. Amazingly, the dome appears to made up of triangles and hexagons when one is under the dome looking slightly to the left, and it appears to be squares when one is under the dome looking slightly to the right. But when one goes to where the triangles are, only pentagonal stars can be seen. As the sun passes overhead, from 10am, to noon, to 2pm the patterns on the floor transform to these computer plots. Every quasicrystal dome, spaceframe, or vault has this two-fold, three-fold, and five-fold symmetry.

3. Although a tradition of studying similar such icosahedrally symmetric structures for use in architecture exists with S. Baer, K. Miyazaki, and H.

Space structures 4. Thomas Telford, London, 1993

Fig. 1.

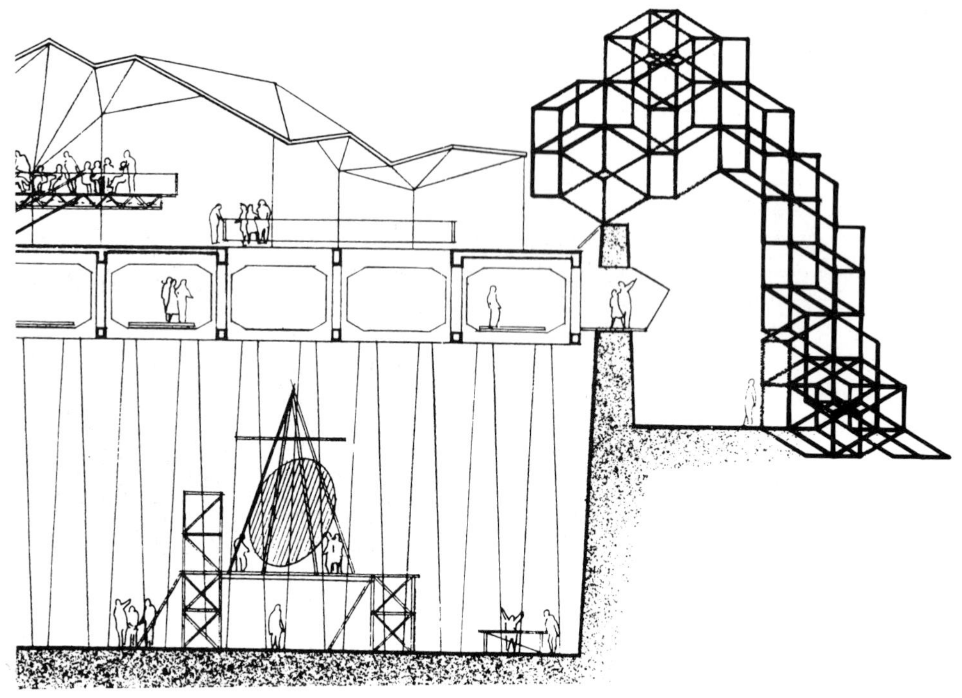

Fig. 2.

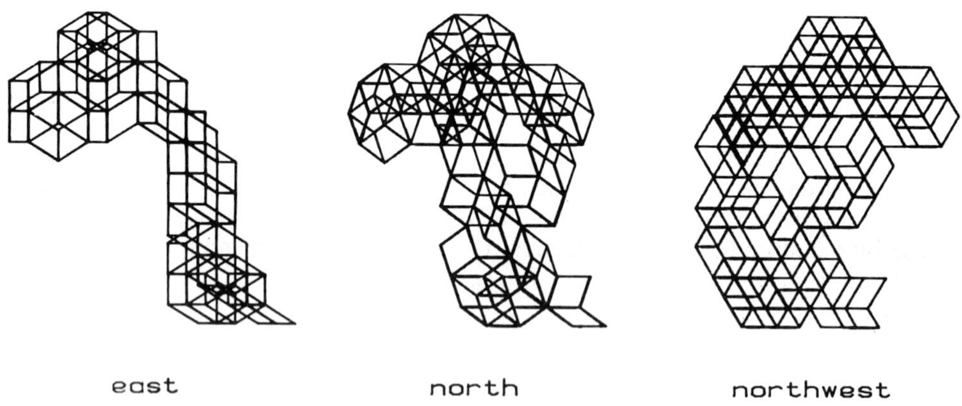

Three views of Coast's Quasicrystal.

Lalvani, it is only when new mathematical tools are borrowed from solid state physics that it is practical to design true, large scale quasicrystal structures, or to discover their unique properties. Baer's dodecahedral nodes are necessary but not sufficient to construct quasicrystals: an algorithm is required to insure that the nodes are correctly connected. True, if one already knows what a quasicrystal looks like, one can use dodecahedral nodes to replicate it and improvise upon it in a way that might be faithful to the mathematics, but Baer did not discover quasicrystals from his nodes and rods and did not make structures with the visual and mathematical properties of true quasicrystals. On the basis of Baer's example, Miyazaki assembled the golden zonohedra from dodecahedral nodes and rods, and speculated about the further assembly of these sub-units into larger groups. Again, the main opportunity of building structures that change their apparent shape as the viewer passes through the structure was missed, probably because Miyazaki had no large unit models or computer models of these structures at this time (the early 1980s). Lalvani has done impressive work in generating large-scale quasicrystals, even inventing new mathematical tools on his own to do this. Like Miyazaki and Robbin, he understands the relationship between quasicrystals structures and the geometry of four spatial dimensions. Yet Lalvani's concentration on studying the families of quasicrystals has not given him the opportunity to build structures or consider their visual properties in detail.

ORIENTATION

4. For COAST, the site for our quasicrystal structure is a north facing wall which will always be in shadow. Behind this wall, COAST's main permanent facility will be built, and in front of the wall is a main pedestrian and automotive street. With Reitzel and COAST architect Soren Koch, Robbin decided that the quasicrystal structure should do four things: 1) It should reflect sunlight onto the stone wall - a moving display on the wall as the sun moves; 2) the structure should reveal its 5 3 2 symmetries as the viewer passes by on the road (Figs 2 and 3) light should be reflected down into the structure so that the structure glows against the dark wall - suggesting the energy and creativity going on behind the wall in the COAST facility; and 4) although we do not emphasise it, the structure should be safe to climb upon, as it may provide an irresistible, though illegal, shortcut into the building.

5. Even at noon in mid-summer the sun does not pass overhead in Lyngby, Denmark, and so a structure oriented primarily for a floor shadow effect would not be effective. On the other hand, most visitors to the University are forced to pass by the structure. How wonderful to have a structure, without moving parts, that would dramatically change its appearance in the 20 seconds or so that it would take to pass by in a bus. Fortunately there is a orientation for quasicrystals that allows for this visual effect, and also allows some planes to be horizontal to the earth (for what become stair treads and perches up the structure) and also allow for some planes to be perpendicular to the earth, so that sunlight can be reflected onto the wall by mirrored and half-mirrored panels. As the sun moves

Fig. 3.

through the day, the patterns on the wall will slide and transform, and the different symmetries of the structure will by emphasised.

STABILITY

6. Although one of us (Robbin) has a theory that with the use of a stretched skin, quasicrystals could be especially efficient structures, the COAST structure is stabilised by the use webs that add bending stiffness to the nodes (Fig. 3) and by using plates in the structure. Such devices are necessary because skeletal quasicrystal structures are inherently deficient in rigidity. The solution to stiffening quasicrystal structures that occurs to most engineers when they first experience them is to triangulate the rhombic faces, and to pass a compression member between opposite corners of the rhombohedral cells that are inevitably engendered. Such a solution, advocated even by Baer, ruins the quasicrystal nature of the structure and defeats the visual opportunities that we have been discussing, and so it was rejected out of hand by both the engineer and the artist.

7. Reitzel has calculated that even with a large node (90mm between parallel faces of a dodecahedron) and rods of 30mm in diameter and 725mm long, the rhombs that make up the structure would have a deflection of 2.3mm for the critical rhombs under the calculated self-weight of 30kN. Wind loads are calculated to be an additional 12.64kN, and snow load adds an estimated 8kN. The 30mm diameter rods were chosen for aesthetic reasons, as it was felt that thinner rods would be visually too light to reveal the geometry of the structure and that thicker rods, which might be more practical from an engineering standpoint, would clutter and obscure the geometry. Robbin's recent experience of building a 5m x 12m x 2m indoor structure at the American Academy for the Advancement of Science with rods 15mm in diameter and 600mm in length reinforced his intuition that 30mm diameter rods would be esthetically appropriate for the COAST structure. Reitzel has specified that stainless steel rod comparable to 18/8 would be sufficient for this application and that each column could have an allowable loading of 24.3kN.

8. At COAST, the stiffening webs between rods are not co-planar with the rods, rather they are normal to the rods; these stiffening webs become partial outer shells of the dodecahedral nodes, further emphasising both the icosahedral symmetry of the structure and, at the same time, making reference to inflation/deflation symmetries of the structures (ref. 10). These stiffening webs are all the same size and shape, and consequently are mass produced. Where they are needed, the pentagon-shaped stainless steel webs are slipped on the rods, by means of a hole in their centre, before the rods are attached to the nodes . The plates are then welded together edge to edge forming a corner resistant to bending. As of this writing, dimensions of these pentagonal plates are under discussion as are some details of the glass plates.

PLANES

9. Additionally rigidity is added to the structure by the planes. The horizontal planes, the stair treads and perches, are metal tracery in deflated quasicrystal

patterns, and these are rigid. Reitzel has proposed substituting 6mm stainless steel plate for mirrored glass at certain key locations. For example, four-cell rhombic dodecaherdra occur frequently as sub-units in the structure, and although six planes are needed to firm up all 15 nodes in this sub-unit, only two planes adds remarkable stiffness to the sub-unit in three perpendicular directions. Finally the coloured glass panels are themselves rigid, and strategically placed, these panels also stiffen the structure. Glass presents a number of challenges and opportunities as a material for the planes. Compared to Lexan, glass is expensive; it cannot be trimmed on site; and it is more dangerous should it break. On the other hand, glass has greater stability and is stronger: it has a static modulus of elasticity of 40,000MPa compared to 700MPa for Lexan, and it has a thermal expansion coefficient of 0.008mm/m/C compared to 0.07mm/m/C for Lexan. The colour of Lexan may not be constant in exterior uses. Reitzel has chosen a Pilkington 10-6 laminated glass, which has the required properties. The laminating layer can be in a wide variety of colours, and patterns can be silk-screened on it, matching the quasicrystal tracery of the floor treads. The glass layers themselves can be coloured, or mirrored, and can be tempered for strength. The glass panels should be attached to the nodes, which would result is both a cleaner look and a better use of the glass's strength. At the time of writing, special modifications of Pilkington's bolt system have yet to be detailed, but it seems such modifications are easy to accomplish.

INFLATION

10. An additional fascination of quasicrystals is their so called inflation/deflation property. Inflation here means that the cells of quasicrystals (either the two dimensional rhombuses that are the unit cells in the two dimensional case or the two rhombohedra that are the cells in three dimensions) can be sub-divided with self-similar units so that a similar but not identical pattern is created. This endless self-similar foliation is not unlike self similarity in chaos theory or fractal theory, except that quasicrystal tessellations fill two dimensional and three dimensional space exactly. In the three dimensional case, which is the one that concerns architects and engineers, one ratio between the edge lengths on the original and deflated cell is Tau cubed to one, where Tau is the golden ratio (1.61803...). This three dimensional inflation/deflation was discovered by the Japanese physicist T. Ogawa. In an attempt to be faithful to the mathematics, we have included smaller unit cells, with edge lengths is this ratio, as non-structural elements in the design. Having cells in two scales and having nodes in three scales, suggests that what we have in front of us at COAST is a manifold - a space with mathematical properties, rather than a singular object.

11. In conclusion, the easy and fruitful collaboration between artist and engineer must be mentioned. Of course, some panels must be where they are because of the angle of the sun at certain times of day; other panels are where they are for structural reasons, and webs must be used where panels interfere with visual effects. Nevertheless, there has been an interplay between engineering and

aesthetics throughout the project, with some of the best engineering ideas coming from the artist and some of the best artistic ideas coming from the engineer.

REFERENCES

1. BAER S. Structural System. U.S. Patent Office, n. 3,722,153, Mar 27, 1973.
2. de BRUIJN N. Algebraic theory of Penrose's non-periodic tilings of the plane. Ned. Akad. Weten Proc. Ser. A .198 1.
3. LALVANI H. Building Structures Based on Polygonal Members and Icosahedral Symmetry. U.S. Patent Office, n. 4,723,382, Feb. 9, 1988.
see also: LALVANI H. Non Periodic Space Structures . Space Structures, vol. 2, no. 2,1 986-7, p.93108. LALVANI H. Non Periodic Space Filling with Golden Polyhedra. Proceedings vol. 1, The First International Conference on Light Weight Structures, Sidney Australia, August 1986, LSA 86 p. 202-211.
4. MIYAZAKI K. and TAKADA, 1. Uniform ant-hills in the world of isozonohedra. Structural Topology 4, 1980.
5. OGAWA T. On the structure of a quasicrystal. Journal of the Physical Society of Japan, vol. 54, No. 9, September 1985.
6. PETERSON I. Shadows and symmetries. Science News, vol. 140 nos. 25 and 26, December 21, 1991.p.408-410.
7. REITZEL E. Kvasikrystaller ved DTH, For project 254-01, October 1992, Reitzel F. R. I, Kobenhaven.
8. ROBBIN T., Fourfield: Computers, Art, & the 4th Dimension, Bulfinch/Little Brown & Co. New York 1992. see also: ROBBIN T. Quasicrystal Architecture, Proceedings of the IASS Symposium, Arkitekskolen Forlag: Copenhagen, 1991.

217. Starman: a space sculpture

I. P. LEW, E. LING, L. PETRELLA and R. GRANT, Lev Zetlin Assoc., USA

SYNOPSIS. The Starman Space Sculpture represents a synthesis of art and structural engineering. The sculpture's diamond shape grid is a progressively increasing pattern that is mathematically generated and automatically analysed and designed by computer. In addition, the computer model provided the artist with the information necessary for the fabrication of the structure.

INTRODUCTION

1. The Starman Space Structure gave the authors the opportunity to explore computer generation of three dimensional structures and the automation of the structural design. The Starman Space Sculpture is the centrepiece of a new plaza in the City of Philadelphia, the birthplace of the American Declaration of Independence. The sculpture was to represent a comet with the head of the comet being a human head, and the tail of the comet being the space sculpture (see Fig. 1).

2. The artist had a series of goals to achieve, but did not have the mathematical ability to realise these goals. These goals were to have a progressively increasing diamond grid on the surface and the thinnest structural members possible. The structural designer's goals were, because of the complexity of the structure, to automate the generation of the geometry, the loading of the structure and the design of the members. The complexity of the structure was due in part to the tilted axis of the structure with respect to the vertical gravitational axis and the constantly increasing member lengths.

GENERATING THE GEOMETRY

3. The base of the Starman Sculpture was a cross section of a head in frontal view. The head's base dimension was 84 in. (2.1 m) from the top of its head to its chin and 52 in. (1.3 m) ear to ear. The base's contour was such that eighteen segments of 12 in. (30 cm) would traverse the perimeter. From this base the space sculpture was to rise 324 in. (8.1 m). In addition to the Space Sculpture, the head itself was 40 in. deep (1 m), plus the podium was 120 in. high (3 m). As a result, the total height of the sculpture was 484 in. (12.1 m).

Space structures 4. Thomas Telford, London, 1993

Fig. 1.

4. The part of the structure that required mathematical generation was the 324 in. (8.1 m) Space Sculpture (see Fig. 2). The space structure was to expand as it grew vertically from its base. On the outer surface of the space structure was to be a diamond pattern that was to be at about a 60 degree angle. The artist desired to start off with equilateral triangles at the base. There would be 18 vertical levels, at which the diamond grid would intersect on the surface of the sculpture.

5. To determine the angle at which the sculpture would expand, an analogy to an equal circumference of a truncated cone was employed. For a perimeter of 18 sections, of 12in. (30 cm) each, to have the same circumference, an equivalent 69 in. (1.75 m) diameter circular base would be required. The objective was to increase the segments at the next level to a 12.45 in. (31 cm) diamond which creates a difference of 8.1 in. (20 cm) in circumference. As a result, the increase in the overall radius (8.1 in. divided by 2π = 1.3in. (3.3 cm)) of the equivalent cone to the first level diamond grid intercepts could be determined. The next goal was to start with an equilateral triangle at the base of 12 in. sides. As a result, the altitude of this triangle would be 12 in. (cos 30 degrees). Knowing the altitude along the incline surface and the increase in horizontal dimension (the radius), the incline of the surface (sine X = increase of radius divided by the inclined altitude of the equilateral triangle of the outer surface) could be found. This turned out to be about 6 degrees with respect to the sculpture's main axis. It was desired that the sides of the diamond grid increase constantly. Expressed mathematically:

$Li = X + (i-1)\,Y$ where Li = length of the ith diagonal,
X = base length
Y = incremental increase in length.

6. Since the diamond grid was to be at a 30 degree angle with respect to altitude, the length of the members was to be increased by one divided by the cosine 30 degrees. In addition, the altitude was at about 6 degrees with respect to the main axis of the sculpture. Therefore, the sum of the altitudes was the vertical height 324 in. (810 cm) divided by the cosine of 6 degrees or 325.8 in. (815 cm). As a result, the total length of diamond lengths on the surface from top to bottom would be 325.8 divided by cos 30 degrees or 376 in. (950 cm).

7. There were to be 18 lengths or modules of the main diamond grid. Each length was to be bigger than the previous element by a set increment or, as previously stated: $Li = X + (i-1)\,Y$.

Therefore,

$$L_{total} = \text{the sum of } Li$$
$$= 18\,X + \text{the sum of } (i-1)Y = 376 \text{ (for } i = 1 \text{ to } 18)$$
$$L_{total} = 18\,X + 153Y = 376.$$

8. If the base X is set at 12 in., then the increment Y is about 1 in. (2.5 cm). Therefore, the first length of the side of the diamond will be 12 in. (30 cm), the second length 13 in. (32.5 cm) and the third length 14 in. (35 cm) and so forth. At the top the skewed length is 29 in. (72.5 cm). At the top of the space sculpture

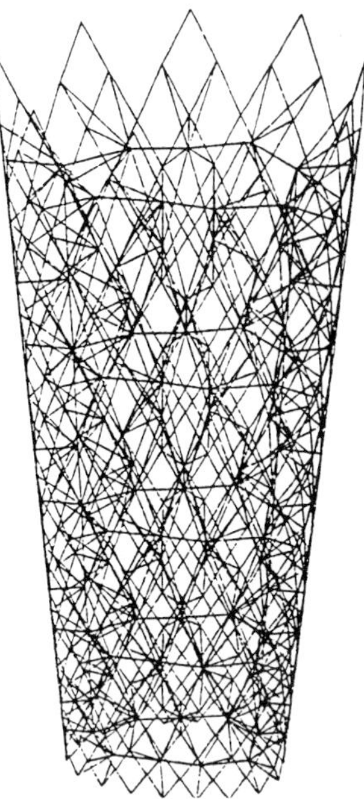

Fig. 2.

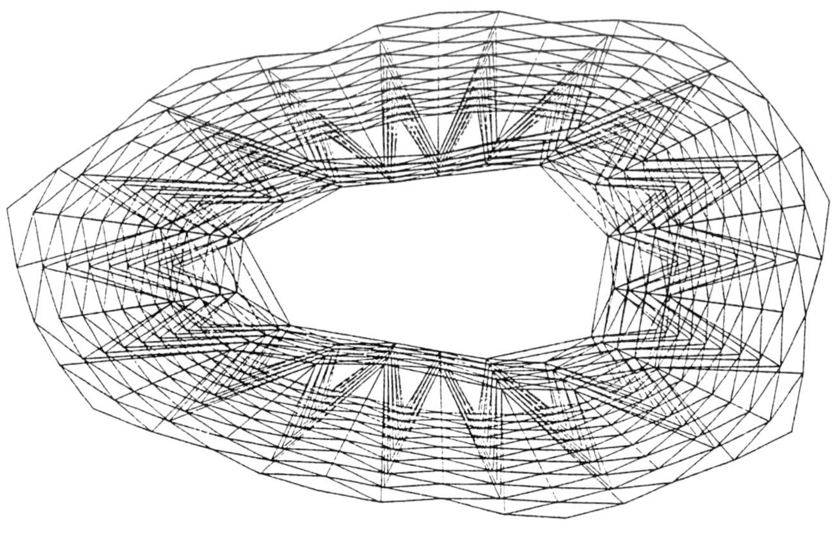

Fig. 3.

the distance from the chin to the top of the head was increased to 152 in. (3.8 m) from 84 in. (4.1 m) and its width to 120 (3 m) from 52 in. (1.3 m). The space structure had an inner layer 10.5 in. (26 cm) inside at its base and 24 in. (60 cm) inside at the top (60 cm).

9. The inner layer diamond grid was twice as large as the outer layer (see Fig. 4). Posts and diagonals between the inner grid intercepts and the outer main grid intercepts created skewed curved trusses along the diamond grid lines of the inner layer grid (see Fig. 5). The intercepts of the inner and outer layer grid were developed as similar shapes using radial generator lines from the base perimeter points using the centroid of the base perimeter as the centre of the radial generator lines and proportional distances (see Fig. 3). Once the joint co-ordinates were generated, the member lengths could be readily found. Based on the member lengths, the loads on the members could be generated.

10. All members were either pipes or rods. The members in the inner and outer grid were pipes from 5/8 in. to 7/8 in. in diameter, whereas the diagonals and posts were rods from 3/8 in. to 5/8 in. in diameter. Each pipe diameter had a corresponding thickness of 1/8 in. (0.3 cm). For each rod and pipe diameter the weight per inch of length was determined based on the area of the member. In addition to the structure's own dead weight, the weight of 1/4 in. (0.6 cm) of ice encasing the pipes and rods was determined and added to the dead weight. Furthermore, these dead weights had to have wind load combined with them. Without ice the design wind velocity by code is 80 mph (128 km/hr), but with a layer of ice on the pipes or rods, the code design wind velocity is reduced to 40 mph (64 km/hr) or, using the power law ($F = K(Velocity^2)$), and the resulting wind pressure with ice encasement is one-quarter the wind pressure of that without ice. All loads were input per unit length and the wind loads were input in the direction of the maximum and minimum tilt of the axis of the space structure in both the positive and negative direction. No sheltering from other members for wind was assumed.

11. The resulting forces were checked by the computer automatically. This checking process was made easier by the fact that the members were pipes or rods and thus axially symmetric. Therefore, bending about any two axes can be combined together vectorially. A unique situation occurred in the design of the members. Under the American Institute of Steel Constructions Code, the slenderness ratio (kl/r) cannot go above 200. The goal was to make the members as small a diameter as possible, but there were practical limitations on how thin the wall of the pipes can be. The smallest thickness wall that could be used was 1/8 in. (0.3 cm) but this was beyond what was required structurally in some members. In addition, the extra thickness reduced the radius of gyration to the point that the slenderness ratio (kl/r) was over 200. If the wall thickness could have been made thinner and the outside diameter kept constant, then the radius of gyration would increase and consequently the slenderness ratio would decrease. As a result, the approach that was taken was to base the structural design only on that portion of the thickness necessary to provide the strength required, and that satisfies the code limitation on slenderness ratios being less than 200. In addition,

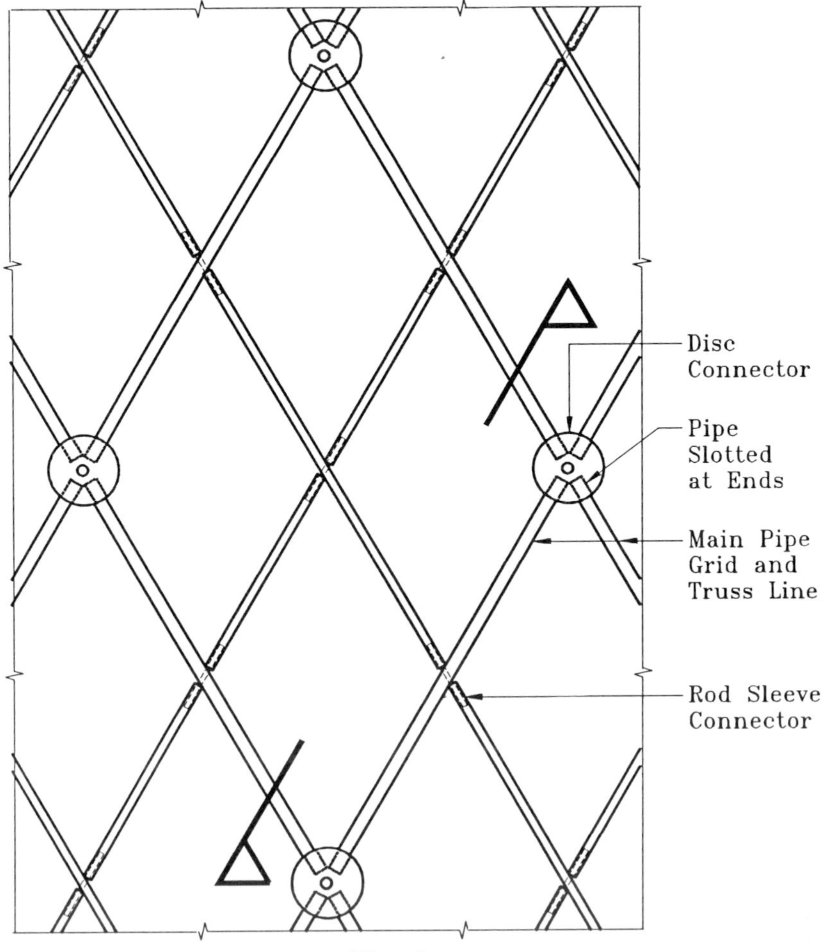
Disc
Connector
Pipe
Slotted
at Ends
Main Pipe
Grid and
Truss Line
Rod Sleeve
Connector

Fig. 4.

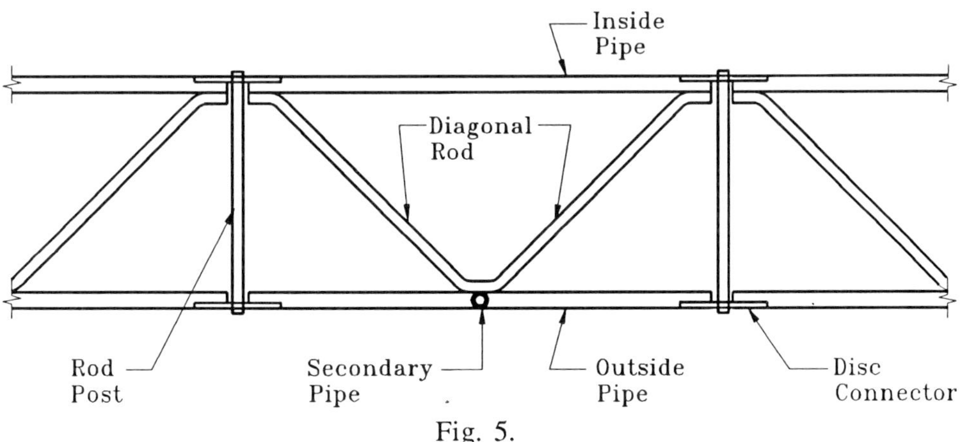
Inside
Pipe
Diagonal
Rod
Rod
Post
Secondary
Pipe
Outside
Pipe
Disc
Connector

Fig. 5.

a comparison of the strength of the full thickness of the pipe by Euler buckling versus the strength of the pipe, based on the reduced thickness meeting the slenderness ratio cut off, showed that the pipe with the full thickness had greater strength. As a result, the reduced thickness approach to meet the code slenderness ratio was conservative.

FABRICATION

12. The space sculpture was made out of stainless steel pipes and rods. Unique details were used to facilitate fabrication. Typical of the details is the circular disc connector at the primary grid intercepts (see Figs 4,5). This connection occurred at the location of the intersection of the inner diamond grid and its corresponding point in the main outer grid. The result is that four major diamond grids made up the height of the space sculpture. The pipe members were continuous between the main intercepts. At these main intercepts the pipe members stopped and started. The ends of the pipes were cut in half to allow the disc connection to fit in. The pipes were then welded to the discs. The outer grid had a secondary diamond grid which intersected at the midpoints of the main diamond grid. Since the pipes were hollow it was possible to put a solid rod sleeve in them to act as a connector for the secondary grid pipes to the main grid pipes or other secondary grid pipes. These rod sleeves fit in the ends of the secondary grid pipes and through the main diamond grid pipes at a 60 degree angle. The sleeve rods were fillet welded to both pipes. The diagonals on the main diamond grid were made out of one continuous rod between the main grid intercises. The rod was bent to form a set of two diagonals in a "V" profile between the two main joints (see Fig. 5). Bending a stainless steel rod increases its strength. The diagonal rod was bent into a curve at the points of intersection which occurred at the inner and outer rings. As result, there were points of tangency between the inner and outer layers' pipe members and the diagonal rod. It is at these points of tangency that the diagonals were welded to the inner and outer layer pipe members. Threaded rod posts connected the inner and outer layer together at the main intercepts and were bolted to the disc connector plates. The result was a simple-to-fabricate space structure using only fillet welds and bolts. The whole space structure (see Fig. 6) was fabricated in the shop and shipped to the field and erected in one piece on the previous elected head structure (see Fig. 7).

13. Other issues that had to be addressed were the dialectic action of the stainless steel space structure to its base support which was structural steel, and the head, which was bronze. A bituminous paint on the support structure connectors provided the isolation.

CONCLUSION

14. The Starman Space Structure is an example of the collaboration between art and structural engineering. The artist sought a spatial sculpture that depended on a mathematical progression to give the sculpture a logic and coherence. This mathematic progression allowed the structural designer to create the mathematical and structural model that allowed the spatial structure to be automatically analysed

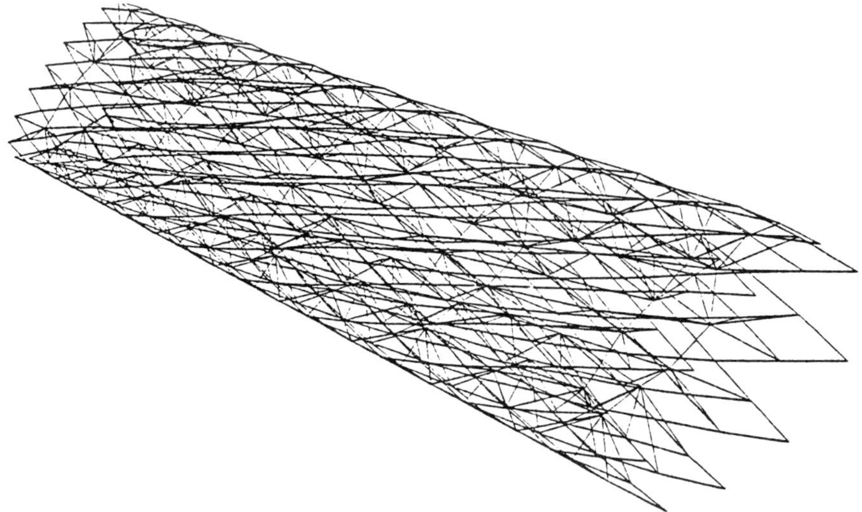

Fig. 6.

Fig. 7.

and designed. In fact, the computer generated model provided the geometry and member lengths that enabled the sculpture to be fabricated. The end product was a very successful collaboration between art and engineering that has become an important outdoor sculpture in the city of Philadelphia.

218. Modular space frame structures

F. ESCRIG, Seville School of Architecture, Spain, and J. VALCARCEL,
La Coruna School of Architecture, Spain

SYNOPSIS. The construction of space frames with prefabricated bars and joints has resulted in a successful method of building roofs and walls, primarily due to the simplicity of connections, the lightness of the structure, the attractive appearance coupled with economy in design. The generalised use of these system confirms their advantages, but there are some restrictions and inconveniences that may make the system inadequate. Modular systems are formed from groups of bars which are easier to assemble compared with single members and consequently can offer an alternative means of constructing space grids. Some constructed structures offering clear advantages are shown to illustrate this aspect.

MODULAR SYSTEMS OF SPACE BAR STRUCTURES

1. One of the greatest advantages in the construction of skeletal space structures is that they can be formed by using members of only a few different lengths.

2. Using bars with lengths which are easy to handle and standard joints, great structures can be constructed, which are light in weight, very rigid with regular geometry and also surprisingly attractive.

3. Spatial structures made out of bars have been successfully developed because of their versatility, economy and rapid erection. The initial difficulty associated with the complex and large analysis of the structures has been overcome with computer techniques that are now used in the automation of the manufacturing process, allowing the construction of structures with very complex geometry's.

4. The advantages of using skeletal space structures, outlined previously, are greatly influenced by the type of joint used in the structure. From earlier proposals, where members were welded directly to each other requiring the complex cutting of the member ends; to more sophisticated nodes, formed in high quality steel, forming almost imperceptible points where many bars meet together; there are numerous different patented node types exhibiting different levels of ingenuity.

5. Quality, strength, appearance, economy and marketing of each patented node have prioritised some nodes in preference to others and indicate why only a few types have become popular.

6. The advantages of skeletal space structures can be summarised as follows:
a) economical in material use resulting in light structures;
b) constructed using relatively few different member types;
c) adaptable to any perimeter and topography;
d) aesthetically pleasing allowing the structures to remain uncovered and visible;
e) easy access for services allowing pipes and tubes unrestricted routing;
f) used to construct roofs and walls and if desired, the whole building;
g) easy transportation of component members because they occupy very little space, when correctly packed.

7. Nevertheless they have some disadvantages too that limit their use, namely:
a) the economy in weight does not necessarily correspond with financial savings;
b) the structures are difficult to analyse and require sophisticated numerical treatment usually undertaken in conjunction with assumptions on loading and support conditions;
c) single layer and foldable structures may be very flexible with large displacements placing limitation on imposed loads;
d) the structures may be highly statically indeterminate but nevertheless stability of the whole structure depends of the stiffness of only few bars, with risk of failure attached to a small number of sections;
e) the simplicity of geometry in practice is complicated by architectural requirements. For instance a flat structure needs some slope for drainage and this means that the depth of the structure may be variable, resulting in different length bars;
f) usually these structures need a complementary structure to support the roof cladding;
g) they may cause access difficulties in their final position when constructing the surrounding structure;
h) nodes are individual elements, usually patented and therefore controlled by specialised builders;
i) nodes are subjected to severe concentration of stresses.

8. This indicates that although these kind of structures may provide the best possible architectural solutions they may not be the panacea desired for and better alternatives may be available.

9. One of these alternatives are "modular systems of bar structures". Space frame structures are not always built from individual bars and nodes. This form of construction eases transportation but complicates the assembly. To build piecemeal requires scaffolding or alternatively building the structure on the ground and lifting it into position, is expensive.

10. However, the modular alternative has serious drawbacks with respect to transportation, primarily due to the large volume, but nevertheless this is compensated by a higher stiffness and simplicity of joints. Both aspects that directly affect the overall economy and aesthetics of the structure.

11. Paradoxically while there exist a wide bibliography on space structures, it is difficult to find publications on structures built from modules. In part, this is due

to the large versatility of solutions but also because it is not a usual technique of structural design.

Fig. 1.

Fig. 2.

12. Because the difficulty of beginning from zero and establishing a guide of possible logical patterns, the authors prefer to explain particular applications designed for construction.

13. The Crystal Palace constructed in London for the International Exhibition of 1851, is usually quoted as the first example of rigorous prefabrication. Although most of the components were planar, the complexity of the whole structure made it desirable to record the behaviour of the rigid modular covered surface. The Crystal Palace was a surprising structure, considering the level of technology at the time of construction (Fig. 1).

14. Modular construction is not only an established method of construction for iron or steel. Reinforced concrete has been used in very complex buildings. Nervi in the Exhibition Palace of Turin (Fig. 2) shows a clever use of repetitive modules, to cover a span of 80 m.

15. Nevertheless, it is in relatively light structures where the use of steel modules is more frequent. Polyhedrical forms are more adequate than other forms to be used repetitively and there are many structural designers using proposals of this kind. For example Candela (Fig. 3), Richter (Fig. 4), or Perez Pinero, the

Spanish architect who built structures, so light, that they could be assembled by hand (Fig. 5).

16. When professional commitments have allowed, the authors have tried to advance proposals with the same aims and goals. The structures constructed are not large projects but they are conceived with vocation of scale.

Fig. 3.

Fig. 4. Fig. 5.

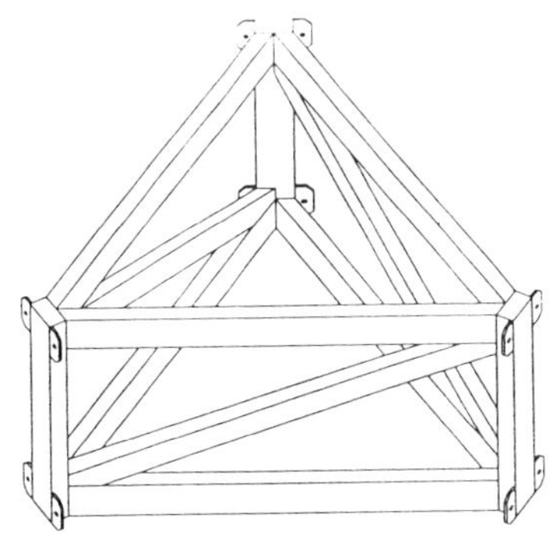

Fig. 6.

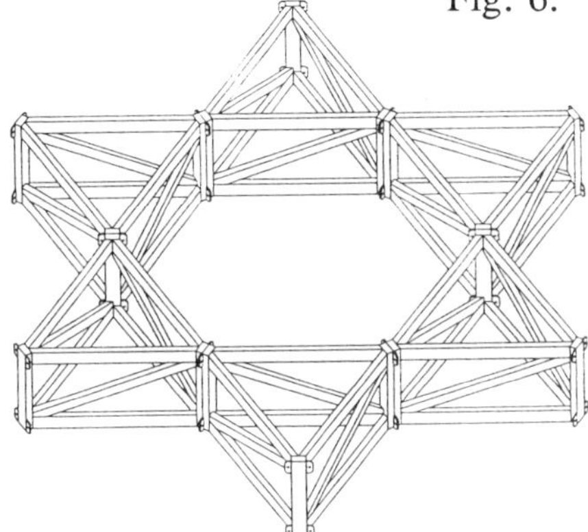

Fig. 8.

Fig. 7.

Fig. 9.

CANTILEVER ROOF FOR THE AUDITORIO OF LA RABIDA, HUELVA.

17. The authors have used triangular modules like those shown in Figs 6 and 7 assembled together to form the structure as shown in Figs 8 and 9. After the initial assembly a diagonal strut has been introduced into the structure to stiffen the hexagonal units so that they could be used to support the roof material.

Fig. 10.

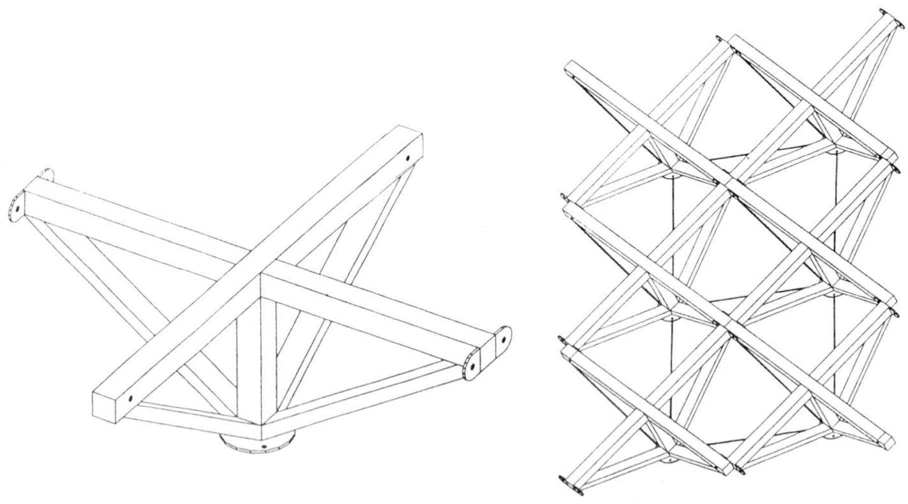

Fig. 11. Fig. 12.

18. As the structure was spanning only a short distance the elements used in the structure are all the same weight, facilitating easy construction, enabling the structure to be erected in two days with the minimum of scaffolding (Fig. 10).

PAVILION OF EXTREMADURA AT EXPO'92 - SEVILLA.

19. A similar system was used to support the glass floor of a building constructed using one basic module (Fig. 11). The modular units were joined together by means of two bolts in the upper layer and a tie in the lower layer. Scaffolding was not used during the assembly of the structure. Each new unit was joined to the previous one, finally forming the entire surface as shown in Figs 12-14.

Fig. 13.

Fig. 14.

SUSPENDED GARDENS FOR EXPO'92 SEVILLA.

20. The aim of this project was to cover 50,000 m^2. of open spaces at EXPO'92 with a repetitive, industrialised, portable structure, capable of supporting very large flowerpots and plants. Economy was another of several conditions necessary to comply with to win the public competition.

21. Fig. 15, shows a scheme of the proposal and Fig. 16 the solution used to achieve it. The authors chose rectangles 12 x 6 m^2 supported on four corners by a collection of four tubes. On each support it was possible to connect between one to four modules at different levels (Fig. 17).

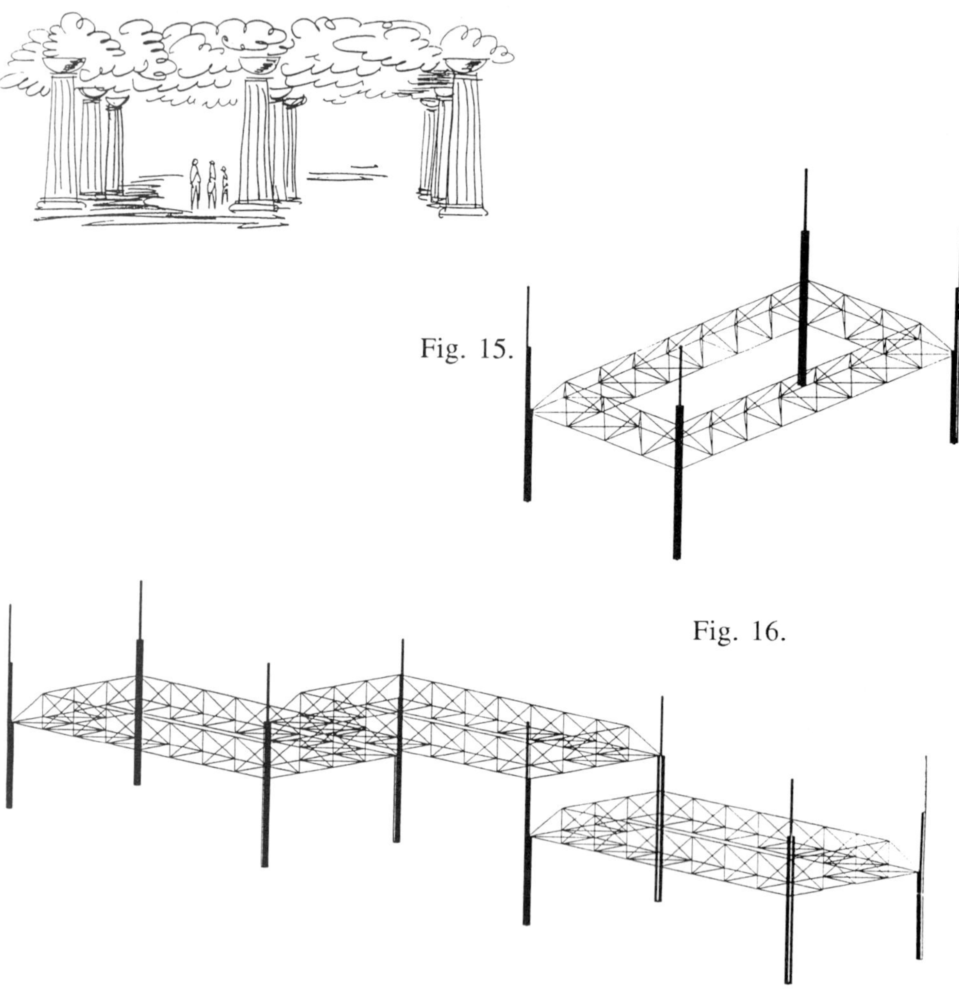

Fig. 15.

Fig. 16.

Fig. 17.

22. Each one of the modules was built in four parts joined together in its final position with bolts (Fig. 18).

23. To reduce the construction costs, joints were formed from bent tubes in such a way that the diagonal web members were made from one length of steel. This method produced problems with the concurrence of the axes of the bars (Fig. 19). Because the radius of curvature must be at least 2·5 times the bar diameter, solution "a" (Fig. 19) was not possible neither solution "b". Solution "c" was chosen giving edges not aligned to one point. The load test showed that the influence of eccentricities was negligible.

24. The connection between the space frames and supports was by a welded element used as a guide introduced into the tubes.

25. To ease the construction, it was not possible to anchor the supports to the ground but instead heavy concrete blocks were used to prevent up lift of the whole structure.

26. Once the structures were assembled it was not possible to erect them directly into their final place because it was first necessary to wait for shrubs and plants placed in containers inside the structures to grow. At the correct time the complete frames together with plants and shrubs were lifted onto a platform, transported and erected on site (Fig. 21).

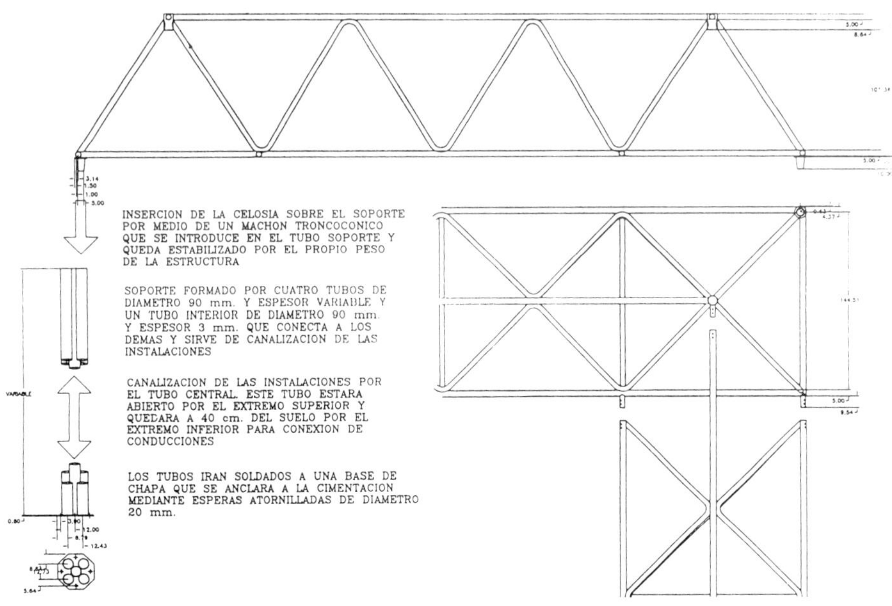

Fig. 18.

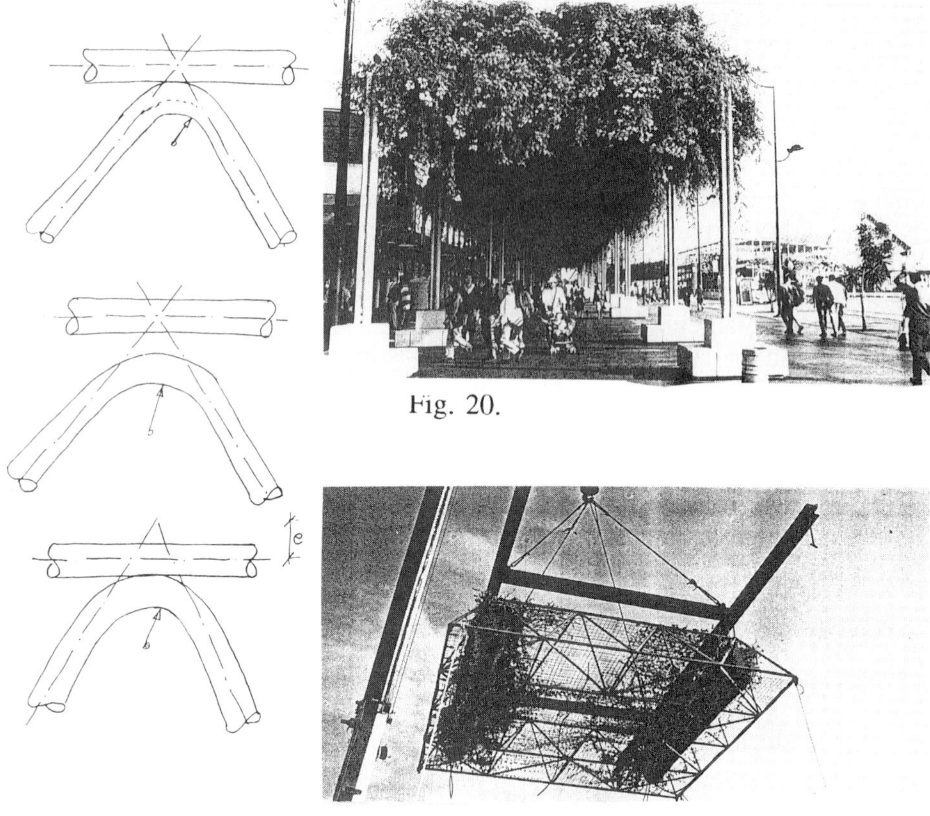

Fig. 20.

Fig. 19.

Fig. 21.

219. Extending Le Corbusier's modulor

A. CATTAN, Consultant, USA

SYNOPSIS. Le Corbusier in his *Modulor,* proposed a proportional system based on the Golden Section, which has a close connection with the Fibonacci series. The present study extends this series into a system of infinite series of numbers and relates them to a broader system of proportions. The Golden Proportion System is a part of this extended system of proportion.

INTRODUCTION TO THE PROPORTION

1. Matila Ghyka in his celebrated book "The Geometry of Art and Life" (ref. 1, p. 1) writes about the notion of proportion: "... *proportion is, in logic and as well as in Aesthetics, one of the most elementary, most important, most difficult to sort out with precision; it is either confused with the notion of ratio or (especially when talking of proportion in the plural) with the notion of a chain of characteristic ratios linked together by a modulus, a common sub-multiple".*

2. Proportion is a relationship of equivalency between two ratios; one element is to a second element as a third element is to a fourth: *a is* to *b* as c is to *d* $(a / b = c / d$. This was called by the Pythagoreans (ref. 2) as *discontinuous proportion* of four terms; *continuous proportion* of three terms is when, one element is to a second element as the second element is to a third: *a is* to *b* as *b is* to c $(a / b = b / d)$.

3. There is only one proportional division which is possible with two terms: when the smaller term is to the larger term in the same way as the larger term is to the smaller plus the larger $(a / b = b / a + b)$. This is known as the *"Golden Proportion"* and is designated by the 21st letter phi (Φ) of the Greek alphabet. The Greek geometers of the Platonic school called it "the section" *par excellence.* Luca Pacioli, Leonardo's friend, called it *"De Divine Proportione"* (ref. 1, p. 4).

4. Leonardo of Pisa, alias Fibonacci, wrote in his book *Liber Abaci* of 1202 A.D. (ref. 3), about a *summation series* (ref. 4), so called by Edward Lucas in 1877 the Fibonacci series (ref. 5, p. 47).

NUMBERS

5. The Fibonacci series is extended here into a system of infinite series of numbers (ref. 6) (Fig. 1). These sequences are generated by the addition of a different number of integers. For example, the Fibonacci series (series 2 in this system) is obtained by adding two successive numbers to get the next number in the sequence, series 3 is obtained by adding three successive numbers to get the next number in the sequence, and so on.

6. In the Fibonacci series (series 2) the quotient between two successive numbers approaches the limit 1.618, in the third series (series 3) the quotient approaches the limit 1.839, in the fourth series (series 4) it approaches 1.927, and the last series (series ∞) it equals 2.

Series 1: 0-1-1-1-1-1-1-1-1-1-1-1-1-1-1-1-1-1-1-1-1-1-1-1-1-1-1-1-1-1-1-1-1-1-1

Series 2: 0-1-1-2-3-5-8-13-21-34-55-89-144-233-377-610-987-1597-2584-4181-6765

Series 3: 0-1-1-2-4-7-13-24-44-81-149-274-504-927-1705-3136-5768-10609-19513

Series 4: 0-1-1-2-4-8-15-29-56-108-208-401-773-1490-2872-5536-10671-20569

....

Series ∞: 0-1-1-2-4-8-16-32-64-128-256-512-1024-2048-4096-8192-16384-32768

Fig. 1. Extension of the Fibonacci series.

7. As in series 2, the ratio between two consecutive terms has the limit Φ, 1.618033988 (b in this system) that when squared equals the number plus one: $b^2 = b + 1$; in series 3 the ratio between two consecutive terms has the limit c (1.839286755), that when cubed equals the number squared plus the number plus one: $c^3 = c^2 + c + 1$. In series 4 the ratio between two consecutive terms has the limit d (1.92756255), where: $d^4 = d^3 + d^2 + d + 1$, and so on. The whole system is represented by the equation $x^{n+1} - 2x^n + 1 = 0$ (ref. 7).

GEOMETRY

8. Each series of numbers in the present system has its own geometrical representation (Fig. 2). The golden rectangle (series 2) is a part of a series of infinite rectangles, which go from the square (series 1) to the limiting case of 1 by 2 rectangle (series ∞).

Series 1

0-1-1-1-1-1-1-1-1-1-1-1-1-1-1-1-1-1-1-1-1-1-1-1-1-1-1-1-1-1-1-1-1-1-1-1-1-1

$a = 1$

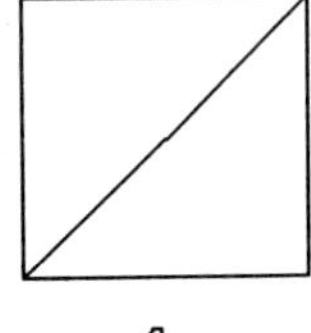

Series 2 (Fibonacci)

0-1-1-2-3-5-8-13-21-34-55-89-144-233-377-610-987-1597-2584-4181-6765

$b^2 = b + 1$
$b^2 / b = b / 1 = 1.618033988$

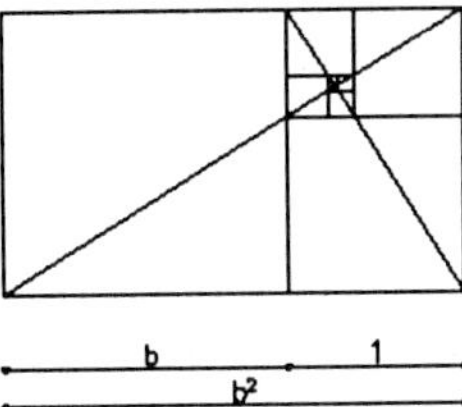

Series 3

0-1-1-2-4-7-13-24-44-81-149-274-504-927-1705-3136-5768-10609-19513

$c^3 = c^2 + c + 1$
$c^3 / c^2 = c^2 / c = c / 1 = 1.839286755$

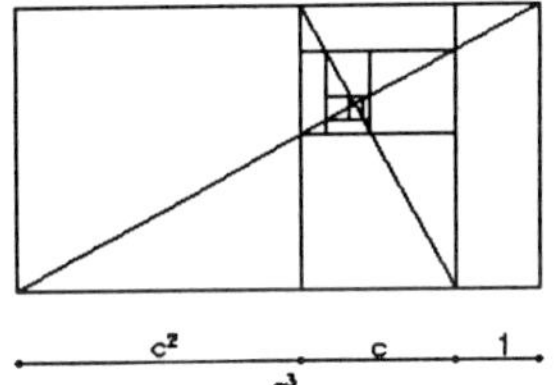

Series 4

0-1-1-2-4-8-15-29-56-108-208-401-773-1490-2872-5536-10671-20569

$d^4 = d^3 + d^2 + d + 1$
$d^4 / d^3 = d^3 / d^2 = d^2 / d = d / 1 = 1.92756255$

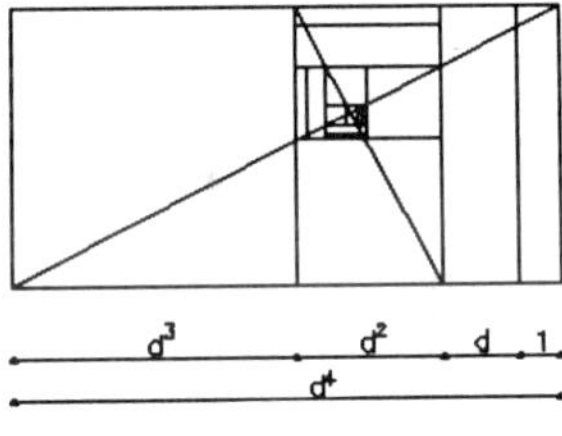

....

Series ∞

0-1-1-2-4-8-16-32-64-128-256-512-1024-2048-4096-8192-16384-32768

$z^n = z^{n-1} + \dots\dots\dots z^3 + z^2 + z + 1$
$z^n / z^{n-1} = z^3 / z^2 = z^2 / z = z / 1 = 2$

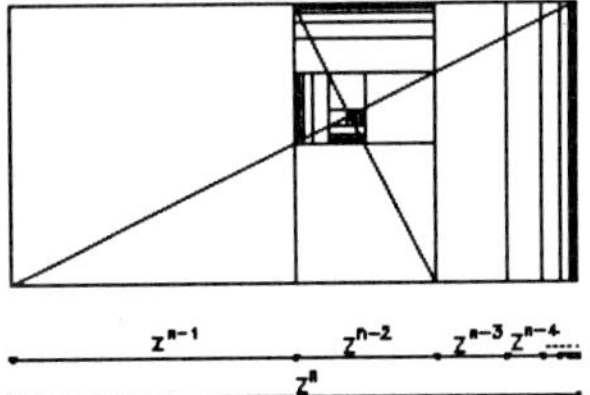

Fig. 2. Geometrical representation of the number sequences.

9. The rectangle of series 3 is selected here, as an example, to show properties common to the system. The rectangle ABCD is proportional to the rectangle EFGH and to the rectangle JILK. This operation can be repeated infinitely, thus getting smaller and smaller rectangles with the same proportion (Fig. 3).

10. As the golden rectangle can be obtained approximately from any additive series of squares (ref. 5, p. 67), closer approximations to the rectangle of series 3 result from any additive series of squares and rectangles (Fig. 4).

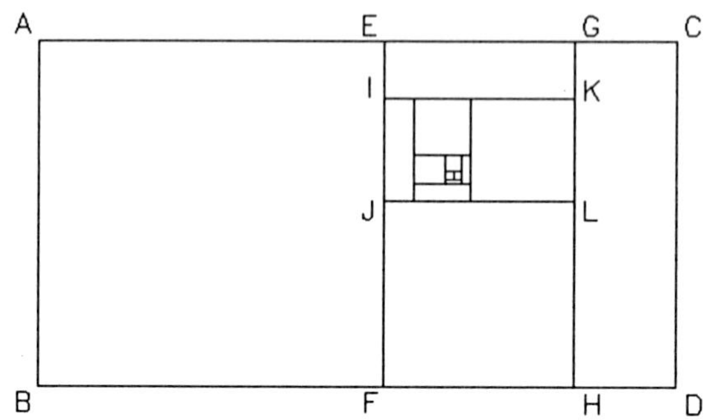

AC / AE = AE / EG = EG / GC = 1.839286755
BD / EF = EF / IK

Fig. 3. Series 3 rectangle and some properties.

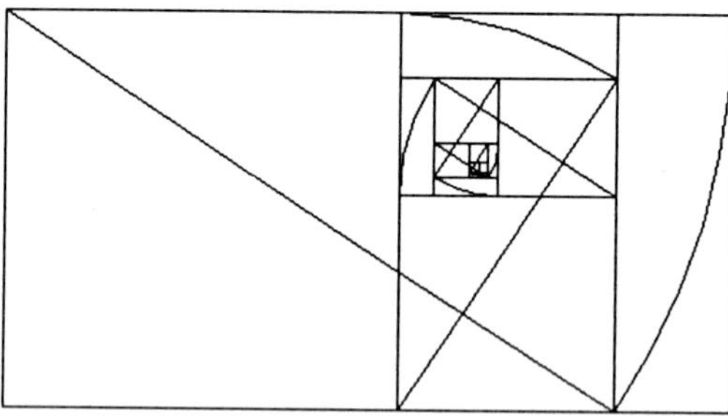

Fig. 4. The additive series of squares and rectangles.

11. In this rectangle, as in the rest of the rectangles of the system, the logarithmic spirals are branching proportionally as fractals (Fig. 5). Based on Anne Tyng's work (ref. 8), the spiral extension of a helix is shown in Fig. 6.

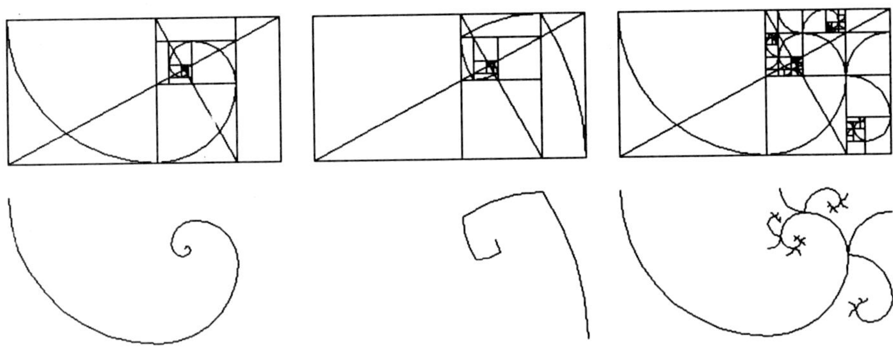

Fig. 5. Logarithmic spiral, and curves in series 3 rectangle.

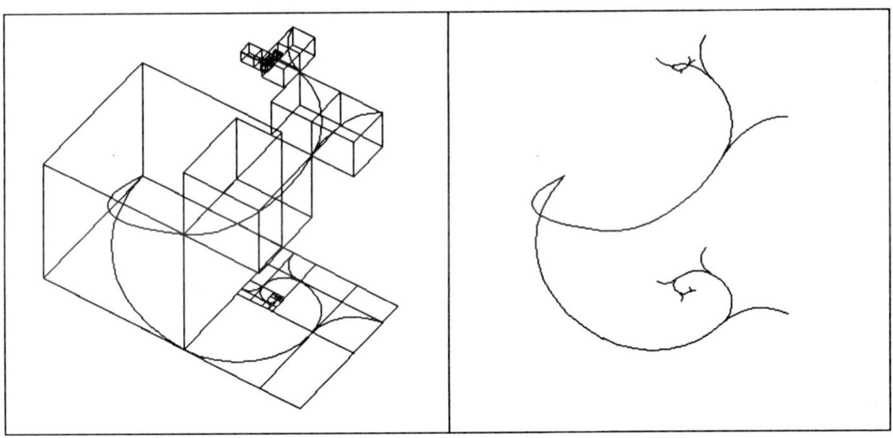

Fig. 6. The spiral extension of helix in series 3.

EXTENDING LE CORBUSIER'S *MODULOR*

12. As the Fibonacci series was extended here into a system of series of numbers, the Golden Series (used by Le Corbusier in his Modulor), where each term is the sum of the two preceding ones in the series, is expanded here into a system of series of progressions. Each of them can be developed as a 2-dimensional and 3-dimensional modular system (Fig. 7) (refs 9,10,11).

Series 1

1, a, a^2, a^3, ..., a^n
$1 = a$

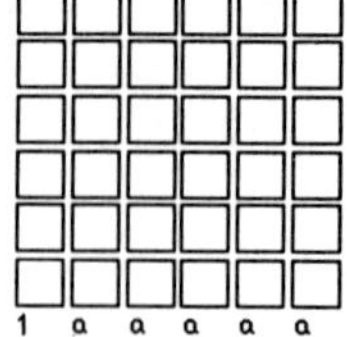

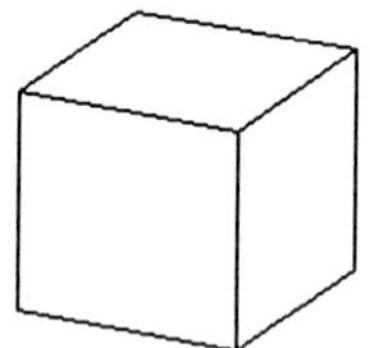

Series 2

1, b, b^2, b^3, ..., b^n
$1 + b = b^2$
$b = 1.6180339$

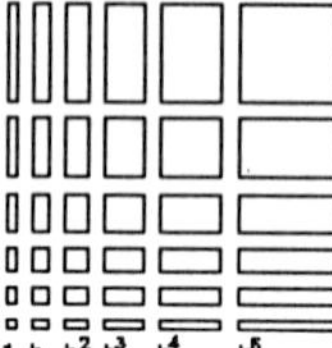

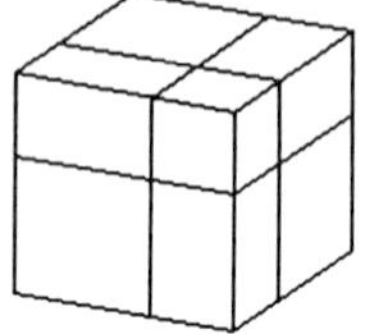

Series 3

1, c, c^2, c^3, ..., c^n
$1 + c + c^2 = c^3$
$c = 1.839286755$

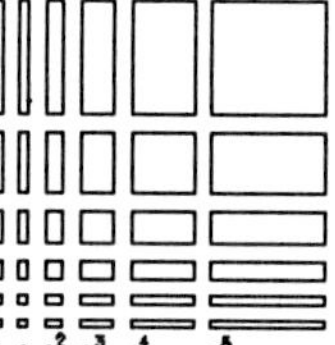

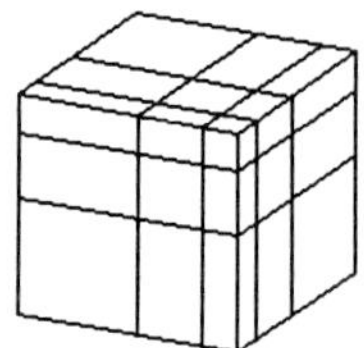

Series 4

1, d, d^2, d^3, ..., d^n
$1 + d + d^2 + d^3 = d^4$
$d = 1.92756255$

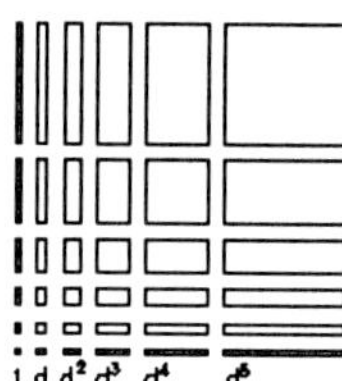

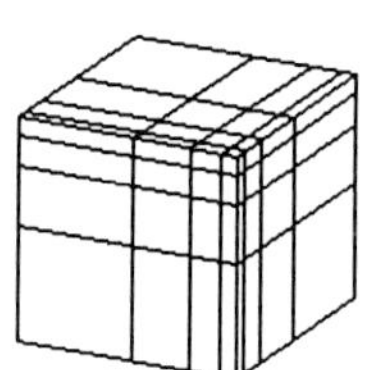

....

Series ∞

1, z, z^2, z^3, ..., z^n
$1 + z + z^2 + ... z^{n-1} = z^n$
$z = 2$

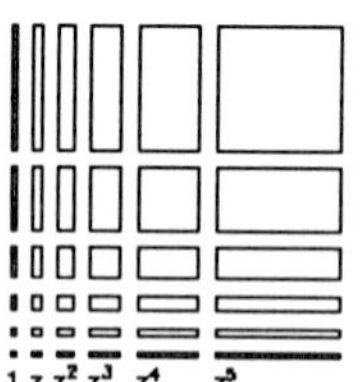

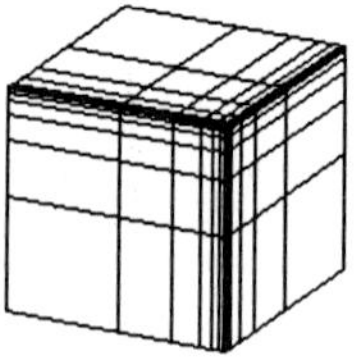

Fig. 7. 2-D and 3-D modular systems based on the number of sequences.

13. This study compares the minimum elements of the Golden Series, progression of series 2 (where b represents the lengths and B represents the areas) in this system, with the minimum elements of the progression of the series 3 (where c represents the lengths and C the areas) (Fig. 8).

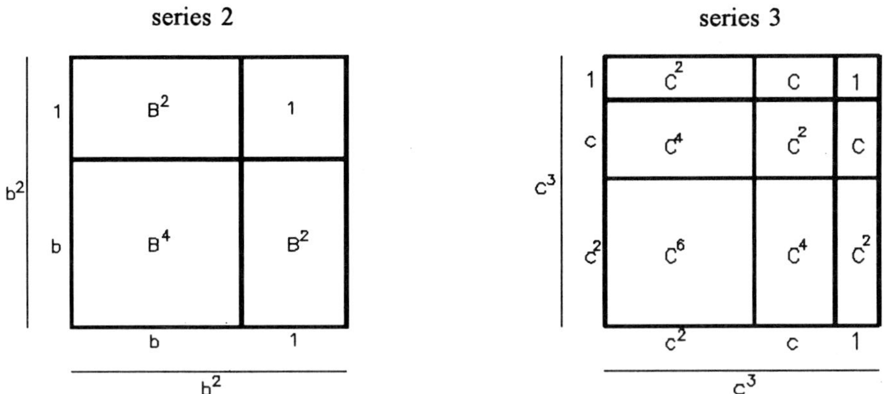

Fig. 8. Minimum elements of the Golden Series and the progressions of series 3.

14. Each of the series of progressions of the system can be combined to give another progression. An example is shown with the progression of series 3 (Fig. 9).

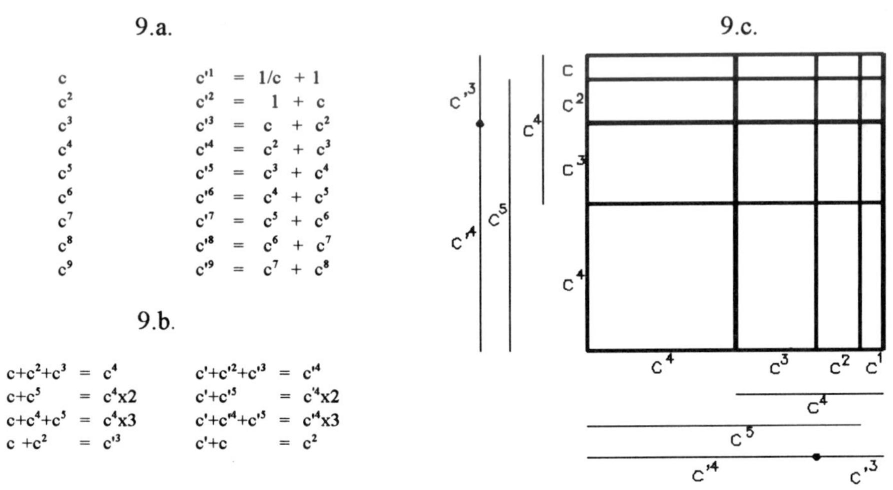

Fig. 9. The combination of two progressions and its relations to produce a combined modular system.

15. Based on this combined system, a series of exercises in two dimensions and three dimensions are shown in Figs 10 and 11, respectively.

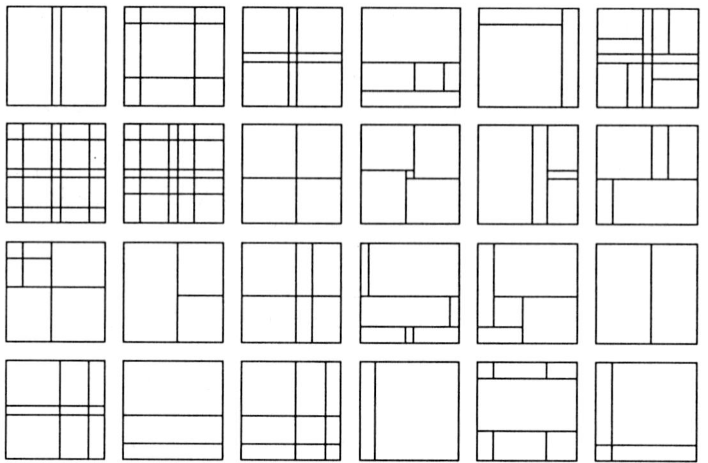

Fig. 10. Panels exercises based on the combined modular system based on Fig. 9.

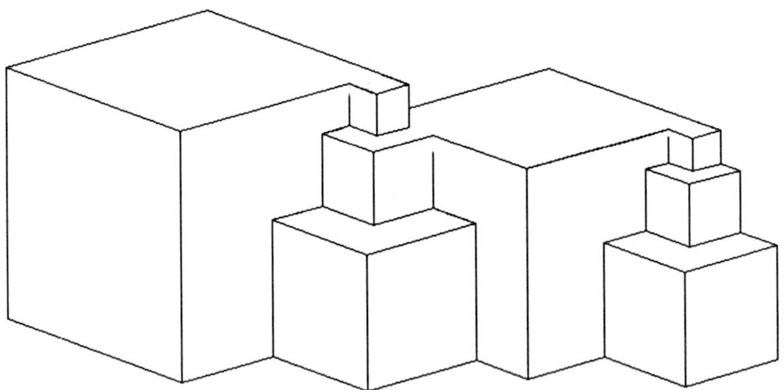

Fig. 11. Modular system in 3-dimensions based on Fig. 9.

16. In the same way that Le Corbusier applied his Modulor to his "Panel Exercise" and eventually to his buildings, the geometries presented here can be extended into a family of 2- and 3-dimensional proportion systems in architecture.

ACKNOWLEDGEMENTS:
My thanks to Professor Haresh Lalvani for his guidance in this project and for his comment on this paper. This work was developed by the author during his M. Arch Thesis in the School of Architecture, PRATT Institute.

REFERENCES

1. GHYKA, MATILA. The Geometry of Art and Life. Dover, New York, 1977.
2. LAWLOR, ROBERT. Sacred Geometry, Philosophy and practice, p. 44. Thomas and Hudson, London, 1982.
3. MENNINGER, KARL. Number Word and Number Symbols, a Cultural History of Numbers, p. 425. Dover, New York, 1992.
4. DOCZI, GYORGY. The Power of Limits, Proportional Harmonies in Nature, Art and Architecture, p. 5. Shambhala, Boston & London, 1985.
5. HUNTLEY, H. E. The Divine Proportion, a Study in Mathematical Beauty. Dover, New York.
6. The concept of the generalisation of the Fibonacci series was developed by the author during his M.Arch. Thesis (1991-92). Earlier work on the k-Fibonacci series has come to the author's attention since (see for example: ref. 12)
7. This equation was shown to the author by Jay Kapraff (personal communication, 1992).
8. TYNG, A. Geometric extensions of man's consciousness, Zodiac 19, Italy, 1969.
9. LE CORBUSIER, Modulor 1 and 2, M.I.T. Press, 1977.
10. WYSOCKI, STANLEY, developed the "form solid" which uses the integer series 1, 2, 3, 4, 5, 6... in a 3-dimensional cubic grid (1970's, unpublished).
11. LALVANI, H. Unpublished (personal communication).
12. SCHROEDER, M.R., Number Theory in Science and Communication, p. 74. Springer-Verlag, 1990.

220. Advanced glass space structures

M. EEKHOUT, Octatube, The Netherlands

SYNOPSIS. This article describes the results of space structures made from structurally loaded, tempered glass panels with steel stabilising and connecting components. These structures have been developed by the Design and Development team of Octatube during the last 4 years, in a continuous product development process stimulated by a number of project-applications.

PROCESS: THE PRODUCT ARCHITECT

1. In the Netherlands the term "project-architect" was introduced by the author in 1988 to distinguish architectural activities around products from those around projects. As product technology becomes more specialised on the one hand and projects become more complex on the other hand, the project-architect generally does not find the time any more, nor has retained the ability to develop new technical components for his building from the many new materials available. The project-architect would rather leave this aspect to a competent producer or to a production colleague.

PRODUCT: STRUCTURAL GLASS

2. Structural glass designs flourish on the current trend from closed structural systems towards open (design responsive) structural concepts and even the positive de-systematising trend in architectural practices nowadays. It is not a complete building product that is applied in architectural projects, but a product-system of which the architect can choose many variables, or even (in the extreme) only a concept like a way of connecting different elements without a fixed prescription of the rest of the built up components. It surely gives architects much more freedom of choice, but it also requires them to be able to participate in the technical set up of the design. Therefore they should be familiar with the basics.

3. As to the topic of structural glass applications, there are three major aspects to be fully aware of, namely:

- material properties of fully tempered glass;
- production technology of glass/metal components;
- application systems for the stabilisation of glass by metal components.

Space structures 4. Thomas Telford, London, 1993

Material Properties

4. The Modulus of Elasticity for glass is 73000 to 75000N/mm^2 while steel has an elastic modulus of 210000N/mm^2 and aluminium 67000 to 73000N/mm^2. The ultimate tensile strength for ordinary glass is 40N/mm^2, 200N/mm^2 for fully tempered and 100120N/mm^2 for half tempered glass (heat strengthened). Some mild steel has an ultimate tensile strength of 360N/mm^2, aluminium 215N/mm^2. The compressive strength of glass (without the effects of buckling) is 800 to 1000N/mm^2. The most deviant property of glass applied as an element of a primary structure, compared with other structural materials, is its brittleness; which turns out to be most decisive in the use of glass as a structural material. The 'Achilles heel', the weakness of glass, is caused by this brittleness. Glass has no capacity to re-distribute high tensions. Small stress concentrations and imperfections in the glass surface can cause concentration factors of over 100. Tempered glass is even more sensitive to imperfections in the surface than normal glass, when the sensitive internal stress balance (compression outside, tension inside) gets disturbed. Very broadly speaking, there are two fracture mechanisms competing to break a material namely: plastic flow and brittle cracking. The material will succumb to whichever mechanism is weaker. If it yields before it cracks, it is ductile. If it cracks before it yields, it is brittle. The potentiality of both forms of failure is always present in most materials. Yielding is a safe and much desired property, spontaneous cracking an undesirable property for a structural material. Thermal conductivity for glass is 0.80W/mk; 40/50W/mk for steel and 200W/mk for aluminium. Polymethylmethacrylate (PMMA) and Polycarbonate (PC) have thermal conductivity values of 0.07 to 0.21W/mk.

5. Thermal expansion is important when combing different materials into one structural entity, subjected as a whole, to temperature loading. Values for glass of 9x10^{-6}W/mk; for steel 12x10^{-6}W/mk; for aluminium 24x10^{-6}W/mk; and for PMMA/PC 7065W/mk, which is a very large expansion coefficient. Specific gravity for glass is 25kN/m^3, for steel 78kN/m^3 for aluminium 27kN/m^3, and for PMMA/PC 12kN/m^3. The maximum workable temperature for tempered glass is 270°C; for ordinary glass 60-110°C; for steel 550°C; for aluminium 270°C; for PMMA/PC 80-115°C and for Borosilicate glass 800°C.

Production Techniques

6. To prevent the glass from cracking it is necessary to subject it to compression. This can be achieved by heating the glass panels again, and chilling the two outer skins of the hot glass panels during manufacture very quickly, so that the two outer skins form with the core, a compression/tension stress state. When the outer surfaces have cooled they solidify, while the core is still hot (700°C). The shrinking during cooling of the core causes the outer skins to be compressed, while the central core will remain in tension. This is a mechanism used to avoid surface cracks, but it hides internal cracks too. After tempering, the final product is a glass panel with higher tensile strengths and a higher impact strength too. Great care has to be taken so that the glass surface is not scratched

by a sharp tool, because if this happens the glass will shatter into thousands of small splinters.

7. In tempered glass panels possible cracks are avoided by the compression mechanism. Consequently, bolt-and-hole type connections can be best achieved by using a pre-tensioning type of bolt connection. In this case the hole edges are not loaded by the bolt when flush, (with the inherent danger of enlarging the micro-cracks around the bolt hole caused by drilling), but the pre-tensioned bolt compresses the two outside metal rings or components onto the glass where mutual friction will provide the connection force. The friction force can be enhanced by grinding or blasting the surface around the bolt hole (before the hardening process). Regarding the size of the glass panels, a possible compression (normal) force introduces the danger of buckling the panel, so that the largest commercially available panel thicknesses of tempered glass (12, 15 and 19mm) will have to be used; these are fairly expensive per volume. It would then be preferable, if at all possible, to load the glass panels in tension instead of compression, suspending the panels rather than stacking them. The description 'curtain wall' would then live up to its full literal meaning! Consequently, with roofs it would be better to combine glass panels into the tensile regions instead of incorporating them into the compression zone.

Statical Systems

8. The static systems applicable to glass structures will all be based on the use of glass panels. The most simple solution is the classical vertical glass panel, capable of spanning vertically with or without metal or glass ribs. One of the first architectural design priorities, is to develop larger scale glass planes with minimal visual disturbances. As a result, visually minimal two-dimensional structures (so-called because of the flat plates) such as guyed structures, are the most logical of structures. Fig. 1 shows the derivation of the guyed glass structure principles, and needs no explanation. The principle behind the guyed glass structure is that short and slender metal compression bars are used along with long thin metal tension bars and glass plates where normal stresses in the form of tension or compression are included. The figure shows the difference between open and closed structural schemes, of importance in regard to the connections with the substructure. An indication too, is given of single-sided and double-sided schemes arising out of architectural considerations, and of the need to stabilise with counter spanning guy trusses when both positive and negative perpendicular loads have to be accommodated.

PROJECTS: FOUR CASES

Glass Music Hall Amsterdam

9. The first major application of Quattro nodes and tensile truss stabilisers are displayed in the walls of the AGA Glass Music Hall in the famous former Exchange of H.P. Berlage in Amsterdam (built in 1903). Project-architect Pieter Zaanen saw in an all glass envelope the only solution to make an acoustical envelope for rehearsals and 200 seat performances by the Dutch Philharmonic

Orchestra, while the old exchange building could house simultaneously a 600 seat orchestra performance in the adjacent Wang Music Hall.

10. A simultaneous set-up by the acoustical advisor Peutz, led to a 2000m^3 volume, a classical form with approximate sizes of 10 by 20m, 10m high, and the minimal requirement of 8mm thick glass with a prescribed level of absorbency.

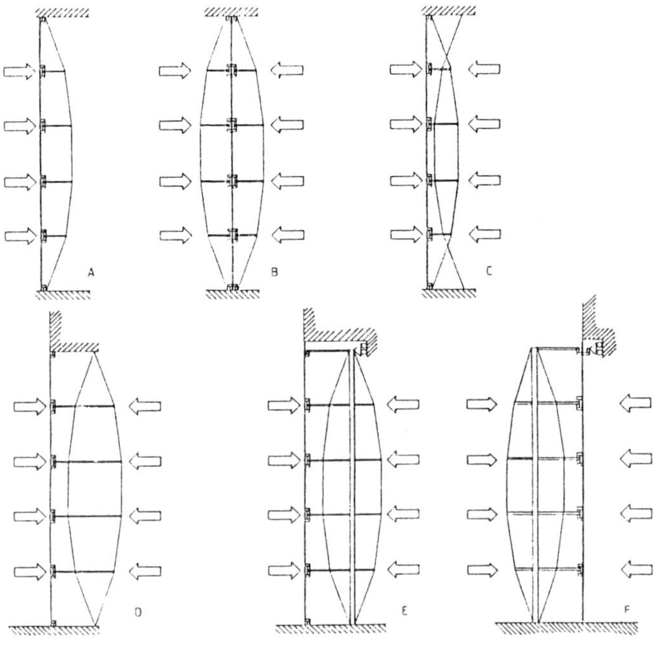

Fig. 1.

Although the targets for the design made simultaneously by the project-architect and the author as a structural designer were set very high to try to achieve both roof and walls in Quattro stabilised glass, experiences failed to convince the fire-prevention officers that the stabilised roof had a 30 minutes fire rating. So a standard space frame was chosen, supported on 6 columns in the characteristically cruciform that Zaanen used as a theme in the renovation of the exchange building. The space frame has been deformed, from a standard space frame with reduced edge-modules in order to get the outside glass panels equal in size, near the corners. The space frame acts as a rigid super structure from which the walls are suspended. The glass volume has been provided with a "cello-belly" long wall in order to prevent acoustical flutter. This curved wall has a different suspension arrangement from the external tubular columns on which the Quattro nodes have been directly welded. The peculiar form of the curved wall has resulted in 12 different Quattro nodes. In the rectilinear walls the glass panels have been suspended from the space frame, directly connected by Quattro nodes.

11. So the highest glass panel carries the dead-weight of the 4 lower panels; the 2 upper bolts carry the dead-weight of 5 glass panels, 1.8 by 1.8m and 8mm thick. The space frame is pre-cambered and is rigid enough not to have a noticeable sag. The space frame had a total length of 21.6m and a width of 9m constructed within a tolerance of 3mm. This overall inaccuracy and local panel tolerances are taken by a special connection. The nodes are prestressed by 2 friction plates placed between node, glass and outside saucer, by tightening the pre-tensioning bolts. This prestress method results in a three fold increase in the allowable forces compared with shear bolts of similar size. The Quattro nodes connect together 4 adjacent glass panels and have been welded on short compression studs, that have been stabilised vertically by 10mm thick solid rods in 2 counter spanning curved lines; one to stabilise against inside over-pressure, one against inside under-pressure. Because of the many different node forms, a suitable production-method was obtained by welding from steel components. Although in the glass industry the general rule is that bolt holes are not to be allocated at 45° from the corners, no disadvantages were noticed in doing just that. The 8mm thick fully tempered glass panels have been produced with the aid of a water laser jet cutter with an extremely high accuracy, enough for the total set-up to be within 0.0 to 1.0mm orthogonally/diagonally panel size and bolt hole position and 0.0 to 0.2mm for the bolt hole diameter.

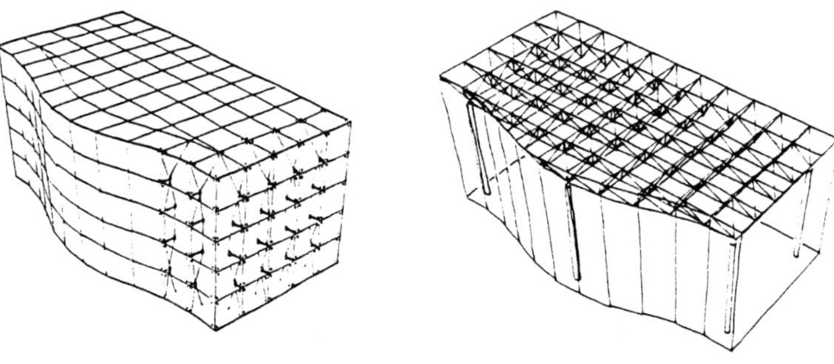

Fig. 2.

Technical data: Wall panels are 1.8 by 1.8m respectively. 1.8/2.1 x 1.8m; 8mm thick fully tempered glass. Roof panels are 1.8 by 1.8m light grey tinted laminated glass 10mm thick. Volume 21.6m long, 9m/13m maximum width, 9m high. (Design by Pieter Zaanen and the author in 1988). Realisation 1989; opening 1990. Engineering + production + building undertaken in 6 months by Octatube.

The roof panels have been used as laminated 5.5.2 annealed glass 1.8 by 1.8m in size. The relative thin glass panels act as flexible bending walls in the acoustical sense. The acoustics of the Glass Music Hall have been called remarkable by its musicians. Mainly due to the many triangular absorbing metal panels, suspended on the inside walls, and the reflective surfaces achieved by suspending acrylic dome lights in the space frame (to disperse sound) as well as white GRP sails, giving remarkably good overall acoustics. The reverberation time is about 1.2 seconds, which makes the hall suitable for live CD-recording.

12. The wall panels of the 3 rectangular walls are tinted grey and the curved wall is clear. Entering the hall over the ramp, with spotlights on the yellow Berlage walls, the hall shows almost black. Entering through the high entrance doors gives a surprise when one suddenly is confronted with the yellow bricks again, visible behind the curved clear wall; a spatial effect of Zaanen, a theatre specialist.

Glass Wall for the sky scraper Garden Road 8888. Hong Kong

13. Adjacent to the China Bank tower in Hong Kong architect Rocco has designed another bank tower. In proportion with the height of these skyscraper buildings the lobby height is more than 30m. The stabilised glass membranes around this atrium are 30 and 32m high, and 10 and 12m wide.

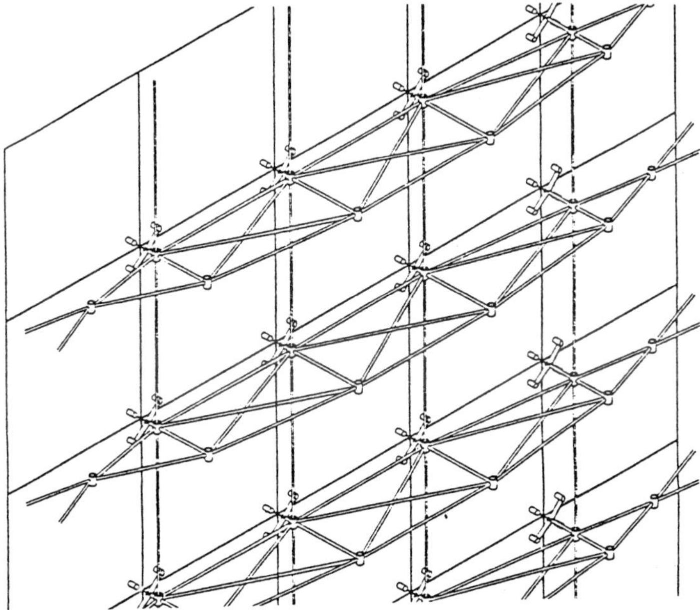

Fig. 3.

Technical details: Glass panels fully tempered 1.950 by 2.0m extra white transparent glass 20mm thick. Total wall area 30 by 10m and 32 by 12m. Design 1990; production and realisation 1991. Design by the author. Engineering and production metalwork by Octatube; glass by Central Glass Japan; erection by Reliance Control, Hong Kong.

Both glass facades have been designed in the Quattro system with the glass panels used 1.950m by 2.0m and 20mm thick in extra white glass manufactured by Central Glass of Japan. The local prevailing wind standards asked for a full size test, which was performed at the Dutch TNO laboratories in Delft, in 1991. A part of the 10m wide facade was tested, 3 panels high and 5 panels wide, under static and dynamic over- and under-pressure, as well as a rainwater test, with a maximum over-pressure of 5250 Pascal (equal to 525 kg/m^2), which is 7 times greater than the value specified in the Dutch codes. The tests were successfully passed. The Quattro maxi nodes and tensile trusses were dimensioned accordingly; the main tensile elements are made of stainless steel with two 27mm diameter rods; the compression studs are 40mm in diameter; the Quattro props are 300 by 300mm apart, to reduce the bending moments and deflections in the glass panels compared with the usual 200mm distance. The connection method is shear bolting, the angle deformation at the corners of the glass panels is taken by neoprene rubber, deforming when loaded.

Flower Gate, Hulst, NL

14. The Flower Gate is a pavilion building in the archetypal form of a gate. The choice of an elementary design concept led to the pillars of the gate (the facades) being closed, unable to receive daylight. The chosen elementary building scheme is based on universal principles of form, and is in itself 'low tech', confronted with a 'high-tech' roof. The origin of architecture is erecting vertical walls and horizontal roofs. The archetypal gate building is giving access symmetrically to a nursery for trees, plants and flowers; at the end of the plot, behind a 3m high 40m wide brickwork wall with a central entrance, is the house of the De Wolf family, floriculturists. The glass roof on the gate shop illustrates the necessity of daylight for flowers.

15. Three alternatives have been investigated, designed and compared. A space frame supporting glass panels, an underspanned metal grid structure supporting glass panels and an underspanned Quattro roof. The outcome in all cases would be a horizontal glass roof. The architect and the client opted for the most intelligent, and most complicated solution; the last proposal. The gate building is composed of two brickwork cubes 6.870 x 6.870 x 6.870m, with an entrance bay of 2.230m wide. The two cubes are entirely enclosed in the outer walls. The glass roof is at a height of 5m, the remaining wall forming a parapet around it. The roof is completely composed of flat double-glass panels (comprising of reflective tempered upper panels of 8mm, a 12mm air cavity and 3.3.1 laminated clear glass for the lower panels). The total size of the roof is 6.090 by 15.150m; the individual panels measure 1.425 x 1.425m. Each panel is supported at the four corners by support brackets designed in the same fashion as the Quattro-nodes of the Glass Music Hall in Amsterdam. All suspended rods are stabilised against downward and upward forces and against asymmetrical loading. The Quattro nodes are glued to the laminated lower glass panels, with a detail allowing the support to be turned slightly under external loading in a vertical direction by means of deflecting rubber buffer cushions. At the same place the upper glass

panel is also supported by a solid prop in the 12mm air cavity, sealed between upper and lower panels. The steel support elements have been glued with a specially tested type of glue to the glass inner pane to give sufficient adherence for horizontal shear and vertical uplift. Each of the glass brackets has been extended with a vertical 20mm thick pole guyed with 12mm steel rods in the node. All tensile rods can be adjusted in length and tension. The whole structure is surrounded by a continuous rectangular hollow section frame section supporting a continuous gutter. Hence this system can still be regarded as halfway between a closed roof system and an open structure with a perimeter ring beam that takes the compression forces omitted in the upper glass panels and converts them into internal bending moments. The slope of the glass panels is 1%. The structure was completed in May 1990.

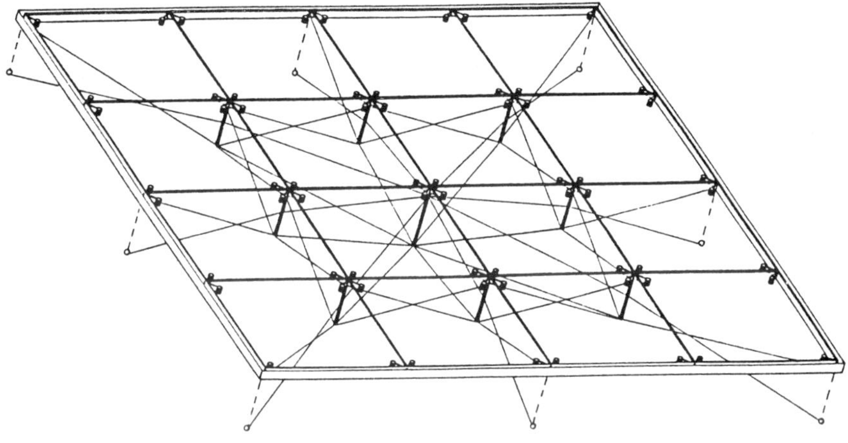

Fig. 4.

Technical data: Roof size 15.15 by 6.09m; panels in double glass 1.425 by 1.425m. Upper pane 8mm light reflective, lower pane 3.3.1 clear glass. Design by project-architect Walter Lockefeer and the author in 1989. Realisation in 1990 by Octatube.

Evoluon Canopy. Eindhoven, NL

16. Inspired by a small sketch of a 4 x 4m glass umbrella in a book by the author "Product Development in Glass Structures", project-architect Gert Grosveld of Philips Architecture office designed with the author a 10 x 10m canopy on a single central steel obelisk-like pole 10m high. In plan, the design resulted in a composition of glass panels 2 by 2m in plan, 10mm thick in single fully tempered glass, stabilised by prop-like adapters, derived from the Quattro

system, stabilised by compression studs and tensile rods, leading via rectangular hollow section cantilevering arms to the central obelisk stem.

17. The stabilisation of the 20 cantilevering outside panels by mean of tensile rods, introduces compression (and buckling) forces in the planes of the glass panels, that are led through to the stem of the obelisk. The panels are analysed to carry both dead-weight up/downward wind-loads and snow-loads, as usual. The peculiar 4 corner tips arose from the spatial arrangement of the Quattro-nodes and the glass panels. The 4 corner panels give a stealth-like secrecy to the 10mm thin wings. The glass plane runs in 4 slightly sloped planes with diagonal subdivisions; the rainwater is drained off by 4 rainwater inlets through the lower part of the obelisk.

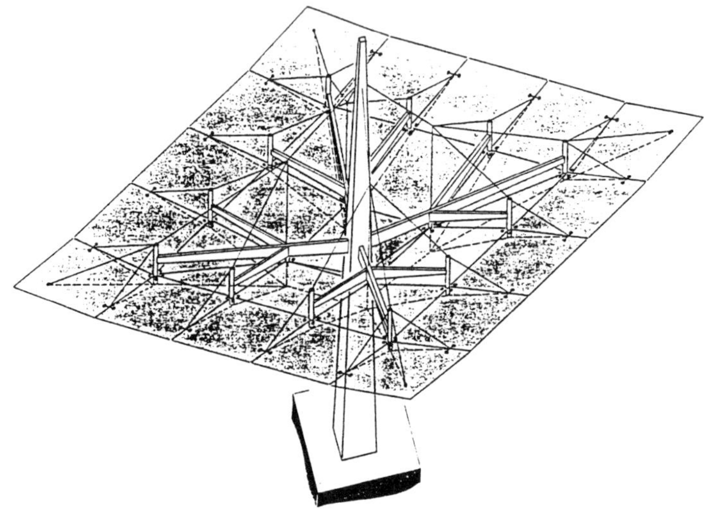

Fig. 5.

Technical data: 21. Surface roof 10 by 10m; height of the obelisk in galvanised and coated steel 10m; glass panels 2 by 2m, 10mm thick. Stabilised by 12mm thick tensile bars. Design by Gert Grosveld and the author in 1991; realisation by Octatube 1991.

221. The architectural potential of polyhedra

J. F. GABRIEL, Syracuse University, USA

SYNOPSIS. Polyhedral buildings have indeed been built, some with brilliant architectural results. While such buildings confirm the potency of polyhedra as space structures and as organising matrices for buildings, the need for a systematic, exhaustive study of polyhedral capabilities in architecture is still unanswered. This paper proposes both an argument for, and a method of, investigation. Three well-known polyhedra are examined. They can be clustered to form infinite structures. The seminal influence that polyhedra could have on architecture depends essentially on a better understanding of spatial changes occurring in infinite structures.

INTRODUCTION

1. Architecture, like the other arts, thrives on renewal. The art of architecture consists of finding new ways to make us enjoy space, lest the spaces to which we are accustomed become stale. New spatial definitions may come from new materials, or from old materials used in new ways. They may also come from the inventiveness of an architect of genius. Geometry is, and has always been, another source of renewal in architecture. Polyhedra, in particular hold an enduring fascination for many of us. Why, then, are there not more buildings erected in the shape of polyhedra?

AN ARGUMENT

2. The beginning of an answer to this question came to us from a long and patient study of polyhedra from the point of view of an architect. It is not that polyhedra are difficult to build. It is not that polyhedra, being so different from conventional building forms, look threatening. It is not that the spaces they contain are impractical. It is simply that the few buildings that have been erected are the products of hasty design, based on a superficial understanding of the complex nature of polyhedra. For this reason, they are sometimes disappointing. The designers of these buildings must nevertheless he commended for their courage and perceptiveness in taking a step in the right direction.

3. What are the problems inherent in the use of polyhedra in architecture, and what might be the solutions?

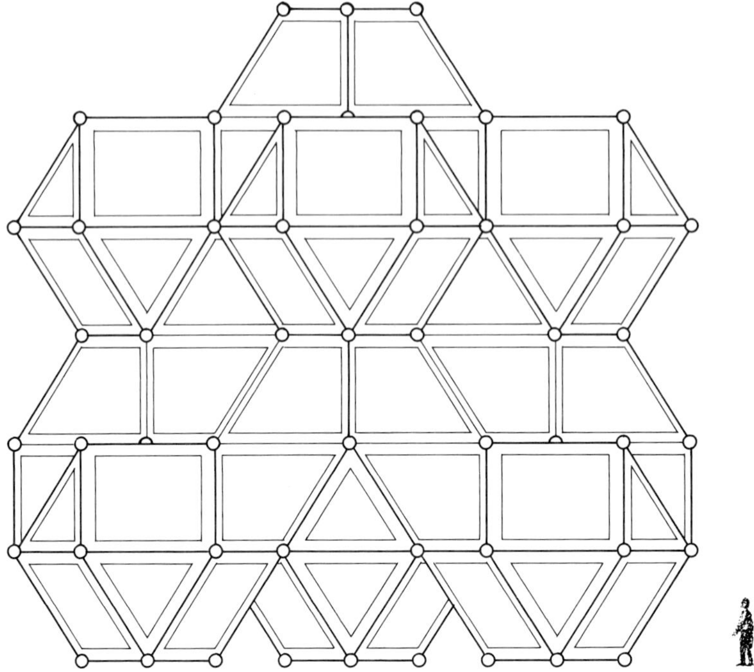

Fig. 1. Simplified elevation of a building whose form duplicates cuboctahedra truncated octahedra and truncated tetrahedra.

DESIGN GUIDELINES

4. A small polyhedron used in isolation can only provide a solution to a simple architectural problem A cluster of polyhedra (also called an infinite structure) is more likely to satisfy varied and complex architectural program It is also in infinite structure that the design problem can he seen more clearly and solution found

5. The following applies principally to the clustering of three polyhedra: the truncated octahedron (T.O.) the cuboctahedron (C.O.) and the truncated tetrahedron (T.T.) (Fig. 1)

6. First the most attractive attribute of polyhedra is probably their geometric perfection. How is this perfection to be retained in the process of transforming polyhedra into buildings? A modification of the faces is unavoidable but every effort should be made to maintain the integrality of the edges when introducing fenestration, decks, balconies and loggias.

7. Second the identity of each polyhedron should match the identity of a specific function as much as possible (Fig. 6)

8. Third the horizontal subdivision of polyhedra into several storeys presents a structural problem caused by varying spans. Intermediate supports might be necessary (ref. 1)

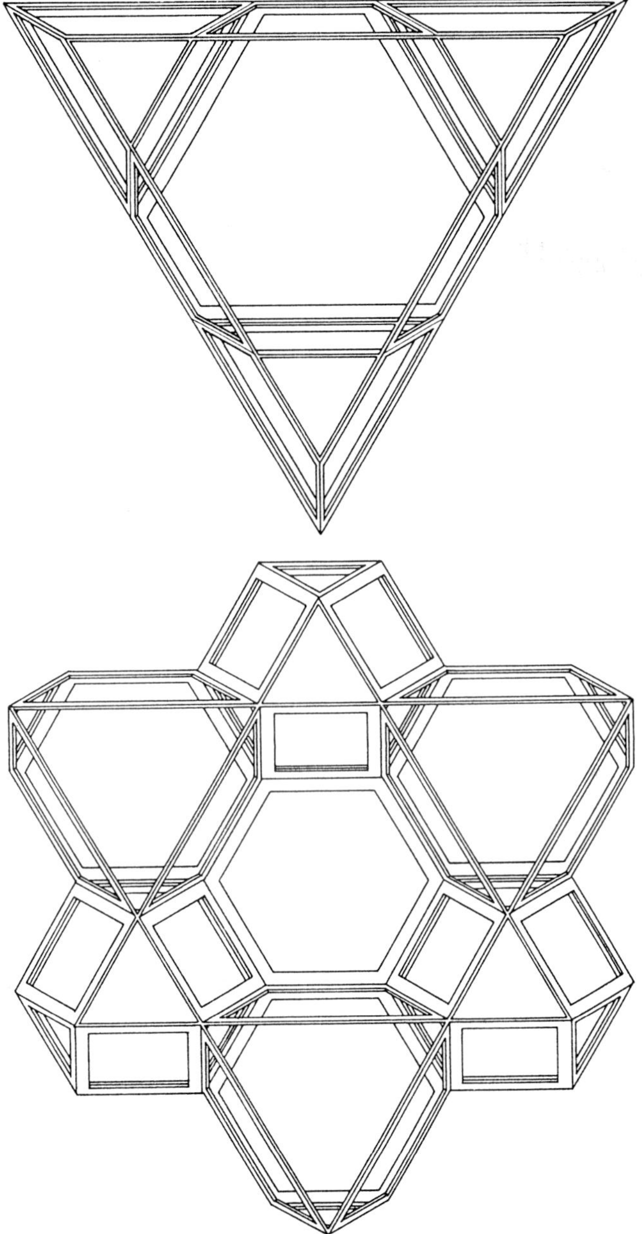

Figs 2 and 3. Spatial relationships in the building form shown in Fig. 1. From the bottom up and are the second and third storeys.

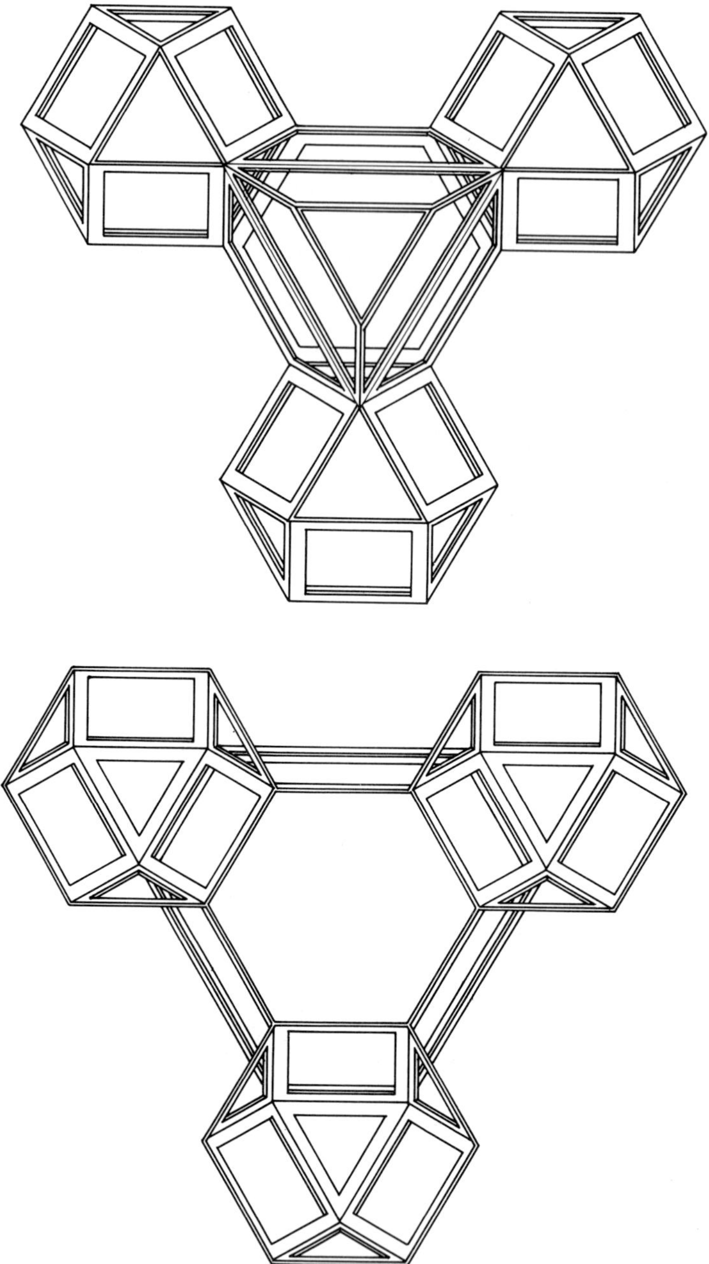

Figs 2 and 3. Spatial relationships in the building form shown in Fig. 1. From the bottom up and are the fourth and fifth storeys.

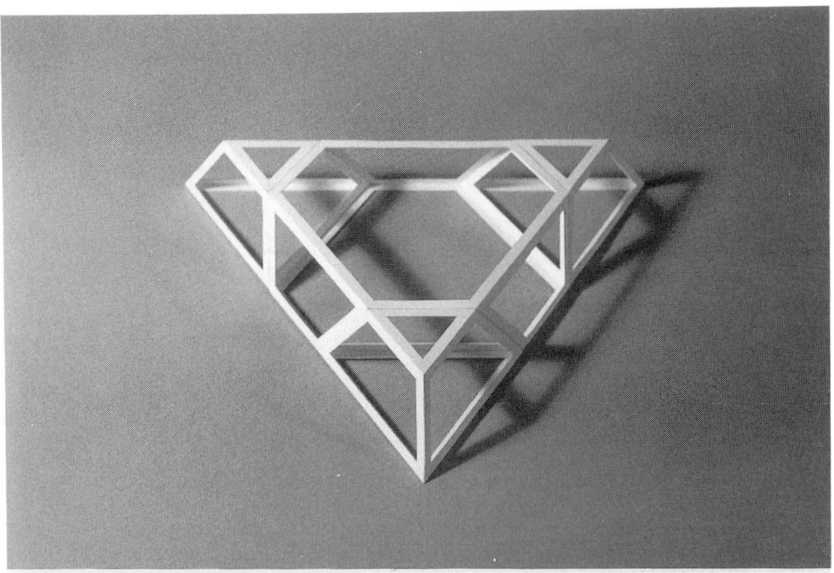

Fig. 4. Represented here in 3-D models, the second and third storeys of the building form shown in Fig. 1. The second storey (above) consists of the lower third of a T.O. surrounded by the upper halves of three C.O.s and the lower section of three T.T.s. The third storey (top) consists of the median section of a T.O. surrounded by the upper portion of three T.T.s. The first or ground storey (not shown), would consist of the lower half of the three C.O.s.

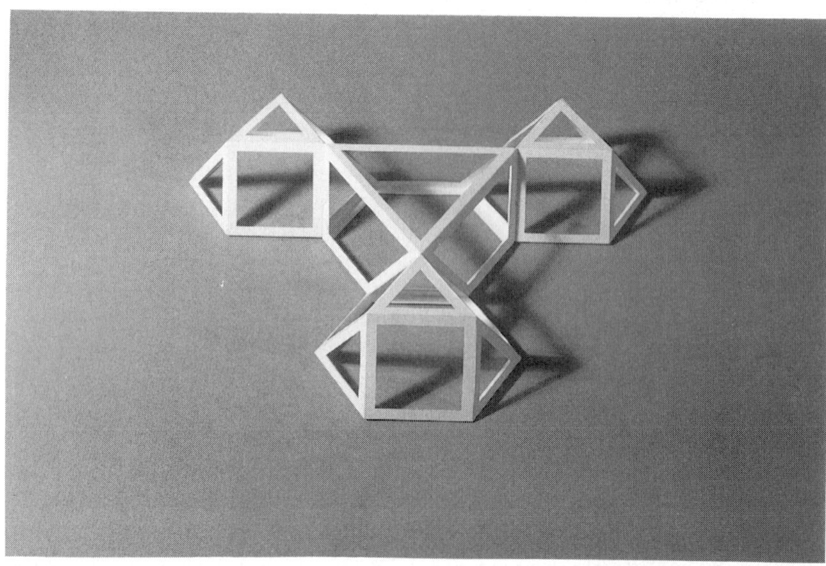

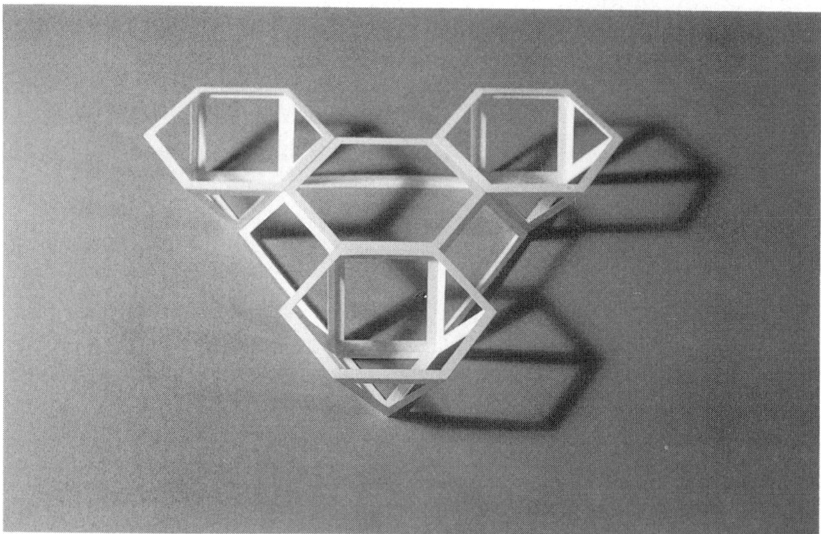

Fig. 5. The fourth storey (above) consists of the upper third of a T.O. surrounded by the lower halves of three cuboctahedra. The fifth storey (top) consists of the lower section of a T.T. surrounded by the upper halves of three cuboctahedra. The sixth storey, not shown, would consist of the upper section of a T.T.

Fig. 6. Student project using T.O.s, T.T.s and C.O.s in the design of an elementary school. Here, T.T.s and C.O.s are one storey high, not two, and T.O.s are a storey and a half.

9. Fourth, the vertical subdivision of polyhedra into smaller rooms presents an opportunity for the introduction of vertical planes. Vertical floor supports can be combined with these.

10. Fifth, infinite structures do not intrude in natural settings as brutally as conventional cubic structures do. Polyhedral buildings have the potential to become metaphorical landscapes (ref. 2).

11. Sixth, in clusters, faces shared by contiguous polyhedra can be eliminated, changing radically the nature of interior space and increasing its flexibility to the advantage of the designer (Figs 2-5).

12. Last, the design of windows requires special care, and this for several reasons. The "walls" of polyhedra are slanting, and bays are either square or triangular. The forms of a cluster being rich and varied, windows should be kept as regular and simple as possible to avoid visual chaos.

13. There are six different contexts for windows. A square bay is leaning out (earthward) or leaning back (skyward). A triangular bay is also leaning out or

leaning back, but the horizontal edge of the triangle is either at the bottom or at the top of the bay.

14. We have already suggested that the geometric purity of polyhedra should he retained and that the best way to do so is to preserve the edges, if not the planes that surround polyhedra. A window should be contained within a specific bay and adopt the shape of that bay. These two rules do not exclude the introduction of vertical glass surfaces in the form of bay windows or oriels, which give the dweller a 180° angle of vision on the outside (ref. 3).

CONCLUSION

15. Anyone attempting to design a polyhedral building will face the seven problems for which we are proposing answers. There are other problems, to which we do not know the answers, or to which answers have already been proposed (refs 4, 5). Materials constitute one such problem. The central issue remains the handling of space, which is always the central issue of architecture. A common perception is that polyhedra are inflexible and that a cluster is made of a series of discrete spaces determined by, and limited to, polygonal faces. Nothing could be further from the truth. Architectural polyhedra can open up onto one another, and the layers of which they are formed create wonderful opportunities for spatial relationships involving several storeys.

16. The more we study polyhedra from the vantage point of an architect, the more attractive the prospect of a polyhedral architecture looks to us.

REFERENCES

l. GABRIEL, J.F. Infinite structures for architecture, Proceedings of the First International Seminar on Structural Morphology, 1992, pp133-143.

2. GABRIEL, J.F. Polyhedra in architecture, Proceedings of the International Conference on the Design and Construction of Non-Conventional Structures, 1987, Vol. 1, pp139-147.

3. GABRIEL, J.F. Megapolyhedra, Proceedings of the International ISS Symposium, 1991, Vol. 2, p354.

4. NOOSHIN, H. (Ed), Studies in space structures, p69, Multi-Science Publishing Co Ltd, Brentwood, Essex, 1991.

5.GABRIEL, J.F. Space frames: an alternative to the architectural cube, International Journal of Space Struetures, 1991, Vol 6, No 4, pp287-295.

ACKNOWLEDGEMENTS

The architectural design illustrated in Fig. 6 is by Hans Graf. The drawings and photographs are by the author. Our gratitude to Bruce Abbey, Dean of the School of Architecture at Syracuse University for his unfailing support .

222. Architectural and structural investigations of a reinforced concrete space sheath

B. S. BENJAMIN, University of Kansas, USA

SYNOPSIS. The author's method uses a thin, reinforced concrete space sheath for vertical extension of old, load-bearing masonry structures. The sheath fits the masonry walls like a surgeon's glove, thereby preserving the architecture of the building. This paper investigates the structural behaviour of the sheath using an IMAGES-3D computer program for varying sheath thickness, base conditions and variations in the size of bands round openings. The application of the research to the rehabilitation of cracked, load-bearing masonry wall structures is investigated. Use of the space sheath for the rehabilitation of masonry wall structures damaged by seismic loads, is also discussed.

INTRODUCTION AND BACKGROUND

1. Overcrowding in cities such as Bombay, India, has led to changes in local municipal regulations permitting old, load bearing, masonry wall structures designed for only two upper storeys to carry an additional two storeys. The common method being used in Bombay is to put reinforced concrete columns, like stilts, around the original structure. This has led to an architecture of stilts, as shown in Figs 1and 2 with complete desecration of the architecture of the building (ref. 1). In some cases, the interaction of the new stilt structure with the original structure leads to the damage of the old structure, as shown in Fig. 3.

THE SPACE SHEATH

2. The author's method provides a thin, reinforced concrete space sheath that fits the old structure like a surgeon's glove, as shown in Fig. 4 (refs 1-3). The space sheath passes round door and window openings and is strengthened by the provision of bands. It passes through continuous cantilever overhangs by the use of vierendeel girders. It is connected to the old masonry structure by wall ties, as shown in Fig. 4. The load from the additional storeys is carried by the thin sheath which is prevented from buckling by the thickness of the original masonry walls. The sheath is provided with independent foundations, which acting together with the foundations of the old structure, transmit the load to the ground.

STRUCTURAL ANALYSIS OF THE SHEATH

3. To understand the structural behaviour of the space sheath, the prototype sheath of Fig. 4, was analysed, without overhangs, using the IMAGES-3D program. The dimensions of the sheath are shown in Fig. 5, with the walls divided, in general, into plates of 1220mm (48in). The load was applied as a line load along the upper edge where the additional two storeys would transmit the load into the plane of the sheath. This load was calculated as 105 N/mm (600 lbs/in). A total of seventeen cases were analysed in all, and these are shown in Table 1.

Figs 1 and 2. Reinforced concrete columns used as stilts, in Bombay.

Fig. 3. Cracking of the old overhang by the new cross member.

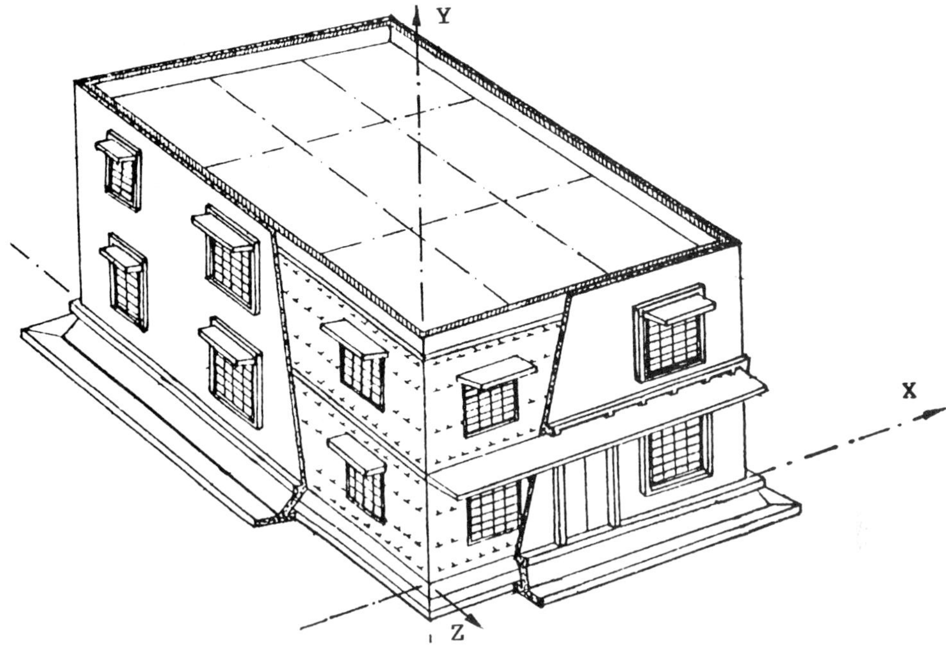

Fig. 4. A reinforced concrete sheath tied to the old masonry.

Table 1. Cases considered for computer analysis.

Case	Thickness	Openings	Band Size(mm)	Base	Notes
A3	76mm	No	---	Pinned	All plates in
B3	76mm	No	---	Fixed	All plates in
C3	76mm	Yes	---	Fixed	No bands, unstable
D3	76mm	Yes	152 x 305	Fixed	Lintels & verticals
E3	76mm	Yes	152 x 305	Pinned	Lintels & verticals
F3	76mm	Yes	152 x 305	Pinned	All bands in
G3	76mm	Yes	152 x 305	Fixed	All bands in
H3	76mm	Yes	152 x 152	Fixed	All bands in
I3	76mm	Yes	76 x 76	Fixed	As C3, but stable
J3	76mm	Yes	76 x 76	Pinned	As C3, but stable
J2	51mm	Yes	76 x 76	Pinned	All bands in
K2	51mm	Yes	51 x 51	Pinned	As C3, but stable
L2	51mm	Yes	51 x 51	Fixed	As C3, but stable
M2	51mm	Yes	152 x 152	Fixed	All bands in
N2	51mm	Yes	152 x 152	Pinned	All bands in
O2	51mm	Yes	152 x 305	Pinned	All bands in
P2	51mm	Yes	152 x 305	Fixed	All bands in

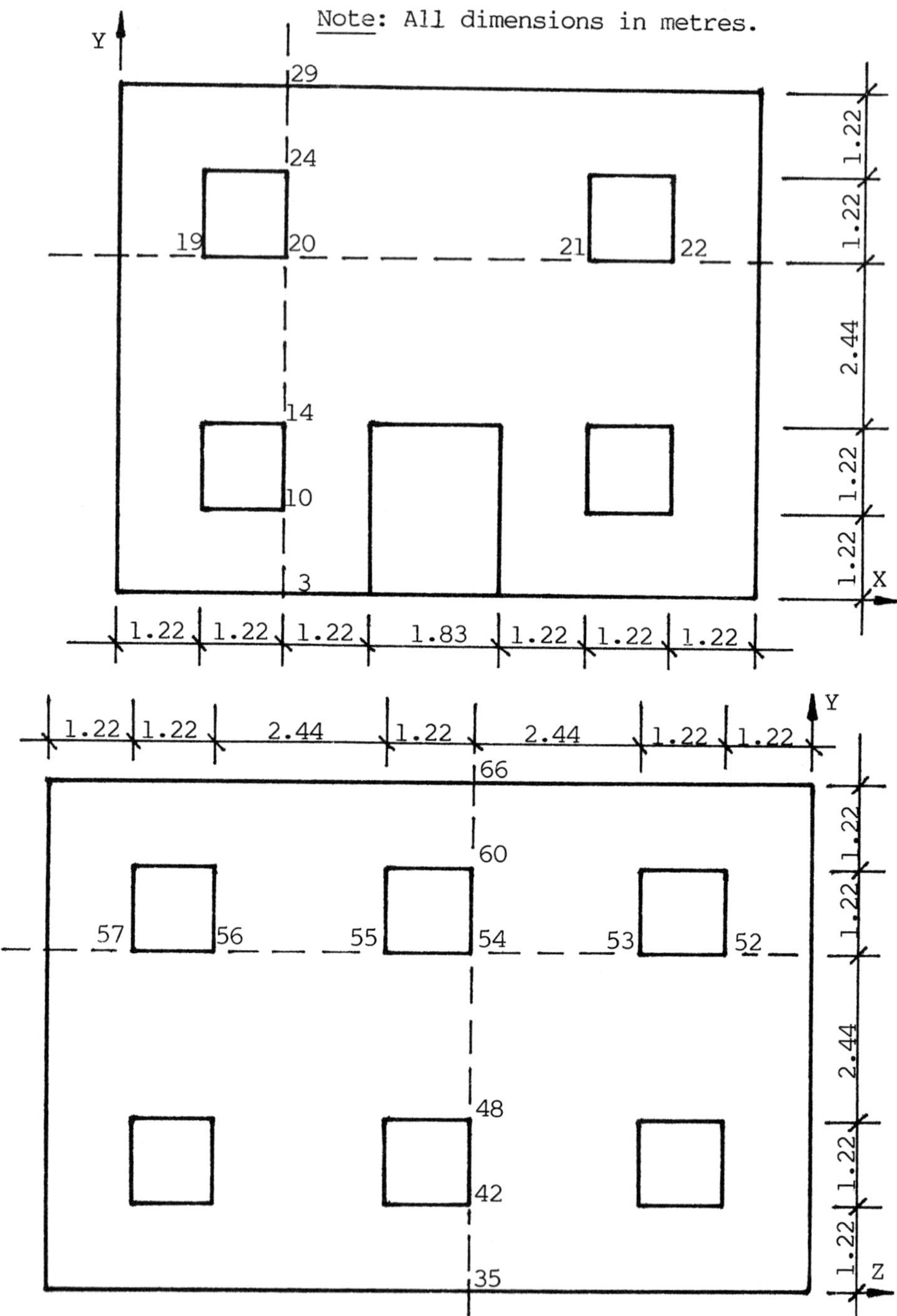

Fig. 5. Front and side elevations of the concrete space sheath.

ANALYSIS OF THE RESULTS

4. The results of the analysis are shown in Figs 6-9. The lines and nodal points along which the results are plotted are shown in Fig. 5. It has not been possible to show all of the results.

5. An analysis of the deflections, Figs 6 and 7, shows that the deflections of the unpunctured sheath are small (A3, B3). Deleting the plates for door and window openings increases deflections considerably (J3, J2, L2). The insertion of sizeable bands, 152mm x 305mm (6in. x 12in.), around openings, however, returns the sheath to a virtually unpunctured condition (D3, E3, F3, G3, O2, P2). The insertion of bands, 152mm x 152mm (6in. x 6in.), reduces the deflections but they are still unacceptable. In Fig. 7, the maximum deflections do not occur at the centre because of plate wrinkling. Values for Pt 52 are shown in Table 2.

6. The variation of the vertical axial force, Fy, along the faces of the sheath are shown in Fig. 8. Here the insertion of bands increases the axial force because of the weight of the bands. The force tapers toward the corners.

7. The variation of the bending moment, Mx, along a vertical line in the short face of Fig. 5, is shown in Fig. 9. Due to the presence of the door opening, the short face moments are somewhat distorted. The end conditions at both top and bottom are modified by the presence of edge beams with torsional resistance. The moments for the 76mm (3 in.) sheath without openings or with sizeable edge bands around openings (A3, B3, D3, E3, F3, G3) are small. However, for the 51mm (2 in.) sheath, in some cases (M2, N2), they are unacceptable.

Table 2. Absolute maximum stress in plates; deflection at Pt. 52.

Case	N/sq mm	Kips/sq in.	mm	in.
A3	2.854	0.414	1.18	0.044
B3	2.868	0.416	1.04	0.041
C3	---	---	--	--
D3	3.013	0.437	0.86	0.034
E3	2.985	0.433	0.94	0.037
F3	2.758	0.400	0.97	0.038
G3	2.765	0.401	0.89	0.035
H3	3.406	0.494	4.90	0.193
I3	4.116	0.597	10.44	0.411
J3	4.130	0.599	14.76	0.581
J2	5.826	0.845	29.57	1.164
K2	5.992	0.869	37.59	1.480
L2	6.198	0.899	29.21	1.150
M2	4.482	0.650	6.48	0.255
N2	4.323	0.627	7.29	0.287
O2	4.357	0.632	0.99	0.039
P2	4.351	0.631	0.91	0.036

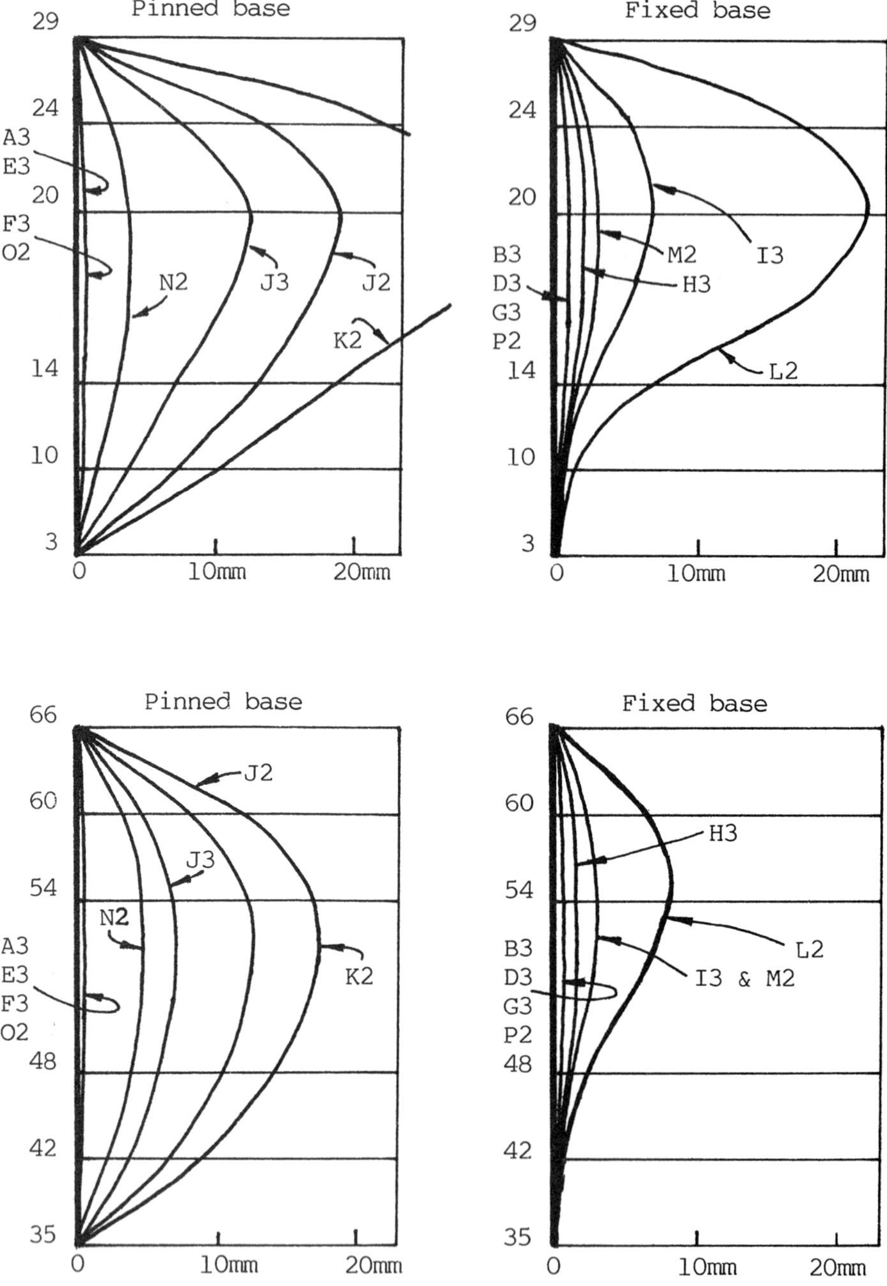

Fig. 6. Horizontal deflections of the sheath on lines 3-29, 35-66.

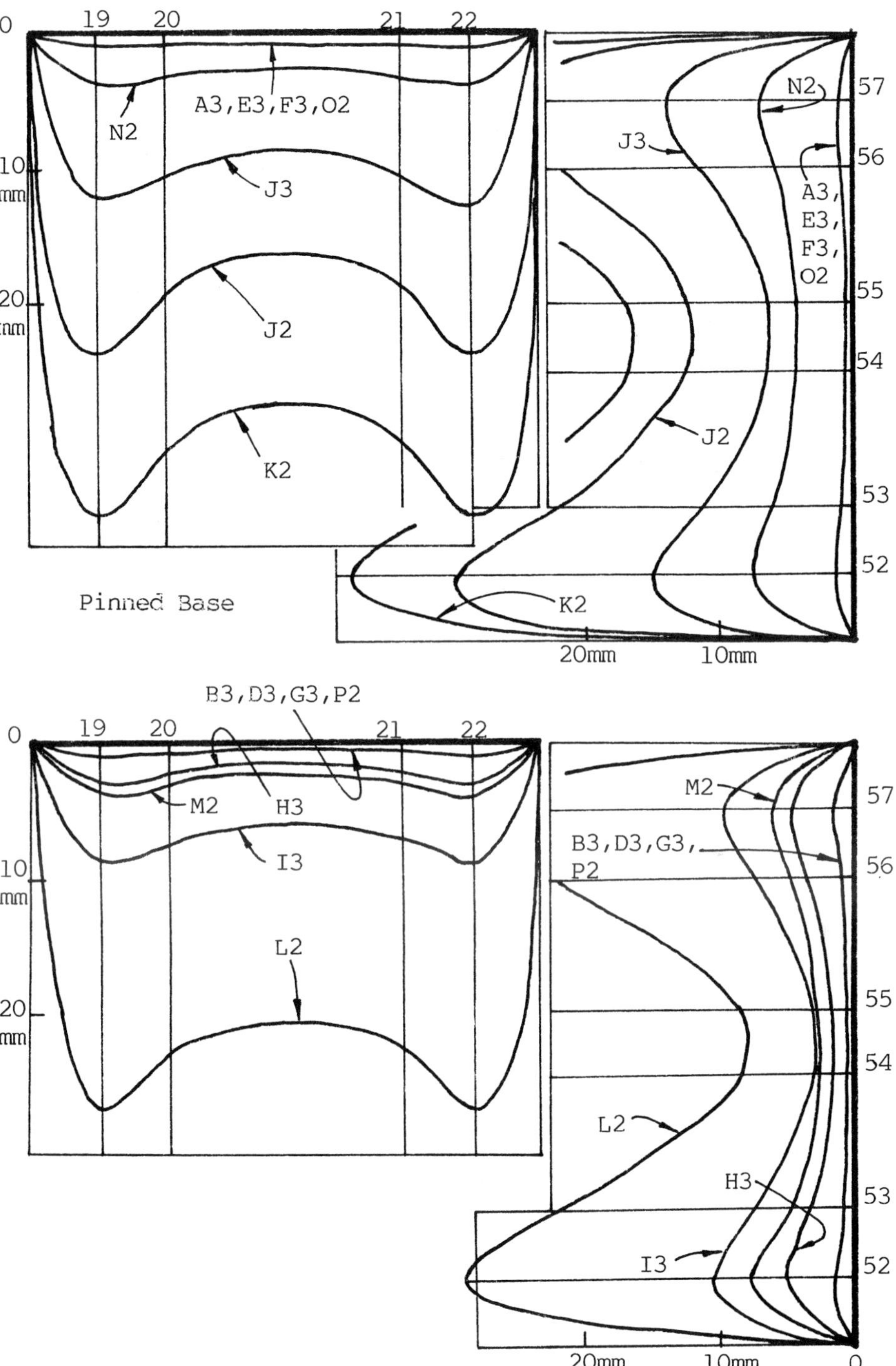

Fig. 7. Horizontal deflections on lines 19-22 and 52-57.

8. From a design point of view, from Table 2, cases H3, I3, J3, J2, K2, L2, M2 and N2 are not acceptable.

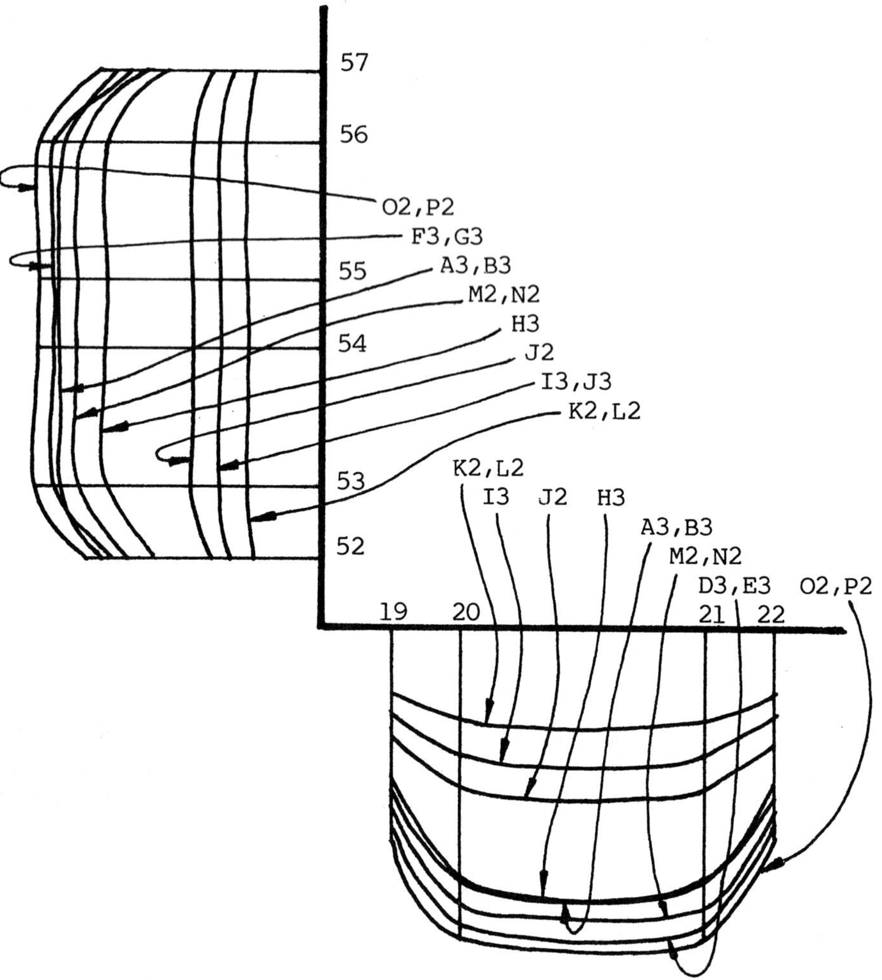

Fig. 8. Variation in Fy along lines 19-22 and 52-57.

APPLICATION OF THE RESEARCH TO THE REHABILITATION OF MASONRY BUILDINGS DAMAGED BY SEISMIC LOADS

9. It becomes apparent that the method has other applications besides those discussed in 1. Masonry buildings severely damaged by earthquakes are written off and demolished. Often they remain unoccupied for years in city centres, like Oakland, California, on prime real estate. The method can be easily applied to the rehabilitation of such structures. It is essential that the thin, space sheath be tied to the old, cracked wall structure and restrained against lateral instability, as shown in Fig. 4.

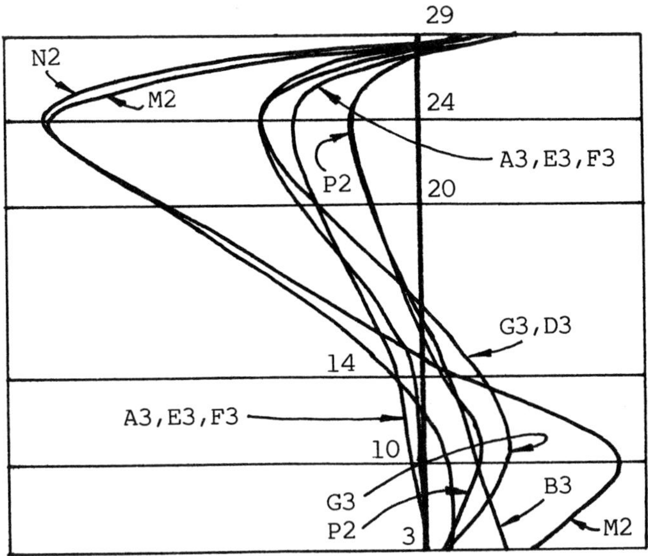

Fig. 9. Variation in bending moment, Mx, along line 3-29.

This has several advantages. The sheath thickness is less. The new construction is tied to the old masonry, thereby making the whole building an integral structure. There is usually considerable reserves of strength still contained within the old construction - after all the building is still standing! - and this can be useful, particularly at points of high stress concentration. It also facilitates transfer of the existing floor loads into the sheath. But how much reserve strength is available in the cracked, masonry wall structure? The author hopes to examine the problem experimentally using models with vertical, horizontal and diagonal cracking. The sheath is tied to the masonry and the model tested to destruction. It is hoped the research will lead to easier rehabilitation of such damaged masonry buildings in city centres - preserving, at the same time, the architecture of the building.

REFERENCES

1. BENJAMIN B. S. A reinforced concrete sheath for old buildings in poor countries. Structural faults and repair conference-87, London. Proc. published by Civil-CompPress, Edinburgh, 1987.

2. BENJAMIN B. S. Futuristic structure: A philosophical enquiry. International Journal of Space Structures, Vol. 4, No. 1, 1989.

3. BENJAMIN B. S. Structural evolution: An illustrated history. A. B. Literary House. P.O.Box 528, Lawrence, Kansas 66044, 1990.

223. Geometry of the spatial folded forms

O. L. TONON, Centro de Estructuras Livianas, Argentina

SYNOPSIS. As an advanced study of the theme, this paper intends to describe the geometric aspects of a specific variable type of spatial shapes. Although this is a theoretical work, focused on the generation of shapes, the final scope of the analysis is to induce their use in real applications into the field of lightweight space structural solutions to the present day problems of covering spaces; this, however, still needs a deeper and wider treatment.

INTRODUCTION

1. Spatial variable shapes, as a product of the folding of laminar planar elements, comprise a very wide spectrum. They include conics, paraboloids, hyperboloids, bubbles, eccentric domes, variable helicoidals and even apparently arbitrary or whimsical configurations, but obeying precise geometrical laws . It is possible, however, through certain particular cases, to understand the origin of a number of different variable spatial shapes - all of them generated on the basis of the folding of planar facets and straightedges - and even to produce different modifications on them just by changing the values of some of their parameters properly. Obviously, the subject will be treated here synthetically.

2. Unlike "constant development" folded shapes (ref. 1), where the pattern that generates them repeats always the same basic geometric figure (BF) without alteration, *the variable ones are generated when the basic figure (BF) repeats itself modifying successively its dimensions or its angles.*

3. To circumscribe the subject from all possible options, only two cases will be considered:

i) that the plane pattern should admit its repetition in two senses (generatrix and directrix) and;

ii) that it could be folded totally as a bellows until it produces a compact packing.

Some patterns only respond partially to these two principles, that is to say, they do not admit a correct repetition in both senses or they cannot be folded completely.

4. In Fig. 1, for example, BF dimensions vary successively, but the pattern only admits repetition in the generatrix sense. In Fig. 2, instead, angles and dimensions vary and the pattern can be repeated in two senses (generatrix and directrix) but it is not possible to fold it completely. Both cases produce variable foldings but they are irrelevant to this study.

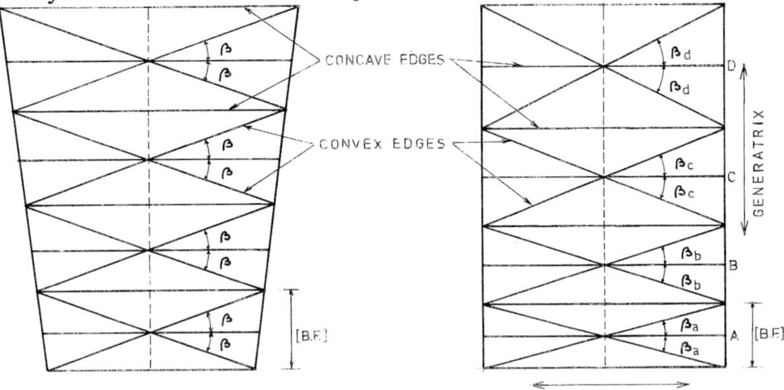

Fig. 1. Plane pattern - variable dimensions.

Fig. 2. Plane pattern-variable dimensions and angles.

5. The basic concept to be introduced, that contemplates both stated options is the following: *an angle Δ which is incorporated to BF marks the angular difference existing between the two facets that converge on one concave edge;* see Fig. 3. When $\Delta = 0°$, there is no angular difference and the same BF repeats itself, generating constant development foldings. If $\Delta \neq 0°$, variations are produced on BF, either on its dimensions or on its angles, which generates variable folded shapes which are interesting to analyse.

6. To guarantee that the successive repetition of BF can be folded completely, the following equation should be fulfilled: $2\beta + \Delta = \text{constant}$, and for the pattern to admit its repetition in both senses (generatrix and directrix), the limits imposed to BF should not be arbitrary, but they should respond to accurate calculations .

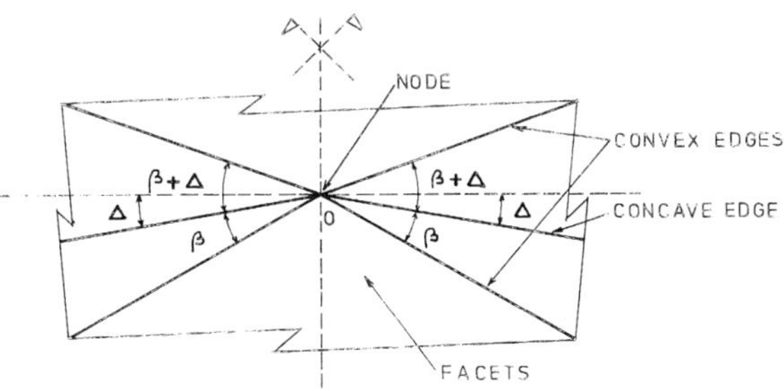

Fig. 3. Basic arrangement of angles for variable shapes.

SINGLE CURVATURE SHAPES

7. To exemplify the above mentioned concept, starting from the simplest case, the BF can be considered to modify only its dimensions and not its angles. If $\beta = 20°$ and $\Delta = 10°$, the plane pattern adopts the form illustrated by Fig. 4, which repeats itself indefinitely in the generatrix sense (5 BF, in this case) and upto a determined number of times in the directrix sense (3 BF, in this case). This pattern can be folded completely, varying the folding angle from $\alpha = 0°$ for the plane non-folded position and up to $\alpha = 90°$ when it is completely folded.

8. If, from all the intermediate positions the silhouette lying flat on a horizontal plane is chosen, each parallel section will respond to a semicircular type polygon and each sector or side of the polygon will be given by the separation between nodes. The spatial shape in its development is conical (Fig .4).

9. The analytic method of calculation is as follows: for the silhouette, 3 BF configurate 6 sectors which compose the whole 180°; whereby each side of the polygon consists of an angle $\phi = 30°$.

$$\tan \beta' = \frac{\cos(\beta+\Delta)}{\cos\beta \cdot \sin\varphi} - \frac{1}{\tan\varphi} = \frac{\cos 30°}{\cos 20° \cdot \sin 30°} - \frac{1}{\sin 30°} = 0{,}111 \qquad \beta' = 6°\,20'$$

$$\tan(\beta+\Delta)' = \frac{\cos\beta}{\cos(\beta+\Delta) \cdot \sin\varphi} - \frac{1}{\tan\varphi} = 0{,}438 \qquad (\beta+\Delta)' = 23°\,40'$$

For the horizontal projection:

$$\sin\alpha_1 = \frac{\cos\beta}{\sin(\beta+\Delta) \cdot \sin\varphi} - \frac{1}{\tan(\beta+\Delta) \cdot \tan\varphi} = 0{,}758 \qquad \alpha_1 = 49°\,20'$$

$$\sin\alpha_2 = \frac{\cos(\beta+\Delta)}{\sin\beta \cdot \sin\varphi} - \frac{1}{\tan\beta \cdot \tan\varphi} = 0{,}305 \qquad \alpha_2 = 17°\,50'$$

DOUBLE CURVATURE SHAPES

10. The next step consists in varying gradually the pattern angles, but only in the generatrix sense. For example, $2\beta + \Delta = 50°$ can be established as a constant, then a different β corresponds to each variation. So, if $\Delta_a = 0°$, $\beta_a = 25°$; if $\Delta_b = 5°$, $\beta_b = 22°\,30'$; if $\Delta_c = 10°$, $\beta_c = 20°$, if $\Delta_d = 15°$, $\beta_d = 17°\,30'$; being the analytic calculation for the silhouette and horizontal projection the same as the former case. According to the repetition of the BF in one way or another, different spatial shapes will be obtained. See Figs 5 and 6.

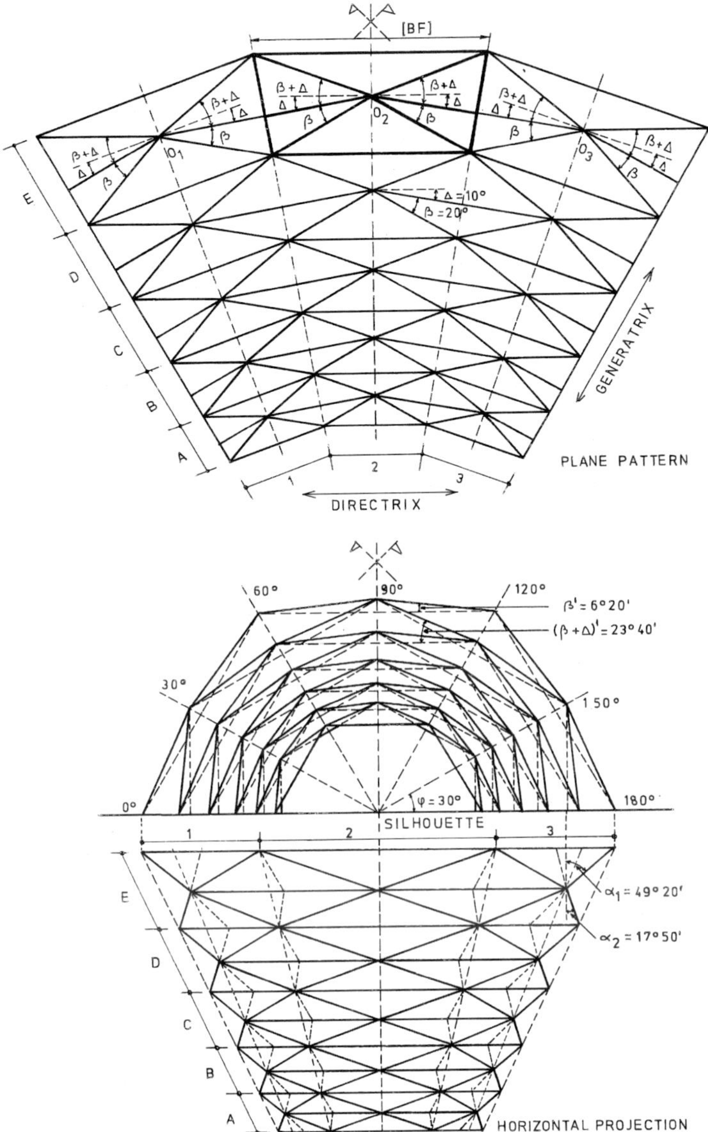

Fig. 4. Single curvature variable shapes (conic. type).

11. The characteristic angle β defines the curvature in the directrix sense and the angular difference defines the curvature in the generatrix sense. By modifying the values of one or another properly, double curvature can be reached and molded at will.

12. A greater difficulty appears when the angle variations in the plane pattern include both directions simultaneously (directrix and generatrix). Even when the resulting shapes are more free and the spatial possibilities practically unlimited, defining previous conditions or laws to be taken into account for each type to be tried, becomes more complex . The three parameters which will determine the shape are:

i) what the edge distribution should be like;
ii) which angles will be suitable and;
iii) how much folding there will be.

At the same time, the premises of repetition and folding established at the beginning should be fulfilled, as well as the equation $2\ \beta + \Delta =$ constant. It is necessary to take into account that with a previous condition given (and the stated premises) only some angles can be fixed *a priori,* the other being unknown and a product of the result of mathematical calculations. Because of this, a skilful mastering of the previous angle selection will be necessary together with their variations to get the final required shape quickly.

13. When a folded double curvature spatial shape is intended to be solved, with a low rise elliptical type silhouette lying flat on a horizontal plane, a previous condition is then being established. In this case, it should also be considered that in the directrix sense that define the silhouette, the angular variation will go from minimum to maximum, from the centre to the extremes. The angle variations produced in the generatrix sense define the double curvature.

14. For each section of the folded silhouette in the directrix sense to lie perfectly flat on a horizontal plane, the concave edges on both extremes of the support should be perpendicular to the BF limits. The possibility exists of considering a parallel section where $\Delta = 0°$ as a starting point, which permits to fix and to know *a priori* the characteristic referential β angles, which duplicated will give the value of the constant in the equation: $2\beta + \Delta =$ constant. All the other angles are variable and unknown, and their calculation exceeds the limitations imposed by this synthesis .

15. For example, in a six sector silhouette whose characteristic fixed angles are: $\beta_{1a} = 16°$, $\beta_{2a} = 21°$, $\beta_{3a} = 27°$ and proposing a variation in the generatrix sense $\Delta_{1a} = 0°$, $\Delta_{1b} = 1°$, $\Delta_{1c} = 2°$, $\Delta_{1d} = 3°$, $\Delta_{1e} = 4°$, the graphic solution will show approximately the true value of each of the angles conforming the plane pattern, starting from the silhouette totally folded (for $\alpha = 90°$). See Fig. 7.

16. Such a silhouette is conducted gradually from the central sector towards the extremes for $\Delta_{1a} = 0°$. Prolonging the extreme limits $\overline{AB}$ and $\overline{EF}$, the vertex $\underline{0}$ is obtained, at which all the nodes of each sector converge so that joining each sector with $\underline{0}$, for each Δ variation given, the location of the different nodes, of other sector, can be obtained. Due to the equation $2\beta + \Delta =$ constant, all the concave edges of each sector are always parallel among themselves, and all the convex edges of each sector differ among themselves at an angle $= \Delta$.

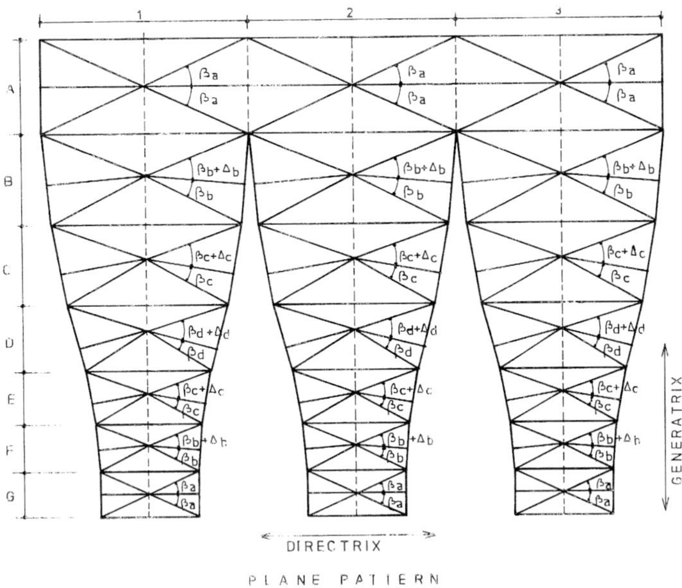

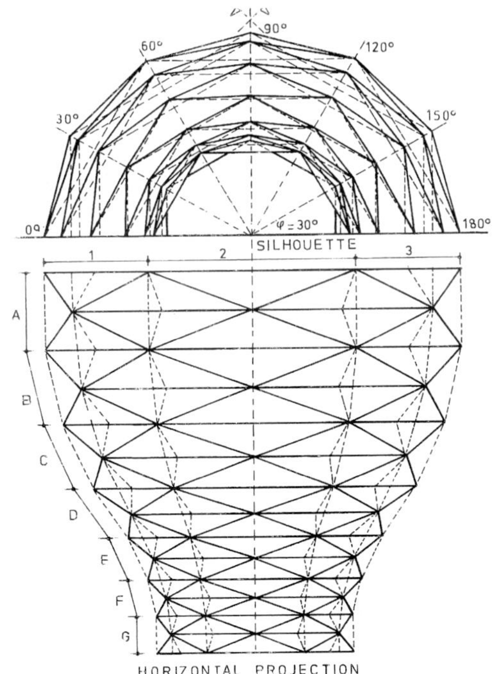

Fig. 5. Double curvature variable shape (bubble type).

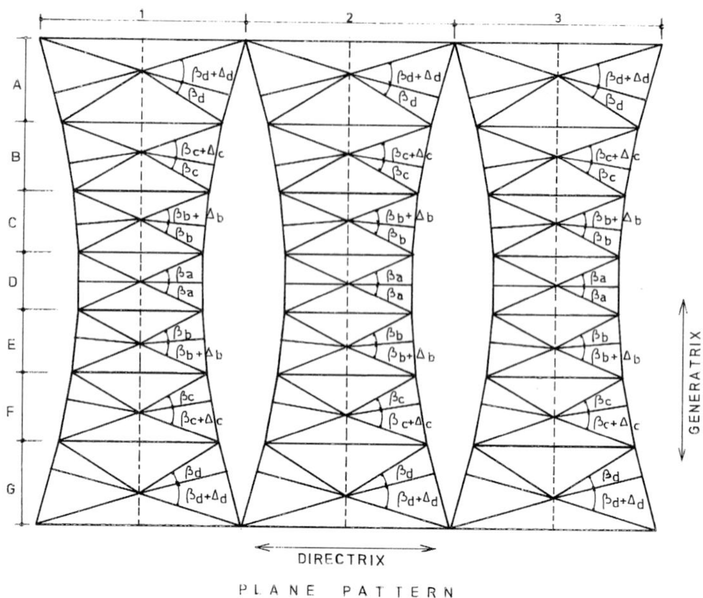

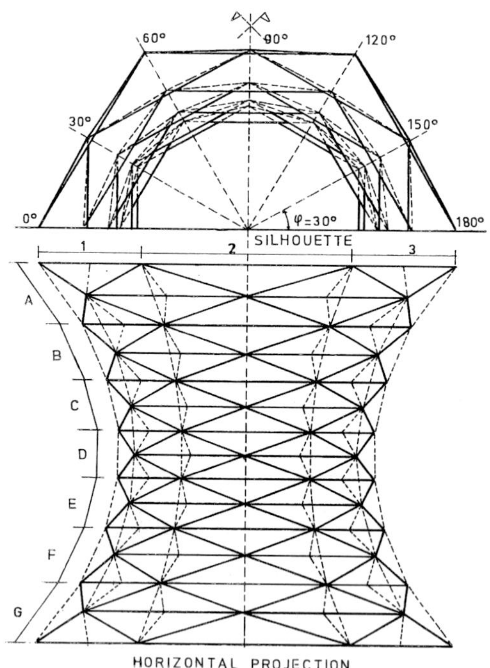

Fig. 6. Double curvature variable shape (hyperbolic type).

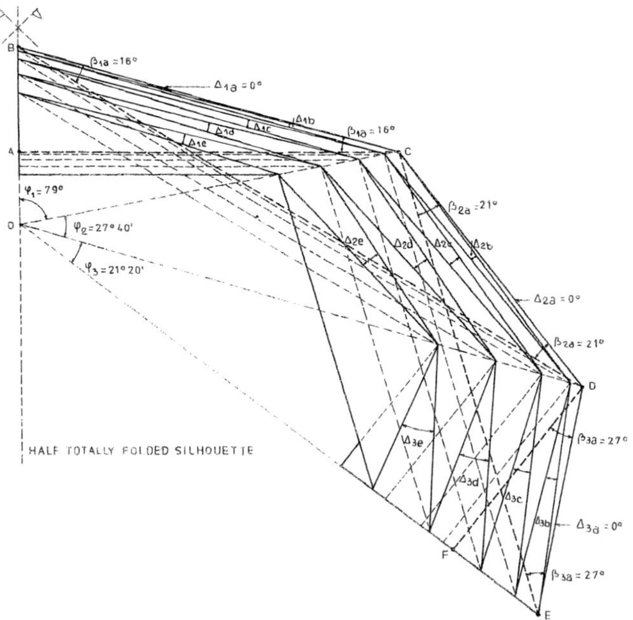

Fig. 7. Graphic method of calculating unknown angles.

If the different Δ_1 values are known, it is relatively easy to rebuild all the other unknown angles in the folded silhouette, resulting, by direct measurement, in:

Sector 1	$\Delta_{1a}=0°$ $\Delta_{1b}=1°$ $\Delta_{1c}=2°$ $\Delta_{1d}=3°$ $\Delta_{1e}=4°$	$\beta_{1a}=16°$ $\beta_{1b}=15°\ 30'$ $\beta_{1c}=15°$ $\beta_{1d}=14°\ 30'$ $\beta_{1e}=14°$	$2\beta_{1a}+\Delta_{1a}=32°$ $\beta_{1b}+\Delta_{1b}=16°\ 30'$ $\beta_{1c}+\Delta_{1c}=17°$ $\beta_{1d}+\Delta_{1d}=17°\ 30'$ $\beta_{1e}+\Delta_{1e}=18°$
Sector 2	$\Delta_{2a}=0°$ $\Delta_{2b}=3°$ $\Delta_{2c}=5°$ $\Delta_{2d}=8°$ $\Delta_{2e}=10°$	$\beta_{2a}=21°$ $\beta_{2b}=19°\ 30'$ $\beta_{2c}=18°\ 30'$ $\beta_{2d}=17°$ $\beta_{2e}=16°$	$2\beta_{2a}+\Delta_{2a}=42°$ $\beta_{2b}+\Delta_{2b}=22°\ 30'$ $\beta_{2c}+\Delta_{2c}=23°\ 30'$ $\beta_{2d}+\Delta_{2d}=25°$ $\beta_{2e}+\Delta_{2e}=26°$
Sector 3	$\Delta_{3a}=0°$ $\Delta_{3b}=6°\ 20'$ $\Delta_{3c}=10°\ 30'$ $\Delta_{3d}=16°\ 30'$ $\Delta_{3e}=19°$	$\beta_{3a}=27°$ $\beta_{3b}=23°\ 50'$ $\beta_{3c}=21°\ 45'$ $\beta_{3d}=18°\ 45'$ $\beta_{3e}=17°\ 30'$	$2\beta_{3a}+\Delta_{3a}=54°$ $\beta_{3b}+\Delta_{3b}=30°\ 10'$ $\beta_{3c}+\Delta_{3c}=32°\ 15'$ $\beta_{3d}+\Delta_{3d}=35°\ 15'$ $\beta_{3e}+\Delta_{3e}=36°\ 30'$

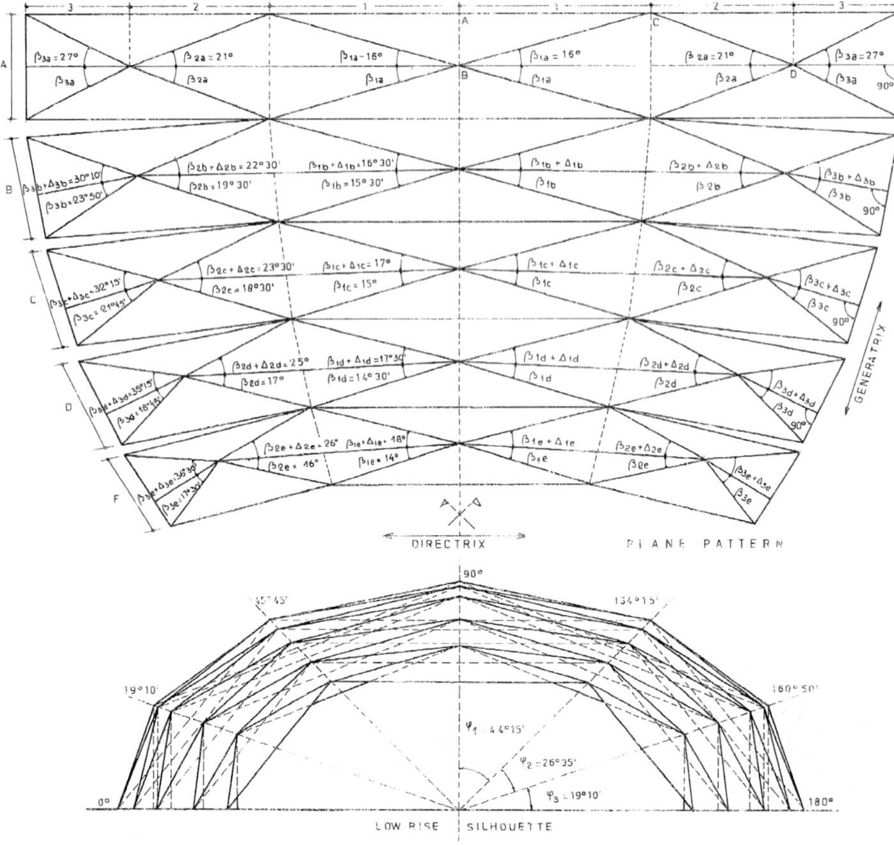

Fig. 8. Double curvature low rise variable shape.

17. This graphic method is an approximation, since its results depend on the accuracy achieved in drawing, but it is good enough to rehearse forms. For greater accuracy, analytical calculation methods should be used. See Fig. 8.

CONCLUSIONS

18. The development of all these variable shapes of single or double curvature, (product of the folding of laminar elements) through plane facets and straight edges, require a skilful mastering of the angles and pattern selection. Existing suitable maximum and minimum limits for their values, if exceeded, cause the distortions when folding and lack or excess of rigidity or even shapes impossible to materialise. With practice, the designer will learn the values of the maximum and minimum which are different for each case. The values of the characteristic β angle, the folding α angle and the Δ angular difference are to be established for any amplitude and variation ranges. The rule is the following: taking into account

that in foldings, the curve is reached through a polygon, the more the polygon approaches a curve (greater number of sectors) the less the variability margin of the angles is and a decreasing of the rigidity of the shape also occurs. On the other hand, the further the polygon is from the curve envelope (lesser number of sectors) the angular variation increases, but an unnecessary excess of rigidity can be reached or even the impossibility to fold the shape completely due to superposition There is an intermediate area of compromise where it is possible to work with in a certain range of variation, without distortions and to obtain spatial shapes of proper rigidity; such an area is the one which imposes the maximum and minimum limits, while not being of precise value, becomes easily evident with experimentation.

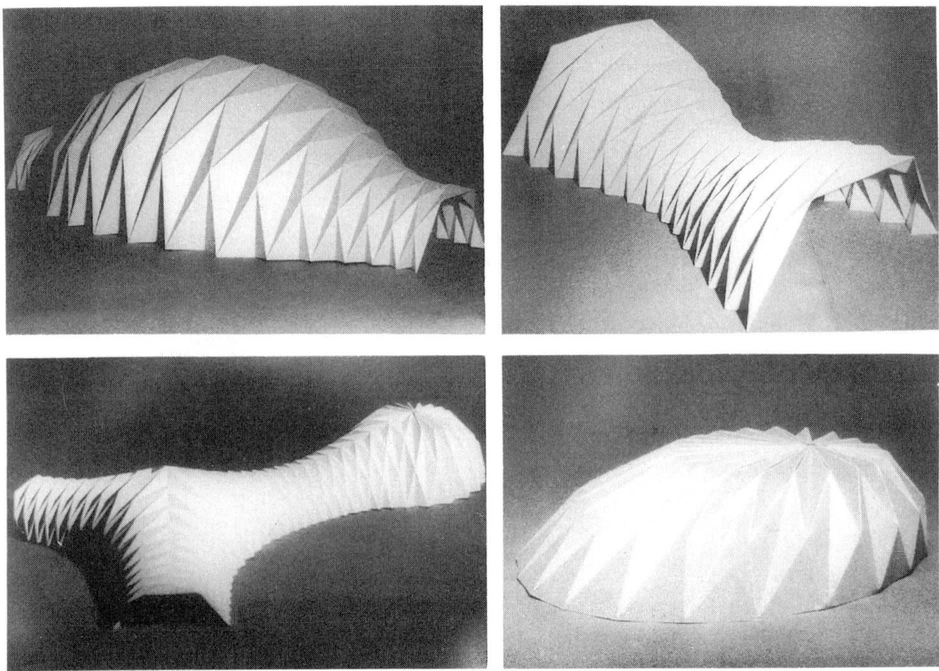

Fig. 9. Different types of variable folded shapes.

TERMINOLOGY:

Facets:	Geometric plane surfaces limited by edges.
Edges:	Lines along which the laminar element folds.
Nodes:	Points of convergence of the edges.
Pattern:	Repetitive arrangement of edges and facets.
Characteristic angle β:	Angle formed by the edges between them.
Folding angle α:	Angular variation of position suffered by the folded facets with respect to their plane original position.

Concave edges:	Those whose facets form valleys on folding.
Convex edges:	Those whose facets form slopes on folding.
Silhouette:	Vertical projection or section in the directrix sense of the folded form.

REFERENCE

1. TONON O. Geometry of spatial folded forms. Int. Journal of Space Structures, Vol. 6, No. 3, 1991.

224. Circular space trusses

R. MEURANT, Institute of Traditional Studies, New Zealand

SYNOPSIS. The archetypal geometry of the centre is given by the regular Star and its counterform the Polar Zonagon Mandala. By monoaxial development, a useful structural geometry for circular space trusses and domes is obtained of a web of equal length members in parallel sets that meet at modular angles of intersection. The projective geometrical nature of the mandala in its metrical expression develops zonahedral properties, which facilitate design and fabrication. Specific applications are indicated in Space, with dynamic structures which contract and expand about an axis. This paper condenses extensive research documented in my Aesthetics of the Sacred (ref. 1).

STRUCTURAL GEOMETRY PRECEDES STRUCTURAL FORM

1. In this paper, and in much of my work generally, I explore structural geometries, and from them deduce structural forms. Geometries which structure space may be intuited and developed as pure geometrical entities, prior to their application as structural forms. A structural geometry is ideal, and the question of it being structurally stable or unstable is not relevant. For example, a cubic lattice has specific harmonic geometrical properties which may be evinced; these apply irrespective of the structural application in which that geometry may be exploited. Only when the geometric structure is physically concretised in a load-resisting application do issues of structural equilibrium and adequate strength become relevant, and achieved through triangulation, stiff connectors, or diaphragms etc.

2. My research has shown that it is beneficial to explore structural geometries fully as theoretical entities (for example in the computer-generated graphic environment). Specific structural applications then become apparent. The virtue of this approach is that the structural geometries respect, and to a degree evince, the natural structural qualities of space prior to their being compromised by the contingent factors of their physical realisation. These geometries by virtue of their projective harmonic nature are very regular, effective and economic in enabling specific structural applications to he conceived and realised.

THE NATURAL GEOMETRY OF THE CENTRE

3. The centre is, I suggest, the most archetypal spatial form. Mircea Eliade shows the centre to be associated with the Sacred; a cosmos comes into being from its centre, and it is re-absorbed back through the centre at the end of its natural life. Spatially, we encounter two- and three-dimensional centres. The geometry of the three-dimensional centre is either approached on a polyaxial or monoaxial basis. The regular polyaxial centre is approached through the symmetries of the regular and semi-regular polyhedra (ref. 2), and their development through configurations such as the centralised zonahedral complexes I develop elsewhere (refs 3 and 4). Alternatively the monoaxial geometry is developed from the geometry of the two-dimensional centre, which is the main concern of this paper. It therefore makes sense to address the natural geometry of the two-dimensional centre (in the plane), and from that develop natural geometries of the monoaxial three-dimensional centre. From these various geometrical expressions, structural applications are deduced.

4. Space structures and trusses are often developed as two-dimensional planar or curved surface forms (as in the R. Buckminster Fuller designed geodesic dome of the World's Fair U.S. Pavilion). Although they have depth, they are essentially surface forms. Further, structures are generally of finite extent. Where these are of planar form, their most fundamental expression is circular (just as in microgravity in three-dimensions the archetypal spatial form of finite structures is spherical). Accordingly, where finite planar space structures are indicated, the natural structural geometry of the two-dimensional centre proves appropriate. Monoaxial developments into three-dimensions are particularly relevant to applications on earth where a gravitational field is encountered which is to all intents and purposes monoaxial, except in structures of very great extent. It is then sensible to align the axis of the structural geometry to that of gravity, as domical forms testify. But such developments are also applicable in Space operations in monodirectional uses such as antennae or telescopes, or where surface structures are required. In particular they are suited for dynamic space structures which expand and contract about an axis.

5. In the two-dimensional geometry of the centre in the plane, the archetypal geometry is given by the regular polygons, more specifically by their development into the infinite regular stars on the one hand, and by their finite counterforms, which I term the polar zonagon mandalas. It proves useful to restrict the frequencies of these, firstly to even frequencies (as odd frequencies are not as symmetrical); secondly for the purposes of illustration of this paper to a frequency of twelve. But their application is envisaged at whatever frequencies are appropriate to the specific use for which they are intended.

6. The regular Star and its counterform the Polar Zonagon Mandala are metrical expressions of archetypal projective geometrical configurations. Through their projective nature, their elements such as nodes, line segments, polygons and circles, firstly, exhibit important projective properties such as coincidence, collinearity, and what I term coconicity (i.e. lying on the same conic). Secondly, their relative proportions, angular and linear, are harmonic. When these projective

forms are expressed in their most regular metrical form, they exhibit metrical properties of equal-length, parallelism, and modularity of angle. Usually these properties are maintained when the geometries are developed into three-dimensions. Harmonic properties are elegantly shown therefore in structural geometries with members of equal length which occur in parallel bundles, and meet at modular angles of intersection, as characteristic of zonahedral geometries.

7. Usually such metrical harmonic geometries generate either equal measure of length, with harmonic angular measure, or equal measure of angle, with harmonic linear measure. The significance of the centralised geometry presented in this paper is that while it is fundamentally an expression of equal measure of angle, it also conveniently develops equal measure of length. Consequently it is very useful for centralised applications.

8. Accordingly, these structural geometries exhibit properties which enable them to be readily conceived, reconfigured in the mind's eye during design, proportioned, modelled in the computer-generated graphic environment, and fabricated as specific structures. Transformations are possible, through circumferential rotation, or radial expansion and contraction, about an axis. These suggest specific applications in Space operations as dynamic structures which can he deployed in various states of openness and closure; or which fold and unfold, and could be delivered in a contracted state before being deployed.

THE REGULAR POLAR ZONAGON MANDALA

9. This form can he visualised and generated in a variety of ways (Figs 1(i)-1(iv):

- From an initial burst of vectors from the centre, adjacent vectors mate to produce parallelograms, and the sequence expands outwards, meeting a natural limit to the expansion which is a two-frequency regular polygon. It then reflects from the boundary, contracts, and eventually implodes and is re-absorbed through the centre.
- The extended regular star is decomposed into concentric star rings as line segments zigzag out and in; these are resized and their order reversed to become mandala rings.
- If a polar inversion of the extended regular star is made about its in-circle, each line inverts into a circle. The polar zonagon mandala is then configured as an eccentric overlapping circular set of circles of equal radius, equally spaced about, and centred on, a central reference circle, each coincident at the centre, their intersections giving the nodes of the mandala. This regular polar inversion is why I regard the mandala as the counterform of the star. Straight line-segments then connect adjacent nodes.
- Thus the mandala can he reconfigured as an eccentric overlapping set of regular polygons centred on the nodes of a central regular polygon, each sharing a central node.
- The mandala can he seen as a checkerboard "hit-and-miss" pattern connecting nodes of intersection of a central ray of equis-paced lines and coincident circles.

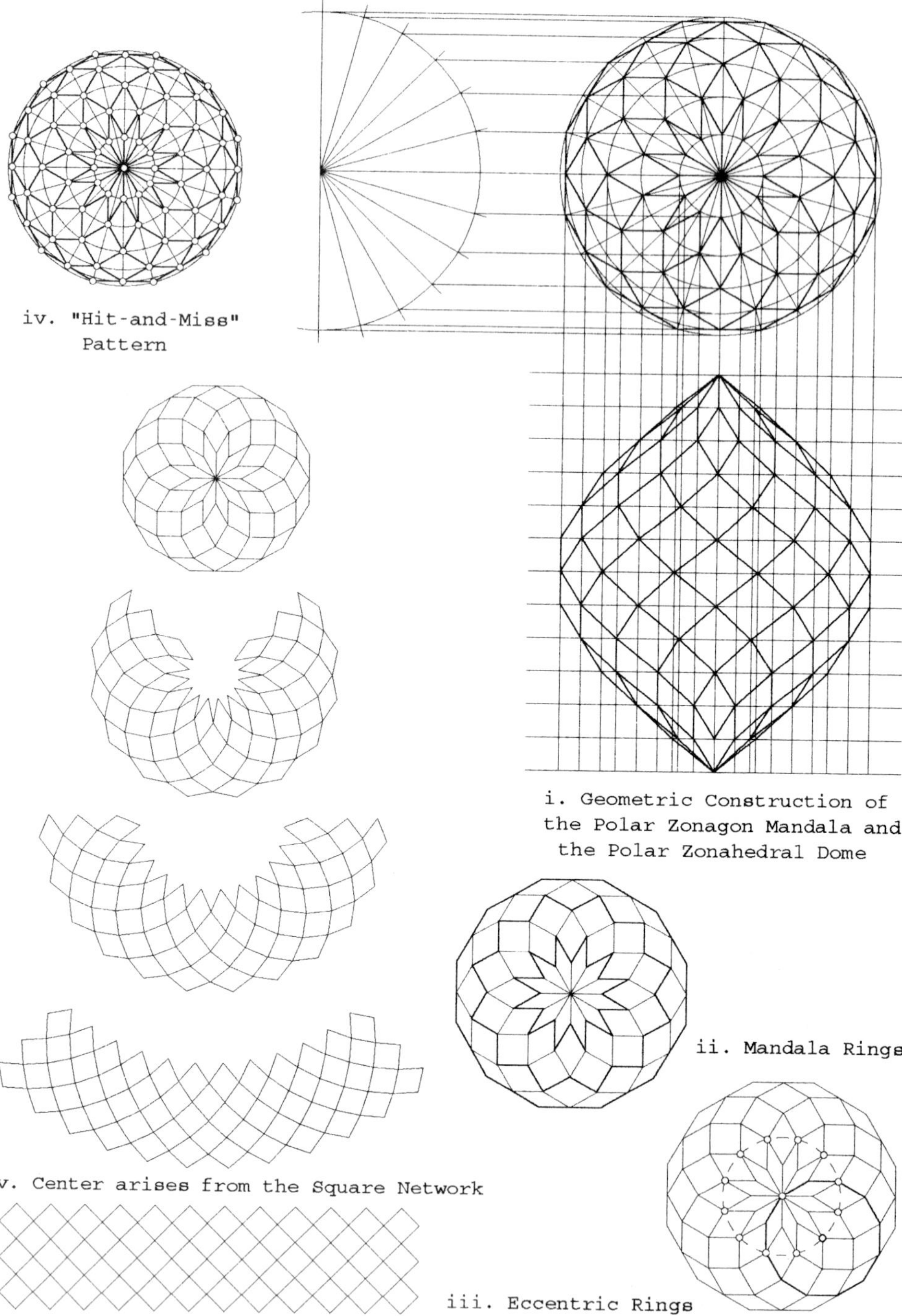

Fig. 1. Construction and properties of the Polar Zonagon Mandala.

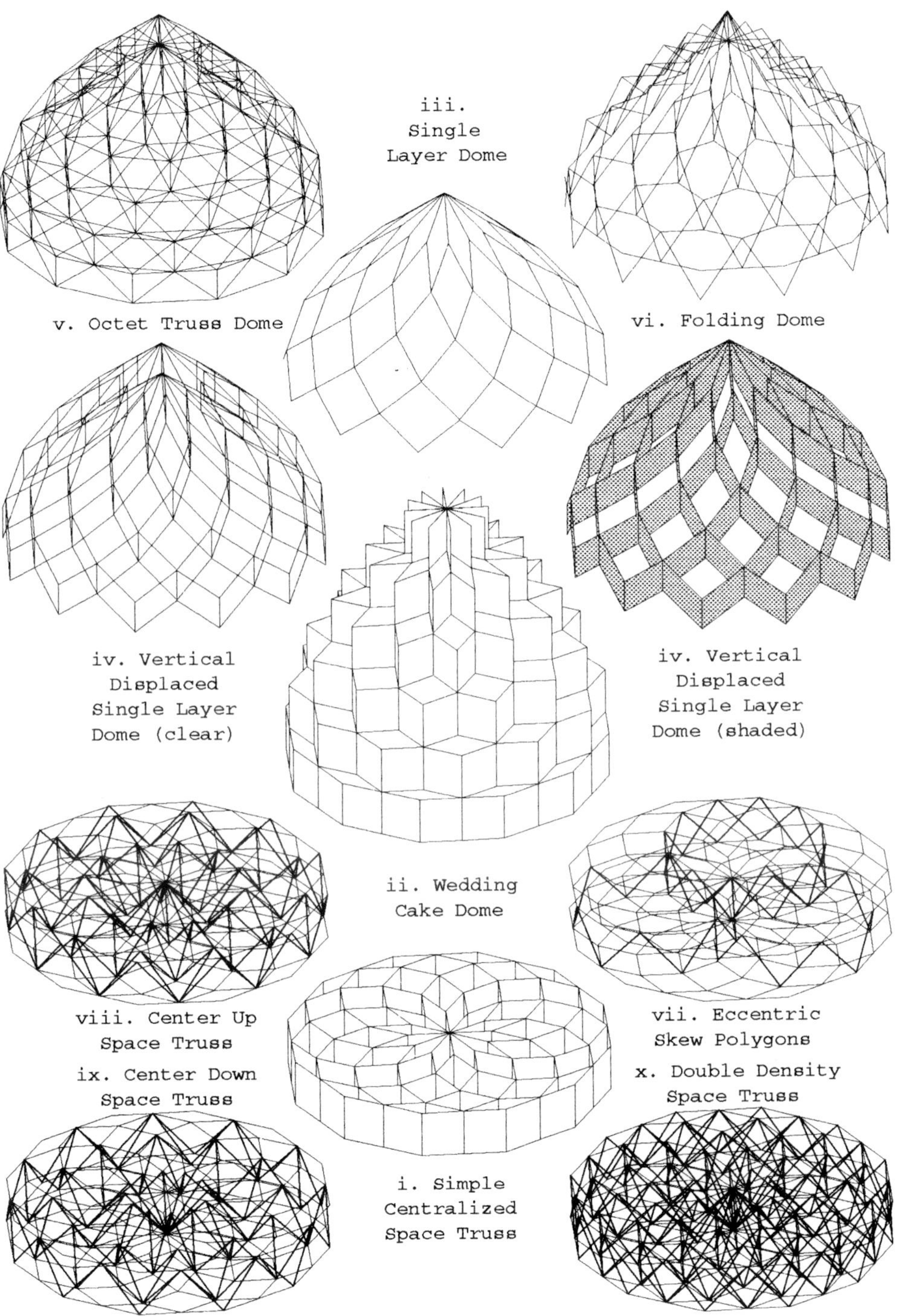

Fig. 2. Monoaxial Development of the Mandala into Trusses and Domes.

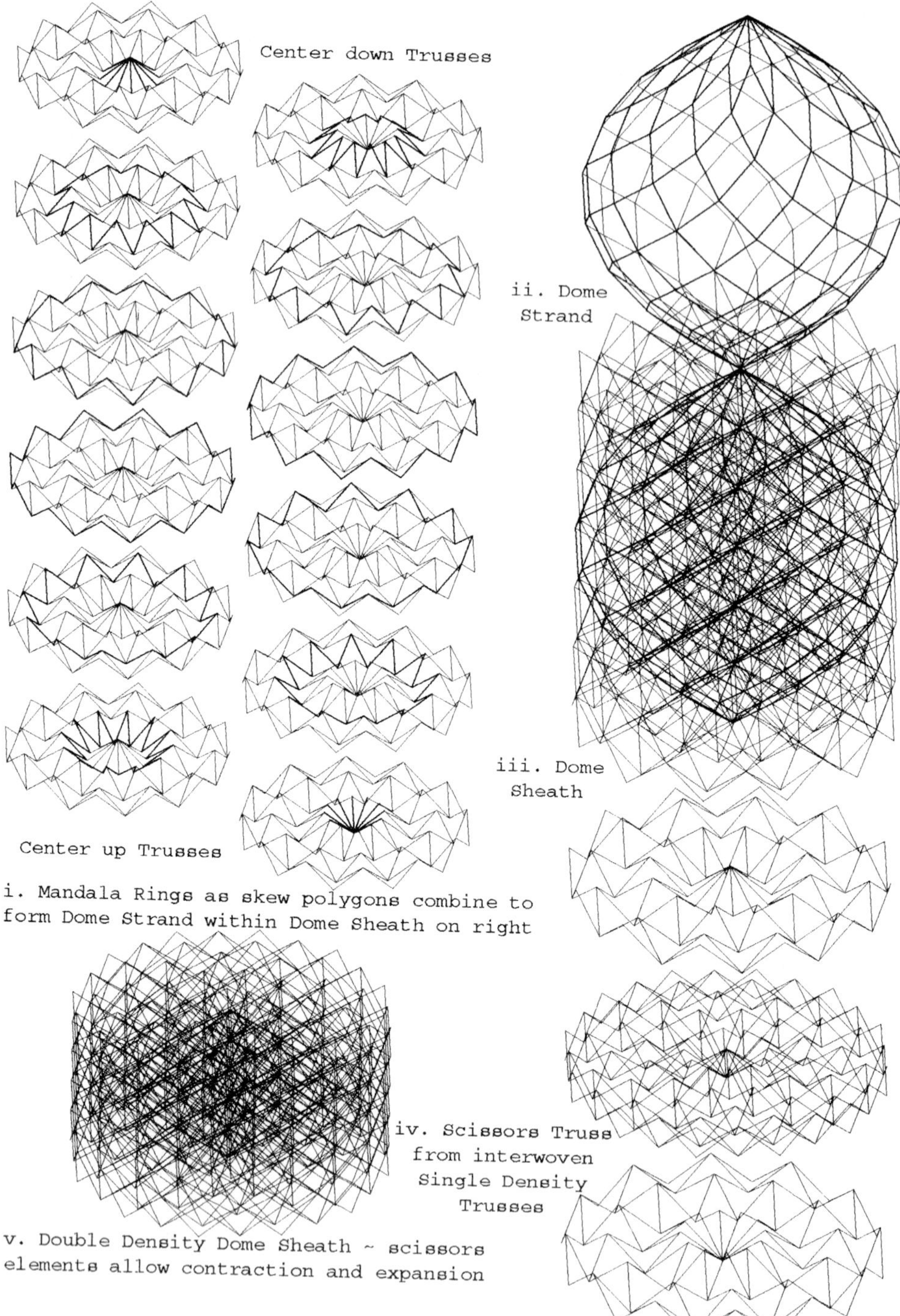

Fig. 3. Single Density (above) and Double Density (below) Trusses and Dome Sheaths.

10. Structural Application of the Planar Mandala. If the outer mandala ring is deleted, the mandala can be constructed as a tensile lattice, fashioned by curving an ordinary tensile net of square mesh. The central ties are doubled. The curvature in Fig. 1(v) of the square net of pin-jointed struts into the circular mandala, forms a structural geometrical transformation of circumferential rotational closure for folding circular trusses and domes.

MONOAXIAL DEVELOPMENT OF THE MANDALA

11. The Dome Strand. If from the plane, line-segments are inclined as vectors of equal slope, a dome formation is generated as in Figs l(i), 2(iii) and 3(ii). Parallel bundles of angled struts are generated about the structure. Individual eccentric polygons develop into regular eccentric polygonal helices. For the case of infinite frequency, the surface of curvature is the surface of rotation of the sine-wave. Nodes of the finite case lie on that surface, and the eccentric circles of the mandala transform into regular eccentric helices lying always on that surface, passing through both centre and periphery. This is an elegant behaviour. The dome structure suggests various structural applications; and the geometry appears to characterise the structural and surface decorative geometry of the Islamic dome (ref. 1).

12. A Simple Centralised Space Truss. The polygonal Mandala is axially displaced and duplicated with corresponding points connected with verticals to give a *linee occulte* grid for a single layer centralized space truss shown in Fig. 2(i). The vertical displacement is made equal to the edge length of the Mandala, hence the vertical faces of the truss are square. (The structure is also formed by curving a single layer of a cubic lattice in the horizontal plane as in Fig. 1(v). Stiff connectors, triangulators, or diaphragms ensure structural stability. If replicated axially, a multi-layer prismatic lattice is obtained.

13. Wedding Cake Dome. In the simple space truss, mandala rings have become star prisms, with square sides folding in and out about their periphery. These recombine in the wedding cake dome of Fig. 2(ii), from which various structural domes are deduced.

14. Octet Truss Dome. For example, using pin joints, transverse elements with surface diamond triangulators are added, to provide an octet truss dome as shown in Fig. 2(v), composed of component octahedra and tetrahedra - the domical transformation of the octet truss. Transverse triangulators between the two skins and horizontal surface diamond triangulators lie in the common horizontal planes of points of the domes.

15. Single-Layer Space Trusses are Generated,which can Contract and Expand. Individual eccentric polygons of the mandala develop into eccentric regular polygonal prisms of the simple space truss. In Fig. 2(vii) diagonals of square side-faces of the prisms connect in either of two ways to generate skew polygons. These combine to generate the webs of single-layer space trusses of either centre-up or centre-down orientation of Figs 2(viii)-2(ix) and 3(i). Each web expands or contracts regularly about its axis, as in Figs 4(i) and 4(ii), as the angle to the horizontal of its members increases on compression or decreases on expansion.

Fig. 4. Expansion and Contraction of Single Density (outside) and Double Density (center) Folding Centralized Space Trusses.

Fig. 5. Some Folded Plate Configurations based on the Single Density Space Truss.

At any stage of expansion the structural geometry is such that web members all have the same degree of slope whilst retaining in plan their orientation.

16. These Combine in the Double-Layer Space Truss, which can Contract and Expand. Both kinds combine in Fig. 2(x) to generate the double-layer space truss of Figs 3(iv)-3(v). Paired diagonals of a square face function as scissor elements which allow the truss to regularly compress and expand transversely about its axis, which is shown in Fig. 4(iii).

17. Dome Sheaths. Figs 3(iii) and 3(v) show trusses axially combined into single and double density sheaths, which can contract and expand. They contain dome strands within them.

18. Folding Vertical Displaced Dome. Scissor elements of two different lengths, given by the diagonals of the parallelograms of Fig. 2(iv), generate the folding dome of Fig. 2(vi) which contracts and expands about its axis. If of low aspect it suggests a Space antenna.

19. Folded Plate Configurations. Folded plate constructions shown in Fig. 5 are derived from the centralised space truss geometry to generate centralised octet trusses for roofs.

REFERENCES

1. MEURANT R.C. The Aesthetics of the Sacred - A Harmonic Geometry and Philosophy of Sacred Architecture (Third Ed.), pp. 5-35 and colour plates A. 1-E.5. A Monograph of the Institute of Traditional Studies, The Opoutere Press, Auckland, 1989. Available from the author P.O. Box 6405, Auckland 1, New Zealand, US$50.

2. MEURANT R.C. A New Order in Space - Platonic and Archimedian Polyhedra and Tilings. International Journal of Space Structures, Vol.6, No.1, University of Surrey, 1991.

3 . MEURANT R.C. The Integral Space Habitation ~ Towards an Architecture of Space (First Ed.). A Monograph of the Institute of Traditional Studies, The Opoutere Press, Boulder and Auckland, 1989 (available from the author, US$15).

4. MEURANT R.C. Structure, Form, and Meaning in Micro-Gravity. The Integral Space Habitation. International Journal of Space Structures, Vol.5, No.2, University of Surrey, 1990.

225. From snow crystals to space enclosure systems: implementing the future of architecture

P. J. PEARCE, Pearce Systems International, USA

SYNOPSIS. During the last ten years Pearce Systems International (formerly Pearce Structures) has built a diverse range of architectural structures. These installations vary in size and configuration with projects in North America, Asia, and the Middle East. During the next ten years, Pearce Systems plans to be a major force in the development of an architecture of the future. This new architecture will incorporate a high-performance vocabulary of space enclosure systems which will celebrate natural light, adaptation, and responsiveness to the human need for diversity and coherence.

INTRODUCTION

1. Developed from over twenty years of theoretical work, Pearce Systems has evolved a highly innovative design/engineering and manufacturing organisation in the field of space structures. Innovative approaches to the geometry of structures, and breakthroughs in product technology, exemplified by the Pearce Mulit-hinge Nodeless Connection System and the Pearce Integral Glazing and Cladding Systems, have enabled Pearce Systems to demonstrate leadership in the field of space structures applications.

2. An underlying design principle that governs the work of Pearce Systems is the concept of *strength of geometry*, in contradistinction to *strength of materials*. In this philosophical framework, structure and form become integrated, design and engineering become one, and a new high-performance architecture for the 21st Century emerges.

STRUCTURE IN NATURE

3. In the early 1960's, inspiration was drawn from Charles Eames, Konrad Wachsmann, and Buckminster Fuller. The evolution of systems design beyond the work of these visionaries seemed to require another step; a broader theoretical view.

4. The observation that a more complete theoretical basis for building systems design was needed led this author to an investigation of the principles of natural

structure. This research ultimately led to the book, *Structure in Nature Is a Strategy for Design*, MIT Press, 1978, 1990.

5. The purpose of this research was an attempt to understand *first principles.* What are the fundamental laws which lead to the rules which govern *form-giving*? Although much of this theoretical work centered around the study of the geometry of crystal structures, repeating networks, and space filling systems, the work of the great natural philosopher, D'Arcy Wentworth Thompson, also played a significant role.

6. In his classic book, *On Growth and Form*, Thompson asserts that all form can be considered a "diagram of forces". He proceeds to give persuasive examples, mostly drawn from the morphology of biological structures, that his assertion is, indeed, accurate. Thompson's view is consistent with many other morphologist regarding formative processes in the world of natural structure. This assertion, that *form is a diagram of forces,* becomes the fundamental basis of an emerging theory of *form-giving* in systemic building structure design.

7. The concept of form as a diagram of forces can be clearly illustrated with the example of the snow crystal. Fig. 1 shows eleven snow crystals of different configurations. There is an infinite variety of snow crystals, with no two exactly alike. However, all snow crystals will exhibit hexagonal symmetry.

Fig. 1. Snow crystals.

8. The water molecule, which is the basic building system of the snow crystal, is the *kit-of-parts*, which determines that all snow crystals have the symmetry of the regular hexagon. This kit-of-parts constrains the form options so that only hexagonal structures are formed. Within that constraint, there are endless nuances of configuration. There are environmental forces which govern the shape of individual snow crystals, in addition to the internal forces of the molecule, which govern the formative process of the snow crystal. The kit-of-parts or building system, which is the water molecule, is a highly responsive and adaptive system, as it diagrams the forces acting within and upon it. It is a prototype for the quintessential building system.

9. An essential attribute in the formative process of the snow crystal is that every configuration that is formed represents an arrangement of *least energy* with respect to the forces that are being *diagrammed*. Each snow crystal takes on the most economical arrangement for the given set of circumstances under which it was formed. This attribute is characteristic of all formative processes in natural structure. Nature's design is always seeking the highest performance for the least input of energy (or material).

10. Natural structure then becomes the ideal model for an architecture of the future. Nature designs highly differentiated patterns of form through adaptation to the forces at play. This is achieved with configurations of minimum energy that are achieved through optimised geometry. What this means for the design of structure and architecture is that performance is achieved through optimising configuration.

STRENGTH OF GEOMETRY

11. The underlying design principle that is implied by the principle of triangulation, and by space frame (truss) structures in general, is the concept of *strength of geometry*. In space frame design, this principle is found to be most deliberate with respect to geometric detail. The overall configuration of space frames, or any overall structural configuration can also be treated from the perspective of strength through geometry. Although there are certainly many examples of structural configurations which exemplify this ideal, structures are commonly designed to rely on *strength of materials* for their integrity. Typically, structural engineers are *given* a preconceived design by a project architect. The expectation is that the engineer will make the proposed design *work* from a structural perspective. At this point, rather than manipulate the configuration, the engineer will adapt to it, and make the project work through brute material force, i.e., strength through materials.

12. Certainly strength of materials is important. It simply should be considered a second order design parameter as a tool for optimisation. A more fundamental parameter is configuration design as a tool for performance optimisation. It has been shown many times that configuration (strength of geometry) is a more fundamental driver of performance than engineering the performance of specific material sizes, shapes, and even material properties.

BUILDING TECHNOLOGY AND PRACTICE

13. The practice of the building arts in our culture has evolved through centuries of history. As we cascade towards the 21st century, the building arts lag disappointingly behind most other major activities of human enterprise. The 20th century has seen many new technologies emerge that, unlike the building arts, had not been known to prior centuries. These technologies include aircraft, automobiles, consumer electronics, computer hardware and software, and a myriad of other enterprises.

14. The rapid evolution of all of these 20th century technologies is characterised by a single design objective: the improvement of performance. This increasing performance has resulted in higher levels of service, usefulness, and creative opportunity at lower cost.

15. When these technologies emerged, none had any real infrastructure in the culture to support them. However, as the benefits of the technologies were understood, new infrastructures of support rapidly evolved.

16. In contradistinction to these technologies, the building arts have been dominated by tradition and habit. The infrastructure which supports the building arts in modern societies is dominated by complex forms of social protocol. Architects are trained to design buildings as a manifestation of style driven by temporal fixations. Engineers are subservient to architects. The general contractors build the projects according to the architects plans with a complex array of subcontractors and tradesmen. The owners pay for them with the banks money.

17. Within the fabric of this social infrastructure, there is, not surprisingly, a lack of integration. Each player has his work scope. There is limited mutual understanding of each -separate scope of work. It is very difficult to optimise building design and construction because of this segmentation of the building process.

18. The building arts are not driven by performance as are all other major socially significant technologies. Engineers attempt to establish the structural viability of the arbitrary form proposed by architects. They do this through scale dependent strength of materials as their primary design tool. There is little integration or true design collaboration.

19. There have been examples in history of the building arts where the architect, engineer, and the builder were one. There was the integration of form, structure, and purpose. This was certainly true for the great cathedrals of the middle ages, and, in a modest way, it is true for most indigenous building. The cathedrals are particularly noteworthy as ambitious work in which form and structure is magnificently integrated.

A NEW ARCHITECTURE

20. In order that a new high-performance architecture emerge for the 21st Century, it will be necessary to set aside the forces of historical accident that drive form-giving in the design of buildings.

21. Such a high-performance architecture will be based upon a systemic vocabulary of space enclosure systems which will be captured in a user-friendly electronic medium. Computer software will become a tool which canonises natural law and concomitant rules, along with libraries of components and solutions. Information will be integrated into a performance driven design tool. This will facilitate a kind of indigenous architecture of the 21st Century.

22. This new architecture will incorporate a high-performance design ethic. It will be a responsive architecture which will adapt to the human need for diversity and coherence. It will provide new levels of physical performance and environmental attributes. Materials will be used efficiently, daylight will be managed to human advantage, the impact of environmental forces such as wind and radiant energy will be addressed in optimum ways, and energy utilisation will be minimised. Structure and form will be integrated to form elegant and user-friendly solutions to complex architectural design problems, and prefinished components will facilitate a long-life, low maintenance architecture, capable of change over time.

SPACE ENCLOSURE SYSTEMS

23. The possibility of implementing a new architecture depends on many factors. Not only do we need a theoretical basis for developing a design vocabulary with certain attributes, we need technical developments and physical systems. With this in mind, we at Pearce Systems have, over a number of years developed an integrated family of building products. Taken as a whole, this family of products is referred to as a Space Enclosure System.

24. These products are currently offered to the building industry for specific project applications. Pearce Systems has been providing space frames for over ten years, and more recently its integral glazing and cladding products. A complementary array of products is thus provided to the building customer. The Pearce space frame provides the primary support structure, and the integral glazing and cladding systems provide the environmental closure solutions.

MULIT-HINGE SPACE FRAME SYSTEM

25. The *Pearce Systems Space Frame System* uses the company's proprietary Mulit-hinge connection technology, which needs no secondary hubs, or fabricated nodes to join tubular steel struts together. In the Pearce System, the connection system is integral to the strut ends. In this system, the struts are simply connected to each other directly, without the intermediary of a node connector. Fig. 2, shows a typical Mulit-hinge connection assembly with eight members meeting, four each in the chordal plane and web planes.

26. A typical strut in the Mulit-hinge System consists of a round steel tube, in which each end has been closed by welding on end caps. Sets of connection elements, called hinge elements, are fillet welded adjacent to each end of the tubular struts. Location of connection elements is controlled by sophisticated modular fixturing. The fabrication costs are very low.

27. The primary attribute of the Mulit-hinge System is that joint stability is guaranteed by its geometry, an example of the philosophy of *strength of geometry*. This joining system incorporates an interposed and spatially distributed family of "hinge type" or pinned connections in which strut ends are bolted directly to each other. This joint assembly is matched to an inherently rigid geometry.

28. This geometric arrangement of hinge connections creates a condition in which all strut members are mutually braced to form an inherently stable array of three-dimensionally linked triangular frames. In such an arrangement, the local instability of the joint assembly is mitigated by the global triangulation which comprises the neighbourhood of each joint. The joint becomes the servant of the globally stable, triangulated space frame, and thus becomes fully stable itself. This localised joint stability is consistent with the global behaviour of the inherently stable, triangulated space frame.

29. The Mulit-hinge System provides a space frame connection strategy offering substantial joint stiffness. This joint stiffness enhances the global load capacity of the space frame assembly through an increase in member ductility. Such increase in ductility in peak regions of force can be of vital significance. It allows for load redistribution following initial buckling, resulting in improved member capacity.

Fig. 2. Pearce Systems Mulit-hinge System connection assembly.

30. The Mulit-hinge System allows complete flexibility of configuration design. With this system, a space frame can be designed to match a broad range of performance requirements and architectural visions. The Mulit-hinge System, because of its inherent design flexibility, offers numerous additional engineering benefits. In addition to the aforementioned inherent stability and ductility, these include: multiple load paths through a given joint, load diffusion, fasteners loaded only in bearing (shear), and variability of connection density at a given connection.

31. The Mulit-hinge System provides high-performance and design/engineering flexibility at low cost. High quality appearance is insured not only through the fine details of the system, but from a fusion-bonded, two-stage powder coat finish.

THE DEVELOPMENT OF THE MULIT-HINGE SYSTEM

32. The development of the Mulit-hinge System was the result of a search for a new paradigm, in an effort to improve the performance, versatility, and economy of space frame connection systems. Engineers and designers continue to accept the primacy of the node or hub as the ideal connection solution. There have been countless attempts to design new kinds of nodes. All these attempts, for the most part, result in minor variations of the original Mero node, and most of these variations are cosmetic. There have been no fundamental improvements in node type systems over the original Mero design.

33. The common paradigm for space frame connection system design clearly derives from the analytical tools of structural engineering. These tools are mathematical and geometric. In the parlance of the mathematical model, space frame structures are built from lines intersecting at points. In this construct the lines become struts and the points become nodes. The analysis places forces in lines which intersect at points.

34. In the real world, the *Law of Spatial Exclusion* becomes a key variable: *More than one object cannot occupy the same space at the same time*. Once lines become physical struts with diameters, it is not possible for struts to meet at literal points without interfering with each other. Therefore a connection solution must mitigate this interference if a structure is to be assembled. The physical implementation of a space frame application clearly compromises the abstract model. The problem becomes: *how do you hold struts together, while establishing a strategy or method for the transfer of forces?* Classically, this would be done with a node as an assumed replica of a point of distribution, as in the mathematical model. Pearce Systems has demonstrated in a wide range of applications over a ten year period, that this problem can be solved without the use of a node.

THE INTEGRAL GLAZING AND CLADDING SYSTEM

35. Historically, structure systems and glazing systems have always been provided by two different organisations. One group presumably understood structure, and the other group presumably understood glazing. The performance of each of these two systems was matched to independent design criteria. There was no real possibility of discovering mutual advantage.

36. Pearce Systems approached the problem differently. The inherent attributes of global space frame behaviour was seen as an opportunity to improve glazing system design. Certain attributes that characterise space frame behaviour have natural advantages for glazing system performance, such as uncommonly small deflections.

37. Furthermore, with these steel space frame structures it is possible to implement configurations which minimise movement caused by thermal gradients. These attributes, in which movements and deflections are minimised, provide substantial benefits for the performance of glazing system weather seals.

38. Since the space frame is steel, the use of steel for glazing system framing insures total thermal congruence. The coefficient of thermal expansion for glass is also very close to steel, which further minimises differential thermal expansion and contraction. This thermal equilibrium creates a benevolent environment within which to provide systemic and reliable weather seals. The steel glazing frames are made from roll-formed steel sections, to which glass is factory-bonded using silicone structural sealants.

39. These factory-bonded unit glass panels are then mechanically attached to the assembled space frame at the job site. Upon completion of this rigid mechanical attachment, a silicone caulk joint is applied between the adjacent glazing panels. Because of the material homogeneity, the use of steel with its comparatively small thermal expansion coefficient and the elastic range of the silicone sealant, there is absolutely no need for any expansion joints within the space frame envelope.

40. In addition to the Integral Glazing System, Pearce Systems has developed a family of other cladding products. These products are designed around the principles of the Integral Glazing System. Indeed, they are merely extensions of the glazing system.

A DECADE OF PROJECTS

41. Over the past decade Pearce Systems has designed, engineered and built over 70 architectural space enclosure projects. These projects have comprised relatively small ornamental canopies as well as large commercial, industrial, and public structures. Indeed, Pearce has designed and built the largest and most complex fully glazed space frame project yet undertaken. Namely, the Biosphere 2 project near Tucson, Arizona.

42. The Biosphere 2 project is comprised of five distinct structures (Fig. 3). These structures are clad with the Pearce designed and produced airtight glazing and metal cladding systems. Five million cubic feet of airtight space is provided. Never before in the history of building has an enclosure of this size been airtight. The project has been completed for nearly two years, and continues its airtight performance.

43. The stepped pyramids of the *Wilderness Biome*, includes a 180 foot clear span structure covering a rain forest. The total building is 550 feet long, and is comprised of 40,000 space frame struts (Figs 4, 5). An *Intensive Agriculture Biome* is comprised of nine intersecting space frame vaults and approximately 15,000 struts. The *Habitat*, with its doubly curved single layer space frame shell, is clad with steel panels and glazing. Two 180 foot steel clad *Lung Domes* are included which contain variable volume membranes designed to regulate internal pressure. These *Lung Domes* are required because the project is entirely airtight. Without the *Lung Domes*, thermal gradients would create dangerously high internal pressures.

44. Pearce Systems helped design the structures, performed all of the engineering of the space frame applications, manufactured the systems, and developed and produced this airtight version of its Integral Glazing System. More than any other project, the Biosphere 2 begins to forecast elements of a high-performance architecture of the 21st century.

45. The Southland Gardens project in Detroit, Michigan, was completed in 1992 (Fig. 6). The project is built with the Mulit-hinge Space Frame System, and includes the Pearce Integral Glazing System. It functions as a large food court and main entrance to a large shopping mall in the Detroit, Michigan area. It creates a winter garden environment and shows off the advantages of the Integral Glazing System.

46. The Fallbrook Square Mall, Canoga Park, California, (Fig. 7) includes a food court atrium and 800 foot mall enclosure. Total plan area is 74,000 square feet. This atrium was the first application of Pearce Systems Integral Glazing System from which the entire food court, including the storefront entrance and food court atrium, was built.

CONCLUSION

47. Of the many projects that the Company has completed, some were truly landmark architectural projects. These projects, in many ways, represent precursors to an architecture of the 21st century. As indicated earlier, system development and product development has been well served by these projects. The opportunity to test building systems in the world of built structures is essential preparation for developing the tools, technology, and vocabulary for a high-performance architecture.

Fig. 3. Biosphere 2 structures.

Fig. 4. Biosphere 2. North end of Wilderness Biome, 180 foot span.

Fig. 5. Biosphere 2. Wilderness Biome interior view.

Fig. 6. Southland Gardens foodcourt.

Fig. 7. Fallbrook Square Mall.

Index